Speicher · Antonarakis · Motulsky
VOGEL AND MOTULSKY'S HUMAN GENETICS
Problems and Approaches

Fourth Edition

Speicher   Antonarakis   Motulsky

VOGEL AND MOTULSKY'S HUMAN GENETICS

Problems and Approaches

Fourth, Completely Revised Edition

With 343 Figures and 76 Tables

Editors
Michael R. Speicher
Institute of Human Genetics
Medical University of Graz
Harrachgasse 21/8
8010 Graz
Austria
michael.speicher@medunigraz.at

Stylianos E. Antonarakis
Department of Genetic Medicine and Development
University of Geneva Medical School
and University Hospitals of Geneva
1 rue Michel-Servet
1211 Geneva
Switzerland
Stylianos.Antonarakis@unige.ch

Arno G. Motulsky
Professor Emeritus
Departments of Medicine (Division of Medical Genetics)
and Genome Sciences
University of Washington
School of Medicine
Box 355065
1705 N.E. Pacific St.
Seattle, WA 98195-5065
USA
agmot@u.washington.edu

ISBN 978-3-540-37653-8   e-ISBN 978-3-540-37654-5
DOI 10.1007/978-3-540-37654-5
Springer Heidelberg Dordrecht London New York

Library of Congress Control Number: 2009931325

© Springer-Verlag Berlin Heidelberg 2010
This work is subject to copyright. All rights are reserved, whether the whole or part of the material is concerned, specifically the rights of translation, reprinting, reuse of illustrations, recitation, broadcasting, reproduction on microfilm or in any other way, and storage in data banks. Duplication of this publication or parts thereof is permitted only under the provisions of the German Copyright Law of September 9, 1965, in its current version, and permission for use must always be obtained from Springer. Violations are liable to prosecution under the German Copyright Law.
The use of general descriptive names, registered names, trademarks, etc. in this publication does not imply, even in the absence of a specific statement, that such names are exempt from the relevant protective laws and regulations and therefore free for general use.

Cover design: Estudio Calamar, Spain

Printed on acid-free paper

Springer is part of Springer Science+Business Media (www.springer.com)

In memory of Friedrich Vogel

The editors

To my wife, Irene, and our children, Alexander and Julia.

Michael R. Speicher

To my parents, my wife, Grigoria, and our children.

Stylianos E. Antonarakis

To the memory of my wife, Gretel, and to my children.

Arno G. Motulsky

Preface

The first edition of *Human Genetics, Problems and Approaches*, was published in 1970 by human geneticists Friedrich Vogel and Arno Motulsky as sole authors. The aim was broad coverage and in-depth analysis of both medical and human genetics with an emphasis on problems and approaches with occasional historical comments. This point of view was fully explained in an introductory chapter of the three previous editions (1970, 1976, 1997). The book acquired an excellent reputation as an advanced text of human genetics and has been translated into Italian, Japanese, Russian, Chinese, and Portuguese. Our general aims for the fourth edition remain similar and together with novel developments are now set out in the Introduction of this new fourth edition.

Around 2004/2005, both Friedrich Vogel and Arno Motulsky, as well as the publishers, felt that the book should be continued with a new fourth edition in the same spirit and coverage as earlier editions, but should now include additional expert authors. After some delay and the death of Friedrich Vogel in the summer of 2006, a new editorial team consisting of Michael R. Speicher of the Medical University of Graz, Austria, Stylianos E. Antonarakis of the University of Geneva Medical School, and Arno G. Motulsky of the University of Washington School of Medicine, was constituted for the fourth edition of the Vogel/Motulsky book in the spirit of the original work.

An outline of the fourth edition's contents was developed and various internationally known geneticists, including the new editors, were selected to write the individual chapters. The resultant titles are listed in the table of contents. Most chapters are entirely new, while only three chapters (1, 5, and 6) utilize the third edition with appropriate, up-to-date revisions. Entirely new chapters include the description of the human genome, epigenetics, pharmacogenetics, genetic epidemiology, human evolution, genetics of mental retardation, autism, alcoholism and other addictions, consanguinity and related matters, gene therapy, cloning and genetic aspects of global health. Multiple chapters of various animal models used in the study of human and medical genetics are novel as are Databases and Genome Browsers as well as Databases Used in Human and Medical Genetics. The final content of the book was the result of many e-mails and conference calls. This new, updated, and totally revised edition does not contain some important and historically interesting chapters on certain topics. These can be found in the third edition of the book published in 1997, which was exclusively authored by F. Vogel and A.G. Motulsky. These topics were: enzymes in Chap. 7 (pp. 258–299); mutation rates in Chap. 9 (pp. 393–413); and mutation induction by ionizing radiation and chemicals in Chap. 11 (pp. 457–493).

The staff of Springer was most helpful in giving us extensive and firm aid in getting the book finished. We thank particularly Doris M. Binzenhöfer-Walker for her work during the early stage of the project and Isabella Athanassiou for her efficient help later. Andrea Pillman was a strict task mistress, who encouraged us to finish the book expeditiously.

The editors of the fourth edition want to express their intellectual indebtedness to Friedrich Vogel for having conceived of and played a major role in the first three editions of this book. Arno Motulsky particularly misses his discussions with Friedrich on human genetics and its role in science and medicine.

The cover illustration portrays a marble statue of Asclepius, the Greek god of healing, grasping a serpent-encircled staff as a symbol of medicine. The double helix of DNA joined to Asclepius symbolizes the applications of basic genetics to medicine.

4th edition
June 28, 2009

Michael R. Speicher, Graz
Stylianos E. Antonarakis, Geneva
Arno G. Motulsky, Seattle

Preface to the First Edition

Human genetics provides a theoretical framework for understanding the biology of the human species. It is a rapidly growing branch of science. New insights into the biochemical basis of heredity and the development of human cytogenetics in the 1950s heightened interest in this field. The number of research workers and clinicians who define themselves as full-time or part-time human and medical geneticists has increased sharply, and detailed well-founded knowledge has augmented exponentially. Many scientists and physicians are confronted with genetic problems and use concepts and methodology of human genetics in research and diagnosis. Methods developed in many different fields of the biologic, chemical, medical, and statistical sciences are being utilized toward the solution of genetic problems. The increasing number and sophistication of well-defined and elegantly solved problems helps to refine an extensive framework of genetic theory. These new conceptual insights in their turn lead to solutions of new questions. To mention only one example, the structure of hemoglobin genes has been elucidated using methods derived from protein chemistry and DNA technology. It is an exciting experience to participate in these developments!

Moreover, scientific progress in genetics has practical implications for human well-being. Improved knowledge of the genetic cause of an increasing number of human diseases helps to refine diagnosis, to find new therapeutic approaches, and above all, to prevent genetic diseases. So far, human genetics has had less of an impact on the behavioral and social sciences. It is possible that genetic differences involved in shaping personality structure, cognitive faculties, and possibly human social behavior may be at least as important as genetic variation affecting health and disease. The data, however, are less clear and more controversial. These problems are discussed in detail in the text. The rapid progress of human genetics in recent decades has attracted – and is still attracting – an increasing number of students and scientists from other fields. Various elementary textbooks, more advanced monographs of various branches of the field, and the original journal literature are the usual sources of introduction to human genetics. What seems to be lacking, however, is a fairly thorough and up-to-date treatise on the conceptual basis of the entire field of human genetics and its practical applications. Often, the absence of a broadly based background in the field leads to misunderstanding of its scope, unclear goals for research, improper selection of methods, and imbalanced theoretical discussions. Human genetics is based on a powerful theory, but this implicit conceptual foundation should be made explicit. This goal is the purpose of this book. It certainly is a formidable and possibly even too audacious task for two sole authors. However, both of us have been active in the field for more than 25 years.

We have worked on various problems and with a variety of methods. Since the early years of our careers, we have met occasionally, followed each other's writings, and were often surprised by the similarity of our opinions and judgments despite quite different early medical and scientific backgrounds. Moreover, our knowledge of the literature turned out to be in part overlapping and in part complementary. Since we are working in different continents, AGM had a better knowledge of concepts and results in the USA, while FV knew more of the continental European literature. Moreover, both of us have extensive experience as editors of journals in human genetics and one (FV) published a fairly comprehensive textbook in Germany some time ago (*Lehrbuch der allgemeinen Humangenetik*, Springer 1961), parts of which were still useful for the new book. We finally decided to take the risk, and, by writing an "advanced" text, to expose our deficiencies of knowledge, shortcomings of understanding, and biases of judgment.

A text endeavoring to expose the conceptual framework of human genetics cannot be dogmatic and has to be critical. Moreover, we could not confine ourselves to hard facts and well-proved statements. The cloud of conjectures and hypotheses surrounding a rapidly growing science had to be depicted. By doing so, we face the risk of being disproved by further results.

A number of colleagues helped by reading parts of the manuscript on which they had expert knowledge and by making useful suggestions: W. Buselmaier, U. Ehling, G. Flatz, W. Fuhrmann, S. Gartler, Eloise Giblett, P. Propping, Laureen Resnick, and Traute M. Schroeder. They should not be held responsible for possible errors. J. Krüger was of supreme help in the statistical parts. Our secretaries, Mrs. Adelheid Fengler and Mrs. Gabriele Bauer in Heidelberg, Mrs. Sylvia Waggoner in Seattle, and Mrs. Helena Smith in Stanford gave invaluable aid. The figures were drawn by Edda Schalt and Marianne Lebküchner. Miriam Gallaher and Susan Peters did an expert job of copy editing. The authors are especially grateful to Dr. Heinz Götze and Dr. Konrad F. Springer, of Springer Publishing Company, for the excellent production. The work could not have been achieved had the two authors not been invited to stay at the Center for Advanced Study in the Behavioral Sciences at Stanford (California) for the academic year of 1976/1977. The grant for AGM was kindly provided by the Kaiser Family Foundation, while the Spencer Foundation donated the grant for FV.

The cover of this book shows the mythical first human couple, Adam and Eve, as imagined by Albrecht Dürer (1504). They present themselves in the full beauty of their bodies, ennobled by the genius and skill of a great artist. The drawing should remind us of the uniqueness and dignity of the human individual. Human genetics can help us to understand humanity better and to make human life happier. This science is a cardinal example of Alexander Pope's statement. "The proper study of mankind is man."

Spring 1979   Friedrich Vogel, Heidelberg
                                                        Arno G. Motulsky, Seattle

Contents Overview

	Introduction	1
1	**History of Human Genetics** Arno G. Motulsky	13
2	**Human Genome Sequence and Variation** Stylianos E. Antonarakis	31
3	**Chromosomes** Michael R. Speicher	55
4	**From Genes to Genomics to Proteomics** Michael R. Speicher	139
5	**Formal Genetics of Humans: Modes of Inheritance** Arno G. Motulsky	165
6	**Linkage Analysis for Monogenic Traits** Arno G. Motulsky and Michael Dean	211
7	**Oligogenic Disease** Jon F. Robinson and Nicholas Katsanis	243
8	**Formal Genetics of Humans: Multifactorial Inheritance and Common Diseases** Andrew G. Clark	263
8.1	**Lessons from the Genome-Wide Association Studies for Complex Multifactorial Disorders and Traits** Jacques S. Beckmann and Stylianos E. Antonarakis	287
9	**Epigenetics** Bernhard Horsthemke	299
10	**Human Gene Mutation: Mechanisms and Consequences** Stylianos E. Antonarakis and David N. Cooper	319
11	**Human Hemoglobin** George P. Patrinos and Stylianos E. Antonarakis	365

12	**Human Genetics of Infectious Diseases**....................................	403
	Alexandre Alcaïs, Laurent Abel, and Jean-Laurent Casanova	
13	**Gene Action: Developmental Genetics**.......................................	417
	Stefan Mundlos	
14	**Cancer Genetics**...	451
	Ian Tomlinson	
15	**The Role of the Epigenome in Human Cancers**........................	471
	Romulo Martin Brena and Joseph F. Costello	
16	**Population Genetic Principles and Human Populations**..........	487
	Emmanouil T. Dermitzakis	
17	**Consanguinity, Genetic Drift, and Genetic Diseases in Populations with Reduced Numbers of Founders**..............	507
	Alan H. Bittles	
18	**Human Evolution**...	529
	Michael Hofreiter	
19	**Comparative Genomics**..	557
	Ross C. Hardison	
20	**Genetics and Genomics of Human Population Structure**........	589
	Sohini Ramachandran, Hua Tang, Ryan N. Gutenkunst, and Carlos D. Bustamante	
21	**Genetic Epidemiology**..	617
	Sophia S. Wang, Terri H. Beaty, and Muin J. Khoury	
22	**Pharmacogenetics**..	635
	Nicole M. Walley, Paola Nicoletti, and David B. Goldstein	
23	**Behavioral Genetics**...	649
23.1	**The Genetics of Personality**..	651
	Jonathan Flint and Saffron Willis-Owen	
23.2	**Mental Retardation and Intellectual Disability**........................	663
	David L. Nelson	
23.3	**Genetic Factors in Alzheimer Disease and Dementia**..............	681
	Thomas D. Bird	
23.4	**Genetics of Autism**...	699
	Brett S. Abrahams and Daniel H. Geschwind	
23.5	**The Genetics of Alcoholism and Other Addictive Disorders**....	715
	David Goldman and Francesca Ducci	
23.6	**Behavioral Aspects of Chromosomal Variants**.........................	743
	Michael R. Speicher	

23.7	**Genetics of Schizophrenia and Bipolar Affective Disorder**	759
	Markus M. Nöthen, Sven Cichon, Christine Schmael, and Marcella Rietschel	
24	**Model Organisms for Human Disorders**	777
	Michael R. Speicher	
24.1	**Mouse as a Model for Human Disease**	779
	Antonio Baldini	
24.2	***Caenorhabditis elegans*, A Simple Worm: Bridging the Gap Between Traditional and Systems-Level Biology**	787
	Morgan Tucker and Min Han	
24.3	***Drosophila* as a Model for Human Disease**	795
	Ruth Johnson and Ross Cagan	
24.4	**Human Genetics and the Canine System**	813
	Heidi G. Parker and Elaine A. Ostrander	
24.5	**Fish as a Model for Human Disease**	827
	Siew Hong Lam and Zhiyuan Gong	
25	**Genetic Counseling and Prenatal Diagnosis**	845
	Tiemo Grimm and Klaus Zerres	
26	**Gene Therapy**	867
	Vivian W. Choi and R. Jude Samulski	
27	**Cloning in Research and Treatment of Human Genetic Disease**	875
	Ian Wilmut, Jane Taylor, Paul de Sousa, Richard Anderson, and Christopher Shaw	
28	**Genetic Medicine and Global Health**	885
	David J. Weatherall	
29	**Genetic Databases**	903
	Introductory Note by Stylianos E. Antonarakis	
29.1	**Databases and Genome Browsers**	905
	Rachel A. Harte, Donna Karolchik, Robert M. Kuhn, W. James Kent, and David Haussler	
29.2	**Ensembl Genome Browser**	923
	Xosé M. Fernández and Ewan Birney	
29.3	**Databases in Human and Medical Genetics**	941
	Roberta A. Pagon, Ada Hamosh, Johan den Dunnen, Helen V. Firth, Donna R. Maglott, Stephen T. Sherry, Michael Feolo, David Cooper, and Peter Stenson	
Subject Index		961

Contents

Introduction		1
1	**History of Human Genetics**	13
	Arno G. Motulsky	
1.1	The Greeks (see Stubbe [83])	14
1.2	Scientists Before Mendel and Galton	15
1.3	Galton's Work	16
1.4	Mendel's Work	17
1.5	Application to Humans: Garrod's Inborn Errors of Metabolism	18
1.6	Visible Transmitters of Genetic Information: Early Work on Chromosomes	20
1.7	Early Achievements in Human Genetics	20
1.7.1	AB0 and Rh Blood Groups	20
1.7.2	Hardy-Weinberg Law	21
1.7.3	Developments Between 1910 and 1930	21
1.8	Human Genetics, the Eugenics Movement, and Politics	21
1.8.1	United Kingdom and United States	21
1.8.2	Germany	22
1.8.3	Soviet Union/Russia (see Harper, Chap. 16 in [38])	23
1.8.4	Human Behavior Genetics	23
1.9	Development of Medical Genetics (1950–the Present)	23
1.9.1	Genetic Epidemiology	23
1.9.2	Biochemical Methods	24
1.9.3	Genetic and Biochemical Individuality	24
1.9.4	Cytogenetics, Somatic Cell Genetics, Prenatal Diagnosis	24
1.9.5	DNA Technology in Medical Genetics	25
1.9.6	The "Industrialization" of Discoveries and the Team Efforts	26
1.9.7	Unsolved Problems	26
	References	27
2	**Human Genome Sequence and Variation**	31
	Stylianos E. Antonarakis	
2.1	The Human Genome	31
2.1.1	Functional Elements	33

2.1.1.1	Protein-Coding Genes	33
2.1.1.2	Noncoding, RNA-Only Genes	36
2.1.1.3	Regions of Transcription Regulation	38
2.1.1.4	Conserved Elements Not Included in the Above Categories	39
2.1.2	Repetitive Elements	40
2.1.2.1	Segmental Duplications	41
2.1.2.2	Special Genomic Structures Containing Selected Repeats	41
2.1.3	Mitochondrial Genome	43
2.2	Genomic Variability	43
2.2.1	Single Nucleotide Polymorphisms	43
2.2.2	Short Sequence Repeats	46
2.2.3	Insertion/Deletion Polymorphisms (Indels)	46
2.2.4	Copy Number Variants	47
2.2.5	Inversions	47
2.2.6	Mixed Polymorphisms	47
2.2.7	Genome Variation as a Laboratory Tool to Understand the Genome	48
	References	48
3	**Chromosomes**	**55**
	Michael R. Speicher	
3.1	History and Development of Human Cytogenetics	56
3.1.1	First Observations on Human Mitotic Chromosomes	56
3.1.2	An Old Error is Corrected and a New Era Begins	57
3.1.3	Birth of Human Cytogenetics 1956–1963	58
3.1.4	Introduction of Banding Technologies from the Late 1960s to the Present	58
3.1.5	The Birth of Molecular Cytogenetics in the Late 1960s	59
3.1.6	Molecular Cytogenetics or Fluorescence In Situ Hybridization 1980 to Date	59
3.1.7	Array Technologies: New Dimensions in Resolution from 1997 to Date	59
3.2	Organization of Genetic Material in Human Chromosomes	60
3.2.1	Heterochromatin and Euchromatin	60
3.2.2	From DNA Thread to Chromosome Structure	60
3.2.2.1	DNA Condensation	60
3.2.2.2	Histone Modifications Organize Chromosomal Subdomains	62
3.2.2.3	Chromatin Diseases	63
3.2.3	Centromeres and Kinetochores	63
3.2.3.1	Function of Centromeres	63
3.2.3.2	Structure of Centromeres	63
3.2.3.3	Centromeres and Human Diseases	65
3.2.4	Chromosome Bands	65
3.2.4.1	G- and R-bands	65
3.2.4.2	In Silico-Generated Bands	68
3.2.5	Telomeres	68
3.2.5.1	Structure of Telomeres (Loop and Proteins)	68
3.2.5.2	The Telomere Replication Problem	69

3.2.5.3	Telomerase	70
3.2.5.4	Telomeres and Human Diseases	71
3.3	Cell Cycle and Mitosis	71
3.3.1	Cell Cycle: Interphase G_1-G_2 and G_0	71
3.3.2	Cell Cycle: Mitosis	72
3.3.2.1	Mitosis and Cytokinesis	72
3.3.2.2	Prophase	73
3.3.2.3	Prometaphase	73
3.3.2.4	Metaphase	74
3.3.2.5	Anaphase	74
3.3.2.6	Telophase	75
3.3.2.7	Cytokinesis	75
3.3.3	Cell Cycle Checkpoints	75
3.3.3.1	Presently Known Checkpoints	75
3.3.3.2	The Spindle-Assembly Checkpoint in Detail	75
3.3.3.3	"Cohesinopathies," Cornelia de Lange and Robert Syndrome	77
3.3.3.4	Possible Additional Checkpoints	77
3.3.3.5	Checkpoint Failures and Human Diseases	77
3.3.4	Cell Cycle Coordinators	77
3.4	Chromosome Analysis Methods	78
3.4.1	Banding Techniques	78
3.4.1.1	Preparation of Mitotic Metaphase Chromosomes	78
3.4.1.2	G-Bands	79
3.4.1.3	R-Bands	79
3.4.1.4	C-Bands	79
3.4.1.5	Ag-NOR Bands	80
3.4.1.6	DA/DAPI Staining	80
3.4.2	Karyotype Description	80
3.4.2.1	Normal Human Karyotype in Mitotic Metaphase Chromosomes	80
3.4.2.2	Description of Normal Human Karyotypes and Karyotypes with Numerical and Structural Changes	81
3.4.3	Fragile Sites	82
3.4.3.1	Fluorescence In Situ Hybridization	83
3.4.3.2	FISH to Metaphase Spreads	84
3.4.3.3	Interphase FISH (Interphase Cytogenetics)	85
3.4.3.4	Fiber FISH	85
3.4.3.5	Detection of Copy Number Changes in the Genome	86
3.4.3.6	Other Array-Based Methods	92
3.5	Meiosis	93
3.5.1	Biological Function of Meiosis	93
3.5.2	Meiotic Divisions	93
3.5.3	Meiotic Recombination Hotspots	95
3.5.4	Differences Between Male and Female Meiosis	96
3.5.4.1	Meiosis in the Human Male	96
3.5.4.2	Meiosis in the Human Female	96
3.5.4.3	A Dynamic Oocyte Pool?	96
3.5.4.4	Maternal Age and Aneuploidy	97

3.5.5	Nonallelic Homologous Recombination During Meiosis Can Cause Microdeletion/Microduplication Syndromes.	98
3.5.6	Molecular Mechanisms Involved in Meiosis	99
3.6	Human Chromosome Pathology in Postnatal Diagnostics	99
3.6.1	Syndromes Attributable to Numeric Anomalies of Autosomes	99
3.6.1.1	Mechanisms Creating Anomalies in Chromosome Numbers (Numerical Chromosome Mutations)	99
3.6.1.2	Down Syndrome	101
3.6.1.3	Other Autosomal Trisomies	104
3.6.1.4	Triploidy	104
3.6.1.5	Mosaics	105
3.6.2	Syndromes Attributable to Structural Anomalies of Autosomes	106
3.6.2.1	Karyotypes and Clinical Syndromes	106
3.6.2.2	Small Deletions, Structural Rearrangements, and Monogenic Disorders: Genomic Disorders and Contiguous Gene Syndromes	113
3.6.3	Sex Chromosomes	115
3.6.3.1	First Observations	115
3.6.3.2	Dosage Compensation for Mammalian X Chromosomes	117
3.6.3.3	X Chromosomal Aneuploidies in Humans	120
3.6.3.4	Chromosomal Aneuploidies in Humans	125
3.6.4	Chromosome Aberrations and Spontaneous Miscarriage	125
3.6.4.1	Incidence of Prenatal Zygote Loss in Humans	125
3.6.4.2	Aneuploidy in Oocytes and Embryos	125
3.7	Chromosome Instability/Breakage Syndromes	126
3.7.1	Fanconi Anemia	126
3.7.2	Instabilities Caused by Mutations of Proteins of the RecQ Family of Helicases: Bloom syndrome, Werner syndrome, and Rothmund-Thomson syndrome	127
3.7.2.1	Bloom Syndrome	127
3.7.2.2	Rothmund–Thomson Syndrome	128
3.7.3	The Ataxia-Telangiectasia Group	128
3.7.3.1	Ataxia-Telangiectasia	128
3.7.3.2	Nijmegen Breakage Syndrome	129
3.7.4	Immunodeficiency, Centromeric Region Instability, and Facial Anomalies Syndrome	129
3.7.5	Roberts Syndrome/SC Phocomelia	129
3.7.6	Mosaic Variegated Aneuploidy	129
	References	130
4	**From Genes to Genomics to Proteomics**	**139**
	Michael R. Speicher	
4.1	Single-Gene Approaches	140
4.1.1	What Is a Gene?	140
4.1.2	Mutations	141
4.1.3	Silent Mutations and Phenotypic Consequences	142
4.1.4	Mutation Detection by Sanger Sequencing	143

4.1.5		Next-Generation Sequencing	144
4.1.6		The Importance of Monogenic Mendelian Disorders	144
4.2		Gene Regulation	145
4.2.1		Genetic Regulation	145
4.2.2		Epigenetic Regulation	146
4.2.3		Regulatory Transcripts of Small RNAs	147
4.3		"-omics" Sciences	147
4.4		Genomics	148
4.4.1		Genomes of Organisms	148
4.4.2		Array and Other Technologies	148
4.4.3		Next-Generation Sequencing	148
4.4.3.1		Roche's (454) GS FLX Genome Analyzer	149
4.4.3.2		Illumina's Solexa IG Sequencer	149
4.4.3.3		Applied Biosystem's SOLiD System	149
4.4.4		Third-Generation Sequencing	152
4.4.5		Personalized Genomics	152
4.4.6		Gene Function	153
4.5		Transcriptomics	154
4.5.1		Capturing the Cellular Transcriptome, Expression Arrays, and SAGE	154
4.5.2		Regulatory Networks	154
4.5.3		Outlier Profile Analysis	155
4.5.4		High-Throughput Long- and Short-Read Transcriptome Sequencing	156
4.5.5		Disease Classification	156
4.5.6		Tools for Prognosis Estimation	156
4.6		Proteomics	156
4.6.1		From Low-Throughput to High-Throughput Techniques	157
4.6.2		Mass Spectrometer-Based Methods	157
4.6.3		Antibody Array-Based Methods	158
4.6.4		Proteomic Strategies	158
4.6.5		Proteomics for Screening and Diagnosis of Disease (Diagnostic and Prognostic Biomarkers)	158
4.7		Conclusions	159
		References	159
5		**Formal Genetics of Humans: Modes of Inheritance**	**165**
		Arno G. Motulsky	
5.1		Mendel's Modes of Inheritance and Their Application to Humans	166
5.1.1		Codominant Mode of Inheritance	166
5.1.2		Autosomal Dominant Mode of Inheritance	167
5.1.2.1		Late Manifestation, Incomplete Penetrance, and Variable Expressivity	169
5.1.2.2		Effect of Homozygosity on Manifestation of Abnormal Dominant Genes	171
5.1.3		Autosomal-Recessive Mode of Inheritance	172
5.1.3.1		Pseudodominance in Autosomal Recessive Inheritance	173

5.1.3.2	Compound Heterozygotes	174
5.1.4	X-Linked Modes of Inheritance	175
5.1.4.1	X-Linked Recessive Mode of Inheritance	175
5.1.4.2	X-Linked Dominant Mode of Inheritance	177
5.1.4.3	X-Linked Dominant Inheritance with Lethality f the Male Hemizygotes [90]	177
5.1.4.4	Genes on the Y Chromosome	179
5.1.5	"Lethal" Factors [32]	179
5.1.5.1	Animal Models	179
5.1.5.2	Lethals in Humans	180
5.1.6	Modifying Genes	180
5.1.6.1	Modifying Genes in the AB0 Blood Group System	180
5.1.6.2	Modifying Genes in Cystic Fibrosis	181
5.1.6.3	Sex-Limiting Modifying Genes	181
5.1.6.4	Modification by the Other Allele	181
5.1.6.5	Modification by Variation in Related Genes	182
5.1.6.6	Modification by a DNA Polymorphism Within the Same Gene	182
5.1.7	Anticipation	182
5.1.8	Total Number of Conditions with Simple Modes of Inheritance Known so far in Humans	184
5.1.8.1	Difference in the Relative Frequencies of Dominant and Recessive Conditions in Humans and Animals?	185
5.1.9	Uniparental Disomy and Genomic Imprinting	185
5.1.9.1	Phenotypic Consequences of UPD	186
5.1.9.2	Human Disorders Involving UPD	187
5.1.10	Diseases Due to Mutations in the Mitochondrial Genome	188
5.1.10.1	Leber Optical Atrophy	188
5.1.10.2	Deletions	190
5.1.10.3	Diseases of Advanced Age	190
5.1.10.4	Interaction Between Nuclear and Mitochondrial Genomes	190
5.1.11	Unusual, "Near Mendelian" Modes of Inheritance	190
5.1.11.1	Digenic Inheritance	190
5.1.11.2	Triallelic Inheritance	193
5.1.12	Multifactorial Inheritance	194
5.2	Hardy–Weinberg Law and Its Applications	194
5.2.1	Formal Basis	194
5.2.1.1	Derivations from the Hardy–Weinberg Law	194
5.2.2	Hardy–Weinberg Expectations Establish the Genetic Basis of AB0 Blood Group Alleies	195
5.2.2.1	Multiple Allelisms	195
5.2.2.2	Genetics of the AB0 Blood Groups	195
5.2.2.3	Meaning of a Hardy–Weinberg Equilibrium	197
5.2.3	Gene Frequencies	198
5.2.3.1	One Gene Pair: Only Two Phenotypes Known	198
5.3	Statistical Methods in Formal Genetics: Analysis of Segregation Ratios	198
5.3.1	Segregation Ratios as Probabilities	198
5.3.2	Simple Probability Problems in Human Genetics	200

5.3.2.1	Independent Sampling and Prediction in Genetic Counseling	200
5.3.2.2	Differentiation Between Different Modes of Inheritance	200
5.3.3	Testing for Segregation Ratios Without Ascertainment Bias: Codominant Inheritance	201
5.3.3.1	Dominance	201
5.3.4	Testing for Segregation Ratios: Rare Traits	202
5.3.4.1	Principal Biases	202
5.3.4.2	Methods for Correcting Bias	203
5.3.5	Discrimination of Genetic Entities Genetic Heterogeneity	204
5.3.5.1	Genetic Analysis of Muscular Dystrophy as an Example	204
5.3.5.2	Multivariate Statistics	205
5.3.6	Conditions without Simple Modes of Inheritance	205
5.3.6.1	Empirical Risk Figures	205
5.3.6.2	Selecting and Examining Probands and Their Families	205
5.3.6.3	Statistical Evaluation, Age Correction	206
5.3.6.4	Example	206
5.3.6.5	Selection of Probands for Genome-Wide Association Studies	207
5.3.6.6	Theoretical Risk Figures Derived from Heritability Estimates?	207
5.4	Conclusions	207
	References	207

6	**Linkage Analysis for Monogenic Traits**	**211**
	Arno G. Motulsky and Michael Dean	
6.1	Linkage: Localization of Genes on Chromosomes	211
6.1.1	Classic Approaches in Experimental Genetics: Breeding Experiments and Giant Chromosomes	212
6.1.1.1	Linkage and Association	213
6.1.2	Linkage Analysis in Humans	213
6.1.2.1	Direct Observation of Pedigrees	213
6.1.2.2	Statistical Analysis	215
6.1.2.3	The Use of LOD Scores	216
6.1.2.4	Recombination Probabilities and Map Distances	217
6.1.2.5	The Sib Pair Method	217
6.1.2.6	Results for Autosomal Linkage, Sex Difference, and Parental Age	219
6.1.2.7	Information from Chromosome Morphology	220
6.1.3	Linkage Analysis in Humans: Cell Hybridization and DNA Techniques	220
6.1.3.1	First Observations on Cell Fusion	220
6.1.3.2	First Observation of Chromosome Loss in Human–Mouse Cell Hybrids and First Assignment of a Gene Locus	221
6.1.3.3	Other Sources of Information for Gene Localization	222
6.1.3.4	DNA Polymorphisms and Gene Assignment	222
6.1.3.5	Gene Symbols to Be Used	223
6.1.3.6	Linkage of X-Linked Gene Loci	223
6.1.3.7	Genetic and Physical Map of the Homologous Segment of X and Y Chromosomes	223

6.1.3.8	The Y Chromosome	224
6.1.3.9	DNA Variants in Linkage	224
6.1.3.10	Practical Application of Results from Linkage Studies	224
6.1.4	Biology and Statistics of Positional Cloning	224
6.2	Gene Loci Located Close to Each Other and Having Related Functions	225
6.2.1	Some Phenomena Observed in Experimental Genetics	225
6.2.1.1	Closely Linked Loci May Show a Cis-Trans Effect	225
6.2.1.2	Explanation in Terms of Molecular Biology	225
6.2.1.3	A Number of Genes May Be Closely Linked	225
6.2.2	Some Observations in the Human Linkage Map	225
6.2.2.1	Types of Gene Clusters That Have Been Observed	225
6.2.2.2	Clusters Not Observed So Far	226
6.2.3	Why Do Gene Clusters Exist?	226
6.2.3.1	They Are Traces of Evolutionary History	226
6.2.3.2	Duplication and Clustering May Be Used for Improvement of Function	226
6.2.4	Blood Groups: Rh Complex (111700), Linkage Disequilibrium	226
6.2.4.1	History	227
6.2.4.2	Fisher's Hypothesis of Two Closely Linked Loci	227
6.2.4.3	Confirmation and Tentative Interpretation of the Sequential Order	228
6.2.4.4	Molecular Basis	228
6.2.4.5	Linkage Disequilibrium	229
6.2.5	Major Histocompatibility Complex [105, 111]	229
6.2.5.1	History	229
6.2.5.2	Main Components of the MHC on Chromosome 6	231
6.2.5.3	Complement Components	232
6.2.5.4	Significance of HLA in Transplantation	232
6.2.5.5	Linkage Disequilibrium	232
6.2.5.6	The Normal Function of the System	234
6.2.6	Unequal Crossing Over	235
6.2.6.1	Discovery of Unequal Crossing Over	235
6.2.6.2	Unequal Crossing Over in Human Genetics	235
6.2.6.3	First Event	236
6.2.6.4	Consequences of Unequal Crossing Over	237
6.2.6.5	Intrachromosomal Unequal Crossing Over	237
6.3	Conclusions	238
	References	238
7	**Oligogenic Disease**	**243**
	Jon F. Robinson and Nicholas Katsanis	
7.1	The Limitations of Mendelian Concepts	244
7.2	PKU and Hyperphenylalaninemia: Genetic Heterogeneity	245
7.3	Cystic Fibrosis: Genetic Modifiers	246
7.4	Lessons Learned from Established Oligogenic Disorders	246
7.4.1	Bardet–Biedl Syndrome	247

7.4.2	Determining Oligogenicity	247
7.4.3	Cortisone Reductase Deficiency	248
7.4.4	Hemochromatosis	248
7.4.5	Hirschsprung Disease	249
7.5	Establishing Oligogenicity: Concepts and Methods	252
7.5.1	Heritability	252
7.5.2	Mouse Models of Oligogenic Inheritance: Familial Adenomatous Polyposis	252
7.5.3	Multigenic Models	253
7.5.4	Linkage Analysis	253
7.6	Molecular Mechanisms of Oligogenic Disorders	254
7.7	Modular or Systems Biology	257
7.8	Conclusions	258
	References	259

8 Formal Genetics of Humans: Multifactorial Inheritance and Common Diseases ... 263
Andrew G. Clark

8.1	Genetic Analysis of Complex Traits	264
8.1.1	Variation in Phenotypic Traits	264
8.1.2	Familial Resemblance and Heritability	264
8.1.3	The Special Case of Twins	267
8.1.4	Embedding a Single Measured Gene Influencing a Continuous Trait	269
8.1.5	A Model for Variance Partitioning	269
8.1.6	Relating the Model to Data	270
8.1.7	Mendelian Diseases Are Not Simple	271
8.2	Genetic Polymorphism and Disease	271
8.2.1	Finding Genes Underlying a Complex Trait	271
8.2.2	Limitations of Pedigree Analysis	271
8.2.3	A Prevailing Model: Common Disease Common Variants	272
8.2.4	Affected Sib-pairs	273
8.2.5	Transmission Disequilibrium Test	273
8.2.6	Full-Genome Association Testing	274
8.3	LD Mapping and Genome-Wide Association Studies	275
8.3.1	Theory and How It Works: HapMap and Genome-Wide LD	275
8.3.2	Technology: The Fantastic Drop in Genotyping Costs	276
8.3.3	Case-Control Studies	277
8.3.4	Statistical Inference with Genome-Wide Studies	277
8.3.5	Replication and Validation	278
8.3.6	Age-Related Macular Degeneration and Complement Factor H	279
8.4	Admixture Mapping and Population Stratification	279
8.4.1	How to Quantify Admixture	279
8.4.2	Using Admixture for Mapping	280
8.4.3	The Perils of Population Stratification	280
8.4.4	How to Correct for Hidden Population Stratification	281

8.5	Complications	282
8.5.1	Genotype by Environment Interaction	282
8.5.2	Epistasis	284
8.6	Missing Heritability: Why is so Little Variance Explained by GWAS Results?	284
8.7	Concluding Remarks	285
	References	285

8.1	**Lessons from the Genome-Wide Association Studies for Complex Multifactorial Disorders and Traits**	**287**
	Jacques S. Beckmann and Stylianos E. Antonarakis	
8.1.1	Lessons from Current GWAS	292
8.1.2	Genomic Topography of Trait-Associated Variants	292
8.1.3	How Important Is the Identified Genetic Contribution to the Variance of the Traits Studied?	292
8.1.4	Predictive Power and Clinical Utility of the Trait-Associated Variants	293
8.1.5	Pathophysiological Dissection of Complex Traits	294
8.1.6	Concluding Remarks	294
	References	295

9	**Epigenetics**	**299**
	Bernhard Horsthemke	
9.1	History, Definition, and Scope	300
9.2	Chromatin-Marking Systems	300
9.2.1	DNA Methylation	301
9.2.1.1	DNA Methyltransferases	302
9.2.1.2	Methyl-Cytosine-Binding Proteins	303
9.2.2	Histone Modification	303
9.2.2.1	Histone Acetylation	304
9.2.2.2	Histone Methylation	304
9.2.2.3	Histone Phosphorylation and Other Histone Modifications	305
9.2.3	Chromatin Remodeling	305
9.2.4	Synergistic Relations Between the Different Chromatin-Marking Systems	305
9.3	Specific Epigenetic Phenomena	306
9.3.1	Genomic Imprinting	307
9.3.2	X Inactivation	310
9.3.3	Allelic Exclusion in the Olfactory System	312
9.4	Epigenetic Variation and Disease	312
9.4.1	Obligatory Epigenetic Variation	313
9.4.2	Pure Epigenetic Variation	313
9.4.3	Facilitated Epigenetic Variation	315
9.5	Transgenerational Epigenetic Inheritance and Evolution	315
	References	316

10	**Human Gene Mutation: Mechanisms and Consequences**	**319**
	Stylianos E. Antonarakis and David N. Cooper	
10.1	Introduction...	319
10.2	Neutral Variation/DNA Polymorphisms.....................	320
10.3	Disease-Causing Mutations	321
10.3.1	The Nature of Mutation	321
10.3.1.1	Nucleotide Substitutions..	322
10.3.1.2	Micro-Deletions and Micro-Insertions	323
10.3.1.3	Expansion/Copy Number Variation of Trinucleotide (and Other) Repeat Sequences	325
10.3.1.4	Gross Deletions...	327
10.3.1.5	Large Insertions (Via Retrotransposition)..................	329
10.3.1.6	Large Insertion of Repetitive and Other Elements	331
10.3.1.7	Inversions..	331
10.3.1.8	Duplications..	332
10.3.1.9	Gene Conversion...	332
10.3.1.10	Insertion-Deletions (Indels)	333
10.3.1.11	Other Complex Defects	333
10.3.1.12	Molecular Misreading..	333
10.3.1.13	Germline Epimutations..	334
10.3.1.14	Frequency of Disease-Producing Mutations..............	334
10.3.1.15	Chromosomal Distribution of Human Disease Genes ..	335
10.3.1.16	Mutation Nomenclature	335
10.3.1.17	Mutations in Gene Evolution	335
10.3.2	Consequences of Mutations	336
10.3.2.1	Mutations Affecting the Amino Acid Sequence of the Predicted Protein, but not Gene Expression	336
10.3.2.2	Mutations Affecting Gene Expression........................	337
10.3.2.3	Transcription (Promoter) Mutations...........................	337
10.3.2.4	mRNA Splicing Mutants	338
10.3.2.5	RNA Cleavage-Polyadenylation Mutants....................	342
10.3.2.6	Mutations in miRNA-Binding Sites	342
10.3.2.7	Cap Site Mutations	343
10.3.2.8	Mutations in 5' Untranslated Regions	343
10.3.2.9	Mutations in 3' Regulatory Regions	343
10.3.2.10	Translational Initiation Mutations	343
10.3.2.11	Termination Codon Mutations...................................	344
10.3.2.12	Frameshift Mutations...	344
10.3.2.13	Nonsense Mutations..	344
10.3.2.14	Unstable Protein Mutants	345
10.3.2.15	Mutations in Remote Promoter Elements/Locus Control Regions	345
10.3.2.16	Cellular Consequences of Trinucleotide Repeat Expansions	346
10.3.2.17	Mutations Producing Inappropriate Gene Expression	346
10.3.2.18	Position Effect in Human Disorders	346
10.3.2.19	Position Effect by an Antisense RNA.........................	347
10.3.2.20	Abnormal Proteins Due to Fusion of Two Different Genes	347

10.3.2.21	Mutations in Genes Involved in Mismatch Repair Associated with Genomic Instability in the Soma	348
10.3.2.22	Mosaicism	348
10.3.2.23	Sex Differences in Mutation Rates	348
10.3.2.24	Concepts of Dominance and Recessiveness in Relation to the Underlying Mutations	349
10.4	General Principles of Genotype-Phenotype Correlations	349
10.5	Why Study Mutation?	351
	References	351

11 Human Hemoglobin ... 365
George P. Patrinos and Stylianos E. Antonarakis

11.1	Introduction	366
11.2	Historical Perspectives	366
11.3	Genetics of Hemoglobins	368
11.3.1	Hemoglobin Molecules	368
11.3.2	Hemoglobin Genes	369
11.3.3	Regulatory Elements	370
11.3.4	Molecular Control of Globin Gene Switching	374
11.3.5	DNA Polymorphisms at the Globin Genes	377
11.4	Molecular Evolution of the Human Globin Genes	377
11.5	Molecular Etiology of Hemoglobinopathies	380
11.5.1	Thalassemias and Related Conditions	380
11.5.2	β-Thalassemia	380
11.5.3	Dominantly Inherited β-Thalassemia	382
11.5.4	δβ-Thalassemias and Hereditary Persistence of Fetal Hemoglobin	383
11.5.5	α-Thalassemia	383
11.5.6	Other Mmutation Types Leading to Hemoglobinopathies	384
11.5.7	Hemoglobin Variants	385
11.6	X-Linked Inherited and Acquired α-Thalassemia	387
11.6.1	Regulatory SNPs and Antisense RNA Transcription Resulting in α-Thalassemia	388
11.6.2	β-Thalassemia Attributable to Transcription Factor Mutations	388
11.7	Population Genetics of Hemoglobin Genes	388
11.8	Diagnosis of Hemoglobinopathies	389
11.8.1	Carrier Screening	389
11.8.2	Hematological and Biochemical Methods	391
11.8.3	Molecular Diagnostics of Hemoglobinopathies	391
11.8.4	Prenatal and Preimplantation Genetic Diagnosis of Hemoglobinopathies	392
11.8.5	Genetic Counseling	393
11.9	HbVar Database for Hemoglobin Variants and Thalassemia Mutations	393
11.10	Therapeutic Approaches for the Thalassemias	394
11.10.1	Hematopoietic Stem Cell Transplantation	394
11.10.2	Pharmacological Reactivation of Fetal Hemoglobin	394
11.10.3	Pharmacogenomics and Therapeutics of Hemoglobinopathies	395

11.10.4	Gene Therapy	395
	References	396

12 Human Genetics of Infectious Diseases — 403
Alexandre Alcaïs, Laurent Abel, and Jean-Laurent Casanova

12.1	Introduction	404
12.2	Mendelian Predisposition to Multiple Infections	406
12.3	Mendelian Predisposition to Single Infections	407
12.4	Mendelian Resistance	408
12.5	Major Genes	409
12.6	Multigenic Predisposition	410
12.7	Concluding Remarks	411
	References	412

13 Gene Action: Developmental Genetics — 417
Stefan Mundlos

13.1	Genetics of Embryonal Development	417
13.1.1	Basic Mechanisms of Development	418
13.1.2	Mechanisms of Morphogenesis	419
13.2	The Stages of Development	422
13.3	Formation of the Central Nervous System	425
13.4	The Somites	431
13.5	The Brachial Arches	433
13.6	Development of the Limbs	435
13.7	Development of the Circulatory System	438
13.8	Development of the Kidney	440
13.9	Skeletal Development	442
13.10	Abnormal Development: Definitions and Mechanisms	445
13.11	Malformations	446
13.12	Disruptions	447
13.13	Deformations	448
13.14	Dysplasias	448
13.15	Terminology of Congenital Defects	449
	References	449

14 Cancer Genetics — 451
Ian Tomlinson

14.1	Inherited Risk	452
14.1.1	Introduction	452
14.1.2	Identification of Mendelian Cancer Susceptibility Genes	452
14.1.3	Unidentified Mendelian Cancer Susceptibility Genes	457
14.1.4	Germline Epimutations	458
14.1.5	A Heterozygote Phenotype in Mendelian Recessive Tumor Syndromes?	458
14.1.6	Some Cancer Genes are Involved in Both Dominant and Recessive Syndromes	458

14.1.7	Phenotypic Variation, Penetrance, and Rare Cancers in Mendellan Syndromes	459
14.1.8	Predisposition Alleles Specific to Ethnic Groups	460
14.1.9	Germllne Mutations can Determine Somatic Genetic Pathways	460
14.1.10	Non-Mendellan Genetic Cancer Susceptibility	460
14.1.11	Concluding Remarks	462
14.2	Somatic Cancer Genetics	462
14.2.1	Mutations Cause Cancer	462
14.2.2	Chromosomal-Scale Mutations, Copy Number Changes, and Loss of Heterozygosity	463
14.2.3	Activating Mutations and Oncogenes; Inactivating Mutations and Tumor Suppressor Genes	464
14.2.4	Epimutations	465
14.2.5	How Many Mutations are Needed to Make a Cancer?	465
14.2.6	The Role of Genomic Instability and Molecular Phenotypes	466
14.2.7	The Conundrum of Tissue Specificity	467
14.2.8	Clinicopathological Associations, Response and Prognosis	467
14.2.9	Posttherapy Changes	468
14.2.10	Concluding Remarks	468
	References	468

15 The Role of the Epigenome in Human Cancers ... 471
Romulo Martin Brena and Joseph F. Costello

15.1	Introduction	472
15.2	DNA Methylation in Development and Cellular Homeostasis	472
15.3	DNA Methylation is Disrupted in Human Primary Tumors	473
15.4	Genome-Wide DNA Methylation Analyses	474
15.5	Gene Silencing Vs. Gene Mutation	476
15.6	Discovery of Cancer Genes via Methylome Analysis	477
15.7	Computational Analysis of the Methylome	478
15.8	Histone Modifications and Chromatin Remodeling in Cancer	478
15.9	Eepigenome–Genome Interations in Human Cancer and Mouse Models: Gene Silencing Vs. Gene Mutation	480
15.10	Epigenetics and Response to Cancer Therapy	480
	References	481

16 Population Genetic Principles and Human Populations ... 487
Emmanouil T. Dermitzakis

16.1	Introduction	487
16.2	Processes Shaping Natural Variation	488
16.2.1	Mutation and Polymorphism	488
16.2.2	Historical Perspective of DNA Polymorphisms and Their Use as Genetic Markers	489
16.2.3	Demography	490

16.2.3.1	Effective Population Size	490
16.2.3.2	Genetic Drift and Isolated Populations	491
16.2.3.3	Migration	491
16.2.3.4	Inbreeding/Nonrandom Mating	492
16.2.4	Recombination	492
16.2.5	Natural Selection	492
16.3	Patterns of Genetic Variation	494
16.3.1	Hardy–Weinberg Equilibrium	494
16.3.2	Coalescent Theory	495
16.3.3	Population Differentiation	495
16.3.4	Patterns of Single-Nucleotide Variation in the Human Genome	496
16.3.5	Patterns of Structural Variation in the Human Genome	498
16.3.6	Haplotype Diversity—Linkage Disequilibrium	499
16.3.7	Detecting Natural Selection	500
16.3.8	Historical Perspective of Population Genetic Studies	501
16.4	Current Themes	501
16.4.1	International Efforts to Detect and Describe Sequence Variation	501
16.4.2	Prospects for Mapping Functional and Disease Variation	502
16.5	Summary	504
	References	504

17 **Consanguinity, Genetic Drift, and Genetic Diseases in Populations with Reduced Numbers of Founders** ... 507
Alan H. Bittles

17.1	Genetic Variation in Human Populations	508
17.1.1	Random Mating and Assortative Mating	509
17.1.2	Genetic Drift and Founder Effects	509
17.2	Consanguineous Matings	509
17.2.1	Coefficient of Relationship and Coefficient of Inbreeding	510
17.2.2	Global Prevalence of Consanguinity	510
17.2.3	Specific Types of Consanguineous Marriage	514
17.2.4	The Influence of Religion on Consanguineous Marriage	514
17.2.5	Civil Legislation on Consanguineous Marriage	515
17.2.6	Social and Economic Factors Associated with Consanguinity	515
17.3	Inbreeding and Fertility	516
17.3.1	Genetically Determined Factors Influencing Human Mate Choice	516
17.3.2	Inbreeding and Fetal Loss Rates	516
17.3.3	Comparative Fertility in Consanguineous and Nonconsanguineous Couples	517
17.4	Inbreeding and Inherited Disease	518
17.4.1	Consanguinity and Deaths in Infancy and Childhood	518
17.4.2	Consanguinity and Childhood Morbidity	518
17.4.3	Consanguinity and Adult Mortality and Morbidity	519
17.5	Incest	519
17.5.1	Mortality and Morbidity Estimates for Incestuous Matings	520

17.6	Genetic Load Theory and Its Application in Consanguinity Studies..	520
17.7	Genomic Approaches to Measuring Inbreeding at Individual and Community Levels	521
17.8	The Influence of Endogamy and Consanguinity in Human Populations..	521
17.8.1	The Finnish Disease Heritage	521
17.8.2	Inter- and Intra-population Differentiation in India......................	523
17.8.3	Consanguinity and the Distribution of Disease Alleles in Israeli Arab Communities..	523
17.9	Evaluating Risk in Consanguineous Relationships	524
17.10	Concluding Comments	525
	References..	525

18 Human Evolution 529
Michael Hofreiter

18.1	Historical Overview...	530
18.1.1	Before and Around Darwin	530
18.1.2	Sarich and Wilson..	530
18.1.3	From a Straight Line to a Bush of Hominid Species and Beyond.	531
18.2	The Fossil Record..	532
18.2.1	Palaeoanthropology	532
18.2.2	Neanderthals and Ancient DNA	534
18.3	The Genetics of Human Evolution	536
18.3.1	The Genomes of Humans and their Relatives.............................	536
18.3.2	Diversity Within the Human Genome	538
18.3.3	Positive and Negative Selection in the Human Genome	541
18.4	Recent Events in Human Evolution...	545
18.4.1	Out of Africa into New-Found Lands..	545
18.4.2	Domestication..	549
18.4.3	Modern Human Population Structure..	550
	Glossary	551
	References..	551

19 Comparative Genomics... 557
Ross C. Hardison

19.1	Goals, Impact, and Basic Approaches of Comparative Genomics	557
19.1.1	How Biological Sequences Change Over Time...........................	558
19.1.2	Purifying Selection	559
19.1.3	Models of Neutral DNA	560
19.1.4	Adaptive Evolution..	562
19.2	Alignments of Biological Sequences and Their Interpretation......	563
19.2.1	Global and Local Alignments..	563
19.2.2	Aligning Protein Sequences...	563
19.2.3	Aligning Large Genome Sequences	563
19.3	Assessment of Conserved Function from Alignments	565

19.3.1	Phylogenetic Depth of Alignments	561
19.3.2	Portion of the Human Genome Under Constraint	568
19.3.3	Identifying Specific Sequences Under Constraint	569
19.4	Evolution Within Protein-Coding Genes	570
19.4.1	Comparative Genomics in Gene Finding	571
19.4.2	Sets of Related Genes	572
19.4.3	Rates of Sequence Change in Different Parts of Genes	574
19.4.4	Evolution and Function in Protein-Coding Exons	574
19.4.5	Fast-Changing Genes That Code for Proteins	575
19.4.6	Recent Adaptive Selection in Humans	576
19.4.7	Human Disease-Related Genes	578
19.5	Evolution in Regions That Do Not Code for Proteins or mRNA	579
19.5.1	Ultraconserved Elements	579
19.5.2	Evolution Within Noncoding Genes	580
19.5.3	Evolution and Function in Gene Regulatory Sequences	581
19.5.4	Prediction and Tests of Gene Regulatory Sequences	582
19.6	Resources for Comparative Genomics	583
19.6.1	Genome Browsers and Data Marts	583
19.6.2	Genome Analysis Workspaces	583
19.7	Concluding Remarks	584
	References	585

20 Genetics and Genomics of Human Population Structure ... 590
Sohini Ramachandran, Hua Tang, Ryan N. Gutenkunst, and Carlos D. Bustamante

20.1	Introduction	590
20.1.1	Evolutionary Forces Shaping Human Genetic Variation	590
20.2	Quantifying Population Structure	592
20.2.1	F_{ST} and Genetic Distance	592
20.2.2	Model-Based Clustering Algorithms	593
20.2.3	Characterizing Locus-Specific Ancestry	594
20.3	Global Patterns of Human Population Structure	595
20.3.1	The Apportionment of Human Diversity	595
20.3.2	The History and Geography of Human Genes	596
20.3.3	Genetic Structure of Human Populations	598
20.3.4	A Haplotype Map of the Human Genome	600
20.4	The Genetic Structure of Human Populations Within Continents and Countries	602
20.4.1	Genetic Differentiation in Eurasia	603
20.4.2	Genetic Variation in Native American Populations	605
20.4.3	The Genetic Structure of African Populations	606
20.5	Recent Genetic Admixture	606
20.5.1	Populations of the Americas	606
20.5.2	Admixture Around the World	608
20.6	Quantiative Modeling of Human Genomic Diversity	609
20.6.1	Demographic History	609
20.6.2	Quantitative Models of Selection	610
	References	613

21	**Genetic Epidemiology**	617
	Sophia S. Wang, Terri H. Beaty, and Muin J. Khoury	
21.1	Introduction	618
21.2	Scope and Strategies of Genetic Epidemiology in the Twenty-First Century	619
21.3	Fundamentals of Gene Discovery: Family Studies	619
21.4	Fundamentals of Gene Discovery: Population Studies and GWAS	623
21.5	Beyond Gene Discovery: Epidemiologic Assessment of Genes in Population Health	626
21.6	Beyond Gene Discovery: Epidemiologic Assessment of Genetic Information in Medicine and Public Health	629
21.7	Policy, Ethical and Practice Considerations: the Emergence of Public Health Genomics	631
	References	632
22	**Pharmacogenetics**	635
	Nicole M. Walley, Paola Nicoletti, and David B. Goldstein	
22.1	Introduction	636
22.1.1	The Goal of Pharmacogenetics	636
22.1.2	Why Pharmacogenetic Studies Are Necessary	636
22.1.1.1	Efficacy	636
22.1.1.2	Safety	636
22.1.1.3	Dose	637
22.2	Important Pharmacogenetic Discoveries to Date	637
22.2.1	CYP2D6 Polymorphism and Pharmacogenetics	637
22.2.2	Classic/Paradigmatic Studies	638
22.2.2.1	Isoniazid, Azathioprine, Mercaptopurine and Other Poor Metabolizer Phenotypes	638
22.2.2.2	Efficacy of Asthma Treatment	638
22.2.2.3	Abacavir Hypersensitivity	638
22.2.2.4	Tranilast and Clinical Pharmacogenetic Investigations	638
22.2.2.5	Cancer Pharmacogenetics and Somatic/Acquired Polymorphisms	639
22.3	Pharmacogenetic Methodology: Clinical Practice	639
22.3.1	Phenotype	639
22.3.2	Sample and Patient Recruitment	640
22.4	Pharmacogenetic Methodology: Genomics	640
22.4.1	Candidate Gene Studies	640
22.4.2	Whole-Genome SNP Analyzes	642
22.4.3	Rare Variants and Whole-Genome Sequencing	643
22.5	Challenges in Pharmacogenetics	643
22.5.1	Polygenic Inheritance	643
22.5.2	Pharmacogenetics in the Clinic	643
22.5.3	Pharmacogenetics Across Ethnic Groups	644
22.5.4	Drug Classes of Urgent Interest	645
2.6	Conclusions	646
	References	646

23	**Behavioral Genetics**	649
	Introductory Note by Michael R. Speicher	

23.1	**The Genetics of Personality**	651
	Jonathan Flint and Saffron Willis-Owen	
23.1.1	Neuroticism	653
23.1.1.1	Genetic Association Studies	653
23.1.1.2	Gene by Environment Effects	654
23.1.1.3	Linkage Studies	656
23.1.1.4	Genome-Wide Association Studies	656
23.1.2	Extraversion	657
23.1.3	Implications for Future Research	657
	References	658

23.2	**Mental Retardation and Intellectual Disability**	663
	David L. Nelson	
23.2.1	Introduction	663
23.2.2	Definition of ID	664
23.2.3	History of ID	664
23.2.4	Frequency of ID	664
23.2.5	Gene Dosage and ID	665
23.2.6	Down Syndrome	666
23.2.7	Recurrent Deletions and Duplications and ID	667
23.2.8	Prader-Willi and Angelman Syndromes	668
23.2.9	Future Directions in Genome Rearrangements and ID	668
23.2.10	Single-Gene Mutations and ID	668
23.2.10.1	Autosomal Recessive	668
23.2.10.2	Autosomal Dominant	669
23.2.11	X-Linked ID	670
23.2.11.1	Fragile X Syndrome	671
23.2.11.2	Rett Syndrome	675
23.2.12	Future Directions	676
	References	676

23.3	**Genetic Factors in Alzheimer Disease and Dementia**	681
	Thomas D. Bird	
23.3.1	Clinical Manifestations of Alzheimer Disease	681
23.3.1.1	Establishing the Diagnosis	682
23.3.1.2	Prevalence	682
23.3.2	Causes	683
23.3.2.1	Environmental	683
23.3.2.2	Heritable Causes [6, 7]	683
23.3.2.3	Unknown	685
23.3.2.4	Molecular Genetic Testing	686
23.3.2.5	Early-Onset Familical AD	688
23.3.3	Genetic Counseling	688
23.3.3.1	Mode of Inheritance	688

23.3.3.2	Risk to Family Members: EOAD	688
23.3.3.3	Related Genetic Counseling Issues	688
23.3.4	Management	689
23.3.4.1	Treatment of Manifestations	689
23.3.4.2	Therapies Under Investigation	689
23.3.4.3	Other	689
23.3.5	Other Causes of Dementia	690
23.3.5.1	Frontotemporal Dementia	690
23.3.5.2	Familial Prion Disorders	691
23.3.5.3	CADASIL	691
	References	692

23.4 Genetics of Autism — 699
Brett S. Abrahams and Daniel H. Geschwind

23.4.1	Background	699
23.4.2	Cytogenetic Findings	701
23.4.3	Linkage	703
23.4.4	Syndromic ASDs	703
23.4.5	Re-sequencing	704
23.4.6	Copy Number Variation	705
23.4.7	Common Variations	707
23.4.8	Towards Convergence	707
	References	710

23.5 The Genetics of Alcoholism and Other Addictive Disorders — 715
David Goldman and Francesca Ducci

23.5.1	Introduction	716
23.5.2	Definition of Substance Use Disorders and Other Addictions	716
23.5.3	Epidemiology and Societal Impact of Addiction	717
23.5.4	Genetics: Family and Twin Studies	718
23.5.4.1	The Heritability of Addictions	719
23.5.4.2	Do Genetic Factors Moderating Risk Differ in Men and Women?	719
23.5.4.3	Developmental Dependence of Genes and Environment in Risk	720
23.5.4.4	What Is the Nature of the Inheritance of Addictions?	720
23.5.4.5	Agent-Specific and Nonspecific Genetic and Environmental Factors	723
23.5.4.6	Are Genetic and Environmental Risk Factors Independent of Each Other?	724
23.5.4.7	Gene by Environment Correlation	726
23.5.4.8	Gene by Environment Interaction	727
23.5.5	Progress Through Intermediate Phenotypes	728
23.5.6	Finding the Specific Genes Underlying Vulnerability to Addiction	728
23.5.6.1	Candidate Genes	729

23.5.6.2	Genetic Mapping	734
23.5.7	Treatment of Addictions	735
23.5.8	Conclusion	736
	References	736

23.6 Behavioral Aspects of Chromosomal Variants ... 743
Michael R. Speicher

23.6.1	Introduction: Human Chromosome Aberrations and Behavior, Possibilities, and Limitations	744
23.6.2	Numeric Autosomal Aberrations	744
23.6.2.1	Down Syndrome	744
23.6.3	Copy Number Variations Associated with Behavioral Disorders	745
23.6.3.1	Autistic Spectrum Disorder	745
23.6.3.2	Schizophrenia	745
23.6.3.3	Bipolar Disorders	746
23.6.4	Aberrations of the X Chromosome	746
23.6.4.1	Turner Syndrome	746
23.6.4.2	Klinefelter Syndrome	746
23.6.4.3	Triple-X Syndrome	748
23.6.5	Aberrations of the Y Chromosome	748
23.6.5.1	XYY Syndrome	748
23.6.5.2	Higher Prevalence Among "Criminals"	748
23.6.5.3	Intellectual Dysfunction or Simply Stature?	749
23.6.5.4	Behavioral Aspects of XYY Men	749
23.6.5.5	Association of Criminal Behavior and Lowered Intelligence in XYY Men	750
23.6.5.6	Social and Therapeutic Consequences	750
23.6.5.7	XXYY Syndrome	751
23.6.6	Other Chromosomal Variants	751
23.6.6.1	22q11.2 Deletion Syndrome	751
23.6.6.2	Smith–Magenis Syndrome	752
23.6.6.3	Prader-Willi and Angelman Syndrome	753
23.6.6.4	Williams–Beuren Syndrome	753
23.6.6.5	Cri-du-chat Syndrome	753
23.6.6.6	Wolf–Hirschhorn Syndrome	754
	References	754

23.7 Genetics of Schizophrenia and Bipolar Affective Disorder ... 759
Markus M. Nöthen, Sven Cichon, Christine Schmael, and Marcella Rietschel

23.7.1	Schizophrenia	759
23.7.1.1	Prevalence	760
23.7.1.2	Environmental Risk Factors	760
23.7.1.3	Formal Genetic Studies	761
23.7.1.3.1	Family Studies	761
23.7.1.3.2	Twin Studies	761

23.7.1.3.3	Adoption Studies	761
23.7.1.4	Gene–Environment Interaction	761
23.7.1.5	The Evolutionary Paradox of Schizophrenia	762
23.7.1.6	Molecular Genetic Studies	762
23.7.1.6.1	Linkage Studies	762
23.7.1.6.2	Candidate Gene Studies	763
23.7.1.6.3	Genome-Wide Association Studies	763
23.7.1.6.4	Submicroscopic Chromosomal Aberrations	764
23.7.1.7	Endophenotypes	764
23.7.2	Bipolar Disorder	765
23.7.2.1	Prevalence	765
23.7.2.2	Environmental Risk Factors	765
23.7.2.3	Formal Genetic Studies	766
23.7.2.3.1	Family Studies	766
23.7.2.3.2	Twin Studies	766
23.7.2.3.3	Adoption Studies	766
23.7.2.4	Molecular Genetic Studies	766
23.7.2.4.1	Linkage Studies	767
23.7.2.4.2	Candidate Gene Studies	767
23.7.2.4.3	Genome-Wide Association Studies	767
23.7.2.5	Endophenotypes	768
23.7.3	Schizophrenia and Bipolar Disorder: Approaches Beyond a Diagnostic Dichotomy	768
23.7.4	Outlook	768
	References	770

24	**Model Organisms for Human Disorders**	**777**
	Introductory Note by Michael R. Speicher	
	References	777

24.1	**Mouse as a Model for Human Disease**	**779**
	Antonio Baldini	
24.1.1	Introduction	779
24.1.2	Gene and Genome Manipulation Strategies	780
24.1.2.1	Knockout	780
24.1.2.2	Knockin and the Use of Site-Specific Recombinases	780
24.1.2.3	Conditional Mutations	781
24.1.2.4	Multigene Deletions and Duplications	782
24.1.2.5	Transgenics	782
24.1.3	Generating Genetically Accurate Models	783
24.1.3.1	Genetic Diseases Caused by Loss of Function Mutations	783
24.1.3.2	Gene Dosage Mutations	783
24.1.3.3	Gain of Function Mutations	784
24.1.3.4	Segmental Aneuploidies	784
24.1.4	Future Perspectives	784
	References	784

24.2	***Caenorhabditis elegans*, A Simple Worm: Bridging the Gap Between Traditional and Systems-Level Biology**.................. Morgan Tucker and Min Han	787
24.2.1	A Primer..	787
24.2.1.1	A Short History..	787
24.2.1.2	The Worm, Its Life Cycle, and Its Cultivation.......................	788
24.2.2	The How and Why of Screening..	789
24.2.2.1	Genetic Screens: A Traditional Single-Gene Approach.........	789
24.2.2.2	RNAi Screens: A High-Throughput Approach......................	790
24.2.2.3	Compound Screens: Identifying Therapeutics.......................	791
24.2.3	Beyond the Simple Screen...	791
24.2.3.1	A Systems Approach: Protein Interaction Maps....................	791
24.2.3.2	Exploring Complex Traits: Aging..	792
24.2.3.3	Modeling Human Disorders: Alzheimer's Disease................	793
24.2.4	Conclusions and Perspectives..	793
	References..	794
24.3	***Drosophila* as a Model for Human Disease**............................ Ruth Johnson and Ross Cagan	795
24.3.1	Why Flies?..	795
24.3.2	Cancer...	796
24.3.2.1	Cell Cycle...	797
24.3.2.2	Cell Death..	797
24.3.2.3	Hyperplasia and Neoplasia...	798
24.3.2.4	Models of Metastasis...	799
24.3.2.5	Models of Specific Cancers...	801
24.3.3	Neurodegenerative Diseases...	802
24.3.3.1	Parkinson's Disease..	803
24.3.3.2	Alzheimer's Disease...	803
24.3.3.3	Triplet-Repeat Diseases..	804
24.3.4	Heart Disease...	804
24.3.5	Diabetes and Metabolic Diseases...	805
24.3.5.1	Body Size..	806
24.3.5.2	Models of Diabetes..	806
24.3.6	Addiction..	807
24.3.6.1	The Genetics of Addiction: Sensitivity and Tolerance..........	808
24.3.7	Sleep Disorders..	808
24.3.8	Conclusions..	808
	References..	809
24.4	**Human Genetics and the Canine System**................................. Heidi G. Parker and Elaine A. Ostrander	813
	Abbreviations...	814
24.4.1	Introduction to the Canine System..	814
24.4.1.1	Origins of the Domestic Dog...	814
24.4.1.2	Dog Breeds...	814

24.4.1.3	Variation Between Breeds	814
24.4.1.4	Lack of Variation Within Breeds	815
24.4.1.5	Benefits of Mapping in a Breed-Based System	816
24.4.2	Navigating the Canine Genome	816
24.4.2.1	Maps	816
24.4.2.2	Sequence	816
24.4.3	Canine Disease Gene Studies	816
24.4.3.1	Canine Disease Mirrors Human Disease	817
24.4.3.2	Canine Disease and Mechanisms of Mutation	818
24.4.3.2.1	SINE Insertions	818
24.4.3.2.2	Simple Repeats	818
24.4.3.2.3	Single Base Mutations	818
24.4.4	Genome Structure in the Domestic Dog	819
24.4.4.1	Linkage Disequilibrium	820
24.4.4.2	Haplotype Structure	820
24.4.4.3	Single Nucleotide Polymorphisms	820
24.4.5	Population Structure in the Domestic Dog	820
24.4.5.1	Canine Breed Clusters Facilitate Mapping Efforts	820
24.4.5.2	Combining Breeds to Improve Mapping	821
24.4.5.3	Homozygosity and Population Bottlenecks	821
24.4.6	Mapping Multigenic Traits in the Dog	822
24.4.6.1	Quantitative Trait Loci	822
24.4.6.2	Establishing a Cohort	822
24.4.6.3	Complex Disease	823
24.4.7	Conclusion	823
	References	823

24.5 Fish as a Model for Human Disease 827
Siew Hong Lam and Zhiyuan Gong

24.5.1	Introduction	828
24.5.1.1	Brief Historical Background and Current Status	828
24.5.1.2	Why Use Fish to Model Human Disorders?	829
24.5.2	Strategies for Modeling Human Disorders in Fish	830
24.5.2.1	Forward Genetics	830
24.5.2.1.1	Chemical Mutagenesis	830
24.5.2.1.2	Insertional Mutagenesis	831
24.5.2.2	Reverse Genetics	831
24.5.2.2.1	Morpholinos	832
24.5.2.2.2	Reverse Genetic Screening by TILLING	832
24.5.2.2.3	Transgenesis	832
24.5.2.3	Physical Manipulation: Chemical Treatment, Environmental Stressor, and Infection	833
24.5.3	Fish Models of Human Disorders	834
24.5.3.1	Blood Disorders	834
24.5.3.2	Heart Disorders	837
24.5.3.3	Cancers	838
24.5.4	Modeling Human Disorders in Fish for Drug Discovery	840
	References	841

25	**Genetic Counseling and Prenatal Diagnosis**	845
	Tiemo Grimm and Klaus Zerres	
25.1	Genetic Counseling	845
25.1.1	Origins and Goals of Genetic Counseling	845
25.1.1.1	Definition of Genetic Counseling	846
25.1.1.2	Origins of Genetic Counseling	846
25.1.1.3	Indications for Genetic Counseling	846
25.1.2	Genetic Diagnosis	848
25.1.2.1	Molecular Diagnosis	848
25.1.2.2	Heterozygote Detection	850
25.1.3	Recurrence Risk	851
25.1.3.1	Recurrence Risk for Multifactorial Conditions	851
25.1.3.2	Reproductive Options and Alternatives	851
25.1.3.3	Predictive Testing	851
25.1.4	Communication and Support	852
25.1.5	Directive Vs. Nondirective Genetic Counseling	852
25.1.6	Assessment of Genetic Counseling and Psychosocial Aspects	853
25.2	Prenatal Diagnosis	854
25.2.1	Indications for Prenatal Diagnosis	854
25.2.1.1	Risk Assessment for Chromosomal Disorders	854
25.2.1.2	Previous Aneuploidy	855
25.2.1.3	Parental Chromosomal Rearrangements	855
25.2.1.4	Family History of a Monogenic Disorder	856
25.2.1.5	Neural Tube Defects	856
25.2.1.6	Psychological Indications	856
25.2.2	Techniques of Prenatal Diagnosis	856
25.2.2.1	Investigations Prior to Implantation	856
25.2.2.2	Noninvasive Diagnosis During Pregnancy	858
25.2.2.3	Invasive Diagnostics During Pregnancy	859
25.2.2.4	Special Problems with Prenatal Diagnosis	860
25.3.2.5	Consultation and Aftercare	861
25.3.2.6	Aftercare	862
25.3.2.7	Psychological Aspects and Outlook	862
25.3	Conclusions	863
25.4	Appendix Example of a Bayesian Table	863
	References	864

26	**Gene Therapy**	867
	Vivian W. Choi and R. Jude Samulski	
26.1	Advent of Gene Therapy	868
26.2	Diseases Considered Suitable for Gene Therapy	868
26.3	Gene Therapy Vectors	868
26.3.1	Viral Vectors	868
26.3.1.1	Adenovirus	868
26.3.1.2	Retrovirus	869
26.3.1.3	Adeno-associated Virus	869
26.3.1.4	Herpes Simplex Virus	869

26.3.2	Nonviral Vectors	869
26.3.2.1	Naked DNA	870
26.3.2.2	Liposomes	870
26.3.2.3	Polymers	870
26.4	Factors Affecting the Success of Gene Therapy	870
26.4.1	Choice of Gene Delivery Vector	870
26.4.2	Route of Administration	870
26.4.3	Integrating or Nonintegrating Vectors	871
26.4.4	Therapeutic Means	871
26.4.4.1	Gene Complementation	871
26.4.4.2	Gene Knockdown	871
26.4.4.3	Gene Correction	871
26.4.5	Regulation of Gene Expression	871
26.5	Safety Issues of Gene Therapy	872
26.6	Difficulties in Achieving Successful Gene Therapy	872
26.7	Conclusions	873
	References	873

27 Cloning in Research and Treatment of Human Genetic Disease ... 875
Ian Wilmut, Jane Taylor, Paul de Sousa, Richard Anderson, and Christopher Shaw

27.1	Introduction	875
27.2	The Value of Cell Lines of Specific Genotype	876
27.2.1	Studies of Inherited Disease	876
27.2.2	Cells for Therapy	877
27.3	Somatic Cell Nuclear Transfer	878
27.3.1	Procedure for Nuclear Transfer	878
27.3.2	Present Successes and Limitations	878
27.4	Cells from Cloned Embryos	879
27.4.1	Cells from Cloned Mouse Embryos	879
27.4.2	Derivation of Cells from Cloned Human Embryos	879
27.4.3	Interspecies Nuclear Transfer	880
27.5	Direct Reprogramming of Somatic Cells	880
27.6	Looking to the Future	881
	References	881

28 Genetic Medicine and Global Health ... 885
David J. Weatherall

28.1	Introduction	886
28.2	Environmental Factors and Genetic Disease	886
28.2.1	Poverty and the Epidemiological Transition	886
28.2.2	Why Are Genetic Disease and Congenital Malformation Commoner in the Developing Countries?	887
28.2.2.1	Selection	887
28.2.2.2	Consanguineous Marriage	887
28.2.2.3	Parental Age	887

28.2.2.4	Population Migration	887
28.2.2.5	Poverty and Dysfunctional Healthcare Systems	888
28.3	Global Burden of Genetic Disease and Congenital Malformation	888
28.4	Monogenic Disease	889
28.4.1	Inherited Disorders of Hemoglobin	889
28.4.1.1	Global Distribution	889
28.4.1.2	Frequency	890
28.4.1.3	Population Genetics and Dynamics	892
28.4.1.4	Clinical Load and Cost–Benefit Issues of Control and Management Posed by the Inherited Hemoglobin Disorders	892
28.4.2	Other Monogenic Diseases in the Developing Countries	894
28.5	Communicable Disease	895
28.5.1	Infectious Agents	895
28.5.2	Vectors	896
28.5.3	Varying Susceptibility	896
28.5.4	Pharmacogenetics and Treatment	896
28.6	Genetic Components of Other Common Diseases: A Global View	897
28.7	Global Control of Genetic Disease	898
28.7.1	Transcultural, Ethical, and Counseling Issues	898
28.7.1.1	Ethnic Differences in Interpreting the Nature of Disease	898
28.7.1.2	Gender Issues	898
28.7.1.3	Patient Discrimination	898
28.7.1.4	Informed Consent	898
28.7.1.5	Lack of Regulatory or Ethical Bodies	899
28.7.1.6	Biobanks and Biopiracy	899
28.7.2	Genetic Services in Developing Countries	899
28.7.3	Organizing International Help for Developing Genetic Programs in Poorer Countries	900
	References	900
29	**Genetic Databases**	**903**
	Introductory Note by Stylianos E. Antonarakis	
29.1	**Databases and Genome Browsers**	**905**
	Rachel A. Harte, Donna Karolchik, Robert M. Kuhn, W. James Kent, and David Haussler	
29.1.1	Historical Background	906
29.1.2	Database Organization	907
29.1.3	Genome Annotations	907
29.1.3.1	Overview of Tracks in the Genome Browser	907
29.1.3.1.1	Mapping and Sequencing Tracks	910
29.1.3.1.2	Phenotype and Disease Associations	911
29.1.3.1.3	Gene and Gene Prediction Tracks and mRNA and EST Tracks	911
29.1.3.1.4	Expression and Regulation	911

29.1.3.1.5	Variation and Repeats	911
29.1.3.2	Comparative Genomics Tracks	912
29.1.3.2.1	Chains, Nets, and Conservation	912
29.1.3.2.2	Cross-Species Protein Alignments	912
29.1.3.3	UCSC Genes Set	912
29.1.4	Displaying and Sharing Data Using Custom Annotation Tracks	913
29.1.5	Example Analysis Using the Genome Browser	913
29.1.6	Using BLAT for Genome-Wide Alignments	916
29.1.7	Table Browser	916
29.1.7.1	Overview	916
29.1.7.2	Example Using the Table Browser	917
29.1.8	Tools	917
29.1.8.1	Introduction	917
29.1.8.2	In Silico PCR	918
29.1.8.3	Lifting Coordinates Between Assemblies	918
29.1.8.4	Gene Sorter	918
29.1.8.5	Proteome Browser	918
29.1.8.6	Genome Graphs	919
29.1.9	Further Information	919
29.1.10	Future Directions	919
	References	920

29.2 Ensembl Genome Browser ... 923
Xosé M. Fernández and Ewan Birney

29.2.1	Genomes Galore	924
29.2.1.1	Genomes in Context	924
29.2.1.2	ENCODE: Shifting the Paradigm	926
29.2.2	Ensembl Annotation	926
29.2.3	Region in Detail: Introduction	927
29.2.3.1	Region in Detail: Features	927
29.2.3.2	Region in Detail: Repeats, Decorations, and Export	927
29.2.4	DAS Sources	927
29.2.5	Comparative Genomics	929
29.2.6	GeneView	930
29.2.7	Variation	932
29.2.7.1	Variation Image	932
29.2.7.2	Comparison Image	932
29.2.8	Ensembl: An Example	932
29.2.9	BioMart Overview	933
29.2.10	Customizing Ensembl	934
29.2.10.1	Displaying User Data on Ensembl	935
29.2.11	Archive	935
29.2.12	*Pre!* Site	935
29.2.13	Further Information	935
29.2.14	Outlook	935
	References	936

	29.3	**Databases in Human and Medical Genetics**	941

Roberta A. Pagon, Ada Hamosh, Johan den Dunnen, Helen V. Firth, Donna R. Maglott, Stephen T. Sherry, Michael Feolo, David Cooper, and Peter Stenson

29.3.1		GeneTests	942
29.3.1.1		Name and URL	942
29.3.1.2		Background	942
29.3.1.3		Purpose and Target Audiences	943
29.3.1.4		Location	943
29.3.1.5		Funding and Governance	943
29.3.1.6		Contents	943
29.3.1.7		Search Mechanisms and Search Results	944
29.3.1.8		Data Maintenance	945
29.3.1.9		Usage	945
29.3.1.10		Future Issues	945
29.3.2		Online Mendelian Inheritance in Man	945
29.3.2.1		Name and URL	945
29.3.2.2		Background	945
29.3.2.3		Purpose and Target Audiences	945
29.3.2.4		Location	946
29.3.2.5		Funding and Governance	946
29.3.2.6		Content	946
29.3.2.7		Search Mechanisms and Search Results	947
29.3.2.8		Data Maintenance	947
29.3.2.9		Usage	947
29.3.2.10		Future Issues	947
29.3.3		HGVS, Locus-Specific Databases	947
29.3.3.1		Name and URL	948
29.3.3.2		Background	948
29.3.3.3		Purpose and Target Audiences	948
29.3.3.4		Location	948
29.3.3.5		Funding and Governance	949
29.3.3.6		Content	949
29.3.3.7		Search Mechanisms and Search Results	949
29.3.3.8		Data Maintenance	949
29.3.3.9		Usage	949
29.3.3.10		Future Issues	949
29.3.4		Decipher	950
29.3.4.1		Name and URL	950
29.3.4.2		Background	950
29.3.4.3		Purpose and Target Audiences	950
29.3.4.4		Location	950
29.3.4.5		Funding and Governance	950
29.3.4.6		Contents	950
29.3.4.7		Search Mechanisms and Search Results	951
29.3.4.8		Data Maintenance	951
29.3.4.9		Usage	952
29.3.4.10		Future Issues	953

29.3.5		Entrez Gene	953
29.3.5.1		Name and URL	953
29.3.5.2		Background	953
29.3.5.3		Purpose and Target Audiences	953
29.3.5.4		Location	953
29.3.5.5		Funding and Governance	953
29.3.5.6		Contents	954
29.3.5.7		Search Mechanisms and Search Results	954
29.3.5.8		Data Maintenance	954
29.3.5.9		Usage	954
29.3.5.10		Future Issues	954
29.3.6		Database of Genotype and Phenotype	954
29.3.6.1		Name and URL	954
29.3.6.2		Background	954
29.3.6.3		Purpose and Target Audiences	955
29.3.6.4		Location	955
29.3.6.5		Funding and Governance	955
29.3.6.6		Contents	955
29.3.6.7		Search Mechanisms and Search Results	955
29.3.6.8		Data Maintenance	956
29.3.6.9		Usage	956
29.3.6.10		Future Issues	956
29.3.7		The Human Gene Mutation Database	956
29.3.7.1		Name and URL	956
29.3.7.2		Background	956
29.3.7.3		Purpose and Target Audiences	956
29.3.7.4		Location	957
29.3.7.5		Funding and Governance	957
29.3.7.6		Contents	957
29.3.7.7		Search Mechanisms and Search Results	958
29.3.7.8		Data Maintenance	958
29.3.7.9		Usage	958
29.3.7.10		Future Issues	958
		References	959

Subject Index 961

Contributors

Laurent Abel
Laboratory of Human Genetics of Infectious Diseases, Necker Branch,
Institut National de la Santé et de la Recherche Médicale, U550,
Necker Medical School, Paris, France
laurent.abel@inserm.fr

Brett S. Abrahams
Program in Neurogenetics and Department of Neurology
Semel Institute for Neuroscience and Behavior at the David Geffen School of Medicine
University of California at Los Angeles, Los Angeles, CA 90095-1769, USA
brett.abrahams@gmail.com

Alexandre Alcaïs
Laboratory of Human Genetics of Infectious Diseases, Necker Branch,
Institut National de la Santé et de la Recherche Médicale, U550,
Necker Medical School, Paris, France
alexandre.alcais@inserm.fr

Richard Anderson
Centre for Reproductive Biology, The Queen's Medical Research Institute,
47 Little France Crescent, Edinburgh, EH16 4TJ, UK

Stylianos E. Antonarakis
Department of Genetic Medicine and Development,
University of Geneva Medical School and University Hospitals of Geneva
1 rue Michel-Servet, 1211 Geneva
Switzerland
Stylianos.Antonarakis@unige.ch

Antonio Baldini
Institute of Genetics and Biophysics, National Research Council,
University of Naples Federico II, and the Telethon Institute of Genetics
and Medicine, Via Pietro Castellino 111, Naples 80131, Italy
baldini@cnr.igb.it

Terri H. Beaty
Department of Epidemiology, The Johns Hopkins Bloomberg
School of Public Health, 615 North Wolfe Street, Baltimore,
MD 21205, USA
tbeaty@jhsph.edu

Jacques S. Beckmann
Service and Department of Medical Genetics, Centre Hospitalier Universitaire
Vaudois and University of Lausanne, Lausanne, Switzerland
Jacques.Beckmann@chur.ch

Thomas D. Bird
University of Washington, Geriatric Research Education and Clinical Center
(182), VA Puget Sound Health Care System, 1660 S Columbian Way,
Seattle, WA 98108, USA
tomnroz@u.washington.edu

Ewan Birney
EMBL-European Bioinformatics Institute, Wellcome Trust Genome Campus,
Hinxton, Cambridge, CB10 1SD, UK

Alan H. Bittles
Edith Cowan University and Murdoch University, Perth, Perth, WA,
Australia; Centre for Comparative Genomics, Murdoch University,
South Street, Perth, WA 6150, Australia
abittles@ccg.murdoch.edu.au

Romulo M. Brena
Department of Surgery, University of Southern California Norris Comprehensive
Cancer Center, Los Angeles, CA 91101, USA

Carlos D. Bustamante
Department of Biological Statistics and Computational Biology,
Cornell University, Ithaca, NY, USA
cdb28@cornell.edu

Ross Cagan
Department of Developmental and Regenerative Biology,
Mount Sinai School of Medicine, 1 Gustave L. Levy Place,
Annenberg Building 25-40, New York, NY 10029, USA
Ross.Cagan@mssm.edu

Jean-Laurent Casanova
Laboratory of Human Genetics of Infectious Diseases, Rockefeller Branch,
The Rockefeller University, New York, NY, USA
Jean-Laurent.Casanova@mail.rockefeller.edu

Vivian W. Choi
Department of Pharmacology, Gene Therapy Center,
University of North Carolina, Chapel Hill, NC 27599, USA
Vivian_Choi@alumni.unc.edu

Sven Cichon
Institute of Human Genetics and Department of Genomics,
Life & Brain Center, University of Bonn, Sigmund-Freud-Str. 25,
53127 Bonn, Germany
sven.cichon@uni-bonn.de

Andrew G. Clark
Department of Molecular Biology and Genetics, Cornell University,
Ithaca, NY 14853, USA
ac347@cornell.edu

David N. Cooper
Institute of Medical Genetics, School of Medicine,
Cardiff University, Heath Park, Cardiff, CF14 4XN, UK
cooperdn@cardiff.ac.uk

Joseph F. Costello
Department of Neurological Surgery, University of California San Francisco,
Diller Family Comprehensive Cancer Center, San Francisco, CA 94143, USA
jcostello@cc.ucsf.edu

Michael Dean
National Cancer Institute, Frederick, MD 21702, USA
deanm@mail.nih.gov

Emmanouil T. Dermitzakis
Department of Genetic Medicine and Development,
University of Geneva Medical School, 1 Rue Michel-Servet, Geneva 1211,
Switzerland
emmanouil.dermitzakis@unige.ch

Francesca Ducci
Division of Psychological Medicine,
P063 Institute of Psychiatry, De Crespigny Park, London, SE5 8AF, UK
Francesca.Ducci@kcl.ac.uk

Johan T. den Dunnen
Leiden University Medical Center, Human and Clinical Genetics,
S4-030, PO Box 9600, 2333RC Leiden, The Netherlands
ddunnen@HumGen.nl

Michael Feolo
National Center for Biotechnology Information, Bldg 45,
4AN.12B, Bethesda, MD 20894, USA
feolo@ncbi.nlm.nih.gov

Xosé M. Fernández
EMBL-European Bioinformatics Institute, Wellcome Trust Genome Campus,
Hinxton, Cambridge, CB10 1SD, UK
xose@ebi.ac.uk

Helen V. Firth
Consultant Clinical Geneticist, Addenbrooke's Hospital, Box 134,
Cambridge, CB2 2QQ, UK
helen.firth@addenbrookes.nhs.uk

Jonathan Flint
Wellcome Trust Centre for Human Genetics, Roosevelt Drive,
Oxford, OX3 7BN, UK
jf@well.ox.ac.uk

Daniel H. Geschwind
Programs in Neurogenetics and Neurobehavioural Genetics,
Neurology Department, and Semel Institute for Neuroscience and Behavior,
David Geffen School of Medicine, University of California at Los Angeles,
Los Angeles, CA 90095-1769, USA
dhg@ucla.edu

David Goldman
Laboratory of Neurogenetics, NIAAA, NIH, 5625 Fishers Lane,
Room 3S32, MSC 9412, Rockville, MD 20852, USA
davidgoldman@mail.nih.gov

David B. Goldstein
Center for Human Genome Variation, Institute for Genome Sciences and Policy,
Duke University, Durham, NC 27708, USA
d.goldstein@duke.edu

Zhiyuan Gong
Department of Biological Sciences, 14 Science Drive 4,
National University of Singapore, Singapore 117543
dbsgzy@nus.edu.sg

Tiemo Grimm
Institute of Human Genetics,
Division of Medical Genetics, University of Würzburg, Biozentrum,
Am Hubland, 97074 Würzburg, Germany
tgrimm@biozentrum.uni-wuerzburg.de

Ryan N. Gutenkunst
Theoretical Biology and Biophysics, and Center for Nonlinear Studies,
Los Alamos National Laboratory,
Los Alamos, NM, USA
ryang@lanl.gov

Ada Hamosh
Institute of Genetic Medicine, Johns Hopkins University School of Medicine,
Blalock 1012D, 600 N. Wolfe St., Baltimore, MD 21287-4922, USA
ahamosh@mail.jhmi.edu

Min Han
Department of Molecular, Cellular, and Developmental Biology,
Howard Hughes Medical Institute, Campus Box 0347,
University of Colorado at Boulder, Boulder, CO 80309, USA
mhan@colorado.edu

Ross C. Hardison
Department of Biochemistry and Molecular Biology,
Center for Comparative Genomics and Bioinformatics,
304 Wartik Laboratory, The Pennsylvania State University,
University Park, PA 16802, USA
rch8@psu.edu

Rachel A. Harte
Center for Biomolecular Science and Engineering, Engineering 2,
University of California Santa Cruz, 1156 High Street, Santa Cruz,
CA 95064, USA
hartera@soe.ucsc.edu

David Haussler
Center for Biomolecular Science and Engineering, Engineering 2,
University of California Santa Cruz, 1156 High Street, Santa Cruz, CA 95064, USA
haussler@soe.ucsc.edu

Michael Hofreiter
Max Planck Institute for Evolutionary Anthropology, Deutscher Platz 6,
04103 Leipzig, Germany
michi@palaeo.eu

Bernhard Horsthemke
Institut für Humangenetik, Universitätsklinikum Essen,
Hufelandstrasse 55, 45122 Essen, Germany
bernhard.horsthemke@uni-due.de

Ruth Johnson
Department of Molecular, Cell and Developmental Biology,
Mount Sinai School of Medicine, 1 Gustave L. Levy Place,
Annenberg Building 18-92, New York,
NY 10029, USA
ruth.johnson@mssm.edu

Donna Karolchik
Center for Biomolecular Science and Engineering, Engineering 2,
University of California Santa Cruz, 1156 High Street, Santa Cruz,
CA 95064, USA
donnak@soe.ucsc.edu

Nicholas Katsanis
Department of Cell Biology, Duke University, 466 A Nanaline Duke Bldg,
Durham, NC 27710, USA
katsanis@jhmi.edu

W. James Kent
Center for Biomolecular Science and Engineering, Engineering 2,
University of California Santa Cruz, 1156 High Street, Santa Cruz,
CA 95064, USA
kent@soe.ucsc.edu

Muin J. Khoury
Office of Public Health Genomics, Centers for Disease Control
and Prevention, 1600 Clifton Road, Atlanta, GA 30333, USA
muk1@cdc.gov

Robert M. Kuhn
Center for Biomolecular Science and Engineering, Engineering 2,
University of California Santa Cruz, 1156 High Street, Santa Cruz,
CA 95064, USA
kuhn@soe.ucsc.edu

Siew Hong Lam
Department of Biological Sciences, 14 Science Drive 4,
National University of Singapore, Singapore 117543
dbslsh@nus.edu.sg

Donna R. Maglott
Natl Ctr Biotechnology Info, 45 Center Dr, MSC 6510,
Bethesda, MD 20892-6510, USA
maglott@ncbi.nlm.nih.gov

Arno G. Motulsky
Departments of Medicine (Division of Medical Genetics)
and Genome Sciences, University of Washington School of Medicine
Box 355065, 1705 N.E. Pacific St., Seattle, WA 98195-5065, USA
agmot@u.washington.edu

Stefan Mundlos
Institute for Medical Genetics,
Charité, Universitaetsmedizin Berlin, Augustenburger Platz 1,
13353 Berlin, Germany; Forschungsgruppe Development & Disease,
Max Planck Institute for Molecular Genetics, Ihnestraße 73,
14195 Berlin, Germany
stefan.mundlos@charite.de

David L. Nelson
Molecular and Human Genetics, Baylor College of Medicine,
One Baylor Plaza, Rm. 902E, Houston, TX 77030, USA
nelson@bcm.tmc.edu

Paola Nicoletti
Center for Human Genome Variation, Institute for Genome Sciences and Policy,
Duke University, Durham, NC, USA; Department of Clinical and Preclinical
Pharmacology, University of Florence, Florence, Italy

Markus M. Nöthen
Institute of Human Genetics and Department of Genomics,
Life & Brain Center, University of Bonn, Sigmund-Freud-Str. 25, 53127 Bonn,
Germany
markus.noethen@uni-bonn.de

Elaine A. Ostrander
National Human Genome Research Institute, National Institutes of Health,
50 South Drive, MSC 8000, Building 50, Room 5351, Bethesda,
MD 20892-8000, USA
eostrand@mail.nih.gov

Roberta A. Pagon
University of Washington, GeneTests, 9725 Third Ave NE, Suite 602,
Seattle, WA 98115-2024, USA
bpagon@u.washington.edu

Heidi G. Parker
National Human Genome Research Institute, National Institutes of Health,
50 South Drive, MSC 8000, Building 50, Room 5334, Bethesda,
MD 20892-8000, USA
hgparker@mail.nih.gov

George P. Patrinos
Department of Pharmacy, School of Health Sciences, University of Patras,
University Campus, Rion, 26504 Patras, Greece;
MGC-Department of Cell Biology and Genetics, Faculty of Medicine and Health
Sciences, Erasmus University Medical Center, P.O. Box 2040, 3000 CA,
Rotterdam, The Netherlands
g.patrinos@erasmusmc.nl;
g.patrinos@upatras.gr

Sohini Ramachandran
Society of Fellows, Harvard University, Cambridge, MA, USA;
Department of Organismic and Evolutionary Biology, Harvard University,
Cambridge, MA, USA
sramach@fas.harvard.edu

Marcella Rietschel
Department of Genetic Epidemiology in Psychiatry,
Central Institute of Mental Health Mannheim,
University of Heidelberg J5, 68159 Mannheim,
Germany
Marcella.Rietschel@zi-mannheim.de

Jon F. Robinson
McKusick-Nathans Institute of Genetic Medicine,
Department of Molecular Biology and Genetics,
Johns Hopkins University School of Medicine, Baltimore, MD 21205, USA

R. Jude Samulski
Department of Pharmacology,
Gene Therapy Center, University of North Carolina,
Chapel Hill, NC 27599, USA
rjs@med.unc.edu

Christine Schmael
Department of Genetic Epidemiology in Psychiatry,
Central Institute of Mental Health Mannheim,
University of Heidelberg J5, 68159 Mannheim,
Germany
c.schmael@gmx.de

Christopher Shaw
Department of Neurology, King's College London School of Medicine,
London, UK

Stephen T. Sherry
Building 38 – Natl Library of Medicine, B1E09B, 38 Center Dr, Bethesda,
MD 20892-6510, USA
sherry@ncbi.nlm.nih.gov

Paul de Sousa
Scottish Centre for Regenerative Medicine, Chancellor's Building,
University of Edinburgh, 49 Little France Crescent, Edinburgh, EH16 4SB, UK

Michael R. Speicher
Institute of Human Genetics, Medical University of Graz,
Harrachgasse 21/8, 8010 Graz, Austria
michael.speicher@medunigraz.at

Peter Stenson
Institute of Medical Genetics, School of Medicine,
Cardiff University, Heath Park, Cardiff,
CF14 4XN, UK
stensonpd@cardiff.ac.uk

Hua Tang
Department of Genetics, Stanford Medical School, Stanford, CA, USA
huatang@stanford.edu

Jane Taylor
Scottish Centre for Regenerative Medicine, Chancellor's Building,
University of Edinburgh, 49 Little France Crescent, Edinburgh, EH16 4SB, UK

Ian Tomlinson
Molecular and Population Genetics,
Wellcome Trust Centre for Human Genetics, Oxford, OX3 7BN, UK
iant@well.ox.ac.uk

Morgan Tucker
Howard Hughes Medical Institute, Department of Molecular,
Cellular, and Developmental Biology, Campus Box 0347,
University of Colorado at Boulder, Boulder, CO 80309, USA
mtucker@colorado.edu

Nicole M. Walley
Department of Biology, Duke University, Durham, NC, USA; Center for Human
Genome Variation, Institute for Genome Sciences and Policy, Duke University,
Durham, NC, USA
nicole.walley@duke.edu

Sophia S. Wang
Division of Cancer Epidemiology and Genetics, National Cancer Institute,
National Institutes of Health, 6120 Executive Boulevard, Room 5104, Rockville,
MD 20852, USA
wangso@mail.nih.gov

David J. Weatherall
University of Oxford,
Weatherall Institute of Molecular Medicine, John Radcliffe Hospital,
Headington, Oxford, OX3 9DS, UK
liz.rose@imm.ox.ac.uk

Saffron Willis-Owen
Molecular Genetics, National Heart and Lung Institute, Guy Scadding Building,
Dovehouse Street, London, SW3 6LY, UK
s.willis-owen@imperial.ac.uk

Ian Wilmut
Scottish Centre for Regenerative Medicine, Chancellor's Building,
University of Edinburgh, 49 Little France Crescent, Edinburgh, EH16 4SB, UK
Ian.Wilmut@ed.ac.uk

Klaus Zerres
Institute of Human Genetics,
Medical Faculty, RWTH University
Pauwelsstr. 30, 52074 Aachen, Germany
kzerres@ukaachen.de

Introduction

Human Genetics as Fundamental and Applied Science

Human genetics is both a fundamental and an applied science. As a fundamental science, it is part of genetics – the branch of science that examines the laws of storage, transmission, and realization of information for development and function of living organisms. Within this framework, human genetics concerns itself with the most interesting organism – the human being. This concern with our own species makes us scrutinize scientific results in human genetics not only for their theoretical significance but also for their practical value for human welfare. Thus, human genetics is also an applied science. Its value for human welfare is bound to have repercussions for theoretical research as well, since it influences the selection of problems by human geneticists, their training, and the financing of their research. Because of its continued theoretical and practical interest, human genetics offers fascination and human fulfillment unparalleled by work in fields that are either primarily theoretical or entirely practical in subject matter.

Science of Genetics

Genetics is based on a powerful and penetrating theory. The profundity of a theory depends on the depth of the problems that it sets out to solve and can be characterized by three attributes: the occurrence of high-level constructs, the presence of a mechanism, and high explanatory power [1]. In genetics, the high-level "construct" is the gene as a unit of storage, transmission, and realization of information. Since the rediscovery of Mendel's laws in 1900, genetic mechanisms have been worked out step by step to the molecular level – deciphering of the genetic code, analysis of transcription and translation, the function of gene-determined proteins, the fine structure of the genetic material, and DNA sequences outside of genes. The problems of regulation of gene activity in the development and function of organisms are currently a principal goal of fundamental research. So far, the explanatory power of the theory has not nearly been exhausted.

How Does a Science Develop?

Kuhn (1962) [10] described the historical development of a science as follows: In the early, protoscientific stage, there is substantial competition among various attempts at theoretical foundation and empirical verification. Basic observations suggest a set of problems that, however, is not yet visualized clearly. Then, one "paradigm" unifies a group within the scientific community in the pursuit of a common goal, at the same time bringing into sharper focus one or a few aspects of the problem field, and suggesting a way for their solution. If the paradigm turns out to be successful, it is accepted by an increasing part of the scientific community, which now works under its guidance, exploring its possibilities, extending its range of application, and developing it into a scientific theory.

This concept of a paradigm has three main connotations:

1. It points to a piece of scientific work that serves as an "exemplar," suggesting ways in which a certain problem should be approached.
2. It delimits a group of scientists who try to explore this approach, expand its applicability, deepen its

theoretical basis by exploration of basic mechanisms, and enhance its explanatory power.
3. Finally, while an elaborate theory must not – and, in most cases, does not – exist when a paradigm is initiated, its germ is already there, and a successful paradigm culminates in the elaboration of this theory.

This process of developing a science within the framework of a paradigm has been described by Kuhn as "normal science." The basic theory is taken increasingly for granted. It would be sterile at this stage to doubt and reexamine its very cornerstones; instead, it is applied to a variety of problems, expanded in a way that is comparable to puzzle solving. From time to time, however, results occur that, at first glance, defy explanation. First, this leads to attempts at accommodating such results within the theoretical framework by additional ad hoc hypotheses. These attempts are often successful; sometimes, however, they fail. If in such a situation an alternative paradigm is brought forward that explains most of the phenomena accounted for by the old theory as well as the new, hitherto unexplained phenomena, a scientific "revolution" may occur. The new paradigm gains support from an increasing majority of the scientific community, it soon develops into a new – more explanatory – theory, and the process of normal science begins anew.

This portrayal of scientific development has been criticized by some philosophers of science [11]. The concept of "normal" science as outlined above does not appeal to some theorists. Working within the framework of a given set of concepts has been denounced as dull, boring, and in any case not as science should be. According to these philosophers, scientists ought to live in a state of permanent revolution, constantly questioning the basic foundations of their field, always eager to put them to critical tests and, if possible, to refute them [15–18]. Many scientists actively involved in research, on the other hand, have readily accepted Kuhn's view; he has apparently helped them to recognize some important aspects in the development of their own fields.

Central Theory of Genetics Looked at as a Paradigm

While Kuhn's concepts were developed on the basis of the history of the physical sciences, his description well fits the development of genetics. Up to the second half of the nineteenth century, the phenomena of heredity eluded analysis. Obviously, children were sometimes – but by no means always – similar to their parents; some diseases were shown to run in families; it was possible to improve crops and domestic animals by selective breeding. Even low-level laws were discovered, for example Nasse's law that hemophilia affects only boys but is transmitted by their mothers and sisters (Chap. 5, Sect. 5.1.4). However, a convincing overall theory was missing, and attempts at developing such at theory were unsuccessful. In this situation, Mendel, in his work *Versuche über Pflanzenhybriden* (1865) [12] first improved a procedure; he complemented the breeding experiment by counting the offspring. He then interpreted the results in terms of the random combination of basic units; by assuming these basic units, he founded the gene concept – the nuclear concept underlying genetic theory (Chap. 1, Sect. 1.4).

Since the rediscovery of his work in 1900, Mendel's insight has served as a paradigm in all three connotations: it provided an exemplar as to how breeding experiments should be designed and evaluated, it resulted in the establishment of a scientific community of geneticists, and it led to the development of a deep and fertile scientific theory. A special problem that has not been answered satisfactorily, in our opinion, concerns the question of why acceptance of Mendel's paradigm had to wait for as long as 35 years after these experiments were published. It would be too simplistic to blame academic arrogance and shortsightedness of contemporary biologists who did not want to accept the work of a "nonacademic" outsider, even if this factor may indeed have been one of the components for this neglect. We believe rather that the many new biological discoveries in the 35 years following Mendel's discovery were of such a revolutionary nature as to qualify as a scientific crisis in the Kuhnian sense and therefore required a completely new approach.

Soon after the rediscovery of Mendel's laws in 1900, however, an initially small, but quickly growing group of scientists gathered who developed genetics in an interplay between theory and experiment and launched the major scientific revolution of the twentieth century in the field of biology.

Human Genetics and the Genetic Revolution

Meanwhile, the biological revolution of the nineteenth century – evolutionary theory – had been accepted by the scientific community. One major consequence was the realization that human beings had evolved from other, more "primitive" primates, that humans are part of the animal kingdom, and that the laws of heredity which had been found to apply for all other living beings are also valid for our species. Hence, Mendel's laws were soon applied to traits that were found in human pedigrees – primarily hereditary anomalies and diseases. Analyzing the mode of inheritance of alkaptonuria – a recessive disease – Garrod (1902) [5] clearly recognized the cardinal principle of gene action: genetic factors specify chemical reactions (Chap. 1, Sect. 1.5). This insight also required 30 years before being incorporated into the body of "normal" science.

Elucidation of inheritance in humans did not begin with Mendel's paradigm. Many relevant observations had been reported before, especially on various diseases. Moreover, another paradigm had been founded by F. Galton in his work on *Hereditary Talent and Character* (1865) [6] and in later works: to derive conclusions as to inheritance of certain traits such as high performance, intelligence, and stature, one should measure these traits as accurately as possible and then compare the measurements between individuals of known degree of relationship (for example, parents and children, sibs, or twins) using statistical methods. This approach did not contain the potential for elucidating the mechanisms of heredity. On the other hand, it seemed to be much more generally applicable to human characteristics than Mendelian analysis; pedigree analysis in terms of Mendel's laws was hampered by the fact that most human traits simply could not be classified as alternate characteristics, as could round and shrunken peas. Human characteristics are usually graded and show no alternative distribution in the population. Moreover, the phenotypes are obviously determined not only by the genetic constitution but by external, environmental influences as well – the result of an interaction between "nature and nurture" (Galton). Therefore, naive attempts at applying Mendel's laws to such traits were doomed to failure. For traits that are regarded as important, such as intelligence and personality, but also for many diseases and mental retardation, there was only the choice between research along the lines suggested by Galton or no research at all. Investigations on genetic mechanisms would have to await elucidations of the genetics of other, more accessible organisms. Under these circumstances, scientists chose to follow Galton. This choice had not only theoretical reasons; it was strongly influenced by the desire to help individuals and families by calculating risk figures for certain diseases, thereby creating a sound basis for genetic counseling. More important, however, was the concern of some scientists about the biological future of the human species, which they saw threatened by deterioration due to relaxation of natural selection. The motives for their research were largely eugenic: it seemed to provide a rational foundation for measures to curb reproduction of certain groups who were at high risk of being diseased or otherwise unfit.

History of Human Genetics: A Contest Between Two Paradigms

The two paradigms – Mendel's gene concept and Galton's biometric approach – have developed side by side from 1900 up to the present; many present-day controversies, especially in the field of behavior genetics but also those concerning strategies in the genetic elucidation of common diseases, are immediately understandable when the history of human genetics is conceived as a contest between these two paradigms. This does not mean that the two paradigms are mutually exclusive; in fact, correlations between relatives as demonstrated by biometric analysis were interpreted in terms of gene action by Fisher in 1918 [4]. Some human geneticists have worked during some part of their career within the framework of the one paradigm, and during another within the framework of the other paradigm. By and large, however, the two streams of research have few interconnections and may even become further polarized because of highly specialized training for each group, epitomized by the biochemical and molecular genetic laboratories for the one and the computer for the other group.

In the first decades of the last century the biometric paradigm of Galton appeared to be very successful. Genetic variability within the human population was believed to be established for normal traits such as stature or intelligence as well as for a wide variety of pathologic conditions such as mental deficiency and psychosis, epilepsy, and common diseases such as diabetes, allergies, and even tuberculosis. Mendelian analysis, on the other hand, seemed to be confined to rare hereditary diseases; the ever repeated attempts at expanding Mendelian explanation into the fields of normal, physical characteristics and common diseases were usually undertaken without critical assessment of the inescapable limitations of Mendelian analysis. The first major breakthrough of Mendelian genetics was the establishment of the three-allele hypothesis for the AB0 blood groups by Bernstein in the 1920s [2] (Chap. 5, Sect. 5.2.2). Further progress, however, had to await the development of genetic theory by work on other organisms such as *Drosophila,* bacteria, and viruses, especially bacteriophages.

The advent of molecular biology in the late 1940s and 1950s had a strong influence on human genetics and, indeed, brought the final breakthrough of Mendel's paradigm. A major landmark was the discovery by Pauling et al. in 1949 [14] that sickle cell anemia is caused by an abnormal hemoglobin molecule.

The foundation of human chromosome research in the late 1950s and early 1960s (Chap. 3, Sect. 3.1) came as a second, important step. At present, most investigations in human genetics have become a part of mainstream research within the framework of genetic theory. The human species, regarded by most early experimental geneticists as a poor tool for genetic research, is now displaying definite advantages for attacking basic problems. Some of these advantages are the large size of available populations, the great number and variety of known mutants and chromosome anomalies, and the unparalleled detailed knowledge of human physiology and biochemistry in health and disease. The improved understanding of human genome structure and its variability (Chap. 2) by the completion of the human genome project, by new sequencing and array technologies, and by efforts to identify all functional elements in the human genome sequence (Chap. 4), further facilitates both basic and applied research in human genetics.

One would expect that such breakthroughs have led to the establishment of Mendel's paradigm as the only leading paradigm in human genetics. This, however, is not the case. In spite of the fact that genetic theory is now pervading many fields that seemed to be closed to it, the paradigm of Galton – biometric analysis – has attained an unsurpassed level of formal sophistication over the past decades. The availability of software tools has greatly facilitated the development and application of biometric techniques. Moreover, in some fields, such as behavior genetics, the application of genetic theory – Mendel's paradigm – is still hampered by severe difficulties (Chap. 23), and here biometric methods have dominated for a long time. In the same field, however, they are most severely criticized and subject to controversial discussions about ethical issues and possible discrimination.

Progress in Human Genetics and Practical Application

The achievements of molecular biology and chromosome research have not only altered human genetics as a pure science, but have also brought marked progress in its application for human welfare. At the beginning, this progress did not appear very conspicuous; the diagnosis of hereditary diseases was improved, and many, hitherto unexplained malformations were accounted for by chromosome aberrations. The first practical success came in the early 1950s when the knowledge of enzyme defects in phenylketonuria (Chap. 1) and galactosemia led to successful preventive therapy by a specific diet. However, a breakthrough on a much larger scale was achieved when the methods of prenatal diagnosis for chromosome aberrations and for some metabolic defects were introduced in the late 1960s and early 1970s (Chap. 25, Sect. 25.2). Suddenly, genetic counseling could now be based not only on probability statements but, in an increasing number of cases, on certainty of individual diagnoses. This scientific development coincided with a growing awareness in large parts of the human population that unlimited human reproduction must not be accepted as a natural law but can – and should – be regulated in a rational way. Introduction of oral contraceptive agents signaled this awareness. The chance to avoid the births of severely handicapped children is now accepted by a rapidly increasing proportion of the population. At the same time, better knowledge of pathophysiological pathways is improving the chances for individual therapy of hereditary diseases, including the promise somatic gene therapy by introduction of genes into cells of functional tissues (see Chap. 27). Applications of human genetics as a practical tool to prevent suffering and disease have found wide resonance and have now one of the most rewarding approaches in preventive medicine. In many countries, the politically responsible bodies have already created, or are now creating the institutions for widespread application of the new tools.

Effects of Practical Applications on Research

These practical applications have led to a marked increase in the number of research workers and the amount of work within the past decades. From the beginning of the twentieth century up to the early 1950s, human genetics had been the interest of a mere handful of scientists for most of whom it was not even a full-time occupation. Many of the pioneers were trained and worked much of their lifetime as physicians in special fields of medicine, such as Waardenburg and Franceschetti in ophthalmology, and Siemens in dermatology. Others were interested in theoretical problems of population genetics and evolution and chose problems in human genetics as the field of application for their theoretical concepts, most notably J.B.S. Haldane and R.A. Fisher. Still others had their point of departure in physical anthropology. This heterogeneous group of scientists did not form a coherent scientific community. For a long time, there was almost no formal infrastructure for the development of a scientific specialty. There were almost no special departments, journals, and international conferences. This lack of focus resulted in a marked heterogeneity in quality and content of scientific contributions.

All this has changed. Departments and units of human and medical genetics are now the standard in many countries; universities and medical schools offer special curricula, many journals and other publications exist, and numerous congresses and conferences are being held. Human genetics is now an active and vigorous field which continues to grow exponentially.

Dangers of Widespread Practical Application for Scientific Development

This development, however, satisfactory as it is, has also a number of potentially undesirable consequences:

(a) Research is promoted primarily in the fields of immediate practical usefulness related to hereditary diseases; fields of less immediate practical importance may be neglected.
(b) Initially the contact with fundamental research in molecular genetics and cell biology was not intensive enough. This may have led to a slowdown in the transfer of scientific concepts and experimental approaches from these fields. Fortunately, this has changed with the advent of recombinant DNA techniques and many other methods. The speed with which results of basic research are being transferred into practical application has increased significantly.
(c) As in other sciences, certain topics may evolve to a mainstream research where vast human and financial resources are being invested, drawing it off other areas, which are then neglected in spite of their great importance. For example, at present the immense activities to unravel complex disorders by high-throughput assays have resulted in a decreased interest in studying monogenic Mendelian disorders although their detailed analyses may provide invaluable insights into consequences of mutations and their associated pathophysiology (see Chap. 4, Sect. 4.1).
(d) Much medical research applies established methods to answer straightforward questions. Many studies collect data with new techniques. Individual results are often not of great import, but the ensemble of such data are the essential building blocks for the future progress of normal science. Much of such work is being carried out in human and medical genetics and is quite essential for many medical and anthropological applications. However, there is continued need in human genetics to develop testable hypotheses and try to test their consequences from all viewpoints.

Human geneticists must not neglect the further development of genetic theory. Basic research is needed in fields in which the immediate practical application of results is not possible but might in the long run be at least as important for the future of the human species as current applications in diagnostic and preventive medicine.

Advantages of Practical Application for Research

The needs of medical diagnosis and counseling have also given strong incentives to basic research. Many phenomena that basic research tries to explain would simply be unknown had they not been uncovered by study of diseases. We would be ignorant regarding the role of sex chromosomes in sex determination had there not been patients with sex chromosomal anomalies. Phenomena such as spontaneously enhanced chromosome instability in Fanconi's anemia or Bloom's syndrome with all its consequences for somatic mutation and cancer formation (Chap. 3, Sect. 3.7) were discovered accidentally in the process of examining certain patients for diagnostic reasons. Genetic analysis of the "supergene" determining the major histocompatibility complex in humans contributes much to our fundamental understanding of how the genetic material above the level of a single gene locus is structured, and how the high genetic variability within the human population can be maintained (Chap. 6, Sect. 6.2.5). However, research in this field would certainly be much less active had there not been the incentive of improving the chances of organ transplantation.

Whether we like it or not, society pays increasing amounts of money for research in human genetics because we want to have practical benefits. Hence, to promote basic research, we must promote widespread practical applications. To guarantee progress in practical application for the future as well – and not only in the field of medicine – basic research needs to be supported. This is also the only way to attract good research workers and to maintain – or even improve – scientific standards. This paradox creates priority problems for all those concerned with research planning.

Human Genetics and the Sociology of Science

The discussion above should have demonstrated that human genetics – as all other sciences – has not developed in a sociological vacuum, following only the inherent logical laws of growth of theory and experimental testing. Human genetics is the work of social groups of human beings who are subject to the laws of group psychology and are influenced by the society at large in their attitudes toward research and their selection of problems. Unfortunately, sociological investigations of group formation and structure in human genetics have not been carried out. Another group active in the foundation of molecular biology, that which introduced the bacteriophages of *Escherichia coli* into the analysis of genetic information, has been studied extensively [3].

We know from this and from other examples that, during a phase in which a new paradigm is being founded, the group that shares this paradigm establishes close within-group contacts. The normal channels of information exchange such as scientific journals and congresses are superseded by more informal information transfer through telephone calls, e-mail communications, preprints, and personal visits. Within the group, influential personalities serve as intellectual and/or organizational leaders. Outside contacts, on the other hand, are often loose. When the acute phase of the scientific revolution is over, the bonds within the group are loosened, and information is again exchanged largely by normal channels of publication.

Similar developments can be observed in the field of human genetics. For example, in Chap. 6 (Sect. 6.2.5) we sketch the groups active in the elucidation of the major histocompatibility complex and in the assignment of gene loci to chromosome segments (Sect. 6.1).

Of similar influence on population genetics has been the first "big science" research project in human genetics – the Atomic Bomb Casualty Commission (ABCC, now the Radiation Effects Research Foundation, RERF; www.rerf.or.jp) project that was launched in the late 1940s in Japan by American and Japanese research workers to examine the genetic consequences of the atomic bombs in Hiroshima and Nagasaki (Chap. 10). In later years, this project led, for example, to comprehensive studies of the genetic effects of parental consanguinity. The second endeavor of this type is the "Human Genome Project" – the attempt at analyzing and sequencing the entire human genome by coordinated international cooperation (see Chap. 44). Today many research efforts are being conducted and can only be accomplished in large, international consortia, as for example the ENCyclopedia Of DNA Elements (ENCODE) project (www.genome.gov) or the Functional Annotation of the Mammalian Genome (FANTOM) project (fantom.gsc.riken.jp).

Many, if not most of the more interesting developments in the field were not initiated by investigators who would declare themselves human geneticists, or who worked in human genetics departments. They were launched by research workers from other fields such as general cytogenetics, cell biology, molecular biology, biochemistry, and immunology, but also from clinical specialties such as pediatrics, hematology, ophthalmology, and psychiatry. A common theme running through many recent developments has been the application of nongenetic techniques from many different fields such as biochemistry and immunology to genetic concepts. On the other hand, techniques originally developed for solving genetic problems, especially for molecular studies of DNA, are being introduced at a rapidly increasing rate into other fields of research, for example in both medical research and practical medicine. In fact, most recent progress in human genetics comes from such interdisciplinary approaches. The number of research workers in the field has increased rapidly. Most did not start as human geneticists but as molecular biologists, medical specialists, biochemists, statisticians, general cytogeneticists, etc. They were drawn into human genetics in the course of their research. This very variety of backgrounds makes discussions among human geneticists stimulating and is one of the intellectual assets of the present state of our field. However, such diversity is also a liability as it may lead to an overrating of one's small specialty at the expense of a loss of an overview of the whole field [8]. With increasing complexity of research methods, specialization within human genetics has become inevitable. However, this brings with it the danger that the outlook of the scientist narrows, whole fields are neglected, and promising research opportunities remain unexploited.

Human Genetics in Relation to Other Fields of Science and Medicine

The rapid development of human genetics during recent decades has created many interactions with other fields of science and medicine. Apart from general and molecular genetics and cytogenetics, these interactions are especially close with cell biology, biochemistry, immunology, and – with many clinical specialties. Until recently, on the other hand, there have been few if any connections with physiology. One reason for this failure to establish fruitful interactions may be a difference in the basic approach: genetic analysis attempts to trace the causes of a trait to its most elementary components. Geneticists know in principle that the phenotype is produced by a complex net of interactions between various genes, but they are interested more in the components than in the exact mechanism of such interactions. At present, genetic analysis has reached the level of gene structure and the genetic code; a final goal would be to explain the properties of this code in terms of quantum physics. A malevolent observer might compare the geneticist with a man who, to understand a book, burns it and analyzes the ashes chemically.

The physiologist, on the other hand, tries to read the book. However, he often presupposes that every copy of the book should be exactly identical; variation is regarded as a nuisance. To put it differently, physiology is concerned not with the elements themselves but with their mode of interaction in complicated functional systems (see Mohr [13]). Physiologists are more concerned with the integration of interacting systems than with the analysis of their components. The analysis of regulation of gene activities by feedback mechanisms, for example, the Jacob-Monod model in bacteria, and some approaches in developmental genetics of higher organisms have taught geneticists the usefulness of thinking in terms of systems. On the other hand, methods for molecular analysis of DNA have been introduced into physiology at an increasing scale. Genes for receptors and their components, for example for neurotransmitters, and genes for channel proteins are being localized in the genome and analyzed at the molecular level. Hence, the gulf between physiology and genetics is now being bridged. With the increasing interest of human geneticists in the genetic basis of common diseases and individual genetic variation in response to influences such as nutrition and stress, genetic concepts are increasingly influencing the many branches of medicine that, in the past, have profited relatively little from genetic theory. Molecular biology is developing increasingly into a common basis for many branches of science, and most biomedical scientists are nowadays becoming better acquainted with the principles of genetics. A field of molecular medicine is emerging.

Fields of Human and Medical Genetics

The field of human genetics is large, and its borders are indistinct. The development of different techniques and methods has led to the development of many fields of subspecialization. Many of these overlap and are not mutually exclusive. The field of *human molecular genetics* has its emphasis in the identification and analysis of genes at the DNA level. Methods such as DNA digestion by restriction endonucleases, Southern blotting, polymerase chain reaction (PCR), sequencing and many others are being applied. *Human biochemical genetics* deals with the biochemistry of nucleic acids, proteins, and enzymes in normal and mutant individuals. Laboratory methods of the biochemist are being used (e.g., chromatography; enzyme assays). *Human cytogenetics* deals with the study of human chromosomes in health and disease. *Immunogenetics* concerns itself largely with the genetics of blood groups, tissue antigens such as the HLA types, and other components of the immune system. *Formal genetics* studies segregation and linkage relationships of Mendelian genes and investigates more complex types of inheritance by statistical techniques.

Clinical genetics deals with diagnosis, prognosis, and to some extent treatment of various genetic diseases. Diagnosis requires knowledge of etiological heterogeneity and acquaintance with many disease syndromes. *Genetic counseling* is an important area of clinical genetics and requires skills in diagnosis, risk assessment, and interpersonal communication. *Population genetics* deals with the behavior of genes in large groups and concerns the evolutionary forces of drift, migration, mutation, and selection in human populations. The structure and gene pool of human populations are studied by considering gene frequencies of marker genes. In recent years population geneticists have become interested in the epidemiology of complex genetic disease that require biometric techniques for their studies. *Behavioral genetics* is a science that studies the hereditary factors underlying behavior in health and disease. Behavior geneticists attempt to work out the genetic factors determining personality and cognitive skills in human beings. The genetics of mental retardation and various psychiatric diseases are also considered. The field of sociobiology tries to explain social behavior by using biological and evolutionary concepts.

Somatic cell genetics is the branch of human genetics that studies the transmission of genes at the cellular level. Cell hybridization between different species is an important tool for the cartography of human genes. *Developmental genetics* studies genetic mechanisms of normal and abnormal development. This field employs to a large extent model organisms and has a strong emphasis on animal experimentation. *Reproductive genetics* is the branch of genetics that studies details of gamete and early embryo formation by genetic techniques. This area is closely related to reproductive physiology. Due to the growing application of assisted reproductive technologies in couples with infertility disorders this field has recently grown significantly. *Pharmacogenetics* deals with genetic factors governing the disposal and kinetics of drugs in the organism. Special interest in human pharmacogenetics relates to adverse drug reactions. *Ecogenetics* is an extension of pharmacogenetics and deals with the role of genetic variability affecting the response to environmental agents.

Clinical genetics has grown very rapidly in recent years because of the many practical applications of diagnosis and counseling, intrauterine diagnosis, and screening for genetic disease. Most research in human genetics is currently carried out in clinical genetics, cytogenetics, molecular and biochemical genetics, somatic cell genetics, and immunogenetics under medical auspices. Research in formal and population genetics has benefited enormously from the increasing knowledge about genome structure and its variation and the availability of new, cheaper high-throughput sequencing approaches.

Future of Human Genetics

Research methods in science are becoming ever more complicated and expensive, and human genetics is no exception. As a necessary consequence mastering of these methods increasingly requires specialization in a narrow field. Purchase of big instruments creates financial difficulties. Hence, the selection of research problems is often directed not by the intrinsic scientific interest in the problems or the conviction that they could, in principle, be solved, but by the availability of research methods, skilled coworkers, and instruments. Many research projects require large patient cohorts and complex, genome-wide analyses, tasks of a magnitude that can only be performed within international consortia. Such efforts are greatly facilitated by web-based databases (Sects. 29.1–29.3) which provide an easy means for distributing results to the genetic community. Furthermore, such databases ensure that new evolving information can easily be utilized by other persons in the field. For example, data on copy number variation in the human genome and possible consequences for the phenotype are now rapidly assessable in databases (Sects. 29.2 and 29.3) and are thus available for genetic counselors who can use this knowledge to provide their patients with detailed up-to-date information.

However, the tendency toward specialization will inevitably continue, and it is possible that, in this process, important parts of human genetics will be resolved into fields mainly defined by research methods, such as biochemistry, chromosome research, immunology, molecular biology [see 12], or into certain clinical areas. For example, hereditary metabolic diseases or syndromes associated with dysmorphic features and developmental delay are often studied and treated by pediatricians with little genetic training. Several departments of neurology have established their own neurogenetics branches, which are often independent from the respective department of human genetics. However, despite this tendency toward subspecialization, it is important to note that a laboratory performing genetic diagnostic procedures needs trained and experienced personnel, up-to-date equipment, and has to fulfill internationally defined quality standards, which are regulated by law in many countries. Therefore, it is probably not cost-effective to perform genetic diagnostics in small laboratories that offer only a few tests. Therefore, large laboratories performing all important human genetics diagnostic procedures may evolve to organizational structures in which human genetics remains united.

Survival of an established field of science has no value in itself. If a field dies because its concepts and accomplishments have been accepted and are being successfully integrated into other fields, little is lost. In human genetics, however, this state has not been reached yet and it may never get to this point. Many concepts of molecular biology, often in combination with "classical" methods such as linkage analysis, are now being applied to humans. A few decades ago human genetics was a medical field mainly dealing with rare syndromes and prenatal diagnostics. This picture has completely changed as the genetic contributions to common diseases are increasingly being unraveled. For example, genetic counseling is now an integral part of care in families with hereditary cancer diseases (Chap. 14) or neurologic disorders. In addition, data evolving from genome-wide association studies (GWAS) have identified numerous new loci in the genome that may change the susceptibility for diseases or phenotypic features. The effect of these loci may often be only moderate (Sect. 8.1), however, the evolving knowledge may further increase requests for genetic counseling. In future, genetic counseling provided by professionals in the field may have to compete with "direct-to-consumer genetic testing" over the Internet that is already offered by several companies. Such developments are accompanied by growing options for predictive genetic diagnosis, which require standardized procedures for both the counseling session and the molecular genetic testing and which often involve difficult ethical issues. Thus, the tasks in human genetics have changed tremendously over the past decades and new challenges are constantly arising in this rapidly evolving field. Newly evolving technologies, such as whole-genome sequencing (see below), will further expand the future of human genetics. In fact, it can be predicted that human genetics will change medicine, as it has the potential to identify persons with an increased risk for certain diseases and it may provide information about treatment options. These aspects are now often referred to as "personalized medicine" (Chap. 4, Sect. 4.4) and they will likely dominate medicine in upcoming years.

Unsolved and Intriguing Problems

With the rapid increase in knowledge over recent years new and often unexpected problems have arisen. At a time when hereditary traits were defined by their modes of inheritance, the relationship between genotype and phenotype appeared relatively simple. This straightforward relationship seemed correct when some hereditary diseases were shown to be caused by enzyme defects, and when hemoglobin variants turned out to be due to amino acid replacements caused by base substitutions. With increasing knowledge of the human genome, however, many hereditary traits with phenotypes that had been considered identical turned out to be heterogeneous. These were caused either by mutations in different genes or by different mutations within the same genes. However, even mutations that are identical by the strictest molecular criteria sometimes have striking phenotypic differences. Analysis of such genotype-phenotype relationships by the study of genetic and environmental modifiers poses intriguing future problems in human genetics.

The establishment of genotype–phenotype relationships was recently further complicated by two new findings. The first finding represents the unanticipated variation within the human genome (Chap. 2). Future research will have to elucidate how copy number variants (CNVs) contribute to human phenotypic diversity and disease susceptibility. CNVs are also of interest for a better understanding of the evolution of the genome, as they provide the raw material for gene duplication and gene family expansion. However, in addition to numerical variation there are extensive structural variations, such as inversions or insertions. Their impact on gene function remains to be elucidated. The second finding was the characterization of functional elements by the ENCODE consortium. To date, only 1% of the human genome has been analyzed by various high-throughput experimental and computational techniques; however, the findings revealed an unexpected number and complexity of the RNA transcripts that the genome produces. These findings have challenged traditional views about regulatory elements in the genome and added new insights into the complexity of human genetics, revealing that our understanding of the genome is still far from being complete. In order to address this, the National Human Genome Research Institute (NHGRI) launched two complementary programs in 2007: an expansion of the human ENCODE project to the whole genome (http://www.genome.gov/ENCODE) and the model organism ENCODE (modENCODE) project to generate a comprehensive annotation of the functional elements in the *Caenorhabditis elegans* and *Drosophila melanogaster* genomes (http://www.modencode.org; http://www.genome.gov/modENCODE). These efforts will likely contribute to a better understanding of genome complexity and gene regulation.

At present our understanding of somatic genome variability is very incomplete. Current concepts suggest that erroneous DNA repair and incomplete restoration of chromatin after damage may be resolved and may produce mutations and epimutations. Both mutations and epimutations have been shown to accumulate with age and such an increased burden of mutations and/or epimutations in aged tissues may increase cancer risk and adversely affect gene transcriptional regulation. This may in turn result in a progressive decline in organ function, a phenomenon frequently observed in aging. With the demographic trend of prolonged life expectancy, a better understanding of somatic genome variability and the stability of the genome may grow in importance.

Other problems may arise from new technologies, such as next-generation or third-generation whole-genome sequencing (Chap. 4, Sect. 4.4), which will make sequencing of entire genomes possible and affordable within in a short period of time. These possibilities will require new bioinformatic tools and interpretation of sequencing results will greatly depend on whether we understand better the aforementioned relevance of structural and copy number variation and whether we can make sense of the various transcriptionally active regions in the genome. If we succeed, there is no doubt that whole-genome sequencing will change human genetics tremendously. They will, for example, contribute to a better understanding of modifier genes in monogenic diseases and thus explain the frequently observed phenotypic variability. Furthermore, they will contribute significantly to further propel research on complex diseases. However, although the new possibilities of human genetics are fascinating they raise at the same time new ethical issues. For example, in prenatal diagnostic settings tests can now be offered not only for devastating diseases but also for common phenotypic traits. Thus, the consequences of the new technologies and new insights do not have consequences only for human geneticists but also for the entire society.

Possible Function of a Textbook

In his book on *The Structure of Scientific Revolutions,* Kuhn in 1962 [2] described the function of textbooks not very flatteringly: they are "pedagogic vehicles for the perpetuation of normal science" that create the impressions as if science would grow in a simple, cumulative manner. They tend to distort the true history of the field by only mentioning those contributions in the past that can be visualized as direct forerunners of present-day achievements. "They inevitably disguise not only the role but the very existence of . . . revolutions . . ."

Below we shall proceed in the same way: we shall describe present-day problems in human genetics as we see them. The result is a largely affirmative picture of normal science in a phase of rapid growth and success. Anomalies and discrepancies may exist, but we often do not identify them because we share the "blind spots" with most other members of our paradigm group. The "anticipation" phenomenon in diseases such as myotonic dystrophy is one example (Chap. 5, Sect. 5.1.7). This disease tends to manifest more severely and earlier in life with each generation. Obviously, this observation did not appear to be compatible with simple mendelism. Therefore, it was explained away by sophisticated statistical arguments which we cited in earlier editions of this book. In the meantime, however, anticipation has been shown to be a real phenomenon, caused by a novel molecular mechanism. What we can do is to alert the reader that human genetics, as all other branches of science, is by no way a completed and closed complex of theory and results that only needs to be supplemented in a straightforward way and without major changes in conceptualization. Our field has not developed – and will not develop in the future – as a self-contained system. Rather, human genetics, as all other sciences, is an undertaking of human beings – social groups and single outsiders – who are motivated by a mixture of goals such as search for truth, ambition, desire to be acknowledged by one's peer group, the urge to convince the society at large to allocate resources in their field – but also the wish to help people and to do something useful for human society.

Therefore, we shall emphasize the history and development of problems and approaches. Occasionally, we shall ask the reader to step back, reflecting with us as to why a certain development occurred at the time it did, why another development did not occur earlier, or why a certain branch of human genetics did not take the direction that one would have expected logically. Inevitably, this implies much more criticism than is usually found in textbooks. Such criticism will – at least partially – be subjective, reflecting the personal stance of the authors. Our goal is to convince the reader that a critical attitude improves one's grasp of the problems and their possible solutions – it is not our intention to convince him that we are always right.

We would have liked to give more information on the ways in which sociological conditions within the field and – still more important – the developments in the society at large have influenced the development of human genetics, and the ways in which thinking on these problems has in turn influenced the societies. The eugenics movement in the United States and the *Rassenhygiene* ideology in Germany have had a strong – and sometimes devastating – influence on human beings as well as on the social structure of society at large. Too little systematic research has been carried out, however, to justify a more extended discussion than that presented in Chap. 1 (Sect. 1.8) [17]. Much more historical research along these lines is all the more urgent, as many of the ethical problems – inherent, for example, in the sterilization laws of many countries during the first decades of the twentieth century – are now recurring with full force in connection with prenatal diagnosis, selective abortion and the possibility of germinal gene therapy (Chaps. 25 and 26). Scientists and physicians working in human genetics were actively involved in and sanctioned ethically abhorrent measures in the past such as killing severely malformed newborns and mentally defectives in Nazi Germany – and how will future generations judge our own activities? These are intriguing questions. They show the Janus face of human genetics: it is a fundamental science – guided by a fertile theory and full of fascinating problems. It is also an applied science, and its applications are bound to have a strong impact on society, leading to novel and difficult philosophical, social, and ethical problems.

References

1. Bunge M (1967) Scientific research, vols 1, 2. Springer, Berlin Heidelberg New York
2. Bernstein F (1924) Ergebnisse einer biostatistischen zusammenfassenden Betrachtung über die erblichen Blutstrukturen des Menschen. Klin Wochenschr 3:1495–1497
3. Cairns J, Stent GS, Watson JD (1966) Phage and the origin of molecular biology. Cold Spring Harbor Laboratory, New York
4. Fisher RA (1918) The correlation between relatives on the supposition of Mendelian inheritance. Trans R Soc Edinb 52:399–433
5. Garrod AE (1902) The incidence of alcaptonuria: a study in chemical individuality. Lancet 2:1616–1620
6. Galton F (1865) Hereditary talent and character. Macmillans Magazine 12:157
7. Harwood J (1993) Styles of scientific thought: the German genetics community 1900–1933. University of Chicago Press, Chicago
8. Hirschhorn K (1996) Human genetics: a discipline at risk for fragmentation. Am J Hum Genet 58:1–6 William Allan Award Address 1995
9. Kevles DJ (1985) In the name of eugenics. Genetics and the uses of human heredity. Knopf, New York
10. Kuhn TS (1962) The structure of scientific revolutions. University of Chicago Press, Chicago
11. Lakatos I, Musgrave A (1970) Criticism and the growth of knowledge. Cambridge University Press, New York

12. Mendel GJ (1865) Versuche über Pflanzenhybriden. Verhandlungen des Naturforschenden Vereins, Brünn
13. Mohr H (1977) The structure and significance of science. Springer, Berlin Heidelberg New York
14. Pauling L, Itano HA, Singer SJ, Wells IC (1949) Sickle cell anemia: a molecular disease. Science 110:543
15. Popper KR (1963) Conjectures and refutations. Rutledge and Kegan Paul, London
16. Popper KR (1970) Normal science and its dangers. In: Lakatos I, Musgrave A (eds) Criticism and the growth of knowledge. Cambridge University Press, New York, pp 51–58
17. Popper KR (1971) Logik der Forschung, 4th edn. Mohr, Tübingen Original 1934
18. Watkins JWN (1970) Against "normal science". In: Lakatos I, Musgrave A (eds) Criticism and the growth of knowledge. Cambridge University Press, New York, pp 25–38

History of Human Genetics*

Arno G. Motulsky

Revised by A.G. Motulsky, S.E. Antonarakis, and M.R. Speicher

Abstract Theories and studies in human genetics have a long history. Observations on the inheritance of physical traits in humans can even be found in ancient Greek literature. In the eighteenth and nineteenth centuries observations were published on the inheritance of numerous diseases, including empirical rules on modes of inheritance. The history of human genetics as a theory-based science began in 1865, when Mendel published his Experiments on Plant Hybrids and Galton his studies on Hereditary Talent and Character. A very important step in the development of human genetics and its application to medicine came with Garrod's demonstration of a Mendelian mode of inheritance in alkaptonuria and other inborn errors of metabolism (1902). Further milestones were Pauling's elucidation of sickle cell anemia as a "molecular disease" (1949), the discovery of genetic enzyme defects as the causes of metabolic disease (1950s, 1960s), the determination that there are 46 chromosomes in humans (1956), the development of prenatal diagnosis by amniocentesis (1968–1969) for the detection of chromosomal defects such as Down syndrome, and the large-scale introduction of molecular methods during the last 25 years. Concepts appropriated from human genetics have often influenced social attitudes and introduced the eugenics movement. Abuses have occurred, such as legally mandated sterilization, initially in the United States and later more extensively in Nazi Germany, where the killing of mentally impaired patients was followed by the genocide of Jews and Romani (Gypsy) people.

Contents

1.1	The Greeks (see Stubbe [83])	14
1.2	Scientists Before Mendel and Galton	15
1.3	Galton's Work	16
1.4	Mendel's Work	17
1.5	Application to Humans: Garrod's Inborn Errors of Metabolism	18
1.6	Visible Transmitters of Genetic Information: Early Work on Chromosomes	20
1.7	Early Achievements in Human Genetics	20
	1.7.1 AB0 and Rh Blood Groups	20
	1.7.2 Hardy-Weinberg Law	21

A.G. Motulsky (✉)
Departments of Medicine (Division of Medical Genetics) and Genome Sciences
University of Washington, School of Medicine, Box 355065, 1705 N.E. Pacific St. Seattle, WA 98195-5065, USA
e-mail: agmot@u.washington.edu

*This chapter is a revised and updated version of the third edition's history chapter originally authored by Friedrich Vogel (died summer 2006) and Arno Motulsky.

1.7.3	Developments Between 1910 and 1930	21
1.8	Human Genetics, the Eugenics Movement, and Politics	21
1.8.1	United Kingdom and United States	21
1.8.2	Germany	22
1.8.3	Soviet Union/Russia (see Harper, Chap. 16 in [38])	23
1.8.4	Human Behavior Genetics	23
1.9	Development of Medical Genetics (1950–the Present)	23
1.9.1	Genetic Epidemiology	23
1.9.2	Biochemical Methods	24
1.9.3	Genetic and Biochemical Individuality	24
1.9.4	Cytogenetics, Somatic Cell Genetics, Prenatal Diagnosis, Clinical Genetics	24
1.9.5	DNA Technology in Medical Genetics	25
1.9.6	The "Industrialization" of Discoveries and the Team Efforts	26
1.9.7	Unsolved Problems	26
References		27

The history of human genetics is particularly interesting since, unlike in many other natural sciences, concepts of human genetics have often influenced social and political events. At the same time, the development of human genetics as a science has been influenced by various political forces. Human genetics because of its concern with the causes of human variability has found it difficult to either remain a pure science or one of strictly medical application. Concerns regarding the heritability of IQ and the existence of inherited patterns of behavior again have brought the field into public view. A consideration of the history of human genetics with some attention to the interaction of the field with societal forces is therefore of interest. We will concentrate our attention on historical events of particular interest for human genetics and refer to landmarks in general genetics only insofar as they are essential for the understanding of the evolution of human genetics.

Recently, an excellent history of medical genetics was published by the medical geneticist Peter Harper in 2008 [38]. This highly readable book with many photographs presents critical assessments of various developments in the field since its beginnings. Many tables document major discoveries and a detailed timeline of both human and medical genetics presents important developments ranging from early discoveries to recent findings. This book is currently the only major comprehensive text devoted to the history of human/medical genetics.

A 30-page "History of Medical Genetics" by Victor McKusick was published as Chap. 1 in Emery and Rimoin's *Principles and Practice of Medical Genetics*, 5th edition, 2007 [59]. This remarkably comprehensive chapter emphasizing clinical aspects starts with a brief description of pre-Mendelian concepts and ends with a broadly conceived assessment of current and future trends of medical genetics.

1.1 The Greeks (see Stubbe [83])

Prescientific knowledge regarding inherited differences between humans has probably existed since ancient times. Early Greek physicians and philosophers not only reported such observations but also developed some theoretical concepts and even proposed "eugenic" measures.

In the texts that are commonly ascribed to Hippocrates, the following sentence can be found:

> Of the semen, however, I assert that it is secreted by the whole body – by the solid as well as by the smooth parts, and by the entire humid matters of the body ... The semen is produced by the whole body, healthy by healthy parts, sick by sick parts. Hence, when as a rule, baldheaded beget baldheaded, blue-eyed beget blue-eyed, and squinting, squinting; and when for other maladies, the same law prevails, what should hinder that longheaded are begotten by longheaded?

This remarkable sentence not only contains observations on the inheritance of normal and pathological traits but also a theory that explains inheritance on the assumption that the information carrier, the semen, is produced by all parts of the body, healthy, and sick. This theory became known later as the "pangenesis" theory. Anaxagoras, the Athenian philosopher (500–428 B.C.), had similar views (see Capelle [15]).

A comprehensive theory of inheritance was developed by Aristotle (see [6]). He also believed in a qualitatively different contribution by the male and the female principles to procreation. The male gives the impulse to movement whereas the female contributes the matter, as the carpenter who constructs a bed out of wood. When the male impact is stronger, a son is born who, at the same time, is more like his father, when the female, a daughter, resembling the mother. This is the reason why sons are usually similar to their fathers and daughters are similar to their mothers.

Barthelmess (our translation) [6] writes: "Reading the texts from this culture, one gets the overall impression that the Greeks in their most mature minds came closer to the theoretical problems than to the phenomena of heredity." Aristotle's assertion even provides an early example of how observation can be misled by a preconceived theoretical concept. Sons are not more similar to their fathers, nor daughters to their mothers.

Plato, in the *Statesman (Politikos)* [71], explained in detail the task of carefully selecting spouses to produce children who will develop into bodily and ethically eminent personalities. He wrote:

> They do not act on any sound or self-consistent principle. See how they pursue the immediate satisfaction of their desire by hailing with delight those who are like themselves and by disliking those who are different. Thus they assign far too great an importance to their own likes and dislikes.
>
> The moderate natures look for a partner like themselves, and so far as they can, they choose their wives from women of this quiet type. When they have daughters to bestow in marriage, once again they look for this type of character in the prospective husband. The courageous class does just the same thing and looks for others of the same type. All this goes on, though both types should be doing exactly the opposite …
>
> Because if a courageous character is reproduced for many generations without any admixture of the moderate type, the natural course of development is that at first it becomes superlatively powerful but in the end it breaks out into sheer fury and madness …
>
> But the character which is too full of modest reticence and untinged by valor and audacity, if reproduced after its kind for many generations, becomes too dull to respond to the challenges of life and in the end becomes quite incapable of acting at all.

In the *Republic* [70], Plato not only requires for the "guards" (one of the highest categories in the social hierarchy of his utopia) that women should be common property; children, should be educated publicly but the "best" of both sexes should beget children who are to be educated with care. The children of the "inferior," on the other hand, are to be abandoned. Democritus, on the other hand, writes: "More people become able by exercise than by their natural predisposition." Here (as in other places), the nature–nurture problem appears already.

1.2 Scientists Before Mendel and Galton

The literature of the Middle Ages contains few allusions to heredity. The new attitude of looking at natural phenomena from an empirical point of view created modern science and distinguishes modern humans from those in earlier periods. This approach succeeded first in investigation of the inorganic world and only later in biology. In the work *De Morbis Hereditariis* by the Spanish physician Mercado (1605) [66], the influence of Aristotle is still overwhelming, but there are some hints of a beginning emancipation of reasoning. One example is his contention that both parents, not only the father, contribute a seed to the future child. Malpighi (1628–1694) [83, p 77] proposed the hypothesis of "preformation," which implies that in the ovum the whole organism is preformed in complete shape, only to grow later. Even after the discovery of sperm (Leeuwenhoek et al. 1677) [3, pp 72-73], the preformation hypothesis was not abandoned altogether, but it was believed by some that the individual is preformed in the sperm, only being nurtured by the mother. The long struggle between the "ovists" and the "spermatists" was brought to an end only when C.F. Wolff [99] attacked both sides and stressed the necessity of further empirical research. Shortly thereafter experimental research on heredity in plants was carried out by Gärtner (1772–1850) [33] and Kölreuter (1733–1806) [48]. Their work prepared the ground for Mendel's experiments [60].

The medical literature of the eighteenth and early nineteenth centuries contains reports showing that those capable of clear observation were able to recognize correctly some phenomena relating to the inheritance of diseases. Maupertuis [57], for example, published in 1753 an account of a family with polydactyly in four generations and demonstrated that the trait could be equally transmitted by father or by mother. He further showed, by probability calculation, that chance alone could not account for the familial concentration of the trait. Probably the most remarkable example, however, was Joseph Adams (1756–1818) (see [1,23,62,64]), a British apothecary who, in 1814, published a book with the title *A Treatise on the Supposed Hereditary Properties of Diseases* [1]. The following findings are remarkable:

(a) Adams differentiated clearly between "familial" (i.e., recessive) and "hereditary" (i.e., dominant) conditions.
(b) He knew that in familial diseases the parents are frequently near relatives.

(c) Hereditary diseases need not be present at birth; they may manifest themselves at various ages.
(d) Some disease predispositions lead to a manifest disease only under the additional influence of environmental factors. The progeny, however, is endangered even when the predisposed do not become ill themselves.
(e) Intrafamilial correlations as to age of onset of a disease can be used in genetic counseling.
(f) Clinically identical diseases may have different genetic bases.
(g) A higher frequency of familial diseases in isolated populations may be caused by inbreeding.
(h) Reproduction among persons with hereditary diseases is reduced. Hence, these diseases would disappear in the course of time, if they did not appear from time to time among children of healthy parents (i.e., new mutations!).

Adams' attitude toward "negative" eugenic measures was critical. He proposed the establishment of registries for families with inherited diseases. Weiss [96] recently pointed out that Adams in the same book also hinted at the existence of evolution stressing the concept of adaptive selection saying that environments such as climate put constraints on people: "By these means a race is gradually reared with constitutions best calculated for the climate" [1].

C.F. Nasse, a German professor of medicine, correctly recognized in 1820 one of the most important formal characteristics of the X-linked recessive mode of inheritance in hemophilia and presented a typical comprehensive pedigree [83, p 180]. He wrote (our translation):

> All reports on families, in which a hereditary tendency towards bleeding was found, are in agreement that the bleeders are persons of male sex only in every case. All are explicit on this point. The women from those families transmit this tendency from their fathers to their children, even when they are married to husbands from other families who are not afflicted with this tendency. This tendency never manifests itself in these women. …

Nasse also observed that some of the sons of these women remain completely free of the bleeding tendency.

The medical literature of the nineteenth century shows many more examples of observations, and attempts to generalize and to find rules for the influence of heredity on disease can be found. The once very influential concept of "degeneration" should be mentioned. Some features that older authors described as "signs of degeneration" in the external appearance of mentally deficient patients are now known to be characteristic of autosomal chromosomal aberrations or various types of mental retardation.

In the work of most of the nineteenth century authors, true facts and wrong concepts were inextricably mixed, and there were few if any criteria for getting at the truth. This state of affairs was typical for the plight of a science in its prescientific state. Human genetics had no dominant paradigm. The field as a science was to start with two paradigms in 1865: biometry, which was introduced by Galton, and Mendelism, introduced by Mendel with his pea experiments. The biometric paradigm was influential in the early decades of the twentieth century, and some examples and explanations in this book utilize its framework. With the advent of molecular biology and insight into gene action, the pure biometric approach in genetics is on the decline. Nevertheless, many new applications in behavioral or social genetics, where gene action cannot yet be studied, rely on this paradigm and its modern elaborations. The laws that Mendel derived from his experiments, on the other hand, have been of almost unlimited fruitfulness and analytic power. The gene concept emerging from these experiments has become the central concept of all of genetics, including human genetics. Its possibilities have not been exhausted.

1.3 Galton's Work

In 1865, F. Galton published two short papers with the title "Hereditary Talent and Character." He wrote [29]:

> The power of man over animal life, in producing whatever varieties of form he pleases, is enormously great. It would seem as though the physical structure of future generations was almost as plastic as clay, under the control of the breeder's will. It is my desire to show, more pointedly than – so far as I am aware – has been attempted before, that mental qualities are equally under control.
>
> A remarkable misapprehension appears to be current as to the fact of the transmission of talent by inheritance. It is commonly asserted that the children of eminent men are stupid; that, where great power of intellect seems to have been inherited, it has descended through the mother's side; and that one son commonly runs away with the talent of the whole family.

He then stresses how little we know about the laws of heredity in man and mentions some reasons, such as long generation time, that make this study very difficult. However, he considers the conclusion to be justified that physical features of humans are transmissible because resemblances between parents and offspring are obvious. Breeding experiments with animals, however, had not been carried out at that time, and direct proof of hereditary transmission was therefore lacking even in animals. In humans, "we have ... good reason to believe that every special talent or character depends on a variety of obscure conditions, the analysis of which has never yet been seriously attempted." For these reasons, he concluded that single observations must be misleading, and only a statistical approach can be adequate.

Galton evaluated collections of biographies of outstanding men as to how frequently persons included in these works were related to each other. The figures were much higher than would be expected on the basis of random distribution.

Galton himself was fully aware of the obvious sources of error of such biological conclusions. He stressed that "when a parent has achieved great eminence, his son will be placed in a more favorable position for advancement, than if he had been the son of an ordinary person. Social position is an especially important aid to success in statesmanship and generalship ... ."

"In order to test the value of hereditary influence with greater precision, we should therefore extract from our biographical list the names of those that have achieved distinction in the more open fields of science and literature." Here and in the law, which in his opinion was "the most open to fair competition," he found an equally high percentage of close relatives reaching eminence. This was especially obvious with Lord Chancellors, the most distinguished lawyers of Great Britain.

Galton concluded that high talent and eminent achievement are strongly influenced by heredity. Having stressed the social obstacles that inhibit marriage and reproduction of the talented and successful, he proceeded to describe a utopic society,

> In which a system of competitive examination for girls, as well as for youths, had been so developed as to embrace every important quality of mind and body, and where a considerable sum was yearly allotted. ... to the endowment of such marriages as promised to yield children who would grow into eminent servants of the State. We may picture to ourselves an annual ceremony in that Utopia or Laputa, in which the Senior Trustee of the Endowment Fund would address ten deeply-blushing young men, all of twenty-five years old, in the following terms....

In short, they were informed that the commission of the endowment fund had found them to be the best, had selected for each of them a suitable mate, would give them a substantial dowry, and promised to pay for the education of their children.

This short communication already shows human genetics as both a pure and an applied science: on the one hand, the introduction of statistical methods subjects general impressions to scientific scrutiny, thereby creating a new paradigm and turning prescience into science. Later, Galton and his student K. Pearson proceeded along these lines and founded biometric genetics. On the other hand, however, the philosophical motive of scientific work in this field is clearly shown: the object of research is an important aspect of human behavior. The prime motive is the age-old inscription on the Apollo temple at Delphi ("know yourself").

Hence, with Galton, research in human genetics began with strong eugenic intentions. Later, with increasing methodological precision and increasing analytic success, such investigations were removed from this prime philosophical motive. This motive helps to understand the second aspect of Galton's work: the utopian idea to improve the quality of the human species by conscious breeding. During the Nazi era in Germany (1933–1945) we saw how cruel the perverted consequences of such an idea may become (Sect. 1.8.2). The question first posed by Galton remains, even more than ever, of pressing importance: What will be the biological future of mankind?

1.4 Mendel's Work

The other leading paradigm was provided by Mendel in his work *Experiments in Plant Hybridization,* which was presented on 8 February and 8 March 1865 before the *Naturforschender Verein* (Natural Science Association) in Brünn (now Brno, Czech Republic) and subsequently published in its proceedings [60]. It has frequently been told how this work went largely unnoticed for 35 years and was rediscovered independently by Correns, Tschermak, and de Vries in 1900 (see [16, 84, 20]). From then on, Mendel's insights triggered the development of modern genetics, including human genetics. A book by Stern and Sherwood [82], which reprints these and a variety of other articles regarding Mendel's paper, is most helpful to assess the impact of this classic work.

Mendel was stimulated to carry out his experiments by observations on ornamental plants, in which he had tried to breed new color variants by artificial insemination. Here he had been struck by certain regularities. He selected the pea for further experimentation. He crossed varieties with differences in single characters such as color (yellow or green) or form of seed (round or angular wrinkled) and counted all alternate types in the offspring of the first generation crosses and of crosses in later generations. Based on combinatorial reasoning, he gave a theoretical interpretation: the results pointed to free combination of specific sorts of egg and pollen cells. In fact, this concept may have occurred to Mendel before he carried out his studies. He may have verified and illustrated his findings by his "best" results, since agreement between the published figures and their expectation from the theoretical segregation ratios is too perfect from a statistical point of view (Fisher [27]). The interpretation of this discrepancy remains controversial [82, 90]. In any case, there is no question that Mendel's findings were correct.

Mendel discovered three laws: the law of uniformity, which states that after crossing of two homozygotes of different alleles the progeny of the first filial generation (F_1) are all identical and heterozygous; the law of segregation, which postulated $1:2:1$ segregation in intercrosses of heterozygotes and $1:1$ segregation in backcrosses of heterozygotes with homozygotes; and the law of independence, which states that different segregating traits are transmitted independently.

What is so extraordinary in Mendel's contribution that sets it apart from numerous other attempts in the nineteenth century to solve the problem of heredity? Three points are most important:

1. He simplified the experimental approach by selecting characters with clear alternative distributions, examining them one by one, and proceeding only then to more complicated combinations.
2. Evaluating his results, he did not content himself with qualitative statements but counted the different types. This led him to the statistical law governing these phenomena.
3. He suggested the correct biological interpretation for this statistical law: The germ cells represent the constant forms that can be deduced from these experiments.

With this conclusion Mendel founded the concept of the gene, which has proved so fertile ever since. The history of genetics since 1900 is dominated by analysis of the gene. What had first been a formal concept derived from statistical evidence has emerged as the base pair sequence of DNA, which contains the information for protein synthesis and for life in all its forms.

1.5 Application to Humans: Garrod's Inborn Errors of Metabolism

The first step of this development is described in this historical introduction: A. Garrod's [30] paper on "The Incidence of Alkaptonuria: A Study in Chemical Individuality." There are two reasons for giving special attention to this paper. For the first time, Mendel's gene concept was applied to a human character, and Mendel's paradigm was introduced into research on humans. Additionally, this work contains many new ideas set out in a most lucid way. Garrod was a physician and in later life became the successor of Osler in the most prestigious chair of medicine at Oxford [8]. His seminal contribution to human genetics remained unappreciated during his lifetime. Biologists paid little attention to the work of a physician. Their interest was concentrated more on the formal aspects of genetics rather than on gene action. The medical world did not understand the importance of his observations for medicine. Garrod first mentioned the isolation of homogentisic acid from the urine of patients with alkaptonuria and stated the most important result of the investigations carried out so far:

> As far as our knowledge goes, an individual is either frankly alkaptonuric or conforms to the normal type, that is to say, excretes several grammes of homogentisic acid per diem or none at all. Its appearance in traces, or in gradually increasing or diminishing quantities, has never yet been observed....

As a second important feature "the peculiarity is in the great majority of instances congenital...." Thirdly: "The abnormality is apt to make its appearance in two or more brothers and sisters whose parents are normal and among whose forefathers there is no record of its having occurred." Fourthly, in six of ten reported families the parents were first cousins, whereas the incidence of first-cousin marriages in contemporary England was estimated to be not higher than 3%. On the other hand, however, children with alkaptonuria are observed in a very small fraction only of all first-cousin marriages.

There is no reason to suppose that mere consanguinity of parents can originate such a condition as alkaptonuria in their offspring, and we must rather seek an explanation in some peculiarity of the parents, which may remain latent for generations, but which has the best chance of asserting itself in the offspring of the union of two members of a family in which it is transmitted.

Then, Garrod mentioned the law of heredity discovered by Mendel, which "offers a reasonable account of such phenomena" that are compatible with a recessive mode of inheritance as pointed out by Bateson [37]. He cited another remark of Bateson and Saunders (Report to the Evolution Committee of the Royal Society) [7] with whom he had discussed his data:

> We note that the mating of first cousins gives exactly the conditions most likely to enable a rare, and usually recessive, character to show itself. If the bearer of such a gamete mates with individuals not bearing it the character will hardly ever be seen; but first cousins will frequently be the bearers of similar gametes, which may in such unions meet each other and thus lead to the manifestation of the peculiar recessive characters in the zygote.

After having cited critically some opinions on the possible causes of alkaptonuria, Garrod proceeded:

> The view that alkaptonuria is a "sport" or an alternative mode of metabolism will obviously gain considerably in weight if it can be shown that it is not an isolated example of such a chemical abnormality, but that there are other conditions which may reasonably be placed in the same category.

Having mentioned albinism and cystinuria as possible examples, he went on: "May it not well be that there are other such chemical abnormalities which are attended by no obvious peculiarities [as the three mentioned above] and which could only be revealed by chemical analysis?" And further:

> If it be, indeed, the case that in alkaptonuria and the other conditions mentioned we are dealing with individualities of metabolism and not with the results of morbid processes the thought naturally presents itself that these are merely extreme examples of variations of chemical behavior which are probably everywhere present in minor degrees and that just as no two individuals of a species are absolutely identical in bodily structure neither are their chemical processes carried out on exactly the same lines.

He suggested that differential responses toward drugs and infective agents could be the result of such chemical individualities. The paper presents the following new insights:

(a) Whether a person has alkaptonuria or not is a matter of a clear alternative – there are no transitory forms. This is indeed a condition for straightforward recognition of simple modes of inheritance.

The condition is observed in some sibs and not in parents.

The unaffected parents are frequently first cousins.

This is explained by the hypothesis of a recessive mode of inheritance according to Mendel. The significance of first-cousin marriages is stressed especially for rare conditions; this may be a precursor to population genetics.

(b) Apart from alkaptonuria several other similar "sports" such as albinism and cystinuria may exist. This makes alkaptonuria the paradigm for the "inborn errors of metabolism." In 1909 Garrod published his classic monograph on this topic [31].

(c) These sports may be extreme and therefore conspicuous examples of a principle with *much more widespread applicability*. Lesser chemical differences between human beings are so frequent that no human being is identical chemically to anyone else.

From these concepts Garrod drew more far-reaching conclusions, which are often overlooked. In a book published in 1931 [32] and reprinted with a lengthy introduction by Scriver and Childs [80], Garrod suggested that hereditary susceptibilities or diatheses are a predisposing factor for most common diseases and not merely for the rare inborn errors of metabolism. These concepts were precursors of current work to delineate the specific genes involved in the etiology of common disease. A valuable biography of Garrod was published by A. Bearn [8], who was a pioneer of human biochemical genetics in the 1950s and later.

Throughout this book the principle of a genetically determined individuality will govern our discussions. Garrod's contribution may be contrasted with that of Adams [23, 62, 64]. Apart from the "familial" occurrence of some hereditary diseases, Adams observed a number of phenomena that were not noted by Garrod, such as the late onset of some diseases, the intrafamilial correlation of age of onset, and the genetic predisposition leading to manifest illness only under certain environmental conditions. However, Adams did not have Mendel's paradigm. Therefore, his efforts could not lead to the development of an explanatory theory and coherent field of science. Garrod did have this paradigm and used it, creating a new area of research: human biochemical genetics.

1.6 Visible Transmitters of Genetic Information: Early Work on Chromosomes

Galton's biometric analysis and Mendel's hybridization experiments both started with visible phenotypic differences between individuals. The gene concept was derived from the phenotypic outcome of certain crossings. At the time when Mendel carried out his experiments nothing was known about a possible substantial bearing of genetic information in the germ cells. During the decades to follow, however, up to the end of the nineteenth century, chromosomes were identified, and mitosis and meiosis were analyzed. These processes were found to be highly regular and so obviously suited for orderly distribution of genetic information that in 1900 the parallelism of Mendelian segregation and chromosomal distribution during meiosis was realized, and chromosomes were identified as bearers of the genetic information [18].

Many research workers contributed to the development of cytogenetics [5,6]. O. Hertwig [41] first observed animal fertilization and established the continuity of cell nuclei: *omnis nucleus e nucleo*. Flemming (1880–1882) discovered the separation of sister chromatids in mitosis [83, p 247]; van Beneden (1883) [85] established the equal and regular distribution of chromosomes to the daughter nuclei. Boveri (1888) [5] found evidence for the individuality of each pair of chromosomes. Waldeyer (1888) (see [18]) coined the term "chromosome."

Meanwhile, Naegeli (1885) [77] had developed the concept of "idioplasma," which contains – to use a modern term – the "information" for the development of the next generation [67]. W. Roux [77] seems to have been the first to set out by logical deduction which properties a carrier of genetic information was expected to have. He also concluded that the behavior of cell nuclei during division would perfectly fulfill these requirements. The most important specific property of meiotic divisions, the ordered reduction of genetic material, was first recognized by Weismann.

These results and speculations set the stage for the identification of chromosomes as carriers of the genetic information, which followed shortly after the rediscovery of Mendel's laws and apparently independently by different authors [16, 20, 84].

Chromosome studies and genetic analysis have remained intimately connected in cytogenetics ever since. Most basic facts were discovered and concepts developed using plants and insects as the principal experimental tools. The fruit fly *Drosophila* played a particularly important role.

The development of human cytogenetics was delayed until 1956 when the correct number of human chromosomes was established as 46 by use of rather simple methods. It should be stressed that this delay could not be explained by the introduction of new cytological methods at that time. In fact, this discovery could have been made many years earlier. The delay was probably related to the lack of interest in human genetics by most laboratory-oriented medical scientists. Human genetics did not exist as a scientific discipline in medical schools since the field was not felt to be a basic science fundamental to medicine. Hereditary diseases were considered as oddities that could not be studied by the methodology of medical science as exemplified by the techniques of anatomy, biochemistry, physiology, microbiology, pathology, and pharmacology. Thus, most geneticists worked in biology departments of universities, colleges, or in agricultural stations. They were usually not attuned to problems of human biology and pathology, and there was little interest to study the human chromosomes. The discovery of trisomy 21 as the cause of Down syndrome and the realization that many problems of sex differentiation owe their origin to sex chromosomal abnormalities established the central role of cytogenetics in medicine. Further details in the development of cytogenetics are described in Chap. 3.

1.7 Early Achievements in Human Genetics

1.7.1 AB0 and Rh Blood Groups

The discovery of the AB0 blood group system by Landsteiner in 1900 [50] and the proof that these blood types are inherited (von Dungern and Hirschfeld [87]) was an outstanding example of Mendelian inheritance applied to a human character. Bernstein in 1924 [11] demonstrated that A, B, and 0 blood group characters are due to multiple alleles at one locus. The combined efforts of Wiener, Levine, and Landsteiner 25–30 years later led to discovery of the Rh factor and established that hemolytic disease of the newborn owes its origin to immunological maternal–fetal incompatibility. The stage was set for the demonstration in the 1960s that Rh hemolytic disease of the newborn can be prevented by administration of anti-Rh antibodies to mothers at risk [73,100].

1.7.2 Hardy-Weinberg Law

Hardy [36], a British mathematici,an, and Weinberg [92], a German physician, at about the same time (1908) set out the fundamental theorem of population genetics, which explains why a dominant gene does not increase in frequency from generation to generation. Hardy published his contribution in the United States in *Science*. He felt that this work would be considered as too trivial by his mathematics colleagues to be published in the United Kingdom. Weinberg was a practicing physician who made many contributions to formal genetics. He developed a variety of methods in twin research [91] and first elaborated methods to correct for biased ascertainment in recessive inheritance [93].

1.7.3 Developments Between 1910 and 1930

The years between 1910 and 1930 saw no major new paradigmatic discoveries in human genetics. Most of the data in formal genetics (such as linkage, nondisjunction, mutation rate) as well as the mapping of chromosomes were achieved by study of the fruit fly, largely in the United States. Many scientists tried to apply the burgeoning insights of genetics to humans. British scientists exemplified by Haldane excelled in the elaboration of a variety of statistical techniques required to deal with biased human data. The same period saw the development of the basic principles of population genetics by Haldane, Fisher, and Penrose [69] in England and by Wright in the United States. This body of knowledge became the foundation of population genetics and is still used by workers in that field. In 1918, Fisher was able to resolve the bitter controversies in England between the Mendelians, on the one hand, and followers of Galton (such as Pearson) on the other, by pointing out that correlations between relatives in metric traits can be explained by the combined action of many individual genes [26]. Novel steps in the development of medical genetics during this period were the establishment of empirical risk figures for schizophrenia and affective disorders by the Munich school of psychiatric genetics.

1.8 Human Genetics, the Eugenics Movement, and Politics

1.8.1 United Kingdom and United States

The first decade of the century saw the development of eugenics in Europe and in the United States [2,19,21,45, 55,76]. Many biological scientists were impressed by their interpretation of an apparently all-pervasive influence of genetic factors on most normal physical and mental traits as well as on mental retardation, mental disease, alcoholism, criminality, and various other sociopathies. They became convinced that the human species should be concerned with encouragement of breeding between persons with desirable traits (positive eugenics) and discourage the sick, mentally retarded, and disabled from procreation (negative eugenics).

A recent reprint of Davenport's 1911 book, *Heredity in Relation to Eugenics*, is accompanied by thoughtful reflections from contemporary geneticists on Davenport's eugenic concepts and recommendations almost one hundred years later [98]. Various eugenic study units were established in the United States (Eugenics Record Office at Cold Spring Harbor) and the United Kingdom. Much of the scientific work published by these institutions was of poor quality. Particularly, many different kinds of human traits such as "violent temper" and "wandering trait" were forced into Mendelian straightjackets. Most serious geneticists became disenchanted and privately disassociated themselves from this work. For various reasons, including those of friendship and collegiality with the eugenicists, the scientific geneticists did not register their disagreement in public. Thus, the propagandists of eugenics continued their work with enthusiasm, and the field acquired a much better reputation among some of the public than it deserved. Thus, many college courses on eugenics were introduced in the United States.

These trends had several important political consequences. Eugenics sterilization laws were passed in many states in the United States, which made it possible to sterilize a variety of persons for traits such as criminality for which no good scientific basis of inheritance existed. The attitude that led to the introduction of these laws is epitomized by United States Supreme Court Justice Holmes' statement that "three generations of imbeciles are enough."

Eugenic influences also played an important role in the passing of restrictive immigration laws in the United States. Using a variety of arguments the proponents of eugenics claimed to show that Americans of northwestern European origin were more useful citizens than those of southern European origin or those from Asia. Since such differences were claimed to be genetic in origin, immigration from southern and eastern European countries and from Asia was sharply curtailed. Similar trends were also operative in the United Kingdom. While solid work in human genetics was carried out by a few statistical geneticists, there was also much eugenic propaganda, including that by the distinguished statistician Pearson, the successor to Galton's academic chair in London.

Kevles [46] has published a wide-ranging and insightful history of eugenics and human genetics in the Anglo-Saxon countries. His book is a most carefully researched and exhaustive study of the uses and abuses of eugenic concepts.

1.8.2 Germany

In Germany [9, 10, 34, 94, 95] eugenics took the name of *Rassenhygiene* from a book of that title published in 1895 by Ploetz [72]. The *Rassenhygiene* movement became associated with mystical concepts of race, Nordic superiority, and the fear of degeneration of the human race in general and that of the German *Volk* in particular by alcoholism, syphilis, and increased reproduction of the feebleminded or persons from the lower social strata. Often representatives of this movement became associated with a dangerous type of sociopolitical prejudice: antisemitism. They warned the public against contamination of German "blood" by Jewish influences. Most followers of the racial hygiene concept were nationalistic and opposed the development of an open society that allows individual freedom and democratic participation. They shared this attitude with a significant segment of the educated classes in Germany. General eugenic ideas divorced from racism and other nationalist notions were often espoused by intellectuals who were concerned about the biological future of mankind. Thus, socialists publicized such views in Germany [34]. In 1931, two years before Hitler's coming into power, the German Society of Racial Hygiene added eugenics to its name. However, all efforts in this area soon became identified with the Nazi ideology.

Prominent German human geneticists identified themselves with the use of human genetics in the service of the Nazi state. Recognized scientists, such as Fischer, F. Lenz, Rüdin, and von Verschuer, accepted Nazi leadership and Nazi philosophy. While most of the propaganda for the new racial hygiene was not formulated by scientists but by representatives of the Nazi party, men such as Fischer and von Verschuer [95] participated in spreading Nazi race ideology. Jews were declared foreign genetic material to be removed from the German *Volk*. A eugenic sterilization law was already passed in 1933 that made forced sterilization obligatory for a variety of illnesses thought to be genetic in origin [74]. Heredity courts were established to deal with interpretation of the sterilization law. This law was hailed by some eugenicists in the United States even at the end of the 1930s [47]. Sterilization laws for eugenic indications were also passed in some Scandinavian countries around the same time but allowed voluntary (in contrast to forced) sterilization [74].

The exact role of the German human geneticists in the increasing radicalization and excesses of the application of Nazi philosophy has been assessed [65, 74, 95]; von Verschuer's role in sponsoring twin and other genetic research by his former assistant Mengele in the Auschwitz concentration and extermination camp is clear. We have no record that any voices were raised by these men in protest against "mercy killings" of the mentally retarded and newborn children with severe congenital defects nor against the mass killings of Jews. Evidence suggests that von Verschuer must have had some idea of such events, since he had continued contact with Mengele when the mass killings at Auschwitz were at their height. The "final solution" to the "Jewish problem" resulted in the murder of about 6 million Jews in the early 1940s [75]. While there is no record that human geneticists favored this type of "solution," their provision of so-called "scientific" evidence for a justification of Nazi antisemitism helped to create a climate in which these mass murders became possible [88]. This episode is one of the most macabre and tragic chapters in the history of man's inhumanity to man in the name of pseudoscientific nationalism. Yet, despite

1 History of Human Genetics

their racist publications, several such "scientists" (including von Verschuer) were given academic positions in post-World War II West Germany.

1.8.3 Soviet Union/Russia (see Harper, Chap. 16 in [38])

Eugenics was initiated in the Soviet Union [21,34] in the 1920s by the establishment of eugenics departments, a eugenic society, and a eugenics journal. Eugenic ideals soon clashed with the official doctrine of Marxism-Leninism, however, and these efforts were abandoned by the late 1920s. Scientists who had become identified with eugenics left the field to work with plants and animals.

Remarkable work in early human cytogenetics was carried out between 1931 and 1936, such as using hypotonic solutions for spreading of chromosomes, analysis of cultured embryonic cells, chromosome analysis of human oocytes, and cytogenetic studies of leukemia and other cancers [3,4]. These studies were published in international journals and later taken up by American and European scientists some 20 years later. Would the critical chromosome-related discoveries of the 1950s have been made by Russian scientists if such work on human genetics had not been terminated by Soviet antigenetic policies? [38]

Interest in the medical application of human genetics nevertheless persisted. A large institute of medical genetics, with 200 physicians, was established in Moscow during the 1920s. Its director, the physician S.G. Levit, made notable contributions [54], but was executed in 1938 (Chap. 16 in [38]), and human genetics was officially declared a Nazi science. The later ascendance of Lysenko [45] stifled all work in genetics, including that of human genetics, and no work whatever was carried out in this field until the early 1960s, after Lysenko's domination ceased (pp. 435–450 in [38]). The reintroduction of human genetics into the Soviet Union occurred by way of medical genetics. A textbook of medical genetics was published by Efroimson in 1964 [22]. A new institute of medical genetics was established in 1969 under the directorship of the cytogeneticist Bochkov, who had been trained by the well-known *Drosophila* geneticist, Timofeeff-Ressovsky [38]. Work in many areas of medical genetics, similar to that carried out elsewhere, is now done in Russia.

1.8.4 Human Behavior Genetics

Vigorous discussion continues regarding the role of genetic determinants in behavior, IQ, and personality. Some observers entirely deny genetic influences on normal behavior or social characteristics such as personality and intellect. This attitude toward genetics is shared by some psychologists and social scientists and even a few geneticists who are concerned about the possible future political and social misuse of studies in human behavioral genetics that claim to show genetic determinants of intelligence and social behavior.

We do not agree with those who deny any genetic influence on behavior or social traits in humans. However, we also caution against a too ready acceptance of results from comparison of twins and other relatives, which claim high heritabilities for many of these traits. Genetic data and pseudodata may be seriously misused by political bodies. However, as biologists and physicians impressed by biological variation under genetic control, we would be surprised if the brain did not also show significant variation in structure and function. Such variation is expected to affect intellect, personality, and behavior, and usually will interact with environmental factors. The extent to which genetic variation contributes to such traits, and especially the biological nature of such variation, will have to await further studies.

1.9 Development of Medical Genetics (1950–the Present)

1.9.1 Genetic Epidemiology

In the 1940s and 1950s a number of institutions pioneered in research on epidemiology of genetic diseases. T. Kemp's institute in Copenhagen, J.V. Neel's department in Ann Arbor, Michigan, and A.C. Stevenson's in Northern Ireland and later in Oxford contributed much to our knowledge on prevalence, modes of inheritance, heterogeneity, and mutation rates of various hereditary diseases. Recent years have seen a renaissance in this area, with special attention to analysis of common complex diseases (see Chap. 8.1). Utilization of new laboratory methods, including DNA techniques, together with more powerful methods of association studies, and the search for rare mutations and structural chromosome changes, provide powerful new approaches in this area.

1.9.2 Biochemical Methods

The years after World War II brought a rapid expansion in the field of human genetics by the development of biochemical, molecular, and cytological methods. Human genetics, which had been the concern largely of statistically oriented scientists, now entered the mainstream of medical research. The demonstration by Pauling et al. [68] that sickle cell anemia is a molecular disease was a key event in this area. The hemoglobins allowed detailed study of the consequences of mutation. The genetic code was found to be valid for organisms as far apart as viruses and humans. Many detectable mutations were found to be single amino acid substitutions, but deletions of various sorts and frameshift mutations similar to those discovered in micro-organisms were discovered. The nucleotide sequences of the hemoglobin genes were worked out using techniques developed in biochemistry and molecular genetics. Many inborn errors of metabolism were shown to originate in various enzyme deficiencies, often caused by a genetic mutation that changes enzyme structure. Methemoglobinemia due to diaphorase deficiency and glycogen storage disease were the first enzyme defects to be demonstrated.

1.9.3 Genetic and Biochemical Individuality

Work on hemoglobin and variants of the enzyme glucose-6-phosphate-dehydrogenase and other enzymes helped to establish the concept of extensive mutational variation. Biochemical individuality explained some drug reactions and led to the development of the field of pharmacogenetics [61, 86, 63, 35]. Marked biochemical heterogeneity of human enzymes and proteins was shown [39]. The uniqueness of humans, which is apparent by the physiognomic singularity of each human being, was shown to apply at the biochemical and immunological level as well. Here, as in several other fields (such as the hemoglobin variants and the mechanism of sex determination), studies in humans led the way to generally valid biological rules. The significance of polymorphism for the population structure (including that of humans) is being widely studied by population geneticists. The hypothesis that some expressed polymorphisms are the genetic substrate against which the environment acts to determine susceptibility and resistance to common disease led to the development of the field of ecogenetics [13,17]. The histocompatibility gene complex has become an important paradigm for the understanding of why several genes with related function occur in closely linked clusters. This locus appears to be of great importance to understand susceptibility to autoimmune diseases. An enormous amount of apparently unexpressed genetic variation has been demonstrated at the DNA and chromosomal level.

1.9.4 Cytogenetics, Somatic Cell Genetics, Prenatal Diagnosis, Clinical Genetics

After cytogenetic techniques became available, they were applied to detect many types of birth defects and intersex states. A specific type of malignancy, chronic myelogenous leukemia, was shown to be caused by a unique chromosomal translocation [78]. Banding techniques developed by Caspersson in 1969 made it possible to visualize each human chromosome and gave cytogenetic methods added powers of resolution.

Soon, biochemical and cytogenetic techniques were combined in somatic cell genetics. Specific enzyme defects were identified in single cells grown in tissue cultures. The development of methods to hybridize human with mouse cells by Henry Harris and Watkins [40] and Ephrussi and Weiss [25] soon allowed the assignment of many genes to specific chromosomes and the construction of a human linkage map.

The developments in somatic cell genetics led to the introduction of prenatal diagnosis in the late 1960s, when amniocentesis at the beginning of the second trimester of pregnancy was developed. This allowed tissue cultures of amniotic cells of fetal origin, permitting both cytogenetic and biochemical characterization of fetal genotypes, assignment of sex, and the diagnosis of a variety of disorders in utero. In the early 1980s chorion villus biopsy – a procedure done during the first trimester of pregnancy – was introduced, and is being widely used. The discovery that neural tube defects are associated with increases in α-fetoprotein of the amniotic fluid permits intrauterine diagnosis of an important group of birth defects [14]. Ultrasound methods to visualize the placenta and to diagnose fetal abnormalities added to the diagnostic armamentarium. This noninvasive method allows phenotypic diagnosis of a variety of fetal defects more frequently.

Clinical Genetics. The field of clinical genetics was initiated in the 1970s [58] and has been growing rapidly. Many medical schools and hospitals are establishing special clinics in which genetic diseases can be diagnosed and genetic counseling provided. The heterogeneity of genetic disease has been increasingly recognized. Genetic counseling – often by specially trained genetic counselors – is now intensified to provide patients and their families with information on the natural history of the disease, recurrence risks, and reproductive options. Screening programs of the entire newborn population for diseases such as phenylketonuria are being introduced in many countries, and other screening programs such as those to detect carriers of Tay-Sachs disease and other conditions more common among Ashkenazi Jews have undergone extensive trials [81].

With the advent of novel biochemical and DNA techniques (Chap. 4), basic work in human genetics is now performed increasingly by biochemists, cell biologists, molecular biologists, and others, who do not necessarily have training in human genetics. However, human genetics is identified with medical genetics in many of its activities. The scientific developments of the past decades are thus being widely applied in practical medicine.

1.9.5 DNA Technology in Medical Genetics

Advances in molecular genetics and DNA technology are being applied rapidly to practical problems of medical genetics. Since understanding of the hemoglobin genes was more advanced than that of other genetic systems, the initial applications related to the diagnosis of hemoglobinopathies (Chap. 11). Several methods are now being utilized. Inherited variation in DNA sequence that is phenotypically silent was found to be common, supplying a vast number of DNA polymorphisms for study. Just as everyone's physiognomy is unique, each person (except for identical twins) has a unique DNA pattern. DNA variants are being used in family or association studies as genetic markers to detect the presence of closely linked genes causing diseases. Direct detection of genetic disease has been achieved by utilizing nucleotide probes that are homologous to the mutations that are searched for. The polymerase chain reaction, together with rapidly increasing knowledge on human DNA sequences, has opened up new opportunities for direct diagnosis at the DNA level. Occasionally, a specific restriction enzyme may detect the mutational lesion. Different DNA mutations at the same locus frequently cause an identical phenotypic disease. This finding makes direct DNA diagnosis without family study difficult unless the specific mutation that causes the disease is known.

Completion of the human gene map and human gene sequence was achieved at the beginning of this century. Several hundred DNA markers and SNPs that are spaced over all chromosomes provide the necessary landmarks for detection of the genes for monogenic diseases and are beginning to hint at the contribution of specific genes to common diseases.

Using normal DNA carried by innocuous viruses to treat patients with genetic diseases carried by defective DNA has been under study for the last 15 years (Chap. 26). Such gene transfer aims to repair affected somatic cells (somatic gene therapy). Human studies have been done but no definitive cures have been reported. However, acute leukemia developed in several children treated for hereditary antibody syndrome presumably due to activation of oncogenes. Germinal gene therapy, i.e., insertion of normal genes into defective germ cells (or fertilized eggs) for treatment of human genetic disease, has never been carried out and is not considered ready for safe study. Such an approach is highly controversial, and is even prohibited by law in some countries.

McKusick [59] described a variety of paradigm shifts in the study of human and medical genetics in recent years. These included an emphasis from structural to functional genomics, from map-based to sequence-based gene discovery, from monogenic disease diagnosis to detection of common disorder susceptibility, from the search for etiology to exploration of mechanisms, from an emphasis on single genes to approaches on systems pathways and gene families, from genomics to proteomics and from "old-fashioned" medical genetics to "genetic medicine," implying that genes may be involved in all diseases. McKusick (p. 28 in [59]) further pointed out that human genetics in recent years has been "medicalized," "subspecialized," "professionalized," "molecularized," "commercialized," and even "consumerized."

1.9.6 The "Industrialization" of Discoveries and Team Efforts

The technological advances, the enormous amount of data generated, the size of the genomes, the impressive variability of individual genomes, the necessary specialized expertise in several disciplines, and the revolution in communication technologies all resulted in the organization and execution of mega-projects related to human genetics in the last 15 years in order to achieve results freely available to the community that provide genome-wide answers to the objectives. These projects, mostly international and funded by different funding agents, often included more than 50 different laboratories and 200 scientists. This paradigm shift is similar to the evolution of experimentation in physics, and underscores the importance of international cooperation in genomic discoveries. In addition, it is remarkable that most of the funding was provided by public sources. The completion of the human genome sequence was the first example of such international projects [44,49]. Other examples include the sequence of the genomes of other organisms and comparative genome analysis [89], the identification of the common genomic variation in a number of human population groups (HapMap project [28,42,43]), the ENCODE project to identify the functional elements in the human genome and that of selected model organisms [12], and the genome-wide association studies to identify common risk variants for the common complex phenotypes [56,79,97] (Chap. 8.1). More recently, the 1000 Genomes Project (http://www.1000genomes.org) and other related efforts aim to identify all genetic variation in the genomes of individuals. The major challenge in the future is to provide causative links between genomic variants and phenotypic variation.

1.9.7 Unsolved Problems

Human genetics had been most successful by being able to guide work that was made possible by the development of techniques from various areas of biology using Mendelian concepts. Important basic frontiers that are still being extended concern problems of gene regulation, especially during embryonic development, control of the immune system and of brain function. Human genetics is likely to contribute to these problems by imaginative use of the study of genetic variation and disease applying novel concepts and techniques. In medical genetics, the problem of common diseases including many birth defects requires study of the specific genes and their interactions involved in such diseases. Insights into the mechanisms of gene action during the aging process remain to be elucidated.

As shown by the many advances in description of genomic anatomy (see Chap. 2) where function is not yet fully understood, there is much need for research in both basic and translational approaches in order to elucidate the role of genomic biology and post-genomic interactions in health and disease. The remarkable similarity of humans and other mammals (and even of more primitive organisms) in both gene number and gene function had not been entirely expected, demonstrating that both new concepts and technical methods will be required to understand and utilize our current and future knowledge for applications in prevention and treatment of disease.

At first glance, the history of human genetics over the past 50 years reads like a succession of victories. The reader could conclude that human geneticists of the last generation pursued noble science to the benefit of mankind. However, how will posterity judge current efforts to make use of our science for the benefit of mankind as we understand it? Will the ethical distinction between selective abortion of a fetus with Down syndrome and infanticide of severely malformed newborns be recognized by our descendants? Are we again moving down the "slippery slope?"

Issues such as selective termination of pregnancy due to disadvantageous genomic variation need to be re-discussed and re-debated due to the ability to diagnose genomic variants with low-penetrance phenotypic consequences. As the dividing line between "severe phenotype" alleles and "low burden" alleles becomes blurred and individualized, consensus criteria and compromised solutions are fluid and constantly revised. Genetic medicine gradually becomes a central preoccupation of health professionals, the patients and their families, and presymptomatic healthy clients.

References

1. Adams J (1814) A treatise on the hereditary properties of disease. J Callow, London, p 33
2. Allen GE (1975) Genetics, eugenics and class struggle. Genetics 79:29–45
3. Andres AH, Jiv BV (1936) Somatic chromosome complex of the human embryo. Cytologia 7:317–388
4. Andres AH, Shiwago PI (1933) Karyologische Studien an myeloischer Leukämie des Menschen. Folia Haematol 49:1–20
5. Baltzer F (1967) Theodor Boveri: Life and work of a great biologist (1862-1915). Berkeley and Los Angeles, University of California Press
6. Barthelmess A (1952) Vererbungswissenschaft. Alber, Freiburg
7. Bateson W, Saunders E (1902) Experimental studies in the physiology of heredity. Reports to the Evolution Committee. R Soc 1:133–134
8. Bearn AG (1993) Archibald Garrod and the individuality of man. Clarendon, Oxford
9. Becker PE (1988) Zur Geschichte der Rassenhygiene (Wege ins Dritte Reich, vol 1). Thieme, Stuttgart
10. Becker PE (1990) SozialDarwinismus, Rassismus, Antisemitismus und völkischer Gedanke (Wege ins Dritte Reich, vol 2). Thieme, Stuttgart
11. Bernstein F (1924) Ergebnisse einer biostatistischen zusammenfassenden Betrachtung über die erblichen Blutstrukturen des Menschen. Klin Wochenschr 3:1495–1497
12. Birney E et al (2007) Identification and analysis of functional elements in 1% of the human genome by the ENCODE pilot project. Nature 447:799–816
13. Brewer GJ (1971) Annotation: human ecology and expanding role for the human geneticist. Am J Hum Genet 23:92–94
14. Brock DJH (1977) Biochemical and cytological methods in the diagnosis of neural tube defects. Prog Med Genet 2:1–40
15. Capelle W (1953) Die Vorsokratiker. Kröner, Stuttgart
16. Correns C (1900) G. Mendel's Regel über das Verhalten der Nachkommenschaft der Rassenbastarde. Ber Dtsch Bot Ges 18:158–168
17. Costa LG and Eaton DL (eds) (2006) Gene-environmental interaction: Fundamentals of ecogenetics. J Wiley and Sons, Hoboken, New Jersey
18. Cremer T (1986) Von der Zellenlehre zur Chromosomentheorie. Springer, Berlin Heidelberg New York
19. Davenport CB (1911) Heredity in relation to eugenics. Henry Holt, New York
20. De Vries H (1889) Intracellulare Pangenesis. Fisher, Jena
21. Dunn LC (1962) Cross currents in the history of human genetics. Am J Hum Genet 14:1–13
22. Efroimson VP (1964) Vvedenie v medicinskuju genetiku (Introduction to medical genetics). Gos Izd Med Lit, Moscow
23. Emery AEH (1989) Joseph Adams (1756–1818). J Med Genet 26:116–118
24. ENCODE Project Consortium (2004) The ENCODE (ENCyclopedia Of DNA Elements) Project. Science 306:636–640
25. Ephrussi B, Weiss MC (1965) Interspecific hybridization of somatic cells. Proc Natl Acad Sci USA 53:1040
26. Fisher RA (1918) The correlation between relatives on the supposition of mendelian inheritance. Trans R Soc Edinb 52:399–433
27. Fisher RA (1936) Has Mendel's work been rediscoverd? Ann Sci 1:115–137
28. Frazer KA et al (2007) A second generation human haplotype map of over 3.1 million SNPs. Nature 449:851–861
29. Galton F (1865) Hereditary talent and character. Macmillans Mag 12:157
30. Garrod AE (1902) The incidence of alkaptonuria: a study in chemical individuality. Lancet 2:1616–1620
31. Garrod AE (1909) Inborn errors of metabolism. Frowde, Oxford University Press, London
32. Garrod AE (1931) The inborn factors of disease. Clarendon Press, Oxford
33. Gärtner CF (1849) Versuche und Beobachtungen über die Bastarderzeugung im Pflanzenreich. Hering, Stuttgart
34. Graham LR (1977) Political ideology and genetic theory: Russia and Germany in the 1920's. Hastings Cent Rep 7:30–39
35. Gurwitz D, Motulsky AG (2007) Drug reactions, enzymes, and biochemical genetics: 50 years later. Pharmacogenomics 8:1479–1484
36. Hardy GH (1908) Mendelian proportions in a mixed population. Science 28:49–50
37. Harper PS (2005) William Bateson, Human genetics and medicine. Human Genet 118:141–151
38. Harper PS (2008) A short history of medical genetics. Oxford University Press, Oxford 557 pp
39. Harris H (1969) Enzyme and protein polymorphism in human populations. Br Med Bull 25:5
40. Harris H, Watkins JF (1965) Hybrid cells from mouse and man: artificial heterokaryons of mammalian cells from different species. Nature 205:640
41. Hertwig O (1875) Beiträge zur Kenntnis der Bildung, Befruchtung und Theilung des tierischen Eies. I Abh Morph Jb 1:347–434
42. International HapMap Consortium (2003) T. I. H. The International HapMap Project. Nature 426:789–796
43. International HapMap Consortium (2005) A haplotype map of the human genome. Nature 437:1299–1320
44. International Human Genome Sequencing Consortium (2004) Finishing the euchromatic sequence of the human genome. Nature 431:931–945
45. Joravsky D (1970) The Lysenko affair. Harvard University Press, Boston
46. Kevles DJ (1985) In the name of eugenics. Genetics and the uses of human heredity, Knopf, New York
47. Köhl S (1994) The Nazi connection. Eugenics, American racism, and German national socialism. Oxford University Press, Oxford
48. Kölreuter JG (1761-1766) Vorläufige Nachricht von einigen das Geschlecht der Pflanzen betreffenden Versuchen und Beobachtungen, nebst Fortsetzung 1,2 und 3. Ostwalds Klassiker der Exakten Wissenschaften, no 41. Engelmann, Leipzig
49. Lander ES, Linton LM, Birren B et al (2001) Initial sequencing and analysis of the human genome. Nature 409: 860–921

50. Landsteiner K (1900) Zur Kenntnis der antifermentativen, lytischen und agglutinierenden Wirkungen des Blutserums und der Lymphe. Zentralbl Bakteriol 27:357–362
51. Landsteiner K, Wiener AS (1940) An agglutinable factor in human blood recognized by immune sera for rhesus blood. Proc Soc Exp Biol 43:223
52. Levine P, Burnham L, Katzin EM, Vogel P (1941) The role of isoimmunization in the pathogenesis of erythroblastosis fetalis. Am J Obstet Gynecol 42:925–937
53. Levine P, Stetson RE (1939) An unusual case of intragroup agglutination. JAMA 113:126–127
54. Levit SG (1936) The problem of dominance in man. J Genet 33:411–434
55. Ludmerer K (1972) Genetics and American society. Johns Hopkins University Press, Baltimore
56. Manolio TA, Brooks LD, Collins FS (2008) A HapMap harvest of insights into the genetics of common disease. J Clin Invest 118:1590–1605
57. Maupertuis PLM (1753) Vénus Physique. The Earthly Venus (trans: Brangier Boas S) (1966). Johnson Reprint Corporation, New York
58. McKusick VA (1975) The growth and development of human genetics as a clinical discipline. Am J Hum Genet 27: 261–273
59. McKusick VA (2007) History of medical genetics. In: Emery AEH, Rimoin DL (eds) Principles and practice of medical genetics, 5th edn. Churchill Livingstone, Philadelphia, Pennsylvania, pp 3–32
60. Mendel GJ (1865) Versuche über Pflanzenhybriden. Verhandlungen des Naturforschenden Vereins, Brünn
61. Motulsky AG (1957) Drug reactions, enzymes and biochemical genetics. JAMA 165:835–837
62. Motulsky AG (1959) Joseph Adams (1756–1818). A forgotten founder of medical genetics. Arch Intern Med 104: 490–496
63. Motulsky AG (2002) From pharmacogenetics and ecogenetics to pharmacogenomics. Medicina nei Secoli Arte e Scienza (Journal of History of Medicine) 14:683–705
64. Motulsky AG (2002) The work of Joseph Adams and Archibald Garrod: possible examples of prematurity in human genetics. In: Hook EB (ed) Prematurity and scientific discovery. University of California Press, Berkeley
65. Müller-Hill B (1988) Murderous science: Elimination by scientific selection of Jews, Gypsies, and others, Germany 1933–1945. Oxford University Press, Oxford English trans of Tödliche Wissenschaft, Rowohlt, Hamburg, 1984
66. Musto DF (1961) The theory of hereditary disease of Luis Mercado, chief physician to the Spanish Hapsburgs. Bull Hist Med 35:346–373
67. Nägeli C (1884) Mechanisch-physiologische Theorie der Abstammungslehre. R Oldenbourg, München and Leipzig
68. Pauling L, Itano HA, Singer SJ, Wells IC (1949) Sickle cell anemia: a molecular disease. Science 110:543
69. Penrose LS (1967) The influence of the English tradition in human genetics. In: Crow JF, Neel JV (eds) Proceedings of the Third International Congress of Human Genetics. The Johns Hopkins University Press, Baltimore, pp 13–25
70. Plato (360 BCE) The republic
71. Plato (360 BCE) The statesman
72. Ploetz A (1895) Die Tüchtigkeit unserer Rasse und der Schutz der Schwachen: ein Versuch über Rassenhygiene und ihr Verhältnis zu den humanen Idealen, besonders zum Sozialismus. Fischer, Berlin
73. Pollack W, Gorman JG, Freda VJ (1969) Prevention of Rh hemolytic disease. Prog Hematol 6:121–147
74. Proctor R (1988) Racial hygiene: Medicine under the Nazis. Harvard University Press, Cambridge, Massachusetts, pp 95–117
75. Reitlinger G (1961) The final solution. Barnes, New York
76. Rosenberg CE (1976) No other gods. On science and American social thought Johns Hopkins University Press, Baltimore
77. Roux W (1883) Uber die Bedeutung der Kerntheilungsfiguren. W Engelmann, Leipzig
78. Rowley JD (1973) A new consistent chromosomal abnormality in chronic myelogenous leukaemia identified by quinacrine fluorescence and Giemsa staining. Nature 243: 290–293
79. Saxena R et al (2007) Genome-wide association analysis identifies loci for type 2 diabetes and triglyceride levels. Science 316:1331–1336
80. Scriver CR, Childs B (1989) Garrod's inborn factors in disease. Oxford University Press, Oxford
81. Scriver CR, Beaudet AL, Sly WS, Valle D (eds) (2005) The metabolic and molecular bases of inherited disease, 8th edn. McGraw-Hill, New York
82. Stern C, Sherwood E (eds) (1966) The origin of genetics: a Mendel source book. WH Freeman, San Francisco
83. Stubbe H (1972) History of genetics, from prehistoric times to the rediscovery of Mendel's laws. MIT Press, Cambridge, Massachusetts, pp 12–50
84. Tschermak E (1900) Über künstliche Kreuzung bei *Pisum sativum*. Ber Dtsch Bot Ges 18:232–239
85. Van Beneden E (1883) Recherches sur la maturation de l'oeuf et la fecondation. Arch Biol IV:265
86. Vogel F (1959) Moderne Probleme der Humangenetik. Ergeb Inn Med Kinderheilk 12:52–125
87. Von Dungern E, Hirszfeld L (1911) On the group-specific structures of the blood. III. Z Immunitats 8:526–562
88. Von Verschuer O (1937) Was kann der Historiker, der Genealoge und der Statistiker zur Erforschung des biologischen Problems der Judenfrage beitragen? Forsch Judenfrage 2:216–222
89. Waterston, R. H. et al (2002) Initial sequencing and comparative analysis of the mouse genome. Nature 420: 520–562
90. Weiling F (1994) Johann Gregor Mendel. Forscher in der Kontroverse. V Med Genet 6:35–50
91. Weinberg W (1901) Beiträge zur Physiologie und Pathologie der Mehrlingsgeburten beim Menschen. Arch Gesamte Physiol 88:346–430
92. Weinberg W (1908) Über den Nachweis der Vererbung beim Menschen. Jahreshefte des Vereins für vaterländische Naturkunde in Württemberg 64:368–382
93. Weinberg W (1912) Weitere Beiträge zur Theorie der Vererbung. IV. Über Methode und Fehlerquellen der

Untersuchung auf Mendelsche Zahlen beim Menschen. Arch Rass Ges Biol 9:165–174
94. Weindling PF (1989) Health, race and German politics between national unification and nazism 1870–1945. Cambridge University Press, Cambridge
95. Weingart P, Kroll J, Bayertz K (1988) Rasse. Blut und Gene, Suhrkamp, Frankfurt
96. Weiss KM (2008) Joseph Adams in the judgment of Paris. Evol Anthropol 17:245–249
97. Wellcome Trust Case Control Consortium (2007) Genome-wide association study of 14, 000 cases of seven common diseases and 3,000 shared controls. Nature 447:661–678
98. Witkowski SA, Inglis JR (eds) (2008) Davenport's dream. 21st century reflections on heredity and eugenics. Cold Spring Harbor Laboratory Press, New York
99. Wolff CF (1759) Theoria Generationis, Halle
100. Zimmerman D (1973) Rh. The intimate history of a disease and its conquest. Macmillan, New York, p 371

Human Genome Sequence and Variation

Stylianos E. Antonarakis

Abstract The knowledge of the content of the individual human genomes has become a *sine qua non* for the understanding of the relationship between genotypic and phenotypic variability. The genome sequence and the ongoing functional annotation require both comparative genome analysis among different species and experimental validation. Extensive common and rare genomic variability exists that strongly influences genome function among individuals, partially determining disease susceptibility.

Contents

2.1	The Human Genome	31
	2.1.1 Functional Elements	33
	2.1.2 Repetitive Elements	40
	2.1.3 Mitochondrial Genome	43
2.2	Genomic Variability	43
	2.2.1 Single Nucleotide Polymorphisms	43
	2.2.2 Short Sequence Repeats	46
	2.2.3 Insertion/Deletion Polymorphisms (Indels)	46
	2.2.4 Copy Number Variants	47
	2.2.5 Inversions	47
	2.2.6 Mixed Polymorphisms	47
	2.2.7 Genome Variation as a Laboratory Tool to Understand the Genome	48
References		48

2.1 The Human Genome

In order to be able to understand the biological importance of the genetic information in health and disease (assign a particular phenotype to a genome variant) we first needed to know the entire nucleotide sequence of the human genome. Thus an international collaborative project has been undertaken named "The Human Genome Project" to determine the nucleotide sequence of the human genome. The project was initiated on 1 October 1990 and was essentially completed in 2004. The potential medical benefits from the knowledge of the human genome sequence were the major rationale behind the funding of this international project. In addition, the involvement and contributions of the biotechnology company Celera may have provided the necessary competition for the timely completion of the project. The last (third) edition of this book was published in 1997 before the knowledge of the human genome sequence; thus, this fourth ("postgenome") edition of the book proudly begins with the discussion of "genome anatomy," as the genomic sequence was named by Victor McKusick.

The goals of the different phases of the Human Genome Project were to: (1) determine the linkage map of the human genome [1, 60]; (2) construct a physical map of the genome by means of cloning all fragments and arrange them in the correct order [32, 69]; (3) determine the nucleotide sequence of the genome; and (4) provide an initial exploration of the variation among human genomes.

As of October 2004 about 93% of the human genome (which corresponds to 99% of the euchromatic portion of the genome) had been sequenced to an accuracy of better than one error in 100,000 nucleotides

S.E. Antonarakis (✉)
Department of Genetic Medicine and Development, University of Geneva Medical School, and University Hospitals of Geneva, 1 rue Michel-Servet, 1211 Geneva, Switzerland
e-mail: Stylianos.Antonarakis@unige.ch

[3, 84, 137]. The DNA that was utilized for sequencing from the public effort came from a number of anonymous donors [84], while that from the industrial effort came from five subjects of which one is eponymous, Dr. J.C. Venter [85, 137]. The methodology used was also different between the two participants: the public effort sequenced cloned DNA fragments that had been previously mapped, while that of Celera sequenced both ends of unmapped cloned fragments and subsequently assembled them in continuous genomic sequences. Detailed descriptions of the genome content per chromosome have been published; the first "completed" chromosome published was chromosome 22 in 1999, chromosome 21 was published in 2000, and all other chromosomes followed in the next 6 years [38, 39, 45, 46, 55, 57, 62, 63, 66, 67, 70, 91, 97, 98, 101, 102, 111, 120, 121, 125, 131, 152, 153]. Figure 2.1 shows the parts of the genome (mainly the heterochromatic fraction) that have not yet been sequenced: the pericentromeric regions, the secondary constrictions of 1q, 9q, 16q, the short arms of acrocentric chromosomes (13p, 14p, 15p, 21p, 22p), and the distal Yq chromosome.

The total number of nucleotides of the finished sequence is 2,858,018,193 while the total estimated length that includes the current gaps is ~3,080,419,480 nucleotides (see Table 2.1, taken from the last hg18 assembly of the human genome http://genome.ucsc.edu/goldenPath/stats.html#hg18). The length of the human chromosomes ranges from ~46 Mb to ~247 Mb. The average GC content of the human genome is 41%. This varies considerably among the different chromosomes and within the different bands of each chromosome. Chromosomal bands positive for Giemsa staining have lower average GC content of 37%, while in Giemsa-negative bands the average GC content is 45%. Interestingly, Giemsa-negative bands are gene-rich regions of DNA (see Chap. 3, Sect. 3.2.4).

Figure 2.2 shows the current status of the "completion" of the human genome sequence [3]. Red bars above the chromosomes represent the sequence gaps. The DNA content of the red blocks (heterochromatin) is still unknown. Heterochromatic regions of chromo-

Table 2.1 Taken from http://genome.ucsc.edu/goldenPath/stats.html#hg18, showing the number of nucleotides per chromosome in the reference genome. Chromosome "M" is the DNA of the mitochondrial genome (see Sect. 2.1.3)

```
NCBI Build 36.1, Mar. 2006 Assembly (hg18)

Chr     Assembled Size   Sequenced    Total Gap    Non-Euch.
Name    (inc. Gaps)      Size         Size         Gap Size
-------------------------------------------------------------
1           247249719    224999719    22250000     20240000
2           242951149    237712649     5238500      4200000
3           199501827    194704827     4797000      4490000
4           191273063    187297063     3976000      3010000
5           180857866    177702766     3155100      3083000
6           170899992    167273992     3626000      3008000
7           158821424    154952424     3869000      3184000
8           146274826    142612826     3662000      3000000
9           140273252    120143252    20130000     18000000
10          135374737    131624737     3750000      2380000
11          134452384    131130853     3321531      3257000
12          132349534    130303534     2046000      1471000
13          114142980     95559980    18583000     17933000
14          106368585     88290585    18078000     18078000
15          100338915     81341915    18997000     18260000
16           88827254     78884754     9942500      9805000
17           78774742     77800220      974522       220000
18           76117153     74656155     1460998      1363998
19           63811651     55785651     8026000      8016000
20           62435964     59505253     2930711      1773661
21           46944323     34171998    12772325     12769767
22           49691432     34851332    14840100     14430000
X           154913754    151058754     3855000      3000000
Y            57772954     25652954    32120000     30500000
M               16571        16571           0            0
-------------------------------------------------------------
Overall
Chrom      3080436051   2858034764   222401287    205472426
```

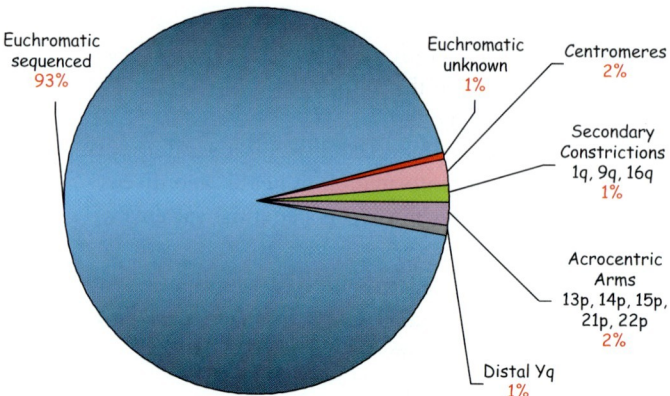

Fig. 2.1 Pie chart of the fractions of the genomes sequenced (*blue*) and not sequenced (*non-blue*)

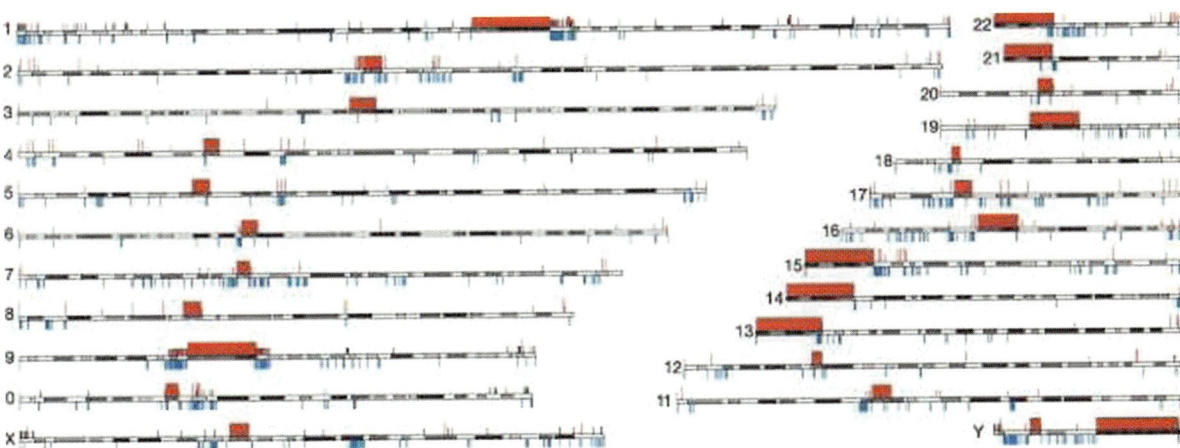

Fig. 2.2 Schematic representation of the completion of the human genome per chromosome. *Red regions* represent areas not sequenced; *blue regions below the chromosomal line* represent gaps in the sequences. The major blocks of unknown sequence include the short arms of acrocentric chromosomes, the pericentromeric sequences, and the large heterochromatic regions (From [3])

somes are those that remain highly condensed throughout the cell cycle (see Chap. 3, Sect. 3.2.1); it is thought that transcription is limited in these regions that contain a considerable number of repetitive elements that renders the assembly of their sequence almost impossible.

The sequence of the human genome is freely and publicly available on the following genome browsers, which also contain many additional annotations (see also Chap. 29):

(a) http://genome.ucsc.edu/
(b) http://www.ensembl.org/
(c) http://www.ncbi.nlm.nih.gov/genome/guide/human/

Representative pages of two of these browsers are shown in Fig. 2.3.

There is now a considerable effort internationally to identify all the functional elements of the human genome. A collaborative project called ENCODE (ENcyclopedia Of DNA Elements) is currently in progress with the ambitious objective to identify all functional elements of the human genome [2, 19].

The genome of modern humans, as a result of the evolutionary process, has similarities with the genomes of other species. The order of genomic elements has been conserved in patches within different species such that we could recognize today regions of synteny in different species, i.e., regions that contain orthologous genes and other conserved functional elements. Figure 2.4 shows a synteny map of conserved genomic segments in human and mouse.

The current classification of the functional elements of the genome contains:

1. Protein-coding genes
2. Noncoding, RNA-only genes
3. Regions of transcription regulation
4. Conserved elements not included in the above categories

2.1.1 Functional Elements

2.1.1.1 Protein-Coding Genes

The total number of protein-coding genes is a moving target, since this number depends on the functional annotation of the genome, the comparative analysis with the genomes of other species, and the experimental validation. The so-called CCDS set (*consensus coding sequence*) is built by consensus among the European Bioinformatics Institute (http://www.ebi.ac.uk/), the National Center for Biotechnology Information (http://www.ncbi.nlm.nih.gov/), the Wellcome Trust Sanger Institute (http://www.sanger.ac.uk/), and the University of California, Santa Cruz (UCSC; http://www.cbse.ucsc.edu/). At the last update (5 July 2009; genome build 36.3) CCDS contains 17,052 genes. This is the minimum set of protein-coding genes included in all genomic databases. The reference sequence (RefSeq) collection of genes of the NCBI contains 20,366 protein-coding gene entries (http://www.ncbi.nlm.nih.gov/

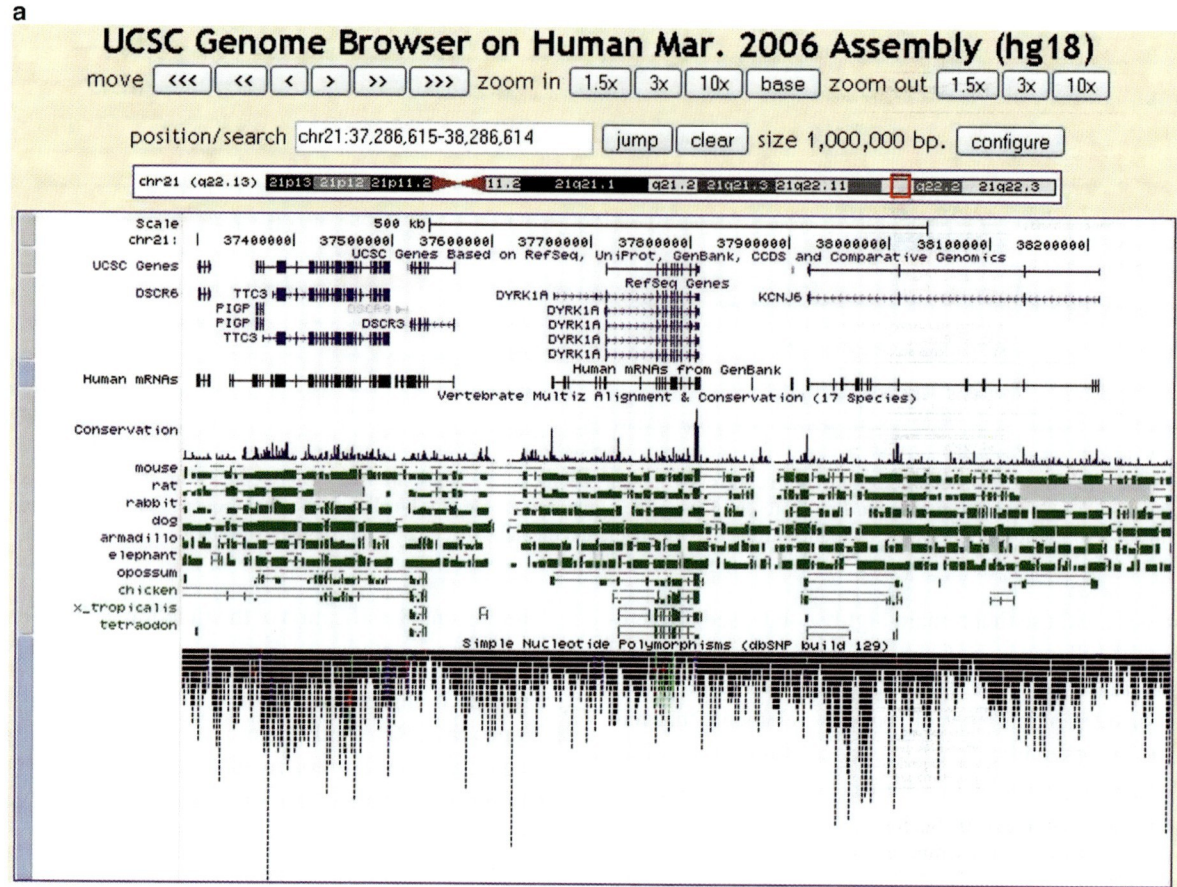

Fig. 2.3 (a) Screenshot of the UCSC genome browser (http://genome.ucsc.edu/) for a 1-Mb region of chromosome 21 (21: 37,286,615–38,286,614). Among the many features that could be displayed, the figure shows genes, sequence conservation in 17 species, and single nucleotide polymorphisms (SNPs) that map in this 1-Mb region. The tracks shown from *top to bottom* include: a scale for the genomic region, the exact location in nucleotides, schematic representation of genes included in the UCSC database, the mRNAs from GenBank, the conservation in the species shown, and the location of SNPs. The *color* of some SNPs corresponds to synonymous and nonsynonymous substitutions.

RefSeq/); the UCSC collection of genes contains 23,008 entries (http://genome.ucsc.edu/); the Ensembl browser contains 21,416 entries (23 June 2009; build 36; http://www.ensembl.org/Homo_sapiens/Info/StatsTable). The total number of annotated exons listed in the Ensembl database is 297,252 (23 June 2009; build 36). The discrepancy among the databases reflects the ongoing and unfinished annotation of the genome.

Table 2.2 lists the number of protein-coding and other genes in humans taken from different databases.

The human genes are not equally distributed in the chromosomes. In general, Giemsa pale bands are gene rich, and this results in unequal numbers of genes per size unit for the different chromosomes. Figure 2.5 from [84] displays the gene density per megabase for each chromosome and the correlation with CpG-rich islands.

Chromosomes 22, 17, and 19 are unusually gene-rich, while chromosomes 13, 18, and X are relatively gene-poor (interestingly, trisomies for chromosomes 13 and 18 are among the few human trisomies at birth). The average number of exons per gene is nine, and the average exon size is 122 nucleotides. Thus, the total number of annotated exons range from 210,000 to 300,000 (depending on the database), and the total exonic genome size is up to 78 Mb.

The mapping position of the genes can be seen in the genome browsers, and their names can be found in the gene nomenclature Web site, which contains 28,182 entries (http://www.genenames.org/; 30 June 2009).

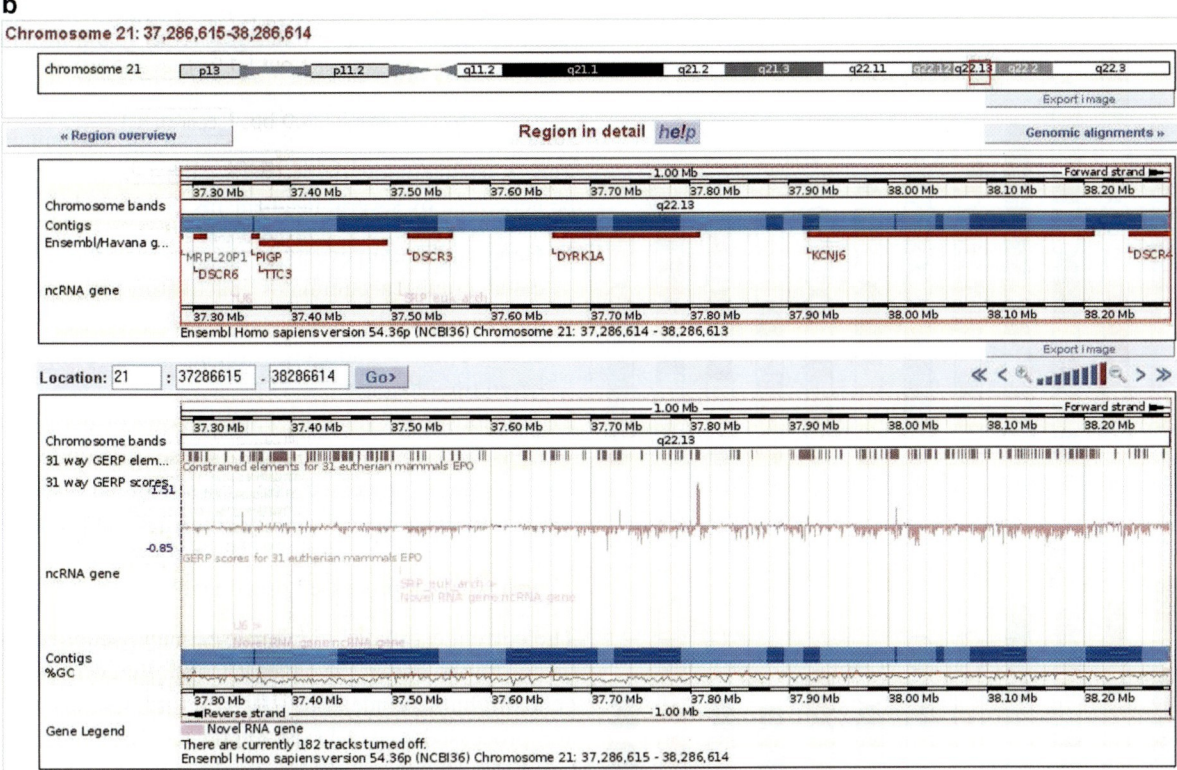

Fig. 2.3 (continued) (**b**) Screenshot of the Ensembl genome browser (http://www.ensembl.org) for a 1-Mb region of chromosome 21 (21: 37,286,615–38,286,614). Among the many features that could be displayed, the figure shows genes (Ensembl/Havana gene track), noncoding RNAs (ncRNA gene track), sequence conservation in 31 species (31-way GERP track), and GC content in this 1-Mb region. The different browsers have similarities and differences, and some features could only be displayed in one browser (for details see Chaps. 29.1 and 29.2)

A single gene may have different isoforms due to alternative splicing of exons, alternative utilization of the first exon, and alternative 5′ and 3′ untranslated regions. There are on average 1.4–2.3 transcripts per gene according the different databases (Table 2.2); this is likely an underestimate since, in the pilot ENCODE 1% of the genome that has been extensively studied, there are 5.7 transcripts per gene [19, 61]. The average number of exons per gene, depending on the database, ranges from 7.7 to 10.9.

The size of genes and number of exons vary enormously. The average genomic size of genes (according to the current annotation) is 27 kb. There are, however, small genes that occupy less than 1 kb, and large genes that extend to more than 2,400 kb of genomic space. There are intronless genes (e.g., histones) and others with more than 360 introns (e.g., titin).

The initial results of the ENCODE and other similar projects provided evidence for additional exons to the annotated genes; these exons could be hundreds of kilobases away (usually 5′) to the annotated gene elements [19, 40, 44]. In addition, there is evidence for chimeric transcripts that join two "independent" genes [103]. The investigation of these complicated transcripts is ongoing, and the functional significance of them is unknown.

Protein-coding genes can be grouped in families according to their similarity with other genes. These families of genes are the result of the evolutionary processes that shaped up the genomes of the human and other species. The members of the gene families could be organized in a single cluster or multiple clusters, or could be dispersed in the genome. Examples of gene families include the globin, immunoglobulin, histones, and olfactory receptors gene families. Furthermore, genes encode proteins with diverse but recognizable domains. The database Pfam (http://pfam.sanger.ac.uk/, http://www.uniprot.org/) is a comprehensive collection of protein domains and families [48]; the current release

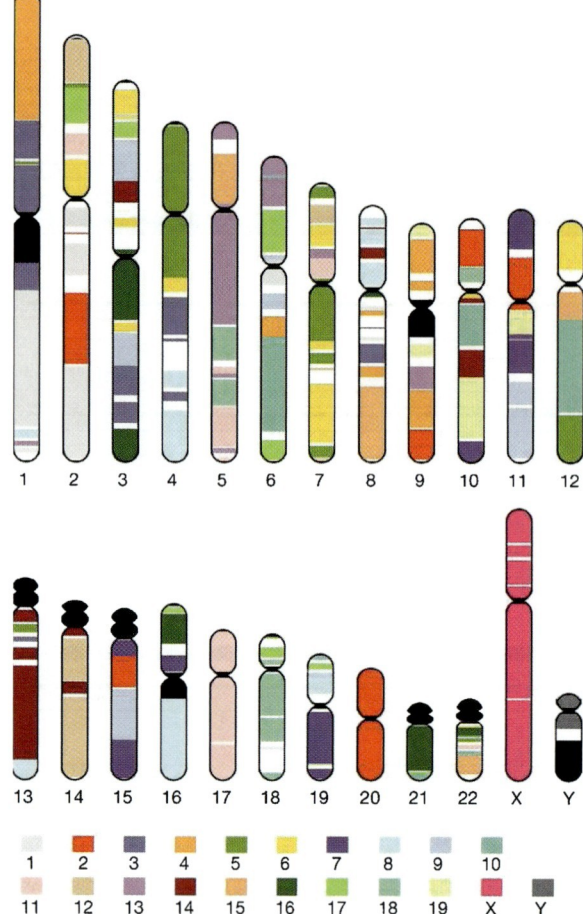

Fig. 2.4 Schematic representation of the observed genomic segments between the human and mouse genomes. The color code of the human chromosomes corresponds to the different mouse chromosomes shown on the bottom. For example, human chromosome 20 is all homologous to mouse chromosome 2; human chromosome 21 is homologous to mouse chromosomes 16, 17, and 10. Centromeric, and heterochromatic regions, and acrocentric p-arms are in *black*. (From [84])

(PF00046) has 430 genes. The identification of domains helps in the prediction of the function and structure of a protein.

Pseudogenes are "dead" nonfunctional genes. These sequences that could be transcribed and spliced contain mutations that render them inactive. Pseudogenes could be generated by several mechanisms that include:

1. Gene duplication events in which one of the duplicated copies accumulates inactivating mutations; alternatively, the duplicated genes may be truncated. These pseudogenes are also called nonprocessed pseudogenes.
2. Transposition events in which a copy of cDNA is reinserted into the genome. These pseudogenes, also called "processed," are not functional, usually because they lack regulatory elements that promote transcription. In addition, inactivating mutations also occur in processed pseudogenes.

The current estimated number of human pseudogenes (according to one of the databases http://www.pseudogene.org/human/index.php) [151] is 12,534 (~8,000 are processed and ~4,000 duplicated pseudogenes; build 36); while according to the Ensembl browser the number is 9,899 (build 36; 23 June 2009). These pseudogenes belong to 1,790 families; e.g., the immunoglobulin gene family has 1,151 genes and 335 pseudogenes, while the protein kinase gene family has 1,159 genes and 159 pseudogenes (http://pseudofam.pseudogene.org/pages/psfam/overview.jsf).

The total number of human genes is not dramatically different from that of other "less" complex organisms. Figure 2.6 depicts the current estimate of the protein-coding gene number for selected species.

2.1.1.2 Noncoding, RNA-Only Genes

Besides the protein-coding genes, there is a growing number of additional genes (transcripts) that produce an

of Pfam (23.0) contains 10,340 protein families. For example, the WD40 domain family (PF00400) includes 609 human genes, while the homeobox domain family

Table 2.2 Human gene, exon, and transcript counts from various databases

Database (June 2009)	Protein-coding genes	RNA-only genes	Total genes	Total number of transcripts	Total number of exons	Average exons per gene	Average transcripts per gene
CCDS	17,052			45,428			2.7
Ensembl	21,416	5,732	27,148	62,877	297,252	10.9	2.3
UCSC	23,008	9,155	32,163	66,802	246,775	7.7	2.1
RefSeq	20,366	2,044	22,410	31,957	211,546	9.4	1.4

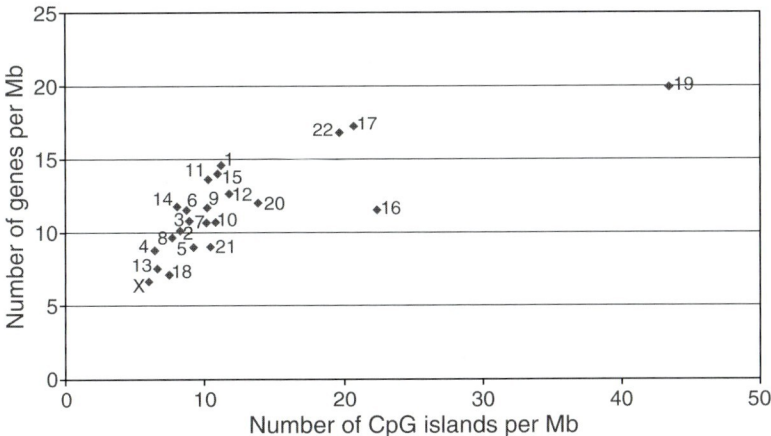

Fig. 2.5 Gene density per chromosome, and correlation with CpG-rich islands of the genome. Chromosome 19 for, example, has the highest gene content and the highest CpG island content. (From [84])

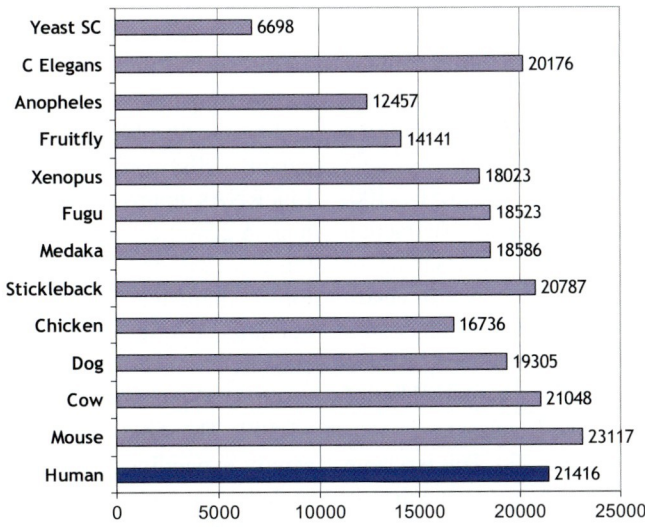

Fig. 2.6 Histogram of the current (5 July 2009) estimate of the number of protein-coding genes in different selected species from the Ensembl browser. These numbers are subject to change

RNA that is not translated to protein (see the databases http://biobases.ibch.poznan.pl/ncRNA/, http://www.ncrna.org/frnadb/search.html, http://www.sanger.ac.uk/Software/Rfam/ and [56]). Table 2.2 contains the current number of these genes, which ranges from 2,044 in RefSeq to 9,155 in the UCSC browser.

The different classes of RNA-only genes are briefly discussed below:

Ribosomal RNA (rRNA) Genes [53, 82, 84]: ~650–900. These are genes organized in tandemly arranged clusters in the short arms of the five acrocentric chromosomes (13, 14, 15, 21, and 22). The transcripts for 28 S, 5.8 S, and 18 S rRNAs are included in one transcription unit, repeated 30–50 times per chromosome. These tandemly arranged genes are continuously subjected to concerted evolution, which results in homogeneous sequences due to unequal homologous exchanges. The transcripts for the 5 S rRNAs are also tandemly arranged, and the majority map to chromosome 1qter. There exist also several pseudogenes for all classes. The total number of these genes is polymorphic in different individuals. The best estimates of the number of rRNA genes are:

28 S (components of the large cytoplasmic ribosomal subunit)	~150–200
5.8 S (components of the large cytoplasmic ribosomal subunit)	~150–200
5 S (components of the large cytoplasmic ribosomal subunit)	~200–300
18 S (components of the small cytoplasmic ribosomal subunit)	~150–200

Transfer RNA (tRNA): ~500 (49 Types). At the last count there are 497 transfer RNA genes (usually 74–95 nucleotides long) encoded by the nucleus and transcribed by RNA polymerase III (additional tRNAs are encoded by the mitochondria genome). There are also 324 tRNA pseudogenes [84]. The tRNA nuclear genes form 49 groups for the 61 different sense codons. Although the tRNA genes are dispersed throughout the genome, more than 50% of these map to either chromosomes 1 or 6; remarkably 25% of tRNAs map to a 4-Mb region of chromosome 6.

Small Nuclear RNA (snRNA) [84, 87, 105]: ~100. These are heterogeneous small RNAs. A notable fraction of these are the spliceosome [139] RNA genes many of which are uridine-rich; the U1 group contains 16 genes, while U2 contains six, U4 4, U6 44, and the other subclasses are represented by one member. Some of these genes are clustered, and there is also a large number of pseudogenes (more than 100 for the U6 class).

Small Nucleolar RNA (snoRNA): ~200. This is a large class of RNA genes that process and modify the tRNAs and snRNAs [135, 147]. There are two main families: C/D box snoRNAs that are involved in specific methylations of other RNAs; and H/ACA snoRNAs, mostly involved in site-specific pseudouridylations. Initially, there were 69 recognized in the first family and 15 in the second [84]; however, the total number is probably larger. A cluster of snoRNAs maps to chromosome 15q in the Prader–Willi syndrome region (at least 80 copies); deletions of which are involved in the pathogenesis of this syndrome [26, 117]. Another cluster of snoRNAs maps to chromosome 14q32 (~40 copies). The majority of snoRNAs map to introns of protein-coding genes and can be transcribed by RNA polymerase II or III.

Micro RNAs (miRNA): (706 Entries on 26 June 2009). These are single-stranded RNA molecules of about 21–23 nt in length that regulate the expression of other genes. miRNAs are encoded by RNA genes that are transcribed from DNA but not translated into protein; instead they are processed from primary transcripts known as pri-miRNA to short stem-loop structures called pre-miRNA and finally to functional miRNA. Mature miRNA molecules are complementary to regions in one or more messenger RNA (mRNA) molecules, which they target for degradation. A database of the known and putative miRNAs, and their potential targets, can be found in http://microrna.sanger.ac.uk/. miRNAs have been shown to be involved in human disorders.

Large Intervening Noncoding RNAs (LincRNAs): ~1,600. This new class has been recently identified using trimethylation of Lys4 of histone H3 as a genomic mark to observe RNA PolII transcripts at their promoter, and trimethylation of Lys36 of histone H3 marks along the length of the transcribed region [95] to identify the spectrum of PolII transcripts. Approximately 1,600 such LincRNA transcripts have been found across four mouse cell types (embryonic stem cells, embryonic fibroblasts, lung fibroblasts, and neural precursor cells) [59]. Among the "exons" of these LincRNAs, approximately half are conserved in mammalian genomes, and are thus present in human. Since this class was described in 2009, further work is needed for its characterization and validation, as well as the potential overlap of its members with the other classes.

Other Noncoding RNAs [7, 75, 113, 126, 136]: ~1,500. The field of noncoding RNA series is constantly expanding. Some of these RNA genes include molecules with known function such as the telomerase RNA, the 7SL signal recognition particle RNA, and the XIST long transcript involved on the X-inactivation [23]. There are also numerous antisense noncoding RNAs, and the current effort to annotate the genome suggests that a substantial fraction of the transcripts are noncoding RNAs.

2.1.1.3 Regions of Transcription Regulation

The genome certainly contains information for the regulation of transcription. The current list of these regulatory elements includes promoters, enhancers, silencers, and locus control regions [92]. These elements are usually found in *cis* to the transcriptional unit, but there is growing evidence that there is also *trans* regulation of transcription. The discovery of the regulatory elements, their functional interrelationship, and their spatiotemporal specificity provides a considerable challenge. A systematic effort during the pilot ENCODE project has provided initial experimental evidence for genomic regions with enriched binding of transcription factors [19, 80, 86, 133]. A total of 1,393 regulatory genomic clusters were, for example, identified in the pilot ENCODE regions; remarkably only ~25% of these map to previously known regulatory regions and only ~60% of these regions overlap with evolutionarily constrained regions. These results suggest that many novel regulatory regions will be recognized in the years to come, and also that there exist regions of transcriptional regulation that are not conserved and thus novel for different clades and species. The use of model organisms facilitates the experimental validation of regulatory elements, and there are systematic efforts underway for the exploration of conserved elements ([106] and http://enhancer.lbl.gov/).

2.1.1.4 Conserved Elements Not Included in the Above Categories

Since it is assumed that functional DNA elements are conserved while nonfunctional DNA diverges rapidly, it is expected that all other conserved elements are of interest and should be studied for potential pathogenic variability. How much of the human genome is evolutionarily conserved? The answer to this question depends on the species compared and the time of their common ancestor. Comparative genome analysis between human and mouse, for example, is particularly instructive, since the time of the common ancestor between these two species is estimated to be ~75 million years ago, and thus the conserved elements are likely to be functional. Approximately 5% of the human genome is conserved compared to mouse [145] (and to several other mammalian genomes). Of this, ~1–2% are the coding regions of protein-coding genes, and ~3% are conserved non-coding DNA sequences (CNCs; Fig. 2.7) [41, 42]. The function of the majority of CNCs is unknown. Please note that this 5% conserved fraction between human and mouse is an underestimate of the functional fraction of the human genome, which is likely to be bigger and to contain additional sequences not conserved with the mouse.

The ENCODE pilot project [19, 90], with data from 1% of the human genome and sequences from the orthologous genomic regions from 28 additional species, also estimated that the constrained portion of the human genome is at least ~4.9%; remarkably, 40% of this genomic space is unannotated and thus of unknown function (Fig. 2.8).

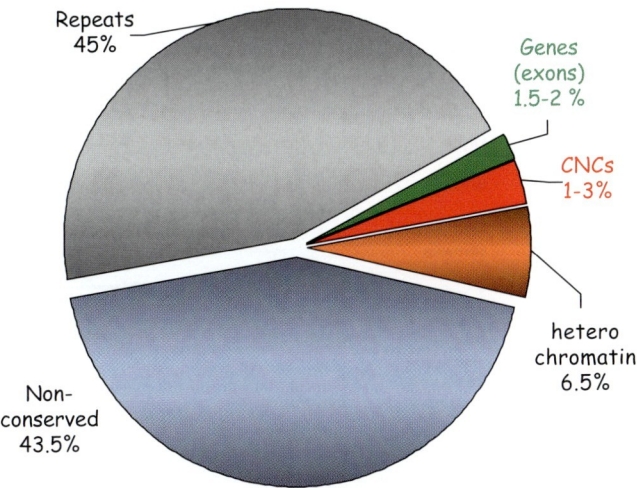

Fig. 2.7 The pie-chart depicts the different fractions of the genome. *CNCs*, conserved noncoding sequences

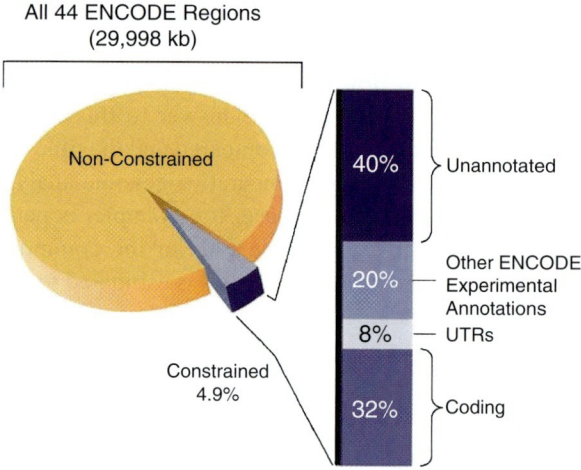

Fig. 2.8 The fractions of different genomic annotations among the 4.9% of constrained sequences in the human genome. Data from the pilot ENCODE project; figure taken from [19]. *UTR*, untranslated region; other ENCODE experimental annotations refers to the fraction of the genome that has been identified using a variety of experimental techniques for transcription, histone modifications, chromatin structure, sequence specific factors, and DNA replication. More information on these experiments is included in Table 1 of [19]

2.1.2 Repetitive Elements

The function of the majority of the human genome is unknown. Remarkably, ~45% of the genome is composed of repetitive elements, and another ~43% is not conserved and does not belong to the functional categories mentioned above. The different interspersed repeats of the human genome are shown in the Fig. 2.9 (from [84]):

- LINEs (long interspersed nuclear elements [76, 77]) are autonomous transposable elements, mostly truncated nonfunctional insertions (average size of 900 bp). More than 20% of the human genome is polluted by LINEs. Transposable elements are mobile DNA sequences which can migrate to different regions of the genome. Autonomous are those that are capable of transposing by themselves. A small fraction of LINEs (~100) are still capable of transposing. The full LINE element is 6.1 kb long, has an internal PolII promoter, and encodes two open reading frames, an endonuclease, and a reverse transcriptase. Upon insertion a target site duplication of 7–20 bp is formed. There are a few subclasses of LINEs according to their consensus sequence. The subfamily LINE1 is the only one capable of autonomous retrotransposition (copy itself and pasting copies back into the genome in multiple places). These LINEs enable transposition of SINEs (defined below), processed pseudogenes, and retrogenes [76, 77]. LINE retrotransposition has been implicated in human disorders [78]. LINEs are more abundant in G-dark bands of human chromosomes.
- SINEs (short interspersed nuclear elements [18]) mainly include the Alu repeats, which are the most abundant repeats in the human genome, occurring on average in every 3 kb. Thus, 13% of the genome is polluted by Alu sequences and other SINEs. They are inactive elements originated from copies of tRNA or from signal recognition particle (SRP; 7SL) RNA. The full-length element is about 280 nt long and consists of two tandem repeats each ~120 nt followed by polyA.

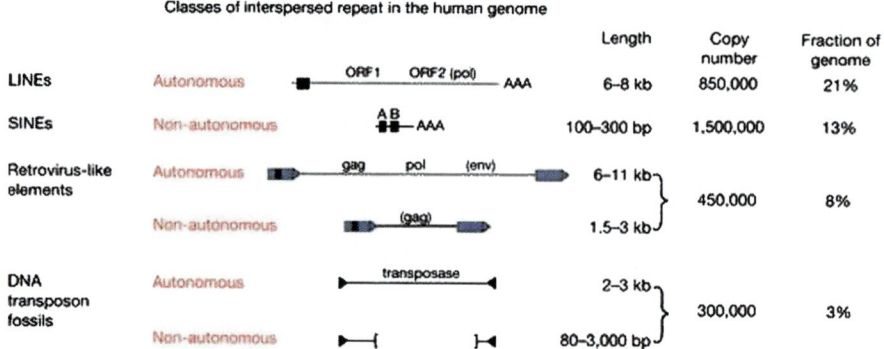

Fig. 2.9 Depicts some basic characteristics of the classes of interspersed repeats in the human genome. For more explanations, see text. (From [84])

Alu sequences are transcriptionally inactive, and are GC-rich. SINEs can retrotranspose in a non-autologous way, since they use the LINE machinery for transposition. Because of their abundance, they could mediate deletion events in the genome that result in human disorders [37]. SINEs are more abundant in G-light bands of human chromosomes (see Sect. 3.2.4).

- Retrovirus like (LTR transposons) are elements flanked by long terminal repeats. Those that contain all the essential genes are theoretically capable of transposition, but that has not happened in the last several million years. Collectively they account for 8% of the genome. Most are known as HERV (human endogenous retroviral sequences) and are transposition defective. Transcription from the HERV genes may modulate the transcriptional activity of nearby protein-coding genes [22].
- DNA transposon fossils [127] have terminal inverted repeats and are no longer active; they include two main families, MER1 and MER2, and comprise 3% of the genome.

More update information about repeats can be found in http://www.girinst.org/server/RepBase/.

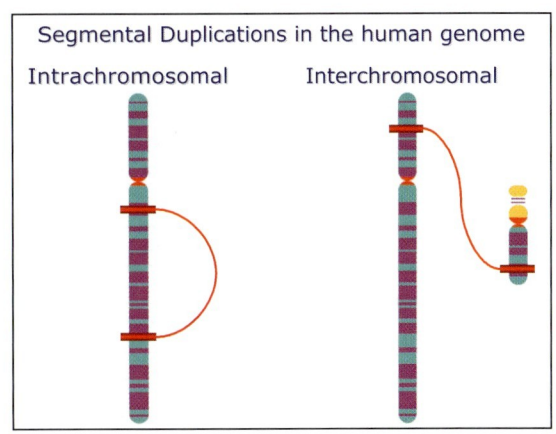

Fig. 2.10 Schematic representation of intra- and inter-chromosomal segmental duplications. The repeat element is shown in *red*, and there is a *connecting line* indicating the highly homologous sequences

2.1.2.1 Segmental Duplications

Approximately 5.2 % of the human genome consists of segmental duplications or duplicons, i.e., regions of more than 1 kb, with greater than 90% identify, that are present more than once in the genome. Segmental duplications are either intrachromosomal (on the same chromosome, 3.9%), or interchromosomal (on different chromosomes, 2.3%; Fig. 2.10). Most of the "duplicons" are in the pericentromeric regions.

Figure 2.11 shows the distribution of intrachromosomal duplicons in the human genome [16, 118]. These duplications are important in evolution and as risk factors for genomic rearrangements that cause human disorders because of unequal crossing-over in meiosis (pathogenic microdeletions and microduplications). Some examples of these include cases of α-thalassemia [65] on chromosome 16p, Charcot–Marie–Tooth syndrome [104] on chromosome 17p, and velo-cardiac-facial syndrome [96] on chromosome 22q, Williams–Beuren syndrome [107] on chromosome 7q, and Smith–Magenis syndrome [29] on chromosome 17p.

2.1.2.2 Special Genomic Structures Containing Selected Repeats

2.1.2.2.1 Human Centromeres

Human centromeres consist of hundreds of kilobases of repetitive DNA, some chromosome specific and some nonspecific [114, 122, 124]. Actually, most of the remaining sequence gaps in the human genome are mapped near and around centromeres. The structure of human centromeres is unknown, but the major repeat component of human centromeric DNA is an α-satellite or alphoid sequence [30] (a tandem repeat unit of 171 bp that contains binding sites for CENP-B, a centromeric-binding protein; see also Chap. 3, Sect. 3.2.3). Figure 2.12 shows an example of the structure of two human centromeres [3].

2.1.2.2.2 Human Telomeres

Human telomeres [109] consist of tandem repeats of a sequence $(TTAGGG)_n$ that spans about 3–20 kb, beyond which at the centromeric side there are about 100–300 kb of subtelomeric-associated repeats [3] before any unique sequence is present.

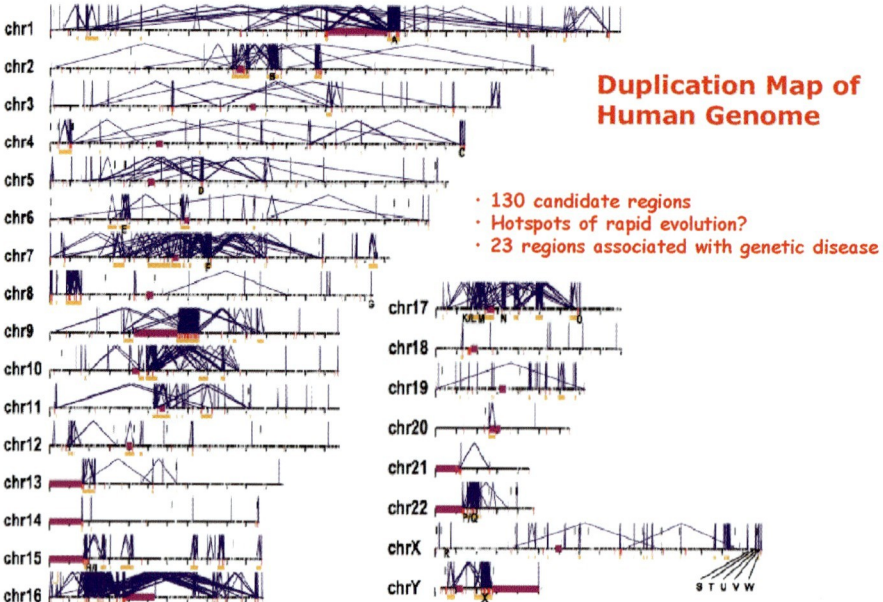

Fig. 2.11 Schematic representation of the intrachromosomal segmental duplications (from [16]). In each chromosome a *blue line* links a duplication pair. For example, on chromosome 21 there is only one duplicon shown; in contrast, on chromosome 22 there is a considerable number of duplications. *Richly blue areas* are considered susceptible to microduplication/microdeletion syndromes

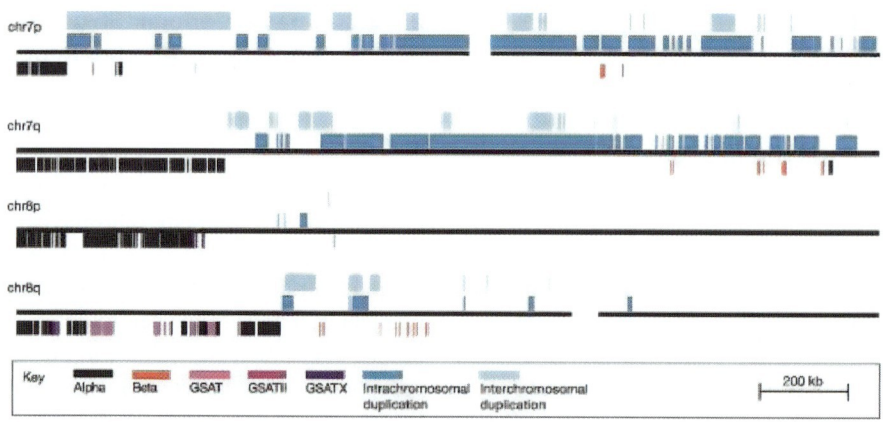

Fig. 2.12 Examples of sequence organization of two human centromeres (chromosomes 7 and 8, (from [3]). Alphoid repeats are the major component of this special chromosomal structure; in addition, several other repetitive elements border the alphoid sequences. The length of these regions is also polymorphic in different individuals

Figure 2.13 schematically shows the sequence organization of six human subtelomeric regions.

2.1.2.2.3 Short Arms of Human Acrocentric Chromosomes

The finished sequence of the human genome does not include the short arms of acrocentric chromosomes (13p, 14p, 15p, 21p, and 22p). Cytogenetic data show that the p arms contain large heterochromatic regions of polymorphic length [35, 138]. Molecular analysis revealed that they are composed mainly of satellite and other repeat families, including satellites I (AT-rich repeat of a monomer of 25–48 bp [73]), II (monomer repeat 5 bp [68]), III (monomer repeat also 5 bp [31]), β-satellite (a tandem repeat unit of 68 bp of the Sau3A family [94, 146]), and repeats ChAB4 [36], 724 [83], and D4Z4-like [89]. These repeats have a complex pattern and are often organized in subfamilies shared

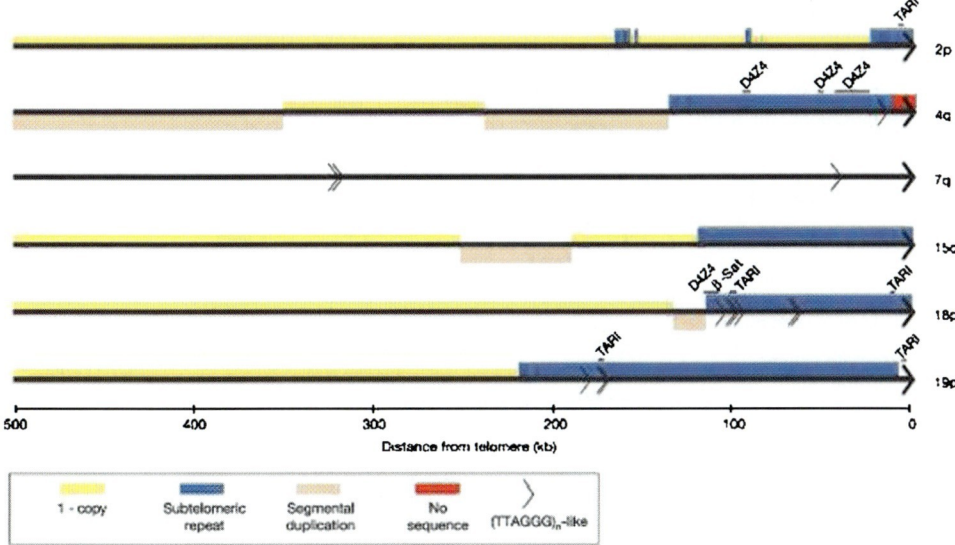

Fig. 2.13 Examples of the sequence organization of six human telomeres (chromosomes 2pter, 4qter, 7qter, 15qter, 18pter, and 19pter, taken from [3]). The *arrows* represent the TTAGGG repeat, while the *blue regions* depict the subtelomeric repeats that mainly consist of TAR1 (telomere associated repeat 1 family [24]), D4Z4 (a 3.3-kb tandem repeat, each copy of which contains two homeoboxes and two repetitive sequences, LSau and hhspm3 [64]) and β-satellite sequences (a tandem repeat unit of 68 bp of the Sau3A family [94])

between different acrocentric chromosomes. The p arms encode the ribosomal (RNR) gene [53, 82] but may also encode other genes [88, 130]. Currently there is an initiative to sequence the short arm of chromosome 21 and thus extrapolate on the structure of the additional p arms of the other acrocentrics [88].

The most common chromosomal rearrangements in humans are Robertsonian translocations (~1 in 1,000 births), which involve exchanges between acrocentric p arms. Three to five percent of these translocations are associated with phenotypic abnormalities [143].

2.1.3 Mitochondrial Genome

In human cells there is also the mitochondrial genome, which is 16,568 nucleotides long and encodes for 13 protein-coding genes, 22 tRNAs, one 23 S rRNA, and one 16 S rRNA ([140–142]; http://www.mitomap.org). The mitochondria genome-encoded genes are all essential for oxidative phosphorylation and energy generation in the cell. Each cell has hundreds of mitochondria and thousands (10^3–10^4) of mitochondria DNA (mtDNA) copies. Human mtDNA has a mutation rate ~20 times higher than nuclear DNA. The inheritance of mtDNA is exclusively maternal (the oocyte contains 10^5 mtDNA copies). Several human phenotypes are due to pathogenic mutations in the mitochondrial genome [140] (Fig. 2.14).

2.2 Genomic Variability

The human genome is polymorphic, i.e., there are many DNA sequence variants among different individuals. These variants are the molecular basis of the genetic individuality of each member of our species. In addition, this genetic variability is the molecular substrate of the evolutionary process. Finally, this variability causes disease phenotypes or predispositions to common complex or multifactorial phenotypes and traits.

2.2.1 Single Nucleotide Polymorphisms

The majority of the DNA variants are single nucleotide substitutions commonly known as SNPs (single nucleotide polymorphisms). The first SNPs were identified in 1978 in the laboratory of Y.W. Kan 3′ to the β-globin

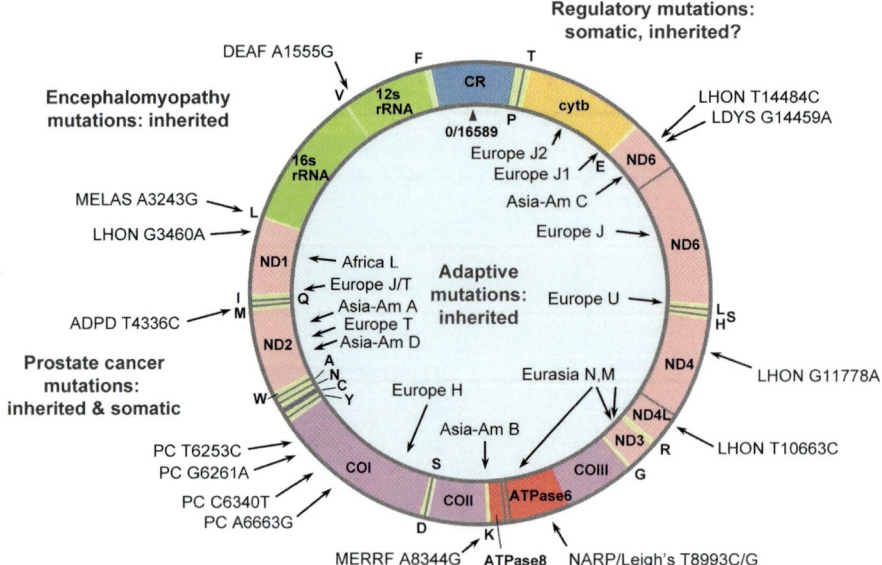

Fig. 2.14 Schematic representation of the circular mtDNA, its genes, its clinical relevant mutations, and certain polymorphic markers. *Letters within the ring* depict the genes encoded. *Letters on the outside* indicate amino acids of the tRNA genes. *CR*, the control of replication region that contains promoters for the heavy and light strands. *Arrows outside* show the location of pathogenic mutations. (From [142]

Fig. 2.15 Schematic representation of a single nucleotide polymorphism. Allele 1 has a C in the sequence, while allele 2 contains a T in the same position

gene [74] (at the time, these DNA polymorphisms were detected by restriction endonuclease digestion of DNAs and were called RFLPs, restriction fragment length polymorphisms). These polymorphic sites have two alternative alleles. In the example shown in Fig. 2.15, the depicted SNP has two alleles in the population: the blue C allele and the red T allele. The frequency of each allele could vary in different populations.

There is on average one SNP in ~1,000 nucleotides between two randomly chosen chromosomes in the population. Many of these SNPs are quite common. A common SNP is that in which the minor allele frequency (MAF) is more than 5%. On average two haploid genomes differ in ~3,000,000 SNPs. In addition, there is a large number of rare (MAF<1%) or near-rare (MAF between 1% and 5%) SNP variants that could be identified by the genome sequencing of various individuals. The majority of heterozygous SNPs in the DNA of a given individual are relatively common in the population; on the other hand, most of the SNPs discovered in a population are more likely to be rare. The NCBI SNP database contains 25 million common and rare SNPs (http://www.ncbi.nlm.nih.gov/

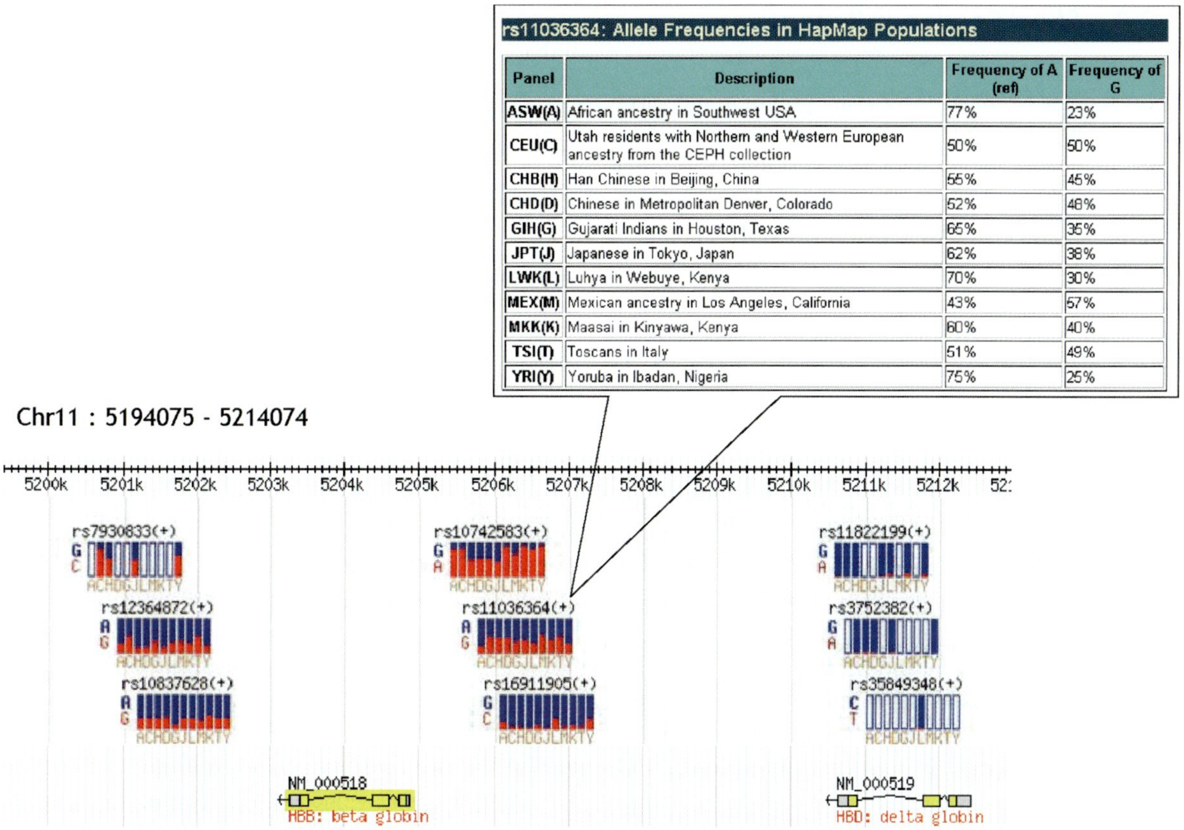

2 Human Genome Sequence and Variation   45

Fig. 2.16 The genomic region of Chr11: 5,194,075–5,214,074 is shown. For each of the nine SNPs shown in the bottom, the frequency of the two alternative alleles is shown in different populations. For example, for SNP rs11036364 that maps between the HBB and HBD globin genes, the allele frequencies are shown in the *callout*. The four original populations of the HapMap project were EUR, YRI, JPT, and CHB, while the other populations were added in a later stage. Modified from http://www.hapmap.org/

SNP/snp_summary.cgi; version 130; July 2009; Fig. 2.16). Of those, ~301,000 are in the protein-coding regions of genes, and ~188,000 result in amino acid substitutions (nonsynonymous substitutions). An international project known as HapMap (http://www.hapmap.org/) [6, 34, 50] has completed the genotyping of ~4,000,000 common SNPs in individuals of different geo-ethnic origins (4,030,774 SNPs in 140 Europeans; 3,984,356 in 60 Yoruba Africans; 4,052,423 in 45 Japanese and 45 Chinese; http://www.hapmap.org/downloads/index.html.en). Additional samples from further populations have been added recently.

The information content of SNPs (and polymorphic variation in general) is usually measured by the number of heterozygotes in the population (homozygotes are individuals that contain the same variant in both alleles; heterozygotes are individuals that contain two different variants in their alleles). The number of heterozygotes is a function of MAF based on Hardy–Weinberg principles (see Chap. 10). The pattern of DNA polymorphisms in a single chromosome is called haplotype (a contraction of "haploid genotype"; allelic composition of an individual chromosome). In the example shown in Fig. 2.17 the haplotype of polymorphic sites for the paternal (blue) chromosome is CGAATC while for the maternally inherited red chromosome it is GACGAT.

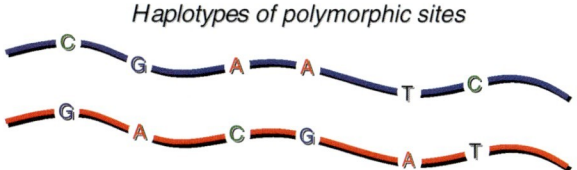

Fig. 2.17 Schematic representation of haplotype of polymorphic variants in a segment of the genome. The parental origin is shown as the *blue* (paternally-inherited) and *red* (maternally-inherited) *lines*. SNPs are shown as *letters interrupting the lines*. The haplotype is defined as the combination of SNP alleles per haploid genome

2.2.2 Short Sequence Repeats

Short sequence repeats (SSRs) are polymorphic variations due to a different number of short sequence repeat units, first described by Wyman and White [150] (then called VNTRs, variable number of tandem repeats[99]), and further elucidated by Jeffreys [72]. Most common are the dinucleotide repeats (described after the introduction of polymerase chain reaction amplification), but SSRs could be tri-, tetra-, or penta- repeats (often called microsatellites where the repeat unit $n = 1–15$ nucleotides). SSRs with longer repeat units ($n = 15–500$ nucleotides) are often termed minisatellites. These sequences comprise ~3% of the genome and there is ~1 SSR per 5 kb [84]. The most frequent dinucleotide SSR is the $(GT)_n$ with an occurrence in the genome of ~28 times per megabase, followed by the $(AT)_n$ SSR with ~19 times per megabase. The most common trinucleotide SSR is the $(TAA)_n$ that occurs approximately four times per megabase. The major advantage of SSRs (or microsatellites) is that there are more than two alleles per polymorphic site, and a large fraction of the human population is heterozygous for each SSR. Therefore, SSRs are extremely useful in linkage mapping and subsequent positional cloning for monogenic disorders [12, 17, 33] and other marking studies of the genome including the development of genomic linkage maps [43, 144]. In addition, SSRs are extensively used in forensic studies [15]. Figure 2.18 shows an example of an SSR with three alleles in the population.

2.2.3 Insertion/Deletion Polymorphisms (Indels)

This variation is due to the presence or absence of certain sequences. These sequences could be a few nucleotides, but they could also be transposons or interspersed repeats such as LINE or SINE elements [18, 112, 149]; alternatively, they could be pseudogenes [8] or other elements. Note that this category of variants is not completely separate from the next one; the arbitrary distinction is just the size of the variation in terms of base pairs. There are usually biallelic polymorphisms, which are not as common as SNPs but are useful for evolutionary studies and for the understanding of the dynamic structure of the human genome. In the example shown in Fig. 2.19, the blue sequence was inserted in the DNA and created a variant with two alleles: the blue allele 1 with insertion and the black allele 2 without.

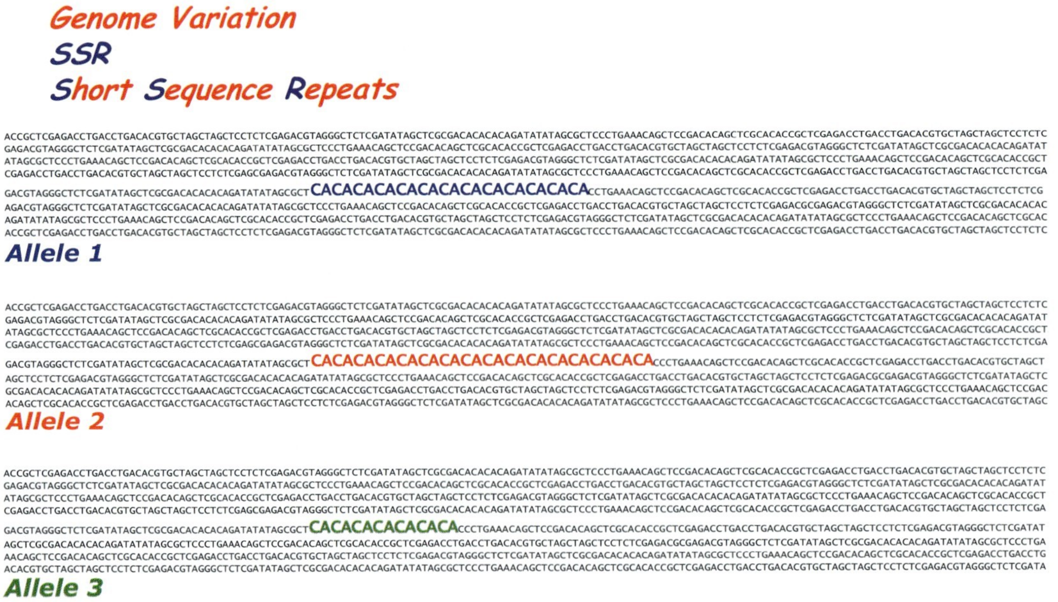

Fig. 2.18 An example of a dinucleotide SSR with three alleles in the population: the *blue* allele with $(CA)_{13}$ repeats, the *red* allele with $(CA)_{16}$, and the *green* allele with $(CA)_7$

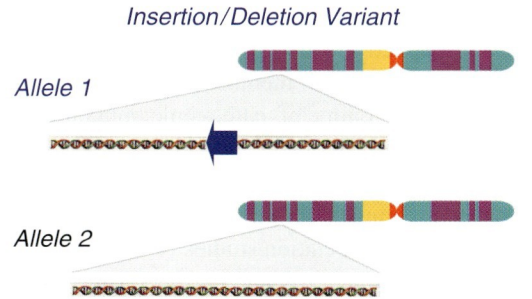

Fig. 2.19 Schematic representation of a polymorphic locus due to insertion deletion of a genomic element, shown as a *blue arrow*

2.2.4 Copy Number Variants

Copy number variant (CNV) refers to large-scale structural variation of our genome in which there are large tandem repeats of 50 kb to 5 Mb long that are present in a variable number of copies. This type of polymorphic variant includes large-scale duplications and deletions [123] (see also Chap. 3, Sect. 3.4.4). These have been known since studies of the α-globin genes in humans [54]. In the Fig. 2.20 example, allele 1 contains three copies and allele 2 five copies of a large repeat. The phenotypic consequences of some of these variants that may contain entire genes is unknown. A CNV map of the human genome in 270 individuals has revealed a total of 1,440 such CNV regions which cover some 360 Mb (~12% of the genome [79, 108]). More recent estimates using more accurate methods for precise mapping of the size of CNVs suggest that ~6% of the genome contains CNVs. A list of these variants can be found at http://projects.tcag.ca/variation/. The extent of CNV in the human genome is certainly underestimated since there are numerous additional CNVs of less than 50 kb. The current methodology for the detection of CNVs is using comparative genomic hybridization (CGH) on DNA microarrays [25]. A further improvement of this method will allow us to detect small CNVs. The most detailed currently available CNV map of the human genome was recently established by the Genome Structural Variation Consortium. This consortium conducted a CNV project to identify common CNVs greater than 500 bp in size in 20 female CEU (European ancestry) and 20 female YRI (African ancestry) samples of the HapMap project. By employing CGH arrays that tile across the

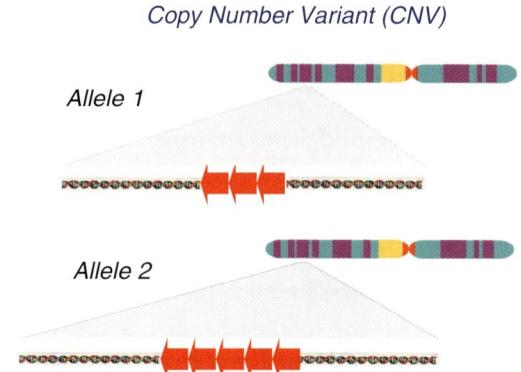

Fig. 2.20 Schematic representation of copy number variation in the human genome. For explanation, see text. Allele 1 in the population contains three copies of a sequence (*red arrowheads*), while allele 2 contains five copies

assayable portion of the genome with ~42 million probes from the company NimbleGen, this consortium could map 8,599 copy number variant events. Parts of these data have been provisionally released to the scientific community and can be viewed at http://www.sanger.ac.uk/humgen/cnv/42mio/.

2.2.5 Inversions

Large DNA segments could have different orientation in the genomes of different individuals. These inversion polymorphisms (Fig. 2.21) predispose for additional genomic alterations [9]. An example of a common inversion polymorphism involves a 900-kb segment of chromosome 17q21.31, which is present in 20% of European alleles but it is almost absent or very rare in other populations [129]. These variants are difficult to identify and most of them have been detected by sequencing the ends of specific DNA fragments and comparing them with the reference sequence [79, 134].

2.2.6 Mixed Polymorphisms

There are combinations of repeat size variants and single nucleotide variants. Figure 2.22 depicts such an example; the repeat units of an SSR contain a SNP and, thus, even alleles with the same repeat number

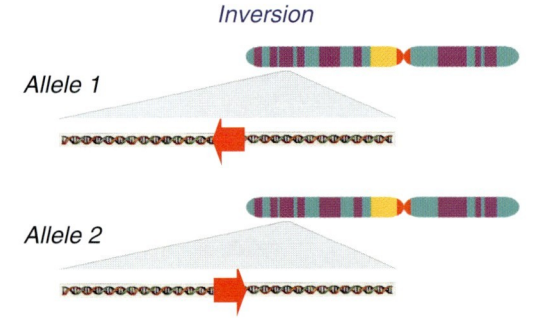

Fig. 2.21 Schematic representation of a polymorphic inversion shown as a *red arrowhead*

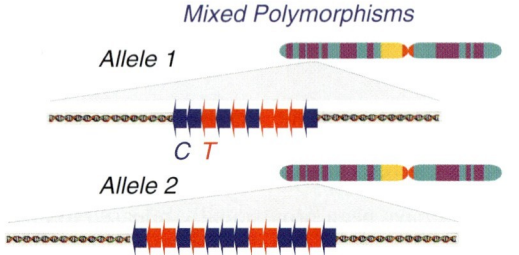

Fig. 2.22 Schematic representation of a highly polymorphic region of the genome with a mixed polymorphism that includes SNPs in the copies of CNVs or SSRs. The copies of the repeat are shown as *arrowheads*; the *blue/red* color of the repeats designates the SNP in them (*blue* for C and *red* for or T)

could be distinguished based on their exact DNA sequence [71]. These highly polymorphic systems could serve as "recognition barcodes" in humans.

2.2.7 Genome Variation as a Laboratory Tool to Understand the Genome

DNA variants, besides their functional importance in health and disease, are very useful in human genetics research because they serve as genomic markers for a variety of studies. Some of the uses of DNA variants are to:

1. Create linkage (genetic) maps of human chromosomes [1, 148]. This has allowed the initial mapping of the human genome and it was a prerequisite for the sequence assembly.
2. Map the genomic location of monogenic phenotypes to human chromosomes by linkage analysis [58, 81]. A large number of such phenotypes have been mapped to small genomic intervals because of the genotyping of members of affected families. Positional cloning of pathogenic mutations was subsequently possible.
3. Map the genomic location of polygenic phenotypes to human chromosomes by genomewide linkage and association studies [4, 20, 119].
4. Allow fetal diagnosis and carrier testing by linkage analysis of the cosegregation of a polymorphic marker and the phenotype of interest [10, 21].
5. Perform paternity and forensic studies [52]. A whole field was developed mainly with the use microsatellite SSR variants [49, 51].
6. Study genome evolution and origin of pathogenic mutations [115, 116].
7. Study the recombination rate and properties of the human genome [28, 93].
8. Study the instability of the genome in tumor tissues [5].
9. Identify loss-of-heterozygosity in human tumors [27, 47].
10. Study uniparental disomy and thus help with understanding genomic imprinting [100, 128].
11. Study parental and meiotic origin, and decipher the mechanisms of nondisjunction [11, 13, 14].
12. Study population history and substructure [110, 132].

The chapters that follow include further discussions on different aspects (including evolution, phenotypic consequences, and disease susceptibility) related to the most precious human genome variability.

Acknowledgments I thank the members of the laboratory, past and present, for discussions, ideas, debates, and data. I also thank my mentors and my students for the learning process.

References

1. No authors listed (1992) A comprehensive genetic linkage map of the human genome. NIH/CEPH Collaborative Mapping Group. Science 258:67–86
2. ENCODE Project Consortium (2004) The ENCODE (ENcyclopedia Of DNA Elements) Project. Science 306:636–640
3. International Human Genome Sequencing Consortium (2004) Finishing the euchromatic sequence of the human genome. Nature 431:931–945
4. Wellcome Trust Case Control Consortium (2007) Genome-wide association study of 14, 000 cases of seven

common diseases and 3, 000 shared controls. Nature 447: 661–678
5. Aaltonen LA, Peltomäki P, Leach FS, Sistonen P, Pylkkänen L et al (1993) Clues to pathogenesis of familial colorectal cancer. Science 260:812–816
6. Altshuler D, Brooks LD, Chakravarti A, Collins FS, Daly MJ, Donnelly P (2005) A haplotype map of the human genome. Nature 437:1299–1320
7. Amaral PP, Dinger ME, Mercer TR, Mattick JS (2008) The eukaryotic genome as an RNA machine. Science 319:1787–1789
8. Anagnou NP, O'Brien SJ, Shimada T, Nash WG, Chen MJ, Nienhuis AW (1984) Chromosomal organization of the human dihydrofolate reductase genes: dispersion, selective amplification, and a novel form of polymorphism. Proc Natl Acad Sci USA 81:5170–5174
9. Antonacci F, Kidd JM, Marques-Bonet T, Ventura M, Siswara P et al (2009) Characterization of six human disease-associated inversion polymorphisms. Hum Mol Genet 18:2555–2566
10. Antonarakis SE (1989) Diagnosis of genetic disorders at the DNA level. N Engl J Med 320:153–163
11. Antonarakis SE (1991) Parental origin of the extra chromosome in trisomy 21 as indicated by analysis of DNA polymorphisms. Down Syndrome Collaborative Group. N Engl J Med 324:872–876
12. Antonarakis SE (1994) Genome linkage scanning: systematic or intelligent? Nat Genet 8:211–212
13. Antonarakis SE, Avramopoulos D, Blouin JL, Talbot CC Jr, Schinzel AA (1993) Mitotic errors in somatic cells cause trisomy 21 in about 4.5% of cases and are not associated with advanced maternal age. Nat Genet 3:146–150
14. Antonarakis SE, Petersen MB, McInnis MG, Adelsberger PA, Schinzel AA et al (1992) The meiotic stage of nondisjunction in trisomy 21: determination by using DNA polymorphisms. Am J Hum Genet 50:544–550
15. Armour JA, Jeffreys AJ (1992) Biology and applications of human minisatellite loci. Curr Opin Genet Dev 2:850–856
16. Bailey JA, Gu Z, Clark RA, Reinert K, Samonte RV et al (2002) Recent segmental duplications in the human genome. Science 297:1003–1007
17. Ballabio A (1993) The rise and fall of positional cloning. Nat Genet 3:277–279
18. Batzer MA, Deininger PL (2002) Alu repeats and human genomic diversity. Nat Rev Genet 3:370–379
19. Birney E, Stamatoyannopoulos JA, Dutta A, Guigo R, Gingeras TR et al (2007) Identification and analysis of functional elements in 1% of the human genome by the ENCODE pilot project. Nature 447:799–816
20. Blouin JL, Dombroski BA, Nath SK, Lasseter VK, Wolyniec PS et al (1998) Schizophrenia susceptibility loci on chromosomes 13q32 and 8p21. Nat Genet 20:70–73
21. Boehm CD, Antonarakis SE, Phillips JA 3 rd, Stetten G, Kazazian HH Jr (1983) Prenatal diagnosis using DNA polymorphisms. Report on 95 pregnancies at risk for sickle-cell disease or beta-thalassemia. N Engl J Med 308:1054–1058
22. Brady T, Lee YN, Ronen K, Malani N, Berry CC et al (2009) Integration target site selection by a resurrected human endogenous retrovirus. Genes Dev 23:633–642
23. Brown CJ, Hendrich BD, Rupert JL, Lafreniere RG, Xing Y et al (1992) The human XIST gene: analysis of a 17 kb inactive X-specific RNA that contains conserved repeats and is highly localized within the nucleus. Cell 71:527–542
24. Brown WR, MacKinnon PJ, Villasante A, Spurr N, Buckle VJ, Dobson MJ (1990) Structure and polymorphism of human telomere-associated DNA. Cell 63:119–132
25. Carter NP (2007) Methods and strategies for analyzing copy number variation using DNA microarrays. Nat Genet 39:S16–S21
26. Cavaille J, Seitz H, Paulsen M, Ferguson-Smith AC, Bachellerie JP (2002) Identification of tandemly-repeated C/D snoRNA genes at the imprinted human 14q32 domain reminiscent of those at the Prader-Willi/Angelman syndrome region. Hum Mol Genet 11:1527–1538
27. Cavenee WK, Dryja TP, Phillips RA, Benedict WF, Godbout R et al (1983) Expression of recessive alleles by chromosomal mechanisms in retinoblastoma. Nature 305:779–784
28. Chakravarti A, Buetow KH, Antonarakis SE, Waber PG, Boehm CD, Kazazian HH (1984) Nonuniform recombination within the human beta-globin gene cluster. Am J Hum Genet 36:1239–1258
29. Chen KS, Manian P, Koeuth T, Potocki L, Zhao Q et al (1997) Homologous recombination of a flanking repeat gene cluster is a mechanism for a common contiguous gene deletion syndrome. Nat Genet 17:154–163
30. Choo K, Vissel B, Nagy A, Earle E, Kalitsis P (1991) A survey of the genomic distribution of alpha satellite DNA on all the human chromosomes, and derivation of a new consensus sequence. Nucleic Acids Res 19:1179–1182
31. Choo KH, Earle E, McQuillan C (1990) A homologous subfamily of satellite III DNA on human chromosomes 14 and 22. Nucleic Acids Res 18:5641–5648
32. Cohen D, Chumakov I, Weissenbach J (1993) A first-generation physical map of the human genome. Nature 366:698–701
33. Collins FS (1990) Identifying human disease genes by positional cloning. Harvey Lect 86:149–164
34. International HapMap Consortium (2003) The International HapMap Project. Nature 426:789–796
35. Craig-Holmes AP, Shaw MW (1971) Polymorphism of human constitutive heterochromatin. Science 174: 702–704
36. Cserpan I, Katona R, Praznovszky T, Novak E, Rozsavolgyi M et al (2002) The chAB4 and NF1-related long-range multisequence DNA families are contiguous in the centromeric heterochromatin of several human chromosomes. Nucleic Acids Res 30:2899–2905
37. Deininger PL, Batzer MA (1999) Alu repeats and human disease. Mol Genet Metab 67:183–193
38. Deloukas P, Earthrowl ME, Grafham DV, Rubenfield M, French L et al (2004) The DNA sequence and comparative analysis of human chromosome 10. Nature 429:375–381
39. Deloukas P, Matthews LH, Ashurst J, Burton J, Gilbert JG et al (2001) The DNA sequence and comparative analysis of human chromosome 20. Nature 414:865–871
40. Denoeud F, Kapranov P, Ucla C, Frankish A, Castelo R et al (2007) Prominent use of distal 5′ transcription start

41. Dermitzakis ET, Reymond A, Antonarakis SE (2005) Conserved non-genic sequences—an unexpected feature of mammalian genomes. Nat Rev Genet 6:151–157
42. Dermitzakis ET, Reymond A, Lyle R, Scamuffa N, Ucla C et al (2002) Numerous potentially functional but non-genic conserved sequences on human chromosome 21. Nature 420:578–582
43. Dib C, Faure S, Fizames C, Samson D, Drouot N et al (1996) A comprehensive genetic map of the human genome based on 5, 264 microsatellites. Nature 380:152–154
44. Djebali S, Kapranov P, Foissac S, Lagarde J, Reymond A et al (2008) Efficient targeted transcript discovery via array-based normalization of RACE libraries. Nat Methods 5:629–635
45. Dunham A, Matthews LH, Burton J, Ashurst JL, Howe KL et al (2004) The DNA sequence and analysis of human chromosome 13. Nature 428:522–528
46. Dunham I, Shimizu N, Roe BA, Chissoe S, Hunt AR et al (1999) The DNA sequence of human chromosome 22. Nature 402:489–495
47. Fearon ER, Vogelstein B, Feinberg AP (1984) Somatic deletion and duplication of genes on chromosome 11 in Wilms' tumours. Nature 309:176–178
48. Finn RD, Tate J, Mistry J, Coggill PC, Sammut SJ et al (2008) The Pfam protein families database. Nucleic Acids Res 36:D281–D288
49. Foster EA, Jobling MA, Taylor PG, Donnelly P, de Knijff P et al (1998) Jefferson fathered slave's last child. Nature 396:27–28
50. Frazer KA, Ballinger DG, Cox DR, Hinds DA, Stuve LL et al (2007) A second generation human haplotype map of over 3.1 million SNPs. Nature 449:851–861
51. Gill P, Ivanov PL, Kimpton C, Piercy R, Benson N et al (1994) Identification of the remains of the Romanov family by DNA analysis. Nat Genet 6:130–135
52. Gill P, Jeffreys AJ, Werrett DJ (1985) Forensic application of DNA 'fingerprints'. Nature 318:577–579
53. Gonzalez IL, Sylvester JE (2001) Human rDNA: evolutionary patterns within the genes and tandem arrays derived from multiple chromosomes. Genomics 73:255–263
54. Goossens M, Dozy AM, Embury SH, Zachariades Z, Hadjiminas MG et al (1980) Triplicated alpha-globin loci in humans. Proc Natl Acad Sci USA 77:518–521
55. Gregory SG, Barlow KF, McLay KE, Kaul R, Swarbreck D et al (2006) The DNA sequence and biological annotation of human chromosome 1. Nature 441:315–321
56. Griffiths-Jones S, Bateman A, Marshall M, Khanna A, Eddy SR (2003) Rfam: an RNA family database. Nucleic Acids Res 31:439–441
57. Grimwood J, Gordon LA, Olsen A, Terry A, Schmutz J et al (2004) The DNA sequence and biology of human chromosome 19. Nature 428:529–535
58. Gusella JF, Wexler NS, Conneally PM, Naylor SL, Anderson MA et al (1983) A polymorphic DNA marker genetically linked to Huntington's disease. Nature 306:234–236
59. Guttman M, Amit I, Garber M, French C, Lin MF et al (2009) Chromatin signature reveals over a thousand highly conserved large noncoding RNAs in mammals. Nature 458:223–227
60. Gyapay G, Morissette J, Vignal A, Dib C, Fizames C et al (1994) The 1993–94 Genethon human genetic linkage map. Nat Genet 7:246–339
61. Harrow J, Denoeud F, Frankish A, Reymond A, Chen CK (2006) GENCODE: producing a reference annotation for ENCODE. Genome Biol 7([Suppl 1]:S4):1–9
62. Hattori M, Fujiyama A, Taylor TD, Watanabe H, Yada T et al (2000) The DNA sequence of human chromosome 21. Nature 405:311–319
63. Heilig R, Eckenberg R, Petit JL, Fonknechten N, Da Silva C et al (2003) The DNA sequence and analysis of human chromosome 14. Nature 421:601–607
64. Hewitt JE, Lyle R, Clark LN, Valleley EM, Wright TJ et al (1994) Analysis of the tandem repeat locus D4Z4 associated with facioscapulohumeral muscular dystrophy. Hum Mol Genet 3:1287–1295
65. Higgs DR, Weatherall DJ (2009) The alpha thalassaemias. Cell Mol Life Sci 66:1154–1162
66. Hillier LW, Fulton RS, Fulton LA, Graves TA, Pepin KH et al (2003) The DNA sequence of human chromosome 7. Nature 424:157–164
67. Hillier LW, Graves TA, Fulton RS, Fulton LA, Pepin KH et al (2005) Generation and annotation of the DNA sequences of human chromosomes 2 and 4. Nature 434:724–731
68. Hollis M, Hindley J (1988) Satellite II DNA of human lymphocytes: tandem repeats of a simple sequence element. Nucleic Acids Res 16:363
69. Hudson TJ, Stein LD, Gerety SS, Ma J, Castle AB et al (1995) An STS-based map of the human genome. Science 270:1945–1954
70. Humphray SJ, Oliver K, Hunt AR, Plumb RW, Loveland JE et al (2004) DNA sequence and analysis of human chromosome 9. Nature 429:369–374
71. Jeffreys AJ, MacLeod A, Tamaki K, Neil DL, Monckton DG (1991) Minisatellite repeat coding as a digital approach to DNA typing. Nature 354:204–209
72. Jeffreys AJ, Wilson V, Thein SL (1985) Hypervariable 'minisatellite' regions in human DNA. Nature 314:67–73
73. Kalitsis P, Earle E, Vissel B, Shaffer LG, Choo KH (1993) A chromosome 13-specific human satellite I DNA subfamily with minor presence on chromosome 21: further studies on Robertsonian translocations. Genomics 16:104–112
74. Kan YW, Dozy AM (1978) Polymorphism of DNA sequence adjacent to human beta-globin structural gene: relationship to sickle mutation. Proc Natl Acad Sci USA 75:5631–5635
75. Katayama S, Tomaru Y, Kasukawa T, Waki K, Nakanishi M et al (2005) Antisense transcription in the mammalian transcriptome. Science 309:1564–1566
76. Kazazian HH Jr (2004) Mobile elements: drivers of genome evolution. Science 303:1626–1632
77. Kazazian HH Jr, Goodier JL (2002) LINE drive. retrotransposition and genome instability. Cell 110:277–280
78. Kazazian HH Jr, Wong C, Youssoufian H, Scott AF, Phillips DG, Antonarakis SE (1988) Haemophilia A resulting from de novo insertion of L1 sequences represents a novel mechanism for mutation in man. Nature 332:164–166

79. Kidd JM, Cooper GM, Donahue WF, Hayden HS, Sampas N et al (2008) Mapping and sequencing of structural variation from eight human genomes. Nature 453:56–64
80. King DC, Taylor J, Zhang Y, Cheng Y, Lawson HA et al (2007) Finding *cis*-regulatory elements using comparative genomics: some lessons from ENCODE data. Genome Res 17:775–786
81. Knowlton RG, Cohen-Haguenauer O, Van Cong N, Frezal J, Brown VA et al (1985) A polymorphic DNA marker linked to cystic fibrosis is located on chromosome 7. Nature 318:380–382
82. Kuo BA, Gonzalez IL, Gillespie DA, Sylvester JE (1996) Human ribosomal RNA variants from a single individual and their expression in different tissues. Nucleic Acids Res 24:4817–4824
83. Kurnit DM, Roy S, Stewart GD, Schwedock J, Neve RL et al (1986) The 724 family of DNA sequences is interspersed about the pericentromeric regions of human acrocentric chromosomes. Cytogenet Cell Genet 43:109–116
84. Lander ES, Linton LM, Birren B, Nusbaum C, Zody MC et al (2001) Initial sequencing and analysis of the human genome. Nature 409:860–921
85. Levy S, Sutton G, Ng PC, Feuk L, Halpern AL et al (2007) The diploid genome sequence of an individual human. PLoS Biol 5:e254
86. Lin JM, Collins PJ, Trinklein ND, Fu Y, Xi H et al (2007) Transcription factor binding and modified histones in human bidirectional promoters. Genome Res 17:818–827
87. Lindgren V, Ares M Jr, Weiner AM, Francke U (1985) Human genes for U2 small nuclear RNA map to a major adenovirus 12 modification site on chromosome 17. Nature 314:115–116
88. Lyle R, Prandini P, Osoegawa K, ten Hallers B, Humphray S et al (2007) Islands of euchromatin-like sequence and expressed polymorphic sequences within the short arm of human chromosome 21. Genome Res 17:1690–1696
89. Lyle R, Wright TJ, Clark LN, Hewitt JE (1995) The FSHD-associated repeat, D4Z4, is a member of a dispersed family of homeobox-containing repeats, subsets of which are clustered on the short arms of the acrocentric chromosomes. Genomics 28:389–397
90. Margulies EH, Cooper GM, Asimenos G, Thomas DJ, Dewey CN et al (2007) Analyses of deep mammalian sequence alignments and constraint predictions for 1% of the human genome. Genome Res 17:760–774
91. Martin J, Han C, Gordon LA, Terry A, Prabhakar S et al (2004) The sequence and analysis of duplication-rich human chromosome 16. Nature 432:988–994
92. Maston GA, Evans SK, Green MR (2006) Transcriptional regulatory elements in the human genome. Annu Rev Genomics Hum Genet 7:29–59
93. McVean GA, Myers SR, Hunt S, Deloukas P, Bentley DR, Donnelly P (2004) The fine-scale structure of recombination rate variation in the human genome. Science 304:581–584
94. Meneveri R, Agresti A, Della Valle G, Talarico D, Siccardi AG, Ginelli E (1985) Identification of a human clustered G+C-rich DNA family of repeats (Sau3A family). J Mol Biol 186:483–489
95. Mikkelsen TS, Ku M, Jaffe DB, Issac B, Lieberman E et al (2007) Genome-wide maps of chromatin state in pluripotent and lineage-committed cells. Nature 448:553–560
96. Morrow B, Goldberg R, Carlson C, Das Gupta R, Sirotkin H et al (1995) Molecular definition of the 22q11 deletions in velo-cardio-facial syndrome. Am J Hum Genet 56:1391–1403
97. Mungall AJ, Palmer SA, Sims SK, Edwards CA, Ashurst JL et al (2003) The DNA sequence and analysis of human chromosome 6. Nature 425:805–811
98. Muzny DM, Scherer SE, Kaul R, Wang J, Yu J et al (2006) The DNA sequence, annotation and analysis of human chromosome 3. Nature 440:1194–1198
99. Nakamura Y, Leppert M, O'Connell P, Wolff R, Holm T et al (1987) Variable number of tandem repeat (VNTR) markers for human gene mapping. Science 235:1616–1622
100. Nicholls RD, Knoll JHM, Butler MG, Karami S, Lalande M (1989) Genetic imprinting suggested by maternal heterodisomy in non-deletion Prader-Willi syndrome. Nature 342:281–285
101. Nusbaum C, Mikkelsen TS, Zody MC, Asakawa S, Taudien S et al (2006) DNA sequence and analysis of human chromosome 8. Nature 439:331–335
102. Nusbaum C, Zody MC, Borowsky ML, Kamal M, Kodira CD et al (2005) DNA sequence and analysis of human chromosome 18. Nature 437:551–555
103. Parra G, Reymond A, Dabbouseh N, Dermitzakis ET, Castelo R et al (2006) Tandem chimerism as a means to increase protein complexity in the human genome. Genome Res 16:37–44
104. Patel PI, Lupski JR (1994) Charcot-Marie-Tooth disease: a new paradigm for the mechanism of inherited disease. Trends Genet 10:128–133
105. Pavelitz T, Liao D, Weiner AM (1999) Concerted evolution of the tandem array encoding primate U2 snRNA (the RNU2 locus) is accompanied by dramatic remodeling of the junctions with flanking chromosomal sequences. EMBO J 18:3783–3792
106. Pennacchio LA, Ahituv N, Moses AM, Prabhakar S, Nobrega MA et al (2006) In vivo enhancer analysis of human conserved non-coding sequences. Nature 444:499–502
107. Perez Jurado LA, Peoples R, Kaplan P, Hamel BC, Francke U (1996) Molecular definition of the chromosome 7 deletion in Williams syndrome and parent-of-origin effects on growth. Am J Hum Genet 59:781–792
108. Redon R, Ishikawa S, Fitch KR, Feuk L, Perry GH et al (2006) Global variation in copy number in the human genome. Nature 444:444–454
109. Riethman H (2008) Human telomere structure and biology. Annu Rev Genomics Hum Genet 9:1–19
110. Rosenberg NA, Pritchard JK, Weber JL, Cann HM, Kidd KK et al (2002) Genetic structure of human populations. Science 298:2381–2385
111. Ross MT, Grafham DV, Coffey AJ, Scherer S, McLay K et al (2005) The DNA sequence of the human X chromosome. Nature 434:325–337
112. Roy-Engel AM, Carroll ML, Vogel E, Garber RK, Nguyen SV et al (2001) Alu insertion polymorphisms for the

study of human genomic diversity. Genetics 159:279–290
113. Royo H, Cavaille J (2008) Non-coding RNAs in imprinted gene clusters. Biol Cell 100:149–166
114. Rudd MK, Willard HF (2004) Analysis of the centromeric regions of the human genome assembly. Trends Genet 20:529–533
115. Sabeti PC, Reich DE, Higgins JM, Levine HZ, Richter DJ et al (2002) Detecting recent positive selection in the human genome from haplotype structure. Nature 419:832–837
116. Sabeti PC, Varilly P, Fry B, Lohmueller J, Hostetter E et al (2007) Genome-wide detection and characterization of positive selection in human populations. Nature 449:913–918
117. Sahoo T, del Gaudio D, German JR, Shinawi M, Peters SU et al (2008) Prader-Willi phenotype caused by paternal deficiency for the HBII-85 C/D box small nucleolar RNA cluster. Nat Genet 40:719–721
118. Samonte RV, Eichler EE (2002) Segmental duplications and the evolution of the primate genome. Nat Rev Genet 3:65–72
119. Saxena R, Voight BF, Lyssenko V, Burtt NP, de Bakker PI et al (2007) Genome-wide association analysis identifies loci for type 2 diabetes and triglyceride levels. Science 316:1331–1336
120. Scherer SE, Muzny DM, Buhay CJ, Chen R, Cree A et al (2006) The finished DNA sequence of human chromosome 12. Nature 440:346–351
121. Schmutz J, Martin J, Terry A, Couronne O, Grimwood J et al (2004) The DNA sequence and comparative analysis of human chromosome 5. Nature 431:268–274
122. Schueler MG, Sullivan BA (2006) Structural and functional dynamics of human centromeric chromatin. Annu Rev Genomics Hum Genet 7:301–313
123. Sharp AJ, Cheng Z, Eichler EE (2006) Structural variation of the human genome. Annu Rev Genomics Hum Genet 7:407–442
124. She X, Horvath JE, Jiang Z, Liu G, Furey TS et al (2004) The structure and evolution of centromeric transition regions within the human genome. Nature 430:857–864
125. Skaletsky H, Kuroda-Kawaguchi T, Minx PJ, Cordum HS, Hillier L et al (2003) The male-specific region of the human Y chromosome is a mosaic of discrete sequence classes. Nature 423:825–837
126. Sleutels F, Zwart R, Barlow DP (2002) The non-coding Air RNA is required for silencing autosomal imprinted genes. Nature 415:810–813
127. Smit AF (1996) The origin of interspersed repeats in the human genome. Curr Opin Genet Dev 6:743–748
128. Spence JE, Perciaccante RG, Greig GM, Willard HF, Ledbetter DH et al (1988) Uniparental disomy as a mechanism for human genetic disease. Am J Hum Genet 42:217–226
129. Stefansson H, Helgason A, Thorleifsson G, Steinthorsdottir V, Masson G et al (2005) A common inversion under selection in Europeans. Nat Genet 37:129–137
130. Tapparel C, Reymond A, Girardet C, Guillou L, Lyle R et al (2003) The TPTE gene family: cellular expression, subcellular localization and alternative splicing. Gene 323:189–199
131. Taylor TD, Noguchi H, Totoki Y, Toyoda A, Kuroki Y et al (2006) Human chromosome 11 DNA sequence and analysis including novel gene identification. Nature 440:497–500
132. Tishkoff SA, Reed FA, Friedlaender FR, Ehret C, Ranciaro A et al (2009) The genetic structure and history of Africans and African Americans. Science 324:1035–1044
133. Trinklein ND, Karaoz U, Wu J, Halees A, Force Aldred S et al (2007) Integrated analysis of experimental data sets reveals many novel promoters in 1% of the human genome. Genome Res 17:720–731
134. Tuzun E, Sharp AJ, Bailey JA, Kaul R, Morrison VA et al (2005) Fine-scale structural variation of the human genome. Nat Genet 37:727–732
135. Tycowski KT, You ZH, Graham PJ, Steitz JA (1998) Modification of U6 spliceosomal RNA is guided by other small RNAs. Mol Cell 2:629–638
136. Umlauf D, Fraser P, Nagano T (2008) The role of long non-coding RNAs in chromatin structure and gene regulation: variations on a theme. Biol Chem 389:323–331
137. Venter JC, Adams MD, Myers EW, Li PW, Mural RJ et al (2001) The sequence of the human genome. Science 291:1304–1351
138. Verma RS, Dosik H, Lubs HA (1977) Size variation polymorphisms of the short arm of human acrocentric chromosomes determined by R-banding by fluorescence using acridine orange (RFA). Hum Genet 38:231–234
139. Wahl MC, Will CL, Luhrmann R (2009) The spliceosome: design principles of a dynamic RNP machine. Cell 136:701–718
140. Wallace DC (1999) Mitochondrial diseases in man and mouse. Science 283:1482–1488
141. Wallace DC (2005) A mitochondrial paradigm of metabolic and degenerative diseases, aging, and cancer: a dawn for evolutionary medicine. Annu Rev Genet 39:359–407
142. Wallace DC (2007) Why do we still have a maternally inherited mitochondrial DNA? Insights from evolutionary medicine. Annu Rev Biochem 76:781–821
143. Warburton D (1991) De novo balanced chromosome rearrangements and extra marker chromosomes identified at prenatal diagnosis: clinical significance and distribution of breakpoints. Am J Hum Genet 49:995–1013
144. Warren AC, Slaugenhaupt SA, Lewis JG, Chakravarti A, Antonarakis SE (1989) A genetic linkage map of 17 markers on human chromosome 21. Genomics 4:579–591
145. Waterston RH, Lindblad-Toh K, Birney E, Rogers J, Abril JF (2002) Initial sequencing and comparative analysis of the mouse genome. Nature 420:520–562
146. Waye JS, Willard HF (1989) Human beta satellite DNA: genomic organization and sequence definition of a class of highly repetitive tandem DNA. Proc Natl Acad Sci USA 86:6250–6254
147. Weinstein LB, Steitz JA (1999) Guided tours: from precursor snoRNA to functional snoRNP. Curr Opin Cell Biol 11:378–384
148. Weissenbach J, Gyapay G, Dib C, Vignal A, Morissette J et al (1993) A second generation linkage map of the human genome. Nature 359:794–801
149. Woods-Samuels P, Wong C, Mathias SL, Scott AF, Kazazian HH Jr, Antonarakis SE (1989) Characterization of a nondeleterious L1 insertion in an intron of the human

factor VIII gene and further evidence of open reading frames in functional L1 elements. Genomics 4:290–296
150. Wyman AR, White R (1980) A highly polymorphic locus in human DNA. Proc Natl Acad Sci USA 77:6754–6758
151. Zhang Z, Harrison PM, Liu Y, Gerstein M (2003) Millions of years of evolution preserved: a comprehensive catalog of the processed pseudogenes in the human genome. Genome Res 13:2541–2558
152. Zody MC, Garber M, Adams DJ, Sharpe T, Harrow J et al (2006) DNA sequence of human chromosome 17 and analysis of rearrangement in the human lineage. Nature 440:1045–1049
153. Zody MC, Garber M, Sharpe T, Young SK, Rowen L et al (2006) Analysis of the DNA sequence and duplication history of human chromosome 15. Nature 440:671–675

Chromosomes

Michael R. Speicher

Abstract The study of human chromosomes started in the late 19th century with first observations on cytological preparations and the area of cytogenetics evolved. Cytogenetics is the study of the structure, function and evolution of chromosomes. Cytogenetics is indispensable for both routine diagnostics in clinical genetics and research. To date cytogenetics is characterized by rapidly growing application of molecular biological techniques which have unraveled many mechanisms involved in fundamental steps during meiosis and mitosis. Furthermore, especially array-technologies allowed a spectacular increase in resolution, which has dramatically changed our view about genome plasticity. As a consequence of this new spectrum of methods cytogeneticists have now new options to improve diagnostics and to address basic research issues. Here basic principles of chromosome structure and biology, latest developments and future perspectives are described.

Contents

3.1	History and Development of Human Cytogenetics	56
	3.1.1 First Observations on Human Mitotic Chromosomes	56
	3.1.2 An Old Error is Corrected and a New Era Begins	57
	3.1.3 Birth of Human Cytogenetics 1956–1963	58
	3.1.4 Introduction of Banding Technologies from the Late 1960s to the Present	58
	3.1.5 The Birth of Molecular Cytogenetics in the Late 1960s	59
	3.1.6 Molecular Cytogenetics or Fluorescence In Situ Hybridization 1980 to Date	59
	3.1.7 Array Technologies: New Dimensions in Resolution from 1997 to Date	59
3.2	Organization of Genetic Material in Human Chromosomes	60
	3.2.1 Heterochromatin and Euchromatin	60
	3.2.2 From DNA Thread to Chromosome Structure	60
	3.2.3 Centromeres and Kinetochores	63
	3.2.4 Chromosome Bands	65
	3.2.5 Telomeres	68
3.3	Cell Cycle and Mitosis	71
	3.3.1 Cell Cycle: Interphase G_1-G_2 and G_0	71
	3.3.2 Cell Cycle: Mitosis	72
	3.3.3 Cell Cycle Checkpoints	75
	3.3.4 Cell Cycle Coordinators	77
3.4	Chromosome Analysis Methods	78
	3.4.1 Banding Techniques	78
	3.4.2 Karyotype Description	80
	3.4.3 Fragile Sites	82
3.5	Meiosis	94
	3.5.1 Biological Function of Meiosis	94
	3.5.2 Meiotic Divisions	94
	3.5.3 Meiotic Recombination Hotspots	96
	3.5.4 Differences Between Male and Female Meiosis	96
	3.5.5 Nonallelic Homologous Recombination During Meiosis Can Cause Microdeletion/ Microduplication Syndromes	98
	3.5.6 Molecular Mechanisms Involved in Meiosis	99
3.6	Human Chromosome Pathology in Postnatal Diagnostics	99
	3.6.1 Syndromes Attributable to Numeric Anomalies of Autosomes	99

M.R. Speicher (✉)
Institute of Human Genetics, Medical University of Graz,
Harrachgasse 21/8, 8010 Graz, Austria
e-mail: michael.speicher@medunigraz.at

3.6.2 Syndromes Attributable to Structural Anomalies of Autosomes............................ 106
3.6.3 Sex Chromosomes... 116
3.6.4 Chromosome Aberrations and Spontaneous Miscarriage 125

3.7 Chromosome Instability/Breakage Syndromes......... 126
3.7.1 Fanconi Anemia 126
3.7.2 Instabilities Caused by Mutations of Proteins of the RecQ Family of Helicases: Bloom syndrome, Werner syndrome, and Rothmund-Thomson syndrome............. 127
3.7.3 The Ataxia-Telangiectasia Group 128
3.7.4 Immunodeficiency, Centromeric Region Instability, and Facial Anomalies Syndrome 129
3.7.5 Roberts Syndrome/SC Phocomelia.............. 129
3.7.6 Mosaic Variegated Aneuploidy 129

References 130

3.1 History and Development of Human Cytogenetics

The chromosome theory of Mendelian inheritance was launched in 1902 by Sutton and Boveri. In the same year Garrod, establishing the autosomal-recessive mode of inheritance for alkaptonuria and commenting on metabolic individuality in general, created the paradigm of "inborn errors of metabolism." Simple modes of inheritance were soon established for many other human disorders. A few years later, Bridges [33] examined in *Drosophila* the first case of a disturbance in chromosome distribution during meiosis and named it "nondisjunction." Cytogenetics of animals and plants flourished during the first half of the twentieth century, and many important phenomena in the field of cytogenetics were discovered during this period. Moreover, cytogenetic methods helped to elucidate many basic laws of mutation.

However, the age of human cytogenetics did not begin until the 1950s, when Tjio and Levan [235] and Ford and Hamerton [79] established the diploid human chromosome number of 46. Lejeune et al. [142] discovered trisomy 21 in Down syndrome, while Ford et al. [80] and Jacobs and Strong [116] established that Turner and Klinefelter syndromes were caused by X-chromosomal anomalies.

The late arrival of human cytogenetics is usually ascribed to shortcomings in the methods used to prepare chromosomes. Indeed, the jumbled masses of chromosomes in old illustrations demonstrate the difficulties encountered by the pioneers who tried to count human chromosomes. Still it is hard to conceive that the development of more adequate methods would have been delayed for such a long period had the cytogeneticists realized there are human anomalies awaiting explanation, and few human geneticists considered the possibility that certain anomalies might be due to chromosomal aberrations.

3.1.1 First Observations on Human Mitotic Chromosomes

It could be said that research on human cytogenetics began with the work of Arnold [13] (Fig. 3.1), Flemming [78], and Hansemann [100], who were the first to examine human mitotic chromosomes. Owing to their affinity for certain stains, Waldeyer [244] dubbed the threadlike structures "chromosomes," which is derived from the Greek words chroma (=color) and soma (=body) and means "colored body."

A report with a strong impact lasting several decades was that of Painter in the 1920s [176]. Painter was the leading cytogeneticist of his time. When he examined chromosome preparations derived from testicles of three individuals in first meiotic divisions he was able to demonstrate that the sex was bivalent, consisting of the X and Y chromosomes, which at anaphase migrated to

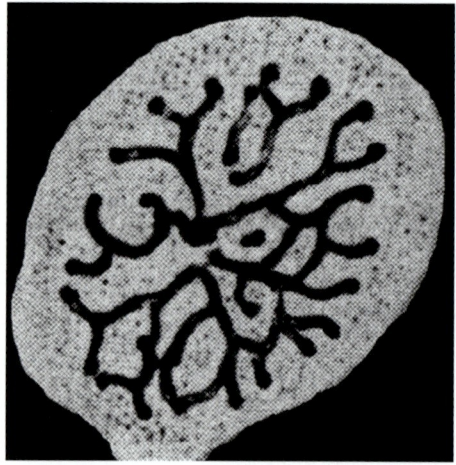

Fig. 3.1 One of the first images of human chromosomes made by the German pathologist J. Arnold in 1879. Arnold examined carcinoma and sarcoma cells because their voluminous nucleus facilitated analysis. The drawing shows a human sarcoma cell. From [13]

opposite poles. In a preliminary report he had described the chromosome number as 46 or 48, but in the definitive report [176] he had decided in favor of 48 chromosomes (Fig. 3.2). Surprisingly, in subsequent decades, a chromosome number of 48 in humans was supported in a number of reports [85]. The problem of the continuing incorrect chromosome count following Painter has been ascribed to a "preconception" [133]. The number was thought to be 48, so subsequent investigators did their utmost possible to make their counts 48.

However, two technical difficulties impeded further progress:

1. Sectioning by the usual histological techniques often disturbed mitoses.
2. The chromosomes tended to lie on top of each other and even to clump together.

These difficulties were ultimately overcome by:

(a) The use of suspensions of intact cells rather than of histological sections.
(b) The use of spindle poisons such as colchicine. Colchicine is an alkaloid derived from the autumn crocus, *Colchicum autumnale*. Because spindle formation is inhibited in cells during mitosis, chromosomes cannot separate during anaphase. The use of colchicines was introduced by Blakeslee and Avery [28] and Levan [145]. By arresting cells in metaphase, colchicine increases the number of metaphase spreads. Furthermore, colchicine increases chromosome condensation. Hence, varying the exposure time of cells to colchicine allows determination of the length of chromosomes.

(c) The subjection of cells to a brief treatment with a hypotonic solution, causing them to swell and burst, thus spreading out the chromosomes for better definition. The hypotonic treatment was described by a Russian group as long ago as in 1934 [262]. However, this discovery was abandoned and the treatment had to be rediscovered in 1952. The hypotonic shock technique paved the way for easy chromosome counting [110, 111]. Interestingly enough, even 30 years later Painter's estimate of 48 was so strongly imprinted on investigators' minds that in the first study on human chromosomes using the new technique the human chromosome number was reported as 48 [110].

3.1.2 An Old Error is Corrected and a New Era Begins

In the summer of 1955 Levan, a Swedish cytogeneticist visited Hsu in New York and learned the technique of squash preparation using hypotonic shock. He and Tjio then improved the technique by shortening the hypotonic treatment and by adding colchicine. They examined lung fibroblasts of four human embryos. To their surprise they found a chromosome number of 46 in most of 261 metaphases [235]. Figure 3.3 shows one example. This evidence was soon supplemented by Ford and Hamerton [18].

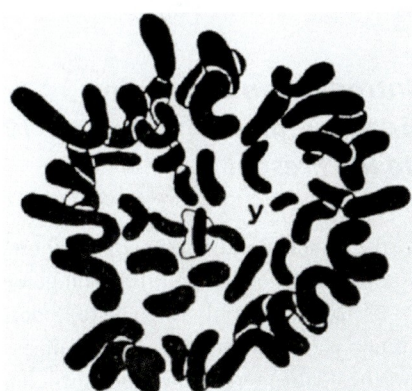

Fig. 3.2 Camera lucida drawing of a human spermatogonial metaphase made by Theopilus S. Painter. From this drawing Painter concluded that humans have 48 chromosomes in their cells. From Springer-Verlag 1979

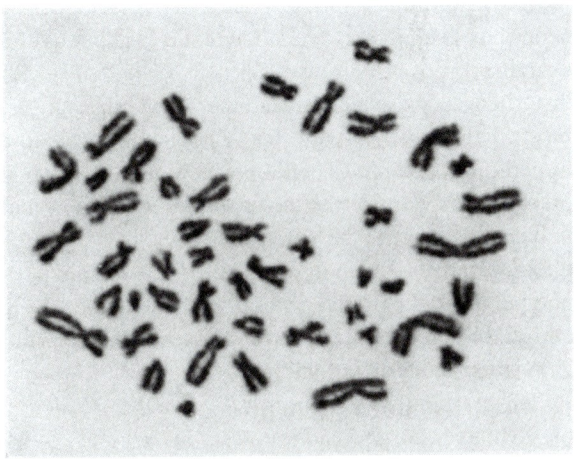

Fig. 3.3 A metaphase of a human embryonic lung fibroblast grown in vitro. The image is from the first report in which the human chromosome number was established as 46. From [235], with permission from Wiley-Blackwell, from the 3rd edition of this work

With these results the stage was set for the development of clinical cytogenetics. Still, it was almost another 3 years before the first abnormal karyotypes in humans were reported.

3.1.3 Birth of Human Cytogenetics 1956–1963

Lejeune et al. [142] reported chromosome studies from fibroblast cultures in nine children with Down syndrome. Fifty-seven diploid cells were regarded as technically perfect. In all of them the chromosomes numbered 47. The supernumerary chromosome was described as small and "telocentric." Meiotic nondisjunction was suggested as the most likely explanation for the additional chromosome. More aspects of Down syndrome are described in Sect. 3.6.1 and in Chap. 23.6.

As soon as in 1949 Barr and Bertram discovered the "X chromatin," an intranuclear body 0.8–1.1 µm in size, which is commonly located at the periphery of the interphase nuclei of females and is not present in males [22]. The discovery was incidental, since it originated in an investigation of the effects of fatigue on the central nervous system in cats. What first seemed to be a sex difference only in the neurons of cats turned out to be a normal finding characteristic of the nuclear inheritance of female mammals, including human females. Corresponding structures, the drumsticks, were found by Davidson and Smith [56] in polymorphonuclear neutrophil leukocytes. The obvious next step was the examination of X chromatin in cells of patients with disturbances in sexual development. In this sample, most male patients with the Klinefelter syndrome turned out to be X chromatin positive in spite of their predominantly male phenotype [171], whereas most female patients with the Turner syndrome were X chromatin negative – again in contrast to their female phenotype. If the X chromatin was directly related to the X chromosomes, these findings pointed to X chromosome anomalies in these two syndromes.

This suspicion was reinforced when the frequency of X-linked color vision defects in patients with Klinefelter syndrome was found to be lower than in normal males but much higher than would be expected in XX females [190].

This situation was explained when Jacobs and Strong [116], examining the chromosomes from bone marrow mitoses in Klinefelter patients, found 47 chromosomes, although both parents had a normal karyotype. The supernumerary chromosome belonged to the group of chromosomes including X chromosomes; the karyotype was tentatively identified as XXY.

Soon after this first report, the XXY karyotype in Klinefelter syndrome was confirmed in many more cases, and it is now known as the standard karyotype in this condition. At the same time, the result in the Klinefelter syndrome was complemented by chromosome examinations in another syndrome in which a discrepancy between phenotypic and nuclear sex seemed to exist: Turner syndrome. Ford et al. [80] showed that the karyotype had only 45 chromosomes, obviously with only one X and no Y chromosome. A third anomaly with 47 chromosomes and three X chromosomes was soon described in a slightly retarded woman with a dysfunction of the sexual organs [117]. Both Klinefelter and Turner syndromes will be discussed further in Sect. 3.6.3.3 and in Chap. 23.6.

Two autosomal trisomies, later identified as trisomies 13 and 18, were described by Patau et al. [178] and Edwards et al. [68]. Furthermore, in 1960 Peter Nowell [172] discovered that phytohemagglutinin stimulated white blood cells to divide, another tremendous stimulus to human cytogenetics. In the same year he and David Hungerford discovered a minute chromosome, later named the Philadelphia chromosome, which was regularly found in peripheral blood in human chronic myeloid leukemia [173]. The first deletion syndrome, cri-du-chat syndrome, caused by a deletion on the short arm of chromosome 5, was observed by Lejeune et al. [143].

3.1.4 Introduction of Banding Technologies from the Late 1960s to the Present

However, the exact identification of the homologous chromosome pairs or of structural rearrangements was still not possible. The only means of "identifying" chromosomes was by noting size differences and the position of the centromere, which is insufficient to distinguish chromosomes with a similar morphology. More detailed chromosome analyses had to await the discovery of chromosome banding. When fluorochromes coupled to an alkylating agent, such as quinacrine mustard, were used a highly characteristic fluorescence pattern for each individual chromosome could be achieved and a complete human karyotype could be presented [40, 41]. However, owing to its practical simplicity, Giemsa banding rapidly replaced quinacrine banding. To this day, Giemsa banding remains the most widely used banding procedure in routine chromosome analyses in most laboratories throughout the world. The improvement in the resolution of chromosome analyses achieved by banding analysis is reflected in the fact that 13 years after the first description of the Philadelphia chromosome, banding analysis revealed that the Philadelphia chromosome was the result not of a deletion in chromosome 22, but of a translocation between chromosome 22 and chromosome 9 [199].

3.1.5 The Birth of Molecular Cytogenetics in the Late 1960s

The first application of molecular techniques to chromosome cytology was based on the perception that sequences that were complementary to each other could anneal, or hybridize, and form much more stable complexes than noncomplementary sequences. The first in situ hybridization, done by Joe Gall and Mary Lou Pardue [83], applied DNA-RNA hybridization to locate the genes coding for ribosomal RNA. A similar in situ hybridization technique had been developed independently in the laboratory of Max Biernstiel [123]. However, these early in situ hybridizations depended on radioactive detection, which was prone to high background and was slow, the film exposure often taking several days or even weeks. Attempts to overcome these problems included the use of fluorescently labeled antibodies to recognize specific RNA-DNA hybrids [200].

3.1.6 Molecular Cytogenetics or Fluorescence In Situ Hybridization 1980 to Date

A more straightforward approach employed the chemical coupling of a fluorophore to an RNA probe for rapid and direct visualization. Such a "fluorescent *in situ hybridiza*tion" (FISH) was first realized in 1980 [24]. The coupling of fluorochrome to a DNA or RNA probe is often referred to as "direct labeling." In contrast, "indirect labeling" means the enzymatic or immunological detection of tags incorporated into a probe. The synthesis of modified nucleotide derivatives containing a biotin label, which could be incorporated by polymerases into probes, was instrumental for the development of indirect labeling techniques [140].

Nonetheless, the successful hybridization of complex probes was restricted by the presence of repetitive sequences, which occur ubiquitously in the genome and which are usually present in complex probes. These repetitive sequences were finally suppressed by the addition of an excess of unlabeled genomic DNA (in first experiments) or Cot-1 DNA to the hybridization mix [138]. With use of the suppression technique the painting of entire chromosomes rapidly became possible [52, 146, 187] and FISH became applicable in clinical cytogenetics. A distinct advantage of FISH is that DNA probes can be visualized in intact interphase nuclei, an approach referred to as "interphase cytogenetics" [51]. In the years since then, a clear aim of the continuing development of cytogenetic methods has been to increase the resolution at which chromosome rearrangements can be identified. This has been achieved by advances in the two crucial elements of cytogenetic analysis, i.e., the probe and the target. Target resolution has improved from metaphase chromosomes (resolution ~5 Mb), through interphase nuclei (50 kb to 2 Mb) and DNA fibers (5–500 kb) to the use of DNA microarrays offering resolutions to a single nucleotide. Simultaneously, probe development has also advanced, to best utilize the improvements in target resolution [222].

3.1.7 Array Technologies: New Dimensions in Resolution from 1997 to Date

The introduction of array technologies increased resolution – depending on the array platform – to the single nucleotide level and is definitively blurring the traditional distinction between cytogenetics and molecular genetics. Since the first array applications [188, 220], a multitude of various array platforms has been developed. One of the most significant findings elucidated by array techniques is the wide scope and prevalence of copy number changes (CNVs) in the human genome [196]. The unearthing of association of CNVs with biological function, recent human evolution, and common and complex human disease is at present one of the most fascinating areas in human genetics.

3.2 Organization of Genetic Material in Human Chromosomes

Chromosomes consist of a number of different building blocks, such as heterochromatin and euchromatin, centromeres, and telomeres. The main features of these building blocks are described below.

3.2.1 Heterochromatin and Euchromatin

Chromosomes are usually not visible in the nuclei of nondividing cells, because the chromatin in interphase nuclei is so densely packed that single chromatin threads are not directly detectable. However, some parts of chromatin may become visible after staining, giving rise to the distinction between "heterochromatin" and "euchromatin" [105]. Heterochromatin refers to chromosomal segments that remain compact and can be stained and thus remain visible during the entire cell cycle. In contrast, euchromatin decondenses in interphase, to the extent that it becomes invisible during late telophase and subsequent interphase. From a functional point of view, heterochromatin represents chromosomal segments with few active genes or none at all. These regions are dominated by repetitive DNA sequences. Our understanding of the establishment and maintenance of these functional domains has improved recently. This distinction is not only based on DNA sequence but also on epigenetic mechanisms, such as methylation patterns and histone modifications, as discussed below.

Heterochromatin can be more finely categorized as constitutive or facultative, depending on whether or not there is a consistent relationship between the DNA sequence involved and a compact organization across cell types and differentiation states. For example, usually all cells of an individual will package the same regions of DNA in *constitutive heterochromatin*. Constitutive heterochromatin can be observed in human cells close to the centromeres of chromosomes 1, 9, and 16. In addition, male cells have a large block of heterochromatin on the long arm of the Y chromosome. Constitutive heterochromatin is usually encountered around the chromosome centromere and near telomeres.

In contrast to constitutive heterochromatin, which is identical in all cells of a body, *facultative heterochromatin* refers to sequences that may be densely packaged in one cell, thus forming heterochromatin, but packaged in euchromatin in another cell. The best-known example of facultative heterochromatin is X-chromosome inactivation in female cells (see also Sect. 3.6.3.2): One X chromosome is packaged in facultative heterochromatin and thus to a large extent, but not completely, silenced. The other X chromosome is packaged in euchromatin. Such mechanisms allow heritable but reversible changes in gene expression without alterations in DNA sequence. Thus, epigenetic "on-off" transcriptional states are largely dependent on the position of a gene within an accessible (euchromatic) or an inaccessible (heterochromatic) chromatin environment (for further discussion see the next section).

In summary, while heterochromatin was first defined on morphological grounds [105], it is now defined in terms of a "histone code" of posttranslational modifications that influence transitions between chromatin states and the regulation of transcriptional activity (see below for further details).

3.2.2 From DNA Thread to Chromosome Structure

3.2.2.1 DNA Condensation

The human genome has a size of approximately 3.4×10^9 base pairs and harbors about 25,000 genes. The total length of DNA in the haploid chromosome complement in nondividing human cells is about 1 meter. However, the total length of the human haploid chromosome complement in metaphase cells is merely 115 µm.

As a consequence, chromosomes are visible as individual structures only during mitosis, with individual chromosomes ranging in size from 3 to 7 µm. These numbers illustrate that considerable packing and unpacking occurs during cell division. Furthermore, there is a direct relationship between higher order chromatin folding and transcriptional control. Hence,

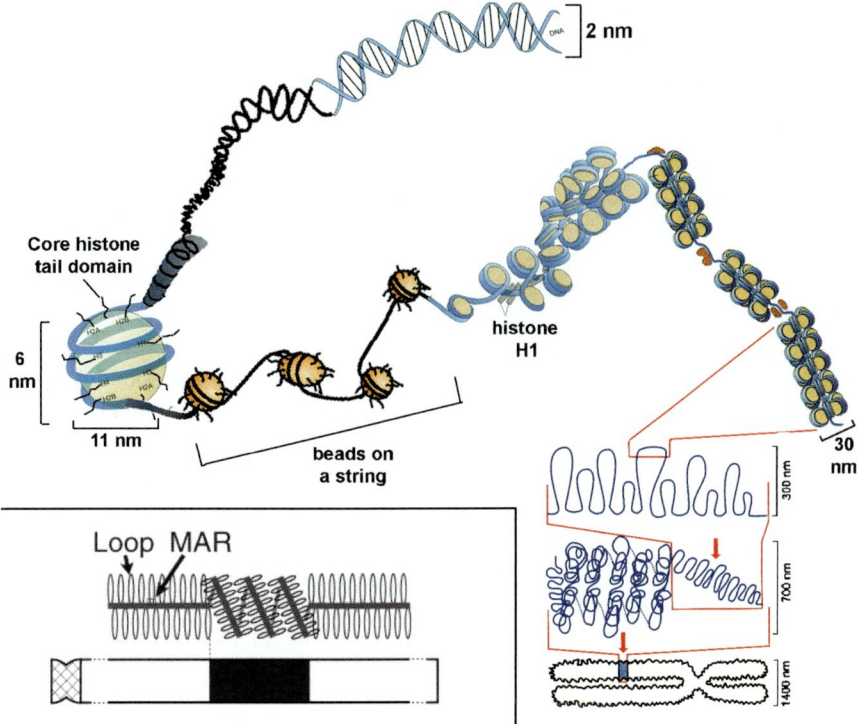

Fig. 3.4 Folding and packing of chromosomal DNA. Scheme of different levels of packing of DNA from the double helix to a metaphase chromosome. (Modified from [109]). *Inset:* Model of a metaphase chromatin structure. Adapted from [202]

the basic building blocks of chromatin determine the interplay between chromatin structure and transcription. Several levels of packing units of DNA can be distinguished. These various building blocks of chromatin are briefly reviewed here (Fig. 3.4).

The thickness of the DNA double helix is estimated at about 2 nm. The fundamental, basic subunit of chromosome structure is the nucleosome, which is made of DNA and histones and which divides the DNA into units of approximately 200 bp in length. At the molecular level, each chromosome is a repetition of nucleosomes and shorter segments of DNA that link the individual nucleosomes. Histones are proteins with a high proportion of positively charged amino acids, allowing them to attach themselves firmly to the negatively charged DNA double helix. A nucleosome consists of 147 bp of DNA wound 1.75 times around a core histone octamer. Such a core histone octamer consists of two copies of each histones H2A, H2B, H3, and H4. Every core histone contains two separate functional domains: a signature "histone-fold" motif sufficient for both histone-histone and histone-DNA contacts within the nucleosome and NH_2-terminal and COOH-terminal "tail" domains containing sites for posttranslational modifications (such as acetylation, methylation, phosphorylation, and ubiquitination). It is these posttranslational modifications, largely located in the N-terminal domains of the histone proteins, which encode most of the epigenetic information specifying chromatin structure and function. The histone octamer forms a cylinder 11 nm in diameter and 6 nm in height. Without histones the chromosome skeleton, surrounded by numerous threads corresponding to the DNA double helix, becomes visible under the electron microscope [180] (Fig. 3.5).

Nucleosomes are organized on a continuous DNA helix in linear strings separated by 10–60 bp of linker DNA. On electron-microscopic photographs they have a "string of beads" appearance. Between nucleosomes the DNA is bound to a fifth histone, histone H1, often referred to as linker histone, which binds DNA as it enters and exits the nucleosome to stabilize two complete turns of the DNA around the histone octamer. Histone tail-mediated nucleosome-nucleosome interaction leads to formation of the 30-nm fiber, which represents a secondary level of compaction [109].

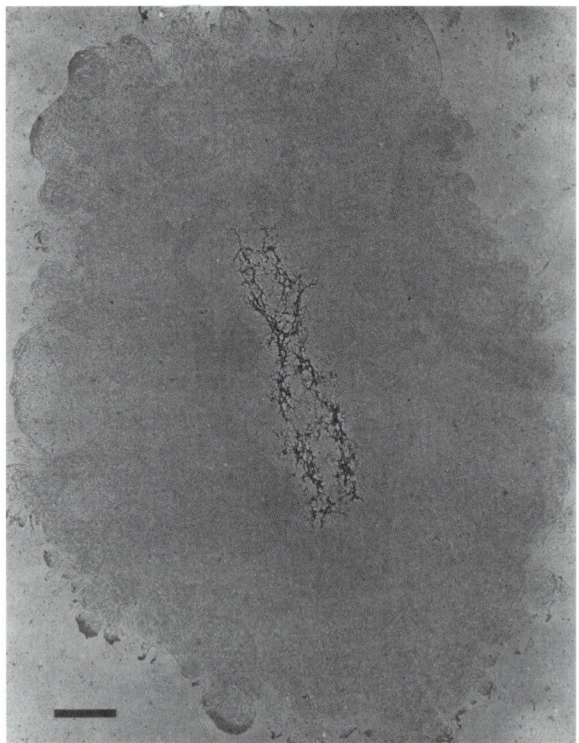

Fig. 3.5 Electron micrograph of a histone-depleted metaphase chromosome from HeLa. The depicted chromosome consists of a central, densely staining scaffold or core surrounded by a halo of DNA extending 6-9 μm outward from the scaffold. The low magnification makes it difficult to see the individual DNA strands except along the edge of the DNA halo. *Bar* 2 μm. From [120], reprinted by permission from Macmillan Publishers Ltd: *Nature Genetics*, copyright 2001

3.2.2.2 Histone Modifications Organize Chromosomal Subdomains

Besides being involved in nucleosome formation, histones are crucial for the functional organization of chromosomal subdomains, e.g., differentiation between euchromatin and heterochromatin. A "histone code" [225] hypothesis has been proposed. This predicts that modifications of histone N-termini, such as acetylation, phosporylation, or methylation, are fundamental mechanisms for the induction and stabilization of distinct chromosomal subdomains (reviewed in [121]). This is shown schematically in Fig. 3.6. In particular, the histone H3 lysine 9 methylation is highly characteristic for the pericentric heterochromatin and differs from the H3-K9 methylation present in other chromosomal regions [183, 184]. This will be discussed in further detail in the next section on centromeres.

Thus, histone modifications result in condensed chromatin, i.e., heterochromatin, or in extended chromatin representing euchromatin. The level of condensation also reflects the transcriptional activity of the corresponding DNA segment: DNA wound around a nucleosome is inactive and unreactive. Sequence-specific DNA-binding proteins are only found between nucleosomes. There are hundreds of sequence-specific DNA-binding proteins that recognize short DNA segments.

Fig. 3.6 (**a, b**) Models for euchromatin and heterochromatin histone tail modifications. (From [121]) (**a**) Schematic representation of euchromatin and heterochromatin as accessible or condensed nucleosome fibers containing acetylated (*Ac*), phosphorylated (*P*), and methylated (*Me*) histone NH_2 termini. (**b**) Model for adding euchromatic (*EU*) or heterochromatic (*HET*) modification marks onto a nucleosomal template. The position of a gene in an accessible (euchromatic) or an inaccessible (heterochromatic) chromatin environment has also been referred to as position-effect variegation (*PEV*). PEV modifiers can enhance variegation [*E(var)*] or suppress variegation [*Su(var)*]. The *left-hand side* depicts an example in which a region is made accessible, while the *right-hand side* shows the reverse, i.e., transfer into inaccessible heterochromatin. Both processes are reversible

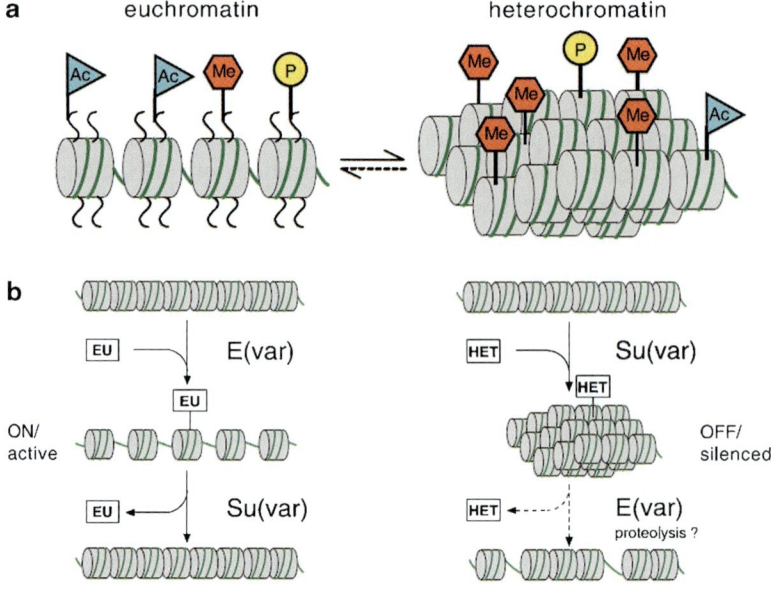

The 30-nm fibers form part of a chromosome segment about 300 nm in diameter. A further packing on metaphase chromosomes is represented by a thickened segment, also called a condensed chromosomal segment, with a diameter of 700 nm. A multitude of these segments make up the chromatids of metaphase chromosomes (Fig. 3.4).

3.2.2.3 Chromatin Diseases

Mutations in genes encoding proteins that control the structure of chromatin have effects on the expression of a potentially large number of genes and may therefore cause diseases. An example are mutations in the gene *ATRX* on the X chromosome at Xq13, which cause in males a syndrome characterized by a mild form of alpha-thalassemia (HbH disease), mental retardation, facial dysmorphisms, and microcephaly. *ATRX* is a helicase that can unwind DNA double helices and is part of large multiprotein complexes controlling the local structure of chromatin. It is likely that *ATRX* is involved in establishing and maintaining repressive chromatin structures [89]. Another example for a chromatin disease is the neurological disorder Rett syndrome, caused by mutations in the human methyl-CpG-binding protein gene *MECP2*, which closely interacts with *ATRX* [164].

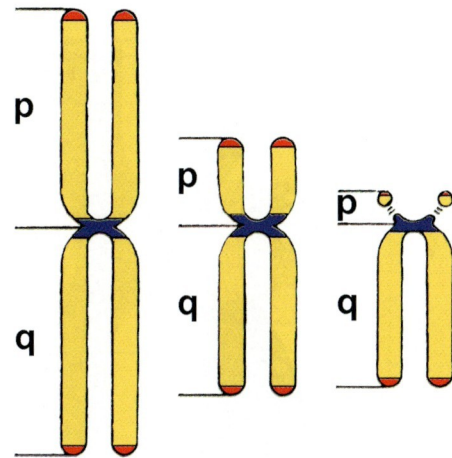

Fig. 3.7 Human metaphase chromosomes: metaphase chromosomes have two chromatids (also termed sister chromatids; *yellow*), which are held together at the centromere (*blue*). In metaphase chromosomes the centromere is usually readily visible as a constriction. The centromere divides each of the chromatids into two chromosome arms. The short arm is referred to as p-arm, and the long arm, as q-arm. Chromosomes with centromeres close to the middle as shown on the *left hand side* have arms of equal length and are known as metacentric chromosomes. If the centromere is not centrally located resulting in arms, which are unequal in length (as shown in the *middle*), the chromosome is termed submetacentric. A chromosome with a centromere very close to one end is called acrocentric. The p-arm of acrocentric chromosomes is often referred to as satellite. The regions at both ends of the chromosome are the telomeres (*red*)

3.2.3 Centromeres and Kinetochores

Each chromosome contains a specialized region known as its centromere. The position of each centromere divides the chromosome into two arms, and its location is characteristic for a given chromosome (Fig. 3.7).

3.2.3.1 Function of Centromeres

Centromeres are essential for normal segregation of chromosomes in both mitotic and meiotic cells. In mitosis, a proteinaceous structure, the kinetochore, assembles at the surface of the centromeres. The kinetochore serves as the attachment site for spindle microtubules and the site at which motors generate forces to power chromosome movement (Fig. 3.8a). The inner kinetochore forms the interface with chromatin, while the laminar outer kinetochore domain (frequently misnamed a "plate") forms the interaction surface for spindle microtubules [43]. Unattached kinetochores are also the signal generators for the mitotic checkpoint, which arrests mitosis until all kinetochores have correctly attached to spindle microtubules, thereby representing the major cell cycle control mechanism protecting against loss of a chromosome (aneuploidy). Thus, centromeres are active components in microtubule capture, stabilization, and empowerment of chromosome movements essential to proper segregation. More than that, they are the signaling elements for controlling cell cycle advance through mitosis.

3.2.3.2 Structure of Centromeres

Human centromeres have sizes approaching 10 Mb. At DNA level centromeres consist of so-called alpha satellite DNA, which is defined by a 171-base pair motif repeated in a tandem head-to-tail fashion [246]. Human

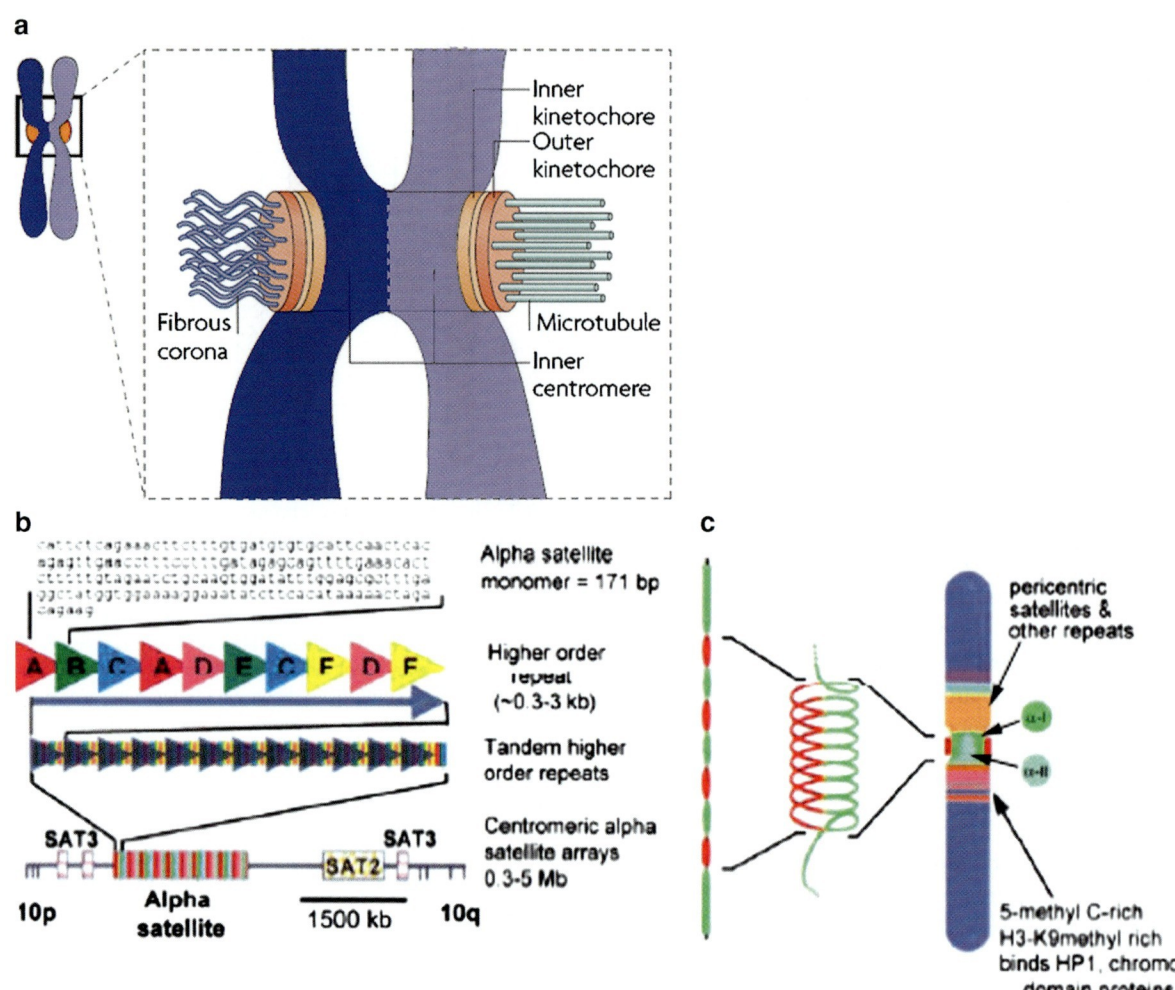

Fig. 3.8 (a) A schematic of a mitotic chromosome with paired sister chromatids and localization of centromere and kinetochore. The right chromatid is attached to microtubules and the chromatid on the *left* is unattached. Without microtubule attachment, the outer kinetochore shows a dense array of fibers, called the fibrous corona. (From [43], reprinted by permission from Macmillan Publishers Ltd: *Nature Reviews Molecular Cell Biology*, copyright 2008) (b) Organization of centromeric DNA. The hierarchic organization of an α satellite DNA is illustrated, with a 171-bp monomer sequence shown at the *top*. Monomer sequences are iterated with nucleotide sequence variations to form a higher order repeat (*colored arrows*), which itself is tandemly iterated with high (>99%) sequence conservation to form extensive arrays of higher order repeats. At the *bottom* is a diagram of the centromere region of chromosome 10, illustrating the ~2-Mb α-satellite array with surrounding pericentric satellite arrays (SAT2 and SAT3). (From [47], with permission from Elsevier) (c) Folding centromeric chromatin in mitotic chromosomes. CENP-A is interspersed with H3-containing nucleosomes. The centromeric chromatin fiber is folded or coiled to achieve the polarized distribution of CENP-A and H3 sites in mature mitotic kinetochores [47]. In this idealized centromere region CENP-A (*red*) assembles onto regular α-I satellite sequences (*green*). α-II satellite sequences (*teal blue*) are localized in the interior or central domain of the centromere. The centromere is flanked by pericentric satellite sequences (*orange, purple*). From [47], with permission from Elsevier

and great ape chromosomes contain alpha satellite organized hierarchically into higher order repeat arrays in which a defined number of alpha satellites have been homogenized as a unit to yield large chromosome-specific arrays that span several megabases [248]. The alpha satellite DNA is flanked by heterochromatin (Fig. 3.8b). Primate centromeres frequently have two major α-satellite families adjoining each other: a highly regular α-I and an α-II, which varies widely in monomer sequences and repeat structure.

The centromere challenges the classic view of a genetic locus. Usually a chromosomal locus is defined by its DNA sequence and its function is contained with the information content present in the primary sequence, e.g., recognition sites for DNA-binding proteins. However, the nature and specification of the centromere is epigenetically determined. As a consequence, chromatin in the centromere differs biochemically from the remainder of the genome in some very fundamental ways.

Chromatin is the key feature, and the centromere domain is built on a distinct type of nucleosome found nowhere else in the genome. Histone H3 is replaced by a divergent (50% identity) homolog usually referred to as CENP-A [218]. CENP-A is interspersed with H3-containing nucleosomes, and both domains are required for complete centromere function (Fig. 3.8b). H3-containing nucleosomes within centromeric chromatin are hypoacetylated, which is typical of heterochromatin, and enriched in dimethylated lysine 4, a modification typically associated with potentiated regions of chromatin [228]. Centromeric heterochromatin containing CENP-A is flanked by chromatin enriched in dimethylated lysine 9, which separates the centromeric heterochromatin from the pericentromeric constitutive heterochromatin. Constitutive heterochromatin is demarcated by enrichment in trimethylated lysine 9 [136].

CENP-A is bound primarily to α-I satellite sequences, at the surface of the chromosome (Fig. 3.8c). In contrast, α-II satellite is localized in the interior or central domain of the centromere, where components such as INCENP, aurora B, and cohesin are concentrated. A core domain built around CENP-A and centromere-specific chromatin-binding proteins establishes the kinetochore-forming component of the centromere, while flanking domains are enriched in proteins involved in chromatid cohesion.

3.2.3.3 Centromeres and Human Diseases

Centromeres are important for accurate chromosome segregation. Centromere impairment may significantly increase the number of aneuploid cells, which in turn is implicated in both aging and cancer. A paradigm for a centromere disease is the ICF (*i*mmunodeficiency, *c*entromeric instability and *f*acial anomalies) syndrome, which is a rare autosomal recessive disease caused by mutations in the DNA methyltransferase gene DNMT3B. These mutations cause demethylation of cytosine residues in classical satellites 2 and 3 at juxtacentromeric regions of these chromosomes, which causes centromeric instability visible during chromosome analysis as formation of radiated chromosomes, especially of chromosomes 1, 9, and 16 (see also Sect. 3.7.4).

3.2.4 Chromosome Bands

3.2.4.1 G- and R-bands

Giemsa staining is the banding technique used for identifying individual human chromosomes in most laboratories. G-Bands are obtained by digesting the chromosomes with the proteolytic enzyme trypsin (Fig. 3.9). This technique is commonly described as GTG (G-bands by trypsin using Giemsa). After Giemsa staining chromosomes display an alternating pattern of Giemsa-dark (G) and -light (R) bands.

In part, these bands reflect different higher order structures and functions at various levels and therefore differ in a number of respects. The light G-bands (R-bands) are rich in GC and replicate early. Furthermore, these bands are gene rich and contain most housekeeping genes and a large number of CpG islands. In addition, R-bands largely contain Alu- and SINE (short interspersed repetitive DNA sequences) repeats. In contrast, dark G-bands are rich in AT and replicate late. G-bands are gene-poor and preferentially contain tissue-specific genes. The most abundant family of repeats in G-bands is that of LINE (long interspersed repetitive sequences) repeats. Moreover, the distribution of G- and R-bands in interphase nuclei follows a specific pattern. G- and R-bands form separate domains, and G-bands are preferentially localized at the nuclear periphery, whereas R-bands are rather positioned in the interior of a nucleus (reviewed in [50]).

Chromatin DNA is composed of loops and matrix-associated regions (MARs), the regions of DNA attaching to nuclear scaffolds [202] (inset in Fig. 3.4). According to metaphase chromatin models, G-bands are regions where AT-strings are tightly folded, whereas R bands are regions where AT-strings are unfolded and located along a longitudinal axis of a chromatin (Fig. 3.4, inset). According to this model, MARs are frequently present in G-bands and sparsely in R-bands [202].

G-Banded chromosomes are used for a standardized and common description in human cytogenetics. At the Paris Conference in 1971 [177] a basic system for designation not only of individual chromosomes but also of chromosome regions and bands was proposed. A system of human cytogenetic nomenclature evolved from further subsequent meetings. Each of the autosomes is numbered, from 1 to 22. The sex chromosomes are X and Y. Within each chromosome, the short arm is the p-arm, and the long arm is the q-arm. Each

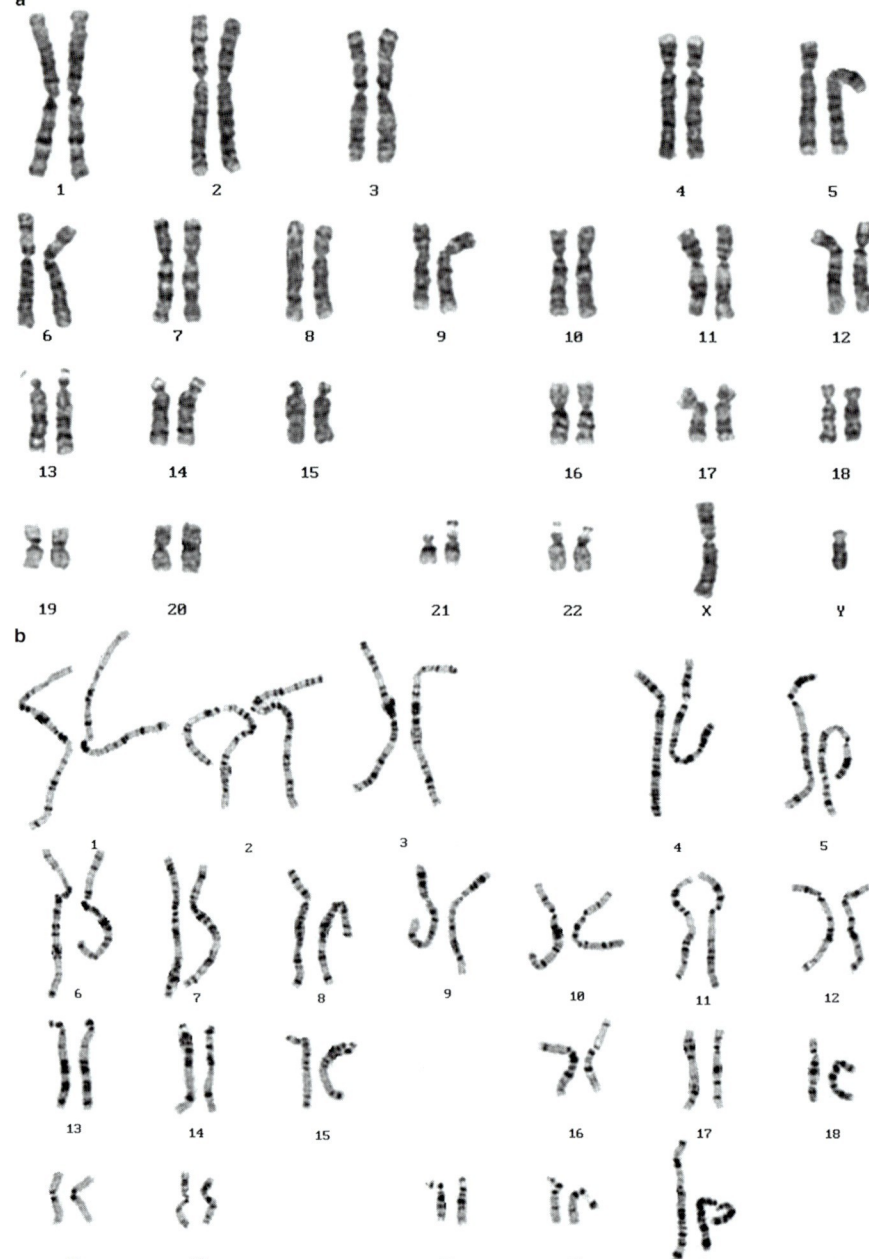

Fig. 3.9 (**a**, **b**) GTG-banded normal metaphase spreads with different resolution. (**a**) Normal male metaphase spread with approximately 450 bands; (**b**) normal female metaphase spread with approximately 600 bands

arm is divided into numbered regions. Within each region, the bands are designated by a number (Fig. 3.10). Such pictograms are in widespread use, in spite of some shortcomings deriving from the fact that the location and width of bands reflect the real chromosome morphology only in part. At present, pictograms created by Uta Francke, who used actual measurements and different gray values to mirror the staining intensities of G-dark bands, are used by most investigators [81] (Fig. 3.10).

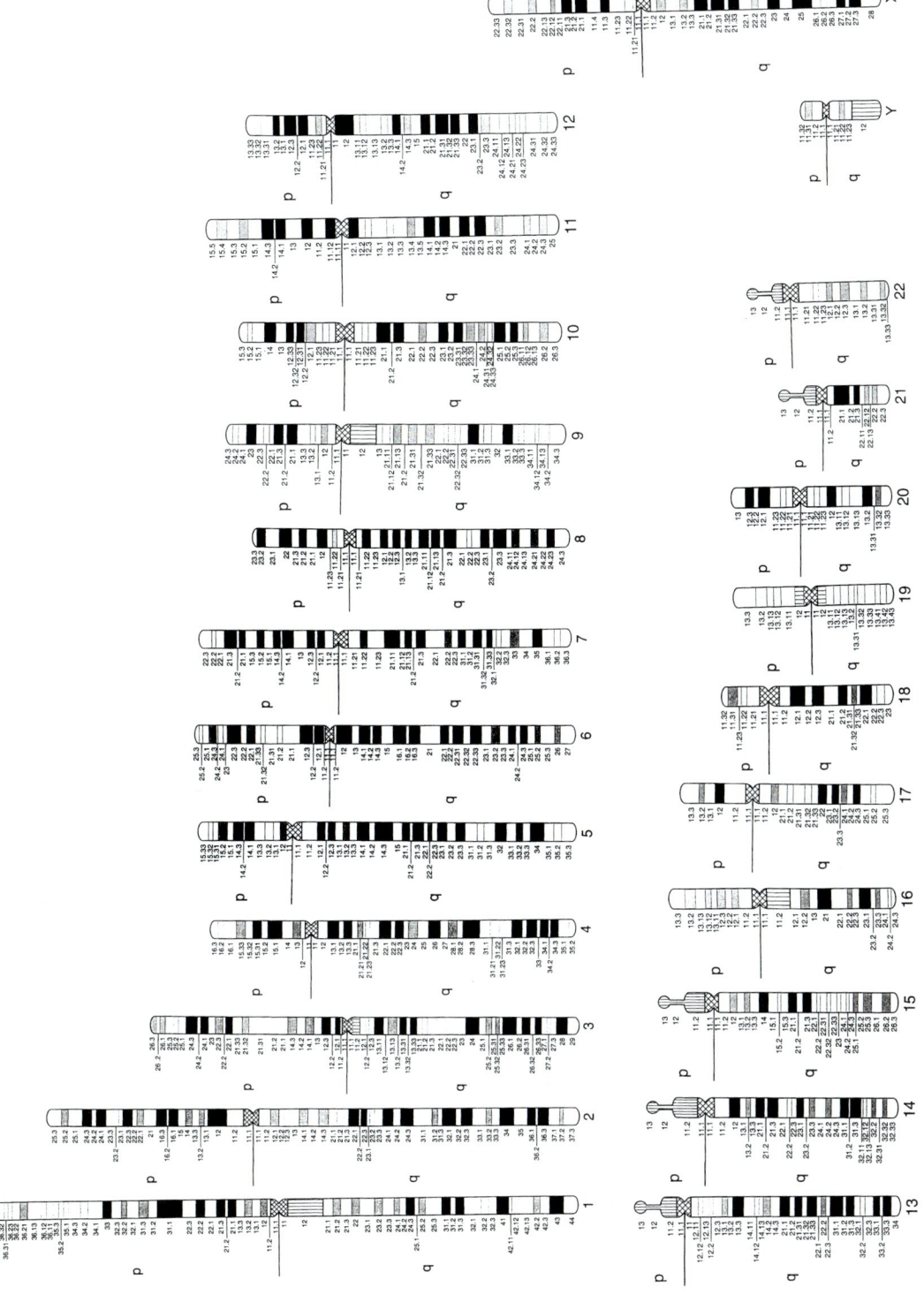

Fig. 3.10 An 850-band pictogram in which the relative widths of euchromatic bands are based on measurements and the staining intensities reflect GTG bands. (From [81], with permission from S. Karger AG, from the 3rd edition of this work)

3.2.4.2 In Silico-*Generated Bands*

More recently, new approaches have been used to achieve *in silico* chromosome staining. Using the DNA sequences of the draft human genome and sophisticated computer software which assigns gray values to chromosomal regions depending on the percentage of the GC content, it has proved possible to achieve successful reconstruction of bands resembling Giemsa bands [170] (Fig. 3.11). Such *in silico* approaches will probably improve with further refinements of draft human genome sequences and should result in the most reliable schematic representations of chromosomes.

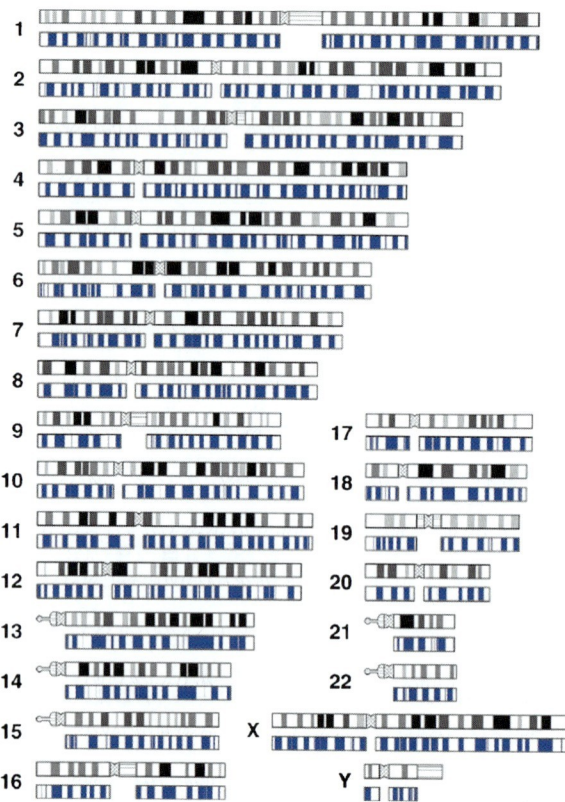

Fig. 3.11 Giemsa and *in silico* bands for all chromosomes. Short (p) and long (q) arms are positioned *left* and *right*, respectively. Giemsa bands obtained from Francke [81] are shown in *black pictograms*. The bands depicted in *black*, *gray*, and *white* represent euchromatins, and the darkness of each band reflects the shading. Pericentromeric heterochromatin and heteromorphic regions of chromosomes 1, 3, 9, 16, 19, and Y are depicted by *crosshatched* and *horizontal lines*, respectively. *In silico* bands constructed by using windows of 2.5 and 9.3 Mb are shown in *blue*. The *thin lines* between Giemsa and *in silico* bands denote aligned G-bands. From [170], copyright 2002, *National Academy of Sciences*, USA

In silico chromosome staining provided the solution to an old problem. As already mentioned, Giemsa-dark and -light bands are generally thought to correspond to GC-poor and GC-rich regions, respectively; however, several experiments have shown that the correspondence is quite poor. *In silico* banding clearly shows that Giemsa-dark bands are *locally* GC-poor regions compared with the flanking regions, but not compared with the entire genome. These findings are consistent with the model that MARs, which are known to be AT rich, are present more densely in Giemsa-dark bands than in Giemsa-light bands (as shown for example in the Fig. 3.4 inset). In fact, G-bands are the regions in which the GC content is only lower relative to the surrounding regions, and not relative to the entire genome [170].

3.2.5 Telomeres

The linear chromosomes of eukaryotes are "sealed" by a specialized region, the telomere, which stabilizes them at both ends. Telomeres protect chromosome ends from being recognized and processed as DNA double-strand breaks and, therefore, from triggering DNA-damage-induced responses and checkpoint activations. Furthermore, telomeres act to prevent the end-to-end fusion of chromosomal DNA molecules and, hence, to prevent the fusion of chromosomes with one another. The first evidence for the importance of telomeres in chromosomal integrity came from a cytogenetic analysis in the 1940s by McClintock on breakage and fusion of maize chromosomes, where loss of telomeric sequences renders DNA ends recombinogenic [155].

Telomeres are essential protein–DNA complexes. Two features characterize the telomere: telomeric DNA loop formation to stabilize the chromosome ends and telomerase activity to compensate for replication-related loss of nucleotides at the chromosome ends. Telomerase was discovered by Elizabeth H. Blackburn together with Carol W. Greider, who was a doctoral student in her laboratory at the time [96]. The first observation that telomeres form large duplex loops was made in the laboratory of Titia de Lange [98].

3.2.5.1 Structure of Telomeres (Loop and Proteins)

The telomere repeats consist of 250–1,500 G-rich tandem sequences (5′-TTAGGG-3′), which are highly conserved

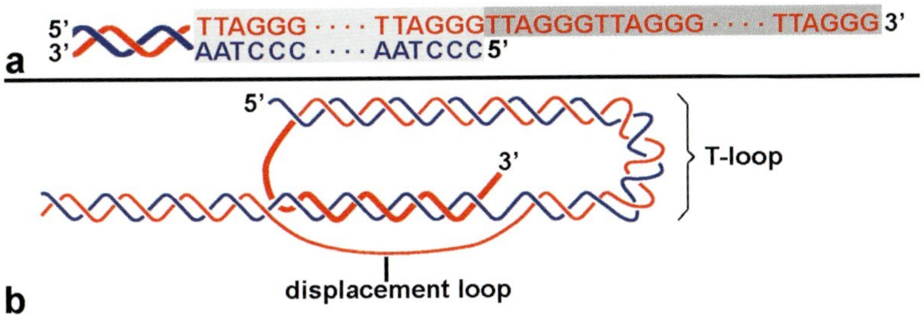

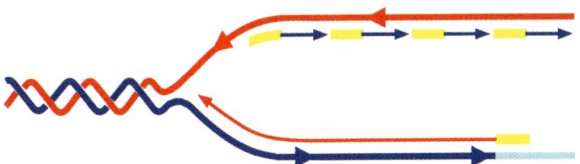

Fig. 3.12 (**a**) The G-rich strand of telomeric DNA (*red*) extends beyond the C-rich strand (*blue*). This creates a 3′ overhang (*dark gray shaded area*). This single stranded 3′ overhang may have a length of several hundred nucleotides. The double-stranded portion of telomeric DNA (*light gray shaded area*) has a length of 5–10 kb. (**b**) The single-stranded 3′ overhang is annealed to a small region of the C-rich strand, causing the formation of a displacement loop while the telomeric duplex DNA forms the T-loop. Adapted from [250]

among different species. Telomeric DNA of mammalian cells is formed from the repeating hexanucleotide sequence 5′-TTAGGG-3′ in one strand (the "G-rich" strand) and the complementary 5′-CCCTAA-3′ in the other (the "C-rich" strand). The G-rich strand is longer by one hundred to several hundred nucleotides, resulting in a long 3′ single-strand overhang (Fig. 3.12a), thus creating a "sticky end." The end of the single-stranded overhanging region has to be put away in order to avoid initiation of the cellular repair machinery. The 3′ overhanging end of the G-rich strand is annealed to a small region of the C-rich strand, causing the formation of a displacement loop, while the telomeric duplex DNA forms a loop (T-loop) [98] (Fig. 3.12b). The T-loop together with the displacement loop helps to protect the ends of linear DNA molecules.

Both the relatively long double-stranded telomeric DNA and the short overhanging end are bound by specific proteins. These proteins participate in the loop formation and are essential for functional telomeres [91, 128]. In fact, these telomere-binding proteins and the telomeric DNA together form the nucleoprotein complexes referred to as "telomeres."

3.2.5.2 The Telomere Replication Problem

The replication machinery has great difficulty in copying sequences at the very ends of linear DNA molecules, because DNA is synthesized in the 5′-to-3′ direction only. Therefore, the two templates of the parent molecule differ with respect to the continuity of synthesis. On the 3′-to-5′ template strand, copying is initiated by a primer and can proceed continuously as it occurs in the same direction as the fork movement (leading strand synthesis) (Fig. 3.13). In contrast, on

Fig. 3.13 During DNA replication, the parental DNA double helix is unwound. The replication process proceeds on the leading strand in leftward direction (leading strand synthesis) whereas, on the 5′ to 3′ template strand, synthesis occurs in the reverse direction relative to the fork movement (lagging strand synthesis). The synthesis of the new "lagging strand" occurs in short fragments of several hundred nucleotides involving short primers (Okazaki fragments, illustrated in *yellow*). The end of the leading strand template cannot be synthesized by DNA polymerase because the primer responsible for initiating leading strand synthesis may bind a significant number of nucleotides away from the 3′ end of the parental strand and the primer itself is lost when it is degraded after maturation of the recently synthesized DNA. The part of the parental strand, which will therefore be lost, is indicated in *light blue*. Adapted from [250]

the 5′-to-3′ template strand, synthesis occurs in the reverse direction relative to the fork movement (lagging strand synthesis; Fig. 3.13). Here, DNA is synthesized in short fragments of about 1,000–2,000 nucleotides in eukaryotes.

The telomere replication problem is caused because the end of the leading strand template cannot be synthesized by DNA polymerase, since the required primer cannot be attached beyond the end of the template strand. In fact, the primer responsible for initiating leading strand synthesis may bind a significant number of nucleotides away from the 3′ end of the parental strand. Furthermore, the primer itself is lost when it is degraded after maturation of the recently synthesized DNA. Therefore, the leading strand synthesis usually results in underreplication of one of the parental strands of DNA. In addition to this underreplication of telomeric DNA ends, cellular exonucleases may contribute to further telomere erosion. Owing to the combined action of underreplication of the leading strand and these exonucleases, it is estimated that in normal human cells telomeres lose 50–100 base pairs of DNA during each cell generation [91].

Thus, in the absence of special telomere maintenance mechanisms, linear chromosomes shorten progressively with every round of DNA replication, eventually leading to cellular senescence or apoptosis. Telomere erosion is inexorably linked to cell division. Chromosomes with critically short telomeres trigger the activation of the p53 and Rb tumor suppressor pathways, providing the signal for replicative senescence [253]. These repetitive elements play a major role in senescence, as progressive shortening of the telomeres occurs with each cell division [3]. The limited replicative potential of primary human cells, also referred to as the Hayflick limit [103], is due to the progressive shortening with each successive round of cell division, leading to significant telomere attrition [91].

3.2.5.3 Telomerase

Because the length of telomeric DNA inexorably shortens during replication, there is a need for compensating mechanisms to preserve genome integrity and telomere functions. In many organisms a specialized reverse transcriptase named telomerase catalyzes the addition of short and simple repeats in a process that is tightly connected with replication.

Telomerase is a specialized polymerase that adds telomere repeats to the ends of chromosomes. It has two essential components: one subunit is a DNA polymerase, i.e., a reverse transcriptase to synthesize DNA from a RNA template. The telomerase holoenzyme provides its own RNA template, which is the second essential subunit. The polymerase is the catalytic component, named human telomerase reverse transcriptase (hTERT), while the "template provider" for nucleotide addition by hTERT, the RNA component is referred to as hTR. The addition of telomeric repeats onto the ends of the chromosome partly triggers the shortening that occurs during DNA replication (Fig. 3.14). In normal human cells telomerase activity is too low to prevent progressive telomere erosion. Substantial telomerase activity in normal cells is found in stem cells, in germ cells and, peri-

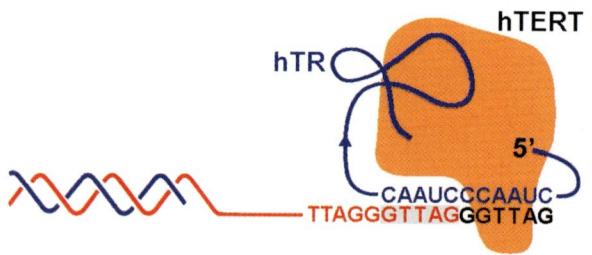

Fig. 3.14 Illustration of the two essential subunits of the human telomerase holoenzyme: the hTERT catalytic subunit (*orange*) and the associated hTR subunit (*blue*), which can elongate telomeres. The 3′ end of chromosomes (*red letters*) is extended by reverse-transcribing the template region of the telomerase RNA. Near the RNA 5′ end are sequences complementary to telomeric DNA repeat sequences (*blue letters*). A short nucleotide sequence of this RNA pairs with terminal DNA sequences (*gray shaded red letters*). The adjacent RNA nucleotides provide the template for adding nucleotides to the 3′ end of the chromosome (*black letters*). Repetition of this process in an iterative fashion makes it possible for telomeres to be elongated. Adapted from [250]

odically, in lymphocytes when they become functionally activated. However, activation of telomerase is a common finding in many tumor cells [59].

In cells expressing telomerase, telomeric DNA trimming still occurs but can be counterbalanced either partially or completely by the elongation of the G-rich strand and by its subsequent complementary replication.

In the absence of telomerase, telomeric DNA loss can also be compensated for by alternative lengthening of telomeres (ALT) mechanisms. The ALT mechanisms are still unclear and appear to rely on homologous recombination, rolling-circle replication, extrachromosomal circle integration, and break-induced replication. ALT-positive cells have a number of characteristics dissimilar from those in cells that use telomerase for telomere maintenance. Telomeres in ALT-positive cells are typically quite long and heterogeneously sized compared with the shorter, more homogeneous population of telomeres usually present in telomerase-positive cells. Based on differences in telomere structure, mechanisms that amplify subtelomeric repeats and mechanisms that lengthen the simple telomeric repeats alone can be distinguished. The ALT pathways can be considered as backups of telomere maintenance in organisms that normally exploit the telomerase system. For instance, whereas ALT is inhibited in normal human cells, some human tumors maintain their telomeres using ALT.

3.2.5.4 Telomeres and Human Diseases

Telomeres shorten with each cell division and ultimately activate a DNA damage response that leads to apoptosis or cell-cycle arrest. Telomere length thus limits the replicative capacity of tissues and has been implicated in age-related diseases. However, ectopic expression of hTERT can reconstitute functional telomerase activity [29]. In fact, in most epithelial tumors, telomere maintenance is accomplished by telomerase, which appears to be an important mechanism involved in the unlimited replicative potential of immortal cancer cells.

Mutations in the essential components of telomerase, i.e., hTR and hTERT, can cause dyskeratosis congenita, a rare hereditary disorder characterized by a triad of mucocutaneous manifestations, including skin hyperpigmentation, oral leukoplakia, and nail dystrophy. Heterozygous mutations underlie the defect in families with dominant inheritance, indicating that half the usual dose of telomerase is inadequate for telomere maintenance in tissues of high turn-over, such as the bone marrow [243]. Furthermore, telomerase mutations have been implicated in families with idiopathic pulmonary fibrosis [10].

3.3 Cell Cycle and Mitosis

3.3.1 Cell Cycle: Interphase G_1-G_2 and G_0

Almost all types of normal cells proliferate only if they receive appropriate signals to divide, which initiates a complex cycle of growth and division, referred to as the "cell cycle."

The mammalian cell cycle can be divided into four phases (Fig. 3.15). The mitotic phase (M) is a relatively short period which alternates with the much longer interphase where the cell prepares itself for the next cell division. Interphase is divided into three phases, G_1 (first gap), S (synthesis), and G_2 (second gap). During interphase cells remaining in the active growth-and-division cycle prepare for the next division. This preparation includes, for example, the duplication of macromolecular constituents which will later be equally distributed to the two daughter cells. The production of proteins and cytoplasmic organelles occurs during all three interphase stages (i.e., G_1, S, and G_2). For example, the centrosomes represent an important component, as

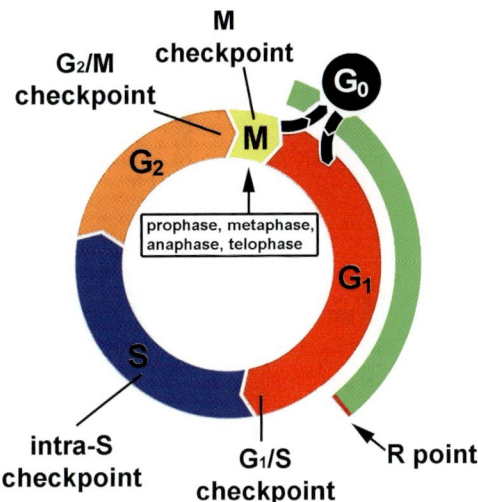

Fig. 3.15 Summary of the mammalian cell cycle and the cell cycle checkpoints: The mammalian cell cycle consists of four phases: G_1 (gap one), S (synthesis/replication of DNA), G_2 (gap two), and M (mitosis). The mitosis can be subdivided into prophase, metaphase, anaphase, and telophase. G_0 indicates a fifth, resting, nonproliferative state of cells, which have withdrawn from the active cell cycle. It is unclear when during G_1 exit into G_0 occurs. Cells may respond to extracellular mitogens and inhibitory factors only during a certain time period that begins at the onset of G_1 and ends before the end of G_1 (indicated in *green*). The end of this time window is designated the restriction (*R*) point. After this time point, the cell is committed to advance through the remainder of the cell cycle through M phase. A checkpoint is a regulatory pathway that controls the order and timing of cell-cycle transitions by checking at the beginning of each new step whether the previous one is completed. The image depicts various DNA-damage checkpoints: the G_1/S checkpoint blocks entrance into S-phase if the genome is damaged; the intra-S checkpoint halts replication in the case of any replication errors or other damages within the genome; the G_2/M checkpoint blocks entrance into M-phase if DNA replication is not completed; and the mitotic checkpoint prevents progress in mitosis if not all chromosomes are properly assembled on mitotic spindle

they organize a microtubule meshwork throughout the cell cycle, thereby influencing both tissue architecture and the accuracy of chromosome segregation [169]. The centrosomes are duplicated in several steps, whereas chromosomes are replicated only during the S phase (Fig. 3.16). The centrosome cycle and the chromosome cycle have to be tightly coordinated. In the G_2 phase, every chromosome has doubled into two identical elements, called sister chromatids. As a consequence, the material of every chromosome is now present twice ($2\times2=4n$). During or after replication,

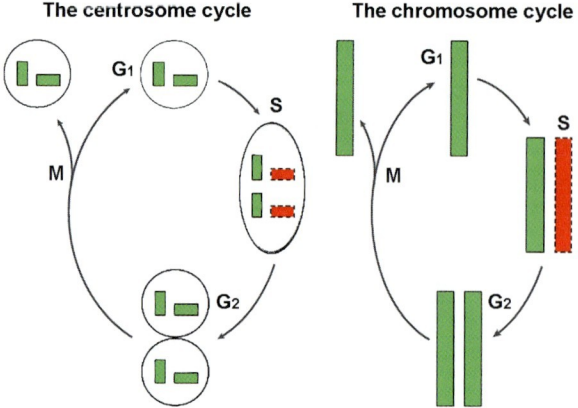

Fig. 3.16 Schematic drawing of the centrosome cycle and the chromosome cycle. Adapted from [169], reprinted by permission from Macmillan Publishers Ltd: *Nature Reviews Cancer*, copyright 2002

An increased SCE rate is observed in cases of defects in the homologous repair defect.

The duration of the different phases varies. G_1 lasts for 12–15 h, the S-phase 6–8 h, but may be much shorter in certain cell types, such as rapidly dividing embryonic cells or lymphocytes, and the G_2 phase may require about 3–5 h. The shortest phase is mitosis, which with its five subphases (prophase, prometaphase metaphase, anaphase, and telophase) takes about 1 h.

The decision on whether a cell will advance through another growth-and-division cycle is made in late G_1 at a transition called the restriction point or R point (Fig. 3.15). At this R point a cell commits itself to proceed beyond G_1 into S phase and then to complete the entire S, G_2, and M phases.

In the absence of mitogenic growth factors a cell can proceed from mitosis into the G_0 quiescent stage. This G_0 state can be reversible if the cell is again exposed to mitogenic growth factors. However, some cells leave the active cell cycle irreversibly without ever re-initiating active growth and division. Thus, some cells in a body have entered such a postmitotic, differentiated state from which they will never re-emerge and resume proliferation.

3.3.2 Cell Cycle: Mitosis

Mitosis is traditionally subdivided into five consecutive and morphologically distinct phases: prophase, prometaphase, metaphase, anaphase, and telophase (Fig. 3.18).

3.3.2.1 Mitosis and Cytokinesis

Mitosis starts with the condensation of chromatin and ends with the separation of the sister chromatids and their drawing to the opposite poles by the spindle fibers. It is followed by cytokinesis, which divides the nuclei, cytoplasm, organelles, and cell membrane into two daughter cells containing roughly equal shares of these cellular components. Both mitosis and cytokinesis represent the mitotic (M) phase of the cell cycle.

As mentioned above, mitosis is divided into five phases, i.e., prophase, prometaphase, metaphase, anaphase, and telophase.

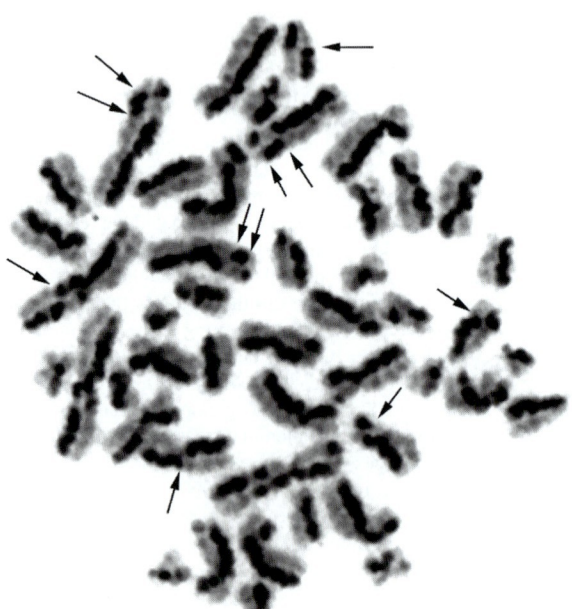

Fig. 3.17 Male metaphase spread after visualization of sister chromatid exchanges. The *arrows* point to some, but not all, locations of sister strand exchanges

the two sister chromatids exchange segments repeatedly, so that the two chromatid arms of a mitotic chromosome have parts of both chromatids. This process is called sister chromatid exchange (SCE) and can be made visible by using a specific staining technique after treatment with bromodeoxyuridine (Fig. 3.17). There are normally about 6–10 SCEs during each cell division.

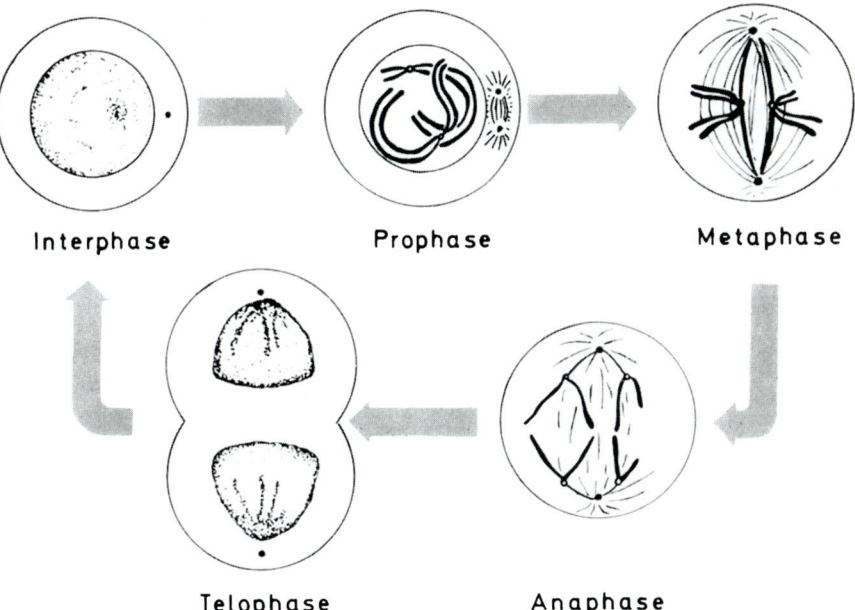

Fig. 3.18 Schematic representation of mitosis. Only 2 of the 46 chromosomes are drawn. For details see text. (Courtesy of Dr. W. Buselmaier, from the 3rd edition of this work)

3.3.2.2 Prophase

At the onset of prophase, chromosomes initiate condensation and become visible through a light microscope.

Each centrosome acts as a coordinating centre for the cell's microtubules. The two centrosomes nucleate microtubules (which may be thought of as cellular ropes or poles) by polymerizing soluble tubulin present in the cytoplasm. Molecular motor proteins create repulsive forces that will push the centrosomes to opposite side of the nucleus.

3.3.2.3 Prometaphase

The nuclear envelope and the interphase microtubule array disassemble and the fully compacted chromosomes spill into what was the cytoplasm to produce prometaphase.

The attachment of microtubules to kinetochores is a stochastic "search and capture" (Fig. 3.19a) [189]. The process begins with unattached sister kinetochores. Initial capture occurs frequently by the binding of one kinetochore of a duplicated chromosome pair along the side of a spindle microtubule allowing rapid (up to 1 μm/s) poleward translocation along that microtubule. This is followed by an attachment of additional microtubules, in humans up to 30. At the spindle pole, additional microtubules bind the captured kinetochore in an end-on fashion to create a microtubule fiber. A probing microtubule from the opposite pole then interacts with the remaining unattached kinetochore. Finally the sister chromatids gather around the center of the spindle, where the sister kinetochores achieve full microtubule occupancy.

The correct, bipolar or bioriented, attachment needed for correct chromosome segregation is called amphitelic (Fig. 3.19b). Here sister kinetochores are oriented to opposite poles and thus bind microtubules from the adjacent pole. However, correct and incorrect attachments can also occur during mitosis, resulting in several kinetochore-microtubule arrangements. Syntelic describes kinetochore-microtubule attachment in which both sister kinetochores face only one of the two poles and consequently attach only to microtubules from that pole (Fig. 3.19b). Monotelic refers to a situation in which both sister kinetochores face opposite poles but only one kinetochore is attached to microtubules. A merotelic kinetochore-microtubule attachment is present when both sister kinetochores face opposite poles but one kinetochore

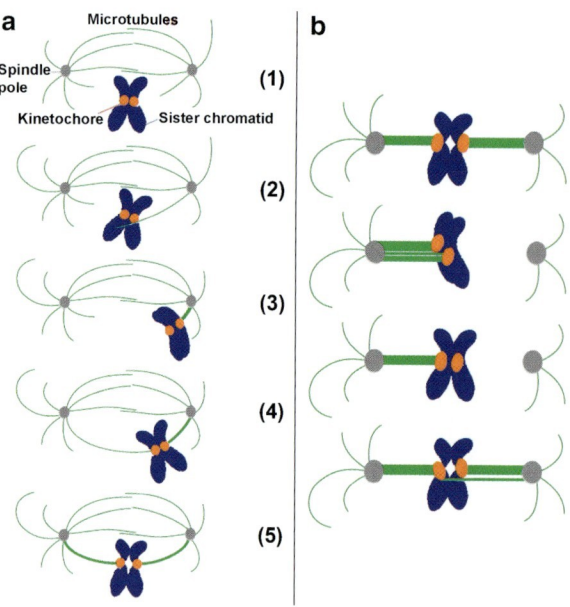

Fig. 3.19 (**a**, **b**) Kinetochore-microtubule attachments. (**a**) The dynamic probing of microtubules for kinetochore attachments represents a stochastic process, dubbed "search and capture." *1:* The *upper panel* shows unattached kinetochore. *2:* During the search and capture process single kinetochores may be captured by spindle microtubules. *3:* The sister chromatids are then pulled to the pole where more microtubules can bind. *4:* The sister kinetochore interacts with microtubules from the opposite pole. *5:* Finally microtubules from both poles have bound resulting in a correct, bipolar or bioriented attachment. (**b**) Types of kinetochore-microtubule attachments. From *above*: amphitelic, syntelic, monotelic, and merotelic. Adapted from [189], with permission from Elsevier

3.3.2.4 Metaphase

The centromeres of the chromosomes convene along the *metaphase plate* or *equatorial plane*, a line usually equidistant from the two centrosome poles. This even alignment is due to the counterbalance of the pulling powers generated by the opposing kinetochores, analogous to a tug-of-war between equally strong people.

Proper chromosome separation requires every kinetochore to be attached to a bundle of microtubules (spindle fibers), and unattached kinetochores generate a signal to prevent premature progression to anaphase without all chromosomes being aligned. The signal creates the mitotic spindle checkpoint (for more details see Sect. 3.4.3.3).

3.3.2.5 Anaphase

The cell proceeds to anaphase when every kinetochore is attached to a cluster of microtubules and the chromosomes have lined up along the metaphase plate. Anaphase ensues about 20 min after the last kinetochore has attached itself to the spindle, and repeated detachment of a chromosome from a spindle by micromanipulation has delayed anaphase indefinitely [197].

The proteins that bind sister chromatids together, i.e., cohesin, are cleaved, allowing them to separate. This converts sister chromatids into sister chromosomes, which are pulled apart by shortening kinetochore microtubules and move towards the respective centrosomes to which they are attached. Subsequently, the nonkinetochore microtubules elongate, pushing the centrosomes (and the set of chromosomes to which they are attached) apart to opposite ends of the cell.

These two stages are sometimes called early and late anaphase. Early anaphase is usually defined as the separation of the sister chromatids, while late anaphase is the elongation of the microtubules and the microtubules being pulled farther apart. At the end of anaphase, the cell has succeeded in separating identical copies of the genetic material into two distinct populations.

Research of essential mitotic processes is ongoing and still reveals new, surprising findings. For example, it was recently demonstrated that alphoid centromeric DNA persists as thin threads connecting separating

binds to microtubules from both poles. Chromosomes with syntelic, monotelic, or merotelic attachments have to be corrected, as otherwise they will segregate improperly.

Mechanisms exist to prevent incorrect chromosome inheritance, which would occur if improper attachment persisted until anaphase. Monotelic attachment is a normal condition during prometaphase before bi-orientation. In syntelic attachment both sisters in a pair connect to the same pole. Merotelic attachment occurs quite frequently. Both syntelic and merotelic attachments are corrected by the aurora-B kinase.

When proper bipolar attachments are formed, the poleward forces of the kinetochore microtubules are opposed by the cohesion between the sister chromatids, putting the chromosomes under tension. When all the chromosomes are aligned properly the cell is in metaphase.

chromosomes even during anaphase. These findings were achieved by the identification of PICH (Plk1-interacting checkpoint helicase), which represents an essential component of the spindle assembly checkpoint and is localized to kinetochores (inner centromeres) [25]. Topoisomerase activity is required during anaphase for the resolution of these alphoid centromere PICH-positive threads, implying that the complete separation of sister chromatids occurs later than previously assumed [245].

3.3.2.6 Telophase

During telophase the cell continues to elongate owing to further lengthening of nonkinetochore microtubules, and sister chromosomes reach the opposite ends of the cell. A new nuclear envelope evolves around each set of separated sister chromosomes. Both sets of chromosomes, now surrounded by new nuclei, decondense and form chromosome territories in their respective nucleus.

3.3.2.7 Cytokinesis

Cell division is finally completed by cytokinesis. A cleavage furrow containing a contractile ring develops where the metaphase plate has been separating the two new nuclei.

3.3.3 Cell Cycle Checkpoints

Cells deploy a series of surveillance mechanisms that monitor each step in cell cycle progression. If certain steps in the execution of a process fail, these monitors rapidly stop further advance through the cell cycle until these problems have been successfully resolved. Another task of such monitors is to ensure that once a particular step of the cell cycle has been completed, it is not repeated until the cell passes through the next cell cycle. These monitoring mechanisms have been termed checkpoints or checkpoint controls [101]. Checkpoints impose quality control to ensure that a cell has properly completed all the requisite steps of one phase of the cell cycle before it is allowed to advance into the next phase.

3.3.3.1 Presently Known Checkpoints

The presently best defined checkpoints are in the G_1/S, intra-S, or G_2/M phases (Fig. 3.15). These checkpoints examine the genome integrity at specific points in the cell cycle and if they sense damage, mechanisms are activated that arrest cell-cycle progression to allow time for repair. If the damage is irreparable checkpoints can activate programmed cell death (apoptosis) or replicative senescence to eliminate the affected cell or prevent it from further replication. If the restoration has been successful, checkpoints re-initiate the continuation of the cell cycle.

3.3.3.2 The Spindle-Assembly Checkpoint in Detail

The molecular mechanisms involved in each checkpoint are beginning to emerge, and some basic plans of signaling cascades are fairly well characterized. Here, only the spindle-assembly checkpoint (SAC) during mitosis will be discussed in more detail. The SAC is a ubiquitous safety device ensuring the fidelity of chromosome segregation in mitosis. The SAC prevents chromosome mis-segregation and aneuploidy (Fig. 3.20), and its dysfunction is implicated in a constitutional disorder, termed mosaic variegated aneuploidy (MVA; further discussed below and in Sect. 3.7.6) and in tumorigenesis.

At the beginning of the M-phase the two sister chromatids of each chromosome are connected by cohesin, which needs to be removed from chromosomes to allow sister chromatid separation in mitosis (Fig. 3.20). Sister chromatid cohesion is essential for bi-orientation of sister kinetochores during mitosis and hence for propagation of the genome during cell proliferation. Cohesin is composed of four core subunits, called SMC1, SMC3, SCC1 (also known as MDC1 and RAD21), and SCC3 (also known as SA2 and STAG2) [148]. These proteins have been proposed to mediate cohesion by embracing sister chromatids as a ring [165]. During early mitosis the bulk of cohesin is already removed from the chromosome arms via a process that requires the WAPL (WAPAL) protein and involves the polo-like kinase PLK1 and Aurora B but does not require proteolytic cleavage of the cohesin subunit SCC1 [185]. However,

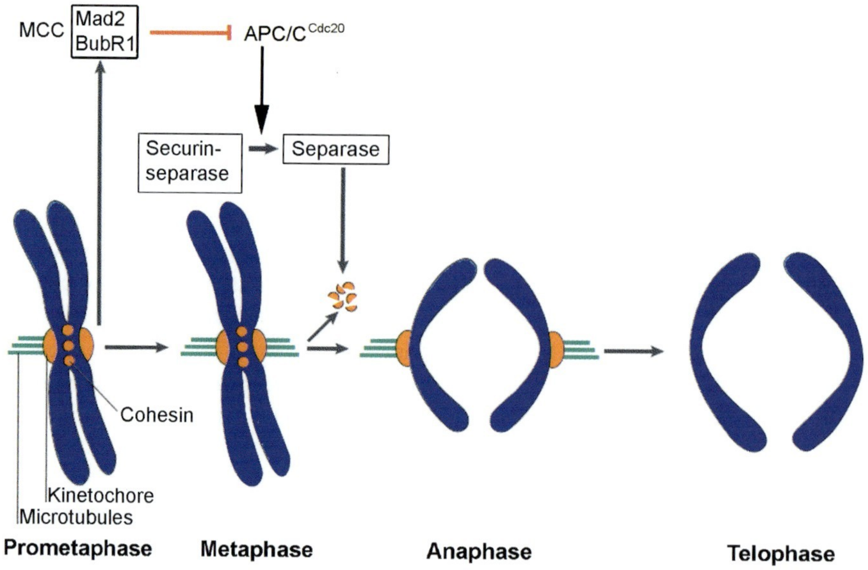

Fig. 3.20 At the beginning of metaphase, both sister chromatids are linked by cohesion. Sister-chromatid cohesion at the centromeres persists until the onset of anaphase. Removal of the centromeric cohesion is performed by a protease named separase. Therefore, separase is kept inactive prior to anaphase. This inactivation is done by binding of the protein securin. Unattached kinetochores contribute to the formation of the mitotic checkpoint complex (MCC), which consists of several kinases (i.e. BUBR1, BUB1) and the MAD1/MAD2 complex. The MCC inhibits CDC20 and thus prevents the activation of the APC/C. The attachment of all sister-kinetochore pairs to kinetochore microtubules and their bi-orientation results in release of CDC20, which can now activate the APC/C. This causes polyubiquitination of securin and therefore activation of separase, which can now remove the centromeric cohesion rings to separate the sister chromatids. Adapted from [185], reprinted by permission from Macmillan Publishers Ltd: *Nature Reviews Molecular Cell Biology*, copyright 2006

chromosomes remain connected to each other by centromeric cohesin until the onset of anaphase, because centromeric cohesion is protected by the shugoshin protein SGO1. Thus, for chromosome segregation to proceed it is essential that the centromeric cohesin is removed by a protease termed separase [185]. Separase is a protease whose activity is required to remove sister-chromatid cohesion at the metaphase-to-anaphase transition. However, separase should not accomplish this task before all kinetochores are connected to the mitotic spindle machinery and before all chromosomes are properly aligned in the metaphase plate. Hence, prior to anaphase, separase is kept inactive by the binding of a protein known as securin (Fig. 3.20).

Furthermore, signals are required to indicate whether kinetochores have attached to microtubules or not. Kinetochores that are not, or not fully, attached with microtubules are capable of binding and activating a collection of mitotic checkpoint components. These include kinases, such as BUB1 (*B*udding *u*ninhibited by *b*enomyl), BUBR1 (also referred to as BUB1B) and MAD2 (*m*itotic *a*rrest-*d*eficient homologue-2), which are part of the mitotic checkpoint complex (MCC), which creates a "wait anaphase" signal by inhibiting the ability of CDC20 to activate the anaphase-promoting complex/cyclosome (APC/C). The levels of MAD2 are high at unattached kinetochores and moderately high at attached kinetochores in a monotelic pair. Owing to the inactivated APC/C^{Cdc20} securin cannot be ubiquitylated, so that separase remains in an inactive state and thereby anaphase and mitotic exit are prevented.

After microtubule attachment, bi-orientation depletes MAD2 and BubR1 from kinetochores and promotes the acquisition of tension in the centromere area. As a consequence, checkpoint signaling is silenced and anaphase ensues after the decay of the previously made inhibitor.

The attachment of all sister-kinetochore pairs to kinetochore microtubules and their bi-orientation, which produces congression to the spindle equator, negatively regulates the SAC signal. This releases CDC20, which can now activate the APC/C. This results in the polyubiquitylation of securin and the subsequent proteolytic destruction. The degradation of securin results in the activation of separase, which targets the cohesin ring by proteolytic cleavage, especially the cohesin subunit SCC1 (RAD21, MCD1), which is holding the centromeric sister chromatids together, thus causing the loss of sister-chromatid cohesion and the separation of sister chromatids (reviewed in [162, 185]. When anaphase onset is delayed by the spindle-assembly checkpoint, the complete removal of cohesin from chromosome arms but not from centromeres generates typical X- or V-shaped chromosomes.

There is a continuing controversy as to whether the mitotic checkpoint is silenced by microtubule attachment or by the tension exerted between bioriented kinetochore pairs after attachment and whether activities of subsets of the known components are selectively silenced by one or the other [47].

Thus, the primary mission of this checkpoint is to prevent errors in chromosome segregation.

3.3.3.3 "Cohesinopathies," Cornelia de Lange and Robert Syndrome

Failure of components of the SAC may cause human diseases. Cornelia de Lange syndrome (CdLS) is characterized by growth and mental retardation, craniofacial anomalies, and microcephaly. This disease can be caused by mutations in a protein that is required to load cohesin onto DNA, called SCC2 (also known as NIPBL and delangin), or by mutations in SMC1 or SMC3. A related disease, Roberts/SC phocomelia syndrome (RBS/SC), has been linked to mutations in ESCO2, a protein implicated in the establishment of cohesion [62] (see also Sect. 3.7.5).

3.3.3.4 Possible Additional Checkpoints

In addition to the aforementioned checkpoints, there may be other checkpoints, which will not be explained in detail. For example, the existence of a topoisomerase II (decatenation) checkpoint has been proposed. DNA topoisomerase II is a highly conserved enzyme that is needed to remove catenations that form between sister DNA molecules during replication. As a consequence, sister chromatids are physically linked and these catenations have to be removed prior to mitosis. Thus, this putative topoisomerase II (decatenation) checkpoint operates during G_2 to prevent cells from entering mitosis with entangled DNA [45].

3.3.3.5 Checkpoint Failures and Human Diseases

Mutations in *BUB1B*, encoding the mitotic checkpoint protein BUBR1, were identified in individuals with mosaic variegated aneuploidy (MVA) [99]. MVA is a rare recessive disease characterized by growth retardation, microcephaly, childhood cancer, and constitutional mosaicism for whole chromosomal gains and losses. At present, MVA is the only human disease related to germline mutations in a spindle checkpoint gene.

On a somatic level mitotic checkpoint defects may promote aneuploidy and tumorigenesis, so that mitotic checkpoints might play a crucial role in cancer development [119]. However, mutations in spindle checkpoint genes are only rarely found in epithelial cancer, and therefore their real impact in tumorigenesis is still a matter of debate at present.

3.3.4 Cell Cycle Coordinators

As described in the previous section, progression into anaphase and beyond depends on the anaphase-promoting complex/cyclosome (APC/C). However, all other events of the eukaryotic cell-division cycle also require control and coordination. The central components of this system are the cyclin-dependent kinases (Cdks). Distinct cyclin-Cdk complexes form at specific cell-cycle stages and initiate the events of the S and M phases.

For example, cyclin B (cyclin B1, CCNB) is expressed predominantly in the G_2/M phase of cell division and is thought to be essential for the induction and coordination of M-phase events (Fig. 3.21). Cyclin B acts together with Cdk1. Therefore, the activated APC/C^{Cdc20} does not only ubiquitilate securing, but also cyclin B, thereby also inactivating Cdk1. Inactivation of cyclin B-Cdk1 enables exit from mitosis, and mitotic cyclin-Cdk complexes drive the events of early mitosis: chromosome condensation and resolution, nuclear envelope breakdown, and assembly of the mitotic spindle [229].

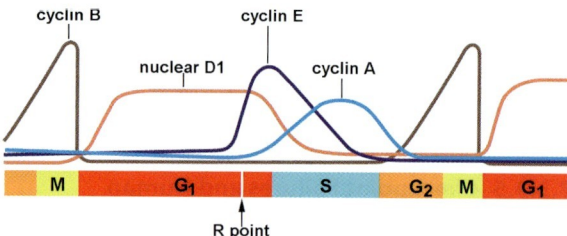

Fig. 3.21 The levels of most mammalian cyclins fluctuate in a tightly coordinated fashion as cells progress through the various phases of the cell cycle. Adapted from [250]

Similarly, levels of the majority of the cyclins fluctuate during the various phases of interphase (Fig. 3.21).

3.4 Chromosome Analysis Methods

A wide spectrum of chromosome analysis methods exists, ranging from traditional banding analyses to sophisticated, molecular array-based technologies. Detailed chromosome analysis has become possible both in metaphase spreads and in interphase nuclei. As the resolution of chromosome analyses has steadily increased, the traditional distinction between cytogenetics and molecular biology is continuously becoming blurred.

3.4.1 Banding Techniques

Banding techniques can generally be divided into differential and selective staining techniques. Differential staining techniques generate dark and light bands along the lengths of all chromosomes and therefore allow assessment of chromosome structure, morphology, and number for all chromosomes. During recent decades these techniques have been indispensable for the unequivocal identification of chromosomes and for standard routine diagnostics. The first differential staining technique introduced employed quinacrine mustard, resulting in bright and dull fluorescence gradations using ultraviolet light. This method was named the QFQ technique (Q-bands by fluorescence using quinacrine). Nowadays, the most commonly used differential staining techniques comprise the G- and the R-bands. In contrast, selective staining techniques do not aim at the staining of all chromosomal regions, but instead at the visualization of distinct chromosomal regions within the genome, such as centromeres or the p-arms of acrocentric chromosomes. These technologies are mainly used to clarify possible chromosomal polymorphisms. Among these selective staining techniques, the most important are the C-bands and AgNOR bands and DA/DAPI staining. In addition to the aforementioned approaches there are a number of other staining and banding technologies. Detailed explanations along with protocols are available [239].

3.4.1.1 Preparation of Mitotic Metaphase Chromosomes

In principle, chromosome preparations can be made from all tissues. For practical reasons, chromosome analysis from peripheral blood, which is usually easily available, is the material most commonly used. However, depending on the clinical question, other tissues may provide more information. For example, in patients with leukemia chromosome status is often assessed from a bone marrow culture. In patients with certain dysmorphic features it may be necessary to perform a fibroblast culture in addition to the blood culture, in order to establish the presence of a possible chromosomal mosaicism.

Mononuclear cells in the blood of healthy individuals usually do not divide; however, chromosomes are always prepared from dividing cells. Therefore, cell divisions must be stimulated artificially. This is usually accomplished by addition of phytohemagglutinin (PHA) to the cell culture. In routine cultures the first cell divisions appear at about 40 h of incubation at 37 °C after stimulation. Thus, a direct preparation of chromosomes immediately after blood is taken is not possible. Most laboratories have adopted a 70- to 76-h incubation protocol, since a higher number of cells entering the second mitosis yield chromosomes with better morphology than those of the first mitosis. To arrest as many cells as possible in prometaphase or metaphase, spindle formation is prevented by a drug with colchicine-like effect, preferably colcemid.

In order to obtain preparations in which the chromosomes are spread out in one plane, the cells are treated for a short period of time (10–30 min) with a hypotonic solution. The cells are then fixed with methanol and acetic acid; a drop of the cell suspension is subsequently placed on the slide, air-dried, and stained. As chromosomes are prepared from dividing cells in which the DNA has already been replicated the investigations consist in metaphase spread analysis of two identical sister chromatids, joined at the centromeres.

3.4.1.2 G-Bands

Giemsa bands (G-bands) can be achieved by different means. G-bands obtained by digesting the chromosomes with proteolytic enzyme trypsin are the most commonly used bands in clinical laboratories for routine chromosome analysis. The composition of G-bands is discussed in more detail in Sect. 3.2.4 (Chromosome Bands; see also Figs. 3.9 and 3.10).

3.4.1.3 R-Bands

The opposite of G-bands are the R-bands (reverse bands). To produce R-bands slides are treated at high temperatures in various buffers, followed by staining with either acridine orange or Giemsa [66]. The bands produced are the reverse of Q- and G-bands.

3.4.1.4 C-Bands

C-bands [15] are applied to stain, specifically, the constitutive heterochromatin in the centromeric regions (Fig. 3.22). Chromosomes 1, 9, and 16 have the largest heterochromatin blocks close to their centromeres, which results in an intensive staining of these regions (Fig. 3.22). In male karyotypes the distal half of the long arm of the Y-chromosome is also C-band positive. These regions often show polymorphisms visible as considerable size differences between the two homologous autosomes, or in the case of the Y-chromosome in significant size difference of the long arm between different males. These size differences are often difficult to interpret by G- or R-banding alone, so that C-bands often complement cytogenetic diagnostics. In fact, the most common chromosomal variant in the human race is the placement of 9q heterochromatin into 9p immediately adjacent to the centromere, which is usually best visible by C-banding.

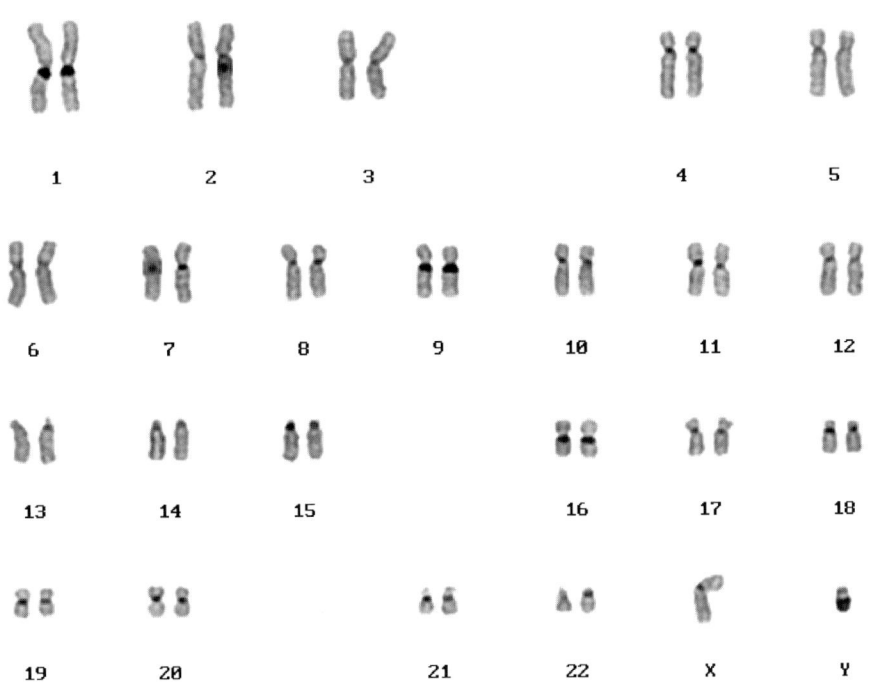

Fig. 3.22 C-Banded human male metaphase spread; C-banding produces selective staining of constitutive heterochromatin. C-Bands are located at the centromere of all chromosomes. They are best visible at chromosomes 1, 9, and 16 and at the long arm of the Y-chromosome

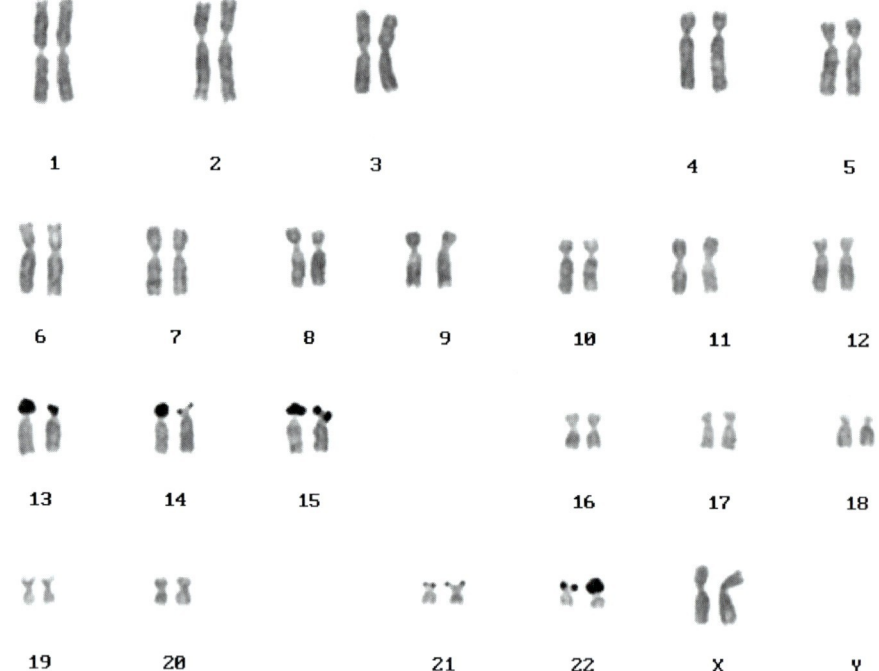

Fig. 3.23 AgNOR-banding of a normal female metaphase. Staining is usually visible at most but not at all p-arms of the acrocentric chromosomes

3.4.1.5 Ag-NOR Bands

Ag-NOR banding [154] aims at the specific staining of the chromosomal regions that form and maintain the nucleoli in interphase nuclei, the so-called nucleolus-organizing regions (NORs) (Fig. 3.23). These regions consist of multiple copies of DNA sequences or genes for ribosomal RNA and are located at the short arms of acrocentric chromosomes. The "Ag" indicates that the staining is done by silver impregnation. Ag-NOR staining reflects the transcriptional activity of the NORs. Thus, frequently not all p-arms of the acrocentric chromosomes show staining. Most individuals have four to seven active NORs per cell.

3.4.1.6 DA/DAPI Staining

DA/DAPI staining involves exposure to the nonfluorescent counterstain distamycin A (DA), followed by staining with fluorescent 4′-6-diamidino-2-phenylindole (DAPI) [209]. As a result, certain heterochromatic regions will appear brightly fluorescent.

3.4.2 Karyotype Description

3.4.2.1 Normal Human Karyotype in Mitotic Metaphase Chromosomes

Karyotypes are described by the number of chromosomes, the sex chromosome constitution, and – if present – any abnormalities (the description of numerical and structural chromosomal abnormalities is explained in detail in the next section). Locations on chromosomes are described according to the Paris Convention, whose basic principles were defined at a conference in 1971 [177]. Bands are numbered counting outwards from the centromere (see Fig. 3.10). Resolution of chromosome analysis depends on the chromosome condensation. In highly extended chromosomes bands split into sub-bands and sub-sub-bands. In routine chromosome analysis a resolution of approximately 500 bands per haploid chromosome set is achieved. At this resolution it is estimated that structural rearrangements with a size of up to 5–10 Mb can be detected. A resolution of less than 400 bands is considered to be inappropriate for the identification of structural chromosomal changes.

An autosome is any chromosome that is not the X or Y sex chromosome. Human cells have 22 pairs of autosomes, which are placed in a karyotype according to their size and position of the centromere. Traditionally, chromosomes in karyotypes are arranged in the following groups: A (chromosomes 1-3), B (4 and 5), C (6-12), D (13-15), E (16-18), F (19 and 20), and G (21 and 22).

3.4.2.2 Description of Normal Human Karyotypes and Karyotypes with Numerical and Structural Changes

Chromosomes are described according to the International System for Human Cytogenetic Nomenclature, abbreviated to ISCN. The latest version is the ISCN 2009 [115]. This publication combines and extends the now classic system of human cytogenetic nomenclature and provides the basic language for the description of chromosomes, especially of chromosomal disorders. For more details the reader should consult the ISCN 2009; here we only describe some basic, general principles for a karyotype description:

The first item to be recorded is the number of chromosomes, followed by a comma (,): the constitution of the sex chromosomes is given next. Thus, a normal female karyotype is designated as 46,XX and a normal male karyotype as 46,XY. Autosomes are only explicitly specified when an abnormality is present.

In the case of chromosome abnormalities sex chromosome aberrations are presented first, followed by abnormalities of the autosomes listed in numerical order irrespective of aberration type. Each abnormality is separated by a comma from the next. For each chromosome numerical abnormalities are listed before structural changes. There are a number of letter designations for the specification of rearranged (i.e., structurally altered) chromosomes. In single chromosome rearrangements the chromosome involved in the change is specified within parentheses immediately after the symbol identifying the type of rearrangement. For example, an inversion has the letter designation "inv," an inversion on chromosome 9 is therefore designated as inv(9). If two or more chromosomes have been altered, a semicolon is used to separate their designations. For example, a translocation has the designation "t," so that a translocation involving chromosomes 9 and 22 is described as t(9;22).

The following list summarizes some of the most important letter designations:

Add	
	Additional material of unknown origin
arrow (->)	From - to, in detailed system
brackets, square ([])	Surround the number of cells
cen	Centromere
colon, single (:)	Break, in detailed system
colon, double (::)	Break and reunion, in detailed system
comma (,)	Separates chromosome numbers, sex chromosomes and chromosome abnormalities
decimal point (.)	Denotes sub-bands
del	Deletion
de novo	Designates a chromosome abnormality which has not been inherited
der	Derivative chromosome
dic	Dicentric
dup	Duplication
fra	Fragile site
h	Heterochromatin, constitutive
hsr	Homogeneously staining region
i	Isochromosome
ins	Insertion
inv	Inversion
mar	Marker chromosome
mat	Maternal origin
minus sign (-)	Loss
p	Short arm of chromosome
parentheses	Surround structurally altered chromosomes and breakpoints
pat	Paternal origin
plus sign (+)	Gain
q	Long arm of chromosome
question mark (?)	Questionable identification of a chromosome or chromosome structure
r	Ring chromosome
rec	Recombinant chromosome
s	Satellite
sce	Sister chromatid exchange
semicolon (;)	Separates altered chromosomes and breakpoints in structural rearrangements involving more than one chromosome
slant line (/)	Separates clones
t	Translocation
ter	Terminal (end of chromsome)
upd	Uniparental disomy

The correct description of karyotypes according to the ISCN 2009 nomenclature represents an especial challenge in the presence of complex chromosomal rearrangements. It is impossible to summarize all

ISCN regulations here. Therefore, interested readers are referred to the ISCN manual [115].

3.4.3 Fragile Sites

Fragile sites represent heritable specific chromosome loci exhibiting an increased frequency of gaps, constrictions, or breaks when chromosomes are exposed to partial DNA replication inhibition. Chromosomal fragile sites are specific loci that preferentially exhibit gaps and breaks on metaphase chromosomes following partial inhibition of DNA synthesis. Fragile sites are present on all human chromosomes. They are named according to the chromosome band they are observed in, e.g., fra(X)(q27.3). In addition, the HUGO nomenclature committee assigns an official symbol. For example, the fra(X)(q27.3) site was called FRAXA (fragile site, X chromosome, A site), because this was the first fragile site detected on the X chromosome.

Fragile sites are classified as either common or rare, depending on their frequency in the population. Whereas common fragile sites are present in all individuals, rare fragile sites are observed in only a small proportion of the population, with a maximal frequency of 1/20, and are inherited in a Mendelian manner. To date, 31 rare and 87 common fragile sites have been described [57, 65, 231] (Tables 3.1, 3.2). A further subdivision is made based on the type of inducing chemicals. The majority of the common fragile sites are induced by aphidicolin, an inhibitor of DNA polymerase, while a smaller group of common fragile sites are induced by BrdU or 5-azacytidine, an inhibitor of DNA methylation. In contrast, the majority of the rare fragile sites are folate sensitive, i.e., they are expressed when cells are grown in folic acid-deficient medium.

Some rare fragile sites, such as FRAXA in the *FMR1* gene, are associated with human genetic disorders, and their study led to the identification of nucleotide-repeat expansion as a frequent mutational mechanism in humans. The FRAXA CGG repeat is located in the 5′ UTR of the fragile X mental retardation 1 (*FMR1*) gene [238]. The *FMR1* gene product (FMRP) is an RNA-binding protein with a high expression in neurons and gonads. However, in the case of so-called full mutations (i.e., >200 CGG repeats), the CpG island in the promotor of the *FMR1* gene is hypermethylated, causing transcriptional silencing of *FMR1* and preventing the synthesis of FMRP. This results in the fragile X syndrome, the most common form of inherited mental retardation [23].

Table 3.1 Classification of fragile sites (from [65])

Class	Number of loci	Sequence
Rare fragile sites	31	
Folate-sensitive	24	(CGG)n
Distimycin A	5	AT-rich repeat[a]
BrdU	2	AT-rich repeat
Common fragile sites	87	
Aphidicolin	76	AT-rich
BrdU	7	AT-rich?[b]
5-Azacytidine	4	?[b]

[a]Fra16B induced by both BrdU and distimycin A
[b]The BrdU- and 5-azacytidine–induced common fragile sites have not been characterized at molecular level

Table 3.2 Classification of rare fragile sites

Subgroup	Fragile site	Location
Folate sensitive ($n=24$)	FRA1M	1p21.3
	FRA2A	
	FRA2B	2q11.2
	FRA2K	2q13
	FRA2L	2q22.3
	FRA5G	2p11.2
	FRA6A	5q35
	FRA7A	6p23
	FRA8A	7p11.2
	FRA9A	8q22.3
	FRA9B	9p21
	FRA10A	9q32
	FRA11A	10q23.3
	FRA11B	11q13.3
	FRA12A	11q23.3
	FRA12D	12q13.1
	FRA16A	12q24.13
	FRA18C	16p13.11
	FRA19B	18q22.1
	FRA20A	19p13
	FRA22A	20p11.23
	FRAXA	22q13
	FRAXE	Xq27.3
	FRAXF	Xq28
		Xq28
Distamycin A-inducible ($n=3$)	FRA8E	8q24.1
	FRA11I	11p15.1
	FRA16E	16p12.1
Distamycin A/BrdU-inducible ($n=2$)	FRA16B	16q22.1
	FRA17A	17p12
BrdU requiring ($n=2$)	FRA10B	10q25.2
	FRA12C	12q24.2

Other fragile sites that have been linked to mental retardation are FRAXE and FRA12A. FRA11B is associated with Jacobsen syndrome, a rare distal 11q deletion syndrome characterized by mental retardation, delayed growth, and specific malformations [57].

Furthermore, common fragile sites have taken on novel significance as regions of the genome that are particularly sensitive to replication stress and that are frequently rearranged in tumor cells. The most prominent example is FRA3B located at the chromosome band 3p14.2. This fragile site maps within the tumor suppressor gene *FHIT*, frequently deleted in cancers, including gastrointestinal tract, cervical, lung, and breast cancers [57].

3.4.3.1 Fluorescence In Situ Hybridization

A clear aim of the continuing development of cytogenetic methods has been to increase the resolution at which chromosome rearrangements can be identified. Crucial to this development was the area of molecular cytogenetics, which is often also referred to as fluorescence in situ hybridization (FISH). In this technique a labeled DNA probe is hybridized to cytological targets such as metaphase chromosomes, interphase nuclei, extended chromatin fibers or, in more recent developments, DNA microarrays (Fig. 3.24). The aim of improving resolution has been achieved by advances in the two crucial elements of cytogenetic analysis,

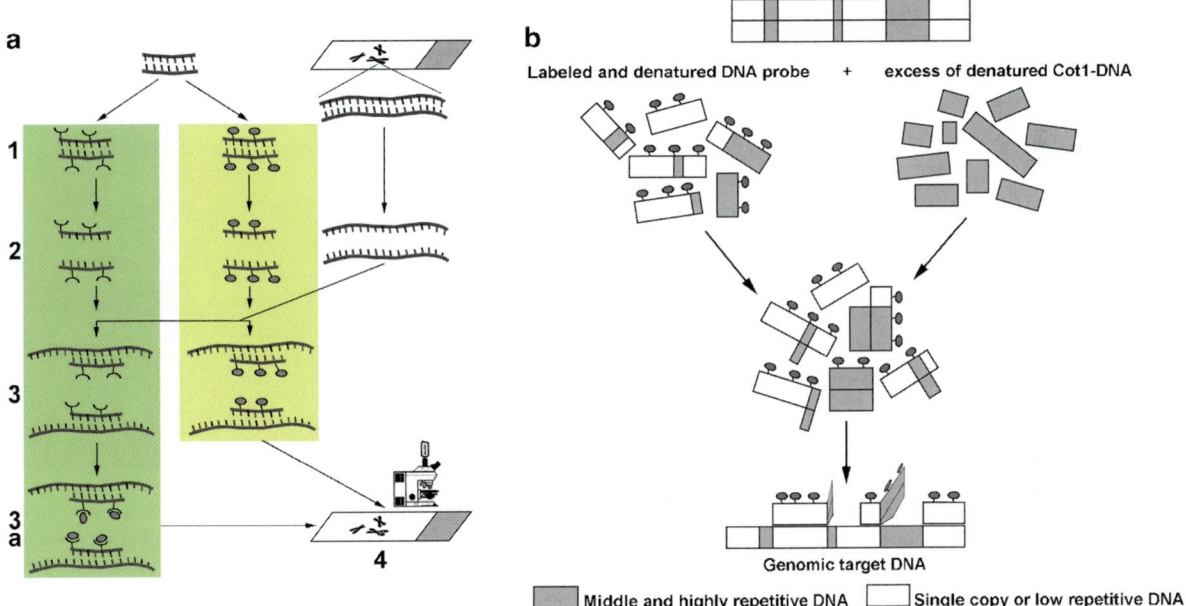

Fig. 3.24 (a) The basic elements of a FISH experiment involve a DNA probe and a target sequence to which the DNA probe is hybridized. Prior to the hybridization, the DNA probe is labeled (*step 1*) by various means such as nick translation, random primed labeling, and PCR. Two different labeling strategies are commonly employed, i.e. indirect labeling (*green box*) and direct labeling (*yellow box*). For indirect labeling, probes are labeled with modified nucleotides containing a hapten, while direct labeling employs incorporation of directly fluorophore-modified nucleotides. The labeled probe and the target DNA are then denatured to yield single-stranded DNA (*step 2*) and brought together allowing the hybridization of complementary DNA sequences (*step 3*). If indirect labeling of the probe has been employed, an additional step is required (*3a*) for visualization of the nonfluorescent hapten using an enzymatic or immunological detection system. While FISH is faster, with directly labeled probes, indirect labeling offers the advantage of signal amplification by using several layers of antibodies and may therefore produce a brighter signal above background. Finally (*step 4*), the signals are evaluated with a fluorescence microscope. Most fluorescence microscopes are equipped with camera systems allowing the imaging of fluorescence signals for digital image processing. (**b**) Complex DNA probes (*top panel*) contain repeat sequences, such as *Alu* and LINES, which are present throughout the genome. The direct use of these probes would result in hybridization signals across the genome and thus a high background level. Therefore, most FISH protocols include a pre-incubation of the denatured complex probe with excess unlabelled Cot-1 DNA. Cot-1 DNA is enriched for the highly repetitive DNA genomic sequences, such as the *Alu* and LINE-1 repeats. Labeled repeat sequences hybridize preferentially to the excess of unlabeled sequences in the Cot-1 DNA, become double stranded, and thus are unavailable for hybridization to the target

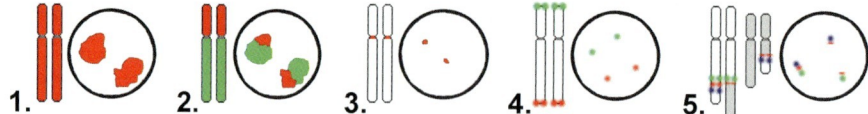

Fig. 3.25 A variety of different DNA probe types are available for hybridization to both metaphase chromosomes and interphase nuclei. The panel shows some examples: *1:* A painting probe stains the entire chromosome. In a normal diploid interphase nucleus two chromosome territories are visible. *2:* Microdissection allows the generation of probes specific for any region within the genome. Differentially labeled chromosome arm-specific probes are shown in this example. In interphase nuclei, the differently labeled chromosome arms within a chromosome territory can be distinguished. *3:* For almost all human chromosomes, probes specific for the centromere are available. Hybridization of these highly repetitive probes does not require suppression with Cot-1 DNA. Owing to their ease of use and high signal intensities, these probes are very popular for the counting of chromosome copy number in interphase nuclei. *4:* For almost any region within the genome, region-specific large insert clones are available. The example shows subtelomeric probes, which are often used to screen for cryptic translocations. Clones for other regions can easily be obtained from publicly available resources. *5:* For known structural rearrangements, special probe sets can be designed to facilitate diagnosis. In this example, the probe set includes a breakpoint-spanning probe and two breakpoint-flanking probes. Use of this probe set allows the structural rearrangement to be detected even in interphase nuclei

i.e., the probe and the target. Target resolution has advanced from metaphase chromosomes (resolution ~5 Mb), through interphase nuclei (50 kb to 2 Mb) and DNA fibers (5–500 kb) to the use of DNA microarrays offering resolutions to a single nucleotide. Probe development has also advanced simultaneously to best utilize the improvements in target resolution.

3.4.3.2 FISH to Metaphase Spreads

The analysis of metaphase spreads has largely benefited from greater probe availability and probe development (Fig. 3.25). As a result of the sequencing of the human genome in the public domain, mapped and sequenced large insert clones are now easily obtainable for virtually any region within the genome [44]. Probes are easily selected by the use of internet browsers such as Ensembl Cytoview (www.ensembl.org) [112], Map-Viewer (www.ncbi.nlm.nih.gov/mapview) [206], or the UCSC genome browser(http://genome.cse.ucsc.edu)[135].Chromosome painting probes which stain an entire chromosome or a chromosome region are also now widely available.

Another important development is the increase in number of differentially labeled probes that can be hybridized and analyzed. The discrimination of many more targets than the number of available spectrally resolvable fluorochromes can be achieved by certain labeling strategies allowing the simultaneous visualization of all 24 human chromosomes, each in a different color in a single hybridization [73]. These 24-color karyotyping technologies are known as multiplex-FISH (M-FISH) [223] (Fig. 3.26), spectral karyotyp-

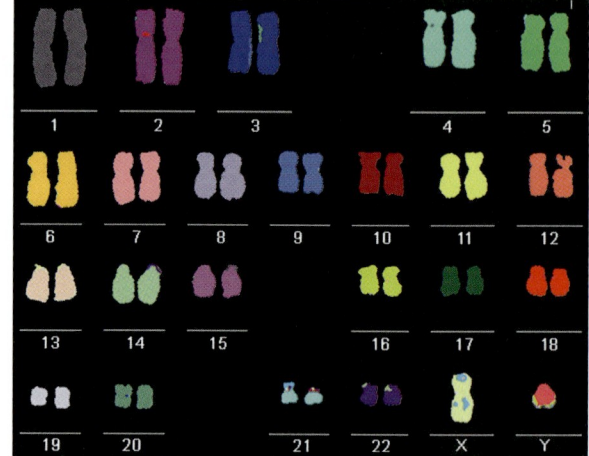

Fig. 3.26 Multicolor classified karyotype of a normal male metaphase, generated by an adaptive spectral classification approach for seven fluorochromes. The automated analysis is assisted by computer-generated pseudocolors. The *light blue colors* on the Y chromosome are caused by cross-hybridization of the Y chromosome to the X chromosome and represent the first pseudoautosomal region at chromosome Xp22.3 (2.6 Mb) and the XY-homologous region at chromosome Xq21.3 (4 Mb). The second pseudoautosomal region at the distal tip of the long arm of the X-chromosome is usually not visible as an extra color (cf. Fig. 3.52)

ing (SKY) [208], and combined binary ratio labeling (COBRA) [232] and have a wide range of uses [73]. The identification of intrachromosomal rearrangements may notably be facilitated by multicolor banding technologies [222].

Together with probe technology there has been considerable improvement in both hardware and software used for the analysis of FISH images. Cooled charge-coupled device (CCD) cameras and more specific and efficient fluorescence filter sets have improved the sensitivity and resolution of imaging on the microscope and sophisticated software facilitates image acquisition and processing.

3.4.3.3 Interphase FISH (Interphase Cytogenetics)

One important feature of FISH-based assays is their ability to yield information about chromosomes or chromosomal subregions in intact interphase nuclei enabling a technology termed "interphase cytogenetics" [51]. Interphase cytogenetics is useful in diagnostic applications where metaphase spreads cannot be obtained, where only small cell numbers are available, or where large cell numbers are to be screened with a particular probe set. Interphase FISH enables rapid screening of large numbers of cells, such as the screening of tumor cell nuclei with centromere-specific probes to establish the presence of chromosomal instability [144] (Fig. 3.27). With careful selection of probes even structural rearrangements, such as translocations and inversions, can be visualized in interphase nuclei [14, 236].

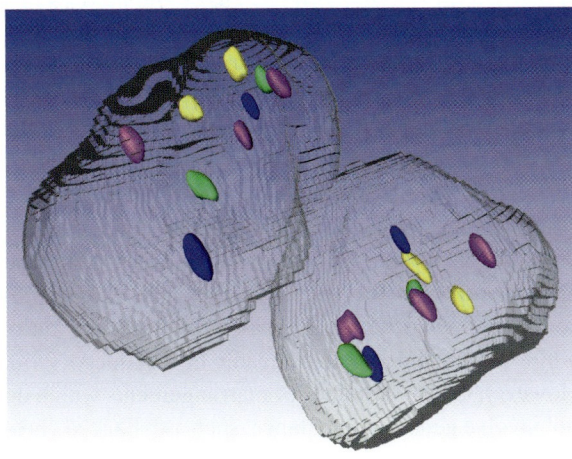

Fig. 3.27 Interphase FISH signals within two nuclei of colon adenoma cells after simultaneous hybridization of four differently labeled chromosome specific centromere probes for chromosomes 7 (*purple*), 8 (*blue*), 11 (*green*), and 17 (*yellow*). Serial optical section in *z*-direction were captured through these nuclei and subjected to 3D reconstruction programs. Each nucleus has three signals for the *purple* chromosome 7-specific probe, suggesting a trisomy 7. In contrast, all other probes yield the expected normal number of two signals

Interphase analysis has also found utility in basic research applications. Interphase FISH, both on fixed nuclei and in living cells, offers the opportunity to analyze how the genome is functionally organized and the dynamic interplay between the genome and its regulatory factors in gene regulation [137]. Furthermore, interphase FISH allows the study of higher order chromatin architecture, which represents an important part of the epigenetic control of gene expression patterns. The current view of higher order chromatin architecture in the cell nucleus is that the cell nucleus has a compartmentalized structure consisting of chromosome territories (CTs) and an interchromatin compartment (IC). The radial arrangement of chromosome territories may be dependent on gene density-related differences in many tissue types [50]. However, in fibroblasts the distribution of chromosome territories showed a probabilistic, highly nonrandom correlation with chromosome size, with small chromosomes being significantly closer to the center of the nucleus, while large chromosomes were located closer to the nuclear or rosette rim (Fig. 3.28) [31] Whether the spatial proximity of chromosomes in interphase nuclei determines the occurrence of possible translocations will be discussed in Sect. 3.6.2.1.

The development of new in vivo labeling techniques has allowed fluorescent labeling of DNA and proteins in living cells (reviewed in [53]). In order to study chromosome territories in living cells, labeling can be performed by incorporation of fluorescent nucleotides in DNA during S-phase. Clonal growth of labeled cells results in segregation of labeled and nonlabeled chromatids during subsequent mitoses.

3.4.3.4 Fiber FISH

While interphase chromatin is less condensed than metaphase chromatin, the highest resolution target for FISH studies is provided by the preparation of released chromatin fibers on microscope slides. This method, known as fiber FISH, provides ordering and structural resolution between 1 kb and 500 kb. A number of alternative techniques for the release of chromatin for fiber FISH have been developed, such as the use of an alkaline lysis buffer or high-salt treatment in an SDS-containing lysis buffer [249] to generate fibers. These approaches produce fibers of varying length and compaction and accordingly are not ideal for the quantification of signal length. For this purpose very uniform, evenly stretched DNA fibers

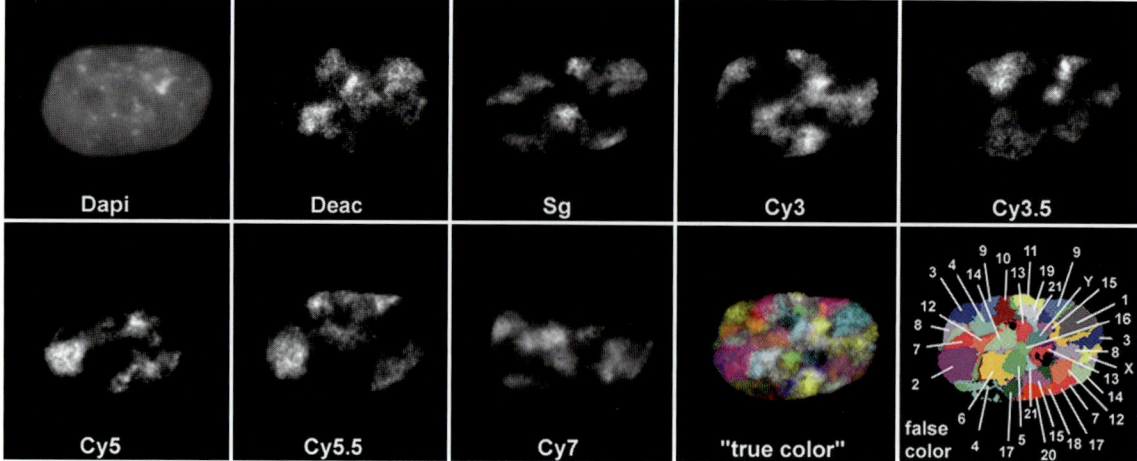

Fig. 3.28 All chromosome domains, each in a different color, can be simultaneously visualized in intact interphase nuclei to study the higher order 3D organization of the genome. The image illustrates a deconvoluted mid-plane nuclear section through a male human G_0 fibroblast nucleus. DAPI is applied to stain the nucleus. For the labeling of chromosome painting probes seven different fluorochromes (diethylaminocoumarin [*DEAC*], spectrum green [*SG*], and the cyanine dyes Cy3, Cy3.5, Cy5, Cy5.5, and Cy7) were used. Here, each channel represents the painting of a subset of chromosome painting probes with the respective fluorochrome. A RGB image of the 24 differently labeled chromosomes (1-22, X, and Y) was produced by superposition of the seven channels. As in 24-color karyotyping, each chromosome has a combinatorial label of a unique fluorochrome combination. Thus, each chromosome territory is amenable to an automated classification so that appropriate software assigns the corresponding chromosome number to a territory. If a stack of these images is collected through a nucleus, a simultaneous 3D reconstruction of all chromosomes is possible. For further details see [31]

are required, as provided by the process of molecular combing [158]. In molecular combing, DNA in solution is stretched at the meniscus as a glass slide is removed at a constant rate, generating fields of very even DNA fibers all parallel to each other.

3.4.3.5 Detection of Copy Number Changes in the Genome

3.4.3.5.1 Conventional CGH on Metaphase Chromosomes

The preparation of high-quality metaphase spreads from clinical and tumor cell samples, and in particular from solid tumors, is often difficult. To overcome this problem comparative genomic hybridization (CGH) was developed [63, 127]. In CGH, DNA is extracted directly from the test sample, e.g., DNA from a patient or a tumor, thus avoiding any culturing artifacts. In a second step the test DNA and normal reference DNA are labeled differentially, for example the test DNA with a green and the reference DNA with a red fluorochrome, and co-hybridized to normal metaphase spreads (Fig. 3.29). The two DNAs compete for the hybridization sites on the target metaphases in such a way that if a region is amplified in the tumor the corresponding region on the normal metaphases becomes predominantly green, and if a region is deleted in the test the corresponding chromosome region becomes red (Fig. 3.29). The actual test to reference DNA fluorescence ratios along all chromosomes is quantified by using digital image analysis systems. DNA gains and amplifications in the test DNA are seen as chromosomal regions with an increased fluorescence ratio, while losses and deletions result in a reduced ratio. However, CGH is limited for rearrangements which do not involve genomic imbalance, such as balanced chromosome translocations and inversions and whole-genome copy number changes (ploidy), which cannot be detected by CGH. Furthermore, CGH does not provide information on the way in which chromosome segments involved in gains and losses are arranged.

In conventional CGH using DNA extracted from hundreds or thousands of cells, chromosomal imbalances are only detected if they are present in most of

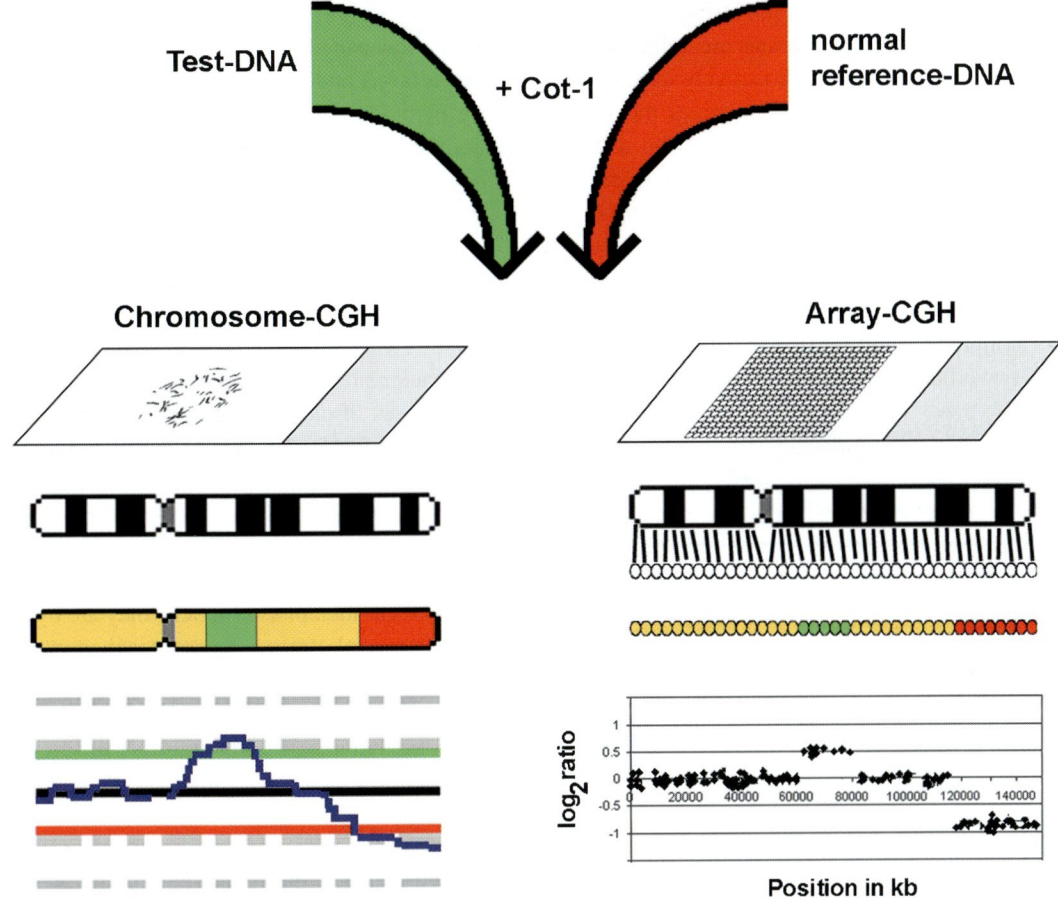

Fig. 3.29 Outline of comparative genomic hybridization (CGH) experiments. CGH maps readily regions of gains and losses in a test-DNA, which is in most cases retrieved from a patient or from a tumor sample. In a CGH analysis, test DNA and normal reference DNA are labeled with different fluorochromes. Here, a test DNA is labeled with a *green* fluorochrome, whereas the reference DNA is labeled with a *red* fluorochrome. In conventional CGH a mixture of these two DNA samples is together with an excess of unlabeled Cot-1 DNA hybridized to normal chromosome spreads (*left-hand side*). In array-CGH/matrix-CGH the hybridization target can consist of large insert clones or oligonucleotides arrayed onto glass slides (*right-hand side*)

the cells in the specimen. Thus, information about heterogeneity may get lost. Furthermore, rare cell events, such as minimal residual disease, where only a single cell or a few cells are available for study, initially could not be analyzed by conventional CGH. However, protocols were developed for the analysis of single cells by CGH. A prerequisite for single-cell CGH analyses are methods for unbiased single-cell DNA amplification yielding amplification products that retain the copy number differences of the original genome [129, 241, 252]. Such approaches have been applied to prenatal diagnostics or for the analyses of minimal residual disease.

3.4.3.5.2 Other Approaches for Evaluation of Copy Number Changes

Copy number alterations of specific genomic sequences can also be measured without the need for chromosomes using multiplex amplifiable probe hybridization (MAPH) [11] and multiplex ligation-dependent probe amplification (MLPA) [207]. MAPH is based on the hybridization of specific probes with uniform linkers to denatured genomic DNA immobilized to a nylon membrane. Probes for different regions vary in length and hybridize to the immobilized DNA in proportion to the copy number of the corresponding sequence in the genome. After hybridization and stringent washing the probes bound to the membrane are released and amplified by PCR using a radioactively or fluorescently labeled primer pair that recognizes the linker sequence. After size separation on a gel the relative intensity of the peaks between a test and reference sample is compared to determine the copy number of the target DNA sequence [11]. MPLA is similar to MAPH, but hybridization and amplification take place in solution with no need to immobilize the DNA onto a nylon membrane (Fig. 3.30). Each region is represented by two adjacent tailed probes that are joined by a ligation reaction on the target DNA. Subsequent amplification can then only take place from ligated probes as target and not from any other sequence [207]. These methods are very rapid and cost-effective and have found favor for specific diagnostic procedures, such as exon deletion screening, but are not easily scaled for whole-genome scanning.

3.4.3.5.3 Array-CGH

The replacement of the metaphase chromosomes with large numbers of clones spotted onto a standard glass slide as the target for CGH has significantly increased the resolution of screening for genomic imbalance. In array-CGH, the test and normal reference genomes are labeled with different fluorochromes, as in conventional CGH. However, as opposed to hybridizing the DNA probes to metaphase spreads, they are hybridized to a microarray platform. The array is then imaged and the relative fluorescence intensities calculated for each mapped clone, where the resulting intensity ratio is proportional to the DNA copy number difference. The resolution of the analysis is only restricted by the clone size and clone density on the array. An additional advantage is the ease with which array-CGH can be automated for high-throughput applications.

The first descriptions of array-CGH using large insert clones were published in the late 1990s [188, 220] and were rapidly followed by the development of whole-genome arrays with one clone every megabasepair (e.g., [75, 219]). The density of clones on the slide has continued to increase, and the highest resolution for array-CGH is now provided by spotted and synthesized oligonucleotide arrays containing as many as 500,000 or even more elements. The inherent noise of hybridizations of these arrays requires statistical analysis for the confident identification of small gains and losses.

Thus, the main advantages of array-CGH over conventional array comprise the increase in resolution, which also includes the capability of breakpoint mapping (Fig. 3.31). At present a number of different array platforms are available for both whole-genome analysis and detailed investigation of selected regions (Fig. 3.32). Array-CGH is extensively used for the analysis of gains and losses in tumors, but is also applied to the analysis of patients with constitutional rearrangements.

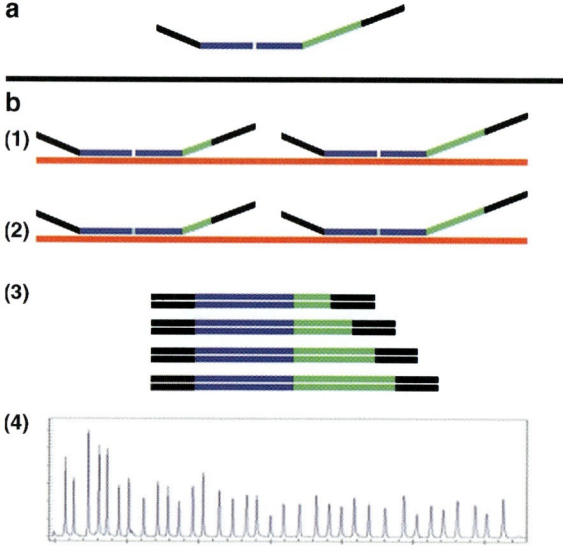

Fig. 3.30 (**a**, **b**) The multiplex ligation-dependent probe amplification (MLPA) procedure. (**a**) Design of a MLPA probe: MLPA probes consist of two separate oligonucleotides, each containing the target sequence (shown in *blue*), a "stuffer sequence" (*green*), and the PCR primer sequences (*black*). The PCR primer sequences are identical for all probes, so that all probes can be later amplified with the same primer set. The stuffer sequence has a different size for each MLPA probe, which allows multiplexing, i.e., multiple probes can be hybridized simultaneously. Different probes are then distinguished by their size. (**b**) The MLPA procedure consists of several steps: *1:* In a first step the DNA to be analyzed is denatured and MLPA probes are hybridized to the DNA. Here, two different MLPA probes are hybridized; the right probe has a larger stuffer sequence than the left probe. *2:* In the second step the ligation reaction takes place. The two probe oligonucleotides hybridize to immediately adjacent target sequences and only then can they be ligated during the ligation reaction. *3:* The ligated probes are amplified by PCR. Probe oligonucleotides that are not ligated only contain one primer sequence and cannot be amplified exponentially. *4:* In the next step amplification products are separated by electrophoresis; and the data can be analyzed. (Adapted from www.mlpa.com)

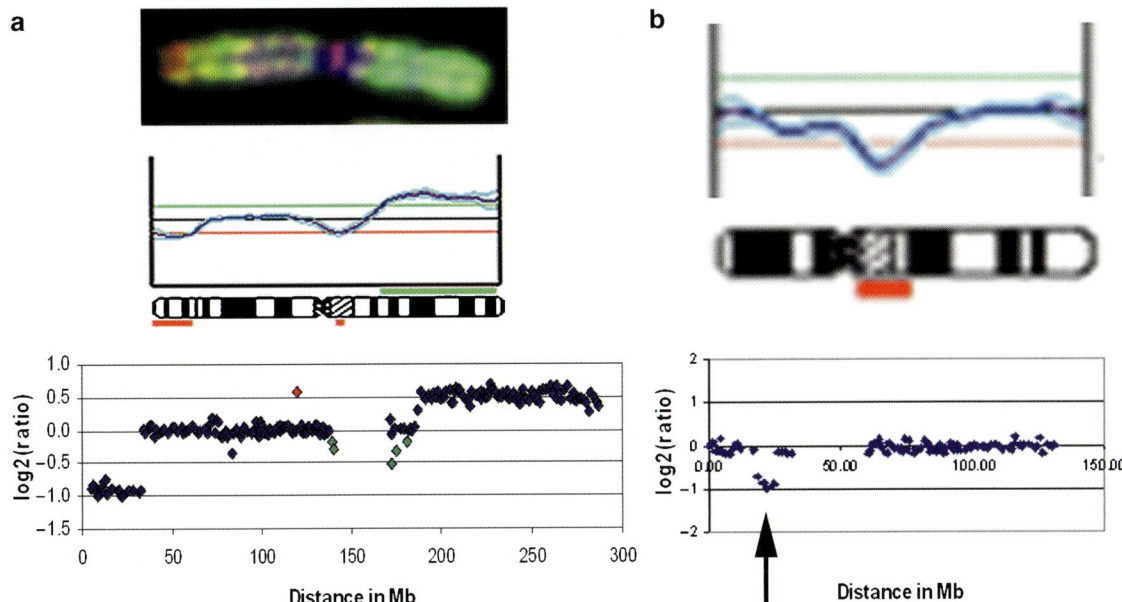

Fig. 3.31 (**a, b**) Comparison of the different resolution limits of conventional and array-CGH. In this example, DNA from a primary renal cell adenocarcinoma line (769P, ATCC No. CRL-1933) was hybridized to both metaphase spreads and a large-insert clone, i.e., BAC array. (**a**) The *top panel* illustrates the hybridization pattern on chromosome 1. The distal tip of the p-arm (*left*) appears *red*, which indicates a loss of this region. In contrast, the q-arm (*right*) is extensively stained with *green*, which suggested overrepresentation of this region. The center of chromosome 1 has very low fluorescence signals as this region represents the large chromosome 1 heterochromatin block. As this region consists only of repetitive sequences, hybridization is blocked by the addition of an excess of unlabeled Cot-1 DNA to the hybridization mix. Interpretation of CGH-hybridization patterns does not depend on visual inspection. Instead computer programs are employed, which calculate the intensities of both the *green* and the *red* fluorochrome and the ratio values between these fluorochromes. The result is displayed in a graph, shown in the *center*. The ratio profiles are usually calculated as a mean value of several metaphase spreads. The *three horizontal lines* above the chromosome 1 pictogram represent different values of the fluorescence intensities between the tumor and the reference DNA. The *black line* represents balanced fluorescence intensities, while the *right line* is the threshold for loss and the *green line*, the threshold for a gain of DNA material. The *lower panel* shows the same result obtained on a BAC array. The respective gains and losses are identified with ease. In contrast to conventional CGH, the breakpoints of lost and gained regions can be accurately mapped as they appear as sharp transitions in the ratio profile. (**b**) Cell line 769P also has a small single-copy deletion on chromosome 9p of about 6.3 Mb. The resolution limits for the detection of deletions or duplications with conventional CGH were estimated to be in the range of about 10 Mb. Consequently, this deletion is not identified with conventional CGH. The "deletion" shown in the *upper panel* is caused by the large heterochromatic block on chromosome 9, which owing to the suppression conditions has no or only very low fluorescence intensities. Such chromosomal regions, rich in repetitive sequences, are prone to resulting in artifacts in chromosome CGH. In array-CGH this heterochromatin block is visible as a large gap around chromosome position 50 Mb because these regions are not represented on arrays. However, the 6.3-Mb deletion is readily visible in array-CGH (*arrow*)

3.4.3.5.4 Copy Number Changes in Healthy Individuals

After completion of the Human Genome Project it was considered that the DNA sequence of essentially every human being was known, as the genomes of healthy individuals were 99.9% identical. The major genetic differences believed to exist between individuals were in the form of scattered single-base pair changes, i.e., single nucleotide polymorphisms (SNPs), accounting for 0.1% of the genome. However, when array-CGH was applied to the genomes of unrelated, healthy individuals hundreds of genomic regions that varied were identified [114, 210]. Many of these copy number polymorphisms are common in gene-rich regions of the genome and are associated with segmental duplication. Using an *in silico* sequence comparison method for identifying not only copy number polymorphism but other structural variants, such as inversions, multiple sites of putative structural variation involving inser-

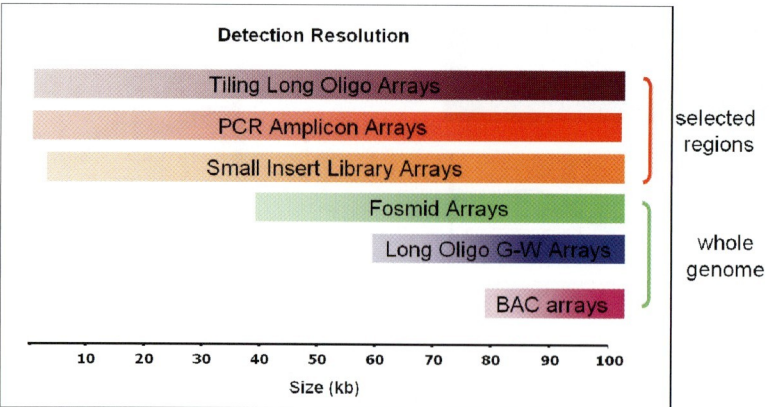

Fig. 3.32 There is now a multitude of different array-platforms for CGH available. In principle, these arrays can be distinguished between arrays to scan the entire genome for gains and losses or in arrays for selected regions. For example, at present many laboratories employ in diagnostic setting for the analysis of DNA samples derived from patients with unknown syndrome long oligo arrays. For more detailed analyses of individual chromosomes there exist chromosome specific tiling long oligo arrays, which provide a resolution in the low kb range

tions, deletions, and inversion breakpoints with the rearrangement ranging in size from 8 kb to as much as 0.9 Mb, were identified [237]. Thus, a surprising and unexpected finding yielded by these high-resolution approaches was that even the genome of normal, healthy individuals has a high number of copy number changes. When several individuals were screened for CNVs, a total of 1,447 copy number variable regions (CNVRs) covering 360 Mb (12% of the genome) were identified. These CNVRs contained hundreds of genes, disease loci, functional elements, and segmental duplications. Notably, the CNVRs encompassed more nucleotide content per genome than SNPs, underscoring the importance of CNVs in genetic diversity and evolution. The data obtained delineate linkage disequilibrium patterns for many CNVs and reveal marked variation in copy number among populations [196].

Submicroscopic CNVs are both intriguing and of particular concern to clinical cytogeneticists, because they can no longer rely on a "standardized" genome – represented at the cytogenetic level as the human karyotype – to identify "abnormal" chromosomal alterations that can be implicated in the etiology of a disease or disorder. In fact, with array-CGH a new terminology was introduced (e.g., structural variants, structural abnormality, CNV, copy-number polymorphism, segmental duplication, low-copy repeat):

Structural variants are operationally defined as genomic alterations that involve segments of DNA larger than 1 kb and can be microscopic or submicroscopic. There is no implication of their frequency, their association with disease or phenotype, or lack thereof. Other alterations that can be considered structural variants include heteromorphisms, fragile sites, ring and marker chromosomes, isochromosomes, double minutes, and gene-conversion products.

The term *structural abnormality* is often used if a structural variant is thought to be disease causing or is discovered as part of a disease study. Here we generally refer to smaller (<1 kb) variations or polymorphisms involving the CNV of a segment of DNA as insertions or deletions (indels).

A *copy-number variant (CNV)* is a segment of DNA that is 1 kb or larger and is present in a variable copy number in comparison with a reference genome. Classes of CNVs include insertions, deletions, and duplications. This definition also includes large-scale CNVs, which are variants involving segments of DNA ≥50 kb in size, allowing them to be detected by clone-based array comparative genome hybridization (array-CGH).

Copy-number polymorphism is the term used for a CNV that occurs in more than 1% of the population. Originally, this definition was used to refer to all CNVs [74].

Segmental duplication or low-copy repeat means a segment of DNA >1 kb in size that occurs in two or more copies per haploid genome, with the different copies sharing >90% sequence identity. They often vary in copy number and can therefore also be CNVs [74].

It is difficult to determine whether a CNV might contribute to phenotypic effects. In general, larger CNVs are probably more likely to cause developmen-

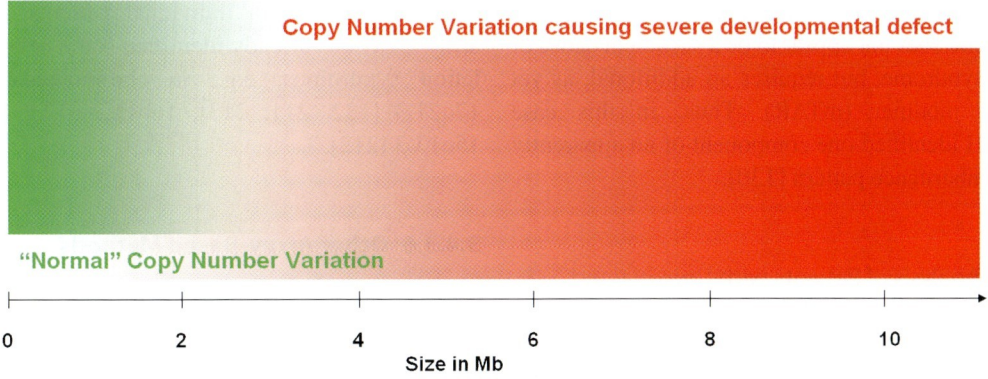

Fig. 3.33 Relationship between CNV size and possible phenotypic effects

tal defects; however, there is no strict threshold to distinguish between "normal" CNVs and disease-causing CNVs (Fig. 3.33).

Association analyses of expression levels of 14,925 transcripts with SNPs and CNVs in individuals who are part of the International HapMap project were performed in order to determine the overall contribution of CNVs to complex phenotypes. SNPs and CNVs captured 83.6 and 17.7% of the total detected genetic variation in gene expression, respectively, but the signals from the two types of variation had little overlap [226].

Interrogation of the genome for CNVs is also an effective way to elucidate the causes of complex phenotypes and disease in humans. In fact, CNVs have already been linked to a variety of different diseases and disease susceptibility, including autism, Crohn disease, AIDS susceptibility, glomerulonephritis, and many others.

Using array-CGH, it was even possible to identify an example of positive selection on a copy number-variable gene, because the copy number of the salivary amylase gene (*AMY1*) is positively correlated with salivary amylase protein level and individuals from populations with high-starch diets have, on average, more AMY1 copies than those with traditionally low-starch diets [181]. This provides a striking example of the role of CNVs in adaptive evolution and of diet in producing selective pressures.

In addition, it even proved possible to identify a novel autosomal-dominantly inherited syndrome associated with a CNV on 4p. This syndrome is characterized by microtia, eye coloboma, and imperforation of the nasolacrimal ducts associated with a CNV on 4p, and represents the first example of an amplified CNV associated with a Mendelian disorder [20].

Furthermore, CNVs may be helpful in population genetics because genome-wide patterns of variation across individuals provide a powerful source of data for uncovering the history of migration, range expansion, and adaptation of the human species [118].

3.4.3.5.5 Array-CGH Identifies Causes of Known Syndromes

Using a 1-Mb resolution array-CGH a deleted region of ~5 Mb at 8q13 was identified in a patient with CHARGE (coloboma of the eye, heart defects, atresia of the choanae, retarded growth and developmental anomalies, genital hypoplasia and/or urinary tract anomalies, and ear anomalies and/or hearing loss) syndrome [240]. Subsequent use of a tiling clone array of chromosome 8 narrowed the CHARGE region to 2.3 Mb containing nine annotated genes. Sequencing of these genes identified mutations in the *CHD7* gene, which were later also found in other patients with CHARGE. Thus, in this case array-CGH pinpointed the region of the disease-causing gene in this disease entity.

Another example was the discovery by array-CGH that patients with Pitt-Hopkins syndrome, a rare syndromic mental disorder, are associated with a microdeletion on 18q21.2 resulting in haploinsufficiency of *TCF4*. Further analysis revealed that Pitt-Hopkins syndrome may also be caused by autosomal dominant mutations in TCF4 [4, 35, 260].

Array-CGH also identified a deletion in chromosome 1q21.1 in patients with TAR (*t*hrombocytopenia-*a*bsent *r*adius) syndrome, which is characterized by hypomegakaryocytic thrombocytopenia and bilateral radial aplasia in the presence of both thumbs.

Interestingly, presence of the deletion on chromosome 1q21.1 alone does not appear to be a sufficient cause of the TAR syndrome, but requires an additional, as yet unknown, modifier (mTAR). Thus, in this case array-CGH identified one component of an apparently complex inheritance pattern [130].

3.4.3.5.6 Analysis of Patients with Unexplained Developmental Delay and Dysmorphic Features

Array-CGH is increasingly being used for diagnosis of constitutional genomic imbalance. It has rapidly evolved to become an indispensable tool for the work-up of patients with unexplained developmental delay and dysmorphic features. Multiple studies related to this group of patients have already been published, suggesting that in up to 20% of cases of mental retardation and multiple congenital anomalies a deletion or duplication that is causative for the phenotype may be found [157]. At present, our knowledge of the phenotypic effects of most gains and losses is minimal. Accordingly, it may be difficult to distinguish between a disease-causing copy number change, a genomic imbalance of unknown clinical significance, and a polymorphism. In general, an imbalance occurring de novo, i.e., which is not present in one of the parents, is more likely to be causative for a conspicuous phenotype. Another indication for a possible causative role is the size of a deletion or duplication: the larger it is the more likely are phenotypic consequences. Owing to the difficulties in establishing whether a copy number is disease associated or not, there are international efforts for a systematic collection of CNVs and their association with specific phenotypes in publicly accessible databases, such as DECIPHER (*D*atabas*e* of *C*hromosomal *I*mbalance and *P*henotype in *H*umans using *E*nsembl *R*esources) (www.sanger.ac.uk/PostGenomics/decipher/) [76] or ECARUCA (*E*uropean *C*ytogeneticists *A*ssociation *R*egister of *U*nbalanced *C*hromosome *A*berrations) (www.ecaruca.net).

3.4.3.5.7 Identification of New Microdeletion Syndromes

The vast majority of imbalances are "private" deletions or duplications, i.e., they are not recurrent and are observed just in individual patients. However, array-CGH has already identified some new microdeletion syndromes, e.g., in chromosomal regions 17q21.3 [132, 211, 213], 16p11.2-p12.2 [21], and 15q13.3 [212].

3.4.3.6 Other Array-Based Methods

3.4.3.6.1 Array Painting: Structural Aberrations at High Resolution

Array-CGH is of little use for the analysis of patients with rearrangements not involving copy number gains or losses, such as inversions or balanced translocations. For balanced translocations, array painting, a modification of the array-CGH method, has been developed which uses flow sorting to separate the derivative chromosomes from the rest of the genome for hybridization to the array [75]. In a balanced translocation, differential labeling of the two derivatives results in the sequences proximal to the breakpoint on a chromosome being labeled in one color while sequences distal to the breakpoint are labeled in the other color. When the two derivatives are hybridized to the array, the measured ratio for the chromosomes involved in the rearrangement switches from high or low (or vice versa) at the breakpoint. If a clone on the array spans the it will report an intermediate value. Array painting will also identify other rearrangements associated with the translocation such as inversion and deletion within the derivative chromosomes [75] and the involvement of additional chromosome regions in the rearrangement [97]. The combined use of array-CGH and array painting provides a comprehensive analysis of genomic rearrangements in patients with apparently balanced translocation and has uncovered a surprisingly high frequency of complexity in these cases [97].

3.4.3.6.2 ChIP on CHIP: Analysis of Chromatin Structure

It is now well established that chromatin organization and modification of associated proteins and complexes have a major functional role in such fundamental processes as transcription, recombination, replication, and

DNA repair. Cytogenetic methods, particularly when combined with chromatin fractionation are proving to be useful tools in the study of chromatin structure and function. Gilbert et al. [90] separated compact and open chromatin structures using sucrose gradients and analyzed the genomic distribution of the differing chromatin states by hybridization of the enriched DNA onto metaphase chromosomes and by array-CGH. They found that regions of open chromatin correlated with high gene density but not necessarily with gene expression, as inactive genes were found in regions of open chromatin and active genes in regions of closed chromatin. Perhaps the most powerful approaches to the study of chromatin structure and function use DNA fractionation and enrichment by chromatin immunoprecipitation (ChIP). In ChIP, chromatin associated with protein-DNA interactions or modifications is specifically enriched by precipitation with an antibody directed against the protein of interest. The enriched DNA sequences can then be mapped and quantified using, in particular, DNA microarrays which provide the resolution required. This combined methodology has been termed ChIP on CHIP. For example, ChIP against phosporylated H2AX, a marker of double-strand breaks in DNA, was used to demonstrate that DNA damage occurring at chromosome ends as a consequence of telomere shortening is associated with cellular senescence [55].

3.5 Meiosis

3.5.1 Biological Function of Meiosis

The somatic cell is diploid, containing both members of a pair of homologous chromosomes (2 n), whereas the germ cell is haploid, containing only one of each pair (n). Thus, in the usual type of somatic cell division, or mitosis, the number of chromosomes in daughter cells remains constant. In contrast, the meiotic process is designed to reduce the number of chromosomes from the diploid number (46 in humans) to one half of this number (23 in humans). Fertilization of two germ cells, each with the haploid number, reconstitutes the diploid number of 46 in the zygote and in all of its descendant cells. Furthermore, meiosis generates genetic diversity, because chance alone determines which of two homologous chromosomes ends up in a given germ cell. Beside this independent assortment of chromosomes, physical exchange of chromosomal regions by homologous recombination during prophase I results in new genetic combinations within chromosomes.

3.5.2 Meiotic Divisions

Complete meiosis consists of two cell divisions, meiosis I and II. Meiosis I is a reductional cellular division involving the segregation of homologous chromosomes. Meiosis II is an equational division with the segregation of sister chromatids. Perhaps the most relevant events occur during the first – and most complicated – phase of meiosis I. In this phase homologous chromosomes pair, become intimately associated (synapsis), and exchange genetic material (crossing over or meiotic recombination; Fig. 3.34).

Meiotic Division I. Prophase I: The complex series of prophase events is subdivided into five stages (leptotene, zygotene, pachytene, diplotene, and diakinesis) reflecting the progression of synapsis and recombination.

Leptotene: Long chromosome threads become visible, but the two sister chromatids are still tightly bound, making them indistinguishable from one another. The chromosomes show a specific arrangement with telomeres oriented towards the nuclear membrane. This stage is called the "bouquet stage."

Zygotene: During this stage homologous chromosomes pair, frequently starting at the chromosome ends. The paired homologous chromosomes are referred to as bivalent, or also as tetrad as they consist of four sister chromatids. The paired homologous chromosomes are connected by the so-called synaptonemal complex, in a process known as synapsis. The synaptonemal complex is a tripartite protein structure between homologous chromosomes, which is essential for meiotic recombination to take place.

Pachytene: After completion of pairing, the chromosomes become shorter through contraction. At this stage, nonsister chromatids of homologous chromosomes may randomly exchange segments of genetic information over regions of homology. This exchange occurs at sites where recombination nodules or chiasmata have formed. Sex chromosomes exchange occurs only over

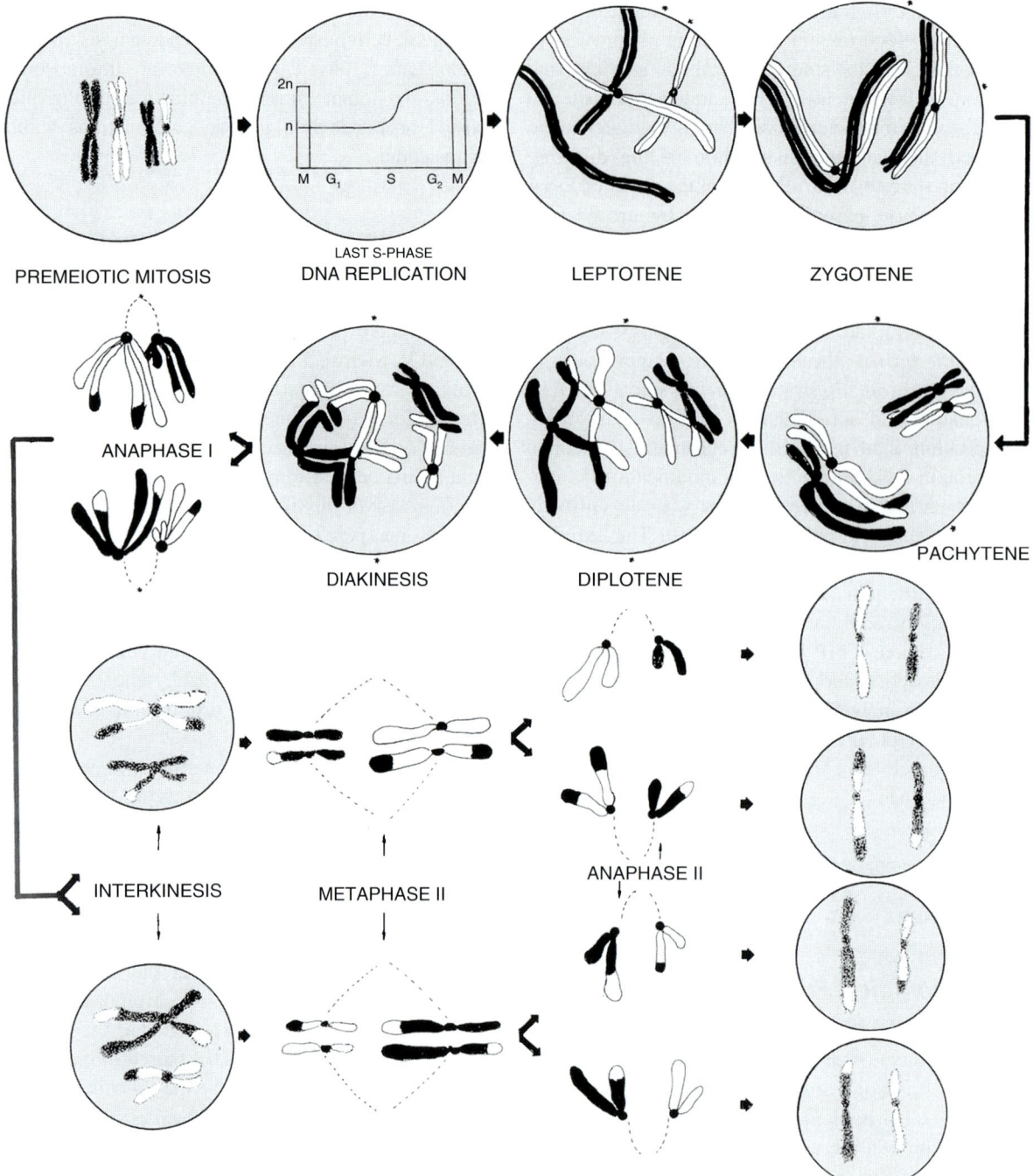

Fig. 3.34 The stages of meiosis. *Black* paternal, *white* maternal chromosomes. Male meiosis is depicted; in female meiosis, polar body formation occurs. (From the 3rd edition of this work)

a small region of homology, the "pseudo-autosomal region" (explained in Sect. 3.6.3.2).

Diplotene: A longitudinal cleft in each pair of chromosomes becomes visible as the synaptonemal complex degrades. This allows some separation of the homologous chromosomes; nonsister chromatids are separated, while sister chromatids remain paired. However, the homologous chromosomes of each bivalent remain connected to each other at the chiasmata – the regions where crossing-over occurred – and become visible between nonsister chromatids. The chiasmata persist until anaphase I. In mammalian females meiosis stops at a modified diplotene stage (dictyate) and can remain in this state for decades.

Diakinesis: During this phase the four chromatids of each kind become visible and appear side by side. The chiasmata are now clearly visible. Much of this stage resembles prometaphase of mitosis as the nucleoli disappear, the nuclear membrane disintegrates and the meiotic spindle begins to form.

Metaphase I: The chromosomes are ordered in the metaphase plane as the centromeres are drawn to the poles. The homologous chromosomes are drawn somewhat apart but are still kept together by the chiasmata.

Anaphase I: As microtubules shorten, pulling homologous chromosomes apart, the chiasmata are resolved. As a consequence, whole chromosomes are pulled toward opposing poles, forming two haploid sets.

Telophase I: The centromeres arrive at the poles and each daughter cell now has half the number of chromosomes with each chromosome consisting of a pair of chromatids. The microtubules disappear and a new nuclear membrane surrounds each haploid set. The chromosomes uncoil back into chromatin. Cytokinesis completes the creation of two daughter cells.

Meiotic Division II: This is in principle a mitotic division of the replicated haploid set of chromosomes. As such, four phases can be distinguished (prophase II, metaphase II, anaphase II, and telophase II). As noted above, meiosis begins after replication. The genetic material which during division I has become fourfold (2×2 homologous chromosomes) is, at the completion of division II, ordinarily distributed to four cells.

A second important aspect in meiosis is the random distribution of nonhomologous chromosomes, which leads to a very large number of possible combinations in germ cells. In humans with 23 chromosome pairs, the number of possible combinations in one germ cell is $2^{23} = 8,388,608$. The number of possible combinations of chromosomes in an offspring of a given pair of parents is $2^{23} \times 2^{23}$ and is further enhanced by crossing over during pairing of homologous chromosomes. The morphological counterpart of crossing over is chiasma formation. Every chiasma corresponds to one crossing over event involving two nonsister chromatids.

3.5.3 Meiotic Recombination Hotspots

Each chiasma corresponds to a locus in which crossing over has occurred. The result of crossing over is an exchange between two chromatids of homologous chromosomes. Any one pair of homologous chromosomes normally experiences at least one crossing-over event during the first meiotic division in order to assure proper chromosomal segregation. A single meiotic crossover event produces two chromosomes, each divided into one portion of grandmaternal and one portion of grandpaternal origin.

Recombination by crossing over also defines the genetic distance between two loci. The *physical distance* is the number of nucleotide base pairs between DNA loci. In contrast, the *genetic distance* between two DNA loci on a chromosome is given as the amount of crossing over between these two markers. The unit of the genetic distance is morgan. Thus, one morgan, M, is the unit of map distance between linked genes; this unit measures recombination frequencies. One centimorgan (cM) means 1% recombination and equals the crossover frequencies between genes.

The distribution of crossing-over events within a chromosome was originally thought to be uniform. However, newer technologies, such as sperm-typing technologies, i.e., screening sperm for recombinant DNA molecules [120], linkage studies, mapping crossovers in large pedigrees by high-density SNP typing [48], and computational inferences from population genetic data have provided detailed high-resolution maps of the sites of crossing over. Together the data suggest that the sites of recombination are not randomly distributed along the chromosomes, and this is likely due to the presence of numerous hotspots and coldspots of recombination [186]. Thus, recombination occurs preferentially in hotspots. Linkage disequilibrium (LD) data, e.g., from the International HapMap Project, can be used to infer the recombination history. Similarities between LD maps of different populations presumably reflect common localizations of recombination hotspots, and these hotspots are frequently referred to as "historical" hotspots. Computational approaches have inferred more than 25,000 historical hotspots, and it is estimated that as many as 50,000 hotspots may exist [163]. Historical hotspots have a median width of 4 kb. Thus, there are many narrow and intense hotspots throughout the genome. It is estimated that 80% of any recombination occurs in only 10–20% of the sequence [12, 163].

3.5.4 Differences Between Male and Female Meiosis

There are two principal aspects by which meiosis differs in males and females:

1. In males all four division products develop into mature germ cells, whereas in females only one of them becomes a mature oocyte, while the others are lost.
2. In males, meiosis immediately follows a long series of mitotic divisions; it is completed when spermatids start developing into mature sperms. In females, meiosis begins at a very early stage of development, immediately after a much smaller series of mitotic divisions. It is then arrested for many years and is only finished after fertilization.

These sex differences are important in human genetics. The fact that only one of the four division products develops into a mature oocyte, and the three polar bodies contain little or no cytoplasm, enables this oocyte to transmit to the new zygote a full set of cytoplasmic constituents, such as mitochondria and messenger RNA. These differences in cell kinetics are probably responsible for sex differences in mutations rates for trisomies, on the one hand, and point mutations, on the other.

3.5.4.1 Meiosis in the Human Male

From the beginning of puberty human spermatocytes continuously undergo meiosis. After the second meiotic division, DNA is densely packed during sperm development and the sperm acquires the ability to move actively.

Genes located on a common segment of the X and Y chromosome can freely recombine with each other, and this region has therefore been dubbed the "pseudoautosomal region" to indicate that it behaves as autosomes do in terms of recombination.

3.5.4.2 Meiosis in the Human Female

In all mammals, oogenesis differs substantially from spermatogenesis. Meiosis is initiated in the human fetal ovary at 11–12 weeks of gestation. On completion of recombination, the oocyte progresses to diplotene of prophase and enters a protracted arrest stage known as dictyate. Around the time of arrest, oocytes become surrounded by somatic cells (pregranulosa cells), forming primordial follicles. Many oocytes are lost during follicle formation, and the newborn ovary contains only a fraction of the total oocytes that entered meiosis in the fetal ovary.

In the sexually mature female one oocyte, on average, completes growth each month and is ovulated in response to a midcycle surge of luteinizing hormone (LH). In response to the LH surge, the oocyte resumes meiosis, as outlined above. Whereas one group of chromosomes remains in the oocyte, the other is segregated to the first polar body. Meiosis I (MI) is immediately followed by meiosis II. The metaphase II-arrested cell is known as an egg, and it remains in arrest until it is fertilized or degenerates. Fusion of the sperm and egg plasma membranes at fertilization triggers the resumption and completion of MII so that sister chromatids segregate, with one group of chromosomes remaining in the egg and the other segregated into a second polar body. After the division, separate nuclear envelopes form around the remaining egg chromosomes and the chromosomes contributed by the sperm, forming a zygote. The chromosomes in the male and female pronuclei undergo DNA replication and condense in preparation for the first mitotic cleavage division.

Thus, in females only one of the four meiotic products develops into an oocyte, while the others become polar bodies that are not fertilized.

3.5.4.3 A Dynamic Oocyte Pool?

The view that the pool of oocytes in the ovary is established during fetal development and all eggs ovulated by the adult female initiate meiosis in the fetal ovary was recently challenged [124, 125]. Evidence for the existence of germline stem cells capable of giving rise to new oocytes in the adult was provided. At present there is a controversial debate about whether such "adult oocytes" with the capability to mature and ovulate exist beside "fetal oocytes."

3.5.4.4 Maternal Age and Aneuploidy

Chromosome anomalies are extraordinarily common in human gametes, with approximately 21% of oocytes and 9% of spermatozoa abnormal. The types of abnormalities are quite different, since most abnormal oocytes are aneuploid, whereas the majority of abnormalities in spermatozoa are structural. Chromosomes 21 and 22 (the smallest chromosomes) are overrepresented in aneuploid gametes in both oocytes and sperm. Chromosome 16 is also frequently observed in aneuploid oocytes, whereas the sex chromosomes are particularly predisposed to nondisjunction in human sperm. Maternal age is clearly the most significant factor in the etiology of aneuploidy, and most aneuploidy derives from errors in maternal meiosis I (Table 3.3). Paternal age does not have a dramatic effect on the frequency of aneuploid sperm; there is some evidence for a modest increase in the frequency of sex chromosomal aneuploidy. Meiotic recombination has a significant effect on the genesis of aneuploidy in both females and males. New techniques allowing the analysis of recombination along the synaptonemal complex have yielded interesting new information in healthy and infertile individuals, establishing a link between infertility and the genesis of chromosome abnormalities. Future studies will unravel more of the underlying causal factors.

At least 7–10% of clinically recognized pregnancies start with an abnormal set of chromosomes [102]. The incidence of chromosomal errors increases exponentially with advancing maternal age. Among women in their early twenties, the risk of trisomy in a clinically recognized pregnancy is 2–3%, but among women in their forties the risk increases to 30–35%. Studies of the origin of human trisomies demonstrated that age influences the likelihood of errors at both meiosis I (MI) and meiosis II (MII) [257]. However, despite considerable research effort, the effect of maternal age remains the "black box" of human aneuploidy. To date, no single hypothesis has provided a satisfactory explanation for this phenomenon. Instead of rather simple terms of cause and effect a more complex picture has emerged, suggesting that female fertility is influenced by a series of events occurring at different stages of egg development. Three vulnerable stages of oogenesis can be distinguished: first, the meiotic prophase events of synapsis and recombination occurring in the fetal ovary; secondly, follicle formation during the second trimester of fetal development; thirdly, oocyte growth in the adult ovary [113]. In fact, eggs may be vulnerable at each of these stages, and environmental influences are likely to have an additional impact [113]. Thus, future research efforts aiming at a better understanding of each of these phases

Table 3.3 Summary of studies of the origin of human trisomies (from [102])

Trisomy	N	Maternal MI (%)	MII (%)	Paternal MI (%)	MII (%)	PZM (%)
Acrocentrics						
13	74	56.6	33.9	2.7	5.4	1.4
14	26	36.5	36.5	0.0	19.2	7.7
15	34	76.3	9.0	0.0	14.7	0.0
21	782	69.6	23.6	1.7	2.3	2.7
22	130	86.4	10.0	1.8	0.0	1.8
Nonacrocentrics						
2	18	53.4	13.3	27.8	0.0	5.6
7	14	17.2	25.7	0.0	0.0	57.1
8	12	50.0	50.0	0.0	0.0	50.0
16	104	100	0.0	0.0	0.0	0.0
18	150	33.3	58.7	0.0	0.0	8.0
XXX	46	63.0	17.4	0.0	0.0	19.6
XXY	224	25.4	15.2	50.9	0.0	8.5

MI meiosis I, *MII* meiosis II, *PZM* postzygotic mitotic

may unravel factors involved in the close relationship between maternal age and increasing aneuploidy.

3.5.5 Nonallelic Homologous Recombination During Meiosis Can Cause Microdeletion/ Microduplication Syndromes

DNA rearrangements during meiosis may occur owing to homologous recombination involving region-specific, low copy repeats. This is also referred to as "nonallelic homologous recombination" (NAHR) between similar sequences (repeats) present at more than one site in the genome. Nonallelic homologous crossover was first described by Sturtevant [227] and Bridges [34] in the bar locus in *Drosophila*. The NAHR mechanism was described and documented in alpha thalassemias, beta thalassemia Lepore, and the growth hormone cluster (see Sect. 3.2.7 on unequal crossing over). More recently, genomic rearrangements attributable to NAHR have been referred to as genomic disorder [150], because they are caused by special conditions that result from genome architecture.

NAHR occurs during meiotic crossing over between homologous chromosomes each carrying two repeats and separated by interrepeat DNA. Recombination between direct repeats results in deletion and/or duplication, whereas recombination between inverted repeats results in an inversion (Fig. 3.35). NAHR can also occur between repeats on the same chromosome or between repeats on different chromosomes, and the specific genetic outcome of NAHR (duplication, deletion, inversion, or chromosomal translocation) depends on the chromosomal location of the repeats and whether the repeats are oriented head to tail, head to head, or tail to tail with respect to one another (reviewed in [224]). Large literature exists showing that many human diseases and syndromes arise by NAHR (reviewed in [224]). Variations in DNA sequence copy

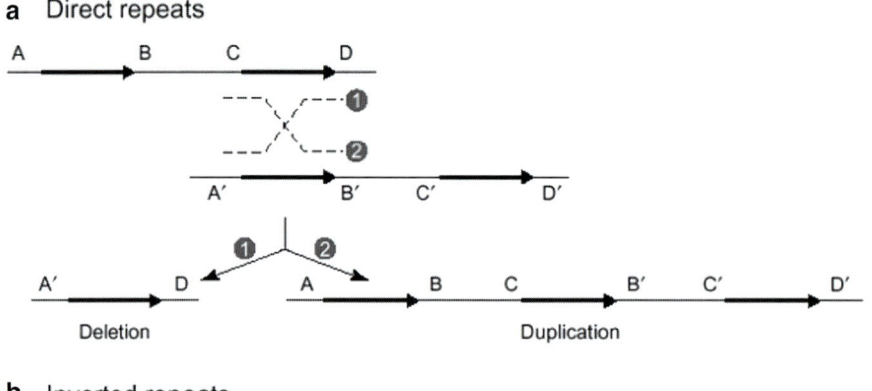

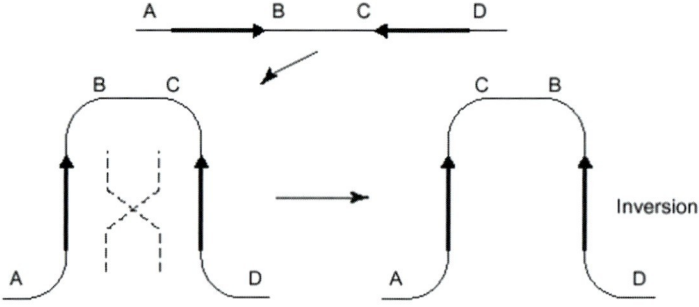

Fig. 3.35 Duplication and deletion formation by NAHR. The *black arrows* represent repeated sequences, and the orientation of these repeats is indicated by the direction of the *arrowhead*. The inter-repeat DNA is depicted as *thin lines*; *capital letters* point to the flanking unique sequences and illustrate consequences of genomic rearrangements. Direct repeats (*a*) result in deletion and/or duplication, whereas inverted repeats (*b*) result in an inversion. From [150], with permission from Elsevier

number in the normal population may also arise by NAHR (see also Sect. 3.5.5).

3.5.6 Molecular Mechanisms Involved in Meiosis

During meiosis spindle formation and chromosome segregation have to proceed in a highly reliable fashion. The spindle assembly checkpoint (SAC) (see Sect. 3.3.3) monitors attachment to microtubules and tension on chromosomes not only in mitosis but also in meiosis. Thus, many components involved in mitosis are also important during meiosis. At the same time differences exist. It would be beyond the scope of this chapter to review the entire current knowledge about the signaling cascade of the SAC for normal chromosome segregation and meiosis here, but some important components will be summarized.

During meiosis two consecutive rounds of nuclear division are required to first segregate homologous chromosomes (at anaphase I) followed by the segregation of sister chromatids (at anaphase II) for the formation of a haploid gamete.

Meiotic prophase I is an especially long and complex phase, because homologous recombination occurs between homologous chromosomes. Formation of chiasmata, which hold homologous chromosomes together until the metaphase I to anaphase I transition, is critical for proper chromosome segregation. Sites of double-stranded DNA breaks (DSBs) have to be generated that are thought to be the starting points of homologous recombination. Processing of these sites of DSBs requires sophisticated repair mechanisms involving the function of RecA homologs, such as RAD51, DMC1, and others. Failure to repair these meiotic DSBs results in abnormal chromosomal alternations, leading to disrupted meiosis.

To pull homologous chromosomes to opposite spindle poles during meiosis I, both sister kinetochores of a homolog must establish attachment to the same pole (syntelic attachment). Sister chromatids of meiotic chromosomes are held together all along the chromosome arms and centromeres by the meiosis-specific cohesin complexes containing REC8. The kinase PIK1 marks REC8 at chromatid arms by phosphorylation for degradation by separase. After silencing of the SAC, at the metaphase I-to-anaphase I transition, separase cleaves REC8 to release cohesin between sister chromatid arms while centromeric cohesin are kept unphosporylated by activity of the phosphatase PP2A. A complex of Shugoshin (SGO) and PP2A is recruited by BUB1 to centromeres at meiosis I to protect Rec8 from phosphorylation by PLK1. The spindle assembly checkpoint (SAC) is constitutively active during prometaphase of meiosis I when sister kinetochores of homologous chromosomes are either co-oriented to the same pole or not fully attached to spindle microtubules and under tension from kinetochore microtubules.

The SAC is also reactivated and halts cell cycle progression when attachment to the spindle is lost at metaphase I by inhibiting the APC/C, as is also characteristic for mitosis. Upon chromosome congression, when chromosomes are under full tension from spindle fibers at metaphase I, the SAC is inactivated and cyclin B and securin become degraded after ubiquination by APC/C. Separase cleaves meiotic cohesin REC8 marked by phosphorylation by PLK1 along chromosome arms.

MPF becomes active again at prometaphase II. However, in the presence of CSF, the mammalian oocyte arrests at metaphase II in spite of aligned chromosomes until fertilization triggers progression into anaphase II for completion of meiosis (reviewed in [182]).

3.6 Human Chromosome Pathology in Postnatal Diagnostics

3.6.1 Syndromes Attributable to Numeric Anomalies of Autosomes

3.6.1.1 Mechanisms Creating Anomalies in Chromosome Numbers (Numerical Chromosome Mutations)

Anomalies in chromosome numbers may be caused by various mechanisms:

(a) The most important mechanism is nondisjunction. Chromosomes that should normally be separated during cell division stick together and are transported in anaphase to one pole. This may occur at mitotic division and during meiosis. Meiotic nondisjunction was discovered by Bridges in 1916 in *Drosophila* [33].). For every gamete with one additional chromosome, another one is formed with one chromosome fewer (Fig. 3.36). After fertilization with a normal gamete the zygote is either trisomic or monosomic. Somatic nondisjunction in mitotic cell division during early development may lead to mosaics with normal, trisomic, and monosomic cells.

(b) A second mechanism leading to numerical abnormalities is loss or gain of single chromosomes. Such alterations in whole chromosomes may be facilitated by errors in the mitotic spindle checkpoints. Chromosome losses can become visible as "anaphase lagging" when one chromosome may lag behind the others. Chromosome loss leads to mosaics with one euploid and one monosomic cell population. Data from preimplantation aneuploidy screening studies suggest that the pronucleus stage and the first cleavage stages are especially susceptible to chromosome segregation errors.

(c) A third mechanism is polyploidization. Here all chromosomes are present more than twice in every cell. For example, in triploid cells the chromosome number is 3n=69. In prenatal diagnosis, triploidy may be observed in early pregnancy (1–3% of recognized pregnancies). However, the vast majority (more than 99%) are lost as first-trimester miscar-

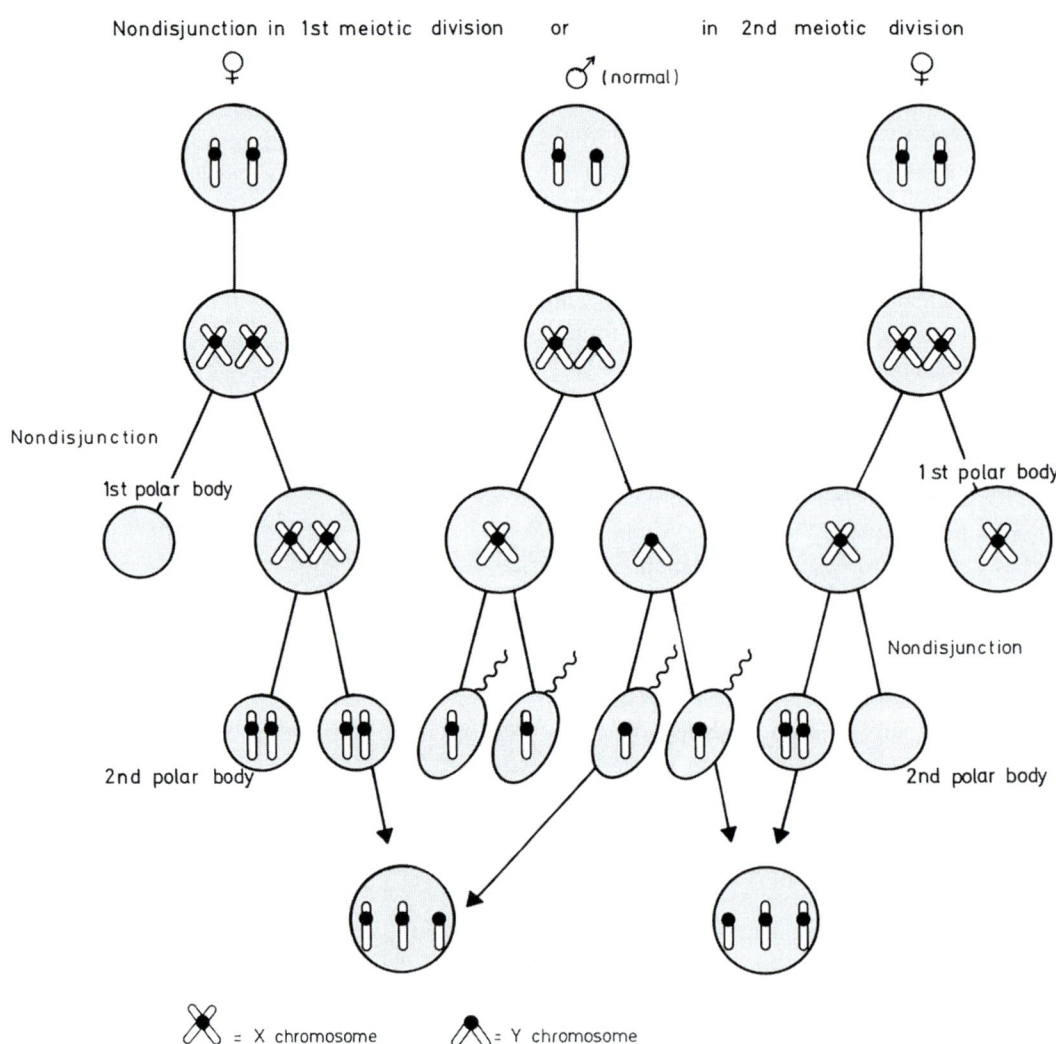

Fig. 3.36 Nondisjunction of the X chromosome in the first (*left*) and second (*right*) meiotic division in a woman. Fertlization by a normal Y sperm. An XXY individual can result from either first or second meiotic division nondisjunction. (From the 3rd edition of this work)

riage or second-trimester fetal death in utero. In some human somatic cell types, such as liver cells, heart muscle cells, or megakaryocytes, polyploidization may occur physiologically. Furthermore, polyploidization is also often observed in tumor cells.

An abnormal number of chromosomes in a cell (aneuploidy) increases the risk of further irregularities. Almost all conditions that result from chromosome imbalances affect multiple systems and produce both structural and functional defects.

There are only three autosomal trisomy syndromes which can be present in all somatic cells and result in the live birth of children, and these are the trisomies 21 (Down syndrome), 13 (Patau syndrome), and 18 (Edwards syndrome). All other autosomal trisomies are assumed to be invariably lethal, and they can be observed by applying chromosome studies in spontaneous abortion material. If autosomal trisomies other than 13, 18, or 21 are observed postnatally, they are usually present only as mosaics, i.e., together with a normal cell line in live newborns. One reason why chromosomes 13, 18, or 21 may result in a live birth is likely their low gene density. For example, chromosomes 18 and 19 have about the same size. However, chromosome 18 has approximately 5.3 genes/Mb, whereas chromosome 19 has approximately 25 genes/Mb. Thus, chromosome 18 is one of the gene-poorest human chromosomes and similar considerations are also true for chromosomes 13 and 21.

In contrast, there is no autosomal monosomy which can result – again with the exception of possible mosaic constellations – in a live birth.

3.6.1.2 Down Syndrome

With an incidence at birth of 1–2/1,000, Down syndrome is the most frequent chromosome aberration syndrome in humans, and a common condition encountered in genetic counseling services. Down syndrome was named after the British physician J. Langdon Down, who recognized the phenotype as a clinical entity in 1866, i.e., long before trisomy 21 was discovered as the genetic basis. In fact, the association between trisomy 21 and Down syndrome was only established in 1959 by Lejeune [142].

Paradigm for Meiotic Nondisjunction: In the majority of cases trisomy 21 is caused by meiotic maternal nondisjunction, which is related to the age of the mother. This nondisjunction is possibly associated with errors in recombination and age-dependent loss of cohesion of meiotic chromosomes [7, 8, 102, 247]. In fact, about 68% of nondisjunctions leading to the conception and birth of children with Down syndrome occur during maternal meiosis I, about 20% during maternal meiosis II, about 4% during paternal meiosis II, and about 3% during paternal meiosis I. The rest (approximately 5%) are postzygotic events, i.e., represent mitotic chromosome segregation errors [7, 8, 102, 247].

Standard Karyotype in Down Syndrome: Trisomy of the entire chromosome is present in about 95% of cases, whereas in a small fraction of cases trisomy of only a part of chromosome 21 leads to Down syndrome. A standard karyotype of a human with Down syndrome is illustrated in Fig. 3.37. Such a free translocation will be found in the vast majority (about 92.5%) of humans with Down syndrome. In about 5% of cases the additional chromosome 21 will participate in formation of a Robertsonian translocation with another acrocentric chromosome, with chromosomes 14 and 21 being the most frequent translocation partners. Very occasionally (<1%) chromosome 21 is translocated to another chromosome. In about 3–5% chromosome 21 is found in mosaic constellations, these cases are typically the results of postzygotic segregation errors.

Phenotypic Variability of Down Syndrome: Some aspects of the phenotype appear to occur in every person with Down syndrome, whereas other traits are more highly variable [9]. However, even for phenotypes that occur in every individual there is variability in expression. Cognitive impairment is apparent in all persons with Down syndrome, but ranges from mild to moderate. By contrast, congenital heart defects occur in only 40%, and its severity also varies. In spite of this appreciable variability of signs, the clinical diagnosis is rarely in doubt for experienced clinicians. The clinical diagnosis can be made at birth based on the presence of marked hypotonia and several minor dysmorphic features, which are especially visible in the craniofacies, hands, and feet (Fig. 3.38). Importantly, there is a temporal dimension to the evolution of the phenotype, which changes with age [6] and which is summarized in Table 3.4. Life expectancy of Down syndrome individuals is reduced. However, and especially because of more efficient treatment of congenital heart diseases by surgery and by compensating the compromised immune system by immuniza-

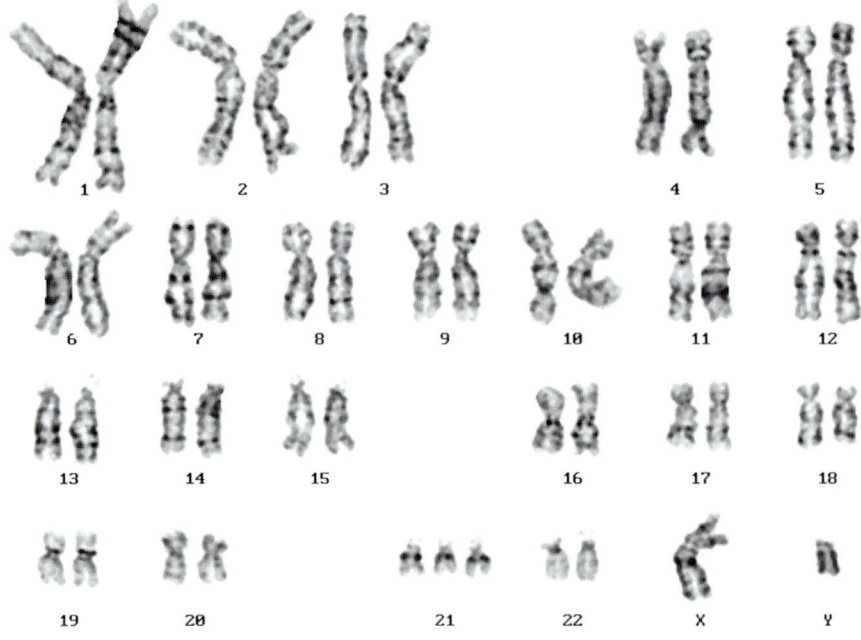

Fig. 3.37 G-Banded chromosome with a free trisomy 21

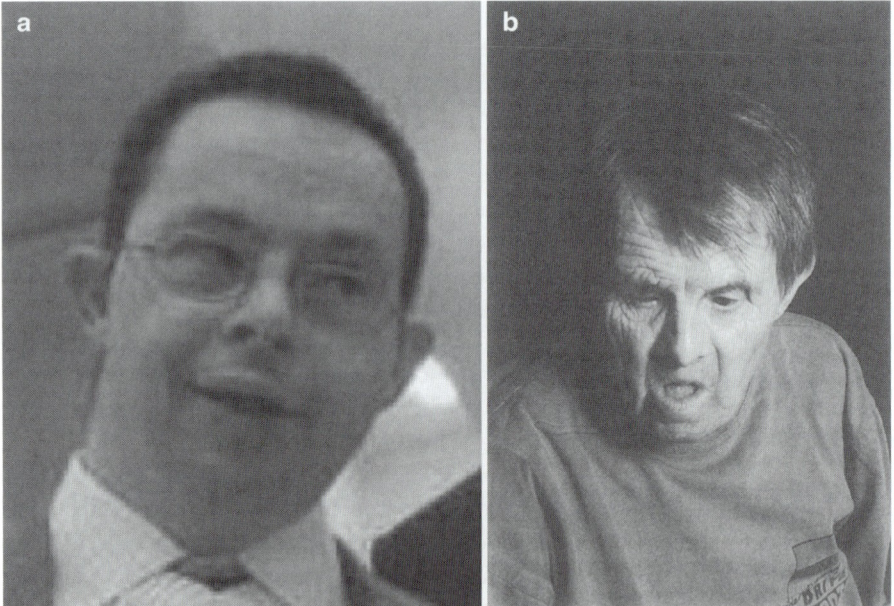

Fig. 3.38 Adults with Down syndrome. (**a**) This person was one of the original cases described by Lejeune [142]. (From [6], with permission from Elsevier) (**b**) Man with Down syndrome at the age of 64. (Courtesy of Dr. G. Tariverdian, from the 3rd edition of this work)

tions and antibiotics, the life expectancy has been rapidly increasing over the last 20 years.

Males with Down syndrome usually have no children and are infertile, likely as a result of defective spermatogenesis. However, there are a few case reports of confirmed paternity in offspring of males with apparently nonmosaic trisomy 21 [192, 215]. These occasional reports are likely very rare exceptions from

Table 3.4 Characteristics of Down syndrome (from [6], with permission from Elsevier)

At birth	Infancy and childhood	Adulthood
Structural		
Dysmorphic features	Growth retardation and obesity	
Congenital heart disease		
Duodenal stenosis or atresia		
Imperforate anus		
Hirschsprung disease		
Central nervous system		
Hypotonia	Developmental and mental retardation	Decrease in cognitive function
	Decreased sensitivity to pain	Alzheimer disease
Immune and hematopoietic systems		
Transient myeloproliferative disorder	Leukemia	
	Immune defects and/or infection	
Other	Thyroid dysfunction	Male sterility
		Reduced longevity

the rule. By contrast, several women with this syndrome have been reported to have offspring. The subfertility in Down syndrome females can likely be attributed to the reduced number of ovarian follicles and the increased rate of atresia. A review summarizing 29 pregnancies in 26 nonmosaic trisomy 21 females reported that 10 offspring also had Down syndrome, 18 offspring were chromosomally normal but 2 of these were mentally retarded, and 4 had other congenital abnormalities. The other pregnancies ended in spontaneous abortions or premature birth followed by death of the babies [215].

Down Syndrome as a Multigene Disorder: The length of the long arm (21q) of human chromosome 21 is 33.6 Mb and represents about 1% of the total euchromatic genome sequences. Current estimates are that 21q harbors 253 protein-coding genes (www.ensembl.org; Ensembl release 52, December 2008); however, this number may be amended as our knowledge of chromosome 21 progresses. Current research is directed at the question of whether only a few of the chromosome 21 genes with a major phenotypic effect are involved in determining the phenotype of Down syndrome or whether the phenotype results from the interaction of several genes of modest effect [6]. At present, based on studies in animal models, the former option may be correct, suggesting that the number of genes involved in Down syndrome is likely considerably lower than their total number on the chromosome. Hence, Down syndrome is likely a disorder of gene dosage caused by increased amounts of the products of the genes on chromosome 21. An example for such a gene-dosage hypothesis is the amyloid precursor protein (*APP*) gene, whose increased dosage is likely involved in the development of Alzheimer disease, which is invariably present in humans with Down syndrome from 35 years of age onward. In addition to protein-coding genes, functional non-protein-coding DNA elements may contribute to phenotypic features, but the exploration of their roles is in their early infancy.

Down Syndrome and Malignant Diseases: Infants with Down syndrome have a 20-fold increased risk of developing acute leukemia, most commonly acute megakaryoblastic (M7) leukemia (AMKL). This is often preceded by a neonatal leukemoid reaction (transient myeloproliferative disorder), which might be a form of transient leukemia. This observation gave rise to the hypothesis that humans with Down syndrome may have a higher risk of aneuploidy owing to mitotic disturbances in blood stem cells. However, and in contrast to such a hypothesis that cells with an autosomal trisomy may be more prone to chromosome segregation errors, Down syndrome individuals are less prone to developing solid tumors [205, 256]. Thus, trisomy 21 appears to increase the risk for a special form of leukemia, while it apparently has a protective effect against other malignancies. Gene dosage alterations caused by the trisomy 21 alone appear not to be sufficient to cause AMKL, as current transformation models suggest that in addition somatic mutations in an X-chromosomal gene, i.e., *GATA1* (GATA-binding protein 1), which encodes an essential transcriptional regulator of normal megakaryocytic differentiation, are required [106]. This unique nonactivating mutation in the *GATA1* gene is associated with both the leukemia and the leukemoid reaction. To explain the reduced incidence of many cancer types in individuals with Down's syndrome, innovative combinations of mouse models were used and these models revealed that the expression level of only one gene, *Ets2*, correlated directly with its copy number and inversely with tumor number in a model for colon cancer [230]. Furthermore, another gene, Down's syndrome candidate region-1 (*DSCR1*, also known as *RCAN1*), which encodes a protein that suppresses vascular endothelial growth factor (VEGF)-mediated angiogenic signaling by the calcineurin pathway, was shown to be increased in Down syndrome tissues. In a mouse model a modest increase in expression afforded by a single extra transgenic copy of *Dscr1* was sufficient to confer significant suppression of tumour growth in mice. This resistance may be a consequence of a deficit in tumor angiogenesis arising from suppression of the calcineurin pathway. Thus, increased expression of *DSCR1* could be another explanation for the reduced cancer risk [17].

3.6.1.3 Other Autosomal Trisomies

Patau et al. [178] described the first case of an autosomal trisomy, i.e., trisomy 13, other than trisomy 21. At about the same time, trisomy 18 was also discovered by Edwards et al. [68]. These trisomies are much less frequent than Down syndrome (about 1:3,000 and 1:5,000, respectively), and both show a maternal age effect. Both syndromes are clinically severe conditions; about 85–90% of liveborns do not survive beyond 1 year of life. For both syndromes there are some case reports describing long-term survival. The main signs and symptoms of both trisomies are shown in Figs. 3.39 and 3.40.

3.6.1.4 Triploidy

In triploidy (3n = 69) there is a double (2n) chromosomal contribution to the conceptus from one parent. Diandry may be due to either two sperms simultane-

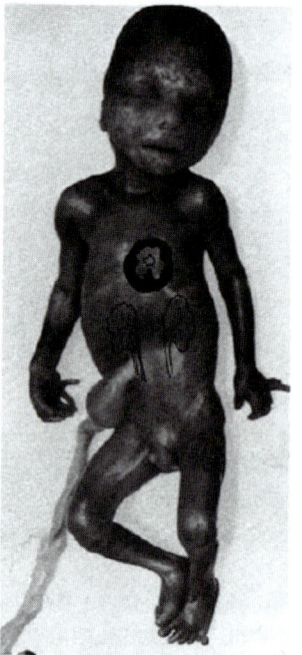

Frequent findings:

Growth retardation
Cardiac malformations (80%)
Holoprosencephaly (60-70%)
Microphthalmia/anophthalmia (60-70%)
Cleft lip/palate (60-70%)
Postaxial polydactyly (60-70%)
Cutis aplasia (scalp defects)
Omphalocele
Kidney malformations
Severe mental retardation

Fig. 3.39 Main clinical findings of trisomy 13. (From the 3rd edition of this work)

Frequent findings:

Growth retardation
Occipital-frontal circumference < 3rd percentile
Cardiac malformations (>90%)
Short sternum with reduced number of ossification centers
Prominent occiput
Low-set, malformed auricles
Clenched hand, tendency for overlapping of index finger over third, fifth finger over fourth
Hypoplasia of nails
Inguinal or umbilical hernia and/or diastasis
Severe mental retardation

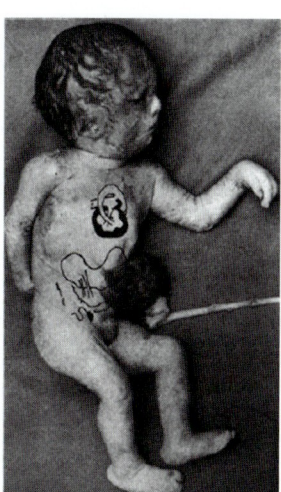

Fig. 3.40 Main clinical findings of trisomy 18. (From the 3rd edition of this work)

ously fertilizing the ovum or of a diploid sperm from a complete nondisjunction in spermatogenesis. Digyny may be the consequence of a complete nondisjunction at either the first or the second meiotic division in oogenesis, of retention of a polar body, or of the fertilization of an ovulated primary oocyte. More than 99% of pregnancies starting with a triploid chromosome set are lost during the first or second trimester. There are only a few anecdotal case reports of live births of triploid newborns, and one exceptional report of a triploid infant who survived for 10.5 months [214].

3.6.1.5 Mosaics

Individuals with two or more genetically different cell populations are referred to as mosaics. This is found relatively often in numerical chromosome aberrations of the sex chromosomes, but also in autosomal aberrations. A mosaic may be formed either by mitotic nondisjunction or by loss of single chromosomes caused by anaphase lagging.

An individual with an aneuploid cell line in only some of her or his tissues is likely to have a less severe phenotype than someone with a nonmosaic aneuploidy. For example, mosaic Down syndrome (i.e., the coexistence of both 47,N,+21 and 46,N cell lines) can result in a less obvious phenotype and with a lesser compromise of intellectual function than in standard trisomy 21. Most aneuploidies of autosomes can only exist in mosaic state because the nonmosaic forms are lethal in utero. If the distribution of the aneuploid cell line is asymmetric, the body shape may be asymmetric in consequence. In sex chromosome mosaicism, fertility can exist when otherwise infertility is the rule, which often represents a likely explanation for fertility reported in individuals with Turner syndrome (45,X/46,XX mosaic instead of 45,X) or with Klinefelter syndrome (47,XXY/46,XY instead of 47,XXY). The diagnosis of mosaics is often difficult, and those with a more obvious phenotypic defect are more likely to be detected. Furthermore, diagnosis is hampered by the obvious fact that many tissue sources are not readily available for a detailed chromosomal analysis.

3.6.2 Syndromes Attributable to Structural Anomalies of Autosomes

3.6.2.1 Karyotypes and Clinical Syndromes

3.6.2.1.1 Breaks, Gaps, and Rearrangements

Chromosomes must first be broken to form any kind of rearrangement. A chromosome may break at any stage of the cell cycle. If a chromosome breaks during G_1 and remains unrepaired through S-phase, the result will be visible in both chromatids as a chromosome break in the next metaphase (Fig. 3.41). If the fragment is not displaced, a gap may become visible. A break not affecting the centromere produces a shorter chromosome with a centromere and an acentric fragment. This fragment may or may not form a small ring (Fig. 3.41) but, lacking a centromere, it runs a high risk of being lost during the subsequent mitosis. Hence, chromosome breakage often leaves behind a cell deficient in a chromosome segment. In most cases, such a loss will be lethal for the respective cell.

When a break occurs during G_2, it usually involves only one of the two chromatids and is therefore called a chromatid break. A single break yields a deleted chromatid and an acentric fragment. Chromatid breaks in two chromosomes can lead to chromatid exchanges and result in quadriradial configurations, which may give rise to interchromosomal rearrangements, which are discussed below.

3.6.2.1.2 Interchromosomal Rearrangements (Interchanges)

In many cases, joining occurs between different chromosomes, homologous or nonhomologous. If breakage occurs in the G_1 phase, joining follows in the G_1 (or early S)-phase before DNA replication. If each of the resulting chromosomes happens to have one centromere, the translocation chromosomes may pass through the next mitosis without difficulties. If one of the resulting chromosomes happens to get two centromeres, a dicentric chromosome is formed. Depending on the exact mode of replication, it may be able to pass the next mitosis, under the following conditions: (a) the two centromeres of the dicentric chromosome migrate to the same pole and (b) replication and sister chromatid exchange between the two centromeres has not led to intertwining of the two chromatids (Fig. 3.42a). However, if the centromeres migrate to opposite poles, anaphase bridges are formed, and the chromosomes may break (Fig. 3.42b, c).

If two breaks occur during G_2, with the subsequent formation of a quadriradial configuration, there are two types of segregation: one in which alternate chromatids will segregate to opposite poles and one in which adjacent chromatids will segregate to opposite poles (Fig. 3.42d). The mitotic anaphase proceeds without further difficulties if the two centromeres happen to be located on different elements (Fig. 3.42d; classes I, III, and V). If the centromeres are located on the same configuration, the resulting daughter cell is in any case aneuploid. Either the centromeres migrate to different poles, in which case an "anaphase bridge" is formed, and the chromosome finally breaks again, or the two centromeres migrate to the same pole, which can only happen with nonhomologous reunions (Fig. 3.42d; classes V, VI, and VII). In the latter case the problem is postponed to the next mitosis, in which the chromosome appears to be dicentric. It may or may not survive this mitosis. In any case, under the conditions mentioned above interchanges cause a great deal of cell loss owing to aneuploidy or mitotic disturbance.

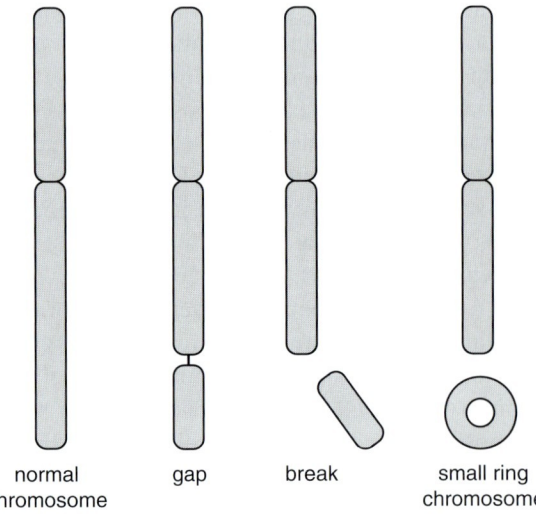

Fig. 3.41 Possible results of G_1 breaks in one chromosome: from *left* to *right*: normal chromosome, gap, break visible as small acentric fragment, and formation of a small ring

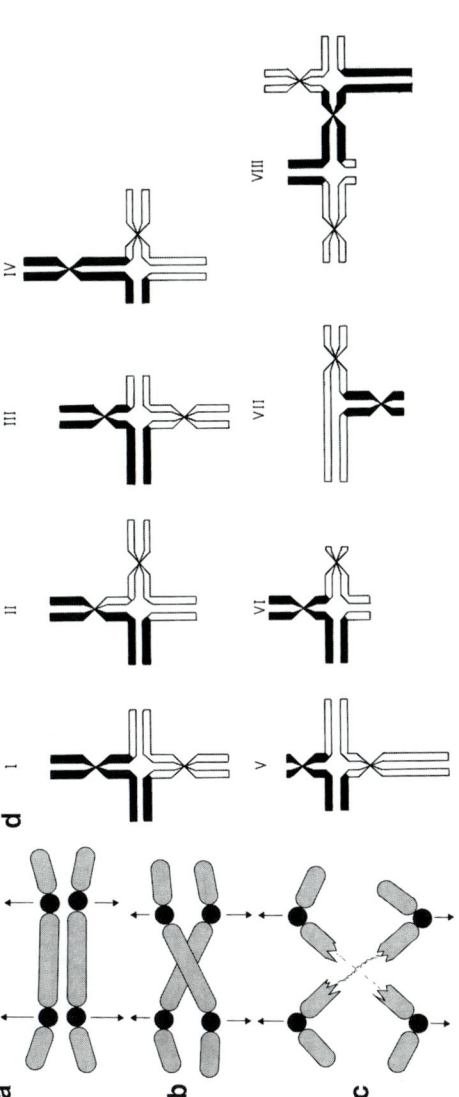

Fig. 3.42 (**a–d**) Mitotic anaphase of a dicentric chromosome (**a–c**) and classes of interchanges found after translocation during G_2-phase (**d**). (**a**) If both centromeres migrate to the same pole the dicentric chromosomes will remain intact. (**b**) If the centromeres are pulled to opposite poles anaphase bridges are formed. (**c**) The chromosomes are broken. (**d**) Illustrations of possible involvements of two homologous chromosomes: *I* Alternate position of the centromeres; exchange of fragments of equal length. *II* Adjacent position of the centromeres; exchange of fragments of equal length. *III* Alternate position of the centromeres; exchange of fragments of different length. *IV* Adjacent position of the centromeres; exchange of fragments of unequal length. Involvement of two non-homologous chromosomes. *V* Alternate position of the centromeres. *VI* Triradial configuration (loss of fragments required). Complex of non-homologous chromosomes (more than two). *VIII* One example of a figure with three chromosomes involved. (From the 3rd edition of this work)

3.6.2.1.3 The Impact of Spatial Proximity on Translocations

Genomes are nonrandomly arranged within the cell nucleus. In higher eukaryotes, each chromosome occupies a distinct, spatially limited space within the interphase nucleus, referred to as "chromosome territory." The position of each territory within the nuclear space is nonrandom. In human cells, the location of chromosomes has been linked to their gene density, with gene-rich chromosomes preferentially located toward the interior of the nucleus, whereas gene-poor chromosomes accumulate at the nuclear periphery. Similar preferential localization patterns have been detected for single genes [50, 54, 221].

A role for spatial proximity in the formation of chromosomal translocations seems obvious considering that the generation of a chromosomal translocation requires the physical interaction of the two translocation partners. Given the nonrandom spatial positioning of genes and chromosomes within the nucleus, the closer two genome regions are on average to each other, the higher is their probability of translocating with each other (Fig. 3.43). Thus, there is evidence for a role of spatial proximity in formation of translocations. Although chromosomes clearly occupy distinct territories, their edges intermingle and chromatin loops from one chromosome invade the territory of its neighbors, bringing genome regions from distinct chromosomes in intimate spatial proximity. The intermingling chromatin loops may be more susceptible to DSBs owing to their decondensed nature, and the zone between chromosome territories may be a preferential site of formation of chromosomal translocations among neighboring chromosomes [32].

Two fundamentally different models for how chromosomal translocations form have been developed (Fig. 3.44). In the contact-first model, translocations can only occur between DSB that are located in close spatial proximity at the time of breakage. In this model, physical association of translocating regions occurs before DSB formation. On the other hand, in a breakage-first model, translocations occur between DSBs located far apart. In this model, broken chromosome ends undergo large-scale motions to relocalize within the nucleus in search of suitable translocation partners and

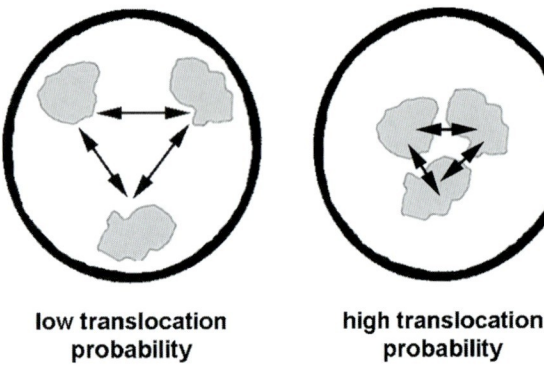

Fig. 3.43 The concept of spatial proximity as a contributor to translocation frequency: chromosome and genome regions that are preferentially located in close spatial proximity owing to the nonrandom organization of the genome in the cell nucleus have a higher probability of undergoing translocations with each other than chromosomes that are located distantly from each other. Adapted from [221], by permission of Oxford University Press

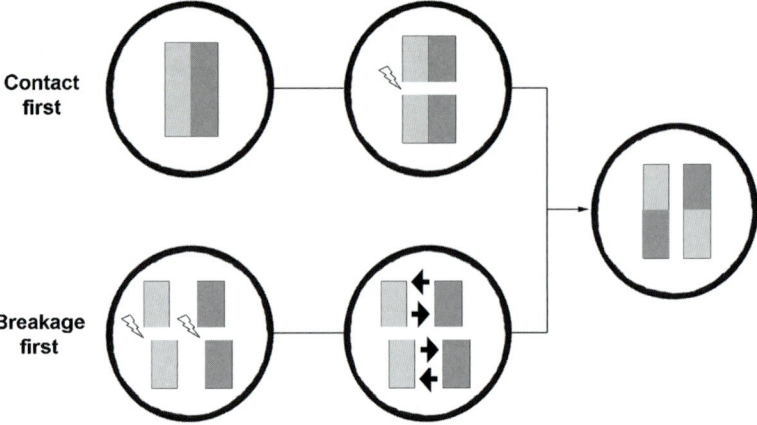

Fig. 3.44 The two models for formation of chromosome translocations: In the contact-first model, translocations occur among proximally positioned DSBs. In the breakage-first model, DSBs roam the cell nucleus in search of potential translocation partners. Adapted from [221], by permission of Oxford University Press

physical association between broken chromosome ends on different chromosomes occurs only after chromosomes suffer DSBs. At present there are experimental data supporting both models [221].

3.6.2.1.4 Segregation and Prenatal Selection of Translocations

The majority of constitutional balanced translocations have no consequences for the carrier. However, a translocation carrier may have an increased risk for a child with mental or physical abnormalities caused by segmental aneuploidy. Such a segmental aneuploidy consists typically of a segment of one of the participating translocation chromosomes, which is duplicated, and a segment of the other translocation chromosome, which is deleted. This results in both a partial trisomy and a partial monosomy. The reason is that at meiosis I, the four chromosomes with common segments form a special configuration, a "quadrivalent," which is best visible during pachytene stage (Fig. 3.45). Depending on which spindle attaches to which centromere, the distribution of the four homologous chromosomes to the daughter cells may vary. In fact, in theory 16 possible chromosomal combinations could be produced in the gametes of a translocation carrier (Fig. 3.45). The various modes of chromosome segregation can be described by the number of chromosomes, which go to the respective cells (i.e., a 2:2 segregation: two chromosomes go to one cell at a time; 3:1 segregation: three chromosomes are moved to one cell and one to the other; 4:0 segregation: four chromosomes go to one and none to the other cell). In addition, the terms "alternate" or "adjacent" describe which centromeres are moved to a daughter cell. "Alternate" describes the situation where these centromeres go alternately to one or the other pole. In contrast, "adjacent" signifies that centromeres that are next to each other travel together. In adjacent-1 segregation chromosomes with nonhomologous centromeres go to the same daughter cell, whereas in adjacent-2 segregation homologous centromeres are pulled to the same daughter cell.

As depicted in Fig. 3.45, only the alternate 2:2 segregation results in balanced daughter cells. In all other cases the gametes are in an imbalance, with the aforementioned double segmental aneuploidy. This may result in a miscarriage or in a child with mental retardation and dysmorphic features. A few translocations, especially those with small translocated chromosomal segments, are associated with a high risk of having an abnormal child.

3.6.2.1.5 Intrachromosomal Rearrangements (Intrachanges)

A single chromosome may break at two different sites, and the intermediate part may rejoin upside down. Inversions can be diagnosed by use of banding methods and/or in situ hybridization. A shift in centromeric position readily identifies pericentric inversions, whereas inversions in which the centromere is not included result in paracentric inversions (Fig. 3.46). Inversion heterozygotes are not particularly rare in human populations. There may be difficulties in chromosome pairing at meiosis, because crossing-over follows the reversed loop model, which allows optimal alignment and pairing of matching segments (Fig. 3.47). If in the case of a paracentric inversion a recombinant gamete is formed after a crossing over in the inverted segment, the chromosome would be either acentric (without centromere) or dicentric (two centromeres) (Fig. 3.47, left panel). Thus, in the case of a paracentric inversion, a recombination within the inverted segment cannot usually produce a viable unbalanced progeny. Few exceptions have been reported. For example, if the crossing-over within the inversion loop reverses upon itself, a "U-loop" is formed, which may result in either a duplication or a deletion. In contrast, a crossing over in the inversion loop of a pericentric inversion may lead to the production of two complementary recombinant chromosomes (Fig. 3.47, right panel).

Another type of intrachange is the ring chromosome. Here two telomeres are usually lost as fragments, and the open ends rejoin. Rings of almost every chromosome have been reported. Ring chromosomes often give rise to new variants. For example, a sister chromatid exchange may lead to a continuous double ring with one centromere. When the centromere divides in anaphase, the daughter centromeres may go to the same pole, with the result that one daughter cell has the double-sized ring while the other daughter cell has no ring. In contrast, if the centromeres move to opposite poles the ring may break randomly, and if the broken ends rejoin the daughter cells have rings of unequal size.

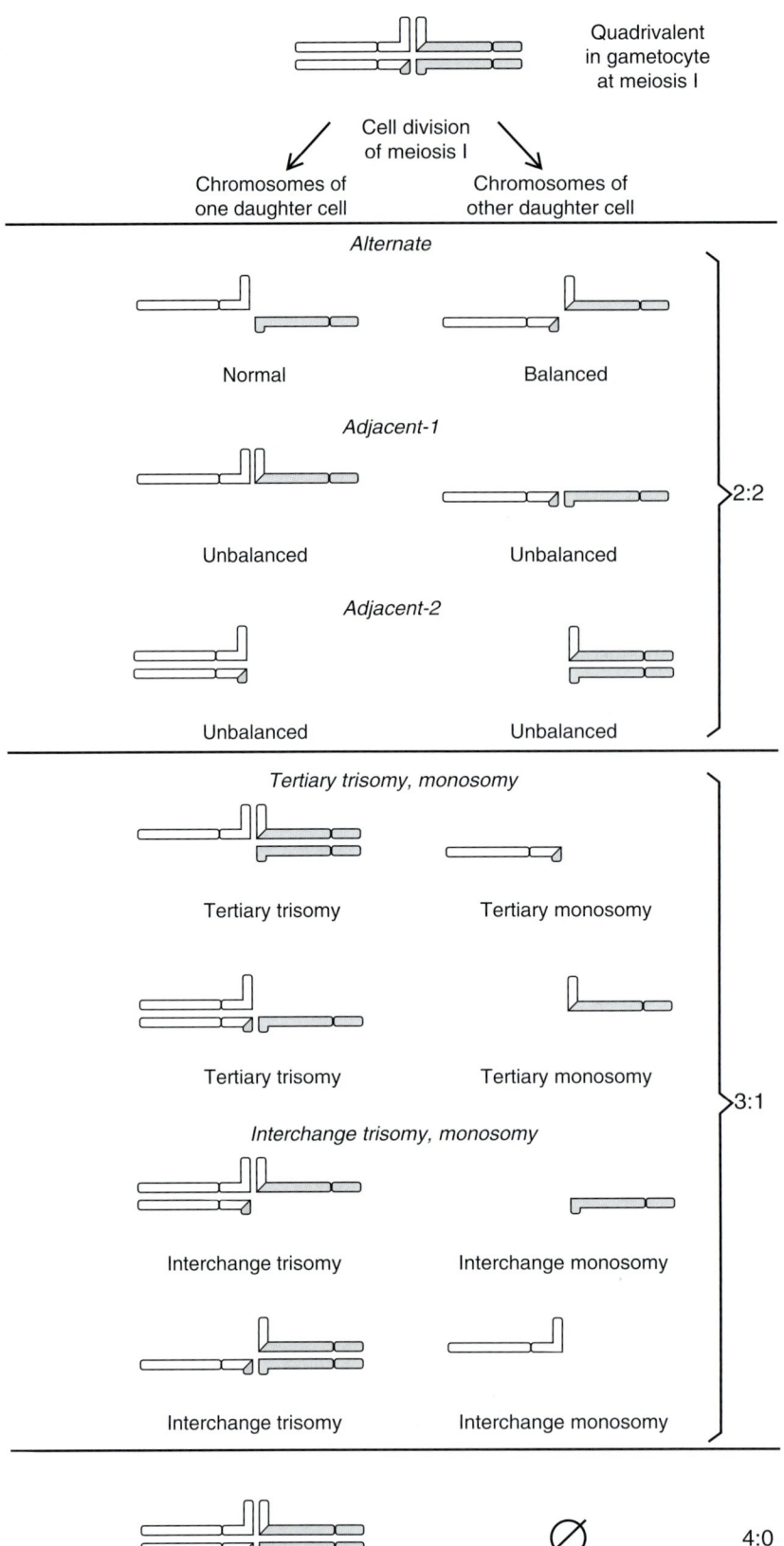

Fig. 3.45 At meiosis I, the four chromosomes with common segments form a quadrivalent. The distribution of the four homologous chromosomes is determined by which spindle attaches to which centromere. This determines different modes of segregation, such as 2:2 segregation (two chromosomes go to one cell at a time); 3:1 segregation (three chromosomes are moved to one cell and one to the other), or 4:0 segregation (four chromosomes go to one and none to the other cell). Adapted from [84]. By permission of Oxford University Press

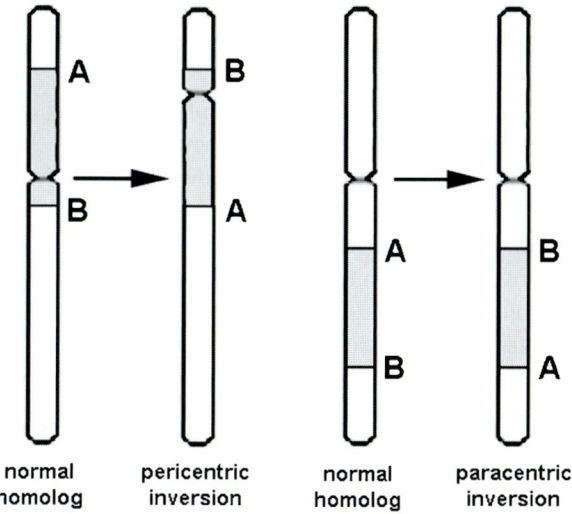

Fig. 3.46 Structure of pericentric (*left*) and paracentric (*right*) inversions

3.6.2.1.6 Deletion Syndromes

An individual who is heterozygous for a deletion is monosomic for a part of the chromosome. The term "deletion syndrome" usually refers to chromosomal losses that are large enough to be seen in standard banding analysis. In contrast, "microdeletions," as discussed in Sect. 3.5.5, are beyond the resolution limits of banding analysis and can therefore only be detected by other means, e.g., FISH or array-CGH.

De Grouchy et al. [58] were apparently the first to publish a case with del 18p-, in 1963. The first deletion syndrome was established by Lejeune et al. [143], also in 1963. They described three children with a deletion of the short arm of chromosome 5 (del 5 p-). In addition to the usual signs of autosomal chromosome aberration, such as developmental retardation and low birth weight, the children showed a moonlike face with hypertelorism. Their appearance was not extraordinarily peculiar, but they had a striking cry that resembled that of a cat (cri du chat = cat cry), which resulted in the name cri-du-chat syndrome.

There are various mechanisms by which a deletion may be formed: (a) true terminal deletion, (b) interstitial

Therefore, ring chromosomes are often unstable in both mitosis and meiosis and are frequently lost. Furthermore, rings can generate mosaics in which different cells contain different derivatives of the original ring.

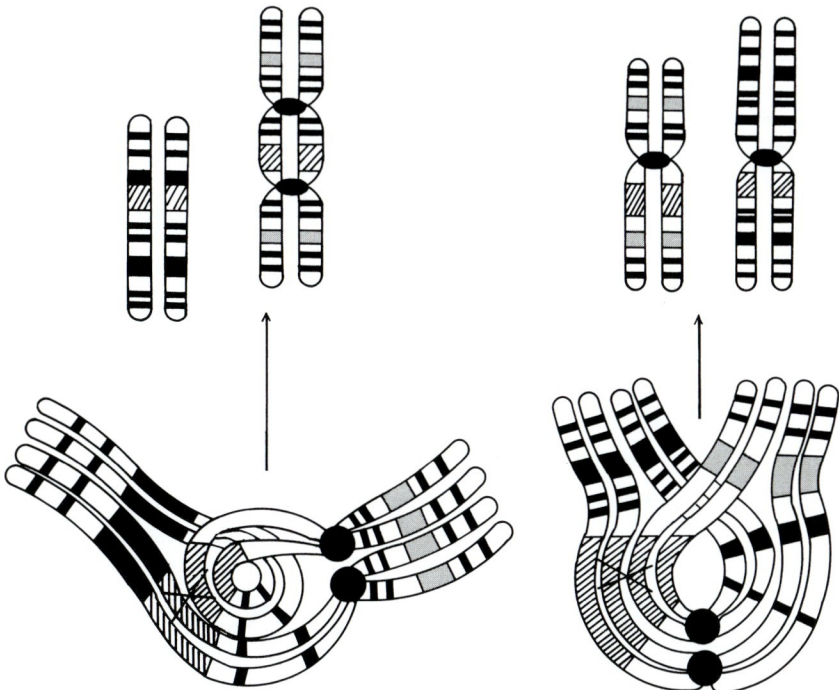

Fig. 3.47 Theoretical recombinant products from classical crossing-over in paracentric (*left*) and pericentric (*right*) inversions. In the two figures, crossing-over is assumed in the segments marked by an *X*. As a consequence, abnormal chromosomes are produced that lead to aneuploidy of zygotes in the next generation. (From the 3rd edition of this work)

deletion, and (c) unbalanced translocation. Array-CGH revealed that in patients with only 5p deletions different deleted regions exist, each with differing effect on retardation. Depending on size and location of the deletion, the level of mental retardation may range from moderate to profound [261].

There are a number of well-defined deletion syndromes, such as De Grouchy I (18p-), De Grouchy II (18q-), Wolf-Hirschhorn (4p-), Jacobsen syndrome (11q-), 1p36 deletion syndrome, 9q subtelomere syndrome, and so on.

3.6.2.1.7 Isochromosomes

Chromosomes consisting of two identical arms are occasionally found. Such chromosomes are known as isochromosomes and presumably originate by abnormal division of metaphase chromosomes, as shown in Fig. 3.48. Isochromosomes for the short arm or isochromosomes for the long arm may result.

Isochromosomes are observed relatively frequently for the X chromosome; an isochromosome of the long arm of the X, i(Xq) leads to the Turner syndrome, since this chromosome is always inactivated and only the normal X is active (Sect. 3.6.3.3).

3.6.2.1.8 Centric Fusions (Robertsonian Translocations)

Centric fusion is the most frequent type of chromosome rearrangement in human populations. The first reported cases of a translocation Down syndrome were due to centric fusion between the long arm of chromosome 21 and one chromosome of the group 13–15 or 21–22 (D or G group). Similar cases have since been observed repeatedly.

Thus, a Robertsonian translocation chromosome comprises the long arm elements of two different acrocentric chromosomes. These translocations are named after the American insect cytogeneticist Robertson, who first described translocations of chromosomes resulting from the fusion of two acrocentrics in his study of insect speciation in 1916.

Only the five acrocentric pairs undergo centric fusion. In the interphase nucleus the short arms and centromeric regions of these chromosomes are located close to the nucleolus, the short arms containing the nucleolus organizers, which carry genes for rRNA. This constellation appears to make acrocentric chromosomes prone to fusions. In fact, Robertsonian translocations are among the most common balanced structural rearrangements, with a frequency in newborn surveys of about 1 in 1,000. The great majority of Robertsonian translocations

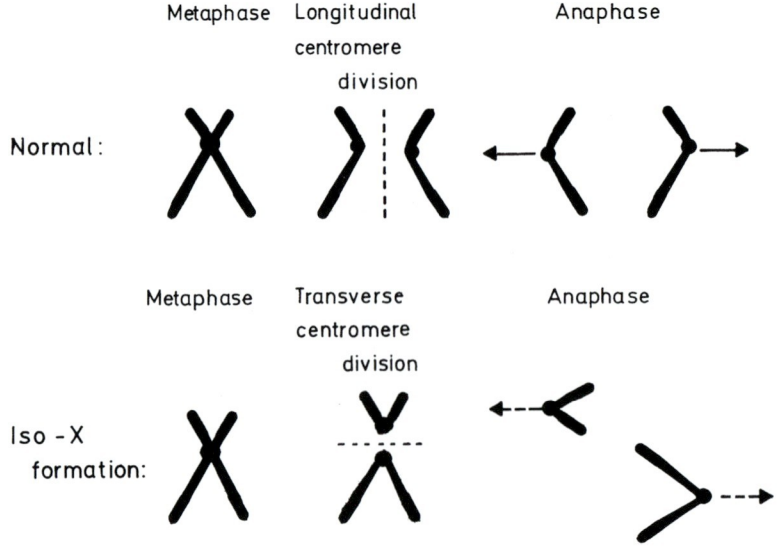

Fig. 3.48 Formation of an isochromosome by abnormal cleavage of the centromere. (From the 3rd edition of this work)

involve two different chromosomes (a heterologous or nonhomologous translocation). Translocations involving the fusion of homologs are very rare. The rob(13q14q) and the rob(14q21q) are predominant.

3.6.2.2 Small Deletions, Structural Rearrangements, and Monogenic Disorders: Genomic Disorders and Contiguous Gene Syndromes

3.6.2.2.1 Genomic Disorders

As pointed out in Sect. 3.5.5, the term "genomic disorder" was coined by Jim Lupski [150] and is typically used to describe a gain (duplication) or loss (deletion) of a specific chromosomal region, associated with a clinical genetic syndrome that may present with congenital anomalies, or with impairment in neurological and cognitive function (Fig. 3.35). A synonym for genomic disorders is "partial aneuploidies," which may be a more descriptive name [9]. Genomic regions associated with known human deletion and duplication syndromes are characterized by the presence of chromosome-specific, low copy repeats, or segmental duplications. Many of these segmental duplications flank the genomic regions that are prone to deletions and duplications and are the underlying basis for genomic disorders. Segmental duplications share a high level of sequence identity, predisposing the regions they occupy to nonallelic homologous recombination (NAHR), which may result in deletions, duplications, and inversions. NAHR may even allow intrachromosomal recombination events, which may result, for example, in translocations [150].

This mechanism explains why identical recurrent deletions/duplications may be observed in various, unrelated individuals and why these rearrangements often have the same size. Therefore phenotypic consequences of genomic disorders are usually very similar, often enabling clinical geneticists to suspect the presence of a specific microdeletion or microduplication based on the observed spectrum of clinical symptoms.

3.6.2.2.2 Contiguous Gene Syndromes

Contiguous gene syndromes refer traditionally to any deletions, i.e., deletions based on the aforementioned NAHR mechanism or deletions occurring randomly somewhere in the genome. Often these are interstitial chromosomal deletions, which contain several genes. The resulting phenotype may be the consequence of the combined haploinsufficiency of all genes in the deleted region, or of haploinsufficiency of only a few deleted genes, or even just of one of the deleted genes. It may be an especial challenge to assign deleted genes to specific phenotypic features. The extent and nature of phenotypic consequences depend on the genes encompassed by the deletion. Often, but not always, the main clinical features of microdeletions are mental retardation, growth retardation, craniofacial dysmorphy, and various congenital defects. Such a combination of symptoms should be clarified by high-resolution approaches, such as microarray analyses.

3.6.2.2.3 22q11.2 Deletion Syndrome

The 22q11.2 deletion syndrome has been characterized as one of the most frequent of the genomic disorders. Estimates about the prevalence of live birth are in the range of 1 in 2,000–4,000 [131]. Low copy repeats mediate genomic instability on chromosome 22 and result in recurrent deletion endpoints. While most patients have a 3-Mb heterozygous deletion, some rare patients have a nested 1.5- or 2-Mb deletion [70] (Fig. 3.49).

The 22q11.2 deletion can cause velo-cardio-facial syndrome (VCFS; MIM: 192430) [216] / DiGeorge syndrome (DGS; MIM: 188400) [61]. VCFS/DGS is characterized by multiple developmental anomalies occurring with varying severity, including craniofacial (cleft palate, velo-pharyngeal insufficiency; 69–100%), thymic, and parathyroid defects, including hypocalcemia (17–60%) and also cardiovascular malformations (70%) [131]. The disorder is fully penetrant, and everyone harboring the deletion is affected with some of the main clinical findings. However, variability is considerable and phenotypes may vary from life-threatening cardiovascular anomalies to mild craniofacial defects and learning disabilities. In some deletion carriers minor symptoms may be detected only after a careful physical examination. In addition, there is overwhelming evidence that children and adults with the 22q11.2 deletion have a characteristic behavioral phenotype, which is discussed in Chap. 23.6.

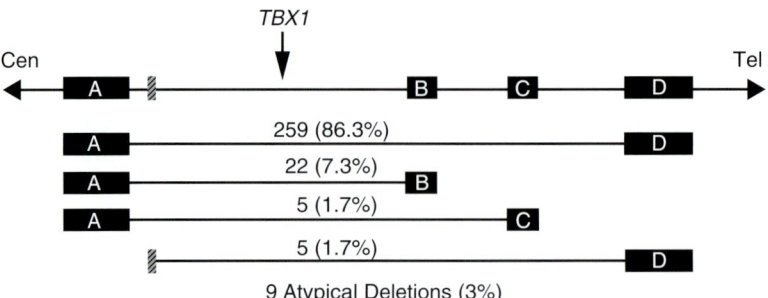

Fig. 3.49 Diagrammatic representation of the deletion region on 22q11.2 and deletion size in 300 patients. In this region there are four low copy repeats (*LCRs*), designated as A, B, C, and D, which are illustrated as *filled boxes*. The figure summarizes results from 300 patients. In 86.3% of these 300 individuals the deletions span the same 3-Mb region from LCR A-D. About 7.3% had a smaller 1.5-Mb deletion (between LCRs A and B), while 1.7% had a 2-Mb deletion (between LCRs A and C). In 3% atypical deletions were found. Adapted from [70]; data from [64]. Reprinted with permission of John Wiley & Sons, Inc.

All the defects in VCFS/DGS derive from the pharyngeal apparatus, which is a temporary embryological structure lateral to the developing head that contributes to diverse tissues of the head, neck, and thorax. Many birth defects, including a large fraction of those in congenital heart disease, derive from developmental problems of the pharyngeal system. Thus, the research interest in this syndrome is driven not only by the obvious clinical significance of the disease but also by a broader biological importance.

Because the 22q11.2 deletion is a contiguous gene syndrome, research is directed at elucidating the contribution of genes within the deleted region to the phenotype. A candidate gene within the deletion region, which may have the highest impact in the etiology of VCFS/DGS, is *TBX1*. This gene is a member of the T-box family of transcription factors, expressed in the pharyngeal apparatus. Inactivation of *Tbx1* in the mouse results in a phenocopy of the syndrome [122]. Furthermore, inactivating mutations in *TBX1* have been found in some patients with VCFS/DGS who were not carriers of the 22q11.2 deletion [255]. However, given the variable expressivity seen in the syndrome, it is obvious that modifying loci for VCFS/DGS must exist. Stochastic, environmental, and genetic factors likely modify the phenotype, and current research efforts are directed at the attempt to elucidate such genetic modifiers. Some potential modifier candidates were identified in the genetic pathway of *TBX1* or in pathways required for the development of the face, thymus, and parathyroid glands and heart [2].

3.6.2.2.4 22q11.2 Duplication Syndrome

The same mechanism as results in the 22q11.2 deletion syndrome, i.e., meiotic homologous recombination after misalignment of LCRs, should not only result in one deleted chromosome, but also in one chromosome containing a duplication of the deleted sequence (cf. Fig. 3.35). Accordingly, 22q11.2 deletions and duplications might be expected to occur with approximately the same frequency in the population. However, although individuals with 22q11 duplications have been identified, they seem to be fewer than anticipated on the basis of the prevalence of the deletion. One explanation for the relative lack of recognized 22q11.2 duplications compared with deletions may be the technical difficulties involved in detecting duplications by the standard cytogenetic analysis for 22q11.2 deletions (metaphase FISH). Similar to the various deletion sizes there is also a typical common approximately 3-Mb microduplication, approximately 1.5-Mb nested duplication, and smaller microduplications within and distal to the DiGeorge/velo-cardio-facial syndrome region.

The first report of a 22q11 duplication described a 4-year-old girl with failure to thrive, marked hypotonia, sleep apnea, and seizure-like episodes in infancy, who later showed delay of gross motor development with poor fine-motor skills, velo-pharyngeal insufficiency, and a significant delay in language skills. Her facial features were mildly dysmorphic, with a narrow face and down-slanting palpebral fissures. Hearing and vision were normal, and there were no detectable cardiac abnormalities [67].

In general, the phenotypes seen in these individuals are usually mild and highly variable, familial transmission is frequently observed. For example, in one family eight individuals over three generations who carried a 3-Mb duplication were identified. The duplication carriers showed intrafamilial phenotypic variation including heart defect, submucous cleft, intellectual disability, speech delay, behavior problems, and brachydactyly [259]. In another family, two duplication carriers were completely normal with high intellect [49]. Thus, it was noted that the delineation of a 22q11.2 microduplication syndrome may be due to ascertainment bias when seeking microdeletions of this region, and the authors of this study suggested that 22q11.2 microduplication could be either a nonpathogenic polymorphism or a syndrome with reduced penetrance [49].

3.6.2.2.5 Other Microdeletion/Microduplication Syndromes

Given the NAHR mechanism, which may cause recurrent deletions and duplications, the obvious question is how many regions there are in the human genome that may be similar to the aforementioned 22q11.2 region prone to being lost or duplicated during meiosis. Methods were developed to analyze *in silico* the public sequence database at the clone level for overrepresentation within a whole-genome shotgun sequence. This test had the ability to detect duplications larger than 15 kb irrespective of copy number, location, or high sequence similarity. This *in silico* approach mapped 169 large regions flanked by highly similar duplications [19]. When this report was published, i.e., in August 2002, only 24 of these hotspots of genomic instability had been associated with genetic disease.

The known number of microdeletion syndromes has dramatically changed over the past few years, because array-CGH has revolutionized the cytogenetic testing. Screening large patient cohorts with mental retardation and some dysmorphic features by array-CGH has recently led to the characterization of many novel microdeletion and microduplication syndromes. Several new recurrent microdeletion/microduplication syndromes were identified, which are flanked by LCRs and whose existence was predicted by the landmark study of Bailey et al. [19]. Such new recurrent microdeletion syndromes were identified for example for chromosomal regions 17q21.31, 15q13.3, 15q24, 1q41-1q42, 16p11-12.1, 2p15-16.1, and 9q22.3[217]. In addition, there are a steadily growing number of other microdeletion and microduplication syndromes, which are apparently not associated with LCRs. As a consequence, the breakpoints of these copy number changes are not recurrent, i.e., they are observed only in single individuals, or if similar regions are deleted in unrelated persons they usually do not have identical breakpoints. Thus, many aberrations are novel or extremely rare, making clinical interpretation problematic and genotype-phenotype correlations uncertain.

Therefore, a new challenge in human genetics is the generation of an appropriate infrastructure facilitating the utilization of the immense wealth of information generated by array-CGH. Aims of such efforts should be the identification of patients sharing a genomic rearrangement and with phenotypic features in common, which may lead to greater certainty in the pathogenic nature of the rearrangement, and the definition of new syndromes. To facilitate the analysis of these rare chromosomal events, interactive web-based databases, such as DECIPHER or ECARUCA, were developed. These databases are explained in detail in chapter 29.

3.6.3 Sex Chromosomes

3.6.3.1 First Observations

3.6.3.1.1 Nondisjunction of Sex Chromosomes and Sex Determination in Drosophila

Meiotic nondisjunction was discovered by Bridges [33] in the sex chromosome of *Drosophila melanogaster*. Morgan [159] had earlier described the X-linked mode of inheritance, and at the same time had elucidated the X–Y mechanism of sex determination in *Drosophila*. In his experiments a few exceptions had occurred that did not conform to the predictions of X-linkage. Bridges explained them by an anomaly in the mechanism of meiosis.

Drosophila has four chromosome pairs, three pairs of autosomes, and two sex chromosomes. Just as in humans, the males have the complement XY, the females

XX. Hence, each normal male germ cell has either one X or one Y chromosome; all female germ cells have an X. In crosses between an affected homozygote for the X-linked recessive trait white and a wild type or normal male, all male offspring would be expected to have white eyes as their mothers do. All daughters should be heterozygous and have normal red eyes. As a rule, this expectation was fulfilled. In exceptional cases, however, male offspring had normal red eyes, and some females were white-eyed. This was shown by Bridges to be due to nondisjunction of the maternal X chromosome, leading to an oocyte with either two X chromosomes or none. Fertilization with sperm from a wild type male was expected to lead to four different types of zygotes: XXX, XXY, XO, and YO. YO was not observed at all; apparently, zygotes without an X chromosome cannot survive. The other three types were observed and gave evidence regarding the mechanism of sex determination: sex chromosome constellations XXX and XXY both resulted in a female phenotype, while XO resulted in a male, sterile phenotype. Hence, the phenotypic sex in this fruit fly depends on the number of X chromosomes. One X chromosome makes a male, while more than one X chromosome makes a female. The Y is also involved in sex determination, as XO males are sterile.

3.6.3.1.2 XO Type in the Mouse

The X-linked mutation scurfy (sf) appeared first by spontaneous mutation, and affected animals have scurfy skin. The *scurfy* mutation causes loss of function of the *FoxP3* gene, which is essential for development and maintenance of naturally occurring regulatory CD4$^+$CD25$^+$ T cells. The hemizygous males are sterile; therefore, the strain can be maintained only by crossing heterozygotes (X^{sf}/X^+) with normal males (X^+/Y). From this mating, scurfy and normal males are expected in a segregation ratio of 1:1; all females should be normal. From time to time, however, an exceptional sf female is observed. As with male hemizygotes, they are sterile. However, their ovaries can be transplanted to normal females, which have been mated with wild-type males. The sons are all sf; the daughters are all normal but fall into two groups, those transmitting sf and those not transmitting it. Further analysis showed that these daughters have two different karyotypes, X^+/O and X^+/X^{sf}; the first group does not transmit sf, but the second does. This experiment showed that, contrary to the findings in *Drosophila*, XO is a fertile female in the mouse [201]. Hence, in this animal, the Y and not the X chromosome is decisive for the phenotypic sex. Subsequently, the XO types of the mouse have been found to be fairly frequent. In most cases the condition is caused not by meiotic nondisjunction but by chromosome loss after fertilization. Not long after the XO type, the XXY type was also discovered in the mouse. It is a sterile male, in contrast to *Drosophila*, where the XXY type was female.

3.6.3.1.3 First X Chromosomal Aneuploidies in Humans

XXY, XO, XXX. Jacobs and Strong [116] studied a 42-year-old man with the typical features of Klinefelter syndrome (Fig. 3.50), including gynecomastia, small testicles, and hyalinized testicular tissue. X Chromatin was found in cells of buccal smears and drumsticks, in granulocytes. Chromosome examination from bone marrow revealed an additional, submetacentric chromosome "in the medium size range." The authors felt that the patient very probably had the constitution XXY. However, "The possibility can not be excluded … that the additional chromosome is an autosome carrying feminizing genes." The patient's parents both had nor-

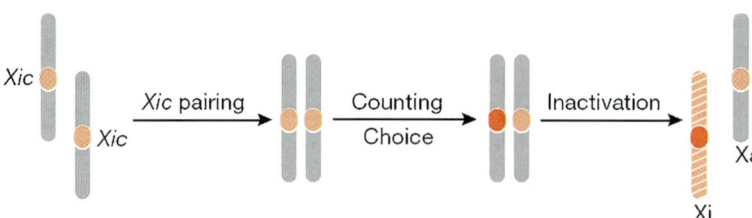

Fig. 3.50 Proposed scheme of X chromosome inactivation. After establishment of an X chromosome-to-autosome ratio one X chromosome is selected to become inactivated. This involves pairing of the two XICs. This may activate Xist transcription on one chromosome which is the one committed to become silenced. The other X chromosome remains active. From [166], reprinted by permission from Macmillan Publishers Ltd: *EMBO Reports*, copyright 2007

mal karyotypes with 46 chromosomes; hence, nondisjunction had occurred in one of their germ cells. Shortly afterward, the XXY status for Klinefelter syndrome was confirmed in many other cases.

At the same time, the XO type was discovered by Ford et al. [80]. Their patient, a 14-year-old girl, presented clinically as a Turner syndrome patient (Fig. 3.53) and was X chromatin negative. The modal number of chromosomes in bone marrow cells was 45; there were only 15 "medium length metacentric chromosomes," as in normal males. The evidence strongly suggested a chromosome constitution XO. The authors, comparing this result with that known from *Drosophila*, concluded that, in contrast to the fly, the XO type in humans may lead to an "agonadal" individual with female phenotype.

3.6.3.2 Dosage Compensation for Mammalian X Chromosomes

3.6.3.2.1 X Inactivation as the Mechanism of Gene Dosage Compensation: Lyon's Hypothesis

Because of their different sex chromosome composition (XX in female cells and XY in male cells), female and male cells have different numbers of genes, which could result in large-scale genetic imbalances between the sexes. Therefore, mechanisms are needed to equalize gene dosage between XX and XY cells. In 1949, Barr and Bertram described a unique nuclear structure present only in XX female cat neurons [22]. However, they did not associate this structure with an X chromosome or a dosage compensation mechanism. This realization came later with the publication of Mary Lyon's hypothesis of X inactivation [151]. Lyon made the step from morphological evidence to function, concluding that the heteropyknotic X chromosome may be either paternal or maternal in origin and is functionally inactive [151]. With this, she formulated one of the most fertile hypotheses in mammalian genetics. Lyon also tentatively explained an observation on a human X-linked disease in the same way: In X-linked ocular albinism, male hemizygotes lack retinal epithelial pigment and have a pale eye fundus. Heterozygous females have irregular retinal pigmentation, with patches of pigment and patches lacking pigment, so that the fundus has a stippled appearance. Lyon also predicted that mosaicism would be demonstrable in other X-linked genes, among them the variants of the enzyme glucose-6-phosphate dehydrogenase (G6PD). However, another human geneticist, Beutler, who should be given credit for having formulated the X-inactivation hypothesis independently of Lyon, using human G6PD as his experimental material [27].

Since then, the phenomenon of X-chromosome inactivation (XCI) has been analyzed in detail, and many features distinguishing the active X (Xa) from the inactive X (Xi) have been described.

In eutherians, X-chromosome inactivation (XCI) affects the paternal or maternal X chromosome randomly during early development. Once silenced, epigenetic mechanisms ensure the maintenance of silencing in future cell divisions, so that the inactive state is then stably inherited, giving rise to adults that are mosaics for two cell types, expressing one or the other X chromosome. Random XCI is achieved in three genetically separable events (Fig. 3.51). First, the X chromosome-to-autosome ratio is counted to ensure that one X chromosome is inactivated per female diploid nucleus. Next, one X chromosome is "chosen" in a mutually exclusive fashion to be the future inactive X (Xi). Lastly, silencing is initiated. The initiation of X inactivation is controlled in mammals by a region on the X chromosome, called the X-inactivation center (Xic), which regulates the different steps of XCI. The Xic produces the noncoding *Xist* transcript responsible for triggering silencing in *cis*. This is achieved by coating of the future Xi by the noncoding Xist RNA, recruitment of silencing factors, and condensation of the X-chromatin.

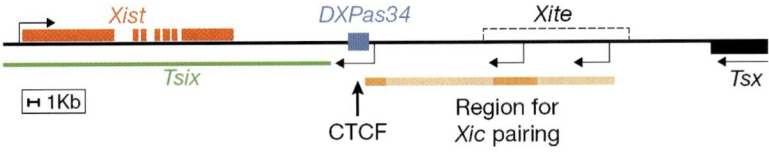

Fig. 3.51 Map of the regulatory elements of the mouse X-inactivation center. These regulatory elements are associated with X-chromosome counting and inactivation choice. From [166], reprinted by permission from Macmillan Publishers Ltd: *EMBO Reports*, copyright 2007

The *Xic* is comprised of several genetic elements that make long noncoding RNAs (ncRNAs), including *Xist*, *Tsix*, *Xite*, *DXPas34*, and *Jpx/Enox* (Fig. 3.51). As explained in detail below, the ncRNAs *Xist* and *Tsix* are choice regulators, which are critically involved in designating the Xa and Xi, respectively. *Tsix* is regulated by enhancers contained in the *Xite* and *DXPas34* elements. Differential methylation of *Xite* and the CCCTC-binding factor (CTCF)-binding sites on *DXPas34* correlate with X chromosome choice in mice [166].

Xist is negatively regulated by its antisense gene partner, *Tsix*. *Tsix* is like *Xist* a choice regulator, which overlaps with the *Xist* gene and is transcribed in the antisense direction. *Tsix* is initially expressed on both X chromosomes and is down-regulated on the Xi before inactivation; conversely, *Tsix* expression persists longer on the Xa. On the X chromosome, which was selected to become inactivated (the future Xi), loss of *Tsix* permits up-regulation of *Xist* and silencing of the *cis* chromosome. In parallel on the other X chromosome, which will be active (Xa), persistence of *Tsix* expression prevents up-regulation of *Xist* and thereby prevents silencing on that chromosome. *Tsix* is in turn regulated by *Xite*, a gene located ~10 kb upstream of *Tsix* and which bears a *Tsix*-specific enhancer. As a consequence, *Xist* is expressed exclusively from the Xi. The *Xist* RNA physically associates with the X chromatin and nuclear matrix around the X and coats the Xi [37] (reviewed in [71, 166]). The observation of Xist spreading along the Xi and maintaining the inactive status instigated the idea of "way stations" or "boosters" at intervals along its length. Mary Lyon proposed long interspersed repeat elements, i.e. line-1 (L1) as a candidate for way stations [152]. In fact there is evidence that L1 sequences are enriched near inactivated genes, which supports the proposed role of L1 elements as way stations. However, the search of sequence features which may be involved in the spread of silencing is still ongoing.

However, XCI includes another challenge because the two X chromosomes in the female must adopt mutually exclusive fates of Xa and Xi, and the cell must ensure that neither both nor neither X be inappropriately inactivated. Recent reports have demonstrated evidence that this task may be accomplished by physical interactions in trans of the Xic loci. In fact, the *Xic* region of the two X chromosomes appear to touch or "pair" just prior to the onset of XCI [16, 254] (Fig. 3.51). According to this pairing model, the two X chromosomes are epigenetically equivalent before the onset of XCI. The pairing of the Xci loci is then critically involved in the asymmetric localization of factors upon separation of the two X chromosomes. Relatively small and diverse 1- to 2-kb DNA elements of very low complexity which lie within *Tsix* and *Xite,* such as the 34mer repeat of *DXPas34* or CTCF, are sufficient to establish ectopic pairing. In addition, ncRNA might also be involved.

In their early studies, Barr and Bertram had already noted that the unique nuclear structure present in female cells, which later became known as the Barr body, often resides near the nucleolus [22]. In fact, the Xi co-localizes with the perinucleolar compartment during mid-late S-phase at a time when the Xi is undergoing DNA replication. This compartment is enriched for Snf2h, a chromatin-remodeling factor known to be required for replication of heterochromatic sequences [261]. Furthermore, XCI involves significant chromosomal reorganization of the Xi, because the core of the Xi consists of silenced nongenic sequences involving centromeric and other repetitive DNA elements, whereas genes that escape X inactivation lie outside or at the edge of the Xi core [42, 46].

Once the silent state of the Xi is created, the repressed status is maintained throughout subsequent cell divisions. *Xist* only has a minor role in the maintenance of the Xi, whereas multiple epigenetic marks, including DNA methylation, late replication and hypoacetylation of histone H4, act synergistically in the maintenance of X inactivation. To achieve this Polycomb group (PcG), proteins are recruited to the Xi to establish specific epigenetic marks at the histones H3 and H2 [104].

3.6.3.2.2 Many Genes Escape Silencing on the Human Xi

Not all, but most, genes on the Xi are transcriptionally silenced. However, approximately 15% of human X-linked genes escape inactivation and are therefore expressed from both the Xa and Xi [39]. Genes expressed from the Xi rarely express to the extent of the Xa. In fact, gene expression data performed on microarrays identified only a small number of X-linked genes with overexpression in females relative to males, so that the X chromosome dosage compensation is virtually complete. The genes that escape inactivation are nonrandomly distributed along the X chromosome, with the majority clustered on the short arm of the X chromosome.

3.6.3.2.3 Pseudoautosomal Regions (PAR1 and PAR2)

Between the X and the Y chromosome there are two limited regions of identical sequence, which are located at the tips of the short and long arms of the X and Y chromosomes. During meiosis, pairing and crossover in men takes place in these two regions, and they have therefore been termed "pseudoautosomal regions" or "PARs." The region at the distal Xp arm is the first pseudoautosomal region (PAR1), with a physical length of approximately 2.7 Mb, while the second pseudoautosomal region (PAR2) is located at the distal Xq arm and has a size of about 0.33 Mb (Fig. 3.52). In men, the PAR1 region exhibits the highest recombination frequencies of the genome. Crossover activity in PAR1 is much higher in men than in women, and also higher than for each of the autosomes. It is estimated that during male meiosis on average at least one crossover occurs. The rate of recombination in PAR2 is much lower than in PAR1, but still higher than the average rate of the remainder of the X chromosome. To date, 24 genes have been reported in PAR1 and 5 genes in PAR2. Possible connections with clinical disorders such as short stature, asthma, psychiatric disorders, and leukemia have been suggested in the past, but only one pseudoautosomal gene, *SHOX* (short stature homeobox), has been unambiguously associated with various short stature conditions and disturbed bone development [77]. Loss of the *SHOX* gene is likely associated with phenotypic features in women with Turner syndrome (Sect. 3.6.3.3).

3.6.3.2.4 Skewed X Chromosome Inactivation

Because inactivation of one X in early development generally occurs in a random manner, most females will have some cells with a maternally derived X and some with a paternally derived X inactivated. However,

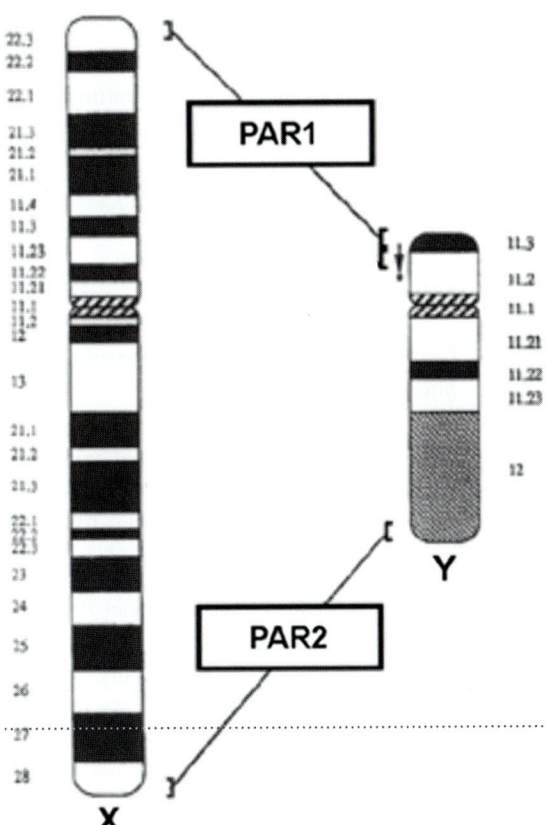

Fig. 3.52 Scheme showing the locations of the pseudoautosomal regions 1 (PAR1) and 2 (PAR2)

owing to the limited number of embryo precursor cells at the time of marking or commitment of a particular X to inactivation, most females have a detectable bias in terms of the proportion of cells that have inactivated one or the other X chromosome. An extremely skewed X-chromosome inactivation, which is defined using an arbitrary cutoff of ≥90% of cells with the same X inactivated, is less common. Such a skewed X chromosome inactivation may be a consequence of various factors, such as (1) biases in the choice of which X to inactivate, (2) X-chromosome mutations or rearrangements, which affect the viability of cells with one or the other X active, or (3) stochastic factors [36]. The frequency of extremely skewed X chromosome inactivation in unselected female persons of reproductive age is approximately 7% when measured using methylation-based assays and DNA derived from whole blood [26]. Females with X chromosome structural abnormalities usually selectively inactivate their abnormal X to avoid the occurrence of genetic imbalance that could result in an abnormal phenotype. Furthermore, X;autosome translocations also usually result in skewed X chromosome inactivation. When the Xic is translocated onto autosomes, spreading of the silent chromatin structure into autosomal chromatin can occur, but is frequently attenuated and incomplete [191].

3.6.3.3 X Chromosomal Aneuploidies in Humans

3.6.3.3.1 Difference Between X Chromosomal and Autosomal Aneuploidies

Soon after these first discoveries a great number of other aneuploidies of sex chromosomes were described. As a group, they show some remarkable differences from the autosomal aneuploidies discussed before.

(a) Mean intelligence can be reduced, but the extent of mental retardation is not nearly as pronounced as in the autosomal conditions; many probands have normal intelligence, and in some it is even above average (Chap. 23.6).
(b) The phenotypic disturbances most severely affect development of the sexual organs and sex-hormone-dependent growth. Other malformations do occur – mainly in Turner syndrome, but except for the small stature of Turner patients they usually are less frequent and less severe.

In brief, X chromosomal aneuploidy does not disturb embryonic development nearly as much as does autosomal aneuploidy. The reason is that normal women have two, normal men only one, X chromosome. This difference led to the development in evolution of a powerful mechanism of gene dosage compensation that happened to benefit carriers of X aneuploidies (Sect. 3.6.3.2).

Clinical Classification of X Chromosomal Aneuploidies: Mosaics. In general, the number of additional X chromosomes enhances the severity of mental retardation. The number of X chromatin bodies is one less than the number of X chromosomes.

Theoretically, zygotes with 45,X should be somewhat more frequent than any other types, since they can be produced by nondisjunction in both sexes and both meiotic divisions. This expectation does not fit in with the observed data, as all the karyotypes together that lead to Turner syndrome are much rarer than XXX or XXY. This finding points to strong selection against

germ cells without the X chromosome and/or to strong intrauterine selection against zygotes with 45,X. The latter expectation is corroborated by observations on aborted fetuses, among which the 45,X chromosome constellation is, indeed, frequent. Another line of evidence points in the same direction: the risk of nondisjunction in general increases with the age of the mother. For XXY and XXX karyotypes this increase can be clearly demonstrated; but not for the 45,X karyotypes. Hence, it is assumed that surviving 45,X zygotes are the result not of meiotic, but of mitotic, nondisjunction or of early chromosome loss. The relatively greater proportion of mosaics in this group compared with XXX and XXY fits this hypothesis.

XYY zygotes, on the other hand, can be formed only by nondisjunction during the second meiotic division in males. Nevertheless, they are about as frequent as XXY zygotes. Therefore, the probability of nondisjunction of Y chromosomes appears to be much higher than the combined probabilities for X chromosome nondisjunction. Mosaics have been observed for all types.

3.6.3.3.2 Intersexes

Sex determination depends on the sex-chromosome complement of the embryo and multiple molecular events involved in the development of germ cells and their migration to the urogenital ridge. In the presence of the Y chromosome (46,XY) testes are formed, whereas the absence of a Y chromosome together with the presence of a second X chromosome (46,XX) results in the formation of ovaries. Many genes have been identified that contribute to the process of sex determination and differentiation. The best-defined gene involved in gonadal differentiation is *SRY,* which is located on the short arm of the Y chromosome and induces the bipotential gonad to differentiate into a testis. *SRY* induces the *SOX9* gene, and both genes have major roles in this process, together with steroidogenic factor 1 (SF-1) and opposition from *DAX1*. Other genes, such as *WT1*, have smaller contributions [153]. Mutations in these genes, which are involved in sex determination and development, may result in intersex anomalies, independently of the constitution of the sex chromosomes.

For example, true hermaphroditism is characterized by ambiguous genitalia, and both ovarian and testicular tissue is present either in the same or in a contralateral gonad. The karyotype is predominantly 46,XX, although testes, ovaries, and ovotestes may be present in various combinations. The molecular events have not been elucidated, but some cases were associated with translocation of a fragment containing the *SRY* gene to the X chromosome [153].

Furthermore, failure to produce testosterone or mutations in the testosterone receptor can result in 46,XY phenotypic females or phenotypic males with various degrees of diminished masculinization (male pseudohermaphroditism). On the other hand, patients with congenital adrenal hyperplasia produce an excess of adrenal androgens, which can cause female pseudohermaphroditism in 46,XX patients. A number of specific mutations have been identified, which are associated with such intersex anomalies. Examples are mutations in *SRD5A2*, located on chromosome 5p15 or *CYP17* (10q24-25), which may cause male pseudohermaphroditism, or mutations in *CYP21* (6q21.3), which may result in female pseudohermaphroditism. There are many other genes that are associated with intersex anomalies [153].

3.6.3.3.3 Turner Syndrome

Since the description of Turner syndrome by Henry H. Turner in 1938, a wealth of information has been added and improved our current understanding of the syndrome.

For a girl or woman to be diagnosed with Turner syndrome, she must be missing all or part of one copy of the second sex chromosome, as confirmed by a chromosome analysis. In addition, mosaicism with two or more cell lines may be present. The first cases described had the "classic" karyotype 45,X. However, this classic karyotype only accounts for 50% of cases; the remaining cases comprise mosaic karyotypes (i.e., cells with 45,X and cells with 46,XX), karyotypes with an isochromosome of X – for example i(Xq) or i(Xp) – or karyotypes with an entire or part of a Y chromosome [18]. All individuals who have Turner syndrome have a female phenotype. Approximately 1 in 2,500 live female births is affected by Turner syndrome.

The Turner syndrome physical phenotype is characterized by abnormalities in three basic systems: the skeletal, the lymphatic, and the reproductive systems. Short stature is the cardinal finding in girls with Turner syndrome, affecting 95–99% of them. Other skeletal defects include cubitus valgus or an unusual carrying angle of the elbows and arms (see Fig. 3.53), as well as a short 4th metacarpal, micrognathia, and a high-arched palate. Without treatment, females with Turner syndrome are typically more than 2 standard deviations below their peers in height and achieve a final height of about 140 cm. Since the advent of growth hormone therapy, as well as of other hormones to augment growth, heights within the bottom end of the normal range can be achieved. Because their facial abnormalities can cause an abnormal orientation of the ear canal, children with Turner syndrome are at high risk of ear infections [107, 198].

The lymphatic system defect is caused by abnormal lymphatic clearance. The consequences are frequently visible in utero as brain hygroma during ultrasound examination. While permanent neck webbing can result after the hygroma recedes, in many cases the hygroma is so severe as to cause fetal demise. Many neonates present with severe edema, which is often the reason for the diagnosis of Turner syndrome. Most postnatal diagnoses are made at birth (15%), during teenage years (26%), and in adulthood (38%), with the remainder being diagnosed during childhood, and

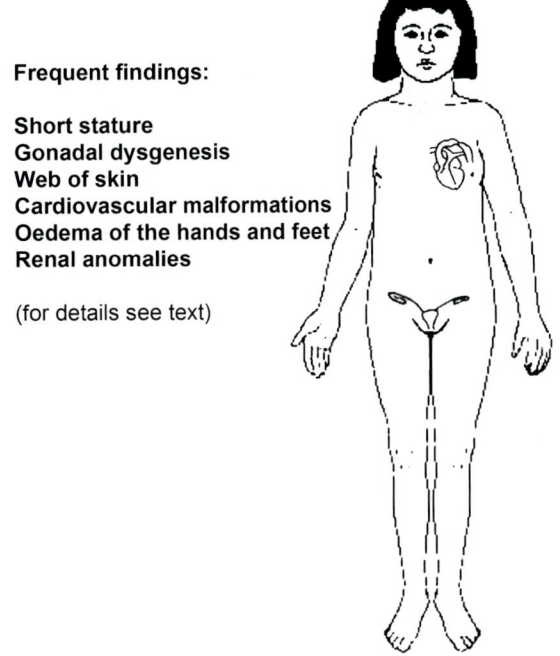

Frequent findings:

**Short stature
Gonadal dysgenesis
Web of skin
Cardiovascular malformations
Oedema of the hands and feet
Renal anomalies**

(for details see text)

Fig. 3.53 Main clinical findings in Turner syndrome

therefore there is considerable delay in diagnosing girls and adolescents. The key to diagnosis was lymphedema in 97% during infancy, and short stature in 82% during childhood and adolescence [107, 198].

Regarding reproductive system defects, most females with Turner syndrome have ovarian dysgenesis owing to streak ovaries containing no ova. As a result, they lack endogenous estrogen and have reduced androgen production. Unless they receive hormonal replacement therapy during adolescence, they remain sexually infantile throughout life. Although the majority of females with Turner syndrome are infertile, a few individuals do spontaneously produce estrogen and undergo normal pubertal development. There are also a handful of women with karyotypes other than 45,X who have successfully reproduced.

In addition, individuals with Turner syndrome are at risk for cardiac abnormalities owing to coarctation of the aorta. Turner syndrome is associated with mainly left-sided CV malformations, such as elongated transverse arch of the aorta seen in 50% of the women, bicuspid aortic valves in 13–43 vs 1–2% in the general population, and coarctation of the aorta (4–14% in Turner syndrome). Less commonly, right-sided malformations such as persistent left vena cava superior and partial anomalous venous return are seen. Rarely atrial and ventricular septal defects, persistent ductus arteriosus, aortic and mitral stenosis and insufficiencies, and hypoplastic left heart syndrome appear [108]. Less frequent other symptoms include renal abnormalities from horseshoe kidneys, multiple pigmented nevi, and nail dysplasia. In addition to the congenital structural cardiac malformations such as bicuspid aortic valve and coarctation of the aorta, hypertension is also thought to be a major factor shown by the relatively high incidence of aortic dissection and rupture of 40 per 100,000 Turner syndrome-years versus 6 per 100,000 general population-years. It also, strikingly, affects Turner syndrome patients at a median age of 35 years, as opposed to 71 years in the general population [95]. Aortic dilation normally precedes dissection, and rupture is seen in 3–42% of randomly selected Turner syndrome women, where aortic diameter correlates significantly to systolic blood pressure but, surprisingly, is not associated with vascular atherosclerotic indices such as aortic stiffness or plasma lipids. An intrinsic arterial defect is therefore likely, as part of the generalized vasculopathy in Turner syndrome.

Morbidity and mortality are increased, which is due especially to the risk of dissection of the aorta and other cardiovascular diseases, as well as the risk of type 2 diabetes, osteoporosis, and thyroid disease [107, 198].

Despite the consistency of these physical features, there is wide variability among affected individuals, and few if any have every abnormality. Generally, a more severe presentation is associated with complete loss of a single X chromosome or the ring X condition, while the least severe presentation is associated with a mosaic karyotype involving a normal 46,XX cell line. Deletions, rearrangements, or translocations of the X chromosome represent intermediary conditions. Part of the explanation of the reduced final height relates to the action of the *SHOX* gene located in the PAR1 region of the X and Y chromosome. Haploinsufficiency of the *SHOX* gene explains the reduction in final height, changes in bone morphology, sensorineural deafness, and other features [194]. However, additional genes are thought to be involved in the pathogenesis of Turner syndrome, but await discovery.

The treatment of Turner syndrome entails biosynthetic recombinant human growth hormone to increase height, estrogen to initiate puberty and maintain normal female functioning, and androgens to advance linear bone growth.

Ideally, the timing of endocrine therapy should allow onset of puberty at the same time as the peers of the patient to avoid social problems at school because of delayed physical and psychological development. This would also allow optimal bone mineralization to take place. Estrogen therapy should be coordinated with the use of growth hormone. This should be individualized for each patient, so as to optimize both growth and pubertal development.

During adulthood, infertility is rated as the most prominent problem of the syndrome. Oocyte donation is an option in many countries. The most recent studies show good results comparable with those of oocyte donation in other groups of patients, although better preparation of the uterus for implantation (uterine size and endometrial thickness) with prolonged treatment with high daily doses of estradiol may improve results [107].

The average intellectual performance is within the normal range. Behavioral aspects of Turner syndrome are described in Chap. 23.6.

3.6.3.3.4 Klinefelter Syndrome

Klinefelter syndrome is the most common sex-chromosome disorder and the most common numerical chromosomal aberration among men, with an estimated frequency of 1:500–1:1,000 of live deliveries [139].

Klinefelter syndrome is characterized by X chromosome polysomy with X disomy being the most common variant (47,XXY). Ninety percent of men with Klinefelter syndrome have nonmosaic X chromosome polysomy [139]. The 47,XXY karyotype of Klinefelter syndrome arises spontaneously by nondisjunction in stage I or II of meiosis. While most human trisomies originate from errors at maternal meiosis I, Klinefelter syndrome is a notable exception, as nearly one-half of all cases derive from paternal nondisjunction [234]. Postfertilization nondisjunction is responsible for mosaicism, which is seen in approximately 10% of Klinefelter syndrome patients. Men with mosaicism are less affected and often are not diagnosed. Advanced maternal age and possibly paternal age have been linked to an elevated risk of Klinefelter syndrome [149]. In fact, increased paternal age may be responsible for an increasing prevalence of Klinefelter syndrome [160].

The X chromosome carries genes that have roles in many body systems, including testis function, brain development, and growth. The "prototypic" man with Klinefelter syndrome has traditionally been described as tall, with narrow shoulders, broad hips (i.e., tall eunuchoid body proportions), sparse body hair, gynecomastia, small hard testicles, micropenis, androgen deficiency, sterility, azoospermia, and decreased verbal intelligence. It is now well known that this original description is not accurate and that men with Klinefelter syndrome represent a broad spectrum of phenotypes, professions, incomes, and socioeconomic status [139]. The classic descriptions of men with Klinefelter syndrome, as shown for example in Fig. 3.54, are based on the most severe cases of phenotypic abnormalities. Klinefelter syndrome is an underdiagnosed condition; only 25% of the expected number of patients are diagnosed, and of these only a minority are diagnosed before puberty.

Most commonly, men with Klinefelter syndrome will present to their urologist with infertility: azoospermia or severe oligospermia, low testosterone, and complications of low testosterone, such as erectile dysfunction and poor libido. Boys will present with

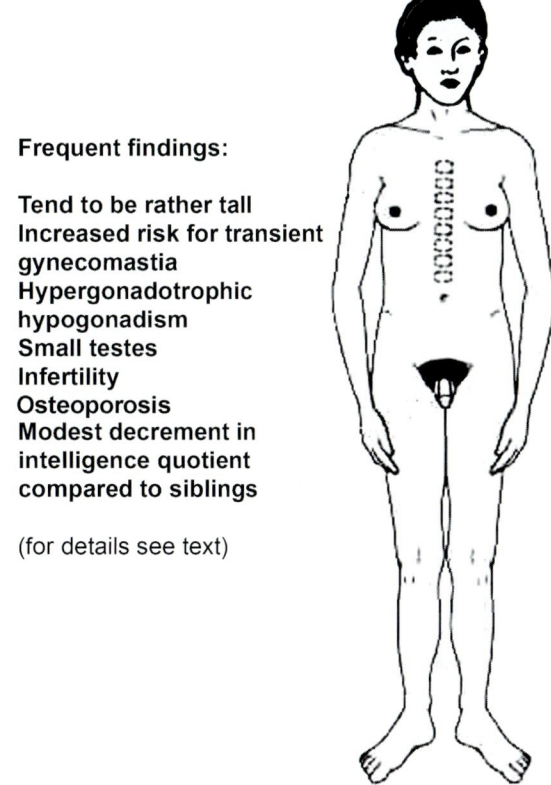

Frequent findings:

Tend to be rather tall
Increased risk for transient gynecomastia
Hypergonadotrophic hypogonadism
Small testes
Infertility
Osteoporosis
Modest decrement in intelligence quotient compared to siblings

(for details see text)

Fig. 3.54 Main clinical findings in Klinefelter syndrome

concerns about genital and pubertal development [139]. Spermatogenic and steroidogenic dysfunction are cardinal and the most prevalent signs of Klinefelter syndrome. A typical patient with Klinefelter syndrome will present with low serum testosterone, high LH and FSH levels, and often elevated estradiol; however, the decline in testosterone production is progressive over the life span, and not all men suffer from hypogonadism.

Patients with Klinefelter syndrome should be treated with lifelong testosterone supplementation that begins at puberty, to secure proper masculine development of sexual characteristics, muscle bulk, and bone structure and to prevent the long-term deleterious consequences of hypogonadism.

The fact that sperm can be found in the testes of men with Klinefelter syndrome has challenged the previous assumption that men with Klinefelter syndrome are always sterile. This has raised the mechanistic questions of whether children with Klinefelter syndrome are born with a severely depleted number of spermatogonia and whether there is a period in life when the spermatogonia undergo massive apoptosis that results in depletion of the spermatogonial population and subsequent azoospermia [30, 175]. Based on the current data, it is reasonable to assume that most men with Klinefelter syndrome are born with spermatogonia. However, during early puberty – most likely after initiation of the first wave of spermatogenesis – the spermatogonia appear to undergo massive apoptosis.

Sperm found in testes of men with Klinefelter syndrome have only a slightly increased frequency of sex chromosome polysomies, and most boys born of fathers with Klinefelter syndrome have a normal karyotype. These findings indicate that during early stem cell proliferation or meiotic division, the checkpoint mechanisms are able to overcome X chromosome polysomy resulting in sperm with a single X or Y chromosome [30, 175].

In the last decade, developments in microsurgical techniques and advances in artificial reproductive technologies allowed over 50% of men with Klinefelter syndrome to have their own children through the combination of microsurgical testicular sperm extraction and use of freshly retrieved sperm for in vitro fertilization.

Men with Klinefelter syndrome are at a higher risk of autoimmune diseases; diabetes mellitus, leg ulcers, osteopenia and osteoporosis, and tumors (breast and germ cells), and historically they have an elevated mortality rate. It is not known whether the morbidity associated with Klinefelter syndrome is a result of hypogonadism and hyperestrogenism or rather due to abnormal function of X chromosome-linked genes [30, 175].

Severe intellectual deficits are rare. Often, the auditory processing delay and language dysfunction seen in men with Klinefelter syndrome are misdiagnosed as cognitive deficits [203]. More behavioral aspects will be discussed in Chap. 23.6.

3.6.3.3.5 Triple-X Syndrome

The triple-X syndrome (47,XXX) was first described by Jacobs et al. [117] in a woman of average intelligence who had secondary amenorrhea. In fact, this finding was described as "Evidence for the existence of the human 'super female'". The triple-X syndrome is one of the most frequent aneuploid variations in female infants, occurring in about 1 per 1,000 newborn females [168]. The incidence increases with maternal age, and nondisjunction appears to occur mainly during maternal meiosis.

To date, several hundred 47,XXX women have been reported. Much of the data is based on isolated case studies of women ascertained through the presence of another condition. Perhaps more unbiased information can be achieved through chromosomal screening of newborns and following them longitudinally. Such studies have revealed that 47,XXX girls have no distinguishing physical features and their pubertal development and reproductive capacity are normal [72]. Most 47,XXX girls present with a normal phenotype at birth [147], but their average birth weight may be slightly lower than that of girls with normal chromosomes. The final height of triple X syndrome females usually exceeds that of controls, and the proportion of their leg length to their overall height increases to a significant level [194]. Bone age is below normal in early childhood, but is within normal limits by 7–10 years of age. Triple X females usually have normal fertility, and almost all their children have normal karyotypes.

Head circumference in 47,XXX infants is often smaller than in control girls. In fact, the presence of an extra X chromosome may cause a reduction in craniofacial growth and reflects on the overall length of the calvaria, the anterior and posterior cranial bases, and the facial complex [134].

The average intelligence of triple X girls seems to be slightly lower than that of siblings or a population control group [167], and an interesting correlation was found between head circumference at birth and IQ at 7 years of age [195]. However, IQ distribution has the potential to improve to within normal limits. Such an improvement has been attributed to providing the parents with appropriate information and counseling, which increases the chance that resources for stimulation of the children, including both parental resources and kindergarten, school, and social system resources, will be available [167]. These observations reinforce the notion that parents who have a child with a sex chromosome abnormality usually need information, counseling, and assistance. The type and magnitude of this assistance depend on the individual child, the specific sex chromosome anomaly, and the parents' own resources, psychologically, socially, and otherwise.

3.6.3.4 Chromosomal Aneuploidies in Humans

3.6.3.4.1 47,XYY Karyotype

The first description of a 47,XYY karyotype was made in 1961 [204]. 47,XYY has an estimated incidence of 1 per 1,000 live births [1]. The clinical features of patients with 47,XYY karyotype can be subtle and variable. Generally, the patients are physically normal in infancy, with phenotypic characteristics becoming increasingly apparent over time. Patients are usually tall and thin and may have delayed speech, lower cognitive function, hyperactivity, learning disabilities, and other central nervous system (CNS) abnormalities (such as intentional tremor and hypotonia). However, several patients with an XYY constitution and a normal phenotype have also been described. Possible behavorial aspects, such as a potential association with an increased risk of criminal and antisocial behaviour, will be discussed in detail in Chap. 23.6.

Males with an extra Y chromosome are mostly fertile and have normal gonadal function. The premeiotic loss of the extra Y chromosome, frequently observed in XYY human males, permits the achievement of spermatogenesis and normal sperm production [82].

3.6.3.4.2 Other Y Polysomies

While the 47,XYY karyotype is relatively frequent, cases with trisomy and tetrasomy Y are rare, having been reported in the literature 20 and 10 times, respectively. No pentasomy Y has been reported. While the small case numbers do not allow reliable descriptions of the phenotype, psychomotor delay ranging from speech delay to severe mental retardation, skeletal abnormalities, and facial dysmorphism have been reported in every case [60].

3.6.3.4.3 XXYY Syndrome

XXYY syndrome was first described in 1960 [161] and occurs in approximately 1:18,000 to 1:40,000 males. Approximately 100 cases have been reported to date. Initially the XXYY syndrome was described as a variant of 47,XXY Klinefelter syndrome because of a shared physical and endocrinological phenotype, such as tall stature, hypergonadotropic hypogonadism, and infertility. However, XXYY differs in its medical, neurodevelopmental and behavioral characteristics.

A review of 95 males with XXYY syndrome reported that the mean age at diagnosis was 7.7 years. Developmental delays and behavioral problems were the most common primary indication for genetic testing (68.4%). Across all age groups physical and facial features such as hypertelorism, clinodactyly, pes planus, and dental problems were common. In both adolescents and adults tall stature was a prominent feature. The mean adult stature was in the range of 192 cm. Frequent medical problems included allergies and asthma (>50%), congenital heart defects (19.4%), radioulnar synostosis (17.2%), inguinal hernia and/or cryptorchidism (16.1%), and seizures (15%). In adulthood, such medical problems as hypogonadism (100%), intention tremor (71%), and deep venous thrombosis and type II diabetes (each 18.2%) were encountered. Brain MRI, done in 35 cases, showed white matter abnormalities in 45.7% of patients and enlarged ventricles in 22.8% [233]. Behavioral patterns of the XXYY syndrome will be discussed in Chap. 23.6.

3.6.4 Chromosome Aberrations and Spontaneous Miscarriage

3.6.4.1 Incidence of Prenatal Zygote Loss in Humans

About 15% of all pregnancies in humans end in recognizable spontaneous abortion, defined as pregnancy loss before the 22nd week (body weight of the embryo: 500 g or less). However, there is good evidence in both humans and other mammals that many more zygotes are lost at an earlier stage of development; they are often severely malformed. It appears that almost 50% of all conceptuses may be lost within the first 2 weeks of development, before the pregnancies are recognized. In humans this early zygote loss usually goes unnoticed.

3.6.4.2 Aneuploidy in Oocytes and Embryos

The high prevalence of aneuploidy in human oocytes and embryos has long been recognized, and the developmental impact of these anomalies, especially those of meiotic origin, is well documented. Present data based on molecular cytogenetic techniques suggest that the incidence of aneuploidy in first trimester spontaneous miscarriages may be as high as 65% [156]. There is a dramatic increase in chromosomally abnormal pregnancies with advancing maternal age. These data are mirrored by findings from direct cytogenetic analysis of human oocytes, which demonstrates oocyte aneuploidy rates in excess of 50% for many women over 40 years of age [102].

Studies applying CGH to the analysis of embryos have provided fascinating data on the variety and frequency of chromosome abnormalities in human embryos and confirmed that aneuploidy can affect any chromosome during human preimplantation development, including the largest chromosomes. In fact, published CGH data reveals that 20–40% of embryos carry chromosome abnormalities [242, 251].

The great majority of these chromosomally abnormal pregnancies end in abortion between 8 and 16 weeks of gestation. A few may remain beyond this time and be lost as a later abortion (which may present as intrauterine fetal death). As among liveborn babies only 1 in 200–250 has an unbalanced chromosome abnormality, there must be very effective natural selection against those large numbers of gametes and conceptions that were abnormal.

3.7 Chromosome Instability/Breakage Syndromes

There are several syndromes that are associated with chromosomal instability. The instability refers to the predisposition of the chromosomes to undergo rearrangement or to display other abnormal cytogenetic patterns. Often multiple random, nonclonal chromosomal aberrations can be observed, which result in a high cell-to-cell variability. Thus, diagnosis often requires analyses of multiple metaphase spreads.

Defects of DNA repair often underlie such syndromes, which frequently exhibit characteristic clinical features and may cause serious clinical results. Although genetic defects in each repair or checkpoint pathway are associated with clinically distinct entities, as a group they are characterized by developmental abnormalities, cancer predisposition, and accelerated aging.

"Classic" chromosomal instability syndromes are Fanconi anemia, Bloom syndrome, and ataxia-telangiectasia. Rarer chromosomal instability syndromes include Nijmegen breakage syndrome, Werner syndrome, Rothmund-Thomson syndrome (RTS), ICF syndrome, Roberts syndrome, mosaic variegated aneuploidy (MVA).

3.7.1 Fanconi Anemia

Fanconi anaemia (FA) is a rare autosomal recessive disease with a prevalence of 1–5 per million. FA is characterized by diverse clinical features that often include developmental anomalies affecting the skeleton (absent or abnormal thumbs and radii), kidneys, heart, or any other major organ system. The life expectancy of FA patients is reduced to an average of 20 years (range 0–50 years), primarily because these individuals develop life-threatening bone marrow failure and are susceptible to developing malignancies, especially acute myeloid leukemia and, to a lesser extent, solid tumors, in particular squamous cell carcinomas.

FA is caused by biallelic mutations in at least 13 different genes (*FANCA, B, C, D1/BRCA2, D2, E, F, G, I, J/BRIP1, L, M,* and *N/PALB2*) interacting with others (e.g., *ATM, RAD51, NBS1*) in cellular DNA damage recognition and repair network. Eight of the FA proteins form a nuclear core complex with a catalytic function involving ubiquitination of the central FANCD2 (Fanconi anemia complementation group D2) protein that interacts with BRCA1 (breast cancer 1, early onset) in microscopically visible subnuclear foci. At present, FA patients are assigned to one of seven complementation groups (FA-A to FA-G; on FA nomenclature), and most (60–80%) fall into group A.

At the cellular level, manifestations of genetic instability include chromosomal breakage, cell cycle disturbance, and increased somatic mutation rates. Because of the extreme variability of the clinical features of this disease, as well as variations in the level of spontaneous chromosome aberrations, a cytogenetic test that quantifies crosslinker-induced chromosomal breakage has become the gold standard for diagnosing

FA [126, 179]. Therefore, genetic testing for FA is routinely based on conventional chromosome breakage analyses. Mainly chromatid breaks and radial figures are counted and compared with those of normal control cells.

3.7.2 Instabilities Caused by Mutations of Proteins of the RecQ Family of Helicases: Bloom syndrome, Werner syndrome, and Rothmund-Thomson syndrome

Homologous recombination (HR) is a process that repairs DNA double-strand breaks and restores productive DNA synthesis following disruption of replication forks. Although HR is indispensable for maintaining genome integrity, it must be tightly regulated to avoid harmful outcomes, because excessive HR can generate DNA damage. For example, the genomes of higher eukaryotes contain many tandem and dispersed repeat sequence DNA. Genetic exchange between nonhomologous copies of these repeat sequences (illegitimate recombination) can generate translocations, deletions, or inversions. Illegitimate recombination events are more likely to occur when the cell is replicating its DNA because the DNA is exposed. If replication forks have been stalled, for example because the polymerase has encountered a damaged base, special structures can be formed, which could be converted into DSBs by up-regulated HR.

To oppose these potentially adverse effects of HR, the cell holds the recombination machinery in check with proteins that have antirecombination activity. The specialized DNA helicases of the RecQ family are intimately involved in the regulation of HR and can inhibit HR through the stabilization of stalled replication forks [174].

The human RecQ helicases have been a major source of interest in the DNA repair field because of their links to human disease. Humans possess five distinct RecQ helicases – RECQL1, BLM, WRN, RECQL4, and RECQL5. Mutations in three of these helicases, *BLM*, *WRN*, and *RECQL4*, result in the genomic instability disorders Bloom syndrome (BS), Werner syndrome (WS), and Rothmund–Thomson syndrome (RTS), respectively. Cells derived from persons with these syndromes display varying types of genomic instability, as evidenced by the presence of different kinds of chromosomal abnormalities and different sensitivities to DNA-damaging agents. Persons with these syndromes exhibit a variety of developmental defects and are predisposed to a wide range of cancers. WS and RTS are further characterized by premature aging. Although these three human disorders share the general characteristics of genomic instability and cancer predisposition syndromes, each syndrome possesses unique clinical, cellular, genetic, and biochemical features that point to nonoverlapping roles in the maintenance of genome integrity [174].

3.7.2.1 Bloom Syndrome

BS is a rare autosomal recessive disorder caused by bi-allelic loss-of-function mutations in the *BLM* gene. Clinically it is characterized by proportionately short stature, sun-sensitive facial erythema, hypo- and hyperpigmented skin lesions, immune deficiency, infertility in males and subfertility in females, lack of subcutaneous fat, and susceptibility to type 2 diabetes. A prominent feature of BS is a marked predisposition to all types of cancers [88], which is notable for its high incidence, broad spectrum (including leukemia, lymphomas, and carcinomas), early diagnosis relative to the same cancer in the general population, and the development of multiple cancers in single individuals [86].

Analysis of metaphases from cultured cells from persons with BS reveals striking chromosome instability, visible as chromatid breaks and gaps, dicentric and ring chromosomes, acentric fragments, pulverized metaphases, telomere associations, and anaphase bridges [87]. BS cells exhibit special cytogenetic signs of dys-regulated HR, which are pathognomic for BS. The first of these abnormalities is the classic quadriradial configuration, which is a symmetric, four-armed arrangement composed of a pair of homologous chromosomes that have apparently undergone somatic crossing over. Such quadriradials are very rarely observed in metaphases from normal lymphocytes (<1 per 1,000) but in BS they are relatively common (1–2 per 100). Another diagnostic cytogenetic finding in BS is a markedly increased level of spontaneous sister chromatid exchange (SCE), which is an exchange event between the sister chromatids. In metaphases prepared from normal persons, the average number of SCEs ranges from 5 to 10 per 46 chromosomes, whereas in BS it is more than 50 per cell.

At the molecular level, the consequences of this excessive HR are increased rates of loss of heterozygosity, unequal SCEs, chromosome deletions and rearrangements – all of which are seen in BS cells – along with an elevated mutation rate. BS cells are hypersensitive to various genotoxic agents, such as ultraviolet light, mitomycin C, and topoisomerase inhibitors, especially when these and other agents are administered to cells synchronized in S-phase [174].

Werner Syndrome: WS is a rare autosomal recessive disorder caused by bi-allelic loss-of-function mutations in the *WRN* gene [258]. In this disorder premature development with features resembling aging, often already shortly after adolescence, is typically observed [93]. Premature aging features include the development of alopecia and graying hair, arteriosclerosis, atherosclerosis, osteoporosis, hypogonadism, cataracts, and type II diabetes. Persons with WS are generally smaller than average. Furthermore, there is an increased cancer predisposition, leading primarily to rare cancers of mesenchymal origin [94].

WS cells exhibit a distinctive cytogenetic abnormality referred to as "variegated translocation mosaicism." Lymphocytes show frequent, nonclonal translocations, and fragile sites appear to constitute preferred sites for chromosomal translocations. Fibroblast cultures are frequently pseudodiploid and exhibit clonal expansions of cells containing different structural rearrangements. In addition, fibroblast cultures also enter premature replicative senescence, implicating WRN in telomere function.

At the molecular level, WS cells accumulate mutations at a higher rate, consisting predominantly of large spontaneous deletions. Cells are hypersensitive to a number of DNA damaging agents [174].

3.7.2.2 Rothmund–Thomson Syndrome

RTS is a rare autosomal recessive genetic disorder characterized by chromosome instability and clinical heterogeneity with growth deficiency, skin and bone defects, premature aging symptoms, and cancer susceptibility, and patients present clinically heterogeneous symptoms. A subset of RTS patients presents mutations of the *RECQL4* gene. Two forms of RTS have been defined based on clinical and molecular analysis: Type I RTS is associated with a characteristic poikiloderma and type II RTS with poikiloderma and an increased risk of osteosarcoma related to deleterious mutations in the *RECQL4* gene [38].

3.7.3 The Ataxia-Telangiectasia Group

In this group of chromosomal instability syndromes the basic pathogenetic process represents a failure in one of the DNA damage monitoring and repair systems. The group comprises ataxia-telangiectasia (AT), Nijmegen breakage syndrome (NBS), and AT-like disorder (ATLD). The genes for AT (*ATM*), NBS (*NBS1*), and ATLD (*MRE11*) encode proteins, which belong to a complex which senses aberrant DNA structures and monitors postreplication DNA repair.

3.7.3.1 Ataxia-Telangiectasia

Ataxia-telangiectasia (AT) is a rare autosomal recessive genetic disorder characterized by progressive neurodegeneration, a high risk of cancer, and immunodeficiency. Humans with AT are also hypersensitive to radiation. No curative strategy for this disease exists.

The gene product defective in this syndrome, ATM (ataxia-telangiectasia mutated), normally recognizes DNA damage. ATM is a protein kinase, which is activated immediately after a DNA double-strand break (DSB) occurs, and the resulting signal cascade generated in response to cellular DSBs is regulated by posttranslational protein modifications, such as phosphorylation and acetylation. ATM signals to the DNA repair machinery and the cell cycle checkpoints to minimize the risk of genetic damage.

ATM also responds to physiological breaks in DNA during the development and differentiation of B and T cells and in the resolution of DNA DSBs that are generated during V(D)J recombination. ATM is involved in supporting efficient Vα–Jα coding. Consistent with this, increased accumulation of unrepaired coding ends during different steps of antigen receptor-gene assembly, for both immunoglobulin and T-cell receptor loci, has been reported in ATM-deficient B and T lymphocytes [141].

The involvement in V(D)J recombination likely explains the cytogenetic hallmarks of AT, which include frequent nonrandom rearrangements of chromosomes 7 and 14 in T lymphocytes, nonspecific chromosomal breaks in fibroblasts, and normal chromosomes. The breakpoints in the lymphocyte rearrangements are at 7p14, 7q35, 14q12, and 14q32, involving the T-cell receptor and immunoglobulin heavy chain genes. In addition, ataxia-telangiectasia homozygotes have markedly increased rates of nonspecific spontaneous translocations.

3.7.3.2 Nijmegen Breakage Syndrome

Nijmegen breakage syndrome (NBS) is a rare recessive genetic disorder. Characteristics include bird-like facial appearance, congenital microcephaly, early growth retardation, immunodeficiency, and high frequency of malignancies.

Upon chromosome analysis NBS cells display frequently spontaneous chromosomal aberrations and are hypersensitive to DNA double-strand break-inducing agents, such as ionizing radiations. NBS shares with AT cytogenetic features because preferentially chromosomes 7 and 14 are involved in rearrangements. The gene underlying the disease, *NBS1*, interacts with the *ATM* gene. NBS1 forms a multimeric complex with MRE11/RAD50 nuclease at the C terminus and retains or recruits them in the vicinity of sites of DNA damage by direct binding to histone H2AX, which is phosphorylated in response to DNA damage. Thereafter, the NBS1 complex proceeds to rejoin double-strand breaks predominantly by homologous recombination repair in vertebrates. NBS cells also show to be defective in the activation of intra-S-phase checkpoint [5].

3.7.4 Immunodeficiency, Centromeric Region Instability, and Facial Anomalies Syndrome

The immunodeficiency, centromeric region instability, and facial anomalies syndrome (ICF) results from a mutated DNA methyltransferase gene, *DNMT3B*. The phenotype, physical and cytogenetic changes, can be regarded as secondary to a failure of methylation. The defective methylation results in chromatin decondensation, which becomes apparent at the juxtacentromeric heterochromatin of chromosomes 1, 9, and 16. As a consequence, these chromosomes can form "windmill" multiradials by interchange within heterochromatic regions [69].

3.7.5 Roberts Syndrome/SC Phocomelia

Roberts syndrome/SC phocomelia (RBS) is an autosomal recessive disorder of symmetric limb defects, craniofacial abnormalities, and pre- and postnatal growth retardation. Intellect is normal. Most affected individuals (about 80%) show a chromosomal phenomenon known as premature centromere separation, sometimes also referred to as "heterochromatin repulsion." This is due to lack of cohesion at the heterochromatic regions around centromeres and the long arm of the Y chromosome, which results in reduced growth capacity, and hypersensitivity to DNA-damaging agents. RBS is caused by mutations in *ESCO2*, which encodes a protein involved in regulating sister chromatid cohesion [92] (see also Sect. 3.3.3).

3.7.6 Mosaic Variegated Aneuploidy

Mosaic variegated aneuploidy (MVA) is a recessive condition characterized by mosaic aneuploidies, predominantly trisomies and monosomies, involving multiple different chromosomes and tissues. The proportion of aneuploid cells varies, but is usually >25% and is substantially greater than in normal individuals. Affected individuals typically present with severe intrauterine growth retardation and microcephaly. Eye anomalies, mild dysmorphism, variable developmental delay, and a broad spectrum of additional congenital abnormalities and medical conditions may also occur. The risk of malignancy is high, with rhabdomyosarcoma, Wilms tumor, and leukemia reported in several cases.

In five families with this disorder, truncating and missense mutations of *BUB1B*, which encodes BUBR1, a key protein in the mitotic spindle checkpoint, were found. These data suggest that germline mutations in a spindle checkpoint gene may increase the rate of aneuploidy and strongly support a causal link between aneuploidy and cancer development [99].

References

1. Abramsky L, Hall S, Levitan J, Marteau TM (2001) What parents are told after prenatal diagnosis of a sex chromosome abnormality: Interview and questionnaire study. Br Med J 322:463–466
2. Aggarwal VS, Morrow BE (2008) Genetic modifiers of the physical malformations in velo-cardio-facial syndrome/DiGeorge syndrome. Dev Disabil Res Rev 14:19–25
3. Allsopp RC, Vaziri H, Patterson C, Goldstein S, Younglai EV, Futcher AB, Greider CW, Harley CB (1992) Telomere length predicts replicative capacity of human fibroblasts. Proc Natl Acad Sci USA 89:10114–10118
4. Amiel J, Rio M, de Pontual L, Redon R, Malan V, Boddaert N, Plouin P, Carter NP, Lyonnet S, Munnich A, Colleaux L (2007) Mutations in TCF4, encoding a class I basic helix-loop-helix transcription factor, are responsible for Pitt-Hopkins syndrome, a severe epileptic encephalopathy associated with autonomic dysfunction. Am J Hum Genet 80:988–993
5. Antoccia A, Kobayashi J, Tauchi H, Matsuura S, Komatsu K (2006) Nijmegen breakage syndrome and functions of the responsible protein, NBS1. Genome Dyn 1:191–205
6. Antonarakis SE, Epstein CJ (2006) The challenge of Down syndrome. Trends Mol Med 12:473–479
7. Antonarakis SE, Petersen MB, McInnis MG, Adelsberger PA, Schinzel AA, Binkert F, Pangalos C, Raoul O, Slaugenhaupt SA, Hafez M, Cohen MM, Roulson D, Schwartz S, Mikkelsen M, Tranebjaerg L, Greenberg F, Hoar DI, Rudd NL, Warren AC, Metaxotou C, Bartsocas C, Chakravarti A (1992) The meiotic stage of nondisjunction in trisomy 21: determination by using DNA polymorphisms. Am J Hum Genet 50:544–550
8. Antonarakis SE, Avramopoulos D, Blouin JL, Talbot CC Jr, Schinzel AA (1993) Mitotic errors in somatic cells cause trisomy 21 in about 4.5% of cases and are not associated with advanced maternal age. Nat Genet 3:146–150
9. Antonarakis SE, Lyle R, Dermitzakis ET, Reymond A, Deutsch S (2004) Chromosome 21 and Down syndrome: from genomics to pathophysiology. Nat Rev Genet 5:725–738
10. Armanios MY, Chen JJ, Cogan JD, Alder JK, Ingersoll RG, Markin C, Lawson WE, Xie M, Vulto I, Phillips JA 3 rd, Lansdorp PM, Greider CW, Loyd JE (2007) Telomerase mutations in families with idiopathic pulmonary fibrosis. N Engl J Med 356:1317–1326
11. Armour JA, Sismani C, Patsalis PC, Cross G (2000) Measurement of locus copy number by hybridisation with amplifiable probes. Nucleic Acids Res 28:605–609
12. Arnheim N, Calabrese P, Tiemann-Boege I (2007) Mammalian meiotic recombination hot spots. Annu Rev Genet 41:369–399
13. Arnold J (1879) Beobachtungen über Kernteilungen in den Zellen der Geschwülste. Virchows Arch Pathol Anat 78:279
14. Arnoldus EP, Wiegant J, Noordemeer IA, Wessels JW, Beverstock GC, Grosveld GC, van der Ploeg M, Raap AK (1990) Detection of the Philadelphia chromosome in interphase nuclei. Cytogenet Cell Genet 54:108–111
15. Arrighi FE, Hsu TC (1971) Localization of heterochromatin in human chromosomes. Cytogenetics 10:81–86
16. Bacher CP, Guggiari M, Brors B, Augui S, Clerc P, Avner P, Eils R, Heard E (2006) Transient colocalization of X-inactivation centres accompanies the initiation of X inactivation. Nat Cell Biol 8:293–299
17. Baek KH, Zaslavsky A, Lynch RC, Britt C, Okada Y, Siarey RJ, Lensch MW, Park IH, Yoon SS, Minami T, Korenberg JR, Folkman J, Daley GQ, Aird WC, Galdzicki Z, Ryeom S (2009) Down's syndrome suppression of tumour growth and the role of the calcineurin inhibitor DSCR1. Nature 459:1126–1130
18. Baena N, De Vigan C, Cariati E, Clementi M, Stoll C, Caballín MR, Guitart M (2004) Turner syndrome: evaluation of prenatal diagnosis in 19 European registries. Am J Med Genet 129A:16–20
19. Bailey JA, Gu Z, Clark RA, Reinert K, Samonte RV, Schwartz S, Adams MD, Myers EW, Li PW, Eichler EE (2002) Recent segmental duplications in the human genome. Science 297:1003–1007
20. Balikova I, Martens K, Melotte C, Amyere M, Van Vooren S, Moreau Y, Vetrie D, Fiegler H, Carter NP, Liehr T, Vikkula M, Matthijs G, Fryns JP, Casteels I, Devriendt K, Vermeesch JR (2008) Autosomal-dominant microtia linked to five tandem copies of a copy-number-variable region at chromosome 4p16. Am J Hum Genet 82:181–187
21. Ballif BC, Hornor SA, Jenkins E, Madan-Khetarpal S, Surti U, Jackson KE, Asamoah A, Brock PL, Gowans GC, Conway RL, Graham JM Jr, Medne L, Zackai EH, Shaikh TH, Geoghegan J, Selzer RR, Eis PS, Bejjani BA, Shaffer LG (2007) Discovery of a previously unrecognized microdeletion syndrome of 16p11.2–p12.2. Nat Genet 39:1071–1073
22. Barr ML, Bertram LF (1949) A morphological distinction between neurones of the male and the female and the behavior of the nucleolar satellite during accelerated nucleoprotein synthesis. Nature 163:676–677
23. Bassell GJ, Warren ST (2008) Fragile X syndrome: loss of local mRNA regulation alters synaptic development and function. Neuron 60:201–214
24. Bauman JG, Wiegant J, Borst P, van Duijn P (1980) A new method for fluorescence microscopical localization of specific DNA sequences by in situ hybridization of fluorochromelabelled RNA. Exp Cell Res 128:485–490
25. Baumann C, Körner R, Hofmann K, Nigg EA (2007) PICH, a centromere-associated SNF2 family ATPase, is regulated by Plk1 and required for the spindle checkpoint. Cell 128:101–114
26. Beever CL, Stephenson MD, Peñaherrera MS, Jiang RH, Kalousek DK, Hayden M, Field L, Brown CJ, Robinson WP (2003) Skewed X-chromosome inactivation is associated with trisomy in women ascertained on the basis of recurrent spontaneous abortion or chromosomally abnormal pregnancies. Am J Hum Genet 72:399–407
27. Beutler E, Yeh M, Fairbanks VF (1962) The Normal human female as a mosaic of X-Chromosome activity:

studies using the gene for G-6-PD-Deficiency as a marker. Proc Natl Acad Sci USA 48:9–16
28. Blakeslee AF, Avery AG (1937) Methods of inducing doubling of chromosomes in plants. J Hered 28:392–411
29. Bodnar AG, Ouellette M, Frolkis M, Holt SE, Chiu CP, Morin GB, Harley CB, Shay JW, Lichtsteiner S, Wright WE (1998) Extension of life-span by introduction of telomerase into normal human cells. Science 279:349–352
30. Bojesen A, Gravholt CH (2007) Klinefelter syndrome in clinical practice. Nat Clin Pract Urol 4:192–204
31. Bolzer A, Kreth G, Solovei I, Koehler D, Saracoglu K, Fauth C, Müller S, Eils R, Cremer C, Speicher MR, Cremer T (2005) Three-dimensional maps of all chromosome positions indicate a probabilistic order in human male fibroblast nuclei and prometaphase rosettes. PLoS Biol 3:e157
32. Branco MR, Pombo A (2006) Intermingling of chromosome territories in interphase suggests role in translocations and transcription-dependent associations. PLoS Biol 4(5):e138
33. Bridges CB (1916) Non-disjunction as proof on the chromosome theory of heredity. Genetics 1:1–52
34. Bridges CB (1936) The bar "gene" a duplication. Science 83:210–211
35. Brockschmidt A, Todt U, Ryu S, Hoischen A, Landwehr C, Birnbaum S, Frenck W, Radlwimmer B, Lichter P, Engels H, Driever W, Kubisch C, Weber RG (2007) Severe mental retardation with breathing abnormalities (Pitt-Hopkins syndrome) is caused by haploinsufficiency of the neuronal bHLH transcription factor TCF4. Hum Mol Genet 16:1488–1494
36. Brown CJ, Robinson WP (2000) The causes and consequences of random and non-random X chromosome inactivation in humans. Clin Genet 58:353–363
37. Brown CJ, Hendrich BD, Rupert JL, Lafrenière RG, Xing Y, Lawrence J, Willard HF (1992) The human XIST gene: analysis of a 17 kb inactive X-specific RNA that contains conserved repeats and is highly localized within the nucleus. Cell 71:527–542
38. Cabral RE, Queille S, Bodemer C, de Prost Y, Neto JB, Sarasin A, Daya-Grosjean L (2008) Identification of new RECQL4 mutations in Caucasian Rothmund-Thomson patients and analysis of sensitivity to a wide range of genotoxic agents. Mutat Res 643:41–47
39. Carrel L, Willard HF (2005) X-inactivation profile reveals extensive variability in X-linked gene expression in females. Nature 434:400–404
40. Caspersson T, Farber S, Foley GE, Kudynowski J, Modest EJ, Simonsson E, Wagh U, Zech L (1968) Chemical differentiation along metaphase chromosomes. Exp Cell Res 49:219–222
41. Caspersson T, Zech L, Johansson C (1970) Analysis of human metaphase chromosome set by aid of DNA-binding fluorescent agents. Exp Cell Res 62:490–492
42. Chaumeil J, Le Baccon P, Wutz A, Heard E (2006) A novel role for Xist RNA in the formation of a repressive nuclear compartment into which genes are recruited when silenced. Genes Dev 20:2223–2237
43. Cheeseman IM, Desai A (2008) Molecular architecture of the kinetochore-microtubule interface. Nat Rev Mol Cell Biol 9:33–46
44. Cheung VG, Nowak N, Jang W, Kirsch IR, Zhao S, Chen XN, Furey TS, Kim UJ, Kuo WL, Olivier M, Conroy J, Kasprzyk A, Massa H, Yonescu R, Sait S, Thoreen C, Snijders A, Lemyre E, Bailey JA, Bruzel A, Burrill WD, Clegg SM, Collins S, Dhami P, Friedman C, Han CS, Herrick S, Lee J, Ligon AH, Lowry S, Morley M, Narasimhan S, Osoegawa K, Peng Z, Plajzer-Frick I, Quade BJ, Scott D, Sirotkin K, Thorpe AA, Gray JW, Hudson J, Pinkel D, Ried T, Rowen L, Shen-Ong GL, Strausberg RL, Birney E, Callen DF, Cheng JF, Cox DR, Doggett NA, Carter NP, Eichler EE, Haussler D, Korenberg JR, Morton CC, Albertson D, Schuler G, de Jong PJ, Trask BJ, BAC Resource Consortium (2001) Integration of cytogenetic landmarks into the draft sequence of the human genome. Nature 409:953–958
45. Clarke DJ, Vas AC, Andrews CA, Díaz-Martínez LA, Giménez-Abián JF (2006) Topoisomerase II checkpoints: universal mechanisms that regulate mitosis. Cell Cycle 5:1925–1928
46. Clemson CM, Hall LL, Byron M, McNeil J, Lawrence JB (2006) The X chromosome is organized into a gene-rich outer rim and an internal core containing silenced nongenic sequences. Proc Natl Acad Sci USA 103:7688–7693
47. Cleveland DW, Mao Y, Sullivan KF (2003) Centromeres and kinetochores: from epigenetics to mitotic checkpoint signaling. Cell 112:407–421
48. Coop G, Wen X, Ober C, Pritchard JK, Przeworski M (2008) High-resolution mapping of crossovers reveals extensive variation in fine-scale recombination patterns among humans. Science 319:1395–1398
49. Courtens W, Schramme I, Laridon A (2008) Microduplication 22q11.2: a benign polymorphism or a syndrome with a very large clinical variability and reduced penetrance? Report of two families. Am J Med Genet 146A:758–763
50. Cremer T, Cremer C (2001) Chromosome territories, nuclear architecture and gene regulation in mammalian cells. Nat Rev Genet 2:292–301
51. Cremer T, Landegent J, Brückner A, Scholl HP, Schardin M, Hager HD, Devilee P, Pearson P, van der Ploeg M (1986) Detection of chromosome aberrations in the human interphase nucleus by visualization of specific target DNAs with radioactive and non-radioactive in situ hybridization techniques: diagnosis of trisomy 18 with probe L1.85. Hum Genet 74:346–352
52. Cremer T, Lichter P, Borden J, Ward DC, Manuelidis L (1988) Detection of chromosome aberrations in metaphase and interphase tumor cells by in situ hybridization using chromosome-specific library probes. Hum Genet 80:235–246
53. Cremer T, Küpper K, Dietzel S, Fakan S (2004) Higher order chromatin architecture in the cell nucleus: on the way from structure to function. Biol Cell 96:555–567
54. Croft JA, Bridger JM, Boyle S, Perry P, Teague P, Bickmore WA (1999) Differences in the localization and morphology of chromosomes in the human nucleus. J Cell Biol 145:1119–1131
55. d'Adda di Magagna F, Reaper PM, Clay-Farrace L, Fiegler H, Carr P, Von Zglinicki T, Saretzki G, Carter NP, Jackson SP (2003) A DNA damage checkpoint response in telomere-initiated senescence. Nature 426:194–198
56. Davidson WM, Smith DR (1954) The nuclear sex of leucocytes. In: Overzier C (ed) Intersexuality. Academic, New York, pp 72–85

57. Debacker K, Kooy RF (2007) Fragile sites and human disease. Hum Mol Genet 16(Spec No.2):R150–R158
58. de Grouchy J, Lamy M, Thieffry S, Arthuis M, Salmon C (1963) Dysmorphie complexe avec oligophrénie: délétion des bras courts d'un chromosome 17-18. CR Acad Sci 256:1028–1029
59. de Lange T (1994) Activation of telomerase in a human tumor. Proc Natl Acad Sci USA 91:2882–2885
60. DesGroseilliers M, Lemyre E, Dallaire L, Lemieux N (2002) Tetrasomy Y by structural rearrangement: clinical report. Am J Med Genet 111:401–404
61. DiGeorge AM (1968) Congenital absence of the thymus and its immunologic consequences: concurrence with congenital hypoparathyroidism. Birth Defects Orig Art Ser IV(1):116–121
62. Dorsett D (2007) Roles of the sister chromatid cohesion apparatus in gene expression, development, and human syndromes. Chromosoma 116:1–13
63. du Manoir S, Speicher MR, Joos S, Schröck E, Popp S, Döhner H, Kovacs G, Robert-Nicoud M, Lichter P, Cremer T (1993) Detection of complete and partial chromosome gains and losses by comparative genomic in situ hybridization. Hum Genet 90:590–610
64. Dunham I, Shimizu N, Roe BA, Chissoe S, Hunt AR, Collins JE, Bruskiewich R, Beare DM, Clamp M, Smink LJ, Ainscough R, Almeida JP, Babbage A, Bagguley C, Bailey J, Barlow K, Bates KN, Beasley O, Bird CP, Blakey S, Bridgeman AM, Buck D, Burgess J, Burrill WD, O'Brien KP et al (1999) The DNA sequence of human chromosome 22. Nature 402:489–495
65. Durkin SG, Glover TW (2007) Chromosome fragile sites. Annu Rev Genet 41:169–192
66. Dutrillaux B, Lejeune J (1971) Sur une novelle technique d'analyse du caryotype humain. C R Acad Sci Hebd Seances Acad Sci D 272:2638–2640
67. Edelmann L, Pandita RK, Spiteri E, Funke B, Goldberg R, Palanisamy N, Chaganti RSK, Magenis E, Shprintzen RJ, Morrow BE (1999) A common molecular basis for rearrangement disorders on chromosome 22q11. Hum Mol Genet 8:1157–1167
68. Edwards JH, Harnden DG, Cameron AH, Crosse VM, Wolff OH (1960) A new trisomic syndrome. Lancet 1:787
69. Ehrlich M, Sanchez C, Shao C, Nishiyama R, Kehrl J, Kuick R, Kubota T, Hanash SM (2008) ICF, an immunodeficiency syndrome: DNA methyltransferase 3B involvement, chromosome anomalies, and gene dysregulation. Autoimmunity 41:253–271
70. Emanuel BS (2008) Molecular mechanisms and diagnosis of chromosome 22q11.2 rearrangements. Dev Disabil Res Rev 14:11–18
71. Erwin JA, Lee JT (2008) New twists in X-chromosome inactivation. Curr Opin Cell Biol 20:349–355
72. Evans JA, MacDonald K, Hamerton JL (1990) Sex chromosome anomalies: prenatal diagnosis and the need for continued prospective studies. Birth Defects Orig Artic Ser 26:273–281
73. Fauth C, Speicher MR (2001) Classifying by colors: FISH-based genome analysis. Cytogenet Cell Genet 93:1–10
74. Feuk L, Carson AR, Scherer SW (2006) Structural variation in the human genome. Nat Rev Genet 7:85–97
75. Fiegler H, Carr P, Douglas EJ, Burford DC, Hunt S, Scott CE, Smith J, Vetrie D, Gorman P, Tomlinson IP, Carter NP (2003) DNA microarrays for comparative genomic hybridization based on DOP-PCR amplification of BAC and PAC clones. Genes Chromosomes Cancer 36:361–374
76. Firth HV, Richards SM, Bevan AP, Clayton S, Corpas M, Rajan D, Van Vooren S, Moreau Y, Pettett RM, Carter NP (2009) DECIPHER: database of chromosomal imbalance and phenotype in humans using ensembl resources. Am J Hum Genet 84:524–533
77. Flaquer A, Rappold GA, Wienker TF, Fischer C (2008) The human pseudoautosomal regions: a review for genetic epidemiologists. Eur J Hum Genet 16:771–779
78. Flemming W (1882) Beiträge zur Kenntnis der Zelle und ihrer Lebenserscheinungen III. Arch Mikrosk Anat 20:1
79. Ford CE, Hamerton JL (1956) The chromosomes of man. Nature 178:1020–1023
80. Ford CE, Miller OJ, Polani PE, de Almeida JC, Briggs JH (1959) A sex-chromosome anomaly in a case of gonadal dysgenesis (Turner's syndrome). Lancet 1:711–713
81. Francke U (1994) Digitized and differentially shaded human chromosome ideograms for genomic applications. Cytogenet Cell Genet 65:206–219
82. Gabriel-Robez O, Delobel B, Croquette MF, Rigot JM, Djlelati R, Rumpler Y (1996) Synaptic behaviour of sex chromosome in two XYY men. Annales de Génétiques 39:129–132
83. Gall JG, Pardue ML (1969) Formation and detection of RNA-DNA hybrid molecules in cytological preparations. Proc Natl Acad Sci USA 63:378–383
84. Gardner RJM, Sutherlnad GR (2004) Chromosome abnormalities and genetic counseling, vol 3. Oxford University Press, Oxford
85. Gartler SM (2006) The chromosome number in humans: a brief history. Nat Rev Genet 7:655–660
86. German J (1997) Bloom's syndrome. XX. The first 100 cancers. Cancer Genet Cytogenet 93:100–106
87. German J, Archibald R, Bloom D (1965) Chromosomal breakage in a rare and probably genetically determined syndrome of man. Science 148:506–507
88. German J, Sanz MM, Ciocci S, Ye TZ, Ellis NA (2007) Syndrome-causing mutations of the BLM gene in persons in the Bloom's Syndrome Registry. Hum Mutat 28:743–753
89. Gibbons RJ, Higgs DR (2000) Molecular-clinical spectrum of the ATR-X syndrome. Am J Med Genet 97:204–212
90. Gilbert N, Boyle S, Fiegler H, Woodfine K, Carter NP, Bickmore WA (2004) Chromatin architecture of the human genome: gene-rich domains are enriched in open chromatin fibers. Cell 118:555–566
91. Gilson E, Géli V (2007) How telomeres are replicated. Nat Rev Mol Cell Biol 8:825–838
92. Gordillo M, Vega H, Trainer AH, Hou F, Sakai N, Luque R, Kayserili H, Basaran S, Skovby F, Hennekam RC, Uzielli ML, Schnur RE, Manouvrier S, Chang S, Blair E, Hurst JA, Forzano F, Meins M, Simola KO, Raas-Rothschild A, Schultz RA, McDaniel LD, Ozono K, Inui K, Zou H, Jabs EW (2008) The molecular mechanism underlying Roberts syndrome involves loss of ESCO2

acetyltransferase activity. Hum Mol Genet 17: 2172–2180

93. Goto M (1997) Hierarchical deterioration of body systems in Werner's syndrome: implications for normal ageing. Mech Ageing Dev 98:239–254

94. Goto M, Miller RW, Ishikawa Y, Sugano H (1996) Excess of rare cancers in Werner syndrome (adult progeria). Cancer Epidemiol Biomarkers Prev 5:239–246

95. Gravholt CH, Landin-Wilhelmsen K, Stochholm K, Hjerrild BE, Ledet T, Djurhuus CB, Sylvén L, Baandrup U, Kristensen BØ, Christiansen JS (2006) Clinical and epidemiological description of aortic dissection in Turner's syndrome. Cardiol Young 16:430–436

96. Greider CW, Blackburn EH (1985) Identification of a specific telomere terminal transferase activity in Tetrahymena extracts. Cell 43:405–413

97. Gribble SM, Prigmore E, Burford DC, Porter KM, Ng BL, Douglas EJ, Fiegler H, Carr P, Kalaitzopoulos D, Clegg S, Sandstrom R, Temple IK, Youings SA, Thomas NS, Dennis NR, Jacobs PA, Crolla JA, Carter NP (2005) The complex nature of constitutional de novo apparently balanced translocations in patients presenting with abnormal phenotypes. J Med Genet 42:8–16

98. Griffith JD, Comeau L, Rosenfield S, Stansel RM, Bianchi A, Moss H, de Lange T (1999) Mammalian telomeres end in a large duplex loop. Cell 97:503–514

99. Hanks S, Coleman K, Reid S, Plaja A, Firth H, Fitzpatrick D, Kidd A, Méhes K, Nash R, Robin N, Shannon N, Tolmie J, Swansbury J, Irrthum A, Douglas J, Rahman N (2004) Constitutional aneuploidy and cancer predisposition caused by biallelic mutations in BUB1B. Nat Genet 36:1159–1161

100. Hansemann D (1890) Über asymmetrische Zelltheilung in Epithelkrebsen und deren biologische Bedeutung. Arch Pathol Anat 119:299–326

101. Hartwell LH, Weinert TA (1989) Checkpoints: controls that ensure the order of cell cycle events. Science 246:629–634

102. Hassold T, Hall H, Hunt P (2007) The origin of human aneuploidy: where we have been, where we are going. Hum Mol Genet 16:R203–R208

103. Hayflick L, Moorhead PS (1961) The serial cultivation of human diploid cell strains. Exp Cell Res 25:585–621

104. Heard E, Disteche CM (2006) Dosage compensation in mammals: fine-tuning the expression of the X chromosome. Genes Dev 20:1848–1867

105. Heitz E (1928) Das Heterochromatin der Moose. I Jahrb Wiss Bot 69:762–818

106. Hitzler JK, Zipursky A (2005) Origins of leukaemia in children with Down syndrome. Nat. Rev. Cancer 5:11–20

107. Hjerrild BE, Mortensen KH, Gravholt CH (2008) Turner syndrome and clinical treatment. Br Med Bull 86:77–93

108. Ho VB, Bakalov VK, Cooley M, Van PL, Hood MN, Burklow TR, Bondy CA (2004) Major vascular anomalies in Turner syndrome: prevalence and magnetic resonance angiographic features. Circulation 110:1694–1700

109. Horn PJ, Peterson CL (2002) Chromatin higher order folding: wrapping up the transcription. Science 297: 1824–1827

110. Hsu TC (1952) Mammalian chromosomes in vitro I. The karyotype of man. J Hered 43:167

111. Hsu TC, Pomerat CM (1953) Mammalian chromosomes in vitro II. A method for spreading the chromosomes of cells in tissue culture. J Hered 44:23–29

112. Hubbard TJ, Aken BL, Ayling S, Ballester B, Beal K, Bragin E, Brent S, Chen Y, Clapham P, Clarke L, Coates G, Fairley S, Fitzgerald S, Fernandez-Banet J, Gordon L, Graf S, Haider S, Hammond M, Holland R, Howe K, Jenkinson A, Johnson N, Kahari A, Keefe D, Keenan S, Kinsella R, Kokocinski F, Kulesha E, Lawson D, Longden I, Megy K, Meidl P, Overduin B, Parker A, Pritchard B, Rios D, Schuster M, Slater G, Smedley D, Spooner W, Spudich G, Trevanion S, Vilella A, Vogel J, White S, Wilder S, Zadissa A, Birney E, Cunningham F, Curwen V, Durbin R, Fernandez-Suarez XM, Herrero J, Kasprzyk A, Proctor G, Smith J, Searle S, Flicek P (2009) Ensembl 2009. Nucleic Acids Res 37(Database issue):D690–D697

113. Hunt PA, Hassold TJ (2008) Human female meiosis: what makes a good egg go bad? Trends Genet 24:86–93

114. Iafrate AJ, Feuk L, Rivera MN, Listewnik ML, Donahoe PK, Qi Y, Scherer SW, Lee C (2004) Detection of large-scale variation in the human genome. Nat Genet 36:949–951

115. ISCN (2009) In: Shaffer LG, Slovak ML, Campbell LJ (eds) An International System for Human Cytogenetic Nomenclature (2005): Recommendations of the International Standing Committee on Human Cytogenetic Nomenclature. S. Karger AG, Basel

116. Jacobs PA, Strong JA (1959) A case of human intersexuality having a possible XXY sex-determining mechanism. Nature 183:302–303

117. Jacobs PA, Baikie AG, Brown WM, MacGregor TN, MacLean N, Harnden DG (1959) Evidence for the existence of the human "superfemale". Lancet 2:423–425

118. Jakobsson M, Scholz SW, Scheet P, Gibbs JR, VanLiere JM, Fung HC, Szpiech ZA, Degnan JH, Wang K, Guerreiro R, Bras JM, Schymick JC, Hernandez DG, Traynor BJ, Simon-Sanchez J, Matarin M, Britton A, van de Leemput J, Rafferty I, Bucan M, Cann HM, Hardy JA, Rosenberg NA, Singleton AB (2008) Genotype, haplotype and copy-number variation in worldwide human populations. Nature 451:998–1003

119. Jallepalli PV, Lengauer C (2001) Chromosome segregation and cancer: cutting through the mystery. Nat Rev Cancer 1:109–117

120. Jeffreys AJ, Kauppi L, Neumann R (2001) Intensely punctate meiotic recombination in the class II region of the major histocompatibility complex. Nat Genet 29:217–222

121. Jenuwein T, Allis CD (2001) Translating the histone code. Science 293:1074–1080

122. Jerome LA, Papaioannou VE (2001) DiGeorge syndrome phenotype in mice mutant for the T-box gene, Tbx1. Nat Genet 27:286–291

123. John HA, Birnstiel ML, Jones KW (1969) RNA-DNA hybrids at the cytological level. Nature 223:582–587

124. Johnson J, Canning J, Kaneko T, Pru JK, Tilly JL (2004) Germline stem cells and follicular renewal in the postnatal mammalian ovary. Nature 428:145–150

125. Johnson J, Bagley J, Skaznik-Wikiel M, Lee HJ, Adams GB, Niikura Y, Tschudy KS, Tilly JC, Cortes ML, Forkert R, Spitzer T, Iacomini J, Scadden DT, Tilly JL (2005) Oocyte generation in adult mammalian ovaries by putative

germ cells in bone marrow and peripheral blood. Cell 122:303–315
126. Kalb R, Neveling K, Nanda I, Schindler D, Hoehn H (2006) Fanconi anemia: causes and consequences of genetic instability. Genome Dyn 1:218–242
127. Kallioniemi A, Kallioniemi OP, Sudar D, Rutovitz D, Gray JW, Waldman F, Pinkel D (1992) Comparative genomic hybridization for molecular cytogenetic analysis of solid tumors. Science 258:818–821
128. Karlseder J, Smogorzewska A, de Lange T (2002) Senescence induced by altered telomere state, not telomere loss. Science 295:2446–2449
129. Klein CA, Schmidt-Kittler O, Schardt JA, Pantel K, Speicher MR, Riethmüller G (1999) Comparative genomic hybridization, loss of heterozygosity, and DNA sequence analysis of single cells. Proc Natl Acad Sci USA 96:4494–4499
130. Klopocki E, Schulze H, Strauss G, Ott CE, Hall J, Trotier F, Fleischhauer S, Greenhalgh L, Newbury-Ecob RA, Neumann LM, Habenicht R, König R, Seemanova E, Megarbane A, Ropers HH, Ullmann R, Horn D, Mundlos S (2007) Complex inheritance pattern resembling autosomal recessive inheritance involving a microdeletion in thrombocytopenia-absent radius syndrome. Am J Hum Genet 80:232–240
131. Kobrynski LJ, Sullivan KE (2007) Velocardiofacial syndrome, DiGeorge syndrome: the chromosome 22q11.2 deletion syndromes. Lancet 370:1443–1452
132. Koolen DA, Vissers LE, Pfundt R, de Leeuw N, Knight SJ, Regan R, Kooy RF, Reyniers E, Romano C, Fichera M, Schinzel A, Baumer A, Anderlid BM, Schoumans J, Knoers NV, van Kessel AG, Sistermans EA, Veltman JA, Brunner HG, de Vries BB (2006) A new chromosome 17q21.31 microdeletion syndrome associated with a common inversion polymorphism. Nat Genet 38:999–1001
133. Kottler MJ (1974) From 48 to 46: cytological technique, preconception, and the counting of human chromosomes. Bull Hist Med 48:465–502
134. Krusinskiene V, Alvesalo L, Sidlauskas A (2005) The craniofacial complex in 47, XXX females. Eur J Orthod 27:396–401
135. Kuhn RM, Karolchik D, Zweig AS, Wang T, Smith KE, Rosenbloom KR, Rhead B, Raney BJ, Pohl A, Pheasant M, Meyer L, Hsu F, Hinrichs AS, Harte RA, Giardine B, Fujita P, Diekhans M, Dreszer T, Clawson H, Barber GP, Haussler D, Kent WJ (2009) The UCSC genome browser database: update 2009. Nucleic Acids Res 37(Database issue): D755–D761
136. Lam AL, Boivin CD, Bonney CF, Rudd MK, Sullivan BA (2006) Human centromeric chromatin is a dynamic chromosomal domain that can spread over noncentromeric DNA. Proc Natl Acad Sci USA 103:4186–4191
137. Lanctôt C, Cheutin T, Cremer M, Cavalli G, Cremer T (2007) Dynamic genome architecture in the nuclear space: regulation of gene expression in three dimensions. Nat Rev Genet 8:104–115
138. Landegent JE, Jansen in dewal N, Dirks RW, Baas F, van der Ploeg M (1987) Use of whole cosmid cloned genomic sequences for chromosomal localization by non-radioactive in situ hybridization. Hum Genet 77:366–370
139. Lanfranco F, Kamischke A, Zitzmann M, Nieschlag E (2004) Klinefelter's syndrome. Lancet 364:273–283
140. Langer PR, Waldrop AA, Ward DC (1981) Enzymatic synthesis of biotin-labeled polynucleotides: novel nucleic acid affinity probes. Proc Natl Acad Sci USA 78:6633–6637
141. Lavin MF (2008) Ataxia-telangiectasia: from a rare disorder to a paradigm for cell signalling and cancer. Nat Rev Mol Cell Biol 9:759–769
142. Lejeune J, Gautier M, Turpin MR (1959) Etude des chromosomes somatiques de neuf enfants mongoliens. C R Acad Sci (Paris) 248:1721–1722
143. Lejeune J, Lafourcade J, Berger R, Vialatte J, Roeswillwald M, Seringe P, Turpin R (1963) Trois cas de délétion partielle du bras court d'un chromosome 5. C R Acad Sci (Paris) 257:3098–3102
144. Lengauer C, Kinzler KW, Vogelstein B (1997) Genetic instability in colorectal cancers. Nature 386:623–627
145. Levan A (1938) The effect of colchicine on root mitosis in Allium. Hereditas 24:471–486
146. Lichter P, Cremer T, Borden J, Manuelidis L, Ward DC (1988) Delineation of individual human chromosomes in metaphase and interphase cells by in situ suppression hybridization using recombinant DNA libraries. Hum Genet 80:224–234
147. Linden MG, Bender BG, Harmon RJ, Mrazek DA, Robinson A (1988) 47, XXX. What is the prognosis? Pediatrics 82:619–630
148. Losada A, Hirano M, Hirano T (1998) Identification of Xenopus SMC protein complexes required for sister chromatid cohesion. Genes Dev 12:1986–1997
149. Lowe X, Eskenazi B, Nelson DO, Kidd S, Alme A, Wyrobek AJ (2001) Frequency of XY sperm increases with age in fathers of boys with Klinefelter syndrome. Am J Hum Genet 69:1046–1054
150. Lupski JR (1998) Genomic disorders: structural features of the genome can lead to DNA rearrangements and human disease traits. Trends Genet 14:417–422
151. Lyon MF (1961) Gene action in the X-chromosome of the mouse (Mus musculus L.). Nature 190:372–373
152. Lyon MF (1998) X-chromosome inactivation: a repeat hypothesis. Cytogenet Cell Genet 80:133–137
153. MacLaughlin DT, Donahoe PK (2004) Sex determination and differentiation. N Engl J Med 350:367–378
154. Matsui S, Sasaki M (1973) Differential staining of nucleolus organisers in mammalian chromosomes. Nature 246:148–150
155. McClintock B (1941) The stability of broken ends of chromosomes in Zea mays. Genetics 26:234–282
156. Menasha J, Levy B, Hirschhorn K, Kardon NB (2005) Incidence and spectrum of chromosome abnormalities in spontaneous abortions: new insights from a 12-year study. Genet Med 7:251–263
157. Menten B, Maas N, Thienpont B, Buysse K, Vandesompele J, Melotte C, de Ravel T, Van Vooren S, Balikova I, Backx L, Janssens S, De Paepe A, De Moor B, Moreau Y, Marynen P, Fryns JP, Mortier G, Devriendt K, Speleman F, Vermeesch JR (2006) Emerging patterns of cryptic chromosomal imbalance in patients with idiopathic mental retardation and multiple congenital anomalies: a new series of 140 patients and review of published reports. J Med Genet 43:625–633

158. Michalet X, Ekong R, Fougerousse F, Rousseaux S, Schurra C, Hornigold N, van Slegtenhorst M, Wolfe J, Povey S, Beckmann JS, Bensimon A (1997) Dynamic molecular combing: stretching the whole human genome for high-resolution studies. Science 277:1518–1523
159. Morgan TH (1910) Sex-limited inheritance in drosophila. Science 32:120–122
160. Morris JK, Alberman E, Scott C, Jacobs P (2008) Is the prevalence of Klinefelter syndrome increasing? Eur J Hum Genet 16:163–170
161. Muldal S, Ockey CH (1960) Double male: new chromosome constitution in Klinefelter's syndrome. Lancet 2:492–493
162. Musacchio A, Salmon ED (2007) The spindle-assembly checkpoint in space and time. Nat Rev Mol Cell Biol 8:379–393
163. Myers S, Bottolo L, Freeman C, McVean G, Donnelly P (2005) A fine-scale map of recombination rates and hotspots across the human genome. Science 310:321–324
164. Nan X, Hou J, Maclean A, Nasir J, Lafuente MJ, Shu X, Kriaucionis S, Bird A (2007) Interaction between chromatin proteins MECP2 and ATRX is disrupted by mutations that cause inherited mental retardation. Proc Natl Acad Sci USA 104:2709–2714
165. Nasmyth K, Haering CH (2005) The structure and function of SMC and kleisin complexes. Annu Rev Biochem 74:595–648
166. Ng K, Pullirsch D, Leeb M, Wutz A (2007) Xist and the order of silencing. EMBO Rep 8:34–39
167. Nielsen J (1990) Follow-up of 25 unselected children with sex chromosome abnormalities to age 12. Birth Defects Orig Artic Ser 26:201–207
168. Nielsen J, Wohlert M (1991) Chromosome abnormalities found among 34, 910 newborn children: results from a 13-year incidence study in Arhus, Denmark. Hum Genet 87:81–83
169. Nigg EA (2002) Centrosome aberrations: cause or consequence of cancer progression? Nat Rev Cancer 2:815–825
170. Niimura Y, Gojobori T (2002) In silico chromosome staining: reconstruction of Giemsa bands from the whole human genome sequence. Proc Natl Acad Sci USA 99:797–802
171. Nowakowski H, Lenz W, Parada J (1959) Diskrepanz zwischen Chromatinbefund und genetischem Geschlecht beim Klinefelter-Syndrom. Acta Endocrinol (Copenh) 30:296–320
172. Nowell PC (1960) Phytohemagglutinin: an initiator of mitosis in cultures of normal human leukocytes. Cancer Res 20:462–466
173. Nowell PC, Hungerford DA (1960) A minute chromosome in human chronic granulocytic leukemia. Science 132:1497
174. Ouyang KJ, Woo LL, Ellis NA (2008) Homologous recombination and maintenance of genome integrity: cancer and aging through the prism of human RecQ helicases. Mech Ageing Dev 129:425–440
175. Paduch DA, Fine RG, Bolyakov A, Kiper J (2008) New concepts in Klinefelter syndrome. Curr Opin Urol 18:621–627
176. Painter TS (1923) Studies in mammalian spermatogenesis II. The spermatogenesis of man. J Exp Zool 37:291–321
177. Paris Conference (1971) Standardization in human cytogenetics. Birth Defects: Original article series, Vol 8, No 7 (The National Foundation, New York 1972); also in Cytogenetics 11:313-362
178. Patau K, Smith DW, Therman E, Inhorn SL, Wagner HP (1960) Multiple congenital anomaly caused by an extra chromosome. Lancet 1:790–793
179. Patel KJ, Joenje H (2007) Fanconi anemia and DNA replication repair. DNA Repair 6:885–890
180. Paulson JR, Laemmli UK (1977) The structure of histone-depleted metaphase chromosomes. Cell 12:817–828
181. Perry GH, Dominy NJ, Claw KG, Lee AS, Fiegler H, Redon R, Werner J, Villanea FA, Mountain JL, Misra R, Carter NP, Lee C, Stone AC (2007) Diet and the evolution of human amylase gene copy number variation. Nat Genet 39:1256–1260
182. Pesin JA, Orr-Weaver TL (2008) Regulation of APC/C activators in mitosis and meiosis. Annu Rev Cell Dev Biol 24:475–499
183. Peters AHFM, O'Carroll D, Scherthan H, Mechtler K, Sauer S, Schofer C, Weipoltshammer K, Pagani M, Lachner M, Kohlmaier A, Opravil S, Doyle M, Sibilia M, Jenuwein T (2001) Loss of the Suv39h histone methyltransferases impairs mammalian heterochromatin and genome stability. Cell 107:323–337
184. Peters AHFM, Mermoud JE, O'Carroll D, Pagani M, Schweizer D, Brockdorff N, Jenuwein T (2002) Histone H3 lysine 9 methylation is an epigenetic imprint of facultative heterochromatin. Nat Genet 30:77–80
185. Peters JM (2006) The anaphase promoting complex/cyclosome: a machine designed to destroy. Nat Rev Mol Cell Biol 7:644–656
186. Petes TD (2001) Meiotic recombination hot spots and cold spots. Nat Rev Genet 2:360–369
187. Pinkel D, Landegent J, Collins C, Fuscoe J, Segraves R, Lucas J, Gray JW (1988) Fluorescence in situ hybridization with human chromosome-specific libraries: detection of trisomy 21 and translocations of chromosome 4. Proc Natl Acad Sci USA 85:9138–9142
188. Pinkel D, Segraves R, Sudar D, Clark S, Poole I, Kowbel D, Collins C, Kuo WL, Chen C, Zhai Y, Dairkee SH, Ljung BM, Gray JW, Albertson DG (1998) High resolution analysis of DNA copy number variations using comparative genomic hybridization to microarrays. Nat Genet 20:207–211
189. Pinsky BA, Biggins S (2005) The spindle checkpoint: tension versus attachment. Trends Cell Biol 15:486–493
190. Polani PE, Bishop PMF, Lennox B, Ferguson-Smith MA, Stewart JSS, Prader A (1958) Color vision studies in the X-chromosome constitution of patients with Klinefelter's syndrome. Nature 182:1092–1093
191. Popova BC, Tada T, Takagi N, Brockdorff N, Nesterova TB (2006) Attenuated spread of X-inactivation in an X;autosome translocation. Proc Natl Acad Sci USA 103:7706–7711
192. Pradhan M, Dalal A, Khan F, Agrawal S (2006) Fertility in men with down syndrome: a case report. Fertil Steril 86(1765):e1–e3
193. Rao E, Weiss B, Fukami M, Rump A, Niesler B, Mertz A, Muroya K, Binder G, Kirsch S, Winkelmann M, Nordsiek G, Heinrich U, Breuning MH, Ranke MB, Rosenthal A,

Ogata T, Rappold GA (1997) Pseudoautosomal deletions encompassing a novel homeobox gene cause growth failure in idiopathic short stature and Turner syndrome. Nat Genet 16:54–63
194. Ratcliffe SG, Pan H, McKie M (1994) The growth of XXX females: population-based studies. Ann Hum Biol 21:57–66
195. Ratcliffe SG, Masera N, Pan H, McKie M (1994) Head circumference and IQ of children with sex chromosome abnormalities. Dev Med Child Neurol 36:533–544
196. Redon R, Ishikawa S, Fitch KR, Feuk L, Perry GH, Andrews TD, Fiegler H, Shapero MH, Carson AR, Chen W, Cho EK, Dallaire S, Freeman JL, González JR, Gratacòs M, Huang J, Kalaitzopoulos D, Komura D, MacDonald JR, Marshall CR, Mei R, Montgomery L, Nishimura K, Okamura K, Shen F, Somerville MJ, Tchinda J, Valsesia A, Woodwark C, Yang F, Zhang J, Zerjal T, Zhang J, Armengol L, Conrad DF, Estivill X, Tyler-Smith C, Carter NP, Aburatani H, Lee C, Jones KW, Scherer SW, Hurles ME (2006) Global variation in copy number in the human genome. Nature 444:444–454
197. Rieder CL, Cole RW, Khodjakov A, Sluder G (1995) The checkpoint delaying anaphase in response to chromosome monoorientation is mediated by an inhibitory signal produced by unattached kinetochores. J Cell Biol 130:941–948
198. Rovet J (2004) Turner syndrome: a review of genetic and hormonal influences on neuropsychological functioning. Child Neuropsychol 10:262–279
199. Rowley JD (1973) A new consistent chromosomal abnormality in chronic myelogenous leukemia identified by quinacrine fluorescence and Giemsa staining. Nature 243:290–293
200. Rudkin GT, Stollar BD (1977) High resolution detection of DNA-RNA hybrids in situ by indirect immunofluorescence. Nature 265:472–473
201. Russell WL, Russell LB, Gower JS (1959) Exceptional inheritance of a sex-linked gene in the mouse explained on the basis that the x/o sex-chromosome constitution is female. Proc Natl Acad Sci USA 45:554–560
202. Saitoh Y, Laemmli UK (1994) Metaphase chromosome structure: bands arise from a differential folding path of the highly AT-rich scaffold. Cell 76:609–622
203. Samango-Sprouse C (2001) Mental development in polysomy X Klinefelter syndrome (47, XXY; 48, XXXY): effects of incomplete X inactivation. Semin Reprod Med 19:193–202
204. Sandberg AA, Koepf GF, Ishihara T, Hauschka TS (1961) An XYY human male. Lancet 2:488–489
205. Satgé D, Sommelet D, Geneix A, Nishi M, Malet P, Vekemans M (1998) A tumor profile in Down syndrome. Am J Med Genet 78:207–216
206. Sayers EW, Barrett T, Benson DA, Bryant SH, Canese K, Chetvernin V, Church DM, DiCuccio M, Edgar R, Federhen S, Feolo M, Geer LY, Helmberg W, Kapustin Y, Landsman D, Lipman DJ, Madden TL, Maglott DR, Miller V, Mizrachi I, Ostell J, Pruitt KD, Schuler GD, Sequeira E, Sherry ST, Shumway M, Sirotkin K, Souvorov A, Starchenko G, Tatusova TA, Wagner L, Yaschenko E, Ye J (2009) Database resources of the National Center for Biotechnology Information. Nucleic Acids Res 37(Database issue):D5–D15
207. Schouten JP, McElgunn CJ, Waaijer R, Zwijnenburg D, Diepvens F, Pals G (2002) Relative quantification of 40 nucleic acid sequences by multiplex ligation-dependent probe amplification. Nucleic Acids Res 30(12):e57
208. Schröck E, du Manoir S, Veldman T, Schoell B, Wienberg J, Ferguson-Smith MA, Ning Y, Ledbetter DH, Bar-Am I, Soenksen D, Garini Y, Ried T (1996) Multicolor spectral karyotyping of human chromosomes. Science 273:494–497
209. Schweizer D (1976) Reverse fluorescent chromosome banding with chromomycin and DAPI. Chromosoma 58:307–324
210. Sebat J, Lakshmi B, Troge J, Alexander J, Young J, Lundin P, Månér S, Massa H, Walker M, Chi M, Navin N, Lucito R, Healy J, Hicks J, Ye K, Reiner A, Gilliam TC, Trask B, Patterson N, Zetterberg A, Wigler M (2004) Large-scale copy number polymorphism in the human genome. Science 30:525–528
211. Sharp AJ, Hansen S, Selzer RR, Cheng Z, Regan R, Hurst JA, Stewart H, Price SM, Blair E, Hennekam RC, Fitzpatrick CA, Segraves R, Richmond TA, Guiver C, Albertson DG, Pinkel D, Eis PS, Schwartz S, Knight SJ, Eichler EE (2006) Discovery of previously unidentified genomic disorders from the duplication architecture of the human genome. Nat Genet 38:1038–1042
212. Sharp AJ, Mefford HC, Li K, Baker C, Skinner C, Stevenson RE, Schroer RJ, Novara F, De Gregori M, Ciccone R, Broomer A, Casuga I, Wang Y, Xiao C, Barbacioru C, Gimelli G, Bernardina BD, Torniero C, Giorda R, Regan R, Murday V, Mansour S, Fichera M, Castiglia L, Failla P, Ventura M, Jiang Z, Cooper GM, Knight SJ, Romano C, Zuffardi O, Chen C, Schwartz CE, Eichler EE (2008) A recurrent 15q13.3 microdeletion syndrome associated with mental retardation and seizures. Nat Genet 40:322–328
213. Shaw-Smith C, Pittman AM, Willatt L, Martin H, Rickman L, Gribble S, Curley R, Cumming S, Dunn C, Kalaitzopoulos D, Porter K, Prigmore E, Krepischi-Santos AC, Varela MC, Koiffmann CP, Lees AJ, Rosenberg C, Firth HV, de Silva R, Carter NP (2006) Microdeletion encompassing MAPT at chromosome 17q21.3 is associated with developmental delay and learning disability. Nat Genet 38:1032–1037
214. Sherard J, Bean C, Bove B, DelDuca V Jr, Esterly KL, Karcsh HJ, Munshi G, Reamer JF, Suazo G, Wilmoth D, Dahlke MB, Weiss C, Borgaonkar DS, Reynolds JF (1986) Long survival in a 69, XXY triploid male. Am J Med Genet 25:307–312
215. Sheridan R, Llerena J Jr, Matkins S, Debenham P, Cawood A, Bobrow M (1989) Fertility in a male with trisomy 21. J Med Genet 26:294–298
216. Shprintzen RJ, Goldberg RB, Lewin ML, Sidoti EJ, Berkman MD, Argamaso RV, Young D (1978) A new syndrome involving cleft palate, cardiac anomalies, typical facies, and learning disabilities: velo-cardio-facial syndrome. Cleft Palate J 15:56–62
217. Slavotinek AM (2008) Novel microdeletion syndromes detected by chromosome microarrays. Hum Genet 124:1–17
218. Smith MM (2002) Centromeres and variant histones: what, where, when and why? Curr Opin Cell Biol 14:279–285

219. Snijders AM, Nowak N, Segraves R, Blackwood S, Brown N, Conroy J, Hamilton G, Hindle AK, Huey B, Kimura K, Law S, Myambo K, Palmer J, Ylstra B, Yue JP, Gray JW, Jain AN, Pinkel D, Albertson DG (2001) Assembly of microarrays for genome-wide measurement of DNA copy number. Nat Genet 29:263–264
220. Solinas-Toldo S, Lampel S, Stilgenbauer S, Nickolenko J, Benner A, Döhner H, Cremer T, Lichter P (1997) Matrix-based comparative genomic hybridization: biochips to screen for genomic imbalances. Genes Chromosomes Cancer 20:399–407
221. Soutoglou E, Misteli T (2008) On the contribution of spatial genome organization to cancerous chromosome translocations. J Natl Cancer Inst Monogr 2008:16–19
222. Speicher MR, Carter NP (2005) The new cytogenetics: blurring the boundaries with molecular biology. Nat Rev Genet 6:782–792
223. Speicher MR, Ballard SG, Ward DC (1996) Karyotyping human chromosomes by combinatorial multi-fluor FISH. Nat Genet 12:368–375
224. Stankiewicz P, Lupski JR (2002) Genome architecture, rearrangements and genomic disorders. Trends Genet 18:74–82
225. Strahl BD, Allis CD (2000) The language of covalent histone modifications. Nature 403:41–45
226. Stranger BE, Forrest MS, Dunning M, Ingle CE, Beazley C, Thorne N, Redon R, Bird CP, de Grassi A, Lee C, Tyler-Smith C, Carter N, Scherer SW, Tavaré S, Deloukas P, Hurles ME, Dermitzakis ET (2007) Relative impact of nucleotide and copy number variation on gene expression phenotypes. Science 315:848–853
227. Sturtevant AH (1925) The effects of unequal crossing over at the bar locus in drosophila. Genetics 10:117–147
228. Sullivan BA, Karpen GH (2004) Centromeric chromatin exhibits a histone modification pattern that is distinct from both euchromatin and heterochromatin. Nat Struct Mol Biol 11:1076–1083
229. Sullivan M, Morgan DO (2007) Finishing mitosis, one step at a time. Nat Rev Mol Cell Biol 8:894–903
230. Sussan TE, Yang A, Li F, Ostrowski MC, Reeves RH (2008) Trisomy represses Apc(Min)-mediated tumours in mouse models of Down's syndrome. Nature 451:73–75
231. Sutherland GR (2003) Rare fragile sites. Cytogenet Genome Res 100:77–84
232. Tanke HJ, Wiegant J, van Gijlswijk RPM, Bezrookove V, Pattenier H, Heetebrij RJ, Talman EG, Raap AK, Vrolijk J (1999) New strategy for multi-colour fluorescence in situ hybridisation: COBRA: COmbined Binary RAtio labelling. Eur J Hum Genet 7:2–11
233. Tartaglia N, Davis S, Hench A, Nimishakavi S, Beauregard R, Reynolds A, Fenton L, Albrecht L, Ross J, Visootsak J, Hansen R, Hagerman R (2008) A new look at XXYY syndrome: medical and psychological features. Am J Med Genet 146A(12):1509–1522
234. Thomas NS, Hassold TJ (2003) Aberrant recombination and the origin of Klinefelter syndrome. Hum Reprod Update 9:309–317
235. Tjio HJ, Levan A (1956) The chromosome numbers of man. Hereditas 42:1–6
236. Tkachuk D, Westbrook C, Andreef M, Donlon T, Cleary M, Suranarayan K, Homge M, Redner A, Gray J, Pinkel D (1990) Detection of bcr-abl fusion in chronic myelogeneous leukemia by in situ hybridization. Science 250:559–562
237. Tuzun E, Sharp AJ, Bailey JA, Kaul R, Morrison VA, Pertz LM, Haugen E, Hayden H, Albertson D, Pinkel D, Olson MV, Eichler EE (2005) Fine-scale structural variation of the human genome. Nat Genet 37:727–732
238. Verkerk AJMH, Pieretti M, Sutcliffe JS, Fu YH, Kuhl DPA, Pizzuti A, Reiner O, Richards S, Victoria MF, Zhang F, Eussen BE, van Ommen GJB, Blonden LAJ, Riggins Gregory J, Chastain JL, Kunst CB, Galjaard H, Caskey CT, Nelson DL, Oostra BA, Warren ST (1991) Identification of a gene (FMR-1) containing a CGG repeat coincident with a breakpoint cluster region exhibiting length variation in fragile X syndrome. Cell 65:905–914
239. Verma RS, Babu A (1995) Human Chromosomes, Principles and Techniques, 2nd edn. McGraw-Hill, Inc
240. Vissers LE, van Ravenswaaij CM, Admiraal R, Hurst JA, de Vries BB, Janssen IM, van der Vliet WA, Huys EH, de Jong PJ, Hamel BC, Schoenmakers EF, Brunner HG, Veltman JA, van Kessel AG (2004) Mutations in a new member of the chromodomain gene family cause CHARGE syndrome. Nat Genet 36:955–957
241. Voullaire L, Wilton L, Slater H, Williamson R (1999) Detection of aneuploidy in single cells using comparative genomic hybridization. Prenat Diagnosis 19:846–851
242. Voullaire L, Slater H, Williamson R, Wilton L (2000) Chromosome analysis of blastomeres from human embryos by using comparative genomic hybridization. Hum Genet 106:210–217
243. Vulliamy T, Marrone A, Goldman F, Dearlove A, Bessler M, Mason PJ, Dokal I (2001) The RNA component of telomerase is mutated in autosomal dominant dyskeratosis congenita. Nature 413:432–435
244. Waldeyer W (1888) Über Karykinese und ihre Beziehung zu den Befruchtungsvorgängen. Arch Mikrosk Anat 32:1–112
245. Wang LH, Schwarzbraun T, Speicher MR, Nigg EA (2007) Persistence of DNA threads in human anaphase cells suggests late completion of sister chromatid decatenation. Chromosoma 117:123–135
246. Warburton PE, Willard HF (1995) Interhomologue sequence variation of alpha satellite DNA from human chromosome 17: evidence for concerted evolution along haplotypic lineages. J Mol Evol 41:1006–1015
247. Warren AC, Chakravarti A, Wong C, Slaugenhaupt SA, Halloran SL, Watkins PC, Metaxotou C, Antonarakis SE (1987) Evidence for reduced recombination on the nondisjoined chromosomes 21 in Down syndrome. Science 237:652–654
248. Waye JS, Willard HF (1987) Nucleotide sequence heterogeneity of alpha satellite repetitive DNA: a survey of alphoid sequences from different human chromosomes. Nucleic Acids Res 15:7549–7569
249. Weier HU (2001) DNA fiber mapping techniques for the assembly of high-resolution physical maps. J Histochem Cytochem 49:939–948
250. Weinberg RA (2007) The biology of cancer. Garland Science. Taylor & Francis Group, London
251. Wells D, Delhanty JD (2000) Comprehensive chromosomal analysis of human preimplantation embryos using whole genome amplification and single cell

comparative genomic hybridization. Mol Hum Reprod 6:1055–1062
252. Wells D, Sherlock JK, Handyside AH, Delhanty JD (1999) Detailed chromosomal and molecular genetic analysis of single cells by whole genome amplification and comparative genomic hybridisation. Nucleic Acids Res 27:1214–1218
253. Wright WE, Shay JW (1995) Time, telomeres and tumours: is cellular senescence more than an anticancer mechanisms? Trends Cell Biol 5:293–297
254. Xu N, Tsai CL, Lee JT (2006) Transient homologous chromosome pairing marks the onset of X inactivation. Science 311:1149–1152
255. Yagi H, Furutani Y, Hamada H, Sasaki T, Asakawa S, Minoshima S, Ichida F, Joo K, Kimura M, Imamura S, Kamatani N, Momma K, Takao A, Nakazawa M, Shimizu N, Matsuoka R (2003) Role of TBX1 in human del22q11.2 syndrome. Lancet 362:1366–1373
256. Yang Q, Rasmussen SA, Friedman JM (2002) Mortality associated with Down's syndrome in the USA from 1983 to 1997: a population-based study. Lancet 359:1019–1025
257. Yoon PW, Freeman SB, Sherman SL, Taft LF, Gu Y, Pettay D, Flanders WD, Khoury MJ, Hassold TJ (1996) Advanced maternal age and the risk of Down syndrome characterized by the meiotic stage of chromosomal error: a population-based study. Am J Hum Genet 58:628–633
258. Yu CE, Oshima J, Fu YH, Wijsman EM, Hisama F, Alisch R, Matthews S, Nakura J, Miki T, Ouais S, Martin GM, Mulligan J, Schellenberg GD (1996) Positional cloning of the Werner's syndrome gene. Science 272:258–262
259. Yu S, Cox K, Friend K, Smith S, Buchheim R, Bain S, Liebelt J, Thompson E, Bratkovic D (2008) Familial 22q11.2 duplication: a three-generation family with a 3-Mb duplication and a familial 1.5-Mb duplication. Clin Genet 73:160–164
260. Zhang LF, Huynh KD, Lee JT (2007) Perinucleolar targeting of the inactive X during S phase: evidence for a role in the maintenance of silencing. Cell 129:693–706
261. Zhang X, Snijders A, Segraves R, Zhang X, Niebuhr A, Albertson D, Yang H, Gray J, Niebuhr E, Bolund L, Pinkel D (2005) High-resolution mapping of genotype-phenotype relationships in cri du chat syndrome using array comparative genomic hybridization. Am J Hum Genet 76:312–326
262. Zhivago P, Morosov B, Ivanickaya A (1934) Über die Einwirkung der Hypotonie auf die Zellteilung in den Gewebekulturen des embryonalen Herzens. Dokl Akad Nauk USSR 3:385–386
263. Zweier C, Peippo MM, Hoyer J, Sousa S, Bottani A, Clayton-Smith J, Reardon W, Saraiva J, Cabral A, Gohring I, Devriendt K, de Ravel T, Bijlsma EK, Hennekam RC, Orrico A, Cohen M, Dreweke A, Reis A, Nurnberg P, Rauch A (2007) Haploinsufficiency of TCF4 causes syndromal mental retardation with intermittent hyperventilation (Pitt-Hopkins syndrome). Am J Hum Genet 80:994–1001

From Genes to Genomics to Proteomics

Michael R. Speicher

Abstract In human genetics many initial research initiatives focused on single genes or were performed on a gene-by-gene basis. However, recent findings, especially those about the extensive transcriptional activity of the genome, changed the concept of what a gene is supposed to be. In addition, novel high-throughput approaches and numerous innovative technologies, such as gene and expression microarrays, mass spectrometry, new sequencing methods, and many more, now enable us to address complex diseases and to unravel underlying involved regulatory patterns. These high-throughput assays resulted in a shift from studying Mendelian disorders towards multifactorial diseases, although monogenic diseases still provide a unique opportunity for elucidating gene function. This chapter describes current concepts about the definition of a gene, possible consequences of mutations and the latest developments in the areas of genomics, transcriptomics, and proteomics and their potential to add to a better understanding of factors contributing to phenotypic features.

Contents

4.1	Single-Gene Approaches	140
	4.1.1 What Is a Gene?	140
	4.1.2 Mutations	141
	4.1.3 Silent Mutations and Phenotypic Consequences	142
	4.1.4 Mutation Detection by Sanger Sequencing	143
	4.1.5 Next-Generation Sequencing	144
	4.1.6 The Importance of Monogenic Mendelian Disorders	144
4.2	Gene Regulation	145
	4.2.1 Genetic Regulation	145
	4.2.2 Epigenetic Regulation	146
	4.2.3 Regulatory Transcripts of Small RNAs	147
4.3	"-omics" Sciences	147
4.4	Genomics	148
	4.4.1 Genomes of Organisms	148
	4.4.2 Array and Other Technologies	148
	4.4.3 Next-Generation Sequencing	148
	4.4.4 Third-Generation Sequencing	152
	4.4.5 Personalized Genomics	152
	4.4.6 Gene Function	153
4.5	Transcriptomics	154
	4.5.1 Capturing the Cellular Transcriptome, Expression Arrays, and SAGE	154
	4.5.2 Regulatory Networks	154
	4.5.3 Outlier Profile Analysis	155
	4.5.4 High-Throughput Long- and Short-Read Transcriptome Sequencing	156
	4.5.5 Disease Classification	156
	4.5.6 Tools for Prognosis Estimation	156
4.6	Proteomics	156
	4.6.1 From Low-Throughput to High-Throughput Techniques	157
	4.6.2 Mass Spectrometer-Based Methods	157
	4.6.3 Antibody Array-Based Methods	158
	4.6.4 Proteomic Strategies	158
	4.6.5 Proteomics for Screening and Diagnosis of Disease (Diagnostic and Prognostic Biomarkers)	158
4.7	Conclusions	159
References		159

M.R. Speicher (✉)
Institute of Human Genetics, Medical University of Graz,
Harrachgasse 21/8, 8010 Graz, Austria
e-mail: michael.speicher@medunigraz.at

4.1 Single-Gene Approaches

Prior to the era of high-throughput analyses, typical research initiatives focused on single genes or were performed on a gene-by-gene basis. However, even research focusing on a single gene may already represent a very complex challenge. Some principles of working with single genes are described below. A particular focus will be on the limitations which have propelled the development of numerous innovative technologies, such as gene and expression microarrays, mass spectrometry, and proteomics, and many more, which now allow investigators to reveal underlying complex regulatory patterns.

4.1.1 What Is a Gene?

Before discussing the steps from genes to proteomics we should reflect on what a "gene" is actually supposed to be. In 1909 the term "gene" was used for the first time by Wilhelm Johannsen. Ever since, the concept of a gene has been under constant development, and numerous gene definitions have been proposed and adjusted as our knowledge of genes has evolved over the past decades. A somewhat surprising result is that although the term "gene" is one of the most commonly used expressions in genetics and although genes are constantly being characterized and more and more mutations in genes are being linked to diseases, the term itself in fact remains poorly defined. An excellent history of operational definitions of a gene over the past decades together with an attempt at an updated definition was recently provided by Gerstein et al. [25]. The authors rightfully argue that the provocative findings of the ENCODE Project [17], which elucidated the complexity of the RNA transcripts produced by the genome, have to change previous definitions of a gene. The preceding views of a gene were centered on protein coding (Fig. 4.1) and did not take the extensive transcriptional activity of the genome into account, most likely because the full extent of transcriptional activity was unknown prior to the ENCODE Project.

Based on the knowledge derived from the ENCODE Project, Gerstein et al. [25] proposed the following, updated definition for a gene: "The gene is a union of genomic sequences encoding a coherent set of potentially overlapping functional products." An illustration of how to apply this definition is provided in Fig. 4.2.

Another implication of this definition is that 5′ and 3′ untranslated regions (UTRs), despite their importance for translation, regulation, stability, and/or localization of mRNAs, would not be part of a gene because they do not participate in encoding the final product of a protein-coding gene. In order to compensate for this, Gerstein et al. [25] suggested a new "category" for regulatory and untranslated regions playing an important part in gene expression, by naming these regions "gene-associated." This terminology may help to acknowledge that additional DNA sequences outside

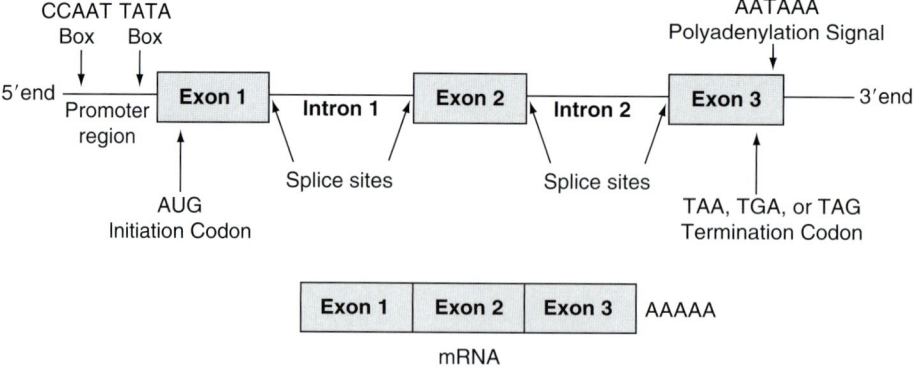

Fig. 4.1 Representative classic view of a gene. Transcription may be initiated from the promoter region located at the 5′ side of a gene. The promoter region often contains a TATA or a CCAAT box and is enriched for the paired nucleotides cytosine and guanine (CG islands). Genes consist of translated (exons) and noncoding (introns) portions. The open reading frame (*ORF*) is situated between the initiation codon (*AUG*) and the termination codon (*TAA*, *TGA*, or *TAG*). Sequences encoding the polyA tail of the protein are located at the end of a gene. The precursor RNA is spliced so that intronic sequences are removed and messenger RNA (mRNA) is formed

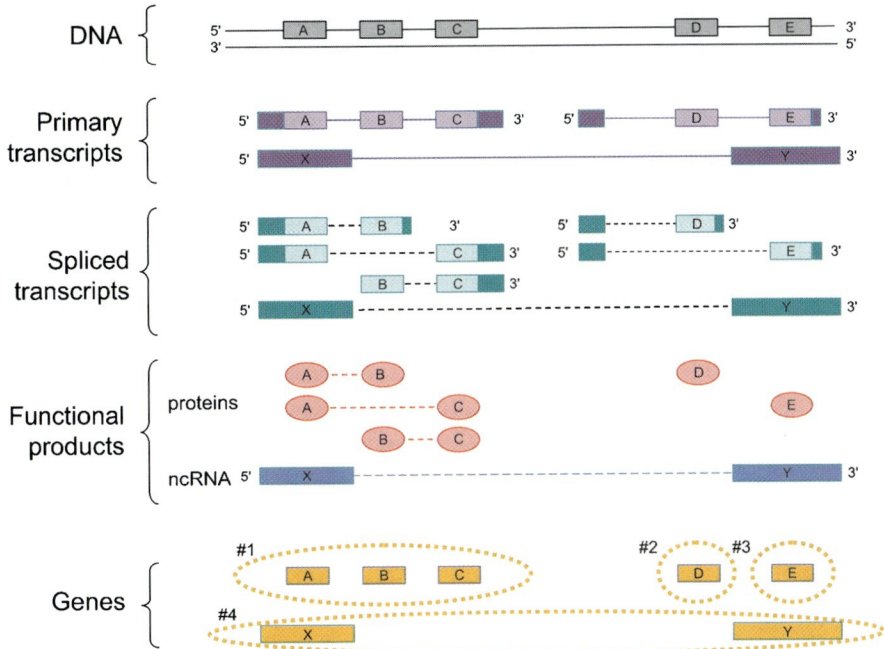

Fig. 4.2 Gerstein et al. [25] proposed a new definition for genes, and this figure illustrates how this definition should be applied. In this region the *gray rectangles* correspond to exonic/protein-coding sequences. Three primary transcripts originate from this genomic region. Two of these transcripts consist, in addition to the 5′ and 3′ un-translated regions, of some of the exons (A, B, C or D, and E); intronic sequences are represented by *solid lines*. The third transcript (X and Y) does not encode a protein but is a noncoding RNA (ncRNA) product. Therefore, such a transcript may share its genomic sequences with protein-coding segments; however, it usually does not exactly correspond to exons. The result of alternative splicing in this example is that the first two transcripts encode five protein products (A-B, A-C, B-C, D, and E; the *dashed lines* illustrate connectivity between the RNA sequences). Thus, exons A, B, and C generate transcripts, each derived from two of these DNA segments. In contrast, the products originating from D and E share a 5′ untranslated region, but their translated regions do not overlap after alternative splicing. The noncoding RNA product is not a coproduct of the protein-coding genes. The functional products are 5 different proteins shown in *ovals* (connected by *dashed lines*) and one RNA product (*rectangles*, also connected by a *dashed line*). As a consequence this region harbors four genes indicated at the bottom within the *orange dashed lines*. Sequence segments A, B, and C comprise gene 1, whereas gene 2 contains D, gene 3 E and gene 4 X and Y. From [25]

of the respective gene themselves have important roles in contributing to gene function.

From this new definition it follows that only continuous DNA sequences coding for a protein or RNA product without overlapping products correspond to the classic and most commonly used view of gene. In fact, the vast majority of our knowledge about "genes" and their functions centers on this subclass of genes. Thus, with these new evolving concepts it is obvious that even "monogenic" disorders are at present incompletely explored and a lot remains to be discovered.

As the updated definition emphasizes the final products of a gene, it disregards intermediate products originating from a genomic region that may happen to overlap. This implies that the number of genes in the human genome is going to increase significantly when the survey of the human transcriptome has been completed.

4.1.2 Mutations

The aforementioned summary of the complexity of a gene and its possible transcripts also suggests that the distinction between pathogenic and nonpathogenic mutations is often very difficult. In general, there are three different types of mutations. *Deletions* involve the loss of at least one nucleotide, whereas *insertions* represent the addition of at least one nucleotide. Both deletions and insertions cause a shift of the reading frame and are therefore also referred to as frameshift mutations. Usually the resulting sequences no longer code for a functional gene product and are thus dubbed "nonsense mutations." Since insertions and deletions usually disturb the gene function significantly, they are often associated with diseases and are therefore frequently pathogenic.

In contrast, a contribution to specific phenotypic features of the *substitution* or *exchange* of a single nucleotide is often very difficult to establish. An exchange of one purine for another purine or of one pyrimidine for another is called transition, whereas an exchange of a purine for a pyrimidine or vice versa is a transversion. A nucleotide substitution does not result in a shift of the reading frame, and possible consequences depend on how a codon has been altered. For example, a substitution may alter a codon so that a wrong amino acid will be present at this site, which is referred to as a "missense mutation." Such missense mutations may have consequences ranging from no changes to severe functional changes, and it is often very difficult to establish the outcome of such mutations. A nucleotide substitution is called a "silent mutation" if the resulting codon still corresponds to the same amino acid. This is possible because of the redundancy of the genetic code, as different nucleotide sequences may code for the same amino acid sequence. For example, the four nucleotide base pairs GCC, GCG, GCT and GCA all code for the amino acid alanine. If GCC represented a codon within an open reading frame a substitution at the third position from C to G or from C to A would still represent a codon with the nucleotide sequence for alanine. Much has been learnt about the phenotypic consequences of mutations, but there are many examples of missense mutations, variants in DNA elements of unknown function, and silent changes in coding regions for which pathogenicity is questionable. Thus, another difficult challenge is to prove that an altered allele is causal to the disease in question.

For example, silent mutations frequently have no consequences for the phenotype. However, in order to illustrate the often enormous difficulties in determining the significance of mutations, two striking examples demonstrating that even "silent" mutations may have severe consequences for a phenotype will be discussed below.

4.1.3 Silent Mutations and Phenotypic Consequences

Two particularly fascinating "silent" mutations with significant consequences for the phenotype are described here.

Hutchinson–Gilford progeria syndrome (HGPS) is a rare genetic disorder. Affected individuals show very early signs of aging, such as loss of hair, lipodystrophy, scleroderma, decreased joint mobility, osteolysis, and facial features resembling those of aged persons, and they die at an average age of 13. In the vast majority (90%), progressive atherosclerosis of the coronary and cerebrovascular arteries is the cause of death [30]. HGPS belongs to a group of conditions called laminopathies, which affect nuclear lamins. The lamins belong to the multiprotein family of intermediate filaments and can be regarded as the main determinants of the nuclear architecture. HGPS is caused by mutations in *LMNA*, resulting in an abnormally formed lamin A. In the majority of progeria patients a classic p.G608G (c.1824C>T) mutation in exon 11 can be found. It is predicted that this mutation is a silent mutation, as it does not cause any change at the amino acid level. However, this change improves the match to a consensus splice donor, activating a cryptic splice site [14, 18]. Owing to this activation of a cryptic splice site, 150 nucleotides, up to the start codon of exon 12, are removed [14, 18]. The last step in the posttranslational processing of prelamin A cannot occur without these nucleotides, so that the mutant prelamin A persists. The mutant prelamin A is called progerin, and it is the presence of progerin, and not the lack of normal lamin A, that causes the phenotype [55].

The second example is the identification of a synonymous *single-nucleotide polymorphism* (SNP), which did not produce altered coding sequences in the *Multidrug Resistance* 1 (*MDR*1) gene [35]. The *MDR*1 gene product is a P-glycoprotein multiple-transmembrane protein pump contributing to the pharmacokinetics of drugs, which is associated with the multidrug resistance of cancer cells. Although *MDR1* harbors many SNPs, some SNPs have been associated with reduced functionality of the pump. This was observed for two SNPs (e.g., C1236T and C3435T) even though neither changes the amino acid sequence of P-glycoprotein. For example, the C1236T polymorphism changes at amino acid position 412a GGC codon to GGT and both encode glycine, whereas the C3435T polymorphism changes at position 1145 ATC to ATT, which in each case encodes isoleucine. However, Kimchi-Sarfaty et al. [35] were able to demonstrate that both polymorphisms resulted in changes from frequent to infrequent codons. As a consequence, ribosome trafficking is slowed down at the corresponding mRNA regions. These alterations likely affect the cotranslational folding pathway of P-glycoprotein, resulting in a different final conformation and eventually in altered substrate specificity. Thus, silent mutations of synonymous codons (changing from frequent to infrequent) in certain genes may alter translation kinetics of mRNA, which might in turn affect final protein conformation.

4.1.4 Mutation Detection by Sanger Sequencing

In 1977 Fred Sanger published three seminal method papers on the rapid determination of DNA sequences [53, 54, 60], for which he received his second Nobel prize in Chemistry in 1980. This technology, which besides Sanger sequencing is also referred to as dideoxynucleotide sequencing, provided a tool for deciphering complete genes and later entire genomes. In fact, Sanger sequencing evolved into the only DNA sequencing method used for three decades after it was first described. DNA can be prepared for sequencing by two approaches: for targeted resequencing, which is done in most diagnostic routine applications: primers flanking the target regions are used to amplify the respective region. In contrast, for shotgun de novo sequencing, DNA is randomly fragmented and cloned into a plasmid, which is subsequently used to transform *Escherichia coli* (Fig. 4.3a). The latter approach played a pivotal role in deciphering the human genome. The

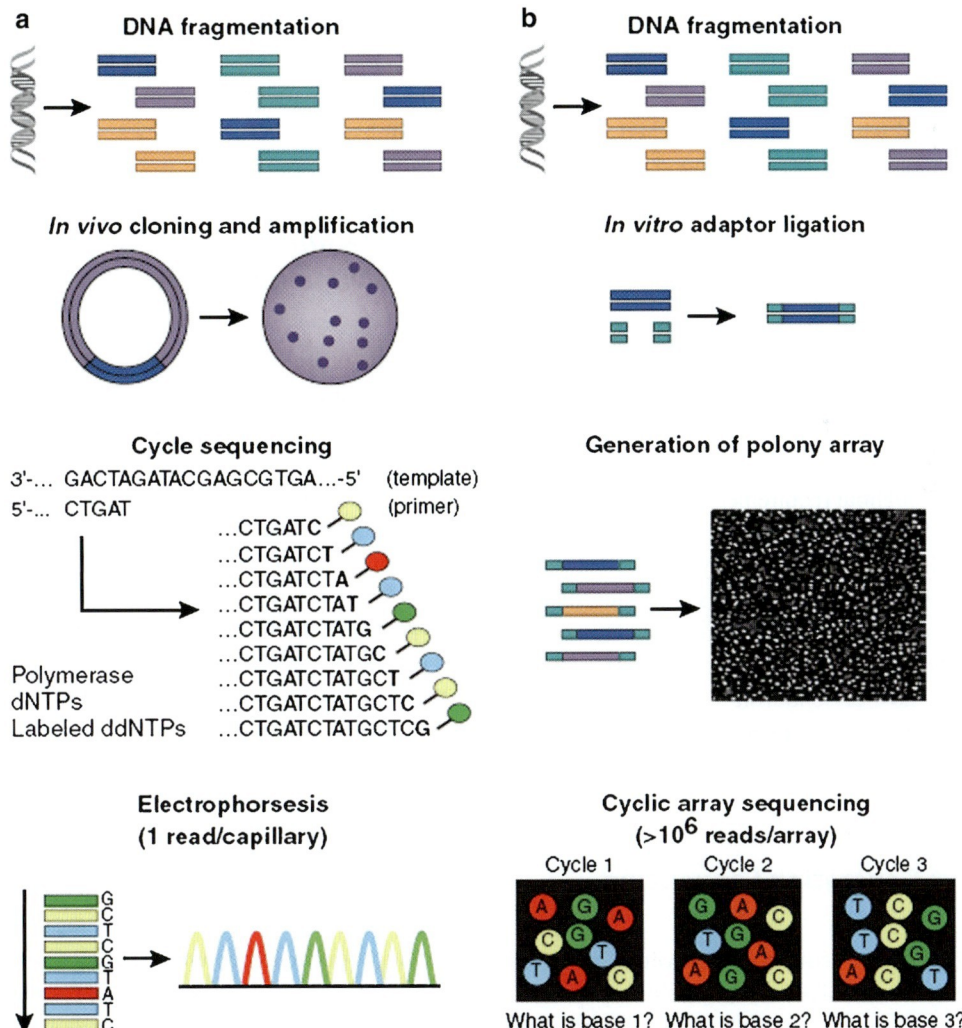

Fig. 4.3 Comparison between (**a**) Sanger and (**b**) next-generation sequencing. (Reprinted by permission from Macmillan Publishers Ltd: Nature Biotechnology [58], copyright 2008)

results of both approaches are multiple templates, which are then subjected to the sequencing reactions, consisting of repeated rounds of template denaturation, primer annealing, and primer extension. In each cycle of the sequencing reaction the primer extension is stochastically terminated by the integration of dideoxynucleotides (ddNTPs), which are labeled with a fluorochrome. This results in a mixture of extension products of different lengths, and the label of the respective terminating ddNTPs reflects the nucleotide identity of its terminal position. Subsequently the sequence can be determined by high-resolution electrophoretic separation of the single-stranded, end-labeled extension products in a capillary-based polymer gel. The DNA sequence is deciphered by analysis of the fluorescent labels at the end of the fragments. The discrete lengths of the fragments determine the nucleotide position, and the nucleotide itself is encoded by laser excitation of the fluorescent labels and a four-color detection of the emission spectra, which are then translated into DNA sequence by appropriate software (Fig. 4.3a).

The application of Sanger sequencing for deciphering the entire human genome represented particularly large-scale sequencing efforts, which were conducted in factory-like environments called sequencing centers. These centers had a specialized and dedicated infrastructure consisting of hundreds of DNA-sequencing instruments, robotics, bioinformatics, computer databases, instrumentation, and a large number of personnel. The aim of deciphering the entire human genome dramatically changed the throughput requirements of DNA sequencing and propelled developments such as automated capillary electrophoresis. Many capillary-based sequencing systems have 96 or more capillaries, meaning that 96 sequence reads can be processed in parallel. However, a simple increase in the number of capillaries was not sufficient for the new enduring tasks in genomics, which required the development of entirely new technologies, as summarized in the next paragraph.

4.1.5 Next-Generation Sequencing

The next-generation sequencing revolution started in 2005 with two seminal papers describing a sequence-by-synthesis technology [47] and a multiplex polony-sequencing protocol [59]. The parallel sequencing throughput capacity is perhaps the most important feature setting next-generation sequencers apart from conventional capillary-based sequencing. In fact, instead of running 96 capillaries or samples at a time, next-generation sequencing allows the processing of millions of sequence reads simultaneously (Fig. 4.3b). This massive parallel sequencing requires only one or two instruments instead of several hundred Sanger-type DNA capillary sequencers and naturally involves significantly fewer personnel operating the machines. Another important difference is that next-generation sequence reads do not depend on vector-based cloning, but are instead derived from fragment libraries. This alone allows a significant speeding up of sequencing (Fig. 4.4). Another difference is that read lengths are shorter (35–250 bp for next-generation sequencing, as against 650–800 bp for capillary sequencers). Next-generation sequencing, often also referred to as second-generation sequencing, and the evolving third-generation sequencing will be discussed in greater detail in Sect. 4.4.

4.1.6 The Importance of Monogenic Mendelian Disorders

The quest for high-throughput assays is also accompanied by a shift away from the Mendelian disorders towards multifactorial diseases. This neglects the fact that linking naturally occurring pathogenic mutations with monogenic disorders provides a unique opportunity for elucidating gene function [1]. Studies on Mendelian traits reveal irreplaceable insights into mutation processes and their associated molecular pathophysiology. Furthermore, it was the investigations of Mendelian disorders that disclosed the existence of genetic phenomena, such as uniparental disomy or parental imprinting.

Only in-depth analysis of monogenic disorders can unravel the consequences of different mutations within the same gene that can give rise to distinct phenotypes. For example, among the most striking examples are mutations in the aforementioned *LMNA* gene, which can cause not only the Hutchinson–Gilford progeria syndrome but also several other, different phenotypes, which are often summarized as primary laminopathies. Phenotypic consequences of mutations in *LMNA* can be further subdivided into laminopathies with striated muscular atrophy [including

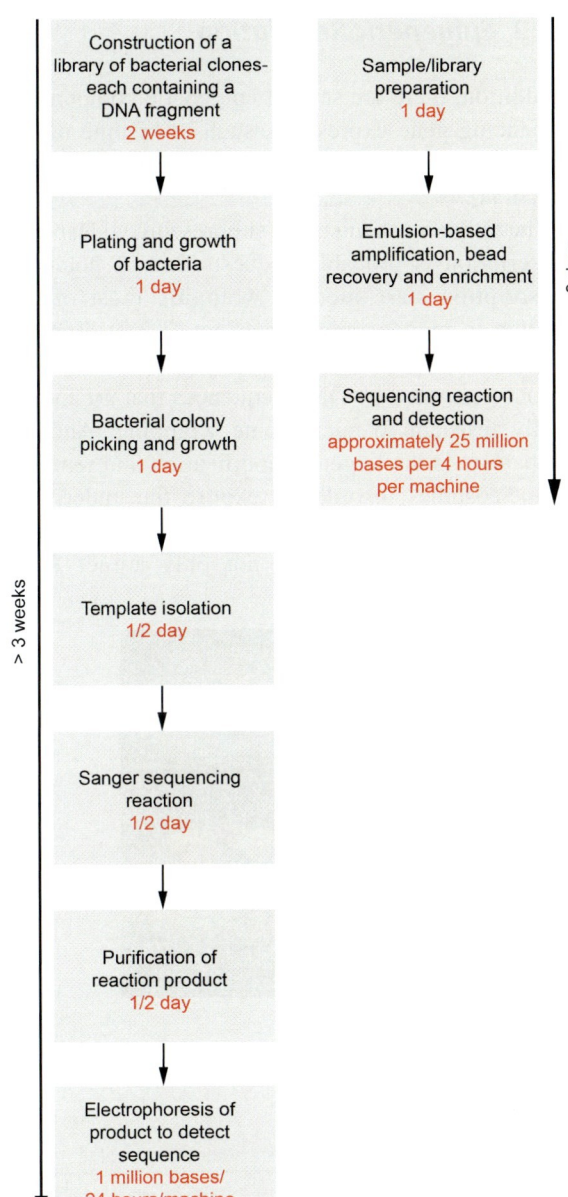

Fig. 4.4 Flow diagrams comparing the time-lapse needed for traditional Sanger sequencing (*left*) and massively parallel sequencing as used for the 454 system (*right*). (Reprinted by permission from Macmillan Publishers Ltd: Nature [52], copyright 2005)

Emery-Dreifuss muscular dystrophy (EDMD2; OMIM 181350), autosomal dominant limb girdle muscular dystrophy 1B (LGMD1B; OMIM 159001) and dilated cardiomyopathy, 1A (DCM1A; OMIM 115200)], laminopathies affecting peripheral nerves [(Charcot-Marie-Tooth disease type 2B1 (CMT2B1; OMIM 605588)] and laminopathies with loss of or reduced adipose tissues [familial partial lipodystrophy, Dunnigan type (FPLD2; OMIM 151660) and congenital generalized lipodystrophy, type 2 (CGL2, OMIM 269700)] [41]. A probably new *LMNA*-associated disease entity that may be classified as a congenital muscular dystrophy (LMNA-related congenital muscular dystrophy, or L-CMD) has recently been described [51] and suggests that even more phenotypes may be caused by mutations in this gene.

Even monogenic diseases have considerable phenotypic complexity, often depending on the genetic background and the status of modifier genes, which may modulate the consequences of specific mutations. In addition, such epigenetic changes as the genomic distribution of 5-methylcytosine DNA and histone acetylation may change the outcome of a mutation, and to make these issues even more complicated, such epigenetic modifications may change as we age [22]. Thus, monogenic disorders are in fact examples of oligogenic inheritance and vary along a continuum from simple to complex disorders [1]. Allelic variation in genes or other functional DNA sequences that modify the phenotypic severity of a monogenic disorder or control variation in gene expression provide links to additional genomic causes related to phenotypic variability.

4.2 Gene Regulation

Genes can be regulated by various means (Fig. 4.5). Obviously there is a "many-to-many" relationship between regulatory regions, epigenetic mechanisms, small RNAs, and genes. In fact, gene expression is a multilevel process, which is controlled by regulatory proteins and DNA sequences (genetic regulation) and by chromatin remodeling and the position of chromosomes in the nucleus (epigenetic regulation). In addition, gene regulation may be affected by complex sets of RNAs that do not produce proteins.

4.2.1 Genetic Regulation

At the beginning of transcription the base sequences of genes are transcribed into RNA by RNA polymerase II. Multiple accessory factors determine the transcriptional start and end points for RNA polymerase II. An important component is the promoter, typically located close to the gene it regulates, which facilitates the transcription of a gene.

Promoters comprise two interacting parts, i.e., the basal promoter elements and the enhancer elements. Basal promoter elements bind accessory transcription initiation factors that position RNA polymerase II in the right place and direction. These basal elements are composed of short, low-complexity sequences (such as the TATA element). Enhancer elements bind regulatory factors that specify the physiological conditions or cell types where the gene will be expressed. The enhancer and basal promoter complexes interact at both functional and physical levels to determine how often an RNA transcript is produced. Enhancers can work over large distances of DNA in both directions.

4.2.2 Epigenetic Regulation

In addition, there are several epigenetic components influencing gene expression, such as histone modifications, DNA methylation, and position effects (Fig. 4.5a).

The major mechanism for suppressing widespread transcription is probably sequestration of potential transcription start sites by wrapping most of the genome in nucleosomes (see Sect. 3.2.2). Typical transcription start sites are found in nucleosome-free regions generated by DNA sequences that are intrinsically resistant to nucleosome wrapping. Another mechanism is the targeted modification and removal of nucleosomes in order to expose the underlying promoter sequences (see Sect. 3.2.2). Thus, functional eukaryotic promoters must not only attract RNA

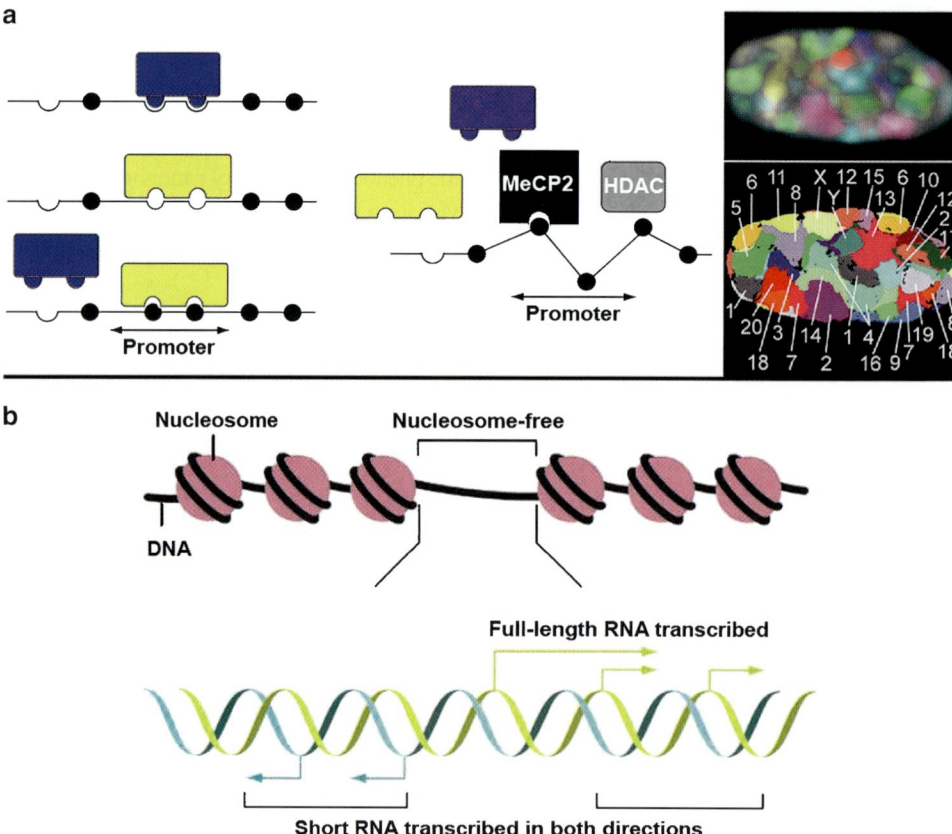

Fig. 4.5 (**a**, **b**) Different means of gene regulation. (**a**) *Left panel*: Consequences of epigenetic regulation by histone methylation: the configuration of the promoter region changes so that transcription factors cannot bind and expression of the respective gene is suppressed; *right panel*: Chromosomes occupy nonrandom positions in cell nuclei, these position effects influence genes expression. From [4]. (**b**) Example for possible gene regulation by small RNAs: promotor-associated transcripts (transcription start sites and transcripts are represented as *bent arrows*) within nucleosome-free DNA close to the promoters may influence gene expression. (From [6]. Reprinted with permission from AAAS)

polymerase II, but also evade nucleosomal repression. The epigenetic modes of gene regulation by histone modifications, DNA methylation, and position effects are discussed in detail in Sects. 3.2.2, 3.4.4.2, and 3.6.2.1.

4.2.3 Regulatory Transcripts of Small RNAs

More recently it has become clear that gene regulation may be affected by complex sets of small (20–30 nucleotides) RNAs that do not produce proteins, i.e., noncoding RNAs (ncRNAs; Fig. 4.5b). In general, effects of small RNAs on gene expression are inhibitory, as small RNAs may bind effector proteins to target nucleic acid molecules through base-pairing interactions. Therefore, activities of small RNAs are frequently summed up as "RNA silencing."

In humans the two main categories of small RNAs – among several classes – are short, interfering RNAs (siRNAs) and microRNAs (miRNAs) [2, 8]. These small RNAs are important regulators of gene expression that control both physiological and pathologic processes (e.g., development and cancer) miRNAs are regulators of endogenous genes, whereas siRNAs are defenders of genome integrity in response to foreign or invasive nucleic acids such as transposons and transgenes. An important distinction between miRNAs and siRNAs is whether or not they silence their own expression. Almost all siRNAs silence the same locus as they were derived from, and they only sometimes have the ability to silence other loci as well. In contrast, most miRNAs do not silence their own loci but do silence other genes. Both RNAs have double-stranded precursors and depend upon the same two families of proteins: Dicer enzymes to excise them from their precursors and Ago proteins to support their silencing effector functions [2, 8]. Single-stranded forms of both miRNAs and siRNAs associate with effector assemblies, which have been dubbed RNA-induced silencing complexes (RISCs).

The genes to be silenced are determined by the small RNA component, which identifies the respective complementary nucleotide sequence. The silencing can be monitored by increased expression of small RNAs or, conversely, by dilution or removal of old ones.

Furthermore, a new class of short RNA transcripts begins near the expected transcription start sites upstream of protein-encoding sequences (Fig. 4.5b). These RNAs often occur in the direction opposite to that of the protein-coding region [12, 29, 50, 57]. Although the function of these RNAs is presently not well defined, they may have an impact on how promoters delineate transcription start sites. These new RNAs are largely derived from DNA in nucleosome-free regions and may therefore arise from random, weak basal promoter elements that escape suppression [12, 29, 50, 57]. Hence, these short promoter-associated RNAs may simply result from incomplete suppression of cryptic initiation which, however, does not exclude an associated function by affecting the expression of the nearby gene.

4.3 "-omics" Sciences

Single-biomarker analysis is increasingly being replaced by multiparametric analysis of genes, transcripts, or proteins, now subsumed under the term "omics" sciences. The current nomenclature of omics sciences includes genomics for DNA variants, transcriptomics for mRNA, proteomics for proteins, and metabolomics for intermediate products of metabolism. The omics sciences use high-throughput techniques often allowing simultaneous examination of changes in the genome (DNA), transcriptome (messenger RNA [mRNA]), proteome (proteins), or metabolome (metabolites) in a biological sample, with the goal of understanding the physiology or mechanisms of disease. Insights derived from the complementary fields of omics sciences are expected to assist the development of new diagnostic, prognostic, and therapeutic tools. The omics sciences have in common that they require the development of novel informatic applications and sophisticated dimensionality reduction strategies. They have an enormous potential to unravel disease and physiological mechanisms and can identify clinically exploitable biomarkers from huge experimental datasets and offer insights into the molecular mechanisms of diseases.

The characteristics of the individual omics sciences and their integration to systems biology can be summarized as follows:

Genomics: Genomics seeks to define our genetic substrate and describes the study of the genomes of organisms.

Transcriptomics: Transcriptomics refers to the detailed analysis of the entire transcriptome, i.e., of all expressed sequences.

Proteomics: Proteomics explores the structure and function of proteins, which are the end-effectors of our genes. Proteomics has been revolutionized in the past decade by the application of techniques such as protein arrays, two-dimensional gel electrophoresis, and mass spectrometry. These techniques have tremendous potential for biomarker development, target validation, diagnosis, prognosis, and an optimization of treatment in medical care, especially in the field of clinical oncology.

Systems Biology: The integration of omic techniques is called "systems biology." This discipline aims at defining the interrelationships of several, or ideally all, of the elements in a system, rather than studying each element independently. Thus, systems biology will capture information from genomics, transcriptomics, proteomics, metabolomics, etc. and combine it with theoretical models in order to predict the behavior of a cell or organism.

4.4 Genomics

Genomics is the systematic study of the genomes of organisms. The field includes intensive efforts to determine the entire DNA sequence of organisms and finescale genetic mapping efforts. The investigation of single genes does not usually fit the definition of genomics. However, as the function of a single gene may affect many other genes, the border between singlegene analysis and genomics is often blurred.

4.4.1 Genomes of Organisms

A major branch of genomics is still concerned with the sequencing of the genomes of various organisms. The genome of the first free-living organism that was completely sequenced (*Haemophilus influenzae* in 1995) had a size of 1.8 Mb [21]. This was followed by complete sequences for *Mycoplasma genitalium* [23] and *Mycobacterium tuberculosis* [11], and subsequently by many other archeal, bacterial, and eukaryotic genomes. A rough draft of the human genome was presented in 2001 [39, 67], followed by an auspiciously completed version in 2004 [32]. Today sequencing efforts for other genomes continue. However, especially the resequencing of genomes, e.g., of human genomes to establish the variability between the genomes of different individuals, was propelled by the new possibilities of next-generation sequencing, which have added the sequences of other human individuals or of the first entire tumor genomes.

4.4.2 Array and Other Technologies

Genomics has certainly benefited from various array technologies that allowed the systematic analysis of entire genomes with various resolutions. These array technologies and other currently frequently employed important diagnostic tools, such as ChIP on chip and MLPA, are described and discussed in detail in Chap. 3 (Sect. 3.4.4.4). However, perhaps the most important recent development in genomics stems from next-generation (also referred to as second-generation) sequencing and the evolving third-generation sequencing (also referred to as single-molecule DNA sequencing).

4.4.3 Next-Generation Sequencing

Next-generation sequencing has already been introduced briefly in Sect. 4.1. An important issue of the new sequencing technologies is a significant reduction in costs: the public Human Genome Project spent US$ 3×10^9 to sequence the human genome, and the National Human Genome Research Institute at the National Institutes of Health aimed at a reduction of these costs to US$ 10^3 by 2014 (www.genome.gov/12513210). The new DNA-sequencing platforms now available do indeed have the potential to achieve the same sequencing results of the Human Genome Project at perhaps 1% of the cost. However, the data obtained with next-generation sequencing depends heavily on the high-quality reference sequence produced by the Human Genome Project. The key to the increased efficiency of the new methods lies in massive parallelization of the biochemical and measurement steps. The second important issue is a significant increase in DNA sequencing speed.

So far there are several commercial next-generation DNA sequencing systems, such as Roche's (454) Genome Sequencer 20/FLX Genome Analyzer, Illumina's Solexa 1G sequencer, Applied Biosystem's SOLiD system, and the Polonator G.007 (Dover Systems/Harvard).

4.4.3.1 Roche's (454) GS FLX Genome Analyzer

This system was commercially introduced in 2004 [47] and is based on pyrosequencing [49]. The sample preparation starts with fragmentation of the genomic DNA (Fig. 4.6a, b). In a next step, adapter sequences are attached to the ends of the DNA pieces to allow the DNA fragments to bind to beads, which have millions of oligomers attached to their surfaces, each with a complementary sequence to the adapter sequences. This is done under conditions allowing only one DNA fragment to bind to each bead. Subsequently, the DNA strands of the library are amplified by emulsion PCR: the beads, each with a single unique DNA fragment, are encased in droplets of oil, which isolate individual agarose beads and keep them apart from their neighbors to ensure that the amplification is uncontaminated. Each droplet contains all reactants needed to amplify the DNA, so that after some hours each agarose bead surface contains more than 1,000,000 copies of the original annealed DNA fragment. This number of DNA strands is needed to produce a detectable signal in the subsequent sequencing reaction. For this sequencing reaction the DNA template-carrying beads are loaded into picoliter reactor wells, each of which just has space for one bead. In these wells pyrosequencing [49], a sequencing-by-synthesis method, takes place, because DNA complementary to each template strand is synthesized. The pyrosequencing reactions flow through each well, and nucleotide and reagent solutions are delivered into it in a sequential fashion. The nucleotide bases used for sequencing release a chemical group as the base forms a bond with the growing DNA chain. This group drives a light-emitting reaction in the presence of specific enzymes and luciferin. The light from the luciferase activity reflects which templates are adding that particular nucleotide, and the emitted light is directly proportional to the amount of the particular nucleotide incorporated. Average read length per sample (or per bead) is about 250 bp.

4.4.3.2 Illumina's Solexa IG Sequencer

Illumina's Genome Analyzer, also commonly referred to as the "Solexa," was the second system commercially launched, in 2006. It is based on "sequencing by synthesis" [3] and is the only next-generation sequencing system that employs bridge-PCR [19] rather than emulsion-PCR (Fig. 4.7). The system applies high-density clonal single-molecule arrays consisting of genomic DNA fragments immobilized to the surface of a reaction chamber. In a first step, DNA fragments are generated by random shearing, and these are then ligated to a pair of oligonucleotides in a forked adapter configuration (Fig. 4.7a). These products can be amplified with two different oligonucleotide primers, which result in double-stranded DNA fragments with different adapter sequences at either end (Fig. 4.7a). In a next step these DNA fragments are denatured and a microfluid cluster station is used to anneal the single strands to the respective complementary oligonucleotides, which are covalently attached to the surface of a glass flow cell (Fig. 4.7b). A new strand is generated using the original strand as template in an extension reaction with an isothermal polymerase. Accordingly, the original strand is removed by denaturation. The adapter sequence of each newly generated strand is annealed to another surface-bound complementary oligonucleotide. This leads to formation of a bridge, and a new site for synthesis of a second strand is generated (Fig. 4.7b). Repeated cycles of annealing, extension, and denaturation result in growth of clusters, each apparently about 1 µm in diameter (Fig. 4.7c). Approximately 50×10^6 separate clusters can be generated per flow cell. For sequencing, each cluster is supplied with polymerase and four differently labeled fluorescent nucleotides (Fig. 4.7d). Based on the concept of "sequencing-by-synthesis," each base incorporation is followed by an imaging step to identify the incorporated nucleotide at each cluster. This iterative process needs about 2.5 days to generate read lengths of 36 bases. As each flow cell has 50×10^6 clusters, each analytical run generates more than 1 billion base pairs (Gb).

4.4.3.3 Applied Biosystem's SOLiD System

The SOLiD (sequencing by *o*ligo *l*igation and *d*etection) system was commercially released in 2007 and represents a development of work published in 2005 [59]. It

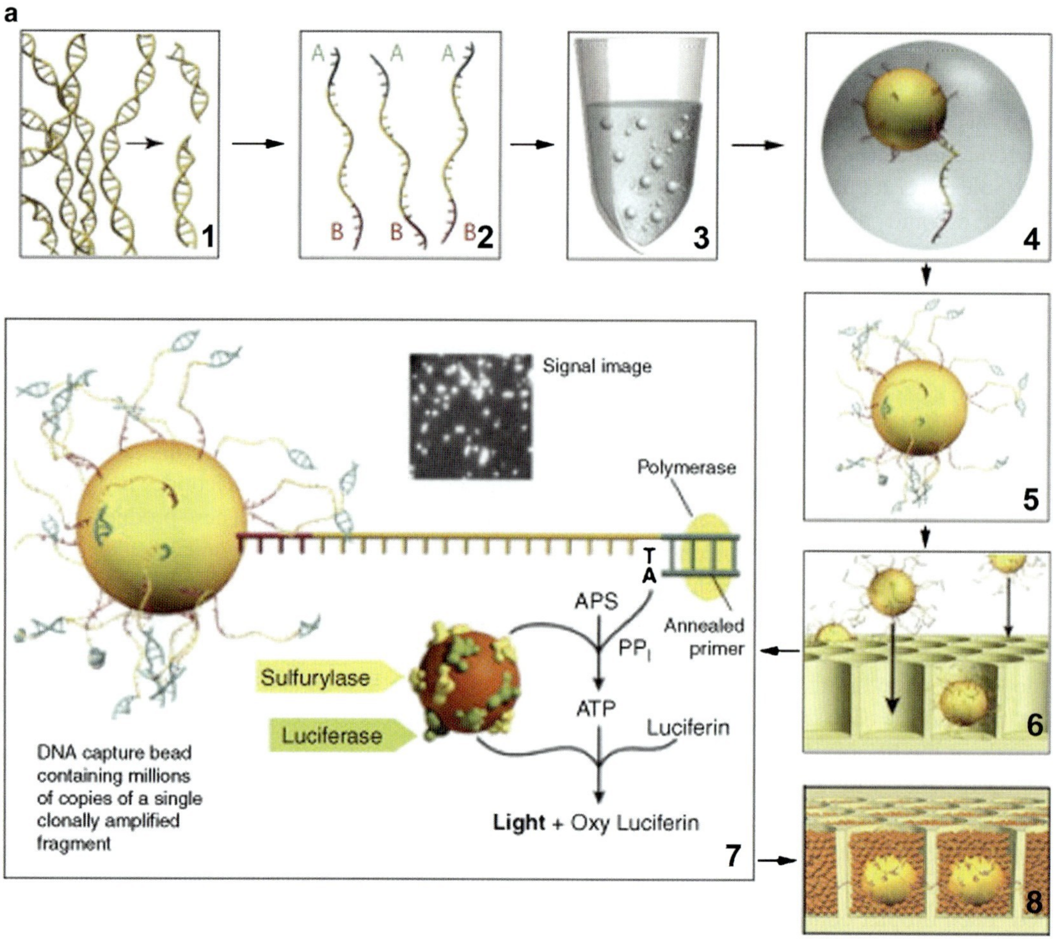

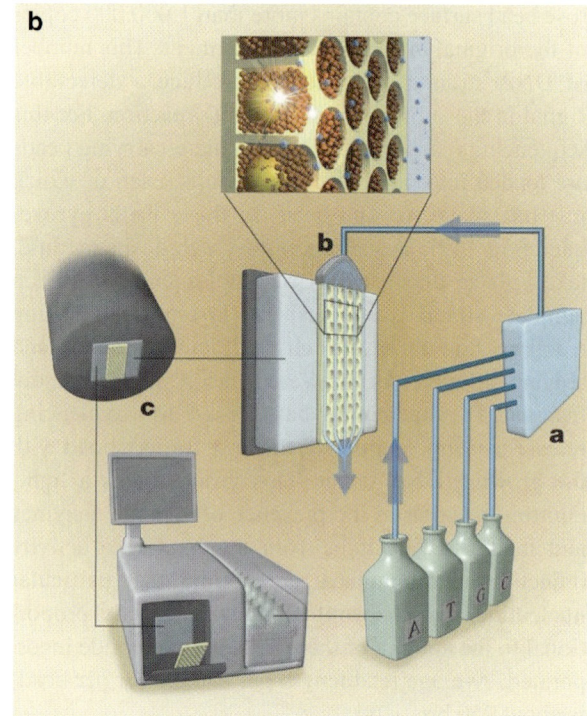

Fig. 4.6 (**a**, **b**) Steps involved in sequencing with the 454 system. (**a**) After isolation genomic DNA is fragmented (*1*) and ligated to adapters (*2*). The DNA is denatured to prepare them for emulsion PCR (*3*). Fragments are bound to beads under conditions that usually allow only one fragment per bead (*4*). The beads are captured in the droplets of a PCR reaction mixture-in-oil emulsion so that a PCR-amplification can be performed within each droplet (*5*). As a result, each bead carries 10 million copies of a unique DNA template. After breaking the emulsion the DNA strands are denatured, and beads carrying single-stranded DNA clones are placed into picotiter plates, i.e., wells of a fiberoptic slide (*6*). In these wells the pyrosequencing reaction takes place (*7* and *8*). (A composite from figures in [46] and [47]) (**b**) Major subsystems of the 454 sequencing instrument: (**b***a*) fluidic assembly; (**b***b*) flow chamber including the well-containing fibre-optic slide; (**b***c*) CCD camera, which captures the light emitted during the pyrosequencing reaction and a computer for instrument control. From [47], reprinted by permission from Macmillan Publishers Ltd: *Nature*, copyright 2005

4 From Genes to Genomics to Proteomics

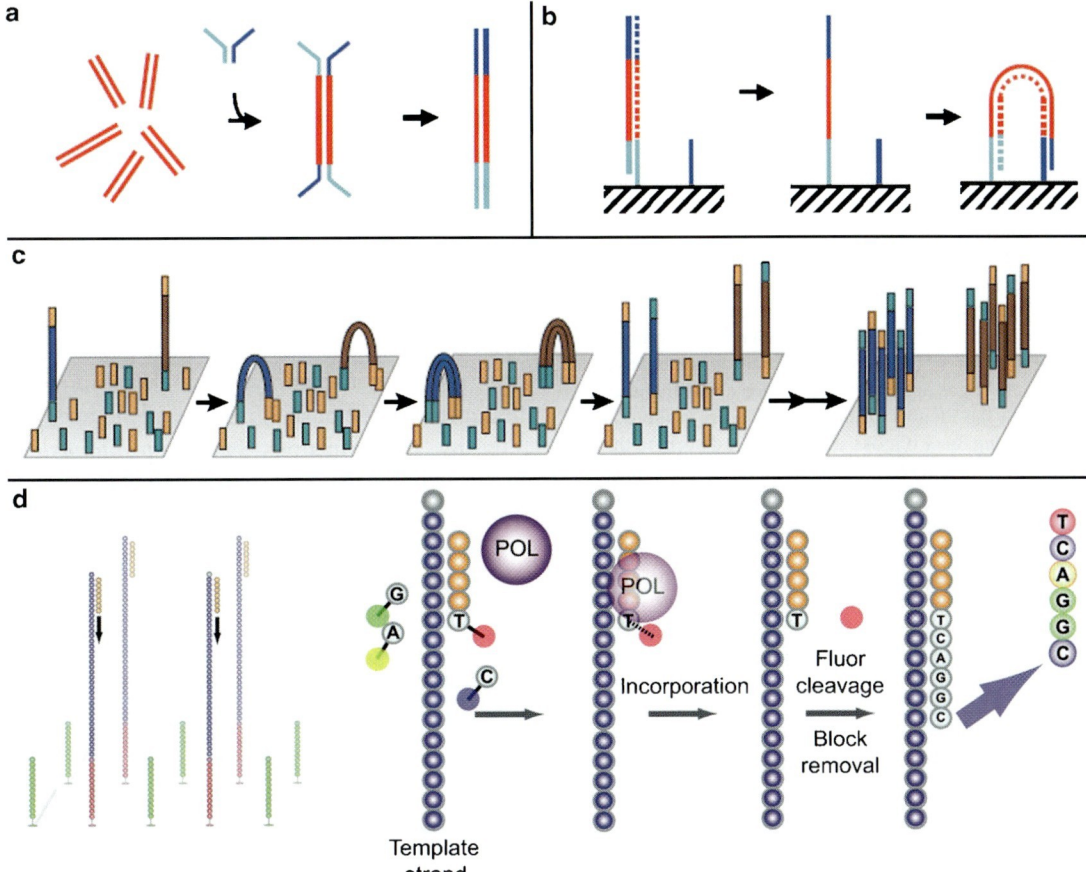

Fig. 4.7 (**a–d**) Steps involved in sequencing with the Illumina system. (**a**) DNA is fragmented by random shearing, and the fragments are then ligated to a pair of oligonucleotides. (**b**) The DNA fragments are denatured and annealed to the respective complementary oligonucleotides, which are covalently attached to the surface of a glass flow cell. A new strand is generated using the original strand as template. The "bridge" amplification relies on captured DNA strands arching over to that they can hybridize to an adjacent anchor oligonucleotide. By this means a bridge is formed and a new site for synthesis of a second strand is generated. (**c**) The adapter sequence of each newly generated strand is annealed to another surface-bound complementary oligonucleotide. Repeated cycles of annealing, extension, and denaturation result in growth of clusters, each appearing about 1 μm in diameter (**c**). (**d**) For sequencing the clusters are denatured, and after a chemical cleavage reaction and wash only forward strands remain for single-end sequencing. Each cluster is supplied with polymerase and four differently labeled fluorescent nucleotides, and each base incorporation is followed by an imaging step to identify the incorporated nucleotide at each cluster. (Reprinted by permission from Macmillan Publishers Ltd: (a,b) Nature [3], (c) Nature Biotechnology [58], copyright 2008)

also employs emulsion-PCR. After amplification the emulsion is broken and beads are covalently attached to the surface of a solid planar substrate, resulting in a dense, disordered array. The ligation-based sequencing process starts with the annealing of a universal primer complementary to the specific adapters on the library fragments. In each sequencing cycle a partially degenerate population of 8mer fluorescently labeled octamers is added (Fig. 4.8). These semi-degenerate oligos are structured in such a way that the label correlates with the identity of the central 2 bp in the octamer ("XX" in Fig. 4.8; the correlation with 2 bp, rather than 1 bp, is the basis of two-base encoding). When an 8mer oligo matches, it can hybridize adjacent to the universal primer 3' end. The DNA ligase can then seal the phosphate backbone. After oligo-ligation a fluorescent readout consisting of imaging in four channels identifies the fixed base with the fluorescence label (the fifth position in Fig. 4.8). Subsequently, a chemical cleavage step removes the sixth through eighth bases ("zzz" in Fig. 4.8) via a modified linkage between bases 5 and 6, which deletes the fluorescent

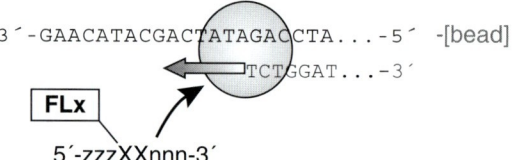

Fig. 4.8 Steps involved in sequencing with the Abi SOLiD system. Sample preparation is similar to that used in 454 technology, because DNA fragments are also ligated to oligonucleotide adapters linked to beads and clonally amplified by emulsion PCR. The ligation-based sequencing process starts with the annealing of a universal primer (5′-zzzXXnnn-3′) complementary to the specific adapters on the library fragments. These semidegenerate oligos are structured in such a way that the label correlates with the identity of the central 2 bp in the octamer ("XX"). Matching 8mer oligos can hybridize adjacent to the universal primer 3′ end and DNA ligase can then seal the phosphate backbone. After oligoligation a fluorescent readout consisting of imaging in four channels identifies the fixed base with the fluorescence label (here the fifth position). Subsequently, a chemical cleavage step removes the sixth through eighth bases (*zzz*), which leaves a free end for another cycle of ligation. For more details see text. (Reprinted by permission from Macmillan Publishers Ltd: Nature Biotechnology [58], copyright 2008)

group and leaves a free end for another cycle of ligation. Several cycles of that kind will iteratively interrogate an evenly spaced, discontinuous set of bases, in this example the sequence of each fragment at five nucleotide intervals. The system is then reset (by denaturation of the extended primer), and the process is repeated with a different offset (e.g., a primer set back from the original position by one or several bases) so that a different set of discontinuous bases is interrogated on the next round of serial ligations. A 6-day instrument run generates sequence read lengths of 35 bases. Placing two flow-cell slides in the instrument per analytical run can produce a combined output of more than 4 Gb of sequence.

A system related to the SOLiD system is the Polonator, which was also developed by the group of George M. Church at Havard [59].

4.4.4 Third-Generation Sequencing

Unlike many of the aforementioned high-speed sequencing technologies currently in use, third-generation sequencing is still under development. This technology is also often referred to as "single-molecule" sequencing, and it reads from individual DNA fragments without the need for amplification, or the risk of introducing errors, or the use of expensive reagents, such as fluorescent tags. As a consequence, third-generation sequencing has the potential to be even faster and cheaper than next-generation sequencing.

There are several different third-generation sequencing approaches, such as exonuclease sequencing, sequencing by synthesis, nanopore sequencing, and transmission electron microscopy [27]. For example, the principle of nanopore sequencing is that DNA can be detected as it passes through a pore by the interruption in the flow of ions through the aperture. The pores, made from a ring of seven α-hemolysin membrane proteins, are the same as those pushed into the membranes of other cells by the infectious bacterium *Staphylococcus aureus* in order to create damaging holes. The identity of each of the four bases traversing the hole might be revealed by distinctive changes in ion flow, which can be read as an electrical signal.

Companies which will likely offer commercial products within the near future include Helicos Bioscience, Complete Genomics, Pacific Biosciences, and Oxford Nanopore.

4.4.5 Personalized Genomics

In April 2008, 454 Life Science sequenced the entire genome of James Watson within 2 months for less than US$1 million [70]. In November 2008, Illumina reported the sequence of the human genome of a person of West African descent [3] and of a person of Han Chinese descent [69], each obtained for about US$250,000 within 8 weeks. At the same time, using the same technology the first complete DNA sequencing of a cytogenetically normal acute myeloid leukemia genome was reported [40]. Thus, the cheap sequencing enabled by next-generation sequencing heralds an era of "personal genomics." In fact, the routine use of whole-genome sequencing as a research tool in human genetics is now possible. At present it actually seems impossible to imagine the potential of third-generation sequencing. For example, Pacific Biosciences, which uses a single-molecule technology with DNA polymerase, aims at producing entire human genomes in less than 3 min by 2013. If these ambitious goals can be realized, the sequencing of an entire genome for about US$1,000 becomes reality and may introduce personalized genomics to the routine work-up in human genetics.

The rapid progress in genetic screening assays and DNA sequencing techniques promises to increase our understanding of the complex relationship between the human genetic make-up (the genotype) and its associated traits (the phenotype). However, what can we expect to learn from the sequences of individual genomes? The first complete genomes demonstrated that it will be extremely difficult to extract medically, or even biologically, reliable inferences from individual sequences. Without any doubt, whole-genome sequencing allows the identification of SNPs, as well as insertion/deletion polymorphisms and structural variations. However, at present they do not accurately define copy-number variants (CNVs, Sect. 3.4.4.4) at the nucleotide level. Thus, next-generation sequencing will improve the catalogue of variants existing in human genomes – SNPs by the million, insertion/deletion polymorphisms by the hundred thousand and structural variants by the thousand. The numbers of these variants will not directly provide information about how such polymorphisms contribute to the wide spectrum of human traits, yet they do provide a necessary step toward accurately defining genomic loci that are likely to be implicated in those traits. Therefore, association studies using complete individual genomes may become the approach of choice for understanding the complexity of human biology and disease.

4.4.6 Gene Function

Regardless of what definition of gene is being used, there is no question that genotype determines phenotype, often together with some environmental factors. At the molecular level, DNA sequences determine the sequences of functional molecules. Thus, an important consequence of the new gene definition as discussed in Sect. 4.1 is that the protein or RNA products must be functional for the purpose of assigning them to a particular gene [25]. This of course results in the important question of, "What is a function?". Many genes remain functionally uncharacterized in the physiological context of disease development. Importantly, the same pathologic mutation may – depending on the genetic background in which it occurs – have different consequences on the phenotype, which is often referred to as expressivity or penetrance. Therefore, high-throughput biochemical and mutational assays, molecular profiling, and interaction studies are needed to define function on a large scale. This is one of the purposes of the -omics sciences.

Gene functions must be clearly defined. This is a tremendous task, considering that biological function has many facets owing to the diversity of cellular activities. Defining the function of a gene is difficult, and it may be influenced by a membership in a specific pathway or a complex network in which the gene product interacts. Depending on this, the function of a gene may have effects across a wide range of spatial and temporal scales.

The gene ontology (GO) database uses a clearly defined and computationally friendly vocabulary for representing the cellular, biochemical, and physiological roles of gene products in a systematic fashion [28]. GO provides a standardized way to assess whether a given number of genes have similar functions. GO terms are organized in a tree-like structure, starting from more general at the root to the most specific at the leaves distributed across three main semantic domains – molecular function, biological process, and cellular location. However, GO describes many, but not all specific biological properties of known genes. In addition to GO, there are many other publicly available data sources, which can be used to get some information about possible gene-product functions (e.g., [31]; Chaps. 29.1–29.3).

Furthermore, there are multiple computational and statistical methods which can be used to deduce the functions of poorly characterized genes from genomic and proteomic datasets via association networks [31].

Many efforts have been made to assign functions to genes computationally. These gene-function predictions are based on parameters such as sequence similarity, the co-occurrence of the protein products in the same macromolecular complex, similarity in mRNA, and protein-expression patterns [71].

A particular challenge in the postgenome era is the deciphering of the biological function of individual genes and gene networks that drive disease. Therefore, at present, alternatives to traditional forward genetics approaches are sought. Such alternatives could consist in the construction of molecular networks defining the molecular states of a system underlying disease. Unlike classic genetics approaches aiming at the indentification of genes underlying genetic loci associated with disease, such approaches seek to identify whole gene networks responding *in trans* to genetic loci driving disease, and in turn leading to variations in the disease traits. The promise of these studies is that investigating how a network of gene interactions affects disease will come to complement more strongly the classic focus of how a single protein or RNA affects disease. Thus, a more detailed picture of the particular network states driving disease may be derived. This in turn may pave the way for more progressive treatments of disease, which may ultimately involve targeting whole networks, as opposed to current therapeutic strategies focused on targeting one or two genes [9].

4.5 Transcriptomics

4.5.1 Capturing the Cellular Transcriptome, Expression Arrays, and SAGE

A detailed analysis of the entire transcriptome requires sophisticated high-throughput approaches. Quantitative real-time PCR (qRT-PCR) represents a very effective technology for gene expression analysis, as it is indeed very quantitative and has a high sensitivity, enabling very accurate measurements of low-abundance transcripts. However, qRT-PCR provides less throughput than the technologies listed in the following paragraphs. Still, many see qRT-PCR as the "gold standard" against which other methods are validated.

Microarray chips have evolved to the most successful and most commonly-used technology for gene expression profiling [13, 56]. Numerous commercially available high-density microarray platforms are accessible, allowing the analysis of more or less entire transcriptomes of complex organisms with relative technical simplicity at low cost.

In parallel with the development of microarrays, computational methods for the analysis of the resulting large data sets were improved and standardized reporting and interpretation guidelines were developed [5]. In principle, two approaches are used for microarray analysis: First, as with CGH (Sect. 3.4.3.5.3), the two differently labeled RNAs are hybridized to the same array and the different fluorescence intensities are compared with one another. In the second approach only one RNA is hybridized to an oligonucleotide platform and stored reference data are being used to derive a comparison.

Microarray-based experiments are performed with RNA isolated from a specific tissue source, which is labeled with a detectable marker. This labeled RNA is then hybridized to arrays comprised of gene-specific probes representing thousands of individual genes.

Each experiment creates a massive amount of data requiring analysis by elaborate computational tools. There are two principle forms of data analysis, i.e., unsupervised and supervised hierarchical clustering analysis. The latter approach detects gene-expression patterns that discriminate tumors on the basis of predefined clinical information [16, 26].

Microarray-based gene expression has propelled our knowledge about transcriptome changes in disease and in physiological conditions. For example, the transcriptome of a normal cell type can be compared with the transcriptome of the same cell type with a specific disease, e.g., after malignant transformation, to elucidate disease-specific alterations. Another frequent application is the analysis of physiological changes, e.g., the comparison of the transcriptome of young versus old cell donors to decipher aging-related changes in the transcriptome [24, 42].

Serial analysis of gene expression (SAGE) [65] represents another approach for gene expression analysis (Fig. 4.9). SAGE is an RNA library-based technology which requires the sequencing of millions of cDNA tags from each library. These tags are then assigned to their genomic location by bioinformatics tools. The main advantage of SAGE is that the transcriptome analysis does not depend on the sequences represented on an array platform. However, SAGE involves significant sequencing efforts, making cost an important issue, so that this technology has not been affordable for many laboratories. Still, the aforementioned new next-generation or third-generation sequencing technologies should significantly decrease costs and may make SAGE even more attractive.

Other, more recent transcriptome analysis approaches are cap analysis of gene expression (CAGE) [7, 36] and polony multiplex analysis of gene expression (PMAGE) [34].

4.5.2 Regulatory Networks

The particular challenges in transcriptomics are to identify every transcript of each cell type and the analysis of regulatory networks for each cell type under different conditions, which may be an important prerequisite for the development of new therapeutic options. As a consequence, system approaches have been developed over the past years to elucidate transcriptional regulatory networks from high-throughput data [61].

The definition of networks includes the identification of all expressed transcripts under any developmental and growth condition. Furthermore, all possible physical interactions between transcriptional regulators and regulatory elements have to be delineated.

As complete transcriptomes of cells are cataloged at increasingly finer levels of detail, we may be able to discern the rules that determine where RNAs are made and how they are processed. However, such rules may change under certain conditions. For example, a cryptic transcription start site upstream of the "correct" initiation site might produce an RNA with additional protein-coding sequence or altered translation efficiency. A minor transcription start site within a gene could produce a truncated protein variant targeted at a different subcellular location. If any of these events provide some selective advantage the cryptic transcription start site could, over the course of time, become an alternative one and eventually the real transcription start site. Such evolutions can only be addressed if entire networks are being analyzed.

4.5.3 Outlier Profile Analysis

A particular challenge in transcriptome analysis could be the inability to extract the essence of recurring specific characteristics that may only be present on a subset of cases within a group. This may be especially true in RNA that has been extracted from cancer samples and which may show heterogeneous patterns of gene amplification, fusion, mutation, or deletion. To overcome these problems, a novel bioinformatics approach dubbed "cancer outlier profile analysis" has been developed as a means of identifying recurring patterns of gene overexpression that may characterize distinct subsets of known cancer types, but may not be detectable with traditional analysis methods (such as t-tests or signal-to-noise ratios) [62]. By using cancer outlier profile analysis, two members of the ETS family of transcription factors, ETV1 and ERG, were identified as outliers in prostate

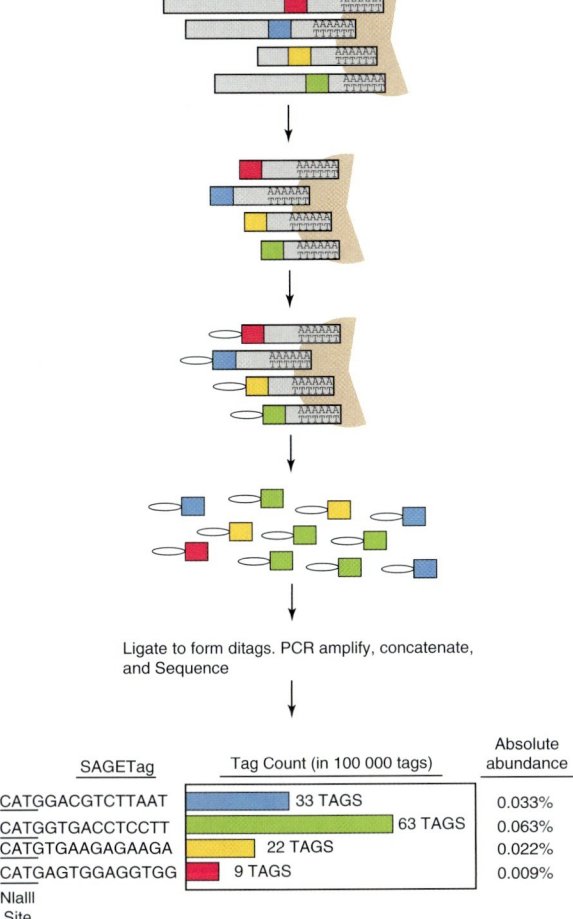

Fig. 4.9 Outline of serial analysis of gene expression (SAGE). In a first step poly-A RNA is captured on oligo-dT-coated beads and subjected to double-stranded cDNA synthesis. The poly-A RNA is cut at defined positions within each transcript by cleavage with an anchoring enzyme (usually NlaIII). Subsequently, linkers are ligated to the immobilized cDNA fragments. These linkers harbor a restriction enzyme type IIs site so that a "tagging enzyme" (usually BsmFI) cuts a short (15-bp) tag from the cDNA. These tags are ligated to form ditags, which can be amplified by PCR. The amplification products are then concatemerized and cloned. Individual tags are then identified by sequencing of concatemere clones. Absolute abundances of tags are calculated by dividing the observed abundance of any tag by the total number of tags analyzed. (Reprinted from [66], with permission from Elsevier)

of how transcription changes during development, with time and space, and especially according to environmental alterations. An integral part of these research efforts is to unravel the control mechanisms which regulate the transcriptome. One aim is the identification cancer. Additional analysis of cDNA transcripts of ERG and ETV1 in prostate cancer cell lines indicated fusion of the 5′ untranslated region of *TMPRSS2* (a prostate-specific, strongly androgen-regulated gene) to either ERG or ETV1. Indeed, cytogenetic analyses performed subsequently confirmed the presence of translocations involving the *TMPRSS2* locus on chromosome 21q22.3 and the corresponding chromosomes harboring one of the ETS family genes. Thus, purely computational manipulation and meta-analysis of existing high-throughput gene expression datasets has eventually led to discovery of a novel group of recurring chromosomal translocations in prostate cancer, which had been neglected by all previously performed cytogenetic or molecular cytogenetic technologies [62].

4.5.4 High-Throughput Long- and Short-Read Transcriptome Sequencing

The same group as initiated outlier profile analysis developed an integrative analysis of high-throughput long- and short-read transcriptome sequencing of cancer cells to discover novel gene fusions [45]. This strategy may represent a powerful tool for the discovery of novel gene chimeras using high-throughput sequencing, opening up an important class of cancer-related mutations for comprehensive characterization [45]. At the same time it becomes obvious that the new sequencing technologies can also be applied to the transcriptome and that they will have a tremendous impact on transcriptomics.

4.5.5 Disease Classification

Interestingly, it has been shown that cancer types can be subclassified based on their *gene* expression patterns. Therefore, gene expression data are often referred to as "signatures" or "molecular portraits," because most tumors show unique expression patterns [10]. Together with appropriate statistical analysis, new or improved classifications have been developed based on expression microarrays for a variety of tumors, such as breast, ovary, prostate, colon, gastric, lung, kidney, brain, leukemia, and lymphoma (reviewed in [10]). These analyses demonstrated that some gene pathways, especially those involved in cell-cycle control, adhesion and motility, apoptosis, and angiogenesis, are frequently affected. Furthermore, these analyses point to pathways, which may represent especially promising targets for therapeutic interventions.

4.5.6 Tools for Prognosis Estimation

Gene expression data have also been used to establish prognostic categories, e.g., in leukemias, breast cancers, and other tumor types [38]. For example, several studies suggest that a panel of 70 genes is sufficient to classify breast cancer into prognostic categories [63, 64]. These analyses resulted in the first multigene panel test approved by the FDA for predicting breast cancer relapse [63].

However, a meta-analysis of seven of the most prominent studies on cancer prognosis based on microarray-expression profiling failed to reproduce the original data in five of these studies [48]. The other two studies yielded much weaker prognostic information than the original data. This suggests that larger sample sizes and careful validation are needed before definite statements about the clinical usefulness of such prognosis predictors can be made. Thus, at present the use of these gene arrays as diagnostic markers cannot yet be recommended [38].

4.6 Proteomics

The proteome is the entire set of proteins encoded by the genome, whereas proteomics is the discipline which studies the global set of proteins and their expression, function, and structure. Proteomics is – after genomics – often considered as a next step in the study of biological systems. Whereas an organism's genome is relatively stable, and therefore more or less constant, the proteome differs from cell to cell and from time to time, making the analysis of the proteome more complicated. Even within a particular cell type, cells may make different sets of proteins at different times or under different conditions. Furthermore, any protein can undergo a wide range of posttranslational modifications, such as phosphorylation, ubiquitination, methylation, acetylation, and so on. As a particular gene can generate multiple distinct proteins, the number of proteins exceeds the number of genes in the corresponding genome by far.

As neither DNA nor mRNA reflects the function of proteins, a number of sophisticated technologies are needed to study individual proteins or the proteome.

4.6.1 From Low-Throughput to High-Throughput Techniques

There are a number of low-throughput techniques which allow testing for the presence of proteins and which can quantify them accurately. These analyses are often performed under certain conditions, e.g., to measure any protein changes during a particular physiological setting or during defined disease stages. Such techniques include Western blot, immunohistochemical staining, and enzyme-linked immunosorbent assay (ELISA). However, in a similar way to DNA or RNA analyses, the study of a protein can quickly become very complex. A frequent aim of proteomics is the identification of biomarkers. This usually requires a detailed understanding of multiple proteins and the complexities of protein-protein interactions. With such an amount of complexity, high-throughput approaches are needed.

At the beginning of proteomics, protein composition studies were performed on two-dimensional gel electrophoresis, which separates proteins in one dimension by molecular weight and in the second dimension by isoelectric point. Spots in the polyacrylamide gel can be cut, and proteins are identified using trypsin digestion and mass spectrometry (MS; Fig. 4.10). The MS tracing provides information on the mass/charge ratio (m/z ratio) of ions. These ratio values can be used to search protein databases. Such a two-dimensional polyacrylamide gel electrophoresis is suitable for high-throughput protein profiling. Basically, procedures to identify biomarkers from clinical specimens can be classified into two principle methodologies: mass spectrometer-based methods and antibody array-based methods, which are similar to DNA microarrays. Mass spectrometry-based approaches are more suitable in cases where the nature of the biomarkers or biosignatures is unknown. In contrast, targeted antibody arrays, which appear to be more cost effective, are more popular for testing proteins for known key pathways.

4.6.2 Mass Spectrometer-Based Methods

The central analytical technique for protein research and for the study of biomolecules is mass spectrometry (MS) [15]. MS is the method most commonly used for the investigation and identification of proteins. MS operates to create ions from neutral proteins, peptides, or metabolites. Therefore, MS depends on effective technologies to softly ionize and to transfer the ionized molecules from the condensed phase into the gas phase without excessive fragmentation. Thus, an MS consists of two main components – an ionization source and a mass analyzer. There are two commonly used techniques to transfer molecules into the gas phase and ionize them prior to mass separation, i.e., electrospray ionization (ESI) [20] and matrix-assisted laser desorption/ionization (MALDI) [33]. After ionization the mass analyzer utilizes the electric charge of the particulates for their separation by speed and/or direction, dependent on the intrinsic m/z of the ion. The types of ion mass separation may include, for example, time-of-flight (ToF), quadrupole electric fields (Q), ion trap (IT), Fourier transform ion cyclotron resonance (FT-ICR) and the Orbitrap [15]. The mass spectrum is characteristic of the molecular mass and/or structure of the metabolite.

Single-stage mass spectrometers are used to evaluate the molecular mass of a polypeptide. However, MS can also provide information about additional structural features, such as amino acid sequence or types of posttranslational modifications. Such analyses are performed after the initial mass determination. Specific ions are selected and fragmented, and structural features of the respective peptides can be deduced from the analysis of these fragments' masses. As two MS analyses are sequentially performed these analyses are usually referred to as tandem MS (MS/MS) [15].

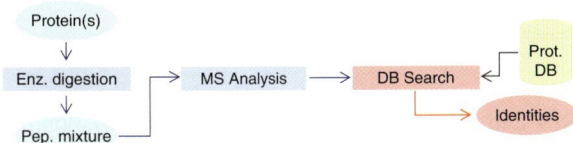

Fig. 4.10 Outline of an experiment in which proteins from 2D gel electrophoresis are identified after enzymatic digestion to create a protein mixture and mass spectrometry of the resulting peptides. The MS tracing provides information on the mass/charge ratio (m/z ratio) of ions, which can be used to search protein databases. (Adapted from [15]. Reprinted with permission from AAAS)

However, like all other approaches, the promising proteomic profiling technologies via MS also have some shortcomings. These include potential artifacts attributable to sample collection and storage, the inherent qualitative nature of mass spectrometers defined by instrument sensitivity, resolution, mass accuracy, dynamic range and throughput, and finally potential artifacts introduced by high-abundance proteins in the serum [38].

4.6.3 Antibody Array-Based Methods

Alternative proteomic strategies include protein microarrays, which depend on immobilization of proteins on a solid support in a way that preserves their folded conformations [44]. For example, antibodies are spotted on the solid surface onto which unmodified proteins are applied. After binding of the proteins to their respective antibodies, a second antibody, which recognizes the same protein and which is labeled for detection by fluorescence, is applied. Such an approach has been referred to as a "sandwich ELISA assay" [37].

Rather like DNA arrays, the direct chemical modification of proteins provides a direct assay mode. Proteins can be labeled with different fluorescent dyes, e.g., as Cy3 and Cy5, and can then be applied to the antibody-spotted slide. This allows the simultaneous analyses of hundreds of target proteins on the same slide. Such an assay is semiquantitative and makes the comparison of two samples applied on the same array, e.g., control versus treated, or normal versus cancer, as in CGH experiments, possible. As in DNA arrays, false-positive or false-negative results have to be excluded, making further validation with other methodologies necessary.

The use of antibody arrays is mainly intended for initial screening of large numbers of proteins to identify candidates for further research. Additional applications include the analysis of posttranslational modifications (such as phosphorylation, acetylation, glycosylation, among others) in complex mixtures of proteins and the analysis of protein/protein interactions [43].

4.6.4 Proteomic Strategies

Several strategies for the analysis of proteins or the proteome have evolved. *MS analysis of substantially purified proteins* corresponds to the aforementioned, classic approach: two-dimensional (2D) gel electrophoresis followed by the mass-spectrometric identification of the protein(s) in a single gel spot. The targeted proteins are digested and identified by mass spectrometry.

In contrast, for *MS analysis of complex peptide mixtures,* also referred to as shotgun proteomics, complex protein samples are digested. The resulting peptide samples are extensively fractionated and analyzed by automated MS/MS. Such an approach allows the analysis of protein samples derived from complete cell lysates or tissue extracts, subcellular fractions, isolated organelles, or other subproteomes.

Furthermore, the establishment of comparative peptide patterns is an important issue. Beside the aforementioned antibody arrays to which two differently labeled protein samples are applied, such a comparison can also be made by 2D gel electrophoresis. For each sample to be analyzed, 2D patterns are generated and the patterns are compared to identify quantitative or qualitative changes. Observed differences can then be further characterized, for example, by sequencing or by determining their posttranslationally modified state.

Future strategies aim at more efficient approaches than those available at present. Such strategies may avoid the situation where the proteome is rediscovered in every experiment. Instead, it would be desirable to use the information from prior proteomic experiments as a guideline for new experiments. This requires the generation of extensive (complete) databases with information to both known and theoretical peptides and their respective proteins to facilitate the targeted, nonredundant analysis of information-rich peptides [15].

4.6.5 Proteomics for Screening and Diagnosis of Disease (Diagnostic and Prognostic Biomarkers)

MS and antibody arrays have evolved into popular platforms for protein screening. It is of special importance that they offer the advantage of multiplexing, can be performed with low sample requirement, and they have the potential for up-scaling using automation.

The availability of methods for measuring the abundance of proteins simultaneously in multiplexed assay formats has opened up opportunities in basic and disease-related research. These technologies can be applied to studies requiring large surveys of changes in protein abundance, to biomarker identification and validation, and to clinical diagnostics using selected targets.

The technologies have matured and can now be used not only for broad protein expression analysis, but also for defining signal transduction pathways, for molecular classification of diseases, for compound profiling and toxicology studies, and for the analysis of patients' individual sensitivities to drugs.

4.7 Conclusions

In human biology the elucidation of gene-product function and regulation is a fundamental objective. In most scenarios a focused single-gene approach is insufficient, making omics sciences indispensible. Owing to its relative stability, the in-depth analysis of the human genome now represents, especially because of new sequencing technologies, an amenable task, although the real extent of genomic variability is so far unknown. The recent completion of the genomic sequences of human and other mammalian species provides researchers with access to a wealth of relevant sequence information necessary for the functional characterization of gene products in a systematic and comprehensive manner. However, analyses of both transcriptome and proteome appear to be significantly more complex than the analysis of the genome. At transcriptome level, the functional characterization of noncoding RNAs represents what will presumably be the greatest challenge. Furthermore, proper biological activity and cellular homeostasis depend on spatially and temporally restricted partitioning of functionally related sets of gene products. A basic and conserved mode of biological control is the organ- and organelle-selective protein accumulation. Therefore, the fundamental biological information encrypted in the human genome can only be understood by the study of the global patterns of protein synthesis and subcellular localization across the major mammalian organ systems. However, at present, much of the human proteome remains poorly annotated in terms of tissue- and organelle-selective expression.

One of the outstanding questions in expression profiling is how well mRNA levels indeed reflect protein abundance and may represent the biological basis for any measurable differences. Although protein synthesis is dependent on mRNA, in many studies often only a modest relationship between mRNA and protein levels was reported. There may be numerous causes for incomplete proteome/transcriptome coverage, such as sample complexity, unknown protein modifications, poor recovery and detection of lower abundance and membrane-associated proteins, and the fact that certain proteins may also be transported between tissues, particularly those associated with circulatory or endocrine functions. This hampers a rigorous definition of the expressed proteome. Hence, the biological significance of differences in mRNA abundance detected among tissues remains to be elaborated at the protein level.

Furthermore, another limitation of transcriptional profiling is that little information is gleaned with respect to the subcellular localization of the translated gene products. Therefore, proteomic methods of examining protein expression and subcellular localization on a genome-wide scale should provide additional insight into the biological context of uncharacterized gene products that can naturally lead to testable hypotheses regarding function.

References

1. Antonarakis SE, Beckmann JS (2006) Mendelian disorders deserve more attention. Nat Rev Genet 7:277–282
2. Bartel DP (2004) MicroRNAs: genomics, biogenesis, mechanism, and function. Cell 116:281–297
3. Bentley DR, Balasubramanian S, Swerdlow HP, Smith GP, Milton J, Brown CG, Hall KP, Evers DJ, Barnes CL, Bignell HR, Boutell JM, Bryant J, Carter RJ, Keira Cheetham R, Cox AJ, Ellis DJ, Flatbush MR, Gormley NA, Humphray SJ, Irving LJ, Karbelashvili MS, Kirk SM, Li H, Liu X, Maisinger KS, Murray LJ, Obradovic B, Ost T, Parkinson ML, Pratt MR, Rasolonjatovo IM, Reed MT, Rigatti R, Rodighiero C, Ross MT, Sabot A, Sankar SV, Scally A, Schroth GP, Smith ME, Smith VP, Spiridou A, Torrance PE, Tzonev SS, Vermaas EH, Walter K, Wu X, Zhang L, Alam MD, Anastasi C, Aniebo IC, Bailey DM, Bancarz IR, Banerjee S, Barbour SG, Baybayan PA, Benoit VA, Benson KF, Bevis C, Black PJ, Boodhun A, Brennan JS, Bridgham JA, Brown RC, Brown AA, Buermann DH, Bundu AA, Burrows JC, Carter NP, Castillo N, Chiara E, Catenazzi M, Chang S, Neil Cooley R, Crake NR, Dada OO, Diakoumakos KD, Dominguez-Fernandez B, Earnshaw DJ, Egbujor UC, Elmore DW, Etchin SS, Ewan MR, Fedurco M, Fraser LJ, Fuentes Fajardo KV, Scott Furey W, George D, Gietzen KJ, Goddard CP, Golda GS, Granieri PA, Green DE, Gustafson DL, Hansen NF, Harnish K, Haudenschild CD, Heyer NI, Hims MM, Ho JT, Horgan AM, Hoschler K, Hurwitz S, Ivanov DV, Johnson MQ, James T, Huw Jones TA, Kang GD,

Kerelska TH, Kersey AD, Khrebtukova I, Kindwall AP, Kingsbury Z, Kokko-Gonzales PI, Kumar A, Laurent MA, Lawley CT, Lee SE, Lee X, Liao AK, Loch JA, Lok M, Luo S, Mammen RM, Martin JW, McCauley PG, McNitt P, Mehta P, Moon KW, Mullens JW, Newington T, Ning Z, Ling Ng B, Novo SM, O'Neill MJ, Osborne MA, Osnowski A, Ostadan O, Paraschos LL, Pickering L, Pike AC, Pike AC, Chris Pinkard D, Pliskin DP, Podhasky J, Quijano VJ, Raczy C, Rae VH, Rawlings SR, Chiva Rodriguez A, Roe PM, Rogers J, Rogert Bacigalupo MC, Romanov N, Romieu A, Roth RK, Rourke NJ, Ruediger ST, Rusman E, Sanches-Kuiper RM, Schenker MR, Seoane JM, Shaw RJ, Shiver MK, Short SW, Sizto NL, Sluis JP, Smith MA, Ernest Sohna Sohna J, Spence EJ, Stevens K, Sutton N, Szajkowski L, Tregidgo CL, Turcatti G, Vandevondele S, Verhovsky Y, Virk SM, Wakelin S, Walcott GC, Wang J, Worsley GJ, Yan J, Yau L, Zuerlein M, Rogers J, Mullikin JC, Hurles ME, McCooke NJ, West JS, Oaks FL, Lundberg PL, Klenerman D, Durbin R, Smith AJ (2008) Accurate whole human genome sequencing using reversible terminator chemistry. Nature 456:53–59
4. Bolzer A, Kreth G, Solovei I, Koehler D, Saracoglu K, Fauth C, Müller S, Eils R, Cremer C, Speicher MR, Cremer T (2005) Three-dimensional maps of all chromosome positions indicate a probabilistic order in human male fibroblast nuclei and prometaphase rosettes. PLoS Biol 3:e157
5. Brazma A, Hingamp P, Quackenbush J, Sherlock G, Spellman P, Stoeckert C, Aach J, Ansorge W, Ball CA, Causton HC, Gaasterland T, Glenisson P, Holstege FC, Kim IF, Markowitz V, Matese JC, Parkinson H, Robinson A, Sarkans U, Schulze-Kremer S, Stewart J, Taylor R, Vilo J, Vingron M (2001) Minimum information about a microarray experiment (MIAME)-toward standards for microarray data. Nat Genet 29:365–371
6. Buratowski S (2008) Transcription. Gene expression--where to start? Science 322:1804–1805
7. Carninci P, Sandelin A, Lenhard B, Katayama S, Shimokawa K, Ponjavic J, Semple CA, Taylor MS, Engström PG, Frith MC, Forrest AR, Alkema WB, Tan SL, Plessy C, Kodzius R, Ravasi T, Kasukawa T, Fukuda S, Kanamori-Katayama M, Kitazume Y, Kawaji H, Kai C, Nakamura M, Konno H, Nakano K, Mottagui-Tabar S, Arner P, Chesi A, Gustincich S, Persichetti F, Suzuki H, Grimmond SM, Wells CA, Orlando V, Wahlestedt C, Liu ET, Harbers M, Kawai J, Bajic VB, Hume DA, Hayashizaki Y (2006) Genome-wide analysis of mammalian promoter architecture and evolution. Nat Genet 38:626–635
8. Carthew RW, Sontheimer EJ (2009) Origins and Mechanisms of miRNAs and siRNAs. Cell 136:642–655
9. Chen Y, Zhu J, Lum PY, Yang X, Pinto S, MacNeil DJ, Zhang C, Lamb J, Edwards S, Sieberts SK, Leonardson A, Castellini LW, Wang S, Champy MF, Zhang B, Emilsson V, Doss S, Ghazalpour A, Horvath S, Drake TA, Lusis AJ, Schadt EE (2008) Variations in DNA elucidate molecular networks that cause disease. Nature 452:429–435
10. Chung CH, Bernard PS, Perou CM (2002) Molecular portraits and the family tree of cancer. Nat Genet 32:533–540
11. Cole ST, Brosch R, Parkhill J, Garnier T, Churcher C, Harris D, Gordon SV, Eiglmeier K, Gas S, Barry CE 3 rd, Tekaia F, Badcock K, Basham D, Brown D, Chillingworth T, Connor R, Davies R, Devlin K, Feltwell T, Gentles S, Hamlin N, Holroyd S, Hornsby T, Jagels K, Krogh A, McLean J, Moule S, Murphy L, Oliver K, Osborne J, Quail MA, Rajandream MA, Rogers J, Rutter S, Seeger K, Skelton J, Squares R, Squares S, Sulston JE, Taylor K, Whitehead S, Barrell BG (1998) Deciphering the biology of Mycobacterium tuberculosis from the complete genome sequence. Nature 393:537–544
12. Core LJ, Waterfall JJ, Lis JT (2008) Nascent RNA sequencing reveals widespread pausing and divergent initiation at human promoters. Science 322:1845–8
13. DeRisi J, Penland L, Brown PO, Bittner ML, Meltzer PS, Ray M, Chen Y, Su YA, Trent JM (1996) Use of a cDNA microarray to analyse gene expression patterns in human cancer. Nat Genet 14:457–460
14. De Sandre-Giovannoli A, Bernard R, Cau P, Navarro C, Amiel J, Boccaccio I, Lyonnet S, Stewart CL, Munnich A, Le Merrer M, Levy N (2003) Lamin A truncation in Hutchinson-Gilford Progeria. Science 300:2055
15. Domon B, Aebersold R (2006) Mass spectrometry and protein analysis. Science 312:212–217
16. Eisen MB, Spellman PT, Brown PO, Botstein D (1998) Cluster analysis and display of genome-wide expression patterns. Proc Natl Acad Sci USA 95:14863–14868
17. ENCODE Project Consortium, Birney E, Stamatoyannopoulos JA, Dutta A, Guigó R, Gingeras TR, Margulies EH, Weng Z, Snyder M, Dermitzakis ET, Thurman RE, Kuehn MS, Taylor CM, Neph S, Koch CM, Asthana S, Malhotra A, Adzhubei I, Greenbaum JA, Andrews RM, Flicek P, Boyle PJ, Cao H, Carter NP, Clelland GK, Davis S, Day N, Dhami P, Dillon SC, Dorschner MO, Fiegler H, Giresi PG, Goldy J, Hawrylycz M, Haydock A, Humbert R, James KD, Johnson BE, Johnson EM, Frum TT, Rosenzweig ER, Karnani N, Lee K, Lefebvre GC, Navas PA, Neri F, Parker SC, Sabo PJ, Sandstrom R, Shafer A, Vetrie D, Weaver M, Wilcox S, Yu M, Collins FS, Dekker J, Lieb JD, Tullius TD, Crawford GE, Sunyaev S, Noble WS, Dunham I, Denoeud F, Reymond A, Kapranov P, Rozowsky J, Zheng D, Castelo R, Frankish A, Harrow J, Ghosh S, Sandelin A, Hofacker IL, Baertsch R, Keefe D, Dike S, Cheng J, Hirsch HA, Sekinger EA, Lagarde J, Abril JF, Shahab A, Flamm C, Fried C, Hackermüller J, Hertel J, Lindemeyer M, Missal K, Tanzer A, Washietl S, Korbel J, Emanuelsson O, Pedersen JS, Holroyd N, Taylor R, Swarbreck D, Matthews N, Dickson MC, Thomas DJ, Weirauch MT, Gilbert J, Drenkow J, Bell I, Zhao X, Srinivasan KG, Sung WK, Ooi HS, Chiu KP, Foissac S, Alioto T, Brent M, Pachter L, Tress ML, Valencia A, Choo SW, Choo CY, Ucla C, Manzano C, Wyss C, Cheung E, Clark TG, Brown JB, Ganesh M, Patel S, Tammana H, Chrast J, Henrichsen CN, Kai C, Kawai J, Nagalakshmi U, Wu J, Lian Z, Lian J, Newburger P, Zhang X, Bickel P, Mattick JS, Carninci P, Hayashizaki Y, Weissman S, Hubbard T, Myers RM, Rogers J, Stadler PF, Lowe TM, Wei CL, Ruan Y, Struhl K, Gerstein M, Antonarakis SE, Fu Y, Green ED, Karaöz U, Siepel A, Taylor J, Liefer LA, Wetterstrand KA, Good PJ, Feingold EA, Guyer MS, Cooper GM, Asimenos G, Dewey CN, Hou M, Nikolaev S, Montoya-Burgos JI, Löytynoja A, Whelan S, Pardi F, Massingham T, Huang H, Zhang NR, Holmes I, Mullikin JC, Ureta-Vidal A, Paten B, Seringhaus M, Church D, Rosenbloom K, Kent WJ, Stone EA, NISC Comparative Sequencing Program; Baylor College of

Medicine Human Genome Sequencing Center; Washington University Genome Sequencing Center; Broad Institute; Children's Hospital Oakland Research Institute, Batzoglou S, Goldman N, Hardison RC, Haussler D, Miller W, Sidow A, Trinklein ND, Zhang ZD, Barrera L, Stuart R, King DC, Ameur A, Enroth S, Bieda MC, Kim J, Bhinge AA, Jiang N, Liu J, Yao F, Vega VB, Lee CW, Ng P, Shahab A, Yang A, Moqtaderi Z, Zhu Z, Xu X, Squazzo S, Oberley MJ, Inman D, Singer MA, Richmond TA, Munn KJ, Rada-Iglesias A, Wallerman O, Komorowski J, Fowler JC, Couttet P, Bruce AW, Dovey OM, Ellis PD, Langford CF, Nix DA, Euskirchen G, Hartman S, Urban AE, Kraus P, Van Calcar S, Heintzman N, Kim TH, Wang K, Qu C, Hon G, Luna R, Glass CK, Rosenfeld MG, Aldred SF, Cooper SJ, Halees A, Lin JM, Shulha HP, Zhang X, Xu M, Haidar JN, Yu Y, Ruan Y, Iyer VR, Green RD, Wadelius C, Farnham PJ, Ren B, Harte RA, Hinrichs AS, Trumbower H, Clawson H, Hillman-Jackson J, Zweig AS, Smith K, Thakkapallayil A, Barber G, Kuhn RM, Karolchik D, Armengol L, Bird CP, de Bakker PI, Kern AD, Lopez-Bigas N, Martin JD, Stranger BE, Woodroffe A, Davydov E, Dimas A, Eyras E, Hallgrímsdóttir IB, Huppert J, Zody MC, Abecasis GR, Estivill X, Bouffard GG, Guan X, Hansen NF, Idol JR, Maduro VV, Maskeri B, McDowell JC, Park M, Thomas PJ, Young AC, Blakesley RW, Muzny DM, Sodergren E, Wheeler DA, Worley KC, Jiang H, Weinstock GM, Gibbs RA, Graves T, Fulton R, Mardis ER, Wilson RK, Clamp M, Cuff J, Gnerre S, Jaffe DB, Chang JL, Lindblad-Toh K, Lander ES, Koriabine M, Nefedov M, Osoegawa K, Yoshinaga Y, Zhu B, de Jong PJ (2007) Identification and analysis of functional elements in 1% of the human genome by the ENCODE pilot project. Nature 447:799–816
18. Eriksson M, Brown WT, Gordon LB, Glynn MW, Singer J, Scott L, Erdos MR, Robbins CM, Moses TY, Berglund P, Dutra A, Pak E, Durkin S, Csoka AB, Boehnke M, Glover TW, Collins FS (2003) Recurrent de novo point mutations in lamin a cause Hutchinson-Gilford progeria syndrome. Nature 423:293–298
19. Fedurco M, Romieu A, Williams S, Lawrence I, Turcatti G (2006) BTA, a novel reagent for DNA attachment on glass and efficient generation of solid-phase amplified DNA colonies. Nucleic Acids Res 34:e22
20. Fenn J, Mann M, Meng C, Wong S, Whitehouse C (1988) Electrospray ionisation for mass spectrometry of large biomolecules. Science 246:64–71
21. Fleischmann RD, Adams MD, White O, Clayton RA, Kirkness EF, Kerlavage AR, Bult CJ, Tomb JF, Dougherty BA, Merrick JM et al (1995) Whole-genome random sequencing and assembly of Haemophilus influenzae Rd. Science 269:496–512
22. Fraga MF, Ballestar E, Paz MF, Ropero S, Setien F, Ballestar ML, Heine-Suñer D, Cigudosa JC, Urioste M, Benitez J, Boix-Chornet M, Sanchez-Aguilera A, Ling C, Carlsson E, Poulsen P, Vaag A, Stephan Z, Spector TD, Wu YZ, Plass C, Esteller M (2005) Epigenetic differences arise during the lifetime of monozygotic twins. Proc Natl Acad Sci USA 102:10604–10609
23. Fraser CM, Gocayne JD, White O, Adams MD, Clayton RA, Fleischmann RD, Bult CJ, Kerlavage AR, Sutton G, Kelley JM, Fritchman RD, Weidman JF, Small KV, Sandusky M, Fuhrmann J, Nguyen D, Utterback TR, Saudek DM, Phillips CA, Merrick JM, Tomb JF, Dougherty BA, Bott KF, Hu PC, Lucier TS, Peterson SN, Smith HO, Hutchison CA 3 rd, Venter JC (1995) The minimal gene complement of Mycoplasma genitalium. Science 270: 397–403
24. Geigl JB, Langer S, Barwisch S, Pfleghaar K, Lederer G, Speicher MR (2004) Analysis of gene expression patterns and chromosomal changes associated with aging. Cancer Res 64:8550–8557
25. Gerstein MB, Bruce C, Rozowsky JS, Zheng D, Du J, Korbel JO, Emanuelsson O, Zhang ZD, Weissman S, Snyder M (2007) What is a gene, post-ENCODE? History and updated definition. Genome Res 17:669–681
26. Golub TR, Slonim DK, Tamayo P, Huard C, Gaasenbeek M, Mesirov JP, Coller H, Loh ML, Downing JR, Caligiuri MA, Bloomfield CD, Lander ES (1999) Molecular classification of cancer: class discovery and class prediction by gene expression monitoring. Science 286:531–537
27. Gupta PK (2008) Single-molecule DNA sequencing technologies for future genomics research. Trends Biotechnol 26:602–611
28. Harris MA, Clark J, Ireland A, Lomax J, Ashburner M, Foulger R, Eilbeck K, Lewis S, Marshall B, Mungall C, Richter J, Rubin GM, Blake JA, Bult C, Dolan M, Drabkin H, Eppig JT, Hill DP, Ni L, Ringwald M, Balakrishnan R, Cherry JM, Christie KR, Costanzo MC, Dwight SS, Engel S, Fisk DG, Hirschman JE, Hong EL, Nash RS, Sethuraman A, Theesfeld CL, Botstein D, Dolinski K, Feierbach B, Berardini T, Mundodi S, Rhee SY, Apweiler R, Barrell D, Camon E, Dimmer E, Lee V, Chisholm R, Gaudet P, Kibbe W, Kishore R, Schwarz EM, Sternberg P, Gwinn M, Hannick L, Wortman J, Berriman M, Wood V, de la Cruz N, Tonellato P, Jaiswal P, Seigfried T, White R, Gene Ontology Consortium (2004) The gene ontology (GO) database and informatics resource. Nucleic Acids Res 32(Database issue): D258–D261
29. He Y, Vogelstein B, Velculescu VE, Papadopoulos N, Kinzler KW (2008) The antisense transcriptomes of human cells. Science 322:1855–1857
30. Hennekam RC (2006) Hutchinson-Gilford progeria syndrome: review of the phenotype. Am J Med Genet 140: 2603–2624
31. Hu P, Bader G, Wigle DA, Emili A (2007) Computational prediction of cancer-gene function. Nat Rev Cancer 7:23–34
32. International Human Genome Sequencing Consortium (2004) Finishing the euchromatic sequence of the human genome. Nature 431:931–945
33. Karas M, Hillenkamp F (1988) Laser desorption ionization of proteins with molecular masses exceeding 10,000 daltons. Anal Chem 60:2299–2301
34. Kim JB, Porreca GJ, Song L, Greenway SC, Gorham JM, Church GM, Seidman CE, Seidman JG (2007) Polony multiplex analysis of gene expression (PMAGE) in mouse hypertrophic cardiomyopathy. Science 316:1481–1484
35. Kimchi-Sarfaty C, Oh JM, Kim IW, Sauna ZE, Calcagno AM, Ambudkar SV, Gottesman MM (2007) A "silent" polymorphism in the MDR1 gene changes substrate specificity. Science 315:525–528
36. Kodzius R, Kojima M, Nishiyori H, Nakamura M, Fukuda S, Tagami M, Sasaki D, Imamura K, Kai C, Harbers M, Hayashizaki Y, Carninci P (2006) CAGE: cap analysis of gene expression. Nat Methods 3:211–222

37. Kopf E, Zharhary D (2007) Antibody arrays–an emerging tool in cancer proteomics. Int J Biochem Cell Biol 39:1305–1317
38. Kulasingam V, Diamandis EP (2008) Strategies for discovering novel cancer biomarkers through utilization of emerging technologies. Nat Clin Pract Oncol 5:588–599
39. Lander ES, Linton LM, Birren B, Nusbaum C, Zody MC, Baldwin J, Devon K, Dewar K, Doyle M, FitzHugh W, Funke R, Gage D, Harris K, Heaford A, Howland J, Kann L, Lehoczky J, LeVine R, McEwan P, McKernan K, Meldrim J, Mesirov JP, Miranda C, Morris W, Naylor J, Raymond C, Rosetti M, Santos R, Sheridan A, Sougnez C, Stange-Thomann N, Stojanovic N, Subramanian A, Wyman D, Rogers J, Sulston J, Ainscough R, Beck S, Bentley D, Burton J, Clee C, Carter N, Coulson A, Deadman R, Deloukas P, Dunham A, Dunham I, Durbin R, French L, Grafham D, Gregory S, Hubbard T, Humphray S, Hunt A, Jones M, Lloyd C, McMurray A, Matthews L, Mercer S, Milne S, Mullikin JC, Mungall A, Plumb R, Ross M, Shownkeen R, Sims S, Waterston RH, Wilson RK, Hillier LW, McPherson JD, Marra MA, Mardis ER, Fulton LA, Chinwalla AT, Pepin KH, Gish WR, Chissoe SL, Wendl MC, Delehaunty KD, Miner TL, Delehaunty A, Kramer JB, Cook LL, Fulton RS, Johnson DL, Minx PJ, Clifton SW, Hawkins T, Branscomb E, Predki P, Richardson P, Wenning S, Slezak T, Doggett N, Cheng JF, Olsen A, Lucas S, Elkin C, Uberbacher E, Frazier M, Gibbs RA, Muzny DM, Scherer SE, Bouck JB, Sodergren EJ, Worley KC, Rives CM, Gorrell JH, Metzker ML, Naylor SL, Kucherlapati RS, Nelson DL, Weinstock GM, Sakaki Y, Fujiyama A, Hattori M, Yada T, Toyoda A, Itoh T, Kawagoe C, Watanabe H, Totoki Y, Taylor T, Weissenbach J, Heilig R, Saurin W, Artiguenave F, Brottier P, Bruls T, Pelletier E, Robert C, Wincker P, Smith DR, Doucette-Stamm L, Rubenfield M, Weinstock K, Lee HM, Dubois J, Rosenthal A, Platzer M, Nyakatura G, Taudien S, Rump A, Yang H, Yu J, Wang J, Huang G, Gu J, Hood L, Rowen L, Madan A, Qin S, Davis RW, Federspiel NA, Abola AP, Proctor MJ, Myers RM, Schmutz J, Dickson M, Grimwood J, Cox DR, Olson MV, Kaul R, Raymond C, Shimizu N, Kawasaki K, Minoshima S, Evans GA, Athanasiou M, Schultz R, Roe BA, Chen F, Pan H, Ramser J, Lehrach H, Reinhardt R, McCombie WR, de la Bastide M, Dedhia N, Blöcker H, Hornischer K, Nordsiek G, Agarwala R, Aravind L, Bailey JA, Bateman A, Batzoglou S, Birney E, Bork P, Brown DG, Burge CB, Cerutti L, Chen HC, Church D, Clamp M, Copley RR, Doerks T, Eddy SR, Eichler EE, Furey TS, Galagan J, Gilbert JG, Harmon C, Hayashizaki Y, Haussler D, Hermjakob H, Hokamp K, Jang W, Johnson LS, Jones TA, Kasif S, Kaspryzk A, Kennedy S, Kent WJ, Kitts P, Koonin EV, Korf I, Kulp D, Lancet D, Lowe TM, McLysaght A, Mikkelsen T, Moran JV, Mulder N, Pollara VJ, Ponting CP, Schuler G, Schultz J, Slater G, Smit AF, Stupka E, Szustakowski J, Thierry-Mieg D, Thierry-Mieg J, Wagner L, Wallis J, Wheeler R, Williams A, Wolf YI, Wolfe KH, Yang SP, Yeh RF, Collins F, Guyer MS, Peterson J, Felsenfeld A, Wetterstrand KA, Patrinos A, Morgan MJ, de Jong P, Catanese JJ, Osoegawa K, Shizuya H, Choi S, Chen YJ, International Human Genome Sequencing Consortium (2001) Initial sequencing and analysis of the human genome. Nature 409:860–921
40. Ley TJ, Mardis ER, Ding L, Fulton B, McLellan MD, Chen K, Dooling D, Dunford-Shore BH, McGrath S, Hickenbotham M, Cook L, Abbott R, Larson DE, Koboldt DC, Pohl C, Smith S, Hawkins A, Abbott S, Locke D, Hillier LW, Miner T, Fulton L, Magrini V, Wylie T, Glasscock J, Conyers J, Sander N, Shi X, Osborne JR, Minx P, Gordon D, Chinwalla A, Zhao Y, Ries RE, Payton JE, Westervelt P, Tomasson MH, Watson M, Baty J, Ivanovich J, Heath S, Shannon WD, Nagarajan R, Walter MJ, Link DC, Graubert TA, DiPersio JF, Wilson RK (2008) DNA sequencing of a cytoenetically normal acute myeloid leukaemia genome. Nature 456:66–72
41. Liu B, Zhou Z (2008) Lamin A/C, laminopathies and premature ageing. Histol Histopathol 23:747–763
42. Ly DH, Lockhart DJ, Lerner RA, Schultz PG (2000) Mitotic misregulation and human aging. Science 287:2486–2492
43. MacBeath G (2002) Protein microarrays and proteomics. Nat Genet 32:526–532
44. MacBeath G, Schreiber SL (2000) Printing proteins as microarrays for high-throughput function determination. Science 289:1760–1763
45. Maher CA, Kumar-Sinha C, Cao X, Kalyana-Sundaram S, Han B, Jing X, Sam L, Barrette T, Palanisamy N, Chinnaiyan AM (2009) Transcriptome sequencing to detect gene fusions in cancer. Nature 458:97–101
46. Mardis ER (2008) The impact of next-generation sequencing technology on genetics. Trends Genet 24:133–141
47. Margulies M, Egholm M, Altman WE, Attiya S, Bader JS, Bemben LA, Berka J, Braverman MS, Chen YJ, Chen Z, Dewell SB, Du L, Fierro JM, Gomes XV, Godwin BC, He W, Helgesen S, Ho CH, Irzyk GP, Jando SC, Alenquer ML, Jarvie TP, Jirage KB, Kim JB, Knight JR, Lanza JR, Leamon JH, Lefkowitz SM, Lei M, Li J, Lohman KL, Lu H, Makhijani VB, McDade KE, McKenna MP, Myers EW, Nickerson E, Nobile JR, Plant R, Puc BP, Ronan MT, Roth GT, Sarkis GJ, Simons JF, Simpson JW, Srinivasan M, Tartaro KR, Tomasz A, Vogt KA, Volkmer GA, Wang SH, Wang Y, Weiner MP, Yu P, Begley RF, Rothberg JM (2005) Genome sequencing in microfabricated high-density picolitre reactors. Nature 437:376–380
48. Michiels S, Koscielny S, Hill C (2005) Prediction of cancer outcome with microarrays: a multiple random validation strategy. Lancet 365:488–492
49. Nyrén P, Pettersson B, Uhlén M (1993) Solid phase DNA minisequencing by an enzymatic luminometric inorganic pyrophosphate detection assay. Anal Biochem 208:171–175
50. Preker P, Nielsen J, Kammler S, Lykke-Andersen S, Christensen MS, Mapendano CK, Schierup MH, Jensen TH (2008) RNA exosome depletion reveals transcription upstream of active human promoters. Science 322:1851–1854
51. Quijano-Roy S, Mbieleu B, Bönnemann CG, Jeannet PY, Colomer J, Clarke NF, Cuisset JM, Roper H, De Meirleir L, D'Amico A, Ben Yaou R, Nascimento A, Barois A, Demay L, Bertini E, Ferreiro A, Sewry CA, Romero NB, Ryan M, Muntoni F, Guicheney P, Richard P, Bonne G, Estournet B (2008) De novo LMNA mutations cause a new form of congenital muscular dystrophy. Ann Neurol 64:177–186
52. Rogers YH, Venter JC (2005) Genomics: massively parallel sequencing. Nature 437:326–327
53. Sanger F, Air GM, Barrell BG, Brown NL, Coulson AR, Fiddes CA, Hutchison CA, Slocombe PM, Smith M (1977) Nucleotide sequence of bacteriophage phi X174 DNA. Nature 265:687–695

54. Sanger F, Nicklen S, Coulson AR (1977) DNA sequencing with chain-terminating inhibitors. Proc Natl Acad Sci USA 74:5463–5467
55. Scaffidi P, Misteli T (2005) Reversal of the cellular phenotype in the premature aging disease Hutchinson-Gilford Progeria syndrome. Nat Med 11:440–445
56. Schena M, Shalon D, Davis RW, Brown PO (1995) Quantitative monitoring of gene expression patterns with a complementary DNA microarray. Science 270:467–470
57. Seila AC, Calabrese JM, Levine SS, Yeo GW, Rahl PB, Flynn RA, Young RA, Sharp PA (2008) Divergent transcription from active promoters. Science 322:1849–1851
58. Shendure J, Ji H (2008) Next-generation DNA sequencing. Nat Biotechnol 26:1135–1145
59. Shendure J, Porreca GJ, Reppas NB, Lin X, McCutcheon JP, Rosenbaum AM, Wang MD, Zhang K, Mitra RD, Church GM (2005) Accurate multiplex polony sequencing of an evolved bacterial genome. Science 309:1728–1732
60. Smith M, Brown NL, Air GM, Barrell BG, Coulson AR, Hutchison CA 3 rd, Sanger F (1977) DNA sequence at the C termini of the overlapping genes A and B in bacteriophage phi X174. Nature 265:702–705
61. Tegner J, Bjorkegen J (2007) Perturbations to uncover gene networks. Trends Genet 23:34–41
62. Tomlins SA, Rhodes DR, Perner S, Dhanasekaran SM, Mehra R, Sun XW, Varambally S, Cao X, Tchinda J, Kuefer R, Lee C, Montie JE, Shah RB, Pienta KJ, Rubin MA, Chinnaiyan AM (2005) Recurrent fusion of TMPRSS2 and ETS transcription factor genes in prostate cancer. Science 310:644–648
63. van de Vijver MJ, He YD, LJ van't Veer, Dai H, Hart AA, Voskuil DW, Schreiber GJ, Peterse JL, Roberts C, Marton MJ, Parrish M, Atsma D, Witteveen A, Glas A, Delahaye L, van der Velde T, Bartelink H, Rodenhuis S, Rutgers ET, Friend SH, Bernards R (2002) A gene-expression signature as a predictor of survival in breast cancer. N Engl J Med 347:1999–2009
64. van't Veer LJ, Dai H, van de Vijver MJ (2002) Gene expression profiling predicts clinical outcome of breast cancer. Nature 415:530–536
65. Velculescu VE, Zhang L, Vogelstein B, Kinzler KW (1995) Serial analysis of gene expression. Science 270:484–487
66. Velculescu VE, Vogelstein B, Kinzler KW (2000) Analysing uncharted transcriptomes with SAGE. Trends Genet 16:423–425
67. Venter JC, Adams MD, Myers EW, Li PW, Mural RJ, Sutton GG, Smith HO, Yandell M, Evans CA, Holt RA, Gocayne JD, Amanatides P, Ballew RM, Huson DH, Wortman JR, Zhang Q, Kodira CD, Zheng XH, Chen L, Skupski M, Subramanian G, Thomas PD, Zhang J, Gabor Miklos GL, Nelson C, Broder S, Clark AG, Nadeau J, McKusick VA, Zinder N, Levine AJ, Roberts RJ, Simon M, Slayman C, Hunkapiller M, Bolanos R, Delcher A, Dew I, Fasulo D, Flanigan M, Florea L, Halpern A, Hannenhalli S, Kravitz S, Levy S, Mobarry C, Reinert K, Remington K, Abu-Threideh J, Beasley E, Biddick K, Bonazzi V, Brandon R, Cargill M, Chandramouliswaran I, Charlab R, Chaturvedi K, Deng Z, Di Francesco V, Dunn P, Eilbeck K, Evangelista C, Gabrielian AE, Gan W, Ge W, Gong F, Gu Z, Guan P, Heiman TJ, Higgins ME, Ji RR, Ke Z, Ketchum KA, Lai Z, Lei Y, Li Z, Li J, Liang Y, Lin X, Lu F, Merkulov GV, Milshina N, Moore HM, Naik AK, Narayan VA, Neelam B, Nusskern D, Rusch DB, Salzberg S, Shao W, Shue B, Sun J, Wang Z, Wang A, Wang X, Wang J, Wei M, Wides R, Xiao C, Yan C, Yao A, Ye J, Zhan M, Zhang W, Zhang H, Zhao Q, Zheng L, Zhong F, Zhong W, Zhu S, Zhao S, Gilbert D, Baumhueter S, Spier G, Carter C, Cravchik A, Woodage T, Ali F, An H, Awe A, Baldwin D, Baden H, Barnstead M, Barrow I, Beeson K, Busam D, Carver A, Center A, Cheng ML, Curry L, Danaher S, Davenport L, Desilets R, Dietz S, Dodson K, Doup L, Ferriera S, Garg N, Gluecksmann A, Hart B, Haynes J, Haynes C, Heiner C, Hladun S, Hostin D, Houck J, Howland T, Ibegwam C, Johnson J, Kalush F, Kline L, Koduru S, Love A, Mann F, May D, McCawley S, McIntosh T, McMullen I, Moy M, Moy L, Murphy B, Nelson K, Pfannkoch C, Pratts E, Puri V, Qureshi H, Reardon M, Rodriguez R, Rogers YH, Romblad D, Ruhfel B, Scott R, Sitter C, Smallwood M, Stewart E, Strong R, Suh E, Thomas R, Tint NN, Tse S, Vech C, Wang G, Wetter J, Williams S, Williams M, Windsor S, Winn-Deen E, Wolfe K, Zaveri J, Zaveri K, Abril JF, Guigó R, Campbell MJ, Sjolander KV, Karlak B, Kejariwal A, Mi H, Lazareva B, Hatton T, Narechania A, Diemer K, Muruganujan A, Guo N, Sato S, Bafna V, Istrail S, Lippert R, Schwartz R, Walenz B, Yooseph S, Allen D, Basu A, Baxendale J, Blick L, Caminha M, Carnes-Stine J, Caulk P, Chiang YH, Coyne M, Dahlke C, Mays A, Dombroski M, Donnelly M, Ely D, Esparham S, Fosler C, Gire H, Glanowski S, Glasser K, Glodek A, Gorokhov M, Graham K, Gropman B, Harris M, Heil J, Henderson S, Hoover J, Jennings D, Jordan C, Jordan J, Kasha J, Kagan L, Kraft C, Levitsky A, Lewis M, Liu X, Lopez J, Ma D, Majoros W, McDaniel J, Murphy M, Newman M, Nguyen T, Nguyen N, Nodell M, Pan S, Peck J, Peterson M, Rowe W, Sanders R, Scott J, Simpson M, Smith T, Sprague A, Stockwell T, Turner R, Venter E, Wang M, Wen M, Wu D, Wu M, Xia A, Zandieh A, Zhu X (2001) The sequence of the human genome. Science 291:1304–1351
68. Voelkerding KV, Dames SA, Durtschi JD (2009) Next-generation sequencing: from basic research to diagnostics. Clin Chem 55:641–658
69. Wang J, Wang W, Li R, Li Y, Tian G, Goodman L, Fan W, Zhang J, Li J, Zhang J, Guo Y, Feng B, Li H, Lu Y, Fang X, Liang H, Du Z, Li D, Zhao Y, Hu Y, Yang Z, Zheng H, Hellmann I, Inouye M, Pool J, Yi X, Zhao J, Duan J, Zhou Y, Qin J, Ma L, Li G, Yang Z, Zhang G, Yang B, Yu C, Liang F, Li W, Li S, Li D, Ni P, Ruan J, Li Q, Zhu H, Liu D, Lu Z, Li N, Guo G, Zhang J, Ye J, Fang L, Hao Q, Chen Q, Liang Y, Su Y, San A, Ping C, Yang S, Chen F, Li L, Zhou K, Zheng H, Ren Y, Yang L, Gao Y, Yang G, Li Z, Feng X, Kristiansen K, Wong GK, Nielsen R, Durbin R, Bolund L, Zhang X, Li S, Yang H, Wang J (2008) The diploid genome sequence of an Asian individual. Nature 456:60–65
70. Wheeler DA, Srinivasan M, Egholm M, Shen Y, Chen L, McGuire A, He W, Chen YJ, Makhijani V, Roth GT, Gomes X, Tartaro K, Niazi F, Turcotte CL, Irzyk GP, Lupski JR, Chinault C, Song XZ, Liu Y, Yuan Y, Nazareth L, Qin X, Muzny DM, Margulies M, Weinstock GM, Gibbs RA, Rothberg JM (2008) The complete genome of an individual by massively parallel DNA sequencing. Nature 452:872–876
71. Zhang B, Horvath S (2005) A general framework for weighted gene co-expression network analysis. Stat Appl Genet Mol Biol 4:Article17

Formal Genetics of Humans: Modes of Inheritance*

Arno G. Motulsky

Revised by S.E. Antonarakis and M.R. Speicher

Contents

5.1 Mendel's Modes of Inheritance and Their Application to Humans 166
    5.1.1 Codominant Mode of Inheritance 166
    5.1.2 Autosomal Dominant Mode of Inheritance 167
    5.1.3 Autosomal-Recessive Mode of Inheritance 172
    5.1.4 X-Linked Modes of Inheritance 175
    5.1.5 "Lethal" Factors [32] 179
    5.1.6 Modifying Genes 180
    5.1.7 Anticipation 182
    5.1.8 Total Number of Conditions with Simple Modes of Inheritance Known So Far in Humans 184
    5.1.9 Uniparental Disomy and Genomic Imprinting 185
    5.1.10 Diseases Due to Mutations in the Mitochondrial Genome 188
    5.1.11 Unusual, "Near Mendelian" Modes of Inheritance 190
    5.1.12 Multifactorial Inheritance 194
5.2 Hardy–Weinberg Law and Its Applications 194
    5.2.1 Formal Basis 194
    5.2.2 Hardy–Weinberg Expectations Establish the Genetic Basis of AB0 Blood Group Alleles 195
    5.2.3 Gene Frequencies 198
5.3 Statistical Methods in Formal Genetics: Analysis of Segregation Ratios 198
    5.3.1 Segregation Ratios as Probabilities 198
    5.3.2 Simple Probability Problems in Human Genetics 200
    5.3.3 Testing for Segregation Ratios Without Ascertainment Bias: Codominant Inheritance 201
    5.3.4 Testing for Segregation Ratios: Rare Traits 202
    5.3.5 Discrimination of Genetic Entities: Genetic Heterogeneity 204
    5.3.6 Conditions Without Simple Modes of Inheritance 205
5.4 Conclusions 207
References 207

> The law of combination of the differing traits, according to which the hybrids develop, finds its foundation and explanation in the proven statement that the hybrids produce germ and pollen cells ... which originate from the combination of the traits by fertilization.
> G. Mendel, Versuche über Pflanzenhybriden, 1865

A.G. Motulsky (✉)
Departments of Medicine (Division of Medical Genetics) and Genome Sciences
University of Washington, School of Medicine, Box 355065, 1705 N.E. Pacific St. Seattle, WA 98195-5065, USA
e-mail: agmot@u.washington.edu

*This chapter is a revised and updated version of the third edition's Modes of Inheritance chapter originally authored by Friedrich Vogel and Arno Motulsky.

5.1 Mendel's Modes of Inheritance and Their Application to Humans

Mendel's fundamental discoveries are usually summarized in three laws:

1. Crosses between organisms homozygous for two different alleles at one gene locus lead to genetically identical offspring (F_1 generation), heterozygous for this allele. It is unimportant which of the two homozygotes is male and which is female (law of uniformity and reciprocity). Such reciprocity applies only for genes not located on sex chromosomes.
2. When these F_1 heterozygotes are crossed with each other (intercross), various genotypes segregate: one-half are heterozygous again, and one-quarter are homozygous for each of the parental types. This segregation 1:2:1 is repeated after crossing of heterozygotes in the following generations, whereas the two types of homozygotes breed pure. As noted previously (Chap. 1), Mendel interpreted this result correctly, assuming formation of two types of germ cells with a 1:1 ratio in heterozygotes (law of segregation and law of purity of gametes).
3. When organisms differing in more than one gene pair are crossed, every single gene pair segregates independently, and the resulting segregation ratios follow the statistical law of independent segregation (law of free combination of genes).

This third law applies only when there is no linkage (Chap. 6). Human diploid cells have 46 chromosomes: the two sex chromosomes and 44 autosomes forming 22 pairs of two homologues each. The pairs of homologues are separated during meiosis, forming haploid germ cells or gametes. After impregnation, paternal and maternal germ cells unite to form the zygote, which is diploid again. Sex is determined genotypically; women normally have two X chromosomes, men have one X and one Y chromosome (Chap. 3).

For an understanding of the statistical character of segregation ratios in humans it is important to realize that the number of germ cells formed is very large, particularly among males. Only a very small sample comes to fertilization. Regarding single gene loci this sampling process can generally be regarded as random.

Two alleles may be termed A and A′. The set of combinations described in Fig. 5.1 are possible. As noted above, these theoretical segregation ratios are probabilities; segregation ratios found empirically should be tested by statistical methods to determine whether they are compatible with the theoretical ratios implied by the genetic hypothesis.

The mating type of identical homozygotes (AA × AA or A′A′ × A′A′) is uninteresting except where it permits conclusions regarding genetic heterogeneity of a recessive condition (Sect. 5.3.5). Mating between the two different homozygous types (AA × A′A′) is usually rare and is therefore of little practical importance. Matings between homozygotes and heterozygotes (AA′ × AA) and between two heterozygotes (A′A × A′A) are most important practically, as explained below.

Mendel found that a genotype does not always determine one distinct phenotype. Frequently heterozygotes resemble (more or less) one of the homozygotes. Mendel called the allele that determines the phenotype of the heterozygote dominant, the other recessive. With more penetrating analysis, some human geneticists have concluded that these terms may be misleading and should be abandoned. In fact, at the level of gene action, genes are not dominant or recessive. At the phenotypic level, however, the distinction is important and useful. Biochemical mechanisms of dominant hereditary diseases usually differ from those of recessive conditions. Hence the mode of inheritance gives a hint regarding the biochemical mechanism likely to be involved.

There are a number of instances in which each of two alleles in a heterozygous state has a distinct phenotypic expression. If both are inherited and phenotypically expressed, this mode of inheritance is sometimes called codominant.

5.1.1 Codominant Mode of Inheritance

The first examples of codominance in man were found in the genetics of blood groups; the MN blood types (111300; numbers refer to identifying numbers of diseases listed in [52]) may serve as an example (Table 5.1). When methods for genetic analysis at the protein level became available, many more examples were soon discovered. The example in Table 5.1 clearly points to a genetic model with two alleles, M and N, the phenotypes M and N being the two homozygotes and MN the heterozygote. This example is used below for a statistical comparison between expected and observed

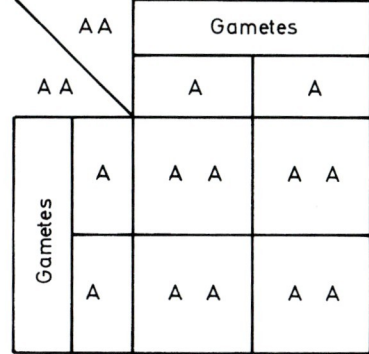
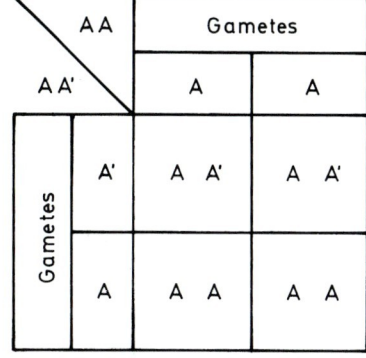
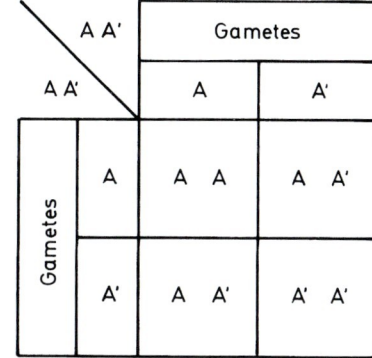

Fig. 5.1 Mating types with two alleles

Table 5.1 Family studies of the genetics of MN blood types from Wiener et al. [92]

Mating type	Number of families	Types of children			Total children
		M	N	MN	
M × M	153	326	0	(1)	327
M × N	179	(1)	0	376	377
N × N	57	0	106	0	106
MN × M	463	499	(1)	473	973
MN × N	351	(3)	382	411	796
MN × MN	377	199	196	405	800
	1,580	1,028	685	1,666	3,379

Parentheses, false paternity.

segregation ratios. The "aberrant" cases in parentheses, which at first glance seem to contradict the genetic hypothesis, were the result of false paternity – a frequent finding in most such investigations.

5.1.2 Autosomal Dominant Mode of Inheritance

The first description of a pedigree showing autosomal dominant inheritance of a human anomaly was Farabee's [22] paper in 1905 on "Inheritance of Digital

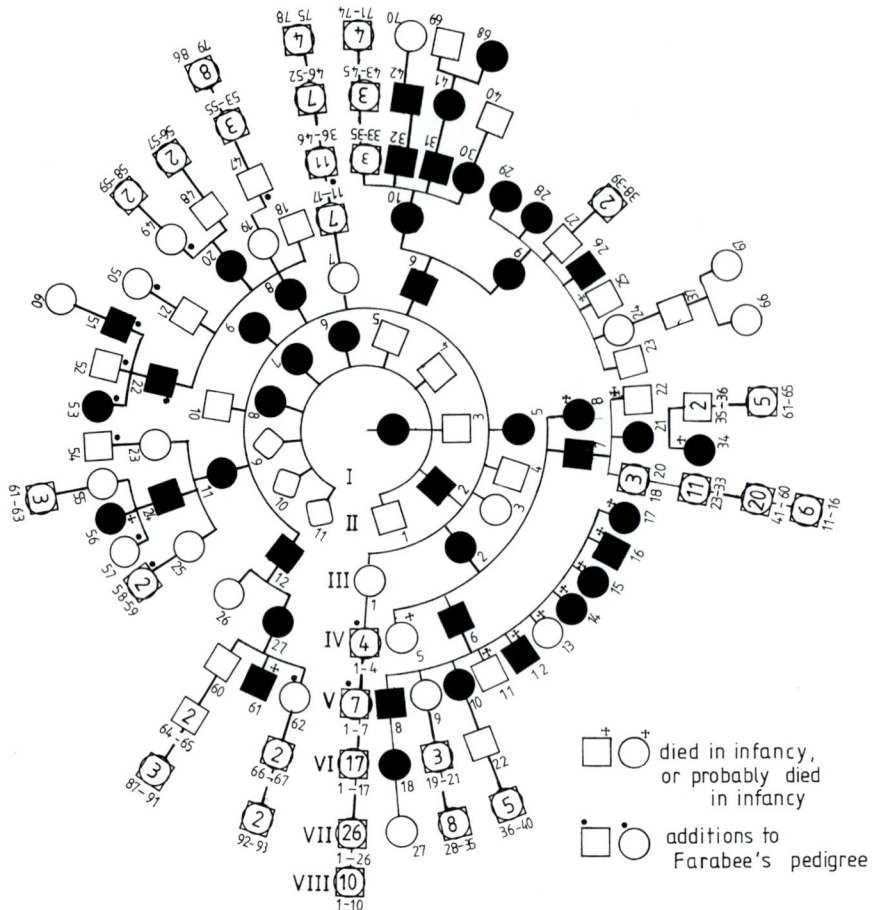

Fig. 5.2 The brachyphalangy pedigree of Farabee [22]. *Black symbols*, affected females (●) and males (■); *numbers*, point to their position in the pedigree

Malformations in Man" (Fig. 5.2). Textbooks usually refer to the condition as brachydactyly (short digits), but from the original paper it is clear that not only were the phalanges of hands and feet shortened, but the number of phalanges was also reduced (Fig. 5.3). In addition, stature was low (average of 159 cm in three males), apparently due to shortness of legs and inferentially also of arms. In every other aspect, Farabee wrote,

> The people appear perfectly normal …and seem to suffer very little inconvenience on account of their malformation. The ladies complain of but one disadvantage in short fingers, and that is in playing the piano; they cannot reach the full octave and hence are not good players.

Figure 5.3 shows the pedigree. There are 36 affected in generations II–V, 13 of which are male and 23 female. Among the unaffected 18 are male and 15 female. The trait is transmitted from one of the parents to about half the children; transmission is independent of sex. Unfortunately, Farabee did not consider the children of the unaffected. Had he done so, he would have found them free from the anomaly. Many other pedigrees have shown absence of the trait among offspring of parents who do not carry the dominant gene. More recently the family has been reexamined [38]. The children of the unaffected family members and some affected family members were added, and X-ray examination confirmed that not only hands and feet were affected but the distal limb bones as well. The basic defect is thought to affect the epiphyseal cartilage.

The condition described by Farabee is now referred to as brachydactyly A-1 (BDA1; OMIM 112500). As pointed out by Farabee, characteristics include shortness of all middle phalanges of the hands and toes,

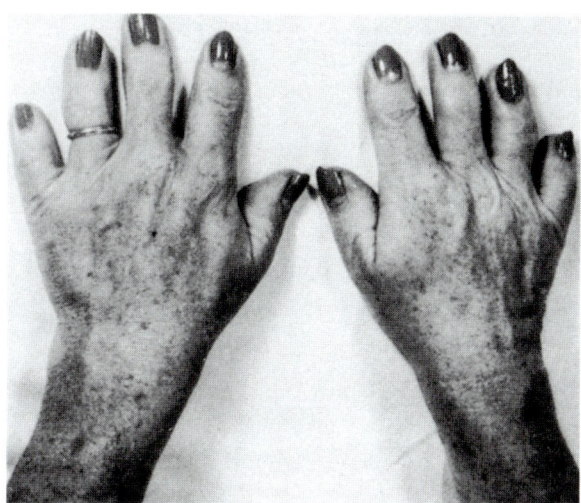

Fig. 5.3 Brachyphalangy in one member of a younger generation of Farabee's pedigree. From Haws and McKusick [38]

occasional terminal symphalangism, shortness of the proximal phalanges of the first digit, and short stature. In 2002 mutations in the Indian Hedgehog gene (*IHH*) were found in descendants of Farabee's family resolving an almost 100-year-old mystery [50, 51]. BDA1 is a heterogeneous condition as an additional locus in another BDA1 family was mapped to 5p13.3-p13.2 [1] (OMIM 607004), and in another BDA1-affected family both the *IHH* locus and the 5p13.3-p13.2 region were excluded [44] suggesting that other, yet unidentified mutations may cause the BDA1 phenotype.

Affected patients are heterozygous for an autosomal allele leading to a clearcut and regular abnormality in the heterozygote. Therefore the trait is, by definition, dominant. The family shows two other characteristics that have since been found to be widespread:

1. The anomalies were described as being almost identical in all family members, and in each person appearing in all four extremities. This is a frequent finding in malformations with a regular mode of inheritance. The reason for the symmetry is evident considering that the same genes act on all four extremities.
2. The anomaly affected the well-being of its bearers only very little. This lack of health impairment is typical for such extended pedigrees. Reproduction is normal. Otherwise the trait would not be transmitted and would soon disappear. This is why, especially in the more serious dominant conditions, extended pedigrees are the exception rather than the rule. Most diseases caused by mutations observed in the present generation have originated rather recently, often even in the germ cell of one of the parents.

5.1.2.1 Late Manifestation, Incomplete Penetrance, and Variable Expressivity

Sometimes a severe dominant condition manifests only during or after the age of reproduction. Here extended pedigrees are usually observed in spite of the severity of the condition. The classic example is Huntington disease (HD) (143100), a degenerative disease of the nerve cells in the basal ganglia (caudate nucleus and putamen) leading to involuntary extrapyramidal movements, personality changes, and a slow deterioration of mental abilities.

Wendt and Drohm [88] carried out a comprehensive study of all cases of HD in the former West Germany. The distribution of ages at onset is presented in Fig. 5.4. The great majority of their patients were married when they developed clinical symptoms. Even among thousands of patients the authors were not able to locate a single case that could be ascribed with confidence to a new mutation. For these reasons and based on results from other early studies, the existence of de novo mutations in HD had long been debated. HD is caused by an increased (CAG) trinucleotide repeat number within the huntingtin gene (*HD*) on 4p16. The unaffected range is $(CAG)_{6-35}$ repeats. Alleles with a length of $(CAG)_{40}$ and above are fully penetrant, i.e., they will cause HD within a normal lifespan. In contrast, alleles of $(CAG)_{36-39}$ confer an increasing risk of developing HD [3]. Analysis of apparently sporadic HD cases revealed that nonpathogenic alleles in the high normal range $((CAG)_{27-35})$ have the potential to expand into the pathogenic range [57]. In fact $(CAG)_{27-35}$ alleles can be unstable during transmission and have a relatively high mutation rate for *HD* of ≥ 10% in each generation [23]. Analysis of the gene is described in Chapter 9.

Another phenomenon occasionally encountered in dominant traits is incomplete penetrance [72]). Penetrance is a statistical concept and refers to the fraction of cases carrying a given gene that manifests a

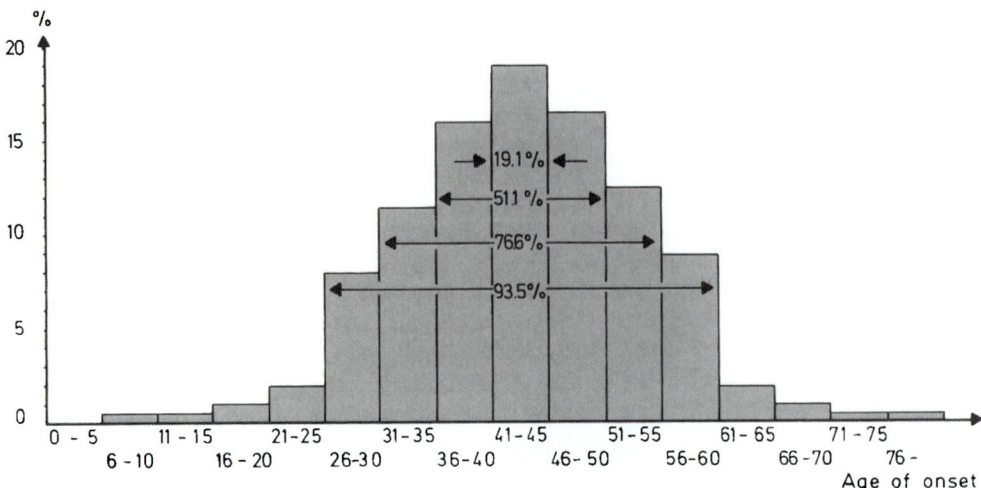

Fig. 5.4 Distribution of ages at onset in 802 cases of Huntington's disease. From Wendt and Drohm [88]

specified phenotype. The transmission seems occasionally to skip one generation, leaving out a person who judging from the pedigree must be heterozygous, or the fraction of those affected among sibs (after appropriate corrections, Sect. 5.3.4) turns out to be lower than the expected segregation ratio. An example is retinoblastoma (180200), a malignant eye tumor of children. Bilateral cases (and cases with more than one primary tumor) are always dominantly inherited, whereas most unilateral, single tumors are nonhereditary, probably being caused by somatic mutation (Chap. 10). Even in pedigrees otherwise showing regular dominant inheritance, however, apparent skipping of a generation is observed occasionally (Fig. 5.5). Calculation of the segregation ratio in a large sample showed that about 45% of sibs were affected instead of the 50% expected in regular dominant inheritance. The penetrance of all cases (unilateral and bilateral) is therefore about 90%. Penetrance in families with bilateral cases is higher than in those with unilateral cases.

In many cases, penetrance is a function of the methods used for examination; higher penetrance is observed with detection methods (clinical or laboratory) that are closer to gene action.

In many dominant conditions the gene may manifest in all heterozygotes, but the *degree of manifestation* may be different. An example is neurofibromatosis (162200). Some cases may show the full-blown picture with many tumors of the skin, café-au-lait spots, and

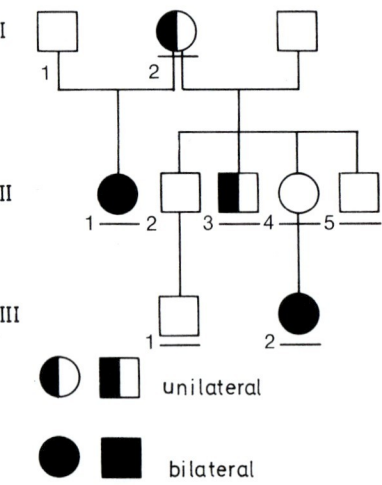

Fig. 5.5 Incomplete penetrance in retinoblastoma. The unaffected woman II,4 must be heterozygous, her mother I,2 and her daughter III,2 being affected; ▫ personally observed. (Personal observation, F. V.)

systemic involvement, whereas other cases – even in the same families – may show only a few café-au-lait spots. The term used to describe this phenomenon is "variable expressivity" [72]. While such terms as "incomplete penetrance" and "variable expressivity" are often needed to convey quick understanding about certain phenomena, they may become dangerous if we forget that they do not explain a biological mechanism but rather are labels for our ignorance.

It is indeed somewhat surprising that so many dominant conditions show such a large interindividual variability in age at onset and severity of manifestation. It would be more understandable if such variability were observed only between different families. Our knowledge of molecular biology (Chap. 10) suggests that the mutational events leading to these conditions are almost always slightly different between families. Indeed, there is usually an intrafamilial correlation between age at onset and severity of manifestation. For HD, for example, Wendt and Drohm [88] calculated a correlation coefficient of +0.57 for age at onset for affected family members. But there usually remains appreciable variability within families, in which the abnormal genes are identical by descent. It is again no more than a label for our ignorance when we invoke the "genetic background" or the action of all other genes for help. In HD, molecular analysis of the gene has provided at least a partial explanation: the number of repeats in the DNA triplet CAG is higher in patients with onset at a very young age. Alleles of $(CAG)_{70}$ repeats or more invariably cause a juvenile onset [3]. Unfortunately, there is no correlation between the number of repeats and age at onset in most patients who develop their clinical disease in the fourth to sixth decades of life.

5.1.2.2 Effect of Homozygosity on Manifestation of Abnormal Dominant Genes

An abnormal gene is called dominant when the heterozygote clearly deviates from the normal. Indeed, almost all bearers of dominant conditions in the human population are heterozygotes. From time to time, however, two bearers of the same anomaly do marry and have children. One quarter of these are then homozygous. This has been observed in several instances, especially when the spouses were relatives. The first example was probably that described by Mohr and Wriedt in [54]. In a consanguineous marriage between two bearers of a moderate brachydactyly (112600) a child was born who not only lacked fingers and toes but also showed multiple malformations of the skeleton and died at the age of 1 year. A sister, however, had only the moderate anomaly, as did her parents [54].

Further examples of homozygosity of dominant anomalies are known. In one family, two parents with hereditary hemorrhagic teleangiectasia had a child showing multiple, severe internal and external telangiectasias who died at the age of 2.5 months [70]. Similarly, a very severe form of epidermolysis bullosa was observed in two of eight children of a couple, both of whom were afflicted with a mild type of this disease.

Another couple, both having a myopathy affecting the distal limb muscles, had 16 children, three of whom showed atypical and especially severe symptoms: the long flexors and the proximal hip muscles were also afflicted, and onset was earlier in life [87].

Epithelioma adenoides cysticum (132700) is a dominant skin disease characterized by multiple nodular tumors. One female patient, whose parents were both affected, had especially severe symptoms, and her eight children all showed this anomaly (Fig. 5.6) [28]. Further examples include achondroplasia (100800), Ehlers-Danlos syndrome (130000), and others. All these cases indicate that homozygotes of dominant anomalies are more severely affected than heterozygotes. It is therefore of interest that there appears to be no clinical difference between heterozygotes and homozygotes for HD, which is therefore a truly dominant disease as defined by Mendel. Clearly a different mechanism must apply to the pathogenesis of such a condition as compared with most other autosomal-dominant diseases, where dose effects are observed [89].

Given what we know about gene action, this is not surprising. In familial hypercholesterolemia (143890) for example, the mechanism of action of a dominant gene is known. A decreased number of receptors for a regulatory substance (low-density lipoprotein) showed the expected differences between heterozygotes and affected homozygotes: 50% decrease and complete absence or very much reduced activity of receptors,

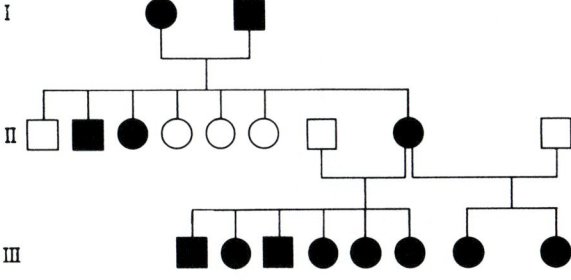

Fig. 5.6 Woman homozygous for epithelioma adenoides cysticum and her progeny in two marriages. From Gaul [28]. The pedigree was complemented in 1958 by Ollendorff-Curth [59]

respectively. Affected homozygotes show massive hypercholesteremia and usually die of myocardial infarction before the age of 30 years.

As noted above, Mendel called a gene dominant when the phenotype of the heterozygote resembled that of one homozygote. The examples of more severe manifestation of dominant genes in the homozygous than in the heterozygous state show that this strict definition is not maintained in human genetics. Here, all conditions are called dominant in which the heterozygote deviates consistently and perceptibly from the normal homozygote – irrespective of the phenotype of the anomalous homozygote. In Mendel's strict definition, most or even all dominant conditions in humans would be "intermediate." However, the more lenient connotation of "dominance" is now in general use.

5.1.3 Autosomal-Recessive Mode of Inheritance

The mode of inheritance is called recessive when the heterozygote does not differ phenotypically from the normal homozygote. In many cases special methods uncover slight detectable differences. Contrary to dominant inheritance, in which almost all crosses are between heterozygotes and homozygous normals (Sect. 5.1.2), the great majority of matings observed in recessive anomalies involve heterozygous and phenotypically normal individuals. Since the three genotypes AA, Aa, and aa occur in the ratio 1:2:1 among the offspring, the probability of a child's being affected is 25%. At the turn of the century when Garrod wrote his paper on alkaptonuria (Chap. 1) the "familial" character of recessive diseases was evident, as family size was large. Today, however, two-children families are generally predominant in industrialized societies. This means that the patient with a recessive disease is very often the only one affected in an otherwise healthy family. However, once an affected child has been born, the genetic risk for any further child of the same parents is 25%. This is important for genetic counseling.

Xeroderma pigmentosum is an autosomal recessive disease (278700). After exposure to ultraviolet light erythema develops, especially in the face, followed by atrophy and telangiectases (Fig. 5.7a). Finally, skin cancers develop that, if untreated, lead to death. Figure 5.7b shows a typical pedigree; here the parents are first cousins. The rate of consanguinity among parents of patients with rare recessive diseases is well above the population average. Usually these parents have inherited this gene from a common ancestor. In Garrod's days this was a powerful tool for recognizing rare recessive diseases; among ten families of alkaptonurics for which this information was available, the parents were first cousins in six cases. Today, however, the consanguinity rate has decreased in most industrialized societies. Hence, even if the rate of consanguinity in families with affected children is substantially increased above the population average, this does not

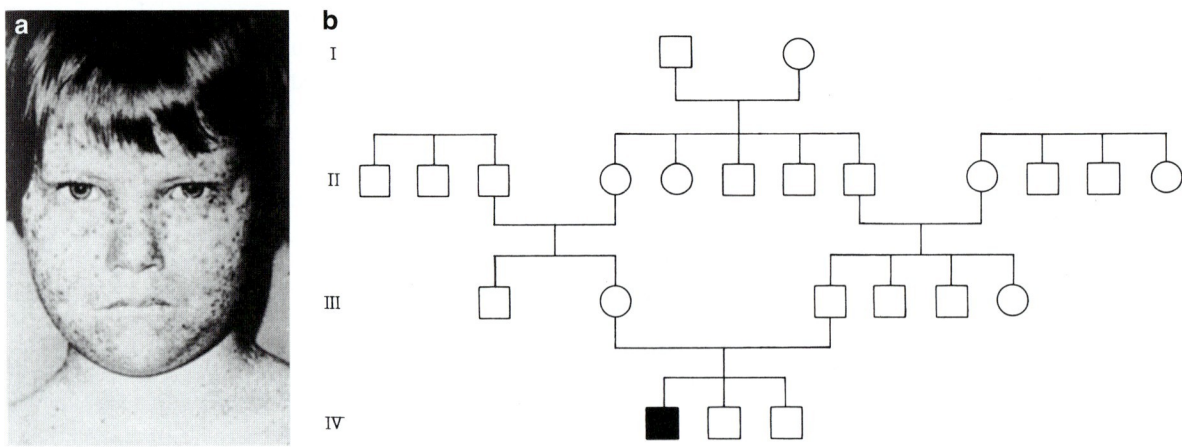

Fig. 5.7 (**a, b**) Xeroderma pigmentosum. (**a**) Girl with this condition (Courtesy of Dr. U. W. Schnyder) (**b**) Pedigree of single case with first-cousin marriage. From Dorn [19]

necessarily lead to the appearance of consanguineous mating when a limited number of families are studied particularly if the abnormal gene is not too rare. This phenomenon together with the small average family size makes it increasingly difficult to recognize an autosomal-recessive mode of inheritance with certainty. Fortunately, however, we no longer need to depend solely on formal genetics. When a rare disease, especially in a child, shows signs of being an inborn error of metabolism, and especially when an enzyme defect can be demonstrated, a recessive mode of inheritance can be inferred in the absence of evidence to the contrary. For purposes of genetic counseling, it must be assumed.

As a rule the vast majority of patients with autosomal-recessive diseases are children of two heterozygotes. Especially decisive for recessive inheritance are the rare matings of two homozygotes with the same anomaly. If both parents are homozygous for the same recessive gene, their mating should exclusively produce affected children. A number of such examples are reported in oculocutaneous albinism (OCA). Some marriages between albinos, however, have produced normally pigmented children [78]. Unless these children are all illegitimate, this proves that the parents must be homozygous for different albino mutations, i.e., more than one albino locus must exist in man. This is the kind of proof that formal genetics can provide to indicate genetic heterogeneity of diseases demonstrating an autosomal recessive mode of inheritance and the same (or a very similar) phenotype. Today, OCA is known as a group of inherited disorders of melanin biosynthesis characterized by a generalized reduction in pigmentation of hair, skin, and eyes. Several types of OCA can be distinguished, with OCA1A (OMIM: 203100) being the most severe type, while OCA1B (OMIM: 606952), OCA2 (OMIM: 203200), OCA3 (OMIM: 203290), and OCA4 (OMIM: 606574) represent milder forms. Each of these four types of OCA is inherited as an autosomal-recessive disorder and at least four genes are responsible for the different types of the disease (i.e., *TYR*, *OCA2*, *TYRP1* and *MATP*) [31].

Another condition for which genetic heterogeneity has been proven in this way is deaf-mutism (Fig. 5.8). Since environmental causes can also cause deafness, it is remarkable that in the pedigree shown here both spouses have an affected sibling, and both parents are consanguineous. Up to date at least 46 genes have been implicated in nonsyndromic hearing loss. The most frequent gene associated with autosomal-recessive nonsyndromic hearing loss is *GJB2*, which is responsible for more than half of cases. Other, relatively frequently implicated genes are *SLC26A4*, *MYO15A*, *OTOF*, *CDH23*, and *TMC1* [39]. Thus, it is likely that in the family shown in Fig. 5.8 the hearing loss was caused by mutations in different genes, e.g., by *GJB2* mutations in one family and *SLC26A4* mutations in the other family. In this scenario the two sons in generation IV would be heterozygous mutation carriers for these two genes; however, this does not result in hearing loss.

5.1.3.1 Pseudodominance in Autosomal Recessive Inheritance

Occasionally matings between an unaffected heterozygote and an affected homozygote are observed. One parent is affected, and the expected segregation ratio among children is 1:1. Since this segregation pattern mimics that found with dominant inheritance,

Fig. 5.8 Pedigree of deaf-mutism showing genetic heterogeneity. Both parents are affected with a hereditary type of deaf-mutism; they have affected sibs and come from consanguineous marriages; however, the two sons are not deaf. They are compound heterozygotes for different deaf mutism genes. From Mühlmann [56]

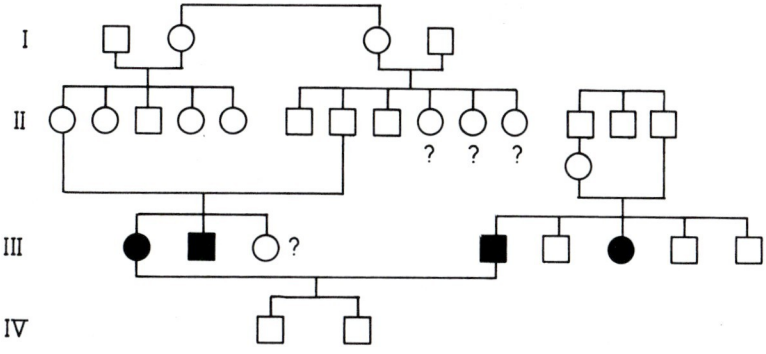

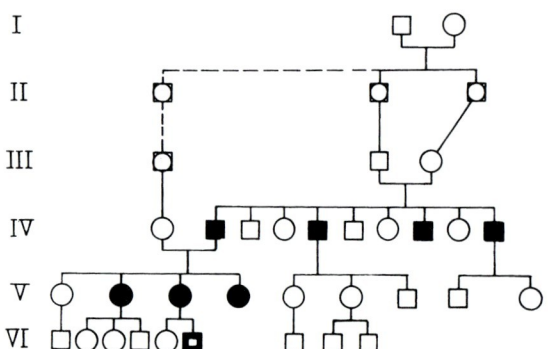

Fig. 5.9 Pedigree of pseudodominance of alkaptonuria, an autosomal-recessive condition. ◼, Suspected alcaptonuric; ◉, sex unknown. (From Milch [53])

this situation is aptly named "pseudodominance." Fortunately for genetic analysis, such matings are very rare.

Garrod's alkaptonuria (203500) provides an example. In all families described since Garrod the autosomal-recessive mode of inheritance had been confirmed until 1956 when a family with a phenotypically similar but apparently dominant form was reported (Fig. 5.9) – a surprising finding. Some years later the authors had to disavow their conclusions: further family investigations had shown typical, recessive alkaptonuria. A number of marriages between relatives (homozygotes × heterozygotes) had led to pseudodominance. If an individual suffering from a recessive disease mates with a normal homozygote, all children are heterozygotes and hence phenotypically normal. As soon as we learn to treat recessive diseases successfully, marriages of affected but treated homozygotes will increase.

Expressivity is generally more uniform within the same family in recessive than in dominant disorders. Incomplete penetrance seems to be rare. Variability between families, however, may be appreciable.

5.1.3.2 Compound Heterozygotes

When a more penetrating biochemical analysis becomes possible, alleles of different origin frequently have slightly different properties. In an increasing number of instances when the gene is analyzed, and the mutations can be identified, such differences can be explained by the properties of the gene-determined proteins and the impairment of their specific functions.

The genes of hemoglobin α and β chains offer an extreme example. Homozygosity of a mutation within the Hbβ gene, for example, may lead to sickle cell anemia or thalassemia major, depending on the precise place of the base substitution. If there are different substitutions within the two alleles, the resulting phenotype might differ from any one of the two true homozygotes. The phenotype of the compound heterozygote who has the sickle cell mutation in one allele and the HbC mutation in the other is different from that of either homozygote (SS or CC). It depends on the population structure how often homozygous patients with a recessive disease are true homozygotes carrying precisely the same mutation twice, and how often they are compound heterozygotes who carry in their two chromosomes different mutations of homologous genes (Fig. 5.10).

We can be reasonably sure that an affected homozygote carries two copies of the same mutation if both copies have a common origin; for example, if his parents are first cousins and if the condition is very rare. Another source of identity by descent are cases from

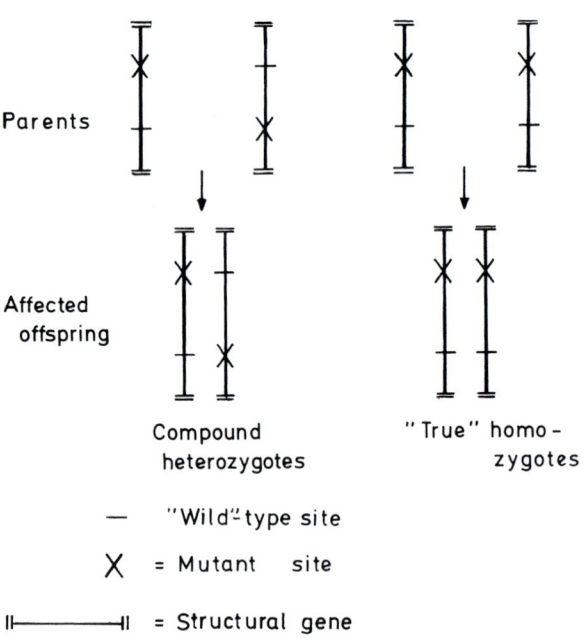

Fig. 5.10 Formation of a compound heterozygote. Each *line* represents the mutant locus on one chromosome in a parent. Among the many possibilities for mutation, two are shown. If parents are heterozygous for mutations that are at identical sites, the affected child is a "true" homozygote; otherwise, he or she is a compound heterozygote

5 Formal Genetics of Humans

an isolate in which a single mutation – which has been introduced by one individual – became frequent, such as the skin disease called Mal de Meleda (OMIM 248300) on the Croatian island of Mljet. Even in a larger and genetically heterogeneous population group, however, the majority of homozygotes may carry the same gene twice. This happens especially when the gene had a selective advantage some time in the past. The *CFTR* (cystic fibrosis) gene is one example: about 60–70% of all abnormal alleles in northwestern European populations are of the type delta 508, meaning that about 40–50% of patients are indeed homozygous for this mutation ($0.7 \times 0.7 = 0.49$). In other diseases the great majority of "homozygous" individuals are in fact compound heterozygotes. With the progress of DNA studies of human genes this question will be answered directly in an increasing number of instances.

5.1.4 X-Linked Modes of Inheritance

In humans, every mating is a Mendelian backcross with respect to the X and Y chromosomes:

		Paternal gametes	
		X	Y
Maternal gametes	X	$1/4$ XX	$1/4$ XY
	X	$1/4$ XX	$1/4$ XY
Total		$1/2$ XX♀ +	$1/2$ XY♂

This implies that on average female and male zygotes are formed at a 1:1 ratio. This, however, is not quite true. The sex ratio at birth (known as the secondary sex ratio in contrast to the primary sex ratio at conception) is slightly shifted in favor of boys (102–106 boys/100 girls). The primary sex ratio is not known exactly, but there are hints that it is also somewhat variable. The formal characteristics of X-linked modes of inheritance can easily be derived from the mode of sex determination. Many studies on the (primary and secondary) sex ratio have been published. Chromosome studies on abortions should reflect the primary sex ratio and point to a value not too far from 100 (boys and girls in a ratio of 1:1). However, the primary and secondary sex ratio also depend on the interval between sexual intercourse and ovulation, frequency of intercourse, general cultural conditions, and even war and peace. After artificial insemination, the fraction of male offspring appears to be appreciably increased.

5.1.4.1 X-Linked Recessive Mode of Inheritance

If we use A for the dominant, normal wild-type and a for the recessive alleles, the following matings are possible:

(a) AA♀ × A♂. All children have the phenotype A. Neither this nor the analogous mating aa × a is useful for genetic analysis.
(b) AA♀ × a♂. All sons have one of the mother's normal alleles. They are healthy. All daughters are heterozygous Aa. They are phenotypically healthy, but carriers of the abnormal allele. In the analogous, very rare mating aa♀ + A♂ all sons are affected (a), and all daughters are heterozygous (Aa).
(c) Aa♀ + A♂. This type is most important. All daughters are phenotypically normal; half are heterozygous carriers. Half of their sons are hemizygous a and affected. The analogous mating Aa♀ × a♂ is extremely rare. There is a 1:1 ratio of affected and heterozygotes among female children and an 1:1 ratio of affected and normals among males.

The principal formal characteristics of X-linked recessive inheritance can be summarized as follows: Males are predominantly – and in rare X-linked conditions almost exclusively – affected. All their phenotypically healthy but heterozygous daughters are carriers. If no new mutation has occurred, and the mother of the affected male is heterozygous, half of his sisters are heterozygous carriers. Among sons of heterozygous women, there is a 1:1 ratio between affected and unaffected.

Strictly speaking, transmission from affected grandfathers via healthy mothers to affected grandsons is helpful, but not altogether decisive for locating the gene on the X chromosome. An autosomal gene with manifestation limited to the male sex could show the same pattern. The fact that all sons of affected men are unaffected, however, is decisive unless the wife is a heterozygous carrier which may not be unusual for common X-linked traits. This criterion can create difficulties in interpretation when a disease is so severe that the patients do not reproduce.

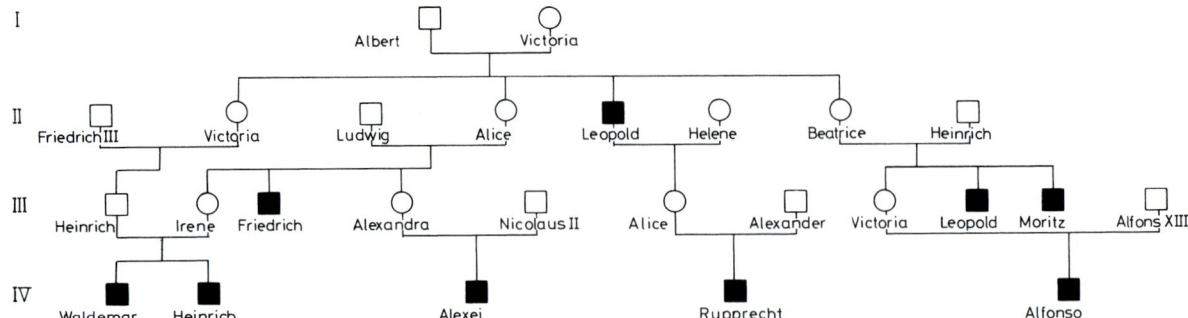

Fig. 5.11 Pedigree of X-linked recessive hemophilia A in the European royal houses. Queen Victoria (I,2) was heterozygous; she transmitted the mutant gene to one hemophilic son and to three daughters

The two most famous and, from a practical standpoint, very important examples are hemophilia A and B (306700, 306900). Due to its alarming manifestations, hemophilia has been known to doctors for a long time and has given rise to the formulation of Nasse's rule (Chap. 1). Figure 5.11 shows the famous pedigree of Queen Victoria's descendants in the European royal houses. One of the hemophilics was the Czarevich Alexei of Russia, and in this case genetic disease influenced politics. Rasputin's power over the imperial couple was based at least partially on his ability to comfort the Czarevich when he was frightened by bleedings. Much larger pedigrees have been described, probably the most extensive being that of hemophilia B in Tenna, Switzerland. As a rule, however, the pedigrees observed in practice are much smaller. Frequently there is only one sibship with affected brothers, or the patient is even the only one affected in an otherwise healthy family. Again, as in dominant conditions (Sect. 5.1.2), this is caused by the reduced reproductive capacity of the patients, which leads to the elimination of most severe hemophilia genes within one or a few generations after they have been produced by new mutation. As expected, almost all hemophilia patients are males. However, there are a few exceptions. Figure 5.12 shows a pedigree from former Czechoslovakia in which a hemophilic had married a heterozygote (who was his double first cousin because in their parents' generation two brothers had married two sisters). The homozygous sisters both had moderately severe hemophilia similar to their affected male relatives.

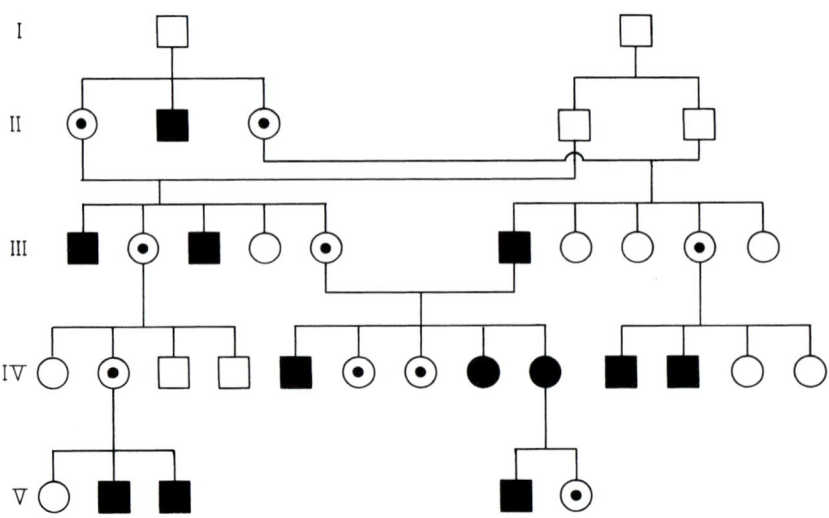

Fig. 5.12 Pedigree of two female homozygotes for X-linked hemophilia. The parents are double first cousins. ⊙, Obligatory heterozygotes. From Pola and Svojitka [64]

Some X-linked conditions have reached considerable frequencies. The most widespread are red-green color vision defects and variants of the enzyme glucose-6-phosphate dehydrogenase, but various types of X-linked mental retardation are also common.

5.1.4.2 X-Linked Dominant Mode of Inheritance

An X-linked dominant condition manifests itself in hemizygous men and heterozygous women. However, all sons of affected males are free of the trait unless their mothers are also affected, and the sons' children are also unaffected. On the other hand, all daughters of affected males are affected. Among children of affected women the segregation ratio is 1:1 regardless of the child's sex, just as in autosomal-dominant inheritance. If affected individuals have a normal rate of reproduction, about twice as many affected females as males are found in the population.

Since only children of affected males provide information in discriminating X-linked dominant from autosomal-dominant inheritance, it is difficult or even impossible to distinguish between these modes of inheritance when the available data are scarce.

The first clearcut example was described by Siemens in [67] in a skin disease that he named "keratosis follicularis spinulosa decalvans (KFSD) cum ophiasi" (308800). The disease manifests follicular hyperkeratosis leading to partial or total loss of eyelashes, eyebrows, and head hair. Severe manifestations were, however, confined to the male members of this family. KFSD is an extremely rare condition as in the last 50 years only 43 additional KFSD cases were identified. A disease-causing gene has not yet been identified [14].

Since then it has been confirmed for all traits with an X-linked dominant mode of inheritance that males are on average more severely affected than females. This finding is no surprise since heterozygous women have a normal allele for compensation, but a satisfactory explanation became possible only when random inactivation of one of the X chromosomes in females was discovered.

Another example of X-linked dominant inheritance is vitamin D-resistant rickets with hypophosphatemia (307800) [93]. In the pedigree shown in Fig. 5.13, all 11 daughters of the affected men suffered from rickets or had hypophosphatemia; all 10 of their sons, however, were healthy. The affected women have both affected and healthy sons and daughters. The probability for the mode of inheritance to be autosomal-dominant and for the affected males to have only affected daughters and only healthy sons is less than 1:10,000. Moreover, in this family male members also tended to be more severely affected than females. Meanwhile, it is established that X-linked hypophosphatemia is caused by mutations in the phosphate-regulating endopeptidase gene (PHEX) [41].

5.1.4.3 X-Linked Dominant Inheritance with Lethality of the Male Hemizygotes [90]

Females with X-chromosomal diseases tend to have milder symptoms than males, as noted above. In some cases the male zygotes may be so severely affected that

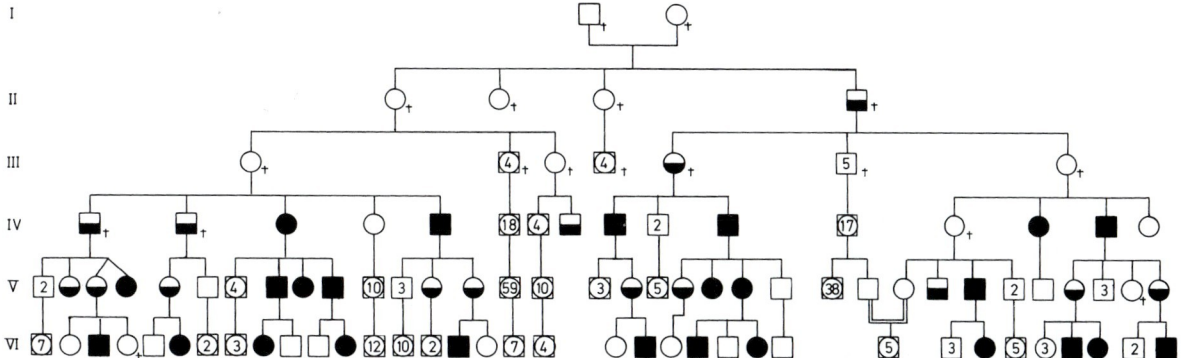

Fig. 5.13 Pedigree of X-linked dominant vitamin D resistant rickets and hypophosphatemia. ■, Hypophosphatemia and rickets; ▬, hypophosphatemia without rickets. From Winters et al. [93]

they die before birth, and only the females survive. This would result in pedigrees containing only affected females, and among their children affected daughters, normal daughters, and normal sons would be found in the ratio of 1:1:1. Among the male hemizygotes who did not die in very early pregnancy, spontaneous abortions (or male stillbirths) would be expected. W. Lenz in [47] was the first to show that this mode of inheritance exists in humans in the condition known as incontinentia pigmenti (Bloch-Sulzberger; 308300).

Around the time of birth the girls affected with this disease develop inflammatory erythematous and vesicular skin disorders. Later, marblecakelike pigmentations appear (Fig. 5.14a). The syndrome additionally comprises tooth anomalies. Figure 5.14b shows a typical pedigree. The alternative hypothesis would be that of an autosomal-dominant mode of inheritance with manifestation limited to the female sex. The two hypotheses would have the following consequences:

a) With autosomal-dominant sex-limited inheritance, and after proper correction (Sect. 5.3.4), there would be a 1:1 ratio of affected to unaffected among sisters of propositae. All brothers would be healthy. If the population sex ratio is assumed to be 1:1, a sex ratio of 2♂:1♀ would be expected among healthy sibs. With X-linked inheritance, on the other hand, the expected number of healthy brothers is much lower, because one-half of the male zygotes are expected to die before birth (possibly leading to an increased rate of spontaneous miscarriages). Among healthy sibs a 1♂:1♀ ratio would be expected.

b) With autosomal-dominant inheritance the abnormal gene may come from the father or from the mother. Therefore more remotely related affected relatives are to be expected among paternal as well as among maternal relatives. With X-linked inheritance, on the other hand, the gene must come from the mother. Considering the rarity of the condition, additional cases would not occur in the father's family.

c) With autosomal-dominant inheritance the loss of mutant genes per generation would be relatively small compared to the total number of these mutations in the population, since the male carriers, being free of symptoms, would reproduce normally. Therefore, assuming genetic equilibrium, the number of new mutations would be small compared to the overall number of cases in the population. With X-linked inheritance, on the other hand, the loss of zygotes is high due to death

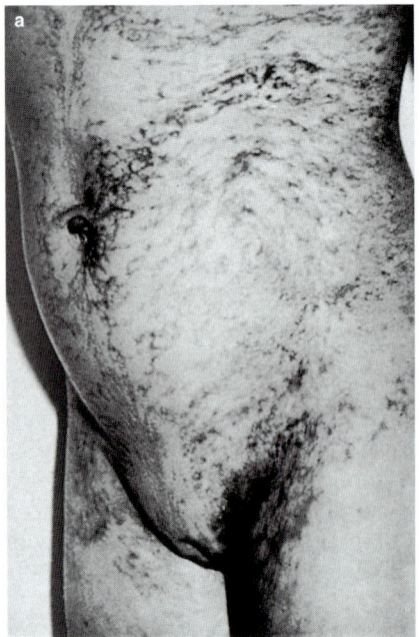
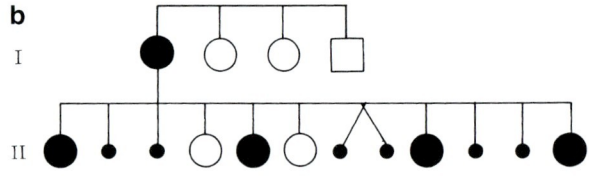

Fig. 5.14 (**a**) Incontinentia pigmenti (Bloch-Sulzberger; courtesy of Dr. W. Fuhrmann). Note the marble cake appearance of skin. (**b**) Pedigree of incontinentia pigmenti. •, Spontaneous abortion; ●, incontinentia pigmenti. From Lenz [47]

of the hemizygote. Hence many of the cases in the population are caused by recent mutation, and extensive pedigrees are rare [7].

The available statistical evidence has consistently supported the hypothesis of an X-linked dominant mode of inheritance with lethality of the male hemizygote. According to Carney et al. [13], 593 female and 16 male cases have been reported. Among the female patients 55% had a positive family history. How can the sporadic males be explained? Of course, the phenomenon of *Durchbrenners* (Hadorn [32] used the term "escapers" – the occasional survival of individuals affected with a lethal genotype) is well known, but Lenz [48] suggested a more specific explanation, assuming, on the basis of a suggestion by Gartler and Francke [27], that a mutation occurs in only one half-strand of the DNA double helix of either the sperm or the oocyte.

Several X-linked syndromes which occur predominantly among females have now been identified. The rareness of affected males in these syndromes is usually attributed to male lethality, which may often occur in the form of early pregnancy loss. About half of the X-linked conditions with predominant expression in females are associated with impairment of cognitive function. Examples include, in addition to the aforementioned incontinentia pigmenti: Aicardi syndrome (OMIM: 304050), focal dermal hypoplasia (Goltz syndrome; OMIM: 305600), Microphthalmia with linear skin defects syndrome (MLS; MIDAS; OMIM: 309801), oral-facial-digital syndrome I (OMIM: 311200), and Rett syndrome (OMIM: 312750). An example for an X-linked syndrome occurring predominantly among females without mental retardation is CHILD syndrome (Congenital hemidysplasia with ichthyosiform erythroderma and limb defects; OMIM: 308050) [74].

5.1.4.4 Genes on the Y Chromosome

Until the 1950s most geneticists were convinced that the human Y chromosome contained genes that occasionally mutate, giving rise to a Y-linked (or holandric) mode of inheritance with male-to-male transmission and males solely being affected. Stern in [72] reviewed the evidence with the result that the time-honored textbook example of Y-linked inheritance of the porcupine man (severe ichthyosis) could no longer be maintained as valid. The only characteristics for which Y-linked inheritance can still be discussed are hairy pinnae, i.e., hair on the outer rim of the ear. A number of extensive pedigrees have been published that show male-to-male transmission. However, the late onset, usually in the third decade of life, and the extremely variable expressivity and high prevalence in some populations (up to 30), makes distinction from a multifactorial mode of inheritance with sex limitation very difficult. Y-linkage can therefore not be fully accepted for this trait.

The Y chromosome contains genes for male differentiation as well as for spermatogenesis.

In experimental animals, segregation ratios deviating from those expected from Mendelian expectations were occasionally reported, one example being the T locus of the mouse [9].

Other cases for which abnormal segregation has been asserted are less well-documented. Since families with many children have become the exception in most industrial societies, the prospect for tracking down and verifying abnormal segregation of pathological genes is becoming more difficult.

5.1.5 "Lethal" Factors [32]

5.1.5.1 Animal Models

Mutations showing a simple mode of inheritance often lead to more or less severe impairment of their bearer's health. There is even evidence (Sect. 5.1.4) that some X-linked conditions prevent the male hemizygote from surviving to birth. It can be assumed that mutations exist which interfere with embryonic development of their carriers so severely as to cause prenatal death.

The first reported case of a lethal mutation in mammalian genetics was the so-called yellow mouse. L. Cuénot [17] reported an apparent deviation from Mendel's law in 1905. A mutant mouse with yellow fur color did not breed true. When yellow animals were crossed with each other, normal gray mice always segregated out. All yellow mice were heterozygous. They all had the same genetic constitution A^Y/A^+; A^Y is a dominant allele of the agouti series, the wild allele of which is termed A^+. When A^Y/A^+ heterozygotes were mated with A^+/A^+ homozygotes, the expected 1:1

ratio between yellow and gray mice was observed. In 1910 it was found that A^Y/A^Y homozygotes are formed but die in utero. Abnormal embryos were later discovered in the expected frequency of 25%.

In this case the allele that is lethal in the homozygous state can be recognized in the heterozygotes by the yellow fur color.

Cases of this sort are exceptional. Generally heterozygotes of lethals are not readily recognizable; therefore lethals occurring spontaneously are difficult to ascertain even in experimental animals and much more so in man.

Usually a lethal mutation kills the embryo in a characteristic phase of its development ("effective lethal phase" [32]). This can easily be explained by the assumption that the action of the mutant gene would be required for further development in this phase.

5.1.5.2 Lethals in Humans

In humans many different types of lethals must occur since many metabolic pathways and their enzymes are essential for survival. It is likely that many still undetected enzyme defects do indeed occur but are not compatible with zygote survival. Moreover, many types of defects of inducer substances needed during embryonic development, and enzymes involved in nucleic acid and protein synthesis, may occur and add to the high incidence of zygote death, which has so far been unexplainable genetically. This problem is discussed from a different standpoint in the context of population genetics (Chap. 16).

According to current estimates, about 15–20% of all recognized human pregnancies end in spontaneous miscarriage. Studies on other mammals suggest that an appreciable number of additional zygote losses go unnoticed, as death occurs during migration through the fallopian tubes. How much of this zygote wastage is due to genetic factors is unknown. A high proportion is caused by numerical or structural chromosome aberrations (Chap. 3). However, there are certainly other maternal causes for abortion as well. While it seemed hopeless to try to relate any proportion of antenatal (or even postnatal) zygote loss to autosomal-dominant or recessive lethals, it appeared more reasonable to speculate about X-linked lethals, as these could influence the sex ratio.

5.1.6 Modifying Genes

So far we have considered phenotypic traits depending on one gene only. However, the phenotypic expression of one gene is usually influenced by other genes. Experiments with animals, especially mammals, show the importance of this "genetic background." One way to overcome analytic difficulties caused by such variation is the use of inbred strains where all animals are genetically alike.

The genetic background is a fairly diffuse concept, but in a number of cases it has been possible to show that penetrance or expressivity of a certain gene can be influenced by another, which is called a "modifier gene" when expressivity is influenced. When penetrance is suppressed altogether, the term "epistasis" (and "hypostasis" of the suppressed gene) is used. In experimental animals cases have been analyzed in which the interaction of two mutations at different loci leads to a completely new phenotype. The classic example is the cross of chickens with "rose" combs and "pea" combs, which leads to the "walnut" comb in homozygotes for both of these mutations. To the best of our knowledge, a similar situation has not been described in man. Modifier genes and epistasis, however, have been demonstrated.

5.1.6.1 Modifying Genes in the AB0 Blood Group System

The best analyzed examples of modifying genes are offered by the AB0 blood group systems. Occurrence of the ABH antigens in saliva (and other secretions) depends on the secretor gene Se. Homozygotes se/se are nonsecretors; heterozygotes Se/se and homozygotes Se/Se are secretors. Hence, se is a recessive suppressor gene. Other rare suppressor genes even prevent the expression of ABH antigens on the surface of erythrocytes.

Bhende et al. [11] discovered a phenotype in 1952 which they called "Bombay" (211100). The erythrocytes were not agglutinated either by anti-A, anti-B or anti-H. The serum contained all three of these agglutinins. Later another family was discovered showing that the bearers of this unusual phenotype did have normal AB0 alleles, but that their manifestation was suppressed (Fig. 5.15; a woman, II, 6, has a Bombay phenotype but

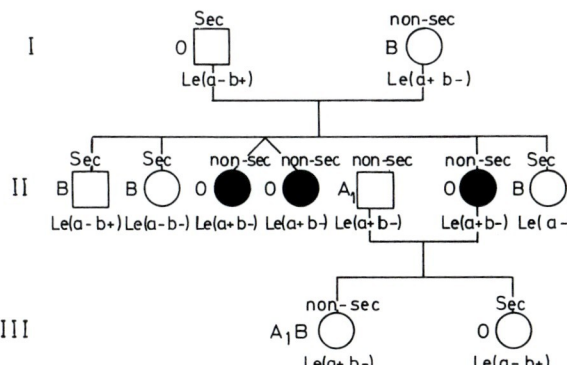

Fig. 5.15 The Bombay blood type. Manifestation of the B antigen is suppressed by a recessive gene x. Note that an O mother (II,6) has an A$_1$B child. From Bhende et al. [11]

transmitted the B allele to one of her daughters). It was further shown that A can also be suppressed, and the available family data suggested an autosomal-recessive mode of inheritance. In the family shown in Fig. 5.15, the parents of the proposita are first cousins.

The locus is not linked to the AB0 locus. The gene pair was named H, h, the Bombay phenotype representing the homozygote, h/h. The gene has been cloned (see [52]). Depending on the nature of the suppressed allele, the phenotype is designated O$_h$A$_1$, O$_h$A$_2$, or O$_h$B. The phenotype has a frequency of about 1 in 13,000 among Maharati-speaking Indians in and around Bombay. A variant with reduced activity is common in the population isolate on Reunion Island [29]. It is caused by the defect of an enzyme that converts a precursor substance into the H antigen, which in turn is a precursor of the A and B antigens [37, 60, 65]. A second gene pair Yy, the rare homozygous conditions of which partially suppresses the A antigen, has been postulated, and subsequently a number of additional families with this condition have been reported.

5.1.6.2 Modifying Genes in Cystic Fibrosis

Cystic fibrosis (CF) is characterized by progressive bronchiectasis, exocrine pancreatic dysfunction, and recurrent sinopulmonary infections. It is a common autosomal recessive disorder with significant morbidity and mortality. The gene, which causes CF, *CFTR*, was already identified 1989; however, the significant phenotypic variation observed in CF suggests that in addition to different mutations in the disease-causing gene and environmental factors, genetic modifiers may contribute to this variability. The identification of such modifiers would have a great potential to improve care for individuals with CF. However, such a modifying effect could to date only be established for a small number of genes. The majority of studies examined the phenotype of lung function and using this parameter, certain alleles of two genes, i.e., transforming growth factor β1 (*TGFβ1*) and mannose binding lectin 1 (*MBL2*), were shown to have an effect on lung function [16]. The efforts of identifying modifier genes show some general problems: First, measurable parameters such as lung function are needed to establish a modifying effect. Second, in a disease affecting multiple organs, such as CF, a modifying effect may have an impact only on one organ but not on others. Third, effects of modifying genes are usually moderate and therefore difficult to identify. Newer tools, such as genome-wide association studies, may further contribute to the elucidation of such modifying genes.

5.1.6.3 Sex-Limiting Modifying Genes

In other, less directly accessible traits the action of modifying genes has been analyzed with statistical methods.

Haldane [33] tried in 1941 to identify such genes in HD, using the family data assembled by Bell in [8]. Harris in [36] examined the problem in a condition called diaphyseal aclasis (133700), which is characterized by multiple exostoses near the cartilaginous epiphyses.

The mode of inheritance is dominant; however, the condition is about twice as common in males as in females. It may be transmitted in some families through unaffected females but not through unaffected males. Statistical analysis of the comprehensive pedigree data collected by Stocks and Barrington [75] suggests in part of the families independent segregation of a factor leading to incomplete penetrance only in females: a sex-limiting modifying gene.

5.1.6.4 Modification by the Other Allele

Phenotypic expression of a gene may be modified not only by genes at other loci but also by the "normal" allele. One example comes from the genetics of the Rh

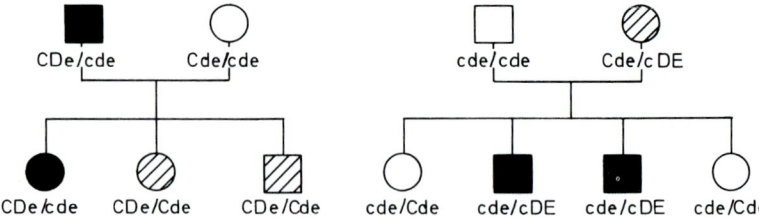

Fig. 5.16 Modification by the homologous allele in the Rh system. ●, D⁺ blood with normal reaction; ◐, weak reaction (Dᵘ variant); ○, D⁻ blood. The haplotype Cde reduces expression of the D factor to Dᵘ. From Ceppellini et al. [15]

factor (Sect. 6.2.4). Occasional blood specimens, when tested with an anti-Rh D serum, give neither a strong positive nor a negative reaction but an attenuated positive reaction. These are called Dᵘ. In most cases a special allele is responsible for this effect, but there are exceptions. In several families the Dᵘ reaction was observed only in family members having Cde as the homologous allele (Fig. 5.16).

5.1.6.5 Modification by Variation in Related Genes

Sickle cell anemia caused by homozygosity for HbS (see Chap. 11) becomes clinically less severe in the presence of several genetic conditions that increase the amount of fetal hemoglobin in the affected red cells. Similarly, the presence of the common alpha thalassemia gene (see Chap. 11) makes for a milder disease manifestation.

5.1.6.6 Modification by a DNA Polymorphism Within the Same Gene

Analysis at the molecular level is revealing new and unsuspected phenomena, including those regarding modification of gene action. Prions are especially interesting proteins. Mutations within the prion gene (176 640) may cause hereditary diseases such as Creutzfeldt-Jakob disease (CJD), Gerstmann-Straussler disease (GSD), or familial fatal insomnia (FFI). The same mutation (Asp → Asn-178) may lead either to CJD or to FFI, depending on a normal polymorphism within the same gene but at a different site: the allele Val 129 segregated in CJD and the allele Met 129 segregated in FFI [30].

Study of various modifying genes and their mechanism is promising to be an important feature for our understanding of the variability of genetic diseases.

The causes of clinical variability in monogenic diseases are:

- Genetic heterogeneity
  - Intra-allelic: different mutations at same locus
  - Inter-allelic: different mutations at other loci
- Modifying genes
  - Additional polymorphisms altering protein conformation
  - Other, as yet unknown mechanisms
- Exposure to various environmental factors required for clinical end result
- Random additional somatic mutations of allele at same locus (e.g., tumors)
- Imprinting (parental origin of mutation)

5.1.7 Anticipation

A time-honored concept popular among physicians in the nineteenth and early twentieth centuries was anticipation. They observed that some hereditary diseases begin earlier in life and follow a more severe course as they progress through generations: the grandfather appeared to be mildly affected; the father was definitely ill, and in the son the disease manifests itself with full force. Anticipation was closely associated with another concept called "degeneration": in some families general, mental, and physical qualities were thought to deteriorate through the generations. These ideas became popular not only among physicians but also among the general public, and were expressed in literary works such as Thomas Mann's novel *Die Buddenbrocks*. In two diseases that tend to manifest during adult life, anticipation seemed to be obvious: HD and myotonic dystrophy (160900) [25]. In the latter, myotonia is associated with relatively mild muscular dystrophy, cataracts, and sometimes mental retardation, or dementia. This disease shows an unusual degree of variability in age at onset, and earlier onset

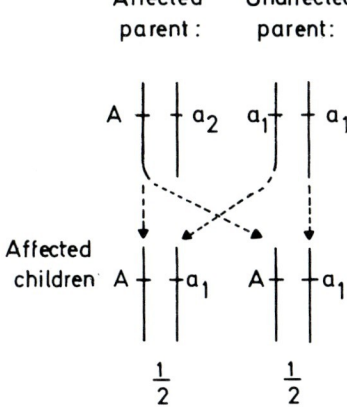

Fig. 5.17 Allelic modification. If manifestation of a dominant, abnormal gene A is modified by the normal allele, and if the allele a_1 causes severe and a_2 milder manifestation of A, there is a correlation in the degree of manifestation between affected sibs but not between affected parent and child. An affected child cannot receive the modifying a_2 allele

as well as a more severe course in some patients of the most recent generation.

When Mendel's laws were rediscovered, anticipation did not fit the new, and otherwise so successful, theory. Therefore scientists interested in genetic problems tried to explain these phenomena away with sophisticated arguments (which were also used in the first two editions of this book). Weinberg [86] pointed out, for example, that anticipation can easily be simulated if families were ascertained directly by patients of the youngest generation who were affected early in life. Their parents and grandparents, on the other hand, who were ascertained through these young probands, could be recognized only if the onset of the disease was so late that they had a chance to have children.

Penrose [63], one of the best human geneticists of his time, explained in great detail that anticipation could be mimicked if ascertainment through the youngest generation combined with dissimilarity of age at onset between parents and children, but similarity between sibs. This would be expected if, in a dominant condition, the normal allele influenced the degree of manifestation of the mutant allele (allelic modification; Fig. 5.17). There can be little doubt that explanations given by Weinberg and Penrose are correct in some instances. However, in HD and myotonic dystrophy molecular analysis revealed specific types of mutations whose effects increase with passage through succeeding generations.

In HD, patients with early onset are more likely to have inherited their mutant genes from the father, whereas late onset is more common when the gene comes from the mother. In myotonic dystrophy, on the other hand, cases of very early onset are less rare; the babies have signs of the disease even at birth. This occurs almost exclusively when the mothers are affected.

Such differences have also been observed in some other monogenic diseases (Table 5.2). On page 169 the huntingtin gene (initially designated "IT15" for "important transcript 15") and the mutations leading to HD were described: amplification of a $(CAG)_n$ repeat beyond 40 copies causes the disease. Moreover, these amplification products are unstable; the predominant tendency appears to be toward an increase in copy numbers by further rounds of amplification; a reduction in copy numbers may occur but is apparently rarer. Higher copy number, on the other side, correlates with earlier onset: a convincing explanation for anticipation.

In myotonic dystrophy an analogous explanation has been found [12, 26, 35]. Here an unstable, amplified sequence was found in the 3′ untranslated region of a gene whose product was predicted to be a member of a protein kinase gene family [12]. It is a $(CTG)_n$ repeat. In normal individuals between 4 and 37 CTG

Table 5.2 Dominant diseases in which parental origin influences the disease (modified from Reik [66])

Disorder	Chromosome	Observations
Huntington disease	4	Early onset frequently associated with paternal transmission
Spinocerebellar ataxia	6	Early onset with paternal transmission
Myotonic dystrophy	19	Congenital form almost exclusively with maternal transmission
Neurofibromatosis I	17	Increased severity with maternal transmission
Neurofibromatosis II	22	Earlier onset with maternal transmission
Wilms tumor	11	Loss of maternal alleles in sporadic tumor
Osteo-sarcoma	13	Loss of maternal alleles in sporadic tumors

repeats are found, 38 to 49 CTG repeats are a premutation. Affected patients may have between 50 and some 2,000 repeats or even more. The repeat number tends to increase over the generations [35]; it is correlated with age at onset and severity of the disease, explaining anticipation.

Thus, the sex of the transmitting parent is an important factor that determines the trinucleotide repeat allele size in the offspring. In the case of myotonic dystrophy it has been speculated that because expansion of the CTG repeat is more rapid with male transmission, negative selection during spermatogenesis may be required to explain the almost exclusive maternal inheritance of severe congenital onset myotonic dystrophy.

5.1.8 Total Number of Conditions with Simple Modes of Inheritance Known So Far in Humans

For many years McKusick has undertaken the task of collecting and documenting known conditions with simple modes of inheritance in man. This extremely valuable resource is now known as OMIM (Online Mendelian Inheritance in Man; www.ncbi.nlm.nih.gov/omim). This web-based full-text, referenced compendium of human genes and genetic phenotypes has the advantage that it can be updated daily, and the entries contain links to other genetics resources. OMIM contains information on all known Mendelian disorders. Table 5.3 provides the number of OMIM entries for autosomal, X-linked, Y-linked, and mitochondrial genes as of 25 May 2009 with information regarding known sequences and phenotypes. Enumeration of dominant and recessive entries was discontinued by OMIM 10 years ago. Note a total number of 19,462 entries, which should be compared with the estimation of about 25,000 human genes based on molecular data. While genetic polymorphisms are included, most conditions listed in this register are rare. Many are rare hereditary diseases. At first glance the list is impressive. However, more detailed scrutiny of the conditions shows that our knowledge of these rare diseases is not nearly as good as it should and could be. There are several reasons:

(a) Most hereditary diseases have become known by occasional observation of affected patients and their families. With rare diseases it is difficult to assess whether they do or do not have a genetic basis. Here, next-generation sequencing or third-generation sequencing may pave the way to finding possible genetic bases in rare diseases.

(b) Some recessive diseases have become known because they happened to be frequent in special populations, primarily in isolates. Isolate studies permit examination of the manifestation of recessive diseases caused by a single mutation. One problem with this approach is that chance determines which genes are studied.

(c) Most human and medical geneticists are working in relatively few industrialized countries. However, genes for rare diseases show a very unequal distribution in different populations. This is particularly true for recessives but has also been shown for dominants with normal or only slightly lowered biological fitness, i.e., when the incidence is not determined by the mutation rate. Hence the developing countries can be expected to abound with hereditary anomalies and diseases that are unclassified to date. Any medical geneticist who has ever walked through, say, an Indian village

Table 5.3 Number of OMIM Entries, 25 May 2009

	Autosomal	X-Linked	Y-Linked	Mitochondrial	Total
* Gene with known sequence	12,111	581	48	37	12,777
+ Gene with known sequence phenotype	347	25	0	0	372
# Phenotype description, molecular basis known	2,293	207	2	26	2,528
% Mendelian phenotype or locus, molecular basis unknown	1,598	141	5	0	1,744
Other, mainly phenotypes with suspected Mendelian basis	1,900	139	2	0	2,041
Total	18,249	1,093	57	63	19,462

knows that this suggestion is not merely a theoretical speculation.

(d) Genetic defects with simple modes of inheritance have a good chance of being detected when they show a clearcut phenotype that is readily recognizable. This is why the inherited conditions of the skin and eye are relatively well known. Other defects, however, may cause anomalies or diseases in some families that are precipitated by environmental factors. Most of such hidden defects are unknown at present.

(e) The real significance of hereditary disease and its total impact can be established only by studies in large populations, using epidemiological methods. Such studies offer the opportunity to detect heterogeneity in etiology and to aid in distinguishing genetic and nongenetic causes. They afford the only basis on which genetic parameters such as mutation rates, biological fitness, and the relative incidence of mild and severe mutations of the same gene can be established. They also help in predicting the long-term and public health effects of medical therapy and of genetic counseling for future generations.

5.1.8.1 Difference in the Relative Frequencies of Dominant and Recessive Conditions in Humans and Animals?

At first glance, there appears to be a difference between humans and experimental animals in the relative frequencies of dominant and recessive conditions. Of the better known mutants of *Drosophila melanogaster* 200 are recessive and only 13 (6.1%) dominant. In the chicken, 40 recessive and 28 dominant mutations have been reported. In the mouse only 17 of 74 mutants are dominant (23%) and the rest recessive. In the rabbit 32 recessive and 6 dominant mutations have been found. (Instances of multiple allelism are counted as one gene locus.) In humans, on the other hand, more dominant than recessive conditions are known. This discrepancy, however, is likely to be caused by diagnostic bias. Our species observes itself most carefully; therefore, defects are detectable that would probably escape observation when present in experimental animals. It would be difficult, for example, to detect brachydactyly in the mouse. This condition, however, leads to a much more severe defect when homozygous.

Hence such a defect, dominant in man, would be counted as recessive in the mouse. Another reason might be that the population of industrialized countries is not in equilibrium for recessive genes. The frequency of consanguineous matings has dropped sharply, and therefore the chance of a recessive gene meeting another mutation in the same gene and becoming homozygous is reduced. A new equilibrium will be reached only in the very distant future when recessive genes could become sufficiently frequent again. In our opinion, there is no significant reason to assume that humans are unique in regard to the ratio of dominant and recessive mutations.

5.1.9 Uniparental Disomy and Genomic Imprinting

In 1980 Eric Engel of the University of Geneva published a paper in which he discussed the possibility of having a chromosomal pair derived from only one parent [21]. He termed this possibility "uniparental disomy" (UPD). The original article included calculations on the potential frequency of UPD; he predicted that as many as 3 individuals out of 10,000 might have UPD for one of the chromosomes involved in common aneuploidies such as 15, 16, 21, 22, and sex chromosomes. Eight years later, the team of Art Beaudet published in the *American Journal of Human Genetics* a case of UPD for chromosome 7 in a female with short stature, cystic fibrosis, and growth hormone deficiency [4]. The authors published a list of possibilities for the mechanism of UPD7 and favored a monosomy 7 conception followed by mitotic nondisjunction or replication of the solitary chromosome 7. The UPD7 in this case was maternal in origin, i.e., there were two chromosomes 7 from the mother and no chromosome 7 contribution from the father. Nonpaternity was obviously convincingly excluded. Isodisomy refers to the case in which the two homologues are identical in sequence (one parental chromosome duplicated); heterodisomy refers to the case in which the two homologues differ (both parental chromosomes inherited). Isodisomy and heterodisomy could be complete, i.e., for the entire chromosome, or partial (segmental) due to recombination events in the parental chromosomes.

The detection of UPD could be done with DNA analysis of the proband and the parents. Single nucleotide polymorphisms, or short sequence repeat polymorphisms, could be used to mark the parental chromosomes, to follow the inheritance, and determine the UPD. In addition the genotyping of the DNA variants could determine the iso- or heterodisomy, either complete of segmental.

UPD has been observed for almost all human chromosomes [20]. The mechanisms resulting in UPD are multiple and include:

(a) "Trisomy rescue" refers to the loss of a chromosome from an initial trisomy (Fig. 5.18). Such reduction from a trisomy to a disomy results from two errors, one meiotic leading to a trisomy state after fertilization by a normal gamete, the other mitotic, removing the supernumerary chromosome by nondisjunction or anaphase lag. Trisomy rescue as a cause of UPD contributes primarily to cases of maternal UPD since most segregation errors occur in oogenesis.
(b) "Gamete complementation" is a mechanism by which a nullisomic gamete meets a disomy gamete. This mechanism implies two errors, one in each sex (Fig. 5.19).
(c) "Rescue of a monosomy" refers to the duplication of a singly inherited chromosome (Fig. 5.20). Such "correction" from a monosomy to a disomy results also from two errors, one meiotic leading to monosomy, the other mitotic duplicating the solitary chromosome.
(d) Somatic recombination (i.e., somatic crossing-over, the symmetrical "trading" of a paternal and maternal homologous chromatid segment) may also be the source of segregants producing cells with segmental UPD (Fig. 5.21).
(e) Chromosomal translocations particularly of acrocentric chromosomes have been found in numerous cases of UPD (Fig. 5.22). Heterologous Robertsonian translocations (of different acrocentrics), or homologous Robertsonian translocations (of the same acrocentric), as well as other translocations provide increased risk for UPD.

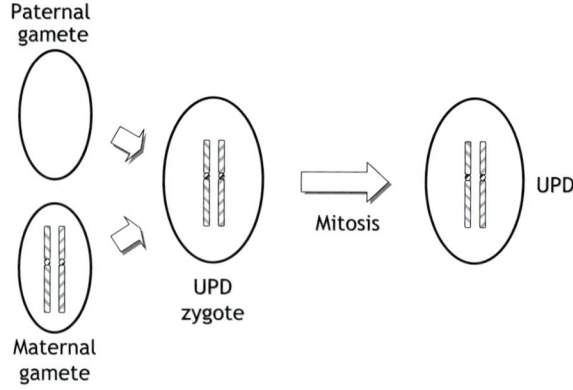

Fig. 5.19 Schematic representation of the mechanism of UPD due to gamete complementation

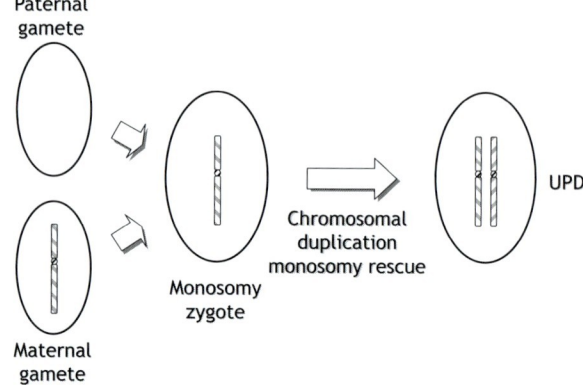

Fig. 5.20 Schematic representation of the mechanism of UPD due to monosomy rescue

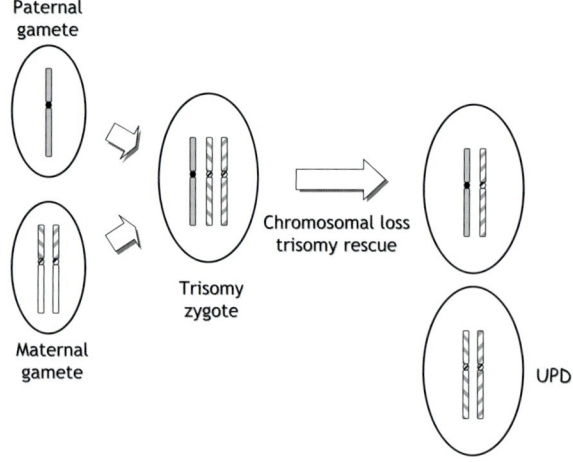

Fig. 5.18 Schematic representation of the mechanism of UPD due to trisomy rescue

5.1.9.1 Phenotypic Consequences of UPD

There are two main reasons for the phenotypic consequences of UPD:

Fig. 5.21 Schematic representation of the mechanism of partial UPD due to somatic–mitotic recombination

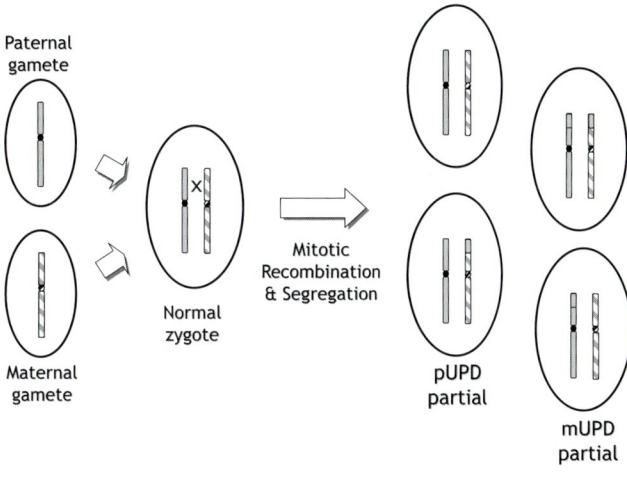

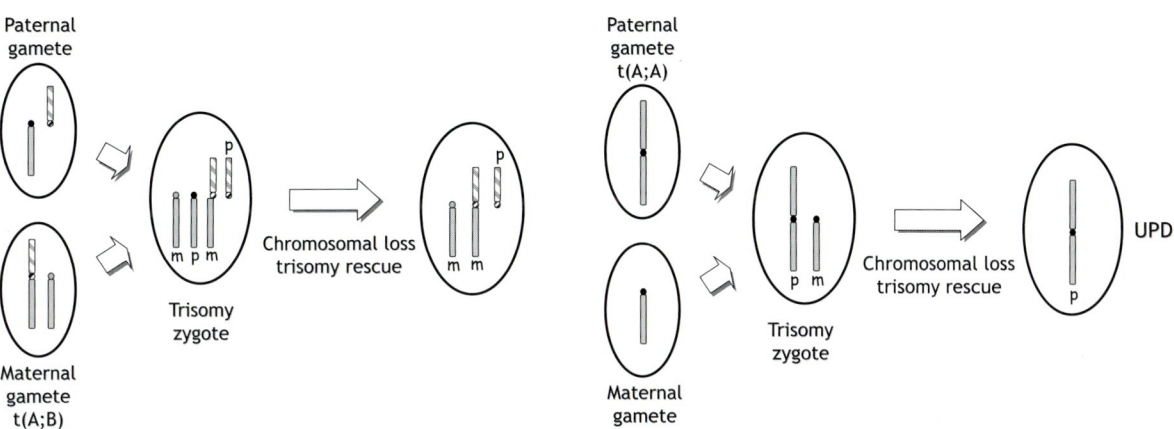

Fig. 5.22 Schematic representation of the mechanism of partial UPD due to translocations of acrocentric chromosomes. The *left panel* depicts a case of a translocation involving two different acrocentrics; the *right panel* shows a case of a translocation involving homologous acrocentrics

1. Duplication of autosomal recessive alleles. In isodisomy, two copies of a mutant allele would result in the disease phenotype. In the originally described case of maternal UPD7, cystic fibrosis was due to two maternally derived copies of the Gly542Ter mutation of the CFTR genes [4] (the mother in that case was a heterozygous carrier of this mutation).

2. Parental imprinting effects. Genomic imprinting refers to parent-of-origin dependent gene expression. Some genes are monoallelically expressed either from the paternally or the maternally derived chromosome. Thus for a paternally-only expressed gene, maternal UPD would result in a null phenotype for this gene. On the other hand, for a maternally-only expressed gene, paternal UPD would result in a similar null phenotype. An example of the former is Prader-Willi syndrome caused by matUPD15; and of the latter is Angelmann syndrome caused by patUPD15 [57].

A considerable number of imprinted genes have been identified in human and mouse [55].

5.1.9.2 Human Disorders Involving UPD

Rare cases of UPD for almost all chromosomes have been identified; the phenotypes are variable. Among them there are some recognizable syndromes which include: (a) Prader-Willi syndrome (matUPD15); (b) Angelmann syndrome (patUPD15); (c) Beckwith-Wiedemann syndrome (patUPD11p15); (d) neonatal transient diabetes mellitus in patUPD6; (e) maternal and paternal UPD14 syndromes; (f) some cases of Russell-Silver syndrome (matUPD7).

5.1.10 Diseases Due to Mutations in the Mitochondrial Genome

As shown in Chapter 2, the mitochondrial genome, mtDNA, consists of a ring-shaped chromosome with 16 596 bp. It encodes a small (12 S) and a large (16 S) rRNA for mitochondrial RNA translation, 22 tRNAs, and 13 genes encoding subunits of the respiratory chain. All these polypeptides are subunits of the mitochondrial energy-generating pathway, oxidative phosphorylation (OXPHOS). OXPHOS encompasses five multiunit enzyme complexes, arrayed within the mitochondrial inner membrane; most of the peptides necessary for building these enzyme complexes are encoded in nuclear genes.

At fertilization the oocyte contains about 200,000 mtDNAs. Once fertilized, the nuclear DNA replicates and the oocyte cleaves, but the mtDNA does not replicate until after the blastocyst is formed. Since the blastocyst cells that are destined to become the embryo proper constitute only a small fraction of all blastocyst cells, and only a fraction of these cells enter the female germ line, few of the oocyte's mtDNA molecules are found in the primordial germ cells. However, it is questionable whether this mechanism is sufficient for creating an mtDNA "population" in human cells that is as homogeneous, as is normally found, especially if we consider the fact that a single mitochondrium contains 5–10 mtDNA molecules.

Most proteins necessary for development of the mitochondria themselves are produced by nuclear genes. Therefore some of the diseases due to malfunction of mitochondria are caused by defects of such genes; they follow classical Mendelian modes of inheritance [82, 84]. On the other hand, diseases due to defects of genes in the mitochondrial genome are transmitted as the mitochondria themselves, i.e., from the mother to all children, irrespective of sex. However, considering the great number of mitochondria that a oocyte contains, and the number of genomes per mitochondrium, it is not surprising that a child may inherit from its mother more than one type of mitochondrial genome; cells containing variable proportions of affected mitochondria are "heteroplasmic." During further development, one genome may become more abundant; different cell lineages may even become "homoplasmic" for different mitochondrial genomes. This may explain in part the enormous phenotypic variation between individuals with the same mitochondrial disease. A heteroplasmic mtDNA mutation may reduce the function of the gene-determined peptide. In most instances this is unimportant, but in a few cells the fraction of mitochondria containing the mutant increases to the extent that OXPHOS enzyme activity decreases until it falls below the cellular or tissue energetic threshold, i.e., the minimum activity necessary to sustain oxidative phosphorylation. Because OXPHOS is necessary for nearly all cells, any organ can be affected in mitochondrial diseases. Thus, respiratory chain deficiencies caused by mitochondrial disorders may generate almost any symptom, in any organ system, and at any stage of life. The heteroplasmy produces marked variability in the severity and symptom patterns of these conditions. The most severe inherited mitochondrial disorders become clinically apparent during infancy, whereas other disorders of mitochondrial function may have an adult onset.

Four categories of diseases due to mutations in the mitochondrial genome may be distinguished (Fig. 5.23) [84]. In the first we find missense mutations with relatively mild phenotypic effects. These are transmitted maternally and appear to be homoplasmic. The second category comprises deleterious point mutations. Of course they can be transmitted maternally only if they are heteroplasmic. The third category, deletion mutants, occur by new mutations during early development, and these are therefore heteroplasmic. In the fourth category of diseases, certain mutations may be present that diminish OXPHOS activity somewhat at onset but not sufficiently to cause functional damage. During life time, however, additional random mutations accumulate in somatic cells, reducing their OXPHOS capacity until the threshold is reached. Then a degenerative disease of advanced age such as Alzheimer or Parkinson disease might ensue.

5.1.10.1 Leber Optical Atrophy

An example of the first category is Leber's hereditary optical neuropathy, (LHON; 308900) [82, 83]. In this disease, rapid vision loss occurs during young adult age; cardiac dysrhythmia is common. Variation in severity of the disease is strong; males are more often and on average more severely affected than females; the proportion of transmitting females in the family is much larger than expected if the mutation were

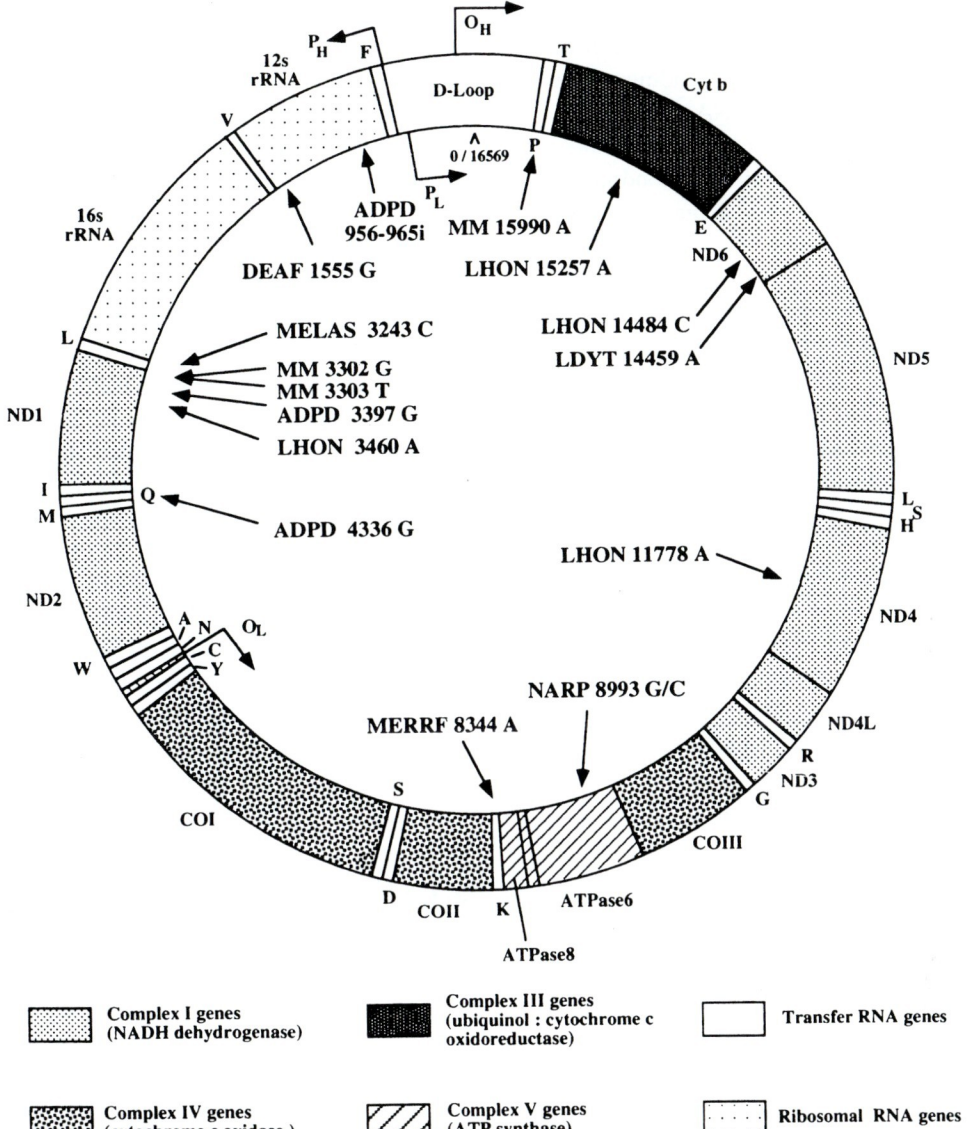

Fig. 5.23 Human tDNA map showing locations of genes and mutations Definitions of gene symbols and mutations, example: MTTK*MERRF8344A.MTTK is the altered mtDNA (MT) gene for tRNA (Lys) (TK); Myoclonic epilepsy and Ragged Red Fiber disease (MERRF) is the most characteristic clinical presentation, 8344 is the altered nucleotide, and A is the pathogenic base. From Wallace [84]

X-linked. Transmission, however, occurs exclusively through females [80]. Molecular analysis revealed a G → A transition (G3460A) leading to an Arg → His replacement in the gene for the NADH subunit 4. The Arg residue must be important for function since it has been conserved in evolution from flagellates and fungi to humans. The mutation is homoplasmic; hence the clinical variability as well as the sex difference must have other causes that are still unknown. Other mitochondrial mutations in closely related genes have occasionally been described [40].

Two diseases apparently belong to the second category – deleterious but heteroplasmic point mutations. In a large kindred, Leber disease was found to be associated with infantile bilateral striatal necrosis. In this family four phenotypes were found: normal, Leber disease, striatal necrosis, and the combination of the two diseases. All members were related through the female line. Since careful analysis has shown no deletion, the disease appears to be due to a deleterious but heteroplasmic point mutation. Depending on the preponderance of the aberrant mtDNA, the clinical signs vary [82]. The second disease of this class is one combining myoclonic epilepsy and mitochondrial myopathy – both conditions with huge interindividual variation [18].

5.1.10.2 Deletions

The third category is that of sporadic and heteroplasmic deletions. These occur as somatic mutations; since all deletions in one individual are identical, they must have arisen by clonal expansion of a single molecular event. Therefore a selective advantage of mutant cells has been suggested [82]. Figure 5.24 shows such deletions. Clinical manifestations again depend on the distribution of mutant mitochondria. A family has been described [94] in which multiple deletions of mtDNA behaved as one autosomal-dominant trait. The affected individuals suffered from progressive external ophthalmoplegia, progressive proximal weakness, bilateral cateract, and precocious death.

5.1.10.3 Diseases of Advanced Age

The fourth category comprises diseases of advanced age that have not found satisfactory explanations so far. In both Alzheimer and Parkinson diseases, for example, pedigrees have been observed in which relatively early onset in middle age is combined with an autosomal-dominant mode of inheritance. In the majority of these cases, however, an accumulation of affected individuals within families is found but no combination of clearcut Mendelian mode of inheritance with onset at more advanced age. Here, mildly to moderately deleterious germ line mutations established in the distant past, and present in a certain proportion of the population in combination with somatic mutations occurring during lifetime of the individual, may lead to such degenerative diseases. For example, a homoplasmic mutation among whites at nucleotide base pair 4,336 leading to a tRNA mutant has been observed in 5% of Alzheimer and Parkinson disease mutations, but appears to be much rarer in the general population. It may contribute to the multifactorial origin of these diseases [76].

In general, mutations within the mitochondrial genome affect mainly organ systems that depend on intact oxidation – central nervous system and muscles. Probably the number of known diseases due to mutations in the mitochondrial genome will increase in future (Table 5.4).

5.1.10.4 Interaction Between Nuclear and Mitochondrial Genomes

Several subunits of the electron transport chain are not encoded within the mitochondrial DNA but by the nuclear DNA. As a consequence, mutations in the nuclear genome can cause secondary mitochondrial DNA information loss. Hence, there are some clinical syndromes in which defects in OXPHOS follow classic Mendelian patterns of dominant-recessive transmission and not the maternal pattern, which is usually associated with this group of disorders. An example is the mitochondrial neurogastrointestinal encephalopathy syndrome (MNGIE; OMIM: 603041) which can be caused by mutations in the gene encoding thymidine phosphorylase (ECGF1).

5.1.11 Unusual, "Near Mendelian" Modes of Inheritance

As a "bridge" between monogenic and polygenic phenotypes it is worth mentioning two concepts that provide an understanding of the increased complexity between genotype and phenotype [2].

5.1.11.1 Digenic Inheritance

In this case, the phenotype is due to one mutant allele in each of two different genes. The first example was that of one form of retinitis pigmentosa published in 1994 from the laboratory of T. Dryja [42]. Individuals with a mutation in the ROM1 gene (OMIM 180721 on chromosome 11q13) *AND* a mutation in the RDS gene (OMIM 179605 on chromosome 6p21) manifest the disease (Fig. 5.25). However, individuals with only heterozygosity of the ROM1 gene mutation, or with only heterozygosity of the RDS gene mutation, were not affected

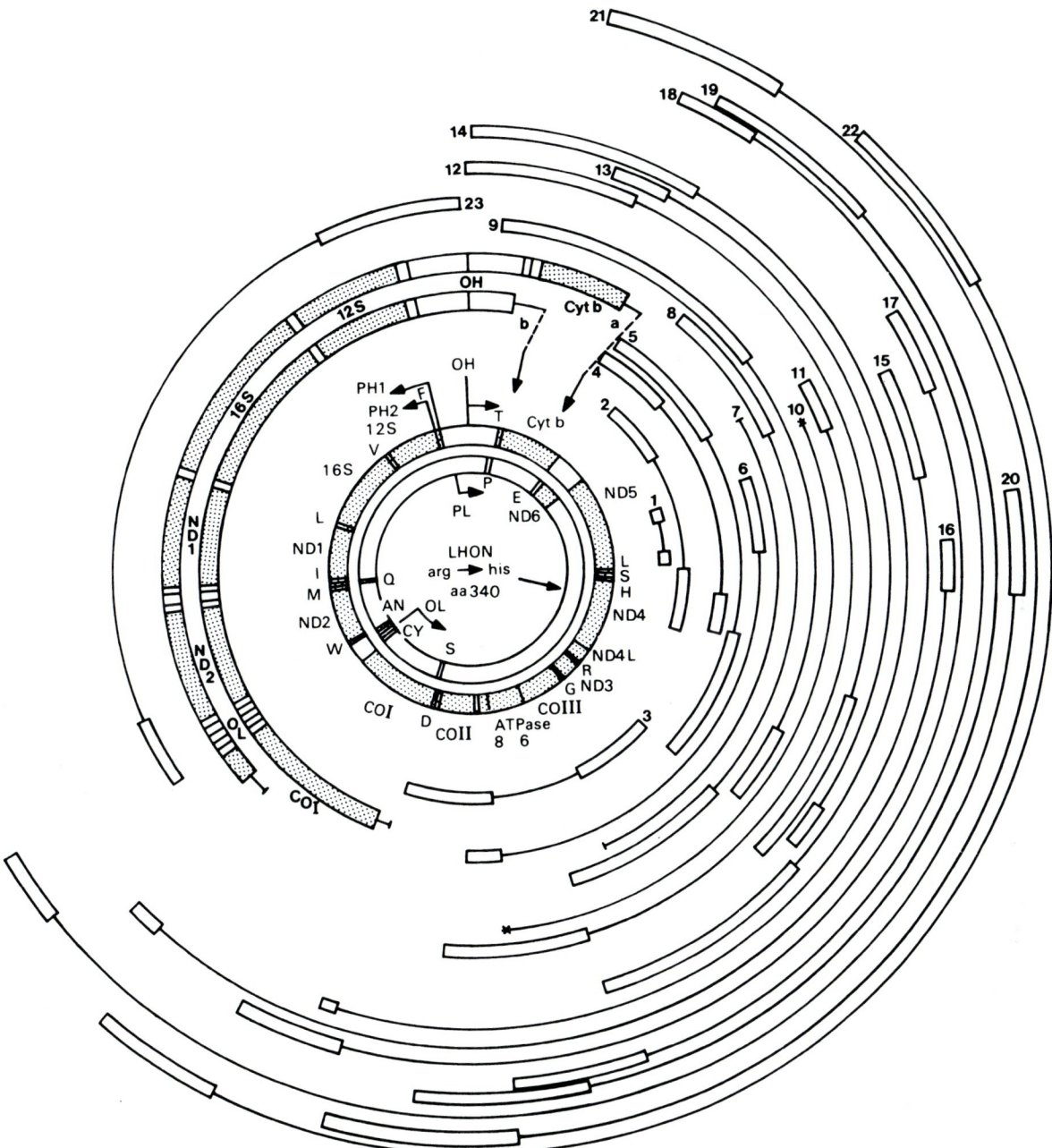

Fig. 5.24 Deletion map of human mtDNA. The *inner circles* show localization of genes and mutations. (See also Fig. 5.23). The arcs no. 1–23 show the mtDNA regions that were lost in various deletions. The *open bars* at the end of the arcs show regions of uncertainty. Deletion 1 was found in a patient with Myoclonic Epilepsy and Ragged Red Fibre Disease (MERRF) together with stroke-like symptoms. Deletions 2–23 were found in ocular myopathy patients with symptoms of varying severity. Deletion 10 was found in about one third of all ocular myopathy patients. The * at the ends of deletion 10 represents the associated 13 base pairs direct repeat. The two partial mtDNA maps labeled "a" and "b" to the left of the function map indicate the regions that were tandemly duplicated in patients with ocular myopathy associated with diabetes mellitus. The insertion sites around MTCYB (cytb) are indicated by *arrows*. From Wallace [83]

Table 5.4 The Mitochondrial Chromosome from McKusick [52]

Location (nt)	Symbol	Title	MIM	Disorder[a]
577–647	MTTF	tRNA phenylalanine	590070	
648–1,601	MTRNR1	12 S rRNA	561000	Deafness, aminoglycoside-induced, 580,000
1,602–1,670	MTTV	tRNA valine	590105	
1,671–3,229	MTRNR2	16 S rRNA	561010	Cloramphenicol resistance, 515,000
3,230–3,304	MTTL1	tRNA leucine 1 (UUA/G)	590050	MELAS syndrome, 540,000; MERRF syndrome, 545,000; Cardiomyopathy; Diabetes-deafness syndrome, 520,000
3,307–4,262	MTND1	NADH dehydrogenase 1	516000	Leber optic atrophy, 535,000
4,263–4,331	MTTI	tRNA isoleucine	590045	Cardiomyopathy
4,400–4,329[b]	MTTQ	tRNA glutamine	590030	Cardiomyopathy
4,402–4,469	MTTM	tRNA methionine	590065	
4,470–5,511	MTND2	NADH dehydrogenase 2	516001	Leber optic atrophy, 535,000
5,512–5,576	MTTW	tRNA tryptophan	590095	
5,655–5,587[b]	MTTA	tRNA alanine	590000	
5,729–5,657[b]	MTTN	tRNA asparagine	590010	Ophthalmoplegia, isolated
5,826–5,761[b]	MTTC	tRNA cysteine	590020	Ophthalmoplegia, isolated
5,891–5,826[b]	MTTY	tRNA tyrosine	590100	
5,904–7,444	MTCO1	cytochrome c oxidase I	516030	
7,516–7,445[b]	MTTS1	tRNA serine 1 (UCN)	590080	
7,518–7,585	MTTD	tRNA aspartic acid	590015	
7,586–8,262	MTCO2	cytochrome c oxidase II	516040	
8,295–8,364	MTTK	tRNA lysine	590060	MERRF syndrome, 545,000
8,366–8,572	MTATP8	ATP synthase 8	516070	
8,527–9,207	MTATP6	ATP synthase 6	516060	Leigh syndrome; NARP syndrome, 551,500
9,207–9,990	MTCO3	cytochrome c oxidase III	516050	Leber optic atrophy 535,000
9,991–10,058	MTTG	tRNA glycine	590035	
10,059–10,404	MTND3	NADH dehydrogenase 3	516002	Leber optic atrophy, 535,000
10,405–10,469	MTTR	tRNA arginine	590005	Leber optic atrophy, 535,000
10,470–10,766	MTND4L	NADH dehydrogenase 4 L	516004	
10,760–12,137	MTND4	NADH dehydrogenase 4	516003	Leber optic atrophy, 535,000
12,138–12,206	MTTH	tRNA histidine	590040	
12,207–12,265	MTTS2	tRNA serine 2 (AGU/C)	590085	
12,266–12,336	MTTL2	tRNA leucine 2 (CUN)	590055	
12,337–14,148	MTND5	NADH dehydrogenase 5	516005	
14,673–14,149[b]	MTND6	NADH dehydrogenase 6	516006	Leber optic atrophy, 535,000

(continued)

5 Formal Genetics of Humans

Table 5.4 (continued)

Location (nt)	Symbol	Title	MIM	Disorder[a]
14,742–14,674[b]	MTTE	tRNA glutamic acid	590025	
14,747–15,887	MTCYB	cytochrome b	516020	
15,888–15,953	MTTT	tRNA threonine	590090	
16,023–15,955[b]	MTTP	tRNA proline	590075	

[a] In addition to the disorders caused by point mutations in individual genes, deletions involving more than one mitochondrial gene have been identified in Pearson syndrome (557,000), early-onset chronic diarrhea with villus atrophy (520,100), and Kearns-Sayre syndrome (530,000), among others [b] Transcribed from light chain (L) in opposite direction from all the other genes which are transcribed from the heavy chain (H)

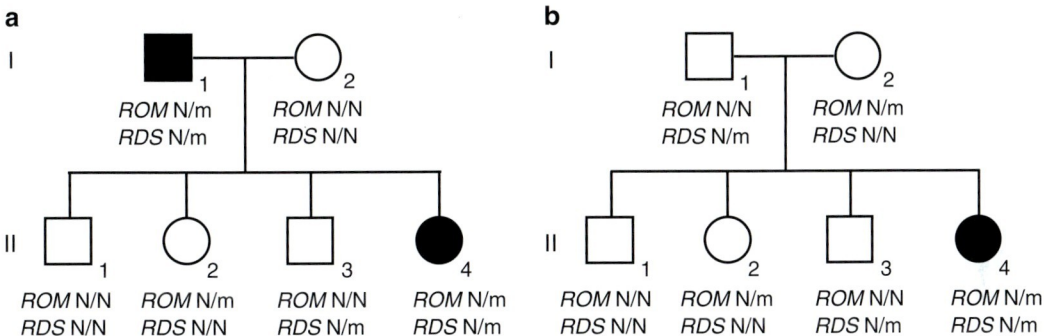

Fig. 5.25 Schematic representation of pedigrees with digenic inheritance; affected individuals are shown in *black*. (**a**) Pedigree with an apparently dominant inheritance. *N* normal allele, *m* mutant allele at the two different genes. (**b**) Pedigree with an apparently recessive mode of inheritance. From [43]

with retinitis pigmentosa. The mode of transmission of this condition resembles an autosomal dominant trait in some pedigrees (vertical transmission, males and females affected) but with 25% affected offspring from an affected parent. In some pedigrees, the transmission resembles a recessive trait when the two parents are heterozygotes for a pathogenic mutation in two different genes. In this case the affected offspring are also 25%.

5.1.11.2 Triallelic Inheritance

A more complicated case is that observed in several forms of Bardet-Biedl syndrome (BBS). The laboratories of N. Katsanis and J.R. Lupski described in 2001 BBS families in which *three* different mutations were necessary to cause the phenotypes of this syndrome [43]: for example, homozygous mutant alleles in the BBS2 gene (OMIM 606151 on chromosome 16q21) AND a heterozygous mutation in the BBS6 gene (OMIM 604896 on chromosome 20p12) needed to be present for the affected status (Fig. 5.26). Homozygous mutations only in the BBS2 gene and normal BBS6 gene were not sufficient for the phenotypic manifestations.

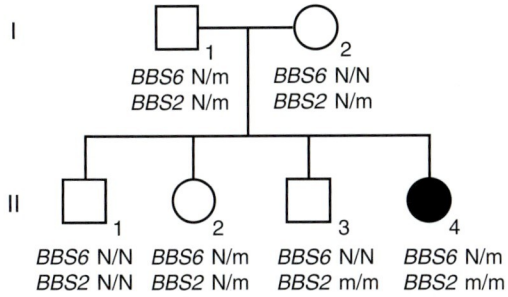

Fig. 5.26 Schematic representation of a pedigree with digenic inheritance; the affected individual II4 is shown in *black*. *N* normal allele, *m* mutant allele Only individuals with three mutant alleles are affected. Note that individual II2 who is heterozygous for mutant alleles in the *BBS6* and *BBS2* genes is not affected; furthermore, individual II3 who is homozygous for mutant alleles in the *BBS2* gene is also not affected. From [43]

Digenic and triallelic inheritance may not be rare in more complex phenotypes, and accumulation of mutations in a few or several genes (also referred to as oligogenic inheritance) may be necessary for complex polygenic disorders.

5.1.12 Multifactorial Inheritance

The majority of human phenotypic features, such as size, weight, intelligence, and others follow multifactorial inheritance. This mode of inheritance is described in Chaps. 4 and 8.

5.2 Hardy–Weinberg Law and Its Applications

5.2.1 Formal Basis

So far the application of Mendel's laws in man has been considered from the standpoint of the single family. What, however, are the consequences for the genetic composition of the population? The field of research that considers this problem is called population genetics. Some basic concepts are introduced here.

These concepts revolve around the so-called Hardy–Weinberg law, discovered by these two authors independently in 1908 [34, 85]. In 1904 Pearson [62] – in the process of reconciling the consequences of Mendel's laws for the population with biometric results – had already derived this law for the special case of equal gene frequencies of two alleles.

The law in its more general form may be formulated as follows: Let the gene frequencies of two alleles in a certain population be p for the allele A and q for the allele B; ($p+q=1$). Let mating and reproduction be random with respect to this gene locus. The gene frequencies then remain the same, and the genotypes AA, AB, and BB in the F_1 generation occur in the relative frequencies p^2, $2pq$, and q^2, the terms of the binomial expression $(p+q)^2$. In autosomal genes, and in the absence of disturbing influences, this proportion is maintained through all subsequent generations.

5.2.1.1 Derivations from the Hardy–Weinberg Law

We assume that at the beginning the proportions of genotypes AA, AB, and BB in the population of both males and females are D, $2H$, and R, respectively. Symbolically, the distribution of genotypes in both sexes may be written as:

$$D \times AA + 2H \times AB + R \times BB \quad (5.1)$$

From this the distribution of mating types for random mating is obtained by formal squaring:

$$(D \times AA + 2H \times AB + R \times BB)^2$$
$$= D^2 \times AA \times AA + 4DH$$
$$\times AA \times AB + 2DR \times AA \times BB + 4H^2$$
$$\times AB \times AB + 4HR \times AB \times BB + R^2$$
$$\times BB \times BB$$

The distribution of genotypes in the offspring of the different mating types is:

AA × AA	AA
AA × AB	$\frac{1}{2}$AA + $\frac{1}{2}$AB
AA × BB	AB
AB × AB	$\frac{1}{4}$AA + $\frac{1}{2}$AB + $\frac{1}{4}$BB
AB × BB	$\frac{1}{2}$AB + $\frac{1}{2}$BB
BB × BB	BB

Inserting these distributions for the mating types into (5.1) yields the distribution of genotypes in the F_1 generation:

$$(D^2 + 2DH + H^2)AA + (2DH + 2DR$$
$$+ 2H^2 + 2HR)AB + (H^2 + 2HR + R^2)BB$$
$$= p^2 AA + 2pq AB + q^2 BB$$

where $p = D + H$, $q = H + R$ are the frequencies of the alleles A and B, respectively, in the parental generation. Thus, the distribution of genotypes in the offspring generation is uniquely determined by the gene frequencies in the parental population:

$$D' = p^2, 2H' = 2pq, R' = q^2.$$

As:

$$p' = D' + H' = p^2 + pq = p,$$
$$q' = H' + R' = pq + q^2 = q,$$

the gene frequencies in the F_1 generation are equal to those in the parental generation. Thus, the genotype distribution in the next generation (F_2) is also the same as in the F_1 generation, and this holds true for all following generations.

This means that in autosomal inheritance these proportions are expected in the first generation and are maintained in the following generations. For X-linked genes the situation is slightly more complicated. At the same time, the concept of gene frequencies $p + q = 1$ was created.

The Hardy–Weinberg law can also be rephrased, indicating that random mating is equivalent to drawing random samples of size 2 from a pool of genes containing the two alleles A and a with relative frequencies p and q. One of the advantages of this law is that frequencies of genetic traits in different populations can be expressed and compared in terms of gene frequencies.

Apart from making it possible to simplify population descriptions, the Hardy–Weinberg law can also help to elucidate modes of inheritance in cases where the straightforward approach through family studies would be too difficult. The classic examples are the AB0 blood types.

5.2.2 Hardy–Weinberg Expectations Establish the Genetic Basis of AB0 Blood Group Alleles

5.2.2.1 Multiple Allelisms

So far, only two different alleles for each locus have been considered. Frequently, however, more than two different states for one gene locus, i.e., more than two alleles, are possible. Examples of such "multiple allelism" in humans and experimental animals abound. Two of the classics are the white series in *Drosophila melanogaster* and the albino series in rabbits.

The formal characteristics can easily be derived:

(a) In any one individual a maximum of only two alleles can be present (unless there are more than two homologous chromosomes, as in trisomics).
(b) Between these alleles, crossing over can be disregarded as they are located at homologous loci. Here the simplest formal model is described, using the AB0 blood groups as an example.

5.2.2.2 Genetics of the AB0 Blood Groups

The AB0 blood groups were discovered by Landsteiner in [46]. Compared to other blood group systems their most important property is the presence of isoantibodies that have led to frequent transfusion accidents. These accidents helped in the discovery of blood groups. The first relevant genetic theory was developed by von Dungern and Hirszfeld in [79]. To explain the four phenotypes A, B, 0, and AB they assumed two independent pairs of alleles (A, 0; B, 0), with dominance of A and B. In 1925 Bernstein [10] tested this hypothesis using the Hardy–Weinberg expectations for the first time. He found their concept to be wrong and replaced it by the correct explanation – three alleles with six genotypes, leading to the four phenotypes due to the dominance of A and B over 0.

The most obvious method to discriminate between these two hypotheses is by family investigation. However, differences between them are to be expected only in matings in which at least one parent carries group AB (Table 5.5). The two-locus hypothesis allows for 0 children while the three-allele hypothesis does not. Although AB is the rarest group, the early literature contained some reports of supposedly 0 children with AB parents; these children were either misclassified or illegitimate. Bernstein, however, was not misled by these observations. His argument goes as follows. It may be assumed that the two-gene pair theory is correct; p may be the gene frequency of A, $1 - p = p'$ of a; q the frequency of B, $1 - q = q'$ of b. The frequencies to be expected in the population are presented in Table 5.6.

Table 5.5 Comparison of the two theories for inheritance of AB0 blood groups (adapted from Wiener [91])

Parents	Children expected from the hypothesis of	
	Two gene pairs	Multiple alleles
0 × 0	0	0
0 × A	0, A	0, A
0 × B	0, B	0, B
A × A	0, A	0, A
A × B	0, A, B, AB	0, A, B, AB
B × B	0, B	0, B
0 × AB	0, A, B, AB	A, B
A × AB	0, A, B, AB	A, B, AB
B × AB	0, A, B, AB	A, B, AB
AB × AB	0, A, B, AB	A, B, AB

Table 5.6 Expectations from multiple allele hypothesis for the AB0 system from Bernstein [10]

Phenotype	Genotype	Frequency	
0	aabb	$(1-p)^2(1-q)^2 = p'^2 q'^2$	
B	aaBB	$(1-p)^2 q^2$	$\left.\right\} = p'^2(1-q'^2)$
	aaBb	$2(1-p)^2 q(1-q)$	
A	AAbb	$p^2(1-q)^2$	$\left.\right\} = (1-p'^2)q'^2$
	Aabb	$2p(1-p)(1-q)^2$	
AB	AABB	$p^2 q^2$	
	AaBB	$2p(1-p)q^2$	$= (1-p'^2)(1-q'^2)$
	AABb	$2p^2 q(1-q)$	
	AaBb	$2p(1-p)2q(1-q)$	

This leads to the following relationships ($\bar{A}, \bar{B}$: frequencies of phenotypes):

$$\bar{0} \times \overline{AB} = \bar{A} + \bar{B}$$

and

$$+A + +AB = 1 - p'^2;\ +B + +AB = 1 - q'^2$$

Thus, it follows:

$$(+A + +AB) \times (+B + +AB) = +AB$$

These identities can be tested. It turned out – and has turned out ever since – that $(\bar{A} + \overline{AB}) \times (\overline{AB} + \overline{AB}) > \overline{AB}$, and $\bar{0} \times \overline{AB} < \bar{A} + \bar{B}$. The differences are so large – and so consistent – that an explanation by chance deviations is inadequate. The first alternative possibility considered by Bernstein was heterogeneity within the examined population. This explanation, however, proved insufficient. On the other hand, it could be shown that the distributions in all populations for which data were available are in perfect agreement with expectations derived from the multiple-allele hypothesis.

To understand Bernstein's argument a fresh look at the Hardy–Weinberg law is necessary. Up to now it has been derived here for the special case of two alleles only. However, it can also be shown to apply for more than two alleles. Assuming n alleles $p_1, p_2, \ldots, p_n$, the relative frequencies of genotypes are given by the terms of the expansion of $(p_1 + p_2 + \ldots p_n)^2$. It follows for the special case of A, B, and 0 with the frequencies p, q, and r that the distribution of genotypes is:

$$p^2(AA) + 2pq(AB) + 2pr(A0)$$
$$+ q^2(BB) + 2pr(B0) + r^2(B0).$$

Now, we follow Bernstein again (our translation): "for the classes" (phenotypes):

$$\bar{0} = 00 \quad \bar{B} = B0 + BB$$
$$\bar{A} = A0 + AA \quad \overline{AB} = AB$$

the following probabilities can be derived:

$$r^2 \quad 2qr + q^2 \quad 2pr + p^2 \quad 2pq$$

It follows:

$$\bar{0} + \bar{A} = (r+p)^2$$
$$\bar{0} + \bar{B} = (r+q)^2$$

and therefore:

$$q = 1 - \sqrt{\bar{0} + \bar{A}}$$
$$q = 1 - \sqrt{\bar{0} + \bar{B}}$$
$$q = 1 - \sqrt{\bar{0}}$$

and the relation:

$$1 = p + q + r = 1 - \sqrt{\bar{0} + \bar{B}} + 1 - \sqrt{\bar{0} + \bar{A}} + \sqrt{\bar{0}}$$

This can be tested using the AB0 phenotype distributions in various populations of the world. The criterion is that the gene frequencies calculated with this formula must add to 1. In addition, expected genotype frequencies can be calculated from these gene frequencies and can be compared with observed frequencies. Apart from the correctness of the genetic hypothesis, however, this result requires still another condition. There must be random mating with regard to this characteristic.

In the data analyzed by Bernstein the agreement already was excellent, and this has proven to hold true for the huge amount of data collected ever since. One example may help in understanding the principle of calculation. The following phenotype frequencies were reported from the city of Berlin ($n = 21{,}104$): 43.23% A ($n = 9{,}123$), 14.15% B ($n = 2{,}987$), 36.60% 0 ($n = 7{,}725$), and 6.01% AB ($n = 1{,}269$).

Using Bernstein's formula, the gene frequencies are:

$$p = 1 - \sqrt{(0.3660 + 0.1415)} = 0.2876$$
$$q = 1 - \sqrt{(0.3600 + 0.4323)} = 0.1065$$
$$r = \sqrt{0.3660} = \frac{0.6050}{0.9991}$$

Thus:

$$p + q + r = 0.9991$$

At first glance, this result agrees well with the expectation, i.e., 1. As a statistical test for examining whether the deviation is significant, the χ^2 method can be applied [73]:

$$\chi_1^2 = 2n\left(1 + \frac{r}{pq}\right)D^2$$

$$D = 1 - (p + q + r)$$

In our example, the result is:

$$\chi_1^2 = 0.88$$

This confirms that the values found are in good agreement with the genetic hypothesis and with the assumptions of random mating for the AB0 system.

In a later paper Bernstein showed how the difference D may be utilized to correct the calculated gene frequencies. The uncorrected gene frequencies may be named p', q', and r', and the following formulas may be used:

$$p = p'(1 + D/2)$$
$$q = q'(1 + D/2)$$
$$r = (r' + D/2)(1 + D/2)$$

and for the example:

$$p = 0.2876(1 + 0.00045) = 0.2877$$
$$q = 0.1065(1 + 0.00045) = 0.1065$$
$$r = (0.6050 - 0.00045)(1 + 0.00045) = 0.6057$$

In the process of testing the two genetic hypotheses for the AB0 system Bernstein developed a method for calculating gene frequencies.

5.2.2.3 Meaning of a Hardy–Weinberg Equilibrium

Populations showing agreement of the observed genotype proportions with the expectations of the Hardy–Weinberg Law are said to be "in Hardy–Weinberg equilibrium." This equilibrium must be distinguished from that between alleles, which is discussed in the contexts of selection and of mutation. The Hardy–Weinberg equilibrium is an equilibrium of the distribution of genes in the population ("gene pool") among the various genotypes. Under random mating this equilibrium is reestablished after one generation, possibly with changed gene frequencies if it is disturbed by opposing forces.

It follows from our discussion, however, that the Hardy–Weinberg law can be expected to be valid only when the following prerequisites are not violated:

(a) The matings must be random with respect to the genotype in question. This can safely be assumed for such traits as blood groups or enzyme polymorphisms. It cannot be assumed for visible characteristics such as stature, and still less for behavioral characteristics such as intelligence. This should be kept in mind when measures used in quantitative genetics, (for example, correlations between relatives), are interpreted in genetic terms.

(b) A deviation from random mating is caused by consanguineous matings. If the consanguinity rate in a population is high, an increase in the number of homozygotes must be expected (Chap. 17). It is even possible to estimate the frequency of consanguinity in a population by means of the deviations from the Hardy–Weinberg proportions.

(c) Recent migrations might disturb the Hardy–Weinberg proportions.

(d) Occasionally selection is mentioned as a factor leading to deviations. This may be true but need not necessarily apply. As a rule, selection tends to cause changes in gene frequencies; selection before reproductive age, for example, in the prenatal period, or during childhood and youth, does not influence the Hardy–Weinberg proportions in the next generation at all. If genotypes are tested among adults in a situation in which a certain genotype had been selected against in children, this genotype is found to decrease in frequency. Even assuming appreciable selection in a suitable age group, ascertainment of statistically significant deviations from Hardy–Weinberg proportions requires large sample sizes – larger than are usually available. Sometimes the absence of significant selection is inferred from the observation that Hardy–Weinberg proportions are preserved in a population. This conclusion, however, unless carefully qualified may easily be wrong. Considering all the theoretical possibilities for disturbance, it is indeed amazing how frequently the Hardy–Weinberg proportions are found to be preserved in the human population.

(e) Formally, a deviation from the Hardy–Weinberg law may be observed if the population is a mixture of subpopulations that do not completely interbreed (random mating only within subpopulations), and consequently the gene frequencies in these subpopulations differ. This was first described by Wahlund in [81], who gave a formula for calculating the coefficient F of the apparent inbreeding from the variance of the gene frequencies between the subpopulations.

(f) Another cause of deviation may be the existence of a hitherto undetected ("silent") allele, a heterozygous carrier of which cannot be distinguished from a homozygous carrier of the usual allele. C.A.B. Smith [69], however, has pointed out that a silent allele causes a significant deviation from the Hardy–Weinberg law only when it occurs at a sufficiently high frequency for the homozygote to be detected.

5.2.3 Gene Frequencies

5.2.3.1 One Gene Pair: Only Two Phenotypes Known

In rare autosomal-recessive diseases only one gene pair is present, and only two phenotypes are usually known when the heterozygotes cannot be identified, or, as is usually the case, when direct data on population frequencies of heterozygotes are not available. This also applies for blood group systems for which only one type of antiserum is available. Here the frequency of homozygotes aa being q^2, the gene frequency is simply. There is no way to test the assumption of random mating.

Table 5.7 [49] is slightly oversimplified; some of the frequencies given vary in different populations. However, the data point out how much more frequent the heterozygotes are, especially for rare conditions. This is important for genetic counseling, and for the much-discussed problem of the number of lethal or detrimental genes for which the average human being might be heterozygous.

5.3 Statistical Methods in Formal Genetics: Analysis of Segregation Ratios

5.3.1 Segregation Ratios as Probabilities

During meiosis – and in the absence of disturbances – germ cells are formed in exactly the relative frequencies expected from Mendel's laws. A diploid spermatocyte heterozygous for alleles A and a produces two haploid sperms with A, and two with a. If all the sperms of a given male come to fertilization, and none of the zygotes die before birth, the segregation ratio among his offspring would be exactly 1:1. There would be no place for any statistics.

Organisms in which such an analysis is indeed possible are yeast and the bread mould *Neurospora crassa,* which has become important in biochemical genetics. In the development of such an organism,

Table 5.7 Differing homozygote and heterozygote frequencies for different gene frequencies (with examples of recessive conditions; adapted from Lenz [49])

Homozygote frequency q^2	Gene frequency q	Heterozygote frequency $2pq$	Approximate homozygote frequencies in European populations
0.64	0.8	0.32	Lp(a-) lipoprotein variant
0.49	0.7	0.42	Acetyl transferase, "slow" variant (Sect. X56)
0.36	0.6	0.48	Blood group 0
0.25	0.5	0.50	Nonsecretor (se/se)
0.16	0.4	0.48	Rh negative (dd)
0.09	0.3	0.42	Lactose restriction (northwestern Germany)
0.04	0.2	0.32	Le(a–b–) negative
0.01	0.1	0.18	β-Thalessemia (Cyprus)
1:2500	1:50	1:25	Pseudocholinesterase (dibucaine-resistant variant), cystic fibrosis; α_1-antitrypsin deficiency
1:4,900	1:70	1:35	Adrenogenital syndrome (Canton Zurich)
1:10,000	1:100	1:50	Phenylketonuria (Switzerland; USA)
1:22,500	1:150	1:75	Albinism; adrenogenital syndrome with loss of NaCl
1:40,000	1:200	1:100	Cystinosis
1:90,000	1:300	1:150	Mucopolysaccharidosis type 1
1:1,000,000	1:1,000	1:500	Afibrinogenemia

there is a phase in which the diploid state has just been reduced to the haploid, and all four meiotic products lie in a regular sequence. They can be removed separately, grown, and examined ("tetrad analysis"). Expected segregation ratios are found with precision.

In higher plants and animals, including humans, only a minute sample of all germ cells comes to fertilization. In the human female about 6.8×10^6 oogonia are formed; the number of spermatogonial stem cells in the male is estimated at about 1.2×10^9; the actual number of sperm is a multiple of this figure. Hence any given germ cell has a very small probability of coming to fertilization. In addition, the sampling process is usually random with respect to a given gene pair A,a. This means that for the distribution of genotypes among germ cells coming to fertilization the rules of probability theory apply, and empirically found segregation ratios may show deviations from their statistical expectations.

Modern humans are fairly accustomed to thinking in statistical terms when solving daily problems. These experiences help us to understand simple applications of probability theory. Everyone, for example, readily recognizes that the following rationale is wrong.

A young mother had always wished to have four children. After the third, however, there was a long pause. The grandmother asked her daughter whether she had now decided differently. Answered the daughter: "Yes, in principle, I would still like four children. But I read in the newspaper that every fourth child born is Chinese. And a Chinese child …there I am reluctant."

In another example, the mistake is less obvious. The parents of two albino children visit a physician for genetic counseling. They wish to know the risk of a third child also being albino. The physician knows that albinism is an autosomal-recessive condition, with an expected segregation ratio of 1:3 among children of heterozygous parents. He also knows that sibships in which all sibs are affected are very rare. Hence, he informs the parents: "As you already have two affected children, the chance that the third child will also be affected is very small. The next child should be healthy." The actual risk, of course, remains 25% (Sect. 5.3.2).

A textbook on human genetics cannot teach probability theory and basic statistics. Therefore, it is assumed that the reader has some knowledge of the basic concepts of probability theory, that he knows the most important distributions (binomial, normal, and Poisson distribution), and has some idea of standard statistical methods. The following presents some applications to problems in human genetics. We are aware of the danger that this section may be used as a "cookbook," without understanding of the basic principles and recommend that the reader become familiar with these principles, for example, in the opening chapters of Feller's *Probability Theory and Its Applications* [24].

5.3.2 Simple Probability Problems in Human Genetics

5.3.2.1 Independent Sampling and Prediction in Genetic Counseling

The physician who gave the wrong genetic counsel to the couple with two albino children did not take into account that the fertilization events leading to the three children are independent of each other, and that each child has the probability of $1/4$ to be affected, regardless of the genotypes of any other children. The probabilities for each child must be multiplied. He was right when he said that illness of all three children is rare in a recessive condition: The probability is $(1/4)^3 = 1/64$ for all three children to be affected; the family to be counseled however, already had two such children and the probability of this occurring was only $(1/4)^2 = 1/16$. It takes only one event with the probability $1/4$ to complete the three-child family with albinism, $1/16 \times 1/4 = 1/64$. It is also intuitively obvious that there is no way for a given zygote to influence the sampling of gametes of the same parents many years later. Chance has no memory!

All possible combinations of affected and unaffected siblings in three-child families can be enumerated as follows (A = affected; U = unaffected):

UUU, AUU, UAU, AAU, \quad UUA, AUA, UAA, AAA

In recessive inheritance, the event U has the probability $3/4$. Thus, the first of the eight combinations (*UUU*) has the probability $(3/4)^3 = 27/64$. This means that of all heterozygous couples having three children $27/64$, or fewer than 50% have only healthy children. On the other hand, all three children are affected in $(1/4)^3 = 1/64$ of all such families. There remain the intermediate groups. Three-child families with one affected child and two healthy ones in that order obviously have the probability $1/4 \times 3/4 \times 3/4 = 9/64$. However, we are not particularly interested in the sequence of healthy and affected children. Therefore the three cases of such families, *UUA, UAU*, and *AUU*, can be treated as equivalent, giving $3 \times 9/64 = 27/64$. The group with two affected can be treated accordingly, giving $3 \times 1/4 \times 1/4 \times 3/4 = 9/64$. As a control, let us consider whether the various probabilities add up to 1:

$$\frac{27+27+9+1}{64}$$

This is a special case of the binomial distribution. There are two consequences for Mendelian segregation ratios one theoretical, the other extremely practical. First, it follows that among all families for which a certain segregation ratio must be expected, an appreciable percentage – 27 of 64 in a three-child family with recessive inheritance – cannot be observed because chance has favored them by not producing any affected homozygotes. Hence, the segregation ratio in the remainder is systematically distorted. Special methods have been devised to correct for this "ascertainment bias" (Sect. 5.3.4). Secondly, and this is a most practical conclusion, with limitation of the number of children to two or three, most parents both of whom are heterozygous for a recessive disease will not have more than one affected child. Since the probability of affected children occurring in another branch of the family is very low – and the rate of consanguinity in current populations of industrialized countries has likewise decreased – almost all affected children represent sporadic cases in an otherwise healthy family; there is no distinct sign of recessive inheritance. Any subsequent child, however, again runs the risk of $1/4$. The layman usually does not know that the condition is inherited. Therefore, genetic counseling must be actively offered to these families.

5.3.2.2 Differentiation Between Different Modes of Inheritance

In Sect. 5.1.4, an X-linked dominant pedigree is shown (Fig. 5.13) for vitamin D-resistant rickets and hypophosphatemia. What is the probability of such pedigree structure if the gene is in fact located on one autosome? Only the children of affected males are informative because among children of affected women a 1:1 segregation irrespective of sex must be expected. The seven affected fathers have 11 daughters, all of whom are affected. The probability of this outcome with autosomal inheritance is $(1/2)^{11}$. The same fathers have 10 sons who are all healthy, giving a probability of $(1/2)^{10}$. Hence, the combined probability of 11 affected daughters and 10 healthy sons is:

$$(1/2)^{21} = \frac{1}{2\,097\,152}$$

This probability is so tiny that the alternative hypothesis of an autosomal-dominant mode of inheritance is convincingly rejected. The only reasonable alternative is the X-linked dominant mode. This hypothesis is corroborated independently by the observation (Sect. 5.1.4) that on average male patients are more severely affected than female.

This is different for a rare skin disease (Brauer keratoma dissipatum). For this condition a Y-chromosomal mode of inheritance has been considered – and indeed all nine sons of affected fathers in a published pedigree show the trait, whereas five daughters in both generations are unaffected. This gives:

$$(1/2)^9 \times (1/2)^5 = (1/2)^{14} = \frac{1}{16\,384}$$

Hence, the probability of this pedigree having occurred by chance as an autosomal-dominant trait is very low indeed. There is an important difference, however, from the example of vitamin D-resistant rickets. Other pedigrees showing autosomal-dominant inheritance are unknown for this type of rickets, and all observations confirm the location of this gene on the X chromosome. For Brauer keratoma dissipatum, on the other hand, some families have been observed exhibiting very similar phenotypes that show clearcut autosomal-dominant inheritance. It is therefore likely that the described pedigree has been selected from an unknown number of observations because of its peculiar transmission. The calculation is misleading as the "universe" from which this sample of observations was drawn (all pedigrees with the same phenotype) is much larger (and ill-defined), and the sample (the pedigree) is biased. The trait seems to be autosomal-dominant.

Another, more obvious example of an error in the definition of the sample space is the mother, above, who did not want a Chinese baby.

5.3.3 Testing for Segregation Ratios Without Ascertainment Bias: Codominant Inheritance

Apart from these limiting cases, calculation of exact probabilities for certain families or groups of families is usually impracticable. Therefore statistical methods are used that are either based on the parameters of the "normal" distribution, which is a good approximation of the binomial distribution (parametric tests), or derive directly from probabilistic reasoning (nonparametric tests). One method that is especially well suited for genetic comparisons is the χ^2 test. This enables us to compare frequencies of observations in two or more discrete classes with their expectations. The most usual form is:

$$\chi^2 = \Sigma \frac{(E-O)^2}{E}$$

(E=expected number; O=observed number). In Farabee's pedigree with dominant inheritance (Sect. 5.1.2), there are 36 affected and 33 unaffected children of affected parents. With dominant inheritance, E is $1/2$ of all children, i.e., 34.5:

$$\chi^2_1 = \frac{(36-34.5)^2}{34.5} = \frac{(33-34.5)^2}{34.5} = 0.13$$

The probability p for an equal or greater deviation from expectation can be taken from a χ^2 table for 1 degree of freedom. The number of degrees of freedom indicates in how many different ways the frequencies in the different classes can be changed without altering the total number of observations. In this case the content of class 2, unaffected, is unequivocally fixed by the content of class 1. Therefore, the number of degrees of freedom is 1. In general the number of degrees of freedom is equal to the number of classes less 1.

A second example is taken from the codominant mode of inheritance (Sect. 5.1.2). Table 5.1 summarizes Wiener's family data for the MN blood types. Are the resultant segregation ratios compatible with the genetic hypothesis? For this problem, matings MM × MM, MM × NN, and NN × NN give no information. Expectations in the matings MM × MN and NN × MN are 1:1, in the mating MN × MN 1:2:1. This leads to Table 5.8 for the χ^2 test: For 4 degrees of freedom we find in the χ^2 table: $p=0.75$. This is very good agreement with expectation.

5.3.3.1 Dominance

The situation becomes slightly more complicated when one allele is dominant and the other recessive. This is the case, for example, in the AB0 blood group system. Here, the phenotype A consists of the genotypes AA and A0. The expected segregation ratios among their offsprings differ. Some of the heterozygous parents A0 can be

Table 5.8 Comparison between expected and observed segregation figures in the MN data of Wiener et al. (Table 4.1 [92])

Mating type	MM	MN	NN	χ^2	Degrees of freedom
MM × MN	$\dfrac{(499-486)^2}{486}$	$\dfrac{(473-486)^2}{486}$	–	0.6955	1
MN × MN	$\dfrac{(199-200)^2}{200}$	$\dfrac{(405-400)^2}{400}$	$\dfrac{(196-200)^2}{200}$	0.1475	2
MN × NN	–	$\dfrac{(411-396.5)^2}{396.5}$	$\dfrac{(382-396.5)^2}{396.5}$	1.0605	1

recognized, for example, in matings with 0 partners by the finding of 0 children. Others have only A children just by chance. Special statistical methods are necessary to calculate correct expectations and to compare empirical observations with these expectations [68].

5.3.4 Testing for Segregation Ratios: Rare Traits

5.3.4.1 Principal Biases

If the condition under examination is rare, families are usually not ascertained at random; one starts with a "proband" or "propositus," i.e., a person showing the condition. This leads to an *ascertainment bias,* which must be corrected. The bias can be of different kinds, depending on the way in which the patients have been ascertained.

(a) Family or truncate selection. All individuals suffering from a specific disease in a certain population at a certain time (or within certain time limits) are ascertained. The individual patients are ascertained independently of each other, i.e., the second case in a sibship would always have been found. Such truncate ascertainment is possible, for example, if the condition always leads to medical treatment, and all physicians report every case to a certain registry – as when an institute carries out an epidemiological study. As a rule, case collections approaching completeness are possible only in ad hoc studies of research workers specializing in a condition or group of conditions.

Here, the ascertainment bias is caused exclusively by the fact that only those sibships are ascertained that contain at least one patient. As noted above, however (Sect. 5.3.3), this leaves out all sibships in which no affected individual has occurred just by chance. Their expected number is:

$$\sum_s q^s n_s \qquad (5.2)$$

(s = number of siblings/sibship; p = segregation ratio; $q = 1 - p$; n_s = number of sibships of size s). In recessive disorders, $p = 0.25$. The smaller the average sibship size, the stronger is the deviation from the 3:1 ratio in the ascertained families.

(b) Incomplete multiple (proband) selection; single selection as limiting case. It is rare that all individuals in a population are ascertained; frequently a study starts, for example, with all patients in a hospital population who have a certain condition. Here an additional bias must be considered: the more affected members a sibship has, the higher is its chance to be represented in the sample. This bias causes a systematic excess of affected persons, which is added to the excess caused by truncate selection as explained above.

Koller [45] gave a simple example that demonstrates the nature of this excess. Let us assume that the probands are ascertained during examination of only a single year's group of conscripts. The population comprises a number of families with three children, at least one of whom has the disease, and one of whom is a member of the current year's group. Ascertainment of the family depends on the presence of an affected child in the 1 year group examined. Thus, all families with three affected siblings but only two-thirds of the families with two affected and one-third of those with only one affected are ascertained.

The methods of correction described below are reliable only if the probability for ascertainment of consecutive siblings is independent of the ascertainment of the first one. In an examination of conscripts, as described above, this may be the case. Most studies, however, begin with a hospital population or some other group of medically treated persons. Here, according to general experience, subsequently affected children are much more frequently brought to a hospital when another child has been treated successfully. The opposite trend, however, is also possible. Becker [3], for example, collected all cases of X-linked recessive Duchenne's muscular dystrophy in a restricted area of southwestern Germany. He had good reason to think that ascertainment was complete for this area. Nevertheless, brothers developing muscular dystrophy as the second or later cases in their sibships were generally not ascertained as probands (i.e., through hospitals and physicians) but through the first proband in the family. In his interviews with the parents Becker found the reason. In the case of the first patient in the sibship the parents usually consulted a physician. Then, however, they discovered that in spite of examinations and therapeutic attempts, the course of the disease could not be influenced. Hence they refrained from presenting a second child to the hospital or the physician.

(c) Apart from these biases, which can be statistically corrected to a certain degree, there are other biases that cannot be corrected. Frequently, for example, a genetic hypothesis is discussed on the basis of families sampled from the literature. Experience shows that such sampling usually leads to reasonable results in autosomal-dominant and X-linked recessive disorders. Autosomal-recessive diseases, however, are more difficult to handle. Families with an impressive accumulation of affected sibs have a higher chance of being reported than those with only one or two affected members. This selection for "interesting" cases was more important early in the twentieth century because families generally had more children. Furthermore, recessive conditions discovered today are usually interesting from a clinical and biochemical point of view as well.

These biases can be avoided only by publishing all cases and by critical interpretation of data from the literature. A statistically sound correction is impossible, as such bias has no simple and reproducible direction.

To summarize, the method of segregation analysis depends on the way in which families are ascertained. It follows that the method of ascertainment should always be described carefully. Above all, the probands should always be fully indicated. It is also of interest whether the author during his case collection has become aware of any ascertainment biases.

These considerations show that complete (truncate) ascertainment of cases in a population, and within defined time limits, is the optimal method of data collection.

5.3.4.2 Methods for Correcting Bias

Two different types of correction are possible: test methods and estimation methods.

In a test method the observed values are compared with the expected values, which have been corrected for ascertainment bias. The first such test method was published by Bernstein in [45]; it corrected for truncate selection. The expected number of affected E_r is:

$$E_r = sn_s \frac{p}{1-q^s} \quad (5.3)$$

in all sibships of size n (definition of symbols as in (5.2)). A similar test method can also be used for proband selection.

Test methods answer a specific question: do the observed proportions fit the expected values according to a certain genetic hypothesis?

In many if not in most actual cases, the question is more general: What is the unbiased segregation ratio in the observed sibships? This is an estimation problem. The earliest method was published in 1912 by Weinberg [86] and was called the sib method. Starting from every affected sib in the sibship, the number of affected and unaffected among the sibs is determined. This method is adequate for "truncate selection," i.e., when each affected person is, at the same time, a proband. The sib method is the limiting case of the "proband method" used when the families are ascertained by incomplete multiple proband selection. The number of affected and unaffected siblings is counted, starting from every proband. A limiting case is single selection. Here each sibship has only one proband, and the counting is done once among the sibs.

These estimates converge with increasing sample size to the parameter p, the true segregation ratio; they are *consistent*. It was realized early, however, that they are not fully *efficient*, except for the limiting case of single selection, i.e., they do not make optimal use of all available information. Therefore improvements have been devised by a number of authors. Today such simple methods are no longer used. Moreover, the problems to be solved by segregation analysis are usually more complex. For example, the families to be analyzed may be a mixture of genetic types with various modes of inheritance; there may be admixture of "sporadic" cases, due either to new mutation or to environmental factors; penetrance may be incomplete, or the simple model of a monogenic mode of inheritance may be inadequate for explaining familial aggregation, and a multifactorial genetic model must be used (for the conceptual basis of such multifactorial models, see Chap. 8). Computer programs are now available for carrying out such analyses; they are available either from their authors' institutions or through an international network of program packages. Some of these also offer programs for comparing predictions from various genetic models.

5.3.5 Discrimination of Genetic Entities: Genetic Heterogeneity

It is a common experience in clinical genetics that similar or identical phenotypes are caused by a variety of genotypes. The splitting of a group of patients with a given disease into smaller but genetically more uniform subgroups has been a major topic of research in medical genetics over recent decades. Frequently such heterogeneity analysis is another aspect of the application of Mendel's paradigm and its consequences: carrying genetic analysis through different levels ever closer to gene action.

It appears at first glance that with modern biological methods discrimination of genetic entities on descriptive grounds, i.e., on the level of the clinical phenotype, would no longer hold interest. In our opinion, however, knowledge of the phenotypic variability of genetic disease in humans is needed for many reasons:

(a) Such knowledge provides heuristic hypotheses for systematic application of the more penetrating methods from biochemistry, molecular biology, immunology, micromorphology, and other fields.
(b) Treatment will often depend upon manipulation of gene disordered biochemistry and pathophysiology of a given disease.
(c) We require insight into the genetic burden of the human population.
(d) Better data are needed for many of our attempts to understand the problems of spontaneous and induced mutation.

5.3.5.1 Genetic Analysis of Muscular Dystrophy as an Example

One group of diseases in which analysis using the clinical phenotype together with the mode of inheritance proved to be successful are the muscular dystrophies. These conditions have in common a tendency to slow muscular degeneration, incapacitating affected patients who often ultimately die from respiratory failure. There are major differences in age at onset, location of the first signs of muscular weakness, progression of clinical symptoms, and mode of inheritance. These criteria were used by medical geneticists to arrive at the following classification of muscular dystrophies:

1. X-linked muscular dystrophies
    a. Severe type (Duchenne) (310200)
    b. Juvenile or benign type (Becker; 310100)
    c. Benign type with early contracture (Cestan-Lejonne and Emery-Dreifuss; 310300)
    d. Hemizygous lethal type (Henson-Muller-de Myer; 309950)
2. Autosomal-dominant dystrophy Facio-scapulo-humeral type (Erb-Landouzy-Déjérine; 158900)
3. Autosomal-recessive muscular dystrophies
    a. Infantile type
    b. Juvenile type
    c. Adult type
    d. Shoulder girdle type

This classification is based on many reports from various populations and, for the rarer variants, on reports of pedigrees. It does not include pedigrees in which affected members showed involvement only of restricted parts of the muscular system, such as distal and ocular types. Congenital myopathies were also excluded. The main criteria for discrimination are obvious from the descriptive terms used in the tabulation; for details, see Becker [6]. At present, various mutations of the X-linked dystrophin gene are known at the molecular level which lead to the Duchenne and Becker types. The gene for Emery-Dreyfus disease has been localized to distal Xq28.

5.3.5.2 Multivariate Statistics

The critical human mind is an excellent discriminator. However, statistical methods for identifying subgroups within a population on the basis of multiple characteristics are now available (multivariate statistics). Such methods can also be applied to the problem of making discrimination of genetic entities more objective.

5.3.6 Conditions Without Simple Modes of Inheritance

The methods discussed so far are used mainly for genetic analysis of conditions thought to follow a simple mode of inheritance. In many diseases, however, especially in some that are both serious and frequent, there are problems:

(a) Diagnosis of the condition may be difficult. There are borderline cases. Expressed more formally: the distribution of affected and unaffected in the population is not an outright alternative (examples: schizophrenia; hypertension; diabetes).
(b) It is known from various investigations, including twin studies, that the condition is not entirely genetic but that certain environmental factors influence manifestation (example: decline of diabetes in European countries during and after World War II).
(c) The condition is so frequent that clustering of affected patients in some families must be expected simply by chance (examples: some types of cancer).
(d) It can be concluded from our knowledge of pathogenic mechanisms that the condition is not a single disease but a complex of symptoms common to a number of different causes (example: epilepsy). In fact, it is becoming apparent that diagnoses such as hypertension and diabetes subsume groups of heterogeneous disease entities.

In no such case can a genetic analysis that starts from the phenotype be expected to lead to simple modes of inheritance. However, for many such conditions, two questions of practical importance arise:

1. What is the risk of relatives of various degrees being affected? Is it higher than the population average?
2. What is the contribution of genetic factors to the disease? Under what conditions does the disease manifest itself?

Familial aggregation can be assessed by calculation of empirical risk figures. Twin studies and comparisons of incidence among relatives of probands with those in the general population are required to answer the questions. Here, we discuss risk figures.

5.3.6.1 Empirical Risk Figures

The expression "empirical risk" is used in contrast to "theoretic risks" as expected by Mendelian rules in conditions with simple modes of inheritance. The early methods were developed largely by the Munich school of psychiatric genetics in the 1920s with the goal of obtaining risk figures for psychiatric diseases.

The basic concept is to examine a sufficiently large sample of affected patients and their relatives. From this material, unbiased risk figures for defined classes of relatives are calculated. These figures are used to predict the risk for relatives in future cases. This approach makes the implicit assumption that risk figures are generally constant "in space and time", i.e., among various populations and under changing conditions within the same population. Considering the environmental changes influencing the occurrence of many diseases such as diabetes, this assumption is not necessarily true but is useful as a first approximation.

The approach can be extended to include the question of whether two conditions A and B have a common genetic component, leading to increased occurrence of patients with disease A among close relatives of patients with disease B.

5.3.6.2 Selecting and Examining Probands and Their Families

In conditions that have simple modes of inheritance, the selection of probands is usually straightforward. The modes of ascertainment are discussed in Sect. 5.3.4. For empirical risk studies the same rules apply. In fairly frequent conditions, complete ascertainment of cases in a population is rarely if ever feasible and is also unnecessary in these investigations. In most situations, a defined sample of probands, such as all cases coming to a certain hospital for the first time during a predefined time period can be used. The mode of ascertainment is single selection, or very close to it. This approach simplifies correction of the ascertainment bias among sibs of probands. The empirical risk figures can be calculated by counting affected and unaffected among the sibs, excluding the proband. Risk figures among children ascertained through the parental generation are unbiased and need no correction.

Frequently, the diagnostic categories are not clearcut. In these cases, criteria for accepting a person as a proband must be defined unambiguously beforehand, and all possible biases of selection should be considered. Are more severe cases normally admitted to the hospital selected for study? Are patients selected from a particular social or ethnic group? Are there any other biases that might influence the comparability of the results? Genuinely unbiased samples are hardly if ever available, but the biases should be known. Most importantly, such biases should be independent of the problem to be analyzed. For example, it would be a mistake to consider only patients who have similarly affected relatives.

The goal of the examinations is to obtain maximal and precise information about the probands and their families as far as possible. Methods for achieving this goal, however, vary. Clinical experience and the study of publications on similar surveys are helpful.

Once the proband and his family are ascertained, the relatives should be noted as completely as possible, and information on their health status must be collected. Here, personal examination by the investigator and historical information provided by the patients and their relatives are indispensible. Such data should be backed by hospital records and various laboratory and radiological studies. Even results of clinical examinations should be regarded with scepticism since not all physicians are equally knowledgeable and careful, and official documents, such as death certificates, are often unreliable regarding diagnostic criteria.

In most cases, the determination of genetic risk figures answers the question of whether the risk is higher than in the average population. Sometimes adequate incidence and/or prevalence data from a complete population in which the study is carried out or a very similar one are available. More often than not, however, a control series must be examined with the same criteria as used for the "test" populations. If possible, examination on normal controls and their relatives should be performed in a "blind" way; i.e., the examiners should be unaware of whether the persons studied come from the patient or the control series. It is a good idea to use matched controls, i.e., to examine for every patient a control person matched in all criteria but not related to the condition to be investigated (such as age, sex, ethnic origin, etc.).

5.3.6.3 Statistical Evaluation, Age Correction

In conditions that manifest at birth, such as congenital malformations affecting the visible parts of the body, further calculations are straightforward: the empirical risk for children is given by the proportion of affected in the sample. In many cases, however, onset occurs during later life, and the period at risk may be extended. Here the question asked is: What is the risk of a person's becoming affected with the condition, provided he or she lives beyond the manifestation period? The appropriate methods of age correction have been discussed extensively in the earlier literature [45]; one much-used is Weinberg's "shortened method." First, the period of manifestation is defined on the basis of a sufficiently large sample (usually larger than the sample of the study itself). Then all relatives who dropped out of the study before the age of manifestation are discarded. The dropping out may be for any of a variety of reasons: death, loss of contact due to change of residence, or termination of the study. All persons dropping out during the age of manifestation are counted as one-half, and all who have survived the upper limit of manifestation age are counted full.

5.3.6.4 Example

Among children of schizophrenics, 50 were affected and 200 unaffected. Of these, 100 have reached the age of 45 and 100 are between the age of 15 and 45 (i.e., the age of manifestation for the great majority of schizophrenic cases).

Thus, the corrected number of unaffected is: $200 - \frac{1}{2} \times 100 = 150$; the empirical risk is:

$$\frac{50}{150+50} = 25\%$$

Chapter 23.7 deals in detail with practical problems, taking schizophrenia and affective disorders as examples.

5.3.6.5 Selection of Probands for Genome-Wide Association Studies

Genome-wide association studies have expanded our possibilities to identify new traits in conditions without simple modes of inheritance. The selection of probands or of big cohorts of individuals with a certain phenotypic feature follows different strategies to the aforementioned examples. This will be explained in detail in Chapters 8.1 and 8.

5.3.6.6 Theoretical Risk Figures Derived from Heritability Estimates?

There are suggestions that empirical risk figures should be replaced by theoretical risk figures computed from heritability estimates for the multifactorial model (Chap. 8), after data are found to agree with expectations from such a model. This could be done when the data compared with a simple diallelic model. Such heritability estimates can be achieved by comparing the incidence of the condition in the general population with that in certain categories of relatives, for example, sibs or, with caution, from twin data. In theory the method permits inclusion of environmental, for example, maternal, effects. Its disadvantage, however, is that it depends critically on the assumption that the genetic model fits the actual situation sufficiently well. Since the genetic model chosen may not apply to the data at hand, there is danger that the sophisticated statistical approach suggests a spuriously high degree of precision of the results.

5.4 Conclusions

The transmission of traits determined by single genes, including hereditary diseases, follows Mendel's laws. Autosomal-dominant, autosomal-recessive, and X-linked modes of inheritance can be identified on the basis of the location of mutant genes on autosomes or on the X chromosome, and noting the phenotypic distinction between homozygotes and heterozygotes. Mutations in mitochondrial DNA are transmitted from the mother to all children. Deviations from the classical Mendelian transmission scheme may occur as a consequence of "genomic imprinting," where the parental origin of the mutation determines the phenotype. "Anticipation," with earlier age of onset in succeeding generations, may owe its origin to unstable mutations. Genotype frequencies in populations follow the Hardy–Weinberg Law, which can be used to estimate gene frequencies. In rare traits, such as those in most hereditary diseases, pedigrees are often ascertained via affected individuals and their sibships; when such pedigrees are used to calculate Mendelian segregation ratios, the resulting "ascertainment bias" in favor of affected persons must be corrected by appropriate statistical methods. New sequencing approaches will now enable researches to find disease-causing genes even in relatively small families. Furthermore, genome-wide association studies have paved the way for identifying genomic loci associated with multifactorial inheritance.

References

1. Armour CM, McCready ME, Baig A, Hunter AG, Bulman DE (2002) A novel locus for brachydactyly type A1 on chromosome 5p13.3–p13.2. J Med Genet 39:186–188
2. Badano JL, Katsanis N (2002) Beyond Mendel: an evolving view of human genetic disease transmission. Nat Rev Genet 3:779–789
3. Bates GP (2005) History of genetic disease: the molecular genetics of Huntington disease – a history. Nat Rev Genet 6:766–773
4. Beaudet AL, Perciaccante RG, Cutting GR (1991) Homozygous nonsense mutation causing cystic fibrosis with uniparental disomy. Am J Hum Genet 48:1213
5. Becker PE (1953) Dystrophia musculorum progressiva. Thieme, Stuttgart
6. Becker PE (1972) Neues zur Genetik und Klassifikation der Muskeldystrophien. Hum Genet 17:1–22
7. Becker PE (ed) (1964–1976) Humangenetik. Ein kurzes Handbuch in fünf Bänden. Thieme, Stuttgart
8. Bell J (1934) Huntington's chorea. Treasury of human inheritance 4. Galton Laboratory
9. Bennett T (1975) The T-locus of the mouse. Cell 6:441–454
10. Bernstein F (1925) Zusammenfassende Betrachtungen über die erblichen Blutstrukturen des Menschen. Z Indukt Abstamm Vererbungsl 37:237
11. Bhende YM, Deshpande CK, Bhata HM, Sanger R, Race RR, Morgan WTJ, Watkins WM (1952) A "new" blood-group character related to the ABO system. Lancet 1:903–904

12. Brook JD, McCurrach ME, Harley HG et al (1992) Molecular basis of myotonic dystrophy. Cell 68:799–808
13. Carney G, Seedburgh D, Thompson B, Campbell DM, MacGillivray I, Timlin D (1979) Maternal height and twinning. Ann Hum Genet 43:55–59
14. Castori M, Covaciu C, Paradisi M, Zambruno G (2009) Clinical and genetic heterogeneity in keratosis follicularis spinulosa decalvans. Eur J Med Genet 52:53–58
15. Ceppellini R, Dunn LC, Turri M (1955) An interaction between alleles at the Rh locus in man which weakens the reactivity of the Rho factor (Du). Proc Natl Acad Sci USA 41:283–288
16. Collaco JM, Cutting GR (2008) Update on gene modifiers in cystic fibrosis. Curr Opin Pulm Med 14:559–566
17. Cuénot L (1905) Les races pures et leurs combinaisons chez les souris. Arch Zool Exp Genet 3:123–132
18. De Vries DD, de Wijs IJ, Wolff G et al (1993) X-linked myoclonus epilepsy explained as a maternally inherited mitochondrial disorder. Hum Genet 91:51–54
19. Dorn H (1959) Xeroderma pigmentosum. Acta Genet Med Gemellol (Roma) 8:395–408
20. Engel E (1993) Uniparental disomy revisited: the first twelve years. Am J Med Genet 46:670–674
21. Engel E (1980) A new genetic concept: uniparental disomy and its potential effect, isodisomy. Am J Med Genet 6:137–143
22. Farabee (1905) Inheritance of digital malformations in man. In: Papers of the peabody museum for american archeology and ethnology, vol 3. Harvard University Press, Cambridge/MA, 69
23. Falush D, Almqvist EW, Brinkmann RR, Iwasa Y, Hayden MR (2000) Measurement of mutational flow implies both a high new-mutation rate for Huntington disease and substantial underascertainment of late-onset cases. Am J Hum Genet 68:373–385
24. Feller W (1970/71) An introduction to probability theory and its applications, 2nd edn. Wiley, New York
25. Fleischer B (1918) Über myotonische Dystrophie mit Katarakt. Grafes Arch Ophthalmol 96:91–133
26. Fu Y-H, Pizzuti A, Fenwick RG et al (1992) An unstable triplet repeat in a gene related to myotonic dystrophy. Science 255:1256–1258
27. Gartler SM, Francke U (1975) Half chromatid mutations: Transmission in humans? Am J Hum Genet 27:218–223
28. Gaul LE (1953) Heredity of multiple benign cystic epithelioma. Arch Dermatol Syph 68:517
29. Gerard G, Vitrac D, Le Pendu J, Muller A, Oriol R (1982) H-deficient blood groups (Bombay) of Reunion island. Am J Hum Genet 34:937–947
30. Goldfarb LG, Petersen RB, Tabaton M et al (1992) Fatal familial insomnia and familial Creutzfeldt-Jakob disease: disease phenotype determined by a DNA polymorphism. Science 258:806–808
31. Grønskov K, Ek J, Brondum-Nielsen K (2007) Oculocutaneous albinism. Orphanet J Rare Dis 2:43
32. Hadorn E (1955) Developmental genetics and lethal factors. Wiley, London
33. Haldane JBS (1941) The relative importance of principal and modifying genes in determining some human diseases. J Genet 41:149–157
34. Hardy GH (1908) Mendelian proportions in a mixed population. Science 28:49–50
35. Harley HG, Rundle SA, Reardon W et al (1992) Unstable DNA sequence in myotonic dystrophy. Lancet 339:1125–1128
36. Harris H (1948) A sex-limiting modifying gene in diaphyseal aclasis (multiple exostoses). Ann Eugen 14:165–170
37. Harris H (1980) The principles of human biochemical genetics, 4th edn. North-Holland, Amsterdam
38. Haws DV, McKusick VA (1963) Farabee's brachydactylous kindred revisited. Johns Hopkins Med J 113:20–30
39. Hilgert N, Smith RJ, Van Camp G (2009) Forty-six genes causing nonsyndromic hearing impairment: which ones should be analyzed in DNA diagnostics? Mutat Res 681:189–196
40. Huoponen K, Lamminen T, Juvonen V et al (1993) The spectrum of mitochondrial DNA mutations in families with Leber hereditary optical neuroretinopathy. Hum Genet 92:379–384
41. HYP Consortium (1995) A gene (PEX) with homologies to endopeptidases is mutated in patients with X-linked hypophosphatemic rickets. Nat Genet 11:130–136
42. Kajiwara K, Berson EL, Dryja TP (1994) Digenic retinitis pigmentosa due to mutations at the unlinked peripherin/RDS and ROM1 loci. Science 264:1604–1608
43. Katsanis N et al (2001) Triallelic inheritance in Bardet-Biedl syndrome, a Mendelian recessive disorder. Science 293:2256–2259
44. Kirkpatrick TJ, Au KS, Mastrobattista JM, McCready ME, Bulman DE, Northrup H (2003) Identification of a mutation in the Indian Hedgehog (IHH) gene causing brachydactyly type A1 and evidence for a third locus. J Med Genet 40:42–44
45. Koller S (1940) Methodik der menschlichen Erbforschung. II. Die Erbstatistik in der Familie. In: Just G, Bauer KH, Hanhart E, Lange J (eds) Methodik, Genetik der Gesamtperson. Springer, Berlin, pp 261–284 Handbuch der Erbbiologie des Menschen, vol 2
46. Landsteiner K (1900) Zur Kenntnis der antifermentativen, lytischen und agglutinierenden Wirkungen des Blutserums und der Lymphe. Zentralbl Bakteriol 27:357–362
47. Lenz W (1961) Zur Genetik der Incontinentia pigmenti. Ann Paediatr (Basel) 196:141
48. Lenz W (1975) Half chromatid mutations may explain incontinentia pigmenti in males. Am J Hum Genet 27:690–691
49. Lenz W (1983) Medizinische Genetik, 6th edn. Thieme, Stuttgart
50. McCready ME, Sweeney E, Fryer AE et al (2002) A novel mutation in the IHH gene causes brachydactyly type A1: a 95-year-old mystery resolved. Hum Genet 111:368–375
51. McCready ME, Grimsey A, Styer T, Nikkel SM, Bulman DE (2005) A century later Farabee has his mutation. Hum Genet 117:285–287
52. McKusick VA (1995) Mendelian inheritance in man, 11th edn. Johns Hopkins University Press, Baltimore
53. Milch RA (1959) A preliminary note of 47 cases of alcaptonuria occurring in 7 interrelated Dominical families, with an additional comment on two previously reported pedigrees. Acta Genet (Basel) 9:123–126
54. Mohr OL, Wriedt C (1919) A new type of hereditary brachyphalangy in man. Carnegie Inst Publ 295:1–64
55. Morison IM, Ramsay JP, Spencer HG (2005) A census of mammalian imprinting. Trends Genet 21:457–465

56. Mühlmann WE (1930) Ein ungewöhnlicher Stammbaum über Taubstummheit. Arch Rassenbiol 22:181–183
57. Myers RH et al (1993) *De novo* expansion of a (CAG)n repeat in sporadic Huntington's disease. Nature Genet 5:168–173
58. Nicholls RD, Saitoh S, Horsthemke B (1998) Imprinting in Prader-Willi and Angelman syndromes. Trends Genet 14:194–200
59. Ollendorff-Curth (1958) Arch Derm Syph 77:342
60. Ott J (1977) Counting methods (EM algorithm) in human pedigree analysis. Linkage and segregation analysis. Ann Hum Genet 40:443–454
61. Pauli RM (1983) Editorial comment: dominance and homozygosity in man. Am J Med Genet 16:455–458
62. Pearson K (1904) On the generalized theory of alternative inheritance with special references to Mendel's law. Philos Trans R Soc 203:53–86
63. Penrose LS (1947/49) The problem of anticipation in pedigrees of dystrophia myotonica. Ann Eugen 14:125–132
64. Pola V, Svojitka J (1957) Klassische Hämophilie bei Frauen. Folia Haematol (Leipz) 75:43–51
65. Race RR, Sanger R (1975) Blood groups in man, 6 th edn. Blackwell, Oxford
66. Reik W (1989) Genomic imprinting and genetic disorders in man. Trends Genet 5:331–336
67. Siemens HW (1925) Über einen, in der menschlichen Pathologie noch nicht beobachteten Vererbungsmodus: dominant geschlechtsgebundene Vererbung. Arch Rassenbiol 17:47–61
68. Smith CAB (1956/7) Counting methods in genetical statistics. Ann Hum Genet 21:254–276
69. Smith CAB (1970) A note on testing the Hardy-Weinberg law. Ann Hum Genet 33:377
70. Snyder LF, Doan CA (1944) Is the homozygous form of multiple teleangiectasia lethal? J Lab Clin Med 29:1211–1216
71. Spence JE et al (1988) Uniparental disomy as a mechanism for human genetic disease. Am J Hum Genet 42:217–226
72. Stern C (1957) The problem of complete Y-linkage in man. Am J Hum Genet 9:147–165
73. Stevens WL (1950) Statistical analysis of the AB0 blood groups. Hum Biol 22:191–217
74. Stevenson RE, Brasington CK, Skinner C, Simensen RJ, Spence JE, Kesler S, Reiss AL, Schwartz CE (2007) Craniofacioskeletal syndrome: an X-linked dominant disorder with early lethality in males. Am J Med Genet 143A:2321–2329
75. Stocks P, Barrington A (1925) Hereditary disorders of bone development. Treasury of human inheritance 3, part 1
76. Stoffel M, Froguel P, Takeda J et al (1992) Human glucokinase gene: isolation, characterization and identification of two missense mutations linked to early-onset non-insulin dependent (type 2) diabetes melitus. Proc Natl Acad Sci USA 89:7698–7702
77. Timoféef-Ressovsky NW (1931) Gerichtetes Variieren in der phänotypischen Manifestierung einiger Generationen von Drosophila funebris. Naturwissenschaften 19:493–497
78. Trevor-Roper PD (1952) Marriage of two complete albinos with normally pigmented offspring. Br J Ophthalmol 36:107
79. Von Dungern E, Hirszfeld L (1911) Über gruppenspezifische Strukturen des Blutes. III. Z Immunitatsforsch 8:526–562
80. Waardenburg PJ, Franceschetti A, Klein D (1961/1963) Genetics and ophthalmology. Blackwell, Oxford vols 1,2
81. Wahlund S (1928) Zusammensetzung von Populationen und Korrelationserscheinungen vom Standpunkt der Vererbungslehre aus betrachtet. Hereditas 11:65–105
82. Wallace DC (1989) Report of the committee on human mitochondrial DNA. Cytogenet Cell Genet 51:612–621
83. Wallace DC (1989) Mitochondrial DNA mutations and neuromuscular disease. Trends Genet 5:9–13
84. Wallace DC (1994) Mitochondrial DNA sequence variation in human evolution and disease. Proc Natl Acad Sci USA 91:8739–8746
85. Weinberg W (1908) Über den Nachweis der Vererbung beim Menschen. Jahreshefte des Vereins für vaterländische Naturkunde in Württemberg 64:368–382
86. Weinberg W (1912) Methoden und Fehlerquellen der Untersuchung auf Mendelsche Zahlen beim Menschen. Arch Rassenbiol 9:165–174
87. Welander L (1957) Homozygous appearance of distal myopathy. Acta Genet (Basel) 7:321–325
88. Wendt GG, Drohm D (1972) Die Huntingtonsche Chorea. Thieme, Stuttgart Fortschritte der Allgemeinen und Klinischen Humangenetik, vol 4
89. Wexler NS, Young AB, Tanzi RE et al (1987) Homozygotes for Huntington's disease. Nature 326:194–197
90. Wettke-Schöfer R, Kantner G (1983) X-linked dominant inherited diseases with lethality in hemizygous males. Hum Genet 64:1–23
91. Wiener AS (1943) Additional variants of the Rh type demonstrable with a special human anti-Rh-serum. J Immunol 47:461–465
92. Wiener AS, di Diego N, Sokol S (1953) Studies on the heredity of the human blood groups. I. The MN types. Acta Genet Med Gemellol (Rome) 2:391–398
93. Winters RW, Graham JB, Williams TF, McFalls VC, Burnett CH (1957) A genetic study of familial hypophosphatemia and viatmin D-resistant rickets. Trans Assoc Am Physicians 70:234–242
94. Zevani M, Serudei S, Gellera C et al (1989) An autosomal dominant disorder with multiple deletions of mitochondrial DNA starting at the D-loop region. Nature 339:309–311

Linkage Analysis for Monogenic Traits

Arno G. Motulsky and Michael Dean

Revised by M.R. Speicher

Abstract Linkage analysis, that is the observation of cosegregation of adjacent genetic markers or traits, is the principal means of constructing genetic maps, and locating genes that cause disease, or genetic traits. Although these methods have been partially supplanted by newer methods such as whole genome sequencing, linkage analysis still has considerable utility. The history of the method and the mathematical basis of linkage analysis are presented as well as specific applications to human genetics. Gene clusters consist of groups of adjacent genes that exist largely though mechanisms of gene duplication. In addition there exist clusters of genes that have related function, the best studied of which is the Major Histocompatibility Complex. The structure and evolutionary history of these clusters provides insight into the history of mammalian genomes.

Contents

6.1	Linkage: Localization of Genes on Chromosomes...	211
	6.1.1 Classic Approaches in Experimental Genetics: Breeding Experiments and Giant Chromosomes............................	212
	6.1.2 Linkage Analysis in Humans......................	213
	6.1.3 Linkage Analysis in Humans: Cell Hybridization and DNA Techniques............	220
	6.1.4 Biology and Statistics of Positional Cloning..................................	224
6.2	Gene Loci Located Close to Each Other and Having Related Functions.................................	225
	6.2.1 Some Phenomena Observed in Experimental Genetics.................................	225
	6.2.2 Some Observations in the Human Linkage Map..	225
	6.2.3 Why Do Gene Clusters Exist?......................	226
	6.2.4 Blood Groups: Rh Complex (111700), Linkage Disequilibrium...............................	226
	6.2.5 Major Histocompatibility Complex [105, 111]......................................	229
	6.2.6 Unequal Crossing Over................................	235
6.3	Conclusions..	238
References..		238

M. Dean (✉)
National Cancer Institute, Frederick, MD 21702, USA
e-mail: deanm@mail.nih.gov

If "to take a possible example, an equally close linkage" (as between the genes for hemophilia and color blindness) "were found between the genes for blood group" and that "determining Huntington's chorea, we should be able, in many cases, to predict which children of an affected person would develop this disease and to advise on the desirability or otherwise of their marriage."

J.B.S. Haldane and J. Bell (1937) The linkage between the genes for colour-blindness and haemophilia in man. Proc. Roy. Soc. B 123, 119.

6.1 Linkage: Localization of Genes on Chromosomes

Genes are located in a linear fashion on the chromosomes. This has the logical consequence that genes located on the same chromosome are transmitted together, i.e., that their segregation is not independent. On the other hand, it is known from cytogenetics that chiasmata are formed during the first meiotic division, and that certain chromosome segments are exchanged between homologous chromosomes (crossing over; see Sect. 3.5). Hence, even genes located on the same chromosome are not always transmitted together; the probability of transmission of two linked genes depends on the distance between them and on the frequency with which they are separated by crossing over. If located on a fairly long chromosome, and if the distance is large enough that numerous crossing over events occur between them, genes located on the same chromosome may even seem to segregate independently. Such genes are syntenic but not linked. It was the great achievement of Morgan and his school in the first two decades of the twentieth century to exploit linkage for localizing genes relative to each other on chromosomes and developing gene maps in the fruit fly *Drosophila melanogaster*.

Studies on linkage and gene mapping in humans lagged behind this development for decades. Sophisticated statistical techniques were developed to get around the difficulty that directed breeding experiments are impossible in humans, and information from naturally occurring families must be used. The application of such techniques, however, was only sparsely rewarded by detection of linkage. A breakthrough occurred only when the new techniques of somatic cell genetics and especially cell fusion were introduced. These techniques permitted the assignment of genes to specific chromosomes and even chromosome segments. Later, methods taken from molecular biology, especially the development of restriction fragment length polymorphisms (RFLPs) revolutionized this progress [9]. Radiation hybrid methods and YAC and BAC clones provided the first high-resolution physical maps and paved the way to the complete sequence of the human genome [50, 104]. Annotation of the human genome sequence has allowed the location of all the known genes to be determined.

In the following we describe first the principle of the classic approach to gene localization as introduced by Morgan and his followers. This provides an opportunity to introduce some general concepts. We then discuss statistical methods for detecting and measuring linkage in humans. The various groups of DNA markers are described next, followed by the principle of cell fusion and its use in localizing genes on chromosomes, as well as the application of radioactive and nonisotopic in situ hybridization for this purpose. Genetic maps are compared to physical maps, and the use of linkage studies as analytical tools in genetic analysis of common diseases with complex etiology and pathogenesis is assessed.

6.1.1 Classic Approaches in Experimental Genetics: Breeding Experiments and Giant Chromosomes

According to Mendel's third law, segregation of two different pairs of alleles is independent; all possible zygotes of two pairs of alleles are formed by free recombination. Mating between the double heterozygote AaBb and the double homozygote aabb leads to:

Paternal gametes	AB	Ab	aB	ab
Maternal gametes ab	$\frac{1}{4}$ AaBb	$\frac{1}{4}$ Aabb	$\frac{1}{4}$ aaBb	$\frac{1}{4}$ aabb

The four genotypes are formed in equal proportions.

Soon after Mendel's laws were rediscovered Bateson et al. [4] found an exception from this rule in the vetch, *Lathyrus odoratus*. Certain combinations were observed more frequently and others less frequently than expected. In some cases, the two parental combinations – in our example AaBb (father) and aabb (mother) – were increased among the progeny; in other cases the two other types Aabb or aaBb were more frequent.

Paternal gametes		AB	Ab	aB	ab
Maternal gametes	ab	AaBb	Aabb	aaBb	aabb
First case (coupling)		$\frac{1}{2} - \Theta$	Θ	Θ	$\frac{1}{2} - \Theta$
Second case (repulsion)		Θ	$\frac{1}{2} - \Theta$	$\frac{1}{2} - \Theta$	Θ

Θ = Recombination fraction Θ.

The alleles of the parental combination seemed either to attract one another or to repel one another. Bateson et al. [4] coined the terms "coupling" for the former phase and "repulsion" for the latter. Morgan in [65] recognized that coupling and repulsion are two aspects of the same phenomenon (i.e., location of two genes on the same or homologous chromosomes). He coined the term "linkage." Coupling occurs when the genes A and B are localized in the doubly heterozygous parent on the same chromosome $\frac{AB}{ab}$, and repulsion occurs when they are localized on homologous chromosomes $\frac{Ab}{aB}$. The terms *cis* and *trans* are more frequently used to refer to genes in coupling or repulsion, respectively. If linkage is complete, only two types of progeny can occur. More frequently, however, all four types are found, albeit two types in smaller numbers. Morgan explained this finding by exchange of chromosome pieces between homologous chromosomes during meiotic crossing over. He also recognized that the frequency of crossing over depends on the distance between two gene loci in one chromosome. Using recombination analysis as an analytic tool, he and his coworkers succeeded in locating a great number of gene loci in *Drosophila* and in establishing chromosome maps. Their results were confirmed in the early 1930s when Heitz, Bauer, and Painter discovered the giant chromosomes of some *Dipterae*. With this experimental tool many gene localizations known from indirect evidence could be confirmed by direct inspection when they were accompanied by small structural chromosomal variation. In the meantime linkage analyses have been carried out in a great number of species.

6.1.1.1 Linkage and Association

It is sometimes assumed that genes which are linked should always show a certain association in the population, i.e., that the chromosomal combinations AB or ab (coupling) occur more frequently than the combinations Ab or aB (repulsion). However, this is not the case in a randomly mating population. Even if the linkage is fairly close, repeated crossing over in many generations causes all four combinations, AB, ab, Ab, and aB, to be randomly distributed in the long run. As a rule, association of genetic traits does not point to linkage. This rule, however, has exceptions. Some combinations of closely linked genes do indeed occur more often than expected with random distribution. Such "linkage disequilibrium" was first postulated in humans for the rhesus blood types (Sect. 6.2.4) and has also been proven for the major histocompatibility complex (MHC), especially the HLA system (Sect. 6.2.5) and for many DNA polymorphisms. It has now been shown that there are blocks of linkage disequilibrium throughout the genome and these blocks vary by location and by population [25] (Sect. 16.3). Linkage disequilibrium may occur for three reasons:

1. The population under examination originated from a mixture of two populations with different frequencies of the alleles A,a and B,b, and the time elapsed since the mixing of the populations was not sufficient for complete randomization (admixture linkage disequilibrium).
2. Two mutants, for example, DNA markers, are located so closely together that an insufficient number of generations has elapsed to separate them by recombination since the two mutations occurred in one chromosome.
3. Certain combinations of alleles at linked gene loci are maintained in high frequency by natural selection.

These problems are discussed in greater detail in connection with the MHC system (Sect. 6.2.5) and in the discussion on association between HLA and disease (Sect. 6.2.5.4).

6.1.2 Linkage Analysis in Humans

6.1.2.1 Direct Observation of Pedigrees

Linkage analysis by classic methods in humans is difficult since no directed breeding occurs. However, in some cases pedigree inspection can provide information. Linkage is excluded, for example, if one of the genes under scrutiny can be localized to the X chromosome while the other is on an autosome. By the same token, there is a high probability of demonstrating formal linkage if both genes are X-linked. Even in this case, however, formal linkage may not be demonstrable since the loci may be so far from each other that crossing over separates them. Similar considerations hold for genes located on a given autosomal chromosome. The term synteny refers to two or more genes being situated on the same chromosome, regardless of whether formal linkage can be demonstrated. Either a large pedigree or a number of smaller pedigrees must be screened to assess the extent of crossing over. Figure 6.1a shows a pedigree with red–green color blindness (303800, 303900) and hemophilia (306700). The males in the sibships at risk either have both conditions or are normal. The genes are in the coupling (or *cis*) state. The pedigree in Fig. 6.1b shows the opposite; here these genes are in the repulsion (or *trans*) phase.

In some exceptional cases linkage between gene loci localized on an autosome can also be established by simple inspection of an extensive pedigree. Figure 6.2 shows a large pedigree in which Huntington disease segregates together with a *Hin*dIII DNA polymorphism detected by a DNA marker, which was named "G 8" [29]. Four allelic variants of this probe are observed in this pedigree, A, B, C, and D. The Huntington gene invariably segregates together with allele C. One individual, VI, 5 (arrow), so far has been unaffected by Huntington disease, but she will be affected later, provided that her father, (who has not been tested), does not happen to have transmitted another chromosome that carries a C allele not linked to the Huntington gene. The pedigree points to close linkage between the locus for this DNA polymorphism and the Huntington gene. Some cross-overs in other such pedigrees have been detected, and the recombination fraction is 4% or less.

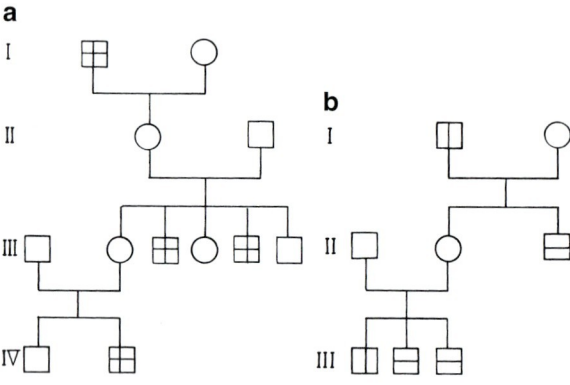

Fig. 6.1 (**a, b**) Pedigrees with red-green color blindness (◒), hemophilia (▨), or both conditions (▩). (**a**) Both abnormal genes in coupling. (From Madlener 1928) (**b**) In repulsion. From Stern [98]

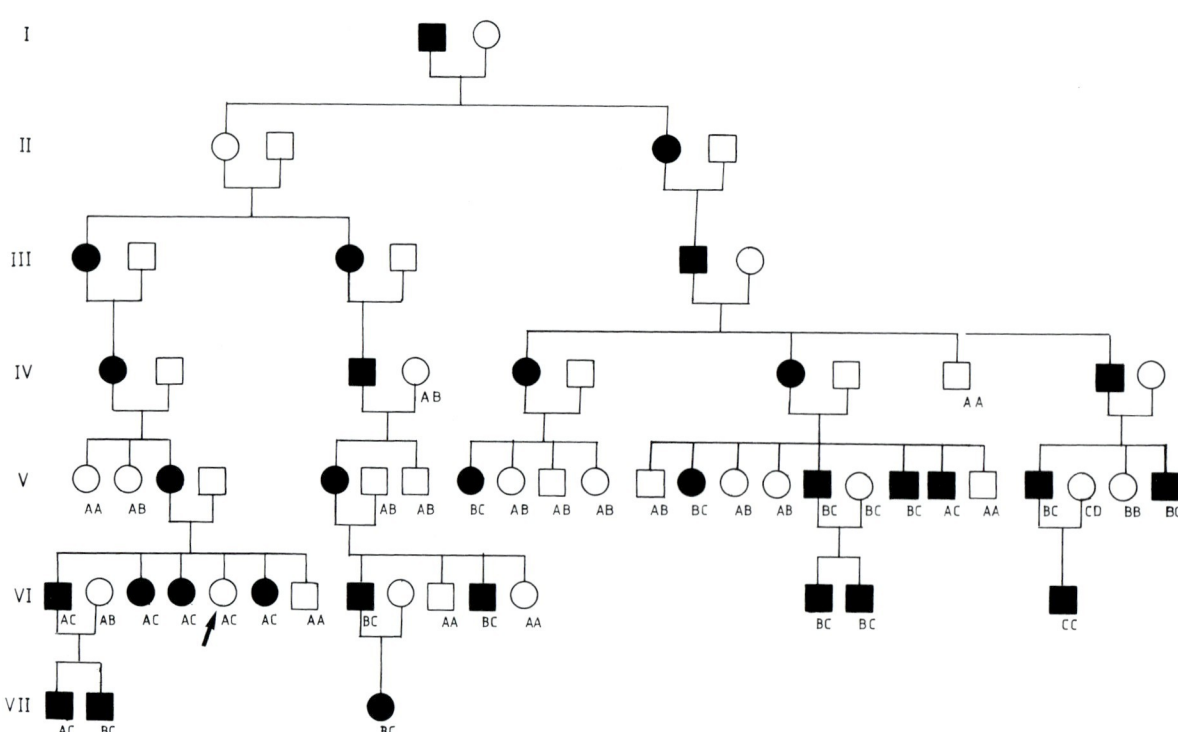

Fig. 6.2 Large pedigree from Venezuela with Huntington disease. *A, B, C, D*, four different "alleles" of a DNA polymorphism. The Huntington gene is transmitted together with "allele" C. One individual, VI,5 (*arrow*) has so far been unaffected. She will most likely be affected later (See text). From Gusella et al. [28]

6 Linkage Analysis for Monogenic Traits

The example of this pedigree shows that the chromosomal phase of alleles at two loci (*cis* or *trans*) can often be ascertained with great precision even in one large pedigree, and that recombinants can be identified if (at least) three generations are available for analysis, and if there are many sibs.

6.1.2.2 Statistical Analysis

In most cases linkage analysis is more difficult. Extensive pedigrees such as that in Fig. 6.2 are exceptional; most available families consist of two parents and their children. Here the problem is that the chromosomal phase is usually unknown: a double heterozygote may be AB/ab (*cis*) or Ab/aB (*trans*). When the alleles are randomly distributed in the population (linkage equilibrium), the two types are expected in about equal frequencies: an AB/ab person forms germ cells in the ratio:

AB	Ab	aB	ab
$\frac{1-\Theta}{2}$	$\frac{\Theta}{2}$	$\frac{\Theta}{2}$	$\frac{1-\Theta}{2}$

whereas a heterozygote Ab/aB forms germ cells in the ratio:

AB	Ab	aB	ab
$\frac{\Theta}{2}$	$\frac{1-\Theta}{2}$	$\frac{1-\Theta}{2}$	$\frac{\Theta}{2}$

Expectations for germ cells are then in any case:

AB	Ab	aB	ab
$\frac{1-\Theta}{2}$	$\frac{\Theta}{2}$	$\frac{\Theta}{2}$	$\frac{1-\Theta}{2}$

or

$\frac{\Theta}{2}$	$\frac{1-\Theta}{2}$	$\frac{1-\Theta}{2}$	$\frac{\Theta}{2}$

which adds up to:

$\frac{1}{4}$	$\frac{1}{4}$	$\frac{1}{4}$	$\frac{1}{4}$

irrespective of Θ. It even remains true if $\Theta=0$ (very close linkage).

All four types of germ cells occur with the same frequencies, regardless of the probability of recombination Θ. Linkage does not lead to any association of alleles A,B or a,b in the population (Exception: linkage disequilibrium; Sect. 6.2). Another criterion for linkage must be found, one that is independent of the phase of the double heterozygote.

Such a criterion would be the *distribution* of children within sibships. In mating of AB/ab persons (*cis* phase) most children show the allele combinations of their parents; in matings of Ab/aB (*trans* phase) most children show these alleles in a new combination. How can these deviations from random distribution within sibships be measured and used for establishing linkage and determining the probability of recombination? Bernstein in [6] was the first to develop such a method. It has now been replaced by the method of "logarithm of differences" (LOD) scores as developed by Haldane and Smith [31] and Morton [65–66] and is generally used to assess linkage. Its principle can be described as follows:

The probability P_2 that the observed family data conform to the behavior of two loci under full recombination without any linkage is calculated and similarly, the probability P_1 that the identical family data are the result of two linked loci under a specified recombination fraction (Θ) is estimated for various families. The ratio of these two probabilities is the likelihood ratio and expresses the odds for and against linkage. This ratio $\dfrac{P_1(F/\Theta)}{P_2(F/(1/2))}$ must be calculated for each family F.

A man may be doubly heterozygous for the gene pairs A,a and B,b. His wife may be homozygous for the two recessive alleles aa, bb. Assume that his two sons, as the father, are doubly heterozygous, i.e., they inherited the dominant alleles A and B from the father. This probability is $1/2 \times 1/2 = 1/4$ for each son if the genes segregate independently. If the gene loci are closely linked without crossing over, the probability for occurrence of this pedigree may be calculated as follows. Either the genes occur in coupling state: AB/ab, then the possibility for common transmission to each of the two sons is $1/2$ (transmission of the combination ab would also have a probability of $1/2$), or the genes occur in repulsion state Ab/aB, where transmission of both dominant alleles to the same son requires crossing over. With close linkage and in absence of crossing over the probability here of common transmission = 0. Hence, the total probability for transmission of the combination aB to either son is $1/2$ and the likelihood ratio is $P_1/P_2 = (1/2)/(1/4) = 2$ in favor of close linkage. Likelihood ratios for the various degree of loose linkage can be calculated in the same way.

For convenience the logarithm of the ratio is used, and a LOD score z (meaning "log odds" or "log probability ratio") is used:

$$z = \log_{10} \frac{P(F|\Theta)}{P(F|(1/2))} \quad (6.1)$$

Here, $P(F|\Theta)$ denotes the probability of occurrence for a family F when the recombination fraction is Θ. Using the logarithms instead of the probabilities themselves has the advantage that the score of any newly found family can be added, giving a combined score $z = S\, z_i$ for all families examined.

Equation (6.1) implies an identical recombination fraction for both sexes. Since sex differences in recombination rates have been described [82] (see below), the z score in actual data should be computed separately for the sexes:

$$z = \log_{10} \frac{P(F/\Theta,\Theta')}{P(F/(1/2,1/2))} \quad (6.2)$$

where Θ is the recombination fraction in females and Θ' in males.

It follows from the definition of the likelihood ratio that the higher its numerator, the stronger is the deviation in the direction of linkage. In terms of logarithms the higher the z score, the better is the evidence for linkage. A LOD score of 3 or higher is generally considered as proof of linkage. Minor corrections for dominance and for ascertainment of pedigrees with rare traits but are not dealt with here [93].

The score $z(\Theta,\Theta')$ for the entire set of data is the sum of the scores of the separate families. For a first approach $\Theta = \Theta'$ is assumed to simplify the calculations. After linkage has been established, a possible sex difference can be looked for.

Numerous computer programs for detection and estimation of linkage are available (for example, see http://linkage.rockefeller.edu/soft, which lists multiple software tools for genetic linkage analysis of human pedigrees). They also allow for testing whether a part only of the observed families show linkage (=linkage heterogeneity). These programs permit to make optimal use of linkage information even in large and sometimes complicated pedigrees. For a detailed account of reasoning on linkage as well as methods of analysis, see Ott [74].

6.1.2.3 The Use of LOD Scores

The ideal mating for linkage studies involves a double heterozygote, i.e., a person heterozygous for two different traits, with a person homozygous for the two genes. The following types of families do not contribute information regarding linkage:

(a) Families in which neither parent is doubly heterozygous
(b) Families in which there cannot be any observable segregation
(c) Families in which the phases of the parents are unknown and there is only one examined child

Most linkage studies involve analysis of two common markers or of a common gene with a gene for a rare genetic disease. Opportunities to study linkage between two rare genes hardly ever exist. The ideal family for linkage studies is a kindred with at least three generations, many matings, and a large number of offspring. Such families are becoming rare in Western societies. An alternative approach involves testing of many small families. This may even have an advantage if more than one gene locus causes a special phenotype. In these instances the study of a single, large pedigree with linkage may create the impression that this gene locus is the only one whose mutations cause the phenotype in question, whereas analysis of many, smaller pedigrees may point to other loci as well, and hence to genetic heterogeneity.

When linkage has been established and a maximum likelihood estimate of Θ achieved, the question of heterogeneity should be examined. If, for example, linkage between the locus for a genetic polymorphism and a rare dominant condition has been established, linkage analysis can help to prove genetic heterogeneity if only part of the family data shows linkage. This occurs very often [61]; the statistical problem posed by such a situation is tricky. Ott [72] has proposed using the χ^2 statistic to compare hypotheses: linkage without heterogeneity vs. nonlinkage, linkage with heterogeneity vs. linkage without heterogeneity, and linkage with heterogeneity vs. nonlinkage. It is also possible to estimate the proportion of families showing linkage in the data set studied.

The human genome is so saturated with genetic markers that one can estimate linkage not only for two loci but for several markers at once (multipoint linkage). Appropriate computer programs for linkage analysis, for example, the LINKAGE package (http://linkage.rockefeller.edu/), have proven to be very useful. In fact, linkage analysis is now often performed using SNP arrays covering several hundred thousand markers which can easily be analyzed with other software tools such as dChip (http://www.biostat.harvard.edu/complab/dchip/).

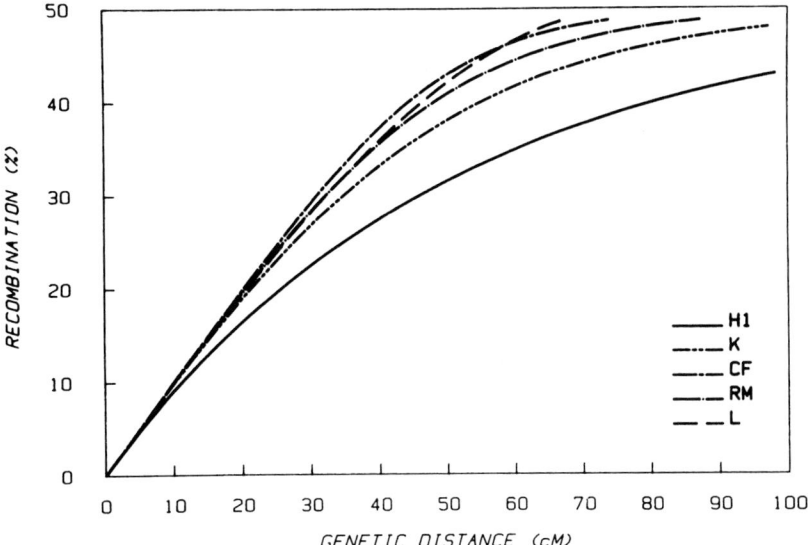

Fig. 6.3 Genetic distance (in centimorgans) in relation to the recombination fraction (in percentage), according to estimates from various authors. *H1* Haldane function with no interference; *K* Kosambi; *CF* Carter and Falconer; *RM* Rao and Morton; *L* Ludwig. From White and Lalouel [110]

6.1.2.4 Recombination Probabilities and Map Distances

Once a number of linkages have been established, the next step is to estimate map distances between these loci. These distances are expressed in morgans and centimorgans, 1 cM (map unit) meaning 1% recombination ($\Theta=0.01$) for small map distances. For larger distances this value must be corrected for double crossing over. Various methods have been proposed. Given a recombination frequency Θ, the map distance (cm) can be read directly from Fig. 6.3.

6.1.2.5 The Sib Pair Method

The use of LOD scores is the ideal method if the mode of inheritance of the two traits to be tested for linkage has been established. Examples include testing of linkage for two genetic markers or for a marker and a clearcut monogenic disease. At least two generations should be available. The analysis becomes more difficult if penetrance of the mutant gene is incomplete, and a definitive phenotype cannot be assigned. While inclusion of a penetrance term in the analysis may be possible, introduction of this and other adjustments can be hazardous since it may lead to false claims, particularly if the data are manipulated in various ways until a "positive" linkage result is obtained.

In general, if the mode of inheritance cannot be established, or when data from only one generation are available, it is preferable to use the sib method first suggested by Penrose in the 1930s [76] (see also [7, 82–84, pp. 90–92]; Fig. 6.4). Its rediscovery has been called by Ott [73] "the cutting of the Gordian knot." (Alexander the Great was challenged to disentangle this knot, which no one had been able to do previously; he cut through it with his sword.) This is because the detection of linkage with this method does not depend on correct assignment of the mode of inheritance but only on the influence of a gene that contributes nonnegligibly to the trait and on linkage of this gene with a marker. Such approaches are termed "nonparametric." Penrose explains, "The method is based on the principle that, when pairs of sibs are taken at random, certain types of sibling pairs will be more frequent if there is linkage than if there is free assortment of the characters studied." The method is used as follows: Codominant genetic markers with several alleles are studied in a series of sib pairs (or other pairs of relatives) both of whom are affected with a disease whose linkage relationship to the marker is to be investigated. If there is *no* linkage to the marker, 25% of affected sib pairs share both maternal and paternal alleles of the

1. Disease and marker locus are not linked

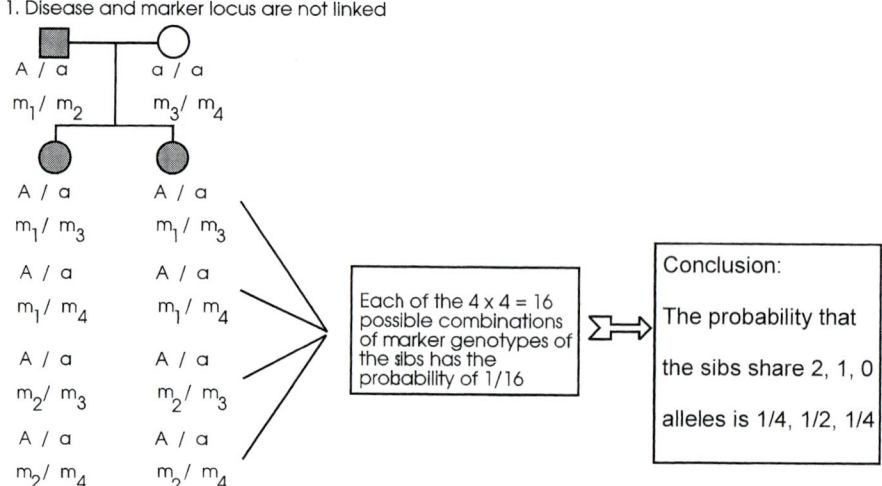

2. Disease and marker locus are linked, for example the following haplotypes:

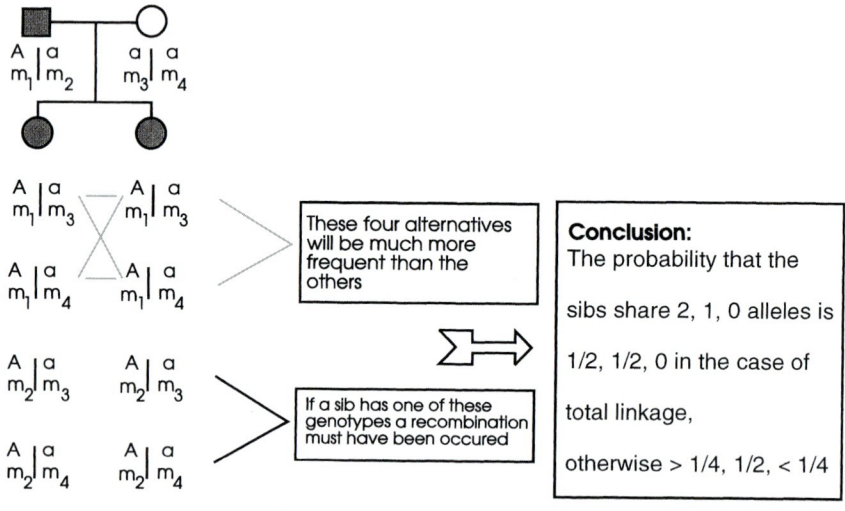

The argument holds true for any parental haplotypes

Fig. 6.4 Principle of the sib pair method for finding linkage. Assume a mating between a parent heterozygous for a dominant disease allele A and a parent homozygous for a normal allele a; each of the four parental chromosomes have different marker alleles at a marker locus. The possible alternatives are presented. Blue bars connect the more common sib pairs

marker, 50% share one of the marker alleles, and 25% differ for both marker alleles. If the marker is linked to a gene that contributes to causing the disease, the proportion of sib pairs with the disease who share one or two marker alleles is increased over the expected 25% or 50%, respectively.

The problem is straightforward when the marker status of both parents is known, and the identity by descent of the marker can be established. However, the method can also be used, albeit with more difficulty, if only marker information on the affected sib pairs is available, and the parents are not investigated (see also [13]). Problems arise here if the affected parent is homozygous for the marker, or if the unaffected parent contributes a marker allele to the child that is identical to the marker that cosegregates with the disease gene [67]. This problem can be avoided in either of two ways: (a) *Unaffected* sib pairs in the same families can be studied. Such pairs are expected to be more similar in the alternative marker alleles. However, the disease must have 100% penetrance so that there is certainty that an unaffected sib really does not carry the disease gene. (b) Another approach is to study haplotypes of several closely linked markers or of multiallelic variable-number tandem repeats (VNTRs) or microsatellites rather than of only a single restriction fragment length polymorphism (RFLP) marker. Under these conditions each parent often has a unique haplotype at the site under study with four haplotypes between the two parents. A child inherits only two haplotypes, one from each parent. With no linkage 25% of sib pairs share two haplotypes, 50% are identical for one, and 25% for none. With linkage, statistically significant increases over the 25% and 50% proportions of shared haplotypes are obtained.

This method has been adapted for pairs of relatives other than sibs as well; computer programs for testing linkage and estimating map distances are available (http://linkage.rockefeller.edu/soft). Risch and Merikangas have shown that the sib pair methods have relatively low power to detect loci with moderate odds ratios [85].

Haplotype analysis is especially useful if general circumstances favor the view that all patients suffer from a genetic disease which can be traced back to one single mutation. (See the discussion of such "founder effects" in Sect. 17.1.2) Here, the time since this mutation occurred may not have been sufficient to randomize marker loci around this mutation by repeated crossing over; it may still exist within the same haplotype in most instances. If the mode of inheritance is autosomal-recessive, patients may be homozygous for this haplotype. In this way, the autosomal-recessive gene for benign recurrent intrahepatic cholestasis (BRIC), an autosomal-recessive condition occurring in the Tyrolian Alps, was mapped to chromosome 18. If its main precondition – evidence for a founder effect – is met, the method is very efficient statistically [35], and the gene can be identified from a single pedigree with a modest number of siblings.

6.1.2.6 Results for Autosomal Linkage, Sex Difference, and Parental Age

The first autosomal linkage in man was found by Mohr [64] between the Lutheran blood groups and the AB0 secretor locus. Some years later linkage between the Rh loci and elliptocytosis (166900) was established and used to detect genetic heterogeneity of elliptocytosis, since not all families with elliptocytosis showed this linkage. A short time later, linkage between the AB0 blood group locus and the dominant nail-patella syndrome (161200) locus was found. This linkage established for the first time a sex difference of recombination probabilities in humans: map distance between these loci was 8 cM for males and 14 cM for females. A great many linkages have since been examined for sex differences; in the majority of instances a higher recombination fraction has been observed in females than in males. The same sex difference had been known for a while in the mouse [80]. It conforms to Haldane's rule [30] that crossing over is generally more frequent in the homogametic than in the heterogametic sex. In humans, however, this rule has exceptions. In the distal portion of 11p, for example, the recombination frequency appears to be higher in males [111]. It appears that such a higher male recombination rate may be characteristic for chromosome parts close to telomeres. There is also good evidence that the absolute exchange rate is higher in chromosome parts close to the telomere [52, 110]. However, the overall recombination rate is definitely higher in females [22, 45]. In *Drosophila*, there is no crossing over at all in the male. Typing of individual sperm shows that the recombination rate in a specific region can be different between individuals [14, 45]. The mechanism controlling this variation in humans has remained elusive. However, recently sequence variants in the 4p16.3 region were identified, which correlated with recombination rates. Interestingly these variants were mapped to the *RNF212* gene. This gene is a putative ortholog of the *ZHP-3* gene that is essential for recombinations and chiasma formation in *Caenorhabditis elegans*. An intriguing finding was that the haplotype formed by two single-nucleotide polymorphisms (SNPs) was associated with the highest recombination rate in males whereas the same haplotype was associated with a low recombination rate in females [46].

There has been considerable discussion as to whether recombination frequency is also influenced by parental age. In the mouse, the data are consistent with decreasing recombination rates with aging in females and increasing rates in males. Weitkamp [109] found a significantly increased incidence of recombinants with increasing birth order in humans for eight closely linked pairs of loci, indicating a parental age effect. There was no difference between males and females for this effect. A similar parental age effect was found for the Lutheran/secretor and Lutheran/myotonic dystrophy (160900) pairs but not for AB0/nail-patella or Rh/PGD pairs.

In a survey of cytogenetically determined chiasma frequencies from 204 males reported in the literature, little or no linear trend with age was found [53]. No cytogenetic data are available for females. The discrepancy between formal recombination data and chiasma frequencies is unexplained [110].

6.1.2.7 Information from Chromosome Morphology

Pairs or clusters of autosomal loci found to be linked (linkage groups) could not be assigned to specific chromosomes by a formal methodology of family study. The first chromosomal localization was accomplished as follows [21, 81].

The long arms of chromosome 1 frequently show a secondary constriction close to the centromere. In about 0.5 % of the population, this constriction appears much thinner and longer than normal. The variant is dominantly inherited. An uncoiler locus (Un-l) appears to be localized on chromosome 1. Linkage studies show close linkage between the blood group Duffy locus and the Un-l trait; $\Theta = 0.05$. Linkage between Duffy and congenital zonular cataract (116200) had been discovered earlier. Hence, a linkage group with three loci, cataract, Duffy, Un-l could be assigned [21].

Another possibility to localize genes on specific chromosomes was afforded by deletions. If a gene locus whose mutation has a dominant effect is lost by deletion, the absence of that gene may occasionally have a phenotype similar to the dominant mutation. More extensive symptoms may also be present, since more genetic material than a single gene would be expected to be lost. In 1963 a retarded child with bilateral retinoblastoma was found to have a deletion of the long arm of one D chromosome [56]. This chromosome was later identified as no. 13, and this 13q14 deletion has been found in a number of other cases with retinoblastoma and additional anomalies. Patients with retinoblastoma without additional symptoms usually have no deletion. The localization of this gene (RB1) has since been confirmed by DNA marker studies and the gene has been cloned [24, 55] (see Sect. 14.1.2).

Another approach, thought to be more generally useful, is the quantitative examination of enzyme activities in cases with chromosome anomalies.

Most enzymes show a clearcut gene dose effect in heterozygotes, i.e., heterozygotes for an enzyme deficiency have approx. 50% of enzyme activity. Therefore a similar gene dose effect might be expected when a gene locus is localized on a chromosome segment that has been lost by deletion.

The results of many early studies of this sort proved disappointing. Later, however, an increasing number of such gene dosage effects have been described in vitro, on trisomic and monosomic cells [48] (Sect. 3.6). To mention only one example, the activity of the enzyme phosphoribosylglycinamide synthetase was studied in several cases of partial monosomy and full and partial trisomy 21, as earlier studies had suggested a gene dosage effect for this enzyme. In regular trisomy 21 an excess was found with a ratio of trisomy 21 to normal of 1.55. A ratio of 0.99 was found in 21q21 → 21pter monosomy; 0.54 in 21q22 → 21qter monosomy; 0.88 in 21q21 → 21pter trisomy; and 1.46 in 21q22.1 trisomy. Therefore the phosphoribosylglycinamide synthetase gene locus could be localized in subband 21q22.1 [15]. Utilization of variants in chromosome morphology (heteromorphisms), such as the secondary constriction on chromosome 1 mentioned above, along with gene dosage studies, slowly opened the way to linkage and gene localization. Another method, using cell fusion, has led to much more rapid progress.

6.1.3 Linkage Analysis in Humans: Cell Hybridization and DNA Techniques

6.1.3.1 First Observations on Cell Fusion

The history of cell fusion is related by Harris [33]. Binucleate cells were observed in 1838 by J. Mueller in tumors, and afterwards by Robin in bone marrow, by Rokitansky in tuberculous tissue, and by Virchow both in normal tissues and in inflammatory and neoplastic lesions. The view that some of these cells were produced by fusion of mononucleate cells derived from the work of de Bary in 1859, who observed that the life cycle of certain myxomycetes involves the fusion of single cells to form multinucleated plasmodia. The earliest reports of multinucleated cells in lesions that can be identified with certainty as being of viral origin appear to be those of Luginbuehl (1873) and Weigert (1874), who described such cells at the periphery of smallpox pustules.

Following the introduction of tissue culture methods, numerous observations were made on cell fusion in cultures of animal tissue (see [32]). Enders and Peebles (1954) found that the measles virus induces cells in tissue culture to fuse to form multinucleated syncytia. Okada (1958) showed that animal tumor cells in suspension can be fused rapidly to form multinucleated giant cells using high concentrations of hemagglutinating parainfluenza virus (Sendai virus).

In 1960 Barski, identified cells generated by spontaneous fusion in a mixed culture of two different but related mouse tumor cell lines. These cells contained the chromosome complements of both parent cells within a single nucleus. This phenomenon was then examined by Ephrussi et al., who concluded that not only closely related mouse cells could be hybridized; even larger genetic differences did not exclude spontaneous cell fusion. However, it soon became obvious that the frequency of spontaneous cell fusion is very low, and that many cell types never fuse spontaneously. Fusion frequency must be increased in some manner. Furthermore, isolation of hybrid cells was possible only when culture conditions gave these cells a selective advantage.

Both problems were soon solved. Littlefield (1964) isolated the rare products of spontaneous fusion in mixed cultures by a technique adopted from microbial genetics. Fusion of two cells deficient in two different enzymes resulted in hybrids that recovered the complete enzyme set by complementation. Only these cells survived selection against the deficient cells.

Harris and Watkins [33] enhanced the fusion rate of various cells by treatment with UV-inactivated Sendai virus. Along with introduction of this method, they showed that fusion can be induced between cells from widely different species, and that the fused cells are viable. With this work, widespread use of the cell fusion method in various branches of cell biology began.

6.1.3.2 First Observation of Chromosome Loss in Human–Mouse Cell Hybrids and First Assignment of a Gene Locus

Weiss and Green [108] fused a stable, aneuploid mouse cell line, a subline of mouse L cells, with a diploid strain of human embryonic fibroblasts. The mouse cell line was deficient in the thymidine kinase (TK) locus and did not grow in hypoxanthine-aminopterin-thymidine (HAT) medium, a culture medium selective for cells containing the human TK locus (188300).

Cultures were initiated by mixing the two types of cells and growing them on standard medium. After 4 days cultures were placed in the selective HAT medium. This led to degeneration of the mouse cells, leaving a single layer of human cells. After 14–21 days hybrid colonies could be detected growing on the human cell monolayer. A number of these colonies were then isolated, grown for a longer time period, and examined. They turned out to maintain the mouse chromosome complement, but 75–95% of the human chromosomes were lost. One human chromosome, however, was present in almost all cells growing in the HAT medium. This suggested that the locus for thymidine kinase is localized on this chromosome. Therefore control experiments were carried out with a bromodeoxyuridine-containing medium. Bromodeoxyuridine, a base analogue for thymine, is accepted by TK in place of thymine and selects against cells containing this enzyme. A special chromosome described as "having a distinctive appearance" was present in almost all HAT cultures but in none of the bromodeoxyuridine cultures. It was concluded that the TK locus is indeed localized to this chromosome. Shortly thereafter the chromosome bearing the TK locus was found to be no. 17 [63] (Fig. 6.5).

This work led to two principles which were later decisive for the use of cell hybridization in linkage work:

1. Hybrids between mouse and human cells tend to lose many human chromosomes. It was later shown that this loss is random, and therefore examining a great number of hybrids one can expect to find a cell that has kept any one specified chromosome.
2. By using an appropriate selective system it is possible to select cells with a certain enzyme activity and to localize the gene loci specifying this enzyme to a specific chromosome.

Whereas genetics has historically been the science of genetic variability within a species, the hybridization method permits the localization of genes that do not show genetic variability in humans, provided only that the gene products of the human and nonhuman cells can be identified. One means of identification is the use of a selective system.

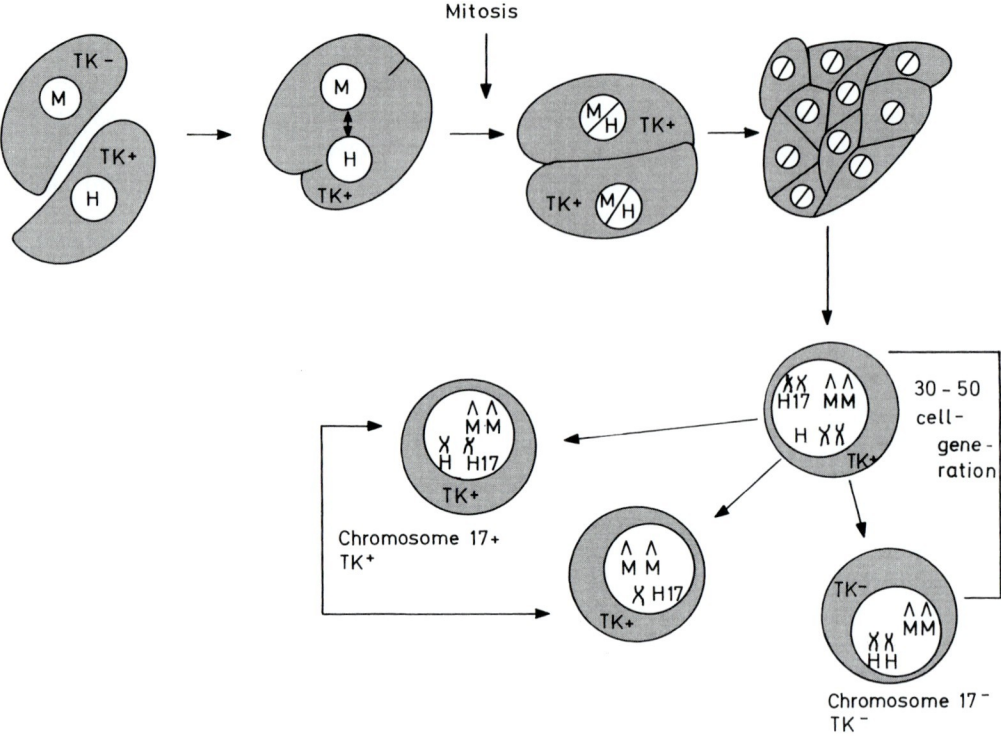

Fig. 6.5 The principle of gene localization on an autosome. Thymidine kinase deficient mouse cells (*M, TK⁻*) are grown in mixed cell culture with normal human cells (*H, TK⁺*). The cells are fused spontaneously, chemically, or by Sendai virus. After 30–50 cell generations the cells have lost part of their human chromosomes. Only cells having kept chromosome 17 show thymidine kinase activity (two cells at *left*). Cells without chromosome 17 show no *TK* activity (cell at *bottom right*)

Since 1967 selective systems have been developed for several enzymes. One example uses the hypoxanthine phosphoribosyltransferase locus on the X chromosome. This system can be used for selection not only of other X-linked loci but also of autosomal loci if a part of the autosome has undergone translocation with the X chromosome. It is also possible to assign loci for which no selective system exists, provided that enzymes produced by the two species have recognizable differences such as electrophoretic variation.

In hybrid cultures chromosome breakage and rearrangement are relatively frequent events. This chromosomal behavior made possible the suitable selection of hybrid clones containing identifiable parts of chromosomes, thereby combining the advantages of deletion mapping and hybridization.

Later the use of irradiation of the donor chromosome led to the development of radiation hybrids. These panels contain many chromosome fragments and were used to map genes and RFLP- and PCR-based markers more precisely [18, 54].

6.1.3.3 Other Sources of Information for Gene Localization

In situ hybridization technologies have also been instrumental for the localization of genes on chromosomes (see Sect. 3.4.4).

6.1.3.4 DNA Polymorphisms and Gene Assignment

The detection of a large number of DNA restriction site polymorphisms and other DNA markers opened up an additional approach to mapping of the human genome. In addition to RFLPs, other markers have been detected, among them the minisatellites [40], short DNA sequences distributed in great number over the human genome and occurring with variable numbers of repeats; this number is different in almost every individual. Therefore the information content for linkage studies is very high. Another such system, that of so-called microsatellites, consists of $(CA)_n$ repeats that occur in great numbers in the genome; the number of repeats per unit is extremely variable, as well. Localization of individual $(CA)_n$ probes in the genome has been achieved by identifying and using specific DNA sequences on both sides of these markers that allows amplification by the polymerase chain reaction (PCR). Genes for many important hereditary diseases could be localized on specific chromosomal sites using such markers.

Immortalized lymphoblastoid cell lines from large three-generation families with known genotypes for many marker loci are available for the study of new markers [111]. These pedigrees, the CEPH families, consist of samples collected in Salt Lake City, Utah, France, and Venezuela [19, 22]. The CEPH families (both parents and a great number of children) have been typed with many genetic markers. DNA from such families is available to the scientific community for further mapping. With increasing numbers of available markers, analysis of linkage relationships not only between two gene loci (e.g., one disease gene and one marker) but also between a disease gene and a set of markers – a haplotype – is routinely carried out. Such haplotypes are used in research studies as well as in genetic counseling and prenatal diagnosis. The proportion of kindreds in which the combination of genotypes is informative for linkage – and hence for prenatal diagnosis – can be enhanced appreciably by using such sets of markers.

6.1.3.5 Gene Symbols to Be Used

In cooperative scientific activities such as mapping of the human genome, certain terminological conventions are necessary. Human gene names are now assigned by the Human Genome Organization (HUGO) gene nomenclature committee. The rules for gene symbols include: only uppercase letters, no hyphens, no more than four or five letters or numbers, etc. For details see http://www.gene.ucl.ac.uk/nomenclature. Standardized nomenclature to describe mutations and genetic variants has also been established [3].

6.1.3.6 Linkage of X-Linked Gene Loci

Assignment of loci to the X chromosome is straightforward when the pedigrees show the typical pattern of X-linked inheritance. The X chromosome is the chromosome most completely saturated with disease loci. Well-known genes for human diseases have been localized. But even with these localizations many X-linked genes need to be mapped; for example, it is estimated that up to 260 genes causing syndromic and nonsydromic X-linked mental retardation (XLMR) may exist. The most common known cause of XLMR is an expansion of a trinucleotide repeat in *FMR1*. Large-scale systematic resequencing of the coding exons of the X chromosome in males with XLMR now allow the identification even of rare, disease-causing sequence variants [101].

6.1.3.7 Genetic and Physical Map of the Homologous Segment of X and Y Chromosomes

When genetic maps of a chromosome were determined, these were complemented by a physical map. The final goal was to identify the DNA sequence of all genes within this area. The following presents an introduction into these methods using the pseudoautosomal region of Xp and Yp as examples.

This region is located in the Giemsa light band Xp22.3 and in Yp11.32 (see Sect. 3.6.3.2). Various authors constructed partial physical maps; and the region comprises about 2,560 kb in its entire length [79]. There is a certain interindividual length variation. A number of families from the CEPH family pool [19] were studied with 11 exactly localized DNA markers; this established very high recombination frequencies in males and lower but clearly elevated frequencies in females. In males, the genetic map is 55 cM long and in females 8–9 cM. Hence the sex difference in this case is opposite to Haldane's rule. Moreover, it is about 20–25 times higher than the average of all chromosomes in the male and 6 times higher in the female. Differences in recombination rates between chromosome regions have also been demonstrated for other chromosomes.

The first step of physical mapping often involved restriction of this area with "rare cutter enzymes" which cut DNA into regions with many CpG islands [49]. As a rule, CpG islands indicate the presence of genes [87]. Closer scrutiny of this terminal area by chromosome jumping revealed at least five regions in which the CpG islands are concentrated. Further analysis, by, constructing of a contig of YAC clones, allowed the first-generation maps to be generated.

6.1.3.8 The Y Chromosome

It is the Y chromosome [105, 113] which determines male sex. Genetic analyses succeeded in localizing specific factors involved in sex determination to certain segments of this chromosome. As in many other instances, analysis of pathological conditions has contributed to understanding of the normal state, such as the study of men with two X but apparently no Y chromosome. As early as 1966, Ferguson-Smith [23] postulated XY translocations which were expected to transfer to the X a small – but for male development decisive – part of the Y chromosome. This expectation has been confirmed by many observations [113]. Since meiotic pairing of X and Y chromosomes occurs in the pseudoautosomal and in adjacent nonhomologous regions, and since pairing errors provide a plausible mechanism for such translocations, the search for the testis-determining factor (TDF) soon concentrated on the short arm. Here the *SRY* (sex reversal gene on Y) gene was finally identified [91]. The mechanism of *SRY* action has been elusive. Recent discoveries shed some light on the role of *SRY*, which is thought to act synergistically with SF1, a nuclear receptor, through an enhancer of *SOX9* to promote Sertoli cell differentiation. *SOX9* is probably the pivotal factor in regulating the gene activity that defines Sertoli cells. Both *SOX9* and SF1 synergize to activate transcription of several downstream genes [88].

The Y chromosome is now known to contain 89 protein-coding genes and at least 27 distinct proteins [92]; www.ensembl.org]. A region within the euchromatic segment of the long arm appears to be important for normal spermatogenesis, since deletions within this region lead to arrested sperm formation either in an early stage, i.e., not even functional spermatogonia are formed, or in postmeiotic stages [105]. The first discovered deletions were so large that they could be recognized by cytogenetic methods [102]. Small deletions were later identified by molecular techniques [105], and their recognition has become important for differential diagnosis of male infertility.

In addition to genes involved in testis development and spermatogenesis, the Y chromosome harbors genes encoding transcription factors, initiation factors, ribosomal proteins, and kinases. Comparisons of the Y chromosomes of human and chimp have provided unique insight into the evolution of Y-chromosome genes [36].

6.1.3.9 DNA Variants in Linkage

The HapMap project has set out to characterize the majority of the SNPs in the human genome and has led to the identification of over 3 million variants [37]. The large number of DNA polymorphisms provide many new markers, and most linkage work is now being carried out with DNA variants often on arrays with 100,000 to 1 million SNPs [59] (Sect. 4.4.2).

Linkage disequilibrium (i.e., failure to demonstrate free assortment; see Sect. 6.2) has frequently been found between the sites of various markers at a given locus. Since these sites are physically very close, crossovers between them are rare, and many generations must pass before linkage equilibrium is reached. Furthermore, current data suggest that recombination rates at closely linked markers may vary considerably between different chromosomal locations. Thus, both "hot" and "cold" spots of recombination appear to exist [26, 68, 69].

6.1.3.10 Practical Application of Results from Linkage Studies

In the past the main interest of linkage studies was theoretical. Practical applications, however, are frequently employed. If, for example, gene A causes a rare hereditary disease manifesting itself later in life, and B is a genetic marker closely linked to A and segregating in the same family, the disease was predicted in a prenatal sample or young individual, and this prediction used in genetic counseling (Sect. 25). Today genetic diagnosis is routinely performed by direct study of the mutant gene itself.

6.1.4 Biology and Statistics of Positional Cloning

For disease loci localized by linkage analysis identifying the gene and mutations involved is termed "positional cloning" [17]. This process involves identifying genes in the interval defined by linkage, and analyzing affected individuals to identify mutations. This process can proceed very rapidly if a gene of obvious biological interest is identified in the interval, and the mutations readily identified. However, the process can be very time consuming if the interval is large, or the disease gene small or poorly expressed, or if there are few affected individuals for mutation screening. The cystic fibrosis gene was localized to an interval now know to be 1.4 Mb in size [106, 112]. It took over 4 years of intense effort to positionally clone the cystic fibrosis gene [86]. With the completion of the sequence of the human genome, this process is greatly aided as most of the genes in an interval are known. Therefore an investigator can sequence all of the coding exons of the positional candidate genes to identify mutations.

To date, most of the genes responsible for the common Mendelian disorders have been identified. However, identification of genes in less common Mendelian disorders with the aforementioned tools will continue to be important [2]. Nowadays, positional cloning is more often applied to complex diseases. This process is similar – a genetic interval is defined via sib pair linkage or whole-genome association and the relevant gene and "disease-causing variant" need to be identified [20]. This process has resulted in identifying genes for such complex diseases as macular degeneration, diabetes, Parkinson disease, obesity, and others. However, many gene haplotypes associated with common disease do not have coding sequence variants, and regulatory effects are proposed.

6.2 Gene Loci Located Close to Each Other and Having Related Functions

6.2.1 Some Phenomena Observed in Experimental Genetics

6.2.1.1 Closely Linked Loci May Show a Cis-Trans Effect

When series of multiple alleles were analyzed in *Drosophila,* crossing over within these series was observed occasionally, indicating that what had been considered as one "gene" can be subdivided by genetic recombination. Such alleles were termed "pseudoalleles" by McClintock in [60]. In some a so-called *cis-trans* effect was shown. When two mutations were located side by side on the same chromosome (*cis* position), the animal was phenotypically normal, but when they were localized on homologous chromosomes (*trans* position), a phenotypic anomaly was seen [58].

6.2.1.2 Explanation in Terms of Molecular Biology

In fungi, bacteria, and phages, genetic recombination is normally observed within functional genes, i.e., DNA regions carrying information for one polypeptide chain. A *cis-trans* effect is now considered to be typical for two mutations that are not able to complement each other functionally, i.e., that are located within the same structural gene. Complementation between two mutations, by the same token, is regarded as an indication that these mutations are located in different functional genes. A gene has many mutational sites and may be subdivided by recombination. Complementation tests are often used to test genetic, biochemically characterized conditions for heterogeneity.

6.2.1.3 A Number of Genes May Be Closely Linked

Close linkage has frequently been described between mutations affecting closely related functions, which are perfectly able to complement each other functionally and show no *cis-trans* effect. In bacteria such as *E. coli,* gene loci for enzymes acting in one sequence have been found to be closely linked and arranged in the sequence of their metabolic pathway. Their activity is subject to a regulating mechanism by a common operator and promoter [44].

6.2.2 Some Observations in the Human Linkage Map

6.2.2.1 Types of Gene Clusters That Have Been Observed

The first impression when examining the human linkage map and DNA sequence is that while most loci are distributed fairly at random, there are a large number of clusters of closely related genes. Here are a few examples:

(a) The loci for human hemoglobins γ, δ, and β are closely linked.
(b) The immune globulin region comprises a number of loci responsible for synthesis of immunoglobulin

chains. The same is true for genes of the T cell receptor (chromosome 14q11). The major histocompatibility complex (MHC) cluster including various components of complement on chromosome 6.
(c) No less than four gene loci involved in the glycolytic pathway are located on chromosome 1.
(d) A number of genes determining closely related enzymes are closely linked, for example, pancreatic and salivary amylase on chromosome 1, and guanylate kinase 1 and 2 on the same chromosome.
(e) The protan and deutan loci for red–green color blindness are located in the same cluster on the X chromosome.

6.2.2.2 Clusters Not Observed So Far

As mentioned above, functionally related genes in bacteria are frequently closely linked; they are subject to common control within an operon. One might predict that, in humans, such operons would also occur, but functionally related genes are rarely clustered. Two genes linked in the same operon in bacteria are those for galactose-1-phosphate uridyltransferase and galactokinase. In humans these genes are located on chromosomes 3 and 17, respectively. Similarly, the gene for G6PD is located on the X chromosome, and that for 6-PGD, the following enzyme in the shunt pathway, is situated on chromosome 1. Genes belonging to one gene family are sometimes but by no means always located close together. For genes involved in the immune system, including those for immunoglobulin synthesis, T cell receptors, and genes in the MHC system, this location has functional significance.

6.2.3 Why Do Gene Clusters Exist?

6.2.3.1 They Are Traces of Evolutionary History

In some cases clustering is simply left over from the evolutionary history of these genes. Early in evolution there was one locus for a given gene. Then gene duplication occurred and offered the opportunity of functional diversification [70]. The first duplication paved the way for further duplications due to unequal crossing over (Sect. 6.2.8) and hence for further functional specialization.

With no further chromosomal rearrangements the gene clusters remain closely linked. It is unknown whether in these cases close linkage is necessary for orderly function. While it may be so in some cases, this explanation is not needed to explain clustering. Evolutionary explanations are sufficient. For example, the red and green color vision genes appear to have arisen by gene duplication.

6.2.3.2 Duplication and Clustering May Be Used for Improvement of Function

The clustering of genes is without obvious functional significance. It would be surprising, however, if evolution were never to take advantage of this situation, combining products of such gene clusters to form higher functional units. This may be the case for the hemoglobin molecule since in the β cluster the ε, γ, β, and δ genes are arranged in the sequence of their successive activation during individual development (Sect. 11.3). In the immunoglobulins and T cell receptors close linkage of a number of genes, possibly a great many, has become important functionally, as their gene products combine to form various classes of functional molecules. In fact, segmental duplications in the human genome are selectively enriched for genes involved in immunity. In this respect, one of the most fascinating, recent discoveries was the identification of interindividual and interpopulation differences in the copy number of a segmental duplication encompassing the gene encoding CCL3L1 (MIP-1alphaP). This gene is a potent human immunodeficiency virus-1 (HIV-1)-suppressive chemokine and ligand for the HIV coreceptor CCR5. Individuals with a CCL3L1 copy number lower than the population average have a markedly enhanced HIV/acquired immunodeficiency syndrome (AIDS) susceptibility [27].

6.2.4 Blood Groups: Rh Complex (111700), Linkage Disequilibrium

The history of the rhesus blood types provides a fascinating illustration of how science develops. First, a new phenomenon was discovered. Scientists soon realized that it eludes explanation by conventional concepts. Then a long-lasting scientific controversy arose as to the most appropriate extension of these concepts. In this controversy, a new explanatory principle was created that survived the controversy in this special case, and that could be applied to an increasing number of other observations. Finally, the problem was solved, and the controversy ended – by new methods.

6.2.4.1 History

In 1939 Levine and Stetson [57] discovered a novel antibody in the serum of a woman who had just delivered a macerated stillborn child and had received blood transfusions from her AB0-compatible husband. Of 101 type 0 bloods only 21 showed a negative reaction with this antibody. There was no association with AB0, MN, or P systems.

The following year Landsteiner and Wiener [51], immunizing rabbits with the blood of rhesus monkeys, obtained an immune serum that gave positive reactions with the erythrocytes of 39 of 45 individuals. Later the antibodies were compared with those of Levine and Stetson and thought to give reactions with the same antigens. This was subsequently found to be not quite true, and now the antigen uncovered by the true anti-rhesus antibody is called LW –, in honor of Landsteiner and Wiener. Rh typing in humans is always carried out with sera of human origin, i.e., according to Levine and Stetson's observation. The following discussion relates only to reactions with these human sera.

The great practical importance of the rhesus system became apparent when transfusion accidents were traced to this antibody, and especially when erythroblastosis fetalis, a common hemolytic disease of the newborn, was explained by Rh-induced incompatibilities between mother and fetus. The red blood cells of about 85% of all whites give positive reactions; family examinations showed that Rh-positive individuals are homozygous Rh/Rh or heterozygous Rh/rh, whereas the rh-negative individuals are homozygous rh/rh.

In 1941 Wiener discovered a different antibody that reacted with the cells of 70% of all individuals and was independent of the basic Rh factor (Rh', according to Wiener). A third related factor was discovered in 1943. These three factors were found in all possible combinations with one another, and the combinations were inherited together. Wiener proposed the hypothesis that these serological "factors" are properties of "agglutinogens," and that these agglutinogens are determined by one allele each of a series of multiple alleles. The agglutinogens were thus thought to determine the factors in different combinations. This descriptive hypothesis is so general that it indeed explains all the complexities discovered later.

6.2.4.2 Fisher's Hypothesis of Two Closely Linked Loci

R.A. Fisher developed a more specific hypothesis. At that time another antibody had been detected, anti-Hr. In 1943 Fisher (see [78]) examined a tabulation prepared by Race, containing the data accumulated so far. He recognized that Rh' and Hr were complementary. All humans have either Rh', Hr, or both. Individuals with both antigens never transmit them together to the same child, and a child always receives one of the two. Fisher explained these findings by proposing one pair of alleles for the two antigens. The pair was named C/c. In analogy, an additional pair of alleles D/d was postulated for the original antigens Rh⁺ and rh⁻, and a third pair of alleles for the third factor that had been discovered. To explain the genetic data close linkage between these three loci was assumed.

Fisher's hypothesis predicted discovery of the two missing complementary antigens d and e. This prediction was later fulfilled for e but not for d. In developing this hypothesis Fisher went one step further. In the British population, there are three classes of frequency of the Rh gene complexes (Figs. 6.6, 6.7). Fisher explained this finding by suggesting that the rare combinations could have originated from the more frequent ones by occasional crossing over. All four combinations of the less common class may have originated from occasional crossing over between the

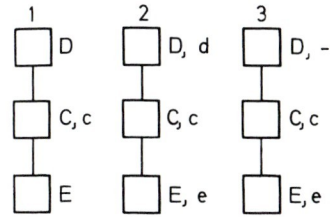

Fig. 6.6 A hypothetical structure of the Rh complex. *1*, On the basis of the evidence known in 1941; *2*, antigens predicted by Fisher and Race; *3*, antigens discovered; antigen *d* was not found

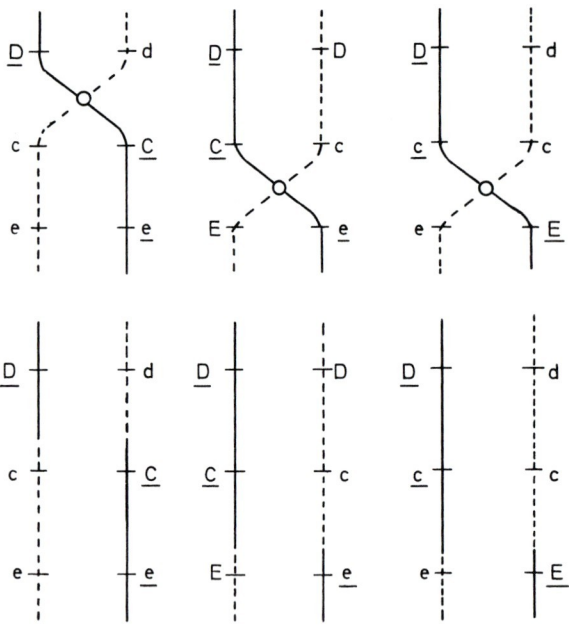

Fig. 6.7 Postulated production of three rare Rh haplotypes from the more common ones by crossing over. Each diagram refers to a different crossing over event. From Race and Sanger [78]

most frequent combinations; not, however, CdE. This complex needs inclusion of a second-order chromosome. Therefore the hypothesis explains why CdE is so rare. Still another prediction is possible. In every crossing over leading, on the one hand, to Cde, CDE, or cdE the complex cDe must also be produced. It follows that the frequency of the three former combinations together should equal the frequency of cDe. Frequencies actually found were: cDe, 0.0257; Cde + cdE + CDE, 0.0241 (in blacks, however, cDe has a high frequency).

Furthermore, Fisher believed the sequence of the three loci to be D-C-E, since cdE – which must have originated by crossing over between D and E from the genotype cDE/cde – is much more frequent in comparison with this genotype than Cde in relation to genotype CDe/cde (crossing over between C and D) and CDE in relation to CDe/cDE (crossing over between C and E).

6.2.4.3 Confirmation and Tentative Interpretation of the Sequential Order

Since Fisher's hypothesis many new observations have been made, the most important for the question of sequence being the combined antigens, for example, ce. These compound antigens were all compatible with the sequence D-C-E, whereas no such antigen suggesting close linkage between D/d and E/e has emerged. Fisher's hypothesis leads to two questions:

1. If the rare types have been formed by occasional crossing over from the more frequent ones, cases of crossing over should occasionally turn up in family studies. One such family has indeed been reported [97]: the father was CDe/cde, the mother cde/cde, four children were cde/cde, and three others CDe/cde, all in concordance with genetic theory. The sixth-born child, however, was Cde/cde. As the discrepancy involved father and child, it could be argued that the child was illegitimate. This, however, was made unlikely by blood and serum groups and by the fact that the family belonged to a religious sect with especially strict rules against adultery.

2. How should we envision the structure of the Rh gene(s) in the light of evidence from molecular genetics? There are two possibilities in principle:

    (a) The Rh complex is one gene with many mutational sites. Mutational changes are expressed as antigenic differences.
    (b) The Rh complex is composed of a number of closely linked genes, possibly three, and the main antigens reflect genetic variability at these genes. One important criterion is the *cis-trans* effect found in mutations affecting the same functional gene. As the ce compound antigen can be found in *cis*-phase CE/ce but not in *trans*-phase Ce/cE, Race and Sanger [77] tentatively concluded that C/c and E/e may be located within the same functional gene.

6.2.4.4 Molecular Basis

The Rh specificities have now been identified as membrane polypeptides. Molecular studies of the gene(s) have shown that in all D+ individuals, two closely related Rh genes in each haploid genome appear to be present. One of these genes is missing in D− individuals [16]. The authors concluded that one of the two genes controls the D polypeptide whereas the C/c and E/e specificities are coded by the second gene, a result confirmed by the molecular cloning [16, 107]. These observations explain at the molecular level why no anti-d serum has been found. They also confirm the sequence D-C-E postulated by Fisher, as well as the above-mentioned conclusion of Race and Sanger [77] that C/c and E/e appear to be located in the same gene product. Thus, both original hypotheses were partially correct: the specificities C,c,E,e are located within the same gene-determined protein, as postulated by Wiener (which does not exclude occasional intragenic crossing over); the D specificity, on the other hand, is located in a second, closely linked gene, as postulated by Fisher. Moreover, attempts at understanding genetics of the Rh system led to the development of a new concept by Fisher that has found widespread application in many fields of human genetics: linkage disequilibrium.

6.2.4.5 Blood Groups: Linkage Disequilibrium

Linkage normally does not lead to association between certain traits in the population (Sect. 6.1.1). Even if initially there is a nonrandom distribution of linkage phases, repeated crossing over randomizes the linkage groups, and in the end the coupling and repulsion phases for two linked loci are equally frequent. There is linkage equilibrium. However, when the population begins with a deviation from this equilibrium, for example, because two populations with different gene frequencies have merged, or because a new mutation has occurred on one chromosome, the time required to reach equilibrium depends on the closeness of linkage: the closer the linkage, the longer the time until equilibrium is reached [12]. It is never reached if certain types have a selective disadvantage.

A selective disadvantage for certain Rh complexes that could lead to their becoming less frequent has not been demonstrated so far; selection works against heterozygotes (Sect. 18.3.3), but this does not mean that a general disadvantage has never existed; neither has a conclusive explanation in terms of population history been postulated. Fisher's hypothesis, by answering some questions, has posed a number of others. However, the concept of linkage disequilibrium proved to be still more important in population genetics and in the genetic analysis of DNA polymorphisms (Sect. 2) and the major histocompatibility (MHC) complex:

6.2.5 Major Histocompatibility Complex [105, 111]

6.2.5.1 History

It had long been known that skin grafts from one individual to another (allotransplants) are usually rejected after a short time. In 1927 K. H. Bauer [5] observed that rejection does not occur when skin is transplanted from one monozygotic twin to the other (isotransplant). Such a transplant is accepted just as a transplant in the same individual (autotransplant). This showed the rejection reaction to be genetically determined. In the following years skin, and later kidney, transplantations between monozygotic twins were occasionally reported. Research on histocompatibility antigens in humans began only when leukocytes were shown to be useful as test cells.

Dausset observed in 1954 that some sera of polytransfused patients contain agglutinins against leukocytes. He later showed that sera from seven such patients agglutinated leukocytes from about 60 % of the French population, but not the leukocytes of the patients themselves. Twin and family investigations soon established that these isoantigens are genetically determined. Other isoantigens (now part of the HLA-B) were discovered by van Rood. Another important achievement was the microlymphocyte toxicity test introduced by Terasaki and McClelland in 1964, which is now the most frequently used method (Figs. 6.8 and 6.9). Subsequently the number of detected leukocyte antigens increased rapidly, and in 1965 it was suggested that most of these antigens were components of the same genetic system. At the histocompatibility workshop in 1967, 16 different teams typed identical samples from Italian families. Here the basic relationships among the different antigens were established. Finally, Kissmeyer-Nielsen [42] proposed the hypothesis of two closely linked loci (now A and B) controlling two series of alleles.

More recently, especially since the PCR technique became available, scientists study MHC genes directly at the DNA level. This has led to a splitting up of serologically defined gene loci, at both the class I and class II antigens (HLA D-DR; see Fig. 6.10) There are over 2,100 alleles described at the MHC locus with the HLA-B gene alone having 728 alleles (http://www.ebi.ac.uk/imgt/hla/stats.html), and several haplotypes have been completely sequenced [34].

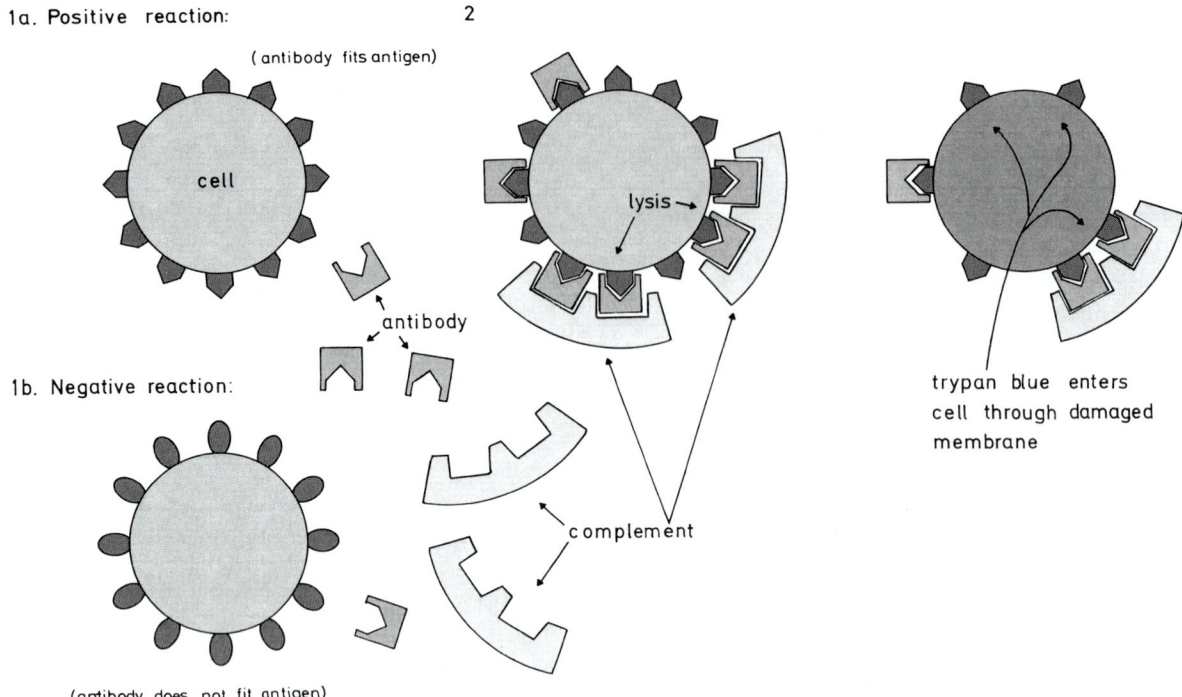

Fig. 6.8 Principle of the lymphocytotoxicity test: A cell having an appropriate antigen reacts with a specific antibody and complement. As a result, trypan blue enters the cell through the damaged membrane and indicates that the cell surface antigen has been recognized by a specific antibody

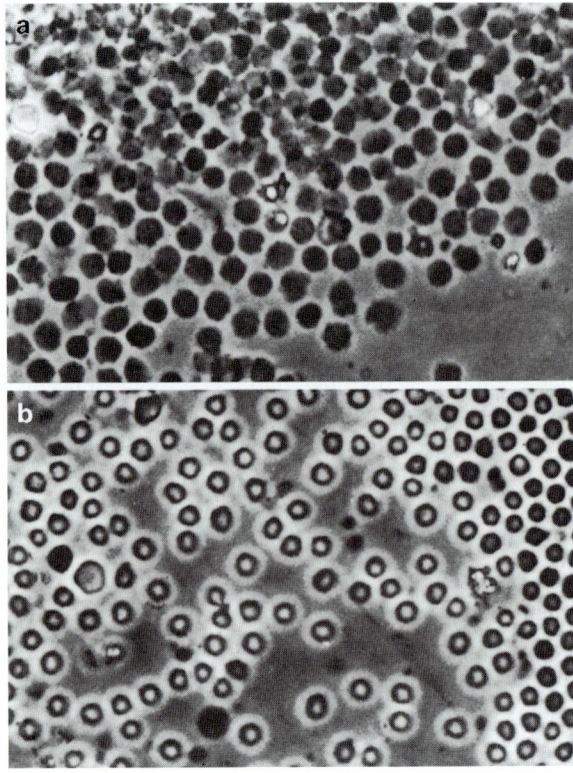

Fig. 6.9 (**a**, **b**) Lymphocytotoxicity test. (**a**) Positive reaction. (**b**) Negative reaction. Positive reaction is indicated by staining of the cells. (Courtesy of Dr. J. Greiner)

a Region of MHC class I gene loci (HLA-A, B, C, E, F, G)

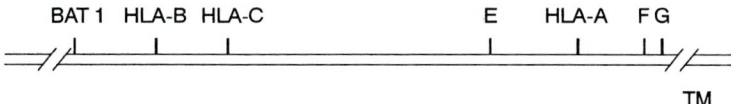

b Region of MHC class II gene loci (HLA-DR, DQ, DO, DN, DP)

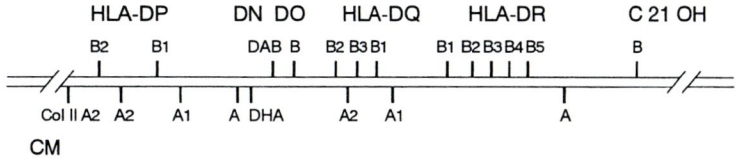

c Region of MHC class III gene loci

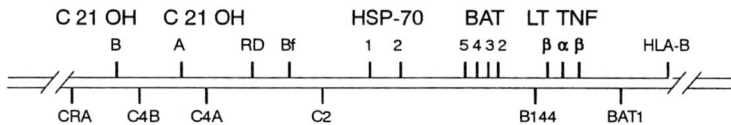

d Overview of the MHC genes

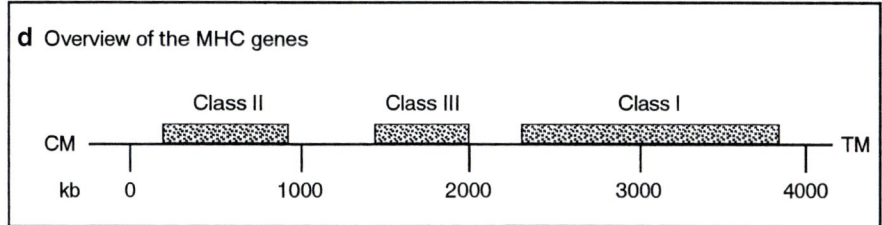

Fig. 6.10 (**a–d**) Sequence MHC class I, II, and III genes on the short arm of chromosome 6. (**a–c**) Detailed view of the three regions, together with their subregions. (**d**) Overview of approx. 4,000 kb. *TM,* Telomere; *CM,* centromere. *Letters, numbers, and their combinations,* the genes and their subregions. Genes of class III (HSP-70, DAT, C2/OH, CT) have no direct functional relationship with the immune response. From Albert [1]

6.2.5.2 Main Components of the MHC on Chromosome 6

The linkage group of the MHC is presented in Fig. 6.10. There are now three classes of MHC antigens. As revealed by studies using mainly molecular methods, each class can be subdivided into a great number of subclasses which are not described here (for details see [103]). Class I comprises the HLA-B, HLA-C, and HLA-A loci (in this order). In class II the HLA-D loci are found together with some other related, transcribed areas. Between these two classes a heterogeneous group of genes is located which have been named MHC class III despite the fact that at least some of these genes, such as those for 21-hydroxylase appear to have no functional relationship to the MHC system, and many non-HLA genes are located between the class I genes.

The function of this system has been elucidated; it plays an important role in the immune response. Here only some genetic aspects are considered.

The concept of four series of alleles is based on the following lines of evidence:

(a) No individual possesses more than two antigens from any of the series.
(b) Recombination between these series has been observed, for example, between the loci for A and B series, 40 crossovers among 4,614 meiotic divisions were described up to 1975, giving a combined ($♀+♂$) recombination frequency of $40/4614=0.0087=0.87$ cM. Ten A-B recombinants

informative for the C series have been reported. In eight of these the C antigen followed B, and in two it followed A. Therefore C is located between A and B, closer to B, a fact confirmed by molecular studies.

(c) When two antigens from the same series are present together in a parent, he or she always transmits one of them – never both or none – to the child. The segregation ratio is 0.5, corresponding to a simple codominant mode of inheritance.

(d) Hardy–Weinberg proportions have been demonstrated for each of the allele series separately in large population samples.

(e) Serological cross reactions occur almost exclusively within the series, not between them. This points to a close biochemical relationship of the antigens within a given series.

(f) Complete sequencing of several MHC haplotypes confirms the presence of four class I genes.

6.2.5.3 Complement Components

Complement consists of a series of at least ten different factors present in fresh serum. The factors are called C1, C2, C3, etc., and C1 is activated by antibodies that react to their corresponding antigens. Then C1 activates C4, this activates C2, and so on. The end result of this "complement cascade" is damage to the cell membrane carrying the antigen and often lysis of the cell. Moreover, activated complement components have a number of other biological properties, such as chemotaxis or histamine release. They are important immune mediators in the body's defense against microbial infection.

The complement system can be activated not only via C1 (the classic pathway) but also via C3 through an alternative pathway involving the "properdin factors." The factor B(BF) acts as "proactivator" for C3.

For some of the complement factors hereditary deficiencies have been described, and polymorphisms are known. BF, C2, C3, and C4, are polymorphic. The loci of C2 and C4 A and B are in class III, together with the properdin factor B with the main alleles BF^F and BF^S. The locus for C3, on the other hand, is located on chromosome 19. Several regulatory factors such as complement factors H and I (CFH, CFI) are also located on autosomes [12, 90].

6.2.5.4 Significance of HLA in Transplantation

One of the main motives for rapid development of our knowledge of HLA antigens has been the hope of improving the survival rate of transplanted organs, primarily kidneys. Indeed, kidneys from HLA-identical and AB0-compatible siblings have a survival rate in the recipient almost equaling that of monozygotic twins. The survival rate is worse in unrelated recipients even if HLA matching is as perfect as possible, and AB0 compatibility is secured. This shows that, apart from the major histocompatibility system – the HLA system – there must be other systems of importance for graft survival. This is not surprising. A great number of such systems are known in the mouse. These systems lead to host-versus-graft reactions in almost all transplantations (Fig. 6.11). These reactions can be managed by immunosuppressive therapy. The chances for survival, and the survival times, of transplanted kidneys have increased substantially. The same is true for transplantation of other organs, such as heart, liver, bone marrow, and pancreas.

Considering the high degree of polymorphism and the low gene frequencies of HLA alleles, successful matching of potential recipients with donor kidneys from others than sibs requires large-scale international organizations. Once kidneys – or other transplantable organs – become available due to the accidental death of an individual, a center is notified in which persons in need of such an organ are registered, together with their HLA status. The donor is typed, and the recipient whose HLA status best fits receives the organ.

6.2.5.5 HLA: Linkage Disequilibrium

One of the most conspicuous properties of the HLA system is that some HLA alleles tend to occur more frequently together than expected by chance. Table 6.1 shows some examples. The A1,B8 haplotype, for example, occurs about five times as often as expected.

Consider two alleles at two linked loci, with frequencies p_1 and p_2. With free recombination between them their combined frequency, i.e., the haplotype frequency h, should be $p_1 \times p_2$. If such a result is obtained, the two loci are said to be in linkage equilibrium. If the haplotype frequency h is higher than expected with free recombination, there is linkage disequilibrium (Δ, deviation from linkage equilibrium), which is often

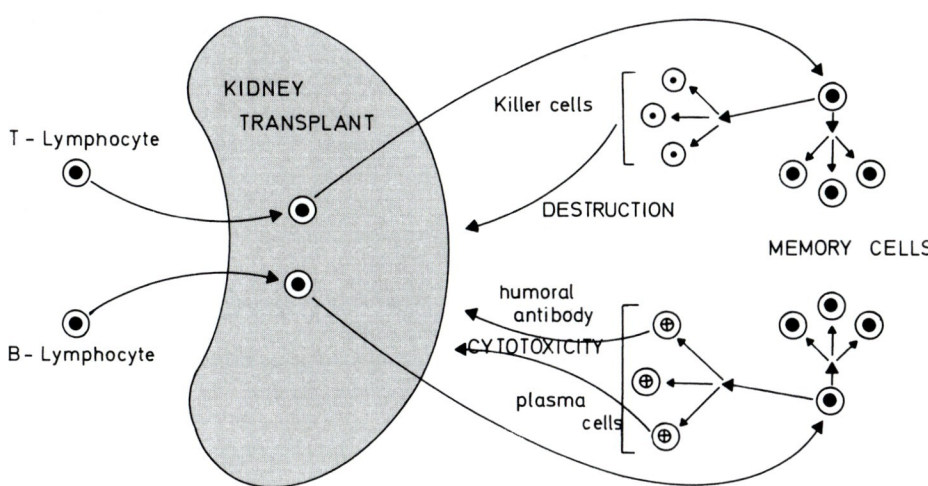

Fig. 6.11 Simplified diagram of the activation of the immune system by a kidney allotransplant. The transplant is recognized as foreign to the host organism by its T and B lymphocytes. This leads to activation of cellular and tumor immune response. From Svejgaard et al. [100]

Table 6.1 Linkage disequilibrium (gametic association; from Svejgaard et al. [100])

Haplotype			Frequency (%)	
A	B	D	Observed	Expected
A1	B8		9.8	2.1
A3	B7		5.4	2.1
	B8	Dw3	8.6	1.4
	B7	Dw2	3.9	1.8

The expected haplotype frequencies were calculated under the assumption of no association

symbolized as $D = h - p_1 p_2$. Haplotype and gene frequencies can be estimated from family and population data. In families the haplotypes of parents can in most cases be derived from those of their children (Table 6.2).

In the HLA system the deviations from linkage equilibrium are indeed striking. The situation is similar to that encountered with the Rh system (Sect. 6.2.4), but there is one important difference: in the Rh system only one case of recombination has been discovered, whereas many cases are known for the HLA system. Hence, genetic data point to much closer linkage in the Rh system than among the MHC genes. This conclusion has been corroborated by molecular studies in both systems (see above).

The observation of linkage disequilibrium – together with identification of immune response (Ir) genes in the mouse – initiated the investigations of HLA associations with diseases.

Table 6.2 Association of HLA-A1 and B8 in unrelated Danes (2×2 table; from Svejgaard et al. [100])

	Number of individuals		
	B8⁺	B8⁻	Total
A1⁺	376	235	611
A1⁻	91	1,265	1,356
Total	467	1,500	1,967

First antigen	Second antigen	+/+ a	+/− b	−/+ c	−/− d	Total n
A1	B8	376	235	91	1,265	1,967

where, for example, +/− means number of individuals possessing the first character (A1) and lacking the second (B8). The χ^2 is:

$$\chi_1^2 = \frac{(ad - bc)^2 N}{(a+b)(c+d)(a+c)(b+d)} = 699.4$$

corresponding to the correlation coefficient:

$$r = \sqrt{\chi^2 / n} = \sqrt{699.4 / 1967} = 0.60$$

Gene frequencies for A1 and B8 can be calculated by Bernstein's formula:

$$p = 1 - \sqrt{1 - \alpha}$$

(where a is the antigen frequency) as 0.170 and 0.127, respectively. The Δ value can be calculated by the formula

$$\Delta = \sqrt{\frac{d}{n}} - \sqrt{\frac{(b+d)(c+d)}{n^2}} = 0.077$$

Thus, the frequency of the HLA-A1, B8 haplotype is
$h_{A1,B8} = p_{A1} p_{B8} + \Delta_{A1,B8} = 0.170 \times 0.127 + 0.077 = 0.099$.

Linkage disequilibrium may have either of two main causes:

1. Two populations homozygous for different haplotypes mixed a relatively short time ago, and repeated

crossing over at a low rate has so far not been sufficient to lead to random distribution of alleles.
2. Certain combinations of alleles on closely linked gene loci caused a selective advantage for their bearers and have therefore been preserved.

To be able to decide between these two possibilities Bodmer [8] calculated how long a linkage disequilibrium would need to disappear in a random mating population.

For these calculations he used the work of Jennings [39], according to which Δ decreases to zero at a rate of $1-\Theta$ per generation, where Θ is the recombination fraction between the two loci. Between the HLA-A and HLA-B loci Θ was found to be of the order of magnitude of 0.008. Taking linkage disequilibrium between HLA-A1 and B8 as an example, Δ values of about 0.06–0.1 have been found in European populations. On the other hand, Δ values between 0.01–0.02 are not statistically significant with reasonable sample sizes. Therefore it is meaningful to examine how many generations are needed to reduce Δ from 0.1 by a factor of 5 to 0.02.

Using the above principle of Jennings, we obtain:

$$(1-\theta)^n (1-0.008)^n = 1/5; n \approx 200$$

This means that Δ would be reduced to an insignificant value within about 200 generations of random mating, i.e., 5,000 years, taking a generation as around 25 years.

This period is approximately the length of time since agriculture first came to parts of northern Europe and is certainly a very short time considering the evolutionary life span of the human species. The fact that such a significant Δ could be eroded in so short a time in the absence of selection suggests at least that this particular combination of HLA-A1, B8 is being maintained at its comparatively high frequency by some sort of interactive selection [50]. We consider it likely that selection will also be found to explain some of the other common cases of linkage disequilibrium and that the effect of recent population mixture will be shown to be of minor importance. Certain haplotypes seem to have a selective advantage that keeps them more frequent than others. This selective advantage, on the other hand, cannot be directly related to the diseases for which associations have been shown so far, as they are too rare. Besides, the onset of most of them is usually delayed until after the age of reproduction. Infectious diseases have probably been the most important selective forces for maintaining the MHC polymorphism as well as linkage disequilibrium. This topic is discussed in Sect. 16).

6.2.5.6 The Normal Function of the System

The HLA determinants are localized at the surface of the cell and are strong antigens. They exhibit the most pronounced polymorphism of expressed genes known so far in humans, with abundant linkage disequilibrium. Disease associations have been shown between HLA antigens and diseases for which an autoimmune mechanism had previously been suspected. Furthermore, similar systems are known in all other mammals examined so far (see [1, 43, 114]). Finally, there is close linkage with other loci concerned with the immune response. All this evidence together is very suggestive of a system that regulates the contact of cells with their environment. In recent years, this function has been elucidated in detail. These genes are important mediators of the immune reaction. Such cell recognition mechanisms may be important in embryonic development and differentiation, especially when they are present on only certain cell types. However, such hypothesis would not explain the selective advantage of the high degree of polymorphism in this system.

Another possible function is protection against viral or bacterial infection. Antigenic material of human origin may be incorporated in the outer membrane of the virus, which is thereby made less recognizable to another human host. However, if the virus contains MHC material from a genetically different individual, it is more readily inactivated by the immune system. Such a mechanism would also explain why the extreme polymorphism of the MHC system has a selective advantage. Further elucidation of the MHC will teach us a great deal about how the organism handles its interaction with the environment. This knowledge is important to our understanding of how natural selection has shaped our genetic constitution in the past, and how recent changes in our environment may influence it in the future.

To broaden the empirical basis for such understanding, however, it may be useful to ask whether there are other examples in nature of such gene clusters with related functions? Can their analysis provide us with hints for a better understanding of the MHC cluster? There is indeed one such example that has been analyzed very carefully – mimicry in butterflies. It cannot

6 Linkage Analysis for Monogenic Traits

be described here for lack of space, since it has no direct relationship with human genetics. But for the reader interested in more general, philosophical aspects of science, it is highly interesting showing how certain general principles may be used by nature in quite different contexts (see also earlier editions of this book).

6.2.6 Unequal Crossing Over

6.2.6.1 Discovery of Unequal Crossing Over

In the early years of work with *Drosophila* some authors observed that the bar mutation, an X-linked dominant character, occasionally reverts to normal, whereas in other cases homozygotes for the allele produce offspring with a new and more extreme allele, later called "double bar." Sturtevant [99] showed that this peculiar behavior is not due to mutations but to unequal crossing over, producing, on the one hand, a chromosome with two bar loci (double bar) and, on the other, a chromosome with no bar locus at all. When the giant salivary chromosomes of *Drosophila* permitted visual testing of genetic hypotheses, Bridges [11] showed that the simple, dominant bar mutation is caused by a duplication of some chromosomal bands. The reversion corresponds to the unduplicated state, whereas double bar is caused by a triplication of that band. Both reversion and triplication can be produced by a single event of unequal crossing over. Bridges did not yet formulate clearly the obvious reason for this event: the mispairing of "structure-homologous" but not "position homologous" chromosome sites (Fig. 6.12).

6.2.6.2 Unequal Crossing Over in Human Genetics

Haptoglobin [10], a transport protein for hemoglobin, is found in the blood serum and shows a polymorphism, the most common alleles being HP^{1F}, HP^{1S} and HP^2. Smithies et al. [94] discovered that the allele HP^2 is almost twice the length of each of the two alleles HP^{1F} and HP^{1S}, as evidenced by the composition of its polypeptide chain. In the HP^2 chain the amino acid sequence of the HP^1 chain is repeated almost completely. They concluded that the HP^2 allele might have been produced by gene duplication. Moreover, they predicted that unequal crossing over might again occur with a relatively high probability between HP^2 alleles, producing, on the one hand, an allele similar to HP^1

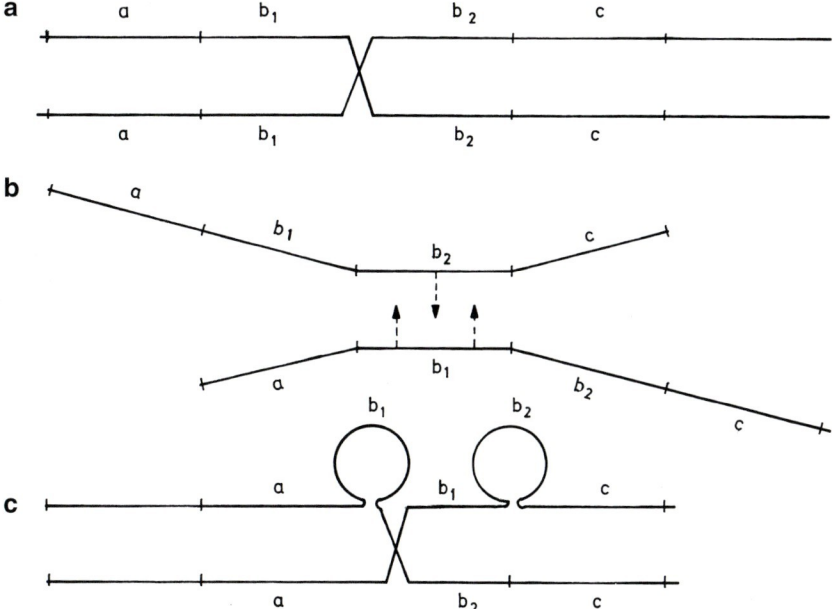

Fig. 6.12 (**a–c**) The principle of unequal crossing over. (**a**) Normal pairing and crossing over. The two genes b_1 and b_2 are assumed to have very similar DNA sequences. (**b**) Genes b_1 and b_2 are pairing. This leads to a shift of the two homologous chromosomes relative to each other. (**c**) Such pairing requires formation of two loops in the upper chromosome

and, on the other, an allele comprising the genetic information almost in triplicate. Repeated occurrence of this event might lead to still longer alleles and hence to a polymorphism of allele lengths in the population. Indeed, such alleles have occasionally been observed and are known as Johnson-type alleles [96].

There is an essential difference between the first unique event that produces the almost double-sized gene (for example, HP^2) from a single gene HP^1, and the unequal but homologous crossing over that becomes possible as soon as the first duplicated allele is present in the population [47].

6.2.6.3 First Event

Given a pair of homologous chromosomes, both partner chromosomes consist of largely identical sequences of nucleotides. Normally these partner chromosomes pair at meiosis, and there can be no unequal crossing over. To allow mispairing and thus unequal crossing over, an initial duplication is necessary. Mechanisms for such a duplication are known in cytogenetics, the simplest being two breaks at slightly different sites in adjacent homologous chromatids during meiosis and subsequent crosswise reunion. Another mechanism would be mispairing due to homology of short base sequences in nonhomologous positions. Our present knowledge of the structure of DNA sequences suggests ample opportunities for such a mispairing (slippage).

If the sites of breakage are separated only by the length of one structural gene, this event results in two gametes that do not contain this gene at all, together with two others containing it in duplicate (Fig. 6.13). The gametes containing a relatively large deletion have a high risk of not being transmitted because of lethality of the ensuing embryo. On the other hand, a gamete with the duplication is likely to develop into a diploid

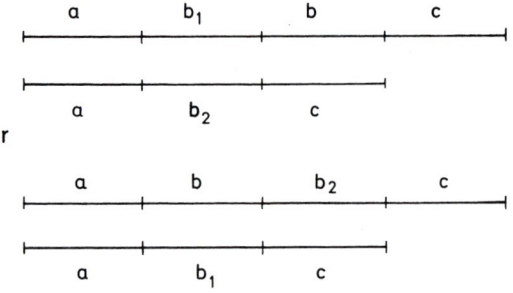

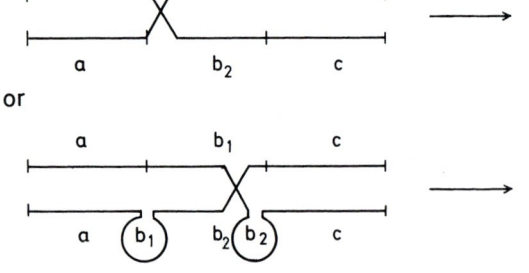

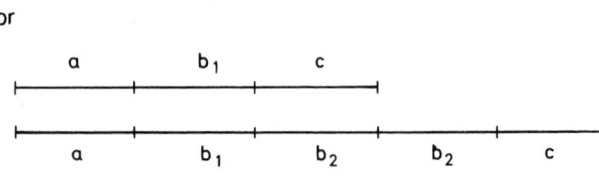

Fig. 6.13 (**a**, **b**) Unequal crossing over between structure-homologous but not position-homologous genes. (**a**) Unequal crossing over always leads to one crossover product with two genes b (b_1b or bb_2) and to another with only one gene (**b**). Formation of larger allele sequences becomes possible if the primary duplication is homozygous. In this case a chromosome with three alleles b (b_1 b_1 b_2 or b_1 b_2 b_2) may be formed. From Krüger and Vogel [47]

individual, providing for the first time a chance for mispairing of homologous sequences and therefore for unequal crossing over.

6.2.6.4 Consequences of Unequal Crossing Over

The consequences are seen in Fig. 6.13. As long as the duplication remains heterozygous, all gametes contain either one or two copies of the duplicated gene. When the duplication becomes homozygous, however, larger allele sequences may be formed. Unequal crossing over may lead, on the one hand, to gametes with only one copy and, on the other, to gametes containing three, and in subsequent generations, more than three copies (Figs. 6.13 and 6.14).

If the probability of unequal crossing over is not too low, high variability is soon found in the number of homologous chromosomal segments that resemble each other in structure but not in position. If selection favors a certain number of such chromosomal segments, which may be as small as a single gene, this number soon becomes the most common one. Selection relaxation leads to an increase in variability in both directions: the proportion of individuals with a very high number of such genes as well as those with a low gene number gradually increases [47]. Another genetic mechanism resembling unequal crossing over in some aspects is gene conversion where nonreciprocal products result.

Other examples besides the haptoglobin genes are the closely linked hemoglobin β- and δ-genes, the color vision pigment locus, and the natural killer cell receptor (KIR) genes [40]. Here the Lepore-type mutants, the X-linked color vision genes, and the diversity of KIR haplotypes are caused by unequal crossing over. Moreover, there are many examples for moderately or highly repetitive DNA sequences within which unequal crossing over should be possible. The presence of short repetitive DNA sequences such as minisatellites (Sect. 2.1.2) provides ample opportunities for pairing "slippage," leading to unequal crossing over. The high mutation rate within such areas (sometimes even a few percent per meiosis (Sect. 3.5) a well as the resulting huge interindividual variability show that this is not merely a theoretical speculation. Other repeated DNA sequences are those coding for the immunoglobulins. Increasing knowledge of the functional significance of repeated DNA sequences will bring a better understanding of the significance of unequal crossing over.

Recently there have been described large DNA segments that are duplicated in tandem or on other chromosomes. These "segmental duplications" can be up to 1 Mb is size or greater. Unequal crossing over between adjacent duplications is the basis for diGeorge/venocardiofacial disorder, Williams syndrome, and several other diseases. This class of diseases has been termed "genomic disorders" [89].

In fact, unequal crossing over, also referred to as "nonallelic homologous recombination," can give rise to numerous, recognizable microdeletion and microduplication syndromes and can significantly contribute to genome plasticity (see also Sect. 3.5.5).

6.2.6.5 Intrachromosomal Unequal Crossing Over

With structure-homologous but not position-homologous genes, such as those found in multigene families (Sect. 3.5.5), unequal crossing over becomes possible not only between homologous chromosomes but also between sister chromatids (intrachromosomal unequal crossing over). Theoretical considerations have shown that this process could have played a role in molecular evolution [41].

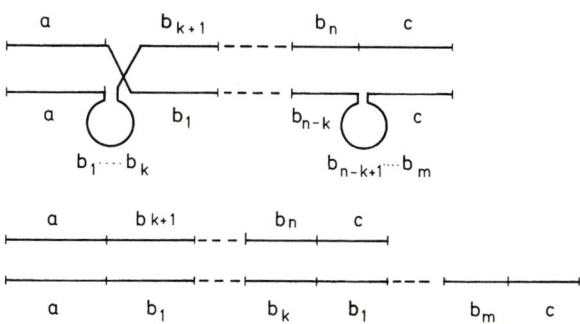

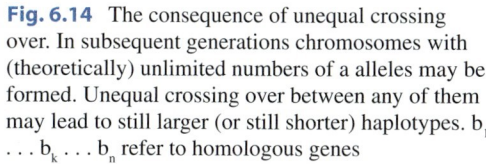

Fig. 6.14 The consequence of unequal crossing over. In subsequent generations chromosomes with (theoretically) unlimited numbers of a alleles may be formed. Unequal crossing over between any of them may lead to still larger (or still shorter) haplotypes. $b_1 \ldots b_k \ldots b_n$ refer to homologous genes

6.3 Conclusions

A few years after the rediscovery of Mendel's laws early in the twentieth century the first exception to Mendel's third law (independent segregation) was discovered: genes located sufficiently close to one other on the same chromosomes often segregate together – they are linked. The frequency of recombination increases with increasing distance between these genes. Genes on the same chromosomes but located far apart from each other, however, may even segregate independently if the distance between them is greater – these are syntenic, but not linked. A great number of genetic markers are available for localizing human genes, and statistical methods for assessing linkage in the human genome and determining the distance between gene loci have been developed. Methods from cell, biochemical, and molecular genetics have helped in localizing genes to specific chromosomes and chromosome segments and led to the molecular isolation of these genes. Such techniques make it possible to localize genes for both normal and abnormal traits and to define the nature of such genes by positional cloning. The identification of genes involved in susceptibilities to common diseases with complex causes by linkage studies remains a major challenge. However, the advent of a human genome sequence, comparative sequence, and a haplotype map, as well as large-scale association studies comparing normal subjects with patients, are leading to some progress in molecular understanding of complex disease.

While genes involved in the same biochemical pathways are seldom located close together, some clusters of closely linked genes exist that have related functions; the genes of the major histocompatibility complex, for example, have been analyzed particularly thoroughly.

References

1. Albert E (1993) Immungenetik. In: Gemsa R (ed) Lehrbuch der Immunologie. Thieme, Stuttgart
2. Antonarakis SE, Beckmann JS (2006) Mendelian disorders deserve more attention. Nat Rev Genet 7:277–282
3. Antonarakis SE (1998) Recommendations for a nomenclature system for human gene mutations. Nomenclature working group. Hum Mutat 11:1–3
4. Bateson W, Saunders PRG (1908) Confirmations and extensions of Mendel's principles in other animals and plants. Report to the Evolution Committee of the Royal Society, London
5. Bauer KH (1927) Homoiotransplantation von Epidermis bei eineiigen Zwillingen. Beitr Klin Chir 141:442–447
6. Bernstein F (1931) Zur Grundlegung der Chromosomentheorie der Vererbung beim Menschen. Z Indukt Abstamm Vererbungsl 57:113–138
7. Bishop MJ (1994) Guide to human genome computing. Academic, London
8. Bodmer WF (1972) Population genetics of the HL-A system: retrospect and prospect. In: Dausset J, Colombani J (eds) Histocompatibility testing. Munksgaard, Copenhagen, pp 611–617
9. Botstein D, White RL, Skolnick M, Davis RW (1980) Construction of a genetic linkage map in man using restriction fragment length polymorphisms. Am J Hum Genet 32:314–331
10. Bowman BH, Kurosky A (1982) Haptoglobin: the evolutionary product of duplication, unequal crossing over, and point mutation. Adv Hum Genet 12:189–261
11. Bridges CB (1936) The bar "gene", a duplication. Science 83:210
12. Briscoe D, Stephens JC, O'Brien SJ (1994) Linkage disequilibrium in admixed populations: applications in gene mapping. J Hered 85:59–63
13. Cantor RM, Rotter JI (1992) Analysis of genetic data: methods and interpretation. In: King A, Rotter JI, Motulsky AG (eds) The genetic basis of common diseases. Oxford University Press, New York, pp 49–70
14. Carrington M, Cullen M (2004) Justified chauvinism: advances in defining meiotic recombination through sperm typing. Trends Genet 20:196–205
15. Chadefaux B, Allord D, Rethoré MO, Raoul O, Poissonier M, Gilgenkrantz S, Cheruy C, Jérôme H (1984) Assignment of human phosphoribosylglycinamide synthetase locus to region 21 q 22.1. Hum Genet 66:190–192
16. Colin Y, Cherif-Zahar B, LeVankim C et al (1991) Genetic basis of the RhD-positive and the RhD-negative blood group polymorphism as determined by Southern analysis. Blood 78:2747–2752
17. Collins FS (1992) Positional cloning: Let's not call it reverse anymore. Nat Genet 1:3–6
18. Cox DR, Burmeister M, Price ER, Kim S, Myers RM (1990) Radiation hybrid mapping: a somatic cell genetic method for constructing high-resolution maps of mammalian chromosomes. Science 250:245–250
19. Dausset J, Cann H, Cohen D et al (1990) Centre d'études du polymorphisme humain (CEPH): collaborative genetic mapping of the human genome. Genomics 6:575–577
20. Dean M (2003) Approaches to identify genes for complex human diseases: lessons from mendelian disorders. Hum Mut 22:261–274
21. Donahue RP, Bias WB, Renwick JH, McKusick VA (1968) Probable assignment of the Duffy blood-group locus to chromosome 1 in man. Proc Natl Acad Sci USA 61:949
22. Donis-Keller H, Green P, Helms C et al (1987) A genetic linkage map of the human genome. Cell 51:319–337
23. Ferguson-Smith MA (1966) X-Y chromosomal interchange in the aetiology of true hermaphroditism and of XX Klinefelter's syndrome. Lancet 2:475–476
24. Friend SH, Bernards R, Rogelj S, Weinberg RA, Rapaport JM, Albert DM, Dryja TP (1986) A human DNA segment with properties of the gene that predisposes to retinoblastoma and osteosarcoma. Nature 323:643–646

25. Gabriel SB, Schaffner SF, Nguyen H, Moore JM, Roy J, Blumenstiel B, Higgins J, DeFelice M, Lochner A, Faggart M, Liu-Cordero SN, Rotimi C, Adeyemo A, Cooper R, Ward R, Lander ES, Daly MJ, Altshuler D (2002) The structure of haplotype blocks in the human genome. Science 296:2225–2229
26. Gerhard DS, Kidd KK, Kidd JR, Egeland JA (1984) Identification of a recent recombination event within the human b-globin gene cluster. Proc Natl Acad Sci USA 81:7875–7879
27. Gonzalez E, Kulkarni H, Bolivar H, Mangano A, Sanchez R, Catano G, Nibbs RJ, Freedman BI, Quinones MP, Bamshad MJ, Murthy KK, Rovin BH, Bradley W, Clark RA, Anderson SA, O'connell RJ, Agan BK, Ahuja SS, Bologna R, Sen L, Dolan MJ (2005) Ahuja SK (2005) The influence of CCL3L1 gene-containing segmental duplications on HIV-1/AIDS susceptibility. Science 307:1434–1440
28. Gusella JF, Wexler NS, Conneally PM, Naylor SL, Anderson MA, Tanzi RE, Watkins PC, Ottina C, Wallace MR, Sakaguchi AY, Young AB, Shoulson I, Bonilla E, Martin JB (1983) A polymorphic DNA marker genetically linked to Huntington's disease. Nature 306:234–238
29. Gusella JF, Tanzi RE, Anderson MA, Hobbs W, Gibbons K, Raschtchian R, Gilliam TC, Wallace MR, Wexler NS, Conneally PM (1984) DNA markers for nervous system diseases. Science 225:1320–1326
30. Haldane JBS (1922) Sex ratio and unisexual sterility in hybrid animals. J Genet 12:101–109
31. Haldane JBS, Smith CAB (1947) A new estimate of the linkage between the genes for colour blindness and haemophilia in man. Ann Eugen 14:10–31
32. Harris H (1970) Cell fusion. Clarendon, Oxford
33. Harris H, Watkins JF (1965) Hybrid cells from mouse and man: artificial heterokyryons of mammalian cells from different species. Nature 205:640
34. Horton R, Gibson R, Coggill P, Miretti M, Allcock RJ, Almeida J, Forbes S, Gilbert JG, Halls K, Harrow JL, Hart E, Howe K, Jackson DK, Palmer S, Roberts AN, Sims S, Stewart CA, Traherne JA, Trevanion S, Wilming L, Rogers J, de Jong PJ, Elliott JF, Sawcer S, Todd JA, Trowsdale J, Beck S (2008) [Variation analysis and gene annotation of eight MHC haplotypes: the MHC Haplotype Project. Immunogenetics 60:1–8
35. Houwen RHJ, Baharloo S, Blankenship K (1994) Genome screening by searching for shared segments: mapping a gene for benign recurrent intrahepatic cholestasis. Nat Genet 8:380–386
36. Hughes JF, Skaletsky H, Pyntikova T, Minx PJ, Graves T, Rozen S, Wilson RK, Page DC (2005) Conservation of Y-linked genes during human evolution revealed by comparative sequencing in chimpanzee. Nature 437:100–103
37. International HapMap Consortium, Frazer KA, Ballinger DG, Cox DR et al (2007) A second generation human haplotype map of over 3.1 million SNPs. Nature 449:851–861
38. Jeffreys AJ, Wilson V, Thein SL (1985) Hypervariable "minisatellite" regions in human DNA. Nature 314:67–73
39. Jennings S (1917) The numerical results of diverse systems of breeding, with respect to two pairs of characters, linked or independent, with special relation to the effects of linkage. Genetics 2:97–154
40. Khakoo SI, Carrington M (2006) KIR and disease: a model system or system of models? Immunol Rev 214:186–201
41. Kimura M (1983) The neutral theory of molecular evolution. Cambridge University Press, Cambridge
42. Kissmeyer-Nielsen F (ed) (1975) Histocompatibility testing 1975. Munksgaard, Copenhagen
43. Klein J (1986) Natural history of the major histocompatibility complex. John Wiley and Sons, New York
44. Knippers R, Philippsen P, Schafer KP, Fanning E (1990) Molekulare Genetik, 5 th edn. Thieme, Stuttgart
45. Kong A, Gudbjartsson DF, Sainz J, Jonsdottir GM, Gudjonsson SA, Richardsson B, Sigurdardottir S, Barnard J, Hallbeck B, Masson G, Shlien A, Palsson ST, Frigge ML, Thorgeirsson TE, Gulcher JR, Stefansson K (2002) A high-resolution recombination map of the human genome. Nat Genet 31:241–247
46. Kong A, Thorleifsson G, Stefansson H, Masson G, Helgason A, Gudbjartsson DF, Jonsdottir GM, Gudjonsson SA, Sverrisson S, Thorlacius T, Jonasdottir A, Hardarson GA, Palsson ST, Frigge ML, Gulcher JR, Thorsteinsdottir U, Stefansson K (2008) Sequence variants in the RNF212 gene associate with genome-wide recombination rate. Science 319:1398–1401
47. Krüger J, Vogel F (1975) Population genetics of unequal crossing over. J Mol Evol 4:201–247
48. Kurnit DM (1979) Down syndrome: gene dosage at the transcriptional level in skin fibroblasts. Proc Natl Acad Sci USA 76:2372–2375
49. Kutsenko AS, Gizatullin RZ, Al-Amin AN, Wang F, Kvasha SM, Podowski RM, Matushkin YG, Gyanchandani A, Muravenko OV, Levitsky VG, Kolchanov NA, Protopopov AI, Kashuba VI, Kisselev LL, Wasserman W, Wahlestedt C, Zabarovsky ER (2002) NotI flanking sequences: a tool for gene discovery and verification of the human genome. Nucleic Acids Res 30:316–3170
50. Lander ES et al (2001) Initial sequencing and analysis of the human genome. Nature 409:860–921
51. Landsteiner K, Wiener AS (1940) An agglutinable factor in human blood recognized by immune sera for rhesus blood. Proc Soc Exp Biol Med 43:223
52. Lange K, Boehnke M (1982) How many polymorphic genes will it take to span the human genome? Am J Hum Genet 34:842–845
53. Lange K, Page BM, Elston RC (1975) Age trends in human chiasma frequencies and recombination fractions. I. Chiasma frequencies. Am J Hum Genet 27:410–418
54. Lawrence S, Morton NE, Cox DR (1991) Radiation hybrid mapping. Proc Natl Acad Sci USA 88:7477–7480
55. Lee WH, Bookstein R, Hong F, Young LJ, Shew JY, Lee EY (1987) Human retinoblastoma susceptibility gene: cloning, identification, and sequence. Science 235:1394–1399
56. Lele KP, Penrose LS, Stallard HB (1963) Chromosome deletion in a case of retinoblastoma. Ann Hum Genet 27:171
57. Levine P, Stetson RE (1939) An unusual case of intragroup agglutination. JAMA 113:126–127
58. Lewis EB (1951) Pseudoallelism and gene evolution. Cold Spring Harbor Symp Quant Biol 16:159–174
59. Maraganore DM, de Andrade M, Lesnick TG, Strain KJ, Farrer MJ, Rocca WA, Pant PV, Frazer KA, Cox DR, Ballinger DG (2005) High-resolution whole-genome association study of Parkinson disease. Am J Hum Genet 77:68–693
60. McClintock B (1944) The relation of homozygous deficiencies to mutations and allelic series in maize. Genetics 29:478–502

61. McKusick VA (1995) Mendelian inheritance in man, 11 th edn. Johns Hopkins University Press, Baltimore
62. Migeon BR, Miller SC (1968) Human-mouse somatic cell hybrids with single human chromosome (group E): link with thymidine kinase activity. Science 162:1005–1006
63. Mittwoch U (1992) Sex determination and sex reversal: genotype, phenotype, dogma and semantics. Hum Genet 89:467–479
64. Mohr J (1954) A study of linkage in man. Munksgaard, Copenhagen Opera ex domo biologiae hereditariae humanae universitatis hafniensis 33
65. Morgan TH (1910) Sex-limited inheritance in drosophila. Science 32:120–122
66. Morton NE (1955) Sequential tests for the detection of linkage. Am J Hum Genet 7:277–318
67. Morton NE (1993) Genetic epidemiology. Annu Rev Genet 27:521–538
68. Murray JC, Mills KA, Demopulos CM, Hornung S, Motulsky AG (1984) Linkage disequilibrium and evolutionary relationships of DNA variants (RFLPs) at the serum albumin locus. Proc Natl Acad Sci USA 81:3486–3490
69. Myers S, Bottolo L, Freeman C, McVean G, Donnelly P (2005) A fine-scale map of recombination rates and hotspots across the human genome. Science 310:32–324
70. Ohno S (1999) Gene duplication and the uniqueness of vertebrate genomes circa 1970–1999. Semin Cell Dev Biol 10:517–522
71. Ohta T (1993) Diversifying selection, gene conversion, and random drift: interactive effects on polymorphism at MHC loci.
72. Ott J (1983) Linkage analysis and family classification under heterogeneity. Ann Hum Genet 47:311–320
73. Ott J (1990) Cutting a Gordian knot in the linkage analysis of complex human traits. Am J Hum Genet 46:219–221
74. Ott J (1991) Analysis of human genetics linkage. Johns Hopkins University Press, Baltimore
75. Page DC, Mosher R, Simpson EM et al (1987) The sex-determining region of the human Y chromosome encodes a finger protein. Cell 51:1091–1104
76. Penrose LS (1934/35) The detection of autosomal linkage in data which consist of pairs of brothers and sisters of unspecified parentage. Ann Eugen 6:133–138
77. Race RR, Sanger R (1969) Xg and sex chromosome abnormalities. Br Med Bull 25:99–103
78. Race RR, Sanger R (1975) Blood groups in man, 6 th edn. Blackwell, Oxford
79. Rappold GA (1993) The pseudoautosomal regions of the human sex chromosomes. Hum Genet 92:315–324
80. Reid DH, Parsons PH (1963) Sex of parents and variation of recombination with age in the mouse. Heredity 18:107
81. Renwick JH (1969) Progress in mapping human autosomes. Br Med Bull 25:65
82. Risch N (1990) Linkage strategies for genetically complex traits. I. Multilocus models. Am J Hum Genet 46:222–238
83. Risch N (1990) Linkage strategies for genetically complex traits. II. The power of affected relative pairs. Am J Hum Genet 46:229–241
84. Risch N (1990) Linkage strategies for genetically complex traits. III. The effect of marker polymorphism on analysis of affected relative pairs. Am J Hum Genet 46:242–253
85. Risch N, Merikangas K (1996) The future of genetic studies of complex human diseases. Science 273:1516–1517
86. Rommens JM, Iannuzzi MC, Kerem BS, Drumm ML, Melmer G, Dean M, Rozmahel R, Cole LJ, Kennedy D, Hidaka N, Zsiga M, Buchwald M, Riordan JR, Tsui LC, Collins FS (1989) Identification of the cystic fibrosis gene: Chromosome walking and jumping. Science 245:1059–1065
87. Sandelin A, Carninci P, Lenhard B, Ponjavic J, Hayashizaki Y, Hume DA (2007) Mammalian RNA polymerase II core promoters: insights from genome-wide studies. Nat Rev Genet 8:424–436
88. Sekido R, Lovell-Badge R (2009) Sex determination and SRY: down to a wink and a nudge? Trends Genet 25:19–29
89. Shaw CJ, Lupski JR (2004) Implications of human genome architecture for rearrangement-based disorders: the genomic basis of disease. Hum Mol Genet 13:R57–R64
90. Shiang R, Murray JC, Morton CC, Buetow KH, Wasmuth JJ, Olney AH, Sanger WG, Goldberger G (1989) Mapping of the human complement factor I gene to 4q25. Genomics 4:82–86
91. Sinclair AH, Berta P, Palmer MS et al (1990) A gene from the human sex-determining region encodes a protein with homology to a conserved DNA-binding motif. Nature 346:240–244
92. Skaletsky H, Kuroda-Kawaguchi T, Minx PJ, Cordum HS, Hillier L, Brown LG, Repping S, Pyntikova T, Ali J, Bieri T, Chinwalla A, Delehaunty A, Delehaunty K, Du H, Fewell J, Fulton L, Fulton R, Graves T, Hou SF, Latrielle P, Leonard S, Mardis E, Maupin R, McPherson J, Miner T, Nash W, Nguyen C, Ozersky P, Pepin K, Rock S, Rohlfing T, Scott K, Schultz B, Strong C, Tin-Wollam A, Yang SP, Waterston RH, Wilson RK, Rozen S, Page DC (2003) The male-specific region of the human Y chromosome is a mosaic of discrete sequence classes. Nature 423:825–837
93. Smith SM, Penrose LS, Smith CAB (1961) Mathematical tables for research workers in human genetics. Churchill, London
94. Smithies O, Connell GE, Dixon GH (1962) Chromosomal rearrangements and the evolution of haptoglobin genes. Nature 196:232
95. Snell GD, Dausset J, Nathenson S (1977) Histocompatibility. Academic, New York
96. Sørensen H, Dissing J (1975) Association between the $C3^F$ gene and atherosclerotic vacuolar diseases. Hum Hered 25:279–283
97. Steinberg AG (1965) Evidence for a mutation or crossing over at the Rh-locus. Vox Sang 10:721
98. Stern C (1973) Principles of human genetics, 3rd edn. Freeman, San Francisco
99. Sturtevant AH (1925) The effects of unequal crossing over at the bar locus in drosophila. Genetics 10:117
100. Svejgaard A, Hauge M, Jersild C, Platz P, Ryder LP, Staub Nielsen L, Thomsen M (1979) The HLA system. An introductory survey, 2nd edn. Karger, Basel
101. Tarpey PS, Smith R, Pleasance E, Whibley A, Edkins S, Hardy C, O'Meara S, Latimer C, Dicks E, Menzies A, Stephens P, Blow M, Greenman C, Xue Y, Tyler-Smith C, Thompson D, Gray K, Andrews J, Barthorpe S, Buck G, Cole J, Dunmore R, Jones D, Maddison M, Mironenko T, Turner R, Turrell K, Varian J, West S, Widaa S, Wray P, Teague J, Butler A, Jenkinson A, Jia M, Richardson D, Shepherd R, Wooster R, Tejada MI, Martinez F, Carvill G, Goliath R, de Brouwer AP, van Bokhoven H, Van Esch H, Chelly J,

Raynaud M, Ropers HH, Abidi FE, Srivastava AK, Cox J, Luo Y, Mallya U, Moon J, Parnau J, Mohammed S, Tolmie JL, Shoubridge C, Corbett M, Gardner A, Haan E, Rujirabanjerd S, Shaw M, Vandeleur L, Fullston T, Easton DF, Boyle J, Partington M, Hackett A, Field M, Skinner C, Stevenson RE, Bobrow M, Turner G, Schwartz CE, Gecz J, Raymond FL, Futreal PA, Stratton MR (2009) A systematic, large-scale resequencing screen of X-chromosome coding exons in mental retardation. Nat Genet 41:535–543
102. Tiepolo L, Zuffardi O (1976) Localization of factors controlling spermatogenesis in the nonfluorescent portion of the human Y chromosome. Hum Genet 34:119–124
103. Trowsdale J, Ragoussis J, Campbell JD (1991) Map of the human MHC. Immunol Today 12:429–467
104. Venter JC et al (2001) The sequence of the human genome. Science 291:1304–1351
105. Vogt P, Keil R, Kirsch S (1993) The "AZF" function of the human Y chromosome during spermatogenesis. In: Sumner AT, Chandley AC (eds) Chromosomes today, vol 2. Chapham and Hall, London, pp 227–239
106. Wainwright BJ, Scambler PJ, Schmidtke J, Watson EA, Law HY, Farrall M, Cooke HJ, Eiberg H, Williamson R (1985) Localization of cystic fibrosis locus to human chromosome 7cen-q22. Nature 318:384–385
107. Wagner FF, Flegel WA (2000) RHD gene deletion occurred in the Rhesus box. Blood 95:3662–3668
108. Weiss MC, Green H (1967) Human-mouse hybrid cell lines containing partial complements of human chromosomes and functioning human genes. Proc Natl Acad Sci USA 58:1104–1111
109. Weitkamp LR (1972) Human autosomal linkage groups. In: Proceedings of the 4th international congress of human genetics, Paris 1971. Excerpta Medica, Amsterdam, pp 445–460
110. White R, Lalouel J-M (1987) Investigation of genetic linkage in human families. Adv Hum Genet 16:121–228
111. White R, Leppert M, Bishop DT, Barker D, Berkowitz J, Brown C, Callahan P, Holm T, Jerominski L (1985) Construction of linkage maps with DNA markers for human chromosomes. Nature 313:101–105
112. White R, Woodward S, Leppert M, O'Connell P, Hoff M, Herbst J, Lalouel JM, Dean M, Vande Woude G (1985) A closely linked genetic marker for cystic fibrosis. Nature 318:382–384
113. Wolf U, Schempp W, Scherer G (1992) Molecular biology of the human Y chromosome. Rev Physiol Biochem Pharmacol 121:148–213
114. Yuhki N, Beck T, Stephens R, Neelam B, O'Brien SJ (2007) Comparative genomic structure of human, dog, and cat MHC: HLA DLA, and FLA. J Hered 98:390–399

Oligogenic Disease

Jon F. Robinson and Nicholas Katsanis

Abstract One of the primary goals of human and medical genetics is to assign predictive value to the genotype – that is to say, to use genetic information to assist in the diagnosis and management of disease. Recent work, originating primarily from disorders thought to be traditionally inherited in a Mendelian fashion, have blurred the boundaries between allele causality in monogenic and complex disease. Studies on genetic variation in disease are now revealing that essentially no disorder is transmitted solely in a Mendelian fashion; rather there are always multiple genetic and environmental factors that cause or modulate a disease phenotype. The focus of this chapter, *oligogenic disorders,* a term describing diseases caused by, or modulated by, a few genes, can provide a conceptual bridge between diseases classically considered monogenic and the poorly understood polygenic or complex disorders.

The inheritance of alleles generally follows Mendelian laws of segregation and independent assortment. However, this axiom does not necessarily hold true when the segregation of disease traits is considered. Mendelian inheritance is founded on the notion that a trait (not exclusively a disease phenotype) is transmitted through a single locus; however, even in the most classic monogenic disorders the 1:1 or 3:1 Mendelian ratio of dominant to recessive phenotypes, respectively, cannot explain the breadth of phenotypic variation found in a clinical setting. Although environment also plays a part, new research is showing that a large amount of the phenotypic variation in "Mendelian" disorders is due to genetic interaction of several genes (Nat Rev Genet 3:779–789, 2002). In that context, most, if not all, disorders should be considered multifactorial; and the main reason they are Mendelized is that the majority of the phenotype can be attributed to variation/mutations at a single locus.

J.F. Robinson
McKusick-Nathans Institute of Genetic Medicine,
Department of Molecular Biology and Genetics, Johns Hopkins
University School of Medicine, Baltimore, MD 21205, USA

N. Katsanis (✉)
Wilmer Eye Institute, Department of Molecular Biology
and Genetics, Johns Hopkins University School of Medicine,
Baltimore, MD 21205, USA
e-mail: katsanis@jhmi.edu

Contents

7.1 The Limitations of Mendelian Concepts 244
7.2 PKU and Hyperphenylalaninemia: Genetic Heterogeneity 245
7.3 Cystic Fibrosis: Genetic Modifiers 246
7.4 Lessons Learned from Established Oligogenic Disorders 246
    7.4.1 Bardet–Biedl Syndrome 247
    7.4.2 Determining Oligogenicity 247
    7.4.3 Cortisone Reductase Deficiency 248
    7.4.4 Hemochromatosis 248
    7.4.5 Hirschsprung Disease 249
7.5 Establishing Oligogenicity: Concepts and Methods 252
    7.5.1 Heritability 252
    7.5.2 Mouse Models of Oligogenic Inheritance: Familial Adenomatous Polyposis 252
    7.5.3 Multigenic Models 253
    7.5.4 Linkage Analysis 253
7.6 Molecular Mechanisms of Oligogenic Disorders 254
7.7 Modular or Systems Biology 257
7.8 Conclusions 258
References 259

One of the primary goals of human and medical genetics is to assign predictive value to the genotype – that is to say, to use genetic information to assist in the diagnosis and management of disease. This includes both Mendelian traits, where interpretation of the relationship between a pathogenic mutation and its phenotypic consequence has been considered more straightforward, and complex traits, where determining the contribution of alleles to the predisposition to disease is far more challenging.

Recent work, originating primarily from disorders thought to be traditionally inherited in a Mendelian fashion, have blurred the boundaries between allele causality in monogenic and complex disease. Studies on genetic variation in disease are now revealing that essentially no disorder is transmitted solely in a Mendelian fashion; rather there are always multiple genetic and environmental factors that cause or modulate a disease phenotype. The focus of this chapter, *oligogenic disorders*, a term describing diseases caused by, or modulated by, a few genes, can provide a conceptual bridge between diseases classically considered monogenic and the poorly understood polygenic or complex disorders.

The inheritance of alleles generally follows Mendelian laws of segregation and independent assortment. However, this axiom does not necessarily hold true when the segregation of disease traits is considered. Mendelian inheritance is founded on the notion that a trait (not exclusively a disease phenotype) is transmitted through a single locus; however, even in the most classic monogenic disorders the 1:1 or 3:1 Mendelian ratio of dominant to recessive phenotypes, respectively, cannot explain the breadth of phenotypic variation found in a clinical setting. Although environment also plays a part, new research is showing that a large amount of the phenotypic variation in "Mendelian" disorders is due to genetic interaction of several genes [5]. In that context, most, if not all, disorders should be considered multifactorial; and the main reason they are Mendelized is that the majority of the phenotype can be attributed to variation/mutations at a single locus.

7.1 The Limitations of Mendelian Concepts

The transition from monogenic to oligogenic models highlights some of the limitations of concepts generated originally to facilitate the identification of pathogenic mutations in Mendelian disease. That is not to say the concepts of monogenic disorders were invalid, especially since they have been instrumental to our progress towards solving a significant proportion of the mutational load in human genetic disease. Nonetheless, an unavoidable side effect is that they have introduced a bias towards monogenic disorders, which will now have to be first recognized and then ameliorated.

Consider a traditional positional cloning approach to identifying a disease locus and the underlying allele(s) under a monogenic model. The three major genetic tools at our disposal are: (a) inheritance data in families; (b) the predicted effect of a change on the target gene and its protein, often coupled to the evolutionarily predicted tolerance for such a genetic lesion; and (c) the prevalence of a candidate disease allele in the general population.

A simple example of linkage analysis highlights some of the biases introduced under an a priori expectation of monogenic disease transmission. In a family in which a trait segregates under an autosomal recessive model (based on both the vertical pattern of inheritance, i.e., transmission, and the horizontal pattern of inheritance, i.e., recurrence risk), affected individuals are expected to carry homozygous or *compound heterozygous* – two different mutant alleles, at the causal locus, whereas normal individuals either have one or no pathogenic mutations (Fig. 7.1). In oligogenic disorders, however, homozygous mutations at the locus of interest may be present in both normal and affected individuals, which, if sufficiently prevalent in the pedigree, might actually mask a statistically significant linkage peak.

Similarly, recognizing the detrimental effect of an allele to the gene and its product is also biased towards a Mendelian paradigm. The majority of mutations reported to date for monogenic disorders are more severe, such as deletions that remove parts of the transcript or cause frameshifts, nonsense mutations, splice site changes or missense codons of highly conserved amino acid sites of a protein. By contrast, in oligogenic disease one can expect that there are several genes contributing to the phenotype, with the average effect of each mutation being less severe in comparison. This in turn poses the risk of misinterpreting true pathogenic alleles as benign variants. Indeed, even in diseases well recognized as monogenic, there are examples of additional alleles within the same locus that can affect the penetrance of the phenotype, even if their predicted effect (from sequence analysis) is at best dubious. For example, a study on erythropoietic protoporphyria (OMIM: 177,000), an autosomal dominant disorder of incomplete penetrance, demonstrated how disease manifestation requires a deleterious mutation at one allele of the *FECH* gene and a common, low-expression polymorphism of the other [38].

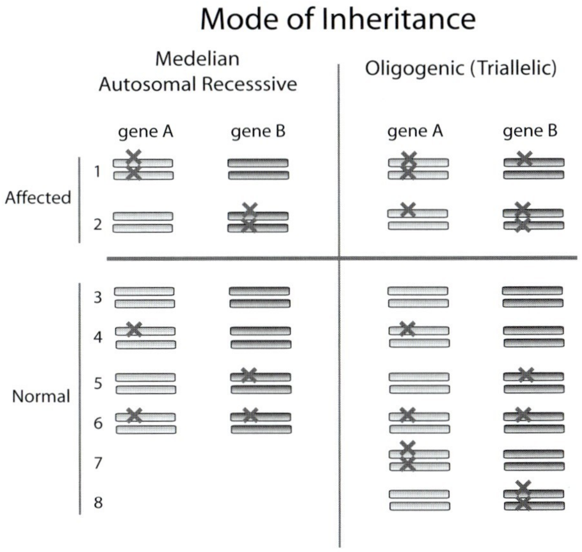

Because most deleterious mutations in monogenic disease have more severe effects on gene/protein function and cell physiology, they are often under negative selection (with a few notable exceptions, such as the hemoglobin sickle cell variant, common in sub-Saharan Africa, which protects against *Plasmodium falciparum* infection of malaria). As such, they are not expected to be in Hardy–Weinberg equilibrium in the population, which typically translates to their absence from a cohort of normal individuals. However, because mutations contributing to oligogenic traits are often expected to be milder and, in some instances, are not sufficient for pathogenesis, they can be found at equilibrium in the general population, which in turn has the potential to mask them as common polymorphisms of no pathogenic potential. This makes discerning these genetic lesions more difficult, because the more common a genetic variant is, the more difficult it becomes to justify its contribution to a particular phenotype. These issues are elaborated on later in the chapter, but overall, as we move from studying monogenic to oligogenic disease it becomes significantly more difficult to establish causal associative relationships between single alleles and the phenotype.

Several disorders have been highlighted to help illustrate how factors such as genetic heterogeneity, genetic modifiers, and oligogenicity can cause deviations from the Mendelian paradigm.

Fig. 7.1 Mendelian vs oligogenic (triallelic) inheritance. Under a Mendelian autosomal recessive model of inheritance, mutations in both alleles of a gene are required to manifest the disease phenotype. In triallelic inheritance, deleterious mutations of three alleles (two at one locus and one at another) are required for disease manifestation. In other forms of oligogenic inheritance single mutations in two different genes can genetically interact to manifest disease (not shown)

7.2 PKU and Hyperphenylalaninemia: Genetic Heterogeneity

Pronounced phenotypic variability in a disease commonly considered monogenic calls into question whether mutations at a single gene are sufficient to account for the variability of the disease phenotype, or whether the involvement of multiple genes should be suspected. This is exemplified by phenylketonuria (PKU; OMIM: 261,600), one of the first genetic disorders for which a biochemical defect was found before the advent of familial genetic analysis. In PKU, mutations in the hepatic (liver) enzyme phenylalanine hydroxylase (PAH) prevent the conversion of the essential amino acid phenylalanine to tyrosine [48]. Prenatal diagnosis of hyperphenylalaninemia in the 1960s provided the opportunity for early treatment; however, ~1% of patients did not respond well to the traditional therapy [5]. This suggested that other genetic factors could be involved, and in 1983 the mapping and cloning of the PAH gene revealed substantial *allelic heterogeneity* [94] – multiple alleles causing the same trait or disease. Allelic heterogeneity can result in variable phenotypes and severity of the disease depending on the severity of mutation in the locus. The discovery of other loci such as *biopterin-synthetase*, also leading to hyperphenylalaninemia (OMIM: 264,070) proved *locus heterogeneity* [16] – multiple loci contributing to a trait or disease – was another reason for the variable response to the traditional therapy. PKU and hyperphenylalaninemia are mostly inherited in a monogenic fashion; however, studies showing extensive phenotypic variability in patients with identical genotypes exemplify how other genetic or environmental factors modulate the phenotypic spectrum.

7.3 Cystic Fibrosis: Genetic Modifiers

Common genetic variation found within functional regions of the genome can often modify the outcome of even the most severe monogenic disorders. Cystic fibrosis (CF; OMIM: 602,421), an autosomal recessive disorder caused by deleterious mutations in the *CFTR* gene (cystic fibrosis transmembrane regulator), is often considered an excellent example of monogenic inheritance. Nearly all patients with classic forms of CF have mutations in the *CFTR* gene, with ~70% of those having the same pathogenic allele, ΔF508 (OMIM: 602,421) [52]. Although homozygous mutations in *CFTR* are almost completely penetrant, there is wide phenotypic variability, especially with regard to the severity of the pulmonary phenotype. Allelic heterogeneity can explain some of this variation, but there are several cases in which patients carrying the same alleles show different pulmonary phenotypes [21]. Environment, which is also a factor, does not explain all the variation, suggesting that there are contributing *genetic modifiers* – genes that modulate the severity of disease but are not necessarily causal. Recent association studies correlating the severity of a particular CFTR phenotype with alleles from candidate genes have found: low-expressing MBL (mannose-binding lectin, OMIM: 154,545), HLA (human leukocyte antigen; OMIM: 142,857) class II, TNFA (tumor necrosis factor-α; OMIM: 191,160) and TGFB1 (transforming growth factor-B1; OMIM: 190,180) all potentially modulate the pulmonary phenotype [5].

It is important to note that many of these modifiers are *polymorphisms*, i.e., genetic variants present in at least 1% of the population, highlighting the point that "polymorphism" and "neutral variant" are not synonymous. Therefore, the severity of a disease is largely dependent upon the presence or absence of certain common variants whose deleterious contribution is context dependent. The effects of these common variants, most of which are functional single nucleotide polymorphisms (SNPs) but can also include copy number variants (CNVs), Mb-sized deletions and duplications, largely go unnoticed in a healthy individual. However, under conditions of cellular or organismal stress that is established by mutations at a primary locus, such as in an individual with homozygous CFTR mutations, the effects of these polymorphisms become unmasked and can be protective or detrimental. For example, in patients with CF, certain *NOS1* (OMIM 163,731) polymorphisms that cause high NO production are associated with a slower decline in lung function [20]. It is thus important always to consider the effects of a disease-causing locus in the context of a person's entire genetic background.

7.4 Lessons Learned from Established Oligogenic Disorders

Even though *epistatic interactions* – genetic interactions between different genes in which the effect of one gene is modulated by one or more genes – between alleles at different loci are expected to be common contributors to the modulation of penetrance and expressivity, there remains a relative paucity of such examples in human genetic disease, especially when considering the fact that more than 1,000 human phenotypes have now been associated with a genetic lesion. This is probably reflective of both the bias towards Mendelizing disease phenotypes, as well as of the innate difficulty of establishing such phenomena. Nonetheless, the development of oligogenic disease models, coupled to higher sequencing capacity and improved computational and experimental protocols for investigating the potential effect of alleles on gene function, are unmasking a rapidly expanding list of oligogenic traits.

7.4.1 Bardet–Biedl Syndrome

Bardet–Biedl syndrome (BBS; OMIM 209,900) is a genetically heterogeneous disorder caused by recessive mutations in 1 or 2 of at least 12 different genes [51, 84]. It is also a clinically heterogeneous or *pleiotropic* disorder, with primary features including mental retardation, postaxial polydactyly, postnatal obesity, hypogenitalism, renal abnormalities, and progressive retinal dystrophy [10]. BBS also exhibits striking phenotypic variability both within and between families, which strongly suggested the presence of epistatic loci.

The oligogenic mode of inheritance now established for BBS illustrates the blurred boundaries between monogenic and multifactorial traits. In some families, homozygous or compound heterozygous recessive mutations at a single BBS locus seem to be sufficient to elicit the disease phenotype [68]. However, there is significant evidence showing that in some families there is a more complex digenic inheritance that requires at least three mutant alleles across two bona fide BBS genes (two at one locus and one at another) [51]. This is exemplified by the pedigree of a US family of European descent (Fig. 7.2), in which haplotype analysis predicted through identity-by-descent that the affected proband and unaffected sibling would carry two mutations in the *BBS2* locus [11]. Identity-by-descent refers to two alleles of the same locus deriving from a common ancestral chromosome and is usually proven by the presence of identical haplotype markers surrounding the gene of interest shared between the ancestor and descendant (i.e., parent and child). The affected sibling was also shown to have an additional mutation in another gene, *BBS6*, which was absent from the unaffected sibling, and the unaffected father was shown to be heterozygous for pathogenic mutations in both *BBS2* and *BBS6* genes (OMIM 209,900).

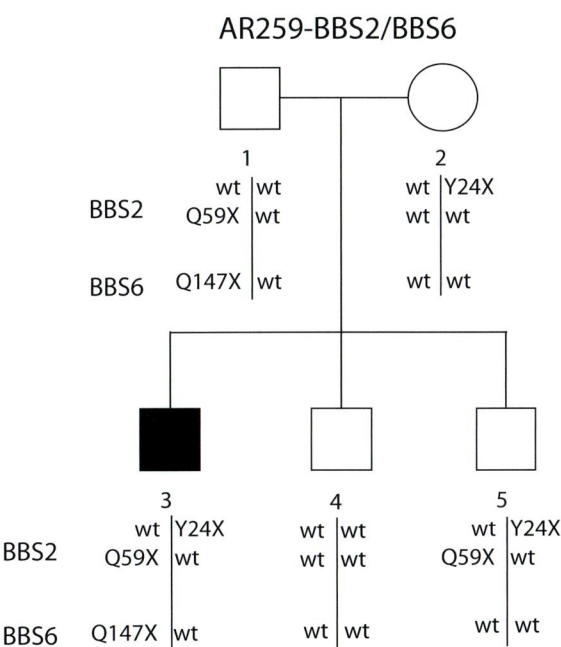

Fig. 7.2 BBS pedigree (triallelic) inheritance. In family AR259, both the affected proband (*3*) and an unaffected sib (*4*) contain compound heterozygous nonsense mutations in BBS2, suggesting other genetic factors were contributing to disease. Sequencing of BBS6 revealed a nonsense mutation in the affected proband that was inherited from the unaffected father but missing in the sibling with BBS2 mutations, suggesting that "triallelic" mutations are required for disease manifestation

This suggested that at least three mutations (two in *BBS2* and one in *BBS6*) were required for manifestation of the BBS phenotype. This type of mutant allele distribution has been observed in many BBS families; however, the genetic contribution of the third allele has been variable. In some families, such as that described in Fig. 7.2, the third BBS mutation can be best described as causal or a modifier of the penetrance of BBS. In other families, however, the third mutation likely acts as a genetic modifier, affecting the severity of the disease (i.e., exacerbating the clinical phenotypes).

7.4.2 Determining Oligogenicity

Recent studies on cohorts of patients with mutations in one of the two major BBS loci, *BBS1* or *BBS10* found approximately 13.3 and 18%, respectively, of patients had potentially three mutations (two at one locus and one at another) [12, 84]. Determining whether these alleles are pathogenic modifiers of penetrance, modifiers of disease severity, or simply benign missense variants can be done by systematically demonstrating three lines of evidence: (1) the presence of unaffected individuals carrying the same two mutations at the locus as affected individuals with three mutations; (2) the carrier frequency of the primary mutant allele is higher than would be expected under a Mendelian recessive model of inheritance. Modification of severity can simply be shown if (3) the allelic variant is correlated with more severe forms of disease. Evidence for oligogenicity in *BBS1* patients came from demonstration of the first two lines of evidence [12] even though a third mutation in another BBS gene had not been found. In a separate study on *BBS10*, the lack of unaffected individuals with homozygous or compound heterozygous *BBS10* mutations suggested that the third allelic variant at a second BBS locus (*BBS1*, *BBS4*, *BBS6*, or *BBS7*) were either severity modifiers or benign variants [84].

7.4.3 Cortisone Reductase Deficiency

Although BBS was one of the first disorders with documented oligogenic inheritance, a number of phenotypes have been described since for which similar genetic models apply. One interesting example is cortisone reductase deficiency (CRD; OMIM 604,931), a disorder characterized by excess of adrenocorticotropin-mediated androgen and phenotypes similar to polycystic ovary syndrome due to inability to convert cortisone to cortisol [29] 11B-HSD (OMIM 218,030) catalyzes interconversion of cortisone to cortisol, making it a natural candidate for harboring mutations that cause CRD. Interestingly, in one study, in patients affected with CRD all mutations were found to localize in intron 3 and none in the six exons of 11B-HSD [29]. Biochemical analysis, in the form of a luciferase reporter assay, showed that the mutant forms of 11B-HSD had 2.5 times lower transcriptional activity than the wild-type gene, suggesting that intron 3 acts as a transcriptional enhancer of 11B-HSD expression. Although 11B-HSD was clearly implicated in the disease, it could not fully explain the CRD phenotype, because 25% of unaffected controls were heterozygous and 3% were homozygous for the intron 3 mutations.

In BBS, the presence of additional genes, mutations in each of which caused the same phenotype, were the catalyst for the establishment of oligogenicity. In CRD, analysis of the biochemical pathway was the route that provided suitable functional candidates. One key observation focused on the requirement of NADPH for 11B-HSD oxo-reductase activity, which in turn led to the sequencing of hexose-6-phosphate dehydrogenase (H6PDH; OMIM 138,090), an enzyme of the pentose phosphate pathway known to generate NADPH [29]. All affected individuals were found to be either homozygous or heterozygous for variants in exon 5 of *H6PDH*, in addition to having mutations in the gene encoding 11B-HSD. Biochemical analysis subsequently demonstrated that this variant had less than 50% the H6PHD activity of wild type. Triallelic inheritance of CRD was then hypothesized because all affected individuals had homozygous mutations in one gene and heterozygous mutations in the other. In support of this model was the presence of unaffected individuals who were solely homozygous for the *H6PDH* variants, solely homozygous for *11B-HSD* mutations, or *doubly heterozygous* – heterozygous for mutations at two different loci – for both *H6PDH* and *11B-HSD* mutations [29]. The suggested model of pathogenicity is noninteracting, nonallelic noncomplementation (see Sect. 7.4.4).

7.4.4 Hemochromatosis

An emerging trend from the earlier examples is that in oligogenic disorders candidate genes for contributing both causal and modifying alleles are often found in the same pathway as the predominant disease locus. A clear example of this can be seen in hereditary hemochromatosis (HH; OMIM 235,200), a typically adult-onset, genetically heterogeneous disorder characterized by abnormal iron deposition in tissues due to increase in duodenal iron absorption and release of iron from macrophages [54]. HH was originally classified as a strictly monogenic autosomal recessive disorder, but recent studies have discovered an autosomal dominant inheritance as well as autosomal recessive oligogenic transmission [54]. A founder mutation in *HFE* (OMIM 235,200), C282Y is the predominant cause of HH as it is homozygous in 64–100% of patients [62] and its iron overload phenotypes have been recapitulated in mouse models [54]. Allelic heterogeneity has been demonstrated by genotype–phenotype correlation studies after the discovery of another common variant with a 2% frequency of homozygosity in European countries causing a histidine-to-aspartate (H63D) substitution in HFE [42]. C282Y/H63D compound heterozygotes represent a large majority, between 74 and 100%, of non-C282Y homozygous patients and are found to have milder iron overload phenotypes [62]. This effect of a common variant modulating the risk for a disease may be a recurring theme in oligogenic disease (see also the RET example in the next section).

Locus heterogeneity has been demonstrated by the presence of rarer, non-HFE-related forms of hemochromatosis due to mutations in genes that encode transferrin receptor 2 (TFR2; OMIM 604,720), ferroportin 1 (FPN1, OMIM: 604,653), hepicidin antimicrobial peptide (HAMP, OMIM: 606,464), and hemojuvelin (HJV, OMIM: 608,374) [42, 54]. Discovery of these other forms, some of which have been associated with early ages of onset and increased severity of disease has led to a classification of hemochromatosis into four different types [75]. However, similar to BBS, where each implicated gene is related by a common system or pathway (the primary cilium), all the genes implicated in hemochromatosis are suggested to modulate HAMP transcriptional response to iron [54]. Interestingly, many of these genes also interact genetically to cause or to modulate hemochromatosis; for example, recent studies have correlated younger age of onset and increased severity of disease in homozygous HFE C282Y patients, with additional heterozygous mutations in one or two of the genes involved in juvenile hemochromatosis, HJV and HAMP [47, 55]. In the oligogenic, or more specifically digenic inheritance, double heterozygous mutations in HFE and HAMP are required for disease manifestation [63]. For this reason, hemochromatosis exemplifies two key paradigms regarding identity and relationships between genetic modifiers and causative genes in oligogenic inheritance: (1) causative genes of a disease are often good candidates for genetic modifiers or contributors in digenic inheritance; (2) genes implicated in an oligogenic disorder are likely involved in the same or similar/converging pathway, suggesting that, not surprisingly, it is the total genetic load in any given physiological system that determines the extent and rate of progression of clinical manifestations.

7.4.5 Hirschsprung Disease

One difficulty when considering oligogenic traits (and modifying alleles in general) is that the epistatic loci can contribute alleles that are sometimes found at high frequency in the general population. This is not surprising, since if all modifying alleles were rare, then epistasis would be infrequent and the recurrence risk in oligogenic traits would potentially be much lower (see subsequent sections for a detailed discussion on recurrence risks). Nonetheless, the effect of common alleles can pose a major challenge and the genetic analysis tools are sometimes insufficient to establish a causal/modifying relationship between the allele and the phenotype.

A prominent example of how a common functional polymorphism can have a significant effect on the penetrance of an oligogenic disorder is seen in Hirschsprung disease (HSCR, OMIM: 142,623) [31]. HSCR is a congenital disorder in which the enteric ganglia are absent along variable regions of the intestine. There are two general modes of inheritance: (1) the most prevalent, a recessive multifactorial mode, characteristic in short segment HSCR is associated with mutations in PMX2B [13] (OMIM: 603,851), and linked to three loci on chromosomes 3p21, 10q11 (most likely RET), and 19q12 [33] in different individuals; (2) an autosomal dominant mode characteristic in long segment HSCR that usually results from severe mutations in the RET proto-oncogene [30] (OMIM: 164,761), or mutations in seven other genes, including EDNRB (OMIM: 131,244), ECE1 (OMIM: 600,423), EDN3 (OMIM: 131,243), GDNF (OMIM: 600,837), NRTN (OMIM: 602,018), SOX10 (OMIM: 602,229), and ZFH1B (OMIM: 605,802) [31]. Under 30% of patients have mutations in one of these eight genes, suggesting the existence of additional HSCR genes, or undiscovered mutations in the known genes, the latter highlighted by the observation that 91% L-HSCR [17] and 88% of S-HSCR [33] families show linkage to the RET locus, yet pathogenic coding RET alleles have been found in only ~50% [4]. Recently, a common sex-dependent variant in a RET enhancer was shown to have a significant contribution to HSCR risk, potentially accounting for 10–20 times the variation of susceptibility in all other known RET mutations [31]. This polymorphic variant, which was shown by luciferase reporter assays to reduce the RET enhancer activity six- to eight-fold, was originally missed in the RET mutational screens because it is located in a conserved noncoding region, RET intron 1, underscoring the importance of including conserved noncoding regions in mutational screens for oligogenic diseases. What is particularly informative in the context of oligogenic models is the high prevalence of this deleterious allele in the general population, with an allele frequency of 0.25 in Europe and 0.45 in Asia, postulated to result from an unknown selective advantage in heterozygotes [31]. This phenomenon is particularly relevant to both oligogenic and polygenic disorders, where many of the contributing genetic factors are in fact common polymorphisms instead of rare mutations. It is conceivable that deleterious polymorphisms are maintained at a high frequency in the population owing to a delicate balance between advantageous and detrimental gene–gene and gene–environment interactions. For example, certain alleles at one locus might not have a differential effect on a particular trait, but in a perturbed state resulting from exposures in the environment or deleterious alleles at other loci these alleles now confer different effects on the trait – a phenomenon referred to as cryptic genetic variation [35].

Overall, these four disorders illustrate how, within the same disease, one can find evidence of monogenic inheritance, effects of genetic modifiers, or oligogenic inheritance; in some ways, the distinction among these terms is somewhat arbitrary. The variability in the mode of inheritance among individuals is ultimately governed by interactions between an individual's genetic background and his environment. Hence, individuals respond to diet, drugs, exercise, climate, infections, and disease differently as a result of common genetic variation, which can take the form of single nucleotide changes, large-scale deletions, duplications, or inversions [78]. A comprehensive list of oligogenic disorders and their implicated genes can be found in Table 7.1.

Table 7.1 Oligogenic disorders

Syndrome or trait	OMIM number	Primary locus	Secondary locus	Effect	Reference
Alagille syndrome	118,450	*Jag1*	*Notch2*	Digenic	[5]
Alzheimer's disease	104,300	*APP*	*TGFB1*	Severity modifier	[5]
Antley–Bixler syndrome	207,410	*FGFR2*	*CYP21* (postulated)	Digenic/severity modifier	[73]
Autosomal dominant exudative vitreoretinopathy	133,780	*FZD4*	*LRP5*	Synergistic	[72]
Autosomal dominant nosyndromic deafness	601,842	*DFNA12*	*DFNA2*	Digenic/additive	[8]
Autosomal-dominant glaucoma	137,750	*MYOC*	*CYP1B1*	Onset/severity modifier	[5]
Bardet–Biedl syndrome	209,900	*BBS6/BBS2*	*BBS2/BBS6*	Digenic "triallelic"	[5]
	209,900	*BBS2/BBS4*	*BBS4/BBS2*	Digenic "triallelic"	[5]
Bartter syndrome type 4	602,522	*CLCNKA*	*CLCNKB*	Digenic	[77]
Becker muscular dystrophy	159,991	*DMD*	*MYF6*	Severity modifier	[5]
Breast and ovarian cancer	113,705	*BRCA1*	*HRAS1*	Penetrance/risk	[5]
Breast cancer	175,100	*BRCA*	*APC*	Modifier/risk	[5]
Congenital disorder of glycosylation type 1a	212,065	*PMM2*	*ALG6*	Severity modifier	[5]
Cortisone reductase deficiency	604,931	*HSD11B1/H6PD*	*H6PD/HSD11B1*	Digenic "triallelic"	[29]
Cystic fibrosis	603,855	*CFTR*	*CFM1*	Severity modifier	[5]
Cystinuria type A/B	220,100	*SLC7A9*	*SLC3A1*	Expressivity	[5]
Deafness: nonsyndromic recessive deafness (DFNB1)	22,090	*GJB2*	*GJB6*	Digenic	[25]
Emery–Dreifuss muscular dystrophy	181,350	*LMNA*	*DESMIN*	Digenic/synergistic	[67]
Epilepsy with febrile seizures plus	604,233	*SCN2A*	*postulated*	Digenic/severity modifier	[45]
Familial adenomatous polyposis	175,100	*APC*	*Mom1, Cox2, cPLA2*	Protective	[5]
Familial amyotrophic lateral sclerosis	147,450	*SOD1*	*CNTF*	Severity modifier	[5]
Familial hypercholesterolemia	143,890	*ARH*	*13q22-q32*	Digenic	[5]
Familial Mediterranean fever	249,100	*MEFV*	*SAA1*	Pleiotropy	[5]
Familial porphyria cutanea tarda	176,100	*UROD*	*HFE*	Digenic	[61]

(continued)

Table 7.1 (continued)

Syndrome or trait	OMIM number	Primary locus	Secondary locus	Effect	Reference
Familial thrombophilia	188,050	Protein C	Factor V	Digenic	[34]
Finnish congenital nephrosis	256,300	NPHS1	NPHS2	Triallelic, severity modifier	[5]
Hemochromatosis	602,390	HFE	HAMP	Digenic	[63]
Hereditary deafness	108,733	PMCA2	CDH23	Digenic/severity modifier	[32]
Hereditary deafness	108,733	CDH23	PCDH1 5	Digenic	[96]
Hirschsprung disease	142,623	RET	NRTN	Digenic, severity modifier	[28]
Holoprosencephaly	142,623	RET	GDNF	Digenic	[5]
Hypertension	142623	RET	3p21; 19p12?	Penetrance/risk	[5]
	236,100	SHH	TGIF	Digenic	[69]
	145,500	ATP1A1	NKCC2	Digenic/synergistic	[36]
Idiopathic hypogonadotropic hypogonadism	146,110	GNRHR	FGFR1	Triallelic/severity modifier	[71]
Junctional epidermolysis bullosa	226,700	FGFR1	NELF	Digenic/severity modifier	[5]
Kallman syndrome	308,700	COL17A1	LAMB3	Triallelic/severity modifier	[27]
		KAL1	PROKR2	Digenic	
Late-onset Fuchs corneal dystrophy	610158	FCD1	postulated	Digenic/onset modifier	[88]
Long QT syndrome	152427	KCNQ1	KCNH2	Digenic	[14]
Maternally inherited deafness	152427	KCNH2	SCN5A	Digenic	[64]
Melanoma	152427	KCNE2	SCN5A	Digenic	[64]
	561000	12 S Ribosomal	D8S277	Penetrance	[5]
	600160	CDKN2A	MC1R	Penetrance/risk	[5]
Midventricular hypertrophic cardiomyopathy	192600	MYLK2	MYH7	Digenic/additive	[24]
Myopathy	610099	8p22-q11	12q13-q22	Digenic	[43]
Osteoporosis	166710	ERS2	NRIP1	Digenic/nonsynergistic	[65]
Parkinson disease	168600	ERS2	ESR1	Digenic/nonsynergistic	[22]
Pheochromocytoma	171300	PRKN	LRRK2	Digenic	[89]
		PINK1	DJ-1	Digenic/synergistic	
Progressive external ophthalmoplegia	157640	2cen	16p13	Digenic	[23]
		POLG	Twinkle	Digenic	[92]
Pseudohypoaldosteronism	264350	MR	ENaC	Digenic	[2]
Refractory auto-inflammatory syndrome	142680	TNFRSF1A	CIAS1	Digenic/severity modifier	[91]
Retinitis pigmentosa	180721	RDS	ROM1	Digenic	[5]
Rett syndrome	312750	MECP2	postulated	Digenic/severity modifier	[74]
Spina bifida occulta (mouse)	182940	PAX1	PDGFRA	Digenic	[44]
Spinal muscular dystrophy	603011	PAX1	E2A	Digenic	[49]
Split hand/foot malformation with long bone deficiency	119100	SMN1	H4F5	Candidate severity modifier	[5]
		SHFLD1 (1q42.2-q43)	SHFLD2 (6q14.2)	Digenic	[70]
Type I von Willebrand disease	601628	VWF	ABO blood group	Penetrance	[5]
Type II diabetes mellitus	125853	VWF	Galgt2	Modifier	[5]
Type II diabetes mellitus	125853	PPARG	PPP1R3A	Digenic/additive	[76]
		TCF1	SHP	Digenic/severity modifier	[90]
Usher syndrome type III	276092	INSR (mouse)	IRS1 (mouse)	Digenic	[19]
		USH3	MYO7A	Severity modifier	[1]
Van der Woude syndrome	604547	VWS	17p11.2	Penetrance	[5]
Waardenburg syndrome type II and ocular albinism	103470	MITF	TYR	Digenic	[5]

7.5 Establishing Oligogenicity: Concepts and Methods

As discussed earlier in this chapter, there are biases in molecular tools and methodology against discovery of genetic factors involved in oligogenic diseases. These biases, however, can be overcome by adopting a methodology that does not ignore mutations with subtle effects, conserved noncoding sequences, variants in Hardy–Weinberg equilibrium, and linkage tests that consider multiple genes in the analysis. There are primarily four lines of evidence that suggest oligogenicity in a disease: (1) poor phenotype-genotype correlations, (2) phenotypic differences in animal models that are dependent on genetic background, (3) identification of a disease trait that is not transmitted in a Mendelian fashion, and (4) establishment of linkage within the same family to more than one locus.

7.5.1 Heritability

The classic Mendelian model used to discover genetic diseases is anchored in the assumption that the traits or clinical features of a disease are transmitted by a molecular defect at a single locus. Inaccuracy of this model is thus evident when there is large phenotypic variability between individuals that have identical alleles at the implicated locus. One example is two individualswith homozygous ΔF508 mutations but drastically different clinical outcomes of CF. To establish oligogenicity, it is necessary to determine whether some of the phenotypic variation is in fact due to genetic variation and not simply environment, or in other words determine the *heritability* – the proportion of phenotypic variation that is due to genetic variation. If one were to simply consider the phenotypic variance (var[P]) as the sum of the genetic variance (var[G]) and environmental variance (var[E]) then heritability could be estimated as: $H^2 = \text{var}[G]/\text{var}[P]$ [86]. Heritability is difficult to measure, but the most common method of measurement is to use twin studies in which the phenotypic variability of monozygotic twins (MZ; genetically identical) and dizigotic twins (DZ; ~50% of alleles shared) is compared. Since variance in environment is generally considered small in this situation and MZ twins share twice as many genes as DZ twins, heritability is approximately twice the difference in correlation between MZ and DZ twins: $[H^2 = 2(R_{MZ} - R_{DZ})]$ [40, 79]. Although heritability can be due to allelic heterogeneity at a single locus, high values suggest that there are multiple genes (locus heterogeneity) which contribute to the phenotypic variation; the heritability in HSCR, for example, is nearly 100%. In addition to high heritability values and poor genotype–phenotype correlation phenotypic variation in animal models can also suggest oligogenicity.

7.5.2 Mouse Models of Oligogenic Inheritance: Familial Adenomatous Polyposis

Mouse models have been powerful tools for recapitulating human phenotypes under a specific, fixed genetic background. This provides a significant advantage when studying a disease with phenotypic variability, because variability attributable to environment can be differentiated from that caused by the genetic background. By keeping the genetic background constant, through successive backcrosses of the mutant to an inbred strain, the genetic heterogeneity between individual mutant mice is virtually eliminated, helping to isolate the phenotypic effects of a single mutant locus. The presence of modifier loci can then be established if large phenotypic variation is present when comparing the mutant phenotype on different genetic backgrounds (backcrossing to two different inbred strains). One such example is elegantly demonstrated in a mouse model of familial adenomatous polyposis (FAP;OMIM 175,100), in which numerous adenomatous polyps develop in the colon and can eventually progress to carcinomas. FAP is caused by dominant mutations in the adenomatous polyposis coli (*APC*) gene [9, 82], which is mutated in the mouse model *Min* (multiple intestinal neoplasia) on chromosome 18 [87]. It was noticed that in *Min/+* mice the number of intestinal neoplasms was heavily affected by genetic background: the *Min/+* mice of the parent congenic strain *C57BL/6 J-Min/+* have an average of 29 tumors, whereas crosses between these mice and the AKR strain resulted in the *Min/+* F1 progeny having an average of six tumors [66]. This phenotypic variation seen in different mouse genetic backgrounds is similar to what is observed in humans where the same mutations in the APC gene can give rise to profound phenotypic differences even among family members.

For example, in some families the number of colonic polyps in affected individuals can range from very few to over 100 [56]. Even more interesting is that individual family members with the same APC mutation can manifest with different types of cancer from colonic polyps, gastric polyps, or nongastrointestinal neoplasms such as, osteomas, sarcomas, and carcinomas [81]. These phenotypic differences between family members are probably due to a combination of differences in genetic background (most likely functional polymorphisms modifying disease severity) and environment; however, it is difficult to differentiate between these two effects in human pedigrees. In contrast, the ability to perform multiple backcrosses and select for genetic and phenotypic markers make mice excellent tools for mapping genetic modifiers. This is illustrated in a subsequent study in which the DNA of the 110 *Min/+* mice from a backcross between two different strains [(AKR×B6−Min)×B6 backcross] were sequenced for 75 simple sequence length polymorphisms (SSLPs) distributed throughout the genome and then analyzed with respect to the quantitative phenotype (number of polyps) [26]. This resulted in the discovery of a dominant modifier that reduced the number of polyps, termed *Mom1* (modifier of Min) on chromosome 4, in a region of synteny to human chromosome 1p36-35. The allelic variation for *Mom1* was suggested to be responsible for approximately 50% of the genetic contribution to phenotypic variation based on assumed allelic differences between mouse strains. A further study found that a candidate gene, *Pla2g2a*, encoding a secretory phosphatase, mapped to the *Mom1* region and had 100% concordance between allele type and tumor susceptibility [59]. Nevertheless, causality of this gene was not proven until a transgenic line expressing *Pla2g2a* (OMIM: 172,411) recapitulated the tumor-resistant effects of the *Mom1* locus in a *Min* mouse line [20]. Interestingly, a later study also identified another locus, *Mom2* (modifier of *Min2*), on chromosome 18, which confers stronger resistance to *Min*-induced tumors than the *Mom1* locus [79].

7.5.3 Multigenic Models

As stated earlier in the chapter, deviations from a Mendelian paradigm noted when looking at disease transmission in familial pedigrees or mouse models can suggest oligogenic models. If a mutation at one locus requires a genetic interaction with a mutation at a second locus to elicit the disease phenotype, then it can easily be overlooked in certain pedigrees if a multigenic model of inheritance is not considered. This is because the mutant allele will be present in both normal and affected individuals (the affected ones carry a mutation at a second locus). In cases where the second locus is unknown, the first locus can be implicated in the disease by showing that the mutant allele segregates more often with the affected individuals than would be expected by its population frequency. If several different loci are later implicated individually with the phenotype it is imperative to go back and resequence the entire cohort – meaning both normal and affected subjects, for mutations in one or more of the implicated loci. As in the case of BBS and other oligogenic diseases, many of the affected patients with known mutations at one locus were found to harbor mutations at a second locus later shown to be implicated in the disease.

7.5.4 Linkage Analysis

Linkage analysis is a powerful tool for identifying loci involved in [the] transmission of a disease. In regard to oligogenic disorders, however, traditional linkage approaches must be modified must be modified to account for multiple segregating loci. A key example can be found in Hirschsprung disease (HSCR). In one study, *parametric linkage analysis* (a linkage test with assumptions such as mode of transmission, penetrance, or allele frequencies) and *nonparametric analysis* (NPL; a linkage test without models or assumptions on trait distribution) were first carried out in a traditional manner on 12 L-HSCR families by comparing segregation of polymorphic microsatellite markers in affected and unaffected subjects [17]. Weak evidence of linkage to chromosome 9q31 was observed (Lod score of 1.51, NPL score of 3.8), in addition to the expected strong linkage to *RET* on chromosome 10q11 (Lod score of 9.2; NPL score of 11.2). Subsequent sequencing and mutational analysis identified a diverse group of mutations at the *RET* locus in 11 of 12 families, reaffirming that *RET* is the major locus in L-HSCR. To circumvent the "monogenic bias" of traditional linkage tests, the approach was then modified by dividing the families into two groups, accepting the hypothesis that weaker *RET* mutations require a second locus to manifest the disease phenotype: group I were *RET* linked with severe nonsense and missense *RET* mutations, and group II were *RET* linked without missense and nonsense mutations. (Note: The introns were not sequenced in this study, suggesting that group II most probably has noncoding *RET* mutations). The reanalysis showed no linkage of group I (Lod = 0, NPL = 0.4), but significant linkage of group II (Lod = 4.8, NPL = 5.3), to 9q31, suggesting that mild *RET* mutations and allelic variants of an unknown locus on chromosome 9 are required for disease pathogenesis in the *RET*-linked families lacking exonic mutations. This study signifies that severity of mutations at the major contributing locus often determines whether additional mutations are required for disease manifestation.

7.6 Molecular Mechanisms of Oligogenic Disorders

Although there are now several examples of genetically determined oligogenic disorders (Table 7.1), few of these studies have functionally characterized the genes involved, resulting in a poor understanding of the underlying molecular mechanisms of oligogenicity. The difficult question to answer is how two mutant alleles at two different loci can act in conjunction to cause or exacerbate the same phenotype, a phenomenon known as *nonallelic noncomplementation*. Although nonallelic noncomplementation has been established in various organisms and physiological processes, such as transcriptional regulation and cytoskeletal motility [95], there are few mechanistic examples for alleles implicated in human genetic disease.

Since mutations and allelic variation contributing to disease ultimately effect expression, function, and/or interaction of proteins (with exceptions attributable to noncoding RNA), understanding protein–protein interaction in the cell is essential to elucidating these oligogenic mechanisms. There are two main models that explain nonallelic noncomplementation: the dosage model and the *poison model*. In the case of the dosage model, a concomitant decrease in dosage (expression or function) of two different interacting proteins is necessary to cause a pathogenic phenotype (Figs. 7.3, and 7.4). One example is found in *Drosophila melanogaster*, where null mutations in a ligand, *slit*, which is required for neuronal migration, fail to complement null alleles of its receptor, robo [18, 53]. By contrast, the poison model posits that a mutant protein disrupts or "poisons" a multimeric protein complex to which it normally binds, but not enough to cause visible pathogenesis (Fig. 7.5). This is most likely, because there are sufficient remaining functional complexes. A second mutation in a different protein of the same complex then lowers the functionality of the multimer beyond a certain threshold, which results in a visible pathogenic phenotype. Therefore, the poison model differs from the dosage model in that at least one mutation does not cause just a reduction of functionality, but a "poisoning" of the complex via sequestration of other components or a novel but disruptive interaction [60, 95]. An example can be seen in *Drosophila* and yeast, where altered α-tubulins act as "poisons" through disruption of microtuble polymerization or sequestering β-tubulin [5]. In humans, a good example of the poison model may also be seen in retinitis pigmentosa (RP), where the mechanisms of digenic inheritance caused by mutations in *ROM1* and *RDS* are well understood. It has been shown that these two proteins first form homodimers, which then combine to form a heterotetramer [37]. Mutations in RDS disrupt the proper formation and function of RDS–RDS homodimers and an additional null mutation in *ROM1* further exacerbates the effect on the tetrameric complex [58]. This digenic interaction is believed to be the cause of photoreceptor degeneration in RP.

In another oligogenic model termed *noninteracting noncomplementation*, mutations in two noninteracting proteins can also give rise to an oligogenic disorder, as seen in the example of cortisone reductase deficiency. In this example, H6PDH and 11β-HSD1 proteins are not thought to interact, but act at different stages of a common pathway – endolumeal H6PDH regenerates NADPH in the endoplasmic reticulum required for the oxo-reductase activity of 11β-HSD1 [29]. Each mutated protein thus contributes quantitatively to the dysfunction of the pathway resulting in an additive reduction of 11β-HSD1 oxo-reductase activity beyond a certain threshold that results in a disease phenotype (Fig. 7.6). Another mechanism of oligogenicity of two non-interacting proteins can be found when they have redundant functionality (Fig. 7.7). An example can be seen between Mitf and Tfe3, two members of the Mitf–Tfe family of basic helix-loop-helix-leucine zipper transcription factors that have redundant roles

7 Oligogenic Disease

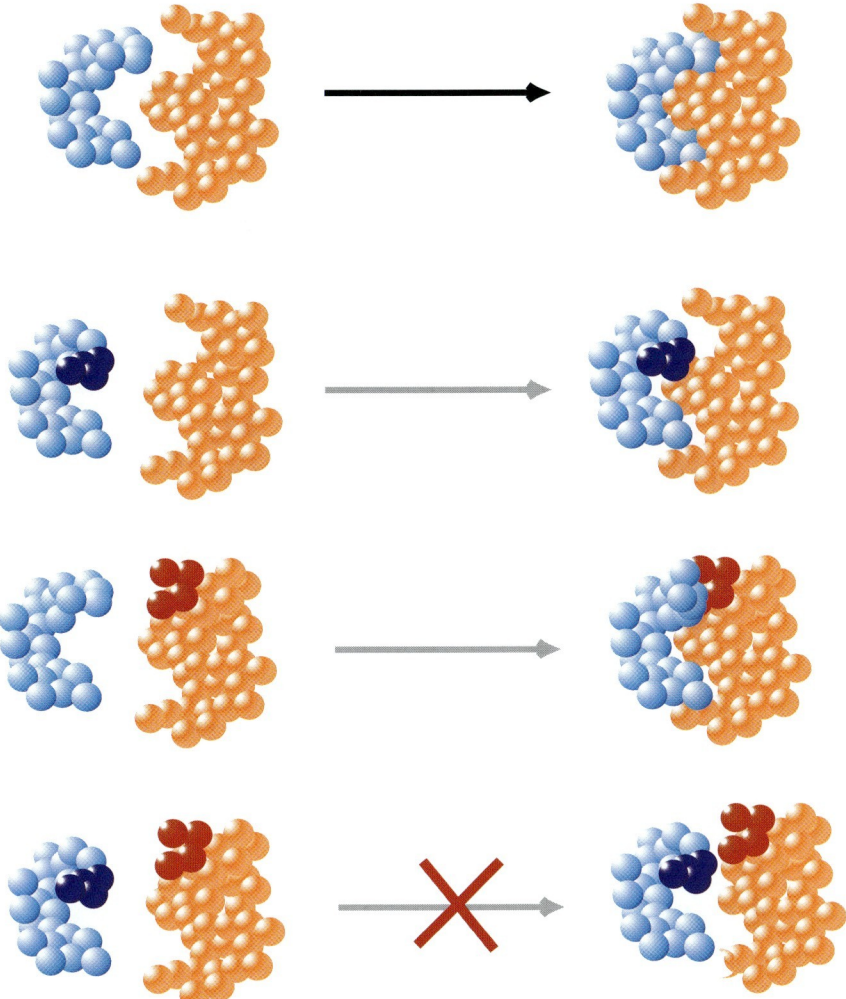

Fig. 7.3 *Dosage model 1:* direct interaction. Mutations at one locus (mutated proteins are indicated by *darker balls*) affect binding but are not enough to disrupt the formation of the complex between the two proteins. Mutations in both loci, however, prevent complex formation or functionality below a critical threshold that results in the disease phenotype

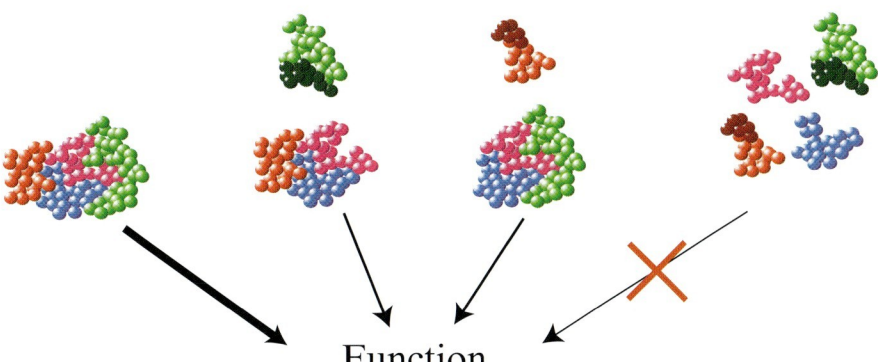

Fig. 7.4 *Dosage model 2:* complex. In a complex between four different proteins, mutations at one locus prevent binding of one protein component and lower (*thinner arrows*), but do not eliminate complex functionality. An additional mutation in another protein that does not directly bind the first one prevents formation or functionality of the entire complex

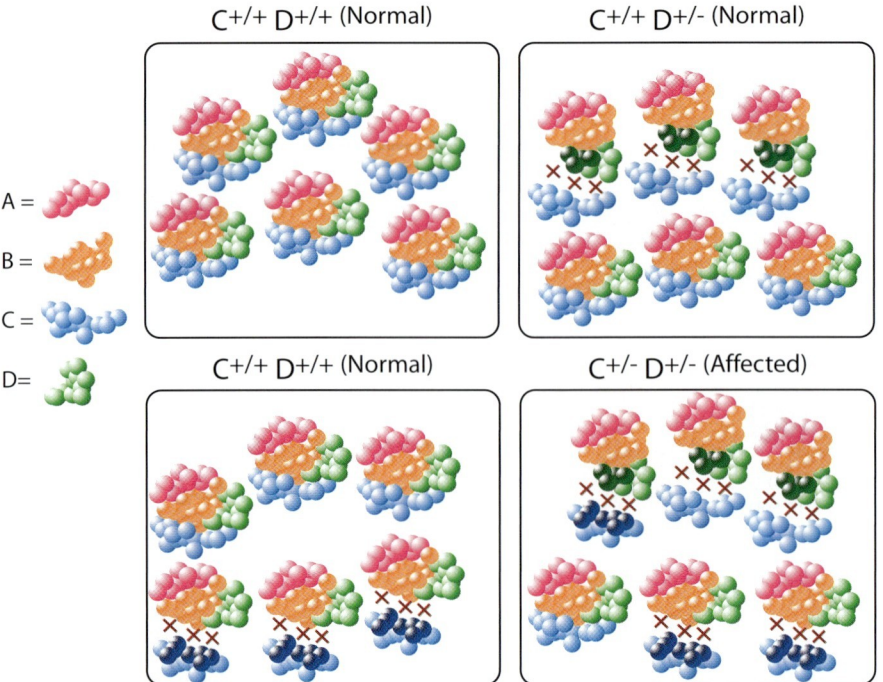

Fig. 7.5 *Poison model.* Mutations in protein C "poison" the complex by binding incorrectly and blocking the binding of another member; however, there are sufficient functional units to prevent a clinical phenotype. Another mutation (or mutations) in protein D disrupt(s) more units of the complex, resulting in very few functional complexes and a disease phenotype

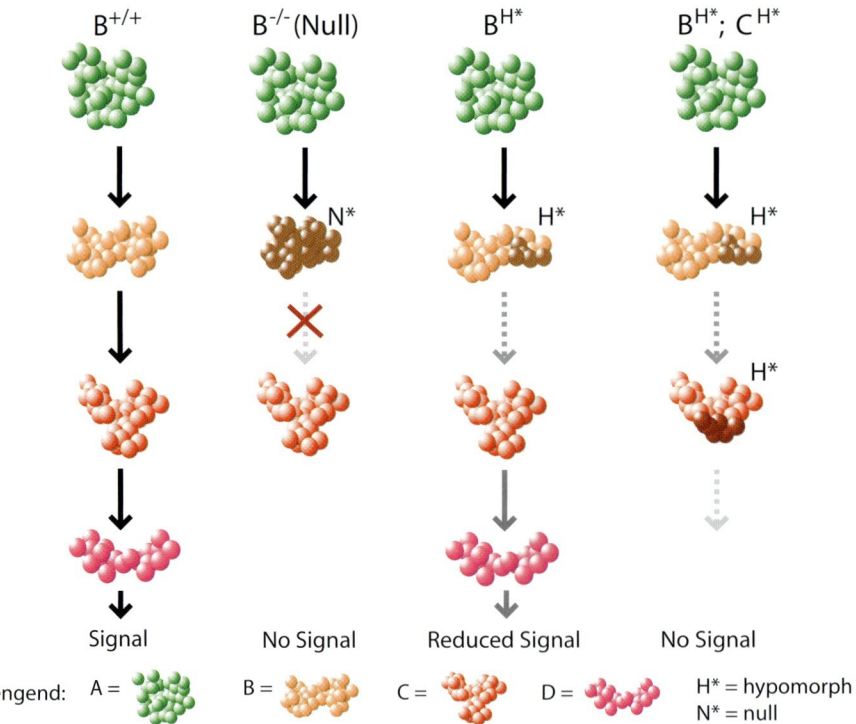

Fig. 7.6 *Noninteracting hypomorphs.* Null alleles of any protein in this pathway (A–D) cause a loss of signal (a null mutation in B that results in loss of signal in the second column). Hypomorphic (H*) mutations in B result in slightly reduced signal that does not cause a phenotype. However, a second hypomorphic mutation in a downstream protein C creates an additive loss of signal that has the same effect as the null mutation

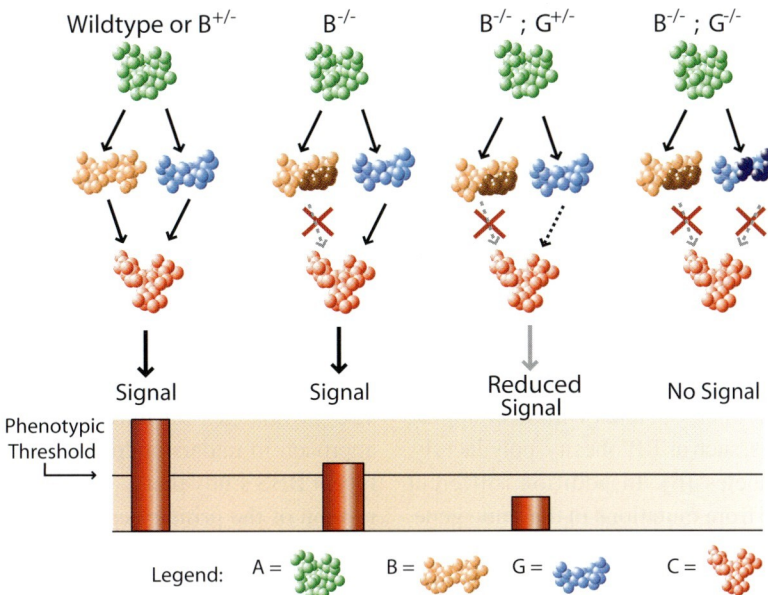

Fig. 7.7 *Noninteracting redundant functionality.* In this signal transduction pathway proteins *B* and *G* have redundant functionality. Homozygous mutations in protein B do not reduce signaling beyond a phenotypic threshold owing to protein *G*'s functional redundancy. However, heterozygous mutations in protein *G* in addition to homozygous loss of function of protein *B* reduces the signal below the critical threshold, resulting in disease

in osteoclast development. In one study, osteoclasts were found to have no abnormalities in *Mitf* or *Tfe3* null mice; however, the combined loss of the two genes resulted in severe osteoporosis [83]. Although Mitf and Tfe3 do interact to form heterodimers, this interaction was shown not to be essential for function or osteoclast development. In contrast, eliminating both the Mitf and Tfe3 homodimers, and thus their individual functions, resulted in the osteoporosis phenotype.

7.7 Modular or Systems Biology

Modular or systems biology is the study of molecular interactions or pathways as a particular cellular system with a discrete function, rather than a focus on the individual functions of proteins. A common theme illustrated by the examples of oligogenic disorders described previously is that the molecular basis of disease usually involves a particular pathway and its individual components. These pathways and their constituents can be better characterized as functional "modules," an evolving concept first described by Hartwell et al., which proposes the existence of discrete subsystems of biological organization. Modules are composed of many different types of molecules, and each module has a discrete function that results from highly regulated interaction of their components: protein, DNA, RNA, and/or small molecules. Oligogenic disorders may result from disruption of two or more components of a particular module. The concept of modules also suggests that both different disorders with similar clinical manifestations and locus heterogeneity within a single disorder probably result from mutations disrupting different components of a single molecular system. For example, cortisone reductase deficiency and polycystic ovary syndrome (PCOS), both disorders of a particular system regulating adrenal steroidogenesis, are due to greater conversion of cortisol to cortisone and impaired biosynthesis of cortisol, respectively, both of which cause adrenal androgen excess and PCOS or PCOS-like phenotypes in women [15, 93]. The four types of hemochromatosis, disorders of abnormal iron deposition, can be characterized into a system or module regulating heme biosynthesis. Extending this line of thought, it is notable that patients with familial porphyria cutanea tarda (OMIM 176,100), a disorder of reduced activity of uroporphyrinogen decarboxylase (UROD), the fifth enzyme in the heme biosynthetic pathway, are also associated with HFE mutations found in hemochromotosis patients [61].

Functional modules, by definition, have a particular function that can be distinguished from other modules through either spatial localization or biochemical specificity. This is exemplified in ciliopathies such as nephronophthisis (OMIM 256,100), Bardet–Biedl syndrome (209,900), Alstrom syndrome (203,800), Meckel–Gruber syndrome (OMIM: 249,000), Joubert syndrome (213,300) [77], and polycystic kidney disease (173,900), where mutated genes are all components of a particular module, viz. the primary cilium [6]. In keeping with the view of the cilium as a functional module, mutations in different ciliary or basal body genes can result in ciliopathies with overlapping clinical manifestations, such as RP, obesity, polydactyly, renal cysts, and diabetes [6]. In addition, different ciliopathies can arise from mutations in the same gene, the severity of which determines the spectrum of clinical manifestations and thus the disease. For example milder mutations in *BBS6* are associated with McKusick–Kaufman syndrome, whereas more severe mutations (or mutations combined with alleles in other BBS genes), lead to BBS [50].

Functional modules are not necessarily rigid or fixed structures, and any given constituent may belong to different modules, which might offer one explanation as to why mutations in the same gene can sometimes lead to unrelated phenotypes. An example is *RET*, which encodes a receptor tyrosine kinase and is expressed in cell lineages derived from neural crests [46]. *RET* is involved in molecular systems that regulate cell proliferation, migration, differentiation, and survival during embryogenesis [3]. Different mutations in *RET* can give rise not only to HCSR, as described earlier, but also to multiple endocrine neoplasia type IIA or type IIB (*MEN2*) [41], or familial medullary thyroid carcinoma (FTC; OMIM 188,470) [39]. Germline missense mutations resulting in constitutively active *RET* can cause the different subclinical cancer types, depending on the tissues affected. There is large phenotypic variability, even between the different cancer types, which can range from medullary thyroid carcinoma, a tumor of the adrenal medulla, and hyperparathyroidism in MEN2A (OMIM: 171,400) [41], to a more complex clinical phenotype including ganglioneuromas on the tongue, lips, and eyelids, intestinal ganglioneuronas, thickened cornea nerves and a marfanoid habitus in MEN2B (OMIM 162,300). HCSR results from mutations that disrupt *RET* function, resulting in a loss of enteric ganglia. This large phenotypic variability resulting from mutations in one gene suggests that the "module-specific" functions of *RET* can be separated temporally (period of development), spatially (tissue or cell type), and/or through chemical specificity (expression of tissue specific ligands or cofactors that interact with *RET*).

The concept of modules or molecular systems in the context of disease can be important in the search for modifier alleles. The involvement of a modular component in a particular disease suggests that mutations disrupting other components of that module may also contribute to the same disease or related disorders. The advantages of having a modular or systems approach in understanding disease are illustrated in a recent BBS study that used the hypothesis that the disruption of the primary cilium or its constituents is the sole basis for BBS and led to the discovery of a BBS modifier, MGC1203 (OMIM 209,900) [7]. Accepting the hypothesis that undiscovered genes involved in BBS would potentially interact with known BBS genes, several yeast two-hybrid screens were performed and identified more than 60 candidates. Since BBS is a ciliopathy, presence of these candidate genes was then screened for in the ciliary proteome – a database of purported ciliary localizing proteins [57]. This resulted in the identification of MGC1203 as the only protein present in both data sets. Further characterization through combined use of biochemical, genetic, and in vivo tools identified MGC1203 as a modifier that has an epistatic effect on BBS mutations, with some allelic variants resulting in greater disease severity. This study underscores the importance of using a variety of molecular and genetic tools under a paradigm of "modularity" of disease components in order to dissect the potentially weak contribution of genes in an oligogenic disorder.

7.8 Conclusions

The study of oligogenic disorders, diseases caused or modulated by a few genes, is elucidating new principles of methodology that are necessary to overcome the inherent limitations of current methods towards classification of monogenic disease. The discovery of genetic modifiers, compounded by the poor genotype–phenotype correlations in even the most quintessential monogenic disorders, is highlighting the vague lines between what should be considered monogenic or oligogenic. In fact, evidence of both modes of inheritance is sometimes found within the same clinically classified disorder.

The highlighted examples of oligogenic disorders suggest that the individual genetic factors are often confined to a similar pathway or module. The alleles of these genes contributing to disease, however, are not necessarily rare mutations, but can be common polymorphisms that only elicit an effect in a particular physiological setting. To disclose these often weak genetic interactions new methodology must be applied, such as considering heritability, using mouse models of varying genetic backgrounds, and modifying traditional linkage approaches. Overall, the future of understanding complex disease lies in analyzing genetic factors from a systems biology viewpoint under the premise of modularity of these factors within the same or between similar disorders.

References

1. Adato A, Kalinski H, Weil D, Chaib H, Korostishevsky M, Bonne-Tamir B (1999) Possible interaction between USH1B and USH3 gene products as implied by apparent digenic deafness inheritance. Am J Hum Genet 65:261–265
2. Arai K, Zachman K, Shibasaki T, Chrousos GP (1999) Polymorphisms of amiloride-sensitive sodium channel subunits in five sporadic cases of pseudohypoaldosteronism: do they have pathologic potential? J Clin Endocrinol Metab 84:2434–2437
3. Arighi E, Borrello MG, Sariola H (2005) RET tyrosine kinase signaling in development and cancer. Cytokine Growth Factor Rev 16:441–467
4. Attie T, Pelet A, Edery P, Eng C, Mulligan LM, Amiel J, Boutrand L, Beldjord C, Nihoul-Fekete C, Munnich A (1995) Diversity of RET proto-oncogene mutations in familial and sporadic Hirschsprung disease. Hum Mol Genet 4:1381–1386
5. Badano JL, Katsanis N (2002) Beyond mendel: an evolving view of human genetic disease transmission. Nat Rev Genet 3:779–789
6. Badano JL, Leitch CC, Ansley SJ, May-Simera H, Lawson S, Lewis RA, Beales PL, Dietz HC, Fisher S, Katsanis N (2006) Dissection of epistasis in oligogenic Bardet-Biedl syndrome. Nature 439:326–330
7. Badano JL, Mitsuma N, Beales PL, Katsanis N (2006) The ciliopathies: an emerging class of human genetic disorders. Annu Rev Genomics Hum Genet 7:125–148
8. Balciuniene J, Dahl N, Borg E, Samuelsson E, Koisti MJ, Pettersson U, Jazin EE (1998) Evidence for digenic inheritance of nonsyndromic hereditary hearing loss in a swedish family. Am J Hum Genet 63:786–793
9. Bapat B, Odze R, Mitri A, Berk T, Ward M, Gallinger S (1993) Identification of somatic APC gene mutations in periampullary adenomas in a patient with familial adenomatous polyposis (FAP). Hum Mol Genet 2:1957–1959
10. Beales PL, Katsanis N, Lewis RA, Ansley SJ, Elcioglu N, Raza J, Woods MO, Green JS, Parfrey PS, Davidson WS, Lupski JR (2001) Genetic and mutational analyses of a large multiethnic Bardet–Biedl cohort reveal a minor involvement of BBS6 and delineate the critical intervals of other loci. Am J Hum Genet 68:606–616
11. Beales PL, Elcioglu N, Woolf AS, Parker D, Flinter FA (1999) New criteria for improved diagnosis of Bardet–Biedl syndrome: results of a population survey. J Med Genet 36:437–446
12. Beales PL, Badano JL, Ross AJ, Ansley SJ, Hoskins BE, Kirsten B, Mein CA, Froguel P, Scambler PJ, Lewis RA, Lupski JR, Katsanis N (2003) Genetic interaction of BBS1 mutations with alleles at other BBS loci can result in non-Mendelian Bardet–Biedl syndrome. Am J Hum Genet 72:1187–1199
13. Benailly HK, Lapierre JM, Laudier B, Amiel J, Attie T, De Blois MC, Vekemans M, Romana SP (2003) PMX2B, a new candidate gene for Hirschsprung's disease. Clin Genet 64:204–209
14. Berthet M, Denjoy I, Donger C, Demay L, Hammoude H, Klug D, Schulze-Bahr E, Richard P, Funke H, Schwartz K, Coumel P, Hainque B, Guicheney P (1999) C-terminal HERG mutations: the role of hypokalemia and a KCNQ1-associated mutation in cardiac event occurrence. Circulation 99:1464–1470
15. Biason-Lauber A, Suter SL, Shackleton CH, Zachmann M (2000) Apparent cortisone reductase deficiency: a rare cause of hyperandrogenemia and hypercortisolism. Horm Res 53:260–266
16. Blau N, Thony B, Heizmann CW, Dhondt JA (1993) Tetrahydrobiopterin deficiency: from phenotype to genotype. Pteridines 4:1–10
17. Bolk S, Pelet A, Hofstra RM, Angrist M, Salomon R, Croaker D, Buys CH, Lyonnet S, Chakravarti A (2000) A human model for multigenic inheritance: phenotypic expression in Hirschsprung disease requires both the RET gene and a new 9q31 locus. Proc Natl Acad Sci USA 97:268–273
18. Brose K, Bland KS, Wang KH, Arnott D, Henzel W, Goodman CS, Tessier-Lavigne M, Kidd T (1999) Slit proteins bind robo receptors and have an evolutionarily conserved role in repulsive axon guidance. Cell 96:795–806
19. Bruning JC, Winnay J, Bonner-Weir S, Taylor SI, Accili D, Kahn CR (1997) Development of a novel polygenic model of NIDDM in mice heterozygous for IR and IRS-1 null alleles. Cell 88:561–572
20. Cormier RT, Hong KH, Halberg RB, Hawkins TL, Richardson P, Mulherkar R, Dove WF, Lander ES (1997) Secretory phospholipase Pla2g2a confers resistance to intestinal tumorigenesis. Nat Genet 17:88–91
21. Cutting GR (2005) Modifier genetics: cystic fibrosis. Annu Rev Genomics Hum Genet 6:237–260
22. Dachsel JC, Mata IF, Ross OA, Taylor JP, Lincoln SJ, Hinkle KM, Huerta C, Ribacoba R, Blazquez M, Alvarez V, Farrer MJ (2006) Digenic Parkinsonism: investigation of the synergistic effects of PRKN and LRRK2. Neurosci Lett 410:80–84
23. Dahia PL, Hao K, Rogus J, Colin C, Pujana MA, Ross K, Magoffin D, Aronin N, Cascon A, Hayashida CY, Li C,

Toledo SP, Stiles CD, Familial heochromocytoma Consortium (2005) Novel pheochromocytoma susceptibility loci identified by integrative genomics. Cancer Res 65:9651–9658

24. Davis JS, Hassanzadeh S, Winitsky S, Lin H, Satorius C, Vemuri R, Aletras AH, Wen H, Epstein ND (2001) The overall pattern of cardiac contraction depends on a spatial gradient of myosin regulatory light chain phosphorylation. Cell 107:631–641

25. del Castillo FJ, Rodriguez-Ballesteros M, Alvarez A, Hutchin T, Leonardi E, de Oliveira CA, Azaiez H et al (2005) A novel deletion involving the connexin-30 gene, del(GJB6–d13s1854), found in trans with mutations in the GJB2 gene (connexin-26) in subjects with DFNB1 non-syndromic hearing impairment. J Med Genet 42:588–594

26. Dietrich WF, Lander ES, Smith JS, Moser AR, Gould KA, Luongo C, Borenstein N, Dove W (1993) Genetic identification of mom-1, a major modifier locus affecting min-induced intestinal neoplasia in the mouse. Cell 75:631–639

27. Dode C, Teixeira L, Levilliers J, Fouveaut C, Bouchard P, Kottler ML, Lespinasse J, Lienhardt-Roussie A, Mathieu M, Moerman A, Morgan G, Murat A, Toublanc JE, Wolczynski S, Delpech M, Petit C, Young J, Hardelin JP (2006) Kallmann syndrome: mutations in the genes encoding prokineticin-2 and prokineticin receptor-2. PLoS Genet 2:e175

28. Doray B, Salomon R, Amiel J, Pelet A, Touraine R, Billaud M, Attie T, Bachy B, Munnich A, Lyonnet S (1998) Mutation of the RET ligand, neurturin, supports multigenic inheritance in hirschsprung disease. Hum Mol Genet 7:1449–1452

29. Draper N, Walker EA, Bujalska IJ, Tomlinson JW, Chalder SM, Arlt W, Lavery GG, Bedendo O, Ray DW, Laing I, Malunowicz E, White PC, Hewison M, Mason PJ, Connell JM, Shackleton CH, Stewart PM (2003) Mutations in the genes encoding 11beta-hydroxysteroid dehydrogenase type 1 and hexose-6-phosphate dehydrogenase interact to cause cortisone reductase deficiency. Nat Genet 34:434–439

30. Edery P, Lyonnet S, Mulligan LM, Pelet A, Dow E, Abel L, Holder S, Nihoul-Fekete C, Ponder BA, Munnich A (1994) Mutations of the RET proto-oncogene in Hirschsprung's disease. Nature 367:378–380

31. Emison ES, McCallion AS, Kashuk CS, Bush RT, Grice E, Lin S, Portnoy ME, Cutler DJ, Green ED, Chakravarti A (2005) A common sex-dependent mutation in a RET enhancer underlies Hirschsprung disease risk. Nature 434:857–863

32. Ficarella R, Di Leva F, Bortolozzi M, Ortolano S, Donaudy F, Petrillo M, Melchionda S, Lelli A, Domi T, Fedrizzi L, Lim D, Shull GE, Gasparini P, Brini M, Mammano F, Carafoli E (2007) A functional study of plasma-membrane calcium-pump isoform 2 mutants causing digenic deafness. Proc Natl Acad Sci USA 104:1516–1521

33. Gabriel SB, Salomon R, Pelet A, Angrist M, Amiel J, Fornage M, Attie-Bitach T, Olson JM, Hofstra R, Buys C, Steffann J, Munnich A, Lyonnet S, Chakravarti A (2002) Segregation at three loci explains familial and population risk in Hirschsprung disease. Nat Genet 31:89–93

34. Gandrille S, Greengard JS, Alhenc-Gelas M, Juhan-Vague I, Abgrall JF, Jude B, Griffin JH, Aiach M (1995) Incidence of activated protein C resistance caused by the ARG 506 GLN mutation in factor V in 113 unrelated symptomatic protein C-deficient patients. The French Network on the behalf of INSERM. Blood 86:219–224

35. Gibson G, Dworkin I (2004) Uncovering cryptic genetic variation. Nat Rev Genet 5:681–690

36. Glorioso N, Filigheddu F, Troffa C, Soro A, Parpaglia PP, Tsikoudakis A, Myers RH, Herrera VL, Ruiz-Opazo N (2001) Interaction of alpha(1)-na, K-ATPase and na, K, 2Cl-cotransporter genes in human essential hypertension. Hypertension 38:204–209

37. Goldberg AF, Molday RS (1996) Subunit composition of the peripherin/rds-rom-1 disk rim complex from rod photoreceptors: hydrodynamic evidence for a tetrameric quaternary structure. Biochemistry 35:6144–6149

38. Gouya L, Puy H, Lamoril J, Da Silva V, Grandchamp B, Nordmann Y, Deybach JC (1999) Inheritance in erythropoietic protoporphyria: a common wild-type ferrochelatase allelic variant with low expression accounts for clinical manifestation. Blood 93:2105–2110

39. Grieco M, Santoro M, Berlingieri MT, Melillo RM, Donghi R, Bongarzone I, Pierotti MA, Della Porta G, Fusco A, Vecchio G (1990) PTC is a novel rearranged form of the ret proto-oncogene and is frequently detected in vivo in human thyroid papillary carcinomas. Cell 60:557–563

40. Griffiths AJF, Miller JH, Suzuki DT, Lewontin RC, Gelbart WM (1999) Introduction to genetic analysis. WH Freeman and Company, New York

41. Hansford JR, Mulligan LM (2000) Multiple endocrine neoplasia type 2 and RET: from neoplasia to neurogenesis. J Med Genet 37:817–827

42. Hanson EH, Imperatore G, Burke W (2001) HFE gene and hereditary hemochromatosis: a HuGE review. Human genome epidemiology. Am J Epidemiol 154:193–206

43. Haravuori H, Siitonen HA, Mahjneh I, Hackman P, Lahti L, Somer H, Peltonen L, Kestila M, Udd B (2004) Linkage to two separate loci in a family with a novel distal myopathy phenotype (MPD3). Neuromuscul Disord 14:183–187

44. Helwig U, Imai K, Schmahl W, Thomas BE, Varnum DS, Nadeau JH, Balling R (1995) Interaction between undulated and patch leads to an extreme form of spina bifida in double-mutant mice. Nat Genet 11:60–63

45. Ito M, Yamakawa K, Sugawara T, Hirose S, Fukuma G, Kaneko S (2006) Phenotypes and genotypes in epilepsy with febrile seizures plus. Epilepsy Res 70(Suppl 1):S199–S205

46. Iwashita T, Kruger GM, Pardal R, Kiel MJ, Morrison SJ (2003) Hirschsprung disease is linked to defects in neural crest stem cell function. Science 301:972–976

47. Jacolot S, Le Gac G, Scotet V, Quere I, Mura C, Ferec C (2004) HAMP as a modifier gene that increases the phenotypic expression of the HFE pC282Y homozygous genotype. Blood 103:2835–2840

48. Jervis GA (1953) Phenylpyruvic oligophrenia deficiency of phenylalanine-oxidizing system. Proc Soc Exp Biol Med 82:514–515

49. Joosten PH, van Zoelen EJ, Murre C (2005) Pax1/E2a double-mutant mice develop non-lethal neural tube defects that resemble human malformations. Transgenic Res 14:983–987

50. Katsanis N (2004) The oligogenic properties of Bardet–Biedl syndrome. Hum Mol Genet 13(Spec No 1):R65–R71

51. Katsanis N, Ansley SJ, Badano JL, Eichers ER, Lewis RA, Hoskins BE, Scambler PJ, Davidson WS, Beales PL, Lupski JR (2001) Triallelic inheritance in Bardet–Biedl syndrome, a Mendelian recessive disorder. Science 293:2256–2259

52. Kerem B, Rommens JM, Buchanan JA, Markiewicz D, Cox TK, Chakravarti A, Buchwald M, Tsui LC (1989) Identification of the cystic fibrosis gene: genetic analysis. Science 245:1073–1080
53. Kidd T, Bland KS, Goodman CS (1999) Slit is the midline repellent for the robo receptor in *Drosophila*. Cell 96: 785–794
54. Le Gac G, Ferec C (2005) The molecular genetics of haemochromatosis. Eur J Hum Genet 13:1172–1185
55. Le Gac G, Scotet V, Ka C, Gourlaouen I, Bryckaert L, Jacolot S, Mura C, Ferec C (2004) The recently identified type 2A juvenile haemochromatosis gene (HJV), a second candidate modifier of the C282Y homozygous phenotype. Hum Mol Genet 13:1913–1918
56. Leppert M, Burt R, Hughes JP, Samowitz W, Nakamura Y, Woodward S, Gardner E, Lalouel JM, White R (1990) Genetic analysis of an inherited predisposition to colon cancer in a family with a variable number of adenomatous polyps. N Engl J Med 322:904–908
57. Li JB, Gerdes JM, Haycraft CJ, Fan Y, Teslovich TM, May-Simera H, Li H, Blacque OE, Li L, Leitch CC, Lewis RA, Green JS, Parfrey PS, Leroux MR, Davidson WS, Beales PL, Guay-Woodford LM, Yoder BK, Stormo GD, Katsanis N, Dutcher SK (2004) Comparative genomics identifies a flagellar and basal body proteome that includes the BBS5 human disease gene. Cell 117:541–552
58. Loewen CJ, Moritz OL, Molday RS (2001) Molecular characterization of peripherin-2 and rom-1 mutants responsible for digenic retinitis pigmentosa. J Biol Chem 276: 22388–22396
59. MacPhee M, Chepenik KP, Liddell RA, Nelson KK, Siracusa LD, Buchberg AM (1995) The secretory phospholipase A2 gene is a candidate for the *Mom1* locus, a major modifier of ApcMin-induced intestinal neoplasia. Cell 81:957–966
60. McCright B, Lozier J, Gridley T (2002) A mouse model of alagille syndrome: Notch2 as a genetic modifier of Jag1 haploinsufficiency. Development 129:1075–1082
61. Mendez M, Sorkin L, Rossetti MV, Astrin KH, Del C Batlle AM, Parera VE, Aizencang G, Desnick RJ (1998) Familial porphyria cutanea tarda: characterization of seven novel uroporphyrinogen decarboxylase mutations and frequency of common hemochromatosis alleles. Am J Hum Genet 63:1363–1375
62. Merryweather-Clarke AT, Cadet E, Bomford A, Capron D, Viprakasit V, Miller A, McHugh PJ, Chapman RW, Pointon JJ, Wimhurst VL, Livesey KJ, Tanphaichitr V, Rochette J, Robson KJ (2003) Digenic inheritance of mutations in HAMP and HFE results in different types of haemochromatosis. Hum Mol Genet 12:2241–2247
63. Merryweather-Clarke AT, Pointon JJ, Jouanolle AM, Rochette J, Robson KJ (2000) Geography of HFE C282Y and H63D mutations. Genet Test 4:183–198
64. Millat G, Chevalier P, Restier-Miron L, Da Costa A, Bouvagnet P, Kugener B, Fayol L, Gonzalez Armengod C, Oddou B, Chanavat V, Froidefond E, Perraudin R, Rousson R, Rodriguez-Lafrasse C (2006) Spectrum of pathogenic mutations and associated polymorphisms in a cohort of 44 unrelated patients with long QT syndrome. Clin Genet 70:214–227
65. Moron FJ, Mendoza N, Vazquez F, Molero E, Quereda F, Salinas A, Fontes J, Martinez-Astorquiza T, Sanchez-Borrego R, Ruiz A (2006) Multilocus analysis of estrogen-related genes in Spanish postmenopausal women suggests an interactive role of ESR1, ESR2 and NRIP1 genes in the pathogenesis of osteoporosis. Bone 39:213–221
66. Moser AR, Dove WF, Roth KA, Gordon JI (1992) The min (multiple intestinal neoplasia) mutation: its effect on gut epithelial cell differentiation and interaction with a modifier system. J Cell Biol 116:1517–1526
67. Muntoni F, Bonne G, Goldfarb LG, Mercuri E, Piercy RJ, Burke M, Yaou RB, Richard P, Recan D, Shatunov A, Sewry CA, Brown SC (2006) Disease severity in dominant emery dreifuss is increased by mutations in both emerin and desmin proteins. Brain 129:1260–1268
68. Mykytyn K, Nishimura DY, Searby CC, Shastri M, Yen HJ, Beck JS, Braun T, Streb LM, Cornier AS, Cox GF, Fulton AB, Carmi R, Luleci G, Chandrasekharappa SC, Collins FS, Jacobson SG, Heckenlively JR, Weleber RG, Stone EM, Sheffield VC (2002) Identification of the gene (BBS1) most commonly involved in Bardet–Biedl syndrome, a complex human obesity syndrome. Nat Genet 31:435–438
69. Nanni L, Ming JE, Bocian M, Steinhaus K, Bianchi DW, Die-Smulders C, Giannotti A, Imaizumi K, Jones KL, Campo MD, Martin RA, Meinecke P, Pierpont ME, Robin NH, Young ID, Roessler E, Muenke M (1999) The mutational spectrum of the sonic hedgehog gene in holoprosencephaly: SHH mutations cause a significant proportion of autosomal dominant holoprosencephaly. Hum Mol Genet 8:2479–2488
70. Naveed M, Nath SK, Gaines M, Al-Ali MT, Al-Khaja N, Hutchings D, Golla J, Deutsch S, Bottani A, Antonarakis SE, Ratnamala U, Radhakrishna U (2007) Genomewide linkage scan for split-hand/foot malformation with long-bone deficiency in a large arab family identifies two novel susceptibility loci on chromosomes 1q42.2-q43 and 6q14.1. Am J Hum Genet 80:105–111
71. Pitteloud N, Quinton R, Pearce S, Raivio T, Acierno J, Dwyer A, Plummer L, Hughes V, Seminara S, Cheng YZ, Li WP, Maccoll G, Eliseenkova AV, Olsen SK, Ibrahimi OA, Hayes FJ, Boepple P, Hall JE, Bouloux P, Mohammadi M, Crowley W (2007) Digenic mutations account for variable phenotypes in idiopathic hypogonadotropic hypogonadism. J Clin Invest 117:457–463
72. Qin M, Hayashi H, Oshima K, Tahira T, Hayashi K, Kondo H (2005) Complexity of the genotype-phenotype correlation in familial exudative vitreoretinopathy with mutations in the LRP5 and/or FZD4 genes. Hum Mutat 26:104–112
73. Reardon W, Smith A, Honour JW, Hindmarsh P, Das D, Rumsby G, Nelson I, Malcolm S, Ades L, Sillence D, Kumar D, DeLozier-Blanchet C, McKee S, Kelly T, McKeehan WL, Baraitser M, Winter RM (2000) Evidence for digenic inheritance in some cases of Antley–Bixler syndrome? J Med Genet 37:26–32
74. Renieri A, Meloni I, Longo I, Ariani F, Mari F, Pescucci C, Cambi F (2003) Rett syndrome: the complex nature of a monogenic disease. J Mol Med 81:346–354
75. Roetto A, Camaschella C (2005) New insights into iron homeostasis through the study of non-HFE hereditary haemochromatosis. Best Pract Res Clin Haematol 18:235–250
76. Savage DB, Agostini M, Barroso I, Gurnell M, Luan J, Meirhaeghe A, Harding AH, Ihrke G, Rajanayagam O, Soos MA, George S, Berger D, Thomas EL, Bell JD, Meeran K, Ross RJ, Vidal-Puig A, Wareham NJ, O'Rahilly S, Chatterjee

VK, Schafer AJ (2002) Digenic inheritance of severe insulin resistance in a human pedigree. Nat Genet 31:379–384

77. Sayer JA, Otto EA, O'Toole JF, Nurnberg G, Kennedy MA, Becker C, Hennies HC et al (2006) The centrosomal protein nephrocystin-6 is mutated in Joubert syndrome and activates transcription factor ATF4. Nat Genet 38:674–681

78. Sebat J, Lakshmi B, Troge J, Alexander J, Young J, Lundin P, Maner S, Massa H, Walker M, Chi M, Navin N, Lucito R, Healy J, Hicks J, Ye K, Reiner A, Gilliam TC, Trask B, Patterson N, Zetterberg A, Wigler M (2004) Large-scale copy number polymorphism in the human genome. Science 305:525–528

79. Silverman KA, Koratkar R, Siracusa LD, Buchberg AM (2002) Identification of the modifier of min 2 (Mom2) locus, a new mutation that influences apc-induced intestinal neoplasia. Genome Res 12:88–97

80. Snieder H, MacGregor AJ, Spector TD, NetLibrary I (2000) Advances in twin and sib-pair analysis. Greenwich Medical Media, London Distributed worldwide by Oxford University Press, Oxford

81. Spirio L, Otterud B, Stauffer D, Lynch H, Lynch P, Watson P, Lanspa S, Smyrk T, Cavalieri J, Howard L (1992) Linkage of a variant or attenuated form of adenomatous polyposis coli to the adenomatous polyposis coli (APC) locus. Am J Hum Genet 51:92–100

82. Spirio L, Olschwang S, Groden J, Robertson M, Samowitz W, Joslyn G, Gelbert L, Thliveris A, Carlson M, Otterud B (1993) Alleles of the APC gene: an attenuated form of familial polyposis. Cell 75:951–957

83. Steingrimsson E, Tessarollo L, Pathak B, Hou L, Arnheiter H, Copeland NG, Jenkins NA (2002) Mitf and Tfe3, two members of the mitf-tfe family of bHLH-zip transcription factors, have important but functionally redundant roles in osteoclast development. Proc Natl Acad Sci USA 99:4477–4482

84. Stoetzel C, Laurier V, Davis EE, Muller J, Rix S, Badano JL, Leitch CC et al (2006) BBS10 encodes a vertebrate-specific chaperonin-like protein and is a major BBS locus. Nat Genet 38:521–524

85. Stoetzel C, Muller J, Laurier V, Davis EE, Zaghloul NA, Vicaire S, Jacquelin C, Plewniak F, Leitch CC, Sarda P, Hamel C, de Ravel TJ, Lewis RA, Friederich E, Thibault C, Danse JM, Verloes A, Bonneau D, Katsanis N, Poch O, Mandel JL, Dollfus H (2007) Identification of a novel BBS gene (BBS12) highlights the major role of a vertebrate-specific branch of chaperonin-related proteins in Bardet–Biedl syndrome. Am J Hum Genet 80:1–11

86. Strachan T, Read AP (1999) Human molecular genetics 2. Garland Science, New York

87. Su LK, Kinzler KW, Vogelstein B, Preisinger AC, Moser AR, Luongo C, Gould KA, Dove WF (1992) Multiple intestinal neoplasia caused by a mutation in the murine homolog of the APC gene. Science 256:668–670

88. Sundin OH, Jun AS, Broman KW, Liu SH, Sheehan SE, Vito EC, Stark WJ, Gottsch JD (2006) Linkage of late-onset fuchs corneal dystrophy to a novel locus at 13pTel-13q12.13. Invest Ophthalmol Vis Sci 47:140–145

89. Tang B, Xiong H, Sun P, Zhang Y, Wang D, Hu Z, Zhu Z, Ma H, Pan Q, Xia JH, Xia K, Zhang Z (2006) Association of PINK1 and DJ-1 confers digenic inheritance of early-onset Parkinson's disease. Hum Mol Genet 15:1816–1825

90. Tonooka N, Tomura H, Takahashi Y, Onigata K, Kikuchi N, Horikawa Y, Mori M, Takeda J (2002) High frequency of mutations in the HNF-1alpha gene in non-obese patients with diabetes of youth in Japanese and identification of a case of digenic inheritance. Diabetologia 45:1709–1711

91. Touitou I, Perez C, Dumont B, Federici L, Jorgensen C (2006) Refractory auto-inflammatory syndrome associated with digenic transmission of low-penetrance tumour necrosis factor receptor-associated periodic syndrome and cryopyrin-associated periodic syndrome mutations. Ann Rheum Dis 65:1530–1531

92. Van Goethem G, Lofgren A, Dermaut B, Ceuterick C, Martin JJ, Van Broeckhoven C (2003) Digenic progressive external ophthalmoplegia in a sporadic patient: recessive mutations in POLG and C10orf2/Twinkle. Hum Mutat 22:175–176

93. White PC, Speiser PW (2000) Congenital adrenal hyperplasia due to 21-hydroxylase deficiency. Endocr Rev 21:245–291

94. Woo SL, Lidsky AS, Guttler F, Chandra T, Robson KJ (1983) Cloned human phenylalanine hydroxylase gene allows prenatal diagnosis and carrier detection of classical phenylketonuria. Nature 306:151–155

95. Yook KJ, Proulx SR, Jorgensen EM (2001) Rules of nonallelic noncomplementation at the synapse in caenorhabditis elegans. Genetics 158:209–220

96. Zheng QY, Yan D, Ouyang XM, Du LL, Yu H, Chang B, Johnson KR, Liu XZ (2005) Digenic inheritance of deafness caused by mutations in genes encoding cadherin 23 and protocadherin 15 in mice and humans. Hum Mol Genet 14:103–111

Formal Genetics of Humans: Multifactorial Inheritance and Common Diseases

Andrew G. Clark

Abstract The study of the genetics of complex traits is made complicated by the fact that the traits themselves are influenced by an interplay of many genes with many environmental factors. In this chapter the historical concepts of quantitative genetics, including additive variance and heritability, will be developed to underscore how important it is to understand that the root of the problem is to explain how genes contribute to the variance in a trait. With molecular genetic markers, such as SNPs, it is possible to test whether there are differences in the measured phenotype among the genotypes at the genetic marker, and this serves as a crude test of association. Many interesting challenges arise when such a test is expanded to 1 million markers spanning the entire chromosome, a design known as a genome-wide association study (GWAS). Complications due to population stratification, admixture, genotype x environment interaction, epistasis, and rare alleles are all considered. Methods that test association by use of excess of allele sharing in siblings (affected sib methods) or other relatives, or by excess cotransmission of alleles and a disease state (transmission disequilibrium test) have their own set of advantages and disadvantages. The chapter closes with some considerations of why the powerful methods presented here nevertheless leave much of the genetic variance in complex traits unexplained.

Content

8.1	Genetic Analysis of Complex Traits	264
8.1.1	Variation in Phenotypic Traits	264
8.1.2	Familial Resemblance and Heritability	264
8.1.3	The Special Case of Twins	267
8.1.4	Embedding a Single Measured Gene Influencing a Continuous Trait	269
8.1.5	A Model for Variance Partitioning	269
8.1.6	Relating the Model to Data	270
8.1.7	Mendelian Diseases Are Not Simple	271
8.2	Genetic Polymorphism and Disease	271
8.2.1	Finding Genes Underlying a Complex Trait	271
8.2.2	Limitations of Pedigree Analysis	271
8.2.3	A Prevailing Model: Common Disease Common Variants	272
8.2.4	Affected Sib-pairs	273
8.2.5	Transmission Disequilibrium Test	273
8.2.6	Full-Genome Association Testing	274
8.3	LD Mapping and Genome-Wide Association Studies	275
8.3.1	Theory and How It Works: HapMap and Genome-Wide LD	275
8.3.2	Technology: The Fantastic Drop in Genotyping Costs	276
8.3.3	Case-Control Studies	277
8.3.4	Statistical Inference with Genome-Wide Studies	277
8.3.5	Replication and Validation	278
8.3.6	Age-Related Macular Degeneration and Complement Factor H	279
8.4	Admixture Mapping and Population Stratification	279
8.4.1	How to Quantify Admixture	279
8.4.2	Using Admixture for Mapping	280

A.G. Clark (✉)
Department of Molecular Biology and Genetics,
Cornell University, Ithaca, NY 14853, USA
e-mail: ac347@cornell.edu

8.4.3	The Perils of Population Stratification	280	8.6	Missing Heritability: Why is so Little Variance Explained by GWAS Results? ... 284
8.4.4	How to Correct for Hidden Population Stratification	281	8.7	Concluding Remarks ... 285
8.5	Complications	282	References	285
8.5.1	Genotype by Environment Interaction	282		
8.5.2	Epistasis	284		

8.1 Genetic Analysis of Complex Traits

A primary goal of genetic analysis is to understand the causal relationships that connect the observed variation in phenotypes to the underlying genetic variation in the population. The simplest case was that observed by Gregor Mendel, where a single gene with two different co-dominant alleles presents a one-to-one correspondence between genotype and phenotype. In this situation, the ability to predict offspring ratios from any given parental phenotypes is very good. Most human traits do not follow these simple rules of transmission, but instead have a more complex association between genotype and phenotype. We become convinced that there is at least some genetic aspect to the transmission, because there is *familial resemblance*. These traits aggregate in families, but do not segregate like a single Mendelian gene. Such traits include stature and body proportions, facial features, skin color, and blood pressure. Many diseases may have a complex nexus of causes, but often the liability may differ between individuals and may be genetic in origin. In earlier years, a Mendelian framework was often superimposed naively on such data, with no testing of the formal requirements for simple modes of inheritance. We will show in this chapter that a fruitful way to approach the genetics of complex traits is to fit the data on individual genotypes and phenotypes to specific models that consider different ways in which the genetic variation may be causing the phenotypic variation. One outcome of this kind of model fitting is to map the genes responsible for the variation. But by the very nature of complex traits, there is also a role of environmental effects on the traits, and the observation that different genotypes respond differentially to environmental pressures means that the inferences about the genotype-phenotype association depends on the environmental context. Let us first consider some basic principles about variation at the phenotypic and genetic levels.

8.1.1 Variation in Phenotypic Traits

A fundamental idea to focus on in considering complex traits is that the primary feature that is of importance is among-individual *variation*. Nearly every trait shows some level of variation among individuals, from overall body size measurements, to the most minute features, such as fingerprints. Most biochemical traits also display variation, including the levels of many components of the blood (cholesterol, hemoglobin) and ranging up to the activities of metabolic enzymes in the liver. It is only because there is variation among individuals that there is an opportunity to identify underlying genes that themselves harbor genetic variation in the form of differences in DNA sequences. These gene variations in turn may mediate the phenotypic variation. Just because we can identify mouse mutants in orthologous genes that have profound effects on a particular phenotype, this does not guarantee that the population will harbor natural polymorphisms in that gene, which in turn will influence trait variability. Similarly, the genes that appear to be most responsible for variation in a trait may play a part in the mechanism for that aspect of biology that seems totally peripheral, or in many cases one has no clue why gene X influences trait Y. This level of decoupling of genetic and phenotypic variation may seem unnerving at first, but for many attributes of profound medical importance there is important phenotypic variability (such as susceptibility to atherosclerosis) and excellent understanding of relevant pathways, but relatively great uncertainty about the causes of variation.

8.1.2 Familial Resemblance and Heritability

Before we make the leap from phenotypic variation to seeking to find the genes responsible for that variation, there is one other attribute of the trait that is of vital importance. The trait might actually not have any genetic variation responsible for the phenotypic variation, but may instead be driven entirely by environmental influences, such as diet or exercise levels. Fortunately there is a rich history of study of the problem of detecting a role of genes in complex traits simply by asking whether the degree of resemblance among relatives is elevated above what one would see by chance.

A fundamental idea in quantitative genetics is that variability in a trait can be partitioned into components that contribute to that variability. We seek to explain the variability in the phenotypic measures in terms of both genetic and environmental factors. Environment is considered as a sort of trash-bin term to encompass all nongenetic factors that influence the phenotypic value. The simplest statement of a model is that the phenotypic value of an individual is composed of the sum of the genotypic value plus the environmental value:

$$P = G + E$$

where P = phenotypic value, G = genotypic value, and E = environmental value.

The phenotypic values of all individuals in a population have a mean and a variance around this mean. The variance is distinguished from other measures of variability by one mathematical property: different variances can be added to give a total variance and, conversely, a total phenotypic variance V_P can be broken down into its components, such as the genotypic variance V_G and the environmental variance V_E:

$$V_P = V_G + V_E$$

The idea that the sum of normally distributed factors yields a normal distribution whose variance is the sum of the variances of the components is true in the limit with many factors, and is a central idea in statistics (indeed, it is called the Central Limit Theorem) (Fig. 8.1).

However, the addition rule for variances applies only if genotypic and environmental values are independent of each other, i.e., when they are not correlated. If there is a correlation between the two, the covariance of G and E must be added:

$$V_P = V_G + V_E + 2\,\text{Cov}_{GE}$$

Let us take an example from the area of genetics that first introduced these concepts – agricultural studies. It is normal practice in dairy husbandry to feed cows

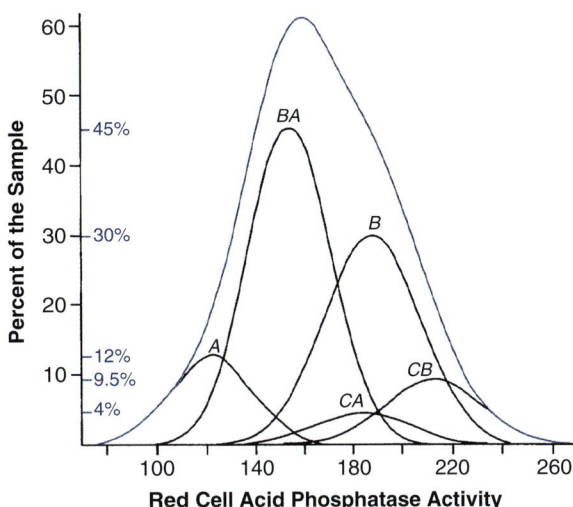

Fig. 8.1 The population distribution of acid phosphatase activity. The bell-shaped curve of total enzyme level (acid phosphatase) may be the sum of enzyme activities for acid phosphatase for genotypes of several polymorphic alleles, each having different overlapping acid phosphatase activity. Most phenotypes have a distribution in a population that results from summing over heterogeneous collections of genotypes (from Harris et al. 1968 Ann N Y Acad Sci. 151:232–242)

according to their milk yield. Cows that produce more milk are given more food. Such correlations of genetic and environmental factors tend to inflate the variance. Whether human societies present environmental perks to individuals in a way that is correlated with genetic proclivity is open to discussion. In any event, any such correlation between genetic and environmental variation should be identified, as it can cause serious problems in the modeling if it is ignored.

Another assumption is that specific differences in environments have the same effect on the various genotypes. When this is not so, there is an interaction between genotype and environment, giving an additional component to the variance V_{GE}. A prime example of genotype by environment interaction occurs with adverse reactions to drugs by a subset of individuals with a susceptible genotype. In the laboratory, where multiple replicate experiments may be run with identical genotypes of plants or animals, genotype × environment interaction is measured by testing the same genotypes across a range of environments.

The genotypic value V_G can be subdivided into several components: an additive component (V_A) and a component (V_D) measuring the deviation attributable to dominance and epistasis (V_I) from the expectation derived from the additive model. The dominance variance is contributed by heterozygotes (Aa) that are not exactly intermediate in value between the corresponding homozygotes (*aa* and *AA*). The variance contributed by epistasis refers to the action of genes that affect the expression of other genes. Hence, the concept of additive variance does not imply the assumption of purely additive action of the genes involved. Even the action of genes showing dominance or epistasis tends to have an additive component. The whole genotypic variance can be written:

Phenotypic variance	Genetic variance	Environmental variance	Genetic × environmental covariance
$V_P =$	$V_A + V_D + V_I$	$+ V_E$	$+ \mathrm{Cov}_{GE}$

To estimate these various components of variance, one measures the phenotypes of individuals that have different known relationships to one another. There are simple algebraic relationships between the correlations of phenotypic measures among relatives and these components of variance. The one that we will focus on is the relationship between parents and offspring. Sir Francis Galton observed a nearly linear relationship between points that represent family groups plotted as follows. Define the *x*-axis as the average of the two parents' phenotypes (also called the midparent), and the *y*-axis as the average of the offspring phenotype. If each point represents a nuclear family, then in a population, such points will fall along a line whose slope is called the "heritability." If the slope is 1.0, this would mean that the average of the offspring is always equal to the average of the parents, and so the resemblance is perfect. More typically, the slope might be about 0.5, meaning that for every increase by a factor of 2 in the phenotype of the midparent, the offspring mean would increase by 1. Galton called this line through the scatter of points (Fig. 8.2) a "regression" line, because the offspring tend to be less deviant from the population mean than do the parents. As Galton put it, the offspring *regress* toward the population mean.

The heritability, as calculated by the midparent-offspring regression can also be written as:

$$h^2 = \frac{V_A}{V_P}$$

When it is written in this way, it is clear that heritability can be considered as the proportion of the total phenotypic variance that is explained by additive genetic effects. This expression varies between 0 and 1, and many morphological traits, such as height, have a heritability in the range of 0.7–0.8, implying that the bulk

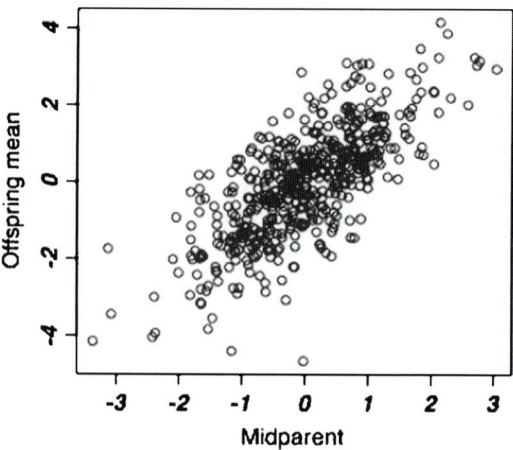

Fig. 8.2 The parent-offspring regression. The *x*-axis plots the average phenotypic measure of the two parents (the midparent) and the *y*-axis is the average phenotype of the offspring from each couple. Thus, each *point* on the plot represents a single nuclear family. The slope of the regression line through these points is the narrow-sense heritability. Francis Galton constructed many such scatter-plots, and inferred the degree of familial resemblance in this way

of the variance is genetic in origin. This very same term, often called the narrow sense heritability, has also been used by plant and animal breeders to predict the outcome of artificial selection for such economically useful traits as milk production in cows and egg laying in chickens. A high proportion of additive genetic variance implies that the trait will respond rapidly to selection. Heritability of many complex disorders, such as diabetes, is more in the range of 30–50%, implying that there clearly is a genetic component, but that this only explains part of the variation in disease risk.

It is worth stating carefully some of the properties of heritability:

(a) Heritability is a ratio. A ratio changes when either the numerator or the denominator changes. There is an increase in h^2 when the numerator (V_G, genotypic, or V_A, additive, variance) increases, or when the denominator (V_P, phenotypic variance) decreases. We could also say that a reduction in environmental variability will actually increase the heritability!

(b) The estimation of heritability is based on theoretical correlations between relatives. These correlations are valid only for random mating. Assortative mating leads to other correlations and, unless taken into consideration, produces systematic errors in the estimation of h^2. The correlations resulting from assortative mating were first calculated by Fisher [7]. These correlations can be used for adjustment of h^2.

(c) An estimation of h^2 is strictly valid only when the assumption is made that covariance and interaction between genotypic and environmental values are 0.

Correlations between relatives do not prove that there is genetic variability; they may also be caused by common environmental influences within families. In animal breeding, where the environment can be controlled, this factor might either be neglected or quantified. In humans, this is almost impossible. One of the major areas of research in human complex trait genetics today is to develop better automated methods for measuring differences in the environments that individuals have experienced. For example, many chemical exposures can be assessed by directly measuring residues in the bloodstream.

8.1.3 The Special Case of Twins

The use of twins has been much more popular in the past as a means for understanding genetic transmission of traits and diseases, but twin studies remain an excellent tool for developing concepts of genetic transmission. Identical or monozygotic (MZ) twins represent a wonderful experiment in nature (Fig. 8.3), since their genetic identity implies that differences between MZ

Fig. 8.3 Monozygotic twins have long fascinated human geneticists, and the questions about their biology change with advancing technologies. Initially interest focused on gross morphological similarities that were easy to measure. Now these questions center on issues of differential epigenetic modifications, differences in somatic mutations, altered patterns of X-inactivation, and similarities in brain activity as measured by functional MRI

twins must be due to accumulated perturbations of the environment [15]. In discussions on methods of quantitative genetics the use of twin data for quantitative assessment of the degree of genetic determination has been mentioned repeatedly. Indeed, twin investigations have played a major role in the history of human genetics. Especially in the field of behavior genetics, much of our current understanding is based on twin data. Therefore critical assessment of the twin method, its advantages, and limitations, is well motivated.

The twin method for assessing heritability is based on the biological observation that MZ twins originate from splitting of one zygote into two identical clones. It follows that any phenotypic differences between MZ twins must be largely caused by environmental influences. Somatic mutations may arise that generate differences between MZ twins, and efforts to quantify differences in somatic mutations between twin pairs using modern genomics technologies are under way in several laboratories. Environmental differences may manifest themselves by altering the epigenetic states of chromosomal regions, and an active area of research is to quantify the magnitude of differences in DNA methylation and histone acetylation between MZ twin pairs [2].

The degree of phenotypic similarity between MZ twins can be contrasted to the similarity between dizygotic (DZ) twins. Assuming that DZ twins are influenced by the same environmental differences but have only one-half of their genes in common by descent, the greater degree of resemblance of MZ twins provides a kind of measure of heritability. This heritability, however, is not the same as the parent-offspring regression approach mentioned above. Instead, the heritability one gets from twins is *broad-sense heritability*:

$$h_B^2 = \frac{V_G}{V_P}$$

where V_G and V_P refer to the total genotypic and phenotypic variance, respectively. This broad-sense heritability can be estimated from MZ and DZ twin pairs by calculating the average correlation between pairs of MZ twins (r_{MZ}) and the average correlation between pairs of DZ twins (r_{DZ}). The broad sense heritability is then $h_B^2 = 2(r_{MZ} - r_{DZ})$. It takes some algebra to show exactly why this is so, and it is of course true only when there is no shared environment effect. If there is a shared environment effect and it is measurable, one can adjust the heritability downward using another formula.

The above model for estimating heritability from twins makes some key assumptions about the biology that deserve to be considered carefully. In particular, twins have a unique shared environment that nontwins do not, and one has to worry whether that shared time in utero may influence their degree of resemblance. Because they have shared nutrition and environmental stresses, this shared environment might be expected to inflate the resemblance of twins. Whether the resemblance is augmented more in MZ twins than DZ twin depends on the details of how the environment is experienced (e.g., in one chorion or in two chorions).

One appreciates the effect of the uterine environment simply by examining medical attributes of twins and nontwins. Twins suffer from a higher frequency of abnormalities during pregnancy and at birth. Their lower birthweight can be attributed only partly to the shorter duration of gestation. The stillbirth rate and infant mortality in early life are considerably higher in multiple births than in single ones; in later years, twins run a higher risk than nontwins of becoming mentally retarded, which is presumably at least partly due to complications during pregnancy and at birth. Even the mean IQ of both MZ and DZ twins is slightly lower than that of control populations.

Some features of twins result in a higher chance that they *differ* in traits. X-inactivation in females occurs at the division of the zygote after X-inactivation (and is a fairly disruptive process). It therefore may happen that all cells in which a certain X-linked gene has been inactivated end up in one twin, while all the cells with active X chromosomes are found in the co-twin. This phenomenon leads to clinical expression of X-linked traits (such as Duchenne muscular dystrophy or color blindness) in only one member of a female twin pair that is heterozygous for the X-linked trait. A striking example is that two of the MZ Dionne quintuplets were color blind! The roles of X-inactivation and of intrauterine effects on epigenetic modifications are two of the many processes that occur during development and result in altered resemblance between twins. Thus, twins remain a fascination for geneticists, although simple calculation of heritability based on twin resemblance is clearly fraught with problems.

8.1.4 Embedding a Single Measured Gene Influencing a Continuous Trait

Consider a trait that has important medical consequences, where the trait has continuous phenotypic variation but we also know about an underlying mechanism for the trait, and we have managed to identify a gene whose variation influences the trait. As an illustration, consider the example of warfarin dose and the *VKORC1* polymorphism. Warfarin is an important anticoagulant drug that is used for heart disease patients and other circumstances where it is important to "thin" the blood to prevent thrombosis (clotting). The problem with warfarin has been that it has a narrow range of dose within which it is effective – too low a dose and it fails to delay clotting time, but at too high a dose it leads to hemorrhagic complications. For each patient there is a period where the optimal dose for that patient must be determined by approximate testing of coagulation status. To make matters worse, there is wide variability among individuals in the best therapeutic dose.

Rieder et al. [17] did a retrospective study on a large cohort of individuals who had been on warfarin therapy. These individuals had been through the battery of tests to determine their correct warfarin dose, and this was the phenotype being considered. The target of warfarin is the vitamin K epoxide reductase complex 1 (*VKORC1*), and a first guess might be that there could be polymorphism in this gene that requires different doses of warfarin for effectiveness. The study was a stunning success, finding mean differences in optimal warfarin dose across genotypes.

Other studies had associated warfarin dose with the cytochrome P450 2C9 (*CYP2C9*) gene, and this raises the question of whether there might be other genes elsewhere in the genome that also contribute to variation in optimal warfarin dose. Cooper et al. [3] did a scan of 181 European warfarin users and a replication sample of 374 individuals. They tested 550,000 SNPs and found that *VKORC1* had by far the strongest association ($P=6.2\times 10^{-13}$) and that a SNP in *Cyp2C9* has moderate significance ($P<10^{-4}$). Because none of the other SNPs attained significance in this study, the conclusion was that common SNPs with large effects on optimal warfarin dose are unlikely to be discovered outside of *VKORC1* and *CYP2C9*. In Sect. 8.1.5 we will see how to partition the variance in a trait, and to determine what fraction of the total variance in a phenotype is attributable to one or two major genes.

8.1.5 A Model for Variance Partitioning

The preceding section showed how the continuously varying phenotype can be thought of as having multiple causal factors that determine the phenotype. If one is lucky and has a handle on one of those factors, it is possible to determine what fraction of the total variation is explained by that one factor. Let us consider a model that seeks to explain variation in continuous traits as the sum of the effects over many loci. We can further assume that, as in the above section, we have a handle on one of the loci. Among a collection of individuals whose genotype is *aa*, we can define the mean phenotype as $-a$. For the *Aa* heterozygotes, let the mean phenotype be d, and for the *AA* homozygote, let the mean be $+a$. If the frequencies of the *A* and *a* alleles are p and q, an the population is in Hardy–Weinberg equilibrium, then the mean phenotype for the whole population is:

$$p^2 a + 2pqd + q^2(-a) = a(p-q) + 2pqd$$

If we were to plot the phenotypes for these three genotypic classes on the *y*-axis, and label the *x*-axis with the genotypes *aa*, *Aa*, and *AA* at coordinates 0, 1, and 2 (think of this as measuring 0, 1, and 2 copies of the *A* allele), then a regression through these points has many useful attributes. The increase in phenotype for each addition of an *A* allele is the "average effect of an allelic substitution" and has the value $\alpha = a + d(q-p)$. The *y*-axis values for the points on the regression line are $-2p\alpha$, $2pqd$, and $2q\alpha$. These are the "breeding values," a term from classic animal breeding analysis. They give the value of each genotype if the allelic substitutions were purely additive. But because there is dominance, we can calculate the deviation of each observed phenotype from this regression line fit (like a residual in a regression). These are the dominance deviations, and they are $-2p^2d$, $2pdq$ and $-2q^2d$, respectively.

From the breeding values and the dominance deviations we can calculate two important attributes of this trait. The additive genetic variance is the variance in breeding values. This is the sum of the squared deviations from the mean, weighted by the population frequencies or:

$$V_A = p^2(2q\alpha)^2 + 2pq[(q-p)\alpha]^2 \\ + q^2(-2p\alpha)^2 = 2pq[a+d(q-p)]^2$$

Recall that one of the definitions of narrow-sense heritability is the additive genetic variance divided by the phenotypic variance. This formula for the additive variance makes it clear that the additive genetic variance depends on allele frequencies, and it drops to zero with either $p=0$ or $p=1$. The variance in dominance deviations is the dominance variance. This is:

$$V_D = p^2(2q^2d)^2 + 2pq(2pqd)^2 + q^2(2p^2d)^2 = (2pqd)^2$$

The above formulae show clearly how the variance components are impacted by allele frequencies, and how heritability itself also varies with allele frequencies. The important point to remember about these measures of quantitative genetics is that these are parameters of a model, and the numbers have meaning only so far as the model explains the data. In many circumstances in plant and animal breeding for agricultural purposes, we have excellent data demonstrating the utility of the models. Human quantitative genetics cannot assess whether the model fits nearly as thoroughly, both because the environment is less well controlled and because the only crosses that can be observed are those drawn from a large, essentially randomly mating population.

In the absence of epistasis or genotype×environment interaction, the total genetic variance is the simple sum of the additive variance and dominance variance: $V_G = V_A + V_D$. This splitting of a variance into two parts is called variance partitioning, and a key part of modern quantitative genetics is to partition variance into components that have biological meaning. For example, in the *VKORC1* example, the total genetic variance in optimal warfarin dose can be partitioned into a component of variance attributable to the *VKORC1* gene, and another component that accounts for the rest of the genome. For further development of the models for partitioning quantitative genetic variation, see [6].

8.1.6 Relating the Model to Data

When there is a measured genotype that it is suspected is involved in a trait, the above model suggests a straightforward way to test what the effects of that gene on the phenotype are. We have to emphasize that this test is valid under the assumption that the effects across genes are additive. If this assumption is not right, then the inferred effects of the measured gene will not be valid – and the estimates can either spuriously overestimate the effects or underestimate them. Thus, a lot hinges on the validity of the assumption of additive effects.

First we bin the individuals in the population into the three bins based on their genotypes at the measured locus. If we plot them as in Fig. 8.4, it can be seen that one way to test the null hypothesis of no effect of this gene would be to perform a linear regression and test whether the regression coefficient (the slope) differs from zero. A nonzero slope indicates that the gene has an additive effect on this trait. A really wonderful aspect of this approach to the problem is that the slope is proportional to the additive effect contributed by this locus. Similarly, if the data are plotted with two x-values, where genotypes *AA* and *aa* are on the left, and *Aa* on the right, then the regression through these points will have zero slope if the heterozygotes are intermediate between the two homozygous classes. This would be true if there were zero dominance. So the estimator for the dominance effect is simply the regression coefficient obtained from the data when arranged in this way.

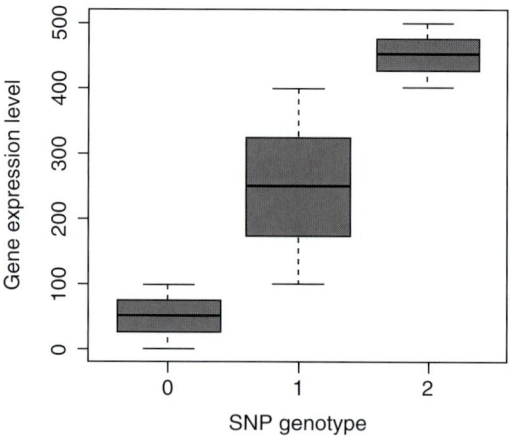

Fig. 8.4 Association of a single SNP with a continuous trait can be assessed by the regression plot depicted here: the *x*-axis has discrete elements for each genotype at one locus, and the *y*-axis is the continuously variable phenotypic measure. The slope of the regression line through these points yields the additive component of variance. If the genotype at this one SNP has no effect on the phenotype, this slope will be zero. The test of significance for whether the regression coefficient is greater than zero is the formal statistic test for whether this SNP shows an association with the phenotype

8.1.7 Mendelian Diseases Are Not Simple

While it is possible to trace the transmission of simple Mendelian traits and show that the trait is consistent with a major gene that is transmitted in the same way as smooth vs wrinkled peas, humans are so acutely aware of subtle phenotypic differences that the full spectrum of phenotypes associated with a major gene is almost never simple. And the departures from the pure Mendelian pattern are not always subtle. There are cases of individuals who are homozygous for a disease-causing allele, but are nevertheless perfectly healthy. Is this an example of a variant allele with reduced penetrance? Or, as is more often the case, are there other genes in the genome conferring a modifying influence, virtually suppressing the disease phenotype in these individuals? Mendelian disorders are not simply composed of two alleles, one healthy and one diseased; rather, a multitude of mutations that knock out function can result in disease, and there is also typically a series of alleles in the healthy group. In this case we have a mutation-selection balance between a fully functional gene and a rainbow series of alleles of reduced function. Heterozygotes may have even more intermediate phenotypes. It should be clear how this presents a situation where, despite the primary role of one major gene, there is nevertheless a continuous spectrum of disease severity in the population.

8.2 Genetic Polymorphism and Disease

Much of our understanding about the genetic basis of complex chronic diseases is based on our knowledge of Mendelian disorders, coupled with experiences in quantitative genetics of agricultural and laboratory organisms. We see that the complex disorders aggregate in families but do not segregate as Mendelian genes do, and so the inevitable conclusion is that the genetic basis involves many genes. In order to find those genes and to better understand the transmission of the disorder, we must construct a model for the genetic architecture. There may, for example, be a single major gene that accounts for most of the disease risk, but a series of modifier genes may temper the expression of this major gene. Or there may be ten genes, each of which is equally important in determining the trait. The frequency of the high-risk alleles may be very low, which may happen if there is natural selection driving them to low frequency in a mutation-selection balance, or they may have more intermediate frequency if they have little influence on reproductive fitness. In the next sections we will examine properties of polymorphisms in human genes and their impact on complex diseases.

8.2.1 Finding Genes Underlying a Complex Trait

In the preceding sections of this chapter we saw the consequences of a gene that has an effect on a trait and how its effect is added to the mix of effects of other genes and environment to add to the among-individual variability in the trait. Now imagine the situation in which you have no information about any of the underlying genes. The only data you have are the measurements of phenotypes of many individuals. You can determine that the trait has a heritable component because of the fact that relatives have correlated phenotypes. It is also clear that, if you did have measurements of the genotypes of a gene that happened to influence the trait (let the genotypes be *AA*, *Aa*, and *aa*), you might be able to see this from the fact that the phenotypes of these three genotypic groups might be different. The challenge is to identify an efficient way to find such genes.

8.2.2 Limitations of Pedigree Analysis

Probably the first approach one would consider for mapping the genes that underlie variation in a trait would be linkage analysis using pedigrees. This is a fundamental approach in human genetics, and it has a long history of success. As soon as one suspects a genetic basis for a syndrome, one has a collection of cases, and so by acquiring DNA samples from relatives, it becomes possible to test the linkage of the syndrome to anonymous marker loci throughout the genome. Methods for performing linkage analysis are provided elsewhere in this text, but there are several attributes of

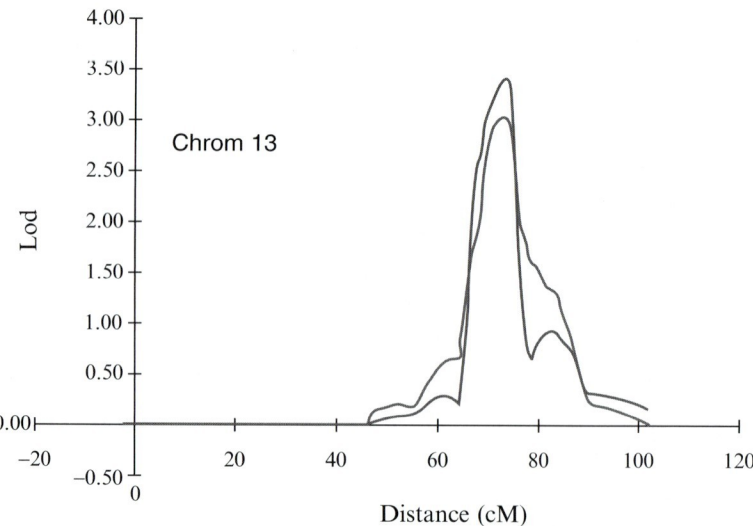

Fig. 8.5 Linkage analysis in pedigrees produces log-odds (LOD) plots like this one. In this example, there were 396 people in 22 families that were identified as having bipolar disorder in at least two members per family (these are called multiplex families). As described in the text, a model is fitted that provides the LOD score, representing the likelihood of obtaining such data given linkage at each position sliding along the chromosome. The salient feature to note is that the width of the LOD peak is nearly 20 cM or 20 Mbp across. This implies that there is relatively little confidence in the location, apart from there being a gene somewhere within that 20-Mbp region. (After [4])

linkage analysis that especially pertain to finding genes for complex traits that render linkage analysis somewhat less than ideal. First, if multiple genes are involved in a trait, the transmission pattern in a pedigree may be highly complex, and we may fail to detect the impact of any single marker through its marginal effect on risk. Even more serious is the fact that the resolution, in terms of accuracy of pinpointing the location of a gene on the genome, is limited by a combination of the sample size, the number of markers, and the number of meiotic exchanges represented by the pedigree. Typical pedigree studies have a mapping resolution of no better than 10 or 20 cM (centiMorgans), which is equivalent to approximately 10–20 Mbp of DNA sequence. This span typically encompasses dozens of human genes, and so one is left with a particularly challenging fine-mapping problem (Fig. 8.5).

8.2.3 A Prevailing Model: Common Disease Common Variants

For genetic association to be found by linkage disequilibrium, a fundamental constraint is that the rare allele must be relatively common (greater than about 10%) or the power to detect the association will be very low. Given that this approach can only find relatively common alleles, one can ask just how badly association mapping works for rare alleles. After all, common alleles, all else being equal, will contribute more to the total population variance in the trait, and will hence have a greater population attributable risk (defined elsewhere). What then are the prospects that common diseases will be caused by these relatively common alleles? Some Mendelian disorders can provide useful insight. If the disease is associated with a change in the environment, such as presence of malarial parasites, then alleles that may cause a disease (sickle cell anemia) may be driven to high frequency by the presence of a worse disease (malaria) against which they confer resistance. This kind of counterselection results in a heterozygote advantage, and any disease associated with alleles showing heterozygote advantage, either now or in the recent past, would be expected to have common alleles. The rapid expansion of the human population and the fact that many human populations have gone through population bottlenecks can also drive deleterious alleles to relatively high frequencies by drift and founder effects. In short, it was plausible that many diseases might have relatively common alleles as an underlying genetic cause. But these arguments do not make it particularly convincing that most complex diseases would be driven mostly by common alleles.

In the end it seems clear that successful identification of the common alleles causing disease would be the most desirable place to begin, since they likely harbor more of the population risk, and diagnostic tests that identify these tests are likely to identify more at-risk individuals than would tests for very rare alleles [1]. Now that more than 300 genome-wide association studies have been completed (http://www.genome.gov/gwastudies/), we can see that in no case was a very large portion of the total variance explained by the associated SNPs. While there are many success stories of finding well-replicated associations between disorders and common SNPs, the effect sizes of those SNPs are all very small. That the common SNPs do not explain much of the variation in risk does not imply that the Common Disease Common Variant Hypothesis is totally in error, however, because it is possible that the variance explained is eroded by the fact that we are looking at effects of marker SNPs, and perhaps not the actual SNPs causing the variation in risk. But the fact that so little of the variance in risk is explained is unfortunate, and it suggests that myriad rare alleles of larger effect might contribute a substantial portion of disease risk in humans.

8.2.4 Affected Sib-pairs

For a brief period in the 1990s, the affected sib-pairs method was very popular, and it met with some success in mapping genes for some traits (more than 600 papers applying affected sib methods appear in PubMed; see [20] for a review of methods). The basic idea is that because full pedigrees are time consuming, expensive, and difficult to collect, one could collect the single kind of relative best matched for age and environment, namely siblings. The principle behind mapping with affected sib-pairs is to score genetic markers throughout the genome in a collection of sibs, and then to scan the genotype data to identify regions of the genome that show an excess of genetic identity between the sibling pairs.

To make sense of affected sib-pair methods we need the concept of *Identity By Descent*. Two alleles sampled from either two individuals or the same individual are said to be identical by descent if they can be traced back to a single ancestor. If two parents have genotypes A_1A_2 and A_3A_4, then a pair of siblings may both be A_1A_3, in which case they share two alleles that are IBD, or they may be A_1A_3 and A_1A_4, in which case they share one allele IBD. Finally, the two siblings may be A_1A_3 and A_2A_4, in which case they share zero alleles IBD. If you consider all possibilities, you find that ¼ of the time they share two alleles IBD, ½ the time they are expected to share one allele IBD, and ¼ of the time they are expected to share no alleles IBD. In table form it looks like this:

	Count of alleles IBD		
	0	1	2
Observed	n_0	n_1	n_2
Expected	$n/4$	$n/2$	$n/4$

where $n = n_0 + n_1 + n_2$ is the total count of sib-pairs in the study. The test of association is to perform a simple Chi-square test. If the null hypothesis is rejected, and if there is an excess count of those sharing one and two alleles, then this SNP shows a positive association with the disorder. It is not so easy to explain the case when the null hypothesis is rejected with an excess of cases sharing zero alleles. It does not imply that the SNP has protective effects. There are many extensions of this simple affected sib-pair test, including use of LOD scoring, application to continuously varying traits, and application to cases where other circumstances result in an empirical deviation from ¼ : ½ : ¼ for the expected allele sharing.

The basic idea of affected sib-pair mapping is to find regions of the genome where affected sibs have an elevated chance of sharing more alleles than this null model. The LOD score equivalent to the Chi-square can be plotted for each SNP as one scans along the chromosome, resulting in plots remarkably like the LOD score plots from full pedigree mapping efforts. Affected sib-pair methods retain the advantage in being much faster and easier to collect than full pedigrees.

8.2.5 Transmission Disequilibrium Test

The problem of hidden population stratification was seen as a serious limitation of direct association testing, because any such stratification could result in false-positive test results that would be difficult to identify without a full independent replication study. The Transmission Disequilibrium Test (TDT) is one of the simplest designs that is immune to the problem of population stratification. Since it was first introduced by Spielman et al. [21], there have been dozens of extensions to allow a similar test approach to apply to other scenarios. We will focus on just the simplest application, since it shows why the test works so well.

Suppose our sample consists of trios, each of parents and an affected offspring. The essence of the TDT is to ask whether the two alleles at a heterozygous SNP are transmitted at a 50:50 ratio to the affected offspring. If the SNP is linked to a mutant allele at a disease-causing gene, then the transmission will be distorted. The test is essentially a Chi-square test for the co-transmission of the SNP and the disease state. If the count of trios where the *A* allele is transmitted is n_A, and the count of trios where the *a* allele is transmitted is n_a, then the Mendelian expectation is that each count would be $(n_A+n_a)/2$, so that the Chi-square is

$$X^2 = \frac{\left[n_A - \left(\frac{n_A+n_a}{2}\right)\right]^2 + \left[n_a - \left(\frac{n_A+n_a}{2}\right)\right]^2}{\left(\frac{n_A+n_a}{2}\right)} = \frac{(n_A-n_a)^2}{(n_A+n_a)}$$

This remarkably simple test has many positive attributes, not the least of which is the virtual immunity to distortions caused by population stratification. Its simplicity and robustness explain in part why it has been applied in nearly 1,200 published studies in human genetics.

8.2.6 Full-Genome Association Testing

In a major paradigm-shifting paper, Risch and Merikangas [18] pointed out the statistical limitations for mapping by determining linkage in pedigrees and carefully showed how we might be able to map in humans purely by association testing. This approach would work if there was relatively little linkage disequilibrium (LD) between SNPs or other genetic variants that are far apart along the chromosome. The hope was that a signature of high LD between a marker and a disease would indicate that the disease had to have risk factors mapping close to the SNP. This strongly motivated the quest for better understanding of LD across the human genome, and eventually led to completion of the human HapMap project [19]. The HapMap project provided us with a map of some 8 million markers and information on the pattern of LD across them in three human population samples. It also stimulated commercial entities to develop methods for genotyping those SNPs with high accuracy and low cost (see Sect. 8.3.2).

Risch and Merikangas [18] made the case for genome-wide association testing by showing that for a given sample size, one could have a greater probability of detecting association (higher power) by doing an association study than by doing a pedigree study. They considered a range of allele frequencies and genotypic relative risks for the disease-causing alleles, and several scenarios for the markers to be scored. It is impressive to see how accurately they foresaw the problems of testing 1,000,000 markers, estimating that a significance level of $a = 5 \times 10^{-8}$ would be needed to have a low probability of false positives. In Table 8.1,

Table 8.1 Sample sizes needed to detect a gene that elevates the risk of a complex disease under different assumptions of frequencies, genotypic relative risks, and testing approaches. (from Risch and Merikangas [18])

Linkage						Association			
						Singletons		SibPairs	
Genotypic risk ratio (γ)	Frequency of disease allele A (p)	Probability of allele sharing (Y)	No. of families required (N)	Probability of transmitting diseases allele A (P(tr-A)	Proportion of heterozygous parents (Het)	(N)	(Het)	(N)
4.0	0.01	0.52	4260	0.800	0.048	1098	0.112	235
	0.10	0.597	185	0.800	0.346	150	0.537	48
	0.50	0.576	297	0.800	0.5	103	0.424	61
	0.80	0.529	2013	0.800	0.235	222	0.163	161
2.0	0.01	0.502	296.71	0.667	0.029	5823	0.043	1970
	0.10	0.518	5382	0.667	0.245	695	0.323	264
	0.50	0.526	2498	0,667	0.5	340	0.474	180
	0.80	0.512	11,917	0.667	0.267	640	0.217	394
1.5	0.01	0.501	4,620.807	0.600	0.025	19,320	0.031	7776
	0.10	0.505	67,816	0.600	0.197	2216	0.253	941
	0.50	0.51	17,997	0.600	0.5	949	0.49	484
	0.80	0.505	67,816	0.600	0.286	1663	0.253	941

From [18], the paper that convinced the human genetics community that by scoring genotypes and phenotypes in direct association tests we ought to be able to identify genetic variants responsible for disease. The genotypic risk ratio (γ) is the ratio of risk of genotypes AA:aa reproduced from their paper, you can see the massive reduction in sample size needed in an association study relative to a pedigree study for the same chance of finding a disease gene.

Note that association testing works by demonstrating a statistical correlation between allelic states of an anonymous marker and a putative risk-elevating locus. This approach is quite distinct from linkage-based mapping methods. The latter rely on identification of recombination events within the sample, and noting that two genes are closely linked if there are relatively few such recombination events. Because linkage methods rely on counting recombination events, the resolution comes from having a large number of such events. Even the largest pedigrees might have only a few thousand recombination events, and this limits the resolution and the statistical confidence in map distances obtained in linkage studies. Association studies seem to depend solely on the statistical correlation of allelic states, but behind this test is the idea that the correlations arise from a combination of low rates of recombination in the ancestral history of the variation and from random genetic drift. Genes that are far apart will have allelic states randomized relative to one another by recombination over a few generations. If the genes are close together, drift can generate LD, and recombination will be very slow to erode it, so at equilibrium there is a tendency for tightly linked genes to display LD.

8.3 LD Mapping and Genome-Wide Association Studies

8.3.1 Theory and How It Works: HapMap and Genome-Wide LD

The basic principle behind LD mapping, also called association mapping, rests on a few key assumptions. Suppose a population is in a state of near equilibrium, with relatively little mixing through migration, so that the resulting genetic variation in the population is in Hardy–Weinberg proportions. In a population that has a steady rain of mutations, there will be a balance between the input of variation by mutation and its loss by random genetic drift. Some of the mutations have a deleterious effect; other mutations have no measurable effect; and very rarely some will be advantageous.

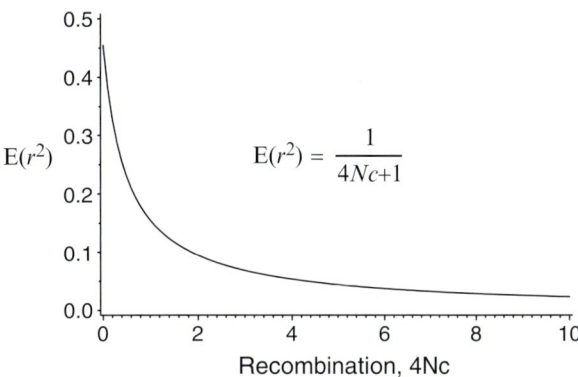

Fig. 8.6 Under the population genetic model, in which there is a balance between mutation, neutral drift, and recombination, there arises an equilibrium level of linkage disequilibrium (LD) as is plotted here. LD is here measured as the correlation coefficient r^2, as described in the text. The theory says that the expected value of r^2, or $E(r^2) = 1/(4N_e c + 1)$, where N_e is the effective population size, and c is the recombination rate. Note that the terms appear as the product $N_e c$, so that one expects the same LD if one doubles the population size and halves the recombination rate. The theory shows that there is a strong inverse relation between $4N_e c$ and LD

Because there is recombination occurring in each generation, the statistical association between mutant alleles will tend to erode over time; however, the effect of random drift is to keep the LD from completely decaying to zero. Instead, there is a balance between mutation, drift, and recombination that produces a steady state level of LD. An approximate relation at steady state is $E(r^2) = 1/(1 + 4N_e c)$, indicating that the expected linkage disequilibrium as measured by r^2 is a simple function related inversely to a term with $4N_e c$, where N_e is the effective population size and c is the recombination rate [14, 22]. According to this theory, one would get the same LD if one halved the recombination rate and doubled the population size, so long as $4N_e c$ is kept the same (Fig. 8.6).

Empirically, the data on human LD support the idea of association mapping very well. In particular, one does find SNPs that are in strong pairwise LD, but basically this only happens if the SNPs are in close physical proximity along the genome (Fig. 8.7). When a pair of SNPs is farther apart than 100 kb or so, they only very rarely have strong LD. This means that a strong association between a disease and an SNP provides fairly convincing evidence that a gene associated with elevating disease risk must reside near the marker SNP.

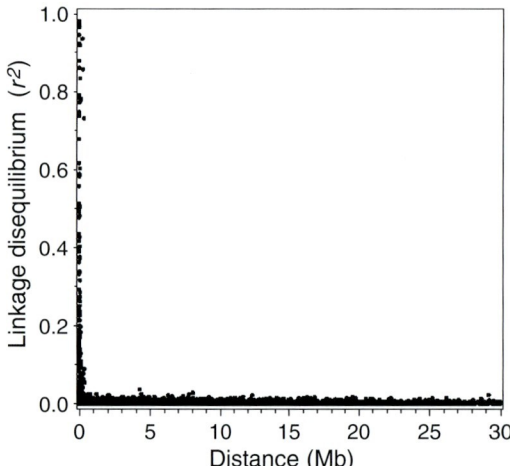

Fig. 8.7 A plot of the pairwise LD for a collection of SNPs from an early SNP study. In this study, the SNP genotypes were determined at several thousand SNPs in a few hundred people, and for each pair of SNPs it was possible to calculate the distance between them (for pairs on the same chromosome) and the level of LD between that SNP pair, showing clearly that SNPs that are far apart almost never have appreciable LD

According to this theory, if one imagines that there are SNPs responsible for disease, then there ought to be a statistical association between case/control status and the genotypes at nearby SNPs. Table 8.2 shows the steps for a genome-wide association test. The quality checking step is particularly vital, because despite the impressive gains in genotyping technologies, artifacts always creep into these studies and any slight perturbation from perfect genotyping calls can and usually does result in false-positive calls. Nearly every GWAS study had a moment of amazement when so many positive signals of association were seen, only for the number to dwindle as quality testing revealed more and more to be artifactual.

8.3.2 Technology: The Fantastic Drop in Genotyping Costs

One cannot overstate the importance of developments in the technology for large-scale molecular biology in accelerating the rate of discovery in human genetics. This is nowhere more true than in the area of genome-wide association testing. As recently as 2002, it cost about 1 U.S. dollar to score the genotype of an individual at one targeted nucleotide in the genome. Just 5 years later, one could score 1 million SNPs for $ 400, a 2,500-fold reduction in cost. This came about through development of mass manufacture of high-quality microarrays and methods to label and hybridize DNA to these arrays that gave highly accurate genotype calls. Competition among multiple manufacturers for competing technologies probably helped to drive the costs down as they drove speed and accuracy up. The next frontier is whole-genome sequencing at costs comparable to those of a CAT scan, and the human genetics community seems to have a consensus that this will happen within the next few years. Returning to the problem of mapping genetic variants that are associated with risk of complex diseases, even if we had complete DNA sequences of all the individuals in the case-control GWAS studies, many of the barriers to identification of genes responsible for inflated risk would still be there.

Table 8.2 Steps for a genome-wide association study

1. Identify the sample. Should be from a homogeneous population. Clearly defined cases and controls matched for gender and age.
2. Score the genotypes. Today this is almost universally done by applying standard commercial SNP genotyping chips from Affymetrix or Illumina.
3. Quality checking. It is necessary to take the genotype calls through rigorous testing for Hardy–Weinberg departures, spurious heterogeneity across runs, clustering of artifacts with cases, etc. Generally poor-quality DNA means removing some individuals, and some SNPs need to be removed.
4. Perform first-pass statistical inference. Nearly everyone starts with single-SNP tests, such as the Cochran–Armitage trends test.
5. Double-check all positives. The vast and overwhelming majority of positive hits seen at the first pass are errors of some sort. Disbelieve them until you fail to prove that they are errors.
6. Perform validation study. Standard practice is to repeat the study in another population to see that the same result is repeated.
7. Perform additional statistical inference. One can check for genotype x environment and epistatic effects, although the power will be low.

8.3.3 Case-Control Studies

Despite the fact that complex disorders are intrinsically embedded in likely interactions with environmental factors, the easiest design to begin genome-wide studies that identify genes associated with the disease is the case-control design. Because these tests entail examination of so many SNPs (typically 500,000 or 1 million SNPs), it is necessary to have large sample sizes so that the *P*-values of tests are sufficiently small, even when effect sizes are moderate, for the statistical tests to retain significance in the face of so many simultaneous tests. For example, the Wellcome Trust Case Control Consortium examined 2,000 cases for each of seven different disorders, and these were each contrasted against 3,000 controls [25]. With a complex disorder it becomes necessary to dichotomize individuals into these two bins, and it is crucial that this be done rigorously and homogeneously across the study., Other variables, such as sex, age, diet, etc. must either be randomized, controlled (e.g., by examining one sex only), or done as matched cases and controls, where the matching is for as many of these ancillary variables as possible. But case-control studies have a solid place in the history of medical research, and the simplicity of their design and ready access to samples stratified in this simple way means they are likely to continue to be useful. In addition, the first-pass statistical tests are very simple indeed.

8.3.4 Statistical Inference with Genome-Wide Studies

If the individuals in the study are placed into discrete bins of ;cases' and "controls," then the simplest way to consider the data is as a 3×2 table:

	AA	*Aa*	*aa*
Cases	n_{11}	n_{12}	n_{13}
Controls	n_{21}	n_{21}	n_{22}

It is legitimate to perform a 3×2 contingency Chi-square test on these data, provided the cell counts are sufficiently large (above 5 or so). For many SNPs one finds that the rare homozygous class has only a few observations, and in these cases one has to be careful about the aberrant behavior of the test statistic with small cell counts. One common way to solve the problem of small cell counts is to perform a permutation test to estimate the probability of a more extreme table. Another approach is to pool cells (e.g., the rarest genotype class, or column, could be pooled with the heterozygotes, yielding a 2×2 table).

Because the three genotypes are not totally independent categories, but rather there is an underlying order to them, a test more appropriate than the 3×2 contingency Chi-square is the Cochran–Armitage trend test. This test assumes that there is a linear trend in the phenotypes as one progresses from *AA* to *Aa* and *aa*, and obtains an asymptotically Chi-square test statistic under this model. Its primary advantage is in statistical power, because it effectively saves a degree of freedom. Just as for the contingency Chi-square, the significance test for the Cochran–Armitage trend test can be based on a permutation, and this allows it to be used even when cell counts are small. One needs to have *P*-values below 10^{-6} to attain significance across the whole study, and the Wellcome Trust Case Control Consortium was successful in achieving this for more than 80 SNPs across the seven disorders they mapped by GWAS (Fig. 8.8).

The genome-wide SNP chips are not successful at producing a reliable genotype call for every SNP in every individual, and the resulting missing data can be a challenge for analysis. One of the interesting features of dense SNP data is that because nearby SNPs are in LD, when one SNP call is missing, there is often some ability to predict the value of the missing genotype by use of the flanking SNPs. This "guessing the missing data" is known in statistics as *imputation* [8, 11]. While it sounds suspicious to fill in the missing data in this way, it is easy enough to test how well it works – simply take a large data set, blind yourself to some of the known genotype calls, and determine whether the imputation procedure gets the correct genotype call. When this is done, the misclassification error rate can be as low as 1%. With genome-wide SNP chips, whose density is one SNP every 3 kb on average, the imputation error rate varies with population but is typically less than 3%. Depending on the analysis, this can make a big difference. For the Wellcome Trust case-control study, use of imputed genotype calls often produced SNPs whose association *P*-values were more significant than the nonimputed SNPs.

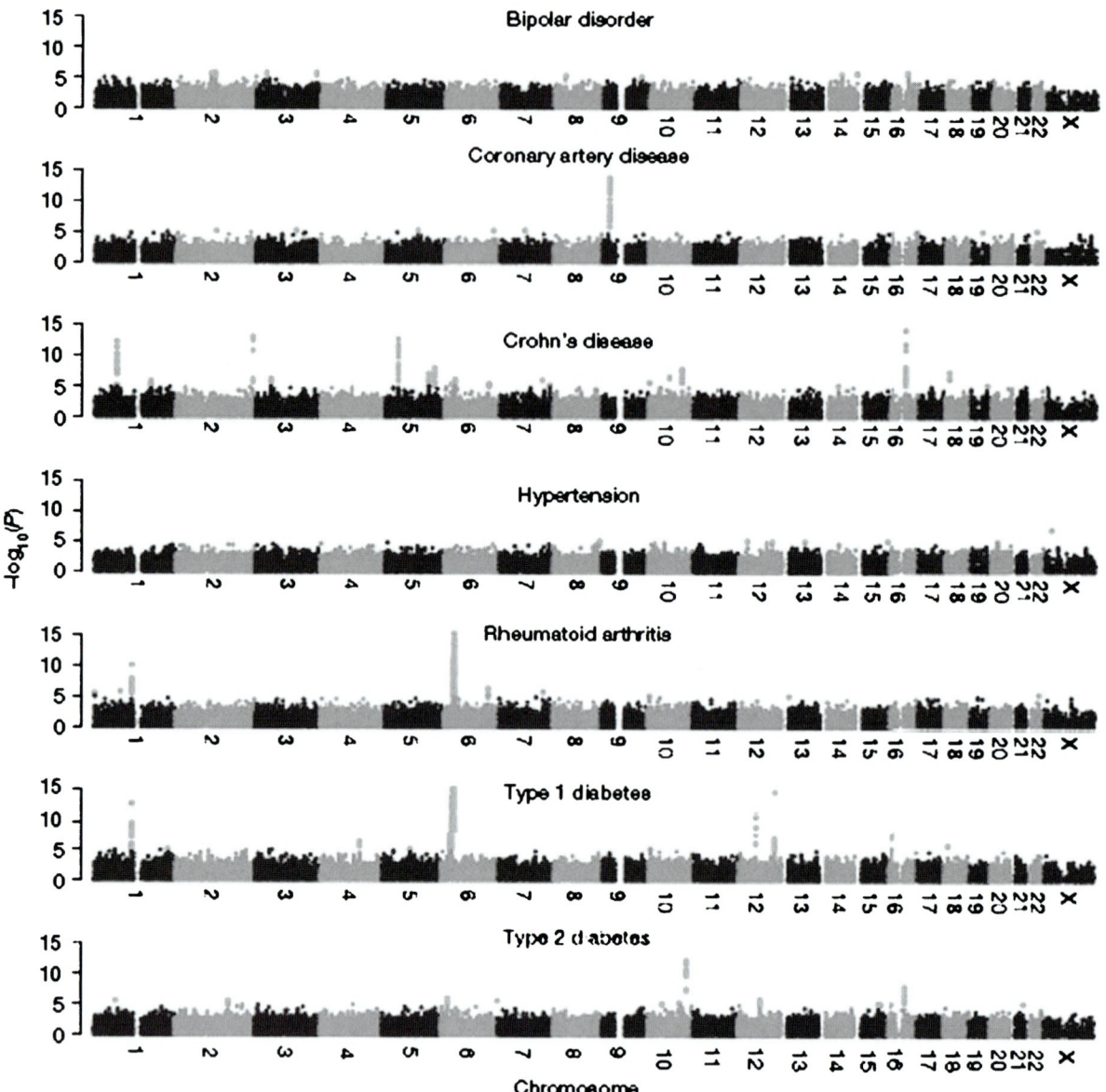

Fig. 8.8 Results observed by the Wellcome Trust Case Control Consortium in a large multi-disorder genome-wide association study. This study examined seven different complex disorders and performed genome-wide association tests for all traits using a common panel of healthy control individuals. Each of these plots (such plots are sometimes called Manhattan plots) shows the results of all 500,000 significance tests for association between each of the 500,000 SNPs and the specified disease. The y-axis of each plot is $-\log_{10}$ (P-value), so that a value of 6 implies a P-value of 10^{-6} (such an event would be seen by chance alone once out of every 1 million trials)

8.3.5 Replication and Validation

A problem with performing 500,000 tests at once is that one expects that 25,000 will be "significant" at $P<0.05$ by pure chance. Even when stringent criteria are applied to control for the false-positive rate, such as Bonferroni correction or use of False Discovery Rate, it is inevitable that if one places all the tests in rank order from the lowest to the highest P-value, that in amongst the significant tests at the lowest P-value range, there will be many tests that are spuriously considered positive. It is felt that the only way around this problem, to distinguish false positives from true posi-

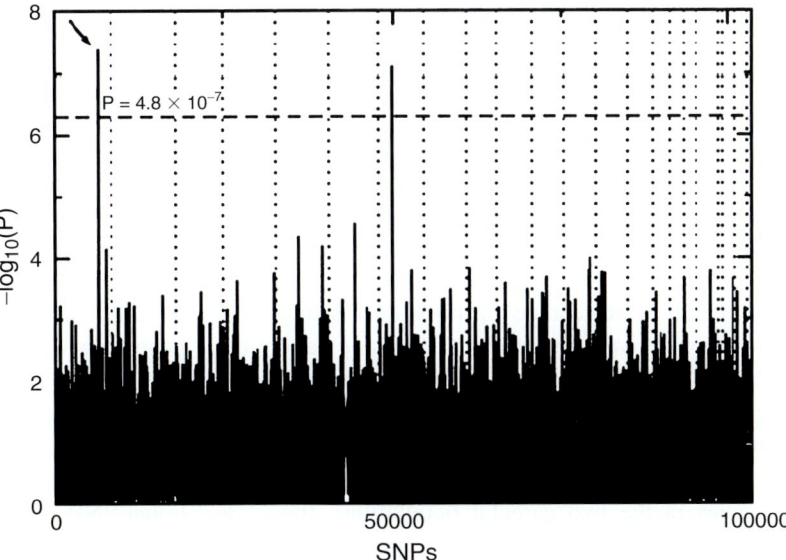

Fig. 8.9 A plot similar to that in Fig. 8.8, showing the outstandingly strong signal from the association of macular degeneration with complement factor H. The *dashed line* is the Bonferroni critical value for $P<0.05$, implying that any point above this line would be expected to occur by chance only 1 out of 20 times even after doing the 100,000 tests. (After [9])

tives, is to "replicate" the study. The word "replicate" is placed in quotes because of course there is no way to truly replicate a human study. Each individual is unique and each set of environmental circumstances is unique. At best, a second study on a similar but independent second population sample might identify overlapping sets of genomic regions harboring variation associated with disease risk. If so, this does indeed lend support to the initial positive result. The rub is that the second population is not identical, and the differences in genotypic and environmental composition between the two studies may in fact account for the difference between the results. That is, it may truly be a positive in the first study and not in the second. For now we hope that this is relatively rare, and are forced to rely on replication as a signature of real and repeatable effects.

8.3.6 Age-Related Macular Degeneration and Complement Factor H

In the early days, when the human genetics community was coming to grips with the idea that genome-wide association studies might actually work, Klein et al. [9] published a paper that showed that it could work far better than anyone could have hoped. The disorder was age-dependent macular degeneration, and they applied a simple case-control design. What was remarkable about the study was that they genotyped only 116,204 SNPs (using one of the early commercial chips) in a ridiculously small sample of 96 cases and 50 controls. To have a test that remains significant in the face of 116,204 tests would require an odds ratio of something like 6.0, and in fact, this is just what they found (Fig. 8.9). The positive hit was in the gene for complement factor H, and the result immediately sent the AMD community scrambling to understand the role of this immunity factor in macular degeneration risk.

8.4 Admixture Mapping and Population Stratification

8.4.1 How to Quantify Admixture

Before considering how to use admixture for mapping purposes, first consider how one might try to determine the degree of admixture of an individual's genome, and whether it is possible to infer which alleles came from which population. If one could identify the "parental" populations from which the admixed population derives, then the first thing to do is to estimate allele frequencies in the parental and admixed populations. In the extreme example where the allele frequencies are 0 and 1 in the parentals, it is easy to see that the allele frequency in the admixed population directly gives an estimate of the proportion of the alleles derived from the second population. If instead the allele frequencies in the two parental populations are p_1 and p_2, and the frequency in the admixed population is p_a, then the admixture proportion, a, giving the proportion of the alleles derived from the second population, is:

$$\alpha = \frac{|p_a - p_1|}{|p_2 - p_2|}$$

It turns out this is a maximum-likelihood estimator for this simple single gene case. The situation gets more interesting when we have genome-wide data. For each region of the genome it is possible to estimate the proportion derived from each parental population, but what we really want is to identify for each individual the population of origin of that individual's two alleles. This is much easier with runs of SNP alleles along the chromosome, or haplotype segments. Based on the frequencies in the two parental populations, there are methods that produce reasonably accurate calls of the stretches of the genotype derived from each parental population. One effective approach applies a Markov hidden Markov model to the genotype data [22].

8.4.2 Using Admixture for Mapping

If two different populations have differing risk of a complex disorder, and there is an admixed population that also manifests the disorder, if one could identify regions of the genome derived from each population for each admixed individual, then a means of mapping might be to look for an association between disease status and population-of-origin of genomic segments. These methods are still being refined, but they appear to be very promising, especially in populations with variation in the degree of mixing of the two genomes [23]. It is good to have large blocks of unrecombined chromosomal segments to attain power, but more finely diced genomic regions are needed in order to map with fine resolution. Also, the method works best when the parental populations are well defined, and when there are only two parental populations that are widely separated from each other historically (to maximize allele frequency differences).

A reasonable target for admixture mapping methods are diseases that differ in incidence between the two parental populations. End-stage kidney disease has a lifetime incidence of about 1.5% in Europeans and about 7.5% in African Americans. At the outset we do not know whether there is a genetic basis for this, but admixture mapping could in principle identify genetic factors if they exist. One particular form of end-stage kidney disease that shows strong familial clustering is focal segmental glomerulosclerosis (FSGS). Relative to Europeans, African Americans have a fourfold increased risk for FSGS and an 18- to 50-fold increased risk for HIV-1-associated FSGS. For this reason, Kopp et al. [10] identified 190 African-American cases and 222 controls for FSGS, obtained genome-wide SNP data and applied admixture mapping. On chromosome 22 they found a region with a LOD score of 9.1, implying that African ancestry for this chromosomal region inflated the risk of FSGS by more than ninefold. Subsequent genotyping of additional SNPs in additional samples narrowed the mapping to the gene *MYH9*. The precise mutation(s) responsible for the elevated risk of African alleles are still not known, but this success and the relative ease of application of admixture mapping in studies of African American population samples, make it likely that we will see many future successes in its application.

8.4.3 The Perils of Population Stratification

Many complex disorders display a wide range of incidences across different human populations. At the outset we cannot say whether the difference in incidence is due to a difference in gene frequencies or whether differences in environmental exposures account for the variation in disease risk. Sometimes a population will face a change in an environmental factor, and then the role of environment can become starkly clear. For example, the increase in saturated fat consumption in the diet of Chinese, especially in large cities, is being accompanied by a sharp increase in cardiovascular disease [24]. The increase in protein content of the diet in post-World War II Japan was accompanied by an astonishing increase in the average stature of that population. But in addition to such clear environmental effects, many genes have allele frequencies that differ among populations, and whenever we try to do association tests when there are differences in disease incidence and allele frequencies, we must be wary of a serious artifact.

Suppose two populations have disease incidences of 4% and 20%. These two populations have been isolated geographically for thousands of years, and many alleles differ in frequency. Suppose one particular gene has allele frequencies of 0.10 and 0.30 in the two populations. Now imagine that there was a large influx of individuals from the second population into the first population, and the population sample consists of a 50/50 mix of individuals from the two populations, but investigators were unable to keep track of the ancestral origin of each individual. The population sample contains hidden stratification of these two population groups. The allele frequency in the sample would be $(0.10+0.20)/2=0.15$, and the disease incidence would likewise be the average of the two populations or 12%. But, assuming that there is zero association between this gene and the disease, the table of genotype and phenotype frequencies would be:

	AA	Aa	aa
Diseased	16	144	320
Healthy	76	856	2,580

This table was constructed by calculating the Hardy–Weinberg proportions in each population (frequencies of 0.01, 0.18, and 0.81 in one population and 0.04, 0.32, and 0.64 in the other), taking the average frequencies across the two populations for each genotypic class, and then calculating the disease incidence for each genotype. The Chi-square test of heterogeneity is $\chi^2 = 10.53$, for which $P<0.005$. We have generated an association that appears significant purely due to the fact that the population with the higher disease incidence happened by chance to have a higher allele frequency for this SNP. In fact, for any SNP having an allele frequency difference of sufficient magnitude between the two populations, there will be this same kind of spurious association. This is why it is so crucial to avoid hidden population stratification in association testing.

8.4.4 How to Correct for Hidden Population Stratification

Fortunately, there are ways to identify the problem of hidden population stratification that allow some degree of correction of the false positives it causes. First note that a mixture of two populations having different allele frequencies results in genotype frequencies that depart from Hardy–Weinberg proportions. The easiest way to see this is to imagine populations with allele frequencies 0 and 1. A mixture of the two would give 50% *AA*, 0% *Aa*, and 50% *aa* individuals. The allele frequency is 50%, but there is a massive deficit of heterozygotes (or excess of homozygotes). One way to tease apart the sample into its original populations is to try to find clusters of individuals each of which form a Hardy–Weinberg population. This is the basic idea behind the program STRUCTURE, which is widely used in heterogeneous population samples to try to understand its partitioning into units [16].

Another approach, first used in 1978 by L.L. Cavalli-Sforza's group [12], is to apply a principal components analysis to the genotype data. This is a multivariate statistical procedure that identifies linear combinations of the SNPs that explain the most among-individual variability (arbitrarily number coded as, for example 0, 1, and 2 for the three genotypes). Generally there are multiple orthogonal sets of "axes" or vectors of SNPs that are needed to describe the variation. What PCA does is provide the weightings for each SNP and each such principal component. In the end, one can simply plot these principal components for each individual, and to the extent that individuals are more genetically similar to each other, they will fall closer together in these plots. If there are separate clusters of individuals, as there might be if there were discrete populations, these would appear as clusters in the PCA plot. Recently this method was applied to a sample of some 7,000 individuals from Europe genotyped at 500,000 SNPs [13], and the plot of the first two principal components produces an astonishingly good reproduction of the geographic map of Europe (Fig. 8.10). What does this imply? Just that there is a measurable isolation by distance among Europeans, and that historically people have tended to marry and settle down not far from their birth place.

To use PCA for association testing, one could identify the discrete clusters and use this as a covariate in the analysis, trying to explain as much of the variance in disease risk by population of origin first, and then explaining the remainder with the allele frequencies. Alternatively, one could directly use the principal components loadings as cofactors in the association analysis. This is an area of active research, and some of the newer approaches for dealing with genetic ancestry

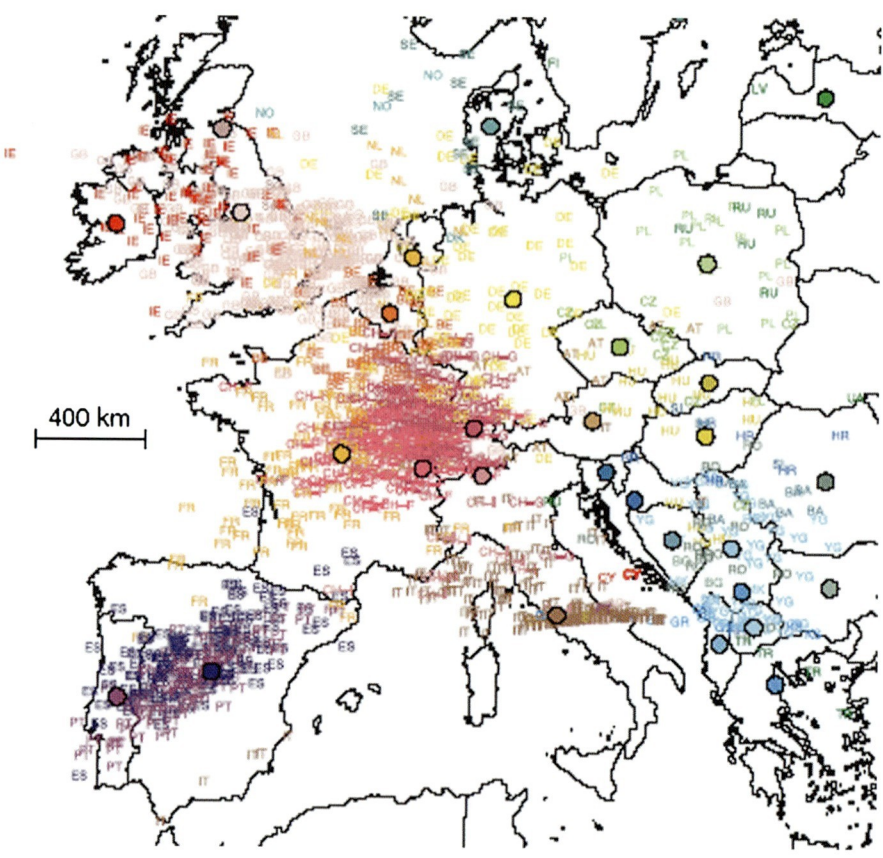

Fig. 8.10 The principal components plot from a study of 500,000 SNPs across a European sample of nearly 7,000 individuals. (From [13]). The raw genotype data were analyzed by Principal Components Analysis to try to find collections of SNPs that explain the most variance. A Principal Component is a combination of weightings of a subset of SNPs, and so after the PCA is run, each individual has a value for each principal component (PC1, PC2, PC3, etc.). If one plots a point (x, y) for the values (PC1, PC2) for each individual, one gets a plot like that shown. Note the impressive correspondence to the map of Europe, indicating that simple geographic distance is well correlated with the degree of genetic difference between individuals living that distance apart

and population structure in association studies are presented in Chap. 20 in this volume.

8.5 Complications

The models that we have presented up to now were purposely simplified so that the principle concepts would be clear. We assumed that the effects of many genes were additive, and proceeded to fit real data to this model without particularly questioning whether the model was correct. In fact, several factors can contribute to departures from this simple additive model, and many people think that these departures are virtually ubiquitous. Departures from additivity do not bring to a halt hopes of finding genes that act on complex traits, but they do make the problem more challenging.

8.5.1 Genotype by Environment Interaction

One of the challenges of studying the genetics of complex traits in humans is that we can never measure the same genotype in more than one controlled environment. Monozygotic twins at least give us some idea of the impact that different environments may have as a zygote undergoes development and eventually manifests mature phenotypes. With model organisms, where it is possible to produce many individuals with the same genotype, a very simple experiment produces a profoundly important result. The experiment is to simply rear the set of genotypes in two or more environments. Figure 8.11 shows an example of one such experiment, where a set of *Drosophila* lines were reared at two different temperatures, and body mass was measured in the resulting adult flies. As you can see, some lines gain weight

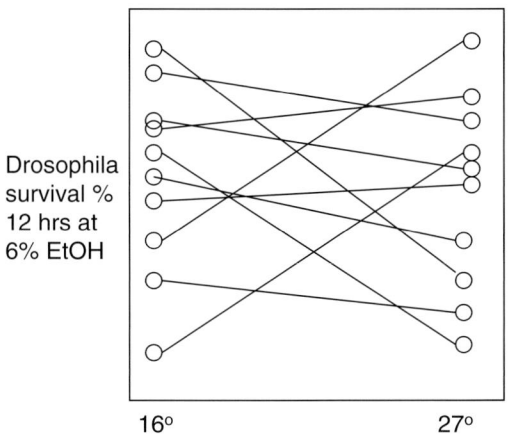

Fig. 8.11 A typical study of genotype × environment interaction obtained from model organism studies where the same genotypes can be reared in two or more environments. This kind of design nearly ubiquitously shows crossing of the mean phenotype lines, indicating a nonlinear effect of the environment attributable to a genotype × environment background. (Data from Kristi Montooth)

when moving from the low to the higher temperature, and other lines do the opposite.

Whenever the lines connecting the mean phenotypes across a range of environments cross, as they do in Fig. 8.11, this is a form of genotype × environment (G×E) interaction. More formally, we could set up an analysis of variance of these data, where the two factors are genotype and environment, and the interaction term in the analysis of variance would quantify the degree of G×E interaction. The impressive feature of this simple experiment is that whenever an experiment of this sort is done having any power at all, the observation of significant G×E is nearly universal.

Human examples of G×E interaction are a bit harder to find, but this is only because one has to define genotypes by particular targeted subsets of SNPs. A clear example of human G×E comes from drug responses. The particular example of *VKORC1* and warfarin response is a prime example. The phenotype of clotting time shows a strong interaction between genotypes at *VKORC1* and the environment of having taken warfarin. Given the ubiquity of G×E in animal and plant studies, one might expect that a differential response to drugs, varying with the genotype of the patient, is probably also nearly universal.

A particularly good example of a genotype by environment is seen in α_1-antitrypsin. The antiproteolytic activity of human serum was detected in 1897, and in 1900 Landsteiner showed this activity to be located in the albumin fraction. Antiproteolytic activity is measured by hydrolysis of artificial substrates by trypsin in the presence of the serum to be tested. The concentration of antiproteolytic activity increases quickly, for example with bacterial infection, after injection of typhoid vaccine, and during pregnancy. Interindividual differences in levels of antiproteolytic activity in the blood were first observed in 1963. A simple recessive mode of inheritance was proposed for low α_1-antitrypsin levels. Many different alleles have been discovered that vary widely in their activity levels. The gene is located on 14q31-32; it spans 10.2 kb, and has five exons. Two variants, Z and S, are especially important because the α_1-antitrypsin level is appreciably reduced relative to the common M type.

Subcutaneous injection of typhoid vaccine and diethylstilbestrol leads to a 100 % increase in activity of subjects with the MM type. Heterozygotes of the MZ type show a moderate increase, whereas in homozygotes of the ZZ type hardly any increase is seen. Many studies have shown that the rate of obstructive pulmonary disease in these ZZ individuals is at least 15 times the rate in the general population. Among ZZ homozygotes only 70–80% develop obstructive emphysema, and in heterozygotes the frequency is much lower. When a patient is exposed to recurrent bronchial irritation, such as that caused by smoking or frequent infections, these enzymes cause digestive damage to the lungs. Tobacco smoking enhances the danger of bronchial infections and hastens the progress of the disease. Once we are in the era of widespread genotyping for medical diagnostics, individuals who are found to be ZZ homozygotes and possibly ZM heterozygotes ought to get extra guidance regarding their exceptional risk of COPD, especially if they smoke.

The a_1-antitrypsin polymorphism is an example in which there is a subset of genotypes with heightened environmental sensitivity. The associated diseased condition can be thought of as one of reduced penetrance, and that penetrance is increased by an environmental trigger. The genetics of COPD appears to be complex, but for individuals with the ZZ genotype of a_1-antitrypsin, the disease is practically Mendelian. This is one of the more hopeful situations motivating the study of genotype × environment interactions – many diseases that we think of as complex and unpredictable may prove to have a simple gene of large effect whose otherwise low penetrance is triggered by an identified environmental factor. Such situations are also highly sought after because they provide a means whereby early genotypic analysis may result in an ability to give advice about environmental hazards that could greatly impact disease prevalence.

8.5.2 Epistasis

In the context of complex traits, epistasis is the situation when the risk of the disorder departs from an additive effect across two or more risk-elevating SNPs. Table 8.3 makes the situation clear. If one locus has marginal phenotypes (means across all other factors) of a_1, a_2, and a_3, and the other locus has marginal phenotypes of b_1, b_2, and b_3, then the two-locus genotypes have phenotypes that might fit the additive pattern as depicted in Table 8.3. Any departure from this additivity is an example of epistasis. One extreme example is where all the genotypes in the table have one phenotype, but the *aabb* genotype in the lower right corner has a radically different genotype. Consider two parallel pathways, where the organism requires the product of one or the other pathway, and the *aa* genotype knocks out one pathway, and the *bb* genotype knocks out the other. In this case, all the genotypes except *aabb* would get the required product, but the *aabb* doubly homozygous mutant would fail in both pathways and would produce the extreme phenotype. This kind of epistasis is rampant in model organisms, but when we try to test for it in human complex traits, it is not so easy to find. The reason is primarily due to the greatly reduce statistical power to detect such interaction effects. Given this low statistical power, it is premature to conclude that epistasis is not very prevalent in humans.

It has been argued that epistasis is especially likely to be found for phenotypes that are closely related to molecular function. The argument is that molecular biology is loaded with intermolecular interactions, and so if there is polymorphism in pairs of molecules that interact in some key pathway, then it is all the more likely that those variants may display an interaction in disease risk. Following this reasoning, Dimas et al. [5] examined pairs of SNPs for possible interactions in driving transcript abundance. They used the genome-wide expression data generated by the Sanger Centre in the 210 cell lines from the unrelated individuals whose genotypes were scored in the HapMap study. Reasoning that coding SNPs might be compensated for by flanking SNPs, they specifically looked for coding-flanking SNP pairs that influenced transcript abundance in nonadditive ways. After identifying nonsynonymous SNPs that affect expression and flanking SNPs that also affect expression, they performed an ANOVA test for each SNP pair to detect main effects and pairwise interactions. At a significance level of $P<0.001$ they expected 331 such interactions by chance, but observed 412. In this set were several cases of strong and highly significant interactions. Although the final conclusion does not overwhelmingly suggest that pairwise interactions are rampant in the human genome, the test had relatively low power given the small sample sizes. As our ability to apply tests of epistasis to larger samples targeted at specific pathways improves, it does seem likely that epistatic interaction among human genetic variants will be seen to play as important a role as has been found in genetic model organisms.

8.6 Missing Heritability: Why is so Little Variance Explained by GWAS Results?

One of the more surprising results from the genome-wide association studies has been that they uniformly find only SNPs of very small effect, and that even the sum of the effects of all the SNP associations that are found only explains a small proportion of the total genetic variance. This implies that if one has the SNP genotype for all the SNPs that impact a trait, one still has rather poor ability to predict the phenotype. This is surprising in light of the density of SNP genotypes obtained (one every 3 kb on average) and the large

Table 8.3 Two-locus genotypes and additive genotypic effects[a]

	BB	Bb	bb
AA	a_1+b_1	a_1+b_2	a_1+b_3
Aa	a_2+b_1	a_2+b_2	a_2+b_3
aa	a_3+b_1	a_3+b_2	a_3+b_3

[a]Define (a_1, a_2, a_3) as the effect of genotypes *AA*, *Aa*, and *aa* on the phenotype, and (b_1, b_2, b_3) as the effects of genotype *BB*, *Bb*, and *bb*, then the matrix below gives the expected genotypic effects for the nine pairwise genotype combinations assuming that the two loci have additive effects. These genotypic effects would be equivalent to the measured phenotypes in the environmental effect is zero

sample sizes (in some studies in excess of 30,000). The most dramatic example of this poor prediction ability is the case of body height (stature). The heritability of stature in humans is approximately 80%, making it one of the more strongly heritable complex phenotypes that we know. Despite this, even the top 20 SNPs found to be associated with stature explain less than 5% of the variance. Because we know from the heritability studies that there are genetic factors explaining the familial resemblance, this problem is sometimes called "missing heritability"; or, by analogy with dark matter in astrophysics, it is also called "dark heritability."

There are several reasons why a GWAS study may fail to explain more of the genetic variance in a complex trait. First, the SNPs that are used as markers are not expected to be the causal factors that drive the phenotype, but instead are correlated with the trait-affecting SNPs. This indirect association would erode the prediction power. Second, the SNPs that are used as markers are only quite common, because they were chosen from the HapMap studies, which specifically sought to catalog common SNPs. If much of the variance in traits is driven by rare SNPs, the correlation between the SNP markers that were used and these rare SNPs could be quite low. Third, it is clear that the complex traits that are studied include an environmental component, and if there are genotype × environment interactions (G×E), each SNP genotype will be averaged across all the environments, so that its effect would appear to be eroded compared with an SNP that had no such G×E interaction. Soon we hope to have the means to directly test for G×E interactions, but the primary challenge that must be tackled is to have accurate and meaningful measurements of the environment. Fourth, the statistical models have only made use of single SNPs at a time, and the trait may instead be driven by interactions among SNPs, or epistasis. It is also possible that there are other sources of heterogeneity, including epigenetic differences among individuals.

8.7 Concluding Remarks

The human genetics community is striving to improve methods for identification of genes that underlie complex genetic disorders and to understand how the effects of genes combine to produce inflated risk of disease. As part of the effort to better understand the role of rare alleles, the 1000 Genomes Project (www.1000genomes.org) was launched to provide the stimulus to accelerate the development of sequencing technologies that reduce the cost while increasing the speed and accuracy of whole-genome resequencing methods. Statistical methods need to be developed that accommodate the known complexities that may connect variation at the genotypic and phenotypic levels. While we can have confidence that methods of genome-wide association testing based on full genome sequence will be developed and improved in the near future, the prediction of an individual's disease risk given only his or her genome sequence may never attain useful accuracy (apart from extreme alleles that are nearly deterministic for some disorders, such as Mendelian disorders), especially if the disorder is heavily impacted by stochastic environmental factors, or by complex interactions between genotype and environment. But prediction of individual risk could make an enormous difference to public health, especially if environmental amelioration of that risk were possible, and so the drive to maximize prediction accuracy will motivate work in this area for years to come.

References

1. Altshuler D, Daly MJ, Lander ES (2008) Genetic mapping in human disease. Science 322:881–888
2. Cheung VG, Bruzel A, Burdick JT, Morley M, Devlin JL, Spielman RS (2008) Monozygotic twins reveal germline contribution to allelic expression differences. Am J Hum Genet 82:1357–1360
3. Cooper GM, Johnson JA, Langaee TY, Feng H, Stanaway IB, Schwarz UI, Ritchie MD, Stein CM, Roden DM, Smith JD, Veenstra DL, Rettie AE, Rieder MJ (2008) A genome-wide scan for common genetic variants with a large influence on warfarin maintenance dose. Blood 112:1022–1027
4. Detera-Wadleigh SD, Badner JA, Berrettini WH et al (1999) A high-density genome scan detects evidence for a bipolar-disorder susceptibility locus on 13q32 and other potential loci on 1q32 and 18p11.2. Proc Natl Acad Sci USA 96:5604–5609
5. Dimas AS, Stranger BE, Beazley C, Finn RD, Ingle CE, Forrest MS, Ritchie ME, Deloukas P, Tavaré S, Dermitzakis ET (2008) Modifier effects between regulatory and protein-coding variation. PLoS Genet 4:e1000244
6. Falconer DS, Mackay TFC (1996) Introduction to quantitative genetics, 4th edn. Longman Group Ltd, London
7. Fisher RA (1918) The correlation between relatives on the supposition of Mendelian inheritance. Trans R Soc Edinb 52:399–433

8. Guan Y, Stephens M (2008) Practical issues in imputation-based association mapping. PLoS Genet 4(12):e1000279
9. Klein RJ, Zeiss C, Chew EY, Tsai JY, Sackler RS, Haynes C, Henning AK, SanGiovanni JP, Mane SM, Mayne ST, Bracken MB, Ferris FL, Ott J, Barnstable C, Hoh J (2005) Complement factor H polymorphism in age-related macular degeneration. Science 308:385–389
10. Kopp JB, Smith MW, Nelson GW, Johnson RC, Freedman BI, Bowden DW, Oleksyk T, McKenzie LM, Kajiyama H, Ahuja TS, Berns JS, Briggs W, Cho ME, Dart RA, Kimmel PL, Korbet SM, Michel DM, Mokrzycki MH, Schelling JR, Simon E, Trachtman H, Vlahov D, Winkler CA (2008) MYH9 is a major-effect risk gene for focal segmental glomerulosclerosis. Nat Genet 40:1175–1184
11. Marchini J, Howie B (2008) Comparing algorithms for genotype imputation. Am J Hum Genet 83:535–539
12. Menozzi P, Piazza A, Cavalli-Sforza L (1978) Synthetic maps of human gene frequencies in Europeans. Science 201:786–792
13. Novembre J, Johnson T, Bryc K, Kutalik Z, Boyko AR, Auton A, Indap A, King KS, Bergmann S, Nelson MR, Stephens M, Bustamante CD (2008) Genes mirror geography within Europe. Nature 456:98–101
14. Ohta T, Kimura M (1969) Linkage disequilibrium due to random genetic drift. Genet Res 13:47–55
15. Poulsen P, Vaag A (2003) The impact of genes and pre- and postnatal environment on the metabolic syndrome. Evidence from twin studies. Panminerva Med 45:109–115
16. Pritchard JK, Stephens M, Donnelly P (2000) Inference of population structure using multilocus genotype data. Genetics 155:945–959
17. Rieder MJ, Reiner AP, Gage BF, Nickerson DA, Eby CS, McLeod HL, Blough DK, Thummel KE, Veenstra DL, Rettie AE (2005) Effect of VKORC1 haplotypes on transcriptional regulation and warfarin dose. N Engl J Med 352:2285–2293
18. Risch N, Merikangas K (1996) The future of genetic studies of complex human diseases. Science 273:1516–1517
19. Sabeti PC, Varilly P, Fry B, Lohmueller J, Hostetter E, Cotsapas C, Xie X, Byrne EH, McCarroll SA, Gaudet R, Schaffner SF, Lander ES, The International HapMap Consortium (2007) Genome-wide detection and characterization of positive selection in human populations. Nature 449:913–918 [Abstract] [PDF]
20. Schaid DJ, Olson JM, Gauderman WJ, Elston RC (2003) Regression models for linkage: issues of traits, covariates, heterogeneity, and interaction. Hum Hered 55:86–96
21. Spielman RS, McGinnis RE, Ewens WJ (1993) Transmission test for linkage disequilibrium: the insulin gene region and insulin-dependent diabetes mellitus (IDDM). Am J Hum Genet 52:506–516
22. Sved JA (1968) The stability of linked systems of loci with a small population size. Genetics 59:543–563
23. Tang H, Coram M, Wang P, Zhu X, Risch N (2006) Reconstructing genetic ancestry blocks in admixed individuals. Am J Hum Genet 79:1–12
24. Wang Y, Mi J, Shan XY, Wang QJ, Ge KY (2007) Is China facing an obesity epidemic and the consequences? The trends in obesity and chronic disease in China. Int J Obes 31:177–188
25. Wellcome Trust Case Control Consortium (2007) Genome-wide association study of 14, 000 cases of seven common diseases and 3, 000 shared controls. Nature 447:661–678

Lessons from the Genome-Wide Association Studies for Complex Multifactorial Disorders and Traits

8.1

Jacques S. Beckmann and Stylianos E. Antonarakis

Abstract Genome-wide association studies (GWAS) between common sequence variation and phenotypic variation were recently performed for a large number of human phenotypes. This was possible due to the discovery of the common variation in human populations, the development of technologies for large-scale and inexpensive genotyping, and the collection of very large number of well-phenotyped samples. GWAS were successful in identifying low risk alleles in candidate genes or loci. More importantly, these studies disclosed unexpected molecular pathways for different common, multifactorial disorders and traits, thereby providing new working hypotheses. Yet, the current clinical utility of these findings remains limited.

Contents

8.1.1	Lessons from Current GWAS	292
8.1.2	Genomic Topography of Trait-Associated Variants	292
8.1.3.	How Important Is the Identified Genetic Contribution to the Variance of the Traits Studied?	292
8.1.4	Predictive Power and Clinical Utility of the Trait-Associated Variants	293
8.1.5	Pathophysiological Dissection of Complex Traits	294
8.1.6	Concluding Remarks	294
	References	295

J.S. Beckmann (✉)
Service and Department of Medical Genetics,
Centre Hospitalier Universitaire Vaudois and
University of Lausanne, Switzerland
e-mail: Jacques.Beckmann@chuv.ch

S.E. Antonarakis
Department of Genetic Medicine and Development
University of Geneva Medical School and University Hospitals
of Geneva, 1 rue Michel-Servet, 1211 Geneva
Switzerland
e-mail: Stylianos.Antonarakis@unige.ch

The recent discovery of the common variation of genomes from different human populations (57, 58) and the availability of technical platforms for massive and low-cost genotyping of hundreds of thousands of polymorphic markers in very large number of samples has made genome-wide association (GWA) studies possible for numerous human complex multifactorial phenotypes (disorders and traits). The latter are complex traits, which are influenced by distinct combinations of multiple genetic and nongenetic factors, contributing together to a graded phenotype.

The results of the first series of GWA studies are very recent, since most of them have been published in the last 18 months (see Table 8.1.1 and Fig. 8.1.1 for the phenotypes studied); several important lessons have emerged from these studies, and a short summary of those is discussed in this addendum.

The prevailing hypothesis underlying this quest for genetic variation contributing to complex traits was that these diseases are essentially caused by a number of common disease variants, each with small to modest effects (the common disease–common variant hypothesis, (32, 51)). One of the few examples of an association of a genetic variation with a disease phenotype in

Table 8.1.1 Alphabetic list of phenotypes studied with GWAS (http://www.genome.gov/GWAStudies, June 1 2009 (20))

Addiction
Adiponectin levels
Age-related macular degeneration
Aging traits
AIDS progression
Alzheimer's disease
Amyotrophic lateral sclerosis
Anthropometric traits
Anti-cyclic citrullianted peptide antibody
APOE*e4 carriers with late-onset Alzheimer disease
Asthma
Asthma (toluene diisocyanate-induced)
Atopic dermatitis
Atrial fibrillation
Atrial fibrillation/atrial flutter
Attention deficit hyperactivity disorder
Attention deficit hyperactivity disorder (time to onset)
Attention deficit hyperactivity disorder and conduct disorder
Attention deficit hyperactivity disorder symptoms (interaction)
Autism
Basal cell carcinoma (cutaneous)
Behcet's disease
Bilirubin levels
Biochemical measures
Biomedical quantitative traits
Bipolar disorder
Black vs blond hair color
Black vs red hair color
Blond vs brown hair color
Blood lipid traits
Blood pressure
Blue vs brown eyes
Blue vs green eyes
Body mass (lean)
Body mass index
Bone mineral density
Bone mineral density (hip)
Bone mineral density (spine)
Brain imaging in schizophrenia (interaction)
Brain lesion load
Breast cancer
Burning and freckling
Celiac disease
Cholesterol, total
Chronic lymphocytic leukemia
Chronic obstructive pulmonary disease
Cognitive test performance
Colorectal cancer
Conduct disorder (interaction)
Coronary artery calcification
Coronary artery disease
Coronary disease
Coronary spasm in women
C-reactive protein
Creutzfeldt-Jakob disease
Crohn's disease
Crohn's disease and sarcoidosis (combined)
Cystatin C
Cystic fibrosis severity
Diabetes-related insulin traits
Diabetic nephropathy
Diastolic blood pressure
Echocardiographic traits
Electrocardiographic conduction measures
Electrocardiographic traits
End-stage renal disease
Environmental confusion in the home
Episodic memory
Essential tremor
Exercise treadmill test traits
Exfoliation glaucoma
Factor VII
Fasting plasma glucose
F-cell distribution
Fetal hemoglobin levels
Folate pathway vitamins
Freckles
Gallstones
General cognitive ability
HDL cholesterol
Hearing impairment
Heart failure
Heart rate variability traits
Height
Hemostatic factors and hematological phenotypes
Hepatitis B
Hip bone size
Hip geometry
Hirschsprung's disease
HIV1 viral setpoint
Hyperactive-impulsive symptoms
Hypertension
Hypertension (young onset)
Idiopathic pulmonary fibrosis
Inattentive symptoms
Incident diabetes
Inflammatory bowel disease
Inflammatory bowel syndrome
Insulin response
Intracranial aneurysm
Iris color
Ischemic stroke
Juvenile idiopathic arthritis
Kawasaki disease
Knee osteoarthritis
Late-onset Alzheimer disease
LDL cholesterol
Lung adenocarcinoma
Lung cancer
Lupus
Major CVD
Major depressive disorder
Male pattern baldness
Mean forced vital capacity from two exams
Mean platelet volume
Melanoma
Memory performance
Menarche (age at onset)

(continued)

8.1 Lessons from the Genome-Wide Association Studies for Complex Multifactorial Disorders and Traits

Table 8.1.1 (continued)

Menarche and menopause (age at onset)
Menopause (age at onset)
Methamphetamine dependence
Morbidity-free survival
Multiple sclerosis
Multiple sclerosis (age of onset)
Multiple sclerosis (severity)
Myeloproliferative neoplasms
Myocardial infarction
Myocardial infarction (early onset)
Myopathy
Narcolepsy
Neuroblastoma
Neuroblastoma (high-risk)
Neuroticism
Nicotine dependence
Nonsyndromic cleft lip with or without cleft palate
Normalized brain volume
Obesity
Obesity (early-onset extreme)
Obesity-related traits
Osteonecrosis of the jaw
Other metabolic traits
Other pulmonary function traits
Other subclinical atherosclerosis traits
Otosclerosis
Pain
Panic disorder
Parkinson disease (familial)
Parkinson's disease
Personality dimensions
Plasma carotenoid and tocopherol levels
Plasma eosinophil count
Plasma level of vitamin B12
Plasma levels of liver enzymes
Plasma levels of polyunsaturated fatty acids
Plasma Lp (a) levels
Progressive supranuclear palsy
Prostate cancer
Protein quantitative trait loci
Psoriasis
Pulmonary function measures
QT interval
QT interval prolongation
Recombination rate (females)
Recombination rate (males)
Red vs non-red hair color
Renal function and chronic kidney disease
Reponse to lithium treatment in bipolar disorder
Response to diuretic therapy
Response to iloperidone treatment (PANSS-T score)
Response to iloperidone treatment (QT prolongation)
Response to interferon beta therapy
Response to ximelagatran treatment
Restless legs syndrome
Rheumatoid arthritis
Sarcoidosis
Schizophrenia
Select biomarker traits
Serum bilirubin levels
Serum IgE levels
Serum markers of iron status
Serum metabolites
Serum urate
Serum uric acid
Skin pigmentation by reflectance spectroscopy
Skin sensitivity to sun
Sleep duration
Sleepiness
Smoking behavior
Smoking cessation
Soluble ICAM-1
Stroke
Subarachnoid aneurysmal hemorrhage
Successful cognitive aging
Systemic lupus erythematosus
Systemic lupus erythematosus in women
Systolic blood pressure
Tanning
Telomere length
Thyroid cancer
Thyroid-stimulating hormone
Tonometry
TP53 carriage
Treatment response for acute lymphoblastic leukemia
Treatment response to TNF antagonists
Triglycerides
Type 1 diabetes
Type 2 diabetes
Type 2 diabetes and 6 quantitative traits
Ulcerative colitis
Urinary albumin excretion
Urinary bladder cancer
Venous thromboembolism
Volumetric brain MRI
Waist circumference and related phenotypes
Waist circumference traits
Warfarin maintenance dose
Weight
Wet age-related macular degeneration
YKL-40 levels

the pre-high-throughput genotyping era was that of the *ApoE* variants and late-onset Alzheimer disease. In this particularly unusual case, a common haplotype of two nonsynonymous codon variants in the *ApoE* gene was strongly associated with Alzheimer disease. The risk allele (*ApoE4* defined by R112 and R158) in homozygosity has repeatedly been shown to confer a 15-fold risk of Alzheimer disease relative to all other allelic combinations (8, 53). After the introduction of high-throughput genotyping, several reports described the strong association of the complement regulatory gene factor H in age-related macular degeneration (ARMD) (11, 17, 19, 28). The successful identification

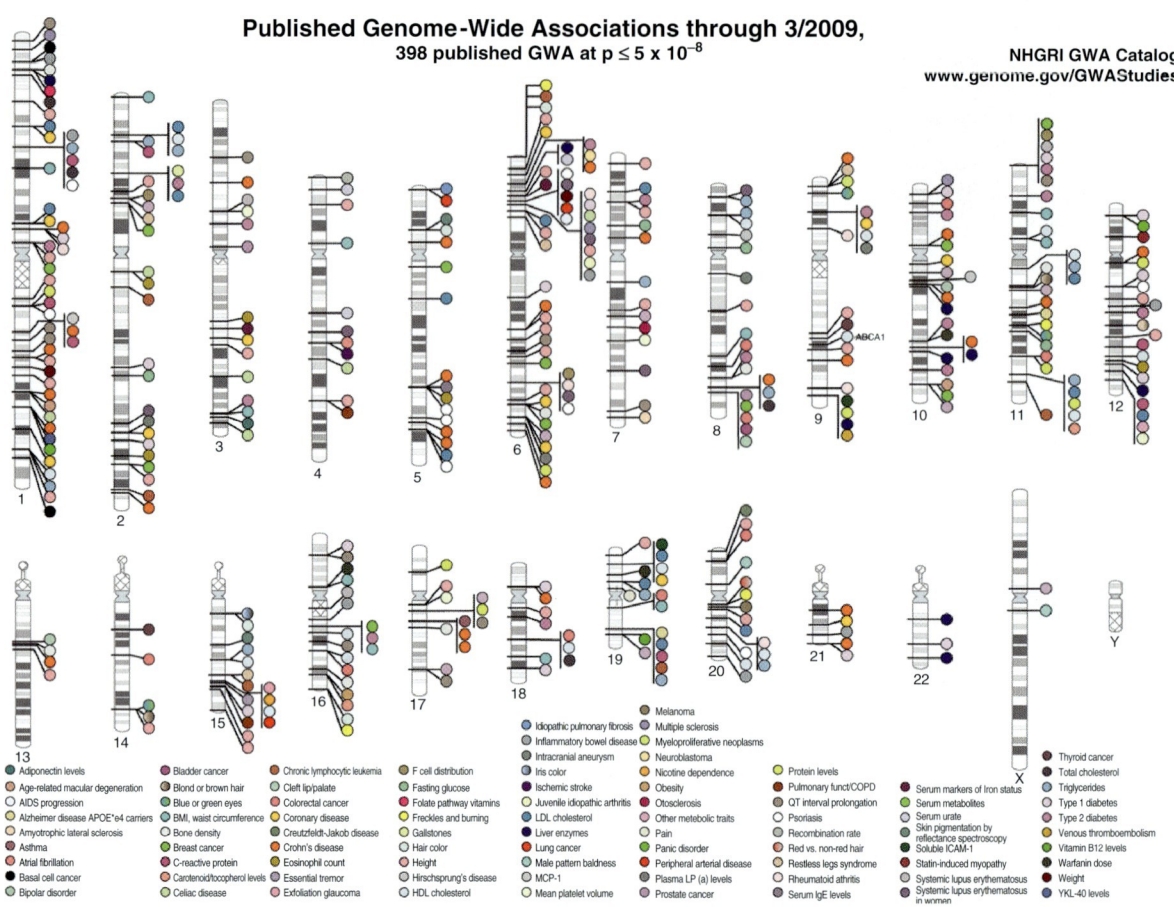

Fig. 8.1.1 Schematic representation of the published genome-wide associations; each *colored dot* represents an associated genomic locus for a given phenotype. (Reproduced from http://www.genome.gov/GWAStudies (20))

of a major locus involved in the etiology of ARMD, a complex phenotype, was heralded as a prelude to a rich harvest of numerous additional strong genetic risk factors for a wide variety of phenotypes. Thus, it took only a couple more years for this genome-wide association study (GWAS) approach to become widespread, and in the following years rapidly growing numbers of such studies were published, mostly in high-impact-factor journals reporting some successes in the quest for susceptibility factors for numerous phenotypes. The GWAS are usually done as follows (Fig. 8.1.2): DNA samples are collected from patients and well-matched controls (for dichotomous traits) or from a population for quantitative traits. Genotypes for several thousand to a million common SNPs are determined in all of these samples. Because of linkage disequilibrium (LD), these SNPs have been selected to extract approximately 80% of the genetic information. Thus, a subset of SNPs – called tagging SNPs – are genotyped, and because of LD the genotypes of nearby SNPs can be inferred, a process referred to as imputation. For each SNP (genotyped or imputed) a p-value is calculated for association of its alleles with the dichotomous or the quantitative trait. Statistical correction for multiple testing is applied to ensure detection of significant SNP-trait associations only. Odds ratios (OR) or relative risk (RR) estimates for the risk alleles are calculated from the data.

Today, it is widely recognized that initial mapping assignments require validation of the GWAS findings in independent cohorts (6, 23, 45). These steps are often accompanied or followed by high-resolution mapping of the associated genomic intervals (Fig. 8.1.2).

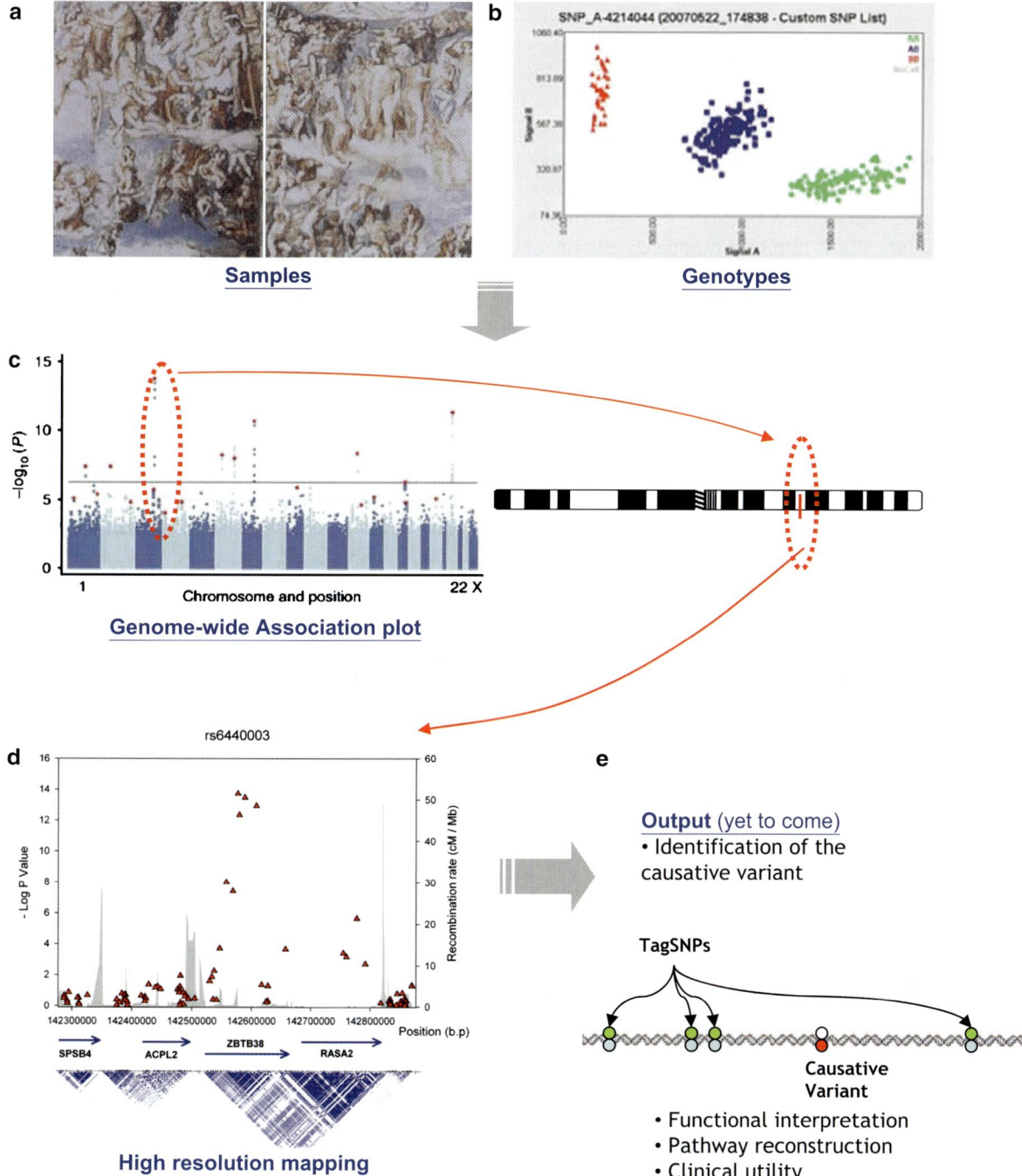

Fig. 8.1.2 Schematic representation of the different steps involved in GWA studies. The starting points for any GWAS are carefully phenotyped (population or case/control) cohorts from which DNA is collected from each participating individual. (**a**) Genotypes for several thousand to a million common SNPs are determined in all of these samples. (**b**) For each SNP (genotyped or imputed) a *p*-value is calculated for association of its alleles with the dichotomous or the quantitative trait. (**c**) The significance levels of the corresponding SNP-phenotype associations are plotted with the SNP order reflecting their corresponding locations along the chromosomes (from 1 to X): so-called Manhattan plots. (**d**) Close-up of about 0.5-Mb-long selected region on chromosome 3 (after (66)) for which SNP significance levels exceed a predefined threshold (*horizontal line* in **C**): *red triangles* represent the significance values of each SNP; *blue arrows* denote the genes falling in this interval and their orientation; *gray curve* shows the inferred recombination rate across this interval; *blue triangles* at the *bottom* show the extent of LD among the SNPs shown (*darker blue* strong LD; *lighter blue* weak LD). This region is subjected to fine mapping and/or replication in an independent validation cohort. (**e**) Once validated and size-restricted, the corresponding candidate intervals are subjected to a variety of analyses, including deep sequencing, and functional assays, in an attempt to eventually identify the causative variant(s) and its/their functional consequences and possibly to translate the resultant findings into clinical and public health utility

8.1.1 Lessons from Current GWAS

Altogether, over 200 GWAS studies have now been published, accounting for the statistical association and subsequent validation between a considerable number of genomic regions defined by SNP variation (more than 500 by June 2009; http://www.hugenavigator.net; http://www.genome.gov/GWAStudies and (20)) and common complex traits (see Table 8.1.1 and Fig. 8.1.1 for selected examples). The database at the NIH as of 06/04/09 includes 334 publications and 1,553 SNPs associated with complex traits. Among the latter there are several strong SNP-phenotype associations with p-values of 10^{-100} or less. These strong associations are in the vicinity of or involve the following genes: *UGT1A1* and serum bilirubin levels (10^{-324}); *VKORC1* and warfarin maintenance dose (10^{-181}); HLA and type 1 diabetes (10^{-129}); *SLC2A9* and serum urate (10^{-168}); *HLA-drb1* and rheumatoid arthritis (10^{-186}); *IRF4*, *HERC2*, and *MC1R* and hair color (10^{-127}, 10^{-103}, and 10^{-142} respectively); and *OCA2* and eye color (10^{-241}).

It should be remembered that these findings represent associated genetic markers that merely point out genomic regions, and the associated markers or regions are usually not the causal etiologic variants and do not point out genes causally responsible for the increased or decreased risk for these phenotypes. One thing of note, however, is that some of these loci include genes coding for known drug targets (e.g., *HMGCR* and statins (26, 69)), and nearly one in five of these association hits map in or near genes known to be mutated in Mendelian syndromes (21) that include the associated phenotype as part of the disease spectrum. For example, markers around the *MC4R* gene have been associated with early-onset obesity (12) or *IFHI1* in type 1 diabetes (46); pathogenic mutations in these genes have been shown to cause, respectively, "monogenic" obesity and type 1 diabetes. Complex SNP-phenotype associations and single-gene disorders may thus be part of a continuous etiologic spectrum on genetic diseases. The elucidation of the causal factors can thus provide an enriched list of candidate genes for both parts of this spectrum (2, 65). Furthermore, what applies to single-gene disorders, i.e., extensive allelic heterogeneity, locus heterogeneity, and incomplete penetrance, is likely to apply equally to common disorders. Thus the study of additional rare Mendelian disorders (beyond the 2,500 for which the molecular basis is known as of June 1 2009, http://www.ncbi.nlm.nih.gov/Omim/mimstats.html), will continue to assist the search for the causative variation on complex common phenotypes (2, 65).

8.1.2 Genomic Topography of Trait-Associated Variants

It is informative to examine the nature and genomic localization of associated SNPs, even if the latter are most likely only proxies to the causal etiologic variant. A recent analysis of the published validated SNP-trait associations revealed that 43% of these associations map to intergenic regions; the rest fall in gene coding regions, 45% in introns; 9% result in a nonsynonymous amino acid change, 2% map to 5′ or 3′ UTRs, and another 2% result in synonymous changes (20).

At first approximation, these initial observations suggest that genomic variants in noncoding genomic regions are likely to substantially contribute to the risk for complex common phenotypes. For these loci, the road from a validated statistical association to the identification of the etiologic variant and of its functional basis is still long, cumbersome, and uncertain (e.g., (43)). However, they also indicate that many associated SNPs are enriched in coding regions. These findings provide some enthusiasm for the identification of the functional causative variant(s) and of the molecular pathway(s) involved in the pathogenesis of the phenotypic trait. Nonsynonymous coding SNPs, which a priori are expected to be more deleterious, are significantly enriched in the associated regions. The corresponding SNPs and genes thus represent interesting candidates for further follow-up studies. A similar trend of enrichment was also seen for SNPs mapping in promoter regions (20).

8.1.3 How Important Is the Identified Genetic Contribution to the Variance of the Traits Studied?

Unfortunately for the overwhelming majority of studies, the OR for the risk allele of each associated SNP to the phenotypic trait is relatively low; in other words, most of the associated common variation explains only a small fraction of the phenotypic variance. Indeed, even after the study of large to huge cohorts (in some extreme cases with more than 30,000 patients and an equal number of controls), with over half a million SNPs for most examined traits, we have only uncovered what can be metaphorically expressed as "the tip of the iceberg": each genomic region identified contributes a modest effect, and collectively all associated regions for a given trait explain only a small fraction (5–10%) of the observed phenotypic variation attributed to genetic elements. With the notable exception of the HLA region on chromosome 6p, most risk factors confer on average a 20% increase in risk (or a RR of disease susceptibility or OR of 1.2). Thus, after the completion of the majority of the GWAS in a large number of appropriate samples most of the risk-contributing genetic variation remains elusive ((1, 14, 21, 43), etc.).

Let us illustrate this with adult human height, a good example of a complex prototypic trait that should be amenable to genetic dissection. Height, a clearly polygenic trait, has several attributes that make it an optimal target for genetic investigations: it has a strong genetic component (heritability in the order of 85–90%): it is easily and accurately measurable, so that height data are available for large cohorts; it is stable over most of adult life and shows an almost normal population distribution; though subject to environmental influences (as evidenced by the height differences in successive generations (7))), the latter play a small overall role. Thus, all these elements combined would suggest that the genetics of height should be easier to elucidate than that of other traits (such as hypertension or type 2 diabetes, T2D). The GWAS harvest was successful. Over 45 "height" loci were identified and confirmed, including genes known to be mutated in Mendelian syndromes of abnormal skeletal growth (e.g., (65)). But cumulatively all these loci explain less than 10% of the genetic variance (i.e., the rest is still unknown), the gene with the strongest effect accounting only for 0.6 cm per contributing allele (15, 34, 55, 65, 66). The situation prevailing in other studies is analogous. *FTO*, for instance, is the locus that has the largest effect on body weight so far, yet it accounts at most for 1% of the overall phenotypic variance (e.g., (4, 37, 70)). Even for the hard-to-dissect hypertension phenotype (59), a first set of genome-wide significant hits was recently obtained, with each susceptibility allele contributing less than 1 mmHg of systolic blood pressure (36, 47). However, for blood pressure, variations in the order of 2 mmHg may explain a substantial burden of cardiovascular disease at population level (56), so that the cumulative effect of multiple genetic variants might still be of public health relevance even if they had low predictive power at the individual level.

We are thus left with an as yet unsolved enigma: where is the rest of the missing heritability (39)? It is of interest to note that methods used to estimate heritability assume that there are no gene-environment interactions, which may be an unreasonable assumption. Where indeed have all the height variants gone, or is it that our premises and hopes are wrong? Numerous scenarios have been put forward to explain the relatively modest contribution of the common variation to the risk of polygenic, complex phenotypes (see, e.g., (1, 13, 14, 21, 29, 44, 61)). These include (1) incomplete marker coverage, and thus the possibility that important areas of the genome have not been examined; (2) allelic heterogeneity at a given locus (see, e.g., (10, 18, 40, 41, 49); (3) the contribution of rare variants (50, 67, 71), including structural (CNVs and smaller) variants (3) as risk factors for complex disorders; initial studies in autism (e.g., (42, 54, 68)) or schizophrenia partially support this possibility (e.g., (64, 72)); (4) epistatic interactions, i.e., the contribution of the combination of alleles at two or more loci to the risk for a given trait; (5) gene–environment interactions or different environmental exposures of subpopulations; (6) epigenetic modifications, such as methylation or transgenerational RNA (i.e., RNA transmitted through the sperm or egg), which might account for a parent-of-origin effect that would be lost under current GWAS investigations; and possibly (7) overestimation of heritability (39). Future studies, including the detection of all variants by high-throughput sequencing, may clarify these issues in the years to come.

8.1.4 Predictive Power and Clinical Utility of the Trait-Associated Variants

As a result of the modest effects contributed by each of the risk alleles tagged by the genotyped SNPs, their risk profiles (individually or even collectively) and the ensuing predictive power are still small. In comparison to the usual clinical factors, estimates relying on known genetic factors are poor predictors of risk, and only marginally improved prognosis (see, e.g., (38)). Thus, their clinical utility is so far limited. At present, this precludes acceptable and useful individual genotypic discrimination, personalized prognosis, and intervention (1, 9, 31, 65), except possibly in the field of pharmacogenetics, as illustrated by the warfarin example (5). It is a matter of current debate whether we will ever be able to provide effective preventive medicine statements based on the genetic profiling of individuals (see (14, 21, 29)). In spite of the current skepticism and the absence of validation of the proposed prediction algorithms, others have already taken the opportunity to venture into direct-to-consumer predictive genetic tests (http://ghr.nlm.nih.gov/handbook/testing/directtoconsumer). The controversies around these possibilities and the tendency for the control of genetic testing to shift from the clinical professionals into the hands of consumers (27) are new challenges for predictive medicine (24, 25, 30, 61). Hence, it is essential that consumers and health providers be made aware of the potentials, but also of the limitations and pitfalls, of many of the promises associated with genetic tests for complex traits. Proper (public and professional) education is thus a necessity, as is adequate genetic counseling of all those seeking to learn more about themselves by knowing their genetic variability profiles.

8.1.5 Pathophysiological Dissection of Complex Traits

Yet even if the associated SNPs do not accurately predict common traits, many of the newly identified loci highlight molecular pathways, including some not previously known to be involved in the corresponding pathophysiological processes (13, 21, 62). GWAS studies of Crohn disease revealed the central role of autophagy and bacterial defense in this disease process (48, 52) in an unbiased/unsupervised fashion, thereby opening new possibilities for translational research. Genes involved in signaling pathways, chromatin, the extracellular matrix or cell cycle, and cancer, among other matters, are implicated in height (22, 65); in T2D we encounter genes involved in pancreatic beta-cell formation and function as well as in pathways affecting glucose levels and obesity (13); furthermore, in obesity, we encounter genes highly expressed in the brain, and particularly in the hypothalamus, and possibly involved in weight regulation (60, 70). Pathway involvement is thus an important finding of GWAS studies and could provide potential new candidate genes, thus enabling further targeted investigations and a new way to categorize diseases.

It is also interesting to note that different GWAS identified common/shared risk SNPs for more than one common disorder revealing either similar or shared pathogenic mechanisms or coincident mapping of etiologic variants (Fig. 8.1.3). In many instances these co-occurrences are not surprising (e.g., that of Crohn disease and T2D and that of prostate, colorectal, or breast cancer), while in others the co-occurrence is unexpected (e.g., for a *TCF2* gene variant in prostate cancer and T2D) (13, 16, 33, 43). Some of the identified variants may have pleiotropic effects, being protective for one disease and a susceptibility factor for another (e.g., *TCF2*; see also Fig. 8.1.3). Discovering functional risk variants for one trait could thus help studies of other traits and diseases.

The discovery of genomic risk loci for complex phenotypes could also help to subclassify these phenotypes, in a similar way to the subclassification of similar phenotypes for Mendelian disorders (2). This may facilitate the study of the pathophysiology of certain phenotypes or identify individuals at risk for adverse reactions to certain medications.

8.1.6 Concluding Remarks

GWAS, based on the hypothesis that there are common risk alleles for common complex phenotypes, have provided a large number of very low risk alleles that are much less impressive than the risk conferred by the well-known HLA system known for several decades. On the other hand, some important risk alleles for a few disorders have been identified (e.g., late-onset macular degeneration (11, 17, 19, 28)); a few unknown regulators of quantitative traits have also been identified (e.g., *BCL11A* and fetal hemoglobin (35, 63)). The results yielded so far by GWAS provide a wealth of hypotheses for identification of molecular pathways involved in the complex phenotypes. However, the majority of the risk alleles identified provide only a very small risk to the phenotypes studied. For most of the studied traits, clinical utility thus remains to be demonstrated. This may be acutely true for those pleiotropic variants that, depending on the disease considered,

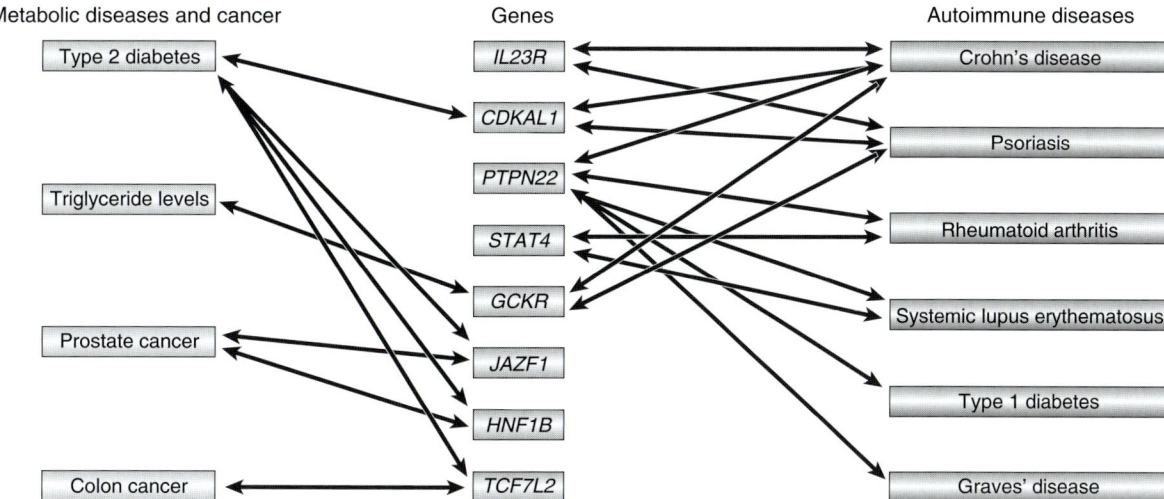

Fig. 8.1.3 Overlap of genetic risk factor loci for common diseases. A surprising finding of genome-wide association studies is that loci can be associated with the risk of developing two or more diseases, probably sharing common molecular causes. For some loci, distinct risk alleles are associated with different diseases or, in other instances, the same alleles can be protective for one disease and confer susceptibility to another. Eight loci are shown here for illustrative purposes with their impact on autoimmune, metabolic diseases or cancer. (After (13))

could assume either a protective or a predisposing role. It seems thus that the predictive value of the risk alleles is doubtful for a given individual. Most of the loci that contribute to the heritability of the common traits remain to be identified.

References

1. Altshuler D, Daly MJ, Lander ES (2008) Genetic mapping in human disease. Science 322:881–888
2. Antonarakis SE, Beckmann JS (2006) Mendelian disorders deserve more attention. Nat Rev Genet 7:277–282
3. Beckmann JS, Estivill X, Antonarakis SE (2007) Copy number variants and genetic traits: closer to the resolution of phenotypic to genotypic variability. Nat Rev Genet 8:639–646
4. Cauchi S, Stutzmann F, Cavalcanti-Proenca C, Durand E, Pouta A et al (2009) Combined effects of MC4R and FTO common genetic variants on obesity in European general populations. J Mol Med 87:537–546
5. Cavallari LH, Limdi NA (2009) Warfarin pharmacogenomics. Curr Opin Mol Ther 11(3):243–251 PMID 19479657
6. Chanock SJ, Manolio T, Boehnke M, Boerwinkle E, Hunter DJ et al (2007) Replicating genotype-phenotype associations. Nature 447:655–660
7. Cole TJ (2003) The secular trend in human physical growth: a biological view. Econ Hum Biol 1:161–168
8. Corder EH, Saunders AM, Strittmatter WJ, Schmechel DE, Gaskell PC et al (1993) Gene dose of apolipoprotein E type 4 allele and the risk of Alzheimer's disease in late onset families. Science 261:921–923
9. Daly AK (2009) Pharmacogenomics of anticoagulants: steps toward personal dosage. Genome Med 1:10
10. Duerr RH, Taylor KD, Brant SR, Rioux JD, Silverberg MS et al (2006) A genome-wide association study identifies IL23R as an inflammatory bowel disease gene. Science 314:1461–1463
11. Edwards AO, Ritter R 3 rd, Abel KJ, Manning A, Panhuysen C, Farrer LA (2005) Complement factor H polymorphism and age-related macular degeneration. Science 308:421–424
12. Farooqi IS, Yeo GS, Keogh JM, Aminian S, Jebb SA et al (2000) Dominant and recessive inheritance of morbid obesity associated with melanocortin 4 receptor deficiency. J Clin Invest 106:271–279
13. Frazer KA, Murray SS, Schork NJ, Topol EJ (2009) Human genetic variation and its contribution to complex traits. Nat Rev Genet 10:241–251
14. Goldstein DB (2009) Common genetic variation and human traits. N Engl J Med 360:1696–1698
15. Gudbjartsson DF, Walters GB, Thorleifsson G, Stefansson H, Halldorsson BV et al (2008) Many sequence variants affecting diversity of adult human height. Nat Genet 40:609–615
16. Gudmundsson J, Sulem P, Steinthorsdottir V, Bergthorsson JT, Thorleifsson G et al (2007) Two variants on chromosome 17 confer prostate cancer risk, and the one in TCF2 protects against type 2 diabetes. Nat Genet 39:977–983
17. Hageman GS, Anderson DH, Johnson LV, Hancox LS, Taiber AJ et al (2005) A common haplotype in the complement regulatory gene factor H (HF1/CFH) predisposes individuals to age-related macular degeneration. Proc Natl Acad Sci USA 102:7227–7232

18. Haiman CA, Patterson N, Freedman ML, Myers SR, Pike MC et al (2007) Multiple regions within 8q24 independently affect risk for prostate cancer. Nat Genet 39:638–644
19. Haines JL, Hauser MA, Schmidt S, Scott WK, Olson LM et al (2005) Complement factor H variant increases the risk of age-related macular degeneration. Science 308:419–421
20. Hindorff LA, Sethupathy P, Junkins HA, Ramos EM, Mehta JP et al (2009) Potential etiologic and functional implications of genome-wide association loci for human diseases and traits. Proc Natl Acad Sci USA 106:9362–9367
21. Hirschhorn JN (2009) Genomewide association studies–illuminating biologic pathways. N Engl J Med 360:1699–1701
22. Hirschhorn JN, Lettre G (2009) Progress in genome-wide association studies of human height. Horm Res 71(Suppl 2):5–13
23. Ioannidis JP, Thomas G, Daly MJ (2009) Validating, augmenting and refining genome-wide association signals. Nat Rev Genet 10:318–329
24. Janssens AC, van Duijn CM (2008) Genome-based prediction of common diseases: advances and prospects. Hum Mol Genet 17:R166–R173
25. Janssens AC, Gwinn M, Bradley LA, Oostra BA, van Duijn CM, Khoury MJ (2008) A critical appraisal of the scientific basis of commercial genomic profiles used to assess health risks and personalize health interventions. Am J Hum Genet 82:593–599
26. Kathiresan S, Melander O, Guiducci C, Surti A, Burtt NP et al (2008) Six new loci associated with blood low-density lipoprotein cholesterol, high-density lipoprotein cholesterol or triglycerides in humans. Nat Genet 40:189–197
27. Kaye J (2008) The regulation of direct-to-consumer genetic tests. Hum Mol Genet 17:R180–R183
28. Klein RJ, Zeiss C, Chew EY, Tsai JY, Sackler RS et al (2005) Complement factor H polymorphism in age-related macular degeneration. Science 308:385–389
29. Kraft P, Hunter DJ (2009) Genetic risk prediction–are we there yet? N Engl J Med 360:1701–1703
30. Kraft P, Wacholder S, Cornelis MC, Hu FB, Hayes RB et al (2009) Beyond odds ratios–communicating disease risk based on genetic profiles. Nat Rev Genet 10:264–269
31. Kruglyak L (2008) The road to genome-wide association studies. Nat Rev Genet 9:314–318
32. Lander ES (1996) The new genomics: global views of biology. Science 274:536–539
33. Lettre G, Rioux JD (2008) Autoimmune diseases: insights from genome-wide association studies. Hum Mol Genet 17:R116–R121
34. Lettre G, Jackson AU, Gieger C, Schumacher FR, Berndt SI et al (2008) Identification of ten loci associated with height highlights new biological pathways in human growth. Nat Genet 40:584–591
35. Lettre G, Sankaran VG, Bezerra MA, Araujo AS, Uda M et al (2008) DNA polymorphisms at the BCL11A, HBS1L-MYB, and beta-globin loci associate with fetal hemoglobin levels and pain crises in sickle cell disease. Proc Natl Acad Sci USA 105:11869–11874
36. Levy D, Ehret GB, Rice K, Verwoert GC, Launer LJ et al. (2009) Genome-wide association study of blood pressure and hypertension. Nat Genet (in press)
37. Loos RJ, Lindgren CM, Li S, Wheeler E, Zhao JH et al (2008) Common variants near MC4R are associated with fat mass, weight and risk of obesity. Nat Genet 40:768–775
38. Lyssenko V, Jonsson A, Almgren P, Pulizzi N, Isomaa B et al (2008) Clinical risk factors, DNA variants, and the development of type 2 diabetes. N Engl J Med 359:2220–2232
39. Maher B (2008) Personal genomes: The case of the missing heritability. Nature 456:18–21
40. Maier LM, Lowe CE, Cooper J, Downes K, Anderson DE et al (2009) IL2RA genetic heterogeneity in multiple sclerosis and type 1 diabetes susceptibility and soluble interleukin-2 receptor production. PLoS Genet 5:e1000322
41. Maller J, George S, Purcell S, Fagerness J, Altshuler D et al (2006) Common variation in three genes, including a noncoding variant in CFH, strongly influences risk of age-related macular degeneration. Nat Genet 38:1055–1059
42. Marshall CR, Noor A, Vincent JB, Lionel AC, Feuk L et al (2008) Structural variation of chromosomes in autism spectrum disorder. Am J Hum Genet 82:477–488
43. McCarthy MI, Hirschhorn JN (2008) Genome-wide association studies: past, present and future. Hum Mol Genet 17:R100–R101
44. McCarthy MI, Hirschhorn JN (2008) Genome-wide association studies: potential next steps on a genetic journey. Hum Mol Genet 17:R156–R165
45. McCarthy MI, Abecasis GR, Cardon LR, Goldstein DB, Little J et al (2008) Genome-wide association studies for complex traits: consensus, uncertainty and challenges. Nat Rev Genet 9:356–369
46. Nejentsev S, Walker N, Riches D, Egholm M, Todd JA (2009) Rare variants of IFIH1, a gene implicated in antiviral responses, protect against type 1 diabetes. Science 324:387–389
47. Newton-Cheh C, Johnson T, Gateva V, Tobin MD, Bochud M, et al (2009) Genome-wide association study identifies eight loci associated with blood pressure. Nat Genet (in press)
48. Parkes M, Barrett JC, Prescott NJ, Tremelling M, Anderson CA et al (2007) Sequence variants in the autophagy gene IRGM and multiple other replicating loci contribute to Crohn's disease susceptibility. Nat Genet 39:830–832
49. Plenge RM, Cotsapas C, Davies L, Price AL, de Bakker PI et al (2007) Two independent alleles at 6q23 associated with risk of rheumatoid arthritis. Nat Genet 39:1477–1482
50. Pritchard JK, Cox NJ (2002) The allelic architecture of human disease genes: common disease-common variant...or not? Hum Mol Genet 11:2417–2423
51. Reich DE, Lander ES (2001) On the allelic spectrum of human disease. Trends Genet 17:502–510
52. Rioux JD, Xavier RJ, Taylor KD, Silverberg MS, Goyette P et al (2007) Genome-wide association study identifies new susceptibility loci for Crohn disease and implicates autophagy in disease pathogenesis. Nat Genet 39:596–604
53. Roses AD (1997) A model for susceptibility polymorphisms for complex diseases: apolipoprotein E and Alzheimer disease. Neurogenetics 1:3–11
54. Sebat J, Lakshmi B, Malhotra D, Troge J, Lese-Martin C et al (2007) Strong association of de novo copy number mutations with autism. Science 316:445–449
55. Soranzo N, Rivadeneira F, Chinappen-Horsley U, Malkina I, Richards JB et al (2009) Meta-analysis of genome-wide

scans for human adult stature identifies novel Loci and associations with measures of skeletal frame size. PLoS Genet 5:e1000445
56. Stamler J, Rose G, Stamler R, Elliott P, Dyer A, Marmot M (1989) INTERSALT study findings. Public health and medical care implications. Hypertension 14:570–577
57. The International HapMap Consortium (2003) The International HapMap Project. Nature 426:789–796
58. The International HapMap Consortium (2005) A haplotype map of the human genome. Nature 437:1299–1320
59. The Wellcome Trust Case Control Consortium (2007) Genome-wide association study of 14, 000 cases of seven common diseases and 3,000 shared controls. Nature 447: 661–678
60. Thorleifsson G, Walters GB, Gudbjartsson DF, Steinthorsdottir V, Sulem P et al (2009) Genome-wide association yields new sequence variants at seven loci that associate with measures of obesity. Nat Genet 41:18–24
61. Todd JA (2006) Statistical false positive or true disease pathway? Nat Genet 38:731–733
62. Torkamani A, Topol EJ, Schork NJ (2008) Pathway analysis of seven common diseases assessed by genome-wide association. Genomics 92:265–272
63. Uda M, Galanello R, Sanna S, Lettre G, Sankaran VG et al (2008) Genome-wide association study shows BCL11A associated with persistent fetal hemoglobin and amelioration of the phenotype of beta-thalassemia. Proc Natl Acad Sci USA 105:1620–1625
64. Walsh T, McClellan JM, McCarthy SE, Addington AM, Pierce SB et al (2008) Rare structural variants disrupt multiple genes in neurodevelopmental pathways in schizophrenia. Science 320:539–543
65. Weedon MN, Frayling TM (2008) Reaching new heights: insights into the genetics of human stature. Trends Genet 24:595–603
66. Weedon MN, Lango H, Lindgren CM, Wallace C, Evans DM et al (2008) Genome-wide association analysis identifies 20 loci that influence adult height. Nat Genet 40: 575–583
67. Weiss KM, Clark AG (2002) Linkage disequilibrium and the mapping of complex human traits. Trends Genet 18:19–24
68. Weiss LA, Shen Y, Korn JM, Arking DE, Miller DT et al (2008) Association between microdeletion and microduplication at 16p11.2 and autism. N Engl J Med 358: 667–675
69. Willer CJ, Sanna S, Jackson AU, Scuteri A, Bonnycastle LL et al (2008) Newly identified loci that influence lipid concentrations and risk of coronary artery disease. Nat Genet 40:161–169
70. Willer CJ, Speliotes EK, Loos RJ, Li S, Lindgren CM et al (2009) Six new loci associated with body mass index highlight a neuronal influence on body weight regulation. Nat Genet 41:25–34
71. Wright AF, Hastie ND (2001) Complex genetic diseases: controversy over the Croesus code. Genome Biol 2: COMMENT2007
72. Xu B, Roos JL, Levy S, van Rensburg EJ, Gogos JA, Karayiorgou M (2008) Strong association of de novo copy number mutations with sporadic schizophrenia. Nat Genet 40:880–885

Epigenetics

Bernhard Horsthemke

Abstract According to its founder, Conrad Hal Waddington, epigenetics studies the causal interactions between genes and their products, which bring the phenotype into being. Waddington was the first to recognize that the developing organism is a nonlinear dynamic system and that development proceeds through a time-ordered set of epigenetic states, which are mitotically stable, but potentially reversible. The metastability of epigenetic states explains why developmental processes are buffered against minor changes in genotype and environment (a phenomenon called canalization), yet one genotype can give rise to more than one phenotype (phenotypic plasticity). Among several epigenetic inheritance systems, chromatin marking is the one that has received most attention from modern molecular biology. It has been recognized that tissue-specific gene expression patterns can be mitotically stable, because the genome is parceled into chromatin states that allow or repress use of the genetic information (permissive and repressive chromatin, respectively). Permissive and repressive chromatin states are characterized by specific patterns of DNA methylation, histone modification, and chromatin configuration. Within a cell, most often both alleles of a gene are either active or inactive. However, there are several examples where the two alleles of a gene, although identical in sequence, are functionally different. The difference can be parent-of-origin specific (as a result of genomic imprinting) or random (X inactivation and allelic exclusion). Epigenetic variation, occurring at random or induced by the environment, can lead to phenotypic variation and, in its most extreme form, to disease. In general, epigenetic states are cleared between generations. It is a matter of debate to what extent epigenetic states can be transmitted through the mammalian germline from one generation to the next.

Contents

9.1 History, Definition, and Scope 300
9.2 Chromatin-Marking Systems 300
    9.2.1 DNA Methylation.. 301
    9.2.2 Histone Modification 303
    9.2.3 Chromatin Remodeling................................... 305
    9.2.4 Synergistic Relations Between the Different Chromatin-Marking Systems 305
9.3 Specific Epigenetic Phenomena 306
    9.3.1 Genomic Imprinting... 307
    9.3.2 X Inactivation.. 310
    9.3.3 Allelic Exclusion in the Olfactory System...... 312
9.4 Epigenetic Variation and Disease............................... 312
    9.4.1 Obligatory Epigenetic Variation..................... 313
    9.4.2 Pure Epigenetic Variation............................... 313
    9.4.3 Facilitated Epigenetic Variation 315
9.5 Transgenerational Epigenetic Inheritance and Evolution 315
References.. 316

B. Horsthemke (✉)
Institut für Humangenetik, Universitätsklinikum Essen,
Hufelandstrasse 55, 45122 Essen, Germany
e-mail: bernhard.horsthemke@uni-due.de

9.1 History, Definition, and Scope

The term "epigenetics" was coined in the 1940s by the British scientist Conrad Hal Waddington. "Epigenetics" is a neologism which combines the words "epigenesis" (embryonic development) and "genetics." Before Waddington, embryology and genetics had been separate disciplines. Embryologists used chemical reagents and surgical techniques to study development, but could not explain the similarity between parents and offspring. Geneticists studied the behavior of genes during inheritance, but could not explain the development of a particular phenotype. Waddington abrogated the distinction between embryology and genetics. He defined epigenetics as: "[T]he branch of biology which studies the causal interactions between genes and their products which bring the phenotype into being" [35]. His concept also led him to expand the classic model of the genotype–phenotype distinction by including the "epigenotype":

Genotype + Epigenotype + Environment
= Particular Phenotype

According to Waddington the epigenotype is the "[S]et of organizers and organizing relations" involved in creating a particular phenotype during development. As the prefix "epi-" implies, this set of organizers and organizing relations operates *on* or *over* the genes.

A fundamental characteristic of the epigenotype is that it is relatively autonomous. It is not strictly determined by the genotype, but also by the environment and subject to random variation. Furthermore, it is relatively stable, but not as stable as the genotype. As a consequence, one genotype can have alternative epigenotypes and hence phenotypes (a phenomenon called phenotypic plasticity), and observed inheritance patterns may deviate from expected Mendelian inheritance patterns. In fact, epigenetics does not only investigate the question of how the genotype brings the phenotype into being, but also why the phenotype often does not correlate with the genotype. Inbred animals and monozygotic twins can be discordant for a certain phenotype because of epigenetic variation.

A very illuminating example of phenotypic plasticity is the bee. In principle, larvae with identical genotypes develop into bees with identical phenotypes. This is because developmental processes are canalized. However, larvae fed with royal jelly develop into queens. Thus, under certain environmental conditions, development follows a different trajectory and leads to a different phenotype.

The field of epigenetics did not see very much activity until the late 1980s, when molecular mechanisms controlling gene activity in higher eukaryotes and the inheritance of cell phenotypes (cellular memory) began to be unraveled. Most importantly, several studies had shown that the methylation of certain cytosine residues within the DNA could control gene expression and that these patterns could be inherited from cell to cell [16]. Concomitant with these studies and the advent of complete genome sequences, the term epigenetics was narrowed down to mean the study of cell-heritable changes in gene activity that are not based on differences in the DNA sequence. To date, most researchers equate epigenetics with the study of chromatin-marking systems (DNA methylation, histone modification, and chromatin remodeling). Although it can be argued that such a narrow definition excludes other important epigenetic inheritance systems, such as steady state systems (e.g., metabolic feedback loops) and structural inheritance systems (e.g., prion-induced protein refolding), the current focus on chromatin-marking systems has greatly advanced our understanding of development and variation in higher organisms.

Although chromatin changes also play a fundamental part in transcriptional control, epigenetics should not be equated with the study of transcriptional control. While the latter deals with the factors and mechanisms involved in switching gene activity on or off, epigenetics deals with the variation and cellular inheritance of gene activity states. It is their relative stability that matters in epigenetics.

9.2 Chromatin-Marking Systems

Although all cells of an organism – with a few rare exceptions – have the same genome, dividing fibroblasts give rise to new fibroblasts only and dividing hepatoblasts give rise to new hepatoblasts only. The tissue-specific gene expression patterns that characterize the fibroblast, hepatoblast, and other cells of the body, are cell heritable, because the genome is parceled into metastable chromatin states that allow or repress the use of the genetic information (permissive and repressive chromatin, respectively). Permissive

Table 9.1 Epigenetic diseases

Disease	Locus	Clinical findings
Diseases caused by a mutation in a gene encoding an epigenetic factor		
ICF syndrome	*DNMT3B*	Immunodeficiency, centromere instability and facial anomalies; recurrent infections
Rett syndrome	*MECP2*	Loss of acquired skills, microcephaly, mental retardation, autistic features, seizures, hand stereotypies, gait apraxia
Rubinstein-Taybi syndrome	*CREBBP*	Mental retardation, downslanting palpebral fissures, hypoplastic alae nasi, broad thumbs and toes
9q34 subtelomeric deletion syndrome	*EHMT1*	Mental retardation, hypotonia, brachycephaly, flat face with hypertelorism, synophrys, anteverted nares, tented upper lip, everted lower lip, prognathism, conotruncal heart defects, behavioral problems
Coffin–Lowry syndrome	*RSK2*	Mental retardation, full lower lip, soft hands with tapering fingers, pectus carinatum, scoliosis
X-linked alpha thalassemia mental retardation	*ATRX*	Mental retardation, tented upper lip, abnormal genitalia, alpha thalassemia
Diseases caused by an epimutation at a specific gene locus		
Prader–Willi syndrome	*MAGEL2, NDN, C15orf2, SNRPN,* snoRNA genes	Neonatal muscular hypotonia, feeding difficulties in infancy, hyperphagia and obesity starting in early childhood, hypogonadism, short stature, small hands and feet, sleep apnea, behavioral problems, mild to moderate mental retardation
Angelman syndrome	*UBE3A*	Microcephalus, ataxia, absence of speech, abnormal EEG pattern, severe mental retardation, frequent laughing
Beckwith–Wiedemann syndrome	*IGF2, CDKN1C*	High birth weight, hypoglycemia, macroglossia, exomphalos, increased risk of Wilms' tumor
Silver–Russell syndrome	*IGF2* and others	Pre- and postnatal growth retardation
Fragile X mental retardation syndrome	*FMR1*	Mental retardation, macroorchidism, long face, large ears, and prominent jaw, behavioral problems
Facioscapulohumeral muscular dystrophy	*D4Z4*	Progressive muscular weakness affecting the face and the scapulae followed by the foot dorsiflexors and the hip girdle

and repressive chromatin states are characterized by specific patterns of DNA methylation, histone modification, and chromatin configuration. Several recognizable syndromes are caused by mutations in genes that encode chromatin-marking proteins (Table 9.1).

9.2.1 DNA Methylation

DNA methylation in the mammalian genome refers to the methylation of cytosine (5-methyl-cytosine, m^5C) within a CpG dinucleotide. CpG is a palindromic sequence, and typically the cytosines in both strands are methylated. It is estimated that approximately 80% of C residues within CpG dinucleotides are methylated. As m^5C can undergo spontaneous deamination to thymine, methylated CpG dinucleotides are hot spots for mutation and are slowly eliminated during evolution. Therefore, in large parts of the mammalian genome, CpG dinucleotides occur at a much lower frequency than expected from the relative frequency of C and G residues. Certain "islands" of 0.5–5 kb, however, are GC rich and have the expected frequency of CpG dinucleotides [2]. In general, these CpG islands are unmethylated and overlap the promoter and exon 1 of a gene. It is likely that the binding of transcription factors or the process of transcription protects them from methylation.

Several techniques have been developed to determine DNA methylation patterns. The gold standard is the sequencing of sodium bisulfite-treated genomic DNA [13]. Sodium bisulfite converts C, but not m^5C, into uracil. During PCR, thymine replaces uracil. The PCR products are then sequenced directly or after subcloning into a plasmid vector. The presence of a CpG dinucleotide in the final sequence indicates that the cytosine residue was originally methylated. A TpG dinucleotide that is not present in the untreated DNA indicates the presence of an unmethylated cytosine in the genomic DNA. Cytosines outside CpG dinucleotides are always unmethylated and should present as thymines if the bisulfite conversion has been complete.

9.2.1.1 DNA Methyltransferases

5-Methyl-cytosine is not incorporated into the DNA by the DNA polymerase, but is the result of posttranscriptional modification of cytosine by DNA methyltransferases. Three active DNA methyltransferases have been identified in human cells: DNMT1, DNMT3A, and DNMT3B (Fig. 9.1). In addition, there is one putative DNA methyltransferase (DNMT2) and a protein that is highly similar to DNMT3A and B, but devoid of catalytic activity (DNMT3L). All methyltransferases use S-adenosyl-methionine as a methyl donor. DNMT1 is ubiquitously expressed and is a maintenance methyltransferase, which methylates hemi-methylated CpG dinucleotides in the nascent DNA strand after replication (Fig. 9.1). It is essential for maintaining DNA methylation patterns in proliferating cells. DNMT3A and DNMT3B are regulated during development. They carry out de novo methylation and thus establish new DNA methylation patterns (Fig. 9.1). DNMT3L cooperates with DNMT3A and DNMT3B to establish methylation imprints (see below). The activity and function of DNMT2 remains undefined.

Targeted mutations of the murine orthologs of the *DNMT1*, *DNMT3A*, and *DNMT3B* genes lead to methylation defects and pre- or post-natal lethality. Targeted mutations in the murine ortholog of *DNMT3L* impair imprinting and gametogenesis. In humans, mutations of the *DNMT3B* gene cause autosomal recessive ICF syndrome (immunodeficiency, *c*entromere instability, and *f*acial anomalies; Table 9.1). Centromere instability correlates with severe hypomethylation of the satellite DNA.

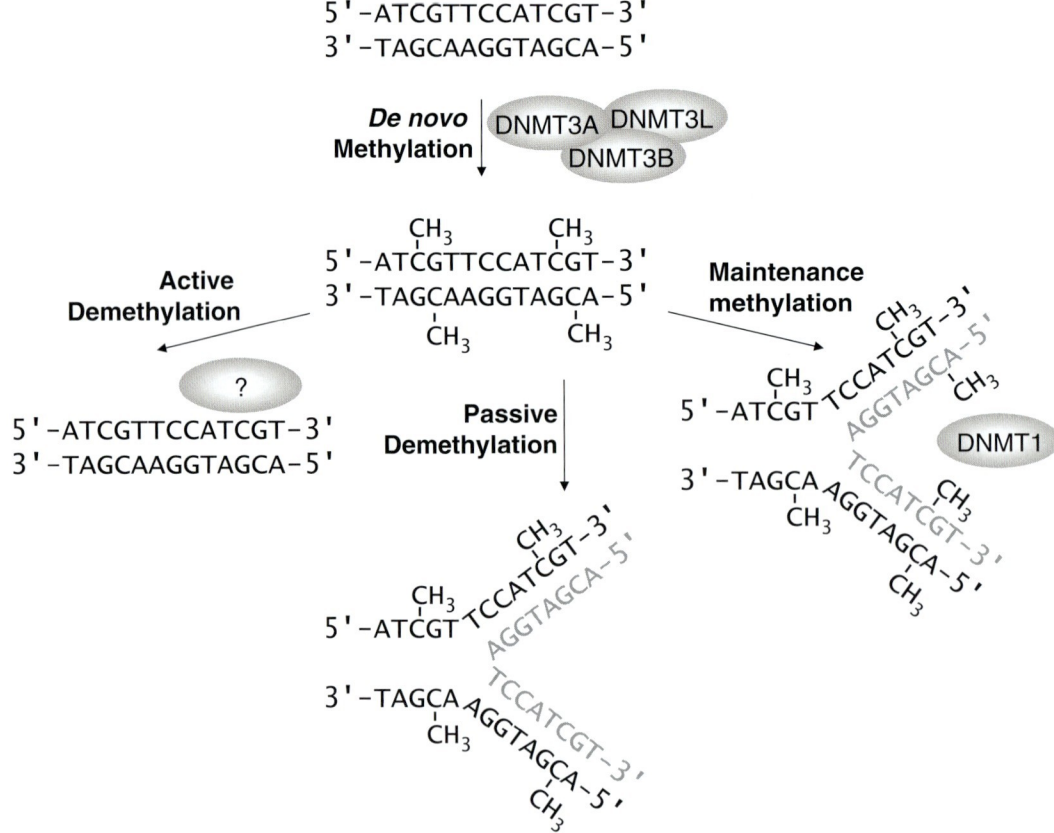

Fig. 9.1 DNA methylation. For details see text

At present, no bona fide DNA demethylase has been identified, although the existence of such an enzyme is likely. Shortly after fertilization, the sperm DNA is actively demethylated by the oocyte. Although there is no convincing evidence for an enzyme that removes the methyl group from m^{5}C, DNA repair enzymes can excise m^{5}C and replace it by C. Passive demethylation occurs by DNA replication in the absence of DNMT1 (Fig. 9.1). In cell culture, passive DNA demethylation can be achieved with DNMT1 inhibitors. The most frequently used compound is 5-aza-cytidine.

9.2.1.2 Methyl-Cytosine-Binding Proteins

Mammalian cells contain several proteins that bind to single m^{5}Cs or clusters of neighboring m^{5}Cs. At present, six methyl-CpG-binding proteins have been identified: MECP2, MBD1, MBD2, MBD3, MBD4, and KAISO. MECP2 contains a methyl-CpG-binding domain (MBD) and a transcription repression domain (TRD). MECP2, MBD1, and MBD2 function as transcription repressors. MBD4 is a DNA glycolase and is involved in DNA mismatch repair. KAISO lacks an MBD domain and binds methylated CGCG through its zinc-finger domain. The CpG-binding proteins recruit chromatin-remodeling and transcription factor complexes to methylated DNA regions in the genome. MECP2, for example, recruits the SIN3A corepressor complex, which contains a histone deacetylase, and sets up repressive chromatin. MBD2 forms a complex with the NuRD complex, which contains an ATP binding-dependent protein as well as a histone deacetylase. This complex represses methylated promoters. These examples also highlight the synergy between the three chromatin-marking systems (see also below).

Targeted mutations in the murine orthologs of the MBD proteins are associated with prenatal lethality (*Mbd3*$^{-/-}$) or neurological and behavioral defects (*Mecp2*$^{-/-}$ and *Mbd2*$^{-/-}$). In humans, *MECP2* mutations are associated with Rett syndrome, which is an X-linked dominant neurodevelopmental disorder. Girls with Rett syndrome have apparently normal development throughout the first 6 months of life, before they lose previously acquired skills (for more clinical details see Table 9.1). Initially it was believed that the loss of MECP2 led to widespread loss of gene repression. This, however, does not appear to be the case. To date, only two genes (*BDNF* and *DLX5*) have been identified as target genes.

9.2.2 Histone Modification

Within the nucleus, the DNA is packaged into chromosomes. Nucleosomes are the basic structural unit of a chromosome. The nucleosome core particle consists of a complex of eight histone proteins – two molecules each of histones H2A, H2B, H3, and H4 – and 146 base pairs of double-stranded DNA wrapped around the histone octamer. Each nucleosome core particle is separated from the next by approximately 80 bp of DNA and the linker histone H1. To a first approximation, the core histones are globular molecules, from which the flexible N-terminus protrudes. The side chains of lysine, arginine, and serine residues within the N-terminal tail of the histones are subject to extensive posttranslational modification (Fig. 9.2). The modifications include

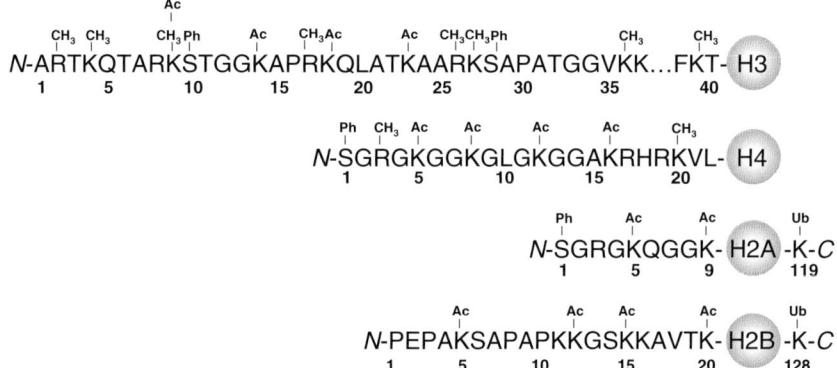

Fig. 9.2 Histone modifications. The amino acid sequences of the N-terminal and C-terminal tails of histones H3, H4, H2A, and H2B are given in the one-letter code. Possible posttranslational modifications of the amino acid side chains of arginine (*R*), lysine (*K*), and serine (*S*) are indicated above the sequence (*CH3* methylation, *Ac* acetylation, *Ph* phosphorylation, *Ub* ubiquitinylation)

the acetylation, ubiquitinylation, and mono-, di-, and trimethylation of lysine, the mono- and dimethylation of arginine, and the phosphorylation of serine. Furthermore, histone ubiquinylation, histone sumoylation, and poly-ADP-ribosylation have been described. The pattern of modification determines the accessibility of promoter and enhancers to transcription factors and thus gene activity.

Many different combinations of histone modifications are possible, providing the cell with a wealth of different possibilities to regulate gene activity. Trimethylation of Lys4 of H3 (H3K4) and acetylation of Lys9 of H3 (H3K9), for example, have been associated with active gene expression, whereas trimethylation of H3K9 and H3K27 has been associated with transcriptional silencing.

The pattern of histone modification at a locus of interest is determined by chromatin immuno-precipitation (ChIP). Living cells are treated with formaldehyde to crosslink DNA and proteins. The chromatin is then fragmented by ultrasound and antibodies are used to precipitate the DNA-protein complexes that contain histones with a specific modification. After reversal of the crosslink, PCR is used to amplify the DNA region of interest and the fold enrichment is calculated. For each modification, a separate experiment has to be done. Instead of PCR, the DNA can also be analyzed by hybridization to microarrays containing oligonucleotides from the region of interest (ChIP on chip) or by massive parallel sequencing (ChIPseq).

9.2.2.1 Histone Acetylation

The degree of histone acetylation, i.e., the acetylation of a lysine side chain, is determined by the activity of histone acetylases (HATs) and histone deacetylases (HDACs). Several transcription factors and co-activators have intrinsic HAT activity. There are three HAT families in humans (the GNAT, MYST, and CBP/p300 families), and more than a dozen other proteins with HAT activity. The acetylation of a lysine residue changes the charge of the histone molecule and leads to an "open," transcriptionally permissive chromatin configuration.

Several mutations affecting HAT genes have been identified in humans. Mutations of the *CREBBP* gene, which encodes the CBP protein, cause Rubinstein–Taybi syndrome (Table 9.1). A fusion of the *CREBBP* and *MOZ* (*MYST3*) genes, resulting from a t(8;16) (p11;p13) translocation, plays a causal role in the development of a subtype of acute myeloid leukemia.

There are at least 18 HDACs in humans, which fall into three classes. Class I and II HDACs have a catalytic domain which includes a zinc ion. Class III HDACs use NAD^+ as a co-factor. Histone deacetylation leads to a "closed," repressive chromatin configuration. As mentioned above, methyl-CpG-binding proteins can recruit HDACs to repress gene transcription.

Histone deacetylation can be inhibited by small compounds such as trichostatin A (TSA). TSA treatment of cells can lead to reactivation of silenced genes. Several *HDAC* genes have been knocked out in mice. A targeted mutation of the murine ortholog of *HDAC1*, for example, leads to intrauterine growth retardation and early embryonic lethality. In humans, *HDAC2* mutations have been found in certain types of cancer.

9.2.2.2 Histone Methylation

In contrast to the acetylation of lysine residues, the methylation of lysine residues does not change the charge of the histone molecule. It does, however, change its basicity and hydrophobicity. Furthermore, histone methylation is considered to be more stable than histone acetylation and thus more relevant for epigenetic inheritance. The degree of histone methylation is determined by the activity of histone methyltransferases (HMTs) and histone demethylases (HDMs).

Like DNA methyltransferases, HMTs use S-adenosyl-methionine as a methyl donor. The catalytic domain of lysine HMTs consists of approximately 130 amino acids and is called the SET domain. The acronym is derived from three drosophila proteins [*su*(var)3-9, *e*nhancer-of zest, *t*rithorax]. At present, more than 300 proteins with a SET domain have been identified. The proteins also have bromo- and chromodomains. It is not clear how mono-, di-, and trimethylation of lysine residues are regulated.

The HMTs differ in their specificity. SUV39, G9A and EHMT1, for example, preferentially methylate H3K9. A targeted mutation of the murine *Suv39* gene leads to loss of H3K9 methylation in heterochromatin and chromosome defects. In $G9a^{-/-}$ mice, loss of H3K9 methylation in euchromatin leads to early embryonic lethality. Haploinsufficiency for *EHMT1* in humans is responsible for the clinical findings associated with the 9q34 subtelomeric deletion syndrome (Table 9.1).

In addition to lysine residues, arginine residues can be methylated. This reaction is catalyzed by protein arginine methyltransferases. At present, seven such enzymes are known in mammals. Similar to the lysine HMTs, the arginine HMTs differ in their specificity.

Only recently, several HDMs have been identified, for example the amine oxidase AOF2 (also called lysine-specific demethylase 1, LSD1) and the peptidyl-arginine-deiminase 4 (PADI4).

9.2.2.3 Histone Phosphorylation and Other Histone Modifications

The phosporylation and dephosphorylation of serin residues within histone molecules is catalyzed by kinases and phosphatases, respectively. Mutations in the *RSK2* gene, which encodes a ribosomal S6 kinase, causes Coffin-Lowry syndrome (Table 9.1). The kinase has several functions, one of which is the phosophorylation of histone H3.

In addition to acetylation, methylation, and phosphorylation, histone ubiquitinylation, histone sumoylation, and poly-ADP-ribosylation have been observed. Ubiquitin is a key component of target protein degradation. The ubiquinylation of histones, however, does not appear to be a marker for degradation, but for active chromatin. Sumoylation appears to have the opposite effect on transcription. SUMO is a *s*mall *u*biquitin-related *mo*difier. Poly-ADP-ribosylation is found in active chromatin.

9.2.3 Chromatin Remodeling

A number of studies have identified protein complexes which bind to DNA, hydrolyze ATP, and use the energy to change the relative position of the nucleosome with respect to the DNA. Chromatin remodeling can make DNA-binding sites accessible or inaccessible to transcription factors (Fig. 9.3). The best characterized complex is SWI/SNF, which contains a number of different proteins. It is of particular relevance to epigenetics that the change in chromatin structure brought about by chromatin remodeling factors persists even after the complex has dissociated from the chromatin.

Loss of function of a chromatin remodeling factor can impair gene expression and cause disease. Mutations

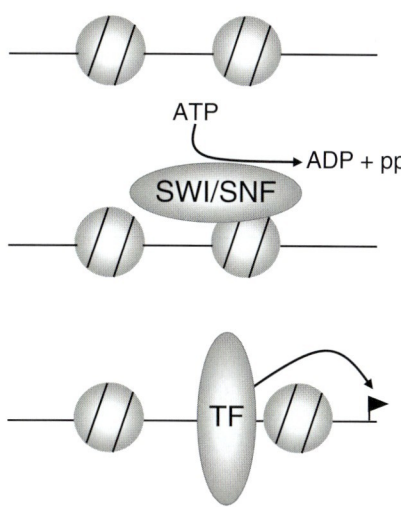

Fig. 9.3 Chromatin remodeling. The DNA (*black line*) is wrapped around nucleosomes (*gray balls*). In an energy-dependent mechanism, the chromatin remodeling complex SWI/SNF changes the relative position of the nucleosomes, so that a transcription factor (*TF*) can bind to its target sequence and activate gene transcription (*arrowhead*)

in the *ATRX* gene, for example, which encodes a member of the SWI/SNF family, lead to X-linked alpha thalassemia mental retardation (ATRX) syndrome. ATRX is a developmental disorder characterized by mental retardation, dysmorphism, and reduced expression of the α-globin genes (Table 9.1).

9.2.4 Synergistic Relations Between the Different Chromatin-Marking Systems

The different chromatin-marking systems cooperate with each other to set, spread, and maintain chromatin states. In some instances, DNA methylation depends on histone methylation, while in other instances the reverse is true. One possible scenario is shown in Fig. 9.4. H3K9 methylation also depends on deacetylation, because this lysine residue can be either acetylated or methylated. Furthermore, histone phosphorylation cooperates with histone acetylation in generating transcriptionally active chromatin. In this case and in other cases, a modification of one amino acid side chain can inhibit or enhance the modification of an amino acid side chain on the same or even a neighboring histone molecule. In addition, many other nonhistone proteins,

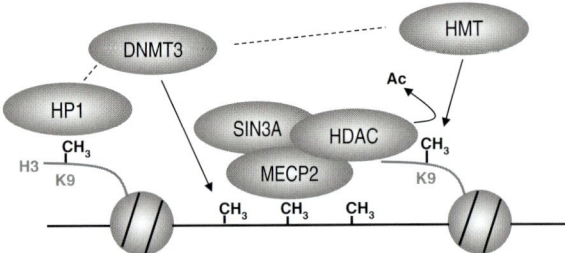
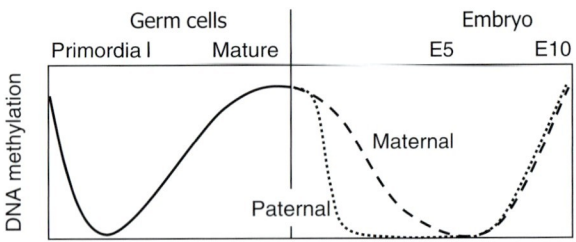

Fig. 9.4 Interaction between different chromatin-marking proteins. One possible scenario is shown. The heterochromatin protein HP1 binds to the methylated lysine residue K9 of histone H3 and triggers DNA methylation (through DNMT3) and histone methylation (through one of several HMTs). The methyl-cytosine-binding protein MECP2 recruits the co-repressor SIN3A and a histone deacetylase (HDAC)

Fig. 9.5 Epigenetic reprogramming during gametogenesis and embryogenesis. DNA methylation is erased in primordial germ cells and newly established during gametogenesis. After fertilization (*vertical line*), the paternal genome is actively demethylated by the oocyte. The maternal genome looses methylation through a passive mechanism. At the blastula stage, de novo methylation starts

e.g., heterochromatic protein 1 (HP1), can be involved. HP1 is methyl-lysine-binding protein localized at heterochromatin sites, where it mediates gene silencing.

The chromatin-marking system is exceedingly complex, although in principle there are only two basic states: permissive and repressive chromatin. So why are there so many different combinations of DNA and histone modifications? It appears that different modification patterns determine the degree of stability of a chromatin state. In some instances, the chromatin state survives many mitotic and even meiotic cell divisions. In other instances, e.g., in embryonic stem cells, certain chromosomal regions are in a semi-permissive state, i.e., the genes in these regions are silent in the stem cells, but easily activated upon differentiation.

9.3 Specific Epigenetic Phenomena

During development, the genome undergoes stage- and tissue-specific epigenetic reprogramming. As a consequence, only a specific set of genes is active in any given cell at any given developmental stage. The most extensive epigenetic reprogramming occurs during gametogenesis and early embryogenesis, when the epigenetic states of almost all loci are erased and newly established. This is most impressive in mice (Fig. 9.5). The clearing of the epigenetic state between generations appears to be necessary to provide a clean state on which differentiation and development can occur.

Within a few hours after fertilization, the paternal genome is actively demethylated by the oocyte [22]. The enzymes involved in this process are still unknown. The maternal genome is protected against this active demethylation, but loses most of its methylation during subsequent rounds of cell division by passive demethylation (see Fig. 9.1).

During development, the developmental potential of the nuclear genome is more and more restricted. The zygote and early blastomeres are totipotent: they can give rise to all embryonic and extraembryonic tissues. Embryonic stem cells can give rise to all embryonic tissues, but not to extraembryonic tissues; they are pluripotent. The developmental potential of tissue-specific stem cells is restricted to certain cell lineages. Hematopoietic stem cells in the bone marrow stem cells, for example, can give rise to all the types of both the myeloid and lymphoid lineages, but not to other cells of the body, at least not under normal conditions. However, cell biologists are learning more and more to extend the developmental potential of such cells. It appears that the epigenetic states of the zygote, embryonic stem cells, tissue-specific stem cells, and terminally differentiated states differ in their relative stability.

The cloning of animals by transfer of somatic nuclei into enucleated oocytes requires clearing of the epigenetic state of the donor nucleus and restoration of cellular totipotency [38]. This process is inefficient and error prone, so that the success rate is rather low. Furthermore, viable offspring often have an unusually large birth weight ("large offspring syndrome") and severe organ malformations. Nevertheless, the successful generation of cloned animals proves that development does not involve genetic changes, but reversible epigenetic changes.

At a given developmental stage, similar cells have similar epigenotypes and therefore similar phenotypes. However, there are interesting exceptions to this rule. One is position-effect variegation in *Drosophila*, which was described first by Muller [25]. He had observed that flies with a certain inversion on the X chromosome had red-white mosaic eyes, although all cells had the same genotype (Fig. 9.6). The inversion placed the *white* gene next to pericentric heterochromatin. Normally, the *white* gene is expressed in every cell of the adult *Drosophila* eye, resulting in a red eye phenotype. In the mutant flies, the *white* gene is expressed in some cells in the eyes and silenced in others. Silencing is due to spreading of the heterochromatin into the *white* locus. Variegation can be suppressed or enhanced by numerous *trans*-acting mutations. The identification of the mutated genes has led to the discovery of important chromatin-marking proteins. The gene affected in the suppressor mutation Su(var)39, for example, encodes a H3K9 histone methyltransferase. The human ortholog (SUV39) has been mentioned above.

Within a cell, most often both alleles of an autosomal gene are either active or inactive. However, there are several examples where the two alleles of a gene, although identical in sequence, are functionally different. The difference can be parent-of-origin specific (as a result of genomic imprinting) or random (X inactivation and allelic exclusion).

9.3.1 Genomic Imprinting

In placental mammals (eutheria) approximately 100 genes are expressed from either the paternal or the maternal allele only. These genes are subject to genomic imprinting, which is an epigenetic process by which the male and the female germline each confer a sex-specific mark (imprint) on certain chromosomal regions. As a consequence, the paternal and the maternal genome are functionally nonequivalent and both are required for normal embryonic development. This was first shown by nuclear transplantation experiments in mice [23,33], but it is also observed in humans. Occasionally, the haploid genome of an unfertilized egg undergoes duplication. Although such an egg has 46 chromosomes, it does not develop into a normal embryo, but into a benign ovarian teratoma, which contains tissues from all three germ layers, but no trophoblast. Similarly, an egg that has been fertilized by two spermatozoa but has lost its maternal genome does not undergo normal development; it develops into a hydatidiform mole, which is degenerated trophoblast tissue.

The nonequivalence of the maternal and paternal genomes is also obvious from uniparental disomies. Uniparental disomy refers to the presence of two copies of a chromosome (or part of a chromosome) from one parent and none from the other [8]. Uniparental disomy per se does not interfere with genomic imprinting or cause clinical problems. However, if the affected chromosome pair carries an imprinted gene, both alleles of this gene will be inactive or active, depending on the parental origin of the chromosomes. The presence of zero or two active copies of an imprinted gene interferes with normal development and growth.

The parental copies of imprinted regions differ with respect to DNA methylation, histone modification and, consequently, gene expression. Despite the identification of parent of origin-specific chromatin differences of imprinted chromatin regions, the nature of the primary imprint is still a matter of debate.

Genomic imprints are erased in primordial germ cells, newly established during later stages of germ cell development, and stably inherited through somatic cell divisions during postzygotic development (Fig. 9.7a).

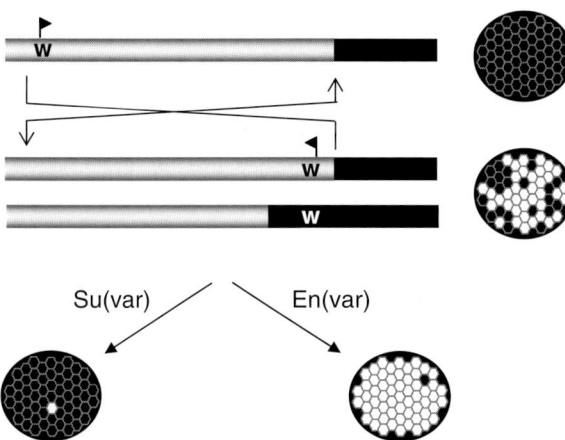

Fig. 9.6 Position effect variegation in *Drosophila*. The diagram shows the *white* gene (*w*) on the X chromosome. Gene transcription is indicated by the *triangular flag*. Pericentromeric heterochromatin is shown in *gray*. A wild-type fly has red eyes (*top right*). A mutant fly has mosaic eyes (*below*). *Trans*-acting mutations can suppress or enhance the spreading of heterochromatin (*bottom left and right*). *Su(var)* suppressor of variegation; *En(var)* enhancer of variegation

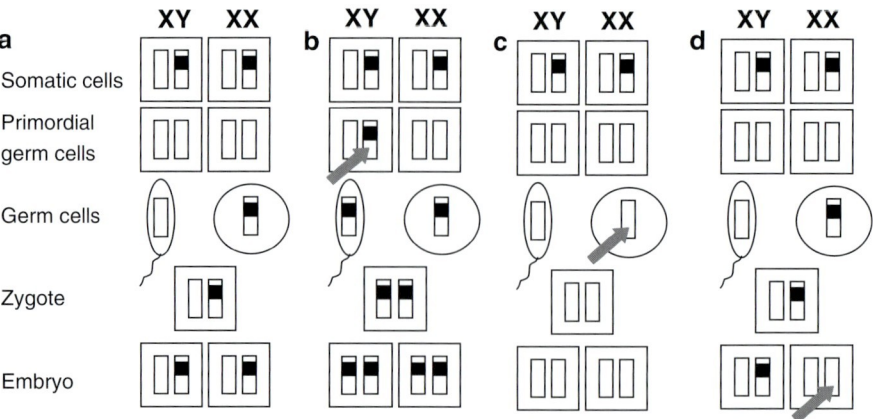

Fig. 9.7 Genomic imprinting. (**a**) Imprints are erased in the primordial germ cells, newly established during gametogenesis and maintained in the zygote and embryo. Imprinting defects can result from an error in imprint erasure (**b**), imprint establishment (**c**) or imprint maintenance (**d**). For clarity, only one pair of homologous chromosomes (*open bars*) and a maternal imprint (*black box*) are shown

They survive the global waves of DNA demethylation and remethylation during early embryonic development (see above), although it is not clear what protects them. In somatic cells, the imprint is read by the transcription machinery and used to regulate parent of origin-specific gene expression so that only the paternal or the maternal allele of a susceptible gene is active.

Many imprinted genes are involved in regulating resource acquisition of the embryo and fetus. In fact, it has been proposed that imprinting co-evolved with the placenta. In eutherian mammals, the fetus grows at the expense of the mother. As proposed by the genetic conflict theory [37], the paternal genome is "interested" in extracting as many resources from the mother as possible. This is because a male can spread his genes through many different females. By contrast, maternally inherited genes protect the mother from being exhausted by the fetus, because a female can spread her genes only through multiple pregnancies.

Imprinted genes are not randomly distributed in the genome, but tend to occur in clusters. In humans, imprinted gene clusters have been found on chromosomes 6, 7, 11, 14, and 15. The clustering of imprinted genes suggests that the primary control of imprinting is not at the single gene level, but at the chromosome domain level. Indeed, several clusters have been found to contain a *cis*-acting imprinting center (IC) which controls imprint establishment and imprint maintenance [4].

The proximal long arm of human chromosome 15 (15q11-q13) contains a cluster of imprinted genes which are affected in the Prader–Willi syndrome (PWS) and the Angelman syndrome (AS; Fig. 9.8, Table 9.1) Paternal-only expression of *MKRN3*, *NDN*, *SNRPN*, and (possibly) *MAGEL2* is associated with differential DNA methylation. Whereas the promoter/exon 1 regions of these genes are unmethylated on the expressing paternal chromosome, the silent maternal alleles are methylated. In addition to DNA methylation, the parental copies of these genes also differ in histone modification.

The *SNRPN* gene encodes two proteins, SNURF and SNRPN, serves as a host for 79 C/D small nucleolar (sno) RNA genes, and overlaps, in an antisense orientation, the *UBE3A* gene. The snoRNAs are encoded within introns of the *SNRPN* gene. They are expressed from the paternal allele only, because they are processed from the paternally expressed *SNRPN* sense/ *UBE3A* antisense transcript during the splice process. Thus, imprinted expression of the snoRNAs is indirectly regulated through *SNRPN* methylation. Unlike other C/D box snoRNAs, the snoRNAs encoded within the *SNRPN* locus do not serve a guide RNAs for 2'-*O*-ribose methylation of nucleotides in rRNA. Their function remains to be determined.

A ~4-Mb de novo interstitial deletion of the paternal chromosome 15 [del(15)(q11-q13)pat], which includes the entire imprinted domain plus several nonimprinted genes, is found in the majority (~70%) of patients with PWS. The second most common genetic abnormality in PWS (~30%) is a maternal uniparental disomy 15 [upd(15)mat], which most often arises from maternal

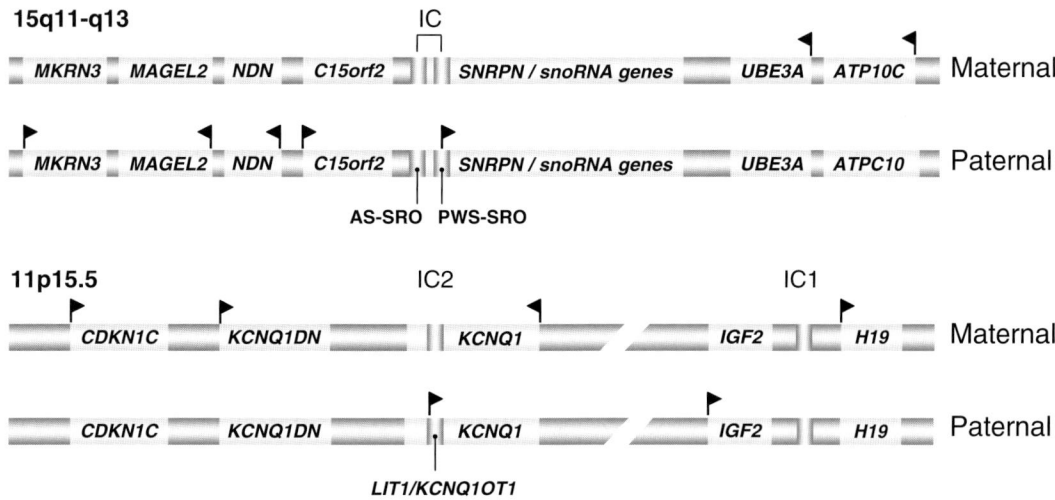

Fig. 9.8 Imprinted chromosome domains. For clarity, the maps (from centromere, *left*, to telomere, *right*) are not drawn to scale, and not all genes are shown. Several genes show imprinted expression in some tissues only. Gene transcription is indicated by *triangular flags* (*IC* imprinting center)

meiotic nondisjunction followed by mitotic loss of the paternal chromosome 15 after fertilization. A few patients with PWS have apparently normal chromosomes 15 of biparental inheritance, but the paternal chromosome carries a maternal imprint (imprinting defect). All three lesions lead to the lack of expression of imprinted genes that are active on the paternal chromosome only. They can easily be detected by DNA methylation analysis. Patients with PWS lack unmethylated alleles of several loci within 15q11-q13.

In contrast to the paternally active genes, the maternally active *UBE3A* gene lacks differential DNA methylation. Another striking difference is that imprinted *UBE3A* expression is tissue-specific. At present it is unclear how tissue-specific imprinting of *UBE3A* is regulated, but the paternally expressed *SNRPN* sense/*UBE3A* antisense transcript may be involved in silencing the paternal *UBE3A* allele.

The loss of function of the *UBE3A* gene leads to AS. Similar to PWS, the major lesion in AS is a common large deletion of 15q11-q13 (~70%), but in AS the deletion is on the maternal chromosome. AS can also result from upd(15)pat (~1% of cases), which most often arises from the postzygotic duplication of a zygote carrying only a paternal chromosome 15, or the lack of a maternal imprint on the maternal chromosome (imprinting defect; ~4% of cases). All three lesions can be detected by DNA methylation analysis. The patients lack a methylated allele of maternally methylated loci within 15q11-q13. A few percent of patients have a mutation in the maternal *UBE3A* gene, and some 10% of patients suspected of having AS have a genetic defect of unknown nature. The latter two classes of patients cannot be detected by methylation analysis.

Imprinting in 15q11-q13 is under the control of a bipartite imprinting center, which overlaps the promoter/exon 1 region of *SNRPN*. The imprinting center (IC) was identified by the mapping of small deletions in patients with an aberrant imprint [4]. An element within the smallest region of deletion overlap in patients with AS (called the AS-SRO) is necessary for establishment of the maternal imprint in the female germline. A deletion of this element prevents maternal imprinting of the mutated chromosome. A child inheriting this chromosome will develop AS, because the maternal *SNRPN* allele is unmethylated and expressed, and the maternal *UBE3A* allele is silenced. Since deletions of the AS-SRO element affect maternal imprinting only, they are silently transmitted through the paternal germline. This explains why in some families only few and distantly related individuals are affected (Fig. 9.9).

An element within the smallest region of deletion overlap in patients with PWS (called the PWS-SRO) is necessary for the postzygotic maintenance of the paternal imprint. A paternally derived deletion of this element leads to an epigenetic state in 15q11-q13 that

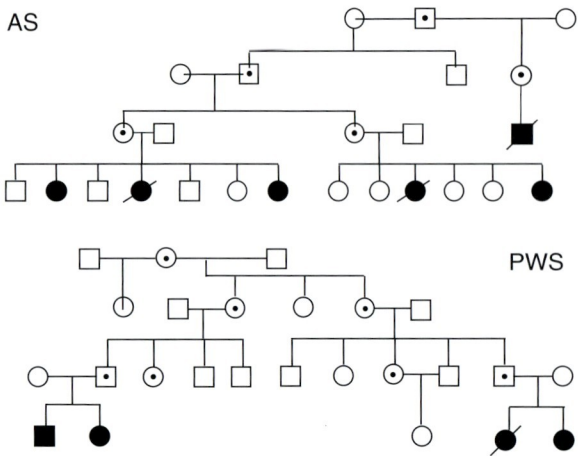

Fig. 9.9 Segregation of imprinting center deletions. Unaffected carriers are indicated by a *dot*. Note that a deletion of the AS-SRO element of the 15q IC in the Angelman syndrome family (*top*) is benign when transmitted through a male. In the female germline it prevents the establishment of the maternal imprint. A deletion of the PWS-SRO of the 15q IC in the Prader–Willi syndrome family (*bottom*) is benign when transmitted through a female. After transmission through a male, the deletion chromosome acquires a maternal methylation pattern during early embryogenesis

resembles the maternal imprint. A child with such a chromosome will develop PWS, because all paternally expressed genes are silent. Since deletions of the PWS-SRO element affect the paternal imprint only, they are silently transmitted through the maternal germline (Fig. 9.9).

Another cluster of imprinted genes with relevance to human disease is located on 11p15.5 (Fig. 9.8). It is affected in patients with Beckwith–Wiedemann syndrome (BWS) and some patients with Silver-Russell syndrome (SRS) (Table 9.1). BWS is caused by overexpression of the paternally active *IGF2* gene and silencing of the maternally expressed *H19* gene or by silencing or mutational inactivation of the maternally active *CDKN1C* gene. These genes map to the short arm of chromosome 11, but are controlled by two different ICs, *IGF2/H19* IC (IC1), and *LIT1/KCNQ1OT1* (IC2), which controls imprinting of *CDKN1C*.

IC1 contains several binding sites for the CCCTC-binding factor (CTCF), which is a multifunctional protein [15]. As a consequence of genomic imprinting, the binding sites on the paternal chromosome (as well as the *H19* promoter) are methylated, while they are unmethylated on the maternal chromosome. Binding of CTCF to the unmethylated IC1 on the maternal chromosome isolates the *IGF2* gene from two enhancers, which it shares with *H19* (Fig. 9.10 a). As a consequence, *IGF2* is silent, whereas *H19* is active. CTCF cannot bind to the methylated IC1 on the paternal chromosome. Therefore, the paternal *IGF2* allele is active, whereas the maternal *H19* allele is inactive.

The function of the IC2 is rather different. It is located within the *KCNQ1* gene and serves as a promoter for an antisense gene (*LIT1/KCNQ1OT1*). The IC2 is methylated on the maternal chromosome and unmethylated on the paternal chromosome. As a consequence, *LIT1/KCNQ1OT1* is expressed only from the paternal chromosome. By an unknown mechanism, *LIT1/KCNQ1OT1* transcription silences *cis* expression of *CDKN1C*.

9.3.2 X Inactivation

In female mammals, one X chromosome is inactivated to ensure that levels of X-linked genes are equal between XY males and XX females. This form of dosage compensation takes place during early development and, once established, is maintained through many cell divisions. In rodent preembryos, X inactivation is initially imprinted, in that the paternal X (X_p) is inactivated. In the inner cell mass of the blastocyst, X_p is reactivated, and then each cell makes a random choice with regard to X inactivation. In extraembryonic tissues, X_p remains inactivated. X inactivation is initiated at a master control locus in Xq13, the X inactivation center (Xic). The Xic is also involved in "counting," by which the cell recognizes whether it has one, two, or more X chromosomes, and "choice," by which the cell determines which X will remain active. Any supernumerary X chromosome will be inactivated.

The Xic harbors the *Xist* gene (X-inactive specific transcript), which is required for the initiation of X inactivation [3]. The gene encodes a 17-kb untranslated RNA. The RNA is transcribed from and coats the X chromosome that has been selected to become inactivated. Within one or two cell cycles gene silencing occurs along the length of this chromosome. The association of *Xist* RNA with chromatin is mediated by sequences that are functionally redundant and dispersed throughout the transcript. The silencing requires

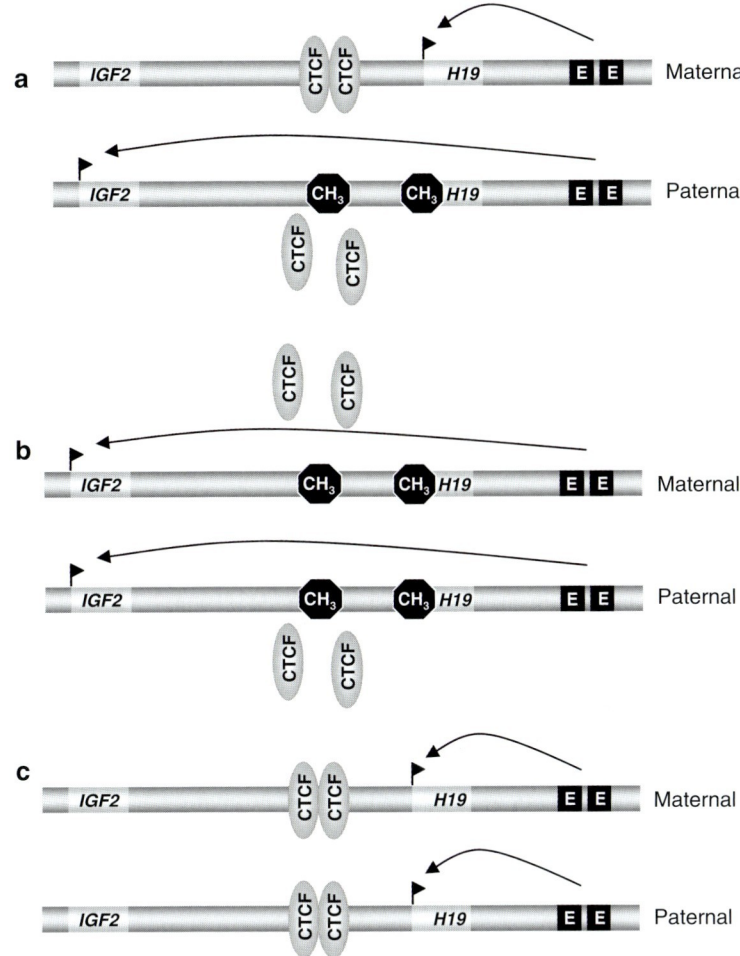

Fig. 9.10 Enhancer competition model. *IGF2* and *H19* share two enhancers (*E*) telomeric to *H19*. Transcription is indicated by the *triangular flags*. (**a**) Normal situation; (**b**) imprinting defect (aberrant methylation of the maternal allele) leading to Beckwith-Wiedemann syndrome; (**c**) imprinting defect (lack of paternal methylation) leading to Silver-Russell syndrome. For details see text

a conserved region at the 5' end. After the initiation of silencing, *Xist* becomes dispensable.

The X-chromosomal genes are inactivated by extensive histone modifications: hypoacetylation of H3 and H4, dimethylation of H3K9, trimethylation of H3K27, and hypomethylation of H3K9. Polycomb group proteins and associated chromatin-modifying enzymes (Eed and Enx1, the latter being a H3K9 and K27 HMT) are mediators of the transition from the reversible inactive, Xist-dependent phase of inactivation, to the irreversible, Xist-independent phase. Later, the incorporation of the histone variant macroH2A and DNA methylation play an important part in maintaining the inactive state.

Although X inactivation in humans is random, not all females have an equal proportion of cells in which X_p and X_m, respectively, are inactivated. There are at least two reasons for this observation. First, the choice appears to be influenced by genetic variation with the Xic, resulting in the preferential inactivation of a certain X chromosome. In the extreme, X inactivation may be heavily skewed. X Inactivation skewing can lead to phenotypic expression of an X-linked recessive trait in a female if the mutation is on the active X chromosome.

Second, if an X chromosome carries a mutation that confers some disadvantage on cell proliferation, cells that have chosen this chromosome to be active will be selected against during the growth or adult life of the female. In fact, it has been observed that the skewing of X inactivation can increase during the life of a woman.

Although X inactivation is rather stable, the chromatin marks are in principle reversible and can be lost during aging. In fact, there is ample evidence for this. Reactivation of the inactive X during aging can have deleterious effects if the reactivated gene carries a mutation.

9.3.3 Allelic Exclusion in the Olfactory System

Although there are more than 1,000 odorant receptor (*OR*) genes in the genome, they are expressed in a mutually exclusive and monoallelic manner in olfactory sensory neurons. The one neuron-one receptor rule is essential for the brain to determine which odorant is present in the environment. Similar to the expression of antigen receptor genes in lymphocytes, DNA rearrangements have long been regarded as a possible mechanism for the allelic exclusion of the *OR* genes. However, mice cloned from the nuclei of mature olfactory sensory neurons expressed the full repertoire of ORs [7,21]. These experiments suggest that *OR* choice and maintenance are epigenetic.

According to a current model, the *OR* genes are initially silent, but in a semipermissive state [31]. By a stochastic and presumably inefficient process, on average a single *OR* allele is activated per cell. At this stage the choice is unstable, but a functional OR protein may mediate a feedback stabilization that commits the cells to the receptor. This commitment involves full activation of the selected *OR* gene by changing its chromatin state from semipermissive to permissive, and repression of all the other *OR* genes by changing their chromatin state from semipermissive to repressive.

It is unclear why *OR* gene expression must be monoallelic. It is possible that this pattern of expression is a consequence rather than a requirement of singular gene choice. A stochastic choice process as described above is more likely to select individual alleles than two copies of the same gene. However, it is also possible that monoallelism is important for some biological process unrelated to *OR* choice.

9.4 Epigenetic Variation and Disease

Epigenetic states are set by developmental processes, but subject to natural variation. Epigenetic variation can result from genetic variation, environmental variation, and errors of the epigenetic machinery. In most cases, all three sources of epigenetic variation play a part, although the relative contribution of each source may vary. It should also be noted that genetic variation, environmental variation, and errors of the epigenetic machinery are not independent of each other. Genetic variants of chromatin-marking proteins, for example, can be associated with an increased error rate of the epigenetic machinery, and the error rate may further be affected by environmental variation.

Normal epigenetic variation is likely to contribute to the phenotypic range of normal individuals. Rare aberrant epigenetic states are called epimutations and have an important role in several diseases [17].

In some cases there is a one-to-one correspondence between genotype and epigenotype (obligatory epigenetic variation) [30] (Fig. 9.11 a). This is seen in certain diseases where a *cis*-acting DNA mutation always leads

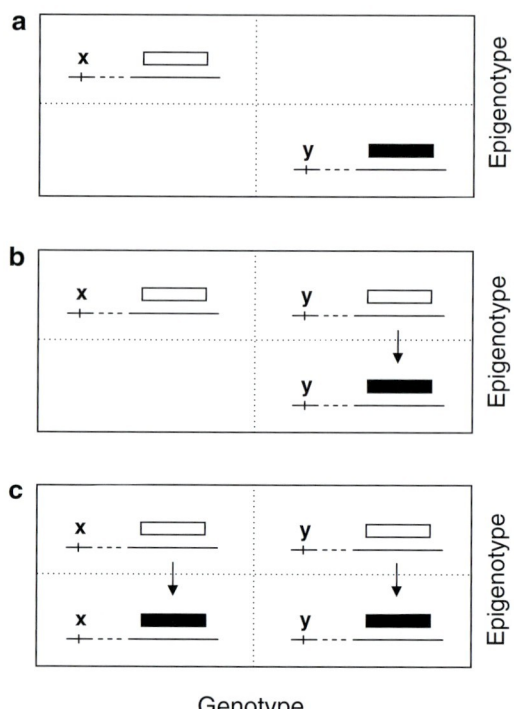

Fig. 9.11 (**a–c**) Three classes of epigenetic variation. The range of autonomy of epigenetic variation in relationship to genotypic context is expressed in three genotype x epigenotype matrices. The *horizontal axis* distinguishes between two genotypes that are represented by alleles *x* and *y*. At a genomic location either in *cis* or in *trans* (*dashed line*), two alternative epigenetic states are depicted as either *open* or *filled boxes*. A In this scenario, the epigenotype of the locus is strictly determined by genotype; so, the epigenotype is an obligatory phenotype of the alternative genotypes. (**b**) The matrix illustrates facilitated epigenetic variation (epiallele formation: *open box–filled box*) that can occur, in a probabilistic manner, only in the context of genotype y. (**c**) In this matrix, stochastic events generate alternative epialleles at some finite frequency regardless of the genotype. (Reprinted from [30], by permission from Macmillan Publishers Ltd: *Nature Reviews Genetics,* copyright 2006)

to an aberrant epigenetic state (secondary epimutation). In other cases, the epigenotype is completely independent of the genotype (pure epigenetic variation; Fig. 9.11 c). This is seen in diseases where an epimutation has occurred without any change in the DNA sequence (primary epimutation), although certain genotypes may increase the risk of such a primary epimutation (facilitated epigenetic variation; Fig. 9.11 b). In practice, it is often impossible to distinguish between pure and facilitated epigenetic variation, because the underlying genotypes are not completely known.

9.4.1 Obligatory Epigenetic Variation

There are at least two monogenic disorders in which an epimutation represents the major pathogenetic mechanism. These are fragile X mental retardation syndrome (FMR1) and facioscapulohumeral muscular dystrophy (FSHD) (Table 9.1). FMR1 is an X-linked dominant disease caused by the expansion of an unstable trinucleotide repeat (CGG) within exon 1 of the *FMR1* gene. It is one of the most common causes of mental retardation. The number of repeats varies in the human population. Repeats with more than 58 copies are unstable and can expand to several hundred copies during proliferation of the diploid oogonia in the fetal ovary. After fertilization of an oocyte carrying an expanded *FMR1* allele, the CGG repeat and *FMR1* promoter are methylated. DNA methylation, histone deacetylation, and the establishment of repressive chromatin in this region silence the *FMR1* gene.

FSHD is an autosomal dominant disorder that has been linked to a 3.3-kb, -tandemly repeated sequence (*D4Z4*) in the subtelomeric region of the long arm of chromosome 4. In normal individuals the number of *D4Z4* repeats varies between 11 and 150 units, whereas FSHD patients have fewer than 11 repeats. It has been proposed that deletion of *D4Z4* is associated with an open chromatin structure in 4q35 and the inappropriate expression of several genes within this region, but the mechanism is still not clear.

Imprinting defects resulting from the deletion of a *cis*-acting imprinting center (IC) are another example of an epimutation resulting from a DNA sequence change. Deletions of the 15q IC impair the imprint in the Prader–Willi/Angelman syndrome region (see Sect. 9.3.1). Approximately 50% of these deletions are familial mutations. Two families are shown in Fig. 9.9. IC deletions have also been described in rare patients with BWS.

A unique epimutation affecting the α-globin gene *HBA2* has been described by Tufarelli et al. [34]. The authors studied an individual with an inherited form of α-thalassemia who has a deletion that results in a truncated, widely expressed gene (*LUC7L*) becoming juxtaposed to the structurally normal α-globin gene *HBA2*. Although it retains all of its local and remote *cis*-regulatory elements, expression of *HBA2* is silenced. *LUC7L* is transcribed from the opposite strand to the α-globin genes. In the patient, RNA transcripts from the truncated copy of *LUC7L* (missing the last three exons) extend into the *HBA2* CpG island, thus generating an antisense transcript with respect to *HBA2*. Antisense RNA transcription appears to mediate methylation of the *HBA2* CpG island during early development and silencing of *HBA2* expression. There are also several examples of chromosomal translocations affecting the epigenetic state of genes adjacent to the breakpoints. This is the case in particular in translocations involving the X chromosome. A very instructive case was published by Jones et al. [20]. The authors studied a male patient with an unbalanced X;13 translocation [46,XY,+der(X;13)(q10;q10),-13] and bilateral retinoblastoma. The patient has an extra copy of Xq, but no signs of Klinefelter syndrome. DNA replication and methylation studies suggested that the extra copy of Xq, which is attached to the long arm of one chromosome 13, was inactivated and that inactivation had spread to chromosome 13 and silenced the *RB1* gene in 13q14. This epimutation is equivalent to a constitutional *RB1* mutation and explains the development of bilateral tumors in this patient.

9.4.2 Pure Epigenetic Variation

As monozygotic twins are assumed to have the same genomic DNA sequence, any phenotypic difference should be attributable to environmental or epigenetic variation. Preliminary studies on global and locus-specific DNA methylation and histone methylation in monozygotic twins suggest that the patterns are highly similar at birth, but less similar in older twins [11]. The differences may be the result of environmental variation and/or random errors of the epigenetic machinery, and these differences appear to accumulate with age. At present it is not clear to what extent epigenetic differences in monozygotic twins explain discordance in disease phenotypes, although several examples are known. DNA methylation analysis of the *AXIN1* locus in monozygotic twins discordant for a caudal duplication anomaly revealed increased methylation in the affected twin [27]. Another example is the Beckwith–Wiedemann syndrome. Interestingly, there is an excess of monozygotic twins among patients with BWS, and almost all of them are female and discordant for the disease. All of the patients with a healthy twin sib are hypomethylated at the maternal *LIT1/KCNQ1OT1* locus. This epigenetic difference is most probably due to a failure of imprint maintenance in a cell of the early embryo. As a consequence, *LIT1/KCNQ1OT1* is expressed from both chromosomes and both alleles of *CDKN1C* are silenced. This defect may then increase the risk of twinning. As suggested by Weksberg et al. [36], it is possible that a group of cells carrying a postzygotic *LIT1/KCNQ1OT1* imprint maintenance defect could preferentially increase the proliferation rate of this group of cells beyond that of normal cells, thereby generating asymmetry of the entire cell mass and increasing the chance of separation of the epigenetically distinct cell clones.

BWS can also be caused by an imprinting defect affecting the *IGF2/H19* IC (Fig. 9.10 b). Methylation of the maternal *CTCF* binding sites and *H19* promoter leads to activation of the maternal *IGF2* allele, two doses of IGF2, and overgrowth. Loss of methylation on the paternal chromosome allows binding of CTCF to the IC. As a consequence, the paternal *IGF2* allele is isolated from the enhancers and silenced (Fig. 9.10 c). Lack of IGF2 results in pre- and postnatal growth retardation and is one cause of Silver-Russell syndrome (Table 9.1).

In genetic syndromes involving imprinted genes the majority of imprinting defects are primary epimutations. They cannot only occur during imprinting maintenance, but can also result from an error of imprint erasure in primordial germ cells or imprint establishment during later stages of gametogenesis. In PWS patients with an imprinting defect not caused by an IC mutation the affected chromosome is always derived from the paternal grandmother [5]. This finding suggests that the (grand)maternal imprint has not been erased in the paternal germline (Fig. 9.7b). In contrast to PWS, imprinting defects in AS result from an error in imprint establishment or an error in postzygotic imprint maintenance (Fig. 9.7c, d). At least 30% of AS patients with a primary imprinting defect are somatic mosaics, and there appears to be some correlation between the percentage of abnormally imprinted cells and the severity of the disease. It is likely that the role of mosaic imprinting defects on chromosome 15 in mental retardation is severely underestimated.

Primary epimutations also have an important role in cancer. In 1983 Feinberg and Vogelstein discovered altered DNA methylation in cancer cells [9]. Subsequently, these authors and others demonstrated that hypomethylation can lead to inappropriate activation of oncogenes. In 1986 Baylin et al. identified hypermethylation of the calcitonin gene in human lung cancers and lymphomas [1], but the role of these changes in tumor development was unknown. Soon after the discovery of the first tumor suppressor gene (the retinoblastoma gene *RB1*), it was found that the *RB1* promoter is methylated in a subset of retinoblastomas [14], suggesting that tumor-suppressor silencing can also occur by way of an epigenetic pathway. Methylation of tumor-suppressor genes has now been found in virtually all tumors, and the field of cancer epigenetics is growing rapidly.

In general, tumor-predisposing epimutations are found only in premalignant or malignant cells. In some instances, however, the epimutation has also been detected in normal somatic cells. Some individuals have constitutional loss of imprinting at the *IGF2* locus [6]. These individuals appear to be at an increased risk for colon cancer.

Primary epimutations also appear to play a part in cardiovascular disease. In a similar way to tumors, atherosclerotic lesions are characterized by global DNA hypomethylation and local DNA hypermethylation. These similarities should not surprise us, because a key step in the atherogenetic process is the proliferation and migration of smooth muscle cells. Once within the intima, the phenotype of the smooth muscle cells switches from contractile to "dedifferentiated." It has been suggested that methylation of estrogen receptor-α gene (*ESR1*) could contribute to these processes [39].

Whereas secondary epimutations result from DNA mutations, it is less clear what triggers primary epimutations. Primary epimutations probably represent errors in the establishment or maintenance of an epigenetic state. An interesting model for tumor-suppressor methylation has been proposed by Clark and colleagues [32]. According to this model, a combination of transient gene silencing and methylation seeding leads to recruitment of the methyl-CpG-binding protein MBD2, histone deacetylase, and DNA methyltransferase. This then leads to the spreading of DNA and histone methylation, and consequently to the establishment of silent chromatin.

The spontaneous epimutation rate can be modified by environmental variation. Cellular levels of the methyl donor *S*-adenosyl-methionine, for example, depend—to a certain degree—on the levels of folic acid, choline, and vitamin B taken up from nutrition. Changes in DNA methylation by folate levels have been observed in various types of cancers and also in animal models. After folate washout in patients with hyperhomocysteinemia, Ingrosso et al. observed biallelic expression of *H19* [18], which normally is expressed from the maternal allele. After folate treatment, they observed a shift back to monoallelic expression.

Since patients with a primary epimutation do not have a DNA sequence mutation, the epimutation should in principle be reversible. As mentioned in the paragraph on chromatin-marking, 5-aza-cytidin and trichostatin A can be used to change DNA methylation and histone acetylation patterns, respectively. Other drugs are being developed to modify histone methylation and phosphorylation patterns. Although these drugs are not locus specific, they appear to have some beneficial effect on cancer development, and clinical trials with these drugs are under way.

9.4.3 Facilitated Epigenetic Variation

Obligatory and pure epigenetic variation are the two extremes of the relation between genotype and epigenotype. In most instances, the genotype has some effect on the epigenotype. For example, certain DNA sequence variants appear to be more susceptible to epimutations than others. There is tentative evidence for a genetic predisposition to epimutations in the Beckwith–Wiedemann syndrome region [26]. Four single nucleotide polymorphisms (SNPs) in a differentially methylation region of the *IGF2* gene occur in 3 out of 16 possible haplotypes: The frequency of one haplotype was significantly higher and that of another haplotype significantly lower in BWS patients than in controls.

Similarly, there appears to be a certain haplotype of the 15q IC that increases the risk of a sporadic imprinting defect leading to Angelman syndrome [40]. The increased risk could be attributed to the alleles of two polymorphisms. It is likely that the IC contains a binding site for a *trans*-acting factor involved in establishing the maternal imprint and that this factor binds with different efficiency to the different alleles. This may increase the risk that the maternal imprint is not, or not completely, established. Genetic variation may also influence epigenetic states in trans, e.g., by affecting proteins that are directly or indirectly involved in the establishment or maintenance of chromatin marks. 5,10-Methylenetetrahydrofolate reductase (MTHFR) is a key enzyme of the one-carbon atom metabolism, which provides DNA and HMTs with the methyl-group donor *S*-adenosyl-methionine. A frequent variant of the *MTHFR* gene (677C>T) encodes a thermolabile protein with reduced enzymatic activity. Homozygosity for the 677C>T variant is associated with reduced global DNA methylation, but only in the presence of low folate levels [12]. This is a good example of the interaction between genetic variation and environmental variation in affecting epigenetic variation. The interaction appears to lead to a shortage of intracellular levels of *S*-adenosyl-methionine, so that postreplicative, hemimethylated DNA is inefficiently methylated by the maintenance methyltransferase DNMT1. As a consequence, methylation is lost during successive rounds of DNA replication. A shortage of *S*-adenosyl-methionine also appears to affect the establishment of the maternal methylation imprint by the de novo methyltransferases DNMT3A and DNMT3B, because mothers who are homozygous for the *MTHFR* 677C>T variant are at increased risk of conceiving a child with an imprinting defect on chromosome 15, which leads to Angelman syndrome [40].

9.5 Transgenerational Epigenetic Inheritance and Evolution

While epigenetic variation in somatic cells has been well documented, there are few data available on epigenetic variation in the human germline. A study on human sperm revealed DNA methylation differences between and within the germlines of normal males [10]. The largest degree of variation was detected within the CpG islands and pericentromeric satellite DNA. It is not clear, however, whether the observed patterns can be transmitted across generations. The inheritance of epigenetic states through mitotic cell divisions of an organism is relatively faithful, but major reprogramming events during gametogenesis and early embryogenesis usually prevent the inheritance of epigenetic states between generations. However, there appear to be exceptions to this rule. At least in plants, epigenetic states have been shown to be transmitted unchanged through the germline. A good example is that of paramutations, which were first described in maize. Paramutations result from an interaction between the two alleles of a gene: at loci susceptible to paramutations a silent allele can silence its active homolog. Paramutations do not involve a change in DNA sequence, but a change in chromatin structure, which is meiotically stable. Phenotypes that are subject to paramutations show deviations from Mendelian inheritance.

Evidence for transgenerational inheritance in mammals is scanty. The best-studied examples are two mouse lines carrying a retrotransposon upstream of a gene, the *agouti viable yellow* (A^{vy}) [24] and the *axin-fused* ($Axin^{Fu}$) mice [29]. In both cases, the transcriptional activity is under the control of a promoter within the retrotransposon. In A^{vy} mice, retrotransposon driven ectopic expression of the *agouti* gene leads to yellow coat color, obesity, diabetes, and increased tumor risk. Methylation of the promoter prevents ectopic gene expression and results in pseudo-agouti coat color. In addition to yellow and pseudo-agouti mice, mottled mice are observed. These mice are somatic mosaics for ectopic *agouti* expression. The phenotype of the mice can be modified by maternal nutrition during pregnancy, specifically by folate supplements. The A^{vy} mice show that genetically identical organisms can have different phenotypes. Furthermore, the phenotypes can be passed on to subsequent generations, although the penetrance is incomplete. For example, yellow dams have more yellow offspring than pseudo-agouti dams have. This suggests that the epigenetic state is not always cleared between generations. Surprisingly, however, the retrotransposon is demethylated during early embryogenesis. Thus, DNA methylation cannot be the inherited mark, and other possibilities, such as histone modifications, have to be considered.

The A^{vy} and $Axin^{Fu}$ mice are rare variants involving a retrotransposon, and it is not clear whether epigenetic states of normal mammalian genes can be transmitted through the germline. In humans there is no direct evidence for transgenerational epigenetic inheritance, although epidemiological studies suggest that it might occur. One such study found that paternal grandfathers' food supply during their prepubertal slow growth phase was linked to the mortality risk ratio of their grandsons [28]. Indirect evidence comes from the study of patients with PWS and an imprinting defect not caused by an IC mutation. As mentioned above, the affected chromosome is always derived from the paternal grandmother, suggesting that the (grand)maternal imprint was not erased in the paternal germline and is transmitted to the child [5].

A few multigenerational families with an epimutation have been described in the literature, but it is not possible in these cases to distinguish between transmission of an epimutation and transmission of a genetic variant predisposing to an epimutation in each generation.

Despite the lack of direct evidence for transgenerational epigenetic inheritance in mammals, especially in humans, it is of great interest for the theory of evolution. Germline transmission of epigenetic states would allow for the inheritance of acquired traits and thus provide a basis for adaptive evolution [19]. Because of random or environmentally induced epigenetic variation, new epigenetic variants appear in the population. If they are transmitted through the germline and are selectively favorable, they will accumulate in the population and allow the genotypes to occupy a new adaptive zone while maintaining the old one. Under continued selection pressure, the favorable epigenetic variants will be fixed by genetic changes, which can be transmitted to the offspring in a more stable way. Genetic fixation of epigenetic changes may also underlie speciation.

References

1. Baylin SB, Hoppener JW, de Bustros A, Steenbergh PH, Lips CJ, Nelkin BD (1986) DNA methylation patterns of the calcitonin gene in human lung cancers and lymphomas. Cancer Res 46:2917–2922
2. Bird AP (1986) CpG-rich islands and the function of DNA methylation. Nature 321:209–13
3. Brown CJ, Ballabio A, Rupert JL, Lafreniere RG, Grompe M, Tonlorenzi R, Willard HF (1991) A gene from the region of the human X inactivation centre is expressed

exclusively from the inactive X chromosome. Nature 349:38–44
4. Buiting K, Saitoh S, Gross S, Dittrich B, Schwartz S, Nicholls RD, Horsthemke B (1995) Inherited microdeletions in the Angelman and Prader–Willi syndromes define an imprinting centre on human chromosome 15. Nat Genet 9:395–400
5. Buiting K, Gross S, Lich C, Gillessen-Kaesbach G, el-Maarri O, Horsthemke B (2003) Epimutations in Prader–Willi and Angelman syndromes: a molecular study of 136 patients with an imprinting defect. Am J Hum Genet 72:571–577
6. Cui H, Horon IL, Ohlsson R, Hamilton SR, Feinberg AP (1998) Loss of imprinting in normal tissue of colorectal cancer patients with microsatellite instability. Nat Med 4:1276–1280
7. Eggan K, Baldwin K, Tackett M, Osborne J, Gogos J, Chess A, Axel R, Jaenisch R (2004) Mice cloned from olfactory sensory neurons. Nature 428:44–49
8. Engel E (1980) A new genetic concept: uniparental disomy and its potential effect, isodisomy. Am J Med Genet 6:137–143
9. Feinberg AP, Vogelstein B (1983) Hypomethylation distinguishes genes of some human cancers from their normal counterparts. Nature 301:89–92
10. Flanagan JM, Popendikyte V, Pozdniakovaite N, Sobolev M, Assadzadeh A, Schumacher A, Zangeneh M, Lau L, Virtanen C, Wang SC, Petronis A (2006) Intra- and interindividual epigenetic variation in human germ cells. Am J Hum Genet 79:67–84
11. Fraga MF, Ballestar E, Paz MF, Ropero S, Setien F, Ballestar ML, Heine-Suner D, Cigudosa JC, Urioste M, Benitez J, Boix-Chornet M, Sanchez-Aguilera A, Ling C, Carlsson E, Poulsen P, Vaag A, Stephan Z, Spector TD, Wu YZ, Plass C, Esteller M (2005) Epigenetic differences arise during the lifetime of monozygotic twins. Proc Natl Acad Sci USA 102:10604–10609
12. Friso S, Choi SW, Girelli D, Mason JB, Dolnikowski GG, Bagley PJ, Olivieri O, Jacques PF, Rosenberg IH, Corrocher R, Selhub J (2002) A common mutation in the 5, 10-methylenetetrahydrofolate reductase gene affects genomic DNA methylation through an interaction with folate status. Proc Natl Acad Sci USA 99:5606–5611
13. Frommer M, McDonald LE, Millar DS, Collis CM, Watt F, Grigg GW, Molloy PL, Paul CL (1992) A genomic sequencing protocol that yields a positive display of 5-methylcytosine residues in individual DNA strands. Proc Natl Acad Sci USA 89:1827–1831
14. Greger V, Passarge E, Hopping W, Messmer E, Horsthemke B (1989) Epigenetic changes may contribute to the formation and spontaneous regression of retinoblastoma. Hum Genet 83:155–158
15. Hark AT, Schoenherr CJ, Katz DJ, Ingram RS, Levorse JM, Tilghman SM (2000) CTCF mediates methylation-sensitive enhancer-blocking activity at the H19/Igf2 locus. Nature 405:486–489
16. Holliday R (1987) The inheritance of epigenetic defects. Science 238:163–170
17. Horsthemke B (2006) Epimutations in human disease. Curr Top Microbiol Immunol 310:45–59
18. Ingrosso D, Cimmino A, Perna AF, Masella L, De Santo NG, De Bonis ML, Vacca M, D'Esposito M, D'Urso M, Galletti P, Zappia V (2003) Folate treatment and unbalanced methylation and changes of allelic expression induced by hyperhomocysteinaemia in patients with uraemia. Lancet 361:1693–1699
19. Jablonka E, Lamb MJ (1989) The inheritance of acquired epigenetic variations. J Theor Biol 139:69–83
20. Jones C, Booth C, Rita D, Jazmines L, Brandt B, Newlan A, Horsthemke B (1997) Bilateral retinoblastoma in a male patient with an X; 13 translocation: evidence for silencing of the RB1 gene by the spreading of X inactivation. Am J Hum Genet 60:1558–1562
21. Li J, Ishii T, Feinstein P, Mombaerts P (2004) Odorant receptor gene choise is reset by nuclear transfer from mouse olfactory sensory neurons. Nature 428:393–399
22. Mayer W, Niveleau A, Walter J, Fundele R, Haaf T (2000) Demethylation of the zygotic paternal genome. Nature 403:501–502
23. McGrath J, Solter D (1984) Completion of mouse embryogenesis requires both the maternal and paternal genomes. Cell 37:179–183
24. Morgan HD, Sutherland HG, Martin DI, Whitelaw E (1999) Epigenetic inheritance at the agouti locus in the mouse. Nat Genet 23:314–318
25. Muller HJ (1930) Types of visible variations induced by X-rays in *Drosophila*. J Genet 22:299–334
26. Murrell A, Heeson S, Cooper WN, Douglas E, Apostolidou S, Moore GE, Maher ER, Reik W (2004) An association between variants in the IGF2 gene and Beckwith–Wiedemann syndrome: interaction between genotype and epigenotype. Hum Mol Genet 13:247–255
27. Oates NA, van Vliet J, Duffy DL, Kroes HY, Martin NG, Boomsma DI, Campbell M, Coulthard MG, Whitelaw E, Chong S (2006) Increased DNA methylation at the AXIN1 gene in a monozygotic twin from a pair discordant for a caudal duplication anomaly. Am J Hum Genet 79: 155–162
28. Pembrey ME, Bygren LO, Kaati G, Edvinsson S, Northstone K, Sjostrom M, Golding J (2006) Sex-specific, male-line transgenerational responses in humans. Eur J Hum Genet 14:159–166
29. Rakyan VK, Chong S, Champ ME, Cuthbert PC, Morgan HD, Luu KV, Whitelaw E (2003) Transgenerational inheritance of epigenetic states at the murine Axin(Fu) allele occurs after maternal and paternal transmission. Proc Natl Acad Sci USA 100:2538–2543
30. Richards EJ (2006) Inherited epigenetic variation–revisiting soft inheritance. Nat Rev Genet 7:395–401
31. Shykind BM (2005) Regulation of odorant receptors: one allele at a time. Hum Mol Genet 14(Spec No 1):R33–39
32. Stirzaker C, Song JZ, Davidson B, Clark SJ (2004) Transcriptional gene silencing promotes DNA hypermethylation through a sequential change in chromatin modifications in cancer cells. Cancer Res 64:3871–3877
33. Surani MA, Barton SC, Norris ML (1984) Development of reconstituted mouse eggs suggests imprinting of the genome during gametogenesis. Nature 308:548–550
34. Tufarelli C, Stanley JA, Garrick D, Sharpe JA, Ayyub H, Wood WG, Higgs DR (2003) Transcription of antisense RNA leading to gene silencing and methylation as a novel cause of human genetic disease. Nat Genet 34:157–165
35. Waddington CH (1968) The basic ideas of biology. In: Waddington CH (ed) Towards a theoretical biology, vol 1. Edinburgh University Press, Edinburgh, pp 1–32
36. Weksberg R, Shuman C, Caluseriu O, Smith AC, Fei YL, Nishikawa J, Stockley TL, Best L, Chitayat D, Olney A, Ives E, Schneider A, Bestor TH, Li M, Sadowski P, Squire

J (2002) Discordant KCNQ1OT1 imprinting in sets of monozygotic twins discordant for Beckwith–Wiedemann syndrome. Hum Mol Genet 11:1317–1325
37. Wilkins JF, Haig D (2003) What good is genomic imprinting: the function of parent-specific gene expression. Nat Rev Genet 45:359–68
38. Wilmut I, Schnieke AE, McWhir J, Kind AJ, Campbell KH (1997) Viable offspring derived from fetal and adult mammalian cells. Nature 385:810–813
39. Ying AK, Hassanain HH, Roos CM, Smiraglia DJ, Issa JJ, Michler RE, Caligiuri M, Plass C, Goldschmidt-Clermont PJ (2000) Methylation of the estrogen receptor-alpha gene promoter is selectively increased in proliferating human aortic smooth muscle cells. Cardiovasc Res 46:172–179
40. Zogel C, Bohringer S, Gross S, Varon R, Buiting K, Horsthemke B (2006) Identification of cis- and trans-acting factors possibly modifying the risk of epimutations on chromosome 15. Eur J Hum Genet 14:752–758

10 Human Gene Mutation: Mechanisms and Consequences

Stylianos E. Antonarakis and David N. Cooper

Abstract A wide variety of different types of pathogenic mutation occur in the human genome, with many diverse mechanisms responsible for their generation. These types of mutation include single base-pair substitutions in coding, regulatory and splicing-relevant regions of human genes, and also micro-deletions, micro-insertions, duplications, repeat expansions, combined micro-insertions/deletions ("indels"), inversions, gross deletions and insertions, and complex rearrangements. A major goal of molecular genetic medicine is to be able to predict the nature of the clinical phenotype through ascertainment of the genotype. However, the extent to which this is feasible in medical genetics is very much disease, gene, and mutation dependent. The study of mutations in human genes is nevertheless of paramount importance for our understanding of the pathophysiology of inherited disorders and for optimizing diagnostic testing, as well as in guiding the design of new therapeutic approaches.

Contents

10.1	Introduction	319
10.2	Neutral Variation/DNA Polymorphisms	320
10.3	Disease-Causing Mutations	321
	10.3.1 The Nature of Mutation	321
	10.3.2 Consequences of Mutations	336
10.4	General Principles of Genotype-Phenotype Correction	349
10.5	Why Study Mutation?	351
References		351

S.E. Antonarakis
Department of Genetic Medicine and Development
University of Geneva Medical School
and University Hospitals of Geneva
1 rue Michel-Servet, 1211 Geneva, Switzerland
e-mail: Stylianos.Antonarakis@unige.ch

D.N. Cooper (✉)
Institute of Medical Genetics, Cardiff University,
Heath Park, Cardiff CF14 4XN, UK
e-mail: cooperdn@cardiff.ac.uk

10.1 Introduction

A wide variety of different types of pathogenic mutation occur in the human genome, with many diverse mechanisms being responsible for their generation. These types of mutation include single base-pair substitutions in coding, regulatory and splicing-relevant regions of human genes, and also micro-deletions, micro-insertions, duplications, repeat expansions, combined micro-insertions/deletions ("indels"), inversions, gross deletions and insertions, and complex rearrangements. A major goal of molecular genetic medicine is to be able to predict the nature of the clinical phenotype through ascertainment of the genotype. However, the extent to which this is feasible in medical genetics is very much disease, gene, and mutation dependent. The study of mutations in human genes is nevertheless of paramount importance for our understanding of the pathophysiology of inherited disorders and for optimizing diagnostic testing, as well as in guiding the design of new therapeutic approaches.

The first description of the exact molecular defect in a human disease (sickle cell mutation, a substitution from Glu to Val at the 6th codon of the β-globin gene) was identified by Ingram in 1956, who found that the difference between hemoglobin A and hemoglobin S lies in a single tryptic peptide (158). His analysis was made possible by methods developed by Sanger for determining the structure of insulin and Edman's stepwise degradation of peptides. Since then, continuous advances have potentiated the identification of numerous disease-related genes and the discovery of thousands of underlying pathologic lesions. Single base-pair substitutions (68%) and micro-deletions (16.4%) are the most frequently encountered mutations in the human genome, the remainder comprising an assortment of micro-insertions (6.6%), indels (1.5%), gross deletions (5.6%), gross insertions and duplications (1.0%), inversions, repeat expansions (0.23%), and complex rearrangements (0.8%). Characterized mutations occur not only in coding sequences, but also in promoter regions, splice junctions, introns and untranslated regions, and any other functional region of the genome. Mutations can interfere with any stage in the pathway of expression from gene activation to synthesis and secretion of the mature protein product. This chapter attempts to provide an overview of the nature of mutations causing human genetic disease and then considers their consequences for the clinical phenotype. The interested reader is also referred to the third edition of this work for an in-depth discussion of mutation rates and factors influencing the generation of mutations. Two online databases contain information on disease-related (pathogenic) mutations: *Mendelian Inheritance in Man* (http://www.ncbi.nlm.nih.gov/Omim/) and the *Human Gene Mutation Database* (http://www.hgmd.org).

10.2 Neutral Variation/DNA Polymorphisms

The term *polymorphism* has been defined (336) as a "Mendelian trait that exists in the population in at least two phenotypes, neither of which occurs at a frequency of less than 1%." Polymorphisms are not rare. Indeed, there is enormous variation in the DNA sequences of any two randomly chosen human haploid genomes. Clearly, not all variations within a gene result in the abnormal expression of protein products. Indeed, single nucleotide substitutions/polymorphisms (SNP) occur in 1/~600–1,200 nucleotides in intervening sequences and flanking DNA (13, 75, 103, 244, 285, 342). These substitutions represent the most common forms of DNA polymorphism that can be used as markers for specific regions of the human genome. Similarly, some single nucleotide substitutions in the coding regions of genes may also be normal (nonpathogenic) polymorphic variants even if they result in nonsynonymous substitutions of the polypeptide product (247). For example, there are three common forms of the β-globin (*HBB*) gene on chromosome 11p. These forms differ at five nucleotides, one of which lies within the first exon of the gene and results in a synonymous codon. The average human gene contains > 120 biallelic polymorphisms, 46 of which occur with a frequency > 5% and 5 within the coding region (78).

Some polymorphisms entail the alteration of an encoded amino acid, e.g., the Lewis *Le* alleles of the *FUT3* gene (245), whereas others may introduce a stop codon that serves to inactivate the gene in question, e.g., the secretor *se* allele of the *FUT2* gene present in 20% of the population (176). However, not all polymorphisms are SNPs. Examples of other types of gene-associated polymorphism in the human genome include triplet repeat copy number (e.g., in the *FMR1* gene; see Sect. 9.2.1.3), gross gene deletion (e.g., *GSTM1* and *GSTT1*) (273), gene duplication (e.g., *HBG2*) (318) intragenic duplication (e.g., *IVL*) (132), micro-insertion/deletion (e.g., *PAI1*) (84), indel (e.g., *APOE*) (120), gross insertion (e.g., the inserted *Alu* sequence in intron 16 of the *ACE* gene) (277), inversion (e.g., the 48 kb Xq28 inversion involving the *EMD* and *FLN1* genes) (307)) and gene fusion (e.g., between the *RCP* and *GCP* visual pigment genes) (241)). It can be seen that the mutational spectrum of polymorphisms in the human genome is qualitatively different to that underlying human disease; they may vary in terms of location and frequency but otherwise they display remarkable similarities indicative of the same underlying mutational mechanisms.

It is likely, however, that some SNPs, both frequent and rare, alter the risk for common complex human phenotypes. A public SNP database now contains more than 10 million entries (dbSNP; http://www.ncbi.nlm.nih.gov/SNP/index.html). An international project has recently been completed, termed the "HapMap project" (8, 68, 144), the goal of which was to define the patterns of common SNP genetic variation in a sample of 270 DNAs from individuals of European, African, Chinese, and Japanese origin (http://www.hapmap.org/). The data obtained from this project constitute approximately 2.8 million SNPs and are publicly available. The results of this project are likely to contribute significantly to our understanding of both common and rare human genetic disorders and traits.

Another form of polymorphic variation in our genome is the presence of variable numbers of tandem repeats (VNTRs). The repeat unit can be 10–60 nucleotides in length and many different alleles may exist at a given locus (164, 358). The combination of a VNTR and single nucleotide substitutions within the repeat unit results in an extremely high level of polymorphic variability that can be used as a unique bar-code to distinguish different individuals (165). The introduction of the polymerase chain reaction (PCR) (286) permitted the rapid detection and analysis of variation in short sequence repeats (SSR), e.g., $(GT)_n$ repeats (212, 347) . These are common polymorphisms that occur on average once every 50 kilobases (kb) of genomic DNA. The SSRs also display many alleles and the repeat unit can be two, three, four, five or more nucleotides. Poly(A) tracts may also be polymorphic, exhibiting variation in the number of A residues (101); many of these polymorphisms are localized at the ends of *Alu* repetitive elements. Another kind of polymorphism in the human genome involves the presence or absence of retrotransposons (i.e., *Alu* or LINE repetitive elements or pseudogenes) at specific locations (10, 71) . Furthermore, duplicational polymorphisms have also been reported for some human genes, e.g., *HBA1, PRB1-4, HBZ, CYP21/C4A/C4B* (39, 71).

The use of comparative genomic hybridization against BAC or oligonucleotide arrays has revealed extensive copy number polymorphisms/variation of sizable genomic regions (CNP or CNV) (156, 294, 300). Details of more than 1,400 such genomic variants may be found in the following databases: *Human Structural Variation Database*, http://paralogy.gs.washington.edu/structuralvariation; *Database of Genomic Variants*, http://projects.tcag.ca/variation. A CNV map of the human genome of the 270 "HapMap" individuals has revealed a total of 1,440 CNV such regions which cover some 360 megabases (12% of the genome) (274). The functional significance, if any, of the majority of these polymorphic variants is however unknown. Copy number variants may predispose to phenotypic variability. For example, it has recently been observed that copy number variation of the orthologous rat and human *Fcgr3/FCGR3B* genes is a determinant of susceptibility to immunologically mediated glomerulonephritis (5). Copy number variants in the *CCL3L1* and *DEFB4* genes have also been found to be associated with increased susceptibility to HIV infection and Crohn's disease, respectively (111, 126).

Deletional polymorphisms are also remarkably frequent in the human genome: a typical individual has been estimated to be hemizygous for some 30–50 deletions >5 kb, spanning >550 kb in total and encompassing >250 known or predicted genes (67, 222). Since such deletions appear to be in linkage disequilibrium with neighboring SNPs, we may surmise that they share a common evolutionary history (145).

Human DNA polymorphisms have proven extremely useful in developing linkage maps, for mapping monogenic and polygenic complex disorders, for determining the origin of aneuploidies and chromosomal abnormalities, for distinguishing normal from mutant chromosomes in genetic diagnoses, for performing forensic, paternity, and transplantation studies, for studying the evolution of the genome, the loss of heterozygosity in certain malignancies, the detection of uniparental disomy, the instability of the genome in certain tumors, recombination at the level of the genome, the study of allelic expression imbalance, and the development of haplotype maps of the genome. However, in studying the role of a candidate gene in a given disorder, it is imperative to distinguish between pathogenic mutations that cause a clinical phenotype and the polymorphic variability of the normal genome.

10.3 Disease-Causing Mutations

10.3.1 The Nature of Mutation

Figure 10.1a depicts the frequencies of the various mutation types responsible for molecularly characterized human genetic disorders, as recorded in the *Human Gene Mutation Database* (HGMD) (http://www.hgmd.org) and elsewhere (196, 312)). HGMD records each mutation *once* regardless of the number of independent occurrences of that lesion. Figure 10.1b shows the frequency of the first mutation per disease recorded in *Mendelian Inheritance in Man* (MIM) (http://www.ncbi.nlm.nih.gov/Omim) and by Antonarakis and McKusick (11). As of March 31, 2009, HGMD contained some 88,317 different mutations in 3,337 human genes, whereas MIM contained examples of allelic variants in 2,514 human genes.

10.3.1.1 Nucleotide Substitutions

Single nucleotide substitutions are the most frequent pathologic mutations in the human genome (Fig. 10.1). Most of these alterations occur during DNA replication, which is an accurate, yet error-prone, multistep process. The accuracy of DNA replication depends on the fidelity of the replicative step and the efficiency of the subsequent error correction mechanisms (214). Analysis of more than 7,000 missense and nonsense mutations associated with human disease has indicated that the most common nucleotide substitution for T (thymine) is to C (cytosine), for C it is to T, for A (adenine) it is to G (guanine) and for G it is to A (195). Transitions are therefore much more common than transversions. Some 61% of the missense/nonsense mutations currently logged in HGMD are transitions (T to C, C to T, A to G, G to A), whilst 39% are transversions (T to A or G, A to T or C, G to C or T, C to G or A).

Among single nucleotide substitutions there is one that clearly predominates, and it represents the most common type of mutational lesion: CpG dinucleotides mutate to TpG at a frequency that is about 5 times that of mutations in all other dinucleotides (15, 195, 361, 363). This substitution, which generates TG when it occurs on one DNA strand and CA ("CG to TG or CA rule") when it is on the other, is a major cause of human genetic disease. This phenomenon was first observed in the factor VIII (*F8*) gene in cases of hemophilia A (361), but it was soon noted in studies of many other genes (74). In hemophilia A, CG to TG or CA mutations account for 46% of point mutations in unrelated patients (14). In the HGMD (312) (http://www.hgmd.org), such mutations currently account for ~20% of the total number of missense/nonsense mutations. Among CpG dinucleotide mutations, transitions to TG or CA account for ~90% of substitutions. The mechanism of this common type of mutation appears to be methylation-mediated deamination of 5-methylcytosine (5mC). In eukaryotic genomes, 5mC occurs predominantly in CpG dinucleotides, most of which appear to be methylated (see (70) for review). 5mC then undergoes spontaneous nonenzymatic deamination to form thymine (Fig. 10.2). There is a bias in terms of the origin of CpG to TpG mutations: most occur in male germ cells

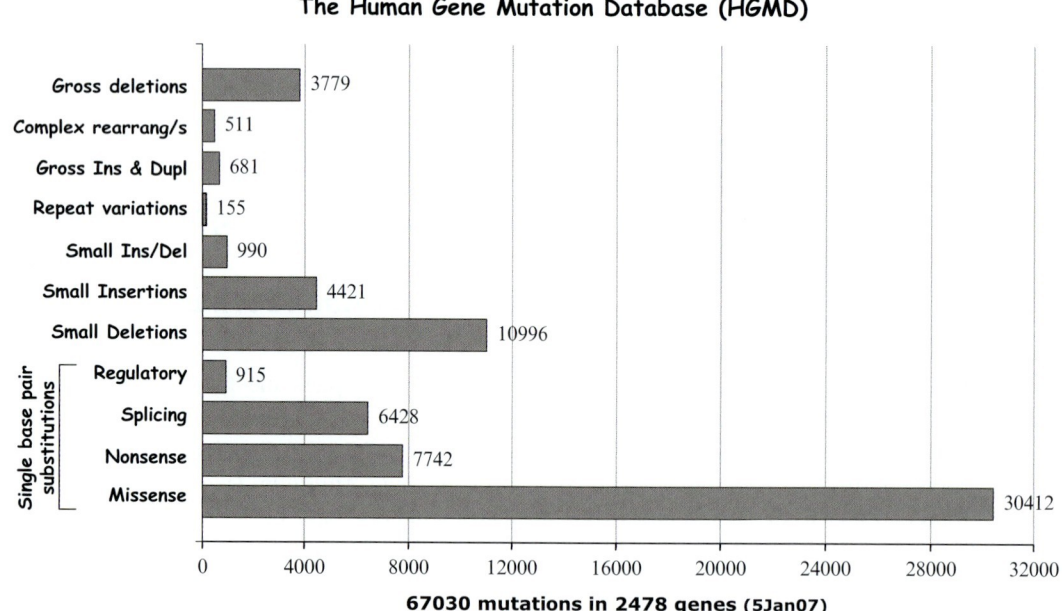

Fig. 10.1 Spectrum of different types of human gene mutations logged in the Human Gene Mutation Database as of January 2007

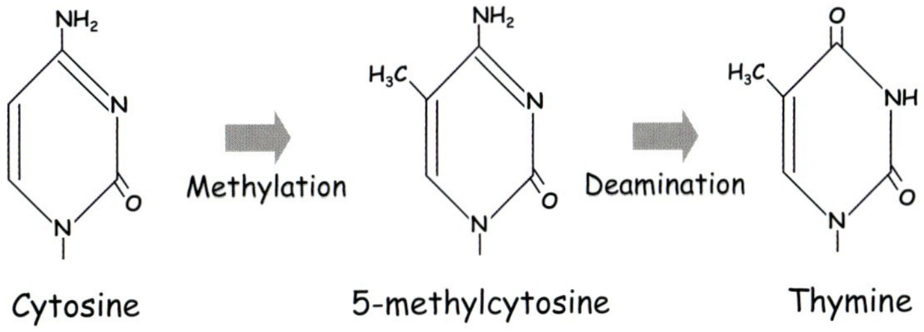

Fig. 10.2 Schematic representation of cytosine, 5′-methylocystosine, and thymine, and the chemical events involved in the mutational transformation of cytosine to thymine

(the male/female ratio is 7 to 1). One reason for this may be that sperm DNA is heavily methylated, whereas oocyte DNA is comparatively undermethylated (99). Another reason may be the considerably higher number of germline cell divisions in males than in females (154).

In a recent analysis, the average direct estimate of the combined rate of all mutations was 1.8×10^{-8} per nucleotide per generation. Single nucleotide substitutions were found to be approximately 25 times more common than all other mutations, whilst deletions were approximately three times as common as insertions; complex mutations were very rare, and the CpG context was found to increase substitution rates by an order of magnitude (185). Rates of different kinds of mutations were also found to be strongly correlated across different loci (185).

10.3.1.2 Micro-Deletions and Micro-Insertions

Deletions or insertions of a few nucleotides are also fairly common as a cause of human inherited disease. Most of these are less than 20 bp in length. Indeed, the majority of micro-deletions involve <5 nucleotides. In HGMD, the deletion of 1 bp accounts for 48% of small deletions, whilst an additional 30% involve 2 or 3 nucleotides. The majority of micro-deletions recorded (78%) result in an alteration of the reading frame. Most micro-deletions occur in regions that contain direct repeats of 2 bp or more. The most common length of direct repeat is 3 bp (48% of direct repeats associated with short deletions (15)). The most plausible mechanism for small deletions mediated by the presence of direct repeats is the slipped mispairing model (104) (Fig. 10.3). In addition, deletions of one or a few nucleotides frequently occur in runs of the same nucleotide, e.g., a poly(T) region (198). Finally, inverted repeats

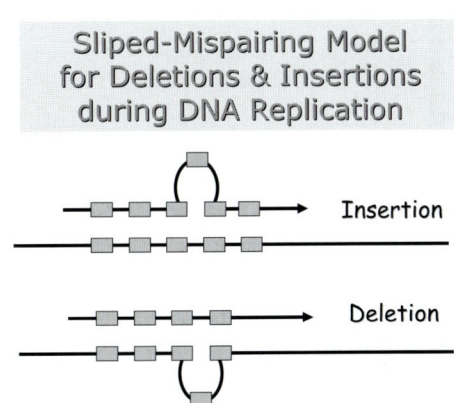

Fig. 10.3 Schematic representation of the slipped mispairing model for deletions and insertions during DNA replication

and "symmetric elements" are also frequently found in the immediate vicinity of micro-deletions (73, 289). Krawczak and Cooper (193) identified a consensus sequence – TG(A/G)(A/G)(G/T)(A/C) – which they claimed represented a deletion hotspot.

Micro-insertions (again up to 20 nucleotides) are rarer than micro-deletions; thus, in HGMD there are three times as many micro-deletions as micro-insertions (Fig. 10.1a). Nearly half of these involve the insertion of only 1 nucleotide (Fig. 10.4). As is the case with micro-deletions, most micro-insertions lead to alterations of the reading frame and are located in regions containing direct or inverted repeats or runs of the same nucleotide. Details of possible mechanisms of generation during replication can be found in elsewhere (72). There are as yet insufficient data available to estimate the frequency ratio of micro-insertions or micro-deletions in male or female germ cells. In the case of such lesions in factor VIII (*F8*) gene, 56% of micro-deletions/-insertions have been reported to occur in DNA regions harboring direct repeats or runs of the same nucleotide (14).

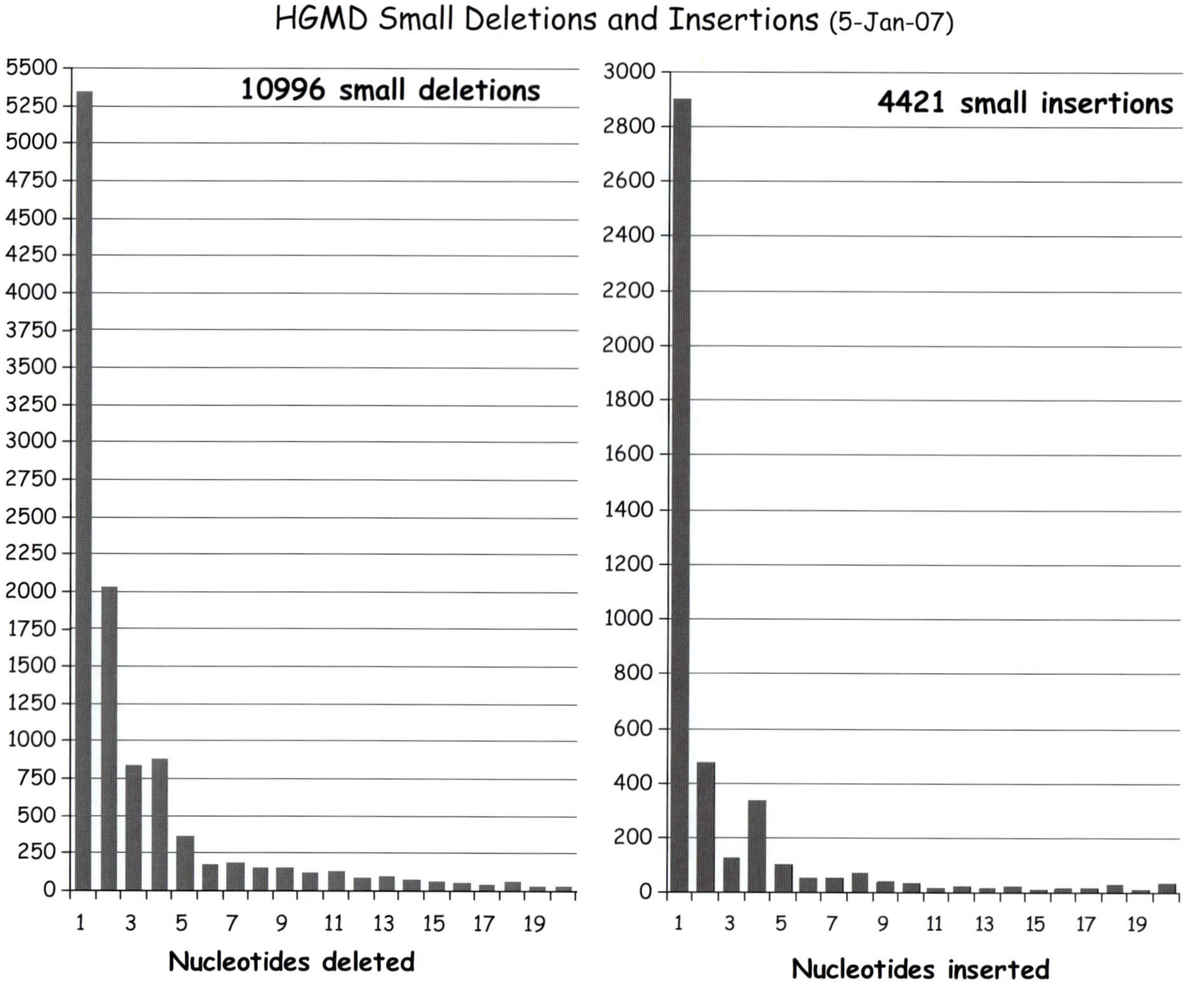

Fig. 10.4 Size distribution of short (<20 bp pathogenic human gene deletions and insertions (HGMD; http://www.hgmd.org, 5 January 2007)

HGMD data (3,767 micro-deletions and 1,960 micro-insertions) were used to perform a meta-analysis of micro-deletions and micro-insertions causing inherited disease, both defined as involving ≤20 bp DNA (23). A positive correlation was noted between the micro-deletion and micro-insertion frequencies for 564 genes for which both micro-deletions and micro-insertions have been reported. This is consistent with the view that the propensity of a given gene/sequence to undergo micro-deletion is related to its propensity to undergo micro-insertion. While micro-deletions and micro-insertions of 1 bp constitute, respectively, 48 and 66% of the corresponding totals, the relative frequency of the remaining lesions correlates negatively with the length of the DNA sequence deleted or inserted. Many micro-deletions and micro-insertions of >1 bp can potentially be explained in terms of slippage mutagenesis, involving the addition or removal of one copy of a mono-, di-, or trinucleotide tandem repeat. The frequency of in-frame 3 and 6 bp micro-insertions and micro-deletions was, however, found to be significantly lower than that of mutations of other lengths, suggesting that some of these in-frame lesions may not have come to clinical attention. Various sequence motifs were found to be overrepresented in the vicinity of both micro-insertions and micro-deletions, including the heptanucleotide CCCCCTG that shares homology with the complement of the 8-bp human minisatellite conserved sequence/

chi-like element (GCWGGWGG). The "indel hotspot" GTAAGT (and its complement ACTTAC) were also found to be overrepresented in the vicinity of both micro-insertions and micro-deletions, thereby providing a first example of a mutational hotspot that is common to different types of gene lesion. Other motifs overrepresented in the vicinity of micro-deletions and micro-insertions included DNA polymerase pause sites and topoisomerase cleavage sites. Several novel micro-deletion/micro-insertion hotspots were noted, and some of these exhibited sufficient similarity to one another to justify terming them "super-hotspot" motifs. Analysis of DNA sequence complexity also demonstrated that a combination of slipped mispairing mediated by direct repeats, and secondary structure formation promoted by symmetric elements, can account for the majority of micro-deletions and micro-insertions. Thus, micro-insertions and micro-deletions exhibit strong similarities in terms of the characteristics of their flanking DNA sequences, implying that they are generated by very similar underlying mechanisms.

A similar analysis of micro-deletions and micro-insertions in 19 human genes yielded evidence for an elevated micro-deletion rate at YYYTG and an elevated micro-insertion rate at TACCRC and ATMMGCC (186). Kondrashov and Rogozin (186) also found that ~45% of micro-deletions led to the removal of a repeated sequence, an event they termed "deduplication" in order to highlight the identity of the deleted sequence and the sequence abutting the site of deletion.

10.3.1.3 Expansion/Copy Number Variation of Trinucleotide (and Other) Repeat Sequences

Another mechanism of human gene mutation causing hereditary disease is the instability of certain trinucleotide repeats and their expansion in affected genes (44, 218, 282). A growing number of disorders (in excess of 150 are now recorded in HGMD), the majority of which involve neuromuscular tissues, have been found to be due to, or associated with, the expansion of repeat sequences; of these, 23 are expansions of triplet repeats. The first such disease was fragile X, a common cause of male mental retardation, which mapped to chromosome Xq27.3. Table 10.1 lists some examples of these disorders, which include Huntington disease, myotonic dystrophy, spinobulbar muscular atrophy, spinocerebellar ataxia 1, spinocerebellar ataxia 3 or Machado-Joseph disease, the fragile E site, and dentatorubral pallidoluysian atrophy. Genetic "anticipation" (the earlier onset and increasingly severe phenotype in successive generations) is a common phenomenon in these disorders (141). The trinucleotide involved is usually either CAG or CGG, but occasionally CTG, GCG, or GAA. It can be located in the 5' untranslated region (UTR), as in the case of the *FMR1* gene underlying fragile X, within the coding region (as in Huntington disease, SCA1, SCA3, and Kennedy disease) encoding poly(Gln), in an intron, as in Friedreich ataxia (*FXN*) and myotonic dystrophy type 2 (*ZNF9*), or in the 3' UTR, as in myotonic dystrophy type 1 (*DMPK*; Table 10.1, Fig. 10.5). The expansion of the triplet repeat either prevents its expression (329), results in a dominant gain–of–function mutation mediated by the longer poly(Gln) peptide (150), or alters the RNA processing of other genes (211, 288).

The trinucleotide repeats are usually polymorphic in human populations. Rarely, however, the number of trinucleotide repeats lies within a high-risk category that is termed "premutation." In such a case, the premutation exhibits a high probability of further expansion (instability) to yield disease-related alleles ("full mutation"). In fragile X, for example, the normal polymorphic alleles of the CGG repeat contain between 10 and 50 triplets, the premutation between 50 and 200, and the full mutation more than 200 triplets (119). Expansion of premutations to full mutations only occurs during female meiotic transmission. The probability of repeat expansion correlates with repeat copy number in the premutated allele. Since the premutation must precede the appearance of a full mutation, all mothers of affected children carry either a full mutation or a premutation (119). Premutation alleles may also be associated with late-onset movement disorders and premature ovarian failure (69, 163)

The precise mechanism of repeat expansion is unclear, although it is known that DNA polymerase progression is blocked by CTG and CGG repeats and the resultant idling of the polymerase could serve to catalyze slippage, leading to repeat expansion (171). In the case of spinocerebellar ataxia 1 (SCA1), interruption of the CAG repeat with a CAT unit is associated with more stable trinucleotide repeat (56). More details about these "dynamic mutations" can be found in the appropriate sections covering individual disorders, and have also been treated by Wells (348). Short

Table 10.1 Various examples of disorders of trinucleotide and other repeat expansions

	Disorder	Inheritance	Gene	Chr	OMIM#	Repeat	Normal	Mutant	Repeat location	Mutation type	Parental gender bias
1	Fragile X syndrome	XLD	FMR1	Xq27.3	309,550	CGG	6–52	60–200 premutation 230–1000 full mut	5'UTR	LOF, FraX	Maternal
2	Fragile E mental retardation	XLD	FMR2	Xq28	309,548	GCC	7–35	130–150 premutation 230–750 full mut	5'UTR	LOF, FraX	ND
3	Myotonic dystrophy	AD	DMPK	19q13	160,900	CTG	5–37	50–3,000	3'UTR	?Dom negative	Maternal
4	Spinobulbar muscular atrophy	XLR	AR	Xq13-21	313,700	CAG	11–33	38–66	Coding	GOF, LOF	ND
5	Huntington disease	AD	HD	4p16.3	143,100	CAG	6–39	36–121	Coding	GOF	Paternal
6	Dentatorubro-pallidoluysian atrophy	AD	DRPLA	12p13.31	125,370	CAG	6–35	51–88	Coding	GOF	Paternal
7	Spinocerebellar ataxia 1	AD	SCA1/ATX1	6p23	601,556	CAG	6–39	41–81	Coding	GOF	Paternal
8	Spinocerebellar ataxia 2	AD	SCA2/ATX2	12q24.1	601,517	CAG	14–31	35–64	Coding	GOF	Paternal
9	Spinocerebellar ataxia 3	AD	SCA3/MJD1	14q32.1	109,150	CAG	12–41	40–84	Coding	GOF	Paternal
10	Spinocerebellar ataxia 6/ Episodic ataxia type 2	AD	CACNA1A	19p13	601,011	CAG	7–18	20–23 EA2 21–27 SCA6	Coding	ND	ND
11	Spinocerebellar ataxia 7	AD	SCA7	3p12-13	164,500	CAG	7–17	38–130	Coding	GOF	Paternal
12	Friedreich ataxia	AR	FRDA1	9q13-21.1	229,300	GAA	6–34	80 premutation 112-1700 full mut	Intron 1	LOF, FraX	Maternal
13	Progressive myoclonus epilepsy 1	AR	CSTB	21q22.3	601,145	CCCCG-CCCCGCG	2–3	35–80	5' flanking	LOF	Paternal
14	Synpolydactyly	AD	HOXD13	2q31-q32	142,989	(GCG)n(GCT)n(GCA)n	15	22–29	Coding	ND	??
15	Oculopharyngeal muscular dystrophy	AD	PABP2	14q11.2-q13	602,279	GCG	6	7–13	Coding	ND	??

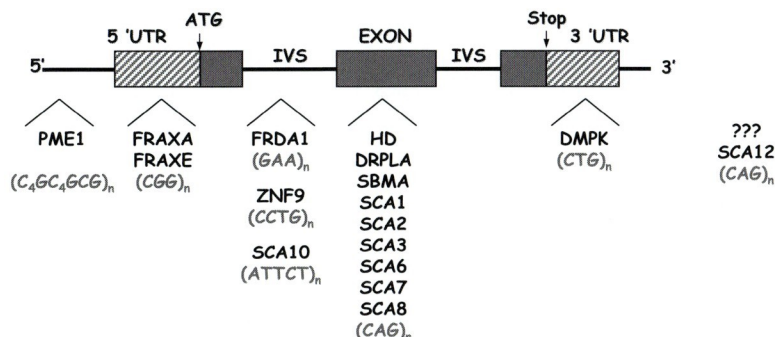

Fig. 10.5 Location of the repeat expansion in selected human disorders

expansions of GCG trinucleotide codons encoding Ala have been observed in the *HOXD13* gene causing dominant polydactyly, and in the *PABP2* gene causing oculopharyngeal muscular dystrophy (37, 237). These mutations may be due to unequal crossing-over rather than polymerase slippage. Generally speaking, it is likely that repeat instability is a consequence of the resolution of unusual secondary structure intermediates during DNA replication, repair and recombination (264).

A repeat expansion of 12 nucleotides (CCCCGCC-CCGCG) in the 5′ flanking region of the *CSTB* gene causes one form of the recessive progressive myoclonus epilepsy (EPM1) (201). This indicates that repeat sequences other than trinucleotides can become expanded and cause human disorders. This particular expansion silences the *CSTB* gene, probably because it alters the spacing of transcription factor binding sites from each other and/or the transcriptional initiation site (202).

A tetranucleotide repeat expansion (CCTG)$_n$ in intron 1 of the *ZNF9* gene causes myotonic dystrophy type 2 (211). This expansion can be between 75 and 11,000 repeats in length. The expansion of the pentanucleotide repeat (ATTCT)$_n$ is responsible for the phenotype of spinocerebellar ataxia 10 (SCA10). The expansion occurs in intron 9 of the *SCA10* gene and can be up to 22.5 kb in length (221). Expansions of even longer repeats have been reported. In Usher syndrome type 1C, for example, there is an expansion of a 45-bp VNTR in intron 5 of the *USH1C* gene (9 tandem repeats instead of the usual less than 6 such repeats); this expansion has been predicted to inhibit transcription of the gene (332). There are also cases in which a large repeat expansion is not associated with a particular phenotype, e.g., the expansion of an AT-rich 33-mer repeat in the dictamycin-sensitive fragile site 16B (364).

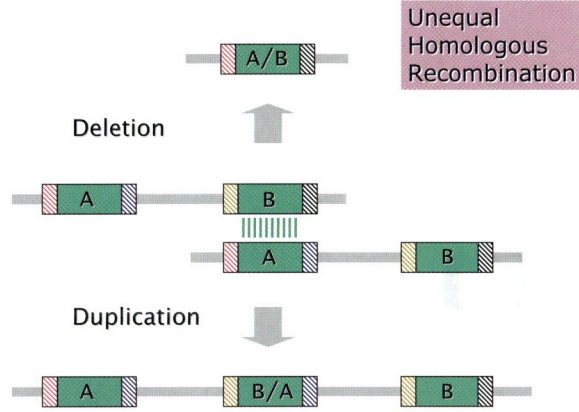

Fig. 10.6 Homologous unequal combination between similar regions of sequences A and B. The recombination events cause either deletions or duplications. In the case of a deletion, a hybrid sequence is generated, with the first part from sequence A and the second, from sequence B. The middle sequence in the duplication product is also a hybrid sequence, with the first part from sequence B and the second, from sequence A

10.3.1.4 Gross Deletions

Gross deletions are common causes of certain disorders and rare in others. In most of the X–linked disorders, for example, large deletions account for about 5% of molecular defects. In other disorders, however, such as steroid sulfatase deficiency, large deletions of the *STS* gene account for 84% of patients (24). The same is true for disorders such as Duchenne muscular dystrophy, growth hormone deficiency and α-thalassemia (92, 242, 335).

A considerable number of large deletions are probably generated by mispairing of homologous sequences and unequal recombination (Fig. 10.6). One of the best examples of homologous unequal recombination is the case of α-globin genes on chromosome 16p. As a result of a recent evolutionary duplication of the α-globin genes, extensive regions of sequence homology exist between the two closely linked α-genes. Unequal crossover results in either deletion of one α-gene or the creation of a fusion hybrid gene (106). The reciprocal product chromosomes carry three α-genes and are not associated with a clinical phenotype (127). Another example of a fusion gene resulting from an unequal crossover is the case of hemoglobin Lepore characterized by a hybrid gene between the δ- and β-globin genes on chromosome 11p (21). In the case of steroid sulfatase deficiency, the deletion can be as large as one megabase (Mb) (299). In Kallmann syndrome, translocation can occur as a result of unequal mispairing of X- and Y-homologous sequences (138).

A number of common genetic disorders are due to large deletions (or duplications) caused by unequal crossing-over of homologous sequences. Figure 10.7 depicts various examples, which include a 1.5-Mb deletion of 17p12 in hereditary neuropathy with liability to pressure palsies (HNPP) (276), deletion of 1.5 Mb of 17q11.2 in neurofibromatosis type 1 (98), deletion of 1.6 Mb of 7q11.23 in Williams syndrome (115), deletion of 5 Mb of 17p11.2 in Smith-Magenis syndrome (169), deletion of either 3 Mb or, more rarely, 1.5 Mb of 22q11 in DiGeorge and velo-cardio-facial syndrome (102, 298), and 4-Mb deletions of 15q in Prader-Willi and Angelman syndromes (54). A recurrent deletion of ~0.5 Mb of 17q21.3, which may be mediated by a common inversion polymorphism, has also been described (188, 301, 305, 311). For a review of chromosomal "duplicons," the low copy repeats that mediate deletions and duplications, see (168). It has been estimated that approximately 5% of the human genome is duplicated either intra- or inter-chromosomally (22). The large deletions or duplications (see below) due to duplicon crossover are also termed "genomic disorders." A recent review of such genomic disorders may be found in (303).

In many cases of large deletion, homologous unequal crossover occurs between repetitive elements such as *Alu* sequences. The *Alu* repeat is the most abundant repetitive element, with about 1.5×10^6 copies in the human genome (86, 203). The element is about 300 bp in length and consists of two similar regions separated by a short A-rich region. Unequal crossover can occur between *Alu* sequences oriented either in opposite directions or in the same direction.

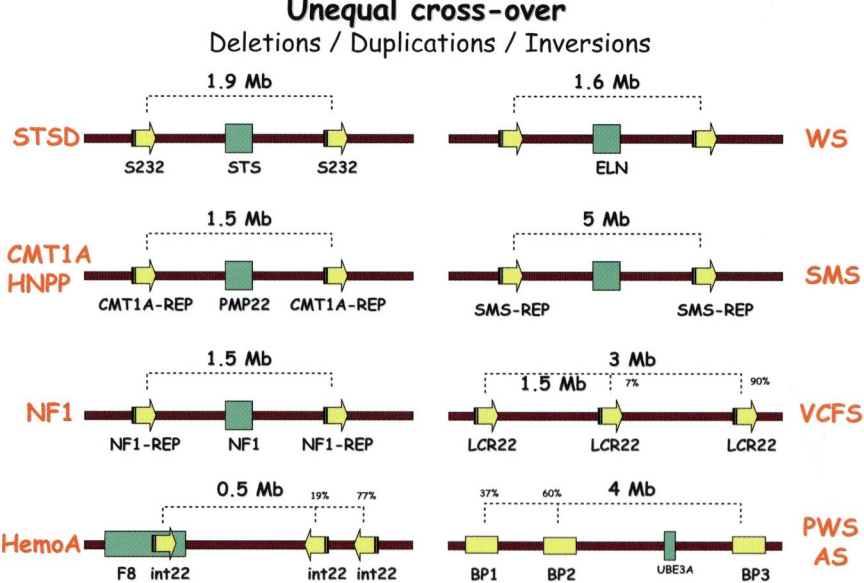

Fig. 10.7 Genes, duplicons, and diseases. Unequal crossover between homologous sequences (duplicons) produce either deletions or duplications of the DNA between the duplicons. The duplicons are shown by *arrows* or *clear boxes*. Genes included in the duplications/deletions are shown as *dark boxes* (*AS* Angelman syndrome *CMTA1* Charcot-Marie-Tooth type 1, *HemoA* hemophilia A, *HNPP* hereditary neuropathy with liability to pressure palsies, *NF1* neurofibromatosis 1, *PWS* Prader-Willi syndrome, *SMS* Smith-Magenis syndrome, *VCFS* velo-cardio-facial syndrome, *STSD* steroid sulfatase deficiency, *WS* Williams syndrome)

In addition, unequal crossings over have been noted between *Alu* elements and nonrepetitive DNA sequences without homology to *Alu*s. The best examples of *Alu-Alu* recombination occur in the genes encoding the low-density lipoprotein receptor (*LDLR*), which underlies familial hypercholesterolemia, and complement component 1 inhibitor (*C1I*) (205, 313)). All but one of the breakpoints associated with *LDLR* gene deletions occur within *Alu* repeats. By contrast, deletions in other *Alu*-rich genes (e.g., *GLA1*) do not necessarily involve *Alu* repetitive elements (189).

Nonhomologous (illegitimate) recombination occurs between two DNA sites that share minimal sequence homology of a few basepairs. This type of recombination during meiosis or alternatively, slipped mispairing during DNA replication mediated by short (2–8) nucleotide direct repeats flanking the deletions is a common finding in many instances of large gene deletions (281). Such deletions have been studied, for example, in hemophilia A; a compilation of 46 junctions from large deletions revealed that about 50% shared 2- to 6-bp homology at the breakpoint junction, as compared with only 17% in which the deletion was due to *Alu-Alu* recombination (356). Similar results have been reported from the intron 7 deletion hotspot in the Duchenne muscular dystrophy (*DMD*) gene; 8/9 deletion breakpoints examined were found to be flanked by DNA sequences with minimal homology (223).

It has also been proposed that alternative DNA conformations may trigger genomic rearrangements through recombination-repair activities. Distance measurements have indicated the significant proximity of alternating purine-pyrimidine and oligo(purine.pyrimidine) tracts to breakpoint junctions in 222 gross deletions and translocations, respectively, involved in human diseases. In 11 deletions analyzed, breakpoints could be explained by non-B DNA structure formation (20).

The Gross Rearrangement Breakpoint Database (GRaBD; http://www.uwcm.ac.uk/uwcm/mg/grabd/). This database was established primarily for the analysis of the sequence context of translocation and deletion breakpoints in a search for characteristics that might have rendered these sequences prone to rearrangement (3). GRaBD, which contains 397 germline and somatic DNA breakpoint junction sequences derived from 219 different rearrangements underlying human inherited disease and cancer, represents a large but not comprehensive collection of sequenced gross gene rearrangement breakpoint junctions. Analysis of these breakpoints has extended our understanding of illegitimate recombination by highlighting the importance of secondary structure formation between single-stranded DNA ends at breakpoint junctions. For example, potential secondary structure was noted between the 5' flanking sequence of the first breakpoint and the 3' flanking sequence of the second breakpoint in 49% of rearrangements, and between the 5' flanking sequence of the second breakpoint and the 3' flanking sequence of the first breakpoint in 36% of rearrangements (58). In addition, deletion breakpoints were found to be AT rich, whereas translocation breakpoints were GC rich. Alternating purine-pyrimidine sequences were found to be significantly overrepresented in the vicinity of deletion breakpoints, while polypyrimidine tracts were over-represented at translocation breakpoints (2).

10.3.1.5 Large Insertions (Via Retrotransposition)

A less common, but nevertheless still fascinating, mechanism of human gene mutation is the de novo insertion of repetitive elements via retrotransposition. The phenomenon was first observed in humans in the factor VIII (*F8*) gene in two unrelated de novo cases of severe hemophilia A (175). Truncated LINE (long interspersed) repetitive elements were introduced into exon 14 of the factor VIII (*F8*) gene, where they caused disruption of the reading frame. The inserted elements contained a poly(A) tract and caused a target site duplication of more than 12 nucleotides. Further analysis of these insertions revealed that, in one case, the inserted element was an exact but truncated copy of a full-length LINE element, with open reading frames found at chromosome 22q11 (97). The master source gene produces an mRNA that is probably reverse transcribed (possibly via a reverse transcriptase encoded by itself) and the double stranded nucleic acid is then reinserted into an A-rich region of the genome (Fig. 10.8). LINEs probably integrate into genomic DNA by a process called target-primed reverse transcription (251). The proposed mechanism of LINE retrotransposition is as follows: an active LINE is transcribed in the nucleus and is subsequently transported to, and translated in, the cytoplasm. The two LINE proteins, ORF1 and ORF2, complex with their encoding LINE transcript in ribonucleoprotein particles. The complex is then transported to recipient DNA sequences where target-primed

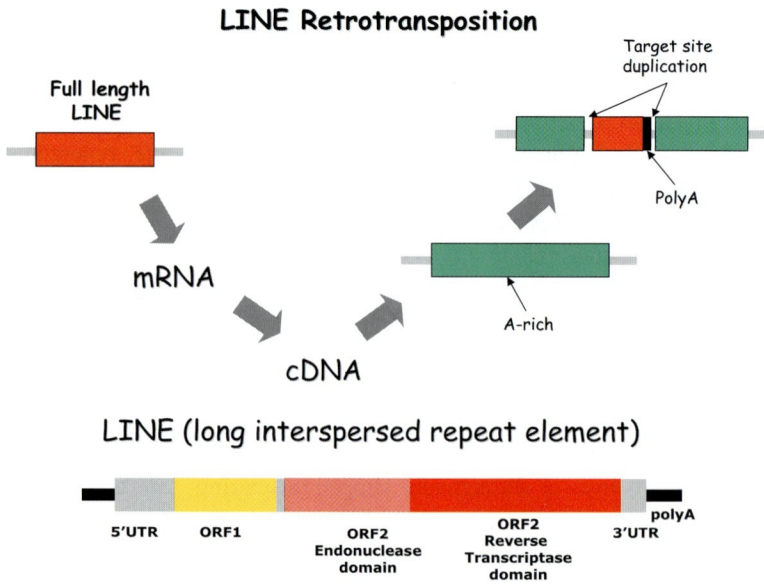

Fig. 10.8 Schematic representation of LINE retrotransposition. A master retrotransposon (full-length LINE from one chromosomal location is transcribed to double-stranded DNA and inserted into an adenine-rich region of another chromosomal location. The transposon has a poly(A) tail and produces a target site duplication

reverse transcription occurs. The new, integrated LINE copy is usually truncated at its 5' end. Over evolutionary time, L1s have shaped mammalian genomes through a number of different mechanisms. First, they have greatly expanded the genome both by their own retrotransposition and by providing the machinery necessary for the retrotransposition of other mobile elements, such as *Alu* sequences or SVA elements (49). Secondly, they have shuffled non-L1 sequence throughout the genome by a process termed transduction. Accidents of retrotransposition can cause disease and a number of such insertions have been reported to date (174, 251). It is noteworthy that insertions of these elements within introns of genes or flanking regions are probably not associated with disease, but instead represent rare, private polymorphisms (355).

Similar retrotranspositions that involve members of the *Alu* sequence family have also been reported in several genes (examples include *Alu* insertions into the *NF1* gene causing type 1 neurofibromatosis, into the factor IX (*F9*) gene causing hemophilia B, and into the cholinesterase (*BCHE*) gene in a case of acholinesterasemia) (238, 333, 341). It is likely that LINEs provide the molecular machinery necessary for the retrotransposition of *Alu*s. One study using mutation analysis of the *F9* gene has estimated the frequency of retrotransposition to be such that it occurs somewhere in the genome of about 1 in every 17 children born (208).

In an analysis of 199 unrelated families with proven mutations in *BTK* X-linked agammaglobulinemia, two families with retrotransposon insertions at exactly the same nucleotide within the coding region of the *BTK* gene have been identified. These insertions, of an SVA element and an *Alu*Y sequence, respectively, occurred 12 bp before the end of exon 9. Both had the typical hallmarks of a retrotransposon insertion, including target site duplication and a long poly A tail. The occurrence of two retrotransposon sequences at precisely the same site suggests that this site may be especially vulnerable to insertional mutagenesis (65).

Some 17% of a collection of gross insertions, all ≥276 bp in length, were due to LINE-1 (L1) retrotransposition involving different types of elements (L1 trans-driven *Alu*, L1 direct, and L1 trans-driven SVA) (49). A meta-analysis of 48 recent L1-mediated retrotranspositional events known to have caused human genetic disease revealed that 26 were L1 *trans*-driven *Alu* insertions, 15 were direct L1 insertions, four were L1 *trans*-driven SVA insertions, and three were associated with simple poly(A) insertions (52). The systematic study of these lesions, when combined with previous in vitro and genome-wide analyzes, allowed several conclusions regarding L1-mediated retrotransposition to be drawn: (a) ~25% of L1 insertions are associated with the 3' transduction of adjacent genomic sequences, (b) ~25% of the new L1 inserts are full length, (c) poly(A) tail length correlates inversely with the age of the element, and (d) the length of target site duplication in vivo is rarely longer than 20 bp. This analysis also suggested that some 10% of L1-mediated retrotranspositional events are associated with significant genomic deletions in humans.

Interestingly, Audrezet et al. (19) reported an indel in the *CFTR* gene that involved the insertion of a short 41-bp sequence with partial homology to a retrotranspositionally-competent LINE-1 element. These authors dubbed such insertions of ultra-short LINE-1 elements "hyphen elements."

10.3.1.6 Large Insertion of Repetitive and Other Elements

The insertion of non-retrotransposons, namely beta-satellite repeats, has been observed in the human genome. The insertion of 18 copies of the 68-bp monomer of the beta satellite repeat in exon 11 of the *TMPRSS3* gene on chromosome 21 caused one form of recessive nonsyndromic deafness, DFNB10 (293). This may have been mediated by invasion of the genomic DNA by a small polydispersed circular DNA (spcDNA).

A patient with a sporadic case of Pallister–Hall syndrome has been shown to have experienced a de novo nucleic acid transfer from the mitochondrial to the nuclear genome. This mutation, a 72-bp insertion into exon 14 of the *GLI3* gene, creates a premature stop codon and predicts a truncated protein product. Both the mechanism and the cause of the mitochondrial-nuclear transfer are however unknown (326). A second example of pathologic mitochondrial-nuclear sequence transfer has been subsequently (and retrospectively) identified in the *USH1C* gene, but appears to have arisen via a novel mechanism, "*trans*-replication slippage" (49).

Gross insertions (>20 bp) comprise <1% of disease-causing mutations. In an attempt to study these insertions in a systematic way, 158 gross insertions ranging in size between 21 bp and approximately 10 kb were identified from the HGMD; their study has revealed extensive diversity in terms of the nature of the inserted DNA sequence and has provided new insights into the underlying mutational mechanisms (49). Some 70% of gross insertions were found to represent sequence duplications of different types (tandem, partial tandem, or complex). In the context of a 26-bp insertion into the *ERCC6* gene, Chen et al. also speculated as to whether they had found evidence for another mechanism of human genetic disease, involving the possible capture of DNA oligonucleotides (49).

10.3.1.7 Inversions

The most common inversion found to date is that associated with the *F8* gene, which occurs via intrachromosomal recombination mediated by a 9.5-kb sequence that is repeated three times in the last megabase of Xqter; once in intron 22 of the *F8* gene and twice about 400 kb telomeric to the first (200, 240) (Fig. 10.9). Most inversions, which are high-frequency independent recurring events, involve the distal sequence. The vast majority of inversions occur in male germ cells (280), perhaps because intrachromosomal recombination is inhibited by the presence of homologous X chromosomes (the male-to-female ratio was estimated

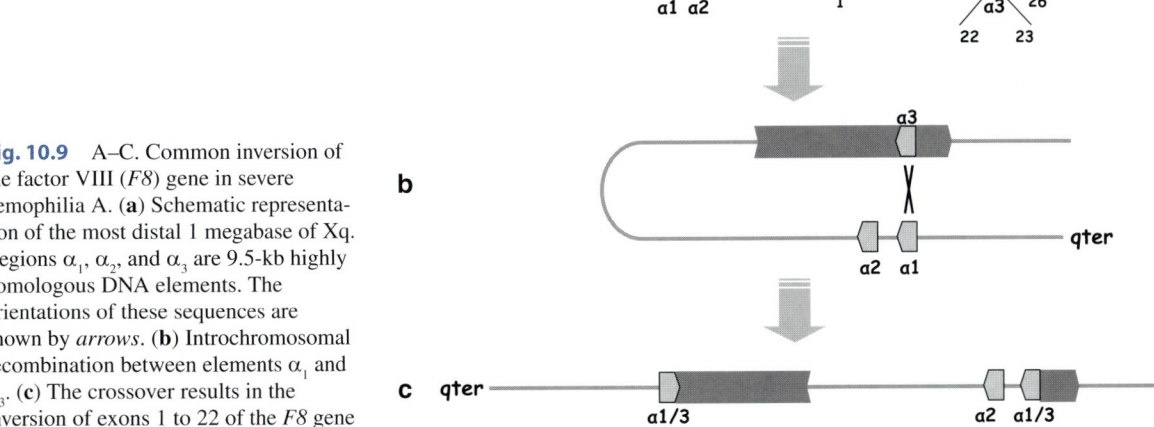

Fig. 10.9 A–C. Common inversion of the factor VIII (*F8*) gene in severe hemophilia A. (**a**) Schematic representation of the most distal 1 megabase of Xq. Regions α_1, α_2, and α_3 are 9.5-kb highly homologous DNA elements. The orientations of these sequences are shown by *arrows*. (**b**) Introchromosomal recombination between elements α_1 and α_3. (**c**) The crossover results in the inversion of exons 1 to 22 of the *F8* gene

to be about 300–1). Almost all mothers of persons with inversion hemophilia A are carriers of the abnormality. DNA diagnosis of the molecular lesion in severe hemophilia A has been greatly facilitated by the frequent occurrence of this common inversion of the *F8* gene (45% of individuals with severe hemophilia A). The frequency of de novo *F8* gene inversion has been estimated at 7.2×10^{-6} per gamete per generation. Another example of inversion has been described in the *IDS* gene (also on Xq) in about 13% of cases of Hunter syndrome (34). Inversions of DNA sequences have also been reported in the β-globin gene cluster on 11p and in the *APOA1-APOC3-APOA4* gene cluster on 11q (167, 172).

A meta-analysis of inversions of ≥5 bp but <1 kb has been performed by Chen et al. (51). Of the 21 mutations studied, 19 were found to be compatible with a model of intrachromosomal serial replication slippage in *trans* (SRStrans) mediated by short inverted repeats. Eighteen (one simple inversion, six inversions involving sequence replacement by upstream or downstream sequence, five inversions involving the partial reinsertion of removed sequence, and six inversions that occurred in a more complicated context) of these were found to be consistent with either two steps of intrachromosomal SRStrans or a combination of replication slippage in *cis* plus intrachromosomal SRStrans. The remaining lesion, a 31-kb segmental duplication associated with a small inversion in the *SLC3A1* gene, was explained in terms of a modified SRS model incorporating the concept of "break-induced replication." This study has therefore lent broad support to the idea that intrachromosomal SRStrans can account for a variety of complex gene rearrangements involving inversions.

10.3.1.8 Duplications

Duplications of whole genes or exons have contributed very significantly to the evolution of the human genome (71). Indeed, most gene clusters (e.g., β-globin, growth hormone, Hox) owe their origin to gene duplications that have occurred during vertebrate evolution. Furthermore, the presence of similar domains in proteins (e.g., immunoglobulin-like domains in many transmembrane proteins) are due to duplications of certain exons.

Occasionally, however, duplications may also be the cause of genetic disorders. The most frequent mechanism of duplication is homologous unequal crossover, as described for large deletions. In fact, most large duplications are generated as the reciprocal product of a deletion resulting from homologous unequal crossover. Duplications are less common, however, than their theoretically reciprocal deletions (see, e.g., (151), for the *DMD* gene). This may be due to the nonpathogenicity of a duplication (e.g., α-globin genes (127)), elimination of duplications as is the case for the *HPRT1* gene, or the fact that not all mechanisms that lead to deletions also produce duplications. A large and common duplication has been identified in cases of Charcot-Marie-Tooth disease type 1A (265). This duplication involves 1.5 Mb of DNA on chromosome 17p containing the peripheral myelin protein 22 (*PMP22*) gene. It results from homologous unequal crossover events between 24-kb repeats that flank the duplicated region. The reciprocal deletion product of this recombination event is responsible for a completely different clinical phenotype: hereditary neuropathy with liability to pressure palsies (Fig. 10.7). Another notable duplication of at least 500 kb that includes the *PLP1* gene is a frequent cause of Pelizaeus-Merzbacher disease (357). The pathogenetic mechanism of these duplications involves unequal crossing-over in meiosis mediated by "duplicons" in the genome (168).

The molecular defect in the majority of cases with ectrodactyly type SHFM3 on chromosome 10q24, is an approximately 0.5-Mb tandem duplication. The exact pathogenetic mechanism of this duplication is unknown (90).

Additional gene duplications causing recognizable syndromes include the *APP* duplication causing early-onset Alzheimer disease (283), the *SNCA* duplication and Parkinson disease (306), and the triplication of an ~605-kb segment containing the *PRSS1* gene in families with hereditary pancreatitis (206).

10.3.1.9 Gene Conversion

Gene conversion is the modification of one of two alleles by the other. It involves the nonreciprocal correction of an "acceptor" gene or DNA sequence by a "donor" sequence, which itself remains physically unchanged. In most known instances of gene conversion as a cause of human genetic disease, the functional gene has been wholly or partially converted to the sequence of a highly homologous and closely linked pseudogene, which therefore acts as the donor sequence. Probable examples include the genes for steroid

21-hydroxylase (*CYP21*) (327), polycystic kidney disease (*PKD1*) (345), neutrophil cytosolic factor p47-phox (*NCF1*) (130), immunoglobulin λ-like polypeptide 1 (*IGLL1*) (230), glucocerebrosidase (*GBA*) (108)), von Willebrand factor (*VWF*) (105), and phosphomannomutase (*PMM2*) (291). These gene/pseudogene pairs are all closely linked with the exception of the *VWF* gene (12p13) and its pseudogene (22q11-q13), and the *PMM2* gene (16p13) and its pseudogene (18p). Together, these two exceptions seem to establish a precedent for the occasional occurrence of gene conversion between unlinked loci in the human genome.

10.3.1.10 Insertion-Deletions (Indels)

A relatively rare type of mutation causing human genetic disease is the insertion-deletion, or *indel*, a complex lesion that appears to represent a combination of microdeletion and micro-insertion. One example is provided by the 9 deleted base-pairs encoding codons 39–41 of the α2-globin (*HBA2*) gene that were replaced by eight inserted bases that served to duplicate the adjacent downstream sequence (250). Indels constitute a fairly infrequent type of lesion causing human genetic disease; some 1.5% of lesions in HGMD fall into this category.

Several indel hotspots have been noted in a meta-analysis of HGMD data on 211 different indels underlying genetic disease (57). A GTAAGT motif was found to be significantly overrepresented in the vicinity of the indels studied. The change in complexity consequent to a mutation was also found to be indicative of the type of repeat sequence involved in mediating the event, thereby providing clues as to the underlying mutational mechanism. The majority of indels (>90%) were explicable in terms of a two-step process involving established mutational mechanisms. Indels equivalent to double base-pair substitutions (22% of the total) were found to be mechanistically indistinguishable from the remainder and may therefore be regarded as a special type of indel.

10.3.1.11 Other Complex Defects

Complex mutational events that involve combined gross duplications, deletions, and/or insertions of DNA sequence have been not infrequently observed and together constitute ~1% of entries in HGMD. One example of this type of gene defect is a 10.9-kb deletion coupled with a 95-bp inversion in the factor IX (*F9*) gene causing hemophilia B (178). The molecular characterization of this type of lesion is often extremely complicated and in most cases the underlying mutational mechanisms could not be readily inferred.

Recently, however, a meta-analysis of 21 complex gene rearrangements derived from the HGMD revealed that all but one could be accounted for by a model of serial replication slippage, involving twin or multiple rounds of replication slippage (50). Thus, of the 20 complex gene rearrangements, 19 (seven simple double deletions, one triple deletion, two double mutational events comprising a simple deletion and a simple insertion, six simple indels that may constitute a novel and noncanonical class of gene conversion, and three complex indels) were compatible with the model of serial replication slippage in *cis*; by contrast, the remaining indel in the *MECP2* gene appears to have arisen via interchromosomal replication slippage in *trans*.

10.3.1.12 Molecular Misreading

Long runs of adenines (and perhaps other mononucleotides or dinucleotides) promote a phenomenon termed "molecular misreading," by which DNA replication/RNA transcription and/or translation result in erroneous products with different numbers of (A)s derived from the original DNA sequence. In a family with hypobetalipoproteinemia, a deletion of one C in the A_5CA_3 coding sequence of the *APOB* gene results in a run of $(A)_8$. The patient, however, did not have severe disease, because some ApoB protein was made. This was the result of molecular misreading, in which ~10% of the resulting mRNAs contained $(A)_9$ instead of the expected $(A)_8$; this partially restored the reading frame, thereby templating the synthesis of low amounts of normal ApoB (210). Similarly, a family with mild to moderately severe hemophilia A with a deletion of one T within the coding A_8TA_2 sequence of the *F8* gene has been reported. The partial "correction" of the phenotype was due to restoration of the reading frame because of molecular misreading in which ~5% of the resulting RNAs contained $(A)_{11}$ instead of the expected $(A)_{10}$. In this family, there was also evidence for ribosomal frameshifting during translation of the mutant RNA (360).

Another example of this phenomenon was observed in the *APC* gene. A T-to-A transversion is present in the coding A_3TA_4 sequence of the *APC* gene in 6% of

Ashkenazi Jews, and in about 28% of Ashkenazim with a family history of colorectal cancer. This mutation creates a small hypermutable region, indirectly causing cancer predisposition because there are many somatic cells in which stretches of $(A)_9$ occur instead of the expected $(A)_8$; the $(A)_9$ results in frameshifting and a truncated dysfunctional APC (199). Interestingly, in the neurofibrillary tangles, neuritic plaques, and neuropil threads in the cerebral cortex of Alzheimer disease and Down syndrome, abnormal forms of β-amyloid precursor protein and ubiquitin B have been observed. These aberrant proteins were produced because of +1 frameshifting that resulted from a deletion of AG in a sequence GAGAG that occurred in the coding regions of both genes (*APP* and *UBB*, respectively). This dinucleotide deletion was again the result of molecular misreading during transcription or posttranscriptional editing of RNA (330). This mechanism is likely to yield a considerable quantity of abnormal RNA molecules and protein products in somatic cells (259).

10.3.1.13 Germline Epimutations

Epimutations are modifications of DNA that constitute clonally heritable (yet potentially reversible) alterations in the transcriptional status of a gene that lead to the abnormal silencing of that gene. Epimutations are not mutations in the strictest sense of the word, since they do not alter the gene's nucleotide sequence. However, germline epimutations of the *MLH1* gene have been reported in individuals with multiple cancers (316) and in the *MLH1* and *MSH2* genes in hereditary nonpolyposis colorectal cancer (147). These heritable inactivating epimutations are characterized by mono-allelic hypermethylation of the *MLH1* gene and, to all intents and purposes, are functionally equivalent to conventional mutations.

10.3.1.14 Frequency of Disease-Producing Mutations

Mutation Frequency Within Genes. The frequency of different molecular defects is not the same for every gene and every disorder. It depends very largely on the DNA sequence characteristics of the gene in question (e.g., the presence of repeat units or homologous sequences) and the function of, and evolutionary constraints experienced by, its encoded protein (314). For some genes, deletions predominate; for others, one particular type of lesion such as an inversion may be especially common. Some genes exhibit mainly frameshifts and stop codons associated with a specific disorder, whereas others manifest mainly missense mutations for a given phenotype, or expansions of trinucleotide repeats.

Disease mutations are nonuniformly distributed within genes (229). Such mutations were found to be statistically overrepresented in conserved domains, and underrepresented in variable regions, even after allowing for the amino acid site variability of domains over long-term evolutionary history. This finding suggests that there is a nonadditive influence of amino acid site conservation on the observed intragenic distribution of disease mutations.

Mutation Frequency Within Human Populations. Population genetic considerations are also likely to be very important in determining why some mutations occur frequently, either within a patient cohort or in the population at large (see *Frequency of Inherited Disorders Database*, http://archive.uwcm.ac.uk/uwcm/mg/fidd/; *FINDbase*, http://www.findbase.org/). Selection, migration and genetic drift are all likely to play a part, as well as the mutation rate (114, 320, 365). Thus, the mutational spectrum of the *PAH* gene underlying phenylketonuria appears to result from a range of different factors including founder effect, range expansion and migration, genetic drift and possibly also heterozygote advantage (368). Selection can also serve to maintain deleterious mutations at high frequencies in particular populations by overdominant selection (heterozygote advantage). Good examples of this phenomenon are provided by a reduction in risk of severe malaria associated with female heterozygotes and male hemizygotes for mutations in the X-linked *G6PD* gene (232, 284), for individuals heterozygous for the β-globin (*HBB*) sickle cell mutation, Glu6Val (4), and for individuals heterozygous and homozygous for α⁺-thalassemia (351). Intriguingly, however, the protection against malaria afforded by sickle cell disease and α⁺-thalassemia when inherited individually is lost when the two conditions are co-inherited (350). Other possible examples of heterozygote advantage include an elevated cortisol response in heterozygous carriers of *CYP21A* mutations (352), higher values for hemoglobin, serum iron and transferrin saturation in women heterozygous for *HFE* gene mutations (82), resistance to prion infection conferred by a common prion protein (*PRNP*) polymorphism (225), resistance to severe sepsis in heterozygous carriers of the factor V Leiden polymorphism, Arg506Gln (177), and increased keratinocyte cell survival in individuals heterozygous for *GJB2* gene mutations (64). Resistance to cholera toxin (122), protection against bronchial asthma (292), and resistance to *Pseudomonas aeruginosa* infection (269) have all been mooted as possible bases for overdominant selection in heterozygous carriers of *CFTR* gene mutations. However, cystic fibrosis heterozygotes have been shown to secrete chloride at the same rate as individuals lacking *CFTR* gene mutations (149).

A number of genetic diseases are known to be particularly prevalent in Jewish populations (236, 252). The presence of four distinct lysosomal storage diseases at significant frequencies among Ashkenazi Jews has often been thought to provide evidence for a selective advantage accruing to heterozygotes in this population. However, evidence in support of the idea of genetic drift appears to be more compelling (117, 278).

Selection may also act at an extremely early stage to boost the frequency of some mutations that are deleterious at a later stage in development. Gain-of-function missense mutations in the fibroblast growth factor receptor 2 (*FGFR2*) gene responsible for Apert syndrome have been shown to confer a selective advantage on spermatogonial cells by promoting the clonal expansion of mutant cells (128, 129)

10.3.1.15 Chromosomal Distribution of Human Disease Genes

Human disease genes are characterized by the greater lengths of their encoded amino acid sequences, larger numbers of longer introns, broader ranges of tissue expression, and wider phylogenetic distributions (187, 216). Human disease genes are also known to be unevenly distributed between human chromosomes (48, 152). Furthermore, synonymous nucleotide substitutions appear to occur at a higher rate in human disease genes, a finding that may reflect increased mutation rates in the chromosomal regions in which disease genes are found (152). It may be that disease genes are more prevalent in genomic regions that experience elevated rates of mutation (55). Another possible explanation is that the disease gene set may contain a disproportionately lower number of genes expressed in the germline (152). This is because mutations in such genes might be expected to be more effectively repaired by transcription-coupled repair (transcription-coupled repair in the germline appears to account for the strand asymmetry that the human genome exhibits in terms of inherited mutations (133, 217). Strand asymmetries with respect to the mutation rate may, however, also arise through the influence of DNA replication origins (321) and recombination (153, 275).

10.3.1.16 Mutation Nomenclature

Some consistency in the way in which mutations are described is essential for the accurate and unambiguous reporting and curation of mutation data. The most recently published set of guidelines on how to describe mutational changes in human genes is to be found in den Dunnen and Antonarakis' work published in 2001 (91).

10.3.1.17 Mutations in Gene Evolution

Mutations in human gene pathology and evolution represent two sides of the same coin in that those same mutational mechanisms that have frequently been implicated in human pathology have also been involved in potentiating evolutionary change (71). Regardless of whether they are advantageous, disadvantageous, or neutral, these mutational changes and their putative underlying causal mechanisms are very similar. It is now clear that the gene has often been a dynamic entity over evolutionary time, and not a static one. Indeed, during vertebrate evolution, many genes have undergone gross rearrangement as a result of the action of a variety of mutational processes, including insertion, inversion, duplication, repeat expansion, translocation, and deletion. What links pathology and evolution is the underlying genomic architecture with its hitherto largely unexplored vocabulary of structural elements, and different types and patterns of repetitive DNA sequences (303). It can thus be seen that the mutational spectra of germline mutations responsible for inherited disease, somatic mutations underlying tumorigenesis, polymorphisms (either neutral or functionally significant), and differences between orthologous gene sequences exhibit remarkable similarities, implying that they are very likely to have causal mechanisms in common.

10.3.2 Consequences of Mutations

10.3.2.1 Mutations Affecting the Amino Acid Sequence of the Predicted Protein, but not Gene Expression

Many missense mutations (i.e., nucleotide substitutions that result in an amino acid substitution) cause hereditary disease in humans. Missense mutations are of importance for understanding the structure or function of a protein, since they usually occur in amino acid residues of structural or functional significance (228). Occasionally, however, not only is the mutated residue not conserved in mouse, but the substituting residue in humans is identical to its wild-type counterpart in the orthologous murine gene (123). It is thought that the most likely explanation for the majority of these cases of fixation of disease mutations in mice is *compensatory mutation*. This hypothesis holds that loss-of-function amino acid substitutions in a protein can be rescued by additional substitutions in the vicinity that compensate structurally for the original change.

It is sometimes difficult to establish a causative link between a missense mutation and a disease phenotype (76). The absence of the mutation in a large sample (usually 200 individuals) from the same ethnic group as the patient serves to exclude the possibility of a common polymorphism. Amino acid substitutions in evolutionarily conserved residues can also be good candidates for true pathogenicity (228). If the function of the protein is known, assessment of the effect of the missense mutation can be performed by in vitro mutagenesis and functional assay. Finally, the introduction of the mutation into an entire organism (e.g., into transgenic mice) and the study of its systemic effects provide one of the best means of assessing its contribution to a particular clinical phenotype. Amino acid substitutions can be shown to reduce or abolish the physiological function of a protein; for example, missense mutations have been identified in factor VIII that abolish thrombin cleavage, which is necessary for its activation (15), interfere with binding to other proteins, such as von Willebrand factor (143), or create or abolish N-glycosylation sites (9). In other proteins, mutations have been identified, e.g., in DNA binding domains, catalytic domains, transmembrane domains, ATP-binding regions, receptor-ligand contact sites, and phosphorylation or other chemical modification sites. Missense mutations may also affect protein folding, causing a dramatic change in secondary and tertiary structure such that the protein can no longer fulfill its physiological function.

A classic example of a missense mutation in the active site of an enzyme is provided by α1-antitrypsin Pittsburgh, found in an individual with a fatal bleeding disorder (253). The underlying mutation in the α1-antitrypsin (*SERPINA1*) gene substituted Arg for Met358 within the active site of the molecule. Substitution by Arg served to alter the substrate specificity of α1-antitrypsin by converting its "bait loop" (which is specific for elastase) to one that was specific for thrombin. In effect, the molecule lost its anti-elastase activity and became a serine protease inhibitor capable of inhibiting thrombin and factor Xa.

Mutations involving gains of glycosylation have generally been considered rare, and the pathogenic role of the new carbohydrate chains has never been formally established. Vogt et al. (337), however, identified three children with Mendelian susceptibility to mycobacterial disease who were homozygous with respect to a missense mutation in the *IFNGR2* gene that created a new N-glycosylation site in the IFNγR2 chain. The resulting additional carbohydrate moiety was found to be both necessary and sufficient to abolish the cellular response to IFNγ. From 10,047 HGMD mutations in 577 genes encoding proteins trafficked through the secretory pathway, 142 candidate missense mutations (~1.4%) in 77 genes (~13.3%) for potential gain of N-glycosylation were identified. Six mutant proteins were shown to bear new N-linked carbohydrate moieties. Thus, it may be that an unexpectedly high proportion of mutations causing human genetic disease do so via the creation of new N-glycosylation sites. Indeed, the pathogenic effects of these mutations may be a direct consequence of the addition of N-linked carbohydrate.

Missense mutations can result in disease by (1) elimination or reduction of the physiological activity/role of the protein; (2) gain of function by which the amino acid substitution creates new functional capabilities of the protein in biochemical and developmental processes in which the protein either does not participate or has a different role; (3) change of the target function of another protein, as in the case of the mutation in the protein C cleavage site at Arg 506 of coagulation factor V, which is associated with thrombophilia (30), or in the case of a mutation in the thrombin cleavage site of factor VIII that eliminates normal activation of factor VIII (16), or in the case of severe obesity from childhood and R236G in the human pro-opiomelanocortin (*POMC*) gene that disrupts the dibasic cleavage site between beta melanocyte-stimulating hormone (beta-MSH) and beta-endorphin (46); and (4) participation of the mutant polypeptide in protein complexes, which renders the entire complex abnormal or nonfunctional, as in the case of the triple helical structure of certain collagens in which incorporation of one abnormal collagen chain results in "protein suicide" or an abnormal structure that degrades rapidly (41).

Missense mutations have a multitude of different effects on protein structure and function including (a) introduction of larger residues within the hydrophobic protein core leading to adverse interactions between residues, (b) introduction of buried charged residues, (c) disruption of protein-protein interactions, (d) disruption of hydrogen bonding, (e) interference with DNA binding, (f) breakage of disulphide covalent linkages, (g) mutation of catalytic residues, (h) perturbation of metal binding, and (i) disruption of quaternary structure.

Without in-depth analytical studies, missense mutations are often difficult to distinguish from polymorphisms with little or no clinical significance. In the "post-genome era," a substantial amount of human genetic variation will become amenable to high-throughput analysis in the form of single nucleotide polymorphisms (SNPs), and many of these SNPs will directly influence the structure, function, or expression of genes and the RNAs/proteins they encode. Prior knowledge as to which SNPs are most likely to be clinically relevant would greatly enhance the power of studies that aim to identify disease genes through the genotypic screening of patients in both families and populations. Inclusion of structural/functional information could be especially important in the elucidation of multifactorial disease, where genetic heterogeneity and complex interactions between genes and environment have so far limited the success of genetic epidemiological studies (146). Recently, several predictive models have been developed that employ a number of different biophysical parameters to estimate the likely impact of an amino acid substitution on the structure and function of a protein (112, 315, 317, 340, 343). These models have been used to distinguish reasonably successfully between pathologic substitutions, functional polymorphisms, and neutral polymorphisms. Vitkup et al. (334) have concluded that mutations at arginine and glycine residues are together responsible for about 30% of cases of genetic disease, whereas random mutations at tryptophan and cysteine have the highest probability of causing disease.

10.3.2.2 Mutations Affecting Gene Expression

Mutations that do not result in amino acid substitution invariably affect gene expression, i.e., transcription, RNA processing and maturation, translation, or protein stability. Total or partial gene deletions, insertions, inversions, and other gross rearrangements obviously result in the loss of gene expression. These types of mutation are usually less frequent unless the genomic sequence environment of specific genes (e.g., presence of repeats) predisposes to such lesions. Disorders with high frequencies of gross rearrangements include α–thalassemia, Duchenne muscular dystrophy, steroid sulfatase deficiency, and hemophilia A. Some partial gene deletions that eliminate one or a few exons in frame result in milder clinical phenotypes because gene expression is not totally eliminated; the resulting protein may lack an amino acid domain that is not critical for its function (362).

10.3.2.3 Transcription (Promoter) Mutations

Mutations in known promoter motifs usually lead to reduced (or occasionally increased) mRNA levels. Such mutations have been studied in the TATA box of the β-globin (*HBB*) gene (12). Other nucleotide substitutions within DNA motifs that bind transcription factors include those located in the CACCC motif of the β–globin (*HBB*) gene influencing transcription factor EKLF binding (248, 266), several motifs in the γ-globin (*HBG*) genes (63), the CCAAT motif of the *F9* gene influencing C/EBP binding (80), the SP1 motif of the *LDLR* gene promoter (183), the HNF-1 binding site in the *PROC* gene (29), and the binding site for the transcription factor Oct-1 in the lipoprotein lipase (*LPL*) gene (359). These few examples are only representative of a total of over 370 known promoter mutations listed in HGMD and causing human genetic disease. The importance of these mutants lies in the specific DNA sequences thereby implicated in binding to transcription factors. Although most of the known mutations reduce the levels of mRNA production, some substitutions actually increase it. Examples include various lesions in the promoters of the Gγ and Aγ globin (*HBG1* and *HBG2*) genes that cause hereditary persistence of fetal hemoglobin due to the inappropriate continuation of γ–globin gene expression into adult life (346). An increase in the distance of promoter elements from the transcriptional start site may also result in gene silencing. Such an example has been found in the promoter elements of the *CSTB* gene in progressive myoclonus epilepsy type 1 (EPM1) (202). Mutations that alter the transcriptional regulation of gene expression have been reviewed elsewhere (295).

The concomitant change in local DNA sequence complexity surrounding a substituted nucleotide is directly related to the likelihood of a regulatory mutation coming to clinical attention (196). This finding is consistent with the view that DNA sequence complexity is a critical determinant of gene regulatory function and may reflect the internal axial symmetry that frequently characterizes transcription factor binding sites.

Polymorphisms in the promoter region that are associated with differential levels of gene expression may predispose to common disorders. For example, a G>A single nucleotide polymorphism (SNP) at nucleotide-6 relative to the transcriptional initiation site of the angiotensin (*AGT*) gene influences the basal level of transcription and may predispose to essential hypertension (159). Listed in HGMD are in excess of 250 disease-associated promoter polymorphisms plus >170 functional promoter polymorphisms that significantly increase or decrease promoter activity but which have not yet been associated with a clinical phenotype.

10.3.2.4 mRNA Splicing Mutants

Single base-pair substitutions in splice junctions constitute at least 10% of all mutations causing human inherited disease. There are, however, a wide variety of mutations within both introns and exons that can affect normal RNA splicing (see (194) for review). The different mechanisms by which disruption of pre-mRNA splicing play a role in human disease were reviewed by Faustino and Cooper in 2003 (110). The most commonly found mutations occur in the invariant dinucleotides GT and AG found at the beginning and end of the donor (5') and acceptor (3') consensus splice sequences (see Fig. 10.10 for the consensus splice elements and Fig. 10.11 for the different kinds of RNA splicing abnormalities). Almost all of these mutations cause either exon skipping or cryptic splice site utilization, resulting in the severe reduction or absence of normally spliced mRNA. In addition, mutations in nucleotides +3, +4, +5, +6, -1 and −2 of the consensus donor splice site have frequently been observed (Fig. 10.12), with variable severity of the RNA splicing defect. Similarly, mutations in positions -3 and the polypyrimidine tract of the consensus acceptor splice site have been noted (Fig. 10.12). In the majority of these cases, some normal splicing occurs and the defect is not severe. Utilization of cryptic splice sites leads to the production of abnormal mature mRNA with premature stop codons or to the inclusion of additional amino acids after translation (see (15) for examples, and references cited therein).

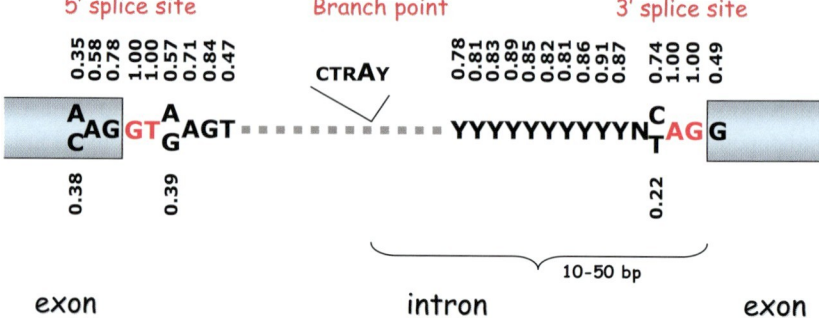

Fig. 10.10 Consensus sequences for the donor (5' splice) and acceptor (3' splice) sites and the branch point. *Numbers* above or below the nucleotides correspond to frequencies of a given nucleotide in a large number of mammalian splice site sequences. Note that the dinucleotides GT and AG (in *red*) at the beginning and end of the intron are invariant

10 Human Gene Mutation: Mechanisms and Consequences

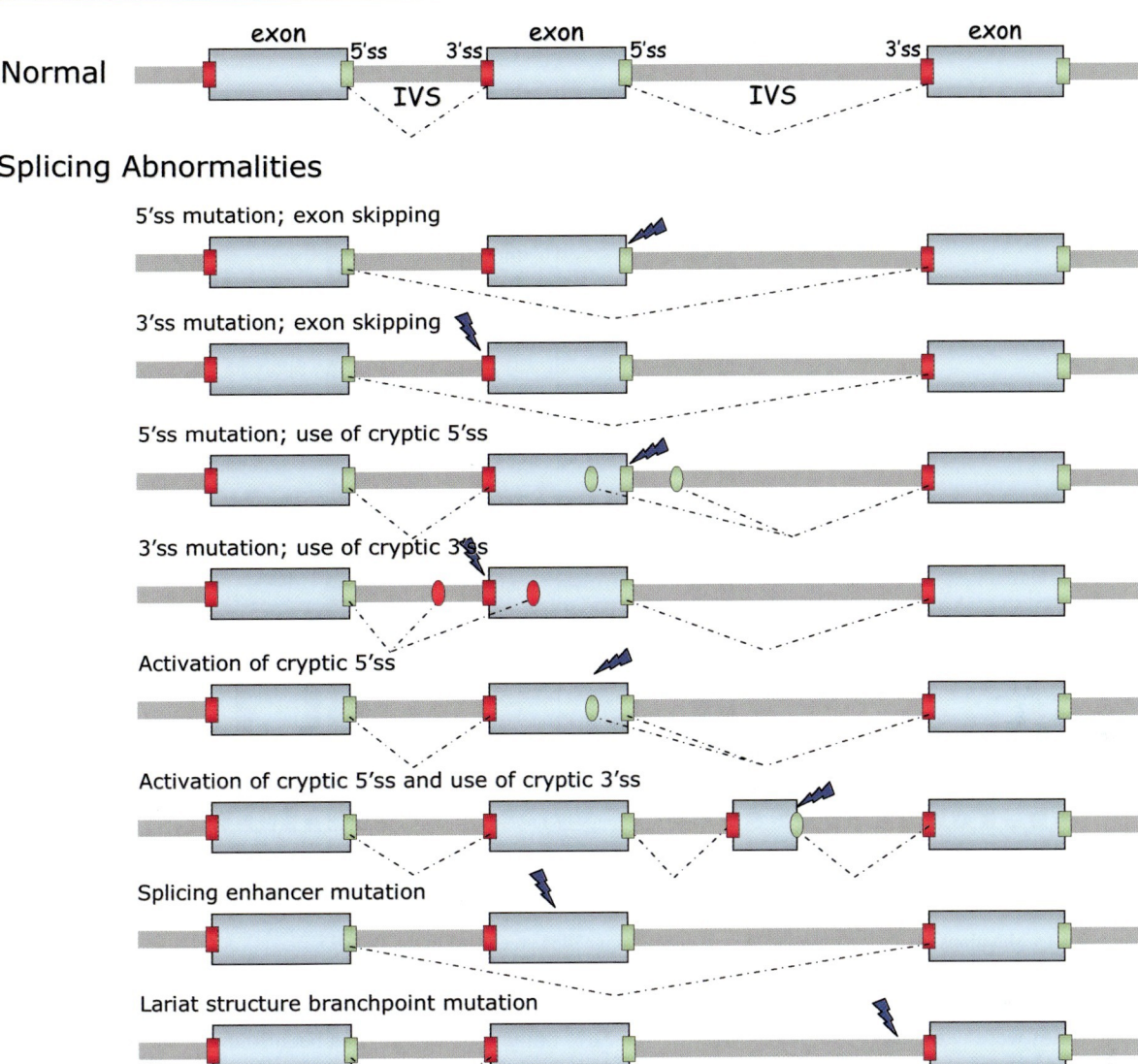

Fig. 10.11 Examples of splicing abnormalities in introns of human genes. Exons are shown as *blue boxes*; introns, as *lines* between exons. *Green squares* denote the normal 5' (donor) splice sites; *red squares* represent the normal 3' (acceptor) splice sites. *Green* and *red circles* denote cryptic 5' and 3' splice sites, respectively. The *broken blue wedge* represents the site of mutation

Employing a neural network for splice site recognition, Krawczak et al. (197) performed a meta-analysis of 478 disease-associated splicing mutations, in 38 different genes, for which detailed laboratory-based mRNA phenotype assessment had been performed. Inspection of the ±50-bp DNA sequence context of the mutations revealed that exon skipping was the preferred phenotype when the immediate vicinity of the affected exon-intron junctions was devoid of alternative splice sites. By contrast, in the presence of at least one such motif, cryptic splice site utilization became more prevalent. This association was, however, confined to donor splice sites. Outside the obligate dinucleotide, the spatial distribution of pathological mutations was found to differ significantly from that of SNPs. Whereas disease-associated lesions clustered at positions -1 and +3 to +6 for donor sites and −3 for acceptor sites, SNPs were found to be almost evenly distributed over all sequence positions considered. When all putative missense mutations in the vicinity of

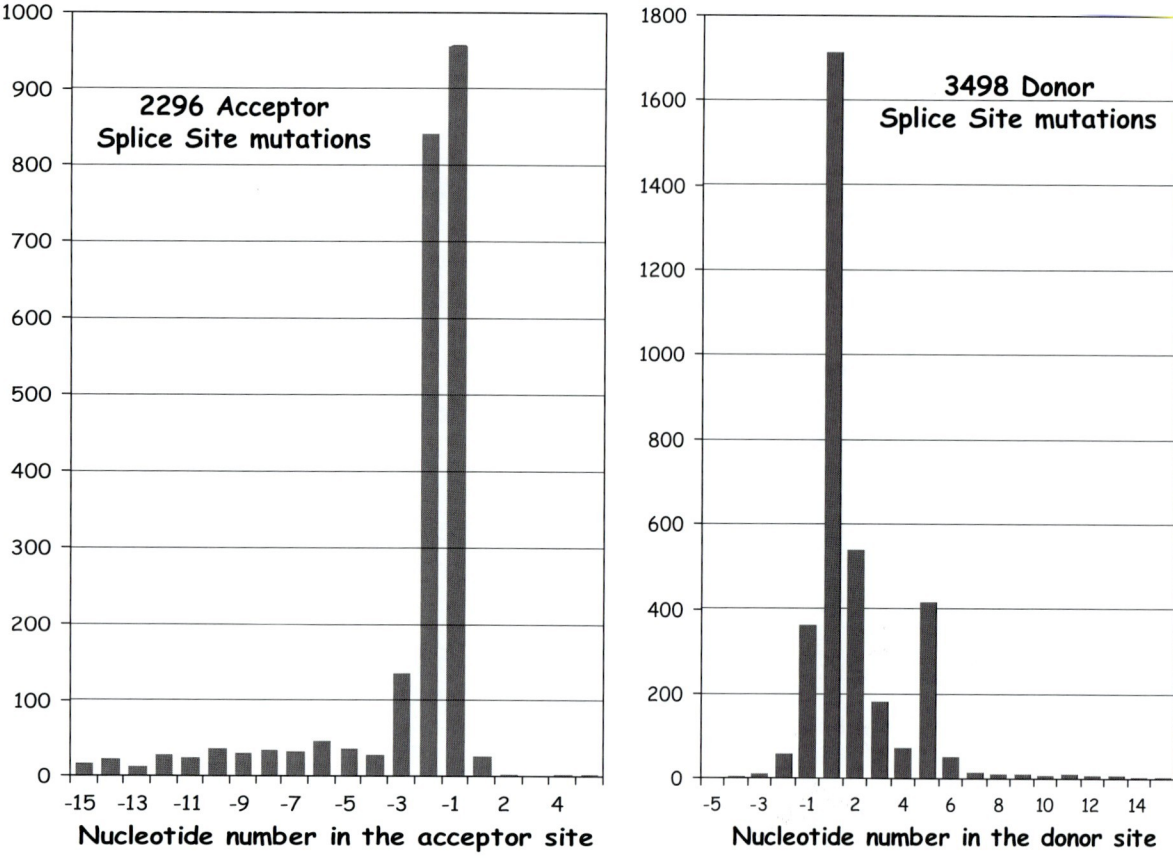

Fig. 10.12 Mutations in the consensus sequences of splice junctions recorded in the HGMD

splice sites were extracted from the HGMD for the 38 studied genes, a significantly higher proportion of changes at donor sites (37/152; 24.3%) than at acceptor splice sites (1/142; 0.7%) was found to reduce the neural network signal emitted by the respective splice site. It is estimated that some 1.6% of disease-causing missense substitutions in human genes are likely to affect the mRNA splicing phenotype.

Other kinds of mutation in introns include those that cause the activation of cryptic splice sites (by altering a sequence so as to make it more similar to an authentic consensus splice site) or by creation of new splice sites (323). In both instances, new intron splice patterns occur with consequent introduction of stop codons or abnormal peptides after translation. These mutations do not completely abolish normal splicing and are therefore not associated with the absence of normal mature mRNA. A mutation in a lariat structure branchpoint (302) has been found in the *L1CAM* gene in a patient with X-linked hydrocephalus (279). By contrast, another mutation in intron 5 of the type 2 neurofibromatosis (*NF2*) gene created a consensus branchpoint sequence and led to the activation of a cryptic exon (87).

Some 98.7% of all splice sites in human genes conform to consensus sequences that include the invariant dinucleotides GT and AG at the 5' and 3' ends of the introns, respectively (40). Noncanonical sequences (e.g., GA-AG, GC-AG, and AT-AC) do, however, occur at human splice junctions, albeit much less frequently (<0.02, 0.69, and 0.05%, respectively. Some of these noncanonical splice sites are nevertheless known to be utilized with high efficiency and may be conserved over quite long stretches of evolutionary time. Such sites have occasionally come to clinical attention when they have harbored mutations causing human inherited disease (304). Moreover, the utilization of a cryptic noncanonical donor splice site within exon 1 of the *HRPT2* gene in a case of familial isolated primary hyperparathyroidism as a consequence of a causative lesion in intron 1 of the gene has been reported. RNA isolated from EBV-transformed lymphoblastoid cell lines derived from the patients was utilized to demonstrate the consequences at the level of the mRNA phenotype (the loss of 30 bases from the mRNA transcript).

Single base-pair substitutions within "splicing enhancer" sequences may also perturb splicing by promoting exon skipping; examples include a mutation in intron 3 of the growth hormone (*GH1*) gene causing short stature (62) and a mutation in exon 5 of the adenosine deaminase (*ADA*) gene causing ADA deficiency (287). In patients with frontotemporal dementia with parkinsonism, three heterozygous mutations in a cluster of 4 nucleotides +13 to +16 of exon 10 of the microtubule-associated protein tau (*MAPT*) gene destabilized a potential stem-loop structure that is probably involved in regulating the alternative splicing of exon 10. This caused more frequent use of the 5' splice site and an increased proportion of tau transcripts that include exon 10. The increase in exon 10+ mRNA increased the proportion of tau protein containing four microtubule-binding repeats, which is consistent with the neuropathology described in families with this type of frontotemporal dementia (155). One mutation found in the *ATM* gene causing ataxia-telangiectasia, was a deletion of four nucleotides (GTAA) in intron 20 within an intron-splicing processing element (ISPE) that is complementary to U1 snRNA. This element mediates accurate intron processing and interacts specifically with U1 snRNP particles (256). Finally, the intronic prothrombin (*F2*) gene 19911A>G polymorphism influences splicing efficiency by altering a known functional pentamer CAGGG motif (338).

Some nonsense mutations cause skipping of one or more exons, presumably during pre-mRNA splicing in the nucleus; this phenomenon has been termed "nonsense-mediated altered splicing" (NAS) but its underlying mechanism is unclear. The first such mutation was described in the *FBN1* gene in Marfan syndrome (95). It is now recognized that any nucleotide substitution within exons (nonsense, missense or translationally silent synonymous point mutation) that disrupts a splicing enhancer or silencer (ESE enhancer splicing element; CERES composite exonic regulatory element of splicing) or creates an exon splicing silencer (ESS) may affect either the pattern or efficiency of mRNA splicing (32, 43, 47, 213) (Fig. 10.13). In exon 12 of the *CFTR* gene, about one quarter of synonymous variations result in exon skipping, and hence lead to the synthesis of an inactive CFTR protein (257). For a review on the effects of exonic variants in splicing, and additional examples of such pathogenic mutations, see (255). It has been estimated that pathogenic effects of ~20% of mutations in the *MSH2* gene result from missense mutations that disrupt ESE sites and perturb splicing. Similarly, the pathogenic effects of ~16% of missense mutations in the *MLH1* gene are thought to be ESE-related (131).

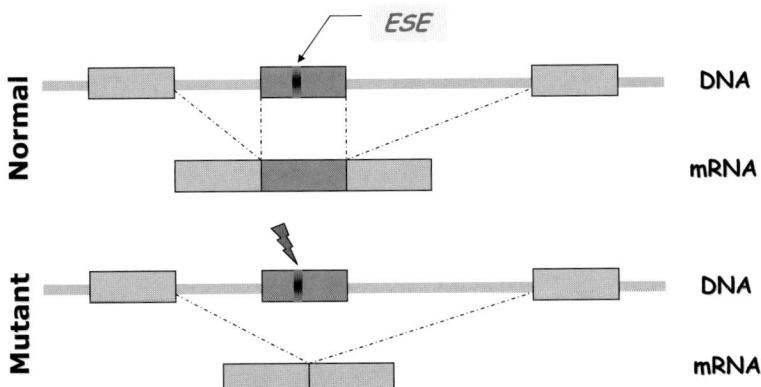

Fig. 10.13 Exon skipping attributable to nonsense, missense, and silent mutations in enhancer splicing elements (*ESE*). This element is shown as a darkened segment of the middle exon

Splice-mediated insertional inactivation involving an *Alu* repeat was first reported by Mitchell et al. in 1991 (231). Analysis of the ornithine δ-aminotransferase (*OAT*) mRNA of a patient with gyrate atrophy revealed a 142 nucleotide insertion at the junction of exons 3 and 4. An *Alu* sequence is normally present in intron 3 of the *OAT* gene, 150 bp downstream of exon 3. The *Alu* sequence found in the cDNA was identical to this one, except that the patient was homozygous for a C→G transversion in the right arm of the *Alu* repeat which served to create a new 5' splice site. This activated an upstream cryptic 3' splice site (the poly(T) complement of the *Alu* poly(A) tail followed by an AG dinucleotide) and a new "exon," containing the majority of the right arm of the *Alu* sequence, was recognized by the splicing apparatus and incorporated into the mRNA. The splice-mediated insertion of an *Alu* sequence in reverse orientation has also been reported in the *COL4A3* gene causing Alport syndrome (182).

A number of "deep intronic" mutations, at some considerable distance from splice sites and known splicing-related sequence elements, have been reported as a cause of human inherited disease (77, 140, 325). Such lesions often create novel splice sites thereby activating cryptic exons ('pseudoexons'). As mutational screening techniques improve, it is anticipated that an increasing number of such lesions will be identified which will turn out to have adverse effects on the mRNA splicing phenotype.

10.3.2.5 RNA Cleavage-Polyadenylation Mutants

A number of examples of RNA cleavage-polyadenylation mutations have now been described (53). Those reported occur in the sequence AAUAAA, which is 10–30 nucleotides upstream of the polyadenylation site and is important for the endonucleolytic cleavage and polyadenylation of the mRNA. Mutation in this sequence of the β-globin (*HBB*) gene results in mild thalassemia (249). In these cases, normal polyadenylation and cleavage occurs at a level about 10% of normal. Alternative AAUAAA sites downstream of the mutated one are used, resulting in larger mRNAs that are highly unstable. Other mutations near the poly(A) cleavage sequence may result in mRNA destabilization; one such mutation has been described 12 bp upstream of the AAUAAA sequence of the *HBB* gene in a patient with β-thalassemia (42).

The G>A mutation at the 3'-terminal nucleotide of the 3' untranslated region (UTR) of the *F2* (prothrombin) gene mRNA gives rise to an elevated prothrombin plasma level and represents a common genetic risk factor for the occurrence of thromboembolic events. This mutation creates an inefficient 3' end cleavage signal and represents a gain-of-function mutation, causing increased cleavage site recognition, increased 3' end processing, and increased mRNA accumulation and protein synthesis (124, 271).

10.3.2.6 Mutations in miRNA-Binding Sites

Micro-RNAs (miRNAs) post-transcriptionally down-regulate gene expression by binding to complementary sequences on the 3' untranslated regions (UTRs) of their cognate mRNAs, thereby inducing either mRNA degradation or translational repression. Over 400 human miRNAs have so far been identified, but many more probably still remain to be discovered. These miRNAs are each likely to down-regulate a large number of different target mRNAs. Mutations in miRNA-binding sites could in principle cause disease, although in practice only one such lesion has so far been reported: a G A transition in a binding site for miR-189 within the 3' UTR of the *SLITRK1* gene of two apparently unrelated Tourette syndrome patients (1). Experimental confirmation of the functional effect of this mutation came from the demonstration that, in the presence of miRNA-189, in vitro constructs bearing the 3' UTR mutation served to increase repression of a reporter gene by comparison with the wild-type.

An instructive pathogenic mutation was recently found in an miRNA target site. A quantitative trait locus with a major effect on muscle mass of Texel sheep was mapped to a chromosome interval encompassing the myostatin (*GDF8*) gene. The *GDF8* allele of Texel sheep is characterized by a G-to-A transition in the 3' UTR that creates a target site for mir1 and mir206, miRNAs that are highly expressed in skeletal muscle. This causes translational inhibition of the myostatin gene and hence contributes to the muscular hypertrophy of Texel sheep (60). A further example of a functional miRNA target site variation involves an SNP in the 3'UTR of the human *AGTR1* gene; the variant allele is not down-regulated by miR155; remarkably, the variant allele has been associated with hypertension in numerous studies (296).

10.3.2.7 Cap Site Mutations

Transcription of the mRNA is initiated at the so-called cap site, which is protected from exonucleolytic degradation by the addition of α-methylguanine. An A-to-C transversion at the cap site of the β-globin (*HBB*) gene was found in a patient with β-thalassemia (354). It is not, however, clear whether this mutation causes reduced transcription or abnormal initiation of transcription since C is found in 6% of transcriptional initiation sites (190) (the most common nucleotide (76%) at position +1 is A). A functional (C/A) polymorphism of the transcriptional initiation site has been noted in the *APOH* gene; the rarer A allele displayed a carrier frequency of 0.12 and was associated with markedly reduced plasma β2 glycoprotein I (226).

10.3.2.8 Mutations in 5′ Untranslated Regions

Sequence motifs in the 5' UTRs of genes are thought to play a role in controlling the translation of the encoding mRNA. The phenotypic effects of lesions in 5' UTRs and their clinical consequences have been reviewed (45). Mutations in the iron response element (IRE) in the 5' UTR of the ferritin (*FTH1*) gene interfere with the post-transcriptional regulation of ferritin synthesis by decreasing the affinity of IRE for IRE-binding protein (125). By contrast, decreases in the steady state level of β-globin (*HBB*) mRNA have been noted in association with a single base deletion at position +10, a G-to-A substitution at position +22, a C-to-G transversion at position +33, and a 4 bp deletion (AAAC) at position +(40-43) in the *HBB* 5' UTR (18, 148, 297).

10.3.2.9 Mutations in 3′ Regulatory Regions

Sequences in the 3' regulatory regions (3' RRs) of genes are known to be involved in controlling mRNA cleavage/polyadenylation and determining mRNA stability, nuclear export, intracellular localization, and translational efficiency. Although such regions are rich in regulatory elements, relatively few pathologic mutations have been reported (53, 66). Although only ~0.2% of mutations currently logged in HGMD are located within 3' RRs, this is likely to represent a rather conservative estimate of their actual prevalence.

A typical example is the G→A transition 69 nucleotides downstream of the polyadenylation site of the δ-globin (*HBD*) gene causing δ-thalassemia (233); the mutation occurs within a GATA motif and serves to increase the binding affinity of the sequence for erythroid-specific DNA binding protein.

In an attempt to study 3' RR mutations systematically, Chen et al. (53) collated 121 3' RR variants in 94 human genes including 17 mutations in the upstream core polyadenylation signal sequence (UCPAS), 79 in the upstream sequence (USS) between the translational termination codon and the UCPAS, 6 in the left arm of the 'spacer' sequence (LAS) between the UCPAS and the pre-mRNA cleavage site (CS), 3 in the right arm of the 'spacer' sequence (RAS) or downstream core polyadenylation signal sequence (DCPAS), and 7 in the downstream sequence (DSS) of the 3'-flanking region. All the UCPAS mutations and the rather unusual cases of *DMPK*, *SCA8*, *FCMD*, and *GLA* mutations were found to exert a significant effect on the mRNA phenotype, and the majority cause monogenic disease. By contrast, most of the remaining variants were polymorphisms, were found to exert a comparatively minor influence on mRNA expression, but may predispose to, protect from, or modify complex clinical phenotypes. The systematic study of these lesions permitted the identification of consistent patterns of secondary structural change that promise to allow the discrimination of nonfunctional USS variants from their functional counterparts.

10.3.2.10 Translational Initiation Mutations

Mutations in the ATG translational initiation codon have been reported in quite a wide variety of disorders (e.g., (270)). Instances of substitutions in all three nucleotides have been observed in β-thalassemia, Norrie disease, albinism, phenylketonuria, McArdle disease, and Albright osteodystrophy, among others. Indeed, a total of 251 mutations within ATG translational initiation codons are recorded in HGMD, representing ~0.6% of all missense and nonsense mutations. Almost invariably, the mutation leads to severe reduction of steady state mRNA levels similar to that associated with nonsense mutations. The mutant mRNA is presumably not translated. The first AUG codon occurs in the context of the so-called Kozak consensus sequence GCCA/GCCAUGG, which is thought to be recognized by the 40S ribosomal subunit (191). Mutations at the initiator methionine ATG may completely abolish translation; however, there are alternative possibilities, viz. utilization of the mutant ATG with much reduced efficiency or translational initiation at the next available ATG codon. A C/T polymorphism immediately 5' to the ATG codon within the Kozak sequence of the *CD40* gene is thought to influence translation efficiency (162).

Some diseases are caused by mutations that perturb the initiation step of translation by changing the context around the start AUG codon or introducing upstream AUG codons (see (192) for a review). The scanning mechanism provides a framework for understanding the effects of these changes in mRNAs. The scanning mechanism refers to the entry of the small ribosomal subunit at the (usually capped) 5' end of the mRNA and linear migration until an AUG codon is encountered. Mutational mechanisms such as: (a) reinitiation at an internal start codon (e.g., thrombopoietin, *TPO*); and (b) leaky scanning (as in the case of the *Rx/rax* gene underlying the mouse eyeless mutation) probably account for such cases.

Naturally occurring mutations in the GCCA/GCCAUGG motif include (for the numbering of the mutant nucleotide, the A of the AUG codon is +1; see references in (192)): (a) +4 G-to-A in the androgen receptor (*AR*) gene in a family with partial androgen insufficiency; (b) -1 C-to-T transition in the α-tocopherol transfer protein (*TTPA*) gene in a family with vitamin E deficiency; (c) a 2nt deletion causes an A-to-C change at position -3 of the α-globin gene (*HBA*) in a patient with α-thalassemia; (d) -3 A-to-T transversion in the mouse *Pax6* gene causes defects in eye development; (e) -3 G-to-C somatic mutation in the *BRCA1* gene in one case of highly aggressive sporadic breast cancer. It is not surprising that most of the naturally occurring mutations involve positions -3 and +4, the positions wherein experimentally induced mutations have the strongest effect.

10.3.2.11 Termination Codon Mutations

The classic example of a termination codon mutant is the case of the α_2-globin Constant Spring, with a mutation in the normal stop codon; this substitution leads to incorporation of an additional 31 amino acid residues in the α_2-globin polypeptide chain (59). The resulting protein is unstable and does not interact properly with the β–globin chains of hemoglobin. Some 81 mutations within Term codons are recorded in HGMD, representing ~0.2% of all missense/nonsense mutations.

10.3.2.12 Frameshift Mutations

A large number of frameshift mutations have been described in numerous disease-related genes. All lead to altered translational termination with abnormal polypeptide chains after the frameshifts; severe phenotypes are usually seen. Frameshifts occur with microdeletions or micro-insertions and exon skipping. The mechanisms underlying these mutations were discussed earlier in this chapter.

10.3.2.13 Nonsense Mutations

Nonsense mutations obviously cause premature termination of translation and truncated polypeptides. Some 48% of nonsense mutations in HGMD are to codon TGA, with 28% being to TAA and 24%, to TAG. About 55% of the newly created TGA codons are CG-to-TG transitions resulting from the methylation-mediated deamination of 5mC described earlier. Many such mutations have been described in a large number of disease-related genes.

Nonsense mutations are usually associated with a reduction in the steady state level of cytoplasmic mRNA (28). This mechanism of "nonsense-mediated mRNA decay" (NMD) is responsible for the degradation of mRNAs that contain a premature termination codon at a position at least 50 nt upstream of an exon-exon boundary (219), but it is not universal (157). One or more parameters could be affected: the transcription rate, the efficiency of mRNA processing or transport to the cytoplasm, or mRNA stability.

Nonsense mutations account for at least 11% of all described gene lesions causing human inherited disease. In the majority of cases, the resulting disorders are recessive in nature as a consequence of the haploinsufficiency resulting from the NMD-induced absence of the truncated proteins (which ensures that such polypeptides do not interfere with the function of the wild-type protein). Nonsense mutations that do not elicit NMD can, however, give rise to a dominant negative condition (e.g., mutations in the *SOX10* gene causing Waardenburg-Shah syndrome (160)). Since, for NMD to be activated, the nonsense mutation must reside at least 50-55 nt upstream of an exon-exon boundary, it follows that the precise location of the nonsense mutation could be an important factor in predicting the pathogenicity of that lesion. By way of example, nonsense mutations within the last exon of the human β-globin (*HBB*) gene do not elicit NMD. As a consequence, the truncated β-globin product has near-normal abundance, fails to associate properly with α-globin, and hence gives rise to a dominantly inherited form of α-thalassemia (318). Different nonsense mutations within the same gene may thus be associated with different clinical phenotypes depending upon whether or not NMD is activated. Another example of this is provided by a nonsense mutation (Q37X) in the *DAX1* gene of an adrenal hypoplasia congenita patient; this lesion is associated with a milder clinical phenotype than expected on account of the expression of a partially functional, amino terminal-truncated DAX1 protein synthesized from an alternative in-frame translational start site at Met83 (254).

In practical terms, the observation of greatly reduced or absent cytoplasmic mRNA associated with nonsense mutations has important implications for mutation screening. Thus, attempts to obtain mRNA for RT-PCR and mutation detection may result in amplification of nucleic acid from only the non-nonsense mutation-bearing allele. Nonsense mutations in the factor VIII (*F8*) gene (hemophilia A) and fibrillin (*FBN1*) gene (Marfan syndrome) have been associated with the skipping of exons containing these mutations (95, 240), and this observation has now been extended to other genes; exon skipping is either complete or partial. The mechanism underlying this phenomenon is unknown although a number of intriguing models have been proposed (118).

10.3.2.14 Unstable Protein Mutants

Missense mutations can cause abnormal protein folding and are therefore associated with reduced expression owing to instability of the protein. Reviews of mutations that affect protein stability can be found in (6, 258). For proteins that circulate in body fluids, most mutations are associated with 'CRM-negative' status in which the amount of protein correlates with the amount of activity or "CRM-reduced" status in which the amount of activity is still lower than the amount of protein produced. Many such mutations have been seen in factor VIII causing mild/moderate hemophilia A (14).

The nature of the biophysical properties of amino acid substitutions in p53 that increase their likelihood of coming to clinical attention has been explored (239); these include solvent inaccessibility, the number of adverse steric interactions introduced and a reduction in H-bond number. This study was extended by modeling *in silico* all amino acid replacements that could potentially have arisen from an inherited single base-pair substitution in five human genes encoding arylsulfatase A (*ARSA*), antithrombin III (*SERPINC1*), protein C (*PROC*), phenylalanine hydroxylase (*PAH*), and transthyretin (*TTR*) (317). A total of 9,795 possible mutant structures were modeled and 20 different biophysical parameters assessed. Comparison with the HGMD-derived spectra of 469 clinically detected mutations indicated that several types of mutation-associated change affected protein function, including the energy difference between wild-type and mutant structures, solvent accessibility of the mutated residue, and distance from the binding/active site. These parameters are considered to be important in protein folding, which adds support to the view that many missense mutations come to clinical attention by virtue of their consequences for protein folding and stability (38, 135).

10.3.2.15 Mutations in Remote Promoter Elements/Locus Control Regions

In the β-globin gene cluster, a regulatory region about 10 kb upstream of the ε-globin (*HBE*) gene has been identified that is capable of directing a high level of position-independent β-globin gene expression (137). This region, termed the locus control region (LCR), is thought to organize the entire 60-kb β-globin gene cluster into an active chromatin domain and to enhance the transcription of individual globin genes (310). A similar LCR is also present in the α-globin gene cluster and other gene clusters (339). Deletions of the LCR in the β-globin gene cluster result in silencing of the β-globin and other genes of the cluster, even though the genes themselves are intact (346). A particular 25-kb deletion, known as Hispanic γδβ-thalassemia, which deletes sequences 9.5–39 kb upstream of the ε-globin gene including the LCR, renders the β-globin gene 60 kb downstream of the deletion nonfunctional (100). This extraordinary effect of the deletion of the LCR is thought to be due to an altered (DNase I-resistant) state of chromatin associated with non-functional genes. Several other examples of similar deletions in the LCR of the α-globin gene cluster have been reported (209).

10.3.2.16 Cellular Consequences of Trinucleotide Repeat Expansions

Trinucleotide repeat expansion has been discussed earlier. In the case of fragile X, the $(CGG)_n$ repeat is located in the 5' UTR of the *FMR1* gene and its expansion to full mutation results in hypermethylation of the promoter region, loss of transcription, and hence silencing of the gene (344). Loss of the encoded protein, FMRP, which is thought to play a role in dendritic mRNA transport and translation, is responsible for the classical fragile X syndrome phenotype. Gene inactivation can also be caused by altering the spacing of promoter elements from the transcriptional start site as in the case of the 12mer repeat expansion in the *CSTB* gene (202).

When the trinucleotide repeat lies within the gene coding region, as in Huntington disease, its expansion results in an abnormal protein with a gain of function owing to the enlargement of the polyglutamine tract. Mutant huntingtin exerts its pathologic effects via abnormal protein aggregation, transcriptional dysregulation, mitochondrial dysfunction, excitotoxicity, and abnormal cellular trafficking, leading to neuronal loss particularly in the dorsal substratum (35).

Another example of a gain-of-function mutation is provided by the expansion of the CTG repeat in the 3' UTR of the *DMPK* gene causing type 1 myotonic dystrophy (DM1). This does not abolish transcription but rather causes nuclear retention of RNA transcripts leading to the transcriptional dysregulation of other genes (83). CTG expansion appears to lead to the sequestration of cellular RNA-binding proteins which in turn gives rise to the abnormal splicing of multiple transcripts. DM1 thus exemplifies a disease whose mechanistic basis lies at the RNA level.

10.3.2.17 Mutations Producing Inappropriate Gene Expression

Hereditary persistence of fetal hemoglobin (HPFH) and hereditary persistence of α-fetoprotein (HPAFP) are two clinical conditions that are prototypes for the inappropriate expression of γ-globin (*HBG1* and *HBG2*) and α-fetoprotein (*AFP*) genes, respectively. Normally the levels of fetal hemoglobin (HbF; α2γ2) in adult life are very low, as there is a switch from fetal to adult hemoglobin during the perinatal period. Similarly, AFP is produced at a high level in fetal liver but declines rapidly after birth. In HPFH and HPAFP, however, the levels of HbF and AFP, respectively, are inappropriately high in adult life. This is often due to single nucleotide substitutions in the promoter regions of the *HBG2*, *HBG1*, or *AFP* genes. A considerable number of mutations that occur in the region -114 to -202 of the γ-globin genes have been characterized and presumably cause persistent expression of their corresponding genes (346). A similar situation has been observed with a -119 mutation in the *AFP* gene (224). These mutations occur within DNA-binding motifs for transcriptional regulators.

A very interesting mutational mechanism has been proposed for facioscapulohumeral muscular dystrophy (FSHD), an autosomal dominant myopathy. This disease is characterized by deletions of a tandem 3.3-kb repeat termed D4Z4 on chromosome 4q35. In the general population, the size of the D4Z4 repeat array may vary between 11 and 150 units, whereas FSHD patients carry fewer than 11 repeats (142). Partial deletion of D4Z4 leads to a local change in chromatin structure (267). As a consequence, genes expressed in muscle and located up to 3 Mb upstream of D4Z4 are inappropriately overexpressed. A multiprotein complex binds D4Z4 and appears to mediate the transcriptional repression of neighboring genes. The deletion of an integral number of D4Z4 repeats below a certain threshold reduces the number of bound repression complexes, and consequently decreases transcriptional repression of 4q35 genes including the *ANT1* gene, an excellent candidate for contributing to the pathogenesis of FSHD (121).

10.3.2.18 Position Effect in Human Disorders

In several instances, a DNA alteration is found well outside the putative gene that is primarily involved with a disease. Mutations acting by "positional effect" are those in which the transcription unit and minimal promoter of the gene remain intact but there is a nearby alteration that influences gene expression (180). These positional effect DNA lesions may involve distal promoter regions, enhancer/silencer elements, or changes in the local chromatin environment. The positional effect could be up to several megabases away from the gene of interest. The examples of the LCR in the β-globin gene cluster and the transcriptional repressor

D4Z4 in FSHD are provided elsewhere in this chapter. Most of the position effects are due to chromosomal rearrangements that frequently lead to alteration of the chromatin environment of the gene. Possible mechanisms that may lead to a positional effect include: (a) separation of the transcription unit from distant *cis*-regulatory elements by the rearrangement (enhancer removal results in gene silencing, whereas silencer removal results in inappropriate gene activation); (b) juxtaposition of the gene with an enhancer element from another part of the genome; (c) removal of an insulator or boundary element may also lead to inappropriate gene silencing; (d) enhancer competition of DNA sequences that were juxtaposed to the gene; (e) positional effect variegation in which the chromosomal rearrangement causes the juxtaposition of an euchromatic gene with a region of heterochromatin.

Some examples of positional effect mutations attributable to translocation breakpoints include genes *PAX6* in aniridia (109), *SOX9* in campomelic dysplasia (268, 331), *POU3F4* in X-linked deafness (88), *HOXD* complex in mesomelic dysplasia (308), *FOXL2* in blepharophimosis/ptosis/epicanthus inversus syndrome (BPES) (31, 79), and the *SHH* gene in preaxial polydactyly (207). In these cases, the translocation breakpoints may be in excess of a megabase away from the inappropriately expressed/silenced gene. Indeed, in one example of campomelic dysplasia, the breakpoint maps ~1.3 Mb downstream of the *SOX9* gene, making this the longest range position effect so far found (331). For a recent review of position effect mutations, see (181).

It is likely that, in the majority of cases, the position effect involves a highly conserved *cis*-acting regulatory element. These *conserved noncoding elements* (CNCs; also termed multiple-species conserved sequences (MCS), conserved non-genic sequences (CNGs); the most highly conserved are also called ultraconserved elements (UCEs)) comprise approximately 1–2% of the human genome and represent potential targets for pathogenic mutations (27, 33, 93, 94, 319). An example of such a lesion is provided by the 52-kb deletion of a large noncoding region downstream of the sclerostin (*SOST*) gene in patients with van Buchem disease, leading to altered expression of the *SOST* gene (215). The deletion disrupts a bone-specific enhancer element that drives *SOST* gene expression.

Pathogenic mutation may also occur in nonconserved elements that could become functional after the introduction of the mutant sequence. This pathogenetic mechanism has been described underlying a variant form of α-thalassemia. Affected individuals from Melanesia have a gain-of-function regulatory single-nucleotide polymorphism (rSNP) in a nongenic region between the α-globin genes and their upstream regulatory elements. The rSNP creates a new promoter-like element that interferes with the normal activation of all downstream α-like globin genes (85).

10.3.2.19 Position Effect by an Antisense RNA

An individual with an inherited α-thalassemia has been described who has a deletion that results in a truncated, widely expressed gene (*LUC7L*) becoming juxtaposed to a structurally normal α-globin (*HBA2*) gene. Although it retained all of its local and remote *cis*-regulatory elements, expression of the *HBA2* gene was nevertheless silenced and its CpG island became completely methylated at an early stage during development. The antisense RNA of the *LUC7L* gene appears to have been responsible for the silencing of the *HBA2* gene (324).

10.3.2.20 Abnormal Proteins Due to Fusion of Two Different Genes

The translation of fusion genes results in novel proteins with different or abnormal properties from their parent polypeptides. Fusion genes are either the result of (1) homologous unequal crossing-over, or (2) junction sequences at breakpoints of chromosomal translocations. Hemoglobin Lepore, a fusion of δ- and β-globin genes, is the prime example of the first mechanism. Other examples of abnormal fusion genes caused by unequal crossover include the case of glucocorticoid-suppressible hyperaldosteronism (GSH), an autosomal dominant form of hypertension, caused by oversecretion of aldosterone (262); some GSH patients have hybrid genes between *CYP11B1* and *CYP11B2*, two highly homologous cytochrome P450 genes on 8q22. The hybrid gene contains the regulatory elements of *CYP11B1*, expressed in the adrenal gland, and the 3' coding region of *CYP11B2*, which is essential for aldosterone synthesis. Another example is the case of abnormalities of color vision resulting from fusion of the green and red color pigment (*RCP*, *GCP*) genes (239). Recombination between the Kallmann gene on Xp22.3 (*KALX*) and its homolog (*KALY*) at Yp11.21

results in a fusion gene that is transcriptionally inactive and is associated with Kallmann syndrome secondary to an X;Y translocation.

A growing number of hematologic malignancies are associated with abnormal fusion proteins, the genes of which are found at the breakpoints of chromosomal translocations. One of the first reported examples was the case of fusion of the *BCR* and *ABL* genes in the t(9;22) known as Philadelphia (Ph) chromosome in chronic myelogenous leukemia. The *BCR* gene is on chromosome 22 and the *ABL* gene is on chromosome 9; after the translocation junction, a fusion gene is created with the promoter elements of the *ABL* gene and the 3' half of the *BCR* gene (25). A new abnormal protein is detected in the leukemia cells, the abnormal function of which probably contributes to the malignant phenotype. Another example is the case of Ewing sarcoma (a solid tumor of bone) in which an 11;22 translocation results in a fusion of the *FLI1* gene on 11q24 with the *EWS* gene on 22q12 (89); for a classic review see (272)). Fusion genes can be readily identified by PCR and can serve either as diagnostic indicators for relapse in the disorders concerned or as indicators of the need for an alternative therapeutic regimen.

10.3.2.21 Mutations in Genes Involved in Mismatch Repair Associated with Genomic Instability in the Soma

The study of somatic mutation is extremely important both for the study of cancer (116) and other diseases such as paroxysmal nocturnal hemoglobinuria (107). Mutations that lead to abnormal or abolished function of genes encoding for proteins involved in DNA mismatch repair are of particular importance because they lead to accumulation of mutations throughout the genome. For example, some forms of hereditary nonpolyposis colon cancer (HNPCC), which may account for up to 10% of colon carcinoma, are due to mutations in genes such as *MSH2* or *MLH1* that encode mismatch repair proteins (113, 204, 260) . In families with mutations in these genes, the DNA of tumor tissue shows considerable instability as detected by the generation of new alleles for numerous DNA polymorphic markers (161). One of the genes affected by the genomic instability is that encoding the type II transforming growth factor-β (TGF-β) receptor (*TGFB2R*), which has a run of 10 adenines in its coding region. This run of As is altered, resulting in a frameshift and absence of the receptor, which in turn releases the cell from TGF-β-inhibitory effects and contributes to malignancy (220). The discovery and further study of genes of the mutation repair system will enhance our understanding both of germline and of somatic mutations.

To date, relatively few studies have attempted to compare germline and somatic mutational spectra for the same genes. This notwithstanding, the mutational mechanisms underlying single base-pair substitutions (290, 328), micro-deletions and micro-insertions (134, 166, 328), and even gross gene rearrangements (184, 246) often appear to exhibit similarities between the germline and the soma.

10.3.2.22 Mosaicism

Germline mosaicism is a relatively frequent mechanism of inherited disease and provides an explanation for the inheritance pattern in cases where multiple affected offspring are born to clinically and phenotypically normal parents (367). It arises through the occurrence of a mutation de novo in a germline cell or one of its precursors during the early embryonic development of the parent. Since mitotic divisions predominate in both spermatogenesis and oogenesis, most germline mutations are likely to be mitotic rather than meiotic in origin. *Somatic mosaicism* results from mutations occurring during mitotic cell divisions in the embryo with subsequent clonal expansion of the affected cells (139). The clinical effect of somatic mosaicism depends critically upon the developmental stage at which the mutation occurs. Thus, a mutation that occurs very early on in embryonic development is likely to affect many somatic tissues. By contrast, mutations occurring rather later may give rise to a phenotype that is confined to a single body region or even to a single organ. Somatic mosaicism arising at a very early embryonic stage can involve both somatic cells and germ cells. Such individuals (*gonosomal mosaics*) are at risk of having affected children.

10.3.2.23 Sex Differences in Mutation Rates

Sex differences in mutation rates may have a variety of different underlying causes. For *premeiotic mutations*, the single most important factors are likely to be the much higher number of cell divisions during spermatogenesis than oogenesis and the fact that the number of male germ cell divisions experienced is age dependent (81). However, the likelihood of a given mutation having originated in a particular parent is often dependent upon the nature of the mutation in question. In general, point mutations tend to display a paternal bias, arising during spermatogenesis, whilst gross deletions tend to occur predominantly in females, having originated during oogenesis (26, 136).

10.3.2.24 Concepts of Dominance and Recessiveness in Relation to the Underlying Mutations

A genetic character is held to be *dominant* if it is manifest in the heterozygous state and *recessive* if it is not. Thus, for a truly dominant condition, homozygotes should be clinically and phenotypically indistinguishable from heterozygotes (349). If this is not so, and the homozygote is more seriously affected, then the respective alleles may be regarded as *semidominant* (366).

In general, most recessive alleles are loss-of-function alleles and include gross gene deletions and rearrangements, frameshift mutations, nonsense mutations, etc. By contrast, dominant alleles are often associated with gain of function, resulting either from dominant negative mutations (which interfere with and hence abrogate the function of the wild-type allele) or from dominant positive mutations (which confer increased, constitutive, novel or toxic activity upon the mutant protein). Examples of dominant negative mutations are to be found in the *GH1* (61) and *KIT* (309) genes, whilst dominant positive mutations have been reported in the *PMP22* (263), *GNAS1* (7), *DMPK* (234), and *SERPINA1* (253) genes. It should be noted that loss-of-function mutations (e.g., *TERT* (17) and *RUNX2* (179)) can also be associated with dominantly inherited conditions in cases where a 50% reduction in the level of the protein product is sufficient to impede function.

For X-linked diseases it is probably inappropriate to use the terms dominant and recessive, since males are hemizygous and females often display variable expressivity of their heterozygous mutations owing to skewed X-inactivation or clonal expansion (96).

10.4 General Principles of Genotype-Phenotype Correlations

Several general principles have emerged as a result of the intensive study of causative mutations in genetic disorders. The following discussion highlights some of these principles. The reader is encouraged to use the Online Mendelian Inheritance in Man (OMIM) database at http://www3.ncbi.nlm.nih.gov/Omim for further information or for specific genes and clinical phenotypes. Wolf's review (353) provides an excellent guide to the complex issues inherent in the study of the relationship between mutant genotype and clinical phenotype.

Mutations in the Same Gene may be Responsible for More than One Disorder. There are many examples to illustrate the principle that mutations in a single gene can cause different and distinct clinical phenotypes ("allelic heterogeneity"). Historically, the first example is that of the β-globin (*HBB*) gene on 11pter. Mutations of this gene cause β-thalassemia, sickle cell disease, and methemoglobinemia. The *L1CAM* gene on Xq28 has been shown to be mutated in hydrocephalus and stenosis of aqueduct of Sylvius, MASA syndrome (*m*ental retardation, *a*phasia, shuffing gait, *a*dducted thumbs), and spastic paraplegia 1. The *COL1A2* gene on 7q21-q22 is involved in four different clinical forms of osteogenesis imperfecta (types II, III, IV, and atypical) and in Ehlers-Danlos syndrome type VII B. The fibroblast growth factor receptor 2 (*FGFR2*) gene is mutated in three different craniosynostosis syndromes, namely Pfeiffer, Crouzon, and Jackson-Weiss. The *COL2A1* gene is implicated in Stickler syndrome type 1, SED congenita, Kneist dysplasia, achondrogenesis-hypochondrogenesis type 2, precocious osteoarthritis, Wagner syndrome type 2, and SMED Strudwick type. In a survey of 1014 genes causing disorders in OMIM, 165 genes were associated with two disorders, 52 genes with three disorders, 24 genes with four disorders, and 19 genes with five or more disorders (11).

One Disorder May Be Caused by Mutations in More than One Gene. There are a plethora of similar clinical phenotypes caused by mutations in different genes. This observation, also known as "nonallelic" or "locus" heterogeneity, is well understood, thanks to linkage analyzes for genetic disorders and the search for mutations in different genes. Thus, tuberous sclerosis, a relatively common autosomal dominant disorder, is caused by lesions in at least two different loci: *TSC1* on 9q34 and *TSC2* on 16p13.3. Approximately 60% of TSC families show linkage to the *TSC2* locus and 40% to the *TSC1* locus. Hereditary nonpolyposis colon cancer has been associated with mutations in five different genes. *MLH1* on 3p, *MSH2* on 2p16, *PMS1* on 2q31-q33, *PMS2* on 7p22 and *MSH6* on 2p16. Retinitis pigmentosa has so far been associated with a total of 23 different genes, and the list is still growing. We expect that disorders of complex or polygenic phenotypes, such as hypertension, atherosclerosis, diabetes, schizophrenia, and manic-depressive illness, will be associated with a considerable number of genes scattered throughout the genome.

One and the Same Mutation May Give Rise to Different Clinical Phenotypes ("Polypheny"). The clinical phenotype does not only depend on the one mutation in the responsible gene; it can be modified by the action of any of the other ~25,000–30,000 genes in the genome (353). The environment can also have an important role in the full development of the clinical phenotype. The classic sickle cell disease mutation in the β-globin (*HBB*) gene (Glu6Val) may be associated with severe or mild sickle cell disease. The amelioration of the severe clinical phenotype in this case can be attributed to the increased expression of γ-globin genes and the presence of high levels of HbF. The genomic environment of the β-globin gene cluster may therefore modify the severity of sickle cell disease, as may genetic variation originating from other loci, e.g., the α-globin genes (73). Another example of this phenomenon has recently been provided by studies of certain craniosynostoses. Both Pfeiffer and Crouzon syndromes can be associated with the same C342Y or C342R mutations in the *FGFR2* gene.

The clinical phenotype associated with the D178N missense mutation in the prion protein (*PRNP*) gene is critically dependent upon the presence of the Met or Val 129 polymorphic allele to which it is coupled. When D178N lies in *cis* to the Met129 allele, fatal familial insomnia (FFI) results, whereas D178N coupled to the Val129 allele is associated with Creutzfeldt-Jakob disease (261). The Met/Val 129 polymorphism also exerts an effect in *trans* through the normal allele, since FFI is more severe and of longer duration in patients homozygous for either the Met or the Val allele.

One of the best examples of the contribution of the environment to the clinical phenotype of single gene disorders is that of phenylketonuria resulting from phenylalanine hydroxylase (PAH) deficiency. Individuals homozygous or compound heterozygous for mutations in the *PAH* gene develop severe mental handicap if fed a normal diet. However, the cognitive status remains normal if these individuals are fed with a special, "phenylalanine-free" diet.

Mutations in More than One Gene May Be Required to Express a Given Clinical Phenotype (Digenic Inheritance; Triallelic Inheritance). Digenic inheritance refers to clinical phenotypes caused by the co-inheritance of mutations in two unlinked genes. Thus one form of retinitis pigmentosa is due to the co-inheritance of mutations in the *RDS* gene on 6p *and* the *ROM* gene on 11q (170). Individuals with either one or the other mutation, do not suffer from the disease. In similar vein, digenic inheritance of mutations in the *MITF* and *TYR* genes has been reported as a cause of Waardenburg syndrome type 2 in conjunction with ocular albinism (235). This phenomenon may be common in polygenic disorders and in disorders with "low penetrance."

Triallelic inheritance refers to clinical phenotypes with apparent recessive mode of inheritance caused by the co-inheritance of three mutant alleles, two in one gene and one in another gene. An example of triallelic inheritance is provided by the Bardet–Biedl syndrome. There are pedigrees in which affected individuals have two mutant alleles in the *BBS6* gene and one mutant allele in the *BBS2* gene. Other pedigrees have two mutant alleles in the *BBS2* gene and one mutant allele in *BBS6* (173). This type of inheritance indicates that some forms of BBS have a complex pattern of inheritance. As above, this phenomenon may be relevant in polygenic disorders and in disorders with "low penetrance."

Different Mutations in the Same Gene May Give Rise to Distinct Dominant and Recessive Forms of the Same Disease. von Willebrand factor (vWF) deficiency is a relatively common monogenic disease of blood coagulation. Many mutations have been studied in the *VWF* gene on chromosome 12p. A proportion of mutations (usually deletions, nonsense codons, or frameshift mutations) cause vWF deficiency with a recessive mode of inheritance; other mutations (mostly missense substitutions), however, are associated with a dominant mode of inheritance of the vWF deficiency (243).

Whereas the majority of hitherto characterized growth hormone (*GH1*) gene lesions (including gross deletions and missense/nonsense mutations) that underlie familial short stature are inherited in autosomal recessive fashion, there is a group of intron 3 splicing mutations that are characterized by a dominant mode of inheritance (62). These lesions result in the in-frame skipping of exon 3 encoding 40 amino acids, including a Cys residue. The dominant negative nature of this mutation is thought to be explicable in terms of the participation of the resulting free unpaired cysteine residue in an illegitimate intermolecular disulfide linkage, leading to dimerization of the mutant molecule with a normal GH molecule and inhibition of GH secretion.

10.5 Why Study Mutation?

The sequencing of the human genome is now essentially complete and its annotation well under way. Full exploitation of the emerging data, specifically in relation to understanding the etiology of inherited disease and disease predisposition, is likely to be hampered by our ignorance of the basic processes underlying inter-individual, inter-population, and inter-species genetic diversity, however. At the population level, such an understanding is seen as essential for any meaningful interpretation of the prevalence/incidence patterns observed for diseases with a genetic basis. Within families, it is a prerequisite for being able to explain how inter-individual variation arises and how variable phenotypic expression can be associated with identical gene lesions. Thus, for human genome sequence data to be useful in the context of molecular medicine, they must eventually be related to the genetic variation underlying human inherited disease. To this end, the meta-analysis of pathological germline mutations in human genes should facilitate:

1. The assessment of the spectrum of known genetic variation underlying human inherited disease.
2. The identification of factors determining the propensity of DNA sequences to undergo germline mutation.
3. The optimization of mutational screening strategies.
4. Improvements in our ability to predict the clinical phenotype from knowledge of the mutant genotype.
5. The identification of disease states that exhibit incomplete mutational spectra, prompting the search for, and detection of, novel gene lesions associated with different clinical phenotypes (227).
6. Extrapolation toward the genetic basis of other, more complex traits and diseases (36).
7. Improvements in our understanding of the function of a given protein.
8. Meaningful comparison between the mechanisms of mutagenesis underlying both inherited and somatic disease.
9. Studies of human genetic diseases in their evolutionary context (303).

Acknowledgments The authors wish to thank Professor Michael Krawczak (Institut für Medizinische Informatik und Statistik, Christian-Albrechts-Universität Kiel, Germany) for his contributions to earlier published versions of this chapter and Peter Stenson for provision of HGMD data.

References

1. Abelson JF, Kwan KY, O'Roak BJ, Baek DY, Stillman AA, Morgan TM, Mathews CA, Pauls DL, Rasin MR, Gunel M, Davis NR, Ercan-Sencicek AG, Guez DH, Spertus JA, Leckman JF, LSt D, Kurlan R, Singer HS, Gilbert DL, Farhi A, Louvi A, Lifton RP, Sestan N, State MW (2005) Sequence variants in SLITRK1 are associated with Tourette's syndrome. Science 310:317–320
2. Abeysinghe SS, Chuzhanova N, Krawczak M, Ball EV, Cooper DN (2003) Translocation and gross deletion breakpoints in human inherited disease and cancer. I nucleotide composition and recombination-associated motifs. Hum Mutat 22:229–244
3. Abeysinghe SS, Stenson PD, Krawczak M, Cooper DN (2004) Gross rearrangement breakpoint database (GRaBD). Hum Mutat 23:219–221
4. Aidoo M, Terlouw DJ, Kolczak MS, McElroy PD, ter Kuile FO, Kariuki S, Nahlen BL, Lal AA, Udhayakumar V (2002) Protective effects of the sickle cell gene against malaria morbidity and mortality. Lancet 359:1311–1312
5. Aitman TJ, Dong R, Vyse TJ, Norsworthy PJ, Johnson MD, Smith J, Mangion J, Roberton-Lowe C, Marshall AJ, Petretto E, Hodges MD, Bhangal G, Patel SG, Sheehan-Rooney K, Duda M, Cook PR, Evans DJ, Domin J, Flint J, Boyle JJ, Pusey CD, Cook HT (2006) Copy number polymorphism in Fcgr3 predisposes to glomerulonephritis in rats and humans. Nature 439:851–855
6. Alber T (1989) Mutational effects on protein stability. Annu Rev Biochem 58:765–798
7. Aldred MA, Trembath RC (2000) Activating and inactivating mutations in the human GNAS1 gene. Hum Mutat 16:183–189
8. Altshuler D, Brooks LD, Chakravarti A, Collins FS, Daly MJ, Donnelly P (2005) A haplotype map of the human genome. Nature 437:1299–1320
9. Aly AM, Higuchi M, Kasper CK, Kazazian HH Jr, Antonarakis SE, Hoyer LW (1992) Hemophilia A due to mutations that create new N-glycosylation sites. Proc Natl Acad Sci USA 89:4933–4937
10. Anagnou NP, O'Brien SJ, Shimada T, Nash WG, Chen MJ, Nienhuis AW (1984) Chromosomal organization of the

human dihydrofolate reductase genes: dispersion, selective amplification, and a novel form of polymorphism. Proc Natl Acad Sci USA 81:5170–5174

11. Antonarakis SE, McKusick VA (2000) OMIM passes the 1,000-disease-gene mark. Nat Genet 25:11
12. Antonarakis SE, Irkin SH, Cheng TC, Scott AF, Sexton JP, Trusko SP, Charache S, Kazazian HH Jr (1984) beta-Thalassemia in American Blacks: novel mutations in the "TATA" box and an acceptor splice site. Proc Natl Acad Sci USA 81:1154–1158
13. Antonarakis SE, Kazazian HH Jr, Orkin SH (1985) DNA polymorphism and molecular pathology of the human globin gene clusters. Hum Genet 69:1–14
14. Antonarakis SE, Kazazian HH, Tuddenham EG (1995) Molecular etiology of factor VIII deficiency in hemophilia A. Hum Mutat 5:1–22
15. Antonarakis SE, Krawczak M, Cooper DN (2001) The nature and mechanisms of human gene mutation. In: Scriver CR, Beaudet AL, Valle D et al (eds) The metabolic and molecular bases of inherited disease. McGraw-Hill, New York, pp 343–377
16. Arai M, Inaba H, Higuchi M, Antonarakis SE, Kazazian HH Jr, Fujimaki M, Hoyer LW (1989) Direct characterization of factor VIII in plasma: detection of a mutation altering a thrombin cleavage site (arginine-372-histidine). Proc Natl Acad Sci USA 86:4277–4281
17. Armanios M, Chen JL, Chang YP, Brodsky RA, Hawkins A, Griffin CA, Eshleman JR, Cohen AR, Chakravarti A, Hamosh A, Greider CW (2005) Haploinsufficiency of telomerase reverse transcriptase leads to anticipation in autosomal dominant dyskeratosis congenita. Proc Natl Acad Sci USA 102:15960–15964
18. Athanassiadou A, Papachatzopoulou A, Zoumbos N, Maniatis GM, Gibbs R (1994) A novel beta-thalassaemia mutation in the 5' untranslated region of the beta-globin gene. Br J Haematol 88:307–310
19. Audrezet MP, Chen JM, Raguenes O, Chuzhanova N, Giteau K, Le Marechal C, Quere I, Cooper DN, Ferec C (2004) Genomic rearrangements in the CFTR gene: extensive allelic heterogeneity and diverse mutational mechanisms. Hum Mutat 23:343–357
20. Bacolla A, Jaworski A, Larson JE, Jakupciak JP, Chuzhanova N, Abeysinghe SS, O'Connell CD, Cooper DN, Wells RD (2004) Breakpoints of gross deletions coincide with non-B DNA conformations. Proc Natl Acad Sci USA 101:14162–14167
21. Baglioni C (1962) The fusion of two peptide chains in hemoglobin Lepore and its interpretation as a genetic deletion. Proc Natl Acad Sci USA 48:1880–1886
22. Bailey JA, Gu Z, Clark RA, Reinert K, Samonte RV, Schwartz S, Adams MD, Myers EW, Li PW, Eichler EE (2002) Recent segmental duplications in the human genome. Science 297:1003–1007
23. Ball EV, Stenson PD, Abeysinghe SS, Krawczak M, Cooper DN, Chuzhanova NA (2005) Microdeletions and microinsertions causing human genetic disease: common mechanisms of mutagenesis and the role of local DNA sequence complexity. Hum Mutat 26:205–213
24. Ballabio A, Carrozzo R, Parenti G, Gil A, Zollo M, Persico MG, Gillard E, Affara N, Yates J, Ferguson-Smith MA et al (1989) Molecular heterogeneity of steroid sulfatase deficiency: a multicenter study on 57 unrelated patients, at DNA and protein levels. Genomics 4:36–40
25. Bartram CR, de Klein A, Hagemeijer A, van Agthoven T, Geurts van Kessel A, Bootsma D, Grosveld G, Ferguson-Smith MA, Davies T, Stone M et al (1983) Translocation of c-ab1 oncogene correlates with the presence of a Philadelphia chromosome in chronic myelocytic leukaemia. Nature 306:277–280
26. Becker J, Schwaab R, Moller-Taube A, Schwaab U, Schmidt W, Brackmann HH, Grimm T, Olek K, Oldenburg J (1996) Characterization of the factor VIII defect in 147 patients with sporadic hemophilia A: family studies indicate a mutation type-dependent sex ratio of mutation frequencies. Am J Hum Genet 58:657–670
27. Bejerano G, Pheasant M, Makunin I, Stephen S, Kent WJ, Mattick JS, Haussler D (2004) Ultraconserved elements in the human genome. Science 304:1321–1325
28. Benz EJ, Forget BG, Hillman DG, Cohen-Solal M, Pritchard J, Cavallesco C, Prensky W, Housman D (1978) Variability in the amount of beta-globin mRNA in beta0 thalassemia. Cell 14:299–312
29. Berg LP, Scopes DA, Alhaq A, Kakkar VV, Cooper DN (1994) Disruption of a binding site for hepatocyte nuclear factor 1 in the protein C gene promoter is associated with hereditary thrombophilia. Hum Mol Genet 3:2147–2152
30. Bertina RM, Koeleman BP, Koster T, Rosendaal FR, Dirven RJ, de Ronde H, van der Velden PA, Reitsma PH (1994) Mutation in blood coagulation factor V associated with resistance to activated protein C. Nature 369:64–67
31. Beysen D, Raes J, Leroy BP, Lucassen A, Yates JR, Clayton-Smith J, Ilyina H, Brooks SS, Christin-Maitre S, Fellous M, Fryns JP, Kim JR, Lapunzina P, Lemyre E, Meire F, Messiaen LM, Oley C, Splitt M, Thomson J, Peer YV, Veitia RA, De Paepe A, De Baere E (2005) Deletions involving long-range conserved nongenic sequences upstream and downstream of FOXL2 as a novel disease-causing mechanism in blepharophimosis syndrome. Am J Hum Genet 77:205–218
32. Blencowe BJ (2000) Exonic splicing enhancers: mechanism of action, diversity and role in human genetic diseases. Trends Biochem Sci 25:106–110
33. Boffelli D, Nobrega MA, Rubin EM (2004) Comparative genomics at the vertebrate extremes. Nat Rev Genet 5:456–465
34. Bondeson ML, Dahl N, Malmgren H, Kleijer WJ, Tonnesen T, Carlberg BM, Pettersson U (1995) Inversion of the IDS gene resulting from recombination with IDS-related sequences is a common cause of the Hunter syndrome. Hum Mol Genet 4:615–621
35. Borrell-Pages M, Zala D, Humbert S, Saudou F (2006) Huntington's disease: from huntingtin function and dysfunction to therapeutic strategies. Cell Mol Life Sci 63:2642–2660
36. Botstein D, Risch N (2003) Discovering genotypes underlying human phenotypes: past successes for mendelian disase, future approaches for complex disease. Nat Genet 33 (Suppl):228–237
37. Brais B, Bouchard JP, Xie YG, Rochefort DL, Chretien N, Tome FM, Lafreniere RG, Rommens JM, Uyama E, Nohira O, Blumen S, Korczyn AD, Heutink P, Mathieu J, Duranceau A, Codere F, Fardeau M, Rouleau GA (1998) Short GCG

expansions in the PABP2 gene cause oculopharyngeal muscular dystrophy. Nat Genet 18:164–167
38. Bross P, Corydon TJ, Andresen BS, Jorgensen MM, Bolund L, Gregersen N (1999) Protein misfolding and degradation in genetic diseases. Hum Mutat 14:186–198
39. Buckland PR (2003) Polymorphically duplicated genes: their relevance to phenotypic variation in humans. Ann Med 35:308–315
40. Burset M, Seledtsov IA, Solovyev VV (2000) Analysis of canonical and non-canonical splice sites in mammalian genomes. Nucleic Acids Res 28:4364–4375
41. Byers P (2001) Disorders of collagen biosynthesis and structure. In: Scriver CR, Beaudet AL, Valle D et al (eds) The metabolic and molecular bases of inherited disease. McGraw-Hill, New York, pp 5241–5286
42. Cai SP, Eng B, Francombe WH, Olivieri NF, Kendall AG, Waye JS, Chui DH (1992) Two novel beta-thalassemia mutations in the 5' and 3' noncoding regions of the beta-globin gene. Blood 79:1342–1346
43. Cartegni L, Chew SL, Krainer AR (2002) Listening to silence and understanding nonsense: exonic mutations that affect splicing. Nat Rev Genet 3:285–298
44. Caskey CT, Pizzuti A, Fu YH, Fenwick RG Jr, Nelson DL (1992) Triplet repeat mutations in human disease. Science 256:784–789
45. Cazzola M, Skoda RC (2000) Translational pathophysiology: a novel molecular mechanism of human disease. Blood 95:3280–3288
46. Challis BG, Pritchard LE, Creemers JW, Delplanque J, Keogh JM, Luan J, Wareham NJ, Yeo GS, Bhattacharyya S, Froguel P, White A, Farooqi IS, O'Rahilly S (2002) A missense mutation disrupting a dibasic prohormone processing site in pro-opiomelanocortin (POMC) increases susceptibility to early-onset obesity through a novel molecular mechanism. Hum Mol Genet 11:1997–2004
47. Chao HK, Hsiao KJ, Su TS (2001) A silent mutation induces exon skipping in the phenylalanine hydroxylase gene in phenylketonuria. Hum Genet 108:14–19
48. Chelala C, Auffray C (2005) Sex-linked recombination variation and distribution of disease-related genes. Gene 346:29–39
49. Chen JM, Chuzhanova N, Stenson PD, Ferec C, Cooper DN (2005) Meta-analysis of gross insertions causing human genetic disease: novel mutational mechanisms and the role of replication slippage. Hum Mutat 25:207–221
50. Chen JM, Chuzhanova N, Stenson PD, Ferec C, Cooper DN (2005) Complex gene rearrangements caused by serial replication slippage. Hum Mutat 26:125–134
51. Chen JM, Chuzhanova N, Stenson PD, Ferec C, Cooper DN (2005) Intrachromosomal serial replication slippage in trans gives rise to diverse genomic rearrangements involving inversions. Hum Mutat 26:362–373
52. Chen JM, Stenson PD, Cooper DN, Ferec C (2005) A systematic analysis of LINE-1 endonuclease-dependent retrotranspositional events causing human genetic disease. Hum Genet 117:411–427
53. Chen JM, Ferec C, Cooper DN (2006) A systematic analysis of disease-associated variants in the 3' regulatory regions of human protein-coding genes II: the importance of mRNA secondary structure in assessing the functionality of 3' UTR variants. Hum Genet 120:301–333
54. Christian SL, Fantes JA, Mewborn SK, Huang B, Ledbetter DH (1999) Large genomic duplicons map to sites of instability in the Prader-Willi/Angelman syndrome chromosome region (15q11–q13). Hum Mol Genet 8:1025–1037
55. Chuang JH, Li H (2004) Functional bias and spatial organization of genes in mutational hot and cold regions in the human genome. PLoS Biol 2:E29
56. Chung MY, Ranum LP, Duvick LA, Servadio A, Zoghbi HY, Orr HT (1993) Evidence for a mechanism predisposing to intergenerational CAG repeat instability in spinocerebellar ataxia type I. Nat Genet 5:254–258
57. Chuzhanova NA, Anassis EJ, Ball EV, Krawczak M, Cooper DN (2003) Meta-analysis of indels causing human genetic disease: mechanisms of mutagenesis and the role of local DNA sequence complexity. Hum Mutat 21:28–44
58. Chuzhanova N, Abeysinghe SS, Krawczak M, Cooper DN (2003) Translocation and gross deletion breakpoints in human inherited disease and cancer II: potential involvement of repetitive sequence elements in secondary structure formation between DNA ends. Hum Mutat 22:245–251
59. Clegg JB, Weatherall DJ, Milner PF (1971) Haemoglobin Constant Spring–a chain termination mutant? Nature 234:337–340
60. Clop A, Marcq F, Takeda H, Pirottin D, Tordoir X, Bibe B, Bouix J, Caiment F, Elsen JM, Eychenne F, Larzul C, Laville E, Meish F, Milenkovic D, Tobin J, Charlier C, Georges M (2006) A mutation creating a potential illegitimate microRNA target site in the myostatin gene affects muscularity in sheep. Nat Genet 38:813–818
61. Cogan JD, Phillips JA 3rd, Schenkman SS, Milner RD, Sakati N (1994) Familial growth hormone deficiency: a model of dominant and recessive mutations affecting a monomeric protein. J Clin Endocrinol Metab 79:1261–1265
62. Cogan JD, Prince MA, Lekhakula S, Bundey S, Futrakul A, McCarthy EM, Phillips JA 3rd (1997) A novel mechanism of aberrant pre-mRNA splicing in humans. Hum Mol Genet 6:909–912
63. Collins FS, Stoeckert CJ Jr, Serjeant GR, Forget BG, Weissman SM (1984) G gamma beta+ hereditary persistence of fetal hemoglobin: cosmid cloning and identification of a specific mutation 5' to the G gamma gene. Proc Natl Acad Sci USA 81:4894–4898
64. Common JE, Di WL, Davies D, Kelsell DP (2004) Further evidence for heterozygote advantage of GJB2 deafness mutations: a link with cell survival. J Med Genet 41:573–575
65. Conley ME, Partain JD, Norland SM, Shurtleff SA, Kazazian HH Jr (2005) Two independent retrotransposon insertions at the same site within the coding region of BTK. Hum Mutat 25:324–325
66. Conne B, Stutz A, Vassalli JD (2000) The 3' untranslated region of messenger RNA: a molecular 'hotspot' for pathology? Nat Med 6:637–641
67. Conrad DF, Andrews TD, Carter NP, Hurles ME, Pritchard JK (2006) A high-resolution survey of deletion polymorphism in the human genome. Nat Genet 38:75–81
68. Consortium. TIH (2003) The International hapmap project. Nature 426:789–796
69. Conway GS, Hettiarachchi S, Murray A, Jacobs PA (1995) Fragile X premutations in familial premature ovarian failure. Lancet 346:309–310

70. Cooper DN (1983) Eukaryotic DNA methylation. Hum Genet 64:315–333
71. Cooper DN (1999) Human gene evolution. Bios Scientific, Oxford
72. Cooper DN, Krawczak M (1991) Mechanisms of insertional mutagenesis in human genes causing genetic disease. Hum Genet 87:409–415
73. Cooper DN, Krawczak M (1993) Human gene mutation. Bios Scientific, Oxford
74. Cooper DN, Youssoufian H (1988) The CpG dinucleotide and human genetic disease. Hum Genet 78:151–155
75. Cooper DN, Smith BA, Cooke HJ, Niemann S, Schmidtke J (1985) An estimate of unique DNA sequence heterozygosity in the human genome. Hum Genet 69:201–205
76. Cotton RG, Scriver CR (1998) Proof of "disease causing" mutation. Hum Mutat 12:1–3
77. Coutinho G, Xie J, Du L, Brusco A, Krainer AR, Gatti RA (2005) Functional significance of a deep intronic mutation in the ATM gene and evidence for an alternative exon 28a. Hum Mutat 25:118–124
78. Crawford DC, Akey DT, Nickerson DA (2005) The patterns of natural variation in human genes. Annu Rev Genomics Hum Genet 6:287–312
79. Crisponi L, Deiana M, Loi A, Chiappe F, Uda M, Amati P, Bisceglia L, Zelante L, Nagaraja R, Porcu S, Ristaldi MS, Marzella R, Rocchi M, Nicolino M, Lienhardt-Roussie A, Nivelon A, Verloes A, Schlessinger D, Gasparini P, Bonneau D, Cao A, Pilia G (2001) The putative forkhead transcription factor FOXL2 is mutated in blepharophimosis/ptosis/epicanthus inversus syndrome. Nat Genet 27:159–166
80. Crossley M, Brownlee GG (1990) Disruption of a C/EBP binding site in the factor IX promoter is associated with haemophilia B. Nature 345:444–446
81. Crow JF (2000) The origins, patterns and implications of human spontaneous mutation. Nat Rev Genet 1:40–47
82. Datz C, Haas T, Rinner H, Sandhofer F, Patsch W, Paulweber B (1998) Heterozygosity for the C282Y mutation in the hemochromatosis gene is associated with increased serum iron, transferrin saturation, and hemoglobin in young women: a protective role against iron deficiency? Clin Chem 44:2429–2432
83. Davis BM, McCurrach ME, Taneja KL, Singer RH, Housman DE (1997) Expansion of a CUG trinucleotide repeat in the 3' untranslated region of myotonic dystrophy protein kinase transcripts results in nuclear retention of transcripts. Proc Natl Acad Sci USA 94:7388–7393
84. Dawson SJ, Wiman B, Hamsten A, Green F, Humphries S, Henney AM (1993) The two allele sequences of a common polymorphism in the promoter of the plasminogen activator inhibitor-1 (PAI-1) gene respond differently to interleukin-1 in HepG2 cells. J Biol Chem 268:10739–10745
85. De Gobbi M, Viprakasit V, Hughes JR, Fisher C, Buckle VJ, Ayyub H, Gibbons RJ, Vernimmen D, Yoshinaga Y, de Jong P, Cheng JF, Rubin EM, Wood WG, Bowden D, Higgs DR (2006) A regulatory SNP causes a human genetic disease by creating a new transcriptional promoter. Science 312:1215–1217
86. Deininger PL, Batzer MA (1999) Alu repeats and human disease. Mol Genet Metab 67:183–193
87. De Klein A, Riegman PH, Bijlsma EK, Heldoorn A, Muijtjens M, den Bakker MA, Avezaat CJ, Zwartoff EC (1998) A G–>A transition creates a branch point sequence and activation of a cryptic exon, resulting in the hereditary disorder neurofibromatosis 2. Hum Mol Genet 7:393–398
88. de Kok YJ, Vossenaar ER, Cremers CW, Dahl N, Laporte J, Hu LJ, Lacombe D, Fischel-Ghodsian N, Friedman RA, Parnes LS, Thorpe P, Bitner-Glindzicz M, Pander HJ, Heilbronner H, Graveline J, den Dunnen JT, Brunner HG, Ropers HH, Cremers FP (1996) Identification of a hot spot for microdeletions in patients with X-linked deafness type 3 (DFN3) 900 kb proximal to the DFN3 gene POU3F4. Hum Mol Genet 5:1229–1235
89. Delattre O, Zucman J, Plougastel B, Desmaze C, Melot T, Peter M, Kovar H, Joubert I, de Jong P, Rouleau G et al (1992) Gene fusion with an ETS DNA-binding domain caused by chromosome translocation in human tumours. Nature 359:162–165
90. de Mollerat XJ, Gurrieri F, Morgan CT, Sangiorgi E, Everman DB, Gaspari P, Amiel J, Bamshad MJ, Lyle R, Blouin JL, Allanson JE, Le Marec B, Wilson M, Braverman NE, Radhakrishna U, Delozier-Blanchet C, Abbott A, Elghouzzi V, Antonarakis S, Stevenson RE, Munnich A, Neri G, Schwartz CE (2003) A genomic rearrangement resulting in a tandem duplication is associated with split hand-split foot malformation 3 (SHFM3) at 10q24. Hum Mol Genet 12:1959–1971
91. den Dunnen JT, Antonarakis SE (2001) Nomenclature for the description of human sequence variations. Hum Genet 109:121–124
92. den Dunnen JT, Bakker E, Breteler EG, Pearson PL, van Ommen GJ (1987) Direct detection of more than 50% of the Duchenne muscular dystrophy mutations by field inversion gels. Nature 329:640–642
93. Dermitzakis ET, Reymond A, Lyle R, Scamuffa N, Ucla C, Deutsch S, Stevenson BJ, Flegel V, Bucher P, Jongeneel CV, Antonarakis SE (2002) Numerous potentially functional but non-genic conserved sequences on human chromosome 21. Nature 420:578–582
94. Dermitzakis ET, Reymond A, Antonarakis SE (2005) Conserved non-genic sequences - an unexpected feature of mammalian genomes. Nat Rev Genet 6:151–157
95. Dietz HC, Valle D, Francomano CA, Kendzior RJ Jr, Pyeritz RE, Cutting GR (1993) The skipping of constitutive exons in vivo induced by nonsense mutations. Science 259:680–683
96. Dobyns WB, Filauro A, Tomson BN, Chan AS, Ho AW, Ting NT, Oosterwijk JC, Ober C (2004) Inheritance of most X-linked traits is not dominant or recessive, just X-linked. Am J Med Genet 129:136–143
97. Dombroski B, Mathias S, Nanthakumar E, Scott A, Kazazian H Jr (1991) Isolation of an active human transposable element. Science 254:1805–1808
98. Dorschner MO, Sybert VP, Weaver M, Pletcher BA, Stephens K (2000) NF1 microdeletion breakpoints are clustered at flanking repetitive sequences. Hum Mol Genet 9:35–46
99. Driscoll DJ, Migeon BR (1990) Sex difference in methylation of single-copy genes in human meiotic germ cells: implications for X chromosome inactivation, parental imprinting, and origin of CpG mutations. Somat Cell Mol Genet 16:267–282

100. Driscoll MC, Dobkin CS, Alter BP (1989) Gamma delta beta-thalassemia due to a de novo mutation deleting the 5' beta-globin gene activation-region hypersensitive sites. Proc Natl Acad Sci USA 86:7470–7474
101. Economou EP, Bergen AW, Warren AC, Antonarakis SE (1990) The polydeoxyadenylate tract of Alu repetitive elements is polymorphic in the human genome. Proc Natl Acad Sci USA 87:2951–2954
102. Edelmann L, Pandita RK, Morrow BE (1999) Low-copy repeats mediate the common 3-Mb deletion in patients with velo-cardio-facial syndrome. Am J Hum Genet 64:1076–1086
103. Editorial (2005) A haplotype map of the human genome. Nature 437:1299–1320
104. Efstratiadis A, Posakony JW, Maniatis T, Lawn RM, O'Connell C, Spritz RA, DeRiel JK, Forget BG, Weissman SM, Slightom JL, Blechl AE, Smithies O, Baralle FE, Shoulders CC, Proudfoot NJ (1980) The structure and evolution of the human beta-globin gene family. Cell 21:653–668
105. Eikenboom JC, Vink T, Briet E, Sixma JJ, Reitsma PH (1994) Multiple substitutions in the von Willebrand factor gene that mimic the pseudogene sequence. Proc Natl Acad Sci USA 91:2221–2224
106. Embury SH, Miller JA, Dozy AM, Kan YW, Chan V, Todd D (1980) Two different molecular organizations account for the single alpha-globin gene of the alpha-thalassemia-2 genotype. J Clin Invest 66:1319–1325
107. Erickson RP (2003) Somatic gene mutation and human disease other than cancer. Mutat Res 543:125–136
108. Eyal N, Wilder S, Horowitz M (1990) Prevalent and rare mutations among Gaucher patients. Gene 96:277–283
109. Fantes J, Redeker B, Breen M, Boyle S, Brown J, Fletcher J, Jones S, Bickmore W, Fukushima Y, Mannens M, Danes S, van Heyningen V, Hanson I (1995) Aniridia-associated cytogenetic rearrangements suggest that a position effect may cause the mutant phenotype. Hum Mol Genet 4:415–422
110. Faustino NA, Cooper TA (2003) Pre-mRNA splicing and human disease. Genes Dev 17:419–437
111. Fellermann K, Stange DE, Schaeffeler E, Schmalzl H, Wehkamp J, Bevins CL, Reinisch W, Teml A, Schwab M, Lichter P, Radlwimmer B, Stange EF (2006) A chromosome 8 gene-cluster polymorphism with low human beta-defensin 2 gene copy number predisposes to Crohn disease of the colon. Am J Hum Genet 79:439–448
112. Ferrer-Costa C, Orozco M, de la Cruz X (2002) Characterization of disease-associated single amino acid polymorphisms in terms of sequence and structure properties. J Mol Biol 315:771–786
113. Fishel R, Lescoe MK, Rao MR, Copeland NG, Jenkins NA, Garber J, Kane M, Kolodner R (1993) The human mutator gene homolog MSH2 and its association with hereditary nonpolyposis colon cancer. Cell 75:1027–1038
114. Flint J, Harding RM, Clegg JB, Boyce AJ (1993) Why are some genetic diseases common? Distinguishing selection from other processes by molecular analysis of globin gene variants. Hum Genet 91:91–117
115. Francke U (1999) Williams-Beuren syndrome: genes and mechanisms. Hum Mol Genet 8:1947–1954
116. Frank SA, Nowak MA (2004) Problems of somatic mutation and cancer. Bioessays 26:291–299
117. Frisch A, Colombo R, Michaelovsky E, Karpati M, Goldman B, Peleg L (2004) Origin and spread of the 1278insTATC mutation causing Tay-Sachs disease in Ashkenazi Jews: genetic drift as a robust and parsimonious hypothesis. Hum Genet 114:366–376
118. Frischmeyer PA, Dietz HC (1999) Nonsense-mediated mRNA decay in health and disease. Hum Mol Genet 8:1893–1900
119. Fu Y-H, Kuhl D, Pizzuti A, Pieretti M, Sutcliffe JS, Richards CS, Verkerk AJMH, Holden J, Fenwick RJ, Warren ST, Oostra BA, Nelson DL, Caskey CT (1991) Variation of the CGG repeat at the Fragile X site results in genetic instability: resolution of the Sherman paradox. Cell 67:1047–1058
120. Fullerton SM, Clark AG, Weiss KM, Nickerson DA, Taylor SL, Stengard JH, Salomaa V, Vartiainen E, Perola M, Boerwinkle E, Sing CF (2000) Apolipoprotein E variation at the sequence haplotype level: implications for the origin and maintenance of a major human polymorphism. Am J Hum Genet 67:881–900
121. Gabellini D, Green MR, Tupler R (2002) Inappropriate Gene Activation in FSHD. A repressor complex binds a chromosomal repeat deleted in dystrophic muscle. Cell 110:339–348
122. Gabriel SE, Brigman KN, Koller BH, Boucher RC, Stutts MJ (1994) Cystic fibrosis heterozygote resistance to cholera toxin in the cystic fibrosis mouse model. Science 266:107–109
123. Gao L, Zhang J (2003) Why are some human disease-associated mutations fixed in mice? Trends Genet 19:678–681
124. Gehring NH, Frede U, Neu-Yilik G, Hundsdoerfer P, Vetter B, Hentze MW, Kulozik AE (2001) Increased efficiency of mRNA 3' end formation: a new genetic mechanism contributing to hereditary thrombophilia. Nat Genet 28:389–392
125. Girelli D, Corrocher R, Bisceglia L, Olivieri O, De Franceschi L, Zelante L, Gasparini P (1995) Molecular basis for the recently described hereditary hyperferritinemia-cataract syndrome: a mutation in the iron-responsive element of ferritin L-subunit gene (the "Verona mutation"). Blood 86:4050–4053
126. Gonzalez E, Kulkarni H, Bolivar H, Mangano A, Sanchez R, Catano G, Nibbs RJ, Freedman BI, Quinones MP, Bamshad MJ, Murthy KK, Rovin BH, Bradley W, Clark RA, Anderson SA, O'Connell RJ, Agan BK, Ahuja SS, Bologna R, Sen L, Dolan MJ, Ahuja SK (2005) The influence of CCL3L1 gene-containing segmental duplications on HIV-1/AIDS susceptibility. Science 307:1434–1440
127. Goossens M, Dozy AM, Embury SH, Zachariades Z, Hadjiminas MG, Stamatoyannopoulos G, Kan YW (1980) Triplicated alpha-globin loci in humans. Proc Natl Acad Sci USA 77:518–521
128. Goriely A, McVean GA, Rojmyr M, Ingemarsson B, Wilkie AO (2003) Evidence for selective advantage of pathogenic FGFR2 mutations in the male germ line. Science 301:643–646
129. Goriely A, McVean GA, van Pelt AM, O'Rourke AW, Wall SA, de Rooij DG, Wilkie AO (2005) Gain-of-function amino acid substitutions drive positive selection of FGFR2 mutations in human spermatogonia. Proc Natl Acad Sci USA 102:6051–6056

130. Gorlach A, Lee PL, Roesler J, Hopkins PJ, Christensen B, Green ED, Chanock SJ, Curnutte JT (1997) A p47-phox pseudogene carries the most common mutation causing p47-phox- deficient chronic granulomatous disease. J Clin Invest 100:1907–1918
131. Gorlov IP, Gorlova OY, Frazier ML, Amos CI (2003) Missense mutations in hMLH1 and hMSH2 are associated with exonic splicing enhancers. Am J Hum Genet 73:1157–1161
132. Green H, Djian P (1992) Consecutive actions of different gene-altering mechanisms in the evolution of involucrin. Mol Biol Evol 9:977–1017
133. Green P, Ewing B, Miller W, Thomas PJ, Green ED (2003) Transcription-associated mutational asymmetry in mammalian evolution. Nat Genet 33:514–517
134. Greenblatt MS, Grollman AP, Harris CC (1996) Deletions and insertions in the p53 tumor suppressor gene in human cancers: confirmation of the DNA polymerase slippage/misalignment model. Cancer Res 56:2130–2136
135. Gregersen N, Bross P, Jorgensen MM, Corydon TJ, Andresen BS (2000) Defective folding and rapid degradation of mutant proteins is a common disease mechanism in genetic disorders. J Inherit Metab Dis 23:441–447
136. Grimm T, Meng G, Liechti-Gallati S, Bettecken T, Muller CR, Muller B (1994) On the origin of deletions and point mutations in Duchenne muscular dystrophy: most deletions arise in oogenesis and most point mutations result from events in spermatogenesis. J Med Genet 31:183–186
137. Grosveld F, van Assendelft GB, Greaves DR, Kollias G (1987) Position-independent, high-level expression of the human beta-globin gene in transgenic mice. Cell 51:975–985
138. Guioli S, Incerti B, Zanaria E, Bardoni B, Franco B, Taylor K, Ballabio A, Camerino G (1992) Kallmann syndrome due to a translocation resulting in an X/Y fusion gene. Nat Genet 1:337–340
139. Hall JG (1988) Review and hypotheses: somatic mosaicism: observations related to clinical genetics. Am J Hum Genet 43:355–363
140. Harland M, Mistry S, Bishop DT, Bishop JA (2001) A deep intronic mutation in CDKN2A is associated with disease in a subset of melanoma pedigrees. Hum Mol Genet 10:2679–2686
141. Harper PS, Harley HG, Reardon W, Shaw DJ (1992) Anticipation in myotonic dystrophy: new light on an old problem. Am J Hum Genet 51:10–16
142. Hewitt JE, Lyle R, Clark LN, Vallely EM, Wright TJ, Wijmenga C, van Deutekom JCT, Francis F, Sharpe PT, Hofker M, Frants RR, Williamson R (1994) Analysis of the tandem repeat locus D4Z4 associated with facioscapulohumeral muscular dystrophy. Hum Mol Genet 3:1287–1295
143. Higuchi M, Wong C, Kochhan L, Olek K, Aronis S, Kasper CK, Kazazian HH Jr, Antonarakis SE (1990) Characterization of mutations in the factor VIII gene by direct sequencing of amplified genomic DNA. Genomics 6:65–71
144. Hinds DA, Stuve LL, Nilsen GB, Halperin E, Eskin E, Ballinger DG, Frazer KA, Cox DR (2005) Whole-genome patterns of common DNA variation in three human populations. Science 307:1072–1079
145. Hinds DA, Kloek AP, Jen M, Chen X, Frazer KA (2006) Common deletions and SNPs are in linkage disequilibrium in the human genome. Nat Genet 38:82–85
146. Hirschhorn JN, Lohmueller K, Byrne E, Hirschhorn K (2002) A comprehensive review of genetic association studies. Genet Med 4:45–61
147. Hitchins M, Williams R, Cheong K, Halani N, Lin VA, Packham D, Ku S, Buckle A, Hawkins N, Burn J, Gallinger S, Goldblatt J, Kirk J, Tomlinson I, Scott R, Spigelman A, Suter C, Martin D, Suthers G, Ward R (2005) MLH1 germline epimutations as a factor in hereditary nonpolyposis colorectal cancer. Gastroenterology 129:1392–1399
148. Ho PJ, Rochette J, Fisher CA, Wonke B, Jarvis MK, Yardumian A, Thein SL (1996) Moderate reduction of beta-globin gene transcript by a novel mutation in the 5' untranslated region: a study of its interaction with other genotypes in two families. Blood 87:1170–1178
149. Hogenauer C, Santa Ana CA, Porter JL, Millard M, Gelfand A, Rosenblatt RL, Prestidge CB, Fordtran JS (2000) Active intestinal chloride secretion in human carriers of cystic fibrosis mutations: an evaluation of the hypothesis that heterozygotes have subnormal active intestinal chloride secretion. Am J Hum Genet 67:1422–1427
150. Housman D (1995) Gain of glutamines, gain of function? Nat Genet 10:3–4
151. Hu XY, Ray PN, Murphy EG, Thompson MW, Worton RG (1990) Duplicational mutation at the Duchenne muscular dystrophy locus: its frequency, distribution, origin, and phenotypegenotype correlation. Am J Hum Genet 46:682–695
152. Huang H, Winter EE, Wang H, Weinstock KG, Xing H, Goodstadt L, Stenson PD, Cooper DN, Smith D, Alba MM, Ponting CP, Fechtel K (2004) Evolutionary conservation and selection of human disease gene orthologs in the rat and mouse genomes. Genome Biol 5:R47
153. Hurles M (2005) How homologous recombination generates a mutable genome. Hum Genomics 2:179–186
154. Hurst LD, Ellegren H (1998) Sex biases in the mutation rate. Trends Genet 14:446–452
155. Hutton M, Lendon CL, Rizzu P, Baker M, Froelich S, Houlden H, Pickering-Brown S et al (1998) Association of missense and 5'-splice-site mutations in tau with the inherited dementia FTDP-17. Nature 393:702–705
156. Iafrate AJ, Feuk L, Rivera MN, Listewnik ML, Donahoe PK, Qi Y, Scherer SW, Lee C (2004) Detection of large-scale variation in the human genome. Nat Genet 36: 949–951
157. Inacio A, Silva AL, Pinto J, Ji X, Morgado A, Almeida F, Faustino P, Lavinha J, Liebhaber SA, Romao L (2004) Nonsense mutations in close proximity to the initiation codon fail to trigger full nonsense-mediated mRNA decay. J Biol Chem 279:32170–32180
158. Ingram VM (1956) A specific chemical difference between the globins of normal human and sickle-cell anaemia haemoglobin. Nature 178:792–794
159. Inoue I, Nakajima T, Williams CS, Quackenbush J, Puryear R, Powers M, Cheng T, Ludwig EH, Sharma AM, Hata A, Jeunemaitre X, Lalouel JM (1997) A nucleotide substitution in the promoter of human angiotensinogen is associated with essential hypertension and affects basal transcription in vitro. J Clin Invest 99:1786–1797

160. Inoue K, Khajavi M, Ohyama T, Hirabayashi S, Wilson J, Reggin JD, Mancias P, Butler IJ, Wilkinson MF, Wegner M, Lupski JR (2004) Molecular mechanism for distinct neurological phenotypes conveyed by allelic truncating mutations. Nat Genet 36:361–369
161. Ionov Y, Peinado MA, Malkhosyan S, Shibata D, Perucho M (1993) Ubiquitous somatic mutations in simple repeated sequences reveal a new mechanism for colonic carcinogenesis. Nature 363:558–561
162. Jacobson EM, Concepcion E, Oashi T, Tomer Y (2005) A Graves' disease-associated Kozak sequence single-nucleotide polymorphism enhances the efficiency of CD40 gene translation: a case for translational pathophysiology. Endocrinology 146:2684–2691
163. Jacquemont S, Hagerman RJ, Leehey M, Grigsby J, Zhang L, Brunberg JA, Greco C, Des Portes V, Jardini T, Levine R, Berry-Kravis E, Brown WT, Schaeffer S, Kissel J, Tassone F, Hagerman PJ (2003) Fragile X premutation tremor/ataxia syndrome: molecular, clinical, and neuroimaging correlates. Am J Hum Genet 72:869–878
164. Jeffreys AJ, Wilson V, Thein SL (1985) Hypervariable 'minisatellite' regions in human DNA. Nature 314:67–73
165. Jeffreys AJ, Neumann R, Wilson V (1990) Repeat unit sequence variation in minisatellites: a novel source of DNA polymorphism for studying variation and mutation by single molecule analysis. Cell 60:473–485
166. Jego N, Thomas G, Hamelin R (1993) Short direct repeats flanking deletions, and duplicating insertions in p53 gene in human cancers. Oncogene 8:209–213
167. Jennings MW, Jones RW, Wood WG, Weatherall DJ (1985) Analysis of an inversion within the human beta globin gene cluster. Nucleic Acids Res 13:2897–2907
168. Ji Y, Eichler EE, Schwartz S, Nicholls RD (2000) Structure of chromosomal duplicons and their role in mediating human genomic disorders. Genome Res 10:597–610
169. Juyal RC, Figuera LE, Hauge X, Elsea SH, Lupski JR, Greenberg F, Baldini A, Patel PI (1996) Molecular analyses of 17p11.2 deletions in 62 Smith-Magenis syndrome patients. Am J Hum Genet 58:998–1007
170. Kajiwara K, Berson EL, Dryja TP (1994) Digenic retinitis pigmentosa due to mutations at the unlinked peripherin/RDS and ROM1 loci. Science 264:1604–1608
171. Kang S, Ohshima K, Jaworski A, Wells RD (1996) CTG triplet repeats from the myotonic dystrophy gene are expanded in Escherichia coli distal to the replication origin as a single large event. J Mol Biol 258:543–547
172. Karathanasis SK, Ferris E, Haddad IA (1987) DNA inversion within the apolipoproteins AI/CIII/AIV-encoding gene cluster of certain patients with premature atherosclerosis. Proc Natl Acad Sci USA 84:7198–7202
173. Katsanis N, Ansley SJ, Badano JL, Eichers ER, Lewis RA, Hoskins BE, Scambler PJ, Davidson WS, Beales PL, Lupski JR (2001) Triallelic inheritance in Bardet-Biedl syndrome, a Mendelian recessive disorder. Science 293:2256–2259
174. Kazazian HH Jr (1998) Mobile elements and disease. Curr Opin Genet Dev 8:343–350
175. Kazazian HH Jr, Wong C, Youssoufian H, Scott AF, Phillips DG, Antonarakis SE (1988) Haemophilia A resulting from *de novo* insertion of L1 sequences represents a novel mechanism for mutation in man. Nature 332:164–166
176. Kelly RJ, Rouquier S, Giorgi D, Lennon GG, Lowe JB (1995) Sequence and expression of a candidate for the human Secretor blood group alpha(1, 2)fucosyltransferase gene (FUT2). Homozygosity for an enzyme-inactivating nonsense mutation commonly correlates with the non-secretor phenotype. Biol Chem 270:4640–4649
177. Kerlin BA, Yan SB, Isermann BH, Brandt JT, Sood R, Basson BR, Joyce DE, Weiler H, Dhainaut JF (2003) Survival advantage associated with heterozygous factor V Leiden mutation in patients with severe sepsis and in mouse endotoxemia. Blood 102:3085–3092
178. Ketterling RP, Ricke DO, Wurster MW, Sommer SS (1993) Deletions with inversions: report of a mutation and review of the literature. Hum Mutat 2:53–57
179. Kim HJ, Nam SH, Kim HJ, Park HS, Ryoo HM, Kim SY, Cho TJ, Kim SG, Bae SC, Kim IS, Stein JL, van Wijnen AJ, Stein GS, Lian JB, Choi JY (2006) Four novel RUNX2 mutations including a splice donor site result in the cleidocranial dysplasia phenotype. J Cell Physiol 207:114–122
180. Kleinjan DJ, van Heyningen V (1998) Position effect in human genetic disease. Hum Mol Genet 7:1611–1618
181. Kleinjan DA, van Heyningen V (2005) Long-range control of gene expression: emerging mechanisms and disruption in disease. Am J Hum Genet 76:8–32
182. Knebelmann B, Forestier L, Drouot L, Quinones S, Chuet C, Benessy F, Saus J, Antignac C (1995) Splice-mediated insertion of an Alu sequence in the COL4A3 mRNA causing autosomal recessive Alport syndrome. Hum Mol Genet 4:675–679
183. Koivisto UM, Palvimo JJ, Janne OA, Kontula K (1994) A single-base substitution in the proximal Sp1 site of the human low density lipoprotein receptor promoter as a cause of heterozygous familial hypercholesterolemia. Proc Natl Acad Sci USA 91:10526–10530
184. Kolomietz E, Meyn MS, Pandita A, Squire JA (2002) The role of Alu repeat clusters as mediators of recurrent chromosomal aberrations in tumors. Genes Chromosomes Cancer 35:97–112
185. Kondrashov AS (2003) Direct estimates of human per nucleotide mutation rates at 20 loci causing Mendelian diseases. Hum Mutat 21:12–27
186. Kondrashov AS, Rogozin IB (2004) Context of deletions and insertions in human coding sequences. Hum Mutat 23:177–185
187. Kondrashov FA, Ogurtsov AY, Kondrashov AS (2004) Bioinformatical assay of human gene morbidity. Nucleic Acids Res 32:1731–1737
188. Koolen DA, Vissers LE, Pfundt R, de Leeuw N, Knight SJ, Regan R, Kooy RF, Reyniers E, Romano C, Fichera M, Schinzel A, Baumer A, Anderlid BM, Schoumans J, Knoers NV, van Kessel AG, Sistermans EA, Veltman JA, Brunner HG, de Vries BB (2006) A new chromosome 17q21.31 microdeletion syndrome associated with a common inversion polymorphism. Nat Genet 38:999–1001
189. Kornreich R, Bishop DF, Desnick RJ (1990) a-galactosidase A gene rearrangements causing Fabry disease. J Biol Chem 265:9319–9326
190. Kozak M (1984) Compilation and analysis of sequences upstream from the translational start site in eukaryotic mRNAs. Nucleic Acids Res 12:857–872

191. Kozak M (1991) Structural features in eukaryotic mRNAs that modulate the initiation of translation. J Biol Chem 266:19867–19870
192. Kozak M (2002) Emerging links between initiation of translation and human diseases. Mamm Genome 13:401–410
193. Krawczak M, Cooper DN (1991) Gene deletions causing human genetic disease: mechanisms of mutagenesis and the role of the local DNA sequence environment. Hum Genet 86:425–441
194. Krawczak M, Reiss J, Cooper DN (1992) The mutational spectrum of single base-pair substitutions in mRNA splice junctions of human genes: causes and consequences. Hum Genet 90:41–54
195. Krawczak M, Ball EV, Cooper DN (1998) Neighboring-nucleotide effects on the rates of germ-line single-base-pair substitution in human genes. Am J Hum Genet 63:474–488
196. Krawczak M, Chuzhanova NA, Stenson PD, Johansen BN, Ball EV, Cooper DN (2000) Changes in primary DNA sequence complexity influence the phenotypic consequences of mutations in human gene regulatory regions. Hum Genet 107:362–365
197. Krawczak M, Thomas NS, Hundrieser B, Mort M, Wittig M, Hampe J, Cooper DN (2006) Single base-pair substitutions in exon-intron junctions of human genes: nature, distribution, and consequences for mRNA splicing. Hum Mutat 28:150–158
198. Kunkel TA (1985) The mutational specificity of DNA polymerases-alpha and -gamma during in vitro DNA synthesis. J Biol Chem 260:12866–12874
199. Laken SJ, Petersen GM, Gruber SB, Oddoux C, Ostrer H, Giardiello FM, Hamilton SR, Hampel H, Markowitz A, Klimstra D, Jhanwar S, Winawer S, Offit K, Luce MC, Kinzler KW, Vogelstein B (1997) Familial colorectal cancer in Ashkenazim due to a hypermutable tract in APC. Nat Genet 17:79–83
200. Lakich D, Kazazian HH Jr, Antonarakis SE, Gitschier J (1993) Inversions disrupting the factor VIII gene are a common cause of severe haemophilia A. Nat Genet 5:236–241
201. Lalioti MD, Scott HS, Buresi C, Rossier C, Bottani A, Morris MA, Malafosse A, Antonarakis SE (1997) Dodecamer repeat expansion in cystatin B gene in progressive myoclonus epilepsy. Nature 386:847–851
202. Lalioti MD, Scott HS, Antonarakis SE (1999) Altered spacing of promoter elements due to the dodecamer repeat expansion contributes to reduced expression of the cystatin B gene in EPM1. Hum Mol Genet 8:1791–1798
203. Lander ES, Linton LM, Birren B, Nusbaum C, Zody MC, Baldwin J, Devon K et al (2001) Initial sequencing and analysis of the human genome. Nature 409:860–921
204. Leach FS, Nicolaides NC, Papadopoulos N, Liu B, Jen J, Parsons R, Peltomaki P, Sistonen P, Aaltonen LA, Nystrom-Lahti M et al (1993) Mutations of a mutS homolog in hereditary nonpolyposis colorectal cancer. Cell 75: 1215–1225
205. Lehrman MA, Goldstein JL, Russell DW, Brown MS (1987) Duplication of seven exons in LDL receptor gene caused by Alu-Alu recombination in a subject with familial hypercholesterolemia. Cell 48:827–835
206. Le Marechal C, Masson E, Chen JM, Morel F, Ruszniewski P, Levy P, Ferec C (2006) Hereditary pancreatitis caused by triplication of the trypsinogen locus. Nat Genet 38:1372–1374
207. Lettice LA, Horikoshi T, Heaney SJ, van Baren MJ, van der Linde HC, Breedveld GJ, Joosse M, Akarsu N, Oostra BA, Endo N, Shibata M, Suzuki M, Takahashi E, Shinka T, Nakahori Y, Ayusawa D, Nakabayashi K, Scherer SW, Heutink P, Hill RE, Noji S (2002) Disruption of a long-range cis-acting regulator for Shh causes preaxial polydactyly. Proc Natl Acad Sci USA 99:7548–7553
208. Li X, Scaringe WA, Hill KA, Roberts S, Mengos A, Careri D, Pinto MT, Kasper CK, Sommer SS (2001) Frequency of recent retrotransposition events in the human factor IX gene. Hum Mutat 17:511–519
209. Liebhaber SA, Griese EU, Weiss I, Cash FE, Ayyub H, Higgs DR, Horst J (1990) Inactivation of human alpha-globin gene expression by a de novo deletion located upstream of the alpha-globin gene cluster. Proc Natl Acad Sci USA 87:9431–9435
210. Linton MF, Pierotti V, Young SG (1992) Reading-frame restoration with an apolipoprotein B gene frameshift mutation. Proc Natl Acad Sci USA 89:11431–11435
211. Liquori CL, Ricker K, Moseley ML, Jacobsen JF, Kress W, Naylor SL, Day JW, Ranum LP (2001) Myotonic dystrophy type 2 caused by a CCTG expansion in intron 1 of ZNF9. Science 293:864–867
212. Litt M, Luty JA (1989) A hypervariable microsatellite revealed by in vitro amplification of a dinucleotide repeat within the cardiac muscle actin gene. Am J Hum Genet 44:397–401
213. Liu HX, Cartegni L, Zhang MQ, Krainer AR (2001) A mechanism for exon skipping caused by nonsense or missense mutations in BRCA1 and other genes. Nat Genet 27:55–58
214. Loeb LA, Kunkel TA (1982) Fidelity of DNA synthesis. Annu Rev Biochem 51:429–457
215. Loots GG, Kneissel M, Keller H, Baptist M, Chang J, Collette NM, Ovcharenko D, Plajzer-Frick I, Rubin EM (2005) Genomic deletion of a long-range bone enhancer misregulates sclerostin in Van Buchem disease. Genome Res 15:928–935
216. Lopez-Bigas N, Ouzounis CA (2004) Genome-wide identification of genes likely to be involved in human genetic disease. Nucleic Acids Res 32:3108–3114
217. Majewski J (2003) Dependence of mutational asymmetry on gene-expression levels in the human genome. Am J Hum Genet 73:688–692
218. Mandel JL (1993) Questions of expansion. Nat Genet 4:8–9
219. Maquat LE (2004) Nonsense-mediated mRNA decay: splicing, translation and mRNP dynamics. Nat Rev Mol Cell Biol 5:89–99
220. Markowitz S, Wang J, Myeroff L, Parsons R, Sun L, Lutterbaugh J, Fan RS, Zborowska E, Kinzler KW, Vogelstein B et al (1995) Inactivation of the type II TGF-beta receptor in colon cancer cells with microsatellite instability. Science 268:1336–1338
221. Matsuura T, Yamagata T, Burgess DL, Rasmussen A, Grewal RP, Watase K, Khajavi M, McCall AE, Davis CF, Zu L, Achari M, Pulst SM, Alonso E, Noebels JL, Nelson DL, Zoghbi HY, Ashizawa T (2000) Large expansion of the ATTCT pentanucleotide repeat in spinocerebellar ataxia type 10. Nat Genet 26:191–194

222. McCarroll SA, Hadnott TN, Perry GH, Sabeti PC, Zody MC, Barrett JC, Dallaire S, Gabriel SB, Lee C, Daly MJ, Altshuler DM (2006) Common deletion polymorphisms in the human genome. Nat Genet 38:86–92
223. McNaughton JC, Cockburn DJ, Hughes G, Jones WA, Laing NG, Ray PN, Stockwell PA, Petersen GB (1998) Is gene deletion in eukaryotes sequence-dependent? A study of nine deletion junctions and nineteen other deletion breakpoints in intron 7 of the human dystrophin gene. Gene 222:41–51
224. McVey JH, Michaelides K, Hansen LP, Ferguson-Smith M, Tilghman S, Krumlauf R, Tuddenham EG (1993) A G–>A substitution in an HNF I binding site in the human alpha-fetoprotein gene is associated with hereditary persistence of alpha-fetoprotein (HPAFP). Hum Mol Genet 2:379–384
225. Mead S, Stumpf MP, Whitfield J, Beck JA, Poulter M, Campbell T, Uphill JB, Goldstein D, Alpers M, Fisher EM, Collinge J (2003) Balancing selection at the prion protein gene consistent with prehistoric kurulike epidemics. Science 300:640–643
226. Mehdi H, Manzi S, Desai P, Chen Q, Nestlerode C, Bontempo F, Strom SC, Zarnegar R, Kamboh MI (2003) A functional polymorphism at the transcriptional initiation site in beta2-glycoprotein I (apolipoprotein H) associated with reduced gene expression and lower plasma levels of beta2-glycoprotein I. Eur J Biochem 270: 230–238
227. Millar DS, Lewis MD, Horan M, Newsway V, Easter TE, Gregory JW, Fryklund L, Norin M, Crowne EC, Davies SJ, Edwards P, Kirk J, Waldron K, Smith PJ, Phillips JA 3rd, Scanlon MF, Krawczak M, Cooper DN, Procter AM (2003) Novel mutations of the growth hormone 1 (GH1) gene disclosed by modulation of the clinical selection criteria for individuals with short stature. Hum Mutat 21:424–440
228. Miller MP, Kumar S (2001) Understanding human disease mutations through the use of interspecific genetic variation. Hum Mol Genet 10:2319–2328
229. Miller MP, Parker JD, Rissing SW, Kumar S (2003) Quantifying the intragenic distribution of human disease mutations. Ann Hum Genet 67:567–579
230. Minegishi Y, Coustan-Smith E, Wang YH, Cooper MD, Campana D, Conley ME (1998) Mutations in the human lambda5/14.1 gene result in B cell deficiency and agammaglobulinemia. J Exp Med 187:71–77
231. Mitchell GA, Labuda D, Fontaine G, Saudubray JM, Bonnefont JP, Lyonnet S, Brody LC, Steel G, Obie C, Valle D (1991) Splice-mediated insertion of an Alu sequence inactivates ornithine delta-aminotransferase: a role for Alu elements in human mutation. Proc Natl Acad Sci USA 88:815–819
232. Mockenhaupt FP, Mandelkow J, Till H, Ehrhardt S, Eggelte TA, Bienzle U (2003) Reduced prevalence of Plasmodium falciparum infection and of concomitant anaemia in pregnant women with heterozygous G6PD deficiency. Trop Med Int Health 8:118–124
233. Moi P, Loudianos G, Lavinha J, Murru S, Cossu P, Casu R, Oggiano L, Longinotti M, Cao A, Pirastu M (1992) Delta-thalassemia due to a mutation in an erythroid-specific binding protein sequence 3' to the delta-globin gene. Blood 79:512–516
234. Mooers BH, Logue JS, Berglund JA (2005) The structural basis of myotonic dystrophy from the crystal structure of CUG repeats. Proc Natl Acad Sci USA 102:16626–16631
235. Morell R, Spritz RA, Ho L, Pierpont J, Guo W, Friedman TB, Asher JH Jr (1997) Apparent digenic inheritance of Waardenburg syndrome type 2 (WS2) and autosomal recessive ocular albinism (AROA). Hum Mol Genet 6:659–664
236. Motulsky AG (1995) Jewish diseases and origins. Nat Genet 9:99–101
237. Muragaki Y, Mundlos S, Upton J, Olsen BR (1996) Altered growth and branching patterns in synpolydactyly caused by mutations in HOXD13. Science 272:548–551
238. Muratani K, Hada T, Yamamoto Y, Kaneko T, Shigeto Y, Ohue T, Furuyama J, Higashino K (1991) Inactivation of the cholinesterase gene by Alu insertion: possible mechanism for human gene transposition. Proc Natl Acad Sci USA 88:11315–11319
239. Nathans J, Piantanida TP, Eddy RL, Shows TB, Hogness DS (1986) Molecular genetics of inherited variation in human color vision. Science 232:203–210
240. Naylor JA, Green PM, Rizza CR, Giannelli F (1993) Analysis of factor VIII mRNA reveals defects in everyone of 28 haemophilia A patients. Hum Mol Genet 2:11–17
241. Neitz M, Neitz J, Grishok A (1995) Polymorphism in the number of genes encoding long-wavelength-sensitive cone pigments among males with normal color vision. Vision Res 35:2395–2407
242. Nicholls RD, Fischel-Ghodsian N, Higgs DR (1987) Recombination at the human alpha-globin gene cluster: sequence features and topological constraints. Cell 49:369–378
243. Nichols WC, Ginsburg D (1997) von Willebrand disease. Medicine (Baltimore) 76:1–20
244. Nickerson DA, Taylor SL, Weiss KM, Clark AG, Hutchinson RG, Stengard J, Salomaa V, Vartiainen E, Boerwinkle E, Sing CF (1998) DNA sequence diversity in a 9.7-kb region of the human lipoprotein lipase gene. Nat Genet 19:233–240
245. Nishihara S, Narimatsu H, Iwasaki H, Yazawa S, Akamatsu S, Ando T, Seno T, Narimatsu I (1994) Molecular genetic analysis of the human Lewis histo-blood group system. J Biol Chem 269:29271–29278
246. Oldenburg J, Rost S, El-Maarri O, Leuer M, Olek K, Muller CR, Schwaab R (2000) De novo factor VIII gene intron 22 inversion in a female carrier presents as a somatic mosaicism. Blood 96:2905–2906
247. Orkin SH, Kazazian HH Jr, Antonarakis SE, Ostrer H, Goff SC, Sexton JP (1982) Abnormal RNA processing due to the exon mutation of beta E-globin gene. Nature 300:768–769
248. Orkin SH, Antonarakis SE, Kazazian HH Jr (1984) Base substitution at position -88 in a beta-thalassemic globin gene. Further evidence for the role of distal promoter element ACACCC. J Biol Chem 259:8679–8681
249. Orkin SH, Cheng TC, Antonarakis SE, Kazazian HH Jr (1985) Thalassemia due to a mutation in the cleavage-polyadenylation signal of the human beta-globin gene. EMBO J 4:453–456
250. Oron-Karni V, Filon D, Rund D, Oppenheim A (1997) A novel mechanism generating short deletion/insertions

following slippage is suggested by a mutation in the human alpha2-globin gene. Hum Mol Genet 6:881–885
251. Ostertag EM, Kazazian IIII Jr (2001) Biology of mammalian L1 retrotransposons. Annu Rev Genet 35:501–538
252. Ostrer H (2001) A genetic profile of contemporary Jewish populations. Nat Rev Genet 2:891–898
253. Owen MC, Brennan SO, Lewis JH, Carrell RW (1983) Mutation of antitrypsin to antithrombin. alpha 1-antitrypsin Pittsburgh (358 Met leads to Arg), a fatal bleeding disorder. N Engl J Med 309:694–698
254. Ozisik G, Mantovani G, Achermann JC, Persani L, Spada A, Weiss J, Beck-Peccoz P, Jameson JL (2003) An alternate translation initiation site circumvents an amino-terminal DAX1 nonsense mutation leading to a mild form of X-linked adrenal hypoplasia congenita. J Clin Endocrinol Metab 88:417–423
255. Pagani F, Baralle FE (2004) Genomic variants in exons and introns: identifying the splicing spoilers. Nat Rev Genet 5:389–396
256. Pagani F, Buratti E, Stuani C, Bendix R, Dork T, Baralle FE (2002) A new type of mutation causes a splicing defect in ATM. Nat Genet 30:426–429
257. Pagani F, Raponi M, Baralle FE (2005) Synonymous mutations in CFTR exon 12 affect splicing and are not neutral in evolution. Proc Natl Acad Sci USA 102:6368–6372
258. Pakula AA, Sauer RT (1989) Genetic analysis of protein stability and function. Annu Rev Genet 23:289–310
259. Paoloni-Giacobino A, Rossier C, Papasavvas MP, Antonarakis SE (2001) Frequency of replication/transcription errors in (A)/(T) runs of human genes. Hum Genet 109:40–47
260. Papadopoulos N, Nicolaides NC, Wei YF, Ruben SM, Carter KC, Rosen CA, Haseltine WA, Fleischmann RD, Fraser CM, Adams MD et al (1994) Mutation of a mutL homolog in hereditary colon cancer. Science 263:1625–1629
261. Parchi P, Petersen RB, Chen SG, Autilio-Gambetti L, Capellari S, Monari L, Cortelli P, Montagna P, Lugaresi E, Gambetti P (1998) Molecular pathology of fatal familial insomnia. Brain Pathol 8:539–548
262. Pascoe L, Jeunemaitre X, Lebrethon MC, Curnow KM, Gomez-Sanchez CE, Gasc JM, Saez JM, Corvol P (1995) Glucocorticoid-suppressible hyperaldosteronism and adrenal tumors occurring in a single French pedigree. J Clin Invest 96:2236–2246
263. Patel PI, Roa BB, Welcher AA, Schoener-Scott R, Trask BJ, Pentao L, Snipes GJ, Garcia CA, Francke U, Shooter EM, Lupski JR, Suter U (1992) The gene for the peripheral myelin protein PMP-22 is a candidate for Charcot-Marie-Tooth disease type 1A. Nat Genet 1:159–165
264. Pearson CE, Nichol Edamura K, Cleary JD (2005) Repeat instability: mechanisms of dynamic mutations. Nat Rev Genet 6:729–742
265. Pentao L, Wise CA, Chinault AC, Patel PI, Lupski JR (1992) Charcot-Marie-Tooth type 1A duplication appears to arise from recombination at repeat sequences flanking the 1.5 Mb monomer unit. Nat Genet 2:292–300
266. Perkins AC, Sharpe AH, Orkin SH (1995) Lethal beta-thalassaemia in mice lacking the erythroid CACCC-transcription factor EKLF. Nature 375:318–322
267. Petrov A, Pirozhkova I, Carnac G, Laoudj D, Lipinski M, Vassetzky YS (2006) Chromatin loop domain organization within the 4q35 locus in facioscapulohumeral dystrophy patients versus normal human myoblasts. Proc Natl Acad Sci USA 103:6982–6987
268. Pfeifer D, Kist R, Dewar K, Devon K, Lander ES, Birren B, Korniszewski L, Back E, Scherer G (1999) Campomelic dysplasia translocation breakpoints are scattered over 1 Mb proximal to SOX9: evidence for an extended control region. Am J Hum Genet 65:111–124
269. Pier GB (2000) Role of the cystic fibrosis transmembrane conductance regulator in innate immunity to Pseudomonas aeruginosa infections. Proc Natl Acad Sci USA 97:8822–8828
270. Pirastu M, Saglio G, Chang JC, Cao A, Kan YW (1984) Initiation codon mutation as a cause of alpha thalassemia. J Biol Chem 259:12315–12317
271. Poort SR, Rosendaal FR, Reitsma PH, Bertina RM (1996) A common genetic variation in the 3'-untranslated region of the prothrombin gene is associated with elevated plasma prothrombin levels and an increase in venous thrombosis. Blood 88:3698–3703
272. Rabbitts TH (1994) Chromosomal translocations in human cancer. Nature 372:143–149
273. Rebbeck TR (1997) Molecular epidemiology of the human glutathione S-transferase genotypes GSTM1 and GSTT1 in cancer susceptibility. Cancer Epidemiol Biomarkers Prev 6:733–743
274. Redon R, Ishikawa S, Fitch KR, Feuk L, Perry GH, Andrews TD, Fiegler H et al (2006) Global variation in copy number in the human genome. Nature 444:444–454
275. Reich DE, Schaffner SF, Daly MJ, McVean G, Mullikin JC, Higgins JM, Richter DJ, Lander ES, Altshuler D (2002) Human genome sequence variation and the influence of gene history, mutation and recombination. Nat Genet 32:135–142
276. Reiter LT, Hastings PJ, Nelis E, De Jonghe P, Van Broeckhoven C, Lupski JR (1998) Human meiotic recombination products revealed by sequencing a hotspot for homologous strand exchange in multiple HNPP deletion patients. Am J Hum Genet 62:1023–1033
277. Rigat B, Hubert C, Alhenc-Gelas F, Cambien F, Corvol P, Soubrier F (1990) An insertion/deletion polymorphism in the angiotensin I-converting enzyme gene accounting for half the variance of serum enzyme levels. J Clin Invest 86:1343–1346
278. Risch N, Tang H, Katzenstein H, Ekstein J (2003) Geographic distribution of disease mutations in the Ashkenazi Jewish population supports genetic drift over selection. Am J Hum Genet 72:812–822
279. Rosenthal A, Jouet M, Kenwrick S (1992) Aberrant splicing of neural cell adhesion molecule L1 mRNA in a family with X-linked hydrocephalus. Nat Genet 2:107–112
280. Rossiter JP, Young M, Kimberland ML, Hutter P, Ketterling RP, Gitschier J, Horst J, Morris MA, Schaid DJ, de Moerloose P, Sommer SS, Kazazian HH, Antonarakis SE (1994) Factor VIII gene inversions causing severe haemophilia A originate almost exclusively in male germ cells. Hum Mol Genet 3:1035–1039
281. Roth DB, Wilson JH (1986) Nonhomologous recombination in mammalian cells: role for short sequence homologies in the joining reaction. Mol Cell Biol 6:4295–4304

282. Rousseau F, Heitz D, Mandel JL (1992) The unstable and methylatable mutations causing the fragile X syndrome. Hum Mutat 1:91–96
283. Rovelet-Lecrux A, Hannequin D, Raux G, Le Meur N, Laquerriere A, Vital A, Dumanchin C, Feuillette S, Brice A, Vercelletto M, Dubas F, Frebourg T, Campion D (2006) APP locus duplication causes autosomal dominant early-onset Alzheimer disease with cerebral amyloid angiopathy. Nat Genet 38:24–26
284. Ruwende C, Khoo SC, Snow RW, Yates SN, Kwiatkowski D, Gupta S, Warn P, Allsopp CE, Gilbert SC, Peschu N et al (1995) Natural selection of hemi- and heterozygotes for G6PD deficiency in Africa by resistance to severe malaria. Nature 376:246–249
285. Sachidanandam R, Weissman D, Schmidt SC, Kakol JM, Stein LD, Marth G, Sherry S et al (2001) A map of human genome sequence variation containing 1.42 million single nucleotide polymorphisms. Nature 409:928–933
286. Saiki RK, Scharf S, Faloona F, Mullis KB, Horn GT, Erlich HA, Arnheim N (1985) Enzymatic amplification of beta-globin genomic sequences and restriction site analysis for diagnosis of sickle cell anemia. Science 230:1350–1354
287. Santisteban I, Arredondo-Vega FX, Kelly S, Loubser M, Meydan N, Roifman C, Howell PL, Bowen T, Weinberg KI, Schroeder ML et al (1995) Three new adenosine deaminase mutations that define a splicing enhancer and cause severe and partial phenotypes: implications for evolution of a CpG hotspot and expression of a transduced ADA cDNA. Hum Mol Genet 4:2081–2087
288. Savkur RS, Philips AV, Cooper TA, Dalton JC, Moseley ML, Ranum LP, Day JW (2004) Insulin receptor splicing alteration in myotonic dystrophy type 2. Am J Hum Genet 74:1309–1313
289. Schmucker B, Krawczak M (1997) Meiotic microdeletion breakpoints in the BRCA1 gene are significantly associated with symmetric DNA-sequence elements. Am J Hum Genet 61:1454–1456
290. Schmutte C, Jones PA (1998) Involvement of DNA methylation in human carcinogenesis. Biol Chem 379:377–388
291. Schollen E, Pardon E, Heykants L, Renard J, Doggett NA, Callen DF, Cassiman JJ, Matthijs G (1998) Comparative analysis of the phosphomannomutase genes PMM1, PMM2 and PMM2psi: the sequence variation in the processed pseudogene is a reflection of the mutations found in the functional gene. Hum Mol Genet 7:157–164
292. Schroeder SA, Gaughan DM, Swift M (1995) Protection against bronchial asthma by CFTR delta F508 mutation: a heterozygote advantage in cystic fibrosis. Nat Med 1:703–705
293. Scott HS, Kudoh J, Wattenhofer M, Shibuya K, Berry A, Chrast R, Guipponi M, Wang J, Kawasaki K, Asakawa S, Minoshima S, Younus F, Mehdi SQ, Radhakrishna U, Papasavvas MP, Gehrig C, Rossier C, Korostishevsky M, Gal A, Shimizu N, Bonne-Tamir B, Antonarakis SE (2001) Insertion of beta-satellite repeats identifies a transmembrane protease causing both congenital and childhood onset autosomal recessive deafness. Nat Genet 27:59–63
294. Sebat J, Lakshmi B, Troge J, Alexander J, Young J, Lundin P, Maner S, Massa H, Walker M, Chi M, Navin N, Lucito R, Healy J, Hicks J, Ye K, Reiner A, Gilliam TC, Trask B, Patterson N, Zetterberg A, Wigler M (2004) Large-scale copy number polymorphism in the human genome. Science 305:525–528
295. Semenza GL (1994) Transcriptional regulation of gene expression: mechanisms and pathophysiology. Hum Mutat 3:180–199
296. Sethupathy P, Borel C, Gagnebin M, Grant GR, Deutsch S, Elton TS, Hatzigeorgiou AG, Antonarakis SE (2007) Human microRNA-155 on chromosome 21 differentially interacts with its polymorphic target in the AGTR1 3' untranslated region: a mechanism for functional single-nucleotide polymorphisms related to phenotypes. Am J Hum Genet 81(2):405–413
297. Sgourou A, Routledge S, Antoniou M, Papachatzopoulou A, Psiouri L, Athanassiadou A (2004) Thalassaemia mutations within the 5'UTR of the human beta-globin gene disrupt transcription. Br J Haematol 124:828–835
298. Shaikh TH, Kurahashi H, Saitta SC, O'Hare AM, Hu P, Roe BA, Driscoll DA, McDonald-McGinn DM, Zackai EH, Budarf ML, Emanuel BS (2000) Chromosome 22-specific low copy repeats and the 22q11.2 deletion syndrome: genomic organization and deletion endpoint analysis. Hum Mol Genet 9:489–501
299. Shapiro LJ, Yen P, Pomerantz D, Martin E, Rolewic L, Mohandas T (1989) Molecular studies of deletions at the human steroid sulfatase locus. Proc Natl Acad Sci USA 86:8477–8481
300. Sharp AJ, Locke DP, McGrath SD, Cheng Z, Bailey JA, Vallente RU, Pertz LM, Clark RA, Schwartz S, Segraves R, Oseroff VV, Albertson DG, Pinkel D, Eichler EE (2005) Segmental duplications and copy-number variation in the human genome. Am J Hum Genet 77:78–88
301. Sharp AJ, Hansen S, Selzer RR, Cheng Z, Regan R, Hurst JA, Stewart H, Price SM, Blair E, Hennekam RC, Fitzpatrick CA, Segraves R, Richmond TA, Guiver C, Albertson DG, Pinkel D, Eis PS, Schwartz S, Knight SJ, Eichler EE (2006) Discovery of previously unidentified genomic disorders from the duplication architecture of the human genome. Nat Genet 38:1038–1042
302. Sharp PA (1987) Splicing of messenger RNA precursors. Science 235:766–771
303. Shaw CJ, Lupski JR (2004) Implications of human genome architecture for rearrangement-based disorders: the genomic basis of disease. Hum Mol Genet 13(Spec No 1):R57–64
304. Shaw MA, Brunetti-Pierri N, Kadasi L, Kovacova V, Van Maldergem L, De Brasi D, Salerno M, Gecz J (2003) Identification of three novel SEDL mutations, including mutation in the rare, non-canonical splice site of exon 4. Clin Genet 64:235–242
305. Shaw-Smith C, Pittman AM, Willatt L, Martin H, Rickman L, Gribble S, Curley R, Cumming S, Dunn C, Kalaitzopoulos D, Porter K, Prigmore E, Krepischi-Santos AC, Varela MC, Koiffmann CP, Lees AJ, Rosenberg C, Firth HV, de Silva R, Carter NP (2006) Microdeletion encompassing MAPT at chromosome 17q21.3 is associated with developmental delay and learning disability. Nat Genet 38:1032–1037
306. Singleton AB, Farrer M, Johnson J, Singleton A, Hague S, Kachergus J, Hulihan M, Peuralinna T, Dutra A, Nussbaum R, Lincoln S, Crawley A, Hanson M, Maraganore D, Adler C, Cookson MR, Muenter M, Baptista M, Miller D,

Blancato J, Hardy J, Gwinn-Hardy K (2003) Alpha-Synuclein locus triplication causes Parkinson's disease. Science 302:841
307. Small K, Iber J, Warren ST (1997) Emerin deletion reveals a common X-chromosome inversion mediated by inverted repeats. Nat Genet 16:96–99
308. Spitz F, Montavon T, Monso-Hinard C, Morris M, Ventruto ML, Antonarakis S, Ventruto V, Duboule D (2002) A t(2;8) balanced translocation with breakpoints near the human HOXD complex causes mesomelic dysplasia and vertebral defects. Genomics 79:493–498
309. Spritz RA, Giebel LB, Holmes SA (1992) Dominant negative and loss of function mutations of the c-kit (mast/stem cell growth factor receptor) proto-oncogene in human piebaldism. Am J Hum Genet 50:261–269
310. Stamatoyannopoulos G (1991) Human hemoglobin switching. Science 252:383
311. Stefansson H, Helgason A, Thorleifsson G, Steinthorsdottir V, Masson G, Barnard J, Baker A et al (2005) A common inversion under selection in Europeans. Nat Genet 37:129–137
312. Stenson PD, Ball EV, Mort M, Phillips AD, Shiel JA, Thomas NS, Abeysinghe S, Krawczak M, Cooper DN (2003) Human Gene Mutation Database (HGMD): 2003 update. Hum Mutat 21:577–581
313. Stoppa-Lyonnet D, Duponchel C, Meo T, Laurent J, Carter PE, Arala-Chaves M, Cohen JH, Dewald G, Goetz J, Hauptmann G et al (1991) Recombinational biases in the rearranged C1-inhibitor genes of hereditary angioedema patients. Am J Hum Genet 49:1055–1062
314. Subramanian S, Kumar S (2006) Evolutionary anatomies of positions and types of disease-associated and neutral amino acid mutations in the human genome. BMC Genomics 7:306
315. Sunyaev S, Ramensky V, Koch I, Lathe W 3rd, Kondrashov AS, Bork P (2001) Prediction of deleterious human alleles. Hum Mol Genet 10:591–597
316. Suter CM, Martin DI, Ward RL (2004) Germline epimutation of MLH1 in individuals with multiple cancers. Nat Genet 36:497–501
317. Terp BN, Cooper DN, Christensen IT, Jorgensen FS, Bross P, Gregersen N, Krawczak M (2002) Assessing the relative importance of the biophysical properties of amino acid substitutions associated with human genetic disease. Hum Mutat 20:98–109
318. Thein SL (2004) Genetic insights into the clinical diversity of beta thalassaemia. Br J Haematol 124:264–274
319. Thomas JW, Touchman JW, Blakesley RW, Bouffard GG, Beckstrom-Sternberg SM, Margulies EH, Blanchette M et al (2003) Comparative analyses of multi-species sequences from targeted genomic regions. Nature 424:788–793
320. Tishkoff SA, Verrelli BC (2003) Patterns of human genetic diversity: implications for human evolutionary history and disease. Annu Rev Genomics Hum Genet 4:293–340
321. Touchon M, Nicolay S, Audit B, BrodieofBrodie EB, d'Aubenton-Carafa Y, Arneodo A, Thermes C (2005) Replication-associated strand asymmetries in mammalian genomes: toward detection of replication origins. Proc Natl Acad Sci USA 102:9836–9841
322.
323. Treisman R, Orkin SH, Maniatis T (1983) Specific transcription and RNA splicing defects in five cloned beta-thalassaemia genes. Nature 302:591–596
324. Tufarelli C, Stanley JA, Garrick D, Sharpe JA, Ayyub H, Wood WG, Higgs DR (2003) Transcription of antisense RNA leading to gene silencing and methylation as a novel cause of human genetic disease. Nat Genet 34:157–165
325. Tuffery-Giraud S, Saquet C, Chambert S, Claustres M (2003) Pseudoexon activation in the DMD gene as a novel mechanism for Becker muscular dystrophy. Hum Mutat 21:608–614
326. Turner C, Killoran C, Thomas NS, Rosenberg M, Chuzhanova NA, Johnston J, Kemel Y, Cooper DN, Biesecker LG (2003) Human genetic disease caused by de novo mitochondrial-nuclear DNA transfer. Hum Genet 112:303–309
327. Tusie-Luna MT, White PC (1995) Gene conversions and unequal crossovers between CYP21 (steroid 21-hydroxylase gene) and CYP21P involve different mechanisms. Proc Natl Acad Sci USA 92:10796–10800
328. Upadhyaya M, Han S, Consoli C, Majounie E, Horan M, Thomas NS, Potts C, Griffiths S, Ruggieri M, von Deimling A, Cooper DN (2004) Characterization of the somatic mutational spectrum of the neurofibromatosis type 1 (NF1) gene in neurofibromatosis patients with benign and malignant tumors. Hum Mutat 23:134–146
329. Van Esch H (2006) The Fragile X premutation: new insights and clinical consequences. Eur J Med Genet 49:1–8
330. van Leeuwen FW, de Kleijn DP, van den Hurk HH, Neubauer A, Sonnemans MA, Sluijs JA, Koycu S, Ramdjielal RD, Salehi A, Martens GJ, Grosveld FG, Peter J, Burbach H, Hol EM (1998) Frameshift mutants of beta amyloid precursor protein and ubiquitin-B in Alzheimer's and Down patients. Science 279:242–247
331. Velagaleti GV, Bien-Willner GA, Northup JK, Lockhart LH, Hawkins JC, Jalal SM, Withers M, Lupski JR, Stankiewicz P (2005) Position effects due to chromosome breakpoints that map approximately 900 Kb upstream and approximately 1.3 Mb downstream of SOX9 in two patients with campomelic dysplasia. Am J Hum Genet 76:652–662
332. Verpy E, Leibovici M, Zwaenepoel I, Liu XZ, Gal A, Salem N, Mansour A, Blanchard S, Kobayashi I, Keats BJ, Slim R, Petit C (2000) A defect in harmonin, a PDZ domain-containing protein expressed in the inner ear sensory hair cells, underlies Usher syndrome type 1C. Nat Genet 26:51–55
333. Vidaud D, Vidaud M, Bahnak BR, Siguret V, Gispert Sanchez S, Laurian Y, Meyer D, Goossens M, Lavergne JM (1993) Haemophilia B due to a de novo insertion of a human-specific Alu subfamily member within the coding region of the factor IX gene. Eur J Hum Genet 1:30–36
334. Vitkup D, Sander C, Church GM (2003) The amino-acid mutational spectrum of human genetic disease. Genome Biol 4:R72
335. Vnencak-Jones CL, Phillips JA 3rd (1990) Hot spots for growth hormone gene deletions in homologous regions outside of Alu repeats. Science 250:1745–1748
336. Vogel F, Motulsky A (1986) Human genetics. Springer, Berlin

337. Vogt G, Chapgier A, Yang K, Chuzhanova N, Feinberg J, Fieschi C, Boisson-Dupuis S et al (2005) Gains of glycosylation comprise an unexpectedly large group of pathogenic mutations. Nat Genet 37:692–700
338. von Ahsen N, Oellerich M (2004) The intronic prothrombin 19911A>G polymorphism influences splicing efficiency and modulates effects of the 20210G>A polymorphism on mRNA amount and expression in a stable reporter gene assay system. Blood 103:586–593
339. Vyas P, Vickers MA, Simmons DL, Ayyub H, Craddock CF, Higgs DR (1992) Cis-acting sequences regulating expression of the human alpha-globin cluster lie within constitutively open chromatin. Cell 69:781–793
340. Wacey AI, Cooper DN, Liney D, Hovig E, Krawczak M (1999) Disentangling the perturbational effects of amino acid substitutions in the DNA-binding domain of p53. Hum Genet 104:15–22
341. Wallace MR, Andersen LB, Saulino AM, Gregory PE, Glover TW, Collins FS (1991) A de novo Alu insertion results in neurofibromatosis type 1. Nature 353:864–866
342. Wang DG, Fan JB, Siao CJ, Berno A, Young P, Sapolsky R, Ghandour G et al (1998) Large-scale identification, mapping, and genotyping of single-nucleotide polymorphisms in the human genome. Science 280:1077–1082
343. Wang Z, Moult J (2001) SNPs, protein structure, and disease. Hum Mutat 17:263–270
344. Warren ST, Nelson DL (1994) Advances in molecular analysis of fragile X syndrome. JAMA 271:536–542
345. Watnick TJ, Gandolph MA, Weber H, Neumann HP, Germino GG (1998) Gene conversion is a likely cause of mutation in PKD1. Hum Mol Genet 7:1239–1243
346. Weatherall DJ, Clegg JB, Higgs DR, Wood WG (2001) The hemoglobinopathies. In: Scriver CR, Beaudet AL, Valle D et al (eds) The metabolic and molecular bases of inherited disease. McGraw-Hill, New York, pp 4571–4636
347. Weber JL, May PE (1989) Abundant class of human DNA polymorphisms which can be typed using the polymerase chain reaction. Am J Hum Genet 44:388–396
348. Wells RD, Warren AC (1998) Genetic instabilities and hereditary neurological disorders. Academic, San Diego
349. Wexler NS, Young AB, Tanzi RE, Travers H, Starosta-Rubinstein S, Penney JB, Snodgrass SR, Shoulson I, Gomez F, Ramos Arroyo MA et al (1987) Homozygotes for Huntington's disease. Nature 326:194–197
350. Williams TN, Mwangi TW, Wambua S, Peto TE, Weatherall DJ, Gupta S, Recker M, Penman BS, Uyoga S, Macharia A, Mwacharo JK, Snow RW, Marsh K (2005) Negative epistasis between the malaria-protective effects of alpha+-thalassemia and the sickle cell trait. Nat Genet 37:1253–1257
351. Williams TN, Wambua S, Uyoga S, Macharia A, Mwacharo JK, Newton CR, Maitland K (2005) Both heterozygous and homozygous alpha+ thalassemias protect against severe and fatal Plasmodium falciparum malaria on the coast of Kenya. Blood 106:368–371
352. Witchel SF, Lee PA, Suda-Hartman M, Trucco M, Hoffman EP (1997) Evidence for a heterozygote advantage in congenital adrenal hyperplasia due to 21-hydroxylase deficiency. J Clin Endocrinol Metab 82:2097–2101
353. Wolf U (1997) Identical mutations and phenotypic variation. Hum Genet 100:305–321
354. Wong C, Dowling CE, Saiki RK, Higuchi RG, Erlich HA, Kazazian HH Jr (1987) Characterization of beta-thalassaemia mutations using direct genomic sequencing of amplified single copy DNA. Nature 330:384–386
355. Woods-Samuels P, Wong C, Mathias SL, Scott AF, Kazazian HH Jr, Antonarakis SE (1989) Characterization of a nondeleterious L1 insertion in an intron of the human factor VIII gene and further evidence of open reading frames in functional L1 elements. Genomics 4:290–296
356. Woods-Samuels P, Kazazian HH Jr, Antonarakis SE (1991) Nonhomologous recombination in the human genome: deletions in the human factor VIII gene. Genomics 10:94–101
357. Woodward K, Kendall E, Vetrie D, Malcolm S (1998) Pelizaeus-Merzbacher disease: identification of Xq22 proteolipid-protein duplications and characterization of breakpoints by interphase FISH. Am J Hum Genet 63:207–217
358. Wyman AR, White R (1980) A highly polymorphic locus in human DNA. Proc Natl Acad Sci USA 77: 6754–6758
359. Yang WS, Nevin DN, Peng R, Brunzell JD, Deeb SS (1995) A mutation in the promoter of the lipoprotein lipase (LPL) gene in a patient with familial combined hyperlipidemia and low LPL activity. Proc Natl Acad Sci USA 92:4462–4466
360. Young M, Inaba H, Hoyer LW, Higuchi M, Kazazian HH Jr, Antonarakis SE (1997) Partial correction of a severe molecular defect in hemophilia A, because of errors during expression of the factor VIII gene. Am J Hum Genet 60:565–573
361. Youssoufian H, Kazazian HH Jr, Phillips DG, Aronis S, Tsiftis G, Brown VA, Antonarakis SE (1986) Recurrent mutations in haemophilia A give evidence for CpG mutation hotspots. Nature 324:380–382
362. Youssoufian H, Antonarakis SE, Aronis S, Tsiftis G, Phillips DG, Kazazian HH Jr (1987) Characterization of five partial deletions of the factor VIII gene. Proc Natl Acad Sci USA 84:3772–3776
363. Youssoufian H, Antonarakis SE, Bell W, Griffin AM, Kazazian HH Jr (1988) Nonsense and missense mutations in hemophilia A: estimate of the relative mutation rate at CG dinucleotides. Am J Hum Genet 42:718–725
364. Yu S, Mangelsdorf M, Hewett D, Hobson L, Baker E, Eyre HJ, Lapsys N, Le Paslier D, Doggett NA, Sutherland GR, Richards RI (1997) Human chromosomal fragile site FRA16B is an amplified AT-rich minisatellite repeat. Cell 88:367–374
365. Zlotogora J (1994) High frequencies of human genetic diseases: founder effect with genetic drift or selection? Am J Med Genet 49:10–13
366. Zlotogora J (1997) Dominance and homozygosity. Am J Med Genet 68:412–416
367. Zlotogora J (1998) Germ line mosaicism. Hum Genet 102:381–386
368. Zschocke J (2003) Phenylketonuria mutations in Europe. Hum Mutat 21:345–356

Human Hemoglobin

11

George P. Patrinos and Stylianos E. Antonarakis

Abstract The hemoglobin molecule can be studied with greater facility than any other human protein. This is because blood can easily be taken from many individuals, hemoglobin is the principal protein of red blood cells, and its extraction does not require complicated biochemical methods. It is therefore not surprising that this protein is the most thoroughly studied and the globin genes the best analyzed in humans. Work on human hemoglobin began with the investigation of sickle cell disease (SCD) in 1910, and in 1925 Cooley and Lee first described a form of severe anemia associated with splenomegaly amd characteristic bone changes. This condition has been known since then as thalassemia – a term derived from the Greek word "$\theta\alpha\lambda\alpha\sigma\sigma\alpha$" (thalassa=sea), referring to the Mediterranean Sea – or Cooley's anemia. Genetically oriented studies of human hemoglobins have proceeded apace, starting with the elucidation of the amino acid sequence and structure of the molecule in the 1960s. The hemoglobin system is currently a paradigm for the understanding of gene action at the molecular level. Hemoglobin research is comparable in human biochemical and molecular genetics to that of research on Drosophila and phage in basic genetics. Most concepts derived from hemoglobin research can be readily applied to other proteins, and it has been possible to teach many conceptual principles of human genetics by means of examples from the hemoglobin system. This chapter summarizes the main aspects of human hemoglobin research that has provided, overall, important insights into: (a) Protein structure and function, (b) Gene structure and expression, (c) Developmental gene regulation, (d) Long-range gene interactions and chromatin structure, (e) Gene evolution, (f) Genotype-phenotype relationships and phenotype modifiers, (g) Mechanisms of action of pathogenic mutations, (h) Polymorphisms and haplotype blocks, (i) Molecular diagnostics of inherited disorders, and (j) Gene therapy of monogenic disorders.

G.P. Patrinos (✉)
Department of Pharmacy, University Campus,
University of Patras, School of Health Sciences, Rion,
26504 Patras, Greece
and
Faculty of Medicine and Health Sciences, MGC-Department
of Cell Biology and Genetics, Erasmus University Medical
Center, P.O. Box 2040, 3000 CA Rotterdam, The Netherlands
e-mail: g.patrinos@erasmusmc.nl; gpatrinos@upatras.gr

S.E. Antonarakis
Department of Genetic Medicine and Development, University
of Geneva Medical School and University Hospitals of Geneva,
1 rue Michel-Servet, 1211 Geneva, Switzerland
e-mail: Stylianos.Antonarakis@unige.ch

Contents

11.1	Introduction...	366
11.2	Historical Perspectives..	366
11.3	Genetics of Hemoglobins..	368
	11.3.1 Hemoglobin Molecules............................	368
	11.3.2 Hemoglobin Genes..................................	369
	11.3.3 Regulatory Elements................................	370
	11.3.4 Molecular Control of Globin Gene Switching...	374
	11.3.5 DNA Polymorphisms at the Globin Genes..................................	377

11.4	Molecular Evolution of the Human Globin Genes	377	11.8	Diagnosis of Hemoglobinopathies	389
				11.8.1 Carrier Screening	389
11.5	Molecular Etiology of Hemoglobinopathies	380		11.8.2 Hematological and Biochemical Methods	391
	11.5.1 Thalassemias and Related Conditions	380		11.8.3 Molecular Diagnostics of Hemoglobinopathies	391
	11.5.2 β-Thalassemia	380			
	11.5.3 Dominantly Inherited β-Thalassemia	382		11.8.4 Prenatal and Preimplantation Genetic Diagnosis of Hemoglobinopathies	392
	11.5.4 δβ-Thalassemias and Hereditary Persistence of Fetal Hemoglobin	383		11.8.5 Genetic Counseling	393
	11.5.5 α-Thalassemia	383	11.9	HbVar Database for Hemoglobin Variants and Thalassemia Mutations	393
	11.5.6 Other Mmutation Types Leading to Hemoglobinopathies	384			
	11.5.7 Hemoglobin Variants	385	11.10	Therapeutic Approaches for the Thalassemias	394
11.6	X-Linked Inherited and Acquired α-Thalassemia	387		11.10.1 Hematopoietic Stem Cell Transplantation	394
	11.6.1 Regulatory SNPs and Antisense RNA Transcription Resultingin α-Thalassemia	388		11.10.2 Pharmacological Reactivation of Fetal Hemoglobin	394
	11.6.2 β-Thalassemia Attributable to Transcription Factor Mutations	388		11.10.3 Pharmacogenomics and Therapeutics of Hemoglobinopathies	395
				11.10.4 Gene Therapy	395
11.7	Population Genetics of Hemoglobin Genes	388	References		396

11.1 Introduction

The hemoglobin molecule can be studied with greater facility than any other human protein. This is because blood can easily be taken from many individuals, hemoglobin is the principal protein of red blood cells, and its extraction does not require complicated biochemical methods. It is therefore not surprising that this protein is the most thoroughly studied and the globin genes the best analyzed in humans. Genetically oriented studies of human hemoglobins have proceeded apace, starting with the elucidation of the amino acid sequence and structure of the molecule in the 1960s. The hemoglobin system is currently a paradigm for the understanding of gene action at the molecular level. Hemoglobin research is comparable in human biochemical and molecular genetics to that of research on *Drosophila* and phage in basic genetics. Most concepts derived from hemoglobin research can be readily applied to other proteins, and it has been possible to teach many conceptual principles of human genetics by means of examples from the hemoglobin system. In particular, hemoglobin research has been instrumental in providing important insights into:

- Protein structure and function
- Gene structure and expression
- Developmental gene regulation
- Long-range gene interactions and chromatin structure
- Gene evolution
- Genotype–phenotype relationships and phenotype modifiers
- Mechanisms of action of pathogenic mutations
- Polymorphisms and haplotype blocks
- Molecular diagnostics of inherited disorders
- Gene therapy of monogenic disorders

11.2 Historical Perspectives

Work on human hemoglobin began with the investigation of sickle cell disease (SCD). In 1910, Herrick observed a peculiar sickle-shaped abnormality of red cell structure in an anemic African-American student [63]. It soon became apparent that this condition is fairly common among African Americans. Affected patients suffer from hemolytic anemia and recurrent episodes of abdominal and musculoskeletal pain. In 1923, Taliaferro and Huck recognized that the condition is hereditary [164]. In 1949 it was shown separately by Neel and Beet that patients with SCD are homozygous for a gene that causes, in heterozygosity, an innocuous condition known as sickle cell trait, which is found in about 8% of the African–American population [105]. In 1925, Cooley and Lee were the first to describe a form of severe anemia associated with splenomegaly and characteristic bone changes [25]. This condition has been known since then as thalassemia or Cooley's anemia [25]. Unlike β-thalassemia, however, the α-thalassemias have taken longer for us to understand, largely because for many years the genomic structure of the α-globin gene cluster, i.e., two α-globin gene copies per haploid genome in normal individuals, was not appreciated. The major clinical syndromes resulting from α-thalassemia (Hb H disease and the Hb Bart's hydrops fetalis syndrome) were first recognized in the mid-1950s and early 1960s through the association of the abnormal hemoglobins (Hb H and Hb Bart's) with hypochromic microcytic anemia in the absence of iron deficiency (for a review see [63]). Identification of β-like globin chain tetramers in patients with these syndromes provided evidence that these conditions result from genetic defects in the α-like globin chains [72].

In 1949, Linus Pauling and his group surmised that a defect of hemoglobin was likely to be the cause of SCD, basing this on evidence indicating that the process of sickling might be intimately associated with the state and the nature of the hemoglobin within the erythrocyte [131]. Therefore, the authors compared the hemoglobins of SCD patients and carriers, and of normal individuals. Using zone electrophoresis, Pauling et al. observed a significant difference between the electrophoretic mobilities of hemoglobin derived from erythrocytes of normal individuals and from those with SCD [131]. In hemoglobin from subjects with the sickle cell trait about 25–40% of the hemoglobin turned out to be identical with that found in SCD, whereas the remainder was indistinguishable from normal. This result was compatible with the genetic data showing that SCD represents the homozygous state of a gene variant for which SCD carriers are heterozygous. Pauling and his co-workers therefore concluded that SCD was indeed the first example of a molecular disorder.

In 1956, Ingram discovered what precisely distinguishes normal from sickle hemoglobin [71]. Using a trypsin-based fingerprinting method, he showed that sickle cell hemoglobin was identical with the normal molecule in all peptides except one. Further analysis showed that sickle cell hemoglobin differed from normal hemoglobin in only one amino acid: glutamic acid was replaced by valine at position 6 of the globin chain. Glutamic acid has two COOH groups and one NH_2 group, whereas valine has only one COOH group, resulting in a charge difference, which explained the electrophoretic differences between normal and sickle hemoglobin.

In the early 1960s, further steps of great importance in hemoglobin research were the establishment and elucidation of the full amino acid sequence of hemoglobin chains and the elucidation of the three-dimensional structure of human hemoglobin [102, 132], which led to the identification of an increasing number of other hemoglobin variants (see also below). Subsequent advances have led to our understanding of structure-function relationships and to detection of various types of mutations, such as deletions and frameshifts. Isolation of the hemoglobin mRNA led to new insights into gene structure and function and opened new paths to the understanding of gene action [138].

Molecular work on the hemoglobins has proceeded at a rapid pace since then. The full DNA sequences of the various hemoglobin genes and their flanking sequences were elucidated in the early 1980s, and the hemoglobin genes and their regulation are probably better understood than any other mammalian gene. Mutations affecting the hemoglobins, particularly the thalassemias [174], have been elucidated, and models for the understanding of gene action at the molecular level have been proposed. In 1982, Orkin et al. demonstrated that a number of sequence variations were linked to specific β-globin gene mutations [111]. These groups of restriction fragment length polymorphisms (RFLPs), termed "haplotypes" (both intergenic and intragenic), provided a first screening approach making it possible to determine which *HBB* gene is mutated [4, 111]. RFLP analysis also made it possible to track a mutant allele for pre- and postnatal diagnosis [14]. At first, the identification of the disease-causing mutation was only possible through the construction of a genomic DNA library from the affected individual, in order to first clone the mutated allele and then determine its nucleotide sequence. Again, many human globin gene mutations have been identified through such approaches [6, 17, 112, 114, 115, 167]. At the same time, in order to provide a shortcut to DNA sequencing, a number of exploratory methods for identifying mutations in patients' DNA were developed. The first methods involved mismatch detection in DNA/DNA or RNA/DNA heteroduplexes [103]. Using these laborious and time-consuming approaches, a number of sequence variations were identified, which made the design of short synthetic oligonucleotides as allele-specific probes for genomic Southern blots possible. This experimental design was quickly implemented for the detection of *HBB* gene mutations [113, 135]. With the advent of the polymerase chain reaction (PCR), characterization of the β-globin gene mutation leading to SCD was possible from a considerably smaller amount of genomic DNA and was also significantly expedited [146, 147]. In addition, the battery of diagnostic tools for globin gene mutation screening was significantly enriched. As with many other genetic disorders, DNA amplification was coupled to a rich repertoire of methodologies, either for the detection of known mutations or for screening for unknown sequence alterations inside the human globin loci [130].

The major discoveries and milestones in hemoglobin research are summarized in Table 11.1.

Table 11.1 Landmark studies and history of human hemoglobin research

Discovery	References
First observation and investigation of sickle cell disease	[62]
First report on thalassemia	[25]
Characterization of sickle cell disease as a molecular disorder	[131]
Identification of the mutation leading to sickle cell disease at the protein level	[72]
Elucidation of hemoglobin's protein sequence and three-dimensional structure	[102, 132]
Elucidation of the full globin genes DNA sequences	[10, 86, 154, 156]
Determination of human β-globin gene cluster haplotypes	[4]
Prenatal diagnosis for sickle cell disease or β-thalassemia using DNA polymorphisms	[14, 75]
Molecular diagnosis of β-thalassemia using allele specific oligonucleotide probes	[112, 135]
Molecular diagnosis of sickle cell disease using restriction site analysis	[146]
Study of developmental regulation of globin gene switching in transgenic mice	[41, 80]
Discovery of the human β-globin Locus Control Region	[56]
Molecular diagnosis of sickle cell disease using PCR-RFLP analysis	[147]
Establishment of the globin gene competition model	[176]
Successful restoration of β-thalassemia phenotype in mice using lentiviral *HBB* gene transfer	[97]
Study of the human β-globin Active Chromatin Hub	[119, 129]

11.3 Genetics of Hemoglobins

11.3.1 Hemoglobin Molecules

Human hemoglobin consists of four globin chains. The general designation of a hemoglobin (Hb) molecule is $\alpha_2\beta_2$ (Hb A), signifying that the four globin chains comprise two pairs of identical polypeptides. Each globin chain carries a heme group, a nonprotein molecule attached at a specific site of the globin molecule. The four globin chains with their respective heme groups constitute the functional hemoglobin molecule that carries oxygen from the lungs to the tissues. A globin chain is either α- or β-*like*, in which case it consists of a string of 141 (α-*like*; ζ-, and α-globin), or 146 amino acids (β-*like*; ε-, γ-, δ-, and β-globin). Most normal human hemoglobins, which are expressed at different developmental stages, have identical α-globin chains, while the non-α (or β-*like*) globin chains differ from each other (see below). The primary DNA and protein sequence of all globin chains is well defined.

The principal hemoglobin of children and adults is Hb A or adult hemoglobin ($\alpha_2\beta_2$). The characteristic subunit of Hb A is the β-globin chain. Additionally, all adults carry a small amount (<3%) of Hb A_2 ($\alpha_2\delta_2$). The δ-globin chains differ in only ten amino acid positions from the β-globin chain. A small amount (<2%) of fetal hemoglobin (Hb F; $\alpha_2\gamma_2$) is also present postnatally in all individuals. The γ-globin chain differs considerably from both the α- and β-globin chains, whereas the α-globin chains of Hb A, Hb A_2, and Hb F are identical.

Several hemoglobins that are characteristic of embryonal and fetal development exist. Hemoglobin molecules of the embryonic erythropoietic stage consist of α-, ζ-, ε-, and γ-globin chains. The ζ- and ε-globin chains disappear after 8–10 weeks of embryonal life (Fig. 11.1a) [178]. The principal hemoglobin of fetal development is Hb F. There are two types of γ-globin chains with very similar properties: those with alanine at position 136 ($^A\gamma$) and those with glycine at the same amino acid position ($^G\gamma$). There is a third type of γ-globin chain, with threonine instead of isoleucine at position 75 in the γ-globin chain; this occurs with a frequency ranging up to 40% and is polymorphic in human populations [148]. Adult hemoglobin can be demonstrated in fetuses as early as at the 6–8-week stage [178].

The embryonic erythropoietic site in humans is the yolk sac, where the α-, ζ-, and ε-globin chains are produced. The transition from primitive to definitive erythropoiesis coincides with a switch in erythropoietic sites from the yolk sac to the fetal liver, where α- and γ-globin chain synthesis largely occurs. Conversely, α-, β-, and δ-globin chains in childhood and later in adulthood are produced in the bone marrow and spleen (Fig. 11.1b).

11.3.2 Hemoglobin Genes

The amino acid sequence of each of the globin chains is specified by a unique globin gene. The human globin genes are arranged in multigene clusters, namely, the α-*like* and β-*like* globin gene cluster, a frequent type of organization of mammalian genes.

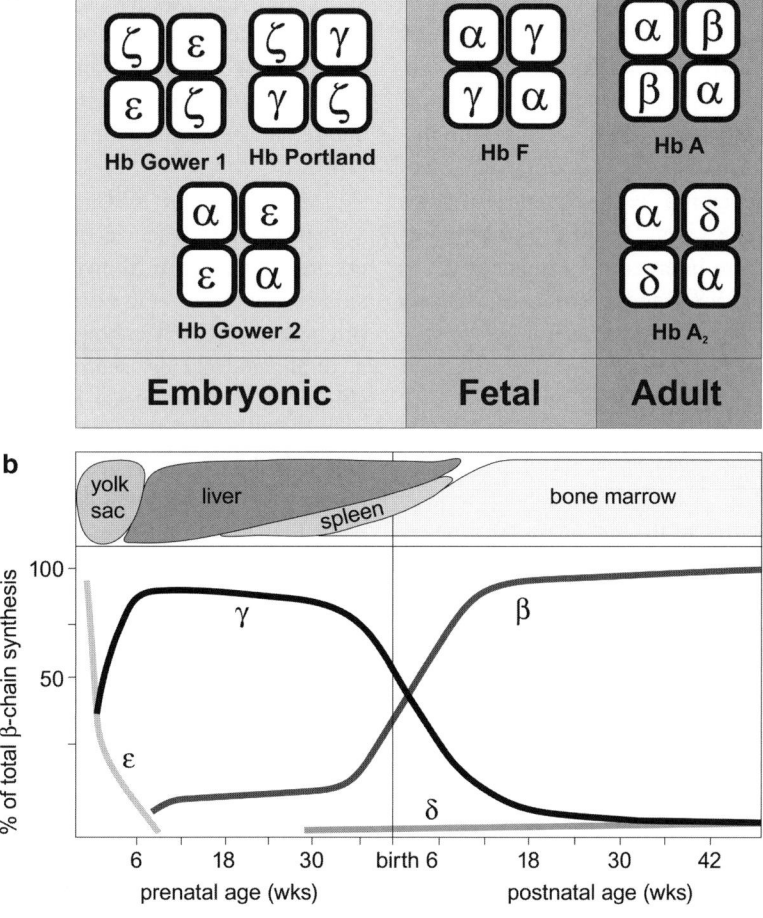

Fig. 11.1 (**a**) Hemoglobin tetramers during the embryonic, fetal and adult developmental stages. (**b**) Globin chain synthesis during development. The various tissues which contribute to hematopoiesis at the various developmental stages are shown above (see also text for details)

The α-globin gene exists in two copies, namely, α2 (*HBA2*)- and α1 (*HBA1*)-globin genes, yielding an identical protein product. The latter, together with the ζ-globin gene (*HBZ*) are resident on chromosome 16. The structure of the α-*like* globin gene cluster [from 5′ (upstream) to 3′ (downstream)] include: the embryonic *HBZ* gene, a *HBZP* pseudogene, two α-globin pseudogenes, namely, *HBAP2* and *HBAP1*, two identical functional α-globin genes, namely, *HBA2* and *HBA1*, and the θ-globin gene of unknown function, (*HBQ*l; Fig. 11.2a) [96]. Similarly, the location of the various genes on the β-globin gene cluster are: the embryonic ε-globin gene (*HBE*), two fetal (γ)-globin genes, namely, $^G\gamma$-(*HBG2*) and $^A\gamma$-(*HBG1*), the ψβ-globin pseudogene (*HBBP*), a δ-globin gene (*HBD*), and a β-globin gene (*HBB*; Fig. 11.2a). The 5′ to 3′ arrangement of the genes in both clusters is in the order of their ontogenetic expression during development. Extensive DNA sequence analysis has been carried out on all hemoglobin genes, and their primary sequence has been fully characterized [10, 24, 86, 154, 156].

All the globin genes have many functional similarities in organization. Three exons or coding sequences code for the unique amino acid sequence of each globin chain. Between exons 1 and 2 and between exons 2 and 3 there are intervening sequences (IVS) or introns known as IVS-I and IVS-II, respectively (Fig. 11.2b).

11.3.3 Regulatory Elements

Transcriptional regulation of the human globin clusters is achieved by proximal and distal *cis*-regulatory elements. Promoter, enhancer, and silencer elements govern the proper developmental expression of individual globin genes. Each proximal regulatory element, whether promoter, enhancer, or silencer, harbors binding sites for erythroid-specific and ubiquitously expressed transcription factors, which interact with other co-factor(s) to regulate gene expression. Each globin gene has a characteristic pattern of transcription factor-binding sites, consisting of a "TATA box," accompanied by two other globin-specific motifs, namely, the "CCAAT box" and the "CACCC box." In addition, major regulatory elements, located at the 5′ end of each cluster, such as the hypersensitive site (HS)-40 in the α-globin gene cluster and the locus control region (LCR) in the β-globin gene cluster, contribute to high-level, tissue-specific, and developmentally controlled expression of the globin genes. The proximal regulatory elements are thought to interact with the LCR in achieving control of individual gene expression at the various developmental stages in a well-defined nuclear territory, termed the "active chromatin hub" (ACH; see also below).

The HBE Gene. The *HBE* promoter has canonical CACCC, CCAAT, and TATA boxes in a 5′>3′ orientation (Fig. 11.2c). Erythroid-specific transcription factor binding, such as GATA-1 and most likely NF-E2, is vital for proper tissue-specific expression of the *HBE* gene [54]. Studies in transgenic mice have shown that the *HBE* gene displays autonomous developmental control, namely, activation and silencing, at the fetal hematopoietic stage. The latter is achieved by a 275-bp silencer element, located 467 bp upstream of *HBE* Cap site. This silencer contains GATA-1- and YY1-binding sites, and when deleted in the context of transgenic mice the *HBE* gene loses its developmental specificity and remains expressed during definitive hematopoiesis [142].

The HBG2 and HBG1 Genes. Like their coding sequences, the sequences of the γ-globin gene promoters are almost identical owing to a recent gene conversion event (see below) and contain (in a 5′>3′ orientation) a CACCC box, two CCAAT boxes within a duplicated 27-bp segment, and the TATA box (Fig. 11.2c). In contrast to the *HBE* promoter and coding regions, in which no mutations have been found [120], both γ-globin gene promoters harbor many mutations, mostly in regulatory motifs, which result in persistent high fetal hemoglobin levels in adults, an inherited condition known as hereditary persistence of fetal hemoglobin (HPFH). These mutants have contributed to the enhancement of our understanding of the molecular mechanisms governing the downregulation of the γ-globin genes in the adult. Although the distal CCAAT box harbors several mutations, leading to HPFH, the proximal CCAAT box is virtually mutation free, probably suggesting that mutations in the proximal CCAAT box are incompatible with fetal life or that these mutations are very mild. Although experimental evidence from transgenic mice suggests that the γ-globin, like the *HBE*, genes silence autonomously during adult life [32], the γ-globin gene silencer region(s) remains to be discovered. Stamatoyannopoulos

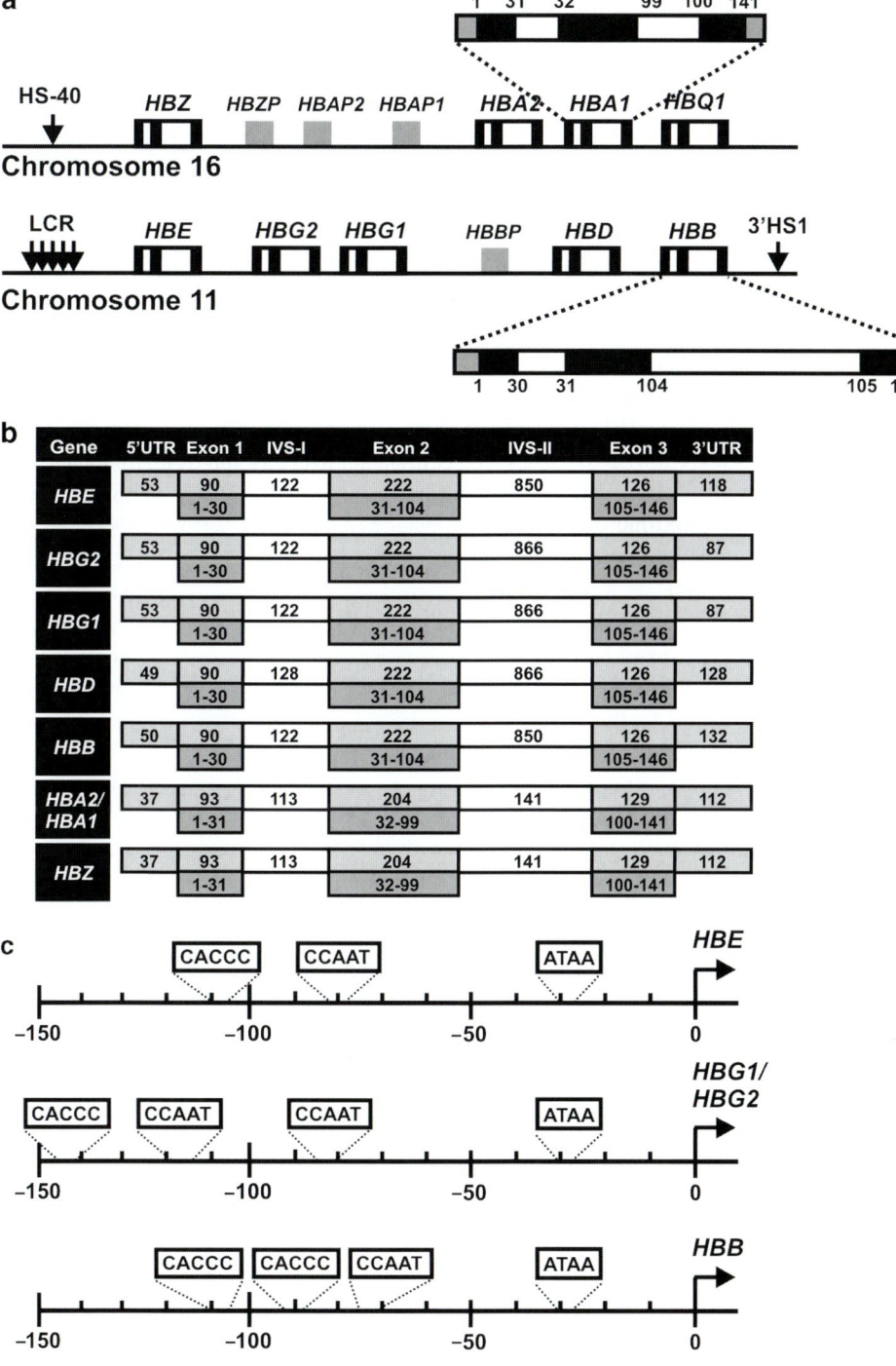

Fig. 11.2 (a) Schematic drawing of the human α-*like* and β-*like* globin gene clusters. Exons and intervening sequences of functional genes are depicted by *black* and *white boxes*, respectively, and pseudogenes as *light gray boxes*. *Dark gray boxes* represent the 5′ and 3′ untranslated regions. *Numbering* corresponds to the first and last amino acid of each exon. (b) Comparison of the sizes of the 5′ and 3′ untranslated regions, exons and intervening sequences of the human globin genes. *Numbers* in the *upper part*, i.e., *light gray and white boxes*, depict the length of each fragment (in bp), while the numbers in the *lower part* (*dark gray boxes*) indicate the amino acid range of each exon. (c) Position of the *cis*-regulatory elements of the human β-*like* globin gene promoters, relative to their Cap site (indicated with an *arrow*)

et al. have demonstrated that a 354-bp region, located −732 bp upstream of the γ-globin genes Cap site, may act as a silencing element and, when deleted, result in the continuing expression of γ-globin genes in adult hematopoiesis [159].

A 750-bp region downstream of the *HBG1* gene has been shown to display enhancer activity, based on cellular expression studies [13]. This fragment harbors eight footprints, three of which are GATA-1-binding motifs [139]. Subsequently, the same fragment has been shown to silence the expression of a reporter gene [91]. Altogether, these data suggest that the function of this "Aγ-enhancer" element remains enigmatic.

The HBB and HBD Genes. The *HBB* gene promoter contains (in a 5′>3′ orientation) two CACCC boxes, a CCAAT box, and a TATA box (Fig. 11.2c). Of the two CACCC boxes, the proximal one appears to play the more important functional role, since naturally occurring mutations in this box, as well as in the TATA box, have been shown to cause mild β-thalassemia. GATA-1 and KLF-1 [or erythroid Krüppel-*like* factor (ELKF)] are the most important erythroid-specific transcription factors that bind to the *HBB* promoter. In particular, targeted deletion of the murine *KLF-1* gene results in a lethal phenotype in the early definitive hematopoietic stage, owing to severe dyserythropoiesis [107]. The human *HBB* gene remains silent during fetal and adult erythropoiesis in *KLF-1*-targeted mice carrying a human β-globin locus transgene and presents with a nonpermissive chromatin configuration [177].

Also, two enhancers, within and downstream of the *HBB* gene, are likely to contol *HBB* gene expression [9]. The intragenic enhancer is located between IVS-II and exon 3, while the downstream enhancer is located just after the *HBB* polyadenylation site. Both enhancer elements contain at least four GATA-1-binding sites, which most likely account for their enhancer properties. Although the *HBB* and the *HBD* genes differ only in ten amino acids in their coding regions, they vary substantially in their promoter regions. *HBD* gene transcription is far less efficient than *HBB* gene transcription [162], which is reflected in the HbA$_2$ and HbA concentrations, respectively.

The HBA2/HBA1 and HBZ Genes. Located in a different chromosome than the β-*like* globin genes, *HBA2/HBA1* and *HBZ* genes have several unique features (Table 11.2). Like the fetal globin genes, *HBA2/HBA1* promoters and coding regions are virtually identical, containing TATA and CCAAT boxes but lacking a CACCC box, despite being GC rich. On the other hand, the *HBZ* gene has the typical globin gene promoter arrangement, with CACCC, CCAAT and TATA boxes in a 5′>3′ orientation. In all cases, GATA-1-binding sites have been identified, along with the ubiquitous transcription factor Sp1 motifs [172].

The Human β-Globin Locus Control Region. The human β-globin locus control region (or LCR) is the major regulatory element governing proper transcriptional regulation of the human β-*like* globin genes. It is located at the 5′ end of the β-globin gene cluster, approximately 20 kb upstream of the *HBE* gene, and consists of five erythroid-specific DNaseI-hypersensitive sites (HS1-5). The first indication of its importance was

Table 11.2 Comparison of the general features of the α- and β-globin gene clusters, suggesting that the chromosomal regions on which the α- and β-globin gene clusters reside are structurally different, reflecting in the differences in transcription, repair, recombination and replication

Features	Gene cluster	
	α-Globin	β-Globin
Location		
Chromosomal location	Telomeric (16p13.3)	Interstitial (11p15.5)
GC content	54%	39.5%
CpG islands	Common	None
Gene density	High	Low
Alu family repeats	25%	~5%
LINE repeats	Rare	Present
Chromatin structure and transcription		
Chromatin	Open	Closed → Open
Matrix attachment sites	None detected	Common
Major regulatory element	Enhancer (HS-40)	Locus Control Region
Effect of deletion of major regulatory element on long-range chromatin structure	No	Yes
Expression in transient expression assays	Enhancer independent	Enhancer dependent
Expression in cell hybrids	Early	Late
Other features		
Replication timing	Early	Late → Early
Predominant mutations	Deletions	Point mutations
Evolution of intergenic regions	Rapid	Slow

obtained when Kioussis et al. [78], identified a deletion mutant in which the *HBB* gene was silent, but structurally normal [79]. Subsequently, two other deletion mutants have been identified, one of which was removing only four of the LCR HSs (HS2-5), leaving the entire β-*like* globin genes, though structurally intact, transcriptionally silent (Fig. 11.3a). Studies in transgenic mice have shown that linkage of the LCR to the *HBB* gene confers high-level copy number-dependent and position-of-integration-independent expression of this gene [56]. This latter feature is what distinguishes the LCR from a classic enhancer element. It is noteworthy that prior to the discovery of the LCR, *HBB* transgenes without LCRs were only expressed in 50% of the transgenic mouse lines and at about 0.1–3% of their expected transcription rate [80].

Each HS contains a 250- to 500-bp core region, which is largely responsible for the HS activities. The core regions contain multiple erythroid-specific and ubiquitous transcription factor-binding sites. From a plethora of experiments it has been shown that HS2-4 display enhancer activities, with HS2 being the strongest enhancer among them, most likely because of the presence of an NF-E2/AP1 transcription-binding site [106]. HS5 functions as a chromatin insulator [44], where a binding site for the protein CCCTC-binding factor (CTCF), which acts as a chromatin insulator [108], has been found. The function of HS1 remains to be defined.

Another property of the LCR is its ability to confer copy number-dependent and position-of-integration-independent expression on a linked gene [56]. Copy number-dependent expression is widely considered to be indicative of permissive chromatin structure, i.e., DNA that is accessible to transcription factors. Involvement of the β-globin LCR in creating open chromatin was first suggested from analysis of β-thalassemia deletional mutants, where part of the LCR was deleted but the β-*like* globin genes were structurally intact [46, 79] (Fig. 11.3a). In keeping with this, only the

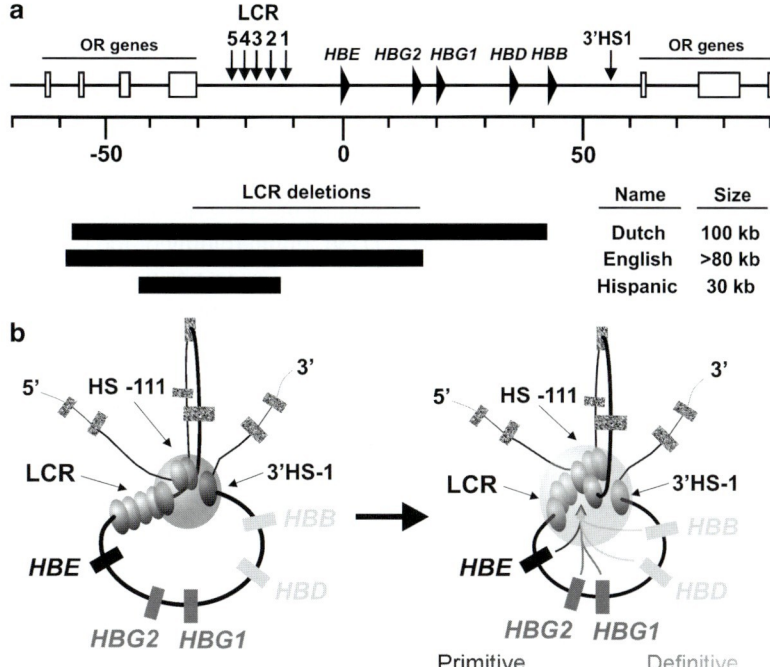

Fig. 11.3 (a) Representation of the large deletions identified to date, removing part or the entire LCR, leading to β-thalassemia. *Black arrows* represent the functional globin genes. (b) Model of the three-dimensional structure of the human β-like globin gene cluster. The chromatin hub (CH), indicated as a *dark gray circle*, is formed from the distant HSs, namely, HS-111 and 3′HS1 together with HS5 from the LCR. The genes are colored depending on their developmental order of expression, namely, embryonic (*black*), fetal (*dark gray*), and adult (*light gray*). Olfactory receptor genes are shown as *hatched gray boxes*. The active chromatin hub (ACH; shown as *light gray circle*) is formed by the remaining HSs of the LCR and the gene proximal regulatory elements, i.e., promoter, at the respective developmental stage, in which the globin gene(s) are active

full LCR (HS1-5) can provide position-independent chromatin-opening activity in transgenic mice carrying the entire β-globin locus [99]. In transgenic mice carrying single copies of gene constructs, including one of the LCR's HSs and a globin gene, only those carrying HS3 were able to confer copy number-dependent gene expression [40], suggesting that HS3 possesses the dominant chromatin-opening activity of the β-globin LCR.

Finally, the β-globin LCR can direct replication timing in a developmentally specific manner in vivo [153], resulting in the human β-globin locus being replicated late in most cell types, but early in erythroid cells [31].

The Human α-Globin Locus Control Region. The α-*like* globin genes also display tissue specificity in their developmental expression. The *HBZ* gene is expressed during the first trimester of embryonic life, whereas the *HBA2/HBA1* genes are expressed during the last two trimesters of gestation and later on in adult erythroid cells. Although tissue specificity and developmental expression pattern are identical to those of β-*like* globin genes, the regulatory mechanisms governing human α-*like* globin gene cluster expression display some distinct features (Table 11.2). The existence of an α-*like* LCR has been established by naturally occurring deletion mutants leading to α-thalassemia, in which the α-globin genes are silenced, albeit structurally intact [60, 144]. In addition, the *HBA2/HBA1* genes alone are expressed in low levels either in cultured erythroid cells or in transgenic mice, but when linked to the β-*like* LCR they are then expressed in high levels and with erythroid specificity [145]. A systematic search for LCR activity revealed an HS 40 kb upstream of the human α-globin gene cluster (termed therefore "HS-40"), displaying properties of a strong erythroid-specific enhancer [64, 151]. The HS-40 appears to function in a very similar way to HS2 of the human β-globin LCR [73].

The Human β-Globin Active Chromatin Hub. The spatial organization of the murine β-globin gene cluster in expressing and nonexpressing cells has recently been studied using the chromosome conformation capture (3C) technology [29] and the RNA TRAP method [18]. It has been shown that in erythroid cells the expressed adult mouse β-globin genes ($β_{maj}$ and $β_{min}$) spatially interact with the HS of the LCR, located 60 kb away. The inactive embryonic β-*like* globin genes (εy and βh1), positioned in between the LCR and adult genes, do not participate in this interaction and loop out [18, 166]. Two sets of HS at either side of the murine β-*like* globin gene cluster, 130 kb apart from each other, are also present in the spatial cluster, namely, 3′HS1 and two additional 5′HS (called HS-60.7 and HS-62.5), with previously unknown function [44]. These HSs form a pre-ACH complex, termed the "chromatin hub" or CH, and are likely to be involved in globin gene regulation in erythroid cells. Most importantly, the transcriptionally silent olfactory receptor (OR) genes located in between the 5′HSs and the LCR, like the inactive embryonic globin genes, do not participate in clustering, and also loop out. The spatial clustering of *cis*-regulatory elements and active genes is termed an "active chromatin hub" or ACH (Fig. 11.3b).

Similar results have been obtained using transgenic mice carrying the human β-globin gene cluster [119], where it was also shown that the formation of the ACH is mediated by multiple regulatory elements [129]. Similar observations in *KLF-1*-targeted mice, in which the functional murine β-globin ACH is lost [37], suggest that KLF-1 is a key player in stabilization of the β-globin ACH in definitive erythroid cells.

It is postulated that the active organization of the human β-*like* globin gene cluster would use similar principles (but on a larger scale) to those used in the folding of an enzyme to create an active site or "pocket" [129]. Initial folding of the locus would be nonspecific in terms of exact position or site, but would provide the template to allow the proper and precise folding of the LCR (which would be analogous to creating an active site of an enzyme) and allow interactions with the globin genes (which, to continue, the analogy, is the substrate). In other words, the function of the CH lies in the fact that it would cause a spatial restriction by forming a loop containing the genes and LCR and hence stimulate ACH formation.

11.3.4 Molecular Control of Globin Gene Switching

Analysis of developmental expression of human globin genes has provided useful insights into the regulation of gene expression. The human β-*like* globin gene cluster had always had a central role in these experiments, since understanding of the molecular mechanisms governing the expression of the β-*like* globin genes would open new perspectives leading toward novel therapeutic strategies for β-thalassemia patients (see also below). The human β-*like* globin gene cluster displays two developmental switches, one from the embryonic to the fetal stage, coinciding from the transition from primitive to definitive hematopoiesis, and a second switch from the fetal to the adult stage during the perinatal period [158]. From different types of experiments it has been established that the regulation of the human β-*like* globin gene is controlled by a dual regulatory mechanism involving (a) *gene competition* for activation from the LCR and (b) *gene silencing* of those genes that are active during the previous developmental stage (Fig. 11.4).

Gene Silencing. Autonomous gene silencing of the globin gene, which is active at the previous developmental stage, has been demonstrated to be one of the two molecular mechanisms involved in human globin gene regulation. Deletion of the putative *HBE*-silencing elements resulted in the abrogation of *HBE* silencing, whose expression persisted during adult erythropoiesis [142]. Similarly, autonomous silencing has also been demonstrated for the *HBZ* gene [155]. However, silencing of the human fetal globin genes has been more difficult to demonstrate. When a 3.3-kb fragment containing the *HBG1* gene was linked to the LCR, it displayed proper developmental control, e.g., was expressed in the embryonic and fetal erythroid stage and not in the adult stage [32]. However, Enver et al. concluded that similar LCR-*HBG1* constructs displayed γ-globin gene expression throughout development [41]. Dissecting the human γ-globin gene promoter suggests that there are two negative regulatory elements [159]. The first is located downstream of position -141, where many mutations leading to nondeletional hereditary persistence of fetal hemoglobin (HPFH) were found (see also below). The second is localized further upstream, between positions -378 and -732, and appears to function as an adult-specific enhancer. In favor of this hypothesis is a mutation leading to increased Hb F levels in the adult [21, 92].

Gene Competition. In erythroid cell cultures or transgenic mice, the *HBB* gene is properly regulated in DNA constructs containing the LCR and the *HBG1* and *HBB* genes in the normal configuration. However, when the *HBG1* gene is removed, *HBB* loses its developmental specificity and remains active throughout development [41]. The model suggests that the LCR

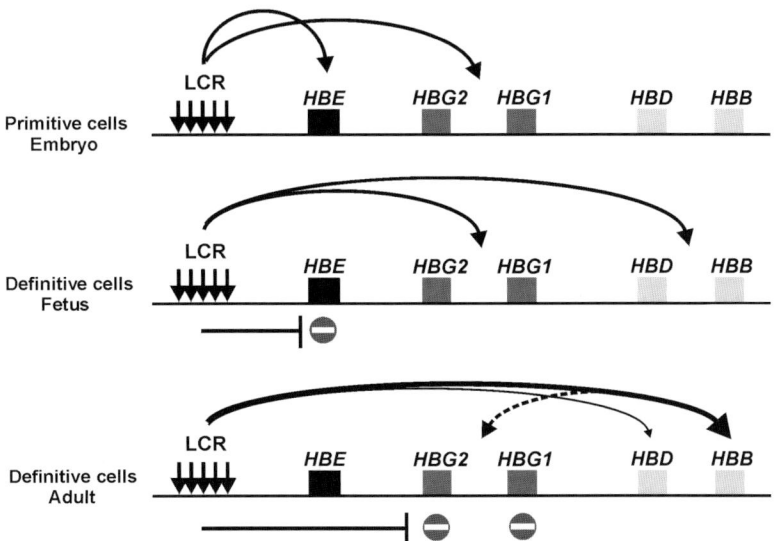

Fig. 11.4 Schematic drawing of the dual regulatory mechanism, which governs human globin gene transcription. The embryonic *HBE* gene is expressed during primitive hematopoiesis. Upon switch from primitive to early definitive hematopoiesis, the *HBE* gene is silenced autonomously and the *HBG1* and *HBG2* genes are activated. Finally, silencing of the human fetal globin genes occurs upon transition of hematopoiesis from early to late definitive erythroid tissues, where the *HBD* and *HBB* genes are transcribed at high levels. At this stage, *HBG1* and *HBG2* genes are expressed at very low levels (indicated with a *dashed line*)

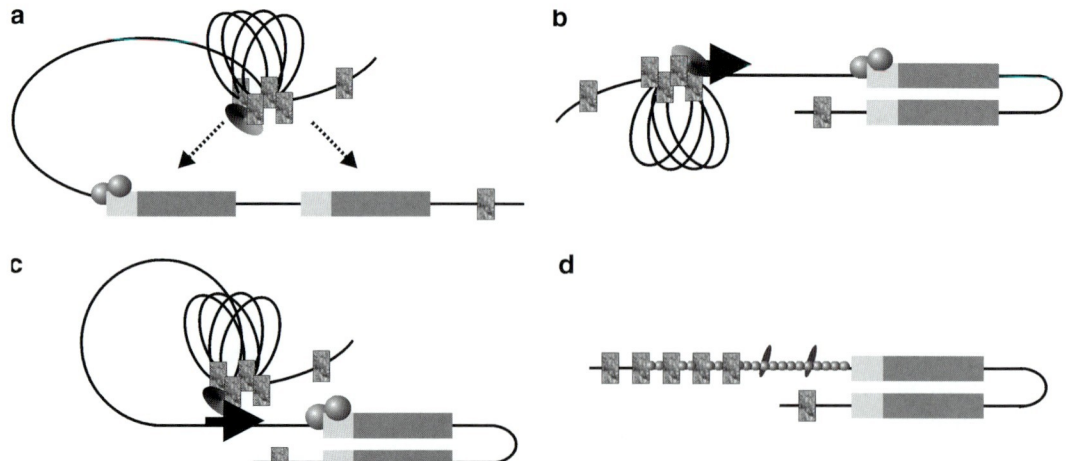

Fig. 11.5 (**a–d**) The four models put forward to explain globin gene competition, namely, looping (**a**), scanning (**b**), facilitated scanning (**c**) and linking (**d**). The outline of each model is provided in the text. *Dark gray boxes* indicate the structural genes, *light gray boxes*, the gene proximal regulatory elements, *large* and *small circles* correspond to the various transcription factors and histone modification proteins, respectively, and *hatched boxes*, to the HSs of the LCR

directly interacts with the gene proximal regulatory elements, e.g., promoters, and this interaction is primarily determined by the transcription factor environment. The gene that successfully interacts with the LCR is expressed, and the unsuccessful one is not. The model has been expanded to explain both switching processes in the human β-*like* globin gene cluster. Four different models have been proposed to explain LCR function in relation to globin gene competition: looping, tracking (or scanning), facilitated tracking, and linking (Fig. 11.5). Available data neither strongly support nor preclude any of them.

The *looping model* suggests that the HSs of the β-globin LCR form a "holocomplex," with the HS core elements creating an active site that binds transcription factors and the core-flanking sequences constraining the holocomplex in the proper conformation (Fig. 11.5a). This structure is looping, so that the LCR comes in close physically proximity to the appropriate gene promoter. Close association with gene-proximal promoter and enhancer elements allows the delivery of LCR-bound transcription proteins and other co-activators that interact with the basal transcription machinery, which is already bound at the promoter to form a stable transcription complex, thus enhanci ng globin gene expression [99, 176]. Further evidence supporting the notion of a holocomplex suggests that the LCR interacts with only one globin gene promoter at any given time and that these interactions may alternate between two or more promoters (also known as "flip-flop"), depending on the developmental stage [176]. The gene order and distance between the LCR and its target gene is a critical parameter in the flip-flop assumption, which has been shown to affect the probability that these two elements, namely, the LCR and the gene promoter, will interact [33]. This probability is constant for a gene at a specific stage of development, determined at each stage from the respective transcription factor environment. A variation of this model suggests that the LCR initially serves as a multiple element receptor that acts as a docking site for transcription factors to initiate chromatin remodeling [55]. Once chromatin-remodeling activity has been completed, the LCR directly interacts with downstream genes to facilitate their expression.

The *tracking* (or *scanning*) *model* suggests that erythroid-specific and ubiquitous transcription factors and cofactors bind to recognition sequences in the LCR and form an activation complex that migrates, or tracks, linearly along the β-*like* globin gene cluster (Fig. 11.5b) [168]. When this transcription complex encounters the basal transcription machinery located at the correct gene promoter, according to the developmental stage, the complete transcriptional apparatus is assembled and transcription of that gene is initiated.

The *facilitated-tracking model* incorporates aspects of both the looping and the tracking models (Fig. 11.5c) [12]. This model suggests that an LCR co-activator complex loops to contact downstream DNA in promoter-distal regulatory regions, where the transcription factor complex is released. This complex then tracks in small steps along the chromatin until it encounters the appropriate promoter with its associated bound proteins. A stable loop structure is then established and gene expression proceeds.

Finally, the *linking model* proposes that chromatin facilitator proteins bound throughout the β-*like* globin gene cluster define the domain to be transcribed and mediate the sequential stage-specific binding of transcription factors (Fig. 11.5d) [16]. Non-DNA-binding facilitator proteins form a continuous protein chain from the LCR to the globin gene to be transcribed, linking proteins bound at a transcriptionally primed gene to one another.

11.3.5 DNA Polymorphisms at the Globin Genes

Gene mapping by restriction enzyme analysis of the β-*like* globin gene cluster led to the identification of numerous single nucleotide polymorphisms (SNPs) between different individuals. Haplotypes, i.e., the pattern of SNPs on a given chromosome, are valuable diagnostic tools, and only a few of them can be found in different ethnicities. Initially, Kan and Dozy reported a DNA polymorphic sequence adjacent to the human *HBB* gene in relation to the sickle cell mutation [78]. Subsequently, at least 17 single nucleotide substitutions have been identified in the β-*like* globin gene cluster (intergenic) and in the *HBB* gene itself (intragenic) [4, 8, 75, 77, 112]; these are symbolized as either present (+) or absent (−) in different individuals. Most of the polymorphic sites are of ancient origin, since they are found in all ethnic groups, but their frequencies vary significantly among different ethnicities and populations (Table 11.3).

A remarkable feature of the β-*like* globin gene cluster SNPs is their linkage disequilibrium. The most reasonable interpretation postulates a recombinational hotspot, leading to high recombination rates, which separates the flanking SNP clusters. If free recombination were involved among these SNPs, one would expect random associations of any two polymorphic sites over many generations, which would have resulted in a very large number of haplotypes [19]. Recent data suggest that SNPs within the β-globin gene cluster comprise two distinct linkage disequilibrium blocks, one extending from the LCR

Table 11.3 Frequency of DNA polymorphic sites in the β-globin gene cluster in different populations (adapted from [8])

Polymorphic sites	Ethnic groups		
	Greeks	African–Americans	Southeast-Asians
Taq I	1.00	0.88	1.00
Hinc II	0.46	0.10	0.72
Hind III	0.52	0.41	0.27
Hind III	0.30	0.16	0.04
Pvu II	0.27	N.D.	N.D.
Hinc II	0.17	0.15	0.19
Hinc II	0.48	0.76	0.27
Rsa I	0.37	0.50	N.D.
Taq I	0.68	0.53	N.D.
Hinf I	0.97	0.70	0.98
Hgi A	0.80	0.96	0.44
Ava II	0.80	0.96	0.44
Hpa I	1.00	0.93	N.D.
Hind III	0.72	0.63	N.D.
Bam HI	0.70	0.90	N.D.
Rsa I	0.37	0.10	N.D.

N.D. Not determined

to the *HBD* gene and the other containing the *HBB* gene [95].

Apart from the main haplotype frame, a handful of SNPs, either within a haplotype context or independently, have been subsequently identified in different regulatory regions within the β-*like* globin gene cluster [34, 84, 110, 121, 122, 127, 137]. These polymorphisms have been correlated with different *HBD* and *HBB* gene mutations, and occasionally with phenotypes of varying severity in β-thalassemia and SCD patients.

11.4 Molecular Evolution of the Human Globin Genes

All of the normal human hemoglobins and their genes have an identical three-dimensional structure and a closely related primary DNA sequence, respectively, suggesting a common evolutionary origin. The closer the resemblance between two globin chains, the more recent the common ancestral sequence. The main evolutionary event in generating a gene family is gene duplication, an essential first step toward the creation of duplicates from a single ancestral gene and expansion of gene clusters. The duplicated copies can subsequently evolve independently to acquire new biological functions. Homologous recombination between repeated elements is the most frequent event leading to gene duplication and was first described by Shen et al. [152] in the human fetal globin genes.

Gene duplication is usually followed by additional recombination events. Unequal crossover can subsequently increase or decrease the number of genes in the family. In the human α-*like* or the β-*like* globin gene clusters, unequal crossover has not only been involved in their evolution but has been also implicated in pathogenic gain or loss of gene copies [67], which leads to phenotypes of various severity, with Hb Lepore [116] and Hb Kenya [68] as the most characteristic examples (see also below).

Gene sequence similarities strongly indicate that the human α-*like* and β-*like* globin genes originate from a single ancestral globin gene. Following gene duplication, the two genes were separated in different chromosomes, approximately 450–500 million years ago (Mya), and have evolved independently ever since (Fig. 11.6). The ancestral α-*like* globin gene has undergone several duplication events since then, leading to the *HBZ* gene (300 Mya), the *HBQ1* gene (260 Mya) and ultimately the *HBA2* and *HBA1* genes (52–72 Mya). Similarly, the ancestral β-*like* globin gene was duplicated some 150–200 Mya, yielding two gene copies of embryonic and adult developmental specificity. Duplication of the ancestral embryonic gene 100–140 Mya resulted in the embryonic *HBE* gene and a fetal-*like* globin gene, whose duplication 47 Mya yielded the *HBG2* and *HBG1* genes. Finally, the ancestral adult gene was duplicated 85–100 Mya, yielding the *HBB* and *HBD* genes.

Following the initial gene duplication, the gene copies accumulate nucleotide substitutions, small or large deletions, and insertions independently, as part of the evolutionary process. Consequently, the two gene copies diverge from each other, and finally each copy has the possibility of gaining an improved function or, on the other hand, of being silenced and becoming a pseudogene. Gene conversion, the nonreciprocal transfer of genetic information between two highly homologous genes, first observed in yeast [141], has been described in the *HBG2*/*HBG1* genes [152, 154], indicating that such events also occur frequently in mammalian cells. Sequence analysis has revealed that a 1.5-kb DNA fragment of the *HBG2* gene, including the promoter up to most of the IVS-II region, is virtually identical to the respective DNA fragment of the *HBG1* gene. This observation indicates that a gene conversion event occurred of such a kind that sequences from the *HBG2* gene converted those of the *HBG1* gene no later than 1 Mya [123].

Similarly, in 1983, Michelson and Orkin showed that, following the α-globin gene duplication event, the duplicated loci underwent a gene conversion event that homogenized their sequences from the 5′ regulatory to the IVS II region [98]. Most interestingly, in both gene copies gene conversion events are restricted to the upstream sequences, leading to homogenization of the 5′ regulatory and coding sequences. On the other hand, the absence of gene correction events at downstream sequences, such as the 3′ UTR, excludes this region from the concerted evolutionary process, leading to the accumulation of specific nucleotide variations. The latter may reflect the differences in the transcriptional and translational efficiency of the *HBA2* and *HBA1* genes [88].

Also, it is interesting that the conversion polarity is defined from the expression level of the participating genes. The gene that is expressed at a higher level ("master" gene) converts the sequence of one expressed

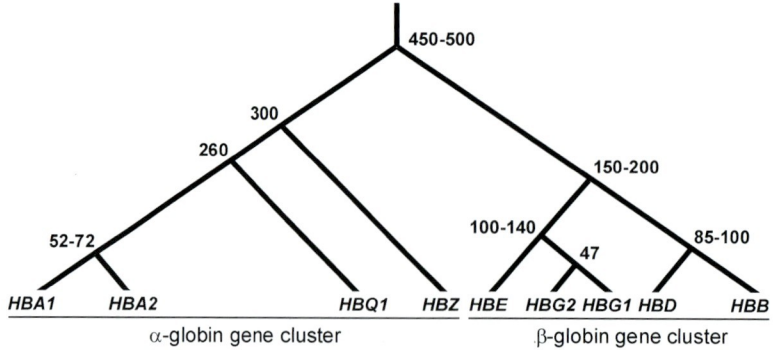

Fig. 11.6 Outline of the human α- and β-globin gene clusters' evolution. *Numbers* indicate the approximate date (in million years) when the respective gene duplication event took place

at a lower level ("slave" gene). This master/slave gene rule, which seems to be involved in the definition of the polarity of the conversion, can be clearly interpreted as an attempt by natural selection to prevent the possible inactivation of the slave gene when two functional gene copies are required [20].

On the other hand gene conversions are less frequent between the *HBD* and *HBB* genes, which has led to their marked sequence divergence, also reflected in their transcriptional efficiency. Nevertheless, interallelic gene conversion seems to have contributed to the spread of β-thalassemic mutations into different chromosomal backgrounds, defined from the respective β-*like* globin haplotypes [123]. Recent data from at least 14 different α-globin chain variants, due to an identical mutation in both the *HBA2* and *HBA1* genes, suggest that such mechanism may be also active in the human α-globin genes [100].

Gene conversion has been shown to have an important role in propagating pathogenic mutations in both globin gene clusters. The most characteristic example is Hb Parchman [1], which is most likely the result of *HBD* gene conversion by the *HBB* gene. Finally, the Cretan type of HPFH, which resulted from a likely gene conversion event identical to the one that originally homogenized the fetal globin gene sequences (Fig. 11.7), [126], further demonstrates the dynamic nature of this mechanism in molecular evolution.

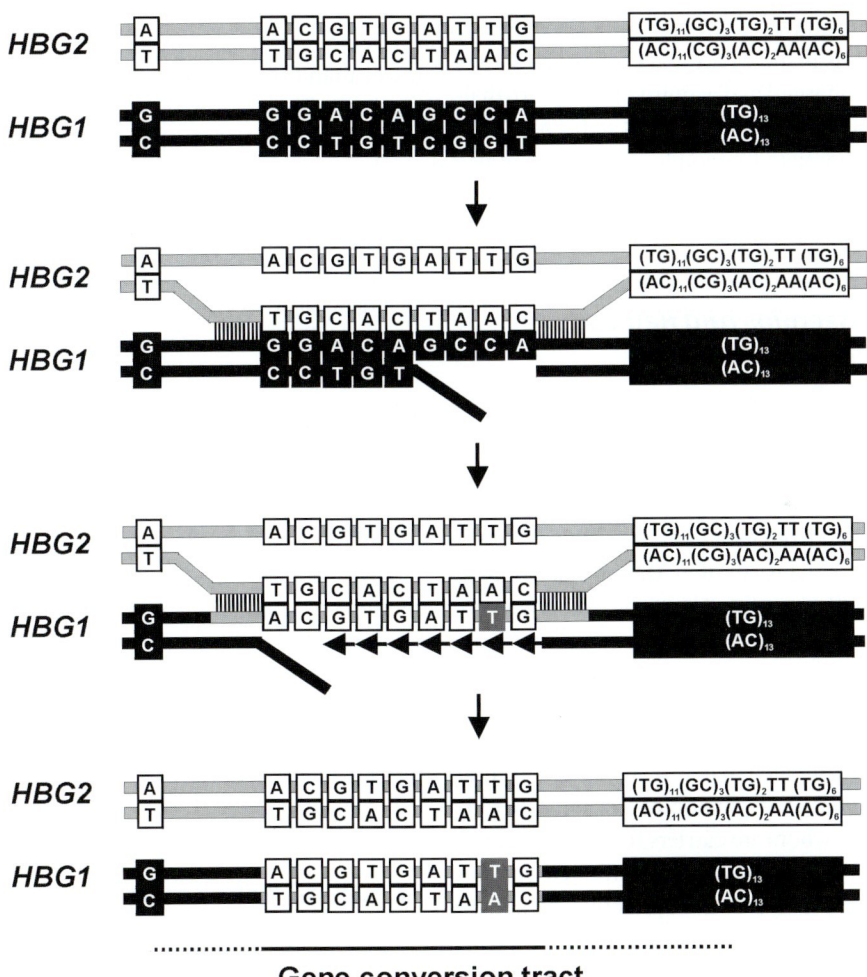

Fig. 11.7 Schematic drawing depicting the convergent evolution of the human fetal globin genes. A gene conversion event resulted in the homogenization of a 1.5-kb fragment of the *HBG1* gene from the *HBG2* paralog. The gene conversion tract length can be deduced from the flanking DNA base differences and spans from 423 to 1,550-bp (indicated as a *dashed line*). An identical de novo gene conversion event has recently been identified, leading to the Cretan type of nondeletional HPFH (shown as a *gray box*), indicative of the dynamic nature of this phenomenon [126]

11.5 Molecular Etiology of Hemoglobinopathies

Hemoglobinopathies refer to a wide range of inherited disorders, including the thalassemias and SCD. The vast majority of hemoglobinopathies result from genetic defects in the human α-*like* or β-*like* globin genes. In other words, the pathogenic mutations lie in *cis* to the globin gene cluster. However, there are a few examples where the mutation resulting in α- or β-thalassemia is located in *trans* of the globin gene clusters. In particular, few mutations in genes enconding for the erythroid-specific transcription factor GATA-1 on chromosome X [180] or the XPD subunit of the general transcription factor TFIIH on chromosome 19 [170] have been shown to cause β-thalassemia. In addition, mutations in the *ATRX* gene have been shown to result in both inherited and acquired forms of α-thalassemia. Interestingly, transcription of antisense RNA or mutations in unrelated genomic regions have also been shown to result in α-thalassemia.

11.5.1 Thalassemias and Related Conditions

A variety of conditions are characterized by genetically determined diminished or absent synthesis of a globin chain. These diseases are known as the thalassemias [25]. This term is derived from the Greek word "θαλασσα" (*thalassa*=sea), referring to the Mediterranean Sea, and was originally used to describe the Mediterranean origin of many gene carriers of these conditions [175].

The thalassemias are among the most common monogenic disorders in the world. The World Health Organization has estimated that as many as 270 million carriers of a globin gene genetic defect exist worldwide, 80 million of whom are carriers for β-thalassemia, and around 300–400,000 severely affected infants are born every year. Because of a selective advantage of heterozygotes against malaria, the frequencies of thalassemia are particularly high in the malarial tropical and subtropical regions of Asia, the Mediterranean, and the Middle East [45]. The carrier frequency of α-thalassemia varies from 1% (e.g., in southern Spain) to 90% (e.g., in tribal populations of India), while the carrier frequency of β-thalassemia varies from 1% (e.g., in northern Italy) to up to 70% (e.g., in some regions of Southeast Asia). Considerable genotypic variation occurs even within a single country. The epidemiology of the disease, however, is changing. In the developing countries, the numbers of affected children are on the increase owing to falling childhood mortality resulting from improved nutrition and better infection control, while in the more developed countries, epidemiology of the disease has been affected by a fall in total birth rate and preventive programs. One of the best-documented control programs involving education, counseling, and prenatal diagnosis that have succeeded in limiting the numbers of new births of affected individuals has been in Cyprus, where thalassemia was first recognized in the 1940s. Implementation of the control program resulted in fewer than 600 severely affected patients with thalassemia in Cyprus in total, with only two or three new cases each year (Fig. 11.8). Furthermore, because of recent population migrations, thalassemia has become an important part of clinical practice in Northern Europe, including the United Kingdom, and in the USA and Australasia [174].

The molecular etiology of the thalassemias is highly heterogeneous. It is noteworthy that advances in understanding the thalassemias at the molecular level have led to better comprehension of the nature and variety of human mutations in general. Depending upon which globin chain is absent or reduced, thalassemias can be classified as α-, β-, γ-, δ- and δβ-thalassemias (the last including exclusively deletion mutants and being further subdivided into ^γδβ-, γδβ- and εγδβ-thalassemias). The commonest and clinically most important are the α- and β-thalassemias (Fig. 11.9), while δ-thalassemia is a clinically silent condition. Elucidation of globin gene regulation has been aided significantly by investigation of various thalassemia mutant alleles, which can be classified in the categories detailed below.

11.5.2 β-Thalassemia

The recessively inherited β-thalassemias are widespread throughout the tropical and subtropical areas of the world and likely owe their frequency to a selective advantage vis-à-vis *Plasmodium falciparum* malaria [173]. The apparent protective effect of thalassemia

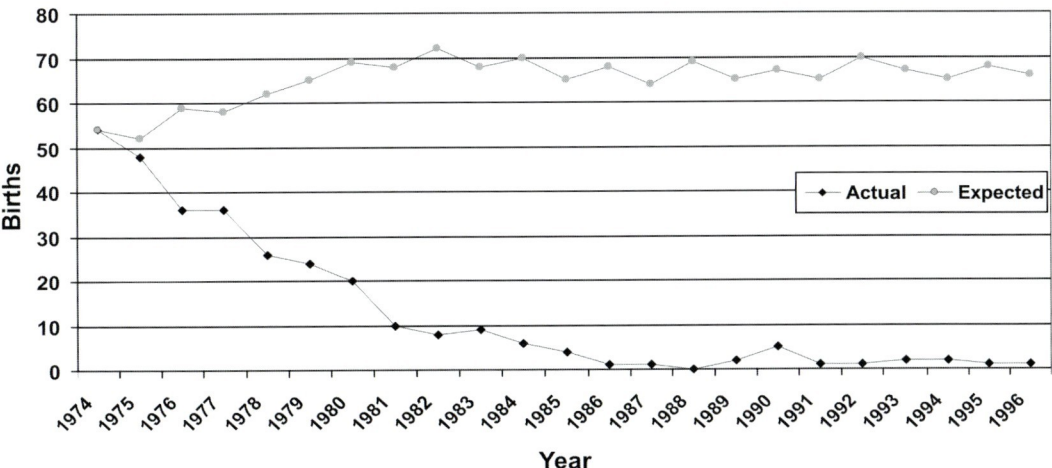

Fig. 11.8 Graph depicting the decline of births of homozygous β-thalassemia children in Cyprus compared with the expected number of homozygous β-thalassemia births. The graph is indicative of the success of the β-thalassemia prevention program in Cyprus

Fig. 11.9 Classification of α- and β-thalassemia depending on the degree of symptoms' severity, in relation to globin chain imbalance and gene expression. (From [130])

α-thalassemia	β-thalassemia	Features
Normal	Normal	Globin gene expression / Globin chain imbalance
α-thalassemia trait	β-thalassemia trait	
Hb H	β-thalassemia intermedia	
Hb Bart's hydrops fetalis	β-thalassemia major	

against malaria may be related to enhanced immune recognition, and hence clearance, of parasitized erythrocytes [93].

Carriers (heterozygotes) of β-thalassemia are clinically normal with mild anemia and are largely unaware of their carrier status. Hb A_2 is slightly elevated, while the red cells are smaller and less filled with hemoglobin, leading to a decrease in the mean cellular hemoglobin (MCH) and mean cellular volume (MCV) values. Heterozygotes usually do not require medical attention or treatment. Severely affected β-thalassemia homozygotes, however, have marked anemia requiring blood transfusions. Hb A is completely absent in β^0 thalassemia homozygotes, and much decreased in β^+ thalassemia homozygotes. The disease is associated with growth failure and often leads to death in adolescence or earlier. Homozygosity for β^0 thalassemia alleles and compound heterozygosity for β^+/β^0 thalassemia alleles are severe hemoglobinopathies and are a serious public health problem in countries where these alleles are common. The simultaneous presence of α-thalassemia ameliorates the clinical severity of homozygous β-thalassemia. Hb S/β^+ thalassemia is common in black populations. Hb E/β-thalassemia is common in Southeast Asia and results in severe anemia, as seen in homozygous β^0 thalassemia. This severity is at least partially explained by the fact that the Hb E mutation itself also causes mild thalassemia (see below). Less severe β-thalassemic alleles include β^{++} and β^{Silent} reflecting the minimal deficit in β-globin chain synthesis.

The remarkable heterogeneity of *HBB* gene mutations explains the frequent finding of compound heterozygotes for β-thalassemia, i.e., affected patients who have inherited a different *HBB* gene mutation from each parent. The frequency of such compound heterozygotes is somewhat lower in population isolates, where a single thalassemia mutant may account for the majority of thalassemia alleles. For example, while the p.Q39X (*HBB*:c.118C>T) nonsense mutant comprises about 27% of all β-thalassemia mutant alleles in general Mediterranean populations (http://www.findbase.org), it accounts for most of the β-thalassemia mutations in Sardinia. Since homozygosity for a given thalassemia mutant may range from only mild reduction to complete absence of globin synthesis, and compound heterozygosity is frequent, a wide spectrum of thalassemias with different clinical severity will be encountered.

A complete list of all pathogenic mutations reported for the human *HBB* gene can be found in the HbVar database for hemoglobin variants and thalassemia mutations (http://globin.bx.psu.edu/hbvar). The commonest mutations leading to β-thalassemia are point mutations or short insertions or deletions (indels) and can be classified in the following categories:

1. Promoter Mutations. Thalassemia mutations that affect the noncoding 5′ upstream regions of the *HBB* gene are regulatory mutations, which affect gene transcription. These mutations have been found at the distal CACCC and TATA boxes, which diminish hemoglobin synthesis and manifest themselves as relatively mild thalassemias [6, 114]. No mutations have yet been found at the *HBB* CAAT box.
2. RNA Cleavage Mutations. These mutations affect the AATAAA polyadenylation signal in the 3′ untranslated region of the *HBB* gene [115].
3. Nonsense Mutations. These mutations result in a new stop codon and hence lead to premature termination of translation, resulting in a shortened and, therefore, nonfunctional globin chain. These mutations therefore lead to β^0 thalassemia. One of the commonest mutations of this category is p.Q39X (*HBB*:c.118C>T), which is mostly found in patients and carriers of Mediterranean origin.
4. Frameshift Mutations. Short indels produce frameshifts with garbled coding, causing effective termination of functional globin synthesis again yielding a shorter globin chain.
5. RNA-Processing Mutations. These mutations affect RNA processing (splicing), mostly altering the GT or AG dinucleotide consensus sequence at the donor or acceptor splicing sites of splice junctions, respectively. These mutations cause β^0-thalassemia. Mutations in the splicing consensus sequence beyond the invariant GT-AG sequence usually cause milder β^+-thalassemia. Often, cryptic splice sites, namely, those that are not used during normal splicing, are sometimes activated by mutations in intervening sequences and cause interference with normal splicing process. Such cryptic splicing sites can also be activated within exons, Hb E and Hb Knossos being the most characteristic examples.

11.5.3 Dominantly Inherited β-Thalassemia

Dominantly inherited β-thalassemias are heterogeneous at the molecular level and are due to mutations at or near the *HBB* gene. Many of these involve mutations of exon 3 of the *HBB* gene. The resulting β-globin chain variants are very unstable, and in many cases the products of the dominantly inherited β-thalassemias are not detectable and the mutant alleles only are deducted from the DNA sequence. The predicted synthesis is supported by the presence of substantial amounts of abnormal β-globin mRNA in reticulocytes, comparable to the amount produced by the normal *HBB* allele. Of the 33 known dominantly inherited β-thalassemia mutations, 22 are located in exon 3 of the *HBB* gene, producing unstable globin and thalassemia intermedia (a term used to characterize a wide clinical spectrum ranging from mild thalassemia conditions to asymptomatic forms that are only slightly more severe than the β-thalassemia trait; see http://globin.bx.psu.edu/hbvar). The mutations leading to dominant β-thalassemia are missense mutations, frameshifts, premature chain termination (nonsense) mutations, and complex rearrangements, often resulting in an early stop codon (6 alleles) or elongated globin chains (7 alleles). Nonsense or frameshift mutations that would produce truncated β-globin chains up to 72 residues in length are usually associated with a mild phenotype in heterozygotes. It is believed that the mRNAs associated with these mutations are not transported to the cytoplasm and hence, no mutant protein is synthesized. On the other hand, mRNAs with mutations in exon 3 are transported and translated normally. They produce long and highly unstable globin gene products that are capable of binding heme, but not combining with α-globin chains to produce any kind of stable Hb tetramer. Hence, these large truncated products tend to precipitate in the red cell precursors, together with excess α-globin chains, to produce large inclusion bodies [165].

11.5.4 δβ-Thalassemias and Hereditary Persistence of Fetal Hemoglobin

Various rare deletions in the human β-globin gene cluster have been described. These conditions are known as δβ-, ᴬγδβ-, γδβ-, and εγδβ-thalassemias, depending on which genes have been removed by the deletion. Such deletions have been important for the identification of the human β-globin LCR (see above). None of these deletions can be recognized cytogenetically, since they are too small for microscopic detection.

Hereditary persistence of fetal hemoglobin (HPFH) is an inherited condition leading to increased Hb F production in adult life. HPFH is caused by deletions in the human β-globin gene cluster (deletion HPFH), removing the *HBD* and *HBB* genes, and is clinically dinstict from δβ-thalassemia. In particular, HPFH is characterized by higher HbF levels (30%), in contrast to δβ-thalassemia (24%), and has pancellular as against heterocellular distribution of Hb F. HPFH can also result from 18 point mutations and 1 short (13-bp) deletion (nondeletion HPFH), residing at the upstream promoter region of either the *HBG1* (10 mutations and 1 deletion) or *HBG2* (8 mutations) genes. These mutations are located in the distal CCAAT box and the upstream GC-rich region (200 bp upstream of *HBG1* and *HBG2* Cap sites) and up-regulate Hb F production, most likely by altering erythroid-specific transcription factor binding, which results in the continuous expression of the (otherwise silenced) fetal globin genes. Studies directed at understanding the regulation of the Hb F switch have far-reaching implications for treatment of thalassemia and SCD, since increased Hb F production in these disorders would be of marked therapeutic benefit.

11.5.5 α-Thalassemia

11.5.5.1 Deletion α-Thalassemia

Most α-thalassemias are caused by gene deletions. Sequence similarity in the 5′ and 3′ regions of the α-globin genes allows incorrect chromosomal alignment, followed by recombination with subsequent deletions and duplications. The crossover chromosomes bearing either a single α-globin gene (-α) or three α-globin genes (ααα) have both been observed. While malarial selection has amplified the frequency of the (-α) alleles among tropical and subtropical populations, the triple (ααα) allele seems to confer neither an advantage nor deleterious effects on its carriers and, hence, is much rarer [65].

Various phenotypes caused by deletions of one, two, three, or four α-globin genes have been documented. Absence of a single Hb α gene (-α/αα) produces little hematological impairment, since three genes remain active. Two principal types of deletion events cause the mild α-thalassemia (-α/αα). The so-called leftward crossover creates a single *HBA2* gene and deletes a 4.2-kb fragment. The so-called rightward crossover derives from misalignment of the *HBA2* and *HBA1* genes, with crossover producing a fusion gene and a 3.7-kb deletion (Fig. 11.10). The rightward single α-globin gene is the most common type of α-thalassemia in Africa and in the Mediterranean countries, while in Asia both leftward and rightward crossovers have been reported. Several deletions that removed the upstream regulatory region (HS-40) of the human α-globin gene cluster were found to silence α-globin gene expression [144].

Deletion of two α-globin genes (-α/-α or --/αα) produces mild anemia, while deletion of three α-globin genes (-α/--) causes a more severe anemia characterized by production of Hb H, a $β_4$ tetramer, owing to α-globin chain deficiency. Co-existing α-thalassemia in patients with SCD is associated with less severe anemia and improved survival. Finally, deletion of all four α-globin genes (--/--) is fatal perinatally and is known as hydrops fetalis, referring to the extensive edema of the stillborn infant (Fig. 11.9). Most of the hemoglobin molecules of such infants consist of a $γ_4$ tetramer (Hb Bart's). Survival of the fetus into late pregnancy is likely to be caused by the presence of functional Hb Portland ($ζ_2γ_2$). The virtual absence of hydrops fetalis from African infants is related to the nonexistence of the chromosome bearing the two α-globin gene deletions in that population.

11.5.5.2 Nondeletion α-Thalassemia

Nondeletion mutations similar to those resulting in β-thalassemia would be expected. A variety of such mutations have in fact been found, most of which involve the *HBA2* gene. No regulatory mutations in the

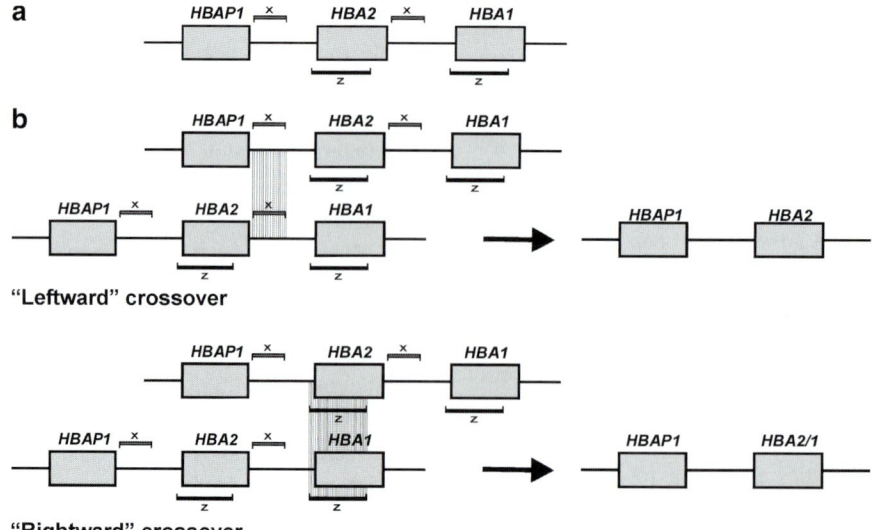

Fig. 11.10 Schematic drawing indicating the mechanism that resulted in the commonest deletion mutants leading to α-thalassemia. (**a**) Regions of homology are shown as "*x*" and "*z*." (**b**) Crossover between the x homology boxes yields a 4.2-kb deletion, termed "leftward," as the homology boxes are on the left side of the gene. Similarly, pairing of the "*z*" homology boxes in the rightward crossover yields a smaller, i.e., 3.7-kb, deletion; both result in a single α-globin gene

upstream region of both α-globin genes besides an LCR deletion (see above) have been detected. Only one splicing mutation, consisting of a 5-bp deletion that abolishes an acceptor site in IVS-I, has so far been found. Few point mutations have altered the human α-globin gene's termination codon. As a result, the α-globin chain is extended by 31 additional amino acids. These variants are unstable, and only small amounts (5%) of them can be detected in the blood. Hb Constant Spring is the most common of these mutants.

11.5.6 Other Mmutation Types Leading to Hemoglobinopathies

11.5.6.1 Fusion Genes

Deletions are frequent events of mispairing between homologous sequences of nucleotides during either meiotic or mitotic divisions in germ cell development. Examination of nucleotide sequences around the areas of deletions for various such mutants shows expected sequence homologies that facilitate mispairing. Recombination or crossover events following mispairing may lead to fusion genes, as the result of mispairing between similar but not identical genes. Therefore, nonhomologous crossover can eventually lead to fusion genes that encode the N-terminal portion of one globin and the C-terminal portion of another. Hb Lepore is the protein product of a *HBD-HBB* fusion gene (Fig. 11.11), with several types of Hb Lepore (http://globin.bx.psu.edu/hbvar), depending upon the site of crossover. Hb Kenya is another example of a protein that results from a fusion gene, owing to the misalignment between *HBG1* and *HBB* genes [109].

11.5.6.2 Duplications

Duplications may affect whole genes, such as the duplications during evolution that led to the various globin genes (see also Sect. 11.4). The commonest examples of globin gene duplication usually involve highly homologous genes, such as the human α- and γ-globin genes. In the former case, α-globin gene triplications are usually found, while in the latter, γ-globin gene triplications, quadruplications, and a quintiplication have been reported [67]. Duplication gene products would also be expected from crossover events as the counterpart of fusion genes. A chromosome with a *HBD*,

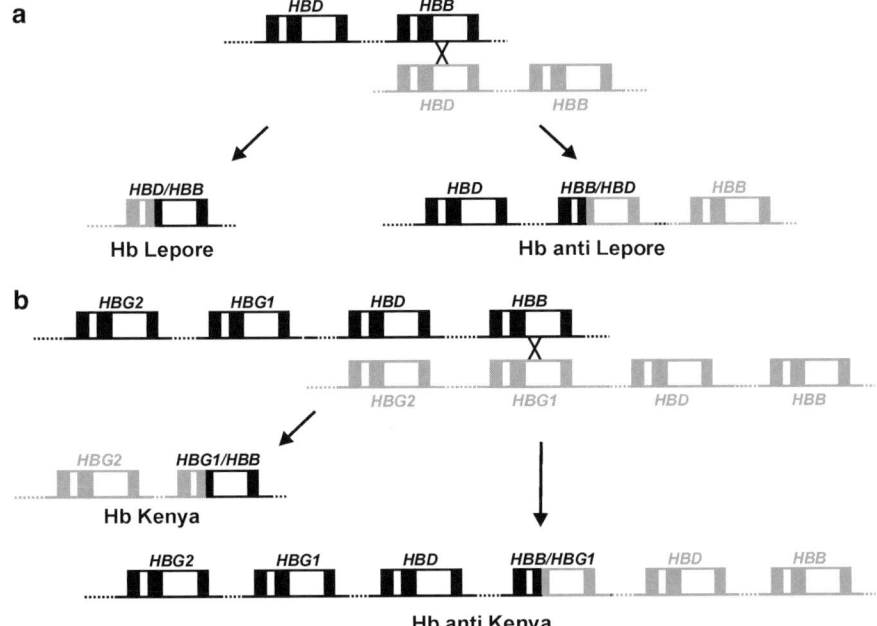

Fig. 11.11 Formation of hemoglobin fusion genes. Mispairing of the *HBD* and *HBB* genes yields a hybrid *HBD/HBB* gene (Hb Lepore) and an alternate product, termed Hb anti-Lepore. Such Hb anti-Lepore molecules have been identified, namely, Hb Miyada, Hb P, Hb Congo). Similarly, mispairing between the *HBG1* and *HBB* genes result in another fusion gene, namely, *HBG1/HBB* (Hb Kenya). The postulated Hb anti-Kenya molecule has never been found

HBB/HBD, *HBB* gene structure or Hb anti-Lepore have in fact been identified several times, while the expected Hb anti-Kenya chromosome (*HBG2*, *HBG1*, *HBD*, *HBB/HBG1*, *HBD*, *HBB*) has not yet been found.

11.5.7 Hemoglobin Variants

Hemoglobin variants are caused by a variety of mutational events affecting a given hemoglobin gene. The most common hemoglobin variants result from single amino acid substitutions of a globin chain, such as the *HBB*:p.E6V mutation, leading to SCD. If the electric charge of the mutant amino acid is different from the normal residue, the variant hemoglobin can be recognized by its altered behavior on electrophoresis or cation exchange chromatography. Mutations that do not change electrophoretic properties are usually detected if they exert a deleterious effect on hemoglobin function and cause disease. The majority of hemoglobin qualitative variants, regardless of charge differences, has a minor effect or none at all on hemoglobin function and is compatible with normal health. In general, amino acid substitutions of the exterior of the hemoglobin chains cause fewer perturbations of function than those replacing amino acids in the chain interior or close to the insertion of the heme group. Substitutions affecting normal helical turns of the chain often cause hemoglobin instability. Amino acid replacements affecting subunit contacts are often associated with abnormalities in oxygen affinity [157]. Most hemoglobin variants are rare, although a few, such as Hb S, Hb C, and Hb E, have reached higher frequencies in certain populations, presumably by positive natural selection.

The results of compromised hemoglobin function caused by hemoglobin variants can produce four different types of disease: (a) SCD due to distortion of the red cell membrane, (b) hemolytic anemia due to unstable hemoglobins, (c) methemoglobinemia attributable to more rapid hemoglobin oxidation and (d) erythrocytosis attributable to abnormal oxygen affinity causing hypoxia with resulting erythropoietin production. Synonymous polymorphisms, i.e., nucleotide variants in codons with no effect on protein's primary sequence also exist.

11.5.7.1 Sickle Cell Disease

Hb S is a recessive disease caused by the substitution of valine for glutamic acid in the sixth codon of the β-globin chain. This particular mutation affects the solubility and crystallization of hemoglobin under conditions of hypoxia. Patients with SCD produce Hb S and lack Hb A. With a relatively low degree of hypoxia, the Hb S of such patients polymerizes into filaments of high molecular weight that associate to form bundles of fibers. These abnormal hemoglobin crystals distort the red cell membrane to its characteristic sickling shape. Some of these cells remain irreversibly sickled and are destroyed prematurely. Sickled cells increase blood viscosity and impede normal circulation in small blood vessels. The resultant hypoxia leads to more sickling, with a vicious cycle of more stagnation and characteristic episodic sickle crises with abdominal and musculoskeletal pain. After several years, necrosis of poorly perfused tissues, such as the spleen, occurs, and this organ atrophies [150].

Carriers of the mutation leading to SCD have only 25–40% Hb S and are clinically normal. Their red cells contain both Hb A and Hb S, and have a normal red cell life span. In vivo sickling occurs only under conditions of severe hypoxia, such as atmospheric conditions at altituds of over 3,000 m [149].

Hb F, when present together with Hb S in a red cell, decreases the extent of sickling [39]. Hb F reduces the crystalization of Hb S so that patients with SCD and large amounts of Hb F have few or no symptoms of SCD. In a few of these instances Hb F is contributed by a HPFH mutation [124]. In general, there is an inverse correlation between the amount of Hb F and the severity of symptoms in SCD. Any manipulation that would increase fetal hemoglobin production would therefore cause clinical improvement in SCD. Co-existing α-thalassemia in patients with SCD is associated with less anemia and improved survival.

Interestingly, there are several genetic factors that modulate the phenotypic outcome of SCD [161]. Hb F plays a key role as a genetic modifier of SCD. The sickle cell mutation arose independently on different chromosomes, as determined by the underlying haplotypes [7, 118]. The latter has been shown to significantly impact on the observed Hb F levels and, hence, on the observed phenotype. In particular, SCD patients bearing the Bantu, Benin, and Cameroon haplotypes present with a severe phenotype with painful occlusive crises, while SCD patients bearing the Senegal and Arab/Indian haplotype have milder symptoms attributable to the increased Hb F levels [83]. Moreover, there are more genetic modifiers that have been found in different chromosomes, influencing the overall Hb F production in SCD patients, namely, on chromosomes Xp [36], 6p [26], and 8q [47], also known as quantitative trait loci (QTL). Similarly, α-thalassemia reduces the concentration of Hb S and, therefore, Hb S polymerization [62].

Hb E/Hb C. Hb E (*HBB*:p.E26K) is the second most common hemoglobin variant, which reaches high frequencies in Southeast Asia, particularly in regions of Laos and Thailand 1102]. The $β^E$ mutation is also only a mildly pathogenic mutation, to the extent that Hb E homozygotes have a clinical picture resembling classic β-thalassemia trait [43]. The $β^E$ mutation affects *HBB* gene expression by improving the efficiency of a normally inactive donor site for RNA splicing between codons 25 and 27 of the *HBB* gene [112]. Through this mechanism, the mutation leads to a mild deficiency in normal β-globin mRNA and to production of small amounts of structurally abnormal mRNA. Hb E homozygotes have microcytic red cells but no significant anemia or other clinical problems. However, co-existence of β-thalassemia with Hb E gives rise to a more severe β-thalassemia-*like* phenotype.

Hb C (*HBB*:p.E6K) is the third most commonly encountered structural hemoglobin variant worldwide. Hb C is frequently found in West Africa [104] and also among American Blacks, though at much lower frequencies. As with Hb S, the *HBB*:p.E6K mutation results in reduced solubility of the Hb C oxy- and deoxygenated forms and the formation of crystals. Crystalization of Hb C can be inhibited in the presence of Hb F [66]. Hb C homozygotes have mild anemia, markedly dehydrated red cells, and moderate spleen enlargement, while Hb C heterozygotes are mostly asymptomatic. Co-existence of Hb S with Hb C result in the same, but less frequent, painful vaso-occlusive crises as are seen in Hb S homozygous patients [171].

11.5.7.2 Unstable Hemoglobins

Over 130 unstable hemoglobins have been described, the majority of which are in the β-globin chain. Many unstable hemoglobins have amino acid substitutions or deletions affecting the heme pocket of the globin chain.

Clinical manifestations vary from mild instability that is not clinically apparent, to severe instability, which causes increased red blood cell destruction. The instability of these hemoglobins is often caused by premature dissociation of the heme from the globin chain. Such heme-depleted globin is precipitated as intracellular material known as Heinz bodies and interferes with cell membrane function. The diagnosis of unstable hemoglobins, if not associated with electrophoretic mobility alterations, is difficult and may require isolation of the precipitated globin chains for further analysis in specialized laboratories. Unstable hemoglobins have been found to result from recurrent mutations, yielding identical hemoglobin molecules (i.e., Hb Köln, Hb Hammersmith).

Methemoglobinemia Attributable to Hb M. Hb M was the very first globin abnormality to be discovered [174]. Methemoglobinemia is caused by the more rapid oxidation of divalent to trivalent iron. Nine different mutations can produce Hb M, six of them producing it by tyrosine replacement of the histidine residues that anchor the heme group in its characteristic pocket of the globin molecule and stabilize the heme iron. Mild hemolysis is common in patients with Hb M. Patients with Hb M mutations of the α-globin chain are cyanotic from birth. Those with Hb M mutations of the β-globin chain do not develop severe cyanosis until 6 months of age, when the γ-globin chains are replaced by β-globin chains. However, Hb M mutations of the γ-globin chain are manifested with cyanosis at birth, which disappears in a few months after β-globin chains have replaced γ-globin chains.

Erythrocytosis due to Hemoglobins with Abnormal Oxygen Affinity. Upon oxygenation, the area of contact between the α/β globin subunits shifts. Deoxyhemoglobin is normally stabilized by a hydrogen bond between α42Y and β99D, while oxyhemoglobin is stabilized by a bond between α94D and β102N [101]. Stabilization of the oxy conformation or destabilization of the deoxy conformation by a mutation may result in increased oxygen affinity. More than 90 hemoglobins with increased oxygen affinity are known to exist. Most hemoglobins with high O_2 affinity have substitutions of the C-terminal of the β-globin chain or at binding sites of diphosphoglycerate (DPG), which are normally involved in maintenance of stability of the deoxy conformation [11]. The increased oxygen affinity reduces oxygen delivery to the tissues, with resultant hypoxia. Hypoxia leads to release of the hormone erythropoietin, which stimulates red cell production with resultant erythrocytosis. Only a few hemoglobins with reduced oxygen affinity have been detected [11]. In these cases, increased oxygen delivery to the tissues caused by the reduced affinity for hemoglobin results in reduced production of erythropoietin and mild anemia.

11.6 X-Linked Inherited and Acquired α-Thalassemia

In contrast to the usual deletion and nondeletion forms of α-thalassemia resulting from *cis*-acting genetic defects, there are two other forms of α-thalassemia, one associated with a variety of developmental abnormalities in a rare, severe form of X-linked mental retardation (ATR-X syndrome) [52] and another acquired abnormality in association with a multilineage myelodysplasia, also known as α-thalassemia myelodysplasia syndrome (ATMDS) [160]. ATR-X syndrome is a rare condition that, to date, has been identified in approximately 100 families from many regions of the world. Patients with ATR-X syndrome carry mutations in the *ATRX* gene [53] residing on the X-chromosome (Xq13.1-q21.1). ATR-X patients present with a strikingly uniform phenotype, comprising severe mental retardation, characteristic dysmorphic facies, genital abnormalities, and an unusual, mild form of Hb H disease, accompanied by reduced *HBA1* and *HBA2* mRNA levels.

Also, although abnormal patterns of hemoglobin synthesis are nearly always inherited, occasionally persons with previously normal hematologic function develop aberrant hemoglobin synthesis as an acquired abnormality, usually within the context of hematologic malignancy. This situation is known as ATMDS. It has been demonstrated that both inherited and acquired mutations in *ATRX* gene cause α-thalassemia (ATR-X syndrome and ATMDS, respectively), suggesting that *ATRX* has a central role in the regulation of α-globin gene expression [52]. The mechanism by which *ATRX* mutations down-regulate α-globin expression is still unknown. The *ATRX* gene encodes a chromatin-remodeling factor, a member of the SWI2/SNF2 family of proteins. Although ATMDS and ATR-X syndrome are rare conditions, they have provided important insights into the general principles underlying more common, clinically important, diseases such as α-thalassemia.

11.6.1 Regulatory SNPs and Antisense RNA Transcription Resulting in α-Thalassemia

A common situation in human molecular genetics is a well-defined clinical phenotype for which the mutation causing the disease cannot be found in the associated gene/s or regulatory sequences. In contrast to the majority of thalassemia patients, in whom the underlying genetic defect is well characterized and located within the structural globin genes or their regulatory elements, there is a fraction of patients whose mutation could not be found in the latter genomic regions.

α-Thalassemia can result from transcription of antisense RNA, leading to an α-globin genes silencing, Tuffareli et al. showed that in an α-thalassemia, transcription of antisense RNA from the juxtaposed, to the *HBA2*, truncated *LUC7L* gene, owing to an 18.3-kb deletion, mediates silencing and methylation of the associated CpG island [169]. In other words, the antisense RNA from the truncated *LUC7L* gene runs through the *HBA2* gene and interferes with normal transcription in the sense direction in *cis*. Hence, although the *HBA2* gene retains all of its local and remote *cis*-regulatory elements, its expression is silenced and its CpG island becomes completely methylated early in development.

Finally, a rare form of α-thalassemia has been shown to occur via a gain-of-function regulatory SNP (rSNP) [28] in a nongenic region between the α-globin genes and their upstream regulatory elements. In particular, this rSNP creates a GATA-1 binding site, nucleates the binding of a pentameric erythroid complex including the transcription factors SCL, E2A, LMO2, and Ldb-1, which are frequently found with GATA-1 at erythroid regulatory elements, and binds RNA polymerase II. The rSNP, therefore, creates a new promoter-*like* element that interferes with normal activation of all downstream α-*like* globin genes. These findings not only demonstrate an additional mechanism causing human genetic disease, but also illustrate that SNPs of functional significance can also be identified in genomic regions, such as multispecies conserved noncoding (CNC) regions, harboring no *cis*-regulatory elements.

11.6.2 β-Thalassemia Attributable to Transcription Factor Mutations

Apart from the plethora of *HBB* mutations leading to β-thalassemia, a few β-thalassemia patients with their phenotype resulting from mutations in genes encoding for general and erythroid-specific transcription factors have been described. Viprakasit et al. have demonstrated that *HBB* gene expression is affected by a mutation in the XPD subunit of the general transcription factor TFIIH [170]. In particular, it has been shown that β-globin chain synthesis is only affected in trichothiodystrophy (TTD), and not in xeroderma pigmentosum (XP), patients. Using standard hematological assays, TTD patients have been identified with substantially elevated Hb A_2 levels, one of the hallmarks of β-thalassemia, reduced mean red cell volume and mean cell hemoglobin values, and no *HBB* genetic defect. In addition, α-globin chain levels were unaffected [170]. These data provide insights into the mechanism that enables a defect in a general transcription factor to affect the expression of specific genes, the *HBB* gene being the first example to confirm this hypothesis. Also, the fact that *XPD* mutations affect *HBB* but not *HBA2/HBA1* gene expression also suggests that *HBB* and *HBA/HBA1* gene promoters may differ in their requirement for TFIIH and, by implication, initiation of transcription from the human *HBB* and *HBA/HBA1* gene promoters may occur via different mechanisms. This would be consistent with many previously described structural and functional differences between the *HBB* and *HBA/HBA1* genes (Table 11.2).

Finally, although mutations in the erythroid-specific transcription factor GATA-1 have been reported to also result in β-thalassemia [180], there has been no mutation found in the erythroid-specific transcription factor KLF-1 in atypical β-thalassemia patients [42].

11.7 Population Genetics of Hemoglobin Genes

The presence of a relatively large number of DNA polymorphisms at the human β-globin gene locus has enabled the elucidation of the geographic origin and spread of several mutations leading to hemoglobinopathies. Initial evidence suggested that the mutation leading

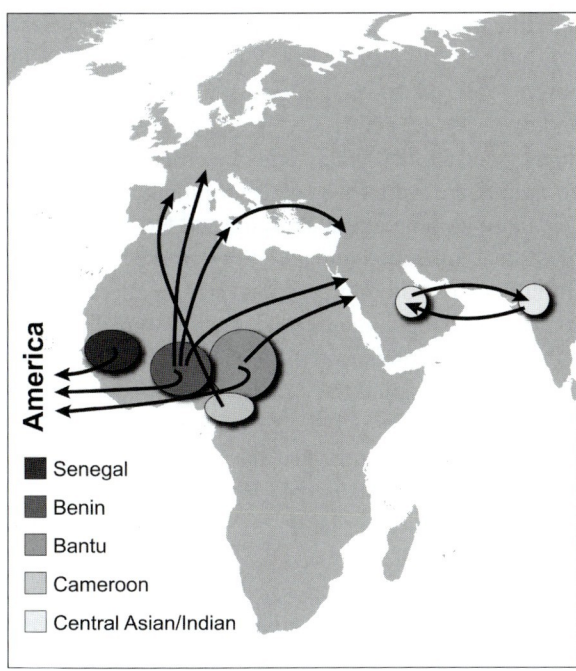

Fig. 11.12 Multicentric origin of the sickle cell mutation in Africa and Asia. *Arrows* indicate the migration routes of each β^S chromosome in various European and Asian populations and the Americas

to SCD (β^S, p.E6V) occurred at least four times in different geographic areas in Africa (Senegal, Benin, Central African Republic and the Bantu-speaking area) [85, 118] and once in Asia (Arab/Indian) [81] and then spread in various regions because of the selective advantage of Hb S carriers vis-à-vis malaria. In total, the β^S mutation has been observed in at least 16 different chromosomal backgrounds [7]. Interestingly, European populations are extremely homogeneous in terms of the β^S mutation origin. In particular, the Benin type of the β^S mutation is predominantly found in Mediterranean populations, namely, Sardinian, Greek, Turkish, and Cypriot, while both the Benin and Senegalese types of the β^S mutation are spread out over the west and northwestern parts of Africa (Fig. 11.12). However, the β^S mutation found in the African-American populations is largely heterogeneous owing to population migrations to the American continent (Fig. 11.2).

As with the β^S, the β^E mutation has been found on three different chromosomal backgrounds in Southeast Asia [5] and on two in European population [76].

While the haplotype data suggest that several mutational events were the origin of Hb S and Hb E chromosomes, it is likely that recurrent mutations are but one of three possible mechanisms, the other two being recombination and gene conversion events (see also Sect. 11.4). The latter events are likely mechanisms to explain the spread of several thalassemia mutations in different chromosomal backgrounds within a single population or ethnic group or closely related ones. In other words, the various common β-thalassemia mutations usually occurred in a unique haplotype with subsequent expansion of such chromosomes because of malarial selection. The p.Q39X nonsense mutation (*HBB*: c.118C>T), having been found in all nine haplotypes in the Sardinian population, stands out as the most characteristic example [136]. Other examples include: (1) *HBB*:c.79G>A [β^E mutation [5], (2) *HBB*:c.20delA [77], (3) *HBB*:c.124_127delTTCT [61] and (4) *HBB*:c.92+5G>T and *HBB*:c.316-197C>T [181].

Interestingly, the chromosomal background of a certain *HBB* mutation leading to hemoglobinopathies is linked to the symptoms' severity. To this end, SCD patients bearing the Senegalese or Asian/Indian haplotypes present with milder symptoms than do patients bearing the Bantu, Benin and CAI haplotypes [83, 118], owing to the higher Hb F levels in the former patient group. Similar data have been recently provided for Hb E/β^0-thalassemia compound heterozygotes in Southeast Asian populations [95].

11.8 Diagnosis of Hemoglobinopathies

Although symptom-free, individuals with thalassemia trait are characterized by a specific hematological profile, which can be indicative for their genotype. A number of comprehensive guidelines for laboratory diagnosis of hemoglobinopathies have been published; all this lies outside the scope of this chapter and will not be discussed in detail.

11.8.1 Carrier Screening

The aim of carrier screening is to identify carriers of hemoglobinopathies in order to assess the risk of a couple having a severely affected child and to provide information on the options available to avoid such an eventuality. Ideally, screening is performed before pregnancy. There are several possible strategies for screening, depending on factors such as disease frequency, heterogeneity of the genetic defects, resources available, and social, cultural and religious factors. Knowing the frequency and heterogeneity of the hemoglobinopathies in a population is critical for the planning of an adequate strategy of carrier identification and for selection of the most suitable laboratory methods.

There are two types of screening: mass screening, provided to the general population before and during childbearing age, and targeted screening, which is restricted to a particular population group, such as couples preparing to marry, before conception or in early pregnancy. Screening may be "retrospective," when couples already have an affected child, or "prospective," when carriers are identified before having an affected child. Screening can be also targeted at different age groups, such as newborn, currently only recommended for sickle cell disease, and adolescent screening [3]. Premarital testing is carried out in several Mediterranean countries (Greece, Italy, and Cyprus). In Cyprus, couples are required to produce a certificate of carrier testing before they can be married. This, together with prenatal diagnosis and genetic counseling, has contributed to the marked decrease in the number of observed homozygous births since the beginning of the program in mid-1970s (Fig. 11.8).

Carrier identification strategies should ensure that no carrier eludes detection. There are two possible methodological approaches for β-thalassemia carrier identification: (a) a primary screen to determine red cell indices, followed by a secondary screen involving hemoglobin analysis in subjects with reduced MCV and/or MCH, recommended in countries with low frequency and limited thalassemia heterogeneity; and (b) complete screening based on determining red cell indices, hemoglobin pattern analysis, and Hb A_2 measurement, recommended in populations where both α- and β-thalassemias are common and where interaction of α- and β-thalassemias could lead to misdiagnoses attributable to the normalization of red cell indices. A flowchart illustrating the strategy used to identify carriers in high-risk populations is shown in Fig. 11.13.

In brief, considering that the iron status (metabolism) is normal, reduced red blood cell (RBC) indices, i.e., mean corpuscular hemoglobin (MCH<28 pg) and mean corpuscular volume (MCV<81fl) levels, are suggestive of thalassemia heterozygosity. If accompanied by elevated Hb A_2 levels, a diagnosis of β-thalassemia trait, co-existing or not with α-thalassemia, is made. Reduced RBC indices accompanied by normal Hb A_2 levels are indicative of α-thalassemia trait, co-existence of α- and δ-thalassemia traits, or co-inheritance of δ- and β-thalassemia alleles. In the latter case (also known as normal Hb A_2-β-thalassemia), caution should be exercised, as the β-thalassemia condition can be overlooked, leading to misdiagnosis.

Reduced MCH and MCV values accompanied with elevated Hb F levels are strong indicators of δβ-thalassemia or compound heterozygosity of β-thalassemia

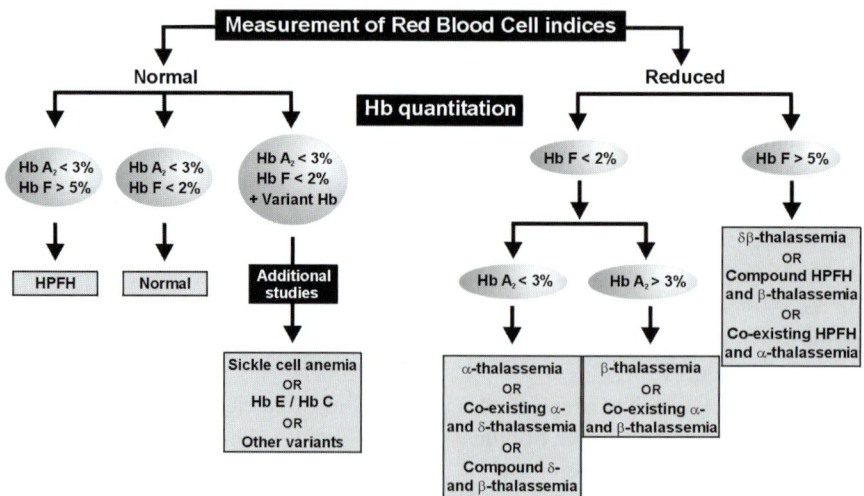

Fig. 11.13 Carrier screening and differential diagnosis of hemoglobinopathies. (From [130])

and HPFH, or co-existence of α-thalassemia with HPFH. In cases where the MCH and MCV values are normal, qualitative, and quantitative hemoglobin analysis is required in order to distinguish between a normal individual, an HPFH hetero- or homozygote (Hb F>5%), and a carrier of Hb S or any other structural hemoglobin variant (Hb C, Hb E, or other).

These cut-off indices are the most widely used; however, appropriate reference values should be independently defined for each population, as there may be slight differences according to the types of thalassemia alleles present. In addition, there should be regular quality control programs in order to monitor the accuracy of laboratory results.

11.8.2 Hematological and Biochemical Methods

With the exception of newborn screening, hematological and biochemical investigations of an individual for a suspected hemoglobinopathy provide useful insights into the hematological phenotype. Routine biochemical investigations mainly consist of electrophoretic and/or chromatographic analysis of an individual's hemoglobin fractions and globin chains.

Hematological Indices. The individual's hematological profile consists of the measurement of the RBC indices and includes hemoglobin concentration, hematocrit, RBC number, MCH, MCV, and red cell distribution (RDW, an indicator of RBC size variation). Routinely, a blood film accompanies the RBC indices. Depending on the severity of the thalassemia condition, minimal or major decrements in most of the RBC indices are observed. Also, quantitation of the hemoglobin fractions, using a variety of electrophoretic and chromatographic techniques, is an essential part of the hematological profile (see below). Iron deficiency alters RBC indices. If necessary, additional investigations can be performed, such as hemoglobin heat stability and sickling tests, oxygen dissociation studies, and in vitro globin chain synthesis, in order to detect and/or better characterize a structural hemoglobin variant.

Biochemical Methods. Electrophoresis has been the method of choice in a traditional hematological laboratory for qualitative and quantitative analysis of the various hemoglobin fractions. Cellulose acetate electrophoresis at alkaline pH (8.2–8.6) and citrate agar or agarose gel electrophoresis at acid pH (6.0–6.2) allows separation of the major hemoglobins, i.e., Hb A, Hb F, Hb S/D, Hb C/E/O-Arab, and a number of less common hemoglobin variants. Because of its simplicity, cellulose acetate electrophoresis remains among the most popular methods for hemoglobin screening. However, apart from being laborious, these electrophoretic techniques have the disadvantages of poor precision and accuracy of hemoglobin quantitation. Also, urea-triton gel electrophoresis provides rapid analysis of very small amounts of hemoglobin from hemolysates and permits examination of globin chain composition as well as globin synthetic ratios [2]. In some cases, mutant globin chains could also be separated with this electrophoretic technique. Finally, isoelectric focusing (IEF) on agarose gels can be used to separate different hemoglobin fractions and globin chains.

Chromatographic methods are also widely used for hemoglobin quantitation and initial screening of hemoglobin variants. Cation exchange-high performance liquid chromatography (CE-HPLC) has become the method of choice to quantify the various normal and abnormal hemoglobin fractions [143]. This method tends to replace electrophoretic techniques for primary screening of hemoglobin of clinical significance and to be at least an additional tool for identification of hemoglobin variants [74]. In recent years, reversed-phase HPLC (RP-HPLC) of globin chains has become an important tool in the study of hemoglobin abnormalities [87]. It has mostly been used for measuring the γ-globin chain ratios in various hemoglobin disorders, but owing to its high sensitivity, the proposed method may be also useful in the diagnosis of hemoglobinopathies and in the detection and study of hemoglobin variants, even those that are indistinguishable through the battery of electrophoretic tests.

11.8.3 Molecular Diagnostics of Hemoglobinopathies

Although a plethora of molecular diagnostic techniques are available for genetic testing of hemoglobin disorders, DNA sequencing is the ultimate method for the definitive identification of unknown sequence alterations, which should always be coupled to additional mutation scanning methods. In the vast majority of analyzes for molecular diagnosis of thalassemias, DNA is extracted from peripheral blood leukocytes, while chorionic villus and amniotic fluid cells are used as DNA source for prenatal diagnosis, fetal cells, and fetal DNA isolated from the maternal plasma and serum [22]. The noninvasive sampling methods are still under development and hence not widely offered for globin genes mutation screening. In preimplantation genetic diagnosis (PGD) polar bodies or blastomeres, either at the cleavage or at the blastocyst stage, are isolated from preimplantation embryos and these cells are used for DNA extraction to perform genetic diagnosis.

The available mutation detection methodology for human globin genes include screening for known mutations, using restriction endonuclease analysis and allele-specific mutation detection or amplification, and whole-gene scanning methods where the mutation in question in not known, using denaturing gradient gel electrophoresis (DGGE) or single-stranded conformation polymorphism (SSCP) analysis, both of which are characterized by their high discriminatory potential [130]. In recent years, denaturing HPLC (DHPLC) has been gradually taken into use by several diagnostic laboratories, as it provides a semiautomated, fast and reliable alternative to DGGE.

Regarding routine detection of large deletion mutants leading to α-thalassemia and a few other β-type hemoglobinopathies, gap-PCR provides an alternative to Southern blot analysis for routine detection of such mutants [23], as it is a rapid, cost-effective approach and simple to use. Today, gap-PCR is the ultimate method for reliable detection of the common deletional α-thalassemia alleles, Hb Lepore, or several deletion mutants in the human β-globin gene cluster, leading to δβ-thalassemia or deletional HPFH. The rationale behind gap-PCR is the generation of wild-type and deletion-specific amplification products from one each, respectively, of two oligonucleotide primer pairs, from which one primer is common. Alternatively, a three-color multiplex ligation-dependent probe amplification (MPLA) approach has been described [59]. This approach utilizes two sets of probes, one of which is α- and the other, β-globin cluster specific, and aims to detect rearrangements in these clusters. This approach is a rapid and sensitive method for high-resolution analysis of the globin gene clusters and an attractive alternative to conventional techniques and gap-PCR. Ultimately, oligonucleotide tiling microarrays could also be used to detect all of these abnormalities.

High-Throughput Globin Gene Mutation Screening. Recent developments in automation and miniaturization technologies have created new standards and changed the practice of molecular diagnostics in the postgenomic era. DNA microarrays have allowed high-throughput mutation detection and large-scale DNA sequencing. There are several microarray-based human globin gene mutation screening methods [27], based on single-base extension, arrayed primer extension, or a microelectronic allele-specific oligonucleotide (ASO)-*like* and multiple ligation probe amplification (MLPA)-*like* approaches. Finally, pyrosequencing is also gaining momentum for resequencing of globin genomic regions.

11.8.4 Prenatal and Preimplantation Genetic Diagnosis of Hemoglobinopathies

Prenatal and preimplantation genetic diagnoses of hemoglobinopathies both aim at reducing the number of affected individuals in at-risk populations.

Prenatal diagnosis became possible in the early 1970s, and in most European countries prenatal diagnosis is available for couples at risk for hemoglobin disorders, with the option of selective termination of pregnancy after genetic counseling [134]. However, it is widely discussed that the latter cannot be considered an optimal or easy solution. Chorionic villus sampling under ultrasound guidance in the first trimester of pregnancy (10–12 weeks of gestation) usually provides sufficient amounts of DNA for further analysis, with a relatively low risk of procedure-related pregnancy loss (0.6%).

Virtually all the DNA diagnostic methods reported for thalassemia diagnosis are suitable for prenatal diagnosis. It is highly recommended, however, that fetal DNA analysis be performed by two independent DNA diagnostic techniques, both yielding unambiguous results before a diagnosis can be made. Maternal contamination of the fetal tissue is the most common problem, which can lead to false-positive or false-negative results. The inheritance of polymorphic markers could always be used to "diagnose" contamination.

PGD is an early form of prenatal diagnosis, in which oocytes or preimplantation embryos are genetically analyzed, so that only those that are judged to be free of the genetic defect under consideration are transferred back to the mother. Detection of mutations in the *HBB* and *HBA2/HBA1* genes, leading to SCA and β- and α-thalassemia, has received considerable attention, because of their high frequency in certain populations. A handful of protocols have been reported for PDG of hemoglobinopathies [130].

11.8.5 Genetic Counseling

Genetic counseling is very important in view of the phenotypic diversity of thalassemia. Generally, the inheritance of two pathogenic *HBB* mutations results in a blood transfusion-dependent thalassemia. However, there are mild β-thalassemia mutations resulting in thalassemia intermedia. For instance, when both parents carry the mild *HBB*:c.-138C>T mutation, the homozygous state generally results in a very mild clinical phenotype. This type of mild thalassemia poses an ethical dilemma, both to parents and to health professionals, as far as prenatal diagnosis and possible termination of pregnancy are concerned. Nevertheless, there are situations in which the couple can be reassured that the phenotype of their affected child will be mild, such as (a) when the parents are silent carriers of β-globin gene mutations, i.e., the *HBB*:c.-151C>T mutation, resulting in very mild clinical phenotype in the homozygous state, (b) when mild and severe β-thalassemia mutations are likely to be interacting in the compound heterozygous state, and (c) when β-thalassemia is co-inherited with α-globin gene triplication (ααα) or an HPFH condition. Prenatal diagnosis is not offered for these cases [133].

Counseling couples who are at risk for SCD is often perceived as relatively simple, but in fact it can be significantly more complex, because of the variation in severity of SCD [150]. In contrast, counseling couples at risk for α-thalassemia is more straightforward, because of the usually severe phenotype for an affected fetus and the likelihood of life-threatening obstetric risks for the mother [133].

11.9 HbVar Database for Hemoglobin Variants and Thalassemia Mutations

In the late 1990s, Titus Huisman published two books that recorded information on hemoglobin variants and thalassemias, entitled *A Syllabus of Human Hemoglobin Variants* (2nd edn) [70] and *A Syllabus of Thalassemia Mutations* [69], which were a rich source of information not only about the mutations, but also about the methods used in detection and analysis, their biochemical properties, associated clinical effects, ethnic distribution, and other data. The sheer amount of information tabulated, the continuous accumulation of new mutation data, and their complexity dictated the need for construction of a globin-specific database as an up-to-date and accessible repository of this information, which has occurred in three discrete stages. HbVar is a comprehensive locus-specific database for the globin genes, realized from a multicenter academic initiative to provide up-to-date information on the various genetic defects leading to hemoglobinopathies [58]. This database, accessible at http://globin.bx.psu.edu/hbvar,), includes detailed information on pathology, hematology, clinical presentation, and laboratory findings (range of hemoglobin levels, hematocrit, etc.) on most of the published and unpublished hemoglobin variants and thalassemia mutations, while considerable biochemical data on the variants is also recorded, including techniques used to identify, isolate, and determine their structure, stability, function, and qualitative distribution in ethnic groups and geographic locations [128]. These data can be easily accessed through summary listings or user-generated queries, which can be highly specific.

Also, HbVar has been linked with the *GALA* database [49], since for several studies the information in HbVar needs to be combined with the wealth of information about features of genomic DNA, such as gene structures, interspecies sequence conservation, and many others, while an online repository for experimental protocols for detecting the globin gene variation has been also developed, as part of HbVar associated resources [50]. Finally, HbVar was one of the locus-specific databases implemented for the PhenCode project, in an effort to combine phenotypic, genotype and clinical data [51].

HbVar is useful not only for the research community, for geneticists and physicians as an aid in diagnosis, but also to other interested individuals, such as patients and their parents, people involved in the provision of genetic services and counseling and pharmaceutical industries. Online Mendelian Inheritance in Man (OMIM; http://www.ncbi.nlm.nih.gov/omim) and the Human Gene Mutation Database (HGMD; http://www.hgmd.cf.ac.uk) also contain information on globin gene mutations. However, OMIM lacks the

11.10 Therapeutic Approaches for the Thalassemias

The gradual elucidation of the pathophysiology of thalassemia has led to the development of new strategies in attempts to cure or mitigate the disease. These potential therapeutic approaches can be divided into three categories, aiming to: (1) address the reduced or absent globin chain synthesis directly; (2) compensate for the reduced or absent globin chain synthesis; and (3) treat thalassemia complications, e.g., by decreasing iron overload and reducing oxidative stress. Apart from these approaches, there are also others, such as RNA interference (RNAi) to abrogate expression of mutant *HBB* alleles, that have only been experimentally used [163]. Although these strategies have not yet been used in practice, they hold promise for their future therapeutic potential [140].

11.10.1 Hematopoietic Stem Cell Transplantation

Hematopoietic stem cell transplantation is conceptually the simplest, and so far the only, approach that may lead to a definitive cure for β-thalassemia. Patients who are likely to benefit most from such treatment are those with early but severe transfusion-dependent disease [48]. However, even in the ideal case scenarios of leukocyte antigen (HLA)-identical family donors, this treatment is associated with 5% mortality and the receipients patients still have a high risk of developing tissue damage with time, such as impaired growth, gonadal failure, and chronic graft-vs-host disease (GVHD) [82]. These disadvantages, namely, donor shortage and GVHD (in the case of allogeneic stem cell transplantation), can be overcome by gene therapy of autologous hematopoietic stem cells (see below).

11.10.2 Pharmacological Reactivation of Fetal Hemoglobin

An alternative therapeutic approach involves pharmacological up-regulation of the fetal globin genes, which are otherwise silent in the adult hematopoietic stage, to compensate for the reduced or absent β-globin chain synthesis in patients with β-type hemoglobinopathies. There are three classes of pharmacological agents that have been shown to be capable of inducing Hb F to therapeutic levels: (1) recombinant human erythropoietin (rhEPO) preparations, acting on human erythropid progenitors cells during erythroid differentiation, (2) short-chain fatty acid derivatives, exerting their effect by promoting cell differentiation and enhancing gene expression, and (3) chemotherapeutic cytotoxic agents, which terminate actively cycling progenitors and, hence, trigger rapid erythroid regeneration and formation of mature red blood cells that contain Hb F. This approach has a number of disadvantages. First of all, the reactivation of human fetal globin genes expression is transient. Secondly, the mechanisms underlying the pharmacological induction of Hb F synthesis remain unclear, complex, and likely to involve multiple pathways, including alteration of the chromatin structure, e.g., by inhibition of histone deacetylase (HDAC) activity, hypomethylation of the human fetal globin promoters, and acceleration of erythroid cell differentiation.

Hydroxycarbamide (hydroxyurea; HU) is the only Federal Drug Authority (FDA)-approved agent used to stimulate Hb F synthesis with proven efficacy in SCD. Despite its use, the mechanism of action by which HU induces Hb F still remains uncertain. HU is cytotoxic; it inhibits ribonucleotide reductase, which interferes with DNA synthesis in the dividing late erythroid progenitors. This leads to a transient arrest of hematopoiesis, enhancing the premature commitment of earlier progenitor cells that still possess the primitive Hb F program [117]. HU has a relatively good safety profile with a low risk of carcinogenicity. Although HU has been shown to be highly effective in reducing the frequency of painful crises in SCD, with increases of Hb F levels, its efficacy in β-thalassemia patients has been less convincing. There have been few reports of its clinical efficacy in the severe transfusion-dependent β-thalassemia patients [15]. The future of HU looks more promising in the thalassemia intermedia syndromes in which sustained modest increases in Hb F can be maintained.

Short-chain fatty acids, such as butyrate and its analogs, are thought to act as HDAC inhibitors. One of the major caveats with the use of butyrates therapeutically is that they have short half-lives; they require continuous intravenous administration for demonstrable efficacy, and they may also inhibit erythroid cell growth. 5-Azacytidine and decitabine are DNA methyltransferase inhibitors, thus acting as hypomethylating agents. These drugs have been shown to have an action that is beneficial to β-thalassemia patients [30, 89].

11.10.3 Pharmacogenomics and Therapeutics of Hemoglobinopathies

The concept that genetic variation contributes to variability in drug responses is widely accepted and has been validated in many research settings. The application of pharmacogenetic testing and pharmacogenomics in hemoglobinopathy therapeutics is particularly attractive, because of the limited therapeutic possibilities presently available and the narrow therapeutic drug index, namely, iron chelation and fetal hemoglobin (Hb F)-inducing agents.

Hb F response of β-hemoglobinopathies patients to HU treatment is variable, particularly in β-thalassemia, with approximately 25% of the patients being poor or nonresponders [117]. Therefore, a capacity to predict a patient's Hb F response to HU and/or other drugs used for the same purpose would aid the selection of patients for treatment and reduce toxicity from unhelpful dose escalation. Polymorphisms in genes regulating Hb F expression, HU metabolism, and erythroid progenitor proliferation might modulate the patient response to Hb F-inducing pharmacological agents.

Correlation of SNPs linked to the human β-globin locus with HbF induction upon HU treatment is a controversial issue. Two association studies investigated β-globin locus-related SNPs, indicating that the most significant modulating factor involved in good and moderate response to HU was positively correlated with the *HBG1*-158 T allele [179], while in another study this correlation was not confirmed [35, 90]. On the other hand, association studies in genomic regions not linked to the human β-globin locus revealed putatively useful pharmacogenetic markers for HU treatment [94].

Data supporting the use of pharmacogenetic testing for HU treatment of hemoglobinopathies are currently very limited [125], and similar studies should also be conducted with more pharmacological agents and different treatment modalities. However, candidate gene studies attempting to associate SNPs with phenotypes of varying severity related to hemoglobinopathies should be interpreted with caution, as a candidate gene approach is necessarily limiting and unlikely to examine all genes affecting a phenotype. Whole-genome association pharmacogenetic studies are only just beginning, facilitated by the most technological advances in microarray-based transcription profiling and genotyping (G.P. Patrinos and F. Grosveld, unpublished work). Such studies may identify different types of genes: (a) genes encoding for putative stage-specific transcription factors that down-regulate the γ-globin genes or sustain high Hb F levels, (b) novel erythroid genes that participate in regulatory pathways involved in erythroid cell differentiation and which can potentially alter the stringent developmental program governing globin gene expression and/or (c) genes involved in the Hb F-inducing drug metabolism. Results from such studies may also better orientate pharmacogenetic marker identification only in those genes that are differentially expressed in good and poor responders to Hb F-inducing therapy. This will in turn facilitate the design of customized high-throughput pharmacogenetic tests for hemoglobinopathies based on the arsenal of the currently available genetic testing methods for globin gene mutation screening.

11.10.4 Gene Therapy

Gene therapy, i.e., the insertion of normal DNA directly into cells to correct a genetic defect of β-thalassemia aims to achieve therapeutic levels of functional *HBB* gene expression leading to more than 20% of total hemoglobin production. Efficient gene therapy requires erythroid lineage specificity and should be carried out by safe and efficient viral or nonviral transfection of autologous hematopoietic stem cells. Early attempts included conventional oncoretroviral systems carrying the *HBB* gene and β-globin LCR "core" elements of the β-LCR and adeno-associated virus (AAV) vectors, but both were unsuccessful because of vector instability, low viral titers, and variable, low β-globin gene expression [38] and position-variegated expression. In 2000, May et al. reported that a lentivirus vector carrying a large human *HBB* gene fragment under the control of the β-globin LCR could yield stable therapeutic levels of *HBB* gene expression [97]. Unlike retrovirus, lentiviral vectors are highly suitable for gene therapy of hemoglobinopathies, as they can transduce nondividing cells, are characterized by relative genomic stability and larger packaging capacity, e.g., can accommodate larger DNA fragments, and are much safer, owing to their self-inactivating design that include the removal of the viral long terminal repeats (LTRs) upon integration. Therefore, lentiviral vectors can stably include larger DNA inserts without rearrangements, allowing the incorporation of larger LCR fragments and chromatin insulator elements to overcome the problem of positional effects with the ability to sustain therapeutic levels of expression. The pioneering study of the Sadelain group [97] paved the way in the field, and preclinical studies, using third-generation lentiviral vectors, are currently under way; provided that the risk of insertional oncogenesis is adequately addressed [57], they hold great promise for clinical trials on gene therapy for hemoglobinopathies.

Acknowledgments We regret that, owing to space limitations, much of the relevant work could not be cited.

References

1. Adams JG III, Marrison WT, Steinberg MH (1982) Hemoglobin Parchman: double crossover within a single human gene. Science 218:291–293
2. Alter BP, Goff SC, Efremov GD, Gravely ME, Huisman TH (1980) Globin chain electrophoresis: a new approach to the determination of the G gamma/A gamma ratio in fetal haemoglobin and to studies of globin synthesis. Br J Haematol 44:527–534
3. Angastiniotis MA, Hadjiminas MG (1981) Prevention of thalassaemia in Cyprus. Lancet 1:369–371
4. Antonarakis SE, Boehm CD, Giardina PJ, Kazazian HH Jr (1982) Nonrandom association of polymorphic restriction sites in the beta-globin gene cluster. Proc Natl Acad Sci USA 79:137–141
5. Antonarakis SE, Orkin SH, Kazazian HH Jr, Goff SC, Boehm CD, Waber PG, Sexton JP, Ostrer H, Fairbanks VF, Chakravarti A (1982) Evidence for multiple origins of the beta E-globin gene in Southeast Asia. Proc Natl Acad Sci USA 79:6608–6611
6. Antonarakis SE, Orkin SH, Cheng TC, Scott AF, Sexton JP, Trusko SP, Charache S, Kazazian HH Jr (1984) beta-Thalassemia in American Blacks: novel mutations in the "TATA" box and an acceptor splice site. Proc Natl Acad Sci USA 81:1154–1158
7. Antonarakis SE, Boehm CD, Serjeant GR, Theisen CE, Dover GJ, Kazazian HH Jr (1984) Origin of the beta S-globin gene in blacks: the contribution of recurrent mutation or gene conversion or both. Proc Natl Acad Sci USA 81:853–856
8. Antonarakis SE, Kazazian HH Jr, Orkin SH (1985) DNA polymorphism and molecular pathology of the human globin gene clusters. Hum Genet 69:1–14
9. Antoniou M, deBoer E, Habets G, Grosveld F (1988) The human beta-globin gene contains multiple regulatory regions: identification of one promoter and two downstream enhancers. EMBO J 7:377–384
10. Baralle FE, Shoulders CC, Proudfoot NJ (1980) The primary structure of the human epsilon-globin gene. Cell 21:621–626
11. Bellingham AJ (1976) Haemoglobins with altered oxygen affinity. Br Med Bull 32:234–238
12. Blackwood EM, Kadonaga JT (1998) Going the distance: a current view of enhancer action. Science 281:61–63
13. Bodine D, Ley TJ (1987) An enhancer element lies 3' to the human A gamma globin gene. EMBO J 6:2997–3004
14. Boehm CD, Antonarakis SE, Phillips JA 3rd, Stetten G, Kazazian HH Jr (1983) Prenatal diagnosis using DNA polymorphisms. Report on 95 pregnancies at risk for sickle-cell disease or beta-thalassemia. N Engl J Med 308:1054–1058
15. Bradai M, Abad MT, Pissard S, Lamraoui F, Skopinski L, de Montalembert M (2003) Hydroxyurea can eliminate transfusion requirements in children with severe beta-thalassemia. Blood 102:1529–1530
16. Bulger M, Groudine M (1999) Looping versus linking: toward a model for long-distance gene activation. Genes Dev 13:2465–2477
17. Busslinger M, Moschonas N, Flavell RA (1981) Beta1thalassemia: aberrant splicing results from a single point mutation in an intron. Cell 27:289–298
18. Carter D, Chakalova L, Osborne CS, Dai YF, Fraser P (2002) Long-range chromatin regulatory interactions in vivo. Nat Genet 32:623–626
19. Chakravarti A, Buetow KH, Antonarakis SE, Waber PG, Boehm CD, Kazazian HH (1984) Nonuniform recombination within the human beta-globin gene cluster. Am J Hum Genet 36:1239–1258
20. Chen JM, Cooper DN, Chukalova N, Ferec C, Patrinos GP (2007) Gene conversion in evolution and disease. Nat Rev Genet 8:762–775
21. Chen Z, Luo HY, Basran RK, Hsu TH, Mang DW, Nuntakarn L, Rosenfield CG, Patrinos GP, Hardison RC, Steinberg MH, Chui DH (2008) A T-to-G transversion at nucleotide -567 upstream of HBG2 in a GATA-1 binding motif is associated with elevated hemoglobin F. Mol Cell Biol 28:4386–4393
22. Chiu RW, Lau TK, Leung TN, Chow KC, Chui DH, Lo YM (2002) Prenatal exclusion of beta thalassaemia major by examination of maternal plasma. Lancet 360:998–1000
23. Chu DC, Lee CH, Lo MD, Cheng SW, Chen DP, Wu TL, Tsao KC, Chiu DT, Sun CF (2000) Non-radioactive Southern hybridization for early diagnosis of alpha-thalassemia with southeast Asian-type deletion in Taiwan. Am J Med Genet 95:332–335
24. Collins FS, Weissman SM (1984) The molecular genetics of human hemoglobins nucleic acid. Prog Res Mol Biol 31:315–462
25. Cooley TB, Lee P (1925) A series of cases of splenomegaly in children with anemia and peculiar bone changes. Trans Am Pediatr Soc 37:29
26. Craig JE, Rochette J, Fisher CA, Weatherall DJ, Marc S, Lathrop GM, Demenais F, Thein SL (1996) Dissecting the

loci controlling fetal haemoglobin production on chromosomes 11p and 6q by the regressive approach. Nat Genet 12:58–64
27. Cremonesi L, Ferrari M, Giordano PC, Harteveld CL, Kleanthous M, Papasavva T, Patrinos GP, Traeger-Synodinos J (2007) An overview of current microarray-based human globin gene mutation detection methods. Hemoglobin 31:289–311
28. De Gobbi M, Viprakasit V, Hughes JR, Fisher C, Buckle VJ, Ayyub H, Gibbons RJ, Vernimmen D, Yoshinaga Y, de Jong P, Cheng JF, Rubin EM, Wood WG, Bowden D, Higgs DR (2006) A regulatory SNP causes a human genetic disease by creating a new transcriptional promoter. Science 312:1215–1217
29. Dekker J, Rippe K, Dekker M, Kleckner N (2002) Capturing chromosome conformation. Science 295:1306–1311
30. DeSimone J, Koshy M, Dorn L, Lavelle D, Bressler L, Molokie R, Talischy N (2002) Maintenance of elevated fetal hemoglobin levels by decitabine during dose interval treatment of sickle cell anemia. Blood 99:3905–3908
31. Dhar V, Skoultchi AI, Schildkraut CL (1989) Activation and repression of a beta-globin gene in cell hybrids is accompanied by a shift in its temporal replication. Mol Cell Biol 9:3524–3532
32. Dillon N, Grosveld F (1991) Human gamma-globin genes silenced independently of other genes in the beta-globin locus. Nature 350:252–254
33. Dillon N, Trimborn T, Strouboulis J, Fraser P, Grosveld F (1997) The effect of distance on long-range chromatin interactions. Mol Cell 1:131–139
34. Dimovski AJ, Adekile AD, Divoky V, Baysal E, Huisman TH (1994) Polymorphic pattern of the (AT)XTY motif at -530 5' to the beta-globin gene in over 40 patients homozygous for various beta-thalassemia mutations. Am J Hematol 45:51–57
35. Dixit A, Chatterjee TC, Mishra P, Choudhry DR, Mahapatra M, Tyagi S, Kabra M, Saxena R, Choudhry VP (2005) Hydroxyurea in thalassemia intermedia-a promising therapy. Ann Hematol 84:441–446
36. Dover GJ, Smith KD, Chang YC, Purvis S, Mays A, Meyers DA, Sheils C, Serjeant G (1992) Fetal hemoglobin levels in sickle cell disease and normal individuals are partially controlled by an X-linked gene located at Xp22.2. Blood 80:816–824
37. Drissen R, Palstra RJ, Gillemans N, Splinter E, Grosveld F, Philipsen S, de Laat W (2004) The active spatial organization of the beta-globin locus requires the transcription factor EKLF. Genes Dev 18:2485–2490
38. Dzierzak EA, Papayannopoulou T, Mulligan RC (1988) Lineage-specific expression of a human beta-globin gene in murine bone marrow transplant recipients reconstituted with retrovirus-transduced stem cells. Nature 331:35–41
39. Eaton WA, Hofrichter J (1995) The biophysics of sickle cell hydroxyurea therapy. Science 268:1142–1143
40. Ellis J, Tan-Un KC, Harper A, Michalovich D, Yannoutsos N, Philipsen S, Grosveld F (1996) A dominant chromatin-opening activity in 5-hypersensitive site 3 of the human beta-globin locus control region. EMBO J 15:562–568
41. Enver T, Raich N, Ebens AJ, Papayannopoulou T, Costantini F, Stamatoyannopoulos G (1990) Developmental regulation of human fetal-to-adult globin gene switching in transgenic mice. Nature 344:309–313
42. Faa V, Meloni V, Moi L, Ibba G, Travi M, Vitucci A, Cao A, Rosatelli MC (2006) Thalassaemia-like carriers not linked to the beta-globin gene cluster. Br J Haematol 132:640–650
43. Fairbanks VF, Oliveros R, Brandabur JH, Willis RR, Fiester RF (1980) Homozygous hemoglobin E mimics beta-thalassemia minor without anemia or hemolysis: hematologic, functional, and biosynthetic studies of first North American cases. Am J Hematol 8:109–121
44. Farrell CM, West AG, Felsenfeld G (2002) Conserved CTCF insulator elements flank the mouse and human beta-globin loci. Mol Cell Biol 22:3820–3831
45. Flintt J, Harding RM, Boyce AJ (1998) Clegg JB (1998) The population genetics of the haemoglobinopathies. Bailliers Clin Haematol 11:1–51
46. Forrester WC, Epner E, Driscoll MC, Enver T, Brice M, Papayannopoulou T, Groudine M (1990) A deletion of the human beta-globin locus activation region causes a major alteration in chromatin structure and replication across the entire beta-globin locus. Genes Dev 4:1637–1649
47. Garner CP, Tatu T, Best S, Creary L, Thein SL (2002) Evidence of genetic interaction between the beta-globin complex and chromosome 8q in the expression of fetal hemoglobin. Am J Hum Genet 70:793–799
48. Gaziev J, Lucarelli G (2003) Stem cell transplantation for hemoglobinopathies. Curr Opin Pediatr 15:24–31
49. Giardine B, Elnitski L, Riemer C, Makalowska I, Schwartz S, Miller W, Hardison RC (2003) GALA, a database for genomic sequence alignments and annotations. Genome Res 13:732–741
50. Giardine B, van Baal S, Kaimakis P, Riemer C, Miller W, Samara M, Kollia P, Anagnou NP, Chui DH, Wajcman H, Hardison RC, Patrinos GP (2007) HbVar database of human hemoglobin variants and thalassemia mutations: 2007 update. Hum Mutat 28:206
51. Giardine B, Riemer C, Hefferon T, Thomas D, Hsu F, Zielenski J, Sang Y, Elnitski L, Cutting G, Trumbower H, Kern A, Kuhn R, Patrinos GP, Hughes J, Higgs D, Chui D, Scriver C, Phommarinh M, Patnaik SK, Blumenfeld O, Gottlieb B, Vihinen M, Valiaho J, Kent J, Miller W, Hardison RC (2007) PhenCode: connecting ENCODE data with mutations and phenotype. Hum Mutat 28:554–562
52. Gibbons RJ, Higgs DR (2000) Molecular-clinical spectrum of the ATR-X syndrome. Am J Med Genet 97:204–212
53. Gibbons RJ, Picketts DJ, Villard L, Higgs DR (1995) Mutations in a putative global transcriptional regulator cause X-linked mental retardation with alpha-thalassemia (ATR-X syndrome). Cell 80:837–845
54. Gong Q, Dean A (1993) Enhancer-dependent transcription of the epsilon-globin promoter requires promoter-bound GATA-1 and enhancer-bound AP-1/NF-E2. Mol Cell Biol 13:911–917
55. Grosveld F (1999) Activation by locus control regions? Curr Opin Genet Dev 9:152–157
56. Grosveld F, van Assendelft GB, Greaves DR, Kollias G (1987) Position-independent, high-level expression of the human beta-globin gene in transgenic mice. Cell 51:975–985

57. Hacein-Bey-Abina S, von Kalle C, Schmidt M, Le Deist F, Wulffraat N, McIntyre E, Radford I, Villeval JL, Fraser CC, Cavazzana-Calvo M, Fischer A (2003) A serious adverse event after successful gene therapy for X-linked severe combined immunodeficiency. N Engl J Med 348:255–256
58. Hardison RC, Chui DH, Giardine B, Riemer C, Patrinos GP, Anagnou N, Miller W, Wajcman H (2002) HbVar: a relational database of human hemoglobin variants and thalassemia mutations at the globin gene server. Hum Mutat 19:225–233
59. Harteveld CL, Voskamp A, Phylipsen M, Akkermans N, den Dunnen JT, White SJ, Giordano PC (2005) Nine unknown rearrangements in 16p13.3 and 11p15.4 causing alpha- and beta-thalassaemia characterised by high resolution multiplex ligation-dependent probe amplification. J Med Genet 42:922–931
60. Hatton CS, Wilkie AO, Drysdale HC, Wood WG, Vickers MA, Sharpe J, Ayyub H, Pretorius IM, Buckle VJ, Higgs DR (1990) Alpha-thalassemia caused by a large (62 kb) deletion upstream of the human alpha globin gene cluster. Blood 76:221–227
61. Hattori Y, Yamane Y, Yamashiro Y, Matsuno Y, Ki Y, Ku Y, Ohba Y, Miyaji T (1989) Characterization of beta-thalassemia mutations among the Japanese. Hemoglobin 13:657–670
62. Herrick JB (1910) Peculiar elongated and sickle-shaped red blood corpuscles in a case of severe anemia. Arch Intern Med 6:517
63. Higgs DR, Vickers MA, Wilkie AO, Pretorius IM, Jarman AP, Weatherall DJ (1989) A review of the molecular genetics of the human alpha-globin gene cluster. Blood 73:1081–1104
64. Higgs DR, Wood WG, Jarman AP, Sharpe J, Lida J, Pretorius IM, Ayyub H (1990) A major positive regulatory region located far upstream of the human alpha-globin gene locus. Genes Dev 4:1588–1601
65. Hill AV, Flint J, Weatherall DJ, Clegg JB (1987) Alpha-thalassaemia and the malaria hypothesis. Acta Haematol 78:173–179
66. Hirsch RE, Lin MJ, Nagel RL (1988) The inhibition of hemoglobin C chrystallization by hemoglobin S. J Biol Chem 263:5936–5939
67. Huisman TH (1987) A short review of human gamma-globin gene anomalies. Acta Haematol 78:80–84
68. Huisman TH, Wrightstone RN, Wilson JB, Schroeder WA, Kendall AG (1972) Hemoglobin Kenya, the product of the fusion of the gamma- and beta-polypeptide chains. Arch Biochem Biophys 153:850–853
69. Huisman TH, Carver MF, Baysal E (1997) A syllabus of thalassemia mutations. Sickle Cell Anemia Foundation, Augusta, GA
70. Huisman TH, Carver MF, Efremov GD (1998) A syllabus of human hemoglobin variants, 2nd edn. Sickle Cell Anemia Foundation, Augusta, GA
71. Ingram VM (1956) A specific chemical difference between the globins of normal human and sickle cell anaemia haemoglobin. Nature 178:792
72. Ingram VM, Stretton AO (1959) Genetic basis of the thalassaemia diseases. Nature 184:1903–1909
73. Jarman AP, Wood WG, Sharpe JA, Gourdon G, Ayyub H, Higgs DR (1991) Characterization of the major regulatory element upstream of the human alpha-globin gene cluster. Mol Cell Biol 11:4679–4689
74. Joutovsky A, Hadzi-Nesic J, Nardi MA (2004) HPLC retention time as a diagnostic tool for hemoglobin variants and hemoglobinopathies: a study of 60000 samples in a clinical diagnostic laboratory. Clin Chem 50:1736–1747
75. Kazazian HH Jr, Antonarakis SE, Cheng T, Boehm CD, Waber PG (1983) Use of haplotype analysis in the beta-globin gene cluster to discover beta-thalassemia mutations. Prog Clin Biol Res 134:91–98
76. Kazazian HH Jr, Waber PG, Boehm CD, Lee JI, Antonarakis SE, Fairbanks VF (1984) Hemoglobin E in Europeans: further evidence for multiple origins of the beta E-globin gene. Am J Hum Genet 36:212–217
77. Kazazian HH Jr, Orkin SH, Markham AF, Chapman CR, Youssoufian H, Waber PG (1984) Quantification of the close association between DNA haplotypes and specific β-thalassemia mutations in Mediterraneans. Nature 310:152–154
78. Kan YW, Dozy AM (1978) Polymorphism of DNA sequence adjacent to human beta-globin structural gene: relationship to sickle mutation. Proc Natl Acad Sci USA 75:5631–5635
79. Kioussis D, Vanin E, deLange T, Flavell RA, Grosveld FG (1983) Beta-globin gene inactivation by DNA translocation in gamma beta-thalassaemia. Nature 306:662–666
80. Kollias G, Wrighton N, Hurst J, Grosveld F (1986) Regulated expression of human A gamma-, beta-, and hybrid gamma beta-globin genes in transgenic mice: manipulation of the developmental expression patterns. Cell 46:89–94
81. Kulozik AE, Wainscoat JS, Serjeant GR, Kar BC, Al-Awamy B, Essan GJF, Falusi AG, Haque SK, Hilali AM, Kate S, Ranasinghe WA, Weatherall DJ (1986) Geographical survey of betaS-globin gene haplotypes: evidence for an independent Asian origin of the sickle-cell mutation. Am J Hum Genet 39:239–244
82. La Nasa G, Argiolu F, Giardini C, Pession A, Fagioli F, Caocci G, Vacca A, De Stefano P, Piras E, Ledda A, Piroddi A, Littera R, Nesci S, Locatelli F (2005) Bone marrow transplantation in adults with thalassemia: treatment and long-term follow-up. Ann NY Acad Sci 1054:196–205
83. Labie D, Pagnier J, Lapoumeroulie C, Rouabhi F, Dunda-Belkhodja O, Chardin P, Beldjord C, Wajcman H, Fabry ME, Nagel RL (1985) Common haplotype dependency of high G gamma-globin gene expression and high Hb F levels in beta-thalassemia and sickle cell anemia patients. Proc Natl Acad Sci USA 82:2111–2114
84. Lanclos KD, Oner C, Dimovski AJ, Gu YC, Huisman TH (1991) Sequence variations in the 5' flanking and IVS-II regions of the Ggamma- and Agamma-globin genes of betaS chromosomes with five different haplotypes. Blood 77:2488–2496
85. Lapoumeroulie C, Dunda O, Ducrocq R, Trabuchet G, Mony-Lobe M, Bada JM, Carnevale P, Labie D, Elion J, Krishnamoorthy R (1992) A novel sickle cell mutation of yet another origin in Africa: the Cameroon type. Hum Genet 89:333–337
86. Lawn RM, Efstratiadis A, O'Connell C, Maniatis T (1980) The nucleotide sequence of the human beta-globin gene. Cell 21:647–651
87. Leone L, Monteleone M, Gabutti V, Amione C (1985) Reversed phase high-performance liquid chromatography of human haemoglobin chains. J Chromatogr 321:407–419

88. Liebhaber SA, Kan YW (1982) Different rates of mRNA translation balance the expression of the two human alpha-globin loci. J Biol Chem 257:11852–11855
89. Lowrey CH, Nienhuis AW (1993) Treatment with azacitidine of patients with end-stage beta-thalassemia. N Engl J Med 329:845–848
90. Lu ZH, Steinberg MH, Multicenter Study of Hydroxyurea (1996) Fetal hemoglobin in sickle cell anemia: relation to regulatory sequences cis to the beta-globin gene. Blood 87:1604–1611
91. Lumelsky NL, Forget BG (1991) Negative regulation of globin gene expression during megakaryocytic differentiation of a human erythroleukemic cell line. Mol Cell Biol 11:3528–3536
92. Luo HY, Mang D, Patrinos GP, Pourfarzad F, Wuc CJY, Eung SH, Rosenfield CG, Daoust PR, Braun A, Grosveld FG, Steinberg MH, Chui DH (2004) A novel single nucleotide polymorphism (SNP), T>G, in the GATA site at nucleotide (nt) -567 5' to the Ggamma-globin gene may be associated with elevated Hb F. Blood 104:145a–146a
93. Luzzi GA, Merry AH, Newbold CI, Marsh K, Pasvol G, Weatherall DJ (1991) Surface antigen expression on Plasmodium falciparum-infected erythrocytes is modified in alpha- and beta-thalassemia. J Exp Med 173:785–791
94. Ma Q, Wyszynski DF, Farrell JJ, Kutlar A, Farrer LA, Baldwin CT, Steinberg MH (2007) Fetal hemoglobin in sickle cell anemia: genetic determinants of response to hydroxyurea. Pharmacogenomics J 7:386–394
95. Ma Q, Abel K, Sripichai O, Whitacre J, Angkachatchai V, Makarasara W, Winichagoon P, Fucharoen S, Braun A, Farrer L (2007) Beta-Globin gene cluster polymorphisms are strongly associated with severity of HbE/beta(0)-thalassemia. Clin Genet 72:497–505
96. Marks J, Shaw J-P, Shen C-KJ (1986) Sequence organization and genomic complexity of primary theta-1 globin gene, a novel alpha-globin-like gene. Nature 321:785–788
97. May C, Rivella S, Callegari J, Heller G, Gaensler KM, Luzzatto L, Sadelain M (2000) Therapeutic haemoglobin synthesis in beta-thalassaemic mice expressing lentivirus-encoded human beta-globin. Nature 406:82–86
98. Michelson AM, Orkin SH (1983) Boundaries of gene conversion within the duplicated human alpha-globin genes. J Biol Chem 258:15245–15254
99. Milot E, Strouboulis J, Trimborn T, Wijgerde M, de Boer E, Langeveld A, Tan-Un K, Vergeer W, Yannoutsos N, Grosveld F, Fraser P (1996) Heterochromatin effects on the frequency and duration of LCR-mediated gene transcription. Cell 87:105–114
100. Moradkhani K, Préhu C, Old J, Henderson S, Balamitsa V, Luo HY, Poon MC, Chui DH, Wajcman H, Patrinos GP (2009) Mutations in the paralogous human alpha-globin genes yielding identical hemoglobin variants. Ann Hematol 88:535–543
101. Morimoto H, Lehmann H, Perutz MF (1971) Moleuclar pathology of human haemoglobin: stereochemical interpretation of abnormal oxygen affinities. Nature 232: 408–413
102. Muirhead H, Perutz MF (1963) Structure of haemoglobin. A three-dimensional fourier synthesis of reduced human haemoglobin at 5–5 A resolution. Nature 199:633–688
103. Myers RM, Lumelsky N, Lerman LS, Maniatis T (1985) Detection of single base substitutions in total genomic DNA. Nature 313:495–498
104. Nagel RL, Ranney HM (1990) Genetic epidemiology of structural mutations of the beta-globin gene. Semin Hematol 27:342–359
105. Neel JV (1949) The inheritance of sickle cell anemia. Science 110:64
106. Ney PA, Sorrentino BP, McDonagh KT, Nienhuis AW (1990) Tandem AP-1-binding sites within the human beta-globin dominant control region function as an inducible enhancer in erythroid cells. Genes Dev 4:993–1006
107. Nuez B, Michalovich D, Bygrave A, Ploemacher R, Grosveld F (1995) Defective haematopoiesis in fetal liver resulting from inactivation of the EKLF gene. Nature 375:316–318
108. Ohlsson R, Renkawitz R, Lobanenkov V (2001) CTCF is a uniquely versatile transcription regulator linked to epigenetics and disease. Trends Genet 17:520–527
109. Ojwang PJ, Nakatsuji T, Gardiner MB, Reese AL, Gilman JG, Huisman TH (1983) Gene deletion as the molecular basis for the Kenya-G gamma-HPFH condition. Hemoglobin 7:115–123
110. Oner C, Dimovski AJ, Altay C, Gurgey A, Gu YC, Huisman TH, Lanclos KD (1992) Sequence variations in the 5' hypersensitive site-2 of the locus control region of betaS chromosomes are associated with different levels of fetal globin in hemoglobin S homozygotes. Blood 79:813–819
111. Orkin SH, Kazazian HH Jr, Antonarakis SE, Goff SC, Boehm CD, Sexton JP, Waber PG, Giardina PJ (1982) Linkage of beta-thalassaemia mutations and beta-globin gene polymorphisms with DNA polymorphisms in human beta-globin gene cluster. Nature 296:627–631
112. Orkin SH, Kazazian HH Jr, Antonarakis SE, Ostrer H, Goff SC, Sexton JP (1982) Abnormal RNA processing due to the exon mutation of beta E-globin gene. Nature 300:768–769
113. Orkin SH, Markham AF, Kazazian HH (1983) Direct detection of the common Mediterranean beta-thalassemia gene with synthetic DNA probes. An alternative approach for prenatal diagnosis. J Clin Invest 71:775–779
114. Orkin SH, Antonarakis SE, Kazazian HH Jr (1984) Base substitution at position -88 in a beta-thalassemic globin gene. Further evidence for the role of distal promoter element ACACCC. J Biol Chem 259:8679–8681
115. Orkin SH, Cheng TC, Antonarakis SE, Kazazian HH Jr (1985) Thalassemia due to a mutation in the cleavage-polyadenylation signal of the human beta-globin gene. EMBO J 4:453–456
116. Ottolenghi S, Giglioni B, Comi P, Gianni AM, Polli E, Acquaye CT, Oldham JH, Masera G (1979) Globin gene deletion in HPFH, delta0beta0 thalassemia and Hb Lepore disease. Nature 278:654–657
117. Pace BS, Zein S (2006) Understanding mechanisms of gamma-globin gene regulation to develop strategies for pharmacological fetal hemoglobin induction. Dev Dyn 235:1727–1737
118. Pagnier J, Mears JG, Dunda-Belkhodja O, Schaefer-Rego KE, Beldjord C, Nagel RL, Labie D (1984) Evidence for

118. ...the multicentric origin of the sickle cell hemoglobin gene in Africa. Proc Natl Acad Sci USA 81:1771–1773
119. Palstra RJ, Tolhuis B, Splinter E, Nijmeijer R, Grosveld F, de Laat W (2003) The beta-globin nuclear compartment in development and erythroid differentiation. Nat Genet 35:190–194
120. Papachatzopoulou A, Menounos PG, Kolonelou C, Patrinos GP (2006) Mutation screening in the human epsilon-globin gene using single-strand conformation polymorphism analysis. Am J Hematol 81:136–138
121. Papachatzopoulou A, Kourakli A, Makropoulou P, Kakagianne T, Sgourou A, Papadakis M, Athanassiadou A (2006) Genotypic heterogeneity and correlation to intergenic haplotype within high HbF beta-thalassemia intermedia. Eur J Haematol 76:322–330
122. Papachatzopoulou A, Kaimakis P, Pourfarzad F, Menounos PG, Pappa M, Kollia P, Grosveld FG, Patrinos GP (2007) Increased fetal hemoglobin levels in beta-thalassemia intermedia patients correlates with a mutation in 3'HS1. Am J Hematol 82:1005–1009
123. Papadakis MN, Patrinos GP (1999) Contribution of gene conversion in the evolution of the human beta-like globin gene family. Hum Genet 104:117–125
124. Papadakis MN, Patrinos GP, Tsaftaridis P, Loutradi-Anagnostou A (2002) A comparative study of Greek non-deletional hereditary persistence of fetal hemoglobin and beta-thalassemia compound heterozygotes. J Mol Med 80:243–247
125. Patrinos GP, Grosveld FG (2008) Pharmacogenomics and therapeutics of hemoglobinopathies. Hemoglobin 32:229–233
126. Patrinos GP, Kollia P, Loutradi-Anagnostou A, Loukopoulos D, Papadakis MN (1998) The Cretan type of non-deletional hereditary persistence of fetal hemoglobin [Agamma -158 C>T] results from two independent gene conversion events. Hum Genet 102:629–634
127. Patrinos GP, Kollia P, Papapanagiotou E, Loutradi-Anagnostou A, Loukopoulos D, Papadakis MN (2001) Agamma-haplotypes: a new group of genetic markers for thalassemic mutations inside the 5' regulatory region of the human Agamma-globin gene. Am J Hematol 66:99–104
128. Patrinos GP, Giardine B, Riemer C, Miller W, Chui DH, Anagnou NP, Wajcman H, Hardison RC (2004) Improvements in the HbVar database of human hemoglobin variants and thalassemia mutations for population and sequence variation studies. Nucleic Acids Res 32:D537–D541
129. Patrinos GP, de Krom M, de Boer E, Langeveld A, Imam AM, Strouboulis J, de Laat W, Grosveld FG (2004) Multiple interactions between regulatory regions are required to stabilize an active chromatin hub. Genes Dev 18:1495–1509
130. Patrinos GP, Kollia P, Papadakis MN (2005) Molecular diagnosis of inherited disorders: lessons from hemoglobinopathies. Hum Mutat 26:399–412
131. Pauling L, Itano HA, Singer SJ, Wells IC (1949) Sickle cell anemia: a molecular disease. Science 110:543
132. Perutz MF (1962) Relation between structure and sequence of haemoglobin. Nature 194:914–917
133. Petrou M (2005) Genetic counseling and ethics in molecular diagnostics. In: Patrinos GP, Ansorge W (eds) Molecular diagnostics. Academic, San Diego, pp 399–407
134. Petrou M, Modell B (1995) Prenatal screening for haemoglobin disorders. Prenat Diagn 15:1275–1295
135. Pirastu M, Kan YW, Cao A, Conner BJ, Teplitz RL, Wallace RB (1983) Prenatal diagnosis of beta-thalassemia. Detection of a single nucleotide mutation in DNA. N Engl J Med 309:284–287
136. Pirastu M, Galanello R, Doherty MA, Tuveri T, Cao A, Kan YW (1987) The same beta-globin gene mutation is present on nine different beta-thalassemia chromosomes in a Sardinian population. Proc Natl Acad Sci USA 84:2882–2885
137. Pissard S, Beuzard Y (1994) A potential regulatory region for the expression of fetal hemoglobin in sickle cell disease. Blood 84:331–338
138. Proudfoot NJ, Shander MH, Manley JL, Gefter ML, Maniatis T (1980) Structure and in vitro transcription of human globin genes. Science 209:1329–1336
139. Purucker M, Bodine D, Lin H, McDonagh K, Nienhuis AW (1990) Structure and function of the enhancer 3' to the human A gamma globin gene. Nucleic Acids Res 18:7407–7415
140. Quek L, Thein SL (2007) Molecular therapies in beta-thalassaemia. Br J Haematol 136:353–365
141. Radding CM (1978) Genetic recombination: strand transfer and mismatch repair. Annu Rev Biochem 47:847–880
142. Raich N, Enver T, Nakamoto B, Josephson B, Papayannopoulou T, Stamatoyannopoulos G (1990) Autonomous developmental control of human embryonic globin gene switching in transgenic mice. Science 250:1147–1149
143. Rogers BB, Wessels RA, Ou CN, Buffone GJ (1985) High performance liquid chromatography in the diagnosis of hemoglobinopathies and thalassemias. Report of three cases. Am J Clin Pathol 84:671–674
144. Romao L, Osorio-Almeida L, Higgs DR, Lavinha J, Liebhaber SA (1991) Alpha-thalassemia resulting from deletion of regulatory sequences far upstream of the alpha-globin structural genes. Blood 78:1589–1595
145. Ryan TM, Behringer RR, Townes TM, Palmiter RD, Brinster RL (1989) High-level erythroid expression of human alpha-globin genes in transgenic mice. Proc Natl Acad Sci USA 86:37–41
146. Saiki RK, Scharf S, Faloona F, Mullis KB, Horn GT, Erlich HA, Arnheim N (1985) Enzymatic amplification of beta-globin genomic sequences and restriction site analysis for diagnosis of sickle cell anemia. Science 230:1350–1354
147. Saiki RK, Chang CA, Levenson CH, Warren TC, Boehm CD, Kazazian HH Jr, Erlich HA (1988) Diagnosis of sickle cell anemia and beta-thalassemia with enzymatically amplified DNA and nonradioactive allele-specific oligonucleotide probes. N Engl J Med 319:537–541
148. Schroeder WA, Huisman TH (1978) Human gamma chains: structural features. In: Stamatoyannopoulos G, Nienhuis A (eds) Cellular and molecular regulation of hemoglobin switching. Grune and Stratton, New York, pp 29–45
149. Sears DA (1978) The morbidity of sickle cell trait. A review of the literature. Am J Med 64:1021–1036
150. Serjeant GR (1993) The clinical features of sickle cell disease. Baillieres Clin Haematol 6:93–115
151. Sharpe JA, Chan-Thomas PS, Lida J, Ayyub H, Wood WG, Higgs DR (1992) Analysis of the human alpha globin upstream regulatory element (HS-40) in transgenic mice. EMBO J 11:4565–4572

152. Shen S, Slightom JL, Smithies O (1981) A history of the human fetal globin gene duplication. Cell 26:191–203
153. Simon I, Tenzen T, Mostoslavsky R, Fibach E, Lande L, Milot E, Gribnau J, Grosveld F, Fraser P, Cedar H (2001) Developmental regulation of DNA replication timing at the human beta globin locus. EMBO J 20:6150–6157
154. Slighton JL, Blechl AE, Smithies O (1980) Human fetal G-gamma- and A-gamma-globin genes: complete nucleotide sequences suggest that DNA can be exchanged in these duplicated genes. Cell 21:627–638
155. Spangler EA, Andrews KA, Rubin EM (1990) Developmental regulation of the human zeta globin gene in transgenic mice. Nucleic Acids Res 18:7093–7097
156. Spritz RA, DeRiel JK, Forget BG, Weissman SM (1980) Complete nucleotide sequence of the human delta-globin gene. Cell 21:639–646
157. Stamatoyannopoulos G (1972) The molecular basis of hemoglobin disease. Annu Rev Genet 6:47
158. Stamatoyannopoulos G, Grosveld F (2001) Hemoglobin switching. In: Stamatoyannopoulos G, Majerus PW, Perlmutter RM, Varmus H (eds) The molecular basis of blood diseases, 3rd edn. WB Saunders and Company, Philadelphia, PA, pp 135–182
159. Stamatoyannopoulos G, Josephson B, Zhang JW, Li Q (1993) Developmental regulation of human gamma-globin genes in transgenic mice. Mol Cell Biol 13:7636–7644
160. Steensma DP, Gibbons RJ, Higgs DR (2005) Acquired alpha-thalassemia in association with myelodysplastic syndrome and other hematologic malignancies. Blood 105:443–452
161. Steinberg MH (2005) Predicting clinical severity in sickle cell anaemia. Br J Haematol 129:465–481
162. Steinberg MH, Adams JG 3rd (1991) Hemoglobin A2: origin, evolution, and aftermath. Blood 78:2165–2177
163. Suwanmanee T, Sierakowska H, Lacerra G, Svasti S, Kirby S, Walsh CE, Fucharoen S, Kole R (2002) Restoration of human beta-globin gene expression in murine and human IVS2-654 thalassemic erythroid cells by free uptake of antisense oligonucleotides. Mol Pharmacol 62:545–553
164. Taliaferro WH, Huck JG (1923) The inheritance of sickle cell anaemia in man. Genetics 8:594
165. Thein SL, Hesketh C, Taylor P, Temperley IJ, Hutchinson RM, Old JM, Wood WG, Clegg JB, Weatherall DJ (1990) Molecular basis of for dominantly inherited inclusion body beta-thalassemias. Proc Natl Acad Sci USA 87:3924–3928
166. Tolhuis B, Palstra RJ, Splinter E, Grosveld F, de Laat W (2002) Looping and interaction between hypersensitive sites in the active beta-globin locus. Mol Cell 10:1453–1465
167. Treisman R, Orkin SH, Maniatis T (1983) Specific transcription and RNA splicing defects in five cloned beta-thalassaemia genes. Nature 302:591–596
168. Tuan D, Kong S, Hu K (1992) Transcription of the hypersensitive site HS2 enhancer in erythroid cells. Proc Natl Acad Sci USA 89:11219–11223
169. Tufarelli C, Stanley JA, Garrick D, Sharpe JA, Ayyub H, Wood WG, Higgs DR (2003) Transcription of antisense RNA leading to gene silencing and methylation as a novel cause of human genetic disease. Nat Genet 34:157–165
170. Viprakasit V, Gibbons RJ, Broughton BC, Tolmie JL, Brown D, Lunt P, Winter RM, Marinoni S, Stefanini M, Brueton L, Lehmann AR, Higgs DR (2001) Mutations in the general transcription factor TFIIH result in beta-thalassaemia in individuals with trichothiodystrophy. Hum Mol Genet 10:2797–2802
171. Warkentin TE, Barr RD, Ali MA, Mohandas N (1990) Recurrent acute splenic sequestration crisis due to interacting genetic defects: hemoglobin SC disease and hereditary spherocytosis. Blood 75:266–270
172. Watt P, Lamb P, Squire L, Proudfoot N (1990) A factor binding GATAAG confers tissue specificity on the promoter of the human zeta-globin gene. Nucleic Acids Res 18:1339–1350
173. Weatherall DJ (1987) Common genetic disorders of the red cell and the 'malaria hypothesis'. Ann Trop Med Parasitol 81:539–548
174. Weatherall DJ, Clegg JG (2001) The thalassemia syndromes, 4th edn. Blackwell, Oxford
175. Whipple GH, Bradford WL (1933) Mediterranean disease-thalassemia (erythroblastic anemia of Cooley); associated pigment abnormalities simulating hemochromatosis. J Pediatr 9:279–311
176. Wijgerde M, Grosveld F, Fraser P (1995) Transcription complex stability and chromatin dynamics in vivo. Nature 377:209–213
177. Wijgerde M, Gribnau J, Trimborn T, Nuez B, Philipsen S, Grosveld F, Fraser P (1996) The role of EKLF in human beta-globin gene competition. Genes Dev 10:2894–2902
178. Wood WG (1976) Haemoglobin synthesis during human fetal development. Br Med Bull 32:282–287
179. Yavarian M, Karimi M, Bakker E, Harteveld CL, Giordano PC (2004) Response to hydroxyurea treatment in Iranian transfusion-dependent beta-thalassemia patients. Haematologica 89:1172–1178
180. Yu C, Niakan KK, Matsushita M, Stamatoyannopoulos G, Orkin SH, Raskind WH (2002) X-linked thrombocytopenia with thalassemia from a mutation in the amino finger of GATA-1 affecting DNA binding rather than FOG-1 interaction. Blood 100:2040–2045
181. Zhang JZ, Cai SP, He X, Lin HX, Lin HJ, Huang ZG, Chehab FF, Kan YW (1988) Molecular basis of beta-thalassemia in South China: strategy for DNA analysis. Hum Genet 78:37–40

Human Genetics of Infectious Diseases

Alexandre Alcaïs, Laurent Abel, and Jean-Laurent Casanova

Abstract A lingering question in the field of infectious diseases is that of the considerable clinical variability between individuals in the course of infection, raising fundamental questions about the actual pathogenesis of infectious diseases. There is increasing epidemiological and experimental evidence to suggest that human genetics plays a major role in susceptibility/resistance to infectious diseases. There seems to be a continuous spectrum of human predisposition to infectious diseases, from monogenic to polygenic inheritance. Many monogenic primary immunodeficiencies have been clinically described and genetically deciphered, and most predispose affected individuals to multiple infections. Other monogenic traits conferring pathogen-specific susceptibility in otherwise healthy individuals are increasingly being described. Examples of Mendelian specific resistance to infectious agents are also being discovered. At the population level, major genes are being identified in a small, but growing number of common infectious diseases. Truly polygenic predisposition to a human infectious disease remains to be definitively demonstrated experimentally, despite the unquestionable identification of individual (but not necessarily interacting) susceptibility genes. Studies of the human genetics of infectious diseases have considerable clinical implications, as improvements in our understanding of the pathogenesis of infectious disease pave the way to both genetic diagnosis and immunological interventions. The genetic investigation of infectious diseases, seen as 'experiments of Nature', also provides a unique approach to definition of the function of host defense *genes in natura* — i.e. in the setting of a natural, as opposed to experimental, ecosystem governed by natural selection.

A. Alcaïs and L. Abel
Laboratory of Human Genetics of Infectious Diseases,
Necker Branch, Institut National de la Santé et de la Recherche
Médicale, U550, Necker Medical School, Paris, France
Université Paris Descartes, Paris, France
e-mail: alexandre.alcais@inserm.fr; laurent.abel@inserm.fr

J.-L. Casanova (✉)
Laboratory of Human Genetics of Infectious Diseases,
Rockefeller Branch, The Rockefeller University,
New York, NY, USA
e-mail: jean-laurent.casanova@mail.rockefeller.edu

Contents

12.1　Introduction 404
12.2　Mendelian Predisposition to Multiple Infections 406
12.3　Mendelian Predisposition to Single Infections 407
12.4　Mendelian Resistance... 408
12.5　Major Genes.. 409
12.6　Multigenic Predispostion 410
12.7　Concluding Remarks.. 411
References.. 412

12.1 Introduction

The determinism of human infectious diseases is still widely misunderstood, with these diseases commonly thought to be purely infectious. As exposure to a microbial agent is obviously required for infection and disease to occur, infectious diseases are often regarded as textbook proof-of-principle examples of purely environmental diseases. Conversely, some inherited metabolic disorders, such as phenylketonuria, are commonly seen as perfect examples of purely genetic traits. However, in the case of phenylketonuria, mental retardation is entirely dependent on the patient's exposure to the amino acid phenylalanine (85). If phenylalanine was absent from our diet, phenylalanine hydroxylase deficiency would be considered merely a neutral polymorphism in human populations. *Conditional* genetic susceptibility to disease, whether Mendelian or more complex, is probably not a trivial exception but a general rule that also applies to infectious diseases. Nonetheless, for most scientists and physicians, the more overt the disease-causing environmental factor, the less other factors, such as human genetics, are taken into account. It is precisely for this reason that infectious diseases are often erroneously considered to be determined solely by the microbial environment. However, even Pasteur's microbial theory of disease states that the microbe is necessary but not sufficient for the development of infectious diseases [66] (Fig. 12.1 and Fig. 12.2).

Once these diseases had been shown to be infectious, the next most important question concerned the astounding level of interindividual clinical variability in populations infected with the same microbe. Speculations about and investigations of natural variability in the development of infectious diseases were boosted in the early twentieth century by Charles Nicolle's discovery of the coexistence of symptomatic

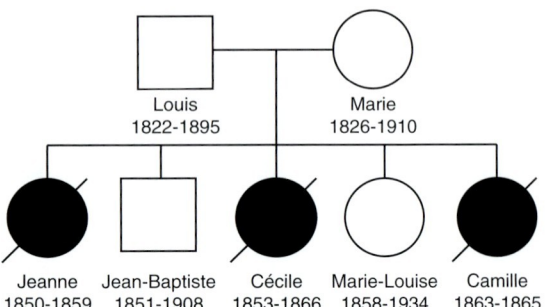

Fig. 12.1 Pasteur's pedigree and the genetic theory of infectious diseases. Louis Pasteur, the founder of the microbial theory of disease, lost three young daughters to "fever" between 1859 and 1866. A few years later, in 1870, he discovered that microbes caused disease in silk worms [65], paving the way for a general microbial theory of disease. Retrospectively, it is clear that his daughters died of infectious diseases. This illustrious family is representative of most families worldwide and throughout history until recent improvements in hygiene and the advent of vaccines and antibiotics, which resulted from the microbial theory. It was not uncommon for at least half the siblings in a family to die of infection. The microbial theory of disease identified the microbial cause of disease, but did not resolve the question of intrafamilial clinical heterogeneity in families exposed to the same microbial environment. As illustrated in Pasteur's own pedigree, one son and one daughter survived into adulthood, despite probable exposure to at least one of the microbes that killed their other siblings. We show here Pasteur's pedigree as it would have been drawn at around 1866 if the genetic theory of infectious diseases had emerged at that time. It is possible that the three children who died carried a Mendelian trait predisposing them to infectious diseases, or at least some form of genetic predisposition

and asymptomatic infections in naive human populations [60]. In addition to microbial variation, three theories have been proposed to account for this heterogeneity. Nonmicrobial environmental factors may be involved, with air temperature or humidity, and the availability of an animal vector particularly crucial (the ecological theory of infection) [28]. Nongenetic host factors, such as age or, since the last century, personal vaccination history may have a key role (the

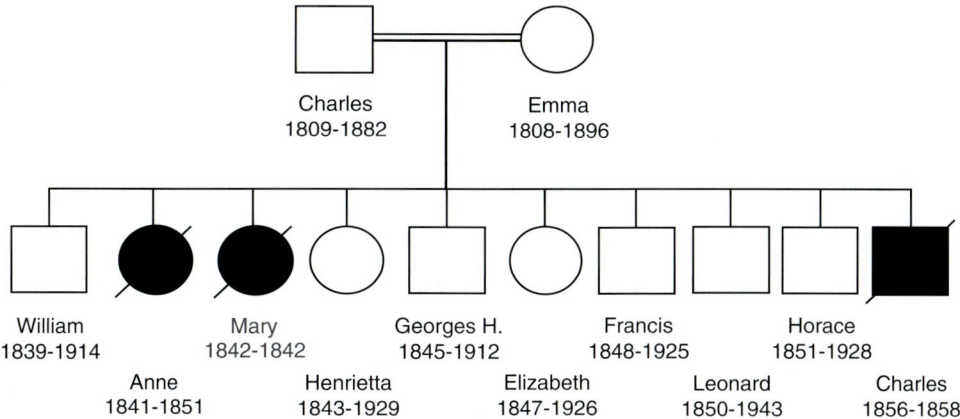

Fig. 12.2 Darwin's pedigree and the genetic theory of infectious diseases. Charles Darwin, the founder of the theory of natural selection, also lost three children to febrile infectious diseases between 1842 and 1858 (Fig. 12.1). In 1859, one year after the last of these three deaths, he published 'On the origin of species by means of natural selection', proposing the theory of natural selection of species. We show here Darwin's pedigree as it would have been drawn in 1858 if the genetic theory of infectious diseases had already emerged. Darwin married a first cousin and it is therefore possible that the three children who died of infection carried a recessive genetic predisposition to infectious diseases

immunological theory of infection) [66]. Finally, epidemiological evidence has accumulated since the 1930s, that shows a particularly important role of human genetic factors in immunodeficiency and susceptibility to infectious diseases (the genetic theory of infection) [20, 46]. The first evidence supporting the genetic theory of infectious diseases came from observations of the ethnic or familial aggregation of both rare and common infections, which even followed a Mendelian (monogenic) pattern of inheritance in some kindreds [4]. Follow-up studies of adoptive children also showed that predisposition to infectious diseases was largely inherited, paradoxically more so than in diseases associated with less well-known environmental risk factors, such as cancer [82]. Finally, the concordance rate of infectious diseases was higher in monozygotic than in dizygotic twins, implicating host genetic background in susceptibility to disease [4, 46].

The field of human genetics of infectious diseases entered the molecular and cellular era in the early 1950s, with the discovery of X-linked agammaglobulinemia by Bruton [16] and that of the protective role of the sickle cell trait for severe forms of *falciparum* malaria by Allison [7]. According to the dominant paradigm following these two landmark discoveries, predisposition to infectious disease segregates in a Mendelian or polygenic pattern of inheritance. An ever-growing number of rare Mendelian syndromes conferring susceptibility to multiple infectious agents is being reported. These syndromes include, in particular, conventional primary immunodeficiencies (PIDs) associated with multiple infections, [63]. For more common infectious phenotypes (such as malaria), genetic predisposition is thought to involve many genes, each of which has a modest marginal effect. However, the distinction between Mendelian predisposition in individuals with rare infections (one gene, multiple infections) and complex predisposition in populations with common infections (one infection, multiple genes) has become somewhat blurred in recent years [19, 22]. First, nonconventional PIDs conferring predisposition to a single type of infection in otherwise healthy individuals are increasingly being recognized [23, 67]. Second, Mendelian resistance to more virulent pathogens has also been described (e.g. [11, 55]). Third, a so-called polygenic susceptibility may primarily reflect the impact of a predominant gene, often referred to as a major gene [1]. Finally, the commonly diffused concept of polygenic inheritance remains to be demonstrated, at least at the individual level. We provide below an overview of the various genetic susceptibilities underlying human infectious diseases (Fig. 12.3), illustrated by key examples.

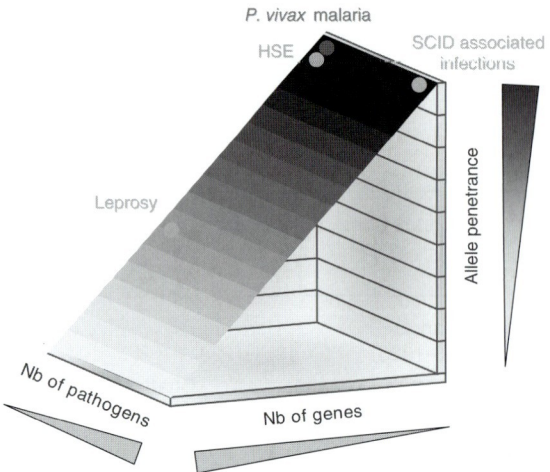

Fig. 12.3 Schematic representation of the continuous genetic models underlying human infectious diseases. This figure summarizes the spectrum of genetic susceptibilities predisposing to infectious diseases at the individual level. Different situations can be distinguished according to the number of genes with an additive impact on genetic susceptibility (in *green*) or resistance (in *red*), the marginal effect of each of these genes and the number of pathogens to which the individual is susceptible. Four textbook examples are shown on the graph: severe combined immune deficiency (SCID)-associated infections (a unique gene with complete penetrance predisposes individuals to a broad spectrum of infectious agents), herpes virus encephalitis (HSE; a single gene with high but incomplete penetrance conferring predisposition to a single infectious agent), *P. vivax* malaria (a single gene with high penetrance conferring resistance to a single infectious agent,) and leprosy (a small number of genes with intermediate penetrance conferring predisposition to a single infectious agent)

12.2 Mendelian Predisposition to Multiple Infections

Perhaps the most compelling evidence that human genetics does indeed determine the development of infectious diseases arises from the group of Mendelian disorders known as PIDs, which were first described as such in 1952 [16, 17]. More than 200 PIDs have been clinically described. Each is individually rare, but the genetic etiology of over 100 of these diseases has been elucidated [61, 63]. A comprehensive review of them is beyond the scope of this chapter, and interested readers are referred to previous reviews and textbooks [61, 63]. Typically, these disorders are monogenic (Mendelian) and confer predisposition to multiple infectious diseases (one gene, multiple infections) – the number and nature of which vary from case to case. They affect immune responses in various ways, in hematopoietic cells, nonhematopoietic cells, or both. The most severe known PID is reticular dysgenesia (RD), which is characterized by a complete lack of leukocytes of both lymphoid and myeloid origin [14]. Severe combined immunodeficiency (SCID) is more common and almost as severe, and is defined as an intrinsic (i.e., hematopoietic) lack of development of autologous T cells [34]. The Di George and Nude syndromes are immunological phenocopies of SCID of extrinsic (i.e., nonhematopoietic) origin. Children with SCID can be further classified according to the underlying genetic etiology, which may or may not affect other lymphoid lineages, such as B and NK lymphocytes. X-linked SCID (typically B+ NK−) is caused by mutations in the cytokine receptor common γ–chain. Other patients with SCID carry deleterious mutations in genes involved in the machinery generating antigen-specific receptors on T and B cells. Children with SCID present with multiple infectious diseases caused by viruses, bacteria, fungi, and parasites in the 1st year of life, typically starting by the age of 2–3 months. A broad range of weakly virulent microorganisms can kill such patients. SCID and RD neatly illustrate the life-threatening impact of Mendelian genetics in terms of infectious diseases.

SCID patients have also proved interesting in terms of the information they provide about the efficacy to save life and the risks of death associated with hematopoietic stem cell transplantation and gene therapy. SCID is invariably fatal when it follows its natural course, even in tertiary hospitals, typically in the 1st year of life. SCID patients progressively succumb to the succession of infections, even if treated with anti-infectious agents. The first successful hematopoietic stem cell transplantation (HSCT) in humans was actually performed in 1968, in a child with SCID [38, 40]. Despite the rarity of graft rejection, it was soon realized that children undergoing HSCT could die of graft-versus-host disease or severe infections. More than 30 years later, a series of ten children with X-linked SCID were treated with gene therapy [25]. A retroviral vector encoding a wild-type copy of the common cytokine γ–chain was introduced into CD34-positive bone marrow cells, which were thought to include committed hematopoietic progenitors and self-renewing stem cells. Remarkably, nine of the ten children with SCID developed normal numbers of functional T cells, which were maintained for up to 7 years in the first child treated [34]. This was the first clinical success of human gene therapy. However, four of the nine children eventually developed acute T-cell leukemia, 3–5 years after gene therapy [34, 42]. Molecular oncogenesis in the first two patients was found to be related to insertion of the retroviral vector into the *LMO2* gene, which encodes a transcription factor normally produced in T-cell progenitors, and an oncogene driving the proliferation of certain T-cell leukemias [42]. The molecular mechanism underlying the T-cell leukemia in the remaining two patients has yet to be reported. However, insertional mutagenesis and aberrant transcription of an oncogene were probably involved in leukemogenesis in all four patients. Sadly, one of these four patients with leukemia died. The other three were treated with chemotherapy, and one also underwent HSCT. The other five patients from the first trial are still doing well. The gene therapy trial was suspended when it was found that at least four of the nine patients had developed T-cell leukemia. One of the patients treated with a similar protocol in the UK has also since developed leukemia [37, 43]. Gene therapy for SCID, by heterologous recombination, has provided the long-awaited proof-of-principle that human gene therapy can be clinically beneficial. It has also revealed that despite the initial success, gene therapy as performed in this trial could eventually be detrimental.

12.3 Mendelian Predisposition to Single Infections

Interestingly, not all PIDs confer predisposition to multiple infections. An increasing number of disorders (summarized in Table 12.1) are known to confer Mendelian predisposition to a single type of infection [23, 67]. Epidermodysplasia verruciformis (EV) was described clinically as early as 1922. The genetic origin of this syndrome was proposed in 1933, and its viral etiology was documented in 1946 [64]. It was therefore probably the first PID ever described – preceding even Bruton's agammaglobulinemia – in that it corresponds to a monogenic predisposition to infection. The lack of an overt immunological phenotype and the extremely narrow spectrum of infections, limited to those caused by certain oncogenic papillomaviruses, probably precluded the use of the term "PID" at this time. The first two EV-causing genes, *EVER1* and *EVER2*, were described in 2002 [72]. An X-linked form of predisposition to lethal infection by Epstein-Barr virus was reported in 1975 [70], and two causal genes have been identified to date [26, 73]. Mendelian predispositions to bacterial infections have also been described. Patients with properdin deficiency or defects in the terminal components of complement (C5–C9, forming the membrane attack complex) display

Table 12.1 Mendelian holes in immunity to infection

Infectious agent	Clinical phenotype	Immunological phenotype	Gene	References
Neisseria	Invasive disease	MAC deficiency	*C5, C6, C7, C8A, C8B, C8G, C9*	Reviewed in [67]
	Invasive disease	Properdin deficiency	*PFC*	
Mycobacteria	MSMDDisseminated tuberculosis	IL-12/23-IFN-γ deficiency	*IFNGR1, IFNGR2, STAT1, NEMO, IL12B, IL12RB1*	Reviewed in [33]
Streptococcus pneumoniae	Invasive disease	IRAK-4 deficiency	*IRAK4, MYD88*	[13, 68]
Epstein-Barr virus	X-linked lymphoproliferative disease	SAP deficiency	*SH2D1A, BIRC4*	[26, 73]
Human papillomavirus	Epidermodysplasia verruciformis	EVER1/EVER2 deficiency	*EVER1, EVER2*	[72]
Herpes simplex virus 1	Encephalitis	Impaired production of type I IFN	*UNC93B1, TLR3*	[24, 89]
Trypanosoma evansi	Febrile episodes	No trypanolytic activity	*APOL1*	[86]
Plasmodium vivax	Natural resistance	Lack of receptor for pathogen	*DARC*	[55]
Human Immunodeficiency virus-1	Natural resistance	Lack of receptor for pathogen	*CCR5*	[6, 51, 78]
Norovirus	Natural resistance	Lack of receptor for pathogen	*FUT2*	[49, 50]

a selective predisposition to invasive meningococcal disease [67]. A predisposition to invasive pneumococcal disease also led to the discovery of IRAK-4 and Myd88 deficiency [13, 68]. The most thoroughly characterized of these syndromes is probably Mendelian predisposition to mycobacterial diseases (MSMD). Despite its clinical description in the 1950s, it was not until 1996 that the first genetic etiology of this syndrome – IFN-γR1 deficiency – was identified [44, 59]. In the last 10 years, up to 13 genetic defects affecting six genes have been reported, including autosomal recessive, autosomal dominant, and X-linked recessive traits, partial and complete defects, and complete defects with a lack of protein or nonfunctional expressed proteins [18, 33]. The six genes are physiologically related, as all are involved in IL-12/23-dependent, IFN-γ-mediated immunity [18, 32, 33]. Finally, Mendelian predisposition to a parasitic infection was recently documented in a patient with APOL1 deficiency who developed trypanosomiasis [86].

These disorders paved the way for the study of herpes simplex virus encephalitis (HSE), the most common form of sporadic viral encephalitis in Western countries. Since its discovery in 1941, it has remained unclear why only a small fraction of otherwise healthy individuals exposed to herpes simplex virus 1 (HSV-1) develop HSE. Moreover, none of the known PIDs, including RD and SCID, increase the risk of HSE. In a genetic epidemiological survey, about 10% of children with HSE were found to have been born to unrelated parents from consanguineous families, strongly suggesting Mendelian inheritance of predisposition to HSE, in at least some patients. Two genetic etiologies of HSE have recently been discovered: autosomal recessive UNC-93B deficiency [24] and autosomal dominant TLR3 deficiency [89]. These disorders lead to impaired recognition of dsRNA intermediates of HSV-1 in the central nervous system, resulting in impaired interferon production, itself resulting in enhanced viral replication and cell death. The virus does not spread to other organs because other cell types seem to be capable of controlling HSV-1 by TLR3-independent processes. Similarly, patients seem to use other means to control other viruses. The incomplete clinical penetrance probably reflects the influence of other factors, such as age at HSV-1 infection or modifier genes. In any event, it probably accounts for HSE being sporadic in the vast majority of kindreds. The identification of a Mendelian basis for HSE provides the first demonstration that a sporadic, life-threatening infectious disease may result from a group of monogenic disorders. It also suggests that other severe infectious diseases, particularly in children, may have a monogenic basis. The field of Mendelian predisposition to infectious diseases thus covers an immense spectrum, ranging from RD, in which patients have no leukocytes and are vulnerable to most (but not all) microbes, to HSE, in which leukocytes are not involved in the immunodeficiency and otherwise healthy patients are vulnerable to primary HSV-1 infection of the central nervous system.

12.4 Mendelian Resistance

Genetically determined resistance to an infectious agent is the obligate mirror of the susceptibility phenotype. Four Mendelian traits have been found to confer resistance to specific infections, as they result in a lack of the receptors used by the invading microbes (Table 12.1). Consequently, individuals carrying the common wild-type alleles are intrinsically susceptible to these particular pathogens, whereas individuals carrying the rare mutant alleles display almost complete and apparently specific protection against these pathogens. Protection against *Plasmodium vivax*, an agent of malaria, is conferred by a lack of erythrocyte expression of the Duffy antigen receptor for chemokines (DARC), a key receptor for the parasite [55]. The resistance trait is recessive, and the single nucleotide mutation affects the GATA-1-binding site in the promoter of the *DARC* gene, thereby selectively preventing gene transcription in erythroid cells [85]. Recessive resistance to human immunodeficiency virus-1 (HIV-1) infection has been found to be conferred by mutations affecting the extracellular domain of another chemokine receptor, CCR5. CCR5 functions with CD4 as a coreceptor for HIV-1 on CD4+ T cells [9]. Subjects homozygous for the most common *CCR5* deleterious mutation, a 32 bp deletion (Δ32), display strong protection against infection by CCR5-tropic HIV-1 [27, 51, 78]. The erythrocyte P antigen is the cellular receptor for parvovirus B19, and the rare people with the p-phenotype, whose erythrocytes do not have this receptor, are resistant to B19 infection, which causes erythema infectiosum [15]. Finally, resistance to norovirus (i.e., Norwalk-like viruses), a leading cause of gastroenteritis, was recently shown to be associated with alleles of the *FUT2* gene [49] encoding an α[1, 2]-fucosyltransferase that regulates the expression of ABH histo-blood group antigens on the surface of epithelial cells and in mucosal secretions [53]. Several inactivating *FUT2* mutations are responsible for the nonsecretor phenotype (Se−), with a lack of expression of ABH antigens on epithelial cells and complete resistance against symptomatic norovirus infection in experimental and natural conditions [50, 84].

Alleles conferring Mendelian resistance to virulent pathogens would be expected to be under strong positive selection pressure. This is clearly the case for the *DARC* mutation, which is not found in Europe but has a frequency of up to 80% in African populations, in which *P. vivax* is endemic [55]. Similarly, variants of erythrocyte disease-causing genes conferring a "major" (but not Mendelian) resistance, as defined in the next section, against *P. falciparum* malaria are much more common in endemic countries. The best example is provided by the worldwide spread of the deleterious hemoglobin S (HbS) allele [47]. Homozygosity for this allele causes life-threatening sickle cell disease (drepanocytosis), but heterozygosity protects against severe *P. falciparum* malaria [7], resulting in heterosis, in which heterozygotes have a selective advantage over both types of homozygote [87]. In contrast, the HbC allele confers recessive, but not dominant, resistance against *P. falciparum* malaria, possibly accounting for the limited spread of this allele in one geographical region (West Africa) [58]. The first population genetics studies of *CCR5* found that the main resistance allele, Δ32, originated from a single ancestor of European origin [83]. The relatively recent estimated date of the mutation event (about 2,000–3,000 years ago), the frequency of the allele in the European population (10% in Western and Central Europe), and the long-range linkage disequilibrium pattern at the *CCR5* locus are highly suggestive of positive selection [62]. However, the intensity [77] and nature [9] of the selective pressure remain to be determined. The situation may be even more complicated, as CCR5-Δ32 homozygosity was recently reported to be associated with symptomatic West Nile virus (WNV) infection [39], indicating that selective pressure may also be negative, depending on the microbial environment. Finally, a single nonsense *FUT2* mutation, G428A, is the most common mutation (>95%) responsible for the Se- phenotype in populations of European and African descent [84].

It remains unclear whether the recurrence of this mutation is due to a hotspot and/or a founder effect under positive selective pressure. Mendelian resistance genes have provided the best overall illustrations of natural selection on the human genome. There are probably many other similar human mutations that have been or are being selected because they confer Mendelian resistance to virulent pathogens.

12.5 Major Genes

The "major gene/locus" concept was developed in the 1960s, following the introduction of the polygenic model by Fisher [35], when clinical geneticists needed a framework that could explicitly specify the effect of single genes in the expression of common diseases [30, 48]. A major gene differs from a Mendelian effect in displaying incomplete penetrance, and its phenotypic expression may be influenced by both environment and other genes in the individual. This concept was first formalized and developed in the context of complex segregation analysis, a statistical method based in its more general expression on a model of inheritance in which a given phenotype may result from the joint effects of a major locus, a polygenic component, and environmental factors [45, 48]. Several major genes identified by segregation analyzes have been reported since the 1970s in a number of complex traits, including infectious disease-related phenotypes in leprosy, malaria, schistosomiasis, and some viral infections [19]. The development of highly polymorphic genetic markers has recently led to the use of the major gene concept for loci identified in the context of genome-wide screening by means of linkage studies. Indeed, the loci detected in genome-wide scans, particularly in those based on the commonly used affected sib-pairs design, should have a substantial influence on the phenotype under study that could be qualified as a major effect [75, 76]. Using this definition, several major loci have been mapped in infectious diseases [19], the first one to chromosome 5q31–q33 and controlling levels of infection with the parasite *Schistosoma mansoni* [55], and the most recent to chromosome 8q12 and conferring a predisposition to pulmonary tuberculosis [10]. However, the major gene concept may evolve further, particularly in the context of technological advances in genomics. Furthermore, major genes may be specific to a given population (depending on ethnic origin, history, age of onset, etc.), and, for an infectious disease (e.g., leprosy), specific for a given phenotype (e.g., paucibacillary leprosy).

The HbS trait may be considered the first major gene identified in a common infectious disease [7], based on both its frequency in some African populations and its estimated effect on severe malaria (relative risk ~10) [47]. However, no genome-wide linkage screen has been conducted for severe malaria, probably because of the rarity of this phenotype and the even greater rarity of families with multiple cases. Consequently, leprosy is the only infectious disease for which a complete successful positional cloning approach, including genome-wide linkage screening followed by refined linkage disequilibrium mapping, has led to the identification of major susceptibility variants. Leprosy is a chronic infectious disease caused by *Mycobacterium leprae* that continues to affect more than 300,000 new subjects per year [88]. Both the development of leprosy per se upon exposure to *M. leprae* and its clinical features (ranging from paucibacillary to multibacillary forms) depend on human genes [6, 18]. The first evidence for this was provided by twin studies carried out in the 1960s, followed by several segregation studies, most detecting the presence of a major gene [2, 3]. Two major genes were only recently mapped by genome-wide linkage studies. The first of these studies focused on paucibacillary leprosy in India and detected a major locus on chromosome 10p13 [80] that has yet to be precisely identified. The second was carried out in Vietnam and mapped a major gene for susceptibility to leprosy per se to chromosome 6q25 [57]. Further linkage disequilibrium studies identified variants of the regulatory region shared by *PARK2*, a gene encoding an E3-ubiquitin ligase called Parkin, and *PACRG* (Parkin coregulated gene) as genetic risk factors for leprosy [56]. In addition to identifying a novel pathway of immunity to *M. leprae* [79], this study was the first to report successful positional cloning of a major locus in a common infectious disease.

12.6 Multigenic Predisposition

It is common to distinguish two patterns of multigenic inheritance, oligogenic and polygenic, according to the number and marginal size effect of the genes influencing the disease. Oligogenicity implies that the phenotype is dependent on two or more major genes, in addition to other factors. Polygenicity implies that the phenotype results from the effect of a large number of genetic loci, each having a small effect. An implicit but fundamental idea underlying multigenic inheritance is that, mechanistically, these definitions apply at the *individual* level. As this idea of a *cumulative* effect of the genes at the individual level is implicit, this point is often overlooked, and it is common practice to describe multigenic predisposition at the *population* level. However, multigenic predisposition at the population level does not necessarily reflect multigenic inheritance at the individual level, as genetic or even phenotypic heterogeneity may be involved, with certain genes acting as major genes in certain individuals or groups of individuals in the population considered. As discussed above, there is no unambiguous definition of a "major gene," and it is even possible to use a mixture of several definitions. Indeed, it would be legitimate to consider tuberculosis to be an example of an oligogenic disease (one major gene on 8q12 derived from genome-wide linkage analysis [10] and another, *NRAMP1*, from candidate gene analysis [12, 41, 52]) if both genes were found to affect the *same* phenotype in the *same* sample. We pointed out above that two major genes conferring a predisposition to leprosy have been identified on chromosomes 10p13 and 6q25. However, it would not be accurate to cite this as an example of oligogenicity, because the first of these two genes seems to affect only the paucibacillary form of the disease, whereas the second affects leprosy per se. Conversely, in the Vietnamese sample used to identify *PARK2/PACRG* [56], *LTA*, encoding lymphotoxin alpha, was recently identified as a second major susceptibility gene for leprosy per se [5]. The study of genetic predisposition to leprosy has therefore led not only to the first successful positional cloning of a major locus in a common infectious disease, but also to the first demonstration of an oligogenic predisposition to a common infectious disease.

The identification of a polygenic predisposition requires a large number of individuals, both because of the small expected effect attributable to each "polygene" and because of the additive nature of these polygenic effects. It is therefore not surprising that the proof-of-principle of such genetic mechanisms at the individual level has been provided by studies of susceptibility to infectious diseases in animal models of experimental infections. As a textbook example, it has been shown in a murine model of malaria caused by infection with *Plasmodium chabaudi* AS (which mimics several pathophysiological aspects of the blood-stage infection in humans, including host response and genetic control of parasitemia and ultimate outcome of infection) that at least five loci control parasitemia and nine control survival [36]. Several genes have often been reported to have a potential influence on the onset of a given infectious disease, but there is currently no proof of polygenic predisposition per se in individual human beings. For example, a number of genes (e.g., *HLA-DR, NRAMP1, IL12RB1*) have been reported to have a role in tuberculosis, but it has never been determined whether these genes act independently and additively on the same phenotype in the same sample. Thus, references to the polygenic nature of predisposition to a human infectious disease have generally been made at population level and should be considered to reflect ignorance of events at the level of individuals. However, with the advent of new chip-based technologies for massive genotyping, it should soon become possible to perform genome-wide scans for association, by capturing a large proportion of the genetic variation through the genotyping of millions of SNPs, in thousands of cases and controls. This need for a large sample for the detection of polygenes is entirely consistent with these genes playing a role in individuals. Indeed, the small size of their effects makes it more likely that the genes detected in such a sample play a role in the vast majority of the individuals, indicating true polygenic susceptibility. This comprehensive approach has started the dissection of the polygenic contribution to several complex diseases in humans such as type I diabetes [81] or Crohn's disease [29, 74]. In infectious diseases, the first whole-genome association was recently published. This study of the genetic control of HIV-1 viral load identified several polymorphisms that explained nearly 15% of the phenotypic variation among asymptomatic individuals [31].

12.7 Concluding Remarks

Infectious diseases are therefore largely genetically determined, probably more so than many other human diseases. The increase in life expectancy observed in the twentieth century occurred despite the retention of poor immunity to particular infectious agents in the genomes of most individuals [21]. There has been no sudden natural selection of high-quality immune system genes worldwide; this persistent immunodeficiency has simply been masked by medical progress. The genetic theory of infectious diseases constitutes a paradigm shift in medicine but does not conflict with the microbial theory of diseases, and Pasteur himself stated in his seminal survey of diseases affecting silk worms, that in the course of "flacherie," which he designated as "hereditary," "[It] is not the microbe that is transmitted from the parents to the offspring, but the predisposition to disease." However, although the genetic theory of infectious diseases has benefited from recent exchanges between different disciplines [21], only a very small fraction of human infectious diseases are understood at the genetic level.

Conventional PIDs have been studied in most detail, over the last 50 years. Association studies have so far identified only a handful of convincing susceptibility alleles (e.g., HbS). Linkage studies have recently identified a set of major genes (leprosy), and novel PIDs have provided a Mendelian basis for predisposition to certain infectious diseases (HSE), bridging the gap between the two fields dealing with genetic predisposition to infection. Much progress is expected in these two fields in the near future. We can also expect more Mendelian resistance genes to be discovered. There is therefore considerable determinism in the human genetics of infectious diseases, and the field as a whole is in its infancy. In addition to the identification of new genes, one key question concerns definition of the proportion of Mendelian and more complex predispositions or resistances in individuals and populations [8, 69]. There is clearly no such thing as a strict Mendelian segregation of phenotypes, because no single-gene organisms exist. All phenotypes are therefore multigenic in essence. It is therefore important to define the hierarchy of the genes involved in predisposition to infection at both the individual and population levels.

This field has considerable clinical implications, as illustrated in this chapter, in terms not only of diagnosis, but also of the development of new therapeutic interventions, such as the life-saving effects of exogenous IFN-γ in patients with mycobacterial disease resulting from insufficient endogenous IFN-γ production, which are of a similar magnitude to those of insulin in diabetic patients. The biological implications have been equally considerable, not only in terms of understanding the pathogenesis of infections but also in terms of immunology, as the human genetics of infectious diseases provides an ideal way to define the function of immune system genes *in natura*, in the setting of a natural ecosystem [71].

References

1. Abel L, Casanova JL (2000) Genetic predisposition to clinical tuberculosis: bridging the gap between simple and complex inheritance. Am J Hum Genet 67:274–277
2. Abel L, Demenais F (1988) Detection of major genes for susceptibility to leprosy and its subtypes. Am J Hum Genet 42:256–266
3. Abel L, Vu DL, Oberti J, Nguyen VT, Van VC, Guilloud-Bataille M, Schurr E, Lagrange PH (1995) Complex segregation analysis of leprosy in southern Vietnam. Genet Epidemiol 12:63–82
4. Alcais A, Abel L (2004) Application of genetic epidemiology to dissecting host susceptibility/resistance to infection illustrated with the study of common mycobacterial infections. In: Bellamy R (ed) Susceptibility to infectious diseases: the importance of host genetics. Cambridge University Press, Cambridge, pp 7–44
5. Alcais A, Alter A, Antoni G, Orlova M, Thuc NV, Singh M, Vanderborght PR, Katoch K, Mira MT, Thai VH, Huong NT, Ba NN, Moraes M, Mehra N, Schurr E, Abel L (2007) Stepwise replication identifies a low-producing lymphotoxin-alpha allele as a major risk factor for early-onset leprosy. Nat Genet 39:517–522
6. Alcais A, Mira M, Casanova JL, Schurr E, Abel L (2005) Genetic dissection of immunity in leprosy. Curr Opin Immunol 17:44–48
7. Allison AC (1954) Protection afforded by sickle cell trait against subtertian malarian infection. Br Med J 1:290–294
8. Antonarakis SE, Beckmann JS (2006) Mendelian disorders deserve more attention. Nat Rev Genet 7:277–282
9. Arenzana-Seisdedos F, Parmentier M (2006) Genetics of resistance to HIV infection: role of co-receptors and co-receptor ligands. Semin Immunol 18:387–403
10. Baghdadi JE, Orlova M, Alter A, Ranque B, Chentoufi M, Lazrak F, Archane MI, Casanova JL, Benslimane A, Schurr E, Abel L (2006) An autosomal dominant major gene confers predisposition to pulmonary tuberculosis in adults. J Exp Med 203:1679–1684
11. Barnwell JW, Nichols ME, Rubinstein P (1989) In vitro evaluation of the role of the Duffy blood group in erythrocyte invasion by Plasmodium vivax. J Exp Med 169:1795–1802
12. Bellamy R, Ruwende C, Corrah T, McAdam KPWJ, Whittle HC, Hill AVS (1998) Variations in the NRAMP1 gene and susceptibility to tuberculosis in west africans. New Engl J Med 338:640–644
13. von Bernuth H, Picard C, Jin Z, Pankla R, Xiao H, Ku CL, Chrabieh M, Mustapha IB, Ghandil P, Camcioglu Y, Vasconcelos J, Sirvent N, Guedes M, Vitor AB, Herrero-Mata MJ, Aróstegui JI, Rodrigo C, Alsina L, Ruiz-Ortiz E, Juan M, Fortuny C, Yagüe J, Antón J, Pascal M, Chang HH, Janniere L, Rose Y, Garty BZ, Chapel H, Issekutz A, Maródi L, Rodriguez-Gallego C, Banchereau J, Abel L, Li X, Chaussabel D, Puel A, Casanova JL (2008) Pyogenic bacterial infections in humans with MyD88 deficiency. Science 321(5889):691–696
14. Bertrand Y, Muller SM, Casanova JL, Morgan G, Fischer A, Friedrich W (2002) Reticular dysgenesis: HLA non-identical bone marrow transplants in a series of 10 patients. Bone Marrow Transplant 29:759–762
15. Brown KE, Hibbs JR, Gallinella G, Anderson SM, Lehman ED, McCarthy P, Young NS (1994) Resistance to parvovirus B19 infection due to lack of virus receptor (erythrocyte P antigen). N Engl J Med 330:1192–1196
16. Bruton OC (1952) Agammaglobulinemia. Pediatrics 9:722–728
17. Bruton OC (1962) A decade with agammaglobulinemia. J Pediatr 60:672–676
18. Casanova JL, Abel L (2002) Genetic dissection of immunity to mycobacteria: the human model. Annu Rev Immunol 20:581–620
19. Casanova JL, Abel L (2007) Human genetics of infectious diseases: a unified theory. Embo J 26:915–922
20. Casanova JL, Abel L (2004) The human model: a genetic dissection of immunity to infection in natural conditions. Nat Rev Immunol 4:55–66
21. Casanova JL, Abel L (2005) Inborn errors of immunity to infection: the rule rather than the exception. J Exp Med 202:197–201
22. Casanova JL, Abel L (2007) Primary immunodeficiencies: a field in its infancy. Science 317:617–619
23. Casanova JL, Fieschi C, Bustamante J, Reichenbach J, Remus N, von Bernuth H, Picard C (2005) From idiopathic infectious diseases to novel primary immunodeficiencies. J Allergy Clin Immunol 116:426–430
24. Casrouge A, Zhang SY, Eidenschenk C, Jouanguy E, Puel A, Yang K, Alcais A, Picard C, Mahfoufi N, Nicolas N, Lorenzo L, Plancoulaine S, Senechal B, Geissmann F, Tabeta K, Hoebe K, Du X, Miller RL, Heron B, Mignot C, de Villemeur TB, Lebon P, Dulac O, Rozenberg F, Beutler B, Tardieu M, Abel L, Casanova JL (2006) Herpes simplex virus encephalitis in human UNC-93B deficiency. Science 314:308–312
25. Cavazzana-Calvo M, Hacein-Bey S, de Saint Basile G, Gross F, Yvon E, Nusbaum P, Selz F, Hue C, Certain S, Casanova JL, Bousso P, Deist FL, Fischer A (2000) Gene therapy of human severe combined immunodeficiency (SCID)-X1 disease. Science 288:669–672
26. Coffey AJ, Brooksbank RA, Brandau O, Oohashi T, Howell GR, Bye JM, Cahn AP, Durham J, Heath P, Wray P, Pavitt R, Wilkinson J, Leversha M, Huckle E, Shaw-Smith CJ, Dunham A, Rhodes S, Schuster V, Porta G, Yin L, Serafini P, Sylla B, Zollo M, Franco B, Bolino A, Seri M, Lanyi A, Davis JR, Webster D, Harris A, Lenoir G, de St Basile G, Jones A, Behloradsky BH, Achatz H, Murken J, Fassler R, Sumegi J, Romeo G, Vaudin M, Ross MT, Meindl A, Bentley DR (1998) Host response to EBV infection in X-linked lymphoproliferative disease results from mutations in an SH2-domain encoding gene. Nat Genet 20:129–135

27. Dean M, Carrington M, Winkler C, Huttley GA, Smith MW, Allikmets R, Goedert JJ, Buchbinder SP, Vittinghoff E, Gomperts E, Donfield S, Vlahov D, Kaslow R, Saah A, Rinaldo C, Detels R, O'Brien SJ (1996) Genetic restriction of HIV-1 infection and progression to AIDS by a deletion allele of the CKR5 structural gene. Science 273:1856–1862
28. Dubos RJ (1950) Louis Pasteur, free lance of science, 1st edn. Little Brown, Boston
29. Duerr RH, Taylor KD, Brant SR, Rioux JD, Silverberg MS, Daly MJ, Steinhart AH, Abraham C, Regueiro M, Griffiths A, Dassopoulos T, Bitton A, Yang H, Targan S, Datta LW, Kistner EO, Schumm LP, Lee AT, Gregersen PK, Barmada MM, Rotter JI, Nicolae DL, Cho JH (2006) A genome-wide association study identifies IL23R as an inflammatory bowel disease gene. Science 314:1461–1463
30. Edwards JH (1969) Familial predisposition in man. Br Med Bull 25:58–64
31. Fellay J, Shianna KV, Ge D, Colombo S, Ledergerber B, Weale M, Zhang K, Gumbs C, Castagna A, Cossarizza A, Cozzi-Lepri A, De Luca A, Easterbrook P, Francioli P, Mallal S, Martinez-Picado J, Miro JM, Obel N, Smith JP, Wyniger J, Descombes P, Antonarakis SE, Letvin NL, McMichael AJ, Haynes BF, Telenti A, Goldstein DB (2007) A whole-genome association study of major determinants for host control of HIV-1. Science 317:944–947
32. Fieschi C, Dupuis S, Catherinot E, Feinberg J, Bustamante J, Breiman A, Altare F, Baretto R, Le Deist F, Kayal S, Koch H, Richter D, Brezina M, Aksu G, Wood P, Al-Jumaah S, Raspall M, Da Silva Duarte AJ, Tuerlinckx D, Virelizier JL, Fischer A, Enright A, Bernhoft J, Cleary AM, Vermylen C, Rodriguez-Gallego C, Davies G, Blutters-Sawatzki R, Siegrist CA, Ehlayel MS, Novelli V, Haas WH, Levy J, Freihorst J, Al-Hajjar S, Nadal D, De Moraes Vasconcelos D, Jeppsson O, Kutukculer N, Frecerova K, Caragol I, Lammas D, Kumararatne DS, Abel L, Casanova JL (2003) Low penetrance, broad resistance, and favorable outcome of interleukin 12 receptor beta1 deficiency: medical and immunological implications. J Exp Med 197:527–535
33. Filipe-Santos O, Bustamante J, Chapgier A, Vogt G, de Beaucoudrey L, Feinberg J, Jouanguy E, Boisson-Dupuis S, Fieschi C, Picard C, Casanova JL (2006) Inborn errors of IL-12/23- and IFN-gamma-mediated immunity: molecular, cellular, and clinical features. Semin Immunol 18:347–361
34. Fischer A, Le Deist F, Hacein-Bey-Abina S, Andre-Schmutz I, Basile Gde S, de Villartay JP, Cavazzana-Calvo M (2005) Severe combined immunodeficiency. A model disease for molecular immunology and therapy. Immunol Rev 203:98–109
35. Fisher RA (1918) The correlation between relatives on the supposition of Mendelian inheritance. Trans R Soc Edinb 52:399–433
36. Fortier A, Min-Oo G, Forbes J, Lam-Yuk-Tseung S, Gros P (2005) Single gene effects in mouse models of host: pathogen interactions. J Leukoc Biol 77:868–877
37. Gaspar HB, Parsley KL, Howe S, King D, Gilmour KC, Sinclair J, Brouns G, Schmidt M, Von Kalle C, Barington T, Jakobsen MA, Christensen HO, Al Ghonaium A, White HN, Smith JL, Levinsky RJ, Ali RR, Kinnon C, Thrasher AJ (2004) Gene therapy of X-linked severe combined immunodeficiency by use of a pseudotyped gammaretroviral vector. Lancet 364:2181–2187
38. Gatti RA, Meuwissen HJ, Allen HD, Hong R, Good RA (1968) Immunological reconstitution of sex-linked lymphopenic immunological deficiency. Lancet 2:1366–1369
39. Glass WG, McDermott DH, Lim JK, Lekhong S, Yu SF, Frank WA, Pape J, Cheshier RC, Murphy PM (2006) CCR5 deficiency increases risk of symptomatic West Nile virus infection. J Exp Med 203:35–40
40. Good RA (1987) Bone marrow transplantation for immunodeficiency diseases. Am J Med Sci 294:68–74
41. Greenwood CM, Fujiwara TM, Boothroyd LJ, Miller MA, Frappier D, Fanning EA, Schurr E, Morgan K (2000) Linkage of tuberculosis to chromosome 2q35 loci, including NRAMP1, in a large aboriginal Canadian family. Am J Hum Genet 67:405–416
42. Hacein-Bey-Abina S, Von Kalle C, Schmidt M, McCormack MP, Wulffraat N, Leboulch P, Lim A, Osborne CS, Pawliuk R, Morillon E, Sorensen R, Forster A, Fraser P, Cohen JI, de Saint Basile G, Alexander I, Wintergerst U, Frebourg T, Aurias A, Stoppa-Lyonnet D, Romana S, Radford-Weiss I, Gross F, Valensi F, Delabesse E, Macintyre E, Sigaux F, Soulier J, Leiva LE, Wissler M, Prinz C, Rabbitts TH, Le Deist F, Fischer A, Cavazzana-Calvo M (2003) LMO2-associated clonal T cell proliferation in two patients after gene therapy for SCID-X1. Science 302:415–419
43. Howe SJ, Mansour MR, Schwarzwaelder K, Bartholomae C, Hubank M, Kempski H, Brugman MH, Pike-Overzet K, Chatters SJ, de Ridder D, Gilmour KC, Adams S, Thornhill SI, Parsley KL, Staal FJ, Gale RE, Linch DC, Bayford J, Brown L, Quaye M, Kinnon C, Ancliff P, Webb DK, Schmidt M, von Kalle C, Gaspar HB, Thrasher AJ (2008) Insertional mutagenesis combined with acquired somatic mutations causes leukemogenesis following gene therapy of SCID-X1 patients. J Clin Invest 118(9):3143–3150
44. Jouanguy E, Altare F, Lamhamedi S, Revy P, Emile JF, Newport M, Levin M, Blanche S, Seboun E, Fischer A, Casanova JL (1996) Interferon-gamma-receptor deficiency in an infant with fatal bacille Calmette-Guerin infection. N Engl J Med 335:1956–1961
45. Khoury MJ, Beaty TH, Cohen BH (1993) Fundamentals of genetic epidemiology. Oxford University Press, New York
46. Kwiatkowski D (2000) Science, medicine, and the future: susceptibility to infection. BMJ 321:1061–1065
47. Kwiatkowski DP (2005) How malaria has affected the human genome and what human genetics can teach us about malaria. Am J Hum Genet 77:171–192
48. Lalouel JM, Rao DC, Morton NE, Elston RC (1983) A unified model for complex segregation analysis. Am J Hum Genet 35:816–826
49. Le Pendu J, Ruvoen-Clouet N, Kindberg E, Svensson L (2006) Mendelian resistance to human norovirus infections. Semin Immunol 18:375–386
50. Lindesmith L, Moe C, Marionneau S, Ruvoen N, Jiang X, Lindblad L, Stewart P, LePendu J, Baric R (2003) Human susceptibility and resistance to Norwalk virus infection. Nat Med 9:548–553
51. Liu R, Paxton WA, Choe S, Ceradini D, Martin SR, Horuk R, MacDonald ME, Stuhlmann H, Koup RA, Landau NR (1996) Homozygous defects in HIV-1 coreceptor accounts for resistance of some multiply-exposed individuals to HIV-1 infection. Cell 86:367–377

52. Malik S, Abel L, Tooker H, Poon A, Simkin L, Girard M, Adams GJ, Starke JR, Smith KC, Graviss EA, Musser JM, Schurr E (2005) Alleles of the NRAMP1 gene are risk factors for pediatric tuberculosis disease. Proc Natl Acad Sci USA 102:12183–12188
53. Marionneau S, Ruvoen N, Le Moullac-Vaidye B, Clement M, Cailleau-Thomas A, Ruiz-Palacios G, Huang P, Jiang X, Le Pendu J (2002) Norwalk virus binds to histo-blood group antigens present on gastroduodenal epithelial cells of secretor individuals. Gastroenterology 122:1967–1977
54. Marquet S, Abel L, Hillaire D, Dessein H, Kalil J, Feingold J, Weissenbach J, Dessein AJ (1996) Genetic localization of a locus controlling the intensity of infection by Schistosoma mansoni on chromosome 5q31–q33. Nat Genet 14:181–184
55. Miller LH, Mason SJ, Clyde DF, McGinniss MH (1976) The resistance factor to Plasmodium vivax in blacks. The Duffy-blood-group genotype, FyFy. N Engl J Med 295:302–304
56. Mira MT, Alcais A, Nguyen VT, Moraes MO, Di Flumeri C, Vu HT, Mai CP, Nguyen TH, Nguyen NB, Pham XK, Sarno EN, Alter A, Montpetit A, Moraes ME, Moraes JR, Dore C, Gallant CJ, Lepage P, Verner A, Van De Vosse E, Hudson TJ, Abel L, Schurr E (2004) Susceptibility to leprosy is associated with PARK2 and PACRG. Nature 427:636–640
57. Mira MT, Alcais A, Van Thuc N, Thai VH, Huong NT, Ba NN, Verner A, Hudson TJ, Abel L, Schurr E (2003) Chromosome 6q25 is linked to susceptibility to leprosy in a Vietnamese population. Nat Genet 33:412–415
58. Modiano D, Luoni G, Sirima BS, Simpore J, Verra F, Konate A, Rastrelli E, Olivieri A, Calissano C, Paganotti GM, D'Urbano L, Sanou I, Sawadogo A, Modiano G, Coluzzi M (2001) Haemoglobin C protects against clinical Plasmodium falciparum malaria. Nature 414:305–308
59. Newport MJ, Huxley CM, Huston S, Hawrylowicz CM, Oostra BA, Williamson R, Levin M (1996) A mutation in the interferon-gamma-receptor gene and susceptibility to mycobacterial infection. N Engl J Med 335:1941–1949
60. Nicolle C (1937) Destin des maladies infectieuses, 3rd edn. Alcan, Paris
61. Notarangelo L, Casanova JL, Conley ME, Chapel H, Fischer A, Puck J, Roifman C, Seger R, Geha RS (2006) Primary immunodeficiency diseases: an update from the International Union of Immunological Societies Primary Immunodeficiency Diseases Classification Committee Meeting in Budapest, 2005. J Allergy Clin Immunol 117:883–896
62. Novembre J, Galvani AP, Slatkin M (2005) The geographic spread of the CCR5 Delta32 HIV-resistance allele. PLoS Biol 3:e339
63. Ochs H, Smith CIE, Puck J (2006) Primary Immunodeficiencies: a molecular and genetic approach, 2nd edn. Oxford University Press, New York
64. Orth G (2006) Genetics of epidermodysplasia verruciformis: insights into host defense against papillomaviruses. Semin Immunol 18:362–374
65. Pasteur L (1870) Etudes sur la maladie des vers a soie. La pébrine et la flacherie (tome I). Gauthier-Villars, Paris
66. Pasteur Vallery-Radot L (1939) Oeuvres complètes de Louis Pasteur, réuniset annotées par Pasteur Vallery-Radot. Masson et Cie, Paris
67. Picard C, Casanova JL, Abel L (2006) Mendelian traits that confer predisposition or resistance to specific infections in humans. Curr Opin Immunol 18:383–390
68. Picard C, Puel A, Bonnet M, Ku CL, Bustamante J, Yang K, Soudais C, Dupuis S, Feinberg J, Fieschi C, Elbim C, Hitchcock R, Lammas D, Davies G, Al-Ghonaium A, Al-Rayes H, Al-Jumaah S, Al-Hajjar S, Al-Mohsen IZ, Frayha HH, Rucker R, Hawn TR, Aderem A, Tufenkeji H, Haraguchi S, Day NK, Good RA, Gougerot-Pocidalo MA, Ozinsky A, Casanova JL (2003) Pyogenic bacterial infections in humans with IRAK-4 deficiency. Science 299:2076–2079
69. Pritchard JK (2001) Are rare variants responsible for susceptibility to complex diseases? Am J Hum Genet 69:124–137
70. Purtilo DT, Cassel CK, Yang JP, Harper R (1975) X-linked recessive progressive combined variable immunodeficiency (Duncan's disease). Lancet 1:935–940
71. Quintana-Murci L, Alcais A, Abel L, Casanova JL (2007) Immunology in natura: clinical, epidemiological and evolutionary genetics of infectious diseases. Nat Immunol 8:1165–1171
72. Ramoz N, Rueda LA, Bouadjar B, Montoya LS, Orth G, Favre M (2002) Mutations in two adjacent novel genes are associated with epidermodysplasia verruciformis. Nat Genet 32:579–581
73. Rigaud S, Fondaneche MC, Lambert N, Pasquier B, Mateo V, Soulas P, Galicier L, Le Deist F, Rieux-Laucat F, Revy P, Fischer A, De Saint Basile G, Latour S (2006) XIAP deficiency in humans causes an X-linked lymphoproliferative syndrome. Nature 444:110–114
74. Rioux JD, Xavier RJ, Taylor KD, Silverberg MS, Goyette P, Huett A, Green T, Kuballa P, Barmada MM, Datta LW, Shugart YY, Griffiths AM, Targan SR, Ippoliti AF, Bernard EJ, Mei L, Nicolae DL, Regueiro M, Schumm LP, Steinhart AH, Rotter JI, Duerr RH, Cho JH, Daly MJ, Brant SR (2007) Genome-wide association study identifies new susceptibility loci for Crohn disease and implicates autophagy in disease pathogenesis. Nat Genet 39:596–604
75. Risch N (1990) Linkage strategies for genetically complex traits. III. The effect of marker polymorphism on analysis of affected relative pairs. Am J Hum Genet 46:242–253
76. Risch N, Merikangas K (1996) The future of genetic studies of complex human diseases. Science 273:1516–1517
77. Sabeti PC, Walsh E, Schaffner SF, Varilly P, Fry B, Hutcheson HB, Cullen M, Mikkelsen TS, Roy J, Patterson N, Cooper R, Reich D, Altshuler D, O'Brien S, Lander ES (2005) The case for selection at CCR5-Delta32. PLoS Biol 3:e378
78. Samson M, Libert F, Doranz BJ, Rucker J, Liesnard C, Farber CM, Saragosti S, Lapoumeroulie C, Cognaux J, Forceille C, Muyldermans G, Verhofstede C, Burtonboy G, Georges M, Imai T, Rana S, Yi Y, Smyth RJ, Collman RG, Doms RW, Vassart G, Parmentier M (1996) Resistance to HIV-1 infection in caucasian individuals bearing mutant alleles of teh CCR5 chemokine receptor gene. Nature 382:722–725
79. Schurr E, Alcais A, de Leseleuc L, Abel L (2006) Genetic predisposition to leprosy: a major gene reveals novel pathways of immunity to Mycobacterium leprae. Semin Immunol 18:404–410
80. Siddiqui MR, Meisner S, Tosh K, Balakrishnan K, Ghei S, Fisher SE, Golding M, Shanker Narayan NP, Sitaraman T, Sengupta U, Pitchappan R, Hill AV (2001) A major suscep-

tibility locus for leprosy in India maps to chromosome 10p13. Nat Genet 27:439–441
81. Sladek R, Rocheleau G, Rung J, Dina C, Shen L, Serre D, Boutin P, Vincent D, Belisle A, Hadjadj S, Balkau B, Heude B, Charpentier G, Hudson TJ, Montpetit A, Pshezhetsky AV, Prentki M, Posner BI, Balding DJ, Meyre D, Polychronakos C, Froguel P (2007) A genome-wide association study identifies novel risk loci for type 2 diabetes. Nature 445:881–885
82. Sorensen TI, Nielsen GG, Andersen PK, Teasdale TW (1988) Genetic and environmental influences on premature death in adult adoptees. N Engl J Med 318:727–732
83. Stephens JC, Reich DE, Goldstein DB, Shin HD, Smith MW, Carrington M, Winkler C, Huttley GA, Allikmets R, Schriml L, Gerrard B, Malasky M, Ramos MD, Morlot S, Tzetis M, Oddoux C, di Giovine FS, Nasioulas G, Chandler D, Aseev M, Hanson M, Kalaydjieva L, Glavac D, Gasparini P, Dean M et al (1998) Dating the origin of the CCR5-Delta32 AIDS-resistance allele by the coalescence of haplotypes. Am J Hum Genet 62:1507–1515
84. Thorven M, Grahn A, Hedlund KO, Johansson H, Wahlfrid C, Larson G, Svensson L (2005) A homozygous nonsense mutation (428G–>A) in the human secretor (FUT2) gene provides resistance to symptomatic norovirus (GGII) infections. J Virol 79:15351–15355
85. Tournamille C, Colin Y, Cartron JP, Le Van Kim C (1995) Disruption of a GATA motif in the Duffy gene promoter abolishes erythroid gene expression in Duffy-negative individuals. Nat Genet 10:224–228
86. Vanhollebeke B, Truc P, Poelvoorde P, Pays A, Joshi PP, Katti R, Jannin JG, Pays E (2006) Human Trypanosoma evansi infection linked to a lack of apolipoprotein L-I. N Engl J Med 355:2752–2756
87. Vogel F, Motulsky AG (1997) Human genetics: problems and approaches, 3 completely revth edn. Springer, Berlin
88. WHO (2006) Global leprosy situation, 2006. Wkly Epidemiol Rec 81:309–316
89. Zhang SY, Jouanguy E, Ugolini S, Smahi A, Elain G, Romero P, Segal D, Sancho-Shimizu V, Lorenzo L, Puel A, Picard C, Chapgier A, Plancoulaine S, Titeux M, Cognet C, von Bernuth H, Ku CL, Casrouge A, Zhang XX, Barreiro L, Leonard J, Hamilton C, Lebon P, Heron B, Vallee L, Quintana-Murci L, Hovnanian A, Rozenberg F, Vivier E, Geissmann F, Tardieu M, Abel L, Casanova JL (2007) TLR3 deficiency in patients with herpes simplex encephalitis. Science 317:1522–1527

Gene Action: Developmental Genetics

Stefan Mundlos

Abstract Developmental genetics studies the mechanisms how genes initiate and control the process by which a single cell can give rise to a mature organism. This includes mechanisms of early patterning, as well as later events that result in the formation and maturation of organ systems. Developmentally active genes exert their effects through many pathways and mechanisms including diffusing morphogens, cell migration, proliferation, and border formation. Transient structures such as the somites, the branchial arches and the apical ectodermal ridge serve as scaffold and signaling centers during embryogenesis. Gene defects frequently result in abnormal development with specific phenotypes that reflect the gene's essential functions during embryogenesis. In many instances this results in a combination of malformations that are characteristic for a specific syndrome.

Contents

13.1	Genetics of Embryonal Development	417
	13.1.1 Basic Mechanisms of Development	418
	13.1.2 Mechanisms of Morphogenesis	419
13.2	The Stages of Development	422
13.3	Formation of the Central Nervous System	425
13.4	The Somites	431
13.5	The Brachial Arches	433
13.6	Development of the Limbs	435
13.7	Development of the Circulatory System	438
13.8	Development of the Kidney	440
13.9	Skeletal Development	442
13.10	Abnormal Development: Definitions and Mechanisms	445
13.11	Malformations	446
13.12	Disruptions	447
13.13	Deformations	448
13.14	Dysplasias	448
13.15	Terminology of Congenital Defects	449
References		449

13.1 Genetics of Embryonal Development

The study of embryonal development focuses on the process by which a single cell can give rise to a mature organism. This involves the early steps of development that set the pattern for the overall bauplan of the body and the development of individual organs studied in model systems, such as the insect eye, the vertebrate limb, or the nervous system. The study of developmental biology, however, goes beyond the study of embryos. It includes the regeneration of lost organs, such as the newt limb, or the lizard tail, and the control of post-embryonic growth, a process that includes metamorphosis and aging. Development is less an adaptation and reaction to certain stimuli and more a programmed set of events in which the transient stages are more

S. Mundlos (✉)
Institute for Medical Genetics,
Charité, Universitaetsmedizin Berlin, Augustenburger Platz 1,
13353 Berlin, Germany
e-mail: stefan.mundlos@charite.de

important than the permanent ones. The information for this process is contained within the genome. The genome holds the code that tells the embryo to develop and controls differentiation of cells, thus directing the entire process. But this is not a one-way process, since the embryo, on the other hand, is able to control its genome, mediating between genotype and phenotype, between the inherited genes, its environment, and the adult organism. Since gene regulation is obviously the key event in the organization of this process, many mechanisms have been proposed and experimentally verified. But how this intricate control system is able to induce and maintain differences between cells, enabling them to differentiate, remains largely unclear. The answer will probably not come from considering merely DNA and its interactions with RNA and proteins. The feedback between the cells, their genome, and the environment will have to be considered. The organism, the regulative pathway including a network of cooperating partners, and the interacting environment have to be studied as a whole for true understanding of this process to be possible. The detection of a mutation in a patient with a heart defect links the function of this gene with heart development, but how this gene contributes to a critical process during development has to be investigated in detail using functional analysis. Mutations in humans can lead to new insights into the factors that contribute to certain developmental processes, but only the study of the entire regulatory network will yield a thorough understanding that will ultimately enable us to comprehend complex human traits and their interaction with environmental stimuli.

Because human experimentation is subject to obvious limitations, appropriate model systems are needed to study the molecular basis of developmental processes. Over the past century developmental biologists have established a wide variety of model systems that have greatly contributed to our understanding of the basic mechanisms of animal and human development. These include the fruit fly *Drosophila melanogaster*, the nematode *Caenorhabditis elegans*, the sea urchin *Lytechinus variegatus*, the zebra fish *Danio rerio*, the frog *Xenopus laevis*, the chick, and the mouse, to mention the most important ones. *Drosophila* was the first animal model to be studied in great detail, because it is genetically amenable and therefore suitable for large-scale mutation screening, genetic analysis, and manipulation. Many of the mechanisms underpinning early embryonic development were established in this species and, surprisingly at the time, were found to be conserved throughout the animal kingdom. The sea urchin has been most instrumental in the study of early development, in particular gastrulation, whereas *C. elegans* is a more recent model system appreciated for its stereotyped developmental program, an almost invariant cell lineage, and easy manipulation. *Xenopus* and the chick have been used as vertebrate model organ systems because both species produce robust embryos that develop outside of the mother, allowing easy manipulation. Both systems, however, have the disadvantage that they cannot be genetically manipulated. In this sense the mouse is the preferred organism, because of its suitability for genetic manipulation (in particular gene targeting) and its closeness to humans. The zebra fish combines many of the above-mentioned advantages owing to its accessibility and genetic amenability.

13.1.1 Basic Mechanisms of Development

The problem of how the embryo determines a pattern is one of the most central questions in biology. How is the cell fate orchestrated in a three-dimensional space by a set of instructions, the genes, that are the same in every cell? This question has captivated biologists and scientists from many other disciplines who have infused the field with their viewpoints. Clearly, the embryo uses a variety of ways to determine its own gestalt and function. At the molecular level several different processes are incorporated that affect the behavior of cells. Cell *proliferation* leads to an increase in cell number and thus expansion and growth of the embryo. New structures can be generated by differentially increasing the rate of proliferation. *Growth* may also be accomplished by the synthesis of macromolecules or an increase in cell size. Programmed cell death or *apoptosis* is an important mechanism to remove transient structures. For example, our fingers and toes are created by the death of interdigital cells in the hand and foot plates. Cell *migration* is the movement of an individual cells or groups of cells with respect to other cells in the embryo, as observed, for example, in neural crest and germ cells that undergo extensive migration and consequently populate parts of the embryo that are very distant from their original locations. During further development, *differentiation* takes place, a process by which cells become structurally and functionally specialized.

Cells become organized in tissues by sticking together, a process called condensation. This can take place through the expression of complementary adhesion molecules on their surfaces, and/or they may form associations with their extracellular matrix. Adhesion molecules function in the same way as a receptor-ligand interaction, except that in this case the adjacent cells carry either the receptor or the ligand. By this mechanism borders can be induced and maintained, ensuring that the cells on either side of the border develop according to a different scheme. The delta/notch pathway represents such a system, where one cell expresses notch whereas the other carries the delta receptor on its surface, each inducing its own set of gene expression. Posttranslational modification of proteins is an important way to modify their interaction with other proteins [13]. For example, fringe, a known modifier of notch signaling, has been shown to function as a fucose-specific *N*-acetylglucosaminyltransferase, demonstrating that notch signaling can be regulated by protein modification. More recently, microRNAs, small RNA molecules that inhibit the translation of specific mRNAs, were shown to be involved in gene regulation in development. All of these mechanisms are instrumental in inducing differentiation, but how a body plan and thus a three-dimensional structure is established cannot be explained by these mechanisms alone. One emerging concept that is likely to be central to the problem is the presence of so-called morphogens, signaling molecules that determine cell fate.

13.1.2 Mechanisms of Morphogenesis

In the oldest sense of the word, a morphogen is a substance that is produced by cells and organizes a pattern by spreading to other cells [1]. Because morphogens are produced at one location, usually referred to as the signaling center, their concentration is thought to decline as a function of distance from the source. Thus, cells that are close to the signaling center will receive a high concentration of morphogen, while those further away receive lower doses. The hypothesis of smoothly declining gradients, originally proposed by Wolpert, assigns positional values to cells that are ultimately translated into cell fate determination (Fig. 13.1). The diffusing morphogen produces a gradient, which is superposed by other morphogen gradients. This results in cross-threshold values at which

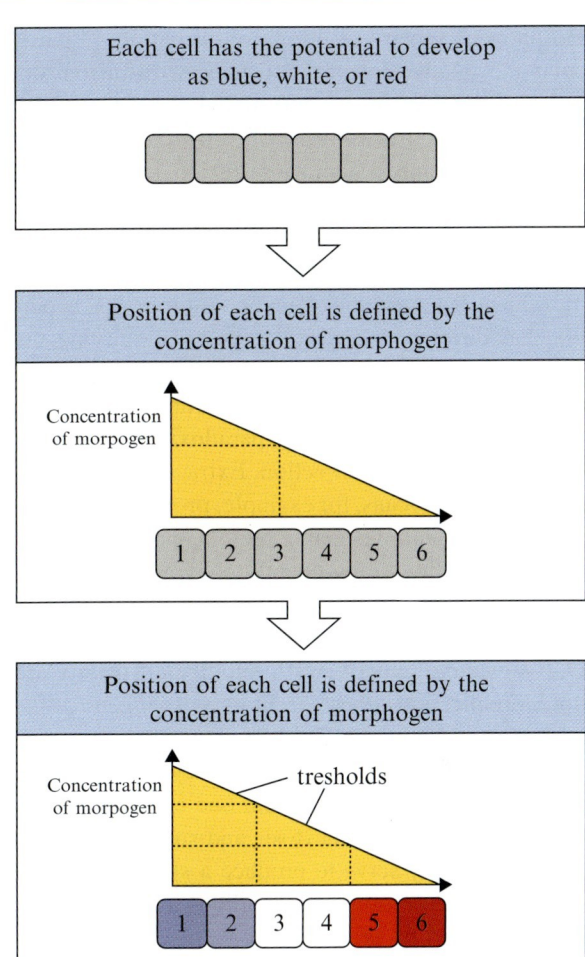

Fig. 13.1 Morphogens determine cell fate. Each cell has the potential to develop as red, white, or blue. A gradient of morphogen produced by a signaling center results in a different concentration at each cell, depending on the distance from the source. A threshold level determines cell fate and identity. A high concentration results in blue cells, medium concentration, in white, and low concentration, in red cells. (From [35], p. 24, Fig. 1.25, by permission of Oxford University Press)

genes are turned on or off. This hypothesis has been supported by the identification of substances such as bicoid and decapentaplegic (Dpp) in *Drosophila* and, subsequently, polypeptides of the fibroblast growth factor (FGF), Wnt, Hedgehog, and transforming growth factor (TGF) families in vertebrates that act as intracellular or extracellular morphogens. During early *Drosophila* development morphogens can diffuse freely because the zygote nucleus undergoes a series of divisions in a common cytoplasm, the syncytial blastoderm. But how do morphogens move in the intercellular space? It is still not clear how gradients of morphogens specify positional information, in particular the relative roles of morphogen diffusion and cell–cell interaction. Several models exist that try to explain how cell identity can be specified over long and short distances. Simple diffusion refers to randomly moving molecules that encounter little impediment from the tortuous intercellular spaces if concentration difference is the driving force of their movement. To direct the effect of a morphogen, evolution has developed intricate ways by modifying the spreading of freely moving molecules. Altering diffusivity allows the morphogen to accumulate to much higher levels near its source, paradoxically resulting in an increased range of action. Extracellular heparan sulfate proteoglycans, for example, promote the transport of *Drosophila* hedgehog protein and are essential for FGF signaling. Consequently, enzymes that change the local content of extracellular matrix heparan sulfate can have a great influence on hedgehog as well as FGF signaling. Other ways to modify diffusivity are lipid modifications; these occur in morphogens of the hedgehog and Wnt families. Morphogens may also be transported through cells by endocytotic vesicles, passing the morphogen on from one cell to the next. Receptor-mediated endocytosis and subsequent rapid degradation may serve to produce a sharp decline in protein concentration. Availability of morphogens may be altered by producing nonactive depot forms, as it is in the case of the transforming growth factors, which are activated upon proteolytic cleavage. Inhibitors may bind to morphogens, thus preventing their diffusion and/or binding to their cognate receptors. The BMP inhibitor Noggin, for example, completely inhibits BMP signaling, but is expressed at distinct sites that only partially overlap with BMP expression, thus directing and diversifying the signal (Fig. 13.2).

A wide variety of different signaling centers have been described that are essential for many developmental processes. In the *Drosophila* egg two "organizing centers" initiate anterior and posterior gradients, each forming its own structures at the poles and interacting with the other gradient to form the central portion of the embryo. Nüsslein–Volhard identified bicoid and hunchback as the proteins crucial for the anterior gradient and thus head and thorax formation, whereas nanos and caudal were found to form the posterior gradient and thus the abdominal segments [6, 7]. The anterior proteins inhibit the translations of the posterior proteins and vice versa, and as a result of this interaction a four-protein gradient is produced in the early embryo, which governs the first steps of *Drosophila* embryogenesis (Fig. 13.3). Signaling centers in vertebrates are, for example, the notochord and the floor plate, both producing the morphogen sonic hedgehog (Shh). As discussed below, different neurons develop in the neural tube depending on their distance from the signaling center and thus the concentration of the morphogen. A similar situation is seen in the limb, where Shh is expressed exclusively in the so-called zone of polarizing activity, a distinct region in the posterior part of the limb bud which controls the asymmetry of our limbs from the thumb to the little finger (Fig. 13.4) [30]. Depending on the distance from the center a pattern is built that determines which cells will finally develop into the individual digits. Duplications of this center to the anterior side result in mirror image duplications producing another set of digits. Many other signaling centers are known, and some of them are discussed below.

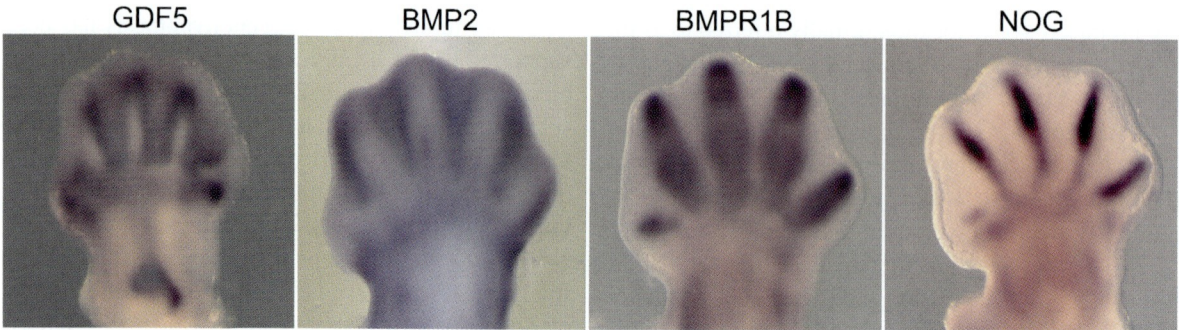

Fig. 13.2 Diverse expression patterns of morphogen, receptor, and inhibitor. Expression pattern of GDF5, BMP2, the receptor BMPR1B, and the inhibitor Noggin during development of the mouse digits. (Courtesy of P. Seemann)

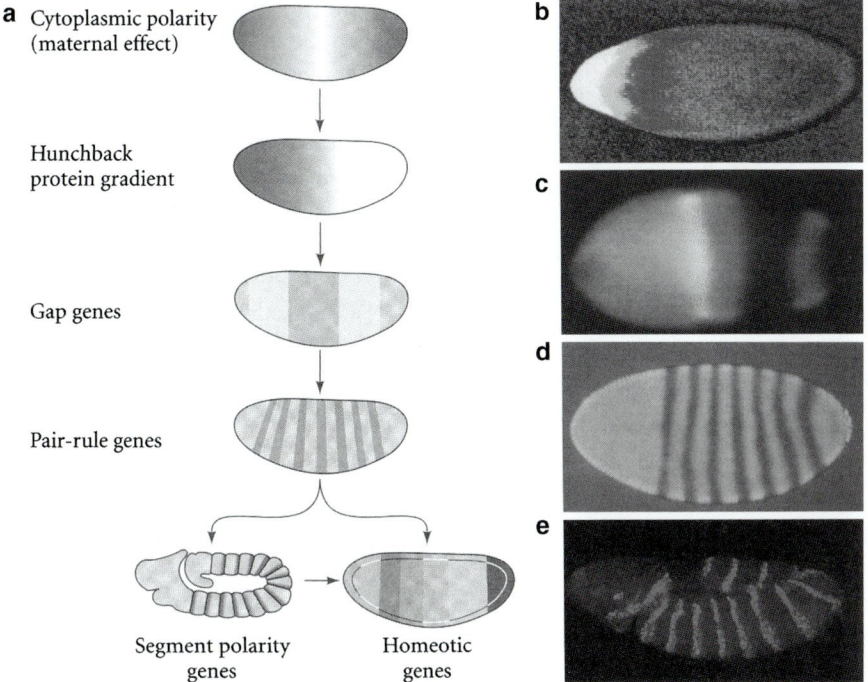

Fig. 13.3 (**a–e**) A generalized model of *Drosophila* pattern formation. (**a**) Anterior-posterior pattern is first generated by maternal effect genes that are located as sequestered mRNAs near the anterior tip (bicoid) and the posterior tip (nanos) of the unfertilized egg. After fertilization, the mRNAs are translated into proteins that can diffuse in the syncytial blastoderm, forming gradients that in turn activate Hunchback protein, which differentially activates gap genes, that define broad regions of the embryo. Gap genes activate pair-rule genes, giving the first indication of segmentation in the fly embryo. Pair-rule genes are expressed in a "zebra-stripe" pattern with an alternating pattern of vertical bands of cells expressing and not expressing a pair-rule gene. Together, these genes control the expression domains of the homeotic genes that define the identity of each segment. (**b**) Maternal effect genes. Bicoid protein concentration is highest at anterior tip (*bright yellow*) and diminishes towards the middle of the embryo (*red*). (**c**) Gap gene expression. Distribution of hunchback (*orange*) and Krüppel (*green*) overlap in the middle part (*yellow*). (**d**) Pair-rule genes. Pair-rule gene fushi tarazu forms seven stripes across the embryo. (**e**) Segment polarity genes. Expression of engrailed dividing the embryo into a repeated series of segmental primordia along the anterior-posterior axis. (From [10], Fig. 9.17a–e)

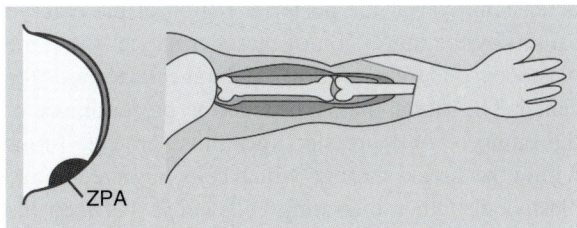

Fig. 13.4 Morphogen controls asymmetry of the limbs. Sonic hedgehog (*Shh*) is expressed in the posterior margin of the developing limb in a region called the zone of polarizing activity (ZPA). Digital identity from thumb to little finger is determined by the proximity to the signaling center and thus the concentration of Shh protein. (From [33], Fig. 2)

Morphogens need effectors, molecules that govern gene expression and by this means determine cell fate. In general this is not accomplished by single molecules but by an entire set of so-called transcription factors that bind to specific DNA sequences in the regulatory regions of target genes, thus resulting in an orchestration of gene expression. One example are the homeotic or *Hox* genes, originally identified in *Drosophila* for their ability to deliver positional identity to cells, meaning the information that tells each cell where it is in the embryo and how it has to behave to generate a regionally appropriate structure. Mutations in *Hox* genes produced a number of flies in which one body

Fig. 13.5 Ultrabithorax mutant. A mutation in the bithorax complex of the *Drosophila Hox* gene cluster results in the homeotic transformation of the posterior halteres (a balancing organ) into wings, resulting in a fly with two pairs of wings instead of one. (From [10], p. 284)

part developed in the likeness of another (called a homeotic transformation). In the mutant *Antennapedia* legs instead of antennae grow out of the head. In the mutant bithorax a second set of wings develops instead of the haltere, a small structure normally used by the fly to keep balance (Fig. 13.5). Molecular analysis has revealed that eight homeobox genes exist in *Drosophila* and that they are arranged in a cluster in the fly genome. Furthermore, the genes are expressed in an overlapping pattern along the head-to-tail axis, dividing the body into discrete zones (Fig. 13.6). The particular combination of genes expressed in each zone appeared to be essential for this the cell's positional information, since manipulating the genes either by mutating them or by overexpressing single *Hox* genes resulted in body part transformation [19, 21]. Very similar clusters of *Hox* genes are found in mammals, but here four clusters are present that are also expressed in an overlapping fashion, with most 3′ genes of a cluster being expressed most anterior and most 5′ in the dorsal region of the embryo (Fig. 13.6) [19]. Mutations in HOX genes in humans do not result in transformation of body parts, but instead lead to a variety of malformations involving the limbs and genitals (Fig. 13.7).

13.2 The Stages of Development

The development of animals actually begins before fertilization of the egg with the production by the female of substances that nourish and control the development of the zygote into a multicellular organism.

After fertilization the zygote divides mitotically to produce the cells of the body, usually without much growth in overall size. In most invertebrates the resulting ball of cells is called a blastula, but in vertebrates the term morula is used. Even at this stage some cells have been determined to form specific tissues in the body. This was shown by experiments in which a piece of blastula is surgically excised and transplanted to a different position or onto another organism. In some cases the cells survive and continue their original path of development irrespective of their new environment. Thus, even before there is any visible distinction, cells are assigned to a specific fate. Cleavage divisions produce a hollow sphere of cells, the fluid-filled blastocele, which is called a blastocyst in mammals.

After the first rapid series of cell divisions, the blastula/morula undergoes a massive reorganization called gastrulation. This process converts an essentially nondescript sphere of cells into an organism with distinct cell layers, often called the primary germ layers. As originally studied in the sea urchin, gastrulation starts with an invagination of a subset of cells located at one side of the blastula into the blastocoel, the central cavity of the blastula. Through extensive cell movements the primary germ layers are formed. As a result, the embryo consists of the outer layer, the ectoderm, which produces the cells of the epidermis and the nervous system, the inner layer, the endoderm, which produces the lining of the digestive tube and its associated organs, and the middle layer, the mesoderm, which gives rise to several organs, including heart, kidney, gonads, and the skeleton. Although similar in principle, gastrulation follows different mechanisms in different species (Fig. 13.8). The major characteristic of avian and mammalian gastrulation is the primitive streak. This structure is visible as a thickening of the outer cell layer at the posterior region of the embryo caused by the ingression of mesodermal cells into the blastocoel and by migration of lateral cells towards the center. This streak marks the anterior–posterior axis of the embryo. A depression (primitive groove) forms within the streak, through which cells migrate into the blastocoel. Other migrating cells move between the two layers and form the mesoderm. At the anterior end of the primitive streak a regional thickening forms, called the primitive knot or Hensen's node. This structure is equivalent to the amphibian's blastopore, where migrating cells from the outside turn inward and travel along the inner surface of the outer cell sheets

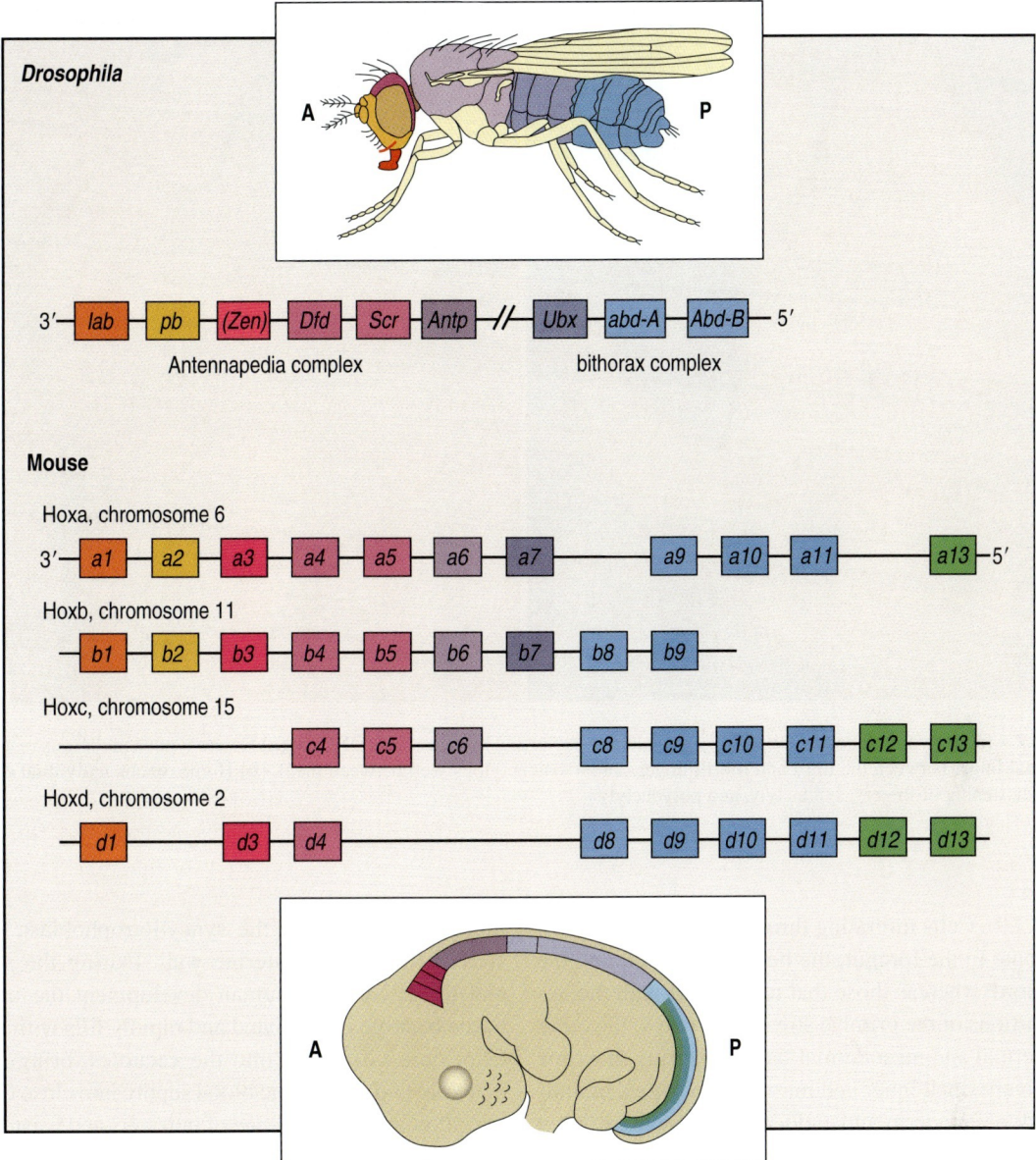

Fig. 13.6 *Hox* genes in *Drosophila* and mammals. *Hox* genes belong to a group of regulatory proteins (transcription factors) that control gene expression. In *Drosophila* there is one *Hox* gene cluster known as HOM-C, which consists of two distinct composites, the Antennapedia and the bithorax complex. In vertebrates *Hox* genes are arranged in four *Hox* clusters called A, B, C, and D, which are likely to have developed from genomic duplications of ancestral *Hox* clusters. In each cluster the order of the genes from 3′ to 5′ corresponds to the sequence in which they are expressed along the anterior-posterior axis of the embryo. Thus, the most 3′ located gene (*red*) is expressed first and furthest in the anterior direction, whereas the most 5′ (*blue/green*) is expressed last and furthest in the posterior direction (a phenomenon called spatial and temporal colinearity, respectively). Vertebrates have four *Hox* gene clusters (*Hox*A, -B, -C, -D), which originate from an ancestral cluster, possibly related to the single *Hox* cluster in the lancelet, a simple chordate. Genes that have arisen by duplication and divergence are referred to as paralogs, and the corresponding genes in each cluster (e.g., *Hoxa9*, *Hoxb9*, *Hoxc9*, *Hoxd9*) are known as a paralogous subgroups. Genes of a paralogous subgroup are more similar than genes within a cluster. (From [35], p. 156, Box 4A, by permission of Oxford University Press)

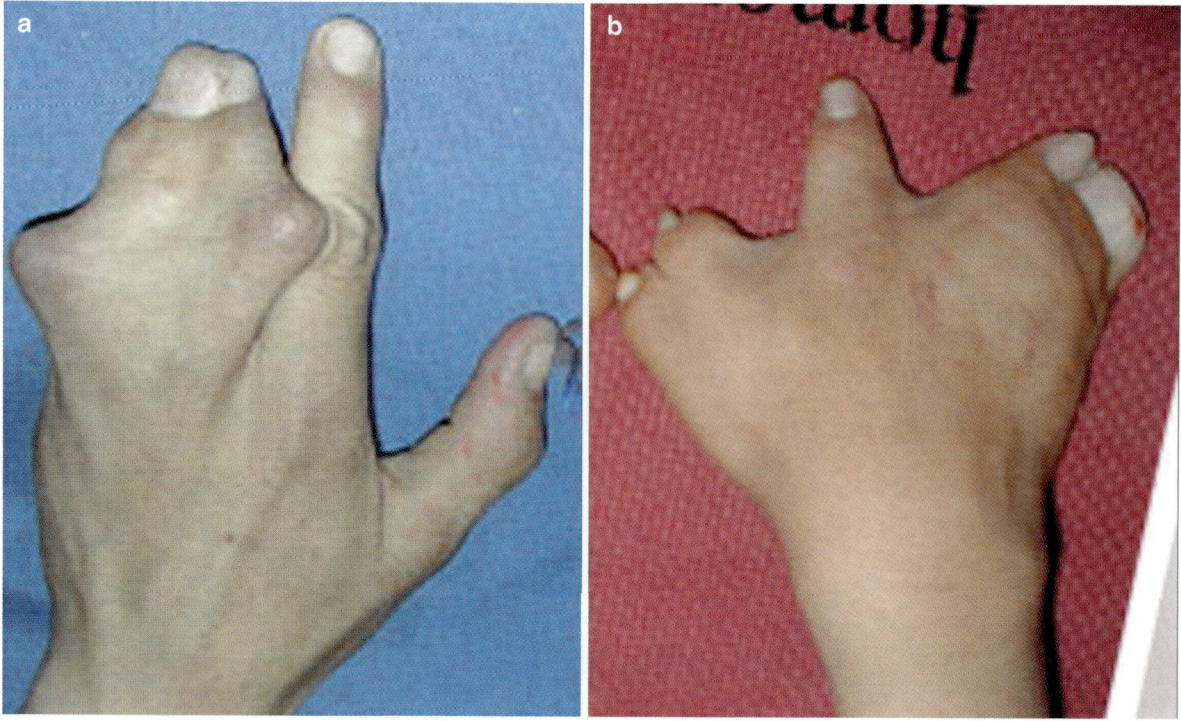

Fig. 13.7 (a, b) Synpolydactyly caused by a polyalanine expansion mutation in HOXD13. (a) Heterozygous mutations result in an additional finger between the third and fourth fingers and a syndactylous web between them. (b) Homozygous individual showing severe shortening of fingers, syndactyly, and polydactyly

(Fig. 13.9). Cells migrating through the primitive knot contribute to the foregut, the head mesoderm, and the notochord, whereas those that migrate through the lateral portions of the primive streak give rise to the other endodermal and mesodermal tissues. During this time a relatively small inner cell mass has become a bilaminar disk of ectoderm and endoderm, each with its own fluid-filled cavity, the amniotic sac, and the yolk sac.

In mammals early development follows the same principle, but not all cells contribute to the embryo, since a major proportion of cells is concerned with establishing tissues that are needed as a life support, namely the extraembryonic membranes and the placenta. Five to six days after conception the human blastocyst arrives in the uterus and attaches to the uterine wall (Fig. 13.10). The blastocyst now consists of an outer cell layer, the trophoblast and an inner layer, the embryoblast. The latter cells congregate at one end of the blastocele to form the inner cell mass. The trophoblast proliferates rapidly and differentiates into an inner layer of cytotrophoblast and an outer multinucleated layer, the syncytiotrophoblast, which starts to invade the uterine wall. During the second and third weeks of human development the invaded tissue becomes vacuolated and rapidly fills with blood. Chorionic villi grow into the vacuoles, bringing the maternal and embryonic blood supply into close contact and allowing the exchange of nutrients and waste products. The fully developed organ consiting of trophoblast tissue and the blood vessels is called the chorion. The chorion fuses with the uterine wall to create the placenta. Besides its role in the exchange of nutrients, the chorion has other important functions as an endocrine organ producing chorionic gonadotropin.

One of the early mesodermal derivatives is the circulatory system, which develops during the third week by vascular channels that arise in the splanchnic mesoderm lining the yolk sac; later a primitive heart that begins pumping by the end of the third week. The narrow connecting stalk that links the embryo to the trophoblast eventually forms the vessels of the umbilical cord. The embryo is now connected to its supply for

13 Gene Action: Developmental Genetics

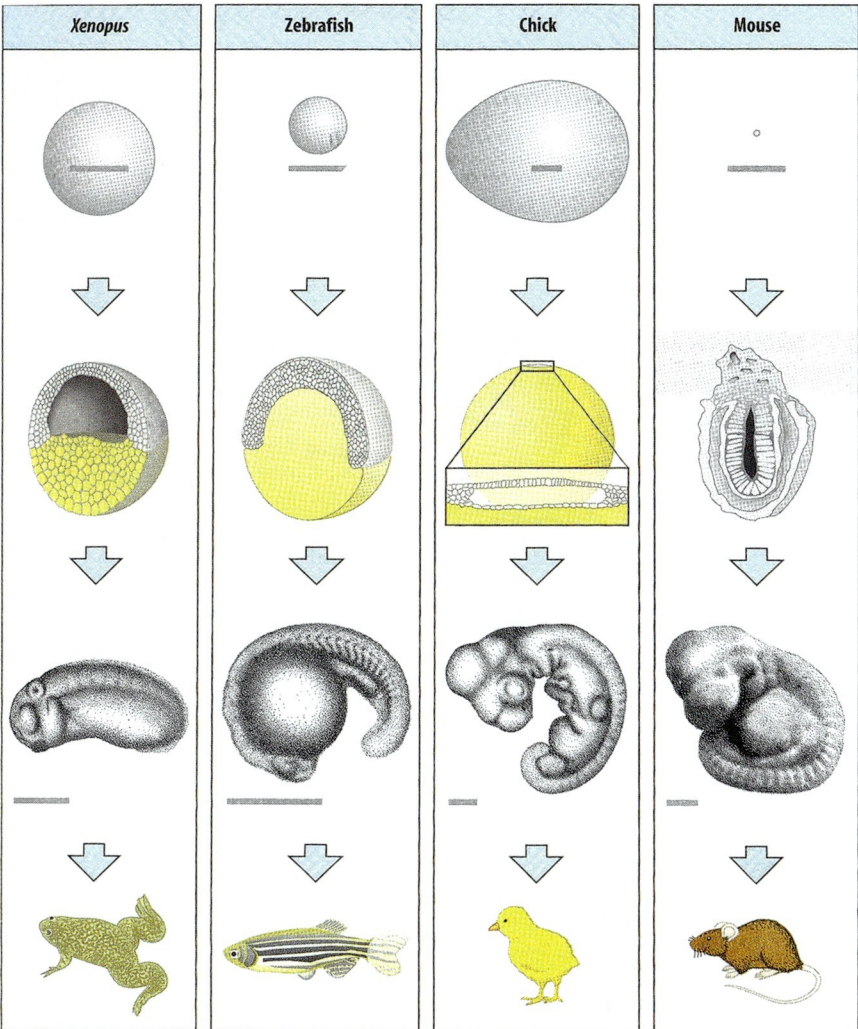

Fig. 13.8 Differences and similarities in development. Vertebrates show considerable differences at the start of their development. The size of the egg is very different (*scale bars* 1 mm except for the chicken egg: 10 mm). The *second row* shows a cross section through the blastula, with the egg yolk shown in *yellow*. At this time the mouse embryo has implanted into the uterine wall and has developed extraembryonic tissues. The embryo itself is the U-shaped structure in the *middle*. Thereafter, gastrulation commences and the embryos develop into polarized bodies consisting of ectoderm, mesoderm, and endoderm. In the following stage (*third row*) all embryos show a certain degree of similarity. The head has formed, and the neural tube, somites, and notochord are present. After this stage their development diverges again, giving rise to such diverse structures as wings, fins, and legs. (From [35], p. 91, Fig. 3.2, by permission of Oxford University Press)

nutrition, hormones and other essential substances, and the stage is set for the period of further patterning and major organogenesis. Early patterning events and the subsequent organ development are extremely complex processes that have been studied in a multitude of model systems. In this overview a few of these systems will be presented in an exemplary way without aiming at a full description of vertebrate organ development.

13.3 Formation of the Central Nervous System

Neurulation is the process by which the embryo forms a neural tube, the rudiment of the central nervous system. The formation of the neural tube is directly related to gastrulation, one of the most important processes in early development. The

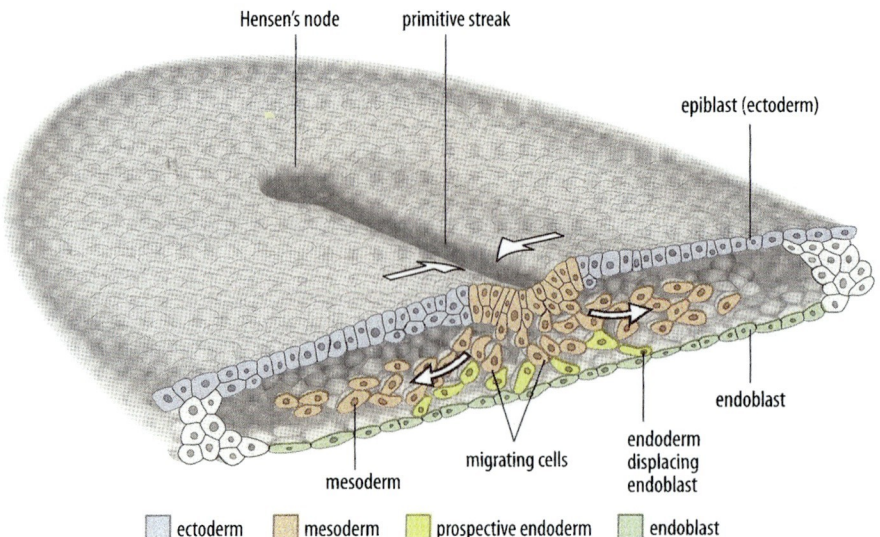

Fig. 13.9 Gastrulation in the chick embryo and Hensen's node. Cells migrate through the primitive streak into the interior of the blastoderm, where they give rise to the endoderm and the mesoderm. At the anterior end of the primitive streak an aggregation of cells known as Hensen's node can be seen. As the streak regresses, the node moves to the posterior end leaving behind the notochord and the first somites. (From [35], p. 102 Fig. 3.15, by permission of Oxford University Press)

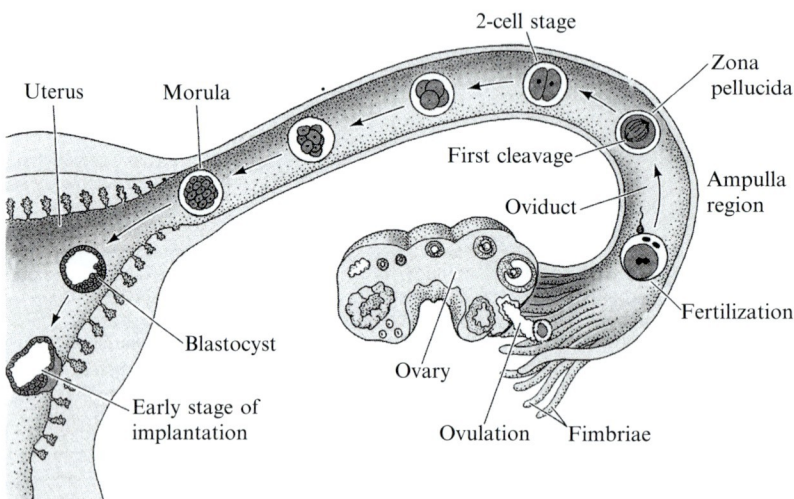

Fig. 13.10 Development of the human embryo from fertilization to implantation. Fertilization of the human oocyte takes place in the ampulla region of the oviduct, and the first cleavage occurs about a day later. The embryo keeps dividing at a slow rate as the cilia of the oviduct push the embryo towards the uterus. In contrast to most other embryos, mammalian blastomeres do not all divide at the same time, and they thus frequently contain odd numbers of cells. The zona pellucida surrounds the embryo and prevents it from attaching to the oviduct. Upon entry into the uterus the blastocyst escapes from the zona pellucida. In the mouse this is accomplished by digesting a small hole in it, through which the blastocyst "hatches." The uterine epithelium (endometrium) secretes a matrix that allows attachment of the blastocyst to the uterine wall. A cocktail of proteases secreted by the trophoblast enables the blastocyst to bury itself within the uterine wall. (From [10], p. 348, Fig. 11.27)

interaction between the dorsal mesoderm and its overlying ectoderm results in the formation of a hollow tube, which will differentiate into the brain and the spinal cord. The ectoderm folds at the most dorsal point, forming an outer epidermis and an inner neural tube. The two layers are connected by a specialized subset of cells, the neural crest. Folding ultimately results in the formation of the spinal cord from the inner layer and closure of the epidermal layer, with the neural crest being situated in between. The neural tube closes as the paired neural folds are brought together at the dorsal midline. This process does not happen at one time point, but is an ongoing process that can best be observed in the chick embryo. While major regions of the neural tube may already be formed in the cephalic (head) region of an embryo, the caudal (tail) region may still be undergoing gastrulation, i.e., the migration of cells from the outer layer inside. Thus, tube formation progresses from head to tail in a zipper like format. In the cephalic region the wall of the tube is broad and thick and a series of swellings and constrictions define the future brain compartments. In those parts of the embryo that form the spinal cord, the neural tube remains a simple tube (Fig. 13.11).

In humans this process follows the same principle. However, the timing and the site of neural tube closure is different. In contrast to the process in the chick, human closure starts in the middle of the embryo and both, the anterior and the posterior neuropores, are open. Closure of the tube progresses from this initial site in both directions, toward the caudal and the cephalic part of the embryo. The cephalic part closes first, followed by the caudal part. Human malformations involving this process are common (Fig. 13.12). They present as spina bifida and anencephaly. In the latter condition there is failure of the neural plate fusion in the cephalic region while spina bifida is observed if the caudal part does not fuse. The reasons for this are complex, caused by genetic as well as environmental factors. Dietary factors such as cholesterol and folic acid appear to be important for normal development. It has been estimated that around 50% of neural tube defects can be prevented when pregnant women take supplemental folic acid. Cholesterol, on the other hand, appears to be necessary for the function of the Sonic hedgehog (Shh) signaling molecule, a morphogen that is essential for neural tube and brain development.

Neural crest cells start a long migration throughout the embryo, contributing to a vast number of tissues, including the bones of the skull, the teeth, the neuronal cells of the gut, and the heart. In humans several syndromes are known that are due to defects in neural crest migration and/or differentiation. For example,

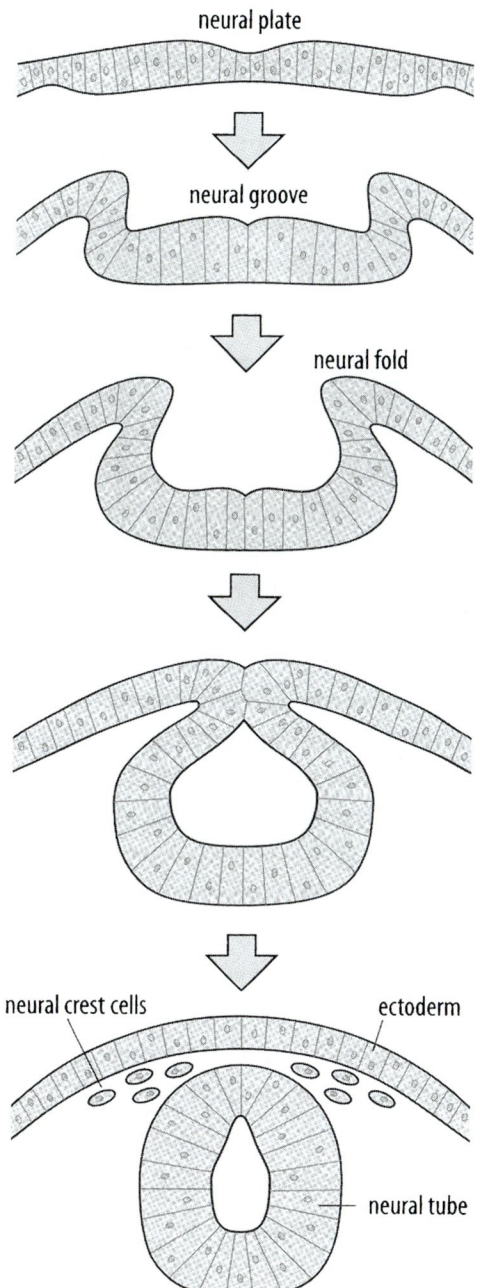

Fig 13.11 Development of the neural tube. Schematic cross sections through the early embryo are shown, representing different stages of neurulation from anterior (head, *top*) to posterior (tail, *bottom*). During neurulation the neural plate bends inwards, creating the neural groove with the neural folds at either side. With further development, the neural folds rise up, extend to the lateral side, and finally form a tube when they meet in the midline. This tube then detaches from the ectoderm, which becomes the epidermis. Neural crest cells which originate from the tip of the folds migrate away towards their distinct destinations. (From [35], p 283, Fig. 7.34, by permission of Oxford University Press)

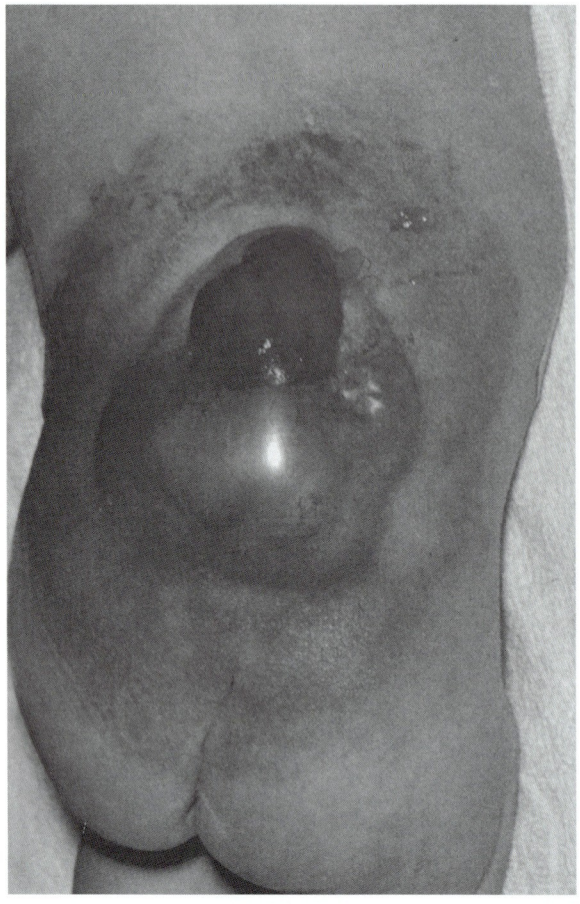

Fig. 13.12 Neural tube defect

mutations in PAX3 result in Waardenburg syndrome, a condition characterized by a specific facial appearance and unpigmented scalp hair. This pigmentation defect is believed to be due to missing neural crest cells, since these cells contribute to the pigment epithelium in the skin and hair.

While the posterior part of the tube is still being formed, the anterior or cephalic part undergoes drastic changes. In humans it subdivides into three primary vesicles, the forebrain (prosencephalon), the midbrain (mesencephalon), and the hindbrain (rhombencephalon). The anterior part of the prosencephalon bulges out laterally, building the two parts of the telencephalon, whereas the more caudal part of the prosencepahlon becomes the diencephalon. Furthermore, secondary bulges, the optic vesicles, extend laterally from each side of the prosencephalon. The mesencephalon does not become subdivided. Its lumen will eventually become the aqueduct connecting the ventricles. The rhomencephalon elongates and becomes subdivided into the metencephalon and the myelencephalon which eventually give rise to the cerebellum and the medulla oblongata, respectively (Fig. 13.13). During this process the early brain increases dramatically in size; however, this increase is primarily due to an increase in cavity size, and not to tissue growth. It has been speculated that increased fluid pressure inside the vesicle is the driving force for

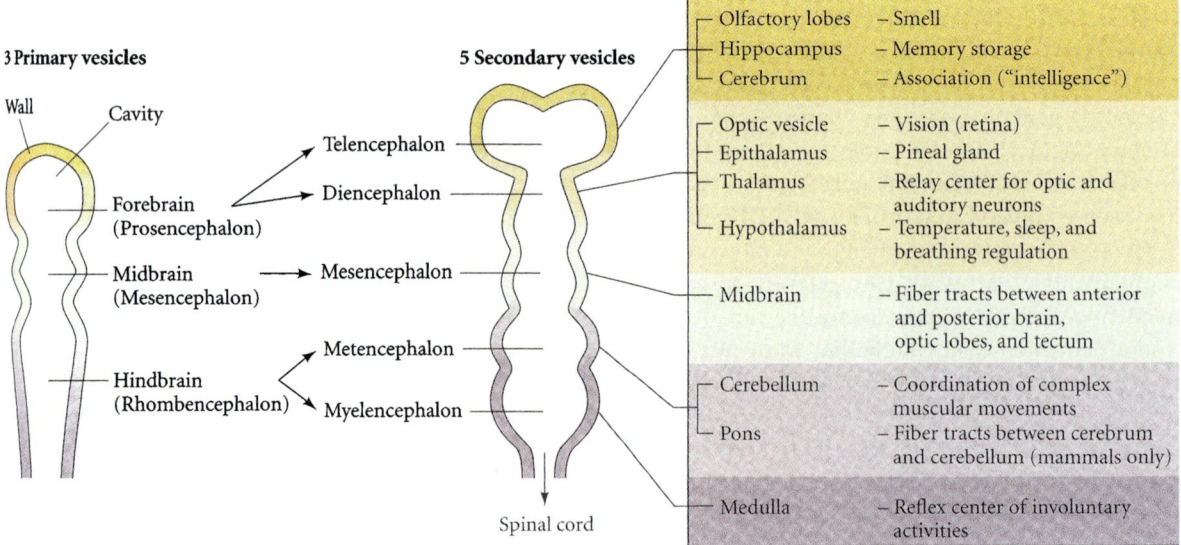

Fig. 13.13 Early human brain development. The three primitive brain vesicles, forebrain, midbrain, and hindbrain, are subdivided as shown on the *right*. These give rise to the adult brain derivatives listed in the *box*. (From [10], p. 381, Fig. 12.9)

this rapid expansion, since reduction of the pressure in chick embryos results in enlargement at a slower rate and the formation of fewer cells.

One important molecule that controls these early patterning events is Sonic hedgehog (Shh)[14]. Hedgehog was first discovered during a screen for mutants in *Drosophila melanogaster*, in which mutations in the single *Hh* gene that presents in this species give rise to an embryo that is covered in spiky cuticular processes called denticles, inspiring the "hedgehog" name. Shh, the vertebrate homologue of Hh, is a secreted molecule that undergoes extensive posttranslational modification including lipid modification, which influences the movement of Hh molecules between cells and its autocatalytic processing in an active and an inactive part. Through its receptor-patched (ptc) and the activation of the transcription factors Gli (cubitus interruptus (Ci) in *Drosophila*) Shh/Hh specifies neuronal identity over short and long distances. Shh's ability to specify identities as a function of its concentration is especially well illustrated by the vertebrate neural tube, where it has a pivotal role in the generation of the diverse types of neurons that are required for the assembly of the spinal cord, the forebrain, and the retina. Shh patterns the neural tube from its two expression sites, the notochord and the floorplate, a triangular wedge of cells located at the ventral midline of the neural tube. The decreasing concentration of Shh from ventral to dorsal establishes distinct progenitor domains which prefigure and predict defined classes of neurons (Fig. 13.14). Without Shh this specification does not takes place and the neural tube consists mainly of dorsal type neurons and, for example, completely misses motorneurons.

How can one signal result in such diverse outcome? Expression of additional factors that modulate Shh signaling may result in a completely different effect. For example, regional differences to Shh signaling have a crucial role in the generation of complexity in the CNS of vertebrates. Shh is expressed in the developing brain in a narrow strip of cells in the vertebrate diencephalon, which is known as the zona limitans intrathalamica. Cells that lie posterior to this structure form the thalamus, whereas those located anterior give rise to the prethalamus. The target genes of *Shh*, *Ptch1* and *Nkx2.2* are expressed on both sides, but other important transcription factors (*Dlx2* on the anterior side and *Gbx2* on the posterior side) are expressed asymmetrically in response to *Shh*. Thus, the response to *Shh* may vary over time and space, further diversifying the response of cells to this important signal.

In humans, heterozygous mutations in SHH have been associated with variable forms of midline facial and brain dysmorphism, in particular holoprosencephaly (HPE) [27]. According to its clinical severity, three types of HPE have been delineated: alobar HPE, the

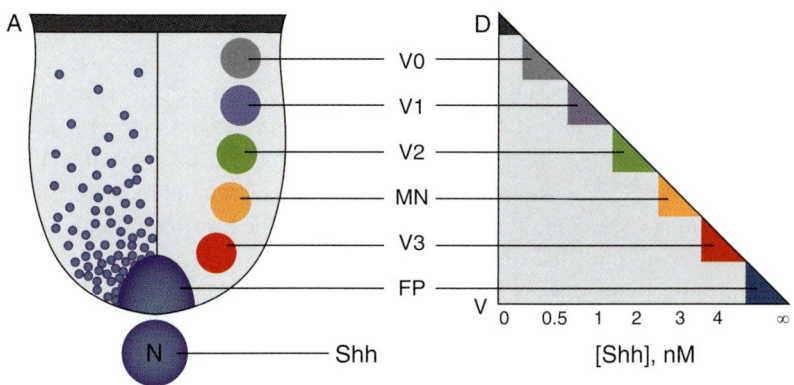

Fig 13.14 *Shh* and the control of spinal cord patterning. *Shh* expressed in the notochord and the floorplate defines the fate of ventral neural progenitors of the spinal cord. The progressively decreasing concentration of Shh protein regulates the expression domains of a series of transcription factors (as indicated by *color code*). *Shh* either represses or induces target genes at different concentration thresholds resulting in a complex pattern of expression combinations. The combinatorial expression of transcription factors in distinct domains determines the type of neuron that arises from each domain. (Reprinted from [15], by permission from Macmillan Publishers Ltd, *EMBO Reports*, copyright 2003)

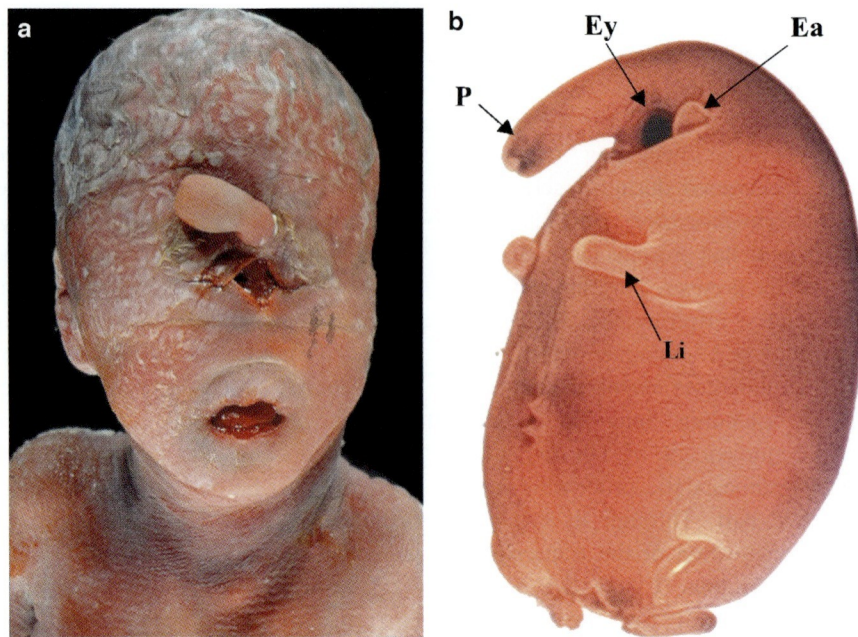

Fig. 13.15 (**a**, **b**) *Shh* regulates many aspects of development. (**a**) Patient with holoprosencephaly. Note nose-like structure in the middle with single eye underneath. (Courtesy of J. Kunze) (**b**) Mouse with inactivated *Shh*. Note single midline eye (cyclops) and nose-like structure above the eye (proboscis) (*P* Proboscis, *Ey* Eye, *Ea* Ear, *Li* Limb)

most severe form, is characterized by complete failure of the forebrain to divide into left and right hemispheres (Fig. 13.15). This form is associated with a single ventricle, and severe anomalies of the face, such as cyclopia, midline cleft, and single nostril are common. In semilobar HPE the interhemispheric fissure is partially present, whereas in lobar HPE the two hemispheres are separated but there is usually microcephaly. All three types, including completely asymtomatic carriers, can be observed within one family. This extreme variability raises the question of whether environmental factors can influence the outcome of this condition. Studies conducted in mammals and birds show that the severity of HPE defects correlates with the stage in which interruption in *Shh* signaling occurs [4]. Furthermore, levels of cholesterol may influence Shh processing and thus its activity. Inactivation of *Shh* in the mouse results in HPE and cyclopia, thus supporting the strong conservation of the pathway.

The initial patterning phase of the human cerebral cortex is paralleled and followed by three overlapping stages that form the structure of the brain. From weeks 5/6 to 6/20, stem cells deep in the forebrain proliferate and differentiate into young neurons or glial cells. During the second phase, between weeks 6/7 and 20/24, neurons migrate away from their place of origin in a radial fashion towards the pial surface, where each successive generation passes the previous one and settles in an inside-out pattern within the cortical plate. When migration is complete, the cortex consists of six layers of neurons that form discrete connections between the neurons and different parts of the CNS. Finally, from week 16 until well into postnatal life organization of the cortical layers takes place, a process associated with synaptogenesis and apoptosis. In this dynamic process more than one stage can occur simultaneously.

Developmental disorders accompanied by brain malformations are important causes of developmental delay, mental retardation, and epilepsy [12]. One of the best-described forms of human brain malformation is lissencephaly, a condition characterized by absent (agyria) or decreased (pachygyria) convolutions, producing a smooth cerebral surface attributable to loss of the folds of the brain, an abnormally thick cortex, and the loss of cortical lamination (Fig. 13.16). Subcortical band heterotopia is a related disorder that can be observed in different regions of the same brain, defining a spectrum of disorders with variable severity, all related to poor cortical architecture in the brain's development. Mutations in the *LIS1* gene cause lissencephaly [26].

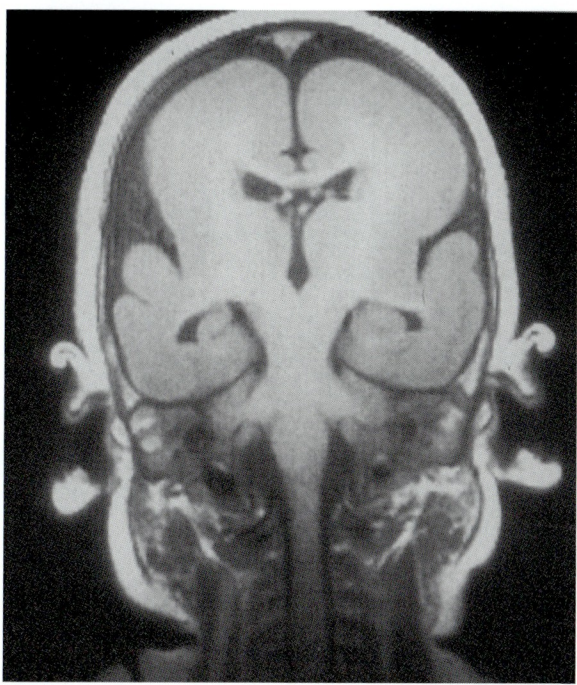

Fig. 13.16 MRI of patient with lissencephaly. Note "smooth" surface of brain owing to absence of gyri

Lis1 has been shown to influence neural progenitor proliferation and migration, possibly by interacting with microtubulin-binding proteins such as dynein and dynamitin. Loss of *Lis1* results in disruption of microtubule function, which interferes with forward translocation of the cell soma during migration and spindle orientation during mitosis. A second gene, DCX, causing an X-linked from of lissencephaly is likely to exert its effect on neuronal migration through its polymerization with microtubules.

Microcephaly vera means a reduction in brain size without marked brain malformation. Affected individuals have very small heads and a variable degree of mental retardation. Mutations in four genes have been shown to cause microcephaly vera; all are associated with the spindle poles of mitotic cells, and the mutations in them are consistent with a centrosomal mechanism in the control of cell division. Loss of microcephalin, for example, leads to premature chromosome condensation in G2 phase and delayed decondensation postmitosis through the condensin proteins. Interestingly, microcephalin has been implicated in human evolution including human brain size based on a strong positive selection in the human evolutionary lineage [8].

13.4 The Somites

Somites are a transient organizational structure of the developing embryo located on both sides of the neural tube consisting of epithelial cells with a periodic structure that originate from the paraxial mesoderm [11]. The formation of new somites and their detachment from the paraxial mesoderm has to occur in a highly ordered fashion simultaneously on both sides of the neural tube in the cranio-caudal direction. A time code for the formation and the budding of new somites is given by oscillations of cycling genes that lead to waves of notch-signaling sweeping up the paraxial mesoderm from the posterior to the anterior pole (Fig. 13.17). Beside this molecular clock, a stable gradient of *Fgf8* expression from the posterior to the anterior pole of the embryo allows a spatial coordination of somite border formation. Dll proteins are notch ligands that reside at the cell surface. Their differential expression determines both the size and the polarity of the somites. Any disturbance in this polarity results in abnormally spaced somites and in fusion of adjacent somites, as exemplified in the mouse mutant pudgy, which carries a mutation in *Dll3*. The phenotype observed in these mice resembles human vertebral malformations.

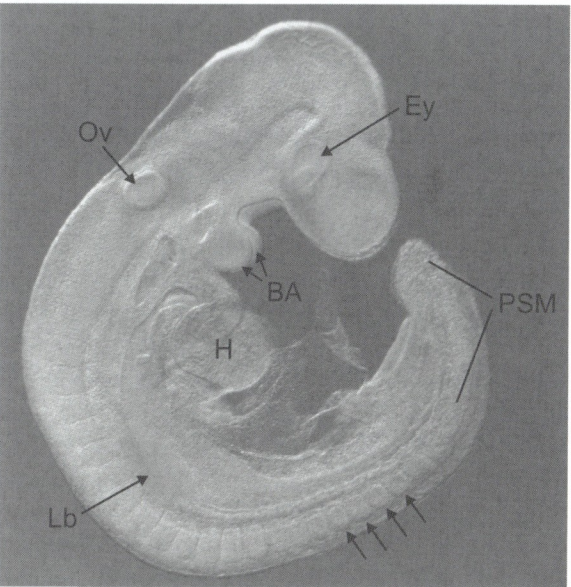

Fig. 13.17 Nine-day mouse embryo. Individual somites (*arrows*) can be seen, as can the paraxial mesoderm (PSM) (*Ey* Eye, *Ov* Otic vesicle, *Lb* Limb bud, *H* Heart, *BA* Brachial arches)

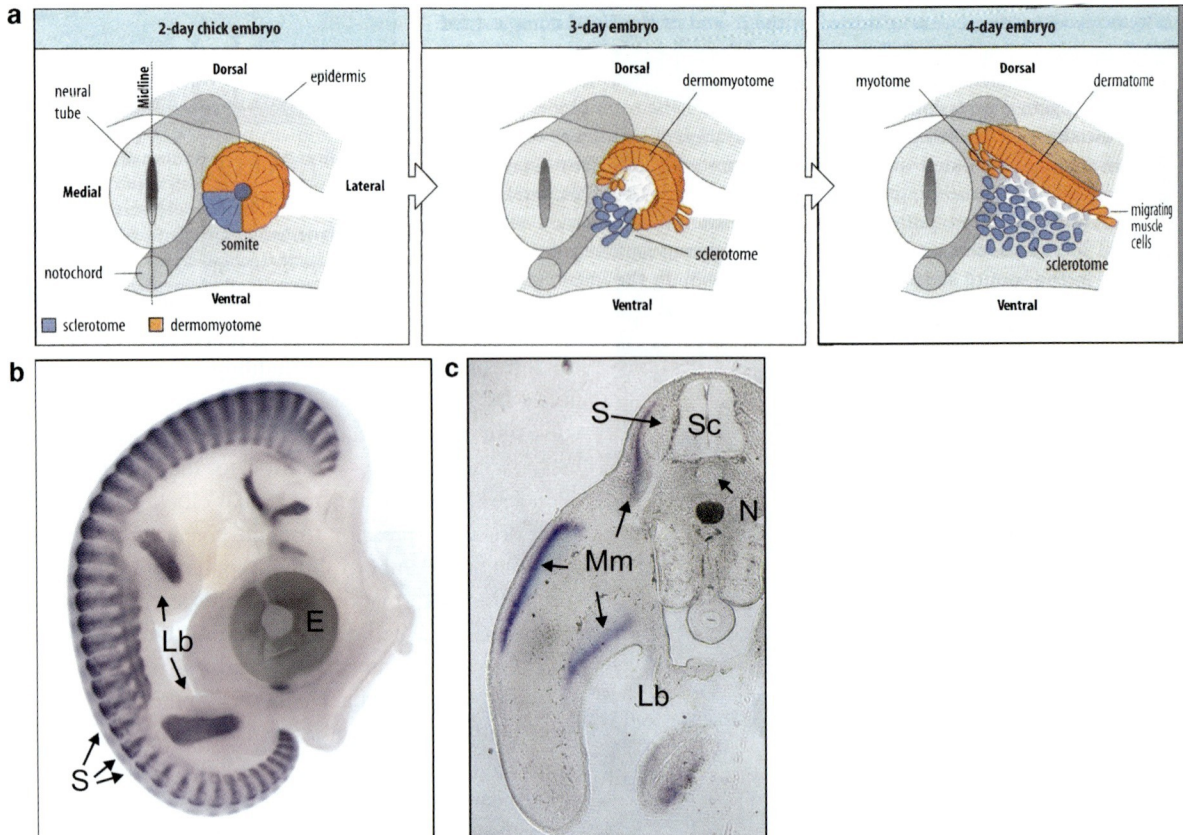

Fig. 13.18 (**a–c**) Development and fate of the somite. (**a**) Somites are transient structures that are formed periodically at each side of the neural tube from presomitic mesoderm. The medial quadrant of each somite gives rise to the sclerotome, whose cells will eventually form the vertebrae and the posterior part of the ribs. The rest of the somite forms the dermatome, which gives rise to the dorsal dermis and the myotome, the origin of all trunk and limb muscles. (**b**) Whole-mount in situ hybridization of a chick embryo showing expression of *MyoD*, a gene expressed during early muscle differentiation. Note periodic expression in somites and in the limb buds. (**c**) Cross section of embryo shown in **B** at anterior limb level. Staining shows migrating muscle cells (*Mm*) lateral to the somite (*S*) and in the limb bud (*Lb*) (*N* Notochord, *Sc* Spinal chord, *E* Eye. (From [35], p 164, Fig. 4.15, by permission of Oxford University Press)

As the somite moves rostrally, it matures and differentiates into the dermatome, the myotome, and the sclerotome (Fig. 13.18). A range of different tissues is generated from these structures. The dermatome originates from the central region of the dorsal layer of the somite (at this stage called the dermatomyotome). It generates the mesenchymal connective tissue of the back skin, the dermis. The myotome originates from the two most lateral portions of the somite. Its cells produce a layer of muscle precursor cells, the myoblasts, which form the epaxial muscles that will give rise to the intercostal and back musculature, and the hypaxial muscles of the body wall, limbs, and tongue. The epaxial part of the myotome is induced by signals from the neural tube (*Wnt1* and *Wnt3a*) and the floorplate (Shh), whereas the hypaxial region is induced by Wnt proteins from the epidermis and bone morphogenetic protein (BMP4) from the lateral plate mesoderm. The latter signal is likely to cause the migration of myoblasts away from the dorsal region into the body wall and the limbs. Further muscle development is established by a cascade of gene activation directed by a set of transcription factors, referred to as the myogenic regulatory factors, including *MyoD* and *Myf5*.

The sclerotome is the most medially located part of the somite and the primary origin of the axial skeleton. *Shh* is the major signal from the notochord that initiates and controls sclerotome formation. Transplantation of parts of the notochord (or *Shh*-producing cells) to other regions of the somite will result in the induction of sclerotome cells at these sites. Sclerotome cells migrate towards the notochord and eventually surround it completely where they condense to form the anlage of the vertebrae. On each side of the neural tube the anterior part of a somite contributes to the caudal part of the vertebral body and the neural arch, whereas the posterior part of the next rostrally located somite is responsible for the rostral part of the vertebral body and the neural arch. It is evident that any disturbance of this process will result in abnormal anlagen of the vertebrae and thus in vertebral malformations. The notochord degenerates by apoptosis but sections of it remain and will eventually form parts of the intervertebral disk, the nucleus pulposus. In addition to the three major compartments of the somite, the sclerotome, the dermatome, and the myotome, two additional regions have been described. The syndetome is another layer of specified cells located between the myotome and the sclerotome. These cells are the precursors of cells that eventually form the tendons, which connect muscle to bone. The fifth compartment is present only in the trunk somites and contains cells that will form the vascular wall of the aorta and the intervertebral blood vessels.

Abnormalities of the ribs and/or vertebrae are a relatively common finding in human malformation syndromes. Those that primarily affect the axial skeleton are summarized under the term spondylocostal dysostoses (SCD) (Fig. 13.19). While the majority of these genetically heterogenous conditions remain unexplained, some types of SCD have been shown to be caused by mutations in members of the notch pathway such as DLL3 and NOTCH2, or in lunatic fringe (LFNG), a secreted protein necessary to maintain oscillation of Notch in the paraxial mesoderm. Affected individuals show a wide variety of vertebral malformations, including fusions and half-vertebrae.

13.5 The Brachial Arches

The brachial or pharyngeal arches are another transient structure in the developing embryo that disappears along with the development of the neck and facial structures. They appear at about 3–4 weeks of human development as a series of bulges found on the lateral surface of the embryo. Four arches are found on each side, separated by clefts on the outside and so-called pharyngeal pouches on the inside (Fig. 13.17). Each

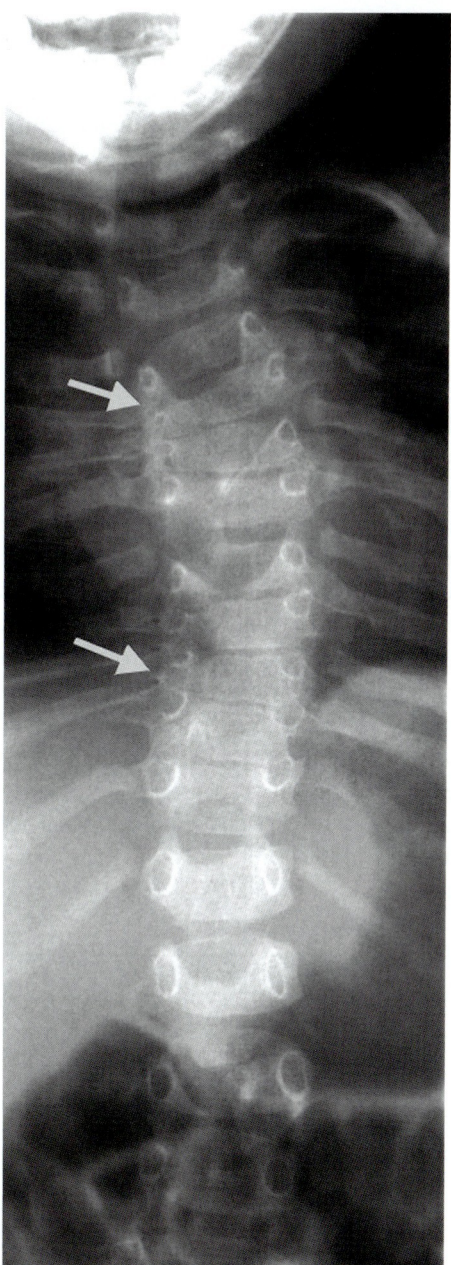

Fig. 13.19 Radiograph of a patient with spondylocostal dysostosis and mutation in DLL3. Organization of the vertebral column is chaotic and many vertebrae are fused or present only as halves

pharyngeal arch is covered externally by ectoderm and internally by endoderm. The core of each arch contains neural crest cells, which surround a central group of mesodermal cells. Each cell type gives rise to a distinct set of tissues. The ectoderm generates the epidermal covering of the arches, whereas the endoderm forms the endothelial lining of the pharynx, the thyroid, the parathyroid, and the thymus. The mesoderm forms both the musculature and the endothelial cells of the arch arteries. The neural crest cells generate the connective tissue of the neck and the skeletal tissues of each arch including Meckel's cartilage and the middle ear (stapes, incus, malleus). Although the arches represent repeated series of similar structures, they have been shown to give rise to distinct parts of the body. The most anterior arch forms the jaw, the malleus, and the incus of the middle ear, the second forms the stapes and the styloid ligament, and the third and fourth arches form the hyoid bone and the thyroid and the cricoid cartilage (Fig. 13.20).

During development, each arch must be patterned to receive its own identity and to be positioned within the embryo along the three axes. A number of mechanisms are known to achieve this, but it is clear that the endoderm plays a prominent role. It is the site where most of the signaling molecules are expressed and the endoderm has been shown to be responsible for the induction of particular arch components including cartilage formation and the precursors of the cranial sensory ganglia. Neural crest cells migrate from the neural tube into the arches. Transplantation studies in both amphibian and avian embryos have demonstrated that crest cells acquire positional information when they are within the neural tube and transfer this information into patterning cues when they have reached the arches. Crest cells migrate in three streams separated by two regions, rhombomeres 3 and 5, which are basically depleted of crest cells. BMP4 have been shown to induce cell death of crest cells in these rhombomeres, while the flanking cells are protected by expression of the BMP inhibitor Noggin. However, other studies point to roles of the mesoderm and the surface ectoderm in patterning the arches, and there is evidence that *Hox* genes play the major part in this process. Inactivation of *Hoxa2*, for example, results in homeotic transformation of elements derived from the second arch into first arch derivatives, including partial duplications of Meckel's cartilage and the ossification centers of the middle ear bones.

The combination of hearing loss (sensorineural and/or conductive), preauricular pits, branchial fistulas, and renal dysplasia characterizes the branchio-oto-renal syndrome. With the exception of the renal problem, these anomalies are considered defects of the brachial arches, and EYA1, the gene mutated in this condition, is expressed in the neural cells of the first arch and in the second, third, and fourth pharyngeal pouches and clefts. Defects of the brachial arches result in

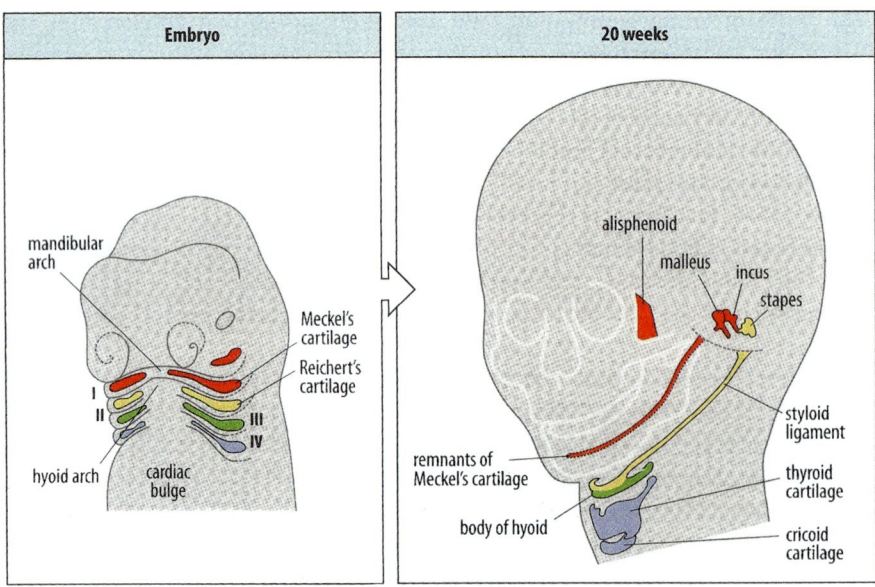

Fig. 13.20 Fate of brachial arch cartilage in humans. Cells from the brachial arches give rise to the three auditory vesicles, the hyoid, and the pharyngeal skeleton. The fate of the various elements is shown by the *color coding* (From [35], p. 502, Fig. 14.6, by permission of Oxford University Press)

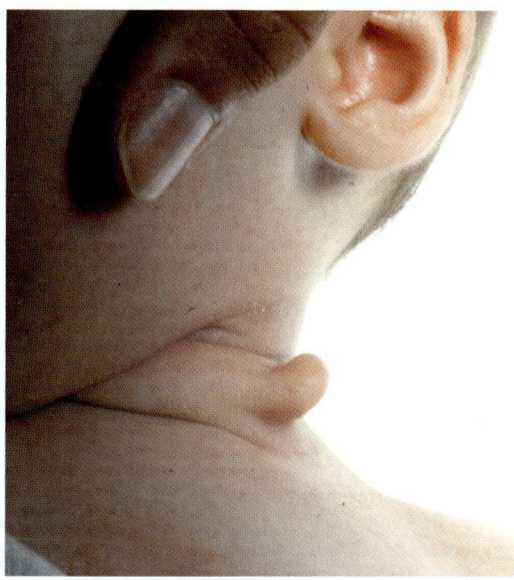

Fig. 13.21 Patient with brachial cyst. (Courtesy of J. Kunze [34], p. 8, Table 4.15)

malformations of the neck, such as brachial fistulas or cysts (Fig. 13.21).

13.6 Development of the Limbs

The limb skeleton originates from the lateral plate mesoderm. All other limb structures, including muscles, nerves, and vasculature, originate from the somitic mesoderm. Outgrowth of the limb bud is the result of a series of interactions between the mesoderm and the overlying ectoderm. As the limb bud grows out, mesenchymal cells begin to differentiate to form the various tissues of the limb in a proximo-distal sequence with structures being laid down progressively from a region of undifferentiated mesenchymal cells at the tip of the limb bud, known as the progress zone. The positional identity, and thus differentiation, of each cell is controlled by a three-dimensional coordinate system consisting of the dorso-ventral, the proximo-distal and the antero-posterior axes. Each axis in controlled by a particular set of signaling molecules/pathways produced by a defined population of cells (signaling centers). The combination of these signals informs the undifferentiated cells in the mesenchyme about their position and their fate in order to form the appropriate structures. Three signaling regions that convey this information have been identified: the apical ectodermal ridge (AER), mediating limb bud outgrowth (proximo-distal axis), ectoderm covering the dorsal sides of the bud governing dorso-ventral pattern, and the zone of polarizing activity (ZPA) controlling antero-posterior pattern (Fig. 13.22). Many of the signaling molecules that are produced by these signaling centers have been identified and characterized and intracellular signaling transduction pathways are being unraveled [32]. Furthermore, mutations in human limb malformations are being identified that help to understand the normal and abnormal mechanisms of limb development [28].

The AER is an anatomical structure consisting of densely packed ectodermal cells located at the very tip of the limb bud. Several different fibroblast growth factors (FGFs) are expressed and secreted by the AER and have been shown to be both, essential and sufficient to initiate and control outgrowth of the limb. FGF signaling is conveyed through the FGF receptors, which are expressed in the underlying mesenchyme. Experimental removal of the AER results in an arrest of limb outgrowth and thus truncations of the limb, depending on the time point of the intervention (Fig. 13.23). Mutations that result in AER loss have a similar outcome, leading to various degrees of limb defects. In humans this mechanism is associated with the ectrodactyly phenotype. This severe limb malformation is characterized by variable degrees of central defects of the digits resulting in cleft hands/feet, or, in the most severe cases, adactyly/monodactyly (Fig. 13.24). Ectrodacyly is genetically heterogeneous, but mutations in TP63L appear to be the most common cause.

The ZPA is a region of mesenchyme located at the posterior limb bud margin. Sonic hedgehog (*Shh*) is expressed in this region and has been shown to be the main mediator of antero-posterior patterning. Implantation of *Shh*-expressing cells can rescue surgical ZPA removal, and expression from the anterior margin of the limb bud results in the formation an anterior ZPA with subsequent mirror duplication of the entire autopod (Fig. 13.25). The restricted expression of *Shh* in the anterior part of the limb bud appears to be regulated by a conserved region approx. 1 Mb away from the *Shh* promotor known as the ZRS. Mutations in ZRS result in polydactyly in humans, mice, and cats by dysregulating *Shh* expression in the limb bud, thus creating an additional anterior *Shh* expression domain

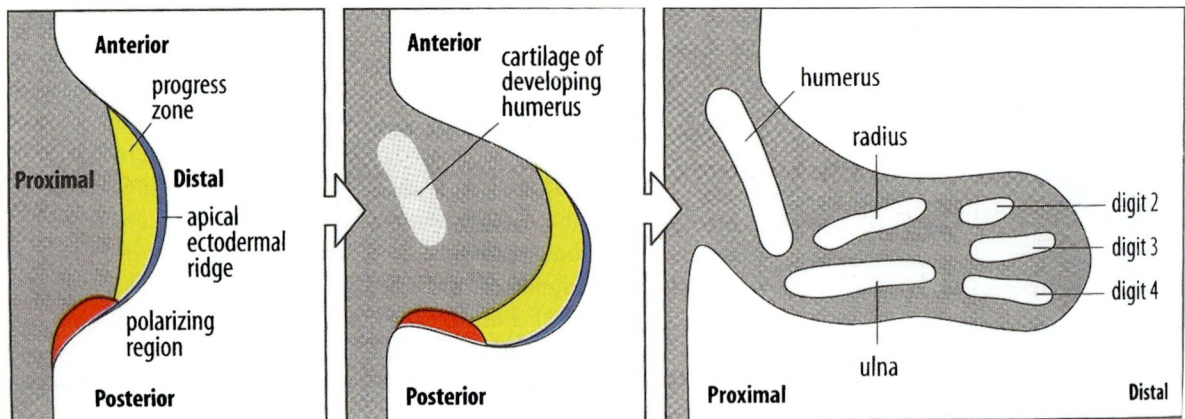

Fig. 13.22 Development of the chick wing with its centers of differentiation and signaling. Cells in the limb bud receive signals from the zone of polarizing activity (ZPA) localized at the posterior margin of the limb, and the apical epidermal ridge (AER), a specialized anatomical structure at the very tip of the bud. The ZPA specifies position along the anterior-posterior axis, whereas the AER controls the proximo-distal axis. Signals from the AER keep cells in the so-called progress zone in an undifferentiated and proliferative state. Once they leave this zone they start to differentiate, giving rise to the mesodermal derivatives of the limb, e.g., the skeleton. Thus, the individual skeletal elements are laid down in a sequential order from proximal to distal. (From [35], p. 142, Fig. 9.6, by permission of Oxford University Press)

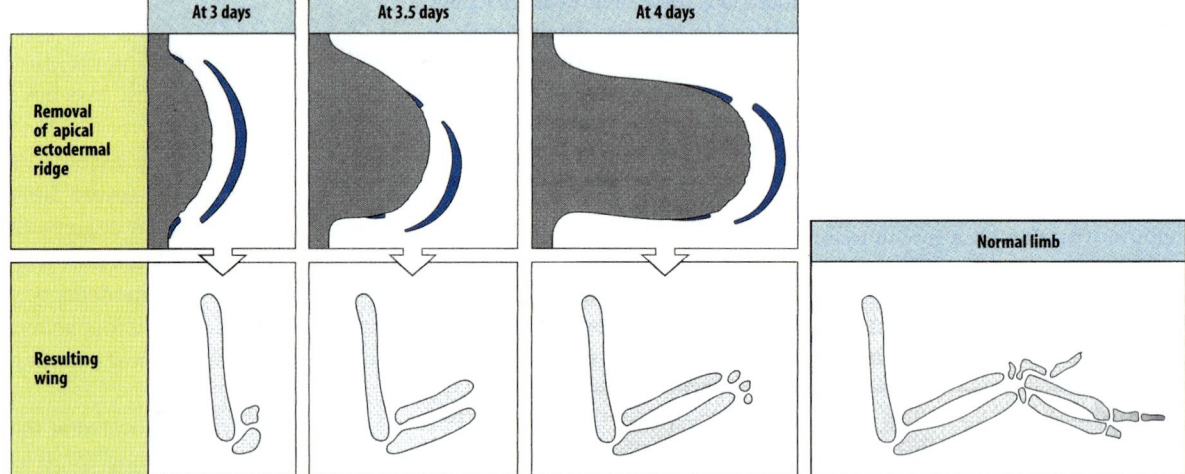

Fig. 13.23 Removal of the AER results in arrest of limb outgrowth and truncation. Surgical removal of the AER results in limb truncations at different levels, depending on the time point of removal. (From [35], p. 341, Fig. 9.7, by permission of Oxford University Press)

(Fig. 13.25b–d). Thus, this regulatory mutation has the same effect as the ZPA transplantation, i.e., activation of ectopic anterior *Shh* expression and subsequent mirror image duplications.

Under the influence of *Shh* the zinc finger transcription factor *Gli3* is converted into the activating form *Gli3A*, whereas otherwise the repressor form *Gli3R* predominates, which in a negative feedback downregulates *Shh*. By this mechanism *Gli3* expression is much stronger in the anterior than in the posterior limb bud, where *Shh* levels are high, thus creating an anteroposterior gradient that has been shown to be important for limb patterning. Mutations in GLI3 result in the Greig and Pallister–Hall syndromes, two conditions characterized by various degrees of polydactyly associated with midline malformations and benign tumors (hamartomas) (Fig. 13.26). In mice it has been shown that *Gli3* deficiency leads to ectopic expression of

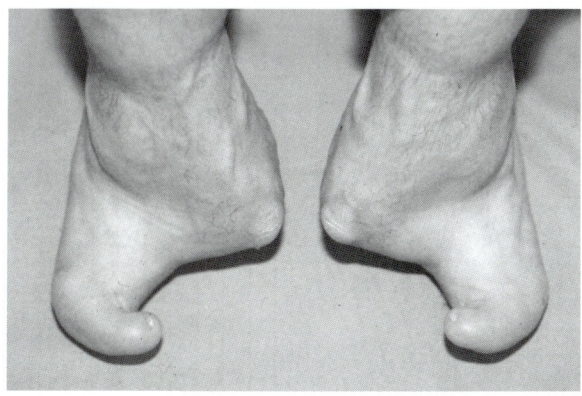

Fig. 13.24 Ectrodactyly associated with mutation in TP63L

Shh in the anterior margin of the limb bud in addition to the normal expression of *Shh* in the ZPA. This results in a double dose of *Shh* and a second ZPA, thus explaining the polydactyly.

A member of the Wnt family of growth factors, Wnt7a, has been shown to be important in dorso-ventral patterning. Expression of Wnt7a in the dorsal ectoderm of the limb bud up-regulates Lmx1b, which belongs to the family of LIM homeodomain transcription factors and forms a dorso-ventral gradient. This close functional relationship explains why Wnt7a-deficient and Lmx1b-deficient mice both develop autopods with a double ventral phenotype. Mutations in

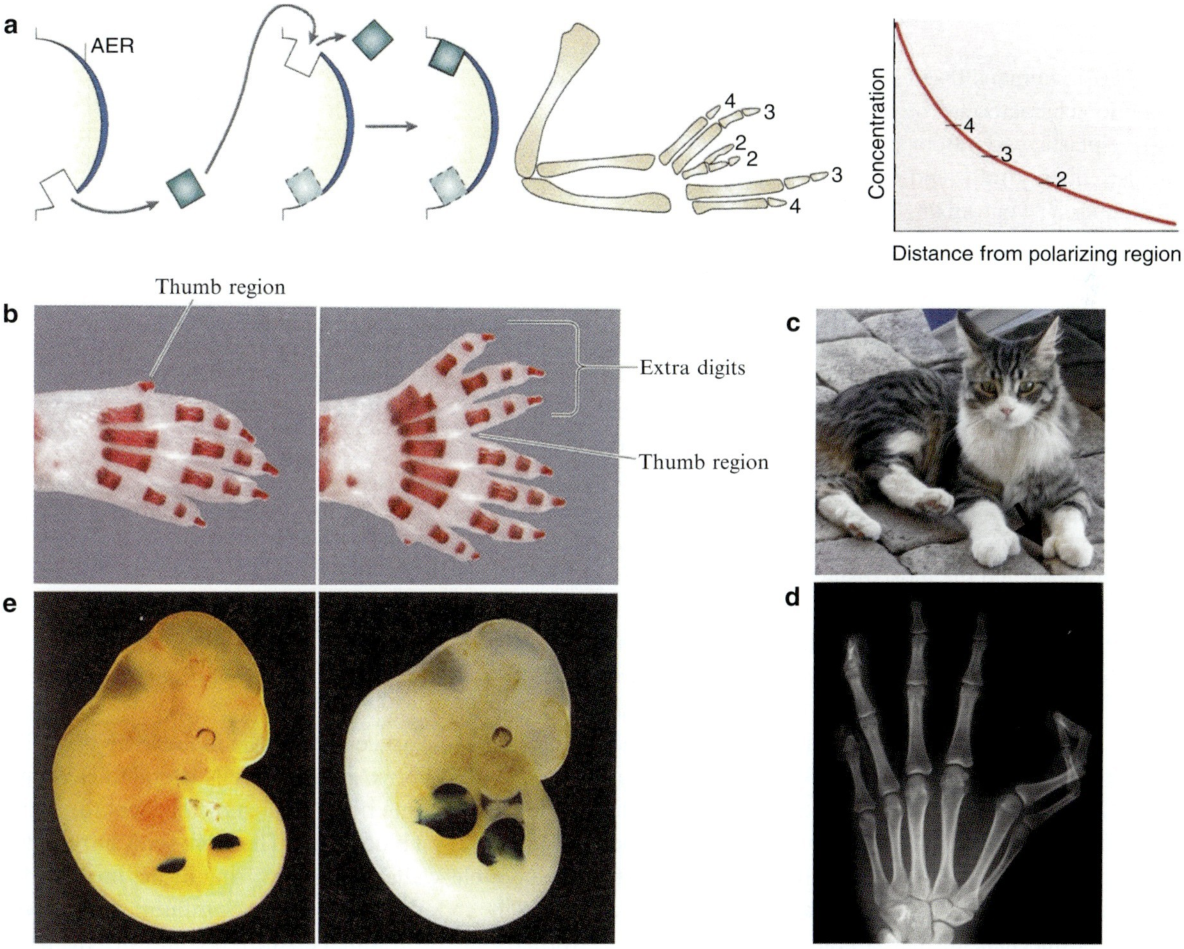

Fig. 13.25 (**a–e**) Digit duplications by AER duplication. (**a**) If an AER is grafted from a donor limb to the anterior region of a recipient a mirror image duplication of the digits is observed. This is due to a double dose of Shh morphogen from both the anterior and the posterior side. Since the distance from the source specifies digit identity, two posterior sides are created. (Reprinted from [32], by permission from Macmillan Publishers Ltd, *Nature Reviews Molecular Cell Biology*, copyright 2005) (**b–d**) Mutations in the regulatory region of the Shh gene in mouse (**b**), cats (**c**), and humans (**d**) result in polydactyly due to ectopic *Shh* expression at the anterior margin of the limb bud. (**e**) Transgenic mouse embryos stage E 12.5 expressing *LacZ* (*blue staining*) under the control of wt (*left*) or mutant (*right*) *Shh* limb regulatory region. Note expression restricted to the posterior limb in embryos expressing the wt construct and expanded expression domain in the mutant. ((**b**, **e**) from [10], p. 517, Fig. 16.17a–d)

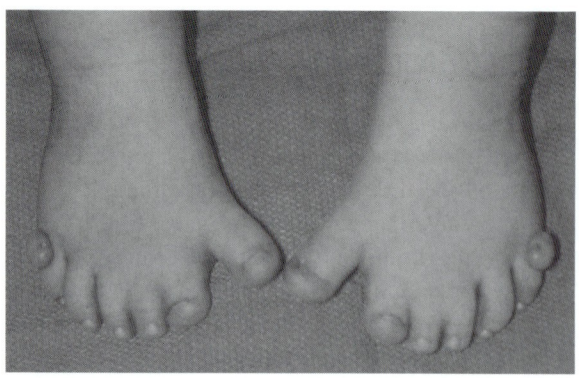

Fig. 13.26 Polydactyly in Pallister–Hall syndrome caused by mutation in GLI3

LMX1B in humans result in nail-patella syndrome, a condition characterized by absent/hypoplastic patellae and dysplastic nails. Both structures represent the dorsal part of the limb, and thus nail-patella syndrome represents a "ventralizing" phenotype. Mutations in WNT7A cause a range of phenotypes, again associated with loss of dorsal structures (nails, patellae), but in most severe cases these mutations are associated with phocomelia, probably because of a complete breakdown of all three signaling centers.

Hox genes from the 5′ region of the A- and D-clusters show characteristic stage-dependent expression patterns that determine the shape and identity of the limb skeletal elements. They are expressed in overlapping patterns, with most 5′ genes having the smallest, the most posterior, and the most distal expression domain. During later stages of development these domains change in a very dynamic way, resulting in an expression of the most 5′ *Hoxd* gene, *Hoxd13* over the entire hand/foot plate. Gene inactivation experiments in the mouse have shown that these domains also correspond to later anatomical regions (Fig. 13.27) in that loss of *Hoxd11* and *Hoxa11* results in absence of radius/ulna and tibia/fibula, whereas loss of *Hoxd11, -12, -13*, and *Hoxa13* results in a severe reduction of hands/feet. This pattern of expression is probably achieved by a repression mechanism that is gradually released during development located in a chromosomal region upstream of *Hoxd13*. Mutations in *Hox* genes result in limb malformations as illustrated by synpolydactyly, a condition characterized by fusion of fingers (syndactyly) together with an additional finger in the syndactylous web (polydactyly) (see also Fig. 13.7).

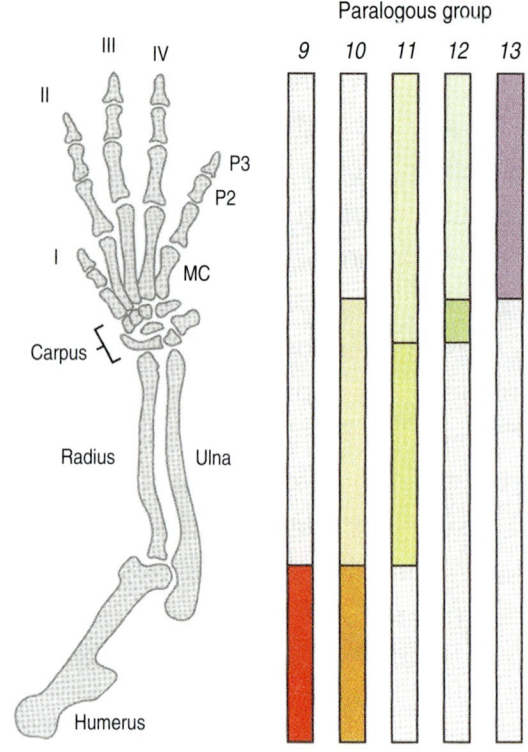

Fig. 13.27 (**a, b**) *Hox* gene patterning of the forelimb. (**a**) *Hox* genes of one paralogous group are expressed in distinct domains of the developing limb that correspond to adult skeletal elements. Whereas *Hoxd13* is exclusively influences patterning of the hand (autopod), *Hoxd11* together with *Hoxd10* are important for pattering of radius and ulna (zeugopod). *Hoxd9* and *Hoxd10* function together to pattern the humerus (stylopod). (From [35], p. 353, Fig. 9.18). (**b**) *Hoxd13* expression in the limb buds and the genital ridge of a mouse embryos stage E12.5. By permission of Oxford University Press

13.7 Development of the Circulatory System

Presumptive heart cells originate in the early primitive streak and form two groups of cells lateral to Hensen's node. At the same time, angiogenic clusters form that contain the first blood cells. These are soon surrounded by a double-walled tube consisting of the inner endocardium and the outer epimyocardium, which will form the endocardium, i.e., the inner lining of the heart and the heart muscles, respectively. As the foregut is closed, the two tubes are brought together and fuse to a single pumping chamber. In humans this occurs at 3 weeks of gestation. By 5 weeks the heart has developed into a two-chambered tube with one atrium and one chamber. The partitioning into the two chambers is accomplished by cells that migrate into a hyaluronate-rich structure, the endocardial cushion, that is later located between the ventricles and the atria. Meanwhile the atrium is partitioned by the development of two septa that grow ventrally towards the endocardial cushion. These septa stay open, thus providing blood flow from the right to the left atrium. In the 7th week of human development the ventricles begin to be separated by the growth of the ventricular septum towards the endocardial cushion [31]. Figure 13.28 shows a schematic of human heart development.

For a proper orientation of the pulmonary (right) and systemic (left) ventricles and for the alignment of the heart chambers with the vasculature the linear heart tube undergoes rightward looping [25]. The direction of cardiac looping is determined by an asymmetric axial signaling system that also affects the position of the lungs, liver, spleen and gut. Before organ formation begins, this signaling cascade directs the asymmetrical expression of Sonic hedgehog (*Shh*) and Nodal, a member of the TGFβ family. In humans mirror image reversal of right–left asymmetry (situs inversus) is often associated with normal organogenesis, but discordance of cardiac and visceral asymmetry (situs ambiguus, also called heterotaxy) is associated with malformations of the heart and other organs. In the latter condition asymmetry in structure and placement of organs still develops, but, owing to the lack of definitive positional information, this happens on a stochastic basis. Cardiac defects typically occurring with situs ambiguus include, but are not limited to, atrial septal defects, ventricular septal defects, transposition of the great arteries, double-outlet right ventricle, anomalous venous return, and aortic arch anomalies. But what determines right–left asymmetry in the first place? Evidence that ciliary function might be involved came from studies of Kartagener syndrome, a condition characterized by situs inversus together with recurrent pulmonary infections attributable to defects in cilia function. Elegant experiments performed with video microscopy have shown that motile cilia are present in the center of the mouse node that propagate directional fluid flow. In addition, it was shown that this flow is abnormal in several mouse mutants with laterality

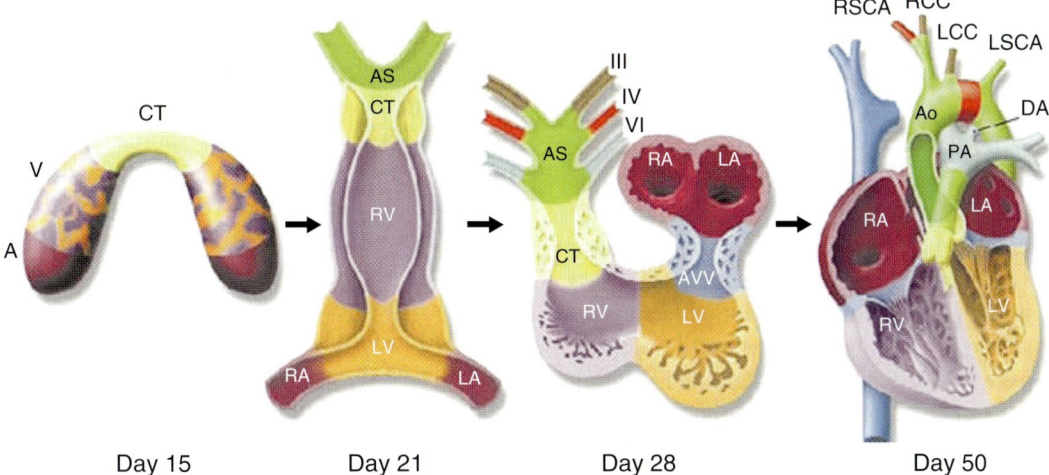

Fig. 13.28 Human heart development. Cardiac development is depicted in four consecutive steps. The heart consists of an inner endocardium, which is the endothelial sheet, and the outer myocardium, which is contractile. The first sign of the developing heart is the formation of a crescent (*left-most panel*), which fuses along the midline to give rise to a tube; this is patterned along the anterior-posterior axis to form the various chambers of the heart. This two-chambered heart is the basic adult form in fish, but in higher vertebrates the tube undergoes looping, a process controlled by the left-right asymmetry of the antero-posterior axis, and further partitioning to form a four-chambered heart. Cells from neural crest origin contribute to the outflow tracts populate the bilaterally symmetrical aortic arch arteries (III, IV, and VI) and aortic sac (AS) that together contribute to specific segments of the mature aortic arch. Mesenchymal cells form the cardiac valves from the conotruncal (CT) and atrioventricular valve (AVV) segments. Corresponding days of human embryonic development are indicated (*A* Atrium, *Ao* Aorta, *DA* Ductus arteriosus, *LA* Left atrium, *LCC* Left common carotid, *LSCA* Left subclavian artery, *LV* Left ventricle, *PA* Pulmonary artery, *RA* Right atrium, *RCC* Right common carotid, *RSCA* Right subclavian artery, *RV* Right ventricle, *V* Ventricle) (Reprinted from [31], by permission from Macmillan Publishers Ltd, *Nature*, copyright 2000)

defects. In the current model, this cilia-directed flow creates an asymmetric accumulation of growth factors such as Nodal and/or Shh that subsequently governs asymmetry. In Kartagener syndrome this is presumably disturbed because the mutations in dynein result in nonfunctional cilia.

The individual segments of the heart can only function if the cardiac valves are properly placed to ensure the unidirectional flow of blood through the heart. Development of the valves starts with the formation of regional swellings, known as cardiac cushions, that form the anlage of the valves. To form the valves a transformation of endocardial cells into mesenchymal cells has to take place, mediated by members of the TGFβ family. Inactivation of Smad6, a transcription factor activated upon TGFβ signaling leads to abnormally thick-ended, gelatinous valves. Transformed cells migrate into the cushions and differentiate into the fibrous tissue of the valves. Atrioventricular canal or endocardial cushion defect represents a developmental abnormality of this process, a congenital heart defect frequently observed in trisomy 21. However, which gene(s) cause this defect is unknown.

Many other genes have been implicated with congenital heart defects. Mutations in TBX5 cause Holt–Oram syndrome, a condition characterized by heart defects, frequently associated with arrhythmia, and limb malformations. TBX5 was shown to be an essential factor for the development of the right ventricle and the outflow tract [3]. In addition, Tbx5 appears to function in the left ventricle and atria by influencing the expression of other transcription factors and proteins that are required for cardiac function. Tbx5 associates with Nkx2.5, another transcription factor shown to be essential for normal heart development. Mutations in NKX2.5 have been identified in families with atrial septal defects and cardiac conduction abnormalities [2]. Sporadic mutations have also been found in individuals with outflow tract alignments defects such as tetralogy of Fallot, or tricuspid valve anomalies. An evolving general theme of congenital heart defects appears to be a considerable variability in the type and severity of cardiac malformations among individuals with mutations in the same gene or even with the same mutation. Mutations in one gene are not predictive for a certain heart defect but rather appear to be responsible for a range of abnormalities, with certain types of defects occurring more often than others.

13.8 Development of the Kidney

The urogenital system, i.e., the kidneys, the gonads, and their respective duct systems, develop from the intermediate mesoderm, the region located between the lateral plate mesoderm (origin of the limbs) and the paraxial mesoderm (origin of the somites). Development of the permanent kidney, the metanephros, is preceded by the formation of a transient structure, the pronephros. The pronephric duct is the first visible sign, a bilateral structure consisting of a single-cell-thick epithelium that extends caudally until it reaches the cloaca. Connected to this tube a linear array of tubules derived from mesenchymal cells adjacent to the primary nephric duct forms. The more caudal tubules contain glomeruli and convoluted proximal tubule-like structures that serve as transient filtration units that degenerate as the development of the permanent kidney takes place [5]. Figure 13.29 shows a summary of kidney development.

The adult kidney or metanephros begins to develop when an outgrowth of the primary nephric duct, termed the ureteric bud, extends into the surrounding metanephric mesenchyme. The ureteric bud branches and first mesenchymal cells begin to aggregate near the tips of the bud. This interaction of mesenchyme with epithelium drives the entire following developmental process. The aggregates first form a polarized renal vesicle, which is still in contact with the bud. Subsequently the so-called comma and S-shaped bodies form and become polarized along a proximo-distal axis as they undergo mesenchyme-to-epithelial conversion. Subsequently the distal end fuses with the ureteric bud to form a single, continuous epithelial tube, which in its distal part will give rise to the proximal tubule, Henle's loop, and the distal tubule. Endothelial cells invade the cleft of the S-shaped body forming the glomerular filtration unit. Interactions between endothelial cells and the mesenchyma-derived glomerular epithelial cells lead to the formation of podocytes and thus the glomerula basement membrane that serves as a filtration barrier. This process of branching, aggregation, mesenchyme-to-epithelial conversion, fusion, and endothelial invasion continues from the inside to the outside in such a way that the oldest nephrons are located closer to the medulla and the youngest ones more peripherally in the nephrogenic zone. In humans this process continues during the

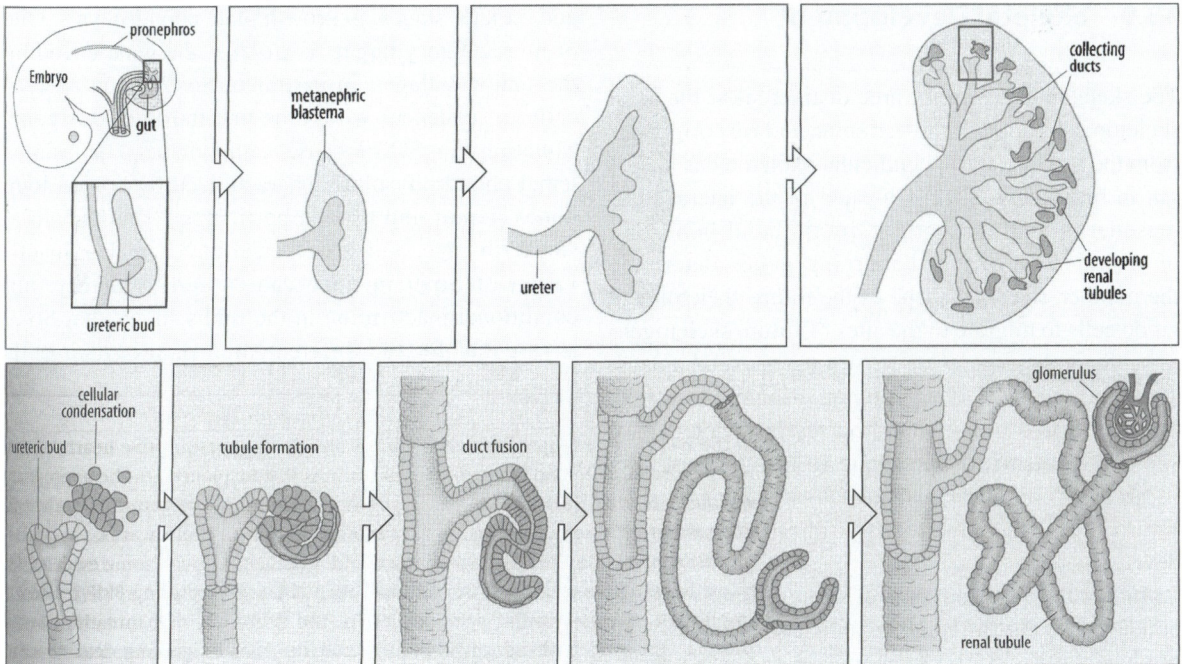

Fig. 13.29 Kidney development. The permanent kidney develops from a loose mass of mesenchyme, the metanephric blastema, which is induced by cells from the ureteric bud. The ureteric bud itself is induced by the mesenchyme to grow and branch. The metanephric blastema eventually develops into the glomerulus and the renal tubule, whereas the ureteric bud will form the collecting ducts that connect the kidney to the ureter. The formation of glomeruli involves several distinct steps, including comma- and S-shaped bodies. The human kidney continues to grow from metanephric blastema until birth. (From [35], p. 377, Fig. 9.46)

entire fetal period until the final size of the kidney is reached.

This complex process in governed by a large number of signaling pathways, including the *Pax/Eya/Six* genes as well as *Lim1* and *Odd1*. Ureteric outgrowth and branching morphogenesis are controlled by the Ret/Gdnf pathway. In the mouse *Wnt9b* and *Wnt4* genes are critical for aggregation and transformation of metanephrogenic mesenchyme into tubular epithelium. If separated from the utereteric bud mesenchymal cells die. FGF2 and BMP7 are two factors secreted by the ureteric bud that prevent apoptosis and promote aggregation. The competence to respond to uteric bud inducers is thought to be regulated, among other genes, by Wt1, a transcription factor originally found to be mutated in a heritable form of childhood kidney tumor, the Wilms tumor. If this factor is missing, the uninduced cells of the metanephrogenic mesenchyme die and no kidney is formed. Later in development Wt1 is found expressed in podocytes of the glomerular basement membrane. Besides leading to tumors, specific mutations in WT1 can result in Denys–Drash syndrome, a condition characterized by Wilms tumors, nephrotic syndrome (severe proteinuria) leading to end-stage renal failure before the age of 3 years, and pseudohermaphroditism with children having either ambiguous external genitalia or a normal female phenotype with an XY karyotype. The latter phenotype points to an important role of Wt1 in genital development. It is similar to the renal phenotype in that Wt1-deficient mice develop no gonads.

The caudal part of the ureteric bud becomes the ureter, which inserts into the bladder and is a part of the urogenital system that, if nonfunctional, will result in hydronephrosis. The bladder develops out of the cloaca, which is divided by a septum into the rectum und the urogenital sinus. The latter will also give rise to the urethra. *Hox* genes appear to be important for this process. Mutations in HOXA13 result in hand-foot-genital syndrome, a condition characterized by short thumbs/toes and abnormalities of the cloaca, the male and female reproductive tracts and the urethra.

13.9 Skeletal Development

The skeleton arises from three distinct sites: the axial skeleton consisting of the vertebrae and ribs originates from the somites, the appendicular skeleton has its origin in mesenchymal cells located in the lateral plate mesoderm, and most of the craniofacial bones are of neural crest origin. Patterning genes determine the number, size, and shape of the future skeleton and guide cells to migrate to the sites of future skeletogenesis, where they aggregate to form the skeletal anlage. The next step is the overt differentiation of these cells into cartilage-forming chondrocytes in endochondral skeletal elements, or into bone forming osteoblasts in areas of membranous bone formation [16, 17]. Figure 13.30 schematically summarizes skeletal development.

In areas of endochondral bone formation, the condensed cells differentiate into chondrocytes that produce cartilage. The transcription factor Sox9 has been shown to be essential for this process. Sox9 is expressed in the early cartilaginous condensations and, at later stages, in growth plate chondrocytes. One of its regulatory targets is *Col2a1*, the gene encoding the major collagen in cartilage. Sox9 binds specifically to sequences within the first intron of this gene. Mutations in SOX9 cause campomelic dysplasia, a lethal chondrodysplasia characterized by bowed long bones (femur and tibia in particular), small scapula, small rib cage, and sex reversal in XY males. Inactivation of Sox9 in mice causes early lethality, but conditional inactivation in the limbs shows that Sox9 is essential for the differentiation of precursor cells into chondrocytes.

The TGFβ/BMP/GDF5 pathway plays an important role in the regulation of condensation and differentiation of precursor cells into chondrocytes. Signaling of the Tgfβ superfamily members requires the binding of the ligand to cell surface receptors consisting of two types of transmembrane serine/threonine kinase receptors classified as type I and type II. The type II receptor transphosphorylates and thus activates the type I receptor. The intracellular substrates of the activated type I receptors are the Smads. Smads 1, 5, and 8 are phosphorylated and then translocated to the nucleus, where

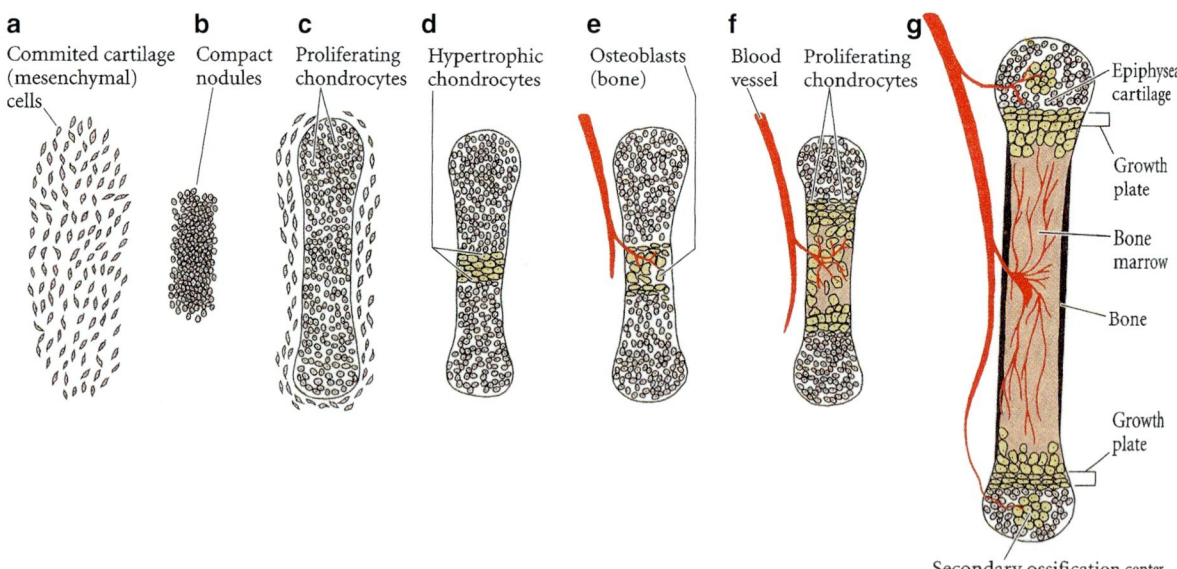

Fig. 13.30 Skeletal development. The first sign of skeletogenesis is the condensation of precursor cells at the site of future bone formation. The condensed cells differentiate into cartilage to form the cartilaginous anlage of the future bone or directly into bone at sites of desmal ossification. Chondrocytes in the center of the cartilaginous anlage begin to hypertrophy, and their matrix begins to mineralize. At this stage the first cortical bone begins to form as a thin layer of osteoid around the shaft of the anlage. In addition, vessels invade the cartilage, introducing monocytic progenitor cells that differentiate into osteoclasts that remove cartilage and bone. A growth plate forms at each end of the bone, in which most of the growth is generated until adulthood. Secondary ossification centers develop in the cartilage heads of the bone. (From [10], p. 456, Fig. 14.14)

they participate in the transcriptional regulation of the expression of genes involved in cartilage and bone formation [22]. BMP signaling is controlled by the binding of BMPs to inhibitors in the extracellular space. One potent inhibitor is Noggin, a gene originally identified for its role to induce head formation in *Xenopus*. Activation of BMP signaling either by overexpression, activation of the receptor, or inhibition of the inhibitor results in grossly enlarged cartilaginous anlagen. The likely mechanism for this effect is the recruitment of mesenchymal precursor cells to the cartilage condensations and to the perichondria, which contribute cells to the anlage by appositional growth. Mutations in GDF5, another member of the BMP/TGFβ family or its receptor BMPR1B result in various limb phenotypes, mainly characterized by shortening of the digits (brachydactyly). Mutations in the inhibitor NOGGIN or activating mutations in GDF5 cause joint fusions (symphalangism, synostosis syndrome), indicating that the tight regulation of BMP signaling is also essential for joint formation [29].

In areas of membranous bone formation, the condensed cells differentiate into osteoblasts which produce bone matrix. Genetic experiments in mice have demonstrated that Cbfa1/Runx2 is essential for this process. Runx2 is a member of a small family of transcription factors that are homologous to the *Drosophila* runt gene. In Runx2 null mice no endochondral or membranous bone is formed owing to an arrest in the early steps of osteoblast differentiation (Fig. 13.31). In contrast, the cartilaginous template is relatively normal. Mutations in RUNX2 cause cleidocranial dysplasia, a skeletal dysplasia characterized by patent fontanelles, aplastic/hypoplastic clavicles, supernumerary teeth, and short stature [24].

In contrast to the bones of the skull, which are formed by a direct transformation of mesenchymal cells into osteoblasts, the major part of the skeleton is formed by endochondral ossification in which a cartilaginous template is formed first, which is subsequently replaced by bone. Central to this process is the formation of a growth plate, a highly organized structure that generates the entire longitudinal growth. Growth plate chondrocytes are invariably arranged in three layers: (1) reserve chondrocytes, (2) proliferating chondrocytes, and (3) hypertrophic chondrocytes. A very complex interplay of different signaling pathways regulates the rate of proliferation and the conversion of proliferating chondrocytes into hypertrophic chondrocytes.

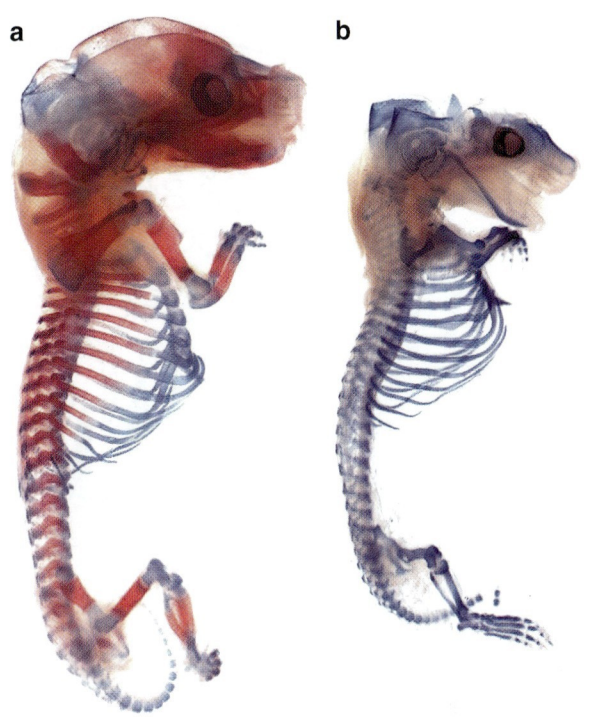

Fig. 13.31 (**a, b**) Role of Runx2 in bone development. Inactivation of Runx2 transcription factor results in complete loss of bone. Mouse skeleton with cartilage stained blue and bone stained red. Note *red* and *blue* staining in control (**a**) but no red staining in Runx2–/– mice (**b**)

During the process of chondrocyte differentiation, the matrix changes dramatically through the production of other components such as collagen type 10, the expression of metalloproteinases, and the calcification of matrix. At the same time, blood vessels begin to penetrate the calcified cartilage, bringing in osteoclasts that remove cartilage and osteoblasts, which build new bone. With further growth, the central and primary centers of ossification expand towards the ends of the bones and secondary centers of ossification form within the cartilage remnants. The growth plate, now localized between the epiphysis (secondary center of ossification) and the metaphysis (distal end of former primary ossification center) generates all longitudinal growth until the end of puberty when primary and secondary ossification centers fuse. At this point, the cartilage of the joints is the only cartilage that remains of the former anlage (Fig. 13.32).

Two major signaling pathways that control proliferation and differentiation of chondrocytes have been identified, the Indian hedgehog (*Ihh*)/parathyroid

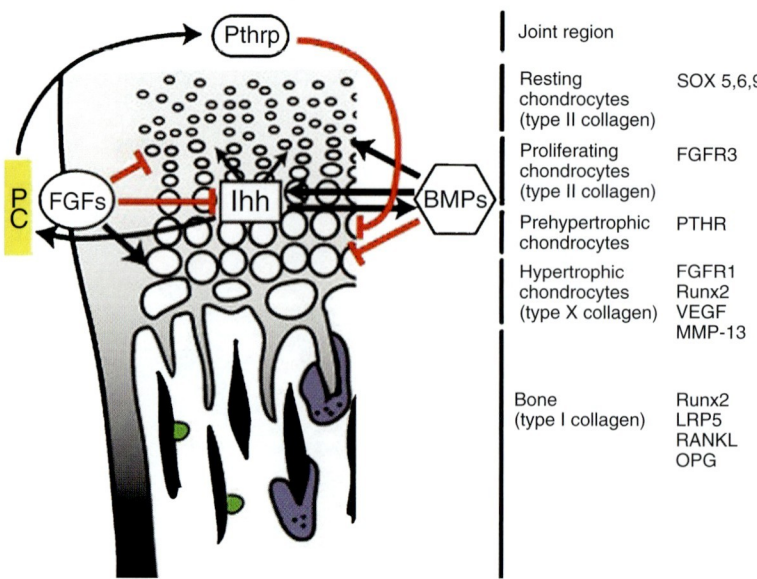

Fig. 13.32 Schematic of growth plate and its regulation. Beginning from the joint region, chondrocytes are arranged in four layers, representing their stages of differentiation: resting, proliferating, prehypertrophic, and hypertrophic chondrocytes. The later are replaced by bone by the joint action of osteoblasts (*green*) and multinuclear osteoclasts (*blue*). Genes that are characteristically expressed within each layer are given on the *right* side. A complex signaling network regulates proliferation and differentiation of chondrocytes. Prehypertrophic chondrocytes express *Ihh*, which regulates PTHrp, which is produced by the joint region, via the perichondrium. PTHrp, in turn, inhibits differentiation of proliferating chondrocytes. FGFs inhibit *Ihh* and chondrocyte proliferation and stimulate differentiation of chondrocytes. BMPs act antagonistically to the FGFs. *Gray shading* symbolizes the degree of calcification of the extracellular matrix

hormone related peptide (PTHrP) pathway and the FGF pathway. Indian hedgehog (*Ihh*), the second mammalian hedgehog ortholog, plays a central role in the regulation of chondrocyte proliferation and hypertrophy. Through a yet unknown cascade of events *Ihh* indirectly regulates the expression of PTHrP. *Ihh*, and PTHrP form a feedback loop whereby *Ihh* up-regulates the synthesis of PTHrP, thereby indirectly slowing down the process of chondrocyte hypertrophy. Mutations in the PTHrP pathway cause pseudohypoparathyroidism (PHP) and pseudopseudoparathyroidism (PPHP). In PHP abnormalities in calcium and phosphate metabolism (hypocalcemia, elevated PTH levels) are accompanied by obesity, mental retardation, short stature, subcutaneous calcifications, and short digits (Albright hereditary osteodystrophy), whereas PPHP patients have osteodystrophy with normal calcium and PTH levels.

FGF signaling plays a major role in the regulation of chondrocyte proliferation and differentiation. More than 20 different FGFs are currently known, and four distinct FGF receptors (FGFR) have been described that bind and are activated by members of the FGF family. FGFs are potent mitogens for chondrocytes as well as for osteoblasts and stimulate bone formation in vitro and in vivo. Inactivation of Fgfr3 in the mouse results in overgrowth of the long bones, whereas expression of an activating mutation results in dwarfism, indicating that Fgfr3 functions as a negative regulator of chondrocyte proliferation and/or hypertrophy. Mutations in FGFR3 cause the most common form of skeletal dysplasia, achondroplasia. Affected individuals are characterized by disproportionate short stature (130 cm adult height). Further activating mutations in FGFR3 cause hypochondroplasia, a less severe variant, and thanatophoric dysplasia, the most severe and lethal form of this dysplasia group.

Cartilage is characterized by a unique extracellular matrix that accounts for around 90% of the tissue volume. The main component is fibrous collagen, which confers tensile strength. The growth plate cartilage contains type II, IX, X, and XI collagen. In the resting and proliferating cartilage type II collagen predominates, which can form fibers together with type IX and type XI. Type X collagen, in contrast, is specific for hypertrophic cartilage. Proteoglycans, and especially aggrecan, are highly abundant. They are giant molecules, which have a gel-like consistency when dissolved in water. Consisting of a core protein to which different kinds of glycosaminoglycans (GAGs) are attached, the proteoglycans are highly sulfated and thus negatively charged. This allows them to bind large amounts of cations and water molecules and to be mutually repellent, which is thought to contribute to the elasticity of the cartilage. Glycoproteins are a third group of matrix components comprising for example perlecan, fibronectin, tenascin, and cartilage oligomeric matrix protein (COMP). Mutations in any of these cartilaginous matrix components result in skeletal dysplasias with growth deficiencies.

In bone, type 1 collagen is the most abundant protein. In contrast to cartilage, bone matrix contains few proteoglycans, but instead consists largely (two thirds) of hydroxyapatite, to ensure that it is rigid. Mutations in either of the two genes that encode for the two chains that contribute to collagen type I (COL1A1 and COL1A2) result in osteogenesis imperfecta, a group of diseases primarily characterized by brittleness of bones and recurrent fractures (Fig. 13.33).

Homeostasis of bone mass and remodeling during skeletal growth are accomplished by an antagonistic action of bone producing osteoblasts and bone resorbing osteoclasts. Osteoclasts originate from mononuclear hematopoietic precursor. When osteoclasts attach to bone they create the so-called ruffled membrane, which contains high levels of V-type H^+-ATPases that pump protons into the tightly sealed resorption lacuna between osteoclast and bone surface. Strong extracellular acidification is crucial for bone resorption, as low pH is needed to dissolve the hydroxyapatite of the bone tissue and to degrade the organic components of the bone matrix, above all type I collagen, by acidic proteases such as cathepsin k. Dysfunction of osteoclasts results in an abnormal accumulation of bone also called osteopetrosis. The recessive forms are life-threatening conditions caused by loss of the bone marrow cavity. They can be caused by defects in the osteoclast acidification machinery, either affecting the osteoclast proton pump TIGR, the chloride transporter CLC-7 [18], or carboanhydrase type II, a protein responsible for synthesizing the protons transported by the H^+-ATPase. Mutations in cathepsin k, which is indispensable for bone resorption since it is the only protease able to cleave the intact collagen triple-helix, result in the sclerosing disorder pycnodysostosis.

13.10 Abnormal Development: Definitions and Mechanisms

The complexities of embryonal development are reflected by the vast number of phenotypes produced by abnormal development. Anomalies that are of prenatal

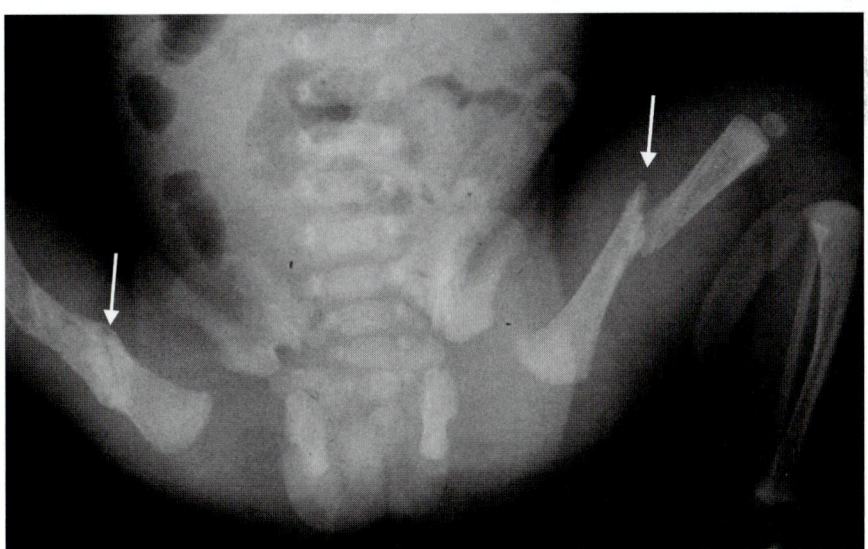

Fig. 13.33 Newborn with type II osteogenesis imperfecta. Note fresh fracture of left femur and old fracture of right femur with callus formation (*arrows*)

onset are generally subsumed under the heading congenital. Such defects are either visible at birth or can be diagnosed at a later time point when they manifest as disease (e.g., deafness, anomalies of teeth). 'Congenital' does not necessarily mean genetic or inherited, because congenital defects can also be caused by maternal disease (e.g., diabetes), intraterine infections (e.g., rubella), or mechanical forces (e.g., uterine constraint caused by oligohydramnion).

Congenital anomalies contribute to a great extent to neonatal morbidity and mortality. It has been estimated that, depending on the methodology of assessment, 4–8% of all newborns are born with a medically relevant anomaly. Thus, approx. 1 in 20 newborns has a recognizable and medically relevant anomaly. This proportion is higher among children who die during the neonatal phase (20%) and is much higher in miscarried babies. In Western countries congenital anomalies are the most frequent cause of neonatal mortality.

Abnormalities of development will have a different outcome depending on the cause, time point, and magnitude of the insult. For example, defects that affect the embryo during the preimplantation phase usually result in either complete restoration or loss of the embryo, whereas deficiencies that occur later can result in either death of the embryo or in birth defects. For the clinician a careful clinical analysis may reveal important information that is of high relevance not only for understanding the condition, but also for successful counseling. As illustrated in Fig. 13.34, all developmental anomalies can be categorized into four subgroups.

13.11 Malformations

Primary malformations are caused by an intrinsic defect of the embryo, i.e., they are genetic or due to an interaction of the embryo's genome with its environment. They can be inherited or may develop as a result of a de novo mutation that has occurred in the oocyte or the sperm, or during early embryogenesis. Malformations are due to inactivation or dysregulation of important developmental genes during the postimplantation phase. These genes regulate early patterning processes and/or organogenesis. Because regulation of early development is their most important role, the developmental defect happens very early and the

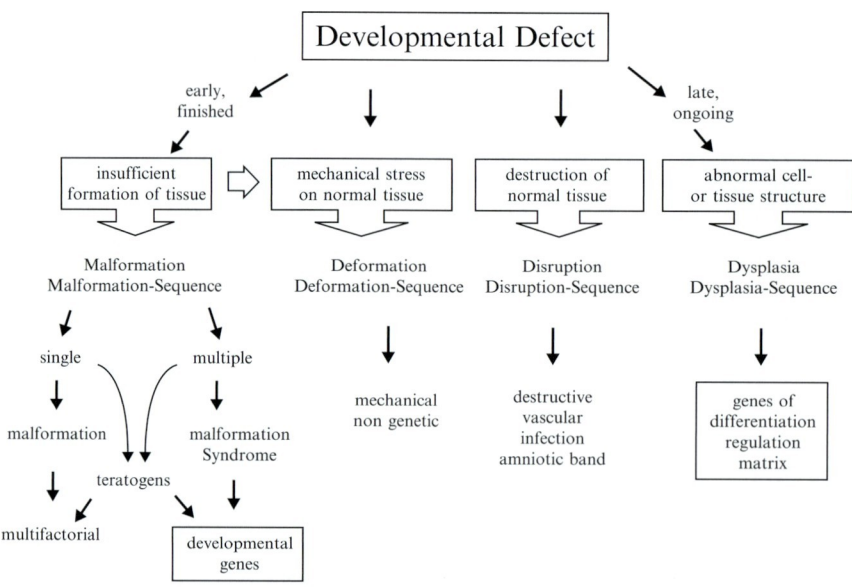

Fig. 13.34 Types of structural defects. Congenital defects can be subdivided in four major groups, malformations, deformations, disruptions, and dysplasias. As shown, their etiology can be genetic, multifactorial, and non-genetic. Each type may induce secondary changes, also called sequence. (From [20], p. 282, Fig. 31.2)

pathogenesis of the condition is thus completed at the time point when it is observed.

The majority of malformations are likely to be caused by a combination of environmental and polygenic factors. Abnormalities in the maternal metabolism, such as diabetes mellitus, are such an environmental factor which, together with yet unknown susceptibility genes, may result in severe malformations, such as caudal regression. A smaller number of birth defects is due to chromosomal aberrations or is caused by single gene defects.

Defects caused by polygenic factors are most commonly "single site defects," i.e., they affect one site or organ of the body while leaving others intact. They occur at relatively high frequency and are either present or absent, i.e., they show little overlap with minor defects. Such malformations include cleft lip and palate, isolated cleft palate, neural tube defects, club foot, congenital heart defects, and pyloric stenosis to mention the most frequent ones. They tend to cluster in families but do not conform to Mendel's laws of gene transmission. Risk calculation has determined that: (1) first-degree relatives have a risk approximating the square root of the population risk; (2) second-degree relatives have a sharply lower risk than first-degree relatives; (3) the higher the number of affected individuals in one family, the higher the risk; (4) consanguinity increases the risk; and (5) the more severe the malformation, the higher the risk. For some defects the influence of environmental factors is larger than for others. For example, neural tube defects can be prevented to a large extent if folic acid is supplemented during early pregnancy. In fact, the folate-neural-tube defect relation represents the only instance in which a congenital malformation can be prevented by changing the environment. In contrast, the frequency of concordance and discordance in monozygotic and dizygotic twins argues against both a single-gene etiology and a major environmental influence in most other conditions.

Cleft lip with or without cleft palate is one of the most common birth defects in the world, with an average prevalence of 1/700. Transforming growth factor (TGFA) a is a growth factor expressed during palatogenesis in the mouse, but mice with inactivation of TGFA have abnormal skin, hair, and eyes, but no clefts. TGFA was found to be associated with human clefts in a first study in 1989 and has been confirmed in several studies since. Taken together, these studies show that TGFA is probably a modifier of clefting in humans, which is consistent with an oligogenic model for clefting in humans. Several causative genes for inherited syndromic forms of cleft lip/cleft palate have been identified and some studies show that they may also contribute to the occurrence of isolated forms.

In contrast to the above-mentioned polygenic conditions, chromosomal imbalances and single-gene disorders are usually associated with a combination of anomalies, hence their name "syndrome" (Greek, meaning "convergence"). For the clinician this means that the search for other associated anomalies is pivotal, since it helps to distinguish single-gene defects from polygenic conditions. As only the combination of symptoms defines a syndrome, a clinical diagnosis cannot be made on the basis of a single defect. In that respect the detection of minor defects may be as helpful as that of major anomalies. The pattern of anomalies defines the condition and, in addition, the functional role of the gene affected. Chromosomal aberrations, including microdeletions, show their "own" pattern, which is usually less defined and broader (e.g., a combination of specific patterning defects with mental retardation) than that of single gene defects because multiple genes are involved that may act in a additive or even epistatic way.

Malformations affecting single sites may result in consecutive changes of other sites, giving rise to a malformation sequence. For example, innervation problems of the tongue resulting in a reduced pressure on the lower jaw result in reduced growth of the mandible, also called micrognathia. The latter is a consequence of the initial defect and thus secondary. Arthrogryposis is a congenital contracture of the fingers with partial joint fusions. In most cases this is not due to a primary defect in joint formation, but to a secondary effect caused by reduced fetal movement either because of reduced muscle mass or reduced muscle innervation. Both examples illustrate that the most obvious abnormality is not necessarily the most informative. Most primary defects lead to secondary effects, and sometimes it is difficult to separate cause from effect.

13.12 Disruptions

Disruptions define destructive problems acting on a normal fetus. The fetus has the ability to develop normally but fails to do so. Such disruptions may be of vascular, infectious, or teratogenic origin. They usually affect several different tissues in a particular anatomic region without adhering to regions defined by developmental processes. For example, an amniotic band can cut through a certain part of an embryo, destroying muscle, skin, bone, and other tissues that are not embryologically related. Probably secondary to amnion rupture, small strands of amnion can encircle developing structures (most frequently the limbs), leading to annular constrictions, edema, disruptive necrosis, amputation, or syndactyly. At birth aberrant strands can be noted and/or remnants of the rolled-up amnion are present at the placental base. In the milder cases this leads to asymmetric amputation-like phenotypes of the digits. In the most severe cases the entire limb may be lost and secondary deformational defects consequent on decreased fetal movement may occur. Other disruptions can be caused by intrauterine infections: rubella, for example, leads to the destruction of certain organ systems (in this case the inner ear) that would have developed in a regular way if the insult had not occurred. Teratogenic substances such as thalidomide or valproate may cause disruptions because they interfere with normal signaling and/or cell proliferation

13.13 Deformations

In contrast to malformations, which are caused by a primary and intrinsic problem in morphogenesis, and disruptions, which represent the breakdown of a previously normal tissue, deformations are abnormalities caused by mechanical forces. They usually occur during the late phase of pregnancy, but can nevertheless have a pronounced influence on fetal development. Deformations may be intrinsic, i.e., due to abnormal development of the fetus itself, or extrinsic, produced by constraint in utero of an otherwise normal fetus. Such extrinsic forces may produce a single localized deformation, such as a deformed foot, or they may cause a deformation sequence. The potter sequence is an example for the complex effects of mechanical forces. In cases of serious oligohydramnion caused by amniotic fluid leakage or due to reduced production of fluid (aplasia of kidneys, urethral valves, etc.) thoracic growth is restrained and the full growth and maturation of the lungs is thwarted, making them incapable of aerobic expansion and oxygen exchange. The nose is flattened and the limbs are in an aberrant position, often with stiff joints owing to insufficient movement. Breech presentation is another important cause of deformation. Owing to the abnormal position in utero, the head is elongated, approaching a scaphocephalic form, and there are redundant skin folds in the posterior neck, presumably due to the constant retroflexion of the head. The legs are usually hyperflexed in front of the fetus, and any gradation of hip dislocation may occur. Most of the deformations have an excellent prognosis once the fetus is released from the constraining environment. However, some need treatment to release contractures, while others are difficult to treat if the deformation interferes with normal organ development.

13.14 Dysplasias

Dysplasias are caused by gene defects that affect the formation and growth of tissues. In contrast to malformations that are due to abnormal events during early embryonic patterning, dysplasias affect the embryo at a later time point, when the patterning phase is completed. Accordingly, the gene defects that cause dysplasias frequently continue to be active after birth. This offers potential treatment options that are hard to envision for malformations. On the other hand, owing to the ongoing gene activity, some dysplasias carry the potential risk for malignant transformation. Mutations in the cartilage-specific COL2A1 gene cause different forms of skeletal dysplasias ranging from lethal types (achondrogenesis), severe dwarfism (SED congenita), to normal stature with premature osteoarthrosis and myopia. Even the most severe cases have normal skeletal patterning, i.e., they have a normal number of fingers, toes, vertebrae, etc., but all bones are very short owing to a defect in tissue (cartilage) formation. Likewise, mutations in one of the collagen type 1 chains (COL1A1, COL1A2) cause brittle bone disease or osteogenesis imperfecta. Here the affected tissue is not cartilage but bone, because COL1A1 and COL1A2 are predominantly expressed in bone and are essential for the mineralization of bone and thus for its strength.

13.15 Terminology of Congenital Defects

Congenital defects come in a bewildering diversity. Further complicating the situation is the extreme variability in some conditions combined with reduced penetrance. Over the past clinical geneticists set out to order these conditions by giving names to diseases that looked similar. Fine discrimination of the phenotype is necessary to distinguish similar entities. For example, achondroplasia was frequently misdiagnosed among individuals who have small stature and chondrodysplasia that only superficially resemble true achondroplasia. Only the clear delineation of signs and symptoms made definite diagnosis on clinical grounds possible. This phenotypic approach proved to be very successful, since most clinically defined conditions were subsequently identified as having similar or identical gene defects. Certain defects that occurred together were defined as a "syndrome," implying that all patients with this condition are likely to have the same gene defect. The identification of mutations in these patients has shown that the theory holds true, at least in principle. However, similar phenotypes may also be caused by different gene defects, a phenomenon known as genetic heterogeneity. It turns out that gene defects that result in similar phenotypes are likely to be linked within a common molecular pathway. The disruption of such a pathway results in a certain phenotype regardless of where it takes place. Noonan syndrome, for example, is a relatively common condition characterized by small stature, characteristic facies, and heart defects. Mutations in KRAS, a proto-oncogene well known from its mutations in tumors, have been shown to be responsible for Noonan syndrome. Mutations in other components of the Ras pathway including PTPN11, SOS1, and HRAS were shown to cause Noonan or Noonan-like phenotypes illustrating how defects in one pathway may lead to overlapping phenotypes [9]. Conditions that are caused by gene mutations within one molecular pathway and that are thus associated with similar phenotypes have also been called "molecular disease families" [23]. The ongoing identification of gene defects associated with specific phenotypes will eventually enable us to link phenotype and genotype and to better understand differences and similarities in human birth defects.

References

1. Ashe HL, Briscoe J (2006) The interpretation of morphogen gradients. Development 133:385–394
2. Benson DW, Silberbach GM, Kavanaugh-McHugh A, Cottrill C, Zhang Y, Riggs S, Smalls O, Johnson MC, Watson MS, Seidman JG, Seidman CE, Plowden J, Kugler JD (1999) Mutations in the cardiac transcription factor NKX2.5 affect diverse cardiac developmental pathways. J Clin Invest 104:1567–1573
3. Bruneau BG, Nemer G, Schmitt JP, Charron F, Robitaille L, Caron S, Conner DA, Gessler M, Nemer M, Seidman CE, Seidman JG (2001) A murine model of Holt–Oram syndrome defines roles of the T-box transcription factor Tbx5 in cardiogenesis and disease. Cell 106:709–721
4. Cordero D, Marcucio R, Hu D, Gaffield W, Tapadia M, Helms JA (2004) Temporal perturbations in sonic hedgehog signaling elicit the spectrum of holoprosencephaly phenotypes. J Clin Invest 114:485–494
5. Dressler GR (2006) The cellular basis of kidney development. Annu Rev Cell Dev Biol 22:509–529
6. Driever W, Nusslein-Volhard C (1988) A gradient of bicoid protein in Drosophila embryos. Cell 54:83–93
7. Driever W, Nusslein-Volhard C (1989) The bicoid protein is a positive regulator of hunchback transcription in the early Drosophila embryo. Nature 337:138–143
8. Evans PD, Gilbert SL, Mekel-Bobrov N, Vallender EJ, Anderson JR, Vaez-Azizi LM, Tishkoff SA, Hudson RR, Lahn BT (2005) Microcephalin, a gene regulating brain size, continues to evolve adaptively in humans. Science 309:1717–1720
9. Gelb BD, Tartaglia M (2006) Noonan syndrome and related disorders: dysregulated RAS-mitogen activated protein kinase signal transduction. Hum Mol Genet 15(Spec No 2):R220–R226
10. Gilbert SF (2006) Developmental biology, 8th edn. Oxford University Press, Oxford
11. Gridley T (2006) The long and short of it: somite formation in mice. Dev Dyn 235:2330–2336
12. Guerrini R, Marini C (2006) Genetic malformations of cortical development. Exp Brain Res 173:322–333
13. Haltiwanger RS (2002) Regulation of signal transduction pathways in development by glycosylation. Curr Opin Struct Biol 12:593–598
14. Ingham PW, McMahon AP (2001) Hedgehog signaling in animal development: paradigms and principles. Genes Dev 15:3059–3087
15. Jacob J, Briscoe J (2003) Gli proteins and the control of spinal-cord patterning. EMBO Rep 4:761–765
16. Karsenty G (2003) The complexities of skeletal biology. Nature 423:316–318
17. Kornak U, Mundlos S (2003) Genetic disorders of the skeleton: a developmental approach. Am J Hum Genet 73:447–474
18. Kornak U, Kasper D, Bosl MR, Kaiser E, Schweizer M, Schulz A, Friedrich W, Delling G, Jentsch TJ (2001) Loss of the ClC-7 chloride channel leads to osteopetrosis in mice and man. Cell 104:205–215
19. Krumlauf R (1992) Evolution of the vertebrate Hox homeobox genes. Bioessays 14:245–252

20. Lentze M, Schaub J, Schulte FJ, Spranger J (2007) Pädiatrie: grundlagen und praxis, 3rd edn. Springer, Heidelberg
21. Maeda RK, Karch F (2006) The ABC of the BX-C: the bithorax complex explained. Development 133:1413–1422
22. Massague J, Chen YG (2000) Controlling TGF-beta signaling. Genes Dev 14:627–644
23. Mundlos S (2009) The brachydactylies - a molecular disease family. Clin Genet 76:123–136
24. Mundlos S, Otto F, Mundlos C, Mulliken JB, Aylsworth AS, Albright S, Lindhout D, Cole WG, Henn W, Knoll JH, Owen MJ, Mertelsmann R, Zabel BU, Olsen BR (1997) Mutations involving the transcription factor CBFA1 cause cleidocranial dysplasia. Cell 89:773–779
25. Ramsdell AF (2005) Left-right asymmetry and congenital cardiac defects: getting to the heart of the matter in vertebrate left-right axis determination. Dev Biol 288:1–20
26. Reiner O, Coquelle FM (2005) Missense mutations resulting in type 1 lissencephaly. Cell Mol Life Sci 62:425–434
27. Roessler E, Muenke M (1998) Holoprosencephaly: a paradigm for the complex genetics of brain development. J Inherit Metab Dis 21:481–497
28. Schwabe GC, Mundlos S (2004) Genetics of congenital hand anomalies. Handchir Mikrochir Plast Chir 36:85–97
29. Seemann P, Schwappacher R, Kjaer KW, Krakow D, Lehmann K, Dawson K, Stricker S, Pohl J, Ploger F, Staub E, Nickel J, Sebald W, Knaus P, Mundlos S (2005) Activating and deactivating mutations in the receptor interaction site of GDF5 cause symphalangism or brachydactyly type A2. J Clin Invest 115:2373–2381
30. Smith JC (1994) Hedgehog, the floor plate, and the zone of polarizing activity. Cell 76:193–196
31. Srivastava D, Olson EN (2000) A genetic blueprint for cardiac development. Nature 407:221–226
32. Tickle C (2002) Molecular basis of vertebrate limb patterning. Am J Med Genet 112:250–255
33. Vincent J-P, Briscoe J (2001) Morphogens. Curr Biol 11(21):R851–854
34. Wiedemann HR, Kunze J, Dibbern H (1989) Atlas der klinischen Syndrome (Atlas of clinical syndromes) fuer Klinik und Praxis, 3rd edn. Schattauer, Stuttgart
35. Wolpert L Principles of development, 3rd edn. Oxford University Press, Oxford

Cancer Genetics

Ian Tomlinson

Abstract This chapter outlines the contribution of inherited genetic variation to cancer susceptibility and of acquired somatic mutations to cancer growth. The chapter deals with all forms of cancer susceptibility from rare, high-penetrance Mendelian syndromes to common alleles with small effects with importance for population risk. The section on somatic cancer genetics deals with underlying processes, such as genomic instability, and the types of functional change that result in the growth of both benign and malignant tumors. Genetic and epigenetic changes in cancers, from chromosomal scale to small mutations, are discussed. Overall, rather than performing an exhaustive survey in this large field, the chapter outlines the principles of cancer genetics using examples from both common and rare tumor types.

Contents

14.1	Inherited Risk		452
	14.1.1 Introduction		452
	14.1.2 Identification of Mendelian Cancer Susceptibility Genes		452
	14.1.3 Unidentified Mendelian Cancer Susceptibility Genes		457
	14.1.4 Germline Epimutations		458
	14.1.5 A Heterozygote Phenotype in Mendelian Recessive Tumor Syndromes?		458
	14.1.6 Some Cancer Genes are Involved in Both Dominant and Recessive Syndromes		458
	14.1.7 Phenotypic Variation, Penetrance, and Rare Cancers in Mendelian Syndromes		459
	14.1.8 Predisposition Alleles Specific to Ethnic Groups		460
	14.1.9 Germline Mutations can Determine Somatic Genetic Pathways		460
	14.1.10 Non-Mendelian Genetic Cancer Susceptibility		460
	14.1.11 Concluding Remarks		462
14.2	Somatic Cancer Genetics		462
	14.2.1 Mutations Cause Cancer		462
	14.2.2 Chromosomal-Scale Mutations, Copy Number Changes, and Loss of Heterozygosity		463
	14.2.3 Activating Mutations and Oncogenes; Inactivating Mutations and Tumor Suppressor Genes		464
	14.2.4 Epimutations		465
	14.2.5 How Many Mutations are Needed to Make a Cancer?		465
	14.2.6 The Role of Genomic Instability and Molecular Phenotypes		466
	14.2.7 The Conundrum of Tissue Specificity		467
	14.2.8 Clinicopathological Associations, Response and Prognosis		467
	14.2.9 Posttherapy Changes		468
	14.2.10 Concluding Remarks		468
References			468

I. Tomlinson (✉)
Molecular and Population Genetics,
Wellcome Trust Centre for Human Genetics, Oxford,
OX37BN, UK
e-mail: iant@well.ox.ac.uk

14.1 Inherited Risk

14.1.1 Introduction

The fact that cancer can aggregate in some families has been known for many years. In some of these families, the phenotype has been striking in terms of early onset, multiple tumors of particular combinations of lesions, including some nontumor features. These observations have allowed the cancers in such families to be distinguished from so-called sporadic lesions. Examples of such kindreds include those with von Hippel-Lindau (VHL) syndrome, in which affected individuals may develop the otherwise very rare tumors cerebellar hemangioblastoma or pheochromocytoma, but may also develop clear-cell renal carcinoma, which is suffered by 0.5–1% of the general population [23]. By contrast, in neurofibromatosis (NF), patients may develop hundreds of benign nerve cell tumors, but malignancy is not common [32]. In familial adenomatous polyposis (FAP), the characteristic lesion is the humble colorectal adenoma, but patients usually develop hundreds or thousands of these tumors, combined with upper-gastrointestinal tumors, desmoid tumors, and congenital hypertrophy of the retinal pigment epithelium [15].

The realization that many of the common cancers might have some genetic basis has been gradual. Occasionally, these cancers present in large families and inheritance appears to be Mendelian. Breast and ovarian cancers caused by *BRCA1* or *BRCA2* mutations and colorectal cancers caused by mismatch repair gene mutations are two examples. Such families tend to present with a few distinguishing features, such as low age at presentation, multi-generational inheritance, and characteristic tumor histology, but these kindreds cannot always be distinguished from those with familial disease resulting primarily from chance. There is, however, evidence from genetic epidemiology to suggest that the relatives of patients with otherwise unremarkable cancers are often at increased risk of the disease themselves. The magnitude of this increased risk and of the genetic contribution to it remain controversial, as will be discussed below.

14.1.2 Identification of Mendelian Cancer Susceptibility Genes

Following naturally from the observation of cancer-prone families, linkage analysis with subsequent positional cloning was for several years the chief method used to identify dominant Mendelian cancer predisposition genes. Here, by following anonymous, polymorphic DNA markers through families and testing for co-segregation with disease, the locations of the cancer genes could be detected. These locations were subsequently refined through further linkage searches to detect critical recombinants, often based on a finer scale map of polymorphic markers, and by allied techniques, such as the identification of large-scale germline changes including microdeletions. Ultimately, a large panel of genes within a particular genomic region usually had to be screened for germline mutations before the culprit was identified. Some of the cancer predisposition genes identified in this fashion are shown in Table 14.1. In many cases, particularly when the phenotype was not highly unusual, a single, relatively large, family provided the crucial localization data.

A relatively recent example of tumor predisposition gene identification in the Mendelian dominant setting is that of succinate dehydrogenase subunit D (SDHD): *SDHD* mutations cause early-onset or multiple paragangliomas and pheochromocytomas (HPGL) [4]. Ascertainment and careful curation of data on families segregating paragangliomas (particularly carotid body tumors) and pheochromocytomas had shown that this condition was likely to be inherited as a Mendelian dominant trait with incomplete, variable penetrance. A genome-wide linkage screen with highly polymorphic markers mapped the gene to chromosome 11q23.1, this location being refined by the use of 16 novel simple tandem repeat polymorphisms generated by the researchers themselves. Genes within the region were identified using available physical maps and additional mapping data, e.g., from the construction of bacterial and yeast artificial chromosome contigs [5]. Screening of candidate genes revealed that affected individuals harbored protein-inactivating mutations in *SDHD*. That linkage searches can be successful despite reduced penetrance has more recently been shown by the successful

14 Cancer Genetics

Table 14.1 Mendelian tumor susceptibility genes

Disease[a]	Gene symbol[b]	Tumor spectrum[c]	Classification	Notes
Ataxia telangiectasia		Leukemia, generally increased risk of malignancy	Tumor suppressor; caretaker	Predisposed to DNA double strand breaks; heterozygotes have increased breast carcinoma risk
Bloom syndrome		Generally increased risk of malignancy, especially hematological	Tumor suppressor; caretaker	Increased mitotic recombination
Hereditary pituitary adenoma	AIP	Pituitary adenoma	Tumor suppressor; gatekeeper	Possible somatotropinoma excess
Familial adenomatous polyposis	APC	Intestinal adenomas and carcinomas, desmoids, thyroid carcinomas, hepatoblastomas, adrenal tumors, brain tumors	Tumor suppressor; gatekeeper	Also attenuated disease variant
Hereditary breast/ovarian carcinoma	BRCA1	Breast carcinoma, ovarian carcinoma, pancreatic cancer, prostate cancer	Tumor suppressor; ?both	
	BRCA2			
Hereditary diffuse gastric carcinoma	CDH1	Diffuse gastric carcinoma; ?lobular breast carcinoma	Tumor suppressor; gatekeeper	
Familial melanoma	CDKN2A/TP16	Cutaneous melanoma; pancreatic carcinoma	Tumor suppressor; gatekeeper	Also known as familial atypical multiple mole melanoma-pancreatic carcinoma (FAMMPC) syndrome
Hereditary mixed polyposis	CRAC1	Colorectal hyperplastic polyps, adenomas, and carcinomas	Not known	Gene not yet cloned
Cylindromatosis; Brooke-Spiegler syndrome	CYLD1	Turban tumors/trichoepitheliomas	Tumor suppressor; gatekeeper	
Fanconi anemia	FANCA	Leukemia, generally increased risk of malignancy	Tumor suppressor; caretaker	Predisposed to DNA double strand breaks
	FANCB			
	FANCC			
	BRCA2			
	FANCD2			
	FANCE			
	FANCF			
	FANCG			
	BRIP1			
	FANCL/PHF9			
	FANCM			

(continued)

Table 14.1 (continued)

Disease[a]	Gene symbol[b]	Tumor spectrum[c]	Classification	Notes
Hereditary leiomyomatosis and renal cell carcinoma	FH	Uterine and cutaneous leiomyomas, type II papillary renal cell carcinoma, uterine leiomyosarcoma; Leydig cell testicular tumor	Tumor suppressor; gatekeeper	
Birt-Hogg-Dube syndrome	FLCN	Fibrofolliculomas, trichodiscomas, acrochordons, renal carcinoma (mostly chromophobe and oncocytoma)	Tumor suppressor; gatekeeper	
Peutz-Jeghers syndrome	LKB1/STK11	Intestinal hamartomas, increased risk of colorectal and several other carcinomas	?	Possible tumor suppressor
Multiple endocrine neoplasia type I	MENIN/MEN1	Pancreatic endocrine tumors, parathyroid adenoma, pituitary adenoma, adrenocortical adenoma	Tumor suppressor; gatekeeper	
Hereditary papillary renal cell carcinoma	MET/HGFR	Papillary renal carcinomas	Oncogene; gatekeeper	
Hereditary non-polyposis colon carcinoma	MSH2	Carcinomas of the colorectum, endometrium, stomach, urothelium and skin (sebaceous), and possibly other sites; brain tumors	Tumor suppressor; caretaker	MLH3 and PMS1 also previously suggested; carcinomas show microsatellite instability
	MLH1			
	MSH6			
	PMS2			
MYH-associated polyposis	MYH/MUTYH	Intestinal adenomas and carcinomas	Tumor suppressor; caretaker	
Neurofibromatosis type I	NF1	Multiple neurofibromas, neuromas of other sites, meningioma, hypothalamic tumor, neurofibrosarcoma, rhabdomyosarcoma, parathyroid adenoma, pheochromocytoma	Tumor suppressor; gatekeeper	
Neurofibromatosis type II	NF2	VIII cranial nerve tumors, meningiomas, schwannomas of spinal cord, sometimes other neuromas or neurofibromas	Tumor suppressor; gatekeeper	
Nijmegen breakage syndrome	NIBRIN	Probable generally increased risk of malignancy, especially hematological	Tumor suppressor; caretaker	
Carney complex	PRKAR1A	Atrial myxoma, endocrine tumors, psammomatous melanotic schwannomas	Tumor suppressor; gatekeeper	MYH8 mutations in variant disease
	MYH8			
Gorlin syndrome	PTCH1	Basal cell nevi and carcinomas	Tumor suppressor; gatekeeper	

14 Cancer Genetics

Disease[a]	Gene symbol[b]	Tumor spectrum[c]	Classification	Notes
Cowden syndrome; Lhermitte Duclos disease; Bannayan-Ruvalcaba-Riley syndrome	PTEN	Multiple hamartomas, including epithelial trichilemmomas, papules, keratoses and verrucous lesions, benign and malignant breast tumors, multiple lipomas, thyroid carcinoma, endometrial carcinoma, cerebellar dysplastic gangliocytoma, intestinal hamartomas	Tumor suppressor; gatekeeper	Considerable, largely unexplained phenotypic heterogeneity
Hereditary retinoblastoma	RB1	Retinoblastoma, osteosarcoma	Tumor suppressor; gatekeeper	
Werner syndrome	RECQL2	Generally increased risk of malignancy, especially sarcoma	Tumor suppressor; caretaker	
Rothmund-Thomson syndrome	RECQL4	Osteosarcoma; possibly generally increased risk of malignancy	Tumor suppressor; caretaker	
Multiple endocrine neoplasia type II	RET	Pheochromocytoma, medullary thyroid carcinoma, parathyroid adenoma, carcinoid, ganglioneuroma	Oncogene; gatekeeper	
Inherited AML	RUNX1/CBFA2	Acute myeloid leukemia	Oncogene?; caretaker	
Hereditary pheochromocytomas and paragangliomas	SDHB	Paraganglioma (including carotid body tumor),	Tumor suppressor;	May also be risk of clear-cell renal cell carcinoma
	SDHC	pheochromocytomas	gatekeeper	
	SDHD			
Juvenile polyposis	SMAD4/MADH4/DPC4	Intestinal polyps, colorectal carcinoma, upper-gastrointestinal carcinoma	Tumor suppressor; gatekeeper	Landscaper suggested for SMAD4
	BMPR1A/ALK3			
Tylosis	TOC	Carcinoma of esophagus	Not known	Gene not yet cloned
Li-Fraumeni syndrome	TP53	Rhabdomyosarcoma, soft tissue sarcomas, osteosarcoma, breast carcinoma, brain tumor, adrenocortical carcinoma	Tumor suppressor; ?both	CHEK2 also a low-penetrance susceptibility gene for breast and other cancers
	CHEK2			
Mosaic variegated aneuploidy	TP53	Rhabdomyosarcoma, Wilms tumor, leukemia	Tumour suppressor; caretaer	
	CHEK2			
Tuberous sclerosis	TSC1	Multiple renal angiomyolipoma, cardiac rhabdomyoma, ependymoma, renal carcinoma, astrocytoma	Tumor suppressor; gatekeeper	
	TSC2			
Von Hippel-Lindau syndrome	VHL	Multiple angiomas, cerebellar hemangioblastoma, pheochromocytoma, clear-cell renal cell carcinoma, pancreatic carcinoma	Tumor suppressor; gatekeeper	

(continued)

Table 14.1 (continued)

Disease[a]	Gene symbol[b]	Tumor spectrum[c]	Classification	Notes
Wilms tumor	WT1	Nephroblastoma	Tumor suppressor; gatekeeper	Associated with Wilms tumor–aniridia–genitourinary anomalies–mental retardation syndrome and Beckwith–Wiedemann syndrome
Xeroderma pigmentosum	XPA, ERCC3, XPC, ERCC2, DDB2, ERCC4, ERCC5, POLH	All types of skin carcinoma	Tumor suppressor; caretaker	Nucleotide excision repair deficiencies

[a]Most commonly used name
[b]"Official" and some unofficial names in common use
[c]Most commonly reported neoplastic lesions; other characteristic lesions are not shown

identification of germline aryl hydrocarbon receptor interacting protein (*AIP*) mutations in pituitary tumor families [55]. Here, a single, very large family was used to map a susceptibility locus based on 16 individuals affected by somatotropinona or mixed adenoma, who were separated by tens of meioses and by many relatives who were obligate gene carriers but showed no evidence of tumor.

Increasingly, linkage analysis and positional cloning have been supplemented by other techniques. Where a Mendelian disease is genetically heterogeneous, screening of sequence or functional homologs has provided an alternative method or a short cut to the gene (see Table 14.1 for details). After *SDHD* was identified as the cause of HPGL, for example, the other components of the SDH heterotetramer were screened for in families with similar phenotypes, and mutations were found in *SDHB* and *SDHC* [3, 36]. The identification of *LKB1/STK11* as the gene for Peutz-Jeghers syndrome relied on linkage analysis that was focused on the short arm of chromosome 19 by the discovery of deletions of this region in polyps from Peutz-Jeghers patients [19].

In addition to the above examples, several other Mendelian tumor syndromes are caused by mutations in more than one gene, each of which encodes a functionally related protein. In many of these cases, the connection between the genes has only become apparent after positional cloning identified the susceptibility loci. In tuberous sclerosis, for example, the proteins hamartin and tuberin were found to have decreased expression after linkage analysis demonstrated that there were two genes implicated, one at chromosome 9q34 and one at chromosome 16p13.3. The genes turned out to be *TSC1* and *TSC2*, respectively, and a mutation in either gene results in a similar phenotype [11]. Both proteins act as tumor suppressors and interact with each other to form a heterodimer that acts within the PI3K/AKT/mTOR signaling pathway. Other examples are provided by the condition of gastrointestinal juvenile polyposis, in which both susceptibility genes (*BMPR1A, SMAD4*) lie in the TGF-beta/BMP signaling pathway, and by neurofibromatosis, where type 2 is caused by mutations in the protein merlin, which is involved in regulation of the Ras/Rac signaling pathway and type 1 by mutations in neurofibromin, which contains a Ras GTPase-activating protein domain and negatively regulates Ras.

The fact that several bioscientific disciplines can contribute to disease gene identification is exemplified by the case of the MMR genes in Lynch syndrome (hereditary nonpolyposis colon cancer, HNPCC). Linkage analysis was intrinsically difficult in this disease, because of the problems in distinguishing the phenotype from sporadic colorectal cancer (see above). Gene identification was, however, clinched by a combination of evidence sources: linkage analysis in large colorectal cancer families; replication errors in familial and sporadic cancers that were reminiscent of similar changes in mismatch repair-deficient yeast; and the identification of germline mutations in MMR genes in HNPCC patients [31]. This approach was taken a step further by the discovery of *MYH/MUTYH* mutations as the cause of a recessively inherited form of colorectal polyposis. Here no linkage analysis or positional cloning was undertaken, and the original study was focused not on *MYH* but on the *APC* gene [1]. However, in examination of polyps from a family with multiple colorectal adenomas, the gene was identified on the basis of a strikingly unusual spectrum of somatic mutations in the *APC* gene. The preponderance of G > T changes within *APC* suggested that oxidative damage to DNA was not being efficiently repaired, perhaps as a result of an underlying defect in base excision repair (BER). This was subsequently confirmed by the finding of bi-allelic mutations in the gene encoding the MYH glycosylase, an enzyme that removes adenine residues which have mispaired with the oxidized base 8-oxo-7,8-dihydroxy-2'-deoxyguanosine.

14.1.3 Unidentified Mendelian Cancer Susceptibility Genes

Linkage analysis, positional cloning, and allied methods cannot succeed in gene identification for all Mendelian tumor syndromes, the chief problems being genetic heterogeneity and the presence of phenocopies within families. Candidate gene screening and searches for mutational signatures (such as in MAP and HNPCC) are not always successful. Therefore, while most Mendelian cancer genes have been identified, there almost certainly remains a small, but important, group of patients who have a single-gene disorder but whose underlying genetic problems have not been found. This contention remains true even if a generous allowance is made for the imperfections of mutation detection methods when screening known genes. In the absence of one or a few large kindreds that would on their own provide power for linkage analysis, these genes may remain unidentified until, by good fortune, a candidate gene search is successful or "something turns up," as it did for MYH.

14.1.4 Germline Epimutations

Some genes can be inactivated by promoter methylation rather than mutation in the soma, but it had generally been thought that stable methylation could not be transmitted through the germline. This reasoning did not, however, exclude acquisition of methylation early enough during development for it to affect sufficient cells to mimic a de novo mutation. This phenomenon, although apparently uncommon, has been reported for the DNA mismatch repair gene *MLH1*, which may acquire promoter methylation relatively easily [47]; the epimutation caused an HNPCC phenotype. There was also evidence that stable methylation might be transmitted through the germline in some unknown fashion. Recently, some cases of *MSH2* epimutation have been shown to result from germline deletions of an upstream gene, *TACSTD1*, which leads to transcript read-through and methylation of the *MSH2* promoter by some unknown mechanism [29]

14.1.5 A Heterozygote Phenotype in Mendelian Recessive Tumor Syndromes?

For recessive tumor predisposition genes, the combined population frequency of disease alleles may be >1%. At this level, the issue of disease risks in heterozygotes potentially becomes important. The mechanism for these effects might, for example, be mild haploinsufficiency or somatic loss of the wild type allele in tumors. The most long-standing example of such an effect comes from heterozygotes for ataxia telangiectasia (*ATM*) mutations. First-degree relatives of ataxia telangiectasia patients have been reported to be at increased risk of breast cancer [48], and recent molecular studies suggest that women with one mutant copy of ATM have an approximately two-fold increased risk of this cancer [39, 51]. Some other heterozygote effects continue to be more controversial. Carriers of single Bloom syndrome (*BLM*) mutations, especially the relatively common Ashkenazi change, a frameshift mutation in exon 10, have been found in some studies to be at increased risk of colorectal cancer [18], but others have found no such effect. There is, similarly, an ongoing debate about the colorectal cancer risk in carriers of single *MYH* mutations; very few studies have found significantly increased risks, but several have found small, nonsignificantly raised risks [10, 49]. Meta-analyses have not resolved this issue to date. Excluding a small risk when mutation carriers are uncommon is extremely difficult, and the case of *MYH* illustrates this problem well.

14.1.6 Some Cancer Genes are Involved in Both Dominant and Recessive Syndromes

For a number of Mendelian dominant cancer predisposition syndromes, there is now clear evidence of an additional, recessive phenotype. *BRCA2* mutant homozygotes or compound heterozygotes have Fanconi anemia [21]; and individuals with bi-allelic MMR mutations have neurofibromatosis, childhood lymphoid malignancies, and brain tumors [12, 16, 31]. Interestingly, even though somatic loss of the wild type allele underlies the dominant versions of these syndromes, the tumor spectrum is not identical to that of the recessive syndrome. For the MMR mutants, for example, perhaps the brain tumors or neurofibromas require multiple additional events that only rarely occur subsequent to allelic loss in mutant/wild type heterozygotes, but additional postulates are required to explain why the mutant homozygotes do not develop very-early-onset bowel tumors. In some cases, the homozygote syndrome is not associated with tumors at all: individuals with heterozygous fumarate hydratase (*FH*) mutations develop leiomyomas and renal cell cancer (HLRCC), but homozygotes die early in life from nonspecific global developmental delay, hypotonia, and acidosis [52].

14.1.7 Phenotypic Variation, Penetrance, and Rare Cancers in Mendelian Syndromes

The Mendelian cancer syndromes vary greatly in the number, aggressiveness, and tissue spectrum of tumors that develop (Table 14.1). In several syndromes, it is common for mutation carriers to reach old age without being affected, whereas others are affected exceptionally early in life. It is not easy to explain these variations, but animal models suggest that modifying genetic loci may sometimes have a role, as, of course, might environmental exposures. Genotype-phenotype associations also occur in Mendelian syndromes. The *APC* gene in FAP provides an excellent example of such an association. Some mutations, especially around codon 1309, cause thousands of colorectal adenomas, probably because these tumors can acquire their "second hits" by LOH, this phenomenon being related to the tumors needing the optimal level of Wnt signaling that is caused by the mutation combination [27]. Most other FAP patients have a few hundred adenomas, and their polyps have "second hits" caused by protein-truncating mutation, but a few so-called attenuated FAP patients have fewer than 100 adenomas, and some have none at all. These AFAP cases have germline mutations in the first four exons of APC, or the second half of the last exon, or in the alternatively spliced exon 9. It seems that many polyps from AFAP patients require "three hits" at APC, including loss or additional mutation of the germline mutant allele [45]. Again, it is likely that a requirement for optimal Wnt signaling is the reason.

Similar genotype-phenotype associations exist for other diseases, including conditions resulting from germline mutations in two or more different genes. In some cases, an explanation is at hand. There are, for example, suggestions that splice-site MMR mutations produce a "weak" phenotype, and it is plausibly argued that this is because they produce some functional protein and hence have less effect on the mutation rate. In other cases, however, the cause of the association is not fully understood. VHL disease (Table 14.1), for example, is often clinically divided into type 1 (without pheochromocytoma) and type 2 (with pheochromocytoma). It can be further subdivided into type 2A (with pheochromocytoma), type 2B (with pheochromocytoma and renal cell carcinoma), and type 2C (with isolated pheochromocytoma without hemangioblastoma or renal cell carcinoma). Protein-truncating mutations are more common in type 1 disease, whereas missense changes predominate in type 2, suggesting that the latter encode proteins that retain some function(s). Intriguingly, type 2C mutations, for example, may encode proteins that retain the ability to down-regulate hypoxia-inducible factor alpha subunit, thought to be critical in the pathogenesis of VHL tumors [26].

While such considerations can help to explain variation among individuals with the same disease, variation between diseases generally requires different explanations. Why are there so many polyps in FAP and neurofibromas in NF, yet so few identifiable lesions in, for example, carriers of E-cadherin mutations? Why do mutations in ubiquitously expressed genes only cause cancer in certain tissue types? Some possible causes for the variation between different tumor syndromes [44] are:

- Our failure to recognize or identify precursor lesions
- The stage of tumorigenesis at which mutation acts (earlier mutation causing more tumors)
- The cell turnover of target organ/tissue/cell
- The importance of the gene at different stages of development or for a particular specialized cell function
- Probabilistic effects, such as the likelihood of cell death unless the next mutation in the pathogenetic pathway happens within a particular time window
- The critical role of a gene, perhaps in relation to the environmental stresses acting on the target organ/tissue/cell

Perhaps even more difficult to explain is the similarity in some phenotypes, such as *MYH*-associated polyposis and FAP, which are caused by mutations in apparently unrelated proteins.

Another phenomenon that is increasingly described is the small absolute risks (although sometimes large relative risks) of certain cancer types in some of the Mendelian syndromes. Examples are shown in Table 14.1. They include modestly increased risks of prostate cancer in carriers of *BRCA2* mutations [54] and of solid malignancies in those affected by several of the syndromes caused by DNA breakage that had previously been strongly associated only with hematological malignancies.

14.1.8 Predisposition Alleles Specific to Ethnic Groups

Increasingly, cancer susceptibility mutations that are specific to particular ethnic groups have been identified. The Ashkenazi population is probably the best studied of these groups, and specific alleles have been identified in genes such as *BLM* (nt2281delATCTGA, insTAGATTC), *BRCA1* (185delAG, 5382insC), *BRCA2* (6174delT), *APC* (I1307K), and *MSH2* (A636P). For *MYH*, particular mutations are overrepresented in the Pakistani (Y90X), Indian (E466X), and Mediterranean (nt1395-7delGGA) populations [8]. Several other examples exist and must be borne in mind when undertaking mutation screening in clinical practice.

14.1.9 Germline Mutations can Determine Somatic Genetic Pathways

The identity of the germline susceptibility gene or allele can determine the somatic tumorigenic pathway that tumors follow. Breast cancers with germline *BRCA1/2* mutations, for example, generally have ductal morphology, high grade, loss of hormone receptors, absence of Her2/Neu amplification, and specific somatic molecular changes (the so-called basal phenotype [53]. HNPCC colorectal cancers are often found in the proximal colon and acquire changes, such as beta-catenin and K-ras mutations, which are very rarely found in their sporadic counterparts that have acquired somatic silencing of *MLH1*. Intriguingly, it has also been shown that genetic pathways of tumorigenesis may vary among ethnic groups, perhaps caused by their different genetic backgrounds [2].

14.1.10 Non-Mendelian Genetic Cancer Susceptibility

For most of the common tumors, there is evidence that first-degree relatives of cases are at a slightly (typically two-fold) increased risk of developing the same disease. These risks are higher for some cancers, such as those of the testis or thyroid. For some cancers, there is also sometimes a smaller increased familial risk of developing cancer in general. These risks are usually higher when the proband has early-onset disease or the family history is more extensive. Whilst some of this risk might be the result of shared environment, twin studies and segregation analyses have shown that genetic factors are highly likely to account for much of the increased risk [28].

The nature of the genetic risk of the "common cancers" remains unknown, although the relative risks associated with the known pathogenic variants in the Mendelian cancer genes are too high to explain the risks observed in the general population. Genes with moderate (say, fivefold) effects on risk may in theory exist, particularly for the more "genetic" cancers, such as those of the colorectum, breast, and prostate. However, efforts to identify such genes have largely been unsuccessful and most studies are currently focused on alleles that have relatively small effects on risk (or, in other words, have low penetrance). In all cases, these alleles may "interact" with other alleles (at the same or different loci), or with the environment, in such a way that risks are much higher (or lower) if an individual has several of them. There are essentially three major, nonexclusive models for the genetic contribution to the common cancers. In the first model – the "common variant-common disease" hypothesis – a relatively small number of frequent polymorphic alleles is supposed to increase disease risk. The second, "rare variant," model assumes that the predisposing alleles are generally uncommon (typically <2% frequency) and might even take the form of multiple different variants within the same gene. The third, "polygenic," model assumes that many, many variants contribute to disease, each conferring a very small risk; such variants may in some cases act via intermediate phenotypes or molecular traits, such as levels of gene expression.

It is arguable that the third model will be extremely difficult to show directly, except by exclusion of the other models, since the relative risks conferred by each allele (and even by a combination of several alleles) will be too low to detect other than by huge studies. Model three may also encompass what might be regarded as normal variants that are under genetic control and associated with cancer risk, such as body mass, skin pigmentation, breast density, and hormone levels. A two-stage search, based on finding the genes for these traits and then associating the traits with cancer, may be appropriate here.

Efforts to demonstrate model one are currently focused on case-control (association) studies involving thousands of patients. The disease-causing variation is generally assumed to take the form of single nucleotide polymorphisms (SNPs) or copy number variants. Platforms are now available to genotype thousands of coding SNPs or millions of noncoding SNPs genome wide. Even within the ranks of those who favor the "common variant-common disease" model, there is a divergence of opinion as to the best method of detecting causal variation, with one group favoring a screen based on known functional variation [50] and the other, a screen based on linkage disequilibrium (using so-called haplotype-tagged SNPs from efforts such as the HapMap study [9]. In the latter case especially, even when a disease-associated allele has been identified and verified by several studies it may prove particularly difficult to pin down the causal variant, given the linkage disequilibrium-based SNP selection involved.

The results of genome-wide SNP screens are at an early stage, but the validity of the common disease-common variant model has been confirmed for all the common cancers and for some rarer cancer types, including childhood solid tumors and hematological malignancies. More than ten SNPs have been found to be associated with each of prostate, breast, and colorectal cancer, although most of these are tagging variants rather than causal for disease. In general, the following conclusions can be drawn from the genome-wide association studies of cancer (Fig. 14.1) [13]:

- Most cancer predisposition SNPs are organ specific, an exception being rs6983267 on chromosome 8q, which is associated with increased risks of prostate, colorectal, and ovarian cancers.
- The relative risks conferred are somewhat lower than expected (typically 1.1–1.3 per allele).
- Power considerations suggest that more, undetected predisposition SNPs exist and that these probably have weak effects.
- SNPs in candidate genes and/or nonsynonymous SNPs are generally absent from those detected.
- Some clinicopathological associations have been found (e.g., with tumor site, stage, or grade).
- Risks at multiple SNPs are additive or log-additive, and no gene-gene interactions have been found to date.
- The most plausible models suggest that many variants have their effects through (long-range) changes in gene expression.
- Although the genes tagged by the SNPs appear to have various functions, there are hints of some pathways being consistently affected (e.g., the bone morphogenetic protein pathway in colorectal cancer).

Intriguingly, there is also evidence for model two, the rare variant' hypothesis [17]. Its proponents argue that variants with all but very small functional effects are unlikely to exist at allele frequencies higher than 1–5%, because natural selection will tend to eliminate them from the population as long as they cause some cancers to develop before reproductive age, or if they have deleterious effects unrelated to cancer. In general, surveys have shown that alleles predicted in silico to have deleterious effects do tend to be rarer, although it is difficult to apply this reasoning a priori to cancer. Rare variants might, moreover, confer higher relative risks than common variants, whilst not violating the

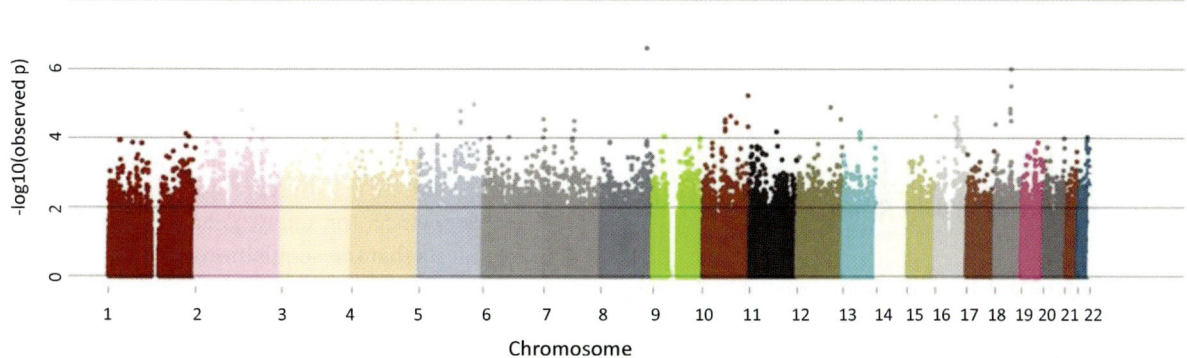

Fig. 14.1 "Manhattan" plot of GWA study data on colorectal cancer. Note the signals on chromosomes 8 and 18, both of which have subsequently been shown to result from true susceptibility loci

observed, population-based relative risk constraint. (Similar arguments can be applied to the effects of recessively acting SNP alleles.) The twin difficulties of identifying rare, disease-associated, variants are, however, discovering them and empowering analysis sufficiently to detect a genuine association (the latter is again also applicable to recessive alleles). Despite these great problems, there are a few potential solutions. Large numbers of rare variants now exist owing to HapMap and related SNP discovery programs, and selection of cases by age of onset and/or family history can greatly increase the power of association-based analyses. The *CHEK2* 1100delC allele is a case in point. This allele, which has a population frequency of ~1%, was originally proposed to be the cause of some Li–Fraumeni syndrome families (see Table 14.1). Although subsequently shown that this was unlikely to be the case, the role of CHEK2 – like BRCA1/2 – in the double strand break (DSB) repair pathway caused the variant to be tested in a set of familial breast cancer cases. It was estimated that carriers of the variant are at a two-fold increased risk of breast cancer [33]. Carriers of other rare *CHEK2* variants may also be at increased breast cancer risk, and perhaps at raised risk of other cancers too, although the latter remains controversial. It is not clear whether the examples of *CHEK2* can be applied generally, in the sense that candidate genes can be selected on the basis of current knowledge of molecular pathways in which susceptibility genes act. For breast cancer, it seems that this model is correct, in that not only are heterozygotes for pathogenic mutations in *ATM* at raised breast cancer risk (see above), but also carriers of rare variants in the related DSB repair genes *BRIP1* [42] and, possibly, *PALB2* [38] have a modestly raised risk of the disease. It is notable, however, that similar screens in the MMR or the Wnt pathways have not yet yielded similar associations between rare variants and colorectal cancer risk.

14.1.11 Concluding Remarks

Many cancers and benign tumors probably have an inherited component. Best characterized are the Mendelian (single-gene) disorders that are associated with a high lifetime tumor risk. There is great heterogeneity in these conditions in terms of the genes involved, tumor site, multiplicity of tumors, molecular pathways involved, and underlying mechanisms of tumorigenesis. Most Mendelian tumor susceptibility genes are dominantly inherited tumor suppressors that normally restrain cell proliferation in some way, or recessively inherited DNA repair or genome integrity alleles. Mendelian syndromes probably account for 1–2% of all human malignancies, and they are best classified on a molecular, rather than a clinical, basis.

The remaining genetic contribution to cancer is "complex," but may explain up to one third of all cases. The alleles involved are likely to have low penetrance, but their successful identification depends on the size of their effects, on their frequencies, and on multiple replication studies. Eventually, however, it may be possible to offer genetic testing for multiple cancers to the general population, based on panels of polymorphisms and/or rarer variants. Some of these alleles may provide a general cancer predisposition, whereas others will be site specific. It will then be possible to offer preventive measures to those at higher genetic risk. In addition, it may be possible to identify germline determinants of tumor aggressiveness or response to treatment (including pharmacogenetics) and of prognosis.

The potential financial burden of population-level genetic testing will be great. It is therefore important to target this testing wherever possible, principally to alleles with relatively large effects on disease risk and to cancers for which effective, selective prophylaxis is possible. There is otherwise a danger that the existing and predicted gains of cancer genetics will not be applied equitably and efficiently.

14.2 Somatic Cancer Genetics

14.2.1 Mutations Cause Cancer

The mutational theory of carcinogenesis has had general acceptance for many years. In short, this theory suggests that the acquisition of changes in the DNA of a susceptible progenitor cell can confer the characteristics of cancer, namely unrestrained growth, invasion, and metastasis. Cancer is thus seen largely as a cell-autonomous condition, whereby tumorigenesis is initiated in a single cell, and as a result of selection, that cell's progeny increase in numbers over their normal counterparts. This process is known as clonal expansion. It is often stated, although rarely formally proven, that the initial mutations must occur within a stem cell in the normal tissue. More recently, it has been proposed that even within a cancer, there are stem and nonstem cell populations, and molecules such as CD133 have been proposed as cancer stem cell markers [34]. The stem cell model of carcinogenesis does not necessarily assume, however, that the cells outside the tumor clone have no part to play in tumor development, and there is indeed increasing evidence that the microenvironment is extremely important in determining tumor behavior.

The nature of the selective advantage that cancer-causing mutations must confer appears to vary among cancer types. For example, the *ras* oncogene family is activated by mutation in many different cancer types, and it causes activation of Pi3-kinase, Mek/Erk, and other signaling pathways. VHL mutations are found in renal and other cancers and cause activation of the hypoxic response pathway. E-Cadherin mutations are found in lobular breast and intestinal gastric cancers and lead to failure of cell adhesion. Retinoblastoma (*RB1*) mutations disrupt the balance between cell cycle arrest/senescence and proliferation.

In many cancer types, the acquisition of mutations tends to follow particular genetic pathways. In other words, while there seems to be no absolute requirement, certain mutations are usually co-selected. Thus, in colorectal cancer, mutations of K-ras and p53 often occur together, whereas mutations of BRAF and p53 are very rarely found together in the same tumor. Moreover, some mutations seem to be selected very early in tumorigenesis – an example being changes in adenomatous polyposis coli (APC) – and others, such as p53, are found in tumors that have already acquired one or more mutations in other genes. In general, the reasons for these findings are currently unknown, but presumably relate to factors such as the interplay between signaling pathways, interactions among tumor cells, and the relationships between tumor and nontumor cells; for example, a large invasive tumor will have very different requirements from a small benign tumor in terms of its many factors, including the availability of oxygen and nutrients, thus causing different mutations to be selected.

14.2.2 Chromosomal-Scale Mutations, Copy Number Changes, and Loss of Heterozygosity

Owing to the relatively long history of cytogenetics, including the examination of tumor karyotypes, chromosomal-scale changes were the first genetic changes to be discovered in cancers. Many, perhaps most, cancers were found to have gross karyotypic abnormalities, comprising a polyploid DNA content (often 3 N or 4 N), multiple aneusomies, structural changes (including large deletions, duplications, inversions, and translocations), and grossly abnormal chromosomes such as double minutes. Newer cytogenetic techniques, such as multicolor-fluorescence in situ hybridization (MFISH; Fig. 14.2), have confirmed and characterized this tendency of cancers to acquire a highly abnormal chromosome complement. Why many tumors are aneuploid and polyploid is, however, unclear, as is the mechanism (e.g., duplication of the whole chromosome complement versus sequential chromosomal gains).

In some tumor types, the chromosomal-scale changes tend to be highly specific. In hematological malignancies, and sarcomas in particular, recurrent translocations tend to place an oncogene under the control of another gene that is highly expressed in the relevant cell type (see http://www.ncbi.nlm.nih.gov/sites/entrez?db = cancer chromosomes). These translocations

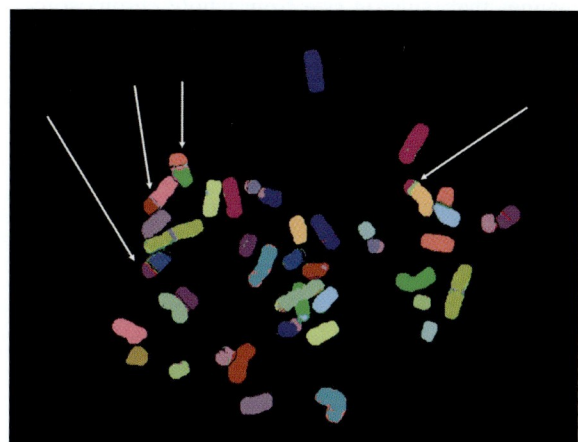

Fig. 14.2 M-FISH analysis of breast tumor cell line AP. The chromosome are pseudo-colored, showing several changes including translocations (examples *arrowed*) and numerical changes

may result in fusion genes or put the oncogene in an inappropriate molecular environment. Whilst each translocation must occur extremely rarely, they are found because they provide a strong selective advantage and, in some cases, may be essential for tumorigenesis. Examples include the classic Philadelphia chromosome t(9;22)(q34;q11) that fuses BCR and ABL in CML and the PAX3-FOXO1 fusion in rhabdomyosarcomas.

In carcinomas, recurrent translocations are relatively uncommon, for reasons that are unclear. Nevertheless, it is usual for a whole variety of chromosomal-scale mutations to be present in carcinomas. Most commonly, these result in copy number changes – gain or deletion of material relative to the overall DNA content of that tumor. Sometimes, however, the copy number changes target specific genes. For example, amplifications may target oncogenes such as EGFR or HER2/ERBB2 [41]; and deletions may target tumor suppressor genes, such as p16 or p53. However, in most cases, chromosomal-scale copy number changes in carcinomas have no obvious gene target. Whether these mutations represent background changes or subtle, selected effects on gene dosage, perhaps at multiple loci on a chromosome, is unclear. However, the fact that some changes appear to be recurrent and to differ in their frequencies among tumor types suggests that many copy number changes are not simply background events. Techniques such as high-density, microarray-based comparative genomic hybridization (CGH) are uncovering copy number changes that range in size from whole chromosomes down to a few kilobases.

A commonly used concept in cancer genetics is loss of heterozygosity (LOH) [7]. LOH can be defined as a loss of or decrease in the intensity of one allele at a polymorphic locus in a tumor, relative to paired normal tissue. The term LOH was introduced as a way of describing molecular observations associated with the inactivation of one copy of a tumor suppressor gene in a diploid cell. However, LOH has rarely been used in this correct context, and its extension to the analysis of cancers in which specimens are impure and nondiploid has created a great deal of confusion. Consider, for example, a perfectly triploid cancer. Every locus in that tumor will show LOH if a threshold of 50% decreased intensity of one allele is used. However, to ascribe functional significance to these observations would clearly be nonsensical. It may be time to accept that the term LOH should be restricted to a copy-neutral situation (i.e., when a chromosomal region becomes homozygous without deletion or gain of genetic material). This process commonly occurs by mitotic recombination (break-induced replication) and seems to be a favored way of inactivating tumor suppressors such as *APC* and *RB1*).

14.2.3 Activating Mutations and Oncogenes; Inactivating Mutations and Tumor Suppressor Genes

Conventionally, genes that undergo somatic mutations (defined so as to include all changes from those at whole-chromosome to single-base level) in cancer have been divided into those that cause gain-of-function (oncogenes) and those that cause loss-of-function (tumor suppressor genes). Perhaps unsurprisingly, this classification is too simple to describe the variety of mutations found in tumors, but it remains true that most mutated genes probably do fall into one of these two broad classes.

The original class of oncogenes was discovered as the normal cellular counterparts of proto-oncogenes found in transforming viruses. Examples include *MYC*, *SRC*, *KIT*, and *ABL*. Since that time, any dominantly acting mutation associated with gain of function has come to be classed as an oncogene. An example of an oncogene without a proto-oncogene equivalent is the Wnt signaling effector, beta-catenin which is mutated in several types of intestinal cancer [35]. In general, oncogenes can be activated by several different mutational mechanisms, including: point mutation or deletion of critical residues; translocation or rearrangement causing overexpression; and amplification (high-level copy number gain). Sometimes, an oncogene can be activated by more than one mechanism in different cancers, or even in the same cancer. For example, the hepatocyte growth factor receptor *MET* is activated by point mutation in some papillary renal cell cancers, but also frequently undergoes copy number gain of the mutant allele as a further "hit" [57].

Tumor suppressor genes conventionally require both copies to be inactivated by two mutational hits for there to be an effect on tumorigenesis. Small-scale mutations that target tumor suppressors are often, although by no means always, nonsense or frameshift changes that result in a truncated or absent protein. Missense changes that target critical residues may also be found if they inactivate protein function. Deletions or copy-neutral LOH are also common. A typical tumor suppressor is a gene such as *VHL*, which is mutated in the great majority of clear-cell renal cancers, mostly by protein-truncating mutation affecting one allele and deletion affecting the other. Loss of function of this ubiquitin ligase adaptor leads to failure to degrade HIF and other substrates, and hence to inappropriate activation of the hypoxia signaling pathway [23].

Although oncogenes can in principle be activated in a similar fashion in both diploid and aneuploid/polyploid cells, tumor suppressors classically require all copies to be inactivated. Thus, three "hits" would be required fully to inactivate a tumor suppressor in a trisomic cell, and four, in a tetrasomic cell. So far, there is little evidence that tumor suppressors harbor several mutations consistent with their inactivation in a polyploid cell, raising the intriguing possibility that tumor suppressor changes are usually relatively early events in tumorigenesis, albeit for largely unknown mutational or selective reasons.

The list of atypical oncogenes or tumor suppressors (or genes that cannot readily be placed into either category) has recently been lengthening. There is a relatively long-standing controversy as to whether p53, apparently one of the archetypal tumor suppressors, has some dominant negative effect in the heterozygous state associated with missense mutations in the protein's DNA-binding domain, despite there usually being two hits found in the gene in tumours [6]. *APC* mutations, found in the great majority of colorectal cancers, are constrained such that there is selection for retention of some protein function owing to a requirement for an optimal resulting level of Wnt signaling [27]. Mutations in the *CDC4/FBXW7* ubiquitin ligase adaptor are usually found in the heterozygous state with the wild type allele, suggesting a dominant negative effect, but may also be present in the conventional two-hit state in some tumors [25]. In addition, several genes mutated in cancer have been proposed to have their effects through haploinsufficiency, especially in mouse models, but further evidence is required before such a model, however plausible, can be accepted.

14.2.4 Epimutations

In the normal genome, CpG dinucleotides outside gene promoters are very often methylated at the cytosine residue. In cancers, there is global hypomethylation of unknown cause, but also a tendency for some promoter CpG islands to become hypermethylated and for expression of the gene concerned to be silenced. In many cases, just like mutations, these methylation changes are background events, but sometimes they target a gene the inactivation of which is selected. Indeed, for some tumor suppressor genes, mutation can affect one allele and hypermethylation the other. Examples of methylated genes with probable functional significance in cancer include the mismatch repair gene *MLH1* (see below), *CDH1*, *MGMT*, and *RUNX3*. In other cases of promoter methylation with proposed functional effects, the patterns and/or role of methylation are complex. At *APC*, for example, only the 1A promoter is methylated and, whilst this affects mRNA levels, it is far from clear that functional consequences result.

Hypermethylation in cancer does not have a known cause, and it is intrinsically different from mutation, in that multiple CpGs within a promoter must usually be methylated for there to be gene silencing. Little is known about whether this is a stochastic process or whether an initiating event triggers regional methylation. A further area of ignorance is why some genes tend to be inactivated by methylation rather than mutation: is this intrinsic to the gene concerned, perhaps reflecting some unknown aspect of normal gene sequence, structure or function?

14.2.5 How Many Mutations are Needed to Make a Cancer?

There is ongoing debate as to the number of mutations, probably acquired in a stepwise fashion, that are required for a malignant tumor to grow. Calculations based on cancer incidence rates have suggested that four to six mutations may be required, but these studies do not take full account of requirements for bi-allelic changes, the molecular processes that underlie complex mutations such as amplification, and the influences of epigenetic changes [20]. Moreover, these studies essentially measure rate-limiting steps rather than total mutation numbers. Recent efforts have enabled the DNA of a few colorectal and breast cancers to be sequenced on a genome-wide scale [56]. Remarkably, these efforts did not turn up large numbers of undiscovered mutations. In colorectal cancer, for example, all of the frequently mutated genes (largely *APC*, K-ras, p53, *SMAD4*, and *CDC4*) were already known, and no good evidence of any metastasis-specific genetic changes was found. Although other, low-frequency genetic changes occur, and may provide a small selective advantage, there is no good evidence that they drive tumorigenesis. Thus, it appears that relatively few (<5) small-scale mutations are required for a cancer to occur. However, this number takes no account of epigenetic changes, mutations in regulatory regions, changes to noncoding sequences such as microRNAs, copy number alterations, and polyploidy. Until the importance of such changes has been determined, the number of mutations required for carcinogenesis must remain uncertain, but it seems likely to be fewer than 20, and perhaps fewer than 10. This number is, of course, several orders of magnitude lower than the actual number of mutations found in any cancer. The processes of normal development and tissue turnover, added to the clonal expansion during tumorigenesis, mean that millions or billions of mutations are expected to occur in the typical cancer, simply as a result of normal rates of DNA damage, replication errors, and imperfect repair.

14.2.6 The Role of Genomic Instability and Molecular Phenotypes

An ongoing and unresolved controversy in cancer genetics is the role of genomic instability, meaning a specific underlying tendency to acquire mutations at a higher rate than normal cells. At one level, a relatively subtle form of genomic instability is probably caused by factors primarily selected for other reasons, such as oncogene activation or inactivation of tumor suppressors (e.g., *PTEN*). However, discussion of genomic instability rarely encompasses this subtle form. Rather, genomic instability is considered in terms of primary defects in processes such as DNA repair, chromosome segregation, cell cycle checkpoint failure, assembly of the mitotic spindle, or telomere structure.

Proponents of the central place of genomic instability in tumorigenesis argue that cancers require too many mutations for them not to require genomic instability [30]. This argument is relatively weak, since the process of clonal expansion can generate sufficient tumor cell progeny to allow five or six mutations to be acquired at normal mutation rates, let alone rates that may be modestly elevated as a result of, say, oncogene activation. More tellingly, it is argued that the observed numbers of mutations in cancers is probably many, many thousands, even though almost all of these are nonselected (or passenger) changes [46]. Others have responded by contending that normal mutation rates can explain even these large numbers of mutations without difficulty, and some authors have even suggested that genomic instability may not provide a selective advantage to tumors under all circumstances, particularly in the early stages of tumorigenesis [43].

What is incontrovertible is that a few cancers do have a specific tendency to genomic instability. Germline mutations in the following pathways, amongst others, cause genomic instability and a greatly increased risk of cancer: base excision repair (*MYH/MUTYH*); nucleotide excision repair (the xeroderma pigmentosum loci); spindle checkpoint (*BUBR1*); homologous recombination repair (*BLM*); mismatch repair (*MLH1, MSH2, MSH6, PMS2*); and double strand break repair (*ATM*). Most of these occur in the context of rare recessive conditions. By contrast, there are very few examples of the equivalent somatic changes in the common cancers. However, bi-allelic methylation of *MLH1* is found in an important minority of cancers, chiefly those of the colorectum and endometrium [24]. Loss of MLH1 expression leads to failure of repair of spontaneously mismatched bases, especially in simple repeat sequences such as microsatellites, and thus to a mutator phenotype termed microsatellite instability (MSI, MIN; Fig. 14.3).

The mutator phenotype in MLH1-deficient tumors raises an issue of tumor classification. MLH1-negative colorectal cancers tend to have a high frequency of mutations in genes such as *BRAF* and *TGFBR1*, and few changes in K-ras and *SMAD4;* MSI+cancers also strongly tend to have a near-diploid karyotype. Thus, it can be said that MSI+colorectal cancers tend to follow a particular genetic pathway. The molecular phenotype of MSI can thus conveniently classify a subset of cancers. By contrast, MSI- negative cancers are frequently aneuploid and polyploid and are said to be chromosomally

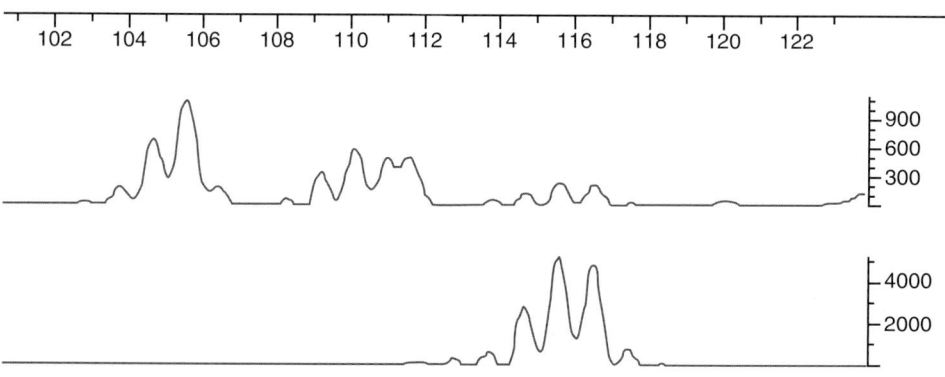

Fig. 14.3 Microsatellite instability in a colorectal tumor. The *lower trace* shows genotype at marker BAT26 in normal cells. The genotype is shifted to the left in the tumor (*above*), indicating deletion of adenine residues within the oligonucleotide tract at this site, and hence microsatellite instability

unstable (CIN+); this is an imprecise term and does not necessarily reflect an underlying tendency to genomic instability. MSI and CIN seem to be (almost) mutually exclusive. There may also be a group of MSI-CIN- cancers, although whether these have some other, undiscovered from of instability remains unknown. A third molecular phenotype, a global tendency to CpG island methylation, is generally, although not universally, accepted to exist; this so-called CIMP phenotype overlaps substantially with the MSI phenotype. Undoubtedly, these molecular phenotypes will become refined for most cancer types in the future.

14.2.7 The Conundrum of Tissue Specificity

Some genes, such as p53 and K-ras, are mutated or otherwise altered in many cancer types, whereas others, such as *APC*, have a very restricted occurrence in tumors of one or a few sites. In general, some tissue specificity of mutations is the norm. In some cases, this phenomenon can readily be explained. For example, changes in the PU.1 transcription factor are specific to hematopoietic cells and associated with the development of AML [40]. However, in most cases, it is not clear why some mutations are strongly associated with certain tumor types. DNA mismatch repair genes are ubiquitously expressed, yet *MLH1* is principally inactivated by promoter methylation in carcinomas of the large bowel and endometrium. There is increasing interest in a model of tumorigenesis in which highly specific levels of dysregulation are required for neoplastic growth. Particular types of cells have Achilles heels, but the specific molecular and general environments generally make these different among tissues. It follows that different genes tend to be altered in different tissue types.

14.2.8 Clinicopathological Associations, Response and Prognosis

Somatic mutations and epimutations are almost certainly the primary determinants of how cancers behave. Many associations between molecular changes (DNA, mRNA or protein) and clinicopathological associations, response and prognosis have been reported, although these are too numerous to list here. In colorectal cancer, for example, MSI+ cancers tend to occur in older individuals in the proximal colon, to show mucinous histology and a high number of tumor-infiltrating lymphocytes. MSI is also associated with a generally good prognosis [37] and perhaps with a poorer response to 5-FU-based therapy [22]. The reasons for these associations are unclear, but may be related to the immune response elicited by MSI+ cancers. With increasing use of molecularly targeted therapeutic agents, the associations between molecular changes and treatment response will become more important. This is already being seen for the use of anti-EGFR therapy in colorectal cancer, for which K-ras mutation is a strong negative predictive marker.

14.2.9 Posttherapy Changes

The use of chemotherapy and radiotherapy enforces strong extrinsic selection on tumors. While this will be strong enough to kill most or all of the tumor cells in many cases, where tumors do recur they often acquire specific resistance as a result of mutations in genes involved in the pathway targeted by the therapy. For example, *BRCA1/2*-mutant breast cancers that show resistance to poly(ADP-ribose) polymerase (PARP) inhibitors have been shown to have acquired reversion mutations that restore near-normal function to the mutant BRCA1 or -2 protein (yet, interestingly, do not cause cancer regression) [14]. Ovarian cancers treated with cisplatin have been reported to acquire MSI, owing to dependence of the effect of the drug on intact DNA mismatch repair and hence strong selection for resistance-conferring mismatch repair gene silencing.

14.2.10 Concluding Remarks

The cancer genome is increasingly becoming tractable to analysis. Methods such as high-density microarrays and massively parallel sequencing promise to deliver all of the important genetic and epigenetic changes in cancers. The challenge then becomes categorization of cancers, functional analysis of these changes, the acquisition of large sample series to test the roles of genetic changes as biomarkers and, ultimately, the design of new anticancer therapies.

References

1. Al-Tassan N, Chmiel NH, Maynard J, Fleming N, Livingston AL, Williams GT, Hodges AK, Davies DR, David SS, Sampson JR, Cheadle JP (2002) Inherited variants of MYH associated with somatic G:C-->T:A mutations in colorectal tumors. Nat Genet 30:227–232
2. Ashktorab H, Smoot DT, Carethers JM, Rahmanian M, Kittles R, Vosganian G, Doura M, Nidhiry E, Naab T, Momen B, Shakhani S, Giardiello FM (2003) High incidence of microsatellite instability in colorectal cancer from African Americans. Clin Cancer Res 9:1112–1117
3. Astuti D, Latif F, Dallol A, Dahia PL, Douglas F, George E, Skoldberg F, Husebye ES, Eng C, Maher ER (2001) Gene mutations in the succinate dehydrogenase subunit SDHB cause susceptibility to familial pheochromocytoma and to familial paraganglioma. Am J Hum Genet 69:49–54
4. Baysal BE, Ferrell RE, Willett-Brozick JE, Lawrence EC, Myssiorek D, Bosch A, van der Mey A, Taschner PE, Rubinstein WS, Myers EN, Richard CW 3 rd, Cornelisse CJ, Devilee P, Devlin B (2000) Mutations in SDHD, a mitochondrial complex II gene, in hereditary paraganglioma. Science 287:848–851
5. Baysal BE, Willoett-Brozick J, Lawrence E (2001) A high-resolution integrated map spanning the SDHD gene at 11q23: a 1.1-Mb BAC contig, a partial transcript map and 15 new repeat polymorphisms in a tumour-suppressor region. Eur J Hum Genet 9(2):121–129
6. Blagosklonny MV (2000) p53 from complexity to simplicity: mutant p53 stabilization, gain-of-function, and dominant-negative effect. FASEB J 14:1901–1907
7. Cavenee WK (1989) Tumor progression stage: specific losses of heterozygosity. Princess Takamatsu Symp 20:33–42
8. Cheadle JP, Sampson JR (2007) MUTYH-associated polyposis–from defect in base excision repair to clinical genetic testing. DNA Repair (Amst) 6:274–279
9. Couzin J (2006) Genomics. The HapMap gold rush: researchers mine a rich deposit. Science 312:1131
10. Croitoru ME, Cleary SP, Di Nicola N, Manno M, Selander T, Aronson M, Redston M, Cotterchio M, Knight J, Gryfe R, Gallinger S (2004) Association between biallelic and monoallelic germline MYH gene mutations and colorectal cancer risk. J Natl Cancer Inst 96:1631–1634
11. Curatolo P, Bombardieri R, Jozwiak S (2008) Tuberous sclerosis. Lancet 372:657–668
12. de Vos M, Hayward B, Bonthron DT, Sheridan E (2005) Phenotype associated with recessively inherited mutations in DNA mismatch repair (MMR) genes. Biochem Soc Trans 33:718–720
13. Easton DF, Eeles RA (2008) Genome-wide association studies in cancer. Hum Mol Genet 17:R109–R115
14. Edwards SL, Brough R, Lord CJ, Natrajan R, Vatcheva R, Levine DA, Boyd J, Reis-Filho JS, Ashworth A (2008) Resistance to therapy caused by intragenic deletion in BRCA2. Nature 451:1111–1115
15. Galiatsatos P, Foulkes WD (2006) Familial adenomatous polyposis. Am J Gastroenterol 101:385–398
16. Gallinger S, Aronson M, Shayan K, Ratcliffe EM, Gerstle JT, Parkin PC, Rothenmund H, Croitoru M, Baumann E, Durie PR, Weksberg R, Pollett A, Riddell RH, Ngan BY, Cutz E, Lagarde AE, Chan HS (2004) Gastrointestinal cancers and neurofibromatosis type 1 features in children with a germline homozygous MLH1 mutation. Gastroenterology 126:576–585
17. Goldstein DB (2009) Common genetic variation and human traits. N Engl J Med 360:1696–1698
18. Gruber SB, Ellis NA, Scott KK, Almog R, Kolachana P, Bonner JD, Kirchhoff T, Tomsho LP, Nafa K, Pierce H, Low M, Satagopan J, Rennert H, Huang H, Greenson JK, Groden J, Rapaport B, Shia J, Johnson S, Gregersen PK, Harris CC, Boyd J, Rennert G, Offit K (2002) BLM heterozygosity and the risk of colorectal cancer. Science 297:2013
19. Hemminki A, Tomlinson I, Markie D, Jarvinen H, Sistonen P, Bjorkqvist AM, Knuutila S, Salovaara R, Bodmer W,

Shibata D, de la Chapelle A, Aaltonen LA (1997) Localization of a susceptibility locus for Peutz-Jeghers syndrome to 19p using comparative genomic hybridization and targeted linkage analysis. Nat Genet 15:87–90

20. Hornsby C, Page KM, Tomlinson IP (2007) What can we learn from the population incidence of cancer? Armitage and doll revisited. Lancet Oncol 8:1030–1038
21. Howlett NG, Taniguchi T, Olson S, Cox B, Waisfisz Q, De Die-Smulders C, Persky N, Grompe M, Joenje H, Pals G, Ikeda H, Fox EA, D'Andrea AD (2002) Biallelic inactivation of BRCA2 in Fanconi anemia. Science 297:606–609
22. Iacopetta B, Kawakami K, Watanabe T (2008) Predicting clinical outcome of 5-fluorouracil-based chemotherapy for colon cancer patients: is the CpG island methylator phenotype the 5-fluorouracil-responsive subgroup? Int J Clin Oncol 13:498–503
23. Kaelin WG (2007) Von Hippel-Lindau disease. Annu Rev Pathol 2:145–173
24. Kane MF, Loda M, Gaida GM, Lipman J, Mishra R, Goldman H, Jessup JM, Kolodner R (1997) Methylation of the hMLH1 promoter correlates with lack of expression of hMLH1 in sporadic colon tumors and mismatch repair-defective human tumor cell lines. Cancer Res 57:808–811
25. Kemp Z, Rowan A, Chambers W, Wortham N, Halford S, Sieber O, Mortensen N, von Herbay A, Gunther T, Ilyas M, Tomlinson I (2005) CDC4 mutations occur in a subset of colorectal cancers but are not predicted to cause loss of function and are not associated with chromosomal instability. Cancer Res 65:11361–11366
26. Kim WY, Kaelin WG (2004) Role of VHL gene mutation in human cancer. J Clin Oncol 22:4991–5004
27. Lamlum H, Ilyas M, Rowan A, Clark S, Johnson V, Bell J, Frayling I, Efstathiou J, Pack K, Payne S, Roylance R, Gorman P, Sheer D, Neale K, Phillips R, Talbot I, Bodmer W, Tomlinson I (1999) The type of somatic mutation at APC in familial adenomatous polyposis is determined by the site of the germline mutation: a new facet to Knudson's 'two-hit' hypothesis. Nat Med 5:1071–1075
28. Lichtenstein P, Holm NV, Verkasalo PK, Iliadou A, Kaprio J, Koskenvuo M, Pukkala E, Skytthe A, Hemminki K (2000) Environmental and heritable factors in the causation of cancer–analyses of cohorts of twins from Sweden, Denmark, and Finland. N Engl J Med 343:78–85
29. Ligtenberg MJ, Kuiper RP, Chan TL, Goossens M, Hebeda KM, Voorendt M, Lee TY, Bodmer D, Hoenselaar E, Hendriks-Cornelissen SJ, Tsui WY, Kong CK, Brunner HG, van Kessel AG, Yuen ST, van Krieken JH, Leung SY, Hoogerbrugge N (2009) Heritable somatic methylation and inactivation of MSH2 in families with Lynch syndrome due to deletion of the 3' exons of TACSTD1. Nat Genet 41:112–117
30. Loeb LA, Christians FC (1996) Multiple mutations in human cancers. Mutat Res 350:279–286
31. Lynch HT, Lynch JF (2004) Lynch syndrome: history and current status. Dis Markers 20:181–198
32. McClatchey AI (2007) Neurofibromatosis. Annu Rev Pathol 2:191–216
33. Meijers-Heijboer H, van den Ouweland A, Klijn J, Wasielewski M, de Snoo A, Oldenburg R, Hollestelle A, Houben M, Crepin E, van Veghel-Plandsoen M, Elstrodt F, van Duijn C, Bartels C, Meijers C, Schutte M, McGuffog L, Thompson D, Easton D, Sodha N, Seal S, Barfoot R, Mangion J, Chang-Claude J, Eccles D, Eeles R, Evans DG, Houlston R, Murday V, Narod S, Peretz T, Peto J, Phelan C, Zhang HX, Szabo C, Devilee P, Goldgar D, Futreal PA, Nathanson KL, Weber B, Rahman N, Stratton MR, Consortium CH-BC (2002) Low-penetrance susceptibility to breast cancer due to CHEK2(*)1100delC in noncarriers of BRCA1 or BRCA2 mutations. Nat Genet 31:55–59
34. Mizrak D, Brittan M, Alison MR (2008) CD133: molecule of the moment. J Pathol 214:3–9
35. Moon RT, Kohn AD, De Ferrari GV, Kaykas A (2004) WNT and beta-catenin signalling: diseases and therapies. Nat Rev Genet 5:691–701
36. Niemann S, Muller U (2000) Mutations in SDHC cause autosomal dominant paraganglioma, type 3. Nat Genet 26:268–270
37. Popat S, Hubner R, Houlston RS (2005) Systematic review of microsatellite instability and colorectal cancer prognosis. J Clin Oncol 23:609–618
38. Rahman N, Seal S, Thompson D, Kelly P, Renwick A, Elliott A, Reid S, Spanova K, Barfoot R, Chagtai T, Jayatilake H, McGuffog L, Hanks S, Evans DG, Eccles D, C. Breast Cancer Susceptibility, Easton DF, Stratton MR (2007) PALB2, which encodes a BRCA2-interacting protein, is a breast cancer susceptibility gene. Nat Genet 39:165–167
39. Renwick A, Thompson D, Seal S, Kelly P, Chagtai T, Ahmed M, North B, Jayatilake H, Barfoot R, Spanova K, McGuffog L, Evans DG, Eccles D, C. Breast Cancer Susceptibility, Easton DF, Stratton MR, Rahman N (2006) ATM mutations that cause ataxia-telangiectasia are breast cancer susceptibility alleles. Nat Genet 38:873–875
40. Rosenbauer F, Wagner K, Kutok JL, Iwasaki H, Le Beau MM, Okuno Y, Akashi K, Fiering S, Tenen DG (2004) Acute myeloid leukemia induced by graded reduction of a lineage-specific transcription factor, PU.1. Nat Genet 36:624–630
41. Sauter G, Lee J, Bartlett JM, Slamon DJ, Press MF (2009) Guidelines for human epidermal growth factor receptor 2 testing: biologic and methodologic considerations. J Clin Oncol 27:1323–1333
42. Seal S, Thompson D, Renwick A, Elliott A, Kelly P, Barfoot R, Chagtai T, Jayatilake H, Ahmed M, Spanova K, North B, McGuffog L, Evans DG, Eccles D, C. Breast Cancer Susceptibility, Easton DF, Stratton MR, Rahman N (2006) Truncating mutations in the Fanconi anemia J gene BRIP1 are low-penetrance breast cancer susceptibility alleles. Nat Genet 38:1239–1241
43. Sieber OM, Heinimann K, Tomlinson IP (2003) Genomic instability–the engine of tumorigenesis? Nat Rev Cancer 3:701–708
44. Sieber OM, Tomlinson SR, Tomlinson IP (2005) Tissue, cell and stage specificity of (epi)mutations in cancers. Nat Rev Cancer 5:649–655
45. Spirio LN, Samowitz W, Robertson J, Robertson M, Burt RW, Leppert M, White R (1998) Alleles of APC modulate the frequency and classes of mutations that lead to colon polyps. Nat Genet 20:385–388
46. Stoler DL, Chen N, Basik M, Kahlenberg MS, Rodriguez-Bigas MA, Petrelli NJ, Anderson GR (1999) The onset and

extent of genomic instability in sporadic colorectal tumor progression. Proc Natl Acad Sci USA 96:15121–15126
47. Suter CM, Martin DI, Ward RL (2004) Germline epimutation of MLH1 in individuals with multiple cancers. Nat Genet 36:497–501
48. Swift M (1997) Ataxia telangiectasia and risk of breast cancer. Lancet 350:740
49. Tenesa A, Campbell H, Barnetson R, Porteous M, Dunlop M, Farrington SM (2006) Association of MUTYH and colorectal cancer. Br J Cancer 95:239–242
50. Terwilliger JD, Hiekkalinna T (2006) An utter refutation of the "fundamental theorem of the HapMap". Eur J Hum Genet 14:426–437
51. Thompson D, Duedal S, Kirner J, McGuffog L, Last J, Reiman A, Byrd P, Taylor M, Easton DF (2005) Cancer risks and mortality in heterozygous ATM mutation carriers. J Natl Cancer Inst 97(11):813–822
52. Tomlinson IP, Alam NA, Rowan AJ, Barclay E, Jaeger EE, Kelsell D, Leigh I, Gorman P, Lamlum H, Rahman S, Roylance RR, Olpin S, Bevan S, Barker K, Hearle N, Houlston RS, Kiuru M, Lehtonen R, Karhu A, Vilkki S, Laiho P, Eklund C, Vierimaa O, Aittomaki K, Hietala M, Sistonen P, Paetau A, Salovaara R, Herva R, Launonen V, Aaltonen LA, Multiple C (2002) Leiomyoma, Germline mutations in FH predispose to dominantly inherited uterine fibroids, skin leiomyomata and papillary renal cell cancer. Nat Genet 30:406–410
53. Turner N, Tutt A, Ashworth A (2004) Hallmarks of 'BRCAness' in sporadic cancers. Nat Rev Cancer 4:814–819
54. van Asperen CJ, Brohet RM, Meijers-Heijboer EJ, Hoogerbrugge N, Verhoef S, Vasen HF, Ausems MG, Menko FH, Gomez Garcia EB, Klijn JG, Hogervorst FB, van Houwelingen JC, van't Veer LJ, Rookus MA, van Leeuwen FE (2005) C. Netherlands Collaborative Group on Hereditary Breast, Cancer risks in BRCA2 families: estimates for sites other than breast and ovary. J Med Genet 42:711–719
55. Vierimaa O, Georgitsi M, Lehtonen R, Vahteristo P, Kokko A, Raitila A, Tuppurainen K, Ebeling TM, Salmela PI, Paschke R, Gundogdu S, De Menis E, Makinen MJ, Launonen V, Karhu A, Aaltonen LA (2006) Pituitary adenoma predisposition caused by germline mutations in the AIP gene. Science 312:1228–1230
56. Wood LD, Parsons DW, Jones S, Lin J, Sjoblom T, Leary RJ, Shen D, Boca SM, Barber T, Ptak J, Silliman N, Szabo S, Dezso Z, Ustyanksky V, Nikolskaya T, Nikolsky Y, Karchin R, Wilson PA, Kaminker JS, Zhang Z, Croshaw R, Willis J, Dawson D, Shipitsin M, Willson JK, Sukumar S, Polyak K, Park BH, Pethiyagoda CL, Pant PV, Ballinger DG, Sparks AB, Hartigan J, Smith DR, Suh E, Papadopoulos N, Buckhaults P, Markowitz SD, Parmigiani G, Kinzler KW, Velculescu VE, Vogelstein B (2007) The genomic landscapes of human breast and colorectal cancers. Science 318:1108–1113
57. Zhuang Z, Park WS, Pack S, Schmidt L, Vortmeyer AO, Pak E, Pham T, Weil RJ, Candidus S, Lubensky IA, Linehan WM, Zbar B, Weirich G (1998) Trisomy 7-harbouring non-random duplication of the mutant MET allele in hereditary papillary renal carcinomas. Nat Genet 20:66–69

The Role of the Epigenome in Human Cancers

Romulo Martin Brena and Joseph F. Costello

Abstract Deregulation of the epigenome is an important mechanism involved in the development and progression of human diseases such as cancer. As opposed to the irreversible nature of genetic events, which introduce changes in the primary DNA sequence, epigenetic modifications are reversible. The conventional analysis of neoplasias, however, has preferentially focused on elucidating the genetic contribution to tumorigenesis, which has resulted in a biased and incomplete understanding of the mechanisms involved in tumor formation. Epigenetic alterations, such as aberrant DNA methylation and altered histone modifications, are not only sufficient to induce tumors, but can also modify tumor incidence and even determine the type of neoplasia that will arise in genetic models of cancer. There is clear evidence that the epigenetic landscape in humans undergoes modifications as the result of normal aging. Thus, it has been proposed that the higher incidence of certain disease in older individuals might be, in part, a consequence of an inherent change in the regulation of the epigenome. These observations raise important questions about the degree to which genetic and epigenetic mechanisms cooperate in human tumorigenesis, the identity of the specific cooperating genes, and how these genes interact functionally to determine the diverse biological paths to tumor initiation and progression. The answers to these questions will partially rely on sequencing relevant regions of the 3 billion nucleotide genome, and determining the methylation status of the 30 million CpG dinucleotide methylome at single nucleotide resolution in different types of neoplasias. Here, we also review the emergence and advancement of technologies to map ever larger proportions of the cancer methylome, and the unique discovery potential of integrating these technologies with cancer genomic data. We discuss the knowledge gained from these large-scale analyses in the context of gene discovery, therapeutic application, and building a more widely applicable mechanism-based model of human tumorigenesis.

R.M. Brena
Department of Surgery, Epigenome Center,
University of Southern California Norris Comprehensive
Cancer Center, Los Angeles, CA 90033, USA

J.F. Costello (✉)
Department of Neurological Surgery, University of California
San Francisco Comprehensive Cancer Center, San Francisco,
CA 94143, USA
e-mail: jcostello@cc.ucsf.edu

Contents

15.1 Introduction 472

15.2 DNA Methylation in Development and Cellular Homeostasis 472

15.3 DNA Methylation is Disrupted in Human Primary Tumors 473

15.4 Genome-Wide DNA Methylation Analyses 474

15.5 Gene Silencing Vs. Gene Mutation 476

15.6 Discovery of Cancer Genes via Methylome Analysis 477

15.7 Computational Analysis of the Methylome 478

15.8 Histone Modifications and Chromat in Remodeling in Cancer 478

15.9 Eepigenome–Genome Interations in Human Cancer and Mouse Models: Gene Silencing Vs. Gene Mutation 480

15.10 Epigenetics and Response to Cancer Therapy 480

References 481

15.1 Introduction

Cancer is typically described in terms of genes that are mutated or deregulated. This gene-based model is derived, in large part, from whole-genome but low-resolution analytical methods, which certainly have biased the process of gene discovery. Higher resolution, high-throughput technical advances in DNA sequencing, genome scanning, and epigenetic analysis have produced an impressive cadre of new cancer gene candidates to fit to the model. A new challenge is thus to distinguish the gene alterations that are active drivers of cancer from those that are passengers, or more passively involved. However, significant portions of the cancer genome and epigenome remain uncharted, suggesting that even more cancer genes and potential targets for diagnosis and therapy remain to be discovered. This realization has stimulated national and international collaborative efforts to fully map various cancer genomes and epigenomes [59, 75, 90, 129], with notable successes in pilot phases. Here we discuss the technologies that have propelled these efforts, the resulting gene discoveries, and the fundamental principles of the pathogenic mechanisms of cancer, with particular emphasis on epigenetic studies of DNA methylation. Because epigenetic mechanisms can cause genetic changes and vice versa, we also review known epigenetic–genetic interactions in the context of an integrated mechanism-based model of tumorigenesis.

15.2 DNA Methylation in Development and Cellular Homeostasis

DNA methylation is essential for normal development, chromosome stability, maintenance of gene expression states, and proper telomere length [18, 23, 43, 64, 67, 95, 96, 109, 114, 123, 160, 163, 180]. DNA methylation involves the transfer of a methyl group to the 5-position of cytosine in the context of a CpG dinucleotide via DNA methyltransferases that create (DNMT3A, 3B) or maintain (DNMT1) methylation patterns. DNMT3A and DNMT3B share sequence similarity with DNMT3L, an enzymatically inactive regulatory factor that interacts with histone tails that are unmethylated at H3K4 and recruits DNMT3A2 to facilitate de novo methylation [86, 125]. Genetic knock-out of *Dnmt1*, *Dnmt3a* or *Dnmt3b* in the mouse embryo results in embryonic or perinatal lethality, underscoring the essential role of DNA methylation in normal developmental processes [109, 124].

In human and mouse, DNA methylation patterns are first established during gametogenesis. However, the genetic material contributed by each of the gametes undergoes profound changes after fertilization. A recent report indicates that the paternal genome is actively demethylated in mitotically active zygotes [72]. This active demethylation phase is followed by a passive and selective loss of DNA methylation that continues until the morula stage [141, 178]. DNA methylation patterns are then reestablished after implantation and maintained through somatic cell divisions [61]. Interestingly, amidst the sweeping genome-wide methylation changes during embryonic development, the methylation status of imprinted genes remains unchanged [164, 178].

The haploid human methylome consists of approximately 29,848,753 CpGs, nearly 70% of which are methylated in normal cells. Just 7% of all CpGs are within CpG islands [136], and most of these are unmethylated in normal tissues. Normally methylated sequences include those few CpG islands associated with the inactive X chromosome and some imprinted and tissue-specific genes, as well as pericentromeric DNA (e.g., Sat2 repeats on chr1 and chr16), intragenic regions and repetitive sequences. In fact, 45% of all CpGs in the genome are in repetitive elements, thus accounting for a large proportion of the total 5-methylcytosine [20, 31, 38, 39, 136, 156]. Normal DNA methylation patterns may vary among individuals [53, 140], potentially stemming from environmental exposure [177], stochastic methylation events [60], or trans-generational inheritance [119]. The importance of interindividual epigenomic variance has been postulated to influence the development of disease, and also the time of disease onset. An intriguing potential example of this phenomenon is illustrated by psychiatric diseases, such as bipolar disorder and schizophrenia in monozygotic twins. In some instances, only one member of the twin pair develops the pathology, while in others the time of disease onset between the twins may differ by several years or even decades. Importantly, molecular studies have failed to identify a genetic component that may account for this phenotypic discordance [16].

Several studies have focused on the influence of nutrition on DNA methylation. Of particular interest is the role played by a set of nutrients directly involved in regenerating or supplying methyl groups. Since methyl groups are labile, chronic deficiency in methyl-supplying nutrients can result in the direct or indirect alteration of SAM-to-S-adenosylhomocysteine (SAH) ratios, consequently reducing the cellular potential for methylation reactions, including DNA methylation [15]. Nutrients that regenerate or supply methyl groups fall into the category of lipotropes, and they include folate, choline, methionine, and vitamin B_{12}. Riboflavin and vitamin B_6 might also contribute to the modulation of DNA methylation processes, since both of these nutrients are integral parts in 1-carbon metabolism [183].

Studies in which rodents were subjected to diets deficient in different combinations of folate, choline, methionine, and vitamin B_{12} showed a reduction in the SAM-to-SAH ratio in those animals. Furthermore, DNA hypomethylation could be detected at the genomic level not only in specific tissues, but also at specific loci [127, 128, 147, 168]. High methyl–donor content in the diet of pregnant agouti mice can partially suppress the phenotypic manifestations of a genetic mutation (an IAP element insertion) in their offspring [171, 175]. Taken together, these results suggest that the mechanisms regulating the epigenome can be influenced by environmental factors and can interact with genetic elements to alter phenotype. Moreover, the modulation exerted by environmental factors on the epigenome can potentially contribute and/or trigger the development or onset of disease. In light of this evidence, high-resolution mapping of the methylome, ideally at single CpG dinucleotide resolution, may provide a new avenue for understanding the disease or susceptibility factors that could be used to detect at-risk individuals.

15.3 DNA Methylation is Disrupted in Human Primary Tumors

DNA methylation patterns are severely altered in primary human tumors. This includes aberrant hypermethylation of CpG islands in promoter regions, which is frequently associated with gene silencing [5, 23, 24, 45, 89], and genome-wide and locus-specific hypomethylation [39, 46–48, 58, 61]. Typically, aberrant CpG island methylation is assessed in genes already known to have a role in tumor development, especially in tumor samples that do not harbor genetic alterations of the gene. This candidate gene approach has identified aberrant methylation-mediated silencing of genes involved in most aspects of tumorigenesis, commonly altering the cell cycle [52, 66, 110, 116, 139, 158], blocking apoptosis [22, 94, 99, 161] and DNA repair [25, 26, 40, 41, 70, 74, 91]. In general, aberrant CpG island methylation tends to be focal, affecting single genes but not their neighbors [3, 185]. Two genomic loci however, are subjected to epigenetic silencing over an entire chromosomal domain of 150 kb in one case and 4 MB in the other [55, 121 154]. It is likely there will be more examples of this type of long-range epigenetic silencing yet to be uncovered.

These and other studies have established an important role for aberrant methylation in tumorigenesis and prognostication, but have focused on only a small number of the estimated 15,000 CpG island-associated promoters in the genome [4], and only on those genes first identified through genetic screens. Among those CpG islands analyzed, many have only been "sampled" for methylation at fewer than 5 of potentially 100 or more CpGs in a single island, and only on one DNA strand. Even more revealing of the early stage of cancer methylome analyses is the fact that, of the roughly 15,000 of non-CpG island-associated promoters that could also be influenced by aberrant methylation at specific CpGs, few have been studied in cancer [153]. Concurrent with promoter hypermethylation, many human tumors exhibit a global decrease in 5-methylcytosine, or genomic hypomethylation, relative to matching normal tissues [38, 47, 49, 57, 58]. In severe cases, hypomethylation can affect more than 10 million CpGs in a single tumor [12]. Three mechanisms by which hypomethylation contributes to malignancy have been proposed, including transcriptional activation of oncogenes, loss of imprinting (LOI), and promotion of genomic instability via unmasking of repetitive elements and by causation of mutations [18, 47, 49]. Most surprisingly, even though hypomethylation has been known about for more than two decades, the vast majority of genomic loci affected by cancer hypomethylation are unknown [46, 49, 83, 142, 143, 179], though presumably a significant proportion of DNA methylation loss occurs in repetitive sequences [76]. A resurgence of interest in hypomethylation, along with newer technologies for assessing hyper- and hypomethylation discussed herein, should address these sizable gaps in our knowledge of the cancer methylome.

15.4 Genome-Wide DNA Methylation Analyses

Analyzing the human genome for changes in DNA methylation is a challenging endeavor. A majority of the approximately 29 million CpG dinucleotides in the haploid genome are located in ubiquitous repetitive sequences common to all chromosomes, which hampers determination of the precise genomic location where many DNA methylation changes occur [101, 134]. In addition, gene-associated CpG islands encompass a minor fraction of all CpG sites, and their hypermethylation therefore has only a limited affect on global 5-methylcytosine levels in cancer cell DNA [35]. However, since changes in CpG island methylation can abrogate gene expression [88], identifying aberrant CpG island methylation often, but not always, identifies genes whose expression is affected during, or because of, the tumorigenic process.

Restriction Landmark Genomic Scanning (RLGS) was the first method to emerge as a genome-wide screen for CpG island methylation and was originally described in 1991 [69, 137]. In RLGS, genomic DNA is digested with the rare-cutting methylation-sensitive restriction enzymes such as *Not*I or *Asc*I. The recognition sequences for these enzymes occur preferentially in CpG islands [27, 111], effectively creating a bias towards the assessment of DNA methylation in gene promoters. Importantly, *Not*I and *Asc*I recognition sequences rarely occur within the same island, effectively doubling the number of CpG islands interrogated for DNA methylation in any given sample [29]. Following digestion, the DNA is radiolabeled and subjected to two-dimensional gel electrophoresis. DNA methylation is detected as the absence of a radiolabeled fragment, which stems from the enzymes' failure to digest a methylated DNA substrate. The main strengths of RLGS are that PCR and hybridization are not part of the protocol, allowing for quantitative representation of methylation levels and a notably low false-positive rate relative to most other global methods for detecting DNA methylation. Additionally, a priori knowledge of sequence is not required [151], making RLGS an excellent discovery tool [30, 105, 152, 184]. However, RLGS is limited to the number of *Not*I and *Asc*I sites in the human genome that fall within the well-resolved region of the profile. In practice, the combinatorial analysis of both enzymes can assess the methylation status of up to 4,100 landmarks [1264].

The success of the Human Genome Project [166] helped stimulate the development of newer methods for genome analysis, which were then adapted for DNA methylation analyses, ranging from single gene, intermediate range and high throughput (e.g., 100–1,000 loci/genes in 200 samples) [8, 37], to more complete methylome coverage (array-based methods, next-generation sequencing) [20, 82, 84, 98, 118, 145, 170, 172]. To allow for more in-depth discussion of these methods, we unfortunately had to exclude discussion of a number of other very effective PCR and array-based methods. Arrays originally designed for genome-wide analysis of DNA alterations have been adapted for methylation analysis. A main advantage of array platforms is their potential to increase the number of CpGs analyzed, and the technically advanced state of array analysis in general. Critical parameters for methylation arrays for analysis of human cancer include effective resolution, methylome coverage (total number of CpGs analyzed), reproducibility, ability to distinguish copy number and methylation events, and accurate validation through an independent method.

Differential methylation hybridization, the first array method developed to identify novel methylated targets in the cancer genome [81], has served as a basis for many newer generation array methods. In this assay, DNA is first digested with *Mse*I, an enzyme that cuts preferentially outside of CpG islands, and then ligated to linker primers. The ligated DNA is subsequently digested with up to 2 methylation sensitive restriction enzymes, such as *Bst*UI, *Hha*I or *Hpa*II. Since these enzymes are 4-base-pair restriction endonucleases, their recognition sequence is ubiquitous in GC-rich genomic regions, such as CpG islands. After the second round of enzymatic digestion, the DNA is amplified by PCR using the ligated linkers as primer binding sites. Detection of DNA methylation is accomplished by fluorescently labeling the PCR product from a test sample, such as tumor DNA and then co-hybridizing it with the PCR products derived from a control sample, such as normal tissue DNA. Aberrantly methylated fragments are refractory to the methylation-sensitive restriction endonuclease digestion, resulting in the generation of PCR products. On the other hand, an unmethylated fragment would be digested, preventing PCR amplification. Therefore, the comparison of signal intensities derived from the test and control samples following hybridization to CpG island arrays provides a profile of sequences that are methylated in one sample and not the other. One potential drawback of most methylation array methods is the need to use potentially unfaithful linker ligation and linker PCR amplification, which is prone to false positives. Nevertheless, massive improvements in oligonucleotide arrays, particularly for allelic methylation analysis, hold promise of even greater methylome coverage to methylation array-based methods in the future [20, 71, 98, 118, 145, 172].

Bacterial artificial chromosome (BAC) arrays have also been successfully introduced as a means of high-throughput DNA methylation analysis [20, 90], and complete tiling path arrays are available now [85]. In one application with BAC arrays, genomic DNA is digested with a rare cutting methylation-sensitive restriction enzyme, the digested sites are filled-in with biotin, and unmethylated fragments are selected on streptavidin beads and then co-hybridized to the BAC array with a second reference genome. In contrast to other array methods, ligation and PCR are not used in this protocol. The use of rare cutting restriction enzymes ensures that most BACs will contain only a single site or single cluster of sites, allowing single-CpG-effective resolution of the methylation analysis and accurate validation. Tiling path BAC arrays can be easily adapted for use with different restriction enzymes to significantly increase the number of analyzable CpGs. However, genome coverage using restriction enzymes is limited by the presence of their recognition sequence in the targets of interest.

The particular combination of array- and methylation-sensitive detection reagent is also critical for tumor methylome analysis. These reagents include methylation-sensitive restriction enzymes, 5-methylcytosine antibody, methylated DNA-binding protein columns, or bisulfite-based methylation detection. Bisulfite is a chemical that allows conversion of cytosine to uracil, but leaves 5-methylcytosine unconverted [56]. This method is a staple of single gene analysis and high-throughput analysis of small sets of genes [73, 107]. However, owing to the significantly reduced sequence complexity of DNA after bisulfite treatment, its use for array application has been more limited [2, 182]. DNA selected through methyl-binding protein columns or by 5-methylcytosine antibody immunoprecipitation has also been applied to microarrays [97, 121, 131, 172, 186, 187]. The effective resolution of methylation using either method is dependent in part on the average DNA fragment size after random shearing, generally 500 bp to 1 kb. It is not yet clear how many methylated CpG residues are needed for productive methylated DNA-antibody binding to occur, or whether the antibody has significant sequence bias. An advantage of this approach is that it is not as limited to specific sequences as restriction enzyme-based approaches. The 5-methylcytosine antibody approach has been used to successfully map the methylome of *Arabidopsis thaliana* [186, 187], with results largely confirmed by shot-gun bisulfite sequencing of the same genome [21, 114]. This approach has also been applied to human cancer cell lines [97, 121].

Methylation-sensitive restriction enzymes, whether rare or common cutters, can theoretically provide single-CpG-precision/effective resolution. In practice, however, common cutters, even when applied to oligonucleotide arrays, will not yield single-CpG resolution because up to ten oligonucleotides spanning multiple common cutter sites are averaged into one value. Additionally, because protocols using common cutters require ligation and PCR [82, 98, 145], the distance and sequence between sites precludes a large proportion of these sites from analysis, reducing genome coverage. The restriction enzyme *McrBc* has also been tested for methylation detection [112, 121], although the resolution of methylation events is undefined owing to the unusual recognition site of *McrBc* (two methylated CpGs separated by 40–3,000 bp of nonspecific sequence).

An innovative large-scale SAGE-like sequencing method has also been employed for methylation analysis of breast cancer and the surrounding stoma cells [80]. Gene expression arrays can also be used to identify methylation-related silencing of genes by focusing on silent genes that are reactivated in tumor cell lines exposed to a DNA demethylating agent [92, 93, 146, 159].

Reduced representation bisulfite sequencing, a large-scale genome-wide shotgun sequencing approach [115], has been successfully employed to investigate loss of DNA methylation in DNMT[1^{kd},$3a^{-/-}$,$3b^{-/-}$] ES cells. An advantage of this method is that it is amenable to gene discovery without preselecting targets, though sites exhibiting heterogeneous methylation might be confounding when represented by only a single sequence read. Substantially increasing the depth of sequencing may mitigate this limitation somewhat. Also, since clone libraries can be constructed, the system can be automated to maximize efficiency.

Human epigenome projects of normal human cells have taken a standard sequencing-based bisulfite strategy, which gives single-CpG resolution of methylation status [36, 85, 130]. While these projects are not primarily designed to determine the methylation status of 29 million CpGs, the efforts to date have been immense and impressive, including different cell type, and interindividual and interspecies comparisons. Combining either bisulfite, the 5-methyC antibody, methyl-binding protein columns, or restriction enzymes with next-generation sequencing also holds great promise. These and other studies are adding to whole new disciplines within epigenetic research, including population epigenetics and comparative epigenetics. In addition to the main goals of these projects, the data will also be of substantial value for comparison with cancer methylome data, whether from arrays or sequencing bisulfite-converted DNA.

15.5 Gene Silencing Vs. Gene Mutation

Tumor suppressor genes are typically discovered through study of familial cancers and through mapping allelic loss of heterozygosity (LOH) in sporadic human tumors [17]. Regions exhibiting recurrent, nonrandom deletion are selected for further identification of a candidate tumor suppressor gene by attempting to identify a second hit involving a point mutation or homozygous deletion [100]. Thus, until recently, surveys for point mutations have been confined largely to regions of recurrent LOH or genomic amplification. Current proposals for sequencing entire cancer genomes aim to identify genes that have escaped detection by lower resolution approaches, to provide new targets for therapy, and to further improve the experimental modeling of cancer. Pilot projects have proven the utility of this approach with great success [32, 33, 150]. A recent zenith in sequencing, including 13,023 genes in 22 tumor cell lines, yielded a wealth of new candidate cancer genes and potential therapeutic targets [150]. A current challenge is to distinguish mutations likely to contribute to the tumorigenic process from the many inconsequential mutations that riddle the cancer genome.

A related hypothesis is being addressed concurrently by taking an unbiased approach to mapping nonrandom and tumor type-specific epigenetic alterations that result in gene silencing [6, 27, 87, 138]. These studies address the hypothesis that there may be tumor suppressor genes that have escaped detection because they are seldom inactivated by genetic lesions, but often silenced by epigenetic mechanisms [185]. Using Restriction Landmark Genome Scanning (RLGS) [69], the first of many large-scale methylation

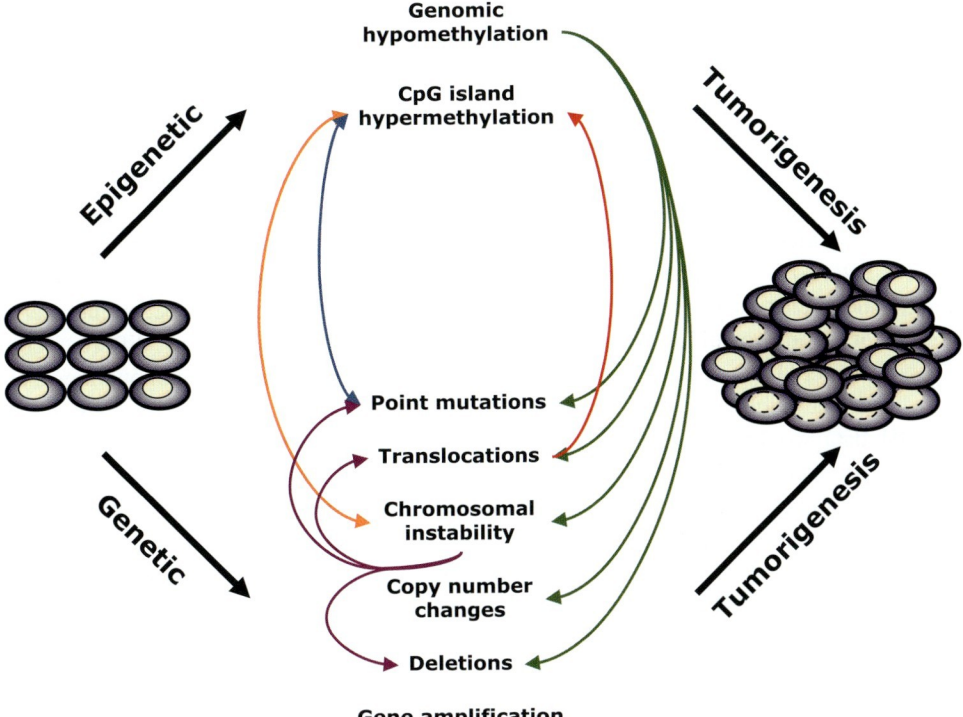

Fig. 15.1 A mechanism-based model of the pathogenesis of human cancer. Data from mouse models of cancer or hereditary human cancer indicate that genetic changes or epigenetic changes alone can initiate tumor formation. Sporadic cancers, which comprise 90–95% of all cancers, almost uniformly exhibit both genetic and epigenetic defects genome-wide, and these mechanisms show substantial interaction (*arrows*). That is, epigenetic events can cause genetic events, and vice versa. Determining the relative contribution of genetic and epigenetic mechanisms to tumor formation is an important goal of current research and should be facilitated by the whole cancer genome and epigenome approaches. Depending on the cancer type, each mechanism can operate early, late, or continuously in the development of the tumor

analysis methods, it was estimated that hundreds of CpG islands may be aberrantly methylated in any given tumor, though the range of methylation across individual tumors varies significantly [27]. Similar to mutation spectra, only a subset of these methylation events are sufficiently recurrent to qualify as nonrandom events, potentially arising through selection of cells harboring a methylation-mediated silencing event that confers a growth advantage. Large-scale integrated genomic and epigenomic tumor profiles have shown that the majority of loci affected by aberrant methylation are in fact independent of recurrent deletions [78, 79, 185]. Genes such as *WNK2*, encoding a serine/threonine kinase that negatively regulates MEK1, are largely subject to epigenetic silencing in one tumor type, but by genetic point mutations in other tumor types [79]. Taken together, these data suggest that genomic and epigenomic approaches are complementary for cancer gene discovery and that their integration could provide an ideal and more comprehensive platform for interrogating the cancer genome (Fig. 15.1).

15.6 Discovery of Cancer Genes via Methylome Analysis

As discussed above, no single current genome-wide DNA methylation approach can assay the entire cancer methylome. Thus, more focused and integrative approaches exploiting the cooperation between genetic and epigenetic mechanisms have been adopted in efforts to identify new cancer genes, many with promising results. Recently, for example, transcription factor 21 (*TCF21*) was identified as a putative tumor suppressor in head and neck and lung cancers by specifically screening a known region of LOH for aberrant DNA methylation [155]. Interestingly, this gene is located in a 9.6-Mb chromosomal domain known to suppress metastasis in melanoma cell lines [174]. However, no candidate gene had been proposed for this region, since mutations in *TCF21* are infrequent [174]. A similar strategy was utilized to identify oligodendrocyte transcription factor 1 (*OLIG1*), a frequently methylated gene and prognostic factor in human lung cancer located in a region of chromosomal loss [11, 108], and also for *HIC1* and others [169]. Like *TCF21*, *OLIG1* also was methylated at a much higher frequency than the existing LOH data would have suggested [107, 155], indicating that aberrant DNA methylation is likely to be the main mode of inactivation for these genes in the tumor types analyzed. Other putative tumor suppressor genes also located in regions of frequent LOH, such as *DLEC1*, *PAX7*, *PAX9*, *HOXB13*, and *HOXB1*, have been identified via the use of affinity columns to enrich methylated DNA sequences [131]. Given their specific technical limitations, these studies indicate that the integration of several experimental strategies will be required in order to maximize the discovery of new cancer-related genes. These studies illustrate the discovery potential of combined approaches, though the current cast of candidate cancer genes derived from methylation screens alone is far larger than can be discussed here.

Sequence-based rules derived from cancer cell methylation data have also been explored as a way to predict the pattern of aberrant methylation in cancer genome-wide [9, 44, 50, 51]. These studies have identified consensus sequences, proximity to repetitive elements, and chromosomal location as potential factors influencing or perhaps determining the likelihood that a CpG island might become aberrantly methylated. If the sequence context in which a CpG island is located influences its likelihood of becoming aberrantly methylated, the convergence of different computational analyses is likely to find commonalities that could help explain this phenomenon. An important goal in these studies will be to distinguish sequence rules that predict pan-cancer methylation from those that predict tumor-type-specific methylation, as these rules could be mutually exclusive. An intriguing and particularly striking association between a subset of genes susceptible to aberrant promoter methylation in adult human cancers and a subset of genes occupied or marked by polycomb group proteins in human embryonic stem cells has been reported independently by three groups [122, 144, 176]. These and earlier studies [135, 167] offer important new insight into possible mechanisms by which certain genes might be susceptible to methylation in cancer, and epigenetic support for the theory that human tumors arise from tissue stem cells (Fig. 15.2). Comparison of the sequences associated with PcG occupancy and those derived from the computational analysis of methylation-prone and methylation-resistant loci described above might be particularly revealing.

15.7 Computational Analysis of the Methylome

Aberrant DNA methylation exhibits tumor-type-specific patterns [27]. However, it is unclear how these patterns are established and why a large number of CpG islands seem to be refractory to DNA methylation, while others are aberrant methylated at high frequency [30, 97, 132, 138, 181]. A functional explanation for this observation could be that all CpG islands may be equally susceptible to DNA methylation, but only a fraction is detected in tumors because of selection pressures. This hypothesis, though probably true for some genes, is unlikely to explain the mechanism responsible for aberrant methylation of all CpG island-associated genes.

15.8 Histone Modifications and Chromatin Remodeling in Cancer

A second epigenetic mechanism of transcriptional regulation and chromosomal functioning (e.g., DNA repair, DNA replication, chromatin condensation) involves reversible histone modifications [63, 102, 103]. Eight histone proteins, two each of histone H2A, H2B, H3 and H4, along with 146 bp of DNA comprise a single nucleosome. Interaction of neighboring nucleosomes can be altered by the complex combinations of covalent modifications on the histones, which may represent a "histone code." Different types of histone modifications include phosphorylation, acetylation,

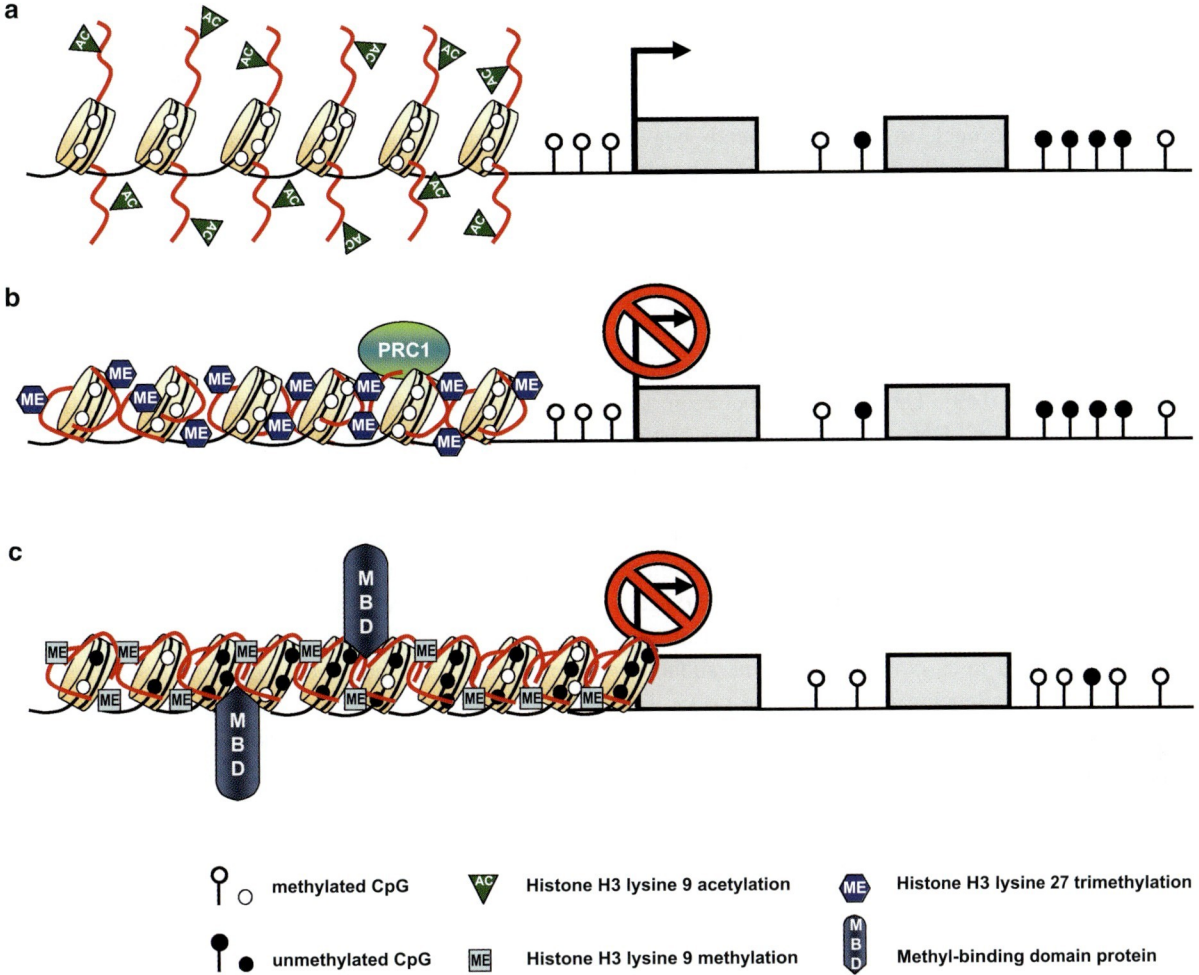

Fig. 15.2 (**a–c**) Role of DNA methylation and histone modifications at promoter CpG islands in normal cells and cancer. In normal cells, most promoter CpG islands do not exhibit DNA methylation. Thus, the expression status of a CpG island containing genes is primarily determined by the presence or absence of transcription factors and by histone modifications around the promoters of such genes. (**a**) In general, transcriptionally active CpG island loci exhibit unmethylated DNA, high levels of histone H3 lysine 9 acetylation, and/or H3K4 trimethylation, which are accompanied by an open chromatin configuration. (**b**) Transcriptionally silent genes, however, are marked by histone H3 lysine 27 trimethylation, a modification catalyzed by EZH2 (enhancer of zeste 2), a member of the Polycomb repressor complex 2 (PRC2). Following H3 lysine 27 trimethylation, these promoters are often bound by members of the Polycomb repressor complex 1 (PRC1), which together prevent transcription initiation by RNA polymerase II. This is also a mechanism of abnormal gene silencing in cancer, in the absence of aberrant DNA methylation. (**c**) In cancer, a large number of CpG islands are hypermethylated at their DNA, which generally correlates with transcriptional repression of the associated genes. These CpG islands generally exhibit a closed chromatin configuration, marked by histone H3 lysine 9 methylation, loss of acetylation, nucleosome occupancy around the transcription start site, and various types of methyl-binding domain proteins. Altogether, these modifications render the chromatin nonpermissive for transcription initiation

methylation (mono-, di-, and trimethylation), ubiquitylation, ADP ribosylation, deimination, proline isomerization, and sumoylation. The modifications may directly alter protein–histone interactions, or can indirectly influence protein–histone or protein–DNA interactions by attracting other proteins that bind specifically to modified histones. The enzymes responsible for these modifications, and for demodification or reversal, have significant specificity for the type of mark, the particular amino acid, and the position of the amino acid in the histone subunit. Histone modifications can be very dynamic in nature, changing rapidly in response to stimuli. Mapping individual histone modifications genome-wide is now possible with chromatin immunoprecipitation applied to tiling path microarray chips, although the resolution is not yet at the level of single nucleosomes and depends heavily on the quality of the antibody that recognizes the modification [7]. Because of the complexity of histone marks on a given nucleosome, new tools and approaches for testing the functional significance of individual modifications will be particularly useful, such as the synthesis of nucleosomes with pure, single modifications added in vitro, as well as new analogues of modified lysines [148, 149]. The interaction between, and inter-dependence of, DNA methylation and histone modifications is the subject of a large number of studies, particularly in cancer [14, 55, 89, 117, 157]. Alterations in the pattern and overall amount of each histone modification have also been reported in human cancers and cancer cell lines [90]. For example, again the silencing mark H3K27 trimethylation in promoters has been reported in association with gene silencing. These and other silencing marks may co-occur with aberrant DNA methylation and function synergistically in gene silencing, and also have been observed in the absence of aberrant DNA methylation. Experimental models using cancer cell lines suggest a relative order of silencing events involving both histone and DNA methylation, but this may be gene and cell type dependent. More globally, two characteristic changes of histone modifications in cancer are a decrease in acetylation of Lys16 and trimethylation of Lys20 on histone H4, in large part from repetitive portions of the genome and in association with hypomethylation of these DNA sequences [54].

15.9 Epigenome–Genome Interations in Human Cancer and Mouse Models: Gene Silencing Vs. Gene Mutation

Genetic and epigenetic mechanisms both contribute to, and probably interact during, tumorigenesis. In genetic mouse models of tumors, disruption of DNA methylation dramatically modifies the incidence of tumor formation and the spectrum of tumor types [107, 133, 165]. Methylation imbalance alone is also sufficient to induce tumors in mice [61, 77]. These studies illustrate a functional role of epigenetic imbalance in tumorigenesis, and also emphasize the interaction of genetic and epigenetic mechanisms in determining tumor incidence and tumor type.

In human tumors, genetic and epigenetic mechanisms can cooperate directly or indirectly. For example, direct cooperation includes complete inactivation of tumor suppressors by methylation of one allele and either deletion or mutation of the other [65, 120]. Epigenetic mechanisms can also cause genetic alterations, and *vice versa*. For example, aberrant methylation-associated silencing of *MLH1* leads to microsatellite instability in colon cancer [74, 91]. Similarly, methylation and silencing of the *MGMT* gene, which encodes a DNA repair enzyme, is significantly associated with G:C to A:T transition mutations in the tumor suppressor gene p53 in colorectal tumors [42]. Indirectly, aberrant loss of methylation in the pericentromeric regions of chromosomes 1 and 16, followed by cell division, is associated with abnormalities of these chromosomes, including loss and gain of whole chromosome arms, in cancer and in ICF (immunodeficiency, centromere instability, and facial anomalies) syndrome patients. Alternatively, translocations of *PML* and *retinoic acid receptor* can create a fusion protein that abnormally recruits the DNA methyltransferase and causes aberrant methylation at specific promoters in leukemia [34]. More global epigenetic defects described as a CpG island methylator phenotype [162, 173] are tightly associated with genetic mutations of the oncogene BRAF, potentially suggesting a common genetic-epigenetic course for these tumors. This candidate gene approach suggests there are important interactions between these two major mechanisms of tumorigenesis, but the extent to which these individual observations can be extrapolated to the whole cancer genome is unknown. Efforts that integrate different technologies, as described above, promise a more complete understanding of the genomic and epigenomic contribution to tumorigenesis.

15.10 Epigenetics and Response to Cancer Therapy

Aberrant methylation of particular CpG islands may also alter the response of a cancer cell to therapeutic agents, or serve as a clinically useful marker of clinical outcome. For example, normal expression of the DNA repair gene, *O*-6-methylguanine DNA methyltransferase (MGMT), is associated with resistance to therapy, whereas aberrant methylation of the MGMT 5′ CpG island, and presumably MGMT silencing [25, 26, 68], is associated with significantly improved anti-tumor response of alkylating agents such as temozolomide [41, 70]. In contrast, cisplatin-resistant cancer cells can be sensitized by relieving repressive histone H3 K27 methylation and DNA methylation, presumably by reactivating silenced tumor suppressors and modulators of cisplatin response [1a]. Efforts directed at finding DNA methylation-based markers for early detection of tumors and predicting tumor response to therapy are underway in research laboratories worldwide [10, 28, 146]. Assays are currently available to detect aberrant DNA methylation in minute samples that are obtained with minimally invasive procedures, such as sputum, blood, feces, urine and nipple aspirates, and which are likely to contain tumor cells and tumor DNA shed from a primary tumor mass [13, 104]. In contrast, loss of methylation from normally methylated promoters of the MAGEA gene family followed by MAGEA gene activation may elicit production of anti-MAGEA antibodies, which are detectable in the blood of patients with melanoma and other cancers [19].

References

1. Abbosh PH, Montgomery JS, Starkey JA, Novotny M, Zuhowski EG et al (2006) Dominant-negative histone H3 lysine 27 mutant derepresses silenced tumor suppressor genes and reverses the drug-resistant phenotype in cancer cells. Cancer Res 66:5582–5591
2. Adorjan P, Distler J, Lipscher E, Model F, Muller J et al (2002) Tumour class prediction and discovery by microarray-based DNA methylation analysis. Nucleic Acids Res 30:e21
3. Akama TO, Okazaki Y, Ito M, Okuizumi H, Konno H et al (1997) Restriction landmark genomic scanning (RLGS-M)-based genome-wide scanning of mouse liver tumors for alterations in DNA methylation status. Cancer Res 57:3294–3299
4. Antequera F, Bird A (1993) Number of CpG islands and genes in human and mouse. Proc Natl Acad Sci USA 90:11995–11999
5. Baylin S, Bestor TH (2002) Altered methylation patterns in cancer cell genomes: cause or consequence? Cancer Cell 1:299–305
6. Baylin SB, Ohm JE (2006) Epigenetic gene silencing in cancer - a mechanism for early oncogenic pathway addiction? Nat Rev Cancer 6:107–116
7. Bernstein BE, Meissner A, Lander ES (2007) The mammalian epigenome. Cell 128:669–681
8. Bibikova M, Lin Z, Zhou L, Chudin E, Garcia EW et al (2006) High-throughput DNA methylation profiling using universal bead arrays. Genome Res 16:383–393
9. Bock C, Paulsen M, Tierling S, Mikeska T, Lengauer T et al (2006) CpG island methylation in human lymphocytes is highly correlated with DNA sequence, repeats, and predicted DNA structure. PLoS Genet 2:e26
10. Brena RM, Plass C, Costello JF (2006) Mining methylation for early detection of common cancers. PLoS Med 3:e479
11. Brena RM, Morrison C, Liyanarachchi S, Jarjoura D, Davuluri RV et al (2007) Aberrant DNA Methylation of OLIG1, a Novel Prognostic Factor in Non-Small Cell Lung Cancer. PLoS Med 4:e108
12. Cadieux B, Ching TT, Vandenberg SR, Costello JF (2006) Genome-wide Hypomethylation in Human Glioblastomas Associated with Specific Copy Number Alteration, Methylenetetrahydrofolate Reductase Allele Status, and Increased Proliferation. Cancer Res 66:8469–8476
13. Cairns P, Esteller M, Herman JG, Schoenberg M, Jeronimo C et al (2001) Molecular detection of prostate cancer in urine by GSTP1 hypermethylation. Clin Cancer Res 7:2727–2730
14. Cameron EE, Bachman KE, Myohanen S, Herman JG, Baylin SB (1999) Synergy of demethylation and histone deacetylase inhibition in the re-expression of genes silenced in cancer. Nat Genet 21:103–107
15. Cantoni GL (1985) The role of S-adenosylhomocysteine in the biological utilization of S-adenosylmethionine. Prog Clin Biol Res 198:47–65
16. Cardno AG, Rijsdijk FV, Sham PC, Murray RM, McGuffin P (2002) A twin study of genetic relationships between psychotic symptoms. Am J Psychiatry 159:539–545
17. Cavenee WK, Dryja TP, Phillips RA, Benedict WF, Godbout R et al (1983) Expression of recessive alleles by chromosomal mechanisms in retinoblastoma. Nature 305:779–784
18. Chen RZ, Pettersson U, Beard C, Jackson-Grusby L, Jaenisch R (1998) DNA hypomethylation leads to elevated mutation rates. Nature 395:89–93
19. Chen YT, Stockert E, Chen Y, Garin-Chesa P, Rettig WJ et al (1994) Identification of the MAGE-1 gene product by monoclonal and polyclonal antibodies. Proc Natl Acad Sci USA 91:1004–1008
20. Ching TT, Maunakea AK, Jun P, Hong C, Zardo G et al (2005) Epigenome analyses using BAC microarrays identify evolutionary conservation of tissue-specific methylation of SHANK3. Nat Genet 37:645–651
21. Cokus SJ, Feng S, Zhang X, Chen Z, Merriman B et al (2008) Shotgun bisulphite sequencing of the Arabidopsis genome reveals DNA methylation patterning. Nature 452:215–219
22. Conway KE, McConnell BB, Bowring CE, Donald CD, Warren ST et al (2000) TMS1, a novel proapoptotic caspase recruitment domain protein, is a target of methylation-induced gene silencing in human breast cancers. Cancer Res 60:6236–6242
23. Costello JF (2003) DNA methylation in brain development and gliomagenesis. Front Biosci 8:S175–S184
24. Costello JF, Plass C (2001) Methylation matters. J Med Genet 38:285–303

25. Costello JF, Futscher BW, Kroes RA, Pieper RO (1994) Methylation-related chromatin structure is associated with exclusion of transcription factors from and suppressed expression of the O-6-methylguanine DNA methyltransferase gene in human glioma cell lines. Mol Cell Biol 14:6515–6521
26. Costello JF, Futscher BW, Tano K, Graunke DM, Pieper RO (1994) Graded methylation in the promoter and body of the O-6-methylguanine DNA methyltransferase (MGMT) gene correlates with MGMT expression in human glioma cells. J Biol Chem 269:17228–17237
27. Costello JF, Fruhwald MC, Smiraglia DJ, Rush LJ, Robertson GP et al (2000) Aberrant CpG-island methylation has non-random and tumour-type-specific patterns. Nat Genet 24:132–138
28. Cui H, Cruz-Correa M, Giardiello FM, Hutcheon DF, Kafonek DR et al (2003) Loss of IGF2 imprinting: a potential marker of colorectal cancer risk. Science 299:1753–1755
29. Dai Z, Weichenhan D, Wu YZ, Hall JL, Rush LJ et al (2002) An AscI boundary library for the studies of genetic and epigenetic alterations in CpG islands. Genome Res 12:1591–1598
30. Dai ZY, Lakshmanan RR, Zhu WG, Smiraglia DJ, Rush LJ et al (2001) Global methylation profiling of lung cancer identifies novel methylated genes. Neoplasia 3:314–323
31. Das R, Dimitrova N, Xuan Z, Rollins RA, Haghighi F et al (2006) Computational prediction of methylation status in human genomic sequences. Proc Natl Acad Sci USA 103:10713–10716
32. Davies H, Bignell GR, Cox C, Stephens P, Edkins S et al (2002) Mutations of the BRAF gene in human cancer. Nature 417:949–954
33. Davies H, Hunter C, Smith R, Stephens P, Greenman C et al (2005) Somatic mutations of the protein kinase gene family in human lung cancer. Cancer Res 65:7591–7595
34. Diala ES, Cheah MS, Rowitch D, Hoffman RM (1983) Extent of DNA methylation in human tumor cells. J Natl Cancer Inst 71:755–764
35. Di Croce L, Raker VA, Corsaro M, Fazi F, Fanelli M et al (2002) Methyltransferase recruitment and DNA hypermethylation of target promoters by an oncogenic transcription factor. Science 295:1079–1082
36. Eckhardt F, Lewin J, Cortese R, Rakyan VK, Attwood J et al (2006) DNA methylation profiling of human chromosomes 6, 20 and 22. Nat Genet 38:1378–1385
37. Ehrich M, Nelson MR, Stanssens P, Zabeau M, Liloglou T et al (2005) Quantitative high-throughput analysis of DNA methylation patterns by base-specific cleavage and mass spectrometry. Proc Natl Acad Sci USA 102:15785–15790
38. Ehrlich M (2000) DNA methylation: normal development, inherited diseases, and cancer. J Clin Ligand Assay 23:144–146
39. Ehrlich M, Gama-Sosa MA, Huang L-H, Midgett RM, Kuo KC et al (1982) Amount and distribution of 5-methylcytosine in human DNA from different types of tissues and cells. Nucleic Acids Res 10:2709–2721
40. Esteller M, Garcia-Foncillas J, Andion E, Goodman SN, Hidalgo OF et al (2000) Inactivation of the DNA-repair gene MGMT and the clinical response of gliomas to alkylating agents. N Engl J Med 343:1350–1354
41. Esteller M, Silva JM, Dominguez G, Bonilla F, Matias-Guiu X et al (2000) Promoter hypermethylation and BRCA1 inactivation in sporadic breast and ovarian tumors. J Natl Cancer Inst 92:564–569
42. Esteller M, Risques RA, Toyota M, Capella G, Moreno V et al (2001) Promoter hypermethylation of the DNA repair gene O(6)-methylguanine-DNA methyltransferase is associated with the presence of G:C to A:T transition mutations in p53 in human colorectal tumorigenesis. Cancer Res 61:4689–4692
43. Fan GP, Beard C, Chen RZ, Csankovszki G, Sun Y et al (2001) DNA hypomethylation perturbs the function and survival of CNS neurons in postnatal animals. J Neurosci 21:788–797
44. Fang F, Fan S, Zhang X, Zhang MQ (2006) Predicting methylation status of CpG islands in the human brain. Bioinformatics 22:2204–2209
45. Feinberg AP (2005) Cancer epigenetics is no Mickey Mouse. Cancer Cell 8:267–268
46. Feinberg AP, Tycko B (2004) The history of cancer epigenetics. Nat Rev Cancer 4:143–153
47. Feinberg AP, Vogelstein B (1983) Hypomethylation distinguishes genes of some human cancers from their normal counterparts. Nature 301:89–92
48. Feinberg AP, Gehrke CW, Kuo KC, Ehrlich M (1988) Reduced genomic 5-methylcytosine content in human colonic neoplasia. Cancer Res 48:1159–1161
49. Feinberg AP, Ohlsson R, Henikoff S (2006) The epigenetic progenitor origin of human cancer. Nat Rev Genet 7:21–33
50. Feltus FA, Lee EK, Costello JF, Plass C, Vertino PM (2003) Predicting aberrant CpG island methylation. Proc Natl Acad Sci USA 100:12253–12258
51. Feltus FA, Lee EK, Costello JF, Plass C, Vertino PM (2006) DNA motifs associated with aberrant CpG island methylation. Genomics 87:572–579
52. Ferguson AT, Evron E, Umbricht CB, Pandita TK, Chan TA et al (2000) High frequency of hypermethylation at the 14-3-3 sigma locus leads to gene silencing in breast cancer. Proc Natl Acad Sci USA 97:6049–6054
53. Fraga MF, Ballestar E, Paz MF, Ropero S, Setien F et al (2005) Epigenetic differences arise during the lifetime of monozygotic twins. Proc Natl Acad Sci USA 102:10604–10609
54. Fraga MF, Ballestar E, Villar-Garea A, Boix-Chornet M, Espada J et al (2005) Loss of acetylation at Lys16 and trimethylation at Lys20 of histone H4 is a common hallmark of human cancer. Nat Genet 37:391–400
55. Frigola J, Song J, Stirzaker C, Hinshelwood RA, Peinado MA et al (2006) Epigenetic remodeling in colorectal cancer results in coordinate gene suppression across an entire chromosome band. Nat Genet 38:540–549
56. Frommer M, McDonald LE, Millar DS, Collis CM, Watt F et al (1992) A genomic sequencing protocol that yields a positive display of 5-methylcytosine residues in individual DNA strands. Proc Natl Acad Sci USA 89:1827–1831
57. Gama-Sosa MA, Wang RY, Kuo KC, Gehrke CW, Ehrlich M (1983) The 5-methylcytosine content of highly repeated sequences in human DNA. Nucleic Acids Res 11:3087–3095
58. Gama-Sosa MA, Slagel VA, Trewyn RW, Oxenhandler R, Kuo KC et al (1983) The 5-methylcytosine content of DNA from human tumors. Nucleic Acids Res 11:6883–6894

59. Garber K (2006) Momentum building for human epigenome project. J Natl Cancer Inst 98:84–86
60. Gartner K (1990) A third component causing random variability beside environment and genotype. A reason for the limited success of a 30 year long effort to standardize laboratory animals? Lab Anim 24:71–77
61. Gaudet F, Hodgson JG, Eden A, Jackson-Grusby L, Dausman J et al (2003) Induction of tumors in mice by genomic hypomethylation. Science 300:489–492
62. Gaudet F, Rideout WM 3rd, Meissner A, Dausman J, Leonhardt H et al (2004) Dnmt1 expression in pre- and postimplantation embryogenesis and the maintenance of IAP silencing. Mol Cell Biol 24:1640–1648
63. Goldberg AD, Allis CD, Bernstein E (2007) Epigenetics: a landscape takes shape. Cell 128:635–638
64. Gonzalo S, Jaco I, Fraga MF, Chen T, Li E et al (2006) DNA methyltransferases control telomere length and telomere recombination in mammalian cells. Nat Cell Biol 8:416–424
65. Grady WM, Willis J, Guilford PJ, Dunbier AK, Toro TT et al (2000) Methylation of the CDH1 promoter as the second genetic hit in hereditary diffuse gastric cancer. Nat Genet 26:16–17
66. Greger V, Passarge E, Höpping W, Messmer E, Horsthemke B (1989) Epigenetic changes may contribute to the formation and spontaneous regression of retinoblastoma. Hum Genet 83:155–158
67. Hansen RS, Wijmenga C, Luo P, Stanek AM, Canfield TK et al (1999) The DNMT3B DNA methyltransferase gene is mutated in the ICF immunodeficiency syndrome. Proc Natl Acad Sci USA 96:14412–14417
68. Harris LC, Remack JS, Brent TP (1994) In vitro methylation of the human O6-methylguanine-DNA methyltransferase promoter reduces transcription. Biochim Biophys Acta 1217:141–146
69. Hatada I, Hayashizaki Y, Hirotsune S, Komatsubara H, Mukai T (1991) A genomic scanning method for higher organisms using restriction sites as landmarks. Proc Natl Acad Sci USA 88:9523–9527
70. Hegi ME, Diserens AC, Gorlia T, Hamou MF, de Tribolet N et al (2005) MGMT gene silencing and benefit from temozolomide in glioblastoma. N Engl J Med 352:997–1003
71. Hellman A, Chess A (2007) Gene body-specific methylation on the active X chromosome. Science 315:1141–1143
72. Hemberger M, Dean W, Reik W (2009) Epigenetic dynamics of stem cells and cell lineage commitment: digging Waddington's canal. Nat Rev Mol Cell Biol 10:526–537
73. Herman JG, Graff JR, Myohanen S, Nelkin BD, Baylin SB (1996) Methylation-specific PCR: a novel PCR assay for methylation status of CpG islands. Proc Natl Acad Sci USA 93:9821–9826
74. Herman JG, Umar A, Polyak K, Graff JR, Ahuja N et al (1998) Incidence and functional consequences of hMLH1 promoter hypermethylation in colorectal carcinoma. Proc Natl Acad Sci USA 95:6870–6875
75. Higgins ME, Claremont M, Major JE, Sander C, Lash AE (2007) CancerGenes: a gene selection resource for cancer genome projects. Nucleic Acids Res 35:D721–D726
76. Hoffmann MJ, Schulz WA (2005) Causes and consequences of DNA hypomethylation in human cancer. Biochem Cell Biol 83:296–321
77. Holm TM, Jackson-Grusby L, Brambrink T, Yamada Y, Rideout WM 3rd et al (2005) Global loss of imprinting leads to widespread tumorigenesis in adult mice. Cancer Cell 8:275–285
78. Hong C, Bollen MIA, Costello JF (2003) The contribution of genetic and epigenetic mechanisms to gene silencing in oligodendrogliomas. Cancer Res 63:7600–7605
79. Hong C, Maunakea A, Jun P, Bollen AW, Hodgson JG et al (2005) Shared epigenetic mechanisms in human and mouse gliomas inactivate expression of the growth suppressor SLC5A8. Cancer Res 65:3617–3623
80. Hong C, Moorefield KS, Jun P, Aldape KD, Kharbanda S et al (2007) Epigenome scans and cancer genome sequencing converge on WNK2, a kinase-independent suppressor of cell growth. Proc Natl Acad Sci USA 104: 10974–10979
81. Hu M, Yao J, Cai L, Bachman KE, van den Brule F et al (2005) Distinct epigenetic changes in the stromal cells of breast cancers. Nat Genet 37:899–905
82. Huang TH, Laux DE, Hamlin BC, Tran P, Tran H et al (1997) Identification of DNA methylation markers for human breast carcinomas using the methylation-sensitive restriction fingerprinting technique. Cancer Res 57:1030–1034
83. Iacobuzio-Donahue CA, Maitra A, Olsen M, Lowe AW, van Heek NT et al (2003) Exploration of global gene expression patterns in pancreatic adenocarcinoma using cDNA microarrays. Am J Pathol 162:1151–1162
84. Ishkanian AS, Malloff CA, Watson SK, DeLeeuw RJ, Chi B et al (2004) A tiling resolution DNA microarray with complete coverage of the human genome. Nat Genet 36:299–303
85. Jeltsch A, Walter J, Reinhardt R, Platzer M (2006) German human methylome project started. Cancer Res 66:7378
86. Jia D, Jurkowska RZ, Zhang X, Jeltsch A, Cheng X (2007) Structure of Dnmt3a bound to Dnmt3L suggests a model for de novo DNA methylation. Nature 449:248–251
87. Jones PA (2005) Overview of cancer epigenetics. Semin Hematol 42:S3–S8
88. Jones PA, Baylin SB (2002) The fundamental role of epigenetic events in cancer. Nat Rev Genet 3:415–428
89. Jones PA, Baylin SB (2007) The Epigenomics of Cancer. Cell 128:683–692
90. Jones PA, Martienssen R (2005) A blueprint for a human epigenome project: the AACR human epigenome workshop. Cancer Res 65:11241–11246
91. Kane MF, Loda M, Gaida GM, Lipman J, Mishra R et al (1997) Methylation of the hMLH1 promoter correlates with lack of expression of hMLH1 in sporadic colon tumors and mismatch repair-defective human tumor cell lines. Cancer Res 57:808–811
92. Karpf AR, Jones DA (2002) Reactivating the expression of methylation silenced genes in human cancer. Oncogene 21:5496–5503
93. Karpf AR, Peterson PW, Rawlins JT, Dalley BK, Yang Q et al (1999) Inhibition of DNA methyltransferase stimulates the expression of signal transducer and activator of transcription 1, 2, and 3 genes in colon tumor cells. Proc Natl Acad Sci USA 96:14007–14012
94. Katzenellenbogen RA, Baylin SB, Herman JG (1999) Hypermethylation of the DAP-Kinase CpG island is a common alteration in B-cell malignancies. Blood 93:4347–4353

95. Kawai J, Hirotsune S, Hirose K, Fushiki S, Watanabe S et al (1993) Methylation Profiles of Genomic Dna of Mouse Developmental Brain Detected By Restriction Landmark Genomic Scanning (Rlgs) Method. Nucleic Acids Res 21:5604–5608
96. Kazazian HH, Moran JV (1998) The impact of L1 retrotransposons on the human genome. Nat Genet 19:19–24
97. Keshet I, Schlesinger Y, Farkash S, Rand E, Hecht M et al (2006) Evidence for an instructive mechanism of de novo methylation in cancer cells. Nat Genet 38:149–153
98. Khulan B, Thompson RF, Ye K, Fazzari MJ, Suzuki M et al (2006) Comparative isoschizomer profiling of cytosine methylation: the HELP assay. Genome Res 16:1046–1055
99. Kissil JL, Feinstein E, Cohen O, Jones PA, Tsai YC et al (1997) DAP-kinase loss of expression in various carcinoma and B-cell lymphoma cell lines: possible implications for role as tumor suppressor gene. Oncogene 15:403–407
100. Knudson AG (1971) Mutation and cancer: statistical study of retinoblastoma. Proc Natl Acad Sci USA 69:820–823
101. Kochanek S, Renz D, Doerfler W (1993) DNA methylation in the Alu sequences of diploid and haploid primary human cells. EMBO J 12:1141–1151
102. Kouzarides T (2007) Chromatin modifications and their function. Cell 128:693–705
103. Kouzarides T (2007) SnapShot: histone-modifying enzymes. Cell 128:802
104. Krassenstein R, Sauter E, Dulaimi E, Battagli C, Ehya H et al (2004) Detection of breast cancer in nipple aspirate fluid by CpG island hypermethylation. Clin Cancer Res 10:28–32
105. Kuromitsu J, Kataoka H, Yamashita H, Muramatsu M, Furuichi Y, Sekine T et al (1995) Reproducible alterations of DNA methylation at a specific population of CpG islands during blast formation of peripheral blood lymphocytes. DNA Res 2:263–267
106. Laird PW (2003) The power and the promise of DNA methylation markers. Nat Rev Cancer 3:253–266
107. Laird PW, Jackson-Grusby L, Fazeli A, Dickinson SL, Jung WE et al (1995) Suppression of intestinal neoplasia by DNA hypomethylation. Cell 81:197–205
108. Lee EB, Park TI, Park SH, Park JY (2003) Loss of heterozygosity on the long arm of chromosome 21 in non-small cell lung cancer. Ann Thorac Surg 75:1597–1600
109. Li E, Bestor TH, Jaenisch R (1992) Targeted Mutation of the Dna Methyltransferase Gene Results in Embryonic Lethality. Cell 69:915–926
110. Liang GN, Robertson KD, Talmadge C, Sumegi J, Jones PA (2000) The gene for a novel transmembrane protein containing epidermal growth factor and follistatin domains is frequently hypermethylated in human tumor cells. Cancer Res 60:4907–4912
111. Lindsay S, Bird AP (1987) Use of restriction enzymes to detect potential gene sequences in mammalian DNA. Nature 327:336–338
112. Lippman Z, Gendrel AV, Black M, Vaughn MW, Dedhia N et al (2004) Role of transposable elements in heterochromatin and epigenetic control. Nature 430:471–476
113. Lister R, O'Malley RC, Tonti-Filippini J, Gregory BD, Berry CC et al (2008) Highly integrated single-base resolution maps of the epigenome in Arabidopsis. Cell 133:523–536
114. Maraschio P, Zuffardi O, Dalla Fior T, Tiepolo L (1988) Immunodeficiency, centromeric heterochromatin instability of chromosomes 1, 9, and 16, and facial anomalies: the ICF syndrome. J Med Genet 25:173–180
115. Meissner A, Gnirke A, Bell GW, Ramsahoye B, Lander ES et al (2005) Reduced representation bisulfite sequencing for comparative high-resolution DNA methylation analysis. Nucleic Acids Res 33:5868–5877
116. Merlo A, Herman JG, Mao L, Lee DJ, Gabrielson E et al (1995) 5' CpG island methylation is associated with transcriptional silencing of the tumour suppressor p16/CDKN2/MTS1 in human cancers. Nat Med 1:686–692
117. Millar DS, Paul CL, Molloy PL, Clark SJ (2000) A distinct sequence (ATAAA)(n) separates methylated and unmethylated domains at the 5'-end of the GSTP1 CpG island. J Biol Chem 275:24893–24899
118. Misawa A, Inoue J, Sugino Y, Hosoi H, Sugimoto T et al (2005) Methylation-associated silencing of the nuclear receptor 1I2 gene in advanced-type neuroblastomas, identified by bacterial artificial chromosome array-based methylated CpG island amplification. Cancer Res 65:10233–10242
119. Morgan DK, Whitelaw E (2008) The case for transgenerational epigenetic inheritance in humans. Mamm Genome 19:394–397
120. Myöhänen SK, Baylin SB, Herman JG (1998) Hypermethylation can selectively silence individual p16ink4A alleles in neoplasia. Cancer Res 58:591–593
121. Novak P, Jensen T, Oshiro MM, Wozniak RJ, Nouzova M et al (2006) Epigenetic inactivation of the HOXA gene cluster in breast cancer. Cancer Res 66:10664–10670
122. Ohm JE, McGarvey KM, Yu X, Cheng L, Schuebel KE et al (2007) A stem cell-like chromatin pattern may predispose tumor suppressor genes to DNA hypermethylation and heritable silencing. Nat Genet 39:237–242
123. Okano M, Bell DW, Haber DA, Li E (1999) DNA methyltransferases Dnmt3a and Dnmt3b are essential for de novo methylation and mammalian development. Cell 99:247–257
124. Okano M, Takebayashi S, Okumura K, Li E (1999) Assignment of cytosine-5 DNA methyltransferases Dnmt3a and Dnmt3b to mouse chromosome bands 12A2–A3 and 2H1 by in situ hybridization. Cytogenet Cell Genet 86:333–334
125. Ooi SK, Qiu C, Bernstein E, Li K, Jia D et al (2007) DNMT3L connects unmethylated lysine 4 of histone H3 to de novo methylation of DNA. Nature 448:714–717
126. Plass C, Weichenhan D, Catanese J, Costello JF, Yu F et al (1997) An arrayed human not I-EcoRV boundary library as a tool for RLGS spot analysis. DNA Res 4:253–255
127. Pogribny IP, Basnakian AG, Miller BJ, Lopatina NG, Poirier LA et al (1995) Breaks in genomic DNA and within the p53 gene are associated with hypomethylation in livers of folate/methyl-deficient rats. Cancer Res 55:1894–1901
128. Pogribny IP, James SJ, Jernigan S, Pogribna M (2004) Genomic hypomethylation is specific for preneoplastic liver in folate/methyl deficient rats and does not occur in non-target tissues. Mutat Res 548:53–59
129. Qiu J (2006) Epigenetics: unfinished symphony. Nature 441:143–145
130. Rakyan VK, Hildmann T, Novik KL, Lewin J, Tost J et al (2004) DNA methylation profiling of the human major his-

tocompatibility complex: a pilot study for the human epigenome project. PLoS Biol 2:e405
131. Rauch T, Li H, Wu X, Pfeifer GP (2006) MIRA-Assisted Microarray Analysis, a New Technology for the Determination of DNA Methylation Patterns, Identifies Frequent Methylation of Homeodomain-Containing Genes in Lung Cancer Cells. Cancer Res 66:7939–7947
132. Raval A, Lucas DM, Matkovic JJ, Bennett KL, Liyanarachchi S et al (2005) TWIST2 demonstrates differential methylation in immunoglobulin variable heavy chain mutated and unmutated chronic lymphocytic leukemia. J Clin Oncol 23:3877–3885
133. Reilly KM, Broman KW, Bronson RT, Tsang S, Loisel DA et al (2006) An imprinted locus epistatically influences Nstr1 and Nstr2 to control resistance to nerve sheath tumors in a neurofibromatosis type 1 mouse model. Cancer Res 66:62–68
134. Rein T, DePamphilis ML, Zorbas H (1998) Identifying 5-methylcytosine and related modifications in DNA genomes. Nucleic Acids Res 26:2255–2264
135. Reynolds PA, Sigaroudinia M, Zardo G, Wilson MB, Benton GM et al (2006) Tumor suppressor p16INK4A regulates polycomb-mediated DNA hypermethylation in human mammary epithelial cells. J Biol Chem 281:24790–24802
136. Rollins RA, Haghighi F, Edwards JR, Das R, Zhang MQ et al (2006) Large-scale structure of genomic methylation patterns. Genome Res 16:157–163
137. Rush LJ, Plass C (2002) Restriction landmark genomic scanning for DNA methylation in cancer: past, present, and future applications. Anal Biochem 307:191–201
138. Rush LJ, Dai ZY, Smiraglia DJ, Gao X, Wright FA et al (2001) Novel methylation targets in de novo acute myeloid leukemia with prevalence of chromosome 11 loci. Blood 97:3226–3233
139. Sakai T, Toguchida J, Ohtani N, Yandell DW, Rapaport JM et al (1991) Allele-specific hypermethylation of the retinoblastoma tumor-suppressor gene. Am J Hum Genet 48:880–888
140. Sandovici I, Kassovska-Bratinova S, Loredo-Osti JC, Leppert M, Suarez A et al (2005) Interindividual variability and parent of origin DNA methylation differences at specific human Alu elements. Hum Mol Genet 14:2135–2143
141. Santos F, Hendrich B, Reik W, Dean W (2002) Dynamic reprogramming of DNA methylation in the early mouse embryo. Dev Biol 241:172–182
142. Sato N, Maitra A, Fukushima N, van Heek NT, Matsubayashi H et al (2003) Frequent hypomethylation of multiple genes overexpressed in pancreatic ductal adenocarcinoma. Cancer Res 63:4158–4166
143. Sato N, Fukushima N, Matsubayashi H, Goggins M (2004) Identification of maspin and S100P as novel hypomethylation targets in pancreatic cancer using global gene expression profiling. Oncogene 23:1531–1538
144. Schlesinger Y, Straussman R, Keshet I, Farkash S, Hecht M et al (2007) Polycomb-mediated methylation on Lys27 of histone H3 pre-marks genes for de novo methylation in cancer. Nat Genet 39:232–236
145. Schumacher A, Kapranov P, Kaminsky Z, Flanagan J, Assadzadeh A et al (2006) Microarray-based DNA methylation profiling: technology and applications. Nucleic Acids Res 34:528–542
146. Shames DS, Girard L, Gao B, Sato M, Lewis CM et al (2006) A genome-wide screen for promoter methylation in lung cancer identifies novel methylation markers for multiple malignancies. PLoS Med 3:e486
147. Shivapurkar N, Poirier LA (1983) Tissue levels of S-adenosylmethionine and S-adenosylhomocysteine in rats fed methyl-deficient, amino acid-defined diets for one to five weeks. Carcinogenesis 4:1051–1057
148. Shogren-Knaak M, Ishii H, Sun JM, Pazin MJ, Davie JR et al (2006) Histone H4-K16 acetylation controls chromatin structure and protein interactions. Science 311:844–847
149. Simon MD, Chu F, Racki LR, de la Cruz CC, Burlingame AL et al (2007) The site-specific installation of methyl-lysine analogs into recombinant histones. Cell 128:1003–1012
150. Sjoblom T, Jones S, Wood LD, Parsons DW, Lin J et al (2006) The consensus coding sequences of human breast and colorectal cancers. Science 314:268–274
151. Smiraglia DJ, Fruhwald MC, Costello JF, McCormick SP, Dai Z et al (1999) A new tool for the rapid cloning of amplified and hypermethylated human DNA sequences from restriction landmark genome scanning gels. Genomics 58:254–262
152. Smiraglia DJ, Rush LJ, Fruhwald MC, Dai ZY, Held WA et al (2001) Excessive CpG island hypermethylation in cancer cell lines versus primary human malignancies. Hum Mol Genet 10:1413–1419
153. Smith JF, Mahmood S, Song F, Morrow A, Smiraglia D et al (2007) Identification of DNA methylation in 3' genomic regions that are associated with upregulation of gene expression in colorectal cancer. Epigenetics 2:161–172
154. Smith JS, Costello JF (2006) A broad band of silence. Nat Genet 38:504–506
155. Smith LT, Lin M, Brena RM, Lang JC, Schuller DE et al (2006) Epigenetic regulation of the tumor suppressor gene TCF21 on 6q23–q24 in lung and head and neck cancer. Proc Natl Acad Sci USA 103:982–987
156. Song F, Smith JF, Kimura MT, Morrow AD, Matsuyama T et al (2005) Association of tissue-specific differentially methylated regions (TDMs) with differential gene expression. Proc Natl Acad Sci USA 102:3336–3341
157. Song JZ, Stirzaker C, Harrison J, Melki JR, Clark SJ (2002) Hypermethylation trigger of the glutathione-S-transferase gene (GSTP1) in prostate cancer cells. Oncogene 21:1048–1061
158. Stirzaker C, Millar DS, Paul CL, Warnecke PM, Harrison J et al (1997) Extensive DNA methylation spanning the Rb promoter in retinoblastoma tumors. Cancer Res 57:2229–2237
159. Suzuki H, Gabrielson E, Chen W, Anbazhagan R, van Engeland M et al (2002) A genomic screen for genes upregulated by demethylation and histone deacetylase inhibition in human colorectal cancer. Nat Genet 31:141–149
160. Takizawa T, Nakashima K, Namihira M, Ochiai W, Uemura A et al (2001) DNA methylation is a critical cell-intrinsic determinant of astrocyte differentiation in the fetal brain. Dev Cell 1:749–758
161. Teitz T, Wei T, Valentine MB, Vanin EF, Grenet J et al (2000) Caspase 8 is deleted or silenced preferentially in childhood neuroblastomas with amplification of MYCN. Nat Med 6:529–535

162. Toyota M, Ahuja N, Ohe-Toyota M, Herman JG, Baylin SB et al (1999) CpG island methylator phenotype in colorectal cancer. Proc Natl Acad Sci USA 96:8681–8686
163. Trasler JM, Trasler DG, Bestor TH, Li E, Ghibu F (1996) DNA methyltransferase in normal and Dnmtn/Dnmtn mouse embryos. Dev Dyn 206:239–247
164. Tremblay KD, Duran KL, Bartolomei MS (1997) A 5' 2-kilobase-pair region of the imprinted mouse H19 gene exhibits exclusive paternal methylation throughout development. Mol Cell Biol 17:4322–4329
165. Trinh BN, Long TI, Nickel AE, Shibata D, Laird PW (2002) DNA methyltransferase deficiency modifies cancer susceptibility in mice lacking DNA mismatch repair. Mol Cell Biol 22:2906–2917
166. Venter JC, Adams MD, Myers EW, Li PW, Mural RJ et al (2001) The sequence of the human genome. Science 291:1304–1351
167. Vire E, Brenner C, Deplus R, Blanchon L, Fraga M et al (2006) The Polycomb group protein EZH2 directly controls DNA methylation. Nature 439:871–874
168. Wainfan E, Poirier LA (1992) Methyl groups in carcinogenesis: effects on DNA methylation and gene expression. Cancer Res 52:2071s–2077s
169. Wales MM, Biel MA, Eldeiry W, Nelkin BD, Issa P et al (1995) P53 Activates Expression of Hic-1, a New Candidate Tumour Suppressor Gene On 17p13.3. Nat Med 1:570–577
170. Wang Y, Hayakawa J, Long F, Yu Q, Cho AH et al (2005) "Promoter array" studies identify cohorts of genes directly regulated by methylation, copy number change, or transcription factor binding in human cancer cells. Ann NY Acad Sci 1058:162–185
171. Waterland RA, Jirtle RL (2004) Early nutrition, epigenetic changes at transposons and imprinted genes, and enhanced susceptibility to adult chronic diseases. Nutrition 20:63–68
172. Weber M, Davies JJ, Wittig D, Oakeley EJ, Haase M et al (2005) Chromosome-wide and promoter-specific analyses identify sites of differential DNA methylation in normal and transformed human cells. Nat Genet 37:853–862
173. Weisenberger DJ, Siegmund KD, Campan M, Young J, Long TI et al (2006) CpG island methylator phenotype underlies sporadic microsatellite instability and is tightly associated with BRAF mutation in colorectal cancer. Nat Genet 38:787–793
174. Welch DR, Chen P, Miele ME, McGary CT, Bower JM et al (1994) Microcell-mediated transfer of chromosome 6 into metastatic human C8161 melanoma cells suppresses metastasis but does not inhibit tumorigenicity. Oncogene 9:255–262
175. Whitelaw NC, Whitelaw E (2006) How lifetimes shape epigenotype within and across generations. Hum Mol Genet 15(Spec No 2):R131–R137
176. Widschwendter M, Fiegl H, Egle D, Mueller-Holzner E, Spizzo G et al (2007) Epigenetic stem cell signature in cancer. Nat Genet 39:157–158
177. Wong AH, Gottesman II, Petronis A (2005) Phenotypic differences in genetically identical organisms: the epigenetic perspective. Hum Mol Genet 14(Spec No 1):R11–R18
178. Wood AJ, Oakey RJ (2006) Genomic imprinting in mammals: emerging themes and established theories. PLoS Genet 2:e147
179. Wu H, Chen Y, Liang J, Shi B, Wu G et al (2005) Hypomethylation-linked activation of PAX2 mediates tamoxifen-stimulated endometrial carcinogenesis. Nature 438:981–987
180. Xu GL, Bestor TH, Bourc'his D, Hsieh CL, Tommerup N et al (1999) Chromosome instability and immunodeficiency syndrome caused by mutations in a DNA methyltransferase gene. Nature 402:187–191
181. Yan PS, Chen CM, Shi HD, Rahmatpanah F, Wei SH et al (2001) Dissecting complex epigenetic alterations in breast cancer using CpG island microarrays. Cancer Res 61:8375–8380
182. Yan PS, Wei SH, Huang TH (2004) Methylation-specific oligonucleotide microarray. Methods Mol Biol 287:251–260
183. Yi P, Melnyk S, Pogribna M, Pogribny IP, Hine RJ et al (2000) Increase in plasma homocysteine associated with parallel increases in plasma S-adenosylhomocysteine and lymphocyte DNA hypomethylation. J Biol Chem 275:29318–29323
184. Yoshikawa H, de la Monte S, Nagai H, Wands JR, Matsubara K et al (1996) Chromosomal assignment of human genomic NotI restriction fragments in a two-dimensional electrophoresis profile. Genomics 31:28–35
185. Zardo G, Tiirikainen MIA, Hong C, Misra A, Feuerstein BG et al (2002) Integrated genomic and epigenomic analyses pinpoint biallelic gene inactivation in tumors. Nat Genet 32:453–458
186. Zhang X, Yazaki J, Sundaresan A, Cokus S, Chan SW et al (2006) Genome-wide high-resolution mapping and functional analysis of DNA methylation in arabidopsis. Cell 126:1189–1201
187. Zilberman D, Gehring M, Tran RK, Ballinger T, Henikoff S (2007) Genome-wide analysis of Arabidopsis thaliana DNA methylation uncovers an interdependence between methylation and transcription. Nat Genet 39:61–69

Population Genetic Principles and Human Populations

16

Emmanouil T. Dermitzakis

Abstract Human genetic variation has been at the core of recent interest for geneticists. The initial studies of human variation were done using mitochondrial DNA in a few individuals, but recent technologies produced large amounts of genetic data in large numbers of individuals. In this chapter we will discuss the nature and structure of genetic variation in humans, and the causes of this variation. We review some recent advances in the detection of human genetic variation, the caveats and issues when studying genetic data from human populations, and the prospects of applying such information for identification of functional genetic variants in the human genome.

Contents

16.1	Introduction	487
16.2	Processes Shaping Natural Variation	488
	16.2.1 Mutation and Polymorphism	488
	16.2.2 Historical Perspective of DNA Polymorphisms and Their Use as Genetic Markers	489
	16.2.3 Demography	490
	16.2.4 Recombination	492
	16.2.5 Natural Selection	492
16.3	Patterns of Genetic Variation	494
	16.3.1 Hardy–Weinberg Equilibrium	494
	16.3.2 Coalescent Theory	495
	16.3.3 Population Differentiation	495
	16.3.4 Patterns of Single-Nucleotide Variation in the Human Genome	496
	16.3.5 Patterns of Structural Variation in the Human Genome	498
	16.3.6 Haplotype Diversity—Linkage Disequilibrium	499
	16.3.7 Detecting Natural Selection	500
	16.3.8 Historical Perspective of Population Genetic Studies	501
16.4	Current Themes	501
	16.4.1 International Efforts to Detect and Describe Sequence Variation	501
	16.4.2 Prospects for Mapping Functional and Disease Variation	502
16.5	Summary	504
References		504

E.T. Dermitzakis (✉)
Department of Genetic Medicine and Development,
University of Geneva Medical School, 1 Rue Michel-Servet,
Geneva 1211, Switzerland
e-mail: emmanouil.dermitzakis@unige.ch

16.1 Introduction

In the early days of human genetics much of the research into levels and patterns of genetic variation in human populations was done using highly polymorphic genetic markers such as mitochondrial DNA [93] and microsatellites [101]. This was due to technological limitations, and only a few individuals could be interrogated; markers with high information content were necessary for the elucidation of recent human population history. With the completion of the human genome [59, 97] and the almost complete availability of annotated genes much of the focus has shifted towards the interrogation and characterization of complete genetic variation within human populations. Natural phenotypic variation can be measured at various levels and dimensions, such as variation in whole-organism phenotypic characteristics, levels of gene expression or protein expression, cellular response to stimuli, and more. However, the heritable component of phenotypic diversity in human populations is naturally based on the differences of the DNA sequence among individuals [12, 71]. Although we have a very good understanding of the four-letter code of DNA and now have advanced technologies to measure this genetic variation on a large scale, the properties and the fine structure of functional genetic variation as well as its causes are not clear to date.

The key forces that shape patterns of genetic variation are:

- *Mutation*: small and large changes of the nucleotide sequence.
- *Recombination*: exchange of genetic information between homologous chromosomes, mostly during meiosis.
- *Genetic drift*: the variance in sampling of genetic information from generation to generation that has particularly large effect in small populations.
- *Natural selection*: forces that increase or decrease the probability of inheritance of certain variants due to difference in fitness.
- *Migration*: exchange of genetic information between partly or fully isolated subpopulations.
- *Other demographic parameters*: structure of genetic variation in a population depends on a number of complex and frequently unknown demographic models such as inbreeding, assortative mating and others.

DNA sequence variation can come in various flavors and forms, from fine scale single nucleotide differences [12, 13, 71] to large-scale variations that affect millions of nucleotides at once [31]. There are, of course, many intermediate level variants in the human genome that can affect genome function also to various degrees of severity and form and under certain conditions. Another aspect of genomic variation is the degree to which populations share it. Some of this genetic variation is responsible for phenotypic variability present in all populations or phenotypic variability that is responsible for differences between populations [82]. The distribution of phenotypic variability in human populations and evolutionary properties of human genetic variation inform us about the history of the human species, the forces that have shaped its current phenotypic state of the human species and the basis for our differentiation from other closely related species.

Although genetic variation in human populations is an intriguing topic in itself and worthy of proper and in-depth interrogation from the basic science perspective, its contribution to disease, disease susceptibility or protection, and drug response has attracted a lot of the attention [2, 96]. In recent years, new technologies and methodologies have been developed specifically to explore the role of genetic variation to health and disease. The deep and proper understanding of the properties of human genetic variation is essential for the correct interpretation of the link between DNA sequence variation and disease [80, 81] and how this information can assist in the development of diagnostic tests as well as effective treatments.

Some of the key parameters to estimate in human populations are:

- Levels of nucleotide and structural variation (e.g., heterozygosity, density, and frequency of variants)
- Patterns and correlation of sequence variation (e.g., linkage disequilibrium)
- Degree of population subdivision (see also Chap. 20) and consequences for the flow of genetic variation
- Impact of natural selection on patterns of sequence variation

The aim of this chapter is to provide the framework and basic principles of human population genetics. Some of the issues raised here will be addressed in more detail in other chapters of this book. We will approach the topic from the points of view both of mutational processes and the forces that shape the history of such variation, such as natural selection and genetic drift, and of population structure and behavior. We will explore the range of sequence variants, the technological and statistical methodologies used to reveal its history and functional consequences, and the recent efforts to develop the infrastructure for the efficient exploration of large numbers of variants in large population samples.

16.2 Processes Shaping Natural Variation

16.2.1 Mutation and Polymorphism

Genetic variation in human as well as other species is primarily generated by mutational mechanisms that introduce new variants to the population. These variants can come in various forms from single nucleotide changes to multi-megabase-pair (Mbp) deletions, insertions, inversions, and translocations. The different types of variants also have different underlying mechanisms of mutation and therefore have different probabilities of occurring and also different regions in which they are more prevalent, and this depends on the DNA sequence context.

The most common and best-studied variants in populations are:

1. *Single nucleotide mutations (or point mutations)*: These are mutations that change the nucleotide base at a particular position in the genome without affecting the length of the chromosome (Fig. 16.1). The average mutation rate is 10^{-8} per generation per site. These kinds of mutations are categorized as transitions (purine to purine or pyrmidine to pyrimidine or A–G or C–T mutations) and transversions (all others). Owing to structural and molecular similarity, transitions are more likely to occur with approximately two-fold higher probability than transversions [73]. Other factors that have an effect on the rate of mutation are the local nucleotide composition, overall nucleotide compositions of the genome and other context-dependent effects (e.g., CpG sites). Models that aim to estimate substitution rates of sequences frequently attempt to estimate and account for these parameters [35]. In most cases there is an assumption of the infinite sites model, which theoretically states that every new mutation in the population will affect a new site not previously mutated [54]. This model becomes impractical when one looks at complete genome sequence data in large numbers of individuals.

2. *Small insertions and deletions*: These are mutations that contribute to the addition or deletion of small numbers of nucleotides to the sequence (Fig. 16.1). The mutation rates and patterns of variation of such mutations have been studied to some extent in inter-species comparisons, but they remain largely unexplored in terms of intrapopulation variability. Projects that currently explore complete sequencing of multiple human genomes, such as the 1,000 Genomes Project, will reveal many aspects of their pattern and potentially allow the proper estimation of their mutation rates.

3. *Variable number tandem repeats and short tandem repeats*: See below.

4. *Large structural variants (SVs) including copy number variants, inversions, and translocations*: These are large insertions or deletions, which for operational purposes are defined as larger than 1 kb or rearrangements and translocations of large segments of genomic DNA that alter the genomic landscape of a chromosome [31] (Fig. 16.2).

16.2.2 Historical Perspective of DNA Polymorphisms and Their Use as Genetic Markers

Historically there have been a number of different types of DNA markers used for this purpose. Various efforts [51, 58] showed that a set of several hundred such polymorphisms, distributed over the entire genome, would allow the mapping of genes on all chromosomes provided that a sufficient number of informative families were available. Over the years several such markers were used and are still being used in some settings:

1. Restriction length fragment polymorphisms (RFLPs). These polymorphisms were discovered first, and they are usually the result of single-base-pair changes in the DNA sequences inherited by Mendelian transmission. Such mutations lead either to removal or, less commonly, to the introduction of a recognition site for a restriction enzyme, causing increases or decreases in the length of restriction

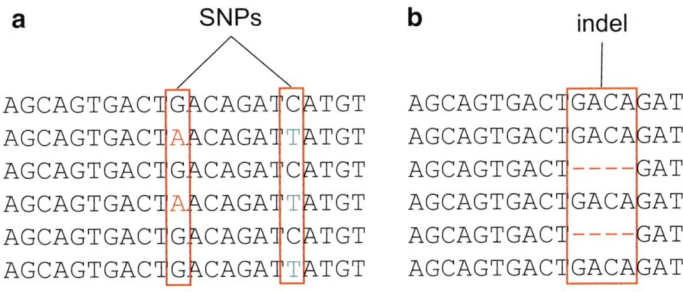

Fig. 16.1 Example of SNP and indel variations. On the *left panel* the two SNPs appear to be in perfect linkage disequilibrium with respect to historical recombination (D¢ =1), but the correlation coefficient *r*<1; there a large number of short-length differences between sequences denoted as insertions or deletions (indels). The *panel* on the *right* shows an example of small insertion/deletion (indel) variation

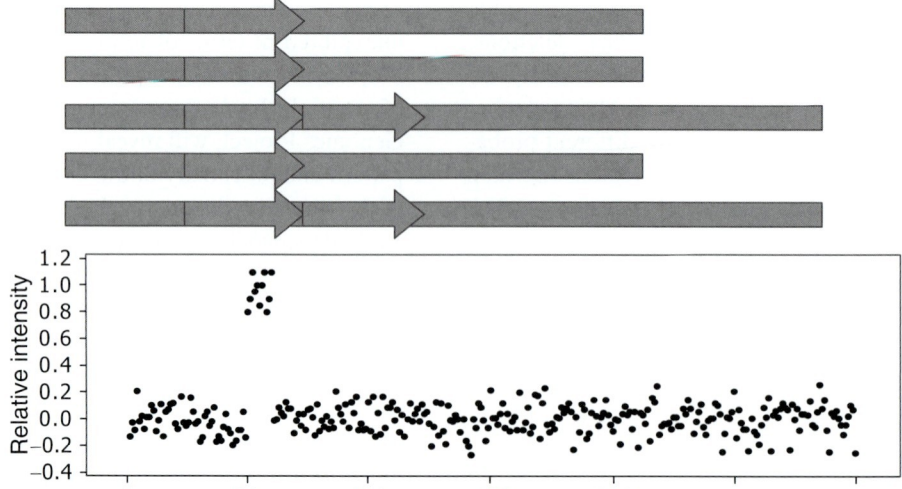

Fig. 16.2 Example of copy number variation. The *top panel* indicates the different haplotypes observed in the population with one or two copies of a piece of DNA. The *bottom panel* shows the hybridization-relative intensity in $\log_2$ units (Y-axis) for a sample homozygous for one copy vs a sample homozygous for two copies of the variant

fragments that have been previously amplified with PCR. These DNA variants are therefore referred to as restriction fragment-length polymorphisms. They are caused by a difference in the number of cleavage sites that are cut by a certain restriction endonuclease in different areas of the genome. They have previously served as proxies for single-nucleotide changes and in some sense represent the first genotyping assay of SNPs. Their use is not very frequent now, although they can still be used as a validation for the detection of a variant or for a high-quality assay in a medical setting.

2. Minisatellites (variable number of tandem repeats, VNTR). These were discovered by Jeffreys et al. [45–47] and are found very frequently in noncoding, repetitive areas of the genome. Sequences of about 9–60 base pairs are repeated in tandem order and may be present in a varying number of repeats per chromosome. One property, however, diminishes their utility for some applications: each of these polymorphisms has a relatively high mutation rate—a small percentage per generation. This is not entirely surprising. The similarity of base composition between multiple sequences creates ideal conditions for meiotic pairing of structure homologous but not position-homologous DNA segments, leading to unequal crossing over.

3. Microsatellites (short tandem repeats, STRs). These were discovered by Weber and May [101].

For example, pairs of two bases $(GT)_n$ may be repeated from a few to very many times. There may be up to about 30 different alleles in a population for one such polymorphism. The best method for their study is the PCR reaction. Two primers attached to the DNA on both sides of the polymorphism are required; this means that short base sequences outside the polymorphism must be known. Information on primers can now be gathered from a genome data base. In addition to the two-base-pair STRs, others comprising variable repeats of three to five base pairs have been described. The common STR markers have become the DNA variants of choice in the study of human and mammalian linkage, as well as in the early of human population studies [94]. The availability of the human genome sequence has made the identification and use of such sequences very common and a standard process.

16.2.3 Demography

16.2.3.1 Effective Population Size

An important parameter in population genetics is the effective population size, usually symbolized as N_e. This does not refer to the actual size of a population under study. The effective population size is: "[T]he number of breeding individuals in an idealized population that would show the same amount of dispersion of allele frequencies under random genetic drift or the same amount of inbreeding as the population under consideration" [104]. In other words, the effective population size represents the idealized population size that has the same genetic parameters as the observed population. The effective population size is usually lower than the observed size. The discrepancy between the effective and observed population size is due to many demographic phenomena: population expansion, bottlenecks (periods of radical reduction of population size), natural selection, and others.

16.2.3.2 Genetic Drift and Isolated Populations

There are specific assumptions and expectations behind the dynamics of polymorphisms in populations. If one considers a population of N diploid individuals (i.e., $2N$ chromosomes) in a population in which most theoretical assumptions above are met, then a new mutation in the population will have a frequency of $1/2N$. Assuming no selection in favor of or against such alleles, most of the mutations will oscillate around values close to their initial frequency ($1/2N$) and eventually disappear (go to zero frequency) from the population. Occasionally, such variants can increase in frequency and eventually either disappear or go to fixation (go to 100% frequency) in the population (Fig. 16.3). It can be shown that under assumptions of neutrality and random mating the probability of fixation of a new mutation is its initial frequency $1/2N$ [36]. In fact, it can be shown that at all times the frequency of a variant at a given time point is its probability of fixation. In populations with large size a lot of mutations are introduced in each generation (owing to the large number of chromosomes available to accumulate mutations), but each one of them has a low probability of fixation owing to a very low initial frequency. On the other hand, a population with small sample size can introduce a small number of mutations, but each one of them has a high probability of fixation owing to its high initial frequency. However, if one accounts for both mutation rate and probability of fixation after the mutation has occurred (and assuming no selection) the probability and rate of fixations are the same in small and large populations. This overall

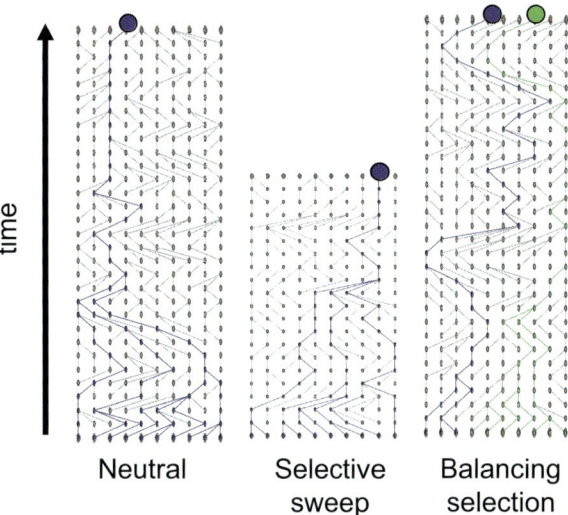

Fig. 16.3 Three models and their impact on coalescent tree structure. The *left panel* is the neutral pattern where the MRCA is the *blue dot* and the *lines* indicate transmission of the allele from one generation to the other. In the *middle panel* we show an example of a selective sweep scenario where the MRCA is more recent than in the neutral model (*blue dots* and *lines*). In the *right panel* we show an example of balancing selection, where even at the same number of generations as in the neutral model two major alleles (*blue* and *green dots* and *lines*) are segregating in the population, suggesting the MRCA is older than in the neutral model

phenomenon is essentially a result of the sampling variance effect from generation to generation and is referred to as "genetic drift."

Genetic drift is an important consideration in human populations because of the nature of human population dynamics. In particular, this is has become relevant in the studies of isolated and small founding populations such as those in Iceland and Finland. The fact that such populations have small effective population sizes (a metric that also represents the degree of genetic variation segregating in the population) makes it more likely that variants that are otherwise rare in the parent population can become quite common by chance, even in the presence of selection acting against them, making these population very useful for the mapping of common and rare disease variants [37, 72].

16.2.3.3 Migration

The segregation and transmission of genetic information does not only occur vertically from generation to generation, but also horizontally between population subgroups that are partly isolated. When the exchange of genetic information between subpopulations is completely free the two subpopulations will eventually become a single panmictic population within a few generations. However, in most cases the exchange of genetic information between subpopulations happens with low rates of migration of individuals either in one direction or in a bidirectional way. The rate by which new mutations or old variants brought to high frequency or fixation by way of either genetic drift or natural selection combined with the rate of migration between subpopulations determines the degree of differentiation between subpopulations after a number of generations.

16.2.3.4 Inbreeding/Nonrandom Mating

Even if individuals live in the same geographic region, the size of the population and also behavioral processes and choices result in distinct patterns of variation. When the population size is small or individuals tend to mate with related individuals more than reflected by the average pairwise degree of relatedness observed, then this leads to increased levels of homozygosity. The degree of this effect can be described by the inbreeding coefficient.

A more generalized pattern of nonrandom mating involves choices that individuals make that are either based on phenotypic similarities between the pair of individuals (assortative mating) or even more general choices based on phenotypic characteristics, such as height, eye color and others. If these phenotypic characteristics have a genetic basis this can lead to deviation from patterns of genetic variation beyond that caused by genetic drift.

16.2.4 Recombination

Another key property of the genome that was mentioned above is the landscape of recombination. In the past, recombination was considered a property of the genome that varied between large genomic regions but was pretty uniform on the fine scale. Pedigree studies have revealed the degree of variation in recombination rates between large genomic regions, and these were largely correlated with properties of genomic sequence, such as GC content, location relative to the centromere, and others [9]. Such recombination rates were based on the observation of recombination events from one generation to the next. There have also been fine-scale experimental studies on recombination, and those have hinted that recombination rates do not only vary across large genomic regions but also on a finer scale [48, 49]. These studies have observed recombination events by sperm typing and have shown that there exist hotspots of recombination within short multi-kilobase-pair regions of the genome. More recently, inferential methodologies, using dense population variation data and appropriate modeling under the coalescence model (see below), have revealed that recombination is distributed in a highly nonuniform way, with 80% occurring in hotspots of recombination, which comprise only a small fraction (about 20%) of the genome [60, 66, 69] (Fig. 16.4). This observation has become common knowledge, changing the way we view genomic variation and correlations between variants. As discussed above, the assumption that recombination is uniform in genomic regions is violated and increases the expected variance of statistics being used for population genetics inference.

16.2.5 Natural Selection

All the above considerations about human populations were not only necessary to understand the specific characteristics of human populations, but also to introduce the reader to some basic principles useful to the interpretation of patterns of sequence variation. The neutral theory of molecular evolution [53] had been proposed as the main process by which DNA sequences evolve. However, it is widely accepted that for a fraction of a genome one of the main forces that influence patterns and levels of sequence variation is natural selection. Natural selection makes some variants in the population have higher or lower probability of fixation relative to what could be estimated from their population frequency [3].

There are three main forms of selection with many versions and alternatives, but we will focus here only on the main points. We will try to avoid detailed statements and we advise the reader to consult the primary literature, since this is an ever-developing area (see Fig. 16.5). The first and most intuitive form of natural selection is purifying selection. Purifying selection

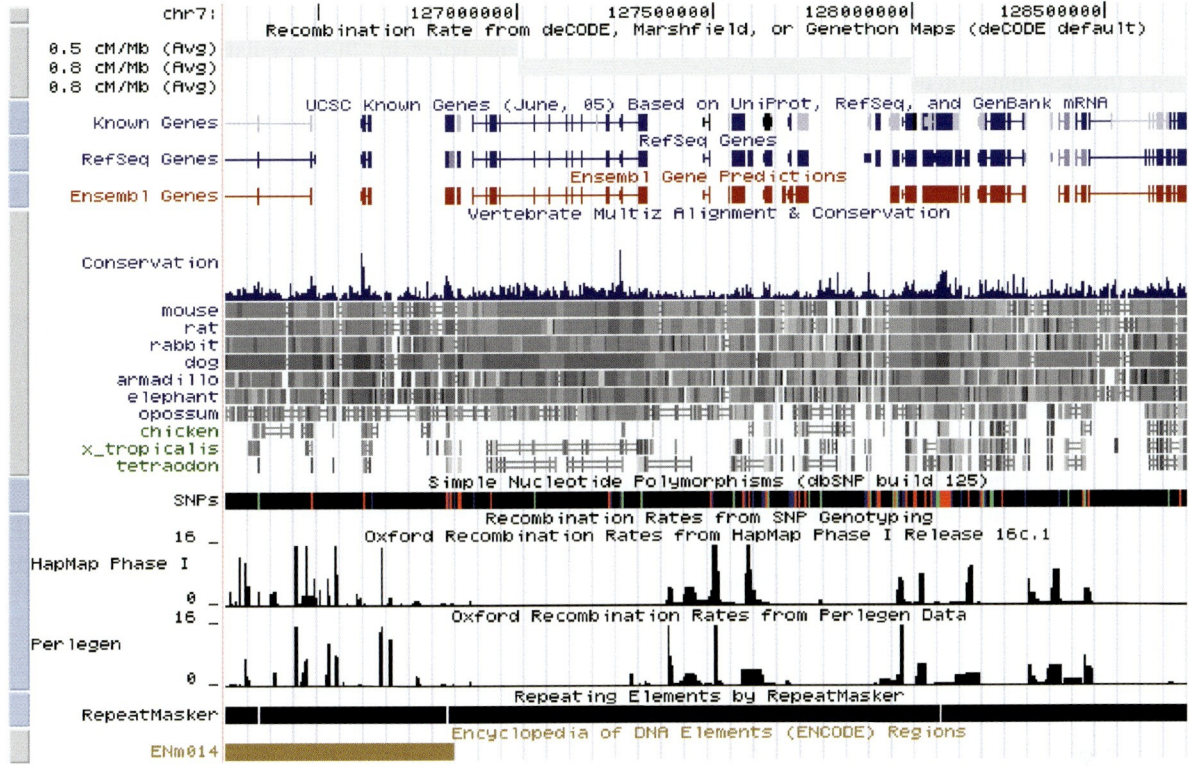

Fig. 16.4 Rates of recombination along the genome. A UCSC genome browser (genome.ucsc.edu) shot in which many genomic elements are shown. At the *bottom* of the *panel* two tracks with population-based estimated recombination rates are shown, based on HapMap SNP data and Perlegen SNP data

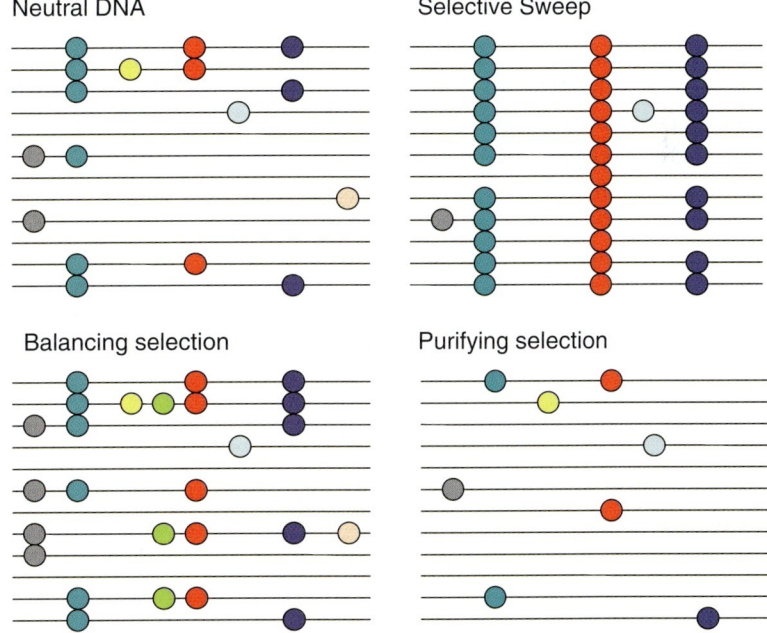

Fig. 16.5 Schematic of patterns of SNP variation. *Colored dots* indicate the new (derived) allele in the population in regions (each *color* indicates an independent mutation) undergoing neutral evolution, positive selection (selective sweep), balancing selection, and purifying selection

(or background selection) is based on the principle that when a new mutation occurs in a functional sequence, and if this new mutation disrupts the pre-existing function, then the individual carrying the new mutation may be at a disadvantage. Mutations that undergo purifying selection tend to disappear from the population quite rapidly. However, depending on whether the effect is co-dominant (or dominant in some cases) or recessive, this "waiting" time can be very short or long. Co-dominant mutations will directly confer a disadvantage to the individual even when in heterozygosity, so that they will disappear faster. Recessive mutations will only start being impacted by purifying selection when in homozygosity, which means that the mutation has to reach appreciable frequency to have a decent probability of being in homozygosity in individuals. They will likely have a long asymptotic process of elimination similar to that of neutral mutations, because of the normal phenotype in heterozygous individuals. Regions that are under purifying selection have an excess of rare variants in their site frequency spectrum relative to neutrally evolving sequences, mainly resulting from the suppression of new mutations reaching high frequency [14, 15].

It is possible that some of the mutations occurring mostly in functional sequences confer an advantage on the individual carrying them. This mode of selection is called positive selection and can occur either if a new mutation is advantageous or if an extant mutation becomes advantageous following changes in the environment. The effect of such an advantageous mutation is that it goes to fixation rapidly and can drag many other new, mostly neutral, closely linked mutations to fixation or near-fixation as well [4, 27–29]. This process is called "selective sweep," [91] and the allele frequency spectrum after a sweep is characterized by an excess of high-frequency derived alleles as a result of new rare variants in the vicinity of the positively selected variant reaching high frequency [26]. Patterns can become quite complex and less interpretable when multiple variants in the same genomic region undergo selective sweep at the same time or within short evolutionary time periods.

Finally, another mode of selection is balancing selection, where a region maintains high levels of variation and a large number of variants at intermediate frequencies. This type of selection is caused by conditional effects of selection of alleles to the frequency of other alleles. The presence of antagonizing advantages between alleles leads to an optimal equilibrium point of frequencies in the population. Examples are heterozygote advantage or frequency-dependent selection. A very common example of heterozygote advantage are the mutations of sickle-cell anemia and the thalassemias [103]. These were kept at high frequency because, although disadvantageous in homozygosity, they made their heterozygote carriers more fit than the wild type homozygotes, because they conferred protection from malaria. Balancing selection is a complex type of selection, and one of the characteristic patterns is that there is an excess of intermediate frequency variants and the tree of the population for such regions has very deep branches (see coalescent theory below), representing very old variation. Methodologies used for detecting possible natural selection acting on certain regions of the genome will be discussed below.

16.3 Patterns of Genetic Variation

16.3.1 Hardy–Weinberg Equilibrium

One of the key principles of population genetics is the Hardy–Weinberg equilibrium (HWE). If one considers a locus with two alleles, A1 and A2, then the frequency of allele A1 can be denoted as p and that of allele A2 as q. If one assumes random mating and no selection acting on the alleles (which in most cases is an approximation), the probability of observing an individual will genotype AA is the product of the frequency of allele A or p^2 and that of aa is q^2. The frequency of the heterozygote individuals Aa is $2pq$ (which stems from the fact that there are two ways to sample a heterozygote, Aa and aA). Of course, owing to finite population size this expectation is never realized exactly and there is always some variance around it. In order to test for deviations from HWE, the standard method is to perform a Chi-square test to compare the frequencies of observed vs expected counts of the three genotypes. Deviations from HWE can exist for many reasons, including population subdivision, recent migration at a high rate, strong natural selection, and others.

Under HWE, the genotypic frequencies of the three different genotypes are described by the equations below:

$$P_{AA} = p^2 \qquad (16.1)$$

$$P_{Aa} = 2pq \tag{16.2}$$

$$P_{aa} = q^2 \tag{16.3}$$

where p is the allele frequency of A and q is the allele frequency of a and $p+q=1$.

16.3.2 Coalescent Theory

In most studies we sample a set of individuals from a population with the main goal of inferring the properties and population genetic parameters of the whole population from this sample under study, with certain assumptions. The history of a set of $2N$ chromosomes that we sampled from the population can be modeled within the coalescent framework that predicts that all these chromosomes have a common ancestor at some point in the past. This is called the most recent common ancestor (MRCA), and the tree leading back to MRCA represents the likely relationships of sequences observed at present into the past (Fig. 16.3). This was first described and formalized by Kingman [55] and was further developed by Hudson [38]. The coalescent theory states that the sample of chromosomes can be viewed as a phylogenetic tree of sequences from the same species, and the distribution of branch lengths of the tree can be estimated by the expected model by which the population has evolved. Current methodologies can accommodate recombination between the sequences, selection, various population events such as bottlenecks, or population expansions as well as gene conversion.

Coalescent models have become popular in the last two decades to describe population genetic processes and infer parameters. There are two key applications of coalescent models. One of them is to facilitate the interpretation of potential processes that have contributed to an observed pattern of genetic variation in a population sample. In such cases, thousands of coalescent simulations are run using parameters estimated from the sample. Different processes that could have led to the observed pattern are tested and followed by an assessment of which of the models tested fits best with the observed pattern. The coalescent model can also be used to estimate certain parameters from the population sample, such as population recombination rates (see below). Overall, the coalescent model has been established as one of the most reliable ways to model the relationships of sampled sequences from a population.

Coalescent models are also being used to look for the impact of natural selection. The pattern of branch lengths is correlated with the type of selection that has acted on the sequence. Let us consider a sample of chromosomes and a neutral coalescent tree for these chromosomes. Under a scenario of purifying selection (or background selection) the coalescent tree of an allele will on average have shorter branches, since many branches will have been lost as a result of deleterious mutations. With a scenario of positive selection, sampled chromosomes will converge to a MRCA faster, since the extant sample will represent a chromosome that was rapidly fixed. Finally, in the case of balancing selection the MRCA will converge much further back in the past owing to the maintenance of old polymorphism at intermediate frequency (see Fig. 16.3, and [3] for a review).

16.3.3 Population Differentiation

There are currently more than 6 billion people on the planet, but the amount of genetic variation we carry as a species is much lower than would be expected from a panmictic population of 6 billion people in equilibrium. The human population has undergone a radical expansion very recently (approximately the last 100,000 years), and this has not allowed for genetic variation to catch up. For this reason, in humans the effective population size has been estimated to be approximately 10,000, a number very small relative to the observed population size [95].

In addition to the small effective population size, human populations have undergone bottlenecks; they have been subdivided for long periods of time; and they have undergone admixture afterwards. This has created a very complex pattern of variation and patterns of correlations of variants in the genome (see Sect. 16.3.6, below) that appear to be unique and potentially a result of natural selection. However, these are artifacts of population structure and can be mistakenly inferred as natural selection. The main point is that the history of human populations has violated most of the standard assumptions made when we apply standard population genetic models, so that great caution is needed when one applies those models in human population data [62].

Methodologies have been developed to deal with the degree of population differentiation and substructure to facilitate genome-wide association studies. One of them uses a two-step scheme to correct for population structure in case-control studies by detecting associations that are significant within population subgroups detected. This methodology is implemented in the softwares STRAT and STRUCTURE [76–78]. Other methodologies estimate the degree of inflation of association P-values attributable to population substructure and apply a genome-wide correction to the distribution of P-values, otherwise known as genomic control [23, 24]. Finally, recent methods make use of the idea of describing genetic data with principal component analysis (Fig. 16.6), as was originally proposed by Menozzi et al. [67], with some modifications, and they use the pattern to correct for population substructure. This methodology has been implemented in the tool EIGENSTRAT [74, 75]. Although these methods can be very useful, such procedures may lead to loss of power, which results in population genetic inferences being more difficult and less accurate in human populations. The only thing that compensates for this is that there is a lot more data generated for human populations, usually in the thousands of individuals, so power and accuracy are recovered from the large amounts of data we use for the estimates.

16.3.4 Patterns of Single-Nucleotide Variation in the Human Genome

The complexity of the issues in population genetics becomes higher once we start evaluating multiple variable positions, especially if they are physically linked. One of the predominant types of sequence variation in the human genome is the large number of single-nucleotide polymorphisms (SNPs), also found in the literature as "segregating sites" or "polymorphic sites." These are single-base differences (Fig. 16.1) between chromosomes and can exist in various frequencies in human populations. According to the infinite sites model [54], each of these SNPs has only two alleles (two alternative nucleotides). Although this is largely the case, it is an approximation and obviously a function of how deep our interrogation of sequence variation is. The availability of nearly complete sequencing of hundreds of individuals from the 1,000 Genomes Project and large-scale sequencing from other projects allows detection of sites for which three alleles segregate in human populations. Conventionally, SNPs are the polymorphisms whose minor allele is present in at least 1% of the chromosomes of a given population. Most of population genetic theory has been developed around the pattern, density, and frequency distribution of SNPs. The average number of differences one can expect when comparing two homologous human chromosomes has been estimated at around 8–10 per 10,000 bps, with these levels being lower in functionally constrained sequences and higher in hypermutable regions (e.g., CpG sites). One can derive many of the historical properties of a given genomic region using the characteristics of SNP data within that region.

There are a number of methods that are being used to assess the degree of nucleotide diversity in sequences. Some of these metrics rely on the number of observed variable (segregating) sites in the sample of haplotypes

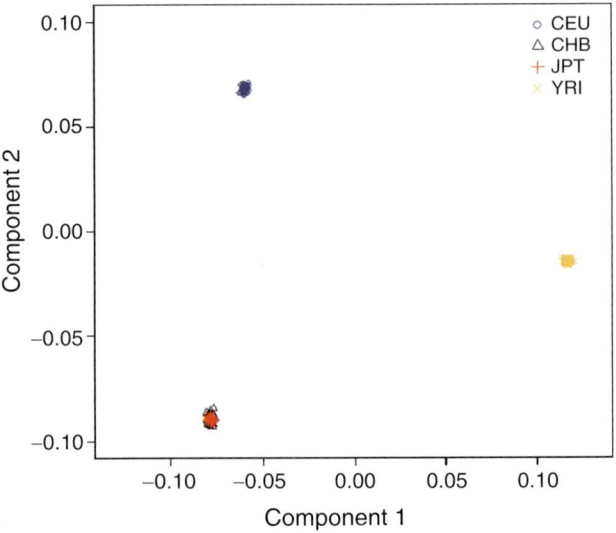

Fig. 16.6 Principal component analysis of SNP data among population samples. This figure shows the multidimensional scaling of SNP data for the four Hapmap populations for the first two principal components using EIGENSTRAT. The figure shows that the Caucasian (*CEU*), African (*YRI*), and Asian (*CHB* and *JPT*) samples cluster away from each other. The differences between CHB and JPT are too small to be detectable in the scale of the first two principal components

from the population and derive assumptions about the expected diversity assuming no selection. One of them is Waterson's q [100], estimated by the following equation:

$$\theta = \frac{S}{\sum_{n=1}^{i=1} 1-i} \quad (16.4)$$

where S is the number of variable sites and i is the number of chromosomes compared.

Other metrics directly count the number of mismatches between any two haplotypes and derive the average expected heterozygosity per nucleotide [70].

$$\pi = \sum_{ij} x_i x_j \pi i = \sum_{i=1}^{n} \sum_{j=1}^{i} x_i x_j \pi ij \quad (16.5)$$

where x_i and x_j are the respective frequencies of the i-th and j-th sequences, pij is the number of nucleotide differences per nucleotide site between the i-th and j-th sequences, and n is the number of sequences in the sample.

In fact, the difference in inference between these two metrics was used by Tajima to detect patterns of variation that deviate from neutral patterns of natural selection with the statistic D [92].

$$D = \frac{\theta - \pi}{Var} \quad (16.6)$$

However, one has to be very cautious about such deviations from expected patterns with metrics using nucleotide diversity inferences in order to detect selection. Such metrics are very sensitive to small deviations in expectations about the population structure, such as random mating and also levels and uniformity of recombination [87], and this leads to false interpretations.

Before we go deeper into the interpretation of signals derived from SNP data it is essential to discuss some basic assumptions that population genetic theory makes. When analyzing data of this type there are caveats one needs to be aware of when using "out-of-the-box" methodologies developed under theoretical assumptions in human data. Many of the standard population genetics models assume panmixia (equal probability of each individual to mate with any other individual), constant population size across generations, nonoverlapping generations, and uniform recombination rates across a region or no recombination at all. Given these assumptions one can contrast patterns of variation expected under a certain set of predefined parameters with those observed in real data and infer realistic population genetic parameters. As discussed previously and below, almost none of the above assumptions are true for human populations. Therefore it is essential to implement more sophisticated models that allow for these parameters to vary in order to interpret human variation data. In this chapter we do not attempt to explore the "deep" statistical modeling necessary for such analysis, but to give pointers to the caveats and suggest methodologies that will either bypass such effects or take them into account.

The frequency distribution of SNPs, otherwise called the site frequency spectrum, is usually informative of the forces or processes that have acted on a given genomic region. In general, it is expected that when a sequence evolves under a neutral model there is always a large number of rare SNPs. This is mainly due to the large number of variants that are new in the population, some of which are single observations (singletons) on individual chromosomes. The degree of this excess relative to the neutral expectation may indicate forces that have distorted the patterns of variation, such as natural selection or population bottlenecks. These issues have been discussed earlier in this chapter.

All the above discussion was based on the assumption that we have full sequencing data for a number of individuals for a given genomic region. However, this has not been possible until recently for human populations. Sequencing was very expensive and only recently new technologies (also called second-generation sequencing technologies) have provided the appropriate framework for rapid and low-cost resequencing of large genomic regions or full genomes [6, 100]. Much of the data and inferences about population-genetic parameters had to rely on genotyping data, i.e., the interrogation of the allelic state of known polymorphic position in the human genome. Why is genotyping not as good as sequencing? The main reason is that in order for a genotyping assay to be designed one needs to know that a nucleotide site is variable in at least one population sample. This means that variation is interrogated and inferred in a small number of individuals and then assayed in a larger number. This biases the frequency spectrum to common variants, since we only know about the rare variants of the individuals who were sequenced and are missing all the rare and singleton variants of the individuals who were just genotyped and not sequenced. This bias can become even worse if the ascertainment of variable sites is not uniform and is highly variable across regions. This was the case in the HapMap project discussed below [41–43]. Methodologies have been developed that can correct for the ascertainment bias if the ascertainment scheme is known [17]. However, these problems and the correction required highlight the fact that we urgently need resequencing data in order to elucidate some of the historical parameters of human population variation, and until we do this many of the signals will be confounded by ascertainment biases.

16.3.5 Patterns of Structural Variation in the Human Genome

Until a few years ago, much of the attention paid to human variation had been devoted to single nucleotide variation data, assuming that each haploid human genome has the same genome size and internal organization (i.e., gene order and number). However, if we want to get a full and complete picture of variants in the human genome that may have functional effects, we should not ignore insertion/deletion (indel) polymorphisms (Fig. 16.1). In this section, and for purely operational reasons, we mean the indels that are small (from one to a few tens of base pairs) and can be identified by simply comparing sequencing reads. Other, large-scale, variants are discussed below. There is a small number of studies that have carefully characterized the patterns and distribution of indels in the human genome. The few that have performed such analysis have shown that indels behave in a similar way to SNPs, with their density being lower in functional portions of the genome. Studies that have used interspecific distribution of indels have shown that one can detect strong signals of selective constraint [61], but such studies have not been applied in human variation on a large scale.

In the last 4–5 years it has become obvious that a newly rediscovered type of variation is also very common in human populations and is likely to contribute to phenotypic diversity and disease. These are the copy number variants (CNVs) and other structural variants (SVs, e.g., inversions, translocations), which are currently attracting a lot of attention [18, 40, 64, 84]. Once again, for operational reasons CNVs are defined as large (more than 1 Kb) [31] regions of the genome that are present in more or less than the expected copies in the genome (more or less than the two copies expected in a diploid individual), and this copy number is variable among individuals (Fig. 16.2). The mutational nature and population genetic properties of CNVs has not yet been fully elucidated, and their contribution to complex disease is still largely unexplored, but the examples we have in monogenic disorders suggest that such variants will be important for the understanding of human diversity and it is therefore worth exploring in human populations. Some initial studies have suggested that CNVs may be single mutations and therefore tractable by LD with SNPs in human populations [64], but some other CNVs are more complex in nature and likely a result of recurrent mutations, and these are unlikely to be easily tractable by LD [79]. CNVs and other structural variants also have consequences for the pattern of SNP variation in the human genome. Inversion and translocations lead to suppression of recombination and create artifacts of increased correlation structure in the human genome. Similarly, CNVs can create distorted patterns of SNP variation owing to naïve assumptions about the genotyping assays.

Two major recent studies have looked closely at patterns of copy number as well as some degree of other structural variation in human populations. The two approaches employed are quite different in resolution and confidence in inference and provide both overlapping and complementary results. The Genome Structural Variation Consortium initially employed a BAC CGH (comparative genome hybridization) array approach by which they assayed 270 individuals from the HapMap project in order to detect regions of the genome that are variable in copy number [79]. They detected more than 1,200 CNV regions in these populations. These were large regions of the genome (owing to resolution of the array), and they turned out to be a result of structural changes in dynamic regions of the genome such as segmental duplications. Another approach employed was that of sequencing both ends of medium-sized clones (fosmids) and mapping them back to the genome [52]. Using the size distribution that these clones were selected for, they were able to find regions of the genome in the eight assayed individuals that are either larger or smaller than the reference and therefore may contain insertions or deletions. In addition, by making use of the orientation of the sequencing reads they were able to assign potential inversions and translocations. A similar approach has been applied using second-generation sequencing technologies [56]. These and other studies have made it very clear that the study of structural variants in the genome is not going to be as straightforward as studies of SNP variation. Nevertheless, there is substantial evidence that CNVs and other structural variants contribute to common diseases from HIV resistance [34] to psychiatric disorders [19, 85, 99], and the drive to properly interrogate the degree and amount of copy number variation in human populations has been strong.

16.3.6 Haplotype Diversity—Linkage Disequilibrium

Variants in genomic regions of the human genome are correlated, because of common history and lack of historical recombination between them. Mutations land on a given haplotype and within a given framework of variants, and unless recombination is given enough time to shuffle them they co-segregate from generation to generation. The correlation of variants is called linkage disequilibrium (LD), to indicate the nonrandomness of alleles linked on the physical space of the chromosome (Fig. 16.7).

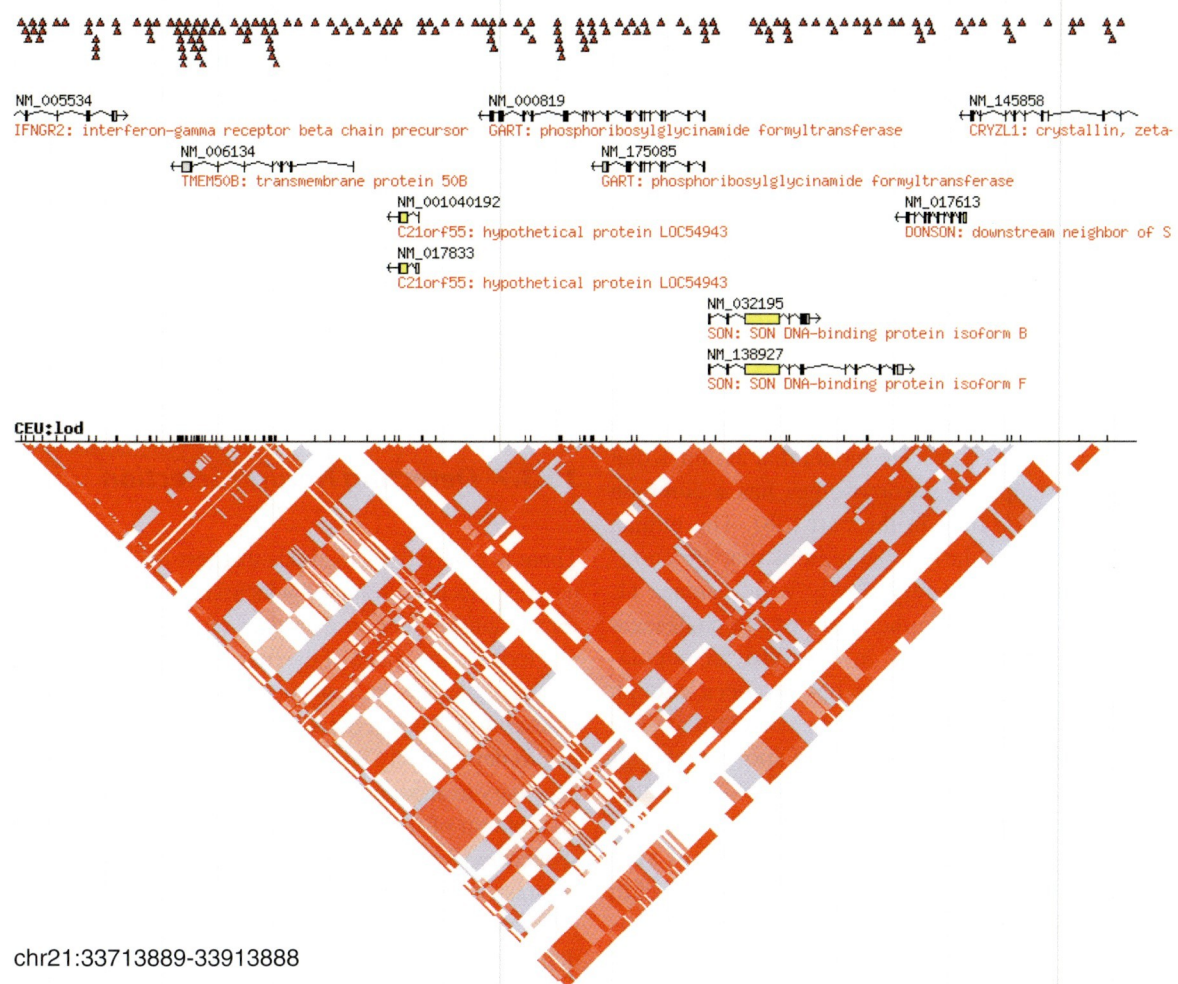

Fig. 16.7 Linkage disequilibrium plots in the context of annotation from the HapMap website browser (www.hapmap.org). The figure shows the annotation and SNP density for a short region of human chromosome 21. The *large inverted triangle below* indicates the pairwise linkage disequilibrium between SNPs. *Strong red* indicates significant linkage disequilibrium, *light red* LD of intermediate significance and *light gray* and *white*, low LD

There are various metrics for pairwise LD that have different properties, such as D, D', and r. D is an absolute measure of LD, D' is a measure that corrects for the frequencies of the variants considered, and r is the statistical correlation between the variants.

$$D = P_{AB} = p_A * p_B \quad (16.7)$$

where p_A, q_a, p_B and q_b are the allele frequencies of alleles and A, a, B and b and P_{AB} are the frequencies of AB haplotypes.

D_{max} is the maximum D value two variants can obtain if there has been no historical recombination between them.

$$D' = \frac{D}{D_{max}} \quad (16.8)$$

$$r = \frac{D}{\sqrt{p_A p_B q_a q_b}} \quad (16.9)$$

where p_A, q_a, p_B and q_b are the allele frequencies of alleles, A, a, B and b.

As discussed in a previous section, we tend to consider the history of human chromosomes in the framework of the coalescent model that assumes that any extant sample of chromosomes has a common ancestor at some time in the past [38]. The human genome has particularly large regions that are correlated, and LD decays very slowly with distance [1, 20, 21]. This is because the depth of the coalescent tree is shallow, especially in non-African populations, which means that there has not been enough time for recombination to break the correlations. The combination of shallow history and spotty recombination (owing to recombination hotspots) has created a pattern of blocky LD in the human genome where variants in large segments of the genome show very high LD within the segment but very low LD with neighboring regions [33]. This blocky structure has led to some ideas in the past that the human genome is divided into "haplotype blocks" that are discrete genomic segments. Although the general principle that the genome is blocky with respect to LD structure is true, it has now become obvious, especially since publication of the HapMap project, that the correlation structure of variants in the genome is much more complex and is not simply defined by physical boundaries on chromosomes [42, 43]. Nevertheless, this pattern of high LD between variants has proved very valuable, since one can interrogate only a subset of genetic markers (in most cases SNPs) to capture the majority of genetic variation of a human population. It is estimated that depending on the origin and history of the population we need 500,000 to 1 million SNPs to capture 70–80% of common genetic SNP variation [57].

A methodological issue when we work with diploid genetic data is to determine the arrangement of alleles on homologous chromosomes, otherwise called "phasing." Most of the data we usually obtain in human populations is in the form of genotyped data, so that the phasing is not known. Sophisticated statistical methods have been developed that make use of prior information, pedigree data, and LD to infer the phasing of SNP or other variants [88, 89]. Some of these methodologies and their performance are reviewed in a paper that compares their performance and efficiency [63]. Such methodologies have become essential in human genetic data analysis. Extensions of such data to accommodate variants such as microsatellites or CNVs have been developed and already being extensively used.

16.3.7 Detecting Natural Selection

There is a variety of methods that test for footprints of natural selection, some of which are more appropriate for human population data than others. Standard tests that look for the shape of the frequency spectrum (distribution of variants in frequency classes) [27, 32, 92] can be useful for human data, but these tests are also sensitive to demography and more empirical approaches are necessary to account for these effects by considering the distribution of the statistics in observed data of neutral sequences. Some other methodologies make use of the contrast of divergence to polymorphism data (HKA [39], MK [65]), but these are also sensitive to human demography. Finally, recent studies have taken the approach of using empirical distributions of such statistics to sample the tail of the distribution of the statistic and then take the regions or genes to another orthogonal (independent) set of analysis. All the methods above require re-sequencing data that is not currently available for human sequences but will soon become so under the framework of the 1,000 Genomes Project and other projects using new sequencing technologies.

Given the absence of full re-sequencing data and the availability of SNP genotype data, haplotype-based tests have become useful in human data. These tests are designed to partly account for human demography and are also fit for analyzing genotyping data of SNPs from a given region [83, 98]. Haplotype-based tests are particularly sensitive to recent events of positive selection. The main principle behind these tests is that recently positively selected variants take other neutral variants to high frequency, generating specific patterns of haplotypic diversity and variant correlation. The signal these tests detect is extended haplotype homozygosity around a potential region that has undergone recent positive selection.

16.3.8 Historical Perspective of Population Genetic Studies

Applications of DNA Marker Studies. The use of DNA markers extended the theoretical and practical applications of genetic linkage work considerably. For example, the high degree of individuality of DNA patterns together with the fact that DNA can be extracted from all nucleated cells, and even minute amounts can be amplified with the PCR reaction, makes DNA polymorphisms excellent tools for identifying individuals even if very little material is available. Thus, forensic applications for the identification of blood and sperm residues have come into common use [45]. While there is no controversy about the conceptual basis of this DNA technique, much discussion has been devoted to statistical issues that arise in calculating the probabilities that a suspect's DNA pattern comes from the same person. It is almost certain that using multiple markers appropriately makes it possible to demonstrate a unique DNA pattern for every person, except for identical twins.

Mitochondrial DNA Polymorphisms. Mitochondria are transmitted only from mothers to all of each mother's sons and daughters; there is no diploidy, no meiosis, and no recombination. Polymorphisms of mitochondrial DNA are especially useful in population genetics, mainly for the analysis of relationships between population groups and population history and most of the mutations do not appear to be subject to selection pressures. Therefore comparison of maternally inherited mtDNA restriction patterns between population groups gives an unbiased picture of the population's genetic history [11].

Y-Chromosome Polymorphisms. The other side of mitochondrial DNA is the Y-chromosome. The Y-chromosome is transmitted only by the father to a son. There is only a small portion of the Y-chromosome, called the pseudoautosomal region, that undergoes recombination with the X-chromosome. Variations on the Y-chromosome are very useful, in particular when combined with mtDNA analysis to reveal patterns of population history and behavioral aspects of ancestral populations [50].

16.4 Current Themes

16.4.1 International Efforts to Detect and Describe Sequence Variation

Cataloguing and proper recording of human variation is necessary to support the infrastructure of the search for human disease variants. A major international effort called the International HapMap Project was launched in 2003 to formally record this information in a number of human populations [41–43]. The project initially aimed at the genotyping of a few tens of thousands of SNPs stored in the SNP database of NCBI (dbSNP) in four chosen populations, a Caucasian population of Northern European origin from Utah (CEU), an African population from Ibadan, Nigeria (YRI), a Han Chinese population from Beijing (CHB), and a Japanese population from Tokyo (JPT). The need for large-scale genotyping technologies and the obvious breakthroughs that such technologies would make in disease gene hunting allowed the project to genotype more than 1 million SNPs in the first phase of the project and almost 4 million SNPs after the second phase. The datasets from the HapMap have some ascertainment problems, which are mainly due to rapid change of SNP selection during the project, and it is not clear yet how much the LD information obtained from the four populations will be applicable to other global populations. However, it is a resource that is beginning to serve the scientific community in a substantial way (Fig. 16.7), and we have already experienced the impact in disease genetics with associations of common variants with common diseases [102].

Another project that is expected to have a major impact on the exploration of variation in human populations is the 1,000 Genomes Project (www.1000genomes.org). This project aims at the re-sequencing of thousands of individuals from a number of diverse global locations, in order to elucidate variants that were previously undetected, in particular those with low frequencies [44]. This has been the result of major advances in genome sequencing over the last 4–5 years, and it is considered only the beginning of genome exploration at this level of detail. Accessibility to rare variants and availability of full genome sequencing is expected to spawn a new set of methodological advances that will provide a lot of new insights into population history as well as functional variation.

16.4.2 Prospects for Mapping Functional and Disease Variation

As described above, mutations can occur in any place in the genome, and depending on where they occur their fate in the population may be different. The majority of the genome is neutral DNA, with a high content of repeats and other nonfunctional sequences. But a substantial fraction is functional, and some of this is shared among all mammals or among vertebrates. In the past, much of the attention was devoted to variation in protein-coding DNA, and most of the functional variation was thought to lie within protein-coding sequences. In recent years, attention has been focused on noncoding DNA and it has been inferred that more than two thirds of the human functional DNA is noncoding [5, 22, 68, 86]. The consequences of nucleotide variants in coding DNA are pretty well understood, with nucleotides changing amino acid sites carrying an overall higher probability of having a functional effect than silent (non-amino acid-changing) sites [8, 10]. However, the code of noncoding DNA is not known, and even if we interrogate all sequence variants in the human genome it will be hard to assign a priori probabilities of which ones may have a functional effect. One way to deal with this is to use interspecific sequence conservation to partition variants within and outside of conserved sequences. This is a logical categorization and has been informative in some disease studies [25], but even then we rarely know the exact functional effect, because we do not understand the biochemical reasons underlying the conservation of the sequence. Furthermore, many of the functional variants in the human genome are not found in conserved sequences and the elucidation of such variants will be more difficult.

A very ambitious project, called the ENCODE (ENCyclopedia Of Dna Elements) project, has been initiated and is aimed at the identification of all functional elements in the human genome [30]. In its pilot effort, this project has generated large catalogs of functional elements in 30 Mb (1%) of the human genome. In addition, the HapMap Project has generated re-sequencing data for 48 individuals for 5 of the 30 Mbp. There was a great opportunity to intersect these two types of data and for the first time describe the patterns and levels of variation in a large set of functional elements (Fig. 16.8). The results of this effort [7] will provide a substantial framework for the interpretation of functional variation in the human and other mammalian genomes.

Another way of interrogating functional variation is to condition it on phenotypic effects. We can choose a phenotype with large enough genetic and heritable variation in human populations, such as cellular functions and responses. By using standard QTL mapping approaches we can map the location of the variants that explain the phenotypic variation. These approaches are generally used in humans for disease mapping in case-control studies, but in this case we apply it to a continuous trait. Such approaches have been used to map functionally variable regulatory regions by studying gene expression phenotypes from EBV-transformed lymphoblastoid cell lines of HapMap populations and using the SNPs genotyped in those populations as markers. Some studies have managed to map such variation [16, 90], but the resolution of the methods is low owing to the high LD in human populations (Fig. 16.9). Additional, first-order, assays are necessary (e.g., binding assays chip-seq, open chromatin assays) to multiple variable haplotypes in humans to reveal the amount, nature and degree of functional noncoding variation in humans.

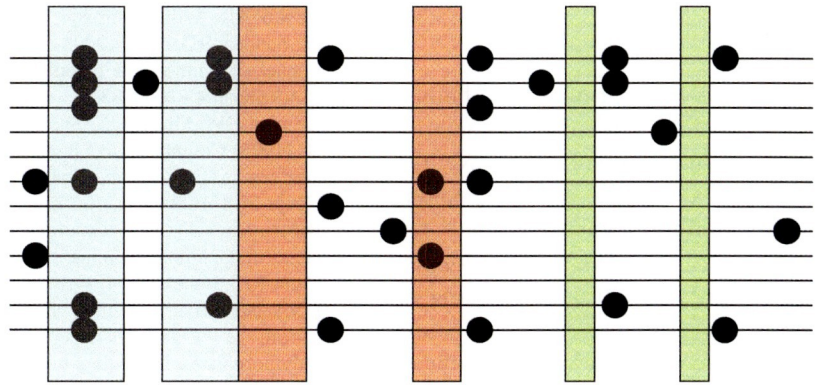

Fig. 16.8 Differential frequency spectrum depending on the type of functional element of a SNP. Patterns of SNP variation in functional DNA elements (*colored boxes*). Note the different patterns of SNP variation in each of them (both density and frequency), indicating the different modes of selection acting on the different types of functional elements

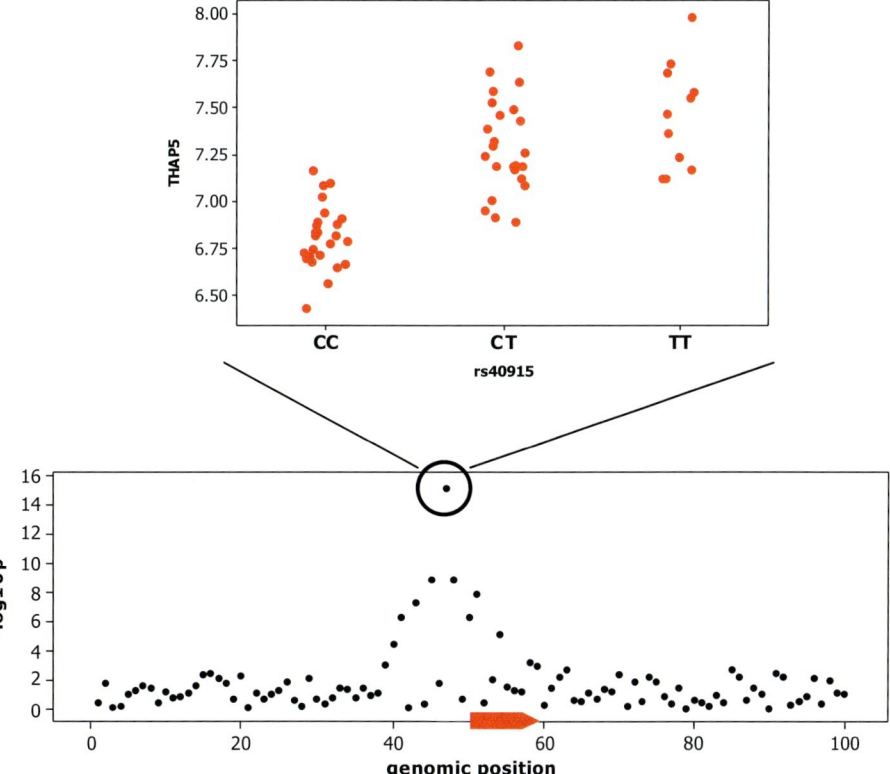

Fig. 16.9 *Cis*-association of gene expression variation in a population. The *panel below* shows the $-\log_{10}$pvalue (*Y*-axis) of the association test of SNPs with gene expression values of gene THAP5 in a given sample of individuals. The *X*-axis is the genomic region, and the *red arrow* is the gene. The *panel above* shows the scatterplot of $\log_2$ expression values (*Y*-axis) for each of the three genotypic classes (*X*-axis)

16.5 Summary

We have discussed some basic principles of human genetic variation, its nature and some characteristics that are specific to human populations. We like to think of our species as special, and our population genetics are not straightforward, but this is not a property of our species only. The study of human genetic variation requires a deep understanding of the population history. Although we have some incomplete historical records we are still not able to develop models that account for all the historical effects, and this is a serious obstacle to our getting a complete picture of our history, what has shaped our genetic diversity, and how different we are from our ancestors. This knowledge extends to the understanding of the causes of disease, some of which are a combination of our genetic makeup and our recent changes in lifestyle. The development of genotyping and sequencing technologies promises that we will soon have steadily improving data to address these issues.

Acknowledgments I would like to thank Stephen Montgomery, Barbara Stranger, and Antigone Dimas for their help and for critically reading earlier versions of this manuscript, and all the members of the Dermitzakis lab for intellectual contributions over the years.

References

1. Abecasis GR (2001) Extent and distribution of linkage disequilibrium in three genomic regions. Am J Hum Genet 68:191–197
2. Altshuler D, Daly MJ, Lander ES (2008) Genetic mapping in human disease. Science 322:881–888
3. Bamshad M, Wooding SP (2003) Signatures of natural selection in the human genome. Nat Rev Genet 4:99–111
4. Begun DJ, Aquadro CF (1992) Levels of naturally occurring DNA polymorphism correlate with recombination rates in D. melanogaster. Nature 356:519–520
5. Bejerano G, Siepel AC, Kent WJ, Haussler D (2005) Computational screening of conserved genomic DNA in search of functional noncoding elements. Nat Methods 2:535–545
6. Bentley DR, Balasubramanian S, Swerdlow HP, Smith GP, Milton J et al (2008) Accurate whole human genome sequencing using reversible terminator chemistry. Nature 456:53–59
7. Birney E, Stamatoyannopoulos JA, Dutta A, Guigo R, Gingeras TR et al (2007) Identification and analysis of functional elements in 1% of the human genome by the ENCODE pilot project. Nature 447:799–816
8. Boyko AR, Williamson SH, Indap AR, Degenhardt JD, Hernandez RD et al (2008) Assessing the evolutionary impact of amino acid mutations in the human genome. PLoS Genet 4:e1000083
9. Broman KW, Murray JC, Sheffield VC, White RL, Weber JL (1998) Comprehensive human genetic maps: individual and sex-specific variation in recombination. Am J Hum Genet 63:861–869
10. Bustamante CD (2005) Natural selection on protein-coding genes in the human genome. Nature 437:1153–1157
11. Cann RL, Stoneking M, Wilson AC (1987) Mitochondrial DNA and human evolution. Nature 325:31–36
12. Cargill M (1999) Characterization of single-nucleotide polymorphisms in coding regions of human genes. Nat Genet 22:231–238
13. Chakravarti A (1999) Population genetics–making sense out of sequence. Nat Genet 21:56–60
14. Charlesworth B (1993) The effect of deleterious mutations on neutral molecular variation. Genetics 134:1289–1303
15. Charlesworth B, Nordborg M, Charlesworth D (1997) The effects of local selection, balanced polymorphism and background selection on equilibrium patterns of genetic diversity in subdivided populations. Genet Res 70:155–174
16. Cheung VG (2005) Mapping determinants of human gene expression by regional and genome-wide association. Nature 437:1365–1369
17. Clark AG, Hubisz MJ, Bustamante CD, Williamson SH, Nielsen R (2005) Ascertainment bias in studies of human genome-wide polymorphism. Genome Res 15:1496–1502
18. Conrad DF, Andrews TD, Carter NP, Hurles ME, Pritchard JK (2006) A high-resolution survey of deletion polymorphism in the human genome. Nat Genet 38:75–81
19. Cook EH Jr, Scherer SW (2008) Copy-number variations associated with neuropsychiatric conditions. Nature 455:919–923
20. Daly M, Rioux JD, Schaffner DF, Hudson TJ, Lander ES (2001) High-resolution haplotype structure in the human genome. Nat Genet 29:229–232
21. Dawson E (2002) A first-generation linkage disequilibrium map of human chromosome 22. Nature 418:544–548
22. Dermitzakis ET (2002) Numerous potentially functional but non-genic conserved sequences on human chromosome 21. Nature 420:578–582
23. Devlin B, Roeder K (1999) Genomic control for association studies. Biometrics 55:997–1004
24. Devlin B, Roeder K, Wasserman L (2000) Genomic control for association studies: a semiparametric test to detect excess-haplotype sharing. Biostatistics 1:369–387
25. Emison ES, McCallion AS, Kashuk CS, Bush RT, Grice E et al (2005) A common sex-dependent mutation in a RET enhancer underlies Hirschsprung disease risk. Nature 434:857–863
26. Fay JC, Wu CI (2000) Hitchhiking under positive Darwinian selection. Genetics 155:1405–1413
27. Fay JC, Wu CI (2000) Hitchhiking under positive Darwinian selection. Genetics 155:1405–1413
28. Fay JC, Wu CI (2001) The neutral theory in the genomic era. Curr Opin Genet Dev 11:642–646
29. Fay JC, Wyckoff GJ, Wu CI (2001) Positive and negative selection on the human genome. Genetics 158:1227–1234

30. Feingold EA, Good PJ, Guyer MS et al (2004) The ENCODE (ENCyclopedia Of DNA Elements) Project. Science 306:636–640
31. Feuk L, Carson AR, Scherer SW (2006) Structural variation in the human genome. Nat Rev Genet 7:85–97
32. Fu YX, Li WH (1993) Statistical tests of neutrality of mutations. Genetics 133:693–709
33. Gabriel SB (2002) The structure of haplotype blocks in the human genome. Science 296:2225–2229
34. Gonzalez E, Kulkarni H, Bolivar H, Mangano A, Sanchez R et al (2005) The influence of CCL3L1 gene-containing segmental duplications on HIV-1/AIDS susceptibility. Science 307:1434–1440
35. Graur D, WH L (2000) Fundamentals of molecular evolution. Sinauer Associates, Sunderland, MA
36. Hartl DL, Clark AG (2007) Principles of Population Genetics. Sinauer, Sunderland, MA
37. Helgason A, Sigurethardottir S, Nicholson J, Sykes B, Hill EW (2000) Estimating Scandinavian and Gaelic ancestry in the male settlers of Iceland. Am J Hum Genet 67:697–717
38. Hudson RR (1991) Gene genealogies and coalescent process. Oxf Surv Evol Biol 7:1–44
39. Hudson RR, Kreitman M, Aguade M (1987) A test of neutral molecular evolution based on nucleotide data. Genetics 116:153–159
40. Iafrate AJ (2004) Detection of large-scale variation in the human genome. Nat Genet 36:949–951
41. International HapMap Consortium (2003) The International HapMap Project. Nature 426:789–796
42. International HapMap Consortium (2005) A haplotype map of the human genome. Nature 437:1299–1320
43. International HapMap Consortium (2007) A second generation human haplotype map of over 3.1 million SNPs. Nature 449(7164):851–861
44. Ionita-Laza I, Lange C, ML N (2009) Estimating the number of unseen variants in the human genome. Proc Natl Acad Sci USA 106:5008–5013
45. Jeffreys AJ, Brookfield JF, Semeonoff R (1985) Positive identification of an immigration test-case using human DNA fingerprints. Nature 317:818–819
46. Jeffreys AJ, Wilson V, Thein SL (1985) Hypervariable 'minisatellite' regions in human DNA. Nature 314:67–73
47. Jeffreys AJ, Wilson V, Thein SL (1985) Individual-specific 'fingerprints' of human DNA. Nature 316:76–79
48. Jeffreys AJ, Ritchie A, Neumann R (2000) High resolution analysis of haplotype diversity and meiotic crossover in the human TAP2 recombination hotspot. Hum Mol Genet 9:725–733
49. Jeffreys AJ, Kauppi L, Neumann R (2001) Intensely punctuate meiotic recombination in the class II region of the major histocompatibility complex. Nature Genet. 29:217–222
50. Jobling MA, Tyler-Smith C (2003) The human Y chromosome: an evolutionary marker comes of age. Nat Rev Genet 4:598–612
51. Kan YW, Dozy AM (1978) Polymorphism of DNA sequence adjacent to human beta-globin structural gene: relationship to sickle mutation. Proc Natl Acad Sci USA 75:5631–5635
52. Kidd JM, Cooper GM, Donahue WF, Hayden HS, Sampas N et al (2008) Mapping and sequencing of structural variation from eight human genomes. Nature 453:56–64
53. Kimura M (1968) Evolutionary rate at the molecular level. Nature 217:624–626
54. Kimura M (1969) The number of heterozygous nucleotide sites maintained in a finite population due to steady flux of mutations. Genetics 61:893–903
55. Kingman J (1982) The coalescent. Stoch Proc Appl 13:235–248
56. Korbel JO, Urban AE, Affourtit JP, Godwin B, Grubert F et al (2007) Paired-end mapping reveals extensive structural variation in the human genome. Science 318:420–426
57. Kruglyak L (1999) Prospects for whole-genome linkage disequilibrium mapping of common disease genes. Nat Genet 22:139–144
58. Lander ES, Botstein D (1989) Mapping Mendelian factors underlying quantitative traits using RFLP linkage maps. Genetics 121:185–199
59. Lander ES, Linton LM, Birren B, Nusbaum C, Zody MC et al (2001) Initial sequencing and analysis of the human genome. Nature 409:860–921
60. Li N, Stephens M (2003) Modeling linkage disequilibrium and identifying recombination hotspots using single-nucleotide polymorphism data. Genetics 165:2213–2233
61. Lunter G, Ponting CP, Hein J (2006) Genome-wide identification of human functional DNA using a neutral indel model. PLoS Comput Biol 2:e5
62. Marchini J, Cardon LR, Phillips MS, Donnelly P (2004) The effects of human population structure on large genetic association studies. Nat Genet 36:512–517
63. Marchini J, Cutler D, Patterson N, Stephens M, Eskin E et al (2006) A comparison of phasing algorithms for trios and unrelated individuals. Am J Hum Genet 78:437–450
64. McCarroll SA (2006) Common deletion polymorphisms in the human genome. Nat Genet 38:86–92
65. McDonald JH, Kreitman M (1991) Adaptive protein evolution at the ADH locus in Drosophila. Nature 351:652–654
66. McVean GA (2004) The fine-scale structure of recombination rate variation in the human genome. Science 304:581–584
67. Menozzi P, Piazza A, Cavalli-Sforza L (1978) Synthetic maps of human gene frequencies in Europeans. Science 201:786–792
68. Mouse Genome Sequencing Consortium (2002) Initial sequencing and comparative analysis of the mouse genome. Nature 420:520–562
69. Myers S, Bottolo L, Freeman C, McVean G, Donnelly P (2005) A fine-scale map of recombination rates and hotspots across the human genome. Science 310:321–324
70. Nei M, Li WH (1979) Mathematical model for studying genetic variation in terms of restriction endonucleases. Proc Natl Acad Sci USA 76:5269–5273
71. Nickerson DA (1998) DNA sequence diversity in a 9.7-kb region of the human lipoprotein lipase gene. Nat Genet 19:233–240
72. Peltonen L (2000) Positional cloning of disease genes: advantages of genetic isolates. Hum Hered 50:66–75
73. Petrov DA, Hartl DL (1999) Patterns of nucleotide substitution in Drosophila and mammalian genomes. Proc Natl Acad Sci USA 96:1475–1479
74. Price AL (2006) Principal components analysis corrects for stratification in genome-wide association studies. Nat Genet 38:904–909

75. Price AL (2008) Discerning the ancestry of European Americans in genetic association studies. PLoS Genet 4:e236
76. Pritchard JK, Donnelly P (2001) Case-control studies of association in structured or admixed populations. Theor Popul Biol 60:227–237
77. Pritchard JK, Stephens M, Donnelly P (2000) Inference of population structure using multilocus genotype data. Genetics 155:945–959
78. Pritchard JK, Stephens M, Rosenberg NA, Donnelly P (2000) Association mapping in structured populations. Am J Hum Genet 67:170–181
79. Redon R (2006) Global variation in copy number in the human genome. Nature 444:444–454
80. Risch N, Merikangas K (1996) The future of genetic studies of complex human diseases. Science 273:1516–1517
81. Risch NJ (2000) Searching for genetic determinants in the new millennium. Nature 405:847–856
82. Rosenberg NA (2002) Genetic structure of human populations. Science 298:2381–2385
83. Sabeti PC (2002) Detecting recent positive selection in the human genome from haplotype structure. Nature 419:832–837
84. Sebat J (2004) Large-scale copy number polymorphism in the human genome. Science 305:525–528
85. Sebat J, Lakshmi B, Malhotra D, Troge J, Lese-Martin C et al (2007) Strong association of de novo copy number mutations with autism. Science 316:445–449
86. Siepel A, Bejerano G, Pedersen JS, Hinrichs AS, Hou M et al (2005) Evolutionarily conserved elements in vertebrate, insect, worm, and yeast genomes. Genome Res 15:1034–1050
87. Spencer CC (2006) The influence of recombination on human genetic diversity. PLoS Genet 2:e148
88. Stephens M, Donnelly P (2003) A comparison of Bayesian methods for haplotype reconstruction from population genotype data. Am J Hum Genet 73:1162–1169
89. Stephens M, Smith NJ, Donnelly P (2001) A new statistical method for haplotype reconstruction from population data. Am J Hum Genet 68:978–989
90. Stranger BE (2005) Genome-wide associations of gene expression variation in humans. PLoS Genet 1:e78
91. Strobeck C, Smith JM, Charlesworth B (1976) The effects of hitchhiking on a gene for recombination. Genetics 82:547–558
92. Tajima F (1989) Statistical method for testing the neutral mutation hypothesis by DNA polymorphism. Genetics 123:585–595
93. Thomas R, Zischler H, Paabo S, Stoneking M (1996) Novel mitochondrial DNA insertion polymorphism and its usefulness for human population studies. Hum Biol 68:847–854
94. Tishkoff SA (1996) Global patterns of linkage disequilibrium at the CD4 locus and modern human origins. Science 271:1380–1387
95. Tishkoff SA, Verrelli BC (2003) Patterns of Human Genetic Diversity: Implications for Human Evolutionary History and Disease. Ann Rev Genomics Hum Genet 4:293–340
96. Tishkoff SA, Williams SM (2002) Genetic analysis of African populations: Human evolution and complex disease. Nat Rev Genet 3:611–621
97. Venter JC (2001) The sequence of the human genome. Science 291:1304–1351
98. Voight BF, Kudaravalli S, Wen X, Pritchard JK (2006) A map of recent positive selection in the human genome. PLoS Biol 4:e72
99. Walsh T, McClellan JM, McCarthy SE, Addington AM, Pierce SB et al (2008) Rare structural variants disrupt multiple genes in neurodevelopmental pathways in schizophrenia. Science 320:539–543
100. Watterson GA (1975) On the number of segregating sites in genetical models without recombination. Theor Popul Biol 7:256–276
101. Weber JL, May PE (1989) Abundant class of human DNA polymorphisms which can be typed using the polymerase chain reaction. Am J Hum Genet 44:388–396
102. Wellcome Trust Case Control Consortium (2007) Genome-wide association study of 14,000 cases of seven common diseases and 3,000 shared controls. Nature 447:661–668
103. Wiesenfeld SL (1968) Selective advantage of the sickle-cell trait. Science 160:437
104. Wright S (1931) Evolution in Mendelian Populations. Genetics 16:97–159

17 Consanguinity, Genetic Drift, and Genetic Diseases in Populations with Reduced Numbers of Founders

Alan H. Bittles

Abstract In western countries, consanguineous marriage often arouses curiosity and prejudice in approximately equally measure, despite the fact that until the mid-nineteenth century cousin marriages were quite common in Europe and North America. Attitudes to consanguinity remain very different in other parts of the world, in particular north and sub-Saharan Africa, the Middle East, Turkey and central Asia, and south Asia, where between 20% and over 50% of current marriages are contracted between biological relatives, with first-cousin unions especially common. Besides intra-familial marriage, in these regions a large majority of marriages also occur within long-established male lineages, e.g., clans and tribes in Arab societies and castes in India. Through time these lineages effectively become separate breeding pools, with founder effect, mutation, genetic drift and bottle-necking separately and collectively influencing gene pool composition. The present chapter first considers the concepts of random and assortative mating and then examines demographic, social, economic, and religious variables that influence the prevalence of preferred types of consanguineous marriage. The effects of consanguinity on human mate choice, reproductive success, and reproductive compensation are identified, and the impact of consanguinity on morbidity and mortality in infancy, childhood and adulthood are discussed and quantified. Three detailed case studies are then used to illustrate the influence of endogamy and consanguinity on human genetic variation and genetic disease: the Finnish Disease Heritage; inter- and intra-population genetic differentiation in India; and the distribution of specific disease alleles in Arab Israeli communities. The scale of global migration during the last two generations, with many millions of individuals, families, and occasionally entire communities moving within and between continents, has created an entirely new scenario in human population genetics. Against this background, consanguinity has re-emerged both as an important feature of community and public health genetics, and as a topic of general interest.

A.H. Bittles(✉)
Centre for Comparative Genomics, Murdoch University, South
Street, Perth, WA 6150, Australia
e-mail: abittles@ccg.murdoch.edu.au

17

Contents

17.1 Genetic Variation in Human Populations 508
    17.1.1 Random Mating and Assortative Mating 509
    17.1.2 Genetic Drift and Founder Effect 509

17.2 Consanguineous Matings 509
    17.2.1 Coefficient of Relationship and Coefficient of Inbreeding 510
    17.2.2 Global Prevalence of Consanguinity 510
    17.2.3 Specific Types of Consanguineous Marriage 514
    17.2.4 The Influence of Religion on Consanguineous Marriage 514
    17.2.5 Civil Legislation on Consanguineous Marriage 515
    17.2.6 Social and Economic Factors Associated with Consanguinity 515

17.3 Inbreeding and Fertility 516
    17.3.1 Genetically Determined Factors Influencing Human Mate Choice 516
    17.3.2 Inbreeding and Fetal Loss Rates 516
    17.3.3 Comparative Fertility in Consanguineous and Nonconsanguineous Couples 517

17.4 Inbreeding and Inherited Disease 518
    17.4.1 Consanguinity and Deaths in Infancy and Childhood 518
    17.4.2 Consanguinity and Childhood Morbidity 518
    17.4.3 Consanguinity and Adult Mortality and Morbidity 519

17.5 Incest 519
    17.5.1 Mortality and Morbidity Estimates for Incestuous Matings 520

17.6 Genetic Load Theory and Its Application in Consanguinity Studies 520

17.7 Genomic Approaches to Measuring Inbreeding at Individual and Community Levels 521

17.8 The Influence of Endogamy and Consanguinity in Human Populations 521
    17.8.1 The Finnish Disease Heritage 521
    17.8.2 Inter- and Intra-population Differentiation in India 523
    17.8.3 Consanguinity and the Distribution of Disease Alleles in Israeli Arab Communities 523

17.9 Evaluating Risk in Consanguineous Relationships 524

17.10 Concluding Comments 525

References 525

17.1 Genetic Variation in Human Populations

The concepts of race and ethnicity often are highly controversial topics, and the use of supposed racial characteristics in differentiating between human populations has been strongly censured. At the same time, genomic microarray studies have convincingly demonstrated significant differences between major human populations living in different parts of the world, with common genetic variants playing an important role in inter-ethnic gene expression [86]. However, microarray studies also have shown that 93–95% of the total genetic variation was intrapopulation rather than interpopulation in origin [75]. While the proportionally minor genetic differences between populations and the attendant race/ethnicity/ancestry controversy are widely discussed and argued, an obvious and potentially more significant question arises with respect to the origins and causes of the very high level of intra-population genetic variation.

How and why did this variation arise, how and why is it maintained, and what, if any, are the consequences in terms of biological fitness, and more especially genetic disease?

Throughout recorded human history, marriage between a male and female has been the predominant institution within which procreation occurred and genes were transmitted. Therefore a key initial step in investigating intra- and inter-population genetic differences is to examine how and why marriage partners are chosen in different societies. Virtually all traditional societies are divided into long-established communities, with limited inter-community marriage. Indeed, genome-based association studies conducted in industrialized Western societies have revealed similar, if less pronounced sub-divisions, and even in countries with large immigrant communities, such as the USA, Canada and Australia, recent arrivals typically marry within their own ethnic and/or religious community during the first and second post-migration generations. Although offering strong social advantages, this tradition has important genetic implications, since it is probable that couples from the same national, ethnic or religious sub-community will have a significant proportion of their genes in common, and therefore that their progeny are more likely to be homozygous for a detrimental recessive disorder [14].

17.1.1 Random Mating and Assortative Mating

One of the theoretical cornerstones of human population genetics, the Hardy–Weinberg principle, incorporates the provisos of infinite population size and random mating. Even cursory consideration of the growth rate of the global human population through time would indicate that blanket assumptions of this nature are seriously flawed. Thus it has been estimated that the total global population in 1,000 AD was some 310 million, increasing approximately 20-fold during the course of the second millennium to 6,070 million, with an additional 4,420 million humans in the twentieth century alone.

Likewise, rather than random mating, in many Western countries first cousin unions were both popular and highly prized up to the mid-nineteenth century and, for example, not only did Charles Darwin marry his first cousin Emma Wedgewood, Darwin's sister Caroline married Emma Wedgwood's brother Josiah, following intermarriage between the Darwin and Wedgwood families in the previous generation. However, in modern Western societies there is a strong belief that marriage between close biological kin is genetically disadvantageous, which has led to a marked decline in the prevalence of consanguineous marriage in these populations.

This does not mean that marriage partner choice has become an essentially random process, and even in societies where consanguinity is regarded with disfavor, positive assortative mating is the rule rather than the exception. Thus despite greater personal mobility, the choice of a marriage partner remains strongly influenced by geography and ethnicity, and by essentially social factors, such as religion, education, economic status, and political beliefs. Under these circumstances the strict concept of random mating does not apply, since it is probable that the marriage partners will have inherited identical alleles at a proportion of gene loci.

17.1.2 Genetic Drift and Founder Effect

The phenomenon of genetic drift is most simply defined as the influence of chance on gene frequencies in successive generations, and the probability of genetic drift is greatest in communities with small effective population sizes, i.e., with restricted numbers of potential mating couples. In evolutionary terms this situation can arise in several ways, for example, through founder effect, when a subgroup of a population establishes a new breeding colony; via a demographic bottleneck following major disease- or disaster-related mortality; and in subdivided populations with multiple, strictly endogamous subcommunities.

Where there is restricted marriage partner choice, genetic drift can lead to random inbreeding, with unions contracted between individuals not known to be biological relatives but drawn from the same confined gene pool. The net effect is similar to positive assortative mating, and the main outcome is a higher probability of homozygosity at some gene loci, resulting in an increased likelihood of recessive gene expression. This is important from a medical genetics perspective, since a recessive founder or de novo mutation can rapidly increase in frequency within a small community by chance alone, resulting in the birth of an affected child whether the parents are known to be consanguineous or believe themselves to be nonrelatives [104].

17.2 Consanguineous Matings

The origin of the term consanguineous is the Latin *consanguineus*, meaning 'of the same blood.' In a human genetics context, a couple are said to be consanguineous if they share one or more common ancestors. Since most pairs of individuals living in the same location will have a common ancestor somewhere in their family trees, for practical purposes the search for a shared ancestor generally does not extend back more than three or four generations. In medical genetics, the definition of consanguinity is usually restricted to a preferential union between a couple related as second cousins or closer, although as discussed in Sects. 17.2.1 and 17.4, important exceptions can and do arise.

17.2.1 Coefficient of Relationship and Coefficient of Inbreeding

Two basic measures are employed to quantify genetic relationships. The first is the coefficient of relationship (r), which is the proportion of genes identical by descent (IBD) shared by two individuals. The coefficient of relationship is calculated from the formula:

$$r = \{(1/2)^n\}$$

where n is the number of steps apart on a pedigree for these two individuals via their common ancestor. Thus for two persons related as first cousins:

$$r = \{(1/2)^4\} + \{(1/2)^4\} = 1/8$$

The coefficient of inbreeding (F) is the proportion of gene loci at which an individual is homozygous by descent (Table 17.1). Incestuous relationships, i.e., between father–daughter, mother–son or brother–sister are the closest form of human mating, with the partners sharing half of their genes ($r=0.5$), and so any offspring would be homozygous at 1/4 of gene loci ($F=0.25$). The closest legally permissible consanguineous unions are between an uncle and niece, which occur mainly in South Indian Hindu communities, or between double-first cousins, as in Muslim populations in the Middle East and Pakistan. In both of these types of marriage the partners share one fourth of their genes ($r=0.25$) and the coefficient of inbreeding in their progeny is $F=0.125$. Double-first cousins have both sets of grandparents in common, whereas in first-cousin marriage the couple shares two common grandparents (Fig. 17.1).

Second cousins have inherited 1/32 of their genes from a common ancestor ($r=0.0313$), and so the offspring of a second-cousin union would be expected to be homozygous (or more strictly autozygous) at 1/64 of their gene loci, i.e., $F=0.0156$. In populations with restricted marriage partner choice, couples who are not second cousins may be related through multiple pathways involving more remote ancestors. Under such circumstances the coefficient of inbreeding for an individual is calculated by summing each of the known pathways of inheritance. Thus, for an individual whose parents are third, fourth and fifth cousins ($r=0.0078$, 0.0039 and 0.00195), the corresponding coefficient of inbreeding is ($F=0.0039+0.00195+0.00098$), i.e., a composite coefficient of inbreeding of $F=0.00683$.

In many societies, specific subcommunities or families have a long and unbroken tradition of consanguineous marriage, resulting in a cumulative coefficient of inbreeding that can greatly exceed the genetic influence of consanguinity in a single generation. To quantify this situation a correction term can be applied using the formula:

$$F = \Sigma(1/2)^n (1 + F_A)$$

where F_A is the ancestor's coefficient of inbreeding, n is the number of individuals in the path connecting the parents of the individual, and the summation (Σ) is taken over each path in the pedigree that goes through a common ancestor. In small endogamous communities with limited numbers of marriage partners, cumulative inbreeding via multiple consanguineous pathways can result in a significant build-up of homozygosity, even within a few generations.

17.2.2 Global Prevalence of Consanguinity

From a global perspective the lowest rates of consanguinity are found in Western Europe, North America and Oceania, where less than 1% of marriages are consanguineous, i.e., they are contracted between couples related as second cousins or closer ($F \geq 0.0156$). In some parts of Southern Europe, South America and Japan approximately 1–5% of current marriages are consanguineous, depending on local geography and

Table 17.1 Human genetic relationships

Biological relationship	Genetic relationships	Coefficient of relationship	Coefficient of inbreeding
Incest[a]	First degree	0.5	0.25
Uncle-niece Double first cousin	Second degree	0.25	0.125
First cousin	Third degree	0.125	0.0625
First cousin once removed Double second cousin	Fourth degree	0.0625	0.0313
Second cousin	Fifth degree	0.0313	0.0156
Second cousin once removed Double third cousin	Sixth degree	0.0156	0.0078
Third cousin	Seventh degree	0.0078	0.0039

[a]Incest is defined as a sexual relationship between father–daughter, mother–son or brother–sister

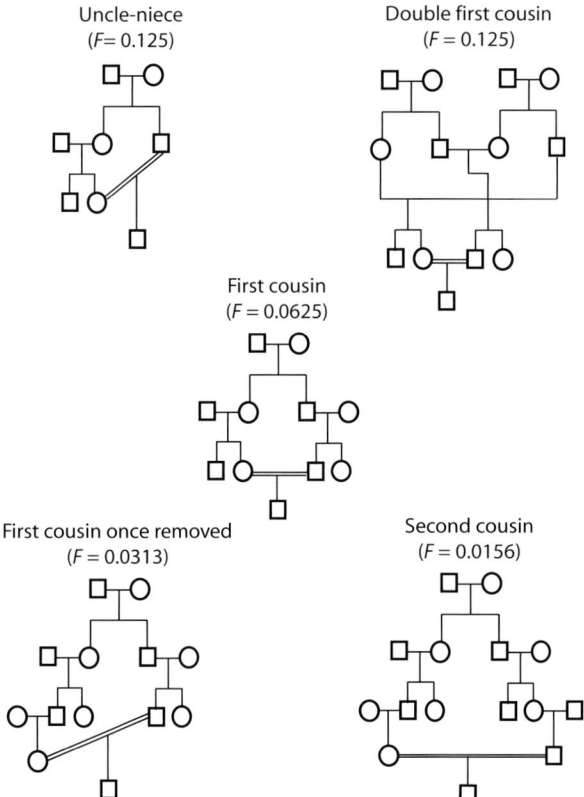

Fig. 17.1 Consanguineous pedigrees

social customs. The highest recorded rates of consanguinity are in North and sub-Saharan Africa, the Middle East, Turkey and Central Asia, and parts of South Asia, where unions between couples related as second cousins or closer account for 20% to over 50% of all marriages (www.consang.net). While a recent decline in the prevalence of consanguineous marriage has been reported in some Middle Eastern countries, such as Jordan [40], increases have been reported in the neighboring Arab states of Qatar [10], and the United Arab Emirates [5]. In the major South Asian countries of India [22], Pakistan [4], and Iran [78] little change appears to have occurred in the prevalence of consanguinity during the latter half of the twentieth century, although there is some evidence that attitudes towards consanguineous marriage are influenced by contemporary political regimes.

Data on consanguinity remains at best partial for many populous countries in Asia, including Bangladesh and Indonesia. Anthropological research in Africa has indicated that cousin marriage is common in many specific communities, but there is little information on its prevalence or the particular types of cousin union that are favored. Although consanguinity has been rare in Western societies since the early twentieth century, most Western countries are now home to large migrant communities which traditionally have contracted consanguineous unions, with all evidence pointing to continued preference for intrafamilial marriage in their newly adopted countries [11, 68]. For this reason, the summary country and regional data on consanguineous marriage presented in Fig. 17.2 are best considered as lower bound estimates of the overall global picture.

If the specific types and frequencies of consanguineous marriage are known, the mean coefficient of inbreeding (α) can be calculated to provide a measure of the intensity of inbreeding in the population, according to the formula:

$$\alpha = \Sigma\, p_i F_i$$

where Σ is the summation of the proportion of individuals p_i in each consanguinity category F_i. As indicated in Table 17.2, the values for α vary widely between

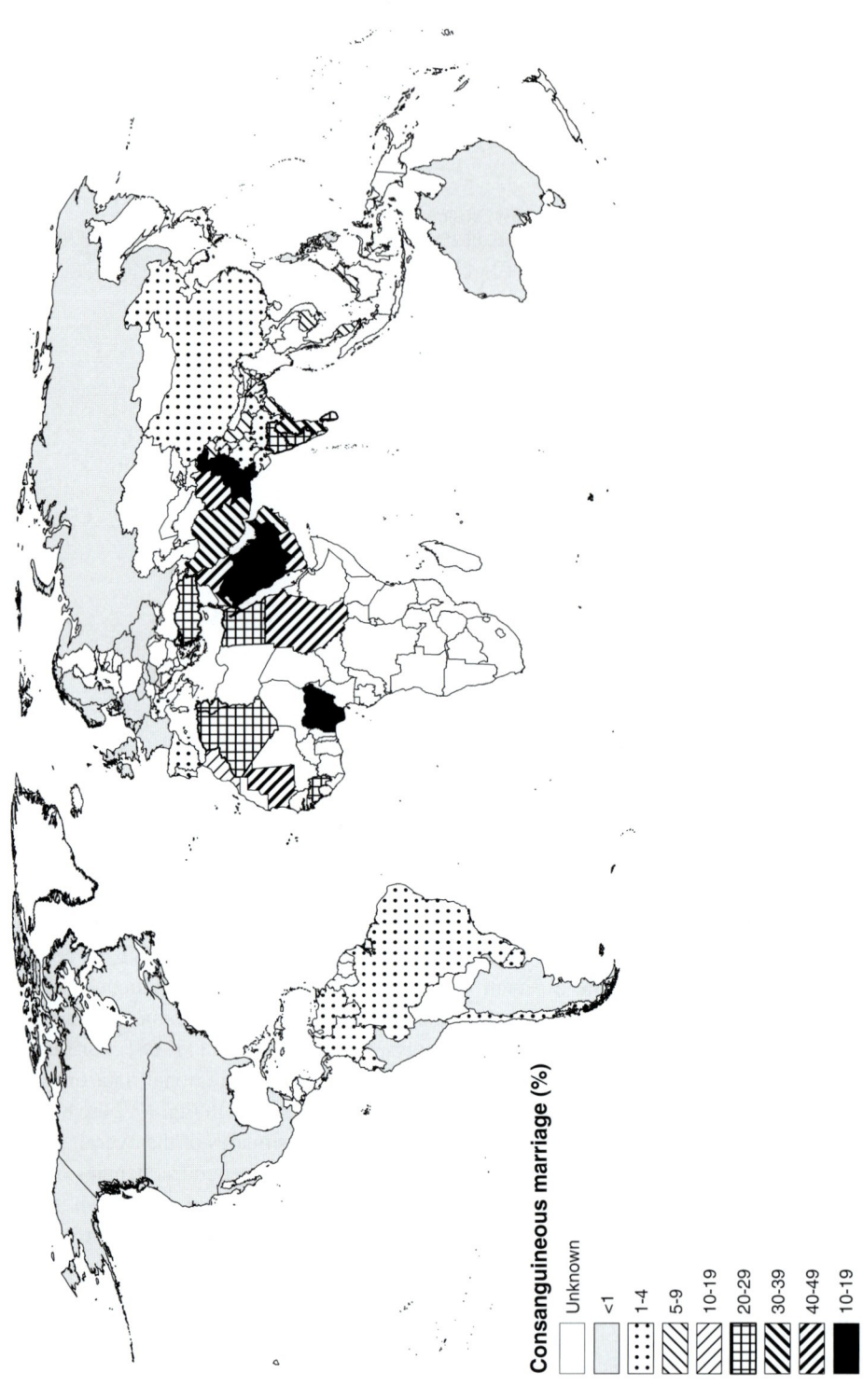

Fig. 17.2 Global distribution of consanguineous marriages

Table 17.2 Prevalence (%) and types of consanguineous marriage in different regions and populations

Region	Study setting	Participants	Number	%	Relationships[a]	Inbreeding (α)	Reference
Western Europe							
Great Britain	Birmingham	Obstetric inpatients	2,431	0.2	1C,2C	0.0001	[23]
Norway	All-Norway	Civil registration	893,941	0.7	1C,2C	0.0002	[56]
Spain	Sigüenza-Guadalajara	RC dispensation	5,315	12.6	1C,1¹/²C,2C	0.0029	[24]
North America							
Canada	Québec	RC dispensation	21,874	1.3	UN,1C,1¹/²C,2C	0.0003	[31]
USA	National	RC dispensation	133,228	0.2	1C,1¹/²C,2C	0.0001	[36]
South America							
Argentina	All-Argentina	Civil registration	212,320	0.4	1C	0.0002	[26]
Brazil	Rio Janeiro	RC dispensation	4,070	1.5	UN/AN,1C,1¹/²C,2C	0.0008	[30]
	Paraiba	RC dispensation	9,521	12.8	UN/AN,1C,1¹/²C,2C	0.0058	[30]
Northern Africa							
Tunisia	North	Obstetric inpatients	5,767	26.9	>1C,1C,1¹/²C,2C	0.0213	[73]
Egypt	National	Household/school/school/workplace	26,554	29.0	D1C,1C,1¹/²C,2C	0.0101	[39]
Sudan	Khartoum	Blood donors	4,833	52.0	1C,2C	0.0302	[79]
West Asia							
Saudi Arabia	National	Household survey	3,212	40.6	1C,2C	0.0241	[34]
UAE	Al Ain	Health centre	1,502	37.4	D1C,1C,1¹/²C,2C	0.0245	[5]
Turkey	National	Household survey	5,257	21.1	1C,2C	0.0096	[92]
South Asia							
India North	Lucknow (Hindu)	Obstetric inpatients	7,955	0.1	UN,1C	0.0001	[2]
- South	Karnataka (Hindu)	Obstetric inpatients	86,448	33.5	UN,1C,2C	0.0333	[22]
Pakistan	National	Household survey	6,611	61.2	1C,2C	0.0332	[4]
East Asia							
China - *Han*	Zejiang	Household survey	15,762	1.1	UN,1C,1¹/²C,2C	0.0006	[100]
Kirgiz	Xinjiang	Household survey	2,863	45.2	D1C,1C,1¹/²C,2C	0.0274	[99]
Japan	National	Household survey	9,225	3.9	1C,1¹/²C,2C	0.0013	[45]
Oceania							
Australia	Western Australia	Civil marriages	62,376	0.2	1C,1¹/²C,2C	0.0001	[67]

[a]Categories of consanguineous marriage: *UN* Uncle-niece, *AN* Aunt-nephew, *D1C* Double first cousin, $F = 0.125$; *1C* first cousin, $F = 0.0625$; *1¹/²C* first cousin once removed, $F = 0.0313$; *2C* second cousin, $F = 0.0156$

populations and regions, from 0.0001 in Western Europe and North America to 0.0241 in Saudi Arabia, 0.0332 in Pakistan and 0.0333 in South India, where consanguineous marriage is widely preferential. In the two latter countries the average level of inbreeding is thus equivalent to all marriages being contracted at the level of first cousin once removed ($F = 0.0313$).

Detailed global estimates of consanguineous marriage in the current generation are available by continent, country, region and population at www.consang.net. However, as indicated in Table 17.2, interpopulation comparability is difficult to achieve because of differences in the numbers of subjects recruited and in the types of study population, e.g., whether based on dispensations granted by the Roman Catholic church for consanguineous couples to marry, compulsory civil marriage registration, or records relating to obstetric inpatients. Likewise, the levels at which data have been collected vary, with some studies counting only first- or second-cousin marriages, while in other populations uncle-niece and aunt-nephew, double-first-cousin, and first-cousin once removed marriages also were recorded. What is, however, clearly apparent from Fig. 17.2 is that consanguineous marriage is not restricted to geographically remote communities or to specific religious, ethnic, or social isolates, as has been popularly believed.

17.2.3 Specific Types of Consanguineous Marriage

The specific patterns of consanguineous marriage contracted in different populations largely reflect their traditional customs and beliefs. The highest levels of consanguineous marriage so far reported in a single generation are from the former French colony of Pondicherry in South India, with 54.9% consanguinity (mean coefficient of inbreeding, α, 0.0449) [70], and among army recruits in the province of Punjab, Pakistan with 77.1% consanguinity (α = 0.0414) [41]. The fact that the mean coefficient of inbreeding was actually higher in Pondicherry than Punjab despite a lower total percentage consanguinity is explained by the fact that most consanguineous marriages in Punjab were between first cousins ($F = 0.0625$), whereas in Pondicherry uncle-niece marriages ($F = 0.125$) predominated.

Local custom also dictates the specific types of first-cousin unions, so that in Arab Muslim communities a marriage between a man and his father's brother's daughter (FBD) is preferred, as opposed to the mother's brother's daughter (MBD) pattern of first-cousin marriage found in such disparate populations as Dravidian Hindus of South India, Han Chinese, and the Tuareg of North Africa [13]. A further factor to be considered is that in communities where consanguinity is preferential, couples in marriages categorized as nonconsanguineous very probably have inherited a significant proportion of their genes from one or more common ancestor, even though they themselves are unaware of any close genetic relationship.

Although the coefficient of inbreeding for FBD and MBD offspring is the same at autosomal loci ($F = 0.0625$), at X-chromosome loci $Fx = 0$ for FBD progeny but 0.125 for children born to MBD couples. Therefore, the specific forms of first cousin union favored and contracted within particular populations can have an important influence on the expression of X-linked disease genes.

17.2.4 The Influence of Religion on Consanguineous Marriage

The major world religions exert a strong influence on consanguineous marriage, both directly in terms of the types of marriages permitted and via the enactment of civil legislation. As indicated in Table 17.3, most of

Table 17.3 Religious attitudes towards consanguineous marriages

Religion	Subcommunity	Attitude
Judaism	Sephardi	Permissive
	Ashkenazi	Permissive
Christianity	Greek and Russian Orthodox	Proscribed
	Roman Catholic	Diocesan approval required
	Protestant	Permissive
Islam	Sunni	Permissive
	Shia	Permissive
Hinduism	Indo-European	Proscribed
	Dravidian	Permissive
Buddhism		Permissive
Sikhism		Proscribed
Confucianism/ Taoism		Partially permissive
Zoroastrian/Parsi		Permissive

the major religions sanction consanguineous unions, although there are quite marked differences within each religious tradition. Judaism and Islam largely follow the guidelines provided in Leviticus 18:7–18, but there is a Quranic prohibition on uncle-niece marriages, which are permitted within Judaism. Despite the Judaic core of Christianity, the Orthodox and Roman Catholic Churches restrict close-kin marriage. However, the strictures requiring dispensation for consanguineous unions were substantially relaxed by the Roman Catholic Church in the early twentieth century and now apply only to couples related as first cousins or closer [12]. By comparison, as part of the sixteenth century Reformation, the Protestant denominations basically reverted to the Levitical proscriptions on marriage, with first-cousin unions permissible. A similar divergence occurs within Hinduism, and while many Dravidian South Indians regard consanguineous marriage as preferential, in North India consanguinity is prohibited under the Indo-European Hindu tradition. A further detailed description of Hindu marriage practices and prohibitions is given in Sect. 17.8.2.

17.2.5 Civil Legislation on Consanguineous Marriage

While consanguineous unions are largely avoided in regions such as Western Europe and Oceania, first-cousin marriage is permissible under civil law in virtually all countries, and since 1987 marriage between half-sibs ($F=0.125$) may be permitted in Sweden under specific circumstances. The situation is quite different in the USA. Until 1861 first-cousin marriage was legal, but through time legislation to ban different types of consanguineous marriage was gradually introduced at state level, the most recent example being a ban on first-cousin marriage adopted by the state of Texas in 2005. This means that first-cousin unions are a criminal offense in 10 states, and are illegal in a further 22 states, despite a Federal recommendation in 1970 that all state laws on first-cousin marriage should be rescinded [15]. The USA is one of the few countries to have enacted legislation of this type, alongside the People's Republic of China and the Democratic People's Republic of Korea.

17.2.6 Social and Economic Factors Associated with Consanguinity

Consanguineous unions have been most frequently reported within the ruling classes and land-owning families of Western societies, and also within powerful mercantile dynasties, such as the Rothschilds, whose family members have worked cohesively across national boundaries for many generations. The picture is quite different in other less economically developed parts of the world, with the highest rates of consanguineous marriage commonly reported among poor, rural, and largely illiterate communities in societies throughout Asia and North Africa [12, 43].

As indicated in Table 17.4, the preference for consanguineous marriage is primarily social in nature, since it is believed that family ties will be strengthened, family honor will be optimally maintained, and health or financial uncertainties that may be encountered following marriage with a partner from another family or community are avoided [15]. Also, in societies where males and females are segregated from late childhood, potential marriage partners are more likely to know each other if they are biological relatives, since they would have been able to meet at family social functions. Premarital arrangements are simplified in a consanguineous union, and the relationship of a couple and their in-laws is expected to be more congenial, which is particularly important for female autonomy in the patrilocal societies typical of most Asian countries (Table 17.4).

As in Western societies, economic considerations are an important facet of marriage partner choice, and in

Table 17.4 Social and economic advantages of consanguineous marriage

The assurance of marrying within the family and the strengthening of family ties
The assurance of knowing one's spouse prior to marriage
Simplified premarital negotiations, with conditions and marriage arrangements agreed in the partners' early or late teens
Greater social compatibility of the bride with her husband's family, in particular her mother-in-law, who also is a relative
Lower risk of undeclared health problems in the intended spouse
Reduced requirement for dowry or bridewealth payments, with consequent maintenance of the family goods and monies
For land-owning families, maintenance of the integrity of family land-holdings, which otherwise might be subdivided by inheritance

countries in which dowry payments are the norm, marriages within the family reduce or even negate the potential financial costs to the bride's family [12, 57]. Problems arising from marriage to a close relative have been cited in a minority of cases, especially where there is a large age gap between spouses. But in most instances marital stability appears strong and divorce is uncommon, possibly reflecting the family disunity that could arise if a marriage between cousins were to fail [13].

17.3 Inbreeding and Fertility

The prevailing suspicion of consanguineous unions in Western societies is centered on the belief that the offspring of a close-kin marriage will be physically and/or mentally disadvantaged. However, it also has been suggested that consanguineous relationships are less fertile than unions between nonrelatives. An influential early example of spiritual guidance on the inadvisability of consanguineous marriage was provided by Pope Gregory I in the late sixth century. Besides rather dubiously citing Leviticus 18:6 as the basis in Holy Scripture for the avoidance of cousin marriage, 'None of you shall approach to any that is near of kin to him, to uncover their nakedness,' and thus avoiding the specific guidelines provided in Leviticus 18:7–18, the Pope also claimed that unions between consanguineous spouses were infertile [38]. Where empirical information has been collected in human populations, the studies often have relied on small sample numbers, a shortcoming that makes the results difficult to assess [21]. However, in general, reduced levels of pathologic sterility have been reported among consanguineous couples [33, 72] with no evidence of an increase in fetal loss rates [21, 48], and indirect indicators of fetal survival, such as multiple birth rates and the secondary sex ratio, also failed to show an adverse inbreeding effect.

17.3.1 Genetically Determined Factors Influencing Human Mate Choice

It has been proposed that an olefactory mate-choice system operates in humans and other mammals. For example, in studies on the Hutterites, a highly endogamous Anabaptist sect resident in the USA, there was a lower than expected incidence of HLA haplotype matches between spouses, which was interpreted as evidence for the instinctive avoidance of partners with similar human leukocyte antigen (HLA) haplotypes [64].

The phenomenon of mate choice also was investigated in Swiss university students, with female students asked to conduct blind smell-testing of cotton T-shirts previously worn by male students and score the resultant body odors in terms of pleasantness and 'sexiness.' It was claimed that level of attractiveness of the male body odors was proportional to the degree of major histocompatibility complex (MHC) dissimilarity between the male subjects and the female testers, although follow-up studies indicated a high level of scoring variance [94]. Quite different results were obtained when a similar experiment was conducted with unmarried Hutterite women. In this case, the women were more likely to favor the odor of a 'donor' with whom they shared an intermediate number of HLA alleles. Furthermore, the positive preference appeared to be based on HLA alleles inherited from the subject's father but not her mother [49].

From the viewpoint of actual marriage partner choice, it can be convincingly argued that the findings of each of these studies have limited relevance in communities where consanguineous unions are strongly preferential, since in these more traditional societies marriage contracts are generally subject to parental decision-making [21].

17.3.2 Inbreeding and Fetal Loss Rates

Enhanced genetic compatibility would be expected between mother and fetus in consanguineous unions owing to their greater proportion of shared maternal and paternal genes. In keeping with this premise it has been claimed that intrauterine mortality is reduced in the pregnancies of consanguineous couples, with lower rates of conditions such as Rhesus (Rh) incompatibility [87] and pre-eclamptic toxemia [88]. Conversely, the fetal allograft hypothesis proposes that antigenic disparity between mother and fetus is beneficial to fetal development [29, 62], which would suggest higher losses in consanguineous pregnancies.

Table 17.5 Average number of live-born children by coefficient of inbreeding (F) (from [21])

	Coefficient of inbreeding (F)[a]				
	Uncle-niece/ double first cousin	First cousin	First cousin once removed	Second cousin	Nonconsanguineous
	$F=0.125$	0.0625	0.0313	0.0156	0
Mean number of live-born children	3.26	3.43	3.18	2.96	2.57
Number of studies	17/30	30/30	19/30	20/30	30/30

[a]The patterns of consanguineous marriage assessed varied between individual studies, with comparative data on fertility in first-cousin and nonconsanguineous matings available for all 30 studies analyzed, for uncle-niece/double first cousin unions in 17 studies, for first cousins once removed in 19 studies, and for second cousins in 20 studies

A positive association between parental HLA sharing at allele loci and recurrent abortion has been reported, with negative selection against individuals homozygous at HLA loci [51]. Unfortunately, retrospective data on pregnancies and prenatal losses may be subject to significant levels of recall bias, resulting in data of dubious reliability and significant underestimation of the levels of prenatal losses [95, 96].

Studies based on sequential human chorionic gonadotrophin (hCG) assays are more reliable, with on average some 40+% of all post-implantation conceptions lost. This figure rises with advancing maternal age, and in a hCG-based study in Bangladesh, while 45% of the pregnancies detected among women at 18 years of age spontaneously miscarried, among women aged 38 years fetal losses increased to 92% [65]. Since these levels of spontaneous abortion/miscarriage are very much higher than generally reported in inbreeding studies, there must be a strong suspicion that early pregnancy losses have been undetected or were underreported in investigations that relied solely on women's recall.

17.3.3 Comparative Fertility in Consanguineous and Nonconsanguineous Couples

A majority of comparative studies into fertility have shown a positive correlation between consanguinity and the number of live-born children. Data analyzed in a meta-analysis of 30 studies conducted in Asian and African countries are summarized in Table 17.5, with a higher mean number of children born in all categories of consanguineous marriage when compared with nonconsanguineous couples. Since the structure of each study varied according to the locally preferred types of consanguineous marriage, complete data were available only for first-cousin and nonconsanguineous couples, with first cousins showing the highest mean number of children ($n=3.43$). But even among the uncle-niece and double-first-cousin marriages ($F=0.125$), information on the numbers of live-born children had been published for 17 of the 30 populations, with mean fertility ($n=3.26$) higher than among nonconsanguineous spouses ($n=2.57$).

Typically, maternal age at marriage is negatively associated with consanguinity, resulting in a younger maternal age at first birth [22]. In addition, a higher mean age of motherhood has been reported among consanguineous couples [91], which supports the belief that early marriage, the earlier commencement of reproduction, and maximization of the maternal reproductive span by consanguineous couples are critical biosocial factors in determining family size.

The uptake of contraception may be lower in consanguineous couples [44], and reproductive compensation has been advanced as an additional explanation for the positive association between consanguinity and fertility, with infants dying at an early age rapidly replaced [63, 83]. Reproductive compensation could involve a conscious decision by parents to achieve their desired family size, but at the same time a further pregnancy following the death of a breast-fed infant may mainly be a consequence of the cessation of maternal lactational amenorrhea. The relationship between consanguinity, fertility, and reproductive compensation is however complicated, since the greater the number of children born to parents who are carriers of one or more detrimental recessive alleles, the higher the expectation that at least some of their progeny will be affected and so could die in early childhood.

17.4 Inbreeding and Inherited Disease

A significant positive association has been repeatedly demonstrated between consanguinity and early mortality, with disorders involving the expression of detrimental recessive genes especially involved. But since the poorest sections of all populations are most disadvantaged in terms of health and health care provision, overrepresentation of poorer and less educated families among consanguineous couples creates problems in assessing the effects of consanguinity on morbidity and mortality.

The first structured study into the medical effects of inbreeding was organized by Dr. Samuel Bemiss of Louisville, Kentucky [9], who in 1858 examined reports forwarded by medical colleagues on the health outcomes of unions ranging from incest ($F=0.25$) to third-cousin marriages ($F=0.0039$). Hundreds of further studies have been undertaken since that time, based on a variety of sampling techniques including pedigree analysis, household surveys and questionnaires administered to hospital in- and outpatients. In populations where uncle-niece or double-first-cousin and first-cousin marriages are preferential, unions beyond second cousins ($F<0.0156$) are of limited medical significance [13]. By comparison, where consanguineous unions generally are rare, biologically remote relationships in the present generation, such as third cousins and beyond ($F \leq 0.0039$) may nevertheless be of clinical importance in families where cumulative inbreeding at differing levels of consanguinity has occurred through time, with a consequent build-up of homozygosity. A similar phenomenon can arise in communities in which close-cousin unions have been proscribed on religious grounds but marriages between couples who are related to a lesser degree are permissible.

17.4.1 Consanguinity and Deaths in Infancy and Childhood

Data on the relationship between consanguinity and birth measurements have been mixed, with some studies suggesting that babies born to consanguineous parents are smaller and lighter, and therefore less likely to survive, whereas others have failed to detect any significant difference. By comparison, there is a general consensus that postnatal mortality and morbidity are higher among the progeny of consanguineous unions, and the rarer the frequency of a deleterious recessive gene in a population, the greater the proportional disadvantageous effect of inbreeding on its expression [13]. Estimates of the overall adverse effects of consanguinity have been highly variable, and it is generally accepted that earlier surveys may have produced spuriously high values due to inadequate control for important non-genetic variables that are known to influence childhood health, including maternal age and education, birth order, and birth intervals.

In developing countries, excess consanguinity-associated deaths are largely concentrated during the 1st year of life, but in many cases no specific cause of death is determined because of inadequate diagnostic facilities and parental reluctance to sanction prenatal diagnosis or autopsy examinations [16, 68]. Where a diagnosis has been possible, a clear link between consanguinity and autosomal recessive disorders is apparent, with multiple deaths reported in a proportion of consanguineous families, the effect being proportional to the level of parental genetic relatedness [13, 90].

17.4.2 Consanguinity and Childhood Morbidity

By definition, studies into the prevalence of birth defects are dependent on the diagnostic criteria employed and, in less developed countries, recognition of the symptoms of congenital disorders can often overlap with and reflect late fetal and neonatal survival rates. In developed countries, on average 4–5% of newborns have some form of birth defect [28]. A significant excess of major congenital defects has been diagnosed in consanguineous offspring, especially disorders with a complex etiology and a higher rate of recurrence, but the reported rates of birth defects associated with consanguinity have varied quite widely. Thus, in an Arab community in Israel first-cousin progeny had 3.8% excess major malformations [47], whereas a 26-year study based on the Medical Birth Registry of Norway reported 1.9% excess birth defects in Norwegian first-cousin couples and 2.4% among Pakistani migrant couples [89]. According to the Latin America Collaborative Study of Congenital

Malformations (ECLAMC) which examined 34,102 newborn infants for congenital anomalies, a significant association with consanguinity was found only for hydrocephalus, postaxial polydactyly, and bilateral cleft lip with or without cleft palate [74].

From these data it is difficult to identify major categories of disease that are specifically overrepresented in consanguineous progeny. Cognitive impairment is more common in consanguineous offspring, and a study of Arab schoolchildren in Israel indicated a 0.8- to 1.3-point decrease in mean IQ scores among first-cousin progeny by comparison with the children of unrelated parents, with a 2.6- to 5.9-point decline in the mean IQ scores of double-first-cousin progeny [8]. There also was a significantly higher level of variance in the IQ scores of the double-first-cousin progeny, suggesting the expression of detrimental recessive genes in some of these children. In Pakistan mild and severe intellectual and developmental disability also have been associated with consanguinity [32], although as with cognitive impairment poor social conditions may play significant causative roles in such cases.

As large, inbred pedigrees offer a cost- and time-effective strategy to locate disease mutations, the technique of homozygosity mapping in consanguineous families [52] has been widely adopted to identify the causative loci for disorders such as autosomal recessive nonsyndromal hearing loss, and blindness caused by early onset retinal dystrophies and childhood glaucoma, each of which has been reported at increased prevalence in specific consanguineous communities.

17.4.3 Consanguinity and Adult Mortality and Morbidity

Although potentially the most intriguing and challenging age range during which the adverse effects of consanguinity on health could be expressed, morbidity in adulthood has been underinvestigated. There is some preliminary evidence that certain cancers, especially breast cancer [54, 85], and specific forms of early-onset cardiovascular disease [46] are more prevalent in consanguineous individuals. The adult progeny of consanguineous unions also are overrepresented in institutions caring for persons with intellectual disability [13].

A major difficulty in assessing many of the findings obtained with adult-onset diseases is that they were derived from composite studies based on investigations conducted across discrete breeding populations, with little control for sociodemographic variables. Because of a lack of precise information on the composition and structure of the consanguineous and nonconsanguineous study groups, and appropriate matching for nongenetic variables, the comparisons drawn often prove to be irreproducible. An exception is the high prevalence of Alzheimer disease diagnosed in an Israeli Arab community, with more than one-third of the cases diagnosed members of a single clan (*hamula*) [35]. This supports an earlier study from the demographically well-characterized Saguenay area of Québec, Canada, which found that cases of late-onset cases of Alzheimer disease associated with the apolipoprotein (APOE) ε4 allele were significantly more inbred than controls [93].

Long-term studies conducted on the Dalmatian Islands, Croatia have suggested that inbreeding is a strong predictor for a wide range of late-onset disorders, including hypertension, coronary heart disease, stroke, cancer, uni-/bipolar depression, asthma, gout and peptic ulcer [76, 77]. At least in the short term, studies which concentrate on subcommunities of this type are more likely to provide information on disease-predisposing alleles than ethnically mixed populations. Although, even in population isolates with extensive pedigree data, failure to allow for the influence of distant genealogical loops can result in false positives in homozygosity mapping [55].

17.5 Incest

Incest is the most extreme example of human inbreeding, with the partners having a coefficient of relationship, r, of 0.5, so that any progeny born of an incestuous union would be expected to have a coefficient of inbreeding (F) of 0.25. Incest also differs from all other forms of inbred union since, in contemporary societies, it is universally regarded as both a criminal and a moral offense. Brother–sister marriages were recorded in Pharaonic and Ptolemaic Egypt, Zoroastrian Persia, the Inca Empire, and other historical dynasties, and they also were noted among nonroyal families in Roman Egypt from the first to the fourth centuries AD [81, 82]. Perhaps because of the high level of disapproval that incest attracts in modern societies, there are very few credible data sets on the outcomes of incestuous pregnancies. Yet the numbers of reported prosecutions on grounds of incest, usually involving father–daughter relationships, suggest that incest may be more common than is generally supposed, with brother–sister incest especially underreported.

In many instances where a child is born to a very young mother, the father of the child is not identified, even though incest may be suspected. If the child is healthy it is probable that no further action will be taken even if the child is offered for adoption. But when a sick child is born there is a greater imperative to investigate the cause of the illness, which in turn may lead to incest being identified. Under these circumstances significant overestimation of the adverse outcomes of incest could result, suggesting that considerable caution needs to be applied in the interpretation of incest data.

17.5.1 Mortality and Morbidity Estimates for Incestuous Matings

As shown in Table 17.6, according to data on 213 children collated from the four best-known studies of incest, conducted in the USA, UK, Czechoslovakia, and Canada over some 50 years [1, 6, 25, 84], only 46.0% of incestuous pregnancies resulted in the birth of a healthy infant. Follow-up ranged from 0.5 to 37 years, and among the incestuous offspring 39.4% had a recognized autosomal recessive disorder or a congenital malformation, had succumbed to sudden infant death, or had severe nonsyndromic intellectual disability, with deaths in 14.1% of cases. A further 14.6% of subjects had a mild disorder, including intellectual and developmental disability. By comparison, just 8.0% of the 113 nonincestuous controls died or were diagnosed with a serious defect, suggesting a mean level of excess mortality or serious defect in the incestuous progeny of 31.4%.

It should be stressed that in many cases the incestuous mothers were very young, with gynecological immaturity a possible adverse factor in the pregnancy, and in a substantial percentage of these cases either the mothers or the fathers, and sometimes both, had serious pre-existing physical or mental disorders [17]. Therefore, it is probable that the adverse pregnancy outcomes may, in part, have been due to causes other than detrimental recessive gene expression. Clarification of this issue will be dependent on additional data becoming available, but as already observed, the collection of unbiased information on the health sequelae of incest is extremely difficult.

17.6 Genetic Load Theory and Its Application in Consanguinity Studies

All humans are heterozygous for a number of detrimental recessive genes, and the term 'genetic load' refers to the decrease in the average fitness of a population caused by the expression of genes which reduce survival. Lethal gene equivalents are defined as the number of detrimental recessive genes carried by an individual in the heterozygous state which, if homozygous, would result in death. Therefore, by comparing death rates in the progeny of consanguineous and unrelated couples, it is possible to estimate the numbers of lethal gene equivalents in a community or population.

The number of lethal gene equivalents in a population can be calculated according to the formula:

$$-\log_e S = A + BF$$

where S is the proportion of survivors in the study population, A measures all deaths that occur under random mating, B represents all deaths caused by the expression of recessive genes via inbreeding, and F is the coefficient of inbreeding [58]. By plotting a

Table 17.6 Mortality and morbidity estimates for incestuous progeny. (From [55, 77, 81, 82])

Number studied	Follow-up (yr)	Autosomal recessive disorders	Congenital malformations/ sudden infant deaths	Nonspecific severe intellectual disability	Others, including mild intellectual disability	Normal
213	0.5–37	11.7%	16.0%	11.7%	14.6%	46.0%

weighted regression of the log proportion of survivors (S) at different levels of inbreeding (F), A can be determined from the intercept on the Y-axis at zero inbreeding (F = 0), and B (the number of lethal gene equivalents) is given by the slope of the regression.

Since consanguineous individuals have a greater probability of inheriting the same mutant allele(s) from a common ancestor, their progeny will be at a higher risk of expressing one or more recessive disorders. By calculating the number of lethal gene equivalents, the results of inbreeding surveys could be transformed into a meaningful and reproducible format, which then could be comparatively applied to the results of surveys in different populations. A multinational meta-analysis conducted on over 600,000 pregnancies and live births collated from 38 study populations indicated 4.4% excess prereproductive mortality in first-cousin progeny (measured from approximately 6 months gestation to a median age of 10 years) [20]. This level of excess mortality equates to 1.4 lethal equivalents per zygote, and a subsequent study of first cousin versus nonconsanguineous marriages in Italy from 1911-1964 produced equivalent results, with 3.5% excess deaths at $F = 0.0625$, i.e., 1.2 lethal equivalents per zygote [27].

17.7 Genomic Approaches to Measuring Inbreeding at Individual and Community Levels

The direct estimation of an individual's inbreeding coefficient by reference to genomic data offers many advantages, since it can include the influence of historical levels and patterns of inbreeding that may not be identifiable within a pedigree. A maximum-likelihood method of analysis has been developed using simulated whole genome data which permits inference of the identity by descent (IBD) status of both alleles of an individual at each marker along the genome. The method also provides a variance measure for the estimates and, for example, it was shown that while the mean value for IBD status for first cousins was 0.0625, at individual loci the calculated values ranged from 0.03 to 0.12 [53].

Microsatellite analysis of DNA samples obtained from UK ethnic migrants showed that in the Pakistani Muslim community, in which consanguineous marriage is widely favored and practiced, the observed F values were much higher than in a co-resident Indian Sikh community which avoided consanguinity [66]. This study also indicated significant genetic substructuring, which could interfere with estimates of the frequency of recessive disease genes. Using both SNP and microsatellite analysis, a subsequent study of UK Pakistani consanguineous individuals with a range of autosomal recessive diseases showed that, on average, persons whose parents were first cousins ($F = 0.0625$) were actually homozygous at 11% of the loci tested, with a range of 5-20% [98].

The findings of these studies indicate the influence of cumulative inbreeding on genome structure at both individual and community levels. In addition, they confirm the desirability of a prior understanding among researchers and clinicians of the social structure of communities, in particular their marriage patterns, since information of this nature could have a major role in determining the patterns and frequencies of specific genetic disorders.

17.8 The Influence of Endogamy and Consanguinity in Human Populations

Inter- and intrapopulation fluctuations in the frequencies of coding genes are well recognized and documented, and it seems probable that similar variations will be demonstrated in the control of gene expression. Three quite different, representative human populations, Finland, India, and Israeli Arabs, will be used to illustrate the impacts of founder effect, random drift and consanguinity on genetic structure and the prevalence and expression of recessive disease genes.

17.8.1 The Finnish Disease Heritage

Finland is a small and formerly quite isolated country with a unique genetic history. The original inhabitants are thought to have been arctic northern European Uralic speakers who settled the territory of Finland some 6,500 years ago after the decline of the last Ice Age.

Somewhat later arrivals included peoples from southeastern and western Europe between 5,000–6,000 and 4,500 years ago, respectively, with later minor waves of German, Scandinavian, and Baltic peoples [60].

The initial population settlement was concentrated in the south and west of the country, and at the start of the twelfth century the total number of inhabitants was less than 50,000. Commencing in the sixteenth century there was internal migration northward, and by the mid-seventeenth century the total numbers had increased to 400,000–450,000. But in the Great Famine of 1696–1697 approximately 25–33% of the inhabitants died, and additional major population losses occurred owing to plague at the beginning of the 1700s, and famine following crop failures in 1866–1868 [60].

The concept of the Finnish Disease Heritage (Table 17.7) was introduced in 1973 to describe some 36 mostly autosomal recessive diseases that are typical of the Finnish population while rare in other populations [59, 61]. Conversely, disorders which are common in most other northern European populations, such as cystic fibrosis and phenylketonuria, are very rare in Finland. The causative genes have been identified for 29 of the Finnish diseases [61], and four main groups of disorders can be categorized according to their patterns of geographic distribution in the current population of 5.3 million (www.findis.org).

For the most common diseases, such as congenital nephrosis, cartilage hair hypoplasia, and aspartylglucosaminuria, a lysosomal storage disease which causes intellectual and developmental disability, the birthplaces of the grandparents of affected individuals in the present generation are widely distributed throughout the country. With a second larger group of disorders, e.g., Mulibrey nanism, Usher syndrome type 3, and nonketotic hyperglycinemia, there is clustering in geographic subregions, usually areas initially populated from the sixteenth century onward. A third group, typified by Meckel syndrome and diastrophic dysplasia, is found predominantly in the western early settlement area. While the fourth group of disorders, comprising Northern epilepsy and the Finnish variant of late infantile neuronal ceroid lipofuscinosis, originated locally in the Kainuu region close to the eastern border and in Southern Ostrobothnia, respectively. Seven other autosomal recessive, autosomal dominant, and X-linked disorders have been included in the Finnish Disease Heritage, and a further five diseases are under investigation and may be incorporated in future years [61].

Given the dispersed nature of much of the population, and the small numbers of individuals, it might

Table 17.7 The Finnish disease heritage. (From [61])

Disease	Incidence in Finland
Autosomal recessive	
Congenital nephrosis	1:8,000
Infantile neuronal ceroid lipofuscinosis	1:14,000
Meckel syndrome	1:15,000
Unverricht-Lundborg disease	1:17,000
Aspartylglucosaminuria	1:18,000
Cartilage-hair dysplasia	1:18,000
Spielmeyer–Sjögren disease	1:19,000
Hydrolethalus syndrome	1:22,000
Diastrophic dysplasia	1:22,000
Autoimmune polyendocrinopathy-candidiasis-ectodermal dystrophy	1:27,000
Lethal congenital contracture syndrome (Herva)	1:29,000
Congenital chloride diarrhea	1:33,000
Mulibrey nanism	1:37,000
Usher syndrome type 3	1:42,000
Salla disease	1:42,000
Cornea plana congenita	1:46,000
Congenital lactase deficiency	1:48,000
Muscle-eye-brain disease	1:52,000
Nonketotic hyperglycinemia	1:52,000
Lethal arthrogryposis with anterior horn cell disease (Vuopala)	1:53,000
Jansky–Bielschowsky disease variant	1:59,000
Hyperornithinemia with gyrate atrophy of choroid and retina	1:63,000
GRACILE syndrome (Fellman)	1:64,000
Selective malabsorption of vitamin B_{12}	1:68,000
Nasu–Hakola disease	1:71,000
Lysinuric protein intolerance	1:76,000
PEHO syndrome	1:78,000
IOSCA syndrome	1:90,000
Cohen syndrome	1:105,000
Rapadilino syndrome	1:105,000
Follicle stimulating hormone-resistant ovaries (Aittomäki)	1:127,000
Northern epilepsy	1:176,000
Autosomal dominant	
Meretoja disease	~1:6,000
Tibial muscular dystrophy	~3/year
X-chromosome	
Choroideremia	~2/year
Retinoschisis	~1:17,000

have been expected that consanguinity contributed substantially to the prevalence of the various recessive disorders. In fact, except for some parishes in northern Finland with a substantial Sami minority, first-cousin marriage was historically rare in the country, in part because of the dispensation requirement for such marriages that remained in place until 1872, with fees payable to the King. Thus, even among the Swedish-speaking Lutheran minority of Finland the attitudes towards consanguinity differed from those in neighboring Sweden, where first-cousin marriage was freed from civil law restrictions in 1844, leading to an increase in first-cousin unions during the remainder of the nineteenth century [19]. Instead, the historical population profile of Finland was characterized by conditions under which founder effects, genetic drift, and demographic bottlenecks occurred, and it is these factors that have shaped and determined the present-day national and regional profiles of genetic disease [60].

17.8.2 Inter- and Intra-population Differentiation in India

The present-day population of India is estimated at some 1,200 million, having increased from 271 million in the year 1900 and 361 million in 1950, resulting in a greatly enlarged overall effective population size. From a genetic perspective a further significant aspect of the Indian population is that, in common with Middle Eastern and North African populations, and neighboring Pakistan and Afghanistan, where tribal and clan marriage boundaries are in place, marriage in India is contracted within highly endogamous castes [18].

Caste membership is hereditary and defines an individual's position within Indian society. The caste system is believed to have been in existence for at least 2,000 years, and in the past it appears to have been somewhat more flexible, with the emergence of new castes and subcastes recorded during the eighteenth and nineteenth centuries. As an example of the current level of demographic and genetic complexity within India, there are seven major religions, and 299 different languages spoken by 4,635 officially recognized ethnic communities, which in turn are composed of an estimated 50,000–60,000 highly endogamous subpopulations [37].

The majority Hindu population, which accounts for approximately 80% of the national population and so currently numbers over 1,000 million, is structured into four major hierarchical groups (*varna*), Brahmins, Kshatriyas, Vaishas, and Sudras, The four *varna* in turn are subdivided into numerous castes (*jati*) and subcastes, and virtually all Hindu marriages continue to be contracted within hereditary caste boundaries. As these institutions are reputed to have been in existence for some 2,000 years, there has been ample opportunity for intercaste genetic differentiation to have occurred via founder effect and genetic drift, especially given the much smaller, and multiply subdivided, population of India in historical times. Therefore, it would be expected that, through time, caste-specific genetic disease profiles would have developed.

There also is a major dichotomy between the majority Indo-Europeans of north India, who avoid consanguineous marriage, and the Dravidian Hindus of south India, where first-cousin and uncle-niece marriage is widely popular and in many communities preferential (Fig. 17.2). This subdivision is believed to date back to the Codes of Manu compiled around 200 BC [50], and it continues to the present day. Given the long-term preference for close consanguineous marriage it was proposed that the endogamous and largely consanguineous populations of south India would have purged lethal recessive genes from their gene pools [80]. Empirical evidence of the range and prevalence of genetic disorders in the current South Indian population has indicated that this outcome is improbable [71], probably due to reproductive compensation, which would effectively delay if not nullify the elimination of deleterious recessives from the gene pool(s).

17.8.3 Consanguinity and the Distribution of Disease Alleles in Israeli Arab Communities

Arab populations in Israel typify a third major form of human genetic organization. For some 500 years prior to the early twentieth century, Arab communities in the Holy Land were part of the Ottoman Empire. As such, members of these communities were able to mix freely with other neighboring Arab populations, although in most cases marriages were contracted within tribes and frequently at the level of the clan (*hamula*). In addition, consanguineous unions were widely favored, in particular father's brother's daughter-first-cousin marriage (termed *ibn amm*), with double-first-cousin and second-cousin unions also quite common. Conditions favoring village endogamy increased markedly in the years following establishment of the state of Israel in 1948, with the initial movement of an estimated 700,000 people to other neighboring countries and the effective closure of the borders between Israel and the surrounding Arab states. Since 1948, the Arab population of Israel has undergone very rapid natural expansion and now totals over 1.2 million.

Autosomal recessive disorders were found to be more common in the progeny of consanguineous parents [101], and the prevalence rates of congenital malformations were higher in Palestinian Arab and Druze communities, where clan endogamy and consanguinity were strongly favored, than in the more exogamous Jewish and Christian communities [102]. As in Finland, some diseases, such as β-thalassemia, familial Mediterranean fever, and deafness are frequent in the whole Arab population, whereas others are restricted to specific regions or villages. For a specific, rare inherited disease, a single founder mutation would ordinarily be expected within a small geographic area. However, in the case of the lysosomal storage disorder metachromatic leukodystrophy caused by a deficiency of arylsulfatase A, multiple causative mutations were identified within a restricted region, suggesting the occurrence of a number of founder mutations at this disease locus [42]. Subsequent studies have further demonstrated some 19 mostly chronic autosomal recessive disorders in a village with 8,600 inhabitants, i.e., a prevalence for these disorders of approximately 1/70 [103].

A detailed investigation of 12 recessive mutations affecting the inhabitants of a single village has indicated both founder effects and de novo mutations, and the transfer of mutations between families via marriage. Under such circumstances, a single family with one or more family members diagnosed with a specific recessive disorder would usually indicate a recent event, whereas a rare disease affecting members of several families would be more convincingly interpreted as an older mutation [104]. But in all such cases, a thorough understanding of past and present marriage patterns is an essential prerequisite, and given the demographic history of founder effects, migration, population bottlenecking, and rapid expansion, in combination with clan endogamy and preferential consanguinity, the resultant overall picture becomes kaleidoscopic.

17.9 Evaluating Risk in Consanguineous Relationships

The three preceding examples illustrate some of the unexpected complexities that can be encountered when dealing with actual human populations, and the importance of at least a basic knowledge of the demographic structure of a population. They also highlight the difficulties that may be encountered in some populations in differentiating between random inbreeding, brought about by founder effect, endogamy and genetic drift, and preferential consanguinity. Yet this differentiation is critical in accurately assessing the outcomes of consanguineous unions, and in providing risk estimates in settings such as a genetic counseling clinic.

The importance of recognizing and controlling for remote levels of consanguinity in gene association studies has already been noted [55], as have the combined roles of consanguinity and population subdivision in many clinical situations [18]. Although it has been claimed that statistical methods such as principal components analysis can be employed to correct for population stratification in genome-wide studies [69], their successful application in societies as multiply subdivided as India would be a very major challenge. Greater care is therefore warranted in the selection of cases and controls for gene association studies and, if properly conducted, greater reproducibility in their outcomes should follow.

From a practical perspective, the ability to purvey risk in an unambiguous and readily understood manner is an all-important issue in human and medical genetics. Risk estimates expressed as relative risks, odds ratios, or attributable risks, i.e., the fraction of cases in a population that can be attributed to a particular risk factor, are very useful in epidemiological studies. However, in a genetic counseling setting the probability of an adverse outcome needs to be presented in as uncomplicated a manner as possible, taking into account factors such as the background population risk, degree of consanguinity, and relevant family history [11].

When dealing with a topic as potentially sensitive as consanguineous marriage, the avoidance of any potential misunderstanding or misinterpretation by clients and their families is critical.

17.10 Concluding Comments

The development of society-compatible education programs on consanguineous marriage, in combination with evidence-based screening and genetic counseling guidelines, is of paramount importance. It has been proposed that communities in which consanguineous marriage is preferential can be at an advantage when screening for deleterious mutations, for example, in the case of β-thalassemia in Pakistan [3], as there is a high probability that all family members will be homozygous for the same mutation. Under these circumstances, identification of the specific mutation in an affected individual can serve as a diagnostic marker for an extended family group at high genetic risk. Some caution is, however, needed in this approach, since past intermarriage with other unrelated families and communities could have led to significant gene admixture and hence increased the likelihood of compound heterozygosis. But in general the concept is valid and useful, as indicated by a successful screening program for autosomal recessive nonsyndromic intellectual disability in an Israeli Arab community [7].

The future status of consanguineous marriage is a matter of conjecture. Currently an estimated 1,000 million people live in countries where from 20% to over 50% of marriages are consanguineous [18], and it seems highly improbable that a form of marriage which remains so widely popular would rapidly decline in popularity. But strenuous semiofficial efforts are being made to lessen the appeal of consanguinity in many developing countries, often to the distress and embarrassment of consanguineous couples. The situation for migrant communities in western societies is different again, since there is both an attraction to continue with a form of marriage that has been undertaken for many generations in their countries of origin, and at the same time a desire among some younger members of migrant families to adopt the social mores of their new homeland, including exogamous marriage customs.

Within migrant communities there is a much greater awareness of genetic disease than would have been the case in their homeland, and of the increased risk of an affected child being born to a consanguineous couple [13]. Ultimately, declining family sizes and a consequent reduction in the availability of potential marriage partners within the immediate family may prove to be the major factor in determining the future prevalence of consanguineous unions. However, it also has to be acknowledged that the presence of several family members with a major disabling disorder may severely limit the marriage opportunities of other family members, thus increasing the probability of further intrafamilial unions [7, 18].

While medical genetics is a relatively new subject, consanguineous marriage has been, and remains, a core feature of many successful human societies. A recent World Health Organization Report on Medical Genetics Services in Developing Countries advised that: 'Preference for consanguineous marriage is a feature of the socio-cultural context within which medical genetic services must work' [97]. Adoption of this eminently sensible and nonjudgmental approach should ensure that the health needs of families and communities in both the developing and developed world can best be met.

References

1. Adams MS, Neel JV (1967) Children of incest. Pediatrics 40:55–62
2. Agarwal SS, Singh U, Singh PS, Singh SS, Das V, Sharma A et al (1991) Prevalence and spectrum of congenital malformations in a prospective study at a teaching hospital. Indian J Med Res 94:413–419
3. Ahmed S, Saleem M, Modell B, Petrou M (2002) Screening extended families for genetic hemoglobin disorders in Pakistan. N Engl J Med 347:1162–1168
4. Ahmed T, Ali SM, Aliaga A, Arnold F, Ayub M, Bhatti MH et al (1992) Pakistan demographic and health survey 1990/91. Pakistan National Institute of Population Studies and Macro International, Islamabad
5. Al-Gazali LI, Bener A, Abdulrazzaq YM, Micallef R, Al-Khayat AI, Gaber T (1997) consanguineous marriages in the United Arab Emirates. J Biosoc Sci 29:491–497
6. Baird PA, McGillivray B (1982) Children of incest. J Pediatr 101:854–857
7. Basel-Vanagaite L, Taub E, Halpern GJ, Drasinover V, Magal N, Davidov B, Zlotogora J, Shohat M (2007) Genetic screening for autosomal recessive nonsyndromic mental retardation in an isolated population in Israel. Eur J Hum Genet 15:250–253

8. Bashi J (1977) Effects of inbreeding on cognitive performance. Nature 31:440–442
9. Bemiss SM (1858) Report on influence of marriage of consanguinity upon offspring. Trans Am Med Assoc 11:319–425
10. Bener A, Alali KA (2006) Consanguineous marriage in a newly developed country: the Qatari population. J Biosoc Sci 38:239–246
11. Bennett R, Motulsky AG, Bittles AH, Hudgins L, Uhlrich S, Lochner Doyle D, Silvey K, Scott RC, Cheng E, McGillivray B, Steiner RD, Olson D (2002) Genetic counseling and screening of consanguineous couples and their offspring: recommendations of the National Society of Genetic Counselors. J Genet Couns 11:97–119
12. Bittles AH (1994) The role and significance of consanguinity as a demographic variable. Popul Dev Rev 20:561–584
13. Bittles AH (2001) Consanguinity and its relevance to clinical genetics. Clin Genet 60:89–98
14. Bittles AH (2002) Endogamy, consanguinity and community genetics. J Genet 81:91–98
15. Bittles AH (2003) The bases of Western attitudes to consanguineous marriage. Dev Med Child Neurol 45:135–138
16. Bittles AH (2003) Consanguineous marriage and childhood health. Dev Med Child Neurol 45:571–576
17. Bittles AH (2005) Genetic aspects of inbreeding and incest. In: Wolf AP, Durham WH (eds) Inbreeding, incest, and the incest taboo. Stanford University Press, Stanford, pp 38–60
18. Bittles AH (2008) A Community Genetics perspective on consanguineous marriage. Community Genet 11:324–330
19. Bittles AH, Egerbladh I (2005) The influence of past endogamy and consanguinity on genetic disorders in northern Sweden. Ann Hum Genet 69:1–10
20. Bittles AH, Neel JV (1994) The costs of human inbreeding and their implications for variations at the DNA level. Nat Genet 8:117–121
21. Bittles AH, Grant JC, Sullivan SG, Hussain R (2002) Does inbreeding lead to decreased human fertility? Ann Hum Biol 29:111–131
22. Bittles AH, Mason WM, Greene J, Appaji Rao N (1991) Reproductive behavior and health in consanguineous marriages. Science 252:789–794
23. Bundey S, Alam H, Kaur A, Mir S, Lancashire RJ (1990) Race, consanguinity and social features in Birmingham babies: a basis for a prospective study. J Epidemiol Community Health 44:130–135
24. Calderón R, Peña JA, Delgado J, Morales B (1998) Multiple kinship in two Spanish regions: new model relating multiple and simple consanguinity. Hum Biol 70:535–561
25. Carter CO (1967) Risk to offspring of incest. Lancet 289:436
26. Castilla EE, Gomez MA, Lopez-Camelo JS, Paz JE (1991) Frequency of first-cousin marriages from civil marriage certificates in Argentina. Hum Biol 63:203–210
27. Cavalli-Sforza LL, Moroni A, Zei G (2004) Consanguinity, inbreeding, and genetic drift in Italy. Princeton University Press, Princeton, p 292
28. Christianson A, Howson CP, Modell M (2006) Global report on birth defects. March of Dimes, White Plains, NY
29. Clarke B, Kirby DRS (1966) Maintenance of histocompatibility polymorphisms. Nature 211:999–1000
30. da Fonseca LG, Freire-Maia N (1970) Further data on inbreeding levels in Brazilian populations. Soc Biol 17:324–328
31. De Braekeleer M, Ross M (1991) Inbreeding in Saguenay-Lac-St-Jean (Québec, Canada): a study of Catholic Church dispensations. Hum Hered 41:379–384
32. Durkin MS, Hasan ZM, Hasan KZ (1998) Prevalence and correlates of mental retardation in Karachi, Pakistan. Am J Epidemiol 147:281–288
33. Edmond M, De Braekeleer M (1993) Inbreeding effects on fertility and sterility: a case-control study in Saguenay-Lac-Saint-Jean (Québec, Canada) based on a population registry. Ann Hum Biol 20:545–555
34. El-Hazmi MAF, Al-Swailem AR, Warsy AS, Al-Swailem AM, Sulaimani R, Al-Meshari AA (1995) Consanguinity among the Saudi Arabian population. J Med Genet 32:623–626
35. Farrer LA, Bowirrat A, Friedland RP, Waraska K, Kprczyn AD, Baldwin CT (2003) Identification of multiple loci for Alzheimer disease in a consanguineous Israeli-Arab community. Hum Mol Genet 12:415–422
36. Freire-Maia N (1968) Inbreeding levels in American and Canadian populations: a comparison with Latin America. Eugen Q 15:22–33
37. Gadgil M, Joshi NV, Prasad UVS, Manoharan S, Patil S (1998) Peopling of India. In: Balusubramanian D, Appaji Rao N (eds) The Indian human heritage. Hyderabad Universities Press, Hyderabad, pp 100–129
38. Goody J (1985) The development of the family and marriage in Europe. Cambridge University Press, Cambridge, pp 134–136
39. Hafez M, El-Tahan H, Awadalla M, El-Khayat H, Abdel-Gafar A, Ghoneim M (1983) Consanguineous matings in the Egyptian population. J Med Genet 20:58–60
40. Hamamy H, Jamhawi L, Al-Darawsheh J, Ajlouni K (2005) Consanguineous marriages in Jordan: why is the rate changing with time? Clin Genet 67:511–516
41. Hashmi MA (1997) Frequency of consanguinity and its effect on congenital malformation – a hospital based study. J Pak Med Assoc 47:75–78
42. Heinisch U, Zlotogora J, Kafert S, Gieselmann V (1995) Multiple mutations are responsible for the high frequency of metachromatic leukodystrophy in a small geographic area. Am J Hum Genet 56:51–57
43. Hussain R, Bittles AH (1998) The prevalence and demographic characteristics of consanguineous marriages in Pakistan. J Biosoc Sci 30:261–275
44. Hussain R, Bittles AH (1999) Consanguinity and differentials in age at marriage, contraceptive use and fertility in Pakistan. J Biosoc Sci 31:121–138
45. Imaizumi Y (1986) A recent survey of consanguineous marriages in Japan. Clin Genet 30:230–233
46. Ismail J, Jafar TH, Jafary FH, White F, Faruqui AM, Chaturvedi N (2004) Risk factors for non-fatal myocardial infarction in young South Asian males. Heart 90:259–263
47. Jaber L, Merlob P, Bu X, Rotter JI, Shohat M (1992) Marked parental consanguinity as a cause for increased major malformations in an Israeli Arab community. Am J Med Genet 44:1–6
48. Jaber L, Merlob P, Gabriel R, Shohat M (1997) Effects of consanguineous marriage on reproductive outcome in an Arab community in Israel. J Med Genet 34:1000–1002
49. Jacob S, McClintock MK, Zelano B, Ober C (2002) Paternally inherited HLA alleles are associated with women's choice of male odor. Nat Genet 30:175–179

50. Kapadia KM (1958) Marriage and the family in India, 2nd edn. Oxford University Press, Calcutta, pp 117–137
51. Kostyu DD, Dawson DV, Elias S, Ober C (1993) Deficit of HLA homozygotes in a Caucasian isolate. Hum Immunol 37:135–142
52. Lander ES, Botstein D (1987) Homozygoisty mapping: a way to map human recessive traits with the DNA of inbred children. Science 236:1567–1570
53. Leutenegger A-L, Prum B, Genin E, Verny C, Lemainque A, Clerget-Darpoux F, Thompson EA (2003) Estimation of the inbreeding coefficient through use of genomic data. Am J Hum Genet 73:516–523
54. Liede A, Malik IA, Aziz Z, Rios L, Kwan E, Narod SA (2002) Contribution of BRAC1 and BRAC2 mutations of breast and overian cancer in Pakistan. Am J Hum Genet 71:595–606
55. Liu F, Elefante S, van Duijn CM, Aulchenko YS (2006) Ignoring distant genealogic loops leads to false-positives in homozygosity mapping. Ann Hum Genet 70:965–970
56. Magnus P, Berg K, Bjerkedal T (1985) Association of parental consanguinity with decreased birth weight and increased rate of early death and congenital malformations. Clin Genet 28:335–342
57. Modell B, Darr A (2002) Genetic counselling and customary consanguineous marriage. Nat Rev Genet 3:225–229
58. Morton NE, Crow JF, Muller HJ (1956) An estimate of the mutational damage in man from data on consanguineous marriages. Proc Natl Acad Sci USA 42:855–863
59. Norio R (2003) Finnish disease heritage. I: characteristics, causes, background. Hum Genet 112:441–456
60. Norio R (2003) Finnish disease heritage. II: population prehistory and genetic roots of Finns. Hum Genet 112:457–469
61. Norio R (2003) Finnish disease heritage. III: the individual diseases. Hum Genet 112:470–526
62. Ober C (1998) HLA and pregnancy: the paradox of the fetal paradox. Am J Hum Genet 62:1–5
63. Ober C, Hyslop T, Hauck WW (1999) Inbreeding effects in humans: evidence for reproductive compensation. Am J Hum Genet 64:225–231
64. Ober C, Weitkamp LR, Cox N, Dytch H, Kostyu D, Elias S (1997) HLA and mate choice in humans. Am J Hum Genet 16:497–504
65. O'Connor KA, Holman D, Wood JW (1998) Declining fecundity and ovarian ageing in a natural fertility population. Maturitas 30:127–136
66. Overall ADJ, Ahmad M, Thomas MG, Nichols RA (2003) An analysis of consanguinity and social structure with the UK Asian population using microsatellite data. Ann Hum Genet 67:525–537
67. Port KE, Bittles AH (2001) A population-based estimate of the prevalence of consanguineous marriage in Western Australia. Community Genet 4:97–101
68. Port KE, Mountain H, Nelson J, Bittles AH (2005) The changing profile of couples seeking genetic counseling for consanguinity in Australia. Am J Med Genet 132A:159–163
69. Price AL, Paterson NJ, Plenge RM, Weinblatt ME, Shadick NA, Reich D (2006) Principal components analysis corrects for stratification in genome-wide association studies. Nat Genet 38:904–909
70. Puri RK, Verma IC, Bhargava I (1978) Effects of consanguinity in a community in Pondicherry. In: Verma IC (ed) Medical genetics in India, vol 2. Auroma, Pondicherry, pp 129–139
71. Radha Rama Devi A, Appaji N, Bittles AH (1987) Consanguinity and the incidence of childhood genetic disease in Karnataka, South India. J Med Genet 24:362–365
72. Rao PSS, Inbaraj SJ (1977) Inbreeding effects on human reproduction in Tamil nadu of South India. Ann Hum Genet 41:87–98
73. Riou SE, Younsi C, Chaabouni H (1989) Consanguinité dans la population du Nord de la Tunisie. La Tunisie Médicale 67:167–172
74. Rittler M, Liascovich R, López-Camelo J, Castilla EE (2001) Parental consanguinity in specific types of congenital anomalies. Am J Med Genet 102:36–43
75. Rosenberg NA, Pritchard JK, Weber JL, Cann HM, Kidd KK, Zhivotovsky LA, Feldman MW (2002) Genetic structure of human populations. Science 298:2381–2385
76. Rudan I, Smolej-Narancic N, Campbell H, Carothers A, Wright A, Janicijevic B, Rudan P (2003) Inbreeding and the genetic complexity of human hypertension. Genetics 163:1011–1021
77. Rudan I, Rudan D, Campbell H, Carothers A, Wright A, Smolej-Narancic N, Janicijevic B, Jin L, Chakraborty R, Deka R, Rudan P (2003) Inbreeding and risk of late onset complex disease. J Med Genet 40:925–932
78. Saadat M, Ansari-Lari M, Farhud DD (2004) Consanguineous marriage in Iran. Ann Hum Biol 31:263–269
79. Saha N, El Sheikh FS (1988) Inbreeding levels in Khartoum. J Biosoc Sci 20:333–336
80. Sanghvi LD (1966) Inbreeding in India. Eugenics Quarterly 13:291-301
81. Scheidel W (1997) Brother-sister marriage in Roman Egypt. J Biosoc Sci 29:361–371
82. Scheidel W (2005) Ancient Egyptian sibling marriage and the Westermarck effect. In: Wolf AP, Durham WH (eds) Inbreeding, incest, and the incest taboo. Stanford University Press, Stanford, pp 93–108
83. Schull WJ, Furusho T, Yamamoto M, Nagano H, Komatsu I (1970) The effects of parental consanguinity and inbreeding in Hirado. Japan. IV Fertility and reproductive compensation. Humangenetik 9:294–315
84. Seemanová E (1971) A study of incestuous matings. Hum Hered 21:108–128
85. Shami SA, Qaisar R, Bittles AH (1991) Consanguinity and adult morbidity in Pakistan. Lancet 338:954–955
86. Spielman RS, Bastone LA, Burdick JT, Morley M, Ewens WJ, Cheung VG (2007) Common genetic variants account for differences in gene expression among ethnic groups. Nat Genet 39:226–231
87. Stern C, Charles DR (1945) The Rhesus gene and the effect of consanguinity. Science 101:305–307
88. Stevenson AC, Davison BCC, Say B, Ustuoplu S, Liya D, Abul-Einem M, Toppozada HK (1971) Contribution of feto-maternal incompatibility to aetiology of pre-eclamptic toxaemia. Lancet 2:1286–1289
89. Stoltenberg C, Magnus P, Lie RT, Dalveit AK, Irgens LM (1997) Birth defects and parental consanguinity in Norway. Am J Epidemiol 145:439–448

90. Stoltenberg C, Magnus P, Skrondal A, Lie RT (1999) Consanguinity and recurrence risk of stillbirth and infant death. Am J Public Health 89:517–523
91. Tunçbilek E, Koç I (1994) Consanguineous marriage in Turkey and its impact on fertility and mortality. Ann Hum Genet 58:321–329
92. Tunçbilek E, Ulusoy M (1989) Consanguinity in Turkey in 1988. Turk J Popul Stud 11:35–46
93. Vézina H, Heyer É, Fortier I, Ouellette G, Robitaille Y, Gauvreau D (1999) A genealogical study of Alzheimer disease in the Saguenay region of Québec. Genet Epidemiol 16:412–425
94. Wedekind C, Seebeck T, Bettens F, Paepke AJ (1995) MHC-dependent mate preference in humans. Proc R Soc Lond Ser B 260:245–249
95. Wilcox AJ, Horney LF (1984) Accuracy of spontaneous abortion recall. Am J Epidemiol 120:727–733
96. Wilcox AJ, Weinberg CR, O'Connor JF, Baird DD, Schlatterer JP, Canfield RE, Armstrong EG, Nisula BC (1988) Incidence of early loss of pregnancy. N Engl J Med 319:189–194
97. WHO (2006) Medical genetics in developing countries: the ethical, legal and social implications of genetic testing and screening. World Health Organization, Geneva
98. Woods CG, Cox J, Springell K, Hampshire DJ, Mohamed MD, McKibbin M, Stren R, Raymond FL, Sandford R, Sharif SM, Karbani G, Ahmed M, Bond J, Clayton D, Inglehearn CF (2006) Quantification of homozygosity in consanguineous individuals with autosomal recessive disease. Am J Hum Genet 78:889–896
99. Wu L (1987) Investigation of consanguineous marriages among 30 Chinese ethnic groups. Hered Dis 4:163–166
100. Zhan J, Qin W, Zhou Y, Chen K, Yan W, Yu W (1992) Effects of consanguineous marriages on hereditary diseases: a study of the Han ethnic group in different geographic districts of Zejiang province. Natl Med J China 172:674–676 In Chinese
101. Zlotogora J (1997) Genetic disorders among Palestinian Arabs: 1 Effects of consanguinity. Am J Med Genet 68:472–475
102. Zlotogora J, Haklai Z, Rotem N, Georgi M, Berlovitz I, Leventhal A, Amitai Y (2003) Relative prevalence of malformations at birth among different religious communities in Israel. Am J Med Genet 122A:59–62
103. Zlotogora J, Hujerat Y, Barges S, Shalev SA, Chakravarti A (2006) The fate of 12 recessive mutations in a single village. Ann Hum Genet 71:202–208
104. Zlotogora J, Shalev SA, Habibullah H, Barjes S (2000) Genetic disorders among Palestinian Arabs: 3 Autosomal recessive disorders in a single village. Am J Med Genet 92:343–345

Human Evolution

Michael Hofreiter

Abstract The study of human evolution is as old as evolutionary biology itself. Despite this long history, progress was slow in many fields. Until the 1980s, neither the closest living relative of our species nor the geographical origin of modern humans was known. However, since then a flood of new data has provided detailed insights into many aspects of human evolution. Thus, population genetic analyses of DNA sequences unequivocally identified Africa as the continent of modern human origin and provided information about the colonization of the whole globe by modern humans. A whole plethora of recently discovered fossil remains of hominid species show a detailed picture – albeit not yet well understood – of modern human ancestors and side branches. And ancient DNA analyses have revealed our relationship to our closest relatives, the extinct Neanderthals, with the prospect of the complete Neanderthal genome being sequenced soon. Finally, first candidate genes have been identified that may have been of critical importance in the evolutionary process of becoming human. At the same time, traits such as cultural tradition and tool use have been discovered in other primate species, especially our closest relatives, the great apes, leaving few traits that may be exclusively human. Identifying these traits, revealing their genetic basis and understanding the evolutionary forces that lead to their selection will be the challenges to research in human evolution with the aim of eventually understanding what makes us human.

Contents

18.1	Historical Overview	530
18.1.1	Before and Around Darwin	530
18.1.2	Sarich and Wilson	530
18.1.3	From a Straight Line to a Bush of Hominid Species and Beyond	531
18.2	The Fossil Record	532
18.2.1	Palaeoanthropology	532
18.2.2	Neanderthals and Ancient DNA	534
18.3	The Genetics of Human Evolution	536
18.3.1	The Genomes of Humans and Their Relatives	536
18.3.2	Diversity Within the Human Genome	538
18.3.3	Positive and Negative Selection in the Human Genome	541
18.4	Recent Events in Human Evolution	545
18.4.1	Out of Africa into New-found Lands	545
18.4.2	Domestication	549
18.4.3	Modern Human Population Structure	550
References		551

M. Hofreiter (✉)
Max Planck Institute for Evolutionary Anthropology,
Deutscher Platz 6, 04103 Leipzig, Germany
e-mail: hofreiter@eva.mpg.de

18.1 Historical Overview

18.1.1 Before and Around Darwin

Humans are a unique species – there can be little doubt about this statement, as it is true for any of the some million species living on earth. However, humans are indeed special in some ways. For example, contrary to all other species, for the human species *Homo sapiens*, no type specimen, i.e., a specimen that uniquely identifies the species exists [59]. Moreover, no other species lives in such complex societies or has such a sophisticated culture. However, maybe one of the traits defining humans best is the fact that humans are capable of thinking about their origin – and do so extensively. This is testified by the fact that the origin of humans has interested people in all parts of the world; almost every culture has some creation myth, trying to explain how humans originated. In the Christian societies of western civilization humans were seen as directly created by God as described in the Bible. Interestingly, this belief did not keep Carl von Linné in his *Systemae Naturae*, published as early as 1758, from putting humans, great apes, and monkeys together in the order Primates. Yet it was another 100 years before this formal placement was provided with a theoretical underpinning by the joint publication of Charles Darwin's and Alfred Russel Wallace's essays on the causes of evolution [28, 153]. Although its principle was laid out in these publications, the theory of evolution – and its implications for humans' place in nature – were not widely recognized for another year, when Darwin published *The Origin of Species* in 1859 [29]. Although the book was, as Darwin himself wrote in the Introduction, just "a long argument" to support evolution as a concept, Darwin did not fail to mention that all his arguments certainly also applied to the human species. Thus, humans did not only look similar to other primates, but this similarity could be explained by the fact that they were distant relatives. This idea dethroned humans from the special position they had occupied in western thinking until then, making them the simple product of a process that also applied to any other living species.

However, its application proved extremely fruitful. Just 3 years before Darwin's publication, the Neanderthal type specimen had been discovered [49] and the theory of evolution now allowed it to be placed in a meaningful context [61]. Thus, in 1863 Huxley, "Darwin's bulldog" wrote an entire book [61] on human evolution, and 8 years later, Darwin himself published *The Descent of Man* [30], in which he not only lays out anatomical similarities but also argues for the differences in mental capabilities between humans and animals being only gradual rather than principal nature. Although erroneous in details, the book is an impressive collection of facts showing beyond reasonable doubt that humans were the product of a long evolutionary process.

18.1.2 Sarich and Wilson

After the publication of *The Descent of Man*, progress in the study of human evolution was quite slow for almost 100 years. A close relationship between humans and great apes was undisputed, and some important fossils such as *Homo erectus* and *Australopithecus africanus* were discovered. However, an important question remained unresolved. This concerned the time-scale of human evolution. In many aspects humans are so different from great apes that estimates of when humans had split from their relatives varied from 5 to 25 million years. This lack of knowledge effectively prevented any meaningful discussion about the process of human evolution. At the same time the fossil record was too sketchy and the dating of fossils too unreliable to provide more detailed information. However, this situation changed radically with the work published by Sarich and Wilson in 1967 [125]. They had used an immunological comparison to put a time-scale on human evolution by immunizing rabbits against human blood serum and then using the antibodies obtained to test their cross-reactivity to serums from different ape and monkey species. Quantification of the reactivity and the use of a calibration point, with the baboon assumed to have split from the human lineage some 30 million years before present, allowed them to estimate that humans and chimpanzees had separated probably as recently as only about 5 million years ago. This result had major implications for the study of human evolution. There was no extended time-span with a lack of fossils as believed by many researchers at that time. Rather the differences between humans and chimpanzees had evolved within a comparatively short period. In a way this result was also the beginning of a different view on these differences, which we know today, are nowhere near as big as many researchers liked to believe only a few decades ago. Thus, the work of Sarich and Wilson opened the door to a completely new understanding of human evolution that has developed during the four decades since their discovery and is based on many new fossils and a whole range of new techniques and data.

18.1.3 From a Straight Line to a Bush of Hominid Species and Beyond

Twelve years after the work by Sarich and Wilson one of the last dogmas in human evolution was destroyed. In a seminal paper Johanson and White [69] finally buried the idea that since the divergence of humans and chimpanzees from a common ancestor a straight line had led to modern humans by unequivocally showing – although strong evidence pointing in this direction had been brought forward some years earlier – that during the Pleistocene at least two hominid species had existed contemporaneously (Fig. 18.1). Since then an ever-growing number of hominid species has been discovered (see Sect. 18.2), with sometimes four or even more species living at the same time (Fig. 18.2; e.g., [15]). Consequently, the history of human evolution now looks like a bushy tree with many dead ends and unclear relationships between the various species – and in fact there is a lot of discussion about which fossils to recognize as different species at all. However, progress has not only been made in palaeoanthropology. The sequencing of DNA made it possible not only to investigate the relationship of humans to their primate relatives but also to investigate the population history of the human species in great detail, including the age of human diversity, the geographical origin of modern humans [14, 62, 148], and their various migrations (e.g., [117, 119]). More recently sequence analyses also resulted in the discovery of genes that were positively selected during human evolution, such as *FOXP2*, a gene involved in language evolution, or the gene coding for human sarcomeric myosin *(MYH16)*, a muscle protein, which may have influenced human skull morphology. Studies of these genes allowed first clues to which genetic changes are responsible for specific human traits (e.g., [40, 51, 134]). Tremendous progress in DNA sequencing has lead to the launch of

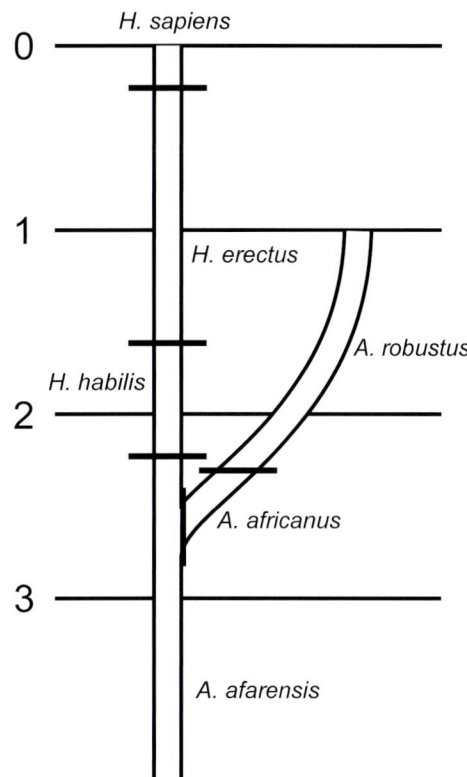

Fig. 18.1 Phylogenetic tree of human evolution from 1979. (After [69])

a Neanderthal genome project [52], which has the potential to reveal how we differed genetically from our closest, albeit extinct, relatives. New techniques have also allowed first insights into not only differences in DNA sequences but also the extent to which the expression of genes differs between or has evolved in parallel in humans and their relatives [77]. Finally, progress has been made in understanding to what extent other animal species show behavioral features which were so far thought to be unique to humans. It has been shown that tool use in free-living animals not only exists in chimpanzees – a fact that was already known to Darwin [30] – but also in evolutionarily distant animals, such as crows and dolphins. Even more strikingly, cultural traditions, long thought to be a unique human trait, have been discovered in both chimpanzees [155] and orangutans [146].

Progress in the study of human evolution has been substantial during the last 20 years and the future is likely to yield further insights to eventually reveal not only what makes us human but also how we got there.

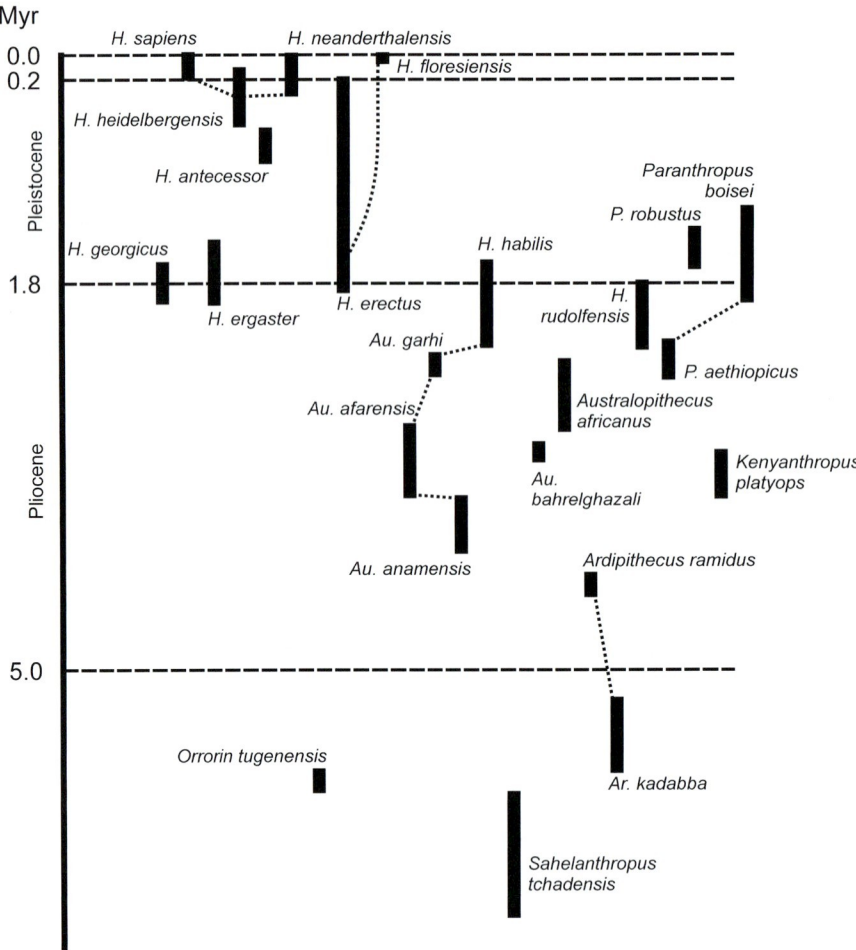

Fig. 18.2 Fossil record of human evolution. *Dotted lines* indicate tentative relationships. (After [15] and other sources)

18.2 The Fossil Record

18.2.1 Palaeoanthropology

18.2.1.1 The African Record

The fossil record testifying to human evolution has become almost fantastically diverse. The first fossil hominid to be described was the Neanderthal type specimen discovered in 1856 and scientifically described as *Homo neanderthalensis* in 1864 – although a Neanderthal skull from Engis in Belgium had already been discovered in 1829, but was not recognized as such until 1936. The next species to be discovered was *Homo erectus* in 1891, followed by *A. africanus* in 1925. In 1979, when the concept of the existence of two contemporaneous hominid lineages had gained substantial support, altogether six hominid species had already been described [69]. Today about 20 different hominid species are recognized. Many, such as the so far oldest hominid fossil *Sahelanthropus tchadensis* [12] and the Late Pleistocene dwarf species *Homo floresiensis* [11], have been discovered during the last 15 years. The former has extended the fossil record of human evolution back to almost 7 million years, bringing the fossil close to the divergence between humans and chimpanzees as calculated from molecular data. Despite the exceptionally good record of hominid fossils, the relationships between the various species are far from being clear (e.g., [157]). In part, this is due to the fact that many fossils, such as *Kenyanthropus platyops* or *Orrorin tugenensis*, are represented by only fragmentary material from few individuals. Despite many fossils, which have to span about 7 million years and a vast geographic area, the fossil record is still quite sketchy. Nonetheless, it is in many ways informative. The discovery of early hominid fossils from both East (*Orrorin tugenensis*, 5.7–6 million years and *Ardipithecus kaddaba*, 5.5–5.8 million years old) and West Africa (*Sahelanthropus tchadensis*, 6–7 million years old) shows that soon after the divergence from the chimpanzee, hominids had already spread widely across Africa. Interestingly, even these early fossils show evidence for bipedal locomotion although a recent finding of a juvenile *Australopithecus afarensis* skeleton suggests that arboreal behavior may still have played a part about 3.3 million years ago [2]. Thus, bipedal locomotion evolved very early in human evolution, much earlier than for example increased brain size, a pattern known as mosaic evolution [15]. Moreover, individual traits did not evolve in a linear fashion. For example, *Sahelanthropus tchadensis*, the earliest fossil species on the human lineage discovered to date, had a brain volume of 320–380 cm³, similar to that in living chimpanzees. *A. africanus*, which lived at least 3 million years later, had a brain volume only 30% larger. However, within the next 3 million years, the brain volume tripled to an average of 1,355 in modern humans ([15] and refs therein; see also Fig. 18.3). Thus, the evolution of human-specific traits took place in both a mosaic and a nonlinear fashion.

Although it is not yet clear which fossils represent direct human ancestors and which represent side branches of the hominid tree that did not leave descendants, such as the robust Australopithecines *Paranthropus robustus, P. boisei, P. aethiopicus* and possibly *K. platyops*, there is no doubt that hominids proliferated in Africa for millions of years, occupying different ecological niches [69].

18.2.1.2 Fossils from Asia and Europe

In addition to the African fossil record, there have been a number of discoveries outside Africa, beginning with the recovery of a number of *Homo erectus* (or *Homo georgicus* as claimed by some authors) fossils from the site of Dmanisi in Georgia, dating back about 1.8 million years [50]. These findings, together with dating of the Indonesian *H. erectus* fossils to a similar age [138], showed that hominids left Africa quite early and were successful enough to settle in geographically distant places. Only slightly younger

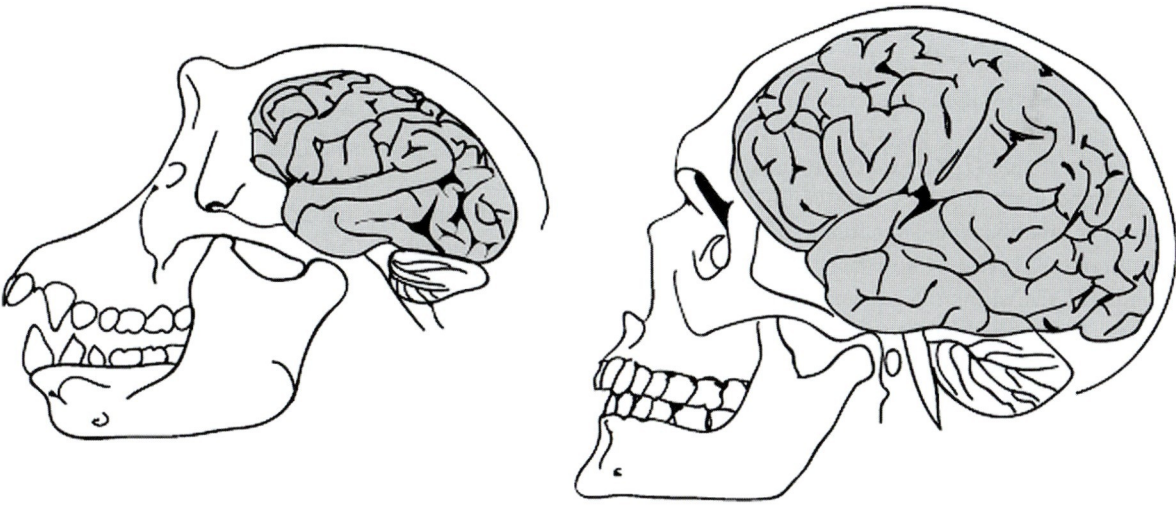

Fig. 18.3 Comparison of the brain of a chimpanzee (*left*) and that of a modern human (*right*). Note also the different shapes of the skulls in the two species. (After [15] and other sources)

artifacts from Majuangou at 40° north in China dating back between 1.66 and 1.32 million years show that hominids could survive in northern Asia over prolonged time periods as soon as the Early Pleistocene [165]. Moreover, discoveries in Spain and Italy showed that hominids occurred in Europe already as early as about 800,000 years ago. Although no hominid fossils of comparative age have been discovered north of the Alps, dating of flint artifacts from the UK indicates that hominids probably reached Europe north of the Alps as much as 700,000 years before present (B.P.) [108].

It now seems that before the dispersal of modern humans there were at least three waves of emigrations out of Africa between 1.8 and 0.5 million years ago and descendants of the different migration waves probably survived for quite a long time in both Europe and Asia. Thus, it has been argued that the Middle Pleistocene hominids from Sima de los Huesos in Spain were the ancestors of the Neanderthals [5], which disappeared about 30,000 years ago [99], and that *H. erectus* in Asia also survived until the Late Pleistocene.

Asia, with the recent discovery of *H. floresiensis*, with an estimated age of about 18,000 years, in fact yielded not only the youngest extinct hominid species to date [11] but also one of the most spectacular and controversial fossils in recent years [4, 66]. *H. floresiensis* has not only a substantially reduced body size; but it also has a very small brain size, a fact that has resulted in arguments about whether the individual discovered was healthy or rather microcephalic. This argument has been extended to the point where the species status of *H. floresiensis* has been questioned and the fossil has been interpreted as simply a microcephalic human pygmy [66]. However, most authors agree that *H. floresiensis* is indeed a separate species and most probably a descendant of *H. erectus* whose small size and specific features evolved as a result of the special evolutionary pressures on small islands [8]. This illustrates beautifully that hominids evolve according to the same evolutionary mechanisms as other animal species, a fact already known to Darwin [30].

Overall, the fossil record has been quite informative with regard to human evolution, but owing to the rapid decay of DNA nothing is known about genetic differences between us and our ancestors or cousins – with the one exception of the Neanderthals.

18.2.2 Neanderthals and Ancient DNA

18.2.2.1 Mitochondrial Sequences

In many ways Neanderthals (*H. neanderthalensis*) occupy a special position in the study of human evolution. They were the first extinct hominid species to be discovered; they are the closest relative of modern *H. sapiens*; and they are the first – and so far only – extinct hominid from which DNA sequences have been obtained [82]. Since the initial publication of DNA sequences from the Neanderthal type specimen almost 10 years ago, mitochondrial (mt) DNA sequences have been obtained from 13 additional specimens (Fig. 18.4) [81] all showing that with respect to mtDNA, Neanderthals fell outside the variation of modern humans, with the two lineages having diverged about 600,000 years ago [83]. While for some time it looked as though if Neanderthals carried very low levels of sequence diversity, several recent publications report somewhat more divergent mtDNA sequences suggesting that sequence diversity in Neanderthals was not that depleted and was probably similar to that in modern humans. Moreover, the identification of Neanderthal mtDNA sequences from the Altai Mountains also extended the geographical range of this extinct human group [81].

These data have not only been used to determine the phylogenetic relationship of Neanderthals and modern humans, they also made it possible – together with comparative analyses on fossils of modern humans – to estimate the maximum amount of mtDNA gene flow that could have been from Neanderthals to modern humans. Using a very conservative approach, Serre et al. [130] concluded that this contribution could not have been larger than 25%. With a more realistic migration model for modern human colonization of Europe a follow-up publication using the same data even concluded that no more than 0.1% gene flow could have occurred [27].

18.2.2.2 Palaeogenomics

While further work on more individuals would be desirable to clarify the extent and possible geographical structure of mtDNA sequence variation in Neanderthals, all the above analyses have the disadvantage that

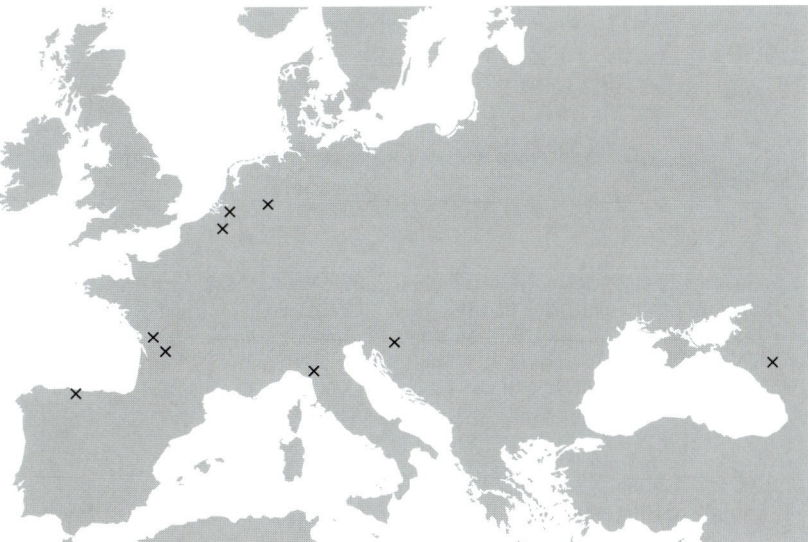

Fig. 18.4 Sites in Europe from where Neanderthal mtDNA sequences have been obtained. Recently Neanderthal mtDNA sequences were also found in fossils from Teshik-Tash, Uzbekistan, and Okladnikov Cave, Russian Altai [81], not depicted on this map

mtDNA represents just a single genetic locus, which is, moreover, inherited only through the maternal lineage. Although for a long time analyses of nuclear (nu) DNA from Neanderthal fossils seemed impossible, progress in the analysis of ancient DNA analyses has made it possible. Thus, two recent publications [52, 106] have reported sequencing and analyses of nuDNA. These two studies used rather different techniques. The first used ancient DNA extracts taken from a Neanderthal specimen from Vindija cave, Croatia, and put the DNA into the bacterium *Escherichia coli*. In this way a genomic library is constructed which can then be sequenced. The second study used a new technique allowing direct sequencing of hundreds of thousands of DNA fragments, which was applied to extracts from the same Neanderthal individual. However, what the two techniques have in common is that they are shotgun techniques, i.e., they do not target any specific sequence region but simply give a random cross section of the DNA present in the ancient DNA extracts. Unfortunately, only the minority of this DNA originates from Neanderthals. Thus, only about 6% of the sequences obtained were of hominid origin, with the remaining ones representing other organisms living in and on the bone, such as fungi, bacteria, and many unknown sequences, which are typically found in such metagenomics studies. Even so, the two studies reported 65,000 bp [106] and about 1 million bp [52] of Neanderthal nuDNA sequences, respectively. These data allowed some interesting insights. First, both techniques showed the Neanderthal sequences evenly distributed along the chromosomes, suggesting there is little bias with respect to different parts of the genome represented in the sequence reads. This is an important insight, as it shows that it is in principle possible to sequence the complete genome of *H. neanderthalensis*. Second, in both studies the divergence time between human and Neanderthal DNA sequences was estimated with somewhat different, albeit overlapping, results. Noonan et al. give a best guess of about 700,000 years (with the confidence interval spanning 450,000 to 1 million years), whereas Green et al., with their larger data set, arrive at 516,000 years with a smaller confidence interval of between 470,000 and 570,000 years. Strikingly, this is hardly older than the average sequence divergence between two modern humans. In addition, Noonan et al. also estimate the population divergence time between humans and Neanderthals, which with 370,000 years is even younger, as sequence divergence always predates population divergence (Fig. 18.5). Although Green et al. did not estimate population divergence, their younger sequence divergence date shows that they would have arrived at an even younger date. These results have several important corollaries.

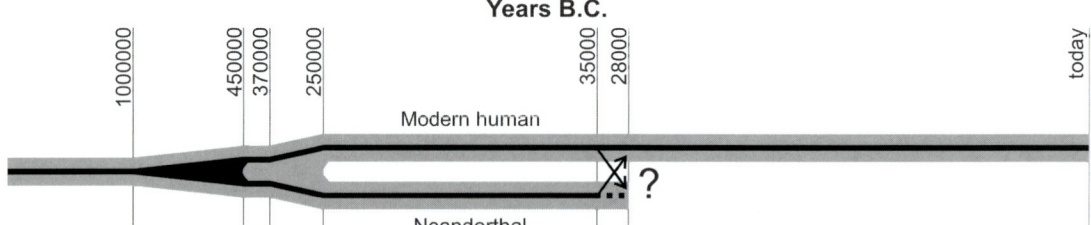

Fig. 18.5 Scheme of modern human – Neanderthal relationships. The *black lines* indicate DNA sequences and the *gray shades*, populations. Between 450,000 and 1 million years ago the nuclear DNA sequences separated, while Neanderthal and modern human populations split only about 370,000 – 250,000 years ago, according to data published by Green et al. [52] and Noonan et al. [106]. Modern humans entered Europe at least 35,000 years ago, while Neanderthals survived until 28,000 years ago or later, leaving at least 7,000 years for potential gene flow

If true, many hominid fossils, such as the older Atapuerca findings from Spain or the El Ceprano fossil from Italy, are too old to represent Neanderthal ancestors. Moreover, there will be few fixed differences between humans and Neanderthals. In fact, Green et al. found that at many DNA sequence positions where modern humans are polymorphic, Neanderthals show the derived sequence state, which the authors interpret as possible evidence for gene flow from modern humans to Neanderthals. This is extremely interesting as the Neanderthals from Vindija have previously been described as unusually gracile and as possible examples for hybridization between humans and Neanderthals. An alternative explanation would be that some sequences represent contamination of the experiments with modern human DNA, a common problem in ancient DNA studies (e.g., [58, 130]). Although in both studies the authors tested for contamination of the extracts with modern human mtDNA, which was found to be very low, it is not clear whether these results are also representative for nuclear DNA. Only more data if possible from several fossils, will clarify how much time and how many sequence differences separate us from Neanderthals.

However, it is noteworthy that population genetic studies on modern humans also found evidence for admixture of DNA sequences from archaic hominids to the gene pool of modern humans [111]. For this to be true, admixture would have to have occurred during a relatively restricted time frame. Although the presence of modern humans and Neanderthals in Europe overlapped temporarily [101], this overlap was probably shorter than previously believed. The oldest modern human remains are from Pesterca de Oase in Romania and have been directly dated to about 35,000 years [143], although there is evidence for the presence of modern humans at other sites, e.g., in south-western Germany around the same time [28]. Conversely, the youngest evidence for the presence of Neanderthals in Europe is from Gibraltar and suggests a terminal age for the existence of Neanderthals of 28,000 years B.P. or younger. Although this allows for 7,000 years of interaction between modern humans and Neanderthals, the sites for the oldest modern human fossils and the youngest Neanderthal fossils are geographically quite distant and the youngest dates for the geographically closer Vindija Neanderthals have recently been revised to about 31,000 years. Despite these revisions, there is little doubt that modern humans and Neanderthals not only met but also interacted ([99, 101] and refs therein). However, whether gene flow occurred during this process, and if so, to what extent and in which direction can only be revealed by future studies.

18.3 The Genetics of Human Evolution

18.3.1 The Genomes of Humans and Their Relatives

Phylogenetically, humans belong to the African great apes, as do gorillas, which represent the first branch diverging from a common ancestor some 8 million years ago, and chimpanzees and bonobos, two closely related sister species that diverged from the human lineage probably 6 – 7 million years ago (Fig. 18.6).

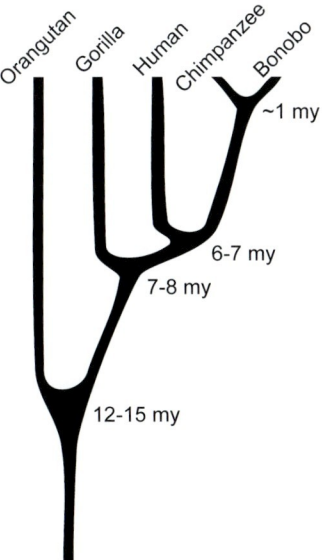

Fig. 18.6 A schematic phylogenetic tree showing the relationships and approximate time-scale for the evolution of humans and great apes. The time estimates for the different divergence events are averages from various studies. It can be seen that humans are evolutionarily as distant from bonobos as they are from common chimpanzees

Ultimately, human evolution is simply a long chain of changes in the genome that have taken place since our own lineage and that of the chimpanzee separated some million years ago. Thus, in a way, the key to human evolution lies in genome sequences. The first published genome sequence was that of the human mitochondrial genome [3] with a length of about 17 kb. This was followed by the homologous sequences from the closest relatives of humans, chimpanzee, gorilla, and orangutan, which allowed scientists to identify the chimpanzee as the human's closest relative [56, 57], a result soon corroborated by nuclear DNA sequence data [57]. However, later analyses have shown that the picture is more complex. An analysis of 53 nuclear loci showed that although for most regions humans and chimpanzees are most closely related, the authors found trees incongruent with that phylogeny for 22 of the 53 loci (42%, [22]). Thus, at many parts of the genome, gorillas are more closely related to either humans or chimpanzees than humans and chimpanzees are to each other. This is possible because in the nuclear genome recombination is a regular event, which breaks up the genetic linkage between regions on a chromosome. Thus, the history of different regions in the nuclear genome may be different. Moreover, the common ancestor of two species always harbors sequence variation. If two populations separate, and eventually evolve into two different species, for some time some of the sequence variants in one species may be more closely related to sequence variants in the other species than to different sequence variants within the same species. Eventually, each species will become fixed for one lineage and its descendants, and the sequence variation between the two species becomes reciprocally monophyletic, a process called lineage sorting. However, if one of the species splits again into two daughter species shortly after the first divergence, lineage sorting may not yet be complete and some sequence variation predating the first population divergence can persist. In this case lineage sorting may occur in a way that the genetic tree for certain regions does not reflect the population divergence tree (Fig. 18.7). As the speciation events of humans, chimpanzees, and gorillas took place within a short time, the ancestral population of humans and chimpanzees still harbored some sequence diversity that predated the divergence of the gorilla, allowing for some genetic regions being more closely related between gorillas and either of the two other species than between these two. Although a recent study [109] showed that the proportion of such regions is somewhat lower than previously assumed, with about 30%, it is still a substantial part of the genome. Even more interestingly, the authors found the divergence between human and chimpanzee sequences to vary substantially among

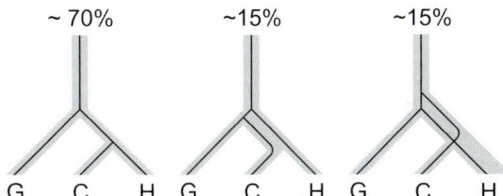

Fig. 18.7 Three possible ways in which genetic lineages can be related and their approximate proportions in genomes of humans, chimpanzees, and gorillas. *Black lines* represent genetic lineages; the *gray outline* shows the population divergence pattern. At each point in time, the populations harbor genetic diversity, which is not shown in the figure. Therefore, when the time between two divergence events is short, some of the genetic diversity in the population at the second divergence may actually predate the first divergence event resulting in the lineage sorting pattern shown in the *middle* and to the *right*. That way, there are three possible ways in which genetic lineages can be related, although there is only one pattern for the population divergence between the three species

different regions of the genome, corresponding to a time range of as much as 4 million years. They explained these results by a complex speciation history of humans and chimpanzees, including secondary hybridization between the two lineages [109]. However, the data could also be explained by a model with strictly allopatric speciation in combination with a large ancestral population size [7]. Although it is not yet clear which interpretation is correct, this example shows the power of genomic analyses to infer events going back to the roots of human evolution.

So far, complete genomes have been sequenced from a number of mammalian species, including mouse, rat, dog, human, and chimpanzee, and more are soon to come. The human genome was the first mammalian genome for which a draft sequence became available [89]. In 2004, the completion of the euchromatic genome sequence was published [35], which contained only 341 gaps, as compared to about 150,000 gaps in the draft sequence. However, the finished genome confirmed, with an estimate of only 20,000–25,000 protein coding genes, the conclusion from the draft sequence that the human genome contains much fewer protein coding genes than previously believed. Completion of the draft sequence of the chimpanzee genome [37] made it possible to estimate the sequence divergence between us and our closest living relatives with great accuracy. Surprisingly, only 1.06% or less of the positions represent fixed differences between the two species [37]. However, owing to insertions or deletions on either lineage, 1.5% (or 40 million basepairs) of the genome sequences are unique in each species. Segmental duplications on one of the lineages change the genomic landscape between humans, with 2.7% difference, more than nucleotide substitutions, with 1.2% difference. They also influence the expression of certain genes [23]. Moreover, a comparison of humans, chimpanzees, bonobos, orang-utans, and gorillas showed that more than 1,000 genes are not present in all species but are unique to one or several of these species and that 134 of these have been created by duplication on the human lineage [48]. How many of these new genes are under positive selection is not yet clear, but for a gene family called Morpheus this has been clearly shown [70]. This gene family expanded in both humans and great apes, with as many as 15 copies present in the human genome and between 25 and 30 copies in the chimpanzee genome. Strikingly, some of these new copies show extremely high Ka/Ks ratios, a strong indication of positive selection acting on these genes. Another gene family shows an even more extreme pattern. One of the 134 human specific genes found in the above-mentioned study encodes a primate-specific sequence domain (DUF1220) that has become extremely amplified on the human lineage with at least 212 copies present in the human genome [112]. In a similar way to Morpheus, many of these sequences show high Ka/Ks values indicating that positive selection has influenced these sequences. Moreover, many of the genes carrying a DUF1220 domain are expressed in neuronal cells in brain regions associated with higher cognitive functions [112]. Thus, gene duplications are likely to have contributed substantially to adaptive processes during human evolution.

There are also large-scale chromosomal differences between humans and great apes, the best known ones being the fusion that created human chromosome 2 and the human-specific pericentric inversions on chromosomes 1 and 18. A recent study probing 12% of the human genome found 63 large-scale, yet cytogenetically undetectable, sites of copy number variations among humans and the great ape species [92]. These rearrangements range in length between 40 and at least 175 kbp, and most of them were found to be located in euchromatic, gene-rich regions of the genome, adding substantially to the differences between humans and their closest relatives.

Thus, humans differ from their closest relatives in many aspects of their genome, ranging from the about 35 million single nucleotide substitutions that separate humans and chimpanzees [37] through small-scale insertion/deletion events and larger sequence duplications to the cytogenetically visible differences, such as the fusion of two ape chromosomes that created human chromosome 2. Yet, besides these differences between the species, there is also ample variation within species, including humans.

18.3.2 Diversity Within the Human Genome

The first studies investigating diversity within humans were done using proteins [16]. Even these early studies showed substantial genetic diversity within the human species. However, the extent of genetic diversity within the human genome could only be properly assessed when analyses of DNA sequences became possible. One of the first surveys investigated restriction fragment length polymorphisms in the beta-globin region from a world-wide sampling of 601 humans [152] and found greater diversity within African populations than within populations outside Africa. A key study with respect to such analyses was published in 1987, when Cann and colleagues presented restriction fragment analyses data from mitochondrial DNA of a world-wide sample of 147 humans [14]. In their survey, they found that 195 of 467 restriction sites were polymorphic in humans. Again, African populations showed the highest diversity. However, as a restriction site contains several nucleotides, where only one needs to be different to create a polymorphism, these data did not allow estimation of mtDNA sequence diversity within humans. The next step in this direction was taken when 189 humans were sequenced for 610 bp of the mitochondrial control region [148]. These samples yielded 135 different sequences, with 179 positions that were polymorphic, i.e., almost a third of the positions studied. As the control region is the most variable part of the mitochondrial genome, these data were also not representative for the mitochondrial genome. Eventually, in 2002, Ingman et al. [62] sequenced complete mitochondrial genomes of 69 humans from a world-wide sampling and found that 657, or about 4%, of the approximately 16,500 bp of the mitochondrial genome were variable in modern humans. Owing to this high within-species diversity, mtDNA has been used extensively in the study of modern human migrations (see Sect. 18.4.1).

Compared with mtDNA sequences, nuDNA sequences show much lower levels of diversity. The first world-wide study of human genetic diversity in the nuclear genome used the same sampling set as the above study on complete mtDNA genomes and surveyed 10 kb on the X-chromosome [72]. Strikingly, only 33 positions, i.e., 0.33% of the analyzed sequence, were found to be variable. Additional studies on the X-chromosome (e.g., [159]) and also on autosomes [163, 164] found similar, albeit somewhat higher levels of variation with between 44 and 75 variable positions within about 10 kb of nuclear sequence. Generally, all these studies indicate a low amount of mean pairwise sequence difference between two human chromosomes, averaging about 0.1% [163].

This low amount of sequence diversity within humans is in stark contrast to the extent of sequence diversity within our closest relatives, the great apes. Sequencing the same 10-kb region of the X-chromosome as in humans, Kaessmann et al. [73] found, with 84 variable positions, more than twice as many variable positions in only 30 chimpanzees than in the above 69 humans. On the same locus more variation than in humans was also detected in 11 gorillas (41 variable positions) and 14 orang-utans (78 variable positions; [74]). Similar results were found for the Y-chromosome, with both chimpanzees and bonobos showing higher levels of sequence variation than humans [135]. However, autosomal sequences show a somewhat different picture. Using about 23 kb from 50 different autosomal loci, Yu and colleagues [160] found bonobos to have a mean pairwise sequence difference (pi) that is even lower, at 0.078%, than that in the corresponding sequences in humans (0.088%). Although chimpanzees show a somewhat higher pi-value (0.132%), the difference between humans and chimpanzees in within-species diversity is much smaller in this study than in the above studies on mtDNA, X-chromosome, and Y-chromosome. Although western lowland gorillas show somewhat higher sequence diversity for these loci, with 0.158% it is only about twice as high as in humans [161]. Taken together, sequence data currently indicate that diversity within the human nuclear genome is somewhat lower than in some great ape species, but comparable to that within bonobos and the western chimpanzee subspecies [160].

However, the term 'low diversity' should not be misunderstood. An average sequence diversity of 0.1% between two individuals sums up to 3 million differences across the complete genome. As polymorphic positions can be used for various purposes, including the mapping of genetic diseases, efforts are under way to create extensive data bases of such single-nucleotide polymorphisms (SNPs). In a recent study report the results of the first phase of the HapMap project, a collection of high-quality SNPs in the human genome, was published [36]. This study involved more than 1 million SNPs with a frequency of the minor allele higher than 5%. In addition, 10 regions with 500 kb in length were sequenced in 48 humans, indicating that the overall frequency of SNPs is about 10-fold higher, with one SNP occurring every 279 bp [36]. In fact, given the number of about 6.5 billion people currently living on earth and the substitution rate of nuclear DNA, every mutation that is compatible with life should be present in at least one individual of the global human population.

In addition to single nucleotide variation there are several other forms of within-human genetic diversity. For example, there are repetitive elements in the human genome that occur in extremely large copy numbers. Two classes of these repetitive elements are SINEs (short interspersed elements, less than 500 bp in length) and LINEs (long interspersed elements). Both LINEs and SINEs increase in numbers by retrotransposition (see, e.g., [8, 132] for reviews). In other words a copy is transcribed into RNA, which is then reverse-transcribed into DNA and inserted somewhere in the genome. The most common SINE element in the human genome is the primate specific Alu family, a 300-bp-long sequence element with altogether more than 1 million copies, which comprise about 10% of the human genome [8]. Alu elements tend to insert in gene-rich regions and can therefore both influence the expression of genes and interrupt genes, resulting in genetic disorders. Owing to their high sequence similarity, Alu elements can also cause recombination between two Alu elements in different regions of the genome, resulting in deletions, duplications, or translocations, which may among other things cause cancer (for a comprehensive review see [8]).

As only a few Alu elements are multiplied at a time, Alu elements can be divided into different families. Although most of these insertions are fixed in humans, there are also families that are quite polymorphic, such as the subfamilies Alu Yc1, Yc2, and Yb9 [122], for which about a third of the members within each family have been found to be polymorphic for presence / absence in humans. Altogether about 5,000 Alu elements inserted into the human genome after the divergence between humans and chimpanzees, but not all of these are polymorphic [8]. Polymorphic Alu elements were also identified on the human X- and Y-chromosomes and represent excellent markers for population genetic studies [122], as Alu element insertions at a specific site in the genome are almost certainly unique events, with very little chance of reversal. Thus, parallel or back-mutations hardly ever occur [8], making them unusually robust markers for inferring evolutionary events.

LINE elements are in many ways similar to SINEs, but they are present in lower copy numbers of about 100,000 in the human genome [132]. However, as they are longer, they also comprise about 15% of the total human genome. LINEs have expanded in numbers in the human genome during the last 100 million years of mammalian evolution, but analogously with Alu elements, there are LINEs that are human specific, such as the L1 Ta subfamily, for which 12% of 249 loci were found to be polymorphic in humans [150]. Together with polymorphic Alu elements, LINE elements of the L1 subfamily are likely to comprise as many as 2,000 polymorphic markers in the human genome, comprising a large amount of genetic diversity within humans and at the same time providing a rich source of information for population studies [132].

Finally, another source of within-species diversity has recently been discovered in the human genome, so-called large-scale copy number polymorphisms (CNPs; reviewed in [21, 47]). These regions range in size from about 1 kb to as much as 3 Mb [47] and have been found to be surprisingly common in the human genome. Thus, an initial screening of 20 individuals revealed 221 CNPs, with an average length of 465 kb, and individuals differing by an average of 11 CNPs. Other studies found similarly high levels of structural variation, indicating that this type of variation may contribute substantially to genetic diversity in humans. As the pieces involved are large, the similarity between two humans may indeed be less than the usually assumed 99.9% [25]. Moreover, as these duplications contain large numbers of genes, they are of substantial clinical relevance [21, 47]. A recent genome-wide study on copy number variations (CNV) in 270 individuals from a world-wide sampling found 1,447 CNVs en-compassing a total of 360 Mb, or 12% of the genome [115]. Interestingly, CNVs encompass, genome-wide, more nucleotides than SNPs [115]. Thus, in a way, CNVs represent more genetic diversity among humans than do SNPs. Moreover, there is evidence that certain regions are hotspots for CNVs, as CNVs have been observed at corresponding loci in both humans and chimpanzees.

Overall, the human genome contains substantial variations between individuals, ranging from individual nucleotides to large-scale structural variations. During human evolution, this variation has been subject to both positive and negative selection.

18.3.3 Positive and Negative Selection in the Human Genome

A key feature of Darwinian evolution lies in the fact that selection works on random variation, i.e., mutations occur in the genome and these are either selectively neutral or positively or negatively selected. In 1968, Kimura proposed that most amino acid substitutions that become fixed in a species are selectively neutral [78], i.e., they have neither a positive nor a negative effect on the fitness of their carrier. It is important to note that the neutral theory does not claim that most mutations are neutral – in fact Kimura assumed that most newly arising mutations are deleterious, but will be removed quickly from the population via negative selection and therefore have no chance of getting fixed. The critical point of the neutral theory is that it assumes that very few substitutions become fixed due to positive selection. To what extent this is true is still a matter of debate. For example, Fay and colleagues estimated that as many as 35% of the amino acid substitutions that were fixed on the human lineage were driven by positive selection [46], a value similar to that estimated for the *Drosophila* genome, for which up to 50% of the substitutions are estimated to become fixed as a result of positive selection [44]. Extrapolated to the human genome, this would result in about 70,000 amino acid substitutions having been under positive selection at some time in human history [15]. If this number is correct, it will be extremely difficult to identify genes that have been under positive selection during human evolution, as each protein would on average have accumulated two or more positively selected amino acid substitutions during the last 5–7 million years. At the same time this number translates into one adaptive substitution getting fixed on the human lineage every 100 years. However, more recent estimates have led to quite different results. Some studies have put the proportion of genes in the human genome that have been under positive selection close to zero [37, 162], and the highest recent estimate for genes under positive selection is about 6% [13]. Thus, the neutral theory seems to apply well to the human genome. Still, if we take a range of estimates between 0.4 [105] and 6% [13] for the proportion of substitutions fixed as a result of positive selection, the number of adaptive substitutions in the human genome still ranges between 800 and 12,000.

Therefore, even with these reduced numbers, searching for genes that have been under positive selection is somewhat like searching for a needle in a haystack. Yet several candidate genes have been identified. Among the first genes claimed to be under positive selection were several proteins critical for male reproduction [158]. Although this claim has been questioned, it does not seem too far fetched, as male reproductive genes are under positive selection in many species and more recent studies have confirmed that male reproductive genes are a primary target of positive selection in humans [124]. Thus, this does not represent an example for selection on a human-specific trait such as upright walking, increased brain size, or language capability, to name just a few (see [15] for more examples). Research focusing on the genetic basis of such human-specific traits is of great interest, as it has the potential to provide insights into what makes us human.

One of the most widely recognized findings along these lines was the identification of a gene that, when defect, causes language disorders, the forkhead protein FOXP2 [86]. Strikingly, even though FOXP2 is a highly conserved transcription factor comprising only one amino acid difference between chimpanzee and mouse, two additional amino acid changes have occurred in the human lineage [40]. Sequencing of 14 kb around the two amino acid substitutions differentiating humans and chimpanzees showed evidence for a recent selective sweep sometime during the last 200,000 years having led to their fixation [40]. While it is unlikely that a complex trait such as language evolved via only two amino acid changes in a single transcription factor, further evidence has been found that FOXP2 does indeed play a crucial part in vocalization. First, additional mutations in FOXP2 resulting in language disorders have been identified in humans [95], and the regions of expression of FOXP2 in human and mouse brains during embryonic development correlate well with the regions of pathology in adult human brains of individuals carrying a defect allele of FOXP2 as suggested by neuroimaging [87]. Second, and much more intriguing, FOXP2 seems to play a critical part in song learning in birds. The first evidence that FOXP2 may also have a role in vocalization in birds' was provided by the fact that its regional expression in the bird's brain is highly similar to that in human brains. Moreover, in song-learning birds such as zebra finches, FOXP2 expression in brain regions critical for song learning varies over time [55]. Thus, in zebra finches it is up-regulated at 35–50 days after hatching, when song learning occurs, and in canary birds FOXP2 expression varies seasonally [55]. These results are strong evidence that FOXP2 has a critical role in vocal learning and may indeed have played an important part in language evolution in humans.

Another human-specific trait that is of prime interest is brain size. Brain size increased only moderately during early human evolution but started to increase rapidly with the appearance of the genus *Homo* in the fossil record about 2.4 million years ago [15]. This increase in brain size was paralleled by a substantial decrease in masticatory muscles even in early representatives of the genus *Homo*. Intriguingly, Stedman and colleagues identified a myosin gene that is expressed only in the muscles of the head and contains an inactivating mutation in humans [134]. Application of molecular clock dating yielded an age of about 2.4 million years for the inactivating mutation, temporally almost exactly coinciding with the appearance of *Homo* in the fossil record and the beginning of accelerated brain size increase in human evolution. However, whether these results indeed provide evidence that inactivation of this gene removed a selective constraint on the increase of brain size, as speculated by the authors, is not clear.

In contrast, there is little doubt that genes influencing the nervous system, and especially the nervous system development, have played a critical part in human evolution. A faster rate of protein evolution has been found for such genes for primates than for rodents, and the effect is most pronounced on the lineage leading to humans [33]. Genes influencing brain development and size seem to show particularly rapid protein evolution on the human lineage [51], although some of these genes also show similarly rapid evolution in other primate lineages, such as the gene ASPM on the gorilla lineage [80]. Thus, the first steps in identifying the genetic changes underlying human-specific traits have been taken, and more results can be expected in the future.

However, research on positive selection in humans has not been restricted to differences between humans and the great apes; the large amounts of SNP data [36] available for different human populations have also resulted in renewed interest in local adaptive evolution in humans, i.e., selection of certain traits in some geographical regions but not in others. Even given these large data sets, identifying candidate genes for local adaptive selection is not straightforward, and even if candidate genes are identified, understanding their function and the selective forces driving their evolution is not a trivial task (reviewed in [124]). To screen these data for possible signs of recent positive selection, several new tests have been developed. One of the most commonly used ones is the extended haplotype homozygosity test (EHH;[123]) and its variants. This test relies on the assumption that recent positive selection leaves a signature of a selective sweep in the form of reduced diversity around a selected site. Thus, if selection has been acting on one population but not on others, the former population is expected to have significantly extended homozygosity around the selected position relative to the later ones. Using the HapMap data [36] it has been shown that recent local adaptation has been taking place in all three major continental groups (Asia, sub-Saharan Africa, and Europe) generally studied for SNPs [151].

The example for local adaptation that is probably best known and understood is sickle-cell anemia which occurs at high frequency in certain regions of Africa where malaria is also prevalent. Despite strong selection against the sickle cell allele in its homozygous state, heterozygous individuals have a sufficient selective advantage in regions with high malaria prevalence to keep the sickle cell allele at high frequency via balancing selection [84]. Another genetic change conferring malaria resistance, a mutation at the *G6PD* gene, was recently found to have been under balancing selection long before malaria became prevalent in Africa, and it may thus originally have had a different adaptive function [147]. Another trait that has long been suspected of being under local positive selection in humans is skin color. There is wide geographical variation in human skin color, but whether these differences are adaptive has long been a matter of debate. In several studies Jablonski and Chaplin systematically investigated the extent of exposure to UV radiation and of skin reflectance in a particular population [64]. In their studies they found a strong correlation between UV radiation and skin reflectance of the exposed population. They explained these results by a trade-off between vitamin D synthesis – for which UV radiation is necessary – and photolysis of folate by UV radiation. Thus, in regions of high UV radiation, vitamin D synthesis is still possible with comparatively dark skin, which at the same time limits folate photolysis, whereas in regions of low UV radiation skins have to be light to ensure sufficient vitamin D synthesis and folate photolysis is not a severe problem. Although additional factors are likely to influence skin color in human populations, Jablonski's and Chaplin's data clearly provide a solid basis for studies on the evolution of skin color in different human populations. Recently progress has also been made in understanding the genetic basis of skin color differences in humans. Skin color in mammals is influenced by a number of genes [65] and differences in skin color can therefore be due to various genetic changes. It has been known for some time that the gene *MC1R* influences both hair and skin color and is under negative selection for full activity in Africa, whereas a number of loss-of-function mutations were found in Europeans at a high frequency, due to either loss of constraint or to positive selection for loss-of-function [116]. A stronger case for positive selection for light skin color in Europeans was made for a gene called *SLC24A5*. Initially found to be locally selected in humans in a whole genome screen [36], its function was at that time unclear. However, the same gene was later identified as a key pigmentation gene in zebra fish, and analyses of human populations have shown that it accounts for about a third of skin color variation between Europeans and West Africans [88]. Moreover, this gene shows signals of recent positive selection in Europeans, supporting the hypothesis that light skin color was indeed selected for in Europe.

Another classic example for recent local selection in humans is lactose tolerance. Most humans, like most other mammals, express the enzyme lactase, which allows digestion of the milk sugar lactose, only during infancy and lactase expression rapidly declines after weaning. However, it has been known for a long time that some humans continue to express lactase also as adults, and the frequency of this trait closely correlates with the consumption of unprocessed milk [137]. Thus, lactase persistence, and consequently lactose tolerance, is high in northern Europe and declines toward the south (Fig. 18.8). Yet, formal genetic evidence for positive selection on this trait had not been provided for a long time. By typing more than 100 SNPs in a region of 3.2 Mb around the lactase gene, Bersaglierie and colleagues finally found strong evidence that one lactase allele was indeed positively selected in Europeans [9]. Moreover, they found the

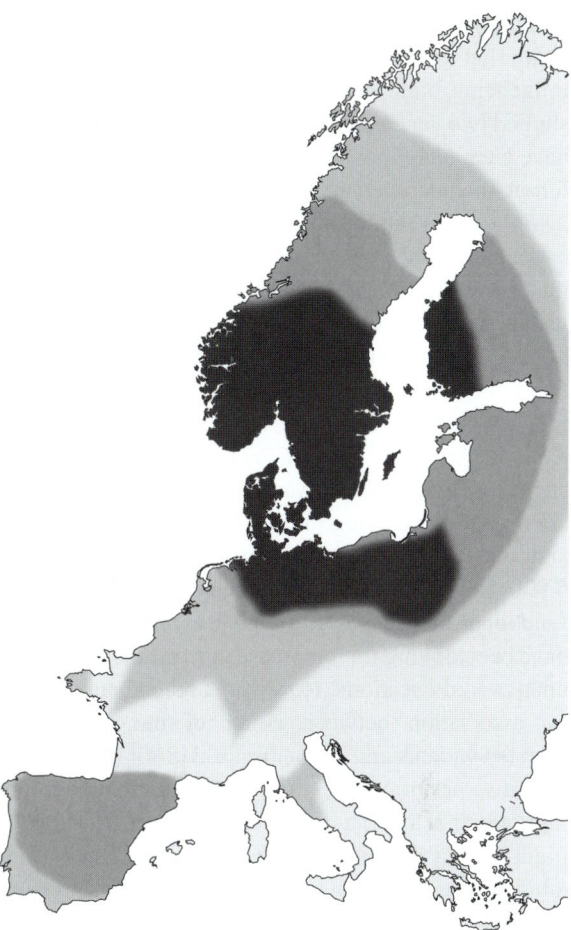

Fig. 18.8 Differences in allele frequency of lactose tolerance in Europe. *Darker shading* indicates higher frequencies of lactose tolerance

selective signal to be one of the strongest observed for any gene in the human genome. Also, molecular dating suggests that strong selection on this allele started about 5,000–10,000 years ago, which is consistent with dairy farming being the cause of a selective advantage of lactase persistence. Apart from Europeans, several other populations in the world show lactase persistence, but the nucleotide position likely to be responsible for this trait in Europeans (C13910T) is found at very low frequency in these populations. Intriguingly, two independent studies in 2006 identified several SNPs that are associated with lactase persistence in Western African populations [63, 140]. Thus, lactase persistence seems to have evolved at least two times independently in human populations, following the same evolutionary pressure.

Finally, some studies have led to the provocative claim that local adaptive selection also affected genes influencing brain development [42, 98]. The authors studied two genes known to influence brain size and to have been under positive selection during human evolution, the microcephalin gene (*MCPH1*; [42]) and the *ASPM* gene (abnormal spindle-like microcephaly associated; [98]). For both genes, they found alleles that showed strong evidence of having been under recent positive selection, such as extended haplotype homozygosity. Intriguingly, these haplotypes are not fixed in the human population and their frequency differs between regions from 0 to 60% for *ASPM* and from 3.3 to 100% for *MCPH1*. Additional analyses suggest that the selected variant of *MCPH1* arose only about 37,000 years ago in the human population [42] and that of *ASPM* even more recently, i.e., about 5,800 years ago [98]. Whether these alleles confer any functional differences and which evolutionary forces have been driving their increase in frequency are so far unanswered questions. However, a recent study found no association between either of the supposedly selected variants and neither brain size nor measures of cognitive performance, calling into question the initial interpretation that the two genes have been selected for brain-related effects [139]. Yet the *MCPH1* story contains another interesting twist. Although the presumably selected haplotypes have a coalescent age of only 37,000 years, their divergence to the nonselected haplotypes dates back as early as 1.1 million years [43]. Applying simulations to obtain data similar to the ones observed, Evans and colleagues concluded that the best explanation for the observed pattern is admixture from an archaic hominid population such as the Neanderthals. However, whether this is true and modern humans thus may have benefited from genetic contributions of archaic hominids needs to be tested in further studies.

18.3.3.1 Human Evolution and Medical Genetics

It has long been hoped that the knowledge about human evolution, and especially human genetics, will improve the possibilities of treating or curing human diseases, especially widespread ones, such as cancer, heart diseases, or type 2 diabetes. Enormous hopes have been set in the decoding of the human genome. As mentioned above, the genetic factors known to influence susceptibility to malaria have long been known and studied in great detail. Yet, a widely effective treatment for malaria has still not been found and malaria remains one of the most deadly diseases in tropical regions, with more than 1 million victims every year. Thus, knowledge about genetic risk factors does not necessarily result in better treatment, and this is true not only for infectious diseases but also for cancer and heart disease to name just two groups.

With the advent of ever-cheaper DNA sequencing and the resources of millions of SNPs (see Sect. 18.3.2) in the human genome, two areas of medical genetics have been the focus of increased research efforts: personalized (sometimes also called Darwinian) medicine and genome-wide association studies searching for risk factors for common diseases. As these two fields are especially linked to human evolution only these topics are discussed in this chapter. However, it should be noted that there are many more research areas in medical genetics, which are beyond the scope of this chapter.

The investigation of DNA sequences has been completely transformed during the last 2 years. Instead of ~70,000 bp that are obtained in a single run on a capillary sequencer using standard Sanger DNA sequencing, new sequencing techniques allow between 20 million and 1 billion basepairs to be sequenced in a single run [60]. Thus, the idea that complete genomes may be sequenced from individual patients to allow medical treatment to be specifically tailored, a concept known as "personalized" medicine, may soon become reality [17, 60]. However, despite increasing interest in the interplay between genetic differences among individuals and both disease susceptibility and drug response [25, 85], there are major obstacles on the road to this aim. First, sequencing techniques are still too expensive to allow the so-called $1,000 genome, which is commonly seen as a prerequisite for clinical use of individual genome sequences. Second, the interplay between individual genetic differences and disease susceptibility or drug response is likely to be complex, owing both to modifying genes elsewhere in the genome and to environmental factors that are different for every patient [128]. Thus, even if genome sequencing becomes cheap enough for complete genome sequences to become a realistic option for every patient, it will be important to ensure that other factors are not simply discounted and that patients rather than genomes are treated [128]. However, if it does indeed become possible to obtain complete genome sequences from each and every patient such information will be of great value for both understanding and treating diseases, including major complex diseases such as cancer or heart disease.

While complete genome sequences for many individuals are still out of reach, technical progress has already made typing of thousands of individuals for large numbers (up to 1 million) of single nucleotide polymorphisms (SNPs) possible. Applied to cohorts of patients and controls, such data can be used to search for genetic regions that are associated with certain diseases [20]. A recent large-scale study identified potential disease linked genetic regions for seven common diseases (bipolar disorder, coronary artery disease, Crohn's disease, hypertension, rheumatoid arthritis, type 1 diabetes, and type 2 diabetes; [38]). While such studies are becoming increasingly popular among medical geneticists, they are unfortunately fraught with problems [20]. First, owing to the high number of tests made in such a study without correction for multiple testing a large number of false positives will be detected. This problem was recently highlighted by an international study that failed to replicate any of 13 previously detected SNPs claimed to be linked to Parkinson's disease [39]. Although in other cases it was possible to confirm linkage between a SNP and disease susceptibility, in many cases replicate studies failed to confirm the results of initial studies, leading to the proposition that association between a SNP and a certain disease should only be considered if it has been replicated in at least two independent studies [20]. Another problem with association studies investigating complex diseases lies in the fact that many loci detected in large-scale studies may increase the individual risk only marginally, as found for a number of new breast cancer susceptibility loci [34]. Thus, only the combination of a large number of genetic loci in combination with environmental factors is likely to increase individual disease risk to a measurable effect, which is not an unexpected result for so-called complex diseases [133]. Finally, even if a SNP is associated with a certain disease, this by no means indicates that the SNP is in any way causative. Establishing a causal relationship between a genetic variant and a disease requires extensive biochemical experiments and further investigations. And once this is established it is still a long way until a medical treatment for a certain genetic susceptibility will be available.

Owing to the complexity of the genetics of disease, another approach, called "The genetics of health" has recently been suggested. This idea is based on the fact that it has been estimated that each individual inherits ~300 mutations with deleterious effects [104]. Despite this, there are individuals who stay healthy until late in life, which is possibly due to the effect of protective alleles for certain genes. A possible alternative route for using evolutionary knowledge in medical genetics would therefore be to search for such protective alleles and modifier genes and – if possible – use the corresponding proteins in medical treatments. As these proteins occur in healthy individuals, adverse side effects are rather unlikely. While this is an interesting and potentially promising alternative idea, as with the other uses of evolutionary insights in medical genetics, applications of this approach are not yet in sight.

Despite all these problems, it is clear that a better understanding of the evolution of the genes that make humans susceptible or resistant to diseases is likely also to lead to better treatments in the long run.

18.4 Recent Events in Human Evolution

18.4.1 Out of Africa into New-found Lands

18.4.1.1 Out of Africa

One of the earliest and maybe most significant findings from human genetics is the insight that modern humans share a recent common origin in Africa [14, 148, 152]. Also controversial in the beginning, it has been supported by numerous studies using markers across the whole genome. The first studies on mtDNA using restriction fragment length polymorphism (RPLPs) [14] and control region sequences [148] have relatively recently been supported by sequencing of a worldwide sample of complete mtDNA genomes [62]. This study not only supported the early studies with respect to a topology of the human mtDNA tree with a lot of diversity in Africa and the rest of the world displaying

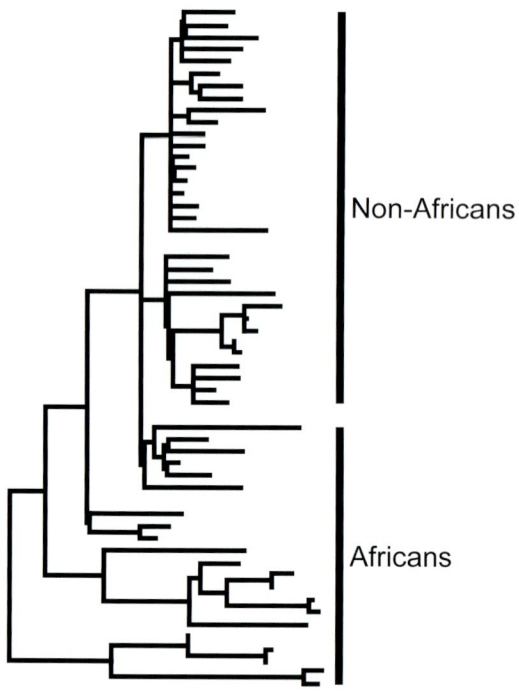

Fig. 18.9 Phylogenetic tree of a worldwide sample of complete human mtDNA sequences. As can be seen, all non-African sequences are nested within African sequence diversity. (After [62])

a subset of this diversity (Fig. 18.9), but also showed that the early studies were surprisingly accurate with respect to the most recent common ancestor (MRCA) of human mtDNA. Thus, while Cann and Vigilant postulated ages for the MRCA of 200,000 and 166,000–249,000 years, respectively, Ingman et al. arrived at a very similar age of 172,000 ± 50,000 years. Interestingly, the oldest modern human fossils have been dated to around 195,000 years B.P. [97]. However, this is a mere coincidence as most sequence regions have far older coalescent dates and should not be taken as evidence that modern humans are direct descendants of these fossils. Apart from determining the relationship of modern mtDNA sequences and the date of their MRCA, Ingman and colleagues were also able to estimate the time when modern humans started to expand in population size to about 38,500 years ago, close to the time when substantial cultural changes occurred in human evolution. This notion of a recent human population expansion was supported by a number of studies using nuclear DNA (e.g., [163, 164]). However, other work indicates that the picture of human population history is more complex and may not be explained by a single population expansion [18, 159]. For example, data from the X-chromosome indicate that before the population expansion detected using mtDNA sequences, there had already been another population expansion in modern humans 100,000–200,000 years ago [74]. A general problem with these analyses lies in the fact that often several alternative models describe the data equally well, so distinguishing between models of long-term growth and those assuming a bottleneck followed by rapid population growth is difficult. Moreover, population expansions seem to have taken place in different geographical regions at different times. Thus, there is evidence for an expansion in Africa as much as 100,000 years ago, whereas the expansions in Asia and Europe date to 52,000 and 23,000 years ago, respectively, [140].

Independently of these signals of population growth, nuclear DNA sequences also support the notion of a recent human origin in Africa (reviewed in [140]). One of the earliest studies using long sequences was on a 10 kb region on the X-chromosome and showed that also for nuclear DNA more sequence diversity is found in Africa [72]. This result, together with the conclusion that humans emigrated relatively recently, has been confirmed by numerous studies, using DNA sequences from autosomes, X- and Y-chromosomes, and other markers, such as Alu insertions from all parts of the genome (e.g., [8, 18, 163, 164]). Most notably, genome-wide analyses of both microsatellites [113] and haplotype structures [26] show a strong negative correlation between the geographical distance of a population from Africa and its genetic diversity. This pattern is best explained by a recent emigration of modern humans out of Africa, with limited gene flow between populations afterwards [113]. Dating this emigration has been somewhat more problematic. However, the molecular estimates center more and more on a date of 60,000–65,000 years B.P. (see e.g., [94, 140, 142]). Intriguingly, Y-chromosomal analyses arrive at a somewhat younger date for the MRCA for all non-African Y-chromosomes, with an age of 44,000 years [145]. This discrepancy can either be explained by Y-chromosomal introgression into Africa subsequent to the initial emigration or, as seems more likely, by uncertainties in the molecular dating. Extensive studies of mtDNA sequences recently also indicated that the initial emigration of modern humans out of Africa took place along a southern coastal route and not through the Levant, as

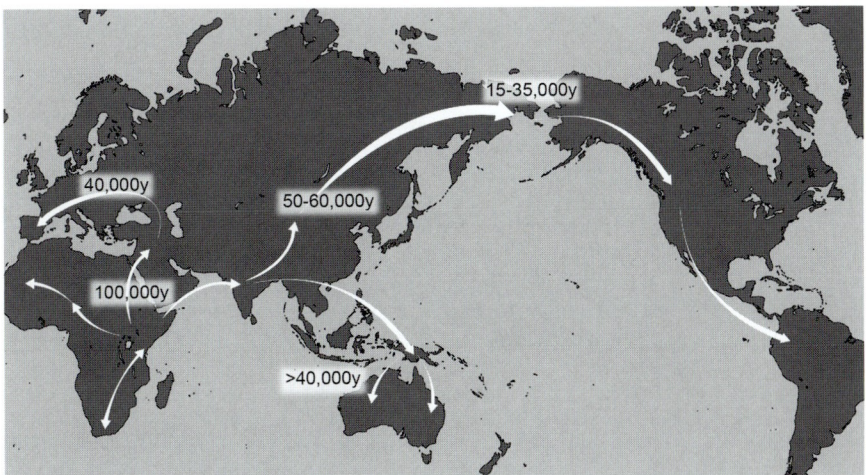

Fig. 18.10 Migration of modern humans out of Africa. The *arrows* indicate suspected routes and the numbers estimated years before present when the various areas were colonized by modern humans. Note that some of the time estimates, especially the timing of the settling of the Americas, are still widely debated

previously believed [94]. This idea is supported by archeological data that indicate an occupation of the Red Sea region together with use of marine food by modern humans around 125,000 years ago [154]. Thus, humans may have emigrated out of Africa from this region along the southern coast of Asia, reaching Australia (Fig. 18.10), as evidenced by archeological findings, as much as 50,000–60,000 years ago [136], which is consistent with the molecular data that hint at an emigration out of Africa about 65,000 years ago [94]. These data also indicate that human migration must have been relatively rapid at about 4 km/year [94]. It is still a matter of debate what triggered the emigration of modern humans out of Africa. Klein [79] has argued for an Upper Paleolithic revolution with some major cognitive or behavioral change having taken place in humans about 45,000–50,000 years ago. However, this notion has been contested with the argument that the archeological evidence provides no strong indication for such a rapid revolution, but rather shows a continuous development [96]. Recently, Mellars [100] argued that rapid climatic changes in Africa about 70,000–80,000 years ago resulted in similarly rapid economic and social changes in African societies, resulting in a population increase that then triggered the emigration out of Africa. Although all these models are intriguing, it should be kept in mind that hominids had emigrated out of Africa several times before, in the absence of sophisticated technology and without any evidence for major sociocultural changes in the African hominid populations preceding these events.

Despite the emigration out of Africa about 65,000 years ago, it was some time before modern humans started colonizing Europe (Fig. 18.10). The earliest modern human fossils in Europe are from Romania, dating back to about 35,000 radiocarbon years B.P. [143]. However, the archeological evidence and differences between uncalibrated carbon dates and true age indicate that modern humans settled in Europe about 40,000 years ago (e.g., [24], reviewed in [101]). Intriguingly, in the Ural region, humans had also managed to settle at the Arctic Circle as early as almost 40,000 years B.P. [110], suggesting that humans reached the Arctic only shortly after they had settled in Europe.

18.4.1.2 Migrations to Europe

Since the initial colonization, several subsequent events have shaped human genetic diversity in Europe, such as recolonization of middle and Northern Europe after the last glacial maximum (e.g., [142]) and the expansion of Neolithic farmers from the Near East north-westward into Europe [19, 103]. To what extent Neolithic farmers really contributed to the modern human gene pool in Europe has been a matter of

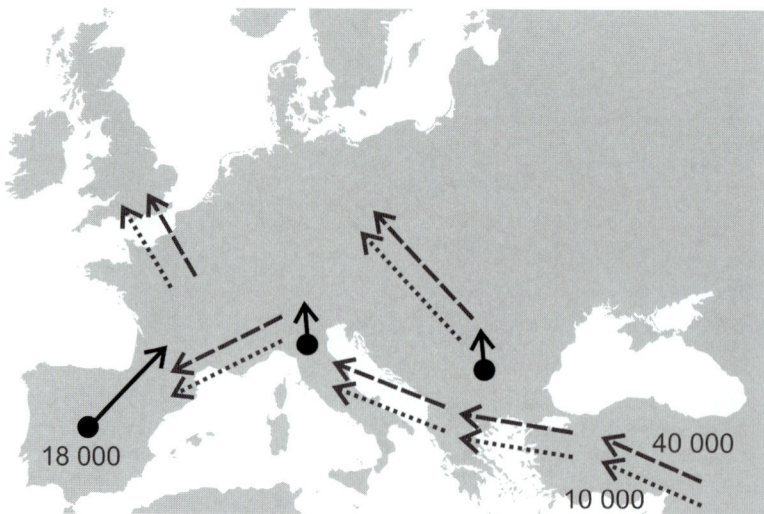

Fig. 18.11 Timing and direction of the three major migration waves that formed the structure of genetic diversity in modern Europeans. (After [6])

debate, ranging from the notion that the contribution was major [19] to the claim that it was less than 25% [117, 142]. However, there is a strong signal of an allele frequency cline from the Near East to northwestern Europe both for autosomal markers [19] and for Y-chromosomal ones [121, 129], making it the predominant pattern of genetic diversity within Europe [6]. At the same time, there is evidence of a postglacial expansion from the Iberian Peninsula [1]. Altogether, the data indicate that at least three major migrations influenced the structure of the modern European gene pool (Fig. 18.11): the initial immigration some 40,000 years ago, a postglacial expansion from southern refugia about 18,000 years ago and the demic diffusion of Neolithic farmers about 10,000 years ago [6].

Analyses of ancient DNA from Neolithic human fossils have added another twist to this complex picture. Haak and colleagues [54] studied a large number of skeletons from Linear pottery culture (LPC) sites and, surprisingly, found a group of haplotypes (N1a) at a frequency of 25%, although this haplogroup is rare (~0.2%) in modern populations all around the world. From these data they concluded that the contribution of Neolithic settlers to the modern European gene pool was only minor. However, it should be noted that LPC is not the same as Neolithic and that the spread of the Neolithic across Europe was not a homogenous event [142].

18.4.1.3 Colonizing America and Oceania

Finally, there are two major regions of the world that have been colonized by humans relatively recently, the Americas and Oceania. The debate about the timing of the settling of the Americas can be summarized in the question whether the initial immigration took place before the appearance of Clovis lithic sites around 13,000 years B.P. or not [127]. Proponents of the Clovis-first hypothesis still contend that there is no solid evidence for the presence of humans in the Americas before the Clovis people. However, there is increasing evidence that human presence in the Americas predated the appearance of the Clovis lithic culture [45]. One of the most important sites in this debate is Monte Verde in Chile, which has been dated to about 14,500 years B.P., clearly predating the Clovis culture [102, 127]. Late Pleistocene sites of human occupation have also been found in Amazonia, and there is an increasing number of sites that predate Clovis culture [45]. Moreover, Y-chromosomal data indicate an entry of humans into the Americas anywhere between 10,000 and 20,000 years B.P. [68, 127], most likely predating the appearance of the Clovis culture but at the same time making a very early entry into the Americas (>25,000 years B.P.) rather unlikely. Further data on both archeological sites and from population genetic studies will hopefully resolve this controversy in the near future.

Oceania was the last major region of the world to become colonized by modern humans, with the earliest human migrations dating to only 800–3,200 years ago [114]. Linguistic evidence points to an Asian origin for Polynesians, whereas archeological evidence argues for a Melanesian origin. The spread of Polynesians is associated with the Lapita cultural complex, but the timing and process of this expansion has been controversial and several competing hypotheses have been presented. Among these is the "slow-boat" hypothesis [75], which proposes that Polynesians originated around the region of Taiwan but moved through Melanesia only slowly, interacting with the population there and also picking up a substantial amount of genetic diversity. This hypothesis is based on the fact that the majority of Polynesian Y-chromosomal haplotypes originated in Melanesia [75]. Further work confirmed this conclusion and at the same time found more than 90% of the mtDNA haplotypes to be of Asian origin [76], suggesting a dual origin of Polynesians. The much larger Melanesian contribution to the Y-chromosome than to mtDNA is hypothesized to be due to the generally matrilocal social structure of Polynesian societies [76]. Other studies have also confirmed a Taiwanese origin for Polynesian mtDNA, and the slow-boat hypothesis is now widely accepted [156]. Moreover, the mtDNA and Y-chromosomal data also show a west-to-east decrease in genetic diversity, which not only indicates a west-to-east direction of settlement in Polynesia but also argues for regular voyaging rather than haphazard settlements [76]. This result is in line with results on the genetic diversity from New Zealand Maoris, which is relatively wide and therefore indicates that settlement of New Zealand was intentional rather than the result of shipwrecks [156]. However, the success story of human settlements in Polynesia also has a dark side. Several Polynesian societies, such as those on Easter Island or Magareva, collapsed because they completely deforested their islands. A comparison of a large number of environmental factors showed that deforestation or lack thereof was not dependent on the respective societies but rather on environmental factors that support or slow down reforestation [118]. In other words, human societies may have a tendency to cause their own collapse by degrading their environment, and the authors suggest that environmental factors may also explain similar deforestation-associated collapses (e.g., Fertile Crescent, Maya, and Anasazi) or the lack thereof (Japan and highland New Guinea) in other parts of the world [118].

18.4.2 Domestication

Possibly the most important event in human history during the last 13,000 years was the beginning of domestication of animals and plants [31]. Domestication was initiated in a maximum of nine regions of the world [31] and resulted in a replacement of hunter-gatherer societies by farmers all around the world although it is not clear to what extent this happened, owing to population replacement, or to what extent assimilation of the farming life-style was involved. The rapid adoption of a farming life style was due to several advantages that farmers have over hunter-gatherer societies. First, farming can support much higher population densities than is possible for hunter-gatherer societies. Second, farming societies are sedentary, which allows accumulation of food stocks, and this is essential for complex technology and centralized states, two hallmarks of modern complex societies [32].

Some of the earliest domestications took place in the Fertile Crescent in the Near East, dating back to about 11,000 years, as evidenced by both archeological and genetic data. Here, several crop species, such as wheat, rye, and barley were domesticated. Moreover, at least one of two independent cattle domestications [10, 93] also took place in this region [144] resulting in modern European cattle (*Bos taurus*). Interestingly, genetic data also show that European aurochs (*Bos primigenius*) were probably never domesticated as a number of ancient DNA sequences from fossil British bones differed substantially in their sequence from all modern European cattle sequences [144]. A second cattle domestication took place in India, leading to the zebu lineage (*Bos indicus* [10, 93]), whereas African cattle are a hybrid population that have a zebu-like appearance, which is consistent with genetic data from their nuclear genome, but have derived their mitochondrial DNA from European cattle. Deep divergences between groups of sequences have been observed for many domesticated species (apart from cattle also for sheep, goat, and donkey) and have usually been interpreted as evidence for independent domestication events. However, this interpretation has to be treated with caution, as deeply diverging lineages may occur in a single wild population, as shown for the yak [53] and are thus in themselves not convincing evidence for independent domestication events. The same is true for high levels of genetic diversity in a domesticated species which has been taken as evidence for domestication from multiple populations of the horse [67]. This interpretation may be correct, but in the absence of knowledge about the genetic diversity of the underlying wild population it is also possible that domestication took place from a single, highly diverse population, as shown for the yak. So far the highest number of domestication events has been claimed for the pig with about a dozen [90]. It remains to be seen whether this is true or rather an artifact of subsequent – intentional and unintentional – back-crossing of domesticated animals with their local wild relatives, which it is suggested, has been a rather frequent practice [149].

Interestingly, the first animal to be domesticated was not a life-stock species but the dog, whose domestication took place about 15,000 years ago, probably in China [126]. Moreover, studies of fossil dogs from America, predating European contact, showed that these Native American dogs were not independently domesticated but rather brought over by humans immigrating from Asia [91]. Thus, domestic animal species may be transported rapidly and over large distances by humans, possibly blurring signals of their geographic origin. A very early (10,000 years B.P.) transportation from Asia to America has also been postulated for another utility species, the bottle gourd [41].

Finally, it should be noted that the effects of domestication on human societies have not all been positive. For example, it has been argued that the African malaria parasite *Plasmodium falciparum* rapidly increased in population size about 10,000 years ago with the onset of agriculture [71]. Thus, domestication did not only improve the nutritional situation for human societies, but also exposed them to new parasite loads. However, domestication was clearly a major step in human evolution and a prerequisite for modern human societies.

18.4.3 Modern Human Population Structure

Human evolution was clearly a dynamic process and the current human population is the result of numerous migrations and population size changes, as shown above. So, the question of how we are to view modern human genetic diversity remains open. As noted before, humans carry relatively little genetic diversity compared with their closest relatives, the great apes, and all modern humans outside Africa share a recent African origin, so that we all seem to be Africans, either living on that continent or in recent exile" [107]. What is a matter of debate is how much structure the modern human gene pool contains. In 2002 Rosenberg and colleagues [119] argued that given a sufficient number of markers, humans can be placed into clusters that correspond to a geographical origin, implying – albeit most probably inadvertently – that major genetic differences exist between continental groups such as Africans, Asians, and Europeans. This conclusion has been contested (e.g., [131]) with the argument that the result of Rosenber et al. [119] was an artifact of the study design, and human genetic diversity is best explained by a clinal model of isolation by distance, without any major jumps in genetic distance over short geographical distances. Thus, rather than being made up of several distinct human groups that are genetically well separated, human genetic diversity changes continuously, with humans living in geographical proximity also being more closely related genetically and humans living at greater geographical distance also being genetically more different. This later view has been confirmed by additional studies that detected a strong signal for isolation by distance in both the data set of Rosenberg et al. [113] and in other data sets [30]. However, expanding their data set, Rosenberg and colleagues reaffirmed that human populations indeed cluster by geographical origin [120]. At the same time, they do find a pattern of isolation by distance, with genetic distance generally increasing with geographical distance. The reason why they nevertheless detect clusters lies in the fact that small jumps in genetic distances occur across short distance of geographical barriers. Thus, the clusters are real but they explain only a small proportion of the genetic differences between humans, a result also seen in other data sets. For example, in a compilation of about 1 million SNPs most are shared between Africans, Europeans, and Asians and only a very few represent fixed differences between the continental groups. So despite some phenotypically obvious differences between human populations, such as skin color, across the whole genome there is very little genetic differentiation between human populations. Given that modern humans originated in Africa only about 200,000 years ago and humans are a notoriously migratory species, it is not surprising that we are all very close relatives.

Acknowledgements I thank the Max Planck Society for financial support, Knut Finstermeier and Sandra Michaelis for help with the figure design, and my research group for their patience with me while I have been mentally absent during writing. I also want to apologize to all those colleagues whose work I have not cited because of lack of space.

Glossary

Clinal model
A model describing the change in genetic diversity across geographical distance. In the clinal model, genetic differences among populations increase more or less constantly with increasing geographical distance.

Clovis culture
Clovis describes a Native American cultural horizon, dated to ~ 11,000 radiocarbon years B.P. Characteristic for this culture are the beautiful bifacial spear points made from stone, also known as Clovis points.

Hominid
A member of the mammalian family Hominidae, which includes humans and the great apes (orangutans, chimpanzees and gorillas).

Lapita culture
A Pacific Ocean culture, dated to ~ 3,500 to 2,500 B.P., which might be ancestral to cultures in Polynesia, Micronesia, and Melanesia.

Linear pottery culture
The linear pottery culture is a Neolithic archeological horizon from Central Europe, dated to about 7,500–6,500 B.P.

Melanesia
The part of Oceania north and north-east of Australia.

Neolithic
The last part of the Stone Age, starting approximately 10,000 years ago. The beginning of the Neolithic also marks the start of farming. It ended when the use of metal tools became common, at different times in different regions.

Oceania
The island region in the Pacific east and north-east of Australia.

Pericentric inversion
An inversion of a part of the chromosome including the centromere.

Polynesians
Inhabitants of Polynesia, the island region roughly in the triangle from Hawaii, New Zealand and Easter Island.

References

1. Achilli A, Rengo C, Magri C, Battaglia V, Olivieri A et al (2004) The molecular dissection of mtDNA haplogroup H confirms that the Franco-Cantabrian glacial refuge was a major source for the European gene pool. Am J Hum Genet 75:910–918
2. Alemseged Z, Spoor F, Kimbel WH, Bobe R, Geraads D et al (2006) A juvenile early hominin skeleton from Dikika, Ethiopia. Nature 443:296–301
3. Anderson S, Bankier AT, Barrell BG, de Bruijn MH, Coulson AR et al (1981) Sequence and organization of the human mitochondrial genome. Nature 290:457–465
4. Argue D, Donlon D, Groves C, Wright R (2006) Homo floresiensis: microcephalic, pygmoid, Australopithecus, or Homo? J Hum Evol 51:360–374
5. Arsuaga JL, Martinez I, Gracia A, Carretero JM, Carbonell E (1993) Three new human skulls from the Sima de los Huesos Middle Pleistocene site in Sierra de Atapuerca, Spain. Nature 362:534–537
6. Barbujani G, Bertorelle G (2001) Genetics and the population history of Europe. Proc Natl Acad Sci USA 98:22–25
7. Barton NH (2006) Evolutionary biology: how did the human species form? Curr Biol 16:R647–R650
8. Batzer MA, Deininger PL (2002) Alu repeats and human genomic diversity. Nat Rev Genet 3:370–379
9. Bersaglieri T, Sabeti PC, Patterson N, Vanderploeg T, Schaffner SF et al (2004) Genetic signatures of strong recent positive selection at the lactase gene. Am J Hum Genet 74:1111–1120
10. Bradley DG, MacHugh DE, Cunningham P, Loftus RT (1996) Mitochondrial diversity and the origins of African and European cattle. Proc Natl Acad Sci US A 93:5131–5135
11. Brown P, Sutikna T, Morwood MJ, Soejono RP, Jatmiko, et al. (2004) A new small-bodied hominin from the Late Pleistocene of Flores, Indonesia. Nature 431:1055–1061
12. Brunet M, Guy F, Pilbeam D, Mackaye HT, Likius A et al (2002) A new hominid from the Upper Miocene of Chad, Central Africa. Nature 418:145–151
13. Bustamante CD, Fledel-Alon A, Williamson S, Nielsen R, Hubisz MT et al (2005) Natural selection on protein-coding genes in the human genome. Nature 437:1153–1157
14. Cann RL, Stoneking M, Wilson AC (1987) Mitochondrial DNA and human evolution. Nature 325:31–36
15. Carroll SB (2003) Genetics and the making of Homo sapiens. Nature 422:849–857
16. Cavalli-Sforza LL (1966) Population structure and human evolution. Proc R Soc Lond B Biol Sci 164:362–379
17. Cavalli-Sforza LL (2007) Human evolution and its relevance for genetic epidemiology. Annu Rev Genomics Hum Genet 8:1–15

18. Cavalli-Sforza LL, Feldman MW (2003) The application of molecular genetic approaches to the study of human evolution. Nat Genet 33(Suppl):266–275
19. Cavalli-Sforza LL, Menozzi P, Piazza A (1993) Demic expansions and human evolution. Science 259:639–646
20. Chanock SJ, Manolio T, Boehnke M, Boerwinkle E, Hunter DJ et al (2007) Replicating genotype-phenotype associations. Nature 447:655–660
21. Check E (2005) Human genome: patchwork people. Nature 437:1084–1086
22. Chen FC, Li WH (2001) Genomic divergences between humans and other hominoids and the effective population size of the common ancestor of humans and chimpanzees. Am J Hum Genet 68:444–456
23. Cheng Z, Ventura M, She X, Khaitovich P, Graves T et al (2005) A genome-wide comparison of recent chimpanzee and human segmental duplications. Nature 437:88–93
24. Conard NJ (2003) Palaeolithic ivory sculptures from southwestern Germany and the origins of figurative art. Nature 426:830–832
25. Conrad B, Antonarakis SE (2007) Gene duplication: a drive for phenotypic diversity and cause of human disease. Annu Rev Genomics Hum Genet 8:17–35
26. Conrad DF, Jakobsson M, Coop G, Wen X, Wall JD et al (2006) A worldwide survey of haplotype variation and linkage disequilibrium in the human genome. Nat Genet 38:1251–1260
27. Currat M, Excoffier L (2004) Modern humans did not admix with Neanderthals during their range expansion into Europe. PLoS Biol 2:e421
28. Darwin C (1858) On the Variation of Organic Beings in a state of Nature; on the Natural Means of Selection; on the Comparison of Domestic Races and true Species. Proc Linnean Soc 3:46–53
29. Darwin C (1859) The origin of species. John Murray, London
30. Darwin C (1871) The descent of man and selection in relation to sex. John Murray, London
31. Diamond J (2002) Evolution, consequences and future of plant and animal domestication. Nature 418:700–707
32. Diamond J, Bellwood P (2003) Farmers and their languages: the first expansions. Science 300:597–603
33. Dorus S, Vallender EJ, Evans PD, Anderson JR, Gilbert SL et al (2004) Accelerated evolution of nervous system genes in the origin of Homo sapiens. Cell 119:1027–1040
34. Easton DF, Pooley KA, Dunning AM, Pharoah PD, Thompson D et al (2007) Genome-wide association study identifies novel breast cancer susceptibility loci. Nature 447:1087–1093
35. Editorial (2004) Finishing the euchromatic sequence of the human genome. Nature 431:931–945
36. Editorial (2005) A haplotype map of the human genome. Nature 437:1299–1320
37. Editorial (2005) Initial sequence of the chimpanzee genome and comparison with the human genome. Nature 437: 69–87
38. Editorial (2007) Genome-wide association study of 14,000 cases of seven common diseases and 3,000 shared controls. Nature 447:661–78
39. Elbaz A, Nelson LM, Payami H, Ioannidis JP, Fiske BK et al (2006) Lack of replication of thirteen single-nucleotide polymorphisms implicated in Parkinson's disease: a large-scale international study. Lancet Neurol 5:917–923
40. Enard W, Przeworski M, Fisher SE, Lai CS, Wiebe V et al (2002) Molecular evolution of FOXP2, a gene involved in speech and language. Nature 418:869–872
41. Erickson DL, Smith BD, Clarke AC, Sandweiss DH, Tuross N (2005) An Asian origin for a 10, 000-year-old domesticated plant in the Americas. Proc Natl Acad Sci USA 102:18315–18320
42. Evans PD, Gilbert SL, Mekel-Bobrov N, Vallender EJ, Anderson JR et al (2005) Microcephalin, a gene regulating brain size, continues to evolve adaptively in humans. Science 309:1717–1720
43. Evans PD, Mekel-Bobrov N, Vallender EJ, Hudson RR, Lahn BT (2006) Evidence that the adaptive allele of the brain size gene microcephalin introgressed into Homo sapiens from an archaic Homo lineage. Proc Natl Acad Sci USA 103:18178–18183
44. Eyre-Walker A (2006) The genomic rate of adaptive evolution. Trends Ecol Evol 21:569–575
45. Falk T (2004) Archaeology Wisconsin dig seeks to confirm pre-clovis Americans. Science 305:590
46. Fay JC, Wyckoff GJ, Wu CI (2001) Positive and negative selection on the human genome. Genetics 158:1227–1234
47. Feuk L, Carson AR, Scherer SW (2006) Structural variation in the human genome. Nat Rev Genet 7:85–97
48. Fortna A, Kim Y, MacLaren E, Marshall K, Hahn G et al (2004) Lineage-specific gene duplication and loss in human and great ape evolution. PLoS Biol 2:E207
49. Fulrott J (1859) Menschliche Überreste aus einer Felsengrotte des Düsselthals. Ein Beitrag zur Frage über die Existenz fossiler Menschen. Verhandl. naturhist. Verein Pruess Rheinl Westphal 16:131–153
50. Gabunia L, Vekua A (1995) A Plio-Pleistocene hominid from Dmanisi, East Georgia, Caucasus. Nature 373:509–512
51. Gilbert SL, Dobyns WB, Lahn BT (2005) Genetic links between brain development and brain evolution. Nat Rev Genet 6:581–590
52. Green RE, Krause J, Ptak SE, Briggs AW, Ronan MT et al (2006) Analysis of one million base pairs of Neanderthal DNA. Nature 444:330–336
53. Guo S, Savolainen P, Su J, Zhang Q, Qi D et al (2006) Origin of mitochondrial DNA diversity of domestic yaks. BMC Evol Biol 6:73
54. Haak W, Forster P, Bramanti B, Matsumura S, Brandt G et al (2005) Ancient DNA from the first European farmers in 7500-year-old Neolithic sites. Science 310:1016–1018
55. Haesler S, Wada K, Nshdejan A, Morrisey EE, Lints T et al (2004) FoxP2 expression in avian vocal learners and non-learners. J Neurosci 24:3164–3175
56. Hasegawa M, Kishino H, Yano T (1985) Dating of the human-ape splitting by a molecular clock of mitochondrial DNA. J Mol Evol 22:160–174
57. Hasegawa M, Kishino H, Yano T (1987) Man's place in Hominoidea as inferred from molecular clocks of DNA. J Mol Evol 26:132–147
58. Hofreiter M, Serre D, Poinar HN, Kuch M, Pääbo S (2001) Ancient DNA. Nat Rev Genet 2:353–359
59. Hughes DA, Cordaux R, Stoneking M (2004) Humans. Curr Biol 14:R367–R369

60. Hutchison CA 3 rd (2007) DNA sequencing: bench to bedside and beyond. Nucleic Acids Res 35:6227–6237
61. Huxley T (1863) Evidence as to man's place in nature. Summer-field Press, London
62. Ingman M, Kaessmann H, Paabo S, Gyllensten U (2000) Mitochondrial genome variation and the origin of modern humans. Nature 408:708–713
63. Ingram CJ, Elamin MF, Mulcare CA, Weale ME, Tarekegn A et al (2006) A novel polymorphism associated with lactose tolerance in Africa: multiple causes for lactase persistence? Hum Genet 120:779–788
64. Jablonski NG, Chaplin G (2000) The evolution of human skin coloration. J Hum Evol 39:57–106
65. Jackson IJ (1997) Homologous pigmentation mutations in human, mouse and other model organisms. Hum Mol Genet 6:1613–1624
66. Jacob T, Indriati E, Soejono RP, Hsu K, Frayer DW et al (2006) Pygmoid Australomelanesian Homo sapiens skeletal remains from Liang Bua, Flores: population affinities and pathological abnormalities. Proc Natl Acad Sci USA 103:13421–13426
67. Jansen T, Forster P, Levine MA, Oelke H, Hurles M et al (2002) Mitochondrial DNA and the origins of the domestic horse. Proc Natl Acad Sci USA 99:10905–10910
68. Jobling MA, Tyler-Smith C (2003) The human Y chromosome: an evolutionary marker comes of age. Nat Rev Genet 4:598–612
69. Johanson DC, White TD (1979) A systematic assessment of early African hominids. Science 203:321–330
70. Johnson ME, Viggiano L, Bailey JA, Abdul-Rauf M, Goodwin G et al (2001) Positive selection of a gene family during the emergence of humans and African apes. Nature 413:514–519
71. Joy DA, Feng X, Mu J, Furuya T, Chotivanich K et al (2003) Early origin and recent expansion of Plasmodium falciparum. Science 300:318–321
72. Kaessmann H, Heissig F, von Haeseler A, Paabo S (1999) DNA sequence variation in a non-coding region of low recombination on the human X chromosome. Nat Genet 22:78–81
73. Kaessmann H, Wiebe V, Paabo S (1999) Extensive nuclear DNA sequence diversity among chimpanzees. Science 286:1159–1162
74. Kaessmann H, Wiebe V, Weiss G, Paabo S (2001) Great ape DNA sequences reveal a reduced diversity and an expansion in humans. Nat Genet 27:155–156
75. Kayser M, Brauer S, Weiss G, Underhill PA, Roewer L et al (2000) Melanesian origin of Polynesian Y chromosomes. Curr Biol 10:1237–1246
76. Kayser M, Brauer S, Cordaux R, Casto A, Lao O et al (2006) Melanesian and Asian origins of Polynesians: mtDNA and Y chromosome gradients across the Pacific. Mol Biol Evol 23:2234–2244
77. Khaitovich P, Hellmann I, Enard W, Nowick K, Leinweber M et al (2005) Parallel patterns of evolution in the genomes and transcriptomes of humans and chimpanzees. Science 309:1850–1854
78. Kimura M (1968) Evolutionary rate at the molecular level. Nature 217:624–626
79. Klein RG (1989) The human career. University of Chicago Press, Chicago
80. Kouprina N, Pavlicek A, Mochida GH, Solomon G, Gersch W et al (2004) Accelerated evolution of the ASPM gene controlling brain size begins prior to human brain expansion. PLoS Biol 2:E126
81. Krause J, Orlando L, Serre D, Viola B, Prufer K et al (2007) Neanderthals in central Asia and Siberia. Nature 449:902–904
82. Krings M, Geisert H, Schmitz RW, Krainitzki H, Paabo S (1999) DNA sequence of the mitochondrial hypervariable region II from the neandertal type specimen. Proc Natl Acad Sci USA 96:5581–5585
83. Krings M, Stone A, Schmitz RW, Krainitzki H, Stoneking M, Paabo S (1997) Neandertal DNA sequences and the origin of modern humans. Cell 90:19–30
84. Kwiatkowski DP (2005) How malaria has affected the human genome and what human genetics can teach us about malaria. Am J Hum Genet 77:171–192
85. Lacana E, Amur S, Mummanneni P, Zhao H, Frueh FW (2007) The emerging role of pharmacogenomics in biologics. Clin Pharmacol Ther 82:466–471
86. Lai CS, Fisher SE, Hurst JA, Vargha-Khadem F, Monaco AP (2001) A forkhead-domain gene is mutated in a severe speech and language disorder. Nature 413:519–523
87. Lai CS, Gerrelli D, Monaco AP, Fisher SE, Copp AJ (2003) FOXP2 expression during brain development coincides with adult sites of pathology in a severe speech and language disorder. Brain 126:2455–2462
88. Lamason RL, Mohideen MA, Mest JR, Wong AC, Norton HL et al (2005) SLC24A5, a putative cation exchanger, affects pigmentation in zebrafish and humans. Science 310:1782–1786
89. Lander ES, Linton LM, Birren B, Nusbaum C, Zody MC et al (2001) Initial sequencing and analysis of the human genome. Nature 409:860–921
90. Larson G, Dobney K, Albarella U, Fang M, Matisoo-Smith E et al (2005) Worldwide phylogeography of wild boar reveals multiple centers of pig domestication. Science 307:1618–1621
91. Leonard JA, Wayne RK, Wheeler J, Valadez R, Guillen S, Vila C (2002) Ancient DNA evidence for Old World origin of New World dogs. Science 298:1613–1616
92. Locke DP, Segraves R, Carbone L, Archidiacono N, Albertson DG et al (2003) Large-scale variation among human and great ape genomes determined by array comparative genomic hybridization. Genome Res 13:347–357
93. Loftus RT, MacHugh DE, Bradley DG, Sharp PM, Cunningham P (1994) Evidence for two independent domestications of cattle. Proc Natl Acad Sci USA 91:2757–2761
94. Macaulay V, Hill C, Achilli A, Rengo C, Clarke D et al (2005) Single, rapid coastal settlement of Asia revealed by analysis of complete mitochondrial genomes. Science 308:1034–1036
95. MacDermot KD, Bonora E, Sykes N, Coupe AM, Lai CS et al (2005) Identification of FOXP2 truncation as a novel cause of developmental speech and language deficits. Am J Hum Genet 76:1074–1080
96. McBrearty S, Brooks AS (2000) The revolution that wasn't: a new interpretation of the origin of modern human behavior. J Hum Evol 39:453–563
97. McDougall I, Brown FH, Fleagle JG (2005) Stratigraphic placement and age of modern humans from Kibish, Ethiopia. Nature 433:733–736

98. Mekel-Bobrov N, Gilbert SL, Evans PD, Vallender EJ, Anderson JR et al (2005) Ongoing adaptive evolution of ASPM, a brain size determinant in Homo sapiens. Science 309:1720–1722
99. Mellars P (2004) Neanderthals and the modern human colonization of Europe. Nature 432:461–465
100. Mellars P (2006) Going east: new genetic and archaeological perspectives on the modern human colonization of Eurasia. Science 313:796–800
101. Mellars P (2006) A new radiocarbon revolution and the dispersal of modern humans in Eurasia. Nature 439:931–935
102. Meltzer DJ (1997) Anthropology – Monte Verde and the Pleistocene peopling of the Americas. Science 276: 754–755
103. Menozzi P, Piazza A, Cavalli-Sforza L (1978) Synthetic maps of human gene frequencies in Europeans. Science 201:786–792
104. Nadeau JH, Topol EJ (2006) The genetics of health. Nat Genet 38:1095–1098
105. Nielsen R, Bustamante C, Clark AG, Glanowski S, Sackton TB et al (2005) A scan for positively selected genes in the genomes of humans and chimpanzees. PLoS Biol 3:e170
106. Noonan JP, Coop G, Kudaravalli S, Smith D, Krause J et al (2006) Sequencing and analysis of Neanderthal genomic DNA. Science 314:1113–1118
107. Paabo S (1999) Human evolution. Trends Cell Biol 9:M13–M16
108. Parfitt SA, Barendregt RW, Breda M, Candy I, Collins MJ et al (2005) The earliest record of human activity in northern Europe. Nature 438:1008–1012
109. Patterson N, Richter DJ, Gnerre S, Lander ES, Reich D (2006) Genetic evidence for complex speciation of humans and chimpanzees. Nature 441:1103–1108
110. Pavlov P, Svendsen JI, Indrelid S (2001) Human presence in the European Arctic nearly 40, 000 years ago. Nature 413:64–67
111. Plagnol V, Wall JD (2006) Possible ancestral structure in human populations. PLoS Genet 2:e105
112. Popesco MC, Maclaren EJ, Hopkins J, Dumas L, Cox M et al (2006) Human lineage-specific amplification, selection, and neuronal expression of DUF1220 domains. Science 313:1304–1307
113. Prugnolle F, Manica A, Balloux F (2005) Geography predicts neutral genetic diversity of human populations. Curr Biol 15:R159–R160
114. Redd AJ, Takezaki N, Sherry ST, McGarvey ST, Sofro AS, Stoneking M (1995) Evolutionary history of the COII/tRNALys intergenic 9 base pair deletion in human mitochondrial DNAs from the Pacific. Mol Biol Evol 12:604–615
115. Redon R, Ishikawa S, Fitch KR, Feuk L, Perry GH et al (2006) Global variation in copy number in the human genome. Nature 444:444–454
116. Rees JL (2000) The melanocortin 1 receptor (MC1R): more than just red hair. Pigment Cell Res 13:135–140
117. Richards M, Macaulay V, Hickey E, Vega E, Sykes B et al (2000) Tracing European founder lineages in the Near Eastern mtDNA pool. Am J Hum Genet 67:1251–1276
118. Rolett B, Diamond J (2004) Environmental predictors of pre-European deforestation on Pacific islands. Nature 431:443–446
119. Rosenberg NA, Pritchard JK, Weber JL, Cann HM, Kidd KK et al (2002) Genetic structure of human populations. Science 298:2381–2385
120. Rosenberg NA, Mahajan S, Ramachandran S, Zhao C, Pritchard JK, Feldman MW (2005) Clines, clusters, and the effect of study design on the inference of human population structure. PLoS Genet 1:e70
121. Rosser ZH, Zerjal T, Hurles ME, Adojaan M, Alavantic D et al (2000) Y-chromosomal diversity in Europe is clinal and influenced primarily by geography, rather than by language. Am J Hum Genet 67:1526–1543
122. Roy-Engel AM, Carroll ML, Vogel E, Garber RK, Nguyen SV et al (2001) Alu insertion polymorphisms for the study of human genomic diversity. Genetics 159:279–290
123. Sabeti PC, Reich DE, Higgins JM, Levine HZ, Richter DJ et al (2002) Detecting recent positive selection in the human genome from haplotype structure. Nature 419: 832–837
124. Sabeti PC, Schaffner SF, Fry B, Lohmueller J, Varilly P et al (2006) Positive natural selection in the human lineage. Science 312:1614–1620
125. Sarich VM, Wilson AC (1967) Rates of albumin evolution in primates. Proc Natl Acad Sci USA 58:142–148
126. Savolainen P, Zhang YP, Luo J, Lundeberg J, Leitner T (2002) Genetic evidence for an East Asian origin of domestic dogs. Science 298:1610–1613
127. Schurr TG, Sherry ST (2004) Mitochondrial DNA and Y chromosome diversity and the peopling of the Americas: evolutionary and demographic evidence. Am J Hum Biol 16:420–439
128. Scriver CR (2002) Why mutation analysis does not always predict clinical consequences: explanations in the era of genomics. J Pediatr 140:502–506
129. Semino O, Passarino G, Oefner PJ, Lin AA, Arbuzova S et al (2000) The genetic legacy of Paleolithic Homo sapiens sapiens in extant Europeans: a Y chromosome perspective. Science 290:1155–1159
130. Serre D, Langaney A, Chech M, Teschler-Nicola M, Paunovic M et al (2004) No evidence of Neandertal mtDNA contribution to early modern humans. PLoS Biol 2:E57
131. Serre D, Paabo S (2004) Evidence for gradients of human genetic diversity within and among continents. Genome Res 14:1679–1685
132. Sheen FM, Sherry ST, Risch GM, Robichaux M, Nasidze I et al (2000) Reading between the LINEs: human genomic variation induced by LINE-1 retrotransposition. Genome Res 10:1496–1508
133. Shriner D, Vaughan LK, Padilla MA, Tiwari HK (2007) Problems with genome-wide association studies. Science 316:1840–1842
134. Stedman HH, Kozyak BW, Nelson A, Thesier DM, Su LT et al (2004) Myosin gene mutation correlates with anatomical changes in the human lineage. Nature 428:415–418
135. Stone AC, Griffiths RC, Zegura SL, Hammer MF (2002) High levels of Y-chromosome nucleotide diversity in the genus Pan. Proc Natl Acad Sci USA 99:43–48
136. Stringer C (2000) Palaeoanthropology Coasting out of Africa. Nature 405(24–5):7
137. Swallow DM (2003) Genetics of lactase persistence and lactose intolerance. Annu Rev Genet 37:197–219

138. Swisher CC 3rd, Curtis GH, Jacob T, Getty AG, Suprijo A, Widiasmoro (1994) Age of the earliest known hominids in Java, Indonesia. Science 263:1118-1121
139. Timpson N, Heron J, Smith GD, Enard W (2007) Comment on papers by Evans et al. and Mekel-Bobrov et al. on evidence for positive selection of MCPH1 and ASPM. Science 317:1036
140. Tishkoff SA, Williams SM (2002) Genetic analysis of African populations: human evolution and complex disease. Nat Rev Genet 3:611–621
141. Tishkoff SA, Reed FA, Ranciaro A, Voight BF, Babbitt CC et al (2006) Convergent adaptation of human lactase persistencein Africa and Europe. Nat Genet . doi:doi:10.1038/ng946 published online 10 December 2006
142. Torroni A, Achilli A, Macaulay V, Richards M, Bandelt HJ (2006) Harvesting the fruit of the human mtDNA tree. Trends Genet 22:339–345
143. Trinkaus E, Moldovan O, Milota S, Bilgar A, Sarcina L et al (2003) An early modern human from the Pestera cu Oase, Romania. Proc Natl Acad Sci USA 100:11231–11236
144. Troy CS, MacHugh DE, Bailey JF, Magee DA, Loftus RT et al (2001) Genetic evidence for Near-Eastern origins of European cattle. Nature 410:1088–1091
145. Underhill PA, Shen P, Lin AA, Jin L, Passarino G et al (2000) Y chromosome sequence variation and the history of human populations. Nat Genet 26:358–361
146. van Schaik CP, Ancrenaz M, Borgen G, Galdikas B, Knott CD et al (2003) Orangutan cultures and the evolution of material culture. Science 299:102–105
147. Verrelli BC, McDonald JH, Argyropoulos G, Destro-Bisol G, Froment A et al (2002) Evidence for balancing selection from nucleotide sequence analyses of human G6PD. Am J Hum Genet 71:1112–1128
148. Vigilant L, Stoneking M, Harpending H, Hawkes K, Wilson AC (1991) African populations and the evolution of human mitochondrial DNA. Science 253:1503–1507
149. Vila C, Seddon J, Ellegren H (2005) Genes of domestic mammals augmented by backcrossing with wild ancestors. Trends Genet 21:214–218
150. Vincent BJ, Myers JS, Ho HJ, Kilroy GE, Walker JA et al (2003) Following the LINEs: an analysis of primate genomic variation at human-specific LINE-1 insertion sites. Mol Biol Evol 20:1338–1348
151. Voight BF, Kudaravalli S, Wen X, Pritchard JK (2006) A map of recent positive selection in the human genome. PLoS Biol 4:e72
152. Wainscoat JS, Hill AV, Boyce AL, Flint J, Hernandez M et al (1986) Evolutionary relationships of human populations from an analysis of nuclear DNA polymorphisms. Nature 319:491–493
153. Wallace AR (1858) On the Tendency of Varieties to Depart Indefinitely From the Original Type. Proc Linnean Soc 3:53–62
154. Walter RC, Buffler RT, Bruggemann JH, Guillaume MM, Berhe SM et al (2000) Early human occupation of the Red Sea coast of Eritrea during the last interglacial. Nature 405:65–69
155. Whiten A, Goodall J, McGrew WC, Nishida T, Reynolds V et al (1999) Cultures in chimpanzees. Nature 399: 682–685
156. Whyte ALH, Marshall SJ, Chambers GK (2005) Human evolution in Polynesia. Hum Biol 77:157–177
157. Wood B, Richmond BG (2000) Human evolution: taxonomy and paleobiology. J Anat 197(Pt 1):19–60
158. Wyckoff GJ, Wang W, Wu CI (2000) Rapid evolution of male reproductive genes in the descent of man. Nature 403:304–309
159. Yu N, Fu YX, Li WH (2002) DNA polymorphism in a worldwide sample of human X chromosomes. Mol Biol Evol 19:2131–2141
160. Yu N, Jensen-Seaman MI, Chemnick L, Kidd JR, Deinard AS et al (2003) Low nucleotide diversity in chimpanzees and bonobos. Genetics 164:1511–1518
161. Yu N, Jensen-Seaman MI, Chemnick L, Ryder O, Li WH (2004) Nucleotide diversity in gorillas. Genetics 166:1375–1383
162. Zhang L, Li WH (2005) Human SNPs reveal no evidence of frequent positive selection. Mol Biol Evol 22: 2504–2507
163. Zhao Z, Jin L, Fu YX, Ramsay M, Jenkins T et al (2000) Worldwide DNA sequence variation in a 10-kilobase noncoding region on human chromosome 22. Proc Natl Acad Sci USA 97:11354–11358
164. Zhao Z, Yu N, Fu YX, Li WH (2006) Nucleotide variation and haplotype diversity in a 10-kb noncoding region in three continental human populations. Genetics 174: 399–409
165. Zhu RX, Potts R, Xie F, Hoffman KA, Deng CL et al (2004) New evidence on the earliest human presence at high northern latitudes in northeast Asia. Nature 431: 559–562

Comparative Genomics

19

Ross C. Hardison

Abstract Comparative genomics harnesses the power of sequence comparisons within and between species to deduce not only evolutionary history but also insights into the function, if any, of particular DNA sequences. Changes in DNA and protein sequences are subject to three evolutionary processes: drift, which allows some neutral changes to accumulate, negative selection, which removes deleterious changes, or positive selection, which acts on adaptive changes to increase their frequency in a population. Quantitative data from comparative genomics can be used to infer the type of evolutionary force that likely has been operating on a particular sequence, thereby predicting whether it is functional. These predictions are good but imperfect; their primary role is to provide useful hypotheses for further experimental tests of function. Rates of evolutionary change vary both between functional categories of sequences and regionally within genomes. Even within a functional category (e.g. protein or gene regulatory region) the rates vary. A more complete understanding of variation in the patterns and rates of evolution should improve the predictive accuracy of comparative genomics. Proteins that show signatures of adaptive evolution tend to fall into the major functional categories of reproduction, chemosensation, immune response and xenobiotic metabolism. DNA sequences that appear to be under the strongest evolutionary constraint are not fully understood, although many of them are active as transcriptional enhancers. Human sequences that regulate gene expression tend to be conserved among placental mammals, but the phylogenetic depth of conservation of individual regulatory regions ranges from primate-specific to pan-vertebrate.

Contents

19.1	Goals, Impact, and Basic Approaches of Comparative Genomics........................ 557		19.1.3	Models of Neutral DNA 560
			19.1.4	Adaptive Evolution 562
	19.1.1	How Biological Sequences Change Over Time............................... 558	19.2	Alignments of Biological Sequences and Their Interpretation 563
	19.1.2	Purifying Selection 559	19.2.1	Global and Local Alignments................ 563
			19.2.2	Aligning Protein Sequences................... 563
			19.2.3	Aligning Large Genome Sequences 564
			19.3	Assessment of Conserved Function from Alignments 565
			19.3.1	Phylogenetic Depth of Alignments 566
			19.3.2	Portion of the Human Genome Under Constraint................................... 568
			19.3.3	Identifying Specific Sequences Under Constraint................................... 569

R.C. Hardison (✉)
Center for Comparative Genomics and Bioinformatics,
Huck Institutes of Life Sciences,
Department of Biochemistry and Molecular Biology,
The Pennsylvania State University,
PA 16802, USA
e-mail: rch8@psu.edu

19.4	Evolution Within Protein-Coding Genes 560		19.5.1	Ultraconserved Elements 579	
	19.4.1	Comparative Genomics in Gene Finding.. 570	19.5.2	Evolution Within Noncoding Genes 580	
	19.4.2	Sets of Related Genes 572	19.5.3	Evolution and Function in Gene Regulatory Sequences 581	
	19.4.3	Rates of Sequence Change in Different Parts of Genes 574	19.5.4	Prediction and Tests of Gene Regulatory Sequences 582	
	19.4.4	Evolution and Function in Protein-Coding Exons........................... 574	19.6	Resources for Comparative Genomics................... 583	
	19.4.5	Fast-Changing Genes That Code for Proteins... 575		19.6.1	Genome Browsers and Data Marts..................................... 583
	19.4.6	Recent Adaptive Selection in Humans.. 576		19.6.2	Genome Analysis Workspaces............... 583
	19.4.7	Human Disease-Related Genes............. 578	19.7	Concluding Remarks... 584	
19.5	Evolution in Regions That Do Not Code for Proteins or mRNA 579		References... 585		

19.1 Goals, Impact, and Basic Approaches of Comparative Genomics

Comparative genomics uses evolutionary theory to glean insights into the function of genomic DNA sequences. By comparing DNA and protein sequences between species or among populations within a species, we can estimate the rates at which various sequences have evolved and infer chromosomal rearrangements, duplications and deletions. This evolutionary reconstruction can then be used to predict functional properties of the DNA. Sequences that are needed for functions common to the species being compared are expected to change little over evolutionary time, whereas sequences that confer an adaptive advantage when altered are expected to have greater divergence between species. Furthermore, sequence comparisons can help in predicting what role is played by a particular functional region, e.g., coding for a protein or regulating the level of expression of a gene.

These insights from comparative genomics are having a strong impact on medical genetics, and their role is expected to become more pervasive in the future. When profound mutant phenotypes lead to the discovery of genes in model organisms (bacteria, yeast, flies, etc.), the human genome is immediately searched for homologs, which frequently are discovered to be involved in similar processes. Control of the cell cycle [76] and defects in DNA repair associated with cancers [24, 47] are particularly famous examples. In studies of the noncoding regions of the human genome, conservation has become almost a proxy for function [20, 26, 64], and we will explore the power and limitations of this approach more in this chapter. The mapping and genotyping of millions of polymorphisms in humans [32] coupled with the availability of genome sequences of species closely related to humans [14, 70] has stimulated great interest in discovering genes and control sequences that are adaptive in humans, which may provide clues to the genetic elements that make us uniquely human (see Chaps. 8 and 16). As more and more loci are implicated in disease and susceptibility to diseases, identifying strong candidates for the causative mutations becomes more challenging. Research in comparative genomics is helping to meet this challenge by generating estimates across the human genome of sequences likely to be conserved for functions common to many species as well as sequences showing signs of adaptive change. Finding disease-associated markers in either type of sequence could rapidly narrow the search for mutations that cause a phenotype.

19.1.1 How Biological Sequences Change Over Time

All DNA sequences are subject to change, and these changes provide the fuel for evolution. Replication is highly accurate but not perfect, and despite the correction of many replication errors by repair processes during S-phase, a small fraction is retained as altered sequences. Mutagens in the environment can damage DNA, and some of these induced mutations escape repair. In addition, DNA bases can change spontaneously, for example, oxidative deamination of cytosine to produce uracil. The *mutation rate* is the number of sequence changes escaping correction and repair that accumulate per unit of time. The average mutation rate in humans has been estimated to be about 2 changes in 10^8 sites per generation [43, 57]. Thus for a diploid genome of 6×10^9 bp, about 120 new mutations arise in each generation. As will be discussed later in more detail, the mutation rate varies among loci and depends on the context, with transitions at CpG dinucleotides occurring about ten times more frequently than other mutations.

Mutations can be substitutions of one nucleotide for another, deletions of strings of nucleotides, insertions of nucleotides, or rearrangements of chromosomes, including duplications of DNA segments. Substitutions are about ten times as frequent as the length-changing alterations, with transitions greatly favored over transversions.

Mutations occur in individuals, and it is instructive to consider how an alteration in a single individual can eventually lead to a sequence difference between two species, which we call a *fixed difference*. Of course, only mutations arising in the germ-line can be passed along to progeny and have some possibility of fixation. Initially, the allele carrying a mutation has a low frequency in the population, i.e., $1/(2N_e)$ for a diploid organism, where N_e is the effective population size. All the mating individuals in a population contribute to the pool of new alleles. Mutant alleles that are disadvantageous will be cleared out of the population quickly, whereas those that confer a selective advantage rapidly will go to fixation (occurrence in most members of a population). However, many of the new mutations will have no effect on the individual; we call these mutations with no functional consequence *polymorphisms* or *neutral changes*. The frequency of these polymorphisms will increase or decrease depending on the results of matings and survival of progeny. The vast majority will be transitory in the population, with most headed for loss. However, the stochastic fluctuations in allele frequencies will allow some to eventually increase to a high frequency. Thus, some of the neutral changes lead to fixed differences. In fact, Kimura [40] and others have argued that such neutral changes are the major contributors to the overall evolution of the genome.

In order for a sequence change to have an effect on an organism, the change has to occur in a region that is involved in some function. Examples of such regions are an exon encoding part of a protein or a promoter or enhancer involved in gene regulation. The rapid removal of disadvantageous alleles results from *negative* or *purifying selection* (Fig. 19.1). The rapid fixation of advantageous alleles is *adaptive evolution* resulting from positive selection. Biological function is inferred from evidence of selection. Thus, the aim of comparative genomics to identify functional sequences can be stated as a goal of finding DNA sequences that show significant signs of positive or negative selection.

In addition to mutations of single bases, strings of nucleotides can be inserted or deleted as a result of replication errors or recombination. Often, the direction

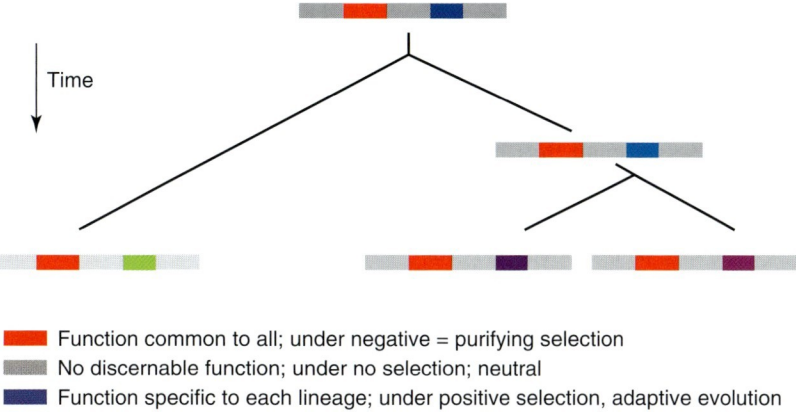

Fig. 19.1 Three modes of evolution, two of which are associated with function. The red line indicates a functional DNA sequence whose role has remained the same from ancestor to contemporary sequences, and thus it has been subject to purifying selection. The blue line represents a sequence that was functional in the ancestor, but changes in separate lineages (illustrated by different shades of blue, green, and purple) are adaptive and hence are subject to positive selection. The gray lines represent sequences of no known function, i.e., neutral DNA

of the event is not known because it is inferred from a gap in an alignment of only two sequences. In these cases the event is called an *indel*. Adding a third sequence to the alignment as an outgroup allows one to conclude with some confidence whether the event is an insertion or a deletion. Indels are less frequent than nucleotide substitution, and their frequency declines sharply with the size of the insertion or deletion. However, a single insertion or deletion can involve tens of thousands of nucleotides. Thus, they account for the majority of the nucleotides that differ between closely related species.

Rearrangements of chromosomes, such as intrachromosomal duplications and inversions or interchromosomal translocations, also lead to large-scale changes both in contemporary populations and over evolutionary time. Some chromosomal rearrangements are associated with human disease (see Chap. XX). In comparisons over evolutionary time, e.g., between mammalian orders, the history of chromosomal rearrangements can be reconstructed with some accuracy.

19.1.2 Purifying Selection

DNA sequences that encode the same function in contemporary species and in the last common ancestral species have been subject to *purifying* selection. The DNA sequence carried out some function in the ancestor, and any changes to this successful invention are more likely to break it than to improve it. Mutations in the sequence tend to work less well than the original one, and those mutations are cleared from the population. Hence the selective pressure to maintain a function prevents the DNA sequence from accumulating many changes, and the selection is referred to as purifying. This type of selective pressure tends to decrease the number of changes observed, and thus it is also called *negative* selection. The sequence under purifying selection is *constrained* by its function to remain similar to the ancestor. Saying that a sequence is subject to constraint is the equivalent of saying that it is subject to purifying selection. Examples of sequences under constraint include most protein-coding regions and many DNA sequences that regulate the level of expression of a gene.

In this chapter, we distinguish between conserved and constrained elements. A feature (e.g., a segment of DNA, a protein, an anatomical structure) that is found in contemporary species and that is inferred as being derived from a similar feature in the last common ancestor is conserved. In particular, a DNA sequence that reliably aligns between two species is considered to be conserved. That does not necessarily mean that it is functional. Evidence of *constraint*, i.e., alignment with a level of similarity greater than expected for neutral DNA, is taken as an indicator of function common to the two species.

The hallmark of purifying selection is a rate of change that is slower than that of neutral DNA. The next section (Sect. 19.2) will delve more deeply into how rates of evolution are determined, but for now assume that we can align related sequences with reasonable accuracy and can use that alignment to measure how frequently mismatches occur. Then the problem of finding sequences under purifying selection becomes one of determining the substitution rate in a segment that is a candidate for being functional and comparing it to the rate in neutral DNA. DNA segments whose inferred rate of evolutionary change is significantly lower than neutral will show a peak of similarity for comparisons at a sufficient phylogenetic distance (e.g., human versus mouse in Fig. 19.2).

In order to distinguish neutral from constrained DNA, sequences of divergent species must be compared. The choice of species to compare will depend on the questions being examined, but enough sequence change must have occurred to distinguish signal from noise. In practical terms, human comparisons with chimpanzee are too close (too similar) to effectively find constrained sequences, but multiple alignments among many primates do have considerable power [8]. Many studies have used comparisons between mammalian orders, such as primate (human) with rodent (mouse), to see the constrained sequences (Fig. 19.2).

19.1.3 Models of Neutral DNA

Although the concept of DNA that has no function is very useful and has led to much insight in molecular evolutionary genetics, it is difficult to establish that any DNA is truly neutral. Several models for neutral DNA are in common use. One of the earliest is the set of nucleotides in protein-coding regions that can be altered without changing the encoded amino acid [41]. The nucleotides are called *synonymous* or *silent* sites. They are neutral with respect to coding capacity, but alterations in particular synonymous sites can affect

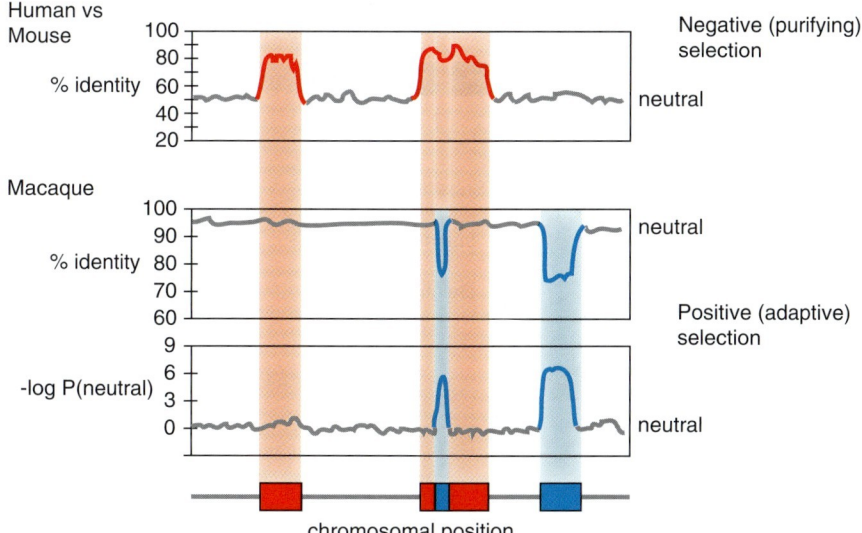

Fig. 19.2 Ideal cases for interpretation of sequence similarity. Idealized graphs of levels of sequence similarity (as percent identity) for a segment of a human chromosome compared with mouse (top) and rhesus macaque (middle), and of the likelihood that the DNA interrogated by the human-macaque comparison is not neutral (negative logarithm of the probability that the sequence similarity comes from the distribution of values for comparisons of neutral DNA, third graph). In the graphs, values that are close to those observed for a model of neutral DNA are shown in gray, those that indicate the action of negative selection are red, and those that indicate positive selection are blue. The bottom map is an interpretation of the graphs as discrete segments of DNA either under negative (red boxes) or positive (blue boxes) selection on a background of neutral DNA (gray line). Note that one segment shows evidence of negative selection since the separation of primates from rodents (red in top graph) but positive selection since the separation of human and Old World monkey (macaque) lineages (blue in middle and bottom graphs)

translation efficiency, splicing, or other processes. The latter appear to be a minority of synonymous sites, and as a group the synonymous sites are the most frequently used neutral model.

Another useful model for neutral DNA are *pseudogenes*. These are copies of functional genes, but the copies no longer code for protein because of some disabling mutation, such as a frameshift mutation or a substitution that generates a translation termination codon. For the period of time since the inactivating mutation, the pseudogene has likely been under little or no selective pressure. The rate of divergence of pseudogenes after inactivation is clearly higher than that of the homologous functional genes, and they have been used successfully as neutral models in many studies of particular gene families (e.g., [48]). One limitation of using pseudogenes as a neutral model is the uncertainty of determining when the inactivating mutation(s) occurred. Also, they are rather sparse for genome-wide studies.

For comparisons in mammalian genomes, *ancestral repeats* (Fig. 19.3) have proved effective, albeit imperfect, models for neutral DNA [27, 85]. The interspersed DNA repeats in the genomes of humans and other mammals are derived from transposable elements, mostly *retrotransposons* that move via an RNA intermediate. Members of an interspersed repeat family generated by recent transposition (on an evolutionary time-scale) are quite similar to each other because they have not had sufficient time to diverge. These are restricted to particular clades, such as the *Alu* repeats that are prevalent in primate genomes. Considerably more differences are observed among members of repeat families that are derived from transposons active in an ancestral species because of the longer divergence time. The members of these older repeat families are present in all the descendant species. Examples include *LINE2* and *MIR* repeats, which are present in the genomes of all eutherian mammals examined. Interestingly, all the members of these ancestral repeat families are quite divergent from each other, indicating that they have not been actively transposing since the separation of the descendant species. Thus, most ancestral repeats appear to be relics of ancient transposable elements, and they are not active even for

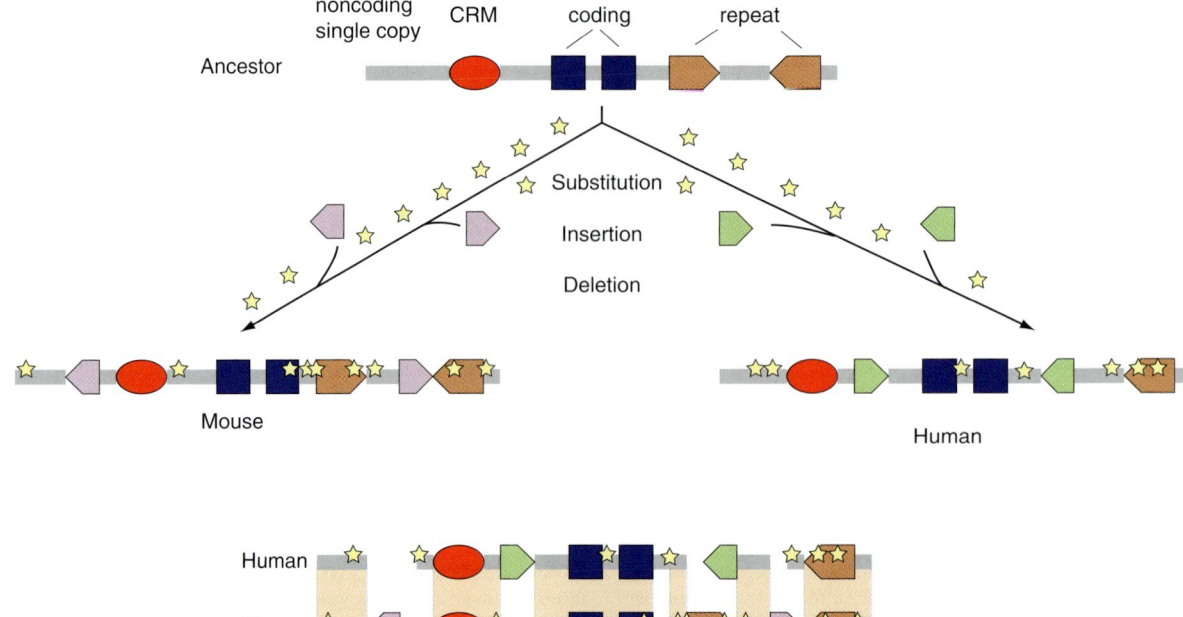

Fig. 19.3 Substitutions, insertions of transposable elements and deletions in the evolution of genomes. (**a**) Illustration of functional regions such as protein-coding exons (blue boxes), cis-regulatory modules (CRMs, red ovals), such as enhancers and promoters, and ancestral repeats (brown pointed boxes). After divergence of rodents and primates, sequences diverge by substitutions (gold stars), insertion of lineage-specific transposable elements (purple and green pointed boxes), and deletions. (**b**) Alignments of the contemporary species allow some of the evolutionary history to be reconstructed, including deletions inferred from the nonaligning portions

transposition. The vast bulk of these ancestral repeats have no apparent function. They are found frequently in eutherian mammals, and thus provide a neutral model with many sites.

When interpreting any measurement or study involving a comparison with a neutral model, it is important to keep in mind that the deduced absence of function is limited by contemporary knowledge. Experimental tests and molecular evolutionary studies have shown that some individual synonymous sites and ancestral repeats are not neutral. They do not constitute the bulk of the sites in these neutral models, and of course the known functional sites can be removed from the neutral set. However, future studies could reveal additional function, which will affect interpretations based on these neutral models.

19.1.4 Adaptive Evolution

The functions of some DNA segments and proteins have changed along the evolutionary lineages to contemporary species. Some sequence changes confer a new function on the DNA or protein that helps the organism adapt to a new environment or condition. These advantageous mutations increase in frequency in a population, leading to fixation (i.e., becoming the predominant allele in the population). The selective pressure favoring these changes is called *positive* selection, since it tends to increase the frequency of changes. This leads to *adaptive evolution*, i.e., a change in a DNA or protein sequence that favors survival and procreation of an organism. The positive selection for new functionality is also referred to as *Darwinian* selection.

The hallmark of adaptive evolution is a rate of sequence change that is faster than that of neutral DNA. Sequences subject to adaptive evolution may change so much that they will not align reliably at greater phylogenetic distances (Fig. 19.1). Also, the selective pressure leading to adaptive changes may apply only recently or in limited clades, such as among humans or among humans and great apes. Thus, sequence comparisons to find adaptive changes are usually done for closely related, recently diverged sequences (Fig. 19.2). The signal for positive selection may be captured as a significant decrease in similarity between species or an increase in the probability that a sequence has not evolved neutrally (Fig. 19.2).

19.2 Alignments of Biological Sequences and Their Interpretation

Biological sequence comparisons are most commonly done with protein sequences (strings of amino acids) or DNA sequences (strings of nucleotides). The comparisons begin with an alignment, which is a mapping of one sequence onto another with insertions of gaps (often indicated by a dash) to optimize a similarity score (Fig. 19.3). The score can be determined in a variety of ways, but in all cases matching symbols (for amino acids or nucleotides as appropriate) are favored, whereas mismatches are not favored and gaps are penalized. The gap penalty frequently takes the form of a gap-open penalty plus an additional, smaller penalty for each position included in the gap. The latter are referred to *as affine gap penalties*.

19.2.1 Global and Local Alignments

A *global* alignment maps each symbol in one sequence onto a corresponding symbol in another sequence. The result is an alignment of the two (or more) sequences from their beginnings to their ends, with any length differences accommodated by gaps that are introduced. This is an appropriate strategy for sequences are related to each other over their entirety. That is the case for many proteins and many mRNAs. The earliest computer program for aligning two biological sequences, written by Needleman and Wunsch [58], generates global alignments. Popular contemporary programs for aligning proteins, such as *ClustalW* [80], also compute global alignments. Global aligners for DNA sequences include *VISTA* [54], *MAVID* [9], and *LAGAN* [10].

A frequent task in comparative genomics is to find matches between two or more sequences that are not related over their entire lengths. For instance, two protein sequences may be related only in one or a few domains, but be different in other parts. The protein-coding portions of genes are frequently divided into short exons that are separated by introns. Exons tend to be under constraint, whereas much of the intronic DNA may be neutral, and thus at a sufficient phylogenetic distance introns can be so divergent that they no longer align, whereas exons will match well. The most common use of comparative genomics is to search a large database of all compiled DNA or protein sequences with a query sequence of interest. In this case, the goal is to find a match that may comprise only one part in billions of the database. When a match between only a portion of two or more sequences is desired, then a *local* alignment should be generated. One of the earliest computer programs for finding local alignments came from Smith and Waterman [75]. The *blast* family of programs (Basic Local Alignment Search Tool, [1]) is used for database searches. One variant, called *blastZ*, has been adapted to compute local alignments of long genomic DNA sequences [72].

19.2.2 Aligning Protein Sequences

Proteins are composed of 20 amino acids, so that for any position in one sequence the possibilities for alignment with a position in a comparison sequence are 1 match, 19 mismatches, or a gap. However, the likelihood for each of the 19 mismatches is not the same. Replacement of an amino acid by a chemically similar amino acid occurs much more frequently than does replacement with a distinctly different amino acid. These different frequencies of amino acid substitutions can be captured as a *scoring matrix*, in which matches are given the highest similarity score and mismatches that occur frequently in protein sequences are given positive scores, decreasing with declining frequencies of the substitution. These scoring matrices are determined by the frequency with which mismatches are observed in well-aligned sequences. Several effective matrices have been generated, beginning with the pioneering work of Dayhoff et al. [18] and continuing on to the BLOSSUM matrices of Henikoff [29].

Alignments can be used to organize relationships among the large number of sequenced proteins. Large compilations of aligned protein sequences are analyzed to find clusters of proteins that appear to share a common ancestor and to find blocks of aligned sequences that are distinctive for various protein domains. Indeed, when genes and their encoded proteins are predicted or identified in genome sequences, the primary basis for making inferences about their function is sequence similarity to known proteins.

Sequence similarity between proteins can be found with considerably greater sensitivity than can be found using a DNA sequence. The reason is that the 20 amino acids found in proteins constitute a much more complex group of characters, or alphabet, than the four nucleotides found in DNA. Thus, alignments between distantly related proteins may only match at a very small percentage of positions, but these are still statistically significant and they can be biologically meaningful.

19.2.3 Aligning Large Genome Sequences

The smaller alphabet for DNA sequences, consisting of only four nucleotides (A, C, G, T), means that the threshold for statistical significance is considerably higher than that used for protein sequences. For random sequences of equal nucleotide composition, any position in one sequence should have a 25% chance of matching any position in the other. However, sufficiently long runs of matching sequences are much less likely, and reliable alignment can be generated between related sequences. Just like for alignments of protein sequences, some substitutions are more likely to occur than others. For example, transitions are much more frequent than transversions. These preferences can be incorporated into the alignment process by using scoring matrices that were deduced from the empirical frequencies of matches and substitutions in reliable alignments, similar to the process that generated scoring matrices for protein alignments.

The portions of DNA sequences that code for proteins tend to be more similar and to have many fewer indels than the rest of a genome for comparisons at a sufficient phylogenetic distance. Hence these are relatively easy to align and different alignment strategies tend to give similar results for coding regions. Other parts of the genome are more likely to have mismatches or to have undergone insertion or deletion, which requires introduction of gaps into the alignment. In these noncoding regions, choice of an alignment strategy is expected to have an impact on the result. Global aligners are expected to have somewhat greater sensitivity, but they may include more inaccurate alignments.

Local aligners will not align sequences that are too dissimilar, even if they occur in analogous positions in the two genomes. More calibration of the various methods is needed to clarify these issues, but at this point there is no consensus on whether the regions that fail to align by local aligners are not homologous, or whether they are homologs that have changed so much that the similarity is not recognizable by these programs [53].

Chromosomal rearrangements complicate the construction of comprehensive alignments between genomes. Genes that are on the same chromosome in one species are *syntenic*. Groups of genes that are syntenic in humans are frequently also syntenic in mouse, and thus these groups of genes display *conserved synteny*. In addition, they frequently maintain a similar order and orientation, indicating *homology*, which is similarity because of common ancestry. The homologous segments between distantly related species rarely extend for entire chromosomes, but rather one human chromosome will align with several homology blocks in mouse, many of which are on different chromosomes in mouse (Fig. 19.4). For genome comparisons, the goal is to find all the reliable alignments within the homology blocks and deduce how the various homology blocks are connected in the genomes of the species being compared. This requires additional steps to the alignment procedure. For local aligners, it means that the large number of individual alignments needs to be organized along chromosomes. For global aligners, it means that homology blocks must be identified prior to execution of a global alignment.

Local alignments are restricted to the DNA segments between rearrangement breakpoints. A collection of local alignments can be organized into *chains* to maintain the order of DNA segments along the chromosome. In this case, local alignment A is connected to local alignment B in a chain if the beginnings of the aligned sequences in B follow the ends of the aligned sequences in A. The chains can be nested in a group, called a *net* [39], and these are used to navigate local alignments through rearrangements (Fig. 19.4). On a large scale, these nets can be used to illustrate chromosomal rearrangements between species, and on a smaller scale they can reveal multiple events associated with rearrangement breakpoints.

Global aligners can be used in genomic regions that have not been rearranged. In practice, for whole-genome alignments, homology blocks are initially identified using a rapid local alignment procedure. Then a global

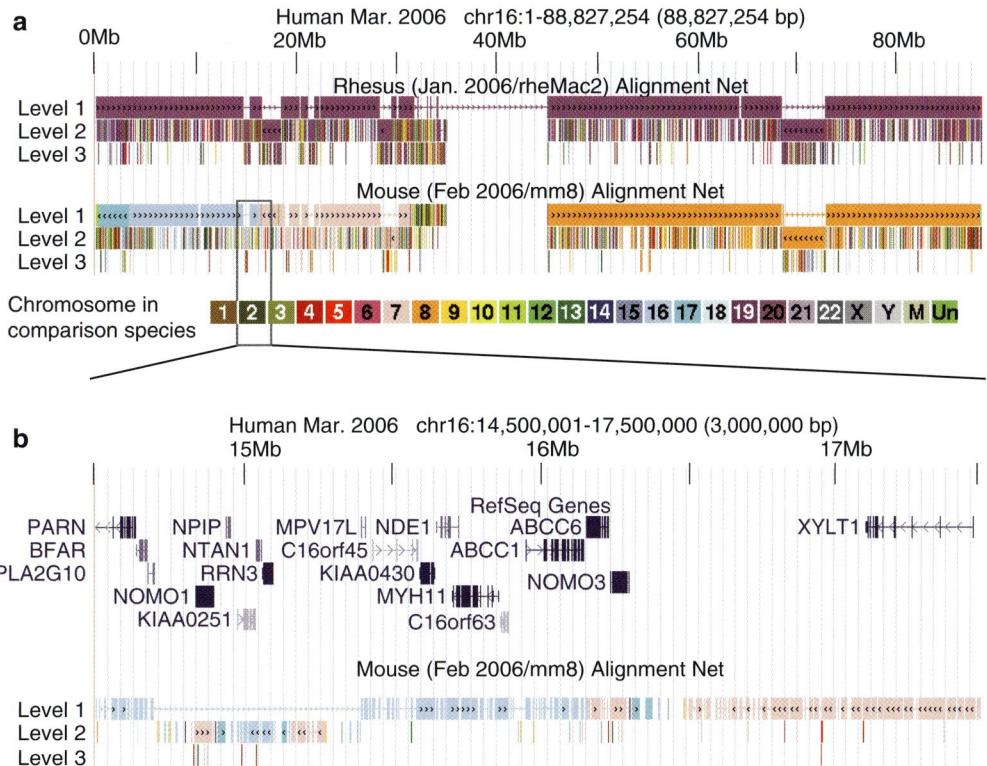

Fig. 19.4 Blocks of conserved synteny and chromosomal rearrangements with human chromosome 16 as the reference sequence. (**a**) Almost all of human chromosome aligns with rhesus chromosome 20, indicated by the purple boxes, but portions of human chromosome 16 align to different chromosomes in mouse, which are color coded by the aligning chromosome in the comparison species. For both comparisons, levels 1, 2, and 3 of a nested set of chained alignments (called a net) are shown. Local alignments form a chain when the start positions of the sequences in one alignment follow the end positions of the sequences in the preceding alignment. The level 1 chain is the highest scoring (usually longest) set of local alignments; the level 1 chain with rhesus covers almost all of rhesus chromosome 20. Gaps in the level 1 chain are filled with the highest scoring additional chains to make level 2 chains, and so on for up to six levels. Inversions are evident by changes in the directions of the arrowheads on the chain maps. (**b**) A higher resolution view of a portion of human chromosome 16 that encompasses a major change in conserved synteny from mouse chromosome 16 (light blue) to mouse chromosome 7 (pink). The diagram illustrates the results of a complex rearrangement history, including an inversion and interlacing of matches to the two mouse chromosomes. Many genes are present in this region despite the complex rearrangements of the chromosome between human and mouse

aligner such as *LAGAN* is run on the sequences in the regions that have not been rearranged [10].

Several powerful Web-servers are available for running these alignment programs on chosen sequences. Often it is prudent to use precomputed alignments because of the complexity of these alignment pipelines and the need for careful adjustment of alignment parameters for different comparisons. Nets and chains of local alignments generated by *blastZ* are available from the UCSC Genome Browser [45] and Ensembl [31]. Precomputed alignments of whole genomes generated by pipelines using *LAGAN* and *VISTA* are also available. As discussed in the next section, analyzes of these alignments can be used to predict function in genomic DNA sequences. Table 19.1 lists a selection of network servers for making and viewing alignments.

19.3 Assessment of Conserved Function from Alignments

Many of the sequences that are conserved between species can be found in the portions of genomes that align. As discussed above, alignment algorithms are good, but imperfect, and no one can guarantee that all

Table 19.1 Selected network servers for making and viewing alignments of genome sequences

Program or pipeline	Name	URL
blastZ, nets and chains	UCSC Genome Browser	http://genome.ucsc.edu/
blastZ, nets and chains	Ensembl	http://www.ensembl.org/
VISTA	VISTA Tools	http://genome.lbl.gov/vista/index.shtml
LAGAN	LAGAN alignment toolkit	http://lagan.stanford.edu/lagan_web/index.shtml
MAVID	MAVID Server	http://baboon.math.berkeley.edu/mavid/
blastZ and others	DCODE.org NCBI	http://www.dcode.org/
blastZ	PipMaker	http://pipmaker.bx.psu.edu/pipmaker/

the conserved sequences will align, especially as the phylogenetic distance between the species increases. Nevertheless, the portions that align should have much of the conserved DNA. Within that conserved DNA is a subset that has a function common to the species being compared; that is the portion that shows evidence of constraint, i.e., purifying selection. Thus, searching genome alignments for evidence of constraint is a major, powerful approach for finding functional DNA sequences.

19.3.1 Phylogenetic Depth of Alignments

The longer two species have been separated, the more divergent their genomes become, and thus one indicator of constraint operating on a sequence is that it aligns with sequences in distantly related species. Several insights can be gleaned by examining the phylogenetic distance at which a particular sequence or class of genomic features continues to align.

As expected, most of the human genome aligns with the genomes of our closest relative, the chimpanzee, and an Old World monkey (the rhesus macaque). The genomes of the comparisons species are not finished for the most part, and thus the values for portion aligning (Table 19.2) will be underestimated, but

Table 19.2 Portions of the human genome conserved and constrained between various species

	Distance from human		Fraction of human aligning to comparison species[d]			
Comparison species[a]	Divergence time (Myr)[b]	Substitutions per synonymous site[c]	Total genome[e]	Coding exons[f]	Regulatory regions[g]	UCEs[h]
Chimpanzee	5.40	0.015	0.95	0.96	0.97	0.99
Macaque	25.0	0.081	0.87	0.96	0.96	0.99
Dog	92.0	0.35	0.67	0.97	0.87	0.99
Mouse	91.0	0.49	0.43	0.97	0.75	1.00
Rat	91.0	0.51	0.41	0.95	0.70	1.00
Opossum	173	0.86	0.10	0.82	0.32	0.95
Chicken	310	1.2	0.037	0.67	0.06	0.95
Zebrafish	450	1.6	0.023	0.65	0.03	0.76
Number			2.858×10^9 nucleotides	250,607	1,3	481

Notes:
[a]Sources of genome sequences are: human: [33]; chimpanzee: [14]; macaque: [70]; dog: [49]; mouse: [85]; rat: [25]; opossum: Broad Institute; chicken: [30]; zebrafish: Zebrafish Sequencing Group at the Sanger Institute
[b]Divergence times for separation from the human branch to the branch leading to the indicated species are from [46]
[c]Estimated substitutions per synonymous site are from [53]
[d]The human genomic intervals in each dataset were examined for whether they aligned with DNA from each comparison species in whole-genome blastZ alignments [42]. An interval that is in an alignment for at least 2% of its length was counted as aligning, but in the vast majority of cases the entire interval was aligned.
[e]The number of nucleotides in the human genome that align with each species was divided by the number of sequenced nucleotides in human (given on the last line)
[f]Coding exons are from the RefSeq collection of human genes [68]
[g]Putative transcriptional regulatory regions were determined by high-throughput binding assays and chromatin alterations in the ENCODE regions [79]; the set compiled by King et al. [42] was used here
[h]Ultraconserved elements (UCEs) are the ones with at least 200 bp with no differences between human and mouse [4]

they are still informative. Since almost all of the genome aligns, of course virtually all known functional regions align between human and apes or Old World monkeys. This includes coding exons [68] and putative transcriptional regulatory regions, which are deduced from high-resolution studies on occupancy of DNA by regulatory proteins [79].

When the comparison is made with genomes of eutherian mammals outside the primate lineage, considerably less of the human genome aligns (Table 19.2). Within the 37–57% of the genome that does align, however, we find almost all of the coding exons (95–97%) and putative regulatory regions (74–89%). Even less of the genome aligns with the marsupial opossum (about 13%). At this phylogenetic distance, the alignments of coding exons tend to persist, but only 39% of the putative regulatory regions still align. Only a small fraction of the human genome aligns to more distant species, such as chickens and fish. At this distance, the estimated substitution rate in neutral DNA (synonymous sites) is so high that a segment of neutral DNA is no longer expected to align, and thus it is highly likely that all the alignments between human and chicken or fish are in functional regions.

The insights about conservation of functional elements are easier to visualize when presented as a function of phylogenetic distance (Fig. 19.5). No single comparison is adequate for all goals. Some are particularly good for one purpose, such as using human-opossum alignments for examining coding regions. Almost all the coding regions still align at this distance, but only 13% of the genome aligns. Most comparisons involve a trade-off between sensitivity (the ability to find the desired feature) and specificity (the ability to reject undesired sequences). One may want to examine alignments at a sufficient distance such that no neutral DNA is aligning, but at that distance (e.g., human-chicken) a third of the coding exons and about 90% of the putative regulatory regions no longer align. This means that the specificity is excellent but the sensitivity is lower than usually desired. In practice, it is common to examine comparisons among multiple species that have given good sensitivity, such as alignments among eutherian mammals, and to apply some discriminatory function to better ascertain the regions that are constrained or show some other evidence of function. Alignments to more distant species can be included as well, but they should not be used as an exclusive filter.

The utility and limitations of examining multiple eutherian species has been studied extensively. About 1,000 Mb align among human, mouse, and rat [25], illustrated by the central portion of the Venn diagram in Fig. 19.6. A similar study of human, dog, and mouse revealed about 812 Mb conserved in all three [49]. This approximately 1 Gigabase of genome sequence found in common can be considered the core of the genome of placental mammals. The DNA sequences needed for functions common to all eutherians are expected to be in this core, and indeed virtually all coding exons and putative regulatory regions are found in it (Table 19.2). However, it seems unlikely that this entire core is under constraint. About 162 Mb of the core consists of repetitive DNA

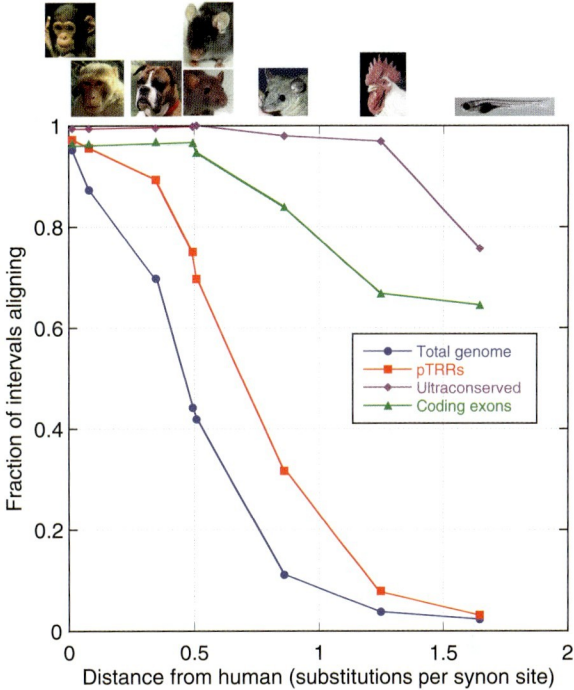

Fig. 19.5 The fraction of genomic intervals that align with comparison species at increasing phylogenetic distance. The fractions of intervals in putative regulatory regions (pTRRs, red squares), coding exons from RefSeq (green triangles) and ultraconserved elements (purple diamonds) substantially exceed the fraction of the human genome (blue circles) that aligns with each species in almost all comparisons. The comparison species in increasing order of distance from human are chimpanzee, rhesus macaque, dog, mouse, rat, opossum, chicken, and zebrafish (pictured above the graph). The distance is the estimated number of substitutions per synonymous site along the path in a tree from human to each species [53]. This measures takes into account faster rates on some lineages, and thus it places mouse and rat more distant from human than dog, despite the earlier divergence of carnivores

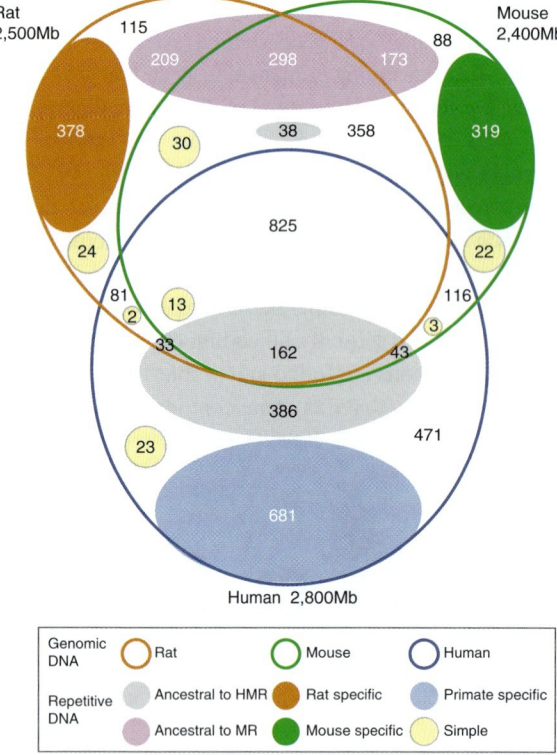

Fig. 19.6 Venn diagram showing common and distinctive sequences in humans and two rodents. As summarized in the key (box under the diagram), the outlined ellipses represent the DNA in each genome, and the overlaps show the amount of sequence aligning in all three species (rat, mouse, and human) or in only two species. Portions of the ellipses that do not overlap represent sequences that do not align. Different types of repetitive DNA are shown as colored disks, and are classified by their ancestry. Those that predate the divergence between rodents and primates are gray, and those that arose on the rodent lineage before the divergence between rat and mouse are lavender. Disks for repeats specific to each species are colored orange for rat, green for mouse, and blue for human; and disks for simple repeats are colored yellow. The disks for the repeats are placed to illustrate the approximate amount of each type in each alignment category. Uncolored areas represent nonrepetitive DNA; the bulk is assumed to be ancestral to the human–rodent divergence. The numbers of nucleotides (in Mb) are given for each sector (type of sequence and alignment category). (Reprinted from Gibbs et al. [25], with permission from Nature Publishing Group)

discrimination of constrained sequences from those that are conserved but are apparently neutral. Figure 19.6 also shows that the rat and mouse genomes share many DNA sequences that are not in human, and about 358 Mb are nonrepetitive. One may expect to find rodent-specific functional sequences in these portions of the mouse and rat genome. Genomic DNA sequences that are found only in rat or only in mouse are dominated by lineage-specific interspersed repeats.

19.3.2 Portion of the Human Genome Under Constraint

Within the subset of the human genome that aligns with other species, we want to know what fraction of it appears to be under constraint (covered in this section), and then to be able to identify the constrained sequences (covered in the next section). One way to estimate the portion of the human genome under constraint is to evaluate all the segments that align with a comparison species for a level of similarity higher than that seen for neutral DNA. This would be a straightforward approach if we knew all the neutral DNA (which we do not; see Sect. 19.1.3), and if the neutral DNA diverged at the same rate at all positions in the chromosome (illustrated by the ideal case in Fig. 19.2). However, the estimated neutral rates show substantial local variation across the human genome (Fig. 19.7). This has been seen for comparison of the human genome with mouse [27, 85], dog [49], and chimpanzee [14]. Thus, estimates of constraint need to take into account the local rate variation.

For comparison of the human and mouse genomes [85], alignments throughout the genomes were evaluated for a level of similarity that exceeds the similarity expected from the amount of divergence in ancestral repeats in the vicinity. The distribution of similarity scores in ancestral repeats is normal, and many similarity scores in the bulk of the genome overlap with those in the neutral distribution (Fig. 19.8). Notably, a pronounced shoulder of alignments presents a score higher than the scores for a vast majority of ancestral repeats. The broad distribution of alignment scores through the genome can be interpreted as the combination of two distributions, one for neutral DNA and one for DNA that is under constraint. Various models lead to the conclusion that about 5% of the human genome falls into the latter distribution. A similar estimate has been obtained for alignments of the human and dog genomes [49]. In support of the idea

that is ancestral to primates and rodents (Fig. 19.6). As discussed above, most of this ancestral repetitive DNA can be considered neutral. Granted that some of these ancestral repeats may indeed be functional, it is unlikely that all of them are. Hence, even in the approximately 800 Mb of the core that is nonrepetitive, it is expected that some, and maybe much, also lack a function conserved in all eutherians. This illustrates the need for further of a conserved eutherian core genome that encompasses the sequences with common function, the human sequences inferred to be under constraint are the same whether the comparison is with dog or mouse [49].

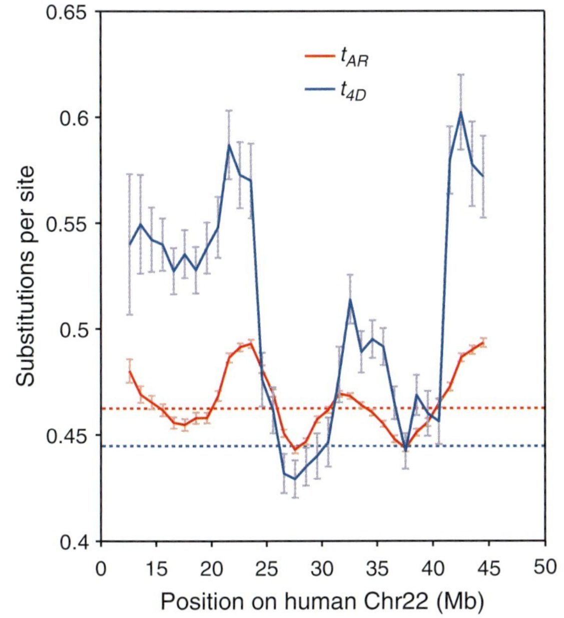

This result tells us that about 5% of the human genome has been under continuous purifying selection since the divergence of primates from carnivores and rodents, approximately 85–100 million years ago. The functions that would be subject to the continuous selection are those that were present in a eutherian ancestor and continue to play those roles in contemporary primates, rodents, and carnivores (and likely all eutherians). This is a lower bound estimate of the portion of the human genome that is functional. DNA sequences that have diverged for new functions in different lineages are not included in this estimate, nor are sequences that have acquired function recently through adaptive evolution. Thus, the portion of the human genome that is functional is certainly higher than 5%, but it is not possible with current knowledge to place an upper bound on the estimate.

The lower bound estimate of the portion under continuous constraint is a remarkable number. The portion of the human genome needed to code for proteins has been estimated at about 1.2%, with another 0.7% corresponding to untranslated regions of mature mRNA [33], giving an estimate of about 2% of the genome devoted to coding for mRNA. This leaves about 3% of the human genome with sequences that do not code for protein but still carry out functions common to eutherian mammals. Among these additional sequences under constraint should be genes for noncoding RNAs and DNA sequences that regulate the level of expression of genes. It is striking that the fraction of the genome devoted to the conserved noncoding functions is greater than the fraction needed to code for proteins.

19.3.3 Identifying Specific Sequences Under Constraint

In order to find particular functional sequences, it is necessary to identify specific sequences whose alignments are likely to be in the portion under constraint. In principle, it is a matter of finding segments with a similarity score above the neutral background (Fig. 19.2). Of course, it is important to adjust the analysis for variation in local substitution rate, as just discussed. For example, from the distribution of S scores in ancestral repeats (Fig. 19.8) based on pairwise human–mouse

Fig. 19.7 Variation in the rate of human-mouse divergence in neutral DNA along human chromosome 22. The substitutions per site in ancestral repeats (t_{AR}, red) and in and in the subset of synonymous sites that are fourfold degenerate (t_{4D}, blue) were estimated in 5 Mb windows, overlapping by 4 Mb. The horizontal dotted lines indicate the estimates of t_{AR} and t_{4D} across the entire human genome. The confidence intervals are shown as brackets; the places where the confidence interval lies outside the genome-wide estimate are those with significant differences in evolutionary rate. (Reprinted from Waterston et al. [85], with permission from Nature Publishing Group)

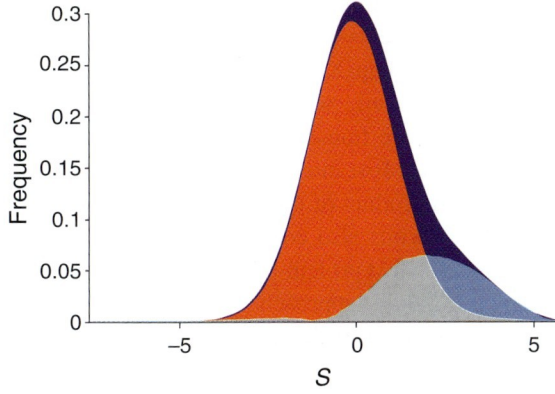

Fig. 19.8 Decomposition of conservation score into neutral and likely selected portions. S is the conservation score adjusted for variation in the local substitution rate. The frequency of the S scores for all 50 bp windows in the human genome, after alignment with mouse, is shown as the blue distribution. The frequency of S scores for ancestral repeats is shown in red. The inferred distribution of scores for regions under constraint is shown in light gray and light blue. This represents about 5% of the human genome. (Reprinted from Waterston et al. [85], with permission from Nature Publishing Group)

alignments, one can compute a probability that a given alignment could result from the locally adjusted neutral rate. Those that are unlikely to result from neutral evolution between humans and nonprimates are likely to be under constraint.

Other measures have been developed to utilize the greater amount of information in multiple sequence alignments to identify constrained sequences. One measure is based on modeling the genome as having two states of "conservation," one that is effectively neutral and one that is the slowly changing, constrained state. By combining phylogenetic models with Hidden Markov models of those states, a score called *phastCons* is computed, which gives the posterior probability that any aligned position came from the constrained state [74]. This measure is routinely computed genome-wide for several sets of genome alignments, and is accessed as the "Conservation" track on the UCSC Genome Browser (Fig. 19.9). Note that it has a form similar to the idealized case in Fig. 19.2, with higher peaks associated with a greater likelihood of being constrained.

A constrained sequence is one that had an opportunity to change because it was mutated in an individual in a population, but the mutation was not fixed in the genome sequence of the species because of selective pressure against the change. Thus, there could have been a substitution, but purifying selection rejected it. Another measure of constraint, called genomic evolutionary rate profiling or GERP [16], explicitly models this process and estimates the number of "rejected substitutions" (Fig. 19.9). Another method, binCons, models the substitution frequency as a binomial distribution, with the contribution of alignments of different species weighted according to their phylogenetic distance from the reference species [52].

In a region evaluated by these methods, some segments are identified as being under constraint by all three, and others are found by only one. Each approach has value, and each has some unique advantages and some idiosyncratic problems. Thus, it is useful to combine the output of each to generate sets of "multispecies conserved sequences" [53, 79]. The strict, moderate, and relaxed sets correspond to the MCSs found by intersection, inclusion in at least two, or the union of the three sets. The example shown in Fig. 19.9 illustrates strong constraint not only in the coding exons but also in the introns. Experimental tests on two of these intronic constrained elements show that they affect the level of expression from a linked promoter [71].

19.4 Evolution Within Protein-Coding Genes

Comparative analysis of protein-coding genes requires several steps. First, a set of protein-coding genes must be defined in each species, and then a set of orthologous genes shared among the species is examined. With this, the rates of change among proteins can be computed and then one can study how those differences in rates correlate with function. Most protein-coding genes are under significant constraint over the course of mammalian evolution. However, genes whose products have roles in reproduction, chemosensation, immunity, and metabolism of foreign compounds are found consistently to be changing more rapidly than other genes. Thus, these are some of the functional classes that determine species-specific functions.

19.4.1 Comparative Genomics in Gene Finding

One of the most important tasks in genomics is to identify the segments of DNA that code for a protein. As covered in Chap. XX, most eukaryotic genes are composed of *exons*, which code for mRNA, and *introns*, which are transcribed but spliced out of the mature mRNA. Most internal exons encode a portion of the protein product of the gene, whereas the initial and terminal exons also contain untranslated regions of the mRNA. Most protein-coding exons can be identified by a variety of approaches. However, combining the exons into genes, including accurate determination of the initial exon (or multiple initial exons), is more of a challenge.

The several approaches for finding exons and genes can be divided into two categories: evidence-based and *ab initio*. Evidence-based methods find genomic DNA segments that align almost exactly with known protein sequences (after translating the genomic sequence) or complete mRNA sequences. Most evidence-based methods also incorporate data on

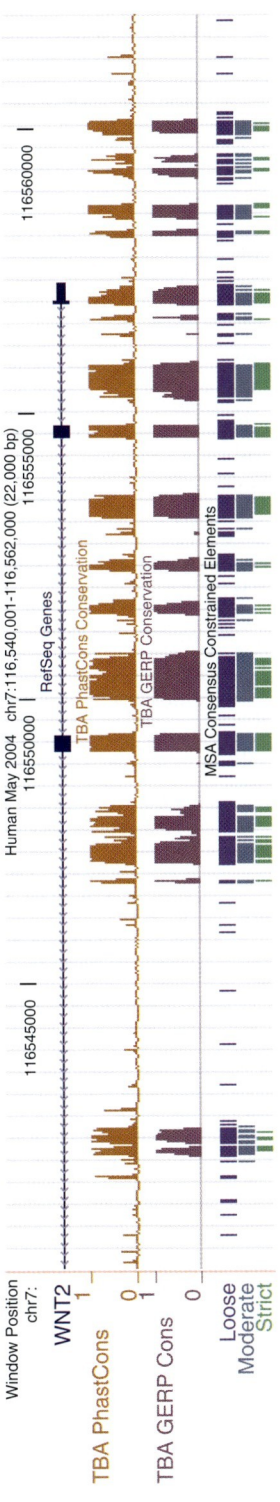

Fig. 19.9 PhastCons and GERP in a portion of ENCODE region ENm001. The first three exons (blue boxes) and introns (lines with arrows showing the direction of transcription from right to left) of the gene *WNT2* are shown on the top line. The next two panels plot the *phastCons* and *GERP* scores, respectively, with higher values indicating a higher probability that a sequence is under constraint. The bottom panel shows the levels of multispecies conserved sequences (see text)

expressed sequence tags (ESTs), which are short sequences containing portions of a very large number of mRNAs, and tags of sequence derived from the 5' capped ends of mRNAs. The mRNA-coding segments of genomic DNA are grouped, using rules about pre-mRNA splicing signals, to find strings of exons that after splicing gives the mRNA sequence, or after splicing and translation gives the protein sequence. In order to find likely exons of genes whose mRNA sequences are not in the databases, *ab initio* methods based on models derived from basic knowledge about gene structure are applied. The genetic code and rules for splice junctions (Chaps. XX) provide the rules that make up the basic grammar for encoding proteins. Hidden Markov models such as those in the programs *genscan* [11] and *genmark* [28] are used to find likely exons and likely arrangements for these exons in genes.

Adding alignments of sequences of other species can improve gene prediction. Two commonly used methods are *Twinscan* [89] and *SGP* [87]; these build on the models in *genscan* but also apply rules from comparative approaches, such as allowing mismatches at degenerate sites in the genetic code. Another program, *exoniPhy* [73], uses the grammar of protein coding and a phylogenetic analysis of multispecies alignments to improve exon finding.

Often the initial and final exons do not code for protein, and thus the *ab initio* predictors no longer benefit from the well-known rules for encoding proteins. Furthermore, it is not uncommon for a gene to have multiple initial exons, with some used at particular times of development or in certain tissues. Thus, the accuracy of fully assembling genes from exons is enhanced by evidence such as mRNA sequences and tags derived from the 5' ends of mRNA. Powerful pipelines for gene annotations have been developed that combine both evidence-based and *ab initio* methods; one of the most widely used is the Ensembl automatic gene annotation system [17].

In the current assembly of the human genome (NCBI build 36, March 2006, hg18), the Ensembl pipeline predicts 270,239 exons. These are arranged into 44,537 mRNAs from 21,662 genes. Most genes code for multiple mRNAs, thereby greatly increasing the diversity of proteins encoded in the human genome. Of these exons and genes, how many are found in other species, and which contribute to lineage-specific characteristics?

19.4.2 Sets of Related Genes

When discussing genes that are shared among species, we usually want to find the genes that are derived from the same gene in the last common ancestor. Homologous genes that separated because of a speciation event are *orthologous*. When there is a simple 1:1 relationship between orthologous genes, such as for *RRM1* in Fig. 19.10a, then any differences between the genes can be interpreted as changes since the time of divergence of the species.

When homologous genes are members of multigene families, then it is important to distinguish genes that have separated as a result of gene duplication (*paralogous* genes) from the orthologous genes, which separated by speciation events (Fig. 19.10a). For instance, the beta-like globin genes in humans arose by duplication in mammals. Within this gene family, each gene is paralogous to the other. For example, *HBE1* and *HBB* are paralogs that resulted from an earlier duplication, whereas *HBG1* and *HBG2* are paralogs that duplicated recently. Each of the four beta-like globin genes in chickens is paralogous to the other three, again because of the duplication history.

When gene duplications have occurred independently in both lineages, then all the duplicated genes in one species are orthologous to each of the genes in the other lineage. This is a many-to-many orthologous relationship. The human *HBB* gene is equally distant from each of the chicken beta-like globin genes, and it is orthologous to each.

Frequently a comparison will involve multigene families in species that share a duplication history, such as the beta-like globin gene clusters in human and macaque (Fig. 19.10b). The gene duplications outlined in panel A pre-date the catarrhine ancestor (ancestor to Old World monkeys, apes and humans). Thus, the *HBB* gene in humans is orthologous to the *HBB* gene in macaque, but it is paralogous to the other macaque beta-like globin genes, such as *HBD*, *HBG1*, etc. Likewise, the human *HBE1* gene is orthologous to the *HBE1* gene in macaque, but paralogous to the others. Comparisons between the orthologs reflect changes that have occurred since the separation of Old World monkeys and humans, whereas comparisons between the paralogs will reflect changes over a much greater phylogenetic distance, i.e., back to the gene duplications that generated the ancestors to the genes being

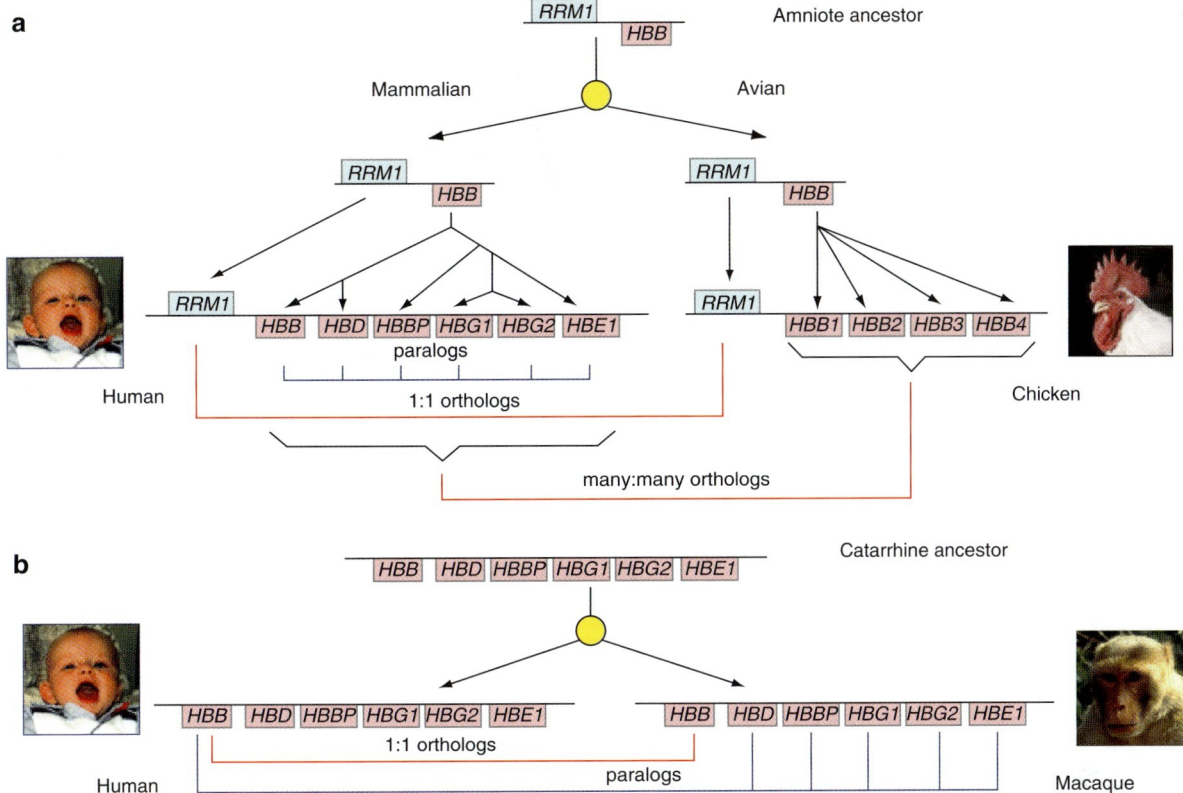

Fig. 19.10 Orthologous and paralogous relationships among genes. Speciation events are shown as yellow disks, and gene duplications are denoted by bifurcating arrows or multiple arrows with a single source. Red lines between genes in contemporary species connect orthologous genes, whereas blue lines connect paralogous genes. (**a**) Illustration of the phylogenetic history of the *RRM1* gene (encoding ribonucleotide reductase M subunit) and the *HBB* gene (encoding beta-globin) and genes related to it by duplication since the divergence of mammalian and avian lineages from the amniote ancestor. The gene duplications in the beta-like globin gene family occurred separately in the mammalian and avian lineages, leading to paralogous relationships within a species and many-to-many orthologous relationships between the species. (**b**) Illustration of the the phylogenetic history of the beta-like globin gene cluster over the much shorter time since humans and macaques (an Old World monkey) diverged from the catarrhine ancestor. The gene duplications predate the ancestor, and thus the speciation event resulted in 1:1 orthologous relationships between human and macaque *HBB*, human and macaque *HBD*, etc. Other relationships, e.g., between human *HBB* and macaque *HBD* are paralogous

compared. In this situation, correct assignments of paralogous and orthologous relationships are particularly important. For instance, an incorrect assignment of paralogous genes as being orthologous between human and macaque would lead to a conclusion of greater sequence change since speciation than would a truly orthologous comparison.

Once gene sets have been defined in two or more species, then orthologous gene sets can be determined. For the cases of 1:1 orthologs, reciprocal highest similarity is a good guide to orthologous relationships. The more complicated cases for multigene families can be summarized as many-to-many orthologous relationships. Figure 19.11 shows the results of comparisons of protein-coding genes among human (*Homo sapiens*), chicken (*Gallus gallus*), and the teleost fish *Fugu rubripes* [30]. Of the almost 22,000 genes annotated in humans in this study, about a third are in 1:1:1 orthologous relationships with chicken and *Fugu*, and about 5% are in many-to-many relationships. About a third of the genes have clear homologs but cannot be definitively assigned as orthologous. Intriguingly, about 4,000 human genes do not have a clear homolog in either chicken or fish. These may encode mammal-specific functions.

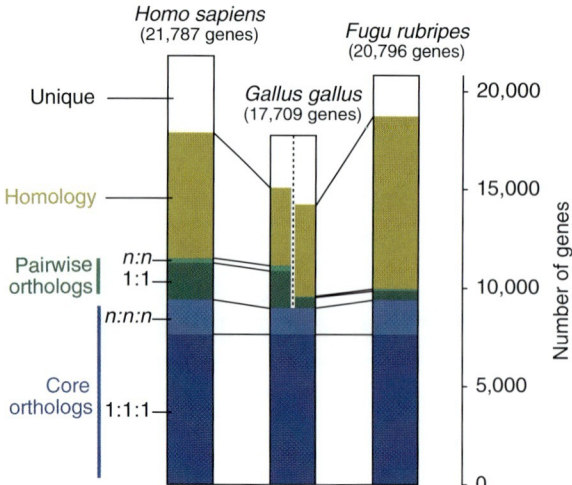

Fig. 19.11 Homology relationships among protein-coding genes in human (*Homo sapiens*), chicken (*Gallus gallus*), and the fish *Fugu rubripes*. Genes in the three species are grouped by their orthology relationships among the three species (1:1:1 or *n:n:n* for many:many:many) or between two species if the gene is not detected in a third species. Genes that are clearly related between species but for which clear orthology relationships cannot be determined are placed in the 'Homology' class. Genes not falling in the orthology or 'homology' classes are considered 'Unique'. (Reprinted from Hillier et al. 2004, with permission from Nature Publishing Group)

19.4.3 Rates of Sequence Change in Different Parts of Genes

Within the set of 1:1 orthologous genes, the amount of sequence similarity can be determined in each of the basic parts of a gene. One of the first genome-wide studies in mammals compared human genes with mouse genes [85], and it confirmed many insights from smaller scale studies. The protein-coding exons are the most similar between human and mouse, showing about 85% identity (Fig. 19.12). The regions adjacent to the splice junctions show peaks of higher identity, reflecting the selection on both coding potential and on splicing function. The introns have the lowest similarity, but they are considerably more similar than is DNA in ancestral repeats (the neutral model in this study), which are about 60% identical. The untranslated regions of exons are about 75% identical. The higher percent identity in the untranslated regions and introns, than in the neutral model, indicate that some portion of these sequences is under constraint. Intronic regions that provide important functions include splicing enhancers and transcriptional enhancers. In the 3′ untranslated region can be found targets for regulation by miRNAs as well as the polyadenylation signals. These short segments can be subject to stringent constraint. If all the intronic and untranslated sequences were subject to such stringent constraint, then their overall percent identity would be closer to that of the coding regions. Thus, one interpretation of these results is that intronic and untranslated regions contain short constrained segments interspersed within larger regions with little or no signature of purifying selection.

19.4.4 Evolution and Function in Protein-Coding Exons

From the earliest comparisons of homologous protein sequences, it was recognized that some proteins change little between species. A classic example is histone H4,

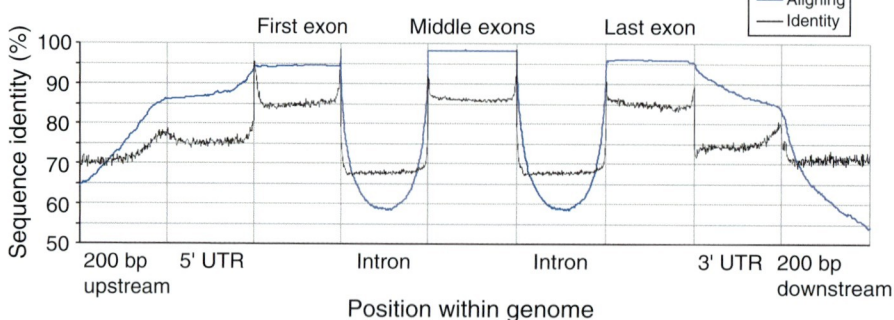

Fig. 19.12 Sequence identity between human and mouse in a generic gene. Within a group of 3,165 RefSeq genes that aligned between the mouse and human genomes, 200 evenly spaced bases across each of the variable-length regions were sampled between human and mouse. The blue line shows the average percentage of bases aligning and the black line shows the average base identity. (From Waterston et al. [85], with permission from Nature Publishing Group)

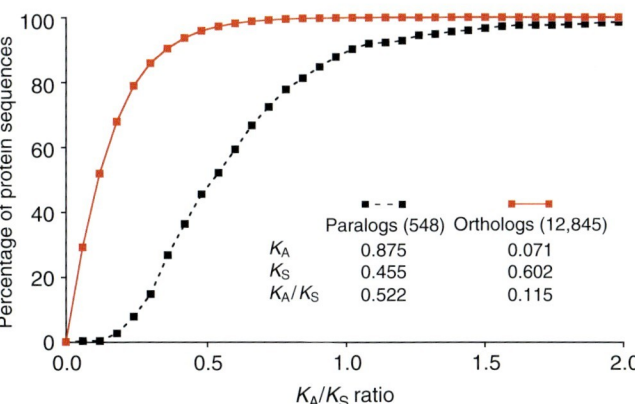

Fig. 19.13 Cumulative distribution of K_A/K_S values for mouse proteins compared with human homologs. The distribution of scores for proteins that are clearly orthologous between human and mouse is shown by the red points and line. The distribution of scores for proteins encoded by locally duplicated, paralogous mouse-specific gene clusters is shown by the black points and line. (From Waterston et al. [85], with permission from Nature Publishing Group)

which has only one amino acid replacement between peas and cows. Other proteins change rapidly. Among the most rapidly changing proteins are the fibrinopeptides, which are segments of fibrinogen molecules that are cleaved off by thrombin during blood clotting. It appears that the amino acid sequence of the fibrinopeptides is not critical for their function, and they are under little or no selective pressure. Interspecies comparisons of even a modest number of proteins showed that the rate of changes in amino acids ranged over 100-fold [60]. Some proteins, such as histones, are under stringent selection over most of their sequence, whereas others seem to be free to change extensively – or have been adapted to new function.

Comparisons of the protein-coding genes for entire mammalian genomes provide the opportunity to examine these issues more comprehensively. The sets of related genes between species can be analyzed to show which genes are under strong purifying constraint and which show signs of adaptive evolution. For protein-coding genes, it is common to consider substitutions at synonymous sites to be neutral. The number of synonymous substitutions per synonymous sites in two species is called K_S. This can be used as an estimate of the neutral rate. Then the number of nonsynonymous substitutions per nonsynonymous site, or K_A, can be compared with K_S to obtain an estimate of the stringency of the purifying selection or the strength of adaptive evolution. As a rule of thumb, a K_A/K_S ratio of 0.2 for human–mouse comparisons is indicative of constraint, whereas ratios of 1 or greater indicate adaptive evolution.

In a study of orthologous genes aligned between mouse and human [85], about 80% show an overall signal for constraint (Fig. 19.13). Very few show evidence of positive selection over their entire length. Thus, at the phylogenetic distance of mouse and human, evolution of protein-coding sequences in orthologous genes is dominated by constraint. This result indicates that the matching, orthologous segments code for proteins that provided a function in the ancestor, and their descendant sequences provide a similar function in contemporary species. Many changes in the encoded amino acid sequences have been selected against because they did not improve the function of the protein. We note that short segments or single codons under positive selection would not be detected in this test.

In contrast, the set of paralogous genes compared between mouse and human are shifted to higher K_A/K_S ratios. Thus, the paralogous genes are more likely to be undergoing adaptive evolution (positive or diversifying selection) than are the orthologous genes. The multigene families are major contributors to lineage-specific function. Duplication of genes leaves at least one copy free to accumulate changes that can provide an adaptive advantage. In contrast, genes that remain as single copies are constrained to fulfill the role that they have played since they arose in some distant ancestor.

19.4.5 Fast-Changing Genes That Code for Proteins

The families of fast-changing genes appear to be adapting to new pressures in a lineage-specific manner. An examination of the types of gene families with this property should provide insights into the types of pressures that lead to adaptive changes. A remarkably consistent result has been found in multiple studies

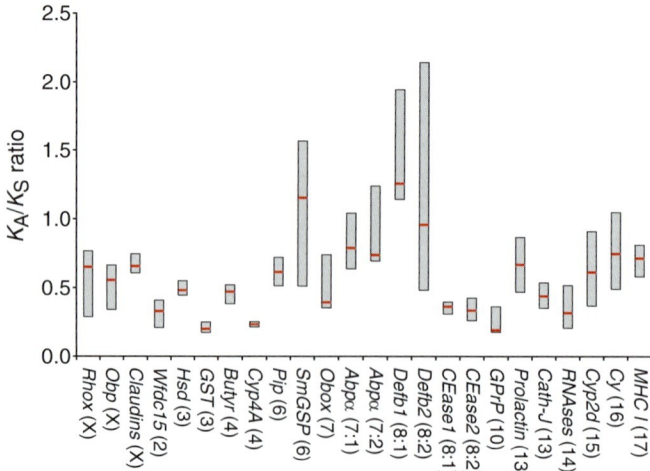

Fig. 19.14 Distributions of K_A/K_S values for duplicated mouse-specific gene clusters. The chromosome on which the clusters are found is indicated in brackets after the abbreviated cluster name. The K_A/K_S values for each sequence pair in the cluster were calculated from aligned sequences. The box plots summarize the distributions of these values, with the median indicated by the red horizontal line and the boxes extending from the 16th and 83 rd percentiles and hence covering the middle 67% of the data. (From Waterston et al. [85], with permission from Nature Publishing Group)

of this question. The four general categories of reproduction, chemosensation, immune response, and xenobiotic metabolism (breakdown of drugs, toxins, and other compounds not produced in the body) encompass many of the genes and gene families subject to positive selection. Thus, these are the major physiological functions in which rapid sequence change leads to adaptive evolution.

For example, the locally duplicated gene families with relatively high K_A/K_S values fall into distinct functional classes (Fig. 19.14). Members of the major categories for adaptive evolution (reproduction, chemosensation, immune response, and xenobiotic metabolism) are apparent. For example, the mouse *Rhox* genes on chromosome X are homeobox genes expressed in male and female reproductive tissue, and targeted disruption of the *Rhox5* gene leads to reduced male fertility [51]. Another example is the oocyte-specific homeobox gene *Obox* on mouse chromosome 7. The *Obp* gene cluster encodes odorant-binding proteins such as lipocalins and aphrodisin, involved in both chemosensation and reproduction. Immune response genes include the *MHC I* genes on chromosome 17, which regulate the immune response, the *Wfdc15* gene, which encodes an antibacterial protein, and the *Defb* genes on chromosome 8 encoding beta-defensins. Several adaptive genes are involved in xenobiotic metabolism, including members of the cytochrome P450 gene family, *Cyp4a* and *Cyp2d*, and a glutathione-*S*-transferase gene (*GST*).

Additional studies of lineage-specific expansions of gene families in comparisons of rat and mouse [25] and of humans and chickens [30] identify the same general categories of reproduction, chemosensation, immune response, and xenobiotic metabolism. Thus, along multiple lineages, these gene families are implicated in adapting to unique pressures on each species. Enrichment of these functional categories for genes implicated in adaptive evolution can be readily rationalized. Changes in genes involved in reproduction and chemosensation could lead to or maintain the differences that cause divergence of species. Adaptation of immune function and the ability to metabolize foreign compounds are important for survival in the distinctive environment of each species. Other families with rapid changes between species include keratins, which are involved in making feathers in birds but hair in mammals.

19.4.6 Recent Adaptive Selection in Humans

In addition to improving our understanding of the evolution of humans within the context of other vertebrates, comparative genomics also provides insights into recent adaptive changes that may eventually tell us what genome sequences make us distinctively human. Comparisons to close relatives such as the chimpanzee and analysis of human polymorphisms drive these new studies.

As was the case for human–mouse comparisons discussed above, the K_A/K_S ratio was computed in genome-wide comparison of the human and chimpanzee gene sets [12, 14, 15, 61]. The ratio for human–chimpanzee comparisons is significantly higher than that seen for mouse–rat comparisons, showing more changes in amino acids in proteins (normalized to synonymous substitutions) in the hominid lineages than in rodents. This does not, however, indicate an overall stronger positive selection in hominids, but rather it reflects the relaxation of purifying selection in species with a small population size. Estimates of effective population size for rodents far exceed those for humans and chimpanzees, and it is well recognized that the severity of selection increases with population size. However, despite this relaxed selection, examination of the orthologous genes with the most extreme ratios of amino acid-changing substitutions to presumptive neutral changes reveals interesting candidates for hominid-specific adaptive evolution. One is the gene for glycophorin C, which is the membrane protein used for invasion of the malarial parasite *Plasmodium falciparum* into human erythrocytes. Others include granulysin, which is needed for defense against intracellular parasites, and semenogelins, which are involved in reproduction. A stronger signal for positive selection can be observed when genes are grouped together, either by physical proximity (often as duplicated genes) or by functional category. For human-chimpanzee comparisons, the sets of genes changing most rapidly include the now-familiar categories of reproduction (e.g., spermatogenesis, fertilization, and pregnancy), chemosensation (olfactory receptors, taste receptors), immunity (immunoglobulin lambda, immunoglobulin receptors, complement activation), and xenobiotic metabolism, plus additional categories such as inhibition of apoptosis.

The distribution of human polymorphisms along chromosomes and their frequency in populations can be analyzed for insights into very recent selection (reviewed in [5, 44]). Positive selection is expected to drive mutations quickly to fixation, so loci under positive selection should be characterized by a skew in the allele frequency distribution toward rare alleles. One measure of that skew is Tajima's D [77].

Also, the rapid fixation of an advantageous allele will bring along linked polymorphisms. These polymorphisms will not have had time to be separated from the selected allele by recombination, and thus linkage disequilibrium will extend further around positively selected alleles than is expected from neutral evolution. Various tests of properties such as these have been developed, and have traditionally been applied to a small number of loci. A major limitation to these studies is that changes in population demographics can generate the same signals. For example, recent expansion in population size, such as that experienced by humans, will also lead to an excess of rare alleles or extended linkage disequilibrium. Thus, it is difficult to disentangle the confounding effects of population demographics and positive selection when only a few genetic loci are examined. However, the recent availability of genome-wide data on polymorphisms [32] provides one solution. Changes in population size should affect all loci in the genome, whereas selection should act on only a few. Thus, when the distribution of values for Tajima's D, long-range haplotype, or related measures are examined for a large number of loci, then it is likely that the outliers are undergoing adaptive evolution [5].

Recent genome-wide studies have identified significant outliers based on frequency of rare alleles (Tajima's D, [13, 37]) and linkage disequilibrium [82, 83]. For example, Carlson et al. [13] calculated Tajima's D in sliding windows across the human genome for populations descended from Africans, Europeans, or Chinese. Several extended regions with consistently negative values for Tajima's D were identified, with most observed in only one of the populations (Fig. 19.15). Negative values for Tajima's D are associated with positive selection if population expansion is not a factor, and the study design to identify outliers in a genome-wide analysis should greatly reduce the confounding effect of such an expansion. Thus, results such as those in Fig. 19.15 indicate that at least one genetic element in the roughly one megabase region with reduced Tajima's D has been under positive selection in humans of European ancestry. Resequencing of targeted genes within these regions has supported the conclusion of positive selection, and in some cases (e.g., *CLSPN* in Fig. 19.15) it has revealed a polymorphism that alters the encoded amino acid sequence [13]. Such a change in amino

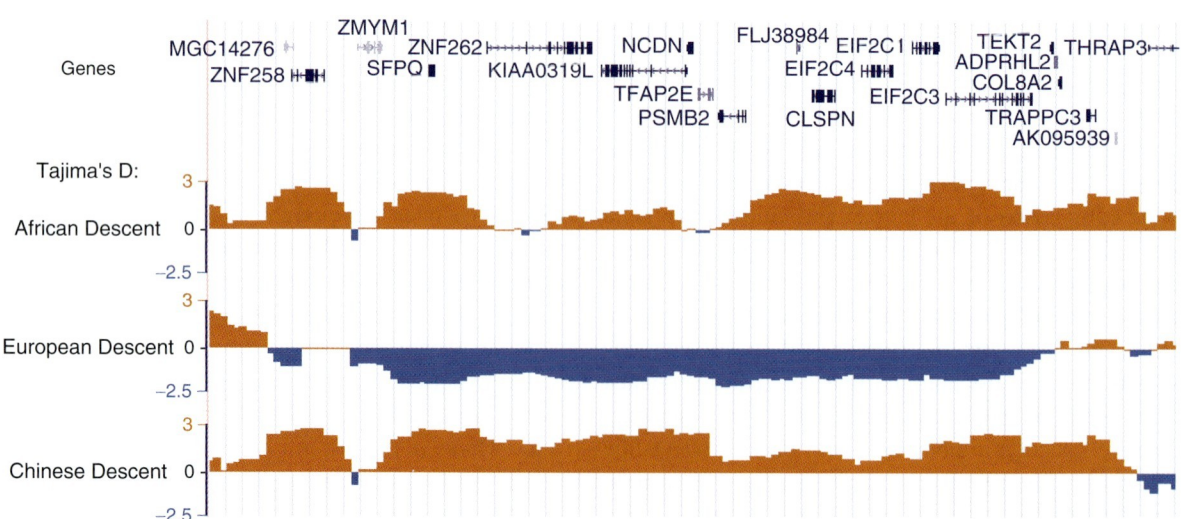

Fig. 19.15 An extended region with an excess of rare alleles indicative of positive selection. The region from human chromosome 1 is one of several identified in the study by Carlson et al. [13] showing an excess of rare alleles in at least one of three human populations (those of European descent in this case) as measured by Tajima's D [77]. Negative values of Tajima's D can be explained by positive selection or population expansion; the design of genome-wide studies favors the former explanation. The full data from the study are available on the UCSC Genome Browser [39]; this figure was generated from the Browser output

acid sequence is a candidate for the functional variant under selection.

A third type of test for recent selection utilizes both human polymorphism data and interspecies divergence between human and close relative, such as chimpanzee. The McDonald-Kreitman [55] test compares the ratio of polymorphisms to divergence (r_{pd}) at nonsynonymous sites (leading to amino acid changes in the protein product) with that ratio in synonymous sites, which do not change the amino acid sequence and are expected to be largely neutral. If the changes in nonsynonymous sites had no selective advantage or disadvantage, then r_{pd} at these sites would not be significantly different from r_{pd} at neutral sites. Deviation from neutral expectation can be evaluated with a chi-square or related statistic. Bustamante et al. [12] applied this test to over 11,000 human genes (with polymorphsims determined in three different populations) compared with chimpanzee. They found that 9% had a significant signal for positive selection and 14% had a significant signal for negative selection.

Each method for finding loci under recent selection in humans has its distinctive strengths and weaknesses. Much effort is currently devoted to examining overlaps and differences in the results. Among the several studies reviewed by Biswas and Akey [5], a total of 2,316 human genes were found to have at least one signature for positive selection. Almost a third of these, including *EDAR*, *SLC30A9*, and *HERC1*, are found in more than one genome-wide study. Other candidate genes for positive selection are found by only one approach, such as *TRPV5* and *TRPV6*. At least to some extent, the failure to overlap reflects the different types of selective events being assayed in the different tests. The features examined by one approach, such as low frequency alleles, are not contributing to other tests, such as linkage disequilibrium measurements based on common alleles [5].

Some genes that are candidates for human-specific selection lead to intriguing and exciting possibilities, such as alterations in *FOXP2* implicated in language acquisition [22] and *MCPH1* and *ASPM* implicated in brain size [23, 56]. Further studies of recent selection in humans should lead to critical new insights into human biology and disease.

19.4.7 Human Disease-Related Genes

Comparative genomics can be used to study the origins and implications of genetic variants associated with human disease. Disadvantageous mutations should be cleared from a population quickly, so why are some genetic diseases rather common?

One factor is the relaxed selection against mildly deleterious alleles resulting from population expansion. A common estimate of the effective population size of humans is about 10,000 individuals, and of course the population has expanded dramatically to the current level of over 6 billion. This would tend to favor the persistence of some deleterious mutations, and the results of a McDonald-Kreitman test [12] indicate that many of the amino acid polymorphisms in humans are moderately deleterious.

Another factor is positive selection in one region of the world driving an allele to high frequency, but that allele is pathogenic in other regions of the world. A classic example is the *HBB-S* allele of the gene encoding beta-globin. This allele encodes a mutant beta-globin that in combination with alpha-globin and heme constitutes HbS. This is the hemoglobin variant that causes red blood cells to form a sickled, inflexible morphology when deoxygenated, and thus leads to sickle cell disease. However, the *HBB-S* allele in heterozygotes reduces the susceptibility of humans to malaria, and thus it is a protective allele in regions of the world in which malaria is endemic. In fact, haplotype analysis has shown that the *HBB-S* allele has arisen independently multiple times in recent human history [2, 63]. This indicates a strong positive selection in the presence of the malarial parasite. Unfortunately, the negative consequence is that people who are homozygous for the *HBB-S* allele are highly prone to sickle cell disease.

A third factor is that some disease-associated variants were protective in the more distant past but are now detrimental for most contemporary human lifestyles. In the "thrifty genotype" hypothesis [59], the limited caloric intake and need for high activity levels in ancestral humans would have favored a thrifty genotype that made efficient use of food. However, many contemporary humans live in an environment with an excess of available food. Being "too thrifty" with energy metabolism could lead to problems such as diabetes. Disease-associated variants that were advantageous in the past should match the amino acid at that position in ancestor, and some of these will still be seen in related species. Indeed, human disease-related variants match with the amino acid in the corresponding position of chimpanzee [14] and rhesus macaque [70] in about 16 and 200 cases, respectively. Further studies of these candidates are needed, but the results suggest that retention of an ancestral state is also contributing to human disease alleles.

19.5 Evolution in Regions That Do Not Code for Proteins or mRNA

Despite the importance of protein-coding regions to genome function, these sequences account for about one-third of the sequences that have been under selection for a common function in eutherian mammals. Accounting for the remaining selection in noncoding regions is a major on-going effort in genomics and genetics. Two functional categories are the focus of much attention: genes that do not code for proteins, such as microRNA (miRNA) genes, and gene regulatory regions. An equally important question is to what phylogenetic depth functional noncoding regions are conserved. These issues will be examined in this section.

19.5.1 Ultraconserved Elements

The level of constraint on genomic sequences spans a wide range, and it likely that different functions are subject to distinctive levels of constraint. The most intense constraint is revealed in the human DNA segments called *ultraconserved elements*, or UCEs [4]. These are the 481 human DNA segments that are identical to mouse DNA for at least 200 nucleotides. Sequences that code for proteins have frequent mismatches between human and mouse at synonymous sites, so these UCEs are under stronger purifying selection than most exons. This pattern of conservation indicates that all nucleotides in the identical segment are critical for some function. The UCEs are broadly conserved in vertebrates, and they show the slowest rate of divergence over the period of vertebrate evolution of any known elements in the genome (Table 19.2, Fig. 19.5).

Determining the roles for the UCEs is currently a matter of intense interest. Only a small fraction (23%) overlaps with mRNA for known protein-coding genes. Thus, the majority is associated with some noncoding function. About half of those tested serve as tissue-specific enhancers in transgenic mouse embryos [64]. A small number are related to each other, and examination of these has revealed a family of sequences derived from an ancient transposable element that have been recruited for activity as a distal enhancer for one gene and part of an exon for another [3]. Another subset of very slowly changing regions (across most eutherians) was examined for rapid change along the human lineage since divergence from chimpanzee. These *human accelerated regions* include a gene that encodes an RNA that may function in cortical development [66]. A full explanation of the stringent constraint on each nucleotide within the UCEs remains elusive. Not only is the intensity of constraint beyond that seen for almost all protein-coding regions, but even RNAs with considerable secondary structure rarely show this resistance to substitution.

Another enigmatic aspect to UCEs is their restriction to vertebrates. Protein sequences, which evolve faster than UCEs in vertebrates, frequently show significant similarity between vertebrate and invertebrates species. Sometimes the similarity extends from vertebrates to eubacteria. In contrast, no homolog to a UCE sequence has been observed outside vertebrates. Worms (and possibly other invertebrates) have analogous highly constrained noncoding sequences, but they differ in sequence from the vertebrate UCEs [81]. Thus, this stringent constraint on noncoding sequences may have evolved in parallel in vertebrates and invertebrates. Finding the sources of the UCEs and explaining how they could be under such intense constraint are important goals for future work. Answers to these questions may reveal aspects of genome function that have yet to be imagined. The fact that the roles and origins of the most stringently constrained sequences in vertebrates are still unknown illustrates how much still needs to be accomplished in comparative genomics.

19.5.2 Evolution Within Noncoding Genes

Many genes do not code for protein, and these must account for some of the noncoding DNA that is under constraint. However, some of the better-known noncoding genes do not help explain the fraction under constraint, but for technical reasons. Consider the genes for RNAs utilized in the mechanics of protein synthesis, such as ribosomal RNAs (rRNA) and transfer RNAs (tRNAs). The rRNA genes are clustered in highly duplicated regions on the short arms of chromosomes 13, 14, 15, 21, and 22. These regions are not included in the assemblies of the human genome, and thus they do not contribute to the minimal estimate of 5% of the genome under constraint in mammals. The tRNA genes are small and contribute little to the selected fraction. Other RNAs, such as snRNAs involved in splicing and processing of precursors to mRNA, also tend to be encoded on small genes. Multiple copies of sequences related to the snRNA genes are present in the human genome, some of which may no longer be active. The contribution of snRNA genes to the fraction of the human genome under constraint needs further study.

The miRNAs do not code for protein, but they negatively regulate mRNA function or abundance. Hybridization of an miRNA to its mRNA target to generate a duplex with some mismatches leads to inhibition of translation of the mRNA. Hybridization of an miRNA to its target to generate a perfect duplex leads to degradation of the target mRNA (see Chap. XX).

The known miRNA genes are constrained, with many conserved from humans to chickens. However, the full set of miRNA genes is not known, and information is limited about the structure and conservation of genes encoding the precursors to miRNAs. Thus, the miRNAs clearly are important contributors to the fraction of the genome under purifying selection, and they could account for substantially more of the constraint that is currently known.

Members of another class of RNA that apparently does not code for protein are detected by hybridization of copies of cytoplasmic RNA to high-density tiling arrays of nonrepetitive human genomic DNA. These results show transcription of protein-coding genes as expected, but about half the transcribed regions are not associated with known genes [34]. These unannotated transcripts, referred to as *transfrags*, are often of low abundance and are expressed in a limited set of tissues. The contribution of transfrags to constrained sequences in human is a matter of current study (e.g., [67, 79]).

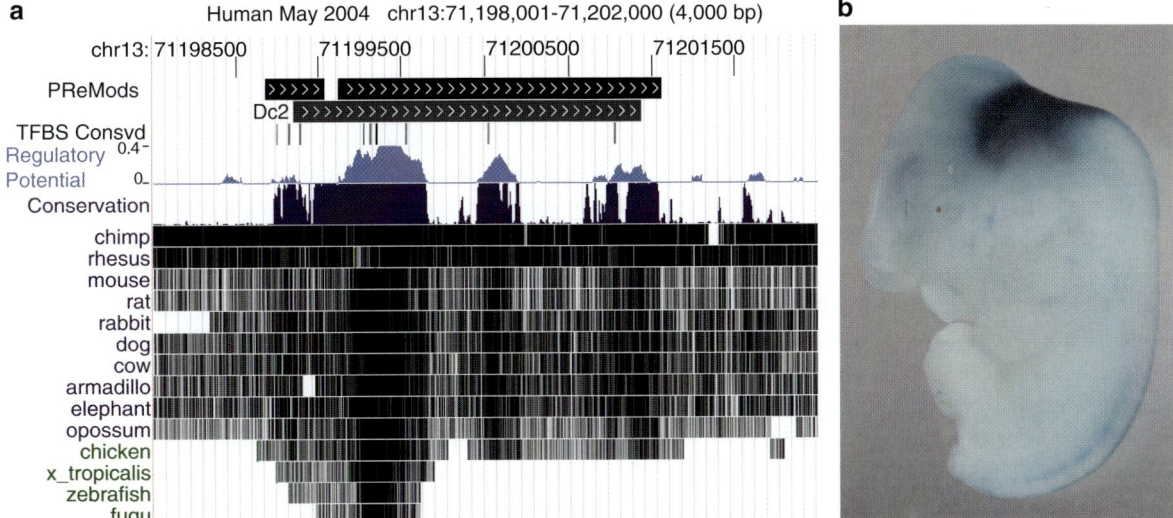

Fig. 19.16 An enhancer of the *DACH1* gene predicted by comparative genomics. This human gene is homologous to the *Drosophila* gene *dachshund*, and it is needed for development of the central nervous system and other organs. Within one of the very large introns of *DACH1* are some deeply conserved DNA segments. (**a**) Several features of the Dc2 region, including its conservation from humans to fish, high regulatory potential [78], and prediction as a regulatory module by the PReMod pipeline [6]. Examples of conserved matches to transcription factor binding site motifs are also shown. (**b**) This DNA segment is sufficient to enhance expression of a beta-galactosidase reporter gene in the hindbrain of a transgenic mouse embryo. The blue stain is a marker for beta-galactosidase activity. The Dc2 region was shown to be an enhancer by Nobrega et al. [62]; the image is from the Enhancer Browser (Table 19.3)

19.5.3 Evolution and Function in Gene Regulatory Sequences

DNA sequences needed to regulate the level, developmental timing, and tissue-specificity of gene expression include promoters that designate the correct start site for transcription, enhancers that increase the level of expression, silencers that decrease the level of expression, and insulators that separate genes and regulatory regions from the effects of neighboring regulatory regions. Many but not all of these regulatory regions are conserved among mammals [26, 64]. Some of the DNA sequences that regulate genes encoding developmental regulatory proteins are conserved from mammals to fish, indicative of strong constraint [62, 88]. One example is shown in Fig. 19.16. However, other regulatory regions show more rapid evolution, e.g., replacing one motif for binding a transcription factor with a similar sequence in another location [19, 50] or being present in only one lineage. Despite numerous studies of the extent of conservation of regulatory regions in individual loci, no clear consensus had emerged on the dominant pattern of conservation.

A major limitation to previous studies has been the small number of regulatory regions that have been identified experimentally. Establishing the role of a segment of DNA in regulation requires multiple experiments, and traditionally these were done in a highly directed manner that did not lend itself to high throughput. Now it is possible to enrich DNA for sites occupied by transcription factors (by chromatin immunoprecipitation or ChIP) and then hybridize this enriched DNA to high-density tiling arrays of genomic DNA (DNA chips). This ChIP-chip experiment [69] reveals sites bound by transcription factors in a high-throughput manner. Experiments by the ENCODE Project Consortium [79] evaluating sites occupied by several transcription factors have yielded a large set (over 1,000) of putative transcriptional regulatory regions in about 1% of the human genome. This large set of DNA intervals implicated in transcriptional regulation was identified by experiments that are agnostic to interspecies sequence conservation, and thus it is an ideal set in which to determine the phylogenetic depth of conservation [42]. As shown in Fig. 19.5 and Table 19.2, about two out of three of these putative transcriptional regulatory regions are conserved from humans to other placental mammals (but no further), and about one out of three are conserved to marsupials. Less than 10% are conserved from humans to birds. An equal fraction, about 3%, is found at the two extremes of conservation, viz., found only in primates or conserved from humans to fish. Thus, the bulk of the regulatory regions are conserved in placental mammals, and we expect that comparisons among these species will continue to be effective at finding and better understanding these regulatory regions. However, a particular phylogenetic depth of conservation is not a consistent property of gene regulatory sequences. Rather, the depth of conservation is a property that varies among the regulatory sequences. Ongoing studies may reveal whether particular functions of regulatory regions or their targets correlate with the depth of conservation.

Although it is not a property shared by all putative regulatory regions, many do have a significant signal for purifying selection. A small majority (about 55%) overlap at least in part with DNA segments that are in the 5% of the human genome that is under strong selection [79]. However, only about 10% of the nucleotides in the putative regulatory regions are under strong constraint, suggesting that small subregions of enhancers and promoters, e.g., binding sites for particular transcription factors, are under purifying selection. Thus, the putative regulatory sequences identified in the ENCODE project contribute only a small amount to the 5% under strong constraint [79].

19.5.4 Prediction and Tests of Gene Regulatory Sequences

Effective use of comparative genomics to find gene regulatory sequences is challenging for at least two reasons. The variation in phylogenetic depth of conservation is a major complication; some human regulatory regions will be observed only in alignments of primates, whereas others align with species as distant as fish. Although the large majority of regulatory regions are conserved in multiple placental mammals, even some apparently neutral DNA aligns reliably at this phylogenetic distance. Thus, the ability to align at this distance is not a property that identifies regulatory regions with good specificity.

Most efforts to detect candidate gene regulatory regions from aligned sequences also use some form of pattern information. For example, the known regulatory regions are clusters of binding sites for transcription factors. The binding sites are short (about 6–8 bp) and many allow degeneracy (e.g., either purine or either pyrimidine works equally well at some sites). Therefore, the binding site motifs themselves do not confer strong specificity. However, in combination with clustering and conservation, this set of criteria has good power to detect novel regulatory regions [6]. A set of about 200,000 regions, called *PReMods*, has been identified as predicted regulatory regions in the human genome using this approach.

The motifs for binding sites in regulatory regions are not known completely. These currently unknown motifs can be incorporated into the prediction of regulatory regions by using machine-learning procedures to find distinctive patterns of alignment columns that are common in a training set of alignments in known regulatory regions, but are less abundant in a set of alignments from likely neutral DNA. The statistical models describing these distinctive patterns are then used to score any alignment for its *regulatory potential*. One implementation of this approach has generated a set of about 250,000 regions of human DNA with a high regulatory potential [78]. Many of these overlap with the PReMods discovered as conserved clusters of transcription factor-binding motifs. Regions with high regulatory potential and a conserved binding site for an erythroid transcription factor are validated at a good rate as enhancers in erythroid cells [84].

In summary, several methods based on comparative genomics can be used with some success to predict gene regulatory sequences, but none achieves the level of reliability desired. Deep conservation of noncoding sequences, e.g., from human to chicken or human to fish, can be used without additional information about patterns such as binding site motifs. However, this approach will miss the majority of gene regulatory regions. For noncoding sequences conserved among placental mammals, clustering of pattern information should be incorporated. The pattern information can either be based on prior knowledge (such as binding motifs) or learned from training sets. Currently, in vivo occupancy of DNA segments by transcription factors is being determined comprehensively by ChIP-chip and related methods. Integration of this information with the comparative genomics should add considerable power to the identification of regulatory regions [21].

19.6 Resources for Comparative Genomics

The large amount and wide variety of data on comparative genomics of mammals and other species can be daunting to those who wish to use them. Also, as discussed throughout this chapter, the level of conservation of functional regions tends to vary from region to region. Detailed information needs to be readily accessible for individual regions and for classes of features across a genome. These needs are accommodated by genome browsers and data marts. Computational tools for further analysis of the data are also available, and one workspace for such tools will be described here.

19.6.1 Genome Browsers and Data Marts

Genome browsers show tracks of user-specified information for a designated locus in a genome. The major browsers for mammalian genomes are the UCSC Genome Browser [45], Ensembl [31], and MapView at NCBI [86] (Table 19.3). Comparative genomics tracks showing results of whole-genome alignments are available at the UCSC Genome Browser and Ensembl. As illustrated in Fig. 19.4, the regions of the human genome aligning with a comparison species can be seen as nets and chains. Inferences about severity of constraint are captured on the "Conservation" track (similar to that in Fig. 19.9), based on phastCons [74].

Often it is desirable to collect and analyze all members of a feature set across a genome or large genomic intervals. This requires the ability to query on the databases of features that underly the browsers. Two such "data marts" are the UCSC Table Browser [35] and BioMart at Ensembl [36]. Both provide interactive query pages to provide access to the data.

19.6.2 Genome Analysis Workspaces

Once the data have been obtained, users frequently need to analyze them further. Different data sets may need to be combined or compared. The level of constraint or regulatory potential may be needed. Estimates of evolutionary rates may be desired. Different tasks will require distinct sets of tools. Considerable progress can be made by acquiring the necessary computer programs and executing them on the user's computer system. However, this leaves it to the user to find or write the needed tools.

An alternative is to connect versatile data acquisition with integrated suites of computational tools in a common workspace such as Galaxy [7] (Table 19.3). This resource allows users to import data from various sources, such as the UCSC Table Browser, BioMart, or files from the user's computer. Once imported, a wide variety of operations can be performed on the data sets, such as edits, subtractions, unions, and intersections. Summary statistics can be computed and distributions can be plotted. Various evolutionary genetic analyzes can be performed.

Table 19.3 Data resources and analysis workspaces for comparative genomics

Name	Description	URL
UCSC Genome Browser	Sequences, comparative genomics, annotations	http://genome.ucsc.edu
Ensembl	Sequences, comparative genomics, annotations	http://www.ensembl.org/
NCBI MapViewer	Gene, EST and other maps of chromosomes	http://www.ncbi.nlm.nih.gov/mapview/
UCSC Table Browser	Query for genomic features	http://genome.ucsc.edu/cgi-bin/hgTables
BioMart	Query for features of genes	http://www.ensembl.org/biomart/martview/
VISTA Enhancer Browser	Data on conserved noncoding regions tested as developmental enhancers	http://enhancer.lbl.gov/
Galaxy	Interactive workspace for analysis of genome sequences, alignments and annotation	http://main.g2.bx.psu.edu/

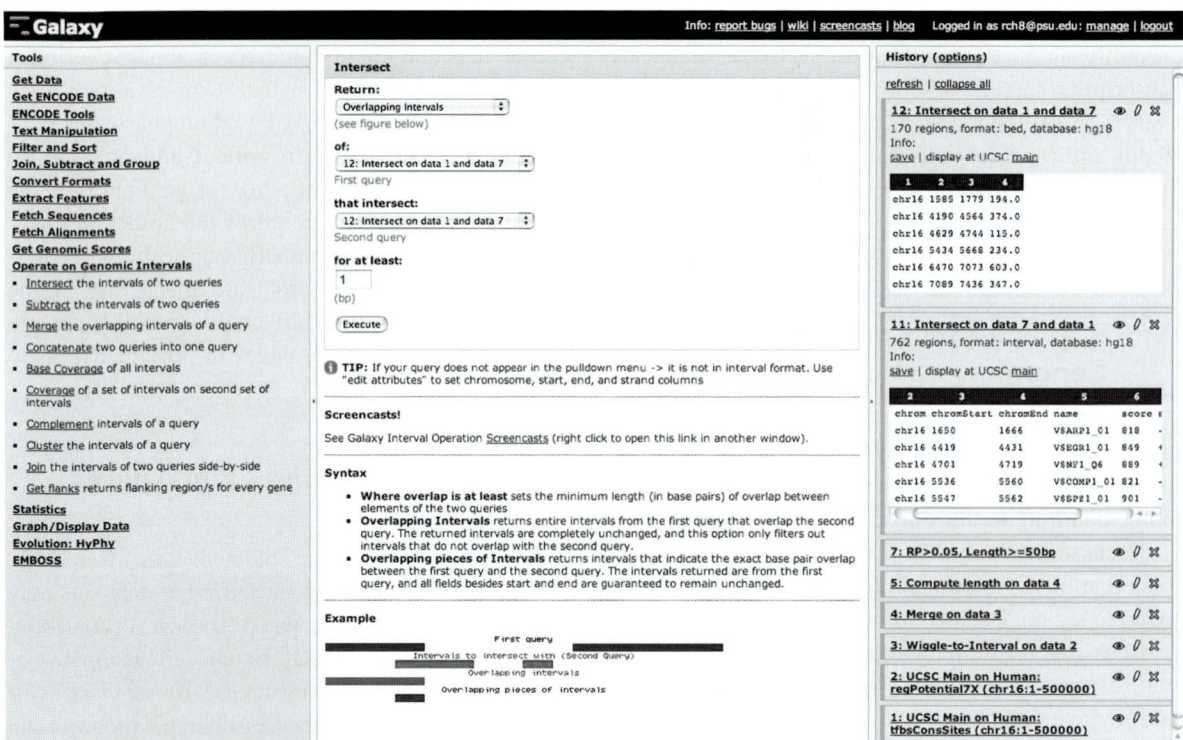

Fig. 19.17 Using Galaxy to find predicted regulatory regions. The user interface for Galaxy has three panels. Tools for obtaining and analyzing data are selected from the left panel, and the user selects input data and other parameters in the central panel. A history of previous results is maintained on the right panel. In this example, candidates for gene regulatory modules in a 500 kb region of human Chromosome 16 are obtained by queries to the UCSC Table Browser to obtain conserved matches to transcription factor binding motifs (query 1) and regions of of high regulatory potential (score >=0.05 in query 2; these results were converted to intervals, merged and filtered for length >= 50 bp to obtain the results in query 7). Intersections reveal conserved motifs that are in regions of high regulatory potential (query 11) and *vice versa* (query 12)

Precomputed scores such as phastCons and regulatory potential can be aggregated on specified intervals. The interface at Galaxy for a series of operations that can predict gene regulatory regions is shown in Fig. 19.17.

19.7 Concluding Remarks

Comparative genomics brings considerable power but daunting challenges to the study of human genetics. No aspect of comparative genomics has been perfected; even the commonly used methods of aligning sequences and predicting protein-coding genes have room for improvement. However, considerable insight and functionality can be gleaned from the predictions and comparisons that are currently available. Real biological variation, for example, in the rate of evolutionary change at different loci or the phylogenetic depth of conservation of a feature class, means that no single threshold for a conservation-based score will be adequate to find all the features of interest. However, as the variation is better understood and as functional correlates of the variation are established, then the potential power of comparative genomics will be better harnessed. Current data can be readily accessed and evaluated. Additional types of data, such as genome-wide ChIP-chip results, coupled with tools for better integration of disparate data types, should lead to considerable future progress in the functional annotation of the human genome.

References

1. Altschul SF, Madden TL, Schaffer AA, Zhang J, Zhang Z, Miller W, Lipman DJ (1997) Gapped BLAST and PSI-BLAST: a new generation of protein database search programs. Nucleic Acids Res 25:3389–3402
2. Antonarakis SE, Boehm CD, Serjeant GR, Theisen CE, Dover GJ, Kazazian HH Jr (1984) Origin of the beta S-globin gene in blacks: the contribution of recurrent mutation or gene conversion or both. Proc Natl Acad Sci USA 81:853–856
3. Bejerano G, Lowe CB, Ahituv N, King B, Siepel A, Salama SR, Rubin EM, Kent WJ, Haussler D (2006) A distal enhancer and an ultraconserved exon are derived from a novel retroposon. Nature 441:87–90
4. Bejerano G, Pheasant M, Makunin I, Stephen S, Kent WJ, Mattick JS, Haussler D (2004) Ultraconserved elements in the human genome. Science 304:1321–1325
5. Biswas S, Akey JM (2006) Genomic insights into positive selection. Trends Genet 22:437–446
6. Blanchette M, Bataille AR, Chen X, Poitras C, Laganiere J, Lefebvre C, Deblois G, Giguere V, Ferretti V, Bergeron D et al (2006) Genome-wide computational prediction of transcriptional regulatory modules reveals new insights into human gene expression. Genome Res 16:656–668
7. Blankenberg D, Taylor J, Schenck I, He J, Zhang Y, Ghent M, Veeraraghavan N, Albert I, Miller W, Makova K et al (2007) A framework for collaborative analysis of ENCODE data: making large-scale analyses biologist-friendly. Genome Res in press
8. Boffelli D, McAuliffe J, Ovcharenko D, Lewis KD, Ovcharenko I, Pachter L, Rubin EM (2003) Phylogenetic shadowing of primate sequences to find functional regions of the human genome. Science 299:1391–1394
9. Bray N, Pachter L (2003) MAVID multiple alignment server. Nucleic Acids Res 31:3525–3526
10. Brudno M, Do CB, Cooper GM, Kim MF, Davydov E, Comparative Sequencing Program NISC, Green ED, Sidow A, Batzoglou S (2003) LAGAN and Multi-LAGAN: efficient tools for large-scale multiple alignment of genomic DNA. Genome Res 13:721–731
11. Burge C, Karlin S (1997) Prediction of complete gene structures in human genomic DNA. J Mol Biol 268:78–94
12. Bustamante CD, Fledel-Alon A, Williamson S, Nielsen R, Hubisz MT, Glanowski S, Tanenbaum DM, White TJ, Sninsky JJ, Hernandez RD et al (2005) Natural selection on protein-coding genes in the human genome. Nature 437:1153–1157
13. Carlson CS, Thomas DJ, Eberle MA, Swanson JE, Livingston RJ, Rieder MJ, Nickerson DA (2005) Genomic regions exhibiting positive selection identified from dense genotype data. Genome Res 15:1553–1565
14. Chimpanzee Sequencing and Analysis Consortium (2005) Initial sequence of the chimpanzee genome and comparison with the human genome. Nature 437:69–87
15. Clark AG, Glanowski S, Nielsen R, Thomas PD, Kejariwal A, Todd MA, Tanenbaum DM, Civello D, Lu F, Murphy B et al (2003) Inferring nonneutral evolution from human-chimp-mouse orthologous gene trios. Science 302:1960–1963
16. Cooper GM, Stone EA, Asimenos G, Green ED, Batzoglou S, Sidow A (2005) Distribution and intensity of constraint in mammalian genomic sequence. Genome Res 15:901–913
17. Curwen V, Eyras E, Andrews TD, Clarke L, Mongin E, Searle SM, Clamp M (2004) The Ensembl automatic gene annotation system. Genome Res 14:942–950
18. Dayhoff MO, Schwartz RM, Orcutt BC (1978) A model of evolutionary change in proteins. In: Dayhoff MO (ed) Atlas of Protein Sequence and Structure. National Biomedical Research Foundation, Washington, DC, pp 345–352
19. Dermitzakis E, Clark A (2002) Evolution of transcription factor binding sites in mammalian gene regulatory regions: Conservation and turnover. Mol Biol Evol 19:1114–1121
20. Dermitzakis ET, Reymond A, Antonarakis SE (2005) Conserved non-genic sequences -an unexpected feature of mammalian genomes. Nat Rev Genet 6:151–157
21. Elnitski L, Jin VX, Farnham PJ, Jones SJ (2006) Locating mammalian transcription factor binding sites: a survey of computational and experimental techniques. Genome Res 16:1455–1464
22. Enard W, Przeworski M, Fisher SE, Lai CS, Wiebe V, Kitano T, Monaco AP, Paabo S (2002) Molecular evolution of FOXP2, a gene involved in speech and language. Nature 418:869–872
23. Evans PD, Gilbert SL, Mekel-Bobrov N, Vallender EJ, Anderson JR, Vaez-Azizi LM, Tishkoff SA, Hudson RR, Lahn BT (2005) Microcephalin, a gene regulating brain size, continues to evolve adaptively in humans. Science 309:1717–1720
24. Fishel R, Lescoe MK, Rao MR, Copeland NG, Jenkins NA, Garber J, Kane M, Kolodner R (1993) The human mutator gene homolog MSH2 and its association with hereditary nonpolyposis colon cancer. Cell 75:1027–1038
25. Weinstock GRA, Metzker ML Muzny DM Sodergren EJ Scherer GM, Scott S, Steffen G, Worley KC Burch PE D et al (2004) Genome sequence of the Brown Norway rat yields insights into mammalian evolution. Nature 428:493–521
26. Hardison RC (2000) Conserved noncoding sequences are reliable guides to regulatory elements. Trends Genet 16:369–372
27. Hardison RC, Roskin KM, Yang S, Diekhans M, Kent WJ, Weber R, Elnitski L, Li J, O'Connor M, Kolbe D et al (2003) Covariation in frequencies of substitution, deletion, transposition and recombination during eutherian evolution. Genome Res 13:13–26
28. Henderson J, Salzberg S, Fasman KH (1997) Finding genes in DNA with a Hidden Markov Model. J Comput Biol 4:127–141
29. Henikoff S, Henikoff JG (1992) Amino acid substitution matrices from protein blocks. Proc Natl Acad Sci USA 89:10915–10919
30. Hillier LW, Miller W, Birney E, Warren W, Hardison RC, Ponting CP, Bork P, Burt DW, Groenen MA, Delany ME et al (2004) Sequence and comparative analysis of the chicken genome provide unique perspectives on vertebrate evolution. Nature 432:695–716
31. Hubbard TJ, Aken BL, Beal K, Ballester B, Caccamo M, Chen Y, Clarke L, Coates G, Cunningham F, Cutts T et al (2007) Ensembl 2007. Nucleic Acids Res 35:D610–D617
32. International Hapmap Consortium (2005) A haplotype map of the human genome. Nature 437:1299–1320
33. International Human Genome Sequencing Consortium (2004) Finishing the euchromatic sequence of the human genome. Nature 431:931–945

34. Kampa D, Cheng J, Kapranov P, Yamanaka M, Brubaker S, Cawley S, Drenkow J, Piccolboni A, Bekiranov S, Helt G et al (2004) Novel RNAs identified from an in-depth analysis of the transcriptome of human chromosomes 21 and 22. Genome Res 14:331–342
35. Karolchik D, Hinrichs AS, Furey TS, Roskin KM, Sugnet CW, Haussler D, Kent WJ (2004) The UCSC table browser data retrieval tool. Nucleic Acids Res 32:D493–D496
36. Kasprzyk A, Keefe D, Smedley D, London D, Spooner W, Melsopp C, Hammond M, Rocca-Serra P, Cox T, Birney E (2003) EnsMart – a generic system for fast and flexible access to biological data. Genome Res 14:160–169
37. Kelley JL, Madeoy J, Calhoun JC, Swanson W, Akey JM (2006) Genomic signatures of positive selection in humans and the limits of outlier approaches. Genome Res 16:980–989
38. Kent WJ, Sugnet CW, Furey TS, Roskin KM, Pringle TH, Zahler AM, Haussler D (2002) The human genome browser at UCSC. Genome Res 12:996–1006
39. Kent WJ, Baertsch R, Hinrichs A, Miller W, Haussler D (2003) Evolution's cauldron: duplication, deletion, and rearrangement in the mouse and human genomes. Proc Natl Acad Sci USA 100:11484–11489
40. Kimura M (1968) Evolutionary rate at the molecular level. Nature 217:624–626
41. Kimura M (1977) Preponderance of synonymous changes as evidence for the neutral theory of molecular evolution. Nature 267:275–276
42. King DC, Taylor J, Cheng Y, Martin J, ENCODE Transcriptional Regulation Group, ENCODE Multispecies Alignment Group, Chiaromonte F, Miller W, Hardison RC (2007) Finding cis-regulatory elements using comparative genomics: some lessons from ENCODE data. Genome Res 17:799–816
43. Kondrashov AS (2002) Direct estimates of human per nucleotide mutation rates at 20 loci causing mendelian diseases. Hum Mutat 21:12–27
44. Kreitman M (2000) Methods to detect selection in populations with applications to the human. Annu Rev Genomics Hum Genet 1:539–559
45. Kuhn RM, Karolchik D, Zweig AS, Trumbower H, Thomas DJ, Thakkapallayil A, Sugnet CW, Stanke M, Smith KE, Siepel A et al (2007) The UCSC genome browser database: update 2007. Nucleic Acids Res 35:D668–D673
46. Kumar S, Hedges SB (1998) A molecular timescale for vertebrate evolution. Nature 392:917–920
47. Leach FS, Nicolaides NC, Papadopoulos N, Liu B, Jen J, Parsons R, Peltomaki P, Sistonen P, Aaltonen LA, Nystrom-Lahti M et al (1993) Mutations of a mutS homolog in hereditary nonpolyposis colorectal cancer. Cell 75:1215–1225
48. Li WH, Gojobori T, Nei M (1981) Pseudogenes as a paradigm of neutral evolution. Nature 292:237–239
49. Wade Lindblad-Toh K, Mikkelsen CM, Karlsson TS, Jaffe EK, Kamal DB, Clamp M, Chang M, Kulbokas EJ 3 rd, Zody MC et al (2005) Genome sequence, comparative analysis and haplotype structure of the domestic dog. Nature 438:803–819
50. Ludwig MZ, Bergman C, Patel NH, Kreitman M (2000) Evidence for stabilizing selection in a eukaryotic enhancer element. Nature 403:564–567
51. Maclean JA 2nd, Chen MA, Wayne CM, Bruce SR, Rao M, Meistrich ML, Macleod C, Wilkinson MF (2005) Rhox: a new homeobox gene cluster. Cell 120:369–382
52. Margulies EH, Blanchette M, Comparative sequencing program NISC, Haussler D, Green ED (2003) Identification and characterization of multi-species conserved sequences. Genome Res 13:2507–2518
53. Margulies EH, Cooper GM, Asimenos G, Thomas DJ, Dewey CN, Siepel A, Birney E, Keefe D, Schwartz AS, Hou M et al (2007) Analyses of deep mammalian sequence alignments and constraint predictions for 1% of the human genome. Genome Res 17:760–774
54. Mayor C, Brudno M, Schwartz JR, Poliakov A, Rubin EM, Frazer KA, Pachter LS, Dubchak I (2000) VISTA: visualizing global DNA sequence alignments of arbitrary length. Bioinformatics 16:1046–1047
55. McDonald JH, Kreitman M (1991) Adaptive protein evolution at the Adh locus in Drosophila. Nature 354:114–116
56. Mekel-Bobrov N, Gilbert SL, Evans PD, Vallender EJ, Anderson JR, Hudson RR, Tishkoff SA, Lahn BT (2005) Ongoing adaptive evolution of ASPM, a brain size determinant in Homo sapiens. Science 309:1720–1722
57. Nachman MW, Crowell SL (2000) Estimate of the mutation rate per nucleotide in humans. Genetics 156:297–304
58. Needleman SB, Wunsch CD (1970) A general method applicable to the search for similarities in the amino acid sequence of two proteins. J Mol Biol 48:443–453
59. Neel JV (1962) Diabetes mellitus: a "thrifty" genotype rendered detrimental by "progress"? Am J Hum Genet 14:353–362
60. Nei M (1987) Molecular evolutionary genetics. Columbia University Press, New York
61. Nielsen R, Bustamante C, Clark AG, Glanowski S, Sackton TB, Hubisz MJ, Fledel-Alon A, Tanenbaum DM, Civello D, White TJ et al (2005) A scan for positively selected genes in the genomes of humans and chimpanzees. PLoS Biol 3:e170
62. Nobrega MA, Ovcharenko I, Afzal V, Rubin EM (2003) Scanning human gene deserts for long-range enhancers. Science 302:413
63. Pagnier J, Mears JG, Dunda-Belkhodja O, Schaefer-Rego KE, Beldjord C, Nagel RL, Labie D (1984) Evidence for the multicentric origin of the sickle cell hemoglobin gene in Africa. Proc Natl Acad Sci USA 81:1771–1773
64. Pennacchio LA, Rubin EM (2001) Genomic strategies to identify mammalian regulatory sequences. Nat Rev Genet 2:100–109
65. Pennacchio LA, Ahituv N, Moses AM, Prabhakar S, Nobrega MA, Shoukry M, Minovitsky S, Dubchak I, Holt A, Lewis KD et al (2006) In vivo enhancer analysis of human conserved non-coding sequences. Nature 444:499–502
66. Pollard KS, Salama SR, Lambert N, Lambot MA, Coppens S, Pedersen JS, Katzman S, King B, Onodera C, Siepel A et al (2006) An RNA gene expressed during cortical development evolved rapidly in humans. Nature 443:167–172
67. Ponjavic J, Ponting CP, Lunter G (2007) Functionality or transcriptional noise? Evidence for selection within long noncoding RNAs. Genome Res 17:556–565
68. Pruitt KD, Maglott DR (2001) RefSeq and LocusLink: NCBI gene-centered resources. Nucleic Acids Res 29:137–140
69. Ren B, Robert F, Wyrick JJ, Aparicio O, Jennings EG, Simon I, Zeitlinger J, Schreiber J, Hannett N, Kanin E et al (2000) Genome-wide location and function of DNA binding proteins. Science 290:2306–2309
70. Rhesus Macaque Genome Sequencing and Analysis Consortium (2007) Evolutionary and biomedical insights from the rhesus macaque genome. Science 316:222–234

71. Schwartz S, Elnitski L, Li M, Weirauch M, Riemer C, Smit A, NISC Comparative Sequencing Program, Green ED, Hardison RC, Miller W (2003) MultiPipMaker and supporting tools: alignments and analysis of multiple genomic DNA sequences. Nucleic Acids Res 31:3518–3524
72. Schwartz S, Kent WJ, Smit A, Zhang Z, Baertsch R, Hardison RC, Haussler D, Miller W (2003) Human-mouse alignments with Blastz. Genome Res 13:103–105
73. Siepel A, Haussler D (2004) Computational identification of evolutionarily conserved exons. In: Proceeding of the 8th annual international conference on research in computational biology (RECOMB '04), pp. 177–186
74. Siepel A, Bejerano G, Pedersen JS, Hinrichs AS, Hou M, Rosenbloom K, Clawson H, Spieth J, Hillier LW, Richards S et al (2005) Evolutionarily conserved elements in vertebrate, insect, worm, and yeast genomes. Genome Res 15: 1034–1050
75. Smith TF, Waterman MS (1981) Identification of common molecular subsequences. J Mol Biol 147:195–197
76. Spurr NK, Gough A, Goodfellow PJ, Goodfellow PN, Lee MG, Nurse P (1988) Evolutionary conservation of the human homologue of the yeast cell cycle control gene cdc2 and assignment of Cd2 to chromosome 10. Hum Genet 78:333–337
77. Tajima F (1989) Statistical method for testing the neutral mutation hypothesis by DNA polymorphism. Genetics 123:585–595
78. Taylor J, Tyekucheva S, King DC, Hardison RC, Miller W, Chiaromonte F (2006) ESPERR: Learning strong and weak signals in genomic sequence alignments to identify functional elements. Genome Res 16:1596–1604
79. The ENCODE Project Consortium (2007) The ENCODE pilot project: identification and analysis of functional elements in 1% of the human genome. Nature 447:799–816
80. Thompson JD, Higgins DG, Gibson TJ (1994) CLUSTAL W: improving the sensitivity of progressive multiple sequence alignment through sequence weighting, position-specific gap penalties and weight matrix choice. Nucleic Acids Res 22:4673–4680
81. Vavouri T, Walter K, Gilks WR, Lehner B, Elgar G (2007) Parallel evolution of conserved non-coding elements that target a common set of developmental regulatory genes from worms to humans. Genome Biol 8:R15
82. Voight BF, Kudaravalli S, Wen X, Pritchard JK (2006) A map of recent positive selection in the human genome. PLoS Biol 4:e72
83. Wang ET, Kodama G, Baldi P, Moyzis RK (2006) Global landscape of recent inferred Darwinian selection for Homo sapiens. Proc Natl Acad Sci USA 103:135–140
84. Wang H, Zhang Y, Cheng Y, Zhou Y, King DC, Taylor J, Chiaromonte F, Kasturi J, Petrykowska H, Gibb B et al (2006) Experimental validation of predicted mammalian erythroid cis-regulatory modules. Genome Res 16:1480–1492
85. Waterston RH, LindbladToh K, Birney E, Rogers J, Abril JF, Agarwal P, Agarwala R, Ainscough R, Alexandersson M, An P et al (2002) Initial sequencing and comparative analysis of the mouse genome. Nature 420:520–562
86. Wheeler DL, Barrett T, Benson DA, Bryant SH, Canese K, Chetvernin V, Church DM, DiCuccio M, Edgar R, Federhen S et al (2007) Database resources of the National Center for Biotechnology Information. Nucleic Acids Res 35:D5–D12
87. Wiehe T, Gebauer-Jung S, Mitchell-Olds T, Guigo R (2001) SGP-1: prediction and validation of homologous genes based on sequence alignments. Genome Res 11:1574–1583
88. Woolfe A, Goodson M, Goode DK, Snell P, McEwen GK, Vavouri T, Smith SF, North P, Callaway H, Kelly K et al (2005) Highly conserved non-coding sequences are associated with vertebrate development. PLoS Biol 3:e7
89. Wu JQ, Shteynberg D, Arumugam M, Gibbs RA, Brent MR (2004) Identification of rat genes by TWINSCAN gene prediction, RT-PCR, and direct sequencing. Genome Res 14:665–671

Genetics and Genomics of Human Population Structure

Sohini Ramachandran, Hua Tang, Ryan N. Gutenkunst, and Carlos D. Bustamante

Abstract Recent developments in sequencing technology have created a flood of new data on human genetic variation, and this data has yielded new insights into human population structure. Here we review what both early and more recent studies have taught us about human population structure and history. Early studies showed that most human genetic variation occurs within populations rather than between them, and that genetically related populations often cluster geographically. Recent studies based on much larger data sets have recapitulated these observations, but have also demonstrated that high-density genotyping allows individuals to be reliably assigned to their population of origin. In fact, for admixed individuals, even the ancestry of particular genomic regions can often be reliably inferred. Recent studies have also offered detailed information about the composition of specific populations from around the world, revealing how history has shaped their genetic makeup. We also briefly review quantitative models of human genetic history, including the role natural selection has played in shaping human genetic variation.

Contents

20.1 Introduction 590
    20.1.1 Evolutionary Forces Shaping Human Genetic Variation 590
20.2 Quantifying Population Structure 592
    20.2.1 F_{ST} and Genetic Distance 592
    20.2.2 Model-Based Clustering Algorithms 593
    20.2.3 Characterizing Locus-Specific Ancestry 594
20.3 Global Patterns of Human Population Structure 595
    20.3.1 The Apportionment of Human Diversity 595
    20.3.2 The History and Geography of Human Genes 596
    20.3.3 Genetic Structure of Human Populations ... 598
    20.3.4 A Haplotype Map of the Human Genome 600
20.4 The Genetic Structure of Human Populations Within Continents and Countries 602
    20.4.1 Genetic Differentiation in Eurasia 603
    20.4.2 Genetic Variation in Native American Populations 605
    20.4.3 The Genetic Structure of African Populations 606
20.5 Recent Genetic Admixture 606
    20.5.1 Populations of the Americas 606
    20.5.2 Admixture Around the World 608
20.6 Quantiative Modeling of Human Genomic Diversity 609
    20.6.1 Demographic History 609
    20.6.2 Quantitative Models of Selection 610
References 613

S. Ramachandran (✉)
Society of Fellows, Harvard University, 78 Mount Auburn Street, Cambridge, MA 02138, USA
e-mail: sramach@fas.harvard.edu

H. Tang
Department of Genetics, Stanford Medical School, Stanford, CA, USA
e-mail: huatang@stanford.edu

R.N. Gutenkunst
Theoretical Biology and Biophysics, and Center for Nonlinear Studies, Los Alamos National Laboratory, Los Alamos, NM, USA
e-mail: ryang@lanl.gov

C.D. Bustamante
Department of Biological Statistics and Computational Biology, Cornell University, Ithaca, NY, USA
e-mail: cdb28@cornell.edu

20.1 Introduction

Technological developments arising from the International Human Genome Sequencing and the International Haplotype Map (The International HapMap Consortium, 2003, 2005, 2007) [20–22] projects are transforming the study of human population genetics by dramatically reducing the cost of sequencing and genotyping. For example, as of early 2009, it costs about U.S. $500 per sample to genotype a million variable DNA sites (i.e., SNPs) and structural variants in the human genome and between $50,000 and $100,000 to sequence a human genome de novo. Recalling that the first human genome cost on the order of $1 billion dollars to sequence, this is a 10^4 gain in efficiency over less than a decade. Furthermore, by the time this book is published the costs we quote above may be reduced by another factor of two or three. In the next 5–10 years, therefore, we will likely see hundreds of thousands (if not millions) of human genomes sequenced, and the vast majority of variation within and among human populations cataloged and analyzed to answer fundamental questions in human and medical genomics.

The purpose of this chapter is to lay the groundwork for thinking about how we will begin to make use of this tremendous abundance of data. While these data will dwarf all that has come before, we will see that many of the questions we wish to answer are actually quite old – some as old as the field of human genetics, itself.

20.1.1 Evolutionary Forces Shaping Human Genetic Variation

Quantifying patterns of human genetic variation serves several important roles in genetics. First, it helps us understand human history and often gives us insights into time periods that have left no written record. For example, global patterns of human genetic variation suggest an African origin of modern humans approximately 150,000–200,000 years ago and are consistent with a "serial" founder model (see Sect. 20.5.1) for subsequent colonization and peopling of the world. Second, it helps us understand human *evolutionary* history. For example, patterns of human genetic variation allow us to delineate what genomic changes are unique to our species (i.e., shared by all humans to the exclusion of other apes), and which may be shared ancestrally (or recurrently) with other species. Likewise, patterns of human genetic variation can give us insight into regions of the human genome that may have experienced recent positive, negative, or balancing selection (see Nielsen et al. [39] for a recent review).

Understanding patterns of human genetic variation is also fundamental for the proper design of medical genomic studies, since population structure can often be a confounding variable in genome-wide association mapping. As the density of markers queried for association with disease increases and we begin to look at rare variants that may show limited geographic distributions, quantifying population structure at ever finer scales will be critical to the interpretation and analysis of experiments which aim to correlate patterns of genetic and phenotypic variation. In order to properly set the stage for our discussion, we will briefly review some key concepts from population genetics, anthropology, and genetics that may be unfamiliar to some readers.

The evolutionary dynamics of natural populations (be they human, plant, animal, or otherwise) are governed by a confluence of different evolutionary forces.

Chief among these is *mutation*, which is the ultimate source of variation. As this book illustrates, the process of mutation is a heterogeneous category of changes in DNA that come about through myriad pathways and ultimately induce changes ranging from single base pair alterations (i.e., single nucleotide polymorphisms or SNPs) to small insertion and deletions to large-scale structural rearrangments or even the addition or deletion of whole chromosomes. Most of the variation we will discuss in this chapter will be of the "small scale" variety, with a particular emphasis on understanding patterns of microsatellite, SNP, and haplotype variation.

We limit ourselves to these marker types largely due to practicality: assaying SNP and microsatellite variation has become standardized, and there are now a plethora of studies – such as those cited later on in this chapter – that have undertaken surveys using these markers across diverse human populations. Our hope is that, as the world of personalized genomics becomes a reality, large and micro-scale structural variation becomes cataloged and standardized in similar ways.

The second key force shaping patterns of human genetic variation is *genetic drift*. As you will recall from Chap. 16, genetic drift is a stochastic force that apportions variation by randomly subsampling variation from one generation to the next. Traditionally, we model genetic drift as simple binomial sampling of alleles. That is, if we consider a biallelic locus under no selection and represent the frequency of an allele A at time t in a given population of size $2N$ as x_t, the frequency in the next generation (x_{t+1}) is binomially distributed with probability of success x_t and sample size $2N$. (It turns out this binomial distribution can be generalized and there is a rich treatment of this subject in theoretical population genetics.) This random sampling from generation to generation induces what is known as a "random walk," such that the collection of allele frequencies from the start of the population history until the current time ($x_0, x_1,..., x_t$) as well as the distribution of long-term average frequencies across different sites can be modeled using a litany of theoretical tools.

For our purposes, we will focus on several qualitative impacts of this neutral evolutionary model. First, for a given population, the dynamics of genetic drift will be governed by the magnitude of $2N$, so that populations with a large number of individuals will "drift" more slowly or take smaller steps in frequency space from generation to generation than small populations. This model also suggests that if we were to follow lines-of-descent (i.e., the number of offspring left some time in the future by a given lineage today) with no difference in average offspring number among lineages, then the probability of a given lineage eventually overtaking the population is simply given by its current frequency. (For example, a lineage or allele at 20% frequency has a 20% chance of eventually getting fixed in the population, and an 80% chance of eventually getting lost.) Likewise, the model predicts that frequency is often a good proxy for age (at least for neutral alleles) so that a mutation at 25% frequency in the population is very likely to be older than a mutation at 5% frequency. For this reason, the distribution of SNP frequencies or the so-called allele-frequency spectrum contains a fair amount of information regarding the history of the population. Mathematically, we would define this quantity using an equation such as the following for a population with sample size of n_i individuals:

$$Y_i := \{\text{the number of SNPs where the sample frequency is } i/(2n_i)\}. \quad (20.1)$$

We will return to Y_i later in the chapter and discuss methods for inferring demographic history and selection from these frequencies. (Note: in the equation above we are assuming directionality as to which allele is the ancestral form and which is the derived. In practice, we infer this information from comparative genomic data, ideally, with correction for multiple mutations occurring at the site. See [17, 18] for a discussion of this problem).

The third force that will affect patterns of human genetic variation is *migration* or, more generally, *demographic history*. By this we mean that a given population (certainly for humans) is unlikely to reproduce as a fully endogamous unit. Rather, there is some probability every generation that new migrants from other populations may enter and contribute to the gene pool of the next generation. We also know that a given population is unlikely to remain the exact same size from generation to generation; it may increase or decrease in size, or go through boom/bust cycles. The number of demographic models one can construct is staggering, but certain general properties of models are described below.

For example, populations that have a closely shared evolutionary history – say they are exchanging migrants often or split from a common ancestral population a short time ago – will show a strong and positive correlation in allele frequencies both over time (i.e., the two populations' x_t values for a specific SNP will be correlated over time) as well as across the genome (i.e., the observed Y_i values will be correlated). We can also define a quantity such as the "joint allele frequency spectrum" to help us quantify this correlation and gauge the impact of different evolutionary forces on sets of populations. Mathematically, for a pair of populations i and j with sample size n_i and n_j this might take the form of a quantity Y_{ij} such that:

$$Y_{ij} = \{\text{the number of SNPs where the sample frequency is } i/(2n_i) \text{ in population } i \text{ and } j/(2n_j) \text{ in population } j\}. \quad (20.2)$$

As we will see throughout this chapter, the allele frequency spectrum both of a single population (i.e., Y_i) and for a pair (Y_{ij}) or more ($Y_{ijkl...}$) contains a fair amount of information regarding the evolutionary history of the populations in question. Many of the commonly used statistics in population genetics such as Wright's F-statistics, defined in Sect. 20.2.1, are

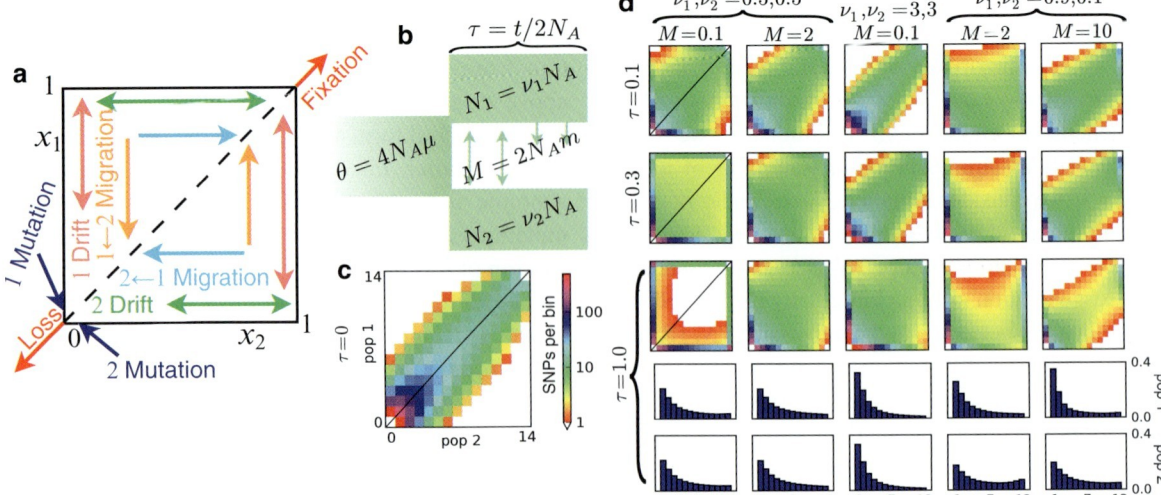

Fig. 20.1 Frequency spectrum gallery (adapted from Gutenkunst et al., manuscript submitted). (**a**) Impact of different evolutionary forces on shared patterns of genetic variation for a pair of populations, as defined by the density of alleles at relative frequencies x_1 and x_2 in populations 1 and 2. (**b**) Graphical description of an evolutionary model in which a pair of populations diverge and continue to exchange migrants. Specifically, an equilibrium population of effective size N_A diverges into two populations $2N_A \tau$ generations ago. Populations 1 and 2 have effective sizes $v_1 N_A$ and $v_2 N_A$, respectively. Migration is symmetric at $m = M/(2N_A)$ per generation, and the scaled mutation rate $\theta = 1,000$. (**c**) The allele frequency spectrum (AFS) at $\tau = 0$. Each entry is colored according to the logarithm of the number of SNPs with a given pairwise sample frequency, ranging from 0 to 14 copies of an allele in each population. (**d**) The AFS at various times for various demographic parameters, on the same scale as **c**. From the two-dimensional spectra, note that increased migration leads to more correlated SNP frequencies, and differences in population size lead to asymmetric genetic drift and thus an asymmetric AFS. For $\tau = 1$, the single-population spectra are also shown, where the scale is the fraction of polymorphisms observed at a given sample frequency. In these, note in particular that when populations experience growth, the spectrum is skewed toward rare alleles, particularly for the middle scenario.

summaries of these quantities. In Fig. 20.1 we show how different demographic forces acting on a population can impact their marginal (Y_i) and joint (Y_{ij}) site-frequency spectra.

The fourth force which contributes to the distribution of human genomic variation is natural selection. As was discussed in the chapter on population genetics (Chap. 16), selection works to decrease the frequency of deleterious alleles, increase the frequency of positively selected variants, and stabilize the frequency of variants subject to balancing selection. In human populations, it appears that selection is a much weaker force than genetic drift or demographic history in shaping global patterns of genomic variation. Nonetheless, there are some clear examples of positive and balancing selection on the human genome which have been recently reviewed (see [39]). Here, we will discuss selection briefly and mostly in light of selection against deleterious alleles, since this is the most prevalent form of selection operating on the human genome (see Sect. 20.6.2).

20.2 Quantifying Population Structure

In this section, we introduce several methods for quantifying and detecting population structure. We begin by introducing the classic F-statistics, which measure the degree of genetic differentiation among pre-defined and discrete subpopulations. We will then focus on model-based clustering methods, which aim to characterize latent and possibly nondiscrete population structure.

20.2.1 F_{ST} and Genetic Distance

Nonrandom mating in a population with substructure has two consequences: first, preferential mating between individuals from the same subpopulation is a form of *inbreeding*, and has the effect of reducing genetic diversity (measured as, say, heterozygosity) in the overall population; second, as the subpopulations experience independent genetic drift, allele frequencies at genetic markers tend to diverge. Originally introduced by Wright in 1921 [69] to quantify the inbreeding effect of population substructure, F_{ST} has become one of the most widely used measures of genetic differentiation between predefined subpopulations.

Consider the simple setting, in which a population consists of several subpopulations. F_{ST} is defined as the decrease in heterozygosity among subpopulations (H_S), relative to the heterozygosity in the total population (H_T):

$$F_{ST} = \frac{H_T - H_S}{H_T}, \quad (20.3)$$

where H_S is the *expected* heterozygosity, computed under the assumption that mating is random within each subpopulation (Hardy-Weinberg equilibrium), while H_T is analogously computed assuming random mating in the entire population without population structure.

Alternatively, F_{ST} is often loosely interpreted as the proportion of variance in allele frequencies at a locus that is explained by the subpopulation level of organization. For example, suppose the frequency of an allele is 0 and 1 in two subpopulations, respectively, then $F_{ST}=1$, meaning the variance in allele frequency is completely explained by the population division. Under this framework, F_{ST} at a biallelic single nucleotide polymorphism (SNP) marker can be computed based on the allele frequencies:

$$F_{ST} = \frac{\sigma_p^2}{\bar{p}(1-\bar{p})}, \quad (20.4)$$

where σ_p^2 is the variance of allele frequencies among subpopulations and $\bar{p}$ denotes the average allele frequency in the pooled population. It can be shown that (20.3) and (20.4) are mathematically equivalent for biallelic markers, but (20.4) is often computationally more convenient.

F_{ST} is often taken as a genetic distance measure, with higher values of F_{ST} reflecting a greater level of genetic divergence. However, both (20.3) and (20.4) define F_{ST} for a specific locus; F_{ST} can vary considerably from locus to locus. Moreover, a locus that is under population- or environment-specific selection can also exhibit unusually high F_{ST}. For example,

across globally-distributed human populations, functional polymorphisms in genes related to skin pigmentation show unusually high levels of F_{ST} (i.e., population differentiation) as compared to the genome-wide distribution (see [45]). To reduce the variance across the markers and the bias due to a small number of strongly selected loci, when F_{ST} is reported as an index for genetic distance among subpopulations, it is often calculated by averaging both the numerator and the denominator in (20.3) or (20.4) across loci.

When one is interested in quantifying the degree of substructure among predefined populations, F_{ST} is a simple and useful measure of genetic distance. However, it is often the case that we are interested in using the genetic data itself to define the populations. In particular, if we are interested in detecting cryptic or hidden population structure, then we need to resort to other approaches (see Sect. 20.4). One method for detecting latent population structure, principal component analysis (PCA), was introduced in Sect. 6.4.4. In the next section, we explain a complementary approach, which defines subpopulations based on statistical genetic models for the data.

20.2.2 Model-Based Clustering Algorithms

Cluster analysis refers to a large family of approaches, whose goal is to simultaneously define subsets (called clusters) and to assign observational units into these clusters, so that members in the same cluster are similar by some criteria. For a comprehensive survey of clustering approaches, readers are referred to Mardia et al. [32] or Hastie et al. [15].

In the context of inferring genetic structure, the data usually consist of individuals genotyped at multiple genetic markers (e.g., restriction fragment length polymorphisms RFLPs, microsatellites, or SNPs). In the discrete population model, all alleles in an individual are assumed to be drawn randomly from one of the subpopulations, according to a set of allele frequencies that are specific to each subpopulation. The goal of the analysis is to simultaneously estimate subpopulation allele frequencies and group membership (i.e., which individuals are drawn from which subpopulation). However, for many human populations, there is often no single group from which individuals derive their ancestry. That is, recent migration gives rise to genetically admixed individuals, whose genomes represent a mixture of alleles from multiple "ancestral" populations (see Sect. 20.5).

Mathematically, this means that an individual may have partial membership in more than one cluster. These clusters are biologically interpreted as ancestral populations for the admixed individuals. For example, African Americans in the United States are a recently admixed group, deriving ancestry from European and West African ancestral populations [65]. Under the admixture model, an African American individual's population membership is characterized by the *individual ancestry* (IA) proportion, which is a vector representing the probability that a randomly selected allele from this individual originates from a European (or alternatively, an African) ancestor.

Under either the discrete or the admixture model, individuals' memberships (or IA values) are jointly inferred with the allele frequencies in each subpopulation, using either maximum likelihood or Bayesian methods. We begin by explaining the maximum likelihood approach for the discrete subpopulation model, as this model illustrates the principles that underlie most of the model-based approaches [63]. Let $G_i^m = (a(i,m), b(i,m))$ denote the genotype of individual i at marker m, with $a(i,m)$ and $b(i,m)$ being the unordered pair of alleles. Let $Z_i \in (1,...,k)$ indicate the subpopulation membership for individual i, and $P = \{p_{m_l}^k\}$ be the frequency of allele l at marker m in population k. Under the assumption that genotypes among markers are independent conditioning on an individual's membership, and that all markers are in Hardy-Weinberg equilibrium within each subpopulation, the likelihood function, treating Z and P as parameters, is simply the product of the probability of observing each allele:

$$L(P, Z; G) \propto \prod_i \prod_m p_{m_{a(i,m)}}^{z_i} p_{m_{b(i,m)}}^{z_i}. \quad (20.5)$$

For the admixture model, one can substitute Z_i by $(Z_{i,m}^a, Z_{i,m}^b)$, the population origin of each allele, and model $Z_{i,m}^a$ and $Z_{i,m}^b$ as independent draws from the multinomial probability vectors of individual ancestry. The inference of population structure amounts to the inference on Z_i, or the genome-wide average of $(Z_{i,m}^a, Z_{i,m}^b)$.

In the maximum likelihood approach, the expectation maximization (EM) algorithm can be used to find the maximum likelihood estimates for the parameter values, (P, Z) [57, 63, 70, 74]. Alternatively, Bayesian approaches incorporate prior distributions into the likelihood, in order to evaluate the posterior distribution. The Bayesian methods offer a flexible framework for incorporating more complex population history models. For example, one of the widely used Bayesian programs, STRUCTURE, includes useful features such as modeling linkage among loci, and the ability to model correlated allele frequencies between evolutionarily related ancestral populations [14, 49].

20.2.3 Characterizing Locus-Specific Ancestry

For admixed populations, methods described in the preceding section can be used to infer individual ancestry, which represents the genome-wide average ancestry proportions in an individual. If admixture has occurred recently, the genome of an admixed individual resembles a mosaic of fairly long chromosomal blocks derived from one of the ancestral populations. With high-density genotype data, it is now feasible to delineate these ancestry blocks with relatively high accuracy. Figure 20.2 illustrates how ancestry blocks can be reconstructed. While numerous statistical methods have been proposed (e.g., [60, 62]), it is important to realize that the source of information underlying all methods is the different allele and haplotype frequencies among the ancestral populations. As such, the accuracy with which one can infer locus-specific ancestry depends on the genetic divergence between the ancestral populations. The distribution of the ancestry blocks also depends on the admixing history: ancient admixing events result in smaller ancestry fragments, while recent admixing events give rise to extended blocks. With any method, the ability to identify a switch in ancestral state deteriorates when the blocks are very small. Therefore, the accuracy of locus-specific ancestry inference depends on (at least) two aspects of the population history: the divergence between the ancestral populations, and the time of the admixing events. Simulation studies using HapMap data suggest that current high-density genotype data harbor sufficient information for accurate ancestry inference for African-Americans or Hispanics [62]. Locus-specific ancestry can provide information regarding the population history of an admixed

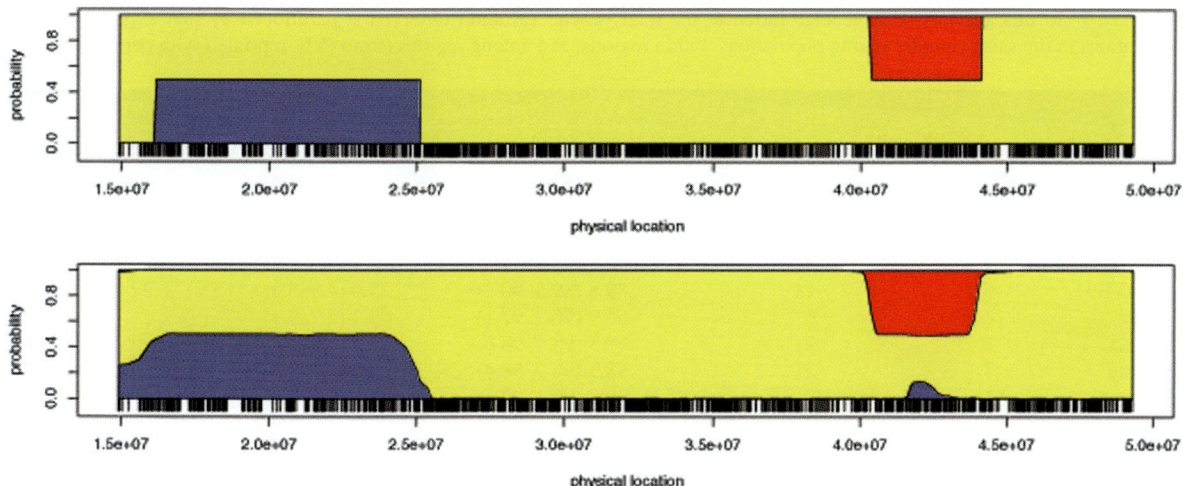

Fig. 20.2 Estimating ancestry along a chromosome. The top panel shows ancestry blocks along a simulated chromosome (*red*: African, *blue*: European; *yellow*: Asian). The bottom panel shows the reconstructed ancestry using high-density SNP markers, which are indicated by the black ticks at the bottom of each panel

population, as well as the finer-scale genetic structure within admixed groups. These are topics we will discuss in greater detail in Sect. 20.5.

20.3 Global Patterns of Human Population Structure

In this and the subsequent section, we begin a detailed exploration of empirical studies of human population genetic structure. First we explore major studies and datasets, now paradigms in the field of human population genetics, that compare human genetic variation at the level of multiple continents; the title of each subsection in this section is the title of a major paper or book in human population genetics. More recent studies of high-density genotyping data reveal patterns in genetic variation at fine geographic scales, as will be discussed after this section's historical perspective is presented.

20.3.1 The Apportionment of Human Diversity

Most studies of human population genetics begin by citing a seminal 1972 paper by Richard Lewontin bearing the title of this subsection [29]. Given the central role this work has played in our field, we will begin by discussing it briefly and return to its conclusions throughout the chapter. In this paper, Lewontin summarized patterns of variation across 17 polymorphic human loci (including classical blood groups such as ABO and M/N as well as enzymes which exhibit electrophoretic variation) genotyped in individuals across classically defined "races" (Caucasian, African, Mongoloid, South Asian Aborigines, Amerinds, Oceanians, Australian Aborigines [29]). A key conclusion of the paper is that 85.4% of the total genetic variation observed occurred within each group. That is, he reported that the vast majority of genetic differences are found within populations rather than between them. In this paper and his book *The Genetic Basis of Evolutionary Change* [30], Lewontin concluded that genetic variation, therefore, provided no basis for human racial classifications.

Lewontin's argument is an important one, and separates studying the geographic distribution of genetic variation in humans from searching for a biological basis to racial classification. His finding has been reproduced in study after study up through the present: two random individuals from any one group (which could be a continent or even a local population) are almost as different as any two random individuals from the entire world (see proportion of variation within populations in Table 20.1 and [20]).

An important point to realize is that Lewontin's calculation (and later work that confirms his finding) are based on the F-statistics introduced in Sect. 20.2.1 (see

Table 20.1 In this analysis of molecular variance, the total genetic variation observed is partitioned by that explained within populations in the same sample, among populations within regions, and among regions (from [53], reprinted with permission from AAAS)

Sample	Number of regions	Number of populations	Variance components and 95% confidence intervals (%)		
			Within populations	Among populations within regions	Among regions
World	1	52	94.6 (94.3, 94.8)	5.4 (5.2, 5.7)	
World	5	52	93.2 (92.9, 93.5)	2.5 (2.4, 2.6)	4.3 (4.0, 4.7)
World	7	52	94.1 (93.8, 94.3)	2.4 (2.3, 2.5)	3.6 (3.3, 3.9)
World-B97	5	14	89.8 (89.3, 90.2)	5.0 (4.8, 5.3)	5.2 (4.7, 5.7)
Africa	1	6	96.9 (96.7, 97.1)	3.1 (2.9, 3.3)	
Eurasia	1	21	98.5 (98.4, 98.6)	1.5 (1.4, 1.6)	
Eurasia	3	21	98.3 (98.2, 98.4)	1.2 (1.1, 1.3)	0.5 (0.4, 0.6)
Europe	1	8	99.3 (99.1, 99.4)	0.7 (0.6, 0.9)	
Middle East	1	4	98.7 (98.6, 98.8)	13 (1.2, 1.4)	
Central/South Asia	1	9	98.6 (98.5, 98.8)	1.4 (1.2, 1.5)	
East Asia	1	18	98.7 (98.6, 98.9)	1.3 (1.1, 1.4)	
Oceania	1	2	93.6 (92.8, 94.3)	6.4 (5.7, 7.2)	
America	1	5	88.4 (87.7, 89.0)	11.6 (11.0, 12.3)	

[67] for a discussion) averaged across single genetic loci. While it is an undeniable mathematical fact that the amount of genetic variation observed within groups is much larger than the differences among groups, this does not mean that genetic data do not contain discernable information regarding genetic ancestry. In fact, we will see that minute differences in allele frequencies across loci when compounded across the whole of the genome actually contain a great deal of information regarding ancestry. Given current technology, for example, it is feasible to accurately identify individuals from populations that differ by as little as 1% in F_{ST} if enough markers are genotyped. (See discussion below for a detailed treatment of the subject.) It is also important to note that when one looks at correlations in allelic variation across loci, self-identified populations and populations inferred for human subjects using genetic data correspond closely [12, 53].

20.3.2 The History and Geography of Human Genes

For more than 40 years, Luigi Luca Cavalli-Sforza and colleagues have worked to document and interpret patterns of human genetic variation. Along the way they have developed and perfected many of the statistical methods used to visualize and quantify patterns of variation and interpret their findings in light of human history and evolution. Their canonical book, *The History and Geography of Human Genes*, summarizes much of what they have learned about the pattern and process of human genetic variation across 1,800 indigenous populations.

Before delving into their findings, it is important to define two important concepts that permeate their work and those of the field a whole. The first is *treeness*, a concept introduced by Cavalli-Sforza and Piazza [5] to summarize population structure across multilocus data. Statistically, we can think of *treeness* as a way of summarizing "block structures" seen in matrices of pairwise genetic distances between populations. Specifically, block structures emerge when populations descended from a common ancestor are grouped together in these matrices, since closely related populations (say sister populations) will show similar levels of differentiation to a distant pair of closely related populations, the matrix will appear to show nearly duplicated rows and columns (or "blocks" of relatedness). By summarizing the blocks as arising from bifurcating trees, one can in theory build up a history of the population splitting events that gave rise to the sampled groups. It is important to emphasize that population trees are somewhat different from traditional phylogenetic (or species) trees since they are summarizing a reticulated history with often a great deal of gene flow among the terminal branches. The second concept that is important to discuss is the technique of principal component analysis (PCA). As we have already seen in Sect. 6.4.4, PCA is a general tool for exploratory data analysis that has found wide application in genetics. Cavalli-Sforza and colleagues were among the first to use PCA of population allele frequency matrices to identify major axes of variation in the data and interpret these axes in light of human history, as we will discuss below. One important distinction to emphasize is that much of the PCA work they carried out was done at the *population* level while much of the PCA that is carried out today is done at the individual level. (That is, PCA analysis of genotype value matrices where the entries are "0," "1," or "2" depending on how many copies of the "A" allele vs. the "a" allele, a given individual carries at a locus).

Using PCA Cavalli-Sforza and colleagues deeply explored human population genetics structure in *The History and Geography of Human Genes*. (A representative example of the PCA plots they generated is given in Fig. 20.3, which summarizes major axes of variation across the sampled populations they studied). A key emphasis of their work was on understanding how or whether language presented barriers to gene flow (i.e., quantifying how much of nonrandom mating in human populations is attributable to language) (see Fig. 20.4). The idea that languages and genes may evolve at similar rates and that a similarity in linguistic markers between two languages may likely reflect a recent shared genetic history among speakers of those languages remains controversial in the field of linguistics. However, Cavalli-Sforza et al. [8] underscored that human evolutionary genetics studies can rely on data and results from other fields – such as anthropology, archaeology, and linguistics – to synthesize inferences about human history.

The book by Cavalli-Sforza and colleagues is known for its numerous *synthetic maps* [34]. Synthetic

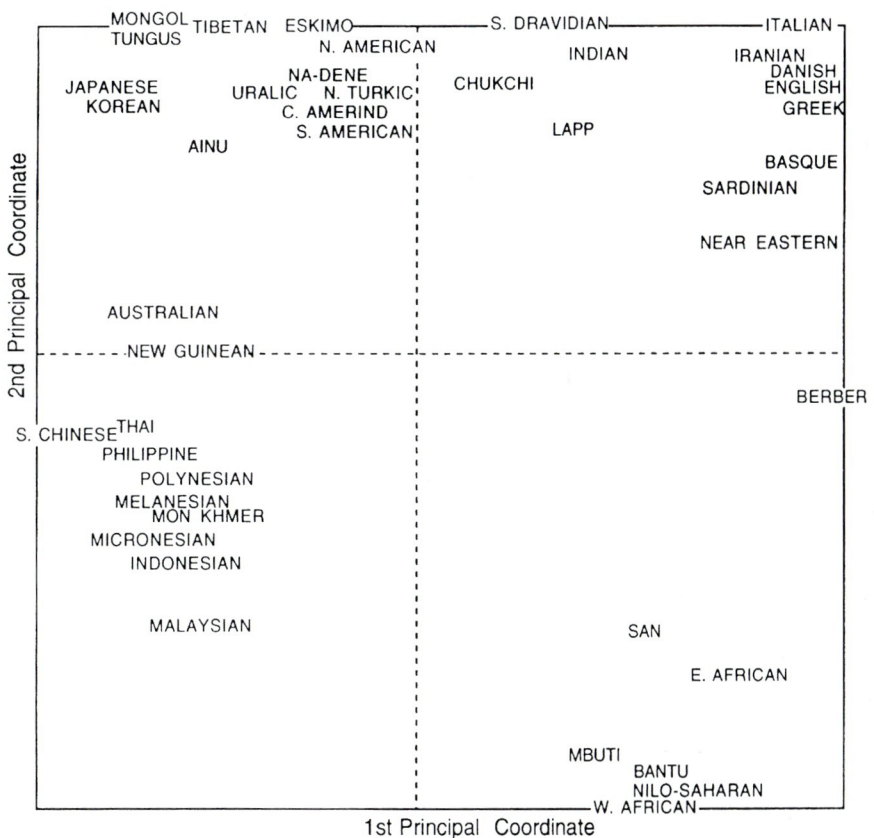

Fig. 20.3 Principal component map of 42 population studies by [6]. The first two PCs summarize 27% and 16% of the variation, respectively. Africans cluster in the lower right quadrant, with Europeans in the upper right, Southeast Asians in the lower left, Northeast Asians and Americans in the upper left. The first PC separates Africans and Europeans from the rest; the authors propose that the first PC does not separate Africans from non-Africans because there are only 6 African populations compared to 36 other populations. From [7]. From Cavalli-Sforza L., The History and Geography of Human Genes, copyright 1994 Princeton University Press. Reprinted by permission of Princeton University Press

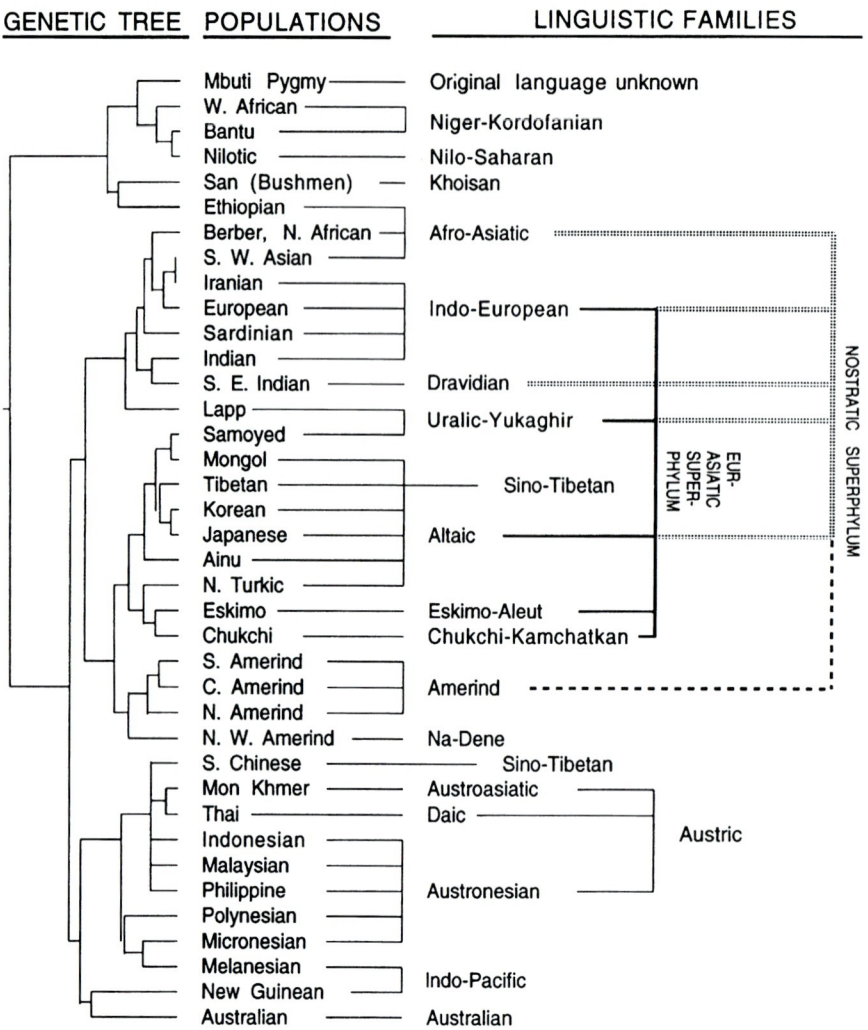

Fig. 20.4 Aligning the genetic tree with linguistic families and superfamilies, as in [8]. From [7]. From Cavalli-Sforza L., The History and Geography of Human Genes, copyright 1994 Princeton University Press. Reprinted by permission of Princeton University Press

maps overlap a single principle component onto a geographic map and are interpreted as revealing migration routes taken through the frequency and geographic spread of the allelic variant (see last three columns of Fig. 20.5). Many synthetic maps reveal north-south and east-west gradients in genetic variation, which might be interpreted as linking variation in a particular gene to climate or ecology. Recent work shows that PCA is expected to reveal axes of genetic variation that are orthogonal [41] and that multiple interpretations may be consistent with a given PCA representation of genetic variation.

20.3.3 Genetic Structure of Human Populations

An important and influential resource for studying human genetic variation has been the Human Genome Diversity Panel. Spearheaded by Howard Cann, Luca Cavalli-Sforza, and Jim Weber [4] (see Box 20.1), the HGDP is a collection of immortalized lymphoblastoid cell lines of over 1,000 individuals from 51 populations. By creating a renewable resource of DNA, the panel has afforded deep inferences on human evolutionary history, especially genomic signatures of

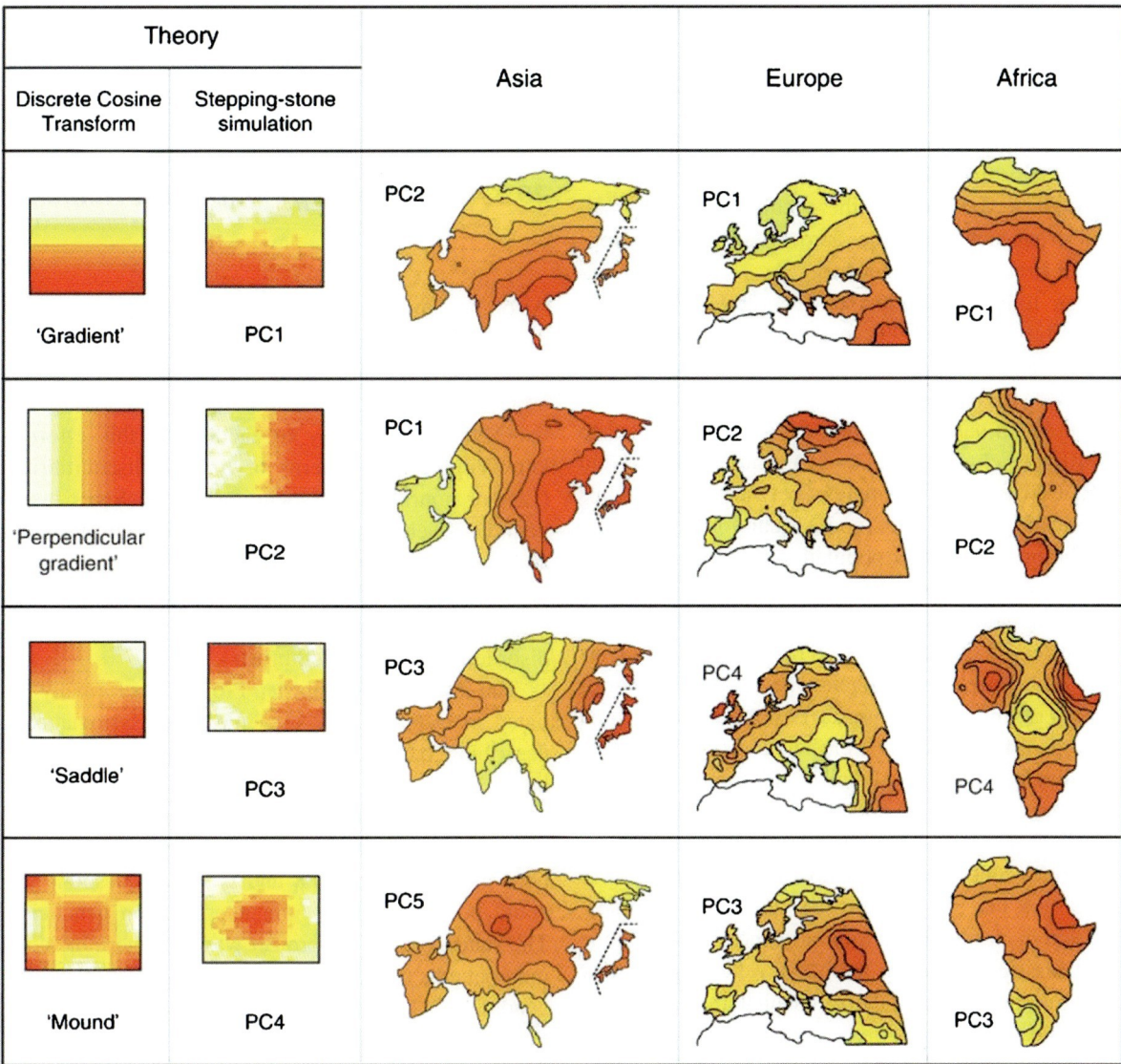

Fig. 20.5 Comparing synthetic maps from [34] with theoretical and empirical expectations. Menozzi et al. [34] performed principal component analyses on frequencies of 38 genes in various populations; the last three columns of this figure depict their original results. In the panel displaying PC1 in Europe, for example, the frequencies of certain alleles decrease (shown by more yellow colors) away from the Middle East; authors state that this pattern parallels the arrival of agriculture, which originated in the Middle East and then spread northward to Europe [6]. The Menozzi et al. [34] figures have been arranged to correspond with the shapes seen in the first two columns, which are based on theoretical and simulation results from [41]. Novembre and Stephens [41] simulated populations evenly-spaced in two-dimensional habitats with homogeneous migration rates aross time and space. Their PCA of these simulations found large-scale orthogonal gradients and "saddle" and "mound" patterns (see first column) when visualizing principal components even under this homogenous migration scheme. The second column displays PCA results from their simulations. The first column shows common structures seen in covariance matrices of population allele frequencies where genetic similarity decreases with geographic distance in a two-dimensional habitat, known as a stepping stone model. The regularity with which they observe these patterns runs counter to Menozzi et al.'s [34] claim that their PCA results are indicative of specific migration events.

historical relationships between populations, by providing a means to genotype (and ultimately sequence) genomes from diverse human populations. The first study of genetic variation in the HGDP scored polymorphism across 377 autosomal microsatellite loci in the panel [53] and recapitulated Lewontin's [29] result that the vast majority of human population genetic variation is found within local populations. However, the study also demonstrated individuals could be assigned to their continent of origin, and in some cases their population of origin using the model-based clustering algorithm STRUCTURE [49]. The authors reported "it was only in the accumulation of small allele-frequency differences across many loci that population structure was identified."

In a follow-up study (see Fig. 20.6), Rosenberg et al. [52] genotyped 993 total markers in the HGDP and demonstrated increased resolution of population structure as a result of increasing the amount of genetic data used. In particular, when the method is asked to identify two clusters (i.e., $K=2$), the authors found that STRUCTURE differentiates between indigenous American (purple cluster) and African (orange cluster) populations, with other populations having a gradient of membership in the African cluster that drops off with geographic distance from Africa. As the number of clusters K used in the STRUCTURE analysis increased, correlations in genotype data within continents of origin allowed Eurasia, East Asia, and Oceania to be identified as separate clusters as well. The structure that is identified is that of differences between continents, with a few notable exceptions. For example, the orange Africa cluster membership in the Mozabites reflects the gene flow this Middle Eastern population has had with Africa, due to the samples' location in North Africa. Similarly, membership in the blue Eurasian cluster in the Maya reflects gene flow with Europe during colonization that this American population experienced to a greater extent than other American populations in the HGDP.

The genotyping of 650,000 SNPs in these populations [31] allowed the detection of individual ancestry and population substructure with very high resolution within continents as well as across them (more examples of analyses with dense SNP maps will be discussed in Sect. 20.4). Li et al. [31] were further able to examine the distribution of ancestral alleles (nucleotides observed in chimpanzee) in HGDP populations by genotyping two chimpanzee samples at the same markers. The ancestral allele-frequency spectrum across loci can yield clues to the history of individual populations, because we expect populations with a small effective size and/or populations that have experienced a bottleneck to have more pronounced genetic drift, which can result in a relatively rapid increase in derived allele frequencies compared to populations with larger effective sizes or populations that have experienced expansion.

20.3.4 A Haplotype Map of the Human Genome

A comprehensive search for genetic causes of common diseases, such as type II diabetes or macular degeneration, requires examining genetic differences between a large number of affected individuals (i.e., cases) and matched controls. Key to facilitating this effort is knowledge about patterns of linkage disequilibrium (LD) or nonrandom association among SNPs in the genome. Such correlations between causal mutations and their haplotypes have long been used in human genetic research of disease (e.g., in studies of the HLA region, and in the identification of causes of Mendelian disorders such as cystic fibrosis).

The International HapMap Project was formally initiated in October 2002 as a means of systematically describing patterns of linkage disequilibrium in the human genome in order to catalyze medical genetic research into the heritable basis of common disease. It also represented the beginning of a paradigm shift in both the amount of data and types of questions that could be answered by human population geneticists. The stated goal of the project was to "determine common patterns of DNA sequence variation in the human genome" [21], with the goal of typing over one million SNPs in 270 individuals including 60 *trios* (samples of two parents and one of their biological children). The project has far surpassed that goal, as seen in Box 20.1.

The HapMap data provide insight into LD patterns across three populations. Chief among the concepts developed around the project is the notion of *tag SNPs* or representative SNPs in a region that can serve as "proxies" for other SNPs. That is, using tag SNPs means genetic variation can be efficiently queried for association with disease without genotyping every SNP in a given chromosomal region (therefore drastically reducing the cost of carrying out a genome-wide association study).

Tag selection methods exploit redundancy among SNPs; however, since the HapMap initially sampled only three populations, an issue for association studies is whether tag SNPs chosen from the HapMap dataset

20 Genetics and Genomics of Human Population Structure

Fig. 20.6 Inferred population structure based on 1,048 individuals and 993 markers. Each individual is represented by a thin line partitioned into *K* colored segments that represent the individuals estimated membership fractions in *K* clusters. Black lines separate populations, whose names are to the left of the figure, with continent listed on the right of the figure. The value of *K* indicates how many clusters STRUCTURE was assuming existed in the dataset for a particular set of runs for the method. From [52]

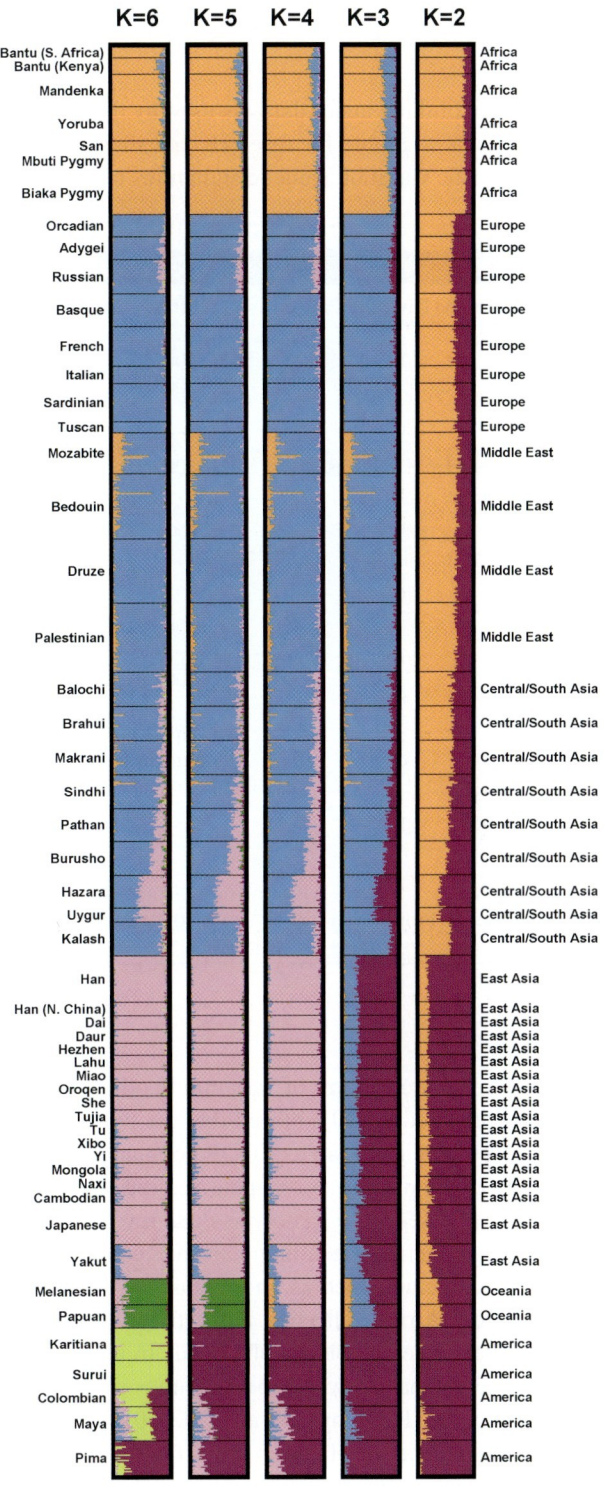

Box 20.1 Examples of publicly available human population genetic datasets

Description of three major datasets used by the human population genetics community. These are not the only large datasets available for research, but illustrate how much data are being generated to better understand human genetic variation, genetic signatures of human history, and the genetic underpinnings of disease.

Dataset Name	Initial reports of data	Amount of data generated
Human Genome Diversity Project	[4] (2002)	lymphoblastoid cell lines from 1,064 individuals in 51 populations
	[53] (2002)	377 autosomal microsatellites
	[52] (2005)	993 markers (microsatellites and insertion/deletion polymorphisms)
	[31] (2008)	650,000 SNPs in 938 of these individuals
International HapMap Project	[21] (2003 paper)	270 people across three populations (30 trios from Yoruba people of Ibadan, Nigeria; 45 unrelated individuals from Tokyo, Japan; 45 unrelated individuals from Beijing, China; 30 United States trios with northern and western European ancestry)
	[22] (2005 paper)	1 million SNPs in these individuals (1 SNP per 5 kilobases)
	[23] (2007 paper)	An additional 2.1 million SNPS (1 SNP per kilobase)
	HapMap Phase III, draft 2 reported online in January 2009	an additional 1.5 million SNPs and an increase to 1184 individuals (populations added: Chinese from Denver; Gujurati from Houston; Luhya from Webuye, Kenya; Mexican ancestry from Los Angeles; Maasai from Kinyawa, Kenya; Toscans from Italy; African ancestry from Southwest USA)
1,000 Genomes Project	www.1000genomes.org	Sequencing the genomes of approximately 2,000 people from around the world

adequately capture patterns of variation in other populations. Conrad et al. [10] showed that the portability of tag SNPs from HapMap to HGDP populations was quite good within large geographic regions such as continents. These dense genotype data reveal other important patterns resulting from continental population structure, such as an increase in LD with distance from Africa, reflecting that African lineages have smaller preserved blocks of LD due to increased time for recombination events to break up correlations (also seen in the HGDP by Conrad et al. [10]).

Data from the initial HapMap project do not enable much inference about evolutionary relationships between populations, so the genotyping of individuals from additional populations has become a priority in human population genetics. As the cost of SNP genotyping lowers, studies allow for across- and within-continental pictures of population structure to emerge. Dense genotype data from multiple populations allow the inference of both continental differentiation and the fine-scale study of within-region relationships among individuals. It is these finer-scale patterns that we will explore in the next section.

20.4 The Genetic Structure of Human Populations Within Continents and Countries

Large-scale human population genetic studies like the Human Genome Diversity Panel and International HapMap Project discussed in Sect. 20.3 initially had to choose between sampling densely geographically and sampling densely genomically. In just the last 2–3 years, improvements in genotyping technologies have allowed studies to report analyses of hundreds of thousands of SNPs genotyped in individuals from many populations. These datasets reveal the genetic signatures of historical events like migrations and conquests in more detail than geneticists could have hoped for when the field began. Here we explore how the history of Eurasia, the Americas, and Africa has shaped patterns of genetic variation of its inhabitants. (Note: the reason we have chosen to start with a discussion of Eurasia is simply that these are the populations that, to date, have been studied most intensively genetically. We believe the next few years will bring fine-scale studies of population structure across global human populations and strongly advocate these studies be undertaken, particularly in parts of the world currently understudied.)

20.4.1 Genetic Differentiation in Eurasia

Instead of grouping individuals into populations a priori (as early population genetic analyses necessitated), today we can let the data speak for themselves and tell us which individuals naturally cluster together based on genetic distance. A convenient means of accomplishing this is undertaking PCA on individual genotype scores (i.e., the "0," "1," "2" matrices mentioned above). Often when this is done, individuals from the same population tend to cluster together in PCA space. In fact, many PCA plots of globally distributed population structure seem to resemble geographical maps of the world with individuals from contiguous geographic regions clustering near each other in PCA space and revealing a close relationship between geographic distance and genetic differentiation (see Figs. 6.4 and 20.7).

Specifically, multicontinental studies of genomic diversity often find a clustering of populations according to their respective continents in the first few principal components, followed by differentiation between regions within continents. When sampling is dense, principal components can often serve as proxies for geographic axes [41, 42], separating Northern from Southern populations or Eastern from Western. For example, multiple studies observe North-to-South clines in European genetic differentiation, as seen using haplotype diversity in Fig. 20.8. Principal components also reveal evidence of genetic admixture between populations that can often be interpreted based on historical events such as colonization or slave trade; these signatures of admixture are discussed at length in Sect. 20.6.

As Fig. 6.4 shows, the European geographic map is an efficient summary of the first two principal components – or, put another way, dimensions – of European genetic variation. Novembre et al. [42] and Heath et al. [16] also showed that individual genotypes, despite low differentiation among populations in Europe as measured by F_{ST} (see Table 20.1), can be used to predict an individual's geographic origin within a few hundred kilometers (when that individual's geographic origin is representative of their ancestry). Likewise, several recent studies using high-density genotyping arrays have demonstrated the ability to reliably distinguish individuals of Ashkenazi Jewish ancestry from those without Ashkenazi Jewish ancestry in both European and European-American populations [36, 47, 64]. The ability to detect fine-scale geographic structure will only improve as whole-genome sequencing data become available; the studies discussed here are based on SNPs whose minor allele frequency is usually greater than 5%. Newer sequencing technologies will call lower frequency alleles more accurately, and low frequency alleles likely reflect recent mutations and may account for much differentiation between neighboring populations.

The larger mean heterozygosity and smaller mean linkage disequilibrium observed in Southern Europe compared to Northern Europen might be explained by an expansion in Europe from the South to the North [28]. South-to-North movement in Europe occurred during the first Paleolithic settlement of the continent by anatomically modern humans, and during the Neolothic expansion [2]. Thus we might expect to see a genetic signature of such movement, although another important controversy in historical anthropology is whether technologies such as agriculture traveled via demic diffusion (the movement of people and their genes) or cultural diffusion (the spread of technologies without a concomitant genetic signature) (see, for example, [46]).

Genetic variation in specific countries has been studied as well, such as that of Finland [24]. Studying a specific population may give insights into inbreeding or homozygosity patterns, as well as the genetic signatures of founder effects or multiple waves of migration

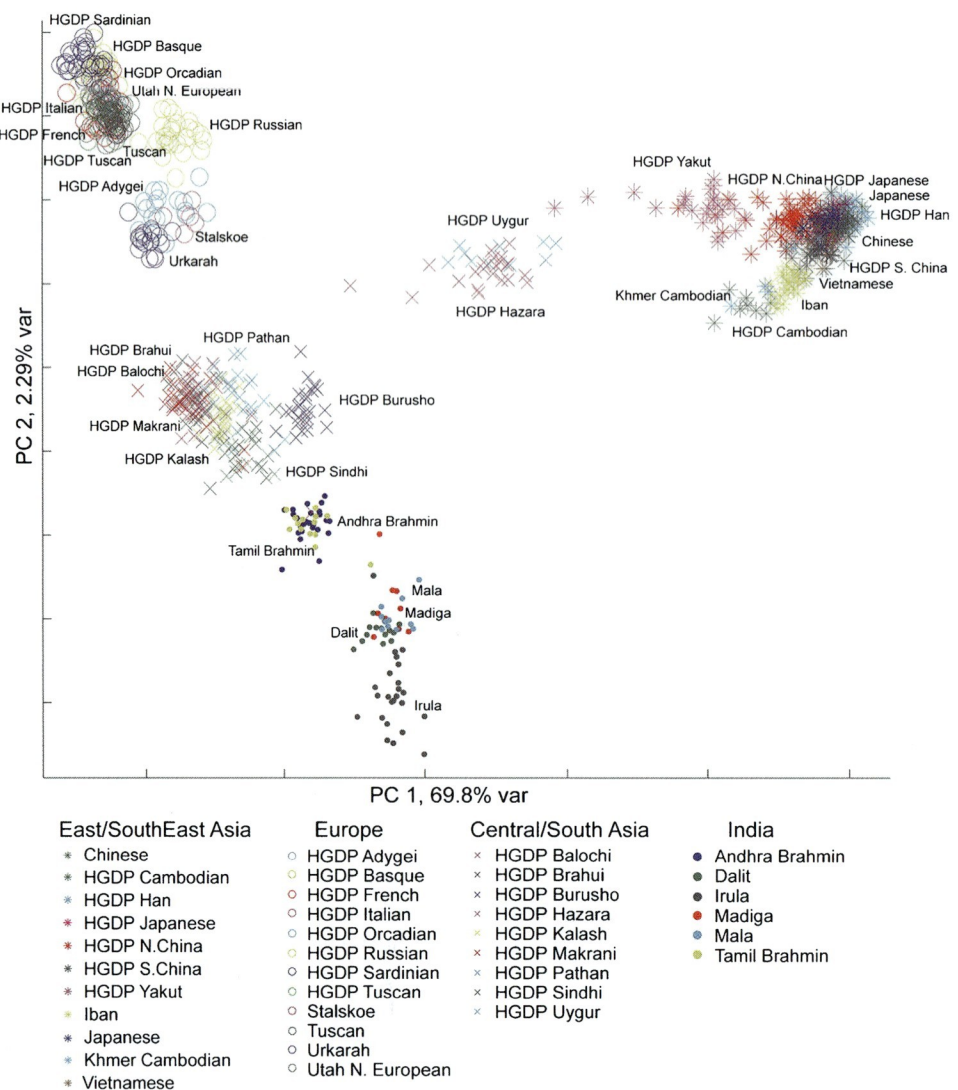

Fig. 20.7 PCA for 815 Eurasian individuals using nearly 50,000 SNPs. Individuals group closely with others in their self-identified population of origin, and the populations are differentiated in a way that mirrors a geographic map of Eurasia, much like the close relationship seen in Fig. 6.4 between genes and geography in Europe. HGDP denotes populations from the Human Genome Diversity Panel [4]. From [71]

in, for example, linkage disequilibrium patterns or admixture blocks. Jakkula et al. [24] found genetic signatures supporting multiple historical bottlenecks resulting from consecutive founder effects, in keeping with Finland's history of two major migration waves (a western one from 4,000 years ago, and a southern and western one from 2,000 years ago). A study of 7,003 Japanese individuals also shows that local regions in Honshu Island, the largest island of Japan, are genetically differentiated despite frequent migration within Japan during the last century [73].

Single-population studies are of great interest, as populations that experienced bottlenecks and subsequent low levels of immigrations (like the Finnish, Askenazi Jewish, or Icelandic peoples) may see a rise in Mendelian disease frequencies, and it has been proposed that gene mapping for complex traits may be easier in these populations than others. The Finns, for

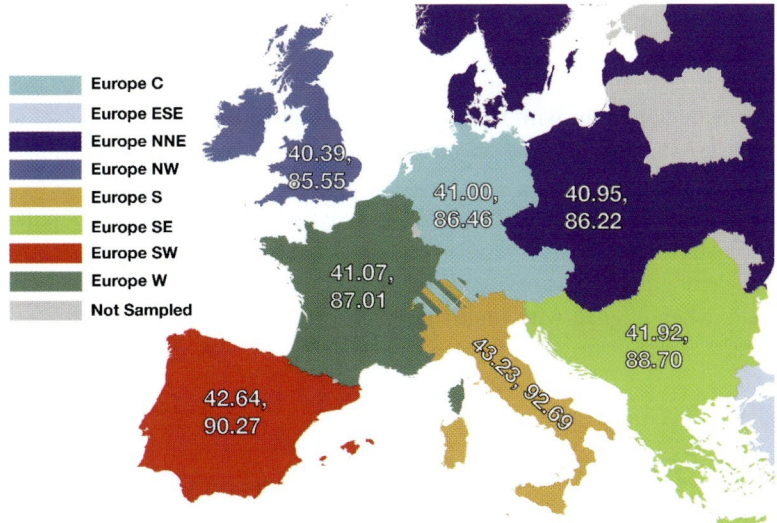

Fig. 20.8 Haplotype diversity within Europe. Two numbers are shown in each region; the first shows the mean number of distinct haplotypes in a region in a genomic window containing 10 SNPs, the second reflects haplotype diversity in a 25-SNP window. The authors find that haplotype diversity, as reflected by the numbers displayed, is higher in southern Europeans countries than in northern European countries, indicating that southern populations have larger effective sizes than northern ones and that the original peopling of Europe happened with migrations from the South to the North. From [1]

example, exhibit a substantial degree of homozygosity due to their population history, although not the amount of homozygosity seen in cultures with consanguineous marriages [24]. In such a population, the tagging of recessive variants in complex disorders may be done with common SNPs; indeed, the use of homozygous segments to identify rare alleles associated with Mendelian mutations has been successful in Finland (Meckel syndrome) [61]. Interestingly, Ashkenazi Jewish populations exhibit very similar patterns of linkage disequilibrium (and, in fact, *less* LD) than the CEPH European populations genotyped as part of HapMap, but approximately 20% higher levels of homozygosity [43]. Studies such as these address the power of genome-wide association datasets to demonstrate history-related stratification even within apparently homogeneous genetic populations, and shed light on the importance of rare variants in fine-scale genomic studies.

20.4.2 Genetic Variation in Native American Populations

During an expansion from a parent population via serial bottlenecks, sometimes called a "serial founder effect," linkage disequilibrium will increase and heterozygosity will decrease with distance from the origin of the expansion in the absence of selection [50]. Large linkage disequilibrium blocks were observed in the five Native American populations genotyped in the HGDP [10]. However, additional population samples from the Americas are important to help us understand the peopling of the Americas via the Bering Strait and what signature colonization, in this case by Europe, might have on genetic data. Wang et al. [66] studied 24 newly sampled populations of Native Americans from Canada, Meso- and South America.

The study found lower heterozygosity at microsatellite loci in indigenous Americans than in the non-American HGDP populations, and also observed a greater variance in heterozygosity among American populations. This could be a signature of a quick initial peopling of the Americas, followed by subsequent isolation of populations in the continent. A rapid coastal migration followed by a slower inland migration was supported by a higher level of genetic diversity in western South America compared to eastern South America. Wang et al. [66] also tested for correlations in differences between linguistic stocks or families and genetic distance between populations, finding that genetic distance and linguistic distance are more highly correlated within linguistic families than between families.

The study found support for an East Asian origin for Native American genetic variation, with relatively higher similarity to East Asian genetic variation in North Americans than South Americans, and also observed a private allele in the Native American samples. Wang et al. [66] showcase how a variety of hypotheses regarding demographic history can be tested with genetic data, when aligned with linguistic and archaeological data.

20.4.3 The Genetic Structure of African Populations

Africa and African populations play an important role in human evolutionary history given the African origin for anatomically modern humans and the amount of our genomic variation shaped by the out-of-Africa bottleneck [11]. However, it is important to recognize that African populations have been evolving since the human diaspora. Tishkoff et al. [65] sampled 121 African populations at over 1,000 microsatellite loci to study the demographic history of Africans as inferred from genetic data.

The investigators identified 14 ancestral clusters in Africa; these clusters approximately correspond to linguistic families, self-identified ethnicities, and/or cultural practices such as hunting-and-gathering. There was also a great deal of mixed ancestry in most populations, a signature of recent migrations in the African continent.

Three hunter-gatherer populations in the study were among the five most genetically diverse populations in the African sample, and African and Middle Eastern populations were found to share a number of alleles not observed elsewhere. Within Africa, the most private alleles were seen in click-speaking populations.

The spatial distribution of heterozygosity was used to pinpoint the origin of the modern human migration within the African continent in the same manner as by Ramachandran et al. [50]. Tishkoff et al.'s [65] analysis places this origin in southwestern Africa near the border of Namibia and Angola, which corresponds to the current San homeland. This lends support to the San being a genetically ancient population, although perhaps their current geographic origin does not reflect

their ancestors' geographic location 100,000 years ago. As geographic analyses become more refined, genetic variation appears to be more clinal than clustered [50]. This is because, at within-continental geographic distances, migration levels may be high and levels of admixture across populations will increase. We cannot study population genetics within continents without understanding recent genetic admixture, the subject of the next section.

20.5 Recent Genetic Admixture

The ultimate cause for population structure is nonrandom mating. For example, if individuals from geographically distant populations are less likely to mate than individuals from the same population, over time, discernible differences in allele frequencies will accumulate (as explained in Sect. 20.1). Compounding these differences across the genome provides power for reliably differentiating individuals from different populations even if, overall, the degree of population differentiation is low [31, 42]. On the other hand, migration facilitates gene flow. Recent global exploration and colonization have led to a rapid increase in gene flow among individuals from different continents. Their offspring are referred to as admixed and we can mathematically model chromosomes from admixed individuals as mosaics of segments derived from different ancestral populations. This section summarizes variation in several admixed populations, with the goal of illustrating the power genetic data have to shed light on a population's recent history.

20.5.1 Populations of the Americas

The population history and genetic structure among the Native American groups was surveyed in Sect. 20.4.2. The first significant wave of European influence arrived in the New World with Christopher Columbus' voyages of 1492–1504. During the sixteenth to nineteenth century, between 9.4 and 12 million Africans (mostly from West and Central Africa) were transported to the New World through the transatlantic slave trade. The arrival of Europeans and Africans in the New World gave rise to numerous admixed populations. In this section, we focus on two such groups: African Americans and Hispanics.

According to the 2007 U.S. Census, 41 million U.S. residents self-identify as having some degree of direct African ancestry (i.e., identify as "black" or African-American). Studies of genetic variation among African-Americans suggest that, on average, 80% of their genetic ancestry is West African, although individual ancestry proportions vary substantially as do the average ancestry proportions for different sampling localities within the United States [44]. Based on chromosomal block lengths, the admixing time between Europeans and Africans is estimated to have occurred 7–14 generations ago [14, 74]. At 25 years per generation, this places admixture as occurring between 175 and 350 years ago. Historical records indicate that the largest sources of the African slaves were the coastline in West and West Central Africa. However, locating the precise African ancestral populations for the African Americans has been challenging, in particular due to the lack of genetic data in geographically and ethnically diverse African populations. The recent study by Tishkoff et al. [65], discussed in Sect. 20.4.3, fills in this gap by genotyping over 2,000 Africans from 113 populations; in the near future, genetic data will likely be used to characterize admixture patterns within the African component of the African Americans.

Hispanics derive their ancestry from European, African, and Native American individuals. The term Hispanic describes populations that share a common language and cultural heritage, including Mexicans, Puerto Ricans, and Cubans, to name a few. However, these groups do not constitute a uniform ethnicity with a similar genetic background. At present, Hispanics represent the largest and fastest-growing minority population in the United States. Although genetic studies characterizing the population structure in the Hispanic population have been limited both in the marker density and in subgroup representation, evidence is mounting that individual ancestry proportions vary tremendously among subgroups that are identified as Hispanic. At a population level, Puerto Ricans have higher African ancestry compared to the Mexicans [56]; even within Mexico, the Native American ancestry proportions vary among States, ranging from 35% in Sonora in the North to 65% in Guerrero in center-Pacific [59].

The history of admixed populations is clearly discernible in patterns of genetic variation as summarized by PCA. Consider, for example, the GlaxoSmithKline POPRES sample [1, 37, 42] consisting of 3,875 individuals of varying ethnic backgrounds from over 80 countries genotyped on the Affymetrix 500 K platform. In Fig. 20.9, we reproduce key results from Nelson et al. [37] on PCA analysis of PopRes and the core HapMap populations (i.e., Yoruba from Ibadan, Nigeria, CEPH with ancestry from Northern Europe, Japanese from Tokyo, and Han Chinese from Beijing). This sample contains representation from several major continental

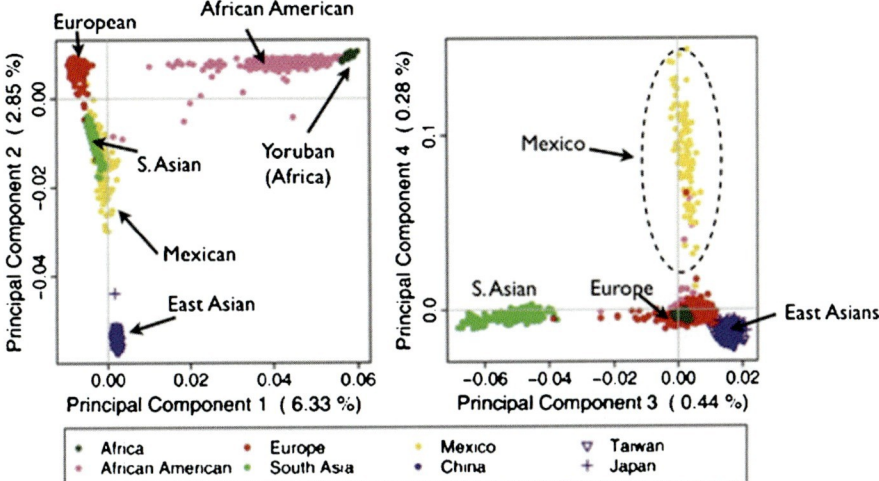

Fig. 20.9 Genetic structure in the PopRes data. Subject scores are colored by continental and/or ethnic origin (see legend). Percent of variation explained by each component is given in parentheses on each axis label. Reprinted from [37], with permission from Elsevier

populations as well as a large sample of African-Americans from the United States and Hispanics from Mexico. We note that the first principal component can be interpreted as an "African-American admixture" principal component (or, equivalently, Africa vs. Europe+Asia); principal component two corresponds to "East Asia vs. Europe"; principal component three corresponds to an "East Asia vs. South Asia" axis of variation; and principal component 4 (PC4) to a "Mexican admixture" axis. Importantly, individuals of admixed ancestry appear on the PCA map as in between the centroids of their putative ancestral populations.

An important feature of this analysis is that along the "Mexican admixture" PC, individuals show varying degrees of admixture between Europeans and a presumably "unsampled" population which likely corresponds to Native Americans. Analyzing just the European, East Asian, and Mexican samples (Fig. 20.10), we find that PC1 is an East Asia vs. Europe principal component and PC2 separates the East Asian sample from the (unsampled) Native American sample (so that the least admixed individuals are furthest away from the East Asian samples along PC2). This suggests that there is substantial genetic differentiation between East Asian and Native American populations so that the former is likely a poor proxy for the later (and vice versa).

20.5.2 Admixture Around the World

Genetic admixture is a worldwide phenomenon and is not limited to the North America. An example of an admixed population in the Eurasian continent is the Uyghur population living in the Xinjiang province in western China. Because of its proximity to the Silk Road – the historically important trade route connecting East Asia with the West, Central Asia, and the Mediterranean world – the Uyghur population derives ancestry from East Asian, European and the Middle East ancestral populations [31]. Using high-density SNP markers, a recent genetic study estimated approximately equal ancestral contributions from the European and East Asian populations to the Uyghur population [72]. In central Algeria in Northern Africa, the Mozabites originated from a Berber ethnic group in the Middle East that had close cultural contact with diverse African and European populations. Not surprisingly, SNP analysis of the Mozabite individuals in HGDP detects substantial European, African, and Middle Eastern ancestry (see Fig. 20.6).

Owing to the availability of high-throughput genotyping technologies, our ability to detect finer-scale population structure and more ancient admixture has dramatically increased during the past few years. The 600,000 SNP markers typed in the HGDP samples revealed genetic structure that was not detected previously using 300 microsatellites: most prominently, the detection of the Middle Eastern populations as a separate cluster [31, 54]. Moreover, the SNP data suggest that the impact of genetic mixture is more profound than has been previously implicated. Many Middle Eastern individuals appear admixed, perhaps because of the continuous migration in this area. Finally, in the analysis of the European populations, while the genetic structure matches geography, there is undeniable continuity between populations. Thus, it is more appropriate to consider the structure within the European continent as continuous clines rather than as a discrete cluster. In summary, the increasing amount of the genetic data will allow us to characterize population structure at a higher resolution within continents.

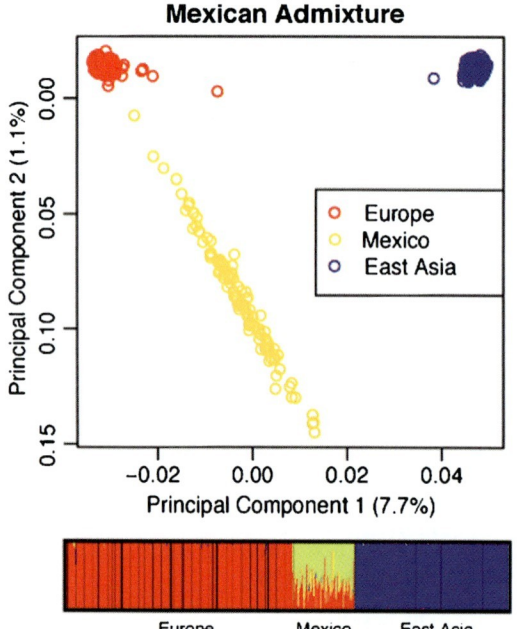

Fig. 20.10 Principal component and STRUCTURE analysis of the PopRes Mexicans from Guadalajara, Europeans from throughout Europe, and East Asians from Japan and China. From [1]

20.6 Quantiative Modeling of Human Genomic Diversity

The primarily qualitative studies described above have given great insight into both the global and local patterns of human genetic history. Quantitative models can offer additional insights; for example, we can use quantitative models to ask how severe particular bottlenecks were, or exactly when populations diverged.

Much quantitative modeling relies on resequencing data. SNP genotyping chips provide a genome-wide picture of variation at low cost, but they can be difficult or impossible to use for quantitative inference. Primarily, this is because the sites assayed on a chip are not a random sample of the genome; they are typically chosen because they are known to be polymorphic in some smaller "discovery" population. This ascertainment process biases the resulting data [9], particularly the allele frequency spectrum. Although this bias can, in some cases, be controlled for (e.g., [25]), it is unfeasible in general [38]. In Fig. 20.11, for example, we report the joint and marginal allele frequency spectra (AFS) for the PopRes and HapMap populations. We note that the joint allele frequency spectra reproduce qualitative patterns of the PCA analysis in Fig. 20.9, such as a stronger correlation in allele frequency and lower F_{ST} for closely related populations (e.g., East Asia from PopRes and JPT+CHB from HapMap or Europe from PopRes and CEPH from HapMap). Nonetheless, the marginal (or one dimensional) spectra are quite skewed toward intermediate frequency alleles as a result of the ascertainment bias in the Affymetrix 500K chip, which favored middle frequency variants for use in GWAS. The goal of this section is to describe how (unbiased) AFS data can be used for quantitative demographic inference of both demographic history and selection.

20.6.1 Demographic History

The demographic history of a set of populations encompasses the order and timing of any divergence or admixture events, as well as changes in population sizes and rates of gene flow over time. In principle, the greatest statistical power for inferring such a model from genetic data would arise from calculating the full likelihood of the data given the model [19]. However, at present such calculations are very difficult at the genomic scale. Thus many methods for inferring demographic events rely on modeling summaries of the data. The allele frequency spectrum is a particularly popular summary. As seen in Fig. 20.1, the frequency spectrum encodes substantial information about demographic history. (Although Myers et al. [35] have shown that it does not alone uniquely determine demographic history.) For example, the center column of part D of Fig. 20.1 shows that the one-dimensional (i.e., single population) allele frequency spectrum is skewed toward rare alleles in situations of population growth, while the right two columns show that asymmetric population sizes yield an asymmetric AFS.

An early study by Marth et al. [33] introduced an analytic method for calculating the allele frequency spectrum for a single population with piecewise constant population size. Using this method to fit models for several global populations revealed signatures of ancient population growth in African-Americans (presumably occurring in their African ancestors), and bottlenecks in the history of both European-American and East-Asian populations. These historical events have been well supported by subsequent genetic studies.

Considering the joint history of multiple populations substantially complicates the models, as divergence and gene flow must be incorporated. Consequently, the computational methods become more demanding. In a ground-breaking study, Schaffner et al. [58] used extensive coalescent simulations to replicate both summaries of the allele frequency spectrum and patterns of LD for West African, European, and East Asian populations, developing the first quantitative model for their joint genetic history. The computationally intensive nature of their analysis, however, precluded them from statistically assessing the confidence of their inferences or testing multiple models.

Recent theoretical and computational advances in the simulation of the frequency spectrum with diffusion theory have enabled more comprehensive statistical characterization of such models (Gutenkunst et al., in press). Figure 20.12 shows an illustrative model of human history, and the resulting expected frequency spectra. Within this model, parameters such as divergence times, migration rates, admixture proportions, and bottleneck sizes have been quantitatively inferred.

Models have also reached further back in time, before the emergence of modern humans. In particular, a recent analysis by Fagundes et al. [13] compared several models of early human history, including the possibility of interbreeding with other hominids. The analysis showed

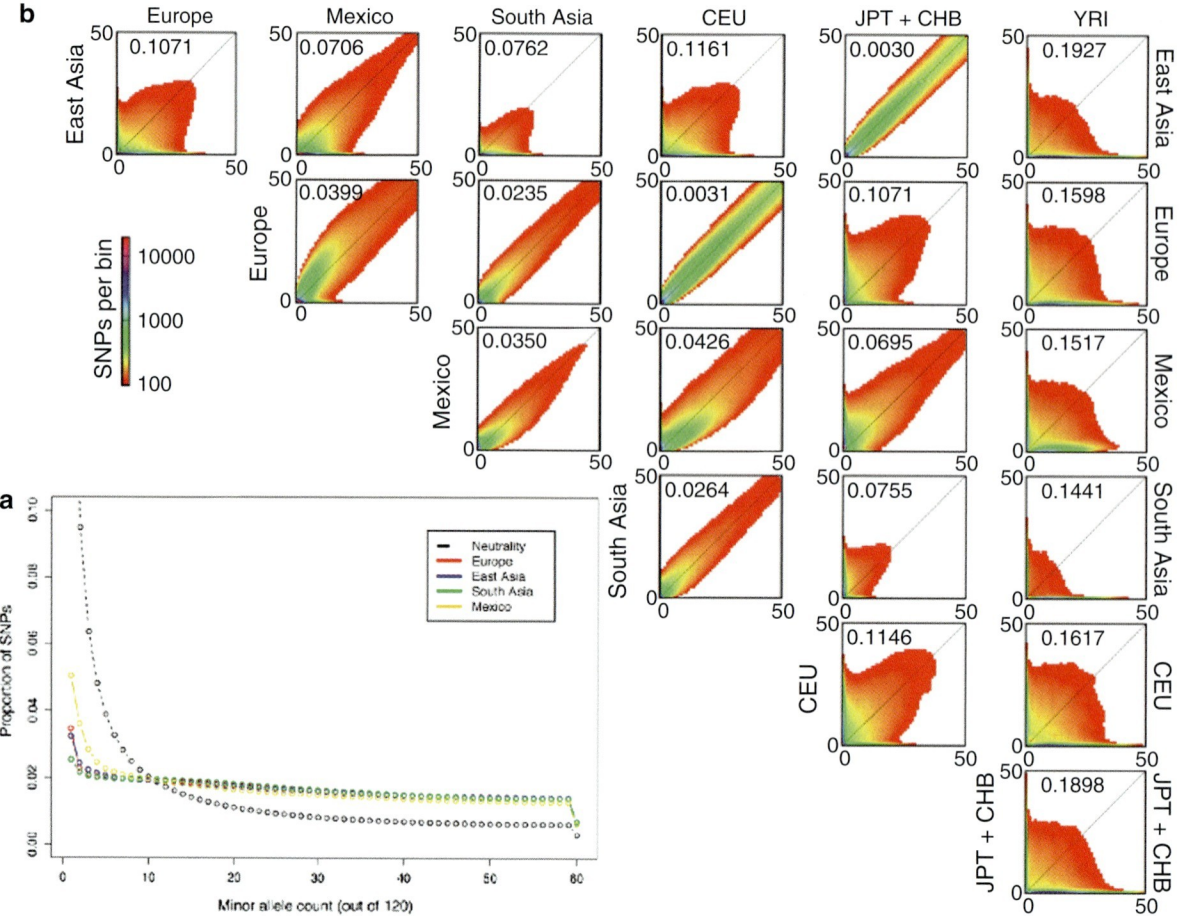

Fig. 20.11 Frequency spectra of the POPRES populations and HapMap samples (*CEU*: CEPH Utah residents with ancestry from northern and western Europe; *CHB* Han Chinese in Beijing, China; *JPT* Japanese in Tokyo, Japan; *YRI* Yoruba in Ibadan, Nigeria). (**a**) Minor Allele Frequency Spectra for the four sub-continental populations. The spectrum expected under neutrality is also shown in black. (**b**) Two-dimensional joint frequency spectra for each pairwise sub-continental population comparison. Colors represent the number of SNPs within each bin. Entries in the spectra containing less than 100 SNPs are shown in white. Autosomal estimates of F_{ST} for each comparison are shown in the upper left hand corner of each figure. From [1]

that a model in which modern humans simply replaced other hominids was best supported by the data.

20.6.2 Quantitative Models of Selection

Quantitative demographic models also play an important role in the search for evidence of selection acting on the genome. In particular, scans for selection seek genomic regions with unusual patterns of genetic variation, and demographic models define the null expectation of how unusual a region must be to be statistically significant when testing the hypothesis that it is under selection [40].

Beyond the search for unusual patterns of genetic variation, quantitative modeling has also given insight into the general signatures left by selection on the human genome. For example, an early analysis of the allele frequency spectrum for different classes of polymorphism revealed strong negative selection on mutations that change amino acid sequence. Furthermore, within those mutations computational algorithms such as Polyphen can predict which changes are most damaging [68].

More recently, the distribution of the selective effects of new mutations has been inferred from the allele frequency spectrum [3]. The selection coefficient s of a mutation is defined as the relative reproductive advantage conferred upon carriers of the mutation. As seen in

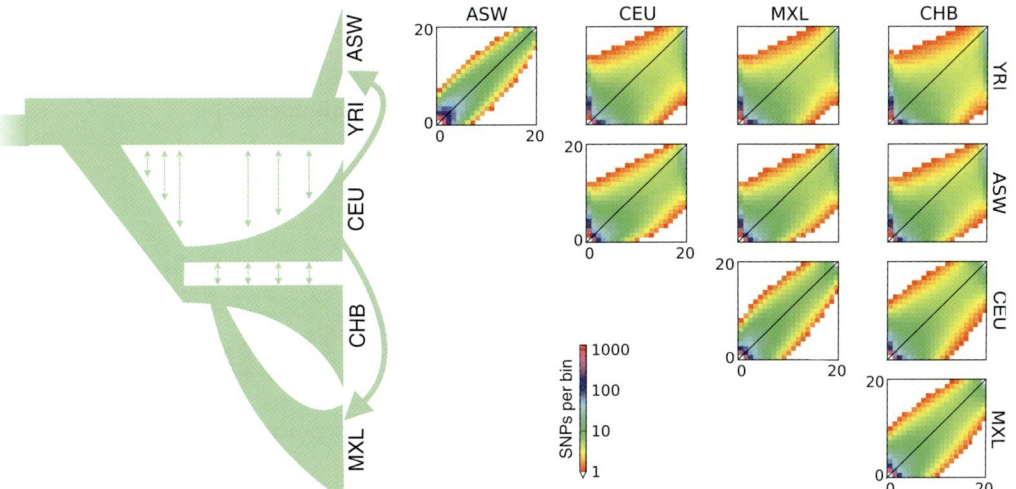

Fig. 20.12 Illustrative model of human expansion out of Africa and across the globe. The model includes African-American (*ASW*), West African (*YRI*), European (*CEU*), East Asian (*CHB*), and Mexican (*MXL*) populations. Using quantitative estimates for divergence times, population sizes, migration rates, and admixture proportions, the expected frequency spectrum under the model can be calculated using either diffusion or coalescent theory. Similar to Fig. 20.1.d., the resulting marginal spectra are shown for each pair of populations. Each spectrum shows distinct signatures of genetic history. For example, the recent European admixture into African-American and Mexican populations results in very highly correlated allele frequencies between populations pairs CEU-ASW (2nd row, 1st column) and CEU-MXL (3rd row, 1st column). Further, the Out-of-Africa bottleneck means that 2D spectra between African and non-African populations are asymmetric. When observed in real data, it is these sorts of signatures that guide quantitative modeling of human history

Fig. 20.13, the frequency spectra of synonymous and nonsynonymous variants differ dramatically. After correcting for demographic history using the synonymous mutations, it was found that the distribution of negative selection coefficients on newly arising amino-acid changing mutations possesses a very long tail. Roughly a third of amino acid substitutions are nearly-neutral ($|s|<0.01\%$), another third are moderately deleterious ($0.01\%<|s|<0.1\%$), and nearly all the remainder are highly deleterious or lethal ($|s|>1\%$). Knowledge of this distribution lets one calculate that very few of the fixed differences between human and chimp are selectively deleterious so that most are neutral or nearly, and that 10–20% of them result from positive selection. As the flood of data from the next generation of sequencing endeavors becomes available (e.g., the 1,000 Genomes Project and associated enterprises), we expect these preliminary estimates to be further refined, along with a quantitative understanding of human demographic history.

Acknowledgments We thank Dr. Sean Myles for helpful comments on an earlier draft of this chapter. Research is supported by GM073059 (to HT) and CDB was supported by NIH R01GM83606)

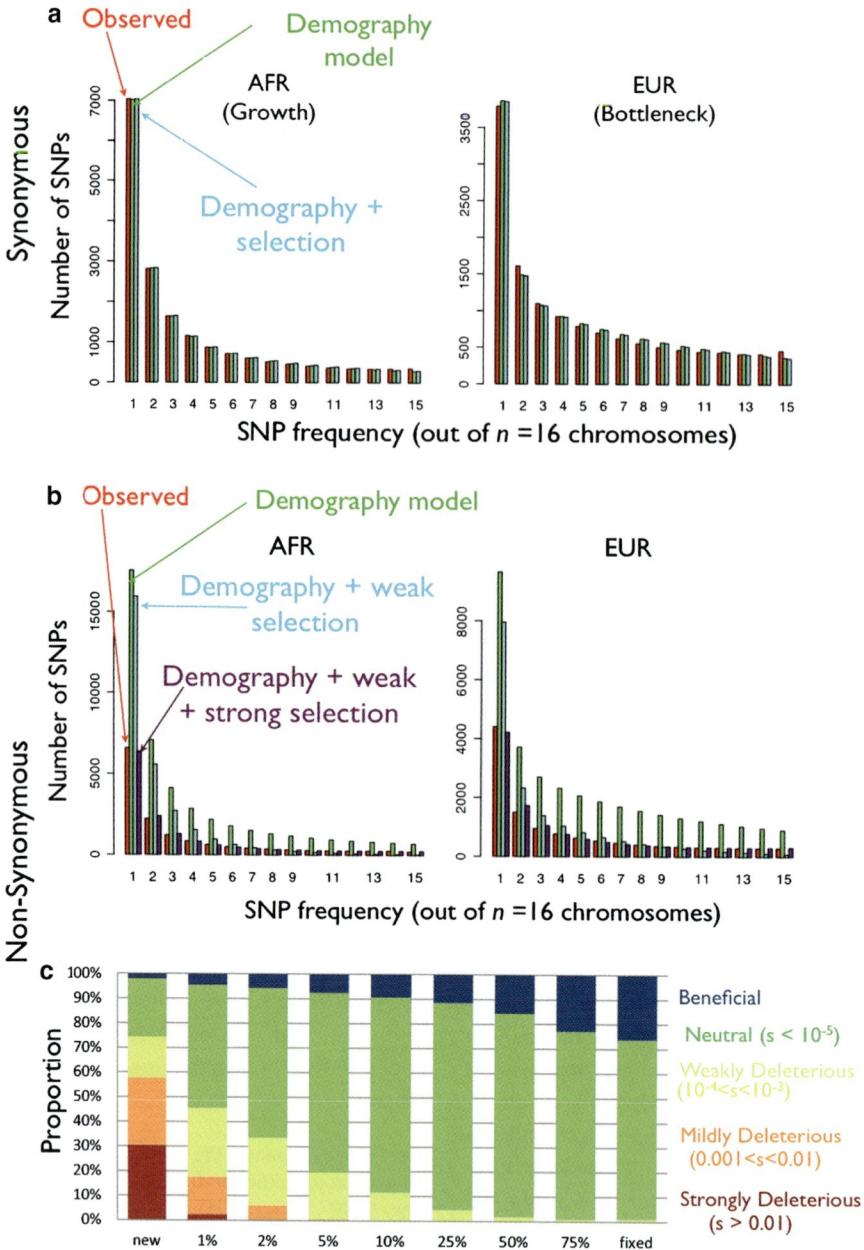

Fig. 20.13 Analysis of site-frequency spectra from over 30,000 coding SNPs found by resequencing of 11,000 genes in 20 European-Americans (*EUR*) and 15 African-Americans (*AFR*) yields estimates of demographic model and distribution of fitness effects of newly arising mutations and SNPs [3]. (**a**) Comparison of observed and predicted SFS for synonymous sites. Predictions are from two different types of models: a demographic model with growth in the AFR and a bottleneck in EUR (*green*) and for a model with weak negative selection on silent sites (*blue*). (**b**) Analogous comparison for nonsynonymous SNPs (nsSNPs) demonstrates that strong purifying selection, weak negative selection, and demographic history are all needed to accurately model the observed distribution of nsSNPs. (**c**) Estimated distribution of fitness effects for newly arising mutations in the human genome as well as SNPs at different population frequencies

References

1. Auton A, Bryc K, Boyko A, Lohmueller K, Novembre J, Reynolds A, Indap A, Wright M, Degenhardt J, Gutenkunst R, King K, Nelson M, Bustamante CD (2009) Global distribution of genomic diversity underscores rich complex history of continental human populations. Genome Res 19:795–803
2. Belle EM, Landry PA, Barbujani G (2006) Origins and evolution of the Europeans' genome: evidence from multiple microsatellite loci. Proc Biol Sci 273:1595–1602
3. Boyko AR, Williamson SH, Indap AR, Degenhardt JD, Hernandez RD, Lohmueller KE, Adams MD, Schmidt S, Sninsky JJ, Sunyaev SR, White TJ, Nielsen R, Clark AG, Bustamante CD (2008) Assessing the evolutionary impact of amino acid mutations in the human genome. PLoS Genet 4:e1000083
4. Cann HM et al (2002) A human genome diversity cell line panel. Science 296:261–262
5. Cavalli-Sforza LL, Piazza A (1975) Analysis of evolution: evolutionary rates, independence and treeness. Theor Popul Biol 8:127–165
6. Cavalli-Sforza LL, Menozzi P, Piazza A (1994) The history and geography of human genes. Princeton University Press, Princeton, NJ
7. Cavalli-Sforza LL, Menozzi P, Piazza A (1996) The history and geography of human genes. Princeton University Press, Princeton, NJ Abridged Paperback edition
8. Cavalli-Sforza LL, Menozzi P, Piazza A, Mountain J (1998) Reconstruction of human evolution; bringing together genetic, archaeological, and linguistic data. Proc Natl Acad Sci USA 85:6002–6006
9. Clark AG, Hubisz MJ, Bustamante CD, Williamson SH, Nielsen R (2005) Ascertainment bias in studies of human genome-wide polymorphism. Genome Res 15:1496–1502
10. Conrad DF, Jakobsson M, Coop G, Wen X, Wall JD, Rosenberg NA, Pritchard JK (2006) A worldwide survey of haplotype variation and linkage disequilibrium in the human genome. Nat Genet 38(11):1251–1260
11. Coop G, Pickrell JK, Novembre J, Kudaravalli S, Li J, Absher D, Myers RM, Cavalli-Sforza LL, Feldman MW, Pritchard JK (2009) The role of geography in human adaptation. PLoS Genetics 5: e1000500
12. Edwards AWF (2003) Human genetic diversity: Lewontin's fallacy. Bioessays 25:798–801
13. Fagundes NJ, Ray N, Beaumont M, Neuenschwander S, Salzano FM, Bonatto SL, Excoffier L (2007) Statistical evaluation of alternative models of human evolution. Proc Natl Acad Sci USA 104(45):17614–17619
14. Falush D, Stephens M, Pritchard JK (2003) Inference of population structure using multilocus genotype data: linked loci and correlated allele frequencies. Genetics 164(4):1567–1587
15. Hastie T, Tibshirani R, Friedman J (2009) The elements of statistical learning: data mining, inference, and prediction, 2nd edn. Springer, Berlin
16. Heath SC, Gut IG, Brennan P, McKay JD, Bencko V, Fabianova E, Foretova L, Georges M, Janout V, Kabesch M, Krokan HE, Elvestad MB, Lissowska J, Mates D, Rudnai P, Skorpen F, Schreiber S, Soria JM, Syvänen A-C, Meneton P, Herçberg S, Galan P, Szeszenia-Dabrowska N, Zaridze D, Génin E, Cardon LR, Lathrop M (2008) Investigation of the fine structure of European populations with applications to disease association studies. Eur J Hum Genet 16:1413–1429
17. Hernandez RD, Williamson SH, Zhu L, Bustamante CD (2007) Context dependent mutation rates may cause spurious signatures of a fixation bias favoring higher GC-content in humans. Mol Biol Evol 24(10):2196–2202
18. Hernandez RD, Williamson SH, Bustamante CD (2007) Context dependence, ancestral misidentification, and spurious signatures of selection. Mol Biol Evol 24(8): 1792–1800
19. Hey J, Nielsen R (2004) Multilocus methods for estimating population sizes, migration rates and divergence times, with applications to the divergence of Drosophila pseudoobscura and D. persimilis. Genetics 167:747–760
20. The Human Genome. Nature 2001;409:following p 812. (series of articles in Nature on the draft genome sequence)
21. The International HapMap Consortium (2003) The International HapMap project. Nature 426:789–796
22. The International HapMap Consortium (2005) A haplotype map of the human genome. Nature 437:1299–1320
23. The International HapMap Consortium (2007) A second generation human haplotype map of over 3.1 million SNPs. Nature 449:851–861
24. Jakkula E, Rehnström K, Varilo T, Pietiläinen OPH, Paunio T, Pedersen NL, deFaire U, Järvelin M-R, Saharinen J, Freimer N, Ripatti S, Purcell S, Collins A, Daly MJ, Palotie A, Peltonen L (2008) The genome-wide patterns of variation expose significant substructure in a founder population. Am J Hum Genet 83:787–794
25. Keinan A, Mullikin JC, Patterson N, Reich D (2007) Measurement of the human allele frequency spectrum demonstrates greater genetic drift in East Asians than in Europeans. Nat Genet 39:1251–1255
26. Kryukov GV, Shpunt A, Stamatoyannopoulos JA, Sunyaev SR (2009) Power of deep, all-exon resequencing for discovery of human trait genes. Proc Natl Acad Sci USA 106(10):3871–3876
27. Lander ES, Schork NJ (1994) Genetic dissection of complex traits. Science 265:2037–2048
28. Lao O, Lu TT, Nothnagel M, Junge O, Freitag-Wolf S, Caliebe A, Balascakova M, Bertranpetit J, Bindoff LA, Comas D, Holmlund G, Kouvatsi A, Macek M, Mollet I, Parson W, Palo J, Ploski R, Sajantila A, Tagliabracci A, Gether U, Werge T, Rivadeneira F, Hofman A, Uitterlinden AG, Gieger C, Wichmann H-E, Rüther A, Schreiber S, Becker C, Nürnberg P, Nelson MR, Krawczak M, Kayser M (2008) Correlation between genetic and geographic structure in Europe. Curr Biol 18:1241–1248
29. Lewontin RC (1972) The apportionment of human diversity. In: Dobzhansky T, Hecht MK, Steere WC (eds) Evolutionary biology 6. Appleton-Century-Crofts, New York, pp 381–398
30. Lewontin RC (1974) The genetic basis of evolutionary change. Columbia University Press, New York
31. Li JZ, Absher DM, Tang H, Southwick AM, Casto AM, Ramachandran S, Cann HM, Barsh GS, Feldman M, Cavalli-Sforza LL, Myers RM (2008) Worldwide human relationships inferred from genome-wide patterns of variation. Science 391:1100–1104

32. Mardia KV, Kent JT, Bibby JM (1980) Multivariate analysis. Academic, London
33. Marth GT, Czabarka E, Murvai J, Sherry ST (2004) The allele frequency spectrum in genome-wide human variation data reveals signatures of differential demographic history in three large world populations. Genetics 166:351–372
34. Menozzi P, Piazza A, Cavalli-Sforza LL (1978) Synthetic maps of human gene frequencies in Europe. Science 201:786–792
35. Myers S, Fefferman C, Patterson N (2008) Can one learn history from the allelic spectrum? Theor Popul Biol 73:342–348
36. Need AC, Kasperaviciute D, Cirulli ET, Goldstein DB (2009) A genome-wide genetic signature of Jewish ancestry perfectly separates individuals with and without full Jewish ancestry in a large random sample of European Americans. Genome Biol 10(1):R7
37. Nelson MR, Bryc K, King KS, Indap A, Boyko AR, Novembre J, Briley LP, Maruyama Y, Waterworth DM, Waeber G, Vollenweider P, Oksenberg JR, Hauser SL, Stirnadel HA, Kooner JS, Chambers JC, Jones B, Mooser V, Bustamante CD, Roses AD, Burns DK, Ehm MG, Lai Eric H (2008) The population reference sample (POPRES): a resource for population, disease, and pharmacological genetics research. Am J Hum Genet 83(3): 347–358
38. Nielsen R, Hubisz MJ, Clark AG (2004) Reconstituting the frequency spectrum of ascertained single-nucleotide polymorphism data. Genetics 168:2373–2382
39. Nielsen R, Hellmann I, Hubisz M, Bustamante MCD, Clark AG (2007) Recent and ongoing selection in the human genome. Nat Rev Genet 8(11):857–868
40. Nielsen R, Hubisz MJ, Hellmann I, Torgerson D, Andrés AM, Albrechtsen A, Gutenkunst R, Adams MD, Cargill M, Hu X, Boyko A, Indap A, Bustamante CD, Clark AG (2009) Darwinian and demographic forces affecting human protein coding genes. Genome Res 19:838–849
41. Novembre J, Stephens M (2008) Interpreting principal component analyses of spatial population genetic variation. Nat Genet 40:646–649
42. Novembre J, Johnson T, Bryc K, Kutalik Z, Boyko AR, Auton A, Indap A, King KA, Bergmann S, Nelson MR, Stephens M, Bustamante CD (2008) Genes mirror geography within Europe. Nature 456:98–101
43. Olshen AB, Gold B, Lohmueller KE, Struewing JP, Satagopan J, Stefanov SA, Eskin E, Kirchhoff T, Lautenberger JA, Klein RJ, Friedman E, Norton L, Ellis NA, Viale A, Lee CS, Borgen PI, Clark AG, Offit K, Boyd J (2008) Analysis of genetic variation in Ashkenazi Jews by high density SNP genotyping. BMC Genet 9:14
44. Parra EJ, Marcini A, Akey J, Martinson J, Batzer MA, Cooper R, Forrester T, Allison DB, Deka R, Ferrell RE, Shriver MD (1998) Estimating African American admixture proportions by use of population-specific alleles. Am J Hum Genet 63(6):1839–1851
45. Pickrell JK, Coop G, Novembre J, Kudaravalli S, Li JZ, Absher D, Srinivasan BS, Barsh GS, Myers RM, Feldman MW, Pritchard JK (2009) Signals of recent positive selection in a worldwide sample of human populations. Genome Res 19(5):826–837
46. Pinhasi R, Fort J, Ammerman AJ (2005) Tracing the origin and spread of agriculture in Europe. PLoS Biol 3:e410
47. Price AL, Butler J, Patterson N, Capelli C, Pascali VL, Scarnicci F, Ruiz-Linares A, Groop L, Saetta AA, Korkolopoulou P, Seligsohn U, Waliszewska A, Schirmer C, Ardlie K, Ramos A, Nemesh J, Arbeitman L, Goldstein DB, Reich D, Hirschhorn JN (2008) Discerning the ancestry of European Americans in genetic association studies. PLoS Genet 4(1):e236
48. Pritchard JK, Rosenberg NA (1998) Use of unlinked genetic markers to detect population stratification in association studies. Am J Hum Genet 65:220–228
49. Pritchard JK, Stephens M, Donnelly P (2000) Inference of population structure using multilocus genotype data. Genetics 155:945–959
50. Ramachandran S, Deshpande O, Roseman CC, Rosenberg NA, Feldman MW, Cavalli-Sforza LL (2005) Support from the relationship of genetic and geographic distance in human populations for a serial founder effect originating in Africa. Proc Natl Acad Sci USA 102:15942–15947
51. Ramachandran S, Rosenberg NA, Feldman MW, Wakeley J (2008) Population differentiation and migration: coalescence times in a two-sex island model for autosomal and X-linked loci. Theor Popul Biol 74:291–301
52. Rosenberg NA, Mahajan S, Ramachandran S, Zhao C, Pritchard JK, Feldman MW (2005) Clines, clusters, and the effect of study design on the inference of human population structure. PLoS Genet 1:e70
53. Rosenberg NA, Pritchard JK, Weber JL, Cann HM, Kidd KK, Zhivotovsky LA, Feldman MW (2002) Genetic structure of human populations. Science 298:2381–2385
54. Jakobsson M, Scholz SW, Scheet P, Gibbs JR, VanLiere JM, Fung H-C, Szpiech AZ, Degnan JH, Wang K, Guerreiro R, Bras JM, Scymick JC, Hernandez DG, Traynor BJ, Simon-Sanchez J, Matarin M, Britton A, van de Leemput J, Rafferty I, Bucan M, Cann HM, Hardy JA, Rosenberg NA, Singleton AB (2008) Genotype, haplotype and copy-number variation in worldwide human populations. Nature 451:998–1003
55. Sabeti PC, Reich DE, Higgins JM, Levine HZP, Richter DJ, Schaffner SF, Gabriel SB, Platko JV, Patterson NJ, McDonald GJ, Ackerman HC, Campbell SJ, Altshuler D, Cooper R, Kwiatkowski D, Ward R, Lander ES (2002) Detecting recent positive selection in the human genome from haplotype structure. Nature 419:832–837
56. Salari K, Choudhry S, Tang H, Naqvi M, Lind D, Avila PC, Coyle NE, Ung N, Nazario S, Casal J, Torres-Palacios A, Clark S, Phong A, Gomez I, Matallana H, Pérez-Stable EJ, Shriver MD, Kwok PY, Sheppard D, Rodriguez-Cintron W, Risch NJ, Burchard EG, Ziv E (2005) Genetic admixture and asthma-related phenotypes in Mexican American and Puerto Rican asthmatics. Genet Epidemiol 29(1):76–86
57. Satten GA, Flanders WD, Yang Q (2001) Accounting for unmeasured population substructure in case-control studies of genetic association using a novel latent-class model. Am J Hum Genet 68(2):466–477
58. Schaffner SF (2004) The X chromosome in population genetics. Nat Rev Genet 5:43–51
59. Silva-Zolezzi I, Hidalgo-Miranda A, Estrada-Gil J, Fernandez-Lopez JC, Uribe-Figueroa L, Contreras A, Balam-Ortiz E, del Bosque-Plata L, Velazquez-Fernandez D, Lara C, Goya R, Hernandez-Lemus E, Davila C, Barrientos E, March S, Jimenez-Sanchez G (2009) Analysis of genomic diversity in Mexican Mestizo populations to

develop genomic medicine in Mexico. Proc Natl Acad Sci USA 106(21):8611–8616

60. Sundquist A, Fratkin E, Do CB, Batzoglou S (2008) Effect of genetic divergence in identifying ancestral origin using HAPAA. Genome Res 18(4):676–682

61. Tallila J, Jakkula E, Peltonen L, Salonen R, Kestila M (2008) Identification of CC2D2A as a Meckel syndrome gene adds an important piece to the ciliopathy puzzle. Am J Hum Genet 82(6):1361–1367

62. Tang H, Coram M, Wang P, Zhu X, Risch N (2006) Reconstructing genetic ancestry blocks in admixed individuals. Am J Hum Genet 79(1):1–12

63. Tang H, Peng J, Wang P, Risch NJ (2005) Estimation of individual admixture: analytical and study design considerations. Genet Epidemiol 28(4):289–301

64. Tian C, Plenge RM, Ransom M, Lee A, Villoslada P, Selmi C, Klareskog L, Pulver AE, Qi L, Gregersen PK, Seldin MF (2008) Analysis and application of European genetic substructure using 300 K SNP information. PLoS Genet 4(1):e4

65. Tishkoff SA, Reed FA, Friedlaender FR, Ehret C, Ranciaro A, Froment A, Hirbo JB, Awomoyi AA, Bodo J-M, Doumbo O, Ibrahim M, Juma AT, Kotze MJ, Lema G, Moore JH, Mortensen H, Nyambo TB, Omar SA, Powell K, Pretorius GS, Smith MW, Thera MA, Wambebe C, Weber JL, Williams SM (2009) The genetic structure and history of Africans and African Americans. Science 324:1035–1044

66. Wang S, Lewis CM Jr, Jakobsson M, Ramachandran S, Ray N, Bedoya G, Rojas W, Parra MV, Molina JA, Gallo C (2007) Genetic variation and population structure in Native Americans. PloS Genet 3:e185

67. Weir B (1996) Genetic data analysis II. Sinauer Press, Sunderland, MA

68. Williamson SH, Hernandez R, Fledel-Alon A, Zhu L, Nielsen R et al (2005) Simultaneous inference of selection and population growth from patterns of variation in the human genome. Proc Natl Acad Sci USA 102:7882–7887

69. Wright S (1921) Systems of mating. I. The biometric relations between offspring and parent. Genetics 6:111–123

70. Wu B, Liu N, Zhao H (2006) PSMIX: an R package for population stratification inference via maximum likelihood method. BMC Bioinformatics 7:317

71. Xing J, Watkins WS, Witherspoon DJ, Zhang Y, Guthery SL, Thara R, Mowry BJ, Bulayeva K, Weiss RB, Jorde LB (2009) Fine-scaled human genetic structure revealed by SNP microarrays. Genome Res 19:815–825

72. Xu S, Jin L (2008) A genome-wide analysis of admixture in Uyghurs and a high-density admixture map for disease-gene discovery. Am J Hum Genet 83(3):322–336

73. Yamaguchi-Kabata Y, Nakazono K, Takahashi A, Saito S, Hosono N, Kubo M, Nakamura Y, Kamatani N (2008) Japanese population structure, based on SNP genotypes from 7003 individuals compared to other ethnic groups: effects on population-based association studies. Am J Hum Genet 83:445–456

74. Zhu X, Zhang S, Tang H, Cooper R (2006) A classical likelihood based approach for admixture mapping using EM algorithm. Hum Genet 120(3):431–445

75. Gutenkunst RN, Hernandez RD, Williamson SH, Bustamante CD (in press) Inferring the joint demographic history of multiple populations from multidimensional SNP data PLoS Genetics; arXiv:0909.0925

Genetic Epidemiology

21

Sophia S. Wang, Terri H. Beaty, and Muin J. Khoury

Abstract In this chapter, we describe both the historical and contemporary terminologies that reflect the evolving field of genetic epidemiology. We discuss the conduct of family-based studies to identify high-penetrance disease genes, along with traditional genetic analyses of human pedigrees to assess Mendelian transmission (segregation analysis) or to locate causal genes (linkage analysis/gene mapping). We also describe epidemiologic approaches used to study gene-disease associations (including genome-wide association studies) plus gene-gene and gene-environment interactions. We review analytic and methodologic issues applicable to each of these studies and emerging, nontraditional epidemiologic methods that can be used as an adjunct to traditional approaches, particularly for the simultaneous study of hundreds of thousands of data points per person. We further discuss the challenging nature of analysis, synthesis, and dissemination of these genetic data, and the value of systematic reviews, meta-analyses and consortia in evaluating large bodies of scientific evidence. Finally, we describe the need for follow-up of these results to identify causal variants and the need for translational research efforts to apply these gene discoveries to personalized medicine, such as evaluation of genetic testing in clinical practice (in terms of analytic validity, clinical validity, clinical utility), and to population health including determining the disease risk and burden in populations (e.g., absolute and attributable risks). We also consider the ethical, legal, and social implications of these discoveries.

S.S. Wang (✉)
Division of Cancer Epidemiology and Genetics, National
Cancer Institute, National Institutes of Health, 6120 Executive
Boulevard, Room 5104, Rockville, MD 20852, USA
e-mail: wangso@mail.nih.gov
and
Division of Cancer Etiology, Department of Population
Sciences, City of Hope National Medical Center,
1500 East Duarte Road, Duarte, CA 91010, USA

T.H. Beaty
Department of Epidemiology, The Johns Hopkins Bloomberg
School of Public Health, 615 North Wolfe Street, Baltimore,
MD 21205, USA
e-mail: tbeaty@jhsph.edu

M.J. Khoury
National Office of Public Health Genomics, Centers for
Disease Control and Prevention, 1600 Clifton Road
Mailstop E-61, Atlanta, GA 30333, USA
e-mail: muk1@cdc.gov

Contents

21.1 Introduction 618

21.2 Scope and Strategies of Genetic
      Epidemiology in the Twenty-First Century 619

21.3 Fundamentals of Gene Discovery:
      Family Studies 619

21.4 Fundamentals of Gene Discovery:
      Population Studies and GWAS 623

21.5 Beyond Gene Discovery:
      Epidemiologic Assessment
      of Genes in Population Health 626

21.6 Beyond Gene Discovery: Epidemiologic
      Assessment of Genetic Information
      in Medicine and Public Health 629

21.7 Policy, Ethical and Practice Considerations:
      the Emergence of Public Health Genomics 631

References 632

21.1 Introduction

Historically, the term "genetic epidemiology" has been used to denote the study of genetic factors in the occurrence of disease in populations [40]. A primary focus of genetic epidemiology has been on studying familial aggregation of disease and on statistical methods for gene discovery in family-based studies. Developments in genetic epidemiology and statistical genetics have largely centered on methods for discovering disease susceptibility genes [8]. These approaches have been successful in identifying mutations in more than 3,000 genes associated with Mendelian disorders (e.g., cystic fibrosis), as well as high-penetrance genetic variants associated with certain common diseases (e.g., hereditary breast and ovarian cancer) in families with multiple affected members [55]. We now know, however, that variants typically identified in such high-risk families explain only a small proportion of all cases; for example, *BRCA1* and *BRCA2* mutations account for an estimated 1–2% of all breast cancer cases in the general population [52].

In the last decade, the definition of genetic epidemiology has broadened, particularly with completion of the human genome project [14, 30] and with rapid developments in testing for numerous genetic variants with increased efficiency (e.g., reduced time and cost) [74]. These advances have accelerated discovery efforts particularly with genome-wide association studies (GWAS) and now make possible the evaluation of numerous gene associations and interactions. (both gene-gene and gene-environment). In an effort to capture these developments, the term "human genomics" was introduced to define "the study of the functions and interactions of all the genes in the genome" [29], and the term "human genome epidemiology" was creatd to further define epidemiological approaches used from gene discovery to applications in medicine and public health [44]. Finally, the term "public health genomics" was coined to reflect "a multidisciplinary field concerned with the effective and responsible translation of genome-based knowledge and technologies to improve population health" [6].

In this chapter, we describe both the historical and contemporary terminologies that reflect the evolving field of genetic epidemiology. We discuss traditional genetic analyses of human pedigrees to assess Mendelian transmission (segregation analysis) or to locate human genes (linkage analysis/gene mapping). We describe traditional epidemiologic approaches used to study gene-disease associations plus gene-gene and gene-environment interactions. We review analytic and methodologic issues applicable to these studies as well as emerging, nontraditional epidemiologic methods that can be used as an adjunct to traditional approaches. As the field becomes increasingly complex with the simultaneous study of hundreds of thousands of data points per person, the importance of rigorous approaches to study design, analysis and interpretation will be greatly magnified [42]. We therefore discuss the challenging nature of analysis, synthesis, and dissemination of this rapidly accumulating evidence, and the value of systematic reviews, meta-analyses and consortia. Finally, we discuss research translation efforts to apply these discoveries in personalized medicine and population health, including ethical, legal, and social implications.

21.2 Scope and Strategies of Genetic Epidemiology in the Twenty-First Century

Genetic epidemiology will continue to broaden its scope in response to continued scientific and technological advances. Monogenic diseases provide a basic understanding of biologic mechanisms in disease causation and traditional family-based studies will therefore remain critical for identifying high-penetrance genes in diseases and disorders for which there is a strong genetic component. For polygenic diseases, genome-wide association studies will continue to play an important role in identifying common genetic variants that contribute to disease. However, we believe that this, too, will evolve as newer technologies such as genome sequencing will supplant current methods. Regardless of the technology, however, most of these discovery efforts will be conducted using traditional epidemiologic studies (either cohort or case-control studies). Though studies are largely based on principles of linkage disequilibrium (LD) mapping, sequencing of the human genome will become feasible in the coming years. Critical to these efforts will be the follow-up and translation of these discoveries to identify the actual causal variants and to determine the disease risk and burden in populations. Whether intervention or prevention is feasible based on these data has yet to be demonstrated. Epidemiologic efforts to identify common variants causing disease are now largely led by multi-institutional consortia and networks intended to maximize statistical power to identify modest but true associations. Consortia will also be critical in genetic epidemiology for identifying and replicating causal associations (Table 21.1).

Although the current focus of gene discovery is on single nucleotide polymorphisms, expansion to other types of genetic variants and to other genetic changes will occur as technologies make broader efforts possible. In this century, the greatest challenge will be to translate these scientific discoveries. To this end, expansion of the field to human genome epidemiology will be critical for full understanding and characterization of the effects of genetic factors in populations [44]. Only by knowing the prevalence of genetic polymorphisms in well-defined populations, by characterizing genotype-phenotype associations, investigating gene-environment interactions, and evaluating the clinical validity and utility of genetic tests can we know how on-going discovery efforts will be useful and potentially utilized in medical practice and public health programs.

21.3 Fundamentals of Gene Discovery: Family Studies

Since human genetics was founded on observational studies of families, it is not surprising that methods of gene discovery first exploited classic linkage analysis to map causal genes for Mendelian diseases. Linkage analysis tests for co-segregation between an observed marker and an unobserved causal locus and evidence of reduced recombination between a genetic marker and a genetic disease constitute compelling evidence that a gene does control the phenotype, while simultaneously locating the causal gene within the genome. In fact, evidence of linkage used to be considered the highest level of statistical evidence that a gene does control a phenotype, since nongenetic factors can easily mimic segregation patterns of a single gene in families but it seems very unlikely that they would co-segregate with a genetic marker [20].

Although the principles of linkage analysis were developed in the 1950s, their application to families was quite limited and did not expand to a genome wide scale until the number of available genetic markers expanded with discovery of anonymous

Table 21.1 Continuum of human genome epidemiology from gene discovery to applications (adapted from [41])

Field	Application	Types of studies
Genetic epidemiology	Gene discovery	Linkage analysis, family-based association studies
Genetic/molecular epidemiology	Gene characterization	Population studies to characterize gene prevalence, gene–disease associations, and gene-gene and gene–environment interaction
Applied epidemiology and health services research	Determining health impact	Studies to evaluate validity and utility of genetic information in clinical trials or observational clinical settings

DNA markers in the 1980s. The very first genetic markers included blood groups and protein variants which were quite limited in their numbers, and their true location in the genome remained unknown until a comprehensive map of the genome could be generated. Markers in DNA itself were first discovered in the 1980s using bacterial restriction enzymes that cut DNA at specific sequences scattered throughout the genome, generating biallelic markers called restriction fragment length polymorphisms (RFLP). Later, single tandem repeat polymorphisms (STRP) or microsatellite markers were discovered to be common in regions outside of genes, and these multi-allelic markers were very informative for linkage analysis. Later still, single nucleotide polymorphisms (SNP) were found to be even more common in the genome (averaging 1 per 1,000 base-pairs), and as technology for mass genotyping improved genome wide linkage studies became more affordable; fixed panels of informative SNPs are now used to provide adequate coverage of the entire genome.

Linkage analysis is most effective for diseases with a known mode of inheritance. Parametric linkage (also called model-based linkage analysis) is a maximum likelihood approach that estimates the recombination fraction (θ), a measure of genetic distance, and uses the log-odds or LOD score to compare the hypothesis of no linkage (independent assortment of markers at two loci) with the hypothesis that linkage does exist at some recombination fraction less than the expected 50%. This parametric or model-based approach estimates the strength of linkage and tests for its statistical significance simultaneously. Statistical power to detect linkage reflects the ability to reconstruct meiotic events in informative matings (i.e., those matings involving a double-heterozygote parent who is heterozygous at both the trait locus and the marker locus). This becomes a function of the information content for meiosis in the families at hand. Marker allele frequencies are also critical in determining information content, and therefore highly polymorphic multi-allelic markers are more informative for linkage because more parents will be heterozygous, so that STRP markers (also called microsatellite markers) became popular for linkage analysis. A minimum of ~400 highly polymorphic STRPs can provide adequate coverage of the entire genome at a resolution of ≤10 centiMorgans (cM) between markers, which means on average there would be no more than 5 cM (equivalent to 5% recombination) between an unknown causal gene and a marker. This relatively loose-scale genome wide linkage array can identify chromosomal regions of interest, but would require follow-up using a separate round of genotyping with more closely spaced markers (typically more SNPs spaced 1–2 cM apart) for fine mapping. Currently SNP panels of 5,000–6,000 markers (with an average spacing of 1 cM between markers) are used for genome-wide linkage studies, although the high correlation among nearby SNPs (which tend to be in tight linkage disequilibrium or LD) must be considered in the statistical analysis. For Mendelian diseases (even those with some incomplete penetrance or possible locus heterogeneity), this strategy is effective in providing evidence sufficient to map a gene down to a small region of a chromosome. These larger SNP panels for linkage can be also used for fine mapping, since there is an average of 1 cM between markers, although even large collections of multiplex families rarely contain enough informative meiotic events to give resolution down to this level. Furthermore, even small chromosomal regions of interest can contain many genes, and the resolution of linkage signals is often limited, so that further sequencing studies will be required to identify specific mutations in genes thought to control a given phenotype.

The parametric or model-based LOD score approach has been used to detect genes exerting a "major effect" on risk (i.e., the disorder may not be strictly Mendelian, but it has limited locus heterogeneity and consistent levels of incomplete penetrance). When there is evidence of locus heterogeneity, i.e., some families show evidence of linkage to a single marker or chromosomal region and others do not, statistics such as the heterogeneity LOD score (HLOD) can be used to consider locus (or linkage) heterogeneity (see Table 21.2 for definition). This HLOD is based on the admixture heterogeneity test [71] and allows maximum-likelihood estimations of both the recombination fraction (θ) and another parameter for the proportion of linked families (α) simultaneously. Thus, parametric tests for linkage can be applied to complex diseases if one is willing to assume a specific model of inheritance. However, errors in the specified model of inheritance will diminish statistical power to detect linkage and will inflate the estimated θ, making fine mapping even less precise and less reliable.

Table 21.2 Definition of terms for linkage analysis

Term	Definition
Recombination fraction (θ)	Proportion of recombinant (nonparental) gametes formed during meiosis. For 2 unlinked loci, alleles will recombine 50% of the time ($\theta = 0.5$), and this becomes the null hypothesis of no linkage. For 2 linked loci, alleles will show <50% recombination and will tend to "co-segregate" in informative matings. Conventionally, tight linkage is considered to be $\theta = 0.05$, and $\theta = 0.0$ would imply complete linkage (no recombination) where alleles at the 2 loci are always inherited together
Informative matings	Matings involving 1 or 2 double heterozygotes which allow meiotic events to be reconstructed by examining genotypes of offspring. The simplest informative mating is between a double heterozygote and a double homozygote where phase is known, because each offspring can be directly classified as having received a "parental" or "nonparental" haplotype from the informative parent and these represent a direct count of nonrecombinant and recombinant gametes formed by meiosis in that parent
Gametic phase	Haplotype combinations of alleles at different loci, which can be inferred from observed genotypes (sometimes with certainty, but not always). In particular, the double-heterozygous individuals who are informative for linkage are ambiguous regarding gametic phase, i.e., the "AaBb" genotype could represent two possible gametic phases: AB‖ab or Ab‖aB, arbitrarily termed "coupling" and "repulsion" in the genetics literature. In family studies, phase can often be established by genotyping parents of such individuals; thus, three generational families can provide "phase-known matings," which provide more statistical power for parametric or model-based linkage analysis. "Phase-unknown matings," on the other hand, must consider the two possible phases and result in reduced power
Parametric linkage or model-based linkage analysis	A statistical approach testing the null hypothesis of no linkage versus linkage at some estimated θ based on a prespecified model of inheritance. This approach used maximum likelihood methods to both estimate θ and compute a log-odds statistic (or LOD score) to summarize the evidence for or against linkage. By convention, a LOD score ≥ 3.0 is taken as evidence for linkage (at that θ), and a LOD score < -2.0 is taken as evidence of no linkage for any single marker. Widely considered the most powerful statistical method for testing for linkage. If, however, the model of inheritance is wrong, the statistical power to detect linkage will be diminished and the estimated θ will be inflated
Nonparametric linkage or model-free linkage analysis	A statistical approach testing for excess allele sharing between affected relatives, which would reflect linkage between an observed marker locus and an unobserved causal locus. In general, the most informative type of affected relative pair is affected sib pairs, and this approach tests for excess sharing of marker alleles that are identical by descent (IBD) by comparing the null hypothesis values of 25% sharing 0 alleles IBD, 50% sharing 1 alleles IBD and 25% sharing 2 alleles between two affected sibs with the observed or estimated proportion of IBD sharing. This approach is nonparametric or model free in the sense it does not assume any particular model of inheritance (recessive, dominant, etc.) for the trait locus, although evaluating significance from the various test statistics often implies specific distributions
Identical by descent	Two alleles are considered identical by descent (IBD) if they were inherited from a common ancestor. Since humans are diploid, it is possible to share 0, 1, or 2 alleles IBD at any autosomal locus, and the prior probabilities of these sharing states are specified solely by the biological relationship (full sibs, parent-offspring, etc.). In affected sib-pair analysis testing for linkage, prior probabilities of IBD sharing are 0.25:0.50:0.25 represent the null hypothesis of no linkage and the observed probabilities are computed for each marker locus or based on a prespecified framework map of multiple markers. The genotypes of the parents are critical in specifying observed patterns of IBD sharing and the most informative type of mating is between two double-heterozygous parents carrying four different alleles, i.e., an AB×CD mating would allow all possible combinations of IBD sharing between pairs of full sibs to be classified unambiguously
Locus or linkage heterogeneity	Heterogeneity between families in evidence for linkage where some families give evidence of linkage to a given marker and other families appear to be unlinked to this same marker. In parametric or model-based linkage analysis, this introduces a second parameter to be estimated (α), which represents the proportion of linked families. In terms of testing the null hypothesis of no linkage, this second parameter requires a sequential testing strategy and an alternative test statistic, the heterogeneity LOD score or HLOD, which summarizes the evidence for linkage in the a proportion of linked families. Thus, this HLOD statistic cannot give evidence against linkage (i.e., it cannot be negative) because there it becomes impossible to distinguish between the (1-α) unlinked families and no linkage to the marker

In the absence of knowledge of any specific model of inheritance, however, more robust nonparametric or model-free methods can be used to test for linkage (although these methods do not estimate the genetic distance between a marker and the unknown trait locus). Nonparametric linkage methods are based on sharing of alleles that are "identical by descent" (IBD), i.e., marker alleles inherited from a common ancestor. Since humans are diploid, any two individuals can share 0, 1, or 2 alleles IBD at a given marker. The prior probabilities of these three sharing states are determined by the relationship between two individuals. For example, parents and offspring share exactly 1 allele IBD, but cannot share 0 or 2 alleles IBD, and therefore will not be informative for linkage. On the other hand, full siblings have a prior probability of sharing 0, 1, or 2 alleles IBD equal to 25, 50, and 25, respectively, and are therefore potentially informative for linkage. Penrose pointed out that full sibs with the same phenotype (e.g., affected sib pairs) should share marker alleles IBD more than these expected prior probabilities *if* the marker is tightly linked to a gene controlling the phenotype [60]. Testing for excess IBD sharing between pairs of affected sibs therefore provides a nonparametric test for linkage, where no model of inheritance need be specified. Other types of relative pairs can also be informative about excess IBD sharing, but to a lesser degree. For example, second-degree relatives (including avuncular pairs, grandparent-grandchild pairs, and half sibs) can share 1 allele IBD with a prior probability of 50% (and a corresponding prior probability of 50% for sharing 0 alleles IBD). If pairs of affected second-degree relatives show estimated IBD sharing above this expected value, this too constitutes evidence for linkage.

This nonparametric or model-free approach allows smaller families (i.e., those with only an affected sib pair or other relative pair) to be used to test for linkage in a robust fashion that makes sense for complex diseases (where the model of inheritance is rarely known). Several statistical methods for nonparametric linkage have been developed [21, 34], but all involve estimating IBD allele sharing between relatives for either individual markers or for multiple markers in a fixed framework map. Generalized methods for estimating IBD sharing based on multiple markers are available where hidden Markov chain algorithms are used to estimate IBD sharing [46]. These can be used to estimate IBD sharing for multiple markers in a fixed-framework map using the Lander–Green algorithm (see [73], and Chap. 7 for a discussion). Given these estimated IBD sharing probabilities for a family, then either parametric or nonparametric methods of linkage analysis can be used for multipoint linkage analysis.

For quantitative traits, parametric linkage models would require complete specification of genotypic means (and their variances) at the trait locus and allele frequencies at both the trait and marker loci, and these could be used to estimate the recombination fraction between the trait locus and the marker (and test its significance). However, good estimates of these parameters are rarely available; therefore, more robust methods are preferred. Again, estimated IBD sharing can also be used to test for linkage by using regression models to relate phenotypic differences between relatives to the observed (or estimated) IBD sharing. For individual markers, simple regression models were proposed where the squared trait difference between two sibs was regressed on the estimated probabilities of sharing alleles IBD at a marker locus [4, 34]. Later extensions showed more information could be obtained by using the squared sum of a mean corrected phenotype, as long as the sample of sib pairs is representative of the general population [21]. Sham et al. proposed regressing the estimated IBD sharing on the measures of phenotype differences between sibs and suggested this was more appropriate for ascertained samples of sib pairs [69]. These various regression approaches for quantitative phenotypes are reviewed by Schaid et al. [65] who point out opportunities for incorporating covariates, and testing for interaction.

Techniques for nonparametric linkage analysis for quantitative phenotypes are complemented by extensions to general linear models, which underlie variance components approaches. Variance components models are the foundation of the original biometrical models for estimating heritability of quantitative phenotypes, and can be extended to families of arbitrary structure [22]. By adding a term that represents an unobserved quantitative trait locus (QTL) completely linked to a marker, it becomes possible to estimate the variance attributable to sharing marker alleles IBD [1]. Using the patterns of IBD sharing among all family members, it is possible to partition the total phenotypic variance into a component attributable to a QTL, to background "polygenes," and residual environmental factors. Thus, evidence of linkage can be obtained as a proportion of the phenotypic variance attributable to IBD sharing among relatives a given marker or to a region of chromosome in a multipoint analysis. Once again, however, when this approach is applied to a quantitative phenotype where both genetic and nongenetic factors contribute to the etiology (the quantitative equivalent of a complex disease) there are limits in resolution (i.e., the peak regions of evidence for linkage are very broad) and evidence across studies is typically inconsistent (suggesting locus heterogeneity).

During the last two decades, considerable efforts have been invested in genome-wide linkage studies of complex diseases using both classic parametric and nonparametric linkage methods in both sib-pair studies and extended pedigrees. While these studies have met with some success, they were often plagued by weak and sometimes contradictory evidence of linkage, even when multiple studies are combined into a meta-analysis of LOD scores (e.g., see Marazita et al. for a meta-analysis of families ascertained through cleft lip with/without cleft palate [50]). More importantly, however, it has proven extremely difficult to narrow regions yielding evidence of linkage, because even large families cannot provide resolution much below 10 cM of genetic distance.

To summarize, linkage studies can be conducted on a genome-wide level to provide an unbiased search for causal genes using families of almost any size. However, linkage analysis always requires families informative enough to either reconstitute meiotic events within families (as part of parametric linkage analysis) or at least informative enough to document excess sharing of marker alleles between relatives in nonparametric linkage analysis. Moreover, all linkage studies of a qualitative disease phenotype (i.e., affected vs. nonaffected) require multiplex families with more than two affected members (at a minimum affected sib-pairs), and these may be difficult to recruit, so that the question arises of whether multiplex families can adequately represent all cases of a given disease and thus all of the genetic causes of that disease. Furthermore, the track record of genome-wide linkage searches is mixed at best. Many studies have failed to identify regions yielding clear evidence of linkage at the genome-wide level, and following the standard linkage approach of a course genome-wide search followed by fine mapping regions of interest does not always narrow regions of linkage for complex diseases. These failures may represent intrinsic limitations of the linkage approach (i.e., linkage alone cannot achieve resolution on a fine scale) or the presence of locus heterogeneity (multiple genes causing disease, plus some non-genetic causes) simply diminishes both power and resolution.

21.4 Fundamentals of Gene Discovery: Population Studies and GWAS

Genome-wide association studies (GWAS) have recently become a technically feasible approach for searching the human genome for genes influencing risk for complex disorders [59]. These studies typically rely on conventional epidemiologic study designs, especially the case-control design or its variations, and utilize high-throughput marker assays where 100,000–1,000,000 SNP markers are typed on each study subject (from population samples or family members). The principles underlying association studies differ from those underlying linkage studies, but both can be used to identify genes for complex disorders [51]. While linkage analysis is designed to test for co-segregation (or co-inheritance) between an unobserved causal gene and a marker within a family, association analysis relies on detecting differences in marker allele frequencies between groups of unrelated individuals such as cases and controls [15]. Close linkage can easily lead to such differences in frequencies because it creates "linkage disequilibrium" (LD), which manifests as significant correlation between alleles at two different loci (e.g., here a marker locus and a high-risk allele at an unobserved causal gene) at the population level. However, while statistical evidence of linkage results in a definitive interpretation that the marker is inherited with a marker, statistical evidence of association is more ambiguous because LD is not the only possible explanation for a difference in frequencies between unrelated cases and controls. For example, a statistically significant association could be due to an artifact created by heterogeneity in the population from which cases and controls were sampled. This phenomenon is termed "confounding" by epidemiologists and "population stratification" by geneticists. It can occur when two subpopulations which differ in both their marker allele frequencies and their disease risk are combined and treated as samples from a single, homogeneous population, creating a completely spurious statistical association between the marker allele and case/control status [58]. Although the statistical evidence from association studies is less definitive than evidence from linkage analysis, the case-control design has the advantage of being able to identify genes with more modest effects on risk that would be missed completely by linkage analysis [63]. Furthermore, because the underlying LD between markers and a potential causal gene spans a much smaller physical distance than can be detected in standard linkage analysis, finding convincing evidence of association gives greater resolution about where a causal gene lies in the genome, i.e., the resolution of association studies is greater.

The conventional case-control design requires a sound epidemiologic study design that avoids introducing various forms of bias (see Table 21.3). Ideally, cases should be a representative sample of affected individuals, preferably incident cases to minimize selection bias caused by survival. Similarly, controls should be a representative sample of at-risk but unaffected individuals. For genetic association studies, extreme caution should be exerted to ensure that the cases and controls are derived from the same base population. Since no human population can be considered completely homogeneous, there is considerable controversy about how to match cases and controls or how to adjust for possible confounding between disease status (case vs. control) and genetic background. Several options are available: (1) careful matching for genetic background (e.g., racial or ethnic background); (2) adjusted analysis using genomic control markers which would reflect effects of confounding [16]; (3) extensive quantification of the genetic makeup of cases and controls through data reduction techniques such as principal components analyzes and possibly incorporating these into the analysis as a covariate [61]; and (4) use of case-family study designs (see below). The matching of cases and controls can be as simple as recording racial/ethnic background and stratifying by these broadly defined social norms. There are also several methods to estimate population membership for individuals [62] and using these to identify and possibly exclude outlier individuals. These Bayesian methods for estimating admixture at the individual level can help identify when significant population heterogeneity or "cryptic substructure" exists in a sample of unrelated individuals [23]. Both the substructure approach and the genomic control approach require typing a substantial number of unlinked markers that are not associated with the phenotypic definition of cases and controls. Genomic control markers are used to adjust the case-control analysis for average confounding. Another adjustment approach is to use a large number of markers to quantify the genetic variance in samples of cases and controls and to condense the genetic diversity to a small number of summary measures. This usually takes the form of principal components analysis, where linear functions are fit to the data on many markers and condensed into orthogonal summary measures that account for successively smaller fractions of the total genetic variance. These linear functions can then be computed for each individual case and control and can serve either to

Table 21.3 Types of biases that can affect epidemiologic studies of gene–disease associations (adapted from [49])

Biases	
Survival bias	Case selection based on those currently available for study may miss fatal and/or quick cases or mild cases
Participation bias	Differential rates of refusal or nonresponse for study participation between cases and controls
Diagnosis bias	Knowledge of a subject's exposure to a putative cause of disease can influence both intensity and outcome of the diagnostic process
Referral bias	Factors related to the probability of referral. Cases who are more likely to receive advanced care or to be hospitalized—such as those with greater access to health care or with co-existing illnesses—can distort associations with other risk factors
Surveillance bias	Differential detection of cases for those under frequent medical surveillance
Recall bias	Cases might recall more intensively potential causative exposures
Family information bias	Information about exposures or illnesses can be stimulated by, or directed to, a new case within a family

graphically document population substructure of the entire sample or can be treated as covariate in the conventional case-control analysis. One option for conventional case-control designs is to use "population" or "universal" controls where the controls are carefully documented to represent the population of reference, but are not necessarily free of a particular phenotype. This strategy was used in a well-recognized English GWAS of seven different complex diseases, and the control group for each case group was the same group of controls selected to carefully represent the English general population but without phenotype information [19]. This can be viewed as an incomplete case-control design, where genotype and phenotype information was available on cases but only genotype information was available on controls. In the United States, such a universal control group can potentially be derived from a sample of the population through carefully conducted representative surveys. One example of such a survey is the National Health Examination and Nutrition Survey (NHANES), which is a stratified weighted random sample of the population with overrepresentation of ethnic and racial minorities. Recently, prevalence estimates of over 90 markers were published [11]. Work is ongoing to measure genome-wide profiles from the same sample population. Nelson et al. recently suggested this type of incomplete design might be very useful for situations where the number of cases is small but the number of controls can be much larger (e.g., 10 controls matched to 1 case through multiple principal components) [57].

Finally, a variation on the conventional case-control study design is the case-family design, where an affected case and his/her relatives are sampled. One version of the case-family design is the case-sib control, where an unaffected sib is used as a control. Case-sib control designs tend to be overmatched for genetic background and thus provide less statistical power to detect effects of genes (G) than are case-unrelated control designs, but they circumvent the problem of confounding, because both cases and their sibs share the same genetic background [80]. However, given the small sibship sizes common in modern populations, case-sib control studies becomes less efficient in a sampling sense also, because not all cases will have an available sib.

Another version of case-family designs is the case-parent trio or "triad" design, where parents are genotyped and used to contrast alleles transmitted to cases to those not transmitted. In more general terms, this design tests the null hypothesis that marker genotypes are completely independent of phenotypes. Case-parent trios are particularly attractive for studies of birth defects or childhood disorders, because parents are generally available when the disorder is detected. Conversely, this design is not practical for late-onset disorders, because the parents will typically not be available. Furthermore, this type of family-based study design has the advantage of using all available cases, and not just those from multiplex families. Since there may be etiologic heterogeneity between cases from simplex versus multiplex families, it will be important to evaluate family history in GWAS, as it may be an indicator of severity. The case-parent trio design uses genotypes of the parents to contrast marker alleles in affected cases to those expected under strict Mendelian transmission. Therefore, rejecting the null hypothesis of independence gives evidence that the marker is both linked to a causal gene and in LD with a high-risk allele at that gene. Because case-family designs focus on distribution of alleles or genotypes within a family, these designs are more robust to confounding or population stratification than the original case-control design. Trio or triad designs can also be adapted to test for interaction between genes and environmental exposures (gene-environment or G×E interaction), which may be very important for many complex diseases where environmental risk factors are already recognized.

In addition, the trio or triad design provides the opportunity to test for "parent-of-origin" effects where marker alleles can give different levels of statistical evidence depending on whether they were transmitted from the mother or the father to the affected child. Evidence of parent-of-origin effects could reflect the effects of maternal genotypes on risk, presumably through control of the in utero environment, or they could reflect "genomic imprinting," where genes are turned on or off depending on if they are inherited through mothers or fathers. This represents a major advantage of the case-parent trio design, because conventional case-control designs (or even case-sib controls without parental information) cannot address the issue of parent-of-origin at any level. However, case-control designs can still address gene effects, G×E interaction and even gene-gene interaction (G×G) with potentially greater power, so that either is a viable choice when designing a GWAS.

To summarize, there are several methods for searching for genes influencing the risk for complex diseases and they should be viewed as complementary not competitive. Neither linkage nor association approaches are perfect for all circumstances. The challenge is to recognize their different strengths and use them appropriately for the question at hand.

21.5 Beyond Gene Discovery: Epidemiologic Assessment of Genes in Population Health

Numerous genome-wide association studies have been conducted to date, and there is little doubt that new susceptibility loci will be identified in the years to come. In cancer epidemiology alone, nearly 50 susceptibility loci have been identified with genome-wide significant (e.g., $P<10^{-7}$) across five major cancers (breast, prostate, lung, colorectal, melanoma) in the last 3 years [18]. ~20 have been identified for prostate cancer alone through GWAS, including SNPs in the 8q24 region. GWAS efforts include both disease associations and associations with various phenotypes and exposures/behavior. For example, GWAS in lung cancer and melanoma have identified SNPs associated with relevant environmental exposures (e.g., *CHRNA3/5* with smoking). GWAS have also identified SNPs associated with other phenotypes such as height, hair color, and skin pigmentation.

To assess the impact of genes in population health, further understanding of the distribution of genetic variants in the population and the joint effect of multiple genes or between genes and the environment (also known as gene–gene and gene–environment interactions) will be critical to explain the occurrence of most human diseases. As genetic and environmental risk factors are identified, determining whether they act independently of one another to alter disease risk, or jointly in a biological (or multiplicative) way, will determine their overall affect on disease risk. Evaluation of gene–environment interactions at the simplest level requires only displaying raw data in a 2-by-4 table (Table 21.4). In a 2-by-4 table, both exposure and the underlying susceptibility genotype are dichotomized as present or absent. The genotype could reflect a dominant model (e.g., heterozygote or homozygote variant) or a combination of alleles at multiple loci if multiple genotypes are considered. Odds ratios are computed for each stratum, using subjects who are unexposed and have no susceptibility genotype as the reference group. Though a simplistic way to present the data, there are several advantages to this approach [5]. Foremost, the role of each factor (e.g., exposure without genotype, genotype without exposure) is independently assessed both in terms of the association and of the potential attributable fraction. The presentation also underscores sample size issues because cell sizes are delineated. This approach also favors effect estimation over model

Table 21.4 Gene–environment interaction analysis in a case-control study (+ present, - absent)

Exposure	Genotype	Cases	Controls	Odds ratio[a] (OR)
−	−	A_{00}	B_{00}	$OR_{00}=1.0$
−	+	A_{01}	B_{01}	$OR_{01}=A_{01}B_{00}/A_{00}B_{01}$
+	−	A_{10}	B_{10}	$OR_{10}=A_{10}B_{00}/A_{00}B_{10}$
+	+	A_{11}	B_{11}	$OR_{11}=A_{11}B_{00}/A_{00}B_{11}$

[a]Case-only OR = $A_{11}A_{00}/A_{10}A_{01} = (OR_{11}/OR_{10}OR_{01})OR_{co}$, where $OR_{co}=B_{11}B_{00}/B_{10}B_{01}$ (control-only odds ratio)
Example: Analysis of oral contraceptive (OC) use, presence of Factor V Leiden mutation (a risk for venous thromboembolism)*.
Case-only odds ratio: 25*36/10*84 = 1.1; control-only odds ratio: 2*100/4*63 = 0.8 (*AF – Pop (%)* attributable fraction (percent) in the population, *AF – Exp(%)* attributable fraction (percent) among exposed). (From [6]; data from [77])

Factor V Leiden	OC	Cases	Controls	Odds ratio	95% CI			AF-Exp (%)
−	−	36	100	Ref	Ref			
−	+	84	63	ORe	3.7	2.18	−	6.32	73.0
+	−	10	4	ORg	6.9	1.83	−	31.80	85.6
+	+	25	2	ORge	34.7	7.83	−	310.0	97.1
Total		155	169					

testing where logistic regression is used to evaluate departure from multiplicative effects. The odds ratios themselves can be used to determine whether departure from multiplicative or additive models of interactions exists. Finally, the 2–by-4 table is advantageous in that it provides the distribution of the exposures among controls.

An alternative method for evaluating gene-environment interaction is to use a case-only study [27, 41]. This study design, where case series are used and for which there are no controls, is limited in its ability to evaluate independent effects of the genotype or exposure. However, among cases, multiplicative interaction can be detected. Specifically, assuming a multiplicative model of interaction where the genotype and exposure are independent, a departure of the case-only odds ratio from 1.0 indicates the presence of gene-environment interaction (Table 21.5).

In the context of GWAS, however, 2-by-4 tables, the use of stratified analyses and even logistic regression analysis will be limited [36]. For complex datasets, novel statistical strategies are emerging, such as hierarchical regression and Bayesian methods [17, 28]. Bayesian methods are attractive to epidemiologists because they allow scientists to integrate a priori expectations and hypotheses. A novel multilocus test of genetic association based on Tukey's 1-*df* model of interaction has been proposed [12]. This method exploits LD patterns among SNPs within a gene and simultaneously accounts for gene-gene and gene-environment interactions. Alternatively, the combinatorial partitioning method (CPM), which represents an extension of analysis of variance between and within genotypes at one locus, can be used to conduct joint analysis of multiple genes for quantitative traits [56]. An association between the gene(s) and a trait is represented in excess variability between the genotypes, relative to that within genotypes. The multifactor dimensionality reduction (MDR) method is an extension of the CPM where genotypes from multiple loci are grouped as high- and low-risk groups [33, 64]. The MDR would potentially allow persons to be classified into two or more distinct groups based on their phenotype or based on the underlying biology of the loci under consideration. Statistical methods continue to evolve, and further research and application are needed.

Ultimately, to assess the role of genes and their interactions with other genes and the environment in population health will require large representative

Table 21.5 Example of a case-only epidemiologic study design (adapted from [27])

I. Scanning for complex genotypes that could have a significant attributable fraction				
Gene variants at N loci			Cases (T)	
1	2	3.....N		
−	−	−.....−	A	
+	−	−.......−	B	
.	.	....	..	
+	+	+......+	X	
For complex genotypes at N loci, the expected proportion of the population with such a combination will decrease markedly with increasing number of loci, even if each variant is common in the population. For example, if we have genetic variants at 10 loci each with 50% prevalence in the population, about 1 in 1,000 or fewer people are expected to be positive for all 10. Therefore, we can use the ratio of X to T to derive an upper bound of population attributable fraction for complex genotypes even in the absence of controls (for more details, consult [5]				
II. Assessing etiologic heterogeneity and genotype–phenotype correlation among cases				
		Cases		
Risk factor (exposure/genotype)		Phenotype 1	Phenotype 2	
Yes		A	C	
No		C	D	
Odds Ratio = AD/BC = 1 if homogeneous subgroups				
III. Screening for multiplicative gene–environment or gene–gene interaction				
		Cases		
Risk factor (exposure/genotype)		Genotype 1	Genotype 2	
Yes		A	B	
No		C	D	
Odds Ratio = AD/BC = 1 if joint effects are multiplicative and > 1 if supramultiplicative				

samples of populations. For published data, systematic reviews and meta-analyses provide a valuable tool for summarizing genetic effects and for identifying and explaining the underlying differences and observed discrepancies between studies [47]. Meta-analyses of gene-disease association studies are accepted as an important method for establishing the genetic components of complex diseases [37, 48]. A set of criteria (the Venice criteria) have been published as one way to assess the credibility of cumulative evidence on genetic associations derived from multiple sources [35]. These criteria take into consideration the amount of evidence, replication, and methodologic issues surrounding the quality of the evidence and potential effects from various biases.

Understanding how gene discoveries may impact population health further requires assessing various epidemiologic measures of disease outcomes, and in particular, absolute and attributable risks. *Absolute risk* is defined as the probability that persons with a particular characteristic, such as a specific genotype, will in fact develop disease (a concept similar to *penetrance* in genetics). To calculate direct estimates of absolute risk requires cohort studies, as it is estimated from the cumulative incidence of disease in a well-defined population. In case-control studies, absolute risk can be imputed if the representative population can be reconstructed and appropriate weights applied to recreate the sampling fraction.

The proportion of cases that would not have occurred within a certain time period had the risk factor (e.g., genotype) been absent is calculated with the *population attributable fraction*. It is the overall contribution of a particular risk factor (e.g., genotype) to the occurrence of disease in a given population. The formula for calculating attributable fraction as proposed by Miettinen is:

Population Attributable Fraction = $f_c(R-1)/R$ where f_c is the fraction of cases with the risk factor and R is the measure of relative risk [54]. Interpreting attributable fraction in the context of potential genetic and environmental interactions is currently poorly understood [78]. The concept of attributable fraction in environmental epidemiology is intuitive since it denotes a reduction in disease occurrence by removal of putative exposures from the population. Though genes cannot be removed, the concept of attributable fraction in gene-disease associations is still valid, as it approximates the influence of the genetic variant on disease occurrence in the population. We note that the genetic concept of heritability is not directly useful for determining the genetic or environmental contributions to disease or for estimating attributable fraction [8]. Heritability is population specific and depends on specific assumptions about additive effects of genetic components and lack of correlation with environmental exposures among family members.

Finally, for any gene discovery to be translated to population health for clinical application, such as genetic testing, parameters for tests in clinical practice (analytic validity, clinical validity, clinical utility) need to be applied (defined in detail in Tables 21.6, and 21.7), as first recommended by the Secretary's Advisory Committee on Genetic Testing [31, 68]. Briefly, such parameters include *analytic validity* which refers to the accuracy of a genetic test. This criterion evaluates the ability of a test to accurately "measure or detect the analyte it is intended to measure or detect" [68]. It is evaluated by calculating the test sensitivity, which provides the probability that a positive test is truly positive (e.g., will detect the analyte when it is in fact present), and test specificity, which provides the probability that the test will not detect the analyte when it is not present. *Clinical validity* com-

Table 21.6 Evaluating the analytic and clinical validity and utility of genetic markers

Type of evaluation[a]	Terms/variables	Types of studies
Analytic validity	Analytic sensitivity, specificity and predictive values	Laboratory studies "Transitional" studies
Clinical validity	Risk of current or future clinical outcomes with and without markers	Traditional epidemiologic study designs
Clinical utility	Risk of disease with and without using biomarker and accompanying interventions	Controlled clinical trials, observational clinical epidemiologic studies

[a]Adapted from the Task Force on Genetic Testing (1997) and the Secretary's Advisory Committee on Genetic Testing (2000)

Table 21.7 Defining analytic and clinical validity and utility of genetic markers*

I. Analytic validity

	Genetic marker	
Test	Present	Absent
+	A	B
-	C	D

Analytic sensitivity: A/A+C
Analytic specificity: D/B+D

II. Clinical validity of biomarkers as risk factors and as clinical tests

	Clinical outcome	
Genetic marker	Present	Absent
+	A	B
-	C	D

Odds ratio = AD/BC (for case-control study)
For cohort study:
Risk ratio = (A/A+B) / (C/C+D)
Clinical sensitivity: A/A+C
Clinical specificity: D/B+D
Positive predictive value (PPV)^a = A/A+B
Negative predictive value (NPV)* = D/C+D
^{a}Assuming total population tested

III. Clinical utility (an example) among persons with a genetic marker; comparing two hypothetical interventions in context of controlled clinical trial: risk ratio = (A/A+B) / (C/C+D)

	Outcome	
Intervention	Disease	No disease
1	A	B
2	C	D

^{a}Tables are for illustrative purposes only and apply to dichotomous biomarkers (present/absent). Additional analyses may involve stratification, person-time analysis in cohort studies, adjustment for confounding and assessing for effect modification

*Adapted from the Task Force on Genetic Testing (1997) and the Secretary's Advisory Committee on Genetic Testing (2000)

prises the sensitivity, specificity, and predictive values of a test to measure its intended clinical (or subclinical) endpoint. Specifically, clinical validity is defined as the "probability that a person with disease, or who will get a disease, will have a positive test result" [68]. Measures of sensitivity and specificity are now evaluated in a representative sample of the population for whom the test is intended. Further, positive predictive value, the "probability that a person with a positive test result has, or will get, the disease for which the analyte is used as a predictor" [68] is determined. Epidemiologic study designs are critical in establishing clinical validity. Finally, *clinical utility* is ideally evaluated in clinical trials to determine the positive and negative predictive values of a test and "to demonstrate the benefits and risks from both positive and negative results" [68]. All values for analytic and clinical validity and utility should ideally be evaluated prior to implementing testing into practice.

21.6 Beyond Gene Discovery: Epidemiologic Assessment of Genetic Information in Medicine and Public Health

Translating genetic information to medicine and public health remains challenging. As in other areas of medicine and public health, translation can be delineated into four overlapping phases as described by Khoury et al. [45]. Briefly, phase 1 translational research is defined by research that seeks to move a basic genome-based discovery into a candidate health application (e.g., genetic test/intervention). Examples of phase 1 translational research are epidemiologic studies aimed at identifying and confirming gene-disease associations, such as ongoing GWAS efforts, and also characterizing gene-disease biology and gene-environment interactions. Phase 2 translational research assesses the value of a genomic application for health practice, leading to the development of evidence-based guidelines. An example of phase 2 translational research is determining the positive predictive value of confirmed gene-disease associations, such as the positive predictive value of 8q24 SNPs in prostate cancer or of *BRCA* mutations among women at high risk for breast cancer. Phase 3 translational research attempts to move evidence-based guidelines into health practice, through delivery, dissemination, and diffusion research. Though phase 3 translational research is premature for GWAS results at the current time of writing, there are successful models for phase 3 translational research, such as studies determining the proportion of women with a family history of breast or ovarian cancer who are tested for *BRCA,* and determining barriers to testing and best approaches to implementation in practice and dissemination of evidence guidelines. Finally, phase 4 translational research seeks to evaluate the "real world" health outcomes of a genomic application in practice. Identifying whether *BRCA* testing in asymptomatic women reduces breast cancer incidence or improves survival outcomes is an example of phase 4 translational research. At present, the vast majority of genomics research can be considered phase 1, as we are still in the early stages of applying genomic information to public health applications. Epidemiologic methods and approaches are critical to all phases. In phases 2 and beyond, these methods are often considered in the domains of clinical epidemiology, applied epidemiology, or health services research (Table 21.1).

"Personalized medicine" has been promised as an end-result of genomics research. However, to attain this goal will require multidisciplinary translational research efforts as described above. To date, the utility of testing for common polymorphism in predicting disease outcomes has yet to be demonstrated. Further, risk estimates and subsequent predictability resulting from joint effects of polymorphisms with other risk factors (e.g., personal and family history, environmental exposures, and behavioral risk factors) remain unmeasured [3, 32, 43, 70]. There have been varying claims regarding the utility and predictability of genetic markers as risk factors. Although the combination of each newly discovered SNP association provides high absolute risk and population attributable fraction, a recent systematic evaluation of seven common SNPs identified in breast cancer showed that the discriminatory value of all seven SNPs had less accuracy than currently known risk factors for breast cancer (e.g., based on ages at menarche and first live birth, family history of breast cancer, and history of breast biopsy examinations). Even the addition of the seven SNPs to the current risk assessment module contributed less to the discriminatory accuracy than did additional results from mammographic density, thus suggesting much larger numbers of SNPs will be required to achieve high discriminatory accuracy [26]. Similar evaluations must be taken for other diseases to determine the true predictability and contribution of SNPs to disease. Importantly, to accurately calculate the predictive values of genetic information will require consideration of gene-gene and gene-environment interactions. Therefore, additional well-conducted epidemiologic studies will be needed, both case-control and cohort studies with long-term follow-up. Only with meticulous analysis from well-conducted studies can the complex relationships between genes, exposures and other population characteristics truly be understood and used to improve population health and to achieve the goals of personalized medicine.

Several on-going initiatives reflect such efforts. The largest effort being conducted is the UK Biobank, which aims to collect information on the health and lifestyle of 500,000 volunteers aged between 40 and 69. The framework for enrolling and following up the Biobank cohort is the UK National Health Service, which provides health care to the entire UK population. Planned follow-up is for 20 or more years, upon which DNA samples and population information will be available for scientifically and ethically approved research [75]. Though no comparable infrastructure exists in the United States, with its highly decentralized health system and mobile population, the Northern California Kaiser Permanente Health Maintenance Organization has recently launched the Research Program on Genes, Environment and Health, which similarly aims to enroll 500,000 volunteers in Northern California with a questionnaire and DNA samples for future research on etiologic and pharmacogenetic outcomes [39]. The American Cancer Society is similarly aiming to enroll 500,000 men and women in their Cancer Prevention Study-3 (CPS-3) to allow evaluation of environmental, lifestyle, and genetic factors as related to cancer. Though unquestionably valuable, these population-based biobanks will require long-term follow-up to ascertain sufficient numbers of persons with selected

Table 21.8 Examples of large-scale population-based genomics studies

Study	Sample size	Population	Study objectives
Decode genetics	>100,000	Iceland	"To identify genetic causes of common diseases and develop new drugs and diagnostic tools" Measures genes, health outcomes, and link with genealogical database
UK Biobank	500,000	Population sample of persons 45-69 years	"To study the role of genes, environment, and lifestyle" Link with medical records
CartaGene[a] (Quebec)	>60,000 persons 25–74 years	Population sample	"To study genetic variation in a modern population." Link with health care records, and genealogical databases
Estonia Genome Project[a]	>1,000,000	Estonian population	"To find genes that cause and influence common diseases" Link with medical records
GenomeEUtwin[a]	~800,000 twin pairs	Twin cohorts from seven European countries and Australia	"To characterize genetic, environmental and lifestyle components in the background of health problems"
Kaiser Permanente Research Program on Genes, Environment and Health (RPGEH)	500,000	Population sample	"To identify genetic and environmental factors that can lead to disease or affect how a person reacts to medications."

[a]Part of the global P3G collaboration (Public Population Project in Genomics); complete list of P3G participants found in http://www.p3gconsortium.org/memb.cfm

health outcomes for scientific analysis [7]. In the meantime, complementary approaches have gained traction. he Wellcome Trust Case Control Consortium and the Genetic Association Information Network (GAIN) are poised to mine genetic association data within previously conducted, large-scale epidemiologic studies [25, 79]. Some on-going efforts are summarized in Table 21.8.

21.7 Policy, Ethical and Practice Considerations: the Emergence of Public Health Genomics

The potential application of genetic information in public health practice raises a number of ethical, legal, and social issues, also referred to commonly with the acroynym ELSI [53, 66, 67]. As the field of genetic epidemiology moves towards big science and biospecimen-intensive collection in large population samples, ethical issues arise concerning enrollment of study participants and collection of biological specimens used for evaluation of genetic data. The specific ethical, legal, and social issues around generating genetic information will vary with time, as they are directly tied to the clinical utility of the genetic data. For example, present specific issues of concern include language on informed consent documents, recruitment of subjects, sending study results to patients who request it, and the potential for discrimination or stigmatization of individuals or groups. The creation of biobanks and large population-based studies relies on language in informed consent documents that specify the risks and benefits of participating in the study are directly related to the meaning of the results. Because we are in the early stages of understanding the role of genetic information in health and disease, the impact and potential risks remain relatively small. However, this may change during the duration of a study, as new results are published regarding potential clinical applications for medicine and public health. In consultation with a multidisciplinary group, the Centers for Disease Control and Prevention (CDC) has published an online consent form template and supplemental information that can be adapted by researchers for population-based genetic epidemiologic studies [2, 9, 10]. Informed consent will surely evolve as additional information is known about the public health significance of genes or panels of genes and gene-environment interactions.

In the meantime, there is an alarming trend of direct-to-consumer advertising of whole-genome analysis by several companies [38]. These companies offer "genetic profile" tests directly to practitioners and the public, accompanied by personalized lifestyle recommendations, such as advice on dietary supplements. This phenomenon can contribute to misunderstanding of the relationship between cause and effect at the individual and population levels [78]. Ethical, legal, and social issues would demand that these issues be explained explicitly and that transparency and rigor in genetic epidemiologic research be promoted. Unfortunately, as these tests are not currently regulated (in the United States), there is no forum for addressing ethical, legal, and social issues in such instances. However, this practice has been reviewed by Congress, and further scrutiny is likely, given their recognized potential for misleading consumers [24, 76].

Before translation of genomic information is truly integrated into clinical and public health practice, it is clear that appropriate ethical, legal, and social issues will need to be addressed. The most common concerns regarding genetic information pertain to consumer access to health insurance, employment, education, and loans [13]. Combating these concerns will require informing health providers as well as patients and the general public of the risks and benefits of genetic data and addressing current misconceptions about genetic information, such as the notion of genetic determinism. Importantly, clarifying complex issues of confidentiality and privacy will be required. As the field of genetic epidemiology and the translational aspects of research evolve, additional legislative and regulatory responses will certainly be needed.

References

1. Amos CI (1994) Robust variance-components approach for assessing genetic linkage in pedigrees. Am J Hum Genet 54(3):535–543
2. Beskow LM, Burke W, Merz JF et al (2001) Informed consent for population-based research involving genetics. JAMA 286(18):2315–2321
3. Bianchi MT, Alexander BM (2006) Evidence based diagnosis: does the language reflect the theory? BMJ 333(7565):442–445
4. Blackwelder WC, Elston RC (1985) A comparison of sib-pair linkage tests for disease susceptibility loci. Genet Epidemiol 2(1):85–97
5. Botto LD, Khoury MJ (2001) Commentary: facing the challenge of gene-environment interaction: the two-by-four table and beyond. Am J Epidemiol 153(10):1016–1020
6. Burke W, Khoury MJ, Stewart A, Zimmern RL (2006) The path from genome-based research to population health: development of an international public health genomics network. Genet Med 8(7):451–458
7. Burton PR, Hansell AL, Fortier I et al (2009) Size matters: just how big is BIG?: quantifying realistic sample size requirements for human genome epidemiology. Int J Epidemiol 38:274–275
8. Burton PR, Tobin MD, Hopper JL (2005) Key concepts in genetic epidemiology. Lancet 366(9489):941–951
9. Centers for Disease Control and Prevention. Informed consent template for population-based research involving genetics. Centers for Disease Control and Prevention 2008 Available from: URL: http://www.cdc.gov/genomics/population/publications/consent.htm
10. Centers for Disease Control and Prevention. Supplemental brochure for population-based research involving genetics. Centers for Disease Control and Prevention 2008 Available from: URL: http://www.cdc.gov/genomics/population/publications/brochure.htm
11. Chang MH, Lindegren ML, Butler MA et al (2009) Prevalence in the United States of selected candidate gene variants: third national health and nutrition examination survey, 1991–1994. Am J Epidemiol 169:54–66
12. Chatterjee N, Kalaylioglu Z, Moslehi R, Peters U, Wacholder S (2006) Powerful multilocus tests of genetic association in the presence of gene-gene and gene-environment interactions. Am J Hum Genet 79(6):1002–1016
13. Clayton EW (2003) Ethical, legal, and social implications of genomic medicine. N Engl J Med 349(6):562–569
14. Collins FS, Morgan M, Patrinos A (2003) The human genome project: lessons from large-scale biology. Science 300(5617):286–290
15. Cordell HJ, Clayton DG (2005) Genetic association studies. Lancet 366(9491):1121–1131
16. Devlin B, Roeder K (1999) Genomic control for association studies. Biometrics 55(4):997–1004
17. Dunson DB (2001) Commentary: practical advantages of Bayesian analysis of epidemiologic data. Am J Epidemiol 153(12):1222–1226
18. Easton DF, Eeles RA (2008) Genome-wide association studies in cancer. Hum Mol Genet 17(R2):R109–R115
19. Editorial (2007) Genome-wide association study of 14,000 cases of seven common diseases and 3,000 shared controls. Nature 447(7145):661–678
20. Elston RC (1992) Segregation and linkage analysis. Anim Genet 23(1):59–62
21. Elston RC, Buxbaum S, Jacobs KB, Olson JM (2000) Haseman and Elston revisited. Genet Epidemiol 19(1):1–17
22. Falconer DC, MacKay TFC (1994) Introduction to quantitative genetics, 4th edn. Chapman and Hall, New York

23. Falush D, Wirth T, Linz B et al (2003) Traces of human migrations in Helicobacter pylori populations. Science 299(5612):1582–1585
24. Food and Drug Administration. Draft guidance for industry, clinical laboratories and FDA staff on in vitro diagnostic multivariate index assays. Food and Drug Administration 2008Available from: URL: http://www.fda.gov/OHRMS/DOCKETS/98fr/ch0641.pdf
25. Foundation for the National Institutes of Health. Genetic Association Information Network (GAIN). Foundation for the National Institutes of Health 2006Available from: URL: http://www.fnih.org/GAIN/GAIN_home.shtml
26. Gail MH (2008) Discriminatory accuracy from single-nucleotide polymorphisms in models to predict breast cancer risk. J Natl Cancer Inst 100(14):1037–1041
27. Gatto NM, Campbell UB, Rundle AG, Ahsan H (2004) Further development of the case-only design for assessing gene-environment interaction: evaluation of and adjustment for bias. Int J Epidemiol 33(5):1014–1024
28. Greenland S (1999) Multilevel modeling and model averaging. Scand J Work Environ Health 25(Suppl 4):43–48
29. Guttmacher AE, Collins FS (2002) Genomic medicine – a primer. N Engl J Med 347(19):1512–1520
30. Guttmacher AE, Collins FS (2003) Welcome to the genomic era. N Engl J Med 349(10):996–998
31. Haddow JE, Palomaki GE (2004) ACCE: a model for evaluating data on emerging genetic tests. In: Khoury MJ, Little J, Burke W (eds) Human genome epidemiology: a scientific foundation for using genetic information to improve health and prevent disease. Oxford University Press, New York, pp 217–233
32. Haga SB, Khoury MJ, Burke W (2003) Genomic profiling to promote a healthy lifestyle: not ready for prime time. Nat Genet 34(4):347–350
33. Hahn LW, Ritchie MD, Moore JH (2003) Multifactor dimensionality reduction software for detecting gene-gene and gene-environment interactions. Bioinformatics 19(3):376–382
34. Haseman JK, Elston RC (1972) The investigation of linkage between a quantitative trait and a marker locus. Behav Genet 2(1):3–19
35. Hoh J, Ott J (2003) Mathematical multi-locus approaches to localizing complex human trait genes. Nat Rev Genet 4(9):701–709
36. Ioannidis JP, Ntzani EE, Trikalinos TA, Contopoulos-Ioannidis DG (2001) Replication validity of genetic association studies. Nat Genet 29(3):306–309
37. Ioannidis JP, Boffetta P, Little J et al (2008) Assessment of cumulative evidence on genetic associations: interim guidelines. Int J Epidemiol 37(1):120–132
38. Janssens AC, Gwinn M, Valdez R, Narayan KM, Khoury MJ (2006) Predictive genetic testing for type 2 diabetes. BMJ 333(7567):509–510
39. Kaiser Permanente Division of Research. Research Program on Genes, Environment and Health. Kaiser Permanente Division of Research 2008Available from: URL: http://www.dor.kaiser.org/studies/rpgeh/index.html
40. Khoury MJ, Beaty TH, Cohen BH (1993) Fundamentals of genetic epidemiology. Oxford University Press, Oxford
41. Khoury MJ, Flanders WD (1996) Nontraditional epidemiologic approaches in the analysis of gene-environment interaction: case-control studies with no controls. Am J Epidemiol 144(3):207–213
42. Khoury MJ, Millikan R, Little J, Gwinn M (2004) The emergence of epidemiology in the genomics age. Int J Epidemiol 33(5):936–944
43. Khoury MJ, Yang Q, Gwinn M, Little J, Dana FW (2004) An epidemiologic assessment of genomic profiling for measuring susceptibility to common diseases and targeting interventions. Genet Med 6(1):38–47
44. Khoury MJ, Little J, Burke W (2004) Human genome epidmiology: a scientific foundation for using genetic information to improve health and prevent disease. Oxford University Press, New York
45. Khoury MJ, Gwinn M, Yoon PW, Dowling N, Moore CA, Bradley L (2007) The continuum of translation research in genomic medicine: how can we accelerate the appropriate integration of human genome discoveries into health care and disease prevention? Genet Med 9(10): 665–674
46. Kruglyak L, Daly MJ, Reeve-Daly MP, Lander ES (1996) Parametric and nonparametric linkage analysis: a unified multipoint approach. Am J Hum Genet 58(6):1347–1363
47. Lau J, Ioannidis JP, Schmid CH (1997) Quantitative synthesis in systematic reviews. Ann Intern Med 127(9): 820–826
48. Lohmueller KE, Pearce CL, Pike M, Lander ES, Hirschhorn JN (2003) Meta-analysis of genetic association studies supports a contribution of common variants to susceptibility to common disease. Nat Genet 33(2):177–182
49. Manolio TA, Bailey-Wilson JE, Collins FS (2006) Genes, environment and the value of prospective cohort studies. Nat Rev Genet 7(10):812–820
50. Marazita ML, Murray JC, Lidral AC et al (2004) Meta-analysis of 13 genome scans reveals multiple cleft lip/palate genes with novel loci on 9q21 and 2q32–35. Am J Hum Genet 75(2):161–173
51. McCarthy MI, Abecasis GR, Cardon LR et al (2008) Genome-wide association studies for complex traits: consensus, uncertainty and challenges. Nat Rev Genet 9(5): 356–369
52. McClain MR, Palomaki GE, Nathanson KL, Haddow JE (2005) Adjusting the estimated proportion of breast cancer cases associated with BRCA1 and BRCA2 mutations: public health implications. Genet Med 7(1):28–33
53. Meslin EM, Thomson EJ, Boyer JT (1997) The Ethical, legal, and social implications research program at the national human genome research institute. Kennedy Inst Ethics J 7(3):291–298
54. Miettinen OS (1974) Proportion of disease caused or prevented by a given exposure, trait or intervention. Am J Epidemiol 99(5):325–332
55. National Library of Medicine. National Library of Medicine, Online Inheritance in Man (OMIM). National Library of Medicine, Online Inheritance in Man (OMIM) 2008Available from: URL: http://www.ncbi.nlm.nih.gov/entrez/query.fcgi?db=OMIM
56. Nelson MR, Kardia SL, Ferrell RE, Sing CF (2001) A combinatorial partitioning method to identify multilocus genotypic partitions that predict quantitative trait variation. Genome Res 11(3):458–470

57. Nelson MR, Bryc K, King KS et al (2008) The Population Reference Sample, POPRES: a resource for population, disease, and pharmacological genetics research. Am J Hum Genet 83(3):347–358
58. Palmer LJ, Cardon LR (2005) Shaking the tree: mapping complex disease genes with linkage disequilibrium. Lancet 366(9492):1223–1234
59. Pearson TA, Manolio TA (2008) How to interpret a genome-wide association study. JAMA 299(11):1335–1344
60. Penrose LS (1938) Genetic linkage in graded human characters. Ann Eugen 8:233–237
61. Price AL, Patterson NJ, Plenge RM, Weinblatt ME, Shadick NA, Reich D (2006) Principal components analysis corrects for stratification in genome-wide association studies. Nat Genet 38(8):904–909
62. Pritchard JK, Stephens M, Donnelly P (2000) Inference of population structure using multilocus genotype data. Genetics 155(2):945–959
63. Risch N, Merikangas K (1996) The future of genetic studies of complex human diseases. Science 273(5281):1516–1517
64. Ritchie MD, Hahn LW, Roodi N et al (2001) Multifactor-dimensionality reduction reveals high-order interactions among estrogen-metabolism genes in sporadic breast cancer. Am J Hum Genet 69(1):138–147
65. Schaid DJ, Olson JM, Gauderman WJ, Elston RC (2003) Regression models for linkage: issues of traits, covariates, heterogeneity, and interaction. Hum Hered 55(2–3):86–96
66. Schulte PA (2004) Some implications of genetic biomarkers in occupational epidemiology and practice. Scand J Work Environ Health 30(1):71–79
67. Schulte PA, Lomax GP, Ward EM, Colligan MJ (1999) Ethical issues in the use of genetic markers in occupational epidemiologic research. J Occup Environ Med 41(8):639–646
68. Secretary's Advisory Committee on Genetic Testing. Task force on genetic testing: Orinitung safe and effective genetic testing in the United States. Final report 1997. Secretary's Advisory Committee on Genetic Testing 2002Available from: URL: http://www.genome.gov/10001733
69. Sham PC, Purcell S, Cherny SS, Abecasis GR (2002) Powerful regression-based quantitative-trait linkage analysis of general pedigrees. Am J Hum Genet 71(2):238–253
70. Sipkoff M (2005) Predictive modeling & genomics: marriage of promise and risk. Manag Care 14(5):60 63–64, 66
71. Smith M, Smalley S, Cantor R et al (1990) Mapping of a gene determining tuberous sclerosis to human chromosome 11q14–11q23. Genomics 6(1):105–114
72. Task Force on Genetic Testing (1997) In: Holtz NA, Watson MS (eds) National Institutes of Health-Department of Energy Working Group on Ethical, Legal and Social Implications of Human Genome Research
73. Thomas DC. Statistical methods in genetic epidemiology. Oxford University Press, Oxford, 2004.
74. Thomas DC, Haile RW, Duggan D (2005) Recent developments in genomewide association scans: a workshop summary and review. Am J Hum Genet 77(3):337–345
75. UK Biobank. UK Biobank: improving the health of future generations. UK Biobank 2006Available from: URL: http://www.ukbiobank.ac.uk
76. United States Government Accountability Office. Testimony before the Special Committee on Again, U.S. Senate: Nutrigenetic testing – tests purchased from four web sites mislead. United States Government Accountability Office 2008Available from: URL: http://www.gao.gov/new.items/d06977t.pdf
77. Vandenbroucke JP, Koster T, Briet E, Reitsma PH, Bertina RM, Rosendaal FR (1994) Increased risk of venous thrombosis in oral-contraceptive users who are carriers of factor V Leiden mutation. Lancet 344(8935):1453–1457
78. Vineis P, Kriebel D (2006) Causal models in epidemiology: past inheritance and genetic future. Environ Health 5:21
79. Wellcome Trust Case Control Consortium. Wellcome Trust Case Control Consortium 2008Available from: URL: http://www.wtccc.org.uk
80. Witte JS, Gauderman WJ, Thomas DC (1999) Asymptotic bias and efficiency in case-control studies of candidate genes and gene-environment interactions: basic family designs. Am J Epidemiol 149(8):693–705

Pharmacogenetics

22

Nicole M. Walley, Paola Nicoletti, and David B. Goldstein

Abstract The goal of pharmacogenetics is to identify genetic differences among patients that influence treatment response. Pharmacogenetics also represents a tractable target for systematic studies, because key treatment responses are both clinically important and amenable to genetic investigation. Despite a number of paradigmatic studies, pharmacogenetics has to date only had a minimal impact in medicine. Here we outline several methodological considerations for implementation in future studies and also detail how we can best apply the lessons learned from extensive genetic studies on disease predisposition in order to advance the state of pharmacogenetic knowledge and to expedite the translation of pharmacogenetic findings into clinical practice. We conclude that while the study of common variation may provide some insight, like the study of disease predisposition itself, pharmacogenetics will ultimately require characterization of rare human gene variants.

Contents

22.1 Introduction 636
    22.1.1 The Goal of Pharmacogenetics 636
    22.1.2 Why Pharmacogenetic Studies Are Necessary 636

22.2 Important Pharmacogenetic Discoveries to Date 637
    22.2.1 CYP2D6 Polymorphism and Pharmacogenetics 637
    22.2.2 Classic/Paradigmatic Studies 638

22.3 Pharmacogenetic Methodology: Clinical Practice 639
    22.3.1 Phenotype 639
    22.3.2 Sample and Patient Recruitment 640

22.4 Pharmacogenetic Methodology: Genomics 640
    22.4.1 Candidate Gene Studies 640
    22.4.2 Whole-Genome SNP Analyzes 642
    22.4.3 Rare Variants and Whole-Genome Sequencing 643

22.5 Challenges in Pharmacogenetics 643
    22.5.1 Polygenic Inheritance 643
    22.5.2 Pharmacogenetics in the Clinic 643
    22.5.3 Pharmacogenetics Across Ethnic Groups 644
    22.5.3 Drug Classes of Urgent Interest 645

2.6 Conclusions 646

References 646

N.M. Walley
Department of Biology, Duke University, Durham, NC, USA

P. Nicoletti
Center for Human Genome Variation, Institute for Genome Sciences and Policy, Duke University, Durham, NC, USA

D.B. Goldstein (✉)
Center for Human Genome Variation, Institute for Genome Sciences and Policy, Duke University, Box 91009, Durham NC 27708, USA
e-mail: d.goldstein@duke.edu

22.1 Introduction

22.1.1 The Goal of Pharmacogenetics

Variation amongst patients in their responses to medication is a consistent clinical experience in most if not all therapeutic areas. Some of this variation is due to environmental factors, such as drug–drug interactions or co-morbid disease, and can be anticipated and at least partially avoided. A large proportion of variable drug-response, however, remains unexplained by environmental factors, suggesting that underlying genetic factors may cause or contribute to a patient's response to a particular medication. The goal of pharmacogenetics is to identify genetic differences among patients that influence treatment response.

22.1.2 Why Pharmacogenetic Studies Are Necessary

Pharmacogenetics also represents a tractable target for systematic studies, because key treatment responses are both clinically important and amenable to genetic investigation. The need for better understanding of the genetic determinants of treatment response is particularly essential with respect to three major response phenotypes: efficacy, safety, and optimum dose.

22.1.1.1 Efficacy

Despite the plethora of pharmacological treatments that exist for any given disorder, drug resistance/non-response remains a serious clinical issue. It is estimated, for example, that 30% of patients with epilepsy, and 30–50% of patients with schizophrenia do not respond to any of the currently available drug treatments. For these treatment-resistant patients there are frequently few, if any, treatment options, which results in a significant proportion of patients with a definitive diagnosis of disease but no possibility of medical intervention.

Efficacy issues also become apparent when the current crisis in drug development is considered: starting in 1998, despite an increase in the productivity of drug discovery, the number of new molecular entities submitted to the FDA and EMEA decreased by nearly 50%. At the same time, clinical trial budgets have increased by 58%, while the failure rate of phase III clinical trials is nearly 50% [26]. The incorporation of pharmacogenetic information into trials could be of financial benefit, however, if preclinical studies began to include pharmacogenetic components. The increased ability to predict the success – or failure – of a drug based on pharmacogenetic knowledge prior to large-scale, expensive trials will simplify phase III trials, reduce costs, and increase the chance of success [26]. In addition, pharmacogenetic discoveries that elucidate the underlying molecular bases of treatment failure stand to facilitate the development of new compounds to treat patients that are resistant to current medications.

22.1.1.2 Safety

Drug treatment is frequently associated with adverse drug reactions (ADRs), which can vary in severity from mild to life threatening. Indeed, many medications fail clinical trials because of the incidence of ADRs, and multiple medications have been withdrawn from the market post approval following the emergence of severe ADRs in patients. Torsade de pointes (TdP/prolonged QT interval) is one severe ADR that has led to the withdrawal of several drugs, including astemizole (an allergy medication), grepafloxacin (an antibiotic), and sertindole (an antipsychotic), in the past 10 years. Currently, an estimated 5–13% of hospital admissions are due to adverse reactions even to FDA-approved drugs [2, 8].

In addition to severe ADRs, there are also adverse reactions that, while not immediately life threatening, are nonetheless treatment limiting. For example, ADRs such as hypersomnolence, diplopia, tremor, or dizziness may prohibit the administration of therapeutic doses. Furthermore, marked weight gain is often associated with several drugs, such as the tricyclic antidepressants amitriptyline and nortriptyline, the anticonvulsants valproate and pregabalin, and the atypical antipsychotics clozapine and olanzapine, among others. The increase in body mass index (BMI) as a result of these treatments can be considerable and can lead to such obesity-related comorbidities as type II diabetes, hypertension, and cardiovascular disease.

The frequency of mild ADRs is difficult to estimate, as they commonly go unreported; however, it is not uncommon for these types of ADRs to precipitate treatment discontinuation, regardless of efficacy outcomes, representing a significant limitation to effective treatment.

22.1.1.3 Dose

Optimal drug dosing represents a third clinical area that stands to be improved by pharmacogenetic investigation. Many clinically used drugs have a narrow therapeutic window, so that doses that are too low are not effective, while doses that are too high quickly lead to drug toxicity and dose-related ADRs. For many of these same drugs, different patients also require dramatically different therapeutic doses. For example, this represents a particular clinical challenge with respect to antiepileptic drugs, where patients must be started on low doses and these slowly titrated until seizure control is achieved or until unacceptable ADRs necessitate medication withdrawal. Unfortunately, dose titration can take months, during which time seizures persist. Pharmacogenetic predictors of dose, however, could allow clinicians to safely expedite the determination of optimal dose and, particularly for patients who require extremely high doses, minimize the amount of time a patient remains in the subtherapeutic range.

22.2 Important Pharmacogenetic Discoveries to Date

The concept that treatment outcome may be affected by genetic variants was first posed by Motulsky in 1957 [32], and the term "pharmacogenetics" was coined by Vogel in 1959. In the 1970s twin studies demonstrated that monozygotic (identical) twins demonstrate highly concordant plasma levels of multiple drugs, whereas the plasma levels in pairs of dizygotic (fraternal) twins vary widely [1, 43], indicating a strong genetic influence. The subsequent identification of debrisoquine polymorphism (cytochrome P450 2D6; CYP2D6) motivated many of the early pharmacogenetic studies.

22.2.1 CYP2D6 Polymorphism and Pharmacogenetics

Debrisoquine polymorphism was identified on the basis of drug clearance studies of the antihypertensive drug debrisoquine, in which two distinct phenotypes were identified: extensive metabolizers (EM) who excrete significant amounts of debrisoquine metabolite, and poor metabolizers (PM) who excrete very little metabolite [7, 24]. This phenotype was later found to be caused by defects in a cytochrome P450 gene [13, 14] now known as *CYP2D6*.

Early studies in pharmacogenetics focused considerable attention on *CYP2D6*, because the gene is highly polymorphic with over 70 identified allelic variants (http://www.cypalleles.ki.se/cyp2d6.htm) and the CYP 2D6 enzyme metabolizes approximately 25% of all medications, including a large number of antidepressants and neuroleptics (Table 22.1). Several studies have found a relationship between the PM phenotype and serum concentrations of a number of drugs, including amitriptyline, nortriptyline, mirtazapine, and aripiprazole [14, 15, 18]. Although there is a clear relationship between *CYP2D6* status and serum concentrations, the correlation does not appear to extend predictably to other important clinical outcomes, such as dose (as determined in empirical medication trials) or efficacy. Therefore, the extent of the clinical utility of *CYP2D6* genotype has not been determined, and there

Table 22.1 Some drugs metabolized by CYP2D6

Drug	Drug class
Sparteine	Antiarrhythmic
Amitriptyline	Antidepressant
Desipramine	Antidepressant
Fluoxetine	Antidepressant
Imipramine	Antidepressant
Nortriptyline	Antidepressant
Paroxetine	Antidepressant
Dextromethorphan	Antitussive
Codeine	Analgesic
Tramadol	Analgesic
Metroprolol	b-Adrenoreceptor blocker
Propanolol	b-Adrenoreceptor blocker
Haloperidol	Neuroleptic
Perphenazine	Neuroleptic
Risperidone	Neuroleptic

are currently no guidelines for the incorporation of *CYP2D6* genetic testing into treatment practice.

22.2.2 Classic/Paradigmatic Studies

22.2.2.1 Isoniazid, Azathioprine, Mercaptopurine and Other Poor Metabolizer Phenotypes

As previously discussed, ADRs are a major cause for concern in clinical use of medications. Indeed, some of the early pharmacogenetic studies concerned ADRs occurring in patients exposed to the drugs isoniazid (used to prevent/treat tuberculosis) and the immunosuppressants azathioprine and mercaptopurine. Each of these drugs is associated with significant ADRs (hepatotoxicity and thiopurine toxicity, respectively), and initial candidate gene studies assessed functional genetic variation in the major drug-metabolizing enzymes (DMEs) of these drugs. These studies revealed that genetic changes resulting in reduced/absent activity of each drug's major DME (*N*-acetyltransferase 2 (*NAT2*) and thiopurine *S*-methyltransferase (*TPMT*), respectively), were associated with the incidence of these ADRs [9, 21, 34]. As a result of these important DME studies, one of the few clinical pharmacogenetic applications became standard: testing of *TPMT* genotype or phenotype (TPMT enzyme level) is used clinically to identify those at risk for thiopurine toxicity upon exposure to azathioprine and mercaptopurine, and dosage adjustments are made before treatment to prevent the occurrence of severe ADRs.

22.2.2.2 Efficacy of Asthma Treatment

Another early pharmacogenetic discovery assessed the efficacy of inhaled albuterol for the treatment of asthma. Albuterol is an agonist of the β-2 adrenergic receptor (*ADRB2*) and provides relief of asthma symptoms. The initial study in 1997 [30] showed a significant association between improved response with a coding polymorphism (Arg-16-Gly) in the *ADRB2* gene with analyzes indicating that homozygotes of the Arg-16 allele were about five times as likely to respond to treatment as homozygotes of the Gly-16 allele. Heterozygotes are 2.3 times as likely to respond. These results may be particularly important because the Arg-16-Gly is a common allele [minor allele frequency (G, Gly) is 0.47], but the clinical relevance of this polymorphism is still under investigation.

22.2.2.3 Abacavir Hypersensitivity

Abacavir is a transcriptase inhibitor used to treat infections with human immunodeficiency virus (HIV). After abacavir was released, it emerged that 4.3% of patients exposed to the drug developed a hypersensitivity reaction that presents as fever, rash, and/or GI upset and can ultimately result in death [19]. This idiosyncratic syndrome usually regresses after treatment withdrawal and has been noted to be much more severe upon rechallenge with the drug. Genetic studies were undertaken focusing on the HLA region of the genome, and initial results indicated that the HLA-B*5701 allele is a significant predictor of the occurrence of the hypersensitivity reaction [20, 29]. Prospective clinical use of this pharmacogenetic association appears to be effective, with no cases of the hypersensitivity reaction reported in a study of 260 patients [35], and systematic genetic testing has been posited as a cost-effective method of avoiding the development of this severe ADR [22].

22.2.2.4 Tranilast and Clinical Pharmacogenetic Investigations

Tranilast is a drug that entered clinical trials as an anti-restenosis drug, but during the course of the phase III trials 12% of patients had a rise in their levels of bilirubin. Pharmacogenetic tests were carried out during the trial and identified a polymorphism in the glucuronyltransferase 1A1 gene (*UGT1A1*) that was significantly associated with the incidence of hyperbilirubinemia [6]. This TA-repeat polymorphism is also known to predispose some individuals to Gilbert's syndrome, an inherited variant that causes hyperbilirubinemia [31].

Tranilast was subsequently abandoned as an anti-restenosis drug owing to lack of efficacy; however, the study itself demonstrates the value of including pharmacogenetics in clinical studies and the importance of collecting DNA samples and detailed phenotype

22.2.2.5 Cancer Pharmacogenetics and Somatic/Acquired Polymorphisms

Though pharmacogenetic studies of many medications focus on germline polymorphisms and the roles that they play in treatment response, cancer pharmacogenetics often seeks rather to discover somatic or acquired polymorphisms in tumor cells that affect treatment outcome. Furthermore, pharmacogenetics of cancer therapy is particularly important, as anticancer treatments have some of the lowest efficacy rates among clinical treatments, at 20–80% for first-line treatments and 5–30% for subsequent treatment regimens [25].

Cancer pharmacogenetics already has a few paradigmatic studies in its relatively short research history. For example, trastuzumab (Herceptin) is an anticancer agent and monoclonal antibody that acts on a particular epidermal growth factor receptor *HER2/ERBB2*. The gene for this particular receptor is amplified in 25–30% of human breast cancer tumors, and the resulting receptor overexpression is associated with tumor aggressivity and an increased incidence of relapse and death (for review see [37]). The drug trastuzumab binds to the overexpressed receptor and signals cancer cells for destruction by the immune system. As a result, the use of trastuzumab is specifically indicated as a treatment option for tumors with *HER2/ERBB2* gene amplification and overexpression [44].

The efficacy of another drug, imatinib (Gleevec), which is used to treat chronic myeloid leukemia (CML), has also been found to depend on the genetic make-up of cancer cells. As a first-line treatment, imatinib is a very effective treatment for CML, effectively binding and inactivating the BCR-ABL protein, a tyrosine kinase that activates downstream proteins that drive white blood cell proliferation (the underlying cause of CML). However, imatinib resistance can be caused by acquired mutations in the BCR-ABL protein complex that inhibit drug binding, and among the patients in whom this occurs there is a significant rate of relapse and a much poorer prognosis. Several such somatic mutations have been characterized, though only six mutations are common and these, taken together, account for 60–70% of all mutations [45].

information on efficacy and safety during the course of preclinical and clinical trials.

22.3 Pharmacogenetic Methodology: Clinical Practice

While the above studies and many others have begun to identify variants that affect treatment response to an array of medications, the field of pharmacogenetic research as a whole has been severely limited by a number of methodological issues, including phenotypic characterization, limited sample sizes, and subpar genomic methods.

22.3.1 Phenotype

One of the most difficult components of designing pharmacogenetic studies is the definition of phenotypes, which must be done in a way that is amenable to genetic study but also reflects clinical endpoints. For the study of ADRs, the definition of phenotype is relatively straightforward as it is frequently easy to classify a patient as having experienced or not experienced a particular adverse reaction. ADRs are not always unambiguous, however, and in order to accurately ascertain their presence/absence or classify the severity of an adverse event it is sometimes necessary to conduct specific tests prospectively. For example, cognitive side effects on exposure to several neuropsychiatric medications, including antipsychotic, antiepileptic, and antidepressant medications, are a clinical concern, but are not easy to discern in a purely clinical setting because patients are being treated for diseases that affect brain function, and many of them are concomitantly also being treated with other neuropsychiatric drugs. In such a scenario, a complaint of a cognitive adverse event cannot easily be attributed to exposure to a single medication. To assess these types of effects as accurately as possible, it is necessary to set up a prospective study design that includes testing the cognitive function of the patient before and during the administration of a drug and close monitoring for any other changes in treatment, disease state, or patient behavior that could also have effects on the outcome. Without detailed information and the ability to compare the change in cognitive ability, the presence and the severity of cognitive adverse events cannot be accurately quantified.

Efficacy studies also present a challenge to pharmacogenetics in terms of defining phenotype. To an extent it is possible to classify patients as responsive or nonresponsive to a particular drug, but there are many other issues that complicate the matter. For example, in epilepsy treatment with antiepileptic drugs (AEDs) it is quite common for patients to be treated with more than one AED, as no AED used in monotherapy has proved sufficient to control seizures. The patients are "responsive," but not to any single drug, so that they are phenotypically different from patients who respond well to the first AED to which they are exposed. In addition, it is not uncommon for patients to have a partial response, where symptoms are significantly affected by a drug but not completely controlled.

Efficacy phenotypes are further complicated by the heterogeneity of certain diseases. Continuing with epilepsy as an example, the disease etiology itself is poorly understood, but it is commonly accepted that patients with epilepsy suffer from a number of different pathophysiological causes. The underlying cause of epilepsy may have a significant impact on treatment outcomes, but this cannot always easily be accounted for in pharmacogenetic studies.

In all, there is no obvious way to assign phenotypes for efficacy; however, it is essential that the designation of phenotype is thorough and unambiguous, with careful consideration of disease etiology and past and present medication use.

22.3.2 Sample and Patient Recruitment

Clinical cooperation is, of course, necessary for pharmacogenetic studies, and study design can be manipulated to accommodate the clinical advantages and limitations at every site. The quality of a genetic study reflects the quality of study design and clinical information that is captured and included. In general, clinics can approach genetic studies in three manners: clinical trials, single-hospital studies, and multiple-hospital studies, each having their own advantages and limitations.

Clinical trials by far surpass other settings in their ability to collect rich information and to selectively enroll patients who meet very particular inclusion criteria. However, clinical trials, by necessity, focus only on the most relevant efficacy and adverse reaction phenotypes, and particularly on those that might prevent a medication from being approved and reaching the market. In such a setting, although the clinical data collected are detailed and accurate, it is not statistically feasible to run analyzes on all clinical measures. Such open-ended studies are limited in statistical power owing to the multiple-comparison corrections that must be made with respect to the assessment of statistical significance. Perhaps most importantly, clinical trials have strict inclusion and exclusion criteria, which mean that the medicines are not always used in a way that reflects real clinical practice.

Single-hospital studies involve the recruitment of patients from a single medical center, which helps to be more certain about the homogeneity of clinical assessments, but also makes it possible to study only common phenotypes. Recruitment from a single medical center is likely to result in patient numbers only in the hundreds, and although these cohorts may outnumber many past pharmacogenetic studies, genome-wide association (GWA) studies require still larger cohorts for statistical power. Therefore, for the study of some of the more severe, and rarer, adverse reactions, single-hospital settings are not likely to provide sufficiently large cohorts.

Studies that draw patients from multiple hospitals are far more likely to involve the large numbers of patients necessary for GWA studies. In particular, genetic assessment of rare adverse reactions will likely be possible only through the collaboration of multiple medical centers. Unfortunately, although study size will be increased in this type of study, patient cohorts are also at risk of inconsistent phenotypic classification. Phenotypic heterogeneity is one of the commonly accepted causes of ambiguous results in genetic studies, and therefore great care must be taken to minimize this factor.

While building up large cohorts is a priority for genome-wide association studies aimed at identifying common variation that contributes to treatment response, some pharmacogenetic phenotypes may be due to rare, highly penetrant variants. Such variants can be identified by whole-genome sequencing in small, carefully collected and phenotyped cohorts (see below). Obviously, the design of the genetic study will dictate which method of sample collection is appropriate.

22.4 Pharmacogenetic Methodology: Genomics

22.4.1 Candidate Gene Studies

While it is likely that genetic variation throughout the genome plays a role in influencing treatment outcomes, a priori hypotheses based on the pharmacokinetic and pharmacodynamic properties of a drug implicate a much smaller number of genes in pharmacogenetic processes. Pharmacokinetic processes are those that are involved in the absorption, distribution, metabolism, and excretion of a drug (the action of the body on a drug), and pharmacokinetic candidate genes include genes that encode drug-metabolizing enzymes and drug transporters. Pharmacodynamic processes describe the action of the drug on the body, and pharmacodynamic candidates include genes that encode drug targets and genes within the pathways of drug targets (Fig. 22.1). This philosophy may narrow the genomic space to which many researchers confine their pharmacogenetic investigations; however, the results obtained by such methods are mixed: candidate gene studies on the widely used anticoagulant warfarin have found that dose depends largely on genetic variation in its drug-metabolizing enzyme (CYP2C9) and its drug target (VKORC1, see below), but candidate gene studies on most other drugs have had negative or inconclusive results.

In general, a candidate gene approach has been the method of choice, but candidate gene studies have several important limitations. Even though the effects

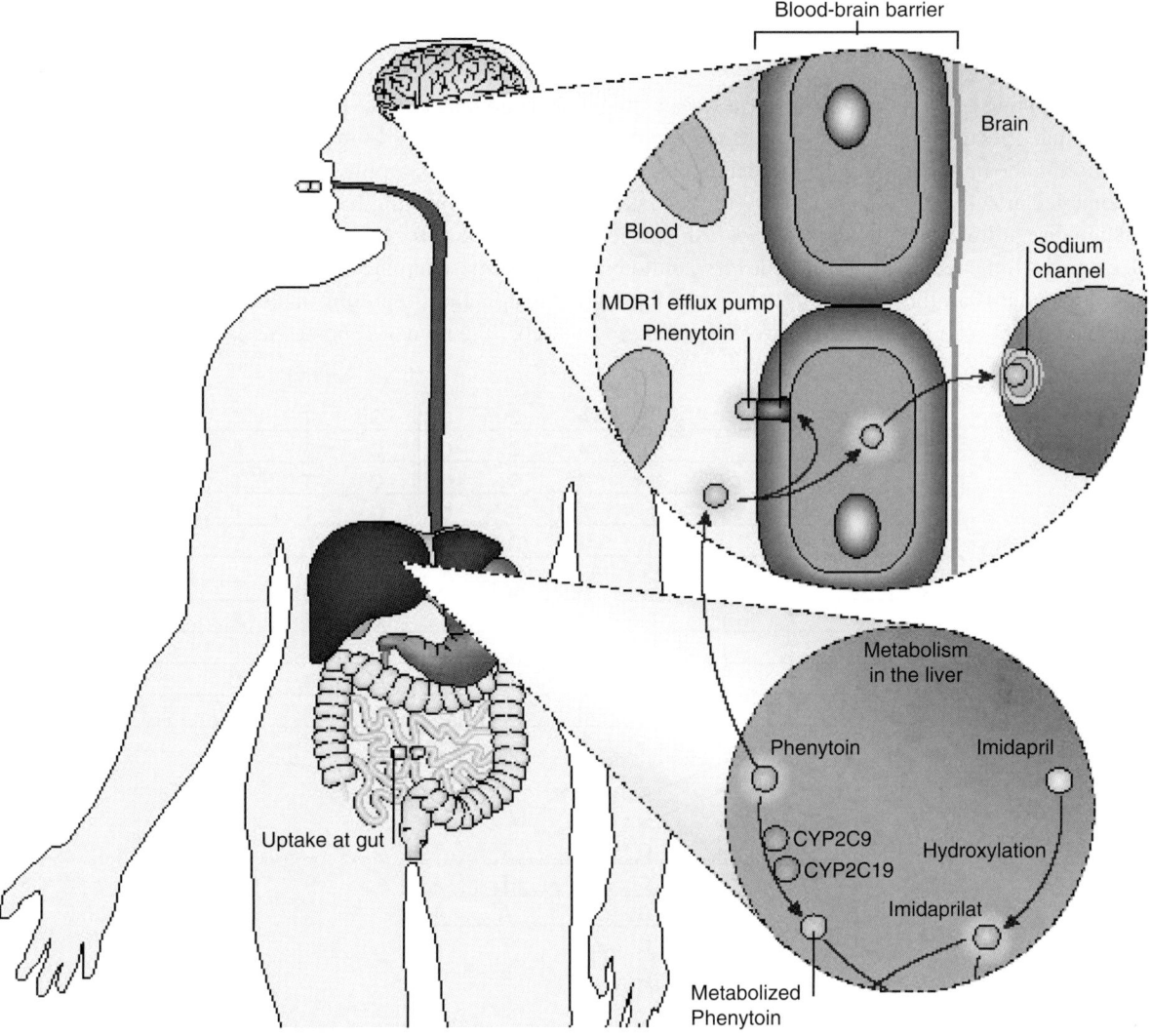

Fig. 22.1 The pharmacokinetic and pharmocodynamic properties of a drug can be used to select candidate genes. Phenytoin (PHT) is illustrated here as an example, PHT is absorbed in the gut, metabolized by two enzymes (CYP2C9 and CYP2C19) in the liver, and transported across the blood-brain barrier. Once in the brain, PHT can act on its drug target, the sodium channel (SCN1A), but is also removed by the MRD1 transporter. All of these genes represent good candidates for genetic studies on PHT

of a drug on the body are thoroughly researched prior to drug approval, it seldom happens that *all* aspects of drug pharmacokinetics and pharmacodynamics are clearly understood. As a result, candidate gene approaches to pharmacogenetics can easily overlook relevant genes. In addition, many studies combine a candidate gene approach with a candidate variant approach, focusing only on single nucleotide polymorphisms (SNPs) or variants that are known to have a specific function. This approach, which fails to consider uncharacterized variation in genes, severely limits the scope of a study.

22.4.2 Whole-Genome SNP Analyzes

It is now routine to carry out genome-wide association studies that effectively represent most of the common variation in the human genome. To understand whole-genome technology, it is first necessary to understand linkage disequilibrium and gene "tagging," the principles on which the technology is based. First, although there are about 10 million common SNPs in the human genome, there are correlations between and among some of these SNPs. The strength of such correlation is measured by the property of r^2, with an r^2 value of 1 indicating a perfect correlation (e.g., when adenosine (A) is found at one SNP, cytosine (C) is always found at a second SNP). Owing to this property, in a genetic study it is unnecessary to type both of these SNPs, because it is known that if SNP 1 is A then SNP 2 is C. In such a scenario, SNP 1 can be used as a "tag" for SNP 2, decreasing the number of SNPs that need to be genotyped (Fig. 22.2).

Recently, the HapMap project [11] has characterized 3.1 million SNPs in 270 individuals from four populations and is estimated to have captured all common variation with an $r^2 \geq 0.9$ (90% certainty). The tagging strategy has been applied to these data, and subsets of tagging SNPs have been identified that essentially represent all common variation across the genome. Technologies have been simultaneously created to genotype up to 1 million tagging SNPs in a single multiplex genotyping array [17]. Although overall cost remains nonnegligible, the cost per genotype is much lower than for single-SNP genotyping. As these platforms continue to become more affordable, and because they represent nearly all common variation across the human genome, whole-genome genotyping

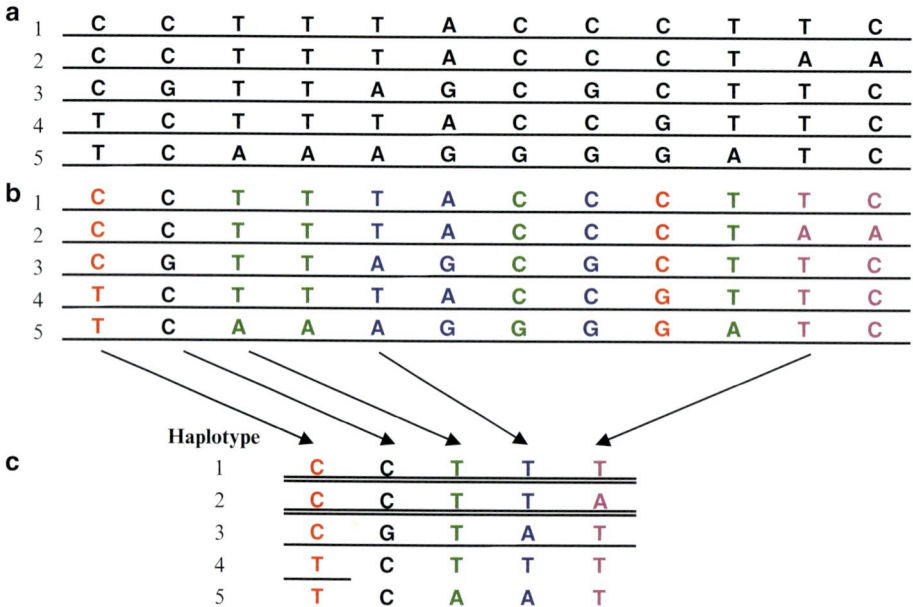

Fig. 22.2 (**a**–**c**) SNP tagging. This diagram represents five different haplotypes defined by 12 SNPs, laid out here in the order in which they occur in the genome (**a**). In this example, the selection of just one SNP from each color-coded group (**b**) will be sufficient to represent all genetic diversity in this area of the genome. Therefore, genotyping only of these five SNPs (**c**) is necessary

is currently the most appropriate technology for pharmacogenetics studies.

Although genome-wide association (GWA) studies currently represent the most comprehensive way to carry out association studies of common variation, they also come with their own suite of problems. Specifically, GWA analyzes require a large number of statistical tests, which significantly decreases the power to detect significant associations unless the cohort and the genetic effect are sufficiently large. In theory, large sample sizes are not a prohibitive issue, but in practice sample size is frequently limited by cost. Sample sizes will almost certainly be inadequate when genetic studies on rare phenotypes are carried out, and the homogeneity of clinical data is often difficult to maintain in large studies.

22.4.3 Rare Variants and Whole-Genome Sequencing

A more fundamental constraint is that after whole-genome scans were completed for most common diseases it became clear that, with a few exceptions such as macular degeneration and Alzheimer's disease, common variants have only a modest impact on disease risk [12]. While there have been few whole-genome studies for drug-response phenotypes, it seems likely that GWA in pharmacogenetics will follow a similar pattern, where a few traits show marked dependence on common variation but many, or perhaps most, do not. These traits will need to be approached by way of whole-genome sequencing.

Rapid advances in sequencing technologies now make it feasible to sequence the entire genome to identify all variants. While the cost per genome remains high, in order to minimize cost an extreme phenotype study design is appropriate. This approach features the selection of patients with only the most severe or definite phenotypes to be whole-genome sequenced. Rare variants that are identified as potential risk factors in the small discovery cohort can then be genotyped in large cohorts to assess their contribution to the response. For example, patients who develop Stevens-Johnson syndrome (SJS) on carbamazepine exposure and patients who have never had allergic reactions or rashes to any medications can be considered extreme phenotype cases and controls. Then, rare variants that are suggestive of being functional can be genotyped in a large cohort of patients exposed to carbamazepine, (a) to corroborate that these rare variants only occur in patients who develop SJS and (b) to assess the contribution of these variants to the development of rash on exposure to carbamazepine. Such a study design is a cost-effective way to include large cohorts of patients in the analysis while only carrying out whole-genome resequencing in a small number of cases and controls.

22.5 Challenges in Pharmacogenetics

22.5.1 Polygenic Inheritance

As with the study of human disease, it remains unclear whether most of the genetic control of drug-response is due to rare variants of relatively large effects or to common variants across multiple genes, with each gene variant having only a modest effect. In the latter case, large GWA studies are necessary, while the former case will require a whole-genome sequencing approach.

Pharmacogenetic studies of warfarin dose illustrate that common variants sometimes have major impacts on complex traits. Warfarin is an anticoagulant with a narrow therapeutic window and with serious consequences (bleeding) in the event of an overdose. The variation in maintenance dose of warfarin is largely attributable to genetic variants in the genes that encode the drug target, vitamin K epoxide reductase complex, subunit 1 (VKORC1), and the major metabolizing enzyme, CYP2C9 [5, 36, 38, 41]. When considered together, these genetic polymorphisms explain 30–40% of the total variation in dose. This not only represents one of the largest genetic predictors of a complex phenotype discovered to date, but also is likely to be implemented in clinical practice on completion of prospective studies.

22.5.2 Pharmacogenetics in the Clinic

Although the goal of pharmacogenetics is to provide guidance to physicians so as to help them know which drugs can be safely used based on a patient's genotype, realization of that goal is proving to be more difficult than initially expected. As previously discussed, there is a lack of well-designed, comprehensive studies in the field, which has resulted in only a modest number of definitive discoveries. In addition, the translation of pharmacogenetic discoveries into clinical practice will only occur after prospective clinical trials have assessed their accuracy and utility and have provided evidence-based guidelines for the clinical use of genetic information. To date, few such studies have been carried out.

Despite these shortcomings in pharmacogenetics, there are a small number of drugs that have had genetic information added to their labeling by the US Food and Drug Administration (Table 22.2). However, the inclusion of genetic warnings on drug labels currently neither requires the incorporation of genetic tests for the clinical use of these drugs nor outlines guidelines for the use of genetic information in clinical practice.

22.5.3 Pharmacogenetics Across Ethnic Groups

Origin of ancestry and ethnicity is an important consideration in genetic studies. In the most general terms, ethnicity was first divided by HapMap into populations of European (CEU), African (YRI), and Asian (CHB + JPT) ancestry. More recently, additional populations have been added to the HapMap databases, which now describe 11 different populations (www.hapmap.org). The inclusion of multiple populations in HapMap has made it quite clear that, across populations, LD patterns and allele frequencies are highly variable (see Table 22.2 for allele frequency differences of pharmacogenetic variants), a fact which has consequences not only for study design, but also for the clinical use of drugs.

With respect to genetic association studies, particularly GWA studies or any study that employs a tagging-based approach, an association relies on LD between a genotyped SNP and a causal genetic variant. When the LD patterns between two populations are different, an association that is observed in one population will not be observed in the other. However, a failure to replicate in a population of different ancestry could be due to several different scenarios that are indistinguishable when the causal variant is not specifically known:

1. The association is not observed in population #2 because of a breakdown in LD
2. The association is not observed in population #2

Table 22.2 Drugs that include pharmacogenetic information in FDA labeling

Drug	Clinical phenotype observed/studied	Associated gene variant	Frequency of risk alleles			
			CEU	YRI	CHB	JPT
Azathiopurine	Thiopurine toxicity	*TPMT* PM	0.036	0.06	0.023	0.003
Mercaptopurine	Thiopurine toxicity	*TPMT* PM	0.036	0.06	0.023	0.003
Irinoetcan	Neutropenia	*UGT1A1*28*	0.32	0.43	0.13	0.113
Erlontinib	Efficacy	EGFR+	n/a	n/a	n/a	n/a
Trastuzumab	Efficacy	HER2 overexpression	n/a	n/a	n/a	n/a
Abacavir	Hypersensitivity	*HLA-B*5701*	0.061	0.0	0.011	0.0
Maraviroc	Efficacy	CCR5	0.951	1.00	0.99	1.00
Warfarin	Dose	*VKORC1*	0.432	0.086[b]	0.924	0.891
		CYP2C9*2, *3	0.1 (*2)	0.0 (*2)	0.0 (*2)	0.0 (*2)
			0.058 (*3)	0.0 (*3)	0.04 (*3)	0.03 (*3)
Atomoxetine	Dose	*CYP2D6* PM[a]	0.07	0.023	0.01	0.01
Thioridazine		*CYP2D6* PM[a]	0.07	0.023	0.01	0.01
Carbamazepine	Stevens-Johnson syndrome	*HLA-B*1502*	0.0	0.0	0.011	0.0
Codeine		*CYP2D6* PM[a]	0.07	0.023	0.01	0.01

Populations: *CEU* Utah residents with Northern and Western European ancestry, *YRI* Yoruba in Ibadan, Nigeria, *CHB* Han Chinese in Beijing, China, *JPT* Japanese in Tokyo, Japan; *PM* poor metabolizer phenotype

[a]Allele frequency is from an African-American population

because the first association is a false positive, or
3. The association is not observed in population #2 because the variant that is causal in the genetic background of population #1 is not causal in the genetic background of population #2.

In fact, of the pharmacogenetic variants that are recognized by the FDA (Table 22.2), at least three are present in only one population (CYP2C9, HLA-B*5701, HLA-B*1502), while the phenotypes that they affect (variation in warfarin dosing, abacavir hypersensitivity, and carbamazepine-related SJS) are relevant concerns in all populations. Therefore, it must be emphasized that pharmacogenetic results are not necessarily applicable across populations and that population-specific studies are necessary to assess the presence and effect of pharmacogenetic variants worldwide.

In addition to population-specific genetic variants, drug administration has also been affected by the observation of population-specific treatment outcomes. For example, it has been noted that ACE inhibitors and beta blockers are not as effective for the treatment of heart failure in African Americans as they are in Caucasians [10]. The FDA has included ethnicity guidelines on ACE inhibitors and at least five other drugs that are currently used (Table 22.3). The disparate effects of these drugs may well be due to pharmacogenetic factors; however these have yet to be determined. Until genetic causes for the inter-ethnic effects of these drugs are elucidated, the administration of some medications will continue to be loosely guided by ethnic considerations.

Another issue of ethnicity that it is important to consider in large-scale population-based studies is the effect of population stratification. Population stratification is caused by differences in allele frequencies across populations, and it is well-established that population stratification can cause both type I and type II error in genetic association studies. While limiting study samples to one ethnic population is good practice for minimizing intercontinental population structure, more subtle allele frequency differences may remain within a continental or ethnic population. Such cryptic population stratification is less easily controlled for methodologically (i.e., by case-control matching), and several statistical methods that identify and correct for population stratification have been developed and validated (reviewed in [42] and in Chap. 20).

The ability to accurately quantify and adjust for population stratification now makes it possible to use large, common population control sample sets in GWA studies that do not have to be carefully matched for ethnicity or geography. This method can enhance the power of a study by increasing the size of the control population and also reduce cost, as it will no longer be necessary for every study to genotype its own large sets of controls. Large, common population control samples will be especially important in cases where the causal variant is rare and large control samples are necessary to demonstrate a statistically significant allele frequency difference between cases and controls. This method has already had some success in disease predisposition studies, resulting in the recent identification of rare copy number variants that cause autism and schizophrenia [33, 39, 46] and is likely to be important in future pharmacogenetics studies when large-effect, but rare, variants influence a drug-response phenotype.

22.5.4 Drug Classes of Urgent Interest

Among all drug classes, a pharmacogenetic focus is specifically indicated where adverse events are the most common and severe and where efficacy rates are low. One drug class that is in urgent need of pharmacogenetic study is that of neuropsychiatric drugs, specifically antipsychotics/antidepressants and antiepileptics.

Antipsychotics and antidepressants have been the most studied of all drugs in pharmacogenetics. As a class, they have efficacy rates of 5–70%, and they are associated with significant ADRs. In general, it has been found that serum levels of both antipsychotics and antidepressants correlate strongly with genotype

Table 22.3 Drugs with FDA warnings concerning the ethnicity of the patient

Drug	Ethnicity information/indication
ACE Inhibitors	Smaller effects in blacks
Isosorbide dinitrate-hydralazine	Indicated for self-identified blacks
Rosuvastatin	Lower dose for Asians
Tacrolimus	Higher dose for blacks
Oseltamivir	Neuropsychiatric events reported mostly in Japan
Warfarin	Lower dose for Asians

at the relevant drug-metabolizing enzymes, and there have been multiple reports of an effect of functional variation in dopamine and serotonin receptors on efficacy outcomes (reviewed in [3, 4]). However, these associations have not yet proved to have any clinical relevance, as serum levels do not strictly correlate with efficacy. A recent retrospective study of the CATIE trial, a large-scale trial designed to estimate the efficacy and incidence of ADRs of antipsychotics showed no significant effect of functional variation in drug-metabolizing enzymes on clinical measures of dose, ADRs, or efficacy [16].

Antiepileptic drugs represent a class of medicines with similarly low efficacy rates (less than 70%), and significant ADR profiles, but very little attention has been devoted to pharmacogenetic studies in antiepileptic drugs. Despite the lack of focus, pharmacogenetic studies have discovered an *HLA-B* allele (*HLA-B*1502*) that is associated with carbamazepine-induced SJS in Asians (Table 22.2) [23, 27]. Unfortunately, aside from the clinical implications of HLA testing for Asian patients beginning treatment with carbamazepine, the field of pharmacogenetics with respect to neuropsychiatric drugs is devoid of discoveries that affect important clinical phenotypes and have important clinical implications.

22.6 Conclusions

As discussed, the next step for pharmacogenetic association studies is to incorporate genome-wide analyzes into study designs that include carefully phenotyped patient cohorts. While this has already become common practice in disease predisposition studies, its application has been limited in pharmacogenetics and there are very few whole-genome association studies of treatment response. This anomaly needs to be corrected, and the examples of warfarin dosing, abacavir hypersensitivity, and statin-induced myopathy [28] illustrate that systematic GWA studies for variable drug responses are likely to generate new findings. What is needed for future success in this research area is a commitment to patient collection, cataloging of common and rare adverse events, exploration of appropriate efficacy and ADR phenotypes, and a transition from candidate gene-based studies to GWA studies. There is, however, every reason to believe that treatment response genetics will follow a similar course to disease genetics, and that even after systematic GWA studies have been performed, much of the variation in treatment response will remain to be explained, even where a strong genetic component is likely. In such cases, discovery genetics will depend on whole-genome sequencing approaches.

References

1. Alexanderson B, Evans DA, Sjoqvist F (1969) Br Med J 4:764–768
2. Alexopoulou A, Dourakis SP, Mantzoukis D, Pitsariotis T, Kandyli A, Deutsch M, Archimandritis AJ (2008) Eur J Int Med 19:505–510
3. Arranz MJ, de Leon J (2007) Mol Psychiatry 12:707–747
4. Binder EB, Holsboer F (2006) Ann Med 38:82–94
5. D'Andrea G, D'Ambrosio RL, Di Perna P, Chetta M, Santacroce R, Brancaccio V, Grandone E, Margaglione M (2005) Blood 105:645–649
6. Danoff TM, Campbell DA, McCarthy LC, Lewis KF, Repasch MH, Saunders AM, Spurr NK, Purvis IJ, Roses AD, Xu, C-F (2003) 4:49–53
7. Eichelbaum M (1984) Fed Proc 43:2298–2302
8. Einarson TR (1993) Ann Pharmacother 27:832–840
9. Evans WE, Horner M, Chu YQ, Kalwinsky D, Roberts WM (1991) J Pediatr 119:985–989
10. Exner DV, Dries DL, Domanski MJ, Cohn JN (2001) N Engl J Med 344:1351–1357
11. Frazer KA, Ballinger DG, Cox DR, Hinds DA, Stuve LL, Gibbs RA, Belmont JW, Boudreau A, Hardenbol P, Leal SM et al (2007) Nature 449:851–861
12. Goldstein DB (2009) N Engl J Med (in press)
13. Gonzalez FJ, Skoda RC, Kimura S, Umeno M, Zanger UM, Nebert DW, Gelboin HV, Hardwick JP, Meyer UA (1988) Nature 331:442–446
14. Gonzalez FJ, Vilbois F, Hardwick JP, McBride OW, Nebert DW, Gelboin HV, Meyer UA (1988) Genomics 2:174–179
15. Grasmader K, Verwohlt PL, Kuhn KU, Dragicevic A, von Widdern O, Zobel A, Hiemke C, Rietschel M, Maier W, Jaehde U et al (2004) Eur J Clin Pharmacol 60: 473–480
16. Grossman I, Sullivan PF, Walley N, Liu Y, Dawson JR, Gumbs C, Gaedigk A, Leeder JS, McEvoy JP, Weale ME et al (2008) Genet Med 10:720–729
17. Gunderson KL, Kuhn KM, Steemers FJ, Ng P, Murray SS, Shen R (2006) Pharmacogenomics 7:641–648
18. Hendset M, Hermann M, Lunde H, Refsum H, Molden E (2007) Eur J Clin Pharmacol 63:1147–1151
19. Hetherington S, McGuirk S, Powell G, Cutrell A, Naderer O, Spreen B, Lafon S, Pearce G, Steel H (2001) Clin Ther 23:1603–1614
20. Hetherington S, Hughes AR, Mosteller M, Shortino D, Baker KL, Spreen W, Lai E, Davies K, Handley A, Dow DJ et al (2002) Lancet 359:1121–1122
21. Huang YS, Chern HD, Su WJ, Wu JC, Lai SL, Yang SY, Chang FY, Lee SD (2002) Hepatology 35:883–889

22. Hughes DA, Vilar FJ, Ward CC, Alfirevic A, Park BK, Pirmohamed M (2004) Pharmacogenetics 14:335–342
23. Hung SI, Chung WH, Jee SH, Chen WC, Chang YT, Lee WR, Hu SL, Wu MT, Chen GS, Wong TW et al (2006) Pharmacogenet Genomics 16:297–306
24. Idle JR, Smith RL (1979) Drug Metab Rev 9:301–317
25. Imyanitov EN, Moiseyenko VM (2007) Clinica Chimica Acta Int J Clin Chem 379:1–13
26. Lesko LJ, Woodcock J (2004) Nat Rev 3:763–769
27. Link E, Parish S, Armitage J, Bowman L, Heath S, Matsuda F, Gut I, Lathrop M, Collins R (2008) N Engl J Med 359:789–799
28. Locharernkul C, Loplumlert J, Limotai C, Korkij W, Desudchit T, Tongkobpetch S, Kangwanshiratada O, Hirankarn N, Suphapeetiporn K, Shotelersuk V (2008) Epilepsia 49:2087–2091
29. Mallal S, Nolan D, Witt C, Masel G, Martin A, Moore C, Sayer D, Castley A, Mamotte C, Maxwell D et al (2002) Lancet 359:727–732
30. Martinez FD, Graves PE, Baldini M, Solomon S, Erickson R (1997) J Clin Invest 100:3184–3188
31. Monaghan G, Ryan M, Seddon R, Hume R, Burchell B (1996) Lancet 347:578–581
32. Motulsky AG (1957) J Am Med Assoc 165:835–837
33. Need AC, Ge D, Weale ME, Maia J, Feng S, Heinzen EL, Shianna KV, Yoon W, Kasperaviciute D, Gennarelli M et al (2009) PLoS Genet 5:e1000373
34. Ohno M, Yamaguchi I, Yamamoto I, Fukuda T, Yokota S, Maekura R, Ito M, Yamamoto Y, Ogura T, Maeda K et al (2000) Int J Tuberc Lung Dis 4:256–261
35. Rauch A, Nolan D, Martin A, McKinnon E, Almeida C, Mallal S (2006) Clin Infect Dis 43:99–102
36. Rettie AE, Haining RL, Bajpai M, Levy RH (1999) Epilepsy Res 35:253–255
37. Ross JS, Fletcher JA, Linette GP, Stec J, Clark E, Ayers M, Symmans WF, Pusztai L, Bloom KJ (2003) Oncologist 8:307–325
38. Rost S, Fregin A, Ivaskevicius V, Conzelmann E, Hortnagel K, Pelz HJ, Lappegard K, Seifried E, Scharrer I, Tuddenham EG et al (2004) Nature 427:537–541
39. Stefansson H, Rujescu D, Cichon S, Pietilainen OP, Ingason A, Steinberg S, Fossdal R, Sigurdsson E, Sigmundsson T, Buizer-Voskamp JE et al (2008) Nature 455:232–236
40. Steimer W, Zopf K, von Amelunxen S, Pfeiffer H, Bachofer J, Popp J, Messner B, Kissling W, Leucht S (2004) Clin Chem 50:1623–1633
41. Takanashi K, Tainaka H, Kobayashi K, Yasumori T, Hosakawa M, Chiba K (2000) Pharmacogenetics 10:95–104
42. Tian C, Gregersen PK, Seldin MF (2008) Hum Mol Genet 17:R143–R150
43. Vesell ES (1978) Twin studies in pharmacogenetics. Human Genet Suppl 1:19–30
44. Vogel CL, Cobleigh MA, Tripathy D, Gutheil JC, Harris LN, Fehrenbacher L, Slamon DJ, Murphy M, Novotny WF, Burchmore M et al (2002) J Clin Oncol 20: 719–726
45. Weisberg E, Manley PW, Cowan-Jacob SW, Hochhaus A, Griffin JD (2007) Nat Rev Cancer 7:345–356
46. Weiss LA, Shen Y, Korn JM, Arking DE, Miller DT, Fossdal R, Saemundsen E, Stefansson H, Ferreira MA, Green T et al (2008) N Engl J Med 358:667–675

Behavioral Genetics

Introductory Note by Michael R. Speicher

The investigation of behavioral genetics is a complicated task, as behavior is caused by combinations of genetic and environmental factors. Some behavioral phenotypes have a strong genetic component. There is, for example, compelling epidemiological evidence indicating that over 50% of the risk for developing alcohol dependence stems from genetic susceptibility, and genetic studies have already identified several risk alleles. However, for many other behavioral traits the degree of genetic contribution is unclear and often a matter of debate. It seems that – in addition to environmental factors – social behavior and social cognition are not the result of variation in a single gene, but are instead modulated by a number of genetic variants, each of which has only a modest effect on behavior. Furthermore, classification and description of a certain behavior is often difficult and standardization is frequently lacking. As a consequence, the linking of specific genetic or environmental risk factors to typical or atypical behaviors represents a challenge and new conceptual tools are being developed in addition to the currently accepted approaches.

These tools include classic, traditional approaches, such as twin studies, family-based studies, and chromosome analysis using biochemical and molecular techniques. Furthermore, animal models involving cross-species trait genetics have made a significant contribution. More recently, genome-wide association studies have elucidated positional candidate genes, which will help to illuminate the often complex etiology of behavioral disorders.

For clinical purposes and for the social aspects of behavioral traits it is very important to understand the neurobiology and neurogenetics of social cognition and behavior. At the same time, behavioral genetics involves highly complex and fluid ethical considerations. Therefore, data on actual risks and benefits of research on behavioral genetics and potential clinical applications has to be monitored carefully and appropriate ethical safeguards need to be developed in parallel.

These aforementioned aspects of behavioral genetics will be addressed for a variety of different behavioral phenotypes in the following chapters (Chaps. 23.1–23.7).

The Genetics of Personality

23.1

Jonathan Flint and Saffron Willis-Owen

Abstract The genetic analysis of human personality, like many other complex traits, has undergone a metamorphosis over the last 10 years. The historically predominant techniques of candidate gene association and linkage are now being replaced by genome-wide association approaches aimed at identifying the small genetic effects that contribute towards individual differences in personality. Like their predecessors, however, these approaches suffer from their own limitations, and as yet there has been no definitive identification of a human gene or variant that robustly contributes to a personality trait (such as "Neuroticism"). In this chapter we discuss the evidence that personality is heritable, consider the main approaches used to identify individual contributory factors, and outline the barriers to success in this rapidly changing field.

Contents

23.1.1 Neuroticism 653
    23.1.1.1 Genetic Association Studies 653
    23.1.1.2 Gene by Environment Effects 654
    23.1.1.3 Linkage Studies 656
    23.1.1.4 Genome-Wide Association Studies 656
23.1.2 Extraversion 657
23.1.3 Implications for Future Research 657
References 658

J. Flint (✉)
Wellcome Trust Centre for Human Genetics, Roosevelt Drive, Oxford, OX3 7BN, UK
e-mail: jf@well.ox.ac.uk

S. Willis-Owen
Molecular Genetics, National Heart and Lung Institute, Guy Scadding Building, Dovehouse Street, London, SW3 6LY, UK
e-mail: s.willis-owen@imperial.ac.uk

Personality, so frequently assessed by self-administered questionnaires, does not at first glance appear to be a favorable target for molecular genetic investigation. Two criticisms have been leveled against the enterprise: that the assessments are state dependent, and therefore meaningless for genetic studies, and second that the answers do not reflect the activity of any meaningful biological process (studying the genetics of personality would be similar to studying the genetics of fever, for instance). Before reviewing the literature on genetic studies of personality, we shall deal with both these points.

One of the most contentious issues in the literature of personality assessment is the degree of state dependence of personality assessment: people may answer the same question in very different ways depending on the circumstances in which they find themselves. Surely, the argument runs, a state-dependent measure can be of little value in genetic studies, since the latter require traits that are, at least in part, free from environmental malleability? One simple refutation of this concern is the observation that questionnaire assessments are remarkably consistent. For example, correlations in excess of 0.9 are found when the same questionnaire is administered even after an interval of more than 2 years [82]. In other words, personality assessments can be considered as stable as many measures of physiological variation (such as blood pressure) during middle adulthood. They do indeed appear to measure traits, not states.

However, response consistency does not guarantee that a personality assessment means very much, if anything at all. Ideally we would like to show that variation in personality reflects the activity of a neurobiological system, as has been claimed by a number of psychologists. To appreciate these arguments it is important to understand how personality factors are generated from questionnaires. It might seem that the value of a completed personality questionnaire depends on how truthfully it has been answered. Hans Eysenck, creator of one the most widely used assessments, gives an example that illustrates the fallacy of this interpretation: "One might see some unfortunate individual sitting down with the questionnaire, his hands trembling and sweating with excitement, his face getting pale and flushed alternately, and his tongue licking his lips, his whole body in a tremor of nervousness; on going over to reassure him, one would find after the question, "Are you generally a nervous sort of person?" he had boldly put the answer, 'No'." [23].

In fact psychologists do not use the individual items: they use factors derived from the pattern of responses. Consider these two questions: "Are your feelings easily hurt" and: "Do ideas run through your head so that you cannot sleep." Although there is no logical reason why the same response should be given to both questions, in fact people who answer yes to the first tend to answer yes to the second, and vice versa. Since this is true for other questions, it is possible to pull out a set of correlated answers. The correlations allow us to recover a relatively small number of correlated responses, or factors.

Over the many years that have passed since the notion of personality was first conceived, evidence has accumulated in favor of a small number of factors. The five-factor model (FFM) divides personality into the five dimensions of Neuroticism, Extraversion, Openness to Experience, Agreeableness, and Conscientiousness [17] (easily remembered by the acronym OCEAN). The FFM is hierarchical; that is to say each broad trait dimension may contain further, more specific, features of personality. For example, Neuroticism contains personality features such as anxiety, angry hostility, depression, self-consciousness, impulsiveness, and vulnerability.

The fact that personality factors are heritable shows that they have, at least in part, a biological basis. Twin studies have been instrumental in demonstrating that about 40% of variation in personality is due to genetic variation (this figure varies between factors, though not substantially) [20, 25, 49, 52].

Heritability does not imply that personality has a biological function. In fact, as Tooby and Cosmides argue, it is consistent with personality being evolutionary noise: "[S]election, interacting with sexual recombination, tends to impose relative uniformity at the functional level in complex adaptive designs, suggesting that most heritable psychological differences are not themselves likely to be complex psychological adaptations. Instead, they are mostly evolutionary by-products, such as concomitants of parasite-driven selection for biochemical individuality" [76]. There are problems with using heritability to validate personality constructs; many valid psychological processes do not appear to be heritable (e.g., the language you speak), while many trivial processes are (such as the sort of car you own).

Does this mean that personality factors have no value in their own right? This has struck a number of investigators as unlikely, since, assuming that the concordance between personality dimensions is meaningful, "[T]he probability is that they are based on the way in which our biology has evolved to cope with the extraordinary range of social structures and physical environments on this planet" [88].

Evidence that other species also have heritable variation in personality traits, homologous to human factors, supports the idea that personality traits have a biological function. Gosling's review of animal personality literature up to 2001 concluded that a number of dimensions appeared repeatedly across multiple species, including a dimension reflecting an individual's reaction to novel stimuli or situations (termed Reactivity, Emotionality, or Fearfulness and possibly homologous to neuroticism) [29]. According to Gosling, "The FFM dimensions of Extraversion, Neuroticism, and Agreeableness showed considerable generality across the 12 species included in their review … The way these personality dimensions are manifested, however, depends on the species. For example, whereas a human scoring low on Extraversion stays at home on Saturday night or tries to blend into a corner at a large party, the octopus scoring low on Boldness stays in its protective den during feedings and attempts to hide itself by changing color or releasing ink into the water" [29].

The combination of heritability, a concordance between personality dimensions measured by different questionnaires, and evidence that other species have similar, if not identical, personality dimensions has been instrumental in promoting a biological interpretation of personality [18]. Gray, arguing for congruence between animal models of trait anxiety and human neuroticism [31-33], put forward the view that there are two interrelated but separable brain systems that subserve both human and rodent traits. The first system, which he terms the fight/flight system, subserves flight, defensive aggression, freezing, and associated autonomic activity. The second system, termed the behavioral inhibition system, subserves the cognitive and information processing aspects of anxiety. Within Gray's theory, neuroticism or emotional stability was seen as a measure of sensitivity to reinforcing events.

Cloninger has argued strongly that personality factors reflect the action of neurobiological systems [14]. He identifies three independent and heritable dimensions of personality: (a) novelty seeking: frequent exploratory activity and excitement in response to novel stimuli; (b) harm avoidance: the tendency to respond intensely to aversive stimuli and to learn to avoid punishment and novelty; (c) reward dependence: the tendency to respond intensely to reward. Cloninger associates each with a different neurobiological system: novelty seeking with low basal dopaminergic activity, harm avoidance with high serotonergic activity, and reward dependence with low basal noradrenergic activity [14]. These views have been important in setting the scene for many of the genetic studies reviewed below because they can be used to justify the choice of candidate genes.

23.1.1 Neuroticism

Neuroticism (N) is a longitudinally and culturally robust measure of emotional stability. Although there are still some divergences in trait designation between measurement instruments, the various manifestations of N have been shown to exhibit considerable overlap in terms of both concept [1, 54, 89] and etiology. A recent longitudinal study of N, for example, revealed large phenotypic, genetic, and environmental correlations (0.57, 0.91, and 0.42, respectively) between N dimensions derived from two independent measurement instruments over a period of 22 years [83]. Likewise, other authors have also shown that N can be detected within a variety of different social strata and cultures [20] and may even be recognized in the behavior of other, less complex, organisms [29, 30].

Neuroticism represents an important construct in the study of human psychiatric disease. Both prospective and cross-sectional studies reveal a close relationship between N and psychiatric phenotypes, including perhaps most notably, major depression. High N scores are robustly associated with an increased risk for depression [2, 36, 42, 44, 79], and experience of a depressive episode yields an elevation in N which persists after recovery (i.e., a scar effect) [62]. Critically, these epidemiological data are also supported by biological evidence of shared causality, which indicate that approximately half of the genetic determinants of these phenotypes are shared (yielding a genetic correlation of 0.49–0.68) with no discernable sex-difference in the magnitude of this correlation [24, 43]. Since N accounts for a substantial proportion of comorbidity between psychiatric disorders – 20–45% of internalizing disorder comorbidity and 19–88% of comorbidity between internalizing and externalizing disorders [47] – it may be hypothesized that N represents a common liability factor for stress and anxiety related disorders.

23.1.1.1 Genetic Association Studies

Historical attempts to identify the genetic determinants of N can primarily be divided into two camps; genome-wide linkage and candidate gene association. While candidate gene studies have been prolific in their production, these studies have focused on relatively few theoretically plausible candidate genes and have been plagued by issues of nonreplication.

The serotoninergic system has been the principal focus in molecular genetic investigations of N. Genetic studies have focused in large part on the role of the serotonin transporter gene (5-HTT), which regulates the re-uptake of 5-HT at synapses. 5-HTTLPR is a polymorphism in the promoter of this gene, which carries two main alleles: a long allele with 16 repeats (*L*) and a short allele with 14 repeats (*S*). The genetically dominant *S* allele results in lower transcriptional activity than the *L* allele, leading to a relative reduction in mRNA levels, serotonin binding, and re-uptake.

In 1996 Lesch et al. demonstrated that the 5-HTTLPR associates significantly with N in a total of 505 individuals [51]. Although a number of subsequent studies have observed a similar association, based both on an N phenotype [19, 51] and on a range of other conceptually related diseases and traits (including Harm Avoidance [40], generalized anxiety disorder [87], and depression [15, 37]), several large investigations have failed to identify any significant effect [26, 82]. Furthermore, where a positive association has been identified its effect size has typically been small, occasionally reversed in direction [8], and sex-specific (although not consistently for the same sex [19, 34]. Similar observations have been made in other complex phenotypes, with the likelihood of successful replication limited by design parameters such as low sample size in the originating study [38].

One method of resolving discrepancies is to apply meta-analytic techniques. Meta-analysis is a method of combining results from independent studies to acquire a much larger data set from which more robust conclusions can be drawn than are obtainable from each of the smaller component studies. The fact that there have been a number of meta-analyzes [55, 57, 66, 67] indicates difficulties in arriving at conclusive decisions about the role of the 5-HTTLPR in personality, and it should be recognized that meta-analyzes are not substitutes for well-powered individual studies [56]. However, it now seems unarguable that if the 5-HTTLPR *does* underlie variation in neuroticism, its impact is likely to be extremely small, contributing much less than 1% of the phenotypic variation.

A number of additional candidate genes have been tested for association to N. However, like the serotonin transporter, most of these genes have been selected using pre-existing evidence of involvement in mood-related neurological systems and/or structures, thereby limiting the potential to detect novel mechanisms of trait causation. Brain-derived neurotrophic factor (BDNF), for example, is known to be expressed in the hippocampus (part of the limbic system), where it moderates neuronal growth, differentiation, and survival, as well as synaptic plasticity. The gene exhibits a transcriptional response to both stressful [69] and antidepressant [28] events and carries a functional mutation in the 5¢ prodomain that influences intracellular BDNF trafficking and activity-dependent secretion [21]. An equivalent mutation has recently been generated in the mouse, and behavioral analysis of these animals suggests that altered BDNF secretion may translate behaviorally to heightened emotionality in anxiogenic contexts [11]. Together, these and other similar data have led researchers to consider the 5¢ functional BDNF mutation (termed val66met) as a candidate for a variety of mood-related phenotypes, including N [50] However, as in the case of the serotonin transporter, evidence of association has been mixed [81].

More recently, a novel class of candidates has emerged; that of the semaphorin axon guidance molecules and their associated co-receptors, the plexins. This family of secreted and transmembrane proteins has been implicated in the etiology of N through several convergent lines of investigation; including both theory- and position-based analyzes. Epidemiologically N is known to predict several clinical phenotypes, including both major depression and schizophrenia. One of the semaphorin receptors, *Plexin A2* (*PLXNA2*) has recently been shown to associate with schizophrenia through genome-wide association [53], and this association appears to replicate successfully between cohorts. Although the mechanism of action is as yet uncharacterized, current evidence indicates that semaphorin-plexin signaling may influence cell migration via an effect on centrosome positioning [63], with a particular emphasis on hippocampal mossy fiber projections [73]. These data fit with current hypotheses regarding mood disorder pathogenesis, which postulate an abnormality in neurogenesis within the adult hippocampus. Consequently, *PLXNA2* has emerged as a potential candidate for N, which may serve as a common susceptibility factor for both clinical phenotypes. Recent analyzes corroborate this hypothesis, revealing significant associations between *PLXNA2* mutations and N, as well as a range of other related phenotypes [84].

23.1.1.2 Gene by Environment Effects

The importance of gene by environment interaction in personality studies is producing a literature no less contentious than that dealing with the main effects of genetic variation on phenotype. We all know that different people, faced with the same stressful situation, react differently, and it is scarcely controversial to assert that this variation has, in part, a genetic origin. Nonetheless, while few doubt that gene by environment interaction exists, its importance has been difficult to assess. The older literature was not optimistic: Jinks and Fulker [39], using the correlation between identical twin intrapair differences and pair sums, found little evidence for genotype environment interaction for cognitive and personality traits.

More recently, the tide has turned. In an influential article published in 1994 Bronfenbrenner and Ceci argue strongly that interaction needs to be taken into account in behavioral genetic studies: "The mechanisms by which genotypes actualize into phenotypes vary as a function of environmental context. When proximal processes are weak, that is when the environment is not conducive to expression of that genotype, heritability is low, as genetic potential is not realized" [7]. A study of cognitive ability in 7-year-old children taking part in the National Perinatal Collaborative Project found that for disadvantaged children, environmental influences accounted for nearly 60% of the variance in IQ, while genetic factors accounted for negligible variance. However, in advantaged children the pattern was almost reversed, good evidence therefore of an interaction [77], a finding that has been replicated in an independent study [35].

In a cross-fostering analysis [13], crime rates in male Swedish adoptees were found to be greatest when both heritable and environmental influences were present, with the interaction accounting for twice as much crime as genetic and environmental influences alone. Cadoret et al. [9] studied adoptees whose parents had either antisocial personality and found that an adverse adoptive home environment interacted with adult antisocial personality in predicting increased aggression in the offspring.

Given that the interactions are there and are important, could it be that molecular variants will not be found unless gene by environment interaction is taken into account? Perhaps modeling the joint effects of genes and environment is necessary to obtain sufficient statistical power to detect the effect. Empirical evidence in favor of this view comes, yet again, from a study of the 5-HTTLPR. Caspi et al. report, that carriers of the 5-HTTLPR short variant are twice as likely to become depressed after stressful events such as bereavement, romantic disasters, illnesses or job loss, and childhood maltreatment significantly increases this probability [10].

Studies of a variant of 5-HTTLPR in nonhuman primates supports this finding. In rhesus monkeys, maternal separation during the first months of life adversely affects later social interaction behavior. As in humans, there is a repeat length variation in the promoter of the serotonin transporter gene (the variant is called rh5-HTTLPR). The rh5-HTTLPR genotype interacts with deleterious early rearing experience to influence attentional and emotional behavior, stress reactivity, and alcohol preference and dependence [5].

The trouble with this argument is that the environmental causes of personality variation are frequently as mysterious as the genetic. Where the environmental effect is known, then increased power might be obtainable, but what happens if we are not so sure? This problem is well known to epidemiologists, who have been struggling for some years to detect subtle environmental effects [74].

Clayton and McKeigue, in a discussion of the value of gene by environment studies, argue as follows: "If we could specify in advance that the effect of the environmental factor on disease risk would be restricted to a subgroup of individuals with a particular genotype, there would, of course, be a gain in power from testing only this subgroup for the effect of the environmental factor. In practice, such an extreme situation is unlikely to be frequently encountered in the study of complex diseases, and entails a level of knowledge of underlying biology which would probably render epidemiological studies redundant. In less extreme situations, and where previous knowledge is more limited, a combined test would need to be done for the main effect of environmental exposure and its interaction with genotype. Since such tests have multiple degrees of freedom, the gain in power is much reduced; indeed, power might even be lost" [12].

As an example, consider the problem of defining environmental effects on depression, a phenotype that is genetically closely related to neuroticism. Here we know that stressful life events (SLE) have an important role in the onset of depression [46], but the temporal relationship between the two is less well characterized. The largest effect is seen in the month succeeding the SLE, but this depends on the type of event [45], raising the possibility that a gene by environment effect will depend on the type of SLE. Some people might be genetically predisposed to weather a marriage break-up better than others, but not to deal so well with the death of a spouse. As important subgroups of each SLE are found we may be faced with an exponentially increasing list of environmental effects to investigate, with consequent disastrous consequences for our power to detect anything at all.

The literature in support of the detection of gene by environment effects at individual loci looks qualitatively very similar to the initial reports of genetic association studies of the main effect of 5HTT [55]: there have been a small number of high profile findings, followed by a mixture of replications and nonreplications [78]. The pattern appears to be pervasive and indicates that the current studies of gene by environment effects are underpowered. It is worth noting that the largest studies of putative gene by environment effects to date have largely produced null results [72], suggesting that findings in smaller studies may represent false positives. In short, we do not yet know whether gene by environment studies will fare any better than other, genetic association studies with simpler designs.

23.1.1.3 Linkage Studies

Linkage studies, on the other hand, while substantially fewer in number, have provided arguably more robust evidence of region involvement. To date five linkage studies have been published describing the genetic basis of N. The first, published in 2003, identified five chromosomal regions. All of these regions exceeded an empirically determined significance threshold [27], and most spanned a region exceeding 30 cM in length, which supports their status as true effects rather than false positives [75]. These loci were positioned on chromosomes 1q, 4q, 7p, 12q, and 13q. In addition to these sites, a number of additional sex-specific loci were also identified, an observation consistent with the elevated phenotype concordance that can be detected in same-sex siblings as opposed to opposite-sex pairs and indicative of partially divergent genetic causality between the sexes.

Two studies included N as a secondary measure, with a primary focus on alcohol [48] and nicotine dependence [60]; two other studies [59, 85] genotyped only extreme-scoring siblings (an extremes design). The joint Australian/Dutch study [85] was particularly impressive in that it combined large sample sets from two populations, providing an opportunity for the investigators to replicate findings. Overall, however, the findings are disappointing. No one locus is consistently found in all studies, although, across all studies, replication can be observed on chromosomes 1, 7, 12, and 18. Additionally, if one considers binary disease traits that are conceptually related to N (e.g., major depression or generalized anxiety disorder), further evidence of genetic overlap can be obtained [80].

Although these data implicate a small number of discrete chromosomal regions in the etiology of N, these regions are extensive, thereby precluding identification of an experimentally tractable number of candidate genes and/or regulatory sequences for further investigation. The N locus originally mapped to chromosome 1 by Fullerton et al. [27] extends over 191 Mb from D1S2667-D1S249, encapsulating over 2,000 known/predicted genes and a multitude of intuitively plausible candidate genes (including two serotonin receptors, a cannabinoid receptor, and a glutamate receptor amongst many others). Since these mapping data were derived from a large, highly selected, sample (consisting of the phenotypic extremes of the N distribution selected from a population in excess of 88,000 individuals) and assessed using a well-validated quantitative measure of liability, it becomes apparent that alternative approaches are likely to be required for the fine-scale dissection of the N phenotype.

23.1.1.4 Genome-Wide Association Studies

The advent of micro-array methods of genotyping, in which hundreds of thousands of genetic variants can be assayed at one time, has made it possible to test all genes in the genome by genetic association. A whole genome association has been used to investigate the genetic basis of N [68]. The results confirm that the genetic effects contributing to heritability of Neuroticism are small. More than 600,000 genetic variants were analyzed and just one, rs702543, replicated in a separate sample. This is hardly robust evidence of genetic association. But the negative evidence is important. Failure to find any genetic variants accounting for more than 1% of the variance means the 40% heritability of N is likely to arise from a large number of loci, each explaining much less than 1% of variance, and/or that N may be determined by rare variants or singleton polymorphisms [41] that are not adequately tagged by current haplotype-based SNP chips in the absence of enrichment (pedigree-based) sampling.

23.1.2 Extraversion

To date no genome-wide positional cloning studies of Extraversion (E) or its close relative novelty seeking conducted either through linkage or association have been reported in the literature. Instead, there has been a predominant focus on candidate gene studies, and as in the case of N, a specific emphasis on a single biochemical pathway and its cognate genes. The dopaminergic system is involved in appetitive and motivational behaviors [16], and pharmacological challenge studies have suggested a relationship between dopaminergic hyperactivity and reward seeking, as well as motivational factors associated with both extraversion and novelty seeking (Netter 2006); all this suggests that these traits may share a common neurobiological basis. On these grounds, dopaminergic genes, in particular the dopamine D4 receptor (*DRD4*) have emerged as plausible candidates for interindividual variation in novelty seeking and extraversion.

Although *DRD4* is highly polymorphic, research evaluating its role in behavioral and psychiatric phenotypes has largely focused on a single variant: a variable number tandem repeat (VNTR) located in exon III, and specifically the presence or absence of a 7-repeat ("long") allele. This variant has been reported to display several functional characteristics, including decreased ligand binding [3], decreased gene expression in vitro, and attenuation of cyclic AMP formation when dopamine is bound to the receptor [4] relative to 6-repeat or shorter allele.s There is, however, some disagreement regarding the optimal grouping of VNTR alleles. The *DRD4* gene also includes a single nucleotide polymorphism in the promoter region (C-521T), which has been reported to be in linkage disequilibrium with the exon III VNTR [22, 71] and is also associated with variation in expression of the D4 receptor, with the *T* allele yielding a ~40% reduction in transcription compared with the *C* allele [61, 64].

By 2008 almost 50 studies had been published reporting data on the association between *DRD4* and extraversion. Many of these studies are assimilated in a meta-analysis published in 2002, which provides support for the association between *DRD4* and trait novelty seeking, both based on the VNTR and C-521T polymorphisms [65]. A further meta-analysis published in 2008 also finds evidence for an effect and shows that the effect size might actually be relatively large by the standards of genetic association studies of behavioral traits (although nevertheless small in absolute terms) [58]. The pooled effect size estimate suggests that the C-521T SNP may account for 2% of the phenotypic variance. Whether this finding will hold up in large-scale individual studies remains to be determined.

Extraversion also associates significantly with drug dependence, with a change of one standard deviation yielding a 24% increase in risk for drug dependence [47]. However, a combination of low E and high N scores also appears to predict several anxiety disorders, including both social phobia and agoraphobia [6]. In keeping with these observations, *Rgs2*, a gene previously shown to underlie murine variation in emotionality by positional cloning [86], has recently been shown to associate with low E scores (introversion) and increased limbic activation during emotion processing in humans [70]. This association cannot be explained by variation in N. These data therefore suggest that genetic variation in *Rgs2* may predispose towards anxiety disorders through the generation of a hyperreactive state in the limbic circuitry, manifesting behaviorally as introversion. These data therefore highlight the potential utility of investigations focused on quantitative personality traits in the genetic dissection of qualitative, diagnosis-based psychiatric phenotypes, and also indicate that cross-species comparisons may be informative in the genetic dissection of human behavioral traits.

23.1.3 Implications for Future Research

The genetic analysis of human complex traits has undergone a metamorphosis over the last 10 years. The predominant techniques of candidate gene association and linkage are now being replaced by genome-wide association approaches aimed at identifying the small genetic effects that contribute to individual differences in the personality factor neuroticism (N). All these approaches, however, suffer from their own limitations, and as yet there has been no definitive identification of a human gene that robustly contributes to N across multiple populations. Alternative approaches are likely to be required, not to replace existing methodologies, but rather to supplement and guide ongoing investigations. For example, the contribution of model organisms, specifically considering the ways in which synteny between human and mouse QTL may be used to guide human association studies, is likely to become increasingly important. In order to progress our knowledge of N and other complex behavioral traits, it may therefore be necessary to adopt a more inclusive, interdisciplinary approach to complex trait analysis, incorporating several convergent sources of information for the identification of clinically relevant genes.

References

1. Aluja A, Garcia O, Garcia LF (2002) A comparative study of Zuckerman's three structural models for personality through the NEO-PI-R, ZKPQ-III-R, EPQ-RS and Goldberg's 50-bipolar adjectives. Pers Individ Dif 33:713–725
2. Angst J, Clayton P (1986) Premorbid personality of depressive, bipolar and schizophrenic patients with special reference to suicidal issues. Compr Psychiatry 27:511–532
3. Asghari V, Schoots O, van Kats S, Ohara K, Jovanovic V, Guan HC, Bunzow JR, Petronis A, Van Tol HH (1994) Dopamine D4 receptor repeat: analysis of different native and mutant forms of the human and rat genes. Mol Pharmacol 46:364–373
4. Asghari V, Sanyal S, Buchwaldt S, Paterson A, Jovanovic V, Van Tol HH (1995) Modulation of intracellular cyclic AMP levels by different human dopamine D4 receptor variants. J Neurochem 65:1157–1165
5. Bennett AJ, Lesch KP, Heils A, Long JC, Lorenz JG, Shoaf SE, Champoux M, Suomi SJ, Linnoila MV, Higley JD (2002) Early experience and serotonin transporter gene variation interact to influence primate CNS function. Mol Psychiatry 7:118–122
6. Bienvenu OJ, Hettema JM, Neale MC, Prescott CA, Kendler KS (2007) Low extraversion and high neuroticism as indices of genetic and environmental risk for social phobia, agoraphobia, and animal phobia. Am J Psychiatry 164:1714–1721
7. Bronfenbrenner U, Ceci SJ (1994) Nature-nurture reconceptualized in developmental perspective: a bioecological model. Psychol Rev 101:568–586
8. Brummett BH, Siegler IC, McQuoid DR, Svenson IK, Marchuk DA, Steffens DC (2003) Associations among the *NEO Personality Inventory*. Revised and the serotonin transporter gene-linked polymorphic region in elders: effects of depression and gender. Psychiatr Genet 13:13–18
9. Cadoret RJ, Yates WR, Troughton E, Woodworth G, Stewart MA (1995) Genetic-environmental interaction in the genesis of aggressivity and conduct disorders. Arch Gen Psychiatry 52:916–924
10. Caspi A, Sugden K, Moffitt TE, Taylor A, Craig IW, Harrington H, McClay J, Mill J, Martin J, Braithwaite A, Poulton R (2003) Influence of life stress on depression: moderation by a polymorphism in the 5-HTT gene. Science 301:386–389
11. Chen ZY, Jing D, Bath KG, Ieraci A, Khan T, Siao CJ, Herrera DG, Toth M, Yang C, McEwen BS, Hempstead BL, Lee FS (2006) Genetic variant BDNF (Val66Met) polymorphism alters anxiety-related behavior. Science 314:140–143
12. Clayton D, McKeigue PM (2001) Epidemiological methods for studying genes and environmental factors in complex diseases. Lancet 358:1356–1360
13. Cloninger CR, Sigvardsson S, Bohman M, von Knorring AL (1982) *Predisposition to petty criminality in Swedish adoptees*. II. Cross-fostering analysis of gene-environment interaction. Arch Gen Psychiatry 39:1242–1247
14. Cloninger CR, Svrakic DM, Przybeck TR (1993) A psychobiological model of temperament and character. Arch Gen Psychiatry 50:975–989
15. Collier DA, Stober G, Li T, Heils A, Catalano M, Di Bella D, Arranz MJ, Murray RM, Vallada HP, Bengel D, Muller CR, Roberts GW, Smeraldi E, Kirov G, Sham P, Lesch KP (1996) A novel functional polymorphism within the promoter of the serotonin transporter gene: possible role in susceptibility to affective disorders. Mol Psychiatry 1:453–460
16. Comings DE, Blum K (2000) Reward deficiency syndrome: genetic aspects of behavioral disorders. Prog Brain Res 126:325–341
17. Costa PT, McCrae RR (1992) Revised NEO Personality Inventory (NEO-PI-R) and NEO Five-Factor Inventory (NEO-FFI) professional manual. Psychological Assessment Resources, Odessa, FL
18. Depue RA, Collins PF (1999) Neurobiology of the structure of personality: dopamine, facilitation of incentive motivation, and extraversion. Behav Brain Sci 22:491–517. doi:discussion 518–569
19. Du L, Bakish D, Hrdina PD (2000) Gender differences in association between serotonin transporter gene polymorphism and personality traits. Psychiatry Genet 10:159–164
20. Eaves LJ, Eysenck HJ, Martin NG (1989) Genes, culture and personality: An empirical approach. Academic, London
21. Egan MF, Kojima M, Callicott JH, Goldberg TE, Kolachana BS, Bertolino A, Zaitsev E, Gold B, Goldman D, Dean M, Lu B, Weinberger DR (2003) The BDNF val66met polymorphism affects activity-dependent secretion of BDNF and human memory and hippocampal function. Cell 112:257–269
22. Ekelund J, Suhonen J, Jarvelin MR, Peltonen L, Lichtermann D (2001) No association of the -521 C/T polymorphism in the promoter of DRD4 with novelty seeking. Mol Psychiatry 6:618–619

23. Eysenck HJ (1957) Sense and nonsense in psychology. Penguin Books, Harmondsworth
24. Fanous A, Gardner CO, Prescott CA, Cancro R, Kendler KS (2002) Neuroticism, major depression and gender: a population-based twin study. Psychol Med 32:719–728
25. Floderus-Myrhed B, Pedersen N, Rasmuson I (1980) Assessment of heritability for personality, based on a short form of the eysenck personality inventory. Behav Genet 10:153–162
26. Flory JD, Manuck SB, Ferrell RE, Dent KM, Peters DG, Muldoon MF (1999) Neuroticism is not associated with the serotonin transporter (5-HTTLPR) polymorphism. Mol Psychiatry 4:93–96
27. Fullerton J, Cubin M, Tiwari H, Wang C, Bomhra A, Davidson S, Miller S, Fairburn C, Goodwin G, Neale MC, Fiddy S, Mott R, Allison DB, Flint J (2003) Linkage analysis of extremely discordant and concordant sibling pairs identifies quantitative-trait Loci that influence variation in the human personality trait neuroticism. Am J Hum Genet 72:879–890
28. Garza AA, Ha TG, Garcia C, Chen MJ, Russo-Neustadt AA (2004) Exercise, antidepressant treatment, and BDNF mRNA expression in the aging brain. Pharmacol Biochem Behav 77:209–220
29. Gosling SD (2001) From mice to men: what can we learn about personality from animal research? Psychol Bull 127:45–86
30. Gosling SD, Kwan VS, John OP (2003) A dog's got personality: a cross-species comparative approach to personality judgments in dogs and humans. J Pers Soc Psychol 85:1161–1169
31. Gray JA (1982) The neuropsychology of anxiety: an enquiry into the function of the septo-hippocampal system. Oxford University Press, Oxford
32. Gray JA (1987) The psychology of fear and stress, 2nd edn. Cambridge University Press, Cambridge, p 387
33.
34. Gray JA, McNaughton N (2000) The neuropsychology of anxiety. Oxford University Press, Oxford
35. Greenberg BD, Li Q, Lucas FR, Hu S, Sirota LA, Benjamin J, Lesch KP, Hamer D, Murphy DL (2000) Association between the serotonin transporter promoter polymorphism and personality traits in a primarily female population sample. Am J Med Genet 96:202–216
36. Harden KP, Turkheimer E, Loehlin JC (2007) Genotype by environment interaction in adolescents' cognitive aptitude. Behav Genet 37:273–283
37. Hirschfeld RM, Klerman GL, Lavori P, Keller MB, Griffith P, Coryell W (1989) Premorbid personality assessments of first onset of major depression. Arch Gen Psychiatry 46:345–350
38. Hoefgen B, Schulze TG, Ohlraun S, von Widdern O, Hofels S, Gross M, Heidmann V, Kovalenko S, Eckermann A, Kolsch H, Metten M, Zobel A, Becker T, Nothen MM, Propping P, Heun R, Maier W, Rietschel M (2005) The power of sample size and homogenous sampling: association between the 5-HTTLPR serotonin transporter polymorphism and major depressive disorder. Biol Psychiatry 57:247–251
39. Ioannidis JP, Ntzani EE, Trikalinos TA, Contopoulos-Ioannidis DG (2001) Replication validity of genetic association studies. Nat Genet 29:306–309
40. Jinks JL, Fulker DW (1970) Comparison of the biometrical genetical, MAVA, and classical approaches to the analysis of human behavior. Psychol Bull 73:311–349
41. Katsuragi S, Kunugi H, Sano A, Tsutsumi T, Isogawa K, Nanko S, Akiyoshi J (1999) Association between serotonin transporter gene polymorphism and anxiety-related traits. Biol Psychiatry 45:368–370
42. Ke X, Taylor MS, Cardon LR (2008) Singleton SNPs in the human genome and implications for genome-wide association studies. Eur J Hum Genet 16:506–515
43. Kendell RE, DiScipio WJ (1968) Eysenck personality inventory scores of patients with depressive illnesses. Br J Psychiatry 114:767–770
44. Kendler KS, Prescott CA (1999) A population-based twin study of lifetime major depression in men and women. Arch Gen Psychiatry 56:39–44
45. Kendler KS, Neale MC, Kessler RC, Heath AC, Eaves LJ (1993) A longitudinal twin study of personality and major depression in women. Arch Gen Psychiatry 50:853–862
46. Kendler KS, Karkowski LM, Prescott CA (1998) Stressful life events and major depression: risk period, long-term contextual threat, and diagnostic specificity. J Nerv Ment Dis 186:661–669
47. Kendler KS, Karkowski LM, Prescott CA (1999) Causal relationship between stressful life events and the onset of major depression. Am J Psychiatry 156:837–841
48. Khan AA, Jacobson KC, Gardner CO, Prescott CA, Kendler KS (2005) Personality and comorbidity of common psychiatric disorders. Br J Psychiatry 186:190–196
49. Kuo PH, Neale MC, Riley BP, Patterson DG, Walsh D, Prescott CA, Kendler KS (2007) A genome-wide linkage analysis for the personality trait neuroticism in the Irish affected sib-pair study of alcohol dependence. Am J Med Genet B Neuropsychiatr Genet 144B:463–468
50. Lake RI, Eaves LJ, Maes HH, Heath AC, Martin NG (2000) Further evidence against the environmental transmission of individual differences in neuroticism from a collaborative study of 45, 850 twins and relatives on two continents. Behav Genet 30:223–233
51. Lang UE, Hellweg R, Gallinat J (2004) BDNF serum concentrations in healthy volunteers are associated with depression-related personality traits. Neuropsychopharmacology 29:795–798
52. Lesch KP, Bengel D, Heils A, Sabol SZ, Greenberg BD, Petri S, Benjamin J, Muller CR, Hamer DH, Murphy DL (1996) Association of anxiety-related traits with a polymorphism in the serotonin transporter gene regulatory region. Science 274:1527–1531
53. Loehlin JC (1992) Genes and environment in personality development. In: Plomin R (ed) Individual differences and development, vol 2. Sage Publications, London
54. Mah S, Nelson MR, Delisi LE, Reneland RH, Markward N, James MR, Nyholt DR, Hayward N, Handoko H, Mowry B, Kammerer S, Braun A (2006) Identification of the semaphorin receptor PLXNA2 as a candidate for susceptibility to schizophrenia. Mol Psychiatry 11:471–478
55. McCrae RR, Costa J, Paul T (1985) Comparison of EPI and psychoticism scales with measures of the five-factor model of personality. Pers Individ Dif 6:587–597
56. Munafo MR, Clark TG, Moore LR, Payne E, Walton R, Flint J (2003) Genetic polymorphisms and personality in

healthy adults: a systematic review and meta-analysis. Mol Psychiatry 8:471–484
57. Munafo MR, Clark T, Flint J (2005) *Promise and pitfalls in the meta-analysis of genetic association studies: a response to Sen and Schinka*. Mol Psychiatry
58. Munafo MR, Freimer NB, Ng W, Ophoff R, Veijola J, Miettunen J, Jarvelin MR, Taanila A, Flint J (2008) *5-HTTLPR genotype and anxiety-related personality traits: a meta-analysis and new data*. Am J Med Genet B Neuropsychiatr Genet
59. Munafo MR, Yalcin B, Willis-Owen SA, Flint J (2008) Association of the dopamine D4 receptor (DRD4) gene and approach-related personality traits: meta-analysis and new data. Biol Psychiatry 63:197–206
60. Nash MW, Huezo-Diaz P, Williamson RJ, Sterne A, Purcell S, Hoda F, Cherny SS, Abecasis GR, Prince M, Gray JA, Ball D, Asherson P, Mann A, Goldberg D, McGuffin P, Farmer A, Plomin R, Craig IW, Sham PC (2004) Genome-wide linkage analysis of a composite index of neuroticism and mood-related scales in extreme selected sibships. Hum Mol Genet 13:2173–2182
61. Neale BM, Sullivan PF, Kendler KS (2005) A genome scan of neuroticism in nicotine dependent smokers. Am J Med Genet B Neuropsychiatr Genet 132:65–69
62. Okuyama Y, Ishiguro H, Toru M, Arinami T (1999) A genetic polymorphism in the promoter region of DRD4 associated with expression and schizophrenia. Biochem Biophys Res Commun 258:292–295
63. Reich J, Noyes R Jr, Hirschfeld R, Coryell W, O'Gorman T (1987) State and personality in depressed and panic patients. Am J Psychiatry 144:181–187
64. Renaud J, Kerjan G, Sumita I, Zagar Y, Georget V, Kim D, Fouquet C, Suda K, Sanbo M, Suto F, Ackerman SL, Mitchell KJ, Fujisawa H, Chedotal A (2008) Plexin-A2 and its ligand, Sema6A, control nucleus-centrosome coupling in migrating granule cells. Nat Neurosci 11:440–449
65. Ronai Z, Szekely A, Nemoda Z, Lakatos K, Gervai J, Staub M, Sasvari-Szekely M (2001) Association between Novelty Seeking and the -521 C/T polymorphism in the promoter region of the DRD4 gene. Mol Psychiatry 6:35–38
66. Schinka JA, Busch RM, Robichaux-Keene N (2004) A meta-analysis of the association between the serotonin transporter gene polymorphism (5-HTTLPR) and trait anxiety. Mol Psychiatry 9:197–202
67. Schinka JA, Letsch EA, Crawford FC (2002) DRD4 and novelty seeking: results of meta-analyses. Am J Med Genet 114:643–648
68. Sen, S., M. Burmeister, and D. Ghosh, *Meta-analysis of the association between a serotonin transporter promoter polymorphism (5-HTTLPR) and anxiety-related personality traits*. Am J Med Genet B Neuropsychiatr Genet 2000.
69. Shifman S, Bhomra A, Smiley S, Wray NR, James MR, Martin NG, Hettema JM, An SS, Neale MC, van den Oord EJ, Kendler KS, Chen X, Boomsma DI, Middeldorp CM, Hottenga JJ, Slagboom PE, Flint J (2008) A whole genome association study of neuroticism using DNA pooling. Mol Psychiatry 13:302–312
70. Smith MA, Makino S, Kvetnansky R, Post RM (1995) Stress and glucocorticoids affect the expression of brain-derived neurotrophic factor and neurotrophin-3 mRNAs in the hippocampus. J Neurosci 15:1768–1777
71. Smoller JW, Paulus MP, Fagerness JA, Purcell S, Yamaki LH, Hirshfeld-Becker D, Biederman J, Rosenbaum JF, Gelernter J, Stein MB (2008) Influence of RGS2 on anxiety-related temperament, personality, and brain function. Arch Gen Psychiatry 65:298–308
72. Strobel A, Lesch KP, Hohenberger K, Jatzke S, Gutzeit HO, Anacker K, Brocke B (2002) No association between dopamine D4 receptor gene exon III and -521C/T polymorphism and novelty seeking. Mol Psychiatry 7:537–538
73. Surtees PG, Wainwright NW, Willis-Owen SA, Luben R, Day NE, Flint J (2006) Social adversity, the serotonin transporter (5-HTTLPR) polymorphism and major depressive disorder. Biol Psychiatry 59:224–229
74. Suto F, Tsuboi M, Kamiya H, Mizuno H, Kiyama Y, Komai S, Shimizu M, Sanbo M, Yagi T, Hiromi Y, Chedotal A, Mitchell KJ, Manabe T, Fujisawa H (2007) Interactions between plexin-A2, plexin-A4, and semaphorin 6A control lamina-restricted projection of hippocampal mossy fibers. Neuron 53:535–547
75. Taubes G (1995) Epidemiology faces its limits. Science 269:164–169
76. Terwilliger JD, Shannon WD, Lathrop GM, Nolan JP, Goldin LR, Chase GA, Weeks DE (1997) True and false positive peaks in genomewide scans: applications of length-biased sampling to linkage mapping. Am J Hum Genet 61:430–438
77. Tooby J, Cosmides L (1990) On the universality of human nature and the uniqueness of the individual: the role of genetics and adaptation. J Pers 58:17–67
78. Turkheimer E, Haley A, Waldron M, D'Onofrio B, Gottesman II (2003) *Socioeconomic status modifies heritability of IQ in young children.* Psychol Sci 14: 623–628
79. Uher R, McGuffin P (2008) The moderation by the serotonin transporter gene of environmental adversity in the aetiology of mental illness: review and methodological analysis. Mol Psychiatry 13:131–146
80. Wetzel RD, Cloninger CR, Hong B, Reich T (1980) Personality as a subclinical expression of the affective disorders. Compr Psychiatry 21:197–205
81. Willis-Owen SA, Flint J (2007) Identifying the genetic determinants of emotionality in humans; insights from rodents. Neurosci Biobehav Rev 31:115–124
82. Willis-Owen SA, Fullerton J, Surtees PG, Wainwright NW, Miller S, Flint J (2005) The Val66Met coding variant of the brain-derived neurotrophic factor (BDNF) gene does not contribute toward variation in the personality trait neuroticism. Biol Psychiatry 58:738–742
83. Willis-Owen SA, Turri MG, Munafo MR, Surtees PG, Wainwright NW, Brixey RD, Flint J (2005) The serotonin transporter length polymorphism, neuroticism, and depression: a comprehensive assessment of association. Biol Psychiatry 58:451–456
84. Wray NR, Birley AJ, Sullivan PF, Visscher PM, Martin NG (2007) Genetic and phenotypic stability of measures of neuroticism over 22 years. Twin Res Hum Genet 10:695–702
85. Wray NR, James MR, Mah SP, Nelson M, Andrews G, Sullivan PF, Montgomery GW, Birley AJ, Braun A, Martin NG (2007) Anxiety and comorbid measures associated with PLXNA2. Arch Gen Psychiatry 64:318–326
86. Wray NR, Middeldorp CM, Birley AJ, Gordon SD, Sullivan PF, Visscher PM, Nyholt DR, Willemsen G, de Geus EJ, Slagboom PE, Montgomery GW, Martin NG, Boomsma DI (2008) Genome-wide linkage analysis of multiple measures

of neuroticism of 2 large cohorts from Australia and the Netherlands. Arch Gen Psychiatry 65:649–658

87. Yalcin B, Willis-Owen SA, Fullerton J, Meesaq A, Deacon RM, Rawlins JN, Copley RR, Morris AP, Flint J, Mott R (2004) Genetic dissection of a behavioral quantitative trait locus shows that Rgs2 modulates anxiety in mice. Nat Genet 36:1197–1202

88. You JS, Hu SY, Chen B, Zhang HG (2005) Serotonin transporter and tryptophan hydroxylase gene polymorphisms in Chinese patients with generalized anxiety disorder. Psychiatr Genet 15:7–11

89. Zuckerman M (1992) What is a basic factor and which factors are basic?Turtles all the way down. Pers Individ Diff 13:675–681

90. Zuckerman M, Kuhlman DM, Joireman J, Teta P, Kraft M (1993) Comparison of three structural models for personality: the big three, the big five, and the alternative five*1. J Pers Soc Psychol 65:757–768

Mental Retardation and Intellectual Disability

23.2

David L. Nelson

Abstract Cognitive or intellectual disability (ID) in humans is a common trait that has a genetic etiology in as many as one-half of cases. Genetic causes range from single-gene defects to trisomies. The discovery of mutations that lead to ID has led to improved diagnosis and the opportunity for reproductive decisions based on prenatal diagnosis. In addition, understanding the gene defects in ID has elucidated molecules and pathways important for normal cognition, assisting with efforts to understand brain function. This chapter touches on the definition and history of ID, then considers in detail several classifications of gene defects that lead to ID. These include unbalanced gene dosage, with attention to Down syndrome and Prader-Willi and Angelman syndromes, and single gene disorders, particularly X-linked disorders Fragile X and Rett syndromes. Each disorder illustrates important principles in human genetics. Future directions in gene discovery, mutation detection and treatment are discussed.

Contents

23.2.1	Introduction	663
23.2.2	Definition of ID	664
23.2.3	History of ID	664
23.2.4	Frequency of ID	664
23.2.5	Gene Dosage and ID	665
23.2.6	Down Syndrome	666
23.2.7	Recurrent Deletions and Duplications and ID	667
23.2.8	Prader-Willi and Angelman Syndromes	668
23.2.9	Future Directions in Genome Rearrangements and ID	668
23.2.10	Single-Gene Mutations and ID	668
23.2.10.1	Autosomal Recessive	668
23.2.10.2	Autosomal Dominant	669
23.2.11	X-Linked ID	670
23.2.11.1	Fragile X Syndrome	671
23.2.11.2	Rett Syndrome	675
23.2.12	Future Directions	676
References		676

D.L. Nelson(✉)
Molecular and Human Genetics, Baylor College of Medicine,
One Baylor Plaza, Rm. 902E, Houston, TX 77030, USA
e-mail: nelson@bcm.tmc.edu

23.2.1 Introduction

Interest in the genetics of human cognition has spanned the period since the rediscovery of Mendel's work. It has been a controversial area, particularly the study of cognitive abilities within the typical (or normal) range, but also in its treatment of patients. The study of the genetics of intelligence is beyond the scope of this chapter. Here, we consider the genetic conditions that can lead to cognitive disabilities.

In recent years, the term "mental retardation" is no longer favored. "Intellectual," and "cognitive," disability have been proposed as preferred terms. Throughout the history of studies of intellectual disability, numerous terms for description of the condition have fallen out of favor as they are adopted by the general public and are viewed as insensitive to those who are disabled. A noteworthy example of this is the use of the terms "idiot," "imbecile," and "moron" to define individuals with varying degrees of mental retardation based on scores on tests that measure intelligence. These terms were once used in the medical literature to classify individuals with intellectual disability, but this is no longer the case, since they became pejorative in the general language. A similar shift away from "mental retardation" has occurred in recent years. In this chapter, the term intellectual disability (ID) will be utilized to conform with current terminology. Nonetheless, the term mental retardation remains widespread, and it is important to note that ID is meant to be synonymous with mental retardation. A useful discussion of issues surrounding this nomenclature can be found elsewhere [100].

23.2.2 Definition of ID

The most widely accepted definition of ID is that it has an age of onset below 18 and is characterized by significant limitations in both intellectual function and adaptive behavior. In this case, "adaptive behavior" refers to conceptual, social, and practical adaptive skills. For intellectual function, disability is typically measured by intelligence test instruments that generate an intelligence quotient, or IQ score. IQ tests are designed to generate a normal distribution in the general population, and scores are distributed around a mean of 100. ID means an IQ score that is less than two standard deviations beneath the mean. Generally, IQ scores below 70 are considered as an indicator of disability, but this is not universally accepted, and it is useful to keep in mind that adaptive aspects of a particular individual's ability can contribute to his or her classification [1].

23.2.3 History of ID

The presence of individuals with significant ID has been noted in the written records of many ancient cultures. It is only in recent centuries that these individuals have received special societal attention, initially in the form of housing in group institutions. Much more recently, a trend toward "mainstreaming" for all children with disabilities (physical and cognitive, as well as those involving senses such as hearing and sight) has taken hold in the United States and elsewhere. This began in the second half of the twentieth century, giving children with ID access to educational opportunities alongside other children and encouraging parents to raise their children at home. The practice has been mandated through federal legislation in the United States that began in 1975, allowing many children who would otherwise have been committed to state institutions to receive appropriate education in the ordinary schools.

A greater appreciation of genetic forms of ID began in the mid-nineteenth century and accelerated with the Darwinian revolution. Francis Galton took a keen interest in understanding variation in the human population, attempting to implicate heritable components with genius [34]. He established the concept of a "normal distribution" for achievement using, for example, academic scores in medical school and appreciated that such a distribution must contain individuals at the low end of the scale. The normal distribution could be applied to many characteristics, and a quantitative measure of "intelligence" was introduced by Binet and Simon in France in 1907 [16], with the first tests designed to evaluate learning potential. Binet's motivation was principally to classify students for appropriate educational levels during schooling, and the score was defined as a "mental age" to help with placing students. Students whose mental age lagged behind their chronological age were classified as mentally retarded. Later, and in contrast to Binet's wishes, these tests were adapted to measure general intelligence, and a scale was developed with a normal distribution of scores with the average set at 100 and a standard deviation of 15, establishing the currently defined IQ score.

23.2.4 Frequency of ID

Today, IQ testing is carried out using a variety of instruments. Among the most common are the Wechsler Intelligence Scales, a series of tests that can be used for individuals within different age groups. These provide subscores in both "verbal" and "performance" in addition to a "full-scale" or combined score. Some forms of genetic ID can result in differences between verbal and performance scores, and the ability to score both can be instructive for some patients. Based on a normal distribution of IQ scores, it is expected that the frequency of ID would be slightly more than 2% in the general population. It has been difficult to define precise frequencies, however a recent study from the US Centers for Disease Control estimated the number of affected individuals among school age children in metropolitan Atlanta, Georgia at 1.2% [75].

Males exceed females with cognitive disability by approximately 50% [110]. This has led many to propose that the increased male vulnerability stems from X-linked mutations or gene variants that result in cognitive problems in hemizygous males, while sparing heterozygous females. We will consider X-linked genetics and intellectual disability in more detail below. In brief, a large fraction of X-linked genes can result in ID when mutated, but these have not explained the enhanced rate of ID in males. ID can result from many causes, including both environmental and genetic [2]. Environmental causes are numerous, and include exposure in utero to many substances such as alcohol, poor prenatal care, birth trauma, an inadequately enriched environment, malnutrition (including iodine deficiency), and infectious disease. It is estimated that the fraction of cases that result from genetic causes approaches 50% and is higher (perhaps 60%) [74] among more severely affected patients (IQ below 50).

The evidence for a genetic contribution to ID is abundant, with many identified causes in both sporadic and familial forms. The best-characterized forms of genetic ID are syndromic; patients exhibit additional symptoms beyond the cognitive disability that allow their disorder to be categorized. Attention has turned more recently to ID without additional symptoms, referred to as nonsyndromic ID, particularly in the case of families with an X-linked inheritance pattern, but also among families without X-linkage and even in sporadic cases. This chapter will discuss some causes in depth. It is not possible to be comprehensive, however, since there are several hundred known genetic lesions that confer cognitive disability. Indeed, based on data from study of X-linked families [114], it is now estimated that mutations in about 10% of all genes can result in ID, emphasizing that normal cognition is dependent upon a very complex interplay of gene products.

23.2.5 Gene Dosage and ID

In his preface to the 1972 edition of Lionel Penrose's classic monograph "The Biology of Mental Defect," J.B.S. Haldane wrote the following [83]:

> The most sensational discovery about mental defect since the last edition [1963] is that not only mongolism, but several other conditions combining mental defect and physical abnormality, are due to cytological aberrations, the commonest being an extra chromosome. All told, they only account for a small fraction of mental defect, but they lead to a conclusion of great importance. A mongoloid is not an imbecile or idiot because he or she possesses abnormal genes in the cell nuclei, like an epiloiac or an amaurotic idiot, but because he or she has too many of certain normal genes. Penrose believes that many unclassified defectives are defective because, though cytologically normal, they have too many genes of a kind which are found in most normal people, and are on the whole desirable in the heterozygous condition.

Haldane's passage and Penrose's insight are remarkable today as the appreciation of gene copy number changes in genetic forms of ID continues to grow as more sensitive means of detection are deployed. The increasing detection of deletions and duplications in patients with developmental delay, particularly de novo events, is one of the more significant advances in the last 5 years of research into the genetics of ID [94].

Changes in gene dosage resulting from deletion or duplication are a common cause of ID and have long been recognized through cytogenetic methods. It has been estimated that as much as 15% of severe ID is accounted for by cytogenetic abnormalities visible in the light microscope with chromosome banding methods [95]. These include whole-chromosome alterations, principally trisomies, and also many deletions and duplications, some involving apparently balanced translocations. Each of the viable trisomies results in ID. Numerous examples of recurrent deletions that can be detected in a karyotype have been described to result in ID.

Down syndrome remains the most frequent form of genetic ID, with an incidence of approximately 1/700. Described as a form of ID that exhibited distinctive physical features in 1866 by J. Langdown Down in England [28] (and by others earlier), Down syndrome was discovered to result in the majority of cases from trisomy of the smallest human chromosome, chromosome 21, in 1959 by multiple groups [17, 31, 49, 63] shortly after the consensus developed regarding the human chromosome number. The other viable trisomies of autosomes also result in ID. Individuals carrying additional copies of the X chromosome are not usually cognitively disabled, but learning disabilities have been described in up to one half of boys with a 47, XYY karyotype [120]. Turner syndrome, or monosomy X, is the only viable human monosomy and can also result in learning disability, but does not typically cause ID [97]. A variety of smaller deletions resulting in partial monosomy and well-defined syndromes including ID are discussed in more detail below. Many partial trisomies can also lead to developmental delay and ID, and a Down phenotype in partial trisomy 21 is perhaps the most common of these. As methods are refined for the detection of smaller trisomies, it is likely that additional syndromes will be described involving recurrent duplications that lead to cognitive disability.

23.2.6 Down Syndrome

Down syndrome is characterized by developmental delay and a constellation of physical features. The epicanthic fold present in these patients' eyelids led to the use of the term "mongolism" to describe individuals with Down syndrome. This term has been discouraged since 1961, but is still encountered [3]. Other prominent features include upslanting palpebral fissures, a flat nasal bridge, a single transverse palmar crease, short stature, short limbs, and poor muscle tone. Individuals are often born with cardiac defects, often ventricular septal defects, requiring surgical correction. Life expectancy has improved with better care, but there is an increased rate of an Alzheimer-like dementia in older individuals, as well as an increased likelihood of leukemia.

In addition to simple trisomy 21, patients with Down syndrome may also exhibit partial trisomies and Robertsonian translocations that lead to triplication of regions or the entire chromosome 21. Approximately 95% of patients have a spontaneous acquisition of an extra chromosome 21 due to nondisjunction. A small fraction of these patients are mosaic for trisomic and normal cells. Robertsonian translocations are commonly inherited as a 14;21 translocation. There is a maternal excess of spontaneous nondisjunction resulting in Down syndrome. The increased incidence of Down syndrome with maternal age was described by Penrose in 1933 [84]. Current data suggest that more than 90% of nondisjunctions are maternal, with three quarters of those occurring in meiosis I. Paternal nondisjunction accounts for some 4% of cases, while another 2% occur in early embryogenesis as a mitotic event [32]. Advanced maternal age is a risk factor for both meiosis I and meiosis II nondisjunction events. Factors that lead to increased nondisjunction in older eggs remain a subject of much research. The long period of time (from fetal development) that eggs remain in stasis after meiosis I is commonly thought to play a role, but other factors, such as patterns of chromosome recombination, are also likely involved [107].

In contrast to ID caused by single-gene disorders, treatment in Down syndrome is particularly daunting owing to the large number of gene aberrations. However, the completion of the DNA sequence of human chromosome 21 provided an important milestone in Down syndrome research, as it made available to investigators the entire catalog of chromosome 21-encoded genes for analysis, particularly for testing hypotheses regarding the gene or genes that might be critical for aspects of the trisomy 21 phenotype [36, 45]. Numerous individual candidates have been considered, and most recently these have been tested in mouse models, where increased copy number of individual genes can be achieved through transgenesis. An example is the DYRK1 gene, which was found (through creation of transgenic mice with additional copies) to alter cognitive function and later shown to induce developmental delay and motor abnormalities [4, 108]. The conservation of gene order between human and mouse chromosomes along with more detailed maps and then sequences from both species, has facilitated the development of trisomy models in the mouse designed to closely approximate the human trisomy. While the mouse chromosome 16 carries many of the human chromosome 21-encoded genes, it is not completely syntenic, with others carried on chromosomes 17 and 10. The Ts65Dn mouse [92] is trisomic for nearly half of the human chromosome 21 genes, and demonstrates a number of phenotypes that resemble those in Down patients. Additional mouse lines with smaller regions of trisomy that overlap the Ts65Dn region have provided the ability to ascribe certain phenotypes in the mouse to subsets of genes [79, 80]. This approach is beginning to provide candidate sets of genes for phenotypes, allowing consideration of specific therapies for patients. Ts65Dn mice, for example, show alterations in neuronal function that suggest the possibility of treatment with specific neurotransmitter antagonists. These have shown efficacy in modifying neuronal phenotypes in the mouse models and led to the possibility of improved medications for Down syndrome patients.

Prenatal screening for trisomy 21 has been indicated for more than 25 years after advances in amniocentesis and chorionic villus sampling. The guidelines have focused on screening pregnancies in mothers of advanced age (mid-30s), since the incidence of Down syndrome rises dramatically from ~0.1% in younger mothers to more than 3% in women aged 45. It is important to note, however, that the majority of Down syndrome children are born to younger women, since so many more children are born to women below age 30 [85]. Advances in screening for Down syndrome in fetuses have continued in recent years, reducing costs and allowing screening in younger populations at reduced risk to the patient. Screening of maternal serum for levels alphafetoprotein, estriol, human chorionic gonadotropin, and inhibin alpha (the Quad Screen) [13] can be utilized as a first-pass screen for the possibility of a Down (or trisomy 18) fetus. Combined with ultrasound detection of nuchal translucency, it is possible to detect up to 95% of trisomies without invasive testing of the fetus. Efforts continue to develop detection of fetal cells in the maternal circulation for improved identification of trisomies and other cytogenetic abnormalities [57].

23.2.7 Recurrent Deletions and Duplications and ID

Numerous syndromes where ID is a feature have been discovered that result from recurrent deletion of chromosomal regions. Many of these were first described using light microscopy and chromosome banding techniques. These include Prader-Willi [60] and Angelman [66] syndromes, Miller-Dieker [27], Williams [38], DiGeorge [23, 39], and Smith-Magenis [40, 72] syndromes, along with 1p [103] and Xp22 [9] deletions. The recognition that deletion of the terminal ends of chromosomes is relatively common among individuals with ID led to the development of "multi-telomere" fluorescence in situ hybridization, using probes that detect subtelomeric regions of each of the chromosomes [55, 59, 90, 102, 125], with a detection rate of more than 2% in all samples studied. More recently, the adoption of DNA-based comparative genome hybridization methods has exposed additional regions that could not be detected with standard cytogenetic methods [6].

Beginning with the completion of a reference chromosome map and subsequently a reference DNA sequence for the human genome, it became possible to contemplate interrogating human samples for copy number changes at increasing fine scale. These analyses began with the efforts of Pinkel and Gray and their coworkers [53], who recognized the capability to carry out comparative hybridization of two fluorescently labeled DNA samples against chromosomes, and later clones immobilized on a surface [87]. Relative hybridization levels allow determination of regions that are under- or overrepresented in one sample relative to the other. These started with large insert-cloned DNAs (e.g., bacterial artificial chromosomes) representing regions of the genome spotted on glass slides. In the initial version, the aim of these studies was to improve understanding of changes in tumor DNAs relative to DNA from normal cells taken from the same individual. However, this became a general method for comparing any sequence to a reference, and the methods have improved to allow use of oligonucleotide probes spotted on slides to vastly increase the resolution of the method. Such arrays that allow comparative genome hybridization (CGH) analysis have quickly became an important adjunct to standard cytogenetics for characterizing patients with a variety of maladies, including developmental delay and ID [5].

The role of repeated DNA sequences in mediating the rearrangements that lead to recurrent deletion or duplication disorders is now appreciated. In particular, low copy repeats have been found to flank regions commonly deleted in a number of well-described common deletion disorders. These repeated sequences are typically uncommon in the genome, usually repeated locally at the site of the deletion, and often polymorphic in the human population. The repeats are often conserved in recent evolution. Lee and Lupski recently reviewed the contribution of repeats of this type to a variety of neurological disorders that involve recurrent deletions or duplications including ID [61]. The principal mechanisms that can explain the generation of rearrangements that involve repeats are nonallelic homologous recombination and nonhomologous end joining [104]. However, some events that appear more complex can be explained by a replication-based mechanism, rather than by recombination [62].

23.2.8 Prader-Willi and Angelman Syndromes

A recurrent deletion in chromosome 15 (15q11.2-q12) was identified in the early 1980s as a cause of Prader-Willi syndrome, a rare, usually sporadic form of mild ID that has other physical features and unusual behaviors, such as hyperphagia [60]. It was noteworthy when it became clear that Angelman syndrome patients were found with the same cytogenetic lesion [66], since the clinical phenotypes are quite distinct, with Angelman patients showing more severe ID along with characteristic language delays, movement disorders, and a happy demeanor with hand flapping. The recognition that deletions from paternally contributed chromosome 15 led to Prader-Willi, while maternal deletions resulted in Angelman syndrome [56], supported the notion that imprinting of the genes in this region was responsible for the differences in phenotype. The recognition that uniparental disomy can cause both Prader-Willi and Angelman syndromes brought significant confirmation of the imprinting hypothesis [76, 77].

In the intervening years since the recognition that imprinted genes in the 15q11-q12 region can result in different forms of ID, intensive efforts by many groups have led to the recognition of individual genes responsible for the bulk of each disorder. Smaller mutations have been found that support the candidates. For Angelman syndrome, a ubiquitin ligase (UBE3A) is silenced on the paternal chromosome and the maternal allele is expressed in brain regions (hippocampus and cerebellum) that support a role in the phenotypic features of the disease [50]. Mice lacking the gene are found to have learning impairments, further supporting the role of UBE3A in the disorder [115]. For Prader-Willi syndrome, the principal candidate is a cluster of small nucleolar RNA genes that are specifically expressed from the paternal chromosome [26, 98].

23.2.9 Future Directions in Genome Rearrangements and ID

The 15q11.2 locus underscores the complexity of genomic rearrangements that lead to ID. It is 20 years since the initial description of the paired deletions in Prader-Willi and Angelman syndromes, and the specific gene lesions have only recently been uncovered. The development of DNA sequence data and reference genomes has allowed such loci to be exhaustively studied. However, regions with recurrent rearrangements tend to have a complex repeat architecture that is often polymorphic in the human population and can confound genome sequence assemblies [8]. Indeed, variation in copy number is a very common feature of the human genome and appears not to be confined to rearrangement-prone regions [54]. The ongoing efforts to characterize these regions in multiple individuals will begin to address the variation in gene and sequence content in the general population, which should in turn provide better assessment of these regions in individuals with ID. Very recent studies using array CGH have pointed to regions with deletions that can result in multiple phenotypic outcomes, reminiscent of the early Prader-Willi and Angelman findings. An example can be found for a region of chromosome 15, which has been found to have copy number variants in individuals with intellectual disability, autism, epilepsy, or psychiatric disorders such as schizophrenia [14, 48, 109]. Whether imprinting, the fine details of the rearrangements (or of the individual's genomic architecture prior to the rearrangement), or effects of variation in the remaining alleles are responsible will likely be different for each locus. Happily, the tools for distinguishing between these possibilities have been developed.

23.2.10 Single-Gene Mutations and ID

23.2.10.1 *Autosomal Recessive*

A large number of single gene mutations can result in ID, although each individually is quite rare. Many of these also result in inborn errors of metabolism and have specific features of these syndromes in addition to the problems with cognition, but it is likely that many more genes have the potential to contribute to

ID. Examples include lysosomal enzyme storage diseases such as Tay-Sachs, Sandhoff, Niemann-Pick, and forms of Gaucher diseases, which are recessive, typically progressive, and cause ID through neurodegeneration. As most of these are autosomal recessive conditions, their frequency is typically quite rare outside of populations that carry founder mutations or cases with consanguinity.

Diseases that result from defects in amino acid metabolism can also result in cognitive disabilities. The example of phenylketonuria (PKU), most frequently caused by mutation in the phenylalanine hydroxylase gene (PAH), is of particular note. Early detection of newborns with PKU provides the ability to place the patient on a special diet with restricted pheylalanine, reducing or eliminating the likelihood of brain damage and ID. The adoption of newborn screening for elevated phenylalanine levels in the blood of newborns has allowed many individuals who would otherwise have been cognitively damaged to escape this consequence. It is estimated that some 250 of cases of ID have been prevented each year in the US through newborn screening and early treatment with diet [2]. This represents a significant success story for research into genetic causes of ID.

Additional genes involved in autosomal recessive ID are being sought. A very useful approach has made use of families with consanguinity, identifying regions of homozygosity, and screening genes within the region(s) for mutations. Syndromic forms with symptoms beyond ID have been more amenable to this approach, but even nonsyndromic forms have been identified [95]. As the cost of DNA sequencing continues to fall, it may be possible to employ mutation detection on a much larger scale for identification of variants in such families, and possibly in sporadic patients. Efforts to identify causative variants in X-linked forms of ID provide insight into the difficulties involved in embarking on such an approach (see below).

23.2.10.2 Autosomal Dominant

Since genetic alterations that lead to moderate or severe ID rarely lead to successful reproduction in individuals who carry them, most autosomal dominant forms of ID result from new mutations. As described above, spontaneous deletions and other chromosome alterations are common forms of new mutation in ID. However, there are some well-described autosomal dominant single gene disorders that feature milder ID with variable expression. These are often described as learning disabilities. It is likely that many more genes will be uncovered and shown to result in ID or learning disability when haploinsufficient, once the genes that underlie the regions with deletions found by array CGH are characterized and tested in additional sets of patients. An example of such a single-gene disorder is neurofibromatosis type I, which is a relatively common (1/4,000) dominant disorder that is inherited in roughly one half of cases, with the remainder appearing de novo. Some 50% of patients are cognitively impaired. In addition to its role as a tumor suppressor and regulator of myelin, the NF1 gene is involved in early brain development through its regulation of the Ras signal transduction pathway [46], and in other aspects of brain function, which may explain the learning issues in patients.

Another dominant disorder involving disturbed signal transduction is tuberous sclerosis. It is caused by dominant loss of function mutations in either TSC1 or TSC2, and patients with this relatively common (1/6,000) disorder can have severe to mild ID or be cognitively unaffected. They typically develop benign tumors in the brain and other organs, and many exhibit seizures. A number of investigators have determined that signaling through the mTOR pathway is altered in TSC and have successfully demonstrated the use of mTOR inhibitors (rapamycin) for modifying phenotypes in cellular and mouse models of the disease. Application of these results to patients in clinical trials is under way, demonstrating the value of understanding basic mechanisms of gene function for potential treatment regimens in disorders involving ID [96, 99].

Myotonic muscular dystrophy (~1/8,000) is typically caused by a large expansion of a trinucleotide repeat sequence in the DMPK gene [18]. This CTG repeat, located in the 3' untranslated region of the gene, can expand to sizes into the thousands of triplets. The repeat has a tendency to expand in subsequent generations and disease severity correlates with repeat length. Very large expansions, inherited almost exclusively from female carriers, can result in a congenital form of the disorder where ID is very common. In typical, adult-onset forms, cognitive difficulties are more often found to be mild, but there can be progressive cognitive difficulties that manifest as dementia, which would not qualify as ID [70, 71]. However, it is clear that the variable and pleiotrophic effects of the DM mutation can lead to cognitive involvement even in individuals without the congenital form. Significant effort in recent years has pointed to a toxic gain of function by mutant mRNA carrying the expanded CUG triplets as a major cause of the disorder. This has the consequence of reducing availability of proteins that normally interact with RNAs bearing this sequence, altering RNA metabolism. One consequence is that changes occur in alternative splicing patterns in a number of genes [81, 82, 89].

23.2.11 X-Linked ID

Our understanding of single-gene forms of ID has benefited significantly from the study of X chromosome-linked forms. The ability to identify families with X-linked inheritance patterns segregating ID allowed linkage mapping to identify regions associated with the disorder in the family. The likelihood of X-linked forms of ID was suggested as early as in the 1930s, and families with an X-linked pattern of inheritance were described shortly thereafter [68]. Numerous groups have collected families of this type and have described both X-linked syndromic forms (those with other phenotypes beyond the ID) and families with ID and no other symptoms, termed "nonsyndromic." A recent tabulation of X-linked conditions with ID discloses a total of 215 defined conditions, with 66 of these classified as nonsyndromic. More than 80 specific genes have been implicated in X-linked ID (XLID), and nearly 100 more have been assigned to chromosomal regions by linkage mapping (Fig. 23.2.1) [21, 93].

A large-scale DNA sequence-based study of some 75% of the known coding exons present on the X chromosome was recently completed in some 200 families with XLID and no known mutation [114]. Mutations could be definitively identified in one quarter of the families, using criteria of multiple mutations in the gene, clear loss of function mutations, segregation of the mutation with disease in the families, and absence of mutation in unaffected individuals. This study added nine genes involved in XLID and provides novel additional mutations and genes for further diagnostic efforts and research into functional contribution to cognition. There were other findings of note, however, which serve as cautions against scaling these types of studies to the whole genome. Approximately 10% of genes studied were found to have clear loss of function mutations in males that did not confer a phenotype and were present in the general population. This suggests that a similar, or higher, frequency of such events can be expected for autosomes, with consequent difficulties in sorting out their relationship to phenotypic effect. In addition, among the families where a clear causative mutation could not be identified numerous unique variants were found (an average of six per individual) that might have a role in disease. These could not be definitively ascribed, even with the benefit of X-linked inheritance patterns and additional affected family members. The variants discovered in this study were confined to the coding sequences, and the complications of analysis can be predicted to be much greater as we begin to consider noncoding variants. This study underscores the difficulties that will be faced as fewer and smaller families, along with singleton cases of ID are considered for mutation analysis.

The large number of genes involved in XLID, and the diverse functions disrupted, have been surprising. These findings have resulted in a reassessment of the role of X-linked genes in the excess of males with ID, since few X-linked loci with a large contribution to ID have been discovered. Instead, many loci with small numbers of patients and families appear to be the pattern, and many males in families with an X-linked pattern of inheritance remain unexplained by mutation. The excess of males with ID may represent a liability conferred by other factors, perhaps hormonal [20, 67]. The families with an X-linked pattern without detectable mutation may represent chance occurrence of multiple members with ID, which is common in the population in general.

Most families examined for potential XLID exhibit the typical X-linked pattern of inheritance, with affected males, no male-to-male transmission, and carrier females. However, some of the X-linked forms of ID manifest in females (e.g., fragile X syndrome), while others are lethal in males in the prenatal period (e.g., Rett syndrome), leaving only affected females. These exceptions to the usual inheritance pattern have been of great interest, and studies of X-linked forms of ID have led to some interesting new principles in human genetics. These studies have also yielded a bountiful number of genes that have impact on cognition.

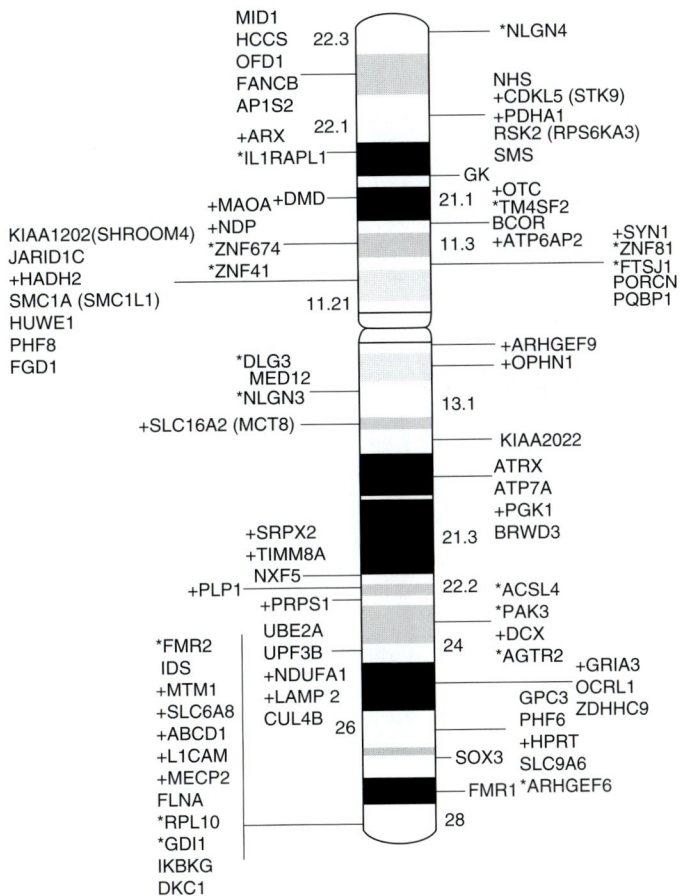

Fig. 23.2.1 Ideogram of the X chromosome with the positions of the 82 known XLMR genes as of 2007. Genes written in *black* cause syndromes, those in *gray* preceded by a *plus sign* cause neuromuscular disorders, and those written in *gray* preceded by an *asterisk* are involved in nonspecific conditions. (Adapted by permission from Macmillan Publishers Ltd: European Journal of Human Genetics, copyright 2008 [21])

23.2.11.1 Fragile X Syndrome

Fragile X syndrome is the most common inherited form of ID, and is second to Down syndrome as a genetic cause of ID. It is estimated to have a prevalence of ~1/3,500 males, and is found in most populations at similar rates [122]. The disorder was first described by Martin and Bell in 1943, who proposed it as an XLID on the basis of a single family with affected males [68]. They described many of the hallmark physical characteristics of the disease, which include a high forehead, large and often protruding ears, large head, and, in postpubescent males, enlarged testicular volume (Fig. 23.2.2). Developmental delay is common, often with speech and language delays, but slow acquisition of motor skills is also typical, with delayed milestones for sitting, standing, and walking. Behavioral characteristics in males are shyness, gaze avoidance, hand flapping, and hypersensitivity to novel environments and intense stimuli. A significant fraction, perhaps as high as 50%, of males with fragile X syndrome can be classified as autistic [35]. The degree of cognitive involvement is highly variable, but ranges from mild to severe. Girls can also be affected, but their cognitive abilities range from normal to moderate levels of ID (Table 23.2.1).

Fragile X syndrome received its name from the discovery in 1969 by Lubs et al. of an unusual chromosomal abnormality present on the X chromosomes of patients with the Martin-Bell phenotype [65]. The "fragile site" was present at the distal end of the long arm, in band q27.3, and manifested as a secondary

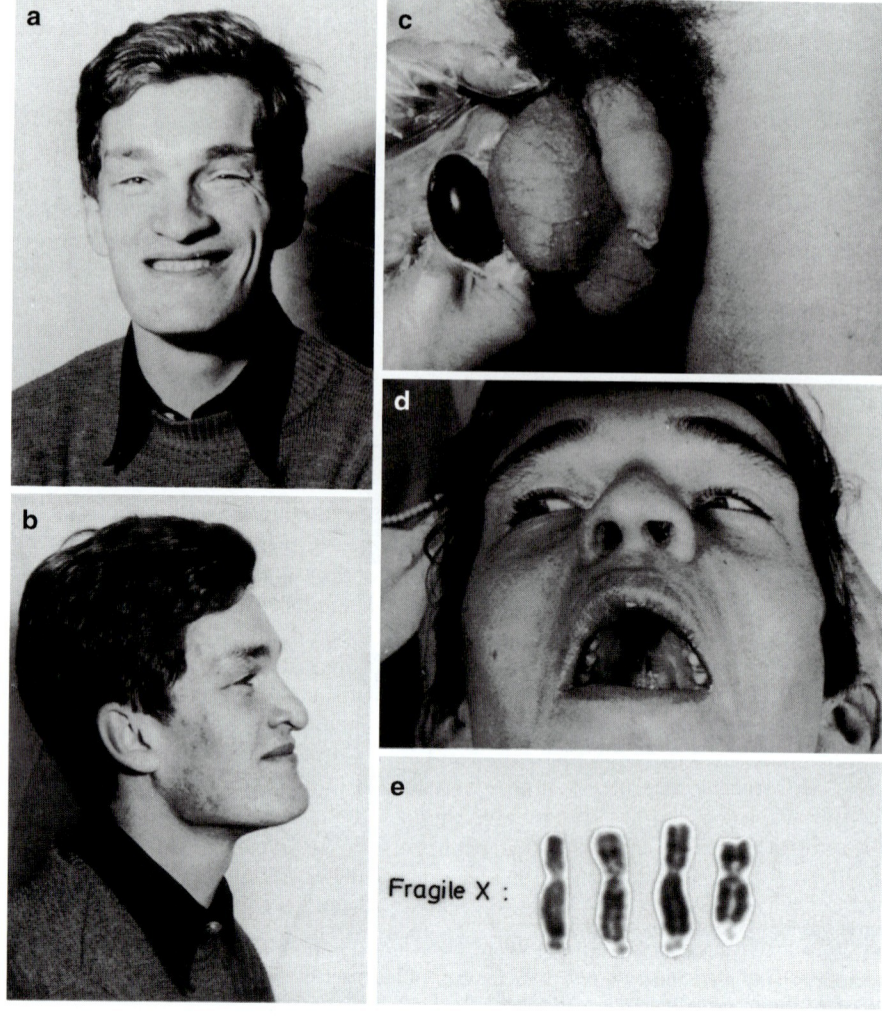

Fig. 23.2.2 Patient with X-linked mental retardation (Martin-Bell or fragile X syndrome): (**a**, **b**) typical face; **c** macro-orchidism; (**d**) high-arched palate; and **e** X chromosomes indicating the fragile X site. From [113]

Table 23.2.1 Clinical features of fragile X syndrome from [119]

Intelligence	IQ range 30–65, sometimes borderline normal or even normal. Occasional hyperactivity or autism in childhood; generally friendly, shy, nonaggressive as teenagers; speech anomaly
Growth	Birth weight normal; usually heavier and taller than normal sibs; head circumference above 50th, sometimes above 97th percentile
Facies	Prominent forehead and jaws, long face, and big ears
Testicles	May be 3–4 cm^3 in childhood (normal 2 cm^3); postpubertal boys 30–60 cm^3 (normal < 25cm^3)
Occasional features	Epilepsy; increased reflexes in lower extremities; gynecomastia, striae, fine skin; thickening of scrotal sac

constriction. The region distal to the fragile site was occasionally broken off in cytogenetic preparations, leading to the constriction being termed a fragile site [111]. A number of reports in the 1970s described this and many more fragile sites in the human genome that could be elicited by a variety of drugs added to cells in culture. The Xq27.3 was found to be induced by low folate media, or by the addition of high levels of folate, and is one of the "folate-sensitive" fragile sites [112]. Interestingly, only a moderate fraction of cells from a patient with fragile X syndrome would exhibit fragile site induction, and rates were never above 50%. The presence of the fragile site was utilized as a diagnostic test for fragile X syndrome for more than a decade.

With a reasonable diagnostic tool, many groups began to collect families with fragile X syndrome and to carry out genetic analyses. These led to a very significant finding: pedigrees segregating fragile X syndrome had unusual transmission characteristics. There were unaffected males in the families, typically grandfathers, who contributed X chromosomes to their grandsons with the disorder. These men were termed normal transmitting males (NTM). The daughters of NTMs were found to have no risk for cognitive problems, but their sons (in aggregate) had a risk that was less than that predicted by Mendelian genetics – just 40%. In some pedigrees, however, it was possible to find generations with typical Mendelian segregation of the disease from carrier females. These typically were found to be in the more recent sibships. These observations were quite puzzling- and were referred to as the Sherman paradox after two papers published by Stephanie Sherman and Patricia Jacobs and their coauthors in the mid 1980s provided the data demonstrating the peculiar pattern of inheritance in the disorder (Fig. 23.2.3) [105, 106].

Linkage mapping using X chromosome markers demonstrated that the fragile X syndrome mutation was in the same region as the cytogenetic abnormality. Using somatic cell genetics, Stephen Warren was able to generate a panel of hybrid cell lines that retained chromosomes with breakpoints apparently at the fragile site, and these were used to determine the locations of fragments of DNA from the X chromosome that were used as markers [123]. Abnormal DNA methylation was detected in the vicinity of the fragile site in male samples using Southern blotting of large restriction fragments separated by pulsed-field electrophoresis, suggesting an unusual mutation [12, 118]. Yeast artificial chromosomes containing the region were identified by several groups and used to characterize the site and identify coding sequences [47]. An expanded, methylated sequence was found to be present in individuals with fragile X syndrome by Southern blotting, but not in controls, or in unaffected individuals carrying the same X chromosome in the pedigree [78, 117, 127]. The expansion was found to arise from a normally polymorphic CGG trinucleotide repeat present in the 5′ untranslated region of the FMR1 gene [117]. The repeats were found to be very unstable within families, and the instability explained the unusual genetics of the disease. NTMs were found to have small expansions, termed premutations. Study of the available families revealed that repeat lengths in individuals with fragile X syndrome exceeded ~200 triplets and were typically several hundred to thousands. Patients showed variation in repeat length and indicated mosaicism, which can include premutation alleles [33, 86].

Discovery of the CGG triplet repeat in fragile X syndrome was followed by that of a large number of unstable triplet repeats that give rise to human genetic disorders. Many of these mutations result in degeneration of neuronal subsets and are present in coding sequences of genes in the form of CAG codons, which encode a stretch of polyglutamines. The first of these to be described was in spinal bulbar muscular atrophy, and was followed by others found in Huntington disease and many spinocerebellar ataxias. Myotonic muscular dystrophy, described above, was also found to result from abnormal expansions of a CTG triplet in the 3′ untranslated region of a gene termed DM protein kinase. The discovery of trinucleotide repeat expansions as mutant alleles in more than one dozen human genetic disorders has been one of the more significant advances in the past two decades [91].

The three allele classes of the CGG repeat in the FMR1 gene are defined by their lengths. Individuals in

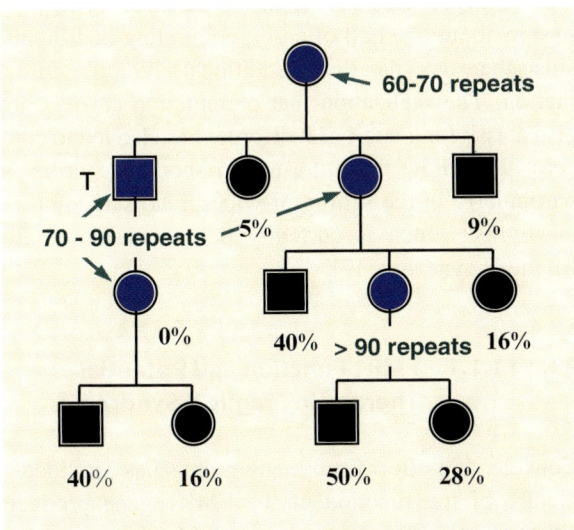

Fig. 23.2.3 Sherman paradox of fragile X inheritance. In this theoretical pedigree, the *black* percentages represent the risk of intellectual disability in a pedigree segregating fragile X syndrome. Unaffected carriers are in *blue*. Note that the risk increases with each generation, and can reach Mendelian expectations for males (*right side*, final generation). The male in the second generation designated by a *T* is a 'normal transmitting male,' who passes the disorder to grandchildren without being cognitively disabled. The *repeat numbers* in *green* indicate the likely CGG repeat lengths that correspond to the risks observed by Sherman et al. The increasing length of repeats in subsequent generations explains the finding that risk is dependent on position in the pedigree

the general population carry alleles that range from as few as 5 to as many as 54 repeats, with the most common alleles being 29 and 30 triplets with AGG triplet interruptions at the 10th and 20th positions. These are transmitted without change to offspring. The moderately expanded premutation alleles are found to range from ~55 to ~230 repeats and are characterized by instability on transmission to the next generation. Premutations alleles typically exhibit long pure CGG sequences, and the development of a premutation from typical alleles appears to often involve loss of one or both of the interrupting AGG triplets. Instability of CGG repeats is found once the length of uninterrupted CGG exceeds approximately 35 triplets [30, 58]. There is a bias toward expansion, with increases outweighing decreases several fold. Interestingly, only female transmissions support the expansion of the premutation to sizes above ~230 repeats, which lead to fragile X syndrome. These larger, full, mutations show methylation of the C residues in the repeat, and this methylation, along with other alterations in chromatin, reflects the loss of FMR1 transcription. While there can be considerable variation among patients owing to mosaicism in methylation and repeat length, the principal result of the full mutation is to eliminate expression of the FMR1 gene.

Premutation alleles are found in all pedigrees with fragile X syndrome; there are no documented examples of a full mutation developing from an allele that would be found in the general population. Moreover, there is no example of a premutation developing from a normal allele. It is estimated that the growth of normal alleles to premutation length takes many generations [73], and the existence of predisposed haplotypes in the general population with unusual patterns of interruption supports the notion of a lengthy process for the development of premutations. The high frequency of fragile X syndrome coupled with the observation of apparently de novo cases led to the prediction that the mutation rate at the locus involved would be extremely high. While this is true in the sense that the premutation to full mutation transition has a high frequency, it is also the case that the high mutation rate is the result of pre-existing premutations, which are primed for additional mutation.

A late-onset neurodegenerative disorder has been described among up to one half of male carriers of premutations [43]. This disorder has been termed fragile X tremor ataxia syndrome (FXTAS) and is characterized by a Parkinsonian tremor, ataxia, cognitive decline, neuronal loss, and the presence of ubiquitin-staining nuclear inclusions in neurons and other cells. Interestingly, patients with fragile X syndrome do not exhibit this disorder, and accumulating evidence points to the expression of CGG-containing RNAs as the toxic agent [44, 51, 52]. This RNA is not expressed in fragile X patients, but is found in premutation carriers, possibly at elevated levels. Some female carriers of premutations have been found with FXTAS, but it appears to be rare in this group.

Counseling in families with fragile X syndrome has benefited from the knowledge of patients' repeat lengths, but it has also introduced considerable complications. The risk of bearing a full-mutation fetus for women carrying a premutation is dependent on the length of the premutation. Risk of an expansion to the full mutation reaches 100% for alleles above ~100 repeats (leading to the expected 50% Mendelian risk), but is very low for those below 60 repeats. Between these two thresholds, the risk escalates with length. An additional complication is presented by the ability to detect female fetuses carrying the full mutation. Since these girls are affected (i.e., have ID defined by IQ testing) in about one half of cases, counseling is difficult, since there is typically no additional predictive information. The realization that premutation carriers are also at risk for a late-onset disorder has also led to concern about how this information should be utilized, particularly in the setting of widespread screening of newborns, which is currently being considered for fragile X syndrome [7].

23.2.11.1.1 FMR1 Function and Potential Therapy in Fragile X Syndrome

Considerable effort from many groups has provided a picture of the function of the FMR1 gene product, although it is far from complete [10]. The protein (FMRP) is a member of a small family (three members) of RNA-binding proteins that have roles in translation of target RNAs. One of the important sites of action for FMRP is in neuronal dendrites, where it is part of a signal transduction cascade that responds to synaptic signaling, FMRP may be involved in transport of specific RNAs to dendrites, as it is found in RNA:protein particles. The protein can be found on polyribosomes, suggesting a role in translational control, and its primary role may be suppression of translation. The absence of FMRP in animal models supports this idea, since overproduction of protein is observed in these models [88].

A number of observations in animal models led to the development of the "mGluR theory of fragile X syndrome" [11]. This theory postulates that glutamate signaling through type I metabotropic receptors (mGluR1 and mGluR5, which are G protein-coupled receptors sensitive to glutamate) is modulated in part by FMRP through translational suppression, and that in FMRP's absence, this system is overstimulated, with a number of consequences, including difficulty for synapses to mature [10]. A prediction of this theory is that reduction of signaling through the mGluRs would ameliorate these effects and might prove therapeutic. Tests of specific inhibitors of mGluR5 proved to modify phenotypes in mouse and fly models [25, 69, 126], and these results have encouraged clinical applications of similar compounds in patients with fragile X syndrome [15]. If these therapies prove effective, this will represent a significant triumph for functional analysis of gene function leading to rational therapy in a common form of genetic ID.

23.2.11.2 Rett Syndrome

Rett syndrome is caused by mutations in the *MECP2* gene in Xq28. The disorder affects girls almost exclusively and is characterized by a progressive loss of function beginning in the 2nd year of life and resulting in severe cognitive disability along with a constellation of physical and behavioral abnormalities (Fig. 23.2.4) [19]. Severity can be highly variable, likely owing to patterns of X-inactivation that differ among patients. The incidence of disease has been estimated to be between 1/10,000 and 1/15,000, and it is found in all populations. Leading up to the description of mutations in the *MECP2* gene, the evidence in favor of a genetic etiology for Rett syndrome was sparse. Most cases appeared to be sporadic, with only a small number of recurrent cases now known to have resulted from mothers who were unaffected owing to favorable lyonization or germline mosaicism. Males carrying typical loss of function mutations in *MECP2* are not viable as embryos, and affected females do not typically bear children, so that, as expected, most cases

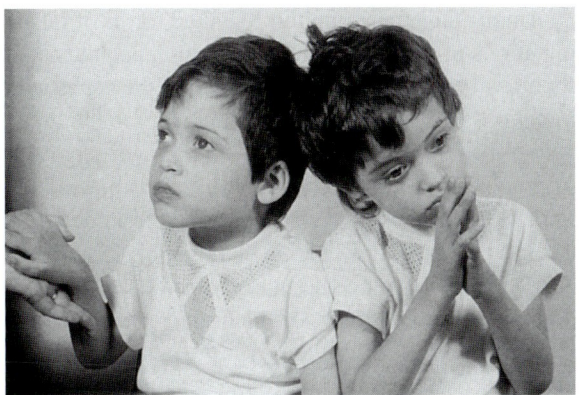

Fig. 23.2.4 Monozygotic twins with Rett syndrome, at the age of 9 years. (Courtesy of Dr. G. Tariverdian)

(~95%) result from de novo mutation. With the focus on the *MECP2* gene, it has become apparent that there are additional disorders that result from *MECP2* mutation, including a relatively common duplication that confers ID in males [24, 116].

The *MECP2* gene product is a nuclear protein that recognizes methylated DNA and, together with other transcription factors, is involved in control of gene transcription. In affected girls, loss of protein function in a portion of cells leads to dysfunction, while overexpression in males is also problematic for normal cognition. A male with a triplication at the locus and a more severe phenotype suggests significant dosage sensitivity for *MECP2* levels [24]. Mice with mutations in *Mecp2* mirror human phenotypes [22, 41, 101]. Of particular note is that adult restoration of *Mecp2* expression can rescue phenotypes [37, 42], offering the possibility that the disorder is dependent on acute levels of *MECP2*, rather than a result of developmental damage. Similar data have been developed in other mouse models for genetic forms of ID, with rescue by gene replacement of drug treatment [29]. These exciting results portend well for potential therapy in Rett syndrome and in other ID disorders.

These findings in Rett sydrome and related disorders underscore the likelihood that the ongoing efforts to identify genomic regions with deletions and duplications will likely be fruitful for further delineation of dosage-sensitive genes involved in ID, and predict that even more subtle alterations in expression levels could prove important for behavioral and cognitive phenotypes, both pathogenic and within the typical range.

Among the other common X-linked forms of ID are mutations in ARX and SLC6A8, which each account for a few percent of XLID [93]. Each of the genes with mutations in XLID has an important story to tell about gene function, mutational mechanism, and genotype/phenotype correlation. Unfortunately, it is beyond the limitations of this chapter to discuss them in detail. A general principle, however, is that the classification of XLID into syndromic and nonsyndromic forms broke down as gene mutations were identified. There are now several examples of mutations in the same gene that can give rise to both syndromic and nonsyndromic forms of ID with X-linked inheritance patterns [93].

23.2.12 Future Directions

The large abundance of genes that can give rise to XLID suggests that a similar fraction of autosomal genes might contribute to cognitive disability. This is likely to be 10% or more of genes and suggests that study of inbred populations will uncover causative homozygous mutations that may be limited to a small number of families. Of possibly more interest will be uncovering the heterozygous mutations that might confer a milder effect, yet could be of appreciable frequency in the general population. The study of autosomal ID in inbred families may point to some of these.

With the large-scale XLID sequencing study [114] and the prior efforts to sequence individual diploid genomes [64, 121, 124], we have come to appreciate that the level of variation between individuals' genomes is large. It will require large numbers of genome sequences from well-characterized individuals to begin to associate sequence variation with phenotype, as many variants will have no effect, while others will be essentially private, making them very difficult to associate with phenotype without family studies or development of models. It is an exciting time, but the accumulation of large data sets of sequence will have to be accompanied by equivalent efforts in determining phenotypes at high precision for the correlations to be worthwhile.

The use of models, principally mouse models, for understanding consequences of mutations on cognition has been very effective. Examples from Rett and fragile X syndromes are noted above. These models have been essential for understanding the functions of the affected proteins, and are being used to devise and test therapies. Nonetheless, the laboratory mouse has significant limitations for study of human cognition. Defects in learning and memory are typically mild if present. This is due in part to the nature of the laboratory mouse, which through inbreeding has been selected for docility. At present, however, there is no viable alternative mammalian model from a cost or convenience viewpoint. The fruit fly *Drosophila* has also been of particular value; the rapid ability to identify modifying genes and even to carry out drug screens has had a very favorable impact for study of mutations of the orthologous fly gene products.

As therapies are developed for disorders that include ID, there will be more interest in developing newborn screening approaches for identifying candidates for treatment. This will be particularly important for disorders such as ID, where early intervention may be of considerable value. The situation in PKU provides an important example, where early diet therapy can prevent a devastating cognitive disability. Testing for mutations in most of the newly discovered genetic forms of ID will require more sophisticated, likely DNA-based, testing, however, which will involve a much greater cost than the current newborn screening for blood analytes. As DNA sequencing costs continue their precipitous decline, it appears likely that complete genome sequencing of newborns may be within reach in the not-too-distant future, providing a comprehensive approach to mutation detection for ID and charting a course of treatment for affected individuals that will preserve or even improve their cognitive capacities. The past 50 years have seen remarkable advances in identification of causes of ID, and the next half century should yield an equivalent series of discoveries that lead to effective therapies for cognitive disability.

References

1. Accardo P, Capute A (1998) Mental retardations. Ment Retard Dev Disabil Res Rev 4:2–5
2. Alexander D (1998) Prevention of mental retardation: four decades of research. Ment Retard Dev Disabil Res Rev 4:50–58
3. Allen G, Benda CE, Book JA, Carter CO, Ford CE, Chu EHY, Hanhart E, Jervis G, Langdon-Down W, Lejeune J et al (1961) Mongolism. Lancet 1:775

4. Altafaj X, Dierssen M, Baamonde C, Marti E, Visa J, Guimera J, Oset M, Gonzalez JR, Florez J, Fillat C et al (2001) Neurodevelopmental delay, motor abnormalities and cognitive deficits in transgenic mice overexpressing DYRK1a (minibrain), a murine model of Down's syndrome. Hum Mol Genet 10:1915–1923
5. Aradhya S, Cherry AM (2007) Array-based comparative genomic hybridization: clinical contexts for targeted and whole-genome designs. Genet Med 9:553–559
6. Aradhya S, Manning MA, Splendore A, Cherry AM (2007) Whole-genome array-cgh identifies novel contiguous gene deletions and duplications associated with developmental delay, mental retardation, and dysmorphic features. Am J Med Genet 143A:1431–1441
7. Bailey DB Jr, Skinner D, Davis AM, Whitmarsh I, Powell C (2008) Ethical, legal, and social concerns about expanded newborn screening: fragile X syndrome as a prototype for emerging issues. Pediatrics 121:e693–e704
8. Bailey JA, Yavor AM, Massa HF, Trask BJ, Eichler EE (2001) Segmental duplications: organization and impact within the current human genome project assembly. Genome Res 11:1005–1017
9. Ballabio A, Bardoni B, Carrozzo R, Andria G, Bick D, Campbell L, Hamel B, Ferguson-Smith MA, Gimelli G, Fraccaro M et al (1989) Contiguous gene syndromes due to deletions in the distal short arm of the human X chromosome. Proc Natl Acad Sci USA 86:10001–10005
10. Bassell GJ, Warren ST (2008) Fragile X syndrome: loss of local MRNA regulation alters synaptic development and function. Neuron 60:201–214
11. Bear MF, Huber KM, Warren ST (2004) The MGLur theory of fragile X mental retardation. Trends Neurosci 27:370–377
12. Bell MV, Hirst MC, Nakahori Y, MacKinnon RN, Roche A, Flint TJ, Jacobs PA, Tommerup N, Tranebjaerg L, Froster-Iskenius U et al (1991) Physical mapping across the fragile X: hypermethylation and clinical expression of the fragile X syndrome. Cell 64:861–866
13. Benn PA, Ying J, Beazoglou T, Egan JF (2001) Estimates for the sensitivity and false-positive rates for second trimester serum screening for down syndrome and trisomy 18 with adjustment for cross-identification and double-positive results. Prenat Diagn 21:46–51
14. Ben-Shachar S, Lanpher B, German JR, Qasaymeh M, Potocki L, Nagamani S, Franco LM, Malphrus A, Bottenfield GW, Spence JE et al (2009) Microdeletion 15q13.3: a locus with incomplete penetrance for autism, mental retardation, and psychiatric disorders. J Med Genet 46:382–388
15. Berry-Kravis E, Hessl D, Coffey S, Hervey C, Schneider A, Yuhas J, Hutchison J, Snape M, Tranfaglia M, Nguyen DV et al (2009) A pilot open label, single dose trial of fenobam in adults with fragile X syndrome. J Med Genet 46:266–271
16. Binet A, Simon T (1907) Les enfants anormaux. A. Colin, Paris
17. Book JA, Fraccaro M, Lindsten J (1959) Cytogenetical observations in mongolism. Acta Paediatr 48:453–468
18. Caskey CT, Pizzuti A, Fu YH, Fenwick RG Jr, Nelson DL (1992) Triplet repeat mutations in human disease. Science 256:784–789
19. Chahrour M, Zoghbi HY (2007) The story of Rett syndrome: from clinic to neurobiology. Neuron 56:422–437
20. Chelly J, Mandel JL (2001) Monogenic causes of X-linked mental retardation. Nat Rev Genet 2:669–680
21. Chiurazzi P, Schwartz CE, Gecz J, Neri G (2008) *Xlmr* genes: update 2007. Eur J Hum Genet 16:422–434
22. Collins AL, Levenson JM, Vilaythong AP, Richman R, Armstrong DL, Noebels JL, David Sweatt J, Zoghbi HY (2004) Mild overexpression of mecp2 causes a progressive neurological disorder in mice. Hum Mol Genet 13:2679–2689
23. de la Chapelle A, Herva R, Koivisto M, Aula P (1981) A deletion in chromosome 22 can cause diGeorge syndrome. Hum Genet 57:253–256
24. del Gaudio D, Fang P, Scaglia F, Ward PA, Craigen WJ, Glaze DG, Neul JL, Patel A, Lee JA, Irons M et al (2006) Increased mecp2 gene copy number as the result of genomic duplication in neurodevelopmentally delayed males. Genet Med 8:784–792
25. de Vrij FM, Levenga J, van der Linde HC, Koekkoek SK, De Zeeuw CI, Nelson DL, Oostra BA, Willemsen R (2008) Rescue of behavioral phenotype and neuronal protrusion morphology in *fmr1* KO mice. Neurobiol Dis 31:127–132
26. Ding F, Li HH, Zhang S, Solomon NM, Camper SA, Cohen P, Francke U (2008) Snorna snord116 (pwcr1/mbii-85) deletion causes growth deficiency and hyperphagia in mice. PLoS ONE 3:e1709
27. Dobyns WB, Stratton RF, Parke JT, Greenberg F, Nussbaum RL, Ledbetter DH (1983) Miller-dieker syndrome: lissencephaly and monosomy 17p. J Pediatr 102:552–558
28. Down J (1866) Observations on an ethnic classification of idiots. Lond Hosp Rep Clin Lect 3:259–262
29. Ehninger D, Li W, Fox K, Stryker MP, Silva AJ (2008) Reversing neurodevelopmental disorders in adults. Neuron 60:950–960
30. Eichler EE, Holden JJ, Popovich BW, Reiss AL, Snow K, Thibodeau SN, Richards CS, Ward PA, Nelson DL (1994) Length of uninterrupted cgg repeats determines instability in the *fmr1* gene. Nat Genet 8:88–94
31. Ford CE, Jones KW, Miller OJ, Mittwoch U, Penrose LS, Ridler M, Shapiro A (1959) The chromosomes in a patient showing both mongolism and the Klinefelter syndrome. Lancet 1:709–710
32. Freeman SB, Allen EG, Oxford-Wright CL, Tinker SW, Druschel C, Hobbs CA, O'Leary LA, Romitti PA, Royle MH, Torfs CP et al (2007) The national Down syndrome project: design and implementation. Public Health Rep 122:62–72
33. Fu YH, Kuhl DP, Pizzuti A, Pieretti M, Sutcliffe JS, Richards S, Verkerk AJ, Holden JJ, Fenwick RG Jr, Warren ST et al (1991) Variation of the cgg repeat at the fragile X site results in genetic instability: resolution of the Sherman paradox. Cell 67:1047–1058
34. Galton F (1869) Hereditary genius: an inquiry into its laws and consequences. Macmillan, London
35. Garber KB, Visootsak J, Warren ST (2008) Fragile X syndrome. Eur J Hum Genet 16:666–672
36. Gardiner K, Davisson M (2000) The sequence of human chromosome 21 and implications for research into down syndrome. Genome Biol 1:REVIEWS0002
37. Giacometti E, Luikenhuis S, Beard C, Jaenisch R (2007) Partial rescue of mecp2 deficiency by postnatal activation of mecp2. Proc Natl Acad Sci USA 104:1931–1936

38. Greenberg F, Ledbetter DH (1988) Chromosome abnormalities and Williams syndrome. Am J Med Genet 30:993–996
39. Greenberg F, Crowder WE, Paschall V, Colon-Linares J, Lubianski B, Ledbetter DH (1984) Familial diGeorge syndrome and associated partial monosomy of chromosome 22. Hum Genet 65:317–319
40. Greenberg F, Guzzetta V, Montes de Oca-Luna R, Magenis RE, Smith AC, Richter SF, Kondo I, Dobyns WB, Patel PI, Lupski JR (1991) Molecular analysis of the smith-magenis syndrome: a possible contiguous-gene syndrome associated with del(17)(p11.2). Am J Hum Genet 49:1207–1218
41. Guy J, Hendrich B, Holmes M, Martin JE, Bird A (2001) A mouse mecp2-null mutation causes neurological symptoms that mimic rett syndrome. Nat Genet 27:322–326
42. Guy J, Gan J, Selfridge J, Cobb S, Bird A (2007) Reversal of neurological defects in a mouse model of rett syndrome. Science 315:1143–1147
43. Hagerman PJ, Hagerman RJ (2004) The fragile-X premutation: a maturing perspective. Am J Hum Genet 74:805–816
44. Hashem V, Galloway JN, Mori M, Willemsen R, Oostra BA, Paylor R, Nelson DL (2009) Ectopic expression of cgg containing mrna is neurotoxic in mammals. Hum Mol Genet 18:2443–2451
45. Hattori M, Fujiyama A, Taylor TD, Watanabe H, Yada T, Park HS, Toyoda A, Ishii K, Totoki Y, Choi DK et al (2000) The DNA sequence of human chromosome 21. Nature 405:311–319
46. Hegedus B, Dasgupta B, Shin JE, Emnett RJ, Hart-Mahon EK, Elghazi L, Bernal-Mizrachi E, Gutmann DH (2007) Neurofibromatosis-1 regulates neuronal and glial cell differentiation from neuroglial progenitors in vivo by both camp- and ras-dependent mechanisms. Cell Stem Cell 1:443–457
47. Heitz D, Rousseau F, Devys D, Saccone S, Abderrahim H, Le Paslier D, Cohen D, Vincent A, Toniolo D, Della Valle G et al (1991) Isolation of sequences that span the fragile x and identification of a fragile X-related cpg island. Science 251:1236–1239
48. Helbig I, Mefford HC, Sharp AJ, Guipponi M, Fichera M, Franke A, Muhle H, de Kovel C, Baker C, von Spiczak S et al (2009) 15q13.3 microdeletions increase risk of idiopathic generalized epilepsy. Nat Genet 41:160–162
49. Jacobs PA, Baikie AG, Court Brown WM, Strong JA (1959) The somatic chromosomes in mongolism. Lancet 1:710
50. Jiang Y, Tsai TF, Bressler J, Beaudet AL (1998) Imprinting in Angelman and Prader-Willi syndromes. Curr Opin Genet Dev 8:334–342
51. Jin P, Zarnescu DC, Zhang F, Pearson CE, Lucchesi JC, Moses K, Warren ST (2003) Rna-mediated neurodegeneration caused by the fragile X premutation rcgg repeats in drosophila. Neuron 39:739–747
52. Jin P, Duan R, Qurashi A, Qin Y, Tian D, Rosser TC, Liu H, Feng Y, Warren ST (2007) Pur alpha binds to rcgg repeats and modulates repeat-mediated neurodegeneration in a drosophila model of fragile X tremor/ataxia syndrome. Neuron 55:556–564
53. Kallioniemi A, Kallioniemi OP, Sudar D, Rutovitz D, Gray JW, Waldman F, Pinkel D (1992) Comparative genomic hybridization for molecular cytogenetic analysis of solid tumors. Science 258:818–821
54. Kidd JM, Cooper GM, Donahue WF, Hayden HS, Sampas N, Graves T, Hansen N, Teague B, Alkan C, Antonacci F et al (2008) Mapping and sequencing of structural variation from eight human genomes. Nature 453:56–64
55. Knight SJ, Regan R (2006) Idiopathic learning disability and genome imbalance. Cytogenet Genome Res 115:215–224
56. Knoll JH, Nicholls RD, Magenis RE, Graham JM Jr, Lalande M, Latt SA (1989) Angelman and prader-willi syndromes share a common chromosome 15 deletion but differ in parental origin of the deletion. Am J Med Genet 32: 285–290
57. Kolialexi A, Tsangaris GT, Papantoniou N, Anagnostopoulos AK, Vougas K, Bagiokos V, Antsaklis A, Mavrou A (2008) Application of proteomics for the identification of differentially expressed protein markers for Down syndrome in maternal plasma. Prenat Diagn 28:691–698
58. Kunst CB, Warren ST (1994) Cryptic and polar variation of the fragile X repeat could result in predisposing normal alleles. Cell 77:853–861
59. Ledbetter DH, Martin CL (2007) Cryptic telomere imbalance: a 15-year update. Am J Med Genet C Semin Med Genet 145C:327–334
60. Ledbetter DH, Riccardi VM, Airhart SD, Strobel RJ, Keenan BS, Crawford JD (1981) Deletions of chromosome 15 as a cause of the Prader-Willi syndrome. N Engl J Med 304:325–329
61. Lee JA, Lupski JR (2006) Genomic rearrangements and gene copy-number alterations as a cause of nervous system disorders. Neuron 52:103–121
62. Lee JA, Carvalho CM, Lupski JR (2007) A DNA replication mechanism for generating nonrecurrent rearrangements associated with genomic disorders. Cell 131:1235–1247
63. Lejeune J, Turpin R, Gautier M (1959) Le mongolisme, maladie chromosomique. (trisomie). Bull Acad Natl Med 143(11-12):256–265
64. Levy S, Sutton G, Ng PC, Feuk L, Halpern AL, Walenz BP, Axelrod N, Huang J, Kirkness EF, Denisov G et al (2007) The diploid genome sequence of an individual human. PLoS Biol 5:e254
65. Lubs HA (1969) A marker X-chromosome. Am J Hum Genet 21:231–244
66. Magenis RE, Brown MG, Lacy DA, Budden S, LaFranchi S (1987) Is Angelman syndrome an alternate result of del(15)(q11q13)? Am J Med Genet 28:829–838
67. Mandel JL, Chelly J (2004) Monogenic X-linked mental retardation: is it as frequent as currently estimated? The paradox of the arX (aristaless X) mutations. Eur J Hum Genet 12:689–693
68. Martin JP, Bell J (1943) A pedigree of mental defect showing sex-linkage. J Neurol Psych 6:154–157
69. McBride SM, Choi CH, Wang Y, Liebelt D, Braunstein E, Ferreiro D, Sehgal A, Siwicki KK, Dockendorff TC, Nguyen HT et al (2005) Pharmacological rescue of synaptic plasticity, courtship behavior, and mushroom body defects in a *Drosophila* model of fragile X syndrome. Neuron 45:753–764
70. Modoni A, Silvestri G, Pomponi MG, Mangiola F, Tonali PA, Marra C (2004) Characterization of the pattern of cognitive impairment in myotonic dystrophy type 1. Arch Neurol 61:1943–1947
71. Modoni A, Silvestri G, Vita MG, Quaranta D, Tonali PA, Marra C (2008) Cognitive impairment in myotonic dystrophy

71. ...type 1 (dm1): a longitudinal follow-up study. J Neurol 255:1737–1742
72. Moncla A, Livet MO, Auger M, Mattei JF, Mattei MG, Giraud F (1991) Smith-magenis syndrome: a new contiguous gene syndrome. Report of three new cases. J Med Genet 28:627–632
73. Morris A, Morton NE, Collins A, Macpherson J, Nelson D, Sherman S (1995) An n-allele model for progressive amplification in the fmr1 locus. Proc Natl Acad Sci USA 92:4833–4837
74. Moser HG (1995) A role for gene therapy in mental retardation. Ment Retard Dev Disabil Res Rev 1:4–6
75. Murphy CC, Boyle C, Schendel D, Decouflé P, Yeargin-Allsop M (1998) Epidemiology of mental retardation in children. Ment Retard Dev Disabil Res Rev 4:6–13
76. Nicholls RD, Knoll JH, Butler MG, Karam S, Lalande M (1989) Genetic imprinting suggested by maternal heterodisomy in nondeletion Prader-Willi syndrome. Nature 342:281–285
77. Nicholls RD, Pai GS, Gottlieb W, Cantu ES (1992) Paternal uniparental disomy of chromosome 15 in a child with Angelman syndrome. Ann Neurol 32:512–518
78. Oberle I, Rousseau F, Heitz D, Kretz C, Devys D, Hanauer A, Boue J, Bertheas M, Mandel J (1991) Instability of a 550-base pair DNA segment and abnormal methylation in fragile X syndrome. Science 252:1097–1102
79. Olson LE, Richtsmeier JT, Leszl J, Reeves RH (2004) A chromosome 21 critical region does not cause specific Down syndrome phenotypes. Science 306:687–690
80. Olson LE, Roper RJ, Sengstaken CL, Peterson EA, Aquino V, Galdzicki Z, Siarey R, Pletnikov M, Moran TH, Reeves RH (2007) Trisomy for the Down syndrome 'critical region' is necessary but not sufficient for brain phenotypes of trisomic mice. Hum Mol Genet 16:774–782
81. Orengo JP, Cooper TA (2007) Alternative splicing in disease. Adv Exp Med Biol 623:212–223
82. Orengo JP, Chambon P, Metzger D, Mosier DR, Snipes GJ, Cooper TA (2008) Expanded ctg repeats within the dmpk 3' utr causes severe skeletal muscle wasting in an inducible mouse model for myotonic dystrophy. Proc Natl Acad Sci USA 105:2646–2651
83. Penrose L (1972) Biology of mental defect, 4th edn. Sidgwick and Jackson, London
84. Penrose LS (1933) The relative effects of paternal and maternal age in mongolism. J. Genet. 27:219–224
85. Penrose LS (1954) Mongolian idiocy (mongolism) and maternal age. Ann NY Acad Sci 57:494–502
86. Pieretti M, Zhang FP, Fu YH, Warren ST, Oostra BA, Caskey CT, Nelson DL (1991) Absence of expression of the fmr-1 gene in fragile X syndrome. Cell 66:817–822
87. Pinkel D, Segraves R, Sudar D, Clark S, Poole I, Kowbel D, Collins C, Kuo WL, Chen C, Zhai Y et al (1998) High resolution analysis of DNA copy number variation using comparative genomic hybridization to microarrays. Nat Genet 20:207–211
88. Qin M, Kang J, Burlin TV, Jiang C, Smith CB (2005) Postadolescent changes in regional cerebral protein synthesis: an in vivo study in the fmr1 null mouse. J Neurosci 25:5087–5095
89. Ranum LP, Day JW (2004) Pathogenic RNA repeats: an expanding role in genetic disease. Trends Genet 20:506–512
90. Ravnan JB, Tepperberg JH, Papenhausen P, Lamb AN, Hedrick J, Eash D, Ledbetter DH, Martin CL (2006) Subtelomere fish analysis of 11 688 cases: an evaluation of the frequency and pattern of subtelomere rearrangements in individuals with developmental disabilities. J Med Genet 43:478–489
91. Reddy PS, Housman DE (1997) The complex pathology of trinucleotide repeats. Curr Opin Cell Biol 9:364–372
92. Reeves RH, Irving NG, Moran TH, Wohn A, Kitt C, Sisodia SS, Schmidt C, Bronson RT, Davisson MT (1995) A mouse model for down syndrome exhibits learning and behaviour deficits. Nat Genet 11:177–184
93. Ropers HH (2006) X-linked mental retardation: many genes for a complex disorder. Curr Opin Genet Dev 16:260–269
94. Ropers HH (2007) New perspectives for the elucidation of genetic disorders. Am J Hum Genet 81:199–207
95. Ropers HH (2008) Genetics of intellectual disability. Curr Opin Genet Dev 18:241–250
96. Rosner M, Hanneder M, Siegel N, Valli A, Fuchs C, Hengstschlager M (2008) The mtor pathway and its role in human genetic diseases. Mutat Res 659:284–292
97. Rovet J (2004) Turner syndrome: a review of genetic and hormonal influences on neuropsychological functioning. Child Neuropsychol 10:262–279
98. Sahoo T, del Gaudio D, German JR, Shinawi M, Peters SU, Person RE, Garnica A, Cheung SW, Beaudet AL (2008) Prader-willi phenotype caused by paternal deficiency for the hbii-85 c/d box small nucleolar rna cluster. Nat Genet 40:719–721
99. Sampson JR (2009) Therapeutic targeting of mtor in tuberous sclerosis. Biochem Soc Trans 37:259–264
100. Schalock RL, Luckasson RA, Shogren KA, Borthwick-Duffy S, Bradley V, Buntinx WHE, Coulter DL, Craig EM, Gomez SC, Lachapelle Y et al (2007) The renaming of mental retardation: understanding the change to the term intellectual disability. Intellect Dev Disabil 45:116–124
101. Shahbazian M, Young J, Yuva-Paylor L, Spencer C, Antalffy B, Noebels J, Armstrong D, Paylor R, Zoghbi H (2002) Mice with truncated mecp2 recapitulate many rett syndrome features and display hyperacetylation of histone h3. Neuron 35:243–254
102. Shao L, Shaw CA, Lu XY, Sahoo T, Bacino CA, Lalani SR, Stankiewicz P, Yatsenko SA, Li Y, Neill S et al (2008) Identification of chromosome abnormalities in subtelomeric regions by microarray analysis: a study of 5, 380 cases. Am J Med Genet 146A:2242–2251
103. Shapira SK, McCaskill C, Northrup H, Spikes AS, Elder FF, Sutton VR, Korenberg JR, Greenberg F, Shaffer LG (1997) Chromosome 1p36 deletions: the clinical phenotype and molecular characterization of a common newly delineated syndrome. Am J Hum Genet 61:642–650
104. Shaw CJ, Lupski JR (2004) Implications of human genome architecture for rearrangement-based disorders: the genomic basis of disease. Hum Mol Genet 13(Spec No 1):R57–R64
105. Sherman SL, Morton NE, Jacobs PA, Turner G (1984) The marker (x) syndrome: a cytogenetic and genetic analysis. Ann Hum Genet 48:21–37
106. Sherman SL, Jacobs PA, Morton NE, Froster Iskenius U, Howard Peebles PN, Nielsen KB, Partington MW,

Sutherland GR, Turner G, Watson M (1985) Further segregation analysis of the fragile X syndrome with special reference to transmitting males. Hum Genet 69: 289–299

107. Sherman SL, Allen EG, Bean LH, Freeman SB (2007) Epidemiology of down syndrome. Ment Retard Dev Disabil Res Rev 13:221–227
108. Smith DJ, Stevens ME, Sudanagunta SP, Bronson RT, Makhinson M, Watabe AM, O'Dell TJ, Fung J, Weier HU, Cheng JF et al (1997) Functional screening of 2 mb of human chromosome 21q22.2 in transgenic mice implicates minibrain in learning defects associated with down syndrome. Nat Genet 16:28–36
109. Stefansson H, Rujescu D, Cichon S, Pietilainen OP, Ingason A, Steinberg S, Fossdal R, Sigurdsson E, Sigmundsson T, Buizer-Voskamp JE et al (2008) Large recurrent microdeletions associated with schizophrenia. Nature 455:232–236
110. Stevenson R, Schwartz C, Schroer R (2000) X-linked mental retardation. Oxford University Press, Oxford
111. Sutherland GR (1977) Fragile sites on human chromosomes: demonstration of their dependence on the type of tissue culture medium. Science 197:265–266
112. Sutherland GR, Ashforth PLC (1979) X-linked mental retardation with macroorchidism and the fragile site at xq27 or 28. Hum Genet 48:117–120
113. Tariverdian G, Weck B (1982) Nonspecific X-linked mental retardation--a review. Hum Genet 62(2):95–109
114. Tarpey PS, Smith R, Pleasance E, Whibley A, Edkins S, Hardy C, O'Meara S, Latimer C, Dicks E, Menzies A et al (2009) A systematic, large-scale resequencing screen of X-chromosome coding exons in mental retardation. Nat Genet 41(5):535–543
115. Tsai TF, Jiang YH, Bressler J, Armstrong D, Beaudet AL (1999) Paternal deletion from snrpn to ube3a in the mouse causes hypotonia, growth retardation and partial lethality and provides evidence for a gene contributing to Prader-Willi syndrome. Hum Mol Genet 8:1357–1364
116. Van Esch H, Bauters M, Ignatius J, Jansen M, Raynaud M, Hollanders K, Lugtenberg D, Bienvenu T, Jensen LR, Gecz J et al (2005) Duplication of the mecp2 region is a frequent cause of severe mental retardation and progressive neurological symptoms in males. Am J Hum Genet 77:442–453
117. Verkerk AJ, Pieretti M, Sutcliffe JS, Fu YH, Kuhl DP, Pizzuti A, Reiner O, Richards S, Victoria MF, Zhang FP et al (1991) Identification of a gene (fmr-1) containing a cgg repeat coincident with a breakpoint cluster region exhibiting length variation in fragile X syndrome. Cell 65:905–914
118. Vincent A, Heitz D, Petit C, Kretz C, Oberle I, Mandel JL (1991) Abnormal pattern detected in fragile-x patients by pulsed-field gel electrophoresis. Nature 349:624–626
119. Vogel F, Motulsky AGM (1997) Human genetics: problems and approaches, 3rd edn. Springer, Berlin
120. Walzer S, Bashir AS, Silbert AR (1990) Cognitive and behavioral factors in the learning disabilities of 47, XXY and 47, XYY boys. Birth Defects Orig Artic Ser 26:45–58
121. Wang J, Wang W, Li R, Li Y, Tian G, Goodman L, Fan W, Zhang J, Li J, Zhang J et al (2008) The diploid genome sequence of an Asian individual. Nature 456:60–65
122. Warren ST, Nelson DL (1994) Advances in molecular analysis of fragile X syndrome. JAMA 271:536–542
123. Warren ST, Zhang F, Licameli GR, Peters JF (1987) The fragile X site in somatic cell hybrids: an approach for molecular cloning of fragile sites. Science 237:420–423
124. Wheeler DA, Srinivasan M, Egholm M, Shen Y, Chen L, McGuire A, He W, Chen YJ, Makhijani V, Roth GT et al (2008) The complete genome of an individual by massively parallel DNA sequencing. Nature 452:872–876
125. Wordsworth S, Buchanan J, Regan R, Davison V, Smith K, Dyer S, Campbell C, Blair E, Maher E, Taylor J et al (2007) Diagnosing idiopathic learning disability: a cost-effectiveness analysis of microarray technology in the national health service of the united kingdom. Genomic Med 1:35–45
126. Yan QJ, Rammal M, Tranfaglia M, Bauchwitz RP (2005) Suppression of two major fragile x syndrome mouse model phenotypes by the mglur5 antagonist mpep. Neuropharmacology 49:1053–1066
127. Yu S, Pritchard M, Kremer E, Lynch M, Nancarrow J, Baker E, Holman K, Mulley JC, Warren ST, Schlessinger D et al (1991) Fragile X genotype characterized by an unstable region of DNA. Science 252:1179–1181

23.3 Genetic Factors in Alzheimer Disease and Dementia

Thomas D. Bird

Abstract Alzheimer disease (AD) is a common and complex disorder affecting several million people world-wide. It is defined clinically as a progressive dementing illness associated with Aβ-amyloid neuritic plaques and neurofibrillary tangles in the brain. From a genetic standpoint it is a heterogeneous disorder. Three separate genes (APP, PSEN1, PSEN2) each cause an autosomal dominant, highly penetrant, early-onset, familial form of the disease. The three proteins encoded by these genes all influence the production of the toxic $A\beta_{1-42}$ form of amyloid. Commercial genetic testing is available for these rare forms of AD. However, mutations in these genes represent less than 2% of all cases of AD. The more common late-onset form of AD is thought to be polygenic and multifactorial. The ε4 allele of apolipoprotein E (ApoE) is a known genetic risk factor for late-onset AD, lowering the average age of onset by unknown mechanisms. Numerous other candidate risk genes are being identified through genome wide association studies, but have been difficult to confirm. Other familial forms of dementia, such as frontotemporal dementia (FTD), prion-associated diseases, and CADASIL, may be caused by autosomal dominant mutations occurring in their respective genes (MAPT, GRN, PRNP, Notch-3).

Contents

23.3.1 Clinical Manifestations of Alzheimer Disease 681
    23.3.1.1 Establishing the Diagnosis 682
    23.3.1.2 Prevalence 682
23.3.2 Causes 683
    23.3.2.1 Environmental 683
    23.3.2.2 Heritable Causes [6, 7] 683
    23.3.2.3 Unknown 685
    23.3.2.4 Molecular Genetic Testing 686
    23.3.2.5 Early-Onset Familial AD 688
23.3.3 Genetic Counseling 688
    23.3.3.1 Mode of Inheritance 688
    23.3.3.2 Risk to Family Members: EOAD 688
    23.3.3.3 Related Genetic Counseling Issues 688
23.3.4 Management 689
    23.3.4.1 Treatment of Manifestations 689
    23.3.4.2 Therapies Under Investigation 689
    23.3.4.3 Other 689
23.3.5 Other Causes of Dementia 690
    23.3.5.1 Frontotemporal Dementia 690
    23.3.5.2 Familial Prion Disorders 691
    23.3.5.3 CADASIL 691

References 692

T.D. Bird (✉)
Geriatric Research Education and Clinical Center (182), VA Puget Sound Health Care System, University of Washington, 1660 S Columbian Way, Seattle, WA 98108, USA
e-mail: tomnroz@u.washington.edu

23.3.1 Clinical Manifestations of Alzheimer Disease

The major clinical manifestation of Alzheimer disease (AD) is dementia that typically begins with subtle and poorly recognized failure of memory and slowly becomes more severe and, eventually, incapacitating. Other common findings include confusion, poor judgment, language disturbance, agitation, withdrawal, and hallucinations. Occasionally, seizures, Parkinsonian features, increased muscle tone, myoclonus, incontinence, and mutism occur [8]. Because the first symptoms of AD are subtle it is difficult to know when the disease begins. Presumably there is brain deposition of Aβ peptide and tau for years before the onset of clinical symptoms. Mild memory loss is common with advancing age. A syndrome called mild cognitive impairment (MCI) refers to a condition where the memory loss is noticeable but not disabling and not associated with other symptoms. Persons with MCI slowly convert to a diagnosis of AD at a rate of approximately 10% per year. However, in some persons MCI does not progress and they do not develop AD. At the present time there is no definitive method of distinguishing between these two groups.

Death usually results from general inanition, malnutrition, and pneumonia. The typical clinical duration of the disease is 8–10 years, with a range of 1–25 years.

23.3.1.1 Establishing the Diagnosis

Establishing the definitive diagnosis of AD relies upon clinical-neuropathologic assessment [104]. Neuropathologic findings on autopsy examination remain the gold standard for diagnosis of AD (Fig. 23.3.1). The clinical diagnosis of AD (prior to autopsy confirmation) is correct about 80–90% of the time [74].

Clinical signs: slowly progressive dementia.

Neuroimaging: CAT and MRI imaging studies will show diffuse cerebral cortical atrophy, often severe in the hippocampal region of the medial temporal lobe [51]. SPECT and PET radionuclide scans may show diffuse decreased cerebral metabolic rates, often beginning in the parietal region. A special PET scan (PIB) using a radioactive ligand for Aβ-peptide shows early increased deposits of material throughout the cortex and basal ganglia [108].

Neuropathologic findings: microscopic extracellular Aβ-amyloid neuritic plaques, intraneuronal neurofibrillary tangles, and amyloid angiopathy at postmortem examination. The plaques should stain positive with Ab-amyloid antibodies and be negative for prion antibodies, which are diagnostic of prion diseases. The numbers of plaques and tangles must exceed those found in age-matched controls without dementia (Fig. 23.3.2). Guidelines for the quantitative assessment of these changes exist [11, 80]. Aggregation of alpha synuclein in the form of Lewy bodies may also be found in neurons in the amygdala [89]. Cerebrovascular disease with microscopic infarcts may also contribute to the pathology [124].

Cerebrospinal fluid (CSF): Analysis shows reduced levels of Aβ-peptide (the metabolic product of the amyloid precursor protein that accumulates in the brains of persons with the disease) and increased levels of tau protein (found in neurofibrillary tangles).

23.3.1.2 Prevalence

AD is the most common cause of dementia in North America and Europe, with an estimated 4 million affected individuals in the US.

The prevalence of AD increases with age. Mild memory loss is often called mild cognitive impairment

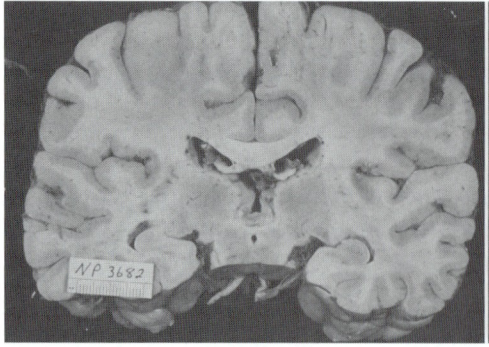
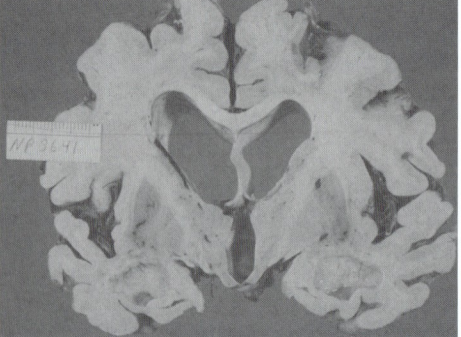

Fig. 23.3.1 Normal adult brain (*top*) compared with Alzheimer brain (*bottom*), showing marked diffuse cortical atrophy and ventricular enlargement. (From [7], with permission)

23.3 Genetic Factors in Alzheimer Disease and Dementia

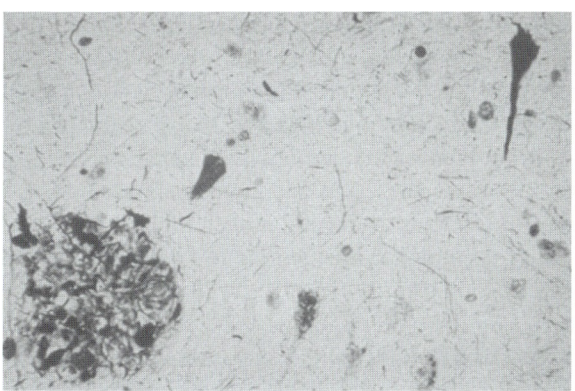

Fig. 23.3.2 Microscopic neuropathology of Alzheimer disease showing an extracellular neuritic plaque (*lower left-hand corner*) and a neurofibrillary tangle containing neuron (*upper right-hand corner*). (From [6], with permission)

(MCI). In many persons MCI is considered an early stage of AD.

The incidence of AD rises from 2.8 per 1,000 person-years (one person living for 1 year = 1 person-year) in the 65- to 69-year age group to 56.1 per 1,000 person-years in the age group older than 90 years [57]. Approximately 10% of persons over age 70 years have significant memory loss, and more than half of these individuals have AD. An estimated 25–45% of persons over age 85 years have dementia.

23.3.2 Causes

About 1–6% of all AD is early onset (before age 60–65 years), and about 60% of early-onset AD is familial, with 13% appearing to be inherited in an autosomal dominant manner [7, 16, 102] (Table 23.3.1).

The distinction between early-onset familial AD (onset before age 60–65 years; EOFAD) and late-onset familial AD (onset after age 60–65 years) is somewhat arbitrary. Early-onset cases can occur in families with generally late-onset disease [14].

Table 23.3.1 Causes of Alzheimer disease

Cause	% of cases
Chromosomal (Down syndrome)	<1%
Familial	~25%
• Late-onset familial (AD2)	• 15–25%
• Early-onset familial AD (AD1, AD3, AD4)	• <2%
Unknown (includes genetic/environment interactions)	~75%

23.3.2.1 Environmental

No environmental agents have been proved to be directly involved in the pathogenesis of AD. It is often speculated that late-onset AD (LOAD) is the result of unknown environmental factors acting on a predisposing genetic background [9]. Twin studies have implicated both genes and environment [30]. The Gatz et al. study found concordance for AD in MZ vs DZ twins of 45% vs. 19% and 61% vs. 41% in males and females, respectively. It was noted that age of onset can vary considerably with twin pair, and heritability in this twin study was estimated at 58–79% [30]. Potential environmental factors include head trauma, viruses, toxins, and low education level. Several investigators have suggested that persons with AD had subtle cognitive or linguistic signs in childhood or early adulthood [120, 132].

23.3.2.2 Heritable Causes [6, 7]

23.3.2.2.1 Chromosomal

Down Syndrome. Essentially all persons with Down syndrome (DS; trisomy 21) develop the neuropathologic hallmarks of AD after age 40 years. More than half of individuals with DS, if carefully observed or tested, also show clinical evidence of cognitive decline [15]. The presumed reason for this association is the lifelong over-expression of the *APP* gene on chromosome 21 encoding the amyloid precursor protein and the resultant overproduction of Aβ-amyloid in the brains of persons who are trisomic for this gene.

The amyloid-β (Aβ) deposition in the brain may begin in the first decade of life in persons with DS [63]. AD was not noted clinically or pathologically in a 78-year-old woman with partial trisomy 21 who did not have an extra copy of the *APP* gene [90]. Two studies have found no association of ApoE genotype with age of onset of dementia in DS [59, 71], but one study did find an association of onset age with a polymorphism in the *APP* gene [71]. Schupf et al. [111] found an unexplained increased risk for AD in mothers who gave birth to children with DS prior to age 35 years (rate ratio = 4.8, 95% CI 2.1–11.2). This study has not been confirmed.

23.3.2.2.2 Single Gene

About 25% of AD is familial (i.e., two or more family members have AD). Familial cases appear to have the same clinical and pathologic phenotypes as non-familial cases (i.e., an individual with AD and no known family history of AD [42, 81]) and are thus distinguished only by family history or by molecular genetic testing. A large volume of research on the molecular and genetic basis of AD has been summarized elsewhere [6, 7, 36, 64, 82, 106, 107, 118].

Late-Onset Familial AD (AD2)

Many families have multiple affected members, most or all of whom have onset of dementia after age 60 or 65 years (Table 23.3.2). Disease duration is typically 8–10 years, but ranges from 2 to 25 years. Investigations have supported the concept that LOAD is a complex disorder that may involve multiple susceptibility genes (reviewed and summarized in [49, 108, 114]. Bertram et al. [5] have performed a meta-analysis on these data. The following information is currently available about genes or loci actually or potentially altering risk for late-onset AD:

There is a very well-documented association of LOAD with the *APOE* e4 allele (19q13). The *APOE* e4 allele, by unclear mechanisms, appears to affect age of onset by shifting the onset toward an earlier age [52, 75, 109].

Apolipoprotein E is a cholesterol transport protein that circulates in plasma and is the major apolipoprotein in brain, where it is synthesized by glia, macrophages and neurons [2]. In humans the three common isoforms differ by a single amino acid: e3 has cysteine at position 112 and arginine at 158, e2 has cysteine at both positions, and e4 has arginine at both positions. Frequencies of ApoE genotpyes in the general population are shown in the first column of Table 23.3.3. Note that 3/3 is most common and 4/4 is the least common. It is not known exactly how ApoE influences the biology of AD [45]. Apo e4 differs from e2 and e3 in lacking a cysteine, having a greater tendency to form a molten globule state, and exhibiting domain interactions [136]. These domain interactions lead to cleavage fragments that may be toxic. An important finding has been that a transgenic mouse model of AD with a mutation in the *APP* gene has a greatly reduced deposition of amyloid plaques when produced on an ApoE knockout background [1]. This demonstrates an important role of ApoE in amyloid plaque formation. The use of APoE genotyping in diagnosis and risk assessment of AD is discussed in Sect. 23.3.3.

- Several other potential genes are under investigation:
- *SORL1* on chromosome 11q23, a protein involved with APP protein trafficking [103].
- *A2M* on chromosome 12 [23]
- *GST01* and *GST02* on chromosome 10 [69]
- *GAB2* on chromosome 11q14 interacting with the *ApoE4* allele [98]
- *PCDH11X* on the X chromosome [17]

Several other potential loci are under investigation on the following chromosomes:

- 12 [24]
- 10 [40, 78, 100]
- 2q, 9p, and 15q [68, 112]
- 19p13 [133]
- 7q36 [93]
- 9q22 (*UBQLN1*) [4, 50, 119]
- Studies of LOAD in a genetically isolated Dutch population have suggested linkage of AD to markers on chromosome 1q22, 3q23, 10q22 and 11q25 [70]
- A genome-wide association study (GWAS) of more than 1,300 cases identified 2 potentially important loci (14q, 6p) in addition to APoE [3].
- A linkage study of >300 families with LOFAD has found several loci of interest [62].

Table 23.3.2 Late-onset familial Alzheimer disease: molecular genetics

Locus name	Gene symbol	Chromosomal locus	Protein name	Test availability
AD2	*APOE*	19q13.2	Apolipoprotein E	Yes

Early-Onset Familial AD (EoFAD)

- *Clinical Features.* Early-onset familial AD (EOFAD) refers to families in which multiple cases of AD occur with the mean age of onset usually before age 65 years, although some studies have used age 60 years or 70 years. Age of onset is usually in the 40s or early 50s, although onset in the 30s and early 60s has been reported. Campion et al. [16] found a prevalence of EOAD in the general population of 41.2 per 100,000 persons at risk (ages 40–59 years). Sixty-one percent of these individuals with EOAD had a positive family history, and 13% met stringent criteria for autosomal dominant inheritance (i.e., affected individuals in three generations). EOFAD cannot be clinically distinguished from nonfamilial AD except on the basis of family history and age of onset. The dementia phenotype is similar to that of LOAD, sometimes with a long prodrome [35, 60, 61].
- *Molecular Genetics.* At least three subtypes of EOFAD (AD1, AD3, AD4) have been identified based on the causative gene. All are autosomal dominant. The relative proportion of each subtype and the causative genes are summarized in Table 23.3.4 [16, 46, 96, 115]. All three types are related to amyloid precursor protein (APP) metabolism. The APP is cleaved by beta and gamma secretases to form the A beta peptide, which is the primary component of the extracellular amyloid plaque deposited in AD (Fig. 23.3.3). (APP may also be cleaved inside the A-beta peptide domain by alpha secretase.) Cleavage at the gamma secretase site may produce Aβ peptide with 40 or 42 amino acids. The Aβ 42 peptide has been demonstrated to be the 'toxic' amyloid plaque-forming peptide. Presenilin 1 (PS1) is part of the gamma secretase complex (and PS2 is a close homolog of PS1). Thus, the three primary genes associated with EOFAD are all related to APP and Aβ amyloid molecular biology. It remains unclear exactly how mutations in these three genes alter APP metabolism and cause AD, whether by increasing production of Aβ 42 or changing the Aβ 40/42 ratio, or some related mechanism [10, 123]. There are transgenic mouse models containing one or more mutations in APP and/or PS1, and even a triple transgenic mouse with mutations in APP, PS1, and MAPT (tau) [83]. It is likely that other genes will be identified as a cause of EOFAD, because kindreds with autosomal dominant FAD with no known mutations in *PSEN1*, *PSEN2*, or *APP* have been described [20, 95].
- Of these three genes, mutations in PS1 are most common. More than 170 mutations have been described in PS1 [61]. Age of onset ranges from 25 to 65 years, usually in the forties. Penetrance is >95%. Very early onset before age 30 has been reported (e.g., mutations I143T, L166P, P436Q). Myoclonus, seizures, aphasia, and cerebellar plaques may occur. Deletions in exon 9 have been associated with spasticity and giant amyloid 'cotton wool' plaques.
- Several different missense mutations have been reported in APP and tend to occur near the secretase cleavage sites (see Fig. 23.3.4). V717I is one of the most common and occurs in the g-secretase site. Mutations near the alpha secretase site may be associated with severe amyloid angiopathy.
- Mutations in PS2 are the least common cause of EOFAD [65]. Only about 12 such mutations have been described. Asn141Ile is a founder mutation that has been found in families with the same Volga German background [65]. Age of onset in PS2 mutations ranges from 40 to 75 years with about 95% penetrance.

Table 23.3.3 Percent of *APOE* genotypes in caucasian controls and individuals with AD (modified from [47])

APOE Genotype	Normal Controls ($n=304$)	All individuals with AD ($n=233$)	Individuals with AD and positive family history of dementia[a] ($n=85$)
e2/e2	1.3%	0%	0%
e2/e3	12.5%	3.4%	3.5%
e2/e4	4.9%	4.3%	8.2%
e3/e3	59.9%	38.2%	23.5%
e3/e4	20.7%	41.2%	45.9%
e4/e4	0.7%	12.9%	18.8%

[a] Most families would be considered to have late-onset familial AD

23.3.2.3 Unknown

Individuals with nonfamilial AD meet the diagnostic criteria for AD and have a negative family history. Onset can be any time in adulthood. The exact pathogenesis of

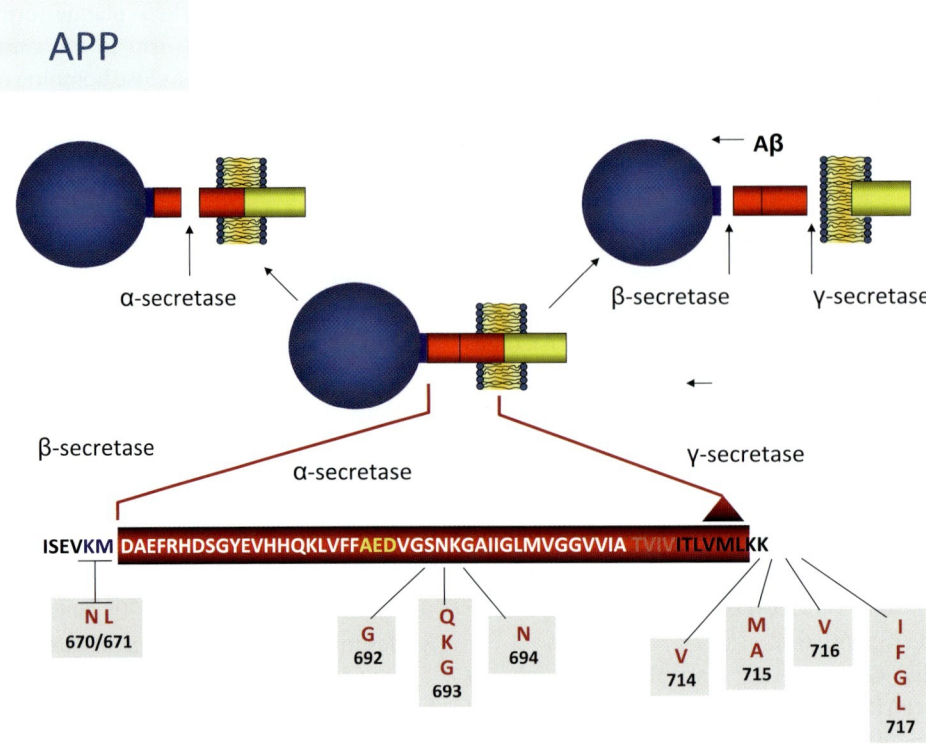

Fig. 23.3.3 Molecular aspects of the APP gene and protein, showing the sites of cleavage by a-, b-, and g-secretases, production of the Ab-peptide (*upper right*) and sites of several disease causing mutations (From [6], with permission)

the disease is unknown. A common hypothesis is that nonfamilial AD is multifactorial and results from a combination of aging, genetic predisposition, and exposure to one or more environmental agents, such as head trauma, viruses, and/or toxins, although no environmental agents have been proved to be directly involved in the pathogenesis of AD [21].

23.3.2.4 Molecular Genetic Testing

23.3.2.4.1 Late-Onset Familial AD

The association of one or two copies of the *APOE* allele e4 (i.e., genotypes e2/e4, e3/e4, e4/e4) with LO AD is well documented (Table 23.3.3) [47, 52, 72].

Table 23.3.4 Early-onset familial Alzheimer disease (EOFAD): molecular genetics

Locus name	Proportion of EOFAD	Gene symbol	Chromosomal locus	Protein N	
name	Test availability				
AD3	20–70%	*PSEN1*	14q24.3	Presenilin-1	Clinical
AD1	10–15%	*APP*	21q21	Amyloid precursor protein	Clinical
AD4	Rare	*PSEN2*	1q31-q42	Presenilin-2	Clinical

- The association between *APOE* e4 and AD is closest when the individual has a positive family history of dementia. The last column of Table 23.3.3 largely represents late-onset familial AD.
- The strongest association between the *APOE* e4 allele and AD, relative to the normal control population, is with the e4/e4 genotype. That genotype occurs in about 1–2% of the normal control population and in nearly 19% of the familial AD population.
- In individuals who have the clinical diagnosis of AD, the probability that AD is the correct diagnosis is increased to about 97% in the presence of the *APOE* e4/e4 genotype [109]. However, note that e4 homozygotes are relatively uncommon in the general population, and the clinical diagnosis of AD by experienced physicians is correct 85–90% of the time without ApoE genotyping.
- The increased risk of AD associated with one *APOE* e4 allele or two *APOE* e4 alleles is also found in African Americans [38] and Caribbean Hispanics [105]. In African Americans the 3/4 genotype is associated with an odds ratio of 2.32 and the 4/4 genotype with an odds ratio of 7.19 for developing AD compared with the 3/3 genotype [77].
- Approximately 42% of persons with AD do *not* have an *APOE* e4 allele. Thus, *APOE* genotyping is not highly sensitive for AD. The absence of an *APOE* e4 allele does not rule out the diagnosis of AD [74]. Thus genotyping is not highly specific [79].
- Breitner et al. [13] have estimated lifetime risks for developing AD based on gender and *APOE* genotype (see Sect. 23.3.4.4.2). This group emphasizes that ApoE genotype primarily affects age at onset of AD, rather than lifetime susceptibility [52]. Figure 23.3.4 shows lifetime risks for AD for various ApoE genotypes.

The usefulness of *APOE* genotyping in clinical diagnosis and risk assessment remains unclear. (See list of "Statements and Policies Regarding Genetic Testing.")

- Although the presence of one *APOE* e4 allele or two *APOE* e4 alleles is neither necessary nor sufficient to establish a diagnosis of AD, *APOE* genotyping may have an adjunct role in the diagnosis of AD because a large proportion of individuals with one *APOE* e4 allele or two *APOE* e4 alleles who are demented have been found to have neuropathologic confirmation of AD at autopsy [47, 73, 78, 131].
- In contrast, *APOE* genotyping was not found to be of significant diagnostic use in identifying AD in a community-based sample with late-onset dementia [125].
- There is some evidence that the *APOE* e2 allele may have a protective effect in regard to risk for AD (Table 23.3.4).

Another way to look at this association between AD and an *APOE* e4 allele is with *APOE* e4 allele frequencies (Table 23.3.5).

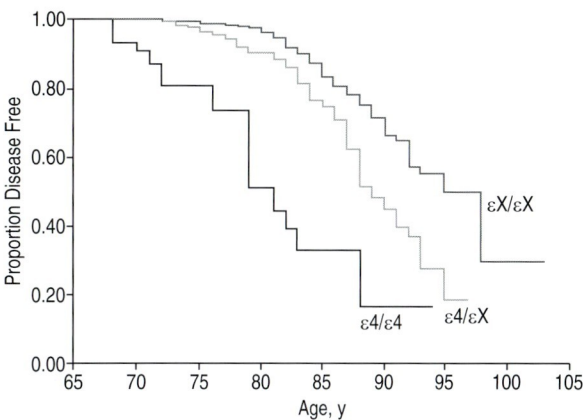

Fig. 23.3.4 Unadjusted disease-free survival by age, stratified by 0, 1, and 2 apolipoprotein E (APOE) e4 alleles, among 3,308 residents of Cache County, Utah, 1995–1997 and 1998–2000. The graph shows three unadjusted product-limit estimates for the three strata of 0 (*hatched line*), 1 (*gray line*), and two (*black line*) APOE e4 alleles (X allele 2 or 3). (From 51, with permission)

Table 23.3.5 *APOE* allele frequencies in controls and individuals with AD (modified from [47])

APOE allele	Normal controls (n=304)	All individuals with AD (n=233)	Individuals with AD and positive family history of dementia[a] (n=85)
e2	9.0%	3.9%	5.9%
e3	76.5%	60.5%	48.2%
e4	13.7%	35.6%	45.9%

[a]Most families would be considered to have late-onset familial AD

23.3.2.5 Early-Onset Familial AD

The three known subtypes of EOFAD, called AD3, AD1, and AD4 [46], can only be distinguished by molecular genetic testing (Table 23.3.4). Genetic testing of individuals who are simplex cases (i.e., a single occurrence of EOAD in a family) is controversial and should be undertaken in the context of formal genetic counseling [127]. A small proportion (<5%) of such cases will have a mutation in PS1.

23.3.3 Genetic Counseling

23.3.3.1 Mode of Inheritance

Because AAD is genetically heterogeneous, genetic counseling of persons with AD and their family members must be tailored to the information available for that family. AD is usually considered polygenic and multifactorial. EOFAD is inherited in an autosomal dominant manner.

23.3.3.1.1 Risk to Family Members: Late-Onset Nonfamilial AD

Genetic counseling for people with nonfamilial AD and their family members must be empiric and relatively nonspecific. It should be pointed out that AD is common and that the overall lifetime risk to any individual of developing dementia is approximately 10–12%.

First-degree relatives of a person with AD have a cumulative lifetime risk of developing AD of about 15–30%, which is typically reported as a 20–25% risk [25, 116]. This risk is about 2.5 times that of the background risk (~27% vs 10.4%) [22, 37].

There is some disagreement as to whether the age of onset of the affected person changes the risk to first-degree relatives. One study found that EOAD increased the risk [116], while another study did not [37].

The number of additional affected family members probably increases the risk to close relatives, but the magnitude of that increase is unclear unless the pattern in the family is characteristic of autosomal dominant inheritance. Having two, three, or more affected family members probably raises the risk to other first-degree relatives in excess of that noted above for nonfamilial cases, although the exact magnitude of the risk is not clear. Heston et al. [43] found a 35–45% risk of dementia in individuals who had a parent with AD and a sib with onset of AD before age 70 years. Jayadev et al. [48] also report data suggesting that offspring of parents with conjugal AD (i.e., both parents affected) had an increased risk of dementia.

23.3.3.2 Risk to Family Members: EOAD

Many individuals diagnosed as having EOAD have another affected family member, although family history is negative 40% of the time [16]. Family history may be 'negative' because of early death of a parent, failure to recognize the disorder in family members, or, rarely, a de novo mutation. The risk to sibs depends upon the genetic status of the affected proband's parent. If one of the proband's parents has a mutant allele, then the risk to the sibs of inheriting the mutant allele is 50%. Individuals with EOFAD (and a mutation in APP, PS1 or PS2) have a 50% chance of transmitting the mutant allele to each child. The risk to other family members depends upon the status of the proband's parents. If a parent is found to be affected, his or her family members are at risk.

23.3.3.3 Related Genetic Counseling Issues

23.3.3.3.1 Use of *APOE* Genotyping for Predictive Testing

In contrast to the utility of *APOE* testing as an adjunct diagnostic test in individuals with dementia, there is general agreement that *APOE* testing has limited value when used for predictive testing for AD in asymptomatic persons. Data suggest that a young asymptomatic person with the *APOE* e4/e4 genotype may have an approximately 30% lifetime risk of developing AD [12]. Further refinement of this risk reveals that females with an *APOE* e4/e4 genotype have a 45% probability of developing AD by age 73 years, whereas males have a 25% risk [13]. These risks are lower – and the likely age of onset later – for persons with only one *APOE* e4 allele (peak age 87 years) or no *APOE* e4 allele (peak age 95 years). These estimates are not generally considered clinically useful; however, a research study to assess the potential utility of *APOE* testing in relatives of individuals with LOAD is under way [37, 101]. The relatively high risk in ε4/ε4 homozygotes is notable, although this is relevant to the smallest number of persons in the general population.

Down Syndrome. Family members of persons with Down syndrome are not at increased risk for AD.

23.3.3.3.2 Testing of At-Risk Asymptomatic EOAD Family Members

- *Testing of At-Risk Asymptomatic Adults.* Testing of asymptomatic adults at risk for EOFAD caused by mutations in the *PSEN1*, *PSEN2*, or *APP* gene is available clinically. Testing results for at-risk asymptomatic adults can only be interpreted after an affected family member's disease-causing mutation has been identified. It should be remembered that testing of asymptomatic at-risk individuals with non-specific or equivocal symptoms is predictive testing, not diagnostic testing. Preliminary results have shown that while relatively few family members choose such testing, they usually cope well with the results, which can affect personal relationships and emotional well-being. [122]. However, significant depression has been reported following such testing [92].
- *Preimplantation Genetic Diagnosis.* Preimplantation genetic diagnosis (PGD) and embryo transfer have been successfully used to achieve pregnancy in a 30-year-old asymptomatic woman with an *APP* disease-causing mutation, resulting in the birth of a healthy child who does not have the *APP* disease-causing mutation identified in the mother and her family [129].

23.3.4 Management

23.3.4.1 Treatment of Manifestations

The mainstay of treatment for AD is necessarily supportive, and each symptom is managed on an individual basis [8]. In general, affected individuals eventually require assisted living arrangements or care in a nursing home.

Although the exact biochemical basis of AD is not well understood, it is known that deficiencies of the brain cholinergic system and of other neurotransmitters are present. Drugs that increase cholinergic activity by inhibiting acetylcholinesterase produce a modest but useful behavioral or cognitive benefit in some affected individuals. The first such drug was tacrine; however, this agent is also hepatotoxic. Newer such drugs with similar pharmacologic action, such as Aricept® (donepezil) [86, 113] (rivastigmine) [26], and galantamine [95], are not hepatotoxic.

Memantine, an NMDA receptor antagonist, has shown some effectiveness in the treatment of moderate to severe AD [99]. Memantine is often added to a cholinesterase inhibitor.

Antidepressant medication may improve associated depression.

23.3.4.2 Therapies Under Investigation

Treatment trials evaluating use of anti-inflammatory agents (NSAIDs), estrogens, nerve growth factors, ginkgo biloba, statins, BACE inhibitors, and antioxidants are under way or have recently been reviewed [54, 66, 67, 73, 85, 134, 135]. Attempts to ameliorate the tau pathology are also underway [41].

23.3.4.3 Other

Vitamins and other over-the-counter medications have been used in the treatment of AD [54, 73, 135].

Some, but not all, reports suggest that affected individuals taking HMG-coenzyme A reductase inhibitors for hypercholesterolemia have a reduced incidence of dementia [66, 67, 134].

Immunization of an AD mouse model with Aβ-amyloid has attenuated the AD pathology and stimulated the search for a possible vaccination approach to the treatment of human AD [110]. A human trial of this approach was halted because of encephalitis in a few subjects [18, 27, 33, 44]. Alternative approaches to immunization therapy have been proposed, including the use of antibodies to A beta peptide [91].

Thus far, treatment of symptomatic AD with estrogens has not proved beneficial [72, 76, 130].

23.3.5 Other Causes of Dementia

Differential diagnosis of AD includes other causes of dementia, especially treatable forms of cognitive decline, such as depression, chronic drug intoxication, chronic CNS infection, thyroid disease, vitamin deficiencies (especially B12 and thiamine), CNS vasculitis, and normal-pressure hydrocephalus [8]. Of these conditions, depression and chronic drug effects are the most common and are treatable with antidepressants and elimination of the drug, respectively.

Other degenerative disorders associated with dementia, such as frontotemporal dementia (FTD), Picks disease, Parkinson disease, diffuse Lewy body disease (LBD), Creutzfeldt–Jakob disease (CJD), and CADASIL, may also be confused with AD [104]. CT and MRI imaging are valuable for identifying some of these other causes of dementia, including neoplasms, normal-pressure hydrocephalus, frontotemporal dementia, and cerebral vascular disease.

23.3.5.1 Frontotemporal Dementia

Frontotemporal dementia (FTD or Pick's disease) is an important cause of dementia that is less common than AD, but typically has an earlier age of onset, such that it is more common than AD in the 5–7th decades of life [28, 39]. The early symptoms of FTD are usually behavioral with a personality change. There may be apathy, loss of social inhibitions, agitation, or inappropriate behavior. Language difficulties are also common. Memory is relatively intact early in the disease, and this helps distinguish it from AD. Brain imaging studies and neuropathological findings demonstrate focal lobar atrophy of the frontal and/or temporal lobes. There is no specific treatment for FTD.

About 30% of FTD cases are familial. Two genes have been discovered that cause familial FTD. One form is produced by a variety of missense or splice mutations in MAPT (microtubule-associated protein tau) [29, 88, 97] (see Fig. 23.3.5). These mutations are autosomal dominant with high penetrance and often associated with tau aggregation, such as neurofibrillary tangles in the brain. Exon 10 may be included or spliced out of the tau mRNA, resulting in 4-repeat or

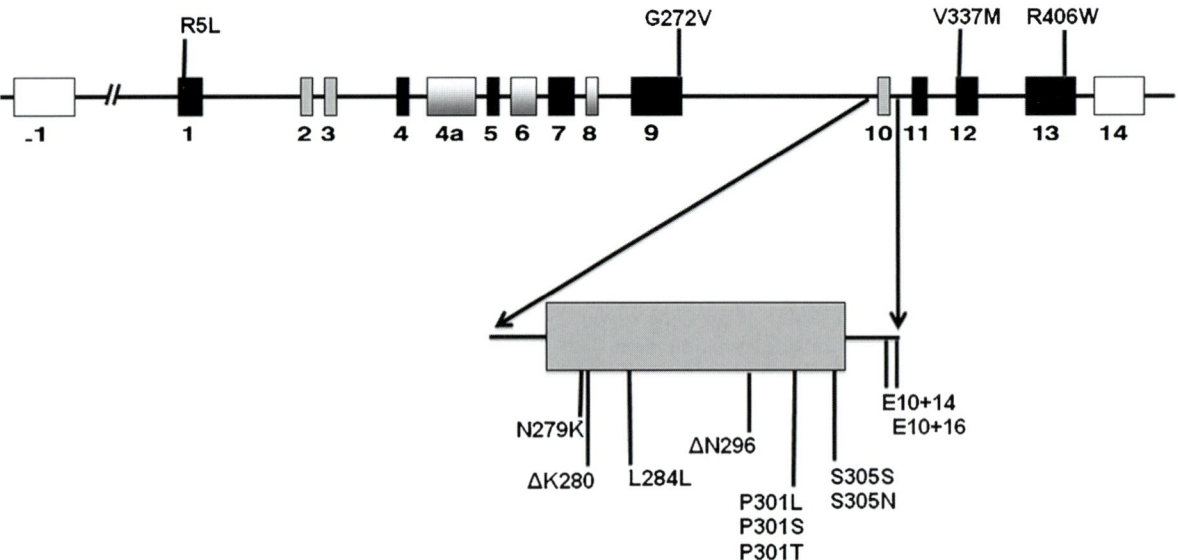

Fig. 23.3.5 MAPT (microtubule-associated protein tau) gene showing sites of several mutations causing hereditary frontotemporal dementia. Exon 10 is a hotspot for mutations and P301L and E10+16 are two of the most common. Splicing of exon 10 determines whether the resulting protein is 3-repeat or 4-repeat tau. *Numbers* below *boxes* in the *top line* refer to exons. Exon 10 is exlarged between the *arrows*. *Numbers* pointing within exons or introns indicate mutations causing FTD designated by codon and amino acid change (e.g., <P301L) or position within the intron (e.g., E10+14). (Courtesy of P. Poorkaj-Navas)

3-repeat tau protein. Many disease-causing mutations are in exon 10 or its adjacent splice site. The mutation may disrupt the function of microtubules or alter the normal 50/50 ratio of the 3-repeat and 4-repeat splicing variants of tau. The most common mutations in MAPT are P301L and E10+16.

The second gene causing FTD is progranulin (GRN) [29, 128]. This variety of FTD is associated with neuronal inclusions of TDP-43 (TARDBP43, a DNA-binding protein). Almost all of the mutations in GRN are prematurely truncating mutations that result in haploinsufficiency of the protein because of nonsense-mediated decay of mRNA [32]. The most common mutation is R493X [94]. Penetrance is reduced with some mutations in GRN, and mutations have been found in occasional sporadic cases. Because of nonsense-mediated decay there are reduced plasma levels of GRN in persons with mutations [31]. Interestingly, mutations in the TARDBP43 gene cause a familial form of ALS (amyotrophic lateral sclerosis), and not FTD [121, 126]. Other rare genetic causes of FTD have occurred with mutations in the valosin-containing protein (VCP, associated with inclusion body myositis and Paget's disease of bone) [53] and CHMP2B genes [117].

23.3.5.2 Familial Prion Disorders

Creutzfeldt–Jakob disease (CJD) is a rare progressive neurodegenerative disease caused by a conformational change in the prion protein with a prevalance of about 1 per million. There are rare familial forms of CJD, representing about 15% of all cases [34]. These familial cases are caused by a variety of missense mutations or insertions in the prion (PrP) gene [55, 58]. These mutations are autosomal dominant but may have reduced penetrance. The clinical syndrome varies considerably including typical CJD (rapidly progressive dementia with rigidity and myoclonus), a more slowly progressive dementia easily confused with AD and the rare but unusual familial fatal insomnia (FFI) [19]. The two most common mutations in PrP are E200K and V210I [56]. Homozygosity for methionine at codon 129 is a risk factor for CJD and also may affect the phenotype in familial cases. When the mutation is D178N and valine is encoded at codon 129 on the same chromosome the phenotype is typical CJD. With the same mutation (D178N) in *cis* with methionine at 129, the phenotype is FFI.

23.3.5.3 CADASIL

CADASIL (*c*erebral *a*utosomal *d*ominant *a*rteriopathy with *s*ubcortical *i*nfarcts and *l*eukoencephalopathy) is caused by missense mutations in the Notch3 gene [87]. A large number of mutations have been found to be spread throughout the 33 exons and cause an autosomal dominant disorder. The clinical characteristics are early migraine headache, followed by recurrent strokes and/or slowly progressive dementia, typically in the 5–7th decades [84]. Brain MRI shows progressive diffuse hyperintensity in cerebral white matter on T2 imaging that may be apparent years prior to any symptoms. Cerebral arterioles show abnormal histopathology of the media with osmophilic granules on electron microscopy. There is no specific treatment.

Published Statements and Policies Regarding Genetic Testing

- American College of Medical Genetics/American Society of Human Genetics Working Group on ApoE and Alzheimer's disease (1995) Statement on use of apolipoprotein E testing for Alzheimer's disease
- American Society of Human Genetics and American College of Medical Genetics (1995) Points to consider: ethical, legal, and psychosocial implications of genetic testing in children and adolescents
- National Institute on Aging/Alzheimer's Association Working Group (1996) Apolipoprotein E genotyping in Alzheimer's disease. Lancet 347:1091–1095 [Medline]
- National Society of Genetic Counselors (1995) Resolution on prenatal and childhood testing for adult-onset disorders
- Post SG, Whitehouse PJ, Binstock RH, Bird TD, Eckert SK, Farrer LA, Fleck LM, Gaines AD, Juengst ET, Karlinsky H, Miles S, Murray TH, Quaid KA, Relkin NR, Roses AD, St George-Hyslop PH, Sachs GA, Steinbock B, Truschke EF, Zinn AB (1997) The clinical introduction of genetic testing for Alzheimer's disease: an ethical perspective. JAMA 277:832–836 [Medline]

References

1. Bales K, Verina T, Dodel RC, Du Y, Altstiel L, Bender M, Hyslop P, Johnstone EM, Little SP, Cummins DJ, Piccardo P, Ghetti B, Paul SM (1997) Lack of Apolipoprotein E Dramatically reduces amyloid b-protein deposition. Nat Genet 17:263–265
2. Bedlack RS, Srittmatter WJ, Morgenlander JC (2000) Apolipoprotein E and neuromuscular disease: a critical review of the literature. Arch Neurol 57:1561–1565
3. Bertram L, Lange C, Mullin K, Parkinson M, Hsiao M, Hogan MF, Schjeide BMM, Hooli B, DiVito J, Ionita I, Jiang H, Laird N, Moscarillo T, Ohlsen KL, Elliot K, Wang X, Hu-Lince D, Ryder M, Murphy A, Wagner SL, Blacker KD, Tanzi RE (2008) Am J Hum Genet 83:623–632
4. Bertram L, Hiltunen M, Parkinson M, Ingelsson M, Lange C, Ramasamy K, Mullin K, Menon R, Sampson AJ, Hsiao MY, Elliott KJ, Velicelebi G, Moscarillo T, Hyman BT, Wagner SL, Becker KD, Blacker D, Tanzi RE (2005) Family-based association between Alzheimer's disease and variants in UBQLN1. N Engl J Med 352:884–894
5. Bertram L, McQueen MB, Mullin K, Blacker D, Tanzi RE (2007) Systematic meta-analyses of Alzheimer disease genetic association studies: the AlzGene database. Nat Genet 39:17–23
6. Bird TD. Alzheimer's disease overview. In: GeneReviews at genetests: medical genetics information resource (database online). Copyright, University of Washington, Seattle. 1997-2008. Available at http://www.genetests.org. Accessed 10/23/2008.
7. Bird T (2008) Genetic aspects of Alzheimer's disease. Gen Med 10(4):231–239
8. Bird TD, Miller BL (2008) Alzheimer's disease and primary dementias. In: Kasper D, Fauci A, Branwald E et al (eds) Harrison's principles of internal medicine, 17th edn. McGraw-Hill, New York, pp 2393–2406
9. Borenstein AR, Copenhaver CI, Mortimer JA (2006) Early-life risk factors for Alzheimer disease. Alzheimer Dis Assoc Disord 20:63–72
10. Bossy-Wetzel E, Schwarzenbacher R, Lipton SA (2004) Nat Med 10(Suppl):S2–S9
11. Braak H, Braak E (1991) Neuropathological stageing of Alzheimer-related changes. Acta Neuropathol (Berl) 82:239–259
12. Breitner JC (1996) APOE genotyping and Alzheimer's disease. Lancet 347:1184–1185
13. Breitner JC, Wyse BW, Anthony JC, Welsh-Bohmer KA, Steffens DC, Norton MC, Tschanz JT, Plassman BL, Meyer MR, Skoog I, Khachaturian A (1999) APOE-epsilon4 count predicts age when prevalence of AD increases, then declines: the cache county study. Neurology 53:321–331
14. Brickell KL, Steinbart EJ, Rumbaugh M, Payami H, Schellenberg GD, Van Deerlin V, Yuan W, Bird TD (2006) Early-onset Alzheimer disease in families with late-onset Alzheimer disease: a potential important subtype of familial Alzheimer disease. Arch Neurol 63:1307–1311
15. Brugge KL, Nichols SL, Salmon DP, Hill LR, Delis DC, Aaron L, Trauner DA (1994) Cognitive impairment in adults with Down's syndrome: similarities to early cognitive changes in Alzheimer's disease. Neurology 44:232–238
16. Campion D, Dumanchin C, Hannequin D, Dubois B, Belliard S, Puel M, Thomas-Anterion C, Michon A, Martin C, Charbonnier F, Raux G, Camuzat A, Penet C, Mesnage V, Martinez M, Clerget-Darpoux F, Brice A, Frebourg T (1999) Early-onset autosomal dominant Alzheimer disease: prevalence, genetic heterogeneity, and mutation spectrum. Am J Hum Genet 65:664–670
17. Carrasquillo MM, Zou F, Pankratz VS, Wilcox SL, Ma L, Walker LP, Younkin SG, Younkin CS, Younkin LH, Bisceglio GD, Ertekin-Taner N, Crook JE, Dickson DW, Petersen RC, Graff-Radford NR, Younkin SG (2009) Genetic variation in PCDH11X is associated with susceptibility in late-onset Alzheimer's disease. Nat Genet 41:92–98
18. Check E (2002) Nerve inflammation halts trial for Alzheimer drug. Nature 415:462
19. Cortelli P, Perani D, Montagna P, Gallassi R, Tinuper P, Federica P, Avoni P, Ferrillo F, Anchisi D, Moresco RM, Fazio F, Parchi P, Baruzzi A, Lugaresi E, Gambetti P (2006) Pre-symptomatic diagnosis in fatal familial insomnia: serial neurophysiological and ^{18}FDG-PET studies. Brain 129:668–675
20. Cruts M, van Duijn CM, Backhovens H, Van den Broeck M, Wehnert A, Serneels S, Sherrington R, Hutton M, Hardy J, St George-Hyslop PH, Hofman A, Van Broeckhoven C (1998) Estimation of the genetic contribution of presenilin-1 and -2 mutations in a population-based study of presenile Alzheimer disease. Hum Mol Genet 7:43–51
21. Cummings JL, Vinters HV, Cole GM, Khachaturian ZS (1998) Alzheimer's disease: etiologies, pathophysiology, cognitive reserve and treatment opportunities. Neurology 51(Suppl):2–17
22. Cupples LA, Farrer LA, Sadovnick AD, Relkin N, Whitehouse P, Green RC (2004) Estimating risk curves for first-degree relatives of patients with Alzheimer's disease: the REVEAL study. Genet Med 6:192–196
23. Depboylu C, Lohmuller F, Du Y, Riemenschneider M, Kurz A, Gasser T, Muller U, Dodel RC (2006) Alpha2-macroglobulin, lipoprotein receptor-related protein and lipoprotein receptor-associated protein and the genetic risk for developing Alzheimer's disease. Neurosci Lett 400:187–190
24. D'Introno A, Solfrizzi V, Colacicco AM, Capurso C, Amodio M, Todarello O, Capurso A, Kehoe PG, Panza F (2006) Current knowledge of chromosome 12 susceptibility genes for late-onset Alzheimer's disease. Neurobiol Aging 27:1537–1553
25. Farrer LA, O'Sullivan DM, Cupples LA, Growdon JH, Myers RH (1989) Assessment of genetic risk for Alzheimer's disease among first-degree relatives. Ann Neurol 25:485–493
26. Feldman HH, Lane R, Study 304 Group (2007) Rivastigmine: a placebo controlled trial of twice daily and three times daily regimens in patients with Alzheimer's disease. J Neuro Neurosurg Psych 78:1056–1063
27. Ferrer I, Boada Rovira M, Sanchez Guerra ML, Rey MJ, Costa-Jussa F (2004) Neuropathology and pathogenesis of encephalitis following amyloid-beta immunization in Alzheimer's disease. Brain Pathol 14:11–20
28. Forman MS, Farmer J, Johnson JK, Clark CM, Arnold SE, Coslett HB, Chatterjee A, Hurtig HI, Karlawish JH, Rosen HJ, Van Deerlin V, Lee VM-Y, Miller BL, Trojanowski JQ, Grossman M (2006) Frontotemporal dementia: clinico-pathological correlations. Ann Neurol 59:952–962

29. Gass J, Cannon A, Mackenzie IR, Boeve B, Baker M, Adamson J, Crook R, Melquist S, Kuntz K, Petersen R, Josephs K, Pickering-Brown SM, Graff-Radford N, Uitti R, Dickson D, Wszolek Z, Gonzalez J, Beach TG, Bigio E, Johnson N, Weintraub S, Mesulam M, White CL 3 rd, Woodruff B, Caselli R, Hsiung GY, Feldman H, Knopman D, Hutton M, Rademakers R (2006) Mutations in prgranulin are a major cause of ubiquitin-positive frontotemporal lobar degeneration. Hum Mol Genet 15:2988–3001
30. Gatz M, Reynolds CA, Fratiglioni L, Johansson B, Mortimer JA, Berg S, Fiske A, Pedersen NL (2006) Role of genes and environments for explaining Alzheimer disease. Arch Gen Psychiatry 63:168–174
31. Ghidoni R, Benussi L, Glionna M, Franzoni M, Binetti G (2008) Low plasma progranulin levels predict progranulin mutations in Frontotemporal lobar degeneration. Neurology 71:1235–1239
32. Gijselinck I, Van Groeckroven C, Cruts M Granulin mutations Hum Mutat 2008, 29:1373–86
33. Gilman S, Koller M, Black RS, Jenkins L, Griffith SG, Fox NC, Eisner L, Kirby L, Rovira MB, Forette F, Orgogozo JM, AN1792(QS-21)-201 Study Team (2005) Clinical effects of Abeta immunization (AN1792) in patients with AD in an interrupted trial. Neurology 64:1553–1562
34. Glatzel M, Stoeck K, Seeger H, Luhrs T, Aguzzi A (2005) Human prion diseases: molecular and clinical aspects. Arch Neurol 62:545–552
35. Godbolt AK, Cipolotti L, Watt H, Fox NC, Janssen JC, Rossor MN (2004) The natural history of Alzheimer disease: a longitudinal presymptomatic and symptomatic study of a familial cohort. Arch Neurol 61:1743–1748
36. Goedert M, Spillantini MG (2006) A century of Alzheimer's disease. Science 314:777–781
37. Green RC (2002) Risk assessment for Alzheimer's disease with genetic susceptibility testing: has the moment arrived? Alzheimer's Care Q 3:208–214
38. Green RC, Cupples LA, Go R, Benke KS, Edeki T, Griffith PA, Williams M, Hipps Y, Graff-Radford N, Bachman D, Farrer LA (2002) Risk of dementia among white and African American relatives of patients with Alzheimer disease. JAMA 287:329–336
39. Grossman M, Libon DJ, Forman MS, Massimo L, Wood E, Moore P, Anderson C, Farmer J, Chatterjee A, Clark CM, Coslett HB, Hurtig HI, Lee VM-L, Trojanowski JQ (2007) Distinct Antemortem Profiles in Patients with Pathologically Defined Frontotemporal Dementia. Arch Neurol 64: 1601–1609
40. Grupe A, Li Y, Rowland C, Nowotny P, Hinrichs AL, Smemo S, Kauwe JS, Maxwell TJ, Cherny S, Doil L, Tacey K, van Luchene R, Myers A, Wavrant-De Vrieze F, Kaleem M, Hollingworth P, Jehu L, Foy C, Archer N, Hamilton G, Holmans P, Morris CM, Catanese J, Sninsky J, White TJ, Powell J, Hardy J, O'Donovan M, Lovestone S, Jones L, Morris JC, Thal L, Owen M, Williams J, Goate A (2006) A scan of chromosome 10 identifies a novel locus showing strong association with late-onset Alzheimer disease. Am J Hum Genet 78:78–88
41. Hattori M, Sugino E, Minoura K, In Y, Sumida M, Taniguchi T, Tomoo K, Ishida T (2008) Different inhibitory response of cyaniding and methylene blue for filament formation of tau microtubule-binding domain. Biochem Biophys Res Comm 374:158–163
42. Haupt M, Kurz A, Pollmann S, Romero B (1992) Alzheimer's disease: identical phenotype of familial and non-familial cases. J Neurol 239:248–250
43. Heston LL, Mastri AR, Anderson VE, White J (1981) Dementia of the Alzheimer type. Clinical genetics, natural history, and associated conditions. Arch Gen Psychiatry 38: 1085–1090
44. Holmes C, Boche D, Wilkinson D, Yadegarfar G, Hopkins V, Bayer A, Jones RW, Bullock R, Love S, Neal JW, Nicoll JAR (2008) Long term effects of $A\beta_{42}$ immunisation in Alzheimer's disease: follow-up of a randomized, placebo-controlled phase I trial. Lancet 372:216–223
45. Irizarry MC, Cheung BS, Rebeck GW, Paul SM, Bales KR, Myman BT (2000) Apolipoprotein E affects the amount, form, and anatomical distribution of amyloid β-peptide deposition in homozygous APP^{V717F} transgenic mice. Acta Neuropathol 100:451–458
46. Janssen JC, Beck JA, Campbell TA, Dickinson A, Fox NC, Harvey RJ, Houlden H, Rossor MN, Collinge J (2003) Early onset familial Alzheimer's disease: mutation frequency in 31 families. Neurology 60:235–239
47. Jarvik G, Larson EB, Goddard K, Schellenberg GD, Wijsman EM (1996) Influence of apolipoprotein E genotype on the transmission of Alzheimer disease in a community-based sample. Am J Hum Genet 58:191–200
48. Jayadev S, Steinbart EJ, Chi Y-Y, Kukull WA, Schellenberg GD, Bird TD (2008) Conjugal Alzheimer disease: risk in children when both parents have Alzheimer disease. Arch Neurol 65(3):373–378
49. Kamboh MI (2004) Molecular genetics of late-onset Alzheimer's disease. Ann Hum Genet 68:381–404
50. Kamboh MI, Minster RL, Feingold E, DeKosky ST (2006) Genetic association of ubiquilin with Alzheimer's disease and related quantitative measures. Mol Psychiatry 11:273–279
51. Kaye JA (1998) Diagnostic challenges in dementia. Neurology 51:45–52
52. Khachaturian AS, Corcoran CD, Mayer LS, Zandi PP, Breitner JC (2004) Apolipoprotein E epsilon4 count affects age at onset of Alzheimer disease, but not lifetime susceptibility: the cache county study. Arch Gen Psychiatry 61:518–524
53. Kimonis VE, Mehta SG, Fulchiero EC, Thomasova D, Pasquali M, Boycott K, Neilan EG, Kartashov A, Forman MS, Tucker S, Kimonis K, Mumm S, Whyte MP, Smith CD, Watts G (2008) Clinical studies in familial VCP myopathy associated with Paget disease of bone and frontotemporal dementia. Am J Med Genet 146A:745–757
54. Klafki HW, Staufenbiel M, Kornhuber J, Wiltfang J (2006) Therapeutic approaches to Alzheimer's disease. Brain 129:2840–2855
55. Kovacs GG, Trabattoni G, Hainfellner JA, Ironside JW, Knight R, Budka H (2002) Mutations of the prion protein gene: phenotypic spectrum. J Neurol 249:1567–1582
56. Kovacs GG, Puopolo M, Ladogana A, Pocchiari M, Budka H, van Duijn C, Collins SJ, Boyd A, Giulivi A, Coulthart M, Delasnerie-Laupretre N, Brandel JP, Zerr I, Kretzschmar HA, de Pedro-Cuesta J, Calero-Lara M, Glatzel M, Aguzzi A, Bishop M, Knight R, Belay G, Will R, Mitrova E (2005) Genetic Prion Disease: the EUROCJD experience. Hum Genet 118:166–174
57. Kukull WA, Higdon R, Bowen JD, McCormick WC, Teri L, Schellenberg GD, van Belle G, Jolley L, Larson EB (2002)

Dementia and Alzheimer disease incidence: a prospective cohort study. Arch Neurol 59:1737–1746
58. Lagogana A, Puopolo M, Poleggi A, Almonti S, Mellina V, Equestre M, Pocchiari M (2005) High incidence of genetic human transmissible spongiform encephalopathies in Italy. Neurol 64:1592–1597
59. Lai F, Kammann E, Rebeck GW, Anderson A, Chen Y, Nixon RA (1999) APOE genotype and gender effects on Alzheimer disease in 100 adults with Down syndrome. Neurology 53:331–336
60. Lampe TH, Bird TD, Nochlin D, Nemens E, Risse SC, Sumi SM, Koerker R, Leaird B, Wier M, Raskind MA (1994) Phenotype of chromosome 14-linked familial Alzheimer's disease in a large kindred. Ann Neurol 36:368–378
61. Larner AJ, Doran M (2006) Clinical phenotypic heterogeneity of Alzheimer's disease associated with mutations of the presenilin-1 gene. J Neurol 253:139–158
62. Lee JH, Cheng R, Graff-Radford N, Foroud T, Mayeux R, NIA LOAD Family Study Group (2008) Analyses of the national institute of aging late-onset alzheimer's disease family study: implication of additional loci. Arch Neurol 65:1518–1526
63. Leverenz JB, Raskind MA (1998) Early amyloid deposition in the medial temporal lobe of young Down syndrome patients: a regional quantitative analysis. Exp Neurol 150:296–304
64. Levy-Lahad E, Bird TD (1996) Genetic factors in Alzheimer's disease: a review of recent advances. Ann Neurol 40:829–840
65. Levy-Lahad E, Wasco W, Poorkaj P, Romano DM, Oshima J, Pettingell WH, Yu C, Jondro PD, Schmidt SD, Wang K, Crowley AC, Fu Y-H, Guenette SY, Galas D, Nemens E, Wijsman EM, Bird TD, Schellenberg GD, Tanzi RE (1995) Candidate gene for the chromosome 1 familial Alzheimer's disease locus. Science 269:973–977
66. Li G, Higdon R, Kukull WA, Peskind E, Van Valen Moore K, Tsuang D, van Belle G, McCormick W, Bowen JD, Teri L, Schellenberg GD, Larson EB (2004) Statin therapy and risk of dementia in the elderly: a community-based prospective cohort study. Neurology 63:1624–1628
67. Li G, Larson EB, Sonnen JA, Shofer JB, Petrie EC, Schantz A, Peskind ER, Raskind MA, Breitner JC, Montine TJ (2007) Statin therapy is associated with reduced neuropathologic changes of Alzheimer disease. Neurology 69:878–885
68. Li Y, Grupe A, Rowland C, Nowotny P, Kauwe JS, Smemo S, Hinrichs A, Tacey K, Toombs TA, Kwok S, Catanese J, White TJ, Maxwell TJ, Hollingworth P, Abraham R, Rubinsztein DC, Brayne C, Wavrant-De Vrieze F, Hardy J, O'Donovan M, Lovestone S, Morris JC, Thal LJ, Owen M, Williams J, Goate A (2006) DAPK1 variants are associated with Alzheimer's disease and allele-specific expression. Hum Mol Genet 15:2560–2568
69. Li YJ, Oliveira SA, Xu P, Martin ER, Stenger JE, Scherzer CR, Hauser MA, Scott WK, Small GW, Nance MA, Watts RL, Hubble JP, Koller WC, Pahwa R, Stern MB, Hiner BC, Jankovic J, Goetz CG, Mastaglia F, Middleton LT, Roses AD, Saunders AM, Schmechel DE, Gullans SR, Haines JL, Gilbert JR, Vance JM, Pericak-Vance MA (2003) Glutathione S-transferase omega-1 modifiesage-at-onset of Alzheimer disease and Parkinson disease. Hum Mol Genet 12:3259–3267
70. Liu F, Arias-Vasquez A, Sleegers K, Aulchnko YS, Kayser M, Sanchex-Juan P, Feng B-J, Bertoli-Avella AM, van Swieten J, Axenovich TI, Heutink P, van Broeckhoven C, Oostra BA, van Duijin CM (2007) A Genomewide screen for late-onset Alzheimer disease in a genetically isolated Dutch population. AJHG 81:17–31
71. Margallo-Lana M, Morris CM, Gibson AM, Tan AL, Kay DW, Tyrer SP, Moore BP, Ballard CG (2004) Influence of the amyloid precursor protein locus on dementia in Down syndrome. Neurology 62:1996–1998
72. Martins CA, Oulhaj A, de Jager CA, Williams JH (2005) APOE alleles predict the rate of cognitive decline in Alzheimer disease: a nonlinear model. Neurology 65:1888–1893
73. Masters CL, Beyreuther K (2006) Alzheimer's centennial legacy: prospects for rational therapeutic intervention targeting the Abeta amyloid pathway. Brain 129:2823–2839
74. Mayeux R, Saunders AM, Shea S, Mirra S, Evans D, Roses AD, Hyman BT, Crain B, Tang MX, Phelps CH (1998) Utility of the apolipoprotein E genotype in the diagnosis of Alzheimer's disease. Alzheimer's disease centers consortium on Apolipoprotein E and Alzheimer's disease. N Engl J Med 338:506–511
75. Meyer MR, Tschanz JT, Norton MC, Welsh-Bohmer KA, Steffens DC, Wyse BW, Breitner JC (1998) APOE genotype predicts when – not whether – one is predisposed to develop Alzheimer disease. Nat Genet 19:321–322
76. Mulnard RA, Cotman CW, Kawas C, van Dyck CH, Sano M, Doody R, Koss E, Pfeiffer E, Jin S, Gamst A, Grundman M, Thomas R, Thal LJ (2000) Estrogen replacement therapy for treatment of mild to moderate Alzheimer disease: a randomized controlled trial. Alzheimer's Disease Cooperative Study. JAMA 283:1007–1015
77. Murrell JR, Price B, Lane KA, Baiyewu O, Gureje O, Ogunniyi A, Unverzagt FW, Smith-Gamble V, Gao S, Hendrie HC, Hall KS (2006) Association of Apolipoprotein E Genotype and Alzheimer Disease in African Americans. Arch Neurol 63:431–434
78. Myers A, Holmans P, Marshall H, Kwon J, Meyer D, Ramic D, Shears S, Booth J, DeVrieze FW, Crook R, Hamshere M, Abraham R, Tunstall N, Rice F, Carty S, Lillystone S, Kehoe P, Rudrasingham V, Jones L, Lovestone S, Perez-Tur J, Williams J, Owen MJ, Hardy J, Goate AM (2000) Susceptibility locus for Alzheimer's disease on chromosome 10. Science 290:2304–2305
79. National Institute on Aging/Alzheimer's Association Working Group (1996) Apolipoprotein E genotyping in Alzheimer's disease. Lancet 347:1091–1095
80. National Institute on Aging/Alzheimer's Association Working Group (1997) Consensus recommendations for the postmortem diagnosis of Alzheimer's disease. Neurobiol Aging 18(Suppl 4):S1–S2
81. Nochlin D, van Belle G, Bird TD, Sumi SM (1993) Comparison of the severity of neuropathologic changes in familial and sporadic Alzheimer's disease. Alzheimer Dis Assoc Disord 7:212–222
82. Nussbaum RL, Ellis CE (2003) Alzheimer's disease and Parkinson's disease. N Engl J Med 348:1356–1364
83. Oddo S, Billings L, Kesslak JP, Cribbs DH, Laferla FM (2004) Ab Immunotherapy leads to clearance of Early, but not late, hyperphosphorylate tau aggregates via the proteasome. Neuron 43:321–332

84. Opherk C, Petes N, Herzog J, Luedtke R, Dichgans M (2004) Long-Term prognosis and causes of death in CADASIL: a retrospective study in 411 patients. Brain 127:2533–2539
85. Overshott R, Burns A (2005) Treatment of dementia. J Neurol Neurosurg Psychiatry 76(Suppl 5):v53–v59
86. Petersen RC, Thomas RG, Grundman M, Bennett D, Doody R, Ferris S, Galasko D, Jin S, Kaye J, Levey A, Pfeiffer E, Sano M, van Dyck CH, Thal LJ (2005) Vitamin E and donepezil for the treatment of mild cognitive impairment. N Engl J Med 352:2379–2388
87. Petes N, Opherk C, Bergmann T, Castro M, Herzog J, Dichgans M (2005) Spectrum of mutations in biopsy-proven CADASIL: implications for diagnostic strategies. Arch Neurol 62:1091–1094
88. Pookaj P, Grossman M, Stienbart E, Payami H, Sadovnick A, Nochlin D, Tabira T, Trojanowski JQ, Borson S, Galasko D, Reich S, Quinn B, Schellenberg G, Bird TD (2001) Frequency of Tau Gene Mutations in Familial and Sporadic Cases of Non-Alzheimer Dementia. Arch Neurol 58:383–387
89. Popescu A, Lippa CF, Lee VM, Trojanowski JQ (2004) Lewy bodies in the amygdala: increase of alpha-synuclein aggregates in neurodegenerative diseases with tau-based inclusions. Arch Neurol 61:1915–1919
90. Prasher VP, Farrer MJ, Kessling AM, Fisher EM, West RJ, Barber PC, Butler AC (1998) Molecular mapping of Alzheimer-type dementia in Down's syndrome. Ann Neurol 43:380–383
91. Qu B, Boyer PJ, Johnston SA, Hynan LS, Rosenberg RN (2006) Abeta42 gene vaccination reduces brain amyloid plaque burden in transgenic mice. J Neurol Sci 244:151–158
92. Quaid KA, Murrell JR, Hake AM, Farlow MR, Ghetti B (2000) Presymptomatic genetic testing with an APP mutation. J Genet Counsel 9:327–345
93. Rademakers R, Cruts M, Sleegers K, Dermaut B, Theuns J, Aulchenko Y, Weckx S, De Pooter T, Van den Broeck M, Corsmit E, De Rijk P, Del-Favero J, van Swieten J, van Duijn CM, Van Broeckhoven C (2005) Linkage and association studies identify a novel locus for Alzheimer disease at 7q36 in a Dutch population-based sample. Am J Hum Genet 77:643–652
94. Rademakers R, Baker M, Gass J, Adamson J, Huey ED, Momeni P, Spina S, Coppola G, Karydas AM, Stewart H, Johnson N, Hsiung GY, Kelley B, Kuntz K, Steinbart E, Wood EM, Yu CE, Josephs K, Sorenson E, Womack KB, Weintraub S, Pickering-Brown SM, Schofield PR, Brooks WS, Van Deerlin VM, Snowden J, Clark CM, Kertesz A, Boylan K, Ghetti B, Neary D, Schellenberg GD, Beach TG, Mesulam M, Mann D, Grafman J, Mackenzie IR, Feldman H, Bird T, Petersen R, Knopman D, Boeve B, Geschwind DH, Miller B, Wszolek Z, Lippa C, Bigio EH, Dickson D, Graff-Radford N, Hutton M (2007) Phenotypic variability associated with progranulin haploinsufficiency in patients with the common 1477C–>T (Arg493X) mutation: an international initiative. Lancet Neurol 6(10):857–868
95. Raskind MA, Peskind ER, Wessel T, Yuan W (2000) Galantamine in AD: a 6-month randomized, placebo-controlled trial with a 6-month extension. The Galantamine USA-1 Study Group. Neurology 54:2261–2268
96. Raux G, Guyant-Marechal L, Martin C, Bou J, Penet C, Brice A, Hannequin D, Frebourg T, Campion D (2005) Molecular diagnosis of autosomal dominant early onset Alzheimer's disease: an update. J Med Genet 42:793–795
97. Reed LA, Wszolek ZW, Hutton M (2001) Phenotypic correlations in FTDP-17. Neurobiol Ageing 22:89–107
98. Reiman EM, Webster JA, Myers AJ, Hardy J, Dunckley T, Zismann VL, Joshipura KD, Pearson JV, Hu-Lince D, Huentelman MJ, Craig DW, Coon KD, Liang WS, Herbert RH, Beach T, Rohrer KC, Zhao AS, Leung D, Bryden L, Marlowe L, Kaleem M, Mastroeni D, Grover A, Heward CB, Ravid R, Rogers J, Hutton ML, Melquist S, Petersen RC, Alexander GE, Caselli RJ, Kukull W, Papassotiropoulos A, Stephan DA (2007) GAB2 Alleles Modify Alzheimer's Risk in APOE e4 Carriers. Neuron 54:713–720
99. Reisberg B, Doody R, Stoffler A, Schmitt F, Ferris S, Mobius HJ (2006) A 24-week open-label extension study of memantine in moderate to severe Alzheimer disease. Arch Neurol 63:49–54
100. Riemenschneider M, Konta L, Friedrich P, Schwarz S, Taddei K, Neff F, Padovani A, Kolsch H, Laws SM, Klopp N, Bickeboller H, Wagenpfeil S, Mueller JC, Rosenberger A, Diehl-Schmid J, Archetti S, Lautenschlager N, Borroni B, Muller U, Illig T, Heun R, Egensperger R, Schlegel J, Forstl H, Martins RN, Kurz A (2006) A functional polymorphism within plasminogen activator urokinase (PLAU) is associated with Alzheimer's disease. Hum Mol Genet 15:2446–2456
101. Roberts JS, Cupples LA, Relkin NR, Whitehouse PJ, Green RC (2005) Genetic risk assessment for adult children of people with Alzheimer's disease: the risk evaluation and education for Alzheimer's disease (REVEAL) study. J Geriatr Psychiatry Neurol 18:250–255
102. Rocca WA, Hofman A, Brayne C, Breteler MM, Clarke M, Copeland JR, Dartigues JF, Engedal K, Hagnell O, Heeren TJ et al (1991) Frequency and distribution of Alzheimer's disease in Europe: a collaborative study of 1980–1990 prevalence findings. The EURODEM-prevalence research group. Ann Neurol 30:381–390
103. Rogaeva E, Meng Y, Lee JH, Gu Y, Kawarai T, Zou F, Katayama T, Baldwin CT, Cheng R, Hasegawa H, Chen F, Shibata N, Lunetta KL, Pardossi-Piquard R, Bohm C, Wakutani Y, Cupples LA, Cuenco KT, Green RC, Pinessi L, Rainero I, Sorbi S, Bruni A, Duara R, Friedland RP, Inzelberg R, Hampe W, Bujo H, Song YQ, Andersen OM, Willnow TE, Graff-Radford N, Petersen RC, Dickson D, Der SD, Fraser PE, Schmitt-Ulms G, Younkin S, Mayeux R, Farrer LA, St George-Hyslop P (2007) The neuronal sortilin-related receptor SORL1 is genetically associated with Alzheimer disease. Nat Genet 39:168–177
104. Rogan S, Lippa CF (2002) Alzheimer's disease and other dementias: a review. Am J Alzheimers Dis Other Demen 17:11–17
105. Romas SN, Santana V, Williamson J, Ciappa A, Lee JH, Rondon HZ, Estevez P, Lantigua R, Medrano M, Torres M, Stern Y, Tycko B, Mayeux R (2002) Familial Alzheimer disease among Caribbean Hispanics: a reexamination of its association with APOE. Arch Neurol 59:87–91
106. Rosenberg RN (2000) The molecular and genetic basis of AD: the end of the beginning: the 2000 Wartenberg lecture. Neurology 54:2045–2054

107. Roses AD, Saunders AM (2006) Perspective on a pathogenesis and treatment of Alzheimer's disease. Alz Dem 2:59–70
108. Rowe CC, Ng S, Ackermann U, Gong SJ, Pike K, Savage G, Cowie TF, Dickinson KL, Maruff P, Darby D, Smith C, Woodward M, Merory J, Tochon-Danguy H, O'Keefe G, Klunk WE, Mathis CA, Price JC, Masters CL, Villemange VL (2007) Imaging b-amyloid burden in aging and dementia. Neurology 68:1718–1725
109. Saunders AM, Hulette O, Welsh-Bohmer KA, Schmechel DE, Crain B, Burke JR, Alberts MJ, Strittmatter WJ, Breitner JC, Rosenberg C (1996) Specificity, sensitivity, and predictive value of apolipoprotein-E genotyping for sporadic Alzheimer's disease. Lancet 348:90–93
110. Schenk D, Barbour R, Dunn W, Gordon G, Grajeda H, Guido T, Hu K, Huang J, Johnson-Wood K, Khan K, Kholodenko D, Lee M, Liao Z, Lieberburg I, Motter R, Mutter L, Soriano F, Shopp G, Vasquez N, Vandevert C, Walker S, Wogulis M, Yednock T, Games D, Seubert P (1999) Immunization with amyloid-beta attenuates Alzheimer-disease-like pathology in the PDAPP mouse. Nature 400:173–177
111. Schupf N, Kapell D, Nightingale B, Lee JH, Mohlenhoff J, Bewley S, Ottman R, Mayeux R (2001) Specificity of the fivefold increase in AD in mothers of adults with Down syndrome. Neurology 57:979–984
112. Scott WK, Hauser ER, Schmechel DE, Welsh-Bohmer KA, Small GW, Roses AD, Saunders AM, Gilbert JR, Vance JM, Haines JL, Pericak-Vance MA (2003) Ordered-subsets linkage analysis detects novel Alzheimer disease Loci on chromosomes 2q34 and 15q22. Am J Hum Genet 73:1041–1051
113. Seltzer B, Zolnouni P, Nunez M, Goldman R, Kumar D, Ieni J, Richardson S (2004) Efficacy of donepezil in early-stage Alzheimer disease: a randomized placebo-controlled trial. Arch Neurol 61:1852–1856
114. Serretti A, Artioli P, Quartesan R, De Ronchi D (2005) Genes involved in Alzheimer's disease, a survey of possible candidates. J Alzheimers Dis 7:331–353
115. Sherrington R, Froelich S, Sorbi S, Campion D, Chi H, Rogaeva EA, Levesque G, Rogaev EI, Lin C, Liang Y, Ikeda M, Mar L, Brice A, Agid Y, Percy ME, Clerget-Darpoux F, Piacentini S, Marcon G, Nacmias B, Amaducci L, Frebourg T, Lannfelt L, Rommens JM, St George-Hyslop PH (1996) Alzheimer's disease associated with mutations in presenilin 2 is rare and variably penetrant. Hum Mol Genet 5:985–988
116. Silverman JM, Li G, Zaccario ML, Smith CJ, Schmeidler J, Mohs RC, Davis KL (1994) Patterns of risk in first-degree relatives of patients with Alzheimer's disease. Arch Gen Psychiatry 51:577–586
117. Skibinski G, Parkinson NJ, Brown JM, Chakrabarti L, Llyod SL, Hummerich H, Neilsen JE, Hodges JR, Spillantini MG, Thusgaard T, Brandner S, Brun A, Rossor MN, Gade A, Johannsen P, Sorensen SA, Gydesen S, Fisher E, Collinge J (2005) Mutations in the endosomal ESCRTIII-complex subunit CHMP2B in frontotemporal dementia. Nat Genet 37:806–808
118. Sleegers K, Van Duijn CM (2001) Alzheimer's disease: genes, pathogenesis and risk prediction. Community Genet 4:197–203
119. Smemo S, Nowotny P, Hinrichs AL, Kauwe JS, Cherny S, Erickson K, Myers AJ, Kaleem M, Marlowe L, Gibson AM, Hollingworth P, O'Donovan MC, Morris CM, Holmans P, Lovestone S, Morris JC, Thal L, Li Y, Grupe A, Hardy J, Owen MJ, Williams J, Goate A (2006) Ubiquilin 1 polymorphisms are not associated with late-onset Alzheimer's disease. Ann Neurol 59:21–26
120. Snowden DA, Kemper SJ, Mortimer JA, Greiner LH, Wekstein DR, Markesbery WR (1996) Linguistic ability in early life and cognitive function and Alzheimer's disease in late life. Findings from the Nun Study. JAMA 275:528–532
121. Sreedharan J, Blair IP, Tripathi VB, Hu X, Vance C, Rogelj B, Ackerley S, Durnall JC, Williams KL, Buratti E, Baralle F, de Belleroche J, Mitchell JD, Leigh PN, Al-Chalabi A, Miller CC, Nicholson G, Shaw CE (2008) TDP-43 mutations in familial and sporadic amyotrophic lateral sclerosis. Science 319:1668–1672
122. Steinbart EJ, Smith CO, Poorkaj P, Bird TD (2001) Impact of DNA testing for early-onset familial Alzheimer disease and frontotemporal dementia. Arch Neurol 58:1828–1831
123. Steiner H, Fluhrer R, Haass C (2008) Intramembrane proteolysis by g-secretase. J Biol Chem 283:30121–30128
124. Troncoso JC, Zonderman AB, Resnick SM, Crain B, Pletnikova O, O'Brien RJ (2008) Effect of infarcts on dementia in the baltimore longitudinal study of aging. Ann Neurol 64:168–176
125. Tsuang D, Larson EB, Bowen J, McCormick W, Teri L, Nochlin D, Leverenz JB, Peskind ER, Lim A, Raskind MA, Thompson ML, Mirra SS, Gearing M, Schellenberg GD, Kukull W (1999) The utility of apolipoprotein E genotyping in the diagnosis of Alzheimer disease in a community-based case series. Arch Neurol 56:1489–1495
126. Van Deerlin VM, Leverenz JB, Berkris LM, Bird TD, Yuan W, Elman LB, Clay D, Wood EM, Chen-Plotkin AS, Martinez-Lage M, Steinbart E, McCluskey L, Grossman M, Neumann M, Wu I-L, Yang W-S, Kalb R, Galasko DR, Montine TJ, Trojanowski JQ, Lee VM-L, Schellenberg GD, Yu CE (2008) TARDBP mutations in amyotrophic lateral sclerosis with TDP-43 neuropathology: a genetic and histopathological analysis. Lancet Neurol 7:409–416
127. van der Cammen TJ, Croes EA, Dermaut B, de Jager MC, Cruts M, Van Broeckhoven C, van Duijn CM (2004) Genetic testing has no place as a routine diagnostic test in sporadic and familial cases of Alzheimer's disease. J Am Geriatr Soc 52:2110–2113
128. Van Swieten JC, Heutink P (2008) Mutations in progranulin (GRN) within the spectrum of clinical and pathological phenotypes of Frontotemporal dementia. Lancet Neurol 7:965–974
129. Verlinsky Y, Rechitsky S, Verlinsky O, Masciangelo C, Lederer K, Kuliev A (2002) Preimplantation diagnosis for early-onset Alzheimer disease caused by V717L mutation. JAMA 287(8):1018–1021
130. Wang PN, Liao SQ, Liu RS, Liu CY, Chao HT, Lu SR, Yu HY, Wang SJ, Liu HC (2000) Effects of estrogen on cognition, mood, and cerebral blood flow in AD: a controlled study. Neurology 54:2061–2066
131. Welsh-Bohmer KA, Gearing M, Saunders AM, Roses AD, Mirra S (1997) Apolipoprotein E genotypes in a neuropathological series from the consortium to establish a registry for Alzheimer's disease. Ann Neurol 42:319–325

132. Whalley LJ, Starr JM, Athawes R, Hunter D, Pattie A, Deary IJ (2000) Childhood mental ability and dementia. Neurology 55:1455–1459
133. Wijsman EM, Daw EW, Yu CE, Payami H, Steinbart EJ, Nochlin D, Conlon EM, Bird TD, Schellenberg GD (2004) Evidence for a novel late-onset Alzheimer disease locus on chromosome 19p13.2. Am J Hum Genet 75:398–409
134. Wolozin B, Kellman W, Ruosseau P, Celesia GG, Siegel G (2000) Decreased prevalence of Alzheimer disease associated with 3-hydroxy-3-methyglutaryl coenzyme A reductase inhibitors. Arch Neurol 57:1439–1443
135. Yaffe K, Clemons TE, McBee WL, Lindblad AS (2004) Impact of antioxidants, zinc, and copper on cognition in the elderly: a randomized, controlled trial. Neurology 63:1705–1707
136. Zhong N, Weisgraber KH (2008) Understanding the association of Apolipoprotein E4 with Alzheimer's disease: clues from its structure. J Biol Chem 284:6027–6031

Genetics of Autism 23.4

Brett S. Abrahams and Daniel H. Geschwind

Abstract We have learned more about the molecular genetics of autism in the last 3 years than in the previous 30. This includes both a new appreciation for the role of rare genetic variation and the identification of the first contributory common variants by genome-wide association. These data show that although the population attributable risk of common variation may be moderate to large, the genotype risk of common variants at the individual level are small. In contrast, a large number of diverse rare mutations of large effect have been identified, but none appear specific to autism. All of these findings point to extreme genetic heterogeneity suggesting complex gene–gene or gene–environment interactions in autism etiology. Available knowledge, reviewed below, also suggests that phenotypic presentation is the result of complex interactions, and that implicated genetic risk factors in many cases cross the boundaries of established clinical diagnostic categories. Acceptance of this complexity and efforts to understand genetic variation in terms of intermediate phenotypes represent important directions for future research.

Contents

23.4.1	Background	699
23.4.2	Cytogenetic Findings	701
23.4.3	Linkage	703
23.4.4	Syndromic ASDs	703
23.4.5	Re-sequencing	704
23.4.6	Copy Number Variation	705
23.4.7	Common Variation	707
23.4.8	Towards Convergence	707
	References	710

B.S. Abrahams (✉) and D.H. Geschwind
Programs in Neurogenetics and Neurobehavioural Genetics, Neurology Department, and Semel Institute for Neuroscience and Behavior, David Geffen School of Medicine, University of California at Los Angeles, Los Angeles, CA 90095-1769, USA
e-mail: brett.abrahams@gmail.com; dhg@ucla.edu

23.4.1 Background

Autistic disorder is characterized by specific deficits in three core domains, language, social behavior, and cognitive flexibility prior to 3 years of age although presentation varies substantially between cases with variable impairment in additional sensory, motor, and medical domains ([56], Table 23.4.1). Current practice does not typically view autistic disorder in isolation, but rather as one of several entities collectively referred to as the autism spectrum disorders (ASDs). Asperger syndrome, where language capability is relatively sparred, would be included here, as would individuals diagnosed with pervasive developmental disorder-not otherwise specified (PDD-NOS), a catch-all used to capture cases showing a subset of relevant behavioral abnormalities but not meeting criteria for autistic disorder as defined in the Diagnostic and Statistical Manual of Mental Disorders (DSM-IV). Childhood disintegrative disorder and Rett syndrome, distinguished by regression (loss of learned skills) and deterioration, respectively, represent a smaller fraction of cases but are likewise included under the ASD umbrella. The development of standardized diagnostic tools including the Autism Diagnostic Interview (ADI-R) and the Autism Diagnostic Observation Schedule (ADOS) have proved critical in characterization of cases and development of uniformity between clinical centers. Recent estimates of frequency for conditions on the spectrum have been discussed in detail elsewhere [54], but are as high as 1/150 when broadest definitions are employed [36].

As highlighted in Fig. 23.4.1, work towards an understanding of the ASDs has covered much ground over the last quarter century. Central to this progress was the shift away from attributing etiology to aspects of parenting, particularly insufficient attention or affection from mothers, a notion popularized (by the prominence of) psychodynamic theories in the 1960s. Subsequent recognition of rare chromosomal aberrations in patients together with the identification of rare syndromes in which autistic symptomatology is elevated was important here and was central for both acceptance of a medical model and the recognition of genetic factors in the ASDs.

Twin studies provide additional important support for a major role for genetic factors in autism susceptibility [11, 123]. Separate studies documenting an increase in the "broader autism phenotype" or subclinical ASD-like behavioral abnormalities amongst first-degree relatives of cases also lend support for genetic factors underlying components of the autistic spectrum [18, 21]. The importance of genetic factors was further highlighted by reports that an autism diagnosis in a first child was associated with a ~10% recurrence rate in subsequent progeny, a significant increase over population levels [108]. This said, early genetic studies were quick to point to significant genetic heterogeneity [72] and question whether genetic factors may govern subcomponents of disease, rather than inheritance of the clinical entity as a whole [53]. Although much has been learned in recent years, such questions remain topical and will be revisited below.

Although the overall focus here is to review current knowledge regarding genetics, consideration of how environmental contributions may likewise impact

Table 23.4.1 Domains of impairment in the autism spectrum disorders (ASD)[a] (reprinted, with permission, from the Annual Review of Medicine, Volume 60 2009, by Annual Reviews, www.annualreviews.org) [56]

Domain	Autism	Asperger	PDD-NOS[b]	ASD
Social communication	Required	Required	Required	
Language	Required	–	Variable	
Repetitive and/or restrictive behaviors	Required	Required	Variable	
Sensory abnormalities[c]	>90%	80%	Variable	94%
Developmental regression[d]	15–40%	?	?	15–40%
Motor signs[e]	60–80%	60%	60%	60–80%
Gross motor delay	10%	?	?	5–10%
Sleep disturbance	55%	5–10%	40%	50%
Gastrointestinal disturbance[f]	45%	4%	50%	4–50%
Epilepsy[g]	10–60%	0–5%	5–40%	6–60%
Comorbid psychiatric diagnosis[h]	70%	60%	>25%	25–70%

[a]Diagnostic features are denoted as required, while those that are not observed are denoted with a dashed line. There have been few large-scale, epidemiologic studies of features associated with the ASDs, so that the frequencies presented above are conservative estimates based on an amalgamation of information from various references [12, 59, 63, 71, 77, 84, 88, 94, 119, 131, 138, 146]. The ASD column at the far right provides estimates for the broad group of related conditions, which includes autism, Asperger syndrome, and PDD-NOS

[b]Pervasive developmental disorder, Not otherwise specified (PDD-NOS) is a defined as a condition in which some, but not all, features of autism or another defined pervasive developmental disorder are present

[c]Responses to sensory stimuli (typically auditory or tactile) different to those observed in typically developing children

[d]Loss of function in either language and social skills (or both)

[e]Motor signs include hypotonia, gait problems, toe walking and apraxia

[f]Six months or more of diarrhea, constipation, reflux, or bloating

[g]The presence of epilepsy varies as a function of other co-morbid features resulting in a relatively large range

[h]Includes mood disorders, conduct disorders, aggression, and attention deficit/hyperactivity disorder (ADHD)

23.4 Genetics of Autism

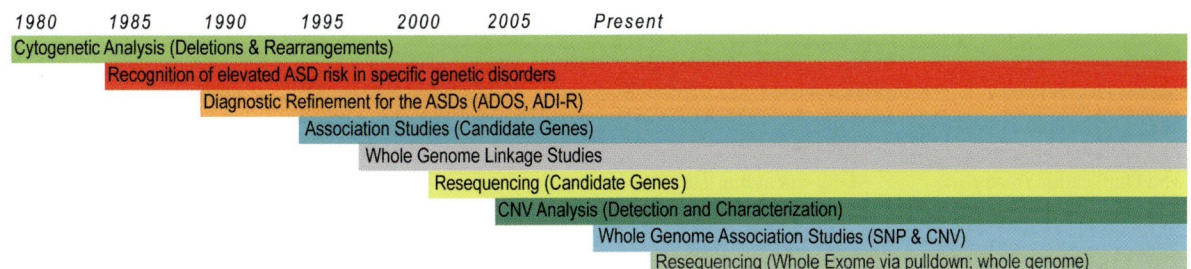

Fig. 23.4.1 Methodological changes have accelerated progress in ASD genetics. Collection of large cohorts via international collaboration, together with array-based technologies enabling genome-wide interrogation of variation, has resulted in major advances. Similar progress will come from massively parallel sequencing of partial and whole genomes. Although such experiments are soon likely to become routine, interpretation of results, particular in the context of diverse phenotype data, will require substantial computational infrastructure. (Reproduced from *Archives of Neurology*, with permission [1])

risk is warranted. Early evidence for a viral link, particularly rubella, suggests that maternal infection in early child development may be contributory [46]. Later work indicates that an array of pre- and perinatal complications is likewise increased in cases [60], but does not highlight specific complications that are obviously increased in cases versus controls. Increased maternal and paternal age have also been implicated [34, 107], although the manner by which having an older parent may come to influence risk remains unclear. It is tempting to speculate that older parents may generate offspring who as a group harbor *de novo* copy number variations (CNVs) (for explanation, see 3, Sect. 3.4.4.4) at an increased frequency but the contribution of maternally derived chromosomal anomalies and paternally derived point mutations must also be considered. Such data are important in the context of an apparent increased prevalence in the ASDs over recent years [54]. Although attributable at least in part to cultural factors – including a broadening of diagnostic criteria and reduced stigma – such effects are unlikely to account for the full extent of the observed increase [66]. Moving forward, our understanding of how risk is modulated and presentation shaped will no doubt benefit from joint consideration of genetics and environment.

Here we review progress in identification of genetic contributions to ASD, ranging from genetic syndromes resulting from rare point mutations and/or copy number variation (CNV) to common genetic variants established to be contributory. One key theme is that although Mendelian mutations clearly play a larger part than previously suspected, some highly associated rare or recurrent mutations show patterns of inheritance more consistent with complex inheritance. Stated otherwise, although major effect alleles exist and play an important part in modulating risk, most if not all of these appear to operate in concert with ancillary factors, environmental or genetic, to shape ultimate presentation.

23.4.2 Cytogenetic Findings

For many years, cytogenetic identification of rare chromosomal abnormalities in cases served as the principal means by which disease-related regions could be isolated [140]. Although typically spanning many megabases, encompassing 10 or more genes, and in many cases complex, such anomalies are estimated to be present in 6–7% of ASD children [90]. This rate increases if ascertainment is limited to individuals with dysmorphic features and intellectual disability but none of these variants or other "syndromic" causes of autism account for more than 1% of ASD, and most are rarer (Table 23.4.2). Although it is well recognized that large maternally derived duplications at 15q11-13 explain the etiology in an estimated 1–2% of cases [41], establishing such relationships for less frequently observed structural variants is challenging.

An additional issue is that because cytogenetically visible events encompass a large number of genes, clear involvement of a particular genomic region does not typically reveal individual molecules which may be contributory. This is important because identification of individual genes that underlie the effects of such variation is necessary to provide insights regarding mechanisms. Within the core 15q11-13 interval, involvement of each of *UBE3A* (a ubiquitin ligase) and

Table 23.4.2 ASD-related syndromes[a,*] (modified from [3])

ASD-related syndrome	Associated gene(s)	Proportion with ASD	Proportion ASD with syndrome	References
1q21 Duplication	Many	50%	~1%?	[91, 128]
3p Deletion / duplication	CNTN4	<50%	~1%	[51, 61, 110]
15q Duplication (maternal)	Many (including UBE3A, GABRB3, SNRPN, and SNURF)	High	~1%	[41]
15q13 Deletion	Many (including CHRNA7)	<50%	Unknown	[15, 118]
16p11 Deletion	Many (including SEZ6L2)	High	~1%	[78, 79, 90, 144]
22q11 Deletion (aka VCFS / DiGeorge)	Many (including TBX1 and COMT)	15–50%	<1%	[52, 139]
22q13 Deletion	SHANK3	High	~1%	[48, 89, 95]
Angelman (15q11-13)	Maternal UBE3A	40–80%	<1%	[22, 102]
Beckwith Weidemann (11p15)	IGF2 and CDKN1C	~7%	Unknown	[73]
Cortical dysplasia focal epilepsy (7q35-36)	CNTNAP2	70%	Negligible	[68, 125]
Cowden/BRRS (10q23)	PTEN	20%	>10% with macrocephaly	[101, 135]
Down (trisomy chr.21)	Many	6–15%	Unknown	[86]
Fragile X (Xq27)	FMR1	25% of males 6% of females	1–2%	[64]
Potocki-Lupski (17p11)	Many (including RAI1)	~90%	Unknown	[106]
Smith–Lemli–Optiz (11q13)	DHCR7	50%	Negligible	[129]
Prader–Willi (15q11-13)	Paternal deletions	20–25%	Unknown	[45]
Rett (Xq26)	MECP2	N/A	~0.5%	[5]
Timothy (12p13)	CACNA1C	60–80%	Negligible	[120]
Tuberous sclerosis (9q34 and 16p13)	TSC1, TSC2	20%	~1%	[10]

*BRRS Bannayan-Riley-Ruvalcaba syndrome, CACNA1C calcium channel voltage-dependent L type alpha 1C subunit, CDKN1C cyclin-dependent kinase inhibitor 1C, CNTN4 contactin 4, CNTNAP2 contactin-associated protein-like 2, DHCR7 7-dehydrocholesterol reductase, FMR1 fragile X mental retardation 1, GABRB3 GABA A Receptor, beta 3 subunit, IGF2, insulin-like growth factor 2, MECP2 methyl CpG-binding protein 2, PTEN Phosphotase and tenoin homolog deleted on chromosome 10; RAI1, retinoic acid-induced 1, SEZL6 seizure-related 6 homolog (mouse)-like 2; SHANK3 SH3 and multiple ankyrin repeat domains 3; SNURF, SNRPN upstream reading frame; SNRPN small nuclear ribonucleoprotein polypeptide N, TSC1 tuberous sclerosis 1, TSC2 tuberous sclerosis 2, UBE3A ubiquitin protein ligase E3A, VCFS velocardiofacial syndrome

[a]The reader should compare values listed above to ASD prevalence in the general public (0.2–0.7%) and among individuals with nonsyndromic MR (~15%) [36]. Other etiologically heterogeneous clinical entities that show unexpectedly high overlap with the ASDs include: bipolar disorder, epilepsy, Joubert syndrome, schizophrenia, specific language impairment, and Tourette syndrome

GABRB3 (an inhibitory neurotransmitter receptor subunit) is well established. Additional GABA receptors (A5 and G3 subunits) and the imprinted SNURF-SNRPN transcripts receive less attention but are also likely to be contributory. This is further complicated by independent structural variants on either side of this core region, which likewise appear to modulate risk. Amongst particularly intriguing candidates here are (CYFIP) the cytoplasmic FMR1 interacting protein 1 [19, 47, 98, 100, 122], MAGEL2 involved in regulation of circadian rhythms [29, 76], the neurexin adaptor molecule APBA2 / MINT2 [16, 75, 99], the nicotinic acetylcholine receptor subunit CHRNA7 [118], and a related hybrid gene CHRFAM7A [44]. Dissection of how individual molecules contribute independently and together to clinical variation in cases represents an important set of problems for the future.

As discussed below, additional variation within such intervals or modifiers elsewhere in the genome are likely to further regulate risk and presentation even in the face of major effect alleles. Along these lines recent work has complicated the relatively simple interpretation that deletions involving the distal portion of 22q are attributable to the postsynaptic scaffolding molecule SHANK3 [48]. Despite the fact that losses encompassing SHANK3 variants are present in an estimated 1% of ASD cases [95], subsequent identification of comparable deletions in typically developing children [61] suggests that additional, as yet unknown, factors must be present to give rise to disease. Similarly, although cytogenetic variation at 7q first identified a possible involvement of the cell adhesion molecule *CNTNAP2* in the modulation of neuropsychiatric phenotypes [136], subsequent work suggests that heterozygous disruption of this gene can be observed in phenotypically normal individuals [14]. As reviewed elsewhere [140], other important regions including 2q37, 5p15, 17p11, and Xp22 shown to harbor cytogenetic lesions in multiple ASD cases represent important targets for further evaluation.

23.4.3 Linkage

Linkage studies have yield mixed results with regards to the identification of ASD loci [3], but the approach was invaluable in the localization of genes underlying ASD-related syndromes (see Table 23.4.2 and below). Genetic heterogeneity likely underlies the observation that increased sample size appears to confer only modest gains in autism linkage studies (Fig. 23.4.2). For example, no locus met criteria for genome-wide significance in the largest linkage study published to date [128]. And yet regions highlighted by this study, including 11p12-p13, may prove to be interesting in prioritization of emerging candidate genes. Other loci have been identified and replicated in multiple studies (e.g., 7q and 17q11–17q21) [9, 33, 130], and although contributory variants have been observed at each of 7q [4, 7, 13, 23, 30, 58, 87, 114] and 17q [126, 127, 143], no single common variant in either region can account for the observed linkage signals. These results suggest that numerous alleles, potentially in multiple different genes, are likely contributory. Strategies designed to enrich for homogeneity amongst cases [28, 96, 124] and employ ASD-related endophenotypes such as cognitive flexibility, social behavior, or language performance [4, 43, 49, 145] are likely to prove useful in future work employing a range of approaches.

23.4.4 Syndromic ASDs

Multiple rare genetic syndromes have been associated with the ASDs [148]. An important distinction here, however, is that in contrast to the cytogenetically recovered loci discussed above, the availability of relatively larger numbers of unrelated probands has permitted the identification of the underlying molecular deficit in a growing number of such entities (Table 23.4.2). Best known is fragile X syndrome, with mutations in the gene for the RNA-binding protein *FMR1* observed in 1–2% of individuals with an ASD. Because only ~25% of boys with *FMR1* mutations meet diagnostic criteria for an ASD, results again suggest that multiple factors are at play. Results are similar for individuals with tuberous sclerosis, which is attributable to autosomal dominant mutations in either *TSC1* or *TSC2*. Although *TSC1/2* mutations are present in an estimated 1% of ASD cases, amongst individuals

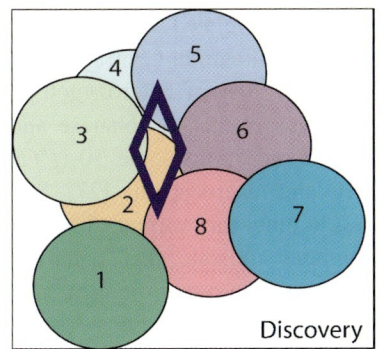
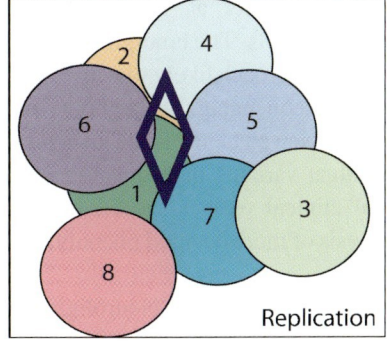

Fig. 23.4.2 Extreme heterogeneity is likely to impede replication of true risk alleles. Known and yet to be identified ASD risk alleles (*colored circles*) each appear to contribute to only a subset of cases. Although allele frequencies are stable across populations (*outer squares*), enormous variation is observed when sampling is incomplete (*blue diamonds*). Thus, alleles that appear to be overrepresented amongst cases in a discovery cohort (*orange; left*) may fail to replicate on follow-up (*orange; right*). Larger cohorts and statistical methods designed to address this heterogeneity will be required. (Concept and execution by Veronia J. Vieland, PhD, Battelle Center for Mathematical Medicine, The Research Institute at Nationwide Children's Hospital)

with tuberous sclerosis only ~20% present with an ASD. Angelman syndrome, most often resulting from the *de novo* loss of the maternal allele at 15q11-13, is likewise observed in ~1% of ASD cases. In contrast to the near 1:1 complete correspondence between maternal *duplications* at this locus and an ASD diagnosis, however, maternal *deletions* appear to give rise to autism in only an estimated 40% of carriers. Additional syndromes, including Down syndrome and neurofibromatosis, show a lower incidence of autistic disorder (~6 and 4%, respectively) but are significantly elevated over baseline levels, which are estimated at ~0.2%. Increased penetrance with regard to a spectrum diagnosis has been reported for less frequently encountered clinical entities, but additional cases will be required here to obtain stable estimates. All told, known syndromic conditions such as those referenced in Table 23.4.2 some of are likely to account for ~15–20% of ASD cases. Identification of the molecules underlying some of these conditions has enabled functional investigation of defined signaling pathways. Discussed below are separate findings obtained through the use of reverse genetics in which genotypes are used to define patient groups with subsequent effort directed towards understanding the associated phenotype.

23.4.5 Re-sequencing

Sequencing of candidate loci, based on linkage or cytogenetic findings [70], along with genome-wide characterization of structural variation [see below] has likewise identified rare variants at multiple loci enriched in cases versus controls. The emerging "genotype first paradigm" is an approach likely to gain even further prominence as sequencing throughput is increased. Also key and reviewed below is the growing recognition that individual variants map only imprecisely onto individual clinical entities, necessitating the collection and analysis of more detailed phenotypic data in cases and controls.

Identification of coding mutations in Neuroligin 3 (*NLGN3*) and Neuroligin 4 (*NLGN4X*) was an important advance that served to focus attention towards the synapse [149]. It was in this context, for example, in which rare missense mutations in the neuroligin interactor Neurexin 1 (*NRXN1*) [25] were identified in cases [50]. Subsequent recognition of de novo deletions at *NRXN1* reinforced the importance of variation at this locus [128] and serve to support an important role for Neuroligin-Neurexin signaling in the ASDs. Such data also helped to make sense of frameshift mutations [68, 125] in the neurexin family member *CNTNAP2* [105]. Recessive inheritance of such mutations in the Amish present with a congenital epilepsy characterized by developmental brain abnormalities, language regression, and an ASD diagnosis in a majority of cases. Likewise, the identification of *SHANK3* as an important modulator of risk [48] was facilitated by the biological plausibility of this candidate given the demonstrated role of other molecules involved in synaptic function. Although these and other rare variants have been important clues in ASD pathophysiology, they are infrequent causes of the ASDs [13, 147]. Given that events in these molecules can also be associated with a diverse number of outcomes including: intellectual disability without features of autism [80], schizophrenia [75, 111], Tourette syndrome [81, 136] and typical development [61, 74], variation at individual loci is unlikely to provide sufficient information for accurate prediction of clinical presentation. The converse – that joint consideration of multiple loci along with relevant environmental risk factors will increase information content – is also true and will be discussed below.

As is hopefully clear from the above discussion, the identification of *NLGN3/4* mutations was an important advance in ASD. At the same time, this discovery has come to so dominate the field that appropriate consideration of molecules operating in distinct signaling pathways has arguably suffered as a result. Support for PI3K-AKT related molecules, for example, merits increased attention. Central here is the finding that an estimated 20% of heterozygous carriers of germline mutations in the phosphatase and tensin homologue deleted on chromosome 10 (*PTEN*) present with an ASD and macrocephaly [27]. Strikingly, and in contrast to other rare variants discussed thus far, available estimates indicate that heterozygous mutations in *PTEN* may be present in as many as 10% of individuals with an ASD diagnosis and macrocephaly [101, 135]. That the receptor tyrosine kinase *MET* (discussed below) and each of *TSC1* and *TSC2* (discussed above) likewise act on PI3K signaling is notable [82]. Recent evidence tying *CACNA1C*, the L-type calcium channel subunit mutated in Timothy syndrome [120], to this same pathway [35] provides additional potential coherence at the molecular level. Whether or not additional calcium channels implicated in the ASDs (e.g., *CACNA1G* [126], and *CACNA1H* [121]) modulate risk by acting on this same pathway remains to be seen. Current notions of these pathways are likely to be primitive but may nevertheless prove useful in organizing emerging results.

MECP2, the gene mutated in Rett syndrome, likewise appears to provide important links between key molecular entities, a somewhat unexpected finding given the phenotypic differences between girls with this disorder and those with autism. That males harboring *MECP2* duplications [134] present with MR suggests that subtle dysregulation of gene expression in either direction can interfere with normal brain development. A relationship to the 15q11-13 locus is supported by the observation that a subset of *MECP2* mutation carriers present clinically with an Angelman-like phenotype [93]. Additional convergence comes from the fact that genes within the Angelman region are dysregulated not only in Rett Syndrome but also in idiopathic autism [115]. Finally, *MECP2* can also be connected to *CADPS2*, a Ca2+ dependent modulator of *BDNF* release, in which rare missense variants are overrepresented in cases relative to controls [114]. The finding that *BDNF* is a direct target of *MECP2* [39], and observed to be an important modulator of disease progression in a mouse model of Rett syndrome [37], provides important additional support for the notion that these independently identified molecules may operate together at a functional level.

Examination of familial segregation of mutations at loci identified by re-sequencing suggests that individual variants are largely insufficient to independently give rise to an ASD. For example, although mutations in axonal initial segment localized sodium channel subunits *SCN1A* and *SCN2A* were seen to be significantly overrepresented in cases relative to controls [142], such variants were observed both in affected children and in their parents. A similar overrepresentation of rare missense variants in the T-type calcium channel subunit *CACNA1H* has likewise been observed in cases compared with controls [121]. Despite the demonstration that individual variants altered channel activity, mutations were again seen both in cases and unaffected relatives. The point here is not to argue that these rare variants are unrelated to disease, but rather underscore the point that many (if not all) risk variants show an imprecise mapping onto affection status. As elegantly noted elsewhere, "[O]nly when penetrances are well above 50% does one approach a familial concentration that begins to look like a standard Mendelian segregation" [20].

23.4.6 Copy Number Variation

Most of the points raised above in the context of re-sequencing are also applicable to analyses of structural variation. Although resolution is much reduced relative to sequencing, the entire genome can be interrogated in parallel using array-based technologies, obviating the need for candidate-driven strategies. Considerable excitement was recently garnered by the identification of recurrent de novo variation at 16p11, observed in an estimated 1% of cases [78, 90, 144]. Subsequent identification of duplication and deletion events in healthy controls [17, 61], however, points again to oligogenic mechanisms, incomplete penetrance, and variable expressivity, consistent with genetic complexity.

Another relatively new finding is that of individuals harboring characteristic 1.5-Mb deletions at 15q13.3 involving *CHRNA7* and the subsequent elucidation of a relationship to intellectual ability and epileptiform abnormalities [118]. Again, however, diverse clinical manifestations of this single variant have been reported, with carriers also showing either schizophrenia [67, 122], autism [15], or typical development [133]. These results underscore the need to understand individual variants in the context of modifiers which may shape presentation.

Other recently published work [26, 29, 61] points to additional loci of interest. Cai et al. [29] determined that duplications encompassing the *ASMT* gene, the last enzyme required for melatonin synthesis, were elevated in cases versus ethnically matched controls (p=0.003 by Fisher's exact test). These results are intriguing in the face of a related study highlighting the relationship between common variation at this locus and each of autism and melatonin levels [92]. Glessner et al. [61] report enrichment of CNVs proximal to each of neuronal cell-adhesion and ubiquitin-degradation molecules. Novel ASD loci were also identified, including a region at 2p24.3 near the uncharacterized cDNA AK123120, which was observed to harbor variants in cases more frequently than in controls (p=3.6×10^{-6}; OR=5.5). Our own investigations [26] prioritized exonic CNVs in an attempt to identify variants most likely to interfere with gene function. These analyses led to case-specific in both *BZRAP1*, an adaptor molecule known to regulate synaptic transmission ($p = 2.3 \times 10^{-5}$) and *MDGA2* ($p = 1.3 \times 10^{-4}$), a cell adhesion molecule with striking structural similarity to Contactin 4 [51, 110]. Each of these results is intriguing and should be explored in additional cohorts. That hundreds of distinct rare variants were each seen only in a single case, however, suggests that massive cohorts will be required to identify the subset of rare alleles relevant to disease.

Unlike re-sequencing studies discussed above, it can be challenging in CNV analyses to define individual genes within regions of interest that may be contributory. This is made all the more complicated by the fact that clinical outcomes may in some cases be attributable to multiple genes within a region. Moreover, multiple distinct variants – likely to differ in effect size and function – can be observed at particular loci of interest (Figs. 23.4.3 and 23.4.4). Results obtained from sequencing candidate genes at 16p11 [79] are consistent with independent contributions from multiple genes including *SEZ6L2*.

International consortia established to make such data available to investigators online will facilitate interpretation of how rare variants come to shape presentation. Such work will likewise benefit from the ongoing collection and characterization of large patient cohorts, particularly when associated data and biomaterials are made available to the entire research community, as is the case with the Autism Genetics Resource Exchange (AGRE) [57].

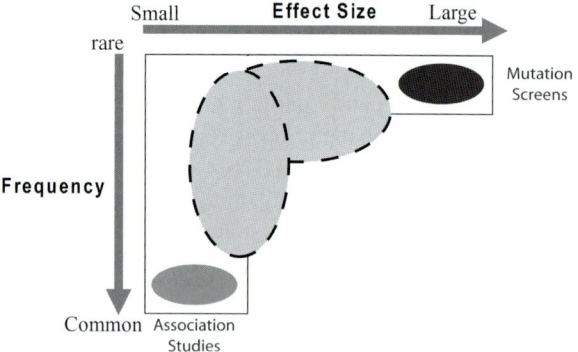

Fig. 23.4.3 Current studies typically attend to two extremes on the allelic spectrum. SNPs with modest effects sizes (*small gray oval*) have been identified through association studies. Rare events that segregate with disease have also been identified through re-sequencing efforts and CNV studies (*small black oval*). Less attention has been directed at events not at either extreme, but also likely to modulate risk. Larger studies will permit consideration of intermediate effect alleles (*ovals with dotted lines*). Although available data do not support the presence of common alleles of large effect for affection status (*empty bottom right corner*) it is possible that such associations may be observed in the future when disease-related endophenotypes, measuring aspects of social behavior or language performance are employed as quantitative endpoints

To round out this discussion of rare variations it should be emphasized again that most individuals harbor multiple rare variants and the presence of individual events in a single patient with a common disease need not be meaningful. That de novo events are observed in ~1% of controls [116] is important here and demands caution in the attribution of observed effects to individual variants. Nevertheless, that such events are increased in frequency in simplex cases (7–10%) suggests that as a group such variants are indeed contributory [90, 116]. The same concern arises in consideration of rare point mutations. For example, it is well established that premature truncation in the

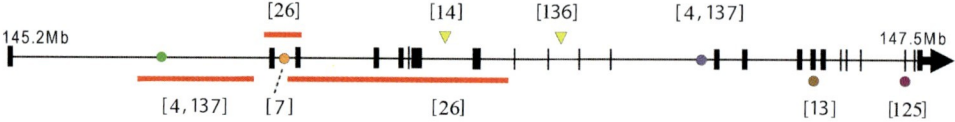

Fig. 23.4.4 Single genes are likely to harbor a variety of functionally distinct alleles. Individuals homozygous for a truncating mutation in *CNTNAP2* (*purple circle*) present with intellectual disability, cortical dysplasia, focal epilepsy, and behavioral abnormalities [125]. Chromosomal translocations at this locus [14, 136] have also been observed (*yellow arrowheads*) as have CNVs (*red lines*) which eliminate exons [26] or an apparently functional *FOXP2*-binding site (*green circle*) [4, 137]. Although potentially contributory, none of these variants are sufficient in the heterozygous state to cause disease, as each has been observed in unaffected parents. Less clear, but of substantial interest, are rare amino acids substitutions (e.g., I869T; *brown circle*) observed in cases [13]. In support of still further complexity, separate SNPs within *CNTNAP2* have been associated with different endpoints. rs7794745 (*orange circle*) is associated with affection status [7], whereas rs2710102 (*blue circle*) is associated with language-related measures [4, 137]. Given that each of these risk alleles is common it may be that the two interact in at least some individuals. That similar effects have been described at other loci suggests that the relationship between genetic variation and outcome is complex even when considering only single genes. Numbers in square brackets identify studies listed in the reference section in which variants were identified or characterized

X-linked calcium channel subunit *CACNA1F* results in congenital night blindness. Less straightforward, however, is whether a gain of function allele at this locus contributes to intellectual disability and autism in a subset of males in from a large family New Zealand [65]. Involvement is plausible, given evidence implicating other related calcium channel subunits, [e.g. 120] but the question is one that requires additional attention. Clear to us is that an integration of results from risk loci across the genome will be required for accurate risk assessment within individuals.

23.4.7 Common Variation

Many candidate genes have been studied, but few replicated. In addition to loci reviewed elsewhere [3], linkage and association at *PRKCB1* support potential involvement in disease risk [103]. Results are particularly compelling given replication in an independent cohort [85]. Additional regionally directed screens point to common variation within *CACNA1G* [126], a T-type voltage gated calcium channel subunit, as well as *DOCK4* [87] observed to regulate dendritic morphogenesis in the hippocampus [132]. As emphasized in these reports and discussed above, however, observed associations are insufficient to account for linkage.

Exciting additional clues are also emerging from initial analyses of recently published GWA data [141]. Wang et al. employed Illumina HumanHap550 data for 2,000 Caucasian cases and contrasted allele frequencies against almost 6,500 unrelated controls of European origin. Not only was a SNP at 5p14.1 found to meet genome-wide significance (rs4307059, $p = 3.4 \times 10^{-8}$), but additional SNPs within a 100 Kb linkage block also showed P-values less than 1×10^{-4}. Importantly, association at this intergenic locus between cadherin 9 and cadherin 10 was confirmed in an additional two independent sets with combined p-values ranging from 7.4×10^{-8} to 2.1×10^{-10}. As is the case with all common variants associated with disease, much work will be required to understand how variation within this intergenic region may act to impact presentation. Of potential utility, however, is the observation that *CDH10* is present at high levels in the developing frontal cortex, a result similar to that observed previously for *CNTNAP2* [2, 4].

Qualitatively similar results at a nonoverlapping locus were obtained from analysis of 1,000 partially overlapping multiplex families genotyped on Affymetrix 5.0 arrays [6]. Although no associations in this initial cohort met genome-wide significance (rs10513025, $p = 7.3 \times 10^{-7}$), focused follow-up in additional families resulted in a genome-wide significant effect ($P = 6 \times 10^{-9}$) at an intergenic region between *TAS2R1* and *SEMA5A*. Although the identity of the underlying causal variant again remains unclear, differential expression of *SEMA5A* in lymphoblasts from cases versus controls supports the hypothesis that the GWAS-identified variant modulates risk by altering transcript levels. This result is of particular interest together with an established interplay between *SEMA5A* and *PLXNB3* [8], in which common variation has been associated with language performance and a reduction in white matter volume [112]. That PLXNB3 likewise interacts with ASD-associated MET [31, 32, 40] suggests that these results may serve to further PBK-AKT refine the discussed above.

Availability of these genome-wide data should also obviate the need for exploratory studies at individual loci. Because different results were obtained from each study despite partially overlapping cohorts, it remains debatable, however, the extent to which these initial results will predict future observations. Differences between the arrays employed and SNPs interrogated are certainly important contributory factors here. Incomplete sampling is also likely to be an important issue (Fig. 23.4.2), and calls for the assembly and analysis of larger cohorts. Although these and other SNPs yet to be identified may only modulate overall risk in subtle ways, the impact on population prevalence is likely to be substantial. For each, identification of the underlying causal variant(s) tagged by the anonymous markers interrogated represents an important next step. If results at *CNTNAP2* (Fig. 23.4.4) generalize to other parts of the genome, individual genes may harbor multiple variants, common and rare, with strongest associations to distinct aspects of presentation [4, 7, 137].

23.4.8 Towards Convergence

It is critical to consider, as we have emphasized above, scenarios in which individual genetic variants are not strongly associated with an ASD diagnosis, but rather to intermediate phenotypes that collectively define ultimate presentation. This paradigm is strongly supported by the observation that first-degree relatives of autistic probands are often enriched for features of the "broader autism phenotype" [21, 83, 104, 145] characterized by subclinical language dysfunction, autistic-like social abnormalities, or increased behavioral rigidity relative to unrelated controls. That study of such intermediate phenotypes [62, 109] has proven useful in the identification of ASD risk loci [38, 117] provides further support for such models. In addition, this model predicts that such variants will contribute both to normal variation and risk for neurodevelopmental conditions that are considered clinically distinct [55]. This is certainly supported by work at the *CNTNAP2* locus which appears modulates language function across in both the ASDS and SLI [4, 137].

From this perspective, the ASDs may best be conceptualized as the end result of multiple rare and common alleles that act in combination to shape different aspects of cognition and behavior (Fig. 23.4.5). Available data are further consistent with the notion that current diagnostic classifications do not adequately capture the underlying etiologies [24, 42, 137]. Careful examination of patient records supports a similar interpretation – that individual genetic risk factors may ultimately predispose to a range of related clinical conditions [113]. As discussed elsewhere [1], this is particularly important in the context of the developing brain, whose circuits should not be expected a priori to show good correspondence to clinically defined disease boundaries.

Thus, genes for which a relationship to ASD is established should be considered as potential candidates for related disorders of cognition including specific language impairment, schizophrenia, bipolar, and mental retardation (and vice versa). Evaluation of how individual variants linked to different disorders are related to underlying endophenotypes will be of criti-

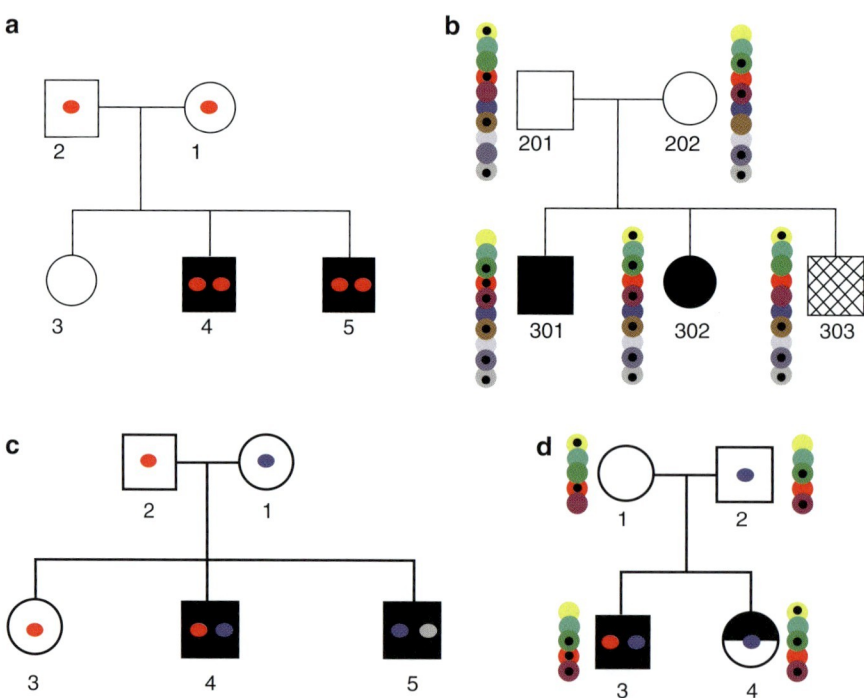

Fig. 23.4.5 Genetic models should conform to available data. Examples of rare alleles which segregate with disease in a Mendelian fashion (**a**) and common variants associated with smaller effects (**b**) are well established. Less widely considered, however, are scenarios in which rare variants of major effect show incomplete penetrance (**c**). Deletions at *NRXN1* and 16p11 appear to fall in this category with at least a subset of carriers showing normal development. Illustrated in **C** is a family in which no rare variant of major effect (*red, blue,* or *gray circle*) appears sufficient to cause disease; instead, all affected individuals carry multiple risk alleles. More complicated still, and most likely with regards to the ASDs, is a scenario in which alleles of varying effect interact to shape presentation and determine affection status (**d**). *Circles* and *squares* correspond to females and males, respectively. *Filled* or *hatched* shapes highlight affected individuals. *Colored circles* alongside or within individuals represent alleles of minor and major effect, respectively

cal importance for explaining the relationships between such conditions. It should be emphasized also that the ever-growing number of genes (Table 23.4.3) makes it increasingly difficult to keep track of the manner by which different kinds of variation contribute to risk. Fortunately, however, it seems that at least a subset of molecules now implicated in pathogenesis can be connected at the level of molecular signaling. Emerging

Table 23.4.3 Evidence scores for *promising* and for *probable* ASD genes[a] (modified from [3], with permission)

Gene	Total score	Syndrome/ mutations	Replicated association	Analysis of variant	Mouse model	Other evidence
Promising						
BZRAP1	1	0	1	0	0	
CACNA1G	1	0	1	0	0	No
DISC1	1	0	0	0	1	No
DOCK4	1	0	1	0	0	Involved in regulation of hippocampal dendrite morphology
ITGB3	1	0	1	0	0	No
MDGA2	1	0	1	0	0	No
PRKCB1	1	0	1	0	0	No
AHI1	2	2	0	0	0	No
ASMT	2	0	1	0	0	ASD-associated variant also associated with melatonin levels
AVPR1A	2	0	0	0	1	Dysregulation in plasma of cases versus controls
CACNA1H	2	0	1	1	0	No
CNTN4	2	2	0	0	0	No
GRIK2	2	0	1	0	0	Homozygous mutation results in nonsyndromic MR
SCN2A	2	0	1	1	0	No
SLC25A12	2	0	1	0	0	Associated with neurite outgrowth; up-regulated ASD brain
Probable						
EN2	3	0	1	1	1	No
MET	3	0	1	1	0	Expression reduced in brains of cases versus controls
NRXN1	3	2	0	0	0	Interacts functionally with neuroligins
OXTR	3	0	1	0	1	Expression reduced in blood of cases versus controls
SHANK3	3	2	0	0	0	Modulates glutamate-dependent reconfiguration of dendritic spines
SLC6A4	3	0	1	1	0	Clinical benefit from inhibitors; variation linked to gray matter volume
CACNA1C	4	2	0	1	0	Linked to PI3K signaling along with *PTEN/TSC1/TSC2*
CADPS2	4	2	0	1	1	No
CNTNAP2	4	2	1	0	0	Downstream target of *FOXP2*
DHCR7	4	2	0	1	0	Hypocholesterolemia in a proportion of probands
FMR1	4	2	0	1	1	No
NLGN3	4	2	0	1	1	No
NLGN4X	4	2	0	1	1	No
GABRB3	5	2	1	0	1	Expression is dysregulated in PDDs
MECP2	5	2	0	1	1	*MECP2* deficiency causes reduced expression of *UBE3A* and *GABRB3*
PTEN	5	2	0	1	1	Linked to PI3K signaling along with *TSC1/TSC2/CACNA1C*
TSC1	5	2	0	1	1	Regulates dendrite morphology and function of glutamatergic synapses
TSC2	5	2	0	1	1	Regulates dendrite morphology and function of glutamatergic synapses
UBE3A	5	2	0	1	1	Expression is dysregulated in PDDs
RELN	6	2	1	1	1	Levels reduced in brains of cases versus controls

(continued)

Table 23.4.3 (continued)

AHI1 Jouberin, ASD autism spectrum disorder, *ASMT* acetylserotonin O-methyltransferase, *AVPR1A* vasopressin V1a receptor, *BZRAP1* benzodiazapine receptor (peripheral), associated protein 1, *CACNA1C* calcium channel, voltage-dependent, L type, alpha 1C subunit, *CACNA1G* calcium channel, voltage-dependent, T type, alpha 1G subunit, *CACNA1H* calcium channel, voltage-dependent, T type, alpha 1H subunit, *CDH9* cadherin 9, *CDH10* cadherin 10, *CADPS2* calcium-dependent secretion activator, 2, *CNTN4* contactin 4, *CNTNAP2* contactin-associated protein-like 2 precursor, *DHCR7* 7-dehydrocholesterol reductase, *DISC1* disrupted in schizophrenia 1, *DOCK4* dedicator of cytokinesis 4, *EN2* homeobox protein engrailed-2, *FMR1* fragile X mental retardation 1 protein, *GABRB3* gamma aminobutyric acid receptor subunit beta-3 precursor, *GRIK2* glutamate receptor, ionotropic kainate 2 precursor, *ITGB3* integrin beta-3 precursor, *MDGA2* MAM domain containing glycosylphosphatidylinositol anchor 2, *MECP2* methyl-CpG-binding protein 2, *MET* met proto-oncogene, *MR* mental retardation, *NLGN3* Neuroligin 3, *NLGN4X* Neuroligin-4, X-linked precursor, *NRXN1* Neurexin-1, *OXTR* oxytocin receptor, PDDs pervasive developmental disorders, *PRKCB1* protein kinase C, beta 1, *PTEN* phosphatase and tensin homolog, *RELN* Reelin precursor, *SCN2A* sodium channel, voltage-gated, type II, alpha subunit, *SEMA5A* sema domain, seven thrombospondin repeats (type 1 and type 1-like), transmembrane domain (TM) and short cytoplasmic domain, (semaphorin) 5A, *SHANK3* SH3 and multiple ankyrin repeat domains protein 3, *SLC6A4* sodium-dependent serotonin transporter, *SLC25A12* calcium-binding mitochondrial carrier protein Aralar1, *TSC1* hamartin, *TSC2* tuberin, *UBE3A* ubiquitin-protein ligase E3A.

ᵃBecause the relationship to disease is most clear for rare variants, we biased our scoring accordingly. Genes associated with an ASD-linked syndrome or mutation resulted in 2 points, whereas other lines of evidence resulted in 1 point. Observation of a rare variant at a particular locus (CNV or bp) was insufficient for inclusion here; the additional requirement of statistical enrichment relative to controls was also necessary. We also excluded regions of clear interest (*e.g.* 1q21) for which the role of individual genes remains ambiguous. Likewise, although compelling evidence supports involvement of regions proximal to each of *AK123120* [61], *CDH9/10* [141], *NHE9* [97] and *SEMA5A* [6] it remains to be determined whether these nearby genes are involved in modulation of risk. To qualify as a mouse model, two out of three "core features" were required to be present. Mean score for the 35 genes included here was 2.97, with a standard deviation of 1.50. We assigned genes with scores 3 or greater ($n=20$) as probable ASD genes and those with scores less than 3 ($n=15$) as possible ASD genes. Although we have done our best to systematically and comprehensively evaluate available data for each of the different molecules discussed, we recognize that these evidence scores are largely arbitrary. Emerging genome-wide data suggests that a large number of molecules are likely to modulate risk in a variety of different ways. Despite unintended omissions of genes likely to be important, we hope this updated table might continue to serve as a starting point for discussion.

data are consistent with a variety of models, although it is increasingly difficult to defend single-gene explanations, even in rare situations. Finally, although the concept of genetic complexity is widely accepted, the logical consequences of this disease architecture, integrating both rare and common variation, must be more widely embraced for necessary progress to occur.

Acknowledgements We gratefully acknowledge the families who have made work in this field possible, as well as the vision and leadership of AGRE and Autism Speaks. We are similarly indebted to the investigators whose work drives this field forward, many of whom we were unable to cite owing to space limitations. Work in the Geschwind laboratory is supported by the National Institute of Mental Health (STAART - U54 MH68172; ACE - P50 HD055784; ACE - MH081754; AGRE R01 MH64547; R37 MH60233).

References

1. Abrahams BS, Geschwind DH (2009) Connecting genes to brain in the autisms. Arch Neurol (in press)
2. Abrahams BS, Tentler D, Perederiy JV, Oldham MC, Coppola G, Geschwind DH (2007) Genome-wide analyses of human perisylvian cerebral cortical patterning. Proc Natl Acad Sci USA 104(45):17849–17854
3. Abrahams BS, Geschwind DH (2008) Advances in autism genetics: on the threshold of a new neurobiology. Nat Rev Genet 9(5):341–355
4. Alarcon M, Abrahams BS, Stone JL et al (2008) Linkage, association, and gene-expression analyses identify CNTNAP2 as an autism-susceptibility gene. Am J Hum Genet 82(1):150–159
5. Amir RE, Van den Veyver IB, Wan M, Tran CQ, Francke U, Zoghbi HY (1999) Rett syndrome is caused by mutations in X-linked MECP2, encoding methyl-CpG-binding protein 2. Nat Genet 23(2):185–188
6. Weiss, Arking DE on behalf of the Gene Discovery Project of John Hopkins and the Autism Consortium (2009) A genome-wide linkage and association scan reveals novel loci for autism. Nature (in press)
7. Arking DE, Cutler DJ, Brune CW et al (2008) A common genetic variant in the neurexin superfamily member CNTNAP2 increases familial risk of autism. Am J Hum Genet 82(1):160–164
8. Artigiani S, Conrotto P, Fazzari P et al (2004) Plexin-B3 is a functional receptor for semaphorin 5A. EMBO Rep 5(7):710–714
9. Badner JA, Gershon ES (2002) Regional meta-analysis of published data supports linkage of autism with markers on chromosome 7. Mol Psychiatry 7(1):56–66
10. Baker P, Piven J, Sato Y (1998) Autism and tuberous sclerosis complex: prevalence and clinical features. J Autism Dev Disord 28(4):279–285

11. Bailey A, Le Couteur A, Gottesman I et al (1995) Autism as a strongly genetic disorder: evidence from a British twin study. Psychol Med 25(1):63–77
12. Baird G, Charman T, Pickles A et al (2008) Regression, developmental trajectory and associated problems in disorders in the autism spectrum: the SNAP study. J Autism Dev Disord 38(10):1827–1836
13. Bakkaloglu B, O'Roak BJ, Louvi A et al (2008) Molecular cytogenetic analysis and resequencing of contactin associated Protein-Like 2 in autism spectrum disorders. Am J Hum Genet 82(1):165–173
14. Belloso JM, Bache I, Guitart M et al (2007) Disruption of the CNTNAP2 gene in a t(7;15) translocation family without symptoms of Gilles de la Tourette syndrome. Eur J Hum Genet 15(6):711–713
15. Ben-Shachar S, Lanpher B, German JR (2009) Microdeletion 15q13.3: a locus with incomplete penetrance for autism, mental retardation, and psychiatric disorders. J Med Genet 46(6):382–388
16. Biederer T, Sudhof TC (2000) Mints as adaptors. Direct binding to neurexins and recruitment of munc18. J Biol Chem 275(51):39803–39806
17. Bijlsma EK, Gijsbers AC, Schuurs-Hoeijmakers JH (2009) Extending the phenotype of recurrent rearrangements of 16p11.2: deletions in mentally retarded patients without autism and in normal individuals. Eur J Med Genet 52:77–87
18. Bishop DV, Maybery M, Maley A, Wong D, Hill W, Hallmayer J (2004) Using self-report to identify the broad phenotype in parents of children with autistic spectrum disorders: a study using the autism-spectrum quotient. J Child Psychol Psychiatry 45(8):1431–1436
19. Bittel DC, Kibiryeva N, Butler MG (2006) Expression of 4 genes between chromosome 15 breakpoints 1 and 2 and behavioral outcomes in Prader–Willi syndrome. Pediatrics 118(4):e1276–e1283
20. Bodmer W, Bonilla C (2008) Common and rare variants in multifactorial susceptibility to common diseases. Nat Genet 40(6):695–701
21. Bolton P, Macdonald H, Pickles A et al (1994) A case-control family history study of autism. J Child Psychol Psychiatry 35(5):877–900
22. Bonati MT, Russo S, Finelli P et al (2007) Evaluation of autism traits in Angelman syndrome: a resource to unfold autism genes. Neurogenetics 8(3):169–178
23. Bonora E, Beyer KS, Lamb JA et al (2003) Analysis of reelin as a candidate gene for autism. Mol Psychiatry 8(10):885–892
24. Boteva K, Lieberman J (2003) Reconsidering the classification of schizophrenia and manic depressive illness–a critical analysis and new conceptual model. World J Biol Psychiatry 4(2):81–92
25. Boucard AA, Chubykin AA, Comoletti D, Taylor P, Sudhof TC (2005) A splice code for trans-synaptic cell adhesion mediated by binding of neuroligin 1 to alpha- and beta-neurexins. Neuron 48(2):229–236
26. Bucan M, Abrahams BS, Wang K (2009) Genome-wide analyses of exonic copy number variants in a family-based study point to novel autism susceptibility genes. PloS Genetics 5:e1000536
27. Butler MG, Dasouki MJ, Zhou XP et al (2005) Subset of individuals with autism spectrum disorders and extreme macrocephaly associated with germline PTEN tumour suppressor gene mutations. J Med Genet 42(4):318–321
28. Buxbaum JD, Silverman J, Keddache M et al (2004) Linkage analysis for autism in a subset families with obsessive-compulsive behaviors: evidence for an autism susceptibility gene on chromosome 1 and further support for susceptibility genes on chromosome 6 and 19. Mol Psychiatry 9(2):144–150
29. Cai G, Edelmann L, Goldsmith JE et al (2008) Multiplex ligation-dependent probe amplification for genetic screening in autism spectrum disorders: Efficient identification of known microduplications and identification of a novel microduplication in ASMT. BMC Med Genomics 1:50
30. Campbell DB, Sutcliffe JS, Ebert PJ et al (2006) A genetic variant that disrupts MET transcription is associated with autism. Proc Natl Acad Sci USA 103(45):16834–16839
31. Campbell DB, D'Oronzio R, Garbett K et al (2007) Disruption of cerebral cortex MET signaling in autism spectrum disorder. Ann Neurol 62(3):243–250
32. Campbell DB, Buie TM, Winter H et al (2009) Distinct genetic risk based on association of MET in families with co-occurring autism and gastrointestinal conditions. Pediatrics 123(3):1018–1024
33. Cantor RM, Kono N, Duvall JA et al (2005) Replication of autism linkage: fine-mapping peak at 17q21. Am J Hum Genet 76(6):1050–1056
34. Cantor RM, Yoon JL, Furr J, Lajonchere CM (2007) Paternal age and autism are associated in a family-based sample. Mol Psychiatry 12(5):419–421
35. Catalucci D, Zhang DH, DeSantiago J et al (2009) Akt regulates L-type Ca2+ channel activity by modulating Cavalpha1 protein stability. J Cell Biol 184(6):923–933
36. CDC (2007) Prevalence of autism spectrum disorders - Autism and developmental disabilities monitoring network. MMWR Surveill Summ 56:1–28
37. Chang Q, Khare G, Dani V, Nelson S, Jaenisch R (2006) The disease progression of Mecp2 mutant mice is affected by the level of BDNF expression. Neuron 49(3):341–348
38. Chen GK, Kono N, Geschwind DH, Cantor RM (2006) Quantitative trait locus analysis of nonverbal communication in autism spectrum disorder. Mol Psychiatry 11(2):214–220
39. Chen WG, Chang Q, Lin Y et al (2003) Derepression of BDNF transcription involves calcium-dependent phosphorylation of MeCP2. Science 302(5646):885–889
40. Conrotto P, Corso S, Gamberini S, Comoglio PM, Giordano S (2004) Interplay between scatter factor receptors and B plexins controls invasive growth. Oncogene 23(30):5131–5137
41. Cook EH Jr, Lindgren V, Leventhal BL et al (1997) Autism or atypical autism in maternally but not paternally derived proximal 15q duplication. Am J Hum Genet 60(4):928–934
42. Craddock N, O'Donovan MC, Owen MJ (2006) Genes for schizophrenia and bipolar disorder? Implications for psychiatric nosology, Schizophr. Bull 32(1):9–16
43. Delorme R, Gousse V, Roy I et al (2007) Shared executive dysfunctions in unaffected relatives of patients with autism and obsessive-compulsive disorder. Eur Psychiatry 22(1):32–38
44. Dempster EL, Toulopoulou T, McDonald C et al (2006) Episodic memory performance predicted by the 2bp deletion in exon 6 of the "alpha 7-like" nicotinic receptor subunit gene. Am J Psychiatry 163(10):1832–1834
45. Descheemaeker MJ, Govers V, Vermeulen P, Fryns JP (2006) Pervasive developmental disorders in Prader–Willi

45. syndrome: the Leuven experience in 59 subjects and controls. Am J Med Genet 140(11):1136–1142
46. Deykin EY, MacMahon B (1979) Viral exposure and autism. Am J Epidemiol 109(6):628–638
47. Doornbos M, Sikkema-Raddatz B, Ruijvenkamp CA (2009) Nine patients with a microdeletion 15q11.2 between breakpoints 1 and 2 of the Prader–Willi critical region, possibly associated with behavioural disturbances. Eur J Med Genet 52(2-3):108–115
48. Durand CM, Betancur C, Boeckers TM et al (2007) Mutations in the gene encoding the synaptic scaffolding protein SHANK3 are associated with autism spectrum disorders. Nat Genet 39(1):25–27
49. Duvall JA, Lu A, Cantor RM, Todd RD, Constantino JN, Geschwind DH (2007) A quantitative trait locus analysis of social responsiveness in multiplex autism families. Am J Psychiatry 164(4):656–662
50. Feng J, Schroer R, Yan J et al (2006) High frequency of neurexin 1beta signal peptide structural variants in patients with autism. Neurosci Lett 409(1):10–13
51. Fernandez T, Morgan T, Davis N et al (2004) Disruption of contactin 4 (CNTN4) results in developmental delay and other features of 3p deletion syndrome. Am J Hum Genet 74(6):1286–1293
52. Fine SE, Weissman A, Gerdes M et al (2005) Autism spectrum disorders and symptoms in children with molecularly confirmed 22q11.2 deletion syndrome. J Autism Dev Disord 35(4):461–470
53. Folstein SE, Rutter ML (1988) Autism: familial aggregation and genetic implications. J Autism Dev Disord 18(1):3–30
54. Fombonne E (2009) Epidemiology of pervasive developmental disorders. Pediatr Res 65(6):591–598
55. Geschwind DH (2008) Autism: many genes, common pathways? Cell 135(3):391–395
56. Geschwind DH (2009) Advances in autism. Annu Rev Med 60:367–380
57. Geschwind DH, Sowinski J, Lord C et al (2001) The autism genetic resource exchange: a resource for the study of autism and related neuropsychiatric conditions. Am J Hum Genet 69(2):463–466
58. Gharani N, Benayed R, Mancuso V, Brzustowicz LM, Millonig JH (2004) Association of the homeobox transcription factor, ENGRAILED 2, 3, with autism spectrum disorder. Mol Psychiatry 9(5):474–484
59. Gillberg C, Billstedt E (2000) Autism and Asperger syndrome: coexistence with other clinical disorders. Acta Psychiatr Scand 102(5):321–330
60. Glasson EJ, Bower C, Petterson B, de Klerk N, Chaney G, Hallmayer JF (2004) Perinatal factors and the development of autism: a population study. Arch Gen Psychiatry 61(6):618–627
61. Glessner JT, Wang K, Cai G et al (2009) Autism genome-wide copy number variation reveals ubiquitin and neuronal genes. Nature 459(7246):569–573
62. Gottesman II, Gould TD (2003) The endophenotype concept in psychiatry: etymology and strategic intentions. Am J Psychiatry 160(4):636–645
63. Hansen RL, Ozonoff S, Krakowiak P et al (2008) Regression in autism: prevalence and associated factors in the CHARGE Study. Ambul Pediatr 8(1):25–31
64. Hatton DD, Sideris J, Skinner M et al (2006) Autistic behavior in children with fragile X syndrome: prevalence, stability, and the impact of FMRP. Am J Med Genet 140(17):1804–1813
65. Hemara-Wahanui A, Berjukow S, Hope CI (2005) A CACNA1F mutation identified in an X-linked retinal disorder shifts the voltage dependence of Cav1.4 channel activation. Proc Natl Acad Sci USA 102(21):7553–7558
66. Hertz-Picciotto I, Delwiche L (2009) The rise in autism and the role of age at diagnosis. Epidemiology 20(1):84–90
67. International Schizophrenia Consortium (2008) Rare chromosomal deletions and duplications increase risk of schizophrenia. Nature 455(7210):237–241
68. Jackman C, Horn ND, Molleston JP, Sokol DK (2009) Gene associated with seizures, autism, and hepatomegaly in an amish girl. Pediatr Neurol 40(4):310–313
69. Jacquemont ML, Sanlaville D, Redon R et al (2006) Array-based comparative genomic hybridisation identifies high frequency of cryptic chromosomal rearrangements in patients with syndromic autism spectrum disorders. J Med Genet 43(11):843–849
70. Jamain S, Quach H, Betancur C et al (2003) Mutations of the X-linked genes encoding neuroligins NLGN3 and NLGN4 are associated with autism. Nat Genet 34(1):27–29
71. Jansiewicz EM, Goldberg MC, Newschaffer CJ, Denckla MB, Landa R, Mostofsky SH (2006) Motor signs distinguish children with high functioning autism and Asperger's syndrome from controls. J Autism Dev Disord 36(5):613–621
72. Jorde LB, Hasstedt SJ, Ritvo ER et al (1991) Complex segregation analysis of autism. Am J Hum Genet 49(5):932–938
73. Kent L, Bowdin S, Kirby GA, Cooper WN, Maher ER (2008) Beckwith Weidemann syndrome: a behavioral phenotype-genotype study. Am J Med Genet B Neuropsychiatr Genet 147B(7):1295–1297
74. Kim HG, Kishikawa S, Higgins AW et al (2008) Disruption of neurexin 1 associated with autism spectrum disorder. Am J Hum Genet 82(1):199–207
75. Kirov G, Gumus D, Chen W et al (2008) Comparative genome hybridization suggests a role for NRXN1 and APBA2 in schizophrenia. Hum Mol Genet 17(3):458–465
76. Kozlov SV, Bogenpohl JW, Howell MP et al (2007) The imprinted gene Magel2 regulates normal circadian output. Nat Genet 39(10):1266–1272
77. Krakowiak P, Goodlin-Jones B, Hertz-Picciotto I, Croen LA, Hansen RL (2008) Sleep problems in children with autism spectrum disorders, developmental delays, and typical development: a population-based study. J Sleep Res 17(2):197–206
78. Kumar RA, KaraMohamed S, Sudi J (2008) Recurrent 16p11.2 microdeletions in autism. Hum Mol Genet 17(4):628–638
79. Kumar RA, Marshall CR, Badner JA, 2 (2009) Association and mutation analyses of 16p11.2 autism candidate genes. PLoS One 4:e4582
80. Laumonnier F, Bonnet-Brilhault F, Gomot M et al (2004) X-linked mental retardation and autism are associated with a mutation in the NLGN4 gene, a member of the neuroligin family. Am J Hum Genet 74(3):552–557
81. Lawson-Yuen A, Saldivar JS, Sommer S, Picker J (2008) Familial deletion within NLGN4 associated with autism and Tourette syndrome. Eur J Hum Genet 16(5):614–618

82. Levitt P, Campbell DB (2009) The genetic and neurobiologic compass points toward common signaling dysfunctions in autism spectrum disorders. J Clin Invest 119(4):747–754
83. Landa R, Piven J, Wzorek MM, Gayle JO, Chase GA, Folstein SE (1992) Social language use in parents of autistic individuals. Psychol Med 22(1):245–254
84. Leekam SR, Nieto C, Libby SJ, Wing L, Gould J (2007) Describing the sensory abnormalities of children and adults with autism. J Autism Dev Disord 37(5):894–910
85. Lintas C, Sacco R, Garbett K et al (2009) Involvement of the PRKCB1 gene in autistic disorder: significant genetic association and reduced neocortical gene expression. Mol Psychiatry 14:705–718
86. Lowenthal R, Paula CS, Schwartzman JS, Brunoni D, Mercadante MT (2007) Prevalence of pervasive developmental disorder in Down's syndrome. J Autism Dev Disord 37(7):1394–1395
87. Maestrini E, Pagnamenta AT, Lamb JA et al (2009) High-density SNP association study and copy number variation analysis of the AUTS1 and AUTS5 loci implicate the IMMP2L-DOCK4 gene region in autism susceptibility. Mol Psychiatry (in press)
88. Malow BA, Marzec ML, McGrew SG, Wang L, Henderson LM, Stone WL (2006) Characterizing sleep in children with autism spectrum disorders: a multidimensional approach. Sleep 29(12):1563–1571
89. Manning MA, Cassidy SB, Clericuzio C et al (2004) Terminal 22q deletion syndrome: a newly recognized cause of speech and language disability in the autism spectrum. Pediatrics 114(2):451–457
90. Marshall CR, Noor A, Vincent JB et al (2008) Structural variation of chromosomes in autism spectrum disorder. Am J Hum Genet 82(2):477–488
91. Mefford HC, Sharp AJ, Baker C (2008) Recurrent rearrangements of chromosome 1q21.1 and variable pediatric phenotypes. N Engl J Med 359(16):1685–1699
92. Melke J, Goubran Botros H, Chaste P et al (2008) Abnormal melatonin synthesis in autism spectrum disorders. Mol Psychiatry 13(1):90–98
93. Milani D, Pantaleoni C, D'Arrigo S, Selicorni A, Riva D (2005) Another patient with MECP2 mutation without classic Rett syndrome phenotype. Pediatr Neurol 32(5):355–357
94. Ming X, Brimacombe M, Wagner GC (2007) Prevalence of motor impairment in autism spectrum disorders. Brain Dev 29(9):565–570
95. Moessner R, Marshall CR, Sutcliffe JS et al (2007) Contribution of SHANK3 mutations to autism spectrum disorder. Am J Hum Genet 81(6):1289–1297
96. Molloy CA, Keddache M, Martin LJ (2005) Evidence for linkage on 21q and 7q in a subset of autism characterized by developmental regression. Mol Psychiatry 10(8):741–746
97. Morrow EM, Yoo SY, Flavell SW et al (2008) Identifying autism loci and genes by tracing recent shared ancestry. Science 321(5886):218–223
98. Napoli I, Mercaldo V, Boyl PP et al (2008) The fragile X syndrome protein represses activity-dependent translation through CYFIP1, a new 4E-BP. Cell 134(6):1042–1054
99. Need AC, Ge D, Weale ME et al (2009) A genome-wide investigation of SNPs and CNVs in schizophrenia. PLoS Genet 5(2):e1000373
100. Nishimura Y, Martin CL, Vazquez-Lopez A et al (2007) Genome-wide expression profiling of lymphoblastoid cell lines distinguishes different forms of autism and reveals shared pathways. Hum Mol Genet 16(14):1682–1698
101. Orrico A, Galli L, Buoni S, Orsi A, Vonella G, Sorrentino V (2009) Novel PTEN mutations in neurodevelopmental disorders and macrocephaly. Clin Genet 75(2):195–198
102. Peters SU, Beaudet AL, Madduri N, Bacino CA (2004) Autism in Angelman syndrome: implications for autism research. Clin Genet 66(6):530–536
103. Philippi A, Roschmann E, Tores F et al (2005) Haplotypes in the gene encoding protein kinase c-beta (PRKCB1) on chromosome 16 are associated with autism. Mol Psychiatry 10(10):950–960
104. Piven J, Palmer P (1997) Cognitive deficits in parents from multiple-incidence autism families. J Child Psychol Psychiatry 38(8):1011–1021
105. Poliak S, Gollan L, Martinez R (1999) Caspr2, a new member of the neurexin superfamily, is localized at the juxtaparanodes of myelinated axons and associates with K+channels. Neuron 24(4):1037–1047
106. Potocki L, Bi W, Treadwell-Deering D (2007) Characterization of Potocki-Lupski syndrome (dup(17)(p11.2p11.2)) and delineation of a dosage-sensitive critical interval that can convey an autism phenotype. Am J Hum Genet 80(4):633–649
107. Reichenberg A, Gross R, Weiser M et al (2006) Advancing paternal age and autism. Arch Gen Psychiatry 63(9):1026–1032
108. Ritvo ER, Jorde LB, Mason-Brothers A et al (1989) The UCLA-University of Utah epidemiologic survey of autism: recurrence risk estimates and genetic counseling. Am J Psychiatry 146(8):1032–1036
109. Ronald A, Happe F, Bolton P et al (2006) Genetic heterogeneity between the three components of the autism spectrum: a twin study. J Am Acad Child Adolesc Psychiatry 45(6):691–699
110. Roohi J, Montagna C, Tegay DH et al (2009) Disruption of contactin 4 in three subjects with autism spectrum disorder. J Med Genet 46(3):176–182
111. Rujescu D, Ingason A, Cichon S et al (2009) Disruption of the neurexin 1 gene is associated with schizophrenia. Hum Mol Genet 18(5):988–996
112. Rujescu D, Meisenzahl EM, Krejcova S (2007) Plexin B3 is genetically associated with verbal performance and white matter volume in human brain. Mol Psychiatry 12(2):190–194 115
113. Rzhetsky A, Wajngurt D, Park N, Zheng T (2007) Probing genetic overlap among complex human phenotypes. Proc Natl Acad Sci USA 104(28):11694–11699
114. Sadakata T, Washida M, Iwayama Y et al (2007) Autistic-like phenotypes in Cadps2-knockout mice and aberrant CADPS2 splicing in autistic patients. J Clin Invest. 117(4):931–943
115. Samaco RC, Hogart A, LaSalle JM (2005) Epigenetic overlap in autism-spectrum neurodevelopmental disorders: MECP2 deficiency causes reduced expression of UBE3A and GABRB3. Hum Mol Genet 14(4):483–492
116. Sebat J, Lakshmi B, Malhotra D et al (2007) Strong association of de novo copy number mutations with autism. Science 316(5823):445–449

117. Schellenberg GD, Dawson G, Sung YJ et al (2006) Evidence for multiple loci from a genome scan of autism kindreds. Mol Psychiatry 11(11):1049–1060
118. Sharp AJ, Mefford HC, Li K (2008) A recurrent 15q13.3 microdeletion syndrome associated with mental retardation and seizures. Nat Genet 40(3):322–328
119. Simonoff E, Pickles A, Charman T, Chandler S, Loucas T, Baird G (2008) Psychiatric disorders in children with autism spectrum disorders: prevalence, comorbidity, and associated factors in a population-derived sample. J Am Acad Child Adolesc Psychiatry. 47(8):921–929
120. Splawski I, Timothy KW, Sharpe LM (2004) Ca(V)1.2 calcium channel dysfunction causes a multisystem disorder including arrhythmia and autism. Cell 119(1):19–31
121. Splawski I, Yoo DS, Stotz SC, Cherry A, Clapham DE, Keating MT (2006) CACNA1H mutations in autism spectrum disorders. J Biol Chem 281(31):22085–22091
122. Stefansson H, Rujescu D, Cichon S et al (2008) Large recurrent microdeletions associated with schizophrenia. Nature 455(7210):232–236
123. Steffenburg S, Gillberg C, Hellgren L et al (1989) A twin study of autism in Denmark, Finland, Iceland, Norway and Sweden. J Child Psychol Psychiatry 30(3):405–416
124. Stone JL, Merriman B, Cantor RM et al (2004) Evidence for sex-specific risk alleles in autism spectrum disorder. Am J Hum Genet 75(6):1117–1123
125. Strauss KA, Puffenberger EG, Huentelman MJ et al (2006) Recessive symptomatic focal epilepsy and mutant contactin-associated protein-like 2. N Engl J Med 354(13):1370–1377
126. Strom SP, Stone JL, ten Bosch JR et al (2009) High-density SNP association study of the 17q21 chromosomal region linked to autism identifies CACNA1G as a novel candidate gene. Mol Psychiatry (in press)
127. Sutcliffe JS, Delahanty RJ, Prasad HC et al (2005) Allelic heterogeneity at the serotonin transporter locus (SLC6A4) confers susceptibility to autism and rigid-compulsive behaviors. Am J Hum Genet 77(2):265–279
128. Szatmari P, Paterson AD, Zwaigenbaum L et al (2007) Mapping autism risk loci using genetic linkage and chromosomal rearrangements. Nat Genet 39(3):319–328
129. Tierney E, Nwokoro NA, Porter FD, Freund LS, Ghuman JK, Kelley RI (2001) Behavior phenotype in the RSH/Smith-Lemli-Opitz syndrome. Am J Med Genet 98(2):191–200
130. Trikalinos TA, Karvouni A, Zintzaras E et al (2006) A heterogeneity-based genome search meta-analysis for autism-spectrum disorders. Mol Psychiatry 11(1):29–36
131. Tuchman R, Rapin I (2002) Epilepsy in autism. Lancet Neurol 1(6):352–358
132. Ueda S, Fujimoto S, Hiramoto K, Negishi M, Katoh H (2008) Dock4 regulates dendritic development in hippocampal neurons. J Neurosci Res 86(14):3052–3061
133. van Bon BW, Mefford HC, Menten B et al (2009) Further delineation of the 15q13 microdeletion and duplication syndromes: A clinical spectrum varying from non-pathogenic to a severe outcome. J Med Genet 46(8):511–23
134. Van Esch H, Bauters M, Ignatius J et al (2005) Duplication of the MECP2 region is a frequent cause of severe mental retardation and progressive neurological symptoms in males. Am J Hum Genet 77(3):442–453
135. Varga EA, Pastore M, Prior T, Herman GE, McBride KL (2009) The prevalence of PTEN mutations in a clinical pediatric cohort with autism spectrum disorders, developmental delay, and macrocephaly. Genet Med 11(2):111–117
136. Verkerk AJ, Mathews CA, Joosse M, Eussen BH, Heutink P, Oostra BA (2003) CNTNAP2 is disrupted in a family with Gilles de la Tourette syndrome and obsessive compulsive disorder. Genomics 82(1):1–9
137. Vernes SC, Newbury DF, Abrahams BS et al (2008) A functional genetic link between distinct developmental language disorders. N Engl J Med 359(22):2337–2345
138. Volkmar FR, Nelson DS (1990) Seizure disorders in autism. J Am Acad Child Adolesc Psychiatry 29(1):127–129
139. Vorstman JA, Morcus ME, Duijff SN (2006) The 22q11.2 deletion in children: high rate of autistic disorders and early onset of psychotic symptoms. J Am Acad Child Adolesc Psychiatry 45(9):1104–1113
140. Vorstman JA, Staal WG, van Daalen E, van Engeland H, Hochstenbach PF, Franke L (2006) Identification of novel autism candidate regions through analysis of reported cytogenetic abnormalities associated with autism. Mol Psychiatry 11(1):18–28 1
141. Wang K, Zhang H, Ma D (2009) Common genetic variants on 5p14.1 associate with autism spectrum disorders. Nature 459(7246):528–533
142. Weiss LA, Escayg A, Kearney JA et al (2003) Sodium channels SCN1A, SCN2A and SCN3A in familial autism. Mol Psychiatry 8(2):186–194
143. Weiss LA, Kosova G, Delahanty RJ et al (2006) Variation in ITGB3 is associated with whole-blood serotonin level and autism susceptibility. Eur J Hum Genet 14(8):923–931
144. Weiss LA, Shen Y, Korn JM (2008) Association between microdeletion and microduplication at 16p11.2 and autism. N Engl J Med 358(7):667–675
145. Wong D, Maybery M, Bishop DV, Maley A, Hallmayer J (2006) Profiles of executive function in parents and siblings of individuals with autism spectrum disorders. Genes Brain Behav 5(8):561–576
146. Xue M, Brimacombe M, Chaaban J, Zimmerman-Bier B, Wagner GC (2008) Autism spectrum disorders: concurrent clinical disorders. J Child Neurol 23(1):6–13
147. Yan J, Oliveira G, Coutinho A et al (2005) Analysis of the neuroligin 3 and 4 genes in autism and other neuropsychiatric patients. Mol Psychiatry 10(4):329–332
148. Zafeiriou DI, Ververi A, Vargiami E (2007) Childhood autism and associated comorbidities. Brain Dev 29(5):257–272
149. Zoghbi HY (2003) Postnatal neurodevelopmental disorders: meeting at the synapse? Science 302(5646):826–830

23.5 The Genetics of Alcoholism and Other Addictive Disorders

David Goldman and Francesca Ducci

Abstract Addictions are common, complex disorders that are to some extent tied together by shared genetic and environmental etiological factors. They are frequently chronic, with a relapsing/remitting course. Addictive disorders, which are in part volitional, in part inborn, and in part determined by environmental experiences, pose the full range of medical, genetic, policy, and moral challenges. Genetic factors account for 40–70% of the variance in addiction liability. There is little evidence for large influences on overall population vulnerability from any single gene. Instead, multiple genetic loci are likely to be involved, each with a small attributable risk. Gene discovery is being facilitated by a variety of powerful approaches and tools, but is in its infancy. Susceptibility loci for addictions include both drug-specific genes (e.g., alcohol-metabolizing genes) and loci moderating neuronal pathways, such as reward, behavioral control, and stress resiliency, that are involved in several psychiatric diseases (e.g., *MAOA* and *COMT*). In recent years, major progress has been made in identification of genes using intermediate phenotypes such as task-related brain activation that confer the opportunity of exploring the neuronal mechanisms through which genetic variation is translated into behavior. Fundamental to the detection of gene effects are also the understanding of the interplay between genes and of genes/environment interactions. The identification of genes altering the liability to addiction and treatment response (e.g., *OPRM1*) could provide new therapeutic targets and an ability to individualize treatment. Although the genetic bases of addiction remain largely unknown, there are reasons to think that more genes will be

Contents

23.5.1 Introduction 716

23.5.2 Definition of Substance Use Disorders and Other Addictions 716

23.5.3 Epidemiology and Societal Impact of Addiction 717

23.5.4 Genetics: Family and Twin Studies 718
    23.5.4.1 The Heritability of Addictions 719
    23.5.4.2 Do Genetic Factors Moderating Risk Differ in Men and Women? 719
    23.5.4.3 Developmental Dependence of Genes and Environment in Risk 720
    23.5.4.4 What Is the Nature of the Inheritance of Addictions? 720
    23.5.4.5 Agent-Specific and Nonspecific Genetic and Environmental Factors 723
    23.5.4.6 Are Genetic and Environmental Risk Factors Independent of Each Other? 724
    23.5.4.7 Gene by Environment Correlation 726
    23.5.4.8 Gene by Environment Interaction 727

23.5.5 Progress Through Intermediate Phenotypes 728

D. Goldman (✉)
Laboratory of Neurogenetics, NIAAA, NIH, 5625 Fishers Lane, Room 3S32, MSC 9412, Rockville, MD 20852, USA
e-mail: davidgoldman@mail.nih.gov

F. Ducci
Division of Psychological Medicine SGDP Centre, PO80 Institute of Psychiatry, UK De Crespigny Park, London, SE5 8AF, UK
e-mail: Francesca.Ducci@iop.kcl.ac.uk

23.5.6 Finding the Specific Genes Underlying
        Vulnerability to Addiction 728
    23.5.6.1 Candidate Genes 729
    23.5.6.2 Genetic Mapping................................. 734
    23.5.7   Treatment of Addictions 735
23.5.8 Conclusion 736
References.. 736

23.5.1 Introduction

Addictions are multistep pathologies associated with maladaptive and destructive behaviors. They share the persistent, compulsive and uncontrolled use of a drug or, more generally, an agent or activity, for example gambling, shopping, or use of the internet.

Once exposure to the addictive agent has occurred, repetitive use induces neuroadaptive changes which promote further agent-seeking behaviors and ultimately lead to an uncontrolled pattern of use. These adaptive changes are the bases for the establishment of tolerance, craving, withdrawal, and affective disturbance that persist long after consumption of the addictive agent ceases. These changes serve as a basis for the cue- and stress-induced triggering of relapse and rapid reinstatement of use. Owing to these progressive and self-maintaining neurobiological mechanisms, addictive disorders are chronic and relapsing in nature [88].

Addictions by definition require exposure to an environmental agent. Both the probability of initiation and the probability of developing a pathologic pattern of use are influenced by individual characteristics (e.g., genetic vulnerability, sex, age, cohort, and psychopathology), environment (e.g., drug availability, social support, exposure to stressful events), and the nature of the addictive agent (e.g., rewarding properties, mode of administration, physiological response, secondary pathology). As genetically influenced diseases, addictions thus fall within the group of diseases that are thought of as complex. Yet, as will be seen, much has been learned about the inheritance of addictions and the influence of specific genes. Gene discovery is clarifying the neurobiological mechanisms of addiction and is being informed and aided by advances in our understanding of those mechanisms, and undoubtedly additional mechanisms will be

discovered in the future. Multiple and complementary approaches will be required to piece together the mosaic of causation.

23.5.2 Definition of Substance Use Disorders and Other Addictions

Addictive disorders are clinically defined by two main systems: the Diagnostic and Statistical Manual of Mental Disorders (DSM) of the American Psychiatric Association, and the International Classification of Disease (ICD) of the World Health Organization. The same diagnostic criteria apply to the variety of addictive agents to which people are exposed. Two categories of addiction are recognized: Abuse (DSM-IV)/harmful use (ICD-10) and dependence. Abuse and dependence are both maladaptive patterns of behavior leading to clinically significant impairment or distress. Diagnostic criteria for substance abuse and dependence are shown in Table 23.5.1.

Table 23.5.1 Diagnostic criteria for substance-use disorders (SUDs) according to the fourth edition of the Diagnostic and Statistical Manual of Mental Disorders (DSM-IV, issued by the American Psychiatric Association)

Substance-use disorders, including abuse and dependence, are maladaptive patterns of substance use that lead to clinically significant impairment or distress. The diagnosis of substance dependence requires at least three of seven criteria and the diagnosis of substance abuse requires one of four criteria.

The seven criteria for substance dependence:
1. The need for markedly increased amounts of the substance to achieve intoxication or desired effect, or diminished effect with continued use of the same amount (tolerance).
2. Withdrawal syndrome or use of the substance to relieve or avoid withdrawal symptoms.
3. One or more unsuccessful efforts to cut down or control use.
4. Use in larger amounts or over a longer period than intended.
5. Important social, occupational or recreational activities are given up or reduced because of substance use.
6. A large amount of time is spent in activities that are necessary to obtain, to use or to recover from the effects of the substance.
7. Continued use despite knowledge of having persistent or recurrent physical or psychological problems that are caused or exacerbated by the substance.

The four criteria for substance abuse:
1. Recurrent use resulting in a failure to fulfill the main obligations at work, school or home.
2. Recurrent use in physically hazardous situations.
3. Recurrent substance-related legal problems.
4. Continued use despite persistent or recurrent social or interpersonal problems that are caused or exacerbated by the substance.

For both disorders, symptoms must occur within the same 12-month period. The abuse diagnosis is excluded in patients who have ever been dependent.

The development of valid and reliable criteria for addictive disorders provided a unifying framework for epidemiologic, treatment, and genetic studies worldwide [51]. However, the nosology of addictions has important limitations. The diagnostic categories are syndromic (based on clusters of symptoms and clinical course) rather than etiologic. The diagnoses are categorical, assuming a cut-off between normal and abnormal behavior, when many of the same problems are found in people who fall below the threshold for diagnosis. Pathologic use (e.g., leading to dangerous driving or problems at work) is underdiagnosed via the categorical DSM approach. Also, the diagnostic categories do not capture features of addiction that are important clinically and in research. For example, binge drinking is a pattern of alcohol use characterized by episodic bouts of intense drinking. In some American Indian Tribes, binge drinking is a common pattern of alcohol use that is generally, but not always, seen in the context of alcohol dependence. Regardless, binge drinking is a strong independent predictor of problems in all four DSM addiction major symptom areas: social, work, physical, and violence/lawlessness [123]. There is also a need to understand, and integrate, etiologic factors that act across diagnostic boundaries. As will be discussed later in this chapter, twin studies detect evidence of etiologic factors shared between addictions and other psychiatric diseases [81] and linking normal (personality) and abnormal (psychopathology) variations [88]. Thus, addiction disease categories are etiologically connected to other psychiatric diseases and to "normality." Future versions of diagnostic classifications may incorporate dimensional indices such as age at onset, and frequency, quantity, and years of use. However, it is improbable that the nosology of addiction will advance until neurobiological indicators, including genotypes, are integrated.

23.5.3 Epidemiology and Societal Impact of Addiction

Substance use and substance use disorders are common According to the World Health Organization (WHO), worldwide there are 2 billion alcohol users, 1.3 billion tobacco users, and 185 million users of illicit drugs (http://www.who.int/substance_abuse/facts/global_burden/en/). In the United States, according to the NESARC Survey (National Epidemiologic Survey on Alcohol and Related Conditions) the lifetime and 1-year point prevalences of alcohol dependence are 12.5 and 3.8%, respectively, [46]. In the same survey, prevalences of 12-month and lifetime drug dependence were 0.6 and 2.6%, respectively, [22]. Finally, the 1-year point prevalence of nicotine dependence is estimated to be 12.8% [48].

From a public health perspective, the cost of substance use and addictions in term of mortality and morbidity is enormous [147] and comparable to that of the worst chronic diseases, including diabetes and cancer (www.drugabuse.gov/about/welcome/aboutdrugabuse/magnitude).

The costs of addictions include teratogenic effects. In the United States, approximately 30% of women consume alcohol during pregnancy, leading to an incidence of fetal alcohol syndrome (FAS) of 0.2–2.0/1,000 live births, a rate comparable to that of Down syndrome. Alcohol crosses the placental barrier and can impair brain development even if the exposure occurs in the third trimester. FAS cognitive disabilities include deficits in memory, attention, behavioral inhibition, and reasoning. Later in life, the affected child is more vulnerable to psychiatric disorders and addictions, perpetuating a cycle of risk. The lifetime medical and social costs of FAS are high: as much as $ 800,000 per child.

Although genetic factors contribute to individual differences in vulnerability to addictions [43], these disorders – at least in theory – can be prevented by environmental intervention [100] and choice not to consume the addictive agent. There is data to validate this view. Alcohol consumption correlates with risk of developing organ damage such as liver cirrhosis at both individual [110, 152] and population levels [119] (Fig. 23.5.1).

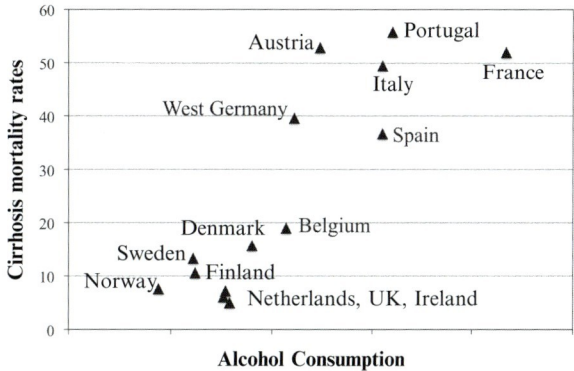

Fig. 23.5.1 Correlation between per capita alcohol consumption and rates of mortality from liver cirrhosis in 13 European countries, adapted from [119]; only data on males reported

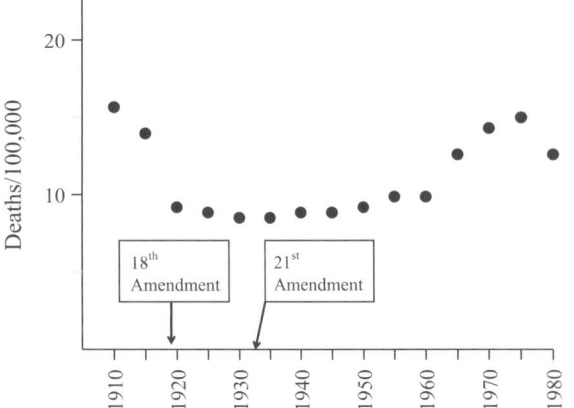

Fig. 23.5.2 Temporal variation in cirrhosis deaths in the U.S in relation to the Prohibition years between the 18th (January 29, 1920) and 21st (December 5, 1933) Amendments to the U.S. Constitution

As shown in Fig. 23.5.2, countries with higher per capita alcohol consumption tend to have higher rates of mortality from cirrhosis. Thus, social policies and customs play a crucial role in determining overall rates of addictions within populations, because they moderate the overall level of exposure. For example, the United States attempted to rid itself of the problem of alcoholism in 1920, when the manufacture, distribution, and sale of alcoholic beverages were prohibited. The "Prohibition," and the Temperance movement that preceded it, appear to have almost halved rates of cirrhosis (see Fig. 23.5.2). However, public demand led to a growth in organized crime, and ultimately the effect of the 18th Amendment to the U.S. Constitution, passed in 1919, was reversed by the 21st Amendment, passed in 1933.

It is important to note that although exposure to legal addictive agents is pervasive, there is an enormous interindividual variation in the pattern of use in terms of quantity, frequency, and duration. In the United States more than 70% of alcohol is consumed by 10% of the population [152]. Similarly, nicotine-dependent and psychiatrically ill individuals consume approximately 70% of cigarettes [48]. Since exposure to addictive agent is widespread, it is important to recognize that a substantial proportion of the exposed population will become addicted, and these vulnerable individuals will account for a large fraction of use of the agent. These statistics highlight the priority of developing targeted preventive strategies focused on individuals who are more vulnerable to developing pathologic use, and they highlight the fact that changes in public policy that are innocuous for most people represent hazards for others.

23.5.4 Genetics: Family and Twin Studies

Family studies have shown that both alcoholism and other addictions cluster in families [8, 77, 101, 104]. First-degree relatives of subjects with substance use disorders have an eightfold increased risk of developing a substance use disorder as compared to relatives of controls [101]. The increased risk among members of the same family might result from either environmental or genetic influences. Twin studies and studies on adoptees have made an enormous contribution to disentangling those contributions. In the classic twin study design the phenotypic resemblance of monozygotic (MZ) twins is compared with that of dizygotic (DZ) twins to estimate three main sources of variation:

1. Heritability (A): Proportion of variance due to additive (A) genetic factors.
2. Shared (or common) environment (C): Proportion of the variance due to factors shared by siblings. C is directly measured in the context of adoption.
3. Unique environment (E): Proportion of the variance due to environmental factors unique to the individual.

Both A and C are sometimes collectively termed familial influences because they contribute to similarities between members of the same family.

It has already been noted that the diagnostic assessment of addictions is imperfect. The usual effect of measurement error is to limit the maximum heritability (A) and to increase the proportion of variance attributable to E. In the same vein, it is important to stress that A, C, and E are latent (unmeasured) influences that should not be interpreted as absolute values. They provide broad estimates of sources of variation in the particular context in which they are evaluated, and do not, positively or negatively, predict effects of intervention to prevent or treat addictions. Furthermore, they do not inform us of sources of variation at the individual level. Nor are they informative of the specific genes and environmental factors involved.

Some of the questions that can be addressed by twin studies of addictions are: Are addictions heritable? Are genetic risk factors moderating vulnerability different in men and women? What is the developmental dependency of genes and environment in risk? What is the nature of inheritance? Are the genetic and environmental risk factors agent-specific or nonspecific? Are genetic and environmental factors independent of each other?

23.5.4.1 The Heritability of Addictions

Results from large, carefully characterized cohorts of twins, including epidemiologically ascertained twins from Virginia, USA) and Australia [15, 32, 53, 71–73, 75, 80, 83, 84, 93, 95, 116, 142], indicate that addictions are among the most heritable psychiatric disorders and most heritable complex traits [Fig. 23.5.3), with heritability estimates ranging from 0.39 (for hallucinogens) to 0.72 (for cocaine) [43].

The moderate to high heritabilities of addictions that were found are paradoxical, because addictions are by definition environmentally mediated, being contingent on exposure and choice. However, it is clear that susceptibilities to several complex diseases, including coronary artery disease, obesity, and diabetes, are genetically influenced, but also depend profoundly on lifestyle choices. Moreover, there is a genetics of choice: individuals vary in their capacity to resist impulse, exploratory behavior, cognitive resources to evaluate risk, and emotional traits that set the stage for vulnerability. In this regard, heritabilities for initiation and use of addictive agents are generally lower than for addictions, but are still significant [31, 79, 93].

It is important to note that in several instances the twins analyzed in studies of the inheritance of addictions were collected using epidemiologic methods such that twins in a particular geographic region and place in time were evaluated systematically and fairly completely (e.g., the Virginia and Australia twin studies), and other twin datasets considered here also represented a systematic sampling within some other general framework (e.g., samples of WWII and Vietnam veterans). Therefore, these do not represent heritabilities within groups of severely affected patients who might be ascertained only in medical settings. However, most of the twin studies were conducted on Caucasian samples from the United States, Europe, and Australia, and thus the results might not be generalizable to other populations.

23.5.4.2 Do Genetic Factors Moderating Risk Differ in Men and Women?

Across populations and on a worldwide basis, addictions are more common in men. Is this because of sex-specific or sex-influenced genetic factors? Alternatively, is it a manifestation of gender-determined differences in opportunities and expectations? Indeed, changing cultural norms have led to disproportionate increases in the risks of addictions among women. However, gender disparities remain. U.S. males have more than twice the risk of alcohol dependence [OR (99%CI): 2.4 (1.75–3.16)] and drug dependence [OR (99% CI): 2.7 (1.16–6.37)] [49].

Twin studies can investigate whether different sets of genes act in males and females. As shown in Fig. 23.5.4, this is accomplished by comparing phenotypic correlations across three types of DZ twin pairs: sex-concordant (male-male and female-female) pairs and opposite-sex (female-male) pairs. If genes act similarly, correlations in these three types of twin pairs will be similar Fig. 23.5.4, scenario A). If the magnitude of genetic influence is higher in males correlations will be higher in male-male DZ pairs (Fig. 23.5.4, scenario B). Finally, if the trait is equally heritable in both sexes but the genes involved are different, there

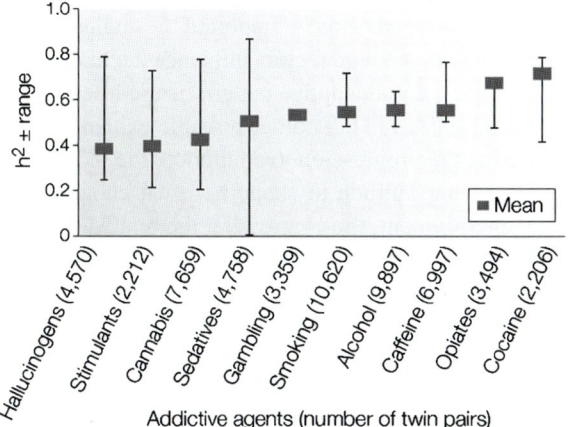

Fig. 23.5.3 The heritability (weighted means and ranges) of ten addictive disorders. These include hallucinogens, stimulants, cannabis, sedatives, gambling, smoking persistence, alcohol dependence, caffeine consumption or heavy use, cocaine dependence or abuse, and opiates. From [43], with permission

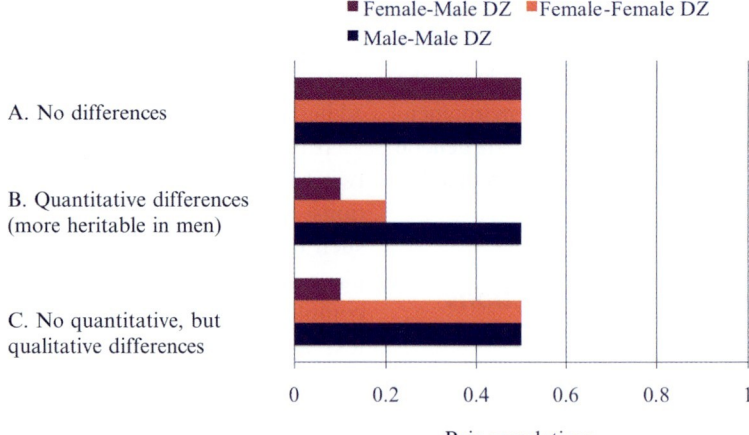

Fig. 23.5.4 (a–c) Phenotypic correlations in gender-concordant (male-male, female-female) and gender-discordant (female-male) dizygotic (DZ) twin pairs, assuming quantitative and/or qualitative differences between the two sexes in genes moderating a hypothetical trait. If gender differences are the same in the two genders, correlations in sex-discordant and sex-concordant twin pairs are similar (a) If the magnitude of genetic influence is higher in men than women, correlations will be highest in male-male DZ pairs (b) Finally, if the trait is equally heritable in both genders but the genes involved are different, there will be a lower correlation in sex-discordant pairs than in both kinds of sex-concordant pairs (c) (see text). Modified from [74], with permission

will be a lower correlation in sex-discordant pairs (Fig. 23.5.4, scenario C).

Twin studies indicate that genetic influences on alcoholism are approximately equivalent in males and females [54, 113]. However, the lesser similarity of opposite-sex pairs provides evidence for qualitatively different influences on risk in the two sexes [113]. At the gene level, gender-specific effects on addiction liability have been described for catechol-*O*-methylransferase (COMT) [35], an enzyme that metabolizes catecholamine neurotransmitters (see Sect. 23.5.6.1.2). Furthermore, animal models have shown sex differences in the regulation of ethanol-stimulated mesolimbic dopamine release by the μ-opioid receptor [67]. However, for nicotine, twin studies do not provide evidence for differences in genetic influences between men and women [96], and for illicit substance use disorders gender differences have not be fully addressed owing to low numbers of affected women in most of the twin studies conducted so far.

23.5.4.3 Developmental Dependence of Genes and Environment in Risk

Longitudinal studies reveal that the impact of genetic factors on addictions changes during development and across the lifespan. Addiction-related behaviors that are not heritable early in life become highly heritable later. Kendler et al. [84] found that the effect of shared environment (C) declines from adolescence into adulthood and disappears by age 35 for nicotine and by age 40 for alcohol (Fig. 23.5.5). In contrast, gene effects (A) are not detectable in early adolescence but gradually grow in importance.

These results can be explained by the action of some genetic factors only after repetitive exposure to the addictive agent, or by genes that act only after the brain has fully developed. This is supported by studies showing that specific genetic factors influence the likelihood of developing a maladaptive pattern of use after initial exposure [1, 78, 111]. Another possible explanation is that during the progression to adulthood, the individual has increasing latitude to shape her own choices and social environment, thus increasing the relative role of genotype [82] (see also active rGE in Sect. 23.5.4.6].

23.5.4.4 What Is the Nature of the Inheritance of Addictions?

Studies of families with addictions do not reveal a Mendelian pattern of inheritance. Factors complicating the inheritance of addictions include:

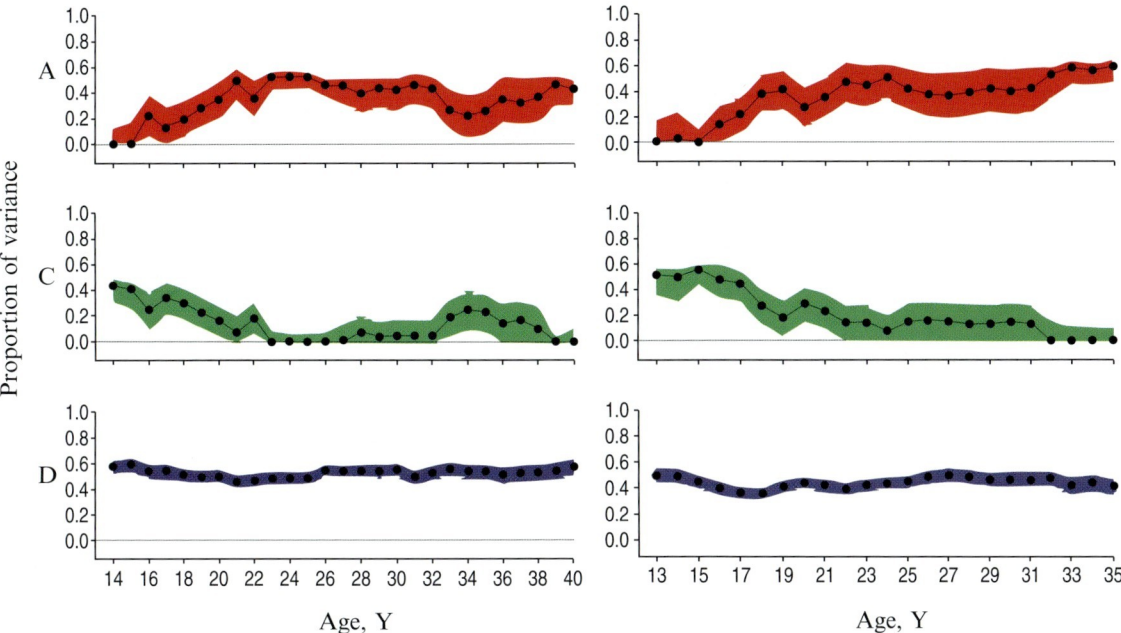

Fig. 23.5.5 Age-dependency of genetic and environmental contributions to variation in liability to alcohol (*left*) and nicotine (*right*) use. Additive genetic effects (*A*), familial environmental factors (*C*) and individual-specific environment (*E*) are represented by age in years (*Y*). The actual parameter estimates for A, C, and E are depicted by the *black lines*. *Colored regions* represent the possible range of estimates ±1 standard error. For both nicotine and alcohol use, genetic influences increase and shared family environmental influences decrease, moving from adolescence to adulthood (see text). Adapted from [84], with permission

Phenocopies: an individual who is not at genetic risk may become at risk because of a severe environment.

Incomplete penetrance: an individual at genetic risk does not become addicted, for example because of a decision to abstain or the intercession of a vigilant spouse.

Assortative mating and bilineal transmission: addicted individuals have a strong tendency to mate [59], enhancing the likelihood of bilineal transmission from both the maternal and paternal sides of the family.

For the addictions and other complex diseases, there is little evidence for large influences on overall population vulnerability from any single gene. Instead, multiple genetic loci are likely to be involved, each with a small attributable risk (for review see [140]). However, there are divergent models involving the action of multiple loci in overall population risk. Two of the most important divergent models, genetic heterogeneity and polygenicity, are contrasted in Fig. 23.5.6.

In the polygenic model, multiple functional alleles act in combination to lead to the phenotype in a fashion that is not merely additive. Therefore, the effect of individual loci would not be independent, a phenomenon also known as gene-by-gene interaction (G×G). In classic American movie terms, the polygenicity model resembles the strategy employed by "The Joker" (Batman's nemesis) to terrorize the citizens of Gotham City, an imaginary American metropolis. The criminal mastermind distributed cosmetics that were harmless when used individually but fatal when used in some combinations. On the left side of Fig. 23.5.6, addicted individuals have different polygenic combinations of alleles that have pushed them over the threshold of vulnerability. In contrast, on the right side of the figure causation via genetic heterogeneity is illustrated. Multiple alleles are predisposing or protective, but the effects of any one can suffice. Genetic heterogeneity can occur at the same gene (within-locus heterogeneity) or at different genes (between-locus heterogeneity) (leading to genocopies of the illness).

Twin concordance ratios make it possible to distinguish between the genetic heterogeneity and polygenicity models to some degree (see Fig. 23.5.7).

As illustrated, polygenic inheritance tends to produce high MZ:DZ concordance ratios, because the

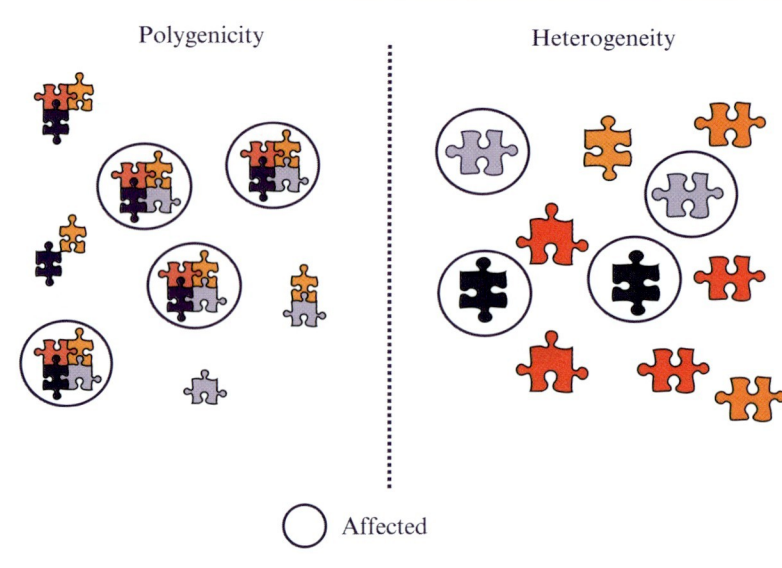

Fig. 23.5.6 Comparison of the polygenicity and heterogeneity models for a hypothetical complex disease in a sample of unrelated individuals. Affected individuals are indicated with a *circle*. Under the polygenicity model (*left*) a combination of different alleles (represented as different *jigsaw puzzle pieces*) is required to determine the disease. Under the heterogeneity model a single allele is sufficient to cause the disease, but different alleles (represented as *dark-blue* and *light-blue pieces of jigsaw puzzle*) are involved in different individuals. In other words, different alleles can cause the same disease (see text). Modified from [43], with permission

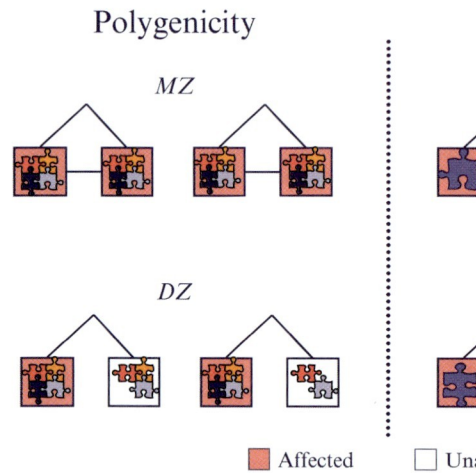
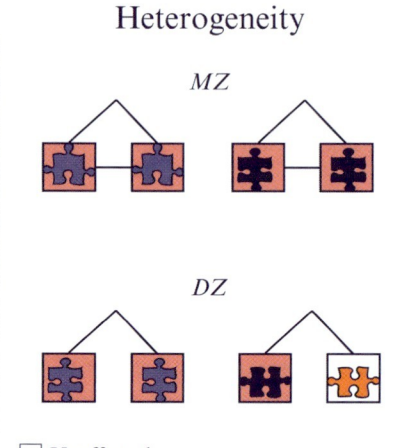

Fig. 23.5.7 Monozygotic (MZ) and dyzygotic (DZ) twin concordances for a hypothetical trait under the polygenicity and heterogeneity models. Under the polygenicity model (*left*) concordances tend to be much higher in MZ pairs than in DZ pairs, because DZ pairs are unlikely to inherit all the alleles (represented by *four pieces of puzzle*) necessary to develop the disease. As a consequence, MZ/DZ phenotypic concordance ratios will be >2:1 (see text). Under the heterogeneity model even just one allele is sufficient to cause the disease (either the *light-blue* or *dark-blue piece of puzzle*). Therefore, concordance ratios between MZ and DZ will tend to be 2:1 (dominant model) or 4:1 (single recessive locus) (see text). Modified from [43], with permission

odds of the DZ twins sharing a large number of alleles is not high. In fact it is $(1/2)^z$ multiplied by the number of such combinations that are available in the twins, where z is the number of alleles necessary to produce vulnerability. On the other hand, monogenic inheritance tends to produce 2:1 MZ:DZ concordance ratios for dominant allele effects and 4:1 ratios for recessive allele effects. As shown in Fig. 23.5.8, the MZ:DZ concordance ratios for addictions approximate the 2:1 ratio expected for monogenic inheritance, with the exception of a 3.7:1 ratio for cocaine. Also arguing against the polygenicity model, in families the degree of phenotypic similarity for addictions tends to fall off in proportion to the decrease in identity by descent, again supporting the heterogeneity model [14].

In line with the heterogeneity model, it has recently been shown that several neuropsychiatric diseases, including schizophrenia and autism [134, 145], are

23.5 The Genetics of Alcoholism and Other Addictive Disorders

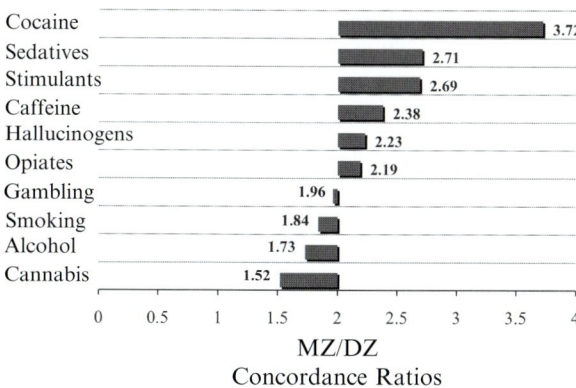

Fig. 23.5.8 MMZDZ twin concordance ratios for ten addictions. MZ/DZ ratios tend to converge on two, consistent with the heterogeneity model (see text). Modified from [43], with permission

sometimes caused by rare, highly penetrant mutations, including large deletions and insertions (copy number variations) that appear to be specific to single cases or families.

23.5.4.5 Agent-Specific and Nonspecific Genetic and Environmental Factors

Co-morbidity (the co-occurrence of different disorders in the same individual) is common amongst the addictions and between addictions and other psychiatric diseases. Both types of co-morbidity are substantially in excess of what would be expected by chance [49, 52, 85]. Such co-morbidity can indicate the existence of etiologic factors that are shared (co-causation) and thus the non-independence of the risk of the two diseases that co-occur. However, co-morbidity can also indicate that one disease tends to lead to the other (e.g., the gateway hypothesis, and self-medication explanations for addictions). Drug use can lead people to alter their environment in such a way that exposure to another drug occurs. For example, 85% of chronic alcoholics smoke cigarettes, but perhaps this is because bars are smoke-filled or because alcohol-consuming peers are more likely to smoke. Twin studies can establish the origins of co-morbidity and evaluate the extent to which genetic and environmental risk factors moderating liability to different diseases are shared or unshared. A common genetic source of co-morbidity between two disorders can be detected by identifying cross-inheritance of vulnerability.

23.5.4.5.1 Co-morbidity Among Addictions

Twin and family studies have revealed that genetic factors acting on different addictive disorders include both substance-specific genetic factors and genetic factors that are shared between different addictions (reviewed in [40, 58]). In a large sample of adult male twins who were veterans of WWII [138], a common genetic factor accounted for more than two-thirds of the total genetic variance in risk for disorders involving marijuana, stimulants, sedatives, and psychedelics, whereas 0.7 of heroin dependence heritability was found to be substance specific. Kendler et al. [83] explored the genetic overlap between several licit and illicit substances, including alcohol, caffeine, nicotine, cannabis, and cocaine, in a portion of a Virginia twin sample consisting of almost 5,000 twins. In this study genetic risk factors for dependence on different psychoactive substances could not be explained by a single factor acting across all substances. Rather, two shared genetic factors were found: one factor mainly explained vulnerability to cannabis and cocaine dependence (illicit drug genetic factor); a second factor mainly explained vulnerability to alcohol, caffeine, and nicotine (licit drug genetic factor). These two factors were not independent but highly correlated. Large substance-specific genetic factors were found mainly for nicotine and caffeine (see Fig. 23.5.9).

23.5.4.5.2 Co-morbidity Between Addiction and Other Psychiatric Disorders

Addictions frequently co-exist with other psychiatric diseases, including both internalizing disorders (disorders marked by anxiety or problems of mood) and externalizing disorders (disorders marked by problems of impulse control) [47, 85]. Twin studies consistently reveal the existence of shared genetic influences between alcoholism and externalizing disorders [59, 81, 88]. Longitudinal studies have shown that externalizing disorders of childhood, such as conduct disorder (CD) and attention deficit hyperactivity disorder (ADHD), are important risk factors for the subsequent development of alcoholism [130]. Evidence from twin

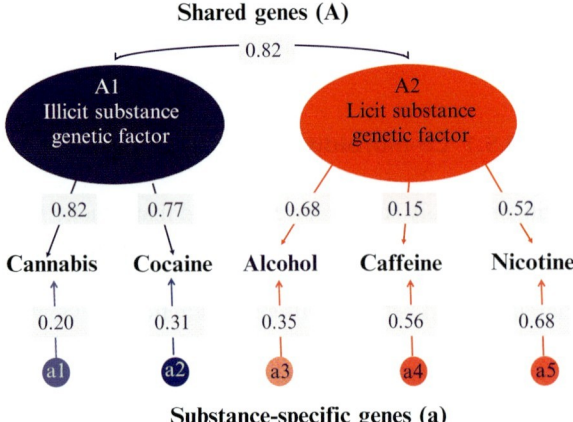

Fig. 23.5.9 Shared and substance-specific additive genetic influences acting on symptom counts for cannabis, cocaine, alcohol, caffeine, and nicotine abuse or dependence. Two common genetic factors are identified, one for illicit substances (*A1*) and one for licit substances (*A2*). Substance-specific genetic influences (*a1-a5*) are represented at the *bottom* of the figure. Path values are standardized loadings and thus need to be squared to reflect the proportion of variation in liability in the observed variable accounted for by the factor. For example, the illicit substance genetic factor (*A1*) accounts for 0.67 (0.82²) of the total variance in cannabis symptoms in this population. The cannabis-specific genetic factor (*a1*) accounts for 0.04 (0.20²) of the variance in this trait. Therefore, the total variance of cannabis symptoms determined by genetic factors – heritability – is 0.71 (0.67+0.04). The remaining variance (1−0.71=0.29) is explained by environmental influences that are not represented here. The *doubleheaded arrow* connecting the illicit and licit substance genetic factors represents the genetic correlation between these factors (see text). Adapted from [83], with permission

studies for shared genetic influences between alcoholism and internalizing disorders are more controversial [76, 81, 114]. However, longitudinal studies have shown that anxiety disorders such as panic disorder and social phobia predict subsequent alcohol problems in adolescents and young adults [154].

Overall, twin studies predict that genes involved in vulnerability to addiction include both substance-specific genes and genes that act on general (common) pathways involved in different diseases. Substance-specific genes include genes involved in pharmacokinetic or pharmacodynamic processes specific to a particular drug. For example, genetic variation in alcohol-metabolizing genes moderates risk to develop alcoholism (see Sect. 23.5.6.1.1). Nonspecific genes include genes affecting neurobiological networks involved in vulnerability to different types of addictions and also genes predisposing to addictions and other psychiatric diseases, such as genes involved in reward, stress resiliency, behavioral control, and personality. For example, the dopamine system is fundamental for the reward effects of all addictive agents [143], and genetic variations in the gene encoding the dopamine two receptor (*DRD2*) have been linked to different types of addictions [23, 148], although with some inconsistencies [42]. As will be discussed, other genes such as monoamine oxidase A (*MAOA*) (see Sect. 23.5.6.1.2), the serotonin transporter (*SLC6A4*) (see Sect. 23.5.6.1.2), and catechol-*O*-methyl transferase (*COMT*) (see Sect. 23.5.6.1.2) have been implicated in the shared genetic liability between addictions and other psychiatric diseases.

23.5.4.6 Are Genetic and Environmental Risk Factors Independent of Each Other?

The simplest, and oldest, model for the etiology of behavioral differences and psychiatric diseases regarded genetic and environmental factors as separate entities contributing to disease liability in an additive fashion. This model polarized discussion, leading to the gene vs environment debate, a false dichotomy.

At the height of the eugenics movement, which in the United States was powerfully embodied at the Cold Springs Harbor Laboratory led by Charles Davenport, Charles Darwin's Malthusian principle of survival of the fittest had been used to develop the concept of Social Darwinism. Those who were psychiatrically ill, cognitively deficient (e.g., illiterate), and poor were thought to be genetically inferior. Social effects included the compulsory sterilization of many thousands of vulnerable individuals. The effect of this movement is encapsulated by the case Buck vs Bell, heard before the U.S. Supreme Court in 1927. Carrie Buck was a ward of the Virginia State Colony for Epileptics and Feebleminded. Although later in her life she became an avid reader, she was one of many thousand proposed for sterilization because she was "feeble-minded" and incorrigibly promiscuous (actually she had given birth to a child after being raped). Carrie Buck's mother had been a prostitute. Writing for an 8–1 majority, Chief Justice Oliver Wendell Holmes, Jr. stated,

It is better for all the world, if instead of executing degenerate offspring for crime, or to let them starve for their imbecility, society can prevent those who are manifestly unfit from continuing their kind. The principle that sustains compulsory vaccination is broad enough to cover cutting the Fallopian tubes.

At the end of his opinion, which greatly accelerated state-sponsored euthanasia for eugenic purposes, Chief Justice Holmes memorably concluded: "Three generations of imbeciles are enough." Carrie Buck, and later her daughter, were among 65,000 individuals compulsorily sterilized in the United States. There were also large eugenic programs in other nations, including Japan, Canada, and Sweden, where in that one nation 21,000 people were forcibly sterilized. Most infamously, Nazi Germany sterilized over 400,000 people and murdered more than 6 million institutionalized individuals, homosexuals, gypsies, and Jews.

Although it is valid to argue that genetic studies on the addictions and other psychiatric diseases have emerged from a different imperative and tradition – namely the need to understand behavior and to prevent and treat psychiatric diseases – the disciplines of psychiatry and psychiatric genetics encouraged and were highly intertwined with the eugenics movement, and with the genocides that were justified on that basis. Therefore, it remains imperative to be alert to the threat of misuse of genetic information, especially where it may be misused to stigmatize whole groups of individuals.

At the opposite pole is the view that addictions and other psychiatric diseases emerge only through upbringing, experience, and choice. Like the genetic view, which was adopted by the Social Darwinists and Nazis to justify policy, the environmentalist viewpoint was also co-opted for political purposes, for example by socialists, whose main point was that all people should not just be treated equally but that there should be equality of outcomes. In the Declaration of Independence, the Founders of the United States uttered the memorable words "All men are created equal"; however, this was a statement of universal rights (for white males!) and not an assertion of equality of ability or vulnerability. Paradoxically, the equal outcomes that are sometimes thought of as a desirable societal goal can only be achieved if individual differences, including the vulnerabilities of individuals, are recognized. However, socialist utopian appeals rested on an assumption of human malleability and ultimately culminated in the excesses of Lysenkoism, a dark time in the former Soviet Union when geneticists studying inheritance were repressed in favor of a Lamarckian view of genomic malleability. On the whole, the unidimensional environmental perspective has also been stigmatizing, in part because since this model of behavioral causation left society with a set of intractable problems that were in some vague sense the fault of the individual, the family, or the community. Indeed, regardless of whether behaviors have a genetic or environmental origin, behaviors will be used for the purposes of stigmatization. American Indians had to contend with the "fire-water myth," though in fact particular American Indian tribes have low rates of alcoholism and within tribes there is differential vulnerability owing to genetic variation. The Irish are frequently characterized as alcoholic although rates of alcoholism are not higher than in several other European countries. Within the Irish, differential vulnerability is again genetically transmitted, as revealed by Prescott et al. [93]. On the other hand, as we understand the genetic and neurobiologic origins of addiction, it is clear that the seeds of these problems are latent in all of us, albeit to different extents, causes are understood, removing the need for over-generalizations, and destigmatization flows from this more sophisticated understanding.

Studies that have examined the combined effect of genes and environment in addiction have revealed that the individual's genotype in probabilistic fashion shapes her relative risk for these common disorders. Genotype also determines reaction range – the range of possible responses to environment. Most people are probably at some risk for an addiction and could become addicted under certain circumstances, and thus the reaction range of most of our genotypes encompasses addiction. However, a shift in level and type of environmental exposure uncovers or covers part of the overall risk density distribution, as shown in Fig. 23.5.10. What are these factors? Several environmental influences powerfully moderate risk of developing addictions, including exposure to maltreatment during infancy [27], age at onset of drinking [45, 46], low socioeconomic status [99], adverse life events, low social support, poor parenting, religiosity [83], peer influences [61], and drug availability.

There are two main types of violations of gene-environment independence: gene by environment correlation and gene by environment interaction.

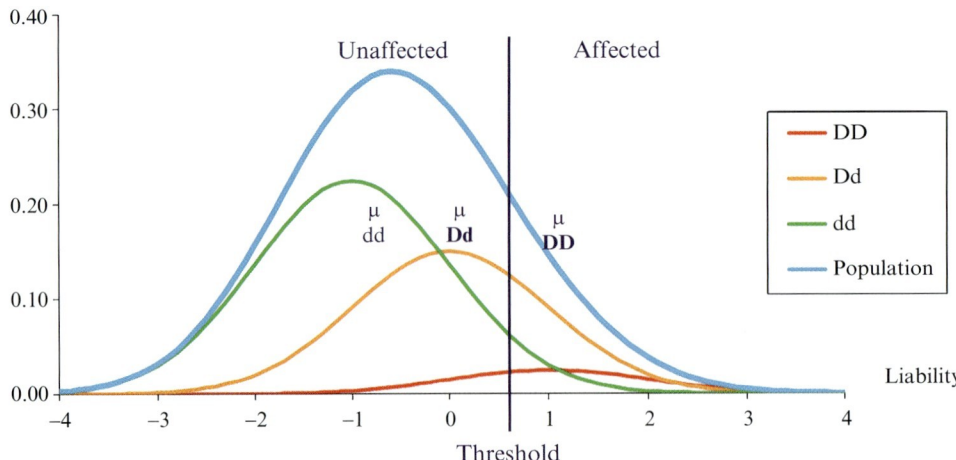

Fig. 23.5.10 Genotype-specific liability and disease threshold. Liability or predisposition to a disease is simplified here as a single continuous dimension. Once liability passes a threshold the discrete phenotype emerges (affected versus unaffected). Given a disease-linked locus with two different alleles (*D* and *d*), carriers of the three different genotypes (*DD, Dd, dd*) have three different liability curves that together compose the overall liability distribution for that population. Factors that change the threshold will therefore change the number of affected and unaffected individuals within each genotype group as well as in the overall population

23.5.4.7 Gene by Environment Correlation

Gene × environment correlation (rGE) occurs when genotype correlates (*r*) with probability of exposure to environmental stressors. There are three main mechanisms leading to rGE: *active rGE, evocative rGE,* and *passive rGE*. Active rGE occurs when an individual's genotype can shape her choice of environment. For example, children with conduct disorder (CD), a precursor to antisocial personality disorder (APD) tend to seek out antisocial peers, exposure to whom increases their risk of developing antisocial behavior and addiction. In evocative rGE, an individual can indirectly shape his/her environment. For example, a child with CD may evoke harsher discipline from her parents, in turn promoting the risk of addictions and other pathologies. In passive rGE, alleles conferring risk in a child also alter the behavior of the parent transmitting the allele. Thus, the children of an addicted parent are at enhanced risk both via transmission risk alleles and via family environment and teratogenic effects of the drug.

Twin studies can address the existence of rGE by measuring the "genetics of the environment." Although it may appear paradoxical, if behavior alters environmental exposures and if the relevant aspect of the behavior is subject to genetic influences, then the environmental measure will be heritable. Inheritance of environmental exposures has been observed. Kendler and Backer [70] reported modest to moderate heritabilities (ranging from 7 to 39%) for several categories of environmental factors that are important or potentially important in addiction vulnerability: stressful life events, parenting, family environment, social support, peer interactions, and marital quality.

Twin studies can help us understand whether the relationship between an environmental variable and an outcome is causal (e.g., directly mediated by an environmental effect) or mediated by genetic/shared environmental influences (as occurs with rGE). The discordant twin design, in which twins discordant for exposure to the environmental factor are studied, is a powerful test of causality. The discordant twin design offers the possibility of testing whether the association between the environmental factor and the disease persists after controlling for genotype and for other shared-environmental factors, such as socioeconomic status and home environment. In this design, the association between the environmental variable and the disease is evaluated in the entire sample, in DZ pairs discordant for exposure, and in MZ pairs discordant for exposure. Figure 23.5.11 shows three possible patterns. If association between the environmental variable and the outcome is entirely mediated by unique environmental factors (case A), the strength of the association (measured by the odds ratios, ORs) will be the same in the whole sample and among MZ and DZ pairs (Fig. 23.5.11a). In case B, the association of the environmental variable is partially mediated by genetic factors. Here the ORs are highest in the total sample (where the association is not controlled for shared environmental and genetic confounds), intermediate in DZ twins (where the association is controlled for shared environment and partially for genetic factors), and lowest among MZ pairs (where the association is fully controlled for both shared environment and genetic factors). Finally, in case C the association is entirely mediated by genetic factors, because it completely disappears among discordant MZ pairs (OR=1).

This approach has been used to deconstruct the origins of an important gene×environment correlation, namely age at first use of the psychoactive substance. Early initiation of alcohol use is associated with an increased risk of developing addiction. According to the NESARC epidemiologic survey, odds of lifetime substance dependence among users are reduced by 4% for illicit drugs and 9% for alcohol for each additional year that onset of drug use is delayed [45, 46]. This association between age at initiation and risk can arise from different mechanisms. It can result from a direct-causal effect (early onset directly increases risk) or might be mediated by genetic/familial influences (early onset and addiction result from a broad shared liability). Results from the Virginia Twin Registry study are more consistent with the second hypothesis. Prescott and Kendler et al. [115] showed that twins with late onset of alcohol use had the same risk as co-twins who experienced early onset of alcohol use. This result indicates that the association between early exposure and alcoholism may result from a shared genetic liability and that early exposure does not independently influence risk of developing alcoholism. This is important, because preventive efforts at delaying drinking are likely to be useful for preventing alcohol-related accidents and injuries, but might not necessarily reduce the risk of developing alcoholism in adulthood. However, the impact of early alcohol exposure on risk of alcoholism remains a critically important issue: few would risk unnecessary early drug and alcohol exposures in children with developing brains and imperfectly developed impulse control.

23.5.4.8 Gene by Environment Interaction

Gene×environment interaction (G×E) occurs when the effect of the environmental exposure on a certain outcome is strongly influenced or contingent upon genotype and vice versa (gene effect on the outcome is contingent on exposure)(for review see [17]).

G×E has been observed for several of the functional alleles identified so far in addictions and psychiatric genetics. By comparison with other complex diseases, including cancer, diabetes, cardiovascular, infectious diseases, and hematologic diseases, where genetic variation moderates resiliency and vulnerability to chemicals, pathogens, nutrients, caloric load, oxidants, and ionizing radiation, in psychiatric genetics much of the progress has been made through the discovery of alleles that influence stress resiliency. Severe childhood stress and neglect both increase vulnerability to addiction and multiple addiction-related psychiatric diseases, including antisocial personality disorder (APD), CD, anxiety disorders, and depression, with the risks of these common diseases being

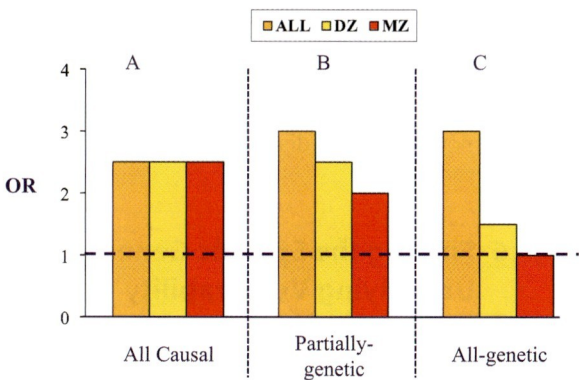

Fig. 23.5.11 (a–c) Causal and noncausal relationship of a hypothetical environmental variable and a hypothetical outcome using MZ and DZ twin pairs that are discordant for exposure. If association between the variable and the outcome is entirely mediated by unique environmental factors, the strength of the association [measured by odds ratio (*OR*)] will be the same in the whole sample and among MZ and DZ pairs (**a**) If the association is partially mediated by genetic factors, the ORs are highest in the total sample, intermediate in DZ twins, and lowest among MZ pairs (**b**) If the association is entirely mediated by genetic factors, it completely disappears among discordant MZ pairs (OR=1) (**c**) (see text). (Modified from [74], with permission)

elevated severalfold in the stress-exposed [122, 144]. However, not all people who are exposed to early life stress develop addiction or other psychiatric diseases, indicating wide variation in resiliency to stress. Functional loci that have been shown to partially account for interindividual differences in stress resiliency include monoamine oxidase A (*MAOA*) (see Sect. 23.5.6.1.2) [18], the serotonin transporter *SLC6A4* [19] (see Sect. 23.5.6.1.2), *COMT* [20] (see Sect. 23.5.6.1.2), the corticotrophin-releasing hormone receptor 1 gene [11], neuropeptide Y (*NPY*) [153], and *FKBP5* [10].

We believe it is important to clarify here the distinction between addictive vs non-additive (or interactive) effects implied in either GXG or GXE. Indeed, misunderstanding of the concept of interaction between factors is frequent. In assessing effects of risk factors on ORs, the ORs attributable to additively acting risk factors are multiplied to predict their combined effect. Thus, if risk factor A confers a relative risk of 2 and risk factor B confers a relative risk of 3, the additive risk (A+B) is $2 \times 3 = 6$. Additivity is frequently mislabeled "interaction," when what has been observed is an additive increase in risk. Also, it is important to recognize that interactions, unless they are large, can be more difficult to detect and accurately quantitate than main effects, thus requiring larger sample sizes or more ingenious sampling frameworks (contrasting populations with different severities of exposure, accessing intermediate phenotypes, etc.).

23.5.5 Progress Through Intermediate Phenotypes

One strategy to discover gene effects in addictions and other etiologically complex diseases is the deconstruction of complex phenotypes into components that are etiologically more homogeneous. Intermediate phenotypes access mediating mechanisms of genes and environmental effects on behavior. Heritable intermediate phenotypes that are disease associated have been termed "endophenotypes" [44]. Several intermediate phenotypes have been specifically associated with addiction. These include alcohol-induced flushing, which is a protective endophenotype, and low response to the effects of alcohol, which is an endophenotype predictive of risk of alcoholism. The genetic origins of alcohol-induced flushing are discussed later (in Sect. 23.5.6.1.1). In humans, the level of response to alcohol is believed to reflect mainly pharmacodynamic variation in response [128] rather than variation in metabolism. A low response to alcohol predicts increased risk of developing alcohol use disorders [55, 124, 129] and has been associated with genetic variation in the serotonin transporter gene (*SLC6A4*) and in the gene encoding the subunit a6 of the g-aminobutyric acid receptor A (*GABRA6*) [62]. Other intermediate phenotypes assist in the exploration of genetic vulnerability to addictions (as well as other psychiatric diseases), and these addiction-relevant intermediate phenotypes include electrophysiologic, neuropsychologic, neuroendocrinologic and, more recently, neuroimaging measures.

Neuroimaging provides access to the neuronal mechanisms underlying emotion, reward, and craving and therefore represents an extraordinary tool to link genes to the neuronal pathways that produce behaviors (for reviews see [102, 149]). For example, amygdala activation after exposure to stressful stimuli predicts anxiety and captures interindividual differences in emotional response and stress resiliency [50]. On the other hand, activation of the prefrontal cortex during working memory performance is used to evaluate prefrontal cognitive function that is impaired in several psychiatric diseases, including addictions. The combination of genetic analysis with brain imaging illustrates the power of cross-disciplinary science.

23.5.6 Finding the Specific Genes Underlying Vulnerability to Addiction

Two main strategies have been used and are increasingly integrated to identify the specific genetic variations influencing addiction: the candidate gene and the genome-wide approaches. In the former, genes known to influence processes involved in the pathogenesis or treatment of addiction are selected. In the latter, the whole genome is interrogated simultaneously in a hypothesis-free fashion. A point of integration between the methods is the study of candidate genes located in chromosome regions implicated by genome-wide scans.

23.5.6.1 Candidate Genes

23.5.6.1.1 Substance-specific Genes Moderating Liability: ADH1B and ALDH2

The alcohol dehydrogenase IB (*ADH1B*) and aldehyde dehydrogenase 2 (*ALDH2*) genes encode for two enzymes catalyzing consecutive steps in alcohol metabolism. In adults, these enzymes have an important role although there are also several other enzymes that can carry out both of these metabolic steps, including catalase, cytochrome P450, and additional enzymes in the ADH and ALDH gene families. In the liver, three ADH genes are expressed at high levels and to some extent at different times of development, and these three enzymes are primarily expressed in hepatocytes. However, it will be seen that despite this complexity of enzyme action in alcohol metabolism, individual functional alleles altering the function of only one enzyme are sufficient to exert a major biochemical effect and an effect on risk. Probably this is because of the relatively greater importance of ADH1B in the adult liver and its lower Km and higher capacity for metabolism than in some of the other enzymes. The ALDH2 enzyme is an ALDH that is encoded in the nuclear genome but translocated to the mitochondrion, where it plays a critical role in the ability of hepatocytes, and other cells throughout the body, to metabolize acetaldehyde. The main roles of the enzymes that have maintained these genes through at least 80 million years of mammalian evolution are in fact somewhat obscure. In their natural environment, mice and rats are not heavy consumers of alcoholic beverages! Yet our distant mammalian cousins possess a full complement of these enzymes. Perhaps the reason for this is that although the liver metabolizes ethanol ingested in beverages it also has to utilize alcohols that are the product of bacterial fermentation in the gut.

The product of ADH is acetaldehyde, a toxic intermediate that may react with a variety of biomolecules. Indeed, acetaldehyde adducts with DNA and both it and alcohol are formally recognized as mutagens. Acetaldehyde is a potent releaser of histamine, triggering the aversive flushing reaction. Symptoms include headache, nausea, palpitations, and flushing of the skin. Ordinarily, acetaldehyde is rapidly converted to acetate, and levels of acetaldehyde remain very low – in the nanomolar range. However, if aldehyde dehydrogenase is blocked by disulfiram (which is used to help alcoholics maintain abstinence) or certain drugs used to treat protozoal infections (e.g., metronidazole) then the flushing reaction is observed after the ingestion of only small quantities of alcohol. Also, if acetaldehyde accumulates the individual is at substantially increased risk of upper gastrointestinal tract cancer, and this can occur either via to pharmacologic blockade of aldehyde dehydrogenase or as a result of natural genetic variation, a factor that physicians may wish to consider in counseling individuals who drink despite carrying the genetic variations that will next be described [12].

Nature has provided two common natural examples of genetic predisposition to alcohol-induced flushing, and it is not surprising that the enzyme variants that lead to flushing are protective against alcoholism. The most important functional loci at *ADH1B* and *ALDH2* are the *ADH1B His47Arg* missense polymorphism, in which Arg47 is a hyperactive allele acting in co-dominant fashion, and *ALDH2 Glu487Lys*, in which the Lys487 allele inactivates ALDH2 dominantly (a manifestation of the tetrameric structure of ALDH2). Higher activity of ADH1B, conferred by Arg47, or lower activity of ALDH2, conferred by Lys487, leads to accumulation of acetaldehyde following alcohol consumption and the flushing reaction. In East Asian populations (e.g., China and Japan), where both His47 and Lys487 are highly abundant, and in Jewish populations, where His47 is abundant, many individuals carry genotypes that are protective against the development of alcoholism. The protective effect seems to vary across environments [139] and shows genotype–genotype additivity [135]. Following up the connection of acetaldehyde to mutation, both polymorphisms have also been associated with enhanced risk of cancers of the oropharynx and esophagus [12, 151]. Both of these functional polymorphisms appear to be ancient in human populations, occurring on characteristic and highly diverged haplotypes. On that basis it is unlikely that either the Arg47 or Lys487 was selected to high frequencies in East Asian populations as protective alleles against alcoholism. One possibility, still speculative, is that the polymorphism alters susceptibility to protozoal infections of the gut, including amebiasis, because an action of metronidazole (an antiprotozoal drug of unknown mechanism) is to inhibit ALDH [41]. However, regardless of the forces responsible for their high frequencies, the pervasive environmental exposure to alcohol that occurs in modern societies has added other dimensions to their effects.

23.5.6.1.2 Genes Moderating Liability to Addiction and Other Diseases: *COMT*, *MAOA*, and *SLC6A4*

Monoamines, including serotonin (5-HT), norepinephrine (NE), and dopamine (DA), are fundamental modulators of emotionality, cognition, reward, and behavioral response to stimuli. Therefore, it is unsurprising that genes regulating monoamines levels such as catechol-*O*-methyltransferase (*COMT*), monoamine oxidase A (*MAOA*), and the serotonin transporter (*SLC6A4*) have been implicated in vulnerability to several psychiatric diseases, including addiction, antisocial personality disorder, depression, and anxiety. In line with these ideas, drugs increasing monoamines in the synaptic cleft or drugs that target receptors of monoamine neurotransmitters are used in the treatment of several psychiatric diseases.

COMT metabolizes DA, NE, and other catecholamine neurotransmitters. COMT plays an important role in the regulation of dopamine levels in the prefrontal cortex because of the paucity of the dopamine transporter in this region [92, 98]. *COMT* knockout mice have increased levels of dopamine in this brain region [37, 150]. In mammals, the COMT enzyme occurs in two distinct forms: as a soluble, cytoplasmic, protein (S-COMT) and as a membrane-bound form (MB-COMT) which – in humans – has 50 additional amino acid residues at the N-terminus. S-COMT predominates in most tissues, accounting for 95% of total COMT activity [66]. However, in brain, the amount of MB-COMT activity is much higher [121]. Val158Met is a common functional single nucleotide substitution of COMT [89], replacing methionine for valine at codon 158 of MB-COMT and at codon 108 of S-COMT. Via its effect on enzyme stability [127, 146] the Met158 allele is three- to fourfold less active than Val158 [21], and the alleles act co-dominantly [138]. Because of its higher activity and the importance of COMT for dopamine metabolism in the frontal cortex, the Val158 allele was predicted to lower dopamine level in that region. Consistent with this idea and with the role of dopamine in tuning frontal cortical function, the Val158 allele has been linked to inefficient frontal lobe function evaluated with different methodologies [29, 39, 97]. Also, in a pharmacogenetic study, the COMT inhibitor tolcapone improved executive function in val/val homozygotes, but not in individuals homozygous for the met allele, a finding consistent with the higher levels of cortical dopamine already expected in individuals with this genotype [38]. On the other hand, the Met158 allele, although associated with better cognitive performance, is associated with decreased stress resiliency and increased anxiety. This allele has been associated with increased anxiety among women populations [33], with increased pain–stress response and a lower pain threshold [26, 155], and with increased amygdala reactivity to unpleasant stimuli [133]. In certain addicted populations, e.g., polysubstance abusers [141], both the Val158 and Met158 alleles have been associated with addictions. The Val158 allele was found to excess among methamphetamine, nicotine, and polysubstance addicts [141]. On the other hand, in addicted populations with high frequencies of internalizing disorders, such as late-onset alcoholics in Finland [137] and Finnish social drinkers [69], increased risk appears to be conferred by the Met158 allele.

This account of multilevel association of a functional locus of the *COMT* gene to behavior illustrates the relative strength of allele effects on intermediate phenotypes such as brain imaging measures of metabolic activity during cognitive tasks or after a painful or emotional challenge test, and the much more modest effects on the common, complex disease. The disease is a more complex phenotype emergent from differences in internal states that are accessed via the intermediate phenotype measures, but the disease state may also emerge for other reasons, or fail to manifest at all. Frustratingly, the addictions are not etiologically defined (people become addicted for different reasons) so that it is illogical to expect gene-consistent effects across different populations of patients (e.g., addicted patients who might differ at age at onset, or whose additions varied in severity, or who had different risk exposures). In the addictions, studies identifying genes such as COMT have had the beneficial effect of focusing more attention on intermediate phenotypes and clinical subgroups, and on the history of environmental exposure, all information with which genetic markers are likely to be used in concert to improve the nosology of these common diseases.

MAOA is an X-linked gene encoding monoamine oxidase A, a mitochondrial enzyme that metabolizes monoamine neurotransmitters including norepinephrine, dopamine and serotonin. *MAOA* knockout mice have higher levels of these neurotransmitters and manifest increased aggressive behavior and stress reactivity [16]. In the human, different *MAOA* genetic variants impair MAOA activity to different degrees, and the reduction in enzyme activity appears to parallel the effect on behavior. In 1993, Brunner et al. [13] reported a Dutch family in which eight males were affected by a syndrome characterized by borderline mental retardation and impulsive behavior including impulsive aggression, arson, attempted rape, fighting, and exhibitionism. The cause was a stop-codon variant in the eighth exon of *MAOA* leading to complete and selective deficiency of MAOA activity, with an X-linked pattern transmission from unaffected mothers carrying the stop-codon (see Fig. 23.5.12).

Discovery of this mutation led to attempts to identify it in other individuals with behavioral dyscontrol, including individuals accused of serious crimes. However, and despite intensive effort, the stop-codon variant was not found in other individuals, and thus represents an example of a rare, private allele. More recently, a common *MAOA* polymorphism influencing MAOA transcription was discovered [126]. This locus, termed the MAOA-linked polymorphic region (*MAOA–LPR*), is a variable-number tandem repeat (VNTR) located approximately 1.2 kb upstream of the *MAOA* start codon and within the gene's transcriptional control region [25, 126]. Alleles at this VNTR have a different number of tandem copies of a 30-bp sequence, with the three- and four-repeat alleles being by far the most common. Alleles with four repeats are transcribed more efficiently than alleles with three copies of the repeat, and therefore lead to higher MAOA enzyme activity [126]. In a longitudinally studied cohort of boys, Caspi et al. [18] found that *MAOA–LPR* moderated the effect of childhood maltreatment on vulnerability to develop antisocial behavior. In this study maltreated boys with the low-activity genotype were more likely to develop antisocial problems later in life than boys with the high-activity genotype. Meta-analysis of several studies that represent attempts at replication revealed a significant pooled G×E effect for *MAOA* and stress. A similar *MAOA* × stress interaction appears to occur in women, although of course a much smaller percentage of women are homozygous for the low expression allele (males being hemizygous for *MAOA*). In a sample of Native American women, the effect of childhood sexual abuse (frequent among women in this and other populations) on risk of developing alcoholism and antisocial personality disorder was contingent upon *MAOA–LPR* genotype [27]. Sexually abused women homozygous for the low-activity *MAOA–LPR* allele had high rates of both disorders, and heterozygous women displayed an intermediate risk pattern. However, in the absence of childhood sexual abuse, there was no relationship between *MAOA* genotype and these disorders.

MAOA G×E has also been studied in animal models. These are useful because of the ability to control stress exposures and many other environmental variables. In the Rhesus macaque (*Macaca mulatta*) early life stress exposure, particularly early separation from the mother, leads to dyscontrolled behavior and enhanced stress response later in life. The behaviors observed in the stressed animals include increased alcohol consumption, higher impulsive aggression, incompetent social behavior and serotonin dysfunction, and increased behavioral and endocrine responsivity to

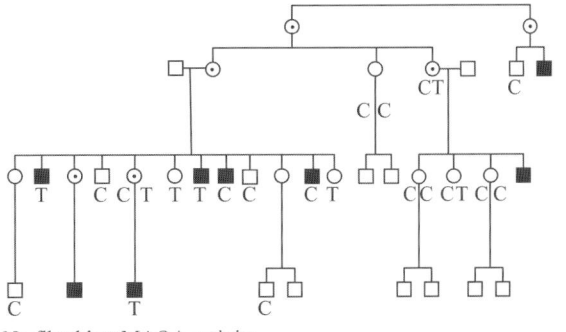

Fig. 23.5.12 Pedigree of a Dutch family with eight males affected by Brunner syndrome, X-linked behavioral dyscontrol caused by a stop codon in monoamine oxidase A. This variant (C936T) leads to complete and selective deficiency of MAOA activity. The X-linked pattern of transmission features unaffected carrier mothers, and both affected (carrier) and unaffected (noncarrier) male offspring

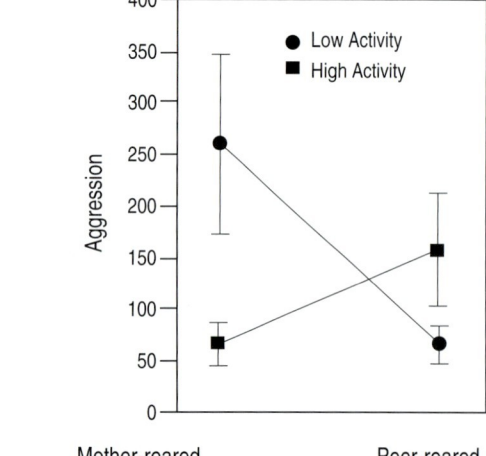
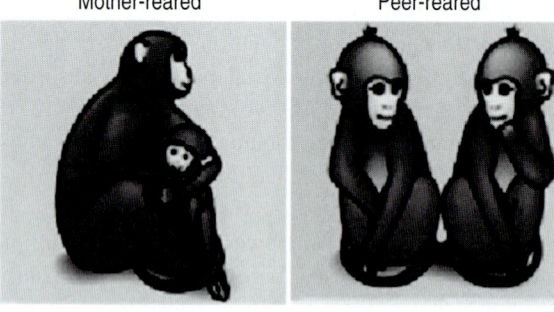

Fig. 23.5.13 Interactive effect between a genetic variant (rhMAOA-LPR) in the promoter of the rhesus monkey MAOA gene and maternal separation on aggressive behavior. The low-activity MAOA allele is associated with increased aggression only among mother-reared monkeys (*left*), and not among peer-reared monkeys (*right*). Adapted from [105], with permission

stress (for review see [5]). Remarkably, an orthologous (same evolutionary origin and same function) VNTR polymorphism is also found in the promoter region of *MAOA* in the Rhesus macaque. Also similar to the human, the lower activity allele predicts aggressive behavior in these animals, and the association is dependent on whether the monkey had been separated from its mother (see Fig. 23.5.13) [105].

Where in the brain does MAOA mediate its effects on behavioral variation? MAOA activity in the hippocampus, a brain region that processes emotional experience and memory, may be critical to the interaction between *MAOA* and childhood trauma. Carriers of low-activity *MAOA–LPR* allele hyperactivate the hippocampus and amygdala during the retrieval of negatively valenced emotional material, but not during the retrieval of neutral material [103]. Therefore, the increased sensitivity to adverse experiences of carriers of the low activity *MAOA* allele might be due to their stronger activation by negative stimuli and their converse impairment in extinguishing adverse memories and conditioned fears.

How do endocrine factors that modulate behavioral control and aggression interact with *MAOA* alleles that modulate the same behaviors? This is a complex question because of the possibility that effects of hormones and *cis*-acting genetic elements could converge on the expression of the gene (in this case *MAOA*). On the other hand, the hormonal environment of the brain could interact with the genotype-influenced availability of monoamine neurotransmitters. In fact, there is a powerful gene × endocrine interaction between *MAOA* and testosterone on the outcome of dyscontrolled and aggressive behavior. This interaction was explored because of the role of testosterone in aggressive behavior, which in part explains high male:female ratios for acts of violence and aggression resulting in criminal convictions. Within males (females have much lower testosterone levels) there is a moderate correlation between testosterone level and lifetime aggression score. However, this relationship is contingent upon the male having the low-expression *MAOA* genotype, as approximately half of males do [132] (Fig. 23.5.14).

The mechanism of the interaction is an open question, because while both low activity and high testosterone lead to behavioral dyscontrol, androgens increase *MAOA* expression through response elements located within the *MAOA* promoter [108].

The serotonin transporter (*SLC6A4*) is a key regulator of the level of serotonin in the synapse. The effects of serotonin are illustrated in part by serotonin transporter blockade, which can be accomplished with drugs such as the serotonin-specific reuptake inhibitors commonly used to treat depression, anxiety, and chronic pain. The serotonin transporter gene *SLC6A4* has a common polymorphism in its promoter region (*5-HTTLPR*). The major alleles, which affect transcription of the gene, involve 16 (*L*) or 14 (*S*) copies of a 20- to 23-bp imperfect repeated sequence [91]. Furthermore there is a relatively common, functional A>G substitution within the *L* allele [63]. The low-transcribing *s* allele has been inconsistently associated with trait anxiety, depression, and alcoholism. However, the effect of this allele on behavior appears to be stronger if stress exposure is taken into account. *5-HTTLPR* moderates the impact of stressful life

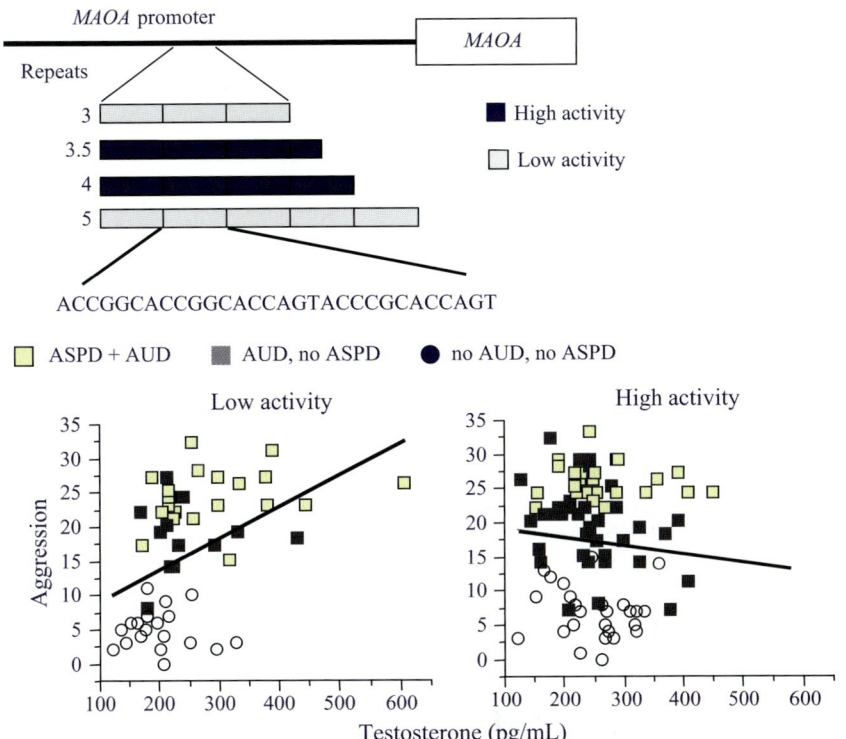

Fig. 23.5.14 Interaction between testosterone level and a genetic variant (MAOA-LPR) in the promoter of the human MAOA gene on lifetime aggression in males from a Finnish population. These include alcoholics with co-morbid antisocial personality disorder (AUD + ASPD), alcoholics without co-morbid antisocial personality disorder (AUD, no ASPD) and controls (no AUD, no ASPD). (**a**) Alleles at the MAOA-LPR locus differ for number of copies of a 30-bp repeated sequence. Alleles with 3.5 and 4 repeats are transcribed more efficiently than alleles with three and five copies of the repeat, and are therefore associated with increased MAOA activity. Indeed, some controversies exist for the activity associated with the five repeats allele (see text). (**b**) Elevated level of testosterone is associated with increased aggression among carriers of the low-activity allele. In contrast, no association between testosterone level and aggression is detected among carriers of the high activity allele. Adapted from [132], with permission

events on risk of depression and suicide [19, 125]. Carriers of the low-transcribing *S* allele exhibit more depression and suicidality following stressful life events than *L* individuals with two copies of the *L* allele [19] (Fig. 23.5.15).

Furthermore, *5-HTTLPR* has been shown to moderate the functions of brain regions, such as the amygdala, that are critical in emotional regulation and response to environmental changes. Carriers of the low-activity allele display increased amygdala reactivity to fearful stimuli [50], reduced amygdala volume [111], and enhanced functional coupling between the amygdala and the ventromedial prefrontal cortex [57], a brain region that ordinarily modulates the activity of the amygdala such that emotional responses are buffered. Closer to the function of the gene, an effect of *5-HTTLPR* genotype on transporter expression in brain in vivo has been reported in some studies [56] but not in others [131]. As was the case for *MAOA*, the Rhesus macaque again has an orthologous polymorphism in the promoter region of its gene. Consistent with findings in humans, the macaque rs-5HTTLPR polymorphism influenced alcohol consumption and stress response, depending on rearing conditions. Carriers of the low-expression serotonin transporter genotype that were separated from their mothers at an early age displayed higher stress reactivity and ethanol preference [6]. The combined effect of *rh-HTTLPR* and environment on stress reactivity suggests that the influence of *HTTLPR* on behavior might be traced to altered regulation of the hypothalamic–pituitary–adrenal (HPA) axis.

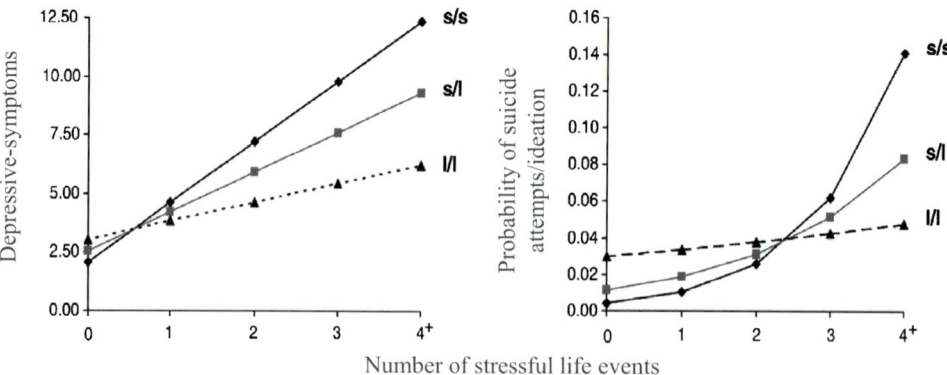

Fig. 23.5.15 Influence of life stress on depression and suicide: moderation by a polymorphism (5HTT-LPR) in the promoter of the serotonin transporter gene (SLC6A4). Individuals homozygous for the low-activity allele (*ss*) who were exposed to more than four stressful events were at higher risk for depression and suicide attempts/ideation than were those homozygous for the high activity allele (*ll*). Differences in risk between the three genotypes emerged progressively in the context of stress life exposure, consistent with a gene by environment interaction. Adapted from [19], with permission

23.5.6.2 Genetic Mapping

Genetic mapping is the localization of genes underlying a trait on the basis of correlation with DNA variation, without the need for prior hypotheses of biological function [3]. For the addictions, genome-wide scans, including whole-genome linkage and whole-genome association (WGA), have implicated several chromosome regions. Here the perspective is the progress of these efforts and the genes thus far identified in what will come to be viewed as the early days of these efforts: in particular a nicotinic receptor gene that has a role in vulnerability to nicotine addiction, as well as lung cancer.

23.5.6.2.1 Whole-Genome Linkage

In whole-genome linkage studies a panel of polymorphisms is tested for meiotic linkage to a disease in family-based samples. This is done by identifying chromosome regions that are shared more often among phenotypically concordant relatives than among phenotypically discordant family members. The implicated chromosomal regions are usually broad, e.g., greater than 10 Mb. Therefore, a more refined search for candidate genes within the disease-linked region is subsequently conducted using association or sequencing.

To perform whole-genome linkage analysis for alcoholism and other addictions, several large family-based data sets have been collected. These include the Collaborative Study on the Genetics of Alcoholism (COGA) [120], the Roscommon study of Irish families [119]; a sample of multiplex families collected in the Pittsburgh area [60]; and samples collected from relatively isolated populations, including Native Americans [28, 30, 94] and Finns [90]. Such isolated populations, and large families within them, are likely to confer the advantage of reduced genetic heterogeneity. A non-exhaustive list of convergent findings across family linkage studies includes a region on chromosome 4q that contains the alcohol dehydrogenase (*ADH*) gene cluster [30, 94, 117, 120] and a chromosome 4p region containing a γ-aminobutyric acid receptor (*GABA$_A$*) gene cluster [94, 120]. In the COGA sample there was also evidence for linkage to chromosomes 1 and 7, and to chromosome 2 at the location of an opioid receptor gene [60]. A region on chromosome 1 was linked to alcoholism and affective disorder in the COGA data set [106], providing more information in support of the existence of a genetic overlap between alcoholism and internalizing disorders. A region on chromosome 7 was linked to alcoholism and/or illicit drug disorders in a subset of COGA families with high density for childhood and adult antisocial behaviors [65]. Linkage analyses have also been conducted with intermediate phenotypes for alcoholism, including low response to alcohol [106], neurophysiological endophenotypes such as P300 [114], and reduced alpha power [28, 36], and chromosome regions identified by these studies overlap partially with those reported for alcoholism.

$GABA_A$ Receptors. γ-Aminobutyric acid (GABA) is the primary inhibitory neurotransmitter in the central nervous system. $GABA_A$ receptor-mediated chloride currents into neurons are facilitated by various drugs including ethanol, benzodiazepines, and barbiturates. Several lines of evidence suggest that GABA is involved in many effects of alcohol, including tolerance, dependence, and cross-tolerance to benzodiazepines and barbiturates. A series of mouse ethanol-related behaviors, including preference, withdrawal severity, and sedation sensitivity, map to quantitative trait loci (QTL) regions where $GABA_A$ receptor-gene clusters are located [24, 87]. In the rat, an Arg100Gln missense variant located in the $GABA_A$ α6 subunit gene (*GABRA6*) was associated with variation in ethanol and benzodiazepine sensitivity [87]. In humans, AUD has been linked to both the chromosome 4 [28, 34] and chromosome 5 [118] GABA clusters. Linkage signals appear to derive from *GABRA6* on chromosome 5 [118] and *GABRA2* [28, 34] and *GABRG1* [33] on chromosome 4. In the COGA sample, the association between *GABRA2* and alcoholism was mainly driven by alcoholics who also abused illicit substances, indicating that this gene might contribute to the shared liability to a different class of addictive disorders [2]. A human variant of *GABRA6* (Pro385Ser), which is located in the chromosome 5 cluster, was associated with sensitivity to alcohol [62] and benzodiazepines [64].

23.5.6.2.2 Whole-Genome Association

Large-scale genotyping techniques have recently become available, and genome-wide analyses for several complex diseases, including diabetes, obesity, bipolar disorder, inflammatory, bowel disease, prostate cancer, breast cancer, colorectal cancer, and rheumatoid arthritis (for review see [3]), have been completed within large data sets of unrelated individuals using dense panels that can include up to 1 million single nucleotides (SNPs). These WGA studies have the advantage of increased power for detecting effects of relatively common alleles (>0.05) and more refined localization of signals to smaller chromosome regions than family-based linkage analyses, which have a reciprocal advantage of being powerful for detecting effects of rare and uncommon alleles that are present in only a small proportion of probands and their families.

A WGA on nicotine addiction was performed using number of cigarettes per day regularly smoked as a phenotype, in two European populations with a total of 7,500 persons. Although no SNP reached genome-wide statistical significance, a trend toward association was found for a common haplotype in the *CHRNA3-CHRNA5* nicotinic receptor subunit gene cluster on chromosome 15. This result has been replicated in a third set of 7,500 additional European individuals [7]. A, a missense mutation in the alpha5 nicotinic cholinergic receptor (*CHRNA5*), has been shown to influence vulnerability to nicotine dependence [9] and lung cancer [136].

For alcoholism, a WGA scan was conducted in a sample of unrelated alcohol-dependent ($n = 120$) and control ($n = 160$) individuals sampled from the COGA pedigrees [68]. This study identified several candidate genes that might moderate vulnerability to alcoholism and whose products are involved in cellular signaling, gene regulation, development, cell adhesion and Mendelian disorders. However, these findings are weakened by the small sample size and the consequent lack of power to identify exhaustively common alcoholism-causing alleles. WGA in large case-control data sets will soon be reported for alcoholism; there are ongoing studies on alcohol consumption including more than 20,000 individuals!

23.5.7 Treatment of Addictions

Treatment of addictive illnesses is enormously beneficial, in the same sense as it is also worthwhile to treat other diseases, such as cancer, where success rates are substantially less than perfect, and where – as in the addictions – volition and lifestyle choices also have a powerful role in etiology and outcome. The maintenance of abstinence for multiyear periods has enormous benefits to the individual, family, and community. Half or more of addicted individuals can make the transition to lifetime freedom from relapse, although this is usually only accomplished through a combination of individual will, lifestyle changes, family support, self-help groups (including Alcoholics Anonymous), and medical care. The multidimensional nature of addictions makes their etiology more difficult to comprehend, but paradoxically increases the range of opportunities for interventions that can sometimes be used in complementary fashion. Effective interventions extend from the spiritual and religious to drug therapies that ease withdrawal, block the action of an addictive drug (antagonist therapies), substitute for the addictive agent (agonist therapies), or reduce symptoms such as anxiety and depression which accompany long-term withdrawal and can trigger relapse. Also, understanding the role of neurobiology and genetic factors in vulnerability to addictions has been crucial in the destigmatization of these illnesses, thus encouraging their diagnosis and treatment. Lastly, the identification of genes altering the liability to addiction and ability to recover are a major focus for genetic studies, because these could provide new therapeutic targets and an ability to individualize treatment (so-called personalized medicine). One of the first examples of pharmacogenetic prediction of treatment response in the addictions is a common functional missense variant of the mu-opioid receptor (*OPRM1* Asn40Asp). In several studies, naltrexone, a mu-opioid receptor antagonist, was observed to augment abstinence and good therapeutic outcome in recovering alcoholics. Carriers of the Asp40 allele appear to be highly likely to show clinical improvement when treated with this drug, encouraging the idea that the treatment of this large clinical population can be better targeted [4, 107].

23.5.8 Conclusion

Addictions are common, complex disorders illustrating the interplay of gene×environment interaction. These disorders, which are in part volitional, in part inborn, and in part determined by environmental experience, pose the full range of medical, genetic, policy, and moral challenges. Gene discovery is being facilitated by a variety of powerful approaches, but is in its infancy. It is not surprising that the genes discovered so far act in a variety of ways: via altered metabolism of drug (the alcohol metabolic gene variants for alcohol), via altered function of the receptor of the drug (the nicotinic receptor for nicotine), and via general mechanisms of addiction (genes such as monoamine oxidase A and the serotonin transporter that modulate stress response, emotion, and behavioral control).

References

1. Agrawal A, Neale MC, Jacobson KC, Prescott CA, Kendler KS (2005) Illicit drug use and abuse/dependence: modeling of two-stage variables using the CCC approach. Addict Behav 30:1043–1048
2. Agrawal A, Edenberg HJ, Foroud T, Bierut LJ, Dunne G, Hinrichs AL, Nurnberger JI, Crowe R, Kuperman S, Schuckit MA, Begleiter H, Porjesz B, Dick DM (2006) Association of GABRA2 with drug dependence in the collaborative study of the genetics of alcoholism sample. Behav Genet 36:640–650
3. Altshuler D, Daly MJ, Lander ES (2008) Genetic mapping in human disease. Science 322:881–888
4. Anton RF, Oroszi G, O'Malley S, Couper D, Swift R, Pettinati H, Goldman D (2008) An evaluation of mu-opioid receptor (OPRM1) as a predictor of naltrexone response in the treatment of alcohol dependence: results from the combined pharmacotherapies and behavioral interventions for alcohol dependence (COMBINE) study. Arch Gen Psychiatry 65:135–144
5. Barr CS, Newman TK, Becker ML, Parker CC, Champoux M, Lesch KP et al (2003) The utility of the non-human primate model for studying gene by environment interactions in behavioral research. Genes Brain Behav 2:336–340
6. Barr CS, Newman TK, Lindell S, Shannon C, Champoux M, Lesch KP et al (2004) Interaction between serotonin transporter gene variation and rearing condition in alcohol preference and consumption in female primates. Arch Gen Psychiatry 61:1146–1152
7. Berrettini W, Yuan X, Tozzi F, Song K, Francks C, Chilcoat H, Waterworth D, Muglia P, Mooser V (2008) Alpha-5/alpha-3 nicotinic receptor subunit alleles increase risk for heavy smoking. Mol Psychiatry 13:368–373
8. Bierut LJ, Dinwiddie SH, Begleiter H, Crowe RR, Hesselbrock V, Nurnberger JI Jr (1998) Familial transmission of substance dependence: alcohol, marijuana, cocaine, and habitual smoking: a report from the Collaborative Study on the Genetics of Alcoholism. Arch Gen Psychiatry 55:982–988
9. Bierut LJ, Stitzel JA, Wang JC, Hinrichs AL, Grucza RA, Xuei X, Saccone NL, Saccone SF, Bertelsen S, Fox L, Horton WJ, Breslau N, Budde J, Cloninger CR, Dick DM, Foroud T, Hatsukami D, Hesselbrock V, Johnson EO, Kramer J, Kuperman S, Madden PA, Mayo K, Nurnberger J Jr, Pomerleau O, Porjesz B, Reyes O, Schuckit M, Swan G, Tischfield JA, Edenberg HJ, Rice JP, Goate AM (2008) Variants in nicotinic receptors and risk for nicotine dependence. Am J Psychiatry 165(9):1163–1171
10. Binder EB, Bradley RG, Liu W, Epstein MP, Deveau TC, Mercer KB, Tang Y, Gillespie CF, Heim CM, Nemeroff CB, Schwartz AC, Cubells JF, Ressler KJ (2008) Association of FKBP5 polymorphisms and childhood abuse with risk of posttraumatic stress disorder symptoms in adults. JAMA 299:1291–1305
11. Blomeyer D, Treutlein J, Esser G, Schmidt MH, Schumann G, Laucht M (2007) Interaction between CRHR1 gene and stressful life events predicts adolescent heavy alcohol use. Biol Psychiatry 63:146–151
12. Brooks PJ, Goldman D, Li TK (2009) Alleles of alcohol and acetaldehyde metabolism genes modulate susceptibil-

ity to oesophageal cancer from alcohol consumption. Hum Genomics 3:103–105
13. Brunner HG, Nelen M, Breakefield XO, Ropers HH, van Oost BA (1993) Abnormal behavior associated with a point mutation in the structural gene for monoamine oxidase A. Science 262:578–580
14. Buster MA, Rodgers JL (2000) Genetic and environmental influences on alcohol use: DF analysis of NLSY kinship data. J Biosoc Sci 32:177–189
15. Carmelli D, Heath AC, Robinette D (1993) Genetic analysis of drinking behavior in World War II veteran twins. Genet Epidemiol 10:201–213
16. Cases O, Seif I, Grimsby J, Gaspar P, Chen K, Pournin S et al (1995) Aggressive behavior and altered amounts of brain serotonin and norepinephrine in mice lacking MAOA. Science 268:1763–1766
17. Caspi A, Moffitt TE (2006) Gene–environment interactions in psychiatry: joining forces with neuroscience. Nat Rev Neurosci 7:583–590
18. Caspi A, McClay J, Moffitt TE, Mill J, Martin J, Craig IW et al (2002) Role of genotype in the cycle of violence in maltreated children. Science 297:851–854
19. Caspi A, Sugden K, Moffitt TE, Taylor A, Craig IW, Harrington H et al (2003) Influence of life stress on depression: moderation by a polymorphism in the 5-HTT gene. Science 301:386–389
20. Caspi A, Moffitt TE, Cannon M, McClay J, Murray R, Harrington H et al (2005) Moderation of the effect of adolescent-onset cannabis use on adult psychosis by afunctional polymorphism in the catechol-O-methyltransferase gene: longitudinal evidence of a gene–environment interaction. Biol Psychiatry 57:1117–1127
21. Chen J, Lipska BK, Halim N, Ma QD, Matsumoto M, Melhem S et al (2004) Functional analysis of genetic variation in catechol-O-methyltransferase (COMT): effects on mRNA, protein, and enzyme activity in postmortem human brain. Am J Hum Genet 75:807–821
22. Compton WM, Thomas YF, Stinson FS, Grant BF (2007) Prevalence, correlates, disability, and comorbidity of DSM-IV drug abuse and dependence in the United States: results from the national epidemiologic survey on alcohol and related conditions. Arch Gen Psychiatry 64:566–576
23. Connor JP, Young RM, Saunders JB, Lawford BR, Ho R, Ritchie TL, Noble EP (2008) The A1 allele of the D2 dopamine receptor gene region, alcohol expectancies and drinking refusal self-efficacy are associated with alcohol dependence severity. Psychiatry Res 160:94–105
24. Crabbe JC, Phillips TJ, Buck KJ, Cunningham CL, Belknap JK (1999) Identifying genes for alcohol and drug sensitivity: recent progress and future directions. Trends Neurosci 22:173–179
25. Deckert J, Catalano M, Syagailo YV, Bosi M, Okladnova O, Di Bella D et al (1999) Excess of high activity monoamine oxidase a gene promoter alleles in female patients with panic disorder. Hum Mol Genet 8:621–624
26. Diatchenko L, Slade GD, Nackley AG, Bhalang K, Sigurdsson A, Belfer I et al (2005) Genetic basis for individual variations in pain perception and the development of a chronic pain condition. Hum Mol Genet 14:135–143
27. Ducci F, Enoch MA, Hodgkinson C, Xu K, Catena M, Robin RW et al (2007) Interaction between a functional MAOA locus and childhood sexual abuse predicts alcoholism and antisocial personality disorder in adult women. Mol Psychiatry 13:334–347
28. Edenberg HJ, Dick DM, Xuei X, Tian H, Almasy L, Bauer LO et al (2004) Variations in GABRA2, encoding the alpha 2 subunit of the GABA(A) receptor, are associated with alcohol dependence and with brain oscillations. Am J Hum Genet 74:705–714
29. Egan MF, Goldberg TE, Kolachana BS, Callicott JH, Mazzanti CM, Straub RE et al (2001) Effect of COMT Val108/158 Met genotype on frontal lobe function and risk for schizophrenia. Proc Natl Acad Sci USA 98:6917–6922
30. Ehlers CL, Gilder DA, Wall TL, Phillips E, Feiler H, Wilhelmsen KC (2004) Genomic screen for loci associated with alcohol dependence in Mission Indians. Am J Med Genet B Neuropsychiatr Genet 129:110–115
31. Ehlers CL, Wall TL, Corey L, Lau P, Gilder DA, Wilhelmsen K (2007) Heritability of illicit drug use and transition to dependence in Southwest California Indians. Psychiatr Genet 17:171–176
32. Eisen SA et al (1998) Familial influences on gambling behavior: an analysis of 3359 twin pairs. Addiction 93:1375–1384
33. Enoch MA, Xu K, Ferro E, Harris CR, Goldman D (2003) Genetic origins of anxiety in women: a role for a functional catechol-O-methyltransferase polymorphism. Psychiatr Genet 13:33–41
34. Enoch MA, Schwartz L, Albaugh B, Virkkunen M, Goldman D (2006) Dimensional anxiety mediates linkage of GABRA2 haplotypes with alcoholism. Am J Med Genet B Neuropsychiatr Genet 141:599–607
35. Enoch MA, Waheed JF, Harris CR, Albaugh B, Goldman D (2006) Sex differences in the influence of COMT Val158Met on alcoholism and smoking in plains American Indians. Alcohol Clin Exp Res 30:399–406
36. Enoch MA, Shen PH, Ducci F, Yuan Q, Liu J, White KV, Albaugh B, Hodgkinson CA, Goldman D (2008) Common genetic origins for EEG, alcoholism and anxiety: the role of CRH-BP. PLoS ONE 3:e3620
37. Giakoumaki SG, Roussos P, Bitsios P (2008) Improvement of prepulse inhibition and executive function by the COMT inhibitor tolcapone depends on COMT Val158Met Polymorphism. Neuropsychopharmacology 33:3058–3068
38. Gogos JA, Morgan M, Luine V, Santha M, Ogawa S, Pfaff D et al (1998) Catechol-O-methyltransferase-deficient mice exhibit sexually dimorphic changes in catecholamine levels and behavior. Proc Natl Acad Sci USA 95:9991–9996
39. Goldberg TE, Egan MF, Gscheidle T, Coppola R, Weickert T, Kolachana BS et al (2003) Executive subprocesses in working memory: relationship to catechol-O-methyltransferase Val158Met genotype and schizophrenia. Arch Gen Psychiatry 60:889–896
40. Goldman D, Bergen A (1998) General and specific inheritance of substance abuse and alcoholism. Arch Gen Psychiatry 55:964–965
41. Goldman D, Enoch MA (1990) Genetic epidemiology of ethanol metabolic enzymes: a role for selection. World Rev Nutr Diet 63:143–160
42. Goldman D, Urbanek M, Guenther D, Robin R, Long JC (1998) A functionally deficient DRD2 variant [Ser311Cys] is not linked to alcoholism and substance abuse. Alcohol 16:47–52

43. Goldman D, Oroszi G, Ducci F (2005) The genetics of addiction: uncovering the genes. Nat Rev Genet 6:521–532
44. Gottesman II, Gould TD (2003) The endophenotype concept in psychiatry: etymology and strategic intentions. Am J Psychiatry 160:636–645
45. Grant BF, Dawson DA (1997) Age at onset of alcohol use and its association with DSM-IV alcohol abuse and dependence: results from the National longitudinal alcohol epidemiologic survey. J Subst Abuse 9:103–110
46. Grant BF, Dawson DA (1998) Age of onset of drug use and its association with DSM-IV drug abuse and dependence: results from the National longitudinal alcohol epidemiologic survey. J Subst Abuse 10:163–173
47. Grant BF, Stinson FS, Dawson DA, Chou SP, Dufour MC, Compton W, Pickering RP, Kaplan K (2004) Prevalence and co-occurrence of substance use disorders and independent mood and anxiety disorders: results from the National epidemiologic survey on alcohol and related conditions. Arch Gen Psychiatry 61:807–816
48. Grant BF, Hasin DS, Chou SP, Stinson FS, Dawson DA (2004) Nicotine dependence and psychiatric disorders in the United States: results from the National epidemiologic survey on alcohol and related conditions. Arch Gen Psychiatry 61:1107–1115
49. Grant BF, Goldstein RB, Chou SP, Huang B, Stinson FS, Dawson DA, Saha TD, Smith SM, Pulay AJ, Pickering RP, Ruan WJ, Compton WM (2008) Sociodemographic and psychopathologic predictors of first incidence of DSM-IV substance use, mood and anxiety disorders: results from the Wave 2 National Epidemiologic Survey on Alcohol and Related Conditions. Mol Psychiatry (in press)
50. Hariri AR, Mattay VS, Tessitore A, Kolachana B, Fera F, Goldman D (2002) Serotonin transporter genetic variation and the response of the human amygdala. Science 297:400–403
51. Hasin DS (2003) Classification of alcohol use disorders. Alcohol Res Health 27:5–17
52. Hasin DS, Stinson FS, Ogburn E, Grant BF (2007) Prevalence, correlates, disability, and comorbidity of DSM-IV alcohol abuse and dependence in the United States: results from the National epidemiologic survey on alcohol and related conditions. Arch Gen Psychiatry 64:830–842
53. Heath AC, Martin NG (1993) Genetic models for the natural history of smoking: evidence for a genetic influence on smoking persistence. Addict Behav 18:19–34
54. Heath AC, Bucholz KK, Madden PA, Dinwiddie SH, Slutske WS, Bierut LJ, Statham DJ, Dunne MP, Whitfield JB, Martin NG (1997) Genetic and environmental contributions to alcohol dependence risk in a national twin sample: consistency of findings in women and men. Psychol Med 27:1381–1396
55. Heath AC, Madden PA, Bucholz KK, Dinwiddie SH, Slutske WS, Bierut LJ et al (1999) Genetic differences in alcohol sensitivity and the inheritance of alcoholism risk. Psychol Med 29:1069–1081
56. Heinz A, Jones DW, Mazzanti C, Goldman D, Ragan P, Hommer D et al (2000) A relationship between serotonin transporter genotype and in vivo protein expression and alcohol neurotoxicity. Biol Psychiatry 47:643–649
57. Heinz A, Braus DF, Smolka MN, Wrase J, Puls I, Hermann D et al (2005) Amygdala–prefrontal coupling depends on a genetic variation of the serotonintransporter. Nat Neurosci 8:20–21
58. Hettema JM, Corey LA, Kendler KSA (1999) Multivariate genetic analysis of the use of tobacco, alcohol, and caffeine in a population based sample of male and female twins. Drug Alcohol Depend 57:69–78
59. Hicks BM, Krueger RF, Iacono WG, McGue M, Patrick CJ (2004) Family transmission and heritability of externalizing disorders: a twin-family study. Arch Gen Psychiatry 61:922–928
60. Hill J, Emery RE, Harden KP, Mendle J, Turkheimer E (2008) Alcohol use in adolescent twins and affiliation with substance using peers. J Abnorm Child Psychol 36:81–94
61. Hill SY, Shen S, Zezza N, Hoffman EK, Perlin M, Allan WA (2004) genome wide search for alcoholism susceptibility genes. Am J Med Genet B Neuropsychiatr Genet 128:102–113
62. Hu X, Oroszi G, Chun J, Smith TL, Goldman D, Schuckit MA (2005) An expanded evaluation of the relationship of four alleles to the level of response to alcohol and the alcoholism risk. Alcohol Clin Exp Res 29:8–16
63. Hu X, Lipsky RH, Zhu G, Akhtar LA, Taubman J, Greenberg BD et al (2006) Serotonin transporter promoter gain-of-function genotypes are linked to obsessive-compulsive disorder. Am J Hum Genet 78:815–826
64. Iwata N, Cowley DS, Radel M, Roy-Byrne PP, Goldman D (1999) Relationship between a GABAA alpha 6 Pro385Ser substitution and benzodiazepine sensitivity. Am J Psychiatry 156:1447–1449
65. Jacobson KC, Beseler CL, Lasky-Su J, Faraone SV, Glatt SJ, Kremen WS, Lyons MJ, Tsuang MT (2008) Ordered subsets linkage analysis of antisocial behavior in substance use disorder among participants in the Collaborative Study on the Genetics of Alcoholism. Am J Med Genet B Neuropsychiatr Genet 147B:1258–1269
66. Jeffery DR, Roth JA (1984) Characterization of membranebound and soluble catechol-0-methyltransferase from human frontal cortex. J Neurochem 42:826–832
67. Job MO, Tang A, Hall FS, Sora I, Uhl GR, Bergeson SE, Gonzales RA (2007) Mu (mu) opioid receptor regulation of ethanol-induced dopamine response in the ventral striatum: evidence of genotype specific sexual dimorphic epistasis. Biol Psychiatry 62:627–634
68. Johnson C, Drgon T, Liu QR, Walther D, Edenberg H, Rice J et al (2006) Pooled association genome scanning for alcohol dependence using 104 268 SNPs: validation and use to identify alcoholism vulnerability loci in unrelated individuals from the collaborative study on the genetics of alcoholism. Am J Med Genet B Neuropsychiatr Genet 141:844–853
69. Kauhanen J, Hallikainen T, Tuomainen TP, Koulu M, Karvonen MK, Salonen JT et al (2000) Association between the functional polymorphism of catechol-*O*-methyltransferase gene and alcohol consumption among social drinkers. Alcohol Clin Exp Res 24:135–139
70. Kendler KS (2007) Baker JH (2007) Genetic influences on measures of the environment: a systematic review. Psychol Med 37:615–626
71. Kendler KS, Prescott CA (1998) Cannabis use, abuse, and dependence in a population-based sample of female twins. Am J Psychiatry 155:1016–1022

72. Kendler KS, Prescott CA (1998) Cocaine use, abuse and dependence in a population-based sample of female twins. Br J Psychiatry 173:345–350
73. Kendler KS, Prescott CA (1999) Caffeine intake, tolerance, and withdrawal in women: a population-based twin study. Am J Psychiatry 156:223–228
74. Kendler KS, Prescott CA (2006) Genes, environment, and psychopathology. The Guilford Press, New York, Understanding the causes of psychiatric and substance use disorders
75. Kendler KS, Heath AC, Neale MC, Kessler RC, Eaves LJ (1992) A population-based twin study of alcoholism in women. JAMA 268:1877–1882
76. Kendler KS, Heath AC, Neale MC, Kessler RC, Eaves LJ (1993) Alcoholism and major depression in women A twin study of the causes of comorbidity. Arch Gen Psychiatry 50:690–698
77. Kendler KS, Davis CG, Kessler RC (1997) The familial aggregation of common psychiatric and substance use disorders in the National comorbidity survey: a family history study. Br J Psychiatry 170:541–548
78. Kendler KS, Neale MC, Sullivan P, Corey LA, Gardner CO, Prescott CA (1999) A population based twin study in women of smoking initiation and nicotine dependence. Psychol Med 29:299–308
79. Kendler KS, Karkowski LM, Neale MC, Prescott CA (2000) Illicit psychoactive substance use, heavy use, abuse, and dependence in a U.S. population-based sample of male twins. Arch Gen Psychiatry 57:261–269
80. Kendler KS, Thornton LM, Pedersen NL (2000) Tobacco consumption in Swedish twins reared apart and reared together. Arch Gen Psychiatry 57:886–892
81. Kendler KS, Prescott CA, Myers J et al (2003) The structure of genetic and environmental risk factors for common psychiatric and substance use disorders in men and women. Arch Gen Psychiatry 60:929–937
82. Kendler KS, Jacobson KC, Gardner CO, Gillespie N, Aggen SA, Prescott CA (2007) Creating a social world a developmental twin study of peer-group deviance. Arch Gen Psychiatry 64:958–965
83. Kendler KS, Myers J, Prescott CA (2007) Specificity of genetic and environmental risk factors for symptoms of cannabis, cocaine, alcohol, caffeine, and nicotine dependence. Arch Gen Psychiatry 64:1313–1320
84. Kendler KS, Schmitt E, Aggen SH, Prescott CA (2008) Genetic and environmental influences on alcohol, caffeine, cannabis, and nicotine use from early adolescence to middle adulthood. Arch Gen Psychiatry 65:674–682
85. Kessler RC, Crum RM, Warner LA et al (1997) Lifetime co-occurrence of DSM-III-R alcohol abuse and dependence with other psychiatric disorders in the National comorbidity survey. Arch Gen Psychiatry 54:313–321
86. Koob GF, Le Moal M (2001) Drug addiction, dysregulation of reward, and allostasis. Neuropsychopharmacology 24:97–129
87. Korpi ER, Kleingoor C, Kettenmann H, Seeburg PH (1993) Benzodiazepine-induced motor impairment linked to point mutation in cerebellar GABAA receptor. Nature 361:356–359
88. Krueger RF, Hicks BM, Patrick CJ, Carlson SR, Iacono WG, McGue M (2002) Etiologic connections among substance dependence, antisocial behavior, and personality: modeling the externalizing spectrum. J Abnorm Psychol 111:411–424
89. Lachman HM, Papolos DF, Saito T, Yu YM, Szumlanski CL, Weinshilboum RM (1996) Human catechol-O-methyltransferase pharmacogenetics: description of a functional polymorphism and its potential application to neuropsychiatric disorders. Pharmacogenetics 3:243–250
90. Lappalainen J, Long JC, Eggert M, Ozaki N, Robin RW, Brown GL et al (1998) Linkage of antisocial alcoholism to the serotonin 5-HT1B receptor gene in 2 populations. Arch Gen Psychiatry 55:989–994
91. Lesch KP, Bengel D, Heils A, Sabol SZ, Greenberg BD, Petri S et al (1996) Association of anxiety-related traits with a polymorphism in the serotonin transporter gene regulatory region. Science 274:1527–1531
92. Lewis DA, Melchitzky DS, Sesack SR, Whitehead RE, Auh S, Sampson A (2001) Dopamine transporter immunoreactivity in monkey cerebral cortex: regional, laminar, and ultrastructural localization. J Comp Neurol 432:119–36
93. Li MD, Cheng R, Ma JZ, Swan GE (2003) A meta-analysis of estimated genetic and environmental effects on smoking behavior in male and female adult twins. Addiction 98:23–31
94. Long JC, Knowler WC, Hanson RL, Robin RW, Urbanek M, Moore E et al (1998) Evidence for genetic linkage to alcohol dependence on chromosomes 4 and 11 from an autosome-wide scan in an American Indian population. Am J Med Genet 81:216–221
95. Lynskey MT et al (2002) Genetic and environmental contributions to cannabis dependence in a national young adult twin sample. Psychol Med 32:195–207
96. Maes HH, Sullivan PF, Bulik CM, Neale MC, Prescott CA, Eaves LJ, Kendler KS (2004) A twin study of genetic and environmental influences on tobacco initiation, regular tobacco use and nicotine dependence. Psychol Med 34:1251–1261
97. Malhotra AK, Kestler LJ, Mazzanti C, Bates JA, Goldberg T, Goldman DA (2002) Functional polymorphism in the COMT gene and performance on a test of prefrontal cognition. Am J Psychiatry 159:652–654
98. Mazei MS, Pluto CP, Kirkbride B, Pehek E (2002) A Effects of catecholamine uptake blockers in the caudate-putamen and subregions of the medial prefrontal cortex of the rat. Brain Res 936:58–67
99. Melchior M, Moffitt TE, Milne BJ, Poulton R, Caspi A (2007) Why do children from socioeconomically disadvantaged families suffer from poor health when they reach adulthood? A life-course study. Am J Epidemiol 166:966–974
100. Merikangas KR, Risch N (2003) Genomic priorities and public health. Science 302:599–601
101. Merikangas KR, Stolar M, Stevens DE, Goulet J, Preisig MA, Fenton B et al (1998) Familial transmission of substance use disorders. Arch Gen Psychiatry 55:973–979
102. Meyer-Lindenberg A, Weinberger DR (2006) Intermediate phenotypes and genetic mechanisms of psychiatric disorders. Nat Rev Neurosci 7:818–827
103. Meyer-Lindenberg A, Buckholtz JW, Kolachana B, Hariri AR, Pezawas L, Blasi G et al (2006) Neural mechanisms of genetic risk for impulsivity and violence in humans. Proc Natl Acad Sci USA 103:6269–6274

104. Midanik L (1983) Familial alcoholism and problem drinking in a national drinking practices survey. Addict Behav 8:133–141
105. Newman TK, Syagailo YV, Barr CS, Wendland JR, Champoux M, Graessle M et al (2005) Monoamine oxidase a gene promoter variation and rearing experience influences aggressive behavior in rhesus monkeys. Biol Psychiatry 57:167–172
106. Nurnberger JI, Foroud T, Flury L, Su J, Meyer ET, Hu K et al (2001) Evidence for a locus on chromosome 1 that influences vulnerability to alcoholism and affective disorder. Am J Psychiatry 158:718–724
107. Oslin DW, Berrettini W, Kranzler HR, Pettinati H, Gelernter J, Volpicelli JR, O'Brien CP (2003) A functional polymorphism of the mu-opioid receptor gene is associated with naltrexone response in alcohol-dependent patients. Neuropsychopharmacology 8:1546–1552
108. Ou XM, Chen K, Shih JC (2006) Glucocorticoid and androgen activation of monoamine oxidase A is regulated differently by R1 and Sp1. J Biol Chem 281:21512–21525
109. Pagan JL, Rose RJ, Viken RJ, Pulkkinen L, Kaprio J, Dick DM (2006) Genetic and environmental influences on stages of alcohol use across adolescence and into young adulthood. Behav Genet 36:483–497
110. Parish MK (1991) Alcohol consumption and the risk of developing liver cirrhosis: implications for future research. J Subst Abuse 3:325–335
111. Pezawas L, Meyer-Lindenberg A, Drabant EM, Verchinski BA, Munoz KE, Kolachana BS et al (2005) 5-HTTLPR polymorphism impacts human cingulate–amygdala interactions: a genetic susceptibility mechanism for depression. Nat Neurosci 8:828–834
112. Porjesz B, Begleiter H, Wang K, Almasy L, Chorlian DB, Stimus AT et al (2002) Linkage and linkage disequilibrium mapping of ERP and EEG phenotypes. Biol Psychol 61:229–248
113. Prescott CA, Aggen SH, Kendler KS (1999) Sex differences in the sources of genetic liability to alcohol abuse and dependence in a population based sample of US twins. Alcohol Clin Exp Res 23:1136–1144
114. Prescott CA, Aggen SH, Kendler KS (2000) Sex-specific genetic influences on the comorbidity of alcoholism and major depression in a population-based sample of US twins. Arch Gen Psychiatry 57:803–811
115. Prescott CA, Kendler KS (1999) Age at first drink and risk for alcoholism: a noncausal association. Alcohol Clin Exp Res 23:101–107
116. Prescott CA, Kendler KS (1999) Genetic and environmental contributions to alcohol abuse and dependence in a population-based sample of male twins. Am J Psychiatry 156:34–40
117. Prescott CA, Sullivan PF, Kuo PH, Webb BT, Vittum J, Patterson DG, Thiselton DL, Myers JM, Devitt M, Halberstadt LJ, Robinson VP, Neale MC, van den Oord EJ, Walsh D, Riley BP, Kendler KS (2006) Genomewide linkage study in the Irish affected sib pair study of alcohol dependence: evidence for a susceptibility region for symptoms of alcohol dependence on chromosome 4. Mol Psychiatry 11:603–611
118. Radel M, Vallejo RL, Iwata N, Aragon R, Long JC, Virkkunen M et al (2005) Haplotype-based localization of an alcohol dependence gene to the 5q34 {gamma}-aminobutyric acid type A gene cluster. Arch Gen Psychiatry 62:47–55
119. Ramstedt M (2001) Per capita alcohol consumption and liver cirrhosis mortality in 14 European Countries. Addiction 96(Suppl 1):S19–S33
120. Reich T, Edenberg HJ, Goate A, Williams JT, Rice JP, Van Eerdewegh P et al (1998) Genome-wide search for genes affecting the risk for alcohol dependence. Am J Med Genet 81:207–215
121. Rivett AJ, Francis A, Roth JA (1983) Distinct cellular localization of membrane bound and soluble forms of catechol-O-methyltransferase in brain. J Neurochem 40:215–219
122. Robin RW, Chester B, Rasmussen JK, Jaranson JM, Goldman D (1997) Prevalence, characteristics, and impact of childhood sexual abuse in a Southwestern American Indian tribe. Child Abuse Negl 21:769–787
123. Robin RW, Long JC, Rasmussen JK, Albaugh B, Goldman D (1998) Relationship of binge drinking to alcohol dependence, other psychiatric disorders, and behavioral problems in an American Indian tribe. Alcohol Clin Exp Res 22:518–523
124. Rodriguez LA, Wilson JR, Nagoshi CT (1993) Does psychomotor sensitivity to alcohol predict subsequent alcohol use? Alcohol Clin Exp Res 17:155–161
125. Roy A, Hu XZ, Janal MN, Goldman D (2007) Interaction between childhood trauma and serotonin transporter gene variation in suicide. Neuropsychopharmacology 32:2046–2052
126. Sabol SZ, Hu S, Hamer D (1998) A functional polymorphism in the monoamine oxidase a gene promoter. Hum Genet 103:273–279
127. Scanlon PD, Raymond FA, Weinshilboum RM (1979) Catechol-O-methyltransferase: thermolabile enzyme in erythrocytes of subjects homozygous for allele for low activity. Science 203:63–65
128. Schuckit MA (1998) Biological, psychological and environmental predictors of the alcoholism risk: a longitudinal study. J Stud Alcohol 59:485–494
129. Schuckit MA, Smith TL, Kalmijn J, Tsuang J, Hesselbrock V, Bucholz K (2000) Response to alcohol in daughters of alcoholics: a pilot study and a comparison with sons of alcoholics. Alcohol Alcohol 35:242–248
130. Sher KJ, Bartholow BD, Wood MD (2000) Personality and substance use disorders: a prospective study. J Consult Clin Psychol 68:818–829
131. Shioe K, Ichimiya T, Suhara T, Takano A, Sudo Y, Yasuno F et al (2003) No association between genotype of the promoter region of serotonin transporter gene and serotonin transporter binding in human brain measured by PET. Synapse 48:184–188
132. Sjöberg RL, Ducci F, Barr CS, Newman TK, Dell'osso L, Virkkunen M et al (2008) A non-additive interaction of a functional MAO-A VNTR and testosterone predicts antisocial behavior. Neuropsychopharmacology 33:425–430
133. Smolka MN, Schumann G, Wrase J, Grusser SM, Flor H, Mann K et al (2005) Catechol-O-methyltransferase val-158met genotype affects processing of emotional stimuli in the amygdala and prefrontal cortex. J Neurosci 25:836–842
134. Stefansson H, Rujescu D, Cichon S, Pietiläinen OP, Ingason A, Steinberg S, Fossdal R, Sigurdsson E, Sigmundsson T,

Buizer-Voskamp JE, Hansen T, Jakobsen KD, Muglia P, Francks C, Matthews PM, Gylfason A, Halldorsson BV, Gudbjartsson D, Thorgeirsson TE, Sigurdsson A, Jonasdottir A, Jonasdottir A, Bjornsson A, Mattiasdottir S, Blondal T, Haraldsson M, Magnusdottir BB, Giegling I, Möller HJ, Hartmann A, Shianna KV, Ge D, Need AC, Crombie C, Fraser G, Walker N, Lonnqvist J, Suvisaari J, Tuulio-Henriksson A, Paunio T, Toulopoulou T, Bramon E, Di Forti M, Murray R, Ruggeri M, Vassos E, Tosato S, Walshe M, Li T, Vasilescu C, Mühleisen TW, Wang AG, Ullum H, Djurovic S, Melle I, Olesen J, Kiemeney LA, Franke B, Sabatti C, Freimer NB, Gulcher JR, Thorsteinsdottir U, Kong A, Andreassen OA, Ophoff RA, Georgi A, Rietschel M, Werge T, Petursson H, Goldstein DB, Nöthen MM, Peltonen L, Collier DA, St Clair D, Stefansson K (2008) Large recurrent microdeletions associated with schizophrenia. Nature 455:232–236

135. Thomasson HR, Edenberg HJ, Crabb DW et al (1991) Alcohol and aldehyde dehydrogenase genotypes and alcoholism in Chinese men. Am J Hum Genet 48:677–681

136. Thorgeirsson TE, Geller F, Sulem P, Rafnar T, Wiste A, Magnusson KP, Manolescu A, Thorleifsson G, Stefansson H, Ingason A, Stacey SN, Bergthorsson JT, Thorlacius S, Gudmundsson J, Jonsson T, Jakobsdottir M, Saemundsdottir J, Olafsdottir O, Gudmundsson LJ, Bjornsdottir G, Kristjansson K, Skuladottir H, Isaksson HJ, Gudbjartsson T, Jones GT, Mueller T, Gottsäter A, Flex A, Aben KK, de Vegt F, Mulders PF, Isla D, Vidal MJ, Asin L, Saez B, Murillo L, Blondal T, Kolbeinsson H, Stefansson JG, Hansdottir I, Runarsdottir V, Pola R, Lindblad B, van Rij AM, Dieplinger B, Haltmayer M, Mayordomo JI, Kiemeney LA, Matthiasson SE, Oskarsson H, Tyrfingsson T, Gudbjartsson DF, Gulcher JR, Jonsson S, Thorsteinsdottir U, Kong A, Stefansson K (2008) A variant associated with nicotine dependence, lung cancer and peripheral arterial disease. Nature 452(7187):638–642

137. Tiihonen J, Hallikainen T, Lachman H, Saito T, Volavka J (1999) Kauhanen J et al Association between the functional variant of the catechol-O-methyltransferase (COMT) gene and type 1 alcoholism. Mol Psychiatry 4:286–289

138. Tsuang MT, Lyons MJ, Eisen SA et al (1996) Genetic influences on DSM-III-R drug abuse and dependence: a study of 3, 372 twin pairs. Am J Med Genet 67: 473–477

139. Tu GC, Israel Y (1995) Alcohol consumption by orientals in North America is predicted largely by a single gene. Behav Genet 25:59–65

140. Uhl GR, Drgon T, Johnson C, Li CY, Contoreggi C, Hess J, Naiman D, Liu QR (2008) Molecular genetics of addiction and related heritable phenotypes: genome-wide association approaches identify "connectivity constellation" and drug target genes with pleiotropic effects. Ann N Y Acad 1141:318–381

141. Vandenbergh DJ, Rodriguez LA, Miller IT, Uhl GR, Lachman HM (1997) High-activity catechol-O-methyltransferase allele is more prevalent in polysubstance abusers. Am J Med Genet 74:439–442

142. van den Bree MB, Johnson EO, Neale MC, Pickens RW (1998) Genetic and environmental influences on drug use and abuse/dependence in male and female twins. Drug Alcohol Depend 52:231–241

143. Volkow ND, Fowler JS, Wang JG, Baler R, Telang F (2008) Imaging dopamine's role in drug abuse and addiction. Neuropharmacology 56(Suppl 1):3–8

144. Walker EA, Gelfand AN, Gelfand MD, Koss MP, Katon WJ (1995) Medical and psychiatric symptoms in female gastroenterology clinic patients with histories of sexual victimization. Gen Hosp Psychiatry 17:85–92

145. Walsh T, McClellan JM, McCarthy SE, Addington AM, Pierce SB, Cooper GM, Nord AS, Kusenda M, Malhotra D, Bhandari A, Stray SM, Rippey CF, Roccanova P, Makarov V, Lakshmi B, Findling RL, Sikich L, Stromberg T, Merriman B, Gogtay N, Butler P, Eckstrand K, Noory L, Gochman P, Long R, Chen Z, Davis S, Baker C, Eichler EE, Meltzer PS, Nelson SF, Singleton AB, Lee MK, Rapoport JL, King MC, Sebat J (2008) Rare structural variants disrupt multiple genes in neurodevelopmental pathways in schizophrenia. Science 320:539–543

146. Weinshilboum R, Dunnette J (1981) Thermal stability and the biochemical genetics of erythrocyte catechol-O-methyl-transferase and plasma dopamine-beta-hydroxylase. Clin Genet 19:426–437

147. World Health Organization (2008) The global burden of disease: 2004 update Geneva, World Health Organization. Available at http://wwwwhoint/evidence/bod

148. Xu K, Lichtermann D, Lipsky RH, Franke P, Liu X, Hu Y, Cao L, Schwab SG, Wildenauer DB, Bau CH, Ferro E, Astor W, Finch T, Terry J, Taubman J, Maier W, Goldman D (2004) Association of specific haplotypes of D2 dopamine receptor gene with vulnerability to heroin dependence in 2 distinct populations. Arch Gen Psychiatry 61:597–606

149. Xu K, Ernst M, Goldman D (2006) Imaging genomics applied to anxiety, stress response, and resiliency. Neuroinformatics 4:51–64

150. Yavich L, Forsberg MM, Karayiorgou M, Gogos JA, Mannisto PT (2007) Site-specific role of catechol-O-methyltransferase in dopamine overflow within prefrontal cortex and dorsal striatum. J Neurosci 27:10196–10209

151. Yokoyama A, Omori T, Yokoyama T, Sato Y, Mizukami T, Matsushita S et al (2006) Risk of squamous cell carcinoma of the upper aerodigestive tract in cancer-free alcoholic Japanese men: an endoscopic follow-up study. Cancer Epidemiol Biomarkers Prev 15:2209–2215

152. Zakhari S, Li TK (2007) Determinants of alcohol use and abuse: impact of quantity and frequency patterns on liver disease. Hepatology 46:2032–2039

153. Zhou Z, Zhu G, Hariri AR, Enoch MA, Scott D, Sinha R, Virkkunen M, Mash DC, Lipsky RH, Hu XZ, Hodgkinson CA, Xu K, Buzas B, Yuan Q, Shen PH, Ferrell RE, Manuck SB, Brown SM, Hauger RL, Stohler CS, Zubieta JK, Goldman D (2008) Genetic variation in human NPY expression affects stress response and emotion. Nature 24(452):997–1001

154. Zimmermann P, Wittchen HU, Hofler M, Pfister H, Kessler RC, Lieb R (2003) Primary anxiety disorders and the development of subsequent alcohol use disorders: a 4-year community study of adolescents and young adults. Psychol Med 33:1211–1222

155. Zubieta JK, Heitzeg MM, Smith YR, Bueller JA, Xu K, Xu Y et al (2003) COMT val158met genotype affects mu-opioid neurotransmitter responses to a pain stressor. Science 299:1240–1243

Behavioral Aspects of Chromosomal Variants

23.6

Michael R. Speicher

Abstract Humans with chromosome aberrations often show, along with many other findings, behavioral abnormalities, which may be mild or severe and may either affect all carriers of an aberration or only some of them. Although chromosome aberrations influence embryonic development in multiple and ill-defined ways, certain genetic syndromes do result in relatively specific patterns of behavior and related attributes. Although sophisticated approaches to examine such "behavioral phenotypes" have been developed, the description and characterization of behavioral aspects is also frequently accompanied by claims and counterclaims which often complicate the subject. Relative to the general population, individuals with intellectual disabilities are at much higher risk of experiencing behavioral, emotional, and psychiatric problems. However, many mental health professionals do not appreciate the co-occurrence of psychiatric problems and intellectual disabilities. Therefore, there are several gaps in the research and treatment of mental health concerns in people with autosomal chromosomal aberrations. Nevertheless, chromosomal aberrations offer the unique opportunity to relate behavioral phenomena to independently ascertained and relatively well-defined genetic causes. In this chapter the impact of various chromosomal rearrangements, ranging from whole-chromosome copy number changes to small deletions or duplications, on behavior is described and discussed.

Contents

23.6.1	Introduction: Human Chromosome Aberrations and Behavior, Possibilities, and Limitations	744
23.6.2	Numeric Autosomal Aberrations	744
	23.6.2.1 Down Syndrome	744
23.6.3	Copy Number Variations Associated with Behavioral Disorders	745
	23.6.3.1 Autistic Spectrum Disorder	745
	23.6.3.2 Schizophrenia	745
	23.6.3.3 Bipolar Disorders	746

23.6.4	Aberrations of the X Chromosome	746
	23.6.4.1 Turner Syndrome	746
	23.6.4.2 Klinefelter Syndrome	746
	23.6.4.3 Triple-X Syndrome	748
23.6.5	Aberrations of the Y Chromosome	748
	23.6.5.1 XYY Syndrome	748
	23.6.5.2 Higher Prevalence Among "Criminals"	748
	23.6.5.3 Intellectual Dysfunction or Simply Stature?	749
	23.6.5.4 Behavioral Aspects of XYY Men	749
	23.6.5.5 Association of Criminal Behavior and Lowered Intelligence in XYY Men	750
	23.6.5.6 Social and Therapeutic Consequences	750
	23.6.5.7 XXYY Syndrome	751
23.6.6	Other Chromosomal Variants	751
	23.6.6.1 22q11.2 Deletion Syndrome	751

M.R. Speicher (✉)
Institute of Human Genetics, Medical University of Graz,
Harrachgasse 21/8, 8010 Graz, Austria
e-mail: michael.speicher@medunigraz.at

23.6.6.2	Smith–Magenis Syndrome	752
23.6.6.3	Prader-Willi and Angelman Syndrome	753
23.6.6.4	Williams–Beuren Syndrome	753
23.6.6.5	Cri-du-chat Syndrome	753
23.6.6.6	Wolf–Hirschhorn Syndrome	754
References		754

23.6.1 Introduction: Human Chromosome Aberrations and Behavior, Possibilities, and Limitations

Humans with chromosome aberrations often show, along with many other findings (Sect. 23.6.4), behavioral abnormalities, which may be mild or severe and may either affect all carriers of an aberration or only some of them. Although chromosome aberrations influence embryonic development in multiple and ill-defined ways, certain genetic syndromes do result in relatively specific patterns of behavior and related attributes. Several approaches to examination of such "behavioral phenotypes" have been developed [25]. A special group that has given immense impetus to the delineation and study of behavioral phenotypes is that of parent support groups. These groups have advocated and raised awareness of major behavioral problems associated with chromosomal aberrations, some of which are often unfamiliar to the clinicians with whom the families have been dealing. However, the description and characterization of behavioral aspects is also frequently accompanied by claims and counterclaims which often complicate the subject.

Most unbalanced autosomal aberrations lead to multiple and severe malformations (Sect. 23.6.4) that also affect the brain, and may therefore cause cognitive impairment ranging from developmental delay to severe mental deficiency. Relative to the general population, individuals with intellectual disabilities are at much higher risk of experiencing behavioral, emotional, and psychiatric problems. However, many mental health professionals do not appreciate the co-occurrence of psychiatric problems and intellectual disabilities. Therefore there are several gaps in the research and treatment of mental health concerns in people with autosomal chromosomal aberrations. Nevertheless, chromosomal aberrations offer the unique opportunity to relate behavioral phenomena to independently ascertained and relatively well-defined genetic causes.

23.6.2 Numeric Autosomal Aberrations

Key findings of behavior and emotional problems in humans with Down syndrome are summarized here.

23.6.2.1 Down Syndrome

Down syndrome is the most common syndrome with a trisomy of an autosome (Sect. 23.6.4). The exact definition of a Down syndrome cognitive phenotype is a particular challenge, as it is influenced by biological development, such as the maturation of specific neural systems over time on the one hand, and experiences and training on the other hand. There are only a few studies following individuals longitudinally, and assessing them at key points in time takes very complex and difficult efforts [11, 51]. As a consequence, it is not surprising that estimates of the intelligence quotient (IQ) attributed to Down syndrome have varied over the years.

With all known shortcomings of IQ estimates it is apparent that cognitive deficits and the IQ range show considerable heterogeneity within the population with Down syndrome. Some may have IQ scores between 50 and 70, indicating moderate intellectual disabilities. However, the vast majority of humans with Down syndrome have scores from over 20 up to 50, corresponding to severe intellectual disability, while a very few have scores of 20 or below [62]. Nevertheless, there is evidence that over the life span performances may decline relatively more frequently in humans with Down syndrome than in the general population [10].

Individuals with Down syndrome may have significant delay in nonverbal cognitive development accompanied by additional specific deficits in speech, language production, and verbal working memory [11, 51]. However, many of these persons can be educated to read and write.

Compared with other groups with specific syndromes or with various causes for their respective disabilities, children with Down syndrome generally show lower rates of significant behavioral or emotional problems [13]. Although rates of psychopathology are relatively low, children with Down syndrome may show behavioral problems. These include – relative to typically developing controls – a higher frequency of externalizing behaviors such as stubbornness, oppositionality, inattention, attention-seeking, and impulsivity [12]. As a consequence, about 6–8% of children with Down syndrome are diagnosed with attention deficit/hyperactivity disorder (ADHD) [13]. In contrast to the prevalent stereotype and view of children with Down syndrome as friendly, sociable, and charming, there is a growing recognition of the co-occurrence of Down syndrome and autism spectrum disorder (ASD) in about 7–10% of the cases [30].

Children with Down syndrome go from having relatively few behavioral problems in their early years to being at markedly increased risks for depression and clinical symptoms of dementia in their adult years. The aforementioned externalizing problems often decline as early as during adolescence. At the same time, internalizing symptoms, such as withdrawal and being more secretive and quiet, increase during this time span, so that persons with Down syndrome often show an age-related withdrawal [14].

Neuropathologic signs of dementia become apparent in most individuals with Down syndrome aged 40 years and older. The full pathology of Alzheimer disease appears to be invariably present from 35 years of age onwards – 50 years earlier than in the normal population. In fact, Down syndrome is unique in conferring a 100% risk of developing early Alzheimer disease [2]. In addition to Alzheimer disease, adults with Down syndrome are particularly prone to depression. This is even true when compared with other groups with intellectual disabilities, and the prevalence rates of depression are estimated to range from 6.1 to 11.4% [34]. In the general population memory changes represent usually preclinical stages of dementia. In contrast, in adults with Down syndrome behavior and personality changes are frequently found at the beginning of dementia, suggesting that frontal lobe dysfunction and dementia of the frontal type may be characteristic of the early course or manifestation of Alzheimer disease in Down syndrome [5].

Other behavioral aspects of adults with Down syndrome include a much lower likelihood of physical aggressiveness than in their counterparts and relatively rare occurrence of bipolar disorder and schizophrenia [13]. Future research efforts will certainly have to focus on the more detailed identification of risk or protective factors for these psycho-behavioral aspects, and also on the efficacy of different interventions and treatment forms.

23.6.3 Copy Number Variations Associated with Behavioral Disorders

Genomic variability is to a large extent caused by structural alterations, such as deletions, duplications or inversions. The widespread presence of copy number variations (CNVs) even in normal individuals was first reported in 2004 [27, 48] and has subsequently been confirmed in numerous other reports (see Sect. 23.6.3). Such structural variations have now been implicated in various behavioral disorders, such as autistic spectrum disorder, schizophrenia, and bipolar disorders. The association between CNVs and these disorders will be described in greater detail in the chapters specifically devoted to them (e.g., autistic spectrum disorder: Chap. 23.4 and schizophrenia: Chap. 23.7) and the treatment of this topic is therefore kept very brief here.

23.6.3.1 Autistic Spectrum Disorder

Array-comparative genomic hybridization (CGH, see Sect. 3.4.3.5.3) performed with genomic DNA of individuals with autistic spectrum disorder and their parents revealed a significant association between de novo CNVs and autism [4, 49]. This observation was confirmed in another study [32]. The CNVs often involve genes critical for CNS development, such as the *SHANK3-NLGN4-NRXN1* postsynaptic density genes, *DPP6-DPP10-PCDH9* (synapse complex), *ANKRD11, DPYD,* or *PTCHD1* [37], supporting the notion that the observed CNVs represent a significant risk factor for autistic spectrum disorders. Based on these results, cytogenetic and array analyses should be included in the standard diagnostic program in persons with suspected autistic spectrum disorders.

23.6.3.2 Schizophrenia

A series of reports also described correlations between CNVs and the occurrence of schizophrenia. Some of these CNVs were also implicated in other psychiatric diseases, such as autistic spectrum disorders and mental retardation. The evolving pattern is that there are some rare CNVs, which contribute to the genetic component of schizophrenia owing to their high penetrance [67]. In particular, deletions in 1q21.1, 15q11.2 and 15q13.3 were found to be significantly associated with schizophrenia [55].

23.6.3.3 Bipolar Disorders

For bipolar disorders there is also increasing evidence for an association with rare structural variants, especially if the age of onset of mania was below 18 years [65, 68].

23.6.4 Aberrations of the X Chromosome

Numeric and structural aberrations of the X and Y chromosomes generally lead to much milder disturbances in embryonic development than autosomal aberrations do (Sect. 3.6.1). Many somatic abnormalities found in these syndromes are related to abnormal sexual development. The psychological disturbances are less overwhelmingly severe and may sometimes be specific.

23.6.4.1 Turner Syndrome

Clinical and chromosomal findings of Turner syndrome are described in Sect. 3.6.3.3.3. The standard karyotype is 45,X; however, many mosaics and structural variations are observed. Turner syndrome is associated with a number of characteristic physical features, such as short stature and absent ovaries, as well as a set of common neuropsychological deficits and social and behavioral features.

Earlier investigations more than 50 years ago reported a significant reduction in mean IQ [23]. Subsequently it has been acknowledged that the lower IQs reflected a significant reduction in Performance IQ, whereas Verbal IQ was normally distributed [17]. Areas in which Turner syndrome females typically score below population norms include Arithmetic, Picture Completion, Coding, and Object Assembly subtests [33, 45]. In fact, visuospatial problems are the cardinal cognitive deficit of women with Turner syndrome. However, selective deficits in attention, memory, and executive processing can also be seen.

Despite the fact that verbal abilities are within normal limits, individuals with Turner syndrome may show reduced fluency, poor articulation, and difficulty processing syntactic structures [61]. In tests of academic achievement individuals with Turner syndrome demonstrate considerable difficulty with arithmetic but perform adequately for age in terms of their reading and spelling and are often described as avid readers. Therefore, at school many girls with Turner syndrome are identified as having a nonverbal learning disability and may have an increased need for special education [45].

In childhood a significant proportion of Turner syndrome patients tend to be hyperactive, while up to 10% have ADHD in adolescence. However, many others are extremely inhibited and shy, especially in adolescence, which is a particularly unhappy time for teenagers with Turner syndrome [45].

Most of the observations seen in children with Turner syndrome persist into adulthood, particularly their visuospatial and visual memory deficits [42, 44]. In contrast, some abilities improve with age, such as weaknesses in perceptual judgment and motor planning skills [41]. A large number of individuals with Turner syndrome demonstrate excellent musical aptitude.

Less than 5% of Turner syndrome women achieve higher professions, although a few individuals with Turner syndrome hold graduate degrees and have become physicians or lawyers. The majority of individuals with Turner syndrome hold clerical or semi-professional positions (i.e., teaching, nursing, early childhood education) and they tend to be overrepresented in child-care positions. There is a high incidence of dependence, with many living at home together with their parents as adults [43].

Psychosocially females with Turner syndrome show a definite female gender identity and assume a typical female gender role. Social immaturity is often described in individuals with Turner syndrome, and their social relations are often difficult [45]. The majority have few friends during adolescence, and even in adulthood difficulties with relationships and coping with Turner syndrome are commonly reported. Owing to their particular set of physical stigmata they generally have a poor body image and low self-esteem [45].

23.6.4.2 Klinefelter Syndrome

Klinefelter syndrome is a relatively common (1/500 to 1/1,000) genetic syndrome caused by an extra X chromosome in males, leading to an XXY karyotype; other karyotypes and also mosaics occur. Adult patients are an average of a few centimeters taller than their normal brothers; in particular, their legs are longer in relation to their overall stature. This growth pattern can be observed as early as in childhood; at the age of puberty the subnormal sexual development becomes obvious: the testicles are small, and there is aspermy (Sect.3.6.3.3.4).

Although Klinefelter syndrome is not rare, many men with Klinefelter syndrome are not aware of their genetic impairment. A recent epidemiologic study suggests that more than 75% of subjects with Klinefelter syndrome are not diagnosed [8]. Furthermore, in many cases diagnosis is significantly delayed, made not during childhood but in adolescence, or even later. Thus, most men with Klinefelter syndrome and their relatives are not aware of the genetic constitution that underlies their cognitive and behavioral problems. As a result, our knowledge of the social behavioral phenotype of Klinefelter syndrome is limited. In fact, cognitive and behavioral dysfunctions in Klinefelter syndrome have generally been underrated relative to the endocrinological and physical features. Previous studies have predominantly assessed global functioning (e.g., academic achievement, occupation, or marital status) rather than specific social abilities. Some studies have specifically focused on social adjustment in adolescents and men with Klinefelter syndrome, yet they primarily collected categorical data (e.g., someone can be either sociable, passive or shy) or involved small sample sizes or lacked control data recorded in individuals from the general population [64].

In most cases, the physical and neurobehavioral characteristics of Klinefelter syndrome are relatively mild, and Klinefelter syndrome is not usually associated with moderate or severe mental retardation. However, Klinefelter syndrome is often associated with significant language-based learning disabilities and executive dysfunction, and there is a general impression that men and boys with Klinefelter syndrome often struggle with social situations (e.g., at school or at work) [19]. Many of the psychological symptoms encountered in males with Klinefelter syndrome can be explained by their diminished androgen production, which is normally required for the expression of male-specific psychological development.

Several studies suggest that individuals with Klinefelter syndrome are at risk for psychosocial and emotional problems such as social withdrawal, social anxiety, shyness, impulsivity, and inappropriate social behavior [19, 52]. In early adulthood a significant portion of XXY men have few or no friends, poor relations with siblings and parents, little energy and initiative, and few or no spare time interests. Their vitality and ability in establishing social contacts is often reduced [35].

The patients show on average slightly reduced intelligence with special difficulties in learning how to read and write. However, IQ values well above average are not rare. On the other hand, Klinefelter syndrome has been found more often in series of mildly mentally subnormal subjects. The literature reports do not unanimously support a well-defined specific defect of mental abilities.

These psychosocial aspects of Klinefelter syndrome can have a significant impact on school performance and learning. School problems are more frequent than expected from intellectual ability and seem to be caused by behavioral problems. Adult patients often hold unskilled jobs; success in higher professional careers has been reported but does not seem to be very common.

A recent study described the psychosocial morbidity in a cohort of young males with hypogonadism attributable to Klinefelter syndrome [52]. One aim of this study was to document the effect of androgen replacement on behavior. Seventeen of 32 postpubertal patients with Klinefelter syndrome required testosterone therapy, while in 11 serum testosterone in the normal adult range was documented. Significant psychosocial and behavioral problems were present in 22 out of 32 of patients with Klinefelter syndrome, including seven who were testosterone replete, with an identifiable pattern of disorder, including marked lack of insight, poor judgment, and impaired ability to learn from adverse experience. Use of long-term replacement testosterone treatment reduced episodes of behavioral indiscretion. This suggests that inadequately treated hypogonadism in Klinefelter syndrome may increase recognized psychosocial morbidity [52].

Therefore, there is a need for prospectively planned and timed support for young men with Klinefelter syndrome, in order to ameliorate current poor psychosocial outcomes.

23.6.4.3 Triple-X Syndrome

This syndrome is described in Sect. 3.6.3.3.5. Many women with the karyotype XXX have developed normally and have children. However, the 47,XXX individual may be at greater risk for poor psychosocial adaptation and early adulthood. As pointed out in Sect. 3.6.3.3.5, much of the available data is based on isolated case studies of women ascertained through the presence of another condition. There are only a few studies in which patients with 47,XXX had been identified through chromosomal screening of newborns and followed up longitudinally [24, 46].

One of these few studies compared 47,XXX women and female sibling controls during adolescence and during early adulthood. The study revealed that 47,XXX woman are less well adapted during both adolescence and young adulthood and they described their lives as more stressful. Furthermore, they had more work, leisure, and relationship problems. Their IQ was lower and they showed evidence of more psychopathology than subjects in the reference group composed of female siblings. Propositi with lower IQs tended to demonstrate poorer psychological adaptation. However, psychiatric status was not determined solely by intelligence; psychological dysfunction occurred even among women with IQs in the average range. Although these women seem to have more difficulties, most of them are self-sufficient and functioning reasonably well, albeit less well than their siblings [24].

Another study reported milder impairment in 47,XXX women [46]. However, a concern of these studies is their representativeness, as all of them have small sample sizes. In summary, there is evidence that children with a supernumerary X chromosome score consistently below controls on Verbal IQ and subtests comprising the Verbal Comprehension factor, but did not differ in Performance IQ, which was in the normal range. Academic achievement is not affected in aneuploid females with higher levels of intelligence.

23.6.5 Aberrations of the Y Chromosome

23.6.5.1 XYY Syndrome

For a description of somatic symptoms of the XYY syndrome see Sect. 3.6.3.4.1. The mean stature of these men is usually taller than that in the population of their origin. Many show normal sexual development and are fertile. 47,XYY males may have delayed speech, lower cognitive function (IQ), hyperactivity, learning disabilities, and other central nervous system (CNS) abnormalities [1]. Several males with an XYY constitution and a normal phenotype have also been reported.

23.6.5.2 Higher Prevalence Among "Criminals"

The XYY syndrome has become widely known since Jacobs et al. [28] carried out a survey of patients who were mentally subnormal and under surveillance in a special institution because of "dangerous, violent, or criminal propensities." Among 196 probands 12 had an abnormal karyotype; 7 with XYY and 1 with XXYY. This frequency was much higher than expected; however, the authors stated that they could not determine whether these men had been institutionalized mainly because of mental subnormality, aggressive behavior, or some kind of combination of these factors. Their results were soon confirmed in a number of studies from institutions for mentally subnormal men with behavior problems, especially among particularly tall inmates. On the basis of such evidence it was concluded that their antisocial behavior was caused by the additional Y chromosome, and that they were genetically predisposed to criminality. The explanation seemed simple. Normal men are more aggressive than normal women; normal men have one Y chromosome, while women do not. Hence, if someone has two Y chromosomes, he should be twice as aggressive as normal men; his aggressiveness may fall outside the socially acceptable range, and he may commit acts of violence.

Gradually, however, some pertinent questions were asked: above all, how frequent is the XYY karyotype in the general population of nonconvicts? Studies on the incidence among male newborns showed a frequency of around 1:1,000, or even higher, similar to that of Klinefelter syndrome [20]. Even in the absence of reliable prevalence studies among the male adult population it was fair to conclude that the prevalence differs little from the incidence at birth, i.e., that there is no preferential mortality. This, however, could just mean that the great majority of XYY men do not come into conflict with the law.

Another question was whether the nature of their crimes revealed a certain pattern and, more specifically, whether acts of violence and sexual aggression prevailed. This was in general not the case: using a population-based sample of men with sex chromosome abnormalities by screening 34,380 infants at birth, Götz et al. [20] compared XYY men, XXY men, and controls for the frequency of antisocial personality disorder and rates of criminal convictions. This study report showed that, with adjustment for the number of years at risk of receiving a criminal conviction the difference in overall delinquency rate ratios was significant at the 1% level ($p = 0.01$), revealing that the XYY men were more likely to have a criminal record than chromosomally normal controls. However, there was no evidence from the sentences imposed on the subjects that the offenses committed by the XYY men were more serious than those committed by the controls. In fact, XYY men committed significantly more offenses within categories such as "breach of the peace" ($p < 0.005$) and theft ($p < 0.01$), yet not in other subcategories, such as "assault," "criminal damage," "alcohol and drug related," or "sexual offenses." Furthermore, there was no evidence from the length of imprisonment or the magnitude of the fines that the delinquent acts of the cases were more extreme than those of controls. In particular, there was no significant difference in crimes of sexual nature, such as indecent assault and shameless indecency. The XYY men received their first conviction at a mean age of 17.6 years, not significantly younger than the controls, at 18.1 years [20]. Thus, while XYY men committed more offenses overall, the offenses were not more serious than those committed by the controls.

Furthermore, the image of XYY men as especially aggressive is also not supported by the behavior of XYY men when they become institutionalized. The question was asked whether they are more aggressive than other men detained in the same institutions. In fact they turned out to be more agreeable; on average they had better relationships with supervisory personnel [56]. Many more psychological and psychiatric studies were carried out. While varying in details, their overall picture seldom differed from that of chromosomally normal inmates of the same institutions with the same range of intelligence.

All these results suggest alternative explanations for the undisputedly higher frequency of XYY probands in institutions for law offenders.

23.6.5.3 Intellectual Dysfunction or Simply Stature?

Many studies have been carried out on convicted and imprisoned law offenders. Their mean IQ is generally low. Intellectually subnormal persons are more often involved in criminal activities – or they run a higher risk of being apprehended. Is the supposedly higher crime rate of XYY men only a result of their reduced average intelligence?

This option appeared to be less likely according to one study, which found an increased rate of criminality in XYY men even after adjusting for social class and intelligence [66]. However, this study was also not conducted on unbiased samples: only men taller than 183 cm were screened. This resulted in a considerable discrepancy between expected and diagnosed numbers of XYY men: assuming an incidence of 1 in 1,000 of the male population the study should have found at least 30 individuals in this cohort of 31,000, whereas in the height-restricted group only 12 XYY men were found. There is the possibility that this resulted in an identification of a subgroup more likely to receive a criminal conviction in court, perhaps because of perceived threat on account of their greater height. In fact, it has been hypothesized that the characteristic tall stature of XYY men may increase the probability of being apprehended [26].

Thus, these important questions can only be addressed by unbiased samples which also look at other behavioral aspects of XYY men.

23.6.5.4 Behavioral Aspects of XYY Men

Broader, less biased studies have been performed on males with XYY syndrome, showing that behavior disorders are not a primary feature in childhood [7]. It is evident that environmental factors play a great role in the development of personality and behavior in males with karyotype 47, XYY as well as in males with a normal chromosome constitution [9].

The aforementioned study by Götz et al. [20] identified no XYY man who would have fitted diagnostic criteria for major psychiatric disorder. However, compared with controls significant differences in antisocial behavior in XYY men were found for unstable occupational history (defined as frequency of job changes, absences from work, and periods of unemployment) and antisocial behavior during adolescence and adulthood (defined as smoothness of school careers; school performance, lying at school and at home, disruptive behavior). Overall, this study suggested a slightly increased liability to antisocial behavior in XYY men [20].

Another study observed longitudinally 38 XYY males, 12 of whom were diagnosed prenatally. XYY males were at a considerably increased risk for delayed language and/or motor development. From birth onward, weight, height, and head circumference were above average values. The majority attended kindergarten in the normal education circuit, although in 50% of these cases psychosocial problems were documented. From primary school age on, there is an increased risk for child psychiatric disorders such as autism. Moreover, although normally intelligent, many of these boys are referred to special education programs [18].

In contrast, 47,XYY boys from families with better socioeconomic status had slightly higher IQs and fewer language problems than those from families of lower socioeconomic status [31].

As a consequence of these studies there can be little doubt that men with the chromosome constitution XYY run a higher relative risk of showing antisocial behavior and coming into conflict with the law than normal XY men.

23.6.5.5 Association of Criminal Behavior and Lowered Intelligence in XYY Men

Whether part of the increased risk of coming into conflict with the law can be traced to the impaired intellectual function of XYY men can only be found if we cultivate a more complex, holistic appreciation of the intervening variables, as opposed to making simple assumptions about genotype-phenotype relationships as has so often been done in the past. While the study conducted by Götz et al. [20] in unselected men confirmed an increase in antisocial and criminal behavior in XYY men, multiple regression analysis showed that this is mediated mainly through their lowered intelligence. Additional background variables, such as the socioeconomic status of the parents, may also account for some of the differences in criminality between the XYY and XY groups [31].

23.6.5.6 Social and Therapeutic Consequences

The evidence shows that the legal consequences for preventing crimes by XYY men as proposed in the heyday of the aggression hypothesis have no basis at all. Still, problems remain. If the XYY status is discovered in a study on newborns, should the parents be informed? Could such information have the effect of a self-fulfilling prophecy in that parents would treat their boy differently, and could this enhance his tendency to deviating behavior? In our opinion, all information should be provided; however, great care is needed in conveying the facts to the parents in a form that causes as little embarrassment as possible and, above all, no damage. The parents should understand that their child might possibly need somewhat more special attention during his education than an XY boy, but that given a stable environment and the same amount of parental protection as other boys enjoy, normal social adjustment is the most likely outcome.

The behavioral problems with XYY individuals (as well as with other persons having deviant sex-chromosomal karyotypes) could probably be alleviated if these conditions had been diagnosed at birth, and if they (and their parents) had received special care during their childhood. Children show marked improvement with appropriate care, e.g., training of their motoric abilities, not only in psychomotoric but also in intellectual development. In an increasing number of countries support groups have been founded to help with these problems.

Inevitably, with the widespread use of antenatal diagnostics XYY, XXY, and XXX karyotypes are discovered by amniocentesis. Parents usually should be fully informed of the findings and the implications of the sex chromosome constitution. The option of abortion as a possibility needs careful discussion; genetic counseling should be nondirective, and the decision should be left to the parents. However, investigations of the rate of pregnancy termination for various fetal aneuploidies suggest that about 57% of pregnancies with a 47,XYY chromosomal constitution are terminated [55].

23.6.5.7 XXYY Syndrome

Phenotypic features of this syndrome are described in Sect. 3.6.4.3. Importantly, the traditional view of the XXYY syndrome as a variant of the Klinefelter syndrome is obsolete, as medical problems are more severe in the former syndrome. Specifically, neurodevelopmental and psychological difficulties are a significant component of the XXYY behavioral phenotype, with developmental delays and learning disabilities universal but variable in severity. A review of 95 males with XXYY syndrome reported that 26% had full-scale IQs in the range of intellectual disability, and adaptive functioning was significantly impaired, with 68% having adaptive composite scores < 70. Overall, rates of neurodevelopmental disorders were elevated to 55.9% and included ADHD (attention-deficit/hyperactivity disorder) (72.2%), autism spectrum disorders (28.3%), mood disorders (46.8%), and tic disorders (18.9%) [59].

23.6.6 Other Chromosomal Variants

Specific behavioral phenotypes exist not only for whole chromosome aneuploidies, but also for numerous other smaller chromosomal aberrations resulting in segmental aneuploidies. The identification of specific cognitive and behavioral associations within a genetic syndrome is important for the characterization of potential etiological pathways of behavior at both the cognitive and the neurobiological level. Therefore, the study of the behavioral phenotype in a known chromosomal disorder provides an important and promising strategy for understanding the genetics and pathogenesis of these disorders in the wider population.

Here only a few syndromes associated with structural chromosomal rearrangements with relatively well-characterized behavioral phenotypes have been selected for discussion. First, the behavioral phenotype of several well-characterized microdeletion syndromes is presented. Microdeletions are so small in size that they usually escape detection in standard banding analysis. However, in many syndromes associated with larger segmental aneuploidies, which are visible by standard chromosome analysis, attempts have been made to identify genotype-phenotype correlations, but cognitive-behavioral aspects of individuals associated with the genotype have not yet been studied systematically. Therefore, data is relatively limited and discussion of behavioral consequences of segmental aneuploidies is confined to two syndromes, i.e., cri-du-chat and Wolf–Hirschhorn syndromes.

23.6.6.1 22q11.2 Deletion Syndrome

The 22q11.2 deletion syndrome (22q11.2DS) refers to a group of related syndromes including velo-cardio-facial syndrome (VCFS), Di George syndrome, and conotruncal anomaly face syndrome. The cause of the deletion is usually nonallelic homologous recombination (NAHR) (Sect. 3.5.5). Many features of this syndrome are discussed in Sect. 3.6.2.2.3. Here the focus will be on the behavioral aspects of this syndrome.

Various behavioral disorders and psychiatric illnesses have been reported within the context of the 22q11.2DS. There is an especially well-documented association between 22q11.2DS and schizophrenia [29], but multiple other behavioral features are linked to this syndrome in addition. Several studies described temperamental and behavioral difficulties, such as poor social skills, problems with social interaction, social withdrawal, and others in early childhood [57]. Other studies found that children with 22q11.2DS exhibit significant attachment to the mother or other caregivers and display clinging behavior and separation anxiety [58]. However, at present longitudinal studies with age-, gender-, and IQ-matched controls are lacking to test whether these early behavioral problems are indicators of future psychiatric disorders.

Attention deficit hyperactivity disorder (ADHD) is the most prevalent psychiatric disorder in children with 22q11.2DS and may occur in up to 40% of affected children [3, 21]. By comparison, the prevalence rates of ADHD in nondeleted school-age children are in the range of 3–5%. When a 22q11.2DS group was compared with another group characterized by similar IQ scores, degree of facial dysmorphism, and cardiac and cleft anomalies, the prevalence of

ADHD was significantly higher in the 22q11.2DS group. This supports the ideas that ADHD in this syndrome may have a genetic basis and that developmental and physical factors may play a smaller role [21]. Furthermore, several studies reported high rates of affective disorders, anxiety disorders, and obsessive-compulsive disorders in children and adults with 22q11.2DS. In addition, a high prevalence (in the range of 14%) of autism spectrum disorders in children with 22q11.2DS was described [40].

Shprintzen et al. [50] were the first to report psychotic symptoms, which they described as resembling "chronic paranoid schizophrenia" in 12 of 90 children and adults with 22q11.2DS [50]. This observation was indeed confirmed by multiple subsequent studies. For example, a longitudinal study comparing adolescents with 22q11.2DS and a control group with idiopathic developmental disability who were matched for age and IQ found that individuals with 22q11.2DS developed psychotic disorders significantly more frequently than control individuals. The available evidence suggests that about 30% of adults with 22q11.2DS have schizophrenia and that the underlying deletion contributes to these high rates [22, 40].

Therefore, the study of 22q11.2DS provides an exciting opportunity to understand the neurobiological basis of psychiatric disorders in both 22q11.2DS and in the wider nondeleted population. Current research is directed towards elucidating the contribution of several susceptibility genes within the 22q11.2 region, such as catechol-o-methyltransferase (*COMT*), proline dehydrogenase (*PRODH*), *GNB1L*, and *TBX1*. In 22q11.2DS mouse models, haploinsufficiency of *Tbx1* and *Gnb1L* is associated with a schizophrenia endophenotype [37]. Thus, *TBX1*, a transcription factor, the mutation of which is likely to be sufficient to cause most of the physical features of 22q11.2DS, may also be associated with the behavioral/psychiatric phenotype. However, further studies in persons with 22q11.2DS will be needed to examine the contribution of *TBX1* and *GNB1L* and other genes and their interaction with other candidate genes to the 22q11.2DS behavioral phenotype.

23.6.6.2 Smith–Magenis Syndrome

Smith–Magenis syndrome (SMS) is generally a sporadic disorder caused by either a 17p11.2 deletion encompassing the retinoic acid-induced 1 (*RAI1*) gene or a mutation of *RAI1* [53, 54]. Approximately 90% of all reported cases with SMS have a 17p11.2 deletion, while the remaining 10% have a mutation in the *RAI1* gene. The 17p11.2 SMS deletions are frequently caused by the NAHR mechanism (Sect. 3.5.5). In fact, chromosome 17p11.2p12 is one of the most recombination-prone regions of the genome and is also associated with hereditary neuropathy with liability to pressure palsies (HNPP) and Charcot–Marie–Tooth disease type 1A. The incidence is estimated at in the range of 1:15,000–25,000; however, this syndrome may be often underdiagnosed.

The physical phenotype is frequently described as consisting of craniofacial anomalies including brachycephaly, frontal bossing, hypertelorism, synophrys, upslanting palpebral fissures, midface hypoplasia, or a broad square face with depressed nasal bridge [15]. However, although the phenotype has some distinctive features, diagnosis is often made because of the behavioral rather than the physical phenotype.

Most SMS individuals have mild-to-moderate mental retardation with IQ ranging between 20 and 78. School-age children with IQs in the low normal range have been identified; however, IQ decreases as the child ages [15]. Sleep disturbance is one of the cardinal features and has been reported in 75–100% of SMS cases. In fact, sleep disturbances are one of the earliest diagnostic indicators of SMS and include reduced 24-h and night sleep, fragmented and shortened sleep cycles with frequent nocturnal and early-morning awakenings, and excessive daytime sleepiness. These abnormal sleep patterns are due to an inverted circadian rhythm of melatonin. An aberrant melatonin synthesis/degradation pathway has been proposed as the underlying cause for the inverted circadian rhythm. Management of sleep disturbances has been one of the challenging tasks. No well-controlled treatment plan has been reported [15, 39].

A number of additional behavioral issues belong to the characteristic features of SMS. Some of these features are unique to SMS, such as onychotillomania (pulling out of fingernails and toenails) and polyembolokoilamania (insertion of objects into bodily orifices). Stereotypical behaviors also unique to SMS include the spasmodic upper body squeeze or "self-hugging," and page-flipping or "lick and flip" behavior often seen in association with excitement. In addition, children with SMS frequently display maladaptive behaviors, including frequent outbursts/temper tantrums, attention seeking, aggression, disobedience, distraction, and self-injurious behaviors. The behavioral phenotype of SMS escalates with age, typically with the onset of puberty [15].

All SMS patients with a 17p11.2 deletion are deleted for *RAI1*, and mutations in *RAI1* are likely to result in a truncated and/or nonfunctional protein, thus leading to haploinsufficiency [55]. While *RAI1* has been shown to be responsible for most SMS features, other genes in the 17p11.2 region may contribute to the variability and severity of the phenotype in 17p11.2 deletion cases.

23.6.6.3 Prader-Willi and Angelman Syndrome

A loss of chromosomal region 15q11-q13 results in one of the two most common microdeletion syndromes, i.e., Prader–Willi (PWS) or Angelman (AS) syndrome. In both syndromes, microdeletion of the respective region on chromosome 15 is observed in about 70% of cases, again caused by NAHR (section 3.5.5). Prader–Willi syndrome results from the absence of certain paternally inherited genes on the long arm of chromosome 15, whereas Angelman syndrome is associated with loss of maternal genes (for a detailed discussion of these syndromes, see Chap. 9: Epigenetics).

PWS is most commonly known for its food-related characteristics of hyperphagia, food-seeking behavior, and consequent obesity. Overall, the behavioral phenotype of Prader–Willi syndrome affects four domains: food-seeking-related behaviors; traits indicating lack of flexibility, oppositional behaviors, and interpersonal problems. Treatment should be offered by a multidisciplinary approach with anticipatory medical and psychiatric care. Importantly, the management requires lifelong dietary restrictive supervision to prevent morbid obesity. Psychopharmacologic management may be exacerbated by metabolic abnormalities [6].

Almost all manifestations of Angelman syndrome seem to be related to lack of *UBE3A* gene expression in the brain. The *UBE3A* gene is an imprinted gene located within the aforementioned 15q11-q13 deletion region. The behavioral phenotype is characterized by a happy demeanor with prominent smiling, poorly specific laughing, and general exuberance, associated with hypermotor behavior and stereotypies. In addition, a number of characteristic features of Angelman syndrome may be seen in the context of the autistic spectrum, including virtual absence of speech, impaired use of nonverbal communicative behaviors (facial expression, body postures/gestures to regulate social interaction and decoding of emotional facial expressions), attention deficits, hyperactivity, feeding and sleeping problems, and delays in motor development [39].

23.6.6.4 Williams–Beuren Syndrome

Williams–Beuren syndrome (WBS) is caused by recurrent de novo microdeletions at 7q11.23, which are also mediated by NAHR (Sect. 3.5.5) between low copy repeats flanking this critical region. WBS is a multisystem disorder with a characteristic dysmorphic face, short stature, particularly typical cardiovascular lesions (e.g., supravalvar aortic stenosis), hypercalcemia, and neurological problems.

Individuals with WBS have mild to moderate intellectual disability or learning difficulties; however, this masks an uneven cognitive profile. The WBS neuropsychological profile is striking, characterized by strengths in certain complex faculties (language, music, face processing, and sociability) alongside marked and severe deficits in visuospatial abilities. Children and adults with WBS also have characteristic personality traits, preferring the company of adults to peers and lacking shyness with strangers, over-friendliness and charismatic speech rich in vocabulary. Approximately 70% also suffer from attention deficit disorder, and many experience anxiety and simple phobias. An interesting feature is the musical creativity observed in WBS individuals [62]. It is likely that dosage-sensitive genes within the region are important for the proper development of human speech and language [37].

23.6.6.5 Cri-du-chat Syndrome

Deletions on chromosome 5p lead to a variety of developmental defects, with most cases classified as cri-du-chat syndrome. These deletions may be terminal or interstitial and occasionally occur in the context of a cytogenetically complex karyotype. Cri-du-chat syndrome has several phenotypic components, including the characteristic cry that gives the syndrome its name, facial dysmorphology, speech delay, and mental retardation (MR). While the physical symptoms have frequently been documented, the developmental and behavioral aspects of the syndrome have not been adequately explored [63].

Array-CGH revealed that in patients with only 5p deletions different deleted regions exist, each having a different effect on retardation. Depending on size and location of the deletion, the level of mental retardation may range from moderate to profound [69].

In a study of 10 children a high rate of distractibility and a low level of object-directed behavior were observed in the play sessions. This demeanor may be an early precursor of hyperactivity, distractibility, and stereotypy, which have been reported to be the characteristic features of the behavioral phenotype of older individuals with 5p-Syndrome [47].

23.6.6.6 Wolf–Hirschhorn Syndrome

Wolf–Hirschhorn syndrome (WHS) is associated with microdeletions in the 4p16.3 region, which are variable in size but may produce similar clinical features in the phenotype that characterizes WHS. A recent study examined the cognitive skills and behavioral repertoire of 12 children, ages 4–17 years, who were diagnosed with WHS and who had some speech and expressive language. It was found that their cognitive deficits ranged from mild to severe MR with a mean IQ score of 44.1 (range: 33–64). Children with WHS exhibited relative strengths in Verbal and Quantitative Reasoning, and relative weaknesses in Abstract/Visual Reasoning and Short-term Memory. The adaptive behavior skills of all the children with WHS we assessed were lower than adequate. However, children with WHS exhibit significant relative strength in socialization compared with their communication and daily living skills. In addition, hyperactivity levels and inattentiveness consistent with a diagnosis of ADD or ADHD were noted. However, ADHD and ADD are frequently observed as comorbid features of individuals with MR [16].

References

1. Abramsky L, Hall S, Levitan J, Marteau TM (2001) What parents are told after prenatal diagnosis of a sex chromosome abnormality: interview and questionnaire study. Br Med J 322:463–466
2. Antonarakis SE, Epstein CJ (2006) The challenge of Down syndrome. Trends Mol Med 12:473–479
3. Antshel KM, Fremont W, Roizen NJ, Shprintzen R, Higgins AM, Dhamoon A, Kates WR (2006) ADHD, major depressive disorder, and simple phobias are prevalent psychiatric conditions in youth with velocardiofacial syndrome. J Am Acad Child Adolesc Psychiatry 45:596–603
4. Autism Genome Project Consortium, Szatmari P, Paterson AD, Zwaigenbaum L, Roberts W, Brian J, Liu XQ, Vincent JB, Skaug JL, Thompson AP, Senman L, Feuk L, Qian C, Bryson SE, Jones MB, Marshall CR, Scherer SW, Vieland VJ, Bartlett C, Mangin LV, Goedken R, Segre A, Pericak-Vance MA, Cuccaro ML, Gilbert JR, Wright HH, Abramson RK, Betancur C, Bourgeron T, Gillberg C, Leboyer M, Buxbaum JD, Davis KL, Hollander E, Silverman JM, Hallmayer J, Lotspeich L, Sutcliffe JS, Haines JL, Folstein SE, Piven J, Wassink TH, Sheffield V, Geschwind DH, Bucan M, Brown WT, Cantor RM, Constantino JN, Gilliam TC, Herbert M, Lajonchere C, Ledbetter DH, Lese-Martin C, Miller J, Nelson S, Samango-Sprouse CA, Spence S, State M, Tanzi RE, Coon H, Dawson G, Devlin B, Estes A, Flodman P, Klei L, McMahon WM, Minshew N, Munson J, Korvatska E, Rodier PM, Schellenberg GD, Smith M, Spence MA, Stodgell C, Tepper PG, Wijsman EM, Yu CE, Rogé B, Mantoulan C, Wittemeyer K, Poustka A, Felder B, Klauck SM, Schuster C, Poustka F, Bölte S, Feineis-Matthews S, Herbrecht E, Schmötzer G, Tsiantis J, Papanikolaou K, Maestrini E, Bacchelli E, Blasi F, Carone S, Toma C, Van Engeland H, de Jonge M, Kemner C, Koop F, Langemeijer M, Hijmans C, Staal WG, Baird G, Bolton PF, Rutter ML, Weisblatt E, Green J, Aldred C, Wilkinson JA, Pickles A, Le Couteur A, Berney T, McConachie H, Bailey AJ, Francis K, Honeyman G, Hutchinson A, Parr JR, Wallace S, Monaco AP, Barnby G, Kobayashi K, Lamb JA, Sousa I, Sykes N, Cook EH, Guter SJ, Leventhal BL, Salt J, Lord C, Corsello C, Hus V, Weeks DE, Volkmar F, Tauber M, Fombonne E, Shih A, Meyer KJ (2007) Mapping autism risk loci using genetic linkage and chromosomal rearrangements. Nat Genet 39:319–328
5. Ball SL, Holland AJ, Hon J, Huppert FA, Treppner P, Watson PC (2006) Personality and behaviour changes mark the early stages of Alzheimer's disease in adults with Down's syndrome: findings from a prospective population-based study. Int J Geriatr Psychiatry 21:661–673
6. Benarroch F, Hirsch HJ, Genstil L, Landau YE, Gross-Tsur V (2007) Prader–illi syndrome: medical prevention and behavioral challenges. Child Adolesc Psychiatr Clin N Am 16:695–708
7. Bender BG, Puck MH, Salbenblatt JA, Robinson A (1984) The development of four unselected 47, XYY boys. Clin Genet 25:435–445
8. Bojesen A, Juul S, Gravholt CH (2003) Prenatal and postnatal prevalence of Klinefelter syndrome: a national registry study. J Clin Endocrinol Metab 88(2):622–626

9. Briken P, Habermann N, Berner W, Hill A (2006) XYY chromosome abnormality in sexual homicide perpetrators. Am J Med Genet B Neuropsychiatr Genet 141B:198–200
10. Carr J (2005) Stability and change in cognitive ability over the life span: a comparison of populations with and without Down's syndrome. J Intellect Disabil Res 49:915–928
11. Chapman RS, Hesketh LJ (2000) Behavioral phenotype of individuals with Down syndrome. Ment Retard Dev Disabil Res Rev 6:84–95
12. Coe DA, Matson JL, Russell DW, Slifer KJ, Capone GT, Baglio C, Stallings S (1999) Behavior problems of children with Down syndrome and life events. J Autism Dev Disord 29:149–156
13. Dykens EM (2007) Psychiatric and behavioral disorders in persons with Down syndrome. Ment Retard Dev Disabil Res Rev 13:272–278
14. Dykens EM, Shah B, Sagun J, Beck T, King BH (2002) Maladaptive behaviour in children and adolescents with Down's syndrome. J Intellect Disabil Res 46:484–492
15. Elsea SH, Girirajan S (2008) Smith–Magenis syndrome. Eur J Hum Genet 16:412–421
16. Fisch GS, Battaglia A, Parrini B, Youngblom J, Simensen R (2008) Cognitive-behavioral features of children with Wolf–Hirschhorn syndrome: preliminary report of 12 cases. Am J Med Genet C Semin Med Genet 148C:252–256
17. Garron DC (1977) Intelligence among patients with Turner's syndrome. Behav Genet 7:105–127
18. Geerts M, Steyaert J, Fryns JP (2003) The XYY syndrome: a follow-up study on 38 boys. Genet Couns 14:267–279
19. Geschwind DH, Dykens E (2004) Neurobehavioral and psychosocial issues in Klinefelter syndrome. Learning Disabil Res Pract 19:166–173
20. Götz MJ, Johnstone EC, Ratcliffe SG (1999) Criminality and antisocial behaviour in unselected men with sex chromosome abnormalities. Psychol Med 29:953–962
21. Gothelf D, Presburger G, Levy D, Nahmani A, Burg M, Berant M, Blieden LC, Finkelstein Y, Frisch A, Apter A, Weizman A (2004) Genetic, developmental, and physical factors associated with attention deficit hyperactivity disorder in patients with velocardiofacial syndrome. Am J Med Genet B Neuropsychiatr Genet 126B:116–121
22. Gothelf D, Feinstein C, Thompson T, Gu E, Penniman L, Van Stone E, Kwon H, Eliez S, Reiss AL (2007) Risk factors for the emergence of psychotic disorders in adolescents with 22q11.2 deletion syndrome. Am J Psychiatry 164:663–669
23. Haddad HM, Wilkins L (1959) Congenital anomalies associated with gonadal aplasia; review of 55 cases. Pediatrics 23:885–902
24. Harmon RJ, Bender BG, Linden MG, Robinson A (1998) Transition from adolescence to early adulthood: adaptation and psychiatric status of women with 47, XXX. J Am Acad Child Adolesc Psychiatry 37:286–291
25. Hodapp RM, Dykens EM (2005) Measuring behavior in genetic disorders of mental retardation. Ment Retard Dev Disabil Res Rev 11:340–346
26. Hunter H (1977) (1977) XYY males: some clinical and psychiatric aspects deriving from a survey of 1811 males in hospitals for the mentally handicapped. Br J Psychiatry 131:468–477
27. Iafrate AJ, Feuk L, Rivera MN, Listewnik ML, Donahoe PK, Qi Y, Scherer SW, Lee C (2004) Detection of large-scale variation in the human genome. Nat Genet 36:949–951
28. Jacobs PA, Brunton M, Melville MM, Brittain RP, McClemont WF (1965) Aggressive behavior, mental subnormality and the XYY male. Nature 208:1351–1352
29. Karayiorgou M, Morris MA, Morrow B, Shprintzen RJ, Goldberg R, Borrow J, Gos A, Nestadt G, Wolyniec PS, Lasseter VK, Eisen H, Childs B, Kazazian HH, Kucherlapati R, Antonarakis SE, Pulver AE, Housman DE (1995) Schizophrenia susceptibility associated with interstitial deletions of chromosome 22q11. Proc Natl Acad Sci USA 92:7612–7616
30. Kent L, Evans J, Paul M, Sharp M (1999) Comorbidity of autistic spectrum disorders in children with Down syndrome. Dev Med Child Neurol 41:153–158
31. Linden MG, Bender BG (2002) Fifty-one prenatally diagnosed children and adolescents with sex chromosome abnormalities. Am J Med Genet 110:11–18
32. Marshall CR, Noor A, Vincent JB, Lionel AC, Feuk L, Skaug J, Shago M, Moessner R, Pinto D, Ren Y, Thiruvahindrapduram B, Fiebig A, Schreiber S, Friedman J, Ketelaars CE, Vos YJ, Ficicioglu C, Kirkpatrick S, Nicolson R, Sloman L, Summers A, Gibbons CA, Teebi A, Chitayat D, Weksberg R, Thompson A, Vardy C, Crosbie V, Luscombe S, Baatjes R, Zwaigenbaum L, Roberts W, Fernandez B, Szatmari P, Scherer SW (2008) Structural variation of chromosomes in autism spectrum disorder. Am J Hum Genet 82: 477–488
33. McGlone J (1985) Can spatial deficits in Turner's syndrome be explained by focal CNS dysfunction or atypical speech lateralization? J Clin Exp Neuropsychol 7:375–394
34. Myers BA, Pueschel SM (1995) Major depression in a small group of adults with Down syndrome. Res Dev Disabil 16:285–299
35. Nielsen J, Johnsen SG, Sorensen K (1980) Follow-up 10 years later of 34 Klinefelter males with karyotype 47, XXY and 16 hypogonadal males with karyotype 46, XY. Psychol Med 10:345–352
36. Osborne LR, Mervis CB (2007) Rearrangements of the Williams–Beuren syndrome locus: molecular basis and implications for speech and language development. Expert Rev Mol Med 9:1–16
37. Paylor R, Glaser B, Mupo A, Ataliotis P, Spencer C, Sobotka A, Sparks C, Choi CH, Oghalai J, Curran S, Murphy KC, Monks S, Williams N, O'Donovan MC, Owen MJ, Scambler PJ, Lindsay E (2006) Tbx1 haploinsufficiency is linked to behavioral disorders in mice and humans: implications for 22q11 deletion syndrome. Proc Natl Acad Sci USA 103:7729–7734
38. Pelc K, Cheron G, Dan B (2008) Behavior and neuropsychiatric manifestations in Angelman syndrome. Neuropsychiatr Dis Treat 4:577–584
39. Potocki L, Glaze D, Tan DX, Park SS, Kashork CD, Shaffer LG, Reiter RJ, Lupski JR (2000) Circadian rhythm abnormalities of melatonin in Smith–Magenis syndrome. J Med Genet 37:428–433
40. Prasad SE, Howley S, Murphy KC (2008) Candidate genes and the behavioral phenotype in 22q11.2 deletion syndrome. Dev Disabil Res Rev 14:26–34
41. Romans SM, Stefanatos G, Roeltgen DP, Kushner H, Ross JL (1998) Transition to young adulthood in Ullrich–Turner

syndrome: neurodevelopmental changes. Am J Med Genet 79:140–147
42. Ross JL, Kushner H, Zinn AR (1997) Discriminant analysis of the Ullrich–Turner syndrome neurocognitive profile. Am J Med Genet 72:275–280
43. Ross J, Zinn A, McCauley E (2000) Neurodevelopmental and psychosocial aspects of Turner syndrome. Ment Retard Dev Disabil Res Rev 6:135–141
44. Ross JL, Stefanatos GA, Kushner H, Zinn A, Bondy C, Roeltgen D (2002) Persistent cognitive deficits in adult women with Turner syndrome. Neurology 58:218–225
45. Rovet J (2004) Turner syndrome: a review of genetic and hormonal influences on neuropsychological functioning. Child Neuropsychol 10:262–279
46. Rovet J, Netley C, Bailey J, Keenan M, Stewart D (1995) Intelligence and achievement in children with extra X aneuploidy: a longitudinal perspective. Am J Med Genet 60:356–363
47. Sarimski K (2003) Early play behaviour in children with 5p- (Cri-du-Chat) syndrome. J Intellect Disabil Res 47: 113–120
48. Sebat J, Lakshmi B, Troge J, Alexander J, Young J, Lundin P, Månér S, Massa H, Walker M, Chi M, Navin N, Lucito R, Healy J, Hicks J, Ye K, Reiner A, Gilliam TC, Trask B, Patterson N, Zetterberg A, Wigler M (2004) Large-scale copy number polymorphism in the human genome. Science 30:525–528
49. Sebat J, Lakshmi B, Malhotra D, Troge J, Lese-Martin C, Walsh T, Yamrom B, Yoon S, Krasnitz A, Kendall J, Leotta A, Pai D, Zhang R, Lee YH, Hicks J, Spence SJ, Lee AT, Puura K, Lehtimäki T, Ledbetter D, Gregersen PK, Bregman J, Sutcliffe JS, Jobanputra V, Chung W, Warburton D, King MC, Skuse D, Geschwind DH, Gilliam TC, Ye K, Wigler M (2007) Strong association of de novo copy number mutations with autism. Science 316:445–459
50. Shprintzen RJ, Goldberg R, Golding-Kushner KJ, Marion RW (1992) Late-onset psychosis in the velo-cardio-facial syndrome. Am J Med Genet 42:141–142
51. Silverman W (2007) Down syndrome: cognitive phenotype. Ment Retard Dev Disabil Res Rev 13:228–236
52. Simm PJ, Zacharin MR (2006) The psychosocial impact of Klinefelter syndrome–a 10 year review. J Pediatr Endocrinol Metab 19:499–505
53. Slager RE, Newton TL, Vlangos CN, Finucane B, Elsea SH (2003) Mutations in RAI1 associated with Smith–Magenis syndrome. Nat Genet 33:466–468
54. Smith AC, McGavran L, Robinson J, Waldstein G, Macfarlane J, Zonona J, Reiss J, Lahr M, Allen L, Magenis E (1986) Interstitial deletion of (17)(p11.2p11.2) in nine patients. Am J Med Genet 24:393–414
55. Stefansson H, Rujescu D, Cichon S, Pietiläinen OP, Ingason A, Steinberg S, Fossdal R, Sigurdsson E, Sigmundsson T, Buizer-Voskamp JE, Hansen T, Jakobsen KD, Muglia P, Francks C, Matthews PM, Gylfason A, Halldorsson BV, Gudbjartsson D, Thorgeirsson TE, Sigurdsson A, Jonasdottir A, Jonasdottir A, Bjornsson A, Mattiasdottir S, Blondal T, Haraldsson M, Magnusdottir BB, Giegling I, Möller HJ, Hartmann A, Shianna KV, Ge D, Need AC, Crombie C, Fraser G, Walker N, Lonnqvist J, Suvisaari J, Tuulio-Henriksson A, Paunio T, Toulopoulou T, Bramon E, Di Forti M, Murray R, Ruggeri M, Vassos E, Tosato S, Walshe M, Li T, Vasilescu C, Mühleisen TW, Wang AG, Ullum H, Djurovic S, Melle I, Olesen J, Kiemeney LA, Franke B, GROUP, Sabatti C, Freimer NB, Gulcher JR, Thorsteinsdottir U, Kong A, Andreassen OA, Ophoff RA, Georgi A, Rietschel M, Werge T, Petursson H, Goldstein DB, Nöthen MM, Peltonen L, Collier DA, St Clair D, Stefansson K (2008) Large recurrent microdeletions associated with schizophrenia. Nature 455:232–236
56. Street DRK, Watson RA (1969) Patients with chromosome abnormalities in Rampton Hospital. In: West DJ (ed) Criminological implications of chromosome abnormalities. Cropwood journal. Institute of Criminology, University of Cambridge, Cambridge, pp 61–67
57. Swillen A, Devriendt K, Legius E, Prinzie P, Vogels A, Ghesquière P, Fryns JP (1999) The behavioural phenotype in velo-cardio-facial syndrome (VCFS): from infancy to adolescence. Genet Couns 10:79–88
58. Swillen A, Vogels A, Devriendt K, Fryns JP (2000) Chromosome 22q11 deletion syndrome: update and review of the clinical features, cognitive-behavioral spectrum, and psychiatric complications. Am J Med Genet 97:128–135
59. Tartaglia N, Davis S, Hench A, Nimishakavi S, Beauregard R, Reynolds A, Fenton L, Albrecht L, Ross J, Visootsak J, Hansen R, Hagerman R (2008) A new look at XXYY syndrome: medical and psychological features. Am J Med Genet 146A(12):1509–1522
60. Tassabehji M (2003) Williams–Beuren syndrome: a challenge for genotype-phenotype correlations. Hum Mol Genet 12(Spec No 2):R229–R237
61. Temple CM (2002) Oral fluency and narrative production in children with Turner's syndrome. Neuropsychologia 40:1419–1427
62. Turner S, Alborz A, Gayle V (2008) Predictors of academic attainments of young people with Down's syndrome. J Intellect Disabil Res 52:380–392
63. Van Buggenhout GJ, Pijkels E, Holvoet M, Schaap C, Hamel BC, Fryns JP (2000) Cri du chat syndrome: changing phenotype in older patients. Am J Med Genet 90:203–215
64. van Rijn S, Swaab H, Aleman A, Kahn RS (2008) Social behavior and autism traits in a sex chromosomal disorder: Klinefelter (47XXY) syndrome. J Autism Dev Disord 38(9):1634–1641
65. Wilson GM, Flibotte S, Chopra V, Melnyk BL, Honer WG, Holt RA (2006) DNA copy-number analysis in bipolar disorder and schizophrenia reveals aberrations in genes involved in glutamate signaling. Hum Mol Genet 15:743–749
66. Witkin HA, Mednick SA, Schulsinger F, Bakkestrom E, Christiansen KO, Goodenough DR, Hirschhorn K, Lundsteen C, Owen DR, Philip J, Rubin DB, Stocking M (1976) Criminality in XYY and XXY males is not related to aggression. It may be related to low intelligence. Science 193:547–555
67. Xu B, Roos JL, Levy S, van Rensburg EJ, Gogos JA, Karayiorgou M (2008) Strong association of de novo copy number mutations with sporadic schizophrenia. Nat Genet 40:880–885
68. Zhang D, Cheng L, Qian Y, Alliey-Rodriguez N, Kelsoe JR, Greenwood T, Nievergelt C, Barrett TB, McKinney R, Schork N, Smith EN, Bloss C, Nurnberger J, Edenberg HJ, Foroud T, Sheftner W, Lawson WB, Nwulia EA, Hipolito M, Coryell W, Rice J, Byerley W, McMahon F, Schulze TG,

Berrettini W, Potash JB, Belmonte PL, Zandi PP, McInnis MG, Zöllner S, Craig D, Szelinger S, Koller D, Christian SL, Liu C, Gershon ES (2008) Singleton deletions throughout the genome increase risk of bipolar disorder. Mol Psychiatry 14:376–380

69. Zhang X, Snijders A, Segraves R, Zhang X, Niebuhr A, Albertson D, Yang H, Gray J, Niebuhr E, Bolund L, Pinkel D (2005) High-resolution mapping of genotype-phenotype relationships in cri du chat syndrome using array comparative genomic hybridization. Am J Hum Genet 76:312–326

Genetics of Schizophrenia and Bipolar Affective Disorder

23.7

Markus M. Nöthen, Sven Cichon, Christine Schmael, and Marcella Rietschel

Abstract Schizophrenia and bipolar affective disorder (bipolar disorder, manic depression) are the paradigmatic illnesses of psychiatry. They profoundly affect thought, perception, emotion, and behavior, and their symptoms cause significant social and/or occupational dysfunction. Schizophrenia and bipolar disorder have been recognized for several millennia, and the WHO (2001) ranks both among the top ten leading causes of the global burden of disease for the age group 15–44 years.

Schizophrenia and bipolar disorder are illnesses with a largely unknown pathophysiology and etiology. Evidence of a clear genetic contribution to the development of these disorders has led to important endeavors to discover the responsible genes. This chapter provides a concise and comprehensive review of the current state of genetic research into schizophrenia and bipolar disorder, and also of its limitations and possible future directions.

Contents

23.7.1 Schizophrenia 759
    23.7.1.1 Prevalence... 760
    23.7.1.2 Environmental Risk Factors 760
    23.7.1.3 Formal Genetic Studies 761
    23.7.1.4 Gene–Environment Interaction 761
    23.7.1.5 The Evolutionary Paradox of Schizophrenia 762
    23.7.1.6 Molecular Genetic Studies 762
    23.7.1.7 Endophenotypes 764

23.7.2 Bipolar Disorder... 765
    23.7.2.1 Prevalence... 765
    23.7.2.2 Environmental Risk Factors 765
    23.7.2.3 Formal Genetic Studies 766
    23.7.2.4 Molecular Genetic Studies 766
    23.7.2.5 Endophenotypes 768

23.7.3 Schizophrenia and Bipolar Disorder: Approaches Beyond a Diagnostic Dichotomy 768

23.7.4 Outlook... 768

References.. 770

M.M. Nöthen (✉) and S. Cichon
Department of Genomics, Life & Brain Center,
University of Bonn, Bonn, Germany
Institute of Human Genetics, University of Bonn, Bonn, Germany
e-mail: markus.noethen@uni-bonn.de;
sven.cichon@uni-bonn.de

C. Schmael and M. Rietschel
Division of Genetic Epidemiology in Psychiatry, Central Institute of Mental Health, Mannheim, Germany
e-mail: c.schmael@gmx.de;
Marcella.Rietschel@zi-mannheim.de

23.7.1 Schizophrenia

Schizophrenia (for diagnostic criteria see Table 23.7.1) is characterized by fundamental and characteristic distortions of thought and perception, inappropriate feelings and/or blunted emotions, and a restricted capacity to act and interact appropriately. The hallmark symptoms of schizophrenia are psychotic phenomena, which include delusions, delusional perceptions, and hallucinations.

Table 23.7.1 Symptoms required for a DMS-IV diagnosis* of major depressive episode, manic episode, and schizophrenia

Major depressive episode	Manic episode	Schizophrenia
Five symptoms during a *2-week* period	*Three (four)* symptoms *during 1 week* of abnormally elevated expansive (or irritable) mood or any duration if hospitalized	*Two* symptoms during a *1-month* period or less if successfully treated
• Depressed mood	• Inflated self-esteem or grandiosity	• Delusions
• Loss of interest or pleasure	• Decreased need for sleep	• Hallucinations
• Change in appetite and or weight	• More talkative than usual or pressure to keep talking	• Disorganized speech
• Insomnia or hypersomnia	• Flight of ideas or racing thoughts	• Grossly disorganized or catatonic behavior
• Psychomotor agitation or retardation	• Distractibility	• Negative symptoms, i.e., affective flattening, alogia, or avolition
• Fatigue or loss of energy	• Increase in goal-directed activity or psychomotor agitation	Only one symptom is required if delusions are bizarre or hallucinations consist of a voice keeping up a running commentary on the person's behavior or thoughts, or two or more voices conversing with each other
• Feelings of worthlessness or excessive or inappropriate guilt	• Excessive involvement in pleasurable activities	Continuous signs of the disturbance present for at least *6 months*, which may include periods of prodromal or residual symptoms
• Diminished concentration or indecisiveness		
• Recurrent thoughts of death/suicidal ideation or suicide attempt		

From [9]

The symptoms are not due to the direct physiological effects of a substance (e.g., a drug of abuse, a medication or other treatment) or a general medical condition

The symptoms cause clinically significant distress or impairment in social, occupational, or other important areas of functioning

Negative symptoms, thought disorders, and neuropsychological deficits, while less striking in nature, are usually more persistent and more indicative of the course of the disorder [50]. Although cognitive deficits and a decline in intellectual capacities are observed in most patients, consciousness and a substantial level of intellectual capacity are maintained. The course of the disorder is often characterized by recurrent episodes and an increased mortality rate. Approximately two-thirds of all affected individuals have persistent or fluctuating symptoms even if they receive optimal treatment [9].

The onset of schizophrenia typically occurs in early adulthood, although premorbid symptoms have often been present for many years [121]. On average, the age at onset is 3–5 years earlier in men than in women, but this gender difference is not observed in patients with a family history of schizophrenia [4, 44, 121].

23.7.1.1 Prevalence

The life-time prevalence of schizophrenia in developed countries is around 0.5–1%. Early studies indicated that schizophrenia occurs at the same rate world-wide, but more recent studies have suggested that the prevalence may vary between countries, a higher prevalence being observed in developed nations [43, 119]. The prevalence in females and males is similar [6, 119].

23.7.1.2 Environmental Risk Factors

A substantial body of epidemiological research has established that there is a set of nongenetic risk factors for schizophrenia. These include being a first- or second-generation migrant, being born or living in an urban area, having had a winter or spring birth, advanced paternal age, prenatal and/or obstetric complications, cannabis use during adolescence, parental unemployment, and socio-economic status [21, 23, 35, 37, 89, 91, 93, 103, 105, 120, 145]. The effect size of individual risk factors is modest; a typical odds ratio is ≤2, except for migration, which appears to confer a higher risk of around 4. The observation that the incidence of schizophrenia may fluctuate over time is consistent with the influence of environmental factors on disease risk [92].

23.7.1.3 Formal Genetic Studies

23.7.1.3.1 Family Studies

A large number of family studies have shown that schizophrenia runs in families. The average risk for the sibling of an affected person is around 9%, that for an offspring is around 13%, and the risk for a parent is around 6% [130]. The lower risk of developing the disorder observed in parents has been explained by the reproductive disadvantage conferred by schizophrenia. Only one study has directly assessed second- and third-degree relatives, and this has reported risks of 3% and 1.5%, respectively [88]. It has been suggested that the relatives of both female patients and patients with an early age at onset may have an increased risk of developing the illness, although findings have been inconclusive [130]. Relatives of schizophrenia patients show an increased risk for the following disorders: schizophrenia-related personality disorders, i.e., schizotypal and paranoid personality disorders; nonschizophrenic psychotic disorders, i.e., schizophreniform disorder, schizoaffective disorder, delusional disorder, and psychotic disorder not otherwise specified; unipolar depression; and bipolar disorder [69, 88]. Symptoms and symptom dimensions showing familiality are age at onset, course of disorder, impairment during disorder, mode of onset, premorbid functioning, psychomotor poverty, disorganization, and manic features [150].

23.7.1.3.2 Twin Studies

Nearly all of the twin studies conducted to date have shown that concordance rates are higher for monozygotic twins than for dizygotic twins. In studies conducted prior to 1980, pooled concordance rates for monozygotic and dizygotic twins were estimated as 53% and 15%, respectively [67]. Reviews of more recent studies, which were conducted using modern methodological standards (blinded and structured recruitment), have found pooled concordance rates of 40–65% for monozygotic and 0–28% for dizygotic twins [24, 25, 79]. Heritability of the liability to schizophrenia is reported as 81% (95% confidence interval: 73–90%) according to meta-analytical estimates from the pooled data of 12 twin studies [137].

23.7.1.3.3 Adoption Studies

Adoption studies have shown that the biological relatives of patients with schizophrenia (adopted-away children of schizophrenia patients, biological relatives of adopted-away children who later develop schizophrenia) have an increased risk of developing the disorder [57, 70, 72–74, 85, 140, 141]. These studies also demonstrate that nongenetic factors influence vulnerability. The adopted-away children of mothers with schizophrenia only developed schizophrenia when placed in adoptive families with psychological abnormalities [140, 141, 153]. Children of healthy parents adopted by a parent who later developed a schizophrenia spectrum disorder did not show increased risk, demonstrating that nongenetic factors alone are not sufficient to cause the disorder [149].

23.7.1.4 Gene–Environment Interaction

Given the importance of environmental risk factors in the development of schizophrenia (see above), it is important to address the issue of whether these factors act with genetic susceptibility in an independent (additive) fashion, or in a synergistic (interactive) fashion in which the effect of one factor is conditional upon the other [139, 145]. Indirect evidence for a gene–environment interaction has been obtained from twin and adoption studies, which have shown, for example, that the disease risk for adopted-away children of mothers with schizophrenia is dependent on the psychological functioning of the family in which such children are placed [140, 141, 153]. Research has also shown that individuals with a familial genetic loading for schizophrenia may be more likely to develop psychosis as a consequence of cannabis use [95].

23.7.1.5 The Evolutionary Paradox of Schizophrenia

Researchers have long been confused by the fact that so debilitating a disorder as schizophrenia has remained so common in the human population despite its negative effect on reproductive fitness. In their in-depth review of the theoretical and empirical evidence for different evolutionary models, Keller and Miller [68] conclude that only a polygenic mutation-selection balance model would appear consistent with the available data on schizophrenia prevalence rates, fitness costs, the probable rarity of susceptibility alleles, and the increased risk of mental disorder associated with brain trauma, inbreeding, and paternal age. While conceding that the alternative models of ancestral neutrality and balancing selection almost certainly play a part in maintaining some susceptibility alleles, they argue that these models lack sufficient convincing support from empirical studies to serve as a general explanation. The authors suggest that these arguments may also be applicable to other neuropsychiatric disorders, such as bipolar affective disorder.

23.7.1.6 Molecular Genetic Studies

The high heritability of schizophrenia has stimulated intense research into the identification of susceptibility genes. Two main approaches are employed in the search for these genes: linkage and association studies using genetic polymorphisms.

In linkage studies, the objective is to identify regions of the genome that are cotransmitted with the disease in families with two or more affected individuals. Using a few hundred evenly spaced genetic markers across the whole-genome, linkage has the potential to locate disease genes by virtue of chromosomal position alone, without any prior knowledge of disease etiology. Linkage studies are ideally suited to detecting genes of large effect, as in monogenic diseases. Their power to detect genes of moderate to small effect is limited, however, and large numbers of families are needed to locate genes for such complex disorders as schizophrenia.

Association studies aim to detect alleles that are more (or less) common in patients than they are within the general population. These studies are more powerful than linkage studies in detecting genes of small effect, assuming that a high proportion of the disease alleles are relatively common in the population and that they are attributable to a common founder [116]. Until recently, one disadvantage of association studies was that they were restricted to candidate genes only, leaving a large proportion of the genome uninvestigated. These candidate genes are selected on the basis that they are either functional candidates, i.e., they encode a protein implicated by an etiological hypothesis, or positional candidates, i.e., they map chromosomal regions implicated by previous linkage studies, or a combination of the two.

The recent introduction of array technology, which permit genome-wide association studies that investigate several hundreds of thousands of markers at a time, has meant that association studies may now be performed in a hypothesis-free fashion, with systematic testing of all genes and most intergenic sequences in the genome for association.

Molecular genetic approaches became widely available in the 1980s, and many groups have since performed linkage and association studies to identify genetic variants conferring susceptibility to schizophrenia. Difficulties in replicating early findings had led to increasing skepticism that such approaches would ever be successful. However, increasing knowledge of the human genome and its variation, as well as the development of new technologies and their successful application to the study of the genetics of schizophrenia, has led to renewed optimism.

23.7.1.6.1 Linkage Studies

The results of early linkage studies into schizophrenia were greeted with some disappointment. Hopes of finding Mendelian forms of the illness, i.e., forms in families showing almost monogenic effects, have not materialized, and most studies have failed to achieve stringent "genome-wide" levels of significance or to replicate pre-existing findings. This is probably attributable to small genetic effects, inadequate sample sizes (<100 families), and marker maps whose size was insufficient for the extraction of genetic information. Replication studies in complex diseases are also hampered by the fact that experimental conditions cannot be reproduced identically, since samples almost certainly differ in genetic architecture as a consequence of ethnicity and recruitment. It is not therefore to be expected that all studies will yield identical findings.

Despite these difficulties, the 20 or more genome-wide linkage studies (genome-scans) published to date have provided some consistent patterns of positive linkage. Three of the best-supported regions are 1q21–q22, 6p24–p22, and 13q32–q34, which have shown genome-wide significance in independent studies, i.e., a linkage value that is expected less than once by chance in 20 complete genome-scans. Further promising regions with evidence for linkage are 1q42, 5q21–q33, 6q21–q25, 8p21–p22, 10p15–p11, and 22q11–q12. Two meta-analyzes of schizophrenia linkage studies have been conducted to address issues of power [10, 84]. Such meta-analyzes enable the identification of genes that make a relatively small, but widespread, contribution to the development of an illness. These studies support the existence of susceptibility genes on 1q, 2q, 3p, 5q, 6p, 8p, 11q, 13q, 14p, 20q, and 22q. Proof that these linkage regions are correct will be obtained when the disease genes are identified.

Studies in isolated populations have been performed as an alternative approach to linkage studies in the hope that founder effects will render the contribution of individual genes larger and thus easier to detect. Some of these studies have suggested new chromosomal regions that may harbor disease genes [146].

23.7.1.6.2 Candidate Gene Studies

The list of candidate gene analyzes of schizophrenia is very comprehensive. Many of these findings are likely to represent false-positive findings since they have not been replicated in independent studies. The convergence of positive linkage findings, however, has led to several detailed mapping studies of linked regions. Some of these positional candidate genes are considered to be the most promising susceptibility genes for schizophrenia, having received the strongest support from independent studies. These include the genes *dystrobrevin-binding protein 1* (*DTNBP1*) at 6p, *neuregulin 1* (*NRG1*) at 8p, *disrupted in schizophrenia 1* (*DISC1*) at 1q, *D-amino acid oxidase activator* (*DAOA, G72/G30*) at 13q, and *catechol-O-methyl-transferase* (*COMT*) at 22q [29, 101, 129, 134, 136]. No consistent evidence of association has been observed across studies for any of these genes, however, and causative mutations still await identification [2, 11, 28, 51, 52, 90, 109, 115, 117, 151].

It is difficult in some cases to judge the replication status of a gene, since different studies report associations with opposite risk alleles for the same marker [2, 106]. Such "flip-flop" phenomena are difficult to interpret, particularly when they are observed in comparably recruited samples from a common population. Differences in linkage disequilibrium (LD) architecture across populations with different ancestral origins may be one plausible theoretical explanation. In populations with a similar ancestral background, however, such differences seem less likely. However, LD patterns may vary even within (sub)populations, either as a consequence of differences in local recombination rates, genetic drift, and population history, or as a result of sampling variation. Lin et al. [86] have applied theoretical modeling to demonstrate that flip-flop associations are possible within samples recruited from the same population when the investigated variants are correlated through interactive effects or LD with a second causal variant. They showed that such flip-flop associations are particularly often observed when the risk allele at the genotyped locus is a relatively common allele in weak LD with an as yet unknown second causal variant. Under such circumstances, the observed direction of allelic association may be especially susceptible to sampling variation, and remarkable variation in LD patterns between the investigated samples is not required.

While further replication of the implicated genes remains the priority for research, the respective contributions of each gene, their relationship to aspects of the phenotype, the possibility of epistatic interactions between genes, and functional interactions between the gene products all await investigation.

23.7.1.6.3 Genome-Wide Association Studies

Genome-wide association studies have recently become possible as a result of enormous technological advances. Such studies involve the use of array technology that can simultaneously genotype up to 1 million single nucleotide polymorphisms (SNPs) per individual. As with complex genetic illnesses in general, this approach holds promise for the identification of common genetic risk variants for schizophrenia. It enables a search for the sites of illness genes on every gene and in most intergenetic regions of the genome in samples of unrelated patients and controls, independent of speculation regarding pathophysiology. In this respect they resemble genome-wide linkage studies (genome-scans), but they are not dependent on the recruitment of families and they have markedly better resolution since, in contrast to linkage, they detect the linkage disequilibrium to the illness-relevant genetic variants.

The first genome-wide association studies for schizophrenia have recently been reported [76, 80, 107, 108]. They demonstrate that schizophrenia is, in principle, amenable to systematic genetic association approaches, but that genetic effect sizes are small (OR in the 1.1–1.3 range), as is the case with other common genetic disorders. The most convincing association reported to date has been for variation in *ZNF804A*, which encodes a zinc finger transcription factor [108]. A recent functional magnetic resonance brain imaging study in healthy individuals demonstrated abnormal functional coupling between hippocampus formation and the dorsolateral prefrontal cortex in carriers of the *ZNF804A* risk variant [38]. Disturbed interaction between these brain areas has been observed in schizophrenia, and this strongly suggests that the identified genetic risk factor of small effect is highly penetrant in the brain and produces pathophysiological patterns observed in overt disease. It is anticipated that future studies involving the use of even larger samples and the pooling of datasets will identify additional specific risk factors for schizophrenia with higher statistical power.

23.7.1.6.4 Submicroscopic Chromosomal Aberrations

The fact that small chromosomal aberrations (copy number variations, genomic imbalances) may confer a risk for schizophrenia is exemplified by the 22q11.2 deletion syndrome (22q11.2DS), which is a common microdeletion syndrome with congenital and late-onset features, including a high risk for neuropsychiatric diseases (up to 25% risk for schizophrenia) [12, 63]. Interestingly, it has not been possible to correlate the extent of the deletion with the occurrence of schizophrenia in 22q11.2 DS patients and there is experimental evidence that altered expression of several genes within the 22q11.2 region may be necessary to increase susceptibility [98, 131]. This may explain why attempts to implicate individual genes from the deletion region as general susceptibility genes for the development of schizophrenia have not led to replicable results [41].

With the use of new technologies, such as comparative genomic hybridization (CGH) or SNP arrays applied in genome-wide association studies, it became possible to identify small chromosomal aberrations on a genome-wide scale. Initial studies found increased overall rates of aberrations in schizophrenia [147, 154], while later studies have been able to implicate specific chromosomal regions [58, 75, 107, 118, 135]. The implicated aberrations include microdeletions in chromosomal regions 1q21.1, 2p16.3, 15q11.2, and 15q13.3, as well as a microduplication in chromosomal region 15q13.1. Although each of these variants is more frequently observed among patients than among controls, the frequency of each individual variant in schizophrenia patients is still low (<1%). Further studies will be required to determine the penetrance, the mutation rate, and the full phenotypic spectrum associated with the aberrations. It has already been shown that some variants occur more frequently in patients with other CNS phenotypes, such as autism, mental disability, and epilepsy [15, 56, 99, 102], suggesting that common etiological factors exist among these disorders.

23.7.1.7 Endophenotypes

Endophenotypes, which are quantitative risk factors that are correlated with disease, have gained importance in molecular genetic analyzes over recent years [8, 14, 45]. It is hypothesized that endophenotypes bear a closer relationship to genetic variation than clinical symptoms and that their use in research may render gene identification more straightforward and successful. In order to be of value in genetic studies it has been suggested that endophenotypes should be (1) associated with illness in the population, (2) heritable, (3) detectable in an individual irrespective of whether or not the illness is active, (4) found to co-segregate with illness in families, and (5) found in unaffected relatives of probands at a higher rate than in the general population [53]. Endophenotypes from a variety of domains have been proposed for schizophrenia, including neurocognition, neurodevelopment, metabolism, and neurophysiology [19, 22]. The fact that only relatively small samples of patients with a specific endophenotype are available may limit their use in gene identification efforts owing to limited power. They will certainly have great potential once disease susceptibility genes are identified and attempts are made to understand the impact of risk variants on specific functions of the human brain.

It may be concluded that molecular genetic studies aiming to identify genes that contribute to the risk for schizophrenia have recently made substantial progress. However, the great majority of genetic risk factors still await identification. A general lesson learnt from genome-wide studies is that the genetic heterogeneity is greater than previously thought, a situation which is, however, similar to that for other common multifactorial disorders.

23.7.2 Bipolar Disorder

Humor, temper, tune, spirit, vibes, sentiment, disposition: the many descriptions of mood attest to its natural unsteadiness. Fluctuations in mood and mood swings are indeed inherent in human nature and are considered normal when they are restricted in intensity and/or duration, but excessive deviations from normal are defined as mood disorders. Mood disorders have been recognized and described for more than 2,000 years [30] and rank among the most common diseases of mankind. The WHO [152] has described mood disorders as a serious global health burden.

Modern classification systems define specific types of mood disorders, depending on the presence of specific patterns of symptoms and signs over specified periods of time (for DSM-IV criteria see Table 23.7.1). This explains the occurrence of many mood syndromes which do not meet diagnostic thresholds. The most common and important mood disorders are major depression (unipolar depression) and bipolar disorder. Mood in a depressed phase is sad and despondent. Other characteristic features are loss of drive, slowness of thought, feelings of guilt, and loss of interest in life. These features are typically accompanied by appetite and sleep disturbances. Depressive phases may occur as single episodes, but tend to recur. Bipolar disorder is characterized by changes in mood between the two poles of depression and mania. An episode of mania is characterized by expansive mood with an exaggerated estimation of ability, increased drive, and rapid thoughts and speech. Marked irritability is another frequent feature of the manic phase. The alternation between depressive and manic episodes may be rapid, taking place within hours or days, or may occur at longer intervals, separated by months or years.

The DSM-IV classification system [9] distinguishes between bipolar I and bipolar II disorder, this distinction being based on the severity of the manic symptoms. Bipolar I disorder has clear manic symptoms and may include delusions. Bipolar II disorder has less distinct symptomatology, with so-called hypomanic symptoms.

In terms of genetic studies, bipolar disorder is the most extensively studied mood disorder to date. The onset of bipolar disorder typically occurs in the mid-teens or twenties, and an episode of major depression or hypomania is usually its first manifestation.

23.7.2.1 Prevalence

Unipolar depression affects women twice as often as men, and has a life-time prevalence of around 10–15%. Bipolar disorder has a life-time prevalence of 0.5–1.5% and affects both sexes equally.

23.7.2.2 Environmental Risk Factors

To date, fewer studies investigating external risk factors have been conducted for bipolar disorder than for schizophrenia, and their findings have been inconclusive. There is some evidence that season of birth may have an influence on the development of bipolar disorder, as has been reported for schizophrenia [143]. Residence in large cities seems to increase the risk for psychotic bipolar disorder, although it does not influence the risk for bipolar disorder per se [65]. No conclusive evidence has been found to suggest that migration [138] or a history of obstetric complications [127] leads to a significant increase in the risk of developing the disorder. A large Danish study found no influence of external factors on the occurrence of bipolar disorder with the exception of parental loss; maternal loss before the age of 5 years was found to increase the risk 4-fold [104].

23.7.2.3 Formal Genetic Studies

Results from a large number of formal genetic studies suggest that genetic factors contribute substantially to the development of bipolar disorder [39, 62, 133].

23.7.2.3.1 Family Studies

All completed family studies have shown that the first-degree relatives of patients with bipolar I disorder have an increased risk of developing the disorder. These studies have used standard diagnostic methods. The relative risk for first-degree relatives is approximately 7. These studies have also shown that first-degree relatives have an increased risk of developing unipolar depression, the risk being approximately double that for the general population. It is not possible in individual cases to determine whether the prevailing depressive illness is related to the familial genetic loading with bipolar disorder, since unipolar depression is very common in the general population. It has been estimated that 70% of the unipolar cases among relatives of bipolar patients share a common genetic background [18]. Other disorders observed in relatives of patients with bipolar I disorder are bipolar II disorder and schizoaffective disorder, raising further suspicions of overlapping etiology.

Many attempts are being made to identify clinical characteristics that influence the risk of illness in families. The identification of such characteristics would be of value in molecular genetic research, since this would make it possible to identify subgroups of patients in whom a higher genetic loading could be assumed. One such characteristic may be early age at onset, for which a large number of studies have reported an increased risk of illness in relatives [83]. Formal segregation analyzes have provided further support for considering age at onset as an important variable in genetic studies of bipolar disorder, with evidence having been obtained for a major gene effect in early-onset families [50, 114].

Other clinical features reported to increase the risk for illness in families include response to lithium [5], a history of psychotic bipolar disorder [112, 113], puerperal manic or hypomanic episodes [63], and co-morbidity with panic disorder [86, 87]. Some of these features have been reported to be familial traits [123], suggesting that they may breed true to some extent.

23.7.2.3.2 Twin Studies

Findings from twin studies also demonstrate the strong contribution of genetic factors to the etiology of bipolar disorder. The average concordance rate for bipolar disorder obtained for monozygotic twins from studies employing a modern concept of bipolar disorder is 50%, as against a rate of 10% for dizygotic twins [7, 17, 25, 71, 78, 142]. For unipolar disorder, the concordance rate is approximately 80% in monozygotic twins and 20% in dizygotic twins [17]. These findings are in accordance with twin studies conducted prior to 1960, which did not distinguish between unipolar and bipolar disorder but which supported the involvement of genes in broadly defined mood disorders (for review see [144]). Heritability estimates range between 60% and 80% for bipolar disorder and 33% and 42% for unipolar disorder [31].

23.7.2.3.3 Adoption Studies

To date there has been only one adoption study from which meaningful conclusions can be drawn concerning bipolar disorder [100]. This study included 29 bipolar and 22 healthy adoptees and their biological and nonbiological parents. In order to test the influence a chronic disease exerts on parental affective status, 31 parents of children with bipolar disorder and 20 parents of children with poliomyelitis were investigated: 31% of parents whose children were affected with bipolar disorder displayed an affective disorder, as opposed to only 12% of those whose children did not suffer from bipolar disorder.

23.7.2.4 Molecular Genetic Studies

The biological mechanisms responsible for the development of bipolar disorder remain largely unknown.

Biological psychiatric research has proposed a multitude of possible mechanisms, including a disturbance of neurotransmitters, intracellular, and neuroendocrine regulation. Whether these mechanisms are genuinely implicated, and to what extent they are involved, is still unclear. The advantage of the molecular genetic approach is that the identification of vulnerability genes reveals causal factors. This in turn allows an understanding of the function of a given gene product and gradually reveals the functional context which ultimately produces the clinical phenotypes.

23.7.2.4.1 Linkage Studies

Linkage studies have proposed a large number of chromosomal regions that are likely to contain genes contributing to bipolar disorder: 4p16, 4q35, 8q24, 10q25–q26, 12q23–24, 13q32–q33, 18p11.2-cen, 18q21–q23, 21q22, and 22q12–q13. Although all of these chromosomal loci have been found by two or more independent research groups, no locus has been consistently replicated by all groups. This reflects both the high degree of locus-heterogeneity and the variability of studies with regard to their definition of phenotype, sample sizes, and the type and density of genetic markers used. The largest linkage study in bipolar disorder combined the original genotype data from 11 genome-wide linkage scans comprising 5,179 individuals from 1,067 families, and established that loci on chromosomes 6q and 8q show genome-wide significance and loci on 9p and 20p show suggestive evidence of linkage [99]. Recently, a first genome-wide interaction linkage scan in bipolar disorder provided evidence of interaction between disease genes on chromosomes 2q22–q24 and 6q23–q24 [3].

Two large meta-analyzes of all published genome-wide linkage studies have been conducted [10, 128]. Since the original studies were not uniform with respect to a number of methodological issues (e.g., diagnostic criteria, genetic markers), the meta-analyzes involved some loss of information. A specific advantage of such meta-analyzes, however, lies in their sensitivity for genes with a relatively small, but widespread, contribution to the development of the illness. These genes may remain undetected in an individual study, but become detectable when results from a large number of studies are combined. It is therefore not surprising that the meta-analyzes, as well as confirming previously implicated loci, have also suggested new loci (10q11–q22, and 14q24–q32). Linkage with schizophrenia has also been reported for some of the chromosomal regions highlighted by the meta-analyzes. Since there is a possible etiological overlap between bipolar disorder and schizophrenia, the genes identified for schizophrenia in these regions are excellent candidate genes for bipolar disorder. Using this strategy, the G72/G30 locus (chr. 13q33) has recently been found to be associated with bipolar disorder [54, 125], although the responsible gene has not yet been unequivocally identified [2].

As with schizophrenia, studies in isolated populations have been performed for bipolar affective disorder, and they have suggested some new chromosomal regions [146].

23.7.2.4.2 Candidate Gene Studies

There have been a large number of studies on candidate genes for bipolar disorder, and a simple interpretation of their findings is difficult. Many candidate gene studies are disadvantaged by small sample sizes and the lack of a systematic approach to the investigation of specific candidate genes (e.g., low number of investigated polymorphisms and insufficient knowledge of haplotype structure). There are also (theoretically) a multitude of biological mechanisms that could be responsible for the illness, and there are therefore a large number of potential candidate genes to be examined. A candidate gene has a higher plausibility when it is located in a chromosomal region that has been shown to have linkage with the disorder.

Among the most discussed candidate genes are the *serotonin transporter gene* (*5-HTT*), the *catechol-O-methyl-transferase gene* (*COMT*), and the *brain-derived neurotrophic factor gene* (*BDNF*) [60]. Reported findings have been inconsistent, and no judgement on their relevance can be made at this time.

23.7.2.4.3 Genome-Wide Association Studies

Although the initial genome-wide association (GWA) studies in bipolar disorder highlighted interesting candidates, none of the findings reached the threshold of genome-wide significance [13, 132, 148]. An important future step towards maximizing the total statistical explanatory power of these samples is likely to be the performance of joint meta-analyzes, as suggested by the results of a first joint analysis of the data from the Wellcome Trust Case-Control Consortium [148] and the Sklar study [132], which were obtained from a total sample of 4,300 bipolar patients and 6,200 controls [40]. The analysis produced a genome-wide significant finding for rs10994336, located in the ankyrinG gene (*ANK3*) (coding for ankyrinG), which was observed to have a *p*-value of 9.1×10^{-9}. The same study also reported a strong association with variability in the gene *CACNA1C* (coding for the alpha 1C subunit of the L-type voltage-gated calcium channel; rs1006737 $p = 7.0 \times 10^{-8}$). An independent study by Schulze et al. [124] has recently replicated the association finding in *ANK3* in independent samples of US American and German origin and found evidence for the presence of independent risk variants for bipolar disorder at this locus. No functional evidence is available for any of the variants, however, to suggest that they are true causative variants.

23.7.2.5 Endophenotypes

It has been suggested that some of the most promising endophenotypes for the study of bipolar disorder are neuropsychological deficits, circadian rhythm instability, dysmodulation of motivation and reward, neuropathological abnormalities, and symptom provocation responses [53].

23.7.3 Schizophrenia and Bipolar Disorder: Approaches Beyond a Diagnostic Dichotomy

In 1896 Kraepelin proposed the categorization of major psychoses into dementia praecox and manic-depressive insanity [77]. Although it has been argued that the two disorders are not distinct entities but should be viewed rather as disorders along a psychosis continuum [33, 34], the distinction made by Kraepelin [77] has influenced all current operational classification systems. There is high diagnostic reliability between psychiatrists, and family studies have shown that relatives of index patients have a higher risk of being assigned the same diagnosis. Psychiatrists are well aware of the limitations of these categorical classification systems, however, since there is substantial overlap in clinical symptoms between the disorders. Around 50% of patients with bipolar disorder display psychotic symptoms, and more than 15% of patients assigned a diagnosis of schizophrenia will develop affective symptoms during the course of the disorder, while around 10% of patients diagnosed as being manic later develop persistent symptoms of schizophrenia [27]. Family and twin studies have also shown that relatives of index patients with bipolar disorder have an increased risk of schizophrenia and vice versa [26, 33]. An increasing amount of molecular genetic evidence over recent years has challenged the validity of the dichotomous classification. Linkage studies have mapped a large number of identical chromosomal loci to schizophrenia and bipolar disorder, and association studies have suggested variants that may increase the risk for both schizophrenia and bipolar disorder [16, 32]. For some of these loci/genes, detailed analysis has revealed that the linkage/ association finding was actually due to subgroups of patients who suffered from symptoms which were common to both diagnostic groups. A linkage finding for schizophrenia and bipolar disorder on chromosome 13p, for example, increased when only those bipolar families who suffered from mood-incongruent psychotic symptoms were included [42]. Similarly, an association found for schizophrenia and bipolar disorder with *G72* markers was found to be due to those bipolar patients with persecutory delusions [122], and an association finding between affective disorder and *BDNF* alleles was also detected in schizophrenia patients with affective symptoms [126]. Analyzes in future genetic studies into bipolar disorder and schizophrenia must therefore include symptoms and symptom dimensions both within and across the categorical diagnoses [36].

23.7.4 Outlook

Genetic epidemiology has demonstrated that modern diagnostic criteria (see Table 23.7.1) define disorders that are highly heritable. It is generally accepted that the inheritance of psychiatric disorders is complex, with multiple genetic as well as environmental factors contributing to the development of a disorder

[1, 20, 46, 55, 64, 81, 94] and with possible interactions occurring among them [139, 145]. When the magnitude of the overall genetic contribution is considered this must be borne in mind.

Any consideration of the magnitude of the overall genetic contribution must take due account of the fact that an unselected sample of schizophrenia or bipolar disorder patients will always include a diverse mixture of individuals whose genetic loading ranges from no or very little genetic contribution to a strong genetic contribution with very little influence from nongenetic factors. It is possible that rare mutations with high penetrance exist in some families in which pronounced clustering and a Mendelian pattern of inheritance is observed, although no such high penetrance mutation has yet been identified.

It has been suggested that complex genetic mechanisms such as imprinting [49, 96], anticipation [97], mitochondrial inheritance [98], and epigenetics [110, 111] account, at least in part, for the irregularities in disease transmission observed within families, but no convincing molecular proof has yet been provided for these hypotheses.

Until recently, systematic genome-wide searches for the genes involved in psychiatric disorders were only possible through the use of a linkage approach. A series of chromosomal regions in which susceptibility genes are likely to be located have been identified through linkage studies. Highly promising association findings have been obtained for some genes in regions with positive evidence for linkage. To date, however, no genetic variant that directly confers a functional effect and is consistently associated with disease across populations has been identified for any of these genes. A similarly cautious conclusion must be drawn for the investigation of numerous candidate genes, although promising findings have been obtained for some of them.

Given the nonreplication of findings that seemed very convincing in the original studies, some authors have challenged the requirement that the true causative variant must show consistent association across populations by the proposal of population specific gene-gene or gene-environment factors. This is pure speculation at present, however, since these factors (if they exist at all) are completely unknown and it is more likely that the true causative variants still await identification [59].

The pace of progress in molecular genetic research is enormous. The first GWA studies in schizophrenia and bipolar disorder have recently been completed, with many more still in progress, and large international meta-analyzes to increase statistical power are being planned. Research findings obtained to date suggest that the variants identified through GWA studies will confer only small individual risks. The major limitation of GWA studies is that they only investigate variants that are common within the population. If a large fraction of the genetic contribution is conferred by rare variants, other approaches will be necessary to identify these factors. A successful first step in this direction has been the identification of rare submicroscopic chromosomal aberrations as causes of schizophrenia. However, owing to methodological restraints, this approach is still limited to aberrations comprising at least several thousand base pairs. Rapid technological developments will provide future studies with increasing resolution. Ultimately, the availability of low-cost whole-genome sequencing technology will make it possible to obtain the complete genomic sequences of large patient samples and compare them with controls. In principle, this will allow the systematic identification of rare variants associated with disease risk, although the existence of a myriad of rare variants in the human genome will render this a complex task. It is hoped that some rare variants confer a larger disease risk, which will facilitate the detection of association in large case-control samples. Rare variants with small disease risk may be extremely difficult to detect because prohibitively large samples sizes may be required to demonstrate significant association.

Once disease susceptibility genes have been identified, future studies will be required to understand the phenotypic dimensions most strongly associated with a specific gene. This will include the analysis of clinical symptoms as well as endophenotypes. The latter may be particularly suited to guiding researchers in the selection of the most promising phenotypes for animal studies [49]. For example, promising endophenotypes for schizophrenia which have already been successfully studied in animals include sensorimotor gating deficits, as indexed by measures of prepulse inhibition of the startle reflex, P50 auditory evoked potential suppression, and antisaccade eye movements.

It is expected that the identification of disease-associated genes will increase our knowledge of the pathophysiology underlying psychiatric disorders in an as yet unforeseen way. The identification of biological pathways has the potential to revolutionize diagnostics and treatment. These developments may well also challenge the existence of the field of psychiatry. A diagnosis of schizophrenia or bipolar disorder cannot be assigned if the disorder is due to the direct physiological effects of a substance or a general medical condition such as organic brain disease, and management of patients has traditionally been taken away from psychiatrists and handed over to specialists from other medical fields when an underlying biological cause for psychiatric symptoms has been identified for which appropriate therapy exists, as in the case of mania due to syphilis, or psychotic symptoms due to hyperthyroidism or porphyria. It will be interesting to see how the field of psychiatry will evolve once genetic research has unraveled the biological pathways responsible for schizophrenia and bipolar disorder. Awareness of the biological causes of psychiatric diseases has increased enormously among psychiatrists over recent decades, and this will continue with the identification of the genetic causes. This may ultimately lead to the development of a very different self-image for the field of psychiatry.

References

1. Abdolmaleky H, Thiagalingam S, Wilcox M (2005) Genetics and epigenetics in major psychiatric disorders: dilemmas, achievements, applications, and future scope. Am J Pharmacogenomics 5:149–160
2. Abou Jamra R, Schmael C, Cichon S, Rietschel M, Schumacher J, Nöthen MM (2006) The G72/G30 gene locus in psychiatric disorders: a challenge to diagnostic boundaries? Schizophr Bull 32:599–608
3. Abou Jamra R, Fuerst R, Kaneva R, Orozco Diaz G, Rivas F, Mayoral F, Gay E, Sans S, Gonzalez MJ, Gil S, Cabaleiro F, Del Rio F, Perez F, Haro J, Auburger G, Milanova V, Kostov C, Chorbov V, Stoyanova V, Nikolova-Hill A, Onchev G, Kremensky I, Jablensky A, Schulze TG, Propping P, Rietschel M, Nöthen MM, Cichon S, Wienker TF, Schumacher J (2007) The first genome-wide interaction and locus-heterogeneity linkage scan in bipolar affective disorder: strong evidence of epistatic effects between loci on chromosomes 2q and 6q. Am J Hum Genet 81:974–986
4. Albus M, Maier W (1995) Lack of gender differences in age at onset in familial schizophrenia. Schizophr Res 18:51–57
5. Alda M, Grof P, Rouleau GA, Turecki G, Young LT (2005) Investigating responders to lithium prophylaxis as a strategy for mapping susceptibility genes for bipolar disorder. Prog Neuropsychopharmacol Biol Psychiatry 29:1038–1045
6. Aleman A, Kahn RS, Selten JP (2003) Sex differences in the risk of schizophrenia: evidence from meta-analysis. Arch Gen Psychiatry 60:565–571
7. Allen MG, Cohen S, Pollin W, Greenspan SI (1974) Affective illness in veteran twins: a diagnostic review. Am J Psychiatry 131:1234–1239
8. Almasy L, Blangero J (2001) Endophenotypes as quantitative risk factors for psychiatric disease: rationale and study design. Am J Med Genet B 105:42–44
9. American Psychiatric Association (1994) Diagnostic and statistical manual of mental disorders, 4th edn. American Psychiatric Association, Washington, DC
10. Badner JA, Gershon ES (2002) Meta-analysis of whole-genome linkage scans of bipolar disorder and schizophrenia. Mol Psychiatry 7:405–411
11. Bartram L (2008) Genetic research in schizophrenia: new tools and future perspectives. Schizophr Bull 34:806–812
12. Bassett AS, Chow EWC, Husted J, Weksberg R, Caluseriu O, Webb GD, Gatzoulis MA (2005) Clinical features of 78 adults with 22q11 Deletion Syndrome. Am J Med Genet 138:307–313
13. Baum AE, Akula N, Cabanero M, Cardona I, Corona W, Klemens B, Schulze TG, Cichon S, Rietschel M, Nöthen MM, Georgi A, Schumacher J, Schwarz M, Abou Jamra R, Höfels S, Satagopan J, Detera-Wadleigh SD, Hardy J, McMahon FJ (2007) A genome-wide association study implicates diacylglycerol kinase eta (DGKH) and several other genes in the etiology of bipolar disorder. Mol Psychiatry 13:197–207
14. Bearden CE, Freimer NB (2006) Endophenotypes for psychiatric disorders: ready for primetime? Trends Genet 22:306–313
15. Ben-Shachar S, Lanpher B, German JR, Qasaymeh M, Potocki L, Nagamani S, Franco LM, Malphrus A, Bottenfield GW, Spence JE, Amato S, Rousseau JA, Moghaddam B, Skinner C, Skinner SA, Bernes S, Armstrong N, Shinawi M, Stankiewicz P, Patel A, Cheung SW, Lupski JR, Beaudet AL, Sahoo T (2009) Microdeletion 15q13.3: a locus with incomplete penetrance for autism, mental retardation, and psychiatric disorders. J Med Genet 46:382–388
16. Berrettini W (2003) Evidence for shared susceptibility in bipolar disorder and schizophrenia. Am J Med Genet (Sem Med Genet) 123:59–64
17. Bertelsen A, Harvald B, Hauge M (1977) A Danish twin study of manic-depressive disorders. Br J Psychiatry 130:330–351
18. Blacker D, Lavori PW, Faraone SV, Tsuang MT (1993) Unipolar relatives in bipolar pedigrees: a search for indicators of underlying bipolarity. Am J Med Genet 48:192–199
19. Braff DL, Freedman R, Schork NJ, Gottesman II (2007) Deconstructing schizophrenia: an overview of the use of endophenotypes in order to understand a complex disorder. Schizophr Bull 33:21–32
20. Burmeister M, McInnis MG, Zöllner S (2008) Psychiatric genetics: progress amid controversy. Nat Rev Genet 9:527–540
21. Byrne M, Agerbo E, Eaton WW, Mortensen PB (2004) Parental socio-economic status and risk of first admission with schizophrenia- a Danish national register based study. Soc Psychiatry Psychiatr Epidemiol 39:87–96
22. Calkins ME, Dobie DJ, Cadenhead KS, Olincy A, Freedman R, Green MF, Greenwood TA, Gur RE, Gur RC, Light GA, Mintz J, Nuechterlein KH, Radant AD, Schork NJ, Seidman LJ, Siever LJ, Silverman JM, Stone WS, Swerdlow NR, Tsuang DW, Tsuang MT, Turetsky BI, Braff DL (2007) The Consortium on the Genetics of Endophenotypes in Schizophrenia: model recruitment, assessment, and endophenotyping methods for a multisite collaboration. Schizophr Bull 233:33–48

23. Cantor-Graae E, Selten JP (2005) Schizophrenia and migration: a meta-analysis and review. Am J Psychiatry 162:12–24
24. Cardno AG, Gottesman II (2000) Twin studies of schizophrenia: from bow-and-arrow concordances to star wars Mx and functional genomics. Am J Med Genet 97:12–17
25. Cardno AG, Marshall EJ, Coid B, Macdonald AM, Ribchester TR, Davies NJ, Venturi P, Jones LA, Lewis SW, Sham PC, Gottesman II, Farmer AE, McGuffin P, Reveley AM, Murray RM (1999) Heritability estimates for psychotic disorders: the Maudsley twin psychosis series. Arch Gen Psychiatry 56:162–168
26. Cardno AG, Rijsdijk FV, Sham PC, Murray RM, McGuffin P (2002) A twin study of genetic relationships between psychotic symptoms. Am J Psychiatry 159:539–545
27. Chen YR, Swann AC, Johnson BA (1998) Stability of diagnosis in bipolar disorder. J Nerv Ment Dis 186:17–23
28. Chubb JE, Bradshaw NJ, Soares DC, Porteous DJ, Millar JK (2008) The DISK locus in psychiatric illness. Mol Psychiatry 13:36–64
29. Chumakov I, Blumenfeld M, Guerassimenko O, Cavarec L, Palicio M, Abderrahim H, Bougueleret L, Barry C, Tanaka H, La Rosa P, Puech A, Tahri N, Cohen-Akenine A, Delabrosse S, Lissarrague S, Picard FP, Maurice K, Essioux L, Millasseau P, Grel P, Debailleul V, Simon AM, Caterina D, Dufaure I, Malekzadeh K, Belova M, Luan JJ, Bouillot M, Sambucy JL, Primas G, Saumier M, Boubkiri N, Martin-Saumier S, Nasroune M, Peixoto H, Delaye A, Pinchot V, Bastucci M, Guillou S, Chevillon M, Sainz-Fuertes R, Meguenni S, Aurich-Costa J, Cherif D, Gimalac A, Van Duijn C, Gauvreau D, Ouellette G, Fortier I, Raelson J, Sherbatich T, Riazanskaia N, Rogaev E, Raeymaekers P, Aerssens J, Konings F, Luyten W, Macciardi F, Sham PC, Straub RE, Weinberger DR, Cohen N, Cohen D (2002) Genetic and physiological data implicating the new human gene G72 and the gene for D-amino acid oxidase in schizophrenia. Proc Natl Acad Sci USA 99:13675–13680
30. Colp R (2000) History of psychiatry. In: Sadock BJ, Sadock VA (eds) Kaplan & Sadock's comprehensive textbook of psychiatry, 7th edn. Lippincott Williams & Wilkins, Philadelphia, PA, p 3301
31. Craddock N, Forty L (2006) Genetics of affective (mood) disorders. Eur J Hum Genet 14:660–668
32. Craddock N, O'Donovan MC, Owen MJ (2005) Genetics of schizophrenia and bipolar disorder: dissecting psychosis. J Med Genet 42:193–204
33. Craddock N, O'Donovan MC, Owen MJ (2006) Genes for schizophrenia and bipolar disorder? Implications for psychiatric nosology. Schizophr Bull 32:9–16
34. Crow TJ (1990) The continuum of psychosis and its genetic origins. The sixty-fifth Maudsley lecture. Br J Psychiatry 156:788–797
35. Dalman C, Allebeck P (2002) Paternal age and schizophrenia: further support for an association. Am J Psychiatry 159:1591–1592
36. Dikeos DG, Wickham H, McDonald C, Walshe M, Sigmundsson T, Bramon E, Grech A, Toulopoulou T, Murray R, Sham PC (2006) Distribution of symptom dimensions across Kraepelinian divisions. Br J Psychiatry 189:346–353
37. El-Saadi O, Pedersen CB, McNeil TF, Saha S, Welham J, O'Callaghan E, Cantor-Graae E, Chant D, Mortensen PB, McGrath J (2004) Paternal and maternal age as risk factors for psychosis: findings from Denmark, Sweden and Australia. Schizophr Res 67:227–236
38. Esslinger C, Walter H, Kirsch P, Erk S, Schnell K, Arnold C, Haddad L, Mier D, Opitz von Boberfeld C, Raab K, Witt SH, Rietschel M, Cichon S, Meyer-Lindenberg A (2009) Neural mechanism of a genome-wide supported psychosis variant. Science 324:605
39. Farmer A, Elkin A, McGuffin P (2007) The genetics of bipolar affective disorder. Curr Opin Psychiatry 20:8–12
40. Ferreira MA, O'Donovan MC, Meng YA, Jones IR, Ruderfer DM, Jones L, Fan J, Kirov G, Perlis RH, Green EK, Smoller JW, Grozeva D, Stone J, Nikolov I, Chambert K, Hamshere ML, Nimgaonkar VL, Moskvina V, Thase ME, Caesar S, Sachs GS, Franklin J, Gordon-Smith K, Ardlie KG, Gabriel SB, Fraser C, Blumenstiel B, Defelice M, Breen G, Gill M, Morris DW, Elkin A, Muir WJ, McGhee KA, Williamson R, MacIntyre DJ, MacLean AW, St CD, Robinson M, Van Beck M, Pereira AC, Kandaswamy R, McQuillin A, Collier DA, Bass NJ, Young AH, Lawrence J, Ferrier IN, Anjorin A, Farmer A, Curtis D, Scolnick EM, McGuffin P, Daly MJ, Corvin AP, Holmans PA, Blackwood DH, Gurling HM, Owen MJ, Purcell SM, Sklar P, Craddock N, Wellcome Trust Case- Control Consortium (2008) Collaborative genome-wide association analysis supports a role for ANK3 and CACNA1C in bipolar disorder. Nat Genet 40:1056–1058
41. Glaser B, Moskvina V, Kirov G, Murphy KC, Williams H, Williams N, Owen MJ, O'Donovan MC (2006) Analysis of ProDH, COMT and ZDHHC8 risk variants does not support individual or interactive effects on schizophrenia susceptibility. Schizophr Res 87:21–27
42. Goes FS, Zandi PP, Miao K, McMahon FJ, Steele J, Willour VL, Mackinnon DF, Mondimore FM, Schweizer B, Nurnberger JI, Rice JP, Scheftner W, Coryell W, Berrettini WH, Kelsoe JR, Byerley W, Murphy DL, Gershon ES, Bipolar Disorder Phenome Group, Depaulo JR, McInnis MG, Potash JB (2007) Mood-incongruent psychotic features in bipolar disorder: familial aggregation and suggestive linkage to 2p11-q14 and 13q21-33. Am J Psychiatry 164:236–247
43. Goldner EM, Hsu L, Waraich P, Somers JM (2002) Prevalence and incidence studies of schizophrenic disorders: a systematic review of the literature. Can J Psychiatry 47:833–843
44. Gorwood P, Leboyer M, Jay M, Payan C, Feingold J (1995) Gender and age at onset in schizophrenia: impact of family history. Am J Psychiatry 152:208–212
45. Gottesmann II, Gould TD (2003) The endophenotype concept in psychiatry: etymology and strategic intentions. Am J Psychiatry 160:636–645
46. Gottesman II, Hanson DR (2005) Human development: biological and genetic processes. Annu Rev Psychol 56:263–286
47. Gould TD, Gottesman II (2006) Psychiatric andophenotypes and the development of valid animal models. Genes Brain Behav 5:113–119
48. Green MF, Kern RS, Braff DL, Mintz J (2000) Neurocognitive deficits and functional outcome in schizophrenia: are we measuring the "right stuff"? Schizophr Bull 26:119–136
49. Grigoroiu-Serbanescu M, Nöthen M, Propping P, Poustka F, Magureanu S, Vasilescu R, Marinescu E, Ardelean V (1995) Clinical evidence for genomic imprinting in bipolar I disorder. Acta Psychiatr Scand 92:365–370

50. Grigoroiu-Serbanescu M, Martinez M, Nöthen MM, Grinberg M, Sima D, Propping P, Marinescu E, Hrestic M (2001) Different familial transmission patterns in bipolar I disorder with onset before and after age 25. Am J Med Genet 105:765–773
51. Guo AY, Sun J, Riley BP, Thiselton DL, Kendler KS, Zhao Z (2009) The dystrobrevin-binding protein 1 gene: features and networks. Mol Psychiatry 14:18–29
52. Harrison PJ (2007) Schizophrenia susceptibility genes and their neurodevelopmental implications: focus on neuregulin 1. Novartis Found Symp 288:246–255 discussion 255-259, 276-281
53. Hasler G, Drevets WC, Gould TD, Gottesman II, Manji HK (2006) Toward constructing an endophenotype strategy for bipolar disorders. Biol Psychiatry 60:93–105
54. Hattori E, Liu C, Badner JA, Bonner TI, Christian SL, Maheshwari M, Detera-Wadleigh SD, Gibbs RA, Gershon ES (2003) Polymorphisms at the G72/G30 gene locus, on 13q33, are associated with bipolar disorder in two independent pedigree series. Am J Hum Genet 72:1131–1140
55. Hattori E, Liu C, Zhu H, Gershon ES (2005) Genetic tests of biologic systems in affective disorders. Mol Psychiatry 10:719–740
56. Helbig I, Mefford HC, Sharp AJ, Guipponi M, Fichera M, Franke A, Muhle H, de Kovel C, Baker C, von Spiczak S, Kron KL, Steinich I, Kleefuss-Lie AA, Leu C, Gaus V, Schmitz B, Klein KM, Reif PS, Rosenow F, Weber Y, Lerche H, Zimprich F, Urak L, Fuchs K, Feucht M, Genton P, Thomas P, Visscher F, de Haan GJ, Møller RS, Hjalgrim H, Luciano D, Wittig M, Nothnagel M, Elger CE, Nürnberg P, Romano C, Malafosse A, Koeleman BP, Lindhout D, Stephani U, Schreiber S, Eichler EE, Sander T (2009) 15q13.3 microdeletions increase risk of idiopathic generalized epilepsy. Nat Genet 41:160–162
57. Heston LL (1966) Psychiatric disorders in foster home reared children of schizophrenic mothers. Br J Psychiatry 112:819–825
58. International Schizophrenia Consortium (2008) Rare chromosomal deletions and duplications increase risk of schizophrenia. Nature 455:237–241
59. Ioannidis JP, Ntzani EE, Trikalinos TA (2004) "Racial" differences in genetic effects for complex diseases. Nat Genet 36:1243–1244
60. Jones I, Craddock N (2001) Candidate gene studies of bipolar disorder. Ann Med 33:248–256
61. Jones I, Craddock N (2002) Do puerperal psychotic episodes identify a more familial subtype of bipolar disorder? Results of a family history study. Psychiatr Genet 12:177–180
62. Jones I, Kent L, Craddock N (2002) Genetics of affective disorders. In: McGuffin P, Owen MJ, Gottesman II (eds) Psychiatric genetics and genomics. Oxford University Press, Oxford, pp 211–245
63. Karayiorgou M, Morris MA, Morrow B, Shprintzen RJ, Goldberg R, Borrow J, Gos A, Nestadt G, Wolyniec PS, Lasseter VK, Eisen H, Childs B, Kazazuan HH, Kucherlapati R, Antonarakis SE, Pulver AE, Housman DE (1995) Schizophrenia susceptibility associated with interstitial deletions of chromosome 22q11. Proc Natl Acad Sci USA 92:7612–7616
64. Kato T, Kuratomi G, Kato N (2005) Genetics of bipolar disorder. Drugs Today 41:335–344
65. Kaymaz N, Krabbendam L, de Graaf R, Nolen W, Ten Have M, van Os J (2006) Evidence that the urban environment specifically impacts on the psychotic but not the affective dimension of bipolar disorder. Soc Psychiatry Psychiatr Epidemiol 41:679–685
66. Keller MC, Miller G (2006) Resolving the paradox of common harmful, heritable mental disorders: which evolutionary genetic models work best? Behav Brain Sci 29:385–452
67. Kendler KS (1983) Overview: a current perspective on twin studies of schizophrenia. Am J Psychiatry 140:1413–1425
68. Kendler KS (2000) Schizophrenia genetics. In: Sadock BJ, Sadock VA (eds) Kaplan & Sadock's comprehensive textbook of psychiatry, 7th edn. Lippincott Williams & Wilkins, Philadelphia, PA, pp 1151–1153
69. Kendler KS, Gruenberg AM (1984) An independent analysis of the Danish Adoption Study of Schizophrenia, VI: the relationship between psychiatric disorders as defined by DSM-III in the relatives and adoptees. Arch Gen Psychiatry 41:555–564
70. Kendler KS, Pedersen N, Johnson L, Neale MC, Mathé AA (1993) A pilot Swedish twin study of affective illness, including hospital- and population-ascertained subsamples. Arch Gen Psychiatry 50:699–700
71. Kendler KS, Gruenberg A, Kinney D (1994) Independent diagnoses of adoptees and relatives as defined by DSM-III in the provincial and national samples of the Danish Adoption Study of Schizophrenia. Arch Gen Psychiatry 51:456–468
72. Kety SS (1983) Mental illness in the biological and adoptive relatives of schizophrenic adoptees: findings relevant to genetic and environmental factors in etiology. Am J Psychiatry 140:720–727
73. Kety SS, Wender PH, Jacobsen B, Ingraham LJ, Jansson L, Faber B, Kinney DK (1994) Mental illness in the biological and adoptive relatives of schizophrenic adoptees. Replication of the Copenhagen Study in the rest of Denmark. Arch Gen Psychiatry 51:442–455
74. Kirov G, Gumus D, Chen W, Norton N, Georgieva L, Sari M, O'Donovan MC, Erdogan F, Owen MJ, Ropers HH, Ullmann R (2008) Comparative genome hybridization suggests a role for NRXN1 and APBA2 in schizophrenia. Hum Mol Genet 17:458–465
75. Kirov G, Zaharieva I, Georgieva L, Moskvina V, Nikolov I, Cichon S, Hillmer A, Toncheva D, Owen MJ, O'Donovan MC (2009) A genome-wide association study in 574 schizophrenia trios using DNA pooling. Mol Psychiatry 14:796–803
76. Kraepelin E (1896) Psychiatrie - Ein Lehrbuch für Studierende und Ärzte. 5th ed. Johann Ambrosius Barth, Leipzig, Germany
77. Kringlen E (1967) Heredity and environment in the functional psychoses. Heinemann, London
78. Kringlen E (2000) Twin studies in schizophrenia with special emphasis on concordance figures. Am J Med Genet 97:4–11
79. Lencz T, Morgan TV, Athanasiou M, Dain B, Reed CR, Kane JM, Kucherlapati R, Malhotra AK (2007) Converging evidence for a pseudoautosomal cytokine receptor gene locus in schizophrenia. Mol Psychiatry 12:572–580
80. Levinson DF (2006) The genetics of depression: a review. Biol Psychiatry 60:84–92

81. Lewis CM, Levinson DF, Wise LH, DeLisi LE, Straub RE, Hovatta I, Williams NM, Schwab SG, Pulver AE, Faraone SV, Brzustowicz LM, Kaufmann CA, Garver DL, Gurling HM, Lindholm E, Coon H, Moises HW, Byerley W, Shaw SH, Mesen A, Sherrington R, O'Neill FA, Walsh D, Kendler KS, Ekelund J, Paunio T, Lönnqvist J, Peltonen L, O'Donovan MC, Owen MJ, Wildenauer DB, Maier W, Nestadt G, Blouin JL, Antonarakis SE, Mowry BJ, Silverman JM, Crowe RR, Cloninger CR, Tsuang MT, Malaspina D, Harkavy-Friedman JM, Svrakic DM, Bassett AS, Holcomb J, Kalsi G, McQuillin A, Brynjolfson J, Sigmundsson T, Petursson H, Jazin E, Zoëga T, Helgason T (2003) Genome scan meta-analysis of schizophrenia and bipolar disorder, part II: Schizophrenia. Am J Hum Genet 73:34–48
82. Lin PI, McInnis MG, Potash JB, Willour V, MacKinnon DF, DePaulo JR, Zandi PP (2006) Clinical correlates and familial aggregation of age at onset in bipolar disorder. Am J Psychiatry 163:240–246
83. Lin PI, Vance JM, Pericak-Vance MA, Martin ER (2007) No gene is an island: the flip-flop phenomenon. Am J Hum Genet 80:531–538
84. Lowing PA, Mirsky AF, Pereira R (1983) The inheritance of schizophrenia spectrum disorders: a reanalysis of the Danish adoptee study data. Am J Psychiatry 140:1167–1171
85. MacKinnon DF, Zandi PP, Cooper J, Potash JB, Simpson SG, Gershon E, Nurnberger J, Reich T, DePaulo JR (2002) Comorbid bipolar disorder and panic disorder in families with a high prevalence of bipolar disorder. Am J Psychiatry 159:30–35
86. MacKinnon DF, Zandi PP, Gershon ES, Nurnberger JI Jr, DePaulo JR Jr (2003) Association of rapid mood switching with panic disorder and familial panic risk in familial bipolar disorder. Am J Psychiatry 160:1696–1698
87. Maier W, Lichtermann D, Minges J, Hallmayer J, Heun R, Benkert O, Levinson DF (1993) Continuity and discontinuity of affective disorders and schizophrenia. Results of a controlled family study. Arch Gen Psychiatry 50:871–883
88. Maier W, Lichtermann D, Franke P, Heun R, Falkai P, Rietschel M (2002) The dichotomy of schizophrenia and affective disorders in extended pedigrees. Schizophr Res 57:259–266
89. Malaspina D, Harlap S, Fennig S, Heiman D, Nahon D, Feldman D, Susser ES (2001) Advancing paternal age and the risk of schizophrenia. Arch Gen Psychiatry 58:361–367
90. Mao Y, Ge X, Frank CL, Madison JM, Koehler AN, Doud MK, Tassa C, Berry EM, Soda T, Singh KK, Biechele T, Petryshen TL, Moon RT, Haggarty SJ, Tsai LH (2009) Disrupted in schizophrenia 1 regulates neuronal progenitor proliferation via modulation of GSK3beta/beta-catenin signaling. Cell 136:1017–1031
91. McDonald C, Murray RM (2000) Early and late environmental risk factors for schizophrenia. Brain Res Rev 31:130–137
92. McGrath J (2006) Variations in the incidence of schizophrenia: data versus dogma. Schizophr Bull 32:195–197
93. McGrath J, Saha S, Welham J, El-Saadi O, MacCauley C, Chant D (2004) A systematic review of the incidence of schizophrenia: the distribution of rates and the influence of sex, urbanicity, migrant status and methodology. BMC Med 28:2–13
94. McGuffin P (2004) Nature and nurture interplay: schizophrenia. Psychiatr Prax 31:189–193
95. McGuire PK, Jones P, Harvey I, Williams M, McGuffin P, Murray RM (1995) Morbid risk of schizophrenia for relatives of patients with cannabis-associated psychosis. Schizophr Res 15:277–281
96. McInnis MG, McMahon FJ, Chase GA, Simpson SG, Ross CA, DePaulo JR Jr (1993) Anticipation in bipolar affective disorder. Am J Hum Genet 53:385–390
97. McMahon FJ, Stine OC, Meyers DA, Simpson SG, DePaulo JR (1995) Patterns of maternal transmission in bipolar affective disorder. Am J Hum Genet 56:1277–1286
98. McQueen MB, Devlin B, Faraone SV, Nimgaonkar VL, Sklar P, Smoller JW, Abou Jamra R, Albus M, Bacanu SA, Baron M, Barrett TB, Berrettini W, Blacker D, Byerley W, Cichon S, Coryell W, Craddock N, Daly MJ, Depaulo JR, Edenberg HJ, Foroud T, Gill M, Gilliam TC, Hamshere M, Jones I, Jones L, Juo SH, Kelsoe JR, Lambert D, Lange C, Lerer B, Liu J, Maier W, Mackinnon JD, McInnis MG, McMahon FJ, Murphy DL, Nöthen MM, Nurnberger JI, Pato CN, Pato MT, Potash JB, Propping P, Pulver AE, Rice JP, Rietschel M, Scheftner W, Schumacher J, Segurado R, Van Steen K, Xie W, Zandi PP, Laird NM (2005) Combined analysis from eleven linkage studies of bipolar disorder provides strong evidence of susceptibility loci on chromosomes 6q and 8q. Am J Hum Genet 77:582–595
99. Meechan DW, Maynard TM, Gopalakrishna D, Wu Y, LaMantia AS (2007) When half is not enough: gene expression and dosage in the 22q11 deletion syndrome. Gene Expr 13:299–310
100. Mefford HC, Sharp AJ, Baker C, Itsara A, Jiang Z, Buysse K, Huang S, Maloney VK, Crolla JA, Baralle D, Collins A, Mercer C, Norga K, de Ravel T, Devriendt K, Bongers EM, de Leeuw N, Reardon W, Gimelli S, Bena F, Hennekam RC, Male A, Gaunt L, Clayton-Smith J, Simonic I, Park SM, Mehta SG, Nik-Zainal S, Woods CG, Firth HV, Parkin G, Fichera M, Reitano S, Lo Giudice M, Li KE, Casuga I, Broomer A, Conrad B, Schwerzmann M, Räber L, Gallati S, Striano P, Coppola A, Tolmie JL, Tobias ES, Lilley C, Armengol L, Spysschaert Y, Verloo P, De Coene A, Goossens L, Mortier G, Speleman F, van Binsbergen E, Nelen MR, Hochstenbach R, Poot M, Gallagher L, Gill M, McClellan J, King MC, Regan R, Skinner C, Stevenson RE, Antonarakis SE, Chen C, Estivill X, Menten B, Gimelli G, Gribble S, Schwartz S, Sutcliffe JS, Walsh T, Knight SJ, Sebat J, Romano C, Schwartz CE, Veltman JA, de Vries BB, Vermeesch JR, Barber JC, Willatt L, Tassabehji M, Eichler EE (2008) Recurrent rearrangements of chromosome 1q21.1 and variable pediatric phenotypes. N Engl J Med 359:1685–1699
101. Mendlewicz J, Rainer JD (1977) Adoption study supporting genetic transmission in manic-depressive illness. Nature 268:327–329
102. Millar JK, Wilson-Annan JC, Anderson S, Christie S, Taylor MS, Semple CA, Devon RS, Clair DM, Muir WJ, Blackwood DH, Porteous DJ (2000) Disruption of two novel genes by a translocation co-segregating with schizophrenia. Hum Mol Genet 22:1415–1423
103. Miller DT, Shen Y, Weiss LA, Korn J, Anselm I, Bridgemohan C, Cox GF, Dickinson H, Gentile J, Harris DJ, Hegde V, Hundley R, Khwaja O, Kothare S, Luedke C, Nasir R, Poduri A, Prasad K, Raffalli P, Reinhard A, Smith SE, Sobeih MM, Soul JS, Stoler J, Takeoka M, Tan WH,

Thakuria J, Wolff R, Yusupov R, Gusella JF, Daly MJ, Wu BL (2009) Microdeletion/duplication at 15q13.2q13.3 among individuals with features of autism and other neuropsychiatric disorders. J Med Genet 46:242–248

104. Mortensen PB, Pedersen CB, Westergaard T, Wohlfahrt J, Ewald H, Mors O, Andersen PK, Melbye M (1999) Effects of family history and place and season of birth on the risk of schizophrenia. N Engl J Med 340:603–608

105. Mortensen PB, Pedersen CB, Melbye M, Mors O, Ewald H (2003) Individual and familial risk factors for bipolar affective disorders in Denmark. Arch Gen Psychiatry 60:1209–1215

106. Murray RM, Jones PB, Susser E, von Os J, Cannon M (2003) The epidemiology of schizophrenia. Cambridge University Press, Cambridge, p 470

107. Mutsuddi M, Morris DW, Waggoner SG, Daly MJ, Scolnick EM, Sklar P (2006) Analysis of high-resolution HapMap of DTNBP1 (Dysbindin) suggests no consistency between reported common variant associations and schizophrenia. Am J Hum Genet 79:903–909

108. Need AC, Ge D, Weale ME, Maia J, Feng S, Heinzen EL, Shianna KV, Yoon W, Kasperaviciute D, Gennarelli M, Strittmatter WJ, Bonvicini C, Rossi G, Jayathilake K, Cola PA, McEvoy JP, Keefe RS, Fisher EM, St Jean PL, Giegling I, Hartmann AM, Möller HJ, Ruppert A, Fraser G, Crombie C, Middleton LT, St Clair D, Roses AD, Muglia P, Francks C, Rujescu D, Meltzer HY, Goldstein DB (2009) A genome-wide investigation of SNPs and CNVs in schizophrenia. PLoS Genet 5:e1000373

109. O'Donovan MC, Craddock N, Norton N, Williams H, Peirce T, Moskvina V, Nikolov I, Hamshere M, Carroll L, Georgieva L, Dwyer S, Holmans P, Marchini JL, Spencer CC, Howie B, Leung HT, Hartmann AM, Möller HJ, Morris DW, Shi Y, Feng G, Hoffmann P, Propping P, Vasilescu C, Maier W, Rietschel M, Zammit S, Schumacher J, Quinn EM, Schulze TG, Williams NM, Giegling I, Iwata N, Ikeda M, Darvasi A, Shifman S, He L, Duan J, Sanders AR, Levinson DF, Gejman PV, Cichon S, Nöthen MM, Gill M, Corvin A, Rujescu D, Kirov G, Owen MJ, Buccola NG, Mowry BJ, Freedman R, Amin F, Black DW, Silverman JM, Byerley WF, Cloninger CR, Molecular Genetics of Schizophrenia Collaboration (2008) Identification of loci associated with schizophrenia by genome-wide association and follow-up. Nat Genet 40:1053–1055

110. Owen MJ, Craddock N, O'Donnovan MC (2005) Schizophrenia: genes at last? Trends Genet 21:518–525

111. Petronis A (2003) Epigenetics and bipolar disorder: new opportunities and challenges. Am J Med Genet C 123C:65–75

112. Petronis A (2004) The origin of schizophrenia: genetic thesis, epigenetic antithesis, and resolving synthesis. Biol Psychiatry 55:965–970

113. Potash JB, Willour VL, Chiu YF, Simpson SG, MacKinnon DF, Pearlson GD, DePaulo JR Jr, McInnis MG (2001) The familial aggregation of psychotic symptoms in bipolar disorder pedigrees. Am J Psychiatry 158:1258–1264

114. Potash JB, Chiu YF, MacKinnon DF, Miller EB, Simpson SG, McMahon FJ, McInnis MG, DePaulo JR Jr (2003) Familial aggregation of psychotic symptoms in a replication set of 69 bipolar disorder pedigrees. Am J Med Genet B 116:90–97

115. Rice J, Reich T, Andreasen NC, Endicott J, Van Eerdewegh M, Fishman R, Hirschfeld RM, Klerman GL (1987) The familial transmission of bipolar illness. Arch Gen Psychiatry 44:441–447

116. Riley B, Kendler KS (2006) Molecular genetic studies of schizophrenia. Eur J Hum Genet 14:669–680

117. Risch N, Merikangas K (1996) The future of genetic studies of complex human diseases. Science 273:1516–1517

118. Ross CA, Margolis RL (2009) Schizophrenia: a point of disruption. Nature 458:976–977

119. Rujescu D, Ingason A, Cichon S, Pietiläinen OP, Barnes MR, Toulopoulou T, Picchioni M, Vassos E, Ettinger U, Bramon E, Murray R, Ruggeri M, Tosato S, Bonetto C, Steinberg S, Sigurdsson E, Sigmundsson T, Petursson H, Gylfason A, Olason PI, Hardarsson G, Jonsdottir GA, Gustafsson O, Fossdal R, Giegling I, Möller HJ, Hartmann AM, Hoffmann P, Crombie C, Fraser G, Walker N, Lonnqvist J, Suvisaari J, Tuulio-Henriksson A, Djurovic S, Melle I, Andreassen OA, Hansen T, Werge T, Kiemeney LA, Franke B, Veltman J, Buizer-Voskamp JE, GROUP Investigators, Sabatti C, Ophoff RA, Rietschel M, Nöthen MM, Stefansson K, Peltonen L, St Clair D, Stefansson H, Collier DA (2009) Disruption of the neurexin 1 gene is associated with schizophrenia. Hum Mol Genet 18:988–996

120. Saha S, Chant D, Welham J, McGrath J (2005) A systematic review of the prevalence of schizophrenia. PLoS Med 141:413–433

121. Saha S, Welham J, Chant D, McGrath J (2006) Incidence of schizophrenia does not vary with economic status of the country: evidence from a systematic review. Soc Psychiatry Psychiatr Epidemiol 41:338–340

122. Schmael C, Georgi A, Krumm B, Buerger C, Deschner M, Nöthen MM, Schulze TG, Rietschel M (2007) Premorbid adjustment in schizophrenia - an important aspect of phenotype definition. Schizophr Res 92:50–62

123. Schulze TG, Ohlraun S, Czerski PM, Schumacher J, Kassem L, Deschner M, Gross M, Tullius M, Heidmann V, Kovalenko S, Jamra RA, Becker T, Leszczynska-Rodziewicz A, Hauser J, Illig T, Klopp N, Wellek S, Cichon S, Henn FA, McMahon FJ, Maier W, Propping P, Nöthen MM, Rietschel M (2005) Genotype-phenotype studies in bipolar disorder showing association between the DAOA/G30 locus and persecutory delusions: a first step toward a molecular genetic classification of psychiatric phenotypes. Am J Psychiatry 162:2101–2108

124. Schulze TG, Hedeker D, Zandi P, Rietschel M, McMahon FJ (2006) What is familial about familial bipolar disorder? Resemblance among relatives across a broad spectrum of phenotypic characteristics. Arch Gen Psychiatry 63:1368–1376

125. Schulze TG, Detera-Wadleigh SD, Akula N, Gupta A, Kassem L, Steele J, Pearl J, Strohmaier J, Breuer R, Schwarz M, Propping P, Nöthen MM, Cichon S, Schumacher J, NIMH Genetics Initiative Bipolar Disorder Consortium, Rietschel M, McMahon FJ (2009) Two variants in Ankyrin 3 (ANK3) are independent genetic risk factors for bipolar disorder. Mol Psychiatry 14:487–491

126. Schumacher J, Abou Jamra R, Freudenberg J, Becker T, Ohlraun S, Otte AC, Tullius M, Kovalenko S, Van den Bogaert A, Maier W, Rietschel M, Propping P, Nöthen MM, Cichon S (2004) Examination of G72 and D-amino-

acid oxidase as genetic risk factors for schizophrenia and bipolar affective disorder. Mol Psychiatry 9:203–207

127. Schumacher J, Jamra RA, Becker T, Ohlraun S, Klopp N, Binder EB, Schulze TG, Deschner M, Schmal C, Hofels S, Zobel A, Illig T, Propping P, Holsboer F, Rietschel M, Nöthen MM, Cichon S (2005) Evidence for a relationship between genetic variants at the brain-derived neurotrophic factor (BDNF) locus and major depression. Biol Psychiatry 58:307–314

128. Scott J, McNeill Y, Cavanagh J, Cannon M, Murray R (2006) Exposure to obstetric complications and subsequent development of bipolar disorder: systematic review. Br J Psychiatry 189:3–11

129. Segurado R, Detera-Wadleigh SD, Levinson DF, Lewis CM, Gill M, Nurnberger JI Jr, Craddock N, DePaulo JR, Baron M, Gershon ES, Ekholm J, Cichon S, Turecki G, Claes S, Kelsoe JR, Schofield PR, Badenhop RF, Morissette J, Coon H, Blackwood D, Curtis D, McInnes LA, Foroud T, Edenberg HJ, Reich T, Rice JP, Goate A, McInnis MG, McMahon FJ, Badner JA, Goldin LR, Ph B, Willour VL, Zandi PP, Liu J, Gilliam C, Juo SH, Berrettini WH, Yoshikawa T, Peltonen L, Lönnqvist J, Nöthen MM, Schumacher J, Windemuth C, Rietschel M, Propping P, Maier W, Alda M, Grof P, Rouleau GA, Del-Favero J, Van Broeckhoven C, Mendlewicz J, Adolfsson R, Spence MA, Luebbert H, Adams LJ, Donald JA, Mitchell PB, Barden N, Shink E, Byerley W, Muir W, Visscher PM, Macgregor S, Gurling H, Kalsi G, McQuillan A, Escamilla MA, Reus VI, Leon P, Freimer NB, Ewald H, Kruse TA, Mors O, Radhakrishna U, Blouin JL, Antonarakis SE, Akarsu N (2003) Genome scan meta-analysis of schizophrenia and bipolar disorder. Part III: bipolar disorder. Am J Hum Genet 73:49–62

130. Shifman S, Bronstein M, Sternfeld M, Pisanté-Shalom A, Lev-Lehman E, Weizman A, Reznik I, Spivak B, Grisaru N, Karp L, Schiffer R, Kotler M, Strous RD, Swartz-Vanetik M, Knobler HY, Shinar E, Beckmann JS, Yakir B, Risch N, Zak NB, Darvasi A (2002) A highly significant association between a COMT haplotype and schizophrenia. Am J Hum Genet 71:1296–1302

131. Shih RA, Belmonte PL, Zandi PP (2004) A review of the evidence from family, twin and adoption studies for a genetic contribution to adult psychiatric disorders. Int Rev Psychiatry 16:260–283

132. Sivagnanasundaram S, Fletcher D, Hubank M, Illingworth E, Skuse D, Scambler P (2007) Differential gene expression in the hippocampus of the Df1/þ mice: a model for 22q11.2 deletion syndrome and schizophrenia. Brain Res 1139:48–59

133. Sklar P, Smoller JW, Fan J, Ferreira MA, Perlis RH, Chambert K, Nimgaonkar VL, McQueen MB, Faraone SV, Kirby A, de Bakker PI, Ogdie MN, Thase ME, Sachs GS, Todd-Brown K, Gabriel SB, Sougnez C, Gates C, Blumenstiel B, Defelice M, Ardlie KG, Franklin J, Muir WJ, McGhee KA, MacIntyre DJ, McLean A, VanBeck M, McQuillin A, Bass NJ, Robinson M, Lawrence J, Anjorin A, Curtis D, Scolnick EM, Daly MJ, Blackwood DH, Gurling HM, Purcell SM (2008) Whole-genome association study of bipolar disorder. Mol Psychiatry 13:558–569

134. Smoller JW, Finn CT (2003) Family, twin, and adoption studies of bipolar disorder. Am J Med Genet 123:48–58

135. Stefansson H, Sigurdsson E, Steinthorsdottir V, Bjornsdottir S, Sigmundsson T, Ghosh S, Brynjolfsson J, Gunnarsdottir S, Ivarsson O, Chou TT, Hjaltason O, Birgisdottir B, Jonsson H, Gudnadottir VG, Gudmundsdottir E, Bjornsson A, Ingvarsson B, Ingason A, Sigfusson S, Hardardottir H, Harvey RP, Lai D, Zhou M, Brunner D, Mutel V, Gonzalo A, Lemke G, Sainz J, Johannesson G, Andresson T, Gudbjartsson D, Manolescu A, Frigge ML, Gurney ME, Kong A, Gulcher JR, Petursson H, Stefansson K (2002) Neuregulin 1 and susceptibility to schizophrenia. Am J Hum Genet 71:877–892

136. Stefansson H, Rujescu D, Cichon S, Pietiläinen OP, Ingason A, Steinberg S, Fossdal R, Sigurdsson E, Sigmundsson T, Buizer-Voskamp JE, Hansen T, Jakobsen KD, Muglia P, Francks C, Matthews PM, Gylfason A, Halldorsson BV, Gudbjartsson D, Thorgeirsson TE, Sigurdsson A, Jonasdottir A, Bjornsson A, Mattiasdottir S, Blondal T, Haraldsson M, Magnusdottir BB, Giegling I, Möller HJ, Hartmann A, Shianna KV, Ge D, Need AC, Crombie C, Fraser G, Walker N, Lonnqvist J, Suvisaari J, Tuulio-Henriksson A, Paunio T, Toulopoulou T, Bramon E, Di Forti M, Murray R, Ruggeri M, Vassos E, Tosato S, Walshe M, Li T, Vasilescu C, Mühleisen TW, Wang AG, Ullum H, Djurovic S, Melle I, Olesen J, Kiemeney LA, Franke B, GROUP, Sabatti C, Freimer NB, Gulcher JR, Thorsteinsdottir U, Kong A, Andreassen OA, Ophoff RA, Georgi A, Rietschel M, Werge T, Petursson H, Goldstein DB, Nöthen MM, Peltonen L, Collier DA, St Clair D, Stefansson K (2008) Large recurrent microdeletions associated with schizophrenia. Nature 455:232–236

137. Straub RE, Jiang Y, MacLean CJ, Ma Y, Webb BT, Myakishev MV, Harris-Kerr C, Wormley B, Sadek H, Kadambi B, Cesare AJ, Gibberman A, Wang X, O'Neill FA, Walsh D, Kendler KS (2002) Genetic variation in the 6p22.3 gene DTNBP1, the human ortholog of the mouse dysbindin gene, is associated with schizophrenia. Am J Hum Genet 71:337–348

138. Sullivan PF, Kendler KS, Neale MC (2003) Schizophrenia as a complex trait: evidence from a meta-analysis of twin studies. Arch Gen Psychiatry 60:1187–1192

139. Swinnen SG, Selten JP (2007) Mood disorders and migration: meta-analysis. Br J Psychiatry 190:6–10

140. Thapar A, Harold G, Rice F, Langley K, O'Donovan M (2007) The contribution of gene-environment interaction to psychopathology. Dev Psychopathol 19:989–1004

141. Tienari P, Wynne LC, Moring J, Läksy K, Nieminen P, Sorri A, Lahti I, Wahlberg KE, Naarala M, Kurki-Suonio K, Saarento O, Koistinen P, Tarvainen T, Hakko H, Miettunen J (2000) Finnish adoptive family study: sample selection and adoptee DSM-III-R diagnoses. Acta Psychiatr Scand 101:433–443

142. Tienari P, Wynne LC, Läksy K, Moring J, Nieminen P, Sorri A, Lahti I, Wahlberg KE (2003) Genetic boundaries of the schizophrenia spectrum: evidence from the Finnish Adoptive Family Study of Schizophrenia. Am J Psychiatry 160:1587–1594

143. Torgersen S (1986) Genetic factors in moderately severe and mild affective disorders. Arch Gen Psychiatry 43:222–226

144. Torrey EF, Miller J, Rawlings R, Yolken RH (1997) Seasonality of births in schizophrenia and bipolar disorder: a review of the literature. Schizophr Res 28:1–38

145. Tsuang MT, Faraone SV (1990) The genetics of mood disorders. The Johns Hopkins University Press, Baltimore
146. van Os J, Rutten BP, Poulton R (2008) Gene-environment interactions in schizophrenia: review of epidemiological findings and future directions. Schizophr Bull 34:1066–1082
147. Venken T, Del-Favero J (2007) Chasing genes for mood disorders and schizophrenia in genetically isolated populations. Hum Mutat 28:1156–1170
148. Walsh T, McClellan JM, McCarthy SE, Addington AM, Pierce SB, Cooper GM, Nord AS, Kusenda M, Malhotra D, Bhandari A, Stray SM, Rippey CF, Roccanova P, Makarov V, Lakshmi B, Findling RL, Sikich L, Stromberg T, Merriman B, Gogtay N, Butler P, Eckstrand K, Noory L, Gochman P, Long R, Chen Z, Davis S, Baker C, Eichler EE, Meltzer PS, Nelson SF, Singleton AB, Lee MK, Rapoport JL, King MC, Sebat J (2008) Rare structural variants disrupt multiple genes in neurodevelopmental pathways in schizophrenia. Science 320:539–543
149. Wellcome Trust Case-Control Consortium (2007) Genome-wide association study of 14,000 cases of seven common diseases and 3,000 shared controls. Nature 447: 661–678
150. Wickham H, Walsh C, Asherson P, Taylor C, Sigmundson T, Gill M, Owen MJ, McGuffin P, Murray R, Sham P (2001) Familiality of symptom dimensions in schizophrenia. Schizophr Res 47:223–232
151. Wickham H, Walsh C, Asherson P, Gill M, Owen MJ, McGuffin P, Murray R, Sham P (2002) Familiality of clinical characteristics in schizophrenia. J Psychiatr Res 36:325–329
152. Williams HJ, Owen MJ, O'Donovan MC (2007) Is COMT a susceptibility gene for schizophrenia? Schizophr Bull 33:635–641
153. World Health Organization (2001) World Health Report 2001 – Mental health: new understanding, new hope. WHO, Geneva www.who.int/whr
154. Wynne LC, Tienari P, Nieminen P, Sorri A, Lahti I, Moring J, Naarala M, Läksy K, Wahlberg KE, Miettunen J (2006) I. Genotype-environment interaction in the schizophrenia spectrum: genetic liability and global family ratings in the Finnish Adoption Study. Fam Process 45:419–434
155. Xu B, Roos JL, Levy S, van Rensburg EJ, Gogos JA, Karayiorgou M (2008) Strong association of de novo copy number mutations with sporadic schizophrenia. Nat Genet 40:880–885

Model Organisms for Human Disorders

Introductory Note by Michael R. Speicher

Given the physical and ethical problems involved in performing experiments on humans, model organisms are vital for our understanding of human biology and disease. Animal models in genetically tractable organisms are indispensible tools for the analysis of the pathogenesis and the development of therapeutic avenues in many human diseases. With the increasingly available number of genomic sequences for multiple organisms spanning the evolutionary tree it is now possible to use comparative genomics to build or select better animal models and to facilitate gene discovery. In addition, for many well-established model systems abundant genetic tools are available.

The most common model organisms are small mammals, usually rats and mice, which offer a variety of tailored model systems with controlled genetic backgrounds. As well as other mammals, e.g., dog, other organisms, such as the nematode *Caenorhabditis elegans*, the fruitfly, *Drosophila melanogaster*, yeast, and the zebrafish, *Danio rerio*, are also extremely important popular model genetic organisms. Their low cost, their smallness, and their external development make these latter organisms excellent tools for development biology and genetic screens for human diseases.

Techniques for large-scale genome mutagenesis and gene mapping, transgenesis, protein overexpression or knockdown, cell transplantation and chimeric embryo analysis, and chemical screens have increased the power of many model organisms immeasurably. It is now possible to rapidly determine the developmental function of a gene of interest in vivo, and then identify genetic and chemical modifiers of the processes involved under normal and pathologic conditions.

In the following chapters a number of model organisms are reviewed and their potential roles for human genetics are discussed. Among various eukaryotic microorganisms, the yeast, *Saccharomyces cerevisiae*, has become legendary for its ease of simple genetic analysis [1]. This organism thus became an important simple model for human genetics. A separate chapter on yeast was therefore planned for the fourth edition of this book, but logistical problems unfortunately prevented its inclusion.

Reference

1. Sherman F (1997) Yeast genetics. In: Mayers RA (ed) The encyclopedia of molecular biology and molecular medicine. VCH Publishers, Weinheim, Germany, pp 302–325

Mouse as a Model for Human Disease

24.1

Antonio Baldini

Abstract Precisely targeted genetic manipulation of the laboratory mouse has been one of the most significant achievements in the field of modeling human genetic diseases. The availability of an ever-increasing number of mutants, and the ability of generating ever-more complex manipulations beyond the "simple" gene knockout, offer unprecedented possibilities for genetic experiments addressing complex questions. Thus, the mouse is getting closer and closer to the most powerful genetic models such as *Drosophila*.

This chapter provides a view of the commonly used or most promising approaches to modeling genetic disease in the mouse. It also provides a discussion on which approach is more suited to modeling different types of mutations.

Contents

24.1.1	Introduction	779
24.1.2	Gene and Genome Manipulation Strategies	780
	24.1.2.1 Knockout	780
	24.1.2.2 Knockin and the Use of Site-Specific Recombinases	780
	24.1.2.3 Conditional Mutations	781
	24.1.2.4 Multigene Deletions and Duplications	782
	24.1.2.5 Transgenics	782
24.1.3	Generating Genetically Accurate Models	783
	24.1.3.1 Genetic Diseases Caused by Loss of Function Mutations	783
	24.1.3.2 Gene Dosage Mutations	783
	24.1.3.3 Gain of Function Mutations	784
	24.1.3.4 Segmental Aneuploidies	784
24.1.4	Future Perspectives	784
References		784

A. Baldini (✉)
Institute of Genetics and Biophysics, National Research Council, University of Naples Federico II, and the Telethon Institute of Genetics and Medicine, Via Pietro Castellino 111, 80131 Naples, Italy
e-mail: baldini@cnr.igb.it

24.1.1 Introduction

Since the development of mouse embryonic stem cell technology [5, 17] combined with homologous recombination [8, 18, 19] was applied to modify genes in those cells, modeling of human genetic diseases in the mouse has been by far the most powerful method used to study the effect of genetic mutations in mammals. With all the limitations related to species-specific characteristics, mouse models remain a virtually "obligatory" step in genetic disease research. Genetically and phenotypically accurate models are essential to understand pathophysiology and experiment therapies.

Gene targeting in the mouse has become routine in many research Institutions, and protocols have become so robust that gene targeting is not only the method of choice for the generation of models of human genetic disease, but also an essential method for studying the function of genes in mammals. The trend of using gene knockout as a tool for mammalian genetics has culminated in multinational initiatives to fund the generation of knockouts for all the mouse genes (the NIH Knockout Project, or KOMP; http://www.nih.gov/science/models/mouse/knockout/index.html and the European EUCOMM initiative; http://www.eucomm.org/info/).

In parallel with all this, the level of sophistication of gene manipulation technologies has increased to a point where a "simple" knockout, i.e., gene inactivation, may be considered a relatively modest research investment for individual research projects if we consider the range of options at our disposal for gene manipulation. More complex strategies, though laborious and time consuming, may, in the long run, provide a significantly greater scientific return.

In this chapter, I will review the tool kit at our disposal to generate mouse models as close as possible to the genetic diseases of interest.

24.1.2 Gene and Genome Manipulation Strategies

24.1.2.1 Knockout

This is a targeted mutation that leads to complete inactivation of the gene. Thus, this method is suitable for modeling diseases caused by loss of function of a gene. Typically, this is obtained by replacing the gene of interest with extraneous DNA, generally a segment encoding a drug resistance protein used as a positive selection cassette (Fig. 24.1.1a). Gene targeting is carried out using a "replacement" type of vector made of two DNA segments identical to genomic DNA flanking the gene of interest (homology arms), separated by a positive selection cassette and followed by a negative selection cassette positioned at one extremity of the vector (Fig. 24.1.1a). After introduction of the vector DNA into ES cells, homologous recombination occurs within the homology arms and the genomic region flanked by the homology arms sequence is replaced by the DNA flanked by the homology arms in the vector. Because the methods used to introduce DNA into ES cells are relatively inefficient, cells that have incorporated vector DNA are selected using a drug to which resistance is conferred by the positive selection cassette carried by the targeting vector. However, only a small percentage of cells that have incorporated vector DNA have actually undergone homologous recombination. Indeed, most of the cells will carry the vector DNA randomly inserted in their genome. The use of a negative selection system is designed to enrich the ES cell population with cells that have undergone homologous recombination. Because the negative selection cassette is positioned externally to the homology arms ("–sel" in Fig. 24.1.1a), it will not be incorporated into the genomic DNA after homologous recombination, but it will if the vector DNA is simply inserted at random in the genomic DNA. Thus, if one uses a drug that kills cells that have incorporated the negative selection cassette, the cell population will become richer in cells that have undergone homologous recombination.

Once ES cells with the desired mutation have been obtained, these are injected into mouse blastocysts and reimplanted into foster mothers. Because the injected ES cells' genome encodes for a different coat color than the blastocyst's genome, the resulting chimeric animals' coats have a characteristic patchy color (Fig. 24.1.2). Germ cells of these animals will also be partly derived from the modified ES cell's genome and from the blastocyst's, wild type genome. These animals are then crossed with wild type animals, and some of the progeny will inherit ES cells' chromosomes, including the one with the targeted mutation. Once established in a mouse, the mutation will be transmitted as a Mendelian trait.

24.1.2.2 Knockin and the Use of Site-Specific Recombinases

Many gene mutations may not be readily classifiable as gain or loss of function. For example, functional interpretation of missense mutations may be difficult. This is just one example of a case in which the use of a knockin gene targeting strategy would be beneficial, perhaps one of the most powerful and flexible approaches to gene manipulation. The method allows the introduction of subtle modifications to the endogenous gene, including point mutations. The basic mechanism is very similar to the one described above, but there are some important differences. In particular, the structure of the targeted gene is not altered, so that it can still be transcribed normally, and no selection cassette is left in the targeted allele (Fig. 24.1.1b). This approach has been made possible thanks to the use of site-specific recombinases. Currently, two recombinases are used in gene manipulation, Cre recombinase, which is by far the more commonly used, and Flp recombinase [6, 13]. For both recombinases, the structure of the target site is made up of two 13-bp palindromic sequences that flank an 8-bp core sequence.

24.1 Mouse as a Model for Human Disease

Fig. 24.1.1 (**a–c**) Schematic representation of gene-targeting approaches to obtain different types of alleles. *Numbers* indicate exons of an hypothetical gene of interest. *PGKneo* is a positive selection cassette, *-sel* is a negative selection cassette. *Small triangles* indicate recombination sites for site-specific recombinases Cre or Flp. (**a**) Generation of a knockout allele; **b** generation of a knockin allele (+ in exon 3 point mutation introduced artificially into the targeting vector); **c** generation of a conditional allele

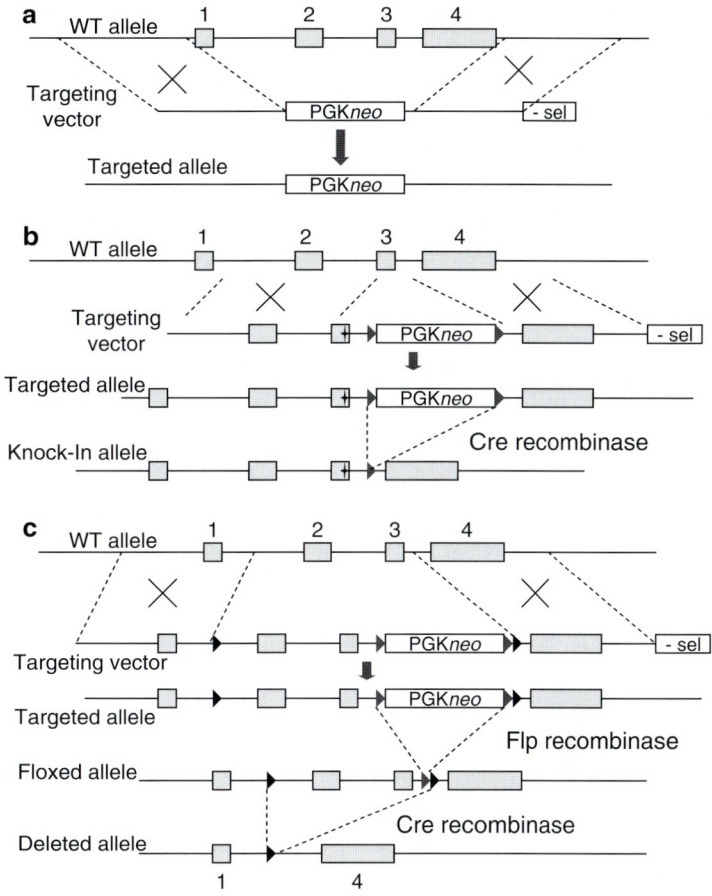

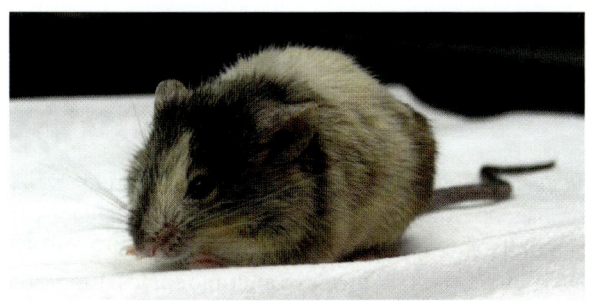

Fig. 24.1.2 A chimeric mouse

However, the sequence of the Cre target site (named *lox*P) is different from that of Flp (named FRT). The target sites are not present in the wild type mouse genome. Therefore, these sites can be introduced artificially at the desired location (using homologous recombination) so that Cre- or Flp-mediated recombination can occur exactly and exclusively at the desired points. In addition, the expression of these recombinases in various tissues in vivo appears to have no phenotypic consequences. The introduction of these site-specific recombinases has been one of the most significant advances in the mouse genetics field, the variety of applications being limited only by the creativity of the investigator.

The knockin strategy uses homologous recombination to target modifications to the endogenous gene, and then uses site-specific recombinases to remove from the targeted allele extraneous sequences that may interfere with the normal function of the allele (e.g., the selection cassette) (Fig. 24.1.1b). This final step, however, leaves behind a *lox*P or FRT site (34 bp) that usually does not harm gene functionality but might do so in some cases, as will be discussed in the next section.

24.1.2.3 Conditional Mutations

The engineered mutations described above, once established in mice, are inherited as Mendelian characters unless they reduce mouse viability. Most autosomal gene mutations are viable in the heterozygous state, but homozygous mutants may be so severe as to make the model practically unusable for studying the pathogenesis of a disease, e.g., if lethality occurs in early embryogenesis (the issue of gene dosage sensitivity in humans and mice will be addressed later in this chapter). In addition, certain genes may have broad expression and their loss may cause a complex phenotype that is difficult to dissect and interpret. Conditional mutations can address these issues, at least partially. Mutations are not present in the germline, but are generated in somatic tissues and at a particular developmental stage.

This strategy, in the great majority of cases, is based on the site-specific recombinase Cre and carefully positioned loxP sites, the target sites of Cre. It requires two genetic components, one of which is a modified allele of the gene of interest, generally referred to as "floxed" (or loxP-flanked) allele. This can be generated as shown in Fig. 24.1.1c. The loxP sites are positioned flanking an essential segment of the gene. The floxed allele must be functional in order to generate truly conditional mutations. While this is often the case, the insertion of loxP sites could have deleterious consequences for gene function even when positioned in apparently "neutral" segments of the gene, for example in nonevolutionary conserved regions of introns. Interference of loxP sites with transcription may be difficult or impossible to predict and may only be discovered after the allele is established in vivo.

The second component is an artificial gene expressed in the desired tissue and time, encoding Cre recombinase. The expression of Cre will cause recombination between the two loxP sites of the floxed allele and delete the intervening DNA, thus effectively causing the loss of function of the gene of interest, but only in the tissues where Cre is expressed. Today there are many mouse lines available that have been engineered to express Cre in different tissues (Cre-drivers) (see, e.g., http://nagy.mshri.on.ca/cre/index.php). These animals can be crossed with mice carrying the floxed allele of the gene of interest, and it is then possible to study the phenotypic consequences of the loss of the gene in specific tissues.

An additional category of Cre drivers uses special types of Cre that normally are not "active" (i.e., are localized in the cytoplasmic compartment of the cell), but they become active on subministration of a drug that causes the Cre protein to translocate to the nucleus, where it can carry out its recombinase activity. These special recombinases are fused to mutant estrogen receptor (mER) sequences that can bind tamoxifen (an estrogen receptor modulator), but they cannot bind the endogenous estrogens. Tamoxifen can be injected into mice to translocate Cre-mER into the nucleus. Because tamoxifen crosses the placenta, it can also be used to activate Cre during embryonic development. The tamoxifen-inducible Cre system is very powerful for testing gene function at specific time points and developmental stages [20].

24.1.2.4 Multigene Deletions and Duplications

Site-specific recombinases can be used to generate large and precisely targeted genomic rearrangements involving many genes [16, 21]. This approach can be used to model segmental aneuploidy syndromes. The strategy includes the positioning loxP sites at the extremities of the chromosomal segments that are to be deleted or duplicated. Cre recombination will cause the deletion of the segment if the two loxP sites are located in the same chromosome, or both deletion and duplication of the segment if they are located one in each homologous chromosome.

24.1.2.5 Transgenics

Transgenics are obtained by injecting nude DNA (transgene) into one-cell mouse embryos. The DNA is generally incorporated (at random) into the genomic DNA of the embryo, in one or multiple copies. Embryos are then reimplanted in a foster mother, and a percentage of the newborn will carry the transgene and transmit it as a Mendelian trait. In contrast to the gene-targeting strategies described above, transgenics are not designed to modify or disable endogenous genes but to force the expression of an additional gene coded by the transgene. The transgene is generally made up of a promoter/enhancer (to drive expression in the desired tissues) and a cDNA encoding the gene of interest or a mutated form. The use of transgenics in modeling human disease will be discussed later. Transgenics, although very useful in several circumstances, have limitations. For example, the expression of the gene carried by the transgene is difficult to control, because of common multiple copy insertions and because the locus of insertion may affect expression; another potential problem is that insertion of the transgene may disrupt an endogenous gene.

24.1.3 Generating Genetically Accurate Models

24.1.3.1 Genetic Diseases Caused by Loss of Function Mutations

This is the simplest scenario, in which ablation (knockout) of the gene of interest is required. Many such attempts may fail (there is no reliable estimate of how many, because negative results are not easily published), either because the loss of function does not have any phenotypic consequences in the mouse or because the consequences are too severe to be informative or to be representative of the disease of interest (e.g., very early embryo lethality). There are a number of strategies that could help in addressing some of these potential problems. The lack of phenotypic consequences of ablation of a human disease gene in a mouse knockout may be due to functional redundancy that is more "effective" in mice than in humans. A possible way to bypass this problem is to ablate the functionally redundant gene (double knockout).

Excessively severe phenotypes in mouse models may occur for different reasons. For example, the mouse gene may have a broader expression of the human gene and broader functions, or there may be partial functional redundancy in humans but not in mice. To get around excessively severe phenotypes, there are at least two approaches. One is to generate hypomorphic alleles (i.e., alleles that have lower functionality than the wild type (wt) allele but not null functionality), while the other is to use a conditional mutation strategy. The latter has been described above (Sect. 24.1.2.3). The generation of hypomorphic alleles is generally done by inserting a positive selection cassette (e.g., PGKneo) into an intron of the gene of interest, using homologous recombination. The insertion of the cassette may affect transcription and or splicing of the targeted gene. The result is a net reduction of mature mRNA available for translation. Unfortunately, the functionality of the modified allele is not easily predictable, so that the outcome of PGKneo insertion may be variable, from total disruption of gene function to very mild reduction of transcription. Nevertheless, hypomorphic alleles are a very important addition to the toolbox for mouse modeling of human diseases.

Another important issue to consider is that the genetic background of the mouse strain used for modeling may have a strong influence on phenotypic presentation. This eventuality may be used to the researcher's advantage, as genetic background-dependent phenotypes may be used to map genetic modifiers.

24.1.3.2 Gene Dosage Mutations

Dominant diseases caused by heterozygous mutations inactivating a gene are relatively common and may pose special challenges to the generation of mouse models. Loss of one of the two copies of an autosomal gene causes phenotypic consequences if the gene is haploinsufficient. Genes that are haploinsufficient in humans are not necessarily so in mice; hence the possibility that a mouse model of this type of mutation may not recapitulate the human phenotype, and this is often the case. Heterozygous mutation in mice may cause a phenotype that only partially resembles the human disease [2, 4, 12, 15], while homozygous mutation may provide a more complete phenotype but it will often be too severe to be an accurate model of the disease. The molecular basis of gene haploinsufficiency is not clear. In general, a gene product needs to be at a sufficient concentration for it to carry out its biological function. Some proteins, e.g., many enzymes, may be able to function even at a much reduced concentration, and other proteins may not function well after a modest reduction of concentration (some transcription factors belong to this category). To complicate the issue further, different biological processes or developmental programs may differ in their sensitivity to the dosage of a given gene, within the same organism. This is why heterozygous mouse mutants of haploinsufficient genes may present with only some of the phenotypic abnormalities. The gene product concentration threshold beyond which certain biological processes start to fail may be different in mouse and humans, which gives rise to the difficulty in accurately modeling disorders caused by gene dosage abnormalities. The use of hypomorphic alleles is the most powerful strategy for modeling haploinsufficiency disorders when heterozygous mutation in the mouse does not recapitulate the disease phenotype. Crossing these alleles with a null allele allows the reduction of gene product dosage below the level afforded by heterozygous mutation.

24.1.3.3 Gain of Function Mutations

Genetic disorders caused by gain of function mutations can be modeled by expressing the mutant form of the gene in the mouse. The accurate strategy that can be used to achieve this objective is to modify the endogenous gene by targeting the mutation into it (see knockin description in Sect. 24.1.2.2). With this strategy, the mutant gene is expressed in the appropriate tissues and at the appropriate transcription level. However, the less laborious transgenic approach is commonly used to model these disorders. Examples of successful use of transgenics are seen in models of disorders caused by expansion of CAG repeats, such as Huntington disease and spinocerebellar ataxia 1 [3, 7].

24.1.3.4 Segmental Aneuploidies

Syndromes caused by deletion or duplication of a chromosomal segment (or an entire chromosome, as in Down syndrome) are relatively frequent [1, 9]. Until recently, genetic disorders of this type could not be modeled accurately in mice. Chromosome engineering has made this possible (see Sect. 24.1.2.4), and the first engineered model of a microdeletion syndrome (the del22q11.2 deletion / velocardiofacial / DiGeorge syndrome) was reported in 1999 [11]. Attempts at modeling Down syndrome, associated with aneuploidy of a much larger region of DNA, have also met with partial success [14], but a new model carrying three copies of most of the region involved in this syndrome has been reported only recently [10].

24.1.4 Future Perspectives

The completion of the sequence of such complex genomes as the human and mouse genomes and many others has initiated the so-called postgenomic phase of biomedical research, when DNA sequence of entire genomes can be analyzed with increasingly sophisticated computational tools. Of course, that does not tell us how those genes work and what they do. By analogy, the generation of inactivating mutations of every mouse gene will inaugurate the post-knockout era. We will be able to look at the phenotypic consequences of the loss of any gene of interest from our desktop computers and develop new hypotheses concerning gene function, possible roles in diseases, interactions, etc. In a way, the post-knockout era has already started for many genes, and, in many cases (especially for developmentally important genes and for genes required in very early development or extra embryonic tissue) the subsequent step has been conditional or other, "specialized" mutations.

Mouse mutants (like any models based on complex organisms) have limitations that are mainly due to species-specific characteristics (e.g., the aforementioned differences in gene dosage sensitivity) or to difficulties in interpreting the consequences that the loss of a critical gene may have in a complex organism. For example, the loss of a gene may perturb multiple genetic networks and have far-reaching consequences. One solution might be to reduce the complexity, either by using a different model (e.g., a tissue culture model) or by restricting the mutation to a tissue or time window.

References

1. Antonarakis SE, Lyle R, Dermitzakis ET, Reymond A, Deutsch S (2004) Chromosome 21 and Down syndrome: from genomics to pathophysiology. Nat Rev Genet 5:725–738
2. Baldini A (2006) The 22q11.2 deletion syndrome: a gene dosage perspective. ScientificWorldJournal 6:1881–1887
3. Bates GP, Mangiarini L, Mahal A, Davies SW (1997) Transgenic models of Huntington's disease. Hum Mol Genet 6:1633–1637
4. Biben C, Weber R, Kesteven S, Stanley E, McDonald L, Elliott DA, Barnett L, Koentgen F, Robb L, Feneley M, Harvey RP (2000) Cardiac septal and valvular dysmorphogenesis in mice heterozygous for mutations in the homeobox gene Nkx2–5. Circ Res 87:888–895

5. Bradley A, Evans M, Kaufman MH, Robertson E (1984) Formation of germ-line chimaeras from embryo-derived teratocarcinoma cell lines. Nature 309:255–256
6. Branda CS, Dymecki SM (2004) Talking about a revolution: the impact of site-specific recombinases on genetic analyses in mice. Dev Cell 6:7–28
7. Burright EN, Orr HT, Clark HB (1997) Mouse models of human CAG repeat disorders. Brain Pathol 7:965–977
8. Doetschman T, Gregg RG, Maeda N, Hooper ML, Melton DW, Thompson S, Smithies O (1987) Targetted correction of a mutant HPRT gene in mouse embryonic stem cells. Nature 330:576–578
9. Fisher E, Scambler P (1994) Human haploinsufficiency – one for sorrow, two for joy [news]. Nat Genet 7:5–7
10. Li Z, Yu T, Morishima M, Pao A, Laduca J, Conroy J, Nowak N, Matsui S, Shiraishi I, Yu YE (2007) Duplication of the entire 22.9 Mb human chromosome 21 syntenic region on mouse chromosome 16 causes cardiovascular and gastrointestinal abnormalities. Hum Mol Genet 16:1359–1366
11. Lindsay EA, Botta A, Jurecic V, Carattini-Rivera S, Cheah Y-C, Rosenblatt HM, Bradley A, Baldini A (1999) Congenital heart disease in mice deficient for the digeorge syndrome region. Nature 401:379–383
12. Mori AD, Bruneau BG (2004) TBX5 mutations and congenital heart disease: Holt–Oram syndrome revealed. Curr Opin Cardiol 19:211–215
13. Nagy A (2000) Cre recombinase: the universal reagent for genome tailoring. Genesis 26:99–109
14. Olson LE, Richtsmeier JT, Leszl J, Reeves RH (2004) A chromosome 21 critical region does not cause specific Down syndrome phenotypes. Science 306:687–690
15. Pu WT, Ishiwata T, Juraszek AL, Ma Q, Izumo S (2004) GATA4 is a dosage-sensitive regulator of cardiac morphogenesis. Dev Biol 275:235–244
16. Ramirez-Solis R, Liu P, Bradley A (1995) Chromosome engineering in mice. Nature 378:720–724
17. Robertson E, Bradley A, Kuehn M, Evans M (1986) Germline transmission of genes introduced into cultured pluripotential cells by retroviral vector. Nature 323:445–448
18. Thomas KR, Capecchi MR (1987) Site-directed mutagenesis by gene targeting in mouse embryo-derived stem cells. Cell 51:503–512
19. Thomas KR, Folger KR, Capecchi MR (1986) High frequency targeting of genes to specific sites in the mammalian genome. Cell 44:419–428
20. Xu H, Cerrato F, Baldini A (2005) Timed mutation and cell-fate mapping reveal reiterated roles of Tbx1 during embryogenesis, and a crucial function during segmentation of the pharyngeal system via regulation of endoderm expansion. Development 132:4387–4395
21. Yu Y, Bradley A (2001) Engineering chromosomal rearrangements in mice. Nat Rev Genet 2:780–790

Caenorhabditis elegans, A Simple Worm: Bridging the Gap Between Traditional and Systems-Level Biology

Morgan Tucker and Min Han

Abstract *Caenorhabditis elegans* is a simple invertebrate roundworm that has served as a model for exploring basic biological questions concerning multicellular organisms and has made numerous contributions to our understanding of human biology. In this chapter we explore some of the approaches C. elegans researchers have employed in an effort to elucidate conserved molecular mechanisms and understand the basic biology underlying many human disorders.

Contents

24.2.1	A Primer	787
	24.2.1.1 A Short History	787
	24.2.1.2 The Worm, Its Life Cycle, and Its Cultivation	788
24.2.2	The How and Why of Screening	789
	24.2.2.1 Genetic Screens: A Traditional Single-Gene Approach	789
	24.2.2.2 RNAi Screens: A High-Throughput Approach	790
	24.2.2.3 Compound Screens: Identifying Therapeutics	791
24.2.3	Beyond the Simple Screen	791
	24.2.3.1 A Systems Approach: Protein Interaction Maps	791
	24.2.3.2 Exploring Complex Traits: Aging	792
	24.2.3.3 Modeling Human Disorders: Alzheimer's Disease	793
24.2.4	Conclusions and Perspectives	793
References		794

M. Tucker and M. Han (✉)
Department of Molecular, Cellular, and Developmental Biology, Howard Hughes Medical Institute, University of Colorado at Boulder, Campus Box 0347, Boulder, CO 80309, USA
e-mail: mtucker@colorado.edu; mhan@colorado.edu

24.2.1 A Primer

The principal goal of this chapter is to introduce the reader to *Caenorhabditis elegans*, a simple roundworm that has been the focus of intense research over the past 40+ years. As a model organism, this small invertebrate has proven to be a valuable tool for exploring basic biological questions concerning multicellular organisms and has also made numerous contributions to our understanding of human biology. We will explore some of the approaches *C. elegans* researchers employ in an effort to elucidate conserved molecular mechanisms and understand the basic biology underlying many human disorders.

24.2.1.1 A Short History

In 2002, the Nobel Prize for Medicine was awarded to Sydney Brenner, Robert Horvitz, and John Sulston for their genetic research on development and programmed cell death in *C. elegans*. The path that led to this award began in the 1960s when Brenner chose this small roundworm as a model for neuronal development and the genetics of behavior. At this time, many of the fundamental processes of molecular biology were just beginning to come to light. The structure of DNA was deciphered a decade earlier, thereby prompting intense investigation of the processes of DNA replication, mRNA transcription, and protein translation. A few pioneering scientists instead focused their attention on new frontiers in biology. These investigators possessed a basic understanding of how the eukaryotic cell works, but their understanding of how these cells propagate and come together to form a complex organism at the molecular level was shadowed in mystery.

Brenner set out to find the perfect model organism in which to study development. His specific interest was centered on how the nervous system was assembled and how this structure subsequently processed intercellular signals to control behavior. *C. elegans* proved to fit the bill. Brenner found that its economy, ease of maintenance in the laboratory, and capacity for long-term storage made it a tractable organism. More importantly, its transparent cuticle allowed each cell to be observed through the light microscope, and the organism was small enough for slices to fit easily under an electron microscope.

In an amazing tour de force, Brenner's colleague John Sulston was able to observe all the cell divisions and map the developmental fate of every single cell in *C. elegans* [22, 23]. The complete cell fate of each of the 959 somatic cells in the adult hermaphrodite was demonstrated to be largely invariant between individuals. Use of the electron microscope was pivotal in generating a complete map of the nervous system, consisting of only 302 neurons, and allowed the connectivity of each of these neurons to be determined [25]. Although this early work was founded in the tradition of descriptive biology, it laid the foundation for more advanced experiments and in many ways made *C. elegans* the powerful tool that it is today. Having a detailed description of every single cell fate and position enables *C. elegans* researchers to link the requirement of a single gene product to the formation of a specific anatomical structure, thus permitting the elucidation of the development processes regulated by a given gene.

C. elegans has provided considerable insights into the mechanisms of basic biology: programmed cell death, organ formation, cell signaling, cell polarity, gene regulation, metabolism, and sex determination have all been studied intensively. However, analyses of the complete *C. elegans* genome sequence suggest that this organism has a broader application as a human disease model. *C. elegans* was the first multicellular organism to have its genome completely sequenced. The sequence of all five autosomes and the single sex chromosome was first published in 1998 [2]. The genome sequence is frequently re-annotated and updated online, thus providing an indispensable bioinformatics resource [3]. The complete sequence consists of 100 million base pairs, encoding approximately 20,000 genes and greater than 1,300 noncoding RNAs. This is comparable to the approximately 23,000 genes predicted in human genome, of which nearly 40% have direct homologs in *C. elegans*. Considering the high degree of structural and functional conservation with mammalian homologs, continued work in *C. elegans* will likely provide considerable insight into the mechanisms of human development and disease at the molecular level.

24.2.1.2 The Worm, Its Life Cycle, and Its Cultivation

C. elegans is a free-living, soil-dwelling nematode that is approximately 1 mm in length – just barely visible to the naked eye. These roundworms have a simple body plan that primarily consists of a gut and a gonad. The intestine is connected at its anterior to the pharynx, a muscled organ that drives the feeding process, and at its posterior to the anal opening (Fig. 24.2.1). The midbody is marked by the vulva, which connects the symmetrical gonad to the external environment. The worm's sinusoidal movement is driven by contraction of the body wall musculature under control of a simple nervous system. The exterior of the worm is covered with a collagenous cuticle, which provides protection from the environment and functions as an external skeleton.

The *C. elegans* life cycle is fairly straightforward. After fertilization of a single oocyte, embryogenesis occurs over the next 13 h. This process consists of standard cell divisions followed by morphogenetic rearrangements and elongation of the embryo just prior to hatching. Under most conditions, larval development consists of four stages (L1–L4), punctuated by cuticle molts and a dramatic increase in overall size, followed by adulthood (Fig. 24.2.2). Alternatively, during conditions of stress, including starvation and/or overcrowding, an alternative third larval stage called

24.2 Caenorhabditis elegans, A Simple Worm

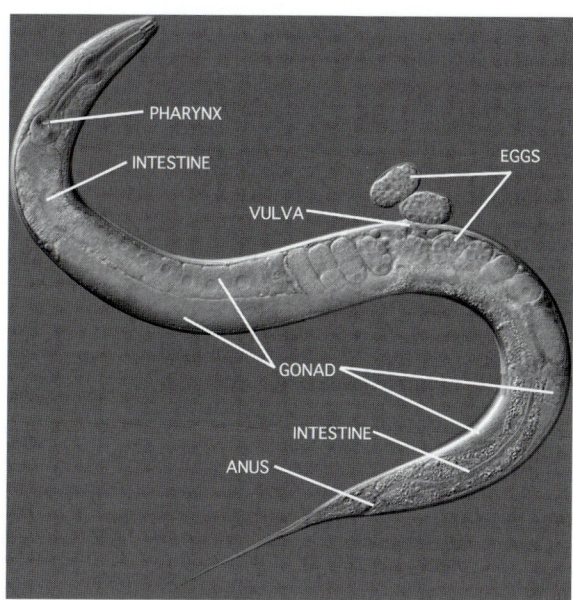

Fig. 24.2.1 Differential interference contrast image of an adult *C. elegans* hermaphrodite. Major anatomical landmarks are indicated. The muscled pharynx pumps food into the intestinal lumen, which is connected at its posterior to the anal opening. The hermaphrodite gonad is comprised of functionally independent anterior and posterior arms. Each U-shaped arm, connected at its proximal end to the vulval opening, produces both sperm and oocytes. *C. elegans* propagate by self-fertilization, and a single hermaphrodite is capable of producing 300+ progeny in only 2–3 days

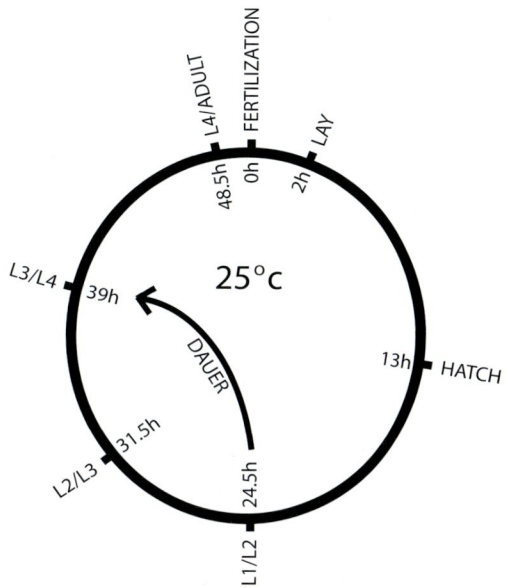

Fig. 24.2.2 *C. elegans* life cycle. In approximately 50 h at 25 C a single fertilized oocyte will complete embryogenesis, grow through four larval stages, and begin laying eggs of its own. At each of the four larval molts, a new cuticle is secreted and the old one is shed. Under conditions of stress or starvation larva can enter diapause, an alternative L3 larval state called dauer. Dauer larva are capable of surviving in the absence of food for as long as 6 months. In the presence of a food supply, dauer larvae will re-enter the development cycle to become fertile adults

the dauer state can be achieved. Dauer larvae do not feed and are highly resistant to stress. These larvae are capable of living for 3–6 months (4–8 times the normal life span) in a dormant state. When growth conditions improve, the larvae can resume the developmental cycle to become fertile adults.

In the wild, *C. elegans* feed on available microorganisms; in the laboratory, they are maintained on agar plates, subsisting on a diet of the common bacterium *Escherichia coli*. Hermaphrodite (XX) worms primarily propagate by self-fertilization. This allows researchers to maintain genetically identical populations over multiple generations with little or no effort. Spontaneous males (XO) do infrequently arise in a population, thus allowing for cross-fertilization between strains and facilitating complementation tests and the construction of double mutants. Transgenic worms containing non-native DNA can be created by microinjecting DNA constructs that are subsequently maintained as extra-chromosomal arrays or that are integrated directly into the *C. elegans* genome [17].

During its life span of approximately 2–3 weeks, an individual hermaphrodite is capable of producing more than 300 progeny. With a generation time of approximately 2–3 days, it is possible for a single worm to produce a large population of genetically identical worms very quickly. This ability means that *C. elegans* is not only suitable for genetic research, but it is also amenable to large-scale genomic experiments and biochemistry.

24.2.2 The How and Why of Screening

24.2.2.1 Genetic Screens: A Traditional Single-Gene Approach

In the postgenomic era, many researchers are quick to dismiss traditional genetic screens as outdated. Nothing could be further from the truth. Forward genetic screens still constitute the most powerful, nonbiased tools biologists have in their quest to assign function to a given gene. In brief, genetic screens are used to identify new genes involved in a given process, either by looking for new mutations that correspond to a previously characterized phenotype or by identifying mutations that can enhance or suppress a phenotype associated with a known gene. This approach allows for identification of gene networks that have a specific relationship, defined based on the screening approach.

C. elegans possesses many advantages that make it a powerful genetic model. Its hermaphroditic lifestyle, rapid generation time, simplicity, ease of manipulation, and fully sequenced genome all facilitate rapid screening and mapping of new mutations. The basic approach to screening is straightforward: mutations are induced in the genome with an appropriate mutagen, and individuals in the population that display the phenotype of interest are subsequently identified. Mapping the mutation of interest is carried out by linkage mapping relative to mutations with known phenotypes, by measuring the recombination frequency between the mutation of interest and single nucleotide polymorphisms (SNPs) found throughout the genome, or by combining these two approaches [7].

This simple approach has been successfully applied in the study of many human disorders. For example, mutations in the human dystrophin–glycoprotein complex (DGC) have been shown to result in muscular dystrophies [1] – one of the most common forms of human genetic disorder. The etiology is characterized by progressive muscle degeneration, although the exact mechanism responsible for such pathogenesis is not well defined. Mutations in the *C. elegans* homologs of the DGC components result in a characteristic hyperactive phenotype and muscle degeneration [10]. An extensive genetic screen for worms that displayed phenotypes identical to those of DGC mutants resulted in the identification of a previously uncharacterized protein, SNF-6 [12]. The *snf-6* gene encodes a novel acetylcholine/choline transporter that is required for the uptake of acetylcholine at neuromuscular junctions. A direct interaction between the DGC and SNF-6 is necessary for proper localization of SNF-6 at neuromuscular junctions [13]. These findings have considerable implications for the treatment of muscular dystrophies. The authors of this study suggest that inhibiting the process of acetylcholine clearing from the neuromuscular junction likely results in prolonged muscle excitation and that this may be the underlying cause of the muscle degeneration that is observed in *C. elegans* mutant for other components of the DGC [11]. The results also suggest that the development of therapeutic agents that affect acetylcholine levels at the neuromuscular junction may be of significant use when treating muscular dystrophy disorders.

24.2.2.2 RNAi Screens: A High-Throughput Approach

In 2006, the Nobel Prize for Medicine was awarded to Andrew Fire and Craig Mello for their research on RNA interference (RNAi) in *C. elegans*. This discovery [10], during the late 1990s, challenged the way gene regulation was conceptualized and also proved to be a powerful tool in many organisms. RNAi is the process by which fragments of double-stranded RNA (dsRNA) interfere with the expression of any genes that share homologous sequences. The molecular mechanism of RNAi has been described in some detail elsewhere [19]. dsRNAs entering a cell, through a mechanism that is not well understood, are bound by the protein Dicer, an RNAse III nuclease, which cleaves the dsRNAs to produce smaller fragments 21–23 bp in length, called small interfering RNAs (siRNAs). These siRNAs are loaded into the RNA-induced silencing complex (RISC), where they guide mRNA degradation or translation silencing, depending on the complementarity of the target. The RNAi process is systemically propagated through transport and replication of the siRNAs. The process is heritable in both plants and *C. elegans* but not, however, in *Drosophila* or mammals, though the exact reason for this difference is unclear.

Researchers have designed a collection of bacteria that individually express dsRNA encoding each gene in the entire *C. elegans* genome [12]. This bacterial library enables worms to be screened for specific phenotypes by essentially knocking out gene function via RNAi. When worms are fed the dsRNA-expressing bacteria, RNA is taken up by the intestinal cells, spreads from cell to cell, and subsequently generates effects that are heritable for multiple generations. The resulting progeny display an approximation of the knockout phenotype for a particular suppressed gene, thereby allowing researchers to quickly assay gene function.

Although RNAi is a powerful screening tool, it is limited in a number of ways. Not all genes are equally responsive to RNAi, and some tissues, notably neuronal cells, are not amenable to RNAi. Also, RNAi only assays for the loss of gene function. Therefore, unlike traditional techniques used in genetics, this method does not make it possible to assay gain of function or dominant phenotypes. With these considerations, RNAi should be regarded as a complement to traditional methods, and not as a complete replacement for genetic screens.

That said, the ease of applying RNAi and the immediate knowledge of the gene target facilitates phenotypic analysis that would simply be unimaginable using traditional methods. Phenotypes that have a very low penetrance or require time-consuming techniques are strong candidates for RNAi-based screens. For example, a technically complicated RNAi-based screen was recently performed in an effort to identify all of the gene products required for the first two rounds of cell division following fertilization of a *C. elegans* oocyte [21]. This screen targeted 98% of all known open reading frames of the *C. elegans* genome via RNAi. Defects in cell division were recorded with the use of differential interference contrast time-lapse microscopy. Approximately 40,000 time-lapse recordings from more than 19,000 individual RNAi experiments were acquired and assayed for phenotypic defects. These experiments identified at least 661 genes that were required for the earliest cell divisions during embryogenesis, about 14% of which had no known prior function. Furthermore, half of the previously uncharacterized gene products had readily identifiable homologs in other organisms, thus demonstrating how experiments that are only possible in *C. elegans* can have broader implications that extend to other organisms.

multicellular animal that is easy to grow and reproduces rapidly; its small size allows it to be readily cultured in a 96-well plate measuring only 5 by 3.5 in, thus facilitating process automation. *C. elegans* also has a large number of conserved genes and disease pathways that have potential therapeutic value when considering the treatment of human disorders. Furthermore, the added ability to carry out traditional genetic screens allows for the rapid identification of molecular targets for potential therapeutics.

The combined ability to screen compound libraries for bioactive small molecules and the ability to efficiently characterize new drug targets makes *C. elegans* a powerful tool for defining the mechanisms of drug action. A recent screen of 14,100 small molecules in *C. elegans* identified 308 compounds that produced discernable phenotypes [14]. Growing worms in the presence of one of these compounds, referred to as nemadipine-A, resulted in reproducible defects in body morphology and inhibited the ability of the worms to lay eggs. Nemadipine-A shares a high degree of structural similarity to the common anti-hypertension drugs 1,4-dihydropyridines (DHPs). DHPs have been shown to function by binding to L-type calcium channels to inhibit their function. A genetic screen for mutations that could suppress the effects of nemadipine-A identified multiple dominant mutations in a single gene, egl-19. egl-19 encodes the alpha subunit of an L-type calcium channel. Sequence analysis of the dominant egl-19 mutations demonstrated amino acid changes in regions of the channel that have been previously shown to be required for interaction with DHPs in mammalian systems. These and other results indicate that the primary target of nemadipine-A is egl-19. This example validates the prospect of using *C. elegans* to rapidly identify active small molecules and their molecular targets, thus facilitating drug discovery.

24.2.2.3 Compound Screens: Identifying Therapeutics

Many researchers are pursuing the identification of new bioactive compounds because of their value as possible pharmaceuticals for treating human disease and their subsequent market potential. *C. elegans* has many features that make it appealing to those interested in large-scale drug screening. Not only is it a

24.2.3 Beyond the Simple Screen

24.2.3.1 A Systems Approach: Protein Interaction Maps

Proteomics refers to the large-scale analysis of all the proteins within a given cell or organism. When and how these proteins are expressed and post-translationally modified and how they interact with each other and with other components within the cellular milieu is information that falls under the proteomics umbrella. In *C. elegans*, many of these areas of systematic research are still in the early stages. However, significant progress has been made in defining the protein–protein interaction map using large-scale two-hybrid analysis. In these experiments, a yeast-based reporter system is used to determine if two proteins are capable of directly interacting with one another [9]. The overall goal is to generate a genome-wide protein interaction map that links uncharacterized proteins to those with known functions in defined biological processes. Although incomplete, mapping projects centered on a subset of metazoan-specific proteins have provided validation of this experimental approach in *C. elegans* [15].

This approach has also been adapted to identify direct interactions between transcription factors and their associated promoter regions, with the overall goal of creating a genome-wide transcription factor-promoter map for *C. elegans*. In theses experiments, a yeast-based reporter system, similar to the two-hybrid assay, is used to determine which proteins within the genome bind to a specific gene promoter and subsequently activate transcription of a reporter gene [9]. In the initial experiments testing this approach, 72 promoters were analyzed [5]. Each was chosen because of its ability to drive gene expression specifically in the gut of the worm. The analysis identified 283 individual protein-DNA interactions involving 117 proteins. Interestingly, a hierarchical nature of the system was observed: overall, there were three or more layers of transcriptional control regulating the genes. A handful of the identified transcription factors were highly connected within the network, binding to many different promoters. These transcription factors were designated 'global regulators.' Another set of transcription factors, termed 'master regulators' bound to fewer, but still multiple, genes. Finally, there were a number of 'specifiers' that only bound to a small number of genes. This hierarchical system, as the authors point out, is similar to the multi-layered transcriptional regulatory networks observed in bacteria, suggesting that the overall structure of the system may be evolutionarily conserved. Further such studies, conducted on a larger scale and coupled with extensive expression data, can provide the first look at the overall developmental transcriptional regulatory cascades of multicellular organisms.

These kinds of network maps will likely lay the foundation for our understanding of more complex systems, including neural development and numerous neurological disorders in humans.

24.2.3.2 Exploring Complex Traits: Aging

Aging is a fundamental and universal process that affects all living organisms. At its core, aging in humans is an incredibly complex trait and is unlikely to fit a simplistic framework of inheritance. Aside from overwhelming environmental influences, the genetic factors affecting longevity are likely multifaceted at best. Remarkably, however, it has been shown that individual mutations can dramatically extend the lifespan of *C. elegans*. These observations, implicate the insulin/insulin growth factor-1 (IGF-1)-like signaling pathway as a critical determinant of lifespan [24]. The insulin/IGF-1-like signaling pathway in *C. elegans* consists of a number of genes, including daf-2 (an insulin/IGF receptor homolog), age-1 (a phosphatidylinositol-3 kinase), daf-18 (a phosphatase and tensin homolog deleted on chromosome X [PTEN] phosphatase), and daf-16 (a Forkhead box class O [FOXO] family transcription factor). Impeding this pathway, by either mutation or RNAi, can significantly extend the *C. elegans* adult lifespan without significantly affecting reproductive fitness. These observations demonstrate that lifespan is clearly subject to regulation and not driven solely by entropy.

Similar results in both flies and mice indicate a level of conservation in the molecular mechanisms affecting longevity, suggesting the possibility of one day developing pharmacological agents to extend the human lifespan. However, the prospect of identifying drugs that can positively influence aging in humans is complicated by the extensive drug-approval guidelines set forth by the Food and Drug Administration (FDA) and the fact that an individual researcher would likely only be able to observe a single cohort of patients, given the time required to complete a clinical trial on aging. With this in mind, a recent study assayed a number of known drug compounds that had already received FDA approval for the treatment of various unrelated human disorders to determine their ability to extend lifespan in *C. elegans* [6]. A class of anticonvulsants traditionally used to modulate neural activity in humans was found to significantly extend both the mean and the maximum lifespan of worms. These results not only serve to support previous findings showing that neural activity in worms can affect longevity; they also suggest a feasible approach for identifying age-extending pharmaceuticals for use in humans by exploiting short-lived model organisms. Testing compounds that have already passed through clinical-trial development mitigates some of the need and cost of this process, although clearly determining the efficacy of these drugs on human longevity will still require multiple generations and some clinical analysis.

24.2.3.3 Modeling Human Disorders: Alzheimer's Disease

Protein misfolding with aberrant protein aggregation is a hallmark of many age-related neurodegenerative disorders, including Alzheimer's, Parkinson's, triplet repeat disorders, and prion-related diseases. Individuals with Alzheimer's disease present with significant and progressive memory loss owing to aberrant cell death in the central nervous system. These clinical traits are usually only observed in individuals in their late fifties or older. The molecular mechanism underlying this age-related neurodegeneration is not completely understood. The histopathology indicates the presence of aberrant protein aggregates, including the presence of amyloid plaques formed by the beta-amyloid peptide (Aβ), a proteolytic fragment of the amyloid precursor protein [20]. The leading hypothesis suggests that Aβ aggregates are neurotoxic and likely play a key part in disease progression. To date, there are no effective therapies to combat the progression of Alzheimer's.

C. elegans has proved to be an influential means for modeling proteotoxic disorders such as Alzheimer's. Ectopic expression of Aβ in the body wall muscles of *C. elegans* results in progressive paralysis [16]. This paralysis is paralleled by Aβ aggregation and deposition along the muscle fibers, as assayed using both amyloid-specific antibodies and dyes [8, 16]. Although *C. elegans* does not encode an amyloid gene homologue, the rationale for constructing such a model is that it allows researchers to utilize the experimental tools available in *C. elegans* to understand how fundamental cellular processes may be affecting the progression of this neurodegenerative disease. As such, this model provides a system in which to address a number of basic questions about aggregate formation and toxicity. For example, it has allowed researchers to address the question of whether aggregate formation is simply a stochastic development that occurs over time during the life of the animal or whether it is a progressive mechanism that is influenced by the aging process. To address this particular question, paralysis of Aβ-expressing worms was measured following down-regulation of insulin signaling via RNAi directed against daf-2 (insulin/IGF receptor homolog) in order to extend the lifespan of these worms [4]. Interestingly, the onset and extent of subsequent paralysis paralleled the altered aging profile of the worms. These results are similar to those showing that slowing the aging process in *C. elegans* can decrease Huntington-associated proteotoxicity [18]. Hence, it appears that there is a direct link between the mechanism of aging and the progression of proteotoxicity. One possible model is that aging reduces an organism's ability to detoxify aggregates, thus suggesting a direct link between longevity and the ability to maintain protein homeostasis. Extrapolating observations such as these to humans may prove useful for identifying potential targets for drug research.

24.2.4 Conclusions and Perspectives

The life sciences have undergone extraordinary advances in the last decade. The dawn of the genomic era in biological research has taken traditional and experimental biology to new heights. No longer are researchers asking what a single component of a cell does at a specific time and place; many now possess the tools to address the interconnections between various components of a cell, how assemblages interact, and how the attributes of living things are derived from the sum of their whole. Recent technological advances are driving this new systems-level approach to biology. In many ways, automated methods based on industrial models have allowed for the development of high-throughput experiments designed to leverage the power of the fully sequenced genome. Despite these advances, it is important to realize that a simple list of genes involved in some aspect of biology is often of little use.

Although such a list can be extremely useful for quickly generating new ideas and research directions, hypothesis validation still requires a return to the basics of experimental biology in the long run. In the end, to fully understand the intricate complexity of a living system, we will need to understand what every single component of the system does at any given time and place. It is with this in mind that we have considered the nematode, *Caenorhabditis elegans*. *C. elegans* research has not only provided a solid foundation in traditional, basic biological inquiry, but is also uniquely suited to modern systems-level approaches. Indeed, many advances in our understanding of human disease mechanisms are likely to revolve around this simple worm over the next decade.

References

1. Batchelor CL, Winder SJ (2006) Sparks, signals and shock absorbers: how dystrophin loss causes muscular dystrophy. Trends Cell Biol 16:198–205
2. C. elegans sequencing consortium (1998) Genome sequence of the nematode *C. elegans*: a platform for investigating biology. Science 282:2012–2018
3. Chen N, Harris TW, Antoshechkin I et al (2005) WormBase: a comprehensive data resource for *Caenorhabditis* biology and genomics. Nucleic Acids Res 33:383–389
4. Cohen E, Bieschke J, Perciavalle RM, Kelly JW, Dillin A (2006) Opposing activities protect against age onset proteotoxicity. Science 313:1604–1610
5. Deplancke B, Mukhopadhyay A, Ao W (2006) A gene-centered *C. elegans* protein-DNA interaction network. Cell 125:1193–1205
6. Evason K, Huang C, Yamben I, Covey DF, Kornfeld K (2005) Anticonvulsant medications extend worm life-span. Science 307:258–262
7. Fay D (2006) Genetic mapping and manipulation: chapter 1-Introduction and basics. In: Wormbook (ed.) The *C. elegans* research community. doi/10.1895/wormbook.1.90.1, http://www.wormbook.org
8. Fay DS, Fluet A, Johnson CJ, Link CD (1998) In vivo aggregation of beta-amyloid peptide variants. J Neurochem 71:1616–1625
9. Fields S, Sternglanz R (1994) The two-hybrid system: an assay for protein-protein interactions. Trends Genet 10:286–292
10. Fire A, Xu S, Montgomery MK, Kostas SA, Driver SE, Mello CC (1998) Potent and specific genetic interference by double-stranded RNA in *Caenorhabditis elegans*. Nature 391:806–811
11. Grisoni K, Martin E, Gieseler K, Mariol MC, Ségalat L (2002) Genetic evidence for a dystrophin-glycoprotein complex (DGC) in *Caenorhabditis elegans*. Gene 294:77–86
12. Kamath RS, Fraser AG, Dong Y et al (2003) Systematic functional analysis of the *Caenorhabditis elegans* genome using RNAi. Nature 421:231–237
13. Kim H, Rogers MJ, Richmond JE, McIntire SL (2004) SNF-6 is an acetylcholine transporter interacting with the dystrophin complex in *Caenorhabditis elegans*. Nature 430:891–896
14. Kwok TCY, Ricker N, Fraser R et al (2006) A small-molecule screen in *C. elegans* yields a new calcium channel antagonist. Nature 441:91–95
15. Li S, Armstrong CM, Bertin N et al (2004) A map of the interactome network of the metazoan *C. elegans*. Science 303:540–543
16. Link CD (1995) Expression of human beta-amyloid peptide in transgenic *Caenorhabditis elegans*. Proc Natl Acad Sci USA 92:9368–9372
17. Mello CC, Kramer JM, Stinchcomb D, Ambros V (1991) Efficient gene transfer in *C. elegans*: extrachromosomal maintenance and integration of transforming sequences. EMBO J 10:3959–7
18. Morley JF, Brignull HR, Weyers JJ, Morimoto RI (2002) The threshold for polyglutamine-expansion protein aggregation and cellular toxicity is dynamic and influenced by aging in *Caenorhabditis elegans*. Proc Natl Acad Sci USA 99:10417–10422
19. Sen GL, Blau HM (2006) A brief history of RNAi: the silence of the genes. FASEB J 20:1293–1299
20. Small SA, Gandy S (2006) Sorting through the cell biology of Alzheimer's disease: intracellular pathways to pathogenesis. Neuron 52:15–31
21. Sönnichsen B, Koski LB, Walsh A et al (2005) Full-genome RNAi profiling of early embryogenesis in *Caenorhabditis elegans*. Nature 434:462–469
22. Sulston JE, Horvitz HR (1977) Post-embryonic cell lineages of the nematode, *Caenorhabditis elegans*. Dev Biol 56:110–115
23. Sulston JE, Schierenberg E, White JG, Thomson JN (1983) The embryonic cell lineage of the nematode *Caenorhabditis elegans*. Dev Biol 100:64–119
24. Vanfleteren JR, Braeckman BP (1999) Mechanisms of life span determination in *Caenorhabditis elegans*. Neurobiol Aging 20:487–502
25. White JG, Southgate E, Thomson JN, Brenner S (1986) The structure of the nervous system of the nematode *C. elegans*. Phil Trans R Soc Lond B Biol Sci 314:1–340

Drosophila as a Model for Human Disease 24.3

Ruth Johnson and Ross Cagan

Abstract *Drosophila melanogaster* has proved a remarkable genetically tractable model organism that continues to provide significant contributions to our understanding of numerous biological processes. In this chapter we discuss insights into a variety of human diseases that have been gained directly from studies conducted in fly labs. These include discoveries relating to the basic biology of diseases (the signaling pathways, for example, that may contribute to disease states), new loci implicated in disease progression or susceptibility (uncovered in large-scale screens and verified in situ using often ingenious assays) and the identification of pharmacological reagents to treat diseases (also identified and tested in well-designed screens and assays). Because genomes, biological processes, and responses have been well conserved, and particularly with the current trend in translational research, studies in flies continue to build a strong foundation for disease studies. Those discussed in this chapter include cancer, neurodegenerative diseases, heart disease, diabetes and metabolic diseases, addiction, and sleep disorders.

Contents

24.3.1	Why Flies?	795
24.3.2	Cancer	796
24.3.2.1	Cell Cycle	797
24.3.2.2	Cell Death	797
24.3.2.3	Hyperplasia and Neoplasia	798
24.3.2.4	Models of Metastasis	799
24.3.2.5	Models of Specific Cancers	801
24.3.3	Neurodegenerative Diseases	802
24.3.3.1	Parkinson's Disease	803
24.3.3.2	Alzheimer's Disease	803
24.3.3.3	Triplet-Repeat Diseases	804
24.3.4	Heart Disease	804
24.3.5	Diabetes and Metabolic Diseases	805
24.3.5.1	Body Size	806
24.3.5.2	Models of Diabetes	806
24.3.6	Addiction	807
24.3.6.1	The Genetics of Addiction: Sensitivity and Tolerance	808
24.3.7	Sleep Disorders	808
24.3.8	Conclusions	808
References		809

24.3.1 Why Flies?

The diminutive fruit fly, *Drosophila melanogaster*, may not be an intuitive model organism for studying human disease. Yet its contributions to this broad topic have been extensive, particularly in the era of genome sequencing, which has simplified the identification of orthologs.

Typically, multiple loci contribute to the susceptibility to or etiology of a single disease. *Drosophila* has a fully sequenced and annotated genome of approximately 14,000 genes and provides a model system that

R. Johnson and R. Cagan (✉)
Department of Developmental and Regenerative Biology,
Mount Sinai School of Medicine, 1 Gustave L. Levy Place,
Annenberg Building 18-92, New York, NY 10029, USA
e-mail: Ross.Cagan@mssm.edu

can identify and manipulate loci with relative ease. This is made possible by a remarkable array of available tools that enable large-scale screening of the *Drosophila* genome for mutations in primary and susceptibility loci. Through the efforts of individual laboratories, the Berkeley Drosophila Genome Project (BDGP), and commercial enterprises such as Exelixis, the community has ready access to fly lines that contain individual targeted mutations that together disrupt nearly 90% of the predicted loci. Deficiencies that cover approximately 70% of the fly genome are available for screening a few dozen genes per fly line. The FLP/FRT and "MARCM" systems are used to create "mitotic clones" of discrete mutants, allowing for the precise study of a mutation within even individual isolated cells [50, 88]. The UAS/GAL4 system and other related systems can be harnessed to express a gene of interest in a tissue- and time-specific manner [54]; with the ease of use of RNA interference in *Drosophila*, these powerful mis-expression systems can also be used to knock down the activity of targeted genes in specific tissues or individual cells [68]. Most of these fly lines are readily available: a key strength of the *Drosophila* community has been a community ethos that emphasizes developing and quickly sharing useful tools and engineered fly lines.

Until recently, most *Drosophila* researchers have focused on broad questions of cell-cell signaling and cell biology as they relate to development. The tiny fruit fly has provided many of the fundamental insights into these issues, and Christiane Nusslein-Volhard, Eric Wieschaus, and Edward Lewis were honored with a Nobel Prize in 1995 for their genetic approaches to development. Recently, an increasing number of *Drosophila* laboratories have turned their attention to specific issues of disease, bringing their powerful tools and a unique vision that emphasizes in situ studies of specific tissues and individual cells.

How similar is *D. melanogaster* to mammals, and how useful is the fruit fly for studying disease? For specific diseases we often do not really know until studies have progressed in both, but the overall signs are encouraging. The fly genome shows a startling degree of homology to that of humans, with better than 60% of disease genes showing a clear ortholog by simple sequence gazing [2] and more careful studies indicating a still higher percentage with functional homology. These similarities extend to diseases that are often thought of in a social context, such as alcoholism, drug abuse, sleep apnea, and also behavioral issues, such as eating habits or aggression. Observing flies reacting to alcohol in ways that mimic human behavior teaches us something about the biological basis of our own responses to intoxication. With a life-span of 2–3 months, a generation time of 10 days, a propensity for each female to produce hundreds of progeny, and easy and inexpensive maintenance, flies provide a cheap, rapid, and powerful approach to the study of specific diseases. And, critically, *Drosophila* are sufficiently complex to permit meaningful generalizations to mammals and humans.

24.3.2 Cancer

Perhaps the first in vivo tumor suppressor identified was the *Drosophila* gene *lethal giant larvae* (*lgl*), which was identified more than 70 years ago by Bridges: recessive loss-of-function mutations led to dramatic overgrowth [13, 14]. *Drosophila* workers bring a somewhat unusual, and potentially very useful, perspective: *Drosophila* is traditionally strong in developmental biology, and the defects observed during oncogenic progression have many aspects that are familiar to *Drosophila* developmental biologists. For these reasons and more, cancer is the disease with the largest efforts in the *Drosophila* field, and the length of this section reflects those efforts.

Of course, *Drosophila* is not a perfect model for human cancer, and a consideration of its similarities and differences is useful. Decades of research have indicated that humans and flies have remarkable similarities in their basic epithelial architecture: junctional proteins are well conserved with some important differences [47, 87]; most major signaling, growth, and death pathways are also well conserved. For example, nearly all of the genes most commonly found altered in tumorigenesis are conserved both structurally and (mostly) functionally in flies, including P53, Ras, Raf, Pten, Src, etc. Perhaps the most important difference lies in the inflammatory response, which has important differences including a lack of the immunoglobulin system. And of course flies do not have bones, which are a major target of breast and prostate metastases, for example. Finally, not all cancer-related genes have clear *Drosophila* orthologs: for example, the BRCA1 and BRCA2 (genes cursive) linked to breast cancer and the mechanisms of telomere maintenance appear to differ.

While potential models of *Drosophila* hematopoietic tumors have been observed (e.g., the emergence of "melanotic tumors" derived from the hematopoietic system), most fly cancer research to date has focused on models of solid tumors. These studies have been extensively reviewed (e.g., [4, 12, 44, 71, 79, 85, 89, 97]), and the purpose below is to introduce the non-fly worker to some of the more notable advances.

24.3.2.1 Cell Cycle

24.3.2.1.1 The Cell Cycle Machinery

The basic cell cycle machinery has been conserved in a broad array of animals from single-cell yeast to mammals, and *Drosophila* shares most of the standard regulators. *Drosophila* has made many of its most important contributions to our understanding of the connections between cell cycle and signaling. A large body of work on *Drosophila* has focused on the spatial regulation of the cell cycle during development (reviewed in [20, 22]), and recent screens have included saturation screens for cell cycle regulators [9]. Recently, *Drosophila* biologists have directed more of their work at aspects that impinge more directly on tissue overgrowth.

Regulation of the Cell Cycle

The cell cycle is regulated at multiple steps by the E2F/Rb pathway, which directly modulates expression of factors such as the cyclins (reviewed in [31]). *Drosophila* has two E2F isoforms: dE2F1 promotes cell proliferation, while dE2F2 opposes it and can also act in other pathways [29, 82]. Both in mammals and in flies, E2Fs act with their heterodimeric partner Dp1; this complex is in turn regulated by physical interactions with Rb. During development, the E2F complex appears to control cell cycle progression through its regulation by a host of other cell cycle-relevant factors (e.g., [19]). As in mammals, deleting the function of the E2Fs indicates they are not essential for cell cycle progression [29]. *Drosophila* Rb acts both to regulate cell cycle and to promote cell death; this balance appears to be at least in part due to regulation by epidermal growth factor receptor (EGFR; [27, 56]), an ortholog of the ErbB family of oncogenes. One goal of E2F/Rb research is to understand the factors that influence Rb to promote cell death vs proliferation and to identify factors that can emphasize the former in advancing tumors that contain activated E2F.

24.3.2.2 Cell Death

Control of cell death is a central aspect of cancer, and *Drosophila* models have provided important contributions to our understanding of cell death pathways, again with the added advantage of being able to examine cell death events in situ.

24.3.2.2.1 Apoptosis Pathway

The primary effectors of apoptotic cell death are the "caspases", cysteine aspases that cleave dozens of downstream targets to initiate the apoptotic process (reviewed in [92]). *Drosophila* has seven caspases, including upstream "initiator" caspases (Dronc, Strica/Dream, and Dredd) and downstream "effector" or "executioner" caspases (Drice, Dcp-1, Damm, Decay) [35]. Regulation of caspase activity represents the central control point of apoptosis. In mammals, caspases are primarily regulated by members of the Bcl-2 superfamily. A second regulatory system is anchored by the IAP (inhibition of apoptosis protein) superfamily, cytoplasmic proteins that inhibit caspase activity through their binding of intermediaries including Smac, Diablo, etc. [35]. Though considered less important as caspase regulators in mammals, IAPs – particularly Survivin and XIAP – represent some of the most strongly overexpressed genes in malignant tumors [70, 100].

In flies, both the Bcl-2 and the IAP systems are active during regulation of apoptosis. Unlike in mammals, IAPs play the dominant role during development [35]. For example, loss of the IAP family member Diap-1 during embryogenesis leads to widespread apoptosis and rapid organism lethality, whereas reduction of either of the characterized Bcl-2 family members (Buffy, Dborg-2) has mild effects. Diap-1 is in turn bound and inhibited by the "H99" group of proteins – Reaper, Hid, and Grim – and other proteins such as Jafrac2, Sickle, and Morgue, in some cases by targeting an

N-terminal IAP-binding motif (IBM). Similar mechanisms appear to be at work in the mammalian IBM motif-containing proteins Smac/DIABLO and HtrA2/Omi, and *Drosophila* has proved a successful model for understanding how this class of proteins acts.

24.3.2.2.2 Neighboring Signals and Cell Death

The Myc complex, composed of Myc, Max, and Mnt family members, regulates cell growth and proliferation in numerous organisms. These are potent oncogenes. Mutations that reduce Myc function lead to small mice and flies, and Myc appears to regulate overall tissue growth. In a similar way to mammals, *Drosophila* Myc (dMyc) represents an important point of cell cycle regulation. For example, the Wg signaling pathway controls Myc expression in the developing wing, and Wingless (Wg) activity at the dorsal-ventral boundary leads to down-regulation of dMyc to create a discrete "zone of nonproliferating cells" (ZNC) at the boundary (e.g., [21]).

Drosophila cells that overexpress dMyc are at a competitive advantage; that is, they grow more quickly than their neighbors. Interestingly, cells neighboring these dMyc overexpressers grow more slowly than expected, indicating that dMyc mediates a signal between cells that keeps overall tissue size within its normal range. In a series of elegant experiments, the Johnston and Basler laboratories used clonal analysis to demonstrate how cells with reduced dMyc activity are cued to die by their wild type neighboring cells [18, 43, 57]. This death signal is mediated at least in part by the Wg and Decapentaplegic (Dpp) pathways, orthologs of mammalian Wnt and BMP signal transduction pathways, respectively. Cell competition and the signals traded between cells are a field of growing interest, holding out promise of improved understanding of some of the local signaling aspects between normal and transformed cells during oncogenesis. These local effects within the epithelium are likely to be better appreciated in mammals as tools improve to permit finer resolution of mammalian tumors.

24.3.2.3 Hyperplasia and Neoplasia

Similar to their mammalian counterparts, *Drosophila* tumor suppressor genes are typically partitioned into "hyperplastic" and "neoplastic." Hyperplastic genes are also known as "growth control" genes. They are thought to be key mediators of setting organ size, and loss of their activity can lead to large imaginal disc epithelia that otherwise develop reasonably normally. Hyperplastic loci include *hippo*, *salvador*, *warts*, *tsc1*, *tsc2*, *pten*, and *src*. Changing a tissue's size is not as simple a task as it might seem: for example, increasing cell proliferation typically leads to compensatory apoptosis, which brings tissue size back to normal. Mutations in hyperplastic loci have the distinct property that they can direct overproliferation and simultaneously block apoptosis. Neoplastic loci (*scribble*, *lgl*, *dlg*) yield a more dramatic overgrowth phenotype: cells lose their apical-basal polarity and masses of expanding mutant tissue are sometimes seen invading other regions. Unlike hyperplastic tissue, neoplastic tissue is poorly constructed.

24.3.2.3.1 Hyperplasia and Growth Control

The primary pathway that appears to mediate size regulation in developing epithelia is the "Hippo pathway" (Fig. 24.3.1) (reviewed in [34]). This pathway couples events at the cell surface – e.g., through the cell surface FERM family proteins Merlin and Expanded and the protocadherin Fat – to regulation of cell growth and cell death. Hippo and Lats/Warts encode Ste-20 and NDR-type serine-threonine kinases, whereas Salvador encodes

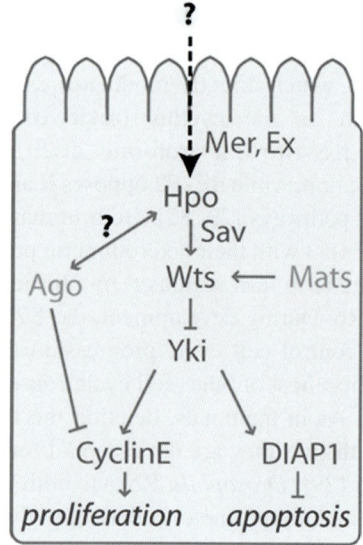

Fig. 24.3.1 The hippo signaling pathway

a scaffolding protein with a WW domain. The kinase activity of Lats/Warts is, in turn, regulated by Mats, an ortholog of the Mob superfamily of tumor suppressors. Reducing activity of any of these proteins leads to activation of the transcription factor Yorkie, which then directs expanded tissues by simultaneously increasing proliferation (via cyclin E) and blocking programmed cell death (via dIAP1). This remarkable ability to simultaneously regulate both aspects of tissue sizing suggests that the Hippo pathway represents a fundamental "ruler" that measures and adjusts tissue size. Critically, human cancer-derived cell lines have germline mutations in the locus encoding the Hippo ortholog *Mst2*, and mutations in murine and human *Lats1* lead to sarcomas and ovarian tumors (e.g., [38, 83]). Similar size-regulation properties have been described in *Drosophila* for the Ste-20 ortholog Slik [37], the F-box protein Archipeligo [55], and the micro-RNA Bantam [10], although the mechanisms by which they act are less clear.

24.3.2.3.2 Neoplasia and Cell Polarity

Neoplastic growth is characterized by uncontrolled proliferation and a failure of cells to enter or maintain terminal differentiation. As carcinomas (malignant epithelial tumors) mature, they are increasingly characterized by a loss of cell morphology (cell structure becomes more relaxed) and tissue polarity (cells pile up on one another). In recent years, *Drosophila* cancer biologists have established models that seek to understand the mechanisms that direct this transition.

Three proteins have taken center stage in *Drosophila* models of neoplasms: Lgl, Dlg, and Scribble. Reducing the activity of any of these proteins leads to massive overgrowth of imaginal disc epithelia and nervous system components (Fig. 24.3.2). These three proteins are associated with apical junctions, and Dlg and Scribble contain multiple protein-protein binding domains that indicate they act in part as molecular scaffolds. Dlg and Scribble act together – likely as a multi-protein complex – to help build and maintain proper apically based polarity [7]. They do so through complex interactions with other apical junction-associated proteins including dPatJ, Crumbs, Stardust, Par-6, Bazooka/Par-3, and Atypical PKC in addition to Lgl [6]. Loss of *lgl*, *dlg*, or *scribble* leads to (a) a loss of morphological and molecular markers of apical cell polarity; (b) a failure of cells to differentiate; and (c) intensive overgrowth.

What is especially fascinating about these loci is that they clearly point to an intimate association between epithelial architecture and growth control. Importantly, all three have mammalian orthologs, although the precise role played by these loci in human tumors is not yet understood.

Curiously, a group of fly proteins that regulate endocytosis also direct neoplastic growth when mutated. Four of these factors have been carefully studied in *Drosophila*: Vps25, Rab5, Tsg101 (also known as Vps23 or Erupted), and Avalanche (reviewed in [33]). Reducing the activity of any of these four factors results in overgrowth that is strikingly similar to mutations in *lgl*, *dlg*, or *scribble*, but their mechanism of action appears to be significantly different. Each of these four factors is a component of the endocytic machinery that regulates protein presence at the cell surface. Reducing activity of any of the four loci leads to a block in proper endocytic trafficking, leaving proteins such as signaling receptors (e.g., Notch, EGF Receptor) and polarity components (e.g., Crumbs, a protein that helps establish a cell's apical domain) stranded inappropriately at the cell surface. Depending on the cellular context, this can lead to overgrowth. Furthermore, reducing activity of Vps25 and Tsg101, components of the ESCRT complex that direct protein sorting from the early endosomes, can lead to overgrowth of surrounding nonmutant cells. These mutants point to the importance of regulating growth factor stability in controlling normal tissue growth and maturation.

24.3.2.4 Models of Metastasis

24.3.2.4.1 Ras

Ras signaling represents a canonical signal transduction pathway that is directly activated in perhaps 20–40% of all solid tumor types. The Ras pathway typically coordinates information from upstream signaling receptors, leading to changes that depend on cell context but include proliferation, migration, and differentiation. Recent work on *Drosophila* has extended previous data demonstrating that Ras can act cooperatively and potently with other factors to direct neoplastic overgrowth and even metastasis.

Strong overexpression of activated isoforms of the *Drosophila* Ras ortholog dRas1 leads to overgrowth, but

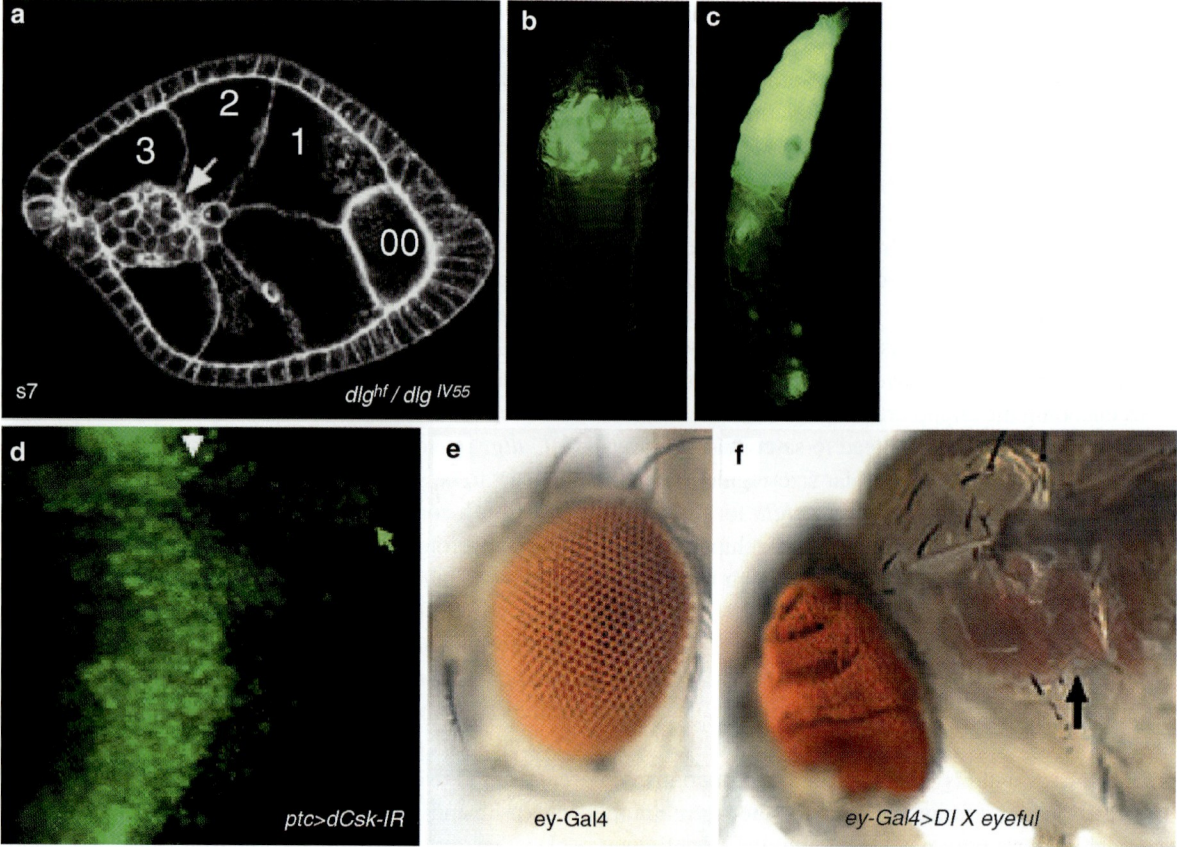

Fig. 24.3.2 (**a–f**) *Drosophila* cancer models. (**a**) dlg Mutant 'follicle cells' expand as a tumor into the egg chamber, a structure that houses and patterns the emerging oocyte. Cells at the anterior and posterior poles of the egg chamber are especially susceptible to transformation by *dlg*. From [30a] (**b**) Expressing the oncogenic Ras isoform dRas1^{G12V} led to low-level, localized overgrowth. (**c**) Reducing *scribble* function in the presence of dRas1^{G12V}, however, was sufficient to produce strongly overgrown tissue that expands away from the head region. Transformed tissue marked with EGFP. From Pagliarini et al. [60] (**d**) At the boundary of a region with reduced dCsk activity, Src activity increases and cells migrate away (*green arrow*) from the original region of the dCsk "tumor." The white arrow indicates the tumor edge; cells are marked with EGFP. (**e, f**) Coupling overexpression of the Notch ligand Delta with the epigenetic silencers Lola and Pipsqueak (which together constitute "eyeful") led to massive overgrowth of the eye (**f**) compared with controls (**e**) in addition to secondary growths of eye-like material (red material indicated by *arrows* in F)

also to compensatory apoptosis (see discussion of this phenomenon above). This triggering of cell death is both autonomous and nonautonomous (that is, neighboring wild type cells are also affected) and often leads to a tissue that is eventually reduced in size. Activated dRas1 (dRas1^{G12V}) strongly cooperated with members of the hyperplastic and neoplastic groups of tumor suppressors. The outcome for each is different in important ways, however. Pairing dRas1^{G12V} with a hyperplastic mutation in *lats* (see above) in the eye led to dramatically increased overgrowth, but the cells remained within the tissue [60]. By contrast, dRas1^{G12V} plus reducing function in *scribble* led to overgrowth and, in addition, degradation of the basement membrane and cell migration in a JNK-dependent manner (Fig. 24.3.2) [11, 40].

24.3.2.4.2 Src

Increase in Src activity has been linked to increased risk of metastasis in dozens of solid tumor types including breast, colorectal, melanomas, etc. While the precise role played by Src in metastasis is not well understood, recent work has linked changes in Src activity to changes in junctions, cell adhesion, cytoskeleton, and cell motility (reviewed in [41]).

24.3 Drosophila as a Model for Human Disease

Drosophila has two Src orthologs, dSrc42A and dSrc64B, and the Src-like Btk ortholog dSrc29 that is a likely downstream Src effector. Reducing activity of the Src repressor *Drosophila* C-terminal Src Kinase (*dCsk*) led to Src activation, tissue overgrowth, an arrest in larval development, and eventual death of the fly. Cells with reduced *dCsk* function showed increased overall proliferation, a block in cell death, and an increase in organ size, placing *dCsk* – and by extension Src – in the category of hyperplastic mutations described above. Indeed, in vitro and genetic studies suggest a link between *dCsk* and Lats.

While reducing dCsk activity throughout a tissue led to an increase in organ size, reducing *dCsk* activity in discrete patches of cells led to basal release of cells and migration away from the *dCsk* "tumor"; these cells eventually died by apoptosis (Fig. 24.3.2) [91]. Interestingly, only cells at the border of the *dcsk* patch showed this behavior, suggesting that interactions with neighboring wild type cells provoke migration. Further genetic studies suggested a pathway – triggered specifically in *dcsk* cells at the "tumor" border – that extends from Src through E-cadherin, P120-catenin, RhoA, JNK, metal metalloproteases, and caspases. While some evidence exists for a change in E-cadherin and P120-catenin at the boundaries of human squamous cell carcinomas [80], further work will be required to determine whether Csk/Src-mediated metastasis commonly utilizes this pathway in human tumors.

24.3.2.4.3 Notch

The Notch signal transduction pathway is broadly utilized during development. Notch is a transmembrane signaling receptor that is activated by Delta and signals through cleavage-dependent translocation of its intracellular domain to the nucleus, where it acts as a transcriptional activator [58]. In humans, activating mutations in Notch can lead to disease, including hematopoetic cancers such as acute lymphoblastic leukemia (e.g., [23]). Indeed, Notch itself is required for proper regulation of cell proliferation in *Drosophila*, and unregulated activation of Notch activity – for example by ectopic expression of its ligand Delta – can lead to uncontrolled proliferation of the larval wing.

Dominguez et al. [25] found a link between Notch and HDAC regulators Lola and Pipsqueak; the latter two act as epigenetic chromosome silencers by regulating HDAC activity. Expressing Delta, Lola, and Pipsqueak together led to a dramatic overgrowth of the eye and, remarkably, "eye tumors" in distant regions such as the abdomen (Fig. 24.3.2). This work provides a wonderful demonstration of how signal transduction pathways can combine with epigenetic elements to drive metastasis-like cell behavior. One surprise is that many of these distant "tumors" retained the ability to develop as eye tissue, suggesting they are not fully transformed or can "revert" once inserted into an epithelium (Fig. 24.3.2).

24.3.2.5 Models of Specific Cancers

While most cancers are polygenic, a few cancers progress through a single monogenic event. Not surprisingly, monogenic cancers have been the first to be examined in *Drosophila*, as the triggering events are easily duplicated.

24.3.2.5.1 Tuberous Sclerosis

In humans, mutations that reduce the activity of Tsc1 or Tsc2 lead to a benign tumor syndrome in which hamartomas emerge in the brain and other organs such as the heart and kidney (reviewed in [61]). Loss-of-function mutations in *Drosophila tsc1* or *tsc2* (originally known as *gigas*) led to an increase in cell size and cell growth due to increased expression of cyclins A, B, D, and E [42, 86], indicating that Tsc activity normally opposes cell proliferation. Work in both mammals and *Drosophila* has linked cell size effects to Tsc-2 GAP activity for the small GTPase Rheb, which in turn acts through the Tor pathway. Tor (target of rapamycin) activity regulates cell metabolism through multiple pathways, including generalized protein biosynthesis [61], and rapamycin is currently in clinical trials in tubular sclerosis patients [28]. Recently, the Tsc/Rheb/Tor pathway has been linked to PI3K signaling, a central mediator of growth response in epithelia. Tuberous sclerosis is an important example of the link between metabolism and oncogenesis, and research on other diseases will no doubt need to account for metabolic processes during the search for effective therapeutics.

24.3.2.5.2 Neurofibromatosis 1

Neurofibromatosis 1 (NF1) is one of the most common inherited nervous system cancer syndromes. Neurofibromas are benign tumors found in peripheral nerves, especially Schwann cells. In addition, *NF1* patients show a palate of problems that include skin pigmentation defects, hamartomas of the iris, optic pathway gliomas, and mental retardation [49]. In a minority of patients tumors advance to become metastatic and can prove fatal.

The *NF1* gene encodes a large protein that includes a functional RasGAP (Ras GTPase activating protein) domain. RasGAP domains stimulate Ras-GTP hydrolysis to inactivate Ras function, and at least some of NF1's activities can be assigned to its regulation of Ras function in mammalian cell culture tumorigenesis models [48]. Work in *Drosophila* provided some of the first evidence that NF1 can act on pathways independently of Ras: mutations in *Drosophila* NF1 lead to small body size that is dependent on cyclic AMP-dependent protein kinase A (PKA). These effects on both Ras and on PKA have also been identified in mammalian systems, leading to the question of whether this large and structurally complex protein acts in other pathways as well.

24.3.2.5.3 Multiple Endocrine Neoplasia Types 1 and 2

Multiple endocrine neoplasia type 1 (MEN1) and type 2 (MEN2) are both cancer syndromes that result in dispersed tumors with strongest representation in hormone-producing endocrine tissues. *MEN1* is an autosomal dominant cancer syndrome characterized by a palate of tumors that include parathyroid, anterior pituitary, and pancreatic islet tumors in addition to nonendocrine tumors such as angiofibroma. The *men1* locus encodes Menin, a large nuclear protein whose the function remains mysterious (reviewed in [98]). Some of the strongest in situ data for Menin's links to both cell cycle and genomic integrity have come from flies. Loss of *Drosophila* Menin (Mnn1) activity led to a viable adult that nonetheless shows poor ability to repair its DNA after ionizing irradiation-induced breaks or handle various physiological stresses and a failure of cells to arrest at the G1-S checkpoint after ionizing irradiation, an observation that was confirmed in mammalian cell lines [15, 62]. Further genetic screens have identified or confirmed functional interactions between *Drosophila* Mnn1 and components of the JNK/AP-1 pathway as well as Ches1, a forkhead/winged helix transcription factor linked to DNA checkpoint regulation. Based on the fly and mammalian data, a picture is beginning to emerge in which Menin is a tumor suppressor that functions as a co-factor for multiple transcription factors and regulators of chromatin integrity during DNA repair.

24.3.2.5.4 Multiple Endocrine Neoplasia 2

MEN2 patients typically have one of several activating mutations in the Ret receptor tyrosine kinase. Patients consistently display medullary thyroid carcinoma (MTC) and can also display pheochromocytomas, parathyroid adenomas, mucosal neuromas, etc. Targeting either *MEN2A*- or *MEN2B*-equivalent Ret mutations to the developing *Drosophila* eye led to a series of tissue defects that mimicked particular aspects of human *MEN2* tumors, including increased cell proliferation, compensatory apoptosis, and disruption of cell fates [63]. A genetic modifier screen identified 140 functionally linked loci; human orthologs of two modifier loci – TNIK and CHD3 – are located within deletions associated with patients displaying secondary pheochromocytomas, suggesting these loci are at least candidate biomarkers of susceptibility to adrenal tumors [63]. This demonstrates how fly genetics can help identify human susceptibility loci.

The anilinoquinazoline ZD6474 inhibited Ret activity in mammalian tumor cell lines. *Drosophila* provided further in situ evidence of its utility: feeding the compound to *MEN2* flies strongly suppressed the eye *MEN2* phenotypes with minimal toxicity to the animal (Fig. 24.3.3) [90]. This compound is being tested in phase III clinical trials to treat patients with medullary thyroid carcinoma (S. Wells, personal communication), pointing to the utility of using *Drosophila* both as a genetic tool and also as a whole-animal screen for therapeutic compounds.

24.3.3 Neurodegenerative Diseases

Neurodegenerative diseases are increasingly prevalent within the aging human population. They have proved especially difficult to treat owing to the complexity of

24.3 Drosophila as a Model for Human Disease

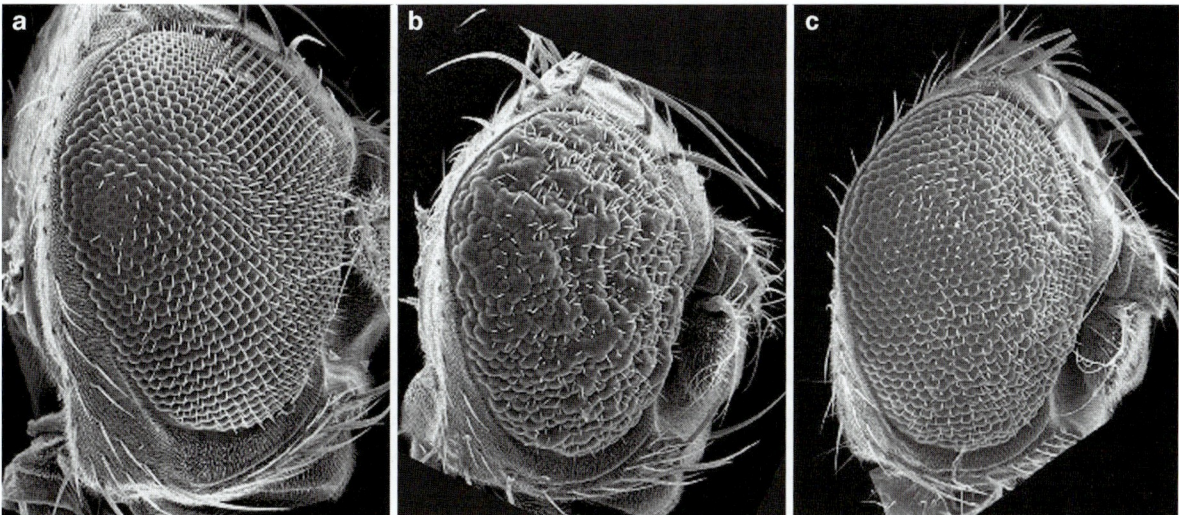

Fig. 24.3.3 Using Drosophila to screen for therapeutic compounds for multiple endocrine neoplasia (MEN). (**a**) Control flies display a smooth eye surface. (**b**) Expression of oncogenic Ret(MEN2) isoforms phenocopies several aspects of tumors including overproliferation, compensatory apoptosis, etc. (**c**) Feeding flies: the compound ZD6474 strongly suppressed Ret(MEN2)-mediated defects

repairing nervous system defects and also to our poor understanding of the biology behind the most common of these diseases. *Drosophila* can help in each of these regards; indeed, some of the first fly models developed for the study of specific human diseases were generated to model neurodegenerative diseases. These include models for Parkinson's disease (PD), Alzheimer's disease (AD) and several triplet-repeat diseases. In humans these late-onset neurodegenerative diseases are often accompanied by the accumulation of aberrant protein deposits: Lewy bodies (PD), insoluble β-amyloid plaques and neurofibrillary tangles (AD), or nuclear inclusions/aggregates of polyQ and other repeat proteins. The relationship between these protein aggregations and neural degeneration is unclear.

24.3.3.1 Parkinson's Disease

Parkinson's disease is characterized by tremors and progressive loss of muscle function owing primarily to loss of dopaminergic neurons within the substantia nigra. At least six genetic loci linked to familial PD in humans have been cloned [24, 95]: *parkin, dj-1, pink1, α-synuclein, leucine rich repeat kinase 2 (LRRK2)*, and *UCH-L1*. In *Drosophila*, mutations in orthologs of many of these loci also result in nervous system defects.

Mutations in *pink1* lead to aberrant mitochondrial morphology and function and reduced ATP levels, male sterility, apoptotic muscle degeneration, and disorganized muscle fibers. These mitochondrial and sterility defects are rescued by ectopic expression of the E3 ubiquitin ligase Parkin. Genetic studies have linked Pink/Parkin to JNK signaling and to *glutathione S-transferase S1*, which may function in the cellular response to oxidation. Also linked to oxidative stress were *Drosophila dj-1*. While *Drosophila* does not have a clear α-Synuclein ortholog, expression of human α-Synuclein in flies led to neuronal degeneration and α-Synuclein positive aggregates a phenotype rescued by overexpression of Parkin or feeding the HSP90-associated chemical geldanamycin.

24.3.3.2 Alzheimer's Disease

Insoluble β-amyloid – in Aβ plaques and neurofibrillary tangles – is commonly found in patients with AD, a late-onset disease also characterized by neural degeneration. The Aβ precursor protein (APP) has been conserved in flies and is encoded by the *Appl* locus, but no endogenous Aβ has been detected in flies. However, expression of a human *Aβ* "minigene" in flies led to neural degeneration, and modifier screens using this disease model yielded a gain of function allele of *neprilysin*, which is able to degrade Aβ and suppress its ability to provoke neuronal cell death [26].

Mammalian presenilins function as essential components of the γ-secretase complex that generates Aβ; mutations in the human *PS1* and *PS2* genes are associated with familial AD [51, 77]. Experiments in *Drosophila* indicated that endogenous presinilin has several processing targets including the transmembrane receptor Notch that is cleaved during signaling activation [84, 99], observations confirmed in mammalian systems. This important observation raises the concern that therapeutic strategies designed to target γ-secretase function would adversely affect Notch signaling, a pathway fundamental to the development and maintenance of many tissues. Expression of another component of the AD-associated tangles, the microtubule-associated protein Tau, also led to neurodegeneration (reviewed in [8, 69]. Fly experiments confirmed that Tau is phosphorylated in vivo by the serine-threonine kinases GSK-3β, PAR-1 and cdk5; cdk5 and GSK-3β associate with neurofibrillary tangles in the vertebrate brain.

24.3.3.3 Triplet-Repeat Diseases

Triplet nucleotide repeat diseases are associated with the further expansion of a domain composed of a tandem string of three amino acids. For example, CAG (glutamine) repeats in the *Huntingtin* locus can expand to 39 or more glutamines in HD [39]. These expanded polyQ-containing proteins form aggregates and nuclear inclusions. In *Drosophila* ectopic expression of either polyQ-containing proteins or even a simple expanded polyQ peptide alone leads to extensive and progressive neural degeneration mirroring the fatal loss of neuronal tissue seen in human patients with different PolyQ diseases (Fig. 24.3.4) [69]. Such experiments suggest that neural toxicity may for the most part be attributed to the polyQ repeat itself and that the mechanism of neural atrophy is likely to be common to the family of triple repeat diseases.

The function of wild type Huntingtin protein (Htt) is largely unknown [39]. In vertebrate systems, expanded Htt isoforms have been detected in nuclear aggregates that localize with the transcriptional co-activator CBP, a histone acetyltransferase. This role for mutant polyQ proteins in transcriptional dysregulation has been confirmed in several independent studies in *Drosophila*. Other molecules shown to modify poly-Q repeat peptides in flies include molecules associated with ubiquitination, chaperone proteins, components of phosphatidylinositol 3-kinase/AKT signaling, ataxin1, and the ATPase VCP. This diverse list illustrates the involvement of many cellular components in mediating or regulating the toxicity of polyQ proteins. A clearer perspective of the biology of triplet-repeat disease biology is just beginning to emerge in flies, representing an exciting new tool in the study of these devastating disease syndromes and demonstrating the power of fly genetics to address complex diseases.

24.3.4 Heart Disease

Heart failure is the most common source of death in Western societies. Although diet and stress can lead to cardiac damage, it is also clear that heart function progressively degrades through simple aging. The prevalence of heart disease has sparked increasing interest in the *Drosophila* community. Although these are early days in studies of heart dysfunction in *Drosophila*, some interesting parallels have already been noted. *Drosophila* has a "single-chambered" heart, really a specialized thickening of the endothelium that consists of just two cell types: the contractile myocardial cells are surrounded by pericardial cells (Fig. 24.3.5). The fly has an open circulatory system (no veins or arteries), and its heart serves to promote blood ("hemolymph") circulation.

Despite these basic differences, the early molecules that define the heart region and direct early heart development are markedly conserved. Another similarity with humans is that the adult fly heart shows an increase in spontaneous arrhythmias and progressively poor response to cardiac stress as it ages [93, 94]. Classic genetic approaches demonstrated that these aspects of cardiac aging could be slowed or suppressed by reducing insulin pathway function, including the insulin receptor and the downstream effector Foxo as well as the downstream metabolic pathway regulator target of rapamycin (TOR [53]). The tools for examining issues of cardiac aging and diseases are advancing rapidly, and though the field is young the study of cardiac dysfunction in *Drosophila* should provide important insights in the coming years.

24.3 Drosophila as a Model for Human Disease

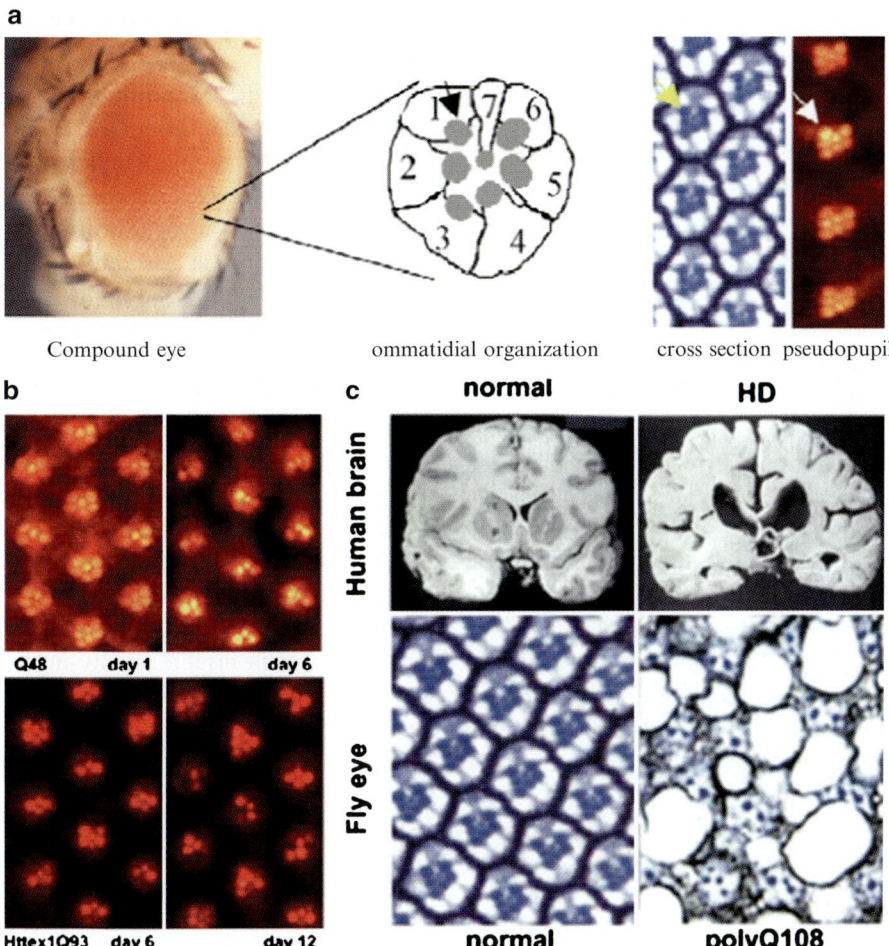

Fig. 24.3.4 (a) The Drosophila eye as a model for neurodegeneration. Structure of the adult Drosophila eye, showing the external eye, a diagram of the structure of the photoreceptor cells in an ommatidiuma, a section of the adult eye showing ommatidia in cross-section or using the pseudopupil technique. Adapted from [52a] (b) The rhabdomeres of flies at different ages expressing a pure polyQ peptide (Q48) and expressing a mutant exon1 fragment of a human Htt gene with 93Qs (Httex1Q93). Note the rhabdomere constellations get progressively worse. (c) Cross-sections through a normal and postmortem HD patient brain demonstrate the dramatic degeneration and loss of neuronal tissue. Cross-sections through the eye of a fly expressing polyQ108 in the photoreceptor neurons show similar significant loss of neuronal tissue

24.3.5 Diabetes and Metabolic Diseases

Drosophila has made important contributions to the study of cellular and organismal metabolic regulation, and interest in disease-related issues such as diabetes is rising. The majority of advances in *Drosophila* to date have emerged from studies of organ and cell size. As in mammals, the insulin signaling pathway plays the primary role in size regulation. Work primarily in yeast, *C. elegans*, flies, and mammals has outlined a basic pathway (Fig. 24.3.6). With the

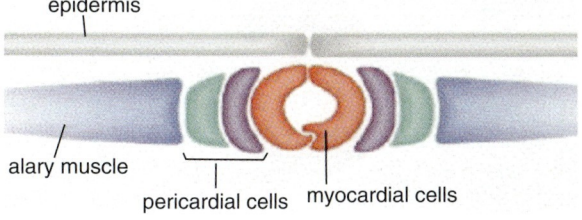

Fig. 24.3.5 (a) Drosophila has a single-chamber heart composed of pericardial and myocardial cells. From [5] (b) M-Mode traces prepared from high-speed movies of dissected flies with exposed hearts at 1 week (*top*) and 5 weeks of age (*bottom*). Hearts were recorded for 10 s. As in humans, the incidence and severity of arrhythmicity increases with age. From [59]

exception of seven insulin-like peptides (Dilp1-7), *Drosophila* contains a single ortholog for each step in this growth-control cascade. This parsimony has simplified the identification and ordering of pathway components.

24.3.5.1 Body Size

The developing *Drosophila* larva needs to gain sufficient weight to advance to the pupal stage while precisely matching the size of each of its organs to make a viable adult. Figure 24.3.4 demonstrates how starvation leads to smaller body size, but overexpression of *Drosophila* insulin-like peptides (see below) can direct larger body size. Insulin itself appears to achieve this growth regulation through its inhibition of the TSC/TOR pathway, a growth pathway known to respond in flies and mammals to amino acids (Fig. 24.3.6) [17, 30]. This inhibition is mediated through the amino acid transporter Slimfast (Fig. 24.3.6) [30], an effector of insulin signaling provided by the "fat body" that serves as the fly's liver and storage of adipose tissue.

24.3.5.2 Models of Diabetes

In addition to studies of nutrition and aging, *Drosophila* is an emerging model system for the study of diabetes and metabolic syndromes. In *Drosophila*, the functional equivalent to beta cells are 14 "insulin-producing cells" (IPCs), which are neurosecretory cells located in the brain that secrete at least three Dilps into the circulatory system ("aorta") and the corpora cardiaca, the source of the glucagon-related hormone adipokinetic hormone. These Dilps carry out the functions of both insulin and IGF1/2, and ablation of the IPCs leads to an interesting model of type I diabetes characterized by hyperglycemia and small body size [67].

Drosophila metabolic controls may differ in some respects from their mammalian counterparts. In particular, ablation of IPCs leads to severe hypoglycemia and lethality, leading to the suggestion [46] that glucagon plays the more central role in regulating carbohydrate metabolism, with insulin activity involved in glucose tolerance. Similar conclusions have been reached based on pancreatectomy of birds, such as geese [45, 78], but the relative importance of glucagons in mammals is less clear. Further, the sulfonylurea receptor subunits of

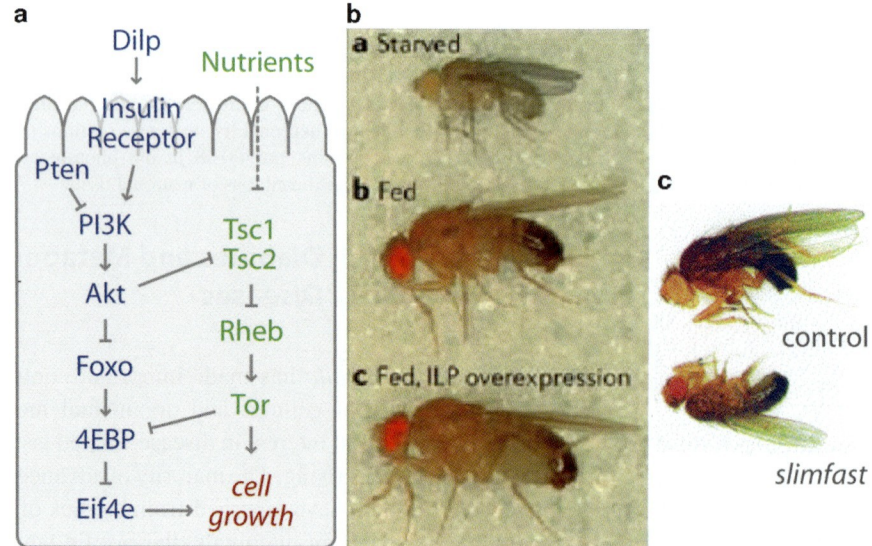

Fig. 24.3.6 (**a–c**) Insulin signaling. (**a**) The insulin and TOR pathways. (**b**) Starvation leads to flies with a smaller body size (*top panel*) compared with controls (*central panel*); overexpression of insulin-like peptides increases body size (*bottom*). (**c**) This reduction in body size is also seen in flies with reduced activity of the amino acid transporter Slimfast

the ATP-sensitive potassium channels – points of insulin regulation in mammalian beta cells – are expressed in the corpora cardiaca in flies [46]. This suggests that glucose acts to regulate itself through glucagon release, although a role has not been ruled out for the IPCs as well. Furthermore, cross-talk between the glucagon and insulin pathways can complicate interpretations of experiments in flies, birds, and possibly mammals.

24.3.6 Addiction

Addiction is a disease of enormous societal and economic cost. Many still equate addiction with a lack of willpower or a weakness of character; addiction research in model organisms such as flies plays a central role not only in elucidating the physiological mechanisms that underlie addiction, but also in overturning such attitudes. *Drosophila* has been used to investigate addiction to alcohol, cocaine, and nicotine [65, 96]. The effects that these drugs have in flies, rodent models and humans are remarkably similar.

In humans, exposure to low doses of alcohol leads to euphoria and loss of inhibition. Increasing doses result in decreased coordination, confusion, and sedation. Similarly, when exposed to ethanol vapor *Drosophila* become hyperactive, then lose coordination and finally become sedated. This behavior translates into a decreased ability of flies to balance on a slanted surface and is used in the design of the somewhat amusingly named "inebriometer" test chamber (Fig. 24.3.7a). The "crackometer" Fig. 24.3.7b) is used to expose flies to free-base cocaine; low doses lead to hyperactivity, moderate doses cause hypokinesis and stereotypic locomotion, and higher doses lead to spasmodic activity, tremors, and finally complete loss of movement (Fig. 24.3.7c). Exposure to volatilized nicotine produces similar hyperactivity, hypokinesis, or akinesia.

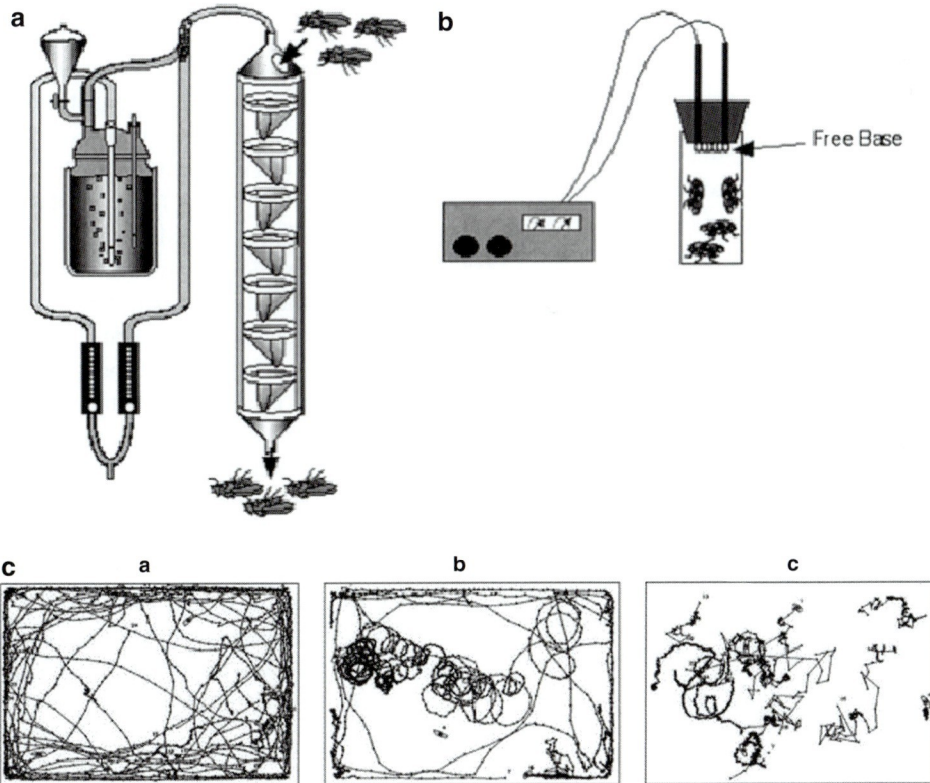

Fig. 24.3.7 (**a–c**) Assaying intoxication in flies. (**a**) The inebriometer. Ethanol vapor is blown on flies at the entrance to the funnel; the time required for flies to fall through the baffles is then assessed. (**b**) The crackometer. Cocaine is volatized and administered to flies. (**c**) Computer-generated traces of the locomotor behavior of a group of five flies exposed to volatilized free-base cocaine. Each panel corresponds to a 1-min period starting 2 min after the end of the cocaine exposure. Ca Mock exposure; Cb exposure to 100 μg of cocaine; Cc exposure to 200 μg of cocaine. Flies have become hyperactive but unresponsive. Adapted from [65, 96]

24.3.6.1 The Genetics of Addiction: Sensitivity and Tolerance

Increasingly sophisticated behavioral tests are being developed to assess the complex movements and behaviors of flies to addictive substances. With these assays, Drosophila provided the first in vivo evidence that cAMP signaling mediates the body's response to ethanol. Mediators involved include the neuropeptide amnesiac (christened "Cheapdate") that acts through adenylate cyclase to increase cAMP levels, adenylate cyclase (Rutabaga), and protein kinase A (PKA). Flies were similarly used to confirm the role of dopamine and serotonin in intoxication responses and, recently, to demonstrate a role for Rho-type GTPases in regulating sensitivity to alcohol intoxication as well as to nicotine and cocaine [66]. These data support a model in which drugs such as alcohol directly affect organization of the cytoskeleton, a structure known to be regulated by small GTPases [32].

Tolerance to narcotics and alcohol is an additional factor contributing to the potency and development of addiction. In flies, adaptation is evident after just one acute exposure to ethanol, termed rapid tolerance [3, 72, 89]. By contrast, prolonged or repeated exposure to small doses of ethanol leads to chronic tolerance: higher levels of alcohol are required to cause intoxication on each consecutive exposure, a hallmark of neuronal habituation. Targeting mutants to particular brain areas permitted mapping of the development of tolerance to a set of neurons within the fly brain's "central complex." Octopamine, the *Drosophila* noradrenaline analog, mediated this tolerance, demonstrating a role similar to noradrenaline in mice. Tolerance also required the nuclear zinc-finger protein *hangover* [73] that mediates a cellular stress response induced on exposure to ethanol. In each case, the majority of mutant fly lines abnormal for ethanol sensitivity or tolerance do not display changes in the rate at which alcohol is absorbed or metabolized [64], indicating that, for the most part, ethanol sensitivity and tolerance are separable from ethanol metabolism.

24.3.7 Sleep Disorders

Though a third of our lifetime is spent in sleep, we understand astonishingly little about why we sleep or what the function of sleep is at the cellular level. Sleep appears to be a common feature of all brain-endowed animals, and flies are proving a useful model system (for reviews see, e.g., [16, 36, 75]). Several drugs that reduce the requirement for sleep in humans, such as caffeine, modafinil, metamphetamine, and antihistamines, also affect the sleep patterns of *Drosophila*. Increasing the release of dopamine and inhibiting its re-uptake stimulates arousal in both humans and flies: for example, feeding flies metamphetamine reduced the frequency and duration of sleep; inhibiting dopamine synthesis led to narcoleptic-like behavior [1]. This is mediated through the voltage-dependent potassium channel Shaker and the dopamine transporter Fumin. Similar to us, sleep-deprived flies tend to sleep more the next day and show reduced performance in several behavior tests. Furthermore, complete sleep deprivation is fatal in flies (which die within 60 h), rats (which survive for 2–3 weeks without sleep), and humans (patients with fatal familial insomnia die 7–36 months after they first present with symptoms) [52, 76]. As we age our need for sleep is reduced and sleep becomes increasingly fragmented, and aging flies display these same sleep pattern changes [48].

Recent work in flies has provided intriguing insights into sleep disruption. The Drosophila Activity Monitoring System (DAMS) instrument enables researchers to simultaneously monitor the activity and sleep patterns of a large number of individual flies; activity is monitored by an infrared light beam bisecting the tube. Devices such as DAMS have allowed researchers to demonstrate roles for homeostatic and circadian mechanisms in regulating sleep in flies. For example, when fed a low dose of the free-radical generator paraquat, aging flies displayed an enhancement of sleep disruption, suggesting a role for oxidative stress in sleep fragmentation; similar effects have been observed in rats [48]. Other factors include serotonin, PKA and cAMP and, interestingly, diet [81].

Finally, recent *Drosophila* work suggests a method to detect biomarkers of sleepiness in humans [74]: researchers found that the levels of amylase mRNA isolated from whole fly heads increased with progressive sleep deprivation, and went on to show an analogous increase in amylase mRNA levels in saliva from sleep-deprived human subjects.

24.3.8 Conclusions

Drosophila offers a host of advantages for studying human disease. A century of study has yielded a dizzying array of tools, giving the fly field perhaps the most powerful, most sophisticated toolset currently available. Although initially focused on development, the shift to human disease reflects both new demands by the public and the natural maturation of the fly field. Once considered pioneers and even outsiders, fly researchers who focus on disease are now considered an important part of the future of fly research. As with every book chapter, some of the data presented here are no doubt out of date, as fly researchers tend to move quickly. Despite their myriad advantages, perhaps the true future of *Drosophila* lies in close cooperation with researchers in mammalian fields. By informing the *Drosophila* community of the details of human diseases, fly researchers can in turn focus on creating more sophisticated models. This is particularly true of polygenic diseases, in which *Drosophila* offers one of the best opportunities to translate the accumulation of complex human mutation data into useful models by altering multiple genes and signaling networks in a manner that few other complex model organisms can match.

References

1. Andretic R, van Swinderen B, Greenspan RJ (2005) Dopaminergic modulation of arousal in Drosophila. Curr Biol 15(13):1165–1175
2. Adams MD, Celniker SE, Holt RA et al (2000) The genome sequence of Drosophila melanogaster. Science 287(5461): 2185–2195
3. Berger KH, Heberlein U, Moore MS (2004) Rapid and chronic: two distinct forms of ethanol tolerance in Drosophila. Alcohol Clin Exp Res 28(10):1469–1480
4. Bier E (2005) Drosophila, the golden bug, emerges as a tool for human genetics. Nat Rev Genet 6(1):9–23
5. Bier E, Bodmer R (2004) Drosophila, an emerging model for cardiac disease. Gene 342(1):1–11
6. Bilder D (2004) Epithelial polarity and proliferation control: links from the Drosophila neoplastic tumor suppressors. Genes Dev 18(16):1909–1925
7. Bilder D, Li M, Perrimon N (2000) Cooperative regulation of cell polarity and growth by Drosophila tumor suppressors. Science 289(5476):113–116
8. Bilen J, Bonini NM (2005) Drosophila as a model for human neurodegenerative disease. Annu Rev Genet 39:153–171
9. Bjorklund M, Taipale M, Varjosalo M et al (2006) Identification of pathways regulating cell size and cell-cycle progression by RNAi. Nature 439(7079):1009–1013
10. Brennecke J, Hipfner DR, Stark A, Russell RB, Cohen SM (2003) bantam encodes a developmentally regulated microRNA that controls cell proliferation and regulates the proapoptotic gene hid in Drosophila. Cell 113(1):25–36
11. Brumby AM, Richardson HE (2003) scribble mutants cooperate with oncogenic Ras or Notch to cause neoplastic overgrowth in Drosophila. Embo J 22(21):5769–5779
12. Brumby AM, Richardson HE (2005) Using Drosophila melanogaster to map human cancer pathways. Nat Rev Cancer 5(8):626–639
13. Bryant PJ, Schubiger G (1971) Giant and duplicated imaginal discs in a new lethal mutant of Drosophila melanogaster. Dev Biol 24(2):233–263
14. Bryant PJ, Levinson P (1985) Intrinsic growth control in the imaginal primordia of Drosophila, and the autonomous action of a lethal mutation causing overgrowth. Dev Biol 107(2):355–363
15. Busygina V, Kottemann MC, Scott KL, Plon SE, Bale AE (2004) Hypermutability in a Drosophila model for multiple endocrine neoplasia type 1. Hum Mol Genet 13(20): 2399–2408
16. Cirelli C, Bushey D (2008) Sleep and wakefulness in Drosophila melanogaster. Ann NY Acad Sci 1129:323–329
17. Colombani J, Raisin S, Pantalacci S et al (2003) A nutrient sensor mechanism controls Drosophila growth. Cell 114(6):739–749
18. de la Cova C, Abril M, Bellosta P, Gallant P, Johnston LA (2004) Drosophila myc regulates organ size by inducing cell competition. Cell 117(1):107–116
19. Dimova DK, Stevaux O, Frolov MV, Dyson NJ (2003) Cell cycle-dependent and cell cycle-independent control of transcription by the Drosophila E2F/RB pathway. Genes Dev 17(18):2308–2320
20. Doroquez DB, Rebay I (2006) Signal integration during development: mechanisms of EGFR and Notch pathway function and cross-talk. Crit Rev Biochem Mol Biol 41(6):339–385
21. Duman-Scheel M, Johnston LA, Du W (2004) Repression of dMyc expression by Wingless promotes Rbf-induced G1 arrest in the presumptive Drosophila wing margin. Proc Natl Acad Sci USA 101(11):3857–3862
22. Edgar BA, Britton J, de la Cruz AF et al (2001) Pattern- and growth-linked cell cycles in Drosophila developmen. Novartis Found Symp 237:3–12 discussion 12-8
23. Ellisen LW, Bird J, West DC et al (1991) TAN-1, the human homolog of the Drosophila notch gene, is broken by chromosomal translocations in T lymphoblastic neoplasms. Cell 66(4):649–661
24. Feany MB, Bender WW (2000) A Drosophila model of Parkinson's disease. Nature 404(6776):394–398
25. Ferres-Marco D, Gutierrez-Garcia I, Vallejo DM et al (2006) Epigenetic silencers and Notch collaborate to promote malignant tumours by Rb silencing. Nature 439(7075):430–436
26. Finelli A, Kelkar A, Song HJ, Yang H, Konsolaki M (2004) A model for studying Alzheimer's Abeta42-induced toxicity in Drosophila melanogaster. Mol Cell Neurosci 26(3): 365–375
27. Firth LC, Baker NE (2005) Extracellular signals responsible for spatially regulated proliferation in the differentiating Drosophila eye. Dev Cell 8(4):541–551
28. Franz DN, Leonard J, Tudor C et al (2006) Rapamycin causes regression of astrocytomas in tuberous sclerosis complex. Ann Neurol 59(3):490–498
29. Frolov MV, Huen DS, Stevaux O et al (2001) Functional antagonism between E2F family members. Genes Dev 15(16):2146–2160
30. Gao X, Zhang Y, Arrazola P et al (2002) Tsc tumour suppressor proteins antagonize amino-acid-TOR signalling. Nat Cell Biol 4(9):699–704

31. Genovese C, Trani D, Caputi M, Claudio PP (2006) Cell cycle control and beyond: emerging roles for the retinoblastoma gene family. Oncogene 25(38):5201–5209
32. Hall A (2005) Rho GTPases and the control of cell behaviour. Biochem Soc Trans 33(Pt 5):891–895
33. Hariharan IK, Bilder D (2006) Regulation of imaginal disc growth by tumor-suppressor genes in Drosophila. Annu Rev Genet 40:335–361
34. Harvey K, Tapon N (2007) The Salvador-Warts-Hippo pathway – an emerging tumour-suppressor network. Nat Rev Cancer 7(3):182–191
35. Hay BA, Guo M (2006) Caspase-dependent cell death in Drosophila. Annu Rev Cell Dev Biol 22:623–650
36. Hendricks JC (2003) Invited review: sleeping flies don't lie: the use of Drosophila melanogaster to study sleep and circadian rhythms. J Appl Physiol 94:1660–1672 discussion 1673
37. Hipfner DR, Cohen SM (2003) The Drosophila sterile-20 kinase slik controls cell proliferation and apoptosis during imaginal disc development. PLoS Biol 1(2):E35
38. Hisaoka M, Tanaka A, Hashimoto H (2002) Molecular alterations of h-warts/LATS1 tumor suppressor in human soft tissue sarcoma. Lab Invest 82(10):1427–1435
39. Huntington's Disease Collaborative Research Group (1993) A novel gene containing a trinucleotide repeat that is expanded and unstable on Huntington's disease chromosomes. Cell 72(6):971–983
40. Igaki T, Pagliarini RA, Xu T (2006) Loss of cell polarity drives tumor growth and invasion through JNK activation in Drosophila. Curr Biol 16(11):1139–1146
41. Ishizawar R, Parsons SJ (2004) c-Src and cooperating partners in human cancer. Cancer Cell 6(3):209–214
42. Ito N, Rubin GM (1999) gigas, a Drosophila homolog of tuberous sclerosis gene product-2, regulates the cell cycle. Cell 96(4):529–539
43. Johnston LA, Prober DA, Edgar BA, Eisenman RN, Gallant P (1999) Drosophila myc regulates cellular growth during development. Cell 98(6):779–790
44. Kango-Singh M, Halder G (2004) Drosophila as an emerging model to study metastasis. Genome Biol 5(4):216
45. Karmann H, Mialhe P (1976) Glucose, insulin and glucagon in the diabetic goose. Horm Metab Res 8(6):419–426
46. Kim SK, Rulifson EJ (2004) Conserved mechanisms of glucose sensing and regulation by Drosophila corpora cardiaca cells. Nature 431(7006):316–320
47. Knust E (2002) Regulation of epithelial cell shape and polarity by cell-cell adhesion (Review). Mol Membr Biol 19(2):113–120
48. Koh K, Evans JM, Hendricks JC, Sehgal A (2006) A Drosophila model for age-associated changes in sleep:wake cycles. Proc Natl Acad Sci USA 103(37):13843–13847
49. Lee MJ, Stephenson DA (2007) Recent developments in neurofibromatosis type 1. Curr Opin Neurol 20(2):135–141
50. Lee T, Luo L (2001) Mosaic analysis with a repressible cell marker (MARCM) for Drosophila neural development. Trends Neurosci 24(5):251–254
51. Levy-Lahad E, Wasco W, Poorkaj P et al (1995) Candidate gene for the chromosome 1 familial Alzheimer's disease locus. Science 269(5226):973–977
52. Lugaresi E, Medori R, Montagna P et al (1986) Fatal familial insomnia and dysautonomia with selective degeneration of thalamic nuclei. N Engl J Med 315(16):997–1003
53. Luong N, Davies CR, Wessells RJ et al (2006) Activated FOXO-mediated insulin resistance is blocked by reduction of TOR activity. Cell Metab 4(2):133–142
54. McGuire SE, Roman G, Davis RL (2004) Gene expression systems in Drosophila: a synthesis of time and space. Trends Genet 20(8):384–391
55. Moberg KH, Bell DW, Wahrer DC, Haber DA, Hariharan IK (2001) Archipelago regulates Cyclin E levels in Drosophila and is mutated in human cancer cell lines. Nature 413(6853):311–316
56. Moon NS, Di Stefano L, Dyson N (2006) A gradient of epidermal growth factor receptor signaling determines the sensitivity of rbf1 mutant cells to E2F-dependent apoptosis. Mol Cell Biol 26(20):7601–7615
57. Moreno E, Basler K (2004) dMyc transforms cells into super-competitors. Cell 117(1):117–129
58. Mumm JS, Kopan R (2000) Notch signaling: from the outside in. Dev Biol 228(2):151–165
59. Ocorr K, Akasaka T, Bodmer R (2007) Age-related cardiac disease model of *Drosophila*. Mech Ageing Dev 128(1):112–116
60. Pagliarini RA, Xu T (2003) A genetic screen in Drosophila for metastatic behavior. Science 302(5648):1227–1231
61. Pan D, Dong J, Zhang Y, Gao X (2004) Tuberous sclerosis complex: from Drosophila to human disease. Trends Cell Biol 14(2):78–85
62. Papaconstantinou M, Wu Y, Pretorius HN et al (2005) Menin is a regulator of the stress response in Drosophila melanogaster. Mol Cell Biol 25(22):9960–9972
63. Read RD, Goodfellow PJ, Mardis ER et al (2005) A Drosophila model of multiple endocrine neoplasia type 2. Genetics 171(3):1057–1081
64. Rodan AR, Kiger JA Jr, Heberlein U (2002) Functional dissection of neuroanatomical loci regulating ethanol sensitivity in Drosophila. J Neurosci 22(21):9490–9501
65. Rothenfluh A, Heberlein U (2002) Drugs, flies, and videotape: the effects of ethanol and cocaine on Drosophila locomotion. Curr Opin Neurobiol 12(6):639–645
66. Rothenfluh A, Threlkeld RJ, Bainton RJ et al (2006) Distinct behavioral responses to ethanol are regulated by alternate RhoGAP18B isoforms. Cell 127(1):199–211
67. Rulifson EJ, Kim SK, Nusse R (2002) Ablation of insulin-producing neurons in flies: growth and diabetic phenotypes. Science 296(5570):1118–1120
68. Ryder E, Russell S (2003) Transposable elements as tools for genomics and genetics in Drosophila. Brief Funct Genomic Proteomic 2(1):57–71
69. Sang TK, Jackson GR (2005) Drosophila models of neurodegenerative disease. NeuroRx 2(3):438–446
70. Schimmer AD, Dalili S, Batey RA, Riedl SJ (2006) Targeting XIAP for the treatment of malignancy. Cell Death Differ 13(2):179–188
71. Schmeichel KL (2004) A fly's eye view of tumor progression and metastasis. Breast Cancer Res 6(2):82–83
72. Scholz H, Ramond J, Singh CM, Heberlein U (2000) Functional ethanol tolerance in Drosophila. Neuron 28(1):261–271
73. Scholz H, Franz M, Heberlein U (2005) The hangover gene defines a stress pathway required for ethanol tolerance development. Nature 436(7052):845–847
74. Seugnet L, Boero J, Gottschalk L, Duntley SP, Shaw PJ (2006) Identification of a biomarker for sleep drive in

flies and humans. Proc Natl Acad Sci USA 103(52): 19913–19918
75. Shaw P (2003) Awakening to the behavioral analysis of sleep in Drosophila. J Biol Rhythms 18(1):4–11
76. Shaw PJ, Tononi G, Greenspan RJ, Robinson DF (2002) Stress response genes protect against lethal effects of sleep deprivation in Drosophila. Nature 417(6886):287–291
77. Sherrington R, Rogaev EI, Liang Y et al (1995) Cloning of a gene bearing missense mutations in early-onset familial Alzheimer's disease. Nature 375(6534):754–760
78. Sitbon G, Khemiss F, Boulanger Y (1982) Effects of total pancreatectomy and amino-acid treatment on plasma amino-acids and glucose in the goose. J Physiol (Paris) 78(3):258–265
79. Song YH (2005) Drosophila melanogaster: a model for the study of DNA damage checkpoint response. Mol Cells 19(2):167–179
80. Soubry A, van Hengel J, Parthoens E et al (2005) Expression and nuclear location of the transcriptional repressor Kaiso is regulated by the tumor microenvironment. Cancer Res 65(6):2224–2233
81. Spiegel K, Knutson K, Leproult R, Tasali E, Van Cauter E (2005) Sleep loss: a novel risk factor for insulin resistance and Type 2 diabetes. J Appl Physiol 99(5):2008–2019
82. Stevaux O, Dimova D, Frolov MV et al (2002) Distinct mechanisms of E2F regulation by Drosophila RBF1 and RBF2. Embo J 21(18):4927–4937
83. St John MA, Tao W, Fei X et al (1999) Mice deficient of Lats1 develop soft-tissue sarcomas, ovarian tumours and pituitary dysfunction. Nat Genet 21(2):182–186
84. Struhl G, Greenwald I (1999) Presenilin is required for activity and nuclear access of Notch in Drosophila. Nature 398(6727):522–525
85. Tapon N (2003) Modeling transformation and metastasis in Drosophila. Cancer Cell 4(5):333–335
86. Tapon N, Ito N, Dickson BJ, Treisman JE, Hariharan IK (2001) The Drosophila tuberous sclerosis complex gene homologs restrict cell growth and cell proliferation. Cell 105(3):345–355
87. Tepass U (2002) Adherens junctions: new insight into assembly, modulation and function. Bioessays 24(8):690–695
88. Theodosiou NA, Xu T (1998) Use of FLP/FRT system to study Drosophila development. Methods 14(4):355–365
89. Vidal M, Cagan RL (2006) Drosophila models for cancer research. Curr Opin Genet Dev 16(1):10–16
90. Vidal M, Wells S, Ryan A, Cagan R (2005) ZD6474 suppresses oncogenic RET isoforms in a Drosophila model for type 2 multiple endocrine neoplasia syndromes and papillary thyroid carcinoma. Cancer Res 65(9):3538–3541
91. Vidal M, Larson DE, Cagan RL (2006) Csk-deficient boundary cells are eliminated from normal Drosophila epithelia by exclusion, migration, and apoptosis. Dev Cell 10(1):33–44
92. Wang ZB, Liu YQ, Cui YF (2005) Pathways to caspase activation. Cell Biol Int 29(7):489–496
93. Wessells RJ, Bodmer R (2004) Screening assays for heart function mutants in Drosophila. Biotechniques 37(1):58–60 62, 64 passim
94. Wessells RJ, Fitzgerald E, Cypser JR, Tatar M, Bodmer R (2004) Insulin regulation of heart function in aging fruit flies. Nat Genet 36(12):1275–1281
95. Whitworth AJ, Wes PD, Pallanck LJ (2006) Drosophila models pioneer a new approach to drug discovery for Parkinson's disease. Drug Discov Today 11(3–4):119–126
96. Wolf FW, Heberlein U (2003) Invertebrate models of drug abuse. J Neurobiol 54(1):161–178
97. Woodhouse EC, Liotta LA (2004) Drosophila invasive tumors: a model for understanding metastasis. Cell Cycle 3(1):38–40
98. Yang Y, Hua X (2007) In search of tumor suppressing functions of menin. Mol Cell Endocrinol 265–266:34–41
99. Ye Y, Lukinova N, Fortini ME (1999) Neurogenic phenotypes and altered Notch processing in Drosophila Presenilin mutants. Nature 398(6727):525–529
100. Zangemeister-Wittke U, Simon HU (2004) An IAP in action: the multiple roles of survivin in differentiation, immunity and malignancy. Cell Cycle 3(9):1121–1123

Human Genetics and the Canine System

24.4

Heidi G. Parker and Elaine A. Ostrander

Abstract With constant advances in canine genomics, the dog has found a permanent position as a source of genetic information for the inheritance of morphologic traits and disease susceptibility. The modern domestic dog is not a typical model organism. They share our environment, our life-styles and often our food. In addition, they experience many of the same diseases that people do and are diagnosed and treated using the same medical procedures and pharmaceuticals. However, unlike humans, the purebred dog maintains a highly structured population organization that, if used correctly, can simplify the genetics of complex traits and disorders. In this chapter, we will discuss the history of canine genomics along with recent advances in resource development. Specific examples will be provided to demonstrate strategies for using population stratification to the best advantage in mapping traits both simple and complex. Together, these data highlight the utility of the canine system for mapping traits and finding mutations important in both human and companion animal science.

Contents

24.4.1 Introduction to the Canine System 814
    24.4.1.1 Origins of the Domestic Dog 814
    24.4.1.2 Dog Breeds 814
    24.4.1.3 Variation Between Breeds 814
    24.4.1.4 Lack of Variation Within Breeds 815
    24.4.1.5 Benefits of Mapping in a Breed-Based System 816
24.4.2 Navigating the Canine Genome 816
    24.4.2.1 Maps 816
    24.4.2.2 Sequence 816
24.4.3 Canine Disease Gene Studies 816
    24.4.3.1 Canine Disease Mirrors Human Disease 817
    24.4.3.2 Canine Disease and Mechanisms of Mutation 818
24.4.4 Genome Structure in the Domestic Dog 819
    24.4.4.1 Linkage Disequilibrium 820
    24.4.4.2 Haplotype Structure 820
    24.4.4.3 Single Nucleotide Polymorphisms 820
24.4.5 Population Structure in the Domestic Dog 820
    24.4.5.1 Canine Breed Clusters Facilitate Mapping Efforts 820
    24.4.5.2 Combining Breeds to Improve Mapping 821
    24.4.5.3 Homozygosity and Population Bottlenecks 821
24.4.6 Mapping Multigenic Traits in the Dog 822
    24.4.6.1 Quantitative Trait Loci 822
    24.4.6.2 Establishing a Cohort 822
    24.4.6.3 Complex Disease 823
24.4.7 Conclusion 823
References 823

H.G. Parker and E.A. Ostrander (✉)
National Human Genome Research Institute, National Institutes of Health, 50 South Drive, MSC 8000, Building 50, Room 5334, Bethesda, MD 20892-8000, USA
e-mail: hgparker@mail.nih.gov; eostrand@mail.nih.gov

24.4

Abbreviations

AKC	American kennel club
BHD	Birt–Hogg–Dube syndrome
DLA	Dog leukocyte antigen
Gb	Gigabases
HLA	Human leukocyte antigen
IBD	Identical by descent
IDID	Inherited diseases in dogs
LD	Linkage disequilibrium
Mb	Mega bases
OMIM	Online mendelian inheritance in man
QTL	Quantitative trait loci (locus)
PC	Principle component
PCA	Principle component analysis
PME	Progressive recessive myoclonic epilepsy
PRA	Progressive retinal atrophy
PWD	Portuguese Water dog
rcd(1,2)	Rod-cone dystrophy
RCND	Canine hereditary multifocal renal cystadenocarcinoma and nodular dermatofibrosis
SINE	Short interspersed nuclear element
SINEC_Cf	Canine-specific short interspersed nuclear element
SNP	Single nucleotide polymorphism

24.4.1 Introduction to the Canine System

Human genetics has found a new partner in an old friend, the domestic dog. While rodent systems are, for the time being, the mainstay of comparative genomics, the sequencing of several mammalian genomes, including the canine genome, offers human geneticists a new set of resources for mapping traits, identifying functional elements, and understanding chromosome structure and evolution. Indeed, as geneticists increasingly exploit the principles of comparative genomics, we can expect the dog to play a significant role in mapping loci for disease susceptibility, morphology, and behavior.

In this chapter, the utility of the canine system for mapping traits and finding mutations important in both human and companion animal science will be discussed. We will describe the population structure of the dog and how it relates to canine genetics along with recent advances in resource development. Moreover, specific examples will be provided to demonstrate the benefits of mapping within and across dog breeds. Finally, the genetics of complex traits such as morphology and behavior will be discussed. In aggregate, the ideas and data presented here support the use of the canine system and the properties of comparative genomics for understanding the genetic basis of traits of interest to mammalian biologists.

24.4.1.1 Origins of the Domestic Dog

The domestic dog is the most recently evolved species in the family Canidae, whose collective history spans nearly 50 million years. Dogs are believed to have evolved from wolves about 40,000 years ago, with the initial site of domestication still under debate [25, 64, 74].

24.4.1.2 Dog Breeds

There are over 400 breeds of dog in the world, about 155 of which are recognized by the American Kennel Club (AKC) [3]. Distinct breeds differ by as much as 100-fold in mass and display amazing levels of morphologic variation, as evidenced by differences in overall body size, skull shapes, leg lengths, and more (Fig. 24.4.1). In fact, the diversity in skeletal size and proportion between dog breeds is greater than that observed within any other terrestrial mammalian species [75, 76]. Equally amazing are the behaviors that characterize specific breeds, such as herding, tracking, retrieving, and guarding.

24.4.1.3 Variation Between Breeds

Because of their unique population structure, domestic dogs are ideal for genetic mapping studies. In a study of 85 dog breeds, Parker et al. showed that while humans and dogs have similar levels of overall nucleotide diversity, i.e., 8×10^{-4} nucleotide substitutions per basepair per generation, the amount of total variation

24.4 Human Genetics and the Canine System

Fig. 24.4.1 Morphologic variation found in AKC recognized dog breeds. (**a**) Examples of gross morphology in skull shape, ear size and carriage, and coat type and patterning. Breeds from *top to bottom*: Basset Hound, Saluki, Skye Terrier, Bulldog. (**b**) Examples of variation in overall body size, coat-type, leg length and proportion. Breeds *clockwise from top-left*: Afghan Hound, Giant Schnauzer, Italian Greyhound, Standard Dachshund, Mastiff, and Chihuahua (*center*) Photographs by Mary Bloom, copyright AKC

that is accounted for by differences between the breeds is much greater than that associated with differences between human populations (27.5% vs. 5.4%, respectively) [50]. Conversely, the degree of genetic variation found within individual dog breeds is much lower than that observed within distinct human populations. For a dog to be a member of a dog breed, therefore, is a more meaningful genetic distinction than for a human to be associated with a particular nationality.

24.4.1.4 Lack of Variation Within Breeds

Dogs are valued and judged in the show ring by the closeness to which each approximates their own breed standard [3]. To be a registered member of a breed, both of a dog's parents must have been documented members of the same breed. Each dog breed effectively represents a closed breeding population defined by small numbers of founders, population bottlenecks, and restricted breeding programs. Owing to the competitive nature of dog shows, sires that perform well in exhibition are bred repeatedly and may contribute excessively to later generations, creating bottlenecks that further reduce genetic variability [50, 51].

The level of intrabreed homozygosity and interbreed heterozygosity is sufficient to allow genetic distinction between breeds based on analysis of small numbers of genetic markers [31, 45]. For example, in the Parker et al. study, data from 96 microsatellite markers spanning all dog autosomes at approximately a 30 MB resolution were tested on 414 dogs to determine the degree to which dogs could be assigned to their breed based solely on genetic data [50]. Using an unsupervised clustering analysis, 95% of the dogs could be correctly assigned to either a single breed or a pair of closely related breeds, such as the Belgian Sheepdog/Belgian Tervuren pairing. Additionally, using the same data set, 99% of the dogs were assigned to the correct breed group by calculating the highest probability of an individual's genotype fitting any of the populations, in a leave-one-out analysis. As a result, the connotation of a "breed" is meaningful at not only the phenotypic level but the genetic level as well.

24.4.1.5 Benefits of Mapping in a Breed-Based System

These results make two functional predictions. First, dog breeds should display not only specific morphologic and behavior traits, but specific disease susceptibilities as well. Within any single breed the number of genes responsible for a trait is apt to be small, reflecting the lack of genetic variation within a breed, which stems from a small number of founders and the ensuing bottlenecks. Thus, mapping disease genes of interest within dog breeds may be a way to simplify the locus heterogeneity that plagues the mapping of many complex human traits. Second, since all dogs share recent common ancestors, the genetic signature of founders used to create related breeds is likely still to exist in the form of common ancestral disease mutations. Thus, one way to move from linked marker to gene may be to take advantage of such common variants. The study of disease susceptibility both within and between breeds supports these hypotheses.

24.4.2 Navigating the Canine Genome

24.4.2.1 Maps

In a few short years the canine mapping community has advanced from the first meiotic linkage map of the dog, composed of 150 markers divided into 30 linkage groups [46] to the development of whole-genome radiation hybrid maps [27], preliminary comparative maps [63], integrated linkage and RH maps [7], and finally, a detailed comparative RH map composed of 10,000 canine gene sequences [30]. These maps provided the foundation for all canine genetics studies, including the much anticipated whole-genome sequencing of the domestic dog.

24.4.2.2 Sequence

The first high-quality draft sequence of the dog, derived from a female boxer, was published in 2005 and comprised over 2.4 Gb estimated to cover ~99% of the euchromatic genome [41]. The sequence was assembled from 31.5 million sequence reads, which was sufficient to cover the genome 7.5 times. The gene count was listed as ~19,000, nearly all of which are orthologs of known human genes. This number is slightly lower than that reported for humans (~22,000 genes), a difference that is likely explained by the frequent occurrence of splice variants and pseudogenes and in some cases, species specific gene gains and losses [21]. The completed canine genome sequence has a critical role in advancing our knowledge of canine disease susceptibility, as discussed in Sect. 24.4.4.

24.4.3 Canine Disease Gene Studies

The domestic dog is second only to human in the number of recorded naturally occurring genetic disorders [4, 53]. Indeed, scientists have appreciated the importance of genetic predisposition in canine diseases for several years [53-55], but only recently have the resources become available to map diseases of interest and identify the underlying DNA variants [30, 36, 41]. Even more recent have been the advances made in studies aimed at identifying genes which regulate truly complex genetic traits, such as morphologic variation [10, 11, 69] and behavior [28, 29].

Careful phenotyping from the veterinary medical community combined with available genetic resources has resulted in the successful mapping of many canine disease genes and, in some cases, identification of the underlying variant (reviewed in [49, 51, 68, 70]). Specific examples include metabolic and endocrine disorders [14, 72], diseases related to the digestive system [16, 60], blindness [1, 26, 65], cancer [43] neurologic disorders [40, 42, 44], disease of the skin and muscle [15], and skeletal and developmental disorders [12, 13, 71]. The largest and most complete listing of documented canine inherited diseases is the Inherited Disease in Dogs database (IDID) [62], which is modeled on OMIM (Online Mendelian Inheritance in Man).

24.4.3.1 Canine Disease Mirrors Human Disease

The above studies demonstrate several principles of interest to human and companion animal scientists. Most studies have focused on diseases that are important in both canine and human medicine and in doing so have expanded our knowledge of the underlying biology of the disease. The most frequently cited example remains the work of Lin et al., who examined a family of Doberman Pinschers in order to show that the sleep disorder, narcolepsy, is caused by the insertion of a short interspersed nuclear element (SINE) in the *hypocretin 2 receptor* gene, resulting in a splicing defect [40]. While inherited narcolepsy is rare in both humans and dogs, the identification of a new class of molecules controlling sleep patterns suggested novel directions for the study of common sleep disorders, such as several forms of insomnia.

Another excellent set of studies are those of progressive retinal atrophy (PRA) in the dog. PRA is a collective term referring to a group of ocular conditions similar to retinitis pigmentosa in humans (reviewed in [57]). Different forms of PRA predominate in different dog breeds. For a number of canine retinal disorders, causative mutations have been found in genes that were previously implicated in clinically similar human diseases (Table 24.4.1). Examination of the disorders as they arise in individual breeds is akin to reading a flowchart of gene pathways and interactions important in eye development. For example, rod-cone dysplasia (rcd) is an early onset form of PRA observed in Irish setters and Cardigan Welsh corgis [58, 66]. In the Irish setter the disease (rcd1) is caused by a mutation in the *PDE6B* gene which encodes the beta subunit of cyclic GMP phosphodiesterase [66]. A similar disease in the corgi, rcd3, is caused by a null mutation in *PDE6A*, the gene encoding the alpha subunit of cyclic GMP phosphodiesterase [58]. Both of these mutations inhibit the enzymatic function of the phosphodiesterase, leading to death of the rod and

Table 24.4.1 Canine retinal disorders

Breed	Canine disease	Human disease	Gene	Reference
Siberian Husky	X-Linked progressive retinal atrophy	X-Linked retinitis pigmentosa	RPGR	[79]
Mastiff	Dominant progressive retinal atrophy	Retinitis pigmentosa 4, autosomal dominant	RHO	[35]
Alaskan Malamute	Cone degeneration	Achromatopsia	CNGB3	[65]
German Shorthaired Pointer	Cone degeneration	Achromatopsia	CNGB3	[65]
Multiple breeds [a]	Progressive rod-cone degeneration	Autosomal recessive retinitis pigmentosa	PRCD[b]	[78]
Irish setter	Rod-cone dysplasia 1	Autosomal recessive retinitis pigmentosa	PDE6B	[66]
Sloughi	Rod-cone dysplasia 1	Autosomal recessive retinitis pigmentosa	PDE6B	[20]
Cardigan Welsh corgi	Rod-cone dysplasia 3	Autosomal recessive retinitis pigmentosa	PDE6A	[58]
Briard	Retinal dystrophy	Leber's congenital amaurosis type-2	RPE65	[2]

[a] Poodle, Cocker spaniel, Labrador retriever, Portuguese water dog, Chesapeake Bay retriever, Nova Scotia duck tolling retriever, and Australian cattle dog all carry the same mutation
[b] Novel gene

cone cells in the retina [57]. In human RP families, mutations in *PDE6A* and *PDE6B* are predicted to account for ~8% of disease [23]. While the phosphodiesterase family of genes had been associated with retinal disorders in humans before they were identified in the dog, another form of rcd (rcd2), has recently been mapped in the collie to a gene of unknown function that is orthologous to human and mouse rd3 [38, 39]. Analysis of splice variations and expression of this gene in the dog may reveal a unique step in the pathway leading to photoreceptor death.

24.4.3.2 Canine Disease and Mechanisms of Mutation

24.4.3.2.1 SINE Insertions

In addition to facilitating our understanding of human disease, canine studies can reveal unique molecular mechanisms involved in genetic disease. Canine narcolepsy was caused by insertion of a canine-specific short interspersed nuclear element (SINEC_Cf) [6, 47, 73]. These elements are retrotransposons derived from a tRNA-Lys that occur frequently throughout the canine genome [6, 18, 36]. As with *Alu* repeats in the human genome, SINE elements are often located in positions affecting gene expression. Other examples include the aberrant insertion of the SINEC_Cf element into the canine *PTPLA* gene, which has been found to cause centronuclear myopathy in the Labrador retriever by causing splicing errors during maturation of the mRNA [56]. In addition, merle coat patterning in several breeds is the result of a SINE element insertion into the *SILV* gene, which plays a part in the biogenesis of premelanosomes, the precursors of pigment organelles [17].

24.4.3.2.2 Simple Repeats

Other diseases have been found to be associated with unique mutagenic events in the dog. For example, Lohi et al. reported that the cause of progressive recessive myoclonic epilepsy (PME) in miniature wirehaired dachshunds was the expansion of a canid-specific dodecamer repeat in the *Epm2b* (*Nhlrc1*) gene [44]. While trinucleotide repeat expansion has been reported in association with several human disorders (reviewed in: [8, 24]), this is the first report of a dodecamer repeat expansion causing a disease in any species. PME is common in multiple dog breeds. An affected basset hound was analyzed, and a shorter repeat expansion mutation was found at the same locus, showing that this mutation is not a single-breed anomaly. This poses an interesting question: will this type of mutation cause disease in humans?

24.4.3.2.3 Single Base Mutations

Finally, the ability to sample multiple generations within canine families has allowed more progress in understanding the role of missense changes in disease susceptibility than has been possible in humans. For instance, canine hereditary multifocal renal cystadenocarcinoma and nodular dermatofibrosis (RCND) is a naturally occurring autosomal dominant form of cancer characterized by bilateral tumors in the kidney and numerous collagen nodules in the skin [32]. This is found exclusively in German shepherd dogs and was mapped to CFA5 in 2000 (Fig. 24.4.2) [32]. Shortly after publication of the canine linkage mapping result, a similar disease in humans, Birt-Hogg-Dube syndrome (BHD), was mapped to chromosome 17q22.1 [34]. Because RCND localized to a portion of CFA5 that corresponded to 17q22.1 it was likely that both the canine and human diseases were due to mutations in the same gene. Indeed, this proved to be the case [43]. The gene implicated in both diseases, *BHD*, encodes a protein called folliculin. Protein-truncating mutations in this gene account for disease in approximately one third of BHD families [48]. However, in the dog, the disease is caused by a single base change in exon 7, creating a conservative missense change, H255R [43]. While it is often difficult to unambiguously determine whether a given missense change is truly disease associated, in this case three lines of evidence support this conclusion [43]. First, only RCND-affected dogs, regardless of country of origin, carried the H255R mutation, while all unaffected German shepherds and dogs of other breeds lacked the variant. Second, a multiple alignment of the folliculin protein sequence in 13 divergent species showed that the histidine !!! residue was completely conserved from dog and human through yeast. Finally, test matings demonstrated that the mutation is embryonic lethal in homozygotes,

24.4 Human Genetics and the Canine System

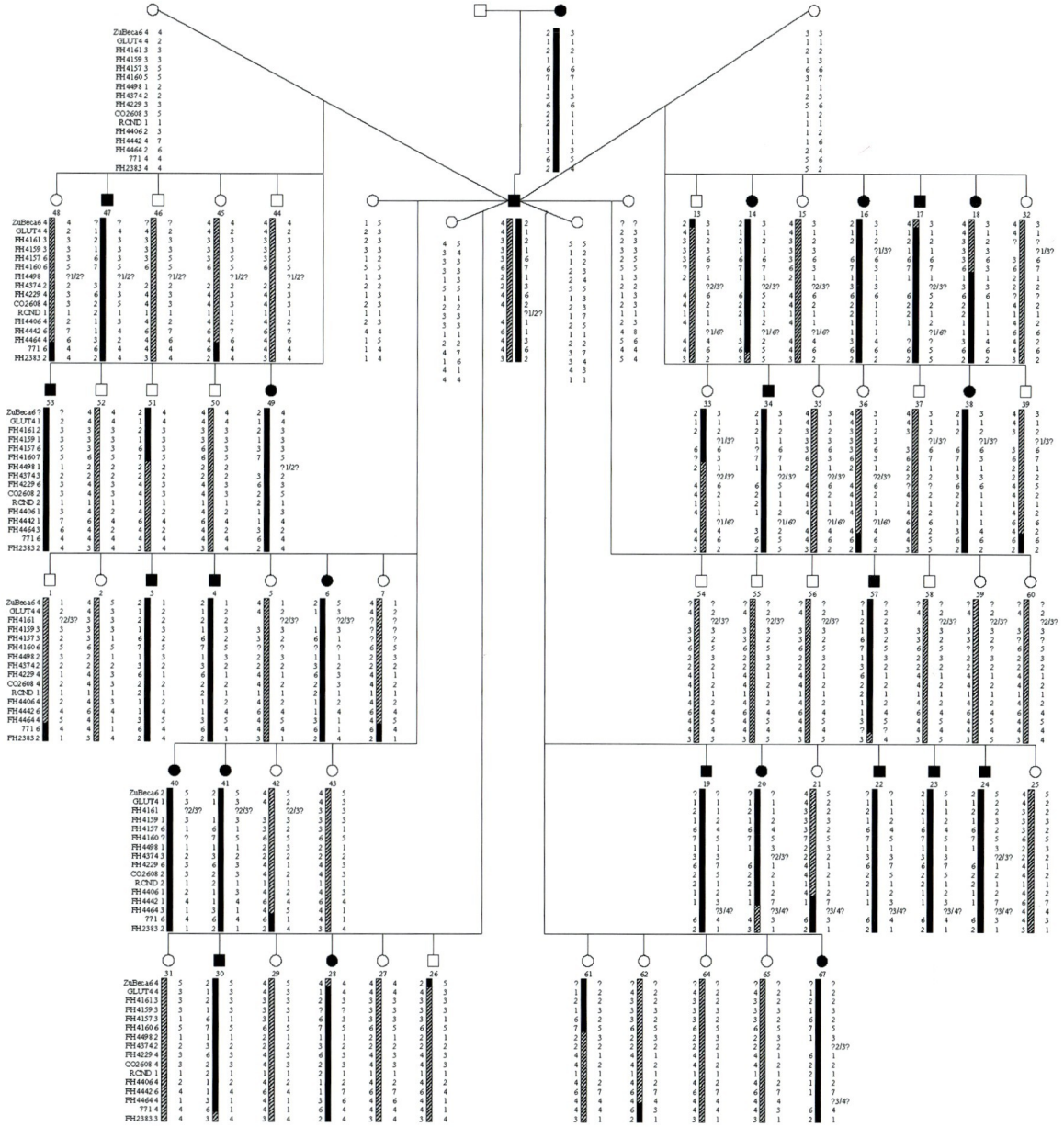

Fig. 24.4.2 The canine pedigree segregating RCND. Affected dogs are represented with black shading and unaffected dogs are unshaded. Marker names are indicated to the left of each row of genotypes. The genotypes of all markers are shown, but the vertical bar representing the haplotypes in the offspring is only shown for the affected proband's side. For the RCND locus, a '1' indicates the wild-type allele (unaffected) and a '2' indicates the mutant allele (affected). (Reprinted from [43], by permission of Oxford University Press)

implying a critical loss of function in the mutant protein. Thus, while it has been nearly impossible to draw conclusions on missense changes in folliculin from studies of humans, it is clear that a single ancestral variant causing the canine disease informs us about a key regulatory portion of the protein.

24.4.4 Genome Structure in the Domestic Dog

While the aforementioned examples (Sect. 24.4.3) illustrate many advantages of the canine system, the future promise of the dog as a genetic system lies in its application to complex traits. Successful mapping of polygenic traits in the dog requires an understanding of linkage disequilibrium (LD) [31, 41, 45, 67]. LD refers to the nonrandom association of two or more usually adjacent loci segregating together through several generations. Two major studies have undertaken the task of describing the extent of LD in the dog [41, 67].

24.4.4.1 Linkage Disequilibrium

In 2004 Sutter et al. examined the extent of LD in five breeds: the akita, Labrador retriever, Bernese Mountain dog, golden retriever, and Pekingese. Five genomic regions were examined and the results averaged. Both this study and the subsequent study of Lindblad-Toh et al., in which ten breeds were examined at ten distinct loci, found that the average length of LD in the dog exceeded 2 Mb [41, 67]. This is 40–100 times greater than the extent of the range of LD found in the human genome [19]. Significant variation was also noted between breeds. Those with long population bottlenecks or small numbers of founders displayed a longer range of LD (e.g., the Irish wolfhound) than did breeds associated with more founders and greater popularity (e.g., the labrador retriever). As we learn more about the canine genome this conclusion will likely bear some modification. We already know that LD varies enormously across the genome, and some of the large differences seen between breeds may reflect the small number of genomic regions examined [41].

24.4.4.2 Haplotype Structure

In addition to LD measurements, both studies looked at haplotypes within and across breeds and found that haplotype diversity was low, while haplotype sharing was high. The data presented by Sutter et al. show that at any given locus, approximately 60% of the chromosomes in each breed carried common haplotypes and 30% carried haplotypes found in every breed studied [67]. Examination of the assembled DNA sequence data from 7.5× whole-genome assembly provides additional insights. Comparison of the two boxer haplotypes, as well as the resequencing and genotyping of nine additional breeds, demonstrated mega-base-sized portions of the genome that are alternatively homozygous and heterozygous. Detailed experiments using 24 dogs from each of ten breeds support these data, suggesting that megabase-sized haplotypes will be common in many dog breeds [35].

The implications of these findings are two-fold. While a typical whole-genome association study in humans may require up to 500,000 SNPs [37], extensive LD suggests that only 10,000-30,000 SNPs will be required for the same study in dogs. In addition, the haplotype results suggest that a single set of SNPs will likely be informative for mapping studies in all breeds and that individual SNP marker sets or SNP chips do not need to be developed for individual breeds or groups of breeds. These facts, coupled with all of the benefits of mapping in the dog discussed in Sect. 24.4.1.4, suggest that for diseases prevalent in both humans and dogs, such as cancer, deafness, blindness, heart disease, epilepsy, cataracts, and motor neuron disorders, it is far easier to conduct initial mapping studies in dogs than in humans.

24.4.4.3 Single Nucleotide Polymorphisms

As a result of the canine sequencing effort approximately 2.1 million single nucleotide polymorphisms (SNPs) have been identified [41] and SNP chips for whole genome analysis have been released (Affymetrix, Santa Clara, CA; Illumina, San Diego, CA). The Affymetrix product supplies approximately 45,000 verified SNPs that are reproducible in multiple dog breeds while the Illumina product comprises approximately 20,000. This density, approximately one SNP every 100 kb, is more than sufficient to cover the range of LD expected within any domestic dog breed, sounds not logical to me in a variety of recent genome wide association studies aimed at finding the genes responsible for both morphologic traits and diseases [5, 22, 33, 61, 77].

24.4.5 Population Structure in the Domestic Dog

24.4.5.1 Canine Breed Clusters Facilitate Mapping Efforts

With the advances described above, the challenge now is to identify methods for mapping disease genes that make full use of the canine population structure as a mechanism for simplifying the mapping process. As an extension of the study described previously (Sect. 24.4.1.2), Parker et al. used microsatellite allele patterns to determine the ancestral relationships between various dog breeds [50]. Using the same data from 96 microsatellite markers analyzed on five unrelated dogs out of 85, breeds they performed an unsupervised clustering analysis using the computer program *structure* [59]. The 85 breeds were ordered into four clusters based on similar patterns of alleles, presumably from a shared ancestral pool. Asian and African breeds grouped together with the gray wolves in what is believed to be the ancient breed cluster. The mastiff-type working dogs formed a distinct cluster. A subset of sighthounds, primarily of European development, grouped with herding dogs of the same regions, and hunting dogs such as hounds, gun dogs, and terriers formed the last cluster (Fig. 24.4.3).

24.4.5.2 Combining Breeds to Improve Mapping

The Parker clusters offered a first look at relationships between breeds that did not rely on historical lore and pedigree records [50]. It is presumed that dog breeds from the same cluster share common ancestors and are therefore more likely to share traits that are identical by descent (IBD). These data, coupled with the findings of the haplotype studies (Sect. 24.4.4.2) suggest a study design for mapping traits that involves multiple breeds [41, 67]. Extensive LD will allow for quick identification of disease-associated regions of the genome within a single breed. However, the extensive LD means that the initial mapping segments will be large, on the order of megabases. Causative mutations will therefore be most easily identified by comparison of haplotypes in affected individuals from multiple breeds [26]. Furthermore, choosing affected dogs from breeds within a single Parker cluster will increase the chance that the breeds in question share an ancestral mutation and improve the likelihood for successful fine mapping [52].

24.4.5.3 Homozygosity and Population Bottlenecks

Analysis of haplotype sharing in the canine genome sequence (Sect. 24.4.4.2) revealed two major bottlenecks in the history of all breeds, a detail that is critical

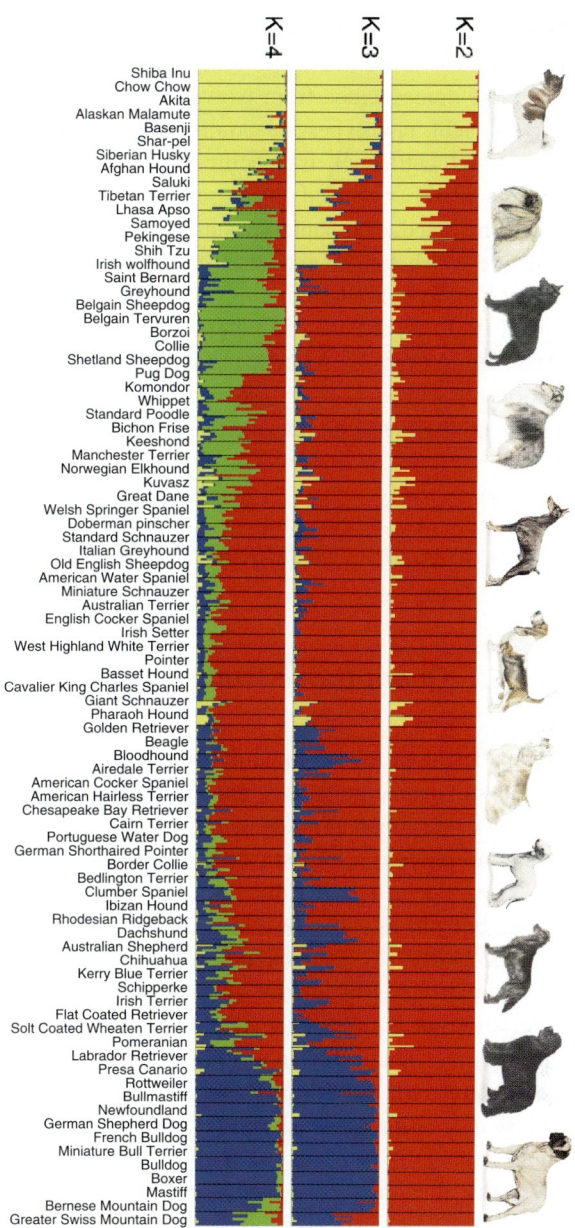

Fig. 24.4.3 Population structure of 85 domestic dog breeds. Each individual dog is represented by a single vertical line divided into K colors where K is the number of clusters assumed. Each color represents one cluster, and the length of the colored segment shows the individual's estimated proportion of membership in that cluster. Thin black lines separate the breeds that are labeled below the figure. Pictures of dogs from representative breeds for each grouping are shown at the top. Results shown are averages over 15 *structure* runs at each value of K. (Reprinted, with permission, from [50], [©AAAS])

for experimental design [41]. The first bottleneck is ancient and is presumed to have occurred at the time dogs were domesticated from wolves. The associated short, ancient haplotypes are common to all breeds. The second bottleneck is specific to the individual breed, and likely occurred during selection for formation of the breed. The haplotypes associated with this event are much longer and often comprise multiple ancient haplotypes. These findings serve as a reminder that all dogs, regardless of modern appearances and breed membership, are related, and the distance of the relationship will be inversely proportional to the length of the haplotypes shared between them.

24.4.6 Mapping Multigenic Traits in the Dog

24.4.6.1 Quantitative Trait Loci

In considering problems to which to apply the above advances in canine genomics, none is more challenging then those aimed at understanding the genetics of canine morphology. Domestic dog breeds differ more than five-fold in height alone and vary far more in over all body mass and appearance (see Sect. 24.4.1.2 and Fig. 24.4.1) [3]. Understanding how members of the same species can tolerate that level of morphologic variation is one of the most interesting questions in the field. Towards this end, studies of quantitative traits in the dog have been initiated that are based on methods developed for studies of inbred plant strains [10].

Quantitative traits are phenotypes controlled by many genes acting in concert not "working in concert". Because each gene contributes only fractionally to the trait, genomic variability in the mapping populations increases the challenge of finding causative genes. The ideal population for quantitative trait loci (QTL) mapping is one in which the phenotypes under question show high levels of variation. Additionally, the population as a whole should ideally derive from a small number of founders, which maximizes mapping power by reducing the number of haplotypes associated with any trait. The Portuguese water dog (PWD) was chosen as the primary focus of the morphology study because it meets the above criteria [11].

24.4.6.2 Establishing a Cohort

In order to identify QTLs for morphologic traits in the dog, investigators at the University of Utah established a cohort of more than several hundred AKC-registered PWDs [11]. DNA samples, X-rays, pedigrees, and detailed health data were collected on each dog. Ninety measurements derived from the X-rays were subjected to principal component analysis (PCA), which groups individual phenotypes into correlated traits. Nine heritable principle components (PCs) relating to overall body size and proportion of the skeleton were identified. The four major PCs are described in Table 24.4.2. Follow-up analysis of the postcranial skeleton has revealed additional PCs related to the trade-off between speed and strength in the canine physique [9]. A genome scan of ~500 microsatellite markers was completed on all dogs and several QTLs were found to be associated with the majority of PCs. Based on these studies, a single gene locus around *insulin-like growth factor 1 (Igf1)* has been identified that is a major contributor to small size in at least 14 diverse small dog breeds [69]. Additional studies are currently under way to find the genes

Table 24.4.2 Top four principle components (PCs) identified from skeletal measurements accounting for 61% of variation [11].

PC	Description	% of total variation	Heritability, %	Heritable variation, %
1	Overall size of skeleton	43.6	23±6	10.0
2	Inverse relationship between the size and strength of the musculoskeletal systems in the pelvis versus the head and neck	8.1	55±8	4.5
3	Inverse relationship between the length of the skull and limbs and the volume of the skull	4.6	24±6	1.1
4	Inverse relationship between length of skull and limbs and the strength of limbs and axial skeleton	4.5	70±6	3.2

responsible for the remaining PCs. Once identified, the data will be invaluable for understanding the genetics of morphologic development. For instance, dogs tend to have either long thin limbs or short squat limbs. Finding the genes controlling growth regulation for these correlated phenotypes will provide a wealth of data about growth regulation during early development.

24.4.6.3 Complex Disease

In addition to finding QTLs for morphology, the PWD study is well poised to identify disease susceptibility loci as well. Because the dogs were selected randomly from the population of living AKC registered PWD, and about 10% of the overall population participated, a representative set of complex genetic disorders is represented in the dataset. Using the already completed genome-wide scan, and health records provided by owners and veterinarians, loci for Addison's disease, osteoarthritis, and hip laxity have been identified [12-14]. Each of these traits has important implications for understanding human disease. For instance, Addison's disease in dogs appears to be immune mediated and, as with humans, occurs late in life with a female to male ratio of 2:1. Two canine loci have been identified to date, one of which is in the canine locus termed DLA (for dog leukocyte antigen) on CFA12 that is comparable to the human HLA locus [14]. The other is in a region of CFA37 that encodes genes for immune suppression. Understanding the exact variation responsible for the disease in dogs will provide insights into the poorly understood human condition as well.

The hip dysplasia/hip laxity study is equally provocative. Nearly 1/1000 human babies are born with hip dysplasia. In the PWD study, two predisposition loci, both on CFA1 have been identified and investigators are currently using haplotype analysis to find the underlying genetic variants. Additionally, investigators working with a mixed population of Greyhounds and Labrador retriever have identified 12 QTLs associated with hip dysplasia, none of which is on CFA1 [71]. The net result of these studies is expected to be a major advancement in both companion animal and human health and to provide insight into mammalian development.

24.4.7 Conclusion

We have argued in recent months that advances in genomic sciences have brought canine genetics into its prime; the dog genome has in effect "come of age". The dog community has now earned the right to take its place alongside other model systems, such as the rat and mouse. In fact, this is a fallacy. The critical experiments in canine genetics were not accomplished through the use of new technologies nor were they designed by scientists in a laboratory. The critical experiments in canine genetics were done by dog breeders and fanciers in the last 200–300 years, in their quest to create dogs of ideal appearance or behavior. Dog breeders are some of the most sophisticated geneticists practicing the craft today. They understand at a very real level the consequences of QTLs, incomplete penetrance, and complex traits. In creating a species that carries within its genome a truly extraordinary level of variation, dog breeders have provided for scientists a mechanism to truly understand the fundamental properties of mammalian biology. It somehow degrades centuries of careful manipulation if the dog is referred to simply as a "model system" for anything. It is an elegant experiment that, in reaching fruition, is of mutual benefit to dogs and humans alike.

Acknowledgements We thank our colleagues who read the article during preparation and made excellent suggestions and the many dog owners, breeders and supporters who provide us with samples and information about their beloved pets. This work is supported by the Intramural Program of the National Human Genome Research Institute, a Burroughs Wellcome Innovation Award, and The American Kennel Club Canine Health Foundation.

During preparation of this article, the 12-year-old Border Collie of one of the authors died after a short and unexpected illness. Only the loss of a parent or child can bring more tears than the loss of a pet. We dedicate this article to the many pet owners who, in similar situations, have shown us how to deal with our loss with grace and dignity.

References

1. Acland GM, Ray K, Mellersh CS et al (1999) A novel retinal degeneration locus identified by linkage and comparative mapping of canine early retinal degeneration. Genomics 59:134–142
2. Aguirre GD, Baldwin V, Pearce-Kelling S et al (1998) Congenital stationary night blindness in the dog: common

1. mutation in the RPE65 gene indicates founder effect. Mol Vis 4:23
2. American KC (1998) The Complete Dog Book, 19th edn. Revised, Howell Book House, New York, NY
3. Association AVM (2002) U.S. Pet Ownership and Demographics Sourcebook. First. American Veterinary Medical Association, Schaumburg, IL
4. Awano T, Johnson GS, Wade CM et al (2009) Genome-wide association analysis reveals a SOD1 mutation in canine degenerative myelopathy that resembles amyotrophic lateral sclerosis. Proc Natl Acad Sci USA 106:2794–2799
5. Bentolila S, Bach JM, Kessler JL et al (1999) Analysis of major repetitive DNA sequences in the dog (Canis familiaris) genome. Mamm Genome 10:699–705
6. Breen M, Hitte C, Lorentzen TD et al (2004) An integrated 4249 marker FISH/RH map of the canine genome. BMC Genomics 5:65
7. Brown LY, Brown SA (2004) Alanine tracts: the expanding story of human illness and trinucleotide repeats. Trends Genet 20:51–58
8. Carrier DR, Chase K, Lark KG (2005) Genetics of canid skeletal variation: size and shape of the pelvis. Genome Res 15:1825–1830
9. Chase K, Adler FR, Miller-Stebbings K, Lark KG (1999) Teaching a new dog old tricks: identifying quantitative trait loci using lessons from plants. J Hered 90:43–51
10. Chase K, Carrier DR, Adler FR et al (2002) Genetic basis for systems of skeletal quantitative traits: Principal component analysis of the canid skeleton. Proc Natl Acad Sci USA 99:9930–9935
11. Chase K, Lawler DF, Adler FR, Ostrander EA, Lark KG (2004) Bilaterally asymmetric effects of quantitative trait loci (QTLs): QTLs that affect laxity in the right versus left coxofemoral (hip) joints of the dog (Canis familiaris). Am J Med Genet 124:239–247
12. Chase K, Lawler DF, Carrier DR, Lark KG (2005) Genetic regulation of osteoarthritis: A QTL regulating cranial and caudal acetabular osteophyte formation in the hip joint of the dog (Canis familiaris). Am J Hum Genet 135: 334–335
13. Chase K, Sargan D, Miller K, Ostrander EA, Lark KG (2006) Understanding the genetics of autoimmune disease: two loci that regulate late onset Addison's disease in Portuguese Water Dogs. Int J Immunogenet 33:179–184
14. Clark LA, Credille KM, Murphy KE, Rees CA (2005) Linkage of dermatomyositis in the Shetland Sheepdog to chromosome 35. Vet Dermatol 16:392–394
15. Clark LA, Wahl JM, Steiner JM et al (2005) Linkage analysis and gene expression profile of pancreatic acinar atrophy in the German Shepherd Dog. Mamm Genome 16:955–962
16. Clark LA, Wahl JM, Rees CA, Murphy KE (2006) Retrotransposon insertion in SILV is responsible for merle patterning of the domestic dog. Proc Natl Acad Sci USA 103:1376–1381
17. Coltman DW, Wright JM (1994) Can SINEs: a family of tRNA-derived retroposons specific to the superfamily Canoidea. Nucleic Acids Res 22:2726–2730
18. Consortium IHM (2003) The international hapmap project. Nature 426:789–796
19. Dekomien G, Runte M, Godde R, Epplen JT (2000) Generalized progressive retinal atrophy of Sloughi dogs is due to an 8-bp insertion in exon 21 of the PDE6B gene. Cytogenet Cell Genet 90:261–267
20. Derrien T, Theze J, Vaysse A et al (2009) Revisiting the missing protein-coding gene catalog of the domestic dog. BMC Genomics 10:62
21. Drogemuller C, Karlsson EK, Hytonen MK et al (2008) A Mutation in hairless dogs implicates FOXI3 in ectodermal development. Science 321:1462
22. Dryja TP, Rucinski DE, Chen SH, Berson EL (1999) Frequency of mutations in the gene encoding the alpha subunit of rod cGMP-phosphodiesterase in autosomal recessive retinitis pigmentosa. Invest Ophthalmol Vis Sci 40:1859–1865
23. Everett CM, Wood NW (2004) Trinucleotide repeats and neurodegenerative disease. Brain 127:2385–2405
24. Germonpre M, Sablin MV, Stevens RE et al (2009) Fossil dogs and wolves from Palaeolithic sites in Belgium, the Ukraine and Russia: osteometry, ancient DNA and stable isotopes. J Archaeol Sci 36:473–490
25. Goldstein O, Zangerl B, Pearce-Kelling S et al (2006) Linkage disequilibrium mapping in domestic dog breeds narrows the progressive rod-cone degeneration interval and identifies ancestral disease-transmitting chromosome. Genomics 88:541–550
26. Guyon R, Lorentzen TD, Hitte C et al (2003) A 1-Mb resolution radiation hybrid map of the canine genome. Proc Natl Acad Sci USA 100:5296–5301
27. Hare B, Brown M, Williamson C, Tomasello M (2002) The domestication of social cognition in dogs. Science 298:1634–1636
28. Hart BL, Hart LA (2005) Breed-specific profiles of canine (*Canis familiaris*) behavior. In: Mills D, Levine E, Landsberg G, Horwitz D, Duxbury M, Mertens P, Meyer K, Huntley LR MR, Willard J (eds) Current issues and research in veterinary behavioral medicine. Purdue University Press, West Lafayette, IN, pp 107–113
29. Hitte C, Madeoy J, Kirkness EF et al (2005) Facilitating genome navigation: survey sequencing and dense radiation-hybrid gene mapping. Nat Rev Genet 6:643–648
30. Hyun C, Filippich LJ, Lea RA et al (2003) Prospects for whole genome linkage disequilibrium mapping in domestic dog breeds. Mamm Genome 14:640–649
31. Jonasdottir TJ, Mellersh CS, Moe L et al (2000) Genetic mapping of a naturally occurring hereditary renal cancer syndrome in dogs. Proc Natl Acad Sci USA 97:4132–4137
32. Karlsson EK, Baranowska I, Wade CM et al (2007) Efficient mapping of mendelian traits in dogs through genome-wide association. Nat Genet 39:1304–1306
33. Khoo SK, Bradley M, Wong FK et al (2001) Birt-Hogg-Dube syndrome: mapping of a novel hereditary neoplasia gene to chromosome 17p12–q11.2. Oncogene 20:5239–5242
34. Kijas J, Cideciyan A, Aleman T et al (2002) Naturally-occurring *rhodopsin* mutation in the dog causes retinal dysfunction and degeneration mimicking human dominant retinitis pigmentosa. Proc Natl Acad Sci USA 99: 6328–6333
35. Kirkness EF, Bafna V, Halpern AL et al (2003) The dog genome: survey sequencing and comparative analysis. Science 301:1898–1903
36. Kruglyak L (1999) Prospects for whole-genome linkage disequilibrium mapping of common disease genes. Nat Genet 22:139–144

38. Kukekova AV, Nelson J, Kuchtey RW et al (2006) Linkage mapping of canine rod cone dysplasia type 2 (rcd2) to CFA7, the canine orthologue of human 1q32. Invest Ophthalmol Vis Sci 47:1210–1215
39. Kukekova AV, Goldstein O, Johnson JL et al (2009) Canine RD3 mutation establishes rod-cone dysplasia type 2 (rcd2) as ortholog of human and murine rd3. Mamm Genome 20:109–123
40. Lin L, Faraco J, Li R et al (1999) The sleep disorder canine narcolepsy is caused by a mutation in the hypocretin (orexin) receptor 2 gene. Cell 98:365–376
41. Lindblad-Toh K, Wade CM, Mikkelsen TS et al (2005) Genome sequence, comparative analysis and haplotype structure of the domestic dog. Nature 438:803–819
42. Lingaas F, Aarskaug T, Sletten M et al (1998) Genetic markers linked to neuronal ceroid lipofuscinosis in English setter dogs. Anim Genet 29:371–376
43. Lingaas F, Comstock KE, Kirkness EF et al (2003) A mutation in the canine BHD gene is associated with hereditary multifocal renal cystadenocarcinoma and nodular dermatofibrosis in the German Shepherd dog. Hum Mol Genet 12:3043–3053
44. Lohi H, Young EJ, Fitzmaurice SN et al (2005) Expanded repeat in canine epilepsy. Science 307:81
45. Lou XY, Todhunter RJ, Lin M et al (2003) The extent and distribution of linkage disequilibrium in a multi-hierarchic outbred canine pedigree. Mamm Genome 14:555–564
46. Mellersh CS, Langston AA, Acland GM et al (1997) A linkage map of the canine genome. Genomics 46:326–336
47. Minnick MF, Stillwell LC, Heineman JM, Stiegler GL (1992) A highly repetitive DNA sequence possibly unique to canids. Gene 110:235–238
48. Nickerson M, Warren M, Toro J et al (2002) Mutations in a novel gene lead to kidney tumors, lung wall defects, and benign tumors of the hair follicle in patients with the Birt–Hogg–Dube syndrome. Cancer Cell 2:157
49. Ostrander EA, Kruglyak L (2000) Unleashing the canine genome. Genome Res 10:1271–1274
50. Parker HG, Kim LV, Sutter NB et al (2004) Genetic structure of the purebred domestic dog. Science 304:1160–1164
51. Parker HG, Ostrander EA (2005) Canine genomics and genetics: running with the pack. PLoS Genet 1:e58
52. Parker HG, Kukekova AV, Akey DT et al (2007) Breed relationships facilitate fine mapping studies: a 7.8 Kb deletion cosegregates with collie eye anomaly across multiple dog breeds. Genome Res 17:1562–1571
53. Patterson D (2000) Companion animal medicine in the age of medical genetics. J Vet Intern Med 14:1–9
54. Patterson DF, Haskins ME, Jezyk PF (1982) Models of human genetic disease in domestic animals. Adv Hum Genet 12:263–339
55. Patterson DF, Haskins ME, Jezyk PF et al (1988) Research on genetic diseases: reciprocal benefits to animals and man. J Am Vet Med Assoc 193:1131–1144
56. Pele M, Tiret L, Kessler JL, Blot S, Panthier JJ (2005) SINE exonic insertion in the PTPLA gene leads to multiple splicing defects and segregates with the autosomal recessive centronuclear myopathy in dogs. Hum Mol Genet 14:1417–1427
57. Petersen-Jones S (2005) Advances in the molecular understanding of canine retinal diseases. J Small Anim Pract 46:371–380
58. Petersen-Jones SM, Entz DD, Sargan DR (1999) cGMP phosphodiesterase-alpha mutation causes progressive retinal atrophy in the Cardigan Welsh corgi dog. Invest Ophthalmol Vis Sci 40:1637–1644
59. Pritchard JK, Stephens M, Donnelly P (2000) Inference of population structure using multilocus genotype data. Genetics 155:945–959
60. Safra N, Schaible RH, Bannasch DL (2006) Linkage analysis with an interbreed backcross maps Dalmatian hyperuricosuria to CFA03. Mamm Genome 17:340–345
61. Salmon Hillbertz NHC, Isaksson M, Karlsson EK et al (2007) Duplication of FGF3, FGF4, FGF19 and ORAOV1 causes hair ridge and predisposition to dermoid sinus in Ridgeback dogs. Nat Genet 39:1318–1320
62. Sargan D (2004) IDID: inherited diseases in dogs: web-based information for canine inherited disease genetics. Mamm Genome 15:503–506
63. Sargan DR, Yang F, Squire M et al (2000) Use of flow-sorted canine chromosomes in the assignment of canine linkage, radiation hybrid, and syntenic groups to chromosomes: refinement and verification of the comparative chromosome map for dog and human [In Process Citation]. Genomics 69:182–195
64. Savolainen P, Zhang YP, Luo J, Lundeberg J, Leitner T (2002) Genetic evidence for an East Asian origin of domestic dogs. Science 298:1610–1613
65. Sidjanin DJ, Lowe JK, McElwee JL et al (2002) Canine CNGB3 mutations establish cone degeneration as orthologous to the human achromatopsia locus ACHM3. Hum Mol Genet 11:1823–1833
66. Suber ML, Pittler SJ, Qin N et al (1993) Irish setter dogs affected with rod/cone dysplasia contain a nonsense mutation in the rod cGMP phosphodiesterase beta-subunit gene. Proc Natl Acad Sci USA 90:3968–3972
67. Sutter NB, Eberle MA, Parker HG et al (2004) Extensive and breed-specific linkage disequilibrium in Canis familiaris. Genome Res 14:2388–2396
68. Sutter NB, Ostrander EA (2004) Dog star rising: the canine genetic system. Nat Rev Genet 5:900–910
69. Sutter NB, Bustamante CD, Chase K et al (2007) A single IGF1 allele is a major determinant of small size in dogs. Science 316:112–115
70. Switonski M, Szczerbal I, Nowacka J (2004) The dog genome map and its use in mammalian comparative genomics. J Appl Genet 45:195–214
71. Todhunter RJ, Mateescu R, Lust G et al (2005) Quantitative trait loci for hip dysplasia in a cross-breed canine pedigree. Mamm Genome 16:720–730
72. van De Sluis B, Rothuizen J, Pearson PL, van Oost BA, Wijmenga C (2002) Identification of a new copper metabolism gene by positional cloning in a purebred dog population. Hum Mol Genet 11:165–173
73. Vassetzky NS, Kramerov DA (2002) CAN – a pan-carnivore SINE family. Mamm Genome 13:50–57
74. Vila C, Savolainen P, Maldonado JE et al (1997) Multiple and ancient origins of the domestic dog. Science 276:1687–1689
75. Wayne RK (1986) Limb morphology of domestic and wild canids: the influence of development on morphologic change. J Morphol 187:301–319
76. Wayne RK (1986) Cranial morphology of domestic and wild canids the influence of development on morphological change. Evolution 40:243–261

77. Wiik AC, Wade C, Biagi T et al (2008) A deletion in nephronophthisis 4 (NPHP4) is associated with recessive cone-rod dystrophy in standard wire-haired dachshund. Genome Res 18:1415–1421
78. Zangerl B, Goldstein O, Philp AR et al (2006) Identical mutation in a novel retinal gene causes progressive rod-cone degeneration in dogs and retinitis pigmentosa in humans. Genomics 26:26
79. Zhang Q, Acland GM, Wu WX et al (2002) Different RPGR exon ORF15 mutations in Canids provide insights into photoreceptor cell degeneration. Hum Mol Genet 11: 993–1003

Fish as a Model for Human Disease

24.5

Siew Hong Lam and Zhiyuan Gong

Abstract As a vertebrate, fish shares many conserved physiological and molecular features with mammals, and its position at the other extreme end of the vertebrate taxon from mammals makes fish an excellent complementary model to existing mammalian disease models for comparative analyses to identify molecular conservation of disorders. Fish, especially small freshwater species that have short generation time, high fecundity, externally and rapidly developing transparent embryos, and low husbandry cost, such as zebrafish (*Danio rerio*), medaka (*Oryzias latipes*), platy-fishes, and swordtails (*Xiphophorus* spp.), are particularly valuable for modeling human disorders at the molecular genetics level. These fish, especially zebrafish and medaka, are amenable to various molecular techniques and are supported by vast genomic resources allowing large-scale mutagenesis screens for the first time in vertebrates, making them highly versatile models of genetic diseases. *Xiphophorus* spp. are particularly valuable for cancer research owing to their ability to generate several spontaneous or induced tumor varieties through interspecies crossing between platyfish and swordtail. In order to fully exploit the fish system for modeling human disorders, various strategies, including both forward and reverse genetics and the use of physical manipulation to induce disease-like states, have been employed. Through a combination of these strategies, these fish are increasingly used to model human genetic diseases caused by both single gene mutation and multiple gene defects, and involving almost all the tissue-organ systems that are found or are homologous in both fish and human. Some of the fish models of human blood or heart disorders, and of cancer are highlighted. The use of these fish models of human diseases for screening of genetic and chemical modifiers of a disease phenotype which can lead to the discovery of drugs and therapeutic targets is also discussed.

Contents

24.5.1 Introduction 828
    24.5.1.1 Brief Historical Background and Current Status 828
    24.5.1.2 Why Use Fish to Model Human Disorders? 829
24.5.2 Strategies for Modeling Human Disorders in Fish 830
    24.5.2.1 Forward Genetics 830
    24.5.2.2 Reverse Genetics 831

S.H. Lam and Z. Gong (✉)
Department of Biological Sciences, National University of Singapore, 14 Science Drive 4, Singapore, 117543
e-mail: dbsgzy@nus.edu.sg

	24.5.2.3	Physical Manipulation: Chemical Treatment, Environmental Stressor, and Infection	833
24.5.3	Fish Models of Human Disorders	834
	24.5.3.1	Blood Disorders	834
	24.5.3.2	Heart Disorders	837
	24.5.3.3	Cancers	838
24.5.4	Modeling Human Disorders in Fish for Drug Discovery	840
References			841

24.5.1 Introduction

24.5.1.1 Brief Historical Background and Current Status

Fish considered as "lower" vertebrates have long been used by biologists to investigate and clarify some of the "complex" physiological processes and pathological problems observed in "higher" vertebrates such as in humans [10]. Researches with fishes provide a conceptual framework and evolutionary reference point for comparative studies. Using fish as a model to unravel some of the fundamental biological mechanisms common in both fish and human could provide insights into normal human biological processes and the pathogenesis of human disorders. Small freshwater species that are amenable to various molecular techniques and have short generation time, high fecundity and low husbandry cost such as zebrafish (*Danio rerio*), medaka (*Oryzias latipes*), platyfishes and swordtails (*Xiphophorus* spp.), are particularly valuable for modeling human disorders at the molecular genetics level.

Medaka (Fig. 24.5.1) came into genetic research as early as 1913 when it was used to demonstrate Mendelian inheritance in vertebrates and in 1921, it became the first vertebrate in which the occurrence of crossing over between X and Y chromosomes and Y-linked inheritance was shown (for reviews see [55, 65]). Medaka was subsequently used to study pigmentation, sex determination, development, and toxicology. It was the first fish species in which stable transgenesis and stable embryonic stem-like cells were established. In 1999, an international consortium, Medaka Genome Initiative (http://www.dsp.jst.go.jp/MGI/), was formed to sequence the medaka genome and to establish genetic and physical mapping resources. The use of *Xiphophorus* spp.

Fig. 24.5.1 (**a**, **b**) Two fish models that are widely used for disease modeling. (**a**) Medaka, *Oryzias latipes*; (**b**) zebrafish, *Danio rerio*

in cancer research can be dated back to 1920s when it was discovered that interspecies hybrids between platyfish *Xiphophorus maculatus* and swordtail *Xiphophorus helleri* develop melanomas spontaneously (for reviews see [37, 62]). The *Xiphophorus* Genetic Stock Center (XGSC; http://www.xiphophorus.org/xgsc.htm) was established in 1939 with the primary interest of developing genetically inbred stock for the identification of genes responsible for cancer. Presently, XGSC maintains 65 pedigreed *Xiphophorus* lines representing 23 species and selected backcross mating between these lines can produced several spontaneous or induced tumor varieties such as melanoma, renal adenocarcinoma, retinoblastoma, fibrosarcoma and schwannoma. Construction of a complete genetic linkage map has been initiated in *Xiphophorus* and development of expressed sequence tags and microsatellite methodologies are ongoing to facilitate genetic studies.

The zebrafish (Fig. 24.5.1) has been used as an experimental model since 1930s to study various environmental stress-related disorders ranging from developmental, anatomical to behavioral, by manipulating the external environment of the fish (for review, see [16]). In 1965, zebrafish became the first fish species used as a chemical carcinogenesis model demonstrating the development of hepatic neoplasia after exposure to carcinogen diethylnitrosamine [57]. However, it was not until the 1970s and 1980s that the zebrafish was developed as a genetic and developmental model. This set the stage for the historic large scale mutagenesis screens in the 1990s which, according to Amsterdam and Hopkins [2], generated more than 6,600 mutations initially but presently only 1,740 genetic mutants with specific embryogenesis defects were maintained and many had been described in the landmark publication of the zebrafish issue in the journal *Development* (volume 123) [9, 19]. Many of these mutant phenotypes resembled human disease conditions and several were found to be affected by orthologues of the human disease genes and serve as zebrafish models of these disorders (see Sect. 24.5.3.1). Within a decade, from a premier vertebrate developmental model, the zebrafish has now positioned itself as a biomedical model supported by vast genomic resources (including the zebrafish genome sequencing project; http://www.sanger.ac.uk/Projects/D_rerio/), a zebrafish stock center (Zebrafish International Resource Center; http://zfin.org/zirc/home/guide.php) and a centralized web-based database Zebrafish Information Network (http://zfin.org/) to facilitate the rapid exchange of genomic data and genetic resources. As a reflection of the growing confidence in the scientific community, Trans-NIH Zebrafish Initiative and European ZF Models consortium were set up to invest and consolidate efforts in developing zebrafish as human disease models for discovering insights into human disorders and novel therapeutics.

Over the span of 100 years, fish has came of age as a species used for model systems; from an experimental animal used for addressing various biological questions it progressed to become a premier model for developmental studies, and now a genetic biomedical model for addressing human disorders, and it even promises to find cures for human diseases. It is now clear that the biology that is observed in fish is likely not restricted to fish alone but may also be found to be applicable to humans. Thus, it is no longer just fish that we are beholding but human lives.

24.5.1.2 Why Use Fish to Model Human Disorders?

As vertebrates, fish and human share most developmental processes, physiological mechanisms, and organ systems. In fact the zebrafish has been described as a "canonical vertebrate" owing to the many similarities shared with mammalian biology [12]. Many of the genes or molecules with essential functions that are found in human are also found in fish. A comparison between the Fugu fish and human genomes revealed that 75% of predicted human proteins have a strong match to Fugu [4]. A number of zebrafish homologues of human disease genes such as those involved in cancer, Alzheimer's disease, muscular dystrophy, diabetes, thrombosis, blood, and heart disorders have been cloned or identified [29, 44, 57]. Thus, many diseases found in human are also observable in fish, allowing it to be used as an experimental animal to probe pathological problems in human.

As the oldest vertebrate, with over 500 million years of evolutionary history, fish is phylogenetically positioned at the opposite extreme end of the vertebrate taxon from human. Given the extreme phylogenetic separation between fish and human, it is reasonable to assume that only important genes, molecules, pathways, and processes that are involved in a particular disease phenotype observed in both fish and human would be conserved. This strategic phylogenetic position in the vertebrate taxon, then, offers unique comparative opportunities for identifying genes and molecules, pathways, and processes that are strongly associated with a disease phenotype. In cases where clinical samples or data are limited, the power of comparative disease modeling between two phylogenetically distant vertebrates such as the fish and rodent is evident; any identified molecular conservation of a disease phenotype between fish and a mammalian model would immediately underscore its fundamental importance in association with the disease and highlight its clinical potential. This comparative advantage was demonstrated when a number of genes that were identified as strongly associated with human and zebrafish liver tumors had also been suggested to have diagnostic, prognostic, or therapeutic values [27].

As the largest and most diverse group of vertebrates, with more than 28,000 species, fish offer a large number of models of adaptation and response to various natural and anthropogenic environments. Some of the specialized physiologies, traits, and adaptations can be exploited to address specific disorders. For example, the short 3-month lifespan of the fish *Nothobranchius furzeri* has been exploited for investigations into aging-related dysfunction of organ systems and drug validation [14].

Fish are very sensitive to change in the environment because of their close physiological contact with the surrounding water. Fish are therefore quick to respond to specific physiological challenges in order to maintain homeostasis, which otherwise will affect susceptible cells, tissues and organs. The intimate physiological-environment relationship and the rapid homeostatic response are more easily defined and their impacts more readily studied in fish than in terrestrial species. Fish are therefore excellent models for studying environment-induced health disorders and for inference of health risk, particularly as a result of environmental pollution, stress, or substance abuse (see Sect. 24.5.2.3).

In addition to the unique advantages mentioned above, the distinct features of some of the small aquarium fishes are what have propelled fish into the forefront of biomedical research. In particular, the zebrafish and medaka are excellent experimental animals owing to their short generation time, high fecundity, externally and rapidly developing transparent embryos, and ease of breeding and maintenance. For example, zebrafish are easily available in large numbers because of their short generation time (3 months), and they spawn throughout the year in a laboratory environment, producing large numbers (100–200) of external transparent embryos that rapidly develop from single-cell stage (with organogenesis occurring within 24 h) to free-feeding larvae with fully functioning organ systems within 5 days. They are inexpensive and easy to maintain in large numbers in a relatively small space. These attributes made large-scale mutagenesis screens feasible for the first time in vertebrates and paved the way for disease modeling in zebrafish and medaka. As a result, the zebrafish and medaka systems are now supported by vast genomic resources and amenable to many molecular tools making them highly versatile for genetic disease modeling (see Sect. 24.5.2).

Moreover, the combination of the above factors placed zebrafish and medaka in a strategic position to bridge the gap between in vitro cell-based models and the in vivo rodent model in biomedical research. The in vitro cell-based model, which is suitable for high-throughput applications, lacks the relevant physiological whole-organism setting, while the rodent model, which provides more relevant in vivo data, is less suited to high-throughput applications. Other established in vivo high-throughput screening systems such as *Drosophila* and *Caenorhabditis elegans* lack important vertebrate organ systems. Zebrafish and medaka are therefore presently the only vertebrate model systems that can be used for medium- to high-throughput bioassays while at the same time providing physiologically relevant data derived from a whole-organism setting. As disease models, these fishes are therefore highly suited for whole-organism-based therapeutic target and small molecule screening for drug discovery [67] (see Sect. 24.5.4).

24.5.2 Strategies for Modeling Human Disorders in Fish

In order for the zebrafish and medaka systems to be fully exploited for modeling human disorders, various genetic manipulations, including both forward and reverse genetics, and many molecular tools have been developed. The amenability of the zebrafish system to these powerful molecular strategies for studying human disorders will be highlighted in this section. For detailed description of each method, the reader is advised to refer to some of the cited literature.

24.5.2.1 Forward Genetics

Forward genetics is a powerful approach to the discovery of dysfunctional genetic control of biological processes that result in diseases without prior knowledge of the genes involved. By generating mutants that phenocopy human diseases, they can serve as experimental models to aid pathological investigations or used for screening of therapeutics (see Sect. 24.5.4). Although there are several ways of performing forward genetic screens, chemical and insertional mutagenesis have been most successful in zebrafish [2, 47].

24.5.2.1.1 Chemical Mutagenesis

Chemical mutagens could induce point mutations in single genes in rapidly dividing premeiotic germ cells at very high efficiency, and *N*-ethyl-*N*-nitrosourea (ENU) has proved to be most effective in zebrafish. Conventionally, chemical mutagenesis is carried out by a three-generation screening approach. Adult males are treated with ENU several weeks before crossing with wild-type females to generate F1 fish, which potentially carry a specific heterozygous mutation. These F1 fish are then crossed either with siblings or with wild-type females to generate F2 families. F2 siblings are then crossed with each other to reveal recessive mutations in the F3 homozygotes. Phenotype screening is then performed in F3 generation. "Targeted" screening may also be performed by focusing on specific organ systems or biological processes. Once a phenotype of interest is detected, positional cloning and candidate gene testing are performed to identify the mutated gene.

ENU screens have successfully generated a number of mutants that are now used for modeling blood and cardiovascular, neurosensory, musculoskeletal, and skin disorders (see Table 24.5.1) [2]. Subsequent cloning of the mutated genes in some of these zebrafish mutants revealed that many of the genes were also known to be mutated in human disease conditions (see Sect. 24.5.3). Likewise, this strategy has also been used to generate mutants for addressing disorders that are more biochemical or pharmacological in nature, such as cocaine addiction [7], d-amphetamine addiction [41], and disorders associated with the endocrine system [36], lipid metabolism [20], and hemostasis [22]. A large-scale ENU mutagenesis screen was also performed in medaka, resulting in 2031 embryonic lethal mutations being identified [13]. These included 312 embryonic lethal mutations causing defects in organogenesis, which were analyzed, and 126 mutations which were characterized genetically and assigned to 105 genes. Seven blood, liver, and thymic mutants have been generated and are potential models for the corresponding disorders in humans.

24.5.2.1.2 Insertional Mutagenesis

Retrovirus-mediated insertional mutagenesis is less efficient in generating mutants than the ENU approach, but it has the advantage of rapid identification of the mutated gene owing to the presence of a molecular tag inserted at the mutated site. Injection of murine retrovirus into a large number of blastula-staged (1,000- to 2,000-cell stage) embryos causes insertion of the viral sequence into the genome of primordial germ cells, causing specific genes to be mutated and transmitted to the next generation. Founder fish are used to produce F1 families, which will be screened for viral inserts. As the inserted viral sequence is known, the mutated genes can be rapidly identified by inverse PCR methods once a phenotype of interest is identified in subsequent generations [2].

By means of this strategy, gene loci or genes involved in cystic kidney disease [59], liver disease [52], and visual system disorders [15] have been identified. In addition, genes associated with craniofacial birth defects and zebrafish models for campomelic dysplasia and Ehlers-Danlos syndrome were also identified using an insertional mutagenesis screen [42]. It has also been found that many insertional mutants with ribosomal protein genes mutated are susceptible to cancer, suggesting similar tumor-suppressing roles of ribosomal proteins in humans that may have escaped detection or been overlooked owing to the many ribosomal proteins present in human [3]. In the same report, a neurofibromatosis type 2 mutant, a known tumor suppressor gene in human, has also been identified with elevated tumor incidence and can therefore be used for modeling tumorigenesis.

24.5.2.2 Reverse Genetics

Reverse genetics is a useful approach to generating disease models when the underlying genetic basis or gene involved in the human disorder is known. However, unlike the mouse model, gene knockout by targeted gene disruption through homologous recombination is still not possible in fish, because there are still no suitable embryonic cell lines. Nevertheless, advances in the derivation of pluripotent germline-competent embryonic cells from medaka [21] and zebrafish [11], and the feasibility of cloning by nuclear transfer in both species [33, 61] are making the prospect of targeted knockout in fish closer to reality. In the meantime, reverse-genetic strategies such as "gene knockdown" of a targeted gene using morpholinos and reverse-genetic screening by TILLING (*t*argeted *i*nduced *l*ocal *l*esion *in* *g*enome) and also tissue-specific overexpression of a foreign gene using transgenic technology have been employed for modeling gene dysfunctions and generating disease models in the zebrafish system.

24.5.2.2.1 Morpholinos

Morpholinos are modified antisense oligonucleotides that are resistant to cellular RNase and are injected into single- or two-cell stage embryos to knock down expression of a target gene product by either blocking translation initiation or interfering with the splicing of a particular exon [39]. Injected morpholinos are unlikely to result in complete loss of function, but usually generate phenotypes with several degrees of severity. This may be seen as an advantage where complete loss of function results in early lethality, and for modeling disorders that display different degrees of severity as a result of sensitivity to the dose-effect of a gene product. Owing to the stability of morpholinos and the rapid development of the zebrafish, phenotypes resulting from the knockdown of the targeted gene can be rapidly observed for several days, hence providing a quick and economical approach for generating disease models.

It has been demonstrated in several cases that injection of morpholinos results in developing embryos/larvae that phenocopy some human disease states. For example, morpholinos against *urod* caused cases of porphyria similar to those seen in humans, including autofluorescence of erythrocytes in ultraviolet light [39], dystrophin morpholinos resulted in muscular degeneration similar to that seen in human muscular dystrophy [18], and *invs* morpholino produced a renal cystic phenotype mimicking cystic kidney disease as observed in children [46]. Similarly, injection of morpholinos targeted against *Pit1* resulted in loss of the lactotroph, somatotroph, and thyrotrophic cells in the developing pituitary and a lack of growth in the juvenile (dwarfism) phenocopying human *Pit1* mutants of combined pituitary hormone deficiencies (CPHD) [40]. Thus, morpholino has become a cost-effective tool for verifying gene function in association with a disorder.

24.5.2.2.2 Reverse Genetic Screening by TILLING

Although the morpholino gene knockdown approach is quick and simple, it is not a germline mutation, so that the effect is only stable for a short period. In addition, some genes may be more difficult to knock down owing to the abundant maternal protein and/or to the late expression during development. On the other hand, mutational screens are efficient in generating mutants, but they are resource intensive in terms of the space, labor, and time required. In order to overcome these limitations, reverse-genetic screening by TILLING has been applied to the zebrafish system. This involves the combination of classic high-saturation ENU screen with high-throughput screening for point mutations in targeted genes of the mutagenized genome [56].

As in ENU screening, adult males are exposed to ENU to induce point mutation in their germ cells and mated with wild-type females to produce F1 generation. DNA can be extracted from tail biopsies of F1 fish maintained in small groups (live library) or from cryopreserved sperm of sacrificed F1 males (frozen sperm library) for scanning of mutation in a targeted gene. TILLING provides a cost-effective high-throughput approach for scanning of ENU-induced point mutation in PCR products of a targeted gene by using the celery mismatched-repair enzyme CEL-1, an endonuclease that cleaves DNA after single-base-pair mismatches, and subsequently allowing for electrophoresis-based detection of the mutation. Once a potential mutant is detected and the nature of the mutation is determined by sequencing, the live fish can immediately be outcrossed to propagate the line, or otherwise cryopreserved sperm is thawed for in vitro fertilization of wild-type eggs to regenerate the line.

Using this approach, Wienholds et al. [63] have identified several *rag1* mutants with amino acid substitutions or with premature stop codon. The mutant with a premature stop codon was intercrossed to produce homozygous fish which are defective in V(D)J recombination, confirming the loss of Rag1 function. In another screen, Wienholds et al. [64] identified 255 mutations, 14 of which had a premature stop codon, 7 had a splice donor/acceptor site mutation, and 119 had an amino acid substitution, and generated 13 potential knockout fish in a few months. This approach has also been used to recover two tp53 mutant lines with missense mutations similar to those found in human cancers, one mutant line being prone to the development of malignant peripheral nerve sheath tumor, which is useful for identifying genes that are associated with the tumor phenotype [5].

24.5.2.2.3 Transgenesis

Overexpressing a gene to mimic or induce a disease state, either by transient overexpression or by transgenesis, can be a useful way to study the gene-disease relationship and to develop a disease model. Transient overexpression of genes in zebrafish embryos is performed by injection of a DNA construct or in vitro-transcribed mRNA into embryos of 1-to 2-cell stage. Transient overexpression is usually performed in parallel with knockdown studies or gain-of-function rescue experiments to decipher gene function. However, it can have distinct advantages over the knockdown approach, as it can be designed to express protein subunits or mutated forms of proteins to decipher structure–function relationships and determine important protein domains associated with a dysfunction. Transient expression of the zebrafish equivalent to a human muscular dystrophy mutant, *CAV3P104L*, which has been identified in human patients with limb girdle muscular dystrophy (LGMD-1C) and rippling muscle disease, causes severe disruption of muscle differentiation in zebrafish and produces a similar dominant phenotype to that in humans [43].

In order to generate stable germline-transmissible foreign gene in zebrafish, DNA-injected embryos are raised to adulthood to produce the next generation. Only a small percentage (1-10%) of the injected embryos will stably integrate the foreign DNA construct into the genome of their germ cells and become transgenic founder fish (F0). Co-injection with an easily screenable reporter, such as a green or red fluorescent protein (GFP or RFP), could facilitate the identification of potential founders. F1 generation is screened for transgene expression by direct fluorescent visualization or other molecular expression. Once a transgenic F1 individual is identified, it can be used to develop a stable transgenic line. Using transgenic technology, several excellent zebrafish leukemia models had been developed [31, 32] (see Sect. 24.5.3.3).

24.5.2.3 Physical Manipulation: Chemical Treatment, Environmental Stressor, and Infection

Another strategy for modeling human disorders in zebrafish is through physical manipulation by using chemical agents, an environmental stressor, or microbial infection to induce disease-like states. Zebrafish are being used to model Parkinson disease by treatment with 1-methyl-4-phenyl-1,2,3,6-tetrahydropyridine (MPTP) and for screening of environmental pollutants, which induce selective loss of dopaminergic neurons in the brain, eliciting symptoms characteristic of Parkinson's disease [6]. MPTP-induced neurodegeneration can be prevented by co-incubation with l-deprenyl (monoamine oxidase-B inhibitor) or nomifensine (dopamine transporter inhibitor), indicating that the mechanism for MPTP-induced dopaminergic neuron toxicity in mammals and zebrafish is conserved [28]. Similarly, zebrafish are being used to model several substance-abuse-related disorders by exposing the fish to cocaine [7], d-amphetamine [41] and ethanol [35]. Likewise, in environmental toxicology, zebrafish liver exposed to arsenic shares a comparable molecular signature with those reported in mammalian systems, suggesting that the zebrafish liver coupled with the available microarray technology presents an excellent in vivo toxicogenomic model for investigating arsenic toxicity [30]. The acute arsenic-induced liver transcriptome changes in zebrafish can be used to infer possible liver damage that could occur in individuals exposed to arsenic in endemic areas where there is a high incidence of liver diseases in these populations. Zebrafish has also been used as a model to investigate molecular mechanisms and signaling molecules that are involved in the physiology and disorders resulting from hypoxic conditioning. In these hypoxic studies, zebrafish were maintained in an artificially induced hypoxic environment by decreasing air saturation in the tank water or by bubbling nitrogen gas in water to reduce oxygen levels. These studies have relevance to human disorders such as stroke, chronic ischemia, tumorigenesis, and intrauterine growth restriction, which increases fetal and neonatal morbidity and mortality [23]. Zebrafish has also been used to model infections, such as mycobacterial, streptococcal, aeromonad, and staphylococcal infections, which are caused by similar bacteria that are common in both human and fish. It has been shown that the zebrafish acute phase responses to *Aeromonas salmonicida* and *Staphylococcus aureus* infections are strikingly similar to the responses in mammals in terms of the type of proteins involved and how they are induced, although additional novel proteins have also been identified in fish only [34].

24.5.3 Fish Models of Human Disorders

Using a combination of strategies, as in sect. 24.5.2, zebrafish are increasingly employed to model human genetic diseases caused by both single gene mutation and multiple gene defects and involving almost all the tissue-organ systems that are found or are homologous in both zebrafish and human. These disease models range from hematopoietic, cardiovascular [44], musculoskeletal [43], and germ cell chromosome disorders [50] to complex neurosensory behavioral [17], cancer [57], environmental health [6] and aging-/degeneration-related disorders [26]. Table 24.5.1 lists some of the selected zebrafish disease models of human disorders, which include mutants with similar genes mutated in corresponding mammalian/human disease states. Some of the fish models of human blood or heart disorders and of cancer will be highlighted in this section (see Fig. 24.5.2).

24.5.3.1 Blood Disorders

The hematopoietic process involving the generation of multi-blood lineages (erythroid, myeloid, lymphoid, platelets), and the expression of many blood-specific genes (*cmyb, gata1, gata2, globin, hhex, ikaros, lmo2, pu1, rag1, runx1, scl,* and *vegf*) are conserved between zebrafish and human, although some hematopoietic sites may differ between the two species [8]. As the developing zebrafish embryo can survive without blood for several days, it allows for the detection of mutant larvae with blood-related defects. More than 50 mutants that affect hematopoiesis have been recovered in large-scale mutagenesis screens (for reviews see [2, 44]). The validity of using zebrafish as a model for human blood disorders is demonstrated by the many blood mutants that phenocopy human blood disease conditions.

The zebrafish *sauternes* (*sau*) mutant is the first animal model for human congenital sideroblastic anemia caused by mutations in δ-aminolevulinate synthase (ALAS2), an erythroid-specific enzyme required to initiate heme biosynthesis. The *sau* mutant is characterized by delayed erythroid maturation and abnormal globin expression, resulting in a microcytic, hypochromic anemia, and positional cloning discloses that the mutant gene encodes ALAS2, as in human conditions. The hypochromic mutant *weissherbst* (*weh*) has significantly reduced cellular hemoglobin and iron levels although the red blood cell count is near normal. Positional cloning of the gene responsible for the *weh* mutant led to the discovery of *ferroportin1*, which encodes a novel iron transporter that is conserved in vertebrates. Soon after the discovery of *ferroportin1* in zebrafish, it was found that similar mutations also occurred in patients suffering from a severe form of hereditary iron overload (hemochromatosis type IV). This is the first demonstration of the potential of using zebrafish to discover a previously unknown gene that is similarly mutated in the corresponding human disease.

Photosensitive blood mutants, such as *yquem* (*yqe*) and *dracula* (*drc*), have erythrocytes that autofluoresce and lyse when exposed to ambient light, phenocopying human congenital erythropoietic porphyrias. By using porphyrin and enzymatic assays, uroporphyrinogen decarboxylase (UROD) is identified as the enzymatic deficiency in the *yqe* zebrafish mutants that prove to be the first animal model for hepatoerythropoietic porphyria in which patients are deficient in UROD. The *drc* mutant has a mutation in *ferrochelatase* which encodes a heme enzyme and is characterized by photosensitive, autofluorescent erythrocytes, accumulation of protoporphyin IX, and liver disease, similar to patients with erythropoietic protoporphyria, a disorder of ferrochelatase. Both the mutants are excellent models for studying the pathogenesis and progression of protoporphyrin-induced liver disease using controlled light conditions.

Several of the blood mutants are characterized by fragile, hemolytic red blood cells attributable to mutations in cytoskeletal proteins. By using positional cloning and candidate gene cloning methods, the *merlot* (*mot*) and *chablis* (*cha*) mutations have been found to be located in the same gene-encoding erythrocyte protein 4.1 (also called band 4.1 or 4.1R), which is a structural membrane protein that anchors the spectrin-actin cytoskeleton to the erythrocyte cell membrane, thus conferring morphological stability to the erythrocyte. Homozygous *mot* or *cha* adults have decreased erythrocyte count, with immature erythrocytes arrested in the basophilic erythroblast stage and erythrocytes displaying abnormal morphology and osmotic fragility suggestive of hemolytic anemia. These mutants provide a model for hereditary elliptocytosis in humans, a rare cause of hemolytic anemia characterized by elliptical red blood cells caused by deficiency of protein 4.1.

Table 24.5.1 Selected zebrafish models of human disorders

Human disorder/disease	Gene product	Fish model: mutant, morphant or transgenic (phenotype)[a]
Blood		
Hypochromic microcytic anemia	SLC11A2	*chardonnay* (Hypochromic blood)
Hereditary elliptocytosis	EBP41	*chablis/merlot* (Hemolytic anemia)
Erythropoietic protoporphyria	Ferrochelatase	*dracula* (Porphyria; blood cell photosensitivity)
Congenital dyserythropoietic anemia type 2	SLC4A1	*retsina* (Hemolytic anemia)
Hereditary spherocytosis; elliptocytosis	beta-spectrin	*riesling* (Hemolytic anemia)
Sideroblastic anemia	Alas2	*sauternes* (Hypochromic anemia, decreasing blood count)
Hemochromatosis type 4	Ferroportin 1	*weissherbst* (Hypochromic anemia)
Hepatoerythropoietic porphyria	Urod	*yquem* (Porphyria)
Dyserythropoietic anemia; thrombocytopenia	Gata1	*vlad tepes* (Fewer blood cells)
Heart		
Holt-Oram syndrome	Tbx5	*heartstring* (Cardiac dysfunction; lacks pectoral fins)
Cardiomyopathy	Titin	*pickwick* (Heart contraction defect)
Cardiomyopathy	Tnnt2	*silent heart* (Absence of heart beat)
Cardiomyopathy	ACTC	*cardiofunk* (Defective endocardial cushion formation)
Cardiomyopathy	Myh6	*weak atrium* (Large and poorly beating atrium)
Timothy syndrome	CACNA1C	*island beat* (Defective heart function)
Long QT syndrome 2	KCNH2	*breakdance; kcnh2* morphant (Defective heartbeat)
Musculo-skeletal/skin (pigmentation)		
Duchenne muscular dystrophy	Dystrophin	*sapje* (Muscle degeneration, decreased motility)
Brody myopathy	ATP2A1	*accordion* (Delayed, prolonged trunk muscle relaxation)
Congenital slow-channel myasthenic syndrome	CHRNA1	*nicotinic receptor* (Nonmotile)
Campomelic dysplasia	Sox9a	*jellyfish* (Cartilage defects)
Progeroid type Ehlers-Danlos syndrome	B4galt7	*b4galt7* (Cartilage defects)
Exostoses, multiple type II	Exostosin 2	*dackel* (Abnormal fin, retinotectal projection, jaw, head, ear)
Oculocutaneous albinism type 1	Tyrosinase	*sandy* (No melanin pigment in body and eyes)
Piebaldism	Kit receptor	*sparse* (Fewer and smaller melanophores than wild type)
Waardenburg-Shah syndrome	sox10	*sox10* (Lacks pigment cells; abnormal ears)
Neuro-/organ-sensory		
Deafness	Myo6b	*satellite* (Uncoordinated swimming owing to vestibular defect)
Usher syndrome type 1b (hearing impairment)	Myo7a	*mariner* (Circular swimming owing to vestibular defect)
Usher syndrome type 1d	Cadherin 23	*sputnik* (Hearing defects, circular swimming, vestibular defect)
Achromatopsia	GNAT2	*no optokinetic response* (Lack optokinetic response)
Congenital cataracts with facial dysmorphy	Ctdp1	*ctdp1* (Small head and eyes)
Pituitary anomalies; holoprosencephaly-like	Gli2a	*you too* (Neural tube, hindbrain, midline axon guidance defects)
Kidney / pancreas / endocrine		
Polycystic kidney disease 2	PKD2	*pkd2* morphant (Cystic kidney)
Glomerulocystic kidney disease	Tcf2	*tcf2* (Cystic kidney)
Maturity-onset diabetes of the young type 5	Tcf2	*tcf2* (Smaller pancreas)
Pancreatic agenesis; neonatal diabetes mellitus	Ptf1a	*ptf1a* morphant (Exocrine pancreas agenesis)
Pituitary-hormone deficiency	Pit1 / POU1F1	*pit1 / pouf1* (Lacks several pituitary cell types)
Adrenal agenesis	Sf-1 / Ff1b	*ff1b* morphant (Impaired interenal development)
Cancer and other conditions		
Neurofibromatosis type 2	NF2	*nf2* (Predisposed to malignant peripheral nerve sheath tumors)
Most commonly mutated in human cancers	p53	*p53* (Predisposed to tumor formation)
Melanoma	hBRAF(V600E)	*mitfa-BRAF (V600E)* in *p53* mutant (Invasive melanoma)
T-cell acute lymphoblastic leukemia	Myc	*rag2-EGFP-Myc* transgenic (T-cell leukemia)
B-cell acute lymphoblastic leukemia	TEL-AML1	*XEF-EGFP-TEL-AML1* transgenic (B-cell leukemia)
DiGeorge syndrome	Tbx1	*van gogh* (Ear, thymus, pharyngeal arches defects)
Telangiectasia, hereditary hemorrhagic type 2	ACVRL1	*violet beauregarde* (Defective circulatory system)

[a] Unless indicated as morphant (morpholino knockdown) or transgenic, all models are mutants generated by forward genetics. The affected genes in these mutants have been cloned and are known to be mutated in human disorders [2, 44]. Only disease-relevant phenotype is indicated

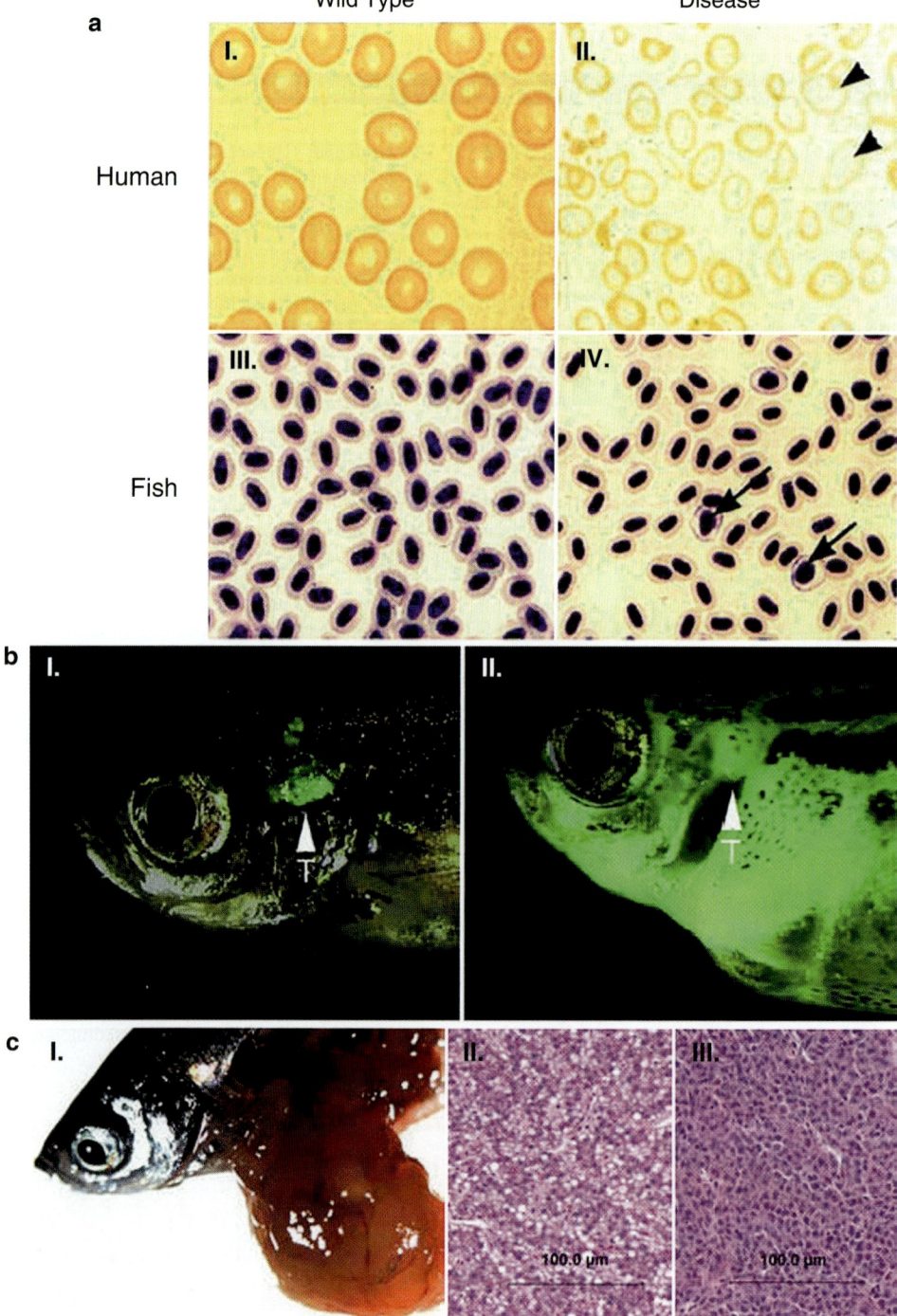

Fig. 24.5.2 (**a–c**) Fish models of human disorders. (**a**) Comparison of the human and zebrafish blood disorders associated with mutations in *ALAS-2*. The red blood cells in the peripheral blood are normally round and uniform (*I*). In human congenital sideroblastic anemia (*II*), cells have abnormal shapes and are hypochromic (owing to a low level of hemoglobin). Red cell precursors, such as reticulocytes, are larger (*arrowheads*). The increase in reticulocytes indicates a response of the marrow to the anemia. A similar hypochromia and increase in precursors is evident in the peripheral blood smear of *sau* homozygous mutants (*arrows* in *IV*) compared to wild-type fish (*III*). (Reproduced with permission from [66]). (**b**) Stable transgenic *rag2-EGFP-mMyc* zebrafish develop GFP-labeled thymic lymphoma, which progresses to T-ALL. Fluorescence microscopic analysis at 50 days of life, showing the thymus of control *rag2-GFP* transgenic fish (*I*) and massive GFP-labeled cellular dissemination of leukemic lymphoblasts in *rag2-EGFP-mMyc* transgenic fish (*II*). *Arrowheads* mark location of the thymus (*T*). (Reproduced with permission from [32]). (**c**) Carcinogen-induced liver tumor in zebrafish. Gross morphology of liver tumor (*I*) and histologic sections showing normal liver (*II*) and the dedifferentiated state of the liver with hepatoblastoma (*III*)

Zebrafish mutant *riesling* (*ris*) is severely anemic owing to membrane instability and rapid hemolysis caused by a mutation in the erythroid *β-spectrin* gene, which encodes the largest cytoskeletal component of the erythrocyte. The *ris* mutant has erythrocytes that are abnormally shaped with large nuclei resembling those seen in the human erythroid disorder hereditary spherocytosis (HS) caused by mutation in human *β-spectrin*. The mutant *retsina* (*ret*) is characterized by the presence immature erythrocytes with bilobed nuclei and is anemic owing to the arrest of erythrocyte maturation at the late erythroblast stage as a result of defective cytokinesis. This is caused by a *band 3* (also called *SLC4A1*) mutation, which produces a dysfunctional cytoskeletal protein responsible for the failure of mitotic chromosomal segregation in developing erythroblast. These features are also observed in patients with congenital dyserythropoietic anemia type 2 and thus *ret* mutant provides a model for investigating the underlying mechanism of the disorder.

It is worth noting that similar blood mutants, such as those with hypochromic anemia, progressive anemia, and erythropoietic porphyrias, had been generated in medaka through ENU mutagenesis screening [53, 60]. For example, the medaka *whiteout* (*who*) mutant identified in an ENU screen displays a normal erythrocyte count initially, but this gradually decreases during the embryonic and larval stages. The erythrocytes in the *who* mutants have elongated morphology and little hemoglobin activity. A missense mutation in a gene for delta aminolevulinic acid dehydratase (ALAD), the second enzyme in the heme synthetic pathway, has been identified in the *who* mutant, which now represents a model for the human disease ALAD deficiency porphyria [53]. There are still many medaka and zebrafish blood mutants that phenocopy human blood disorders and have yet to be cloned, and further characterization of these mutants may yield novel insights into these diseases.

24.5.3.2 Heart Disorders

The anatomy of the two-chambered (atrio-ventricular) zebrafish heart lined by an inner endothelial layer and an outer myocardial layer resembles that of the human heart at 3 weeks of gestation. However ,unlike the situation in mammals, the zebrafish embryonic heart function can be assessed visually and the fish embryo is not dependent on blood circulation for survival during embryogenesis, thus allowing many zebrafish mutants with cardiac defects to be identified through mutagenesis screens (for reviews see [2, 44]).

The mutant *heartstrings* (*hst*), which encodes the human ortholog *TBX5*, a member of the T-box family of transcription factors, does not develop pectoral fins (analogous to mammalian limbs) and has severe heart defects. In humans, mutations in *TBX5* result in Holt-Oram syndrome, a genetic disorder characterized by heart and upper limb defects. Low-level morpholino knockdown of *tbx5* also results in a variety of bilateral and asymmetric fin (limb) deformities similar to haploinsufficiency of *TBX5* in humans. The study demonstrated the versatility of zebrafish to model different degrees of severity of a human disorder that is sensitive to dose-effect of a gene product.

The *silent heart* (*sih*) mutants have a heart that fails to contract although embryonic cardiac development proceeds normally. The mutations causing *sih* are found in the sarcomere component *cardiac troponin T* (*tnnt2*) and were confirmed by morpholino and DNA rescue experiments. The dysfunctional heart muscle of *sih* is caused by failure in cardiac sarcomere assembly attributable to the defects in thin filament stability in the absence of Tnnt2. The *sih* mutants are the first animal model of Tnnt2 deficiency, as they phenocopy human familial hypertrophic cardiomyopathy characterized by sarcomere loss and myocyte disarray as a result of mutations in *TNNT2*. Similarly, the *pickwick* (*pik*) mutant has embryonic, thin, cardiac myofibrils that can contract, but fail to assemble into normal sarcomeres and thus suffer contractility defects resulting in very low systolic pressure and insufficient blood being pumped out from the heart. Use of a positional cloning approach showed that a mutation in cardiac-specific exons of *titin* (*ttn*) caused *pik*. The *pik* mutant phenocopies the defects seen in human heritable dilated cardiomyopathy caused by mutations in *TTN*.

The zebrafish mutant *island beat* (*isl*) has a defective heartbeat, the atrium exhibiting rapid discoordinated contractions that are not propagated to the ventricle. Besides not beating, the ventricle has a lesser number of cardiomyocytes and is small. The mutation of *isl* embryos is in the *alpha-1C L-type calcium channel subunit* (*C-LTCC*) encoding the primary ion-conducting pore-forming subunit of the L-type calcium channel in cardiac tissue. In humans, mutations of the gene (also known as $Ca_v1.2$) cause cardiac arrhythmias that can lead to syncope and sudden death and are part of a multisystem disorder known as Timothy syndrome. The cardiac arrhythmia is characterized by extreme prolongation of the QT interval on electrocardiogram, as a result of prolongation of cardiomyocyte action potentials and delayed secondary depolarizations owing to defective voltage-dependent channel inactivation and abnormal Ca^{2+} signaling. Similarly, cardiomyocytes in zebrafish mutant *isl* are absent from L-type calcium currents owing to a defect in calcium channel function, thus modeling human arrhythmia disorder caused by cardiac L-type calcium channel mutations.

In addition, there are many zebrafish heart mutants that phenocopy human heart disorders, such as cardia bifida (*miles apart*), cardiac valve defects (*jekyll*), cardiac hypertrophy (*liebeskummer*), bradycardia (*slow mo*), coarctation of the aorta (*gridlock*), and atrium and/or ventricle defects (*heart and soul*, *pandora*, *foggy*, *acerebellar*). in which their respect mutation has been identified in zebrafish but a similar gene mutation has yet to be found in the corresponding human disease counterpart. Even so, with the combination of the phenotype and knowledge of the mutated gene, these mutants are useful for understanding the molecular basis of these pathologies and for screening of potential drug candidates that can suppress the disease phenotypes (see Sect. 24.5.4).

24.5.3.3 Cancers

The discovery that melanomas can be generated in the *Xiphophorus* (swordtail and platyfish) hybrid model opened a new frontier for melanoma and tumorigenesis research. A gene encoding a novel receptor tyrosine kinase (Xmrk2), which has homology to the human oncoprotein epidermal growth-factor receptor, has been identified as associated with melanoma formation. Molecular signaling pathways induced by Xmrk consist of essential steps in tumor development, such as preventing cell differentiation, activating unrestricted proliferation, exerting an antiapoptosis effect, and inducing migration [37]. Furthermore, a second genetic locus that is involved in *Xiphophorus* melanoma development has been narrowed down to a tumor suppressor gene called *CDKN2AB* (because of its similarity to both human *CDKN2A* and *CDKN2B* loci), which is mutated in many human cancers, including melanomas [25]. The *Xiphophorus* melanoma, which can be initiated by simple crossings and the well-defined signaling pathways governing its tumor growth and progression, provides a model for a comprehensive study of the molecular changes and regulatory networks underlying tumor formation involving different levels of organization from molecules to where the tumor interacts with healthy organ systems of the whole organism [37]. This should provide further insights into some basic principles of cancer biology and identify new areas for melanoma research.

Zebrafish is susceptible to various chemical carcinogens and produces many neoplasm types in various tissues, showing remarkable histopathologic resemblance to human and mammalian cancers [1]. Comparative expression profile analysis of human and zebrafish tumors revealed that both tumors shared a similar molecular hallmark, having a large number of differentially expressed genes coding for proteins involved in cell cycle/proliferation, apoptosis, DNA replication and repair, cytoskeletal organization, cell adhesion and motility, RNA processing and protein synthesis, and metastasis [29, 30]. Moreover, when microarray data from histopathologically graded liver tumor samples of humans and zebrafish were compared, a large number of genes were observed to share strikingly similar expression profiles correlating with tumor progression (Fig. 24.5.3). This study provides the first validation of a large set of genes that are associated with liver tumor and its progression in fish and human, and highlights the use of comparative expression genomics between two phylogenetically distant species for identifying biomarkers with clinical potential based on the strong association with a disease phenotype such as liver cancer [27].

The zebrafish has several excellent transgenic leukemia models. The transgenic rag2–EGFP-*mMyc* zebrafish line develops GFP-labeled T-cell acute lymphoblastic leukemia (T-ALL) that expresses the zebrafish orthologs of the human T-ALL oncogenes *tal1/sel* and *lmo2*, similar to the most prevalent molecular subgroup of human T-ALL [31]. These fluorescent leukemic cells allow monitoring of the pathogenesis of the disease, as it can be transplanted into irradiated wild-type adult fish and will home to the thymus to produce a pervasive leukemia. A conditional transgenic zebrafish model of myc-inducible T-cell acute lymphoblastic leukemia has also been developed using the Cre/loxP system for the breeding and maintenance of stable transgenic lethal disease models [32]. This

24.5 Fish as a Model for Human Disease

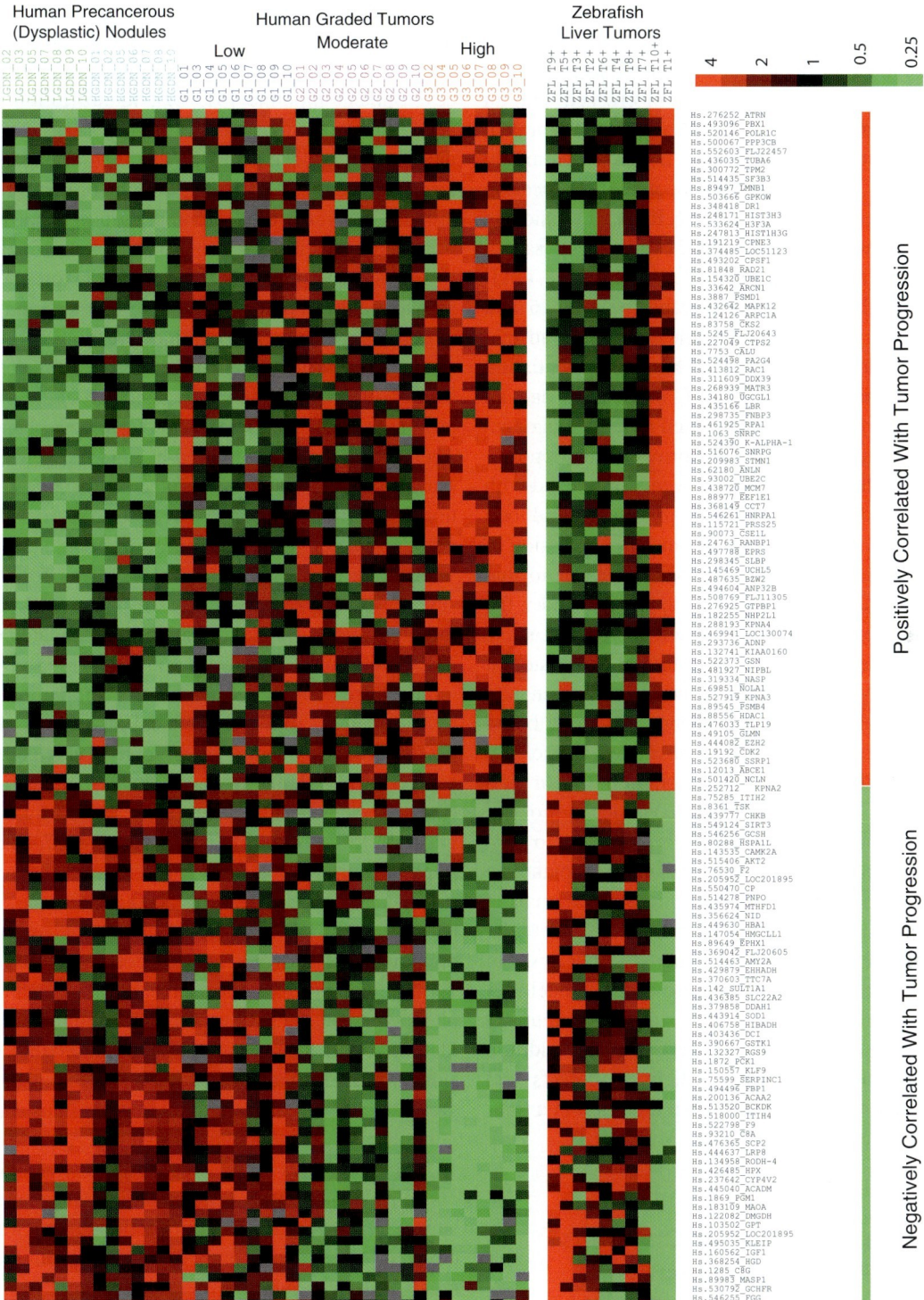

Fig. 24.5.3 Expression profiles of genes showing similar correlations, with tumor progression in both zebrafish and human liver tumors. The *color* in each cell reflects the expression level of the corresponding gene in the corresponding tissue sample relative to its mean expression level across the entire respective set of human and zebrafish tissue samples. *Gray cell* indicates missing or excluded data (*LGDN* low-grade dysplastic nodule, *HGDN* high-grade dysplastic nodule, *G1-G3* grades 1–3)

stable *rag2-loxP-dsRED2-loxP-EGFP-mMyc* transgenic zebrafish line has red fluorescent thymocytes and will not develop leukemia until *Cre* RNA is injected into the single-cell stage embryos or crossed with an activator *Cre* transgenic line. Besides T-ALL, a zebrafish model of human precursor B (pre-B) ALL, commonly caused by a t(p13;q22) chromosomal translocation resulting in a TEL AML1 (ETV6 RUNX1) fusion in pre-B cells, has also been generated by transgenesis [51]. This leukemia was transplantable to irradiated wild-type fish with pathology observable between 6 and 9 weeks after transplantation. The study showed that TEL AML1 induces the arrest of B-cell differentiation and found that the loss of TEL expression and an elevated Bcl1/Bax ratio contributed to leukemia development. This transgenic model provides the opportunity to study the various genetic events associated with *TEL-AML1*-induced leukemia, a prevalent childhood cancer. Apart from stable lines, transient expression of a zebrafish *tel-jak2a* fusion oncogene similar to that seen in human chronic myeloid leukemia, under the control of the *spi1* promoter, which is strongly active in myeloid precursors [45], and transient expression of a human *RUNX1-CBF2T1* [24], were able to generate models useful for the experimental study of blood malignancy. Likewise, a melanoma model is generated when *mitfa-BRAF*^{V600E}, the most common human melanoma BRAF mutant form (V600E) under the control of the melanocyte *mitfa* promoter, is injected into p53-deficient embryos causing invasive and transplantable human-like melanomas to develop in these animals at 4 months of age [48]. Besides transgene expression, there are several zebrafish mutants that exhibit increased tumor incidence that are useful models for cancer research, namely tumor suppressor mutants tp53 [5] and neurofibromatosis type 2 [1], in addition to *crb* mutants that display cell cycle defects and genome instability attributable to a mutation in a transcriptional regulator *bmyb* [54].

24.5.4 Modeling Human Disorders in Fish for Drug Discovery

The zebrafish disease models that are generated using forward and reverse genetics or by a chemical/environmental stressor are not only useful for studying the underlying genetic basis of a disease but also provide excellent systems for screening of genetic and chemical modifiers of a disease phenotype which can lead to the discovery of drugs and therapeutic targets along the drug development pipeline [57, 67]. Identification of the genes that are responsible for the phenotypes of disease models might lead directly to the identification of therapeutic targets. Positional cloning of ENU-induced mutant or inverse PCR cloning of insertional mutant followed by validation using morpholino knockdown or rescue experiment by transient overexpression can help to pin down potential targets. Another approach is to apply large-scale morpholino screening by systematically knocking down genes using a morpholino library in order to identify gene knockdowns that prevent or slow the development of a disease phenotype. Therapeutic targets identified by forward genetic and morpholino screens could become the focus of conventional, target-based drug discovery efforts, including in vitro high-throughput screening.

Alternatively, small-molecule screens to identify suppressors of zebrafish disease phenotypes can be initiated before the identification of a validated target. The feasibility of performing phenotype-based small-molecule screens in zebrafish has been demonstrated by the identification of chemical suppressors of the *gridlock* mutant phenotype, which is a vascular defect caused by a *hey2* mutation [49], and *crb* mutant phenotype, which is a cell cycle defect caused by a *bmyb* mutation [58]. The zebrafish mutation *gridlock* causes impaired aortic blood flow similar to aortic coarctation in humans. After screening 5,000 small molecules, two structurally related novel classes of compounds previously not known to affect blood vessel formation, which completely suppress the *gridlock* coarctation phenotype, hence enabling survival to adulthood, were identified [49]. The *crb* mutation decreased the expression of cyclin B1 that is responsible for driving cell progression from G2 phase through mitosis, thus resulting in mitotic arrest and genome instability. Approximately 16,000 compounds were screened in 16 weeks before one novel compound, persynthamide (psy 1), which suppresses bmyb-dependent mitotic defects, was discovered. Psy-treated embryos showed an S-phase delay and up-regulation of *cyclin B1* mRNA, which promotes the progression of cells through mitosis [57]. Since most cell-cycle genes are conserved between zebrafish and humans, this approach could be applied to a number of zebrafish cell-cycle mutants or cancer models to screen for cancer suppressors. These studies show that the phenotype-based whole-organism screen allows us to approach biological questions that simply cannot be addressed in vitro, such as suppression of the *gridlock* coarctation phenotype and reversal of the cardiac dilation or contractility defects in a zebrafish model of heart failure [67]. Even when an in vitro screen is available, the results obtained by whole-organism screens might be more relevant than those obtained using the in vitro screen, because physiologically relevant organismal setting may be required for the mechanisms of action of the compounds. Likewise, in vitro screens, while suited to high-throughput applications, lacks the relevant physiological whole-organism setting for toxicology, and toxicity or unwanted side effects are major reasons for the high attrition of compounds in the drug development pipeline. By using zebrafish model, in vivo toxicology can be performed earlier, if not simultaneously, in the drug discovery process at a lower cost than in a rodent model. Zebrafish has been noted to show a similar toxic response to that of mammals with regard to chemicals that cause endocrine disruption, reproductive toxicity, behavioral defects, teratogenesis, carcinogenesis, cardiotoxicity, and liver toxicity [67]. Zebrafish also responds to several drugs in a similar manner to that in humans. For example, 100 compounds was screened in a high-throughput assay developed for bradycardia in zebrafish embryos, and 22 of 23 drugs known to cause QT prolongation in humans were also found to cause bradycardia in zebrafish. In addition, known drug-drug interactions between erythromycin and cisapride and also between cimetidine and terfenadine, which lead to QT prolongation, were also detected by the assay, highlighting the fact that complex pharmacokinetic and pharmacodynamic processes occurring in humans can also be reproduced in zebrafish [38]. Moreover, with the availability of more fluorescent transgenic zebrafish, disease models that allow for easy visualization and quantification of the degree of severity of a disease phenotype, the zebrafish will become even more attractive for drug screening.

In summary, fish, especially small species such as zebrafish, medaka, and *Xiphophorus* have proved valuable in disease modeling at the molecular genetics level. Currently, the zebrafish is the most intensively studied fish model for human disorders, and it allows for disease pathogenesis, forward genetic, reverse genetic, small molecule screening, and toxicology to be performed in the same organism at higher throughput and lower cost than in rodent models. As a vertebrate, the zebrafish shares many conserved physiological and molecular features with mammals, and its position at the other extreme end of the vertebrate taxon makes zebrafish an excellent complementary model to existing mammalian disease models for comparative pathology to identify molecular conservation of disorders. Thus, it is not surprising that the zebrafish model is beginning to be used at various preclinical stages of disease modeling and drug discovery processes. Given the strong interest within the scientific community, as reflected by the setting up of the Trans-NIH Zebrafish Initiative and European ZF Models Consortium, and the many genetic screens that are ongoing in several laboratories, it is expected that the zebrafish will gain more prominence for modeling human disorders.

References

1. Amatruda JF, Shepard JL, Stern HM, Zon LI (2002) Zebrafish as a cancer model system. Cancer Cell 1:229–231
2. Amsterdam A, Hopkins N (2006) Mutagenesis strategies in zebrafish for identifying genes involved in development and disease. Trends Genet 22:473–478
3. Amsterdam A, Sadler KC, Lai K, Farrington S, Bronson RT, Lees JA, Hopkins N (2004) Many ribosomal protein genes are cancer genes in zebrafish. PLoS Biol 2:E139
4. Aparicio S, Chapman J, Stupka E, Putnam N, Chia JM, Dehal P, Christoffels A, Rash S, Hoon S, Smit A, Gelpke MD, Roach J, Oh T, Ho IY, Wong M, Detter C, Verhoef F, Predki P, Tay A, Lucas S, Richardson P, Smith SF, Clark MS, Edwards YJ, Doggett N, Zharkikh A, Tavtigian SV, Pruss D, Barnstead M, Evans C, Baden H, Powell J, Glusman G, Rowen L, Hood L, Tan YH, Elgar G, Hawkins T, Venkatesh B, Rokhsar D, Brenner S (2002) Whole-genome shotgun assembly and analysis of the genome of Fugu rubripes. Science 297:1301–1310
5. Berghmans S, Murphey RD, Wienholds E, Neuberg D, Kutok JL, Fletcher CD, Morris JP, Liu TX, Schulte-Merker S, Kanki JP, Plasterk R, Zon LI, Look AT (2005) tp53 mutant zebrafish develop malignant peripheral nerve sheath tumors. Proc Natl Acad Sci USA 102:407–412
6. Bretaud S, Lee S, Guo S (2004) Sensitivity of zebrafish to environmental toxins implicated in Parkinson's disease. Neurotoxicol Teratol 26:857–864
7. Darland T, Dowling J (2001) Behavioral screening for cocaine sensitivity in mutagenized zebrafish. Proc Natl Acad Sci USA 98:11691–11696
8. de Jong JL, Zon LI (2005) Use of the zebrafish system to study primitive and definitive hematopoiesis. Annu Rev Genet 39:481–501
9. Driever W, Solnica-Krezel L, Schier AF, Neuhauss SC, Malicki J, Stemple DL, Stainier DY, Zwartkruis F, Abdelilah S, Rangini Z, Belak J, Boggs C (1996) A genetic screen for

mutations affecting embryogenesis in zebrafish. Development 123:37–46
10. Epstein FH, Epstein JA (2005) A perspective on the value of aquatic models in biomedical research. Exp Biol Med 230:1–7
11. Fan L, Moon J, Crodian J, Collodi P (2006) Homologous recombination in zebrafish ES cells. Transgenic Res 15:21–30
12. Fishman MC (2001) Zebrafish: the canonical vertebrate. Science 294:1290–1291
13. Furutani-Seiki M, Sasado T, Morinaga C, Suwa H, Niwa K, Yoda H, Deguchi T, Hirose Y, Yasuoka A, Henrich T, Watanabe T, Iwanami N, Kitagawa D, Saito K, Asaka S, Osakada M, Kunimatsu S, Momoi A, Elmasri H, Winkler C, Ramialison M, Loosli F, Quiring R, Carl M, Grabher C, Winkler S, Del Bene F, Shinomiya A, Kota Y, Yamanaka T, Okamoto Y, Takahashi K, Todo T, Abe K, Takahama Y, Tanaka M, Mitani H, Katada T, Nishina H, Nakajima N, Wittbrodt J, Kondoh H (2004) A systemic genome-wide screen for mutations affecting organogenesis in Medaka, Oryzias latipes. Mech Dev 121:647–658
14. Genade T, Benedetti M, Terzibasi E, Roncaglia P, Valenzano DR, Cattaneo A, Cellerino A (2005) Annual fishes of the genus Nothobranchius as a model system for aging research. Aging Cell 4:223–233
15. Gross JM, Perkins BD, Amsterdam A, Egana A, Darland T, Matsui JI, Sciascia S, Hopkins N, Dowling JE (2005) Identification of zebrafish insertional mutants with defects in visual system development and function. Genetics 170:245–261
16. Grunwald DJ, Eisens JS (2002) Headwaters of the zebrafish – emergence of a new model vertebrate. Nat Rev Genet 3:717–724
17. Guo S (2004) Linking genes to brain, behavior and neurological diseases: what can we learn from zebrafish? Genes Brain Behav 3:63–74
18. Guyon JR, Mosley AN, Zhou Y, O'Brien KF, Sheng X, Chiang K, Davidson AJ, Volinski JM, Zon LI, Kunkel LM (2003) The dystrophin associated protein complex in zebrafish. Hum Mol Genet 12:601–615
19. Haffter P, Granato M, Brand M, Mullins MC, Hammerschmidt M, Kane DA, Odenthal J, van Eeden FJ, Jiang YJ, Heisenberg CP, Kelsh RN, Furutani-Seiki M, Vogelsang E, Beuchle D, Schach U, Fabian C, Nusslein-Volhard C (1996) The identification of genes with unique and essential functions in the development of the zebrafish, Danio rerio. Development 123:1–36
20. Ho S, Thorpe J, Deng Y, Santana E, DeRose R, Farber S (2004) Lipid metabolism in zebrafish. Methods Cell Biol 76:87–108
21. Hong Y, Schartl M (2006) Isolation and differentiation of medaka embryonic stem cells. Methods Mol Biol 329:3–16
22. Jagadeeswaran P, Gregory M, Johnson S, Thankavel B (2000) Haemostatic screening and identification of zebrafish mutants with coagulation pathway defects: an approach to identifying novel haemostatic genes in man. Br J Haematol 110:946–956
23. Kajimura S, Aida K, Duan C (2005) Insulin-like growth factor-binding protein-1 (IGFBP-1) mediates hypoxia-induced embryonic growth and developmental retardation. Proc Natl Acad Sci USA 102:1240–1245
24. Kalev-Zylinska ML, Horsfield JA, Flores MVC, Postlethwait JH, Vitas MR, Baas AM, Crosier PS, Crosier KE (2002) Runx1 is required for zebrafish blood and vessel development and expression of a human RUNX1-CBF2T1 transgene advances a model for studies of leukemogenesis. Development 129:2015–2030
25. Kazianis S, Morizot DC, Coletta LD, Johnston DA, Woolcock B, Vielkind JR, Nairn RS (1999) Comparative structure and characterization of a CDKN2 gene in a Xiphophorus fish melanoma model. Oncogene. 18(36):5088–5099
26. Keller ET, Murtha JM (2004) The use of mature zebrafish (Danio rerio) as a model for human aging and disease. Comp Biochem Physiol C Toxicol Pharmacol 138:335–341
27. Lam SH, Gong Z (2006) Modeling liver cancer using zebrafish: a comparative oncogenomics approach. Cell Cyle 5:573–577
28. Lam CS, Korzh V, Strahle U (2005) Zebrafish embryos are susceptible to the dopaminergic neurotoxin MPTP. Eur J NeuroSci 21:1758–1762
29. Lam SH, Wu YL, Vega VB, Miller LD, Spitsbergen J, Tong Y, Zhan H, Govindarajan KR, Lee S, Mathavan S, Murthy KR, Buhler DR, Liu ET, Gong Z (2006) Conservation of gene expression signatures between zebrafish and human liver tumors and tumor progression. Nat Biotechnol 24:73–75
30. Lam SH, Winata CL, Tong Y, Korzh S, Lim WS, Korzh V, Spitsbergen J, Mathavan S, Miller LD, Liu ET, Gong Z (2006) Transcriptome kinetics of arsenic-induced adaptive response in zebrafish liver. Physiol Genomics 27:351–361
31. Langenau DM, Traver D, Ferrando AA, Kutok JL, Aster JC, Kanki JP, Shuo L, Prochownik E, Trede NS, Zon LI, Look AT (2003) Myc induced T cell leukemia in transgenic zebrafish. Science 299:887–890
32. Langenau DM, Feng H, Berghmans S, Kanki JP, Kutok JL, Look AT (2005) Cre/lox-regulated transgenic zebrafish model with conditional myc-induced T cell acute lymphoblastic leukemia. Proc Natl Acad Sci USA 102:6068–6073
33. Lee KY, Huang H, Ju B, Yang Z, Lin S (2002) Cloned zebrafish by nuclear transfer from long-term-cultured cells. Nat Biotechnol 20:795–799
34. Lin B, Chen S, Cao Z, Lin Y, Mo D, Zhang H, Gu J, Dong M, Liu Z, Xu A (2007) Acute phase response in zebrafish upon Aeromonas salmonicida and Staphylococcus aureus infection: striking similarities and obvious differences with mammals. Mol Immunol 44:295–301
35. Lockwood B, Bjerke S, Kobayashi K, Guo S (2004) Acute effects of alcohol on larval zebrafish: a genetic system for large-scale screening. Pharmacol Biochem Behav 77:647–654
36. McGonnell IM, Fowkes RC (2006) Fishing for gene function – endocrine modelling in the zebrafish. J Endocrinol 189:425–439
37. Meierjohann S, Schartl M (2006) From Mendelian to molecular genetics: the Xiphophorus melanoma model. Trends Genet 22:654–661
38. Milan DJ, Peterson TA, Ruskin JN, Peterson RT, MacRae CA (2003) Drugs that induce repolarization abnormalities cause bradycardia in zebrafish. Circulation 107: 1355–1358
39. Nasevicius A, Ekker S (2000) Effective targeted gene 'knockdown' in zebrafish. Nat Genet 26:216–220
40. Nica G, Herzog W, Sonntag C, Hammerschmidt M (2004) Zebrafish pit1 mutants lack three pituitary cell types and develop severe dwarfism. Mol Endocrinol 18:1196–1209

41. Ninkovic J, Bally-Cuif L (2006) The zebrafish as a model system for assessing the reinforcing properties of drugs of abuse. Methods 39:262–274
42. Nissen RM, Amsterdam A, Hopkins N (2006) A zebrafish screen for craniofacial mutants identifies wdr68 as a highly conserved gene required for endothelin-1 expression. BMC Dev Biol 6:28
43. Nixon SJ, Wegner J, Ferguson C, Méry PF, Hancock JF, Currie PD, Key B, Westerfield M, Parton RG (2005) Zebrafish as a model for caveolin-associated muscle disease; caveolin-3 is required for myofibril organization and muscle cell patterning. Hum Mol Genet 14:1727–1743
44. North TE, Zon LI (2003) Modeling human hematopoietic and cardiovascular diseases in zebrafish. Dev Dyn 228: 568–583
45. Onnebo SM, Condron MM, McPhee DO, Lieschke GJ, Ward AC (2005) Hematopoietic perturbation in zebrafish expressing a tel-jak2a fusion. Exp Hematol 33:182–188
46. Otto EA, Schermer B, Obara T, O'Toole JF, Hiller KS, Mueller AM, Ruf RG, Hoefele J, Beekmann F, Landau D, Foreman JW, Goodship JA, Strachan T, Kispert A, Wolf MT, Gagnadoux MF, Nivet H, Antignac C, Walz G, Drummond IA, Benzing T, Hildebrandt F (2003) Mutations in INVS encoding inversin cause nephronophthisis type 2, linking renal cystic disease to the function of primary cilia and left-right axis determination. Nat Genet 34:413–420
47. Patton EE, Zon LI (2001) The art of designing genetic screens: zebrafish. Nat Rev Genet 2:956–966
48. Patton EE, Widlund HR, Kutok JL, Kopani KR, Amatruda JF, Murphey RD, Berghmans S, Mayhall EA, Traver D, Fletcher CD, Aster JC, Granter SR, Look AT, Lee C, Fisher DE, Zon LI (2005) BRAF mutations are sufficient to promote nevi formation and cooperate with p53 in the genesis of melanoma. Curr Biol 15:249–254
49. Peterson RT, Shaw SY, Peterson TA, Milan DJ, Zhong TP, Schreiber SL, MacRae CA, Fishman MC (2004) Chemical suppression of a genetic mutation in a zebrafish model of aortic coarctation. Nat Biotechnol 22:595–599
50. Poss KD (2004) A zebrafish model of germ cell aneuploidy. Cell Cycle 3:1225–1226
51. Sabaawy HE, Azuma M, Embree LJ, Tsai HJ, Starost MF, Hickstein DD (2006) TEL-AML1 transgenic zebrafish model of precursor B cell acute lymphoblastic leukemia. Proc Natl Acad Sci USA 103:15166–15171
52. Sadler K, Amsterdam A, Soroka C, Boyer J, Hopkins N (2005) A genetic screen in zebrafish identifies the mutants vps18, nf2 and foie gras as models of liver disease. Development 132:3561–3572
53. Sakamoto D, Kudo H, Inohaya K, Yokoi H, Narita T, Naruse K, Mitani H, Araki K, Shima A, Ishikawa Y, Imai Y, Kudo A (2004) A mutation in the gene for delta aminolevulinic acid dehydratase (ALAD) causes hypochromic anemia in the medaka, oryzias latipes. Mech Dev 121:747–752
54. Shepard JL, Amatruda JF, Stern HM, Subramanian A, Finkelstein D, Ziai J, Finley KR, Pfaff KL, Hersey C, Zhou Y, Barut B, Freedman M, Lee C, Spitsbergen J, Neuberg D, Weber G, Golub TR, Glickman JN, Kutok JL, Aster JC, Zon LI (2005) A zebrafish bmyb mutation causes genome instability and increased cancer susceptibility. Proc Natl Acad Sci USA 102:13194–13199
55. Shima A, Mitani H (2004) Medaka as a research organism: past, present and future. Mech Dev 121:599–604
56. Sood R, English MA, Jones M, Mullikin J, Wang DM, Anderson M, Wu D, Chandrasekharappa SC, Yu J, Zhang J, Paul Liu P (2006) Methods for reverse genetic screening in zebrafish by resequencing and TILLING. Methods 39:220–227
57. Stern HM, Zon LI (2003) Cancer genetics and drug discovery in the zebrafish. Nat Rev Cancer 3:533–539
58. Stern HM, Murphey RD, Shepard JL, Amatruda JF, Straub CT, Pfaff KL, Weber G, Tallarico JA, King RW, Zon LI (2005) Small molecules that delay S phase suppress a zebrafish bmyb mutant. Nat Chem Biol 1:366–370
59. Sun Z, Amsterdam A, Pazour G, Cole D, Miller M, Hopkins N (2004) A genetic screen in zebrafish identifies cilia genes as a principal cause of cystic kidney. Development 131: 4085–4093
60. Tanaka K, Ohisa S, Orihara N, Sakaguchi S, Horie K, Hibiya K, Konno S, Miyake A, Setiamarga D, Takeda H, Imai Y, Kudo A (2004) Characterization of mutations affecting embryonic hematopoiesis in the medaka, Oryzias latipes. Mech Dev 121:739–746
61. Wakamatsu Y, Ju B, Pristyaznhyuk I, Niwa K, Ladygina T, Kinoshita M, Araki K, Ozato K (2001) Fertile and diploid nuclear transplants derived from embryonic cells of a small laboratory fish, medaka (Oryzias latipes). Proc Natl Acad Sci USA 98:1071–1076
62. Walter RB, Kazianis S (2001) Xiphophorus interspecies hybrids as genetic models of induced neoplasia. Inst Lab Anim Res J 42:299–321
63. Wienholds E, Schulte-Merker S, Walderich B, Plasterk R (2002) Target-selected inactivation of the zebrafish rag1 gene. Science 297:99–102
64. Wienholds E, van Eeden F, Kosters M, Mudde J, Plasterk R, Cuppen E (2003) Efficient target-selected mutagenesis in zebrafish. Genome Res 13:2700–2707
65. Wittbrodt J, Shima A, Schartl M (2002) Medaka—a model organism from the Far East. Nat Rev Genet 3:53–64
66. Zon LI (1999) Zebrafish: a new model for human disease. Genome Res 9:99–100
67. Zon LI, Peterson RT (2005) In vivo drug discovery in the zebrafish. Nat Rev Drug Discov 4:35–44

Genetic Counseling and Prenatal Diagnosis

Tiemo Grimm and Klaus Zerres

Abstract The expanding knowledge in human genetics has led to practical applications at an increasing rate – especially in genetic counseling and genetic screening. Conventional and invasive diagnostic procedures have been complemented or entirely replaced by genetic-testing. DNA tests allow us to predict diseases and to modify risk figures. With increasing numbers of both diagnostic and predictive genetic tests available, genetic counseling is becoming more important in virtually all fields of clinical practice. The traditional areas of genetic counseling included pediatrics (assessment of children with developmental delay and dysmorphic features) and obstetrics (prenatal diagnosis). However, as the genetic basis of more and more diseases is unraveled, genetic counseling is now increasingly requested from other disciplines, including neurology, oncology, ophthalmology.

Contents

25.1 Genetic Counseling.. 845
    25.1.1 Origins and Goals of Genetic Counseling 845
    25.1.2 Genetic Diagnosis.. 848
    25.1.3 Recurrence Risk... 851
    25.1.4 Communication and Support 852
    25.1.5 Directive Vs. Nondirective Genetic Counseling.. 852
    25.1.6 Assessment of Genetic Counseling and Psychosocial Aspects................................. 853

25.2 Prenatal Diagnosis 854
    25.2.1 Indications for Prenatal Diagnosis.............. 854
    25.2.2 Techniques of Prenatal Diagnosis............... 856

25.3 Conclusions 863

25.4 Appendix Example of a Bayesian Table................... 863

References... 864

T. Grimm (✉)
Institute of Human Genetics, Division of Medical Genetics,
University of Würzburg, Biozentrum, Am Hubland, 97074,
Würzburg, Germany
e-mail: tgrimm@biozentrum.uni-wuerzburg.de

K. Zerres
Institute of Human Genetics, Medical Faculty, RWTH University,
Pauwelsstr. 30, 52074 Aachen, Germany
e-mail: kzerres@ukaachen.de

25.1 Genetic Counseling

Genetic counseling is an important area of applied human genetics. Patients request advice or are referred by their physicians for counseling to help them to understand biological facts and the medical implications and recurrence risks of genetic diseases. As the public media and the medical literature disseminate more news about genetics, public and medical interest in genetic disease is expanding [25, 30, 31, 53, 55].

25.1.1 Origins and Goals of Genetic Counseling

Genetic counseling is a communication process that deals with human problems associated with the occurrence or the risk of occurrence of genetic disorders in individuals or families. This process involves an attempt by one or more appropriately trained specialists to help an individual or family to:

1. Comprehend the medical facts, including origin, diagnosis, and probable course of a given disorder and the available management

2. Appreciate the way heredity contributes to the disorder and the risk for recurrence in specific relatives
3. Understand the alternatives for dealing with the risk of recurrence

25.1.1.1 Definition of Genetic Counseling

Genetic counseling refers to the totality of activities that:

1. Establish the diagnosis
2. Assess the recurrence risk
3. Communicate the likelihood of recurrence for the patient and family
4. Provide information about the many problems raised by the disease and its natural history, including the potential medical, economic, psychological, and social burdens
5. Provide information regarding potential reproductive options, including prenatal diagnosis
6. Provide for referral of patients to appropriate specialists

The range of problems and questions arising during genetic counseling is wide. Generally, only 30–50% of patients and families turn out to have classic genetic diseases such as monogenic diseases or chromosomal aberrations. Many consultations deal with various birth defects, mental retardation, delayed development, dysmorphic-looking children, short stature, and similar problems, which may or may not have a genetic cause.

Genetic counseling is usually carried out by physicians who have specialist training in medical genetics. They usually work in medical genetics clinics and carry out much of the information gathering, counseling, and follow-up. In several European countries trained social workers, or genetic counselors in the USA, are also involved in genetic counseling. In fact, in the USA nonphysician genetic counselors have an increasingly important role. Genetic counseling services are often provided by post-baccalaureates (largely women) with 2 years' specialized training in medical genetics. These persons usually work as team members with clinical geneticists or with obstetricians, and increasingly with commercial genetic-testing laboratories.

Genetic counseling is increasingly applied in an interdisciplinary manner. Important examples are seen in the case of neuromuscular disorders. A rational diagnostic approach in this field should be discussed with specialist neurologists, pathologists, and geneticists.

In clinical oncology, e.g., in families with familial breast cancer or hereditary nonpolyposis colorectal cancer (HNPCC), close contact of involved experts trained in different fields and regular interdisciplinary exchange are essential. The role of genetic counseling in familial or potentially familial cancer in particular has gained significantly in importance owing to the identification of more and more cancer-predisposing genes. With increasing knowledge about the complex genetic basis of diseases, clinical geneticists attach great importance to the provision of information and to the medical care of affected families.

25.1.1.2 Origins of Genetic Counseling

Genetic counseling is medically oriented. It is considered inappropriate to include eugenic considerations. Couples asking for advice are encouraged to make their own reproductive decisions. Even though under 2% of human afflictions of medical relevance follow a monogenic pattern of inheritance, an increasingly large number of common diseases have a genetic component that can be assessed with genetic tests that in most cases do not yet have clinical utility. There are good medical and genetic reasons for genetic counseling, which needs to be provided in humane and ethically responsible ways.

25.1.1.3 Indications for Genetic Counseling

Important questions in genetic counseling are as follows:

25.1.1.3.1 Birth of a Child with a Congenital or Developmental Disorder

If a child with birth defects or developmental delay is born to healthy parents, the most common question concerns the risk to other children. For example, the recurrence risk for carrier status of balanced chromosomal translocations is an important issue (see Sect. 25.2.1.3). In order to answer this question, a correct diagnosis of the child's condition must be established.

25.1.1.3.2 A Parent Is Affected

The diseases of a parent are a frequent concern leading to the question of recurrence risk in the parents' children. If the consultand is affected or even pregnant, the question of the impact of pregnancy on the course of the disease is another important issue.

25.1.1.3.3 Diseases or Developmental Disorders in Relatives of a Consultand

As a rule, this situation requires communication with multiple members of a family. Formal genetic aspects are often crucial. Careful pedigree analysis and risk calculations frequently yield relatively low risks. Occasionally an increased risk can be ruled out entirely. The issue of predictive testing for late-onset diseases in unaffected but at-risk family members must be considered carefully.

25.1.1.3.4 Age Risks

Owing to an increasing number of pregnancies in older women in western societies, the demand for counseling on this topic is increasing. Most consultands are aware that advanced maternal age increases their risk of having a child with a chromosomal disorder. Even though the age-related risk for trisomy 21 only amounts to less than 1% at age 35, and to less than 5% at age 45, these age-related risks are usually overestimated and considerably lower than perceived by the counselees [3]. Maternal serum screening for aneuploidy and ultrasound (see Sects. 25.2.1.1 and 25.2.2.2) have become part of routine prenatal care in many countries. Information on the limitations and risks of antenatal diagnostic measures should be provided by trained professionals. Elevated paternal age increases the risk for point mutations, but again, this risk is relatively low (1% or less) [71].

25.1.1.3.5 Recurrent Pregnancy Loss or Stillbirths

Approximately every eighth pregnancy ends in miscarriage, the cause of which remains unknown in many cases. Chromosomal aberrations are the most frequent finding among early pregnancy losses, with triploidy, trisomy 16, and monosomy X accounting for around 10% of spontaneous abortions prior to week 12. However, the vast majority are lethal conditions which occur spontaneously, with usually low recurrence risks that mean routine chromosome analysis of tissues from spontaneous pregnancy loss is of limited clinical value. Parental chromosome analysis should, however, be carried out after three unexplained abortions in order to exclude carrier status for balanced chromosomal translocations. Such translocations are found in about 5% of couples with two or more previous abortions. In addition to frequent and recurrent pregnancy losses, the family history in translocation carrier families includes births of both clinically normal individuals and children with birth defects. Male carriers may have oligo-astheno-teratozoospermia syndrome with fertility problems.

25.1.1.3.6 Teratogenic or Mutagenic Effects

The question of a child's risk attributable to exposure to exogenous factors such as drugs, radiation, alcohol, or prenatal infections during pregnancy (teratogenic risks) is a frequent indication for genetic counseling [24]. Again, potential risks for adverse effects are often overestimated. The number of drugs with proven teratogenic or mutagenic effects is rather small [39]. Risks inherent in radiation exposure are also often overestimated.

In order to mitigate excessive concerns of pregnant women, accurate information should be provided. A history of drug and alcohol consumption should be carefully evaluated and discussed. It has become clear that even low doses of alcoholic beverages can have adverse effects on intrauterine development. Because there is no known safe amount of alcohol consumption during pregnancy, the American Academy of Pediatrics recommends abstinence from alcohol for woman who are pregnant or who are planning a pregnancy [1]. Prenatal infections represent complex situations which require interdisciplinary management.

25.1.1.3.7 Consanguinity

First cousins and more remote relatives who are contemplating marriage occasionally ask for advice about the risks of having children with inherited diseases. Marriages between first cousins are illegal in 30 states of the United States of America. Consanguinity definitely increases the risks of disease caused by homozygosity for recessive genes (Chap. 13), but the absolute risks are relatively low in families without evidence for genetic disorders. It has been estimated that the rate of various diseases, birth defects, and mental retardation among offspring of first-cousin matings is at most twice the background rate faced by any given couple; thus, the chance that a child from such a mating will be normal is around 93–95%. These risks are still lower for more remote consanguinity, and thus are difficult to separate from the population background rate for such disorders [15]. There are no additional risks for offspring of a normal person married to an unrelated person when one partner has consanguineous parents. On the other hand, the risks are considerable for children of incestuous matings involving first-degree relatives, such as sib-sib and father-daughter matings; there may be up-to a 50% risk that such a child will be affected by severe abnormality, childhood death, or mental retardation [2, 33, 58]. It is remarkable that defects in offspring of consanguineous matings mostly manifest themselves as nonspecific congenital malformations, childhood death, and mental retardation rather than as well-defined autosomal-recessive disease entities. However, detailed searches for the many different recessive inborn errors of metabolism have rarely been carried out, and it is likely that a significant proportion of childhood deaths involve such unrecognized recessive disorders.

25.1.2 Genetic Diagnosis

Accurate diagnosis of a genetic disease using all the modalities of modern medicine is the cornerstone of genetic counseling. Diagnostic accuracy is emphasized, since similar phenotypes may sometimes have different modes of inheritance or may not be inherited at all. The family history is important, because a clear-cut pattern of inheritance such as that in the case of autosomal-dominant traits often provides the basis for counseling when a definitive diagnosis may not be clear. Previous medical and hospital records are helpful in arriving at a correct diagnosis. Since many genetic diseases are associated with somewhat characteristic facial features, inspection of photographs of family members may be helpful. Chromosomal examinations are frequently required in the diagnosis of complex birth defects. Since many genetic diseases are rare, even trained medical geneticists and specialists in a given field of medicine may have difficulty in arriving at an accurate diagnosis. They cannot be equally knowledgeable about all genetic diseases in every area of medicine, but do need to be aware of recent monographs and computerized expert systems to establish the appropriate diagnosis. The Catalog of Mendelian Traits in Man published by McKusick [43] and its computerized version OMIM are helpful [51]. It is essential in clinical genetics to have a good and up-to-date library and to know how to consult the current literature. Because of the rapid expansion of knowledge, utilization of the journal literature, as opposed to textbooks and monographs, is more important in clinical genetics than in most fields of medicine. This is facilitated today by computerized searches with such syndrome identification programs as POSSUM [52] or the London Dysmorphology Database [42]. However, their proper use requires knowledge, experience, and a critical mind.

Parents whose infants are stillborn or die in the neonatal period often request genetic counseling about recurrence risks. Usually little or no information is available on the specific abnormality of the stillbirth, since no pathological or other diagnostic study has been conducted. It has been recommended that as a minimum a gross autopsy, photographic and radiographic records, and bacterial cultures be performed in all cases of stillbirth or early neonatal death. Such information is essential for subsequent genetic counseling [27]. A definitive diagnosis often cannot be made even by experienced specialists owing to the enormous complexity of development and its possible perturbation by frequently unknown genetic, epigenetic, and environmental factors. Fewer diagnostic uncertainties occur with monogenic diseases than with various birth defects. However, even in this area the growth of the McKusick catalog over the years, i.e., from 866 defined gene loci in 1971 to 11,867 in April 2009 [51] attests to the rapid expansion of knowledge in this field.

25.1.2.1 Molecular Diagnosis

As more genes are cloned and the molecular nature of mutations causing disease becomes known, direct DNA diagnosis of genetic disease is increasingly possible. Unlike indirect diagnosis using linked DNA markers, a family study is not required for direct genetic-testing. However, the exact nature of the mutation to be detected must usually be known. It is good practice to isolate and store DNA from patients with genetic diseases in DNA banks for appropriate future analysis. The resultant information may be of great help in counseling family members in the future.

If the gene in question has been isolated, but the mutation remains unknown, a closely linked or intragenic DNA polymorphism (RFLP, VNTR, CA repeat, SNIP) can be used to track the mutant gene by cosegregation of the marker when the exact nature of the mutation is unknown. The possibility of nonallelic genetic heterogeneity must be kept in mind, however, as this may be a source of error.

Where neither the gene nor its mutations have been defined, linked DNA polymorphism can still be helpful. However, as with intragenic DNA polymorphisms, the exact "phase" of the DNA marker in relation to the disease gene must be known. In such situations, a sufficient number of family members is often not available, precluding the reliable application of indirect genetic-testing. Problems of genetic heterogeneity and possible recombination between the linked DNA marker and the disease gene render the indirect approach less than 100% accurate.

The examples of Duchenne and Becker muscular dystrophy illustrate these principles. About two-thirds of all Duchenne and Becker mutations are caused by deletions and about 7% by duplications of the X-linked dystrophin gene [16, 70]. Direct DNA diagnosis for these deletions and duplications is relatively simple [e.g., by in situ hybridization (see Sect. 3.4.4) and multiplex ligation-dependent probe amplification (MLPA; see Sect. 3.4.3.4)] [56, 57]. These techniques can be used for prenatal diagnosis and for carrier detection. If a deletion cannot be detected, screening for one of the many different missense point mutations that can cause the disease is not always practical. However, indirect DNA diagnosis using DNA markers of the dystrophin gene combined with a family study can be attempted. Since the gene is very large, intragenic crossovers are relatively frequent (about 5–10% [29]). This problem can be overcome by using flanking markers on either side of the disease gene. In view of these complexities, measurement of creatine phosphokinase levels that are elevated in Duchenne muscular dystrophy carriers (but less so in Becker muscular dystrophy carriers) may aid further in carrier detection [35].

Genetic advice concerning multifactorial conditions, such as many birth defects, common diseases of late-onset, and major psychoses, lacks the precision that can be achieved in counseling patients with Mendelian disorders. Empirical risk figures need to be used based on the frequency of recurrence of the disease in many affected families. These recurrence risks are usually lower than those in the Mendelian diseases, ranging between 3 and 5% for many common birth defects, such as neural tube defects, cleft lip, and cleft palate. Risks to first-degree relatives (sibs, parents, and children) for common diseases such as hypertension, schizophrenia, and affective disorders are in the order of 10–15%.

In contrast to monogenic disorders, the risk in multifactorial disorders increases with an increasing number of affected relatives and with increasing severity of the disease; risk is usually negligible for distant relatives.

A careful search for the rare monogenic variety of a disease that appears multifactorial must always be kept in mind. For example, rare patients with gout may have an X-linked monogenic disease caused by hypoxanthine-guanine phosphoribosyl-transferase deficiency. Alternatively, their gout may be caused by autosomal-dominant phosphoribosylpyrophosphate synthetase deficiency. Among male patients under the age of 60 years with coronary heart disease about 5% have familial hypercholesterolemia – an autosomal-dominant trait.

Transmitted chromosome abnormalities, such as translocations, do not segregate by Mendelian ratios, and counseling must be based on empirical risk figures.

Citing percentage figures of absolute recurrence risk is more meaningful to a family than relative risks based on the relative likelihood of the disease compared with that in the general population. A 100-fold increase in risk for a condition that occurs in the population with a frequency of 1:10,000 carries an actual risk of only 1:100 – a small recurrence risk. For Mendelian conditions the recurrence risks are fixed regardless of whether several affected children or none have previously been born. Chance has no memory! In multifactorial diseases such as congenital heart disease or cleft lip and palate, the following considerations apply: if two or more first-degree relatives are affected in a given family, more disease-producing genes are likely to be involved in that family, and the risk for future offspring becomes higher than the usual 3–5% [30]. However, differentiation from an autosomal-recessive variety of the condition with a recurrence risk of 25% may sometimes be difficult. Detailed discussions of the approaches to genetic counseling and risk data for many different types of diseases can be found elsewhere [30].

25.1.2.2 Heterozygote Detection

It is particularly important to detect heterozygotes in sisters of boys affected with X-linked recessive diseases, such as hemophilia and Duchenne- or Becker-type muscular dystrophy. Regardless of their partner's genetic constitution, there is a 50% risk that the sons of female heterozygotes will be affected. In contrast, autosomal-recessive diseases become evident only when both parents are heterozygotes; a heterozygote sib of an affected patient must mate with another heterozygote for the disease to occur. The probability that an unrelated mate of a person who is a carrier of an autosomal-recessive disease carries the same mutation is usually quite low.

Biochemical and functional tests must be carefully standardized on normal subjects and obligate heterozygotes before they are applied for individual carrier identification. The detection of heterozygotes is accurate and simple in the hemoglobinopathies. An increasing number of heterozygote conditions for various autosomal enzyme deficiencies (e.g., hexosaminidase deficiency causing Tay-Sachs disease) can be recognized by enzyme assays as well as by DNA tests [34]. However, measuring mutant enzyme activity in a group of heterozygotes will often demonstrate a wide range of enzyme activity in such subjects. Thus, both low and high enzyme activity may be found in different heterozygotes for the same mutation. Such subjects can reliably be characterized as heterozygotes only by DNA testing. In contrast, enzyme levels, which may be useful genetically and clinically, will not be demonstrated by DNA tests.

If there is overlap in laboratory results of tests for enzyme activity between normal subjects with a low value and carriers with a high value, the significance of an identical laboratory result in various individuals may differ depending upon the a-priori probability of the tested person being a carrier. Tests that are excellent for carrier detection in sisters of males affected with X-linked diseases may give too many false positives in screening studies of extended kindreds or particularly in the general population, where the probability that the tested subject is a carrier is low.

For example, 5% of the average female population would be identified as having a high risk of being carriers of hemophilia using the same standards that identify sisters of hemophilic boys as heterozygotes.

In some of these situations additional statistical techniques may be helpful for refinement of a genetic prognosis. Assume a woman's brother is affected with an X-linked recessive condition; a maternal uncle is also affected. She therefore has a 50% risk of being heterozygous. Assume that she already has two nonaffected sons, and that a test for heterozygote detection is not available. The information that her two sons are normal reduces her chance of being a carrier. Alternately, such a woman may have a negative result for a test that detects 90% of heterozygotes. In this case her risk of being a carrier is very low. Special statistical methods are needed for calculation of the exact recurrence risks in such complex situations [47, 72] (Appendix).

The increasing availability of DNA markers facilitates carrier diagnosis in X-linked diseases. In any given diagnostic problem the simplest and most direct approach should be selected, which increasingly is direct DNA diagnosis. However, use of biochemical tests is often necessary and complementary. Information from DNA markers can be combined with biochemical tests and pedigree information. The fragile X mental retardation syndrome is a good example of how DNA diagnosis has become very helpful. Since the number of CGG triplet repeats responsible for the syndrome can be easily assessed, DNA diagnosis can discriminate between affected males who have a large triplet expansion (i.e., more than 200 CGG repeats) and normal transmitting males who carry a premutation (i.e., between 55 and 200 CGG repeats). The nonaffected carrier daughters of such transmitting males will have only moderate expansion of CGG triplets, if any at all. They are usually not affected, but do carry an increased risk for their sons. For women who have inherited the expansion from their mothers and carry a full mutation there is a 50% risk of at least mild learning difficulties.

Carrier detection for mutations for autosomal-recessive disorders is often required for close relatives of affected children (i.e., healthy siblings, aunts, uncles, nephews, and nieces). Depending on the disease and the detection rate, sensitivity and specificity of possible heterozygote tests, genetic counseling is essential for discussion of the limitations and possible consequences of the available tests and the assessment of possible risks. The availability of a mutational analysis of the affected relative is essential for risk assessment.

Owing to the possibility of compound heterozygosity for mutations a prediction of possible consequences can be difficult or even impossible and might be an important point in genetic counseling. A good example is cystic fibrosis (CF). CF is caused by mutations in the gene on chromosome 7 encoding the CF transmembrane conductance regulator (CFTR) gene, which regulates sodium transport and potassium channels and conducts chloride across the cell membrane. The primary pathophysiology that leads to chronic illness in CF patients is mucosal obstruction of exocrine glands. The most common mutation in the CFTR gene is ΔF508; however, over 1,400 other mutations have also been reported (see www.genet.sickkids.on.ca/cftr/app). The ΔF508 mutation has been reported in a compound heterozygous state with several other mutations, so that reliable genotype-phenotype correlations can be made for some compound heterozygotes. For example, in one study patients with the genotype R117H/ΔF508 significantly differed from age-matched and sex-matched ΔF508 homozygotes because they were older at the time of diagnosis, were more likely to have pancreatic sufficiency, and had lower sweat chloride concentrations [13]. For other compound heterozygotes there are fewer reports in the literature, making prediction of phenotypic consequences difficult.

Population screening, for example for hemoglobinopathies in populations of African origin or Tay-Sachs disease and some other recessive conditions in Jewish communities, addresses difficult ethical issues regarding abortion. Religious tradition can impact on decision making around genetics testing and requires a detailed understanding of culturally sensitive ethical care.

25.1.3 Recurrence Risk

Genetic risks in Mendelian diseases are clearly defined and depend on the specific mode of inheritance. The actual clinical risks to the patient, particularly in autosomal-dominant inheritance, depend upon variable penetrance and expression, especially in late-onset disorders. Patients are more interested in the actual recurrence risk of the clinical feature than in the formal genetic risks alone. In diseases with decreased penetrance the actual recurrence risk is lower than the formal risk of genetic transmission. For example, an offspring's risk of an autosomal-dominant disease with 70% penetrance is 35% rather than 50% ($0.5 \times 0.7 = 0.35$). The risk declines with late-onset diseases, as a person remains unaffected beyond the age at which the disease first becomes manifest.

The McKusick catalog is available as a computerized data base on-line (OMIM) and is updated continuously [51]. In addition, keywords can be introduced, so that more comprehensive cross-searches that also allow access to abstracts of recent articles may be possible and helpful.

25.1.3.1 Recurrence Risk for Multifactorial Conditions

For most complex "common" diseases with a polygenic basis, empirical risks for close relatives exist. These risks usually increase with an increasing number of affected relatives and with worsening severity of the disease. With increasingly remote relationship to the affected person the risk usually decreases. In several disorders with different incidences depending on the sex of affected persons, recurrence risks for close relatives can be different, with the higher risk for persons whose gender has the lower incidence. This phenomenon is known as the Carter effect: the incidence in relatives is higher when the index case is of the less commonly affected sex. For example, in pyloric stenosis the incidence is highest in the sons of affected women and lowest in daughters of affected men.

25.1.3.2 Reproductive Options and Alternatives

If a couple decides that the risks of having children of their own are too high, several options besides contraception should be discussed. Adoption is becoming less practicable because fewer babies are available. Artificial insemination by a donor (heterologous insemination) may be acceptable for some couples to avoid autosomal-recessive disease or autosomal-dominant disease contributed by the male partner, but often raises difficult ethical questions.

25.1.3.3 Predictive Testing

Predictive testing in families in which genetic diseases may not be obvious at birth but may manifest sometime later is an important reason for genetic counseling. For some conditions the identification of gene carriers may be lifesaving if followed by suitable therapy. A sib of a patient with Wilson disease has a 25% chance of being affected, but may be too young to exhibit overt symptoms. Children of patients with the autosomal-dominant condition familial adenomatous polyposis have a 50% chance of being gene carriers, and if they inherit the mutant *APC* gene they have a nearly 100% risk of malignant transformation in one of the many polyps seen in this condition. In general, attempts should be made to examine relatives when a genetic condition causes serious preventable or treatable diseases. Information should always be given to a relative who is at-risk by the family members themselves.

Predictive testing in genetic diseases of late-onset has been performed for many years in Huntington's disease, since the identification of the underlying mutation in 1993. Much literature is available with a focus on the psychological impact for tested persons at-risk and their families [31]. Huntington's disease is an example of a disease with no relevant therapeutic options for persons who test positive, while therapeutic options themselves are in the focus of research in hereditary cancer syndromes such as polyposis coli, Lynch syndrome, and familial breast cancer. Growing data collections including data on affected families and systematic studies about the experience and follow-up with different therapeutic options have made genetic counseling one of the important keystones in family care of these families [66].

An interdisciplinary concept in genetic counseling is useful for many diseases, which has been shown for many years in Huntington disease with detailed guidelines for the application of predictive testing [65].

25.1.4 Communication and Support

The meaning of genetic risks must be conveyed in terms that can be understood by patients. The probability that 3–4% of all children of healthy parents are born with birth defects or possible genetic diseases should be communicated as a baseline risk figure that applies to the general population. There may be problems in communicating the extent of uncertainty. For example, with a sporadic case of an undiagnosable birth defect the risk might be zero if the disease is non-genetic, 2–3% if there is a multifactorial etiology, and 25% if it is caused by an autosomal-recessive trait. Therefore, an empirical risk is based on the probability of the various possibilities. Such a risk figure might be 5% on the assumption that monogenic recessive varieties of this birth defect tend to be rare. Both the physical and the emotional burden of the disease must be discussed. It is well-known that very severe conditions that are invariably fatal in early life often place a less severe burden to the family than do those associated with chronic or slowly progressive diseases. Various reproductive options and alternatives need to be discussed.

Since problems may be complex and may prove to be emotionally difficult for the counselee, it may be necessary to have several counseling sessions. In any case, the counselor should provide a written summary of the counseling session using lay language.

25.1.5 Directive Vs. Nondirective Genetic Counseling

In the tradition of paternalistic medicine, which is now obsolete, occasional physicians are still accustomed to giving directive advice for or against future pregnancies. The practice of medical genetics has clearly evolved in the direction of nondirective counseling. Some of this nondirectiveness may have sociological reasons. When genetic counseling began in the United States some 40 years ago, it was usually carried out by nonphysician geneticists, who lacked the medical profession's tradition of giving directive (paternalistic) advice. Nondirective genetic counseling mirrors recent trends to increasing patient autonomy. Since each family is unique and reactions to risks vary, nondirectiveness fosters mature decision making. However, absolutely neutral advice is rarely possible or even desirable. The person or family requesting advice usually wants and needs more than a computer-like professional who only dispenses facts. The counselor may unconsciously emphasize the more positive or the more frightening aspects of a given disease. These feelings tend to affect the counseling process directly or indirectly, often through nonverbal clues. Not all couples have the necessary educational background and social or emotional maturity to make fully informed decisions. In addition, many couples expect the medical geneticist, whom they consider an experienced expert on their disease, to assist them in arriving at a decision they can live with. "What would you do if you were in our position?" is a frequent question from counselees, regardless of background. However, since a couple's economic situation, religious affiliation, and cultural background (for example) may differ substantially from that of the counselor, the counselor's choices for his/her own circumstances would usually not be appropriate. Reproductive decisions differ widely among individual couples even if the genetic facts and the disease burden are identical.

As predictive testing for late-onset diseases becomes increasingly possible, nondirective advice no longer applies when a disease can be prevented or treated by appropriate measures. The relevant medical recommendations on preventing and treating the disease must be given (including all options).

25.1.6 Assessment of Genetic Counseling and Psychosocial Aspects

Most professionals engaged in genetic counseling agree that counselees should achieve sufficient understanding of the medical significance and social impact of the disease to allow them to make appropriate decisions. Some observers have measured the success of genetic counseling with regard to increased recurrence risks in a further pregnancy by subsequent reproductive behavior. If more couples with a high risk (>10%) than couples with a low risk were deterred from reproduction, genetic counseling would be considered to have been successful. This result has in fact been noted in several studies. Such a narrow end-point is considered an inadequate evaluation of genetic counseling. It would be better to know whether full sharing of information and comprehension of the disease and its recurrence risks have been achieved, and whether all needs for information and psychological and social support have been met.

Various studies of genetic counseling agree that many patients are confused about recurrence risks and do not fully understand the nature of the disease. In the late 1970s a large study was carried out by a group of sociologists in 47 genetic counseling clinics in the United States; 205 counselors and over 1,000 female counselees were involved [61]. Many different conditions were included, and both counselors and counselees were questioned about their experiences and assessment of the counseling process. The results showed that counselors tended to emphasize recurrence rates during the counseling sessions, while counselees were often interested in causation, prognosis, and treatment of the disease – an area which, according to the counselees, was seldom discussed as fully as desired. While both counselors and counselees were generally more interested in the medical and genetic aspects of the consultation, counselees occasionally had psychosocial concerns which were not addressed by the professionals.

This study found that 54% of counselees who were given a risk level and 40% of those given a diagnosis were unable to report these data shortly after counseling. This failure to correctly memorize risk or diagnosis occurred regardless of whether MDs, PhDs, or genetic counselors had carried out the counseling. It was unrelated to the experience of the counselor. Counselors with many years of experience had no better results than more recent graduates. Several other studies have yielded results that are substantially better, but by no means perfect [21]. Usually, but not always, education is found to be correlated with a better level of understanding.

Genetic counseling services have been used more extensively by families with good educational backgrounds than by less advantaged population groups. Couples who are motivated to learn about the disease and the risk of its recurrence are more likely to be affected in their reproductive decisions by the information provided than those who have been referred and are not always certain about the purpose of the genetic consultation. Thus, self-referred patients also tend to have better comprehension of genetic counseling information.

Another study investigated perception of counseling information [40, 41]. Perception of recurrence rates was often not used by the counselees in the probabilistic sense represented by the figures given. Percentage risks were more frequently perceived as binary, i.e., even with lower risks it was believed that the disease either would or would not occur, with all the attendant fears of recurrence. Parents were then overwhelmed by multiple uncertainties, such as how to make reproductive choices, how others would react to their decision, what it would mean to have an affected child, and whether they would be able to fulfill their role as parents. Such perceptions appeared more important for decision making than the actual facts of diagnosis, prognosis, and risk. These data show that there is often a discrepancy between the mind-set of the scientifically oriented counselor and the reflection processes of the counselees, who find it difficult to deal with probabilistic information. Bridging this gap is a real challenge.

Genetic counseling – as currently practiced in most countries – places less specific emphasis on emotional aspects than "counseling" activities in other areas, such as psychological and marriage counseling. Some observers have recommended that more attention be given to the psychodynamic aspects of genetic disease [8, 20, 36]. If there are deep-seated psychological problems, referral to a psychiatrist or psychotherapist is the most appropriate course of action. An empathic and understanding approach to families with awareness of the many social and psychological aspects of the disease and support in these matters, however, needs to be encouraged. Genetic counseling is more than mere diagnosis, risk assessment, and factual dispensation of information.

There are imperfections in the genetic counseling process as currently practiced. Nevertheless, most educated counselees who receive definite information about the matters troubling them usually appear to be satisfied. The majority of counselees with low risks are relieved to find that their actual risks are much lower than they had feared.

The interaction of patients and professionals in any encounter has many variables, and scientific study of this process is difficult. Nevertheless, genetic counseling, as a new field, demands further investigation of the process, its psychosocial effects, and outcomes, so that optimum results can be worked out. Controlled studies comparing patients who received counseling with those with a similar disease who did not would be interesting.

25.2 Prenatal Diagnosis

In 1956 Edwards discussed for the first time the possibility of "antenatal detection of hereditary disorders" [17]. The first use of amniotic-fluid examination in the diagnosis of genetic diseases was reported by Fuchs and Riis in 1956 [26]. They were able to determine the fetal sex from cells found in amniotic-fluid based on the presence or absence of the Barr body, the inactivated X-chromosome.

The improvement in cytogenetics and in somatic cell genetics led to the introduction of prenatal diagnosis in the late 1960s [19]. In 1966 Steele and Breg showed that cultured amniotic-fluid cells were suitable for fetal karyotyping [63]. In 1968 Nadler reported one of the first cases of trisomy 21 diagnosed after amniocentesis [48].

In cases where prenatal diagnosis is being considered in genetic counseling, several risk factors for the unborn child must be examined:

1. Family history of the pregnant woman and her pregnancy history
2. The possibility of consanguinity, depending on the ethnical origin
3. Exposure to mutagenic and teratogenic agents
4. Use or abuse of medicines and drugs

25.2.1 Indications for Prenatal Diagnosis

In most cases prenatal diagnosis is performed with the aim of diagnosing or excluding a serious disease or condition, and may be followed by termination of pregnancy. In this situation prenatal diagnosis is usually requested if there is a known risk for a clearly defined genetic disorder, which can be examined in the fetal cells, in the amniotic-fluid, in the fetal blood, or in the morphology or the skin of the fetus. Also some basic factors should be taken into consideration [30]:

1. Severity of disorder
2. Possibility of treatment
3. Acceptance of termination of an affected pregnancy
4. Availability of an accurate prenatal diagnostic test
5. Assessment of the genetic risk

In a second group, prenatal diagnosis is performed with the aim of diagnosing a fetal condition prenatally in order to start a therapy immediately after birth or even prenatally. An example is adrenal hyperplasia. Prenatal diagnosis may also be helpful to define the risks of prenatal infections.

The different problems of prenatal diagnosis require technical experience and proven authority, and also early and efficient cooperation between medical specialists in many different fields of activity.

25.2.1.1 Risk Assessment for Chromosomal Disorders

25.2.1.1.1 Maternal Age

The most common indication for prenatal diagnosis is above-average maternal age. The chief condition for which children born to women who become pregnant

Fig. 25.1 Approximate incidence of trisomy 21 at time of amniocentesis (16 weeks). (Data after [23])

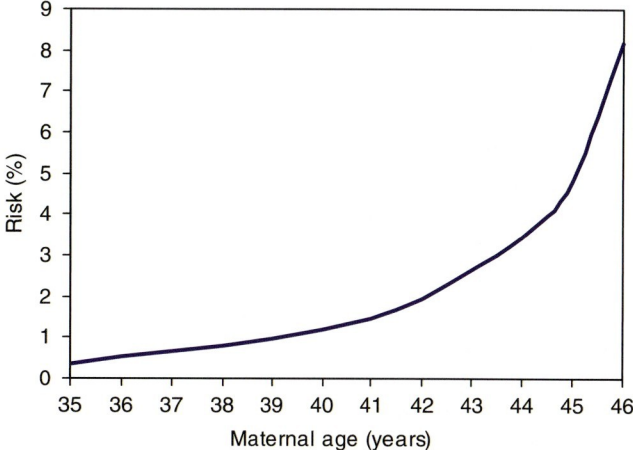

when already at a relatively advanced age are at-risk is Down syndrome (trisomy 21; Fig. 25.1).

There is no standard criterion for determining at what age a pregnant woman should be offered the option of invasive prenatal testing for fetal chromosome analysis. In most countries amniocentesis is offered somewhat arbitrarily to women aged 35 years or over. In the USA courts have considered a physician to be negligent if he fails to offer prenatal diagnostics to older mothers. With the advent of high-resolution ultrasound and first-trimester maternal serum screening, age has become a less important indication for invasive prenatal diagnosis.

About 8.1% of all trisomy 21 cases are caused by paternal nondisjunction in meiosis I or II [32]. However, in these cases no correlation with advanced paternal age has been reported.

25.2.1.1.2 First-Trimester Screening

First-trimester screening facilitates noninvasive risk assessment for certain chromosomal disorders. Assuming a 5% false-positive rate, the detection rate is 30% for age-based screening (age cut-off at 37 years). If nuchal translucency measurement and maternal serum concentrations of free beta-human chorionic gonadotroplin (hCG) and pregnancy-associated plasma protein A (PAPP-A) are included, the detection rate increases to about 90% of all pregnancies with trisomy 21 (Down's syndrome; see also "Maternal Blood Sampling" in Sect. 25.2.2.2). The increased predictive value of these integrated noninvasive tests seems to be resulting in a decline of invasive prenatal diagnosis procedures. At present, most institutions do not perform abortions for Down syndrome based on the results of noninvasive tests only, but require in addition confirmation by amniocentesis.

25.2.1.2 Previous Aneuploidy

A previous child with Down syndrome or other autosomal trisomy slightly increases the risk of recurrence. The recurrence risk for Down syndrome is 0.5–1% higher than the mother's age-related risk. Maternal germline mosaicism may be a possible explanation for this increased risk.

25.2.1.3 Parental Chromosomal Rearrangements

Balanced parental chromosomal rearrangements (translocations or pericentric inversions), which are relatively frequent in the normal population (frequency: 1:500), can lead to variable risks of offspring with chromosomal syndromes. These risks do not correspond to those expected from chromosomal segregation patterns, but are based on empirical data, presumably because of selection against unbalanced gametes [64]. The risk for translocation trisomy 21 is about 15% when the mother is the carrier and only 1–2% if the father is the carrier of a Robertsonian translocation involving chromosome 21 (e.g., t14q21 and t21q22q) [22]. The risk is 100% if one of the parents carries a 21q21q Robertsonian translocation. In reciprocal translocations the risk of future affected offspring is significantly higher (~20%) if ascertainment occurs via an affected live offspring as opposed to ascertainment by way of recurrent abortions (5%) [64]. More extensive unbalanced duplications/deletions (3–6 chromosome bands out of a total of 200) are associated with a lower recurrence risk (9–16%) than those with duplications/deletions affecting only 1 or 2 bands (34%). Presumably larger defects are often not viable and the affected fetuses are aborted spontaneously prior to amniocentesis. Possible recurrence risks have to be assessed individually.

25.2.1.4 Family History of a Monogenic Disorder

In many families monogenic disorders have risks of 25% (autosomal-recessive and X-linked with a 50% risk for boys only) or of 50 % (autosomal-dominant) in sibs of affected children. The accurate diagnosis of a mutation is possible through DNA analysis of fetal cells, in the same way as in postnatal genetic-testing. Therefore molecular prenatal diagnosis is now available for a large and ever-increasing number of monogenic disorders. If direct mutation analysis is not possible, prenatal diagnosis may be offered using linkage analysis with polymorphic DNA markers. However, such analyzes are often complicated by family availability, crossovers, allelic heterogeneity, etc.

It is the exception for prenatal diagnosis to be requested in autosomal-dominant diseases.

25.2.1.5 Neural Tube Defects

Amniocentesis (not chorionic biopsy) for amniotic-fluid AFP is carried out in women at high risk, such as those with previously affected children, or following a confirmed maternal high blood AFP level. The empirical risk for neural tube defects after the birth of an affected child is about 5%. It rises to more than 10% if the parents have two affected children. Ultrasound is very efficient in these circumstances. Other markers in amniotic-fluid (e.g., increases in acetyl cholinesterase) may provide further clues to the presence of neural tube defects and other anomalies.

25.2.1.6 Psychological Indications

Each prenatal investigation can refer only to a defined genetic or teratogenic risk. If no such defined risk is present, from the viewpoint of medical genetics parents' fear of having a handicapped child is not in itself strictly regarded as an indication for an invasive prenatal diagnosis procedure that itself carries a risk of at least 0.5% for a subsequent pregnancy loss. Moreover, a "normal" conventional chromosomal result does not guarantee a healthy child. The genetic counselor has to convey these limitations to the parents. Nevertheless, there may be individual cases in which the genetic advisor comes to the conclusion that a prenatal test may be justified as "psychologically indicated" in order to alleviate exaggerated anxiety and fears.

25.2.2 Techniques of Prenatal Diagnosis

The principle of the different procedures for obtaining fetal tissue is shown in Fig. 25.2 and Table 25.1. Weeks of gestational age are usually counted starting from the first day of the last menstruation.

25.2.2.1 Investigations Prior to Implantation

25.2.2.1.1 Preimplantation Genetic Diagnosis

For some couples at high risk of transmitting a serious genetic disorder conventional prenatal diagnosis may be unacceptable, as it potentially implicates termination of pregnancy at an advanced stage. In this situation preimplantation genetic diagnosis (PGD) may be an acceptable alternative. There is increasing demand for PGD for late-onset autosomal-dominant disorders for which conventional prenatal diagnosis is not requested. Following in vitro fertilization and blastomere biopsy at the 4- to 8-cell stage, molecular or cytogenetic analysis can be performed. After the in vitro analysis only those embryos shown to have an unaffected genotype are transferred into the mother's uterus. The success rate of this procedure is often less than 25% per cycle of treatment [18]. The other group of PGD users is made up of those who have been found to be subfertile or infertile and who wish to combine assisted reproduction with mainly cytogenetic testing of the early embryo.

Fig. 25.2 (**a**) Amniocentesis: puncture of the amniotic cavity through the abdominal wall; (**b**) chorionic villus sampling: puncture of the chorion by the abdominal approach

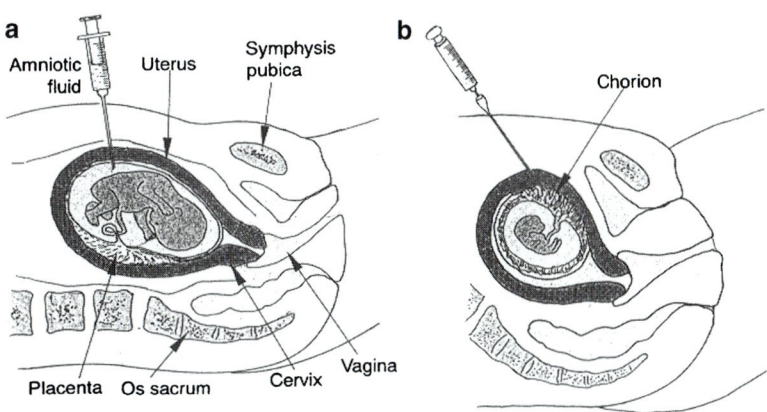

Table 25.1 Techniques used in prenatal diagnosis

Prior to implantation		
Technique	Optimal time (in weeks)	Main indications
Preimplantation genetic diagnosis (PGD)		a) Very high risk for a well-known disease without an effective therapy
		b) For parents who carry a high risk for a chromosomal aberration
Polar body diagnosis		Investigation from second polar body before fertilization of the oocyte for genetic diagnostics of certain maternal monogenic disorders and structural chromosomal aberrations
After implantation		
Noninvasive prenatal diagnosis		
First-trimester screening	11–13	Down syndrome and other chromosomal disorders
Ultrasound		Structural abnormalities, e.g., CNS, heart, kidneys and limbs
Invasive prenatal diagnosis		
Chorionic villus sampling (CVS)	10–12	Chromosome abnormalities, metabolic disorders, molecular defects
Amniocentesis	15–16	Neural tube defects, chromosome abnormalities, metabolic disorders, molecular defects
Cordocentesis		Chromosome abnormalities, hematological disorders, prenatal infections
Fetobiopsy(e.g., skin)		e.g., Hereditary skin disorders

Frequent indications for PGD are:

1. Monogenic disorders
2. Sex selection in X-linked disorders
3. Chromosomal anomalies (e.g., Robertsonian or reciprocal translocation)

Children conceived with the aid of IVF procedures, however, have a slightly higher rate of major birth defects (1–2%) [50]; whether this risk is due to culture conditions or to the mechanical treatment of the embryo cannot be determined at present.

A further restriction is diagnostic accuracy. The probability that a result is obtained on the basis of investigation of the genome of a single isolated cell is about 90–95%. The probability that this result is correct is again 90–95%. The biggest problem is a false-negative test result resulting from contamination with foreign DNA or from allelic dropout, i.e., the analysis limited to a single allele.

PGD is prohibited by law in some countries (e.g., Austria, Germany, Italy, and Switzerland).

25.2.2.1.2 Polar Body Analysis

The symmetrical cell division during female meiosis results in the production of polar bodies. The two polar bodies are relatively small cell structures inside an ovum and can be obtained by zona drilling or laser drilling. Genetic analysis of the polar body can be an alternative to PGD for certain monogenic disorders and maternal structural chromosomal aberrations. This analysis represents an indirect method of genotyping; the unfertilized oocyte and the first polar body arise from each other during meiosis I. They thus contain different members of each pair of homologous chromosomes. For example, if in a carrier of a heterozygous autosomal-recessive mutation the mutation is detected in the polar body, the oocyte will contain the wild-type copy of the gene. Polar body diagnosis has been successfully used in couples at-risk for having a baby with cystic fibrosis and hemophilia, but like PGD, polar body diagnosis is technically difficult and unlikely to become a routine procedure.

25.2.2.2 Noninvasive Diagnosis During Pregnancy

25.2.2.2.1 Ultrasound

Almost all pregnant women are routinely offered an ultrasound scan at around 12 gestational weeks, and repeated ultrasound examination has become part of routine pregnancy surveillance in some countries. Current studies indicate that ultrasound is harmless to the developing fetus. During later stages of pregnancy, ultrasound examination allows the detection of a variety of fetal structural anomalies. Often a detailed ultrasound screen is performed around the 20th gestational week. Below is a list of reasons for prenatal ultrasound and of fetal conditions that can be detected prenatally by this means in most cases [30]:

Obstetric indications:
    Accurate gestational dating
    Multiple pregnancies
    Placental localization before amniocentesis or chorionic villus sampling
CNS disorders:
    Anencephaly
    Hydrocephaly
    Encephalocele
    Meningomyelocele
    Spina bifida
    Holoprosencephaly
    Microcephaly

Skeletal defects:
    Severe neonatal bone dysplasias
    Congenital types of osteogenesis imperfecta
    Severe limb defects
Internal abnormalities:
    Severe congenital heart defects
    Gastroschisis
    Omphalocele
    Duodenal atresia
    Esophageal atresia
    Renal agenesis
    Congenital polycystic kidneys
    Severe obstructive uropathy
    Various fetal tumors

Obscure or pathologic ultrasound findings are often an indication for fetal chromosomal analysis. Many ultrasonographic markers may be of transient and of physiological nature rather than indicative of fetal pathology. Others may indicate abnormality in only a proportion of cases, e.g., nuchal edema and assessment of nasal bone in trisomy 21 (Table 25.2). This increased nuchal fold results from water stored under the skin of the embryo, which is most prominent between the 10th and 14th weeks of gestation.

Table 25.2 Multiplication factor for the modified age risk for trisomy 21 with ultrasonographic markers[a] [4, 60, 62]

Ultrasonographic markers	Time of the occurrence	Multiplication factor for the risk for trisomy 21
Nuchal translucency thickness and fetal nasal bone assessment	10–14 weeks of gestation	Dependent on width of edema, between 3 and 22
Nuchal fold thickening > mm	Second trimester	10
Echogenic bowel	Second trimester	5.5
Shortened femur length	Second trimester	2.5
Dilation of the renal pelvis	Second trimester	1.5
Choroid plexus cysts	Second trimester	1.5

[a]Example: A 29-year-old woman in week 13+2 of gestation (body weight: 56.1 kg) with fetal crown-heel length of 71.1 mm; nuchal transparency of 1.3 mm, PAPP-A (MoM) of 0.45, and free Beta-HCG (MoM) of 1.27 carries an adjusted risk for trisomy 21 of 1:3,500, which is much lower than the age-related risk of 1:730. (PAP-A-MoM and free beta-HCG-MoM are adjusted for body weight.)

25.2.2.2.2 Maternal Blood Sampling

The screening of maternal blood for elevated α-fetoprotein (AFP) levels to detect neural tube defects and some other fetal anomalies is standard practice in many centers [11, 28]. AFP screening as one parameter may also be useful to detect Down syndrome, since fetuses with trisomy 21 have lower AFP levels than normal fetuses [12]. Abnormalities that increase or decrease amniotic-fluid levels of AFP include:

Increased AFP:
  Neural tube defects (anencephaly; open spina bifida)
  Spontaneous intrauterine death
  Abdominal wall defect (omphalocele; gastroschisis)
  Congenital nephrotic syndrome
  Incorrect gestational age
Lowered AFP:
  Down syndrome

Other useful biochemical markers for Down syndrome screening include unconjugated estriol (μE3; decreased), human chorionic gonadotropin (β-hCG; increased), inhibin-A (increased) and pregnancy-associated plasma protein A (PAPP-A; decreased). Multiple marker screening for fetal Down syndrome (2nd trimester) using AFP, μE3, β-hCG in addition to maternal age achieves a detection rate of about 60–70%, with a false-positive rate of approx. 5%. Using ultrasound to detect increased fetal nuchal translucency (NT), the two biochemical markers AFP, β-hCG and the age of the pregnant woman, the detection rate would increase to 86% if 5% of all pregnancies were tested [9, 67, 68].

25.2.2.2.3 Detection of Fetal Cells and Fetal DNA in the Maternal Circulation

Using fetal cells in the peripheral blood of pregnant women to diagnose or screen for fetal chromosome abnormalities is still in the test stage [54, 69]. Analysis of cell-free fetal DNA in the circulation of the pregnant woman is also being tested as a noninvasive prenatal diagnostic method that might be developed [73]. Preliminary examination of principal results with next-generation sequencing suggests that detection of fetal aneuploidies may become possible through the analysis of cell-free fetal DNA in the circulation of pregnant women [10]. The use of fetal cells from transcervical samples collected in early pregnancy may also be a means of noninvasive prenatal diagnosis in the future [7].

25.2.2.3 Invasive Diagnostics During Pregnancy

All invasive diagnostic procedures during pregnancy involve a defined risk and should only be applied after detailed genetic consultation has been offered.

25.2.2.3.1 Amniocentesis

Amniocentesis [49] is the most common form of invasive prenatal diagnostic technique offered to pregnant women. It is carried out at the beginning of the second trimester of pregnancy (13th–15th week of pregnancy; Fig. 25.2a). It involves aspiration of 10–20 ml of amniotic-fluid by transabdominal puncture under ultrasound guidance. The amniotic-fluid is spun down to yield a pellet of cells and supernatant fluid. The fetal cells are derived from amnion, fetal skin, fetal lung, gastrointestinal tract, and urinary tract epithelium. The cell pellet is cultured with fetal calf serum, which stimulates cell growth, and some cells will start to divide and form cell colonies. After 8–12 days of incubation there are usually enough dividing cells for chromosome or DNA analysis. In many centers, a FISH test taking less than 24 h is performed on uncultivated amniotic-fluid cells. This test involves identification of chromosomes 13, 18, 21, X, and Y via fluorescence in situ hybridization (FISH). Alternatively, quantitative polymerase chain reaction (PCR) or multiplex ligation-dependent probe amplification (MLPA) can be used to identify numerical aberrations involving these chromosomes. However, these rapid tests yield only limited information, so that they cannot replace a complete chromosome study as the gold standard using amniocyte cultures.

25.2.2.3.2 Chorionic Villus Sampling

Chorionic villus sampling (CVS) [5, 44, 59] was first developed in China. This procedure may be performed by collecting chorionic villi by the cervical or by the abdominal approach (Fig. 25.2a). The abdominal approach is now generally preferred because of the lower risk of infection. Chorionic villus sampling must be performed under ultrasound guidance. Chorionic tissue of fetal trophoblastic origin can be used for cytogenetic analysis, DNA testing, or biochemical analysis. CVS can be carried out around the 12th week of gestation. The procedure therefore has psychological and practical advantages over amniocentesis, which is not carried out until the 13–16th week of gestation. In addition, cytogenetic results are available much sooner. However, chromosome analysis is less reliable and the risk of miscarriage after chorionic villus sampling is higher than after amniocentesis (1–2% vs 0.5–1%) [37, 46]. There is also evidence that CVS can cause limb abnormalities if carried out before the 12th week of gestation. Placental biopsy is the term used if the puncture is performed at later stages of pregnancy.

25.2.2.3.3 Cordocentesis

Cordocentesis [14] examines blood from the fetus to detect fetal abnormalities. The sample of fetal blood is taken from the umbilical cord. Cordocentesis is usually done when diagnostic information cannot be obtained through amniocentesis, CVS, or ultrasound or these tests have yielded inconclusive results. Cordocentesis is usually performed after the 20th week of gestation. It may be performed to help find whether any of the following concerns is justified:

1. Fetal infection (e.g., toxoplasmosis or rubella)
2. Fetal anemia
3. Chromosome abnormalities

The risk of miscarriage after cordocentesis is between 1 and 2%.

25.2.2.3.4 Fetoscopy

Fetoscopy [38] is usually done after the 18th week of gestation in women who have a high risk of having a child with a major birth defect that can be found only using fetoscopy. Because fetoscopy, which was initially performed to detect congenital birth defects, has been replaced by high-resolution ultrasound scanning, the main use of fetoscopy today is to collect samples of tissue (usually skin) from the fetus. The tissue can then be tested for hereditary skin diseases in rare conditions where a DNA analysis is not yet available. Fetoscopy is associated with a 3–5% risk of miscarriage, and there are very few indications for this procedure.

25.2.2.4 Special Problems with Prenatal Diagnosis

25.2.2.4.1 Failure to Obtain Fetal Material

With each prenatal diagnosis it must be pointed out that there is a <1% probability that it will not be possible to obtain any fetal cells or that any cells obtained will subsequently fail to grow.

25.2.2.4.2 Mosaic Constellation for Chromosome Aberrations

Chorionic villus sampling. In approximately 1% of cases, CVS shows evidence of apparent chromosome mosaicism. This can have various reasons:

1. Contamination of the fetal sample with maternal cells. This arises more frequently from cultured cells than with direct preparations.
2. Mosaicism as a culture artifact. Because of this possibility, at least two separate cultures are usually established for diagnostic use. If the aberrant cell lineage occurs only in one of the culture bottles, it may not be representative of the fetus.
3. Limitation of mosaic to a portion of the placenta. This arises through a replication error during the formation and development of the trophoblast. These mosaics are only of importance for the development of the fetus if they lead to placental insufficiency.
4. True fetal mosaicism. This situation may have effects on the development of the fetus.

Amniocentesis specimens are normally set up as two or three separate cultures. If a single abnormal cell is identified in only one of the cultures, it is highly probable that it represents a culture artifact. This is referred to as level 1 mosaicism or pseudomosaicism. If an aberrant karyotype is present in two or more cells of a single culture bottle, level 2 mosaicism is present. In this situation, 80% of such cases represent culture artifacts and only 20% reflect true fetal mosaicism. If the aberrant cell lineage is present in two or more cells in two separate culture bottles, level 3 mosaicism is present.

This level most likely represents true fetal mosaicism, requiring further diagnostic steps (e.g., high-resolution ultrasound).

25.2.2.4.3 Chromosome Abnormality

The most frequently diagnosed numerical chromosome abnormalities of concern are trisomies 21, 13, and 18, and other autosomal trisomies are found only occasionally. Gonosomal aneuploidies are usually discovered by chance and, with appropriate genetic counseling, are of less concern than autosomal aberrations.

Except for trisomy 21, almost all autosomal trisomies are associated with a poor prognosis as far as the probability of survival of the fetus is concerned. The situation is altogether different in sex chromosomal aneuploidies (47,XXX, 47,XXY, or 47,XYY). The life expectancy of the affected children is normal, malformations are discrete, and the intelligence, while somewhat below sibling scores, is usually within the normal range. The search for sex chromosomal abnormalities is therefore not a standard indication for invasive prenatal diagnostics. Following the coincidental discovery of a gonosomal aberration only a minority of parents opts for termination of pregnancy [6].

25.2.2.4.4 Structural Chromosomal Rearrangements

The incidence of structural chromosomal rearrangements amounts to approx. 1:500 in the newborn population. Half of these are Robertsonian translocations (translocations involving acrocentric chromosomes). Structural chromosomal rearrangements can be balanced or unbalanced. In case of a balanced structural chromosomal rearrangement it is helpful to analyze the parental chromosomes. If one of the parents is a carrier of the rearrangement, this will not usually cause problems in the fetus. If the balanced structural chromosomal rearrangement occurs as a de novo event in the fetus, there is a 5% empirical risk of a developmental handicap, which means that 19 of 20 children will be born without any impairment. The 5% risk is attributed to submicroscopic loss of genetically relevant material at the breakpoints.

25.2.2.4.5 Marker Chromosome

A marker chromosome is a small additional chromosome fragment which carries portions of one or more chromosomes. If the marker is present in one of the parents, it is unlikely that it will interfere with pre- or postnatal development.

If the parents do not have the marker chromosome, a rough estimate of an a-priori risk of 15% for a developmental handicap can be given for a marker with euchromatic material. However, with currently available methods in molecular cytogenetics, including FISH, further analyzes can clarify whether the marker chromosome contains mostly heterochromatic or euchromatic material. The origin of the marker can frequently be determined such that more precise risk information can be given to the parents.

25.3.2.5 Consultation and Aftercare

Prenatal diagnosis should be preceded by genetic counseling, during which the pregnant woman should be given information about the following aspects:

1. Indication for and goal and risks of the investigation
2. Possibilities and limitations of the prenatal diagnostic procedures offered
3. Basis risk, individual risk, and risks inherent in the particular diagnostic procedure under discussion
4. Interpretation and validity of prenatal test results, including possible problems with mosaicism
5. Types and severity levels of diagnosable pathologic conditions
6. Possible therapeutic options if pathologic findings are present
7. Psychological and ethical conflicts

Informed consent of the pregnant women is a prerequisite for any type of prenatal diagnosis. The risks, time of analysis, possible limitations, and possible consequences of this investigation have to be pointed out, and furthermore information about risks, e.g., bleeding, infection, or abortion, of invasive prenatal procedures must be made explicit. Moreover, the geneticist must inform the woman that a normal cytogenetic, molecular-genetic, or biochemical result does not exclude all possible diseases and handicaps. The counselee has to understand that a normal outcome of any of the diagnostic procedures does not guarantee the birth of a normal child. The abrupt confrontation with pathologic serum findings causes undue anxiety in the majority of women. In order to avoid such overreactions, maternal serum screening and ultrasound investigations require the careful provision of information to the patient prior to performance of these procedures. One of the main objectives of prenatal diagnosis is to alleviate fears and anxieties in the pregnant woman rather than provoking such feelings.

Following a pathologic finding, an unhurried counseling session with the pregnant woman, preferably together with her partner, should take place and include the following information:

1. Origin type and natural history (prognosis) of the disorder or developmental disturbance of the unborn child, and possible complications
2. Intrauterine therapy / operational measures (for the pregnant woman, or the fetus)
3. Postnatal therapy and supportive measures
4. Preparation for the special circumstances following the birth of a child with handicap/genetic disorder
5. Aspects of living with a handicapped child (establishment of contact with other parents, and support groups if requested)
6. Modalities and legal/emotional implications of termination of pregnancy
7. Possibilities of psychotherapeutic and psychosocial assistance

Informing pregnant women about pathologic findings requires expertise and experience. The decision of the counselee and the extent of the information provided need to be documented in a written report.

25.3.2.6 Aftercare

In the case of termination of pregnancy or the birth of an affected child, the following points should be considered:

1. Careful documentation of clinical findings (photographic, X-ray pictures and/or fetal pathologies)
2. Confirmation or correction of the result of prenatal diagnosis
3. Genetic counseling of parents (e.g., about possible prenatal diagnostic procedures in subsequent pregnancies; discussion of implications of the findings for other family members)
4. Encouragement of acceptance of findings and provision of psychotherapeutic support

25.3.2.7 Psychological Aspects and Outlook

Antenatal diagnosis is widely used in today's industrialized countries. The possibility of early detection of an affected embryo or fetus often encourages parents to start a pregnancy incircumstances in which the fear of having an affected infant would previously have deterred them from childbearing. While medical termination of pregnancies has become acceptable in many countries and cultures, a significant proportion of the population in the United States and elsewhere strongly opposes prenatal diagnosis for religious or other reasons. Some of the opponents are particularly concerned about the value judgments of human life implicit in these procedures. They feel that such practices are the beginning of a "slippery slope" that would ultimately lead to rejection of relatively minor defects and encourage the search for the "perfect" baby, possible resulting in the resurgence of eugenic and racist ideas. Fears have also been expressed that society would be less inclined to pay large sums of money to take care of children with genetic diseases when abortion could have prevented the birth of the disabled child. Largely because of the realization that most human disabilities arise during postnatal life, utilitarian or eugenic views have not become a driving force in prenatal diagnosis. The fact that in recent years many countries have made provisions to meet their obligations towards handicapped citizens argues against the validity of these fears [45].

New challenges for genetic counseling arise from the recent general availability of DNA testing that provide risk data for a variety of traits and diseases based on DNA genotyping. Several companies offer an "exploration of your genome" at low costs (usually in the range of $400). These companies provide test kits to collect cheek cells in saliva or cells from a buccal swab. The DNA is then genotyped within 4 or 6 weeks, and the user can check results on a designated internet page. Data used for genotyping comes often from published whole-genome association studies and provides information about certain conditions (e.g., risk for obesity) or common diseases (e.g., diabetes, breast cancer). However, the data employed is often incomplete, as it has often not been validated by repeated analysis or not confirmed for populations with different ethnic backgrounds, and it may therefore provide misleading information. The relative risks for phenotypic features or diseases are usually very small (odds ratios are often below 2). Although these companies often have well-prepared and informative websites and now often employ trained genetic counselors, people now have the chance to obtain some information about their genome without any assurance about whether they understand possible implications and whether they can interpret the data and draw correct conclusions from the results.

Further challenges come from new technologies allowing the sequencing of entire genomes at low costs (see Sect. 4.4), which make it likely that whole-genome data will be introduced into routine clinical genetics in the near future. Numerous new bioinformatics tools will be needed for data interpretation, but once such tools exist these technologies will provide a wealth of information, which will include susceptibilities for both common conditions and diseases. Together with the aforementioned web-based genetic-testing offerings, these new technologies will change genetic counseling tremendously. The main focus of genetic counseling is at present concerned about a certain disease in an individual or in the individual's family. However, presymptomatic, predictive testing will become more important as more information about our genome becomes available. Such predictive testing will not only concern diseases that might be prevented by appropriate means, but also such common conditions as the likelihood of obesity, baldness, and so on. Professional genetic counselors have to meet these challenges and have to provide appropriate services. Such services may be in competition with companies that may offer similar information at a different standard.

25.3 Conclusions

Knowledge of human genetics can be applied for genetic counseling of individuals and families at-risk for hereditary anomalies and diseases. Genetic counseling refers to the sum of activities that (a) establish the diagnosis of such diseases, (b) assess the recurrence risk, (c) communicate to the client and family the chance of recurrence, and (d) provide information regarding the many problems raised by the disease, including natural history and variability of the disease. Formal methods applied in risk assessment include the use of genetic algorithms, statistical considerations, and empirically derived risk estimates if the mode of inheritance is unknown. Laboratory techniques involve prenatal diagnosis by ultrasonography, amniocentesis, and chorionic villus sampling and the diagnostic application of techniques such as cytogenetics and molecular genetics. Genetic counseling is not merely concerned with scientific issues that are at the root of genetic disease. It must provide understanding and empathy for the clients´ concerns, and it must deal appropriately with the many psychological aspects of the process. As such, genetic counseling is an important part of comprehensive medical care.

25.4 Appendix Example of a Bayesian Table

Assuming a woman's brother and her maternal uncle are affected by an X-linked recessive condition, her prior probability of being a carrier will be ½, or 50%. If this consultand has two healthy sons, her risk of being a carrier will be reduced to 1/5 or 20% (posterior probability) according to Bayes' theorem.

	Consultand is a carrier	Consultand is not a carrier	Sum: 1/8 + ½ = 5/8
Prior probability of the consultand	½	½	
Two healthy sons	½ × ½ = ¼	1	
Joint probability	= ½ × ¼ = 1/8	= ½ × 1 = ½	
Prior probability of the consultand	½	½	
Two healthy sons	½ × ½ = ¼	1	
Joint probability	= ½ × ¼ = 1/8	= ½ × 1 = ½	
Posterior probability	=1/8 / 5/8 = 1/5	=1/2 / 5/8 = 4/5	
	= 0.2	= 0.8	

	Consultand is a carrier	Consultand is not a carrier	Sum: 1/80+½=41/80
Prior probability of the consultand	½	½	
2 healthy sons	½×½=¼	1	
Negative test result	1/10	1	
Joint probability	=½×¼×1/10=1/80	=½×1×1=½	
Prior probability of the consultand	½	½	
2 healthy sons	½×½=¼	1	
Negative test result	1/10	1	
Joint probability	=½×¼×1/10=1/80	=½×1×1=½	
Posterior probability	=1/80/41/80=1/41	=1/2/41/80=40/41	
	» 0.024	» 0.976	

	Consultand is a carrier	Consultand is not a carrier	Sum: 1/20+½=11/20
Prior probability of the consultand	½	½	
Negative test result	1/10	1	
Joint probability	=½×1/10=1/20	=½×1=½	
Prior probability of the consultand	½	½	
Negative test result	1/10	1	
Joint probability	=½×1/10=1/20	=½×1=½	
Posterior probability	=1/20/11/20=1/11	=1/2/11/20=10/11	
	» 0.09	» 091	

Assuming a woman's brother and her maternal uncle are affected by an X-linked recessive condition, her prior probability of being a carrier will be ½, or 50%. If this consultand has a negative result for a test that detects 90% of heterozygotes (sensitivity=0.9), her risk of being a carrier will be reduced to 1/11 or 9% (posterior probability) according to Bayes' theorem.

Assuming a woman's brother and her maternal uncle are affected by an X-linked recessive condition, her prior probability of being a carrier will be ½, or 50%. If this consultand has two healthy sons and a negative result for a test that detects 90% of the heterozygotes (sensitivity=0.9), her of risk being a carrier will be reduced to 1/41 or 2.4 % (posterior probability) according to Bayes' theorem.

References

1. American Academy of Pediatrics (2000) Fetal Alcohol Syndrome and Alcohol-Related Neurodevelopmental Diaorders. Pediatrics 106:358–361
2. Baird PA, McGillivray B (1982) Children of incest. J Pediatr 101:854–857
3. Bekker HL, Hewison J, Thornton JG (2004) Applying decision analysis to facilitate informed decision making about prenatal diagnosis for Down syndrome: a randomized controlled trial. Prenat Diagn 24:265–275
4. Bianchi DW, Simpson JL, Jackson LG, Elias S, Holzgreve W, Evans MI, Dukes KA, Sullivan LM, Klinger KW, Bischoff FZ, Hahn S, Johnson KL, Lewis D, Wapner RJ, de la Cruz F (2002) Fetal gender and aneuploidy detection using fetal cells in maternal blood: analysis of NIFTY I data. National Institute of Child Health and Development Fetal Cell Isolation Study. Prenat Diagn 22:609–615
5. Brambati B, Simoni G, Danesino C, Oldrini A, Ferrazzi E, Romitti L, Terzoli G, Rossella F, Ferrari M, Fraccaro M (1985) First-trimester fetal diagnosis of genetic disorders: clinical evaluation of 250 cases. J Med Genet 22:92–99
6. Brun JL, Gangbo F, Wen ZQ, Galant K, Taine L, Maugey-Laulom B, Roux D, Mangione R, Horovitz J, Saura R (2004) Prenatal diagnosis and management of sex chromosome aneuploidy: a report on 98 cases. Prenat Diagn 24:213–218
7. Bussani C, Cioni R, Scarselli B, Barciulli F, Bucciantini S, Simi P, Fogli A, Scarselli G (2002) Strategies for the isolation and detection of fetal cells in transcervical samples. Prenat Diagn 22:1098–1101
8. Capron AM, Lappe M, Murray RF, Powledge TM, Twiss SB, Bergsma D (eds) (1979) Genetic counseling: facts, values, and norms. Birth Defects 15(2):187–200

9. Caughey AB, Lyell DJ, Washington AE, Filly RA, Norton ME (2006) Ultrasound screening of fetuses at increased risk for Down syndrome: how many missed diagnoses? Prenat Diagn 26:22–27
10. Chiu RW, Chan KC, Gao Y, Lau VY, Zheng W, Leung TY, Foo CH, Xie B, Tsui NB, Lun FM, Zee BC, Lau TK, Cantor CR, Lo YM (2008) Noninvasive prenatal diagnosis of fetal chromosomal aneuploidy by massively parallel genomic sequencing of DNA in maternal plasma. Proc Natl Acad Sci USA 105:20458–20463
11. Crandall BF, Robertson RD, Lebherz TB, King W, Schroth PC (1983) Maternal serum alpha-fetoprotein screening for the detection of neural tube defects. West J Med 138:524–530
12. Cuckle HS, Wald NJ, Lindenbaum RH (1984) Maternal serum alpha-fetoprotein measurement: a screening test for Down syndrome. Lancet 1:926–929
13. Cystic Fibrosis Genotype–Phenotype Consortium (1993) Correlation between genotype and phenotype in patients with cystic fibrosis. N Engl J Med 329:1308–1313
14. Daffos F, Capella-Pavlovsky M, Forestier F (1985) Fetal blood sampling during pregnancy with use of a needle guided by ultrasound: a study of 606 consecutive cases. Am J Obstet Gynecol 153:655–660
15. de Costa CM (2002) Consanguineous marriage and its relevance to obstetric practice. Obstet Gynecol Surv 57:530–536
16. Den Dunnen JT, Grootscholten PM, Bakker E, Blonden LA, Ginjaar HB, Wapenaar MC, van Paassen HM, van Broeckhoven C, Pearson PL, van Ommen GJ (1989) Topography of the Duchenne muscular dystrophy (DMD) gene: FIGE and cDNA analysis of 194 cases reveals 115 deletions and 13 duplications. Am J Hum Genet 45:835–847
17. Edwards JH (1956) Antenatal detection of hereditary disorders. Lancet 1:579
18. El-Toukhy T, Khalaf Y, Braude P (2006) IVF results: optimize not maximize. Am J Obstet Gynecol 194:322–331
19. Emery AE (1973) Antenatal diagnosis of genetic disease. Williams & Wilkins, Baltimore, MD
20. Epstein CJ, Curry CJR, Packman S, Sherman S, Hall BD (eds) (1979) Risk, communication, and decision making in genetic counseling. Birth Defects 15(5C):335–367
21. Evers-Kiebooms G, van den Berghe H (1979) Impact of genetic counseling: a review of published follow-up studies. Clin Genet 15:465–474
22. Ferguson-Smith MA (1983) Prenatal chromosome analysis and its impact on the birth incidence of chromosome disorders. Br Med Bull 39:355–364
23. Ferguson-Smith MA, Yates JR (1984) Maternal age specific rates for chromosome aberrations and factors influencing them: report on a collaborative European study on 52965 amniocenteses. Prenatal Diag 4:5–44
24. Fisher B, Rose NC, Carey JC (2008) Principles and practice of teratology for the obstetrician. Clin Obstet Gynecol 51:106–118
25. Fletcher JC, Berg K, Tranøy KE (1985) Ethical aspects of medical genetics. A proposal for guidelines in genetic counseling, prenatal diagnosis and screening. Clin Genet 27:199–205
26. Fuchs F, Riis P (1956) Antenatal sex determination. Nature 177:330
27. Fuhrmann W, Vogel F (1983) Genetic counseling, 3rd edn. Springer, Berlin Heidelberg science library 10
28. Fuhrmann W, Weitzel HK (1985) Maternal serum alpha-fetoprotein screening for neural tube defects. Report of a combined study in Germany and short overview on screening in populations with low birth prevalence of neural tube defects. Hum Genet 69:47–61
29. Grimm T, Muller B, Dreier M, Kind E, Bettecken T, Meng G, Muller CR (1989) Hot spot of recombination within DXS164 in the Duchenne muscular dystrophy gene. Am J Hum Genet 45:368–372
30. Harper PS (2004) Practical genetic counseling, 6th edn. Arnold, London
31. Harper PS, Clarke AJ (1997) Genetics society and clinical practice. IOS Scientific Publishers, Oxford
32. Hassold T, Hall H, Hunt P (2007) The origin of human aneuploidy: where we have been, where we are going. Hum Mol Genet 16(2):R203–R208
33. Jancar J, Johnston SJ (1990) Incest and mental handicap. J Ment Defic Res 34:483–490
34. Kaback MM, Zeiger RS, Reynolds LW, Sonneborn M (1974) Approaches to the control and prevention of Tay-Sachs disease. Prog Med Genet 10:103–134
35. Keller H, Emery AEH, Spiegler AWJ, Apacik C, Müller CR, Grimm T (1996) Age effects on serum creatine kinase (SCK) levels in obligate carriers of Duchenne muscular dystrophy (DMD) and Becker muscular dystrophy (BMD) and its implication on genetic counseling. Acta Cardiomiologica 8:27–34
36. Kessler S (1980) The psychological paradigm shift in genetic counseling. Soc Bioi 27:167–185
37. Kozlowski P, Knippel A, Stressig R (2008) Individual risk of fetal loss following routine second trimester amniocentesis: a controlled study of 20,460 cases. Ultraschall Med 29:165–172
38. Lange IR (1985) Congenital anomalies: detection and strategies for management. Semin Perinatol 9:151–162
39. Lemire R, Shepard T (2007) Catalog of teratogenic agents. Johns Hopkins University Press, Baltimore, MD
40. Lippman-Hand A, Fraser FC (1979) Genetic counseling - the postcounseling period. I. Parents' perceptions of uncertainty. Am J Med Genet 4:51–71
41. Lippman-Hand A, Fraser FC (1979) Genetic counseling - the postcounseling period. II. Making reproductive choices. Am J Med Genet 4:73–87
42. London Dysmorphology Database, London Neurogenetics Database & Dysmorphology Photo Library on CD-ROM (2001) In: Michael B, Robin M. Winter (eds.), Oxford Medical Database, 3 rd edn. Oxford University Press, Oxford
43. McKusick VA (1998) Mendelian inheritance in man. A catalog of human genes and genetic disorders, 12th edn. Johns Hopkins University Press, Baltimore, MD
44. Modell B (1985) Chorionic villus sampling. Evaluating safety and efficacy. Lancet 1:737–740
45. Motulsky AG, Murray J (1983) Will prenatal diagnosis with selective abortion affect society's attitude toward the handicapped? In: Berg K, Tranoy KE (eds) Research ethics. Liss, New York, pp 277–291
46. Mujezinovic F, Alfirevic Z (2007) Procedure-related complications of amniocentesis and chorionic villous sampling: a systematic review. Obstet Gynecol 110:687–694

47. Murphy EA, Chase GA (1975) Principles of genetic counseling. Year Book Medical Publishers, Chicago
48. Nadler HL (1968) Antenatal detection of hereditary disorders. Pediatrics 42:912–918
49. Nadler HL, Gerbie AB (1970) Role of amniocentesis in the intrauterine detection of genetic disorders. N Engl J Med 12:596–599
50. Olson CK, Keppler-Noreuil KM, Romitti PA, Budelier WT, Ryan G, Sparks AE, Van Voorhis BJ (2005) In vitro fertilization is associated with an increase in major birth defects. Fertil Steril 84:1308–1315
51. Online Mendelian Inheritance in Man, OMIM (TM). McKusick-Nathans Institute of Genetic Medicine, Johns Hopkins University (Baltimore, MD) and National Center for Biotechnology Information, National Library of Medicine (Bethesda, MD), World Wide Web URL: http://www.ncbi.nlm.nih.gov/omim/
52. POSSUM (2008) World Wide Web URL: http://www.possum.net.au/index.htm
53. President's Commission for the Study of Ethical Problems in Medicine and Biomedical and Behavioral Research (1983) Screening and counseling for genetic conditions. The ethical, social, and legal implications of genetic screening, counseling, and education program. US Government Printing Office, Washington, DC
54. Purwosunu Y, Sekizawa A, Koide K, Okazaki S, Farina A, Okai T (2006) Clinical potential for noninvasive prenatal diagnosis through detection of fetal cells in maternal blood. Taiwan J Obstet Gynecol 45:10–20
55. Resta RG (2006) Defining and redefining the scope and goals of genetic counseling. Am J Med Genet C Semin Med Genet 142:269–275
56. Ried T, Mahler V, Petal V (1990) Direct carrier detection by in situ suppression hybridization with cosmid clones of the Duchenne/Becker muscular dystrophy locus. Hum Genet 85:581–586
57. Schwartz M, Duno M (2004) Improved molecular diagnosis of dystrophin gene mutations using the multiplex ligation-dependent probe amplification method. Genet Test 8:361–367
58. Seemanova E (1971) A study of children of incestuous matings. Hum Hered 21:108–128
59. Simoni G, Brambati B, Danesino C, Terzoli GL, Romitti L, Rossella F, Fraccaro M (1984) Diagnostic application of first-trimester trophoblast sampling in 100 pregnancies. Hum Genet 66:252–259
60. Snijders RJ (1996) Isolated choroid plexus cysts: should we offer karyotyping? Ultrasound Obstet Gynecol 8:223–224
61. Sorenson JR, Swazey JP, Scotch NA (eds) (1981) Reproductive pasts reproductive futures. Genetic counseling and its effectiveness. Birth Defects 17(4):131–144
62. Spencer K, Souter V, Tul N, Snijders R, Nicolaides KH (1999) A screening program for trisomy 21 at 10–14 weeks using fetal nuchal translucency, maternal serum free beta-human chorionic gonadotropin and pregnancy-associated plasma protein-A. Ultrasound Obstet Gynecol 13:231–377
63. Steele MW, Breg WR Jr (1966) Chromosome analysis of human amniotic-fluid cells. Lancet I:383–385
64. Stengel-Rudkowski S, Stene J, Gallano P (1988) Risk estimates in balanced parental reciprocal translocations. Analysis of 1120 pedigrees. Monographie des Annales de Génétique, Paris
65. Tibben A (2007) Predictive testing for Huntington's disease. Brain Res Bull 72:165–171
66. Vogelstein B, Kinzler KW (2002) The genetic basis of human cancer, 2nd edn. McGraw-Hill, New York
67. Wald NJ, Watt HC, Hackshaw AK (1999) Integrated screening for Down's syndrome on the basis of tests performed during the first and second trimesters. N Engl J Med 341:461–467
68. Wald NJ, Bestwick JP, Huttly WJ, Morris JK, George LM (2006) Validation plots in antenatal screening for Down's syndrome. J Med Screen 13:166–171
69. Walknowska J, Conte FA, Grumbach MM (1969) Practical and theoretical implications of fetal-maternal lymphocyte transfer. Lancet I:1119–1122
70. White SJ, Aartsma-Rus A, Flanigan KM, Weiss RB, Kneppers AL, Lalic T, Janson AA, Ginjaar HB, Breuning MH, den Dunnen JT (2006) Duplications in the DMD gene. Hum Mutat 27:938–945
71. Yang Q, Wen SW, Leader A, Chen XK, Lipson J, Walker M (2007) Paternal age and birth defects: how strong is the association? Hum Reprod 22:696–701
72. Young I (2007) Introduction to risk calculation in genetic counselling, 3rd edn. Oxford University Press, Oxford
73. Zimmermann BG, Maddocks DG, Avent ND (2008) Quantification of circulatory fetal DNA in the plasma of pregnant women. Methods Mol Biol 444:219–229

Gene Therapy

26

Vivian W. Choi and R. Jude Samulski

Abstract The integrity of the human genome is essential for maintaining the well being of an individual. Mutations, either caused by extrinsic or intrinsic means, that alters the genetic code can lead to genetic diseases, including, but not limited to, immature aging, developmental disorder, neurological disorder, cancer, etc. Traditional therapeutics may be able to treat or attenuate the condition of the genetic disease, but the mutated gene is still maintained in the patient's genome and the root of the problem is not addressed. Gene therapy, or genetic replacement therapy, aims to treats the disease at the genetic level, either by introducing a wild-type sequence gene over the defective one or a gene that encodes a therapeutic protein. There have been great challenges in developing gene therapy treatment in a safe, effective and specialized way for treating different genetic diseases. We will discuss (1) what diseases are considered to be suitable for gene therapy, (2) currently available methods used in gene delivery, (3) the choice of different viral and non-viral gene delivery tools and how they can be introduced into the target tissues or organs, (4) the safety aspects of gene therapy and (5) challenges yet to be solved in increasing the success rate of gene therapy. We hope to introduce to the readers the latest advancement of the gene therapy field and further discuss challenges and difficulties in having a successful outcome of gene therapy.

Contents

26.1 Advent of Gene Therapy 868
26.2 Diseases Considered Suitable for Gene Therapy 868
26.3 Gene Therapy Vectors 868
    26.3.1 Viral Vectors 868
    26.3.2 Nonviral Vectors 869
26.4 Factors Affecting the Success of Gene Therapy 870
    26.4.1 Choice of Gene Delivery Vector 870
    26.4.2 Route of Administration 870
    26.4.3 Integrating or Nonintegrating Vectors...... 870
    26.4.4 Therapeutic Means 871
    26.4.5 Regulation of Gene Expression 871
26.5 Safety Issues of Gene Therapy 871
26.6 Difficulties in Achieving Successful Gene Therapy 872
26.7 Conclusions 873
References 873

V.W. Choi (✉)
Department of Pharmacology, Gene Therapy Center,
University of North Carolina, Chapel Hill, NC 27599, USA
e-mail: Vivian_Choi@alumni.unc.edu

R.J. Samulski
Room 7119 Thurston Bowles Building, Gene Therapy Center,
CB#7352, University of North Carolina, Chapel Hill,
NC 27510, USA
e-mail: rjs@med.unc.edu

26.1 Advent of Gene Therapy

Gene therapy, or genetic replacement therapy, is one of the most recent medical treatments for previously incurable genetic diseases [33]. The idea of gene therapy is to treat diseases by supplying genetic materials to modulate the pathophysiology caused by a malfunctioning gene in the patient's genome, with the ultimate goal of achieving long-term cure in a single treatment. The completion of the human genome project complemented the field of gene therapy by providing an incredibly vast amount of genetic information for medical research [9, 10], and the better tools for diagnosis and the advancement in molecular biology assisted in accelerating the development of gene therapy in the clinics.

26.2 Diseases Considered Suitable for Gene Therapy

Several types of genetic diseases are considered suitable for gene therapy. Monogenic hereditary disorders, such as cystic fibrosis, adenosine deaminase (ADA) deficiency, hemophilia A and B, familial hypercholestrolemia, Canavan disease, muscular dystrophy, and X-linked SCID [39, 57], are good candidates for gene therapy treatment, because the disease treatment can be carried out by introducing the correct sequence of the mutated gene. With a better understanding of more complicated diseases, the feasibility of using gene therapy in treating inflammatory joint diseases, such as arthritis, neurological diseases, such as Batten's, Parkinson's and Alzheimer's diseases, and various types of cancer, such as glioma, Lewis lung carcinoma, chronic lymphocytic leukemia, and cervical carcinomas [5, 30], has been demonstrated.

26.3 Gene Therapy Vectors

Even though the technology needed to synthesize genetic materials is readily available, the problem of developing gene delivery vehicles sufficiently efficient to transfer the DNA to all the target tissues remains challenging. An efficient gene delivery vector should fulfill the following criteria:

1. High safety level with minimal side effects
2. High efficiency and specificity
3. Large packaging capacity
4. Scalable to produce large quantities
5. Regulable to control gene expression

The two main categories of gene delivery vectors are viral and nonviral, and the most commonly used vectors, are discussed below.

26.3.1 Viral Vectors

Viral vectors are genetically modified viruses that have reduced pathogenicity while retaining the ability to infect cells. They usually consist of a modified viral genome, with a minimum amount of genetic material from the wild-type (wt) virus, packaged inside a capsid. Different viral gene delivery vectors (or viral vectors) offer various advantages that can be exploited to treat different diseases. In general, optimization of tissue tropism can be achieved by capsid mutagenesis, while long-term transgene expression is determined by the nature and the composition of the viral genome. Therefore, choosing the most suitable viral vector for each treatment is crucial to safe and effective gene therapy.

26.3.1.1 Adenovirus

Adenovirus (Ad) is a nonenveloped, double-stranded DNA virus with an icosahedral capsid diameter of 70–100 nm. Its vector genome is nonintegrating but remains episomal (to be discussed in Sect. 26.4.3) [44]. Among the few versions of Ad vectors developed in the past decade, the latest generation, the gutless vector, carries only the 5′ and 3′ inverted terminal repeats (ITRs) and the packaging signal (Ψ) as a minimal required *cis* element to generate the vector [40]. One of the few advantages of Ad vector is its large packaging capacity of up to 36 kb to accommodate either large or multiple transgene expression cassettes, ideal for the treatment of polygenic diseases. However, the major disadvantage of the Ad vector is that its systemic delivery can induce adaptive humoral and innate immune response, thus causing tissue damage and removal of infected cells by macrophages [18, 61]. This prohibits the use of Ad vector in establishing long-term transgene expression for life-long therapy. Another challenge of using Ad vector is the potential contamination of wt or replication competent Ad, causing serious side effects.

26.3.1.2 Retrovirus

Retrovirus is an enveloped virus carrying a single-stranded RNA genome. Moloney murine leukemia virus (MMLV), and human immunodeficiency virus (HIV) are examples of retroviral gene delivery vectors. These vector genomes, with packaging sizes of 8–10 kb, carry the minimal *cis* elements, including the long terminal repeats (LTRs) and the packaging signal from the wt virus and a transgene expression cassette. The other components needed for reverse transcription and packaging are provided in *trans*, including gag/pol, rev, and envelope [49]. Unlike Ad vector, retroviral vector genomes can lead to long-term gene expression by vector genome integration into the host cell chromosomes. This is also a disadvantage for retroviral vectors because the mechanism of integration can cause insertional mutagenesis. For example, it has been well documented that retroviral vectors can cause oncogenic transformation [31]. The newest generation of HIV vector, the self-inactivating vector (SIN), does not carry the viral LTR enhancer/promoter activity and is anticipated to provide a higher safety profile than the prior generations of retroviral vectors [32].

26.3.1.3 Adeno-associated Virus

Adeno-associated virus (AAV) belongs to the family *Parvoviridae*, whose members are among the smallest of the DNA animal viruses. It infects vertebrates and cannot replicate on its own; therefore it is characterized as a *Dependovirus*. An AAV virion is 26 nm in diameter and is composed of 60 subunits of capsid proteins, forming an icosahedral structure encapsidating a single-stranded DNA genome. AAV is regarded as one of the most promising gene therapy vectors for clinical use owing to its lack of pathogenicity in humans, effectiveness in transduction, and inability to self-replicate [46]. Additionally, both naturally occurring serotypes and artificially mutated AAV capsids can infect a wide host-cell range of tissues with high transduction efficiency [7, 15]. AAV can also lead to long-term gene expression owing to episomal persistence of AAV genomes, even though site-specific integration at human chromosome 19 AAVS1 site is possible for wt AAV2 at a very low efficiency [6, 8, 43, 47–59]. In addition, a newer generation of AAV vector, self-complementary AAV (scAAV) is developed as an improved vector, as it carries a genome that resembles a double-stranded DNA template ready for transcription upon infection [34]. Even though the genome packaging size of AAV vector (4.7 kb) is smaller than other viral vectors, improved transgene expression cassette designs and the use of split vector system have overcome this hurdle [8, 28].

26.3.1.4 Herpes Simplex Virus

Herpes Simplex Virus (HSV) is an enveloped and double-stranded DNA virus, which consists of an external envelope surrounding an icosadeltahedral capsid containing a wt genome of 152 kb [2-4]. HSV vectors are used for targeting neuronal tissue (e.g., Parkinson's, malignant gliomas, and cerebral ischemia). In a similar way to Ad and AAV vectors, the HSV genome is also maintained episomally. Despite the high efficiency of targeting central nervous tissues, one of the drawbacks of HSV vector is that wt HSV contamination is highly pathogenic and cerebral injection of such a vector can cause fatal encephalitis [22, 38]. Even though there are concerns in using HSV vector for gene therapy, a replication-defective HSV vector is considered to be one of the most desirable vectors for cancer gene therapy [4].

26.3.2 Nonviral Vectors

Nonviral vectors are chemically assembled vectors made up of lipids, nucleic acids, peptides, and/or inorganic materials. To carry out gene delivery successfully, these vectors need to:

1. Be of biocompatible composition
2. Bind to cell surface and get internalized
3. Perform endosomal escape
4. Traffic through the cytoplasm

5. Deliver the genome to the nucleus

A great advantage of nonviral vectors is that it can possibly carry a transgene expression cassette of no size limitations. In addition, it has been suggested that nonviral vectors induce fewer inflammatory responses or have a higher safety level compared to viral vectors, but this still requires further validation [11]. However, nonviral vectors are far less efficient than viral vectors, because viruses have evolved to infect cells efficiently in order to complete their life cycles. Therefore, further improvements to the nonviral vectors are necessary to generate a gene-targeting vector whose efficiency approaches that of viral vectors.

26.3.2.1 Naked DNA

Naked DNA is the cheapest and easiest material to be produced for gene therapy. By using the hydrodynamic delivery method (large-volume injection into the bloodstream over a short duration), DNA is forced into the bloodstream and then into the internal organs. However, rapid degradation by nucleases and clearance by the mononuclear phagocyte system reduce the inefficiency of in vivo transfection of naked DNA [11].

26.3.2.2 Liposomes

The most popular liposomes are the cationic amphiphile and neutral phospholipids [14]. The plasmid DNA is condensed and packed inside the liposomes through interactions with the polar headgroups. The hydrophobic tail assists the formation of micelles, which fuse with the cellular membrane at the time of transfection, leading to endocytosis [16]. One variant form of improved liposomes is seen in the DNA-ligand conjugates, which guide the complex to bind to specific cellular surface receptors for more efficient gene transfer [25]. Commonly used lipid vector backbones include 2,3-dioleyloxy-N-[2(spermine-carboxamido) ethyl]-N,N-dimethyl-1-propanaminiumtrifluoroacetate (DOSPA) and dioleylphosphatidylethanolamine (DOPE) [11].

26.3.2.3 Polymers

Polymers used for gene therapy are generally classified in two categories: natural and synthetic. Polymer size is determined by the amine-to-phosphate (N/P) ratio. Higher molecular weight polymers are more stable, while lower molecular weight ones have a higher transfection efficiency. Some examples of polymer vector backbones are poly(ethyleneimine) (PEI), poly (β-aminoesters), β-cyclodextrin, poly(amidoamine) dendrimers (PAMAM), poly(2-diethylaminoethyl methacrylate) (PDEAEMA), and polyphosphoesters [11].

26.4 Factors Affecting the Success of Gene Therapy

From diagnosis to clinical treatment, gene therapy requires the involvement of a wide range of expertise. In order to achieve life-long correction of genetic diseases, we need to take many different factors into consideration so that the therapy can be safe and effective.

26.4.1 Choice of Gene Delivery Vector

Careful selection of a delivery vector is crucial for each clinical case, since every vector has its own strengths and weaknesses. The advantages of using nonviral vectors are the low production costs with less complicated procedures and their ability to accommodate any size of gene expression cassette. However, the transfection efficiency of nonviral vectors is significantly lower than that of most of the viral vectors. In addition, the plasmid DNA packaged in the nonviral vectors is readily degraded and the CpG island sequence in the bacterial backbone can initiate immune responses [26, 42]. On the other hand, viral vectors may involve the risk of insertional mutagenesis and of acute immune responses. Nevertheless, the broad tissue tropism, ability to establish long-term gene expression by episomal persistence or chromosomal integration, and capability of infecting dividing and nondividing cells have attracted many investigators in utilizing viral vectors [41].

26.4.2 Route of Administration

The route of administration is another key variable in gene delivery, because the biodistribution of the vectors can directly affect the efficiency of vector delivery. For systemic delivery, intravascular injection is the route most commonly used. Direct muscle injection can be used in muscle-related disease (e.g., muscular dystrophy) or to serve as the tissue to generate secretable protein to be released into the bloodstream (e.g., a-antitrypsin) [5]. Hepatic artery injection is used in targeting the vector to the liver. This route is especially important for many enzyme-dependent therapies, since these molecules are modified in the liver to establish full enzymatic activities [5]. Alternative direct injection protocols have been developed to target other organs. Besides direct introduction of vectors to the patient (in vivo targeting), ex vivo administration is also widely used, particularly when retroviruses are used as the delivery vector. This method is performed by infecting (viral) or transfecting (nonviral) gene delivery vectors to cells that are removed from the patient's body. The cells containing the vector genomes are then reintroduced into the patient.

26.4.3 Integrating or Nonintegrating Vectors

Integration of a gene therapy vector cassette into the target cells chromosome allows long-term correction in both dividing (e.g., bone marrow) and nondividing or slowly dividing (e.g., muscle) cells. In particular, retroviral vectors have been widely used for ex vivo gene therapy based on genome integration into rapidly growing cells [36, 45, 50]. However, random insertional mutagenesis can be dangerous, as it can cause activation of oncogenes or disruption of functional genes. Site-directed integration, which is a more challenging method, has been proposed to minimize the manipulations or changes caused to the host genome by the introduction of a gene therapy vector. For nondividing or slowly dividing cells, vector genomes that can persist episomally in the nucleus of the cell can also establish long-term gene expression. For example, long-term gene expression has been demonstrated with AAV vectors in animals, such as mice, dogs, and primates [43, 55, 59] owing to viral genome persistence as episomal circles and/or concatemers [6, 47, 48, 59]. Preliminary clinical trial data suggested that transgene expression is established in human for at least 4 years [23]. The premise of long-term gene expression without potential insertional mutagenesis into patients' genomes that would cause secondary mutations is optimistic.

26.4.4 Therapeutic Means

Recent advances in molecular biology and cell biology provide new insights on the types of therapeutic means applicable to gene therapy.

26.4.4.1 Gene Complementation

A typical gene therapy vector carries a transgene expression cassette including, but not limited to, the following basic elements: a promoter, a cDNA of the transgene of interest, and a poly-A signal. This cassette supplements the patient with the correct gene and utilizes the protein production mechanism in the patient's body to produce the therapeutic protein.

26.4.4.2 Gene Knockdown

Alternatively, when reducing protein level is therapeutically desirable, siRNA can be used to knock down protein levels by degrading specific mRNAs [13, 20, 62]. Since the specific siRNA sequences are only 21–23 nucleotides, it can easily fit into all vectors that are limited by smaller packaging capacity. Gene knockdown strategy has been mostly used in cancer gene therapy because overexpression of oncogenes is one of the leading causes of tumorigenesis.

26.4.4.3 Gene Correction

Another potential therapy utilizes zinc fingers, a class of DNA-binding proteins with specific recognition sequences. Zinc fingers recognizing a specific DNA sequence can be delivered by a gene therapy vector, and this element targets the repair of the mutated gene in the chromosome [24, 53], so that the patient can produce the therapeutic protein from the endogenous gene.

26.4.5 Regulation of Gene Expression

It is important to design the cassette with regulatory elements so that the expression of the transgene can be controlled in very specific ways. Hyper- or hypotransgene expression would lead to over- or underdosage.

Therefore, a regulation system should be included in the vector DNA cassette. For example, tetracycline-dependent regulable gene expression, which relies on the drug doxycyclin, controls the expression of the transgene through transcriptional regulation [19]. Other approaches include dimerizer, ecdysone-responsive, quorum-sensing and zinc-finger-based regulatory systems [52].

26.5 Safety Issues of Gene Therapy

As for any type of disease treatments, patient safety is critical. For the gene therapy field, the importance of the safety issue has been exemplified by multiple clinical cases, most notably the Gelsinger and the *LMO2* insertional mutagenesis cases.

In 1999, patient Jesse Gelsinger, who suffered from ornithine transcarbamylase deficiency, was treated with a replication-defective Ad vector, which was injected through his hepatic artery into the liver. Four days later, Jesse died of a cascade of organ failures triggered by the immune response to the Ad vector [29, 54]. The three main contributions to his death were identified as his poor liver function, other complications of his health prior to the trial, and the incomplete disclosure of the potential risks of Jesse's treatment to his family [51]. More stringent gene therapy regulations have been established since this case to further protect the safety of patients.

In fall 2002, the Necker Hospital in Paris announced that the two youngest boys among 11 patients enrolled in a gene therapy trial for the treatment of X-linked severe combined immunodeficiency (X-SCID) had developed leukemia [21]. The retroviral vector used in this trial inserted the vector genome close to the promoter of the *LMO2* gene in the host genome, which encodes a transcription factor. The LMO2 protein was then continuously expressed, leading to the development of T-cell acute lymphoblastic leukemia. The trial is ongoing, and all participating patients are closely monitored. Even though it has been speculated that the insertional mutagenesis is due to the retroviral vector used, a recent report suggested that the choice of transgene is equally important. The one used in this trial, *IL2RG*, was demonstrated to be more highly oncogenic than expected [58]. This result illustrates that the safety of transgenes used in other trials should also be assessed.

Despite the setback of several incidences, successful cases of gene therapy have been demonstrated. For example, two patients with SCID are able to live outside of a protective environment 10 years after gene therapy treatment [51]. Similarly, the trial performed in the United Kingdom for four young SCID boys was successful [17]. In addition, researchers in the U.S. started a hemophilia B and a Canavan disease trial in which patients are treated with viral vectors expressing factor IX and aspartoacylase enzyme, respectively. The trials are still ongoing with no signs of toxicity [23, 35]. Several current trials have also demonstrated that gene therapy is a safe treatment.

The impact of the deaths of patients in gene therapy trials led to a meeting of the health recombinant DNA advisory committee (RAC) at the National Institutes of Health (NIH) to assess the safety of gene therapy [1]. This meeting included a comprehensive discussion of the experimental protocols, preclinical data (including large animal experimental data), and ethical issues (including the informed consent document) [1]. The committee released a report suggesting that better communication of scientific findings among the scientific community is needed (safety and toxicity data, purity and integrity of vector preparation, dosage, and route of administration to be used). In addition, clear and in-depth information making informed consent possible should be provided to the patients and their families, with all potential risks and benefits related to the trial detailed, and the health condition of the patient before and after the trial should be monitored carefully.

26.6 Difficulties in Achieving Successful Gene Therapy

1. *Expensive Large-Scale Production of Vectors*. Gene therapy has been considered to be an expensive therapy because the cost of manufacturing high-titer vectors sufficient for clinical uses is extremely high. It is particularly true for generating viral vectors. For example, the traditional triple-transfection method [60] used in generating AAV vectors in mammalian cells has been a valuable method for generating enough high-titer and high-quality AAV vectors for research and small-scale clinical trials. However, in order to generate enough vectors for a large number of patients, alternative methods must

be developed (e.g., an insect cell packaging system) [27]. In addition, the generation of retroviral vectors has been improved by the development of vector-producing cell lines so that higher yield production of vectors can be achieved [12].

2. *Uniqueness of Every Disease*. One of the biggest challenges of gene therapy is that every single disease requires a unique vector for the therapy, and it is therefore impossible to generate a "universal vector" for gene delivery. However, these genetic diseases share one common element – i.e., a mutated gene. As the general principle of gene therapy is to introduce the correct gene to the patient's body, this strategy can be widely used in treating most, if not all, genetic diseases.

3. *Ethical Issues of Gene Therapy*. One of the biggest and most debated topics in gene therapy is the possibility of performing germline or in utero gene correction or modification. One would discern that this is morally incorrect if it created an unfair advantage to those who can afford or have access to the technology in creating "designer babies," leading to segregation of humans into classes. However, if law enforcement can oversee the legal issues of germline correction, it would be possible to prevent the mutated gene(s) from being passed on to subsequent generations and thus eliminate the social and economic tribulations that arise out of genetic diseases. Alternative strategies to prevent or cure diseases should be our priority, but the issue of germline correction should merit continuous discussion both in the government and in the wider community (including patients and scientists) [37, 56].

26.7 Conclusions

In conclusion, we have summarized the basic principle of gene therapy and discussed the types of gene delivery vectors and other criteria in leading to successful gene therapy. This is a newly emerging field, and further basic research and clinical trials would provide us with valuable knowledge that would assist us in making gene therapy a more effective and a safer method of treating diseases. It is certainly a complex process for gene therapy to go from benchtops to clinics, but with all the lessons we have learnt from the past, gene therapy is destined to develop into a successful therapeutic option, leading to better lives for mankind.

References

1. Alba R, Bosch A, Chillon M (2005) Gutless adenovirus: last-generation adenovirus for gene therapy. Gene Ther 12(Suppl 1):S18–S27
2. Argnani R, Lufino M, Manservigi M et al (2005) Replication-competent herpes simplex vectors: design and applications. Gene Ther 12(Suppl 1):S170–S177
3. Berto E, Bozac A, Marconi P (2005) Development and application of replication-incompetent HSV-1-based vectors. Gene Ther 12(Suppl 1):S98–S102
4. Carter BJ (2005) Adeno-associated virus vectors in clinical trials. Hum Gene Ther 16:541–550
5. Chen CL, Jensen RL, Schnepp BC et al (2005) Molecular characterization of adeno-associated viruses infecting children. J Virol 79:14781–14792
6. Choi VW, McCarty DM, Samulski RJ (2005) AAV hybrid serotypes: improved vectors for gene delivery. Curr Gene Ther 5:299–310
7. Choi VW, Samulski RJ, McCarty DM (2005) Effects of adeno-associated virus DNA hairpin structure on recombination. J Virol 79:6801–6807
8. Collins FS, Green ED, Guttmacher AE et al (2003) A vision for the future of genomics research. Nature 422:835–847
9. Collins FS, Morgan M, Patrinos A (2003) The human genome project: lessons from large-scale biology. Science 300:286–290
10. De Laporte L, Cruz Rea J, Shea LD (2006) Design of modular non-viral gene therapy vectors. Biomaterials 27:947–954
11. Dubensky TW Jr, Sauter SL (2003) Generation of retroviral packaging and producer cell lines for large-scale vector production with improved safety and titer. Methods Mol Med 76:309–330
12. Dykxhoorn DM, Palliser D, Lieberman J (2006) The silent treatment: siRNAs as small molecule drugs. Gene Ther 13:541–552
13. Editorial (2002) Assessment of adenoviral vector safety and toxicity: report of the National Institutes of Health Recombinant DNA Advisory Committee. Hum Gene Ther 13:3–13
14. Felgner JH, Kumar R, Sridhar CN et al (1994) Enhanced gene delivery and mechanism studies with a novel series of cationic lipid formulations. J Biol Chem 269:2550–2561
15. Gao G, Vandenberghe LH, Alvira MR et al (2004) Clades of Adeno-associated viruses are widely disseminated in human tissues. J Virol 78:6381–6388
16. Gao X, Huang L (1995) Cationic liposome-mediated gene transfer. Gene Ther 2:710–722
17. Gaspar HB, Parsley KL, Howe S et al (2004) Gene therapy of X-linked severe combined immunodeficiency by use of a pseudotyped gammaretroviral vector. Lancet 364: 2181–2187
18. Gilgenkrantz H, Duboc D, Juillard V et al (1995) Transient expression of genes transferred in vivo into heart using first-generation adenoviral vectors: role of the immune response. Hum Gene Ther 6:1265–1274
19. Goverdhana S, Puntel M, Xiong W et al (2005) Regulatable gene expression systems for gene therapy applications: progress and future challenges. Mol Ther 12:189–211
20. Gura T (2000) A silence that speaks volumes. Nature 404: 804–808

21. Hacein-Bey-Abina S, Von Kalle C, Schmidt M et al (2003) LMO2-associated clonal T cell proliferation in two patients after gene therapy for SCID-X1. Science 302:415–419
22. Izumi KM, Stevens JG (1990) Molecular and biological characterization of a herpes simplex virus type 1 (HSV-1) neuroinvasiveness gene. J Exp Med 172:487–496
23. Jiang H, Pierce GF, Ozelo MC et al (2006) Evidence of Multiyear Factor IX Expression by AAV-Mediated Gene Transfer to Skeletal Muscle in an Individual with Severe Hemophilia B. Mol Ther 14(3):452–455
24. Kaiser J (2005) Gene therapy. Putting the fingers on gene repair. Science 310:1894–1896
25. Khalil IA, Kogure K, Akita H et al (2006) Uptake pathways and subsequent intracellular trafficking in nonviral gene delivery. Pharmacol Rev 58:32–45
26. Klinman DM, Yamshchikov G, Ishigatsubo Y (1997) Contribution of CpG motifs to the immunogenicity of DNA vaccines. J Immunol 158:3635–3639
27. Kohlbrenner E, Aslanidi G, Nash K et al (2005) Successful production of pseudotyped rAAV vectors using a modified baculovirus expression system. Mol Ther 12:1217–1225
28. Lai Y, Yue Y, Liu M et al (2005) Efficient in vivo gene expression by trans-splicing adeno-associated viral vectors. Nat Biotechnol 23:1435–1439
29. Lehrman S (1999) Virus treatment questioned after gene therapy death. Nature 401:517–518
30. Li C, Bowles DE, van Dyke T et al (2005) Adeno-associated virus vectors: potential applications for cancer gene therapy. Cancer Gene Ther 12:913–925
31. Li Z, Dullmann J, Schiedlmeier B et al (2002) Murine leukemia induced by retroviral gene marking. Science 296:497
32. Logan AC, Haas DL, Kafri T et al (2004) Integrated self-inactivating lentiviral vectors produce full-length genomic transcripts competent for encapsidation and integration. J Virol 78:8421–8436
33. Marcum JA (2005) From the molecular genetics revolution to gene therapy: translating basic research into medicine. J Lab Clin Med 146:312–316
34. McCarty DM, Fu H, Monahan PE et al (2003) Adeno-associated virus terminal repeat (TR) mutant generates self-complementary vectors to overcome the rate-limiting step to transduction in vivo. Gene Ther 10:2112–2118
35. McPhee SW, Janson CG, Li C et al (2006) Immune responses to AAV in a phase I study for Canavan disease. J Gene Med 8:577–588
36. Miller AD (1992) Human gene therapy comes of age. Nature 357:455–460
37. Motulsky AG (1983) Impact of genetic manipulation on society and medicine. Science 219:135–140
38. Nunes ML, Pinho AP, Sfoggia A (2001) Cerebral aneurysmal dilatation in an infant with perinatally acquired HIV infection and HSV encephalitis. Arq Neuropsiquiatr 59:116–118
39. O'Connor TP, Crystal RG (2006) Genetic medicines: treatment strategies for hereditary disorders. Nat Rev Genet 7:261–276
40. Palmer D, Ng P (2003) Improved system for helper-dependent adenoviral vector production. Mol Ther 8:846–852
41. Palmer DH, Young LS, Mautner V (2006) Cancer gene-therapy: clinical trials. Trends Biotechnol 24:76–82
42. Pisetsky DS (1996) The immunologic properties of DNA. J Immunol 156:421–423
43. Rivera VM, Gao GP, Grant RL et al (2005) Long-term pharmacologically regulated expression of erythropoietin in primates following AAV-mediated gene transfer. Blood 105:1424–1430
44. Rowe WP, Huebner RJ, Gilmore LK et al (1953) Isolation of a cytopathogenic agent from human adenoids undergoing spontaneous degeneration in tissue culture. Proc Soc Exp Biol Med 84:570–573
45. Sadaie MR, Zamani M, Whang S et al (1998) Towards developing HIV-2 lentivirus-based retroviral vectors for gene therapy: dual gene expression in the context of HIV-2 LTR and Tat. J Med Virol 54:118–128
46. Samulski RJ, Berns KI, Tan M et al (1982) Cloning of adeno-associated virus into pBR322: rescue of intact virus from the recombinant plasmid in human cells. Proc Natl Acad Sci USA 79:2077–2081
47. Schnepp BC, Clark KR, Klemanski DL et al (2003) Genetic fate of recombinant adeno-associated virus vector genomes in muscle. J Virol 77:3495–3504
48. Schnepp BC, Jensen RL, Chen CL et al (2005) Characterization of adeno-associated virus genomes isolated from human tissues. J Virol 79:14793–14803
49. Sinn PL, Sauter SL, McCray PB Jr (2005) Gene therapy progress and prospects: development of improved lentiviral and retroviral vectors–design, biosafety, and production. Gene Ther 12:1089–1098
50. Sorrentino BP, Brandt SJ, Bodine D et al (1992) Selection of drug-resistant bone marrow cells in vivo after retroviral transfer of human MDR1. Science 257:99–103
51. Thompson L (2000) Human gene therapy. Harsh lessons, high hopes. FDA Consum 34:19–24
52. Toniatti C, Bujard H, Cortese R et al (2004) Gene therapy progress and prospects: transcription regulatory systems. Gene Ther 11:649–657
53. Urnov FD, Miller JC, Lee YL et al (2005) Highly efficient endogenous human gene correction using designed zinc-finger nucleases. Nature 435:646–651
54. Verma IM (2000) A tumultuous year for gene therapy. Mol Ther 2:415–416
55. Wang L, Calcedo R, Nichols TC et al (2005) Sustained correction of disease in naive and AAV2-pretreated hemophilia B dogs: AAV2/8-mediated, liver-directed gene therapy. Blood 105:3079–3086
56. Wivel NA, Walters L (1993) Germ-line gene modification and disease prevention: some medical and ethical perspectives. Science 262:533–538
57. Wood KJ, Fry J (1999) Gene therapy: potential applications in clinical transplantation. Expert Rev Mol Med 1999:1–20
58. Woods NB, Bottero V, Schmidt M et al (2006) Gene therapy: therapeutic gene causing lymphoma. Nature 440:1123
59. Xiao X, Li J, Samulski RJ (1996) Efficient long-term gene transfer into muscle tissue of immunocompetent mice by adeno-associated virus vector. J Virol 70:8098–8108
60. Xiao X, Li J, Samulski RJ (1998) Production of high-titer recombinant adeno-associated virus vectors in the absence of helper adenovirus. J Virol 72:2224–2232
61. Yang Y, Wilson JM (1995) Clearance of adenovirus-infected hepatocytes by MHC class I-restricted CD4+ CTLs in vivo. J Immunol 155:2564–2570
62. Zamore PD, Tuschl T, Sharp PA et al (2000) RNAi: double-stranded RNA directs the ATP-dependent cleavage of mRNA at 21 to 23 nucleotide intervals. Cell 101:25–33

Cloning in Research and Treatment of Human Genetic Disease

Ian Wilmut, Jane Taylor, Paul de Sousa, Richard Anderson, and Christopher Shaw

Contents

27.1 Introduction 875
27.2 The Value of Cell Lines of Specific Genotype 876
    27.2.1 Studies of Inherited Disease 876
    27.2.2 Cells for Therapy 877
27.3 Somatic Cell Nuclear Transfer................................ 878
    27.3.1 Procedure for Nuclear Transfer 878
    27.3.2 Present Successes and Limitations 878
27.4 Cells from Cloned Embryos.. 879
    27.4.1 Cells from Cloned Mouse Embryos 879
    27.4.2 Derivation of Cells from Cloned Human Embryos 879
    27.4.3 Interspecies Nuclear Transfer.................... 880
27.5 Direct Reprogramming of Somatic Cells.................. 880
27.6 Looking to the Future.. 881
References... 881

I. Wilmut (✉), J. Taylor, and P. de Sousa
Scottish Centre for Regenerative Medicine, University of Edinburgh, Chancellor's Building 49 Little France Crescent, Edinburgh, EH16 4SB, UK
e-mail: Ian.Wilmut@ed.ac.uk

R. Anderson
Centre for Reproductive Biology, The Queen's Medical Research Institute, 47 Little France Crescent, Edinburgh, EH16 4TJ, UK

C. Shaw
Department of Neurology, King's College London School of Medicine, London, UK

27.1 Introduction

Revolutionary new opportunities to study human disease were heralded by the development of two new cell-based techniques. These are the methods able to obtain normal development after somatic cell nuclear transfer (SCNT) [45] and to derive stem cells from human embryos [38]. Use of these two new techniques together will provide opportunities to study inherited human disease that are not available in any other way. Such cells would provide new opportunities to study the pathophysiology of inherited human diseases and in turn make it possible to use high-throughput screens to identify drugs able to prevent symptoms of the disease.

In the longer term, cells derived from cloned human embryos may be used in treatment of disease. They would offer the potential advantage of being immunologically matched to the patient who donated the nuclear donor cell and so could be used for therapy without the need for immunosuppression to prevent rejection. Cell lines that are homozygous at the major histocompatibility antigens are expected to be particularly valuable for transplantation. Calculations suggest that if these genotypes could be selected a comparatively small number of lines would provide an immune match to a very large proportion of the population [36]. Such a bank of cell lines could be established by somatic cell nuclear transfer (SCNT). None of these objectives have yet been realized, and it is clear that a great deal remains to be learned before these new approaches become available for either research or therapy.

This chapter will describe the potential benefits that may arise from the production of embryo stem cells by SCNT and then describe the present procedures for and outcomes of nuclear transfer, before considering new approaches to the development of effective procedures.

27.2 The Value of Cell Lines of Specific Genotype

27.2.1 Studies of Inherited Disease

There are many different approaches to the isolation of affected cells for the study of inherited disease, and it is unlikely that any one will be optimal for all diseases. In the case of conditions that affect an accessible tissue, such as the skin or the hematopoetic system, then cells may be obtained directly from the patient. Full benefit from this approach will depend upon the isolation of progenitor cells that can be multiplied extensively in culture to allow detailed and repeated experimentation with the same population of cells over a considerable period of time. These methods have been established for some tissues and used for research and therapy [37, 57].

Several other strategies involve derivation of the affected cell type from embryo stem cells that have the causative genotype. If dominant mutations have been identified and the affected cell types can be derived from human embryo stem cells then the introduction of the mutation into an existing stem cell line makes possible a direct comparison between the same cell population with or without the mutated gene.

Alternatively, preimplantation genetic diagnosis may be used by prospective parents to ensure the birth of a child who has not inherited the disease by selection of embryos that do not carry the causative mutation. By contrast, with the consent of the parents those embryos that carry the mutation may be used for research, rather than being destroyed. In this case there is no unaffected control population of cells.

The specific circumstances in which nuclear transfer would provide a unique opportunity apply when the causative allele(s) has/have not been identified and the affected tissue cannot be recovered from the patient. The family of diseases that are variously known as motor neuron disease (MND), amyotrophic lateral sclerosis (ALS), or Lou Gehrig's disease is one such case.

ALS is a relentlessly progressive muscle-wasting disease [8]. Degeneration of motor neurons is the common cause of this fatal condition, but the molecular mechanisms leading to degeneration are not well understood. Several genetic and environmental factors contribute to the pathogenesis of ALS. The great majority of cases are sporadic with no evidence of inheritance in the family, but in a small proportion of cases (approximately 10%) the disease does occur frequently within a family. Penetrance is variable, but in the great majority of cases those with mutations in superoxide dismutase (SOD1) develop the symptoms of ALS. Mutations in SOD1 account for approximately 20% of inherited cases, and genetic analysis has identified other dominant and recessive familial ALS loci on chromosomes 2, 9, 15, 18, and 20 [17].

It was at first assumed that ALS caused by mutations in SOD1 reflected a loss of function of the damaged gene, but this seems not to be the case. No symptoms of ALS were evident in mice in which SOD1 had been deleted [9]. By contrast, mice do develop symptoms of ALS if mutated human genes are overexpressed, suggesting that the effect of the mutation is through a toxic effect of the abnormal protein, rather than a loss of function [9].

While the symptoms of the disease reflect loss of motor neurons, until recently it was not known whether these cells were the site of the primary lesion. New studies in mice carrying mutant SOD1 genes that could be removed from motor neurons or microglia in a tissue-specific manner identified two phases in the development of the disease. Removal of the transgene from microglia had little effect on the development of first symptoms, but dramatically slowed progression through later stages [2]. Conversely, removal of the gene from progenitors of motor neurons delayed early stages of the disease. These observations suggest that the first symptoms reflect changes in the motor neurons, but that later in the development of the disease changes in microglia exacerbate the damage in the motor neurons. The observation that ALS is not cell autonomous has profound implications for cell therapy, because it suggests that healthy glial cells may be able to provide protection for failing motor neurons. In addition, they also show that both glia and neurons should be studied in search of an understanding of the causes of ALS.

Cells suitable for these analyses could be obtained from mouse or human embryo stem cell lines. Cell lines derived from mice carrying a mutant human SOD1 transgene could be studied to identify changes in gene expression and other aspects of cell function. There are also several new ways by which suitable cells derived from human embryo stem cells may be obtained. For those cases in which the mutation has been identified, the mutation can be introduced into existing embryo stem cells or pre-implantation genetic diagnosis used to identify embryos that have the mutation.

To complement this approach, nuclear transfer will make it possible to study familial cases in which the mutation has not yet been identified, which means some 8% of all cases. As resources allow, it will also be possible to study neural lineages from sporadic cases (90%) in search of the factors that have made the patients unusually sensitive to environmental factors.

The aim of all these approaches would be to identify the molecular changes in the cells with a view to establishing a high-throughput screen for small molecules able to prevent degeneration of motor neurons or glial cells. Differences associated with the disease would be sought by profiling gene expression at RNA and protein levels and by observation of protein distribution within cells. At present, candidate drugs may be assessed for their ability to delay development of the disease in mice in which transgenes over express mutant forms of human SOD1. A handful of compounds can be compared each year, at considerable cost. By contrast, a high-throughput screen based upon tissue culture would be able to compare thousands of compounds in the same time and at a similar cost. Candidates identified in this way would be then be evaluated in animal studies before being considered for clinical trials.

Many other human genetic diseases could be studied by the same approach. Other candidate conditions for study include other neurodegenerative diseases, psychiatric conditions, cardiomyopathy, and some forms of cancer. The advantage of using nuclear transfer is greatest if the mutation that causes the disease is not known. It is also essential that the affected cell types can be produced from embryo stem cells in the laboratory. Finally, it must be possible to make observations on the cells in search of the molecular mechanism that causes the disease. These studies need not necessarily concern the physiological function of the cells. In the case of ALS, the demonstration of a harmful effect of corrupted protein in transgenic animals strongly suggests that the fate of the protein in the cells should be an initial subject of study.

27.2.2 Cells for Therapy

When considering cell therapy, the ideal would be to be able to offer a perfect immunological match for every patient. This might be achieved by having a very large network of banks or by using SCNT to produce "patient-specific cells." It is implied that SCNT could be used to produce a cell for each person as required. In terms of the requirement for skilled personnel, high-quality facilities and the supply of oocytes, the sheer scale of that task is almost beyond imagination. In reality it seems probable that SCNT might be used to provide cells for very specific purposes, such as gene therapy or to produce cell lines that could be used to treat large numbers of patients, rather than individuals (see below).

Similarly, the idea of holding a universal bank of cell lines collected at random from donated embryos is also impracticable on several practical grounds, which include cost. The first estimates have been made of the effect of only being able to hold a limited number of lines on the probability of being able to provide cells with an acceptable match for the population of the United Kingdom [36]. To estimate the likely genotype of cell lines provided by donors available at random, the databases of the genotype of 10,000 consecutive organ donors was used. Patient populations were simulated by use of genotype information on 6,577 patients waiting for organ transplantation. The authors considered only HLA-A, HLA-B, and HLA-DR, and calculated the best possible match. A bank of 150 lines was estimated to provide a perfect match for nearly 20% of the population. The probability of having either a perfect match or a mismatch at HLA-A or HLA-B approached 40% (37.9%). A panel of this size was expected to provide at least a match for HLA-DR for 84.9% of the patient population. As the size of the bank was increased beyond 150 lines there was a diminishing return for the additional resource.

By contrast, it was predicted that there would be great advantage in obtaining cells lines that are homozygous for common HLA types [36]. A bank of just 10 lines homozygous for common HLA types was expected to provide a perfect mach for 37.7% and a mismatch only at HLA-A or HLA-B for 67.7% of the patient population. These calculations provide a robust indication of the potential benefit of using SCNT to obtain cell lines of selected homozygous genotypes.

27.3 Somatic Cell Nuclear Transfer

SCNT is just one of a number of methods of producing cells of specific genotype. They may be produced by SCNT using a cell from a person with the desired genotype as the nuclear donor. Alternatively, progress is being made toward the establishment of methods for treating somatic cells in such a way that they acquire many of the characteristics of ES cells. The alternatives will be considered later in the chapter.

27.3.1 Procedure for Nuclear Transfer

Two cells are required for SCNT, an oocyte and a nuclear donor cell [5]. Typically the oocytes are at the second metaphase of meiosis, but a recent publication has demonstrated a benefit following the use of early zygotes as recipient cells [27]. In most procedures, the oocyte is enucleated before the nucleus of a donor cell is introduced. However, in some cases the nucleus is introduced before the same pipette is used to enucleate the oocytes [26]. Nuclear transfer may occur by fusing the entire donor cell to the enucleated oocytes or by rupturing the donor cell and injecting the nucleus and some cytoplasm directly into the oocyte.

If an oocyte has been used as the recipient cell, the reconstructed embryo must then be induced to resume meiosis by parthenogenetic activation. A variety of procedures have been established, and there appear to be differences between species in the specific procedure that is most effective. In addition, there are differences between species in the optimum interval between introduction of the donor nucleus and the activation stimulus, which may be either simultaneous or delayed. In sheep, there was no evidence of an effect of varying this interval [4] and in cattle there was an advantage in delaying activation [42], whereas in the mouse cloned pups were only produced if activation was delayed [40].

Typically, the embryos are cultured for a period before transfer to recipient females. This step provides information on the early development of the embryos, and because only those embryos that are developing normally are transferred to surrogate females it also reduces the number of recipient females that are required.

27.3.2 Present Successes and Limitations

The present procedures for SCNT are widely used, and in that sense repeatable, but they are very inefficient. Viable offspring have been produced after nuclear transfer in a number of species, including mouse, cow, pig, rabbit, rat, horse, cat, and dog. Although these species have been cloned there are no cloned primates at present, despite a considerable research effort.

Similarly, offspring have been obtained after transfer of a variety of different adult somatic cell types (reviewed in [24]). In general, the experiments have not been carried out on a large enough scale to permit meaningful statistical comparisons; this would require several hundred, perhaps thousands, of nuclear transfers for each cell type. There is a suggestion that as cells are taken from later stages of development the efficiency decreases [47], as had been shown previously in amphibians [12], but there are reports suggesting exceptions [32]. This confusion may reflect the fact that in some experiments the procedures being used had not been optimized for each cell type.

The overall inefficiency reflects losses at all stages of development and after birth. The general experience is that offspring derived by SCNT require additional assistance at birth. The birth of calves is often induced to facilitate provision of assistance [43]. Typically, cloned mice are delivered by caesarean section and the pups cross fostered [40]. A variety of different abnormalities have been reported in cloned offspring. These include difficulties in breathing and increased birth weight [43].

At the present time the molecular mechanisms that lead to the birth of cloned offspring are not understood. Following transfer of the nucleus into the cytoplasm of the recipient oocytes or zygote, unknown factors in the oocyte cytoplasm must act to modify the epigenetic mechanisms that regulate gene expression of the transferred nucleus so that the embryo is able to develop. It is possible that offspring are derived from donor cells that are more amenable to reprogramming. Alternatively, the inefficiency may reflect a requirement for a number of key epigenetic changes that may be independent events such that failure for them all to occur appropriately leaves the resulting embryo with some, but not all, of the required changes. In this case it would be expected that embryos produced by nuclear transfer a variable and unusual pattern of gene expression and chromatin organisation. This is indeed the case.

In some of the resulting embryos the pattern is similar to that in embryos produced by fertilization; however, in others the pattern of DNA methylation and histone modification is more like that of a donor nucleus than of an embryo produced by fertilization (reviewed in [50]). Improvements to the efficiency of SCNT probably depend upon methods of assisting in the reprogramming of the nucleus. Techniques to reprogram cells directly may well contribute to attainment of this objective (see Sect. 27.5).

27.4 Cells from Cloned Embryos

Several independent groups have now demonstrated that ES cells can be derived from cloned mouse embryos and that these lines are able to contribute to chimeras with normal efficiency [18, 22, 41]. An intriguing point is that it is possible to derive ES cell lines from many more cloned embryos than would have become offspring (16% vs 2%) had the embryos been transferred to recipients [39].

27.4.1 Cells from Cloned Mouse Embryos

A recent study analyzed a sample from a collection of more than 150 cell lines and demonstrated equivalence of embryo stem cells derived from mouse blastocysts produced by nuclear transfer and fertilization [39]. This was found to be so in regard to ability to form all lineages in vivo, patterns of gene expression as determined by microarray analysis, expression of markers of pluripotency, and presence of tissue-specific patterns of DNA methylation. These studies lend great encouragement to plans to use cells from cloned human embryos in research. However, it should be appreciated that only a very small sample of the lines was assessed.

27.4.2 Derivation of Cells from Cloned Human Embryos

By contrast, progress in the development of methods for the derivation of cells from cloned human embryos has been very limited, and this area of research marred by fraudulent claims. Two groups have reported the production of blastocysts after transfer of nuclei to human oocytes [15, 31]. Neither of these reports led to the derivation of embryo stem cell lines. In both of those studies fresh oocytes were available and results were less encouraging when oocytes that had not been fertilized during an IVF protocol were then used as recipient oocytes.

It is noticeable that despite considerable research effort there are no reported clones of adult primates. Several groups have described development to the blastocyst stage of embryos produced by SCNT, and in some cases pregnancy was established, but these failed to develop beyond 60 days in any case [23]. This limited success may reflect practical difficulties in carrying out research in these species, but it has also been reported that spindle formation is different in primates and evidence has been presented to suggest that critical factors are removed during the nuclear transfer process in primates [28]. As development appeared to be normal after fertilization of oocytes from which the spindle was removed temporarily and then replaced, it seems that the effect is not due to the physical trauma of micromanipulation or exposure to drugs during the procedure, but to actual loss of as yet unknown critical factors during enucleation [28].

The fact that variations in protocol were required for success in nuclear transfer in different species suggests that we should not be surprised to find that further modifications are required to establish effective procedures in a new species.

27.4.3 Interspecies Nuclear Transfer

In an alternative approach to the production of cells from cloned embryos, cells with many of the characteristics of stem cells were derived following transfer of human nuclei into rabbit oocytes [7]. At present this is the only approach that has produced "human" cloned cells in culture.

When it was published this report was a cause of great controversy. Previous experiments had provided very little evidence to suggest that normal development was possible after interspecies nuclear transfer unless the two species were extremely closely related. However, since the initial report, several groups have shown that it is possible to obtain at least early development after transfer between two divergent species. Embryos have developed to the blastocyst stage after transfer of primate nuclei into rabbit [48, 49] and bovine [16] oocytes. Two factors that may limit the value of interspecies SCNT concern the source of mitochondria in the cells derived from these embryos and the existence of differences between species in the epigenetic mechanisms that regulate early development.

A number of studies have monitored the fate of mitochondria after interspecies nuclear transfer and demonstrated that mitochondria derived from both the oocyte and donor cell were present during early cleavage, although the oocyte contributed a greater number [14, 20, 21].

The limited information that is available has revealed differences between species in some of the epigenetic mechanisms that regulate development, despite the fact that in a general sense they are conserved. Demethylation of DNA in the transferred nucleus was compared when nuclei were transferred from pig cells into rabbit oocytes or rabbit cells into pig oocytes [6]. These were associated with the oocytes rather than the transferred nucleus. Similarly, between-species sperm injections revealed differences between the oocytes in demethylation of the male pronucleus, although in this case the origin of the sperm also had an effect [1].

It seems likely that procedures for the derivation of ES cells of a specific genotype through nuclear transfer can be significantly improved. Many factors might be expected to influence development of embryos produced in this way. Is the organization of mitochondrial genomes in the two species compatible? Should the protocols used for production and culture of such embryos be those for the species from which the oocytes were recovered or the nuclear donors? Moreover, a great deal remains to be learned about embryos and cells produced by interspecies nuclear transfer. Are the cells capable of being passaged many times, essentially indefinitely, as is expected of embryo stem cells?

Several reasons are put forward for using of animal oocytes. First, if there are indeed differences in early development between primates and other groups of mammals it may be possible to identify these by comparison of development between embryos produced by transfer of nuclei from the same primate donor population into primate and nonprimate oocytes. Additionally, it may be possible to derive cell lines in this way. If they are obtained, it will be important to discover whether the cells in fact have all of the characteristics of human embryo stem cells. If not, do they have sufficient similarity to allow them to be used in research to study disease? Clearly there would be great benefit in being able to use animal oocytes, most clearly in the numbers available. In fact it will only be possible to pursue this approach to the study of many diseases if such a source of oocytes is identified.

27.5 Direct Reprogramming of Somatic Cells

The birth of offspring following transfer of nuclei from adult somatic cells into enucleated oocytes was the first observation to demonstrate that under appropriate circumstances gene function in such nuclei can be reprogrammed to allow them to control normal development [45]. Since that time several approaches have been used to investigate reprogramming, and more particularly to assess the ability of other methods to achieve similar ends. Most strikingly, the introduction of transcription factors into skin fibroblasts gave a small proportion of the cells many of the characteristics of embryo stem cells (see below).

In the first studies seeking factors that may change epigenetics during SCNT the small volume of cytoplasm in mammalian oocytes imposes strict limits on biochemical analyses, but this has been overcome by use of amphibian oocytes [3, 13, 29, 30] or embryo stem cells (see below).

It is many years since it was first demonstrated that amphibian oocytes are able to reprogram gene expression in mammalian nuclei (reviewed in [13]), providing an opportunity to define the molecular events involved and perhaps in the longer term to identify the active factors. The first change in the nuclei is a pronounced increase in volume of the transferred nucleus [30], and a further 2 days' incubation are required for the initiation of expression of key genes, such as Oct4. However, the oocytes are incubated at 18°C and the authors argue that this is equivalent to 12 h at 37°C [3]. Gene expression is accompanied by demethylation of promoter regions, and key binding sites have been identified [30]. Two proteins have been shown to have a role in reprogramming [19, 35] The identity of other active factors remains to be defined, but in the future this approach may provide another means of obtaining cells of a chosen phenotype from a specific patient.

When adult somatic cells were fused to embryo stem cells it was found that the somatic nucleus was reprogrammed toward ES cell phenotype [33], provided that the nucleus of the embryo stem cell was present [11]. Since that initial experiment a number of groups have studied the phenomenon, with a view either to developing a means of direct reprogramming or to using this approach to identify the active factors responsible for the reprogramming activity [10]. Despite a number of ingenious approaches, so far nobody has devised a method for the effective specific removal of the chromosomes of the embryo stem cell from the heterokaryon. While cell fusion seems unlikely to yield cells suitable for therapy, any new understanding of the nature of the key reprogramming factors would be extremely valuable.

Recently a radically different approach has provided another means of producing cells of specific genotype. The authors speculated that the introduction of key transcription factors would be able to modify cells in such a way that they acquired the characteristics associated with the specific panel of transcription factors. They identified in total 22 factors that are known either to be essential for embryo stem cell maintenance or were at least known to be expressed in embryo stem cells. Retroviral vectors were used to introduce this group of candidate transcription factors into fibroblast cultures [34]. After careful analysis it was found that just four factors were able to make some murine fibroblasts pluripotent. These factors were Oct3/4, Sox2, c-Myc, and Klf4. The cells were able to form tissues of all the major lineages in culture and in chimeras, although in the first study none of the chimeras survived to term. Subsequently, it was shown that cells selected on the basis of their expression of Nanog or Oct4 after introduction of these four factors produced cells that are able to contribute to germline chimeras [25, 44].

This most remarkable demonstration has created an entirely new means of obtaining cells of specific genotype, although it remains to be shown whether an effective procedure can be established for human somatic cells.

27.6 Looking to the Future

It is essential in the development of these methods that a precise assessment is made of the effect of the reprogramming procedure upon the cell lines that are derived. A number of strategies may be considered to clarify the situation. A great deal has been made in both scientific and popular press of the potential value of cells from cloned embryos in therapy, but in their haste to consider this use many people have overlooked the considerable benefits that may be gained by the use of such cells in research into inherited diseases. Such studies would provide opportunities to study diseases that are not available in any other way at present and to perform high-throughput screening of potential therapeutics. This laboratory-based research will also enable us to learn whether or not cells obtained by reprogramming function normally might indeed be suitable for use in therapy.

References

1. Beaujean N, Taylor JE, McGarry M, Gardner JO, Wilmut I, Loi P, Ptak G, Galli C, Lazzari G, Bird A, Young LE, Meehan RR (2004) The effect of interspecific oocytes on demethylation of sperm DNA. Proc Natl Acad Sci USA 101:7636–7640
2. Boillee S, Yamanaka K, Lobsiger CS, Copeland NG, Jenkins NA, Kassiotis G, Kollias G, Cleveland DW (2006) Onset and progression in inherited ALS determined by motor neurons and microglia. Science 312:1389–1392
3. Byrne JA, Simonsson S, Western PS, Gurdon JB (2003) Nuclei of adult mammalian somatic cells are directly reprogrammed to oct-4 stem cell gene expression by amphibian oocytes. Curr Biol 13:1206–1213

4. Campbell KH, McWhir J, Ritchie WA, Wilmut I (1996) Sheep cloned by nuclear transfer from a cultured cell line. Nature 380:64–66
5. Campbell KH, Alberio R, Lee JH, Ritchie WA (2001) Nuclear transfer in practice. Cloning Stem Cells 3:201–208
6. Chen T, Zhang YL, Jiang Y, Liu JH, Schatten H, Chen DY, Sun QY (2006) Interspecies nuclear transfer reveals that demethylation of specific repetitive sequences is determined by recipient ooplasm but not by donor intrinsic property in cloned embryos. Mol Reprod Dev 73:313–317
7. Chen Y, He ZX, Liu A, Wang K, Mao WW, Chu JX, Lu Y, Fang ZF, Shi YT, Yang QZ, Chenda Y, Wang MK, Li JS, Huang SL, Kong XY, Shi YZ, Wang ZQ, Xia JH, Long ZG, Xue ZG, Ding WX, Sheng HZ (2003) Embryonic stem cells generated by nuclear transfer of human somatic nuclei into rabbit oocytes. Cell Res 13:251–263
8. Cleveland DW, Rothstein JD (2001) From Charcot to Lou Gehrig: deciphering selective motor neuron death in ALS. Nat Rev Neurosci 2:806–819
9. Cluskey S, Ramsden DB (2001) Mechanisms of neurodegeneration in amyotrophic lateral sclerosis. Mol Pathol 54:386–392
10. Cowan CA, Atienza J, Melton DA, Eggan K (2005) Nuclear reprogramming of somatic cells after fusion with human embryonic stem cells. Science 309:1369–1373
11. Do JT, Scholer HR (2004) Nuclei of embryonic stem cells reprogram somatic cells. Stem Cells 22:941–949
12. Gurdon JB (1960) Factors responsible for the abnormal development of embryos obtained by nuclear transplantation in Xenopus laevis. J Embryol Exp Morphol. 8:327–340
13. Gurdon JB (2006) From nuclear transfer to nuclear reprogramming: the reversal of cell differentiation. Annu Rev Cell Dev Biol 22:1–22
14. Hua S, Zhang Y, Song K, Song J, Zhang Z, Zhang L, Zhang C, Cao J, Ma L (2008) Development of bovine-ovine interspecies cloned embryos and mitochondria segregation in blastomeres during preimplantation. Anim Reprod Sci 105:245–257
15. Hwang W-S and colleagues as reported by the investigating committee of Seoul National University, Korea 2006
16. Illmensee K, Levanduski M, Zavos PM (2006) Evaluation of the embryonic preimplantation potential of human adult somatic cells via an embryo interspecies bioassay using bovine oocytes. Fertil Steril 85(Suppl 1):1248–1260
17. Jones JM, Vance C, Al-Chalabi A, Smith BN, Hu X, Sreedharan J, Siddique T, Schelhaas HJ, Kusters B, Troost D, Baas F, De Jong V, Shaw CE (2006) Familial amyotrophic lateral sclerosis with frontotemporal dementia is linked to a locus on Chromosome 9p13.2–21.3. Brain 129:868–876
18. Kawase E, Yamazaki Y, Yagi T, Yanagimachi R, Pedersen RA (2000) Mouse embryonic stem (ES) cell lines established from neuronal cell-derived cloned blastocysts. Genesis 28:156–163
19. Koziol MJ, Garrett N, Gurdon JB (2007) Tpt1 activates transcription of oct4 and nanog in transplanted somatic nuclei. Curr Biol 17:801–807
20. Lloyd RE, Lee JH, Alberio R, Bowles EJ, Ramalho-Santos J, Campbell KH, St John JC (2006) Aberrant nucleo-cytoplasmic cross-talk results in donor cell mtDNA persistence in cloned embryos. Genetics 172:2515–2527
21. Ma LB, Yang L, Zhang Y, Cao JW, Hua S, Li JX (2008) Quantitative analysis of mitochondrial RNA in goat-sheep cloned embryos. Mol Reprod Dev 75:33–39
22. Munsie MJ, Michalska AE, O'Brien CM, Trounson AO, Pera MF, Mountford PS (2000) Isolation of pluripotent embryonic stem cells from reprogrammed adult mouse somatic cell nuclei. Curr Biol 24:989–992
23. Ng SC, Chen N, Yip WY, Liow SL, Tong GQ, Martelli B, Tan LG, Martelli P (2004) The first cell cycle after transfer of somatic cell nuclei in a non-human primate. Development 131:2475–2484
24. Oback B, Wells DN (2007) Donor cell differentiation, reprogramming, and cloning efficiency: elusive or illusive correlation? Mol Reprod Dev 74:646–654
25. Okita K, Ichisaka T, Yamanaka S (2007) Generation of germline-competent induced pluripotent stem cells. Nature 448:313–317
26. Peura TT (2003) Improved in vitro development rates of sheep somatic nuclear transfer embryos by using a reverse-order zona-free cloning method. Cloning Stem Cells 5:13–24
27. Schurmann A, Wells DN, Oback B (2006) Early zygotes are suitable recipients for bovine somatic nuclear transfer and result in cloned offspring. Reproduction 132:9–48
28. Simerly C, Dominko T, Navara C, Payne C, Capuano S, Gosman G, Chong KY, Takahashi D, Chace C, Compton D, Hewitson L, Schatten G (2003) Molecular correlates of primate nuclear transfer failures. Science 300:297
29. Simonsson S, Gurdon J (2004) DNA demethylation is necessary for the epigenetic reprogramming of somatic cell nuclei. Nat Cell Biol 6:984–990
30. Simonsson S, Gurdon JB (2005) Changing cell fate by nuclear reprogramming. Cell Cycle 4:513–515
31. Stojkovic M, Stojkovic P, Leary C, Hall VJ, Armstrong L, Herbert M, Nesbitt M, Lako M, Murdoch A (2005) Derivation of a human blastocyst after heterologous nuclear transfer to donated oocytes. Reprod Biomed Online 11:226–231
32. Sung LY, Gao S, Shen H, Yu H, Song Y, Smith SL, Chang CC, Inoue K, Kuo L, Lian J (2006) Differentiated cells are more efficient than adult stem cells for cloning by somatic cell nuclear transfer. Nat Genet 38:1323–1328
33. Tada M, Takahama Y, Abe K, Nakatsuji N, Tada T (2001) Nuclear reprogramming of somatic cells by in vitro hybridization with ES cells. Curr Biol 11:1553–1558
34. Takahashi K, Yamanaka S (2006) Induction of pluripotent stem cells from mouse embryonic and adult fibroblast cultures by defined factors. Cell 126:1–14
35. Tamada H, Van Thuan N, Reed P, Nelson D, Katoku-Kikyo N, Wudel J, Wakayama T, Kikyo N (2006) Chromatin decondensation and nuclear reprogramming by nucleoplasmin. Mol Cell Biol 26:1259–1271
36. Taylor CJ, Bolton EM, Pocock S, Sharples LD, Pedersen RA, Bradley JA (2005) Banking on human embryonic stem cells: estimating the number of donor cell lines needed for HLA matching. Lancet 366:2019–2025
37. Thomas ED, Lochte HL, Lu WC, Ferrebee JW (1957) Intravenous infusion of bone marrow in patients receiving radiation and chemotherapy. N Engl J Med 257:491–496
38. Thomson JA, Itskovitz-Eldor J, Shapiro SS, Waknitz MA, Swiergiel JJ, Marshall VS (1998) Embryonic stem cell lines derived from human blastocysts. Science 282:1145–1147

39. Wakayama S, Jakt ML, Suzuki M, Araki R, Hikichi T, Kishigami S, Ohta H, VanThuan N, Mizutani E, Sakaide Y, Senda S, Tanaka S, Okada M, Miyake M, Abe M, Nishikawa SI, Shiota K, Wakayama T (2006) Equivalency of nuclear transfer-derived embryonic stem cells to those derived from fertilized mouse blastocyst. Stem Cells 24:2023–2033
40. Wakayama T, Perry AC, Zuccotti M, Johnson KR, Yanagimachi R (1998) Full-term development of mice from enucleated oocytes injected with cumulus cell nuclei. Nature 394:369–374
41. Wakayama T, Tabar V, Rodriguez I, Perry AC, Studer L, Mombaerts P (2001) Differentiation of embryonic stem cell lines generated from adult somatic cells by nuclear transfer. Science 292:740–743
42. Wells DN, Misica PM, Tervit HR, Vivanco WH (1998) Adult somatic cell nuclear transfer is used to preserve the last surviving cow of the Enderby Island cattle breed. Reprod Fertil Dev 10:369–378
43. Wells DN, Forsyth JT, McMillan V, Oback B (2004) The health of somatic cell cloned cattle and their offspring. Cloning Stem Cells 6:101–110
44. Wernig M, Meissner A, Foreman R, Brambrink T, Ku M, Hochedlinger K, Bernstein BE, Jaenisch R (2007) In vitro reprogramming of fibroblasts into a pluripotent ES-cell-like state. Nature 448:318–324
45. Wilmut I, Schnieke AE, McWhir J, Kind AJ, Campbell KH (1997) Viable offspring derived from fetal and adult mammalian cells. Nature 385:810–813
46. Wilmut I, Beaujean N, de Sousa PA, Dinnyes A, King TJ, Paterson LA, Wells DN, Young LE (2002) Somatic cell nuclear transfer. Nature 419:583–586
47. Yamazaki Y, Makino H, Hamaguchi-Hamada K, Hamada S, Sugino H, Kawase E, Miyata T, Gawa M, Yanagimachi R, Yagi T (2001) Assessment of the developmental totipotency of neural cells in the cerebral cortex of mouse embryo by nuclear transfer. Proc Natl Acad Sci USA 98:14022–6
48. Yang CX, Han ZM, Wen DC, Sun QY, Zhang KY, Zhang LS, Wu YQ, Kou ZH, Chen DY (2003) In vitro development and mitochondrial fate of macaca-rabbit cloned embryos. Mol Reprod Dev 65:396–401
49. Yang CX, Kou ZH, Wang K, Jiang Y, Mao WW, Sun QY, Sheng HZ, Chen DY (2004) Quantitative analysis of mitochondrial DNAs in macaque embryos reprogrammed by rabbit oocytes. Reproduction 127:201–205
50. Yang X, Smith SL, Tian XC, Lewin HA, Renard JP, Wakayama T (2007) Nuclear reprogramming of cloned embryos and its implications for therapeutic cloning. Nat Genet 39:295–302

Genetic Medicine and Global Health

David J. Weatherall

Abstract Recent estimates suggest that nearly 8 million children are born each year with a serious birth defect of genetic or partial genetic origin, and over 3 million children under the age of 5 years die from birth defects each year. As poorer countries go through the epidemiological transition following improvements in social conditions and health care, many babies with serious genetic diseases who would have died in early life with these conditions unrecognized are now surviving for diagnosis and management. The reasons for the particularly high frequency of some genetic diseases in poorer countries are complex; undoubtedly natural selection, consanguinity, and increased parental age are important factors. Common monogenic diseases such as the hemoglobinopathies are presenting an increasingly severe health burden. DNA diagnostics and modern genomic technologies have an increasingly important part to play in the control of many common communicable diseases, and as countries go through the epidemiological transition many of them are encountering epidemics of diseases of westernization, including cardiac disease, hypertension, and type 2 diabetes, almost certainly reflecting changes in environment associated with variable genetic susceptibility. The introduction of clinical genetics and genetic technology into the developing countries raises many complex ethical and social issues, not to mention the high costs of this technology. Recent reports from the World Health Organization have stressed the potential value of evolving North/South partnerships between centers in genetics in the richer countries and the developing countries; these might be followed by South/South partnerships between emerging countries with expertise gained in this area and adjacent countries where no such skills exist. But none of these developments will occur without a greater recognition of the importance of genomics on the part of the major international health agencies.

Contents

28.1	Introduction		886
28.2	Environmental Factors and Genetic Disease		886
	28.2.1	Poverty and the Epidemiological Transition	886
	28.2.2	Why Are Genetic Disease and Congenital Malformation Commoner in the Developing Countries?	887
28.3	Global Burden of Genetic Disease and Congenital Malformation		888

D.J. Weatherall (✉)
Weatherall Institute of Molecular Medicine, University of Oxford, John Radcliffe Hospital, Headington, Oxford, OX3 9DS, UK
e-mail: liz.rose@imm.ox.ac.uk

28.4 Monogenic Disease 889
    28.4.1 Inherited Disorders
            of Hemoglobin 889
    28.4.2 Other Monogenic Diseases
            in the Developing Countries 894

28.5 Communicable Disease 895
    28.5.1 Infectious Agents 895
    28.5.2 Vectors 896
    28.5.3 Varying Susceptibility 896
    28.5.4 Pharmacogenetics and Treatment 896

28.6 Genetic Components of Other Common
Diseases: A Global View 897

28.7 Global Control of Genetic Disease 898
    28.7.1 Transcultural, Ethical,
            and Counseling Issues 898
    28.7.2 Genetic Services in Developing
            Countries 899
    28.7.3 Organizing International Help
            for Developing Genetic Programs
            in Poorer Countries 900

References 900

28.1 Introduction

During the second half of the twentieth century the development of clinical genetics was restricted largely to the richer, developed countries. This is not surprising, considering that the developing countries were still suffering from high levels of childhood mortality that were due to malnutrition, poor sanitation, and the ravages of communicable disease. In the face of these problems international health agencies have tended to disregard the burden of disease caused by genetic disorders and congenital malformation; even in a recent authoritative review of priorities for research in the developing countries, with the exception of the hemoglobinopathies they are not mentioned [20].

On the other hand, the partial completion of the human genome project in 2001 and subsequent successes in sequencing the genomes of a variety of pathogens and disease vectors suggested that genomics might have an increasing place to play in global health issues. In 2002 the World Health Organization (WHO) published a report entitled *Genomics and World Health* [48], which concluded that DNA technology should be slowly introduced into some of the developing countries, particularly for the control of common monogenic diseases and communicable diseases.

The current state of clinical genetics and the availability of genetic technology varies widely, particularly among the developing countries. Furthermore, their application in these countries raises social and ethical problems that are completely different from those encountered in the developed world. This chapter summarizes a few of the more important aspects of this complex scene.

28.2 Environmental Factors and Genetic Disease

28.2.1 Poverty and the Epidemiological Transition

During the last decade of the twentieth century gross domestic product per head in the developing countries grew by 1.6% a year, and the proportion of people living on less than $1 a day fell from 29 to 23%. Most of this progress was made in Asia, however, and in other parts of the world, notably sub-Saharan Africa, the number of poor people increased. Furthermore, 150 million children living in low- and middle- income economies are still suffering from malnutrition and, unless the situation improves, a similar number are expected to be underweight by 2020. And the plight of the developing countries has become even more difficult because of an inability to control their major killers. About 70% of the 40 million people affected by HIV/AIDS are concentrated in countries with dysfunctional healthcare systems. Tuberculosis, often drug resistant, has re-emerged, with 9 million new cases and 2 million deaths each year, and death rates from malaria are similar [47, 52].

Yet despite this depressing scenario, the latter half of the twentieth century did see some significant improvements in the overall health of many populations in the developing countries. Largely as the result of, improvements in hygiene and social conditions the childhood mortality rate, i.e., the number of children who died in the first 5 years of life, started to decline, in some countries quite dramatically [10]. These improvements in population health were most marked in some of the middle-income countries, notably in South America and the Caribbean, East Asia and the Pacific, the Mediterranean region, the Middle East, and parts of North Africa.

The consequences of this epidemiological transition for the recognition of the importance of genetic disease in many developing countries were quite remarkable. The high frequency of thalassemia in Cyprus is a good example [39]. This condition was not known to occur on the island until 1944, when the clinical findings in 20 patients were reported. This paper highlights the difficulty in identifying this disease in an island population against the background of chronic malaria and other infections; it was published at the end of an extremely successful 3-year program to control anopheline breeding, at which time it was found that anemia secondary to malaria had almost disappeared. Hence, within this remarkably short period it became clear that there was a high frequency of genetic anemia with features of thalassemia in the island population. By the early 1970s it was estimated that, if no steps were taken to control the disease, in about 40 "years" time the blood required to treat all the children with thalassemia would amount to 78,000 units per annum, 40% of the population would need to be donors, and the total cost to the health services would equal or exceed the island's health budget [39]. The same pattern of the increasing realization that many populations of the developing countries have relatively high frequencies of genetic disease has slowly emerged throughout the world over the last 50 years.

In short, the epidemiological transition has improved children's health to the extent that many of those who were born with genetic disease at an earlier time, and whose condition would not have been recognized against a background of severe malnutrition and infection, have begun to survive long enough to present for diagnosis and treatment.

28.2.2 Why Are Genetic Disease and Congenital Malformation Commoner in the Developing Countries?

As discussed in the next section, there is considerable evidence showing that genetic disease and congenital malformation occur more frequently in countries with a low gross national income (GNI) per capita. There are several reasons why this might be the case [10].

28.2.2.1 Selection

As discussed later in this chapter, there is strong evidence to show that the very high frequency of the thalassemias and structural hemoglobin variants in many tropical countries is the result of heterozygote selection by relative resistance to infection by *P. falciparum* malaria. A similar mechanism is responsible for the very high frequency of glucose-6-phosphate dehydrogenase (G6PD) deficiency in parts of the world where malaria has been very common in the past or still is a major health hazard. Although, as discussed below, many problems remain to be clarified, there is no doubt that natural selection working through malaria has been largely responsible for the extraordinarily high gene frequencies for these disorders in many countries of the world [39].

28.2.2.2 Consanguineous Marriage

Although it has been extremely difficult to obtain accurate data, it is clear that consanguineous marriage is still practiced in many parts of the world and may be acceptable to a minimum of 20% of the world's population [10]. This practice is especially common throughout the eastern Mediterranean, North Africa and the Indian subcontinent and, to a lesser extent, parts of South America and sub-Saharan Africa [6, 7, 25]. As well as increasing the birth prevalence of autosomal recessive diseases, the risk of neonatal and childhood death, intellectual disability, and serious birth defects appears to be significantly increased in first-cousin marriages.

28.2.2.3 Parental Age

The percentage of women over the age of 35 years delivering infants is high in middle- and low-income countries, many of which do not have screening, prenatal diagnosis, or related services. For example, the birth prevalence of Down syndrome can reach 2–3 per 1,000 live births in these populations, a range currently double that seen in high-income countries [44, 49].

28.2.2.4 Population Migration

Large-scale population movements from areas of high frequency for single gene defects may introduce these disorders into new populations. As evidenced by the spread of the hemoglobin disorders, notably sickle cell disease, to the Americas, the Caribbean, and Europe by the slave trade and later migrations, this mechanism has led to a particularly high frequency of sickle cell anemia in many developed or developing countries. Other examples include the introduction of Huntington disease to Venezuela, spinocerebellar atrophy to Cuba, and porphyria to South Africa [10].

28.2.2.5 Poverty and Dysfunctional Healthcare Systems

There is an increased rate of birth defects in poorer countries. The reasons are not clear but may reflect maternal malnutrition and increased exposure to alcohol and infection. And since many of these countries have very limited facilities for the correction of structural birth defects, their frequency increases steadily in these communities. Similarly, the frequency of neural defects, fetal alcohol syndrome, congenital syphilis and rubella, and related disorders reflects the lack of prenatal care in many of these populations [11].

28.3 Global Burden of Genetic Disease and Congenital Malformation

For a variety of reasons, including limited diagnostic facilities in many of the developing countries and the extraordinary heterogeneity in the gene frequency of monogenic disorders even within small geographic distances, it has been extremely difficult to obtain accurate data about the global frequency of genetic disease and congenital malformation [45]. The most ambitious attempt to date to produce data of this kind was reported in the recent study supported by March of Dimes [10].

Overall global figures for the frequency of genetic disease and congenital malformation are summarized in Table 28.1. Data relating to individual diseases will be found in the appropriate chapters of this book and in the March of Dimes report [10]. A word of caution about these figures is necessary, however. While reasonably accurate data are available in many developed countries, this is not the case for many parts of the developing world. In countries with high infant and childhood mortalities resulting from malnutrition and infection, congenital malformation or genetic disease may not be recognized and reporting is limited. Furthermore, genetic disease that has reached a high frequency by natural selection tends to be very unevenly distributed even within short geographic distances, a subject to which we will return when we consider the genetic disorders of hemoglobin. Hence, analyzing populations in a few centers does not make it possible to obtain a figure for the overall frequency in a particular country. Hence, much of the data shown in Table 28.1 reflects information obtained from developed countries, augmented with what little is known about these frequencies in the developing world.

Even more serious problems face attempts to determine the economic burden of genetic disease and congenital malformation in many countries. Although there are some reasonable data on the costs of caring

Table 28.1 Minimum estimates for the birth prevalence of infants with serious congenital disorders, by WHO region [46]

WHO region	Population, millions (1996)	Births/ yarrmillions (1996)	Congenital malformations /1,000	Chromosomal disorders /1,000	Single gene disorders /1,000	Total congenital disorders /1,000	Annual affected live births
Eastern Mediterranean	506	18.1	35.7	4.3	27.3	69	1,237,225
Africa	540	23.0	30.8	4.4	25.0	61	1,412,427
SE Asian	1,401	38.2	31.0	3.9	14.7	51	1,946,606
Europe	867	10.8	31.3	3.7	12.4	49	522,832
Americas	782	16.2	30.9	3.8	11.9	48	774,235
Western Pacific	1,650	31.3	30.6	3.5	11.4	47	1,464,067
Total	5,746	137.6	31.5	3.9	16.8	53	7,357,392

for patients with genetic disorders and cost-benefit analyses of different forms of management for these conditions in the developed countries, the position regarding the developing world is much more complex and uncertain. There are widespread differences in the ways these conditions are managed, if they are managed at all, and it has been difficult to obtain accurate information about costs or cost-benefit ratios. A start has been made in this important field in the case of the hemoglobinopathies, and current progress is discussed later in this chapter.

Currently, the economic burden of disease is assessed by converting conditions to Disability Adjusted Life Years (DALYs), a measurement that attempts to include not only mortality but the burden and effectiveness of supporting patients with particular diseases [27]. Using this approach, health economists attempt to compare the economic burden of different diseases with one another. Overall, most work in this field has been directed at the major infectious diseases and the important diseases of the richer countries: heart disease, cancer, stroke, and similar disorders. Genetic disease has received little attention from health economists. However, if cases are to be made to governments and international health organizations about the importance of genetic disease it will be vital to attempt to analyze these conditions in terms of DALYs in the future. We return to this question below, when we consider the genetic disorders of hemoglobin.

28.4 Monogenic Disease

Although monogenic diseases occur in every ethnic group, and owing to founder effects and other mechanisms may occur at unusually high frequencies in localized regions, only the inherited disorders of hemoglobin occur at sufficiently high frequencies to present a major public health problem for a large number of the developing countries. These conditions are described in detail elsewhere in this book. Here, issues relating to their impact on global health are discussed, with particular reference to how limited successes in their control in at least some developing countries offer a model on which programs for the control and management of genetic diseases might be based in these difficult environments.

28.4.1 Inherited Disorders of Hemoglobin

Although there are many different forms of thalassemia, and over 700 structural hemoglobin (Hb) variants, α– and β–thalassemia together with three structural variants, Hbs S, C and E, are the only hemoglobinopathies that reach high enough frequencies to cause a major burden on the healthcare services of the developing countries.

28.4.1.1 Global Distribution

The global distribution of the important hemoglobin disorders [24, 38, 39] is summarized in Figs. 28.1 and 28.2. The β-thalassemias are distributed at varying frequencies right across the tropical belt, ranging from sub-Saharan Africa through the Mediterranean region, the Middle East, the Indian subcontinent, and East and Southeast Asia. The α–thalassemias have a different pattern of distribution, a fact of considerable global health significance. There are two main forms, α$^+$-and α^0-thalassemia. α$^+$-Thalassemia is caused by deletion or inactivation by mutation of one of the pairs of α–globin genes (-α/αα), while α^0-thalassemia is due to the deletion of both pairs of α–genes (--/αα). The α$^+$-thalassemias, the commonest monogenic diseases, are spread at varying and sometimes extremely high frequencies across the region shown in Fig. 28.1, while the α^0-thalassemias are restricted to parts of Southeast Asia and the Mediterranean islands. The result is that the severe, symptomatic forms of α–thalassemia are restricted to the latter regions.

The world distribution of the Hb S and E genes, shown in Fig. 28.2, is quite remarkable [24, 39, 40]. The sickle cell gene is found in populations stretching from sub-Saharan Africa through the oasis regions of the Middle East to localized regions of the Indian subcontinent. It is not observed further east. On the other hand, Hb E is found on the eastern side of the Indian subcontinent and stretches through Burma (Myanmar) to Southeast Asia. In short, the Old World has extremely high frequencies of two β–globin-chain variants, which are separated by a line that runs North-South through the eastern part of the Indian subcontinent.

Hemoglobin C, which is of less clinical importance, occurs only in localized areas of West and North Africa.

Fig. 28.1 World distribution of α- and β-thalassemia [39]

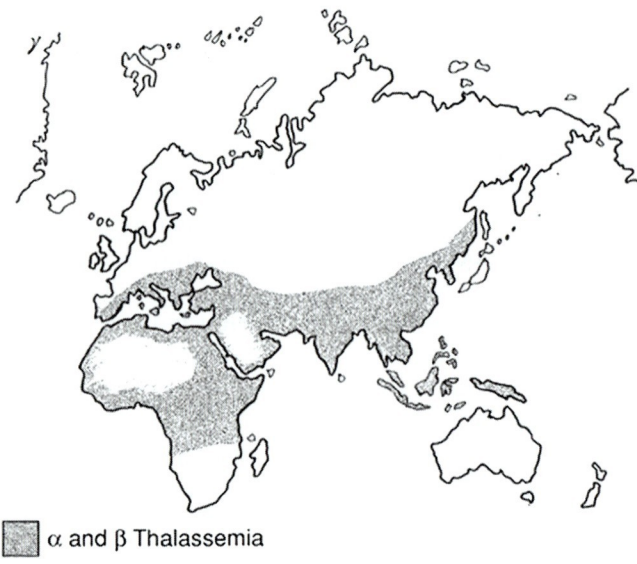

α and β Thalassemia

Fig. 28.2 World distribution of hemoglobin S and E [38]

HbE   HbS

28.4.1.2 Frequency

Estimated frequencies for the different inherited disorders of hemoglobin are discussed in several reviews [10, 24, 39, 40]. Frequencies for each of the designated regions of the World Health Organization are given in Table 28.2. As explained earlier, these are very approximate data, simply because of the extraordinary heterogeneity in the frequency of these gene frequencies within short geographic distances of one another. Even

Table 28.2 Percent carrier frequencies for common hemoglobin disorders by WHO region [24, 38, 39]

Region	Hb S	Hb C	Hb E	β-Thalassemia	α°-Thalassemia	α⁺-Thalassemia
Americas	1–20	0–10	0–20	0–3	0–5	0–40
Eastern Mediterranean	0–60	0–3	0–2	2–18	0–2	1–80
Europe	0–30	0–5	0–20	0–19	1–2	0–12
Southeast Asia	0–40	0	0–70	0–11	1–30	3–40
Sub-Saharan Africa	1–38	0–21	0	0–12	0	10–50
Western Pacific	0	0	0	0–13	0	2–60

Note: many of these data are derived from small population samples

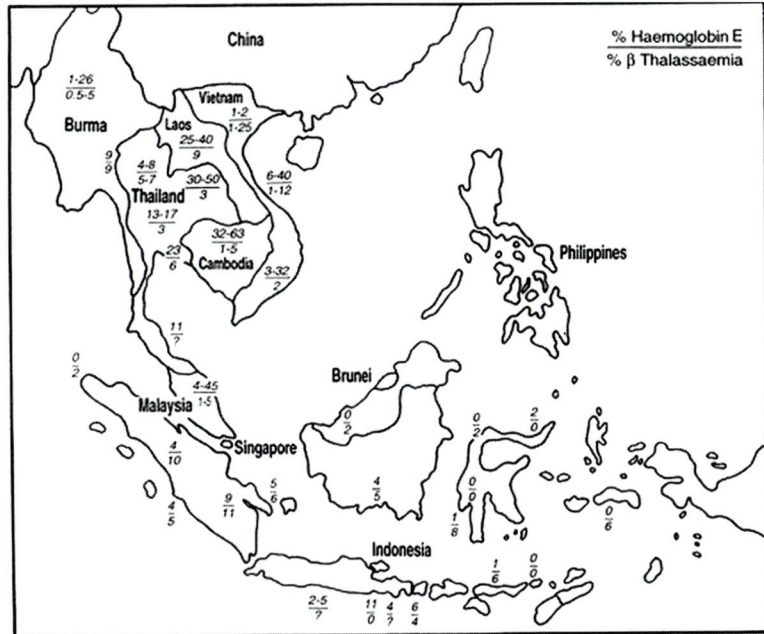

Fig. 28.3 Distribution of Hb E and β thalassemia in SE Asia [13, 39]

in a relatively small island population such as Sri Lanka there are considerable differences between populations resident only a few hundred miles apart [13].

Despite these uncertainties, certain generalizations can be made. The sickle cell and β thalassemia genes, though they occur at varying frequencies in different populations, rarely occur in above 20% of the population; there are a few exceptions, including the particularly high frequencies of the sickle cell gene in some of the oasis populations of the Middle East and localized pockets in India, and the very high frequency of β-thalassemia in the Maldives. On the other hand, the α⁺-thalassemias occur at much higher frequencies, affecting up to 70 or 80% of the population in some regions. The α°–thalassemias occur at similar frequencies to the β-thalassemias. As in the α⁺-thalassemias, the carrier rates for Hb E greatly exceed those of Hb S; in many Asian populations they reach as high as 70%, notably in parts of northern Thailand and Cambodia.

The remarkable diversity of these gene frequencies within populations presents a major problem for developing rational public health programs. The distribution of β-thalassemia and Hb E in the Indonesian islands is shown in Fig. 28.3, and the estimated calculation of new births with serious forms of thalassemia in this population is shown in Table 28.3. Although the latter gives some indication of the likely health burden of these conditions for the future, even after a study carried out at this level of detail the estimated number of new births is beset with considerable uncertainty. Approximate figures for new births with serious hemoglobin disorders for Thailand, a country in which frequencies have been mapped in more detail than most, is given in Table 28.4.

Table 28.3 An estimated population burden of the β-thalassemias in Indonesia

Population 200 million		
Annual births		(750–2,850)
β-Thalassemia major	1,600	(1,000–4,750)
Hb E thalassemia	2,500	
Total annual units of blood for treatment	1.25 million	

From A.S.M. Sofro, J.B. Clegg, D.J. Weatherall, unpublished data

Table 28.4 The public health burden produced by thalassemia in Thailand in 1999

Population 61.5 million (68% rural)	
Infant mortality 25/1.000	
Total patients with thalassemia	523,750
Births of new cases per year	
β-Thalassemia major	625 (6,250)
Hb E/β-Thalassemia	3,250 (97,500)
Hb H Disease	7,000 (420,000)

Data supplied by the Thai Thalassemia Association and Dr. S. Fucharoen; bracketed figures are estimates of current numbers of patients in the community

28.4.1.3 Population Genetics and Dynamics

Although from association studies and other sources of evidence it has been suspected for a long time that the high gene frequencies for the common hemoglobin disorders are attributable to the effects of natural selection mediated through carrier protection against *P. falciparum* malaria, it is only recently that this mechanism has been confirmed directly by case control studies, at least in the case of the sickle cell, Hb C, and α-thalassemia genes [22, 38, 40, 50]. Taking strict World Health Organization criteria for the severity of malaria, particularly the occurrence of coma or profound anemia, the sickle cell trait seems to offer approximately 60–70% protection, while the homozygous state for α^+-thalassemia offers 30–60% protection. Similarly, high protection rates have been found in the case of the homozygous state for Hb C and, although the appropriate case control studies have not been done, there is a considerable body of evidence suggesting that the high carrier states for β-thalassemia and Hb E also reflect protection against malaria.

The eradication of *P. falciparum* malaria would undoubtedly result in a slow decline in the frequency of the important inherited hemoglobin disorders, but this would take many generations before its effect became apparent in clinical practice [38]. Furthermore, there is widespread drug resistance to *P. falciparum* and clear evidence of a recrudescence of the disease in some parts of Africa and Asia. Thus, for public health planning there is no reason to believe that the hemoglobin disorders will show any decline in their current frequency for the foreseeable future. These observations provide further evidence that as countries pass through an epidemiological transition the hemoglobin disorders will present an increasingly serious public health problem. Current estimates for the annual numbers of births of babies with these diseases [10] suggest that the figure is approximately 300,000, though this may well be a considerable underestimate owing to lack of accurate population data in many high-frequency countries.

28.4.1.4 Clinical Load and Cost–Benefit Issues of Control and Management Posed by the Inherited Hemoglobin Disorders

Like most of the inherited disorders of hemoglobin, the common sickle cell disorders, i.e., sickle cell anemia, Hb SC disease, and HbS β thalassemia, all show considerable clinical heterogeneity. The fact that the form of sickle cell anemia which arose independently in Asia is, overall, milder than that seen in Africa suggests that genetic factors play an important part in this clinical heterogeneity. However, while the ability to produce Hb F and the co-inheritance of α-thalassemia have been identified as playing a part, from such twin data as are available it is clear that there must be other genetic factors, and in particular that the environment is a major determinant of the course of the disease [43].

There are limited mortality data for the sickling disorders. Early studies in Africa suggested that very few children with sickle cell anemia survived beyond the first few years of life [15]; however, in some parts of Africa, following the introduction of antimalarial and other public health measures the situation seems to have improved although no accurate mortality data are available [1, 26]. The position is different in some of the developing countries. Recent US data suggest that the median age of death is 42 years for men and 48 years for women [14]. A cohort study in Jamaica has shown how improved social conditions and treatment have increased survival; 70% of those enrolled starting in 1973 survived to age 20 years, as did 80% of those enrolled 3–6 years later [42].

Preliminary attempts at assessing the burden of sickle cell disease in terms of DALYs have been reported recently [42]. Based on birth and scanty survival data it appears that in Africa this disorder may account for between 0.5 and 4.5 million DALYs, or less than 1%, to approximately 2% of the burden for children under 5. Again, although data are extremely limited, it appears that sickle cell anemia may account for at least 1–2.2 million DALYs in low- and middle-income countries. Clearly, we have only the flimsiest of ideas about the total global health burden imposed by the sickle cell disorders. Such as are available probably reflect a major underestimate, and the situation will undoubtedly change as developing countries, particularly those in sub-Saharan Africa and parts of Asia, pass through the epidemiological transition.

Extensive data on the economic aspects of sickle cell anemia have been obtained from the USA [12]. An estimated 75,000 hospital admissions of both children and adults occurred each year from 1989 through 1993. In 1993 the direct costs of management were $575 million. Although it has been reported that the development of centers with expertise in the treatment of sickle cell anemia may significantly lower health,care costs [28, 53], these advantages may have been reduced by the more intensive care of certain subsets of patients over recent years, e.g., the use of chronic transfusion. A pilot study in Benin found that the development of a similar program reduced the frequency and severity of acute complications, with an annual cost per family of $40 and an annual cost for each hospitalization of $100 [33].

To what extent are current approaches to the management of the sickle disorders cost effective? From the limited data that are available it appears that specialized treatment centers and, in particular, neonatal screening programs are effective approaches to the control of sickle cell anemia. It is now clear that neonatal screening followed by the use of appropriate prophylactic treatment with penicillin prevents deaths [37], and in a recent survey of studies of this kind it became quite clear that this approach was highly cost effective [42]. However, although other recent advances in therapy, e.g., the control of strokes by monitoring and transfusion, have been shown to be clinically effective, none of them have yet been exposed to more detailed cost-benefit studies.

Except in a limited number of cases in which the phenotype of the homozygous or compound heterozygous state for β-thalassemia is modified by either a mild β–thalassemia allele or some other genetic modifier/s, these conditions are almost always transfusion dependent, with resulting costs for procuring and processing blood and ensuring it is free from infection, chelating agents to prevent iron overload, and treatment for the many complications that arise. The life expectancy varies widely between different populations. While there has been a major improvement in many of the developed countries, such data as there are from the developing countries suggests that the situation is much worse. In many, there are no government programs for the control and management of thalassemia and, although transfusions may be available, only those who can afford them are able to obtain drugs for removal of iron and treatment of complications. Hence, the median survival for these children in Thailand is approximately 10 years; in many developing countries it is shorter and may not exceed 1–2 years.

Several detailed analyses of the cost of treatment have been reported [3, 42]. Remarkably, the basic costs were similar in Thailand, east European countries, and Canada. Recently, an attempt has been made to convert these data into DALYs [42]. Considering deaths alone, it appears that severe β–thalassemia contributes about 1.5–3 million DALYs to the world burden. Since there are no accurate global estimates of the number of treated survivors it is difficult to assess the true number of DALYs caused by this disease, but from limited regional data it appears that the number of DALYs in parts of Asia may be comparable with that for some of the common communicable diseases [42].

There is another neglected aspect of the global health problem caused by the β-thalassemias. In populations east of the Indian subcontinent there is a very high frequency of both Hb E and β thalassemia, and hence Hb E β-thalassemia is very common. Indeed, in parts of India and in Bangladesh, Burma (Myanmar), Thailand, and Indonesia, the frequency of Hb E β–thalassemia greatly exceeds that of homozygous or compound heterozygous β–thalassemia. Hb E β–thalassemia is remarkable for its clinical heterogeneity; different individuals who received the same β–thalassemia mutations may have phenotypes ranging from those identical to thalassemia major to a mild condition which does not require transfusion and is associated with long survival. Although some of the genetic factors responsible for this heterogeneity have been determined [32, 34, 51] it is clear that many genetic and environmental factors remain to be discovered before the wide phenotypic variation of this disease can be understood. This is a particular challenge for public health measures directed at the prevention and management of the β–thalassemias in many Asian countries. For example, the current inability to predict the phenotype poses considerable ethical and counseling problems for developing prenatal diagnosis programmes and presents equally complex problems for determining the optimal approach to management.

Because of their more localized occurrence and the fact that babies homozygous for α°-thalassemia are stillborn or die shortly after delivery, this condition produces less of a health burden although considerable distress to families. The delivery of these babies is often associated with severe obstetric complications, including postpartum hemorrhage. Although some of these babies have been rescued for life-long transfusion by intrauterine transfusion, this practice has not become widespread, notably because of the high frequency of the associated congenital abnormalities.

While Hb H disease has been thought to be a fairly innocuous condition, there is increasing evidence from such high-frequency countries as Thailand that a significant proportion of the patients may be more severely aemic and become transfusion dependent. So far, it has not been possible to assess the burden of the α thalassemias in terms of DALYs.

As regards cost-benefit analyses of different approaches to managing the β-thalassemias, although, as mentioned earlier, there are some data on costs of standard therapy there is very little published data on the cost-benefit ratios of particular approaches [42]. If matching donors are available, bone marrow transplantation does appear to be cost effective as a treatment for β-thalassemia [5]. There are no comparable data for the α-thalassemias. There is even less information about the costs of managing the thalassemias relative to national health budgets; it has been calculated that, in Sri Lanka, management of the disease will consume approximately 5–8% of the country's health budget, based on 1999 figures [13].

Control by Screening and Prenatal Diagnosis. Screening and prenatal diagnosis programe for the control of β-thalassemia have been established in many countries [8, 38]. In Europe, 80–90% of counseled at-risk couples now request prenatal diagnosis, and in the late 1970s and 1980s in Cyprus and Sardinia the birth rate of new cases of severe β-thalassemia fell by almost 90%; a comparable fall occurred in Greece and Italy and is now occurring in many other countries. There is no doubt that this is the most cost-effective approach to the control of β-thalassemia.

Screening programs that incorporate both α– and β–thalassemia allow the detection of couples who are homozygous for α°-thalassemia and hence risk having a baby with the Hb Bart's hydrops syndrome. While these babies die in utero or at term, screening and termination of pregnancy is still considered to be important because pregnancies associated with this condition show a very high maternal morbidity.

Hitherto, premarital screening and prenatal diagnosis have been much less widely applied for the control of sickle cell anemia. This is probably because, if adequately counseled, couples realize the extraordinary clinical diversity of this condition and many are loth to undergo termination of pregnancy. A much higher uptake is reported from some European countries; whether this reflects more directed counseling or a different parental perception of the level of severity of a disease for which they would be willing to undergo termination, is not known. The same questions apply for Hb H disease and Hb E β thalassemia, both of which are similarly associated with extremely variable phenotypes.

28.4.2 Other Monogenic Diseases in the Developing Countries

Although, because of their high frequency and the fact that they have acted as model diseases for developing clinical genetic programs in the developing countries, this section has focused on the hemoglobinopathies, there are other single-gene disorders that have important implications for developing countries.

The relationship between the high frequency of G6PD deficiency and malaria has been discussed earlier. The increased sensitivity of those with this enzyme deficiency to antimalarial drugs, notably primaquine, is imposing increasingly important problems for the control of *P. vivax* malaria in many Asian countries. Primaquine is currently the only drug that is available to destroy *P. vivax* in the hepatic stage of its life cycle, which gives rise to recurrent attacks of the disease. Although full primaquine resistance has not yet been encountered, there are signs that it is on the way, and already much longer courses of primaquine are being administered for the eradication of *P. vivax* infections. There is an urgent need for a much simpler, cheaper method of identifying those who are G6PD deficient in high-frequency countries. G6PD deficiency is also associated with a high frequency of neonatal jaundice and kernicterus in many countries [10]. Although the precise mechanisms of the high bilirubin levels and the reasons why this phenomenon is restricted to certain countries remain to be determined, neonatal screening for G6PD deficiency has become absolutely essential in many countries.

There is now considerable evidence showing that the 7/7 promoter genotype of *UGT1A1*, which is common in many countries and is associated with Gilbert's disease, is an important hazard for thousands of individuals, particularly in the Indian subcontinent and Africa, with chronic hemolytic disorders such as Hb E β-thalassemia and sickle cell anemia [18, 31]. This genotype is associated with particularly high bilirubin levels in individuals with these conditions and a very marked increase in the likelihood of developing gallstones.

Examples of the growing importance of pharmacogenetics for the developing countries are given in a later section; a brief account of other monogenic disorders in the developing countries, together with common birth defects, is given elsewhere [10].

28.5 Communicable Disease

Progress towards the application of genomics for the control and management of communicable disease has involved studies of the genetic constitution of infectious organisms and of the vectors that transmit them. In addition, much is being learnt about the genetic basis for individual variation in susceptibility to infectious diseases, work that also promises to have important practical implications in the future [41, 48].

28.5.1 Infectious Agents

Since the first establishment of the complete DNA sequence of the genome of a virus in 1977, the sequences of many viruses and bacteria have been determined and a start has been made on obtaining similar information about the much more complex genomes of common parasites. As well as providing invaluable information about some of the basic mechanisms of pathogenicity and the ways in which infectious agents are able to evade the immune system, these studies have provided a wide range of new diagnostic agents and vaccine candidates [48].

DNA technology has also been of great value in subtyping different strains of viruses and in identifying new infectious agents with the potential to produce major epidemics. In practice, DNA diagnostics have to be carefully compared with more standard culture methods for the diagnosis of infectious disease and for the identification of antibiotic-resistant organisms.

DNA-diagnostic agents are turning out to be of particular value for the identification of organisms that are difficult to grow in culture, an approach which is already proving cost effective in some developing countries. By establishing simplified PCR methods and by the use of inventive approaches, including simplification of protocols and bulk preparation of reagents from crude ingredients, combined with recycling, it has been possible to develop routine diagnostic procedures for such diseases as leishmaniasis, dengue, and leptospirosis in several South American countries [17, 48].

Knowledge of the pathogen genome is also being applied to the development of new therapeutic agents and, in particular, of vaccines [23]. The pathogen genome has already yielded vaccine candidates against *Neisseria meningitidis*, group B, and *Mycobacterium tuberculosis* [48]. And although progress has been extremely slow toward developing a vaccine against HIV/AIDS, current drug therapy for this condition is based almost entirely on studies of the virus at the molecular level. Progress toward the development of vaccines against viruses, bacteria, and malarial parasites is the subject of several extensive reviews [21, 36].

In 2002 the complete genome sequence of the human malaria parasite *P. falciparum* was established, and considerable progress is being made towards sequencing the genome of the other important human malaria parasite, *P. vivax* [9, 16]. Apart from providing important information about the biology of these parasites, this work is already leading to some practical applications. Drug resistance has become a major problem in the control and treatment of severe malaria caused by *P. falciparum*; chloroquine resistance is now almost universal, for example. It has been found that mutations in the chloroquine resistance transporter of *P. falciparum*, encoded by the gene *pfcrt*, confer chloroquine resistance in laboratory strains of the parasite. The identification of these mutations has the potential to be of considerable value for public health surveillance of antimalarial resistance [48]. Information collected from the genome of *P. falciparum* has also led to the development of at least one new antimalarial drug [48] and has become an integral part of efforts to produce vaccines against the different stages of the life cycle of the parasite.

Table 28.5 Genetic variation and malarial susceptibility

Red cell
α- and β-Thalassemia. Hb S, C and E
Membrane: ovalocytosis (band 3). Others
Metabolism: G6P-D. PK (murine)
Blood groups: Duffy. GYPA. GYPB (S-s-U) GYPC (Gerbich). P. Se
Receptors: CR1
Immune genes
HLA-DR, IFN-G, IL1A/IL1B, IL10, IL12B, MBL2, NOS2A, TNF, CD40 Lig
Other host receptors
ICAM1. CD36. CD31

GYP Glycophorin; *CR* Complement receptor; *IFN*-G Interferon gamma; *IL* Interleukin; *MBL* Mannose binding lectin; *NOS* Nitric oxide synthase; *TNF* Tumor necrosis factor; *ICAM* Intercellular adhesion molecule, *pk* Pyruvate kinase

28.5.2 Vectors

The complete genome sequence of the malaria-transmitting mosquito *Anopheles gambiae* was reported in 2002 [19]. This has led to a considerable increase in knowledge about the biology of this vector and has also raised the possibility of novel approaches to vector control. For example, the discovery that transposons can be used to introduce genes into the genome of mosquitoes, and hence reduce their ability to transmit malaria, suggests that, despite the inherent difficulties and potential dangers involved, genomics will provide new approaches to the control of malaria transmission [2]. Similar avenues are being explored for the potential modification of other communicable disease vectors.

28.5.3 Varying Susceptibility

As discussed above, there is now strong evidence to indicate that some of the important hemoglobin disorders have risen to their highest frequencies because of relative resistance of heterozygotes to *P. falciparum* malaria. However, many other genes have now been identified that have the same effect. As shown in Table 28.5, as well as blood group antigens and other red cell proteins, they include those that regulate the HLA-DR system and a variety of cytokines. Similarly, progress is also being made to identify genes that modify susceptibility to other infections, including tuberculosis, HIV/AIDS, leprosy, and hepatitis B.

This rapidly moving field may have important practical indications for both prevention and management of the important communicable diseases, particularly those that affect the developing countries. For example, the discovery that individuals who are negative for Duffy blood group are resistant to *P. vivax* malaria, and the subsequent finding that the Duffy chemokine receptor is vital for entry of the parasite into red cells has led to efforts to develop a vaccine directed at this protein [9]. Furthermore, if vaccines are to be tested in field studies in tropical countries, and particularly if they are attenuating vaccines, it may be extremely important to know the frequency of genetic polymorphisms, such as α–thalassemia and the sickle cell gene which, themselves, produce a considerable degree of protection, when designing appropriate protection studies.

28.5.4 Pharmacogenetics and Treatment

The problems of managing *P. vivax* malaria in populations with a high frequency of G6PD deficiency and the identification of chloroquine-resistant malaria parasites have been discussed above. However, there is increasing evidence suggesting that there may be other polymorphisms that may be of significance in the treatment of common infectious diseases. For example, polymorphisms of *MDR1*, a gene that regulates the expression *p*-glycoprotein and which may hence be important in defense mechanisms against potentially toxic agents ingested in the diet, have been shown to be much more common in West African and African-American populations than in those of European or Japanese backgrounds. It has been suggested that a particular variant of *MDR1* is common in Africa because it offers a selective advantage against gastrointestinal-tract infections. However, there is now clear evidence that this variant reduces the efficacy of the protease inhibitors and related agents that are now widely used for the treatment of HIV-1 infections [35].

The application of pharmacogenetic studies as an adjuvant to drug therapy in the developing world will clearly add to the costs of treatment; each case will have to be examined for its cost-effectiveness on an individual basis.

28.6 Genetic Components of Other Common Diseases: A Global View

As shown in Table 28.1, congenital malformations, including heart defects, neural tube defects, and cleft lip with or without cleft palette, constitute an important proportion of children born with genetic disease or congenital malformation. These disorders almost certainly reflect interactions between environmental factors and genetic susceptibility mediated through multiple genes. Some of these conditions are discussed elsewhere in this book, and some of the environmental factors involved have been reviewed recently [10].

Approaches to determining the genetic component of major noncommunicable adult diseases, cardiovascular disease, diabetes, and neurological disease, for example, are also described elsewhere in this book. Here, we review very briefly a few of the global aspects of this growing and particularly complex field, many aspects of which are covered in more detail elsewhere [20, 48].

As countries pass through the epidemiological transition, the burden of disease attributable to malnutrition and infection declines and the pattern of illness starts to assume that of the developed countries [20]. For example, by 2001 cardiovascular disease was responsible for about 30% of deaths worldwide. It is now predicted that it will be the leading cause of death and disability worldwide by 2020, largely because of its increasing frequency in low- and middle-income countries. In 2003 about 194 million people worldwide had diabetes, and it is estimated that by 2025 this figure will increase to 333 million. In 2003 the developing countries accounted for 141 million of those with diabetes. It is believed that the number of people with diabetes will double in three of the six developing regions of the world by 2025; these extraordinary epidemiological changes must reflect a major environmental component in the causation of these conditions.

The major goal in attempting to define the particular susceptibility genes involved in these multigenic disorders is to learn more about their underlying pathophysiology and hence to develop more logical approaches to their treatment. A longer term goal is to define individuals at particularly high risk for these conditions so that public health measures may be focused on these particular groups with the more effective use of man power and facilities.

Several messages are emerging from global epidemiological studies that may have important implications for the search for susceptibility genes for these conditions. First, it is becoming clear that many of these diseases, type 2 diabetes, the metabolic syndrome, and asthma, for example, are not single entities but reflect several different conditions, which may have different genetic and environmental causes. Hence, when whole-genome searches or association studies are launched in attempts to identify the genes involved, the importance of accurate phenotyping and associated epidemiological data cannot be overestimated. Furthermore, there are indications that the pattern of response to environmental factors may vary considerably from race to race. For example, African races seem to be particularly prone to hypertension and cardiovascular disease, whereas Asian populations and some population isolates are more prone to the development of obesity and type 2 diabetes. Again, this raises the possibility of the environmental agents of western life acting in different genetic backgrounds.

Similar types of information relating to differences in age-specific incidence rates of Alzheimer's disease, together with low incidence rates reported from parts of India and Africa, raise the possibility of variation in environmental factors or gene-environment interactions in the causation of this condition [20]. These possibilities point to the importance of ensuring ethnic homogeneity of studies directed at identifying genes involved in susceptibility to common diseases and, equally importantly, to the value of developing partnerships in research of this type between workers in different parts of the world, a topic that we shall return to in the next section.

28.7 Global Control of Genetic Disease

The development of approaches to the global control and management of genetic disease is much too large a topic to be covered fully. It has been reviewed in detail in several reports [4, 10, 48]. Here, a few key ethical and organizational issues are summarized briefly.

28.7.1 Transcultural, Ethical, and Counseling Issues

Ethical issues in genetic research, screening, and testing in developing countries are discussed in detail in a report by the World Health Organization [48] and in reports by the Nuffield Council on Bioethics [29, 30].

While the general principles that have evolved for good ethical and regulatory practice in clinical genetics in the developing countries apply equally to the poorer countries of the world, their application in different social backgrounds raises many problems, a field that is in urgent need of further investigation. A few of these issues are highlighted here.

28.7.1.1 Ethnic Differences in Interpreting the Nature of Disease

It is important to realize that even the best-intentioned efforts to introduce the medical practices of developed countries, such as genetic counseling, into developing countries may fail simply because of the completely different perceptions of the nature of disease. Disease is often viewed as the action of evil spirits or other external forces, a belief that may be confirmed by visits to local healers. In these circumstances even the simplest explanations of the nature of genetic disease may be extremely difficult to communicate; a great deal of time and patience is required to evolve more appropriate approaches and methods to confirm what has been understood.

28.7.1.2 Gender Issues

Many developing countries still have strongly patriarchal societies, and this raises particular problems for women who are carriers of genetic disease or who have had affected children. Despite careful explanation of the mode of inheritance of recessive disorders, husbands very frequently blame their wives for having a child with a serious genetic disease. This often has the effect of breaking up the family, and in some communities there is a high rate of suicide in these families.

Genetic information can also be used to discriminate or stigmatize in the context of other social practices. For example, in countries in which arranged marriages are still common, the fact that a screening program has shown that a woman is a carrier for a genetic disease can make her unmarriageable, or it can subject her to physical and other harms if she gives birth to children with diseases for which she is deemed responsible. In such countries as India, for example, the use of sex selection to avoid the birth of female babies is still practiced and has had a substantial effect on the sex ratio of the population in some regions. As pointed out in the report of the World Health Organization [48], sex selection is the result of deep-seated, entrenched beliefs and values in societies that have long histories of subordinating and devaluing women; long-term public education strategies are needed to combat these beliefs and values effectively.

28.7.1.3 Patient Discrimination

Observations in a number of developing countries have shown that children with severe genetic disease may be stigmatized and even ostracized by their communities. For example, in Sri Lanka children with thalassemia who require regular blood transfusions have been ostracized by their peers, been classified as "blood suckers," and have found difficulties in continuing their education or obtaining employment when they are older.

28.7.1.4 Informed Consent

Although the principles of informed consent for research in the developing countries are similar to those in developed countries, their application may be much more difficult in societies with a limited understanding of the nature of disease. This is particularly relevant to genetic disease. The principles to be applied in research in developing countries have been defined in a discussion paper [30], and these difficult issues are discussed in detail in a further report [48]. Major problems include: explaining how some genetic tests cannot be necessarily followed up by adequate therapeutic interventions; difficulties in communicating the nature of the tests or associated research in populations with limited education; the danger of patients agreeing to being part of a research project as the only way of obtaining medical care; and, in particular, the fact that those performing genetic research in developing countries are often scientists from developed countries or large multinational pharmaceutical or biotechnology companies with research needs that are not directly relevant to the needs of developing countries.

28.7.1.5 Lack of Regulatory or Ethical Bodies

Another problem for the establishment of clinical genetic programs or research in developing countries results from the lack of established ethical or regulatory bodies. While at least some developing countries are now establishing governmental or institutional ethical review bodies, through lack of knowledge it is very difficult for them to assess or regulate the establishment of genetic services or genetic research, particularly if the latter is being carried out by scientists from outside the country. Again, these problems will only be solved by extensive education programs.

28.7.1.6 Biobanks and Biopiracy

Some developing countries offer desirable opportunities for the development of genetic databases, particularly when the population is genetically homogeneous. Although in some cases these are established by public health authorities within the country, in others they are set up by private corporations. These databases raise a number of ethical issues, including profit sharing with the community from which the data are gathered, questions of informed consent, and the participation of individuals in these programs. These important issues are considered in detail in reports from the World Health Organization [48] and the Nuffield Council for Bioethics [29].

28.7.2 Genetic Services in Developing Countries

There have been several reviews of approaches to developing genetic services in general [3, 10, 48] and with particular respect to the requirements of the developing countries [4, 42, 48].

Although the basic principles are similar irrespective of the environments in which genetic services are established, there are problems particular to the developing countries. First, it is essential to try to persuade the particular government involved to accept that there is a clinical need and to appoint a reasonably senior person in government to take responsibility for developing the service. Trained medical personnel must be available to lead the development, and this often requires a period of training in a genetic center in a developed country. Establishing programs for the hemoglobinopathies is different, because this field, probably for historical reasons, has never come under the auspices of clinical genetics and usually requires special training in centers with experience of the field for appropriate pediatricians and hematologists.

Once the basic personnel are in place a prioritized program should be developed, which includes the establishment of the frequency of particular genetic disorders in the community, the training of appropriate counselors, and the development of a major public education program. Next, decisions about whether community control of serious genetic disease will be achieved by prenatal diagnosis or community education alone should be addressed. The final stage involves the development of premarital or prenatal screening programs and the establishment of one or more center with expertise in both the laboratory diagnosis and clinical management of genetic disorders.

While much of this may be self-evident to geneticists in the developed countries there are problems to be overcome at each stage in many developing countries. Their governments often do not think that genetic disease is a high priority, given the many other more pressing problems of communicable disease and the increase in frequency of noncommunicable disease that most of them are experiencing. It may be difficult for them to obtain adequate training for their staff overseas, and even more difficult to provide facilities for them after training. Unless there is strong leadership from government or clinicians it is extremely difficult to organize and synchronize the development of programs of this kind. And, last but by no means least, there are major problems in funding both preventative and management programs.

Currently, it seems unlikely that many developing countries will be able to evolve clinical genetic programs without at least some international help. How can this best be organized?

28.7.3 Organizing International Help for Developing Genetic Programs in Poorer Countries

Although the World Health Organization has accepted that genetic disease and congenital malformation are becoming increasingly serious international health problems, the same is not true of other large international public health agencies or funding organizations. What, therefore, is the best way forward?

In a report published by the World Health Organization [48] and in a follow-up article [41], suggestions were made about how the international medical and scientific community might best help the developing countries in establishing programs such as clinical genetics. These approaches were based in part on the successes in the hemoglobin field over the previous 20 years in developing programs for the control of the thalassemias in many developing countries. They were built on the concept of North/South partnerships, i.e., partnerships between workers in centers with expertise in the developed countries and those in whom these skills were lacking in the developing world. At best, they were sustainable and led to both the training of staff and the transfer of both clinical and laboratory technology to the developing countries. Programs of this type led to the successful application of carrier screening and prenatal diagnosis for the thalassemias in many developing countries, and in some cases were later integrated into the control of other genetic diseases.

The natural successor to a North/South partnership would be the development of a South/South partnership in which those developing countries that have gained skills in the diagnosis and treatment of genetic disorders formed partnerships with countries in which these skills were lacking. Several South/South partnerships along these lines are being planned as this book is being prepared.

There is of course a major problem that will have to be solved for the development of further partnerships of this type. Many North/South partnerships were funded on a research basis by the developed countries and, when successful, at least in some cases were then taken over by governments of the developing countries. But this is not always possible, and for a process of this type to be developed further more sources of funding will have to be obtained, either as part of the aid programs of the Northern partners or from large international agencies and charities, or both.

Another major issue that will have to be faced is the high priority that must be given to determining more accurately the frequency of genetic disease and hence the future health burden that it will present to the governments of developing countries as they go through their epidemiological transition. The more accurate the assessment of this load and its conversion into DALYs, the only language which the international public health community understands, will be a major priority for North/South and South/South partnerships in the immediate future if the true problem of genetic disease for the global community is to be appreciated. In this context announced recently that the next Global Burden of Diseases, Injuries and Risk Factors programme, an extremely influential series of studies that attempt to assess the health burden of all the most important diseases, and which does much to influence the governmental planning of priorities for health care, is to include the hemoglobinopathies and an assessment of genetic risk factors for both communicable and noncommunicable diseases. If successful, these new additions to the program should do much to increase the awareness of the importance of genetic disease and DNA technology among the international health community.

References

1. Akinyanju O (2001) Issues in the management and control of sickle cell disorder. Arch Ibadan Med 2:37–41
2. Alphey L, Beard CB, Billingsley P, Coetzee M, Crisanti A et al (2002) Malaria control with genetically manipulated insect vectors. Science 298(5591):119–121
3. Alexandria Alwan A, Modell B (1997) Community control of genetic and congenital disorders, EMRO technical publication series 24. WHO, Alexandria
4. Alwan A, Modell B (2003) Recommendations for introducing genetics services in developing countries. Nat Rev Genet 4(1):61–68

5. Angelucci E, Lucarelli G (2001) Bone marrow transplantation in thalassemia. In: Steinberg MH, Forget BG, Higgs DR, Nagel RL (eds) Disorders of hemoglobin. Cambridge University Press, New York, pp 1052–1072
6. Bittles AH (1990) Consanguineous marriage: current global incidence and its relevance to demographic research. Research Report No. 90-186, University of Michigan, Population Studies Centre, Detroit, MI
7. Bittles AH, Mason WM, Greene J, Rao NA (1991) Reproductive behavior and health in consanguineous marriages. Science 252(5007):789–794
8. Cao A, Galanello R, Rosatelli MC (1998) Prenatal diagnosis and screening of the haemoglobinopathies. Clin Haematol 11:215–238
9. Carlton J (2003) The Plasmodium vivax genome sequencing project. Trends Parasitol 19(5):227–231
10. Christianson A, Howson CP, Modell B (2006) March of Dimes global report on birth defects. March of Dimes Birth Defects Foundation, New York
11. Christianson AL, Modell B (2004) Medical genetics in developing countries. Annu Rev Genomcs Hum Genet 5:219–265
12. Davis H, Moore RM Jr, Gergen PJ (1997) Cost of hospitalizations associated with sickle cell disease in the United States. Public Health Rep 112(1):40–43
13. de Silva S, Fisher CA, Premawardhena A, Lamabadusuriya SP, Peto TE et al (2000) Thalassaemia in Sri Lanka: implications for the future health burden of Asian populations. Sri Lanka thalassaemia study group. Lancet 355(9206): 786–791
14. Dover GJ, Platt OS (1998) Sickle cell disease. In: Nathan DG, Orkin SH (eds) Hematology in infancy and childhood. W.B. Saunders, Philadelphia, PA, pp 762–801
15. Fleming AF, Storey J, Molineaux L, Iroko EA, Attai ED (1979) Abnormal haemoglobins in the Sudan savanna of Nigeria. I. Prevalence of haemoglobins and relationships between sickle cell trait, malaria and survival. Ann Trop Med Parasitol 73(2):161–172
16. Gardner MJ, Hall N, Fung E, White O, Berriman M et al (2002) Genome sequence of the human malaria parasite Plasmodium falciparum. Nature 419(6906):498–511
17. Harris E, Tanner M (2000) Health technology transfer. BMJ 321(7264):817–820
18. Haverfield EV, McKenzie CA, Forrester T, Bouzekri N, Harding R et al (2005) UGT1A1 variation and gallstone formation in sickle cell disease. Blood 105(3):968–972
19. Holt RA, Subramanian GM, Halpern A, Sutton GG, Charlab R et al (2002) The genome sequence of the malaria mosquito Anopheles gambiae. Science 298(5591):129–149
20. Jamison DT (2006) Investing in health. In: Jamison DT et al (eds) Disease control priorities in developing countries. Oxford University Press and the World Bank, New York, pp 3–34
21. Kaufmann SH, McMichael AJ (2005) Annulling a dangerous liaison: vaccination strategies against AIDS and tuberculosis. Nat Med 11(4 Suppl):S33–S44
22. Kwiatkowski DP (2005) How malaria has affected the human genome and what human genetics can teach us about malaria. Am J Hum Genet 77(2):171–192
23. Letvin NL, Bloom BR, Hoffman SL (2001) Prospects for vaccines to protect against AIDS, tuberculosis, and malaria. JAMA 285(5):606–611
24. Livingstone FB (1985) Frequencies of hemoglobin variants. Oxford University Press, Oxford
25. Modell B, Kuliev AM (1989) Impact of public health on human genetics. Clin Genet 36:286–298
26. Molineaux L, Fleming AF, Cornille-Brogger R, Kagan I, Storey J (1979) Abnormal haemoglobins in the Sudan savanna of Nigeria. III. Malaria, immunoglobulins and antimalarial antibodies in sickle cell disease. Ann Trop Med Parasitol 73(4):301–310
27. Murray CJL, Lopez AD (1996) The global burden of disease: a comprehensive assessment of mortality and disability from diseases, injuries and risk factors in 1990 and projected to 2020. Harvard University Press, Cambridge, MA
28. Nietert PJ, Silverstein MD, Abboud MR (2002) Sickle cell anaemia: epidemiology and cost of illness. Pharmacoeconomics 20(6):357–366
29. Nuffield Council on Bioethics (2002) The ethics of patenting DNA. A discussion paper. Nuffield Council on Bioethics, London
30. Nuffield Council on Bioethics (2002) The ethics of research related to healthcare in developing countries. Nuffield Council on Bioethics, London
31. Premawardhena A, Fisher CA, Fathiu F, de Silva S, Perera W et al (2001) Genetic determinants of jaundice and gallstones in haemoglobin E beta thalassaemia. Lancet 357(9272): 1945–1946
32. Premawardhena A, Fisher CA, Olivieri NF, de Silva S, Arambepola M et al (2005) Haemoglobin E β thalassaemia in Sri Lanka. Lancet 366:1467–1470
33. Rahimy MC, Gangbo A, Ahouignan G, Adjou R, Deguenon C et al (2003) Effect of a comprehensive clinical care program on disease course in severely ill children with sickle cell anemia in a sub-Saharan African setting. Blood 102(3): 834–838
34. Rees DC, Styles J, Vichinsky EP, Clegg JB, Weatherall DJ (1998) The hemoglobin E syndromes. Ann NY Acad Sci 850:334–343
35. Schaeffeler E, Eichelbaum M, Brinkmann U, Penger A, Asante-Poku S et al (2001) Frequency of C3435T polymorphism of MDR1 gene in African people. Lancet 358(9279): 383–384
36. Targett GA (2005) Malaria vaccines 1985–2005: a full circle? Trends Parasitol 21(11):499–503
37. Tsevat J, Wong JB, Pauker SG, Steinberg MH (1991) Neonatal screening for sickle cell disease: a cost-effectiveness analysis. J Pediatr 118(4 (Pt 1)):546–554
38. Weatherall DJ, Clegg JB (2001) The thalassaemia syndromes, 4th edn. Blackwell Science, Oxford
39. Weatherall DJ, Clegg JB (2001) Inherited haemoglobin disorders: an increasing global health problem. Bull WHO 79:704–712
40. Weatherall DJ, Clegg JB (2002) Genetic variability in response to infection Malaria and after. Genes Immun 3:331–337
41. Weatherall DJ (2003) Genomics and global health: time for a reappraisal. Science 302:597–599
42. Weatherall DJ, Akinyanju O, Fucharoen S, Olivieri N, Musgrove P (2006) Inherited disorders of hemoglobin. In: Jamison D et al (eds) Disease control priorities in developing countries, 2nd edn. Oxford University Press and the World Bank, New York, pp 663–680

43. Weatherall MW, Higgs DR, Weiss H, Weatherall DJ, Serjeant GR (2005) Phenotype/genotype relationships in sickle cell disease: a pilot twin study. Clin Lab Haematol 27(6):384–390
44. WHO (1996) Control of hereditary diseases, WHO technical report series 865. WHO, Geneva, Switzerland
45. WHO (1997) Community control of genetic and congenital disorders, Eastern Mediterranean office technical publication series 24. WHO, Alexandria, Egypt
46. WHO (2000) Primary health care approaches for prevention and control of congenital and genetic disorders. World Health Organization, Geneva
47. WHO (2002) The World health report 2002. Reducing risks, promoting healthy life. WHO, Geneva
48. WHO (2002) Genomics and World health. WHO, Geneva
49. WHO/World Alliance for the Prevention of Birth Defects (1999) Services for the prevention and management of genetic disorders and birth defects in developing countries. WHO, Geneva
50. Williams TN (2006) Red blood cell defects and malaria. Mol Biochem Parasitol 149(2):121–127
51. Winichagoon P, Thonglairoam V, Fucharoen S, Wilairat P, Fukimaki Y, Wasi P (1993) Severity differences in β-thalassaemia.haemoglobin E syndromes: implication of genetic factors. Br J Haematol 83:633–639
52. World Development Indicators (2002) World Bank, Washington
53. Yang YM, Shah AK, Watson M, Mankad VN (1995) Comparison of costs to the health sector of comprehensive and episodic health care for sickle cell disease patients. Public Health Rep 110(1):80–86

Genetic Databases

Introductory Note by Stylianos E. Antonarakis

The establishment and maintenance of databases are absolutely essential for access to and exploration of the genome and understanding of genomic variability and how it relates to human phenotypes. It is no longer possible to carry out genetic research or practice clinical genetics without the use of databases. Scientific progress in genetics/genomics virtually always requires the use of databases.

The editors of the new edition of this book felt a compelling need to include chapters on many of the existing important databases. Not all valid databases, however, have been included. The rationale for inclusion of databases was to cover knowledge and information on (a) the human genome and (b) human genetic pathogenic and polymorphic variability and phenotypic, including clinical, variability. For the first goal, we include the genome browsers of UCSC (chapter 29.1) and ENSEMBL (chapter 29.2). The information included in these databases is complementary and partly redundant, but each of the two has specific advantages/particular features, making them attractive to different users. In addition, these two genome browsers collaborate and exchange information, which adds to their value.

We have included several examples of the more medically (phenotype) oriented databases: the historical "OMIM" (the first gene and phenotype database, initiated by the late Victor McKusick), "Gene Tests," the Locus Specific Databases, DECIPHER, the database of copy number variants, "EntrezGene," "dbGAP," a genotype and phenotype database that includes information on genome-wide association studies, and "HGMD," the human genome mutation database. Goals, infrastructure, funding, and likely users of each of these databases are described and differ considerably, but all are extremely useful for our understanding of human genetic phenotypes and clinical manifestations as well as for planning diagnostic and presymptomatic laboratory studies.

Databases and Genome Browsers

29.1

Rachel A. Harte, Donna Karolchik, Robert M. Kuhn, W. James Kent, and David Haussler

Abstract The advent of the human genome project and subsequent projects to sequence genomes of other species and multiple individuals has driven the need for tools that can visualize vast amounts of genomics data. Software for genome browsing has had a vast impact in the arenas of human medical and genetics research, enabling researchers to process and integrate different data types from a large variety of sources. Three major genome browsers are freely accessible online – the University of California, Santa Cruz (UCSC) Genome Browser, the Wellcome Trust Sanger Institute (WTSI)/European Bioinformatics Institute (EBI) Ensembl browser and the National Center for Biotechnology Information (NCBI) MapViewer. The UCSC Genome Browser is a key part of the UCSC Genome Bioinformatics suite of integrated tools that facilitate data mining, together with allowing users to visualize and query their own data in the context of the existing Genome Browser annotations. The chapter provides an overview of the types of annotation data displayed by the Genome Browser, as well as step-by-step examples illustrating how to create custom tracks and query both the Genome Browser and Table Browser. The Genome Browser offers links to several programs: BLAT for performing fast sequence alignment to genomes; the In Silico PCR tool for aligning primers to the genome, and liftOver for converting genomic coordinates from one assembly to another. Other tools in the suite include the Gene Sorter for sorting genes based on their relationships such as expression profiles and genomic proximity; the Proteome Browser, which shows protein-related information in graphical form and links out to external protein-related sites; and Genome Graphs, which allows the user to display genome-wide datasets such as those from SNP association studies, linkage studies and homozygosity mapping. The suite of UCSC Genome Bioinformatics tools, data downloads, extensive documentation and links to further training materials can be found at http://genome.ucsc.edu.

Contents

29.1.1 Historical Background 906
29.1.2 Database Organization 907
29.1.3 Genome Annotations 907
    29.1.3.1 Overview of Tracks in the Genome Browser 907
    29.1.3.2 Comparative Genomics Tracks 912
    29.1.3.3 UCSC Genes Set 912
29.1.4 Displaying and Sharing Data Using Custom Annotation Tracks 913
29.1.5 Example Analysis Using the Genome Browser 913
29.1.6 Using BLAT for Genome-Wide Alignments 916
29.1.7 Table Browser 916

R.A. Harte(✉), D. Karolchik, R.M. Kuhn, W.J. Kent, and D. Haussler
Center for Biomolecular Science and Engineering,
University of California Santa Cruz,
Santa Cruz, CA 95064, USA
e-mail: hartera@soe.ucsc.edu

	29.1.7.1	Overview	916
	29.1.7.2	Example Using the Table Browser	917
29.1.8	Tools		917
	29.1.8.1	Introduction	917
	29.1.8.2	In Silico PCR	918
	29.1.8.3	Lifting Coordinates Between Assemblies	918
	29.1.8.4	Gene Sorter	918
	29.1.8.5	Proteome Browser	918
	29.1.8.6	Genome Graphs	919
29.1.9	Further Information		919
29.1.10	Future Directions		919
References			920

29.1.1 Historical Background

The past 30 years have brought many exciting developments to the field of genomics, culminating in the full sequencing of many organisms and the development of methodologies to further the functional annotation of the genomes. As the vast amount of sequence data accumulated, a major question arose: how should it be visualized together with the available annotations?

This question has been answered by the development of several programs that allow researchers to view genomic data for model organisms on the web. The Saccharomyces Genome Database (SGD) (http://www.yeastgenome.org/) [11,39] was the first such program, designed for the *Saccharomyces cerevisiae* genome assembly. ACEDB [12,26] was adopted by both the *Caenorhabditis elegans* sequencing project [53] and the WormBase project (http://www.wormbase.org/) [44,50], which hosts data on *C. elegans* and related worm species.

The sequencing of the human genome demanded a more robust method of viewing the genome and its annotations, and the ability to link to other relevant biological databases, integrating data from many different sources. The Genome Browser at the University of California, Santa Cruz (UCSC) (http://genome.ucsc.edu/) [19,23,25,30,33,34] is one of three tools that was developed – along with Ensembl (http://www.ensembl.org) [22] and the NCBI MapViewer (http://www.ncbi.nlm.nih.gov/mapview/) [45] – to provide a web-based method of interacting with the human genome. The annotations combined with various data mining tools facilitate many aspects of medical and human genetics research (see Box 29.1.1). All of these browsers have since expanded to include the genomes of a number of model organisms. The UCSC Genome Browser includes genome assemblies from human, chimp, mouse, rat, zebrafish, and other selected vertebrates, *C. elegans*, *S. cerevisiae*, *Drosophila melanogaster*, and a large collection of other flies and insects.

Box 29.1.1. Importance of genome browsers in medical and human genetics research

- Visualization of the results of genome-wide scans such as SNP association studies, linkage studies, and homozygosity studies.
- Convenient definition of candidate regions and candidate genes.
- Browsing a wealth of information on features in a candidate region including protein-coding and non-protein-coding genes, alternative splicing, conserved regulatory regions, and known polymorphisms.
- In-depth information on genes, including Online Mendelian Inheritance in Man (OMIM) [37], Gene Ontology (GO) [54], RefSeq [43], and Swiss-Prot summary information from UniProt [56], expression level in various tissues, and links to orthologs in model organism databases.
- Methods for large-scale data mining to define sets of genes meeting various criteria.
- Facilities for viewing personal data sets side-by-side with public data and combining personal data with public data in the data mining tools.
- Display of simple nucleotide polymorphisms (SNPs) and copy number polymorphisms (CNPs), including deletions of various sizes in the human population and in cancerous cells.
- Access to the evolutionary history of each base from multiple genome alignments and plots of conservation levels.

The UCSC Genome Browser is a web-based CGI (Common Gateway Interface) application adapted from a program implemented in the C programming language to view alternative splice forms of a gene prediction in *C. elegans* [23,31,32]. Extensive changes were made to adapt this program for the display of the human genome, which is 30 times larger than that of the worm. The resulting Genome Browser is fast, easy to use, and freely available through the Internet. The Genome Browser database is extensible to accommodate new genomes and annotation data as they become available.

29.1.2 Database Organization

As a sequencing project for an organism matures, genomic assembly updates are released, each containing improvements in the sequence coverage and/or assembly. The sequence and annotation data for each genome assembly are stored in their own database in the Genome Browser [23], using the publicly available software, My Structured Query Language (MySQL) (http://dev.mysql.com/doc/). As more assemblies are added, selected older assemblies are moved to the archive server (http://genome-archive.cse.ucsc.edu). Each of the web servers for the Genome Browser has its own local copy of the databases to enable rapid access to the data.

Within each assembly database, there are tables containing data sets anchored to the assembly by coordinate positions. These data tables are the basis for the annotation tracks in the Genome Browser. Additionally, a data set may have secondary or tertiary tables containing nonpositional data such as identifiers that link the data to other annotation databases. Programs have been specially developed to facilitate the batch loading of various types of data, including BED (Browser Extensible Data), PSL (Pattern Space Layout), WIG (wiggle format for continuously variable data), microarray and gene prediction data in GFF (Gene-Finding Format) or GTF (Gene Transfer Format). Supplemental databases, such as UniProt [56] or Gene Ontology (GO) [54], are obtained from external sources. The hgFixed database contains data that do not change with time, such as microarray data, and that may be used by several different assemblies as well as by the Gene Sorter tool [29] (see Sect. 29.1.8.4). The Proteome Browser [19,21] (see Sect. 29.1.8.5) and the VisiGene tool [33] also have their own separate databases.

Database tables are optimized to allow rapid display in the Browser graphic. This is achieved by table indexing; some tables also use a binning scheme to speed up table lookups [30]. For very large data sets, such as mRNAs, ESTs and alignments, rapid data lookup and display are improved by creating a data table for each chromosome. Data are stored in Indexed Sequential Access Method (ISAM) tables that may be swapped in and out separately, which allows continuous updating of databases. The ability to modify tables is of particular importance for frequently updated tables such as those tables relating to mRNAs, which are updated nightly, and EST tables, which are updated weekly.

29.1.3 Genome Annotations

To access annotations, open the URL http://genome.ucsc.edu in an Internet browser and then click on the Genomes link, which leads to the Gateway page (Fig. 29.1a). Here, several of the most recent assemblies are available for each genome. Upon selection of an assembly, the "position or search term" text box displays the default position of entry. A new location can be selected by typing in a different genomic position or a search term, such as a gene name, description, accession number or other annotation identifier. The Gateway page presents a set of example search terms supported for the selected assembly. To open the Genome Browser at the requested location, press the "submit" button. If a search term is ambiguous, the user is taken to an intermediate page displaying information about the various results. The search result will be highlighted in the Browser display. The same query method may also be used in the position/search box above the Browser image (Fig. 29.1b).

29.1.3.1 Overview of Tracks in the Genome Browser

The Genome Browser displays annotations as horizontal "tracks" using the genome sequence as the display coordinates. Most features are displayed horizontally, with the exception of "wiggle" tracks (such as the Conservation track), which show a score associated with features on a vertical axis (Fig. 29.1b). There are

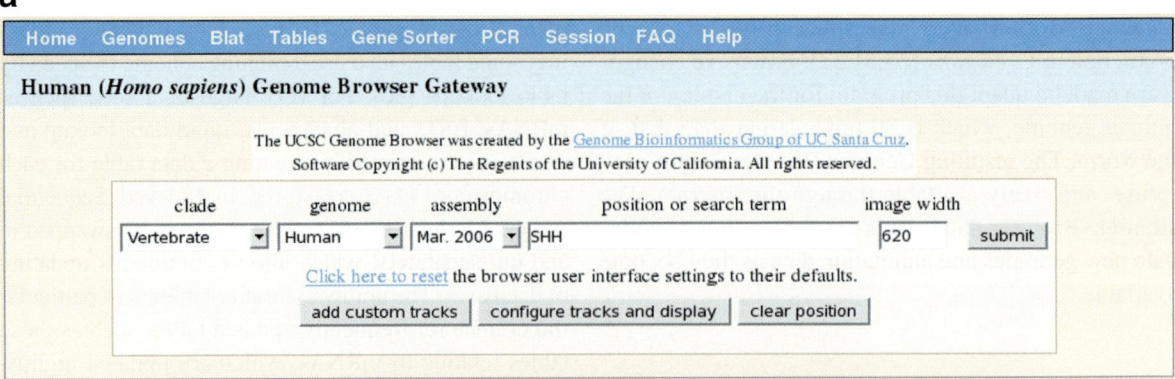

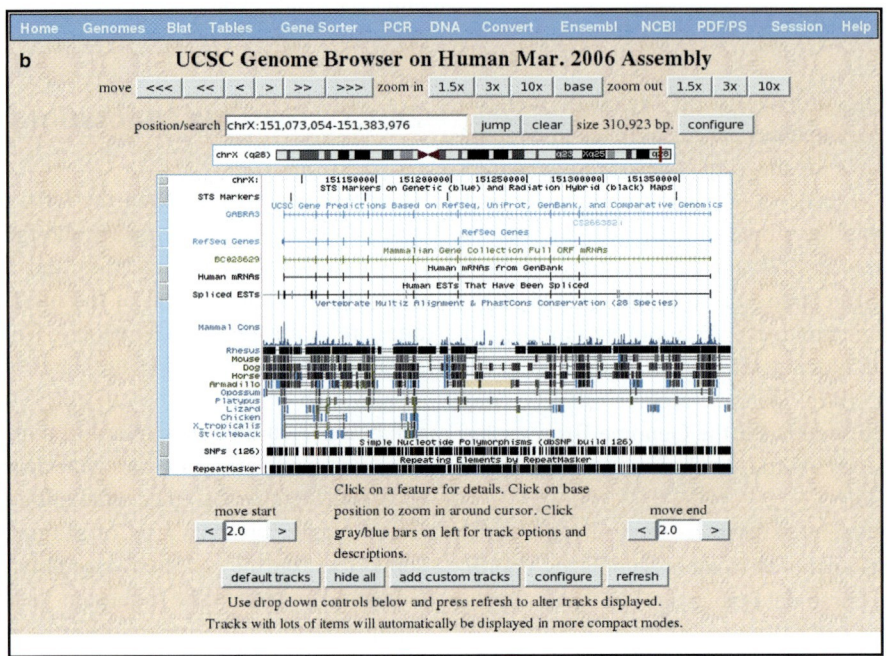

Fig. 29.1 (**a–d**) Browser features. (**a**) Genome Browser Gateway page showing SHH in the search/position box. (**b**) Human hg18 (NCBI Build 36) Genome Browser default view. (**c**) Adding a custom track to the human (hg18) Genome Browser. This interface may be reached via the "add custom tracks" button on the Gateway page (**a**) or underneath the Browser image (**b**).

a wide range of tracks particularly for human, mouse and other model organisms. Many of these tracks are computed at UCSC using public software developed internally or by other institutions; the remaining tracks are contributed by external collaborators. Annotations are organized into groups based on data type (Table 29.1). Track data may be hidden, compressed, or displayed in full using the set of track controls – organized by group – found below the Browser image. The compression feature is useful if there is a large amount of data in the displayed region or if a large number of tracks are visible.

Each track has an associated description page that may be accessed by clicking on its track control or the mini-button to the left of the track in the Browser graphic. Any color-coding conventions or configuration options used in the track are explained on this page. Click on an item in a track to view a details page giving further information related to that item and links to external databases, if appropriate.

Some tracks are organism specific, e.g., the Consensus CoDing Sequence (CCDS) (http://www.ncbi.nlm.nih.gov/projects/CCDS/) [42] annotation for the human and mouse genomes. Newer human

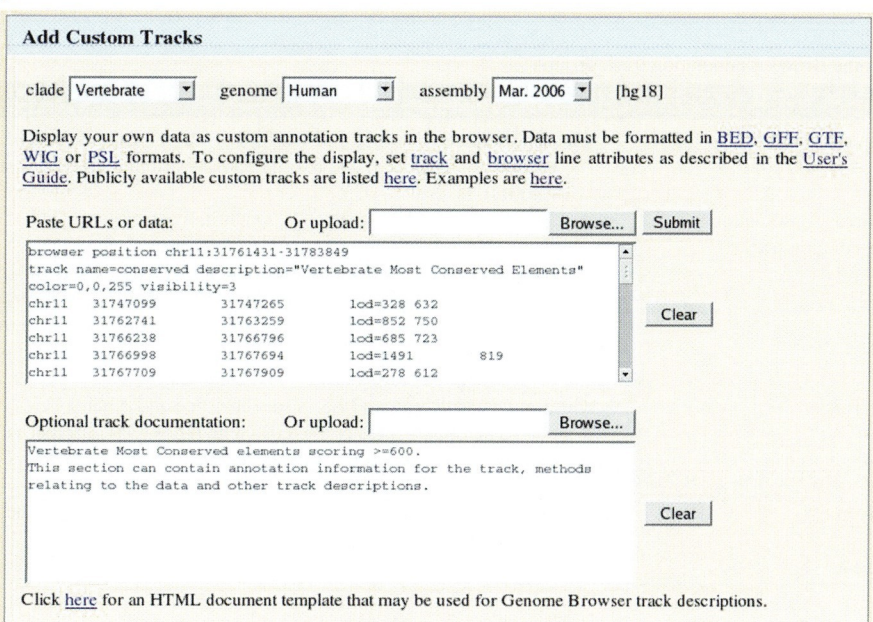

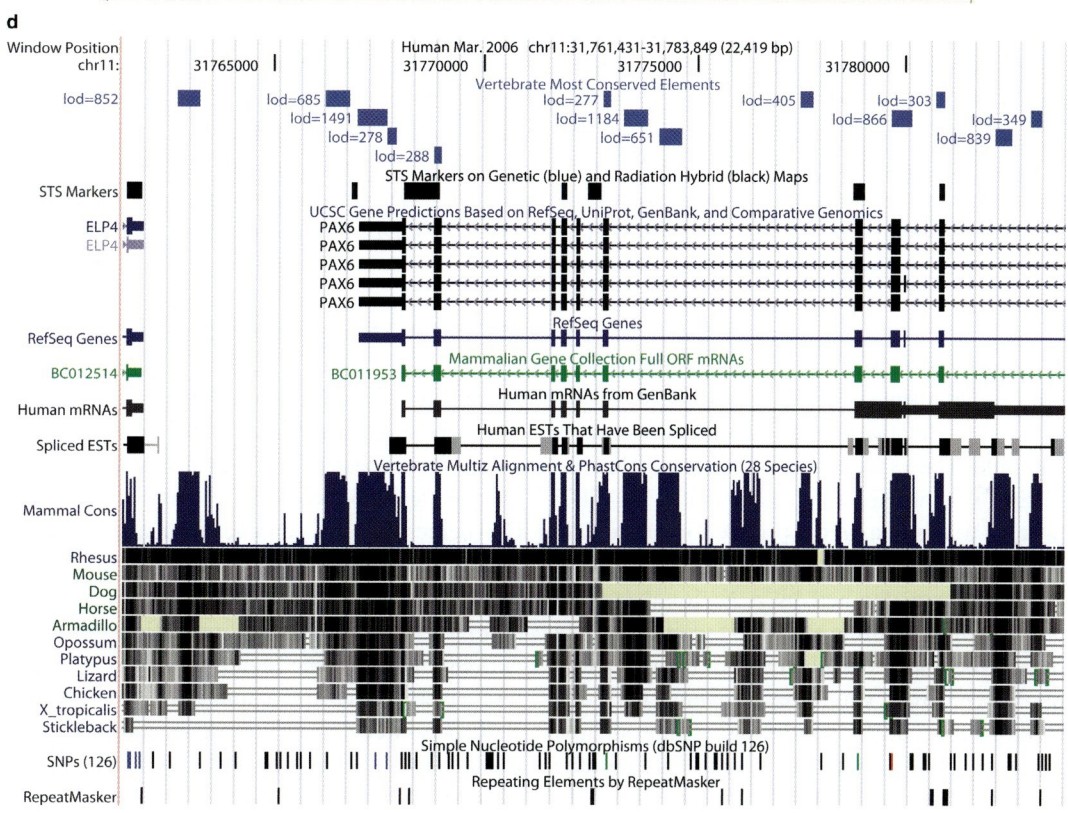

Fig. 29.1 (continued) The human March 2006 assembly is selected. Data are pasted into the box and a description can also be added where indicated. In this example, the custom track specifies the default Browser position, the color of the track features and the initial display visibility (*full*). Data shown are in BED format and they represent highly conserved regions from a multiple alignment of genomes of 28 species. (**d**) Viewing the custom track of highly conserved elements in the human (hg18) Genome Browser. The default position for the track shows part of the *PAX6* gene locus. Data displayed in the custom track represent regions of the Most Conserved track with a score of ≥600. The Most Conserved track is based on data from the multiple alignment of the genomes of 28 vertebrate species displayed in the Vertebrate Multiz Alignment and PhastCons Conservation track. Conservation in this track is displayed in the graph-like "wiggle" format

Table 29.1 Genome Browser annotation track groups

Track group	Example track types
Mapping and sequencing tracks	Genome sequence, contig and scaffold positions, GC percent, bacterial artificial chromosome (BAC), and fosmid ends, restriction enzymes, short match (motifs)
Phenotype and disease associations	GAD View [5], OMIM Genes [2], Quantitative trait loci (QTL)
Gene and gene prediction tracks	Known genes set (organism specific), Consensus Coding DNA Sequencing Project (CCDS) gene annotations [42], RefSeq mRNAs [43] for both reference and other species, Vertebrate Genome Annotation (VEGA) genes [61], Mammalian Gene Collection (MGC) genes [55], Ensembl genes [22], TransMap [34], non-coding RNAs, pseudogenes, gene predictions such as N-SCAN [17], Acembly [57] and Geneid [18]
mRNA and EST Tracks	GenBank [6] mRNAs and ESTs for reference and other species, EST clusters such as UniGene [45] and TIGR Gene Index [35], alternative splice forms
Expression and regulation	In situ hybridization and microarray probes, expression data, *cis*-regulatory regions
Comparative genomics	Multiple alignments, conservation, pairwise alignments with the genomes or proteins of other organisms
Variation and repeats	SNPs and other variation data [59], recombination, repeats, microsatellites

assemblies have additional tracks of data generated by the Encyclopedia of DNA elements (ENCODE) Consortium [13], for which UCSC hosts the official sequence-related data repository [ENCODE Data Coordination Center (DCC)]; microarray data are hosted by GEO [4] and ArrayExpress [40]. The ENCODE project groups provide the Genome Browser with a rich source of data on DNA replication, chromatin regulation, promoter function, gene models, transcription (tiling arrays and RNA-seq), variation, and multiple species comparisons to aid researchers in understanding the functional elements of the genome. Tracks from the pilot phase [14] of the ENCODE project are presented in separate ENCODE track groups; tracks displaying production phase data are integrated directly into the track groups listed in Table 29.1 and are denoted by a National Human Genome Research Institute (NHGRI) icon displayed to the left of the track control label. To organize large amounts of data from similar experiments with different cell lines and/or conditions, many of the ENCODE tracks are composite tracks that include subtracks. For example, the ChIP-chip and the ChIP-seq data include experiments that determine the binding sites of DNA-binding proteins such as transcription factors. These data are often based on experiments in multiple cell lines, which are grouped together into one track. The UCSC ENCODE DCC portal page (http://genome.ucsc.edu/ENCODE/index.html), which is accessed through the ENCODE link on the Genome Browser home page, displays recent project announcements and links to information on the cell types used in the project, a log of recently released tracks, data downloads, contributors, publications, the ENCODE Consortium data policy and an alternative ENCODE pilot project browser portal [58].

29.1.3.1.1 Mapping and Sequencing Tracks

Several tracks contain data pertaining to the genome assembly. Contig and scaffold locations in the genome assembly can be displayed. A Gap track shows the unsequenced gaps between these elements, and the local GC percent in the genome sequence is shown as a graph. The genome sequence itself may be viewed when zoomed in to base level.

Fluorescent in situ hybridization (FISH) clones annotations show the location of FISH-mapped bacterial artificial chromosome (BAC) clones from the BAC Resource Consortium. The Chromosome Bands track shows the approximate locations of Giemsa-stained chromosome bands. The Sequence-Tagged Site (STS) Markers track shows STS markers including markers from NCBI's UniSTS database [45]. Many of these markers are used in the construction of genome-wide genetic and physical maps. Pairs of sequences that form the 5′ and 3′ ends of single BAC clones are aligned using BLAT (BLAST-like alignment tool) [27] (see also Sect. 29.1.6) and displayed in the BAC End Pairs track. Two dynamically created tracks, which have no underlying tables, are Short Match and Restriction Enzymes. The Short Match track indicates the position of short nucleotide motifs of the user's choice in the display; Restriction Enzymes shows the location of the cutting sites of enzymes and may be user limited to a subset of restriction enzymes.

29.1.3.1.2 Phenotype and Disease Associations

This group, which is focused primarily on human assemblies, contains data relating to mutations associated with human diseases and their phenotypes. Data from the Genetic Association Database (GAD) (http://www.grc.nia.nih.gov/branches/rrb/dna/association) [5] is hosted in the GAD View track. GAD is a curated archive for published human genetic linkage data and genetic association studies of complex diseases and disorders. The OMIM Genes track displays genes associated with the Online Mendelian Inheritance in Man (OMIM) database (http://www.ncbi.nlm.nih.gov/sites/entrez?db=omim) [2], a collection of human genes and genetic phenotypes. Quantitative trait loci (QTL) tracks provide both human and rat data from the Rat Genome Database (RGD) (http://rgd.mcw.edu/) and mouse data from Mouse Genome Informatics (MGI) (http://www.informatics.jax.org/). Each locus is linked back to its RGD or MGI report, which includes information such as the trait, associated disease, phenotypes, population and markers. More genetic association tracks will be added as research grows in this field.

29.1.3.1.3 Gene and Gene Prediction Tracks and mRNA and EST Tracks

The mRNA, EST, and gene and gene prediction tracks show physical evidence supporting the positions of genes. GenBank mRNAs and ESTs [6] are aligned to the genome sequence using BLAT [27] (see Sect. 29.1.6). The spliced EST track is created by examining the resulting EST alignments for evidence of splicing. In a similar manner, the RefSeq Genes track is created from RefSeq mRNAs [43]. The Institute for Genomic Research (TIGR) Gene Index [35] is the alignment of clusters of ESTs; UniGene [45] mRNA/EST clusters are aligned by BLAT. The human and mouse UCSC Genes and rat Known Genes tracks consolidate gene prediction information from various lines of evidence and are supported by rich details pages with links to resources at other locations [20] (see Sect. 29.1.3.3).

Vertebrate Genome Annotation (VEGA) genes (from the Wellcome Trust Sanger Institute) [61] provide a set of high-quality manually curated genome annotations. Non-protein-coding RNAs, RNAs with secondary structure such as those predicted by EvoFold [41], pseudogenes, alternative splice form predictions and Ensembl protein superfamilies are also represented in tracks for some organisms.

Tracks for gene prediction annotations are obtained mainly from external sources. The Ensembl [22] and AceView (Acembly) [57] tracks predict genes based on genomic, mRNA and EST evidence. The Exoniphy program finds exons conserved in multiple species based on a phylogenetic hidden Markov model (HMM) [49]. The N-SCAN [17] tool produces gene predictions using information from the genome and from genomic multiple sequence alignments. Geneid [18] is an example of *ab initio* gene predictions.

29.1.3.1.4 Expression and Regulation

This group contains tracks that display experimental data relating to genome elements such as mRNA transcripts, BLAT alignments [27] (see Sect. 29.1.6) of sequences used for microarray probe selection such as those for Affymetrix GeneChip arrays and probes used for *in situ* hybridization experiments. Microarray data tracks exhibit red and green coloring in the probe sequence alignments to show the level of expression of a gene relative to expression over a wide range of tissues as for the Gene Atlas data from the Genomics Institute of the Novartis Research Foundation (GNF) [51,52]. Clicking on an alignment in this type of display leads to a detailed view of the expression for individual probes in that region.

29.1.3.1.5 Variation and Repeats

Repeats are found by RepeatMasker (http://www.repeatmasker.org) (software courtesy of Arian Smit,

Institute for Systems Biology) and Tandem Repeat Finder (TRF) [7] (Simple Repeats track). The Microsatellite track shows di- and trinucleotide repeats that have a highly polymorphic tendency. Segmental Duplications track shows regions that are likely to have been duplicated in the genome. Simple Nucleotide Polymorphisms (SNPs) from NCBI's dbSNP (http://www.ncbi.nlm.nih.gov/projects/SNP/) [45,47] contain single nucleotide polymorphisms and small insertions and deletions. Other SNP-related tracks show commercially available genotyping SNP arrays, recombination rates and hotspots [59].

29.1.3.2 Comparative Genomics Tracks

The UCSC Genome Browser is noted for its wide selection of comparative genomics tracks showing homology with other species. These annotations include pairwise and multiple alignments of genomes, sequence conservation and alignments of proteins from another species. Homology with other species is important for locating putative genes or regions of regulation, for assigning possible functions to genes, and for studying evolutionary aspects of genome development.

29.1.3.2.1 Chains, Nets, and Conservation

Pairwise cross-species whole genome alignments are created using the BLASTZ [46] program, which is very sensitive for alignments on a genome-wide scale. Related alignment fragments are linked together to form larger structures called "chains" [28]. Chains contain homologous genes. From these chains, best-in-genome alignments are selected to form the net tracks. Net tracks are useful for visualizing long-range synteny between organisms.

Multiple alignments are also generated from the best-in-genome pairwise alignments using multiz [9]. Conservation scores and regions of high conservation between genomes of organisms in the multiple alignment are computed using phastCons [48]. The Browser displays these scores in a continuously variable format that provides a visually intuitive view of conserved regions. The comparative alignments and conservation scores may be obtained from the UCSC downloads server.

29.1.3.2.2 Cross-Species Protein Alignments

Amino acid sequences diverge more slowly than nucleic acid sequences, and alignments of proteins from one species to another are therefore particularly useful for finding genes in distantly related or incompletely sequenced organisms and for annotating genes in other organisms for which a rich experimental annotation has not been established. This is of particular importance for a recently sequenced genome that does not have extensive gene annotations of its own. Human Proteins tracks [19] have been created for most such organisms. These human homologs are found by first aligning human known gene proteins (see Sect. 29.1.3.3) to the genome using BLAT [27] (see Sect. 29.1.6) to find the exon boundaries relative to the proteins. The peptide sequences of the putative exons are then aligned to the genome using tBlastn [1], which is a more sensitive alignment program than BLAT. Exon alignments are chained together, and the single best chain is retained after filtering on certain criteria. In a similar manner, the *D. melanogaster* proteins are aligned to other insect genomes. The UCSC Genes details pages (see Sect. 29.1.3.3) show the best Blastp [1] homologs in the gene sets for a number of model organisms.

29.1.3.3 UCSC Genes Set

The UCSC Known Genes protein-coding genes set [20] was produced for the human, mouse and rat assemblies by a fully automated process that combines both mRNA and protein evidence for the existence of protein-coding genes. Through improvements in the Known Genes process, this annotation evolved into the UCSC Genes set for the human and mouse genomes, which includes both protein-coding and putative noncoding genes. A moderately conservative prediction set, UCSC Genes contains 99.9% of the RefSeq Genes and is based on sequence data from GenBank [6], RefSeq [43] and UniProt [56]. To be incorporated into the UCSC Genes set, a transcript must have at least one supporting GenBank mRNA and another line of evidence. RefSeq genes require no additional evidence. If possible, a UniProt protein is found to represent the protein encoded by a transcript. Some transcripts that are annotated as non-coding may actually be protein-coding, but the evidence for the associated protein is weaker. The UCSC Genes set is more inclusive than RefSeq, containing more protein-coding genes, greater coverage of splice variants, and more non-coding genes. UCSC Genes have their own accessions, which remain stable between releases with only a change of the suffix to indicate a change of version. An additional track, Alt Events, is created as part of the UCSC Genes method. This track depicts various types of alternative splicing, alternate promoter usage, and other events that result in the production of multiple transcripts from a gene.

Each UCSC Genes or Known Genes annotation is accompanied by a details page rich in collected information and references from many sources. A gene symbol is assigned to the gene from Entrez Gene at NCBI [31] if it is a RefSeq mRNA; otherwise the Human Gene Organization (HUGO) (http://www.hugo-international.org) [10] gene name is used. Isoforms and variant names are shown, together with a gene description. The Quick Links section connects to views of the gene in UCSC tools and external websites. The sequence section links to the genomic, mRNA and protein sequences. If available, relevant microarray expression data are displayed. Homologs in other species are identified by the best Blastp alignment [1]. For closely related species such as human, mouse and rat, the non-syntenic alignments are removed. For more evolutionarily distant pairs of species, reciprocal-best Blastp is used for ortholog identification. Links to internal and external sites show related homolog information. Links to external pathway databases, protein structures, GO [54] annotations, and further descriptions from various sources such as GenBank [6] mRNAs and UniProt [56] are also displayed if available. The VisiGene tool [33], which provides a virtual microscope for viewing in situ hybridization images, is fully integrated with both the UCSC Genes and with the Gene Sorter (see Sect. 29.1.8.4).

29.1.4 Displaying and Sharing Data Using Custom Annotation Tracks

It is possible to temporarily display personal data alongside the existing genome annotations in the Genome Browser using the custom track feature [25,33,34]. Custom tracks ensure data privacy because they can be viewed only on the machine on which the data reside or via a custom URL that can be shared with collaborators. Data may be uploaded in a number of different formats: BED, GFF, GTF, PSL alignment, and WIG (graph-like display) formats. Data in custom tracks may be used to query the underlying database tables of the Genome Browser (see Table Browser, Sect. 29.1.7). The Genome Browser User's Guide contains instructions for creating and uploading a custom track. To load a custom track, click the "add custom tracks" button on the Gateway page (Fig. 29.1a) or below the Browser image (Fig. 29.1b). Links to several custom tracks submitted by Genome Browser users may be accessed via the "Custom Tracks" link on the Genome Browser home page. See Fig. 29.1c and 29.1d for an example of custom track loading.

29.1.5 Example Analysis Using the Genome Browser

Example: Search for homologs for a human gene in another organism

Suppose you are interested in finding genes belonging to the hedgehog family of signaling proteins. Here are the steps that one would take to find homologs in another species using the human March 2006 assembly (NCBI Build 36, known as hg18 at UCSC) on the UCSC Genome Browser:

1. Go to the Genome Browser home page (http://genome.ucsc.edu) and click on the "Genomes" or "Genome Browser" link on the top or side menu.
2. On the Gateway page, select "Mammal" as the clade, "Human" as the genome, and "March 2006" as the assembly (Fig. 29.1a).
3. Enter SHH – the HUGO gene name for the human sonic hedgehog gene – into the "position or search term" box and press the "submit" button.
4. Several search results are returned. Click on the SHH gene in the RefSeq Genes section. The Browser will display the region of the genome to which this mRNA aligns (Fig. 29.2a).
5. If the Conservation track is not displayed, scroll down to the Comparative Genomics section below the Browser image and open the track by selecting

the "pack" display mode from the Conservation track control and clicking the "refresh" button. This track shows multiple alignments of other species to the human genome and a measure of evolutionary conservation among the species. The exons are in regions of high conservation; conservation levels tend to drop off at the exon/intron boundary. Other regions of conservation may play as yet undiscovered, regulatory roles. In our example, the mouse appears to have high conservation in the coding region of the SHH gene; thus, it would be a good organism to use to look for homologs of SHH.

6. Next, display the Mouse Chain and Mouse Net tracks by setting their track controls (also in the Comparative Genomics section) to "full" visibility. The Mouse Chain track shows regions that are homologous to the human genome; the Mouse Net track can be used to determine which homologs are in the same syntenic location in the two species, and hence likely to be orthologs. Each alignment is color coded to indicate the chromosome on which the aligning region resides in the mouse genome. For example, the first alignment in the chain and the net track is red which, from the color key below the Browser graphic, signifies chromosome 5. The net track shows the best aligning sequence from mouse to this region of the human genome; thus, the best long-range homology to mouse for this region lies on mouse chromosome 5.

7. The top alignment in a Chain track is typically the best long-range alignment. Select the top chain in the Mouse Chain track to view more details about the alignment, including a link to the corresponding sequence in the mouse Genome Browser (Fig. 29.2b). Note that both the UCSC Genes and the RefSeq Genes tracks in this region of the mouse genome show a mouse homolog of the SHH gene (Fig. 29.2c).

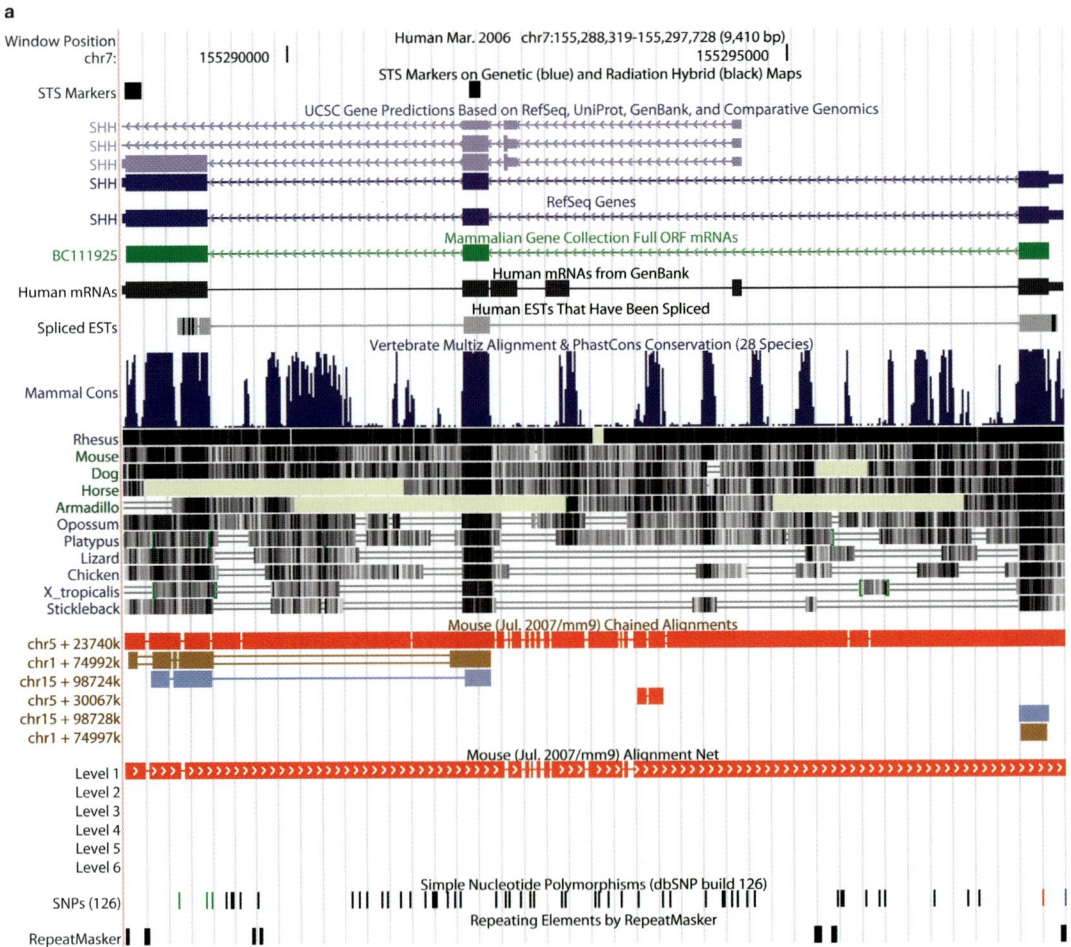

29.1 Databases and Genome Browsers

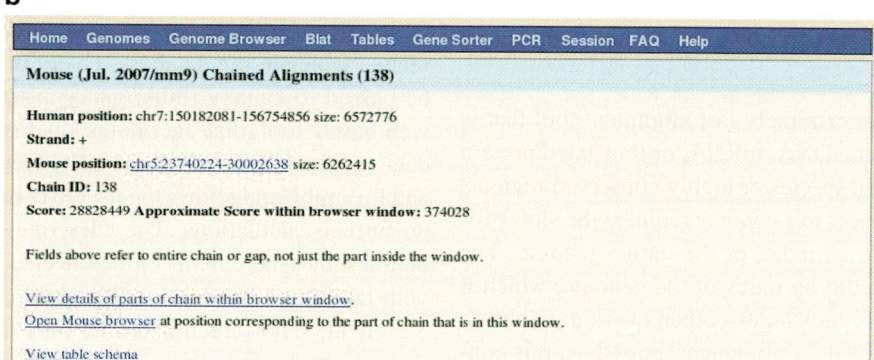

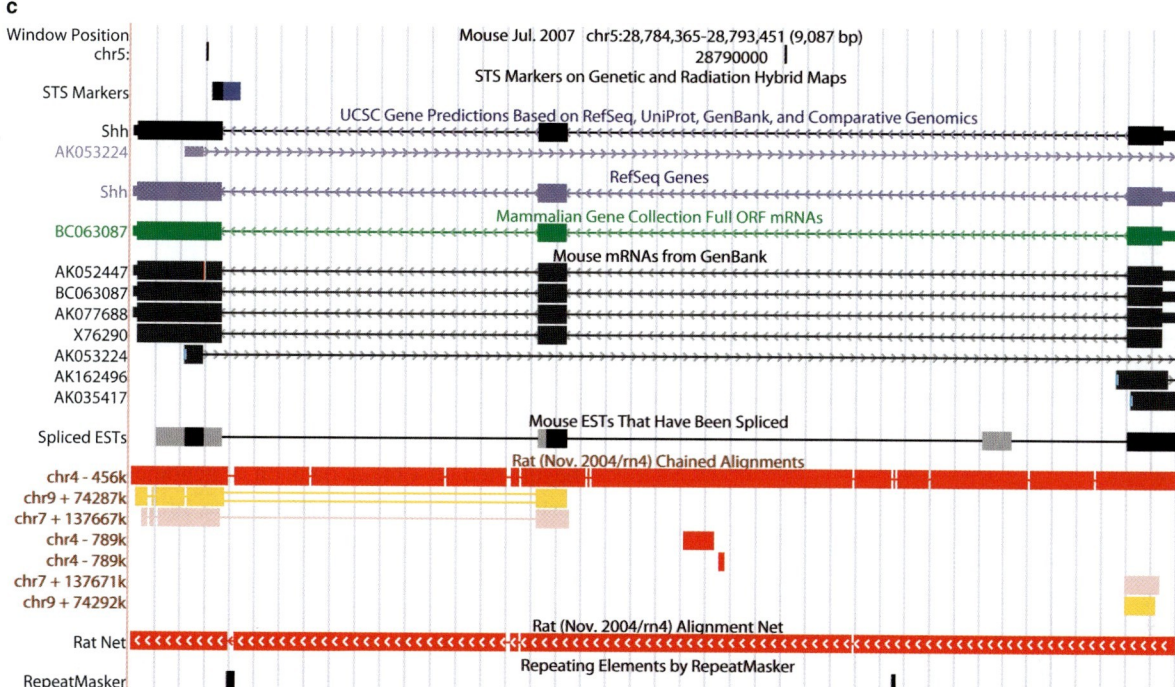

Fig. 29.2 Finding homologs using the Genome Browser. (**a**) Human hg18 (NCBI Build 36) Genome Browser showing the SHH gene with Mouse (mm9, NCBI Build 37) Chain and Mouse Net tracks visible. In the Conservation track (shown at "pack" visibility), scores and pairwise alignments are displayed as a gray-scale density plot. Regions of conservation between genomes of different species are seen for the exonic regions of SHH in this track. The first mouse chain (*red*) aligns to the mouse *Shh* ortholog gene locus; the second mouse chain (*brown*) is an alignment of the mouse genome region containing the Ihh (Indian hedgehog) gene and the third mouse chain (*blue*) is an alignment to the mouse genome region containing the Dhh (Desert hedgehog) gene. Ihh and Dhh are paralogs of the sonic hedgehog gene. Clicking on the first mouse chain (*red*) leads to the view shown in (**b**). (**b**) Details page for the first mouse chain and link to the mouse Genome Browser for the region of the chain in this window. (**c**) Mouse Genome Browser for the mm9 assembly, showing the region of the mouse genome that was aligned in the mouse chain in (**b**). The gene in both the UCSC Genes and the RefSeq Genes tracks is Shh, which is the mouse ortholog of human SHH. Rat Chain and Rat Net tracks are displayed to show that there is a Shh homolog on chromosome 4 of the rat genome (rn4, Baylor v3.4 assembly)

8. Repeating this for the other alignments in the chain track shows that the alignment from mouse chromosome 1 is for Ihh (Indian hedgehog). Similarly, the sequence aligning from mouse chromosome 15 contains the Dhh (desert hedgehog). Dhh and Ihh are members of the same protein family as sonic hedgehog. The chain track therefore identifies homologs of SHH in mouse.

29.1.6 Using BLAT for Genome-Wide Alignments

BLAT [27] is an extremely fast alignment tool that is useful for aligning DNA, mRNA, or translated protein sequences within species or highly conserved regions. Aligning sequences to a genome requires the slow process of building an index of the entire genome. The BLAT server builds an index of the genome, which it holds in memory; this index is then queried with each sequence presented for alignment. For DNA, this consists of all nonoverlapping 11mers, except those from regions that are heavily involved in repeats. For proteins, the index is created using tetramers.

DNA BLAT is designed to find aligning regions of 95% identity or more and of at least 40 bases in length. It can find perfect matches of 30 bp and sometimes down to 20 bp, although it may miss more divergent or shorter sequences. By default, web-based BLAT can find perfect matches as short as 21 bp in nonrepeat regions. Protein BLAT works optimally for alignments of at least 80% identity and of at least 20 amino acids in length.

Many of the Genome Browser pages contain links to a web interface for BLAT. Sequences can be pasted into the text box provided on the BLAT web page or uploaded from a file. A genome-wide search for the sequence typically takes just seconds.

The BLAT program executables and source code are available for download for use on the command line. BLAT is freely obtainable for academic, personal, and nonprofit use, but a license must be sought for commercial purposes. Command-line BLAT settings may be optimized to adjust the sensitivity. If several nearby regions of homology are returned between two sequences as separate alignments, BLAT will stitch these together into a single alignment. It is also able to correctly position splice sites. Therefore, BLAT is used to create mRNA and EST alignments to genomes for display in the Genome Browsers.

29.1.7 Table Browser

29.1.7.1 Overview

The Table Browser [19,24] provides a web-based means of extracting data from the underlying MySQL databases using filtering and free-form query options. Queries can be made on a selected table as well as on related tables in the database. Query results also may be passed to Galaxy (http://galaxy.psu.edu/) [16] – a web-based tool that facilitates queries on multiple data sources, including the UCSC Genome Browser and Ensembl, and allows the user to combine results or do further calculations. The "describe table schema" button shows the schema of the selected table, together with table field descriptions and examples of data in each field. When creating a filter for a database query, if there are related tables, then the option is given to add fields of the related tables into the filter. One of the output formats also allows the selection of fields from the selected and related tables. Other output formats include BED, GTF, custom track, and sequence. In addition, data points can be downloaded for wiggle (graph-like) format tracks, or multiple alignment format (MAF) may be used to obtain alignments from the multiple alignment tracks. Subregions of features can also be specified for the output. For example, the BED format output allows selection of upstream or downstream regions, introns, coding exons, or whole genes.

Certain identifiers or accessions may be provided to allow batch searching of tables. Otherwise, tables can be queried to obtain data on a genome-wide scale, restricted to the ENCODE regions, or for a specified gene, accession, or position.

Custom tracks can be created from Table Browser queries and then viewed in the Genome Browser or loaded as a track into the Table Browser. The custom track data are then available for further Table Browser queries and for intersections with other tracks, thereby allowing compound queries. Data uploaded into the Genome Browser to form a custom track are also available in the Table Browser, making it an excellent tool for comparing one's own data with the Browser's annotations.

Another useful feature is the ability to do correlations. The position/score vectors of two selected tables are then intersected. Scores are retained only if both tables have a score. A linear regression is performed on the two vectors, resulting in a report with several statistics, including the Pearson's correlation coefficient (r), scatterplots of the two data vectors and residuals versus fitted, and value histograms.

29.1.7.2 Example Using the Table Browser

Problem: Find the human UCSC Genes on chromosome one that overlap with genes represented by Affymetrix probe sets in the Affymetrix HG-U133 Plus 2.0 GeneChip.

1. Open the Table Browser by clicking on the Tables link on the Genome Browser home page.
2. Select the genome and assembly of interest, e.g., human (March 2006) (hg18) (Fig. 29.3).
3. Select "Genes and Gene Predictions" as the group, "UCSC Genes" as the track, and "knownGene" as the table.
4. Select "genome" as the region.
5. Press the filter button and type "chr1" into the box adjacent to "chrom does match" then press "submit."
6. Press the intersection button. Select "Expression and Regulation" as the group, "Affy U133Plus2" as the track and "AffyU133Plus2" as the table, then press "submit."
7. Select "BED" as the output format, then press "get output."
8. Select "Whole Gene" and then press "get BED."

The output will list the positions of UCSC Genes that overlap with the sequences used for the Affymetrix HG-U133 Plus 2.0 GeneChip probe design.

29.1.8 Tools

29.1.8.1 Introduction

Several web-based tools are available to supplement the effectiveness of the Genome Browser as a data mining tool. All of the tools mentioned in this section are accessible from either the side menu on the home page or the top menu on the application pages. Some of these tools can be downloaded as executables; the source is available in the Genome Browser source tree, which may also be downloaded. As with BLAT, the source is free for academic, nonprofit, and personal use, but commercial use requires a license.

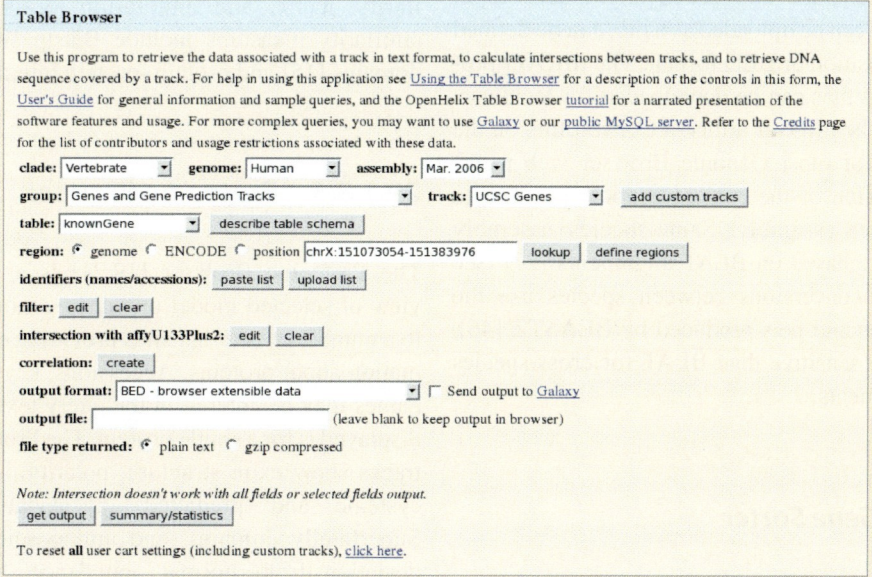

Fig. 29.3 Using the Table Browser. Table Browser interface with human hg18 (NCBI Build 36) chosen as the assembly, and the knownGene table from the UCSC Genes track also selected. The "genome" option is selected for the region setting to search the entire genome. Filtering is set up and an intersection with the Affymetrix U133Plus2 track is created. BED is selected as the output format.

29.1.8.2 In Silico PCR

The in silico PCR tool (isPcr) [19] is available through a web interface, and also as a download for use on a local machine. It indexes the genome for fast sequence searching of an exact match of primer pairs to a genome or to UCSC Genes transcripts on selected human and mouse assemblies. If successful, it returns the sequence that lies between and includes the primer sequences. ExonPrimer (courtesy of Tim Strom), which is accessible through a link on UCSC Genes or Known Genes details pages (see Sect. 29.1.3.3), is a useful tool that can be used in conjunction with isPcr. It is a third-party program that aids in the design of intronic primers with certain criteria for different assemblies.

29.1.8.3 Lifting Coordinates Between Assemblies

The liftOver program [19] has a web interface, reached via the Utilities link on the home page, that allows the conversion of genomic position coordinates or annotations from one assembly version to another. It also retrieves putative homologous regions in other species by using the chained and netted alignments. Input regions, which can be uploaded as a file, can be defined using either position or BED format. The liftOver program and input files can be downloaded for local use. In addition to the liftOver utility, a Convert link on the top menu bar of most Genome Browser web pages allows conversion of the current genome position to that of another assembly. Same-species assembly conversions are based on BLAT alignments [27] (see Sect. 29.1.6); conversions between species use the pairwise chains and nets produced by BLASTZ [46], which is more sensitive than BLAT for cross-species genome alignments.

29.1.8.4 Gene Sorter

Genes evolve together and perform functions together, so that it is important to be able to view their relationships to each other. This is the principle behind the Gene Sorter [29,34], which presents a gene-based view of a genome rather than the chromosome-based view shown in the Genome Browser. The Gene Sorter is available for most of the model organism assemblies. For the human, mouse and rat genomes, the gene set produced for the UCSC Genes or Known Genes track (see Sect. 29.1.3.3) is shown. The gene sets of other organisms are obtained from third-party annotations (Ensembl [22], FlyBase [22,31], WormBase [8], SGD [11,39]). To examine the relationship of a set of genes, enter a gene name or accession in the "search" text box; the Gene Sorter will return a list of related genes sorted by the criteria specified in the "sort by" list. For each gene displayed, the Gene Sorter by default displays the gene name, selected microarray expression data, a Blastp [1] E-value from the alignment of each gene to the selected gene, the genomic position and the mRNA description. The configuration controls allow selection of other types of data for display.

By default, genes are sorted by distance based on expression data. In the Gene Sorter, the expression data are gene-specific and not transcript-specific as in the Genome Browser. For each gene, the best representative probe set representing a single transcript has been selected for that gene. For expression data, the distance between genes is calculated as a weighted sum of the log expression ratio values. Genes may also be sorted by proximity within the genome, protein similarity, name, and annotation terms. The protein similarity measures include Blastp similarity, Pfam domains [15] and protein–protein interactions.

29.1.8.5 Proteome Browser

The Proteome Browser [19,21] is a protein-centric view of selected model organism genomes with links to a number of external sites providing extensive information about proteins. Accessions, keywords, or gene names may be entered in the query text box; the view displayed is for a single protein. The Proteome Browser tracks show exon structure, polarity, hydrophobicity, cysteine and predicted glycosylation locations, Superfamily domains, and amino acids that exhibit deviation from normal abundance. The FAST-All (FASTA) protein sequence is also provided. Histograms show where the protein falls on the genome-wide distribution for protein properties such as isoelectric point, molecular weight, exon number, InterPro domains [38], cysteine locations, hydrophobicity and other characteristics. The Proteome Browser provides links to protein domain information from a wide variety of resources at other locations.

29.1.8.6 Genome Graphs

Genome Graphs [25] is a tool that was developed to support whole-genome linkage, association and homozygosity studies. It allows visualization of these data types in a genome-wide view that also supports user-generated data as custom tracks. Linkage and association studies facilitate the identification of regions of the genome that carry variants associated with disease. Genome Graphs may be used to upload several sets of genome-wide data so that they can be viewed together. The display can be restricted to regions that pass a desired significance threshold. Genes in the regions that pass the threshold can be displayed in the Gene Sorter (see Sect. 29.1.8.4), an area of interest can be viewed in the Genome Browser, and the correlation coefficient (r) among the data sets may also be calculated.

29.1.9 Further Information

An extensive online documentation set accompanies the Genome Browser, which can be accessed via the Help and FAQ links on the top menu bars of most web pages. Further training can be accessed via the Training link of the left blue menu bar of the home page. Data downloads are available through the downloads server (http://hgdownload.cse.ucsc.edu/). A mailing list, genome@soe.ucsc.edu, provides a forum to which users can post questions regarding Genome Browser usage and search previously answered questions. A low-volume mailing list, genome-announce@soe.ucsc.edu, broadcasts important announcements about new software and data releases and problems with the website. Several full and partial mirrors of the Genome Browser are listed on the mirrors page, which is accessible via a link on the side menu bar of the home page; these are useful when the UCSC site is unavailable. Some mirrors do not have all the organism assemblies found on the UCSC site, but they keep at least the two latest assemblies for most of the main organisms. Questions regarding Genome Browser mirrors or the setup of mirrors may be directed to the mailing list, genome-mirror@soe.ucsc.edu.

29.1.10 Future Directions

In future years, the Genome Browser database will continually expand to include newly sequenced and updated vertebrate genomes, as well as selected genomes from other clades, as they become available. The quality of the UCSC Genes set (see Sect. 29.1.3.3) will be further refined. In addition to the existing multiple alignments found in the Comparative Genomics group, protein multiple alignments that mirror these will be explored. Support for high-throughput data and large-scale sequence alignments will be added. Data from published genome-wide association studies will be made available as tracks for Genome Graphs (see Sect. 29.1.8.6). To identify specific disease-associated variants in patients, DNA sequencing of entire regions is becoming increasingly common; medical sequencing and human variation data will become more widely available in future versions of the Genome Browser, along with support for custom tracks that display medical sequencing data and highlight the variations found among different patients. Color-coding will be used to differentiate potentially deleterious mutations from those that may be harmless. Because the UCSC Genome Browser is continually evolving, the software and data may have changed in the interim since this chapter was written.

Acknowledgments The UCSC Genome Browser project has been funded in whole or in part with funds from the National Human Genome Research Institute (5P41 HG002371-09 for the UCSC Center for Genomic Science, and 5U41 HG004568-02 for the UCSC ENCODE Data Coordination Center); Howard Hughes Medical Institute; the National Cancer Institute (Contract No. N01-CO-12400 for the Mammalian Gene Collection) and California Institute for Quantitative Biosciences (QB3). The content of this publication does not necessarily reflect the views of policies of the Department of Health and Human Service, nor does mention of trade names, commercial products or organizations imply endorsement by the U.S. Government.

We would like to thank the Genome Bioinformatics Group at the UCSC Center for Biomolecular Science and Engineering, the many collaborators who have contributed data to our project, our Scientific Advisory Board for their valuable advice and recommendations, and our users for their feedback and support.

References

1. Altschul SF, Madden TL, Schaffer AA et al (1997) Gapped BLAST and PSI-BLAST: a new generation of protein database search programs. Nucleic Acids Res 25:3389–3402
2. Amberger J, Bocchini CA, Scott AF et al (2009) McKusick's Online Mendelian Inheritance in Man (OMIM(R)). Nucleic Acids Res 37:D793–D796
3. Ashburner M, Drysdale R (1994) FlyBase–the Drosophila genetic database. Development 120:2077–2079
4. Barrett T, Troup DB, Wilhite SE et al (2009) NCBI GEO: archive for high-throughput functional genomic data. Nucleic Acids Res 37:D885–D890
5. Becker KG, Barnes KC, Bright TJ et al (2004) The genetic association database. Nat Genet 36:431–432
6. Benson DA, Karsch-Mizrachi I, Lipman DJ et al (2009) GenBank. Nucleic Acids Res 37:D26–D31
7. Benson G (1999) Tandem repeats finder: a program to analyze DNA sequences. Nucleic Acids Res 27:573–580
8. Bieri T, Blasiar D, Ozersky P (2007) WormBase: new content and better access. Nucleic Acids Res 35(Database issue):D506–D510
9. Blanchette M, Kent WJ, Riemer C et al (2004) Aligning multiple genomic sequences with the threaded blockset aligner. Genome Res 14:708–715
10. Bruford EA, Lush MJ, Wright MW et al (2008) The HGNC Database in 2008: a resource for the human genome. Nucleic Acids Res 36:D445–D448
11. Cherry JM, Adler C, Ball C et al (1998) SGD: Saccharomyces Genome Database. Nucleic Acids Res 26:73–79
12. Eeckman FH, Durbin R (1995) ACeDB and macace. Methods Cell Biol 48:583–605
13. ENCODE Project Consortium (2004) The ENCODE (ENCyclopedia Of DNA Elements) Project. Science 306:636–640
14. ENCODE Project Consortium (2007) Identification and analysis of functional elements in 1% of the human genome by the ENCODE pilot project. Nature 447:799–816
15. Finn RD, Tate J, Mistry J et al (2008) The Pfam protein families database. Nucleic Acids Res 36:D281–D288
16. Giardine B, Riemer C, Hardison RC et al (2005) Galaxy: a platform for interactive large-scale genome analysis. Genome Res 15(10):1451–1455
17. Gross SS, Brent MR (2006) Using multiple alignments to improve gene prediction. J Comput Biol 13:379–393
18. Guigó R, Knudsen S, Drake N et al (1992) Prediction of gene structure. J Mol Biol 226:141–157
19. Hinrichs AS, Karolchik D, Baertsch R (2006) The UCSC Genome Browser Database: update 2006. Nucleic Acids Res 34(Database issue):D590–D598
20. Hsu F, Kent WJ, Clawson H et al (2006) The UCSC Known Genes. Bioinformatics 22:1036–1046
21. Hsu F, Pringle TH, Kuhn RM et al (2005) The UCSC Proteome Browser. Nucleic Acids Res 33(Database issue):D454–D458
22. Hubbard TJP, Aken BL, Ayling S et al (2009) Ensembl 2009. Nucleic Acids Res 37:D690–D697
23. Karolchik D, Baertsch R, Diekhans M et al (2003) The UCSC Genome Browser Database. Nucleic Acids Res 31:51–54
24. Karolchik D, Hinrichs AS, Furey TS et al (2004) The UCSC Table Browser data retrieval tool. Nucleic Acids Res 32(Database issue):D493–D496
25. Karolchik D, Kuhn RM, Baertsch R et al (2008) The UCSC Genome Browser Database: 2008 update. Nucleic Acids Res 36:D773–D779
26. Kelley S (2000) Getting started with Acedb. Brief Bioinform 1:131–137
27. Kent WJ (2002) BLAT–the BLAST-like alignment tool. Genome Res 12:656–664
28. Kent WJ, Baertsch R, Hinrichs A et al (2003) Evolution's cauldron: duplication, deletion, and rearrangement in the mouse and human genomes. Proc Natl Acad Sci USA 100:11484–11489
29. Kent WJ, Hsu F, Karolchik D et al (2005) Exploring relationships and mining data with the UCSC gene sorter. Genome Res 15:737–741
30. Kent WJ, Sugnet CW, Furey TS et al (2002) The human genome browser at UCSC. Genome Res 12:996–1006
31. Kent WJ, Zahler AM (2000) Conservation, regulation, synteny, and introns in a large-scale *C. briggsae-C. elegans* genomic alignment. Genome Res 10:1115–1125
32. Kent WJ, Zahler AM (2000) The intronerator: exploring introns and alternative splicing in *Caenorhabditis elegans*. Nucleic Acids Res 28:91–93
33. Kuhn RM, Karolchik D, Zweig AS (2007) The UCSC Genome Browser Database: update 2007. Nucleic Acids Res 35(Database issue):D668–D673
34. Kuhn RM, Karolchik D, Zweig AS et al (2009) The UCSC Genome Browser Database: update 2009. Nucleic Acids Res 37:D755–D761
35. Lee Y, Tsai J, Sunkara S et al (2005) The TIGR Gene Indices: clustering and assembling EST and known genes and integration with eukaryotic genomes. Nucleic Acids Res 33:D71–D74
36. Maglott D, Ostell J, Pruitt KD et al (2007) Entrez Gene: gene-centered information at NCBI. Nucleic Acids Res 35:D26–D31
37. Mckusick VA (2007) Mendeliar Inheritance in man and its online version, OMIM. AM Genet 80:588–604
38. Mulder NJ, Apweiler R, Attwood TK (2007) New developments in the InterPro database. Nucleic Acids Res 35 (Database issue):D224–D228
39. Nash R, Weng S, Hitz B et al (2007) Expanded protein information at SGD: new pages and proteome browser. Nucleic Acids Res 35:D468–D471
40. Parkinson H, Kapushesky M, Kolesnikov N et al (2009) ArrayExpress update–from an archive of functional genomics experiments to the atlas of gene expression. Nucleic Acids Res 37:D868–D872
41. Pedersen JS, Bejerano G, Siepel A et al (2006) Identification and classification of conserved RNA secondary structures in the human genome. PLoS Comput Biol e33:251–262
42. Pruitt KD, Harrow J, Harte RA et al (2009) The consensus coding sequence (CCDS) project: identifying a common protein-coding gene set for the human and mouse genomes. Genome Res 19:1316–1323
43. Pruitt KD, Tatusova T, Klimke W et al (2009) NCBI Reference Sequences: current status, policy and new initiatives. Nucleic Acids Res 37:D32–D36

44. Rogers A, Antoshechkin I, Bieri T et al (2008) WormBase 2007. Nucleic Acids Res 36:D612–D617
45. Sayers EW, Barrett T, Benson DA et al (2009) Database resources of the National Center for Biotechnology Information. Nucleic Acids Res 37:D5–D15
46. Schwartz S, Kent WJ, Smit A et al (2003) Human-mouse alignments with BLASTZ. Genome Res 13:103–107
47. Sherry S, Ward M-H, Kholodov M, Baker J, Phan L, Smigielski EM, Sirotkin K (2001) dbSNP: the NCBI database of genetic variation. Nucleic Acids Res 29:308–311
48. Siepel A, Bejerano G, Pedersen JS et al (2005) Evolutionarily conserved elements in vertebrate, insect, worm, and yeast genomes. Genome Res 15:1034–1050
49. Siepel A, Haussler D (2004) Computational identification of evolutionarily conserved exons. In: Proceedings of the eighth annual international conference on Research in computational molecular biology, 177–186
50. Stein L, Sternberg P, Durbin R et al (2001) WormBase: network access to the genome and biology of *Caenorhabditis elegans*. Nucleic Acids Res 29:82–86
51. Su AI, Cooke MP, Ching KA et al (2002) Large-scale analysis of the human and mouse transcriptomes. Proc Natl Acad Sci USA 99:4465–4470
52. Su AI, Wiltshire T, Batalov S et al (2004) A gene atlas of the mouse and human protein-encoding transcriptomes. Proc Natl Acad Sci 101:6062–6067
53. The C. elegans Sequencing Consortium (1998) Genome sequence of the nematode *C. elegans*: a platform for investigating biology. Science 282:2012–2018
54. The Gene Ontology Consortium (2008) The Gene Ontology project in 2008. Nucleic Acids Res 36:D440–D444
55. The MGC Project Team (2004) The status, quality, and expansion of the NIH full-length cDNA project: the Mammalian Gene Collection (MGC). Genome Res 14: 2121–2127
56. The UniProt Consortium (2009) The Universal Protein Resource (UniProt) 2009. Nucleic Acids Res 37:D169–D174
57. Thierry-Mieg D, Thierry-Mieg J (2006) AceView: a comprehensive cDNA-supported gene and transcripts annotation. Genome Biol 7:S12
58. Thomas DJ, Rosenbloom KR, Clawson H (2007) The ENCODE Project at UC Santa Cruz. Nucleic Acids Res 35(Database issue):D663–D667
59. Thomas DJ, Trumbower H, Kern AD (2007) Variation resources at UC, Santa Cruz. Nucleic Acids Res 35(Database Issue):D716–D720
60. Tweedie S, Ashburner M, Falls K et al (2009) FlyBase: enhancing Drosophila Gene Ontology annotations. Nucleic Acids Res 37:D555–D559
61. Wilming LG, Gilbert JGR, Howe K et al (2008) The vertebrate genome annotation (Vega) database. Nucleic Acids Res 36:D753–D760

Ensembl Genome Browser

29.2

Xosé M. Fernández and Ewan Birney

Abstract Recent years have seen the release of huge amounts of sequence data from genome sequencing centers. However, this raw sequence data is most valuable to the laboratory biologist when provided along with quality annotation of the genomic sequence.

Ensembl provides access to genomic information with a number of visualization tools, becoming one of the world's primary resources for genomic research, a resource through which scientists can access the human genome as well as the genomes of other model organisms. Thus, researchers can download data directly, whether it is the DNA sequence of a genomic contig, or positions of SNPs in a given gene. The key Ensembl web pages are highlighted in this chapter.

Because of the complexity of the genome and the many different ways in which scientists want to use it, Ensembl provides many levels of access with a high degree of flexibility. Through the Ensembl website a wet-lab researcher with a simple web browser can for example perform BLAST searches against the assembly of a genome, download a genomic sequence, or search for all members of a determined protein family. But Ensembl is also an all-round software and database system that can be installed locally to serve the needs of a genomic center or a bioinformatics division in a pharmaceutical company, enabling complex data mining of the genome or large-scale sequence annotation.

Contents

29.2.1	Genomes Galore	924
	29.2.1.1 Genomes in Context	924
	29.2.1.2 ENCODE: Shifting the Paradigm	926
29.2.2	Ensembl Annotation	926
29.2.3	Region in Detail: Introduction	927
	29.2.3.1 Region in Detail: Features	927
	29.2.3.2 Region in Detail: Repeats, Decorations, and Export	927
29.2.4	DAS Sources	927
29.2.5	Comparative Genomics	929
29.2.6	GeneView	930
29.2.7	Variation	932
	29.2.7.1 Variation Image	932
	29.2.7.2 Comparison Image	932
29.2.8	Ensembl: An Example	932
29.2.9	BioMart Overview	933
29.2.10	Customizing Ensembl	934
	29.2.10.1 Displaying User Data on Ensembl	935
29.2.11	Archive	935
29.2.12	*Pre!* Site	935
29.2.13	Further Information	935
29.2.14	Outlook	935
References		936

X.M. Fernández (✉) and E. Birney
EMBL-European Bioinformatics Institute, Wellcome Trust Genome Campus, Hinxton, Cambridge, CB10 1SD, UK
e-mail: xose@ebi.ac.uk

29.2.1 Genomes Galore

The number of species sequenced is continuously increasing[1] owing to the efforts of genome-sequencing centers worldwide. What used to be a trickle is now a flood, as more and more data is obtained with new sequencing technologies [7, 12, 13, 76, 80, 99, 100]. These high-throughput whole-genome shotgun sequencing technologies deliver genomes several orders of magnitude faster than "traditional" electrophoretic methods, but we should remember that any raw sequence is only valuable to most scientists when provided along with quality annotation of the underlying genomic sequence [8].

Nowadays, the generally accepted "gold standard" for annotation of eukaryotic genomes is that made by a human being [5]. Manual annotation is based on information derived from sequence homology searches and the results of various *ab initio* gene prediction methods [14, 15], and also from literature reviews. Annotation of large genomes, such as mouse and human, in this way is slow and labor intensive, taking large teams of annotators years to complete. As a result, the annotation can almost never be entirely up to date [73] and free of inconsistencies (the annotation process usually begins before the sequencing process is complete). Hence, an automated annotation system is desirable, since it provides a relatively rapid way of delivering annotation at genome level and can be updated as new data becomes available from newer (and more complete) assemblies. To address this need, the Ensembl annotation system has been developed by observing how annotators build gene structures and condensing this process into an algorithm that codes this set of rules.

Ensembl provides an automatically annotated gene set alongside a graphical web-based interface for visualization of genomes [39]. Ensembl gene models are based on experimental evidence [9], which is imported from the manually curated protein set provided by UniProt/Swiss-Prot [117], the manually curated NCBI RefSeq [91], and the automatically annotated UniProt/TrEMBL protein records. UTR sequences are annotated to the extent supported by EMBL [67, 112] mRNA records.

29.2.1.1 Genomes in Context

Looking back in time to before Ensembl was available, we can trace the need for such an approach. After the sequencing of the first microbial genome, that of bacteriophage ΦX174 [104], many more genome sequences followed. The first bacterial genome was fully sequenced in 1995, when the sequence of the *Haemophilus influenzae* strain *Rd* was completed [38]; since then hundreds of genomes have been sequenced.

Moving to the human arena, the first human gene was cloned in 1977, but we had to wait 22 years before the first human chromosome was completely sequenced [32]. The pace accelerated [22, 26, 27, 31, 51, 52, 58, 82, 101] toward the ultimate goal: completion of the sequencing of the entire human genome within the Human Genome Project (HGP). When we refer to the HGP we are talking about the international effort that formally started in October 1990 and which delivered the draft sequence of the human genome in 2003 [59, 60]. In addition to determining the complete sequence of the 3 billion DNA base pairs in the human genome, the HGP carried out a number of parallel, preliminary studies, mainly on *Escherichia coli* [11] and mouse [81], in order to develop new approaches to optimize the tools used to handle the amount of information generated for human. This process helped to extend our understanding of human gene function (for a more detailed account see [21]).

The following landmarks highlight the effort undertaken: by February 2001, when the draft human sequence was published, the yeast *Saccharomyces cerevisiae* [46], the microscopic soil worm

---

[1] At the time of writing, the following completed eukaryotic genomes were deposited at EMBL: *Anopheles gambiae, Arabidopsis thaliana, Ashbya gossypii, Aspergillus fumigatus, Aspergillus niger, Bos taurus, Caenorhabditis briggsae, Caenorhabditis elegans, Candida albicans, Candida glabrata, Canis familiaris, Cryptococcus neoformans, Cryptosporidium parvum, Cyanidioschyzon merolae, Danio rerio, Debaryomyces hansenii, Dictyostelium discoideum, Drosophila melanogaster, Drosophila pseudoobscura, Drosophila simulans, Drosophila yakuba, Encephalitozoon cuniculi, Equus caballus, Gallus gallus, Guillardia theta, Hemiselmis andersenii, Homo sapiens, Kluyveromyces lactis, Leishmania braziliensis, Leishmania infantum, Leishmania major, Macaca mulatta, Monodelphis domestica, Mus musculus, Ornithorhynchus anatinus, Oryza sativa, Oryzias latipes, Ostreococcus lucimarinus, Ostreococcus tauri, Pan troglodytes, Paramecium tetraurelia, Pichia stipitis, Plasmodium falciparum, Rattus norvegicus, Saccharomyces cerevisiae, Schizosaccharomyces pombe, Spizellomyces punctatus, Toxoplasma gondii, Trypanosoma brucei, Trypanosoma cruzi, Vitis vinifera,* and *Yarrowia lipolytica.*

29.2 Ensembl Genome Browser

Caenorhabditis elegans [115], and the fruit fly *Drosophila melanogaster* [1, 85] genomes were finished. Drafts were available for the plant *Arabidopsis thaliana* [4] and for one of the chromosomes of the malaria parasite *Plasmodium falciparum* [40]. In addition, complete sequences had been obtained for over 30 microorganisms (including the bacteria *E. coli* [11] and *Mycobacterium tuberculosis* [20]; Table 29.2.1). Work was progressing on many other genomes, some mammalian species (chimpanzee [116], opossum [78], rat [43], mouse [81], rhesus macaque [95]), chicken [53], *Danio rerio* [108], *Takifugu rubripes* [3], *Tetraodon nigroviridis* [98], *Caenorhabditis briggsae* [110], and *Drosophila pseudoobscura* [96], but also plants such as rice [45, 127], amongst others (see Table 29.2.2).

Both human and mouse genomes are assembled by The Genome Reference Consortium (The Wellcome Trust Sanger Institute, The Genome Center at Washington University and The National Centre for Biotechnology Information). The same genomic assembly can be browsed using different genome browsers: UCSC Genome Browser [68] (featured in Chap. 29.1), NCBI Map Viewer [124], and Ensembl [39]. Since different algorithms are used to annotate genomes this can lead to different gene sets, a less than ideal scenario. This is why the Consensus CDS (CCDS) project has been established; members include groups from the European Bioinformatics Institute (EBI), NCBI, Wellcome Trust Sanger Institute, and University of California, Santa Cruz (UCSC), all working toward a unique set of human and mouse protein coding regions consistently annotated by different projects.

Table 29.2.1 A selection of genomes with complete or draft sequence published

Species	Size (Mb)	Divergence (My)[a]	Gene count
Homo sapiens	3,272	–	~25,000
Pan troglodytes (chimpanzee)	2,734	6–14	~25,000
Mus musculus (mouse)	2,932	75	~28,000
Oryctolagus cuniculus (rabbit)	3,400	40[b]	–
Cavia porcellus (guinea pig)	3,400	51[c]	–
Loxodonta africana (African elephant)	3,000	60[d]	–
Echinops telfairi (tenrec)	3,000	70[d]	–
Felis catus (domestic cat)	3,000	75[d]	–
Sorex araneus (common shrew)	3,000	75[d]	–
Erinaceus europeaus (hedgehog)	3,600	80[d]	–
Dasypus novemcinctus (armadillo)	3,000	95[d]	–
Caenorhabditis elegans (soil roundworm)	100	97[e]	21,249
Monodelphis domestica (opossum)	1,571	173 [69]	~21,000
Takifugu rubripes (tiger pufferfish)	329	450[f] [70]	~22,000
Tetraodon nigroviridis (green spotted pufferfish)	402	450[e] [70]	~28,000
Danio rerio (zebrafish)	1,600	360[e] [70]	~25,000
Drosophila melanogaster (fruit fly)	180[g]	250[h] [41]	14,752
Anopheles gambiae PEST	278	250[g] [41]	~13,000
Arabidopsis thaliana (mouse-ear cress)	120	150[i] [17]	~26,000
Oryza sativa (rice)	450	150[h] [17]	~43,000
Saccharomyces cerevisiae (yeast)	12.1	–	7,122
Escherichia coli K12	4.64	–	4,289 [11]
Mycobacterium tuberculosis H37Rv	4.41	–	3,918 [20]
Haemophylus influenzae KW20 Rd	1.83	–	1,737 [38]
ΦX174	5,386 bp	–	12 [104]

[a]Estimated divergence from the last common ancestor (compared with human, unless specified differently)
[b]Although there is some debate in the phylogeny of rodents, the placement of lagomorphs is uncontroversial [57, 83]
[c]Rodent phylogeny still debated, *Caviomorpha* placed according to [57]
[d]Phylogeny based on molecular data [83, 109]
[e]Estimated divergence time between *C. elegans* and *C. briggsae* [18]
[f]Divergence between zebrafish and pufferfish lineages 280 Mya. *Tetraodon* and *Takifugu* are 20–30 Mya apart [98]
[g]*Drosophila*'s genome is made up of 120 Mb euchromatic plus 60 Mb heterochromatic DNA
[h]Divergence between *Drosophila* and *Anopheles* lineages
[i]Divergence between *Arabidopsis* and rice lineages

Table 29.2.2 A selection of genomes currently being sequenced

Species	Size (Mb)	Divergence (Mya)[a]
Homo sapiens neanderthalensis (Neanderthal)	3,000	0.3[b]
Gorilla gorilla (gorilla)	4,000	7[b]
Pongo pygmaeus albelii (orang-utan)	3,000	14[c]
Callithrix jacchus (marmoset)	3,000	35
Tachyglossus aculeatus (echidna)	3,500	180[d]
Macropus eugenii (wallaby)	3,600	80 [83]
Petromyzon marinus (sea lamprey)	2,070	460[e]
Saccoglossus kowalevskii (acorn worm)	1,100	545

[a]Phylogenies are supported either by morphology or molecular data when available
[b]Based on preliminary analysis of 62,250 bp [88]
[c]Estimation based on morphological [47], fossil [44] and molecular data [111]
[d]Phylogeny based on recent molecular, morphological and fossil data [84]
[e]Suggested by molecular maximum likelihood trees [114]

29.2.1.2 ENCODE: Shifting the Paradigm

The ENCyclopedia Of DNA Elements (ENCODE) [34] is an international consortium organized by the National Human Genome Research Institute (NHGRI). NHGRI is an institute within the National Institutes of Health (NIH) that has approached the exploratory study of 1% of the human genome using different techniques in order to assess which ones will be best extended to the rest of the genome later. Initial findings [35] support a reformulation of the traditional view of the genome as a collection of independent genes separated by "junk" DNA [42]. Unexpectedly, it appears that most human DNA is transcribed. Many mRNA transcripts bridge so-called noncoding regions and join established protein-coding genes, and these transcripts are extensively overlapping [118].

29.2.2 Ensembl Annotation

Ensembl has developed a complex algorithm that delivers automatically annotated gene sets [25, 36] based on biological evidence. Ensembl uses biological evidence from various protein sequence and mRNA sequence sources (e.g., UniProt and RefSeq) to deliver transcript models placed on the genome assembly. In addition to protein coding genes, Ensembl also annotates noncoding RNA genes (ncRNAs include siRNA, miRNA, tRNA); in this case the primary source of information is Rfam [48].

Ensembl has developed a comparative approach to annotate these coverage genomes (e.g., bushbaby), where the alignment of complete proteins and cDNAs would not deliver a complete gene set. The approach is based on aligning whole genomes [106] using a reference genome (i.e., projecting a dataset from a well-annotated species, such as human, onto the genome of a less well-understood species, such as bushbaby). In this way, Ensembl can overcome fragmentation, gaps, missing sequence and misassemblies in the underlying genomic assembly.

The comparative method also identifies regions of sequence similarity across species that have persisted through evolution (at least for the last 100 My). These conserved sequences are indicative of some function: we could include here genes and smaller regulatory elements [77] (such as promoters, enhancers, transcription factor-binding sites, and other key players in the determination of activation of genes and pathways in the cell).

At the time of writing, Ensembl incorporates over 50 genomes (mostly vertebrates) into its databases and browser. At the time of its conception, only the human genome, the first vertebrate sequence fully determined, was displayed. Now, using comparative methods and the draft sequences currently available[2] we can estimate gene number (see Table 29.2.1) and locate regulatory elements even of species for which we only have a limited (i.e., low-coverage) assembly.

D. melanogaster, *C. elegans*, and *S. cerevisiae* are manually annotated by FlyBase [30], WormBase [18], and SGD [19], respectively. Ensembl imports these

---
[2]At the time of writing, Ensembl incorporates annotation for human, chimpanzee, orang-utan, macaque, marmoset, gorilla, mouse, rat, dog, cow, and chicken; opossum, platypus, and the frog *Xenopus tropicalis*; fishes such as zebrafish, *Takifugu rubripes*, *Tetraodon nigroviridis*, stickleback, and *Medaka*; *Drosophila melanogaster*, *Aedes* and *Anopheles* mosquitoes; the tunicates *Ciona intestinalis* and *C. savignyi*; *Caenorhabditis elegans* and the yeast *Saccharomyces cerevisiae*. Several organisms were sequenced with low-coverage assemblies: alpaca, armadillo, bushbaby, cat, dolphin, elephant, guinea pig, hedgehog, horse, hyrax, kangaroo rat, lesser hedgehog tenrec, megabat, microbat, mouse lemur, pika, rabbit, shrew, squirrel, tarsier, tree shrew.

datasets to incorporate them into the comparative analysis, but no gene build is run for these genomes.

29.2.3 Region in Detail: Introduction

In order to visualize features on a genome, Ensembl has condensed the annotated assembly into a customizable view. The "Region in detail" page provides a graphical representation of a region, where different panels allow visualization of varying windows of the chromosomal region at different resolutions (Fig. 29.2.1). Gene predictions (e.g., Ensembl genes, EST genes, *ab initio* models) and annotations, such as variations (e.g., SNPs), are displayed along clones, microarray probesets and other genomic features.

Features displayed above the DNA genome assembly (blue bar) are in the forward strand, while features below the DNA are on the reverse strand. Nonstranded features are shown at the bottom of the panel (e.g., SNPs). In Fig. 29.2.1, IL12B_HUMAN (ENSG00000113302) is on the reverse strand [as the arrowhead before the gene ID ("<") indicates], while ENSG0000022160 (a snRNA in the ncRNA track) is on the forward strand. The chromosome in this region is also covered by contig AC011418 (the CTB-9P16 clone from the human tilepath covers this region).

In order to navigate in this view, the zooming ladder allows you to focus on a particular region or to expand the field of view. There is also the possibility of rubberbanding ("click and drag" your mouse) around a region in the "Chromosome" or "Top panel" panel to change the view.

Region in detail allows you to see up to 1 Mb; if you want to zoom out and see a wider region, use Region overview. Syntenic blocks (conserved gene order across species) are available; however, some higher resolution annotations, such as SNPs, are not available in this page. (From the "Configure this page" window go to the "Top panel" tab and select the species for which you want to compare conserved gene order.)

29.2.3.1 Region in Detail: Features

Ensembl provides gene sets (Ensembl genes, Ensembl EST gene models based on EST evidence, along with imports such as tRNA and miRNA genes, manually annotated genes from VEGA, and *ab initio* gene models such as GENSCAN), but is not limited to gene models. Several microarray probe sets (*GeneChip*® from *Affymetrix*, *BeadChip*™ from *Illumina*, *Agilent* and *BioArrays* from *CodeLink*™) are mapped to the assembly along with regulatory elements (CpG islands, models from predictive programs such as Eponine [29] and FirstEF [93], and elements in the *CisRED* [97] database), variations (SNPs, insertions, deletions) and other sequences (from NCBI RefSeq, UniGene [124], and other databases of ESTs, mRNAs, and/or proteins). These annotations can be displayed using the options available in the "Configure this page" menu ("Main panel") of "Region in detail."

29.2.3.2 Region in Detail: Repeats, Decorations, and Export

"Region in detail" is highly customizable: there are over 100 tracks available. Images can be exported in different editable formats (e.g., PDF, SVG). Context menus yield more information about individual features and tracks.

Repeat sequences are also available (as *Repeats* under the Main panel tab on the Configuration page). Genome size does not correlate with organism complexity: our genome is 200 times as large as that of *Saccharomyces cerevisiae*, but that of *Amoeba dubia* is 200 times the size of ours, largely because of different quantities of repetitive sequences. In human we find that the genome is 50% repeat sequences and only 5% coding sequence. In "Region in detail," retrotransposons such as long terminal repeats (LTR), long interspersed elements (LINE), short interspersed elements (SINE, the most thoroughly characterized as belonging to the Alu family [55]), tandem repeats, and other repeats are displayed.

29.2.4 DAS Sources

Ensembl can enrich its annotation by displaying third-party data external to Ensembl databases by means of the Distributed Annotation System (DAS) [28]. DAS is a communication protocol that allows the exchange

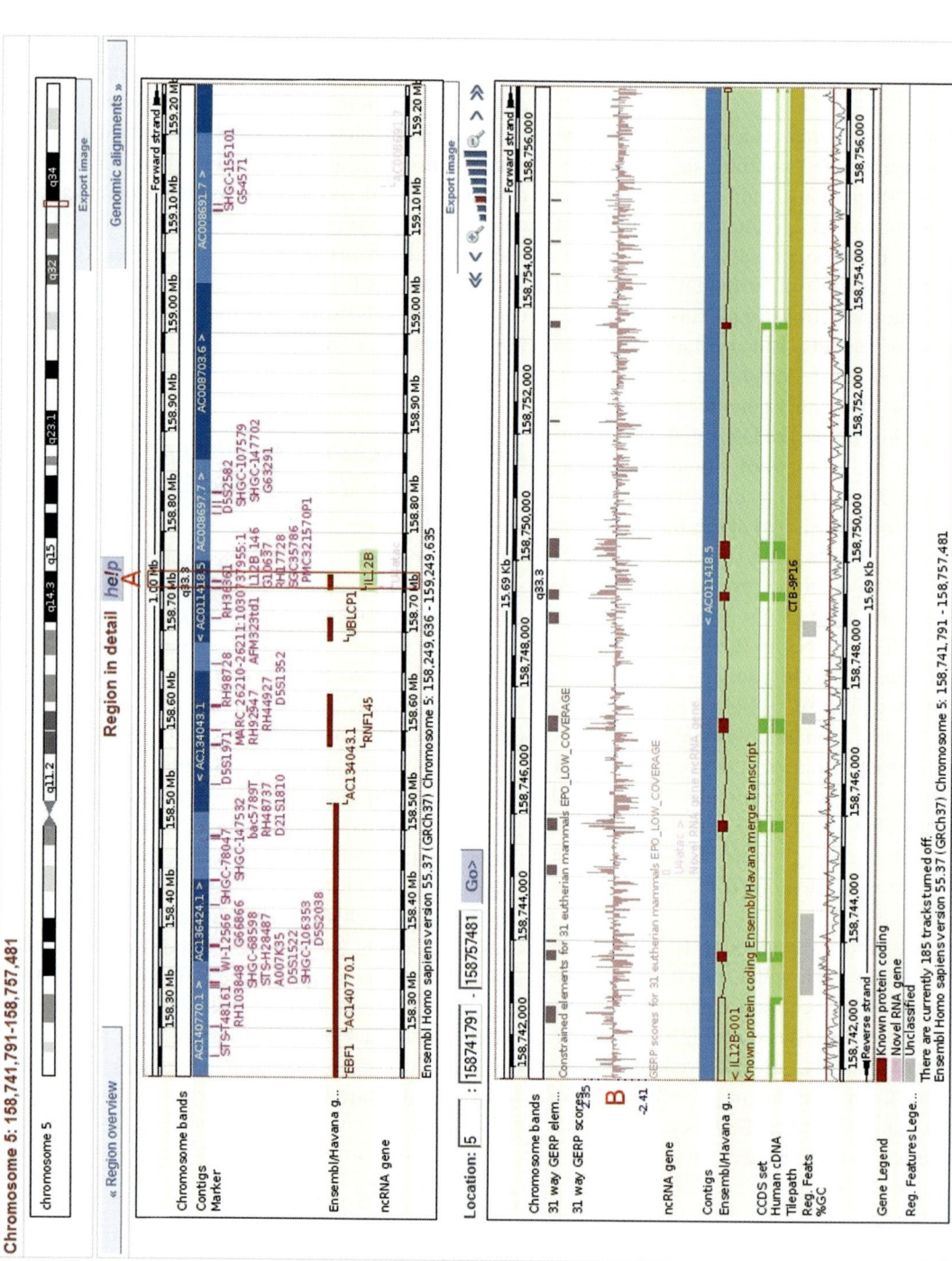

Fig. 29.2.1 'Region in detail' is structured in *three panels* representing different zoom levels of the genome. An ideogram of chromosome 5 appears at the *top*, with chromosomal bands labeled (5q33.3). The following panel (*top panel*) shows a region of 1 Mb corresponding to the *red box* shown on the chromosome (labeled A). Gene models and markers are displayed. A *scale bar* depicts the physical map coordinates for the region, just above the contig track where the individual contig sequences that form the genome sequence assembly are shown, in different shades of blue. The *third panel* (*main panel*) is very rich in features and can be customized (see text)

of biological sequence annotation, removing the need to store it in a single database; in this way resources can be spread over multiple servers in different geographic locations. Furthermore, it provides a way to display data across different databases, enhancing ease of access for the user.

Ensembl acts as a DAS client, gathering and integrating sequence annotation from multiple remote servers and displaying it in "Region in detail" as additional tracks. Note that shortly after a new genome assembly is released some DAS sources might not be available, as it can take some time to map their information to the new reference (Ensembl is not responsible for these remote servers).

The DAS registry (www.dasregistry.org) is a directory of DAS services available. It provides statistics for every source and keeps track of every service (contacting the owner in case the service was down).

Examples of DAS sources are clones (CpG island clones), alternative annotations (NCBI Gnomon, IMGT Genes), new features (CNV, CAGE), data from research projects (DECIPHER), BAC libraries, etc. The DAS sources available vary with species and are accessible for viewing in the configuration panel.

29.2.5 Comparative Genomics

The study of evolution has moved to the genomic era with the availability of whole-genome sequences from several species [102, 119]. Nowadays, we can track how the pieces of the genomic jigsaw have been shuffled during evolution, outlining the importance of gene order conservation, going beyond individual gene and protein comparison across species.

Function is often inferred from coding sequences, but it is not well understood how to decipher functional properties from noncoding sequences. Hence, there is a strong bias toward the analysis and description of events in coding regions [125], such as gene duplications and protein sequence evolution, while noncoding, regulatory sequences often go unnoticed. Since most of the genome is noncoding we are reaching the limit of what can be expected from evolutionary genetics on the sole basis of coding sequences [121]. We should move forward, taking full advantage of the range of organisms sequenced, understanding regulatory sequences as well as the coding sequences.

Cross-species comparison of closely related genomes seems the obvious way to identify functional elements in the human genome (both coding and noncoding) [90]. Evolution has conserved many of the DNA sequences behind coding genes or in the elements that regulate gene expression, making comparisons of genome sequences between species an effective and efficient means of finding new genes and functional sequences [72]. A number of eutherian mammals have been sequenced (Tables 29.2.1 and 29.2.2), providing us with a blueprint or characteristic gene set for the mammalian clade. The next step is to model the effect of mutation on functional elements of the genome, before investigating noncoding regions in order to understand gene expression control [90].

Conserved regions are identified using whole-genome comparisons on the nucleic acid level between two species (e.g., human/mouse) determined with BLASTZ [106]. When comparing more distant species, the translated BLAT [64] algorithm is used to compare genomes at the amino acid level; thus, there would be a bias toward protein-coding regions, although conserved noncoding regions are also detected.

Global multiple alignments are calculated with the Pecan algorithm (B. Paten, unpublished work) and can be displayed in "Region in detail" ("Multiple alignments" in the main panel tab). Conservation scores are calculated using the multiple alignments, and "constrained elements" are determined using GERP [23]. These can be displayed in "Region in detail" ("Conservation" track, labeled B in Fig. 29.2.1).

Ensembl also identifies large regions of conserved sequence (synteny), as previously mentioned in Sect. 29.2.3. These syntenic regions are determined using BLASTZ-net analysis, and they can be viewed using the configuration panel ("Synteny") in "Region overview." Syntenic blocks are also available in the Synteny page (Location | Synteny); conserved regions are displayed by means of a diagram of chromosomes, along with a table of homologous genes between two species for a region.

Gene Trees. Ensembl performs multiple transcript comparisons to identify genes that are paralogous (homology follows a gene duplication; i.e., two genes are descended from one ancestral gene and may have evolved to have different functions) and orthologous (functional equivalent genes diverged in two genomes further to a speciation event). Lack of selection on paralogous genes can result in pseudogenes, i.e., one copy loses its promoter and is not transcribed, so that it can mutate over time with few negative consequences. The transcript set in Ensembl is scanned for signs of processed pseudogenes (genes that have been reverse-transcribed from mRNA and reintegrated at random into the genome) and annotated as such. However, pseudogenes resulting from gene duplications in the genome (complete with introns) are much harder to annotate.

Orthology and paralogy can help us understand genome rearrangements and the evolution of function. These relationships are displayed using phylogenetic gene trees, as shown in Fig. 29.2.2. Maximum-likelihood phylogenetic trees are generated with TreeBeST (Li Heng et al., unpublished work) in order to provide a representation of the evolutionary history of gene families. Duplication and speciation nodes are inferred from the reconciliation with their species tree. The Gene Tree page (Fig. 29.2.2) can be accessed from specific gene pages using the link at the left.

29.2.6 GeneView

Once a gene has been selected, the Gene tab shows annotation focused on a gene (Fig. 29.2.3). At the top of the view there is a summary for the entry with links to external databases (HGNC, CCDS) and to other pages within Ensembl (e.g., "Region in detail" shown in Fig. 29.2.1).

There are several shortcuts to more pages:

- *Splice Variants.* All Ensembl transcripts for the gene. Protein domains and motifs are aligned in the graphical display.
- *Supporting Evidence.* Where the initial proteins and mRNAs aligned to the genome in the Ensembl annotation are shown.
- *Sequence.* Features in the gene structure such as exons or SNPs can be highlighted using the options available in the "Configure this page" menu.
- *External References.* Where external records, or identifiers and information in databases external to Ensembl, are specifically attached to the Ensembl gene.
- *Regulation.* Species-specific regulatory feature information if available. In the case of human, this includes predicted promoters and enhancers from the Ensembl Regulatory Build, which represent a single best guess of regulatory elements and annotations based on statistically significant association with genomics features.
- *Comparative Genomics.* A number of options are available within this category: Gene Trees, orthologues, paralogues, and protein families.
- *Genetic Variation.* Provides information about SNPs and genetic variation in general.
- *ID History.* Ensembl keeps its identifiers (for genes, transcripts, and proteins) stable throughout releases, and can therefore be tracked in case an identifier was retired and a new one assigned (or two identifiers may be merged). This page provides an *ID History Map* showing the release number on the *x*-axis and stable IDs on the *y*-axis. Small squares or nodes correspond to the ID shown on the left and represent an update in the version of the ID. Versions are updated if there has been a change in the gene, transcript, or protein model. Nodes (squares) are connected by a line if the versions are related. This line reflects the score of how well the versions match, for recent releases. If a score is not calculated, the line will be gray (unknown score).

There is a cartoon depicting the structure of the gene (in this case on the reverse strand, drawn under the assembly; this means that this gene's 5′ end is on the right and its 3′ end, on the left; Fig. 29.2.3). In this particular instance Ensembl only annotates one unique transcript; if there were alternative splicing forms, they would be seen here.

UTRs are displayed as unfilled boxes, and coding regions as filled boxes. Manually annotated transcripts (VEGA) are displayed in blue, while Ensembl transcripts are displayed in red; EST-based models are displayed in magenta. In Fig. 29.2.3, only the Ensembl-known protein-coding transcript is displayed. For options to view other annotations, such as VEGA transcripts, choose the option in the "Configuration panel."

When a transcript is selected, there will be a "Transcript" tab, which also offers several shortcuts on the left.

29.2 Ensembl Genome Browser

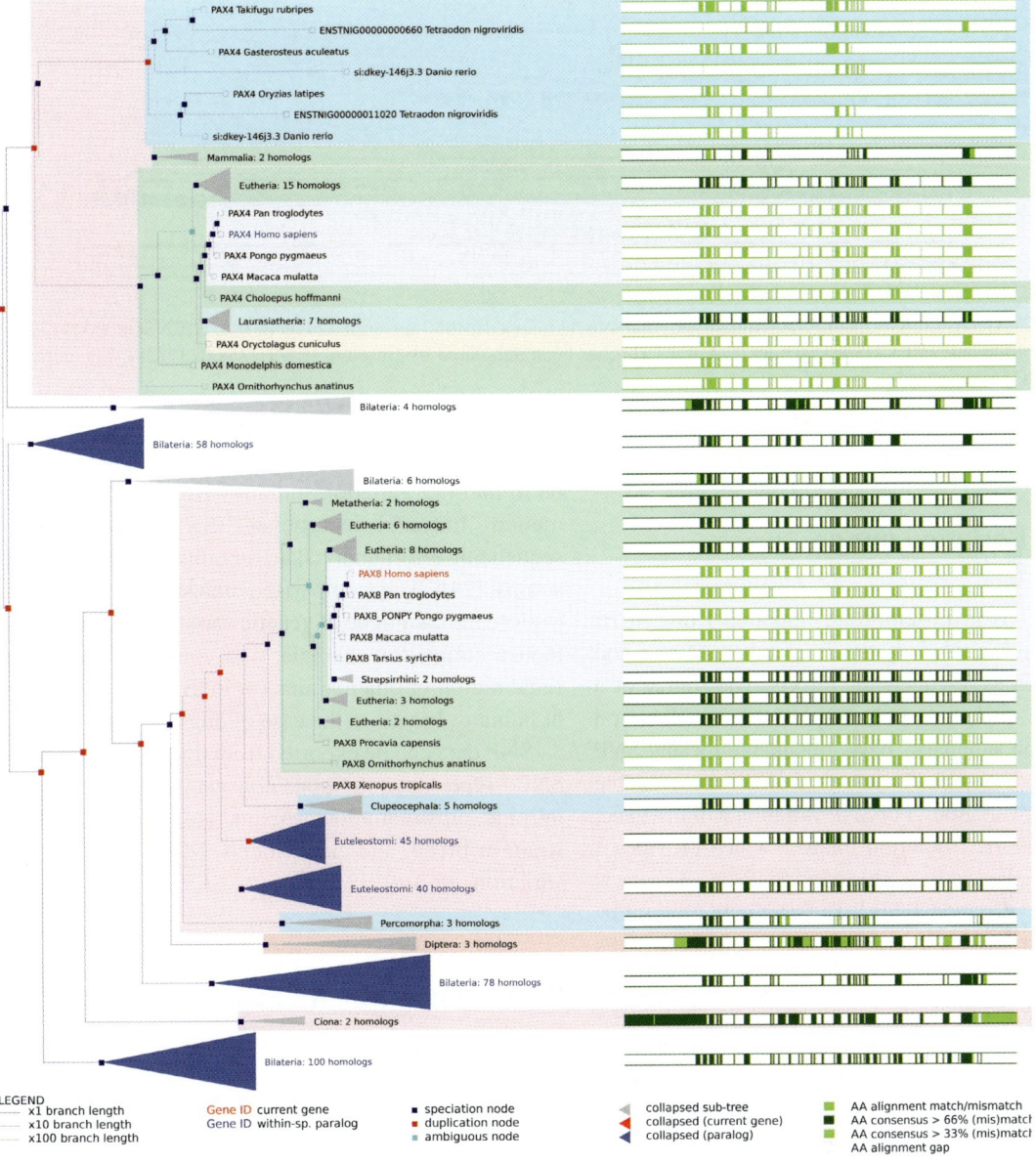

Fig. 29.2.2 Detail from *Gene Trees*, displaying the phylogenetic tree and schematic representation of multiple alignment (in *green*) for the PAX2/PAX5/PAX8 cluster centered on the human PAX8 gene (in *red*). Distinct branches are visible with the fish clade at the top displaying the teleost ancestral duplication [62], and the orthologue gene PAX4 (from chicken to human). Some nodes (associated to orthologues PAX2 and PAX5) appear collapsed (*blue* and *gray* triangles). The *green bars* at the *right* of the tree provide a schematic representation of the multiple alignment of the peptides (*full boxes* indicate matches/mismatches; *open boxes* indicate gaps in the alignment)

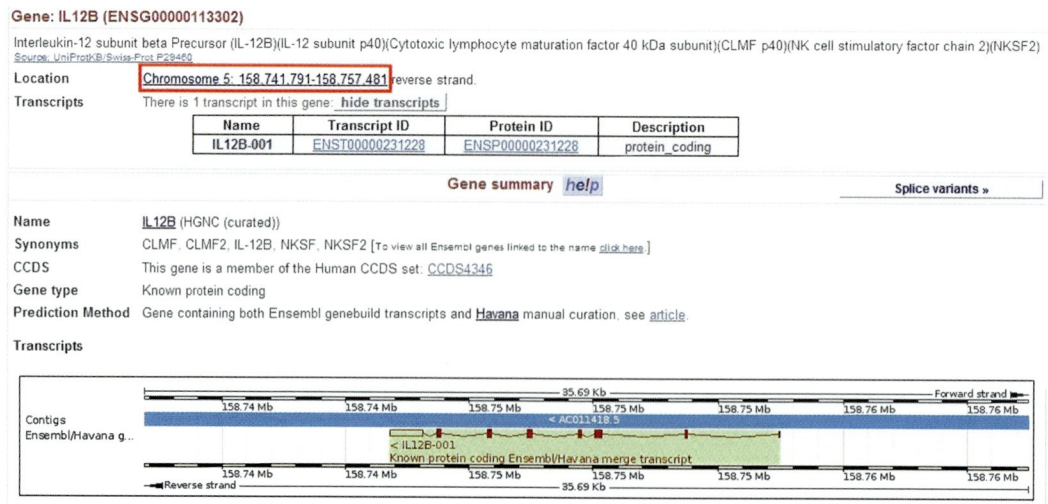

Fig. 29.2.3 The Gene summary incorporates links to external sources mapped to this transcript. This gene IL12B (ENSG00000113302) is a member of the human CCDS set (see text). A transcript has been annotated by Ensembl (ENST00000231228). The chromosomal location is highlighted

29.2.7 Variation

29.2.7.1 Variation Image

Most variations in Ensembl are imported from NCBI dbSNP, though a minority are from other sources and projects, such as resequencing efforts. The source of a SNP (single nucleotide polymorphism) is available on the "Variation summary" page, obtainable for any SNP displayed in Ensembl. A variety of views are available for exploring variations. For SNP variation within a gene, one option is the "Variation image," which provides a graphical display, while a table of all SNPs in one gene is available in "Variation Table." The user may filter information on the basis of the validation method of the SNPs displayed (either by NCBI dbSNP terms: frequency, cluster, or by genotype data from HapMap, etc.). SNPs may also be filtered by class [insertion-deletions (indels), microsatellite repeats, multinucleotide polymorphisms (MNPs) etc.], or by type (intronic, upstream, synonymous or silent mutation, frameshift, etc.).

29.2.7.2 Comparison Image

Comparison image provides a display of the SNPs and variations within the nucleic acid and/or protein sequence in one view. Any variation can then be clicked on to find the Variation summary page for more information. Ensembl incorporates SNPs from the resequencing effort undertaken for multiple inbred mouse strains [24]. The comparison image allows users to explore the catalogue of genetic variants for a particular mouse strain, for example, and compare it with the same region in other strains (or individuals in the case of humans, breeds in the case of dogs).

SNP data can be exported from this page, filtered by SNP type (synonymous/nonsynonymous, coding/intronic, upstream/downstream, etc.), by individual, strain or breed, and class (indels, MNP, heterozygous variation, etc.).

29.2.8 Ensembl: An Example

We will now explore Ensembl, starting with some basic information about the Interleukin-12 beta chain precursor (*IL12B*). This gene encodes a subunit of this cytokine that acts on natural killer cells. Searching for "human gene IL12B" in the text search box found on the home page (www.ensembl.org) takes us to the results page, where we have a link to the Gene summary page for this entry.

On the Gene summary page we find information about orthologues and paralogues across species, whole-genome alignments, splice variants, matches in other databases, and more. For example, in release 55, human IL12B is ENSG00000113302, has no splice variants, and is located on chromosome 5, bp 158,741,791–158,757,481 (as shown in the Location tab). Alignments are shown with other mammals and vertebrates, and it has good correspondence (100% match) to the IL12B_HUMAN entry in UniProtKB/Swiss-Prot. Find these facts on the Gene summary page in the stable archive site:

http://Jul2009.archive.ensembl.org/Homo_sapiens/Gene/Summary?db=core;g=ENSG00000113302

If we follow the link to "Region in detail" from the search results or from the "Location" tab of the Gene summary page (alternatively, click the base pair numbers indicating the location of the gene), we see this gene in its genomic context:

http://Jul2009.archive.ensembl.org/Homo_sapiens/Location/View?db=core;g=ENSG00000113302

At the top of "Region in detail," STS markers (sequence tagged sites) from several databases (thus, there may be synonyms) are displayed along the assembly. They have been mapped to the genome using e-PCR [105], and following any marker link brings up the Marker page, showing the PCR primers that define this unique segment (as well as the expected size of the amplicon).

We can see the IL12B gene is located in chromosome 5 band q33.3, and more specifically, in contig AC011418. If we select clone sets from "Misc regions" menu in the "Configuration panel" we can see clones for this region on this display.

Variations in the form of SNPs can be selected in the "Variation features" menu, but they are only shown when the region displayed is less than 50 kb long. Selectable subsets of these variations are SNPs which have been genotyped, and two GeneChip® Human mapping array sets (100K and 500K) from Affymetrix.

For a more detailed walk-through of the website and information about the pages, please see [37].

29.2.9 BioMart Overview

BioMart is a data "warehouse" originally developed for Ensembl [63], which now has become a joint project between the European Bioinformatics Institute (EBI) and the Ontario Institute for Cancer Research (OiCR), providing a query-oriented data management system to interact with different datasets (Ensembl is just one of many).

The initial query system has now shifted toward a federated approach that has been deployed for several biological databases and therefore provides a gateway to Ensembl from numerous databases, with no need to store those sources locally.

BioMart retrieves information from databases without users having to become familiar with their schema or having any programming expertise. It also provides interactive access to information stored in the Ensembl databases, as well as allowing integration of in-house data (for those ready to set up a local BioMart).

Example Using BioMart. BioMart can be used to convert Entrez Gene IDs [74] to Ensembl identifiers (e.g., ENSG00000113302) and/or RefSeq [91], Entrez Genes or EMBL [67] (GenBank) identifiers.

Select database (Ensembl 51) and dataset (*Homo sapiens* genes), limit your query using "Filters," and in the "Gene" panel enter your ID. Choose "External references" from the "Attributes" in the "External" panel (RefSeq, Entrez Gene ID, and EMBL), and get your "Results" (Fig. 29.2.4).

This tool can also export gene, peptide, UTR (untranslated region), and flanking sequences in FASTA format for a given list of genes, or even all the genes in a genome. Names of clones spanning a particular region can also be quickly obtained.

A powerful capacity of BioMart is the possibility of joining queries from different databases using the same web interface (select the second "Dataset" option in the web interface, Fig. 29.2.4). In this way, for example, information from UniProt can be displayed alongside Ensembl annotation.

In the next example we will retrieve annotation associated to *kinases*. For this purpose we will use the Gene Ontology (GO) [16] term associated with *kinase activity* (GO: 0016301). GO is functional clustering based on a hierarchical vocabulary, providing a description of a gene (molecular function, cellular component, and biological process). From the BioMart Central Server (www.biomart.org) select "Ensembl Genes" and "*Homo sapiens* genes" as your database and dataset, respectively. Amongst the "Attributes" include RefSeq, EMBL (GenBank), and EntrezGene ID. We can filter the results of this query with information retrieved from UniProt by choosing this secondary dataset (Fig. 29.2.5).

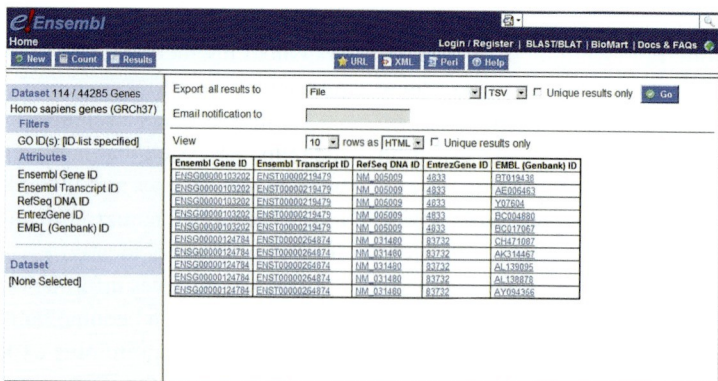

Fig. 29.2.4 The BioMart result preview window showing gene IDs for a specified gene list. For the full table, the view can be changed to 'all' rows, or the file exported using the '*Go*' button. In this example, Ensembl Transcript IDs are mapped to NCBI RefSeq entries, Entrez Gene IDs, and mRNAs in EMBL-Bank

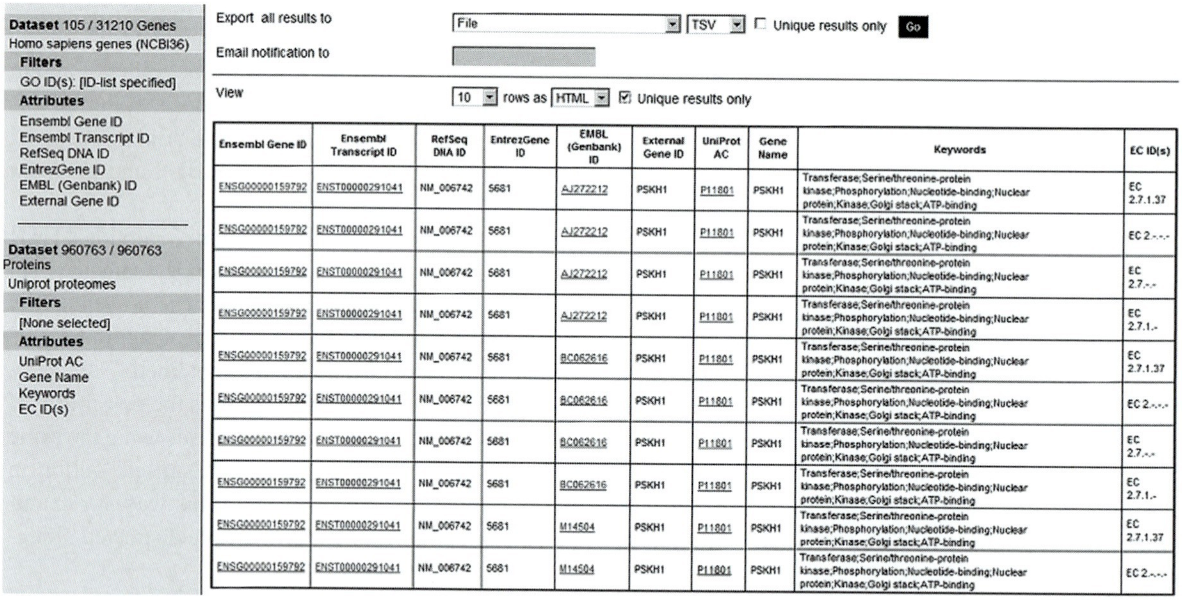

Fig. 29.2.5 Export table from BioMart using the linked Dataset option. As summarized on the *left*, Ensembl *Homo sapiens* genes are chosen as the first dataset and are linked to UniProt proteomes. Attributes (*column headers*) are shown for both datasets in one table

Amongst the "Attributes" available from UniProt, we will select "Gene Name," "Keywords," and "EC (Enzyme Commission) [61] ID" (the last assigned by the International Union of Biochemistry and Molecular Biology, IUMIB).

More tutorials and documentation about BioMart can be found at www.biomart.org, and also in the help and information sections of the Ensembl browser.

29.2.10 Customizing Ensembl

User Accounts. Ensembl has introduced free user accounts, opening up new scope for customization. Normally, customized settings are stored locally as cookies, but when users move to a different computer those settings do not transfer to the new machine. When the user has an account, any customized settings are stored centrally in Ensembl, and when the user logs on these will be taken into account no matter what computer is being used.

An additional advantage of having an account in Ensembl is the possibility of creating user groups (e.g., for a research group or for collaborating consortia). Thus, annotation could be added (e.g., relevant publications/notes about a particular gene) and shared within the group. Furthermore, BLAST results may be stored for an extended period when using one's account.

29.2.10.1 Displaying User Data on Ensembl

Ensembl offers the option of uploading your own information, either with DAS or using a URL-based upload. This data can be temporarily displayed in "Region in detail" and "Region overview" along Ensembl's annotation without the need to set up your own server. Upload of data formatted in the following formats is supported: GFF, GTF, BED, and PSL (see Sect. 29.2.4). Our help documentation (in the "Custom Annotation" section of the Help pages) includes a step-by-step explanation of the procedure to follow so as to visualize your own data in the Ensembl framework.

29.2.11 Archive

Ensembl releases a new version of the browser bimonthly and updates gene builds whenever new assemblies become available. Ensembl has implemented an archive site where researchers can access older versions of Ensembl. This archive (http://archive.ensembl.org) is the appropriate way of referring to a particular region or link in publications, ensuring that other investigators can visualize exactly the same genomic landscape and annotation as is described in the publication. Similarly, old releases of BioMart can be accessed through these archive sites. An archive site was used in Sect. 29.2.6.

29.2.12 *Pre!* Site

Although Ensembl updates gene sets whenever a new genome assembly is made public, this process can take several months. A skeleton site is put together shortly after we can access the new assembly at (http://pre.ensembl.org), so that investigators can peek at the new genomic landscape. These *Pre!* sites include some basic annotation, such as CpG islands, mRNAs and proteins, and GENSCAN transcripts, but no Ensembl gene models.

29.2.13 Further Information

Ensembl links to the EBI, providing cross links to the wealth of information hosted there (e.g., literature, patent information from the European Patent Office, microarray data from ArrayExpress [89], macromolecular structures from MSD [113], and molecular interactions from IntAct [65], amongst many other databases). A sister project, Ensembl Genomes, has extended Ensembl across the taxonomic space providing annotation to bacteria, plants, fungi, protists and metazoa.

Ensembl believes in open source, and therefore all the data generated is freely available [56, 103]. The browser can be installed locally (although this is not a trivial task, as the hardware requirements might exceed those available to the average user), and help is provided through a dedicated helpdesk that can be reached at helpdesk@ensembl.org.

Ensembl provides access to RDBMS systems[3] that can be used for high-level access to our data via direct SQL or via APIs that the Ensembl project provides, as well as via any third-party software that can directly communicate with these RDBMS instances.

29.2.14 Outlook

New species will be incorporated into Ensembl as their genome assemblies are established following the sequencing effort. The analysis of the great ape clade will come into the spotlight as a valuable tool to increase our understanding of human evolution. Genomic sequences from orang-utan *(Pongo pygmaeus albelii),* chimpanzee *(Pan troglodytes),* rhesus macaque *(Macacca mulata),* and marmoset *(Callithrix jacchus,* a New World monkey extensively used in the

---

[3]At ensembldb.ensembl.org (with username "anonymous"), note that this is not an URL but the location of a MySQL instance.

laboratory) should allow comparative analysis, providing more resolution to the picture and helping to identify those features in the human genome that differ among primates [75], with the ultimate goal of defining and better understanding the unique DNA sequences that set primates apart from other mammals and, moreover, humans [50] apart from other primates.

Exciting advances in sequencing and gene detection are revealing more and more about genomes. New "parallel" high-throughput low-cost sequencing platforms deliver short reads, which are the perfect match for ancient DNA, which is fragmented and not amenable to sequencing by traditional methods. New sequencing techniques sacrifice read length for much higher throughput, resulting in high coverage with short reads. Thus, the 454 single-molecule sequencing method could deliver an assembled genome from millions of reads.

Next-generation sequencing paves the way to new information that was unobtainable or prohibitively expensive with traditional methods, e.g., experimental platforms such as microarrays. Epigenome sequencing and genome-wide profiles of DNA-protein interactions become reality with these technologies [79]. Similar methods based on ditag sequencing [87] are being used to profile the methylation status of cancer genomes [122].

Genome sequences of individual humans are already available [71, 120],[4] and there are more to come. More individual genomes will provide additional haplotypes, and Ensembl is committed to providing tools for this future data.

Genome-wide association (GWA) studies [2, 33, 49, 107, 122, 123, 126] should provide new tools for unraveling the genetic basis of many common causes of human morbidity and mortality [54]. The Cancer Genome Atlas is an effort coordinated by the NCI and NHGRI to map the genomic changes involved in cancer, while ClinSeq, a pilot study from the NHGRI, attempts large-scale medical sequencing focusing on common diseases (such as coronary heart disease, with around 200–400 genes associated) and plans to analyze 1,000 individuals over 2 years. Other projects will focus on whole-genome resequencing of hundreds of individuals.

Gains and losses of large segments of DNA sequence (from 10,000 to 5 million base pairs) are known as copy number variation (CNV) [94]. New technologies allow high-resolution characterization of CNV within the framework of GWA studies. Genome browsers such as Ensembl will adapt and expand their scope to incorporate these new types of data in the exciting future ahead.

Acknowledgments The Ensembl project is principally funded by the Wellcome Trust, with additional funding from EMBL, NIH–NIAID, EU, and BBSRC. We are grateful to users of our website and to the developers on our mailing lists for their useful feedback and discussion.

References

1. Adams MD, Celniker SE, Holt RA, Evans CA, Gocayne JD, Amanatides PG et al (2000) The genome sequence of *Drosophila melanogaster*. Science 287(5461):2185–2195
2. Amundadottir LT et al (2006) A common variant associated with prostate cancer in European and African populations. Nat Genet 38:652–658
3. Aparicio S, Chapman J, Stupka E, Putnam N, Chia JM, Dehal P et al (2002) Whole-genome shotgun assembly and analysis of the genome of *Fugu rubripes*. Science 297(5585):1301–1310
4. Arabidopsis Genome Initiative (2000) Analysis of the genome sequence of the flowering plant *Arabidopsis thaliana*. Nature 408(6814):796–815
5. Ashurst JL, Collins JE (2003) Gene annotation: prediction and testing. Annu Rev Genomics Hum Genet 4:69–88
6. Bentley DR (2006) Whole-genome re-sequencing. Curr Opin Genet Dev 16(6):545–552
7. Birney E, Bateman A, Clamp ME, Hubbard TJ (2001) Mining the draft human genome. Nature 409:827–828
8. Birney E et al (2004) An overview of ensembl. Genome Res 14:925–928
9. Blattner FR, Plunkett G 3rd, Bloch CA, Perna NT, Burland V, Riley M et al (1997) The complete genome sequence of *Escherichia coli* K-12. Science 277(5331):1453–1474
10. Böcker S (2003) Sequencing from compomers: using mass spectrometry for DNA *de novo* sequencing of 200+ nt. Lect Notes Comput Sci 2812:476
11. Braslavsky I, Hebert B, Kartalov E, Quake SR (2003) Sequence information can be obtained from single DNA molecules. Proc Natl Acad Sci USA 100(7):3960
12. Burge C, Karlin S (1997) Prediction of complete gene structures in human genomic DNA. J Mol Biol 268(1):78–94
13. Burge CB, Karlin S (1998) Finding the genes in genomic DNA. Curr Opin Struct Biol 8(3):346–354
14. Camon E, Magrane M, Barrell D, Lee V, Dimmer E, Maslen J et al (2004) The gene ontology annotation (GOA) database: sharing knowledge in Uniprot with gene ontology. Nucleic Acids Res 32(1):D262–D266

---

[4] Available http://jimwatsonsequence.cshl.edu, here users can explore the Nobel Prize winner's genome.

15. Chaw SM, Chang CC, Chen HL, Li WH (2004) Dating the monocot-dicot divergence and the origin of core eudicots using whole chloroplast genomes. J Mol Evol 58(4):424–441
16. Chen N, Harris TW, Antoshechkin I, Bastiani C, Bieri T, Blasiar D et al (2005) WormBase: a comprehensive data resource for *Caenorhabditis* biology and genomics. Nucleic Acids Res 33:D383–D389
17. Cherry JM, Adler C, Ball C, Chervitz SA, Dwight SS, Hester ET et al (1998) SGD: saccharomyces genome database. Nucleic Acids Res 26:73–80
18. Cole ST, Brosch R, Parkhill J, Garnier T, Churcher C, Harris D et al (1998) Deciphering the biology of *Mycobacterium tuberculosis* from the complete genome sequence. Nature 393(6685):537–544
19. Collins FS, Morgan M, Patrinos A (2003) The human genome project: lessons from large-scale biology. Science 300(5617):286–290
20. Collins JE, Goward ME, Cole CG, Smink LJ, Huckle EJ, Knowles S et al (2003) Reevaluating human gene annotation: a second-generation analysis of chromosome 22. Genome Res 13(1):27–36
21. Cooper GM, Stone EA, Asimenos G, NISC Comparative Sequencing Program, Green ED, Batzoglou S et al (2005) Distribution and intensity of constraint in mammalian genomic sequence. Genome Res 15(7):901–913
22. Cunningham F, Rios D, Griffiths M, Smith J, Ning Z, Cox T et al (2006) TranscriptSNPView: a genome-wide catalog of mouse coding variation. Nat Genet 38(8):853
23. Curwen V, Eyras E, Andrews TD, Clarke L, Mongin E, Searle SM et al (2004) The Ensembl automatic gene annotation system. Genome Res 14(5):942–950
24. Deloukas P, Matthews LH, Ashurst J, Burton J, Gilbert JG, Jones M et al (2001) The DNA sequence and comparative analysis of human chromosome 20. Nature 414(6866):865–871
25. Deloukas P, Earthrowl ME, Grafham DV, Rubenfield M, French L, Steward CA et al (2004) The DNA sequence and comparative analysis of human chromosome 10. Nature 429(6990):375–381
26. Dowell RD, Jokerst RM, Day A, Eddy SR, Stein L (2001) The distributed annotation system. BMC Bioinformatics 2(1):7
27. Down TA, Hubbard TJ (2002) Computational detection and location of transcription start sites in mammalian genomic DNA. Genome Res 12(3):458–461
28. Drysdale RA, Crosby MA, FlyBase Consortium (2005) FlyBase: genes and gene models. Nucleic Acids Res 33:D390–D395
29. Dunham A, Matthews LH, Burton J, Ashurst JL, Howe KL, Ashcroft KJ et al (2004) The DNA sequence and analysis of human chromosome 13. Nature 428(6982):522–528
30. Dunham I, Shimizu N, Roe BA, Chissoe S, Hunt AR, Collins JE et al (1999) The DNA sequence of human chromosome 22. Nature 402:489–495
31. Easton DF et al (2007) Genome-wide association study identifies novel breast cancer susceptibility loci. Nature 447:1087–1093
32. ENCODE Project Consortium (2004) The ENCODE (ENCyclopedia of DNA elements) project. Science 306(5696):636–640
33. ENCODE Project Consortium (2007) Identification and analysis of functional elements in 1% of the human genome by the ENCODE pilot project. Nature 447(7146):799–816
34. Fernandez-Suarez XM, Searle S, Birney E (2006) Ensembl's annotation pipeline and its use in eukaryotic genomes. In: Mulder N, Apweiler R (eds) *In Silico* genomics and proteomics: functional annotation of genomes and proteins. Nova Science Publishers, New York, pp 109–123
35. Fernandez-Suarez XM, Schuster MK (2007) Using the ensembl genome server to browse genomic sequence data. Curr Protoc in Bioinformatics Unit 1.15, Suppl. 16
36. Fleischmann RD, Adams MD, White O, Clayton RA, Kirkness EF, Kerlavage AR et al (1995) Whole-genome random sequencing and assembly of *Haemophilus influenzae* Rd. Science 269(5223):496–512
37. Flicek P, Aken BL, Beal K, Ballester B, Caccamo M, Chen Y et al (2008) Ensembl 2008. Nucleic Acids Res 36. doi:10.1093/nar/gkm988
38. Gardner MJ, Hall N, Fung E, White O, Berriman M, Hyman RW et al (2002) Genome sequence of the human malaria parasite *Plasmodium falciparum*. Nature 419(6906):498–511
39. Gaunt MW, Miles MA (2002) An insect molecular clock dates the origin of the insects and accords with palaeontological and biogeographic landmarks. Mol Biol Evol 19(5):748–761
40. Gerstein MB, Bruce C, Rozowsky JS, Zheng D, Du J, Korbel JO et al (2007) What is a gene, post-ENCODE? History and updated definition. Genome Res 17(6):669–681
41. Gibbs RA, Weinstock GM, Metzker ML, Muzny DM, Sodergren EJ, Scherer S et al (2004) Genome sequence of the brown Norway rat yields insights into mammalian evolution. Nature 428(6982):493–521
42. Gibbs S, Collard M, Wood B (2002) Soft-tissue anatomy of the extant hominoids: a review and phylogenetic analysis. J Anat 200:3–49
43. Goff SA, Ricke D, Lan TH, Presting G, Wang R, Dunn M et al (2002) A draft sequence of the rice genome (*Oryza sativa* L. ssp. *japonica*). Science 296(5565):92–100
44. Goffeau A, Barrell BG, Bussey H, Davis RW, Dujon B, Feldmann H et al (1996) Life with 6000 genes. Science 274(5287):546 563–567
45. Goodman M, Porter CA, Czelusniak J, Page SL, Schneider H, Shoshani J et al (1998) Toward a phylogenetic classification of Primates based on DNA evidence complemented by fossil evidence. Mol Phylogenet Evol 9(3):585–598
46. Griffiths-Jones S et al (2003) Rfam: an RNA family database. Nucleic Acids Res 31:439–441
47. Gudmundsson J et al (2007) Genome-wide association study identifies a second prostate cancer susceptibility variant at 8q24. Nat Genet 39:631–637
48. Harrison PM, Gerstein M (2002) Studying genomes through the aeons: protein families, pseudogenes and proteome evolution. J Mol Biol 318(5):1155–1174
49. Heilig R, Eckenberg R, Petit JL, Fonknechten N, Da Silva C, Cattolico L et al (2003) The DNA sequence and analysis of human chromosome 14. Nature 421(6923):601–607
50. Hillier LW, Fulton RS, Fulton LA, Graves TA, Pepin KH, Wagner-McPherson C et al (2003) The DNA sequence of human chromosome 7. Nature 424(6945):157–164
51. Hillier LW, Miller W, Birney E, Warren W, Hardison RC, Ponting CP et al (2004) Sequence and comparative analysis of the chicken genome provide unique perspectives on vertebrate evolution. Nature 432(7018):695–716
52. Hirschhorn JN, Daly MJ (2005) Genome-wide association studies for common diseases and complex traits. Nat Rev Genet 6:95–108

53. Houck CM, Rinehart FP, Schmid CW (1979) A ubiquitous family of repeated DNA sequences in the human genome. J Mol Biol 132(3):289–306
54. Hubbard T, Birney E (2000) Open annotation offers a democratic solution to genome sequencing. Nature 403(6772):825
55. Huchon D, Madsen O, Sibbald MJ, Ament K, Stanhope MJ, Catzeflis F et al (2002) Rodent phylogeny and a timescale for the evolution of glires: evidence from an extensive taxon sampling using three nuclear genes. Mol Biol Evol 19(7):1053–1065
56. Humphray SJ, Oliver K, Hunt AR, Plumb RW, Loveland JE, Howe KL et al (2004) DNA sequence and analysis of human chromosome 9. Nature 429(6990):369–374
57. International Human Genome Sequencing Consortium (2001) Initial sequencing and analysis of the human genome. Nature 409:860–921
58. International Human Genome Sequencing Consortium (2004) Finishing the euchromatic sequence of the human genome. Nature 431(7011):931–945
59. International Union of Biochemistry and Molecular Biology. Nomenclature Committee, Webb EC (1992) Enzyme nomenclature 1992: recommendations of the nomenclature committee of the international union of biochemistry and molecular biology on the nomenclature and classification of enzymes. Academic, San Diego
60. Jaillon O, Aury JM, Brunet F, Petit JL, Stange-Thomann N, Mauceli E et al (2004) Genome duplication in the teleost fish *Tetraodon nigroviridis* reveals the early vertebrate proto-karyotype. Nature 431(7011):946–957
61. Kasprzyk A, Keefe D, Smedley D, London D, Spooner W, Melsopp C et al (2004) EnsMart: a generic system for fast and flexible access to biological data. Genome Res 14(1):160–169
62. Kent WJ (2002) BLAT–the BLAST-like alignment tool. Genome Res 12(4):656–664
63. Kerrien S, Alam-Faruque Y, Aranda B, Bancarz I, Bridge A, Derow C et al (2007) IntAct–open source resource for molecular interaction data. Nucleic Acids Res 35:D561–D565
64. Kersey P, Bower L, Morris L, Horne A, Petryszak R, Kanz C et al (2005) Integr8 and Genome Reviews: integrated views of complete genomes and proteomes. Nucleic Acids Res 33:D297–D302
65. Kulikova T, Akhtar R, Aldebert P, Althorpe N, Andersson M, Baldwin A et al (2007) EMBL nucleotide sequence database in 2006. Nucleic Acids Res 35:D16–D20
66. Kumar S, Hedges SB (1998) A molecular timescale for vertebrate evolution. Nature 392(6679):917–920
67. Kumazawa Y, Yamaguchi M, Nishida M (1999) Mitochondrial molecular clocks and the origin of euteleostean biodiversity: familial radiation of perciforms may have predated the Cretaceous/Tertiary boundary. In: Kato M (ed) The biology of biodiversity. Springer, Berlin, pp 35–52
68. Levy S, Sutton G, Ng PC, Feuk K, Halpern AL et al (2007) The diploid genome sequence of an individual human. PLoS Biol 5(10):e254
69. Loots GG et al (2000) Identification of a coordinate regulator of interleukins 4, 13, and 5 by cross-species sequence comparisons. Science 288(5463):136–140
70. Loveland J (2005) VEGA, the genome browser with a difference. Brief Bioinform 6(2):189–193
71. Maglott D, Ostell J, Pruitt KD, Tatusova T (2007) Entrez gene: gene-centered information at NCBI. Nucleic Acids Res 35:D26–D31
72. Margulies EH, Vinson JP, Miller W, Jaffe DB, Lindblad-Toh K, Chang JL et al (2005) An initial strategy for the systematic identification of functional elements in the human genome by low-redundancy comparative sequencing. Proc Natl Acad Sci USA 102(13):4795–4800
73. Margulies M, Egholm M, Altman WE, Attiya S, Bader JS, Bemben LA et al (2005) Genome sequencing in microfabricated high-density picolitre reactors. Nature 437(7057):376–380
74. Maston GA, Evans SK, Green MR (2006) Transcriptional regulatory elements in the human genome. Annu Rev Genomics Hum Genet 7:29–59
75. Mikkelsen TS, Wakefield MJ, Aken B, Amemiya CT, Chang JL, Duke S et al (2007) Genome of the marsupial *Monodelphis domestica* reveals innovation in non-coding sequences. Nature 447(7141):167–177
76. Mikkelsen TS, Ku M, Jaffe DB, Issac B, Lieberman E, Giannoukos G et al (2007) Genome-wide maps of chromatin state in pluripotent and lineage-committed cells. Nature 448(7153):553–560
77. Mitra RD, Shendure J, Olejnik J, Olejnik EK, Church GM (2003) Fluoescent *in situ* sequencing on polymerase colonies. Anal Biochem 320:55–65
78. Mouse Genome Sequencing Consortium (2002) Initial sequencing and comparative analysis of the mouse genome. Nature 420(6915):520–562
79. Mungall AJ, Palmer SA, Sims SK, Edwards CA, Ashurst JL, Wilming L et al (2003) The DNA sequence and analysis of human chromosome 6. Nature 425(6960):805–811
80. Murphy WJ, Eizirik E, O'Brien SJ, Madsen O, Scally M, Douady CJ et al (2001) Resolution of the early placental mammal radiation using Bayesian phylogenetics. Science 294(5550):2348–2351
81. Musser AM (2003) Review of the monotreme fossil record and comparison of palaeontological and molecular data. Comp Biochem Physiol A Mol Integr Physiol 136(4):927–942
82. Myers EW, Sutton GG, Delcher AL, Dew IM, Fasulo DP, Flanigan MJ et al (2000) A whole-genome assembly of *Drosophila*. Science 287(5461):2196–2204
83. Ng P, Tan JJ, Ooi HS, Lee YL, Chiu KP, Fullwood MJ et al (2006) Multiplex sequencing of paired-end ditags (MS-PET): a strategy for the ultra-high-throughput analysis of transcriptomes and genomes. Nucleic Acids Res 34(12):e84
84. Noonan JP, Coop G, Kudaravalli S, Smith D, Krause J, Alessi J et al (2006) Sequencing and analysis of Neanderthal genomic DNA. Science 314(5802):1113–1118
85. Parkinson H, Kapushesky M, Shojatalab M, Abeygunawardena N, Coulson R, Farne A et al (2007) ArrayExpress – a public database of microarray experiments and gene expression profiles. Nucleic Acids Res 35:D747–D750
86. Pennacchio LA, Rubin EM (2001) Genomic strategies to identify mammalian regulatory sequences. Nature Rev Genet 2(2):100–109
87. Pruitt KD, Tatusova T, Maglott DR (2007) NCBI reference sequences (RefSeq): a curated non-redundant sequence database of genomes, transcripts and proteins. Nucleic Acids Res 35:D61–D65
88. Ramana V, Davuluri RV, Grosse I, Zhang MQ (2001) Computational identification of promoters and first exons in the human genome. Nat Genet 29:412–417

89. Redon R, Ishikawa S, Fitch KR, Feuk L, Perry GH, Andrews TD et al (2006) Global variation in copy number in the human genome. Nature 444(7118):444–454
90. Rhesus Macaque Genome Sequencing and Analysis Consortium, Gibbs RA, Rogers J, Katze MG, Bumgarner R, Weinstock GM et al (2007) Evolutionary and biomedical insights from the rhesus macaque genome. Science 316(5822):222–234
91. Richards S, Liu Y, Bettencourt BR, Hradecky P, Letovsky S, Nielsen R et al (2005) Comparative genome sequencing of *Drosophila pseudoobscura*: chromosomal, gene, and *cis*-element evolution. Genome Res 15(1):1–18
92. Robertson G, Bilenky M, Lin K, He A, Yuen W, Dagpinar M et al (2006) cisRED: a database system for genome-scale computational discovery of regulatory elements. Nucleic Acids Res 34:D68–D73
93. Roest Crollius H, Jaillon O, Dasilva C, Ozouf-Costaz C, Fizames C, Fischer C et al (2000) Characterization and repeat analysis of the compact genome of the freshwater pufferfish *Tetraodon nigroviridis*. Genome Res 10(7):939–949
94. Ronaghi M, Uhlén M, Nyrén P (1998) DNA sequencing: a sequencing method based on real-time pyrophosphate. Science 281(5375):363–365
95. Ronaghi M (2001) Pyrosequencing sheds light on DNA sequencing. Genome Res 11(1):3
96. Ross MT, Grafham DV, Coffey AJ, McLay K, Howell GR, Burrows C et al (2005) The DNA sequence of the human X chromosome. Nature 434(7031):325–337
97. Rubin GM, Yandell MD, Wortman JR, Gabor Miklos GL, Nelson CR, Hariharan IK et al (2000) Comparative genomics of the eukaryotes. Science 287(5461):2204–2215
98. Salzberg S, Birney E, Eddy S, White O (2003) Unrestricted free access works and must continue. Nature 422(6934):801
99. Sanger F, Coulson AR, Friedmann T, Air GM, Barrell BG, Brown NL (1978) The nucleotide sequence of bacteriophage phiX174. J Mol Biol 125(2):225–246
100. Schuler GD (1998) Electronic PCR: bridging the gap between genome mapping and genome sequencing. Trends Biotechnol 16:456–459
101. Schwartz S, Kent WJ, Smit A, Zhang Z, Baertsch R, Hardison RC et al (2003) Human-mouse alignments with BLASTZ. Genome Res 13(1):103–107
102. Sladek R et al (2007) A genome-wide association study identifies novel risk loci for type 2 diabetes. Nature 445:881–885
103. Sprague J, Clements D, Conlin T, Edwards P, Frazer K, Schaper K et al (2003) The zebrafish information network (ZFIN): the zebrafish model organism database. Nucleic Acids Res 31(1):241–243
104. Springer MS, Murphy WJ, Eizirik E, O'Brien SJ (2003) Placental mammal diversification and the cretaceous-tertiary boundary. Proc Natl Acad Sci USA 100(3):1056–1061
105. Stein LD et al (2003) The genome sequence of *caenorhabditis briggsae*: a platform for comparative genomics. PLoS Biol 1(2):E45
106. Stewart CB, Disotell TR (1998) Primate evolution – in and out of Africa. Curr Biol 8(16):R582–R588
107. Stoesser G, Moseley MA, Sleep J, McGowran M, Garcia-Pastor M, Sterk P (1998) The EMBL nucleotide sequence database. Nucleic Acids Res 26(1):8–15
108. Tagari M, Tate J, Swaminathan GJ, Newman R, Naim A, Vranken W et al (2006) E-MSD: improving data deposition and structure quality. Nucleic Acids Res 34:D287–D290
109. Takezaki N, Figueroa F, Zaleska-Rutczynska Z, Klein J (2003) Molecular phylogeny of early vertebrates: monophyly of the agnathans as revealed by sequences of 35 genes. Mol Biol Evol 20(2):287–292
110. The C. elegans Sequencing Consortium (1998) Genome sequence of the nematode *C. elegans*: a platform for investigating biology. Science 282:2012–2018
111. The Chimpanzee Sequencing and Analysis Cosortium (2005) Initial sequence of the chimpanzee genome and comparison with the human genome. Nature 437(7055):69–87
112. The UniProt Consortium (2007) The universal protein resource (UniProt). Nucleic Acids Res 35:D193–D197
113. Tress ML, Martelli PL, Frankish A, Reeves GA, Wesselink JJ, Yeats C et al (2007) The implications of alternative splicing in the ENCODE protein complement. Proc Natl Acad Sci USA 104(13):5495–5500
114. Ureta-Vidal A, Ettwiller L, Birney E (2003) Comparative genomics: genome-wide analysis in metazoan eukaryotes. Nat Rev Genet 4(4):251–262
115. Venter JC, Adams MD, Myers EW, Li PW, Mural RJ, Sutton GG et al (2001) The sequence of the human genome. Science 291(5507):1304–1351
116. Waterman MS (1995) Introduction to computational biology: maps, sequences, and genomes: interdisciplinary statistics, 1st edn. Chapman and Hall, Boca Raton, FL
117. Wei CL, Wu Q, Vega VB, Chiu KP, Ng P, Zhang T, Shahab A, Yong HC, Fu Y, Weng Z et al (2006) A global map of p53 transcription-factor binding sites in the human genome. Cell 124:207–219
118. Wellcome Trust Case Control Consortium (2007) Genome-wide association study of 14, 000 cases of seven common diseases and 3, 000 shared controls. Nature 447(7145):661–678
119. Wheeler DL, Barrett T, Benson DA, Bryant SH, Canese K, Chetvernin V et al (2007) Database resources of the national center for biotechnology information. Nucleic Acids Res 35:D5–D12
120. Wootton JC, Federhen S (1996) Analysis of compositionally biased regions in sequence databases. Methods Enzymol 266:554–571
121. Yeager M et al (2007) Genome-wide association study of prostate cancer identifies a second risk locus at 8q24. Nat Genet 39:645–649
122. Yu J, Hu S, Wang J, Wong GK, Li S, Liu B et al (2002) A draft sequence of the rice genome (*Oryza sativa* L. *ssp. indica*). Science 296(5565):79–92

Databases in Human and Medical Genetics 29.3

Roberta A. Pagon, Ada Hamosh, Johan den Dunnen, Helen V. Firth, Donna R. Maglott, Stephen T. Sherry, Michael Feolo, David Cooper, and Peter Stenson

Abstract This chapter provides an introduction to the major, freely available, Internet-accessible databases in human and medical genetics used by healthcare providers in the diagnosis, management, and genetic counseling of persons with inherited disorders and their families, as well as by researchers for gene discovery, recording allelic variants, and cataloging genotype-phenotype relationships. Databases discussed include: GeneTests (view: www.genetests.org); Online Mendelian Inheritance in Man (view: www.ncbi.nlm.nih.gov/Omim); locus specific databases (LSDBs) identified at the Human Genome Variation Society (HGVS) web site (http://www.HGVS.org/dblist.html); DatabasE of Chromosome Imbalance and Phenotype in Humans using Ensembl Resources (view: http://decipher.sanger.ac.uk); Entrez Gene (view: ncbi.nlm.nih.gov/gene); dbGap: Database of Genotype and Phenotype (view: ncbi.nlm.nih.gov/dbgap); and the Human Gene Mutation Database HGMD® (view: http://www.hgmd.org).

Contents

- 29.3.1 GeneTests 942
  - 29.3.1.1 Name and URL 942
  - 29.3.1.2 Background 942
  - 29.3.1.3 Purpose and Target Audiences 943
  - 29.3.1.4 Location 943
  - 29.3.1.5 Funding and Governance 943
  - 29.3.1.6 Contents 943
  - 29.3.1.7 Search Mechanisms and Search Results 944
  - 29.3.1.8 Data Maintenance 945
  - 29.3.1.9 Usage 945
  - 29.3.1.10 Future Issues 945
- 29.3.2 Online Mendelian Inheritance in Man 945
  - 29.3.2.1 Name and URL 945
  - 29.3.2.2 Background 945
  - 29.3.2.3 Purpose and Target Audiences 945
  - 29.3.2.4 Location 946
  - 29.3.2.5 Funding and Governance 946
  - 29.3.2.6 Content 946
  - 29.3.2.7 Search Mechanisms and Search Results 947
  - 29.3.2.8 Data Maintenance 947
  - 29.3.2.9 Usage 947
  - 29.3.2.10 Future Issues 947
- 29.3.3 HGVS, Locus-Specific Databases 947
  - 29.3.3.1 Name and URL 948
  - 29.3.3.2 Background 948
  - 29.3.3.3 Purpose and Target Audiences 948
  - 29.3.3.4 Location 948
  - 29.3.3.5 Funding and Governance 949
  - 29.3.3.6 Content 949
  - 29.3.3.7 Search Mechanisms and Search Results 949
  - 29.3.3.8 Data Maintenance 949
  - 29.3.3.9 Usage 949
  - 29.3.3.10 Future Issues 949
- 29.3.4 Decipher 950
  - 29.3.4.1 Name and URL 950
  - 29.3.4.2 Background 950
  - 29.3.4.3 Purpose and Target Audiences 950
  - 29.3.4.4 Location 950
  - 29.3.4.5 Funding and Governance 950
  - 29.3.4.6 Contents 950
  - 29.3.4.7 Search Mechanisms and Search Results 951
  - 29.3.4.8 Data Maintenance 951
  - 29.3.4.9 Usage 952
  - 29.3.4.10 Future Issues 953
- 29.3.5 Entrez Gene 953
  - 29.3.5.1 Name and URL 953
  - 29.3.5.2 Background 953
  - 29.3.5.3 Purpose and Target Audiences 953
  - 29.3.5.4 Location 953
  - 29.3.5.5 Funding and Governance 953
  - 29.3.5.6 Contents 954
  - 29.3.5.7 Search Mechanisms and Search Results 954
  - 29.3.5.8 Data Maintenance 954
  - 29.3.5.9 Usage 954
  - 29.3.5.10 Future Issues 954
- 29.3.6 Database of Genotype and Phenotype 954
  - 29.3.6.1 Name and URL 954
  - 29.3.6.2 Background 954
  - 29.3.6.3 Purpose and Target Audiences 955
  - 29.3.6.4 Location 955
  - 29.3.6.5 Funding and Governance 955
  - 29.3.6.6 Contents 955
  - 29.3.6.7 Search Mechanisms and Search Results 955
  - 29.3.6.8 Data Maintenance 956
  - 29.3.6.9 Usage 956
  - 29.3.6.10 Future Issues 956
- 29.3.7 The Human Gene Mutation Database 956
  - 29.3.7.1 Name and URL 956
  - 29.3.7.2 Background 956
  - 29.3.7.3 Purpose and Target Audiences 956
  - 29.3.7.4 Location 957
  - 29.3.7.5 Funding and Governance 957
  - 29.3.7.6 Contents 957
  - 29.3.7.7 Search Mechanisms and Search Results 958
  - 29.3.7.8 Data Maintenance 958
  - 29.3.7.9 Usage 958
  - 29.3.7.10 Future Issues 958
- References 959

29.3.1 GeneTests

29.3.1.1 Name and URL

GeneTests (view: www.genetests.org)

29.3.1.2 Background

GeneTests was initially funded in 1992 by the National Institutes of Health as "Helix: A Directory of DNA Diagnostic Laboratories," a stand-alone desktop database accessible via telephone and fax to the data manager. Helix was released in late summer 1993 with listings for 110 diseases offered by 100 laboratories, without distinction between clinical laboratories and research-only laboratories. In 1996, Helix became Internet accessible and was joined in 1997 by an NIH-funded companion web site, "GeneClinics," comprised of full-text disease descriptions relating genetic testing to patient care. The name GeneTests replaced "Helix" in 1998. In the next 3 years, a Clinic Directory and Educational Materials including an Illustrated Glossary were added; GeneClinics was renamed GeneReviews; and, in 2001, both web sites were joined under the name GeneTests.

29.3.1.3 Purpose and Target Audiences

The purpose of GeneTests is to integrate genetic services into patient care by providing information on:

- Genetic testing for inherited disorders that is currently available in clinical laboratories worldwide (GeneTests Laboratory Directory)
- Genetics and prenatal diagnosis clinics (GeneTests Clinic Directory)
- Use of genetic testing in patient diagnosis, management, and genetic counseling (GeneReviews)

The primary intended audience is made up of all healthcare providers. Although genetic counseling and testing terms and concepts are used throughout the GeneTests web site, the companion Education Materials and Illustrated Glossary are intended to help bridge the gap for those clinicians who are not familiar with genetic terminology.

The secondary intended audience is that of researchers, who are served by the ability to list their research laboratories that are investigating a specific phenotype and/or gene(s) in order to ascertain research subjects and their relatives.

Individuals with inherited disease and their families are not an intended audience, as the language used and the required underlying medical knowledge base are too sophisticated for the average health consumer, for whom educational materials generally involve more explanation.

29.3.1.4 Location

The GeneTests and GeneReviews database and staff are located in the Department of Pediatrics at the University of Washington, Seattle, WA, USA.

GeneReviews has also been published since July 2005 on the BookShelf site developed and maintained at the National Center for Biotechnology Information (NBCI) at the National Library of Medicine (NLM) at the NIH.

29.3.1.5 Funding and Governance

GeneTests is funded by a contract from the National Institutes of Health: "Creation and Maintenance of GeneTests Database Records" (N01-LM-4-3503).

An Editorial Board that meets once a year helps decide policy. Contract performance standards are set by the NIH.

29.3.1.6 Contents

The five content areas are: *GeneReviews*, Laboratory Directory, Clinic Directory, Resources, and Educational Materials [14].

GeneReviews comprises more than 490 entries, all authored by experts and peer reviewed. One new entry is added each week; entries are updated every 2–3 years and revised as needed to keep up to date with current genetic test availability. The entries in *GeneReviews* are highly structured disease descriptions with a uniform format that allows clinicians to locate information within each entry quickly. Each entry links directly to the disease-specific clinical laboratory listings in the Laboratory Directory.

The Laboratory Directory provides information on molecular genetic testing, specialized cytogenetic testing [such as FISH (fluorescent in situ hybridization) and array CGH (comparative genome hybridization)], and biochemical genetic testing. About 615 laboratories offer testing for approximately 1,450 inherited disorders; over two-thirds of these tests are offered by clinical laboratories and about one-third, by research laboratories only. Sixty-two percent of laboratories are in the US, and 38% are non-US.

Any laboratory performing "in-house" testing may list its services. The one restriction is that any US laboratories listed as a clinical laboratory must be CLIA certified. In contrast, international laboratories do not need any official approval to be designated as clinical.

The Clinic Directory, comprised of clinics offering genetic evaluation and genetic counseling, provides contact information and information on services offered (i.e., adult genetics, biochemical genetics, cancer genetic counseling/risk assessment, pediatric genetics, preimplantation genetic diagnosis, prenatal genetics, and/or telemedicine).

The Resources database provides links to and contact information for consumer health-oriented national or international resources.

A context-sensitive Illustrated Glossary of more than 200 terms makes genetic counseling and testing terms accessible to a broad healthcare provider audience.

gene produces one phenotype such as Huntington disease
2. A hierarchy in which an alteration in one gene, such as *APC*, produces several phenotypes, each with a different name (familial adenomatous polyposis, Gardner syndrome, and Turcot syndrome) (See Fig. 29.3.1)
3. A hierarchy in which one phenotype, such as hereditary hemorrhagic telangiectasia, is produced by alterations in more than one gene (See Fig. 29.3.2)

In the hierarchical search result, the Testing button corresponds to a name that reflects a change in a gene. Because many disease names are affixed to a discrete phenotype and not the continuum of gene-related phenotypes, the latter names, such as "*APC*-associated polyposis conditions," typically do not exist in MeSH.

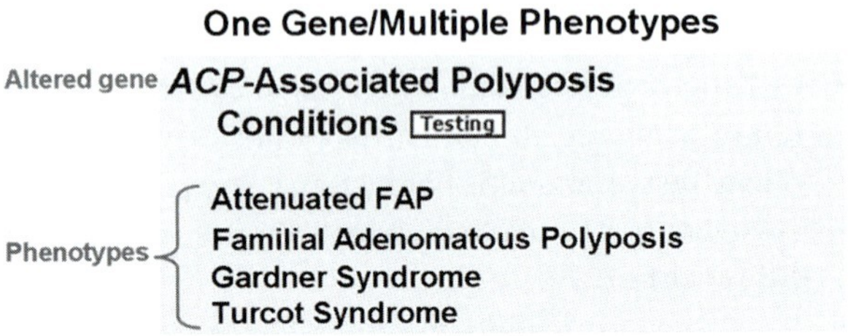

29.3.1.7 Search Mechanisms and Search Results

The GeneTests database is searched by selecting GeneReviews, Laboratory Directory, or Clinic Directory from the navigation bar at the top of the homepage. GeneReviews and the Laboratory Directory can be searched simultaneously by disease name, gene symbol, and protein name. In addition, GeneReviews can be searched by author and title from an alphabetic list and the Laboratory Directory by services, director's name, location, and laboratory name. The Clinic Directory is searched by state and services (US), zip code (US), and country (international). A global search function is not yet available.

Search results for GeneReviews and the Laboratory Directory may be followed by up to four buttons:

1. "Testing" (a listing of laboratories offering clinical testing)
2. "Research" (listing of laboratories conducting research)
3. "Reviews" (*GeneReview*)
4. "Resources"

The shared GeneReviews and Laboratory Directory "Disease" search function yields three possible results:
1. A result for a disorder in which alteration in one

Fig. 29.3.1 GeneTests database hierarchical search result, in which the "parent" term is a associated with a listing of laboratories offering *APC* gene testing (accessed by selecting the "Testing" button) and the "children" are the names of phenotypes (familial adenomatous polyposis, Gardner syndrome, and Turcot syndrome) that a clinician might use as search terms

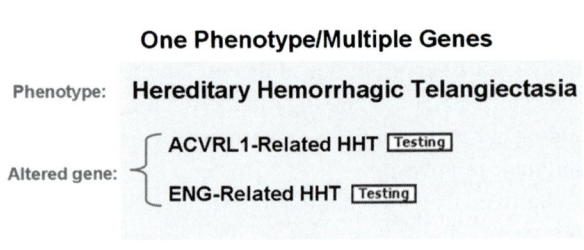

Fig. 29.3.2 GeneTests database hierarchical search result, in which the parent term is a phenotype used by clinicians as a search term and the children are the causative genes associated with a listing of laboratories offering testing for that gene (accessed by selecting the "Testing" button)

29.3.1.8 Data Maintenance

Each entry in *GeneReviews* is revised as needed to reflect currently available testing and is updated in a formal, comprehensive process every 2–3 years. Dates on each entry indicate the initial posting and either last revision or last update, whichever is more recent.

Laboratory and Clinic Directory listings are revised as needed and updated annually in a formal, online process. The date of the last update is displayed. Laboratories and clinics that fail to update are designated as "not current" for 1 year, after which they are deleted.

Resource listings are revised as needed and updated every 2 years.

29.3.1.9 Usage

GeneTests is a recognized resource in the international medical community [4, 13]. Its growth is steady; the number of new disease listings in the Laboratory Directory is about 100 per year; about 50 new expert-authored, peer-reviewed GeneReviews are added each year.

GeneTests and GeneReviews are copyrighted by the University of Washington and are made available to the general public subject to certain restrictions that are explained by the "Terms of Use" link on page footers and by selecting "About GeneTests" on the navigation bar.

As of January 2008, GeneTests had, on average, about 4,000 unique users daily, with an average of 40,000 page views per day.

29.3.1.10 Future Issues

Laboratory Directory and GeneReviews. Use of standard mutation nomenclature to improve data exchange and/or bidirectional links with mutation databases.

Laboratory Directory. Develop international collaboration(s) to exchange data on laboratory listings to reduce redundancy and ensure that only laboratories meeting the quality standards of their governing bodies are listed.

Clinic Directory. Develop a method for listing data about services provided by specialty clinics to permit patients to identify those clinics that provide the services they seek.

29.3.2 Online Mendelian Inheritance in Man

29.3.2.1 Name and URL

Online Mendelian Inheritance in Man (view: www.ncbi.nlm.nih.gov/Omim)

29.3.2.2 Background

OMIM is the online version of *Mendelian Inheritance in Man* (MIM) and is a comprehensive knowledge base of authoritative descriptions of human genes and genetic disorders. Created by Victor McKusick, MIM was first published in 1966 with 1,487 entries. This edition was subtitled "*Catalogs of Autosomal Dominant, Autosomal Recessive and X-linked Phenotypes*." The last print edition, the 12th, was published in 1998 and was subtitled "A Catalog of Human Genes and Genetic Disorders," and it contained 8,587 entries. The MIM catalogs have been maintained in a computer file since late 1963. In 1979, the National Library of Medicine adopted MIM as the experimental basis for the development of a full-text search engine called IRX (Information Retrieval experiment). IRX was made available for the authoring process in 1985 and became generally available in 1987 through funding by the Howard Hughes Medical Institute under the designation Online Mendelian Inheritance in Man or OMIM. From 1989 through 1995, OMIM was distributed along with GDB (Genome DataBase) from Johns Hopkins University. In late 1995, distribution via the World Wide Web and development of OMIM moved to the National Center for Biotechnology Information (NCBI). OMIM became fully integrated with the ENTREZ suite of programs in 2000.

29.3.2.3 Purpose and Target Audiences

OMIM is the descriptive repository of genotype-phenotype interactions. As such, OMIM catalogs all inherited traits, both simple and complex, and the genes that underlie them. As of October 2007, OMIM contained over 18,100 entries. Select mutations implicated in disease or associated with risk of disease (currently over 15,600) are cataloged as allelic variants in the relevant gene entry. OMIM also maintains a synopsis of the gene map with special emphasis on all genes and loci implicated in human disease (OMIM Morbid Map). In addition, in anticipation of phenotypic consequence of dysfunction, OMIM catalogs genes of known function.

OMIM is comprehensive and timely and is targeted at clinicians (especially clinical geneticists), researchers, and teachers. OMIM is not intended for a lay audience.

29.3.2.4 Location

Authoring and editing of OMIM are headquartered at the McKusick-Nathans Institute of Genetic Medicine of the Johns Hopkins University in Baltimore, MD, USA. Some authoring is distributed to authors from around the world. NCBI in Bethesda, MD, stores the electronic files and distributes them on the World Wide Web.

29.3.2.5 Funding and Governance

OMIM is funded by a contract from the National Institutes of Health: "Creation and Maintenance of Online Mendelian Inheritance in Man" (N01-LM43504).

OMIM has an Editorial Board comprised of subject experts and geneticists from around the world. Contract performance standards are set by the NIH.

29.3.2.6 Content

OMIM entries are divided into four main catalogs: autosomal, X-linked, Y-linked, and mitochondrial; these distinctions are made by the MIM number that defines the entry (See Table 29.3.1).

Within these catalogs, the entries are distinguished by type and defined by symbol (see Table 29.3.2):

Table 29.3.1 Numbering system for OMIM entries

MIM number	Catalog entry
1 (100,000)	Autosomal loci or phenotypes (entries created before May 15, 1994)
2 (200,000)	
3 (300,000)	X-linked loci or phenotypes
4 (400,000)	Y-linked loci or phenotypes
5 (500,000)	Mitochondrial loci or phenotypes
6 (600,000)	Autosomal loci or phenotypes (entries created after May 15, 1994)

Table 29.3.2 Symbol system for OMIM entries

Symbol	Entry type
Asterisk (*)	Indicates a gene with known sequence
Plus sign (+)	Indicates a description of a gene plus phenotype
Number sign (#)	Indicates a descriptive entry, usually of a phenotype, and does not represent a unique locus. (The reason for the use of a # sign is stated in the first paragraph of the entry)
Percent sign (%)	Indicates that the entry describes a confirmed Mendelian phenotype or phenotypic locus for which the underlying molecular basis is not yet known
Caret symbol (^)	Indicates that the entry no longer exists because it was removed from the database or moved to another entry as indicated
No symbol	Generally indicates a description of a phenotype for which the Mendelian basis, although suspected, has not been clearly established or that the separateness of this phenotype from that in another entry is unclear

Most entries have subheadings (e.g., description, cloning, gene function, gene structure, molecular genetics, clinical features, genotype-phenotype correlations, animal model). Gene entries may include allelic variants, which are a selected list of the published mutations in a gene. The selection criteria include the first six mutations, common mutations, mutations causing a different phenotype, mutations with an unusual mechanism, and missense mutations, which may give insight into structure/function relationships. OMIM now includes allelic variants that describe increased or decreased risk of a complex trait if that variant is very highly associated, replicated, and/or has demonstrated functional significance.

Phenotype entries may include a clinical synopsis, i.e., a hierarchical, structured description that begins with inheritance and ends with molecular basis (if known). Affected organ systems are included, with a semi-controlled vocabulary allowing ease of search.

The Morbid Map is a tabular listing of phenotypes and their relationship to the genome by locus or gene. The Morbid Map is included in the OMIM Synopsis of the Human Gene Map. The Morbid Map can be viewed alphabetically by disease or by chromosomal location of the disease.

The content of OMIM is greatly enhanced by copious links to other NCBI resources, including Entrez resources such as DNA and protein sequence, PubMed, and outside resources, such as GeneTests, Coriell Cell Repository, Locus Specific Mutation Databases (LSDBs), Human Gene Mutation Database (HGMD), HUGO Nomenclature, and more.

29.3.2.7 Search Mechanisms and Search Results

OMIM may be searched by MIM number, disorder, gene name/symbol (e.g., "Marfan," "FBN1," or "fibrillin"), or perhaps most conveniently by plain English (e.g., "hyponatremia and failure to thrive"). A search can be further restricted by using a "Limits" function to restrict the search to a subset of entries, field(s) within an entry, or date of update. Regardless of the method used, the search engine ranks the retrieval set so that the most relevant entries are in the top ten entries retrieved.

29.3.2.8 Data Maintenance

OMIM is updated daily. The link to the "Update log" in the blue bar on the left margin of the home page displays the number of new and changed entries by month, and within each month lists the specific entries that were added and changed.

29.3.2.9 Usage

OMIM is used extensively by researchers and clinicians around the world. As of September 2006, OMIM had, on average, 13,000 unique users daily with an average of 160,000 page views per day. In addition, data from OMIM are downloaded and displayed in many other resources around the world. OMIM is copyrighted by the Johns Hopkins University and is made available to the general public subject to certain restrictions that are explained by the "Restriction on Use" link on the home page.

29.3.2.10 Future Issues

Managing the triage and synthesis of the ever-burgeoning growth of biomedical research publications remains the greatest challenge going forward. Topics that will be challenging to render in a database include how to represent the complexities of gene and environment interaction on phenotype, the effects of genomic events, such as copy number variation, regulatory elements, and spatial and temporal changes. Specific enhancements to OMIM include collaborations with other researchers to increase links to animal model databases and to map allelic variants to the genomic sequence.

29.3.3 HGVS, Locus-Specific Databases

By definition, locus-specific databases (LSDBs) contain a listing of sequence variants in a specific gene(s) causing a Mendelian disorder or a change in phenotype, curated by (an) expert(s) in that gene [5]. Besides these listings, LSDB-connected web sites often contain a wealth of information related to the gene and the disorder/phenotype studied. Generally, LSDBs are initiated and driven by the interests of the curator, which may be research, clinical, or diagnosis in nature. The Human Genome Variation Society (HGVS) arose out of the HUGO-Mutation Database Initiative (MDI), a group of LSDB curators exchanging ideas on issues related to cataloging (pathogenic) gene variants.

29.3.3.1 Name and URL

Currently, most LSDBs are Internet based – a complete and up-to-date overview can be obtained through the HGVS web site (http://www.HGVS.org/dblist.html).

29.3.3.2 Background

LSDBs were initiated to collect all DNA variants found world-wide, both with and without pathogenic consequences, including both published variants and variants reported directly to the database. LSDBs fill the gap generated by scientific journals and central mutation databases; e.g., OMIM [8] and HGMD [20]. Journals, in general, publish only the first findings of variants in a specific gene, focused on variants that have pathogenic consequences. When more and more reports on a specific gene are published, new variants are unlikely to get published, unless they have not been reported before and/or provide new insights into the underlying disease mechanism. The central databases leave a gap here, because they do not collect all variants in all genes. OMIM generally collects the first variants described and later some with unique characteristics. HGMD collects the first report of every variant, and therefore only the clinical/laboratory phenotype associated with that particular example of the variant.

Initially, one of the most confusing aspects of databases containing information on variants in the DNA/protein sequence of genes was the way that these changes were described. Discussions regarding the uniform and unequivocal description of sequence variants were initiated in 1993 by Beaudet and Tsui [2]. These original suggestions have been widely discussed, modified, and extended, ultimately resulting in the HUGO-MDI/HGVS nomenclature recommendations [6] (http://www.HGVS.org/mutnomen/). These recommendations have now been largely accepted as a consensus and are applied world-wide, giving much more uniformity among databases.

29.3.3.3 Purpose and Target Audiences

Historically, the main purpose of an LSDB is to assist clinical diagnostics. The complete listing of all variants identified in a gene, in combination with the associated phenotype (as complete as possible), means that the LSDB should be the first resource to consult when a variant has been identified in a patient. Based on the presence of the variant and the associated phenotypic consequences as reported by others, the clinician should be helped in drawing a conclusion regarding the associated phenotype, i.e., "pathogenic or not." As such, the result of the LSDB query forms the basis for the diagnostic report. At the same time, through submission of the details of the new case to the LSDB, the clinician updates the database and increases its authority. Especially when a variant is new or has uncertain pathogenic consequences, this submission gives the clinician a unique opportunity to make these findings public immediately, giving colleagues world-wide the opportunity to assist in sorting out the potential pathogenicity that should be connected.

In addition to the list of DNA variants collected, LSDBs often contain a wealth of information on the gene, the disease, its diagnosis, and other related issues. As such, the use of LSDBs goes far beyond the LSDB itself. LSDBs attract the attention of a diverse group of users and, therefore, often provide their information in a format that can be understood by audiences ranging from scientists to interested lay persons.

29.3.3.4 Location

Since most LSDBs are initiated by interested gene experts, they are stored on many different servers world-wide (See http://www.HGVS.org/dblist.html). More recently, the HGVS has instigated initiatives to promote a more standardized LSDB structure, content, and data format, leading to databases that can be accessed from a central location. These initiatives are generally built around specific software packages for LSDB curation, including: dbRBC [3], MUTbase [18], UMD [19], and LOVD [7].

The UMD (Universal Mutation Database; http://www.UMD.be) and LOVD (the Leiden Open-source Variation Database; http://www.LOVD.nl) software are freely available. In addition, LOVD offers assistance in database set up including free-of-charge LSDB hosting.

29.3.3.5 Funding and Governance

An LSDB is generally the initiative of an interested gene expert. As a consequence, with few exceptions, LSDBs have little or no funding; and governance has not been clearly defined. The HUGO-Mutation Database Initiative and currently the HGVS have tried to organize LSDBs and attract financial support, but thus far these initiatives have not been very successful.

29.3.3.6 Content

LSDB content varies widely. In its simplest form an LSDB just lists the sequence variants reported with a reference to its source only. At the other end of the spectrum, an LSDB contains, in addition to the sequence variant and its source, a wealth of data on the characterization of the variant as well as a detailed description of the phenotype of the patient. The lack of standardization among LSDBs can make their efficient use difficult. To improve the quality and overall usefulness of LSDBs, the HGVS is promoting standardization of LSDBs by regular publication of recommendations and guidelines on such topics as the description of DNA variants, LSDB content and structure, and phenotype data reporting and collection (See http://www.HGVS.org). Ultimately, these recommendations should help to achieve more standardization and allow queries across different LSDBs; e.g., identifying specific candidate genes for diagnostic testing based on a detailed phenotypic description.

29.3.3.7 Search Mechanisms and Search Results

Because LSDBs widely vary in structure and data content, the ways to query these databases vary widely. When entries are sorted simply by position in the gene, they all basically generate an overview of all variants that have been identified. In addition, many LSDBs provide static lists summarizing overall database content, including figures on the total number of variants and the type of variants (substitution, deletion, insertion, etc.) collected. Only a few LSDBs provide extensive search options (e.g., using functional or phenotypic characteristics).

More complex queries are supported using Boolean operators per column (AND, OR, NOT) and by combining queries over different columns.

Wider use of LSDBs is seriously hampered by their overall lack of standardization, as explained above. General databases such as OMIM (for phenotype) and HGMD (for DNA mutability) allow some of the more general questions to be answered.

29.3.3.8 Data Maintenance

For most LSDBs, data maintenance is performed on an ad hoc basis; i.e., when the curator has time. A publication describing a set of new variants in the gene of interest is often the trigger for such an update. Only a few larger LSDBs have an active group of curators who regularly scan the literature and/or directly contact diagnostic laboratories with a request to submit all new variants identified. LSDBs using software like LOVD allow web-based data submission by external contributors and include software tools for data curation, error checking, and automatic updating of the database.

29.3.3.9 Usage

LSDBs are widely used by those performing DNA-based diagnostics and also by lay persons searching for information on a disorder of interest to them. Usage of LSDBs largely depends on the frequency of the disorder in the population (i.e., potentially interested users) and by the amount of information available on the LSDB web site. User statistics are usually not mentioned by LSDBs; but, based on our experiences with the Leiden Muscular Dystrophy pages (http://www.DMD.nl), we anticipate that a general LSDB attracts some 500–2,000 unique visitors each month. Collectively, the 700 known LSDBs (http://www.HGVS.org) would thus attract 350,000–1,500,000 unique visitors monthly.

29.3.3.10 Future Issues

Given the myriad possible genetic variations in the estimated 24,000 genes in the human genome, it would be useful to have standard criteria for their collection and storage.

Another important issue is completeness of LSDBs. Since it is not routine practice for researchers and diagnostic laboratories to automatically submit all variants identified, LSDBs are mostly far from complete. This is unfortunate, because all users, especially diagnostic laboratories, would benefit from completeness, thus ensuring interpretation of a sequence variant ("pathogenic or not") based on existing data. It seems that a good step forward would be to make data submission an intrinsic quality control step for certified clinical laboratories.

29.3.4 Decipher

29.3.4.1 Name and URL

DatabasE of Chromosome Imbalance and Phenotype in Humans using Ensembl Resources (view: http://decipher.sanger.ac.uk)

29.3.4.2 Background

It took nearly 10 years from the first paper on subtelomeric FISH analysis for the phenotypes of some of the subtelomeric deletions to be published. Since there are 43 telomeric loci and in excess of 30,000 probes on a high-density genomic microarray, it was clear that a large-scale collaborative approach would be needed to move the process of assigning phenotype to genomic interval forward in at timely fashion.

29.3.4.3 Purpose and Target Audiences

DECIPHER is an innovative collaborative project, based at the Sanger Institute, to map molecularly defined chromosomal rearrangements onto the human genome map. The project has been developed jointly by a clinical geneticist (Helen Firth) and a molecular cytogeneticist (Nigel Carter) and members of the bioinformatics Web team at the Sanger Institute.
DECIPHER has the joint aims of:

- Making information about the gene content of deletions/duplications/translocations available to clinical geneticists to guide their management of patients
- Mapping clinical features to regions of the genome so that these observations can be used to help to define the function of genes whose role in development and disease is currently unknown

The target audiences are staff working in departments of clinical genetics world-wide and research teams studying the function of human genes.

29.3.4.4 Location

DECIPHER is located at the Wellcome Trust Sanger Institute, Hinxton, Cambridge, UK, which is on the same campus as the European Bioinformatics Institute (EBI). DECIPHER interrogates the current version of the human genome assembly displayed in the Ensembl genome browser (currently NCBI-36). Ensembl itself builds all its annotated genes and associated features onto the most recent version of the human genome assembly, which is created by the NCBI. The annotation of genes and their features are continually being reviewed, refined, and updated.

29.3.4.5 Funding and Governance

DECIPHER is funded by the Wellcome Trust Sanger Institute. The project has Multi-centre Research Ethics Committee approval in the UK (MREC 04/mre5/50 until 2014). The ethical, legal, and social implications of the project are monitored by the DECIPHER Advisory Board, which meets once or twice a year to discuss policy issues and strategy.

29.3.4.6 Contents

DECIPHER is an interactive web-based database with three main components:

1. *DECIPHER Syndromes,* a public access compendium of microdeletion/duplication disorders. The Syndromes pages within DECIPHER provide a

29.3 Databases in Human and Medical Genetics

single, curated resource of information and web-links. For each syndrome entity DECIPHER provides:

- A brief clinical synopsis
- Information on the size and origin of a deletion/duplication
- An ideogram of the location of the deletion/duplication on the relevant chromosome
- A list of the genes contained within the aberrant interval
- An up-to-date publication reference list
- Links to support groups and other information sources (e.g., *GeneReviews*)
- A direct, clickable link to visualize the deletion/duplication in Ensembl DECIPHER Syndromes is supported by a panel of Expert Advisors.

2. *DECIPHER Projects*. These are password-protected secure domains into which centers affiliated to DECIPHER can deposit genomic and phenotypic data about individual patients in order to interpret results. As for DECIPHER Syndromes, a DECIPHER patient report generates: Information regarding the size and origin of a deletion/duplication

- An ideogram of the location of the deletion/duplication on the relevant chromosome
- The gene content of the aberrant interval. In DECIPHER, the clinician has a choice of gene lists: (1) HGNC genes (i.e., all genes in the interval recognized by the Human Genome Nomenclature Committee), (2) OMIM Morbid Map genes (i.e., all genes with importance in human disease that are included in the OMIM Morbid Map database), (3) OMIM genes (i.e., all genes known to the OMIM database) and (4) Imprinted genes (i.e., imprinted genes contained in the www.geneimprint.org database). With patient consent, a basic phenotype description using the restricted terminology of the London Neurogenetics database and the genomic location are passed into the publicly accessible Ensembl database. In Ensembl, they are displayed in the DECIPHER track alongside tracks displaying copy number variants identified in the normal population. This allows the clinician to see whether the chromosomal aberration overlies a known region of copy number variation or whether it coincides with a recognized syndrome and, if not, whether it has been seen before by another member of the consortium and, if so, what the phenotype of the patient(s) was.

3. *DECIPHER – Ensembl view.* From the patient report, it is possible with a single click to see the deletion/duplication/translocation in its genomic context within the Ensembl genome browser (See Fig. 29.3.3). Here known syndromes are denoted by a dark red or dark green line for syndromes caused by deletions/duplications, respectively. Single-patient entries are denoted by a pink or pale green line. If clusters of patients with the same or overlapping aberrations are seen, DECIPHER enables the clinical teams to contact each other to share information about the phenotype to facilitate the process of syndrome delineation and understanding of gene function.

29.3.4.7 Search Mechanisms and Search Results

All of the consented data held within the DECIPHER database are searchable. Searches can be made by each of the following, individually or in combination:

- Chromosomal band
- Genomic position
- Chromosomes
- Phenotypes

DECIPHER members can also use this search function to interrogate nonconsented data within their own project groups. The ability to search by a phenotype term enables users to "search the genome by phenotype" and generate a display of genomic intervals associated with particular phenotypes, e.g., cleft palate. (See Fig. 29.3.4)

29.3.4.8 Data Maintenance

Each center maintains its project domain. Syndrome Reports are reviewed once a year by the DECIPHER development team, with input from the Expert Advisor for each syndrome.

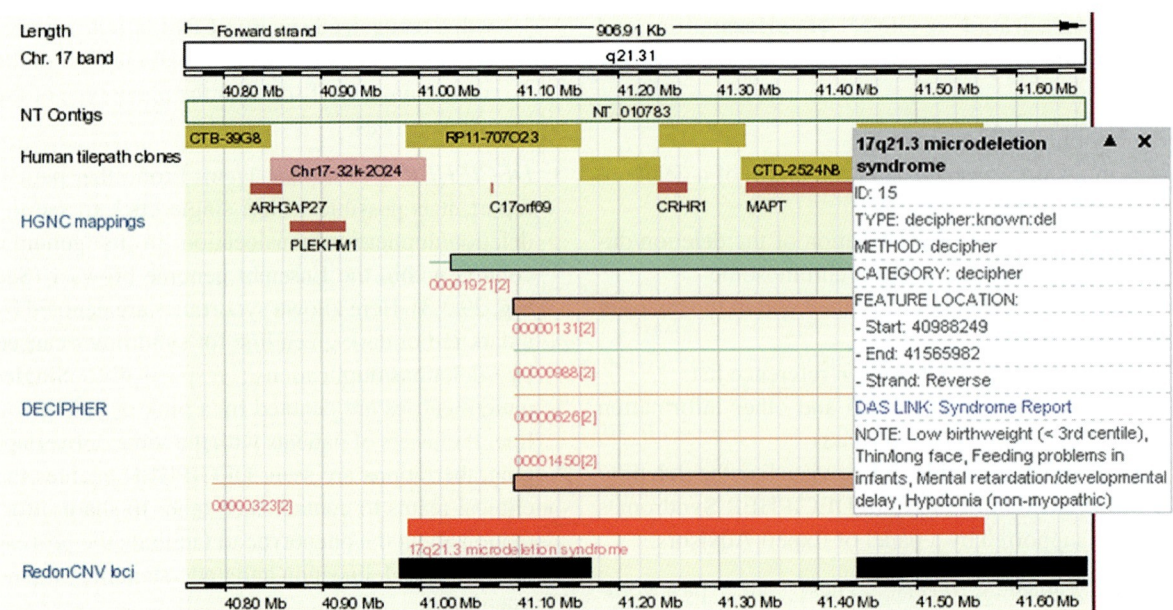

Fig. 29.3.3 A *dark red* denotes a known syndrome caused by a deletion; a *dark green line* denotes a known syndrome caused by a duplication. Single-patient entries are denoted by a *pink* or *pale green line*

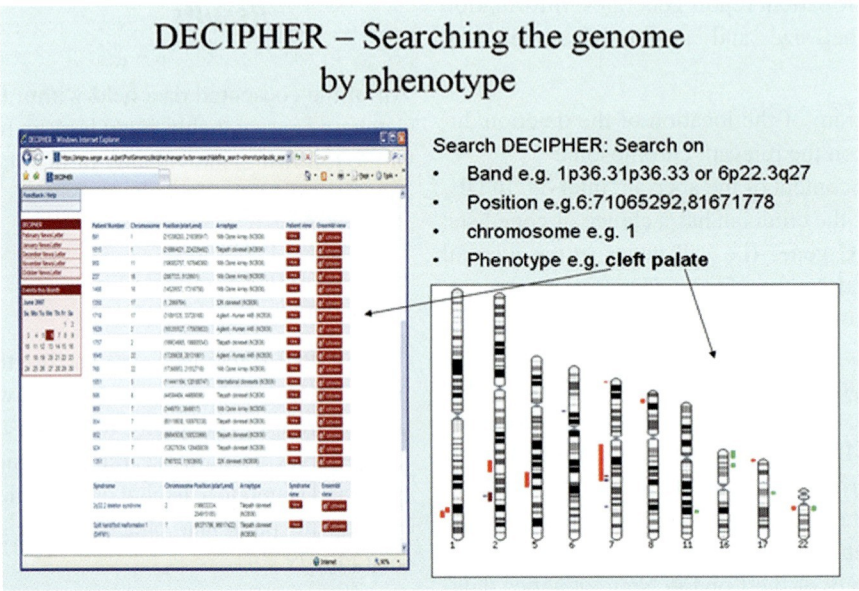

Fig. 29.3.4 Display of genomic intervals associated with the phenotype cleft palate

29.3.4.9 Usage

As a web-based resource, DECIPHER serves its content globally. Since its inception in 2004, when it already served over 31,000 web pages, usage has increased exponentially, with over 240,000 pages viewed in the first 5 months of 2007. Data in the database have risen sharply from 30 patients in the first 6 months of its existence to over 1,000 patients by September 2007. As of September 2007 the database has 48 syndrome reports, 22 array types in use, and 84 projects derived from more than 80 centers worldwide.

29.3.4.10 Future Issues

DECIPHER interacts with the latest version of the Genome Browser, so that each time the database is queried the results are derived from the latest version of the genome assembly (currently NCBI v36). This means that gene lists and positions are always the most up to date available.

As more information about copy number variation becomes available, this will be assimilated by DECIPHER.

One of the biggest challenges for DECIPHER in the coming years will be the analysis and display of complex inheritance in which a phenotype is caused by the interaction of several loci rather than a single highly penetrant locus.

29.3.5 Entrez Gene

29.3.5.1 Name and URL

Entrez Gene (view: ncbi.nlm.nih.gov/gene)

29.3.5.2 Background

The National Center for Biotechnology Information (NCBI) at the US National Library of Medicine (NLM) started to manage gene-specific information in 1999, in conjunction with RefSeq, a project to provide reference sequence records for genomes, RNAs, and proteins [15]. The public database that resulted from that effort was called LocusLink. In 2004, LocusLink was replaced with Entrez Gene [11]. Entrez Gene is a key component in the set of the NCBI databases that supports geneticists, including PubMed, sequence databases, OMIM (X.2), dbSNP, and dbGaP (X.6) [23]. Entrez Gene is the product of multiple collaborations throughout the world.

29.3.5.3 Purpose and Target Audiences

Entrez Gene supports integration of gene-specific information both within NCBI and internationally. By assigning a unique and stable numeric identifier to a gene and connecting that identifier to names, sequences, and publications, Entrez Gene establishes a foundation that all can use to navigate to gene-specific information on the Internet. Some information is reported directly within Gene, some in other databases at NCBI, and some at other databases internationally.

Entrez Gene is intended for diverse users who want to look up information about a particular gene in one species. Key data elements that describe that gene (i.e., summary of the names and functions of a gene and its products, the pathways in which the encoded protein participates, reports of sequences and maps on which the gene is located) are provided. A table of contents facilitates navigation to subcategories of interest. More detailed information is provided via links to more specialized resources. For example, Entrez Gene does not support detailed summaries of genes or diseases; it points to OMIM (X.2) for that function. Entrez Gene does not provide detailed reports of variation known to occur in a gene; for that function it points to dbSNP and Variation Viewer. The site is likely to be of most interest to scientists, but links are provided to resources of broader interest.

29.3.5.4 Location

Entrez Gene is maintained by staff at NCBI, which is part of the National Library of Medicine at the National Institutes of Health (NIH) in Bethesda, MD, USA. Data are available without restriction; thus, many elements from Entrez Gene are integrated into multiple databases world-wide.

29.3.5.5 Funding and Governance

Entrez Gene is produced by staff of NIH and its contractors. It is overseen by a Board of Scientific Counselors to NCBI.

29.3.5.6 Contents

The contents of any Entrez Gene record vary according to the depth of available information. Data in scope for processing include names, an annotated bibliography (GeneRIF or Gene References into Function), public accessions of sequences known to be gene specific, diagrams of gene structure, descriptions of pathways or interactants for a gene's products, and names of diseases or other phenotypes associated with variants of a gene.

29.3.5.7 Search Mechanisms and Search Results

Data in Entrez Gene can be retrieved by text words, by sequence, or by genomic location. Text word searching is provided by NCBI's Entrez interface (http://www.ncbi.nlm.nih.gov/books/bv.fcgi?rid=helpentrez.chapter.EntrezHelp). Searching can be as simple as entering a word or phrase into NCBI's query box and reviewing results. Specificity can be added to the query by using field restriction. Entrez and Entrez Gene provide detailed help documentation online (http://www.ncbi.nlm.nih.gov/books/bv.fcgi?rid=helpbook.TOC).

Records in Entrez Gene can also be retrieved by sequence comparisons. Users of NCBI's BLAST services [1] are directed to records in Gene if the results returned by their queries are sequences associated with records in Gene (indicated by a blue icon with a letter G).

Users can also browse the genome via NCBI's Map Viewer. Clicking on the Gene symbol takes the user to the record in Entrez Gene.

Entrez Gene provides multiple views of gene-specific data. The ones used most frequently are: Full Report (most data elements displayed); Summary (name, species, and location data); and Gene Table (a summary of the sequences of products of protein-coding genes). The LinkOut display is also of particular interest, because it reports resources outside of NCBI that have registered, indicating that they have information related to a record in Gene. Connections to locus-specific databases (LSDBs) (Sect. 29.3.3) are provided by this method.

29.3.5.8 Data Maintenance

Data are maintained by a combination of automated analyses and curatorial review. Entrez Gene is updated daily with information describing each gene's structure, names, function, publications, defining sequence, disease associations, etc. It is also updated with links to gene-specific views of related information, such as variation, related genes in other species (orthologs), and expression. Entrez Gene welcomes input from the scientific community, via the RefSeq/Gene update site (http://www.ncbi.nlm.nih.gov/RefSeq/update.cgi).

In addition to the gradual, daily updates, there are periodical comprehensive revisions of the sequence and/or annotation of a genome. In that case, most records for that species will be changed. This happens about once a year for each species.

29.3.5.9 Usage

Entrez Gene is widely used, both interactively and by multiple databases world-wide that download the files Entrez Gene provides. In September 2007, the site averaged about 8,000 distinct interactive users a day.

29.3.5.10 Future Issues

Entrez Gene will continue to add functionality and content in response to advances in technology and interpretation.

The RefSeqGene project is fostering a detailed review of sequence and annotation of genes for which variants have been shown to increase risk or cause a disease.

29.3.6 Database of Genotype and Phenotype

29.3.6.1 Name and URL

dbGap: Database of Genotype and Phenotype
   Public content: (view: ncbi.nlm.nih.gov/dbgap)
   Authorized access content: (view: ncbi.nlm.nih.gov/dbgap controlled)

29.3.6.2 Background

dbGaP was developed in 2006 to archive and distribute the results of studies that investigate individual-level phenotype, exposure, genotype, and sequence data and the associations between them. Included are genome-wide association studies (GWAS), medical sequencing, and molecular diagnostic assays.

29.3.6.3 Purpose and Target Audiences

The advent of high-throughput and cost-effective methods for genotyping and sequencing has provided powerful tools that allow for the generation of massive amounts of genotypic data. dbGaP provides stable identifiers that make it possible for published study reports to contain discussion and citations of primary data with specificity and uniformity. dbGaP provides access to the large-scale genetic and phenotypic datasets required for GWAS designs, including public access to study documents, summaries of phenotype variables, statistical overviews of the genetic information, position of published associations on the genome, and authorized access to individual-level data.

Two levels of access are provided to dbGaP:

- Open access on the web is available for descriptions of each study, the variables being evaluated, and analyses of genotype-phenotype associations.
- Authorized access is required to download datasets of pedigrees, de-identified phenotypes, genotypes for individual study subjects, and study documents covered by copyright and having restricted terms of distribution. In addition, authorized access is required to obtain additional study data that are covered by a publication embargo and not yet public. Examples include precomputed association data, genotype QC results, cluster plots and sliding window linkage disequilibrium data provided for cases and controls.

The target audience is primarily made up of researchers. Researchers must each be identified as primary investigators in the NIH eRA system to submit requests for authorized access data.

29.3.6.4 Location

dbGaP is maintained by staff at NCBI, which is part of the National Library of Medicine at the National Institutes of Health (NIH) in Bethesda, MD, U.S.A.

29.3.6.5 Funding and Governance

dbGaP is produced by staff of NIH and its contractors. It is overseen by several advisory groups.

29.3.6.6 Contents

dbGaP archives study documents (protocols, questionnaires, data dictionaries, etc.); phenotype measures; genotypes reported from genotyping arrays or sequencing; and details of statistical associations between phenotypes and genotypes [12]. These data are organized in such a way that measurements of phenotype are linked to the study documentation. Measurements are accepted in diverse formats and are converted into a common distribution format without modification or standardization. All content elements are assigned stable, unique public accession identifiers (IDs), allowing specific data or studies to be cited in publications or to be linked to other bioinformatics resources. Variable summaries and document data dictionaries are available via anonymous FTP.

All individual-level data are de-identified by the submitter so that individuals are represented by coded IDs, to which dbGaP does not have access. Each authorized access study has a study configuration report (pdf file) available as an authorized access download component that describes the full set of available download components.

Individual level data are organized into participant sets according to any restrictions on use dictated by the study's informed consent process. Each participant set groups all study individuals who have elected the same set of restrictions. Studies that offer participants choices about the use of their data (e.g., commercial versus noncommercial uses or disease-specific only versus general research use) will possibly have individual-level data partitioned into two or more such sets.

Access to these data is granted by an NIH oversight data access committee (DAC) that reviews each request and proposed use for compliance with the restrictions on use described above.

29.3.6.7 Search Mechanisms and Search Results

Searching of the web site is provided by NCBI's Entrez system. As such, a user can query by any term, with or without field restriction. A tutorial is provided (http://www.ncbi.nlm.nih.gov/entrez/query/Gap/gap_tmpl/dbGaP_HowTo.pdf).

Results of a query are categorized by study, variables, study documents, and analyses. The user can click on any tab to view more information.

Results may be provided in multiple formats, such as .pdf for documents and a web-based browser of genomic regions with significant association to a phenotype.

29.3.6.8 Data Maintenance

Potential submissions are reviewed to ensure compliance with appropriate standards of practice [12]. NCBI staff then work with the submitters to identify and correct any errors.

When submission of phenotypic and genotypic information is complete, NCBI can calculate association data or report associations calculated by others. Links are then provided to NCBI's Map Viewer to represent regions of interest identified by a study.

A versioning system supports periodic (semi-annual) updates by submitters. Additional components may be created when new technologies or methods are applied to current studies; e.g., additional genotypes on a new genotyping platform or sets of imputed genotypes derived from a new algorithm.

29.3.6.9 Usage

Because this resource is very new, usage statistics are not yet available.

29.3.6.10 Future Issues

In addition to representing information from large-scale studies, dbGaP will integrate genotype-phenotype relationships from multiple sources, including publications and OMIM.

29.3.7 The Human Gene Mutation Database

29.3.7.1 Name and URL

The Human Gene Mutation Database HGMD® (view: http://www.hgmd.org)

29.3.7.2 Background

The Human Gene Mutation Database (HGMD®) [20] constitutes a comprehensive core collection of data on germline mutations in nuclear genes underlying or associated with human inherited disease. HGMD does not include either somatic or mitochondrial mutations, because these data are covered by the Catalogue of Somatic Mutations in Cancer (COSMIC; http://www.sanger.ac.uk/genetics/CGP/cosmic/) and the Human Mitochondrial Genome Database (MitoMap; http://www.mitomap.org/), respectively. By March 2008, the database contained almost 80,000 different lesions detected in ~3,000 different genes, with new mutation entries currently accumulating at a rate exceeding 9,000 per annum. Although originally established for the scientific study of mutational mechanisms in human genes, HGMD has since acquired a much broader scope and utility.

The database was first made publicly available in April 1996. A collaborative agreement between HGMD and BIOBASE GmbH (http://www.biobase.de) was signed in January 2006. BIOBASE, based in Wolfenbüttel (Germany) and Beverly, MA, USA, is the leading content provider of biological databases, knowledge tools, and software for the life sciences. This collaborative agreement covers the exclusive world-wide marketing of HGMD to both academic and commercial subscribers.

29.3.7.3 Purpose and Target Audiences

HGMD records the first report of a disease-causing mutation or disease-associated/functional polymorphism and provides these data in a readily accessible form to all interested parties, whether they are from an academic, clinical, or commercial background. In practice, HGMD has become, de facto, the central disease-associated mutation database available to the scientific community.

The main target audiences for HGMD are academic/nonprofit institutions and commercial companies interested in utilizing mutation data. Users include: genetic counselors; clinicians; researchers; medical diagnostic laboratories; and pharmaceutical, biotechnology, and bioinformatics companies.

29.3.7.4 Location

The HGMD database curation staff is based in the Institute of Medical Genetics, part of the School of Medicine at Cardiff University, Cardiff, UK.

29.3.7.5 Funding and Governance

Because HGMD does not receive any public funding to support its upkeep, it has been necessary to develop a sustainable model for both the current and future funding of the database. The ideal model (in the opinion of the curators) would be a mixture of income from both public and private sources. This, in principle, would allow HGMD to provide free database access to academic/nonprofit users alongside a subscription-based distribution for commercial users marketed by a commercial company. With this eventual aim in mind, the HGMD curators opted to market the data in collaboration with BIOBASE GmbH, their commercial partner, while simultaneously continuing to make a core version of the database available as a free service to registered users from academic/nonprofit institutions via the Cardiff web site (http://www.hgmd.org). By insisting that commercial entities pay for access to the latest HGMD data and software tools, while providing a less up-to-date version free of charge to registered users from academic/nonprofit institutions, the HGMD curators believe that they can continue to allow free access to the bulk of their mutation data, while at the same time generating income to support HGMD from its commercial distribution.

HGMD employs a team of curators with expertise in the collection, interpretation, and editing of mutation data. The curatorial/editorial team decides which data are to be included in HGMD, and which ought to be omitted (for inclusion criteria, see Sect. 29.3.7.6).

29.3.7.6 Contents

Mutation data are obtained by means of a combination of both manual and computerized search procedures. Thus, online library screening, the PubMed database, and publicly available, locus-specific mutation databases (LSDBs; see Sect. 29.3.3) are all used to optimize (and maximize) data acquisition.

HGMD data are subdivided into ten primary categories of mutation: missense/nonsense; splicing and regulatory single nucleotide substitutions; microdeletions, micro-insertions and indels of 20 bp or less; gross deletions and gross insertions of greater than 20 bp (including copy number variations); complex rearrangements; and repeat variations. Entries are viewable on a gene-wise basis, and access to the sub-categorized mutation data is available via hypertext link from each gene page. Additional links to complementary data sources are also provided here (GDB, OMIM, HUGO Nomenclature Committee, Entrez Gene, GeneCards, GenAtlas, UniGene, SwissProt, and the Human Protein Reference Database), along with access to over 2,800 annotated cDNA sequences.

Disease-causing mutations and disease-associated/functional polymorphisms are included in HGMD, if adequate evidence pointing to their pathological involvement/relevance is to be found in the published report. Exclusions occur, however, if the published data are deemed to be of insufficient quality (either by virtue of the description provided, or because of a tenuous/nonsignificant association with a clinical or laboratory phenotype). Once identified and validated, the data are entered into HGMD in a standard format.

HGMD Professional (see http://www.biobase-international.com/pages/index.php?id=hgmddatabase) contains additional information including "extended cDNA sequences" that include splice junctions, a mutation viewer/map that superimposes HGMD mutation data over cDNA sequences, links to related genes, mutation data broken down by clinical/laboratory phenotype and substituted/substituting amino acid properties and orthologous protein alignments for missense mutations. *HGMD Professional* also provides an

Table 29.3.3 Number of new records entered into HGMD (by year of entry)

Year	Number of new mutation entries[a]	Number of new gene entries[a]
2001	7,453	189
2002	5,851	197
2003	5,990	214
2004	5,480	257
2005	7,651	241
2006	9,906	287
2007	9,378	441
2008*	2,499	86

[a]Figures accurate to March 15, 2008

Advanced Search feature that has been designed to enhance mutation searching, viewing, and retrieval.

The number of new entries being logged in HGMD has been steadily increasing over the last few years (See Table 29.3.3) and now stands at over 9,000 mutations per annum. The number of new genes being entered into HGMD has also followed this upward trend, with 441 new genes being introduced in 2007 alone.

29.3.7.7 Search Mechanisms and Search Results

The public version of HGMD can be searched by gene symbol, gene name, OMIM number, GDB number, and disease/phenotype.

HGMD Professional contains an expanded search engine with full-text Boolean searching and improved gene and mutation viewing enabled. Users may additionally search for chromosomal location, literature reference, codon number, and HGMD accession number. When utilizing the Advanced Search, users may tailor their queries with more specific criteria, including amino acid change, nucleotide substitution, microdeletion/insertion/indel size and composition, motif searching (both created and abolished), dbSNP number, and article title/abstract/keywords.

Results are generally returned as a list of genes matching the search criteria. Users may then click on a gene symbol to access the gene page and, from there, the mutation data and features associated with that gene. Advanced Search results are returned as a list of mutations matching the user's search parameters and are downloadable in their entirety.

29.3.7.8 Data Maintenance

The public version of HGMD operates with a 2 1/2-year time delay, but mutation data are added automatically once this period has expired. *HGMD Professional* is released by Biobase GmbH to subscribers every 3 months, with newly added mutation data and features. The current (March 31, 2008) release is version 8.1 and contains 79,078 mutations. This is the ninth release since the beginning of 2006.

29.3.7.9 Usage

The public version of HGMD has attracted over 18,000 user registrations from over 140 different countries world-wide. On a monthly basis, there are, on average, ~14,000 queries for genes (with an equal number accessing HGMD genes via external links) from almost 6,000 users, with a total of over 160,000 pages served. *HGMD Professional* is used by both academic and commercial customers worldwide.

HGMD is a registered trademark, and the data are copyrighted by Cardiff University. Users of the public site may not download HGMD data in their entirety without permission. This is, however, generally granted if the data are to be used for non-commercial, collaborative research purposes only. Collaborators who wish to access HGMD data in full must sign a confidentiality agreement. Recent successful collaborations include the projects to sequence the genomes of *Macaca mulatta* [17] (Rhesus Macaque Genome Sequencing and Analysis Consortium 2007) and *Rattus norvegicus* [16] (Rat Genome Sequencing Consortium 2004), as well as a recent study into gains of glycosylation mutations INSERM [22]. HGMD data have also been utilized by researchers in several other recent studies [9, 10, 21].

29.3.7.10 Future Issues

The primary aim of the HGMD curators is to secure sustainable funding for HGMD via a subscription-based model. During 2008, there are plans to incorporate a fully comprehensive, functional/disease-associated polymorphism dataset into HGMD to complement the existing disease-causing mutation data. The provision of full genomic sequences for all HGMD genes and genomic coordinates for as many mutations as possible are also seen as high priorities. The provision of extra supplementary information, including additional clinical phenotypes observed with a given mutation, along with data on the in vitro characterization of specific mutations, will be added to HGMD once resources permit.

References

1. Altschul SF, Gish W, Miller W, Myers EW, Lipman DJ (1990) Basic local alignment search tool. J Mol Biol 215:403–410
2. Burke W (2002) Genetic testing. N Engl J Med 347(23):1867–1875
3. Cotton RG, Auerbach AD, Beckmann JS, Blumenfeld OO, Brookes AJ, Brown AF, Carrera P, Cox DW, Gottlieb B, Greenblatt MS, Hilbert P, Lehvaslaiho H, Liang P, Marsh S, Nebert DW, Povey S, Rossetti S, Scriver CR, Summar M, Tolan DR, Verma IC, Vihinen M, den Dunnen JT (2008) Recommendations for locus specific databases and their curation. Hum Mutat 29(1):2–5
4. den Dunnen JT, Antonarakis SE (2000) Mutation nomenclature extensions and suggestions to describe complex mutations: a discussion. Hum Mutat 15:7–12
5. Fokkema IF, den Dunnen JT, Taschner PE (2005) LOVD: easy creation of a locus-specific sequence variation database using an "LSDB-in-a-box" approach. Hum Mutat 26: 63–68
6. Hamosh A, Scott AF, Amberger JS, Bocchini CA, McKusick VA (2005) Online Mendelian Inheritance in Man (OMIM), a knowledgebase of human genes and genetic disorders. Nucl Acids Res 33:D514–D517
7. Kryukov GV, Pennacchio LA, Sunyaev SR (2007) Most rare missense alleles are deleterious in humans: implications for complex disease and association studies. Am J Hum Genet 80(4):727–739
8. Levy S, Sutton G, Ng PC, Feuk L, Halpern AL, Walenz BP, Axelrod N, Huang J, Kirkness EF, Denisov G, Lin Y, MacDonald JR, Pang AW, Shago M, Stockwell TB, Tsiamouri A, Bafna V, Bansal V, Kravitz SA, Busam DA, Beeson KY, McIntosh TC, Remington KA, Abril JF, Gill J, Borman J, Rogers YH, Frazier ME, Scherer SW, Strausberg RL, Venter JC (2007) The diploid genome sequence of an individual human. PLoS Biol 5(10):e254
9. Maglott D, Ostell J, Pruitt KD, Tatusova T (2007) Entrez Gene: gene-centered information at NCBI. Nucleic Acids Res 35:26–31
10. Mailman MD, Feolo M, Jin Y, Kimura M, Tryka K, Bagoutdinov R, Hao L, Kiang A, Paschall J, Phan L, Popova N, Pretel S, Ziyabari L, Lee M, Shao Y, Wang ZY, Sirotkin K, Ward M, Kholodov M, Zbicz K, Beck J, Kimelman M, Shevelev S, Preuss D, Yaschenko E, Graeff A, Ostell J, Sherry ST (2007) The NCBI dbGaP database of genotypes and phenotypes. Nat Genet 39(10):1181–1186
11. Pagon RA (2005) Uses of databases. In: Korf B, Jorde L (eds) The encyclopedia of genetics, genomics, proteomics, and bioinformatics. Wiley, London Chapter 99
12. Pagon RA (2006) GeneTests: an online genetic information resource for healthcare providers. Med Libr Assoc 94(3):343–348
13. Pruitt KD, Tatusova T, Maglott DR (2007) NCBI reference sequences (RefSeq): a curated non-redundant sequence database of genomes, transcripts and proteins. Nucleic Acids Res 35:61–65
14. Rat Genome Sequencing Consortium, Cooper GM, Batzoglou S, Brudno M, Sidow A, Stone EA, Venter JC, Payseur BA, Bourque G, Lopez-Otin C, Puente XS, Chakrabarti K, Chatterji S, Dewey C, Pachter L, Bray N, Yap VB, Caspi A, Tesler G, Pevzner PA, Haussler D, Roskin KM, Baertsch R, Clawson H, Furey TS, Hinrichs AS, Karolchik D, Kent WJ, Rosenbloom KR, Trumbower H, Weirauch M, Cooper DN, Stenson PD, Ma B, Brent M, Arumugam M, Shteynberg D, Copley RR, Taylor MS, Riethman H, Mudunuri U, Peterson J, Guyer M, Felsenfeld A, Old S, Mockrin S, Collins F (2004) Genome sequence of the Brown Norway rat yields insights into mammalian evolution. Nature 428(6982):493–521
15. Rhesus Macaque Genome Sequencing and Analysis Consortium, Gibbs RA, Rogers J, Katze MG, Bumgarner R, Weinstock GM, Mardis ER, Remington KA, Strausberg RL, Venter JC, Wilson RK, Batzer MA, Bustamante CD, Eichler EE, Hahn MW, Hardison RC, Makova KD, Miller W, Milosavljevic A, Palermo RE, Siepel A, Sikela JM, Attaway T, Bell S, Bernard KE, Buhay CJ, Chandrabose MN, Dao M, Davis C, Delehaunty KD, Ding Y, Dinh HH, Dugan-Rocha S, Fulton LA, Gabisi RA, Garner TT, Godfrey J, Hawes AC, Hernandez J, Hines S, Holder M, Hume J, Jhangiani SN, Joshi V, Khan ZM, Kirkness EF, Cree A, Fowler RG, Lee S, Lewis LR, Li Z, Liu YS, Moore SM, Muzny D, Nazareth LV, Ngo DN, Okwuonu GO, Pai G, Parker D, Paul HA, Pfannkoch C, Pohl CS, Rogers YH, Ruiz SJ, Sabo A, Santibanez J, Schneider BW, Smith SM, Sodergren E, Svatek AF, Utterback TR, Vattathil S, Warren W, White CS, Chinwalla AT, Feng Y, Halpern AL, Hillier LW, Huang X, Minx P, Nelson JO, Pepin KH, Qin X, Sutton GG, Venter E, Walenz BP, Wallis JW, Worley KC, Yang SP, Jones SM, Marra MA, Rocchi M, Schein JE, Baertsch R, Clarke L, Csuros M, Glasscock J, Harris RA, Havlak P, Jackson AR, Jiang H, Liu Y, Messina DN, Shen Y, Song HX, Wylie T, Zhang L, Birney E, Han K, Konkel MK, Lee J, Smit AF, Ullmer B, Wang H, Xing J, Burhans R, Cheng Z, Karro JE, Ma J, Raney B, She X, Cox MJ, Demuth JP, Dumas LJ, Han SG, Hopkins J, Karimpour-Fard A, Kim YH, Pollack JR, Vinar T, Addo-Quaye C, Degenhardt J, Denby A, Hubisz MJ, Indap A, Kosiol C, Lahn BT, Lawson HA, Marklein A, Nielsen R, Vallender EJ, Clark AG, Ferguson B, Hernandez RD, Hirani K, Kehrer-Sawatzki H, Kolb J, Patil S, Pu LL, Ren Y, Smith DG, Wheeler DA, Schenck I, Ball EV, Chen R, Cooper DN, Giardine B, Hsu F, Kent WJ, Lesk A, Nelson DL, O'brien WE, Prufer K, Stenson PD, Wallace JC, Ke H, Liu XM, Wang P, Xiang AP, Yang F, Barber GP, Haussler D,

Karolchik D, Kern AD, Kuhn RM, Smith KE, Zwieg AS (2007) Evolutionary and biomedical insights from the rhesus macaque genome. Science 316(5822):222–234
16. Riikonen P, Vihinen M (1999) MUTbase: maintenance and analysis of distributed mutation databases. Bioinformatics 15:852–859
17. Soussi T, Asselain B, Hamroun D, Kato S, Ishioka C, Claustres M, Beroud C (2006) Meta-analysis of the p53 mutation database for mutant p53 biological activity reveals a methodologic bias in mutation detection. Clin Cancer Res 12:62–69
18. Stenson PD, Ball EV, Mort M, Phillips AD, Shiel JA, Thomas NS, Abeysinghe S, Krawczak M, Cooper DN (2003) Human gene mutation database (HGMD): 2003 update. Hum Mutat 21(6):577–581
19. Subramanian S, Kumar S (2006) Evolutionary anatomies of positions and types of disease-associated and neutral amino acid mutations in the human genome. BMC Genomics 7:306
20. Vogt G, Chapgier A, Yang K, Chuzhanova N, Feinberg J, Fieschi C, Boisson-Dupuis S, Alcais A, Filipe-Santos O, Bustamante J, de Beaucoudrey L, Al-Mohsen I, Al-Hajjar S, Al-Ghonaium A, Adimi P, Mirsaeidi M, Khalilzadeh S, Rosenzweig S, de la Calle Martin O, Bauer TR, Puck JM, Ochs HD, Furthner D, Engelhorn C, Belohradsky B, Mansouri D, Holland SM, Schreiber RD, Abel L, Cooper DN, Soudais C, Casanova JL (2005) Gains of glycosylation comprise an unexpectedly large group of pathogenic mutations. Nature Genetics 37(7):692–700
21. Wheeler DL, Barrett T, Benson DA, Bryant SH, Canese K, Chetvernin V, Church DM, DiCuccio M, Edgar R, Federhen S, Geer LY, Kapustin Y, Khovayko O, Landsman D, Lipman DJ, Madden TL, Maglott DR, Ostell J, Miller V, Pruitt KD, Schuler GD, Sequeira E, Sherry ST, Sirotkin K, Souvorov A, Starchenko G, Tatusov RL, Tatusova TA, Wagner L, Yaschenko E (2007) Database resources of the national center for biotechnology information. Nucleic Acids Res 35:D5–12

Subject Index

Readers searching for information beyond the index should refer to the detailed table of contents (pp. xv-xliv)

A

ab initio gene prediction method 924
AB0 blood group system 20, 180
   genetics 195
   inheritance 196
abacavir 638
abnormal spindle-like microcephaly 544
acetylation 61
acetylcholinesterase 689
acholinesterasemia 330
achondrogenesis 448
achondroplasia 444, 449
acid phosphatase 265
active chromatin hub 370, 374
acute megakaryoblastic leukemia (AMKL) 104
acute myeloid leukemia 455
ADA, see adenosine deaminase
adaptive evolution 559, 562
addiction 715, 807
   agonist therapy 736
   antagonist therapy 736
   co-causation 723
   co-morbidity 723
   maximum heritability 718
   treatment 735
addictive disorder 715
Addison's disease in dogs 823
adenine 322
adeno-associated virus 395, 869
adenomatous polyposis coli (APC) 463
adenosine 642
   deaminase (ADA) 341
      deficiency 341, 867
adenovirus 868
adenylate cyclase 808
adhesion molecule 419
admixture mapping 279, 280
adrenal hypoplasia congenita 345
adverse drug reaction 636
Aeromonas salmonicida 833
affected sib pair method 273
Affymetrix 500 K platform 607, 609
Affymetrix GeneChip 911
agammaglobulinemia 405

age-dependent macular degeneration 279
agglutinogen 227
aggression 748, 749
aging 792
Ag-NOR banding 80
Ago protein 147
agouti gene 316
agyria 430
Aicardi syndrome 179
albinism 19, 200, 343
Albright hereditary osteodystrophy 343, 444
albuterol 638
alcohol 717
   dehydrogenase 729, 734
   dependence 719
alcoholism 715, 718, 723, 727, 735
aldehyde dehydrogenase 729
alignment
   conserved function 564
   genome sequences 566
   large genome sequences 563
   phylogenetic depth 565
   protein sequences 563
   strategy 564
alkaptonuria 3, 18, 19, 172, 174
   possible causes 19
allele
   frequency spectrum 591, 609
   heterogeneity 246, 349
allele-specific oligonucleotide 392
allotransplant 229
alpha satellite 64
alpha-antitrypsin 283
alpha-fetoprotein 346, 667
alpha-globin
   active chromatin hub 374
   gene 367, 374
   triplication 393
Alstrom syndrome 258
alternative lengthening of telomere (ALT) 70
Alu
   element 540
   repeat 562

Alu (cont.)
  sequence 328, 330, 342
    Alu-Alu recombination 329
Alzheimer disease 190, 289, 334, 519, 643, 681, 745, 793, 803, 897
  early-onset 685
  nonfamilial 688
amenorrhea 124, 517
amino acid 563
  substitution 45, 386, 541
amitriptyline 637
amniocentesis 856, 857, 859, 860
amygdala 730, 733
amylase gene (AMY1) 91
amyloid precursor protein (APP)
  gene 103
  metabolism 685
amyloid-beta 683
amyotrophic lateral sclerosis 876
anaphase
  bridge 106
  of mitosis 74
ancestral
  gene 377
  repeat 561, 568
anencephaly 427
aneuploidy 63, 97, 109, 855
  in oocytes and embryos 125
  of sex chromosomes 120
Angelman syndrome 187, 308, 310, 315, 328, 668, 704, 753
*Anopheles gam*biae 896
ANOVA test 284
antenatal diagnosis 862
Antennapedia 422
antidepressant 645
antiepileptic drug 640, 646
antiproteolytic activity 283
antipsychotics 645
antisaccade eye movement 769
antisemitism 22
antisense RNA 347
antisocial personality disorder 726, 727
α-antitrypsin 871
anxiety disorder 656
APC gene in Ashkenazi Jews 333
Apert syndrome 335
aphidicolin 82
apical ectodermal ridge 435
APOE e4 allele 684, 687
apolipoprotein 519
  E 684
    genotyping 687
apoptosis 75, 797
Arabidopsis thaliana 476
Ardipithecus kaddaba 533
aripiprazole 637
Aristotle 14

array CGH 88, 92
array-comparative genomic hybridization 745
arrhythmia 440
arthrogryposis 447
artificial insemination 851
aryl hydrocarbon receptor interacting protein 457
ascertainment bias 202, 207
ASD-related syndrome 702
Ashkenazi Jews 334, 603
aspartoacylase 872
aspartylglucosaminuria 522
AS-SRO 309
astemizole 636
ataxia telangiectasia 128, 341, 453, 458
  homozygotes 129
atherosclerosis 264
atrophy 172
ATR-X syndrome 387
attention deficit /hyperactivity disorder (ADHD) 723, 745, 751, 752
Australopithecines 533
Australopithecus
  afarensis 533
  africanus 530
autism 699, 722
Autism Diagnostic Observation Schedule (ADOS) 700
Autism Diagnostic Interview (ADI-R) 700
autism spectrum disorder 699, 745, 752
autosomal dominant limb girdle muscular dystrophy 145
autosomal trisomy 103
autosomal-dominantly inherited syndrome 91
autotransplant 229
azoospermia 123

B

bacterial artificial chromosome 475, 910
bait loop 336
Bannayan-Riley-Ruvalcaba syndrome 455, 702
Bardet–Biedl syndrome 193, 247, 258, 350
Barr body 118, 854
Bayes' theorem 863, 864
Bayesian method 594, 627
Becker's muscular dystrophy 849, 850
Beckwith–Wiedeman syndrome 187, 310, 314, 315
behavioral
  genetics 8, 649
  phenotype 743, 744
benign recurrent intrahepatic cholestasis (BRIC) 219
benzodiazepine sensitivity 735
Berkeley Drosophila Genome Project 796
Bernstein's formula 196, 197, 203
beta melanocyte-stimulating hormone (beta-MSH) 337
beta-amyloid peptide 793
beta-endorphin 337
beta-globin gene 320, 350
beta-human chorionic gonadotropin 855
biallelic polymorphism 46
bias can 609

binCons 570
binge drinking 717
biobank 631, 899
biochemical genetics 19
biomarker 158
BioMart 583, 933
biometry 16
biopiracy 899
biopterin synthase 246
bipolar disorder 746, 759, 765
    adoption studies 766
    candidate gene studies 767
    family studies 766
    formal genetic studies 766
    genome-wide association studies 767
    linkage studies 767
    twin studies 766
birth defect 447
Birt–Hogg–Dube syndrome 454
blast family 563
blastocyst 422, 424, 780
BLASTZ program 912
BLAT
    algorithm 929
    alignment 911, 916
blood mutant 834
blood-related defect 834
Bloom syndrome 6, 127, 453, 458
Bombay phenotype 180
bone
    formation 442
    morphogenetic protein 432
Bonferroni correction 278
bonobo 536
Bos
    indicus 549
    primigenius 549
    taurus 549
brachial arch 433
brachydactyly 185, 443
brachyphalangy 169
    pedigree of Farahee 168
brain hygroma 121
brain-derived neurotrophic factor (BDNF) 654
branchio-oto-renal syndrome 434
Brauer keratoma dissipatum 201
breakage syndrome 126
break-induced replication 332, 464
breast carcinoma 453, 462, 519
breeding
    experiment 17, 212
    program 816
brittle bone disease 448
broader autism phenotype 700, 708
broadsense heritability 268
bromodeoxyuridine 72, 221
Brooke-Spiegler syndrome 453
Browser Extensible Data 907

Bruton's agammaglobulinemia 407
budding uninhibited by benomyl (Bub1) 76

C

CADASIL 691
cadherin 9 707
Caenorhabditis
    briggsae 925
    elegans 219, 418, 777, 787, 830, 906, 925
        hermaphrodite 789
café-au-lait spot 170
calcium channel subunits 707
campomelic dysplasia 347, 442
Canavan disease 868
cancer
    chromosome 463
    genetics 451
    outlier profile analysis 155
Cancer Prevention Study 630
candidate gene 656, 707, 728
canine
    studies 818
    disease 816, 817
    genomics 813
        maps 816
        sequence 816
cap site mutation 343
carbamazepine 643
cardiac dysrhythmia 188
cardiovascular disease 897
Carney complex 454
carrier identification 390
Carter effect 851
cartilage hair hypoplasia 522
cartilage oligometric matrix protein (COMP) 445
case-control design 277, 624
case-family design 625
case-only study 627
catechol-a-methyltransferase (COMT) 720
CATIE trial 646
cation exchange-high performance liquid chromatography 391
C-band 79
cell
    cycle
        checkpoints 75
        machinery 797
    fusion frequency 221
    hybridization 220, 221
    migration 418
    nucleus 20
    proliferation 418
    therapy 877
cell-cylce arrest 71
CENP-A 65
centimorgan (cM) 95
Central Limit Theorem 265

centromere 41, 63, 106
　instability 65
　region instability 129
centrosome 71
CEPH family 223
cerebellar hemangioblastoma 452
CGG triplet repeat 850
CGH, see comparative genomic hybridization
Charcot–Marie–Tooth disease 41, 145, 332, 752
CHARGE region 91
chemical mutagenesis 830
chemosensation 576
Chi square test 274, 277, 494
　of heterogeneity 281
CHILD syndrome 179
childhood kidney tumor 441
chimpanzee 536, 538
cholesterol 427
cholinesterase 330
chondrocyte 442
　hypertrophy 444
chorionic
　gonadotropin 424
　villus sampling 859, 860
chromatid 106
　break 106
chromatin 65, 93, 300, 310, 316, 674
　analysis 92
　disease 63
　hub 374
　immunoprecipitation (ChIP) 93, 304, 581
　insulator 373
　loop 108
　remodelling 305
chromatin-marking
　protein 301, 312
　system 305
chromosome 20, 23
　aberration 125, 447, 743
　analysis method 78
　anaphase 95
　　lagging 100
　aneuploidy 124
　anomalies 4
　banding 58
　instability 126
　metaphase 95
　mispairing 236
　mitotic 56
　telophase 95
　territories 85, 108
chronic myeloid leukemia 639, 840
Cimp phenotype 467
cis-acting
　DNA mutation 312
　imprinting center 308
cis-trans effect 225
cleft lip/cleft palate 447, 519
clinical genetics 8, 24

cloaca 441
cloned embryo 879
cloning 224
Clovis people 548
ClustalW 563
cluster
　algorithm 600
　analysis 593
CNV, see copy number variant
coalescent theory 495
cocaine addiction 831
Cochrane-Armitage trend test 277
codominance 166, 201
cohesin 75
Colchicum autumnale 57
colon polyps 253
color blindness 268
colorectal cancer 460, 467
combinatorial partitioning method 627
combined binary ratio labeling (COBRA) 84
combined pituitary hormone deficiency 832
comma body 440
common alpha thalassemia gene 182
Common Disease Common Variant
　　Hypothesis 273, 287
comparative genomics 557, 929
　hybridization (CGH) 86, 498, 667, 764
comparison image 932
compensatory mutation 336
complement cascade 232
complex disease 545
compound
　heterozygosity 174
　screen 791
conduct disorder 723, 726
configuration panel 930
confounding 623
congenital
　defect 449
　nephrosis 522
conotruncal anomaly face syndrome 751
consanguinity 507, 524, 847
　adult mortality and morbidity 519
　childhood morbidity 518
　deaths in infancy and childhood 518
　distribution of disease alleles 523
　global prevalence 510
　marriage 513, 514, 887
　　civil legislation 515
　pedigrees 511
　risk evaluation 524
　social and economic factors 515
consensus coding sequence (CCDS) set 33
conservation track 570
conserved
　noncoding elements 347
　noncoding region 388
　synteny 564
Contactin4 706

Index

contiguous gene syndrome 113
cooled charge-coupled device (CCD) 84
Cooley's anemia 367
copy number
    variant (CNV) 10, 47, 59, 91, 498, 540, 701, 705, 745
        mapping 47
        region 90
    polymorphism (CNP) 91
cordocentesis 860
Cornelia de Lange syndrome 77
correcting bias 203
corticotropin-releasing hormone receptor 1 gene 728
cortisone reductase deficiency 248, 257
coupling 213
Cowden syndrome 455
CpG dinucleotide 301
crackometer 807
Cre target site 781
Cre-driver 782
Creutzfeldt–Jakob disease 182, 690, 691
cri-du-chat syndrome 58, 753
criminality 749
Crohn's disease 294, 321, 411
crossing over
    consequences 237
    in human genetics 235
    intrachromosomal 237
cryptic exon 340
cubitus interruptus 429
cyclin-dependent kinase (Cdk) 77
cyclopia 430
cylindromatosis 453
cysteine 684
cystic fibrosis 175, 181, 246, 522, 600, 851, 858, 867
    gene 225
cystic kidney disease 831
cystinuria 19
cytochrome P450 269
cytogenetics 8, 24, 212
cytokinesis 72, 75
cytosine 301, 322
cytotrophoblast 424

D

D chromosome 220
DA/DAPI staining 80
d-amphetamine addiction 831
Danio rerio 418, 777, 827, 828, 925
dark heritability 285
Darwin, Charles 530
data
    access comittee 955
    mart 583
database
    genotype 954
    phenotype 954
De Grouchy syndrome 112
deafness 173, 331, 347, 524
debrisoquine polymorphism 637
Decapentaplegic (Dpp) pathway 798
DECIPHER 92, 115, 950
deduplication 325
deep intronic mutation 342
Defb gene 576
deformation 448
degeneration 16
deletion 190, 558, 704
    syndrome 58, 111, 113, 751, 764
dementia 670, 683, 688, 745
    causes 690
    praecox 768
Democritus 15
denaturing gradient gel electrophoresis (DGGE) 392
Denys–Drash syndrome 441
Dependovirus 869
depression 654, 733
dermatomyotome 432
designer baby 872
developmental genetics 8, 417
diabetes 411, 805
    mellitus 447
Diagnostic and Statistical Manual of Mental Disorders (DSM) 716
diakinesis 95
diaphyseal aclasis 181
Dicer enzyme 147
dideoxynucleotide (ddNTP) 144
diencephalon 428
diethylstilbestrol 283
diffuse Lewy body disease 690
DiGeorge syndrome 113, 237, 328, 406, 751, 784
1,4-dihydropyridine 791
Dilps 806
dinucleotide 333
diphosphoglycerate 387
diploid cell 166
diplotene 95
Dipterae 213
direct labeling 59
direct-to-consumer genetic testing 9
Disability Adjusted Life Years (DALYs) 889, 893
disease-causing mutation 321, 334
disruption 447
Distributed Annotation System 927
    registry 929
disulfiram 729
dizygotic twins 268
DNA 18
    analysis 7
    BLAT program 916
    chips 581
    condensation 60
    demethylation 308
    double-stranded breaks 99, 128
    genotyping 862
    intragenic polymorphism 849
    marker 501
    methylation 118, 301, 302, 305, 309, 472, 478, 480, 675

DNA (cont.)
    methyltransferase 302
    microarray 83
    microsatellite analysis 521
    mitochondrial 487
    mitochondrial polymorphism 501
    noncoding 502
    polymerase 302
    polymorphism 25, 222, 229, 501
    protein-coding 502
    remethylation 308
    repair 93, 558
    segments
        ultraconserved selecting 579
    sequence 1, 144 153, 276, 300, 370
        archaic hominids 536
        from autosomes 546
DNA-binding protein 64
DNA-RNA hybridization 59
dog breeds 814
domestic dog 814
    genome structure 819
    haplotype structure 820
    linkage disequilibrium 820
domestication 549
dominance 201
donezepil 689
dopamine 730, 731
    D4 receptor 657
    transporter 808
    two receptor 724
double bar 235
double knockout 783
double-stranded
    break 462
    DNA breaks 99
    RNA 790
Down syndrome 20, 26, 56, 101–103, 334, 665, 666, 671, 683, 689, 704, 717, 744, 855, 859
    chromosome studies 58
drepanocytosis 409
Drosophila 4, 20, 23, 56, 98, 99, 117, 213, 225, 235, 282, 307, 366, 419, 676, 779, 790, 830
    Activity Monitoring System 808
    Alzheimer disease 803
    cancer models 800
    cell cycle 797
    diabetes 805
    embryogenesis 420
    gene lethal giant larvae 796
    genome 541
    heart disease 804
    hedgehog protein 420
    hyperplasia 798
    melanogaster 115, 185, 195, 212, 254, 418, 429, 777, 795, 906, 925
    Menin 802
    metabolic disease 805

    neoplasia 798
    Parkinson's disease 803
    pseudoobscura 925
drug
    dependence 719
    discovery
        in fish 840
drug-drug interaction 636
drug-metabolizing enzyme 638, 641
drug-resistance protein 780
drug-response phenotype 645
Duchenne's muscular dystrophy 202, 268, 327, 329, 337, 849, 850
Duffy
    chemokine receptos 896
    locus 220
duplication 332, 384, 558, 704
    syndrome 113
duplicon 41
Durchbrenners 179
dynamic mutation 325
dynamitin 431
dynein 431
dysplasia 448
dystrobrevin-binding protein 763
dystrophin-glycoprotein complex 790

E

E-cadherin mutation 462
ECARUCA database 92, 115
ecogenetics 8
ectoderm 426
ectrodactyly 435, 437
Edwards syndrome 101
efficacy phenotype 640
Ehlers–Danlos syndrome 171, 349
electrophoresis 391
electrospray ionization 157
ELISA, see enzyme-linked immunosorbent assay
elliptocytosis 219
embryogenesis 306
    defect 829
embryonic cell 23
Emery-Dreifuss muscular dystrophy 145, 204
emotional stability 653
encephalitis 408, 689
ENCODE project 10, 26, 33, 35, 39, 140, 502, 581, 582, 910, 926
endocardium 438
endocytosis 870
endoderm 434
endogamy 521, 524
endonuclease 40
endophenotype 728, 764, 768
energy metabolism 579
Ensemble 932
    annotation 926
    customizing 934
    Genome Browser 923

Entrez Gene 953
enzyme-linked immunosorbent assay (ELISA) 157
epidermal growth factor receptor 797
epidermodysplasia 407
epidermolysis bullosa 171
epigenetic
   disease 301
   regulation 146
   variation 312
epigenetics 299, 769
epigenome project 476
epilepsy 205
epimutation 313, 334, 458, 465
epimyocardium 438
epistasis 180, 270, 284
epithelioma adenoides cysticum 171
Epstein-Barr virus 407
erythrocytosis 387
erythropoiesis 372
   protophyria 245
erythropoietin 385
Eschericha coli 6, 142, 225, 535, 789, 924
estriol 667
estrogen receptor-α gene 314
ethanol 735
ethical problems 11
euchromatic genome sequence 539
euchromatin 60, 62
eugenics 22, 725
   measure 16
   negative 21
   positive 21
   sterilization laws 21
eutherian core genome 569
Ewing sarkoma 348
exon 571
   protein-coding 574
   splicing silencer (ESS) 341
exonic genome 34
exoniPhy 570
exonuclease 70
   sequencing 152
expectation maximization 594
expressed sequence tag 572
exraversion 657
extended haplotype homozygosity test 542
extensive metabolizer 637
eye coloboma 91

F

facial anomaly 65, 129
facilitated
   epigenetic variation 315
   tracking model 376
factor V Leiden polymorphism 335
facultative heterochromatin 60
false discovery rate 278

familial
   adenomatous polyposis 252, 452, 453, 944
   fatal insomnia 182, 691
   hypercholesterolemia 171, 329, 867
   melanoma 453
   partial lipodystrophy 145
   porphyria cutanea tarda 257
   prion disorder 691
   resemblance 264
Fanconi anemia 6, 126, 453, 458
fascioscapulohumeral muscular dystrophy 313, 346
fast-changing gene 575
F-box protein Archipeligo 799
fetal hemoglobin 294, 375
   hereditary persistence 383
fetal karyotyping 854
α-fetoprotein 25
fetoscopy 860
FGF signaling 420
fiber FISH 85
fibroblast 38, 300
   growth factor 435
fight/flight system 653
Finnish disease heritage 521
fish model of human disorders 834
FISH, see fluorescence in situ hybridization
Fisher's
   exact test 705
   hypothesis 227, 228
five-factor model 652
flip-flop phenomina 763
floxed allele 782
fluorescence in situ hybridization (FISH) 83, 910
   metaphase spreads 84
fluorescent
   leukemic cell 838
   protein 833
fluorochome 58
fluorophore 59
5-fluorouracil (5-FU) 467
focal
   dermal hypoplasia 179
   segment glomerulosclerosis 280
folate photolysis 543
folate-neural-tube defect 447
forced sterilization 22
formal genetics 8
fossil record 532
founder effect 509
FOXP2 541
fragile
   mental retardation syndrome 313
   site 82
   X syndrome 82, 346, 671, 673, 703
   X tremor ataxia syndrome 674
frameshift mutation 344, 382
Friedreich ataxia 325

frontotemporal dementia 341, 690
F-statistics 595
Fugu fish 829
Fugu rubripes 573
full mutation 82

G

galactokinase 226
galactose-I-phosphate uridyltransferase 226
galactosemia 4
Gallus gallus 573
Galton, F. 3, 16
gamete complementation 186
gametogenesis 306
gamma secretase 685
gamma-aminobutyric acid (GABA) 735
gamma-globin 372, 384
   chain 369
Gardner syndrome 944
Garrod, A. 18
gastric polyps 253
gastrointestinal
   juvenile polyposis 457
   tract, infection 897
gastrulation 422, 425, 427
GATA-binding protein 103
gel electrophoresis 391
GenBank 913
gene
   action 3
   assignment 222
   codominant markers 217
   competition 375
   complementation 871
   conversion 332
     tract 379
   correction 871
   definition 140
   delivery vector 870
   discovery
     absolute risk 628
     analytic validity 628
     clinical utility 629
     clinical validity 628
     family studies 619
   dosage 665
     mutation 783
   duplication 226, 378, 537
   encoding thymidine phosphorylase 190
   environment
     correlation 726
     effects 654
   expression 156
   fast-changing 575
   frequencies 198
   function 153
   germinal therapy 25
   human disease-related 578

   knockdown 871
   mapping 25, 212
   noncoding, RNA-only 36
   on chromosomes 211
   ontology 153
   orthologous and paralogous relationship 573
   phenotypic expression 181
   pool 197
   prediction tracks 911
   protein-coding 33, 35, 570
   regulation 145
   regulatory sequence 582
   sequence 25
   silencing 375
   somatic therapy 25
   tagging 642
   targeting 779
   therapy 395, 868, 871
     safety issues 872
     vectors 868
Gene Sorter 918
gene-environment interaction 626, 727
GeneReview 943
GeneTest 945
genetic
   admixture 606, 608
   analysis of human personality 651
   background 180
   cancer 460
   counselling 25, 200, 845
     definition 846
     directive 852
     nondirective 852
     psychosocial aspects 853
     thalassemia 393
   database 903
   diagnosis 848
   differentiation 603
   disease 507
     global control 898
   distance 592
   drift 275
   drift 488, 491, 507, 509, 591, 592
   epidemiology 617
   heterogeneity 204
   heterogeneity 449
   linkage 216
   mapping 734
   medicine 885
   polymorphism 216
   polymorphism 619
   profile test 632
   relationship 510
   revolution 2
   screen 789
   screening assay 153
   services
     in developing countries 899

structure of human population 598
testing 848, 942, 943
  direct-to-consumer 9
tree 598
variation 487
genetics 1
development 2
human evolution 536
of health 545
of the environment 726
GeneView 930
genmark 572
genome
analysis workspace 583
analyzer 149
anatomy 31
annotations 907
browser 583, 905, 906, 913
galore 924
Graph 919
genome-wide association study (GWAS) 9, 206, 276, 284, 287, 618, 623, 763
genomic(s) 147
comparative tacks 912
disorders 113
evolutionary rate profiling (GERP) 570
hybridization (CGH) 47
hypomethylation 474
imprinting 185, 207, 307, 308, 625, 769
marker 48
topography 292
genotype-phenotype
correlation 349
distinction 300
relationship 9, 750
genotyping 558
by environment interaction 282
costs 276
genus Homo 542
germ cell 199
germinal gene therapy 25
germline
epimutation 334, 458
mosaicism 348, 675
mutation 466
susceptibility gene 460
Gerstmann–Straussler disease 182
giant chromosome 212
Giemsa
banding 59, 68, 79
  negative band 32
staining 65
gigas 801
Gilbert's disease 895
global
health 885
regulator 792
α-globin gene 313

glucose-6-phosphate dehydrogenase 24, 887
glutathione-S-transferase gene 576
glycoprotein 896
glycosaminoglycan (GAG) 445
gonosomal
aneuploidy 861
mosaics 348
Gorlin syndrome 454
graft-versus-host disease 394
granulysin 577
Gray's theory 653
Greig syndrome 436
grepafloxacin 636
gridlock
coarctation phenotype 841
mutant 840
gross deletion 327
growth hormone deficiency 327
GWAS, see genome-wide association study

H
Haemophilus influenzae 148, 924
hand-foot-genital syndrome 441
haploid genome 228
haplosufficiency 465
haplotype 45, 233, 280, 548, 822
analysis 219
blocks 500
diversity 499
Map 589
  of human genome 600
haplotype-based test 501
HapMap
population 609
project 26, 47, 91, 224, 274, 285, 320, 498, 500, 502, 600, 602
study 461
haptoglobin 235
Hardy–Weinberg
equilibrium 197, 245, 252, 269, 494, 593
law 21, 194, 195, 207
  derivations 194
principles 45, 509
proportions 198, 232, 275, 281
Hayflick limit 70
HBB-S allele 579
health economist 889
heart
development 439
failure 804
heartstring (hst) mutant 837
Heinz body 387
hemangioblastoma 452
hematopoietic stem cell 306
transplantation (HSCT) 394, 406
hemochromatosis 248, 249, 257
hemoglobin (Hb) 365, 382
A 320
Bart's 383

hemoglobin (Hb) (cont.)
  gene 369
  inherited disorders 889
    global distribution 889
  Lepore 347, 384
  molecule 368
  Portland 383
  variants 385, 393
hemoglobinopathy 380, 388, 391, 894
  carrier screening 389
  molecular diagnosis 391
  pharmacogenomics 395
  prenatal diagnosis 392
hemophilia 176, 214, 322, 330, 868
  A 332, 337
  X-linked 176
Henle's loop 440
Hensen's node 422, 426, 438
hepatoblast 300
hepatotoxicity 638
Herceptin 639
hereditary
  diffuse gastric carcinoma 453
  hemorrhagic telangiectasia 944
  mixed polyposis 453
  nonpolyposis 350
    colon cancer (HNPCC) 348, 846
  pituitary adenoma 453
heritability 252, 264, 266
  estimation 207
hermaphroditism 120, 788, 789
herpes simplex virus 869
  encephalitis 408
HERV gene 41
heterochromatic protein 306
heterochromatin 60, 62, 347
  centromeric 65
heterodisomy 185
heterogeneity LOD score 620
heterokaryon 881
heteromorphism 90, 220
heterotaxy 439
heterozygosity 166, 366, 494
hidden Markov model 570, 911
hidden population stratification 281
high-resolution mapping 291
hip dysplasia/hip laxity study 823
Hippokrates 14
Hirschsprung disease 249, 253
histamine 729
histone 574
  acetylation 304
  code 60, 62, 478
  deacetylase 303
  methylation 304
  modification 303, 480
  phosphorylation 305
  sumoylation 304
  ubiquinylation 304

HLA matching 232
holocomplex 376
holoprosencephaly 429
Holt–Oram syndrome 440
homeobox domain family 36
hominid
  lineage 576
  species 531
Homo
  erectus 530, 532–534
  floresiensis 532, 534
  neanderthalensis 532, 534, 535
  sapiens 530, 573
homologous recombination 127
homozygosity 166, 174, 381, 492, 494, 821
  of dominant anomalies 171
human
  accelerated region 876
  beta-globin 345
  biochemical genetics 19
  brain 541
  centromere 41
  chorionic gonadotropin 667
    assay 517
  chromosome 4
  cytogenetics 8
  disease-related gene 578
  diversity 595
  embryonic fibroblast 221
  evolution 529
    medical genetics 544
  gene
    history and geography 596
  Gene Mutation Database 321, 393, 956
  genetic variation 590
  genetics 1
    database 941
    history 13
    practical applications 5
    probability problems 200
  genome
    diversity 538
    haplotype map 600
    patterns of structural variation 498
  genomics 618
  immunodeficiency virus-1 (HIV-1) 226, 408
  leukocyte antigen haplotype 516
  malformation syndrome 433
  migration 547
  mitochondrial genome 537
  population structure 550, 595
  telomerase reverse transcriptase (hTERT) 70
  telomere 41
Human Genome Diversity Panel 598, 602
Human Genome Organization (HUGO) 223
Human Genome Project 6, 31, 88, 148, 474
Human Genome Variation Society 947

huntingtin
    gene 169, 183
    protein 804
Huntington disease (HD) 169, 170, 213, 325, 346, 784, 852, 888, 944
Hutchinson–Gilford progeria syndrome (HGPS) 141, 144
hydrocephalus 519
hydronephrosis 441
hydroxycarbamide 394
hydroxyurea 394
hypercholesterolemia 171
hyperestrogenism 124
hyperhomocysteinemie 315
hypermethylation in cancer 465
hyperphagia 668, 753
hyperphenylalaninemia 245, 246
hypertelorism 111
hypertension 897
hypocalcemia 444
hypochondroplasia 444
hypogonadism 124
hypophosphatemia 177
hypoplastic left heart syndrome 122
hypostasis 180
hypothalamic-pituitary-adrenal axis 733
hypoxanthine phosphoribosyltransferase 222
hypoxanthine-aminopterin-thymidine (HAT) medium 221
hypoxia 386, 387, 465

I

ichthyosis 179
idendity by descent 273, 510, 521, 622, 821
idiopathic pulmonary fibrosis 71
idioplasma 20
IFN-γ 411
illicit drug genetic factor 723
Illuma's Genome Analyzer 149
immune response 576
immunodeficiency 65, 129, 411
immunogenetics 8
imprinting
    center 309
    defect 314
imputation 277
in silico PCR tool 918
in silico-generated band 68
inborn errors of metabolism 56
inbreeding 514, 519, 520, 592
incest 518, 519
    adverse effects 516
    coefficient 517
    fertility 516
    fetal loss rates 516
    inherited disease 518
incontinentia pigmenti 178
indel, see insertion deletion
Indian Hedgehog gene (IHH) 169
indirect labeling 59
individual ancestry 594

inebriometer test chamber 807
infection, genetic theory 405
informed consent 898
inheritance 3
    autosomal dominant mode 167
    codominant mode 166
    conditions without simple modes 205
    empirical risk figures 205
    pseudodominance 173
    statistical evaluation 206
    triallelic 193
    X-linked 175, 178
inherited disease
    database (IDID) 817
    study 876
inhibin alpha 667
insertion-deletion (indel) 333, 559
    hotspot 325
insulin growth factor 792
insulin-producing cells 806
intellectual dysfunction 664, 749
interbreed heterozygosity 816
interchromatin compartment 85
inter-ethnic gene expression 508
International Classification of Disease (ICD) 716
International HapMap Project 602
International Human Genome Sequencing 590
interphase
    analysis 85
    chromatin 85
    cytogenetics 59, 85
    FISH 85
intersexes 120
interspecies nuclear transfer 880
intrabreed homozygosity 816
intron 570
inversion polymorphism 47
ion mass separation 157
island beat (isl) mutant 837
isochromosome 90, 112
isodisomy 185

J

Jacob-Monod model in bacteria 7
Jacobsen syndrome 83, 112
Johnson-type allel 236
joint allele frequency spectrum 591
Joubert syndrome 258
juvenile polyposis 455

K

Ka/Ks ratio 576
Kallmann syndrome 328, 348
Kartagener syndrome 439, 440
karyotype 80, 81
Kearns–Sayre syndrome 193
Kennedy disease 325
keratosis follicularis spinulosa decalvans cum
    ophiasi 177

kidney
  development 440
  disease 280
  tumor of the childhood 441
kinetochore 63, 73, 76, 99
Klinefelter syndrome 56, 58, 105, 116, 123, 313, 747, 748
knockout 780
K-ras mutation 460, 467
lactase
  allele 543
  persistence 543

L

LAGAN 565
Lander–Green algorithm 622
large offspring syndrome 306
Lathyrus odoratus 212
Latin America Collaborative Study of Congenital Malformations (ECLAMC) 519
law
  of independence 18
  of segregation 18
  of uniformity 18
leading strand synthesis 69
Leber's hereditary optical neuropathy (LHON) 188
leiomyomatosis 454
leprosy 409–411
leptotene 93
lethal
  chondrodysplasia 442
  in humans 180
leukemia 407, 666, 840
  chromosome 78
leukemogenesis 407
leukocyte 229
  antigen 394
leukodystrophy 524
Lewis allel 320
Lhermitte Duclos disease 455
Li-Fraumeni syndrome 455, 462
liftOver program 918
limb
  girdle, muscular dystrophy 833
  progress zone 435
  zone of polarizing activity 435
LINE, see long interspersed nuclear element
lineage sorting 537
linkage 213
  analysis 9, 215, 253, 271, 619
    definition of terms 621
    in humans 213
    nonparametric 253
    parametric 253
  disequilibrium 95, 224, 229, 274, 499, 604, 605, 763
  equilibrium 215, 232, 275
    mapping 275
  map 225
  studies 656

linking model 377
Linné, Carl von 530
liposome 870
lissencephaly 430
liver
  cirrhosis 717
  disease 831
locus
  control region (LCR) 345, 370, 372
  heterogeneity 246
locus-specific database 947
LOD score 216
LOG odds plot 272
logarithm of differences 215
London Dysmorphology Database 848
long interspersed nuclear element (LINE) 40, 65, 118, 540
  retrotransposition 329
Loo Gehrig disease 876
looping model 376
loss of heterozygosity (LOH) 48, 463, 464, 476
luteinizing hormone (LH) 96
lymphoblast 707
lymphocytotoxicity test 230
lymphotoxin alpha 410
Lynch syndrome 457, 852
Lysenkoism 725
Lytechinus variegatus 418

M

Macaca mulatta 731
macular degeneration 643
major depressive episode 760
major gene/locus concept 409
major histocompatibility complex (MHC) 6, 213, 226, 229, 234, 516
Mal de Meleda 175
malaria 272, 334, 380, 381, 389, 409, 494, 544, 887, 892
maltrexone 736
mammalian cell 71
mania 765
manic episode 760
manic-depressive insanity 768
mannose-binding lectin 1 181
MAOA-linked polymorphic region 731
map distance 217
mapping 558, 910
Marfan syndrome 341, 345
marker chromosome 861
Markov model 280
MASA syndrome 349
mass spectrometry 157
maternal blood sampling 859
matrix-assisted laser desorption/ionization 157
McArdle disease 343
McDonald-Kreitman test 578, 579
McKusick–Kaufman syndrome 258
mean cellular
  hemoglobin 381
  volume 381

Meckel syndrome 522, 605
Meckel–Gruber syndrome 258
Meckel's cartilage 434
medaka 828, 830, 837, 841
medical genetics 23, 544
    database 941
Mediterranean fever 524
medullary thyroid carcinoma 802
meiosis 93, 198, 620, 878
    in females 96
    in males 96
    prophase 99
melanoma cell line 478
memantine 689
Mendel, Gregor 17, 166
Mendelian
    backgross 175
    cancer 452
    disorder 144, 225, 244, 271, 618
        recurrence risk 851
    recessive tumor syndrome 458
    resistance 408
    tumor susceptibility gene 453
Mendelism 16
Mendel's laws 1–3, 20, 194, 238
    gene transmission 447
    law of independence 18
    law of segregation 18
    law of uniformity 18
mental retardation 663, 708, 754
mesencephalon 428
mesoderm 434
mesomelic dysplasia 347
metabolic disease 9, 805
metabolome 147
metanephros 440
metaphase
    chromosome 59, 93
    FISH 114
methemoglobinemia 24, 349, 385, 387
methionine 691
methylated DNA 675
methylation 61, 477
    imbalance 480
    tumor-type-specific 478
methylation-sensitive restriction endonuclease
        digestion 475
5-methylcytosine 473, 474
methyl-cytosine-binding protein 303
methylenetetrahydrofolate reductase (MTHFR) 315
methylome 473, 475
metronidazole 729
Mexican admixture 608
microarray
    chips 154
    method 656
microcephalin gene 544
microcephaly vera 431

microdeletion 111, 320, 323, 324
    syndrome 92, 113, 114
microduplication syndrome 114
micrognathia 121
micro-insertion 323, 324
microphthalmia 179
micro-RNA 342, 579
    Bantam 799
microsatellite 46, 223, 490, 593, 606, 608
    marker 620
microtia 91
microtubule 73
Middle Ages 15
midparent 266
migration 488, 591
mild cognitive impairment 682
minisatellite 490
miRNA, hybridization 580
mirtazapine 637
mismatch repair gene 465
missense mutation 141, 336, 345
missing heritability 284
mitochondrial
    DNA 487
    enzyme 730
    genome 38, 43, 188
    inheritance 769
    neurogastrointestinal encephalopathy syndrome 190
mitosis 72, 78
    anaphase 74
    equatorial plane 74
    prometaphase 73, 74
    prophase 73
mitotic
    arrest-deficient homologue-2 (Mad2) 76
    chromosome 56
    clone 796
mixed polymorphism 47
model organism 777, 787
model-based
    clustering algorithm 593
    linkage analysis 620
modeling human disorders in fish 830
molecular
    biology 4
    disease family 449
    misreading 333
    protrait 156
monoamine 730
    oxidase A 724
monogenic
    bias 253
    disease 856, 889
        in developing countries 894
    Mendelian disorder 144
    obesity 292
monomer repeat 42
mononucleotide 333

monosomy 109, 66
    rescue 186
    X 847
monozygotic twins 267
Morbid Map 947
Morpheus gene family 539
morphogen 419
morphogenesis 419
morpholino 832
morphologic trait 822
morula 422
mosaic 104
mosaic variegated aneuploidy (MVA) 77, 129, 455
most recent common ancestor 495
motor neurin disease 876
Mouse chain track 914
Mouse Net track 914
mRNA splicing mutants 338
Mulibrey nanism 522
multicolor-fluorescence in situ hybridization (MFISH) 463
Multidrug Resistance 1 (MDR1) gene 141
multifactor dimensionality reduction 627
multigene disorder 103
multi-megabase-pair deletion 489
multiple
    alignment format (MAF) 916
    allelism 195
    endocrine neoplasia 454, 455, 802
    infectious disease 406
    intestinal neoplasia 252
    ligation probe amplification 392
    regression analysis 750
multiplex
    amplifiable probe hybridization (MAPH) 87, 88
    family 623
    FISH (M-FISH) 84
    ligation-dependent probe amplification (MLPA) 87
multispecies conserved sequenced 570
muscular dystrophy 204
mutant
    estrogen receptor 782
    mitochondria 190
mutation 141, 488, 590
    affecting gene expression 337
    cause cancer 462
    disease-causing 321, 334
    in 3' regulatory region 343
    in 5' untranslated region 343
    in gene evolution 335
    in miRNA binding sites 342
    in remote promoter elements 345
    nomenclature 335
    producing inappropriate gene expression 346
    rate 558
mutator phenotype 466
Myc complex 798
Mycobacterium
    leprae 410
    tuberculosis 148, 895, 925
Mycoplasma genitalium 148
myoblast 432
myoclonic epilepsy 189
myotonic dystrophy 182, 183, 669

N

nail-patella syndrome 219, 438
naked DNA 870
Nasse's laws 2
National Health Examination and Nutrition Survey (NHANES) 625
natural killer cell receptor 237
natural selection 492
    detecting 500
Neanderthal genome project 531
Neanderthals 536
negative
    eugenics 21
    selection 559
Neisseria meningitidis 895
nematode 788
Neolithic 547
    expansion 603
    human fossil 548
nephronophthisis 258
nephrotic syndrome 441
NESARC survey 727
neural
    crest cell 427, 434
    tube defect 428, 856
Neurexin 704
neurodegenerative disease 802
neurofibromatosis 330, 340, 452, 454, 458, 802, 840
neurofibromin 457
Neuroligin 3 704
neuromuscular disorder 846
neuropeptide Y 728
neuropsychiatric
    disease 722
    phenotype 703
neuroticism 652, 653
neurulation 425
neutral DNA 502, 560, 568
next-generation sequencing 144, 148, 184
nicotine 720, 735
nicotinic cholinergic receptor 735
Nijmegen breakage syndrome (NBS) 129, 454
Noggin 420, 443
non-allelic homologous recombination (NAHR) 98, 113
nonallelic noncomplementation 254
noncoding sequence 582
noncomplimentation
    nonallelic 254
    noninteracting 254
nondeletion mutation 383
noninteracting noncomplementation 254
nonketotic hyperglycinemia 522

nonparametric linkage analysis 253
nonpaternity 185
nonsense mutation 141, 341, 344, 382
nonviral vector 869
Noonan syndrome 449
norepinephrine 731
normal science 2
normal transmitting male 673
Norrie disease 343
nortriptyline 637
Norwalk-like virus 408
Notch signal transducing pathway 801
notochord 432
nuchal edema 858
nuclear DNA sequence 537
nuclear transfer 878
nucleolus-organizing region (NOR) 80
nucleosome 61, 303, 480
   wrapping 146
nucleotide 32, 34, 141, 323, 560, 563
   sequence 24
   substitution 322
nucleus pulposus 433
Nude syndrome 406

O

obligatory epigenetic variation 313
Obox gene 576
Obp gene 576
OCEAN 652
oculocutaneous albinism (OCA) 173
oculopharyngeal muscular dystrophy 327
olfactory
   sensory neuron 312
   system
      allelic exclusion 312
oligo-astheno-teratozoospermia syndrome 847
oligodendrocyte transcription factor 478
oligogenic disorder 244, 249, 254
oligogenicity 247–249, 252, 410
oligonucleotide 151, 476
   array 88
oligospermia 123
OMIM, see Online Mendelian Inheritance in Man
oncogene 464
Online Mendelian Inheritance in Man (OMIM) 184, 393, 817, 851, 945
   Genes track 911
onychotillomania 752
oocyte 23, 96, 97, 188
   aneuploidy 126
oogonia 199
oral-facial-digital syndrome 179
organogenesis 425
ornithin transcarbamylase deficiency 872
Orrorin tugenensis 533
Oryzias latipes 827
osteoblast 443

osteogenesis imperfecta 445, 448
osteopetrosis 445
ovarian carcinoma 453
oxidative phosphorylation (OXPHOS) 188

P

5p syndrome 754
pachygyria 430
pachytene 93
palaeoanthropology 531
palaeogenomics 534
Pallister–Hall syndrome 331, 436, 438
pangenesis theory 14
papillary renal cell carcinoma 454
paradigm 1
paraganglioma 452, 455
paralogous gene 572
parametric
   linkage analysis 253
   test 201
paramutation 316
parasitemia 411
paraxial mesoderm 431
parental age 887
parent-offspring regression 266
parent-of-origin effect 625
Parkin coregulated gene 410
Parkinson's disease 190, 545, 690, 803
paroxysmal nocturnal hemoglobinuria 348
Parvoviridae 869
patient discrimination 898
patient-specific cell 877
Pearson syndrome 193
Pecan algorithm 929
Penetrance 169
penicillin 893
pentanucleotide repeat 327
personal genomics 152
personalized medicine 630, 736
pervasive developmental disorder-not otherwise specified 699
Peutz–Jeghers syndrome 454, 457
PGKneo 783
pharmacogenetics 8, 635
   ethnic groups 644
pharyngeal arch 433
phasing 500
phastCons 570, 583, 584
PhenCode project 393
phenocopy 721
phenotypic
   plasticity 300
   trait 264
phenylalanine hydroxylase (PAH) 246, 350, 669
phenylketonuria 4, 25, 245, 246, 343, 350, 522, 669
pheochromocytoma 452, 455, 459
Philadelphia chromosome 58, 59, 348, 464
phocomelia 438
phosphate-regulating endopeptidase gene 177

phosphatidylinositol-3-kinase 792
phosphorylation 61
photosensitive blood mutant 834
phytohemagglutinin (PHA) 78
Pick's disease 690
Pitt-Hopkins syndrome 91
Plasmodium
   chabaudi 411
   falciparum 245, 550, 577, 896, 925
      malaria 380, 405, 409, 887, 892
Plato 15
pleiotropic disorder 247
plexin 654
Plk1-interacting checkpoint helicase (PICH) 75
ploidy 86
pneumonia 682
point mutation 489
poison model 254, 256
Poisson distribution 199
poistional cloning 224
poly(ADP-ribose) polymerase (PARP) inhibitor 468
polyadenylation signal 574
poly-ADP-ribosylation 304
Polycomb repressor complex 479
polycystic
   kidney disease 258
   ovary syndrome 248, 257
polydactyly 437, 438, 519
polyembolokoila 752
polygene 410, 622
polygenecity model 721
polygenic
   disease 619
   trait 293
polyglutamine 673
polymer 870
polymerase 59
polymerase chain reaction (PCR) 223, 321, 368
   amplification 46
polymorphism 320, 488, 539, 558, 577
Polynesians 549
Polyphen 610
polyploidy 466
polyubiquitylation of securin 76
PopREs 607, 609
population
   attributable fraction 628
   cryptic substructure 624
   demographic history 609
   differentiation 495
   ethnic background 624
   genetics 8
   migration 887
   quantitive models of selection 610
   stratification 280, 623, 645
porphyria 88
position effect 346

positive
   eugenics 21
   selection 562
POSSUM program 848
postgenome 31
potter sequence 448
poverty 886
Prader-Willi syndrome 38, 187, 668, 753
preconception 57
predictive testing 851
pre-eclamptic toxemia 516
pregnancy-associated plasma protein 855
pregranulosa cell 96
preimplantation genetic diagnosis (PGD) 392, 856
premeiotic mutation 348
PReMods 582
premutation 325
prenatal
   diagnosis 24, 392, 854, 894, 943
   screening 667
   ultrasound 858
presinilin 804
presomitic mesoderm 432
pressure palsy 752
principal component analysis 596, 822
principle of Jennings 234
prion disease 682, 691
probalitiy in human genetics 200
progranulin 691
progressive
   myoclonus epilepsy 338
   recessive myoclonic epilepsy 818
   retinal atrophy 817
prometaphase of mitosis 74
promoter
   hypermethylation 474
   mutation 337, 382
prophase of mitosis 73
prosencephalon 428
protein 418
   aligning sequences 563
   arginine methyltransferase 305
   chromatin-marking 301
   coding exon 574
   DNA-binding 64
   GATA-binding 103
   interaction map 791
   kinase 808
      gene family 183
   methyl-cytosin-binding 303
   misfolding 793
protein-coding gene 33, 35, 570
protein-histone interaction 479
protein-protein interaction 157
protein-truncating mutation 465
proteinuria 441
proteoglycan 444
proteome 147, 156, 158

Proteome Browser 907, 918
proteomics 148, 156
    antibody array-based method 158
    strategies 158
proteotoxicity 793
proto-oncogene 464
prototypic trait 293
proximo-distal axis 435
pseudo-agouti 316
pseudoallel 225
pseudoautosomal region (PAR) 119
pseudodominance in autosomal recessice inheritance 173
pseudoexon 342
pseudogene 36, 378, 561
    nonprocessed 36
    processed 36
pseudohermaphroditism 441
    female 121
    male 121
pseudohypoparathyroidism 444
Pseudomonas aeruginosa infection 335
pseudouridylation 38
public health genomics 618, 631
pure epigenetic variation 313
purifyfing selection 559
purine 142, 582
PWS-SRO 309
pycnodysostosis 445
pyrimidine 142, 582
pyrosequencing 149

Q
Q-bands by fluorescence using quinacrine (QFO) 78
quantitative trait locus 622, 822

R
radiation hybrid method 212
ragged red fiber disease (MERRF) 189
random walk 591
rapamycin 801, 804
Ras signaling 799
recognition barcode 48
recombinant human erythropoietin (rhEPO) 394
recombination 488, 492
regression 266
regulatory transcripts of small RNA 147
relateddependent macular degeration 289
renal cell carcinoma 459
renal cystadenocarcinoma and nodular dermatofibrosis 818
RepeatMasker 911
replication 279, 558
    break-induced 332
reproduction 576
reproductive genetics 8
repulsion 213
restriction
    fragment length polymorphism (RFLP) 44, 212, 489, 539, 593, 620
    landmark genomic scanning 474
RET mutation 254
reticular dysgenesia 406
retinitis pigmentosa 190, 254
retinoblastoma 170, 455, 463
    gene 314
retinoic acid receptor 480
retrotransposition 329
retrotransposon 561
retrovirus 869
retrovirus-mediated insertional mutagenesis 831
retsina (ret) mutant 837
Rett syndrome 63, 179, 670, 675, 705
reverse
    band (R-band) 79
    genetics 831
    transcriptase 40
reversed loop model 109
RFLP, see restriction fragment length polymorphism
Rh
    complex 228
    factor 20
rheumatoid arthritis 292
rhombencephalon 428
Rhox gene 576
riboflavin 473
ribosomal RNA 580
riesling (ris) mutant 837
ring chromosome 111
RNA 418
    cleavage mutation 382
    cleavage-polyadenylation mutant 342
    interference 394
    polymerase 146
RNA-processing mutation 382
Roberts syndrome 77, 129
Robertsonian translocation 43, 101, 112, 666, 855, 861
Rothmund–Thomson syndrome 127, 128, 455
ruffled membrane 445
Russell–Silver syndrome 187

S
Saccharomyces cerevisiae 777, 906, 924
Saccharomyces Genome Database 906
Sahelanthropus tchadensis 532, 533
SAM-to-S-adenosylhomocysteine 473
Sanger sequencing 143, 144
SC phocomelia syndrome 77, 129
scanning model 376
scar effect 653
Schistosoma mansoni 409
schistosomiasis 409
schizophrenia 206, 293, 473, 636, 654, 668, 705, 708, 722, 746, 752, 759
    adoption study 761
    evolutionary paradox 762
    gene–environment interaction 761
    genome-wide association studies 764

schizophrenia (cont.)
    linkage studies 762
    molecular genetic studies 762
    twin study 761
scientific development 2
sclerotome 432, 433
screening program 894
scurfy mutation 116
segmental
    aneuploidy 111, 784
    duplication 41, 237
segregation
    analysis 203, 617
    ratio 201
selective sweep 494
semaphorin axon 654
semenogelins 577
Sendai virus 221
sequence analysis 245
sequencing
    by oligoligation and detection (SOLiD) 149
    track 910
sequencing-by-synthesis method 149
serial
    analysis of gene expression (SAGE) 154
    founder effect 605
    replication slippage in trans (SRStrans) 332
serine/threonine kinase 477
serotonin 731
    transporter 724, 728, 732
        gene 653
sertindole 636
serum bilirubin 292
severe combined immunodeficiency (SCID) 406
sex chromosome 116
    aberration 81
    mosaicism 105
sex-limiting modifying gene 181
Sherman paradox 673
short interspersed nuclear element (SINE) 40, 540, 817, 818
short sequence repeats (SSR) 46, 321
short stature homeobox (SHOX) 119
    haploinsufficiency 122
short tandem repeat 490
short-chain fatty acid 394
Shugoshin 99
sib pair method 217, 218, 273
sickle cell
    anemia/disease 4, 24, 272, 349, 350, 366, 386, 390, 409, 892
    gene 891
    hemoglobin 367
signature 156
silent
    allel 198
    chromatin 315
    heart (sih) mutant 837
    mutation 141

site 561
    olfactory receptor 312
Silver-Russell syndrome 310
SINE, see short interspersed nuclear element
single base mutation 818
single nucleotide
    mutation 489
    polymorphism (SNP) 43, 89, 142, 337, 338, 377, 461, 496, 539, 545, 590, 620, 642, 790, 820
        haplotype-tagged 461
single restriction fragment length polymorphism 219
single site defect 447
single tandem repeat polymorphism 620
single-biomarker analysis 147
single-molecule sequencing 152
single-stage mass spectrometer 157
single-stranded conformation polymorphism (SSCP) 392
sister chromatid exchange (SCE) 72, 127
situs
    ambiguus 439
    inversus 439
skeletal
    development 442
    dysplasia 444
skeleton
    congenital anomaly 446
    malformations 446
sleep disorder 808
slow-boat hypothesis 549
small deletion 489
small insertion 489
small interfering RNA 790
Smith-Magenis syndrome 328, 752
SNP, see single nucleotide polymorphism
sodium bisulfite 301
Solexa 149
somatic
    cancer genetics 462
    cell 880
        genetics 8, 24
        nuclear transfer 875, 878
    gene therapy 25
    mosaicism 348
    mutation 348
somites 431
Sonic hedgehog (Shh) 429, 435
    signaling molecule 427
spectral karyotyping (SKY) 84
β-*spectrin* 837
spermatogenesis 124
spermatogenesis 224
spina bifida 427
spindle-assembly checkpoint (SAC) 75, 99
spinocerebellar ataxia 325, 784
spinocerebellar atrophy 888
spliceosome RNA gene 38
splicing enhancer sequence 341

spondylocostal dysostosis 433
spontaneous miscarriage 125
sporadic lesion 452
Src activity 800
Staphylococcus aureus 152, 833
startle reflex 769
statistical analysis 215
sten cell model of carcinogenesis 462
sterilization laws 22
steroid sulfatase deficiency 337
steroidogenic factor 1 (SF-1) 120
Stevens-Johnson syndrome 643
STRAT software 496
stressful life event 655
striatal necrosis 189
structural chromosomal rearrangement 861
STRUCTURE software 281, 496, 594, 600
subtelomere syndrome 112
succinate dehydrogenase subunit D (SDHD) 452
sudden infant death 520
suicide 733
supergene 6
super-hotspot 325
superoxide dismutase 876
symphalangism 443
synaptogenesis 430
syncytiotrophoblast 424
syndetome 433
synostosis syndrome 443
synpolydactyly 424, 438
Synteny 929
synthetic map 598, 599

T

Table Browser 916
Tajima's D 577
Takifugu rubripes 925
targeted-induced local lesion in genome (TILLING) 831, 832
TATA box 370
Tau 804
Tay-Sachs disease 25, 850
T-cell leukemia 407, 872
telangiectasia 171, 172
telomerase 70
telomere 41, 68
　erosion 70
　replication problem 69
teratogenic effect 717
termination codon mutation 344
testis-determining factor 224
testosterone 121, 123
tetrad analysis 199
tetranucleotide repeat expansion 327
Tetraodon nigroviridis 925
thalassemia 41, 367, 887, 889, 900
　alpha 98, 347, 383, 387, 894
　　myelodysplasia syndrome 387

beta 342, 343, 349, 367, 375, 380, 524
　Lepore 98
gene 891
thalidomide 448
thanatophoric dysplasia 444
thiamine 690
thiopurine toxicity 638
third-generation sequencing 152
thrifty genotape hypothesis 579
thrombocytopenia-absent adius (TAR) syndrome 91
thymidin kinase 221
thymine 322
Timothy syndrome 704, 838
T-loop 69
topoisomerase 75
tracking model 376
tranilast 638
transcription
　factors 421
　mutation 337
transcriptome 155, 159
　high throughput long- and short read sequencing 156
transcriptomics 148, 154
transfer RNA 580
transforming growth factor 181, 348
transgenesis 832
transgenics 782
translational initiation mutation 343
transmembrane conductance regulator gene 851
transmission disequilibrium test (TDT) 273
transposon 46
trans-replication slippage 331
trastuzumab 639
treeness 596
triad design 625
triallelic inheritance 193
trichothiodystrophy 388
trimethylated lysine 65
trimethylation 38
trinucleotide repeats 325
　allele 184
　expansion 346
triplet nucleotide repeat disease 804
triple-X syndrome 124, 748
triploid cancer 464
triploidy 104, 847
trisomy 110, 663, 665, 855
　13 104, 861
　16 847
　18 105, 861
　21 20, 104, 220, 666, 847, 855, 858, 861
　autosomal 104
　rescue 186
truncate selection 203
trypanosomiasis 408
tuberculosis 638, 886
tuberous sclerosis 455, 457, 669, 703, 801,
tumor methylome analysis 475

tumor suppressor
　gene 314, 464, 466, 476
　mutant 840
tumorigenesis 466, 473, 480
Turcot syndrome 944
Turner syndrome 56, 58, 105, 112, 117, 120, 122, 666
　clinical findings 119
twins 267
　dizygotic 268
　monozygotic 267
tylosis 455
typhoid vaccine 283

U

ubiquitin ligase 668, 701
ubiquitination 61
UCSC
　genes set 912
　Genome Browser 583, 906
U-loop 109
ultrabithorax mutant 422
ultraconserved element 579
ultraconserved elements (UCE) 347, 579
unindentified Mendelian cancer susceptibility gene 457
uniparental disomy 307
uniparental disomy (UPD) 185, 307
unipolar depression 765
unstable
　hemoglobin 386
　protein mutant 345
untranslated region (UTR) 140
uroporphyrinogen decarboxylase (UROD) 257
Usher syndrome 522

V

valproate 448
variable numbers of tandem repeats (VNTR) 219, 321, 657
variance partitioning 270
variation image 932
vascular endothelial growth factor (VEGF) 104
VEGF, see vascular endothelial growth factor
velocardiofacial syndrome 702, 784
Venice criteria 628
Venn diagram 567
venocardiofacial disorder 237
vertebrate 829
Vertebrate Genome Annotation (VEGA) 911
very-early-onset bowel tumor 458
VisiGene tool 907, 913
VISTA 5
visual system disorder 831
vitamin D synthesis 543
von Hippel-Lindau syndrome 452, 455
von Willebrandt factor (vWF) 336, 350

W

Waardenburg syndrome 350, 428
Waardenburg-Shah syndrome 344

warfarin 269, 292, 294, 641, 643, 645
Waterson's q 497
web-based tool 917
Wechsler Intelligence Scale 664
Weinberg's method 206
Wellcome Trust Case Control Consortium 277
Werner syndrome 127, 128, 455
West Nile virus (WNV) infection 409
whole-genome linkage 734
Williams syndrome 237, 328
Williams–Beuren syndrome 41, 753
Wilms tumor 441
Wilson disease 851
Wingless (Wg) activity 798
Wolf–Hirschhorn syndrome 112, 754
Wright's F-statistics 591

X

X chromatin 58, 116, 117
X chromosome 112, 166, 539, 546
　aberration 746
　aneuploidy 120
　inactivation 117
　mapping 223
　polysomy 123, 124
X inactivation 310
　center 117
xenobiotic metabolism 576
Xenopus 443
　laevis 418
Xeroderma pigmentosum 172, 388, 466
Xiphophorus spp. 827, 828, 841
　melanoma 838
X-linked
　agammaglobulinemia 330, 405
　dystrophin gene 204
　hemophilia 176
　hydrocephalus 340
　inheritance 178
　lethal 180
　monogenic disease 849
　recessive disease 850
　recessive inheritance 175
　severe combined immunodeficiency (X-SCID) 868, 872
　syndromic form 670
XXYY syndrome 125, 751
XYY syndrome 748, 749
　behavioral problems 750
XYY zygote 120

Y

Y chromosome 79, 119, 125, 166, 224, 539
　aberration 748
　mapping 223
　polymorphism 501
Y polysomy 125
yolk sac 369

Z

zebra finches 542
zebrafish 828–830, 833, 841
  sauternes 834
zinc finger 871

zona limitans intrathalamica 429
zone of nonproliferating 798
zygote 177
zytogene 94